OPTICAL RESOLUTION PROCEDURES
for
CHEMICAL COMPOUNDS

VOLUME 2
ACIDS
Part I

Paul Newman
Manhattan College

The Section on Asymmetric Synthesis by

Spiro Alexandratos
University of Tennessee

A publication of the

OPTICAL RESOLUTION INFORMATION CENTER
Manhattan College
Riverdale, New York 10471

I S B N 0-9601918-1-X
I S B N Set 0-9601918-3-6

Library of Congress Catalog Card Number
78-61452

Distributed by: Optical Resolution Information Center
Manhattan College
Riverdale, New York 10471

PERMISSIONS

The following journals are copyrighted by the American Chemical Society which has kindly granted permission to reprint the procedures taken therefrom. The specific journal references appear in Section 5 of this volume.

Analytical Chemistry
Journal of Organic Chemistry
Journal of the American Chemical Society
Journal of Medicinal Chemistry
Journal of Agricultural and Food Chemistry
Inorganic Chemistry

All procedures from the *Canadian Journal of Chemistry* are reproduced by permission of the National Research Council of Canada.

All procedures from the *Journal of Pharmaceutical Sciences* are reproduced with permission of the copyright owner, the American Pharmaceutical Association.

I also wish to thank the following publishers for giving me permission to reproduce experimental procedures from their respective journals which are cited in the Reference List in Section 5 of this volume.

Acta Chemica Scandinavica
Munksgaard, International Publishers, Ltd., Copenhagen, Denmark
Acta Chimica (Budapest)
Kultura, Budapest, Hungary
Acta Chimica Academiae Scientiarum Hungaricae
Kultura, Budapest, Hungary
Acta Pharmaceutica Suecica
Swedish Academy of Pharmaceutical Sciences, Stockholm, Sweden
Agricultural and Biological Chemistry
Agricultural Chemical Society of Japan, Tokyo, Japan
Anales de la Asociacion Quimica Argentina
Librart S.R.L., Buenos Aires, Argentina
Angewandte Chemie, International Edition
Verlag Chemie GMBH, Weinheim, West Germany
Annales de Chimie (Paris)
Masson et Cie, Paris, France
Annales Pharmaceutiques Francaises
Masson et Cie, Paris, France
Arkiv foer Kemi
Almqvist and Wiksell International, Stockholm, Sweden
Archiv der Pharmazie (Weinheim)
Verlag Chemie, GMBH, Weinheim, West Germany
Arzneimittel-Forschung
Editio Cantor KG, Aulendorf, West Germany
Biomedical Mass Spectrometry
Heyden & Son, Ltd., London, England
Bioorganic Chemistry
Academic Press, Inc., New York, New York
Bollettino Chimico Farmaceutico
Societa Editoriale Farmaceutica, Milan, Italy
Bulletin de L'Academie Polonaise des Sciences, Serie des Sciences Chimiques
Ars Polona-RUCH, Warsaw, Poland
Bulletin of the Chemical Society of Japan
The Chemical Society of Japan, Tokyo, Japan
Bulletin des Sociétés Chimique Belges
Bibliotheque des Societes Chimiques Belges, Brussels, Belgium
Bulletin de la Société Chimique de France
Masson et Cie, Paris, France
Canadian Journal of Chemistry
National Research Council of Canada, Ottawa, Canada
Chemica Scripta
Almqvist and Wiksell International, Stockholm, Sweden
Chemical and Pharmaceutical Bulletin
Pharmaceutical Society of Japan, Tokyo, Japan
Chemische Berichte
Verlag Chemie GMBH, Weinheim, West Germany
Chemistry and Industry
Society of Chemical Industry, London, England
Chemistry Letters
The Chemical Society of Japan, Tokyo, Japan
Chimia
Chimia, Aarau, Switzerland
Chromatographia
Pergamon Press, Ltd., Oxford, England
Collection of Czechoslovak Chemical Communications
Academia, Prague, Czechoslovakia

Comptes Rendus de l'Academie des Sciences
Gauthier-Villars, Paris, France
Croatica Chemica Acta
Croatian Chemical Society, Zagreb, Yugoslavia
European Journal of Medicinal Chemistry
European Journal of Medicinal Chemistry, Chatenay-Malabry, France
Gazzetta Chimica Italiana
Societa Chemica Italiana, Rome, Italy
Helvetica Chemica Acta
Schweizerische Chemische Gesellschaft, Basel, Switzerland
Heterocycles
Pharmaceutical Institute, Tohoku University, Sendai, Japan
Hoppe-Seyler's Zeitschrift für Physiologische Chemie
Walter de Gruyter and Company, Berlin, West Germany
Il Farmaco
Casella Postale 114, Pavia, Italy
Indian Journal of Chemistry
Council of Scientific and Industrial Research, New Delhi, India
Israel Journal of Chemistry
Weizmann Science Press, Jerusalem, Israel
Izvestiya Akademii Nauk SSSR, Seriya Khimicheskaya (Engl. Translation)
Plenum Publishing Corp., New York, New York
Journal of Biological Chemistry
American Society of Biological Chemists, Baltimore, Maryland
Journal of the Chemical Society
The Chemical Society, London, England
Journal of Chromatographic Science
Preston Publications, Inc., Niles, Illinois
Journal of Chromatography
Elsevier Scientific Publishing Company, Amsterdam, The Netherlands
Journal of Heterocyclic Chemistry
Journal of Heterocyclic Chemistry, Albuquerque, New Mexico
Journal of the Indian Chemical Society
Indian Chemical Society, Calcutta, India
Journal of the Institution of Chemists, Calcutta
S.N. Mitra Institution of Chemists (India), Calcutta, India
Journal of Pharmaceutical Sciences
American Pharmaceutical Association, Washington, D.C.
Journal of Pharmacology and Experimental Therapeutics
The Williams and Wilkins Co., Baltimore, Maryland
Journal of Pharmacy and Pharmacology
Pharmaceutical Society of Great Britain, London, England
Journal für Praktische Chemie
Johann Ambrosius Barth Verlag, Leipzig, East Germany
Journal of the Scientific Research Institute, Tokyo
Tokyo Rikagaku Kenyusho, Tokyo, Japan
Justus Liebigs Annalen der Chemie
Verlag Chemie GMBH, Weinheim, West Germany
La Chemica e l'Industria
Societa Chemica Italiana, Milan, Italy
Life Sciences
Pergamon Press, Ltd., Oxford, England
Magyar Kémiai Folyóirat
Kultura, Budapest, Hungary
Memorias de la Real Academia de Ciencias Exactus Fisicas y Naturales Revista
Academia de Ciencias Exactus, Fisicas y Naturales, Madrid, Spain
Molecular Pharmacology
Academic Press, Inc., New York, New York
Monatshefte für Chemie
Springer-Verlag, Berlin, West Germany
Nippon Kagaku Zasshi
The Chemical Society of Japan, Tokyo, Japan
Proceedings of the Royal Society of London
The Royal Society, London, England
Recueil des Travaux Chimiques des Pays-Bas
The Royal Netherlands Chemical Society, The Hague, The Netherlands
Science
American Association for the Advancement of Science, Washington, D.C.
Tetrahedron
Pergamon Press Ltd., Oxford, England
Tetrahedron Letters
Pergamon Press Ltd., Oxford, England
Zeitschrift für Physikalische Chemie
Akademische Verlagsgesellschaft, Frankfurt, West Germany
Zhurnal Obshchei Khimii (Engl. Translation)
Plenum Publishing Corp., New York, New York
Zhurnal Organischeskoi (Engl. Translation)
Plenum Publishing Corp., New York, New York

The following is a list of the journals scanned in the preparation of these volumes. In most cases they were scanned from the very first volume through 1980.

Acta Chemica Scandinavica
Acta Chimica (Budapest)
Acta Pharmaceutica Suecica
Agricultural and Biological Chemistry
Annales de Chimie (Paris)
Archiv der Pharmazie und Berichte der Deutschen Pharmazeutischen Gesellschaft (Weinheim)
Arkiv foer Kemi
Australian Journal of Chemistry
British Journal of Pharmacology and Chemotherapy
Bulletin of the Chemical Society of Japan
Bulletin des Societes Chimiques Belges
Bulletin de la Societe Chimique de France
Canadian Journal of Chemistry
Chemische Berichte
Chemistry Letters
Chemical and Pharmaceutical Bulletin
Chemica Scripta
Collection of Czechoslovak Chemical Communications
Croatica Chemica Acta
Doklady Akademii Nauk SSSR
European Journal of Medicinal Chemistry - Chimica Therapeutica
Farmaco, Edizione Scientifica
Gazzetta Chimica Italiana
Helvetica Chimica Acta
Heterocycles
Hoppe-Seyler's Zeitschrift fuer Physiologische Chemie
Indian Journal of Chemistry
Inorganic Chemistry
Israel Journal of Chemistry
Izvestiya Akademii Nauk SSSR, Seriya Khimicheskaya
Journal of the American Chemical Society
Journal of the Chemical Society, London
Journal of the Chemical Society, Chemical Communications
Journal of the Chemical Society, Perkin Transactions 1
Journal of the Chemical Society, Perkin Transactions 2
Journal of Heterocyclic Chemistry
Journal of the Indian Chemical Society
Journal of Organic Chemistry
Journal of Organometallic Chemistry
Journal of Pharmacology and Experimental Therapeutics
Journal of Pharmacy and Pharmacology
Journal of Pharmaceutical Sciences
Journal fuer Praktische Chemie
Justus Liebigs Annalen der Chemie
Monatshefte fur Chemie
Nippon Kagaku Zasshi
Recueil des Travaux Chimiques des Pays-Bas
Roczniki Chemii
Tetrahedron
Tetrahedron Letters
Yakugaku Zasshi
Zhurnal Obshchei Khimii
Zhurnal Organicheskoi Khimii

Dissertation Abstracts, International

ACKNOWLEDGMENTS

I wish to express my gratitude to the following individuals: Dr. George Brumlik, an imaginative scientist who first pointed out to me the need for a compilation of optical resolutions, Dr. Ernest Braasch, Dr. Richard Corelli, Dr. Ralph Santa Cruz, Dr. Spiro Alexandratos, Dr. John Hummel, former students who, over the years since 1966, have patiently and with dedication, assisted me in the literature searching. I am indebted to Tatsuo Yamada who made the literature search in the Japanese journals and who carried out all the Japanese translations. Special mention should be made of Fernando Betancourt, a man of many talents, who did the initial layout and drawings. Finally, I wish to express my profound indebtedness to my wife, Anne, for her untiring, exacting work on all phases of the assembly of this compilation, and for her patience, sacrifice and kindness. Without her assistance, this book could never have been completed.

CONTENTS

DEDICATION

It gives me great pleasure to dedicate this volume to two men who are a source of inspiration to me and many others. The first is Professor Kurt Mislow, of Princeton University, Princeton, New Jersey, teacher and research director, who introduced me to the wonders and excitement of stereochemistry in his unique stimulating way and who is a model of what a university professor should be. May he continue to teach, inspire, and innovate.

The second is Professor Arne Fredga, of Uppsala University, Uppsala, Sweden, learned professor and kind, helpful human being, a man I have not had the privilege of meeting but who has been a source of inspiration to me because of his contributions to stereochemistry. He was of great assistance to me in the assembling of material for this volume.

INTRODUCTION

For the past twelve years, with the assistance of a number of devoted students, I have attempted to compile all of the optical resolution procedures published for both *organic* and *inorganic* compounds starting with the first resolution by Louis Pasteur in 1848. This has been accomplished by (1) a page-by-page scan of all of the pertinent chemical journals (some fifty in number),* in most cases from the very first issue to the present,** (2) an exhaustive use of *Chemical Abstracts* and *Chemical Titles* and other sources to locate resolution procedures found in less accessible journals and the patent literature, and (3) a page-by-page scan of *Dissertation Abstracts International* for pertinent theses which might yield resolution procedures. In addition, I am grateful to have received a significant number of unpublished resolution procedures and information about resolutions through the kindness of colleagues from many universities and industrial laboratories throughout the world.

The fruits of these labors will be found in this four volume compendium containing over six thousand resolution procedures of both organic and inorganic compounds divided into the following categories.

Volume 1. Amines and Related Compounds
Volume 2. Acids
Volume 3. Alcohols, Aldehydes and Ketones
Volume 4. Organometallic Compounds, Inorganic Compounds, Compounds Containing Hetero Atoms and Hydrocarbons

It is contemplated that this compilation would be useful in at least the following ways:

1) A laboratory worker who finds it necessary to carry out a resolution can, by perusing a particular volume, determine whether the resolution has already been reported and will find detailed procedures for carrying it out (in many cases with more than one resolving agent), thus saving the time that might be expended in an extensive literature search. The procedures are presented exactly as they appear in the journals since they are photocopies of the pertinent experimental section of a journal paper. In those cases where comments were made by the authors about their resolutions, these are added as notes. In the event that the resolution of a particular compound has not been reported, one can scan the appropriate volume to locate a compound or compounds having general structural similarities which have been resolved. The resolving agent used in these cases might be tried first.

* A list of the journals scanned is given on Page vii
**Literature covered through 1980

2) These volumes contain many resolutions that are located in papers, patents or theses *whose titles give no indication that a particular resolution is contained within.* In these cases, a typical literature search would be futile.
3) These volumes also contain many resolutions from relatively inaccessible journals, private communications, patents or theses, whose acquisition can be troublesome and time-consuming.
4) As of now, an optical resolution carried out in the usual manner is an empirical process. Each volume of this compilation contains a large number of compounds with the same functional group but varying, in many cases subtly, in molecular structure. The data relating molecular structure of compound resolved with resolving agent may be useful to theoreticians attempting to determine principles which will allow for a predictable choice of resolving agent for a new resolution.
5) These volumes contain the most complete collection of resolving agents and specific rotation values for resolved compounds in the chemical literature. It is also the only collection of optical resolution procedures.

Format and Scope

At the beginning of each volume there appears a compilation of resolving agents which may be used to resolve a given class of compounds, together with their structures, melting points, specific rotations, and references to methods for their preparation or resolution. In several instances, the resolution procedure contains the method for preparation of a new synthetic resolving agent.

Since the compounds are arranged in order of increasing number of carbons and hydrogens, with elements other than carbons and hydrogens arranged in alphabetical order, no index is necessary. Starting on page 819, there is found a short addendum, again arranged in order of increasing complexity of molecular formula, which contains both resolutions reported after the main body of the text had been completed, and omissions.

Editor's notes have been added where it was felt it would be useful to include additional information and comments. Where available, information concerning the optical purity of the resolved compound and the method used for its determination is given.

Resolution procedures originally published in English, German, French, Italian and Spanish are reproduced as they appear in the literature. However, procedures in all other languages have been translated into English.

These volumes mainly contain resolution procedures which involve the formation of diastereomeric derivatives of the original mixture of enantiomers in solution, followed by their separation by fractional crystallization and isolation of the separated enantiomers. Also included, however, are a large number of examples of various types of chromatographic procedures (Section 3), examples of resolution by entrainment and enzymatic resolution procedures (Section 4). In addition, because of the increasing importance of asymmetric synthesis which produce a given enantiomer in high enantiomeric excess, it was felt that a section devoted to a survey of practical asymmetric hydrogenation procedures for the preparation of chiral acids would be useful (Section 6). This section has been prepared by Professor Spiro Alexandratos of the Chemistry Department of the University of Tennessee. Finally, there will be found, in Section 5, a summary of literature references to methods for determining optical purities of acids. In addition, there are some examples of partial resolutions when this is the only example of a resolution of a compound, or when this is the only example of the use of a specific resolving agent. Finally, there are reports of unsuccessful resolutions. The aim of these volumes is to make available to the reader examples of all practical methods for producing chiral molecules of high optical purity on a preparative scale. In many instances, the resolution of a given compound has been reported by more than one investigator, using the same resolving agent under similar conditions. In these cases, the procedure which gave the highest specific rotation of the resolved enantiomer, is given.

This compilation has been prepared mainly* by photocopying original sources from the chemical literature, patents or theses, and is published by a photo-offset process to insure maximum accuracy. Every effort has been made to make the photocopies as readable as possible, despite the limitations of age of journal and the quality of reproducing facilities.

I intend to update this compendium every several years by publishing supplements based upon a continuing page-by-page scan of the pertinent chemical literature and the receipt of personal communications which are most welcome.

As a result of current work, most notably by Professor G. Helmchen, Institut fur Organische Chemie, Biochemie und Isotopenforschung of the University of Stuttgart, West Germany; and Professor W. Pirkle, Department of Chemistry, University of Illinois, Urbana, U.S.A., the time is near when one will be able to carry out many resolutions by high performance liquid chromatography of diastereomers obtained using rationally designed and selected chiral derivatizing reagents. These methods will allow one to obtain both enantiomers in high optical purity, and in high yield, on a preparative scale. In addition, with the evolution of models which can rationalize differences in the behavior of diastereomers during liquid chromatography, one should be able to predict the elution order of the diastereomers and hence determine the absolute configuration of the finally obtained enantiomers. This is a most remarkable and exciting development in a laboratory procedure that, as usually practiced, has changed little in 130 years. Some examples of the use of this method will be found in these volumes. As additional details of these procedures become available, they shall be included in future supplements.

*In those cases where the quality of print of primary source was poor, the material was typed.

SECTION 1

RESOLVING AGENTS FOR ACIDS

This section contains compounds reported to have been used as resolving agents for acids together with their melting or boiling points, specific rotations and references to their preparation and/or resolution.

RESOLVING AGENTS FOR ACIDS

N-Acetyl-3,5-dibromo-L-Tyrosine (for amino acids)

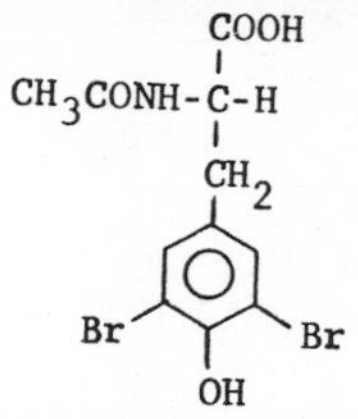

Melting point: 118-120° (dec.) (hemihydrate)
Specific rotation: $[\alpha]_D^{25}$ +34.5° (c 4, methanol)

Reference
F.J. Kearley and A.W. Ingersoll, J. Am. Chem. Soc., 73, 5783 (1951).
H.D. DeWitt and A.W. Ingersoll, J. Am. Chem. Soc., 73, 5782 (1951).

2,3,4,6-tetra-O-Acetyl-β-glucopyranosyl isothiocyanate (for the resolution of amino acids by HPLC)

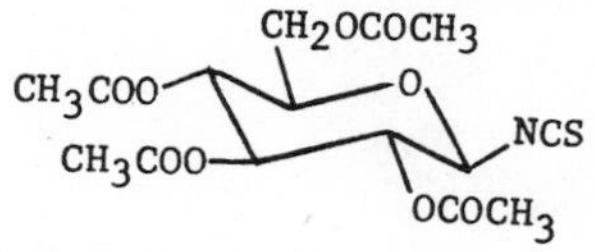

Melting point: 113-115°.

Reference
G. Jager and R. Geiger, Liebigs Ann. Chem., 1535 (1973).
A.M. Kolodziejczyk and A. Arendt, Pol. J. Chem., 53, 1017 (1979).

2-Aminobutane

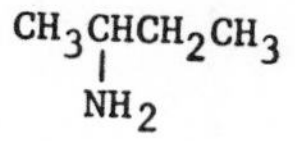

Boiling point: 63°
Specific rotation: $[\alpha]_D^{20}$ 7.48 (neat)*

Reference
L.G. Thomé, Chem. Ber., 36, 582 (1903)
*See also Volume 1, Ref. 12, page 35 of this series.
Available from Chemical Dynamics Corp., South Plainfield, New Jersey 07080

2-Amino-1-butanol

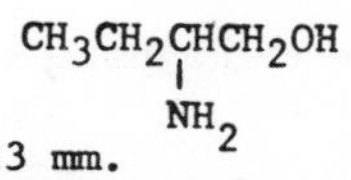

Boiling point: 81°/13 mm.
Specific rotation: $[\alpha]_D^{20}$ -10.1° (neat); $[\alpha]_D^{20}$ 10.1° (neat) — 99.4% optically pure

Reference
D. Pitré and E.B. Grabitz, Chimia, 23, 399 (1969).
Available from Chemical Dynamics Corp., South Plainfield, New Jersey 07080

1-Aminoethyl-4-dimethylaminonaphthalene (for the resolution of carboxylic acids by HPLC)

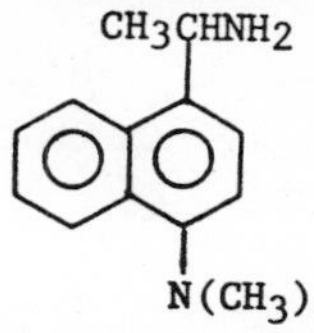

Melting point: 240-241° (decomp.) (for the hydrochloride)
Specific rotation: $[\alpha]_D^{15}$ -17.7° (C, 0.37, CH_3OH)
$[\alpha]_D^{15}$ 17.3° (C, 1.4, CH_3OH)

Reference
J. Goto, N. Goto, A. Hikichi, T. Nishimaki and T. Nambara, Anal. Chim. Acta, 120, 187 (1980).

trans-2-Amino-1-indanol

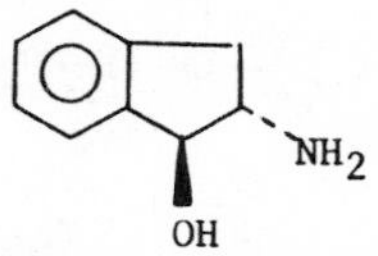

Melting point: 206-209° (as hydrochloride)
Specific rotation: $[\alpha]_D^{20}$ -13.4° (c 0.75, H_2O) [as hydrochloride]

Reference
E. Dornhege, Liebigs Ann., 743, 42 (1971).

threo-2-Amino-1-(4-methylmercaptophenyl)-1,3-propanediol

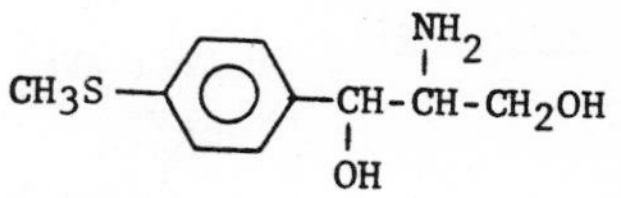

Melting point: 151.9-152.9°
Specific rotation: $[\alpha]_D^{25}$ -21° (c 1, ethanol)
$[\alpha]_D^{25}$ +21° (c 1, ethanol)

Reference
R.A. Cutler, R.J. Stenger and C.M. Suter, J. Am. Chem. Soc., 74, 5475 (1952).

3-Aminomethylpinane

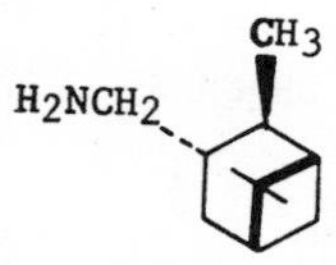

Boiling point: 110°/20 mm.
Specific rotation: $[\alpha]_D^{25}$ 44.5° (c 4, methanol) As hydrochloride

Reference
J. Paust, S. Pfohl, W. Reif and W. Schmidt, Liebigs Ann. Chem., 1024 (1978).
W. Himmele, H. Siegel, S. Pfohl, J. Paust, W. Hoffman and K. von Fraunberg, German Patent 2404306 (1975), U.S. Patent 4081477 (1978).

RESOLVING AGENTS (continued)

threo-2-Amino-1-(4-methylsulfonylphenyl)-1,3-propanediol

$$CH_3SO_2-\text{C}_6\text{H}_4-\underset{OH}{\underset{|}{CH}}-\underset{NH_2}{\underset{|}{CH}}-CH_2OH$$

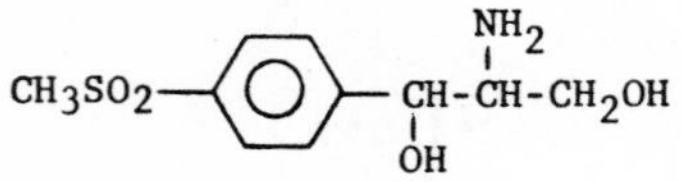

Melting point: 141.4-142.6°
Specific rotation: $[\alpha]_D^{25}$ -19.8° (c 1, 95% ethanol)
$[\alpha]_D^{25}$ +19.7° (c 1, 95% ethanol)

Reference
R.A. Cutler, R.J. Stenger and C.M. Suter, J. Am. Chem. Soc., <u>74</u>, 5475 (1952).

threo-2-Amino-1-phenyl-1,3-propanediol

$$\text{C}_6\text{H}_5-\underset{NH_2}{\underset{|}{\underset{OH}{\underset{|}{CH}}-CH}}-CH_2OH$$

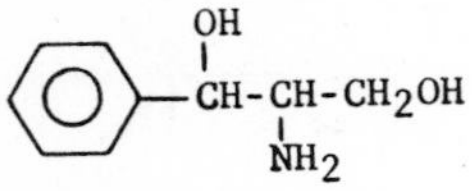

Melting point: 115-117°
Specific rotation: $[\alpha]_D^{27}$ 18° (c 2, H_2O)
$[\alpha]_D^{27}$ -18° (c 2, H_2O)

Reference
M. Nagawa and Y. Murase, Takamine Kenkyujo Nempo, <u>8</u>, 15 (1956), Chem. Ab., <u>52</u>, 308^d (1958).
V. D'Amato and R. Pagani, Brit. Patent 738,064 (1955).

2-Amino-1-propanol

$$CH_3\underset{NH_2}{\underset{|}{CH}}CH_2OH$$

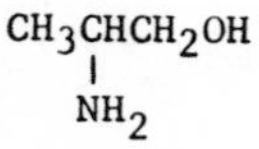

Boiling point: 72-73°/11 mm.
Specific rotation: $[\alpha]_D^{20}$ 15.8° (neat)

Reference
A. Stoll, J. Peyer and A. Hofmann, Helv. Chim. Acta, <u>26</u>, 929 (1943).
Available from Chemical Dynamics Corp., South Plainfield, New Jersey 07080

Amphetamine

$$\text{C}_6\text{H}_5-CH_2\underset{CH_3}{\underset{|}{CH}}NH_2$$

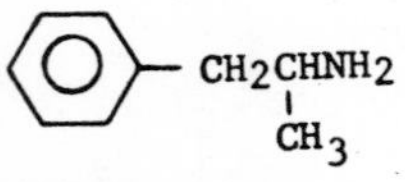

Boiling point: 80°/10 mm.
Specific Rotation: $[\alpha]_D^{15}$ 36.2 (c 10.62, ethanol)

Hydrochloride -
  Melting point: 156°
  Specific rotation: $[\alpha]_D^{15}$ 24.8° (c 9.00, H_2O)

Reference
W. Leithe, Chem. Ber., <u>65</u>, 660 (1932).

2-(Anilinomethyl)-pyrrolidine

$$\underset{H}{\underset{|}{N}}\text{-pyrrolidine-}CH_2NHC_6H_5$$

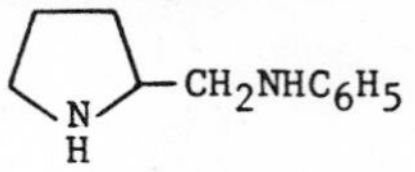

Boiling point: 113°/0.4 mm.
Specific rotation: $[\alpha]_D^{24}$ -19.4° (c 1, ethanol)

Reference
Japanese Patent 5 4079-269 (1979) Sagami Chemical Research Center

Note: Above prepared as shown by the following:

To a mixt. of 5-oxo-2-pyrrolidinecarboxanilide (1.02 g) and dry THF (25 ml) was gradually added LiAlH$_4$ (475 mg) under ice water-cooling. The resultant mixt. was refluxed at 75-80°C (bath temp.) for 1.5 hrs. After cooling, the reaction mixt. was mixed gradually with water and then 2 N NaOH (7 ml) and extracted with methylene chloride. Organic layer was dried, evaporated and distilled to give 2-(anilinomethyl)-pyrrolidine (690 mg) b.pt. 113°C/0.4 mm Hg. $[\alpha]_D^{24}$ -19.4° (c =1, ethanol). Yield = 78%.

L-Arginine (for amino acids)

$$\begin{array}{c} CO_2H \\ | \\ NH_2-C-H \\ | \\ (CH_2)_3 \\ | \\ NH \\ | \\ NH_2-C=NH \end{array}$$

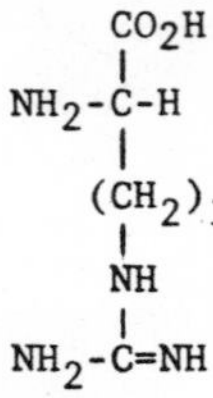

Melting point: 217-218°
Specific rotation: $[\alpha]_D^{25}$ 12.5 (c 2, H_2O)

Reference
S.M. Birnbaum, M. Winitz and J.P. Greenstein, Arch. Biochem. Biophys., <u>60</u>, 496 (1956).
S.M. Birnbaum and J.P. Greenstein, Arch. Biochem. Biophys., <u>39</u>, 108 (1952).

2-Benzylamino-1-butanol

$$CH_3CH_2-\underset{NHCH_2C_6H_5}{\underset{|}{CH}}-CH_2OH$$

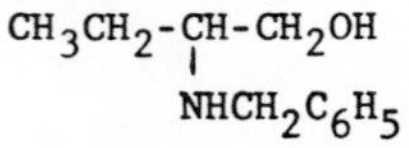

Melting point: 76-78°
Specific rotation: $[\alpha]_D^{23}$ 25.5° (c 4.0, ethanol)
$[\alpha]_D^{21}$ -25.0° (c 4.0, ethanol)

Reference
A. Stoll, J. Peyer and A. Hofmann, Helv. Chim. Acta, <u>26</u>, 929 (1943).
Available from Chemical Dynamics Corp., South Plainfield, New Jersey 07080

2-Benzylamino-1-propanol

$$\text{C}_6\text{H}_5-CH_2NH\underset{CH_2OH}{\underset{|}{CH}}CH_3$$

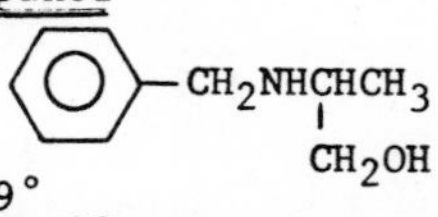

Melting point: 47-49°
Specific rotation: $[\alpha]_D^{22}$ -44.25° (c 4.0, ethanol)
$[\alpha]_D^{23}$ +44 (c 4.0, ethanol)

Reference
A. Stoll, J. Peyer and A. Hofmann, Helv. Chim. Acta, <u>26</u>, 929 (1943).

RESOLVING AGENTS (continued)

cis-N-Benzyl-2-(hydroxymethyl)cyclohexylamine

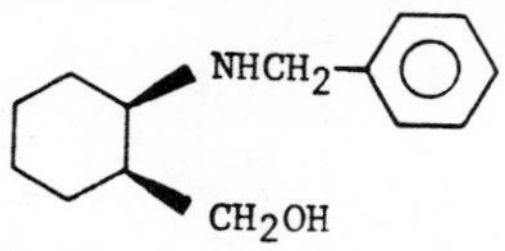

Melting point: 68°
Specific rotation: $[\alpha]_D^{25}$ 40.6 (c 1, ether)
$[\alpha]_D^{25}$ -40.6 (c 1, ether)

Reference
J. Nishikawa, T. Ishizaki, F. Nakayama, H. Kawa, K. Saigo and H. Nohira, Nippon Kagaku Kaishi, 754 (1979).

N-Benzyl-α-phenylethylamine

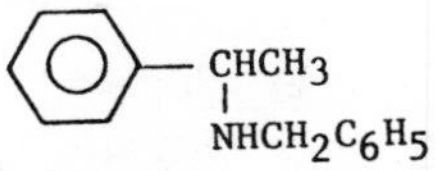

Boiling point: 120°/1 mm.
Melting point: 177° (as hydrochloride)
Specific rotation: $[\alpha]_D^{20}$ 58.6° (C 1, 96% ethanol)

Reference
(Rotation data kindly supplied by Dr. J.P.M. Houbiers, Océ-Andeno B.V., Venlo, Holland)
Chemical Dynamics Corp., South Plainfield, N.J. 07080

Binaphthylphosphoric acid (for amino acids)

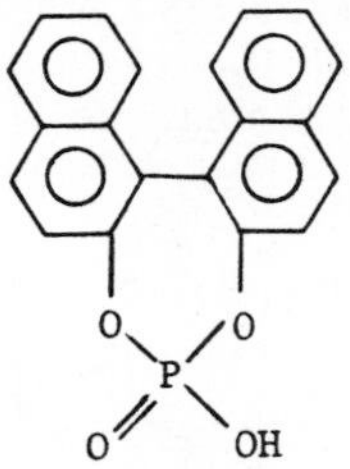

Melting point: 232-234°
Specific rotation: $[\alpha]_D^{18}$ -580° (CH_3OH)
$[\alpha]_D^{22}$ +480° (c 0.3, $CHCl_3$)

Reference
R. Viterbo and J. Jacques, British patent 1,360,946 (1974).

endo-Bornylamine

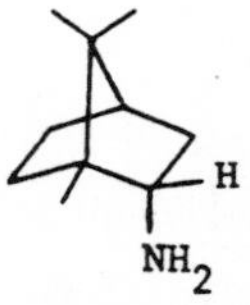

Melting point: 330° (as hydrochloride)
Specific rotation: $[\alpha]_D$ 23° (c 4.4, ethanol)

Reference
Personal communication from Prof. L.A. Paquette, Ohio State University, Columbus, Ohio.

3-Bromocamphor-9-sulfonic acid (for amino acids)

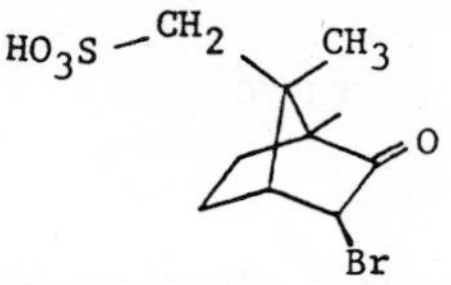

Melting point: 195-196°*
Specific rotation: $[\alpha]_D^{14}$ +88.3° (c 2.577, water)*
*See, however, F.S. Kipping and W.J. Pope, J. Chem. Soc., 63, 548 (1893)

Ammonium salt

Melting point: 284-285°
Specific rotation: $[\alpha]_D^{22}$ +85.25° (c 4.0878, water)
$[\alpha]_D^{22}$ -84.58° (c 4.0260, water)

The free acid may also be obtained from the above salt by passing a solution of the salt through Amberlite IR-120 (H+) and eluting with water [T. Kunieda, K. Koga and S. Yamada, Chem. Pharm. Bull., 15, 337 (1967)].

Reference
W.J. Pope and A.W. Harvey, J. Chem. Soc., 79, 74 (1901).
W.J. Pope and J. Read, J. Chem. Soc., 97, 2199 (1910).
H. Regler and F. Hein, J. Prakt. Chem., 148, 1 (1937).
A.W. Ingersoll and S.H. Babcock, J. Am. Chem. Soc., 55, 341 (1933).

Brucine

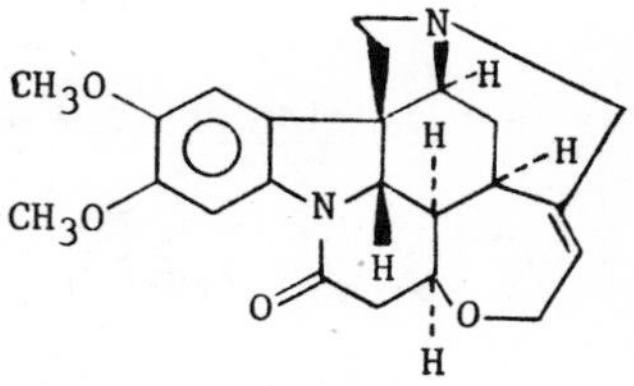

Melting point: 177-178° (anhydrous)
Specific rotation: $[\alpha]_D^{20}$ -121° (c 2.5, $CHCl_3$)
$[\alpha]_{5461}^{17.3}$ -98.7° (c 3, ethanol)

Reference
T.S. Patterson and J.D. Fulton, J. Chem. Soc., 51 (1927).
T.P. Hilditch, J. Chem. Soc., 101, 198 (1912).

cis-Camphoric acid (for amino acids)

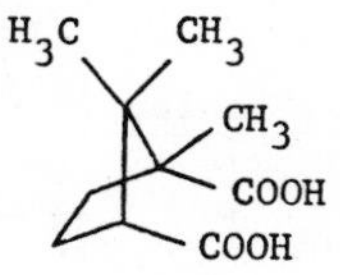

Melting point: 188°
Specific rotation: $[\alpha]_D^{16}$ -48.1° (c 8.24, ethanol)
$[\alpha]_D^{17}$ +47.6° (c 8.24, ethanol)

Reference
J.D.M. Ross and I.C. Sommerville, J. Chem. Soc., 2770 (1926).
C.P. Berg, J. Biol. Chem., 115, 1 (1936).
L. Tschugaeff and W. Budrick, Justus Liebigs Ann. Chem., 388, 280 (1912).

RESOLVING AGENTS (continued)

Camphor-10-sulfonic acid (for amino acids)

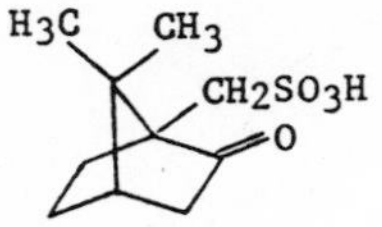

Melting point: 197.4-198.0 (dec.)
Specific rotation: $[\alpha]_D^{20}$ +24.0° (c 5, water)
$[\alpha]_D^{16}$ +20.5° (c 4.5, water) as ammonium salt

For a recently reported resolution of the racemic acid using D- and L-carnitinnitrile as resolving agent, which allows one to obtain both camphor-10-sulfonic acid enantiomers quantitatively and optically pure, and also allows one to recover the resolving agents quantitatively, see D. M. Muller, E. Strack and I. Lorenz, J. Prakt. Chem., 317, 689 (1975). Previously reported resolutions of the acid may be found in the following references: B. Rewald, Chem. Ber., 42, 3136 (1909); H. Burgess and C. S. Gibson, J. Soc. Chem. Ind., 44T, 496 (1925); K. L. Marsi, C. A. VanderWerf and W. E. McEwen, J. Am. Chem. Soc., 78, 3063 (1956); C. Szantay, L. Novak and I. Jelinek, Hung. Pat. 3415, Chem. Ab., 76, 153972u (1972); W. E. Thompson and A. Pohland, U. S. Pat. 2230838, Chem. Ab., 78, 97833r (1973); T. Doi, T. Yushi, K. Hiraoka, and N. Murakami, Jap. Pat. 7343730, Chem. Ab., 81, 37676y (1974).

Reference
T. Hilditch, J. Chem. Soc., 101, 198 (1912).
W. J. Pope and C. S. Gibson, J. Chem. Soc., 97, 2214 (1910).

Carbobenzoxy-L-phenylalanine (for amino acids)

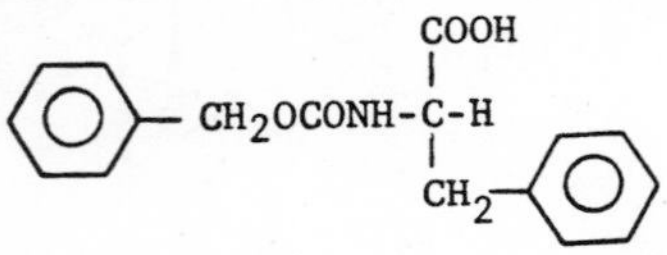

Melting point: 88-89°
Specific rotation: $[\alpha]_D^{20}$ 5.1° (c 2, ethanol)
$[\alpha]_D^{24}$ -4.6° (c 4, acetic acid)

Reference
W. Grassman and E. Wünsch, Chem. Ber., 91, 449, 462 (1958).
W. Grassman, E. Wünsch and A. Riedel, ibid., 91, 455 (1958).
C.S. Smith and A.W. Brown, J. Am. Chem. Soc., 63, 2605 2605 (1941).

Carvomenthoxyacetyl chloride (for amino acids)

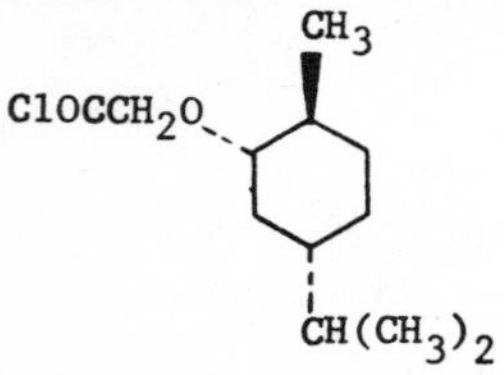

Boiling point: 138-140°/14 mm.
Specific rotation: $[\alpha]_D^{20}$ -108.3 (neat)

Reference
K. Witkiewicz, F. Rulko and Z. Chabudzinski, Pol. J. Chem., 48, 651 (1974).

Δ^4-Cholestene-3-one-6-sulfonic acid (for amino acids)

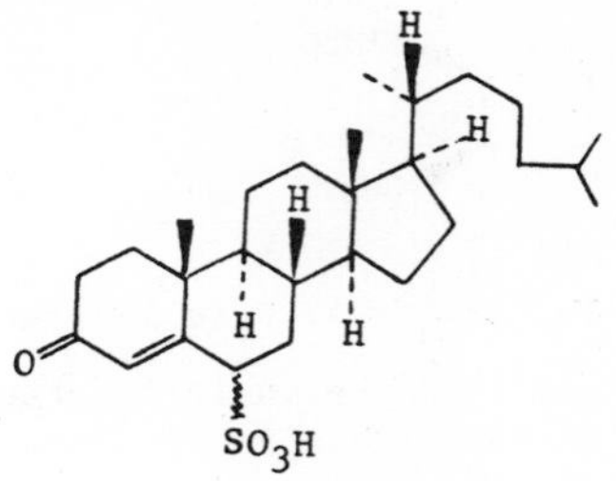

Melting point: 193-195° (decomp.)

Reference
A. Windaus and E. Kuhr, Liebigs Ann., 532, 57 (1937).

Cinchonidine

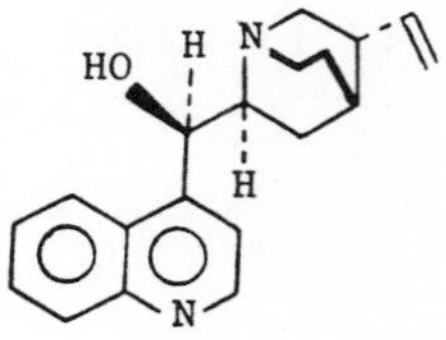

Melting point: 205-206°
Specific rotation: $[\alpha]_D^{17}$ -109.6° (c 1.54, ethanol) [anhydrous compound]

Reference
A.C. Oudemans, Jr., Liebigs Ann., 182, 33 (1876).
O. Hesse, Liebigs Ann., 176, 219 (1875).

Cinchonidine methohydroxide

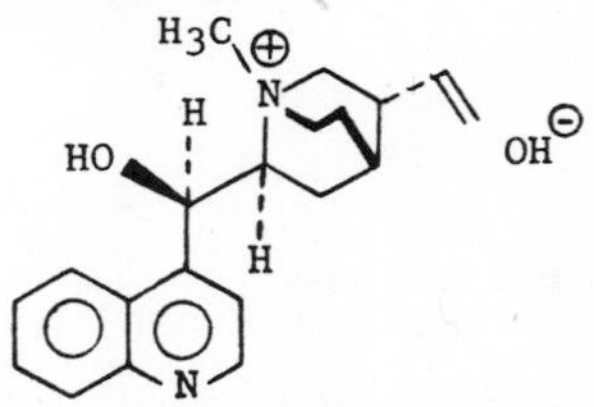

For method of preparation of this compound, see resolving agent quinidine methohydroxide.

Cinchonine

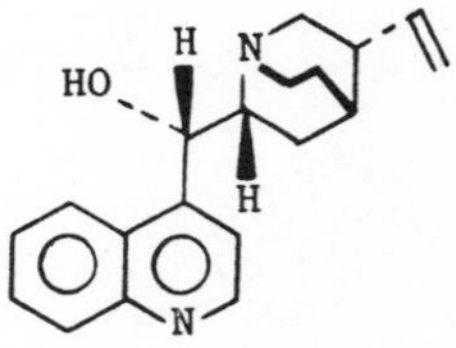

Melting point: 268.8° (corr.)
Specific rotation: $[\alpha]_D^{17}$ 223.3° (c 0.5, ethanol) [anhydrous compound]

Reference
A.C. Oudemans, Jr., Liebigs Ann., 182, 33 (1876).

Cinchonine methohydroxide

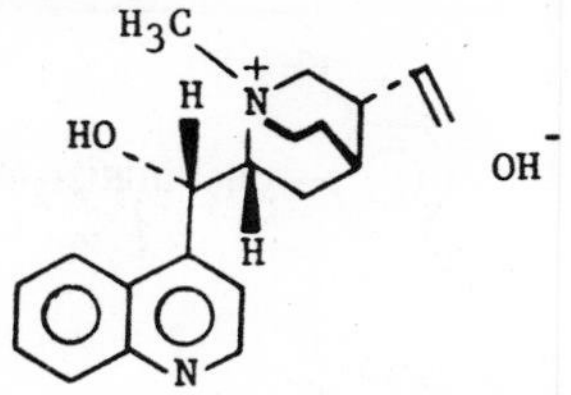

For method of preparation of this compound, see resolving agent quinidine methohydroxide.

Cinchotoxine (cinchonicine)

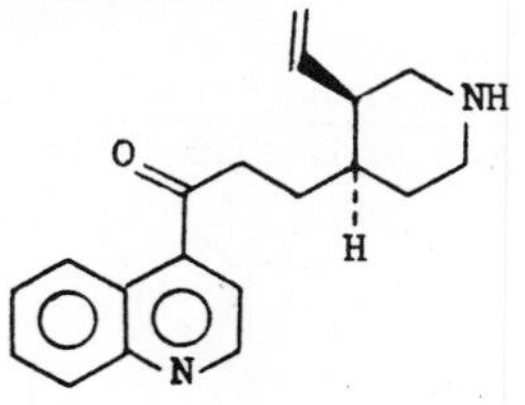

Melting point: 58-59°
Specific rotation: $[\alpha]_D^{15}$ 48° (c 1, 95% ethanol)
$[\alpha]_D^{14}$ 57.7° (c 1, ethanol)

Reference
O. Hesse, Liebigs Ann., 178, 253 (1875).

Codeine

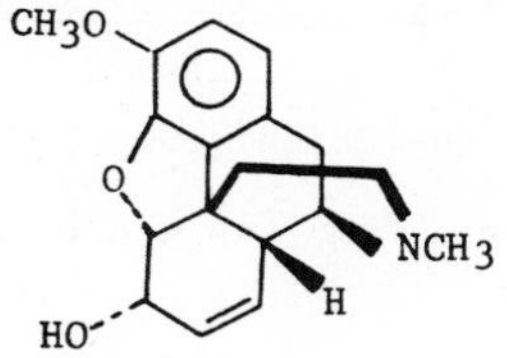

Melting point: 157-158.5° (corr.) monohydrate
Specific rotation: $[\alpha]_D^{25}$ -136° (c 2.8, ethanol) as monohydrate

Reference
M. Gates, J. Am. Chem. Soc., 75, 4340 (1953)

Coniine

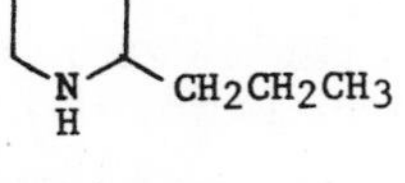

Boiling point: 65-66°/20 mm.
Specific rotation: $[\alpha]_D^{25}$ 8.4° (c 4.0, $CHCl_3$)

Reference
J.C. Craig and A.R. Pinder, J. Org. Chem., 36, 3648 (1971).

trans-1,2-Cyclohexandicarboxylic anhydride

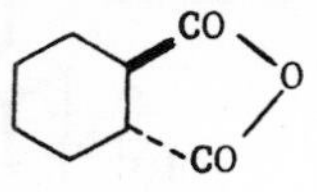

Melting point: 159-164°
Specific rotation: $[\alpha]_{589}^{25}$ + and -87.0° (c 1, dioxane)

Reference
K. Murakami, N. Katsuta, K. Takano, Y. Yamamoto, T. Kakegawa, K. Saigo and H. Nohira, Nippon Kagaku Kaishi, 765 (1979).

N-Cyclohexyl-10-amino-α-pinene

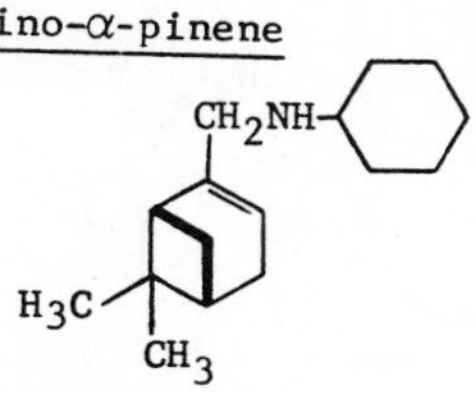

Boiling point: 108°/0.4 Torr.
Specific rotation: $[\alpha]_D^{20}$ -17.4° (neat)

Reference
S.W. Markowicz, Pol. J. Chem., 53, 157 (1979).

Dehydroabietylamine

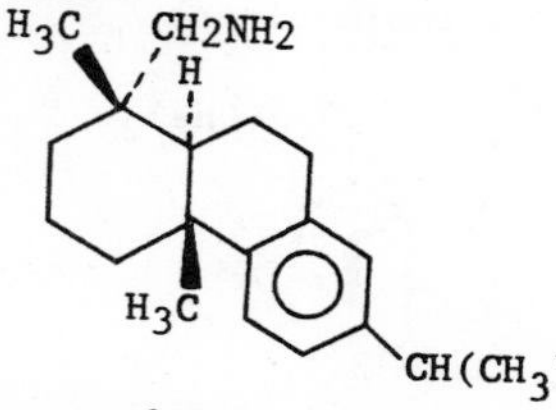

Melting point: 44-45° (free base)
144-146° (as acetate)
Specific rotation: $[\alpha]_D^{20}$ 46° (c 4, CH_3OH) as free base
$[\alpha]_D^{23}$ 31.9° (c 6.4 CH_3OH) as acetate

Reference
W.J. Gottstein and L.C. Cheney, J. Org. Chem., 30, 2072 (1965).
G. Bellucci, G. Berti, A. Borraccini and F. Macchia, Tetrahedron 25, 2979 (1969).
E.J. Corey and S. Hashimoto, Tetrahedron Lett., 22, 299 (1981).

Deoxyephedrine

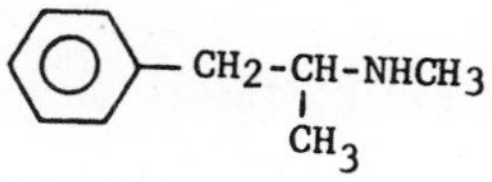

Boiling point: 68°/0.3 mm (Kugelrohr)
Specific rotation: $[\alpha]_D^{26}$ 2.24 ± 0.02 (neat, l 1)

(Deoxyephedrine hydrochloride: melting point 175.5-177°; specific rotation: $[\alpha]_D^{26}$ 17.9 ± 0.1° (c 1.086, H_2O)

Reference
J. Jacobus and T.B. Jones, J. Am. Chem. Soc., 92, 4583 (1970).

RESOLVING AGENTS (continued)

N,N-Diethyl-10-amino-α-pinene

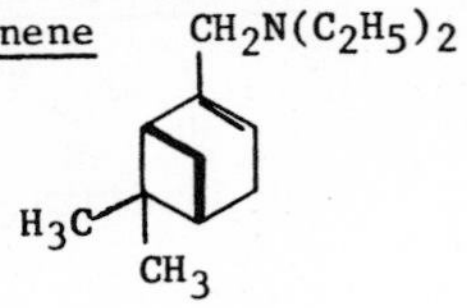

Boiling point: 65-6°/1.5 Torr.
Specific rotation: $[\alpha]_D^{20}$ -21.4 (neat)

Reference
S.W. Markowicz, Pol. J. Chem., 53, 157 (1979).

3-Dimethylaminomethylpinane

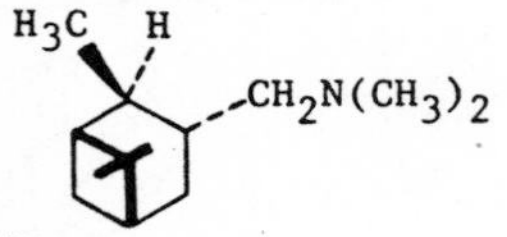

Boiling point: 93-95°/4 mm.
Melting point: 239° (as hydrochloride)
Specific rotation: $[\alpha]_D^{23}$ 51.5° (c 1, methanol)

Reference
W. Hoffman, W. Himmele, J. Paust, K. von Fraunberg, H. Siegel and S. Pfohl, U.S. Patent 4,081,477 (1978).

N,N-Dimethyl-10-amino-α-pinene

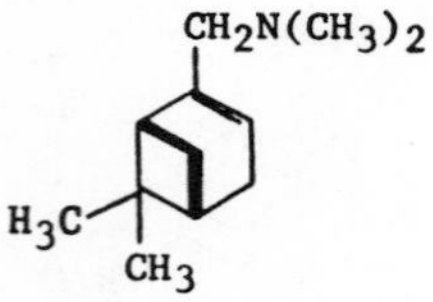

Boiling point: 40-1°/0.6 Torr.
Specific rotation: $[\alpha]_D^{20}$ -78.4° (neat)

Reference
S.W. Markowicz, Pol. J. Chem., 53, 157 (1979).

1-Dimethylamino-2-propanol

$CH_3\text{-}CH(OH)\text{-}CH_2N(CH_3)_2$

Boiling point: 124.5-125.5°
Specific rotation: $[\alpha]_D^{20}$ 16.7° (C 2, H_2O)

Reference
a) T. Dohi, T. Yu and F. Inoue, Japanese Patent 74-19252
b) Chemical Dynamics Corp., South Plainfield, N.J. 07080

2-(2,5-Dimethylbenzylamino)-1-butanol

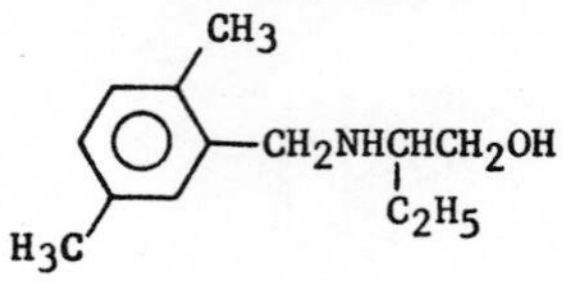

Melting point: 85-88°
Specific rotation: $[\alpha]_D^{25}$ -28.0° (c 2.5, CH_3OH)

Reference
I.A. Halmos, U.S. Patent 4,151,198 (1979) (Resolution of N-Acyl-DL-phenylalanines).

(-)-1,7-Dimethyl-7-norbornyl isothiocyanate
(for resolution of α-amino acid by HPLC)

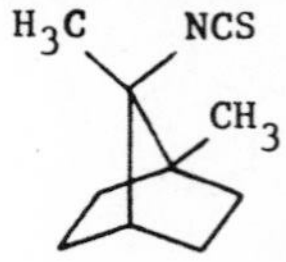

Boiling point: 94°C/5 mm.
Specific rotation: $[\alpha]_D^{20}$ -69.9° (c 1,3, $CHCl_3$)

Reference
T. Nambara, S. Ikegawa, M. Hasegawa and J. Goto, Anal. Chim. Acta, 101, 111 (1978).

cis-Dinitrobis(ethylenediamine)cobaltic bromide (for α-amino acids)

cis-$[Co(en)_2(NO_2)_2]Br$

Specific rotation: $[\alpha]_D$ -44° (c 0.5, H_2O)

Reference
F.P. Dwyer and F.L. Garvan, Inorg. Synth., 6, 195 (1960).
F.P. Dwyer, B. Halpern and K.R. Turnbull, Austr. J. Chem., 16, 510 (1963).

1,2-Diphenyl-1-aminoethane

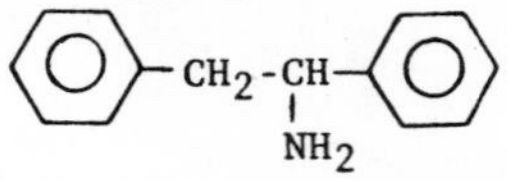

Boiling point: 120-123°/0.3 mm
Specific rotation: $[\alpha]_D^{23}$ 13.5° (neat)

Reference
Information kindly supplied by Dr. Yoshio Suzuki, Sumitomo Chemical Company, Ltd., Institute for Biological Science, Hyogo-Ken, 665, Japan.

threo-1,2-Diphenyl-1,2-diaminoethane

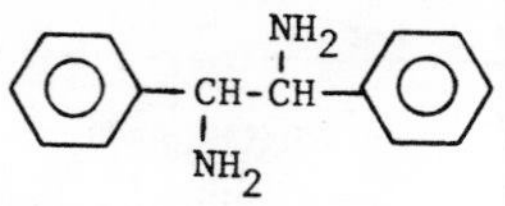

Melting point: 83°
Specific rotation: $[\alpha]^{25}_{547}$ -123.67° (C 1.0, CH_3OH)
$[\alpha]^{25}_{547}$ 123.85° (C 1.0, CH_3OH)
$[\alpha]^{20}_{D}$ -108° (CH_3OH)

Reference
F. Vögtle and E. Goldschmitt, Chem. Ber., 109, 1 (1976).
I. Lifschitz and J.G. Bos, Rec. Trav. Chim. Pays-Bas, 59, 173 (1940).

erythro-α,β-Diphenyl-β-hydroxyethylamine

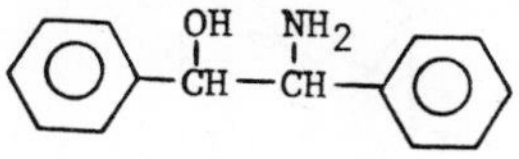

Melting point: 143°
Specific rotation: $[\alpha]^{25}_{D}$ -10.1° (c 0.59, ethanol)
$[\alpha]^{25}_{D}$ 10.2° (c 0.61, ethanol)

Reference
J. Weijlard, R. Pfister,3d, E.F. Swanezy, C.A. Robinson and M. Tishler, J. Am. Chem. Soc., 73, 1216 (1951).

Ephedrine

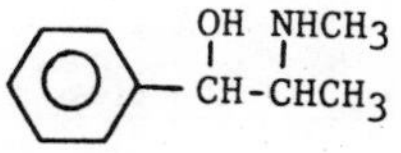

Melting point: 40.0-40.5° (free base)
218-218.5° (hydrochloride)
Specific rotation: $[\alpha]^{24}_{D}$ 35.6° (C, 2, H_2O)
$[\alpha]^{24}_{D}$ -35.5° (C, 2, H_2O) as hydrochloride

Reference
R.H.F. Manske and T.B. Johnson, J. Am. Chem. Soc., 51, 1906 (1929).

Ethyl α-aminophenylacetate

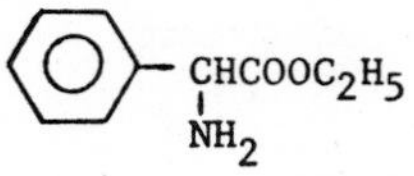

Melting point: 246° (decomp.) (as hydrochloride)
Specific rotation: $[\alpha]^{20}_{D}$ -91° (C, 1, H_2O)
(92% optical purity)

Reference
K. Naumann, German Patent 2,826,952 (1979).

N-Ethyl-10-amino-α-pinene hydrochloride

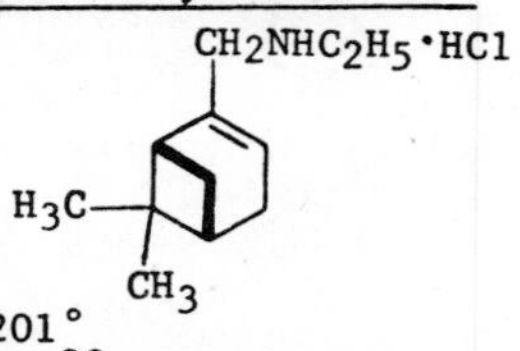

Melting point: 198-201°
Specific rotation: $[\alpha]^{20}_{D}$ -24.7° (c 0.65, $CHCl_3$)

Reference
S.W. Markowicz, Pol. J. Chem., 53, 157 (1979).

Ethylhydrocupreidine

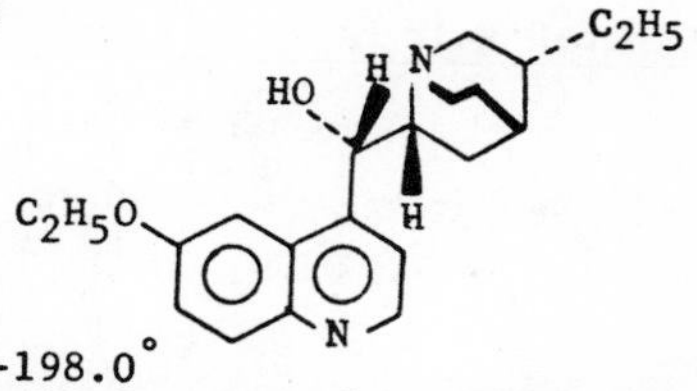

Melting point: 197.5-198.0°
Specific rotation: $[\alpha]^{23.5}_{D}$ +212.8° (c 1.008, ethanol)

Reference
M. Heidelberger and W. A. Jacobs, J. Am. Chem. Soc., 41, 817 (1919).
H. King and A.D. Palmer, J. Chem. Soc., 121, 2577 (1922).

α-Fenchylamine

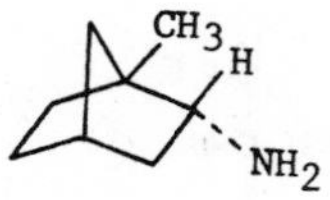

Boiling point: 189-190°
Specific rotation: $[\alpha]^{25}_{D}$ 25.5° (c 5, 95% ethanol)
$[\alpha]^{25}_{D}$ -25.4° (c 4, 95% ethanol)

Reference
L.R. Overby and A.W. Ingersoll, J. Am. Chem. Soc., 73, 3363 (1951).
W.A.H. Huffman and A.W. Ingersoll, J. Am. Chem. Soc., 73, 3366 (1951).

Galactamine

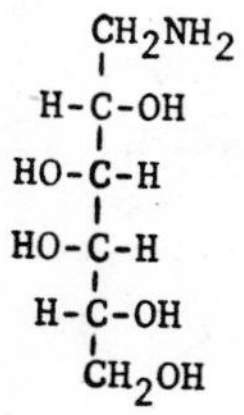

Melting point: 136°
Specific rotation: $[\alpha]^{17}_{D}$ -2.77° (C 8.992, H_2O)

Reference
F. Kagan, M.A. Rebenstorf and R.V. Heinzelman, J. Am. Chem. Soc., 79, 3541 (1957).
M. Roux, Ann. Chim. Phys., [8], 1, 77 (1904).

Glucamine

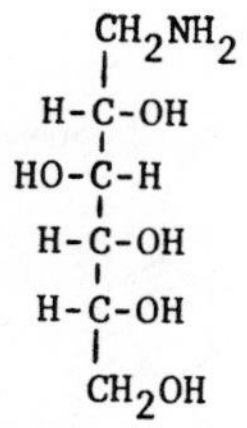

Melting point: 127-128°
Specific rotation: $[\alpha]^{25}_{D}$ -7.5° (c 2.7, H_2O)

Reference
W. Wayne and H. Adkins, 62, 3314 (1940).
F. Kagan, M.A. Rebenstorf, and R.V. Heinzelman, J. Am. Chem. Soc., 79, 3541 (1957).

RESOLVING AGENTS (continued)

2-(D-Gluco-D-gulo-hepto-hexahydroxyhexyl)-benzimidazole

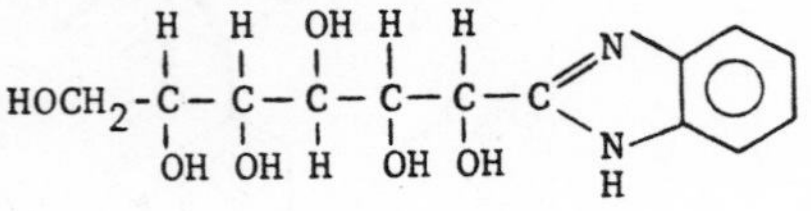

Melting point: 215° (decomp.)
Specific rotation: $[\alpha]_D^{20}$ 14.3 (c 2, 1N HCl)

Reference
W.T. Haskins and C.S. Hudson, J. Am. Chem. Soc., 61, 1266 (1939).
A.F. Holleman, Organic Synthesis, Coll. Vol. 1, page 497 (1941).

L-Glutamic acid (for amino acids)

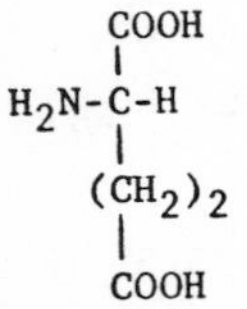

Melting point: 211-213° (dec.)
Specific rotation: $[\alpha]_D^{25}$ 26.0 (c 2, H_2O)

Reference
"Chemistry of The Amino Acids," Volume 3, J.P. Greenstein and M. Winitz, John Wiley and Sons, Inc. (1961).

L-Histidine (for amino acids)

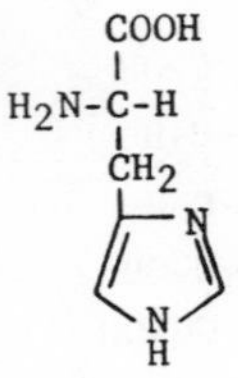

Melting point: 287° (decomp.)
Specific rotation: $[\alpha]_D^{25}$ -38.5 (c 2, H_2O)

Reference
"Chemistry of The Amino Acids," Volume 3, J.P. Greenstein and M. Winitz, John Wiley & Sons, Inc., (1961).

Hydrocinchonidine

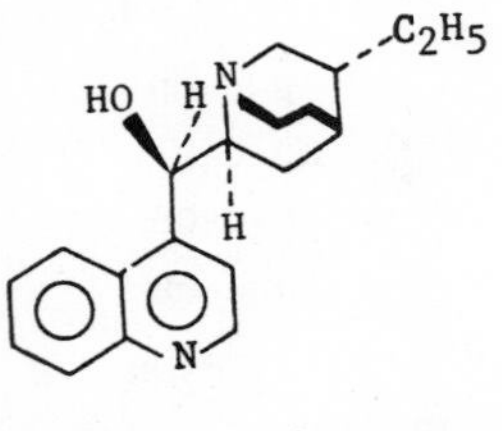

Melting point: 237°
Specific rotation: $[\alpha]_D^{20}$ -95.8 (c 0.3770, 99% ethanol)

Reference
H. Emde, Helv. Chim. Acta, 15, 557 (1932).

1-Hydroxy-2-aminopropane

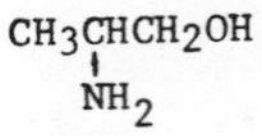

Melting point: 110.3°
Specific rotation: $[\alpha]_D^{25}$ -46.2° (c 2.21, benzene)

Reference
Information kindly supplied by Dr. Tsuneyuki Nagase, Central Research Laboratory, Sumitomo Chemical Co. Ltd., Takatsuki City, Osaka, Japan.

Hydroxymethylenecamphor (for amino acids)

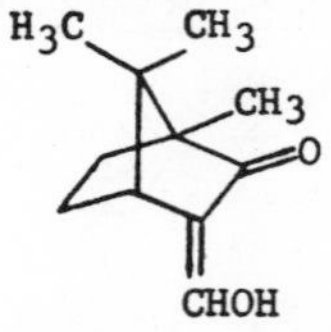

Melting point: 81-82°
Specific rotation: $[\alpha]_D^{20}$ 198° (c 1.009, ethanol)

Reference
W. J. Pope and J. Read, J. Chem. Soc., 103, 444 (1913).

Isobornyl chloroformate (for amino acids)

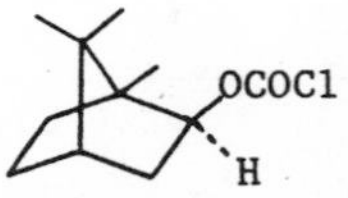

Boiling point: 77-78°/0.4-0.45 Torr.
Specific rotation: $[\alpha]_D^{20}$ -56.6° (C, 1, $CHCl_3$)
$[\alpha]_D^{20}$ 54.3° (C, 1, $CHCl_3$)

Reference
G. Jager and R. Geiger, Liebigs Ann. Chem., 1535 (1973).
A.M. Kolodziejczyk and A. Arendt, Pol. J. Chem., 53, 1017 (1979).

Isopilosine

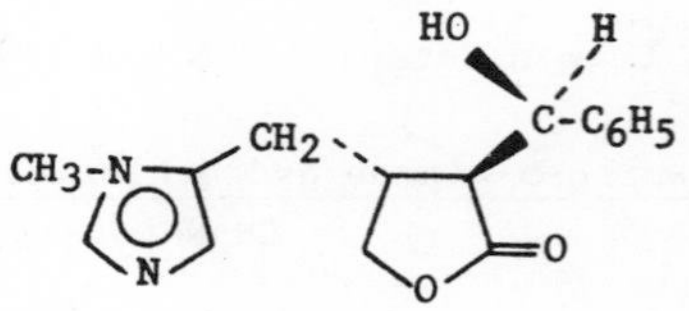

Melting point: 182-182.5°
Specific rotation: $[\alpha]_D^{25}$ 33.5° (c 1, H_2O)

Reference
E. Tedeschi, J. Kamionsky, D. Zeider, S. Fackler, S. Sarel, V. Usieli and J. Deutsch, J. Org. Chem., 39, 1864 (1974).

RESOLVING AGENTS (continued)

L-Leucinamide (for amino acids)

$$H_2N-\underset{\underset{\underset{CH(CH_3)_2}{|}}{\underset{CH_2}{|}}}{\overset{\overset{CONH_2}{|}}{C}}-H$$

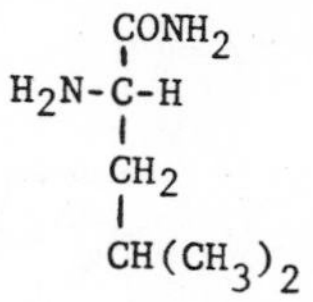

Melting point: 101-102°
Specific rotation: $[\alpha]_D^{19}$ 7.86° (c 5, H_2O)

Reference
T. Kato and Y. Tsuchiya, Agr. Biol. Chem., 26, 467 (1962).

L-Lysine
(for amino acids)

$$H_2N-\underset{\underset{\underset{CH_2NH_2}{|}}{\underset{(CH_2)_3}{|}}}{\overset{\overset{COOH}{|}}{C}}-H$$

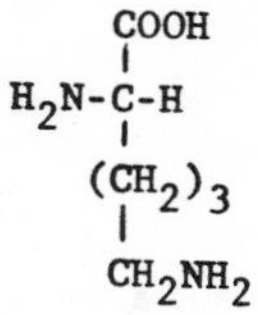

Melting point: 224-5° (decomp.)
Specific rotation: $[\alpha]_D^{25}$ 13.5° (c 2, H_2O)

Reference
"Chemistry of the Amino Acids," Volume 3, J.P. Greenstein and M. Winitz, John Wiley & Sons, Inc., (1961).

Menthol

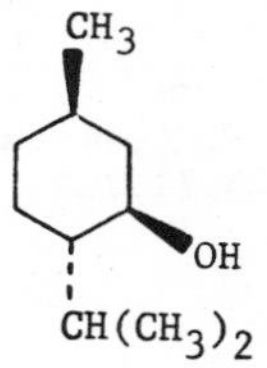

Boiling point: 97-99°/10 mm.
Melting point: 42°
Specific rotation: $[\alpha]_D^{20}$ 49.5° (c 2, 95% ethanol)

Reference
A.W. Ingersoll, Organic Reactions, Vol. 2, 376 (1944).

Menthoxyacetyl chloride (for amino acids)

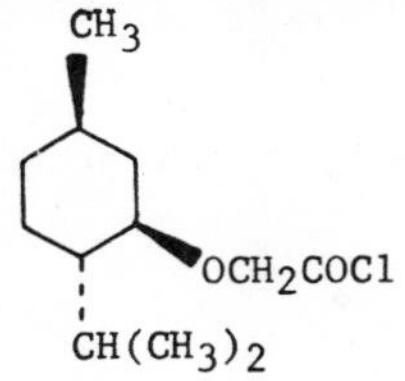

Boiling point: 85-87°/0.4 mm.
Specific rotation: $[\alpha]_D^{20}$ -102° (neat)

Reference
D.F. Holmes and R. Adams, J. Am. Chem. Soc., 56, 2093 (1934).
T.G. Cochran and A.C. Huitric, J. Org. Chem., 36, 3046 (1971).
A.W. Ingersoll, Organic Reactions 2, 376 (1944).

Menthyl phosphate (for basic amino acid)

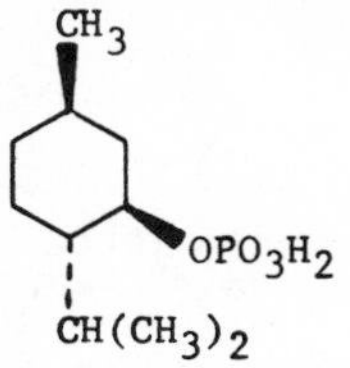

Melting point: 113-114°
Specific rotation: $[\alpha]_D^{20}$ -40.9° (c 2, methanol)

Reference
S. Watanabe and K. Suga, Israel J. Chem., 7, 483 (1969).

α-Methoxy-α-methyl-1-naphthaleneacetic acid
(For resolution of amino acid methyl esters by HPLC)

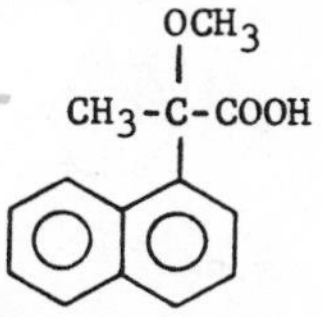

Melting point: 111-112°
Specific rotation: $[\alpha]_D^{13}$ -106.3° (C, 0.16, $CHCl_3$)

Reference
J. Goto, M. Hasegawa, S. Nakamura, K. Shimada and T. Nambara, J. Chromatog., 152, 413 (1978).

α-Methoxy-α-methyl-2-naphthaleneacetic acid
(For resolution of amino acid methyl esters by HPLC)

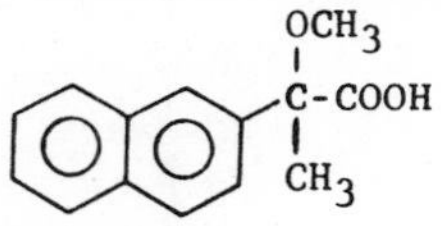

Melting point: 67-68°
Specific rotation: $[\alpha]_D^{17}$ -63.8° (C 0.08, $CHCl_3$)

Reference
J. Goto, M. Hasegawa, S. Nakamura, K. Shimada and T. Nambara, J. Chromatog., 152, 413 (1978).

Erythro-1-O-Methoxyphenyl-1-hydroxy-2-aminopropane

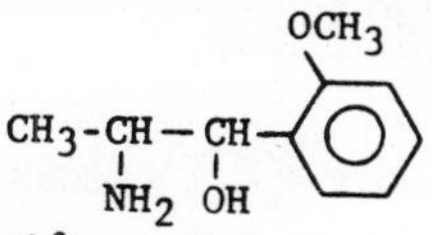

Melting point: 130.4°
Specific rotation: $[\alpha]_D^{25}$ -27.7° (c 0.795, $CHCl_3$)

Reference
Information kindly supplied by Dr. Tsuneyuki Nagase, Central Research Laboratory, Sumitomo Chemical Co., Ltd., Takatsuki City, Osaka, Japan.

RESOLVING AGENTS (continued)

α-Methylbenzylamine

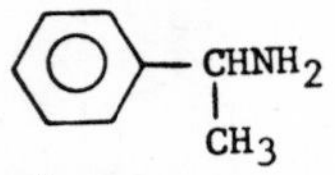

Boiling point: 74°/14° Torr.
Specific rotation: $[\alpha]_D^{22}$ -40.3° (neat)
(d = 0.950)
$[\alpha]_D^{20}$ 40.5° (neat)
(d = 0.940)

Reference
W. Theilacker and H.G. Winkler, Chem. Ber., 87, 690 (1954).
A.P. Terent'ev, G.V. Panova, G.N. Koval and O.V. Toptygina, Zh. Obshch. Khim., 40, 1409 (1970).

N-Methylephedrine

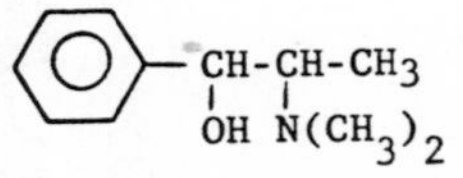

Melting point: 85-86°
Specific rotation: $[\alpha]_D^{20}$ -29.8° (c 4.50, methanol)

Reference
T. Maskiko, S. Terashima and S. Yamada, Yakugaku Zasshi, 100, 319 (1980).

N-Methyl-D-glucamine

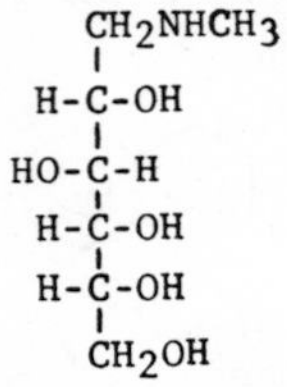

Melting point: 129-131°
Specific rotation: $[\alpha]_D^{20}$ -19.5° (C 2, H_2O)

Reference
Chemical Dynamics Corp., South Plainfield, N.J. 07080

α-Methyl-p-nitrobenzylamine

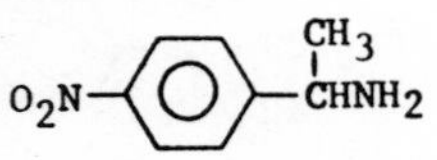

Boiling point: 119-120°/0.5 Torr.
Specific rotation: $[\alpha]_D^{24}$ 17.7° (neat)
$[\alpha]_D^{24}$ 16.9° (c 3.8, $CHCl_3$)

Reference
C.W. Perry, A. Brossi, K.H. Deitcher, W. Tautz and S. Teitel, Synthesis 492 (1977).

Morphine

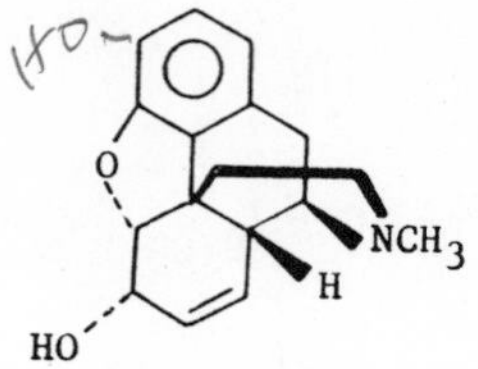

Melting point: 253-254° [decomp.] (anhydrous)
Specific rotation: $[\alpha]_D^{23}$ -130.9° (c 2.3 methanol)

Reference
S.B. Schryver, F.H. Lees, J. Chem. Soc., 77, 1037 (1900).
L. Knorr, H. Hörlein, Chem. Ber., 40, 4890 (1907).

α-(1-Naphthyl)-ethylamine

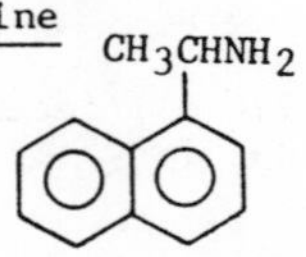

Boiling point: 153°/11 mm; 156°/15 mm.
Specific rotation: $[\alpha]_D^{17}$ 82.8° (neat)
$[\alpha]_D^{19}$ 61.6 (c 2.2, ethanol)

Reference
E. Samuelsson, Ph.D. Thesis, University of Lund (1923).

C.W. Den Hollander, W. Leimgruber and E. Mohacsi, U.S. Patent 3,682,925 (1972).

α-(2-Naphthyl)ethylamine

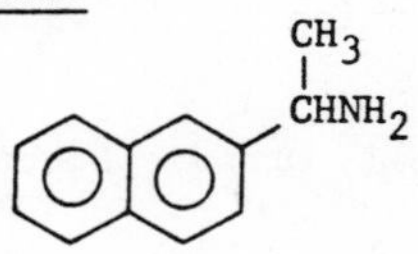

Melting point: 53°
Specific rotation: $[\alpha]_D^{19}$ 19.4° (c 1.86, ethanol)
$[\alpha]_D^{20}$ -18.9° (c 2.38, ethanol)

Reference
E. Samuelsson, Ph.D. Dissertation, University of Lund (1923).

RESOLVING AGENTS (continued)

Erythro-1-(1-Naphthyl)-1-hydroxy-2-aminopropane

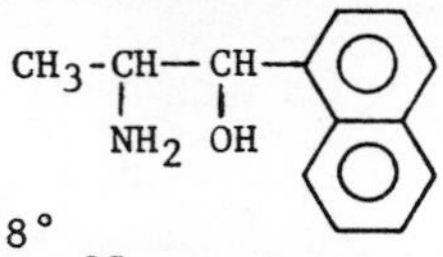

Melting point: 173.8°
Specific rotation: $[\alpha]_D^{25}$ -122° (c 1.00, $CHCl_3$)

Reference
Information kindly supplied by Dr. Tsuneyuki Nagase, Central Research Laboratory, Sumitomo Chemical Co., Ltd., Takatsuki City, Osaka, Japan.

(+)-Neomenthyl isothiocyanate
(for resolution of α-amino acids by HPLC)

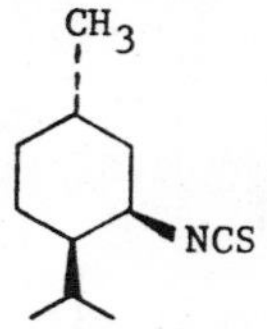

Boiling point: 55°/5 mm.
Specific rotation: $[\alpha]_D^{17}$ 34.5° (c 2.0, $CHCl_3$)

Reference
T. Nanbara, S. Ikegawa, M. Hasegawa and J. Goto, Anal. Chim. Acta, 101, 111 (1978).

Nicotine

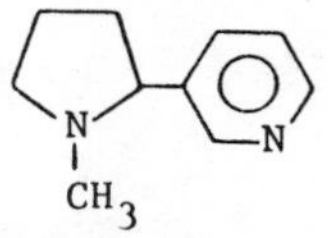

Boiling point: 246°/730.5 mm.
Specific rotation: $[\alpha]_D^{20}$ -166.4° (neat)

Reference
A. Pictet and A. Rotschy, Chem. Ber., 37, 1225 (1904).

threo-1-p-Nitrophenyl-2-amino-1,3-propanediol

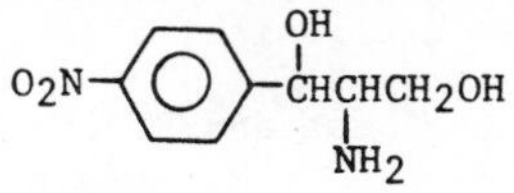

Melting point: 162-163°
Specific rotation: $[\alpha]_D^{24}$ -23.5° (c 1.316, methanol)
$[\alpha]_D^{24}$ +23.4° (c 1.321, methanol)

Reference
P. Pratesi, Farmaco, Ed. Sci., 8, 41 (1953).

threo-1-p-Nitrophenyl-2-N,N-dimethylaminopropane-1,3-diol

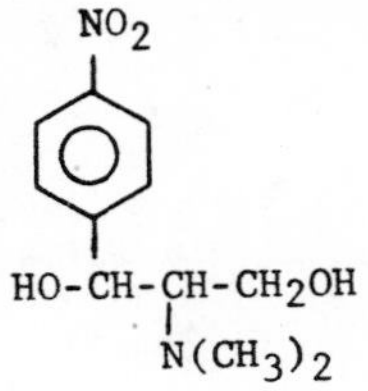

Melting point: 101°
Specific rotation: $[\alpha]_D$ (+) and (-) 26±1° (c 1, ethanol)

Reference
French Patent 1,481,978 (1967).
J. Controulis, M.C. Rebstock and H.M. Crooks, Jr. J. Am. Chem. Soc., 71, 2463 (1949).

1-(p-Nitrophenyl)ethylamine

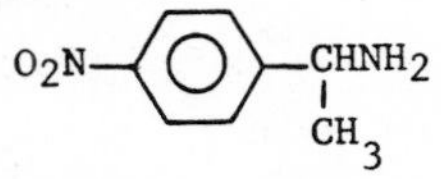

Boiling point: 119-120°/0.5 Torr.
Specific rotation: $[\alpha]_D^{24}$ 17.7° (neat)
$[\alpha]_D^{24}$ 16.9° (c 3.8, $CHCl_3$)

Reference
C.W. Perry, A. Brossi, K.H. Deitcher, W. Tautz and S. Teitel, Synthesis, 491 (1977).

Chemical Dynamics Corp., South Plainfield, N.J. 07080

Nopinylamine

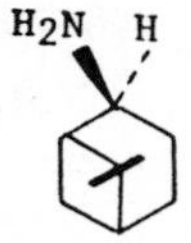

Boiling point: 76°/11 Torr.
Specific rotation: $[\alpha]_D^{25}$ -21.5° (c 4, H_2O) (as hydrochloride)

Reference
J. Paust, S. Pfohl, W. Reif and W. Schmidt, Liebigs Ann. Chem., 1024 (1978).

Norephedrine

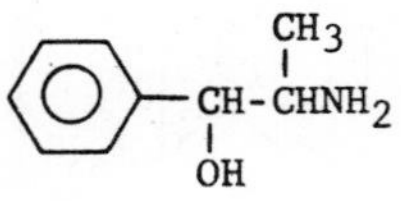

Melting point: 52°
Specific rotation: $[\alpha]_D^{27}$ 14.76° (c 3.75, ethanol)
$[\alpha]_D^{20}$ -14.56° (c 3.43, ethanol)
Hydrochloride Melting point: 171-172°
$[\alpha]_D^{27}$ 33.40 (c 6.45, water)
$[\alpha]_D^{20}$ -33.27° (c 3.43, water)

Reference
W.N. Nagai and S. Kanao, Liebigs Ann., 470, 157 (1929).

RESOLVING AGENTS (continued)

Norepinephrine

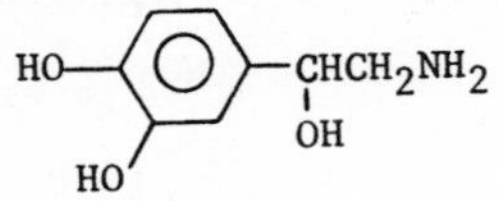

Melting point: 216.5-218° (dec.)
Specific rotation: $[\alpha]_D^{26}$ +37.4 (c 5, H_2O + 1 equiv. HCl)
$[\alpha]_D^{25}$ -37.3° (c 5, H_2O + 1 equiv. HCl)

Norepinephrine hydrochloride: m.p. 146.8-147.4°)
$[\alpha]_D^{25}$ -40° (c 6, H_2O); $[\alpha]_D^{27}$ 39° (c 6, H_2O)

Reference
B.F. Tullar, J. Am. Chem. Soc., 70, 2067 (1948).

Norisoephedrine

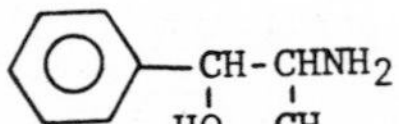

Melting point: 77.5-78.0°
Specific rotation: $[\alpha]_D^{20}$ -32.64° (c 3.48, ethanol)
$[\alpha]_D^{20}$ +33.14° (c 4.84, ethanol)

Hydrochloride melting point: 180-181°
Specific rotation: $[\alpha]_D^{20}$ -42.68° (c 6.94, water)
$[\alpha]_D^{20}$ 42.58° (c 7.01, water)

Reference
W.N. Nagai and S. Kanao, Liebigs Ann., 470, 157 (1929).

L-Ornithine
(for amino acids)

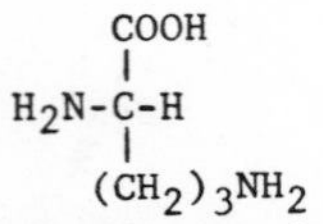

Melting point: 245° (decomp.) as hydrochloride
Specific rotation: $[\alpha]_D^{25}$ 28.3° (c 2,5N HCl)
$[\alpha]_D^{24}$ 23.6° (c 4, 6N (HCl) as hydrochloride

Reference
Chemistry Of The Amino Acids, Vol. 3, J.P. Greenstein and M. Winitz, John Wiley, New York (1961).

Pavine

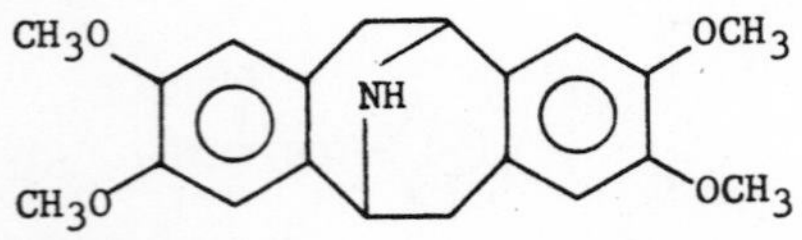

Melting point: 224°
Specific rotation: $[\alpha]_D^{23}$ -150.8° (c 0.99, $CHCl_3$)
$[\alpha]_D^{23}$ +150.3° (c 0.99, $CHCl_3$)

Reference
W.J. Pope and C.S. Gibson, J. Chem. Soc., 97, 2207 (1910).

threo-1-Phenyl-2-dimethylamino-1,3-propanediol

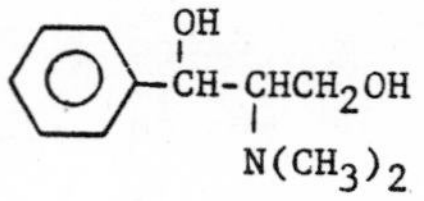

Melting point: 50-53°
Specific rotation: $[\alpha]_D^{20}$ 35.2° (C 10, methanol)

Reference
Physical constants kindly furnished by Dr. Wolfgang Dannenberg, Riedel-de Haen AG, Seelze, West Germany

α-Phenylethylthiouronium acetate

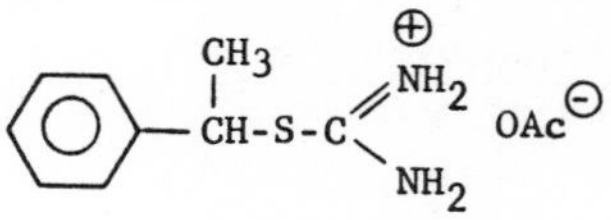

Melting point: 152-154°
Specific rotation: $[\alpha]_D^{20}$ 160° (c 2.5, ethanol)

Reference
W. Klötzer, Monatsch. Chem., 87, 346 (1956).

α-Phenylethylthiuronium chloride

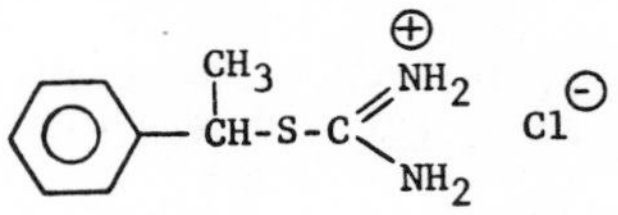

Melting point: 159-162°
Specific Rotation: $[\alpha]_D^{20}$ 170° (c 5, ethanol)

Reference
W. Klötzer, Monatsch. Chem., 87, 346 (1956).

α-Phenylpropylamine

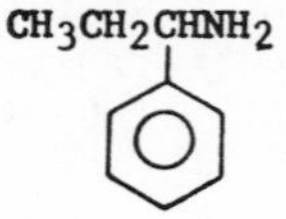

Boiling point: 204-206°; 98°/23 mm.
Specific rotation: $[\alpha]_D^{20}$ -21.2 (neat)
$[\alpha]_D^{20}$ +21.2° (neat)

Reference
M.E. Warren and H.E. Smith, J. Am. Chem. Soc., 87, 1757 (1965).
A.J. Little, J. M'Lean and F.J. Wilson, J. Chem. Soc., 336 (1940).

1-Phenyl-2-p-tolyl-2-aminoethane

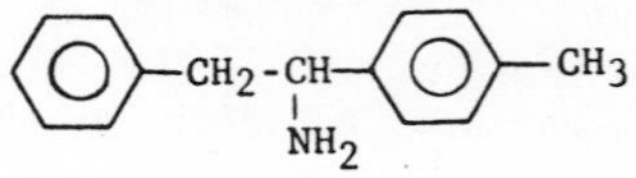

Boiling point: 117-120°/0.1 mm
Specific rotation: $[\alpha]_D^{24}$ 15.2° (neat)

Reference
Information kindly supplied by Dr. Yoshio Suzuki, Sumitomo Chemical Company Ltd., Institute for Biological Science, Hyogo-Ken, 665, Japan.

RESOLVING AGENTS (continued)

Phenyl-β-(p-tolyl)ethylamine

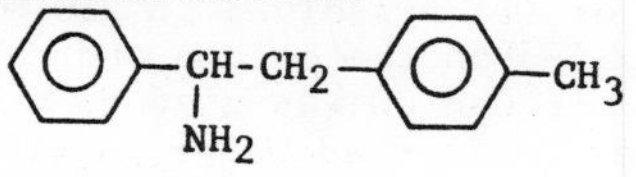

Boiling point: 131°/1 mm.
Specific rotation: $[\alpha]^{23}_{D}$ 6.7° (c 2.2, $CHCl_3$)

Note: can be resolved using L-aspartic acid to give the (+)-amine.

Reference
Data kindly supplied by Dr. Noritada Matsuo, Institute for Biological Science, Sumitomo Chemical Co., Ltd., Takatsukasa, 4-2-1, Takarazuka, Hyogo 665, Japan.

Pilosine

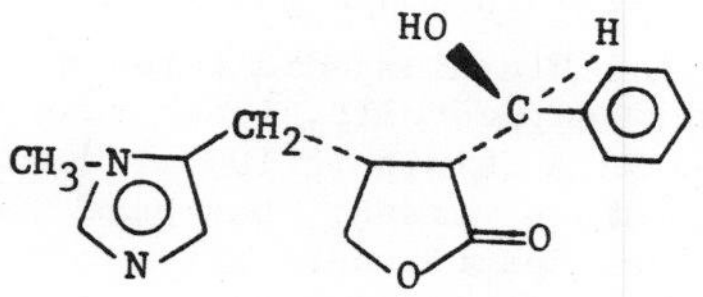

Melting point: 172° (anhydrous)
101-104° (as dihydrate)
Specific rotation: $[\alpha]^{20}_{D}$ 118° (c 1, ethanol) (as dihydrate)
$[\alpha]^{25}_{D}$ 136.8° (c 1, H_2O) (anhydrous)

Reference
H. Link and K. Bernauer, Helv. Chim. Acta, 55, 1053 (1972).
H. Link, K. Bernauer, K. Oberhaensli and E. Willi, Helv. Chim. Acta, 57, 2199 (1974).
E. Tedeschi, J. Kamionsky, D. Zeider, S. Fackler, S. Sarel, V. Usieli and J. Deutsch, J. Org. Chem., 39, 1864 (1974).

3-Piperidinomethylpinane

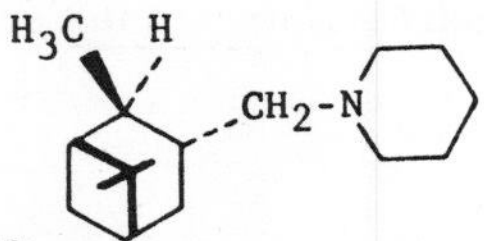

Boiling point: 138-140°/5 mm.
Melting point: 256°
Specific rotation: $[\alpha]^{23}_{D}$ 47.8° (c 1, methanol) (as hydrochloride)

Reference
W. Hoffman, W. Himmele, J. Paust, K. von Fraunberg, H. Siegel and S. Pfohl, U.S. Patent 4,081,477 (1978).

Pseudoephedrine

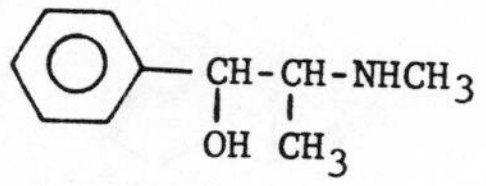

Melting point: 118-118.7°
Specific rotation: $[\alpha]^{22.5}_{D}$ 52.9° (c 4.0, ethanol)

Pseudoephedrine hydrochloride:

Melting point: 182.5-183.5°
Specific rotation: $[\alpha]^{20}_{D}$ 62.8° (c 1.04, H_2O)

Reference
E. Späth and R. Göhring, Monatsh. Chem., 41, 319 (1920).

3-Pyrrolidinomethylpinane

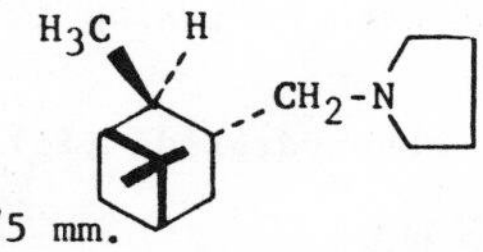

Boiling point: 125-127°/5 mm.
Specific rotation: $[\alpha]^{23}_{D}$ 49.4 (c 1, methanol)

Reference
W. Hoffman, W. Himmele, J. Paust, K. von Fraunberg, H. Siegel and S. Pfohl, U.S. Patent 4,081,477 (1978).

Pyrrolidone-5-carboxylic acid

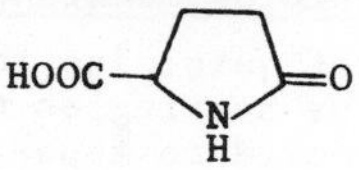

Melting point: 152-160° (sinters at 148°)
Specific rotation: $[\alpha]^{22}_{D}$ -9.9° ± 0.6° (c 7.9, H_2O)

Reference
M. Brenner and H.R. Rickenbacher, Helv. Chim. Acta, 41, 181 (1958).
W.H. Gray, J. Chem. Soc., 1266 (1928).

Quinidine

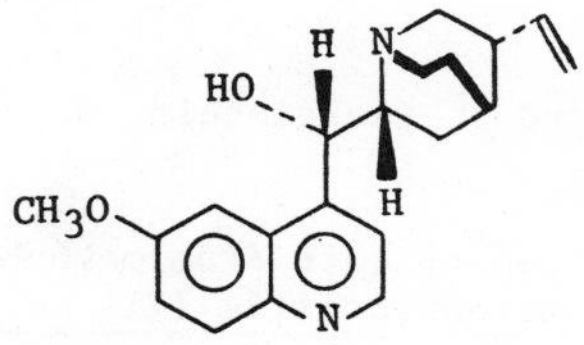

Melting point: 174-175°
Specific rotation: $[\alpha]^{17}_{D}$ 255.4° (c 1.62, ethanol) (anhydrous compound)

Reference
A.C. Oudemans, Jr., Liebigs Ann., 182, 33 (1876).

RESOLVING AGENTS (continued)

Quinine

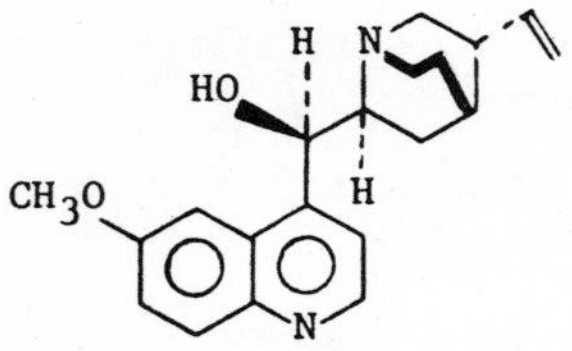

Melting point: 177° (anhydrous)
57° (trihydriate)

Specific rotation: $[\alpha]_D^{17}$ -167.5° (c 1.63, ethanol) [anhydrous compound]

Reference
A.C. Oudemans, Jr., Liebigs Ann. 182, 33 (1876).

Quinidine methohydroxide (I)
or
Quinine methohydroxide (II)

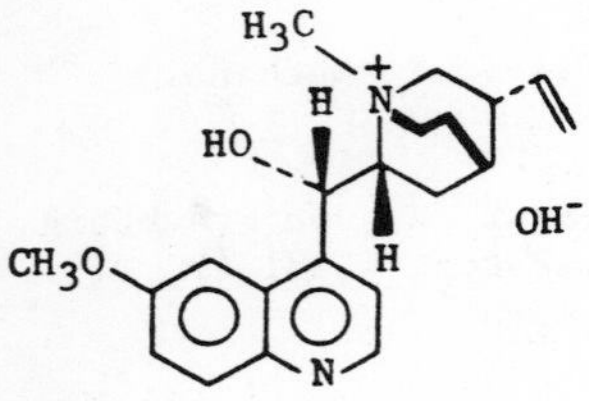

(I)

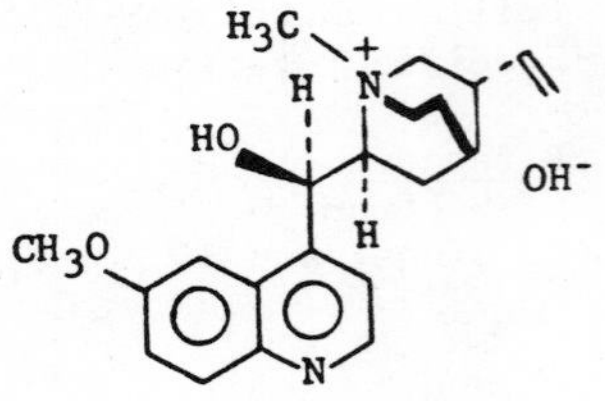

(II)

a) Preparation of quinidine methiodide (A)

A solution of 169 gm. of quinidine in 5 liters of methanol was prepared by heating on the steam bath. When the temperature cooled to about 49°, 85 gm. of CH_3I was added. After stirring to mix intimately, the flask was well stoppered and kept at room temperature. The stopper must be very secure because the reaction mixture warms up on standing. After 24 hours, the solution was concentrated to about 400 cc. Crystallization sets in and when the flask was cooled, the contents turned completely solid. The product was filtered and washed with a small amount of methanol and ether. Then it was recrystallized from hot water and treated with Norit. Long silky needles separated on cooling and were dried at 115° for 2 hours.

Reference
R.T. Major and J. Finkelstein, J. Am. Chem. Soc., 63, 1368 (1941).

b) Preparation of quinidine methohydroxide (I) or quinine methohydroxide (II)

Quinidine methoiodide (A) and quinine methoiodide (B) synthesized as shown above were recrystallized from H_2O. (A) had a m.p. of 236-237° (dec.) and (B) had a m.p. of 229°C (dec.).

A 10 gm. amount of Amberlite IRA-401, after being washed with 500 ml. each of 1N aqueous NaOH, H_2O, 1N HCl, H_2O, C_2H_5OH, and H_2O was conditioned with 10 gm. of NaOH dissolved in 225 ml. of H_2O, followed by 250 ml. each of H_2O and dry C_2H_5OH. A 2.0 mmol. amount of (A) or (B) dissolved in 250 ml. of dry C_2H_5OH was applied on the column and eluted. An additional 250 ml. of C_2H_5OH was used to wash the column. The eluates were collected under N_2 or Ar, concentrated below 25°C under reduced pressure and diluted with dry C_2H_5OH to 25.0 ml. The concentrations of alcoholic solutions of (II) and (I) were determined by titration with 0.0107 m HCl. (II) was found to be 0.08 m. The concentration of (I) was also assumed to be about 0.08 m.

Reference
K. Hermann and H. Wynberg, J. Org. Chem., 44, 2238 (1979).

Alternate Procedure

1. N-Methylchininiumhydroxid

N-Methylchininiumjodid wurde nach Major und Finkelstein[5] dargestellt, indem zu einer gekühlten Lösung von 93 g Chinin in 100 ml Methanol 42 g Methyljodid zugefügt wurden. Das beim Stehen über Nacht ausgefallene Reaktionsprodukt wurde abgesaugt, mit Methanol/Äther (1:1) gewaschen und getrocknet. 30 g N-Methylchininiumjodid wurden in 100 ml Methanol aufgeschwemmt, mit 90 g hydroxidbeladenem stark basischem Anionenaustauscher Merck versetzt und 2 Std. geschüttelt. Danach wurde der Austauscher abgesaugt, mit Methanol nachgewaschen und das Filtrat in einem Messkolben mit Methanol auf 250 ml aufgefüllt, 5.0 ml dieser Lösung wurden nach Verdünnen mit Wasser mit 0,1 n HCl titriert (Tashiro). Verbrauch 12,25 ml 0,1 n HCl.

[5] R.T. Major und J. Finkelstein, J. Am. Chem. Soc., 63, 1368 (1941).

Reference
J. Knabe and R. Kräuter, Arch. Pharm. (Weinheim), 298, 1 (1965).

Quinotoxine (Quinicine)

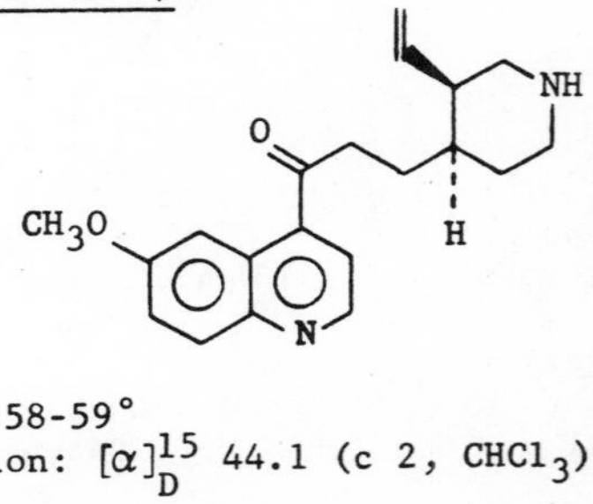

Melting point: 58-59°
Specific rotation: $[\alpha]_D^{15}$ 44.1 (c 2, $CHCl_3$)

Reference
W. Miller and G.Rohde, Chem. Ber., 28, 1056 (1895).
O. Hesse, Liebigs Ann., 178, 244 (1875).

RESOLVING AGENTS (continued)

Sodium-1-hydroxy-p-menthane-2-sulfonate (for amino acids)

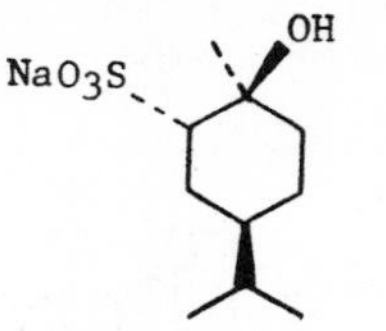

Melting point: 195-200°
Specific rotation: $[\alpha]_D$ 15.5° (C, 10.0, H_2O)

Reference
S.G. Traynor, B.J. Kane, M.F. Betkouski and L.M. Hirschy, J. Org. Chem., 44, 1557 (1979).

Sodium-2-hydroxy-p-menthane-1-sulfonate (for amino acids)

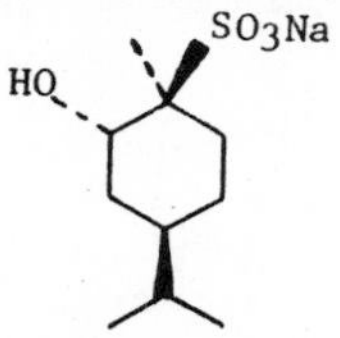

Melting point: 213-216°
Specific rotation: $[\alpha]_D$ -5.2° (C, 20.03, H_2O)

Reference
S.G. Traynor, B.J. Kane, M.F. Betkouski and L.M. Hirschy, J. Org. Chem., 44, 1557 (1979).

Strychnine

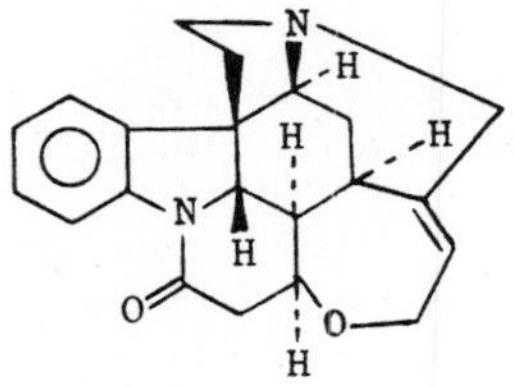

Melting point: 286-288°
Specific rotation: $[\alpha]^{18}_D$ -139.3° (c 2, $CHCl_3$)
$[\alpha]^{18}_D$ -104.3° (c 0.2, ethanol)

Reference
K. Warnat, Helv. Chim. Acta, 14, 998 (1931).
T.S. Patterson and J.D. Fulton, J. Chem. Soc., 51 (1927).

Testosterone

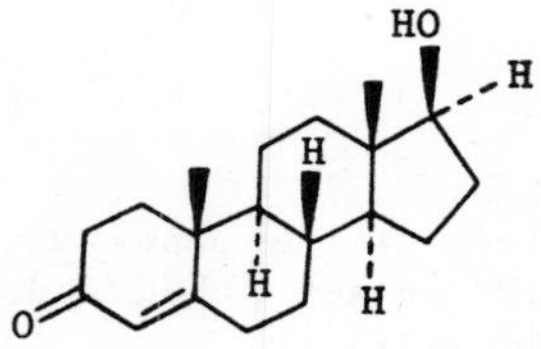

Melting point: 155°
Specific rotation: $[\alpha]^{24}_D$ 109° (C, 4, ethanol)

Reference
R.R. Crenshaw, T.A. Jenks, G.M. Luke and G. Bialy, J. Med. Chem., 17, 1258 (1974).

Tetrahydrofurfurylamine

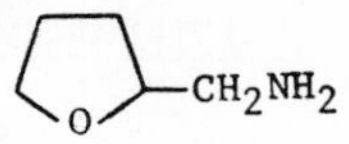

Boiling point: 49-50°/13 mm.
Specific rotation: $[\alpha]^{20}_D$ 3.2° (neat)

Reference
a) E.J. Cragoe,Jr., U.S. 3,577,409 (1971)
b) Chemical Dynamics Corp., South Plainfield, N.J. 07080

Erythro-1-O-Tolyl-1-hydroxy-2-aminopropane

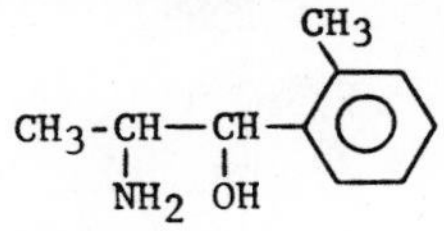

Melting point: 156.9°
Specific rotation: $[\alpha]^{25}_D$ -60.4° (c 0.323, $CHCl_3$)

Reference
Information kindly supplied by Dr. Tsuneyuki Nagase, Central Research Laboratory, Sumitomo Chemical Co., Ltd., Takatsuki City, Osaka, Japan.

Threo-1-O-Tolyl-1-hydroxy-2-aminopropane

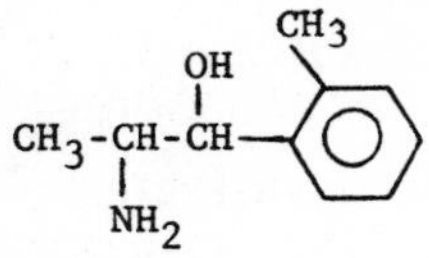

Melting point: 129.4°
Specific rotation: $[\alpha]^{25}_D$ 100.4° (c 0.325, $CHCl_3$)

Reference
Information kindly supplied by Dr. Tsuneyuki Nagase, Central Research Laboratory, Sumitomo Chemical Co., Ltd., Takatsuki City, Osaka, Japan.

1-p-Tolyl-2-phenyl-2-aminoethane

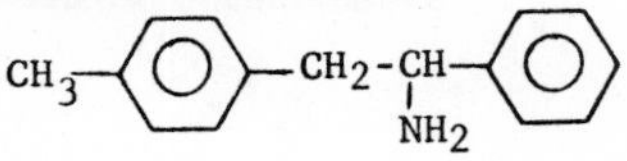

Boiling point: 115-118°/0.1 mm
Specific rotation: $[\alpha]^{20}_D$ 12.7° (neat)

Reference
Information kindly supplied by Dr. Yoshio Suzuki, Sumitomo Chemical Company, Ltd., Institute for Biological Science, Hyogo-Ken, 665, Japan.

RESOLVING AGENTS (continued)

Triethylenediamincobaltibromide

$[Co\ en_3]Br_3$

Specific rotation: $[\alpha]_D$ -115° (c 1, H_2O)
$[\alpha]_D$ +117° (c 1, H_2O)

Reference
A. Werner and M. Basyrin, Chem. Ber., 46, 3229 (1913).

L-Tyrosinamide

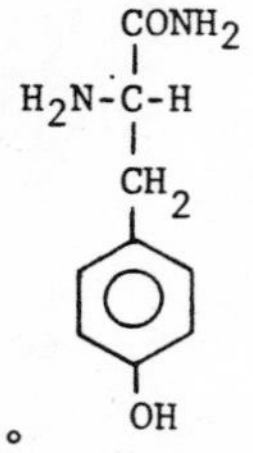

Melting point: 153-154°
Specific rotation: $[\alpha]_D^{21}$ 19.47° (c 2, H_2O)

Reference
T. Kato and Y. Tsuchiya, Agr. Biol. Chem., 26, 467 (1962).

L-Tyrosinhydrazide (for amino acids)

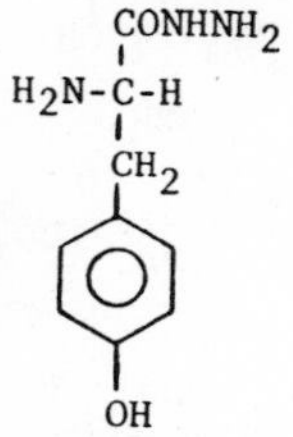

Melting point: 198-200°
Specific rotation: $[\alpha]_D^{23}$ -70.3°(c 2, 3N HCl)
$[\alpha]_D^{23}$ 70.2° (c 2, 3N HCl)

Reference
K. Vogler and P. Lanz, Helv. Chim. Acta, 49, 1348 (1966).

Yohimbine

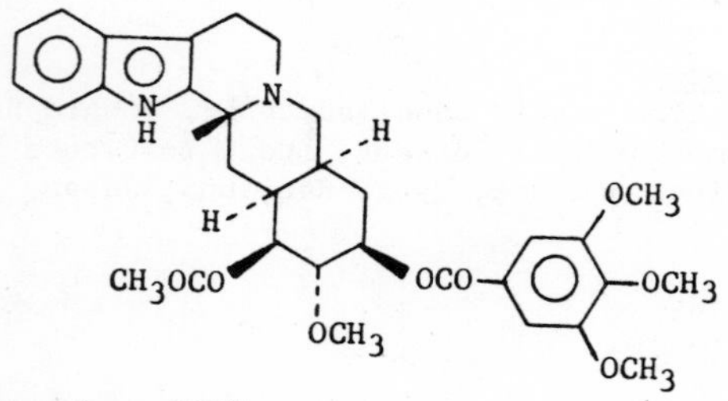

Melting point: 235°
Specific rotation: $[\alpha]_D^{17}$ 54.2° (c 1.56, ethanol)
$[\alpha]_D^{17}$ 107.9° (c 0.93, pyridine)

Reference
G. Hahn, Chem. Ber., 60, 1683 (1927).

A D D E N D U M

N-Butyl-10-amino-α-pinene

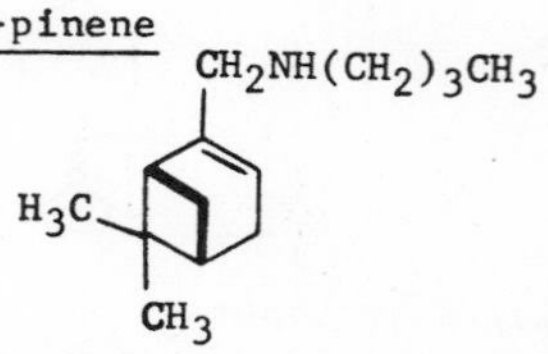

Boiling point: 84°/0.6 Torr.
Specific rotation: $[\alpha]_D^{20}$ -22.6° (neat)

Reference
S.W. Markowicz, Polish J. Chem., 53, 157 (1979).

O.O'-Dibenzoyltartaric acid (anhydrous) (for amino acids)

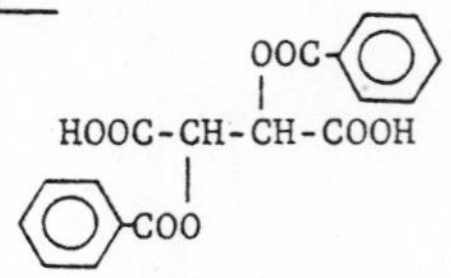

Melting point: 138-140°
Specific rotation: $[\alpha]_D^{20}$ -118.51° (c 4.9011, ethanol)

Reference
F. Zetzsche and M. Hubacher, Helv. Chim. Acta, 9, 291 (1926).

O,O'-Dibenzoyltartaric acid (monohydrate)

Melting point: 90-91°
Specific rotation: $[\alpha]_D^{20}$ -115.5° (c 1.20,ethanol)

Reference
G. Carrara, G.F. Cristiani, V. D'Amato, E. Pace and R. Pagani, Gazz. Chim. Ital., 82, 325 (1952).

Hydrocinchonine

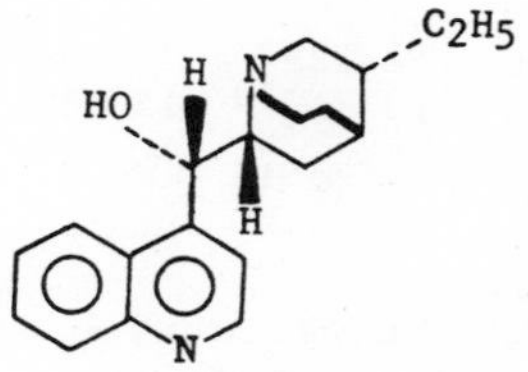

Melting point: 268-269° (free base)
221-223° (as hydrochloride)

Specific rotation: $[\alpha]_D^{15}$ 204.5° (c 0.6 ethanol) free base
$[\alpha]_D^{25}$ 159.3° (c 0.741, H_2O) as hydrochloride

Reference
O. Hesse, Liebigs Ann. 300, 42 (1898).
M. Heidelberger and W.A. Jacobs, J. Am. Chem. Soc., 41, 817 (1919).

SECTION 2

OPTICAL RESOLUTION PROCEDURES FOR ACIDS*

This section contains photocopies of optical resolution procedures for acids as they appear a) in a particular journal, b) in a particular patent, c) in a particular thesis, or d) furnished to the author through correspondence. Where it was felt that it would be useful, comments about the resolution made by the author of the paper are included, as are editor's notes. When the primary source did not lend itself to being photocopied, the resolution procedure was typewritten.

* Including lactones

CH_2ClIO_3S

```
   Cl
   |
I-CHSO3H
```

Ref. 1

Derivatives of d-*Chloroiodomethanesulphonic Acid.*—Optically impure ammonium *d*-chloroiodomethanesulphonate (16·1 g., $[\alpha]_{5461}$ + 9·2°), prepared according to Pope and Read (*loc. cit.*, p. 818), was dissolved in hot H_2O (500 c.c.); 80% of the calc. amount of a hot, conc. aq. solution of brucine sulphate was then added quickly, with stirring. A cryst. separation began in about 2 sec., and after the mixture had cooled the brucine salt was collected and washed with cold H_2O. The brucine was completely eliminated by shaking the salt for a few hrs. at room temp. with a mixture of $CHCl_3$ and the calc. quantity of very dil. aq. NH_3. The aq. solution, when washed with $CHCl_3$ and evaporated to dryness on the water-bath, gave ammonium *d*-chloroiodomethanesulphonate (12·8 g.) having $[\alpha]_{5461}$ + 12·8° (c = 1, in H_2O). Three successive repetitions of this process, starting with the salt having $[\alpha]_D$ +12·8°, yielded further specimens of NH_4 salt with $[\alpha]_{5461}$ + 16·1° (10·2 g.), + 16·7° (8·2 g.), and + 16·7° (6·5 g.). When recryst. from the min. amount of hot abs. EtOH, the last fraction gave large colourless plates with $[\alpha]_{5461}$ + 17·0°; a slight insol. residue consisted of $(NH_4)_2SO_4$. Finally, an alc. solution was evaporated almost to crystallising point and then diluted with hot acetone: small glistening plates separated, having $[\alpha]_{5461}$ + 17·0°. Opalescence on addition of acetone denotes the presence of traces of $(NH_4)_2SO_4$. The rotatory power of the salt was unaffected by further crystn.

Pure *ammonium* d-*chloroiodomethanesulphonate* melts at 229—230° (decomp.). It is moderately sol. in boiling glac. AcOH, from which it crystallises in glistening plates (Found: NH_3, 6·2. $CHO_3ClIS{\cdot}NH_4$ requires NH_3, 6·2%). In 1% aq. solution (l = 2; room temp., *ca.* 15°) it gave $[\alpha]_{5461}$ + 17·0°, $[M]_{5461}$ + 46·5°; $[\alpha]_D$ + 13·2°, $[M]_D$ + 36°. A 1·25% solution in abs. EtOH gave $[\alpha]_{5461}$ + 20·9°, $[M]_{5461}$ + 57°; $[\alpha]_D$ + 16·9°, $[M]_D$ + 46°. When an equiv. of H_2SO_4 was added to 0·2008 g. of the NH_4 salt, the resulting solution (20·0 c.c.) of free *d*-chloroiodomethanesulphonic acid [in presence of $(NH_4)_2SO_4$] gave α_{5461} + 0·25°, $[\alpha]_{5461}$ + 13·3°, $[M]_{5461}$ + 34°; α_D + 0·19°, $[\alpha]_D$ + 10·1°, $[M]_D$ + 26°.

The Ba salt crystallised from hot H_2O in small glistening leaflets; it underwent no loss of weight at 110°, and in 1% aq. solution had $[\alpha]_{5461}$ + 14·0°, $[\alpha]_D$ + 11·0°.

A resolution procedure for the above compound using hydroxyhydrindamine* may be found in Ref. 1a.

*Hydroxyhydrindamine is cis-2-amino-1-hydroxyhydrindene.

```
    Cl
    |
Br-CH-COOH
```

$C_2H_2BrClO_2$

Ref. 2

Resolution of Chlorobromoacetic Acid.—In our preliminary study (*loc. cit.*) the dextrorotatory acid was isolated in the form of its quinine salt. The lævorotatory acid has now been obtained by means of the *brucine* salt, which separates when a solution of sodium chlorobromoacetate (0·1 mol.) and brucine acetate (0·1 mol.) in a total volume of 1 litre are mixed.

The alkaloidal salts may be purified by "cold crystallisation" (*Proc. K. Akad. Wetensch. Amsterdam*, 1925, **28**, 64). This precaution, however, is not necessary, since the tendency to racemisation is not very strong; a careful crystallisation from warm water is sufficient. The solution must not be heated long because of the instability of chlorobromoacetic acid in hot water.

The brucine salt, therefore, is shaken with 2 parts of water at room temperature and mixed with 20 parts of boiling water. The clear solution, obtained within a minute, is quickly cooled until crystallisation sets in. After about five such crystallisations, the brucine salt is optically pure. It crystallises with 2 or 3 molecules of water, which are given off at 110° and in a vacuum over phosphoric oxide (*a*. Found: H_2O, 6·3; calc. for $2H_2O$, 6·0%. *b*. Found: H_2O, 8·5; calc. for $3H_2O$, 8·7%. *c*. Found in salt dried in a vacuum: Cl + Br, 20·4. $C_2H_2O_2ClBr,C_{23}H_{26}O_4N_2$ requires Cl + Br, 20·4%).

In our first experiments the alkaloidal salts were converted into the ammonium salt, whose concentration was estimated by evaporation of its solution. More trustworthy results were obtained by decomposing 2 or 3 g. of the brucine salt with 15 c.c. of water and sodium hydroxide in excess, extracting the brucine with pure chloroform, and titrating the excess of alkali with sulphuric acid (0·2*N*) and phenolphthalein. The solution of the sodium salt (about 20 c.c.), the molecular concentration of which was thus known, was examined in a polarimeter.

In one experiment 200 g. of the brucine salt were recrystallised from 25 times its weight of water. The crystallisation was repeated four times, a sample being used after each crystallisation for polarimetric examination. The brucine salt then weighed only 14 g. and showed no rise of rotatory power.

About 2 g. were decomposed with 15 c.c. of water and 3·47 c.c. of sodium hydroxide (1·035*N*). The solution, after being extracted with chloroform and neutralised with 1·20 c.c. of sulphuric acid (0·198*N*), showed α_D = + 0·21° (l = 2). The rotation of the sodium salt was thus $[M]_D$ = + 6·3°.

$C_2H_2ClFO_2$

```
     Cl
     |
F-CH-COOH
```

Ref. 3

Resolution of **1**. Dehydroabietylamine (**5**) was obtained from commercial Amine 750 (Hercules Powder Co., Inc.) through repeated crystallizations from toluene of its acetate[13] to a m.p. 144–146°, $[\alpha]_D^{25}$ +31° (*c*, 5, MeOH) (lit.,[13] m.p. 141–143·5°, $[\alpha]_D^{25}$ +30·2°); the free base had m.p. 41·5–43° (lit.,[13] m.p. 44–45°).

A soln of **5** (70 g, 0·246 mole) in EtOAc (400 ml) was added to a soln of (±)-**1** (27·5 g, 0·246 mole) in the same solvent (100 ml). After 4 hr at room temp the salt (75 g) was collected, washed with EtOAc (150 ml) and recrystallized from 600 ml EtOAc to give 51 g, m.p. 145–148°, $[\alpha]_D^{25}$ +29·4° (*c*, 0·65, $CHCl_3$); after 4 more crystallizations from EtOAc, m.p. and specific rotation remained constant: m.p. 150·5–151·5°, $[\alpha]_D^{24}$ +36·3° (*c*, 0·595, $CHCl_3$). (Found: C, 66·28; H, 8·50; N, 3·42; F, 4·77; Cl, 9·28. $C_{22}H_{33}ClFNO_2$ requires: C, 66·40; H, 8·36; N, 3·52; F, 4·77; Cl, 8·91%). A soln of this salt (12 g) in water was made alkaline with 5% ammonia, **5** was eliminated by extracting 4 times with 100-ml portions ether, the aqueous layer was acidified with HCl and extracted again with 5 100-ml portions ether; the dried ($MgSO_4$) ether extracts were evaporated and the residue was distilled to give 2·2 g of (+)-**1**, b.p. 65–67°/10 mm, n_D^{23} 1·4140, d_4^{25} 1·532, $[\alpha]_D^{27}$ +24·7° (neat). The specific rotation did not change after a second distillation. Table 1 gives the specific rotations in several different solvents.

The combined mother liquors from the initial precipitation and from the first crystallization were evaporated to dryness and the residue was crystallized twice from 200 ml $CHCl_3$ to give 15·2 g of a salt, m.p. 146–147°, $[\alpha]_D^{25}$ +21·4° (*c*, 0·50, $CHCl_3$). Three more crystallizations led to constant m.p. and specific rotation: 7·2 g, m.p. 148·5–149°, $[\alpha]_D^{25}$ +16·9° (*c*, 0·476, $CHCl_3$). (Found: N, 3·62. $C_{22}H_{33}ClFNO_2$ requires: N, 3·52%). The acid (−)-**1** was isolated as described above: 1·25 g, b.p. 65–67°/10 mm, n_D^{22} 1·4144, $[\alpha]_D^{28}$ −29·9° (neat).

TABLE 1

Solvent	*c*	$[\alpha]_{589}$	$[\alpha]_{578}$	$[\alpha]_{546}$	$[\alpha]_{436}$	$[\alpha]_{365}$
neat		+24·7°	+24·8°	+28·8°	+57·4°	+103·5°
benzene	3·18	+46·5°	+46·5°	+52·7°	+94·3°	+157·4°
ethyl ether	3·48	+24·7°	+25·0°	+27·2°	+52·8°	+91·6°
water	2·03	+16·4°	—	—	—	—
0·46N HCl aq	5·15	+24·2°	+24·2°	+28·4°	+55·0°	+102·1°
ammonium salt in water	2·34	+6·3	+6·3°	+7·8°	+17·8°	+39·4°

[13] W. J. Gottstein and L. C. Cheney, *J. Org. Chem.* **30**, 2072 (1965).

Authors' comment: The optical purity of the acid with $[\alpha]_D^{28}$ -29.9° has a minimum value of 88%; the actual value probably is somewhat above 90%. See body of paper for basis for this statement.

```
I-CH-COOH
  |
  Cl
```

$C_2H_2ClIO_2$

Ref. 4

Resolution of Chloroiodoacetic Acid with l-*Hydroxyhydrindamine*.—Hot methyl-alcoholic solutions of the acid (10 g. in 200 c.c.) and base (6·7 g. in 200 c.c.) were mixed in a shallow dish and evaporated to dryness at once, whilst cooling, in a brisk current of air. The residual mass of almost colourless, silky needles was redissolved in the minimum quantity of hot methyl alcohol, diluted considerably with ethyl acetate, and rapidly concentrated on the water-bath until crystallisation occurred readily when the solution was cooled and stirred. Two successive fractions had $[\alpha]_D$ − 32·4° in aqueous solution (c = 1·00). Upon recrystallisation from the same solvent, separation occurred more readily than from the original solution, and a constant value of $[\alpha]_D$ − 34·0°, $[M]_D$ − 126°, was readily attained for pure l-*hydroxyhydrindamine* l-*chloroiodoacetate* in dilute aqueous solution (c = 1·00). The salt forms brittle transparent prisms, m. p. 149° (decomp.); it dissolves readily in water or in warm methyl alcohol, sparingly in acetone, and is practically insoluble in ethyl acetate or chloroform (Found: Cl + I, 43·6. $C_{11}H_{13}O_3NClI$ requires Cl + I, 44·0%). The solutions in organic solvents darken when kept. Aqueous solutions exhibited no mutarotation, and only a very slight decrease in rotatory power was observed after evaporating such solutions to dryness. A distinct downward mutarotation occurred in methyl-alcoholic solution (c = 1·20), the value of $[M]_D$ declining from − 135° to − 103° in 5 days; at the same time, the solution gradually developed a brown colour, indicative of decomposition. A solution in glacial acetic acid decomposed much more rapidly, with liberation of iodine. In chloroform containing a little methyl alcohol, the salt gave $[\alpha]_D$ − 32·0°, $[M]_D$ − 118° (c = 0·67). Upon treatment with the requisite quantity of brucine acetate in hot methyl alcohol, it yielded brucine *l*-chloroiodoacetate, having $[\alpha]_D$ − 22·0° in methyl alcohol (c = 0·60); when this salt was decomposed in the cold with ammonia, the resulting ammonium *l*-chloroiodoacetate had $[\alpha]_D$ − 24·4°, $[M]_D$ − 58° in aqueous solution (c = 2·75).

The mother-liquor from the original separation of *l*-hydroxyhydrindamine *l*-chloroiodoacetate became very dark; after most of the solvent had evaporated, the residual oily mass was induced to crystallise by treatment with light petroleum. Further treatment with ether yielded two fractions, having $[\alpha]_D$ − 5·9° and − 2·0°, and $[M]_D$ − 22° and − 7°, respectively, in aqueous solution (c = 1·00). The second fraction, composed of minute, feathery needles, appeared to consist of almost pure l-*hydroxyhydrindamine* d-*chloro-*

iodoacetate, the calculated value for the molecular rotation of the corresponding acid ion in dilute aqueous solution being + 53·5°. The brucine salt precipitated from the mixed fractions was fractionally recrystallised from hot water, and the final fraction, having $[\alpha]_D$ + 14·0° in methyl alcohol (c = 0·60), when decomposed in the usual way with dilute ammonia yielded an optically impure ammonium *d*-chloroiodoacetate having $[\alpha]_D$ + 16·0°, $[M]_D$ + 38° in aqueous solution (c = 2·00).

Resolution of dl-*Chloroiodoacetic Acid with Brucine*.—1. The crude brucine salt, obtained by mixing the components in warm methyl alcohol, had $[\alpha]_D$ − 9·7° in methyl alcohol (c = 0·60). Recrystallisation from warm methyl alcohol gave two successive fractions yielding specimens of ammonium chloroiodoacetate with $[\alpha]_D$ − 2° and + 1°, respectively, in dilute aqueous solution.

2. Following preliminary experiments on fractional precipitation with brucine in accordance with the method of Pope and Read (J., 1914, **105**, 817), a warm aqueous solution of ammonium *dl*-chloroiodoacetate (4·4 g. of acid in 300 c.c.) was treated with a similar solution of one-quarter the calculated weight of brucine hydrochloride (2·3 g. of base in 75 c.c.). After the crystalline brucine salt had been separated from the cold liquid, the filtrate was treated thrice in succession in an exactly similar way with fresh quantities of brucine, the volume of the solution being kept practically constant by evaporation on the water-bath after each filtration. The four fractions of brucine salt had roughly the same weight (2·75 g.), and the respective values of $[\alpha]_D$ in methyl alcohol (c = 0·60) were − 22°, − 8°, + 6°, and + 8°. After two recrystallisations from water containing a little methyl alcohol, the first fraction gave, upon decomposition with ammonia, a specimen of ammonium *l*-chloroiodoacetate having $[\alpha]_D$ − 25·1°, $[M]_D$ − 60°, in water (c = 2·11). Further observations indicated that this specimen of ammonium *l*-chloroiodoacetate was optically pure. An aqueous solution of ammonium *l*-chloroiodoacetate (6·53 g. in 250 c.c.) having $[M]_D$ − 32°, prepared from a number of intermediate fractions of the brucine salt which had accumulated in several series of operations, was treated in the usual way with two successive quantities of one-quarter the calculated amount of brucine hydrochloride. The two specimens of ammonium *l*-chloroiodoacetate obtained by decomposing the ensuing fractions of brucine *l*-chloroiodoacetate had the respective values $[\alpha]_D$ − 25·0°, $[M]_D$ − 59°, and $[\alpha]_D$ − 24·4°, $[M]_D$ − 58° in 3% aqueous solution. The two specimens were then mixed and treated once again with one-quarter the calculated amount of brucine hydrochloride under the usual conditions: the resulting brucine salt, when decomposed in the cold with dilute ammonia, yielded ammonium *l*-chloroiodoacetate having $[\alpha]_D$ − 25·0°, $[M]_D$ − 59°. The corresponding pure, air-dried *brucine* l-*chloroiodoacetate* had $[\alpha]_D$ − 22·0° in methyl alcohol (c = 0·60); this salt forms small, glistening prisms, m. p. 158—160° (decomp.), which dissolve very sparingly in hot water and readily in warm methyl alcohol.

After the addition of one-half the calculated amount of brucine hydrochloride to a hot aqueous solution of ammonium *dl*-chloroiodoacetate, the succeeding fractions of the brucine salt usually exhibit a low dextrorotation in methyl alcohol; the fractions are also more readily soluble and consist of large, glistening prisms. A fraction of the brucine salt having $[\alpha]_D$ + 6° in methyl alcohol yielded ammonium *d*-chloroiodoacetate having $[\alpha]_D$ + 17·5°, $[M]_D$ + 42°, in 4% aqueous solution. After the original brucine salt had been recrystallised three times from hot water, the corresponding ammonium *d*-chloroiodoacetate had $[\alpha]_D$ + 15·0°, $[M]_D$ + 36°. Fractional precipitation of this salt with one-half the calculated quantity of brucine gave an ammonium salt having $[\alpha]_D$ + 15·8°, $[M]_D$ + 38°. Prolonged evaporation of aqueous solutions of brucine *d*-chloroiodoacetate on the water-bath appears to bring about partial racemisation of the acid component of the salt.

3. The preparation of optically pure brucine *d*-chloroiodoacetate by the normal methods of fractional crystallisation or fractional precipitation is thus difficult; success was achieved, however, by conducting the second of these operations from a solution having about 3·5 times the concentration noted in paragraph (2) above. Two aqueous solutions were prepared, each of 500 c.c.; these contained, respectively, 22·0 g. of *dl*-chloroiodoacetic acid as ammonium salt and 46·6 g. of brucine in the form of hydrochloride. Upon rapidly adding a portion (150 c.c.) of the hot brucine solution to the whole of the first solution on the water-bath, crystallisation began about one minute after mixing and continued rapidly. The fraction (19·8 g.) collected by filtering the cold mixture consisted of practically pure *brucine* d-*chloroiodoacetate*. This had $[\alpha]_D$ + 21·0° in methyl alcohol (c = 0·60); 0·7647 g. of the corresponding ammonium *d*-chloroiodoacetate contained in 20·0 c.c. of aqueous solution gave α_D + 1·80° in a 2-dcm. tube, whence $[\alpha]_D$ + 23·5° and $[M]_D$ + 56°. A second fraction of brucine salt, yielded similarly by the filtrate from the first fraction, gave a specimen of ammonium *l*-chloroiodoacetate having $[M]_D$ − 32°. Both these fractions of brucine salt were allowed to separate spontaneously without inoculation. Further experiments indicated that the character of the crystalline separation may be controlled to some extent by seeding the supersaturated solution in such instances with specimens of brucine *l*- or *d*-chloroiodoacetate; it is proposed to submit this interesting aspect of the investigation to fuller inquiry.

Ammonium l- *and* d-*Chloroiodoacetate*.—The ammonium salts were prepared by shaking cold aqueous suspensions of the corresponding brucine salts with the calculated amounts of a standard solution of ammonia in presence of chloroform, the ammonia being added in small portions. Traces of brucine were finally removed from the aqueous layer by two further extractions with chloroform. After the value of α_D in a 2-dcm. tube had been observed for the resulting aqueous solution of the ammonium salt, an aliquot portion of the solution was evaporated to dryness on the water-bath in order to determine the concentration. The highest values observed for ammonium *l*- and *d*-chloroiodoacetate were $[\alpha]_D$ − 25·1°, $[M]_D$ − 60°, and $[\alpha]_D$ + 23·5°, $[M]_D$ + 56°, respectively, in dilute aqueous solution as recorded above. The salts form colourless, glistening needles (decomp. 149—150°), which usually become yellow and then brown when kept for a few days in a desiccator (Found: NH_3, 7·2. $C_2H_5O_2NClI$ requires NH_3, 7·2%). It is thus advisable to keep the active acids in the form of their brucine salts. Crystalline ammonium *l*-chloroiodoacetate having $[\alpha]_D$ − 25·1°, $[M]_D$ − 60°, after being kept for 10 days in a desiccator, gave $[\alpha]_D$ − 24·6°, $[M]_D$ − 58° in aqueous solution (c = 0·50) and $[\alpha]_D$ − 24·4°, $[M]_D$ − 57° in methyl alcohol (c = 0·75).

No appreciable racemisation was noticed upon evaporating 5 c.c. of a 3·76% aqueous solution of ammonium *d*-chloroiodoacetate, having $[\alpha]_D$ + 19·0°, to dryness on the water-bath. Sustained evaporation of larger volumes, however, produced a noticeable decline in the rotatory power; a slow decline occurred also on keeping aqueous solutions at the ordinary temperature for long periods.

Upon dissolving 0·2116 g. of ammonium salt having $[M]_D$ − 58° in 20·0 c.c. of water containing 2 c.c. of 2*N*-ammonia and evaporating the solution to dryness on the water-bath, the recovered salt had $[\alpha]_D$ − 10·4°, $[M]_D$ − 25°. Further, 0·2200 g. of ammonium salt having $[\alpha]_D$ − 24·1° was dissolved in water containing an equivalent of sodium hydroxide and made up to 20·0 c.c.; the value of α_D in a 2-dcm. tube after 30 minutes was − 0·50°, and when kept for 24 hours at the ordinary temperature the solution had become optically inactive. No appreciable removal of halogen by hydrolysis was detected at the end of this time. In a similar experiment the value of α_D declined over-night from − 0·68° to zero.

An aqueous solution of 0·1882 g. of ammonium salt having $[\alpha]_D$ + 19·0°, $[M]_D$ + 44°, was made up to 20·0 c.c. with water containing one equivalent of hydrochloric acid. The resulting value of α_D in a 2-dcm. tube was + 0·15°, whence the free acid under these conditions has $[\alpha]_D$ + 0·9°, $[M]_D$ + 2°.

BrCHCOOH
|
SO_3H

$C_2H_3BrO_5S$ Ref. 5

Séparation du composant dextrogyre par „cristallisation froide" du sel de strychnine.

Le sel neutre de strychnine se prépare par double décomposition d'un bromosulfoacétate soluble et de l'acétate de strychnine. Il cristallise avec 4 molécules d'eau en petites aiguilles, qui, essorées, forment une masse feutrée très luisante.

Substance 0.4798 g.; perte à 105° 0.0356 g. Subst. 0.4908 g.; perte 0.0357 g.
Trouvé: H_2O 7.42, 7.27.
$C_2H_3O_5BrS$. 2 $C_{21}H_{22}O_2N_2$. 4 H_2O (959.45). Calculé: „ 7.51.
Substance séchée 1.166 g.; 2.619 m. équiv. de NaOH (phénolphtaléine).
Trouvé: Poids équiv. 445.2.
$C_2H_3O_5BrS$. 2 $C_{21}H_{22}O_2N_2$. Calculé: „ „ 443.7.

Après quelques expériences préliminaires le sel du composant droit pur a été obtenu ainsi:

Une solution de 15.3 g. (40 m.mol.) du bromosulfoacétate de baryum $C_2HO_5BrSBa . 1\frac{1}{2}H_2O$ dans 1 litre d'eau a été mélangée avec une solution de 26.7 g. (80 m.mol.) de strychnine dans 1 litre d'eau et 6 à 7 cm³. d'acide acétique.

Le lendemain 7 g. du sel de strychnine (à peu près un cinquième de la quantité totale) s'étaient déposés.

Une partie de ce produit, transformée en sel ammoniacal, a été soumise à l'examen polarimétrique.

On a recristallisé le sel de strychnine en le décomposant à 0° par l'ammoniaque et en le reprécipitant par l'acétate de strychnine, opération qu'on a répétée deux fois.

La rotation du sel ammoniacal n'a plus changé après la première recristallisation.

Un lavage de ce sel de strychnine à l'eau froide, destiné à l'élimination des parties les plus solubles, n'a pas eu d'influence sur le pouvoir rotatoire du sel ammoniacal, tandis que le même traitement a fait monter la rotation d'un échantillon moins pur jusqu'à celle du produit pur.

```
HO3S-CH-COOH
     |
     Cl
```

$C_2H_3ClO_5S$ Ref. 6

d-*Chlorosulphoacetic Acid.*—Sodium chlorosulphoacetate (0·05 mol.) in 1150 c.c. of water was treated with 33·4 g. (0·1 mol.) of strychnine dissolved in the same volume of water containing 11 g. of acetic acid. Slender needles began to separate within 1 hour, and after 10 hours, at 20°, 9 g. of strychnine salt were collected, corresponding to 20% of the quantity formed (44·8 g.).

For polarimetric examination, weighed quantities of the strychnine salt (0·5—2 g.) were shaken at 0° with the theoretical quantity of dilute ammonia, the volume was brought to 20 c.c., and the solution of the ammonium salt, after being extracted four times with half its volume of chloroform, was examined by means of a polarimeter (Schmidt and Haensch) with monochromator. In some cases, the concentration of the ammonium salt was verified by evaporation of the solution. The rotatory power was measured for $\lambda = 589\ \mu\mu$ (D) and for two arbitrary wave-lengths in the green ($\lambda = 533\ \mu\mu$) and blue ($\lambda = 494\ \mu\mu$). The ammonium salt, prepared from the strychnine salt mentioned above, gave the following figures: Concentration 0·00480 g.-mol. in 100 c.c.; $l = 2$; $\alpha_D = +0{\cdot}18°$; $[M]_D = +18{\cdot}7°$.

For recrystallisation, 4·48 g. (0·005 mol.) of the active strychnine salt were decomposed by ammonia. The separated strychnine was dissolved in dilute acetic acid and added to the solution of the ammonium salt. The total volume being 230 c.c., the concentration was the same as above. 1·3 Grams, or nearly 30%, separated. Rotation of the ammonium salt: Conc. 0·01115 g.-mol. in 100 c.c.; $l = 2$; $\alpha_D = +0{\cdot}45°$, $\alpha_{533} = +0{\cdot}58°$, $\alpha_{494} = +0{\cdot}74°$; $[M]_D = +20{\cdot}2°$, $[M]_{533} = +26{\cdot}0°$, $[M]_{494} = +33{\cdot}2°$. Conc. 0·00560 g.-mol. in 100 c.c.; $l = 2$; $\alpha_D = +0{\cdot}23°$, $\alpha_{533} = +0{\cdot}30°$, $\alpha_{494} = +0{\cdot}37°$; $[M]_D = +20{\cdot}5°$, $[M]_{533} = +26{\cdot}8°$, $[M]_{494} = +33{\cdot}0°$.

Addition of sulphuric acid to liberate the acid (2 mols. for 1 mol. of ammonium salt) increased the rotatory power, but this did not change further on addition of more sulphuric acid. Rotation of the free acid: Conc. 0·005 g.-mol. in 100 c.c.; $l = 2$; $\alpha_D = +0{\cdot}40°$, $\alpha_{533} = +0{\cdot}52°$, $\alpha_{494} = +0{\cdot}62°$; $[M]_D = +40°$, $[M]_{533} = +52°$, $[M]_{494} = +62°$.

l-*Chlorosulphoacetic Acid.*—Sodium chlorosulphoacetate (0·05 mol.) was mixed with 30 g. (0·1 mol.) of cinchonine dissolved in water containing 9 g. of acetic acid. The total volume was 1150 c.c. After 2 days, 5·5 g. of the cinchonine salt, or 14% of the total amount (39 g.), had separated.

The product was decomposed with ammonia like the strychnine salt, the only difference being that the alkaloid was filtered off before extraction of the solution with chloroform. Rotation of the ammonium salt: Conc. 0·00278 g.-mol. in 100 c.c.; $l = 2$; $\alpha_D = -0{\cdot}08°$; $[M]_D = -14{\cdot}4°$.

The active cinchonine salt (3·9 g.; 0·005 mol.) was recrystallised in the same way as the strychnine salt. From 115 c.c., the same concentration as above, there separated 0·95 g. or 25%. Rotatory power of the ammonium salt: Conc. 0·00324 g.-mol. in 100 c.c.; $l = 2$; $\alpha_D = -0{\cdot}12°$, $\alpha_{533} = -0{\cdot}17°$, $\alpha_{494} = -0{\cdot}23°$; $[M]_D = -18{\cdot}5°$, $[M]_{533} = -26{\cdot}2°$, $[M]_{494} = -35{\cdot}5°$.

The acid liberated by an excess of sulphuric acid gave the figures: Conc. 0·00305 g.-mol. in 100 c.c.; $\alpha_D = -0{\cdot}23°$, $\alpha_{533} = -0{\cdot}29°$, $\alpha_{494} = -0{\cdot}38°$; $[M]_D = -37{\cdot}7°$, $[M]_{533} = -47{\cdot}5°$, $[M]_{494} = -62{\cdot}3°$.

Thus the mean values for the rotatory power are:

Chlorosulphoacetic acid: $[M]_D = \pm 39°$, $[M]_{533} = \pm 50°$, $[M]_{494} = \pm 62°$.

Neutral ammonium salt: $[M]_D = \pm 20°$, $[M]_{533} = \pm 26°$, $[M]_{494} = \pm 34°$.

A resolution procedure for the above compound using hydroxyhydrindamine* may be found in Ref. 6a.

*Hydroxyhydrindamine is cis-2-amino-1-hydroxyhydrindene.

```
CH3-CH-COOH
    |
    Br
```

$C_3H_5BrO_2$ Ref. 7

l-Brompropionsäure.

200 g inactive käufliche Brompropionsäure werden zunächst nach Ramberg's Vorschrift in fünf Litern Wasser von 45° gelöst und unter tüchtigem Schütteln 200 g gepulvertes Cinchonin in kleinen Portionen zugegeben. Statt diese Flüssigkeit nach Ramberg verdunsten zu lassen, ist es bequemer, sie unter stark vermindertem Druck aus einem Bade, dessen Temperatur zwischen 40° und 50° liegt, auf die Hälfte einzudampfen. Dabei scheidet sich ein erheblicher Theil des Cinchoninsalzes krystallinisch ab. Nach dem völligen Erkalten wird das Salz filtrirt. Seine Menge beträgt 190—200 g. Falls das Salz sich ölig abscheidet, ist es rathsam, einige Krystalle einzuimpfen, die man sich durch Verdunsten einer kleinen Menge der Lösung bei gewöhnlicher Temperatur ziemlich rasch verschaffen kann. Das so gewonnene Cinchoninsalz wird nun gepulvert, in Wasser von 40° durch Schütteln gelöst und diese Flüssigkeit abermals unter vermindertem Druck eingeengt, bis der grösste Theil ausgeschieden ist, circa 170 g. Diese Art der Krystallisation muss nun sehr häufig (15—20 Mal) wiederholt werden. Zur Isolirung der freien Säure löst man das Salz in überschüssiger verdünnter Salzsäure, äthert die Brompropionsäure aus, trocknet den Aether mit Natriumsulfat, verdampft und fractionirt den öligen Rückstand unter sehr geringem Druck. Bei 0,2—0,4 mm destillirt die Säure aus einem Bade, dessen Temperatur 70—80° beträgt, völlig über. Auf diese Art wurde eine Säure gewonnen, die bei 20° Natriumlicht 45,64° nach links drehte und das spec. Gew. 1,7084 hatte. Daraus berechnet sich $[\alpha]_D^{20}$ —26,7°. Die Analyse desselben Präparates ergab:

0,2035 g gaben 0,2509 AgBr.

	Berechnet für $C_3H_5O_2Br$	Gefunden
Br	52,26	52,47

Die Ausbeute an diesem Producte beträgt ungefähr 5 pC. der inactiven Säure, also 10 pC. der Theorie. Ob die Säure ganz rein gewesen ist, lässt sich schwer sagen. Es verdient aber bemerkt zu werden, dass das Cinchoninsalz, nachdem seine Säure bereits den Drehungswinkel — 42° im 1 dm-Rohre erreicht hatte, noch fünfmal umkrystallisirt war.

The resolution of the above compound using the same resolving agent is found in L. Ramberg, Chem. Ber., 33, 3354 (1900).

$BrCH_2CHCOOH$
 |
 OH

$C_3H_5BrO_3$ **Ref. 8**

Spaltung der Brom-milchsäure.

27 g Morphin werden mit 120 ccm Wasser aufgekocht und durch Zugabe einer lauwarmen Lösung von 30 g reiner Brom-milchsäure in 15 ccm Wasser gelöst. Die Flüssigkeit wird sofort in Eis gebracht; dabei scheidet sich das Morphinsalz der linksdrehenden Brom-milchsäure in rechteckigen, gut ausgebildeten, mikroskopischen Platten ab. Es wird scharf abgesaugt, mit Wasser gewaschen und noch feucht mit 17 g Kaliumbisulfat fein zerrieben, wobei ein dünner Brei entsteht. Diesen vermengt man mit 50 g entwässertem Natriumsulfat, das in einer halben Stunde die Hauptmenge des Wassers bindet. Die erstarrende Masse extrahiert man durch Verreiben mit Äther, den man mehrmals erneuert. Zuletzt wird mit Seesand fein zerrieben und im Soxhlet-Apparat mit Äther der Rest gewonnen. Die Extrakte entwässert man mit Natriumsulfat und verjagt den Äther, zuletzt im Vakuumexsiccator. Der krystallinische Rückstand wird mit der doppelten Gewichtsmenge kalten Chloroforms extrahiert, das die linksdrehende Brom-milchsäure zurückläßt. Ausbeute 10 g.

Für die Analyse wurde das Produkt in wenig Wasser kalt gelöst, von einer Trübung abfiltriert, im Vakuumexsiccator zur Trockne verdampft und schließlich 2 Stunden über Phosphorpentoxyd bei 12 mm auf 65° erwärmt, wobei geringe Mengen der Säure sublimieren.

0.1720 g Sbst.: 0.1915 g AgBr.

$C_3H_5O_3Br$ (168.96). Ber. Br 47.30. Gef. Br 47.38.

0.1629 g Sbst. Gesamtgewicht der wäßrigen Lösung 1.5648 g. d_4^{18} = 1.054. Drehung im 1-dm-Rohr bei 18° und Natriumlicht 0.21° nach links. Mithin $[\alpha]_D^{18} = -1.91°$ (± 0.14).

Eine Probe des Morphinsalzes wurde noch zweimal aus Wasser umkrystallisiert und lieferte eine Säure von $[\alpha]_D^{20} = -1.94°$ (± 0.2). Die Säure schmilzt nach vorherigem Erweichen bei 80° und erweist sich in allen Eigenschaften als der optische Antipode der aus *l*-Isoserin gewonnenen Brom-milchsäure.

Die Mutterlauge vom brom-milchsauren Morphin wird unter vermindertem Druck auf die Hälfte eingeengt und über Nacht im Eisschrank aufbewahrt. Dabei krystallisiert etwa 1 g Morphinsalz von gleicher Krystallform wie die Hauptfraktion aus. Die filtrierte Lösung wird mit 5 g Kaliumbisulfat versetzt und unter vermindertem Druck zum Sirup eingeengt, der ausgeäthert, mit Natriumsulfat verrieben und im Soxhlet-Apparat mit Äther erschöpft wird. Der vereinigte Ätherextrakt wird eingedampft, der Rückstand in 10 ccm lauwarmem Wasser aufgenommen und mit der kalten Lösung von 48 g wasserhaltigem Brucin in 180 ccm 95-proz. Methylalkohol versetzt. Das Brucinsalz der rechtsdrehenden Brom-milchsäure beginnt nach wenigen Augenblicken in langgestreckten Prismen zu krystallisieren und wird nach zweistündigem Stehen bei gewöhnlicher Temperatur abgesaugt. Ausbeute 39 g. Es wird mit der Lösung von 15 g Kaliumbisulfat in möglichst wenig kochendem Wasser übergossen und auf dem Wasserbad digeriert, bis die Masse dünnflüssig geworden ist. Beim Einstellen in Eis bildet sich ein starrer Krystallkuchen, der mit Sand zerrieben und mit Äther gründlich extrahiert wird. Die Ätherlösung wird wie bei der linksdrehenden Säure verarbeitet und liefert 9—10 g rechtsdrehender Säure, die mit der aus *l*-Isoserin bereiteten identisch ist.

0.1163 g Sbst. Gesamtgewicht der wäßrigen Lösung 1.097 g. d_4^{22} = 1.050. Drehung im 1-dm-Rohr bei Natriumlicht und 22° 0.20° nach rechts. Mithin $[\alpha]_D^{22} = +1.80°$ (± 0.14). Eine Probe des Brucinsalzes wurde 6-mal aus Wasser umkrystallisiert und lieferte eine Säure von $[\alpha]_D^{22} = +1.93°$ (± 0.1).

Aus der methylalkoholischen Mutterlauge des Brucinsalzes lassen sich etwa 5 g fast inaktiven Ausgangsmaterials zurückgewinnen.

$ClCH_2CHCOOH$
 |
 OH

$C_3H_5ClO_3$ **Ref. 9**

Spaltung der β-Chlor-milchsäure.

50 g Chlor-milchsäure, nach Richter[4]) hergestellt, werden in 1000 ccm Wasser warm gelöst und unter tüchtigem Umrühren mit 158 g käuflichem Brucin (Kahlbaum) versetzt. Die klare Lösung wird 1 Stde. auf dem Wasserbade auf 60—70° erwärmt. Bei Zimmer-Temperatur scheidet sich das Brucinsalz der *d*-Verbindung in farblosen, meist strahlenförmig gruppierten Nadeln ab, das nach 1-tägigem Stehenlassen abgenutscht und mit kaltem Alkohol und Äther gewaschen wird. Um die Beimengung der isomeren Verbindung zu entfernen, genügt 4-malige Krystallisation, zuerst aus 800 ccm und zuletzt aus 250 ccm heißem Wasser. Ausbeute an reinem *d*-Salz 60 g. In heißem Alkohol leicht, in Äther nicht, in kaltem Wasser bzw. in heißem leichter löslich. Schmp. 130—138°.

d-β-Chlor-milchsäure: 52.4 g Brucinsalz wurden in Wasser warm gelöst, mit 101 ccm *n*-Natronlauge zerlegt, das Brucin nach 1-tägigem Stehen bei 0° möglichst scharf abgenutscht, die Krystalle mit kaltem Wasser angerührt und wieder abgenutscht. Die vereinigten Filtrate wurden mit 101 ccm *n*-Salzsäure angesäuert, unter vermindertem Druck eingeengt und im Extraktionsapparat 10 Stdn. lang mit Äther extrahiert. Der Äther-Rückstand beginnt, im Exsiccator über Schwefelsäure lange Zeit getrocknet, zu krystallisieren. Zur Reinigung wird die Substanz in Wasser gelöst und 12 Stdn. ausgeäthert. Ausbeute 12.0 g. In Wasser, Alkohol und Äther leicht löslich. Schmp. 91.5°.

0.0512 g Sbst.: 0.0551 g CO_2, 0.0188 g H_2O. — 0.0483 g Sbst.: 0.0513 g CO_2, 0.0175 g H_2O.

$C_3H_5O_3Cl$. Ber. C 28.89, H 4.04.
Gef. „ 29.33, 28.96, „ 4.11, 4.05.

$[\alpha]_D^{18} = +0.85° \times 26.3013/4.0763 \times 1.0549 \times 2 = +2.60°$ (in Wasser).
$[\alpha]_D^{20} = +0.68° \times 25.7276/2.4448 \times 1.0350 \times 2 = +3.45°$ (in Wasser).

[4]) Richter, Journ. prakt. Chem. [2] **20**, 193 [1879].

FCH_2-CD-COOH
 |
 NH_2

$C_3H_5DFNO_2$ **Ref. 10**

All optical rotations were determined on a Carl Zeiss photoelectric precision polarimeter Model LEP A1 as a 6% solution in 1 N hydrochloric acid at 25 °C unless otherwise specified. NMR spectra were obtained with a Varian T-60 spectrometer and sodium 3-(trimethylsilyl)propanesulfonate as the internal standard and interpreted by Dr. Alan W. Douglas. Microanalyses were done through the courtesy of Mr. J. P. Gilbert and associates. Melting points were determined on a Thomas-Hoover apparatus and are uncorrected.

Resolution of 3-Fluoro-DL-alanine-*2-d* (1). Racemic 3-fluoroalanine-*2-d*[6] (100 g, 0.935 mol), 2,4-pentanedione (103 g, 1.03 mol), and quinine (320 g, 0.99 mol) in 1870 ml of anhydrous methanol were treated under reflux in a nitrogen atmosphere for 1 h. After cooling and stirring for 1 h at 20 °C the quinine salt of *N*-(1-methyl-2-acetylvinyl)-3-fluoro-L-alanine-*2-d* (**4a**) CH_3OH solvate was collected, washed with cold CH_3OH, and dried in vacuo at room temperature to give 176 g, $[\alpha]^{25}D$ −89.6° (*c* 2.0, 95% EtOH), mp 143–144 °C dec.

Anal. Calcd for $C_{28}H_{36}FN_3O_5 \cdot CH_3OH$: C, 63.83; H, 7.39; F, 3.48; N, 7.70. Found: C, 63.78; H, 7.30; F, 3.69; N, 7.90.

Removal of the solvent from the filtrate and washes, followed by crystallization of the residue from 500 ml of EtOAc, gave after 16 h at 0 °C a second crop of 55.2 g. The yield of analytically pure first and second crops was 96.6%. The anhydrous form was obtained by drying a sample at 50 °C for 2 h.

Anal. Calcd for $C_{28}H_{36}FN_3O_5$: C, 65.48; H, 7.07; F, 3.70; N, 8.18. Found: C, 65.66; H, 7.46; F, 3.77; N, 8.23.

The residue (**4b**) from the second crop was dissolved in 360 ml of water and basified by adding 450 ml of 1.2 N (0.54 mol) sodium hydroxide with stirring at 10–15 °C. The liberated quinine was removed by chloroform extraction and the aqueous phase was acidified and stirred with 500 ml of 2 N hydrochloric acid at 15–20 °C. After 15 min, the protecting group was removed as evidenced by the dissolution of the precipitated *N*-(1-methyl-2-acetylvinyl)-3-fluoro-D-alanine-*2-d*.

After extraction with chloroform to remove 2,4-pentanedione, the aqueous phase was clarified with charcoal and the filtrate was percolated through 1.1 l. of Dowex 50W × 4 (H^+ form). The column was washed free of Cl^- with water, then the product was eluted with 0.5 N ammonium hydroxide, collecting and concentrating in vacuo the ninhydrin-positive fractions until the product crystallized (≈150 ml). The mixture was cooled to 5 °C and aged for several hours, and the crystalline 3-fluoro-D-alanine-*2-d* was collected, washed with cold water, and vacuum dried at 40 °C to give 28.5 g (57%), mp 174–175 °C, $[\alpha]^{25}D$ −10.4°.

Anal. Calcd for $C_3H_6NO_2F$: C, 33.65; H, 5.65; N, 13.08; F, 17.74. Found: C, 33.41; H, 5.74; N, 12.96; F, 17.53.

3-Fluoro-L-alanine-*2-d* was separated from the crystalline quinine *N*-(1-methyl-2-acetylvinyl)-3-fluoro-L-alaninate-*2-d* (231.5 g) as described above for its enantiomer. In this manner 35.5 g (68%) of 3-fluoro-L-alanine-*2-d* was obtained, mp 174–175 °C, $[\alpha]^{25}D$ +10.3°.

Anal. Calcd for $C_3H_6NO_2F$: C, 33.65; H, 5.65; N, 13.08; F, 17.74. Found: C, 33.45; H, 5.78; N, 13.01; F, 17.77.

Resolution of 3-fluoro-DL-alanine was performed starting from potassium *N*-(1-methyl-2-acetylvinyl)-3-fluoro-DL-alaninate.

To a solution of 22.72 g (100 mmol) of potassium *N*-(1-methyl-2-acetylvinyl)-3-fluoro-DL-alaninate (3) in 200 ml of methanol was added 40.5 g (102.5 mmol) of quinine hydrochloride dihydrate. The mixture was heated under reflux for a period of 1 h. Application of the same method as described above for the separation and isolation gave 3-fluoro-D-alanine, $[\alpha]^{25}D$ −10.4°, and 3-fluoro-L-alanine, $[\alpha]^{25}D$ +10.4°, in 54.2 and 64% yield, respectively.

Authors' comments:

β-Diketones have been previously used for protecting α-amino acids as their enamines during peptide synthesis.[2,3] Prior to our report these derivatives were synthesized as their potassium salt[2,4] or the more crystalline dicycohexylamine salt.[3] We explored the possibility of using an optically active base to form their crystalline diastereomers, and found that the quinine salt crystallized easily in good yield and excellent purity. Further advantages of this resolution method include the one-step derivatization–salt formation and the ease of removing the protecting group.

Conversion of 3-fluoro-DL-alanine (1)[5,6] to the quinine *N*-(1-methyl-2-acetylvinyl)-3-fluoro-DL-alaninate (4) is accomplished by warming 1 with quinine and 2,4-pentanedione

Scheme I

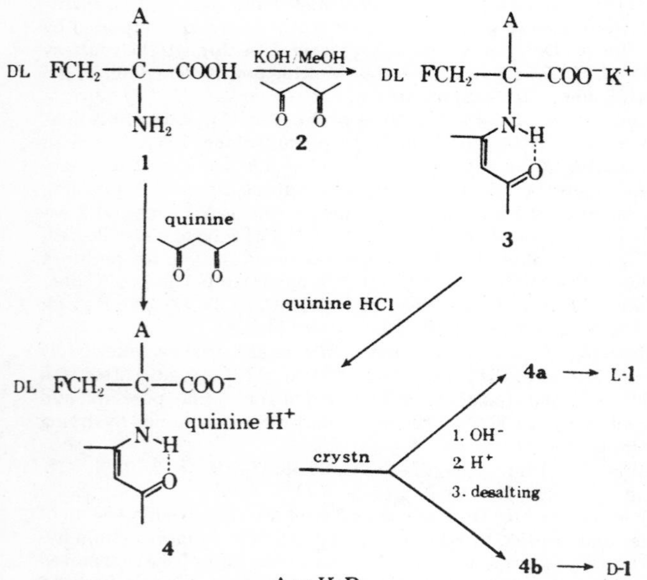

A = H, D

in methanol. The L isomer **4a** crystallizes from the reaction mixture in high (≈99%) optical purity.

The quinine salt **4** can also be made by treatment of the potassium salt **3** with 1 equiv of quinine HCl. From the resolved **4a** or **4b** the quinine is separated by extraction with chloroform of the basified solution. The masking group is readily cleaved from the resolved substrate by mild acid hydrolysis and separated by extraction. In the final step ion exchange is employed for the removal of inorganics. The L-1 or D-1 is eluted with 0.5 N ammonium hydroxide and isolated by crystallization after reducing the volume of eluate in vacuo.

(1) (a) F. M. Kahan and H. Kropp, Abstracts, 15th Interscience Conference on Antimicrobial Agents and Chemotherapy, Washington, D.C., Sept 24–26, 1975, No. 100; (b) H. Kropp, F. M. Kahan, and H. B. Woodruff, *ibid.*, No. 101; (c) J. Kollonitsch, L. Barash, N. P. Jensen, F. M. Kahan, S. Marburg, L. Perkins, S. M. Miller, and T. Y. Shen, *ibid.*, No. 102; (d) F. M. Kahan, H. Kropp, H. R. Onishi, and D. P. Jacobus, *ibid.*, No. 103.

(2) (a) E. Dane and T. Dockner, *Angew. Chem., Int. Ed. Engl.*, **3**, 439 (1964); (b) *Chem. Ber.*, **98**, 789 (1965).

(3) (a) G. L. Southard, G. S. Brooke, and J. M. Pettee, *Tetrahedron Lett.*, 3505 (1969); (b) G. L. Southard, G. S. Brooke, and J. M. Pettee, *Tetrahedron*, **27**, 1359 (1971).

(4) R. G. Hiskey and G. L. Southard, *J. Org. Chem.*, **31**, 3582 (1966).

(5) (a) C. Y. Yuan, C. N. Chang, and Y. F. Yeh, *Acta Pharm. Sinica*, **7**, 237 (1959); *Chem. Abstr.*, **54**, 12096 (1960); (b) H. Lettré and U. Wölcke, *Justus Liebigs Ann. Chem.*, **708**, 75 (1967).

(6) 3-Fluoro-DL-alanine and 3-fluoro-DL-alanine-*2-d* were prepared from fluoropyruvic acid: (a) U. H. Dolling, E. J. J. Grabowski, E. F. Schoenewaldt, M. Sletzinger, and P. Sohar, 7th Northeast Regional Meeting of the American Chemical Society, August 8–11, 1976, Abstract No. 170; (b) U. H. Dolling et al., German Offen. 2550, 109; (c) Belgian Patent 835 358.

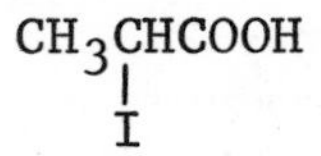

$C_3H_5IO_2$ Ref. 11

1 Tl. Säure wird in einem von Krystallkeimen sorgfältig gereinigten Kolben in 4 Tln. Wasser gelöst und mit der berechneten Menge (0.605 Tln.) aktivem Phenäthylamin (z. B. *d*-Amin) versetzt. Die Lösung wird mit Salz desselben Amins, von einem Vorversuch stammend (am besten reines Homogyrsalz), geimpft, der Kolben verschlossen und bei 15—17° 24 Stdn. sich selbst überlassen. Die dabei in langen, traubenförmig angeordneten Nadeln krystallisierende Salzmenge, Fraktion I, wird abgesaugt und beträgt in der Regel 17—23% der gesamten berechneten Menge. Die Säure aus diesem Salz hat in Äther-Lösung $[\alpha]_D = 30—34°$ mit der Drehungsrichtung des angewandten Amins (+). Fraktion I besteht also hauptsächlich aus Homogyrsalz. Um sich vom Erfolge der Spaltung zu überzeugen, muß man die Säure in Freiheit setzen, da das Salz in wäßriger Lösung nur geringe Aktivität besitzt. Manchmal krystallisiert eine größere Salzmenge als oben angegeben; in diesen Fällen ist die Spaltung weniger vollständig.

Die Mutterlauge von Fraktion I läßt man im offenen Kolben zwei Tage unter Umrühren stehen. Die an den Wänden des Kolbens gebildete Salzkruste dient als Impfmaterial. Die so erhaltene lockere und feinkrystallinische Salzmenge, Fraktion II, wird abgesaugt und getrocknet; sie beträgt meistens 14—18% der gesamten berechneten Salzmenge. Die Säure aus diesem Salz besitzt in Äther-Lösung $[\alpha]_D = 36—40°$ mit der dem angewandten Amin entgegengesetzten Drehungsrichtung. Fraktion II besteht somit hauptsächlich aus Antigyrsalz.

Die Mutterlauge wird mit einem Warmluft-Ventilator bis fast zur Trockne eingeengt, wobei die Lösung nicht über 25—30° erwärmt werden darf, da sonst schnelle Hydrolyse der Säure eintritt. Die geringe Menge Mutterlauge wird abgesaugt, das Salz nach dem Trocknen gewogen und im 2½-fachen seines Gewichts an Wasser von 60—70° gelöst. Nachdem alles in Lösung gegangen ist, wird die Lösung gekühlt, wobei man wie oben zwei neue Fraktionen erhält, die mit I und II vereinigt werden. Mehrere Male gelöstes und durch Eindunsten wiedergewonnenes Salz gibt wegen Störung der Krystallisation durch Verunreinigungen bedeutend kleinere Mengen beider Fraktionen.

Fraktion I wird in Teilen von 20 g 3-mal umkrystallisiert. Die 2.3- bis 2.6-fache Gewichtsmenge Wasser wird auf 75—80° erwärmt und über das Salz gegossen, worauf man ½ Min. schüttelt. Wenn etwas Salz ungelöst bleiben sollte, wird von diesem abgegossen und die Lösung sodann unter fließendem Wasser rasch gekühlt, um Hydrolyse der Säure weitestgehend einzuschränken. Es krystallisieren jedesmal 50—63% der angewandten Salzmenge. Wenn die Drehung der Säure im ursprünglichen Salz 32° betrug, ist sie im allgemeinen nach der ersten Krystallisation ungefähr 43°, nach der zweiten 47—48° und nach der dritten ungefähr 49°. Die an Säure aus durch Umkrystallisieren gereinigtem Homogyrsalz beobachtete höchste Drehung war 49.8°. Das Salz der Mutterlaugen wird auf oben angegebene Weise weiterbehandelt und umkrystallisiert.

Fraktion II wird wegen ihres schlechten Krystallisationsvermögens nicht umkrystallisiert, sondern daraus die Säure mit verd. Schwefelsäure in Freiheit gesetzt und ausgeäthert. Das Phenäthylamin wird aus der schwefelsauren Lösung regeneriert. Nach dem Abdunsten des Äthers wird die Säure in der 3-fachen Gewichtsmenge Wasser gelöst und mit dem optischen Antipoden des vorher benutzten Amins neutralisiert, wobei man das entsprechende Homogyrsalz erhält. Dieses wird in gleicher Weise wie sein aus Fraktion I erhaltener optischer Antipode gereinigt. Schon das ursprüngliche Salz enthält ziemlich stark aktive Säure mit $[\alpha]_D = 44-49^0$.

Löslichkeit und Krystallform eines Homogyr- und eines Antigyrsalzes wurden untersucht. Das Homogyrsalz war *d-d*-Verbindung, wie oben angegeben durch Umkrystallisation erhalten. Das Antigyrsalz war *l*-Phenäthylammonium-*d*-α-Jod-propionat, erhalten durch Neutralisation von Säure aus dem reinsten *d-d*-Salz mit *l*-Amin und einmaliges Umkrystallisieren des gebildeten Salzes.

Die Salze wurden bei der angegebenen Temperatur einige Stunden mit Wasser turbiniert. Die Konzentration der Lösungen wurde durch Jod-Analyse in einer gewogenen Menge Lösung bestimmt.

100 g Wasser lösen g Substanz:

Temperatur	0°	20°
Homogyrsalz	14.14	16.17
Antigyrsalz	11.06	13.60.

Die optisch aktiven Formen der α-Jod-propionsäure.

d-α-Jod-propionsäure wurde mit verd. Schwefelsäure aus ihrem Homogyrsalz in Freiheit gesetzt und ausgeäthert. Die ätherische Lösung wurde getrocknet, der Äther abdestilliert, die durch freies Jod dunkelrote Säure mit Hilfe einer kleinen amalgamierten Kupferdrahtnetzrolle entfärbt und sodann die Drehung bestimmt. Hierauf wurde die Säure bei 0.12 bis 0.22 mm Druck (Temperatur des Bades 75—100⁰) destilliert und neuerdings die Drehung und die Dichte bestimmt. Die Destillation hatte keine merkbare Racemisierung zur Folge.

Aus der Säure wurde sodann das Ammoniumsalz dargestellt und dieses zweimal umkrystallisiert. Die Drehung der Säure aus dem umkrystallisierten Salz wurde wieder vor und nach einer folgenden Hochvakuum-Destillation bestimmt. Wie sich aus den unten wiedergegebenen Werten ergibt, steigt die Drehung der Säure durch Umkrystallisieren des Ammoniumsalzes.

Analyse einer aus *d*-Homogyrsalz erhaltenen destillierten Säure: 0.2968 g Sbst. verbrauchen 14.41 ccm 0.1035-*n*. Barytlösung. — 0.1659 g Sbst.: 0.1941 g AgJ.
$C_3H_5O_2J$ (200.0). Ber. Äquiv.-Gew. 200.0, J 63.46. Gef. Äquiv.-Gew. 199.1, J 63.26.
Dichte: $D_4^{18} = 2.073$.

Aktivität der *d*-α-Jod-propionsäure.

	Säure aus dem Homogyrsalz t = 18.5°	Säure aus dem Ammoniumsalz, t = 17.5°
Drehung im 2.5-cm-Rohr	+21.33 ± 0.05°	+21.75 ± 0.05°
Spez. Drehung in Substanz	+41.16 ± 0.12°	+41.97 ± 0.12°
Spez. Drehung in Äther-Lösung	+49.35 ± 0.3°	+50.73 ± 0.3°
Molekulare Drehung in Substanz	+82.3°	+83.9 ± 0.2°
Spez. Drehung in 0.4-*n*. Wasser-Lösung, 2.5-*n*. in Bezug auf H_2SO_4	—	+41.4°
Spez. Drehung in 0.2-*n*. Wasser-Lösung, 2.5-*n*. in Bezug auf H_2SO_4	—	+40.4°

Wie weiter unten bei den Salzen erwähnt wird, drehen das nicht dissoziierte Molekül der Säure und ihr Ion in entgegengesetzter Richtung. Die molekulare Drehung des Anions der *d*-Säure ist $[M]_D = -30^0$ bis -31^0. Die *d*-Säure konnte weder durch Kohlensäure-Schnee noch durch monatelanges Aufbewahren während des Winters im Freien (−5⁰ bis −40⁰) zur Krystallisation gebracht werden. Ebenso verhält es sich mit der *l*-Säure.

Die *l*-α-Jod-propionsäure aus dem *l*-Phenäthylaminsalz wurde unmittelbar in das Ammoniumsalz übergeführt. Vorversuche ergaben, daß einmaliges Umkrystallisieren genügte, um Salz mit maximaler Drehung zu erhalten (s. unten). Obwohl die spez. Drehung des Ammoniumsalzes ebenso groß wie die des entsprechenden *d*-Salzes war, zeigte die aus demselben in Freiheit gesetzte Säure eine bedeutend geringere Drehung als die *d*-Säure. Es ergab sich, daß die Säure auch nach der Vakuum-Destillation eine Verunreinigung enthielt, die sich beim Lösen der *l*-Säure in Wasser als weiße Trübung zu erkennen gab; die *d*-Säure löst sich vollständig in Wasser

Optische Aktivität der *l*-α-Jod-propionsäure.

Drehung im 2.5-cm-Rohr bei 17.5°	−21.42 ± 0.05°
Spez. Drehung in Substanz $[\alpha]_D^{17.5}$	−41.34 ± 0.12°
Spez. Drehung in Äther-Lösung (ca. 0.3-*n*.)	−49.87 ± 0.5°
Molekulare Drehung in Substanz $[M]_D^{17.5}$	−82.7 ± 0.2°
Drehung im 2.5-cm-Rohr bei 25°	−20.5 ± 0.1°

$C_3H_6O_3$

$CH_3CHCOOH$
$\quad\;|$
$\quad OH$

Ref. 12

50 ccm sirupöse Milchsäure werden mit 600 ccm Wasser 8 Stdn. gekocht. Die eine Hälfte der Lösung wird heiß mit 94 g Morphin versetzt, die andere Hälfte wird mit Soda genau neutralisiert (etwa 18 g wasser-freies Salz) und zu der ersten Hälfte gegeben. Nun wird aus einem Bade von 40⁰ bei Unterdruck eingedampft, bis die Krystallisation beginnt. Die Masse wird in Eis gestellt und nach einigen Stunden scharf abgesaugt. Die Mutterlauge wird durch weitere Konzentrationen noch 2-mal ebenso auf Morphinlactat verarbeitet. Die letzten 80—90 ccm werden beiseite gestellt. Das fast farblose Lactat wird 2-mal aus wenig Wasser umkrystallisiert; jedesmal werden die letzten 30 bis 35 ccm nicht weiter verarbeitet. Unnötiges Erhitzen ist sorgfältig zu vermeiden. Das erhaltene Morphinlactat wird in 400 ccm Wasser von 50⁰ gelöst, das Morphin mit konz. Ammoniak gefällt und nach raschem Abkühlen abgesaugt. Das Filtrat wird zur Entfernung des Ammoniaks im Vakuum ein wenig eingeengt, hierauf mit 10 g reinem Zinkcarbonat bis nahezu zum Verschwinden des Ammoniak-Geruches 2—4 Stdn. gekocht, filtriert und in der Kälte zur Krystallisation gebracht. Das Filtrat wird im Vakuum auf 15 ccm eingeengt und liefert eine zweite Krystallisation. Die vereinigten Fraktionen werden nochmals 1—2-mal umkrystallisiert. Ausbeute 21—22 g; das sind nahezu 60% in Bezug auf das angewendete Morphin und etwa 30% der Milchsäure.

Das Salz zeigt im Ammoniak die l. c. vorgeschriebene Drehung ($[\alpha]_{578}^{20} = +10.95^0$)[18].

Sämtliche Mutterlaugen werden vereinigt, mit konz. Ammoniak versetzt, vom Morphin abfiltriert, wieder eingeengt und schließlich mit konz. Ammoniak mehrere Wochen verschlossen aufbewahrt, damit alles Morphin ausfällt. Wenn das zurückgewonnene Morphin nicht rein weiß ist, muß es über das Hydrochlorid mit Kohle gereinigt werden.

[18]) C. E. Wood, J. E. Such und F. Scarf, Journ. chem. Soc. London **123**, 606 [1923], geben für Natriumlicht $[\alpha]_D^{18} + 10.09^0$ an.

Resolutions of the above compound using the same resolving agent are also found in: J.C. Irvine, J. Chem. Soc., 89, 935 (1906); T.S. Patterson and W.C. Forsyth, J. Chem. Soc. 2263 (1913); C.E. Wood, J.E. Such and F. Scarf, J. Chem. Soc., 600 (1923).

A resolution procedure for the above compound using strychnine as resolving agent may be found in T. Purdie, J. Chem. Soc. 61, 754 (1892).

```
       HOCH2-CH-COOH
             |
             OH
```

$C_3H_6O_4$ Ref. 13

225 g Brucin werden in der gerade nöthigen Menge heissem Alkohol gelöst und mit 50 g Glycerinsäure versetzt. Nach einigem Stehen schied sich in der Kälte glycerinsaures Brucin aus, das abgesaugt und noch viermal aus Alkohol umkrystallisirt wurde. Diese Verbindung wurde in Wasser gelöst und mit Barytwasser in geringem Ueberschuss versetzt. Nach 24-stündigem Stehen wurde das ausgeschiedene Brucin durch Absaugen, der in Lösung befindliche Antheil durch mehrfaches Ausschütteln mit Chloroform entfernt, und Kohlensäure bis zur Sättigung eingeleitet. Die eingeengte Flüssigkeit wurde dann zur Abscheidung gelösten Baryumcarbonats zum Sieden erhitzt, filtrirt und auf ein kleines Volumen verdampft. Beim Versetzen mit Alkohol fiel das glycerinsaure Baryum in festen weissen Krystallen aus, die abgesaugt und aus wässriger Lösung durch Alkohol wieder zur Abscheidung gebracht wurden.

Die Analyse der im Vacuum über Phosphorsäureanhydrid und dann bei 100° zur Gewichtsconstanz getrockneten Substanz ergab:

0.1683 g Sbst.: 0.1287 g CO_2 und 0.0455 g H_2O. — 0.1442 g Sbst.: 0.0973 g $BaSO_4$.

$C_6H_{10}O_8Ba$. Ber. C 20.75, H 2.89, Ba 39.48.
Gef. » 20.87, » 2.98, » 39.52.

Bei der Drehungsbestimmung wurde gefunden für eine Lösung, die in 10.0 ccm 0.7704 g enthielt:

$$[\alpha]_D = -17.38^0$$
$$(\alpha = -1.34^0,\ l = 1).$$

Da das aus Glucuronsäure dargestellte glycerinsaure Baryum unter gleichen Bedingungen ein zwar entgegengesetztes, aber ebenso grosses Drehungsvermögen ($[\alpha]_D = +17.1^0$) aufweist, ist an der Reinheit der Verbindung nicht zu zweifeln.

Die Mutterlauge von jener als *d*-glycerinsaures[2]) Brucin zu bezeichnenden Verbindung schied beim Stehen noch eine kleine Portion unreineren Salzes aus. Die davon abfiltrirte Flüssigkeit ergab beim Einengen eine Fraction, die in das Baryumsalz verwandelt, in 10-proc. Lösung nur noch ca. -0.5^0 drehte. Durch fortgesetzte fractionirte Krystallisation haben wir eine *l*-Glycerinsäure erhalten, deren Baryumsalz im höchsten Falle

$$[\alpha]_D = +8.75^0$$
$$(\alpha = 1.20^0,\ l = 1,\ c = 13.71)$$

zeigte.

[1]) Chem. Soc. 63, 298 [1893].

[2]) Die aus dem reinen linksdrehenden Baryumsalz in Freiheit gesetzte Glycerinsäure ist rechtsdrehend.

A resolution procedure for the above compound using quinine as resolving agent may be found in E. Anderson, Amer. Chem. J., 42, 401 (1909).

```
       HO2Se-CH-COOH
             |
             CH3
```

$C_3H_6O_4Se$ Ref. 14

III. Dédoublement optique de l'acide α-sélénine-propionique.

L'acide α-sélénineproprionique donne un sel bien caractérisé avec deux molécules de quinine. Le sel que forme le composant dextrogyre est beaucoup mieux soluble que le dérivé du composé énantiomorphe.

On peut donc obtenir à l'état pur le sel de quinine de l'acide α-sélénineproprionique lévogyre, renfermant 4 mol. d'eau. Il se ramollit vers 115° et va fondre vers 128° en devenant brun.

Substance 0.2920 g: perte à 80° 0.0231 g.
„ 198.99 mg: Se 17.56 mg.
„ 0.2560 g: baryte 0.5680 équiv.

$C_3H_6O_4Se \cdot 2\,C_{20}H_{24}O_2N_2 \cdot 4\,H_2O$ (905.73). Trouvé: H_2O 7.91: Se 8 82; p. éq. 450.7.
Calculé: „ 7.96; „ 8.75; „ 452.9.

Une solution de 17.8 g (0.05 mol.) d'α-sélénineproprionate de baryum dans 300 cm³ d'eau et une solution de 32.4 g de quinine dans 50 cm³ d'acide chlorhydrique 0.2 n et 500 cm³ d'eau ont été mélangées à 60°. Le mélange filtré et refroidi a donné 27.5 g du sel de quinine.

Pour la recristallisation on l'a dissous, par chauffage modéré, dans le poids double d'alcool et on a versé cette solution dans 10 à 15 fois son volume d'eau tiède (60°).

On a examiné ce sel, en décomposant 0.25 g environ par la quantité équimoléculaire de baryte, en extrayant la quinine à quatre reprises par du chloroforme et en mesurant le pouvoir rotatoire du sel barytique dans un volume de 25 cm³ au moyen d'un tube de 4 dm. Après six recristallisations le sel était optiquement pur. On s'en est assuré par deux cristallisations supplémentaires. Enfin il n'est resté que 0.3 g.

La concentration de la solution du sel barytique était connue d'après le poids du sel de quinine employé; elle fut contrôlée toujours par un dosage du baryum dans 2 portions à 5 cm³ de la solution mesurée. Les autres 15 cm³ de la solution du sel barytique furent additionnés de 10 cm³ d'acide chlorhydrique 2 n pour étudier la rotation de l'acide. Dans ces circonstances l'acide organique, se trouvant en présence d'un excès notable d'acide chlorhydrique, n'est pas dissocié. Un plus grand excès d'acide ne diminue plus la rotation.

Cristaliis.	No. 1	2	3	4	5	6	7	8
Quantité	27.5 g	19.2	11.8	7.5	3.5	2.0	0.7	0.3
$-[M]_D$ Sel		61°	87	109	134	142	141.5	142.5
$-[M]_D$ Acide		44°	58	72	—	100	100	—

Donc on trouve pour la rotation des sels neutres (ion divalente) $[M]_D = -142°$ et pour la rotation de l'acide non-dissocié $[M]_D = -100°$.

Le tableau suivant donne la dispersion rotatoire de l'acide et des sels.

λ (U. A.)	6740	6265	5893	5605	5365	5160
$[M]_D$ Sels neutres	108.5	121.7	142	160.7	177.3	193
$[M]_D$ Acide non-diss.	81.5	89	100	111	120	—

En décomposant une solution du sel barytique par la quantité équimoléculaire de l'acide sulfurique, on obtient l'acide libre, qui est ionisé en partie. Une solution renfermant 2 mol. mg par 100 cm³ a montré la rotation moléculaire $[M]_D = -110°$. La dissociation partielle augmente donc l'activité optique.

```
         SO3H
          |
       H3C-CH-COOH
```

$C_3H_6O_5S$ Ref. 15

Le d-sulfonpropionate acide de strychnine.

L'évaporation de la solution de quantités équimoléculaires de l'acide sulfonpropionique et de la strychnine fait cristalliser le sel acide. [1])

La substance, cristallisée dans l'eau à plusieurs reprises, p. e. cinq fois, constitue, comme nous allons le voir, le sel pur de l'acide dextrogyre.

Voici les résultats de l'analyse.

210.9 mgr. de subst.: 440.7 mgr. de CO_2 et 121.7 mgr. de H_2O
106 mgr. de subst.: 5.3 c.c. de N_2, 19°, 765.3 m.m.
284.5 mgr. de subst. ont perdu à 105°: 10.0 mgr.
Dosage de l'acide sulfonprop.; 506 mgr. de subst.: 2.01 équiv. de baryte (phénolphtaléine).

Trouvé: C 56.99; H 6.45, N 5.74, H_2O 3.52, acide 30.6.
Calculé: „ 56.88; „ 5.97, „ 5.53, „ 3.56, „ 30.4.
Formule: $C_3H_6O_5S + C_{21}H_{22}O_2N_2 + H_2O$.

Le d-sulfonpropionate acide de strychnine, cristallisé dans l'eau, forme des prismes solides, contenant une molécule d'eau. Séchés, ils se décomposent vers 245—250°, en dégageant du gaz et en donnant un liquide brun.

Le pouvoir rotatoire de ce sel ainsi que celui des autres composés a été déterminé pour la lumière du sodium à 20°, à l'aide de tubes de 20 à 50 c.m. de longueur. Les concentrations variaient d'environ 5 à 8 molécules milligrammes par 100 c.c.

Voici la rotation moléculaire: $[M] = -72°$.

[1]) Ce Rec. 39, 692 (1920).

$HO_3SCH_2-CH-COOH$
$\quad\quad\quad\quad\quad | $
$\quad\quad\quad\quad SO_3H$

$C_3H_6O_8S_2$ Ref. 16

Dédoublement optique de l'acide α,β-disulfopropionique.

On fait recristalliser le sel tertiaire de strychnine dans dix fois son poids d'eau. On élimine la strychnine avec la quantité nécessaire de soude caustique (phénolphtaléine) et on examine la solution du sel sodique dans un polarimètre. Après huit cristallisations du sel de strychnine, le pouvoir rotatoire du sel sodique ne change plus.

2.5060 g de sel de strychnine (1.905 m.mol.) ont été décomposés par une lessive de soude caustique. Volume 25 cm^3; rotation du sel sodique dans un tube de 4 dm: —0.40°; rotation après addition de 2 cm^3 d'acide chlorhydrique concentré: +0.325°.

Acide $[M]_D = +11.4°$. Sel sodique: $[M]_D = -13.1°$.

La neutralisation d'une fonction ou de deux fonctions acides (SO_3H) de l'acide tribasique ne change que peu le pouvoir rotatoire. Mais la neutralisation de la troisième fonction acide (CO_2H) provoque l'inversion du pouvoir rotatoire.

Une solution de l'acide actif, chauffée pendant 30 heures au bain-marie, s'est racémisée complètement.

$CH_3-CH-COOH$
$\quad\quad\quad | $
$\quad\quad AsO_3H_2$

$C_3H_7AsO_5$ Ref. 17

1. *Acide α-arsonopropionique.* Une solution neutralisée de 30.5 g d'acide α-arsonopropionique et la quantité bimoléculaire du chlorhydrure de quinine a donné, dans un volume total de 2.5 litres, 120 g du sel secondaire de quinine.

Substance 0.7990 g; perte à 80° 0.0895 g.
" 0.3836 g; $Mg_2As_2O_7$ 0.0623 g.

	Trouvé:	H_2O 11.20;	As 7.84.
$C_3H_7O_5As \cdot 2\,C_{20}H_{24}O_2N_2 \cdot 6\,H_2O$ (954.53).	Calculé:	" 11.32;	" 7.85.

On a dissous le sel à chaud dans 300 cm^3 d'alcool et on a versé cette solution dans 2 litres d'eau bouillante; 86 g de sel ont cristallisé par refroidissement lent.

On examine le sel de quinine en le décomposant en présence de chloroforme par la quantité calculée (équimoléculaire) de baryte et on extrait la solution aqueuse du sel barytique au chloroforme pour éliminer les restes de quinine. Le sel barytique n'a montré encore qu'une rotation négative très faible, mais après acidification la solution a présenté une rotation positive assez notable. On a répété la recristallisation dans l'alcool dilué selon le même procédé jusqu'à ce que la rotation soit restée constante.

Poids	120	86	54	30	22	12	7.2	4.5	2.8 g
$[M]_D$ sel	—	—1.6	—2.9	—6.2	—6.5	—7.1	—7.9	—8.5	—8.5
" acide	—	+9.3	+21.8	+27.9	+32.2	+35.3	+39.1	+41.0	+41.0

Comme la rotation change de signe par neutralisation, nous avons examiné la rotation du sel acide. L'inversion de la rotation s'opère pendant la neutralisation de la seconde fonction acide. Un excès de baryte (sel tertiaire) n'a qu'une faible influence.

Acide α-arsonopropionique $[M]_D = +41°$; sel barytique primaire $[M]_D = +16°$; sel barytique secondaire $[M]_D = -8.5°$.

Dispersion rotatoire.

λ (μμ)	674.0	626.5	589.3	560.5	536.5	516.0
[M] sel second.	—6.5°	—7.3°	—8.5°	—9.7°	—10.5°	—
" acide	+31.5°	+36.1°	+41.0°	+46.0°	+52.2°	+58.9°

$CH_3-CH-COOH$
$\quad\quad\quad | $
$\quad\quad\quad NH_2$

$C_3H_7NO_2$ Ref. 18

Spaltung des *d,l*-Alanin-benzylesters mit Dibenzoyl-*d*-weinsäure: Zu 6 g *d,l*-Alanin-benzylester werden bei 21° 6.6 g Dibenzoyl-*d*-weinsäure, gelöst in 45 ccm absol. Äthanol, gegeben. Die Kristallisation erfolgt sehr langsam in farblosen Nadelbüscheln. Nach 24 Stdn. wird abgesaugt und getrocknet: *d*-Alanin-benzylester-dibenzoyl-*d*-tartrat, Schmp. 146—149°, $[\alpha]_D^{18}$: —33.5° (*c* = 0.2541, in Äthanol). Die Hydrolyse, wie oben beschrieben, liefert *d*-Alanin-Hydrochlorid, $[\alpha]_D^{19}$: —7.0° (*c* = 2.0038, in Wasser). Von E. Fischer[7]) wird für das optisch reine *l*-Alanin-Hydrochlorid $[\alpha]_D$: +10.4° angegeben.

Von der Mutterlauge wird der Alkohol i. Vak. abdestilliert. Der Rückstand wird getrocknet: Schmp. 116—127°; $[\alpha]_D^{22}$: —55.3° (*c* = 0.2531, in Äthanol).

Die Hydrolyse erfolgt wie oben beschrieben. Das Hydrochlorid wird in wenig absol. Äthanol gelöst und mit absol. Äther gefällt. Dabei kristallisieren Prismen neben Nadeln. Letztere werden mit Äther abdekantiert und getrocknet: *l*-Alanin-Hydrochlorid, $[\alpha]_D^{21}$: +10.2° (*c* = 2.0024, in Wasser).

Die Prüfung auf Reproduzierbarkeit ergab folgendes Bild:

		Schmp.	$[\alpha]_D$	*d*-Alanin-Hydrochlorid
1. Fraktion:	1.	137—143°	—38.6°	—6.8°
	2.	149°	—49.7°	—7.4°
	3.	138—146°	—52.6°	—7.1°
	4.	139—145°	—46.3°	—6.2°

		Schmp.	$[\alpha]_D$	*l*-Alanin-Hydrochlorid
2. Fraktion:	1.	118—125°	—56.4°	+ 7.1°
	2.	121—123°	—58.4°	+10.3°
	3.	115—123°	—59.1°	+10.1°
	4.	116—125°	—55.6°	+10.3°

Spaltung des *d,l*-Alanin-benzylesters mit *d*-Weinsäure: Zu 7.3 g *d,l*-Alanin-benzylester wird bei 21° eine Lösung von 6.1 g *d*-Weinsäure in 80 ccm absol. Äthanol gegeben. Die Kristallisation erfolgt langsam unter Abscheidung farbloser Blättchen. Nach 24 Stdn. wird abgesaugt und getrocknet: *l*-Alanin-benzylester-*d*-bitartrat; Schmp. 136—148°, $[\alpha]_D^{23}$: +31.9° (*c* = 0.501, in Äthanol).

Das Ester-bitartrat wird zersetzt, indem man es in eine mit Äther überschichtete gesättigte Kaliumcarbonatlösung einträgt und durchschüttelt. Der Äther, der den freien Ester enthält, wird abgetrennt und abdestilliert. Der zurückbleibende Ester wird mit 10-proz. Salzsäure unter Erwärmen hydrolysiert. Die erkaltete Lösung wird ausgeäthert und i. Vak. eingedampft. Das zurückbleibende Aminosäure-Hydrochlorid wird in wenig absol. Äthanol gelöst, mit Äther ausgefällt und getrocknet: *l*-Alanin-Hydrochlorid, $[\alpha]_D^{24}$: +9.9° (*c* = 1.924, in Wasser).

Von der Mutterlauge wird das Lösungsmittel i. Vak. entfernt. Der krist. Rückstand schmilzt zwischen 124 und 129°, $[\alpha]_D^{20}$: +10.9° (*c* = 0.5037, in Äthanol).

Hydrolyse wie oben beschrieben: *d*-Alanin-Hydrochlorid, $[\alpha]_D^{20}$: —8.2° (*c* = 2.0136, in Wasser).

In vier weiteren Ansätzen wurde auf Reproduzierbarkeit geprüft:

		Schmp.	$[\alpha]_D$	*l*-Alanin-Hydrochlorid	Temp.
1. Fraktion:	1.	136—148°	+22.0°	+ 9.6°	21°
	2.	121—130°	+ 9.9°	+ 8.4°	23°
	3.	120—128°	+11.8°	+ 6.2°	23°
	4.	139—148°	+27.7°	+10.0°	21°

		Schmp.	$[\alpha]_D$	*d*-Alanin-Hydrochlorid	Temp.
2. Fraktion:	1.	122—126°	+11.1°	—9.0°	21°
	2.	129—134°	+11.0°	—4.6°	23°
	3.	125—131°	+ 8.0°	—3.2°	23°
	4.	121—126°	+10.1°	—9.6°	21°

7) Ber. dtsch. chem. Ges. 39, 464 [1906].

A resolution of the above compound using the same resolving agent may be found in: G. Losse and H. Jeschkeit, Chem. Ber., 90, 1275 (1957).

$C_3H_7NO_2$ Ref. 18a

EXAMPLE 1

Resolving of N,N-dibenzyl-DL-alanine

(A) PREPARATION OF THE SALT OF L(+)-THREO-1-p-NITRO PHENYL-2-AMINO PROPANE-1,3-DIOL OF N,N-DIBENZYL-L-(−)-ALANINE

13.5 g. of N,N-dibenzyl-DL-alanine and 11.0 g. of L(+)-threo-1-p-nitro phenyl-2-amino propane 1,3-diol are heated under reflux in 120 cc. of absolute ethanol. 140 cc. of warm water are gradually added thereto. The mixture is allowed to cool slowly to a temperature of 30° C., the precipitated salt is filtered off, washed with a small quantity of aqueous alcohol containing 60% of water, and recrystallized from a mixture of 60 cc. of ethanol and 70 cc. of water. The crystals are filtered off at 25° C. yielding 10.3 g. of the salt of L(+)-threo-1-p-nitro phenyl-2-amino propane-1,3-diol and N,N-dibenzyl-L(−)-alanine. The yield amounts to 42.5% corresponding to a resolving yield of 85%. The salt is obtained in the form of yellow leaflets which are insoluble in water and only slightly soluble in ethanol, ether, acetone, benzene, and chloroform. The salt melts at 177–178° C.; its rotatory power is $[\alpha]_D^{20} = -15° \pm 1°$ (concentration: 1% in methanol).

(B) PREPARATION OF THE SALT OF D(−)-THREO-1-p-NITRO PHENYL-2-AMINO PROPANE-1,3-DIOL AND N,N-DIBENZYL-D(+)-ALANINE

The mother liquors resulting from the preparation and recrystallization of the above described salt of L(+)-threo-1-p-nitro phenyl-2-amino propane-1,3-diol and N,N-dibenzyl-L(−)-alanine are collected and are allowed to stand at room temperature overnight. Thereby, 0.9 g. of a salt rich in the levorotatory diastereoisomer precipitate. Said salt is separated and may be added to a new charge to be resolved.

The resulting filtrate is concentrated in a vacuum to a volume of about 150 cc. and is rendered alkaline by the addition of 28 cc. of N sodium hydroxide solution. L(+)-threo-1-p-nitro phenyl-2-amino propane-1,3-diol crystallizes and is removed from the filtrate. 3.85 g. thereof corresponding to 35% of the amount of resolving agent initially employed are recovered. The alkaline filtrate is then acidified by the addition of 2 cc. of acetic acid. The oily precipitate is extracted three times with 25 cc., 10 cc., and 10 cc. of chloroform. The combined chloroform solutions are washed twice, each time with 10 cc. of water, dried over magnesium sulfate, and evaporated to dryness in a vacuum. 6.2 g. of the impure trihydrate of N,N-dibenzyl-D(+)-alanine are obtained thereby. Said compound is treated in the same manner as described above under (A) with 4.8 g. of D(−)-threo-1-p-nitro phenyl-2-amino propane-1,3-diol and yields thereby the salt of D(−)-threo-1-p-nitrophenyl-2-amino propane-1,3-diol and N,N-dibenzyl-D(+)-alanine. The yield amounts to 70% calculated for the racemic mixture employed initially. Said salt is obtained in the form of yellow leaflets which are insoluble in water and only slightly soluble in ethanol, ether, acetone, benzene, and chloroform. Its melting point is 177–178° C.; its rotatory power is $[\alpha]_D^{20} = +15° \pm 1°$ (concentration: 1% in methanol).

(C) PREPARATION OF N,N-DIBENZYL-L(−)-ALANINE

10.3 g. of the salt of L(+)-threo-1-nitro phenyl-2-amino propane-1,3-diol and N,N-dibenzyl-L(−)-alanine prepared as described hereinabove under (A), are vigorously agitated with 22.5 cc. of N sodium hydroxide solution for 15 minutes. L(+)-threo-1-p-nitro phenyl-2-amino propane-1,3-diol crystallizes and is separated from the solution. 4.4 g. thereof are recovered thereby. The yield of recovered resolving agent is 40% of the amount initially employed in the resolving procedure described hereinabove under (A). The filtrate is acidified by the addition of 1.5 cc. of acetic acid. The oily precipitate is extracted by means of chloroform and the resulting solution is washed with water as indicated hereinabove under (B). The chloroform solution is dried over magnesium sulfate and evaporated to dryness in a vacuum. The resulting residue is again dissolved in one part by volume of chloroform. 30 cc. of water are added thereto. The chloroform is removed therefrom by vacuum distillation in the cold. The resulting crystalline mass is filtered off. 5.8 g. of the trihydrate of N,N-dibenzyl-L(−)-alanine are obtained thereby. It forms white needles which are insoluble in water and soluble in ethanol, ether, and acetone. Its melting point is 64–65° C.; its rotatory power is $[\alpha]_D^{20} = -45° \pm 1°$ (concentration: 2% of the anhydrous compound in methanol). The yield of the anhydrous compound amounts to 70%.

(D) PREPARATION OF THE HYDROCHLORIDE OF L(+)-ALANINE

2 g. of a palladium black catalyst containing 6% of palladium are added to a solution of 4 g. of the trihydrate of N,N-dibenzyl-L(−)-alanine in 80 cc. of 80% ethanol containing 1.5 cc. of concentrated hydrochloric acid. The reaction mixture is heated to 70° C. Hydrogen is introduced into said mixture. 550 cc. of hydrogen are absorbed within 30 minutes. The catalyst is filtered off and the filtrate is evaporated to dryness. On addition of acetone to the residue the hydrochloride of L(+)-alanine crystallizes. It is filtered off, washed with acetone, and dried. 1.05 g. of the optically pure compound, corresponding to a yield of 70%, are obtained thereby. Its melting point is 205° C.; its rotatory power $[\alpha]_D^{20} = +9.5° \pm 0.5°$ (concentration: 3% in water).

$C_3H_7NO_2$ Ref. 18b

Preparation of Purified l(+)-Alanine and d(−)-Alanine

Benzoyl-dl-alanine—A mixture consisting of 267 gm. (3 moles) of technical *dl*-alanine, 500 cc. of distilled water, and 175 cc. of saturated sodium hydroxide solution was cooled to about 0°. 360 cc. (3.12 moles) of benzoyl chloride were added rapidly while the mixture was stirred. Ice and alkali solution (125 cc.) were added as needed to maintain the temperature low and the reaction alkaline to phenolphthalein. 200 cc. of concentrated hydrochloric acid were added, and the mixture was allowed to stand overnight in the refrigerator. The suspension of benzoic acid and benzoyl-*dl*-alanine was filtered and the precipitate washed with 1.5 liters of cold distilled water and 600 cc. of isopropyl ether. The yield of product (m.p. 160–161°, uncorrected), dried in air, was 550 to 557 gm. (95 to 97 per cent of theory). According to Pope and Gibson (18), the melting point is 160° (uncorrected), while Fischer (13) found 162–163° (uncorrected) and 165–166° (corrected).

Because he was not able to get satisfactory yields of benzoyl-*dl*-alanine by the original method of Baum (31), Fischer (13) substituted sodium bicarbonate for sodium hydroxide in the benzoylation. By this means he prepared 6.0 gm. of crude and 4.5 gm. of once recrystallized product in 97 and 73 per cent, respectively, of the theoretical amount. Carter and Stevens (32) found that the yield of the purified benzoyl derivative of *l*-*p*-methoxyphenylalanine was higher when sodium hydroxide was employed in place of sodium bicarbonate and benzoyl chloride was used in the ratio of 2 moles to 1 of the amino acid instead of the 3:1 ratio recommended by Fischer. Adopting this procedure, Levy and Palmer (21) obtained 38 gm. (79 per cent of the theoretical amount) of benzoyl-*dl*-alanine which was stated to have the theoretical titration equivalent and the accepted melting point. By the present authors' relatively simple procedure only a slight excess of benzoyl chloride is required and large amounts of nearly pure benzoylated product are readily prepared in high yield.

Strychnine Benzoyl-l(+)-alanine—The procedure of Pope and Gibson (18) was followed in the resolution of benzoyl-*dl*-alanine. Strychnine benzoyl-*l*(+)-alanine is allowed to crystallize from an aqueous solution containing 2 equivalents of benzoyl-*dl*-alanine and 1 equivalent each of potassium hydroxide and strychnine. It was found, in agreement with Levy and Palmer (21), that this method is as effective and is more economical of alkaloid than that of Fischer (13) as modified by Pacsu and Mullen (20). The last authors crystallized strychnine benzoyl-*l*(+)-alanine from an aqueous solution containing equivalent quantities of benzoyl-*dl*-alanine and strychnine.

From seven lots (109 to 177 gm.) of benzoyl-*dl*-alanine (total, 1064 gm.) there were obtained 1239.7 gm. (79.9 per cent of theory) of recrystallized *l*-strychnine benzoyl-*l*(+)-alanine which, according to Pope and Gibson (18), is the dihydrate. The specific rotation (Sample A, Table I) in water was −10.45° at 24°. That of Pope and Gibson's product (Sample B, Table I) was −10.66° at 20°. Pacsu and Mullen (20) prepared 310.7 gm. (79.5 per cent of theory) of this salt. These authors calculated the yield to be 84.6 per cent of theory based, apparently, on the assumption that the salt is anhydrous.

Brucine Benzoyl-d(−)-alanine—The mother liquor from which the strychnine benzoyl-*l*(+)-alanine had been removed was evaporated to about one-half of its volume. The residual solution was made strongly alkaline with 3 N sodium hydroxide, the mixture was cooled overnight in the refrigerator, the suspension of strychnine was filtered, the precipitate was washed, and the filtrate was acidified to Congo red with concentrated hydrochloric acid. The precipitate of crude benzoyl-*d*(−)-alanine, which formed immediately, was removed by filtration. Two additional crops of crystals were obtained from the filtrate. The yield of crude benzoyl-*d*(−)-alanine was 443.2 gm. (83 per cent of theory).

The crude product (2.06 moles, if 90 per cent pure) was dissolved in 4.5 liters of boiling water and there were added 820 gm. (2.08 moles) of anhydrous *l*-brucine and 12.9 gm. (0.23 mole) of potassium hydroxide dissolved in 100 cc. of water. The resulting solution was cooled, seeded, and let stand for 36 hours in the refrigerator. The suspension of fine crystals was filtered. The crystals were washed twice with ice-cold water and dried at 50°. The yield[1] (brucine benzoyl-*d*(−)-alanine·$4\frac{1}{2}H_2O$, according to Pope and Gibson (18)) was 1059 gm. (69.1 per cent of theory based on 443.2 gm. of crude benzoyl-*d*(−)-alanine and 71 per cent based on 1064 gm. of benzoyl-*dl*-alanine). The yield of twice recrystallized product was 890 gm. (84 per cent recovery). The crude product melted at 84–86° (uncorrected) and the final product at 86–88° (uncorrected). Pope and Gibson (18) found that a product which had been recrystallized four times melted at 89–91°. The specific rotation in water of the authors' product (Sample C, Table I) was −26.53° at 23.0° and Pope and Gibson's product (Sample D, Table I) was −33.6° at 20°.

Benzoyl-l(+)-alanine—A hot aqueous solution containing 1120.9 gm. (1.99 moles) of once recrystallized strychnine benzoyl-*l*(+)-alanine was made strongly alkaline with 3 N sodium hydroxide. The mixture was cooled and the suspension of strychnine was filtered immediately. The filtrate was acidified strongly with concentrated hydrochloric acid, seeded, and let stand overnight in the refrigerator. The resulting crystals were removed by filtration, washed, and dried at 50°. The yield of crude benzoyl-*l*(+)-alanine (m.p. 136–139°, uncorrected) was 379.7 gm. (99 per cent of theory).

The crude benzoyl-*l*(+)-alanine was suspended in 2.5 liters of water heated to 40°, 121 cc. of concentrated sodium hydroxide and 9 gm. of norit were added, and the mixture was stirred for 10 minutes. The suspension was filtered, and the precipitate was washed. Concentrated hydrochloric acid (215 cc.) was added to the filtrate and the suspension which formed after the solution had stood overnight was filtered. The crystals were washed with cold water, sucked as dry as possible, and dried at 50°. There were obtained from 326.3 gm. of crude benzoyl-*l*(+)-alanine 321.5 gm. (98.5 per cent recovery) of recrystallized product which melted at 136–138° (uncorrected). Its titration equivalent was 189.2 (theory, 193.1) and the moisture content was 1.5 per cent. The pure substance melts at 147–148° (uncorrected) or 150–151° (corrected) according to Fischer (13), Pope and Gibson (18), and Pacsu and Mullen (20). The specific rotation (Sample E, Table I) was +33.4° in 1 N sodium hydroxide at 22.5°. That of Pope and Gibson's product (Sample F, Table I) was +34.6° in 0.077 N potassium hydroxide at 20°, and of Pacsu and Mullen's product (Sample G, Table I) was +37.12° in 0.34 N potassium hydroxide at 20°. Although Pope and Gibson, Pacsu and Mullen, and Levy and Palmer (21) have shown that the specific rotation of benzoyl-*l*(+)-alanine in an aqueous solution containing an equivalent of alkali varies with the concentration of the solute, it is difficult to determine from the available data the correct rotation under any particular set of conditions.

Benzoyl-d(−)-alanine[2]—1337 gm. (2.00 moles) of *l*-brucine benzoyl-*d*(−)-alanine·$4\frac{1}{2}H_2O$ were dissolved in 5 liters of water and 480 cc. of saturated sodium hydroxide solution were added. The suspension of brucine, which formed rapidly, was stirred for 10 minutes, cooled to room temperature, and filtered. The brucine was washed with 0.5 N sodium hydroxide solution and the combined filtrate and washings were acidified to Congo red with 340 cc. of concentrated C.P. hydrochloric acid. The suspension was filtered after it had stood overnight in the refrigerator. The crystals were washed with cold water and dried for 2 days at 50°.

The yield of crude benzoyl-*d*(−)-alanine was 364.9 gm. (94.5 per cent of theory). The specific rotation (Sample H, Table I) was −32.5° in 1.05 N sodium hydroxide at 25°. The values, −37.3° and −36.9° at 20°, were found by Fischer (13) and Pacsu and Mullen (20), respectively, under approximately the same conditions.

l(+)-Alanine—292.5 gm. (1.51 moles) of benzoyl-*l*(+)-alanine suspended in 1460 cc. of 6 N hydrochloric acid were heated for 4.5 hours on a boiling water bath. The mixture was cooled overnight in the refrigerator, 1 liter of distilled water was added, and the suspension of crystals was filtered. After the benzoic acid was extracted from the precipitate, about 40 gm. of unchanged benzoyl-*l*(+)-alanine were recovered.

The acid filtrate was distilled to dryness *in vacuo* and, after the addition of 300 cc. of water, this process was repeated twice. The residual *l*(+)-alanine hydrochloride was dissolved in 400 cc. of distilled water, 89 cc. of concentrated ammonium hydroxide and 293 cc. of 95 per cent ethanol were added, and the suspension of crystals which formed after the mixture had stood overnight in the refrigerator was filtered. The crystals were washed with 95 per cent ethanol until the washings were free from chloride. The yield of *l*(+)-alanine, dried at 50°, was first crop, 65.5 gm., second crop, 9.3 gm., and total 74.8 gm. (64 per cent of theory). The specific rotations in 5.97 N hydrochloric acid at 25° were first crop (Sample I, Table I), +13.84°, and second crop (Sample J, Table I), +13.73°.

72 gm. of the foregoing *l*(+)-alanine were dissolved in 214 cc. of boiling distilled water and 520 cc. of 95 per cent ethanol were added. The yield of dried product was 66.9 gm. (93 per cent recovery). It contained less than 0.004 per cent chloride, phosphate, iron, or heavy metal ion, and it was 100.6 per cent pure according to Van Slyke volumetric analysis of amino nitrogen. Its specific rotation (Sample K, Table I) in 6.08 N hydrochloric acid at 25.0° was +13.83°.

53.2 gm. of this product were dissolved in 162 cc. of boiling distilled water and 380 cc. of 95 per cent ethanol were added. The yield of dried product was 49.8 gm. (93.6 per cent recovery). Its specific rotation (Sample L, Table I) in 5.97 N hydrochloric acid at 25.1° was +13.79°.

47.8 gm. of this product were dissolved in 150 cc. of boiling distilled water and 172 cc. of redistilled absolute methanol were added. The yield of dried product containing less than 0.004 per cent chloride, ammonium, iron, phosphate, or heavy metal ion was 40.3 gm. (84.3 per cent recovery). The product was 100.3 (100.2, 100.3, 100.3) per cent pure according to semi-micro-Kjeldahl analysis. Its specific rotation (Sample M, Table I) in 5.97 N hydrochloric acid at 25.0° was +13.82°.

3.0 gm. of this product were dissolved in 9.0 cc. of boiling distilled water. This solution was cooled at 5° for 4 hours, the suspension of crystals was filtered, and the crystals were washed with 33 per cent, 50 per cent, and absolute methanol. The yield of dried product was 1.63 gm. (54.3 per cent recovery). The specific rotation (Sample N, Table I) in 5.97 N hydrochloric acid at 25.0° was +13.70°. About 14 cc. of absolute methanol were added to the alcoholic filtrate and the yield of dried product which crystallized at 5° was 0.80 gm. The specific rotation (Sample O, Table I) in 5.97 N hydrochloric acid at 25.1° was +13.70°.

Since all of the foregoing seven specific rotations are either within, or only slightly beyond, the probable precision of the polarimeter, it was considered that this sample of *l*(+)-alanine was analytically pure, its specific rotation being $[\alpha]_D^{25.0} = +13.77°$ (±0.02°, probable error of the mean) in 6.0 N hydrochloric acid.

d(−)-Alanine—230 gm. (1.19 moles) of recrystallized benzoyl-*d*(−)-alanine were hydrolyzed with concentrated hydrochloric acid, the suspension of benzoic acid filtered, and *d*(−)-alanine isolated from the filtrate essentially by the method employed in the preparation of *l*(+)-alanine from benzoyl-*l*(+)-alanine. The yield of crude *d*(−)-alanine was first crop, 76.2 gm., second crop, 18.7 gm., third crop, 6.4 gm., and total, 101.3 gm. (95.8 per cent of theory).

[1] Pacsu and Mullen (20) prepared 359 gm. of the brucine salt. The yield, 88.2 per cent, calculated by these authors seems to be in error, since it appears to be 74 per cent based on 140 gm. of benzoyl-*d*(−)-alanine and 77.7 per cent based on 268 gm. of benzoyl-*dl*-alanine.

[2] Pacsu and Mullen (20) prepared 77.5 gm. of purified benzoyl-*d*(−)-alanine. The yield was 91 per cent of theory based on brucine benzoyl-*d*(−)-alanine·$4\frac{1}{2}H_2O$. The yield, 81 per cent, given by these authors apparently was calculated on the assumption that brucine benzoyl-*d*(−)-alanine is anhydrous.

TABLE I

Specific Rotations of Substances Referred to in Text of This Paper†*

Sample	Substance	Solvent	Weight of sample	Volume of solution	Temperature	α	$[\alpha]_D$
			gm.	*cc.*	*°C.*	*degrees*	*degrees*
A	*l*-Strychnine benzoyl-*l*(+)-alanine dihydrate	Water	0.2154	50.04	24	−0.180	−10.45
B	" "	"	0.1088	29.94	20	−0.155	−10.66‡
C	*l*-Brucine benzoyl-*d*(−)-alanine·4½ H_2O	"	0.5314	50.01	23.0	−1.127	−26.53
D	" "	"	0.3117	29.94	20	−1.40	−33.6‡
E	Benzoyl-*l*(+)-alanine	1 N NaOH	0.2301	50.01	22.5	+0.614	+33.4
F	"	0.077 N KOH	0.5238	29.94	20	+2.42	+34.6‡
G	"	0.34 " "	0.6681	10	20	+2.48	+37.12§
H	Benzoyl-*d*(−)-alanine	1.05 " NaOH	0.5008	50.0	25	−1.366	−32.5
I	*l*(+)-Alanine	5.97 " HCl	0.5309	26.293	25	+1.118	+13.84
J	"	5.97 " "	0.5352	21.131	25	+1.125	+13.73
K	"	6.08 " "	0.5447	26.044	25.0	+1.157	+13.83
L	"	5.97 " "	0.5281	26.293	25.1	+1.108	+13.79
M	"	5.97 " "	0.5556	26.131	25.0	+1.175	+13.82
N	"	5.97 " "	0.5256	26.206	25.0	+1.099	+13.70
O	"	5.97 " "	0.5422	26.293	25.1	+1.130	+13.70
P	*d*(−)-Alanine	6.08 " "	0.5436	22.044	25.0	−1.333	−13.57
Q	"	5.97 " "	0.5303	26.044	25.1	−1.107	−13.59
R	"	5.97 " "	0.4312	26.126	25.0	−1.111	−13.66
S	"	5.97 " "	0.5255	25.949	25.0	−1.103	−13.62
T	"	5.97 " "	0.5306	26.044	25.2	−1.107	−13.58

* A 1 dm. tube was used to measure the rotation of benzoyl-*l*(+)-alanine, Sample G. In all other cases the tube length was 4 dm.

† The data given were obtained by the present authors with the exceptions noted.

‡ Pope and Gibson (18).

§ Pacsu and Mullen (20).

74.7 gm. of this product were recrystallized from aqueous ethanol solution. The yield of dried product containing less than 0.004 per cent chloride, phosphate, iron, or heavy metal ion was 69.6 gm. (93.2 per cent recovery). This product was 100.8 per cent pure according to Van Slyke volumetric analysis of amino nitrogen. Its specific rotation (Sample P, Table I) in 6.08 N hydrochloric acid at 25.0° was −13.57°.

54.1 gm. of this product were recrystallized from aqueous ethanol solution. The yield of dried product was 50.5 gm. (93.3 per cent recovery). The specific rotation (Sample Q, Table I) in 5.97 N hydrochloric acid at 25.1° was −13.59°.

50.1 gm. of this product were recrystallized from aqueous methanol solution. The yield of dried product containing less than 0.004 per cent chloride, ammonia, iron, phosphate, or heavy metal ion was 42.8 gm. (85.4 per cent recovery). This product was 100.2 (100.3, 100.4, 100.0) per cent pure according to semimicro-Kjeldahl analysis. Its specific rotation (Sample R, Table I) in 5.97 N hydrochloric acid at 25.0° was −13.66°.

3.0 gm. of this product were recrystallized from aqueous methanol solution. The yield of dried product was 1.65 gm. (55.0 per cent recovery). The specific rotation (Sample S, Table I) in 5.97 N hydrochloric acid at 25.0° was −13.62°. 30 cc. of absolute methanol were added to the alcoholic filtrate and the mixture let stand at 5°. The yield of dried product was 0.87 gm. The specific rotation (Sample T, Table I) in 5.97 N hydrochloric acid at 25.2° was −13.58°.

Since all of the foregoing five specific rotations are either within, or only slightly beyond, the probable precision of the polarimeter, it was considered that this sample of *d*(−)-alanine was analytically pure, its specific rotation being $[\alpha]_D^{25.0}$ = −13.60° (±0.01°, probable error of the mean) in 6.0 N hydrochloric acid.

Authors' comments:

The optically active alanines have been prepared most commonly by chemical resolution of *dl*-alanine. The benzoyl (5, 13, 18–22), ethyl ester (23, 24), *p*-toluenesulfonyl (25), α- and β-naphthalenesulfonyl (26), benzenesulfonyl (27), camphorsulfonyl (23), bromocamphorsulfonyl (23), menthoxyacetyl (20, 28), hydroxymethylenecamphor (24), and *o*-, *m*-, and *p*-nitrobenzoyl (29, 30) derivatives of the brucine (5, 13, 18–22, 25, 26, 29, 30), strychnine (5, 13, 18–22, 25, 26, 29, 30), cinchonine (30), cinchonidine (29, 30), quinine (30), and ephedrine (27) salts of *dl*-alanine have been used for this purpose. Some of the alkaloidal derivatives were found to be unsatisfactory because of unfavorable solubility relations of the enantiomorphic complexes and the formation of partial racemates or mixed crystals of the diastereoisomeric salts. It was observed by Colles and Gibson (30) that the benzoyl, *p*-toluyl, cinnamoyl, and *m*- and *p*-nitrobenzoyl derivatives of alanine are readily hydrolyzed when they are refluxed for 3 hours with constant boiling hydrochloric acid, while the benzenesulfonyl and analogous sulfonyl derivatives are incompletely hydrolyzed even under more drastic conditions. It appears, however, that resistance to hydrolysis is not of critical importance, since, according to Pacsu and Mullen (20), *l*(+)-alanine is not racemized appreciably during 24 hours refluxing with 20 per cent hydrochloric acid.

The method described by Pope and Gibson (18) in 1912 for the resolution of *dl*-alanine was finally adopted, since it appeared to be as satisfactory as any other. It was first employed in the authors' laboratory in 1935 in the preparation of 2.5 gm. of purified *l*(+)-alanine and 6.5 gm. of purified *d*(−)-alanine. In this procedure, the strychnine salt of benzoyl-*l*(+)-alanine and the brucine salt of benzoyl-*d*(−)-alanine are isolated in the order given. The precipitation of these complexes in the reverse order, according to the method devised originally by Fischer (13), is not as satisfactory as the procedure of Pope and Gibson, since the strychnine salt of *l*(+)-benzoylalanine is much less soluble in water than the brucine salt of *d*(−)-alanine. Although Pope and Gibson did not extend the resolution beyond the benzoylalanine stage, Pacsu and Mullen (20), as well as Levy and Palmer (21), have reported recently the preparation of both alanine antipodes essentially by the method of Pope and Gibson.

5. Fischer, E., *Ber. chem. Ges.*, **39**, 530 (1906).
13. Fischer, E., *Ber. chem. Ges.*, **32**, 2451 (1899).
18. Pope, W. J., and Gibson, C. S., *J. Chem. Soc.*, **101**, 939 (1912).
19. Heidelberger, M., An advanced laboratory manual of organic chemistry, New York, 78 (1923).
20. Pacsu, E., and Mullen, J. W., 2nd, *J. Biol. Chem.*, **136**, 335 (1940).
21. Levy, M., and Palmer, A. H., *J. Biol. Chem.*, **146**, 493 (1942).
22. Sjollema, B., and Seekles, L., *Rec. trav. chim. Pays-Bas*, **45**, 232 (1926).
23. Colombano, A., and Sanna, G., *Atti accad. Lincei*, **22**, 292 (1913); *Chem. Abstr.*, **8**, 668 (1914).
24. Kipping, F. B., and Pope, W. J., *J. Chem. Soc.*, 494 (1926).
25. Gibson, C. S., and Simonsen, J. L., *J. Chem. Soc.*, **107**, 798 (1915).
26. Colles, W. M., and Gibson, C. S., *J. Chem. Soc.*, **125**, 2505 (1924).
27. Gibson, C. S., and Levin, B., *J. Chem. Soc.*, 2754 (1929).
28. Holmes, D. F., and Adams, R., *J. Am. Chem. Soc.*, **56**, 2093 (1934).
29. Colles, W. M., and Gibson, C. S., *J. Chem. Soc.*, 99 (1928).
30. Colles, W. M., and Gibson, C. S., *J. Chem. Soc.*, 279 (1931).
31. Baum, J., *Z. physiol. Chem.*, **9**, 465 (1885).
32. Carter, H. E., and Stevens, C. M., *J. Biol. Chem.*, **138**, 627 (1941).

$C_3H_7NO_2$ Ref. 18c

Resolution of N-Acetyl-DL-alanine.

N-Acetyl-DL-alanine (65 g, 0.5 mol) and L-leucinamide (71.5 g, 0.55 mol) were dissolved in 1 l of ethanol at 50°C. After being kept overnight at room temperature, separated needle crystals were filtered, washed with a little amount of ethanol and dried in vacuo. Yield, 45.5 g (70% based on L-form); m.p. 152°C; $[\alpha]_D^{20}$−10.50° (c, 2: water).

N-Acetyl-L-alanine was obtained by the use of cation exchanger in the same way as previously described. Yield, 22 g (98%); m.p. 124∼125°C; $[\alpha]_D^{18}$ −63.6° (c, 2: water).

N-Acetyl-L-alanine was hydrolyzed for 5 hrs under refluxing with one and one-half (1.5) equivalents of normal hydrochloric acid. L-Alanine was obtained. Yield, 14.5 g (97%); $[\alpha]_D^{20}$+14.76° (c, 10: 6 N HCl).

L-Leucinamide was recovered by the same method as that described in the resolution of DL-methionine. Yield, 95%; m.p. 102°C; $[\alpha]_D^{19}$+7.78°.

$C_3H_7NO_2$

Ref. 18d

C) Optische Spaltung von Z-DL-Alanin. – 1. *L-Tyrosinhydrazidsalz von Z-D-Alanin.* – a) *Mit der berechneten Menge L-Tyrosinhydrazid.* 44,6 g (0,2 Mol) Z-DL-Alanin [16] wurden in eine siedende Suspension von 39 g (0,2 Mol) pulverisiertem L-Tyrosinhydrazid in 800 ml Alkohol eingetragen und unter Rückfluss gekocht. Man filtrierte von wenig ungelöster Substanz ab, impfte das Filtrat mit dem L-Tyrosinhydrazidsalz von Z-D-Alanin an, rührte 2 Std. während des Abkühlens und bewahrte den Kristallbrei 20 Std. bei 25° auf. Nach dem Abnutschen, sorgfältigem Nachwaschen mit 200 ml Alkohol und 100 ml Äther und Trocknen erhielt man 38,3 g (91%) L-Tyrosinhydrazidsalz von Z-D-Alanin, Smp. 149–150°, $[\alpha]_D^{23} = +55,2°$ ($c = 2$, in H_2O).

$C_{20}H_{26}O_6N_4$ (418,44) Ber. C 57,40 H 6,26 N 13,39% Gef. C 57,51 H 6,33 N 13,47%

Die Mutterlauge, die zur Hauptsache das L-Tyrosinhydrazidsalz von Z-L-Alanin enthielt, wurde gemäss C3) zum Dicyclohexylaminsalz von Z-L-Alanin umgesetzt.

b) *Mit der halben Menge L-Tyrosinhydrazid.* 22,3 g (0,1 Mol) Z-DL-Alanin und 9,75 g (0,05 Mol) pulverisiertes L-Tyrosinhydrazid wurden in 300 ml Alkohol am Rückflusskühler gekocht, heiss filtriert, mit dem L-Tyrosinhydrazidsalz von Z-D-Alanin angeimpft, 2 Std. gerührt und 20 Std. bei 25° aufbewahrt. Nach dem Abnutschen, Waschen mit Alkohol und Essigester sowie Trocknen erhielt man 15,8 g (75%) L-Tyrosinhydrazidsalz von Z-D-Alanin, Smp. 148–150°, $[\alpha]_D^{23} = +54,5°$ ($c = 2$, in H_2O).

2. *Z-D-Alanin.* 50,3 g (0,12 Mol) L-Tyrosinhydrazidsalz von Z-D-Alanin wurden in 200 m 3N Salzsäure suspendiert, 3mal mit je 150 ml Essigester extrahiert, die Essigesterphasen 2mal mit je 40 ml Wasser nachgewaschen, über $MgSO_4$ getrocknet, im Vakuum eingedampft und der Rückstand aus Essigester-Petroläther kristallisiert. Ausbeute 25,8 g (96%) Z-D-Alanin, Smp. 84–86°, $[\alpha]_D^{23} = +13,8°$ ($c = 2$, in Eisessig). Literaturwert [17]: Smp. 84°, $[\alpha]_D = +13,9°$ ($c = 8,5$, in Eisessig).

Optisch noch nicht reines Z-D-Alanin kann notfalls analog C3) über das Dicyclohexylaminsalz gereinigt werden.

3. *Dicyclohexylaminsalz von Z-L-Alanin.* Die Mutterlauge der Spaltung C1) wurde im Vakuum eingedampft und der Rückstand analog C2) mit 3N Salzsäure und Essigester zersetzt. Man erhielt 22,7 g (95%) optisch noch nicht reines Z-L-Alanin, $[\alpha]_D^{23} = -9,4°$ ($c = 2$, in Eisessig). 21,7 g dieses Rohproduktes wurden in 200 ml Alkohol gelöst, mit 17,6 g Dicyclohexylamin versetzt, nach 1 Std. abgenutscht, mit 50 ml Alkohol sorgfältig nachgewaschen und getrocknet. Ausbeute 26,1 g (67%) Dicyclohexylaminsalz von Z-L-Alanin, Smp. 180–182°, $[\alpha]_D^{23} = +5,2°$ ($c = 2$, in Alkohol). Literaturwert [13]: Smp. 180–181,5°, $[\alpha]_D = +5,0°$ ($c = 2,11$, in Alkohol).

Aus der Mutterlauge des Dicyclohexylaminsalzes erhielten wir nach dem Einengen und Ausfällen mit Äther 8 g (20,6%) Salz, $[\alpha]_D^{23} = +1,5°$ ($c = 2$, in Alkohol).

4. *Z-L-Alanin.* Die erste Fraktion des Dicyclohexylaminsalzes von Z-L-Alanin von C3) wurde gemäss [13] in Z-L-Alanin übergeführt; Ausbeute 93%, Smp. 83–85°, $[\alpha]_D^{23} = -13,8°$ ($c = 2$, in Eisessig). Literaturwert [16]: Smp. 84°, $[\alpha]_D^{17} = -14,3°$ ($c = 9$, in Eisessig).

5. *Dicyclohexylaminsalz von Z-DL-Alanin.* Es wurde hergestellt aus Z-DL-Alanin und Dicyclohexylamin gemäss der allgemeinen Vorschrift für das Dicyclohexylaminsalz von Z-L-Alanin [13] und aus wenig Alkohol umkristallisiert: Smp. 172-173°. Das DL-Salz war etwas leichter löslich als das L- bzw. D-Salz. Der Löslichkeitsunterschied war ausreichend für die optische Reinigung von partiell racemischen Z-L-Alanin bzw. Z-D-Alanin.

$C_{23}H_{36}O_4N_2$ (404,53) Ber. C 68,28 H 8,97 N 6,93% Gef. C 68,26 H 9,00 N 6,96%

$C_3H_7NO_2$

Ref. 18e

Resolution of (±)alanine

1.84 g of (±)alanine was dissolved in 5% NaOH solution and 4.8 g of chloride 3 was added in small portions. The formed emulsion, after several minutes of shaking was introduced into 10% HCl solution with crushed ice. Acid solution was decanted from above the congealing emulsion and the residue was treated with acetone. The mixture then separated into an oily layer and a solution from which the crystalline substance came down. The crystals were filtered off and washed with 60% ethanol. The crude product (0.95 g) m.p. 158—173°C after three recrystallizations from 60% ethanol yielded 0.6 g of pure amide; m.p. 172—173°C, $[\alpha]_D^{20} = -52.6°$ (acetone, $c = 2$).

The amide was hydrolyzed by means of HBr in aqueous ethanol for 6 hrs giving 0.2 g of (+)alanine hydrobromide; m.p. 206—208°C, $[\alpha]_D^{20} = +12.1°$ (ethanol, $c = 2$); benzoyl derivative had m.p. 149°C, $[\alpha]_D^{20} = +36°$ (20% NaOH, $c = 1.5$).

Note: The chloride 3 mentioned in the above procedure is (-)-carvomenthoxyacetyl chloride. The authors pointed out in their paper that the use of (+)-carvomenthoxyacetyl chloride allows one to obtain optically pure (-)alanine and the use of the (-)enantiomer as resolving agent allows one to obtain (+)alanine in optically pure form.

$C_3H_7NO_2$ Ref. 18f

Resolution of DL-Alanine. Application of the same method gave quinine *N*-(1-methyl-2-acetylvinyl)-L-alaninate in 93% yield, mp 142–143 °C, $[\alpha]^{25}D$ −73.8°.

Anal. Calcd for $C_{28}H_{37}N_3O_5 \cdot \frac{1}{2}H_2O$: C, 66.64; H, 7.59; N, 8.33. Found: C, 66.45; H, 7.80; N, 8.10.

L-Alanine was obtained from the crystalline enamine–quinine salt as described above for the fluoroalanine in 76% yield, $[\alpha]^{25}D$ +12.8° (*c* 5%, 5 N HCl) (lit.[7] +13°).

(7) Merck Index, 8th ed, 1968, p 27.

The method referred to in the above procedure may be found in Ref. 10.

$C_3H_7NO_2$ Ref. 18g

Note:

Using trans-1,2-cyclohexane dicarboxylic acid anhydride as resolving agent, obtained a less soluble diastereomeric amide with m.p. 204-206° $[\alpha]^{25}_D$ -5.8 (c 1, ethanol). The acid hydrochloride obtained from the amide had a $[\alpha]^{25}_D$ -9.8° (c 1, 2N HCl).

$C_3H_7NO_2$ Ref. 18h

EXPERIMENTAL.

dl-Alanine, prepared by Zelinsky and Stadinov's method, was converted into its ethyl ester, which was then heated for 2 hours on the water-bath with an equimolecular quantity of *d*-hydroxymethylenecamphor. On pouring into water, extracting with light petroleum, and evaporating the petroleum solution after drying with potassium carbonate, an oil was obtained which readily crystallised; the crystalline product was repeatedly crystallised from light petroleum until its melting point became constant at 108—109°. The substance thus obtained in long, white needles is the d-*methylenecamphor*-l-*alanine ethyl* ester, $C_8H_{14}<\begin{matrix}C{:}CH{\cdot}NH{\cdot}CHMe{\cdot}CO_2Et \\ | \\ CO\end{matrix}$ (Found: C, 68·9; H, 8·9. $C_{16}H_{25}O_3N$ requires C, 68·8; H, 9·0%). The substance is so soluble in most organic solvents that it can only be conveniently crystallised from light petroleum. 0·1008 Gram, made up to 20 c.c. with absolute alcohol, gave $\alpha_{Hg\ green} = +2{\cdot}58°$ in a 2-dcm. tube at 20°; whence $[\alpha]_{Hg\ green} = +256°$.

This compound, when treated by the bromine method of Pope and Read, yielded only resinous products from which alanine could not be separated; it is, however, hydrolysed by distillation in steam after admixture with concentrated hydrochloric acid. Hydroxymethylenecamphor distils away and *l*-alanine hydrochloride remains behind; the hydrochloric acid is removed with the aid of lead hydroxide, the excess of the latter by hydrogen sulphide, and the *l*-alanine then separated by crystallisation from the residual aqueous solution. 1·8761 Gram, made up to 20 c.c. with 1·25*N*-hydrochloric acid solution, gave $\alpha_{Hg\ green} = -3{\cdot}22°$ in a 2-dcm. tube at 20°; whence $[\alpha]_{Hg\ green} = -12{\cdot}2°$; this value is numerically identical with that given by *d*-alanine prepared from silk, but is, of course, of opposite sign. This simple type of method will be further studied in its applications to the resolution of racemic aminocarboxylic acids.

$C_3H_7NO_2$ Ref. 18i

Experimental

***l*-Menthoxyacetyl Chloride.**—Menthoxyacetic acid was prepared in a manner similar to that described by Frankland and O'Sullivan except that toluene was substituted for benzene as a solvent.[2] It was purified by vacuum distillation, b. p. 165–175° (7–10 mm.).

The acid chloride was produced by dissolving 90 g. of the acid in 150 cc. of thionyl chloride and warming at 50° for three hours. The excess thionyl chloride was removed by warming in a water-bath under diminished pressure and the acid chloride used directly. For identification and tests of purity a few simple derivatives were prepared.

Menthoxyacetyl Amino Acids.—These were all prepared by the addition of menthoxyacetyl chloride to an equivalent amount of the amino acid dissolved in 5% sodium hydroxide solution and shaking the emulsion at room temperature for several minutes until a clear solution resulted. This solution was then poured very slowly in small portions into dilute hydrochloric acid containing cracked ice, allowing the oil which separated to solidify after each addition. In this manner a solid product was obtainable in every case, and no difficulty was encountered in procuring beautiful crystals thereafter. After filtering, the mixed diastereoisomers were separated by fractional crystallization from either 60% ethyl alcohol or a 75 : 25 mixture of high-boiling petroleum ether and ethyl acetate, the latter usually giving the more satisfactory results. The more soluble form was in each case contaminated with the less soluble and no attempt was made to obtain it in a pure state. The less soluble form was easily obtained pure after a very few crystallizations. In early experiments the amides separated frequently as oils from the reaction mixture, and thus introduced considerable difficulty. By following carefully the procedure just described, the products were obtained solid in every instance.

Hydrolysis of the Menthoxyacetyl Amino Acids.—The amino acid derivative was suspended in a small amount of 20% hydrobromic acid and just enough alcohol added to effect solution when boiling. This solution was refluxed from four to six hours, and the alcohol then removed by boiling in the open. After washing with ether, the acid solution was evaporated to dryness and the amino acid hydrobromide dissolved in absolute ethyl alcohol. The free amino acid was precipitated by making slightly ammoniacal with concentrated aqueous ammonia, filtering, and washing with hot absolute alcohol to remove traces of ammonium bromide.

(2) Frankland and O'Sullivan, *J. Chem. Soc.*, **99**, 2329 (1911).
(3) Rule and Tod, *ibid.*, 1932 (1931).

TABLE I

ROTATIONS OF AMIDE AND AMINO ACIDS

Amino acids	*l*-Menthoxyacetamino acids Made up to 15 cc. With acetone: $l = 2$ Wt., g.	α_D	$[\alpha]_D^{25}$	Amino acids $[\alpha]_D^{25}$ Found	Literature
Glycine	0.4940[a]	−3.65	− 73.7	0	
d-Alanine	0.0840	−0.74	− 50.6	+ 10.6[b]	+ 10.4[4]
l-Alanine	0.2697	−2.69	− 74.8	− 2.2[b]	− 10.3[4]
d-Valine	0.1650	−0.86	− 39.1	+ 6.3[c]	+ 6.4[5]
l-Valine	0.3731	−3.86	− 77.6	− 14.5[d]	− 29.0[5]
				− 2.7[c]	− 5.7[6]
d-Phenylglycine	0.2322	+1.02	+ 32.9	+157.5[e]	+157.8[7]
l-Phenylglycine	0.2638	−4.32	−122.7	− 67.2[e]	
				−140.6[e,f]	−157.8[7]
dl-Phenylalanine[g]	0.6055	−4.42	− 54.7	0	

[a] Made up to 10 cc. with acetone: $l = 1$. [b] Rotation of the hydrochloride in water. [c] Rotation in water. [d] Rotation in 20% hydrochloric acid. [e] Rotation in 2.6% hydrochloric acid. [f] Rotation of more soluble fraction after recrystallization from water. [g] Over half the product was obtained in large crystals of this melting point, yielding an inactive amino acid on hydrolysis, as did the residual oil from the mother liquor. Recrystallization from the solvents aqueous alcohol, petroleum ether and ethyl acetate, petroleum ether and benzene, at various temperatures, did not alter the melting point. Attempts to prepare crystalline brucine and strychnine salts failed.

(4) Fischer, *Ber.*, **39**, 464 (1906).
(5) Fischer, *ibid.*, **39**, 2325 (1906).
(6) Ehrlich, *Biochem. Z.*, **1**, 29 (1906).
(7) Fischer and Weichhold, *Ber.*, **41**, 1291 (1908).

TABLE II

CONSTANTS

OF *l*-MENTHOXYACETAMINO ACIDS

l-Menthoxyacetyl derivative	M. p., °C.
Glycine	155–156
d-Alanine	147–148
l-Alanine	117–118
d-Valine	156–157.5
l-Valine	93– 96
d-Phenylglycine	162
l-Phenylglycine	113–116
dl-Phenylalanine	100–101

Authors' comments:

By the use of an optically active chloride, diastereoisomeric amides of the type RCONHCHR'-COOH are formed, separated by fractional crystallization and then hydrolyzed. *l*-Menthoxyacetyl chloride has been used as the active acid chloride, a reagent which is readily prepared from natural menthol. It reacts with amino acids in a cold alkaline solution to give in excellent yields well-crystallized amides. By fractional recrystallization the two diastereoisomers, especially the less soluble form, could readily be obtained pure. Hydrolysis of the amides in aqueous alcohol by means of hydrobromic acid yielded the pure active amino acid without difficulty and no racemization occurred in the cases studied. The absence of any menthol or menthol derivatives in the hydrolyzed products was proved by the complete inactivity of glycine which had been put through a similar procedure.

$C_3H_7NO_2$ Ref. 18j

Although a number of optical resolutions of DL-amino acids have been reported, most of them have employed chemical or enzymatic procedures, and reports of optical resolution by physicochemical procedure have appeared less often.

If successfully applied, physicochemical resolution, especially the resolution by preferential crystallization, is considered to be one of the most advantageous procedures for the practical production of optically active amino acids[1]. However, satisfactory application of this type of simple procedure has been limited for several amino acids such as histidine[2], threonine[3] and glutamic acid[4], and it is also conceivable that there have been many failures, in spite of all efforts to apply this method to amino acids generally. So far as alanine is concerned, no report has appeared on the successful total optical resolution by preferential crystallization procedure. This communication describes a direct optical resolution of DL-alanine which has been carried out as a first approach to establish the general method for the optical resolution of amino acids.

DL-Alanine itself was proved to be unsuitable for direct resolution due to its properties of forming the racemic compound and also its solubility. Therefore, DL-alanine was converted to readily obtainable salts and derivatives, and the properties of these compounds were investigated. As a result, possibility of direct resolution was indicated in the case of alanine benzene sulphonate. Namely, the solubility of optically active alanine benzene sulphonate was much less than that of DL-modification, although its IR-spectra indicated the formation of racemic compound.

Thus conditions required for optical resolution of DL-alanine benzene sulphonate were studied in detail. As a result, pure optical isomers could be obtained from a supersaturated solution containing excess of desired antipode by seeding the solution with the crystals of the respective antipode and by filtering the precipitated

[1] R. M. SECOR, Chem. Rev. *63*, 297 (1963).
[2] R. DUSCHINSKY, Chemy Ind. *53*, 10 (1934).
[3] L. VELLUZ and G. AMIARD, Bull. Soc. chim. Fr. *20*, 903 (1953).
[4] T. AKASHI, J. chem. Soc. Japan, Pure Chem. Sec. *83*, 417 (1962)

Experiment No.	Amount of addition		Composition of solution		Amount of inoculation (g)	Resolved crystals[a]		
	DL-form (g)	Active form (g)	DL-form (g)	Active form (g)		Yield (g)	$[\alpha]_D^{25}$	Optical purity (%)
1	26.00	(L) 0.90	26.00	(L) 0.90	(L) 0.10	2.16	+ 7.50°	100
2	2.06	0	25.74	(D) 1.16	(D) 0.10	2.68	− 7.30°	97.3
3	2.58	0	25.55	(L) 1.35	(L) 0.10	2.43	+ 7.50°	100

[a] Specific rotation of pure L-alanine benzenesulphonate: $[\alpha]_D^{25} = +7.50°$ (c = 2, ethanol).

crystals when the amount of the separated crystals attained was approximately twice that of the antipode initially in excess.

Alanine benzene sulphonate was prepared by dissolving DL-, D- or L-alanine in an aqueous solution of benzene sulphonic acid.

Typical resolution procedures are as follows: DL-alanine benzene sulphonate, 52.0 g, and D-alanine benzene sulphonate, 1.30 g, were dissolved in 200 ml of 97% aqueous acetone at elevated temperature and cooled slowly to 25°. The solution was seeded with 0.20 g of finely pulverized D-alanine benzene sulphonate and allowed to stand at the same temperature for 16 h. The precipitated crystals were filtered and 4.40 g of D-alanine benzene sulphonate was obtained. Anal. Found: N, 5.62%: calcd. for $C_9H_{13}O_5NS$: N, 5.66%. The product was optically pure, $[\alpha]_D^{25} = -7.50°$ (c = 2, ethanol). The D-alanine benzene sulphonate, 3.00 g, was dissolved in 60 ml of distilled water and passed through a column of Amberlite IR-120 in H-form. The D-alanine absorbed on the resin was eluted with 180 ml of 1 N-NH_4OH and the elute was concentrated to dryness. The residue was crystallized from aqueous methanol to give 0.98 g of pure D-alanine (91% of the theoretical). Anal. Found: N, 15.70: calcd. for $C_3H_7O_2N$: N, 15.72. $[\alpha]_D^{25} = -14.60°$ (c = 2, 5 N HCl).

For further optical resolution, the mother liquor can be used repeatedly to separate the other enantiomorph. Namely, the same amount of DL-modification as that of the enantiomorph previously separated out, is added to the mother liquor and dissolved at an elevated temperature. The supersaturated solution was cooled, seeded and crystallized in the same way as described above. By repeating these procedures, L- and D-alanine benzene sulphonates were successfully obtained. The examples of the first several runs in 100 ml scale are shown in the Table.

Benzene sulphonic acid in the effluent of ion exchangers charged with solution of optically active alanine benzene sulphonate was readily recovered as DL-alanine benzene sulphonate by the addition of the corresponding amount of DL-alanine to the effluent and by further concentration of the solution.

Thus the total optical resolution of DL-alanine benzene sulphonate can be accomplished. This simple procedure is considered to be one of the most advantageous methods for optical resolution of DL-alanine, because the method requires neither optically active resolving agent nor conversion of DL-alanine into complicated derivatives.

$C_3H_7NO_2$ Ref. 18k

Experimental[12]

Synthesis of N-Carbobenzoxy Derivatives.—The compounds were prepared by a generally applicable procedure. The amino acid was exactly neutralized with sodium hydroxide in 5 to 10 parts of water. The solution was cooled to 5° in an ice-bath and stirred mechanically. From dropping funnels one equivalent each of sodium hydroxide solution and benzylchloroformate in toluene[13] were added at rates to maintain a slightly basic solution and a temperature of 5–10°. The mixture was washed twice with ether and the carbobenzoxy derivative precipitated in the cold by a slight excess of hydrochloric acid. The crude product was crystallized from appropriate solvents.

a. **N-Carbobenzoxy-DL-alanine.**—The yield of crude material was almost 100%, m.p. 90–95°. It was recrystallized from benzene, giving a 93% yield in two crops, m.p. 113.5°, reported[14] m.p. 114–115°.

b. **N-Carbobenzoxy-L-alanine** was prepared from natural L-alanine. Upon acidifying the aqueous solution of the sodium salt an opalescent mixture was obtained. Fine needles formed overnight at 4°, yield 86%, m.p. 84–85°, $[\alpha]^{25}D$ −13.59° (c 4, glacial acetic acid) unchanged after recrystallization from benzene. Bergmann and Zervas[14] reported m.p. 84°, $[\alpha]^{25}D$ −14.3° (glacial acetic acid).

c. **N-Carbobenzoxy-DL-phenylalanine.**—The crude product was oily but soon solidified. After drying 24 hours in vacuum it could be powdered in a mortar; yield 96%, m.p. 102–103°. The yield was 90% after recrystallization from benzene, m.p. 102–103°, in agreement with literature values.[14]

d. **N-Carbobenzoxy-DL-tryptophan.**—After addition of all of the reagents the sodium salt of the carbobenzoxy derivative separated from the alkaline solution as a viscous mass. Acidification gave an oil which soon solidified and was treated as described above; yield 97%, m.p. 165° (not sharp). Recrystallizations were in general unsatisfactory, but it was crystallized conveniently from ethyl acetate; over-all yields 85–87%, m.p. 168–169°, literature[15] value 169–170°.

Resolution of N-Carbobenzoxy-DL-alanine. (a) Separation of Diastereoisomeric Salts with (−)-Ephedrine.—N-Carbobenzoxy-DL-alanine (11.7 g.) and (−)-ephedrine (8.7

(12) Melting points are uncorrected and were determined in capillary tubes.

(13) H. E. Carter, R. L. Frank and H. W. Johnston, *Org. Syntheses*, **23**, 13 (1943).

(14) M. Bergmann and L. Zervas, *Ber.*, **65**, 1192 (1932).

(15) E. L. Smith, *J. Biol. Chem.*, **175**, 39 (1948).

g.) were dissolved in 75 cc. of warm ethyl acetate. Cooling at room temperature gave 9.4 g. of the less soluble diastereoisomeric salt of N-carbobenzoxy-D-alanine, $[\alpha]^{25}D$ −23.88° (*c* 2, methanol). This was recrystallized from 275 cc. of ethyl acetate, giving 8.3 g., $[\alpha]^{25}D$ −23.92° (*c* 2, methanol). Thus the first crystallization yielded (92%) a diastereoisomeric salt with maximum rotation. No more crystalline material was obtained from the filtrates containing the more soluble salt.

b. **N-Carbobenzoxy-D-alanine** was obtained from the less soluble salt after acid decomposition by a generally applicable procedure: The salt (7.0 g.) was dissolved in 100 cc. of water and neutralized with a slight excess of dilute hydrochloric acid. Crystallization at 4° yielded 3.7 g. of N-carbobenzoxy-D-alanine, m.p. 84–85°, $[\alpha]^{25}D$ +13.65° (*c* 4, glacial acetic acid). The melting point and rotation were not changed by recrystallization from benzene. The constants agree closely with those of the L-form.

The aqueous acid filtrate containing ephedrine was combined with other similar material for recovery of either the free base or the hydrochloride. The free base was extracted into ethyl acetate after neutralizing the solution with sodium hydroxide. Alternately, the acid solution was evaporated to dryness under reduced pressure and the residue washed with acetone to remove traces of carbobenzoxy derivatives. The ephedrine hydrochloride remained as a fine powder. The products could be reused without further purification.

c. **N-Carbobenzoxy-L-alanine** along with some of the racemic compound was obtained from the more soluble salts in the original ethyl acetate mother liquors. Evaporation of the ethyl acetate, followed by decomposition of the salts with hydrochloric acid as above, gave the partially resolved material, m.p. 83–84°, $[\alpha]^{25}D$ −10.4° (*c* 4, glacial acetic acid). Systematic recrystallization from benzene indicated that the DL-form was less soluble than the active form. The impure N-carbobenzoxy-L-alanine therefore could not be purified by fractional crystallization from benzene. However it was easily obtained in pure form through the diastereoisomeric salts of (+)-ephedrine, made available by resolution of DL-ephedrine with N-carbobenzoxy-L-alanine, as shown below.

d. **D-Alanine** was obtained by hydrogenolysis of the corresponding carbobenzoxy derivative by a generally applicable procedure. Five grams of N-carbobenzoxy-D-alanine was dissolved in 50 cc. of methanol containing 2 cc. of concentrated hydrochloric acid. About 0.2 g. of palladium black[16] was added. The mixture was stirred vigorously and hydrogen passed in at atmospheric pressure. Carbon dioxide was detected almost immediately in the exit gases. After 45 minutes the catalyst was removed by filtration, and the methanol solution evaporated to dryness under reduced pressure. The residue was dissolved in 95% ethanol and the free amino acid precipitated with a slight excess of ammonium hydroxide, and then recrystallized from aqueous ethanol. The recrystallized D-alanine had $[\alpha]^{25}D$ −14.23° (*c* 2, 6 *N* HCl).[17] Dunn and Rockland[18] reported $[\alpha]^{25}D$ +14.47° (*c* 10, 5.97 *N* HCl) for L-alanine.

(18) M. S. Dunn and L. B. Rockland, *Advances in Protein Chem.*, **3**, 296 (1947).
(19) Merck and Co., Inc., Rahway, N. J.

$C_3H_7NO_2$

Ref. 18-1

The resolution of racemic amino acids by the formation of diastereoisomers possesses some difficulties in practice. It is usually possible to isolate the less soluble of the diastereoisomeric pairs and then regenerate from it the optically pure acid. However the more-soluble derivative remaining in the mother liquor is contaminated with a small amount of the less-soluble material and therefore is difficult to obtain pure, since recrystallization is not very effective in removing small amounts of an insoluble material from a soluble material. Marckwald[1] has pointed out that the salts *dA.lB* and *lA.dB* as well as *lA.lB* and *dA.dB* are enantiomorphic and possess the same solubility. Hence if an *acid* (*dlA*) is treated with an optically active base (*lB*), the less-soluble diastereoisomer (say *dAlB*) can be obtained pure. The partially resolved but impure acid, obtained from the more-soluble diastereoisomer (*lAlB*) can then be recombined with the optical antipode of the original active base (*dB*). The enantiomorphic salt (*lAdB*) is now less soluble than its diastereoisomer (*dAdB*) and is the first to crystallize. In order to accomplish resolution by Marckwald's method it is necessary to have supplies of both *d* and *l* forms of the resolving base in the optically pure state. Hitherto, only very few such pairs were available, hence this method has had very little use. For instance, the naturally occurring bases brucine, cinchonidine, cinchonine, morphine, quinine, strychnine, ephedrine, menthylamine, and the like have been used for the resolution of racemic amino acids, but they only occur in one form.[2]

We have now found[3] that metal complex cations such as *cis*-dinitrobis-(ethylenediamine)cobalt(III) ion and oxalatobis(ethylenediamine)cobalt(III) ion, which are easily prepared and readily resolved to give both enantiomorphs,[4,5] can be used successfully for the resolution of amino acid derivatives.

The resolutions are carried out by mixing aqueous solutions of the metal complex acetates with sodium salt solutions of the amino acid derivatives. One diastereoisomer precipitates from the solution according to the following equation.

$$2l[\text{Complex}]^+X^- + 2\text{Na}dA^- \rightarrow l[\text{Complex}]^+dA^- + l[\text{Complex}]^+lA^- + 2\text{Na}^+X^-.$$

The well-defined diastereoisomeric salts crystallize rapidly from solution. The diastereoisomeric pairs have widely different solubilities, and a sharp separation of isomers is obtained (> 90%).

Recovery of the resolved amino acid derivative is effected by slurrying the diastereoisomer in water in the presence of potassium iodide, when the very insoluble complex iodide is precipitated. After acidification of the filtrate the amino acid derivative usually crystallizes out.

From the solution containing the more-soluble diastereoisomeric salt, the partly resolved amino acid derivative can be isolated. This amino acid derivative is obtained optically pure by applying Marckwald's method using the antipode of the resolving agent.

The resolved amino acid derivatives have outstanding optical purity as demonstrated by biochemical assay using D-amino oxidase (< 1/500 of D-isomer) and radioactive tracer experiments.[6]

In general, resolution of the parent racemic amino acids themselves could not be performed using the complex metal cations, since the diastereoisomers were too soluble. Conversion of the amino acids to suitable substituted derivatives, however, modifies the solubilities of the diastereoisomers sufficiently to cause precipitation of the less-soluble diastereoisomer from moderate concentration. Hence the choice of a suitable derivative is important and for the resolution of neutral amino acids, the benzoyl, tosyl, and phthaloyl derivatives were found to be most useful and convenient.

During the resolution of phthaloyl-DL-phenylalanine and phthaloyl-DL-alanine ring opening of the phthaloyl ring was observed. The isolated resolved amino acid derivatives were identified readily as the corresponding *o*-carboxybenzoyl derivatives by infrared spectroscopy. The formation of the *o*-carboxybenzoyl derivatives is almost certainly due to the slightly alkaline reaction conditions during the resolution procedure. The alkali sensitivity of phthaloyl peptides and the isolation of the corresponding *o*-carboxybenzoyl peptides by paper chromatography has already been reported.[7]

* Manuscript received November 23, 1962.
† The John Curtin School of Medical Research, The Australian National University, Canberra, A.C.T.

[1] Marckwald, W. (1896).—*Ber. dtsch. chem. Ges.* 29: 43.
[2] Greenstein, J. P., and Winitz, M. (1961).—"The Chemistry of the Amino Acids." Vol. 1. 723–8. (John Wiley: New York.)
[3] Dwyer, F. P., and Halpern, B. (1962).—*Nature* 196: 270.
[4] Dwyer, F. P., and Garvan, F. L. (1960).—*Inorg. Synth.* 6: 195.
[5] Dwyer, F. P., Garvan, F. L., and Reid, I. K. (1961).—*J. Amer. Chem. Soc.* 83: 1285.
[6] Garnett, J., Halpern, B., Law, S. W., and Turnbull, K. R. (1963), unpublished data.
[7] Hanson, H., and Illhardt, R. (1954).—*Z. Physiol. Chem.* 298: 210.

Resolution of Benzoyl-DL-alanine.—Benzoyl-DL-alanine (7·72 g)[8] was dissolved in sufficient 1N NaOH (40 ml) to give an exactly neutral solution, which was added to a solution of *d-cis*-dinitrobis(ethylenediamine)cobalt(III) acetate [prepared from the *d*-iodide (9·0 g) and silver acetate (3·73 g) in water (25 ml) according to the directions of Dwyer and Garvan[4]]. The solution was evaporated to 30 ml under reduced pressure at 30°C. After cooling to 5°C, the yellow diastereoisomeric salt *d-cis*-dinitrobis(ethylenediamine)cobalt(III)-benzoyl-D-alaninate [A] was filtered, dried, and recrystallized by dissolving in warm water (30 ml), evaporating to 20 ml and cooling to 5°C (5·0 g, 54%), $[\alpha]_D^{20}$ +17°, 1% in water (Found: C, 36·1; H, 5·6; N, 21·0%. Calc. for $C_{14}H_{26}O_7N_7Co$: C, 36·3; H, 5·7; N, 21·2%). The filtrates from the above were combined [B] and set aside for recovery of partly resolved benzoyl-L-alanine. The diastereoisomer [A] (5·0 g) was added in small quantities to a solution of KI (5 g) in water (20 ml), the suspension stirred at room temperature for 15 min and the *d-cis*-dinitrobis(ethylenediamine)cobalt(III) iodide was collected. The filtrate was acidified with 6N HCl, cooled at 5°C for 15 min and the colourless crystals were filtered. Recrystallization from water gave benzoyl-D-alanine (1·9 g) $[\alpha]_D^{20}$ −34°, 1% in 1N NaOH (Dunn[8] reports $[\alpha]_D^{25}$ −32·5°, 1% in 1·05N NaOH).

TABLE 1

AMINO ACID DERIVATIVES RESOLVED

Amino Acid Derivative*	Least-soluble Diastereoisomer	Product	
		Rotation	Literature
Phthaloyl-DL-alanine[10]	*d-cis*-Dinitrobis(ethylenediamine)-cobalt(III)-*o*-carboxybenzoyl-D-alaninate $[\alpha]_D^{20}$ +38°, 0·48% in water; Found: N, 19·6%. Calc. for $C_{15}H_{26}O_9N_7Co$: N, 19·3%	*o*-Carboxybenzoyl-D-alanine $[\alpha]_D^{20}$ +24°, 2·6% in ethanol	
Phthaloyl-DL-phenylalanine[10]	*l-cis*-Dinitrobis(ethylenediamine)-cobalt(III)-*o*-carboxybenzoyl-D-phenylalaninate $[\alpha]_D^{20}$ +72°, 0·3% in water; Found: C, 43·3; H, 5·1; N, 16·5%. Calc. for $C_{21}H_{30}O_9N_7Co$: C, 43·2; H, 5·2; N, 16·8%	*o*-Carboxybenzoyl-D-phenylalanine $[\alpha]_D^{20}$ +213°, 1% in ethanol	
Tosyl-DL-serine[11]	*l-cis*-Dinitrobis(ethylenediamine) cobalt(III)-tosyl-L-serinate $[\alpha]_D^{20}$ +4°, 0·4% in water; Found: C, 31·4; H, 5·4; N, 18·3%. Calc. for $C_{14}H_{28}O_9N_7SCo$: C, 31·8; H, 5·3; N, 18·5%	Tosyl-L-serine $[\alpha]_D^{20}$ −33°, 1% in pyridine	$[\alpha]_D^{25}$ −32·3°, 2% in pyridine[12]
Benzoyl-DL-valine	*l-cis*-Dinitrobis(ethylenediamine)-cobalt(III)-benzoyl-L-valinate $[\alpha]_D^{20}$ +42°, 0·45% in water; Found: C, 39·3; H, 6·1; N, 19·8%. Calc. for $C_{16}H_{30}O_7N_7Co$: C, 39·1; H, 6·2; N, 20·0%	Benzoyl-L-valine $[\alpha]_D^{20}$ +19·6°, 1·1% in ethanol, $[\alpha]_D^{20}$ +53°, 1·0% in chloroform	$[\alpha]_D^{25}$ +21·8°, 4·9% in 95% ethanol[13] For benzoyl-D-valine $[\alpha]_D^{31}$ −51·0°, 1·04% in chloroform[14]

The combined filtrates [B] were evaporated to 20 ml under reduced pressure at 30°C, and KI (5 g) was added. After removal of the precipitated *d-cis*-dinitrobis(ethylenediamine)cobalt(III) iodide, the filtrate was acidified with 6N HCl, cooled at 5°C for 15 min, and the partially resolved benzoyl-L-alanine was collected (4·5 g). This product was re-resolved with *l-cis*-dinitrobis(ethylenediamine)cobalt(III) acetate using the same procedure as described above. Recrystallization from water gave benzoyl-L-alanine (2·0 g), $[\alpha]_D^{20}$ +33°, 1% in 1N NaOH. (Dunn[8] reports $[\alpha]_D^{22.5}$ +33·4°, 0·46% in 1N NaOH.)

Benzoyl-DL-alanine (19·3 g) was dissolved in sufficient 1N NaOH (100 ml) to give an exactly neutral solution which was added to a solution of *d*-oxalatobis(ethylenediamine)cobalt(III) acetate [ex *d*-bromide (36·5 g) and silver acetate (16·5 g) in water (100 ml) prepared according to the directions of Dwyer[5]]. The solution was evaporated to 140 ml under reduced pressure at 30°C. After standing at room temperature for 24 hr, the pink diastereoisomeric salt *d*-oxalatobis(ethylenediamine)cobalt(III)-benzoyl-D-alaninate was filtered, dried, and recrystallized from water. The diastereoisomer (10 g) was added in small quantities to a solution of KI (10 g) in water (50 ml), stirred at room temperature for 15 min and the *d*-oxalatobis(ethylenediamine)-cobalt(III) iodide was filtered. The filtrate was acidified with 6N HCl, cooled at 5°C for 15 min, and the colourless crystals were filtered. Recrystallization from water gave benzoyl-D-alanine (3·5 g).

Recovery of L-Alanine from Benzoyl-L-alanine.—The benzoyl derivative (1·3 g) was dissolved in 6N HCl (10 ml) and hydrolysed by heating under reflux for 3 hr. After cooling, benzoic acid was removed by four extractions with ether. The aqueous layer was concentrated to a solid, dissolved in ethanol, and evaporated to dryness four times. Alanine then was precipitated from a solution of alanine hydrochloride in 95% ethanol by the addition of aniline. The alanine was filtered at 0°C and washed with ethanol and then ether. The L-alanine $[\alpha]_D$ +15° (assayed biochemically with D-amino oxidase[9]) contained less than one part in 500 of the D-antipode.

Several other amino acid derivatives were resolved by procedures similar to those described above. The results are summarized in Table 1.

[8] Dunn, M. S., Stoddard, M. S., Rubin, L. B., and Bovie, R. C. (1943).—*J. Biol. Chem.* **151**: 241.

[9] Negelein, E., and Bromel, H. (1939).—*Biochem. Z.* **300**: 225.

$C_3H_7NO_2$ Ref. 18m

13.1 g. of *dl*-N-acetylalanine and 13.5 g. of *l*-amphetamine are dissolved with heating in 125 ml. of 95% ethanol. The solution is allowed to stand at room temperature for 16–18 hours during which time a feathery precipitate is formed. The precipitate consisting of the *l*-amphetamine salt of *l*-N-acetylalanine is removed by filtration and recrystallized from 95% ethanol;

$$[\alpha]_D^{25} = -33°$$

(2% in water). The recrystallized material is dissolved in water and passed through an ion exchange column containing a sulfonate resin in the acid form (Dowex 50). The effluent is concentrated to dryness. After recrystallization from water the product, *l*-N-acetylalanine, has an optical rotation, $[\alpha]_D^{25}$, of −66.2° (2% in water).

$C_3H_7NO_2$ Ref. 18n

The enzymatic procedure for the resolution of amino acids developed in this Laboratory[2a,b] depends upon the asymmetric enzymatic hydrolysis of the N-acyl or amide derivative of the racemic amino acid. The liberated free L-amino acid is separated from the unhydrolyzed D-derivative by the addition of ethanol; the latter derivative is subsequently converted into the D-amino acid by acid hydrolysis. However, since certain amino acids, *e.g.*, isovaline, are soluble in alcohol, and hence cannot be separated in this manner, another isolation procedure was devised. This alternative procedure, employing ion-exchange chromatography for the separation of the products of the asymmetric enzymatic hydrolysis, was designed to permit the use of the small amounts of material usually involved in the synthesis of isotopically labeled amino acids. The method also has been used for the large scale preparation of the enantiomorphs of isovaline.[3]

If cation-exchange chromatography is applied to the mixture obtained after enzymatic resolution of the N-acyl derivative, the free L-amino acid is retained by the column and the unhydrolyzed D-amino acid derivative is eluted with water. After action of the enzyme upon the amide derivative, however, a weaker cation-exchange resin is used, which permits the free amino acid to pass through the column, while the unhydrolyzed derivative is retained and subsequently removed from the column with weak acid. Cation exchange therefore affords sufficiently mild conditions to avoid hydrolysis of the N-acyl or amide derivative.

Experimental

Preparation of the Ion-exchange Resin Columns.—Dowex-50[4] and Amberlite XE-64[5] in the acid phase were used. The resin as received from the manufacturer was subjected to two cycles of washing with 5 *N* HCl, water, 1 *N* NaOH, water, and followed by a final 5 *N* HCl and water wash. Washing with water in each case was continued to completion as indicated by congo red and phenolphthalein. Resin columns (see Table I for dimensions) were prepared

(1) Presented in part before the Division of Biological Chemistry at the 120th Meeting of the American Chemical Society at Atlantic City, N. J., September, 1952.

(2) (a) S. M. Birnbaum, L. Levintow, R. B. Kingsley and J. P. Greenstein, *J. Biol. Chem.*, **194**, 455 (1952); (b) J. P. Greenstein, S. M. Birnbaum and M. C. Otey, *ibid.*, in press.

(3) C. G. Baker, S-C. J. Fu, S. M. Birnbaum, H. A. Sober and J. P. Greenstein, THIS JOURNAL, **74**, 4701 (1952).

(4) A strong cation-exchange resin with sulfonic acid functional groups (200–400 mesh) obtained from the Dow Chemical Company, Midland, Mich.

(5) A weak cation-exchange resin with carboxylic acid exchange groups obtained from the Resinous Products Division, Rohm and Haas, Philadelphia, Penna.

Table I

Experimental Conditions for Obtaining the Optical Isomers of Amino Acids (See Text)

DL-Amino acid derivative	Wt. of derivative, g.	Wt. of added enzyme, mg.	Time of incubation, hr.	Resin used	Column size height × diam., cm.	Hydrochloric acid eluting agent, N	Agent for converting the hydrochloride into the free amino acid
N-Chloroacetylaspartic acid	3.15	Acylase II 90.0	20.5	Dowex-50	32.5 × 2.5	1	Aniline
N-Acetylhistidine	2.77	Acylase I 194.0[b]	40.0	XE-64	55.0 × 2.5	1	LiOH
Proline amide[a]	[a]	[a]	[a]	XE-64	38.0 × 2.5	0.12	Ag_2CO_3
N-Chloroacetylserine	3.45	Acylase I 3.0	6.5	Dowex-50	57.0 × 2.5	2.5	Aniline
N-Chloroacetylphenylalanine	2.94	Carboxypeptidase 33.0	10.0	Dowex-50	35.0 × 2.0	5	Aniline
N,N′-Dichloroacetylornithine	4.32	Acylase I 115.0[c]	42.0	Dowex-50	43.0 × 2.5	5	.....
N-Acetylalanine	2.94	Acylase I 2.2	13.5	Dowex-50	31.3 × 1.1	2.5	Aniline
N-Acetylmethionine	2.56	Acylase I 0.2	13.5	Dowex-50	31.5 × 2.5	5	Aniline
N-Acetylvaline	2.72	Acylase I 5.0	7.0	Dowex-50	36.4 × 2.5	5	Aniline

[a] Non-enzymatic experiment: 0.880 g. of L-proline amide plus 0.887 g. of L-proline (see text). [b] Added in two portions: 179 mg. initially and an additional 15 mg. at 18 hr. [c] Added in two portions: 105 mg. initially and an additional 10 mg. at 18 hr.

from suspensions of the washed resin, allowed to settle in a glass column and then washed with an additional liter of distilled water. Dowex-50 was used for alanine, methionine, valine, aspartic acid, serine, ornithine and isovaline, and XE-64 was used for proline, histidine and arginine.

Substrates and Enzymes.—Many of the amino acid derivatives used were donated by Dr. J. P. Greenstein. All gave melting points in agreement with those in the literature[2] and the theoretical elemental analyses. The purified hog kidney enzymes, acylase I and II,[2] were prepared by Drs. S. M. Birnbaum and K. R. Rao, respectively. Crystalline carboxypeptidase was obtained from the Worthington Biochemical Sales Company.

Incubation with Enzyme.—N-Acyl-DL-amino acid, equivalent in amount to one gram of each enantiomorph, was dissolved in water and the *p*H adjusted to 7.6 (cresol red) with 6 *N* lithium hydroxide. Sufficient acylase was added to produce the theoretical 50% hydrolysis in one to two hours[2] and the *p*H was readjusted to 7.6. The mixture was diluted to a final substrate concentration of 0.1 *M* and incubated at 37°, the progress of the hydrolysis being followed by manometric-ninhydrin determinations. When 50% hydrolysis had been attained, the incubation mixture was deproteinized at the isoelectric point of the enzyme[2] with Norit, and the solution was reduced *in vacuo*[6] to a small volume.

Chromatography and Isolation of the D-Amino Acid.—The concentrated solution from the enzymatic run was added to the top of the appropriate resin column at a rate of about 0.5 ml. per minute and carefully washed into the resin with several 1-ml. aliquots of water. The N-acyl-D-amino acid was then eluted from the resin by water at a rate of about 10 ml./hr. and collected in fractions of 5–15 ml. The D-amino acid can be detected in the fractions most simply by the drop in *p*H which results from the acidic nature of the derivative.[7] Alternatively, aliquots of each fraction may be spotted on filter paper and examined with: (a) ninhydrin for the appearance of a ninhydrin positive color *only after* preliminary hydrolysis at 100° in 2 *N* HCl,[8] (b) various specific amino acid reagents,[9] or (c) the reagent of Rydon and Smith.[10]

The fractions containing the N-acyl-D-amino acid were combined and evaporated to dryness *in vacuo*. The residue was taken up in about 150 ml. of absolute ethanol or dry acetone and filtered to remove any residual protein. The filtrate was evaporated to dryness *in vacuo*, and the residue was hydrolyzed at 100° for 2 to 3 hours in 2 to 2.5 *N* HCl. The solution was then treated with Norit, filtered and taken to dryness *in vacuo* four times to remove excess HCl. The residue of D-amino acid hydrochloride was then converted to the free amino acid by the reagents indicated in Table I. The compounds were recrystallized from water–ethanol, with the exception of proline, in which case absolute ethanol was employed. Occasionally a second recrystallization was required.

Isolation of the L-Amino Acid.—Several hundred ml. of water was passed through the column subsequent to the complete emergence of the N-acyl-D-amino acid derivative from the column. Elution of the free amino acid was accomplished with hydrochloric acid at concentrations depending upon the resin and the particular amino acid involved. The conditions used are given in Table I. Detection of the fractions containing the L-amino acid was accomplished most conveniently by means of the ninhydrin spot test on paper,[11] although specific reagents also may be used.[9]

The fractions containing the L-amino acid were combined and evaporated to dryness *in vacuo*. The residue was taken up in about 150 ml. of concentrated HCl and filtered to remove salts which usually are eluted with some of the later fractions. The filtrate was diluted and after treatment with Norit the filtered solution was taken to dryness *in vacuo* four times to remove excess HCl. The L-amino acid hydrochloride was converted into the free amino acid in the manner used for its enantiomorph.

Determination of Optical Purity.—Optical rotations were all performed in a two-decimeter tube with a Hilger M375 polarimeter with triple field-type polarizer at 27°. Several of the isomers (*cf.* Table II) were examined for their enantiomorphs by the enzymatic procedure of Meister, Levintow, Kingsley and Greenstein[12] which is sensitive to at least one part of enantiomorph in one thousand parts of the assayed material.

The general procedure given above was followed for the individual amino acids and the specific details are given in Table I. In order to test the feasibility of the column separation, as indicated in Table I, a preliminary non-enzymatic experiment was performed wherein a concentrated solution of L-proline and L-proline amide was put directly on the column. Since proline is not retained by the XE-64 column used, it was eluted by the passage of water through

(6) All concentration procedures were performed *in vacuo* below 40°.

(7) Indicator paper is sufficiently sensitive. This test will apply only when the acyl derivative is used, since with the amide derivatives, the free amino acid emerges first from the XE-64 column.

(8) A 0.25% solution of ninhydrin in acetone is used; *cf.* G. Toennies and J. J. Kolb, *Anal. Chem.*, **23**, 823 (1951). A positive ninhydrin test in the early non-acidic as well as in the acid fractions, except where the amide derivative has been used, should be taken as evidence that the capacity of the resin column for that particular amino acid has been exceeded.

(9) The Sakaguchi test for arginine, the isatin test for proline and hydroxyproline, the Pauly test for histidine and tyrosine, the platinic iodide test for methionine, and the periodate-Nesslerization procedure for hydroxy amino acids; *cf.*, R. J. Block and D. Bolling, "The Amino Acid Composition of Foods. Analytical Methods and Results," 2nd Edition, C. C. Thomas, Springfield, Ill., 1952. The corresponding colors in the tests for proline amide and for the N-acetyl derivatives of histidine, arginine and methionine develop somewhat more slowly. The N-acetyl derivatives of the hydroxy amino acids do not give NH_3 after treatment with periodate.

(10) Except for methionine, tyrosine, cysteine and cystine, a blue starch–iodine color is given in this test with amino acids, peptides, proteins, amides and N-acyl derivatives previously exposed to chlorine; *cf.* H. N. Rydon and P. W. C. Smith, *Nature*, **169**, 923 (1952). A positive test with this reagent in the absence of a positive ninhydrin reaction, locates the N-acyl amino acids.

(11) Because of the acidity of the effluent fractions, the paper containing the aliquot spots should be exposed to an atmosphere of NH_3 for about 5 minutes and the excess NH_3 blown off before the paper is dipped into the ninhydrin solution.[8]

(12) A. Meister, L. Levintow, R. B. Kingsley and J. P. Greenstein, *J. Biol. Chem.*, **192**, 535 (1951).

TABLE II

YIELDS AND PROPERTIES OF RESOLVED AMINO ACIDS

Amino acid	Isomer	Yield, %	Concn. mg. per 2 ml.	Solvent	$[\alpha]_D$ Found, degrees	Nitrogen, % Calcd.	Nitrogen, % Found
Aspartic acid	D[b]	63	38.3	6 *N* HCl	−24.7	10.5	10.4
	L	92	40.0	6 *N* HCl	+25.3		10.6
Histidine	D	55	37.5	H_2O	+38.5	27.1	27.1
	L	49	33.5	H_2O	−38.4		27.1
Proline	L[a]	54	32.0	H_2O	−84.4	12.2	12.2
	L	47	26.8	H_2O	−85.1		12.1
Serine	D	69	82.7	1 *N* HCl	−14.5	13.3	13.2
	L	48	95.6	1 *N* HCl	+14.4		13.2
Phenyl-alanine	D[b]	36	32.4	H_2O	+35.1	8.5	8.7
	L[b]	35	31.9	H_2O	−34.5		8.6
Ornithine·2 HCl	D	46	65.4	5 *N* HCl	−18.2[c]	13.7	13.9
	L	60	62.3	5 *N* HCl	+17.9[c]		13.7
Alanine	D	65	55.9	5 *N* HCl	−14.2	15.7	15.9
	L[b]	61	81.7	5 *N* HCl	+14.2		15.8
Methionine	D	66	52.4	6 *N* HCl	−22.7	9.4	9.4
	L	54	40.0	6 *N* HCl	+23.6		9.4
Valine	D[b]	52	35.5	6 *N* HCl	−27.3	11.9	11.9
	L[b]	74	46.0	6 *N* HCl	+28.2		11.9

[a] Derived from the amide (see text). [b] Examined by the enzymatic test[12] and shown to contain less than 0.1% enantiomorph. [c] The only recorded value in 5 *N* HCl is ±18.2°, S-C. J. Fu, K. R. Rao, S. M. Birnbaum and J. P. Greenstein, *J. Biol. Chem.*, **199**, 207 (1952).

the column. After an additional several hundred ml. of water had passed through, proline amide was eluted with weak acid. In a similar non-enzymatic experiment with L-arginine and N-acetyl-DL-arginine, the acetyl derivative was eluted from the XE-64 column with water and the free amino acid removed with weak acid.

It should be pointed out that although the di-amino acids would be retained by the weak cation exchanger, XE-64, it was necessary to use the strong cation-exchange resin, Dowex-50, for the column separation in the ornithine resolution reported here. Starting with the N,N′-dichloroacetyl-amino acid, only the α-amino group is liberated. The products of the enzymatic hydrolysis are similar in their behavior on the strong exchange resin to those obtained in the resolution of the monoamino-monocarboxylic amino acids.

Results and Discussion

The results obtained with nine representative amino acids are given in Table II. Amino acid enantiomorphs with rotation values in good agreement with those in the literature, with theoretical nitrogen analyses[13] and, where tested with optical purity[12] of at least 99.9%, were obtained in the yields listed in Table II. These yields were calculated on the basis of the acyl or amide derivatives.

The percentage yields obtained with the small amounts employed in this study are similar to those obtained with larger amounts using the usual procedure.[2] A batch procedure with methionine was investigated with Dowex-50 and was considerably more complicated, involved larger volumes of solution and resulted in somewhat lower yields.

It is thus believed that, in general, the chromatographic procedure is more suitable for small amounts of material. Furthermore, in some instances, namely, proline and histidine, the column procedure is simpler than the usual technique.

The amino acids before recrystallization were, in every case tested, free of enantiomorph (better than 99.9%) when examined by the enzymatic test.[12] However, since strong HCl elution from the Dowex-50 column also carried along salts with free amino acid, the amino acid hydrochlorides obtained *directly* from the column were not *analytically* pure, as shown by low nitrogen analyses and specific rotation values one to two degrees low. The yield of both *optically* and *analytically* pure amino acid could no doubt be increased even beyond the recorded values, either by prechromatographic desalting of the enzymatic resolution mixture or by more selective elution of the free amino acid with weaker HCl at a lower amino acid load per resin column.[14]

(13) Analyses by R. J. Koegel and his staff.

(14) S. Moore and W. H. Stein, *Cold Spring Harbor Symposia*, **14**, 179 (1949).

For other examples of enzymatic methods of resolution of racemic amino acids, see H. Zahn and E. Schnabel, Z. Physiol. Chem., 311, 260 (1958); L. Levintow, V.E. Price and J.P. Greenstein, J. Biol. Chem., 184, 55 (1950); and P.J. Fodor, V.E. Price and J.P. Greenstein, J. Biol. Chem., 178, 503 (1949).

$C_3H_7NO_2S$ $HSCH_2CH(NH_2)\text{-}COOH$ Ref. 19

l- bzw. *d*-S-Benzyl-β,β-dimethyl-cysteine (IV).

Wir lösten 49 g *d,l*-S-Benzyl-N-formyl-β,β-dimethyl-cystein in 140 cm^3 absolutem Methanol und gaben eine Suspension von 72 g wasserfreiem Brucin in 100 cm^3 absolutem Methanol zu. Dabei trat erst klare Lösung, nach kurzem Reiben mit dem Glasstab aber Krystallisation ein. Das ausgefallene Brucinsalz des *l*-S-Benzyl-N-formyl-β,β-dimethyl-cysteins (III) wurde in der Kälte abgenutscht und zweimal aus Methanol umkrystallisiert. Die getrocknete Substanz war äusserst hygroskopisch. Smp. 96—100°. $[\alpha]_D^{23} = -6$ bis $-10°$ (c = 1% in Methanol).

$C_{36}H_{43}O_7N_3S$	Ber.	C 65,33	H 6,55	N 6,35%
	Gef.	,, 64,95	,, 6,83	,, 6,23%

58 g krystallisiertes *l*-Brucinsalz III wurden in einem Scheidetrichter mit 150 cm^3 Chloroform und 300 cm^3 n. Ammoniak geschüttelt. Die ammoniakalische Lösung wurde mit Chloroform nachgewaschen und im Vakuum auf 250 cm^3 eingeengt. Zum Rückstand setzten wir 20 cm^3 konz. Salzsäure zu und erhitzten eine ½ Stunde unter Rückfluss zum Sieden, wobei die Fällung in Lösung ging. Nach Neutralisieren mit konz. Ammoniak krystallisierte das *l*-S-Benzyl-β,β-dimethyl-cystein (IV) aus. Nach Umkrystallisieren aus Methanol war sein Smp. 184—185° und die Drehung $[\alpha]_D^{21} = +97°$ (c = 1,085 in n. HCl).

$C_{12}H_{17}O_2NS$	Ber.	C 60,22	H 7,16	N 5,85%
	Gef.	,, 60,46	,, 7,47	,, 6,00%

Die Mutterlaugen des krystallisierten *l*-Brucinsalzes III, welche die *d*-Komponente enthalten, wurden im Vakuum eingeengt und wie oben mit Ammoniak und Chloroform gespalten. Durch Zusatz von 20 cm³ konz. Salzsäure und Kochen am Rückfluss wurde die Formylverbindung verseift und das *d*-S-Benzyl-β,β-dimethyl-cystein (IV) mit konz. Ammoniak gefällt. Es schmolz nach Umkrystallisieren aus Methanol bei 184—185°. $[\alpha]_D^{21} = -96°$ (c = 1,100 in n. HCl).

[1]) Alle Schmelzpunkte sind korrigiert.

Resolution procedures for the above compound using the same resolving agent may be found in: V. du Vigneaud and W.I. Patterson, J. Biol. Chem., 109, 97 (1935) and J. L. Wood and V. du Vigneaud, J. Biol. Chem., 130, 109 (1939).

$C_3H_7NO_2S$

Ref. 19a

9,2 g S-Benzyl-D,L-cysteinäthylester wurden mit 20 cm³ absol. Äthanol verdünnt und das Gemisch mit einer filtrierten Lösung von 16,1 g reiner D-Weinsäure in 124 cm³ Äthanol versetzt. Die Kristallisation des S-Benzyl-L-cystein-äthylester-D-bitartrates setzte bald danach ein und war nach 48 Stunden beendet. Kristallisationsverzögerungen ließen sich dadurch vermeiden, daß eine dem Spaltansatz entnommene Probe mit Äther zur Kristallisation gebracht und mit dem so gewonnenen D,L-Estertartrat die Hauptmenge angeimpft wurde.

Ausbeute an L-Ester-D-bitartrat. 6,3 g entsprechen 86% d. Th. mit einem optischen Reinheitsgrad von 90%. Durch Umkristallisieren aus Äthanol (15 cm³ pro g) ließ sich das Salz völlig optisch rein gewinnen.

Ausbeute: 5 g; Schmp. (korr.) 127—128°; $[\alpha]_D^{18}$: −8,51° (c = 2,0 in Methanol).

$C_{16}H_{23}NO_8S$ (389,4)

Ber.	C 49,36%	H 5,97%	N 3,62%
Gef.	C 49,07%	H 6,10%	N 3,46%.

Aus der 1 Mol überschüssige Weinsäure enthaltenden Mutterlauge wurde mit 200 cm³ absol. Äther im Verlaufe von 48 Stunden 7,1 g (96% d. Th.) S-Benzyl-D-cysteinäthylester-D-hydrogentartrat ausgefällt. Das Salz besaß eine optische Reinheit von 70%. Schmp. (korr.) 103—104°; $[\alpha]_D^{18}$: +18,9° (c = 2,0 in Methanol).

$C_{16}H_{23}NO_8S$ (389,4)

Ber.	C 49,36%	H 5,97%	N 3,62%
Gef.	C 49,11%	H 6,18%.	N 3,82%.

Gewinnung der optisch aktiven S-Benzyl-cysteinäthylesterhydrochloride

S-Benzyl-L-cystein-äthylester-hydrochlorid

Zur Suspension von 7 g L-Esterhydrogentartrat in 100 cm³ absol. Äther wurden 1,5 cm³ absol. Methanol als Lösungsvermittler hinzugefügt und mit trockenem Ammoniak gesättigt. Vom Ammoniumtartrat wurde abgesaugt, mit absol. Äther nachgespült und das Lösungsmittel im Vakuum vertrieben. Der zurückbleibende Ester wurde mit 4 cm³ Methanol versetzt und in diese Mischung trockener Chlorwasserstoff bis zur sauren Reaktion eingeleitet. Mit 200 cm³ absol. Äther versetzt, ließen sich 4,8 g (95% d. Th.) L-Esterhydrochlorid isolieren. Schmp. (korr.) 156°; $[\alpha]_D^{18}$: −19,1° (c = 4,0 in Wasser).

$C_{12}H_{17}NO_2S \cdot HCl$ (275,8)

Ber.	C 52,25%	H 6,57%	N 5,07%
Gef.	C 52,40%	H 6,26%	N 5,05%.

Wie die Verseifung zur freien Aminosäure zeigte, besaß das S-Benzyl-L-äthylesterchlorhydrat eine optische Reinheit von mindestens 97%. K. C. Hooper, H. N. Rydon, J. A. Schofield und G. S. Heaton[17]) geben einen Schmp. von 155° und einen Drehwert von $[\alpha]_D$: −25,4° (c = 2 in Wasser) an. Dieser Drehwert konnte nicht bestätigt werden. Wie wir fanden, zeigte sowohl aus natürlichem Cystin als auch aus der Racematspaltung stammendes S-Benzyl-L-cystein äthylesterhydrochlorid auch nach mehrfacher Umkristallisation aus Alkohol/Äther eine konstante spezifische Drehung von $[\alpha]_D^{18}$: − 19,1°.

[17]) K. C. Hooper, H. N. Rydon, J. A. Schofield u. G. S. Heaton, J. chem. Soc. **1956**, 3151.

S-Benzyl-D-cysteinäthylesterhydrochlorid

Aus dem D-Ester-D-hydrogentartrat wurde wie beim Antipoden beschrieben ein D-Esterchlorhydrat mit 70proz. optischer Reinheit erhalten. Durch langsame fraktionierte Kristallisation aus Methanol mit absol. Äther gewinnt man ein optisch reines Produkt. Ausbeute: 50% d. Th.; Schmp. (korr.) 148°; $[\alpha]_D^{18}$: + 19,0° (c = 4,0 in Wasser).

$C_{12}H_{17}NO_2S \cdot HCl$ (275,8)

Ber.	C 52,25%	H 6,57%	N 5,07%
Gef.	C 52,18%	H 6,67%	N 5,28%.

Aus dem L- und D-Äthylesterhydrochlorid lassen sich mit ammoniakalischem Äther in quantitativer Ausbeute die reinen aktiven Ester freisetzen.

Gewinnung der optisch aktiven S-Benzyl-cysteine

S-Benzyl-L-cystein

S-Benzyl-L-äthylesterhydrochlorid wurde mit der 20fachen Menge konz. Salzsäure 1 Stunde am Rückfluß verseift und im Vakuum zur Trockne eingedampft. In wenig Wasser aufgenommen, wurde die Lösung mit Ammoniak auf p_H 3–4 gebracht. Die freie Säure fiel in der Kälte quantitativ aus. Nach Umkristallisation aus Wasser erhielten wir in 90proz. Ausbeute das reine S-Benzyl-L-cystein, Schmp. 206–207°; $[\alpha]_D^{18}$: + 24,7° (c = 1,0 in 1 n NaOH).

V. du Vigneaud und J. L. Wood[2]) geben für diese Verbindung an: $[\alpha]_D$: + 23,5° (c = 1 in 1 n NaOH).

S. M. Birnbaum und J. P. Greenstein[3]) geben eine spezifische Drehung von $[\alpha]_D^{26}$: + 25,5° bei einer optischen Reinheit von mehr als 99,9% an. Somit betrug die optische Reinheit unseres Produktes wenigstens 97%.

$C_{10}H_{13}NO_2S$ (211,3)

Ber.	C 56,84%	H 6,20%	N 6,63%
Gef.	C 56,98%	H 6,63%	N 6,51%.

S-Benzyl-D-cystein

Die Aufarbeitung erfolgte wie für den L-Antipoden beschrieben; Schmp. 203–204°; $[\alpha]_D^{18}$: – 26,0° (c = 1,0 in 1 n NaOH). S. M. Birnbaum und J. P. Greenstein[3]) geben für ein optisch reines S-Benzyl-D-cystein einen Drehwert von $[\alpha]_D^{26}$: – 26,0° (c = 1,0 in 1 n NaOH) an.

$C_{10}H_{13}NO_2S$ (211,3)

Ber.	C 56,84%	H 6,20%	N 6,63%
Gef.	C 56,90%	H 6,40%	N 6,64%

L- und D-Cystin

Nach V. du Vigneaud und J. L. Wood[2]) durch hydrierende Abspaltung des Benzylesters mit Natrium in flüssigem Ammoniak und anschließender Oxydation des Cysteins in wäßriger Lösung mit $FeCl_3$-Luftsauerstoff.

Ausbeute: 86%; Schmp. 258° und 256–261° (Zers.).

$C_3H_7NO_2S$ Ref. 19b

In eine Lösung von 87,7 kg DL-Cystein-hydrochlorid-monohydrat in 500 l Wasser werden 50 kg Dicyandiamid eingetragen. Nach 40stündigem Rühren bei Raumtemperatur haben sich 79,3 kg = 84,5% der Theorie DL-2-Guanidino-1,3-thiazolincarbonsäure-4 abgeschieden.

In 5400 l Wasser von 45° C werden 33,9 kg DL-Guanidino-thiazolincarbonsäure (180 Mol) und 49,6 kg Kupfer-L-aspartat (180 Mol) eingetragen. Die Lösung wird bei 25 bis 30° C etwa 4 Tage gerührt. Nach Zusatz von 10 kg Kieselgur wird zentrifugiert. Der Rückstand kann auf L-Cystein und L-Asparaginsäure aufgearbeitet werden.

Das Zentrifugat wird bei Raumtemperatur mit 10 l 25%iger Natronlauge und etwa 23,5 kg Kaliumcyanid bis zur Entfärbung versetzt. Mit 7 l Eisessig stellt man auf pH 7,5, kühlt und rührt mindestens 10 Stunden bei 5 bis 10° C nach. Die ausgefallene D-Guanidinothiazolincarbonsäure wird abgeschleudert und getrocknet. Man erhält 7,95 kg Säure, die man in 70 l Wasser und 35 l konz. Salzsäure löst. Nach Stehen über Nacht bei Raumtemperatur wird die Lösung im Vakuum auf etwa 1/4 ihres Volumens eingeengt, abgekühlt und das S-Guanidinocarbonyl-D-cystein-dihydrochlorid abgesaugt. Man wäscht und trocknet an der Luft. Die erhaltene Menge von 10,2 kg des Salzes wird in 48 l Natronlauge (25%ig) von 35° C eingetragen, 2 Stunden bei 35 bis 40° C nachgerührt, unter Kühlung bei max. 40° C mit etwa 25 l konz. Salzsäure neutralisiert bis pH 7-8 und bei 10 bis 12° C 24 Stunden nachgerührt. Man trennt etwa 800 g eines Nebenproduktes (N-Guanidinocarbonyl-D-cystein) ab, versetzt das Filtrat mit 1 l konz. Ammoniak und oxidiert unter Kühlen mit etwa 3 l 16,5%iger Wasserstoffperoxidlösung, bis die SH-Reaktion negativ ist. Mit konz. Salzsäure und Eisessig wird auf pH 5 gestellt, 48 Stunden unter pH-Kontrolle und Kühlung nachgerührt und das D-Cystin abgesaugt. Nach Lufttrocknung erhält man etwa 4 kg D-Cystin einer Reinheit von mindestens 98%. Die spezifische Drehung $[\alpha]_D^{20}$ beträgt + 210° ($c = 3/1$ n-HCl).

$C_3H_7NO_3$ $HOCH_2CH(NH_2)COOH$ Ref. 20

1. *N*-Formyl-*O*-benzyl-DL-serin: 1040 g *O*-Benzyl-DL-serin werden in 4,24 *l* 98—100proz. Ameisensäure gelöst, worauf unter Eiskühlung und kräftigem Rühren 880 *ml* frisch destilliertes Acetanhydrid zugetropft werden. Man läßt unter Kühlung noch 30 Min. nachrühren und darauf 4 Stdn. bei Raumtemperatur stehen. Nach dieser Zeit hat die CO_2-Entwicklung nahezu aufgehört. Nach Zugabe von 880 *ml* Wasser und 4stdg. Aufbewahren in der Kälte fällt der größte Teil des Reaktionsprodukts aus; es wird abgesaugt und im Vak. über Kaliumhydroxyd getrocknet. Aus der Mutterlauge kann der restliche Teil nach Einengen auf etwa $^1/_3$ erhalten werden. Nach Umkristallisieren aus 90proz. Äthanol/Petroläther farblose Kristalle: Schmp. 156—159°. Ausb. 1098 g (92% d. Th.).

$C_{11}H_{13}NO_4$ (223,2)	Ber.	C 59,18	H 5,82	N 6,27
	Gef.	C 59,30	H 6,00	N 6,13

2. *N*-Formyl-*O*-benzyl-D-serin-Brucinsalz: 848,2 g (3,8 Mol) *racem.* *N*-Formyl-Verbindung und 1499 g wasserfreies Brucin werden in 10,5 *l* heißem absol. Äthanol gelöst. Beim Abkühlen auf Raumtemperatur tritt, meist ohne Animpfen, Kristallisation ein. Nach Stehenlassen über Nacht saugt man ab und wäscht den Rückstand mit wenig eiskaltem Äthanol (Frakt. I). Die Mutterlauge wird im Vak. auf die Hälfte eingeengt; nach 1stdg. Aufbewahren im Kühlschrank wird die erhaltene Fällung abgesaugt, wie oben gewaschen und sofort aus 2,5 *l* Äthanol umkristallisiert (Frakt. II).

Nach Umkristallisieren der beiden Fraktionen aus absol. Äthanol erhält man farblose Kristalle vom Schmp. 117—118°, $[\alpha]_D^{20}$: —26,8° ± 1 (*c* = 1, in Wasser), $[\alpha]_{546}^{20}$: —33,6° (*c* = 1, in Wasser). Ausb. 1002 g (85,4% d. Th.).

Alle alkoholischen Mutterlaugen der Brucinsalz-Fällung werden im Vak. zur Trockene eingedampft. Der Rückstand wird in 2500 *ml* Chloroform aufgenommen und mit 1200 *ml* 2*n* NaOH extrahiert. Die wäßrig-alkalische Phase läßt man unter Rühren in eiskalte 2*n* HCl im geringen Überschuß (1250 *ml*) rasch einfließen, wobei Kristallisation eintritt. Die Fällung wird durch kurzes Stehenlassen in der Kühltruhe vervollständigt und schließlich abgesaugt. Den Rückstand verreibt man mit Äther, saugt erneut ab und trocknet an der Luft. Zur Entfernung von Brucinresten wird zweckmäßig mit Chloroform unter Verwendung eines Schnellmischers (Ultra-Turrax) digeriert. Die so erhaltene und lufttrockene Formylverbindung wird im Hochvak. über P_2O_5 getrocknet. Ausb. 465 g.

3. *N*-Formyl-*O*-benzyl-L-serin-Chininsalz: Das aus den alkohol. Mutterlaugen erhaltene Produkt wird zusammen mit 676 g wasserfreiem Chinin in 6 *l* heißem absol. Äthanol gelöst. Nach Abkühlen auf Raumtemperatur und mehrstündigem Stehenlassen saugt man das abgeschiedene Chininsalz ab; eine zweite Fraktion erhält man durch Einengen des Filtrats auf etwa $^1/_3$. Das Rohprodukt wird aus 6 *l* Äthanol umkristallisiert: 711,3 g reines Chininsalz, $[\alpha]_D^{20}$: —93,0° ± 1 (*c* = 1, in Wasser); $[\alpha]_{546}^{20}$: —113,5° (*c* = 1, in Wasser). Aus der Mutterlauge werden nach Einengen auf $^1/_2$ bzw. $^1/_3$ weitere Mengen gewonnen, die nach Umkristallisieren aus Äthanol noch 60,2 g Chininsalz von genanntem spezif. Drehwert geben. Gesamtausb. 771,5 g (etwa 74% d. Th.).

4. a) *N*-Formyl-*O*-benzyl-D-serin: 1002 g Brucinsalz werden in 2,2 *l* Chloroform gelöst und mit 810 *ml* eiskalter 2*n* NaOH zerlegt. Nach Trennen der Phasen wäscht man die Chloroformschicht mit 100 *ml* 1*n* NaOH nach, extrahiert die vereinigten wäßrig-alkalischen Lösungen mit Chloroform und läßt sie dann unter Rühren in eiskaltes Wasser einfließen, das 900 *ml* 2*n* HCl enthält. Nach 2stdg. Aufbewahren im Kühlschrank wird der feinkörnige Niederschlag abgesaugt, mit eiskaltem Wasser gewaschen und gut abgepreßt. Der hohe Wassergehalt der Formylverbindung kann durch Verreiben mit Äther und erneutem kräftigem Absaugen stark vermindert werden. Nach Trocknen über P_2O_5 und Kaliumhydroxyd im Vakuumtrockenschrank: Schmp. 135—136°; $[\alpha]_D^{20}$: —48,7° ± 0,5; $[\alpha]_{546}^{20}$: —58,30° (*c* = 3,5, in 80proz. Äthanol). Ausb. 344 g (81% d. Th.).

$C_{11}H_{13}NO_4$ (223,2)	Ber.	C 59,18	H 5,82	N 6,27
	Gef.	C 59,15	H 6,00	N 6,05

b) *N*-Formyl-*O*-benzyl-L-serin: 771,5 g Chininsalz, gelöst in 2 *l* Chloroform werden mit 750 *ml* 2*n* NaOH zerlegt und wie unter a) beschrieben aufgearbeitet. Schmp. 135—136°; $[\alpha]^{20}$: +48,60° ± 0,5, $[\alpha]_{546}^{20}$: +58,1° (*c* = 3,5, in 80proz. Äthanol). Ausb. 304,5 g (72% d. Th.).

$C_{11}H_{13}NO_4$ (223,2)	Ber.	C 59,18	H 5,82	N 6,27
	Gef.	C 59,13	H 5,96	N 6,19

5. a) *O*-Benzyl-D-serin: 223,2 g Formylderivat werden in 1100 *ml* 1*n* HBr suspendiert und am Wasserbad erwärmt, bis eine klare Lösung entsteht, die beim Abkühlen keine Fällung mehr ergibt (etwa 1,5 Stdn.). Die auf 0° abgekühlte Lösung neutralisiert man nun mit Ammoniak, wobei sich der Benzyläther des Serins in farblosen Nadeln abscheidet; er wird abgesaugt und mit Eiswasser gewaschen. Die Mutterlauge liefert nach Einengen im Vak. noch eine weitere Fraktion. Umkristallisieren aus heißem Wasser ergibt große farblose Nadeln oder Blättchen; Schmp. 212—213°; $[\alpha]_D^{20}$: —22,65° ± 0,5, $[\alpha]_{546}^{20}$: —27,60° (*c* = 2, in 80proz. Essigsäure + 1 Äquiv. HCl). Ausb. 181,4 g (93% d. Th.).

$C_{10}H_{13}NO_3$ (195,2)	Ber.	C 61,50	H 6,70	N 7,17
	Gef.	C 61,33	H 6,81	N 7,14

b) *O*-Benzyl-L-serin: 223,2 g Formylderivat werden wie unter a) mit 1*n* HBr gespalten und das freie *O*-Benzyl-L-serin isoliert. Nadeln oder Blättchen aus Wasser: Schmp. 213—214°; $[\alpha]_D^{20}$: +22,75° ± 0,5, $[\alpha]_{546}^{20}$: +27,50° (*c* = 2 in 80proz. Essigsäure + 1 Äquiv. HCl). Ausb. 183,35 g (94% d. Th.).

$C_{10}H_{13}NO_3$ (195,2)	Ber.	C 61,50	H 6,70	N 7,17
	Gef.	C 61,42	H 6,82	N 7,08

6. a) D-Serin: 3,9 g *O*-Benzyl-D-serin werden in heißem Wasser gelöst, 1 Äquiv. 1*n* HBr zugefügt und in Gegenwart von Palladiumschwarz hydriert. Nach Abfiltrieren vom Katalysator gibt man die äquival. Menge 1*n* NH_4OH hinzu und dampft die Lösung im Vak. zur Trockene. Nach Aufnehmen in möglichst wenig heißem Wasser und Zugabe von absol. Äthanol kristallisiert die Aminosäure in farblosen Nadeln: $[\alpha]_D^{20}$: —14,99° ± 0,2, $[\alpha]_{546}^{20}$: —18,10 (*c* = 5, in 1*n* HCl). Ausb. fast quantitativ.

b) L-Serin: 3,9 g *O*-Benzoyl-L-serin werden wie unter a) beschrieben in die freie Aminosäure übergeführt. Farblose Nadeln aus Wasser/Äthanol: $[\alpha]_D^{20}$: +14,97° ± 0,2, $[\alpha]_{546}^{20}$: +18,05° (*c* = 5, in 1*n* HCl).

$C_3H_7NO_3$ Ref. 20a

1. p-Tosyl-DL-serin. Diese Verbindung ist bereits von *D. W. Woolley*[1]) dargestellt worden, doch fanden wir die hier wiedergegebene Vorschrift vorteilhafter. Zu einer gekühlten Lösung von 8,91 g DL-Serin in 50 cm³ 2-n. Natronlauge und 50 cm³ Dioxan tropfte man unter gutem Rühren eine Lösung von 19 g p-Tosyl-chlorid in 50 cm³ Dioxan, wobei man durch gleichzeitiges Zutropfen von 50 cm³ 2-n. Natronlauge die Lösung ständig alkalisch hielt. Man rührte über Nacht, wobei die Lösung allmählich Zimmertemperatur erreichte; beim Ansäuern mit starker Salzsäure (kongo) und Einengen der Reaktionslösung schied sich p-Tosyl-DL-serin in feinen Blättchen vom Smp. 215—216° ab. Ausbeute 20,5 g, d. s. 79% d. Th.

2. Brucinsalz von p-Tosyl-L-serin. 19,34 g p-Tosyl-DL-serin und 29,5 g Brucin löste man in 250 cm³ heissem Methanol. Beim Abkühlen der methanolischen Lösung kristallisierten 20 g Brucinsalz des p-Tosyl-L-serins aus, welches beim Umkristallisieren aus Methanol 18,8 g des reinen Salzes in Prismen vom Smp. 195—196° lieferte.

3. Chininsalz von p-Tosyl-D-serin. Die Mutterlaugen des unter 2. beschriebenen Brucinsalzes wurden im Vakuum zur Trockne verdampft, der Rückstand in Chloroform aufgenommen und mit etwas mehr als der berechneten Menge 2-n. Natronlauge geschüttelt. Durch Ansäuern der wässerigen Phase wurden 9,79 g p-Tosyl-serin abgeschieden, das zusammen mit 10,3 g wasserfreiem Chinin in 100 cm³ heissem Methanol gelöst wurde. Beim Abkühlen der methanolischen Lösung schieden sich 16 g p-tosyl-D-serin-saures Chinin in Nadeln vom Smp. 193—194° ab.

4. p-Tosyl-L-serin. 20 g des unter 2. beschriebenen Brucinsalzes wurden durch Aufnehmen in Chloroform und Schütteln mit 22 cm³ 2-n. Natronlauge zerlegt und das p-Tosyl-L-serin durch Ansäuern der wässerigen Lösung in Freiheit gesetzt. Man erhielt so 7,54 g des L-Serin-Derivates in Nadeln vom Smp. 230—232°.

Für die Analyse wurde eine Probe der Substanz aus Methanol, worin sie ziemlich schwer löslich ist, umkristallisiert.

$C_{10}H_{13}O_5NS$	Ber. C 46,32	H 5,05	N 5,40%
(259,27)	Gef. ,, 46,06	,, 5,18	,, 5,37%

$[\alpha]_D^{20}$ = + 13,3° (c = 0,4 in Alkohol)

5. p-Tosyl-D-serin. Aus 16 g des unter 3. beschriebenen p-tosyl-D-serin-sauren Chinins wurde die Säure wie unter 4. beschrieben freigesetzt. Man erhielt so 5,7 g p-Tosyl-D-serin in Nadeln vom Smp. 230—232°.

Für die Analyse wurde eine Probe der Substanz aus Methanol umkristallisiert und im Hochvakuum getrocknet.

$C_{10}H_{13}O_5NS$	Ber. C 46,32	H 5,05	N 5,40%
(259,27)	Gef. ,, 46,57	,, 4,98	,, 5,69%

$[\alpha]_D^{20}$ = − 13,2° (c = 0,4 in Alkohol)

Resolution procedures for the above compound using the same resolving agent may also be found in: E. Fischer and W.A. Jacobs, Chem. Ber., 39, 2942 (1906) and E. Fischer and W.A. Jacobs, Chem. Ber. 40, 1060 (1907).

$C_3H_7NO_3$ Ref. 20b

C. Racematspaltung des Serinamides

6,2 g D,L-Serinamid wurden in 30 cm³ absolutem Methanol gelöst, filtriert und mit einer ebenfalls filtrierten Lösung von 23,6 g Dibenzoyl-D-weinsäure in 40 cm³ absolutem Methanol versetzt. Die Kristallisation setzte nach kurzer Zeit ein, nach drei Tagen wurde das L-Serinamid-dibenzoyl-D-hydrogentartrat abgesaugt.

Ausbeute: 13,3 g, entspr. 92% d. Th.; Schmp. 178—180°.
$[\alpha]_D^{20}$: −93,1° (c = 1,37 in Methanol).

$C_{21}H_{22}O_{10}N_2$ (462,4)	ber.	C 54,66	H 4,81	N 6,07
	gef.	54,37	5,10	6,17.

Wie der Abbau des L-Amidtartrates zum L-Amid-hydrochlorid und zur freien L-Aminosäure zeigte, lag es in einer optischen Reinheit von 75% vor.

Nach dem Umkristallisieren aus absolutem Methanol resultierten 8,7 g (60% d. Th.) reines Salz. Schmp. 182,5—183,5°.

$[\alpha]_D^{20}$: −92,4° (c = 1,10 in Methanol).

gef. C 54,41 H 4,93 N 6,38.

Nach Hinzufügen von Äther zum Filtrat des Spaltansatzes kristallisierten noch 1—2 g Salz geringerer optischer Reinheit aus. Das Filtrat wurde im Vakuum bis fast zur Trockne eingedampft und mit viel absolutem Äther das rohe D-Serinamid-dibenzoyl-D-hydrogentartrat ausgefällt.

Ausbeute: 12,4 g entspr. 85% d. Th.; Schmp. 158—161°.
$[\alpha]_D^{20}$: −88,3° (c = 1,16 in Methanol).

Gef. C 54,24 H 4,88 N 6,13.

Gewinnung der optisch aktiven Serinamid-hydrochloride

L-Serinamid-hydrochlorid: In die Suspension des gereinigten L-Serinamid-dibenzoyl-D-hydrogentartrates in abs. Äthanol wurde bis zur Sättigung trockener Chlorwasserstoff eingeleitet, wobei das Salz in Lösung ging. Mit absolutem Äther wurde das L-Serinamid-hydrochlorid ausgefällt.

Ausbeute: 85%, bezogen auf eingesetztes Tartrat.
Schmp. 187—188°. $[\alpha]_D^{20}$: +13,2° (c = 1,10 in Wasser).

$C_3H_8O_2N_2 \cdot HCl$ (140,5)	ber.	C 25,63	H 6,45	N 19,94
	gef.	25,74	6,44	19,76.

Die Verseifung zum L-Serin-hydrochlorid zeigte, daß die Verbindung optisch rein vorlag.

D-Serinamid-hydrochlorid: Das rohe D-Serinamid-dibenzoyl-D-hydrogentartrat wurde in gleicher Weise wie beim L-Antipoden in das D-Serinamid-hydrochlorid überführt.

Ausbeute: 92% d. Th., bezogen auf eingesetztes Tartrat.
Schmp. 184—186°. $[\alpha]_D^{20}$: −8,36° (c = 1,26 in Wasser), entspr. 62% optischer Reinheit.

Gef. N 20,14.

Durch fraktionierte Kristallisation aus Alkohol—Äther wurde das optisch reine D-Serinamid-hydrochlorid erhalten.

Ausbeute: 75%, bezogen auf das rohe Amidhydrochlorid.
Schmp. 183—184°. $[\alpha]_D^{20}$: −13,42° (c = 0,56 in Wasser).

Gef. C 25,84 H 6,77 N 19,89.

Auch hier zeigte die Verseifung zum D-Serinhydrochlorid, daß die Substanz optisch rein vorlag.

Gewinnung der optisch aktiven Serine

L-Serin: L-Serinamid-hydrochlorid wurde mit der 40fachen Menge 5 n HCl eine Stunde im Ölbad auf 120° erhitzt, die Säure dann im Vakuum entfernt, der Rückstand in der eben nötigen Menge Wasser gelöst und die 10fache Menge Alkohol zugefügt. Das L-Serin ließ sich mit dem 1,3fachen der berechneten Menge Anilin ausfällen.

Ausbeute: 90% d. Th., bezogen auf L-Serinamid-hydrochlorid.
Schmp. 224—226°. $[\alpha]_D^{20}$: −6,7° (c = 1,79 in Wasser).
EMIL FISCHER[11]) gibt für L-Serin in wäßriger Lösung den Drehwert von $[\alpha]_D^{20}$: −6,83° ± 0,1 an.

L-Serin-hydrochlorid: Aus L-Serin durch Eindampfen mit Salzsäure. Schmp. 128—131°. $[\alpha]_D^{20}$: +14,64° (c = 0,47 in n HCl). F. SCHNEIDER[12]) fand für das L-Serin-hydrochlorid die Drehung von $[\alpha]_D^{20}$: +14,75° (in n HCl). Die Verbindung lag also optisch rein vor.

D-Serin: Die Verbindung wurde auf dieselbe Weise aus D-Serinamid-hydrochlorid erhalten, wie beim L-Antipoden beschrieben.

Ausbeute: 85% d. Th.; Schmp. 230—231°.
$[\alpha]_D^{20}$: +6,97° (c = 1,43 in Wasser).
E. FISCHER[13]) fand für D-Serin in wäßriger Lösung $[\alpha]_D^{20}$: +6,87°.

D-Serin-hydrochlorid: Schmp. 130—132°.
$[\alpha]_D^{20}$: −14,47 (c = 0,76 in n HCl).
E. FISCHER und W. A. JACOBS[13]) geben für D-Serin-hydrochlorid die spezifische Drehung von $[\alpha]_D^{20}$: −14,32° (in n HCl) an.

[11]) E. FISCHER, Ber. dtsch. chem. Ges. **39**, 2948 (1906).
[12]) F. SCHNEIDER, Liebigs Ann. Chem. **529**, 10 (1937).
[13]) E. FISCHER u. W. A. JACOBS, Ber. dtsch. chem. Ges. **39**, 2946 (1906).

$C_3H_7NO_3$ Ref. 20c

B. Racematspaltung des DL-Serin-benzylesters

6.75 g *DL-Serin-benzylester* wurden mit 10 ccm absol. Methanol verdünnt, filtriert und die Mischung mit der Lösung von 12.6 g *Dibenzoyl-D-weinsäure* in 35 ccm absol. Methanol bei Zimmertemp. versetzt. Nach kurzer Zeit setzte die Kristallisation des *D-Serin-benzylester-dibenzoyl-D-tartrates* ein und war nach 48 Stdn. beendet. Das auskristallisierte Salz (6—7 g) wurde abgesaugt und aus absol. Methanol umkristallisiert. Ausb. 4.8 g (75% d. Th.) optisch reines Salz vom Schmp. 165—168° (Zers., korr.). $[\alpha]_D^{20}$: −65.5° (*c* = 0.244, in Methanol).

$C_{38}H_{40}N_2O_{14}$ (748.7) Ber. C 60.96 H 5.34 N 3.74 Gef. C 60.60 H 5.13 N 3.88

Durch Zugabe von Äther zur Mutterlauge des Spaltansatzes wurden weitere 1—2 g Salz mit geringerer optischer Reinheit isoliert. Die überschüss. Dibenzoylweinsäure enthaltende Lösung wurde i. Vak. bis fast zur Trockne eingeengt und das *L-Serin-benzylester-dibenzoyl-D-tartrat* mit viel absol. Äther ausgefällt. Ausb. 4.0 g (62% d. Th.); Schmp. 143—147° (Zers., korr.) $[\alpha]_D^{20}$: −70.3° (*c* = 0.366, in Methanol).

$C_{38}H_{40}N_2O_{14}$ (748.7) Ber. C 60.96 H 5.34 N 3.74 Gef. C 60.17 H 5.27 N 3.84

C. Gewinnung der optisch aktiven Serin-benzylester-hydrochloride

D-Serin-benzylester-hydrochlorid: In die Suspension von *D-Serin-benzylester-dibenzoyl-D-tartrat* in wenig absol. Methanol wurde bis zur Sättigung trockener *Chlorwasserstoff* eingeleitet, wobei das Salz in Lösung ging. Durch Zugabe von absol. Äther wurde das *D-Serin-benzylester-hydrochlorid* ausgefällt und aus Alkohol/Äther umkristallisiert. Ausb. 80% d. Th., bez. auf das Tartrat. Schmp. 174—175° (Zers., korr.). $[\alpha]_D^{20}$: +4.12° (*c* = 4.60, in Methanol).

$C_{10}H_{13}NO_3 \cdot HCl$ (231.7) Ber. C 51.94 H 6.06 N 6.06 Gef. C 51.96 H 6.30 N 5.99

Wie die Verseifung zum Aminosäure-hydrochlorid zeigte, wurde damit das *D-Serin-benzylester-hydrochlorid* in einer optischen Reinheit von 95.5% erhalten.

L-Serin-benzylester-hydrochlorid: Bei der Überführung von *L-Serin-benzylester-dibenzoyl-D-tartrat* in *L-Serin-benzylester-hydrochlorid* erzielten wir auf dieselbe Weise den optischen Reinheitsgrad 85%. Durch fraktionierte Kristallisation aus Alkohol/Äther gewannen wir die optisch reine Verbindung in 80-proz. Ausb., bezogen auf Tartrat. $[\alpha]_D^{20}$: −4.19° (*c* = 4.53, in Methanol), entspr. 97.2% optischer Reinheit.

$C_{10}H_{13}NO_3 \cdot HCl$ (231.7) Ber. N 6.06 Gef. N 6.12

D. Gewinnung der optisch aktiven Serin-hydrochloride

D-(—)-Serin-hydrochlorid: D-Serin-benzylester-hydrochlorid wurde mit der 20fachen Menge konz. Salzsäure 1 Stde. unter Rückfluß erhitzt, die Lösung mehrmals ausgeäthert und das als Verdampfungsrückstand der Extrakte verbleibende Öl mit absol. Äther zur Kristallisation gebracht. Ausb. 91% d. Th. *D-(—)-Serin-hydrochlorid*; Schmp. 130—133°. $[\alpha]_D^{20}$: —13.6° (c = 2.17, in n HCl).

E. Fischer und W. A. Jacobs[5] geben für D-Serin-hydrochlorid $[\alpha]_D^{20}$: —14.32° (0.5 g in 5.59 g n HCl) an, die optische Reinheit des erhaltenen Produktes betrug somit 95%.

$C_3H_7NO_3 \cdot HCl$ (141.6) Ber. C 25.53 H 5.67 N 9.85 Gef. C 24.81 H 5.68 N 9.81

L-(+)-Serin-hydrochlorid wurde auf dieselbe Weise aus *L-(—)-Serin-benzylester-hydrochlorid* erhalten. Ausb. 86% d. Th., Schmp. 126—129°. $[\alpha]_D^{20}$: +14.02° (c = 2.36, in n HCl), entspr. einem optischen Reinheitsgrad von 95%. F. Schneider[6] gibt für die optisch reine Verbindung $[\alpha]_D^{20}$: +14.75° (in n HCl) an.

$C_3H_7NO_3 \cdot HCl$ (141.6) Ber. C 25.53 H 5.67 N 9.85 Gef. C 25.11 H 5.41 N 9.79

Aus den aktiven Serin-hydrochloriden lassen sich auf üblichem Wege mit Ammoniak die Antipoden der freien Aminosäure gewinnen.

---

[5] Ber. dtsch. chem. Ges. **39**, 2946 [1906]. [6] Liebigs Ann. Chem. **529**, 10 [1937].

$C_3H_7NO_3$ Ref. 20d

EXAMPLE 7

Resolving of N,N-dibenzyl-DL-serine

(A) PREPARATION OF THE SALT OF L(+)-THREO-1-p-NITRO PHENYL-2-AMINO PROPANE-1,3-DIOL AND N,N-DIBENZYL-L(—)-SERINE

14.25 g. of N,N-dibenzyl-DL-serine and 11 g. of L(+)-threo-1-p-nitro phenyl-2-amino propane-1,3-diol are dissolved, while heating, in 300 cc. of aqueous 25% methanol. The solution is kept for 1 hour at 50° C. and for two more hours at 35° C. so as to cause crystallization of the salt. The crystals are filtered, washed with a small quantity of water, and recrystallized from 160 cc. of aqueous 25% methanol. 9.3 g. of the salt of L(+)-threo-1-p-nitro phenyl-2-amino propane-1,3-diol and N,N-dibenzyl-L(—)-serine are obtained thereby. The resolving yield is 75%. The compound is obtained in the form of yellow leaflets which are only slightly soluble in water, ether, acetone, benzene, and chloroform, and are soluble in ethanol. The compound melts at 164–165° C.; its rotatory power is $[\alpha]_D^{20}$=—38°±1° (concentration: 1% in methanol).

(B) PREPARATION OF N,N-DIBENZYL-L(—)-SERINE

9.3 g. of the salt of L(+)-threo-1-p-nitro phenyl-2-amino propane-1,3-diol and N,N-dibenzyl-L(—)-serine, obtained according to preceding Example 7(A), are treated with 20 cc. of N sodium hydroxide solution for 15 minutes while stirring. The crystallized resolving agent is filtered off. 3.8 g. thereof are recovered corresponding to a yield of 47% of the initially employed resolving agent. The alkaline filtrate is heated to 40° C. and acidified by the addition of 1.5 cc. of acetic acid. N,N-dibenzyl-L(—)-serine crystallizes in long needles. The mixture is cooled with ice, the crystals are filtered off, washed with a small quantity of water, and dried in a drying oven. 5.05 g. of N,N-dibenzyl-L(—)-serine are obtained thereby. The yield amounts to 95%. The acid has a melting point of 142–143° C.; its rotatory power is $[\alpha]_D^{20}$=—79°±1° (concentration: 2% in methanol). The compound is soluble in warm water, ethanol, ether, acetone, and chloroform, and only slightly soluble in benzene.

(C) TREATMENT OF THE MOTHER LIQUORS

The mother liquors obtained on preparing and recrystallizing the salt of L(+)-threo-1-p-nitro phenyl-2-amino propane-1,3-diol and N,N-dibenzyl-L(—)-serine are combined and concentrated by evaporation to a volume of about 150 cc. 32 cc. of N sodium hydroxide solution are added thereto. 4.8 g. of the resolving agent, corresponding to a yield of 44% of the initially employed resolving agent are recovered thereby. The alkaline filtrate is acidified by the addition of 3 cc. of acetic acid and extracted with chloroform. The chloroform solution yields, on working up as described hereinbefore in the preceding examples, 8 g. of an optically impure compound which is treated with 6.4 g. of D(—)-threo-1-p-nitro phenyl-2-amino propane-1,3-diol in 250 cc. of aqueous 20% methanol. By working up the reaction mixture in the manner as described hereinbefore in the preceding examples and recrystallizing the resulting salt from 75 cc. of aqueous 20% methanol, 8 g. of the salt of D(—)-threo-1-p-nitro phenyl-2-amino propane-1,3-diol and N,-N-dibenzyl-D(+)-serine are obtained thereby. The resolving yield amounts to 64%. Said compound is obtained in the form of yellow leaflets which are only slightly soluble in water, ether, acetone, benzene, and chloroform, and are soluble in ethanol. The melting point of said salt is 164–165° C.; its rotatory power is $[\alpha]_D^{20}$=+38°±1° (concentration: 1% in methanol).

(D) HYDROGENOLYSIS OF THE N,N-DIBENZYL-L(—)-SERINE

2 g. of N,N-dibenzyl-L(—)-serine in 30 cc. of 80% ethanol are hydrogenated at a temperature of 60–65° C. in the presence of 1 g. of a palladium black catalyst containing 10% of palladium and with the addition of 1 cc. of concentrated hydrochloric acid for 1 hour. The catalyst is filtered off, the filtrate is concentrated to a volume of 10 cc., 1.5 cc. of pyridine are added thereto, and the precipitate is filtered off. Optically pure L(+)-serine is obtained in a yield of 75%. Its rotatory power is $[\alpha]_D^{20}$=+15°±1° (concentration: 4% in N hydrochloric acid).

$C_3H_7NO_3$ Ref. 20e

EXAMPLE 2

Resolution of N-3,5-dinitro benzoyl-DL-serine and preparation of D- and L-serine

(a) FORMATION OF THE SALT OF N-3,5-DINITRO BENZOYL-L-SERINE WITH L(+)-THREO-1-(P-NITRO PHENYL)-2-AMINO PROPANE-1,3-DIOL

Process I.—10 g. of N-3,5-dinitro benzoyl-DL-serine, obtained according to Example 1 are dissolved at 60° C. in 30 cc. of methanol. 4.2 g. of L(+)-threo-1-p-nitro phenyl-2-amino propane-1,3-diol are added to said solution. The reaction mixture is cooled to 10° C. Thereby 6.8 g. of the optically pure salt of N-3,5-dinitro benzoyl-L-serine with L(+)-threo-1-(p-nitro phenyl)-2-amino propane-1,3-diol are obtained. The yield amounts to 80% of the theoretical yield. The melting point of the new salt is 160–161° C., its rotatory power is $[\alpha]_D=+33° \pm 1°$ (concentration: 1% in water).

Process II.—10 g. of N-3,5-dinitro benzoyl-DL-serine are dissolved at 60° C. in 5 cc. of methanol. 4.2 g. of L(+)-threo-1-p-nitro phenyl-2-amino propane-1,3-diol are added thereto. 55 cc. of acetic acid ethyl ester are then introduced into the resulting reaction mixture. The mixture is cooled. The precipitated salt is filtered off, washed, and dried. In this manner 8.2 g. of the optically pure salt of N-3,5-dinitro benzoyl-L-serine and L(+)-threo-1-(p-nitro phenyl)-2-amino propane-1,3-diol are obtained thereby. The yield amounts to 96% of the theoretical yield.

(b) PREPARATION OF N-3,5-DINITRO BENZOYL-L-SERINE

8.2 g. of the above mentioned salt are dissolved in 16.5 cc. of N sodium hydroxide solution. Thereby L(+)-threo-1-(p-nitro phenyl)-2-amino propane-1,3-diol is split off in crystalline form. The crystals are filtered off and washed. In this manner 3.1 g. of the resolving agent, corresponding to 74% of the theoretical yield, are recovered. The filtrate is acidified by means of 1.7 cc. of concentrated hydrochloric acid. Thereby N-3,5-dinitro benzoyl-L-serine crystallizes. It is filtered off, washed, and dried. 4.5 g. of N-3,5-dinitro benzoyl-L-serine are obtained in crystalline form. The yield amounts to 90% based upon N-3,5-dinitro benzoyl-DL-serine used as starting material. The monohydrate of said compound has a melting point of 110–112° C. Its rotatory power is $[\alpha]_D^{20}=+23.5° \pm 1°$ (concentration: 1% in 50% ethanol).

(c) ISOLATION OF L-SERINE

A solution of 4.3 g. of N-3,5-dinitro benzoyl-L-serine in 60 cc. of 5 N hydrochloric acid is heated under reflux for one hour. The resulting solution, after filtration, is evaporated to dryness in a vacuum. The remaining hydrochloride of L-serine is recrystallized from acetone and is dissolved in 1 cc. of water. 1.4 cc. of aniline and thereafter 10 cc. of absolute ethanol are added to said solution. On cooling, 1.3 g. of optically pure crystalline L-serine are obtained. The yield amounts to 90% of the theoretical yield. The melting point of said optically active L-serine is about 228° C. Its rotatory power $[\alpha]_D=+15° \pm 1°$ (concentration: 4% in N hydrochloric acid).

(d) ISOLATION OF D-SERINE

The acetic acid ethyl ester-methanol mother liquor obtained after removing the crystallized salt of N-3,5-dinitro benzoyl-L-serine and of L(+)-thero-1-(p-nitro phenyl)-2-amino propane-1,3-diol obtained according to Example 2*a*, Process II, as described hereinabove, is extracted first with 10 cc. of 0.5 N hydrochloric acid and then with 5 cc. of water. The aqueous solutions are combined and are then rendered alkaline by the addition of sodium hydroxide solution. In this manner 0.4 g. of the resolving agent, corresponding to 10% of the theoretical yield, are recovered by filtration. The organic phase is concentrated and petroleum ether is added thereto. Thereby 4.85 g. of N-3,5-dinitro benzoyl-D-serine corresponding to 97% calculated for N-3,5-dinitro benzoyl-DL-serine used as starting material, are isolated. Said compound can readily be purified by recrystallization from aqueous ethanol 1:1. The crude compound melts at 109–111° C., its rotatory power is $[\alpha]_D^{20}= -22° \pm 1°$ (concentration: 1% in 50% ethanol).

D-serine is obtained from said N-acyl compound in the same manner as described hereinabove under (*c*) for the preparation of L-serine by hydrolyzing N-3,5-dinitro benzoyl-D-serine by means of hydrochloric acid.

EXAMPLE 3

Preparation of N-3,5-dinitro benzoyl-D-serine by means of D(—)-threo-1-p-nitro phenyl-2-amino propane-1,3-diol

The procedure is the same as described hereinabove in Example 2*a*, whereby D(—)-threo-1-p-nitro phenyl-2-amino propane-1,3-diol is reacted with N-3,5-dinitro benzoyl-DL-serine. Thereby the salt of N-3,5-dinitro benzoyl-D-serine with D(—)-threo-1-p-nitro phenyl-2-amino propane-1,3-diol is obtained. The melting point of said new compound is 160–161° C.; its rotatory power is $[\alpha]_D^{20}=-33° \pm 1°$ (concentration: 1% in water). Said salt is converted, by following the same procedure as described hereinabove in Example 2*b*, into optically pure N-3,5-dinitro benzoyl-D-serine. The monohydrate of said compound has a melting point of 110–112° C. Its rotary power is $[\alpha]_D=-23.5° \pm 1°$ (concentration: 1% in 50% ethanol). D-serine is obtained from said N-acylated compound by following the procedure described hereinabove in Example 2*c*. Said D-serine has a rotary power $[\alpha]_D=-15° \pm 1°$ (concentration: 4% in N hydrochloric acid).

$C_3H_7NO_3$ Ref. 20f

EXAMPLE 11

Resolution of N-Benzoyl-DL-Serine

1400 g. of N-benzoyl-DL-serine was dissolved in three liters of methanol by heating to boiling and this solution was added to a boiling solution of 1430 g. of L-threo - 2 - amino-1 - (p - methylmercaptophenyl) - 1,3-propanediol in six liters of methanol. The solution was seeded with a few crystals of the L-threo-2-amino-1-(p-methylmercaptophenyl)-1,3-propanediol salt of N-benzoyl-L-serine and then cooled and stirred for two hours at 25° C. The solid which separated from the solution was collected on a filter and washed with ethanol. There was thus obtained 1415 g. of the L-threo-2-amino-1-(p-methylmercaptophenyl)-1,3-propanediol salt of N-benzoyl-L-serine which melted at 170–175° C. This product was slurried in three liters of boiling 95% ethanol and the resulting solution was cooled to 25° C. There separated from solution 1315 g. of the L-threo-2-amino-1-(p-methylmercaptophenyl)-1,3-propanediol salt of N-benzoyl-L-serine which melted at 176–178° C. Concentration of the filtrate to one-quarter of its original volume by evaporation and cooling the concentrated solution yielded a second crop of crude L-threo-2-amino-1-(p-methylmercaptophenyl) - 1,3 - propanediol salt of N-benzoyl-L-serine weighing 20 g. which melted at 170–175° C. The mother liquor from the second crop of the L-threo-amine-L-acid salt was concentrated under reduced pressure by evaporation to a volume of 2.5 liters. This solution was allowed to stand for twenty-four hours at 5° C. and the solid which separated from solution was collected on a filter, washed with cold ethanol, and dried at 50° C. There was thus obtained 1250 g. of crude L - threo - 2 - amino-1-(p-methylmercaptophenyl)-1,3-propanediol salt of N-benzoyl-D-serine as a hydrate which melted at 88–120° C. By concentrating the filtrate and cooling, there was obtained an additional crop of 275 g. of the same hydrate.

Combination of the crops of L-threo-2-amino-1-(p-methylmercaptophenyl)-1,3-propanediol salt of N-benzoyl-D-serine and recrystallization from five liters of hot water by keeping the solution at 25° C. for several hours yielded as a precipitate 1100 g. of the L-threo-2-amino-1-(p-methylmercaptophenyl)-1,3-propanediol salt of N-benzoyl-D-serine as a hydrate which melted at 90–130° C. (A portion of this hydrate was converted to the anhydrous form, which melted at 143–145° C., by slurrying in a 1:1 mixture of ethanol and ether.) Cooling of the filtrate from the 1100 g. crop of the hydrated salt at 5° C. overnight yielded a second crop of 210 g. of the L-threo-amine-D-acid salt hydrate.

The mother liquor from the crops of hydrated salt was made alkaline by treatment with sodium hydroxide solution, thereby precipitating 54 g. of L-threo-2-amino-1-(p-methylmercaptophenyl) - 1,3 - propanediol, and the filtrate from this product was acidified and saturated with sodium chloride and cooled to yield 30 g. of solid consisting chiefly of N-benzoyl-DL-serine.

N-Benzoyl-L-Serine

1317 g. of the L-threo-2-amino-1-(p-methylmercaptophenyl)-1,3-propanediol salt of N-benzoyl-L-serine was suspended in three liters of hot water and this mixture was mixed with a stoichiometrically equivalent quantity of concentrated hydrochloric acid (265 ml.). The clear solution thus obtained was cooled to 20° C. and was then made alkaline by rapidly adding, with strong stirring, 525 ml. of 35% aqueous sodium hydroxide solution. A heavy precipitate of L-threo-2-amino-1-(p-methylmercaptophenyl)-1,3-propanediol separated from the solution rapidly. After the mixture had been cooled to 10–15° C., the precipitated solid was collected on a filter, washed with 300 ml. of cold water and then with a few ml. of benzene, and dried. There was thus obtained about 650 g. of L-threo-2-amino-1-(p-methylmercaptophenyl)-1,3-propanediol. The aqueous filtrates were combined and acidified by mixing with 270 ml. of concentrated hydrochloric acid, and the solution was saturated with sodium chloride. The solution was then cooled to 5° C. The solid which separated from solution was collected on a filter, washed with a few ml. of ice water, and dried at 50° C. There was thus obtained 575 g. of N-benzoyl-L-serine which melted at 145–148° C. Recrystallization of this product from three parts by weight of isopropyl alcohol yielded 520 g. of N-benzoyl-L-serine which melted at 148–149° C. The filtrate was made alkaline by addition of sodium hydroxide solution, a small crop of L-threo-2-amino-1-(p-methylmercaptophenyl)-1, 3-propanediol which separated from solution was removed by filtration, and the filtrate was acidified, thereby causing the precipitation of another 50 g. of N-benzoyl-L-serine.

N-Benzoyl-D-Serine

880 g. of L-threo-2-amino-1-(p-methylmercaptophenyl)-1,3-propanediol salt of N-benzoyl-D-serine (hydrate) was trated in a manner similar to that described above for the treatment of the N-benzoyl-L-serine salt, and there was thus obtained 355 g. of N-benzoyl-D-serine which melted at 148–150° C.

$C_3H_7NO_3$ Ref. 20g

Separazione degli antipodi ottici a mezzo della cristallizzazione frazionata del suo sale con L(+)-treo-1-p-nitrofenil-2-amino-1,3-propandiolo

In una beuta da litri 2 si pongono cc 725 di H_2O, g 100 di tosil-DL-serina e g 82 di L(+)-treo-1-p-nitrofenil-2-amino-1,3-propandiolo e si portano in soluzione scaldando a 85-90°. Si lascia in riposo a temperatura ambiente per 24 ore e si filtrano i cristalli formatisi lavandoli più volte con acqua fredda.

Si ottengono g 88 di secco (98%) che cristallizzato da acqua dà un prodotto a p.f. 188-189°, $[\alpha]_D^{20}$ = + 41 (1% in dimetilformamide).

All'analisi ha fornito:

	trov. %	:	C 48,2;	H 5,29;	N 9,02;	S 6,95
per $C_{19}H_{22}N_3O_9S$	calc.	:	48,6;	5,34;	8,91;	6,80

Le acque madri che contengono il diastereoisomero si trattano con HCl conc. al rosso Congo. Precipita la tosil-(+)-serina che viene cristallizzata da metanolo.

p.f. = 230-232° $[\alpha]_D^{20}$ = + 54 (5% NaOH N/1)

Al sale della tosil-(—)-serina con il propandiolo, precedentemente isolato, e sciolto in cc 400 di acqua calda, si aggiungono, sotto agitazione, cc 26,4 di HCl conc. Precipita la tosil-(—)-serina pura.

p.f. = 230-232° $[\alpha]_D^{20}$ = — 54 (5% NaOH N/1)

Dalle acque madri di precipitazione sia della tosil-(—)-serina che della tosil-(+)-serina si ricupera il propandiolo alcalinizzando alla fenolftaleina.

Separazione degli antipodi ottici per cristallizzazione differenziale

In un pallone da 2 litri, munito di agitatore, si pongono cc 1000 di acqua, g 4 di NaOH, g 33 di tosil-DL-serina e g 3 di tosil-(—)serina e si scalda a 70°, sotto agitazione, fino a soluzione. Sempre sotto agitazione si inizia il raffreddamento. A 60° circa comincia la cristallizzazione. Si lascia scendere la temperatura per circa 10° e poi si filtra rapidamente lavando con pochissima acqua fredda.

Si ottengono g 6 di tosil-(—)serina che fonde a 227-229°.

$[\alpha]_D^{20}$ = — 50 (5% NaOH N/1)

Il prodotto cristallizzato da metanolo dà $[\alpha]_D^{20}$ = — 54 (5% NaOH N/1) e p.f. 232-233°.

Le acque madri sono addizionate di g 6 di tosil-DL-serina. Si riscalda a 70-75° sino a soluzione completa e si lascia raffreddare con le stesse modalità sopra descritte.

Si ottengono g 6 di tosil-(+)-serina.

p.f. = 228-230° $[\alpha]_D^{20}$ = + 51 (5% NaOH N/1)

In qualche caso si può verificare che la cristallizzazione dell'antipodo ottico inizi a temperatura leggermente superiore o inferiore ai 60°. In tal caso si conserva sempre l'intervallo di 10° prima di raccogliere il cristallizzato.

L'operazione si può ripetere alternativamente per un gran numero di volte, sino a quando le impurezze accumulatesi nelle acque madri le rendono non più idonee.

$C_3H_7NO_3$ Ref. 20h

Varieties of *N*-protecting groups for amino acids have recently been developed by splendid studies of peptide syntheses and have been used well. The *N*-benzyloxycarbonyl group, explored by Bergmann and Zervas in 1932[1], is, however, still one of the most favorable *N*-protecting groups for the sequential synthesis of peptides. The direct resolution of *N*-benzyloxycarbonyl derivatives of racemic amino acids would, therefore, be a much more convenient way of preparing optically active *N*-benzyloxycarbonylamino acids than the route of incorporating the protecting group onto optically active amino acids obtained *via* the appropriate resolution of temporary derivatives of racemic amino acids, *e.g.*, *N*-acetyl ones.

In this paper, the resolution of racemic *N*-benzyloxycarbonylamino acids by using (−)- and (+)-ephedrine alternately and a survey of the resolution will be described. In these experiments, a half equi-amount of ephedrine in relation to racemic *N*-benzyloxycarbonylamino acids was used principally. Although ethyl acetate could be employed as a predominant solvent in many cases, ethyl acetate-petroleum ether (2 : 3v/v) for *N*-benzyloxycarbonyl-DL-leucine and ether for *N*-benzyloxycarbonyl-DL-isoleucine and *N*-benzyloxycarbonyl-DL-phenylalanine were also effectively used in concentrations of *N*-benzyloxycarbonylamino acid and ephedrine of 0.45—4.68 and 0.25—2.58 M respectively. The selection of a suitable solvent for the aforementioned resolution is important, especially in the case of phenylalanine; usually only a racemate or a partially-resolved salt was deposited from a solution in a solvent such as ethyl acetate, benzene, or chloroform, though when ether was used the resolution was favored. The results obtained with salts with natural (−)-ephedrine are listed in Table 1. Of all the amino acids used, *N*-benzyloxycarbonyl derivatives of amino acids without a *β*-methyl side group—*β*-methyl DL-aspartate, DL-leucine, DL-methionine, and *O*-benzyloxycarbonyl-DL-serine—deposited to afford predominantly, if not exclusively, salts of the D-isomer of amino acids with natural (−)-ephedrine. Unlike the above cases, however, the same derivatives of amino acids with a *β*-methyl side group—DL-isoleucine, *erythro*-*β*-methyl-DL-leucine, *threo*-*N*, *β*-dimethyl-DL-leucine, and DL-valine—afforded salts bearing a reverse configuration, L, at the α-carbon atom of the amino acids. Entirely opposite results yielding each antipode were attained when (+)-ephedrine was used instead of (−)-base. Thus, it is possible to choose either antibode of ephedrine before the resolution, because it is possible, as a rule, to predict the configutarion of the amino acid to be obtained as an insoluble salt, depending on the presence of a *β*-methyl side group. This would be especially useful for the resolution of a new amino acid, as we have shown successfully in a preceding paper.[2] In consonance with this view, the preceding observation, made by

TABLE 1

Salt of *N*-Benzyloxycarbonylamino Acid with (−)-Ephedrine	Molecular Formula	Analysis Calcd C	H	N	Found C	H	N	Yield (%)	Mp (°C)	$[\alpha]_D$** (deg)
β-Methyl D-aspartate	$C_{23}H_{30}O_7N_2$	61.87	6.77	6.27	61.84	6.74	6.60	71	141	−30.3
D-Leucine	$C_{24}H_{34}O_5N_2$	66.95	7.96	6.51	66.44	8.06	6.44	78	92—94	−12.3
D-Methionine	$C_{23}H_{32}O_5N_2S$	61.59	7.19	6.25	61.40	7.29	5.96	89	129—130	−28.1
O-Benzyloxycarbonyl-D-serine	$C_{29}H_{34}O_8N_2$	64.67	6.36	5.20	64.87	6.61	5.17	95	133—134	−33.8
L-Isoleucine	$C_{24}H_{34}O_5N_2$	66.96	7.96	6.51	66.68	7.95	6.69	89	138	−17.2
erythro-*β*-Methyl-L-leucine*	$C_{25}H_{36}O_5N_2$	67.54	8.16	6.30	67.50	8.09	6.28	45	124—125	−17.6
threo-*N*, *β*-Dimethyl-L-leucine*	$C_{26}H_{38}O_5N_2$	68.09	8.35	6.11	68.26	8.24	6.27	94	160—161	−75.0
L-Valine	$C_{23}H_{32}O_5N_2$	66.32	7.74	6.73	66.27	7.83	6.43	78	145	−20.3
L-Phenylalanine	$C_{27}H_{32}O_5N_2$	69.80	6.94	6.03	69.97	7.04	6.28	85	139—140	+11.3

* Represented from our preceding paper.[2]

** All optical rotation values were measured in a solution of absolute alcohol, at 22—27°C.

1) M. Bergmann and L. Zervas, *Ber.*, **65**, 1192 (1932).

2) H. Kotake, T. Saito and K. Ōkubo, This Bulletin, **42**, 1367 (1969).

Overby and Ingersoll,[3] that a salt of *N*-benzyloxycarbonyl-D-alanine with (−)-ephedrine was produced as a less soluble product when *N*-benzyloxycarbonyl-DL-alanine was allowed to react with natural ephedrine in ethyl acetate, also supports our above conclusion. An exception to the rule was, however, observed in the case of phenylalanine, which bears a β-aromatic group but no β-methyl group; thus, a salt of the L-isomer with natural (−)-ephedrine was deposited preferentially. In order to determine the effect of a β-aromatic side group on the resolution, further attempts have been made to resolve *N*-benzyloxycarbonyl-DL-tyrosine and -DL-tryptophan into their respective isomers, but these attempts have so far been unsuccessful. Even if the base was used in a half- or equi-molar amount ratio in various solvents, neither *N*-benzyloxycarbonyl-DL-aspartic acid nor *N*-benzyloxycarbonyl-DL-glutamic acid have provided favorable results. A successful resolution of DL-aspartic acid was, however, achieved by protecting β-carboxyl as its methyl ester. An additional attempt to resolve the γ-methyl DL-glutamate was unsuccessful, though. Similarly, the resolution of DL-serine was performed successfully only when its *N,O*-dibenzyloxycarbonyl derivative was employed. Thus, one recommendation, that the third functional groups of amino acids should be protected for the favorable resolution, can be made on the basis of our experiments involving aspartic acid and serine. Each antipode of *N*-benzyloxycarbonylamino acid with an entirely optical purity resulted, when (+)-ephedrine was used alternately, from a solution of the same solvent used in the corresponding case (Table 2), though it could also be produced in an optically active state from the mother liquor obtained after the removal of the salts with natural ephedrine. All the salts obtained herein were treated with 2N hydrochloric acid to remove the basic moieties and to produce optically active *N*-benzyloxycarbonylamino acids. Each amino acid was then regenerated in an optically pure state by the removal of the benzyloxycarbonyl group using hydrogen bromide in glacial acetic acid,[15] followed by neutralization with triethylamine, with the exception of the deprotection of *N,O*-dibenzyloxycarbonyl serine. In the latter case, the protecting groups were removed by catalytic hydrogenation, using palladium-black as a catalyst in ethyl alcohol, in order to avoid any *O*-acylation, which is likely caused by the action of hydrogen bromide in glacial acetic acid.[16] The specific rotations of resolved *N*-benzyloxycarbonylamino acids and free amino acids are listed in Table 3, and their values are compared with those reported in the literature. The specific rotation value of β-methyl *N*-benzyloxycarbonyl-D-aspartate is comparable in magnitude to that of its L-antipode. The value of *N*-benzyloxycarbonyl-L-phenylalanine, along with that of the D-isomer, is relatively higher in magnitude than those reported. A striking difference is found to exist in the optical-rotation values of benzyloxycarbonyl valine reported in the literature; there are only a few values compatible with that of benzyloxycarbonyl valine obtained in this experiment. However, the two isomers of benzyloxycarbonyl valine afforded optically pure L- and D-valine respectively by debrocking.

TABLE 2

Salt of *N*-Benzyloxycarbonylamino Acid with (+)-Ephedrine	Analysis Found C	H	N	Yield (%)	Mp (°C)	$[\alpha]_D$** (deg)
β-Methyl L-aspartate	62.31	6.81	6.29	53	141—142	+31.0
L-Leucine	66.43	8.01	6.47	68	91—93	+12.3
L-Methionine	61.27	7.27	6.09	82	130—131	+28.2
O-Benzyloxycarbonyl-L-serine	64.50	6.39	5.36	91	132	+33.0
D-Isoleucine	67.06	8.00	6.36	83	139	+15.7
erythro-β-Methyl-D-leucine*	67.60	8.14	6.32	50	124—125	+17.7
threo-N, β-Dimethyl-D-leucine*	67.97	8.31	6.16	89	159—160	+73.5
D-Valine	65.98	7.73	6.82	96	145	+19.1
D-Phenylalanine	69.44	7.05	6.28	85	139—140	−11.1

* Represented from our preceding paper.[2]
** All optical rotation values were measured in a solution of absolute alcohol, at 22—27°C.

Experimental

All the melting points are uncorrected. The optical rotation values were measured with a Jasco DIP-SL-type polarimeter.

Starting Materials. *A) (−)- and (+)-Ephedrine Hemihydrate Used Throughout This Work.* Commercially-available (−)-ephedrine hydrochloride, with a specific rotation value of −33.8° in water, was treated with a slight excess of N sodium hydroxide, and the organic base thus liberated was immediately taken up into ethyl acetate. The organic layer was dried over anhydrous sodium sulfate, filtered, and evaporated to dryness *in vacuo* to afford a crystalline material. This was then recrystallized from petroleum ether and dried spontaneously under an atmosphere to give (−)-ephedrine hemihydrate as needles; the melting point was 39—40°C, $[\alpha]_D = -5.8°$ in water. Found: C, 68.80; H, 9.14; N, 8.09%. Calcd for $C_{10}H_{15}ON \cdot 1/2H_2O$: C, 68.93; H, 9.26; N, 8.04%. Similarly, from (+)-ephedrine hydrochloride ($[\alpha]_D = +33.1°$ in water), (+)-ephedrine

3) L. R. Overby and A. W. Ingersoll, *J. Amer. Chem. Soc.*, **82**, 2067 (1960).

15) D. Ben-Ishai and A. Berger, *J. Org. Chem.*, **17**, 1564 (1952); D. Ben-Ishai, *ibid.*, **19**, 62 (1954).

16) K. Okawa, This Bulletin, **30**, 977 (1957); J. Noguchi, T. Saito, T. Hayakawa and Y. Hayashi, *Nippon Kagaku Zasshi*, **80**, 299 (1959).

hemihydrate was obtained; mp 40°C, $[\alpha]_D = +5.9°$ in water. (Found: C, 69.29; H, 9.27; N, 8.21%).

B) N-Benzyloxycarbonyl Derivatives of Racemic Amino Acids. The compounds were prepared, in excellent yields, by the generally-applicable Schotten-Baumann method. Recrystallization from appropriate solvents afforded pure products as follows [name of *N*-benzyloxy-carbonyl-DL-amino acid, mp°C, (lit, mp°C).]: *β*-methyl DL-aspartate, 110—112°C, (112°C).[17] DL-Isoleucine, 64—66°C, (48°C[18]). *erythro-β*-Methyl-DL-leucine, 62—64°C.[2] *threo-N,β*-Dimethyl-DL-leucine, 79°C.[2] DL-Leucine, 53—54°C, (52—55°C[4]). DL-Methionine, 112°C, (112°C[4]) DL-Phenylalanine, 102°C, (103°C[4]). *O*-Benzyloxycarbonyl-DL-serine, 93°C, (94°C[4]). DL-Valine, 74—75°C, (76—78°C[4]).

General Procedure for the Resolution of *N*-Benzyloxycarbonyl-DL-amino Acids. To a solution of 0.100 mol of *N*-benzyloxycarbonyl-DL-amino acid in ethyl acetate at a suitable concentration, depending on the amino acid employed, as will be shown below individually, 0.055 mol of (−)-ephedrine hemihydrate was added; then the solution was allowed to stand at room temperature (or stored in a refrigerator in some cases) until there was an adequate amount of a precipitate

TABLE 3

	N-Benzyloxycarbonyl derivative			Free amino acid regenerated
	Yield (%)	Mp (°C)	$[\alpha]_D$** (deg)	$[\alpha]_D$*** (deg)
β-Methyl D-aspartate	88	98—99	+17.8[a]	
D-Leucine	100	oil		−15.9 (lit, −15.1[11,e])
D-Methionine	95	70—71 (lit, 69—70[4])	+17.2 (lit, +18.2[7])	−26.0 (lit, −23.9[12,e])
O-Benzyloxycarbonyl-D-serine	96	83—84	−4.3, −23.5[b]	−14.0[f] (lit, −14.3[13])
L-Isoleucine	100	oil		+40.8 (lit, +40.7[11,14])
erythtro-β-Methyl-L-leucine*	100	oil		+38.9[e]
threo-N,β-Dimethyl-L-leucine*	92	98—99	−75.9	+38.3[e]
L-Valine	96	66—67 (lit, 66—67[4] 57—61[8])	+ 0.7 (lit, +0.1[4] −4.3[8,c])	+29.5 (lit, +28.8[14])
L-Phenylanine	94	86—89 (lit, 88—89[4])	+ 8.4 (lit, +5.1,[4] +5.3[9,c])	−31.9[g] (lit, −34.8[14,g])
β-Methyl L-aspartate	90	96 (lit, 96—98[5])	−16.5[a] (lit, −17.4[5,a])	
L-Leucine	100	oil		+14.6 (lit, +14.9[14])
L-Methionine	93	69—70 (lit, 68—69[4])	−19.4 (lit, −16.6[10])	+23.4 (lit, +23.5[14])
O-Benzyloxycarbonyl-L-serine	95	83—84	+4.4, +23.6[b]	+13.9[f] (lit, +14.5[14,f])
D-Isoleucine	100	oil		−40.3 (lit, −40.7[11])
erythro-β-Methyl-D-leucine*	100	oil		−39.4[e]
threo-N,β-Dimethyl-D-leucine*	93	98—99	+75.3	−39.3[e]
D-Valine	93	66—67 (lit, 60—62[6])	−0.53 (lit, +5.0,[6,d] +4.2[8,c])	−28.3 (lit, −26.4[12,e])
D-Phenylalanine	91	86—87 (lit, 88—89[4])	−8.3 (lit, −4.6[4,c])	+32.4[g] (lit, +33.9[12,e])

* Represented from our preceding paper.[2]
** Measured in solution of absolute alcohol unless stated otherwise, at 22—26°C.
*** Measured in a solution of 6N hydrochloric acid unless stated otherwise, at 22—25°C.
a) Measured in pyridine. b) Measured in ether. c) Measured in glacial acetic acid. d) Measured in methanol. e) Measured in 5N hydrochloric acid. f) Measured in 1N hydrochloric acid. g) Measured in water.

4) J. P. Greenstein and M. Winitz, "Chemistry of the Amino Acids," Vol. 2 John Wiley & Sons Inc., New York (1961), pp. 892—894.

5) H. Schwarz, F. M. Bumpus and I. H. Page, *J. Amer. Chem. Soc.*, **79**, 5701 (1957).

6) R. Sargers and B. Witkop, *ibid.*, **87**, 2020 (1965).

7) N. F. Albertson and F. C. McKay, *ibid.*, **75**, 5323 (1953).

8) E. Schröder, *Ann. Chem.*, **692**, 241 (1966).

9) D. W. Clayton, J. A. Farrington, G. W. Kenner and J. M. Furner, *J. Chem. Soc.*, **1957**, 1398.

10) K. Hofmann, A. Jöhl, A. E. Furlenmeier and H. Kappeler, *J. Amer. Chem. Soc.*, **79**, 1636 (1957).

11) W. A. H. Huffman and A. W. Ingersoll, *ibid.*, **73**, 3366 (1951).

12) J. R. Parikh, J. P. Greenstein, M. Winitz and S. M. Birnbaum, *ibid.*, **80**, 953 (1958).

13) E. Fischer and W. A. Jacobs, *Ber.*, **39**, 2942 (1906).

14) S. Akabori and S. Mizushima (Eds.), "Protain Chemistry" (Tanpakushitu Kagaku) Vol. 1 Kyoritsu Shuppan K.K., Tokyo (1952), pp. 112.

17) C. H. Bamford, A. Elliot and W. E. Hanby, "Synthetic Polypeptides," Acad. Press, New York (1956), p. 46.

18) Yu. I. Khurgin and M. G. Dmitrieva, *Tetrahedron*, **21**, 2305 (1965).

(this took from several hours to several days, depending on the materials employed.). The precipitate was collected by filtration and washed with ethyl acetate. An almost pure product was recrystallized from ethyl acetate to afford a fine salt of optically active *N*-benzyloxycarbonylamino acid with (−)-ephedrine (Table 1).

The suitable solvents and concentrations for *N*-benzyloxycarbonylamino acid favored for the resolution were as follows [name of *N*-benzyloxycarbonylamino acid, solvent (ethyl acetate was used unless stated otherwise), concentration in mole (M)]: *β*-methyl DL-aspartate, 0.45. DL-Isoleucine, ether, 2.49. *erythro-β*-Methyl-DL-leucine, ether, 2.00. *threo-N,β*-Dimethyl-DL-leucine, 1.00. DL-Leucine, ethyl acetate - petroleum ether (2 : 3 V/V), 2.62. DL-Methionine, 2.46. DL-Phenylalanine, ether, 0.17. *O*-Benzyloxycarbonyl-DL-serine, 1.00. DL-Valine, 4.68.

The above mother liquor was washed with a small amount of N hydrochloric acid and then water. The organic layer was dried over anhydrous sodium sulfate and evaporated to dryness *in vacuo*. The residue thus obtained was redissolved in a half volume of the same solvent as was used in the initial resolution stage, and 0.055 mol of (+)-ephedrine was added. The solution was then allowed to stand. The deposition of salt was usually much faster than in the initial case. The precipitate was filtered and washed with ethyl acetate. Recrystallization from ethyl acetate gave pure antipodal salt (Table 2).

Liberation of Optically Active *N*-Benzyloxycarbonylamino Acids from Salts. A suspension of the salt of optically active *N*-benzyloxycarbonylamino acid with ephedrine in ethyl acetate was treated with a slight excess of 2N hydrochloric acid, and the organic layer was washed with a small amount of water. The solution of ethyl acetate was then dried over anhydrous sodium sulfate. After filtration, the filtrate was evaporated to dryness *in vacuo* to give an almost optically pure material. Recrystallization from an appropriate solvent was performed to afford pure materials (Table 3).

Regeneration of Free Amino Acids Bearing Entirely Optical Activities. *A)* According to the usual way, optically active *N*-benzyloxycarbonylamino acid was dissolved in threefold equi-amounts of 30% hydrogen bromide in glacial acetic acid. After about one hour, much anhydrous ether was added and deposition was collected by fitration. A solution of the resulting hydrobromide in ethyl alcohol was neutralized with triethylamine to deposit free amino acid. This was recrystallized from water or water–ethyl alcohol to give optically pure amino acids (Table 3).

B) In the case of serine, optically active *N,O*-dibenzyloxycarbonylserine was dissolved in ethyl alcohol and a palladium-black catalyst was added. Hydrogenolysis was carried out under slight pressure for 6 hr at room temperature with the aid of an efficient shaking machine. Then the catalyst, with the substance adhering to it, was filtered and the product was separated from the catalyst by extraction with hot water. To the aqueous solution, dioxane was added to complete the precipitation. Recrystallization from water and acetone afforded optically active serine (Table 3).

$C_3H_7NO_3$ Ref. 20i

441 g. of *dl*-N-acetylserine and 405 g. of 1-amphetamine are dissolved with heating and stirring in 2800 ml. of 95% ethanol. Stirring is continued for 16 to 20 hours at 23–25° C. The crystals which have formed are then filtered and dried. The crystalline material consisting of the *l*-amphetamine salt of *l*-N-acetylserine is dissolved in water and the solution is passed through an ion exchange column containing a sulfonate resin (Dowex 50, acid form). The effluent is concentrated and cooled, and the resulting crystalline product, *l*-N-acetylserine, is filtered off and dried; $[\alpha]_D^{28} = -21.9°$ (2% in absolute ethanol).

The filtrate first obtained above is reduced to dryness in vacuo. Two equivalents of 20% sodium hydroxide solution are added to the residue and the *l*-amphetamine is extracted with toluene. The residue is taken up in water and the aqueous solution is passed through an ion exchanger containing a sulfonic acid resin (Dowex 50, acid form) to remove the sodium ions. The effluent is concentrated to 250–300 ml., seeded with *d*-N-acetylserine crystals and allowed to stand overnight at room temperature. The resulting crystalline product, *d*-N-acetylserine, is separated by filtration and is dried; $[\alpha]_D^{25} = +23.25°$ (2% in absolute alcohol).

$C_3H_7NO_3$ Ref. 20j

For experimental details of enzymatic methods of resolution, see Ref. 18 n and the following: 1) S. Akabori, T.T. Otani, R. Marshall, M. Winitz and J.P. Greenstein, Arch. Biochem. Biophys. 83, 1 (1959), 2) S.M. Birnbaum, L.Levintow, R.B.Kingsley and J.P. Greenstein, J.Biol. Chem. 194, 455 (1952); M. Jaeger, S. Iskric and M.Wickerhauser, Croat. Chem. Acta, 28, 5 (1956) and "Chemistry of the Amino Acids, Vol. 3., J.P. Greenstein and M. Winitz, pgs. 2229-2234 (1961), J. Wiley and Sons, Inc.

For resolution of serine by fractional crystallization methods of derivatives, see the following: 1) N. Sugimoto, I.Chibata, S.Yamada and M. Yamamoto, U.S.Patent 3440279 (1969), 2) I. Chibata, S. Yamada, M. Yamamoto, M. Wada and T. Yoshida, Japanese Patent 43-24410 (1968), and 3) I. Chibata, S. Yamada and M. Yamamoto, Japanese Patent 46-29843 (1971).

$C_3H_7N_5$ Ref. 21

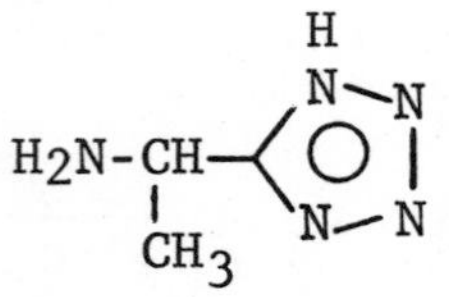

Resolution of tetrazole analogues of N-*benzyloxycarbonyl*-DL-*amino acids*

Separation of diastereomeric salts. Tetrazole analogue of N-benzyloxycarbonyl-DL-amino acid (1 mmole) and hydrazide of L-tyrosine (1 mmole) in MeOH (for volume of solvent see Table 1) were heated to reflux temperature and some undissolved material was filtered off. The filtrate was set aside at −5° in a refrigerator for 24 hr. The separated material (crude salt "A") was collected and recrystallized several times from a minimal amount of MeOH.

The crude salt "B" was obtained from the mother liquor after evaporation of MeOH.

Enantiomers of tetrazole analogues of N-*benzyloxycarbonylamino acids.* A suspension of finally ground diastereomeric salt "A" in 1 N HCl aq (5 ml) was stirred for 15 min and then left for 2–4 hr in a refrigerator. The product was filtered and washed several times with cold water. Results are summarized in Table 2.

The same procedure was applied to crude salts "A" and "B". The results are presented in Table 3.

Table 1. Results of fractionation of diastereomeric salts formed from racemic tetrazole analogues of N-benzyloxycarbonyl amino acids and hydrazide of L-tyrosine

				Precipitating material (salt "A")						Material recovered from mother liquor (crude salt "B")		
				Crude salt "A"		Sterically homogeneous salt "A"						
No	Racemic tetrazolic acid[a]	MeOH[b] ml	Time of crystallization hr	Yield %	M.p. °C	Number of recrystallizations	Yield %	M.p. °C	Configuration	Yield %	M.p. °C	Configuration
1	Z-AlaT	2	13	100	154–159	4	41	166–168	D–L	96	161–164	L–L
2	Z-CysT/BZL/	8	16	102	169–171	2	89	171–172	D–L	93	143–149	L–L
3	Z-LeuT	1·25	20	95	166–169	3	59	174–176	D–L	99	147–150	L–L
4	Z-PheT	6	16	101	148–153	3	60	153–154	L–L	97	156–159	D–L
5	Z-Valt	1·5	18	103	171–175	2	72	183–184	D–L	90	80–89	L–L

[a] Hydrazide of L-tyrosine was used in equimolar amount; proposal of nomenclature and symbolism of tetrazole analogues of amino acids is given in ref. 10; Z– benzyloxycarbonyl; AlaT, CysT(BZL), LeuT, PheT, ValT—tetrazole analogues of alanine, S-benzyl-cysteine, leucine, phenylalanine and valine, respectively.
[b] The given volume of MeOH was used for 1 mmole quantities of tetrazolic acid and resolving base.

Table 2. Enantiomers of tetrazole analogues of N-benzyloxycarbonyl amino acids prepared from sterically homogeneous salts "A"

No	Compound	Yield %	M.p. °C	$[\alpha]_D^{20}$ c=1	Lit. data of reference L-form M.p. °C	$[\alpha]_D^{20}$ c=1	Ref.
1	Z-D-AlaT	89	139–141	+34·5[a]	140–141	−36[a]	3
2	Z-D-CysT(BZL)	92	138–139	+35·5[a]	138–140	−36[a]	3
3	Z-D-LeuT	91	100–102	+39·5[a]	101–102	−40[a]	3
4	Z-L-PheT	96	182–183	−47[b]	181–182	−48[b]	9
5	Z-D-ValT	93	156–158	+33[a]	156–158	−33[a]	This paper

[a] In MeOH [b] In DMF

Table 3. Tetrazole analogues of N-benzyloxycarbonyl amino acids obtained from crude salts "A" and "B"

		Material obtained from crude salt "A"				Material obtained from crude salt "B"			
No	Compound	Yield %	M.p. °C	$[\alpha]_D^{20}$ c=1	Enriched in form	Yield %	M.p. °C	$[\alpha]_D^{20}$ c=1	Enriched in form
1	Z-AlaT	85	124–128	+16·5[a]	D	87	118–124	−11[a]	L
2	Z-CysT(BZL)	93	108–115	+24[a]	D	94	105–108	−25[a]	L
3	Z-LeuT	92	93–96	+22·5[a]	D	98	95–105	−13[a]	L
4	Z-PheT	96	167–171	−27·5[b]	L	96	166–171	+19[b]	D
5	Z-ValT	90	143–146	+20[a]	D	95	135–142	−18[a]	L

[a] In MeOH [b] In DMF

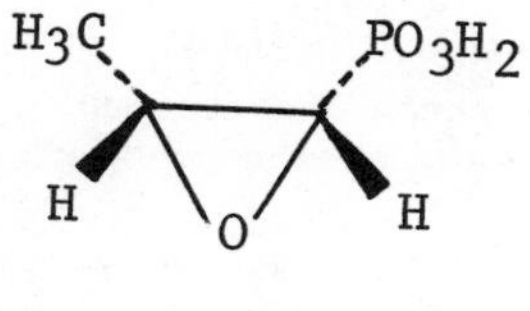

$C_3H_7O_4P$ Ref. 22

Resolution of (±) (cis - *1,2 - epoxy - 1 - propyl) phosphonic acid*

The ammonium salt of (±) (*cis* - 1,2 - epoxy - 1 - propyl) phosphonic acid (5 g., .032 mole), is treated with quinine (6 g., 0.0185 mole) in 200 ml. of methanol. The solution is concentrated to a syrup, and the residue dissolved in 50 ml. of methanol. The mixture is allowed to stand at room temperature, and the crystalline salt settled out of solution. The solid mass is slurried with additional methanol, filtered, and washed with ethanol and acetone to yield 3.2 g. of the crystalline quinine salt having a decomposition point of 150°C. The filtrate is concentrated to a small volume, seeded with crystals of the quinine salt of (−) (*cis* - 1,2 - epoxy - 1 - propyl) phosphonic acid, and allowed to stand at room temperature. The salt settles out of solution and is filtered off, and washed with ethanol and acetone to yield 3.2 g. of quinine salt having a decomposition point of 150°C. The latter salt is recrystallized from methanol. The 3.2 g. obtained from the first crystallization above is also recrystallized. The two recrystallized products are combined and recrystallized again from 30 ml. of methanol to yield the quinine salt of (−) (*cis* - 1,2 - epoxy - 1 - propyl) phosphonic acid. 165 mg. of the quinine salt is then reconverted to the ammonium salt of (−) *cis* - 1,2 - epoxy - 1 - propyl phosphonic acid by suspending the salt in water and adding a small amount of ammonium hydroxide. To the resulting suspension is added a small volume of chloroform, and the aqueous layer which contained the ammonium salt is separated from the organic layer and freeze-dried to obtain the ammonium salt of (−) (*cis* - 1,2 - epoxy - 1 - propyl) phosphonic acid. The salt is dissolved in 2 ml. of water, and the specific rotation is measured in a 0.5 decimeter tube. The product, calculated on the basis of the free acid, had a specific rotation of −16.2° at 405 mμ.

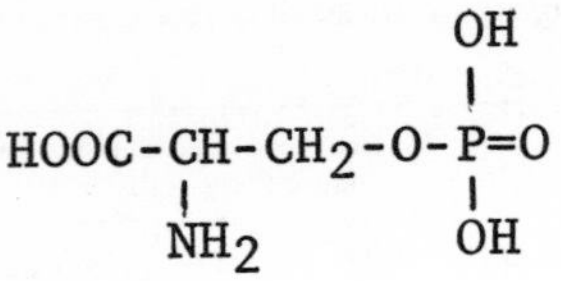

$C_3H_8NO_6P$ Ref. 23

EXPERIMENTAL[5]

Dextro-Serinephosphoric Acid—The *d,l*-serine required for the preparation of *d,l*-serinephosphoric acid was prepared essentially according to the method of Leuchs and Geiger.[6] However, instead of the chloroacetal, the bromoacetal (obtained from paraldehyde with bromine) was treated with sodium ethylate on the steam bath,[7] thereby producing the ethoxyacetal by a much simpler process and with a very good yield. The *d,l*-serine (200 gm.) was phosphorylated[2] in portions of 8 gm. and yielded 106 gm. of the barium salt of *d,l*-serinephosphoric acid.

Two portions of the barium salt (50 gm. each) were each suspended in 400 cc. of water. Sulfuric acid was added in slight excess and the mixture vigorously shaken mechanically during 1 hour. The solutions were combined, filtered, and quantitatively freed from barium and sulfate ions. The clear solution was made faintly alkaline by addition of a concentrated solution of brucine in methyl alcohol. The excess brucine was then removed by extraction with chloroform and the aqueous layer evaporated under diminished pressure to a thick syrup.

The product was dissolved in 750 cc. of boiling methyl alcohol and the solution cooled in ice water during 2 hours. On filtering through a Buchner funnel, there were obtained 175 gm. of nicely crystallized brucine salt. On concentrating and standing in the refrigerator the mother liquor yielded an additional 40 to 50 gm. of the salt.

175 gm. of the first crop of brucine salt were recrystallized four times from absolute methyl alcohol and a small sample then converted into the barium salt. The optical rotation of the barium salt, dissolved in 10 per cent hydrochloric acid, was

$$[\alpha]_D^{25} = \frac{+0.75^\circ \times 100}{2 \times 6} = +6.3^\circ$$

After seven recrystallizations, the rotation under the same conditions was +8.6°. After nine recrystallizations, the rotation was constant (+9.4°) and was not changed by further crystallization. The yield was 40.5 gm.

Dried under diminished pressure at 60°, the dibrucine salt had the following composition.

5.940 mg. substance: 0.382 cc. N_2 (757 mm. at 26°)
4.190 " " : 8.980 mg. molybdate

$C_{49}H_{60}O_{14}N_5P$.*	Calculated.	N 7.20, P 3.19
973.5	Found.	" 7.31, " 3.11

* In our former paper[4] the formula and calculated values for the dibrucine salt should be corrected.

On heating, the salt sinters at 100° and decomposes[8] around 130°. The barium salt, obtained from the brucine salt with barium hydroxide, had the following composition after drying at 110°.

10.580 mg. substance: 0.387 cc. N_2 (762 mm. at 26°)
3.472 " " : 23.180 mg. molybdate

$C_3H_6O_6NPBa$.	Calculated.	N 4.36, P 9.65
320.4	Found.	" 4.18, " 9.69

For the determination of the specific rotation, a 6 per cent solution of the barium salt in 10 per cent hydrochloric acid was used.

$$[\alpha]_D^{25} = \frac{+1.13^\circ \times 100}{2 \times 6.0} = +9.4^\circ$$

Calculated as the free acid, the rotation[3] was

$$[\alpha]_D^{25} = \frac{+1.13^\circ \times 100}{2 \times 3.47} = +16.3^\circ$$

[5] We wish to thank Dr. R. S. Tipson for his kindness in assisting in the preparation of the manuscript.

[6] Leuchs, H., and Geiger, W., *Ber. chem. Ges.*, **39**, 2644 (1906).

[7] Späth, E., *Monatsh. Chem.*, **36**, 4 (1915).

[8] Levene, P. A., and Schormüller, A., *J. Biol. Chem.*, **103**, 537 (1933).

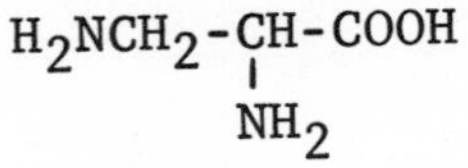

$C_3H_8O_2N_2$ Ref. 24

Melting points were determined in open capillaries on a Tottoli apparatus and are uncorrected. Optical rotations were measured with a Perkin-Elmer model 141 automatic polarimeter, using 1 dm jacketed quartz tubes kept at costant temperature by circulating water.

Optical resolution of 2,3-bis(benzoylamino)propionic acid

A solution containing 47 g (0.15 mol) 2,3-bis(benzoylamino)propionic acid[14], m. p. 197°C, and 18 g (0.15 mol) (*R*)(+)-α-phenylethylamine in 1100 m*l* ethyl acetate was heated to 60°C and then filtered.

The solution was seeded with a few crystals of the pure (*R*)(+)-α-phenylethylamine salt of (*S*)(−)-2,3-bis-(benzoylamino)propionic acid (I) and allowed to reach room temperature overnight. The crystalline solid was collected, washed with dry ether and dried; yield 25.5 g, m. p. 140°C, $[\alpha]_D^{22}$: − 19.98° (c = 2, in H_2O). Recrystallization from 2500 m*l* of ethyl acetate yielded 19.4 g (60%), m. p. 148°C, $[\alpha]_D^{20}$: − 21.3° (c = 2, in H_2O).

Three recrystallizations from ethyl acetate gave $[\alpha]_{589.3}^{20}$: − 21.7° (c = 2, in H_2O).

for $C_{25}H_{27}N_3O_4$ (433.5) Calc. C 69.26 H 6.28 N 9.69
Found. C 68.90 H 6.48 N 9.67

The mother liquors from the optical resolution were concentrated to dryness under reduced pressure. The residue was dissolved in 500 m*l* of water at 40°C, made alkaline with 30% sodium hydroxide and then extracted with three 60-m*l* portions of methylene dichloride. The aqueous layer was acidified with 6N hydrochloric acid to yield 26.3 g of crude (*R*)(+)-2,3-bis(benzoylamino)-propionic acid: m. p. 150—200°C, $[\alpha]_D^{22}$: + 16.2° (c=2, 0.1N NaOH). The crude acid was dissolved with 10.2 g of (*S*)(−)-α-phenylethylamine in 800 m*l* boiling ethyl acetate and allowed to cool overnight. The yield of the (*S*)(−)-α-phenylethylamine salt of (*R*)(+)-2,3-bis-(benzoylamino)propionic acid (II) was 24 g m. p. 146°C, $[\alpha]_D^{22}$: + 21.3° (c = 2, in H_2O).

An analytical sample was obtained by three further crystallizations from ethyl acetate: $[\alpha]_{589.3}^{20}$: 21.3° (c = 2, in H_2O).

for $C_{25}H_{27}N_3O_4$ (433.5) Calc. C 69.26 H 6.28 N 9.69
Found. C 68.92 H 6.31 N 9.67

(S)(−)-2,3-Bis(benzoylamino)propionic acid (III)

A solution of 18.7 g (0.043 mol) of I in 200 m*l* of water at 40°C was made alkaline with concentrated sodium hydroxide solution and extracted with two 50-m*l* portions of ether. The aqueous layer was acidified with HCl, forming a gum, which crystallized to give, after washing and drying, 13.3 g, m. p. 158°C.

Recrystallization from 75 m*l* of 46% aqueous ethanol yielded 12.3 g (90%), m. p. 159°C, $[\alpha]_D^{20}$: − 33° (c = 2, in 0.1N NaOH)[17].

(R)(+)-2,3-Bis(benzoylamino)propionic acid (IV)

The compound was obtained by the same procedure as described for III, but using II as starting material: yield 89%, m. p. 159°C, $[\alpha]_D^{20}$: + 32.8° (c = 2, in 0.1N NaOH).

[14] E. Fischer and W. A. Jacobs, Ber. dtsch. Chem. Ges. **39**, 2942 [1906].

[16] S. M. Birnbaum, R. J. Koegel, S.-C. J. Fu and J. P. Greenstein, J. biol. Chemistry **198**, 335 [1952].

[17] $[\alpha]_D^{20}$: − 35.9° in the conditions given by l. c.[16].

(S)(+)-2,3-Diaminopropionic acid hydrochloride (V)

A stirred suspension of *(S)*(−)-2,3-bis(benzoylamino)-propionic acid (III) (10 g, 0.052 mol) in 150 m*l* of 18% HCl was refluxed for 4.5 h, giving a clear solution after 1.5 h.

Benzoic acid which separated on cooling was filtered and washed with 30 m*l* of water. The filtrate was concentrated to dryness under vacuum. The residue was dissolved in 15 m*l* of water and precipitated with 60 m*l* ethanol, yielding 4 g (86%) of solid, m. p. 232°C decomp., $[\alpha]_D^{20}$: + 14.9 (c = 2, in 1N NCl).

Recrystallization from 19 m*l* of water yielded 2.5 g, m. p. 240°C decomp., $[\alpha]_D^{21}$: + 24.9 (c = 2, in 1N HCl) (l. c.[6] $[\alpha]_D^{25}$: + 25.1°).

(R)(−)-2,3-Diaminopropionic acid hydrochloride (VI)

Starting with 9.5 g *(R)*(+)-2,3-bis(benzoylamino)-propionic acid and saponifying as described for V but using IV as starting material, 2.2 g of *(R)*(−)-2,3-diaminopropionic acid hydrochloride were obtained: m. p. 240°C, $[\alpha]_D^{25}$: − 25.2° (c = 2, in 1N HCl) (l. c.[16] $[\alpha]_D^{25}$: − 25.2°).

$C_3H_8N_2O_2$ Ref. 24a

L-*Diaminopropionic Acid*—15 gm. of α,β-diacetyl-DL-diaminopropionic acid (12) were dissolved in 700 ml. of water and the solution was brought to pH 7.5 with 2 N lithium hydroxide. Water was added to bring the volume to 800 ml. (0.1 M concentration of the compound), and the solution was treated with 1 gm. of hog kidney acylase I powder (1).[3] After 6 hours of incubation at 38°, manometric ninhydrin measurements revealed that exactly 50 per cent hydrolysis of the compound had occurred. The digestion was allowed to proceed for 12 hours more, but no further hydrolysis took place. The mixture was brought to pH 5 with acetic acid, treated with Norit, filtered, and the nearly clear filtrate evaporated *in vacuo* to about 50 ml. The solution was filtered to remove traces of protein and treated with 4 volumes of absolute alcohol. A gelatinous precipitate appeared which was filtered with suction, washed with alcohol, and dried in air. The mother liquor and washings which contained the α,β-diacetyl-D-diaminopropionic acid were combined and set aside.

The dried, precipitated β-acetyl-L-diaminopropionic acid was taken up in boiling water, and the hot solution filtered clear with the aid of Norit. Addition of hot alcohol to the filtrate led to the crystallization of the compound in tufts of long needles. It was filtered with suction, and dried. The yield was 75 per cent of the theoretical. Calculated, C 41.1, H 6.9, N 19.2; found, C 40.9, H 7.0, N 18.9. $[\alpha]_D$ for a 2 per cent solution in 5 N HCl at 25° = −42.5°.[4]

4 gm. of β-acetyl-L-diaminopropionic acid were refluxed 2 hours with 50 ml. of 2.5 N hydrochloric acid. The solution was evaporated *in vacuo* to a syrup, the residue dissolved in a little water, filtered, and the filtrate treated with absolute alcohol to 80 per cent. A crystalline precipitate of α,β-L-diaminopropionic acid monohydrochloride appeared almost immediately. It was filtered with suction and recrystallized from the minimum amount of hot water. The resulting crystals were filtered and washed successively with ice water, alcohol, and ether. The yield was 3 gm. or 70 per cent or the theoretical. Calculated, N 19.9, Cl 25.2; found, N 19.7, Cl 25.1. $[\alpha]_D$ for a 2 per cent solution in 1 N HCl at 25° = +25.2°. Fischer and Jacobs reported $[\alpha]_D$ = +25.09° (±0.1°) at 20° for the enantiomorph which they called *d*-diaminopropionic acid monohydrochloride (7).

[2] The optical enantiomorphs of proline have been obtained by the asymmetric action of a hog kidney amidase preparation on DL-proline amide (11).

[3] The rate of hydrolysis of the L isomer of α,β-diacetyl-DL-diaminopropionic acid by crude hog kidney aqueous homogenate at 38° and at pH 7.0 in 0.1 M phosphate buffer is 19 μM per hour per mg. of N. Under the same conditions, the rate with acylase I is 708 (*cf.* (1) for experimental conditions). In view of the instability of α,β-dichloroacetyl-DL-diaminopropionic acid, the rate values with acylase I described earlier (1) must be modified.

[4] It is interesting to note that the sign of optical rotation is positive for the analogous compounds, ε-chloroacetyl-L-lysine (1, 5) and δ-chloroacetyl-L-ornithine (6). Furthermore, the isolation in good yield of the β-acetyl-L-diaminopropionic acid is additional evidence that acylase I action is restricted to α-substituted acyl radicals (*cf.* (5)).

D-*Diaminopropionic Acid*—The combined alcoholic mother liquor and washings from the resolution mixture should contain all of the α,β-diacetyl-D-diaminopropionic acid plus some unprecipitated β-acetyl-L-diaminopropionic acid. The solution was evaporated *in vacuo* to dryness, and the oily residue taken up in 100 ml. of water. Manometric ninhydrin measurements revealed the presence of 1.3 gm. of an α-amino acid, calculated as β-acetyl-L-diaminopropionic acid.

The customary procedure at this point has generally been to acidify with hydrochloric acid to pH 1.7, and to extract the acidified solution exhaustively with ethyl acetate, thereby bringing the acylated D-amino acid into the non-aqueous solvent while the hydrochloride of the unprecipitated L-amino acid remains in the aqueous layer (1–3). We noted, however, that the α,β-diacetyl-D-diaminopropionic acid did not pass into the ethyl acetate layer under these conditions. It was necessary to remove or destroy the L derivative before the D component could be recovered. For this purpose, we employed a device which worked satisfactorily in the resolution of arginine and histidine (3), namely to oxidize the β-acetyl-L-diaminopropionic acid with snake venom L amino acid oxidase. This compound is readily oxidized by *Crotalus adamanteus* and by *Bothrops jararaca* venoms at a rate of about 1 μM of oxygen consumed per hour per mg. of N.[5] The solution was therefore brought to pH 8.0 by addition of 2 N lithium hydroxide, and to it was added a concentrated aqueous solution of 2.5 gm. of *C. adamanteus* venom. The pH dropped thereby to 7.4. No catalase was employed. The mixture was placed in a bath at 38° and treated with a stream of moist air for 24 hours. At the end of this period, the mixture was diluted with water to 400 ml., treated with acetic acid to pH 5.0, and the protein was coagulated at 95°. After filtration with the aid of Norit, the clear filtrate was evaporated *in vacuo* to 100 ml. Concentrated hydrochloric acid was added to 2 N, and the solution refluxed for 2 hours. The reddish purple solution was briefly boiled with Norit, filtered, and the water-clear filtrate evaporated *in vacuo* to a syrup. This residue was taken up in the minimum amount of water, filtered, and treated with alcohol to 80 per cent. The resulting crystals of D-diaminopropionic acid monohydrochloride weighed 5 gm. The compound was recrystallized from the minimum amount of hot water. The yield of pure material was 4 gm., or 72 per cent of the theoretical. Calculated N 19.9, Cl 25.2; found N 19.5, Cl 25.1. $[\alpha]_D$ for a 2 per cent solution in 1 N HCl at 25° = −25.2°. Fischer and Jacobs reported $[\alpha]_D$ = −24.98° (±0.1°) at 20° for the enantiomorph, which they called *l*-diaminopropionic acid monohydrochloride (7).

At pH 8.2 in the presence of 0.1 M pyrophosphate buffer, D-diaminopropionic acid is nearly completely oxidized by hog kidney D-amino acid oxidase preparations. The rate of oxidation is about half that observed for D-alanine.

Optical Purity of L-*Diaminopropionic Acid*—The routine method employed in this laboratory, to determine the degree of contamination of an optical isomer by its enantiomorph to better than 1 part in 1000, has been described (1–3, 13). The appreciable rate of oxidation of D-diaminopropionic acid by D-amino acid oxidase stands in contrast with the behavior of the higher homologues, namely D-lysine and D-ornithine, which are very little affected by this enzyme. It was possible therefore to test our L-diaminopropionic acid for the presence of the D enantiomorph. 1000 μM of the L isomer were completely inert to D-amino acid oxidase. 1 μM of the D isomer added to 1000 μM of the L isomer was nearly quantitatively oxidized by the enzyme. It was therefore possible to conclude that the L-diaminopropionic acid contained less than 0.1 per cent of D-aminopropionic acid, and hence had an optical purity greater than 99.9 per cent.

In contrast with β-acetyl-L-diaminopropionic acid, the free L-diaminopropionic acid was weakly oxidized by *C. adamanteus* and *B. jararaca* venom L-amino acid oxidase, and therefore the optical purity of the D-diaminopropionic acid isomer could not be tested by the present procedure.[6]

[5] *Crotalus adamanteus* venom was obtained from Ross Allen's Reptile Institute, Silver Springs, Florida. *Bothrops jararaca* venom was obtained through the courtesy of Dr. G. Rosenfeld, of the Instituto Butantan, San Paulo, Brazil.

[6] L-Lysine is also relatively resistant to L-amino acid oxidase, but the ε-substituted derivatives are readily oxidized (14).

```
    COOH
     |
  H-C-Br
     |
  Br-C-H
     |
    COOH
```

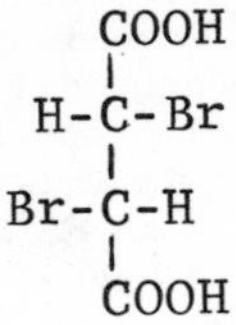

$C_4H_4Br_2O_4$ Ref. 25

Resolution of isoDibromosuccinic Acid.

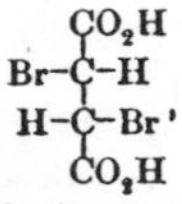

Several alkaloids were tried for the purpose of resolving the acid into its optically active components, and morphine was finally selected as the most suitable. The use of water as the solvent was at first attempted, the morphine being used in the proportion both of 1 mol. and of 2 mols. for 1 mol. of acid and the temperature conditions being varied. Water is, however, an impractical solvent. The iso-acid underwent a certain amount of decomposition into bromofumaric acid under the temperature conditions employed, and, owing to the ease with which hydrogen bromide was eliminated, the crystals which separated consisted mainly of morphine hydrobromide. A small proportion of laevorotatory iso-acid was actually obtained from these crystals, but the method obviously did not appear likely to be successful.

The solvent selected was methyl alcohol, and it was found advantageous to use equimolecular proportions of morphine and acid. The alkaloidal salt which separated when the iso-acid (1 mol.) was added to a solution of morphine (1 mol.) in methyl alcohol could not, however, be submitted to the customary treatment, namely, resolution by crystallisation. It was so very sparingly soluble, not only in methyl alcohol, but also in many other organic solvents, that crystallisation from such solvents was not a practical operation. It dissolved, however, in warm water, but the salt which separated was morphine hydrobromide. An acid of a high optical activity can, however, be obtained under the following conditions.

Morphine (21 grams, 1 mol.) was dissolved in warm methyl alcohol (290 c.c.), and, when the temperature of the solution was 46°, 19 grams (1 mol.) of the powdered iso-acid were added in one instalment, the mixture being shaken vigorously. The solution of the acid was immediately followed by the separation of glassy crystals of a morphine salt. After twenty-four hours at about 15°, the crystals (22 grams) were dissolved in dilute sulphuric acid. The acid obtained by extraction with ether amounted to 6·5 grams, and was strongly laevorotatory in ethyl acetate solution:

$l=2$, $c=8·6$, $a_D -17·32°$, $[a]_D -100·7°$.*

The mother liquor, from which the morphine salt had been removed, gave a dextrorotatory acid. 6·5 Grams of the acid with $[a]_D -100·7°$ were then added to a solution of 7·1 grams of morphine in 100 c.c. of methyl alcohol. The resulting crystals (9·5 grams) gave an acid with the following rotation in ethyl acetate solution:

$l=2$, $c=7·4$, $a_D -16·93°$, $[a]_D -114·4°$.

This acid (2·5 grams) was then acted on by a solution of 2·8 grams of morphine in 80 c.c. of methyl alcohol. The yield of salt was 3·5 grams, and the acid from this gave the following rotation in ethyl acetate solution:

$l=2$, $c=8·47$, $a_D -19·98°$, $[a]_D -118°$.

The acid was not yet homogeneous, and it was also obvious that the preparation of the optically pure acid could not be effected with readiness by this treatment.

Many unsuccessful attempts were made to obtain the pure l-acid by crystallising the strongly laevorotatory acid, which is easily obtained by treating the iso-acid once with morphine according to the method just described; for example, 25 grams of an acid with $[a]_D -116°$ in ethyl acetate solution were crystallised first from benzene (1800 c.c.) and then from a mixture of benzene (200 c.c.) and acetone (6 c.c.), the acid being sparingly soluble in boiling benzene and readily so in cold acetone. The product amounted to 8 grams, and had $[a]_D -128°$ in ethyl acetate solution, but further attempts to obtain the pure acid by crystallising this product from boiling solvents failed owing partly to the tendency of the acid to decompose under the conditions employed.

---

* The behaviour of the dibromosuccinic acid (m. p. 255—256°), obtained from fumaric acid, was examined under conditions similar to the above, 10 grams of acid, 12 grams of morphine, and 153 c.c. of methyl alcohol being used. The morphine salt did not separate at once as in the case of the iso-acid, but, after five days, 4·5 grams had crystallised. The acid obtained from this salt was optically inactive. Other experiments on the action of various alkaloids on this dibromo-acid were made, and no evidence of the possibility of resolving it was obtained (compare Holmberg, *J. pr. Chem.*, 1911, [ii], 84, 145).

The pure l-acid was finally obtained as follows: Morphine (83 grams) was dissolved by heating with methyl alcohol (1150 c.c.), and, when the temperature had sunk to 46°, isodibromosuccinic acid (75 grams) was added. On shaking vigorously, the acid dissolved, and, two minutes after its addition, the morphine salt began to crystallise in glassy prisms. After eighteen hours at the laboratory temperature (summer), the crystals (87 grams) were removed.

Examination of Filtrate.—The filtrate was allowed to evaporate spontaneously. After a week, gritty crystals (8 grams), differing in appearance from those of the main quantity removed, had separated; they were sparingly soluble in dilute sulphuric acid, and they consisted of morphine hydrobromide. After several weeks the residual oily solid was decomposed by dilute sulphuric acid, and the bromo-acid extracted with ether. Thirty-four grams were obtained with the following rotation in ethyl acetate:

$l=2$, $c=5·45$, $a_D +5·07°$, $[a]_D +46·5°$.

By adding a large excess of light petroleum to the concentrated solution of this product in ethyl acetate, a solid was obtained with $[a]_D +7·7°$. On allowing the solvent to evaporate at the ordinary temperature, and on repeating the treatment, a further solid with $[a]_D +21·3°$ resulted, whilst the filtrate contained an oily dextrorotatory acid of high rotation. The separation of the pure d-acid from the latter product was not attempted.

Examination of Solid.—The 87 grams of morphine salt were decomposed by sulphuric acid, and the bromo-acid extracted with ether. The dried ethereal solution was allowed to evaporate spontaneously at the ordinary temperature in order to minimise the risk of decomposition. The resulting acid was at first oily, but ultimately became solid after ten days in a vacuum. Yield, 25 grams. On polarimetric examination in ethyl acetate solution it gave the value:

$l=2$, $c=6·99$, $a_D -15·57°$, $[a]_D -111·4°$.

This product was very soluble in cold ethyl acetate. On dissolving a portion of it in the minimum amount of this solvent, and then adding a large excess of light petroleum or of carbon tetrachloride, the solid which separated was laevorotatory to a less degree than the original, whilst the acid obtained by allowing the filtrate to evaporate spontaneously at the ordinary temperature was more active; for example, when 5 grams of the acid were treated in this manner with light petroleum, 2·3 grams with $[a]_D -82°$ separated, whereas the acid obtained from the filtrate had $[a]_D -128·5°$.

This tedious method was accordingly employed. The acid which separated was neglected, and the acid from the filtrate was examined polarimetrically, its concentrated solution in ethyl acetate then being precipitated afresh. At first light petroleum was used as the precipitant, and afterwards carbon tetrachloride. The acids obtained from the successive filtrates had the values $[a]_D -127°$, $-131·6°$, $-140°$, $-144°$ under similar conditions of concentration in ethyl acetate. The acid with $[a]_D -144°$ was then crystallised twice from boiling benzene, in which it dissolved with great difficulty, the heating not being unduly prolonged. On cooling, the acid separated quickly in beautiful, glassy needles. After two more crystallisations from benzene, the rotation was practically unchanged. The yield of the pure acid was only about 1 gram.

l-αβ-Dibromosuccinic acid,
```
      CO_2H
       |
   Br-C-H
       |
   H-C-Br
       |
      CO_2H
```
is very sparingly soluble in boiling benzene, from which it separates on cooling in glassy needles grouped in rosettes. It is easily soluble in ethyl acetate, acetone, methyl alcohol, or ethyl alcohol, and sparingly so in chloroform, light petroleum, or carbon tetrachloride. Like the r-isomeride, it dissolves very readily in cold water. It melts and decomposes fairly sharply at 157—158° with slight preliminary softening:

0·2013 required 28·1 c.c. of $N/19·18$ baryta for neutralisation.

Equivalent = 137·4 (calc., 137·9).

0·2228 gave 0·3016 AgBr. Br = 57·6.

$C_4H_4O_4Br_2$ requires Br = 57·9 per cent.

Its specific rotation was determined in ethyl acetate solution:

$l=2$, $c=5·788$, $a_D^{12} -17·13°$, $[a]_D^{12} -148·0°$.

The solution used in this determination was allowed to remain in the polarimeter tube for three days at the ordinary temperature; the rotation was then slightly greater than before, namely, $a_D^{12} -17·39°$, whence $[a]_D -150·2°$.

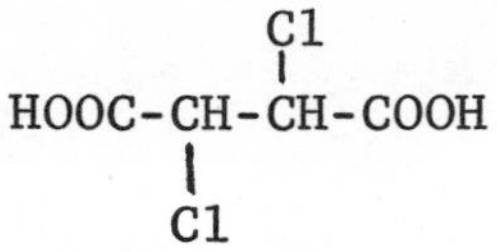

$C_4H_4Cl_2O_4$

Ref. 26

Aus der Mutterlauge nach dem reinen sauren d-dichlorbernsteinsauren d-α-Phenäthylamin wurden 5,5 g. einer Säure erhalten, welche unscharf bei 168-170° unter Gasentwickelung schmolz, und welche in Essigäther $[\alpha]_D$ = -11°,29 zeigte.

4. Saures l-dichlorbernsteinsaures l-α-Phenäthylamin und Darstellung reiner l-Dichlorbernsteinsäure.

Von der Dichlorbernsteinsäure mit $[\alpha]_D$ = -26°,84 wurden 7,4 g. zusammen mit 4.8 g. l-α-Phenäthylamin in 200 g. Wasser von 50° gelost. Bei Kühlung der Lösung mit Eis kristallisierten 5.5 g. weisses Kristallpulver aus. Dieses Salz schmolz wie das d,d-Salz bei 142-142,5.

Die in gewöhnlicher Weise aus dem Salz gewonnene Säure kristallisierte im Exsiccator als zentimeterlange, flache Prismen. Schmelzpunkt 164°-165° unter Gasentwickelung. In Essigäther zeigte die Säure $[\alpha]_D$ = -77°,87.

Nach Lösen in Essigäther und Fällen mit Kohlenstofftetrachlorid wurden farblose Prismen erhalten, welche bei 164-165° unter Gasentwickelung schmolzen.

1,028 g. in Essigäther zu 9,52 ccm. gelöst zeigten in 1-dm. Rohr bei +23°C. α_D = -8°41'. Also $[\alpha]_D^{23}$ = -80°,38 und $[M]_D^{23}$ = -150°,3.

Diese Säure ist also als reine l-Dichlorbernsteinsäure anzusehen.

Aus der Mutterlauge nach dem reinen sauren l-dichlorbernsteinsauren l-α-Phenäthylamin wurden 3,5 g. Säure erhalten, welche bei 171°-173° schmolz und in Essigäther $[\alpha]_D$ = +8°,64 zeigte.

3. Saures d-dichlorbernsteinsaures d-α-Phenäthylamin und Darstellung der reinen d-Dichlorbernsteinsäure.

In 400 ccm. +50°C. warmem Wasser wurden 18,7 g. (= 1/10 Mol.) inaktive Säure und 12,1 g (1/10 Mol.) d-α-Phenäthylamin gelöst. Die klare Lösung wurde mit Eis gekühlt, wobei ein weisses, kristallinisches Pulver abgeschieden wurde. Ausbeute 17 g., also etwas mehr als die Hälfte der ganzen Salzmenge. Dieses Salz, welches ja teilweise racemisch sein musste, schmolz bei 133°-133°,5 und hatte die stöchiometrische Zusammensetzung $C_6H_5CH(CH_3)NH_2$, $(HOCOCHCl)_2$.

Das Salz wurde in Verdünnter Schwefelsäure gelöst und die Lösung viermal mit Äther extrahiert, wonach die Säure wie oben gewonnen wurde, Ausbeute 10,2 g. Diese Säure schmolz unscharf bei 164° unter Gasentwickelung und zeigt in Essigäther $[\alpha]_D$ = +20,60.

Die wässerige Mutterlauge nach dem Salz wurde mit Schwefelsäure versetzt, mit Äther extrahiert u. s. w., wobei 8 g. Säure gewonnen wurden, welche unscharf bei 165° unter Gasentwickelung schmolz und in Essigäther $[\alpha]_D$ = -26°,84 zeigte.

10,0 g. der Rechts-Säure mit $[\alpha]_D$ = +20°,60 wurden zusammen mit 6,5 g. d-α-Phenäthylamin in 300 g. Wasser von 50°C. gelöst, und die Lösung mit Eis gekühlt. Ausbeute an kristallinischem Salz 6,3 g. Schmelzpunkt des Salzes 142°-142°,5.

Aus dem übrig gebliebenen Salz wurde die Säure in gewöhnlicher Weise gewonnen. Im Exsiccator kristallisierte sie aus der zum Schluss zurückbleibenden Wasserlösung als zentimeterlange, schmale Prismen. Ausbeute 3,5 g. Schmelzpunkt 163°-164° unter Gasentwickelung, $[\alpha]_D$ in Essigäther = +79°,05.

3,3 g. dieser Säure wurden in 9 g. Essigäther gelöst und 100 g. Kohlenstofftetrachlorid zugesetzt. Nach vier Stunden hatten sich 1,3 g. kleine, dicke, glasglänzende Prismen abgesetzt, welche bei 164°-165° unter Gasentwickelung schmolzen.

1,178 g. in Essigäther zu 9,52 ccm. gelöst zeigten in 1 dm. Rohr bei +23°C. α_D = +9°57'. Also $[\alpha]_D^{23}$ = +80°,41 und $[M]_D^{23}$ = +150°,4.

Die hier erhaltene Säure darf als reine d-Dichlorbernsteinsäure angesehen werden. Dass der Schmelzpunkt und das Drehungsvermögen durch die Kristallisation aus Essigäther + Kohlenstofftetrachlorid sogar ein wenig erhöht wurden beruht natürlich darauf, dass kleine Spuren von Verunreinigungen dadurch entfernt wurden.

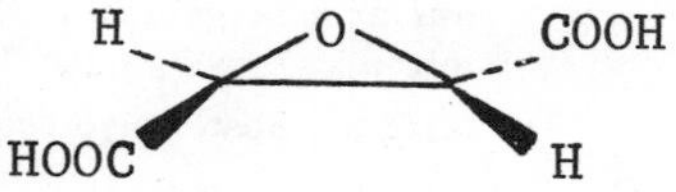

$C_4H_4O_5$

Ref. 27

Optical Resolution of *trans*-DL-2,3-Epoxysuccinic Acid. The epoxy acid, 6.6 g (0.05 mol), and *l*-ephedrine, 8.25 g (0.05 mol), were dissolved in 10 m*l* of hot methanol and 90 m*l* of acetone was added. After cooling, crystallization of the ephedrine salt began by seeding with an authentic specimen which was obtained in another crystallization experiment. After standing 30 min at room temperature, the crystals were filtered and washed with acetone. The crystals, 5.8 g (50%), were recrystallized from methanol and acetone, yield 5.5 g (47.6%), mp 198°C, $[\alpha]_D^{25}$ +2.1° (*c* 5.15, MeOH). The melting point and the specific rotation did not change after further purification.

Found: C, 62.46; H, 7.33; N, 6.18%. Calcd for $C_{24}H_{34}N_2O_7$: C, 62.32; H, 7.41; N, 6.06%.

(+)-*trans*-2,3-Epoxysuccinic Acid (III). Ephedrine salt, 11.0 g (0.024 mol), was dissolved in 10 m*l* of water. The solution was applied to a column of Dowex 50×2 (H-form, 100—200 mesh, 2 cm×40 cm) and washed with water until the effluent became neutral. The effluent was evaporated to dryness under reduced pressure. Crude (+)-epoxy acid, 3.0 g (97.5%) was obtained. This was recrystallized from dioxane and *n*-hexane. (+)-Epoxy acid-dioxane adduct was obtained. This adduct lost dioxane easily at room temperature and thus free (+)-epoxy acid (III) was obtained, mp 188°C, $[\alpha]_D^{25}$ +117.8° (*c* 2.6, EtOH).

Found: C, 36.33; H, 3.14%. Calcd for $C_4H_4O_5$: C, 36.38; H, 3.05%.

A resolution procedure for the above compound using morphine as the resolving agent may be found in R. Kuhn and F. Ebel, Chem. Ber., 58, 919 (1925).

```
      OH Cl
      |  |
HOOC-CH-CH-COOH
```

$C_4H_5ClO_5$ Ref. 28

Die auf 65° erwärmte Lösung von 19.5 g Chlor-äpfelsäure (Schmp. 145°) in 150 ccm Wasser wurde mit einer auf 60° erwärmten Lösung von 30.4 g wasserfreiem Brucin in 225 ccm Alkohol + 750 ccm Wasser versetzt. Das nach einiger Zeit auskrystallisierte Brucin-Salz wog nach dem Trocknen über $CaCl_2$ 33 g und schmolz bei 190° unt. Zers. Aus dem Filtrat schieden sich beim Stehen über Nacht noch weitere 3.5 g ab.

Zur Gewinnung der aktiven Chlor-äpfelsäure aus dem Brucin-Salz eigneten sich das Schütteln des Salzes mit feuchtem, HCl-haltigem Äther, sowie die Zersetzung durch Kaliumbisulfat[6]). Letztere Methode lieferte die (+)-Chlor-äpfelsäure I in sehr reiner Form. Aus 31.3 g Brucin-Salz (bei 110° getrocknet) erhielten wir 4.8 g (+)-Chlor-äpfelsäure. Aus den Mutterlaugen schieden sich beim Stehen über Nacht noch weitere 3.5 g ab. Zur Reinigung wurde auf Ton abgepreßt, in Äther (absol.) gelöst und bis zur eben beginnenden Trübung mit Benzin versetzt. Ausbeute 2.5 g vom Zers.-Pkt. 166—167°.

$[\alpha]_D^{19}$ in Wasser = + 0.48° × 100/3.730 × 1.894 = + 6.80° (c = 3.73);
$[\alpha]_D^{19}$ in Wasser = + 0.25° × 100/1.865 × 1.894 = + 7.07° (c = 1.87).

0.2096 g (+)-Chlor-äpfelsäure wurden in 20 ccm Wasser gelöst und mit $n/_1$-NaOH auf 25 ccm gebracht. Nach 3-stdg. Stehen wurden 20 ccm der Lösung angesäuert und mit $n/_{10}$-$AgNO_3$ nach Volhard titriert.

0.1677 g Sbst.: 9.70 ccm $n/_{10}$-$AgNO_3$. — $C_4H_5O_5Cl$ (168.5). Ber. Cl 21.07. Gef. Cl 20.50.

6) K. Freudenberg, B. **47**, 2027 [1914], u. zw. S. 2035.

```
        Cl
        |
HOOC-CH-CH-COOH
        |
        OH
```

$C_4H_5ClO_5$ Ref. 29

Für die Spaltung sind Cinchonin, Cinchonidin und Brucin in wäßriger und alkoholischer Lösung nicht geeignet. Aus 96-proz. Alkohol fällt das Strychnin-Salz der (+)Säure als Gallerte zuerst aus. Besser ist es, die primären Morphin-Salze zu fraktionieren.

9.0 g wasser-freies Morphin (1/2 Mol.) werden bei 70° in 400 ccm 96-proz. Alkohol gelöst und auf einmal mit einer warmen Lösung von 10 g *d,l*-Chlor-äpfelsäure II (Schmp. 151°) in 20 ccm Alkohol versetzt. Nach etwa 1/2 Min. fällt, noch in der Wärme, das Morphin-Salz der Rechtssäure in harten, weißen Kryställchen aus. Dreimal mit viel Alkohol ausgekocht, beträgt das Drehungsvermögen:

$[\alpha]_D^{20}$ = (— 1.99° × 100) : (1.37 × 2) = — 72.6° (in Wasser).

13.5 g feingepulvertes Morphin-Salz rührt man in kleinen Anteilen in einen mit wenig Wasser bereiteten Brei von etwa 25 g Kaliumbisulfat[15]) ein und gibt nach etwa 1/2 Stde. so viel entwässertes Natriumsulfat zu, bis der Brei steif ist. Nach dem Trocknen über Phosphorpentoxyd wird fein gemahlen und im Soxhlet unter Ausschluß von Feuchtigkeit mit Äther extrahiert. Man gewinnt 3 g rohe Säure, die durch fraktioniertes Lösen in reinem Äther und durch fraktioniertes Fällen der ätherischen Lösungen mit Petroläther (30—50°) gereinigt wird. Bei der Fraktionierung geht beigemengte Racemsäure zuerst in den Äther. In anderen Solvenzien ist die Löslichkeit für *d,l*- und Rechtssäure sehr ähnlich. Aus Essigester erhält man sehr schöne, regelmäßige, vierseitige Pyramiden. Schmp. 166—167° (unkorr., unt. Zers.).

10.869 mg Sbst. verbrauchen 6.688 ccm 0.009734-*n*. Silbernitrat nach Volhard.
$C_4H_5O_5Cl$ (168.5). Ber. Cl 21.07. Gef. Cl 21.26.

$[\alpha]_D^{20}$ = (+ 0.23° × 100) : (2.46 × 1) = + 9.35 ± 1.5° (in Wasser).

Bei 40-stdg. Kochen der wäßrigen Lösung verschwindet das Drehungsvermögen gänzlich. Nach dem Einengen fällt auf Zusatz von Calciumchlorid und Ammoniak *meso*-weinsaures Calcium in den für diese Substanz charakteristischen, regelmäßigen, nahezu quadratischen Krystallen aus.

0.1946 g Sbst. (lufttrocken): 0.1090 g $CaSO_4$.
$C_4H_4O_6Ca + 3 H_2O$. Ber. Ca 16.51. Gef. Ca 16.49.

15) K. Freudenberg, B. **47**, 2027, und zwar S. 2035 [1914].

(—)Chlor-äpfelsäure II[16]).

Die alkohol. Mutterlauge des Morphin-Salzes hinterläßt beim Eindampfen im Vakuum einen Sirup, der in der angegebenen Weise mit Kaliumbisulfat zerlegt und wie bei der Rechtssäure weiterverarbeitet wird. Nach wiederholter Fraktionierung mit Äther-Petroläther erhält man 2 g reine (—)Chlor-äpfelsäure II (aus 5 g Rohprodukt), die wie der Antipode bei 166—167° (unkorr., unt. Zers.) schmilzt und im Gemisch mit der gleichen Menge des Antipoden den von R. Kuhn und F. Ebel angegebenen Schmp. 151° (unkorr.) der *d,l*-Chlor-äpfelsäure II zeigt.

[16]) **Die isomeren Chlor-äpfelsäuren II sind nur bis zur Konstanz des Schmelzpunktes gereiuigt.**

CCl_3CHCH_2COOH (with OH on the CH)

$C_4H_5Cl_3O_3$ Ref. 30

The resolution of the acid proceeds very smoothly, and the following is a description of a typical preparation. Ninety-one grams of quinine (1 mol.) were dissolved by heating with 600 c.c. of ethyl alcohol, and 50 grams of the anhydrous acid (1 mol.) were added. Solution of the acid took place with readiness. On cooling, crystallisation started quickly, and after one day in the ice-chest, the resulting solid was crystallised from 1200 c.c. of ethyl alcohol. The quinine *l*-salt (59 grams) separated in glassy prisms, which melt with decomposition at about 208°. At 25·9°, 100 c.c. of its ethyl-alcoholic solution contain 0·44 gram of salt. After the salt was decomposed by dilute sulphuric acid, the chloro-acid was extracted with ether. The yield amounted to 19 grams, and after two crystallisations from benzene the acid (16·5 grams) was pure.

In another resolution, where 50 grams of the *r*-acid were again employed, the quinine salt was crystallised from alcohol five times, and the resulting acid crystallised three times from benzene. The optical activity of the acid was the same as before, and therefore the prolonged crystallisation of the quinine salt is not necessary.

l-γγγ-*Trichloro*-β-*hydroxybutyric acid* melts at 104—105°. It is readily soluble in ethyl alcohol and in acetone at the ordinary temperature. It is readily soluble in hot benzene, and sparingly soluble in the cold solvent, from which it separates in glassy, hexagonal prisms. It is sparingly soluble in cold toluene, chloroform, or light petroleum (b. p. 40—60°). It is much more soluble in water than is the inactive isomeride, and it differs from the latter in not crystallising with water of crystallisation (Found: C = 23·2; H = 2·3. Calc., C = 23·1; H = 2·4 per cent.). Its specific rotation was determined as follows:

In ethyl alcohol:

$l = 2$, $c = 1·5528$, $\alpha_D^{17°} = -0·92°$, whence $[\alpha]_D^{17°} - 29·6°$.
$l = 2$, $c = 4·0004$, $\alpha_D^{15·5°} = -2·41°$, whence $[\alpha]_D^{15·5°} - 30·1°$.
$l = 2$, $c = 4·0004$, $\alpha_{5461}^{15°} = -2·77°$, whence $[\alpha]_{5461}^{15°} - 34·6°$.

In acetone:

$l = 2$, $c = 1·553$, $\alpha_D^{16·5°} = -0·70°$, whence $[\alpha]_D^{16·5°} - 22·5°$.
$l = 2$, $c = 1·553$, $\alpha_{5461}^{16·5°} = -0·78°$, whence $[\alpha]_{5461}^{16·5°} - 25·1°$.

In the preceding resolution, the first filtrate, after the removal of the crystals containing a preponderance of the *l*B*l*A salt, gave on partial evaporation 26 grams of crystals in which the *l*B*d*A salt was in excess. The acid from this amounted to 8·5 grams which, when crystallised three times from benzene, gave 5 grams of the pure *d*-acid (Found: Cl = 51·4. Calc., Cl = 51·3 per cent.). d-γγγ-*Trichloro*-β-*hydroxybutyric acid* gave values for its specific rotation in agreement with those for the *l*-isomeride.

In ethyl alcohol:

$l = 2$, $c = 1·551$, $\alpha_D^{15·5°} = +0·87°$, whence $[\alpha]_D^{15·5°} + 28°$.
$l = 2$, $c = 1·551$, $\alpha_{5461}^{15·5°} = +1·08°$, whence $[\alpha]_{5461}^{15·5°} + 34·8°$.

In acetone:

$l = 2$, $c = 1·5528$, $\alpha_D^{15°} = +0·71°$, whence $[\alpha]_D^{15°} + 22·9°$.
$l = 2$, $c = 1·5528$, $\alpha_{5461}^{15°} = +0·81°$, whence $[\alpha]_{5461}^{15°} + 26·1°$.

Note:

J.D.M. Ross, J. Chem. Soc., 718 (1936) using quinine as resolving agent obtained active acids with $[\alpha]_D^{17}$ -32.0° (c 1, ethanol) and $[\alpha]_D^{17}$ +32.0° (c 1, ethanol).

$C_4H_5Cl_3O_3$ Ref. 30a

4) *Racematspaltung von γ.γ.γ-Trichlor-β-hydroxy-buttersäure*[7]): Durch fraktionierte Kristallisation eines aus 41.4 g (200 mMol) *d,l-γ.γ.γ-Trichlor-β-hydroxy-buttersäure* (Schmp. 116°) und 24.2 g (200 mMol) *(+)-α-Phenyl-äthylamin* ($[\alpha]_{578}^{21}$: +32.5°) dargestellten Salzes aus Äthanol erhält man in den ersten Fraktionen 25 g (76.3%) negativ drehende Substanz (Schmp. 185—188°, $[\alpha]_{578}^{25}$: −2.5°, Wasser, $c = 1$), die nach Behandlung mit Salzsäure und Ausäthern 12 g (57.8%) *(−)-γ.γ.γ-Trichlor-β-hydroxy-buttersäure*, Schmp. 116° (Benzol/Petroläther), $[\alpha]_{578}^{25}$: −28.9° (Äthanol, $c = 1$) ergibt. Aus der Äthanol-Mutterlauge läßt sich die (+)-drehende Form in 80-proz. optischer Reinheit ($[\alpha]_{578}^{21}$: +25.5°) isolieren.

[7]) Vgl. a. *A. McKenzie* und *H. J. Plenderleith*, J. chem. Soc. [London] **123**, 1093 (1923).

```
     CF
      | 3
CH -C-COOH
  3   |
     OH
```

$C_4H_5F_3O_3$ Ref. 31

Optical Resolution of α-Hydroxy-α-trifluoromethylpropionic Acid.—A solution of the racemic fluoro-acid (5·28 g.) in methyl alcohol was mixed with a solution of brucine (15·59 g.) in methyl alcohol (total volume 350 c.c.), and the mixture was heated to boiling, filtered, and set aside until crystallisation appeared to be complete (3 days). This brucine salt was recrystallised (five times) from methanol until the rotation was constant. The product obtained (1·17 g.) consisted of large needles which showed $[\alpha]_D^{20}$ −30·3° (*c*, 1·0 in water). A portion of this salt (0·583 g.) was treated with sodium hydroxide solution, the precipitated brucine was removed by filtration, the solution was acidified, and the product was extracted with ether and purified as already described. The *lævo*-acid thus obtained (0·124 g.) had $[\alpha]_D^{16}$ −11·3° (*c*, 5·0 in water).

The mother-liquors from which the original precipitate was obtained were concentrated, and the more soluble isomer of the brucine salt was purified by systematic fractional crystallisation until it showed a constant negative rotation. The pure diastereoisomer (0·434 g.) thus produced had $[\alpha]_D^{20}$ −17·4° (*c*, 1·0 in water). Treatment of this brucine salt as above gave the *dextro*-form of the acid (0·092 g.), which showed $[\alpha]_D^{20}$ +11·4° (*c*, 3·7 in water).

```
        CH3
         |
ClCH2-C-COOH
         |
        Cl
```

$C_4H_6Cl_2O_2$ Ref. 32

Quinine Salt of I. This salt was prepared by mixing a hot saturated solution of quinine in ethyl acetate with a hot solution of 2,3-dichloro-2-methylpropionic acid in ethyl acetate. Upon cooling, a 95% yield of quinine 2,3-dichloro-2-methylpropionate was obtained. This salt had an $[\alpha]_D^{30}$ −137° (ethyl alcohol) and contained the acid and alkaloid in a 1 to 1 ratio. Found: Cl, 14.74. Calculated for $C_{24}H_{30}Cl_2N_2O_4$: Cl, 14.73. Three recrystallizations raised the specific rotation to a constant $[\alpha]_D^{30}$ −140° (ethyl alcohol).

Cinchonine Salt of I. Cinchonine was added to a hot solution of 2,3-dichloro-2-methylpropionic acid in carbon tetrachloride, followed by sufficient carbon tetrachloride to accomplish solution while the mixture was heated at gentle reflux. Crystallization occurred within 24 hours after cooling. Cinchonine 2,3-dichloro-2-methylpropionate (found: Cl, 23.08; calculated for $C_{27}H_{34}Cl_4N_2O_5$: Cl, 23.3) melted at 134°–5° C. and had $[\alpha]_D^{30}$ −98° (ethyl alcohol). Two recrystallizations from a 1 to 7 acetone–isopropyl ether mixture gave a salt $[\alpha]_D^{30}$ −141° (ethyl alcohol). Further recrystallization did not alter the rotation.

***d*-α-Methylbenzylamine Salt of I.** *d*-α-Methylbenzylamine was obtained according to the method of Ingersoll (*2*). 2,3-Dichloro-2-methylpropionic acid was added to the amine in hexane to form a salt [m.p. 124–5° C., $[\alpha]_D^{34}$ −11.4° (ethyl alcohol)]. Found: Cl, 25.51. Calculated for $C_{12}H_{17}Cl_2NO_2$: Cl, 25.48. This salt was recrystallized twice from isopropyl ether and three times from a 1 to 8 mixture of acetone and isopropyl ether. Final rotation was $[\alpha]_D^{30}$ −3° (ethyl alcohol).

Note:

A $[\alpha]_D^{27}$ = -15° (methanol) was found for the acid obtained from each of the above salts.

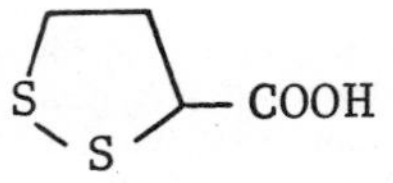

$C_4H_6O_2S_2$ Ref. 33

Preliminary tests on resolution of 1,2-dithiolane-3-carboxylic acid

Racemic acid (0.001 mole) and optically active base (0.001 mole) were dissolved at room temperature with protection against light. If crystals were not formed after several days in a refrigerator, some of the solvent was evaporated. From the crystals formed, the acid was liberated and the optical activity measured in 96% ethanol. The results are given in Table 1.

Table 1.

Base	Solvent	% Salt obtained	$[\alpha]_D^{25}$
Cinchonidine	96 % Ethanol	13	0°
	90 % Ethanol	8	0°
	70 % Ethanol	50	+73°
	Acetone	38	0°
	Methanol	Oil	—
	75 % Methanol	11	0°
	50 % Methanol	30	+23°
	Water	Oil	—
	Chloroform	35	0°
	Benzene	53	+ 5°
	Ethyl acetate	40	+ 7°
Morphine	96 % Ethanol	12	−82°
	Benzene	69	0°
	Ethyl acetate	55	0°
	Acetone	Oil	—
Strychnine	96 % Ethanol	9	+30°
Brucine	96 % Ethanol	38	+ 5°
Cinchonine	96 % Ethanol	Oil	—
Quinine	Benzene	70	0°
	Acetone	87	0°
	Ethyl acetate	75	0°

Optical resolution of 1,2-dithiolane-3-carboxylic acid

Fifteen grams (0.1 mole) of racemic 1,2-dithiolane-3-carboxylic acid in 100 ml of ethanol was added with cooling and mechanical stirring to a suspension of 29 g (0.1 mole) of cinchonidine in 200 ml of 50% ethanol. The cinchonidine dissolved when all of the carboxylic acid solution had been added. Some polymeric disulphide had then precipitated and was removed by filtration. After standing overnight in the refrigerator, 24 g of crystalline salt (m.p. 180–182°) was obtained. Several recrystallizations from about 70% ethanol gave finally 6 g of pure salt. At every recrystallization some polymer was also formed. The acid liberated from the salt had $[\alpha]_D^{25} = +337°$, and further recrystallization did not raise the optical rotation.

(+)-*1,2-Dithiolane-3-carboxylic acid*. One gram of the cinchonidine salt of (+)-1,2-dithiolane-3-carboxylic acid was added to 20 ml of 50% ethanol at 0°. The resulting suspension was stirred and cooled in an ice-bath while 15 ml of hydrochloric acid (1:2) was slowly added. The acidic solution was saturated with potassium chloride and extracted with 6 × 25 ml of methylene chloride. The combined extracts were dried over magnesium sulphate, the solvent removed *in vacuo*, and the residue dissolved in 5 ml of benzene. Polymers were removed by filtration, and the benzene solution was concentrated *in vacuo* to about 1 ml, and 5 ml of cyclohexane was added. From this solvent mixture 35 mg of (+)-1,2-dithiolane-3-carboxylic acid crystallized, m.p. 68–70°, $[\alpha]_D^{25} = +337°$ (c = 0.5, 96% ethanol), equiv. wt. 150.8 (calc. 150.2).

(−)-*1,2-Dithiolane-3-carboxylic acid*. The levorotatory acid was isolated from the mother liquor of the first crystallization of the cinchonidine salt. On recrystallization of the acid from benzene–cyclohexane, two different types of crystals formed, one consisting of pale yellow prisms of the same appearance as those of the racemic acid, and the other consisting of bright yellow needles. These two types of crystals were separated manually. The pale yellow prisms showed very low levorotation, and the bright yellow needles showed levorotation of the same magnitude as the dextrorotatory antipode, $[\alpha]_D^{25} = -313°$ (c = 0.5, 96% ethanol), m.p. 68–70°, equiv. wt. 151.0 (calc. 150.2).

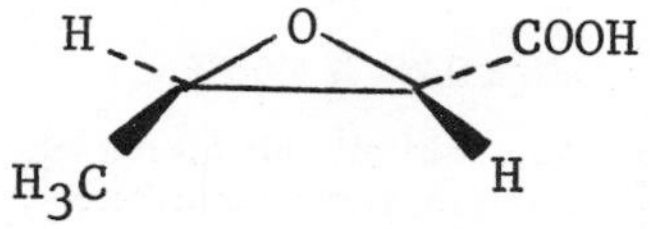

$C_4H_6O_3$ Ref. 34

Brucine Salt of (+)-*trans*-2,3-Epoxybutyric Acid (II).—Epoxy acid (I), 20.4 g. (0.2 mol.), and brucine, 78.8 g. (0.2 mol.), were dissolved in 200 ml. of hot methanol. The solution was filtered and the solvent was evaporated in vacuo. The residual syrup was dissolved in a hot mixture of methanol (20 ml.) and acetone (180 ml.). After cooling, crystallization of brucine salt began by seeding with an authentic specimen which was obtained in another crystallization experiment. After standing overnight at room temperature, the crystals were filtered and were washed with cold acetone. The crystals, 35.0 g., were recrystallized from a hot mixture of methanol (5 ml.) and acetone (95 ml.). Yield, 32.0 g., m. p. 173°C, $[\alpha]_D^{25} = -26.0°$ (*c* 2.1, H_2O).

(−)-*trans*-2,3-Epoxybutyric Acid (III).—Brucine salt (II), 14.3 g. (0.03 mole), was dissolved in 100 ml. of water. To this solution, 1 N sodium hydroxide solution, 33 ml., was added. The precipitated brucine was separated by filtration. The filtrate was extracted with benzene three times to remove the residual brucine. Concentrated hydrochloric acid was used to bring the pH of the aqueous solution to 2.5. After saturation of the solution with sodium chloride, the epoxy acid was extracted with ether five times (100 ml. × 5). The ether solution was dried with anhydrous sodium sulfate. After evaporation of ether, 1.80 g. (60%) of crude III was obtained. This was recrystallized from benzene. 1.20 g. (40%), m. p. 61°C. $[\alpha]_D^{25} = -82.5°$ (*c* 0.59, benzene). Neut. equiv. Found: 102. Calcd. for $C_4H_6O_3$.

$C_4H_6O_3$ Ref. 34a

Resolution of (±)-trans-2,3-*Epoxybutyric Acid* (9a).—(−)-(2*R*,3*S*)-*trans*-2,3-Epoxybutyric acid was obtained by recrystallization of racemic (9a) with brucine according to a published procedure.[10] The recovered (−)-acid (9a) was more easily purified by sublimation at 60—70° and 1 mmHg.

The enantiomer, (+)-(2*S*,3*R*)-*trans*-2,3-epoxybutyric acid (17a), was obtained by the following procedure. Racemic epoxy-acid (9a) (30·2 g; sublimed) and quinine (103 g, 1 equiv.) were dissolved in acetone (400 ml). The crystals which separated after 5 days at 20° were filtered off and recrystallized twice from acetone to constant m.p. 160—163° (decomp.), $[\alpha]_D^{20}$ −135° (*c* 1·05, H_2O) (yield 28 g). This material was dissolved in water (200 ml) and cooled to 10°, then treated with 2N-NaOH (37 ml). The resultant mixture was extracted with ether (3 × 200 ml). The aqueous phase was acidified to pH 7 with HCl, and saturated with NaCl, then further acidified to pH 2·2, and extracted rapidly with ether (10 × 100 ml). The extract was dried (Na_2SO_4) and evaporated, giving crude (17a) which was sublimed at 65° and 1 mmHg (yield 2·83 g), m.p. 62°, $[\alpha]_D^{20}$ +79° (*c* 0·6, benzene).

[10] K. Harada and J. Oh-hashi, *Bull. Chem. Soc. Japan*, **1966**, 39, 2311.

$C_4H_6O_3$ Ref. 34b

Note:

Resolved with (-)-1-α-naphthylethylamine in absolute ethanol. The resolved salt was found to have $[\alpha]_D^{22}$ +11.8° (CH_3OH). The acid was recovered quantitatively from the salt by treatment with 1.0 equiv. of methanesulfonic acid in ether, washing with a small amount of saturated Na_2SO_4 and evaporation of the ether. The acid obtained had a m.p. 61° and $[\alpha]_D^{22}$ +82.1° (benzene).

SH
|
HOOC-CH-CH-COOH
|
SH

$C_4H_6O_4S_2$ Ref. 35

Preliminary experiments on resolution. Gerecke *et al.* [8] have resolved this acid and found that the (+)-form can be obtained from its brucine salt in methanol by further recrystallization from water. The yield was however very low (ca. 2%) and bearing in mind the cost of the acid and its intended use for further syntheses, a reinvestigation was undertaken in hope of finding an improved method. (−)-Acid of good activity was obtained from its quinine and strychnine salts. An attempt on a larger scale with quinine in dil. ethanol was undertaken and two recrystallizations gave an acid of activity $[\alpha]_D^{25} = -128°$ (abs. ethanol). During the recrystallizations however, a distinct smell of hydrogen sulphide was noticed together with darkening of the solution. It was clear that the acid could not be resolved in this way into its optical antipodes in acceptable yield, as it could not stand heating with the alkaloids. Therefore, a method of partial resolution was followed, without recrystallizations. Brucine in dil. acetone gave the best results.

Partially active (+)-*acid*. Brucine (43 g, 0.109 mol) was dissolved in 250 ml boiling acetone. (±)-Acid (20 g, 0.110 mol) in 500 ml hot water (60°) was added and the solution placed at room temperature. Crystals began to separate almost immediately and after 2 hours, 24 g salt were filtered off. The salt was treated with 2-M sulphuric acid at once and the liberated acid taken up in ether which was dried and evaporated. The acid was recrystallized from ether-benzene (1:3) or ethyl acetate–petroleum ether and yielded 5.0 g, m.p. 123–125°, $[\alpha]_D^{25} = +75°$ (abs. ethanol).

Partially active dimethyl (+)-*ester* was prepared from (+)-I, $[\alpha]_D^{25} = +75°$ (abs. ethanol), in the same way as the (±)-ester above, b.p. 145–147°/9 mm, n_D^{25} 1.5130, $[\alpha]_D^{25} = +20.3°$ (methanol).

Partially active (−)-*acid*. The acetone in the mother liquor from the (+)-acid was immediately evaporated. The resulting aqueous solution was acidified and extracted with ether. Further treatment as above gave 6 g of (−)-acid with m.p. 123–125° and $[\alpha]_D^{25} = -43°$ (abs. ethanol).

8. Gerecke, M., Friedheim, E. A. H., and Brossi, A., ibid. *44*, 955 (1961).

$C_4H_6O_4S_2$ Ref. 35a

3. Spaltung von 2,3-Dimercaptobernsteinsäure (IIIb) in optische Antipoden. – (+)-*2,3-Dimercaptobernsteinsäure* (*IIIb*): 60 g der *rac.*-Dicarbonsäure IIIb werden in 2 l Methanol gelöst und zu einer Lösung von 200 g Brucin in 2 l Methanol gegeben. Nach kurzem Stehenlassen wird filtriert. Man erhält 260 g eines Brucinsalzes von IIIb vom Smp. 156–158° (unter vorherigem Sintern); $[\alpha]_D^{22} = -56°$ ($c = 1$ in Dimethylformamid). 130 g davon werden in 4,3 l heissem Wasser gelöst. Nach 2-stdg. Stehen wird filtriert. Es kristallisieren 43 g eines Salzes, das nach nochmaligem Umlösen aus 1,5 l Wasser und 6-stdg. Stehen 22 g des Brucinsalzes der (+)-drehenden Dicarbonsäure IIIb liefert. Smp. 153–155°; $[\alpha]_D^{22} = -53° \pm 2°$ ($c = 1$ in Dimethylformamid).

$C_4H_6O_4S_2$, 2 $C_{23}H_{26}N_2O_4$, 4 H_2O (1043,18) Ber. N 5,37 S 6,15% Gef. N 5,29 S 6,20%

Zur Gewinnung der freien Säure suspendiert man 21 g des obigen Brucinsalzes in 50 ml Wasser, überschichtet mit Äther und versetzt mit 50 ml 3 N Salzsäure. Die Extraktion mit Äther wird dreimal wiederholt. Die vereinigten Ätherextrakte geben nach dem Digerieren mit Benzol 3,1 g einer in Äther (+)-drehenden Säure. Nach dem Auskochen mit Benzol und Umlösen der in Benzol unlöslichen Anteile aus Essigester-Petroläther erhält man 500 mg (+)-2,3-Dimercaptobernsteinsäure (IIIb) als Prismen vom Smp. 124–125°, $[\alpha]_D^{22} = +128° \pm 2°$ ($c = 1$ in Äther). Eine aus Benzol umgelöste Probe kristallisiert in Nadeln, ohne dass sich Smp. und Drehung ändern. Das IR.-Spektrum zeigt die SH-Bande bei 3,92 μ und die CO-Bande bei 5,90 μ.

$C_4H_6O_4S_2$ (182,2) Ber. C 26,38 H 3,32 S 35,19% Gef. C 26,59 H 3,35 S 35,00%

(−)-*2,3-Dimercaptobernsteinsäure* (*IIIb*): Zu einer Lösung von 21,5 g Brucin in 300 ml Methanol gibt man eine Lösung von 13,3 g der Dicarbonsäure Ib in 300 ml 90-proz. wässerigem Methanol. Nach 20-minutigem Stehen wird filtriert. Man erhält 12 g eines Brucinsalzes vom Smp. 129–131°, $[\alpha]_D^{22} = -59° \pm 2°$ ($c = 1$ in Dimethylformamid).

$C_8H_{10}O_6S_2$, $C_{23}H_{26}N_2O_4$, 2 H_2O (696,8) Ber. S 9,20 N 4,02% Gef. S 9,26 N 4,08%

Zur Isolierung der freien Säure löst man in 50 ml Chloroform und schüttelt mit 50 ml 3 N Natronlauge. Die Verseifung der Acetylthio-Gruppierungen macht sich durch Temperatursteigerung bemerkbar. Man trennt die natronalkalische Lösung ab, wäscht mit Äther, stellt mit Salzsäure kongosauer und nimmt in Äther auf. Den nach dem Einengen erhaltenen Rückstand kocht man mit Benzol aus. Die unlöslichen Teile werden filtriert. Aus der Benzollösung kristallisieren 150 mg (−)-drehende Dicarbonsäure IIIb, die nach dem Umlösen aus Essigester-Petroläther 60 mg opt. reine (−)-2,3-Dimercaptobernsteinsäure (IIIb) vom Smp. 124–125° in Form feiner Nadeln geben, die sich aber nach einigem Stehen in Prismen umlagern. $[\alpha]_D^{22} = -128° \pm 2°$ ($c = 1$ in Äther). Das IR.-Spektrum ist mit demjenigen von (+)-IIIb identisch, aber verschieden von *rac.*-IIIb.

$C_4H_6O_4S_2$ (182,2) Ber. C 26,38 H 3,32 S 35,19% Gef. C 26,54 H 3,37 S 35,08%

HOOCCHCH$_2$COOH (with OH on the CH)

$C_4H_6O_5$ Ref. 36

Equimolecular proportions of *l*-malic acid and quinine were taken in ethyl-alcoholic solution and the salt which separated was crystallised twice from ethyl alcohol, in which it is very sparingly soluble at the ordinary temperature. It melted at 177—179° with decomposition (Found: C = 63·0; H = 6·3. $C_{20}H_{24}O_2N_2,C_4H_6O_5$ requires C = 62·9; H = 6·6 per cent.). When the acid and alkaloid were employed in the proportions of 1 mol. of acid to 2 mols. of quinine with ethyl alcohol as solvent, the resulting salt melted at 186—187° [Found: C = 67·3; H = 7·0. $(C_{20}H_{24}O_2N_2)_2,C_4H_6O_5$ requires C = 67·5; H = 7·0 per cent.]. Also, when the acid (1 mol.) and quinine (½ mol.) were taken with water as solvent, the salt, after crystallisation, melted at 186—187° [Found: C = 67·5; H = 7·0. $(C_{20}H_{24}O_2N_2)_2,C_4H_6O_5$ requires C = 67·5; H = 7·0 per cent.]. In each of the above cases, the salts were dried in a desiccator over sulphuric acid.

12·1 Grams of quinine (1 mol.) were dissolved in a solution of 5 grams of *r*-malic acid (1 mol.) in 75 c.c. of ethyl alcohol. The resulting crystals were crystallised three times from ethyl alcohol and then amounted to 6·2 grams. The barium salt prepared from this was inactive. This negative result was probably due to the formation of a partially racemic quinine salt. A similar result was obtained when the proportions employed were 1 mol. of acid to 2 mols. of quinine. Under the following conditions a resolution was, however, effected. 3·63 Grams of quinine (½ mol.) were added to a boiling solution of 3 grams of *r*-malic acid (1 mol.) in 30 c.c. of water. The crystals which separated on cooling were crystallised from water. Yield = 2 grams. The salt was decomposed by the calculated amount of potassium hydroxide, and the precipitated quinine was removed. The filtrate was extracted three times with chloroform, then neutralised with hydrochloric acid, and acted on by the requisite amount of barium chloride. The salt precipitated was removed and washed free from chloride, the residue proving to be pure barium *d*-malate (0·4 gram), as was shown as follows: 0·2362 Gram was decomposed by 1·55 c.c. of dilute hydrochloric acid containing the calculated amount of acid. 3·85 C.c. of potassium hydroxide (0·3251*N*) were added, and then 6·04 c.c. of a standard solution of uranium nitrate (32·936 grams of the hydrated salt made up to 1 litre with water). The solution, when made up to 25 c.c. with water, gave $\alpha_D + 2{\cdot}11°$ ($l = 2$), whereas pure barium *l*-malate gave under similar conditions $\alpha_D - 2{\cdot}12°$.

$C_4H_6O_5$ Ref. 36a

l-Malic Acid from Its α-Phenylethylamine Salt

From 5.0 g of the α-phenylethylamine salt of *l*-malic acid was obtained 2.50 g (theory 2.66 g) of the pure acid, $[\alpha]_D^{25}$ −2.24° (*c*, 20 in H_2O) (lit. $[\alpha]$ −2.4°). Elution with hydrochloric acid gave 2.45 g (theory 3.0 g) of α-phenylethylamine hydrochloride.

Note:

For a description of the experimental details, see Ref. 54a.

$C_4H_6O_5$ Ref. 36b

Cinchonine l-Malate.—Cinchonine (29.4 gm.) and inactive malic acid (13.4 gm.) in molecular proportions are dissolved by heating to boiling with 130 cc. of technical methyl alcohol. The solution was allowed to cool slowly and eventually placed in a cold place overnight. Crystallization takes place very readily in the form of opaque, rather soft masses of crystals which, under the microscope, appear as stout prisms. They are filtered off and washed with a little cold methyl alcohol. The yield of cinchonine *l*-malate varies from 19.0 to 19.7 gm., compared with a theoretical yield of 21.4. As already stated, acetone may be used equally as well as methyl alcohol as solvent. It is well to dissolve the malic acid first in boiling acetone and then add the powdered cinchonine. The *l*-malate separates at once.

The crude product is surprisingly free from impurities and on decomposition it gave *l*-malic acid, which on treatment with uranium acetate showed a specific rotation of −470° compared with −482° for the perfectly pure acid, under similar conditions. The dextro-rotation (+574°) observed after addition of ammonium molybdate was equal to that recorded by McKenzie and Plenderleith for the pure levo acid (+568°), using the conditions prescribed by these authors.

The complete purification of cinchonine *l*-malate may be effected either by crystallization from about 10 parts of boiling methyl alcohol or, even better, by crystallization from boiling water. The salt separates from the latter solvent in clear hard prisms. The salt, crystallized from either solvent, retains no water of crystallization and melts sharply at 197–198° (uncorrected). It dissolves readily in hot water, but requires about 50 parts of cold water (10°) for solution. Unlike the salt of the dextro acid, it is very sparingly soluble in acetone even when boiling. At 7° its solubility in cold acetone is about 0.465 per cent. It is practically insoluble in chloroform and benzene. Its specific rotation was observed in aqueous solution:

$$c = 2.0; \quad l = 2.2; \quad \alpha = +6.37^\circ$$

$$[\alpha]_D^{18} = +146^\circ$$

Cinchonine d-Malate.—The methyl alcoholic or acetone filtrate from the levo salt is concentrated on the water bath, adding water occasionally to replace the alcohol. After all the methyl alcohol or acetone has been removed the residue is dissolved in about 90 cc. of hot water and allowed to crystallize in a cool place. Crystallization of the *d*-malate takes place readily in the form of fine rosettes of clear silky prismatic needles. The salt, after filtering and washing with a little cold water, is practically pure and, after drying in warm air, weighs 22 gm. It retains 2 molecules of water of crystallization so that the yield of anhydrous salt is 20.2 gm.

In order to test the purity of the salt, some of it was converted into malic acid without further purification. On treatment with uranium acetate the acid showed $[\alpha]_D^{18} = -488^\circ$, indicating complete optical purity.

Cinchonine *d*-malate may be best recrystallized from hot water in which it is freely soluble. It is sparingly soluble in cold water, a saturated solution at 10° retaining about 1.7 per cent of the salt.

Its water of crystallization is retained with some tenacity until a temperature of about 95° is reached when it is completely lost.

0.9462 gm. lost 0.0742 gm. H_2O *in vacuo* at 95–97° = 7.84 per cent H_2O.
$C_{19}H_{22}N_2O \cdot C_4H_6O_5 \cdot 2H_2O$ requires 7.76 per cent H_2O.

It is extremely soluble in methyl alcohol and acetone, but sparingly soluble in chloroform and insoluble in benzene. The anhydrous salt may be crystallized from a mixture of acetone and chloroform, but water is a far better solvent for the purpose.

Cinchonine *d*-malate, unlike the levo salt, possesses no definite melting point. The salt, containing water of crystallization, begins to melt indefinitely around 106° while the anhydrous salt melts indefinitely to a turbid wax-like mass at 125–135°, but does not become perfectly clear until about 150°.

The rotation of the salt was observed in aqueous solution:

$$c = 2.31 \text{ (anhydrous salt)}; \quad l = 2.0; \quad \alpha = +7.06^\circ$$

$$[\alpha]_D^{18} = +153^\circ$$

d- and l-Malic Acids.—In order to recover the active malic acids from their cinchonine salts they were dissolved in hot water and the alkaloid was removed by precipitation with ammonia. In the earlier experiments the filtrate was precipitated with basic lead acetate and the lead malate in aqueous suspension decomposed with hydrogen sulfide. This method is by no means quantitative and it was eventually replaced by ether extraction. The ammonium malate solution was concentrated to a thin syrup, acidified with phosphoric acid, and then extracted with ether in an apparatus for rapid continuous extraction. The whole of the acid is readily recovered by 24 hours extraction. The ether residue is best dissolved in a little hot water and concentrated in a desiccator. Crystallization is readily effected. Both forms melted at 99–100°.

The optical rotation in water was observed in 7 per cent solution. The values obtained, $[\alpha]_D^{18} = +2.33^\circ$ and $[\alpha]_D^{18} = -2.31^\circ$, agree closely with previous observations.

The enhanced rotation induced by uranium salts was observed under the following conditions. The active acid (about 0.1 gm.) was exactly neutralized with sodium hydroxide, and then 1 drop of acetic acid together with 0.5 gm. of powdered uranium acetate. The whole was diluted to 20 cc. and allowed to stand 1 hour.

Dextro Acid. $c = 0.5; \quad l = 2.2; \quad \alpha = +5.32^\circ$

$$[\alpha]_D^{18} = +483^\circ$$

Levo Acid. $c = 0.5673; \quad l = 2.2; \quad \alpha = -6.02^\circ$

$$[\alpha]_D^{18} = -482^\circ$$

McKenzie, A., and Plenderleith, H. J., *J. Chem. Soc.*, 1923, cxxiii, 1090.

$C_4H_6O_5$ Ref. 36c

Note:

Using N-benzyl-2-(hydroxymethyl)cyclohexylamine as resolving agent, obtained a less soluble diastereomeric salt after 2 recrystallizations from acetone-ethanol (2:1), having $[\alpha]_{589}^{25}$ -12.3 (c 1, ethanol). (Melting point of authentic material is 153-155°.) The acid obtained from this salt had $[\alpha]_{589}^{25}$ +8.8° (c 1, acetone). Authentic more soluble salt had a m.p. 125-127°, $[\alpha]_{589}^{25}$ +6.2° (c 1, ethanol).

$$\text{HOOC-CH(OH)-CH(OH)-COOH}$$

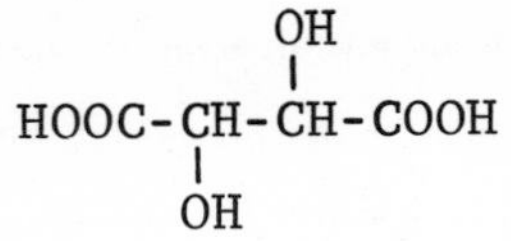

$C_4H_6O_6$ Ref. 37

Resolution of *dl*-Tartaric Acid by *d*-Quinotoxine.—The resolution of *dl*-tartaric acid was readily effected through the *d*-quinotoxine *d*-tartrate hexahydrate described above; it is highly probable that this was the salt with which Pasteur worked in 1853.[48]

The pure (natural) *d*-quinotoxine regenerated from 2.00 g. of the pure anhydrous *d*-tartrate, m. p. 152–155°, was dissolved in a small quantity of benzene and treated with 0.708 g. of *dl*-tartaric acid (monohydrate) and 1.00 cc. of water. The solution was then heated to drive off the benzene, cooled, and seeded with the hexahydrate, m. p. 55–63° (seeding was not necessary to induce crystallization). The yellow needles of the salt which separated were filtered by centrifuging, and twice recrystallized from 1.2 cc. of water. The pure *d*-quinotoxine *d*-tartrate hexahydrate (0.71 g., 58%) was obtained in characteristic stout yellow needles, m. p. 55–63°, which did not depress the melting point of an authentic sample. The salt was soluble in water at room temperature to the extent of *ca.* 0.05 g./cc. The sample from the resolution was converted on crystallization from absolute ethanol into the anhydrous salt, m. p. 151–154°, identical with an authentic sample.

(48) This in spite of the fact that *dl*-tartaric acid is smoothly resolvable by natural *d*-quinotoxine. It is a little known albeit historically an important fact, that this, and the similar resolution by cinchotoxine were the first examples [Pasteur, *Compt. rend.*, **37**, 162 (1853)] of the now universally used method of resolution of a racemic compound by combination with an active material, followed by separation of the resulting diastereomers. Confusion seems to have resulted from the fact that quinotoxine and cinchotoxine (which were discovered by Pasteur [ref. 6]) were in 1853 known as quinicine and cinchonicine, and the resemblance to cinchonidine has led some to believe that the latter alkaloid was the first resolving agent [*cf.* Lowry, "Optical Rotatory Power," Longmans, Green and Co., 1935, p. 34] while others have assumed that the well-known alkaloids quinine and cinchonine were used. Further, Pasteur gave no experimental details of his work; in fact the original note to the French Academy was largely concerned with other matters, and the discovery is mentioned only in passing, there, and later, in the course of a series of lectures summarizing his work on molecular asymmetry ["Researches on the Molecular Asymmetry of Natural Organic Products," (1860); "Alembic Club Reprints," No. 14, p. 41]. Further, the published information elsewhere with regard to the tartrates of quinotoxine is fragmentary at best (*cf.* ref. 7). Even Beilstein contains no reference to Pasteur's work). A description of our experiences with the above resolution and with the characterization of the quinotoxine tartrates will be found in the Experimental Part of this paper.

$C_4H_6O_6$ Ref. 37a

Resolution of Racemic Tartaric Acid.—Twenty grams of racemic tartaric acid[10] was dissolved in 150 ml. of hot water and 35.5 g. of 2-[D-*gluco*-D-*gulo-hepto*-hexahydroxyhexyl]-benzimidazole added with stirring. To the clear solution was added 50 ml. of ethanol and crystallization was readily induced by scratching. After standing a few hours at 5° the crystals were filtered and washed successively with cold 50, 75 and 95% ethanol. The acid L-tartrate salt thus obtained (28.8 g. or 100%) occurs as the dihydrate and may be recrystallized with negligible loss from 5 parts 25% ethanol if desired. Specific rotation[11] of the dihydrate is $-0.5°$ (H_2O, $l = 4$, $c = 0.84$) and m. p.

(10) The racemic acid was made from commercial tartaric acid according to the directions of Holleman, *Org. Syntheses*, **1**, 484 (1921).

(11) All specific rotations were taken at 20° with sodium light; l = length in dcm.; c = grams per 100 ml. of solution.

is 118–125° (corr.). The water of crystallization may be quantitatively removed at 78° *in vacuo*.

Anal. Calcd. for $C_{17}H_{24}O_{12}N_2 \cdot 2H_2O$: C, 42.12; H, 5.83; H_2O, 7.44. Found: C, 42.33; H, 5.70; H_2O, 7.49.

The L-tartaric acid was recovered by decomposing the acid salt with excess ammonium hydroxide and removing the base by filtration. The filtrate was made barely acid with acetic acid and lead L-tartrate was precipitated with lead acetate solution. The filtered and washed lead L-tartrate was suspended in water and the lead removed by precipitation with hydrogen sulfide. The clear filtrate was evaporated to a small volume and allowed to crystallize in a desiccator. Recovery of 96.5% of L-tartaric acid from the original acid salt was obtained; after one recrystallization from water, it showed a specific rotation of $-14.2°$ (H_2O, $l = 4$, $c = 4.05$) and a m. p. of 168–170° (corr.), which are the known values for pure L-tartaric acid.

Preparation of [2] Substituted Benzimidazoles

2 - [D - gluco - D - gulo - hepto - Hexahydroxyhexyl]-benzimidazole.—The preparation of this benzimidazole will be taken as an example. All others were prepared in an analogous manner, except the two cases noted in Table I where 2 moles of hydrochloric acid were added during evaporation to facilitate ring closure.

Fifty grams of D-*gluco*-D-*gulo*-heptonic lactone,[12] 26.6 g. of *o*-phenylenediamine (1:1.02 molecular ratio), and 50 ml. of water were heated on a steam-bath for seven hours with occasional addition of 25-ml. portions of water whenever the cake became dry. The resulting mass of crystals was treated with 100 ml. of ethanol and allowed to stand overnight. After filtering and washing with ethanol 57 g. of the crude product was obtained. One recrystallization from 25 parts of 50% ethanol yielded 53 g. (74%) of a product which gave a specific rotation of $+14.3°$ ($N/1$ HCl, $l = 1$, $c = 2.00$) and a decomposition point of 215° (corr.), these constants remaining unchanged upon further recrystallization.

(12) This lactone is readily prepared from D-glucose by the modification of the Kiliani cyanohydrin synthesis described by Hudson, Hartley and Purves, THIS JOURNAL, **56**, 1248 (1934).

Resolution procedures for the above compound using cinchonine and hydroxyhydrindamine may be found in: J. Read and W.G. Reid, J. Soc. Chem. Ind.,8T,(1928).

Resolution procedures for the above compound using 1) l-menthol as resolving agent may be found in H. Wren and K.H. Hughes, J. Chem. Soc., 125, 1739 (1924) and 2) l-borneol as resolving agent may be found in H. Wren et al, J. Chem. Soc., 117, 191 (1920).

A detailed procedure for the resolution of racemic tartaric acid which affords the levorotatory enantiomer with m.p. 168-170°, $[\alpha]_D^{20}$ -14° (c 4, H_2O) using 2-(D-glycero-D-gulo-hexahydroxyhexyl)-benzimidazole as resolving agent may be found in "Methods in Carbohydrate Chemistry," Vol. 2, pg. 49 [R. L. Whistler and M.L. Wolfrom, Eds.), Academic Press (1963).

```
HOOCCH2-CH-COOH
         |
        SO3H
```

$C_4H_6O_7S$ Ref. 38

V. Dédoublement optique de l'acide sulfosuccinique racémique.

Sulfosuccinate acide de strychnine.

En évaporant une solution de l'acide sulfosuccinique tribasique avec une molécule de strychnine, on obtient un sirop. Le sel primaire de strychnine ne semble donc pas cristalliser.

En dissolvant deux molécules de strychnine dans l'acide et en évaporant la solution, un sel cristallise; cependant, ce n'est pas le sel secondaire avec 2 molécules de strychnine, mais il renferme plus de strychnine.

Lorsqu'on veut préparer le sel tertiaire de strychnine, en chauffant au bain-marie une solution aqueuse de l'acide avec trois molécules de strychnine, il reste toujours environ $^1/_3$ molécule de strychnine qui ne se dissout pas.

La solution refroidie dépose des cristaux d'un sel acide, dont la composition ne varie que peu et est située entre celles des sels secondaire et tertiaire. D'après le titrage, qu'on peut effectuer en présence de phénolphtaléine, ce sel acide renferme environ 8 mol. de strychnine par 3 mol. d'acide. La composition ne change pas en faisant recristalliser le sel dans l'eau. L'analyse mentionnée ci-dessous se rapporte à un sel qu'on a fait recristalliser plusieurs fois (I).

Le même sel s'obtient en faisant cristalliser dans l'eau chaude le sel neutre (tertiaire) de strychnine, qui sera décrit ci-dessous.

1.5 g. de ce sel neutre, digéré avec 20 cm³. d'eau chaude, a déposé une partie de sa strychnine et, par refroidissement de la solution filtrée, 1 g. du même sel acide de strychnine (II). Le sel acide cristallise dans l'eau en petits prismes ou en aiguilles, se décomposant vers 245°.

I. Substance 0.5836 g.; perte de poids à 100° 0.0671 g.; 1.432 m. équiv. de NaOH.
II. „ 0.3606 „ „ „ „ „ 100° 0.0414 „ 0.878 „ „ „ „
Trouvé: H_2O 11.5 (I), 11.48 (II).
$3\,C_4H_6O_7S \cdot 8\,C_{21}H_{22}O_2N_2 \cdot 24\,H_2O$ (3699.6). Calculé: „ 11.68.
Sel anhydre. Trouvé: poids équiv. 361 (I), 363.6 (II).
$3\,C_4H_6O_7S \cdot 8\,C_{21}H_{22}O_2N_2$ (3267.6). Calculé: „ „ 363.1.

Ce sel acide, décomposé à l'ammoniaque, a donné un sel ammoniacal lévogyre. Cependant, les rotations ainsi obtenues étaient trop faibles pour permettre un dédoublement de l'acide.

Sulfosuccinate neutre de strychnine.

Il est possible de faire dissoudre trois molécules de strychnine en présence d'une molécule de l'acide tribasique, en opérant en grande dilution.

Ainsi un demi litre, renfermant $2^1/_2$ m. équiv. d'acide, a dissous complètement $2^1/_2$ m. mol. (0.84 g.) de strychnine. La solution, évaporée à la température ordinaire, a déposé de grands cristaux du sel neutre.

Substance 0.3624 g.; perte de poids à 110° 0.0454 g. (12.5 %); 0.796 équiv. de NaOH.
Sel séché. Trouvé: poids équiv. 398.
$C_4H_6O_7S \cdot 3\,C_{21}H_{22}O_2N_2$. Calculé: „ „ 400.3.

Le même sel neutre se prépare plus aisément en milieu alcoolique. On dilue 10 cm³. d'une solution renfermant 5 m. équiv. de l'acide pour 200 cm³. d'alcool absolu et on y dissout au bain-marie 5 m. mol. (1.67 g.) de strychnine.

La solution refroidie dépose lentement le sel neutre cristallisé. Le sel doit être exposé à l'air pendant plusieurs jours pour atteindre un poids constant. Comme dans ces circonstances il n'est pas exclu que le sel perde une partie de l'eau de cristallisation, un dosage de l'eau n'a pas beaucoup d'importance. Aussi, la teneur en eau était différente pour les diverses préparations.

Voici l'analyse de deux échantillons.

Substance 0.3832 g.; perte à 120° 0.0406 g. (10.6 %); 0.857 m. équiv. de NaOH.
„ 0.3268 „ „ „ 120° 0.0303 „ (9.3 %); 0.074 „ „ „ „
Sel séché. Trouvé: poids équiv. 400.0, 400.6.
$C_4H_6O_7S \cdot 3\,C_{21}H_{22}O_2N_2$. Calculé: „ „ 400.3.

Ce sel neutre a servi au dédoublement optique de l'acide.

Dédoublement optique de l'acide par l'intermédiaire du sel neutre de strychnine.

Le sel neutre de strychnine cristallisant en milieu alcoolique renferme un excès de l'acide dextrogyre.

En décomposant le sel de strychnine très prudemment à froid par la quantité théorique d'ammoniaque, en présence de chloroforme pour faire dissoudre la strychnine, on obtient une solution du sel ammoniacal dont la concentration est connue. L'examen polarimétrique de cette solution, bien extraite au chloroforme, nous apprend donc la rotation moléculaire du sel ammoniacal.

On fait recristalliser le sel de strychnine en le dissolvant dans environ cent fois son poids d'alcool à 90 % en vol., chauffé préalablement à 50°, l'alcool bouillant provoquant une racémisation partielle.

Trois cristallisations successives du sel de strychnine ont p. e. fourni des sels ammoniacaux présentant la rotation moléculaire: $[M]_D$ 9°.5, 15°.4, 22°.5. Comme il peut arriver qu'en négligeant les précautions exigées la recristallisation d'un sel actif fasse diminuer son pouvoir rotatoire, le dédoublement est assez pénible.

La rotation la plus grande, observée pour le sel ammoniacal, a été $[M]_D = 34°$.

Dispersion de la rotation de l'acide et de ses sels.

On obtient l'acide actif en faisant précipiter le sel ammoniacal par le nitrate d'argent et en décomposant le produit à l'hydrogène sulfuré.

On a mesuré la rotation de l'acide pour diverses longueurs d'onde et ensuite on a examiné les sels sodiques, en ajoutant respectivement 1, 2 et 3 molécules de soude caustique à la solution acide.

En rapportant les angles mesurés à la valeur maxima, observée pour la rotation moléculaire de l'acide $[M]_D = 53°$, on obtient les rotations moléculaires du tableau suivant.

Rotation moléculaire de l'acide sulfosuccinique et de ses sels.

Longueur d'onde.	Acide.	Sel primaire.	Sel secondaire.	Sel tertiaire.
6600 u. Å.	47	46	34	29
6400	47	46	34	29
6200	48	47	34	29
6100	49	48	36	30
6000	51	50	37	31
5890 (Na)	53	52	39	33
5700	57	56	41	35
5500	63	62	46	38
5300	69	68	51	43
5100	76	75	55	48
5000	80	79	59	51

```
         SO3H
          |
HOOC-CH-CH-COOH
     |
    SO3H
```

$C_4H_6O_{10}S_2$ Ref. 39

On a ajouté lentement une solution de l'acide disulfosuccinique (276 cm³ 1.8 n = 0.125 mol.) à 83.5 g de strychnine (0.25 mol.) dans 4 litres d'alcool de 89 %, en maintenant la température à 74°. Une partie du sel cristallise au cours de l'addition (A 82 g). En refroidissant, une nouvelle portion se dépose (B 20.8 g). L'eau-mère évaporée à sec abandonne le reste (C 21.8 g). Rendement total 124.6 g au lieu de 125 g.

On examine les fractions cristallines en décomposant 2.5 ou 1.25 g à l'ammoniaque et en éliminant la strychnine par extraction au chloroforme.

Les solutions du sel ammoniacal, diluées à 15 cm³, sont soumises à l'examen polarimétrique, dans des tubes de 20 cm.

On a trouvé à 16° pour la rotation moléculaire $[M]_D$ des sels neutres ammoniacaux dérivant des trois fractions du sel de strychnine: A + 17.2°, B — 18.7°, C + 5.9°.

79 g de A, extraits pendant 3 heures avec 4 litres d'alcool bouillant à 80 %, ont abandonné un résidu (Aa 67 g) donnant un sel ammoniacal de $[M]_D$ = + 9.8°. La solution alcoolique refroidie a laissé cristalliser à la longue une fraction Ab (5.4 g); $[M]_D$ = + 22.3°.

65 g de Aa ont donné de la même manière un résidu, Aaα 54.4 g, $[M]_D$ = + 7.2°, puis après refroidissement Aaβ 5.6 g, [M[= + 25.4°.

La même méthode, appliquée de nouveau, abaisse le pouvoir rotatoire. Le chauffage prolongé avec l'alcool semble causer une racémisation.

En répétant la préparation du sel de strychnine on n'obtient pas les mêmes chiffres, si ce n'est par hasard. Une autre cristallisation, aux dépens des mêmes quantités d'acide et de strychnine, a donné A 87 g, B 18.2 g, C 16 g; $[M]_D$ = + 12.5°, — 26.1°, — 9.5°.

Cette fois, pour éviter l'extraction à l'alcool, nous avons transformé la fraction A du sel de strychnine en acide libre, par l'intermédiaire du sel barytique peu soluble, et nous en avons préparé de nouveau le sel de strychnine en l'ajoutant lentement à une solution alcoolique de strychnine: Aa 55 g, Ab 4 g, Ac 14 g; $[M]_D$ = + 20.0°, + 2.0°, — 7.5°. La première fraction Aa, traitée encore de la même manière, a donné deux fractions Aaα et Aaα1 avec $[M]_D$ + 21.4° et + 22.0°. La rotation n'augmente plus et en répétant la recristallisation on observe parfois un décroissement par racémisation.

Les valeurs les plus élevées que nous ayons observées pour les deux énantiomorphes sont + 25.4° et — 26.1° à la température de 16°. On obtient donc pour la rotation des sels neutres de l'acide disulfosuccinique au moins $[M]^D$ = 26°.

$CH_3CH_2CHCOOH$
|
Br

$C_4H_7BrO_2$ Ref. 40

Versuch 2. Die hier befolgte Methode der Zerlegung lieferte die besten Ergebnisse. In 120 ccm Alkohol von 95 Volumprozent wurden 20 g Brombuttersäure und 40 g Strychnin gelöst. Es wurden 47 g Salz bei 5° C erhalten; diese Menge ging bei schnellem Erwärmen auf 60° mit 115 ccm Alkohol in weniger als einer halben Minute in Lösung. Das Umkrystallisieren wurde in gleicher Weise wiederholt. Aus einem Teil der Mutterlauge der sechsten Umkrystallisation wurde nach Wasserzusatz die Säure mit Schwefelsäure freigemacht und in Äther aufgenommen. Die ätherische Lösung zeigte 1 $\alpha = -0,23°$; 2,00 ccm der Lösung verbrauchten zur Neutralisation 1,67 ccm 0,1023 n-Bariumhydroxydlösung. Hieraus ergibt sich $[\alpha]_D = -16°$ bei etwa 15—16° C. Bei fortgesetztem Umkrystallisieren wurde jedesmal in der Mutterlauge das Drehungsvermögen der Säure in der angegebenen Weise bestimmt. Es wurde gefunden für die freie Säure aus der Mutterlauge beim

sechsten Umkrystallisieren	$[\alpha]_D = -16°$;	auskrystallisierte Salzmenge	9,5 g
siebenten "	$[\alpha]_D = -19°$;	" "	7,3 g
achten "	$[\alpha]_D = -23,5°$;	" "	5,7 g
neunten "	$[\alpha]_D = -32°$;	" "	3,9 g
zehnten "	$[\alpha]_D = -31°$;	" "	2,8 g

Nach diesen Bestimmungen war also mit der achten Umkrystallisation die maximale Drehung für das Salz erreicht. Die Salzmenge war hier 5,7 g. Es wurden 2,8 g zehnmal umkrystallisiertes Salz erhalten und durch systematisches Einengen und Umkrystallisieren aus dem nicht verbrauchten Teil der Mutterlaugen noch 3,6 g Salz von derselben Reinheit.

Analyse von $C_{21}H_{22}O_2N_2$, $HOCOC_3H_6Br$, $3\,H_2O$ (Mol.-Gew. 555).

0,6540, 0,6518 und 0,6415 g Salz, das in der Luft längere Zeit aufbewahrt worden war, verbrauchten nach Kjeldahl 4,74, 4,66 und 4,72 ccm n/2-HCl*);

0,3101 g Salz verbrauchten nach dem Erhitzen mit Kaliumhydroxydlösung 5,55 ccm 0,1002 n-$AgNO_3$.

Eine Salzmenge wurde unmittelbar nach dem Absaugen in einen Vakuumexsiccator über Schwefelsäure gestellt. Nach 20 Minuten betrug das Gewicht 0,3291 g, nach weiteren 3 Stunden 0,3290 und nach 10 Stunden 0,3287. Bei den Wägungen war das Salz zwischen Uhrgläsern aufbewahrt. Trotzdem wurde eine Gewichtszunahme von 0,0002—0,0003 g während der ersten 30 Minuten nach dem Herausnehmen aus dem Exsiccator beobachtet. Das Salz wurde nun an die Luft gelegt und von Zeit zu Zeit gewogen. Hierbei wurde ein anfangs langsames Abnehmen des Gewichtes gefunden, das nach etwa 24 Stunden die größte Geschwindigkeit erreicht hatte, um dann wieder langsamer zu werden. Erst nach etwa 8 Tagen wurde das konstante Gewicht, 0,3106 g, erreicht. — Weder im Exsiccator über Schwefelsäure, noch bei mäßigem Erhitzen konnte eine weitere Gewichtsabnahme konstatiert werden.

	% N			% Br
Gef.	5,07,	5,00	5,13	14,35
Ber.		5,04		14,42

Die Art der Gewichtsveränderung und die Analyse deuten darauf hin, daß das über Schwefelsäure getrocknete Salz noch Krystallalkohol enthält, der aber an der Luft unter Umwandeln der Krystalle und Aufnahme von Krystallwasser abgegeben wird. Hierauf deutet auch der Befund, daß ein Salz, das längere Zeit an der Luft aufbewahrt worden ist, sich langsamer in warmem Alkohol auflöst als frisch abgeschiedenes.

Um zu untersuchen, ob die Säure ohne Destillation in reinem Zustand erhältlich sei, wurden teilweise eingetrocknete Mutterlaugen des Strychninsalzes in Wasser gelöst, die Säure in der oben angeführten Weise freigemacht und in Äther aufgenommen. Nach dem Abdestillieren des Äthers wurde der Rückstand in einen kleinen Destillierkolben übergeführt, der mit einem bis auf den Boden reichenden Siedefaden versehen wurde. Der Kolben wurde mit der Wasserstrahlpumpe verbunden, wodurch ein Luftstrom durch die Capillare gesaugt wurde, dessen Stärke in einer Stunde 2 Liter, bei Atmosphärendruck gemessen, betrug. Die Luft war getrocknet; der Kolben stand in einem Bade von 50—55° C und der Druck im Kolben wurde bei 18—20 mm Hg gehalten. Nach 3 Stunden hatte die Säure das spez. Gew. $D^{18}_{18} = 1,5309$, und nach 5 weiteren Stunden $D^{18}_{18} = 1,5301$, während für racem-Säure $D^{18}_{18} = 1,578$ angegeben ist. Auf diesem Wege ist also eine Reinigung von Säuren, die Zersetzungsprodukte enthalten, undurchführbar. Für eine Säure aus reinem krystallisierten Salz liegt, wie oben angeführt, die Sache anders.

Eine aus Mutterlaugen des Strychninsalzes freigemachte Säure zeigte in 43 prozent. ätherischer Lösung die Drehung $[\alpha]_D = -16,4°$. Nach einer Vakuumdestillation wurde gefunden $\alpha/4 = -5,80°$, $[\alpha]_D = -14,7°$ und nach einer zweiten, 15 Minuten dauernden und bei 15 mm ausgeführten Destillation $\alpha/4 = -5,83°$. Racemisation ist also während dieses Erwärmens nicht eingetreten.

Folgende Bestimmungen des Drehungsvermögens der (—)-Säure sind ausgeführt worden.

Reine Säure: $\alpha/4 = -12,68°$ bei 16,5° C und 20 Tage später $\alpha/4 = -12,71°$ bei 15,5° C. Mit dem spez. Gewicht von 1,572 ergibt sich hieraus

$$[\alpha]_D^{16°} = -32,3° \quad \text{und} \quad [M]_D^{16°} = -54,0°.$$

$C_4H_7BrO_2$ Ref. 40a

(+)-*α-Bromo-n-Butyric Acid* (3)—The resolution of the racemic acid was accomplished by extracting the brucine salt with acetone and with varying mixtures of acetone and chloroform. A typical example of the procedure is the following. The *dl-α*-bromo-*n*-butyric acid was dissolved in a mixture of 5 parts of acetone and 1 part of chloroform, and 1 equivalent of brucine was added. As soon as a crystalline precipitate formed (after several hours), it was filtered off. It is essential not to allow this mixture to stand overnight. The brucine salt was then extracted ten times with acetone at 50°. The operation was followed by extraction three times with a mixture of 9 parts of acetone and 1 part of chloroform at 50°; then three times with a mixture of 8 parts of acetone to 2 parts of chloroform and finally four times with a mixture of 7 parts of acetone and 3 parts of chloroform. After each of the last three extractions portions were taken and converted into the free acids. The optical rotation, $[\alpha]_D^{25} = +39.5°$, $[M]_D^{25} = +66.0°$ (in ether) ($c = 9.1$), was identical for each sample. Levene, Mori, and Mikeska (3) report $[M]_D^{20} = +59°$.

The free acid was distilled. B.p. 66–69°, 0.04 mm.; $d_4^{25} = 1.568$.

$C_4H_7O_2Br$.	Calculated.	C 28.75, H 4.23, Br 47.86
	Found.	" 28.86, " 4.19, " 47.96

$[\alpha]_D^{25} = +36.3°$; $[M]_D^{25} = +60.6°$ ($c = 9.5$); maximum $[M]_D^{25} = +66°$ (in ether)
$[\alpha]_D^{25} = +33.3°$; $[M]_D^{25} = +55.6°$; maximum $[M]_D^{25} = +60°$ (homogeneous)

Thus, racemization to the extent of 8 per cent occurred during distillation.

About 1 year later the rotation of this same substance was again determined; it was 80 per cent of the maximal value, and thus had undergone a further racemization of 12 per cent. Its rotation was $[\alpha]_D^{25} = +27.2°$; $[M]_D^{25} = +45.4°$ ($c = 9.1$); maximum $[M]_D^{25} = +57°$ (in 30 per cent alcohol).

(—)-*α-Bromo-n-Butyric Acid*—The first mother liquor obtained from the brucine salt was liberated from the alkaloid and the free acid was converted into the strychnine salt. This was dissolved in 95 per cent alcohol at 75° and cooled rapidly. The crystalline deposit was recrystallized in the same manner. After eight crystallizations a sample of the acid was obtained which had $[\alpha]_D^{25} = -31.7°$; $[M]_D^{25} = -57.4°$ (in ether) ($c = 14$), *i.e.* 87 per cent of the maximum. The resolution was not continued further.

$C_4H_7BrO_2$ (S)-(−)-2-*Bromobutyric acid* Ref. 40b

Racemic 2-bromobutyric acid[22] (50 g, 0·3 mole) was dissolved in 20 ml Me_2CO and added to a soln of dehydroabiethylamine[23] (84·5 g, 0·3 mole) in 150 ml of the same solvent. After standing overnight at room temp, the salt (122 g) was collected and washed with Me_2CO; it had m.p. 153–156°, $[\alpha]_D^{25}$ +19·28° (*c* 1·6, $CHCl_3$). 50 g of this material was dissolved in 400 ml EtOH by gently warming and allowed to cool at room temp. After 12 hr 18 g of salt separated out, m.p. 159–161°, $[\alpha]_D^{25}$ +16·79° (*c* 1·5, $CHCl_3$). A similar crystallization of another 50 g of the same salt afforded 18 g of product, m.p. 156–159°, $[\alpha]_D$ +16·93° (*c* 1·6, $CHCl_3$). The two fractions were combined and again crystallized from EtOH (250 ml) to obtain 15 g of salt m.p. 158–161°, $[\alpha]_D^{25}$ +16·31° (*c* 1·5, $CHCl_3$). After treatment of all of this salt with 10% NaOHaq, the base was eliminated by extraction with Et_2O, the aqueous layer was acidified with 10% HClaq, extracted with Et_2O, and the dried ($MgSO_4$) extract was evaporated. Distillation of the residue afforded 3·88 g of the (*S*)-(−)-acid, b.p. 105–107°/15 mm, $[\alpha]_D^{25}$ —14·70° (*c* 5·0, Et_2O). Lit.[17] b.p. 66–69°/0·04 mm, $[\alpha]_D$ +39·5° for *R*-(+)-1.

[17] P. A. Levene and M. Kuna, *J. Biol. Chem.* **141**, 391 (1941).

[22] E. Fischer and A. Mouneyrat, *Ber. Dtsch. Chem. Ges.* **33**, 2383 (1900).

[23] W. J. Gottstein and L. C. Cheney, *J. Org. Chem.* **30**, 2072 (1965).

Cl
|
$CH_3CH_2CHCOOH$

$C_4H_7ClO_2$ Ref. 41

(−)-**2-Chlorobutanoic Acid (V).**—The salt (m.p. 148.5–150°) obtained from evaporation of a solution prepared from 353 g. (1.2 moles) of cinchonidine (Fluka AG., Buchs, SG, Switzerland), 147 g. (1.2 moles) of (±)-V, and 300 ml. of methanol was suspended in 1500 ml. of boiling acetone and enough methanol (*ca.* 100 ml.) was added slowly to give complete solution. The solution was allowed to cool to room temperature and stand for 6 hr. The white crystals that separated were collected, and after four additional recrystallizations from acetone–methanol, they weighed 370 g. and had a melting point of 149.5–152°. This material was dissolved in 200 ml. of methanol, and the solution was allowed to stand at 0° for 2 weeks. The solid that separated weighed 107 g. (21%) and melted at 153.5–154.5°. To a mixture of 400 g. of crushed ice and 170 g. of 60% perchloric acid covered with 150 ml. of ether was added 107 g. (1.26 moles) of the above cinchonidine salt. The mixture was shaken vigorously, and after about 10 min., all of the cinchonidine salt had disappeared. The ether phase was separated, and the aqueous phase was extracted three times with 100-ml. portions of ether. The ether extracts were combined, washed twice with 20-ml. portions of water, and dried over magnesium sulfate. The ether was removed by distillation at atmospheric pressure, and distillation of the residual oil yielded 29.3 g. (93%) of (−)-2-chlorobutanoic acid, b.p. 93–94° (9 mm.), $n^{26}D$ 1.4377, $[\alpha]^{25}D$ −9.1° (0.374 g./5 ml. of chloroform); lit.,[10] b.p. 98.5° (14 mm.), $n^{25}D$ 1.4414, $[\alpha]^{25}D$ 17.2° (neat) for (+)-V.

CH_3CHCH_2COOH
|
Cl

$C_4H_7ClO_2$ Ref. 42

20 g *β*-Chlorbuttersäure und 64.8 g Chinin (1 Mol.) wurden in 100 ccm Methylalkohol gelöst und mit 100 ccm Wasser versetzt. Dann wurde die Lösung bei einer 25° nicht übersteigenden Temperatur unter vermindertem Druck etwa auf das halbe Volumen eingedampft. Es schied sich ein Öl ab, das auf Zusatz von vorher gewonnenen Impfkrystallen und gründlichem Reiben krystallinisch erstarrte. Das Produkt wurde abgenutscht und noch zweimal in der gleichen Weise in 75 ccm Methylalkohol gelöst, mit 75 ccm Wasser versetzt und die Hauptmenge des Alkohols unter vermindertem Druck abdestilliert. Die Ausbeute an so gewonnenem, dreimal krystallisiertem Salz betrug 48.5 g. Es wurde nochmals aus 50 ccm Wasser, dann aus 40 cm Wasser, nach Verdampfen des vorher zur Lösung verwandten Methylalkohols umkrystallisiert, wobei 7 g in den Mutterlaugen blieben. Dann wurde in 40 ccm Methylalkohol gelöst und mit Wasser bis zur Trübung versetzt. Das Salz scheidet sich langsam in großen, wohlausgebildeten Krystallen ab, die nach 48-stündigem Aufbewahren im Eisschrank abfiltriert wurden. Die Ausbeute betrug 23 g und nach nochmaligem Umkrystallisieren in der gleichen Weise 16 g. Eine Probe wurde mit kalter, verdünnter Schwefelsäure vorsichtig zerlegt, die Chlorbuttersäure mit Äther extrahiert, die ätherische Lösung getrocknet und der Äther verdampft. In 10-prozentiger Lösung in Toluol betrug $[\alpha]_D = -83.4°$, während die reine *d*-Säure $[\alpha]_D = +46.6°$ hatte. Die Gesamtmenge wurde nun zerlegt.

Die reine *l*-*β*-Chlorbuttersäure krystallisierte in Nadeln aus, sie konnte nach Zugabe von wenig hochsiedendem Ligroin und gutem Abkühlen von der Mutterlauge getrennt werden.

CH_2=CH-CHCOOH
|
NH_2

$C_4H_7NO_2$ Ref. 43

Production of D-*vinylglycine by the action of baker's yeast on the racemate.* DL-Vinylglycine (200 mg) and sucrose (7 g) were dissolved in water (50 ml), and baker's yeast (5 g) was suspended in the mixture by stirring. The suspension was set aside at room temperature for two days, and the liquid was decanted through a 1 cm layer of Filtercel. The filtrate was concentrated and applied to a cation exchange resin (Amberlite IR 120, 1.3 × 17 cm, H^+-form). The column was rinsed with water, and vinylglycine was eluted with aqueous pyridine (0.5 M), 5 ml fractions being collected. Fractions 8 – 12, which contained vinylglycine, were combined and evaporated to dryness. The residue (104 mg) was dissolved in a few ml of water and passed through a filter having a small layer of carbon. The filtrate was evaporated to dryness, leaving a colourless sample of D-vinylglycine (78 mg), m.p. 216 – 218° (decomp.), $[\alpha]_D^{22}$ – 93.8° (*c* 1.5, H_2O). The paper-chromatographic behaviour and IR- and ^1H NMR-spectra were identical with those observed for the racemate. A sample of D-vinylglycine with $[\alpha]_D^{22}$ – 83.1°, produced in a preliminary experiment, revealed, dissolved in hydrochloric acid, an $[\alpha]_D^{22}$ – 95° (*c* 0.3, 2 M HCl).

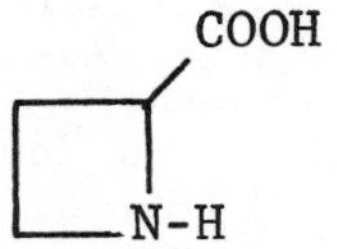

$C_4H_7NO_2$ Ref. 44

A solution of **2** (3.0 g., 0.013 mole) in methanol (60 ml.) was treated with L-tyrosine hydrazide (2.49 g., 0.013 mole) and the suspension was boiled for ten minutes and filtered giving a white crystalline solid. The filtrate was stirred for twenty hours at 25° and the small amount of solid which separated was combined with the original precipitate giving 2.70 g. (98%) of the D salt, m.p. 205-206.5°. Evaporation of the solvent from the mother liquor provided 2.53 g. (92%) of the L salt as a semisolid.

Dissolution of the D salt in 12 ml. of water, addition of 3 ml. of concentrated hydrochloric acid and extraction with ethyl acetate followed by drying (magnesium sulfate) and evaporation of the solvent gave 1.40 g. (95%) of the pure D-form of the *N*-carbobenzoxy derivative (**2**) as an oil, $[\alpha]^{20}_D$ = +98.5° (*c* = 3.9 in chloroform) (6). Treatment of the L salt in a similar manner provided 1.23 g. (89%) of **2** (L) as an oil, $[\alpha]^{20}_D$ = 99.0° (*c* = 3.9 in chloroform) (6).

Catalytic hydrogenolysis of compound **2** (D) in methanol over 10% palladium on charcoal at 4 atmospheres gave 0.35 g. (59%, 55% overall yield) of D-azetidine-2-carboxylic acid, $[\alpha]^{20}_D$ = +107.5° (*c* = 3.5 in water). Similar hydrogenolysis of **2** (L) afforded 0.39 g. (74%, 61% overall yield) of L-azetitine-2-carboxylic acid, $[\alpha]^{20}_D$ = -109° (*c* = 3.6 in water), literature value (7), $[\alpha]^{20}_D$ = -108°.

Previously we had attempted to utilize several of the more common resolving bases such as brucine, strychnine, etc.; however, admixtures of these bases with several different *N*-acyl derivatives of DL-azetidine-2-carboxylic acid gave only noncrystallizable oils.

The procedure described above constitutes an exceedingly facile method of chemical resolution and it is practical as well since the yields are good and most of the resolving agent can be recovered (4).

(4) K. Vogler and P. Lanz, *Helv. Chim. Acta*, **49**, 1348 (1966).

```
HOOCCH2CHCOOH
        |
        NH2
```

$C_4H_7NO_4$ Ref. 45

Standard Aspartic Acid Amine Salts. — *(L-Aspartic Acid D-Amine) Salt.* — L-Aspartic acid (0.67 g., 0.005 mol.) was dissolved in a mixture of 2.0 ml. of water and 0.66 g. (0.005 mol.) of D(−)-methylbenzylamine ($[\alpha]^{27}_D$ = −40.6° in benzene). To this, 10 ml. of methanol and 5 ml. of acetone were added, and the mixture was kept in a refrigerator overnight. The (L-aspartic acid D-amine) salt crystallized; m. p. 260~263°C (decomp.), $[\alpha]^{27}_D$ = −9.9° (*c* 2.04, water). Found: C, 56.48; H, 7.34; N, 11.23. Calcd. for $C_{12}H_{18}N_2O_4$: C, 56.68; H, 7.13; N, 11.02%.

(D-Aspartic Acid L-Amine) Salt. — The salt was obtained from D-aspartic acid ($[\alpha]^{27}_D$ = −24.3°) and L(+) amine, $[\alpha]^{27}_D$ = +10.5° (*c* 2.13 water). Found: C, 56.43; H, 7.14; N, 11.18%.

(D-Aspartic Acid D-Amine) Salt.—The salt did not crystallize under the above conditions, but it did crystallize upon the addition of an extra 10 ml. of acetone to the salt solution; m. p. 258~263°C (decomp.), $[\alpha]^{27}_D$ = +9.2° (*c* 1.04, water). Found: C, 56.66; H, 7.17; N, 10.92%.

The Resolution of DL-Aspartic Acid. — DL-Aspartic acid (6.65 g., 0.05 mol.) was dissolved in a mixture of 200 ml. of water and 6.10 g. (0.05 mol.) of D(−)α-methylbenzylamine ($[\alpha]^{27}_D$ = −40.6° in benzene) and filtered. To this, 50 ml. of methanol and 60 ml. of acetone were added, and the flask was rubbed with a glass rod (or preferably seeded with L-aspartic acid D-amine salt) and kept overnight in a refrigerator. The product, 4.70 g. (71%) of (L-aspartic acid D-amine) salt, was recrystallized by dissolving it in 12 ml. of water and precipitated with 15 ml. of methanol and 35 ml. of acetone. Pure salt (3.60 g.) was obtained; m. p. 260~263°C (decomp.), $[\alpha]^{27}_D$ = −9.7° (*c* 2.18, water). Found: C, 56.68; H, 6.97; N, 11.01. Calcd. for $C_{12}H_{18}N_2O_4$: C, 56.68; H, 7.13; N, 11.02%.

In the same way, (D-aspartic acid L-amine) salt was isolated by the use of L(+)-α-methylbenzylamine ($[\alpha]^{27}_D$ = +39.3° in benzene); 4.50 g. of salt was thereby isolated. After recrystallization, 3.48 g. of pure salt was obtained; m. p. 260~263°C (decomp.), $[\alpha]^{27}_D$ = +10.9° (*c* 2.08, water). Found: C, 56.63; H, 7.03; N, 11.13%.

Optically-Active Aspartic Acid. — *L-Aspartic Acid.*—Three grams of (L-aspartic acid D-amine) salt was dissolved in 15 ml. of water, and the pH was adjusted to about 2.8 by the addition of 3 N hydrochloric acid. Then 15 ml. of ethanol was added, and the mixture was kept in a refrigerator overnight. The crystallized L-aspartic acid was filtered and washed with cold water and absolute alcohol. L-Aspartic acid (1.35 g.) was obtained; $[\alpha]^{27}_D$ = +25.0° (*c* 1.97, 6 N HCl). Found: C, 36.25; H, 5.39; N, 10.48. Calcd. for $C_4H_7NO_4$: C, 36.09; H, 5.30; N, 10.52%. From the mother liquor, optically-impure D-aspartic acid (1.50 g.) was obtained, $[\alpha]^{27}_D$ = −12.3° (*c* 1.07, 6 N HCl). Found: C, 36.21; H, 5.43; N, 10.34%.

D-Aspartic Acid. — D-Aspartic acid L-amine salt (1.20 g.) was treated in the way described above. D-Aspartic acid (0.53 g.) was isolated; $[\alpha]^{27}_D$ = −23.0° (*c* 2.30, 6 N HCl). Found: C, 35.88; H, 5.45; N, 10.37%.

C4H7NO4 Ref. 45a

Zerlegung der racemischen Benzoylasparaginsäure in die optisch-activen Componenten.

Dieselbe lässt sich ebenfalls mit Brucin ausführen und liefert sogar beide optischen Isomeren im reinen Zustand. Denn die Benzoylasparaginsäure bildet mit der Base ein neutrales und ein saures Salz, von welchen das eine bei der *l*-Verbindung und das andere bei der *d*-Verbindung schwerer löslich ist.

20 g wasserhaltige *r*-Benzoylasparaginsäure werden mit 78 g Brucin (2 Mol.) in 200 ccm Wasser heiss gelöst. Beim 12—15-stündigen Stehen der abgekühlten Lösung scheiden sich farblose Nadeln oder Blättchen ab, welche meist zu kugeligen Aggregaten vereinigt sind. Sie bestehen der Hauptmenge nach aus dem Brucinsalz der Benzoyl-*l*-asparaginsäure, welches aber erst durch häufig wiederholtes Umlösen aus heissem Wasser von dem Isomeren getrennt werden kann. Bei der vierten Krystallisation aus 200 ccm Wasser trat schon ein deutlicher Unterschied hervor; denn zuerst schieden sich grosse Prismen oder Tafeln ab, und die abgegossene Mutterlauge lieferte dann beim mehrstündigen Stehen feine, glänzende, meist kugelig vereinigte Nadeln. Diese Erscheinung wiederholte sich beim weiteren Umkrystallisiren der grossen Tafeln, und erst nach dreimaliger Krystallisation war das Salz ganz rein. Die Ausbeute betrug ungefähr 50 pCt. der Theorie.

Zur Isolirung der Benzoylasparaginsäure wurden 25 g des Brucinsalzes in 200 ccm Wasser warm gelöst, mit 50 ccm Normal-Kalilauge versetzt, auf 0° abgekühlt, filtrirt, die Mutterlauge mit 50 ccm Normal-Salzsäure versetzt und unter stark vermindertem Druck eingedampft. Die Ausbeute ist nahezu quantitativ.

Das Product schmolz bei 180—181° (corr. 184—185°) und zeigte unter denselben Bedingungen, wie sie oben beschrieben sind, auch das Drehungsvermögen der Benzoyl-*l*-asparaginsäure (gefunden $[\alpha]^{20}_D$ = + 37.4°).

Verwandlung der Benzoylverbindung in *l*-Asparaginsäure.

Die Spaltung findet vollständig statt, wenn die Benzoylverbindung mit der achtfachen Menge 10-procentiger Salzsäure $2^1/_2$ Stunden auf 100° erhitzt wird. Die abgeschiedene Benzoësäure wird ausgeäthert, und beim Verdampfen der wässrigen Lösung bleibt das Chlorhydrat der Asparaginsäure zurück. Löst man dasselbe in wenig Wasser und versetzt mit der berechneten Menge Normal-Kalilauge, so scheidet sich in der Kälte die Asparaginsäure ab. Zum Vergleich mit der gewöhnlichen, natürlichen Verbindung wurde die optische Untersuchung in alkalischer Lösung ausgeführt.

Eine Lösung vom Gewicht 11.5923 g, welche 0.3853 g Substanz und 3 Mol.-Gew. Natriumhydroxyd enthielt und das spec. Gewicht 1.0394 besass, drehte im 2-Decimeterrohr bei 20° das Natriumlicht 0.155° nach links. Daraus berechnet sich $[\alpha]_D^{20} = -2.24°$.

Zum Vergleich wurde reine, käufliche Asparaginsäure (aus gewöhnlichem Asparagin), welche nochmals aus Wasser umkrystallisirt und deren Reinheit durch eine Stickstoffbestimmung und durch Alkalimetrie controllirt war, zunächst ganz unter den gleichen Bedingungen untersucht:

Eine Lösung von 13.2857 g, welche 0.4454 g Substanz und 3 Mol.-Gew. Natriumhydroxyd enthielt und das spec. Gewicht 1.0398 besass, drehte bei 20° im 2-Decimeterrohr das Natriumlicht 0.165° nach links. Mithin $[\alpha]_D^{20} = -2.37°$.

Dann wurden noch zwei weitere Versuche bei grösserer Concentration und mit 2 Mol.-Gew. Base ausgeführt. Sie zeigen, dass mit steigender Concentration die specifische Drehung abnimmt:

1. Eine Lösung von 20.4814 g, welche 1.2290 g Substanz und 2 Mol.-Gew. Natriumhydroxyd enthielt und das spec. Gewicht 1.0521 besass, drehte bei 20° im 2-Decimeterrohr das Natriumlicht 0.24° nach links. Mithin $[\alpha]_D^{20} = -1.9°$.

2. Eine Lösung von 12.3382 g, welche 1.1382 g Substanz und 2 Mol.-Gew. Natriumhydroxyd enthielt und das spec. Gewicht 1.0811 besass, drehte bei 20° im 2 dm-Rohr das Natriumlicht 0.23° nach links. Mithin $[\alpha]_D^{20} = -1.15°$.

Diese Werthe stimmen leidlich überein mit den Angaben von Pasteur[1]) über die specifische Drehung des asparaginsauren Natriums, welche er für eine 13-procentige Lösung bei 12° und für weisses Licht −2.23° fand. Sie weichen dagegen stark von den Angaben A. Becker's[1]) ab. Letzterer fand unter Bedingungen, wie die oben innegehaltenen, und bei einem Natriumgehalt, welcher von 1—5 Mol.-Gew. schwankte, die specifische Drehung der Asparaginsäure $[\alpha]_D = -9.04$ bis −9.07°.

Da zur Zeit von Becker's Untersuchung für die Darstellung der Asparaginsäure eine bequeme Methode fehlte, so ist zu vermuthen, dass er ein unreines, vielleicht asparaginhaltiges Material für seine Bestimmungen benutzt hat.

[1]) Diese Berichte 14, 1037.

Benzoyl-*d*-asparaginsäure: Dieselbe befindet sich in den Mutterlaugen, aus welchen das benzoyl-*l*-asparaginsaure Brucin auskrystallisirt ist. Zu ihrer Reinigung dient, wie schon erwähnt, das saure Brucinsalz. Um dasselbe darzustellen, ist es aber zunächst nöthig, die Benzoylasparaginsäure aus den Laugen zu isoliren. Das geschah, wie bei dem obigen Versuch, durch Fällen des Brucins mit überschüssiger Kalilauge, Ansäuern des Filtrats mit der entsprechenden Menge Salzsäure und Concentriren durch Eindampfen im Vacuum. Aus den Mutterlaugen des obigen Versuches wurden im Ganzen 11 g trockne Benzoylasparaginsäure zurückgewonnen.

Als 10.3 g davon mit 20.3 g Brucin (1 Mol.) in 100 ccm heissem Wasser gelöst waren, schied sich beim längeren Stehen der erkalteten Flüssigkeit das saure Brucinsalz in kleinen, hübsch ausgebildeten Prismen ab. Das Salz wurde zunächst dreimal aus je 150 ccm Wasser umkrystallisirt. Seine Menge betrug dann 14 g. Da die aus einer Probe regenerirte Benzoyl-*d*-asparaginsäure bei der optischen Untersuchung noch etwas zu schwaches Drehungsvermögen zeigte, so wurde das Salz noch zweimal in der gleichen Art aus Wasser umgelöst. Die jetzt isolirte Benzoyl-*d*-asparaginsäure schmolz bei 180—181° (corr. 184—185°) und hatte auch das Drehungsvermögen des optischen Antipoden.

0.523 g Substanz wurde mit 2 Mol.-Gew. KOH und Wasser gelöst; das Gesammtgewicht der Lösung betrug 5.3669 g und das spec. Gewicht 1.065. Sie drehte im 1 dm-Rohr das Natriumlicht 3.9° nach links. Daraus berechnet sich für Benzoyl-*d*-asparaginsäure in alkalischer Lösung $[\alpha]_D^{20} = -37.6°$. Die Differenz mit dem optischen Antipoden (0.2°) liegt innerhalb der Versuchsfehler.

Zur Umwandlung in die *d*-Asparaginsäure wurde die Benzoylverbindung ebenso behandelt, wie es oben für die optisch-isomere Substanz beschrieben ist. Die so erhaltene Asparaginsäure ist zweifellos identisch mit der von Piutti aus dem rechtsdrehenden Asparagin erhaltenen. Für die optische Bestimmung wurde hier die salzsaure Lösung benutzt.

0.3419 g Substanz wurden mit 3 Mol.-Gew. Salzsäure gelöst; das Gesammtgewicht der Lösung betrug 8.2273 g, das spec. Gewicht 1.032 und die Drehung bei 20° im 1 dm-Rohr bei Natriumlicht 1.09° nach links. Daraus berechnet sich $[\alpha]_D^{20} = -25.5°$.

Zum Vergleich diente natürliche *l*-Asparaginsäure, welche unter denselben Bedingungen folgendes Resultat gab:

0.3455 g Substanz wurden mit 3 Mol.-Gew. Salzsäure gelöst; das Gesammtgewicht der Lösung betrug 8.2749, das spec. Gewicht 1.033 und die Drehung bei 20° im 1 dm-Rohr bei Natriumlicht 1.107° nach rechts. Daraus berechnet sich $[\alpha]_D^{20} = +25.7°$.

Ein zweiter Versuch bei grösserer Concentration (Procentgehalt 8.8173 und spec. Gewicht 1.0705) aber sonst gleichen Bedingungen gab $[\alpha]_D^{20} = +26.47$.

Dieser Werth nähert sich sehr der alten von Pasteur[1]) angegebenen Zahl 27.86, welche ungefähr bei gleicher Concentration der Lösung, aber für weisses Licht (Uebergangsfarbe) gefunden wurde.

Stark abweichend sind dagegen wieder die Werthe, welche Becker[2]) gefunden hat. Ueber die vermuthliche Ursache dieser Differenz habe ich mich oben ausgesprochen.

$C_4H_7NO_4$

Ref. 45b

1-(α-Benzyl N-Benzyloxycarbonyl-D-β- and -L-β-aspartamido)-2-acetamido-1,2-dideoxy-3,4,6-tri-O-acetyl-β-D-glucose——To a solution of 2.0 g. of IV and 1.94 g. of 2-acetamido-2-deoxy-3,4,6-tri-O-acetyl-β-D-glucosylamine in 108 ml. of dry tetrahydrofuran was added 1.3 g. of dicyclohexylcarbodiimide. The solution was stirred for 2 hr. at room temperature and allowed to stand overnight. A few drops of AcOH was added to the reaction mixture and stirred for a further short time. Dicyclohexylurea separated was filtered and the filtrate was evaporated to dryness under reduced pressure. The residue was extracted with a small amount of $CHCl_3$ and filtered from insoluble materials. The extract was washed with cold diluted HCl and water, then with aqueous $NaHCO_3$ and water. The solution was dried over Na_2SO_4 and was evaporated to dryness under reduced pressure.

The residue, 2.9 g., was separated by fractional crystallization from EtOH into two products, which were diastereomeric, 0.63 g. of needles (VIII) and 0.92 g. of small needles (VII), more soluble than the former, respectively.

The former, m.p. 216~217°, $[\alpha]_D^{35}$ +10.2° (c=0.98, $CHCl_3$), was identified with authentic specimen, 1-(α-benzyl N-benzyloxycarbonyl-L-β-aspartamido)-2-acetamido-1,2-dideoxy-3,4,6-tri-O-acetyl-β-D-glucose in the points of melting point, mixed melting point and IR spectra. *Anal.* Calcd. for $C_{33}H_{39}O_{13}N_3$: C, 57.81; H, 5.73; N, 6.13. Found: C, 58.07; H, 5.59; N, 6.25.

The latter (VII), m.p. 194~196°, $[\alpha]_D^{35}$ −13.8° (c=1.44, $CHCl_3$). *Anal.* Calcd. for $C_{33}H_{39}O_{13}N_3$: C, 57.81; H, 5.73; N, 6.13. Found: C, 57.72; H, 5.63; N, 6.15.

$C_4H_7NO_4$ Ref. 45c

15,6 g D, L-Asparaginsäure-diäthylester wurden mit 50 cm³ Methanol vermischt und mit einer Lösung von 31,6 g Dibenzoyl-D-weinsäure in 450 cm³ Methanol bei Zimmertemperatur versetzt. Nach einiger Zeit begann die Kristallisation des D-Asparaginsäure-diäthylester-dibenzoyl-D-hydrogentartrates. Nach 48 Stunden wurden 300 cm³ absol. Äther hinzugefügt und nach weiteren 24 Stunden das auskristallisierte saure Tartrat (25,6 g) abgesaugt und aus Methanol umkristallisiert. Ausbeute: 17 g entspr. 75% d. Th. optisch reines Salz; Schmp. (korr.) 173–174°; $[\alpha]_D^{20}$: −87,8° (c = 2,0 in Methanol).

$C_{26}H_{29}NO_{12}$ (547,5)

Ber.	C 57,04%	H 5,33%	N 2,56%
Gef.	C 56,98%	H 5,45%	N 2,80%.

Die Mutterlauge des Spaltansatzes wurde im Vakuum bei möglichst niederer Temperatur nahezu vollständig eingeengt und L-Asparaginsäure-diäthylester-dibenzoyl-D-hydrogentartrat mit viel Äther ausgefällt. Ausbeute: 17,6 g (78% d. Th.) nahezu optisch reines Salz; Schmp. (korr.) 165°; $[\alpha]_D^{18}$: −73,0°.

$C_{26}H_{29}NO_{12}$ (547,5)

Ber.	C 57,04%	H 5,33%	N 2,56%
Gef.	C 56,84%	H 5,63%	N 2,99%.

Gewinnung der optisch-aktiven Asparaginsäure-diäthylesterhydrochloride

D-Asparaginsäure-diäthylester-hydrochlorid

Das Estertartrat wurde in wenig absolutem Methanol suspendiert und trockener Chlorwasserstoff eingeleitet, bis das Salz gelöst war. Durch Zugabe von viel absol. Äther wurde das D-Esterchlorhydrat in schönen Nadeln ausgefällt. Ausbeute: 90% d. Th. (bezogen auf das Tartrat).

Werden diese Operationen nicht unter vollkommenem Feuchtigkeitsausschluß ausgeführt, so erhält man schlecht kristallisierende Öle, deren Reinigung und Aufarbeitung erhebliche Mühe macht. Schmp. (korr.) 97°; $[\alpha]_D^{8}$: −8,4° (c = 1,0 in Wasser).

$C_8H_{15}NO_4 \cdot HCl$ (225,6)

Ber.	C 42,59%	H 7,13%	N 6,19%
Gef.	C 42,25%	H 7,06%	N 6,31%.

Wie die Verseifung zur freien Aminosäure zeigte, wurde hierbei das D-Esterhydrochlorid in einer optischen Reinheit von 100% gewonnen.

L-Asparaginsäure-diäthylester-hydrochlorid

Das L-Estertartrat wurde auf die gleiche Weise mit methanolischer Salzsäure umgesetzt, wobei in 84proz. Ausbeute (bezogen auf das Tartrat) die nahezu optisch reine Substanz erhalten wurde. Schmp. (korr.) 100–105°; $[\alpha]_D^{18}$: +7,98° (c = 1,0 in Wasser) entspr. einer optischen Reinheit von 98%. R. E. NEUMAN und E. L. SMITH[19]) geben einen Schmp. von 107–110° und einen Drehwert von $[\alpha]_D$: +8,1° für diese Verbindung an.

$C_8H_{15}NO_4 \cdot HCl$ (225,6)

Ber.	C 42,59%	H 7,13%	N 6,19%
Gef.	C 42,15%	H 7,12%	N 6,23%.

Gewinnung der optisch reinen Asparaginsäuren

D-(−)-Asparaginsäure

D-Asparaginsäure-diäthylester-hydrochlorid wurde mit der 20fachen Menge konz. Salzsäure 1 Stunde am Rückfluß erhitzt, dann im Vakuum zur Trockne eingeengt und die überschüssige Salzsäure durch Wasserzusatz und wiederholtes Eindampfen beseitigt. Die Reindarstellung der Säure erfolgte nach A. C. CHIBNALL und Mitarb.[12]). Der Rückstand wurde dazu in wenig Wasser aufgenommen, in der Siedehitze mit einem Überschuß von Kupfercarbonat behandelt und vom Ungelösten schnell abfiltriert. Beim Abkühlen kristallisierte sofort Kupferasparaginat aus. Die Mutterlauge wurde mit Pyridin auf p_H 3–4 gebracht und bei 0° über Nacht stehen gelassen. Dann saugt man weiteres Kupfersalz ab, wäscht dieses mit Eiswasser und Alkohol und fällt aus wäßriger Suspension in der Siedehitze das Kupfer mit Schwefelwasserstoff.

Die wäßrige Lösung wurde im Vakuum eingedampft und der Rückstand mit 95proz. Alkohol ausgezogen. Ausbeute: 80–90% d. Th. (bezogen auf das Esterhydrochlorid); Schmp. (korr.) 345° (Zers.); $[\alpha_D^{18}]$: −25° (c = 1,3 in 1 n HCl)
$[\alpha]_D^{18}$: −24,2° (c = 2,0 in 6 n HCl).

[12]) A. C. CHIBNALL, A. C. REES u. E. F. WILLIAMS, Biochem. J. **37**, 372 (1943).
[19]) R. E. NEUMAN u. E. L. SMITH, J. biol. Chem. **193**, 97 (1951).

C. G. BAKER und H. A. SOBER[20]) fanden für reine D-Asparaginsäure eine spezifische Drehung von $[\alpha]_D^{27}$: − 24,7° (c = 2,0 in 6 n HCl). Die optische Reinheit unserer Substanz betrug damit mindestens 98%

$C_4H_7N_4O$ (133,1)

Ber.	C 36,09%	H 5,27%	N 10,52%
Gef.	C 36,34%	H 5,42%	N 10,53%.

L-(+)-Asparaginsäure

Wurde wie der D-Antipode aus L-Asparaginsäure-diäthylester-hydrochlorid erhalten. Ausbeute: 85% d. Th. (bezogen auf das Esterhydrochlorid); Schmp. (korr.) 331° (Zers.); $[\alpha]_D^{18}$: + 24,2° (c = 1,33 in 1 n HCl) entsprechend einer optischen Reinheit von 96,8%. O. LUTZ und B. JIRGENSONS[21]) fanden für die optisch reine Verbindung $[\alpha]_D$: + 25,0° (c = 1,33 in 1 n HCl); C. G. BAKER und H. A. SOBER[20]). $[\alpha]_D$: + 25,3°.

$C_4H_7NO_4$ (133,1)

Ber.	C 36,09%	H 5,27%	N 10,52%
Gef.	C 36,23%	H 5,47%	N 10,30%.

Aus den Tartraten kann man die Säuren auch direkt auf folgendem Wege isolieren: Das Estertartrat wird im Scheidetrichter in der 10fachen Menge 20proz. Salzsäure suspendiert, mit Äther überschichtet und bis zur völligen Lösung des Salzes geschüttelt. Anschließend extrahiert man die wäßrige Schicht noch zwei Stunden mit Äther und arbeitet, wie beim Esterhydrochlorid beschrieben, zur Asparaginsäure auf.

20) C. G. BAKER u. H. A. SOBER, J. Amer. chem. Soc. **75**, 4060 (1953).

21) O. LUTZ u. B. JIRGENSONS, Ber. dtsch. chem. Ges. **63**, 451 (1930).

$C_4H_7NO_4$ Ref. 45d

Resolution of Racemic Aspartic Acid

1.33 g. of DL-aspartic acid and 2.46 g. of D-threo-2-amino-1-(p-methylsulfonylphenyl)-1,3-propanediol were dissolved together in 5-ml. of water. The solution was seeded, and allowed to stand for one hour at 25° C. The crop of the monobasic D-threo-2-amino-1-(p-methylsulfonylphenyl)-1,3-propanediol salt of L-aspartic acid, having the molecular formula $C_{10}H_{15}NO_4S.C_4H_7NO_4$, which crystallized from solution was collected on a filter, washed with 95% ethanol and dried at 70° C. The salt melted at 198–200° C. By mixing an aqueous solution of this salt with a slight excess of hydrochloric acid, there was obtained a precipitate of L-aspartic acid, M.P. 270° C., $[\alpha]_D^{25}$ +24.5° (c. 2% in 6 N hydrochloric acid).

The monobasic D-threo-2-amino-1-(p-methylsulfonylphenyl)-1,3-propanediol salt of D-aspartic acid, having the molecular formula $C_{10}H_{15}NO_4.C_4H_7NO_4$, was recovered from the filtrate. An aqueous solution of this salt was converted by mixing with a slight excess of hydrochloric acid to yield a precipitate of D-aspartic acid, M.P. 270° C., $[\alpha]_D^{25}$ −24.5° (c. 2% in 6 N hydrochloric acid).

$C_4H_7NO_4$ Ref. 45e

DL-Aspartic acid (10·0 gm.) was dissolved in 750 ml. of 4 per cent acetic acid by warming at 60°. To this was added 16·0 gm. of Baker's cupric acetate monohydrate while warm and the resulting solution was filtered. When the filtrate cooled to approximately 50°, it was inoculated with about 2 mgm. pure L-aspartic acid copper complex. After the L-complex precipitated (precipitation was complete in a period of 10 min.–3 hr.) the crystals were filtered. This filtrate was inoculated with the pure D-complex and the precipitate of D-complex collected in the same way. By inoculation with alternate forms, successive crops were obtained. Six crops each of L-isomer and of D-isomer were obtained in a yield of 7·7 gm. total of L- and 7·6 gm. total of D-isomers.

To the mother liquor warmed to 60°, 10·0 gm. of DL-aspartic acid and 15·0 gm. cupric acetate were added; the resultant solution was filtered. From this supersaturated solution was obtained by seeding 8·3 gm. L-complex and 8·5 gm. D-complex. By addition of 19·5 gm. sodium hydroxide to the mother liquor, 1·57 gm. of L-complex and 1·66 gm. of D-complex were obtained. The total recovery of L-copper complex was 17·6 gm. (85 per cent) and 17·8 gm. (85 per cent) of D-complex.

The isolated copper complex salts were each recrystallized from 1·5 litre 3 per cent acetic acid containing 3·0 gm. cupric acetate. The recoveries were 14·2 and 14·1 gm. of L- and D-forms respectively. Each of these was dissolved in 1 *N* hydrochloric acid and the solution saturated with hydrogen sulphide. The aspartic acid isomers were isolated by adjusting a 200-ml. volume of aqueous solution to *p*H 3. Thus was obtained 7·4 gm. (74 per cent) L-aspartic acid and 7·4 gm. (74 per cent) D-aspartic acid. The respective $[\alpha]_D^{26}$ values in 6 *N* hydrochloric acid were + 24·4° ± 0·0° and − 24·6° ± 0·0°. Nitrogen analyses agreed with the theoretical value.

$C_4H_7NO_4$ Ref. 45f

Typical resolution procedures are as follows: DL-aspartic acid, 14·0 g, was dissolved in a mixture of 25·0 g of ammonium formate and 40 ml. of water by heating at 80° C. The hot solution was filtered and cooled slowly to 43° C. To this solution, 0·50 g of finely pulverized L-aspartic acid was added and mixed with the supersaturated DL-aspartic acid solution. The seeded solution was allowed to stand undisturbed at room temperature (23° C) for 15 min. Precipitated crystals were filtered and washed with a small amount of cold water and ethanol. A weight of 1·57 g of L-aspartic acid (including seed) was isolated. This was recrystallized from 44 ml. of water and 1·28 g of purified L-aspartic acid was obtained, $[\alpha]_D^{11}$ = + 24·6° (c. 2·13, 6 N hydrochloric acid). (Found: C, 36·09; H, 5·34; N, 10·57. Calcd. for

$C_4H_7O_4N$: C, 36·09; H, 5·30; N, 10·52 per cent.) To the filtrate, 0·50 g of pulverized D-aspartic acid was added and stirred. The mixture was kept at room temperature undisturbed for 45 min. A weight of 1·69 g of D-aspartic acid (including seed was) obtained. This was recrystallized from water and 1·39 g of purified D-aspartic acid was obtained, $[\alpha]_D^{25} = -23{\cdot}6°$ (c. 2·03, 6 N hydrochloric acid). (Found: C, 36·11; H, 5·30; N, 10·42 per cent.)

For experimental details of enzymatic methods of resolution of racemic aspartic acid, see Ref. 18-1 and S.M. Birnbaum, L. Levintow, R.B. Kingsley and J.P. Greenstein, Biol. Chem., 194, 455 (1952), and J. Parikh, J.P. Greenstein, M. Winitz and S.M. Birnbaum, J. Am. Chem. Soc., 80, 953 (1958).

There are a number of reports in the chemical literature describing the asymmetric synthesis of L-aspartic acid by the addition of NH_3 to fumaric acid (or fumarate ion) or maleic acid in the presence of the enzyme aspartase as catalyst. Some examples of these are: a) Japanese Patent 74-25189-Chem. Abstr. 81, 8979h (1974), b) M. Kisumi, Y. Ashikaga and I. Chibata, Bull. Agr. Chem. Soc. Japan, 24, 296 (1960), c) S. Kinoshita, K. Nakayama and S. Kitada, J. Ferment. Assoc. Japan, 16, 517 (1958), d) K. Kitahara, S. Fukui and M. Misawa, J. Agr. Chem. Soc. Japan, 34, 44 (1960), e) Y. Takamura, I. Kitamura, M. Ikura, K. Kono and A. Ozaki, Agr. Biol. Chem. 30, 338 (1966).

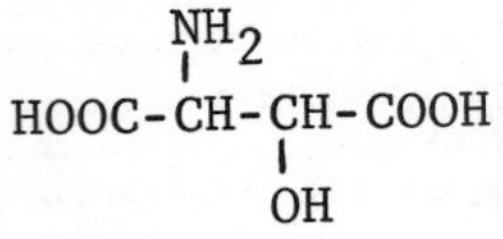

$C_4H_7NO_5$ Ref. 46

L-*Ephedrine Salt of* N-*Benzyl*-threo-β-*hydroxy*-L-*aspartic Acid*.—A mixture of N-benzyl-*threo*-β-hydroxy-DL-aspartic acid [1] (11·95 g.) and L-ephedrine semi-hydrate (8·7 g.) was dissolved by heating under reflux in 80% ethanol (400 ml.). The hot filtered solution was left for 48 hr. at room temperature. The *salt* crystallised quantitatively, m. p. 213° (decomp.) $[\alpha]_D^{25} - 12{\cdot}9°$ (*c* 12·5 in 2N-HCl) (Found: C, 62·2; H, 7·2; N, 6·8. $C_{21}H_{28}N_2O_6$ requires C, 62·3; H, 6·9; N, 6·9%).

N-*Benzyl*-threo-β-*hydroxy*-L-*aspartic Acid*.— The ephedrine salt of *N*-benzyl-*threo*-β-hydroxy-L-aspartic acid (1 g.) was dissolved in water (5 ml.) by heating and concentrated hydrochloric acid was added dropwise to pH 3—4. On cooling the product precipitated practically quantitatively, m. p. 194°; $[\alpha]_D^{16} + 5{\cdot}3°$ (*c* 15 in 5N-HCl).

threo-β-*Hydroxy*-L-*aspartic Acid*.—L-Ephedrine salt of *N*-benzyl-*threo*-β-hydroxy-L-aspartic acid (19·5 g.) was dissolved in glacial acetic acid (200 ml.) and a 30% palladium chloride in Norite catalyst (0·5 g.) was added. Hydrogenolysis was carried out for 5 hr. at 70°. After cooling, the catalyst with the substance adhering to it was filtered off. The product was separated from the catalyst by extraction with hot water. The aqueous solution was concentrated *in vacuo* and on addition of ethanol the product (6·5 g., 90%) precipitated, $[\alpha]_D^{27} + 1{\cdot}6$ (*c* 2·5 in 1N-HCl) [lit.,[2] $[\alpha]_D^{27} + 1{\cdot}3$ (*c* 3·12 in 1N-HCl)] [Found: N (Van Slyke), 9·3. Calc. for $C_4H_7NO_5$: N (Van Slyke). 9·4%].

In order to recover the L-ephedrine the acetic acid solution was evaporated *in vacuo* to dryness and the residue was dissolved in 2N-sodium hydroxide solution (100 ml.), extracted with ether (3 × 100 ml.), and the ether extracts were dried ($MgSO_4$) and evaporated to dryness. To the resulting oil light petroleum was added and on cooling L-ephedrine deposited in colourless crystals.

L-*Ephedrine Salt of* N-*Benzyl*-threo-β-*hydroxy*-D-*aspartic Acid*.—The mother-liquor after the removal of the ephedrine salt of *N*-benzyl-*threo*-β-hydroxy-L-aspartic acid was evaporated *in vacuo* to dryness. Yield, quantitative, $[\alpha]_D^{25} - 17{\cdot}7°$ (*c* 15 in 2N-HCl).

N-*Benzyl*-threo-β-*hydroxy*-D-*aspartic Acid*.—This was obtained from the ephedrine salt in the same manner as its antipode, m. p. 194°; $[\alpha]_D^{20} - 5{\cdot}9°$ (*c* 15 in 5N-HCl).

threo-β-*Hydroxy*-D-*aspartic Acid*.—This was obtained by hydrogenolysis of the L-ephedrine salt of *N*-benzyl-*threo*-β-hydroxy-D-aspartic acid, $[\alpha]_D^{27} - 1{\cdot}44°$ (*c* 2·5 in 1N-HCl) [lit.,[2] $[\alpha]_D^{24} - 1{\cdot}2°$ (*c* 5·16 in 1N-HCl)].

[1] Y. Liwschitz, Y. Rabinsohn, and A. Haber, *J. Chem. Soc.*, 1962, 3589.

[2] T. Kaneko and H. Katsura, *Bull. Chem. Soc. Japan*, 1963, 36, 899.

$C_4H_7NO_5$ Ref. 46a

L-Lysine *threo*-Hydroxy-L-aspartate (L-Lys·L-*t*Hya). A solution of L-lysine monohydrochloride (4.57 g, 25 mmol) in water was placed in a column (2.4 × 20 cm) of Dowex 50 (H^+ form), and the column was washed with water and eluted with 2 N ammonia. The eluate was then evaporated *in vacuo* to dryness. The residue was added in a mixture of DL-*t*Hya (3.73 g. 25 mmol) and water (30 *ml*), and to the solution methanol (25 *ml*) was added. It was allowed to stand overnight at room temperature. The crystals which precipitated were collected by filtration, washed with aqueous methanol, and recrystallized once from water-methanol (the mother liquor and washings were set aside for the isolation of D-*t*Hya); yield of the air-dried product, 2.97 g (76%); mp 182—183°C (decomp.); $[\alpha]_D^{20}$ −5.0° (*c* 1, H_2O), +17.2° (*c* 1, 5 N HCl).

Found: C, 38.54; H, 7.21; N, 13.46%. Calcd for $C_{10}H_{21}O_7N_3{\cdot}H_2O$: C, 38.33; H, 17.40; N, 13.41%.

***threo*-Hydroxy-L-aspartic Acid.** L-Lys·L-*t*Hya·H_2O (2.82 g, 9 mmol) dissolved in water was added to a column (2.4 × 12 cm) of Dowex 1 (acetate form). The column was washed with 0.5 N acetic acid (50 *ml*) to remove the lysine, and then with 2 N acetic acid (200 *ml*) to elute hydroxyaspartic acid. The residue obtained by the evaporation of the eluate was recrystallized from water-ethanol; yield, 1.19 g (88%). The specific rotations are presented in Table 2. (Found: C, 31.95; H, 4.83; N, 9.32%.)

D-Lysine *threo*-Hydroxy-D-aspartate. The mother liquor and washings from L-Lys·L-*t*Hya were evaporated to a small volume, and the solution was treated with a column of Dowex 1 (acetate form), using successively 0.5 and 2 N acetic acid, as has been described for the isolation of L-*t*Hya from L-Lys·L-*t*Hya. The fractions containing hydroxyaspartic acid were evaporated to dryness, and the residue dissolved in water was neutralized with D-lysine which had been prepared from D-lysine monohydrochloride.[23] The solution was then evaporated, and the residue which remained was recrystallized from water-methanol; yield of D-Lys·D-*t*Hya·H_2O, 2.58 g (66%); mp 182—184°C (decomp.); $[\alpha]_D^{20}$ +17.0° (*c* 1, 5 N HCl). (Found: C, 38.61; H, 7.36; N, 13.62%.)

***threo*-Hydroxy-D-aspartic Acid.** D-Lys·D-*t*Hya·H_2O was treated in the same manner as has been described for the preparation of L-*t*Hya; yield, 86%. (Found: C, 31.98; H, 4.83; N, 9.25%.)

23) N. Izumiya, *Nippon Kagaku Zassi (J. Chem. Scc. Japan, Pure. Chem. Sect.)*, **72**, 149, 445 (1951).

TABLE 2. SUMMARY OF SPECIFIC ROTATIONS OF THE FOUR OPTICAL ISOMERS OF HYDROXYASPARTIC ACID

Investigators	Specific rotation, $[\alpha]_D$ *t*Hya L	*t*Hya D	*e*Hya L	*e*Hya D
Dakin[7]	−11.9° (H_2O)	+12.1° (H_2O)		
Sallach *et al.*[1]			+51.2° (N HCl)	
Wieland *et al.*[4]				−54.2° (N HCl)
Kaneko *et al.*[12]	−8.5° (H_2O)	+8.9° (H_2O)	+41.4° (H_2O)	
	+1.3° (N HCl)[a]	−1.2° (N HCl)	+53.0° (N HCl)	−49.2° (N HCl)
Present authors[b]	−8.5° (H_2O)	+8.6° (H_2O)	+47.0° (H_2O)	−46.8° (H_2O)
	+6.4° (5N HCl)	−6.5° (5N HCl)	+52.0° (5N HCl)	−51.8° (5N HCl)
	+2.8° (N HCl)[a]			

a) *c* 3.12 in N HCl.
b) Temperature, 20°C; *c* 1.0 in H_2O or 5N HCl.

NOTE: In the above table, the two rotation values of Dakin should appear in the eHya column instead of the tHya colum.

```
NH2 OH
 |   |
HOOC-CH—CH-COOH
```

$C_4H_7NO_5$ Ref. 47

L-Ornithine *erythro*-Hydroxy-L-aspartate. DL-*e*Hya (3.73 g, 25 mmol) was dissolved in a solution of L-ornithine (25 mmol) in water (30 m*l*). After methanol (20 m*l*) had been added to the solution, it was allowed to stand overnight at room temperature. The crystals thus obtained were recrystallized from water-methanol; yield of the air-dried product (L-Orn·L-*e*Hya·H_2O), 3.14 g (84%); mp 235—236°C (decomp.); $[\alpha]_D^{18}$ +26.6° (*c* 1, H_2O), +40.2° (*c* 1, 5 N HCl). (Found: C, 35.82; H, 7.07; N, 13.88%.)

***erythro*-L-Hydroxy-L-aspartic Acid.** This was obtained from L-Orn·L-*e*Hya·H_2O (2.99 g) by the method described for the separation of L-*t*Hya; yield, 1.43 g (96%). (Found: C, 32.05; H, 4.70; N, 9.31%.)

D-Ornithine *erythro*-Hydroxy-D-aspartate. The mother liquor and washings from L-Orn·L-*e*Hya were treated with a Dowex 1 coulmn, and the portion of hydroxyaspartic acid was neutralized with D-ornithine[14] in the manner described for the preparation of L-Orn·L-*e*Hya·H_2O; yield of D-Orn·D-*e*Hya·H_2O, 2.92 g (78%); mp 233—234° (decomp.); $[\alpha]_D^{18}$ −26.2° (*c* 1, H_2O), −40.2° (*c* 1, 5 N HCl). (Found: C, 35.88; H, 7.02; N, 13.93%.)

***erythro*-Hydroxy-D-aspartic Acid.** D-Orn·D-*e*Hya·H_2O was treated as in the D-*e*Hya preparation; yield, 92%. (Found: C, 32.19; H, 4.71; N, 9.31%.)

14) M. Kondo and N. Izumiya, Abstr. of the 17th Annual Meeting of The Chemical Society of Japan, April, 1964, p. 278.

$C_4H_7NO_5$ Ref. 47a

We required the L-isomer of erythro-β-hydroxy aspartic acid for the synthesis of oligo-peptides and polymeric polypeptides containing this tetra-functional amino acid. On searching the literature for procedures describing the resolution of the racemate we came across Dakin's work [1] using strychnine and quinine for this purpose which gives specific rotations of $[\alpha]_D^{20}$ + 12.1° and -11.9° for the antipodes in water remarking that the optical rotation of the dextro acid is increased about 30% on addition of hydrochloric acid. However, Sallach [2] who had demonstrated the enzymatic formation of the L-isomer by a transamination reaction between glutamate and oxaloglycolate, reported a much higher specific rotation $[\alpha]_D^{25}$ + 51° (in N HCl) [3]. Kaneko and Katsura [4] who prepared it from an optically active trans-epoxysuccinic acid arrived at virtually the same optical rotation $[\alpha]_D^{21}$ + 53.2° (in N HCl). Wieland [5] had isolated the D-isomer from phallacidin, a mushroom poison, giving a specific rotation of $[\alpha]_D^{25}$ -54° (in N HCl). Recently Okai et al. [6] reported the resolution of erythro-β-hydroxy-DL-aspartic acid by means of L- and D-ornithine giving specific rotations of $[\alpha]_D^{20}$ + 52° and -51.8° (in 5N HCl) respectively. These latter authors who tabulated all known specific rotations of the enantiomorphs of both the threo- and erythro-isomers wrongly attributed Dakin's resolution to the threo-series ("para" according to Dakin's nomenclature) whereas Dakin actually dealt with the erythro-series ("anti" according to his nomenclature).

Thus it appeared that Dakin's resolution was either incomplete or he was dealing with a different substance*.

In order to clarify this point we prepared erythro-β-hydroxy-DL-aspartic acid according to Dakin's method [8] starting with fumaric acid according to the following scheme:

* In the book Chemistry of the Amino Acids [7] Dakin's procedure for the resolution of erythro-β-hydroxy-DL-aspartic acid is given in full containing the low specific rotations, although this is juxtaposed on page 2418 with Sallach's value which is over 4 times as high.

$$\begin{array}{c} \text{HC-COOH} \\ \text{HOOC-CH} \end{array} \xrightarrow[\text{Cl}_2]{\text{NaOH}} \begin{array}{c} \text{H} \\ \text{Cl-C-COOH} \\ \text{HOOC-C-OH} \\ \text{H} \end{array} \xrightarrow{\text{NH}_3}$$

$$\left[\begin{array}{c} \text{H} \\ \text{C-COOH} \\ \text{O} \\ \text{HOOC-C} \\ \text{H} \end{array} \right] \longrightarrow \begin{array}{c} \text{H} \\ \text{HO-C-COOH} \\ \text{H}_2\text{N-C-COOH} \\ \text{H} \end{array}$$

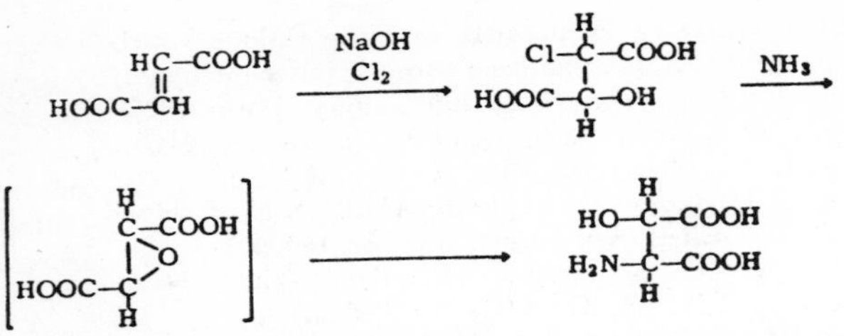

Dakin had stated that the reaction mixture contained about 20% of the para (threo) isomer and between 60-70% of the anti (erythro) isomer which could be separated by fractional crystallisation from water. However, we were unable to isolate any amount of the threo isomer, although there might have been traces of it. In addition to the erythro isomer a substance was isolated from the reaction-mixture, which was much more soluble in water, the constitution of which has not been determined yet.

The erythro isomer thus obtained was identical in all respects (IR-spectrum, TLC, etc.) with an authentical sample prepared according to our method [9] which is strictly stereospecific.

On carrying out the optical resolution by means of strychnine adhering minutely to Dakin's procedure [1] a salt was obtained whose specific rotation conformed approximately to the one given by Dakin, but the β-hydroxy-aspartic acid liberated from it, gave a much higher optical rotation which was moreover of opposite sign ($[\alpha]_D^{20}$ - 33.4° (in N HCl)). A further fraction of the free acid crystallising later from the same mother liquor gave $[\alpha]_D^{20}$ + 4.1° (in N HCl). This peculiar behaviour is in accordance with a statement by Dakin [1] "on recrystallising a mixture of the dextro and inactive acids from water, the mother liquor which was at first dextro-rotatory, on standing in contact with the separated crystals gradually became entirely inactive while the separated dextro acid increased in amount. It is inferred that the inactive acid is a dl-mixture at any rate in solution at room temperature and not truly racemic since under these circumstances the shifting of the equilibrium is readily comprehensible." This alone is proof enough for the fact that the acid after treatment with strychnine still contained some racemate.

The mother liquor from the strychnine salt was treated with ammonia and the crude enantiomer contained in the filtrate was obtained by neutralization with nitric acid. This was then reacted with quinine base whereby a salt was formed with a specific rotation similar to the one given by Dakin [1]. The free erythro-β-hydroxy aspartic acid resulting from this had $[\alpha]_D^{20}$ + 28.4° (in N HCl). From this it may be inferred that Dakin's method does not afford complete optical resolution of erythro-β-hydroxy-DL-aspartic acid and that he was probably misguided by the fact that he obtained, by chance, like but opposite specific rotations after liberation of the strychnine and quinine salts.

We then turned to the N-benzyl derivative of this amino acid previously prepared by us [9] and searched for resolving agents which led to the discovery that L-histidine was eminently suited for this purpose. As in the resolution of N-benzyl-threo-β-hydroxy-DL-aspartic acid with L-ephedrine [10] in which the difference in solubility of the two diastereo-isomers is so pronounced that one precipitates quantitatively and the other remains in solution even when the volume of the solvent is reduced by two-thirds, the same happened here. The histidine salt of N-benzyl-erythro-β-hydroxy-D-aspartic acid crystallised quantitatively and the diastereo-isomeric salt could be isolated in an optically pure state by evaporating the solvent. The free N-benzyl derivatives are then obtained by dissolving the salts in water and adding concentrated hydrochloric acid to pH 2. The optically active erythro-β-hydroxy-aspartic acids finally result by hydrogenolysis of the N-benzyl-derivatives in glacial acetic acid with a palladium chloride on carbon (30%) catalyst at about 70°. The substances which adhere to the catalyst are extracted with hot water. The specific optical rotations of the two enantiomorphs measured in N hydrochloric acid were the highest reported yet in the literature, i.e. $[\alpha]_D^{20}$ - 55.6° and +54.6° respectively.

Experimental

Micro-combustion analyses were carried out by Mrs. M. Goldstein of the Microanalytical Laboratory of the Hebrew University, to whom our thanks are due. Optical rotations were determined in a Perkin-Elmer selfrecording polarimeter 141.

L-Histidine salt of N-benzyl-erythro-β-hydroxy-D-aspartic acid

N-Benzyl-erythro-β-hydroxy-Dl-aspartic acid (4.78 g, 0.02 mole) and L-histidine (free base) (3.1 g, 0.02 mole) were heated under reflux with a mixture of ethanol (100 ml) and water (250 ml) until a clear solution resulted. After being left overnight at room temperature the salt was filtered off and washed with a small volume of ethanol. 3.55 g (90%).
(Found: C, 51.8; H, 5.7; N, 14.3. $C_{17}H_{22}N_4O_7$ requires: C, 51.8; H, 5.6; N, 14.2%). $[\alpha]_D^{20}$ - 3.75° (c 4.6 in 5N HCl).

N-Benzyl-erythro-β-hydroxy-D-aspartic acid

The L-histidine salt (3.5 g) was dissolved in boiling water (100 ml), the pH was adjusted to 2 by addition of concentrated hydrochloric acid. The solution was evaporated in vacuo to about 30 ml. On standing overnight the substance crystallised and was filtered off. 1.9 g (87%).
(Found: N, 5.8. Calculated for $C_{11}H_{13}NO_5$: N, 5.9%).
$[\alpha]_D^{20}$ - 17.1° (c 2.9 in 5N HCl).

erythro-β-Hydroxy-D-aspartic acid

N-Benzyl-erythro-β-hydroxy-D-aspartic acid (1.38 g) was suspended in glacial acetic acid (70 ml) and a palladium chloride on charcoal (30%) catalyst (0.4 g) added. Hydrogenolysis was carried out for 20 hr at 70°.

After cooling, the catalyst with the substance adhering to it was filtered off. Separation of the acid from the catalyst was effected by extraction with hot water, the volume of which was much reduced by evaporation in vacuo. On addition of ethanol the substance crystallised. 0.7 g. (85%).
(Found: N, 9.2; N(Van Slyke), 9.3. Calculated for $C_4H_7NO_5$: N, 9.4; N(Van Slyke), 9.4%). $[\alpha]_D^{20}$ - 55.6° (c 3.2 in N HCl).

L-Histidine salt of N-benzyl-erythro-β-hydroxy-L-aspartic acid

The original mother liquor from the resolution was evaporated to dryness in vacuo. The residue weighed 3.7 g (90%).
(Found: C, 49.2; H, 5.7; N, 13.5. $C_{17}H_{22}N_4O_7 \cdot H_2O$ requires: C, 49.5; H, 5.8; N, 13.6%). $[\alpha]_D^{20}$ + 14.3° (c 2.3 in 5N HCl).

N-Benzyl-erythro-β-hydroxy-L-aspartic acid

This substance was obtained from its L-histidine salt in the same manner as its antipode. (Found: N, 6.1. Calculated for $C_{11}H_{13}NO_5$: N, 5.9%). $[\alpha]_D^{20}$ + 15.2°(c 1.6 in 5N HCl).

erythro-β-Hydroxy-L-aspartic acid

This substance was obtained by hydrogenolysis of the N-benzyl derivative. (Found: N, 9.4; N(Van Slyke), 9.5. Calculated for $C_4H_7NO_5$: N, 9.4; N(Van Slyke), 9.4%). $[\alpha]_D^{20}$ + 54.6° (c 1.9 in N HCl).

REFERENCES

1. H.D. DAKIN, J. biol. Chem., 50, 410 (1922).
2. H.J. SALLACH and T.H. PETERSON, J. biol. Chem., 223, 629 (1956).
3. H.J. SALLACH, J. biol. Chem., 229, 437 (1957).
4. T. KANEKO and H. KATSURA, Bull. chem. Soc. Japan, 36, 899 (1963).
5. T. WIELAND and H.W. SCHNABEL, Ann., 657, 218 (1962).
6. H. OKAI, N. IMAMURA and N. IZUMIYA, Bull. chem. Soc. Japan, 40, 2154 (1967).
7. J.P. GREENSTEIN and M. WINITZ, Chemistry of the Amino Acids, Vol. 3, pp. 2416-2418, Wiley, New York (1960).
8. H.D. DAKIN, J. biol. Chem., 48, 273 (1921).
9. Y. LIWSCHITZ, Y. RABINSOHN and A. HABER, J. chem. Soc., 3589 (1962).
10. Y. LIWSCHITZ, YOLAN EDLITZ-PFEFFERMANN and A. SINGERMAN, J. chem. Soc., (C), 2104 (1967).

$$\begin{array}{c} \text{CH}_3\text{CH}_2\text{-CH-COOH} \\ | \\ \text{SeO}_2 \end{array}$$

$C_4H_7O_4Se$ Ref. 48

Dédoublement optique des acides α-séléninebutyrique et α-séléninevalérique.

Acide α-séléninebutyrique.

A l'aide de quinine on obtient l'énantiomorphe *dextrogyre* de cet acide.

Une solution de 0.125 mol.g de l'α-séléninebutyrate de sodium dans ½ litre d'eau est additionnée d'une solution neutre de 40 g de quinine anhydre (0.125 mol.g) dans ½ litre d'acide chlorhydrique dilué. La solution chaude, décolorée au charbon végétal et filtrée, est abandonnée à elle-même. Après 24 heures le volume total est rempli du sel de quinine, qui cristallise en petites aiguilles. Le sel, essoré et séché à l'air, pèse 35 g. C'est le sel neutre de quinine, quoiqu'on ait employé des quantités équimoléculaires de l'acide dibasique et de la quinine. On le fait recristalliser plusieurs fois en le dissolvant dans l'alcool et en versant cette solution dans une grande quantité d'eau tiède.

On décompose chaque fois un échantillon pesant environ 0.4 g par la quantité équivalente de baryte, on élimine la quinine par quatre extractions au chloroforme et on dilue la solution filtrée jusqu'à 25 cm³.

Après avoir examiné cette solution du sel barytique dans un tube polarimétrique de 4 dm, on en prélève 5 cm³ pour contrôler la concentration du sel et on les remplace par 5 cm³ d'acide chlorhydrique pour déterminer la rotation de l'acide.

Après six recristallisations du sel de quinine, la rotation du sel barytique a atteint sa valeur finale.

Cristallis. No.	1	2	3	4	5	6	7	8
Quantité	35 g	19	10	4.2	3.4	2.0	1.0	0.4
$[M]_D$ Sel neutre	—	41.5°	61	65.5	68	82	81.5	82
$[M]_D$ Acide	—	7.5°	11	15	20	24	25	25

Une solution du sel de baryum, renfermant 1.47 mol.mg dans 100 cm³, a présenté une rotation $\alpha_D = + 0.48°$ pour 4 dm. $[M]_D = +82°$.

20 cm³ de cette solution, additionnés de 5 cm³ d'acide chlorhydrique 5 n, et renfermant ainsi 1.18 mol.mg de l'acide actif à l'état non-dissocié, ont donné le résultat suivant. $\alpha_D = +0.12°$(4 dm). $[M]_D = +25°$.

La *dispersion rotatoire* a été déterminée pour l'acide dextrogyre et son sel de baryum.

λ (A.E.)	6265	5893	5605	5365	5160
[M] Sels	71°	82	94	107	121
[M] Acide	21°	25	29	33.5	39

Acide α-séléninevalérique.

Le dédoublement à l'aide de quinine donne l'énantiomorphe *lévogyre*.

Une solution de 0.15 mol.g de l'α-séléninevalérate de sodium dans ½ litre d'eau a été mélangée à une solution neutre et chaude de 50 g de quinine anhydre (48.6 g = 0.15 mol.g) dans ½ litre d'acide chlorhydrique dilué. En 24 heures 45 g du sel neutre de quinine se sont déposés. Après cinq recristallisations dans l'alcool dilué de la manière décrite le sel fut obtenu à l'état pur. Comme le pouvoir rotatoire est plus faible que dans le cas précédent, on a pris chaque fois au moins 0.8 g du sel de quinine pour examiner sa rotation à l'état de sel barytique, dans un volume de 25 cm³.

Cristallis. No.	1	2	3	4	5	6
Quantité	45 g	35	25	10	4.5	2
$-[M]_D$ Sel neutre	—	9.5°	10	12	12.5	12.5
$-[M]_D$ Acide	—	8°	—	11	11.5	11 5

Une solution de 3.8 mol. mg du sel de baryum dans 100 cm³ a donné la rotation $\alpha_D = -0.19°$ (4 dm); donc $[M]_D = -12.5°$.

20 cm³ de cette solution, mélangés de 5 cm³ d'acide chlorhydrique 5 n et renfermant ainsi 3.04 mol.mg de l'acide non-dissocié dans 100 cm³, ont donné la valeur $\alpha_D = -0.14°$ (4 dm). $[M]_D = -11.5°$.

La faible différence entre les valeurs de l'acide et des sels est frappante.

La dispersion rotatoire présente les valeurs suivantes.

λ (A.E.)	6265	5893	5605	5365	5160
[M] Sels	10.3°	12.5	14.6	17	19
[M] Acide	9 °	11.5	14	16.5	19

```
H2NCOCH2-CH-COOH
         |
         NH2
```

$C_4H_8N_2O_3$ Ref. 49

1. 18.8 g (0.05 mol) of dibenzoyl-d tartaric acid monohydrate are solved 500 ml of water, during continuous heating and stirring, 7,5 g (0,05 mol) f DL-asparagine monohydrate.

The temperature of mixture must not exceed 90 °C. After dissolving e solution is allowed to cool to room temperature, and kept at 0 °C overght. The precipitated crystals were filtered, solved in 50 ml of water during eating, and kept at room temperature overnight, filtered at room temperature. he obtained salt is L-asparagine-dibenzoyl-d-hemitartarate at a yield of 13.6 g.

2. The obtained salt (13.6 g of L—I—VI) is refluxed in 50 ml of anhydrous ethanol for 1 hour, then filtered. The obtained product is crude L—asparagine —(I) of 1.9 g (50.8%)

$[\alpha]_D^{20}$: + 29.3° (c:10; nHCl)

3. The combined mother liqours of 1 and 2 are concentrated in vacuo 50 ml, basified to pH:6 with NH_4OH. The precipitate of crude D-asparagine —I) is recrystallized in 10 ml of water.

ield: 1.95 g 52%

$[\alpha]_D^{20}$: —29.0° (c: 10; nHCl)

4. 7,5 g (0.05 mol) of DL-asparagine (I) monohydrate are suspended in 12.7 ml of distilled water, adding 2.08 ml (0.025 mol) of aq. HCl (c:37%; pw: 1.19 g/cm³), then heating to about 80 °C. 9.58 g (0.025 mol) of dibenzoyl-d-tartaric acid (VI) solved in 12.7 ml of methanol are added. These solutions combined at about 40 °C are allowed to cool during stirring, start crystallization. Left at 10 °C overnight, the other day the crystals are separated by filtration, washed with 2×2 ml of methanol (50%). The obtained salt was L-asparagine-dibenzoyl-d-hemitartarat (L—I—VI).

Yield: 11,1 g

5. The combined filtrates of methanol-water are basified to pH:6 with cc. NH_4OH. The crystallization is completed at 10 °C overnight. The precipitated D-asparagine-monohydrate is filtered, washed with 2×1 ml of methanol and recrystallized in 10 ml of distilled water.

Yield: 3.0 g (80.2%)

$[\alpha]_D^{20}$: —30.2° (c:10; nHCl)

6. 11.1 g of L—I—VI are suspended in 15 ml of distilled water and neutralised to pH:7 with cc. NH_4OH, kept overnight at 10 °C. The precipitated crystals — crude L-asparagine monohydrate — are filtered, recrystallized from 10 ml of distilled water.

Yield: 3.05 g (81.3%)

$[\alpha]_D^{20}$: —31.2° (c: 10; nHCl)

7. Mother liquors 4,5,6 are combined, methanol distilled and the residue stirred and acidified to pH:1 with ccHCl at 10 °C. The precipitated dibenzoyl-d-tartaric acid (VI) is filtered and washed with 2×5 ml of water.

Yield: 8.2 g (85.8%)

mp.: 86—88 °C

The obtained VI can be re-used.

8. The obtained 3 g of D-asparagine monohydrate are boiled in the mixture of 10 ml of distilled water and 4.5 ml of cc HCl for two hours. The obtained solution is basified to pH:3, with NH_4OH, and left overnight at 10 °C.

The precipitated D-asparagine is filtered, washed with 3×1 ml of water.

Yield: 2.38 g (90.0%)

$[\alpha]_D^{20}$: +29.8° (c:10; 6nHCl)

9. The obtained 3.05 g of L-asparagine is hydrolysed as above.

Yield: 2.43 g (90,0%)

$[\alpha]_D^{20}$: —30° (c:10; 6nHCl)

$C_4H_8N_2O_3$ Ref. 49a

Salificazione della carbobenzossi-D.L.α-asparagina con cinconina. — Grammi 5,1 di carbobenzossi-D.L.α.asparagina vengono sciolti, a caldo, in 180 cm³ di acqua ed a questa soluzione si aggiungono g 5,64 di cinconina sciolti in 60 cm³ di alcool bollente. La soluzione idroalcoolica viene evaporata fino a raggiungere un volume di 200 cm³ e quindi lasciata in riposo per 24 ore. Si ottiene in tal modo un deposito cristallino che pesa g 2,2 ed ha p.f. 133-35°. Per ulteriore concentrazione della soluzione fino al volume di 140 cm³ si ottengono altri 2,3 g di deposito che viene riunito al precedente. Per cristallizzazione da acqua si ottengono dei bei cristalli a p.f. 136-138° e $[\alpha]_D = + 109,8°$.

La sostanza posta in stufa a 120° diminuisce lentamente di peso: le determinazioni analitiche indicano che cristallizza con 3 mole di acqua. Analisi:

		C	H	N
	trov.% :	60,40;	6,86;	8,99;
per $C_{31}H_{36}O_6N_4 . 3H_2O$	calc. :	60,57;	6,88;	9,11.

Concentrando ulteriormente le acque madri della salificazione si separa un olio. Si evapora allora completamente il solvente sotto vuoto e il residuo oleoso viene ripreso con poco alcool: dalla soluzione alcoolica, per aggiunta di etere; si ottiene la separazione di un deposito cristallino che risulta essere un miscuglio di sali di cinconina. La soluzione etereo-alcoolica si evapora e si ottiene una massa solida a punto di fusione non netto e, poichè risulta difficilmente cristallizzabile, viene utilizzata tal quale per ottenere la carbobenzossi-asparagina antipodica.

Carbobenzossi-(—)α.asparagina. — Il sale di cinconina a p.f. 136-138° e $[\alpha]_D = + 109,8°$ viene macinato in mortaio con soluzione diluita di soda; si filtra la cinconina separatasi e dalla soluzione alcalina si precipita la carbobenzossi-L.(—)α.asparagina con HCl conc. Il precipitato filtrato, lavato con acqua e seccato fonde a 162-164° e mostra $[α]_D = +6{,}8°$ (c = 0,4 in alcool). Analisi:

trov.%: C 54,23; H 5,48; N 10,46;
per $C_{12}H_{14}O_5N_2$ calc. : 54,13; 5,30; 10,52.

L.(—)α.asparagina. — Grammi 1 di carbobenzossi-L.(—)α.asparagina vengono sciolti in 30 cm³ di alcool etilico; questa soluzione viene aggiunta ad un miscuglio di 8 cm³ di acqua, 2 cm³ di ac. acetico glaciale e 0,1 g di nero di Pd saturato con idrogeno. Si fa gorgogliare attraverso la soluzione una corrente di H_2 per 40-45'. Durante l'operazione si nota la separazione di una sostanza cristallina incolora che alla fine viene disciolta per aggiunta di poca acqua. Si filtra il catalizzatore e si evapora il solvente in essiccatore a vuoto; l'α.asparagina si lava più volte con alcool metilico e si cristallizza da acqua. Contiene una molecola di acqua di cristallizzazione. Il suo potere rotatorio, determinato sotto forma di cloridrato, è $[α]_D = +20{,}0°$ (c = 0,4 in acqua). Non fonde, ma si decompone. Analisi:

trov.%: N 18,78;
per $C_4H_8O_3N_2 \cdot H_2O$ calc. : 18,66.

Carbobenzossi-D.(+)α.asparagina. — Si ottiene spostando con alcali la cinconina dal sale più solubile della carbobenzossi-asparagina, con le stesse modalità descritte per l'antipodo. Mostra p.f. 163-164° e $[α]_D = -6{,}3°$ (c = 0,35 in alcool). Analisi:

trov.%: C 54,27; H 5,52; N 10,41;
per $C_{12}H_{14}O_5N_2$ calc. : 54,18; 5,30; 10,52.

D(+)α.asparagina. — Si ottiene con lo stesso procedimento descritto per la L.(—)α. asparagina. Cristallizza con una molecola di acqua. Per riscaldamento decompone senza fondere. Mostra $[α]_D = -19{,}8°$ (come cloridrato in acqua, c = 0,4). Analisi:

trov.%: N 18,83;
per $C_4H_8O_3N_2 \cdot H_2O$ calc. : 18,66.

For a resolution procedure of racemic asparagine using acetylmandelyl chloride as resolving agent, see S. Berlingozzi, G. Adembri, and G. Bucci, Sperimentale, Sez. chim, biol. **6**, 1 (1955).

Ref. 50

HOOC-CH(NH₂)-CH(NH₂)-COOH

Brucine salt of N,N'-bisbenzyloxycarbonyl-S,S-diaminosuccinic acid (3)

N,N'-Bisbenzyloxycarbonyl-RR,SS-diaminosuccinic anhydride[5] (1) (2.87 g; 7.2 mmole) or N,N'-bisbenzyloxycarbonyl-RR,SS-diaminosuccinic acid[5] (2) (3.0 g; 7.2 mmole) was boiled for 15 min in a mixture of 60 ccm of acetone and 6 ccm of water. To the hot solution brucine dihydrate (6.2 g; 14.4 mmole) was added and boiling was continued for 5 min. The mixture was left overnight in a refrigerator and the crystals (5,5 g; over 100%) were collected. The mother liquors here obtained were evaporated and used in the preparation of acid **14**.

The brucine salt (5.5 g) was boiled for 5 min in a mixture of 60 ccm of acetone and 6 ccm of water and allowed to stand overnight in a refrigerator. The crystals were collected, washed out with a small amount of acetone and dried. The yield was 4.5 g (98.5%) of dibrucine salt tetrahydrate **3**. M.p. 113—161°C.

Analysis:
For $C_{66}H_{72}N_6O_{16} \cdot 4H_2O$ (1276.8) — Calcd.: 62.10% C, 6.33% H, 6.58% N;
found: 62.48% C, 6.10% H, 6.42% N.

5. Biernat J. F., *Roczniki Chem.*, **43**, 427 (1969).

N,N'-Bisbenzyloxycarbonyl-S,S-diaminosuccinic acid (4)

Salt **3** (4.5 g) was treated with 40 ccm of 1 N HCl, free N,N'-bisbenzyloxycarbonyl-S,S-diaminosuccinic acid was extracted with ethyl acetate. The organic layer was washed twice with water and dried over magnesium sulphate. Ethyl acetate was removed under reduced pressure and the oily residue was heated to boiling with 30 ccm of benzene for over 2—5 min. The crystalline compound was collected, washed with benzene, petroleum ether and dried. The yield was 1050 mg (70%) of acid **4** melting at 161°C and showing $[α]_D^{20} = 66 \pm 2°$ (c = 1; dioxane).

Analysis:
For $C_{20}H_{20}N_2O_8$ (416.3) — Calcd.: 6.73% N;
found: 6.88% N.

S,S-Diaminosuccinic acid (7)

Acid **4** (416 mg; 1 mmole), dissolved in a solution of 5 ccm methanol, 2 ccm of water and 0.25 ccm of conc. HCl was hydrogenolyzed in presence of palladized charcoal, until chromatography showed 1 spot of diaminosuccinic acid. Then the catalyst was removed by suction and the filtrate evaporated to dryness under reduced pressure. The residue was suspended in 6 ccm ethanol and neutralized with pyridine. The precipitate was filtered off and washed with ethanol. The yield of the amino-acid dihydrate **7**, decomposing at 217—230°C, was 170 mg (92.5%). The crude amino-acid was dissolved in 1 ccm of 1 N KOH and precipitated with acetic acid. Yield 120 mg (65%) of the pure product melting at 270—275°C with decomposition. $[α]_D^{20} = 54.5 \pm 2°$ (c = 1; 5% HCl) (Lit.[4] $[α]_D^{20} = 59°$ (c = 2; 5% HCl)).

Analysis:
For $C_4H_8N_2O_4 \cdot 2H_2O$ (184.2) — Calcd.: 15.22% N;
found: 15.12% N.

4. Hochstein F. A., *J. Org. Chem.*, **24**, 679 (1959).

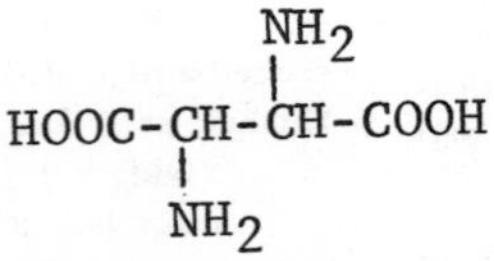

N,N′-Bisbenzyloxycarbonyl-R,R-diaminosuccinic acid (14)

From the evaporated mother liquors, obtained during resolution of racemic N,N′-bisbenzyloxycarbonyl-diaminosuccinic acid, there was isolated the laevorotatory compound **14** (950 mg; 63.5%) in the same manner as described for the dextrorotatory enantiomer. The product melts at 160—162°C. $[\alpha]_D^{20}$ = −66±2° (c = 1; dioxane).

Analysis:
For $C_{20}H_{20}N_2O_8$ (416.3) — Calcd.: 6.73% N;
found: 6.65% N.

$C_4H_8O_3$ $CH_3CH_2CH(OH)COOH$ Ref. 51

The racemic acid (250 g), dissolved in water (1 300 ml), was resolved by means of brucine (966 g of anhydrous alkaloid base), as described by Guye and Jordan [7, 22]. The salt (410 g) was recrystallized from water (700 ml) to give a product (270 g), which possessed rotation values in accord with those reported [7]. To a solution of the salt in water (300 ml), conc. ammonia (50 ml) was added and the precipitated brucine was filtered off and washed. The filtrate was acidified with conc. HCl, saturated with NaCl, and subjected to continuous extraction with ether [9] for 6 h. After removal of the solvent from the dried ether solution, the crystalline (−)-2-hydroxybutyric acid remained (45 g). It was purified by distillation, b. p. 97° at 1.2 mm, and was obtained as hygroscopic, colourless needles (32 g), m. p. 53°, unchanged on recrystallization from a mixture of chloroform and pentane, $[\alpha]_D^{21}$ −15.3° (172 mg of acid in water, containing 1.3 equiv. of ammonia; total volume 5.00 ml). Literature values: m. p. 55—55.5° (recryst. from CCl_4)[10], 52.7—53.5° (from CCl_4: hexane)[12], $[\alpha]_D^{25}$ −15.9° (NH_4-salt in H_2O, c 2.7)[10].

7. Jordan, C. 1re Thèse, *Dédoublement de l'acide butanoloique 2. et recherches sur les dérivés actifs de cet acide.* (Diss.) University, Genève 1895.
10. Fredga, A., Tenow, M. and Billström, I. *Arkiv Kemi, Mineral. Geol.* **16A** (1943) No. 21.
12. Horn, D. H. S. and Pretorius, Y. Y. *J. Chem. Soc.* **1954** 1460.
22. Guye, P. A. and Jordan, C. *Bull. soc. chim. France* [3] **15** (1896) 474.

For a resolution procedure for the above compound using the same resolving agent, see P.A. Guye and C. Jordan, Bull. Soc. Chim. Fr., [3], 15, 474 (1896).

$C_4H_8O_3$ Ref. 51a

An aqueous solution of the free acid was neutralized with morphine; the solution was heated on the steam bath for half an hour and then filtered. The filtrate was concentrated under reduced pressure to a thick syrup. When this was left in the ice box overnight, part of it crystallized. The crystals were removed on a Buchner funnel and then twice recrystallized from 50 per cent alcohol.

The morphine salt was dissolved in water and decomposed by the addition of a slight excess of ammonia. The morphine was removed by filtration and the filtrate converted into the barium salt in the usual way. In water the barium salt had the following rotation.

$$[\alpha]_D^{24} = \frac{+0.43° \times 100}{1 \times 5.6} = +7.7°.$$

A barium salt prepared from the acid obtained on decomposition of the mother liquors in the above resolution yielded the following rotation.

$$[\alpha]_D^{22} = \frac{-0.56° \times 100}{2 \times 5.0} = -5.6°.$$

The free acid was liberated from the second salt in the following manner. 2.0 gm. of barium salt ($[\alpha]_D^{22}$ = −5.6°) were dissolved in cold water and 5.0 cc. of 2.32 N hydrochloric acid added. The volume was made up to 10 cc. and the rotation taken immediately.

$$[\alpha]_D^{22} = \frac{+0.54° \times 100}{2 \times 12} = +2.3°.$$

$CH_3CH(OH)CH_2COOH$

$C_4H_8O_3$ Ref. 52

The resolution of racemic β-hydroxybutyric acid, described in 1902 by McKenzie[2] and repeated by Levene and Haller,[3] depends primarily on inoculation of an aqueous solution of the quinine salts with a crystalline sample of the salt of the L acid obtained from diabetic urine.

In the procedure here described advantage is taken of the hitherto unrecorded great difference in the solubility in acetone of the two quinine salts, the D variety of which requires nearly ten times as much of the solvent as the L isomer for solution. The relationships are illustrated in Table I.

TABLE I

APPROXIMATE PERCENTAGE CONCENTRATION OF SATURATED SOLUTIONS OF THE QUININE SALTS OF D- AND L-β-HYDROXYBUTYRIC ACIDS IN ACETONE AND IN WATER AT VARIOUS TEMPERATURES

Acetone		Water	
D	L	D	L
0.49/1°	4.4/1°	3.5/0°	2.6/0°
1.33/21°	13.2/25°	4.0/25°	5.8/25°
		10/60°	10/36°

EXPERIMENTAL

To a hot solution of 200 meq. of DL-β-hydroxybutyric acid (91.3% by titration) in 500 ml. of acetone, 65 g. (200 mmoles) of anhydrous quinine base was gradually added. When solution was complete the mixture was chilled at 0–1° for 24 hr.; the crystalline salt was collected with suction, washed with 50 ml. of ice cold acetone, and then digested with 300 ml. of boiling acetone for 30 min. The suspension was cooled, held at 0–1° overnight, and filtered with suction; the crystals were washed with 30 ml. of cold acetone, digested as before with 150 ml. of boiling acetone, and dried in air. The yield was 36 g. (81 mmoles, calculated as monohydrate) of quinine D-β-hydroxybutyrate.

The acetone in the combined filtrates and washings was removed as completely as possible by distillation from a steam bath; the sirupy residue was dissolved in 100 ml. of water, the solution was gently warmed until the odor of acetone was no longer perceptible, and was then chilled at 0–1° for 2 days. The needle crystals of quinine L-β-hydroxybutyrate were collected with suction and washed with 10 ml. of ice water in small portions, and the adhering solution was largely displaced by washing with ether. The product was then recrystallized from 80 ml. of water as before, washed with 10 ml. of ice water and finally with ether.[4] After being dried in air, the crystals, which weighed 31.3 g., were dried *in vacuo* to constant weight, 26.1 g. These values correspond to 62 mmoles of the hydrated (4.5 H_2O) and anhydrous salts, respectively.

The free acids were liberated by the gradual addition of 45-ml. quantities of 45% H_2SO_4 to suspensions of the above products in 100 ml. of water. During this operation, quinine sulfate crystallized at first but later dissolved with the formation of the more soluble acid sulfate. The optically active β-hydroxybutyric acids were extracted in a continuous apparatus by a rapid current of ether during 8 hr. and after the removal of solvent the residues were dissolved in water. The solutions were cleared with Norit and aliquots taken for titration and measurement of rotation. The 81 mmoles of D salt yielded 75 mmoles of D acid, $[\alpha]_D^{25} = +23.9°$. The 62 mmoles of L salt yielded 58 mmoles of L acid, $[\alpha]_D^{25} = -24.5°$.

(2) A. McKenzie, *J. Chem. Soc.*, **81**, 1402 (1902).
(3) P. A. Levene and H. L. Haller, *J. Biol. Chem.*, **65**, 49 (1925).

(4) The yields of both salts could no doubt be materially increased by evaporation of the aqueous filtrates to dryness and repetition of the crystallizations from acetone and water.

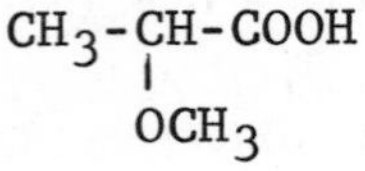

$C_4H_8O_3$ Ref. 53

Resolution of (±)-*α-methoxypropionic acid* (XII) (−)-[S]-*α-Methoxypropionic acid* (−)-XII.

The solution of acid (XII) (13·5 g) and (−)-α-phenylethylamine (15·65 g) in MeOH (10 ml) was heated to boiling, MeOH evaporated, residue (29·15 g, m.p. 89–95°, $[\alpha]_D^{22}$ −9·56° (c, 1, in EtOH) was five times recrystallized from benzene to afford 9·2 g of (−)-amine salt with (−)-[S]-α-methoxypropionic acid, m.p. 124–125°, $[\alpha]_D^{22}$ −34·8° (*c*, 1·5 in EtOH). The salt obtained was dissolved in water, acidified with hydrochloric acid and extracted with ether. The etherial extract was evaporated to afford 3·5 g of (−)-[S]-α-methoxypropionic acid. ((−)-XII, b.p. 88–89 /15 mm, $[\alpha]_D^{20}$ −70·5° (pure liquid)) n_D^{21} 1·4131. Lit.[33], b.p. 106–110°/30 mm $[\alpha]_D^{20}$ −75·47° (pure liquid).

(+)-[R]-*α-Methoxypropionic acid* [(+)-XII]. From the first mother liquor of the above experiment, 1·5 g of (+)-α-methoxypropionic acid were isolated as described above, b.p. 86–88°/15 mm, $[\alpha]_D^{22}$ +24° (pure liquid), n_D^{21} 1·4131. Lit.[21]: b.p. 113–115°/30–32 mm, $[\alpha]_D^{22}$ +72° (pure liquid).

[21] K. Freudenberg and L. Market, *Ber. Dtsch. Chem. Ges.* **60**, 2447 (1927).
[33] T. Purdie and J. C. Irvine, *J. Chem. Soc.* **75**, 486 (1899).

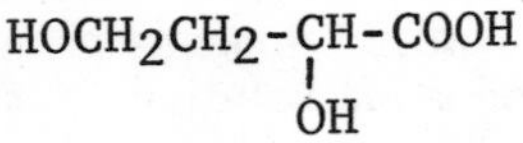

$C_4H_8O_4$ Ref. 54

The Resolution of the *dl*-1,3-Dihydroxy-butyric Acid.—To 36 g. of the acid lactone which had a boiling point of 102° at 3 mm., was added, in one liter of water, 165 g. of brucine. After the usual procedure[8] 182 g. of crude brucine salts was obtained. A sample of this crude salt after a treatment with benzene to remove any free brucine and drying to constant weight over sulfuric acid *in vacuo*, was found to have a specific rotation of −26.73°, *i. e.*, 1 g. in 24 g. of water at approximately 20° gave α = −2.16° in a 2 dcm. tube; sp. gr. 1.010.

The *dl*-brucine salt was now dissolved in twice its weight of hot water. The solution was allowed to stand overnight during which time crystals were deposited (Crop I from water, see below) which were removed by filtration. The filtrate was concentrated until the ratio of water to salt was again 2:1. This solution again yielded a crop of crystals. When no more crystals could be obtained in this way from water, the solution was subjected to complete distillation at 60° and 14 mm. and the residue dissolved in 10 parts of boiling absolute alcohol. When the solution had cooled, the crystals deposited (Crop I from alcohol, see below) were removed by filtration, and the filtrate concentrated until the ratio of salt to alcohol again was 1:10. This process was continued until no more crystals could be obtained. The following table shows the results. The crystals were dried to constant weight *in vacuo* over sulfuric acid, and in each case a 1g. sample was dissolved in 24 g. of water. The specific gravity was found to be 1.010 and readings were taken in a 2dcm tube at approximately 20°.

Crops I, II and III from water were united and recrystallized from 2 parts of water. This process gave 65 g. of crystals. The specific rotation was −20.79. Six subsequent recrystallizations from 2 parts of water gave crops each of which had a specific rotation of −20.79°. This indicates that the substance was the pure brucine salt of one of the optical components of the *dl*-acid. The melting point of this brucine salt was 169°.

FROM WATER. Crop.	Parts of water used.	Weight. G.	α. Observed.	$[\alpha]_D^{20}$.
I	2	49	−1.68°	−20.79°
II	2	8	−1.71°	−21.16°
III	1	20	−1.72°	−21.28°
Residue		104		
FROM ABSOLUTE ALCOHOL.	Parts of alcohol used.	Weight. G.	α. Observed.	$[\alpha]_D^{20}$.
I	10	75	−2.77°	−34.28°
II	10	18	−3.09°	−38.23°
III	10	1	−3.18°	−39.35°
Residue		9.2		

[8] Glattfeld and Hanke, This Journal, **40**, 976 (1918), footnote.

Crops I, II and III from alcohol were united and recrystallized from 10 parts of alcohol, and thus 78 g. of crystals was obtained with a specific rotation of −32.67°. Six subsequent recrystallizations from 10 parts of absolute alcohol gave crops each of which had a specific rotation of −32.67°. This indicates that the substance was the pure brucine salt of the other optical component of the *dl*-acid. The melting point of this acid was 169°.

In all cases of recrystallization from alcohol, the mother liquors when concentrated *in vacuo* at 60° yielded second crops of crystals, the specific rotations of which were much higher than −32.67°, the invariable rotation of the first crops. That this was due to free brucine was shown as follows. When separate samples from Crops I, II and III, from alcohol, $[\alpha]^{20}_{D}$ −34.28, −38.23, −39.35, respectively, were refluxed for 2 hours with benzene, the specific rotations of the extracted brucine salts fell in each case to −32.67° and the benzene was shown to contain brucines. Hydrolysis was not observed in the case of recrystallization of the pure brucine salt of low rotation from water. This salt is apparently perfectly stabel in water solution. Furthermore the specific rotation of this salt was not changed after it had been boiled for 2 hours with benzene under a reflux condenser.

The Optically-active 1,3-Dihydroxy-butyric Acids.—The active acids were obtained by the usual method[9] from the pure brucine salts by treatment with barium hydroxide, etc. The crude acid was distilled at 3 mm. and the distillate in each case was a perfectly colorless, transparent, mobile, odorless liquid with a boiling point of 96°. The yield from 100 g. of brucine salt $[\alpha]^{20}_{D}$ −20.79°, was 18.0 g. and from 100 g. of brucine salt $[\alpha]^{20}_{D}$ −32.67°, was 17.8 g.

The specific rotations of the two acids were now determined with the following results. In each case an exactly 4% solution was used in a 2dcm. tube. The specific gravity was 1.010 and observations were made at approximately 20°.

The figures show that equilibrium between free acid and lactone was reached in water solution at ordinary temperatures in 7 days. Experi-

	α. Obs. 5 min.	α. Obs. 7 days.	α. Obs. 12 days.	$[\alpha]^{20}_{D}$. 5 min.	$[\alpha]^{20}_{D}$. 7 days.	$[\alpha]^{20}_{D}$. 12 days.
Acid from −20.79° brucine salt	+1.64°	+1.21°	+1.21°	+20.29°	+14.97°	+14.97°
Acid from −32.67° brucine salt	−1.62°	−1.20°	−1.20°	−20.05°	−14.86°	−14.86°

ments with these acids showed that equilibrium was reached at 100° in water solution in 4 hours.

[9] Ref. 8, p. 981, footnote.

```
HOCH2-CH-CH2COOH
      |
      OH
```

$C_4H_8O_4$ Ref. 55

Brucine, cinchonine, quinine and strychnine were tried and brucine found to be best for the resolution. The brucine salt was made in the usual way[1] by adding a slight excess of the alkaloid to an aqueous solution of the acid and heating the mixture on the water-bath. The use of an electrically-driven mechanical stirrer greatly hastened the solution of the alkaloid.

Eighty-three g. of the *dl*-acid and 305 g. of brucine were heated in about 2 liters of water on the boiling water-bath until the solution reacted alkaline to litmus. Three g. of brucine did not go into solution. After extraction with benzol and subjection of the extracted solution to complete vacuum distillation, 337 g. of crude brucine salt was obtained. This was treated with 750 cc. of boiling absolute alcohol. Fifteen g. of flocculent material remained undissolved. This material was separated by hot filtration. The clear filtrate was allowed to stand overnight and deposited crystals which weighed 58.4 g. The mother liquor was subjected to complete vacuum distillation at 60°. The residue, weighing 257.5 g., was dissolved in 300 cc. of boiling absolute alcohol. This solution deposited crystals which weighed 13.55 g. The mother liquor from the second crop was subjected to complete vacuum distillation at 70°. The residue weighed 204 g. This was treated with 140 cc. of boiling alcohol. This solution deposited a third crop of crystals which weighed 54.8 g. A fourth crop of crystals which weighed 17 g. was obtained in the same way by treating the residue, which weighed 141 g., with 72 cc. of hot absolute alcohol. The mother liquor from the fourth crop was subjected to complete vacuum distillation. The residue weighed 130 g. No more crystals could be obtained.

The rotations of these 4 crops of brucine salts were taken in exactly 4% aqueous solution. The density of the solution was considered to be 1.011. The following results were obtained.

Crop.	$[\alpha]^{20}_{D}$.	α in one dcm. tube.
I	—29.42	—1.19
II	—27.95	—1.12
III	—27.95	—1.12
IV	—26.95	—1.09

The rotation of the successive crops of brucine salts convinced us that a partial separation of the *dl*-acid into its optical components had been effected. Our next effort was, therefore, to obtain the less soluble brucine salt in its purest form. Another quantity of crude brucine salt was prepared and recrystallized as above. Again the specific rotation of the first crop was —29.42°. The successive crops had approximately the same rotations as the corresponding ones in the first experiment. The salt with the specific rotation of —29.42° was now dissolved in the smallest possible quantity of hot absolute alcohol. The solution deposited a crop of crystals whose specific rotation was taken. This crop was again dissolved in the smallest possible quantity of absolute alcohol and allowed to stand overnight during which time it deposited a crop of crystals, which was separated by filtration and dried to constant weight *in vacuo* over sulfuric acid. The rotation of this crop was taken. This process was repeated 7 times. The rotations of the successive crops were invariably between —29.18° and —29.42°. It was, therefore, concluded that the *pure* brucine salt of one of the optical isomers had the specific rotation of approximately —29.42°.

The Free Acid from the Brucine Salt, $[\alpha]^{20}_{D}$, —29.42°.—One hundred and thirty-seven g. of the brucine salt, $[\alpha]^{20}_{D}$ —29.42°, was dissolved in about 4 liters of hot water and treated with a hot solution of 118 g. of crystallized barium hydroxide in the usual way[1] to remove brucine from its salts. After the removal of brucine by filtration and extraction of the filtrate with benzol, the barium was removed with sulfuric acid and the filtrate from the barium sulfate was subjected to complete vacuum distillation. The residue was taken up in absolute alcohol and the solution was filtered to remove any inorganic salts present. The alcohol was completely removed by distillation *in vacuo* finally at 100°, and the acid was left as a clear yellow and very mobile oil.

The rotation of a portion of this acid was then taken and another portion was titrated. Other portions were converted into the barium and calcium salts and the phenylhydrazid.

Rotation.—The specific rotation of the acid in approximately 4% solution was found to be —8.29°, *i. e.*, 2.32 g. acid dissolved in 45.78 g. water gave α in a one dcm. tube —0.40°. The density of the solution was assumed to be 1.00 and the temperature was approximately 20°.

The Free Acid from the Non-crystallizable Brucine Salt.—The mother liquor from the 4 crops of crystalline brucine salts was subjected to complete vacuum distillation at 60° and gave a residue which weighed 130 g. as mentioned above. The brucine was set free in the usual way and 15.25 g. of a light brown ether-soluble oil was obtained. The specific rotation of this acid was determined as +7.18°, *i. e.*, 1.65 g. of acid in 39.6 g. of water gave α +0.287° in a one dcm. tube. As the specific rotation of the optical isomer of the acid from the brucine salt rotating —29.42° should be +8.29°, the acid now under consideration was evi-

the calculated amount is then added and the mixture heated on the boiling water-bath for 15 minutes. It is then cooled and a measured quantity of 0.1 *N* hydrochloric acid is added to acid reaction and the mixture boiled to expel carbon dioxide. It is again cooled and the excess hydrochloric acid is determined with 0.1 *N* sodium hydroxide. solution. The quantity of sodium hydroxide solution added at first in the cold determines the amount of free acid present and the remainder added determines the quantity of lactone.

dently still contaminated with some of the racemic acid. It was, therefore, converted into the barium salt. The 15.25 g. of acid yielded 14.8 g. of the vacuum-dried barium salt. This salt had a specific rotation of —1.48°, *i. e.*, one g. of salt in 24 g. of water gave α —0.03°. This indicates that the salt was pure, as this rotation is equal and opposite to that of the barium salt from the crystalline brucine salt. The gum was set free from this salt and purified in the usual way. The weight of free acid was 6 g. It was a light yellow mobile oil.

The rotation of a portion of this acid was taken, another portion was titrated and another portion was converted into the phenylhydrazid with the following results.

The Rotation.—The specific rotation of the acid was found to be +8.00°, *i. e.*, 0.83 g. of acid dissolved in 19.92 g. of water gave α +0.32° in a one dcm. tube.

[1] This Journal, 40, 976 (1918). Footnote. [1] *Ann.*, 376, 35 (1910).

$CH_3CH-CH-COOH$
OH OH

$C_4H_8O_4$ Ref. 56

(−)-Erythro-2,3-dihydroxybutyric Acid (1)

20 g (0.165 mole) of (±)-*erythro*-2,3-dihydroxybutyric acid was dissolved in 250 ml of water. The solution was heated to 50 °C and neutralized with 60 g (0.185 mole) of quinine. The cooled solution was filtered to remove undissolved quinine and cooled to 0°. After two days the crystals formed were collected and washed with ice water. The crystals (30 g) were recrystallized twice from water and dried *in vacuo* over H_2SO_4. Yield, 29.0 g; m.p. 195°; $[\alpha]_D^{20}$ −115° (*c*, 5.04, H_2O); [lit. (4), $[\alpha]_D^{20}$ −113° (*c*, 5.0, H_2O)].

Concentration of the original filtrate gave a further 6 g of crystalline material.

The free acid was regenerated by passing an aqueous solution of the resolved quinine salt through a Bio-Rex-70 (H^+) resin column (7) yielding (−)-*erythro*-2,3-dihydroxybutyric acid, $[\alpha]_D^{25}$ −9.5° (*c*, 1.0, H_2O); [lit. (4), $[\alpha]_D^{20}$ −9.30° (*c*, 0.5, H_2O)].

Rotatory dispersion (*c*, 0.5, H_2O): $[\alpha]_{589}$ −9.5°, $[\alpha]_{250}$ −1060°, $[\alpha]_{221}$ −5380° (trough), $[\alpha]_{210}$ −1790°.

NOTE: Because of the poor recovery of the resolved acids from their salts in procedures previously reported, i.e., E. Hoff-Jorgensen, Z. Physiol. Chem., 268, 194 (1941) and R.S. Morrell and E.K. Hanson, J. Chem. Soc. 85, 197 (1904), the authors have devised a new general method for the regeneration of optically active acids and bases from resolved salts which is found in F.W. Bachelor and G.A. Miana, Can. J. Chem., 45, 79 (1967) and is given below.

The resolution of optically active acids and bases is a sufficiently tedious task that extensive losses on regeneration from the resolved salts are frustrating. We have availed ourselves of a technique which essentially circumvents the pitfalls of classical chemical regeneration and yields the resolved acid or base almost quantitatively.

Although ion-exchange resins have been used for many purposes, their utility for the regeneration of simple organic salts into their primary components does not seem to be a common practice. This method is particularly useful when the salt is a resolved salt of an optically active acid or base. We were interested in preparing the optically active enantiomers of the *erythro* and *threo* dihydroxybutyric acids according to methods already reported (1, 2), but found that the chemical regeneration of the acids led to mostly polymeric material and very little crystalline resolved acids. In addition, one of the resolving agents, quinine, could not be recovered (3) and constituted a considerable expense, since we wished to use the process for a large-scale preparative scheme. A similar situation occurs in the resolution of α-phenylethylamine with *l*-malic acid (4), in which the recovery of the relatively expensive acid is tedious and in the hands of the uninitiated frequently leads to naught.

The proper choice of a stable ion-exchange resin readily allows one to recover both the acid and base free from inorganic ions or other hard-to-remove constituents. We chose to use Bio-Rex 70,[2] a carboxylic acid resin, since we were primarily interested in the acids. The passage of either the quinine salt of *l-erythro*-dihydroxybutyric acid or the quinidine salt of *l-threo*-dihydroxybutyric acid, in warm water, through a column of Bio-Rex 70 yielded the free acids in aqueous solution. Removal of the water *in vacuo* gave almost quantitative yields of the acids, which were readily crystallized. A similar sequence yielded *l*-malic acid from its resolved α-phenylethylamine salt.

In addition, the bases could be recovered from the columns by elution with 1 equivalent of hydrochloric acid followed by a sufficient amount of water to wash the hydrochloride salt through the column. The base could then be recovered as the hydrochloride, or the salt could be neutralized with ammonia and the base extracted with a suitable organic solvent. In this manner, quinine was recovered in a 67% yield.

After the base is washed from the ion-exchange resin, the column is again ready for use. This procedure appears to be quite general and should be easily adaptable to large-scale processes.

---

[2]Purchased from Bio Rad Laboratories, 32nd and Griffin Ave., Richmond, California.

1. E. Hoff-Jørgensen. Z. Physiol. Chem. **268**, 194 (1941).
2. J. W. E. Glattfield and J. W. Chittum. J. Am. Chem. Soc. **55**, 3663 (1933).
3. E. L. Eliel. Stereochemistry of carbon compounds. McGraw-Hill Book Co., Inc., New York. 1962. p. 51.
4. A. H. Blatt (*Editor*). Organic synthesis. Collective Vol. II. John Wiley & Sons, Inc., New York. 1943. p. 506.

General Method

An 18 mm glass column is filled with 100 g of Bio-Rex 70 (H^+) in water. It is then washed with distilled water until the effluent is neutral. Five grams of the resolved salt in a minimum amount of water (warmed if necessary) is placed on the column and eluted with distilled water until no more organic acid comes off the column. (This generally requires about 300 ml of water.) The base is then eluted by adding an equivalent amount of 5 *N* hydrochloric acid to the top of the column and eluting with water until no more acidic material is eluted.

Both the acid and base hydrochloride are isolated in an essentially pure state by removing the water *in vacuo*. Alternatively, the free base is isolated by neutralizing the hydrochloride solution and extracting with chloroform.

```
           OH
           |
CH3-CH-CH-COOH
     |
     OH
```

$C_4H_8O_4$ Ref. 57

(−)-Threo-2,3-dihydroxybutyric Acid (2)

To a hot solution of 20 g (0.165 mole) of (±)-*threo*-2,3-dihydroxybutyric acid in 500 ml of water 60 g (0.185 mole) of quinidine was added. The hot solution was filtered and cooled. The crystalline salt (37.3 g) was dried in a vacuum desiccator over H_2SO_4. Concentration of the filtrate gave a further 2.0 g of material. After several recrystallizations from water, a salt with constant rotation was obtained, m.p. 117°; $[\alpha]_D^{20}$ +145.5° (*c*, 3.3, H_2O); [lit. (6), m.p. 114°, $[\alpha]_D^{18}$ +142.2°, (*c*, 3.31, H_2O)].

Regeneration of the free acid was accomplished in the same manner as with the *erythro* acid yielding (−)-*threo*-2,3-dihydroxybutyric acid as a syrup, $[\alpha]_D^{25}$ −17.75° (*c*, 1.0 H_2O); [lit. (6), $[\alpha]_D^{16}$ −13.51° (*c*, 6.0, H_2O)].

Rotatory dispersion (*c*, 0.5, H_2O): $[\alpha]_{589}$ −17.75°, $[\alpha]_{250}$ +360°, $[\alpha]_{221}$ +2640° (peak), $[\alpha]_{210}$ +880°.

6. R. S. Morrell and E. K. Hanson. J. Chem. Soc. **85**, 197 (1904).

```
CH3CH2CHCOOH
        |
        SO3H
```

$C_4H_8O_5S$ Ref. 58

Le *sulfobutyrate neutre de quinine* se dépose aussi bien d'une solution contenant une molécule de quinine, qu'en présence de la quantité théorique de deux molécules. Il se présente sous la forme de jolies rosettes.

Substance 0.2160 g.; perte de poids à 110° 0.0385 g.
Trouvé: H_2O 17.82.
$C_4H_8O_5S \cdot 2\,C_{20}H_{24}O_2N_2 \cdot 10\,H_2O$ (996.7). Calculé: H_2O 18.08.

Dédoublement de l'acide sulfobutyrique à l'aide de strychnine.

La cristallisation du sulfobutyrate acide de strychnine dans l'eau chaude donne un mélange, contenant le sel de l'acide dextrogyre en excès.

Afin de le libérer de l'autre composant, on le fait recristalliser à plusieurs reprises dans huit fois son poids d'eau.

Comme dans les sels neutres l'acide sulfobutyrique possède une rotation plus élevée — quoique de signe opposé — que dans les sels acides, l'état de pureté du sulfobutyrate acide de strychnine peut être contrôlé, en le transformant dans le sel neutre de baryum.

Une méthode plus économiqne et plus commode consiste à examiner la rotation de l'eau-mère de chaque cristallisation après sa décomposition par la baryte.

Dès que la rotation de l'eau-mère est devenue constante, elle doit être égale à celle des cristaux.

Voici, à titre d'exemple, les rotations moléculaires des sels barytiques, préparés par décomposition de six eaux-mères successives: — 12°, — 21°, — 33°, — 33°, — 32°.

Donc la séparation des deux antipodes était complète. En effet, le sel de strychnine de la dernière cristallisation a donné un sel barytique de la même rotation moléculaire.

Afin d'obtenir le sel de l'acide sulfobutyrique antipode, on évapore l'eau-mère de la première cristallisation. Le résidu sirupeux cristallise par refroidissement en grandes rosettes graisseuses.

Ce sel de strychnine est cristallisé plusieurs fois dans l'alcool absolu.

Dans ce cas aussi, les eaux-mères, dont on prépare le sel barytique, rendent d'excellents services au contrôle.

Voici une série de rotations moléculaires des sels barytiques, préparés aux dépens des eaux-mères depuis la troisième cristallisation jusqu'à la sixième: + 15°, + 21°, + 26°, + 30°.

On peut aussi obtenir ce sel de strychnine à l'état pur, libre de l'autre composant, en le précipitant partiellement de sa solution alcoolique par de l'éther.

On prépare les sels barytiques des acides sulfobutyriques actifs, en décomposant les sels de strychnine par un petit excès de baryte, en éliminant l'alcaloïde par filtration et par épuisement au chloroforme, en précipitant l'excès de baryte par l'anhydride carbonique, et en évaporant enfin la solution filtrée.

On peut le faire recristalliser dans l'eau par évaporation partielle ou bien par précipitation à l'alcool.

Dédoublement de l'acide au moyen de quinine.

Cette réaction a réussi de la manière suivante:

21.1 gr. de quinine (0.065 mol.) sont dissous à chaud dans un demi-litre d'une solution d'acide sulfobutyrique contenant 0.065 gr. mol.

Par refroidissement, le sel neutre de quinine cristallise en état volumineux et blanc, l'eau-mère restant verte.

Le sel consistait déjà pour les trois quarts en dérivé de l'acide lévogyre, puisqu' un échantillon, transformé en sel barytique, montrait une rotation moléculaire de + 20°.

On a fait recristalliser le sel de quinine encore cinq fois dans l'eau et ensuite on l'a décomposé à froid par l'eau de baryte.

Après filtration et extraction des restes de quinine par le chloroforme, l'excès de baryte a été enlevé par l'anhydride carbonique.

Dans la solution filtrée et concentrée, l'alcool a précipité le sel barytique de l'acide sulfobutyrique lévogyre. Sa rotation moléculaire dans une solution de 3.3 °/₀ était de + 30.7°.

```
CH3-CH-CH2COOH
     |
     SO3H
```

$C_4H_8O_5S$ Ref. 59

Séparation de l'acide dextrogyre à l'aide de brucine. Une solution d'acide β-sulfobutyrique 0.3 n, chauffée au bain-marie avec deux molécules de brucine, filtrée et refroidie lentement, sépare le sel neutre de brucine en grands cristaux bien formés.

On a fait recristalliser le sel, à plusieurs reprises, dans cinq fois son poids d'eau, et chaque fois on a examiné l'eau-mère de la façon suivante:

L'eau-mère, décomposée par un petit excès de baryte, filtrée, extraite trois fois au chloroforme pour écarter les restes de brucine, traitée à l'acide carbonique, filtrée et concentrée, a donné par addition d'alcool le sel de baryum qu'on a examiné au polarimètre.

Il faut répéter la cristallisation du sel de brucine, jusqu'à ce que le pouvoir rotatoire du sel de baryum ait atteint son maximum.

Pour les sels de baryum, provenant de cinq eaux-mères successives, on a trouvé les rotations spécifiques que voici:

$$[\alpha]_D = -2°.4, +1°, +5°3, +5°.8, +5°.8.$$

Le sel de brucine de la dernière cristallisation s'est montré identique au sel dissous dans l'eau-mère, en donnant un sel de baryum de la même rotation de + 5°.8. Dès lors, le d-β-sulfobutyrate de brucine était déjà pur après la troisième cristallisation.

Séparation de l'acide lévogyre à l'aide de quinine. Une solution de l'acide (0.16 n), dans laquelle on avait dissous une molécule de quinine par molécule d'acide, a déposé le sel dit neutre, renfermant deux molécules de quinine. Cristallisé deux fois dans l'eau chaude, il a donné par décomposition à la baryte un sel de baryum de rotation $[\alpha]_D$ = — 5°.1.

Comme le maximum pour le sel barytique de l'acide droit se trouve à 5°.8, la séparation est donc presque complète. Le seul inconvénient de la méthode consiste en ce que les solubilités du sel de quinine dans l'eau chaude et dans l'eau froide diffèrent peu, de sorte qu'on perd beaucoup de matière.

C'est pour cela que nous nous sommes servis d'une autre méthode pour la préparation de l'acide gauche pur.

Séparation de l'acide lévogyre par cristallisation d'un mélange actif de sels barytiques.

Cette méthode repose sur la connaissance de la solubilité des sulfobutyrates racémique et actif de baryum.

β-Sulfobutyrate racémique de baryum: 100 gr. d'une solution saturée à 29°.8 renferment 22°.7 gr. de sel racémique anhydre.

β-Sulfobutyrate dextrogyre de baryum: 100 gr. d'une solution saturée à 29°.8 renferment 23.7 gr. de sel actif anhydre.

Donc, les solubilités du sel racémique et du sel actif sont presque égales. Il en résulte qu'un mélange dans lequel le sel actif prédomine pourra fournir par cristallisation ce sel à l'état pur.

La première liqueur-mère, obtenue dans la préparation de l'acide dextrogyre, a donné un sel barytique lévogyre, renfermant 30 % du sel racémique ($[\alpha]_D$ = — 4° au lieu de 5°.8).

Ce mélange, cristallisé dans l'eau, a donné de beaux cristaux clairs et rectangulaires du 1-sulfobutyrate de baryum pur ($[\alpha]_D$ = —5°.8).

L'eau-mère de cette cristallisation a enfin servi à la séparation à l'aide de quinine.

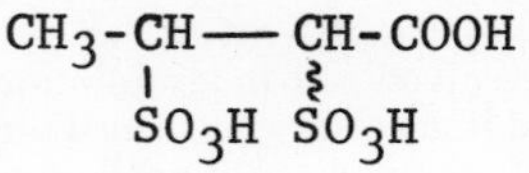

$C_4H_8O_8S_2$ Ref. 60

2. *Dédoublement optique de l'acide α,β-disulfobutyrique.* Après douze recristallisations du sel de strychnine dans l'eau, on a trouvé le même pouvoir rotatoire pour le sel cristallisé et pour le sel dissous.

On décompose le sel de strychnine par la soude caustique et on élimine la strychnine par filtrage et par extraction au chloroforme. Après avoir mesuré la rotation du sel sodique, on ajoute de l'acide hydrochlorique afin de pouvoir examiner l'acide libre.

Nous avons trouvé pour le sel sodique $[M]_D$ = —6° et pour l'acide $[M]_D$ = +16°.

Dispersion rotatoire de l'acide:

λ(μμ) =	656	589	546	486.
[M] =	14°	16°	20°	25°.

Une solution du sel sodique, additionnée d'un excès (3 mol.) de soude caustique et chauffée à 100°, a été examinée à différents intervalles.

heures	0	3½	23	38	62	81	125.
rotation	—0.10	—0.07	—0.06	—0.04	0.00	+0.04	0.00°.

Donc la valeur de la rotation gauche tend vers zéro, pour changer ensuite de signe. Il est probable que l'hydrogène de l'atome de carbone-α subit la racémisation et que la rotation droite due à l'atome-β se maintient. Enfin la rotation devient égale à zéro, mais alors la solution réduit le permanganate et le produit s'est donc décomposé.

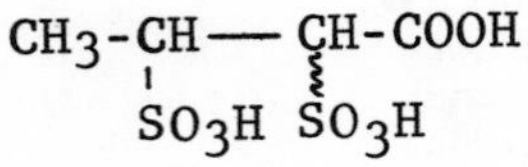

$C_4H_8O_8S_2$ Ref. 61

Dédoublement optique de l'acide α,β-disulfobutyrique..

On fait recristalliser le sel de strychnine plusieurs fois dans de l'eau; la solubilité du produit diminue régulièrement. On décompose le sel au moyen de soude caustique et on examine le pouvoir rotatoire du sel sodique. Après douze cristallisations du sel de strychnine, le pouvoir rotatoire du sel sodique est constant. Nous avons trouvé les valeurs suivantes:

Acide: $[M]_D = +9.4°$. Sel sodique: $[M]_D = +9.1°$.

Racémisation. Une solution de l'acide, chauffée pendant 5 heures à 100°, s'est racémisée complètement.

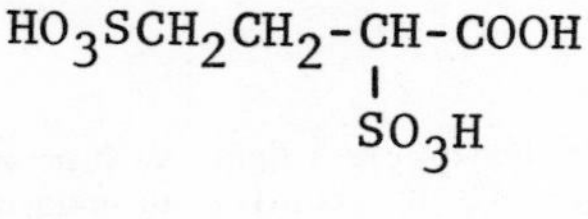

$C_4H_8O_8S_2$ Ref. 62

3. *Dédoublement optique de l'acide α,γ-disulfobutyrique.* Le sel de strychnine, qui cristallise très bien, peut servir au dédoublement. Après sept recristallisations dans huit fois son poids d'eau, le sel cristallin actif est identique à celui dissous dans l'eau-mère. On trouve pour la rotation: $[M]_D$ acide 7.2°; $[M]_D$ sel 9.1°.

Dispersion rotatoire.

λ (μμ)	656	589	546	486.
[M] acide	7°	9°	11°	14°.
[M] sel	6°	7°	10°	12°.

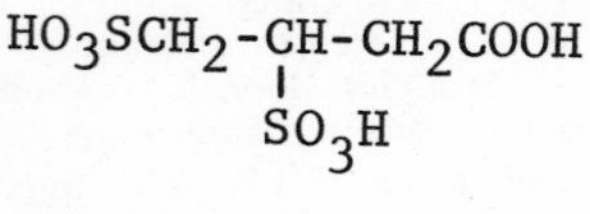

$C_4H_8O_8S_2$ Ref. 63

6. ***Dédoublement optique de l'acide β,γ-disulfobutyrique.*** Le sel de strychnine de l'acide β,γ-disulfobutyrique, préparé aux dépens de l'acide γ-sulfobutyrique (réaction 4), est recristallisé six fois dans l'eau. Le sel cristallisé présente alors la même rotation que le sel dissous. L'acide dextrogyre, dont le sel de strychnine est le moins soluble, s'obtient à l'état pur.

Rotation moléculaire de l'acide et du sel barytique neutre.

λ (μμ)	656	589	546	486.
[M] acide	47°	59°	69°	92°.
[M] sel neutre	41°	51°	59°	81°.

L'acide β,γ-disulfobutyrique, qu'on a obtenu aux dépens de l'acide γ-chlorovinylacétique (5), recristallisé plusieurs fois à l'état de sel de strychnine, donne les mêmes valeurs pour la rotation moléculaire de l'acide et du sel.

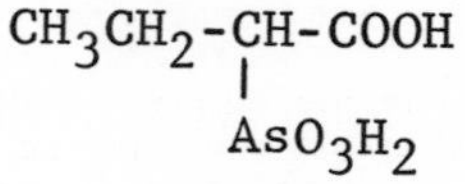

$C_4H_9AsO_5$ Ref. 64

2. *Acide α-arsonobutyrique.* 15 g d'acide α-arsonobutyrique neutralisés par de la soude caustique et la quantité calculée de chlorhydrure de quinine (volume total du mélange 5 litres) ont donné un dépôt de 30 g du sel secondaire de quinine.

Substance 0.4985 g; perte à 80° 0.0469 g.
„ 0.2555 g; $Mg_2As_2O_7$ 0.0406 g.
„ 0.2077 g; 10.74 cm^3 de N_2 (18.5°, 753 mm).
Trouvé: H_2O 9.41; As 7.67; N 5.86.
$C_4H_9O_5As . 2 C_{20}H_{24}O_2N_2 . 5 H_2O$ (950.53). Calculé: „ 9.48; „ 7.88; „ 5.88.

On fait recristalliser le sel plusieurs fois selon la méthode décrite pour l'arsonopropionate de quinine, en versant la solution alcoolique dans de l'eau bouillante.

Poids	30	11	8.5	4.5	2.7	1.5	0.8 g
$[M]_D$ sel	—	—	— 7.6°	— 9.8°	—10.2°	—10.5°	—10.5°
„ acide	—	+12.0°	+18.5°	+21.6°	+24.1°	+25.5°	+25.7°

Dispersion rotatoire.

λ (μμ)	674.0	626.5	589.3	560.5	536.5	516.0
[M] sel	—	— 9.4°	—10.5°	—11.9°	—13.4°	—
„ acide	+19.7°	+22.7°	+25.7°	+29.1°	+32.5°	+36.3°

$CH_3CH_2CHCOOH$
$\qquad\qquad|$
$\qquad\quad NH_2$

$C_4H_9NO_2$ **Ref. 65**

d-Benzoyl-α-Aminobuttersäure.

41 g r-Benzoylaminobuttersäure und 60 g Morphin werden in 125 g heissem Wasser gelöst. Wird dann auf 0° abgekuhlt und die Wandung des Gefässes öfter mit einem Glassstabe gerieben, so beginnt nach einigen Studen die Krystallisation. Sorgt man durch häufiges Umrühren für Vertheilung der Krystalle in der dicklichen Flüssigkeit, so sind bei 0° nach etwa 15 Stunden 39 g derselben abgeschieden. Sie wurden abgesaugt und mit wenig Eiswasser gewaschen. Die Mutterlauge, auf etwa drei Viertel ihres ursprünglichen Volumens eingedampft, giebt eine zweite Krystallisation von etwa 7 g. Zur Reinigung wird das Morphinsalz noch viermal aus heissem Wasser umkrystallisirt, wobei man immer an Wasser 5/4 vom Gewichte des Salzes nimmt. Dabei verliert man ungefähr die Hälfte der ersten Krystallisation, sodass die Ausbeute an reinem Salz etwa 40 pCt. der Theorie beträgt. Das Salz bildet ziemlich grosse, spiessige Krystalle, welche bei 100° getrocknet den Schmp. 145-146° (uncorr.) zeigten.

Um daraus die active α-Benzoylaminobuttersäure zu gewinnen, löst man in der 12-fachen Menge Wasser, kühlt auf 0° ab und fügt so lange eine Lösung von Ammoniumcarbonat hinzu, als noch eine Fällung erfolgt. Das Morphin wird filtrirt und die Mutterlauge mit einem geringen Ueberschuss von Salzsäure versetzt. Dabei fällt der grossere Theil der activen Benzoylaminobuttersäure aus, den Rest gewinnt man durch Eindampfen der Mutterlauge im Vacuum. Das Product wird einmal aus heissem Wasser umkrystallisirt.

Für die Analyse war es bei 100° getrocknet.

0.2104 g Subst.: 12.7 ccm N (20°,762 mm). 0.2016 g Subst.: 0.4717 g. CO_2, 0.1152 g H_2O.

$C_{11}H_{13}O_3N$. Ber. C 63.76, H 6.28, N 6.76.
Gef. " 63.80, " 6.35, " 6.90.

Die Substanz schmilzt bei 120-121° (corr.), mithin erheblich niedriger als der Racemkörper; sie ist auch in Wasser leichter löslich, denn bei 20° verlangt sie davon nur 93 Theile. Desgleichen wird sie von anderen Lösungsmitteln durchgehends leichter aufgenommen. Für die optische Bestimmung diente eine wässrige Lösung, welche mit der für 1 Molekül berechneten Menge Natronlauge hergestellt war.

Gewicht der Lösung 14.3061 g. Gewicht der Substanz 1.1 g, spec. Gewicht 1.0391, Procentgehalt 7.68 pCt., Temperature 20°, Drehung bei Natrium-Licht im 2-Decimeterrohr 4.9°.

Mithin in alkalischer Lösung $[\alpha]_D^{20} = -30.7°$.

Eine zweite Bestimmung unter denselben Bedingungen gab +30.8°.

d-α-Aminobuttersäure.

Die Benzoylverbindung wird mit der fünffachen Menge 10-procentiger Salzsäure 6 Studen am Rückflusskuhler gekocht, die Lösung nach dem Erkalten ausgeäthert und auf dem Wasserbade zur Trockne verdampft, wobei das Chlorhydrat der Aminobuttersäure als Schwach braun gefärbte Krystallmasse resultiert. Dasselbe lässt sich durch Lösen in absolutem Alkohol unter Zusatz von einigen Tropfen alkoholischer Salzsäure und Fällen mit Aether in feinen, farblosen Nadeln gewinnen, welche fur die Analyse bei 100° getrocknet wurden.

0.2032 g Subst.: 0.2095 g AgCl

$C_4H_{10}O_2NCl$. Ber. Cl 25.45, Get. Cl 25.45.

Das Salz ist in Wasser sehr leicht löslich und dreht nach rechts.

Gesammigewicht der Lösung 15.6917 g. Gewicht der Substanz 0.7801 g, spec. Gewicht 1.0201, Procentgehalt 4.97 pCt., Temperatur 20°, Drehung bei Natrium-Licht im 2-Decimeterrohr +1.46°.

Mithin $[\alpha]_D^{20°} = +14.51°$.

Um aus dem Salz die freie Aminosäure darzustellen, löst man in ungefähr 70 Theilen Wasser und kocht mehrere Stunden mit einem grossen Ueberschuss von gelbem Bleioxyd bis eine Probe keine Chlorreaction mehr giebt. Das Filtrat wird dann mit Schwefelwasserstoff gefällt, mit Thierkohle aufgekocht, die farblose Flüssigkeit auf dem Wasserbade bis zur beginnenden Krystallisation verdampft und mit Alkohol versetzt. Dabei fällt die Ami osäure in sehr feinen, farblosen Blättchen. Die Ausbeute ist nahezu theoretisch. Für die Analyse wurde bei 100° getrocknet.

0.2037 g Subst.: 0.3494 g CO_2, 0.1604 g H_2O. - 0.2017 g Subst.: 24 ccm N (24°,766 mm).

$C_4H_9NO_2$. Ber. C 46.60, H 8.73, N 13.59.
Gef. " 46.78, " 8.75, " 13.45.

Im geschlossenen Capillarrohr schmilzt sie unter Zersetzung gegen 303°(corr.), mithin nur wenige Grade niedriger als der Racemkörper. In wässriger Lösung giebt sie mit Kupferacetat ähnlich dem Racemkörper ein schwer lösliches, blaues Kupfersalz. Die wässrige Lösung dreht nach rechts.

0.738 g Subst.: Gesammtgewicht der Lösung 13.6546 g. Mithin Procentgehalt 5.406 pCt. Spec. Gewicht 1.0102. Temperatur 20°. Drehung im 2-Decimeterrohr bei Natriumlicht +0.87°. Mithin spec. Drehung in wässriger Lösung.

$[\alpha]_D^{20°} = +8.0°$.

Zur Prufung auf Reinheit wurde die Substanz in derselben Weise wie der Racemkörper benzoylirt und die Benzoylverbindung, in Wasser mit der äquivalenten Menge Natronlauge gelöst, für die optische Bestimmung benutzt.

Eine Lösung von 4.47 pCt. gab, im Decimeterrohr bei 20° gepruft, die specifische Drehung +29°, während eine Controllprobe mit ganz reiner d-Benzoylaminobuttersäure unter den gleichen Bedingungen +29.35° zeigte. Die kleine Abweichung von dem früher gegebenen Werthe ist wohl durch die geringere Concentration der Lösung bedingt.

l-Benzoyl-α-Aminobuttersäure.

Für die Bereitung der Säure dient entweder der Racemkörper, oder, noch besser, die Mutterlauge von der Krystallisation des Morphinsalzes. Sie wird zunächst zur Entfernung des Morphins mit Ammoniumcarbonat gefällt und das Filtrat mit Salzsäure angesäuert. Dabei fällt ein Gemisch von linksdrehender und racemischer Benzoylaminobuttersäure aus; den in der Lösung verbleibenden Rest gewinnt man durch Verdampfen im Vacuum. Die vereinigten Producte werden durch einmalige Krystallisation aus heissem Wasser gereingt.

Für die Bereitung des Brucinsalzes, welches die Isolirung der Linksverbindung ermöglicht hat, löst man 50 g der gepulverten Säure mit 112 g Brucin in 190 g kochendum Wasser und überlasst die syrupöse und schwach bräunliche Lösung im Eisschrank der Krystallisation. Diese lässt sich durch Impfung sehr beschleunigen und ist dann nach 24 Stunden in der Regel beendet. Das ausgeschiedene Product wird abgesaugt und mit wenig eiskaltem Wasser gewaschen. Zur völligen Reinigung ist nochmalige Krystallisation aus heissem Wasser unter denselben Bedingungen wie beim Morphinsalz nöthig. Dabei entstehen so grosse Verluste, dass die Schliessliche Ausbeute nicht mehr als 25 pCt. der Theorie beträgt. Das reine Salz bildet Ziemlich grosse, durchsichtige Krystalle, welche, uber Schwefelsäure getrocknet, bei 86-87° schmelzen.

Zur Gewinnung der freien Säure löst man 20 g des Salzes in 200 g Wasser, fällt mit 30 ccm Normalnatronlauge, kühlt auf 0° ab, filtrirt und versetzt die Flüssigkeit mit 31 ccm Normalsalzsäure. Der grösste Theil der Benzoylverbindung wird dabei gefällt, den Rest gewinnt man durch Verdampfen der Mutterlauge im Vacuum. Zur völligen Reingung genugt einmaliges Umkrystallisiren aus Wasser.

Die Substanz wurde für die Analyse bei 100° getrocknet.

0.1883 g Subst.: 11 6 ccm N (19°,755 mm). - 0.2018 g Subst.: 0.4719 g CO_2, 0.115 g H_2O.

$C_{11}H_{13}NO_3$. Ber. C 63.76,H 6.28, N 6.76.
Gef. " 63.74 " 6.33 " 7.03.

Die Substanz zeigt denselben Schmelzpunkt, die gleiche Löslichkeit und dieselbe äussere Form der Krystalle, wie der optische Antipode, sie dreht aber in alkalischer Lösung nach links.

Gewicht der Lösung 13.803 g. spec. Gewicht 1.0392, Gewicht der Substanz 1 g, Procentgehalt 7.25, Temperatur 20°. Drehung bei Natriumlicht im 2-Decimeterrohr -4.80°.

Mithin $[\alpha]_D^{20°} = -31.8°$,

während für die d-Verbindung +30.68° gefunden wurde.

l-α-Aminobuttersäure.

Sie wurde genau so dargestellt, wie die d-Verbindung, und zeigte mit Ausnahme der Drehung ganz die gleichen Eigenschaften.

Für die freie Säure wurde gefunden in wässriger Lösung $[\alpha]_D^{20°} = -7.92°$ bein 5.31 pCt. Gehalt.

Gewicht der Lösung 16.643 g, spec. Gewicht 1.009, Gewicht der Substanz 0.8836 g, Temperatur 20°. Drehung bei Natriumlicht im 2-Decimeterrohr -0.85°.

Das salzsaure Salz gab in wässriger Lösung bei 4.77 pCt. Gehalt $[\alpha]_D^{20°} = -14.34°$.

Gewicht der Lösung 11.4952 g. spec. Gewicht 1.0202, Gewicht der Substanz 0.549 g, Temperatur 20°. Drehung bei Natriumlicht im 2-Decimeterrohr -1.40°.

Die Abweichungen von den Werthen, welche bei den d-Verbindungen gefunden wurden, liegen innerhalb der Versuchsfehler.

$C_4H_9NO_2$ Ref. 65a

EXAMPLE 4

Resolving of N,N-dibenzyl-DL-α-amino butyric acid

(A) PREPARATION OF L(+)-THREO-1-p-NITRO PHENYL-2-AMINO PROPANE-1,3-DIOL AND N,N-DIBENZYL-L(−)-α-AMINO BUTYRIC ACID

40 g. of N,N-dibenzyl-DL-α-amino butyric acid and 31 g. of L(+)-threo-1-p-nitro phenyl-2-amino propane-1,3-diol are dissolved, while heating under reflux, in 500 cc. of absolute ethanol. 360 cc. of warm water are gradually added thereto, thereby maintaining the temperature at the boiling point. The mixture is cooled to 30° C. within two hours. The crystallized salt is filtered off, washed with a small quantity of 50% of ethanol, and recrystallized from 500 cc. of 50% ethanol. 30.5 g. of the salt of L(+)-threo-1-p-nitro phenyl-2-amino propane-1,3-diol and the N,N-dibenzyl-L(−)-α-amino butyric acid are obtained thereby after drying at 80° C. in a drying oven. The resolving yield amounts to 87%. Said compound forms yellow leaflets which are insoluble in water, ether and chloroform, and only slightly soluble in ethanol and acetone. Its melting point is 182° C.; its rotatory power is $[\alpha]_D^{20}=-32.5°\pm1°$ (concentration: 1% in methanol).

(B) PREPARATION OF N,N-DIBENZYL-L(−)-α-AMINO BUTYRIC ACID

30.5 g. of the salt of L(+)-threo-1-p-nitro phenyl-2-amino propane-1,3-diol and N,N-dibenzyl-L(−)-α-amino butyric acid obtained as described hereinabove under (A) are treated with 67 cc. of N sodium hydroxide solution for 15 minutes while agitating vigorously. The crystallized L(+)-threo-1-p-nitro phenyl-2-amino propane-1,3-diol is separated from the mother liquor. In this manner, 12.3 g., corresponding to 40% of the initially employed resolving agent, are recovered. The filtrate is acidified by the addition of 4.5 cc. of acetic acid. The acid mixture is extracted with chloroform. The chloroform solution is washed with water, dried over magnesium sulfate, and evaporated to dryness in a vacuum. The oily residue is dissolved in 70 cc. of warm methanol. 35 cc. of water are added to said solution and the mixture is allowed to crystallize for one hour with cooling. The crystals are filtered off, washed with a small quantity of 50% methanol, and dried in a drying oven at 80° C. Thereby, 16.7 g. of optically pure N,N-dibenzyl-L-(−)-α-amino butyric acid, corresponding to a yield of 95% of the theoretical yield, are obtained. Said compound forms white crystals which are insoluble in water, soluble in ethanol, acetone, benzene, and chloroform, and only slightly soluble in ether. Its melting point is 133–134° C.; its rotatory power is $[\alpha]_D^{20}=-34.5°\pm1\%$ (concentration: 2% in 5 N hydrochloric acid) or, respectively,

$$[\alpha]_D^{20}=-96.5°\pm1°$$

(concentration: 2% in methanol).

(C) TREATMENT OF THE MOTHER LIQUORS

The motor liquors obtained on preparing and recrystallizing the salt of L(+)-threo-1-p-nitro phenyl-2-amino propane-1,3-diol and the N,N-dibenzyl-L(−)-α-amino butyric acid are treated as indicated in the preceding examples. Thereby, 12.4 g. of L(+)-threo-1-p-nitro phenyl-2-amino propane-1,3-diol, corresponding to 40% of the initially employed resolving agent, are recovered. 20.2 g. of optically impure N,N-dibenzyl-D(+)-α-amino butyric acid, corresponding to 50.5% of the starting racemic mixture, are obtained. The rotatory power of this acid is $[\alpha]_D^{20}=+75.5°\pm1°$ (concentration: 2% in methanol). Said impure acid is purified by recrystallization from aqueous 75% methanol or by reaction and salt formation with D(−)-threo-1-p-nitro phenyl-2-amino propane-1,3-diol in the same manner as indicated hereinabove in Examples 1(B), 2(B), or 3(B), respectively.

$C_4H_9NO_2$ Ref. 65b

Darstellung von *d*(—)-Leucin: 5 g *racem.* Leucin, 18 g rohe Cholestenon-sulfonsäure und 30 ccm absol. Alkohol werden in der angegebenen Weise vereinigt. Nach wenigen Min. erstarrt die Mischung zu einem Krystallbrei, der abgesaugt, in Alkohol gelöst und filtriert wird. Man engt im Vak. auf 20 ccm ein, läßt über Nacht krystallisieren, saugt ab und wäscht mit absol. Äther nach.

Die alkoholischen Mutterlaugen wurden zur Weiterverarbeitung auf *l*(+)-Leucin aufbewahrt.

Das Salz des *d*(—)-Leucins mit der Cholestenonsulfonsäure stellt weiße Nadeln dar. Schmp. 192—193° unter Gasentwicklung.

0.00759 g Sbst.: 0.158 ccm N (19.1°, 757 mm).

$C_{33}H_{57}O_6NS$. Ber. N 2.45. Gef. N 2.42.

Das Leucin wurde aus dem cholestenonsulfonsauren Salz, wie oben beschrieben, isoliert und gereinigt.

$[\alpha]_D^{20} = -0.74° \times 2.1698/0.0850 \times 1 \times 1.10 = -17.17°$ (in 20-proz. HCl).

Das *d*(—)-Leucin wurde nach dieser Methode also optisch reiner erhalten, als es nach der Methode von E. Fischer möglich ist, wahrscheinlich weil in unserem Arbeitsgang keine Racemisierungsgefahr besteht.

Das aus den alkohol. Mutterlaugen isolierte Leucin zeigte:

$[\alpha]_D^{20} = +0.41° \times 2.3128/0.1138 \times 1 \times 1.1 = +7.6°$ (in 20-proz. HCl).

Die Darstellung von *d*(—)-*n*-α-Amino-buttersäure geschah analog der Darstellung von *d*(—)-Leucin aus 5 g *racem. n.* α-Amino-buttersäure, 23 g Cholestenon-sulfonsäure und 40 ccm absol. Alkohol. Die Krystallisation trat nach wenigen Min. ein. Nach dem Umkrystallisieren stellt das Salz weiße Nadeln dar.

0.00859 g Sbst.: 0.181 ccm N (18.2°, 757 mm).

$C_{31}H_{53}O_6NS$. Ber. N 2.46. Gef. N 2.46.

Die aus dem Salz isolierte Aminobuttersäure zeigte: $[\alpha]_D^{20} = -0.73° \times 2.1064/0,1013 \times 1 \times 1.016 = -14.94°$ (in Wasser unter Zusatz der zur Salzbildung errechneten Menge Salzsäure).

$C_4H_9NO_2$ Ref. 65c

Note:

Using trans-1,2-cyclohexane dicarboxylic acid anhydride as resolving agent, obtained a less soluble diastereomeric amide from dioxane with m.p. 200-201° and $[\alpha]^{25}_{435}$ -8.5° (c 1, methanol). The acid hydrochloride obtained from the amide had a $[\alpha]^{25}_{D}$ -15.0° (c 1, 2N HCl).

$C_4H_9NO_2$ Ref. 65d

EXAMPLE 1

14.5g of N-acetyl-DL-2-aminobutyric acid was dissolved in 1 liter of water and the pH thereof was adjusted to 7.5 with LiOH. To the solution was added 87,000 units of acylase I derived from the kidney of pigs and reacted at 37°C. When the L-2-aminobutyric acid produced was followed by a colorimetric determination using ninhydrin, it was found that about 100% conversion was attained in 17 hours. After the reaction was completed, 1 g of activated carbon was added and the acylase I was removed by filtration. The filtrate was passed through a column of an ion exchange resin Dowex 50-X8 (trade name) to adsorb the L-2-aminobutyric acid thereon and N-acetyl-D-2-aminobutyric acid was separated. The L-2-aminobutyric acid adsorbed on the column was eluted with 2 N NH_4OH and then water was removed under reduced pressure, and thus 4.2 g of L-2-aminobutyric acid was obtained.

The conversion was 81.5% assuming that the theoretical amount of L-2-aminobutyric acid was 5.15 g and the m.p. (decomposition) was 272° to 273°C.

Elemental Analysis:	C	H	N
Calculated (%)	46.6	8.8	13.6
Found (%)	46.7	8.7	13.7

$(\alpha)_D^{33} = +18.5° \pm 0.8°$ (C = 6.225, 6N HCl)

The optical purity was 99.2% on the basis of the value:
$(\alpha)_D^{19} = +18.65°$ (C=4.8, 6N HCl) which is described in Merck Index (7th Edition).

On the other hand, the liquid which was not adsorbed on the ion exchange resin was combined and substantially all of the water was removed under reduced pressure. Thereafter, the unreacted N-acetyl-D-2-aminobutyric acid was extracted with ethyl acetate. After removing the ethyl acetate and a small amount of acetic acid, 5.7 g of N-acetyl-D-2-aminobutyric acid was obtained.

The yield was 78.6% assuming that the theoretical amount of N-acetyl-D-2-aminobutyric acid was 7.25 g and the m.p. was 132.4°C.
$(\alpha)_D^{31} = +40.2° \pm 0.1°$ (C=10.5, H_2O)

Additional enzymatic resolution procedures of racemic 2-aminobutanoic acid may be found in: R. Marshall, S.M. Birnbaum, and J.P. Greenstein, J. Am. Chem. Soc., 78, 4636 (1956), S.M. Birnbaum, L. Levintow, R.B. Kingsley, and J.P. Greenstein, J. Biol. Chem., 194, 455 (1952), and J.P. Greenstein, J.B. Gilbert, and P.J. Fodor, ibid., 182, 451 (1950).

$CH_3CH(NH_2)CH_2COOH$

$C_4H_9NO_2$ Ref. 66

Spaltung des dl-β-Aminobuttersäuremethylesters in die optisch aktiven Komponenten.

Zu einer Lösung von 116 g d-Camphersulfosäure[2]) (0,5 Mol.) in 350 g trocknem Methylalkohol fügten wir unter Kühlung zuerst 58,5 g reinen β-Aminobuttersäuremethylester (0,5 Mol.) und dann unter Umschütteln 1300 ccm trocknen Äther. Nach kurzer Zeit begann die Krystallisation des Camphersulfonats, das sehr leichte mikroskopische Nädelchen bildet. Nach 12 stündigem Stehen im Eisschrank wurde die Krystallmasse, welche die Flüssigkeit ganz durchsetzte, scharf abgesaugt und mit einer auf 0° abgekühlten Mischung von 1 Tl. trocknem Methylalkohol und 3 Tln. trocknem Äther ausgewaschen. Die Ausbeute betrug ungefähr 130 g oder 3/4 der Gesamtmenge des gelösten Salzes. Das Salz enthält den Ester der linksdrehenden Aminosäure im Überschuß, das Filtrat diente dementsprechend zur Darstellung der d-Verbindung. Das krystallisierte Salz wurde von neuem in der doppelten Gewichtsmenge trocknem Methylalkohol gelöst und nach Zusatz des dreifachen Volumens Äther im Eisschrank der Krystallisation überlassen, wobei wieder ungefähr 3/4 der Gesamtmenge ausfielen. Die Trennung der beiden Camphersulfonate ging leider auf diesem Wege so langsam vor sich, daß selbst nach zehnmaligem Umkrystallisieren die optische Aktivität der aus dem Salz isolierten Aminosäure erst 40 Proz. des richtigen Wertes betrug. Wir haben uns deshalb in der Regel mit vier oder fünf Krystallisationen begnügt und die aus dem Salze regenerierte Aminosäure durch Krystallisation aus Methylalkohol gereinigt. Nach der fünften Krystallisation betrug die Menge des Camphersulfonates nur noch 45 g. Selbstverständlich haben wir dann alle Mutterlaugen systematisch aufgearbeitet.

Aus dem Camphersulfonat ließ sich der freie Ester auf folgende Art isolieren. 45 g Salz wurden in etwa 22 ccm warmem Methylalkohol gelöst und hierzu ein geringer Überschuß von methylalkoholischem Ammoniak von bekanntem Titer zugegeben. Das schwer lösliche Ammoniumcamphersulfonat krystallisierte bald und wurde vollständig durch Zusatz des zehnfachen Volumens Äther gefällt. Nach einstündigem Stehen im Eisschrank wurde abgesaugt, mit etwas Äther nachgewaschen und das Filtrat unter vermindertem Druck bei etwa 20° eingedampft. Bei der Destillation des Rückstandes unter 12 mm Druck ging nach einem beträchtlichen Vorlauf der Ester von 53—57° über. Er wurde mit Natriumsulfat getrocknet und zeigte bei abermaliger Fraktionierung bei 13 mm den Siedep. 54—55°.

0,1710 g gaben 0,3204 CO_2 und 0,1440 H_2O.
0,1869 g gaben 19,4 ccm Stickgas bei 17° und 744 mm Druck.

	Ber. für $C_5H_{11}O_2N$ (117,1)	Gef.
C	51,23	51,10
H	9,47	9,42
N	11,96	11,82

Der zweimal destillierte Ester hatte d^{19} = 0,991, er drehte im 1 dm-Rohr bei 19° und Natriumlicht 6,91° (± 0,02°) nach links. Mithin

$$[\alpha]_D^{19} = -6,97° (\pm 0,02°).$$

Wie später auseinandergesetzt wird, ist diese Zahl viel zu klein. Sie beträgt kaum 1/4 des richtigen Wertes.

[2]) A. Reychler, Bull. soc. chim [3] 19, 121 (1898).

Durch Kochen mit Wasser lieferte dieser Ester eine Aminosäure von der spezifischen Drehung $-7{,}9°$.

Aus der Mutterlauge, die bei der oben beschriebenen ersten Krystallisation des d-camphersulfosauren l-β-Aminobuttersäuremethylesters blieb und die noch 44 g Salz enthielt, wurde in der gleichen Weise ein rechtsdrehender β-Aminobuttersäuremethylester dargestellt. Er hatte nach zweimaligem Destillieren denselben Siedepunkt, drehte aber etwas stärker, und zwar bei 20° und Natriumlicht 8,81° (± 0,02°) nach rechts; $d^{20} = 0{,}989$. Mithin

$[\alpha]_D^{20} = +8{,}91°\ (\pm 0{,}02°)$.

0,1828 g gaben 0,3415 CO_2 und 0,1554 H_2O.
0,1755 g „ 16,8 ccm Stickgas bei 15° und 777 mm Druck.

	Ber. für $C_5H_{11}O_2N$ (117,1)	Gef.
C	51,28	50,95
H	9,47	9,50
N	11,96	11,48

Aus diesem Ester wurde durch Verseifung eine β-Aminobuttersäure von $[\alpha]_D^{20} = +10{,}1°$ gewonnen. Nimmt man an, daß die später beschriebene aktive Aminosäure von $[\alpha]_D^{20} = +35{,}3°$ optisch rein gewesen ist und daß bei der Verseifung des Esters keine Racemisation eintritt, so würde sich für den reinen Methylester ungefähr $[\alpha]_D^{20} = +31°$ berechnen.

l-β-Aminobuttersäure.

Zur Gewinnung der Aminosäure aus dem Camphersulfonat ihres Esters ist dessen Isolierung nicht nötig. Man kommt bequemer zum Ziel, wenn man seine ätherisch-methylalkoholische Lösung, die nach dem Auskrystallisieren des camphersulfosauren Ammoniums resultiert, wiederholt mit kleinen Mengen Wasser ausschüttelt, bis dieses nicht mehr alkalisch reagiert. Das ließ sich durch zehnmaliges Ausschütteln leicht erreichen. Die vereinigten wäßrigen Lösungen des Esters wurden dann 4 Stunden am Rückflußkühler gekocht und schließlich die Flüssigkeit unter vermindertem Druck verdampft. Die Ausbeute an Aminosäure war so gut wie quantitativ. Die weitere Verarbeitung dieses Präparates auf optisch reine Aminosäure geschah durch Krystallisation aus trocknem Methylalkohol. Wir wollen den Verlauf der Krystallisation schildern für 8 g Aminosäure von $[\alpha]_D^{20} = -6{,}6°$, die also noch über 80 Proz. inaktive Substanz enthielt. Die 8 g Rohprodukt wurden in etwa 200 ccm trocknem Methylalkohol gelöst und auf 40 ccm eingeengt. Nach 15 stündigem Stehen im Eisschrank waren 4,5 g $[\alpha]_D = -12°$ auskrystallisiert. Die nach Einengen des Filtrats erhaltene zweite Krystallisation von 1,7 g erwies sich als fast inaktiv. Beim weiteren Umkrystallisieren obiger 4,5 g aus der vierfachen Gewichtsmenge Methylalkohol wurden erst 3 g von $-18{,}2°$ und dann 2,1 g von $-26{,}8°$ erhalten. Das Präparat war nun soviel schwerer löslich geworden, daß die zur Lösung erforderliche Menge Methylalkohol relativ erheblich erhöht werden mußte und daß nach dem Einengen auch schon aus der achtfachen Gewichtsmenge Methylalkohol der größere Teil wieder ausfiel. Es wurden so erhalten 1,3 g von $-33{,}6°$, dann 1 g von $-34{,}9°$ und schließlich 0,6 g von $[\alpha]_D^{20} = 35{,}2°$. Da dasselbe Resultat auch bei der rechtsdrehenden Aminosäure erhalten wurde, so scheint hiermit der richtige Wert ganz oder doch nahezu erreicht zu sein. Leider war uns eine weitere Prüfung durch Krystallisation aus anderen Lösungsmitteln nicht möglich, denn das Trennungsverfahren ist nicht allein recht mühsam, sondern auch sehr verlustreich. Aus diesem Grunde haben wir auch für die Umsetzungen der Aminosäure nicht die Präparate vom höchsten optischen Wert, sondern die leichter zugänglichen mittleren Krystallisationen verwendet. Die von uns erhaltene reinste aktive β-Aminobuttersäure unterscheidet sich von dem Racemkörper sehr deutlich durch die Krystallform, die geringere Schmelzbarkeit und die geringere Löslichkeit in Methylalkohol.

Während der Racemkörper aus Methylalkohol in mikroskopischen Nädelchen ausfällt, die meist zu kugeligen Aggregaten vereinigt sind, krystallisiert die aktive Säure aus Methylalkohol in gut ausgebildeten, dicken Prismen, die wir leicht bis zu 1 mm Länge erhielten. Beim langsamen Verdunsten der wäßrigen Lösung im Vakuumexsiccator bekamen wir dünnere, bis zu 1 cm lange Prismen. Der Geschmack ist wenig charakteristisch. Die Aminosäure hat keinen richtigen Schmelzpunkt. Beim raschen Erhitzen im offenen Capillarrohr tritt gegen 220°, also etwa 30° höher als beim Racemkörper, völlige Zersetzung unter Gasentwickelung ein. Die über Schwefelsäure getrocknete Substanz verlor bei 76° und 15 mm über P_2O_5 nicht mehr an Gewicht. Die optisch reinste Aminosäure gab folgende Zahlen:

0,1201 g gaben 0,2059 CO_2 und 0,0964 H_2O.
0,1118 g „ 12,8 ccm Stickgas bei 15° und 772 mm Druck.

	Ber. für $C_4H_9O_2N$ (103,1)	Gef.
C	46,56	46,76
H	8,80	8,98
N	13,59	13,64

0,1290 g Substanz, gelöst in Wasser. Gesamtgewicht 1,2947 g. $d^{20} = 1{,}025$. Drehung im 1 dm-Rohr bei 20° und Natriumlicht 3,59° (± 0,02°) nach links. Mithin

$[\alpha]_D^{20} = -35{,}2°\ (\pm 0{,}2°)$.

Wir führen auch noch die optische Untersuchung der vorletzten Krystallisation an:

0,1290 g Substanz, gelöst in Wasser. Gesamtgewicht 1,2917 g. $d^{20} = 1{,}025$. Drehung im 1 dm-Rohr bei 20° und Natriumlicht 3,57° (± 0,02°) nach links. Mithin

$[\alpha]_D^{20} = -34{,}9°\ (\pm 0{,}2°)$.

d-β-Aminobuttersäure.

Sie wurde aus dem in der ersten methylalkoholischen Mutterlauge verbliebenen Camphersulfonat des rechtsdrehenden Methylesters genau so dargestellt, wie zuvor für die l-Verbindung beschrieben ist. Das Rohprodukt hatte hier schon $[\alpha]_D^{20} = +10{,}1°$. Es gelang dementsprechend auch durch Krystallisation aus Methylalkohol rascher, die hoch drehenden Präparate zu erhalten. Die vorletzte Krystallisation zeigte $[\alpha]_D^{20} = +34{,}9°\ (\pm 0{,}4°)$. Für die letzte Krystallisation geben wir die vollen Daten.

0,1520 g gaben 0,2597 CO_2 und 0,1214 H_2O.
0,1146 g „ 13,2 ccm Stickgas bei 19° und 762 mm Druck.

	Ber. für $C_4H_9O_2N$ (103,1)	Gef.
C	46,56	46,60
H	8,80	8,94
N	13,59	13,32

0,1297 g Substanz. Gesamtgewicht der wäßrigen Lösung 1,3561 g. $d^{20} = 1{,}023$. Drehung im 1 dcm-Rohr bei 20° und Natriumlicht 3,45° (± 0,02°) nach rechts. Mithin

$[\alpha]_D^{20} = +35{,}3°\ (\pm 0{,}2°)$.

Die Substanz zeigte in Krystallform, Löslichkeit, Geschmack und Verhalten in der Hitze Übereinstimmung mit dem Antipoden.

Von diesem Präparat haben wir auch noch die Drehung in salzsaurer und in alkalischer Lösung bestimmt.

0,0454 g Substanz, gelöst in n-Salzsäure. Gesamtgewicht 0,4843 g. $d^{20} = 1{,}04$. Drehung in ½ dm-Rohr bei 20° und Natriumlicht 1,45° (± 0,02°) nach rechts. Mithin

$[\alpha]_D^{20} = +29{,}7°\ (\pm 0{,}4°)$.

0,0343 g Substanz, gelöst in n-Natronlauge. Gesamtgewicht 0,3805 g. $d^{20} = 1{,}06$. Drehung im ½ dm-Rohr bei 20° und Natriumlicht 0,70° (± 0,02°) nach rechts. Mithin

$[\alpha]_D^{20} = +14{,}7°\ (\pm 0{,}4°)$. *

*See note on following page.

Note:

K. Balenovic, D. Cerar and Z. Fuks, J. Chem. Soc., 3316 (1952) reported the following physical constants for 3-aminobutanoic acid: m.p. 212° $[\alpha]_D^{18}$ +38.8° ± 1° (c 0.48 in water), $[\alpha]_D^{19}$ +37.07 ± 1° (c 6.0 in water).

$$H_2NCH_2\underset{\displaystyle CH_3}{\underset{|}{C}}HCOOH$$

$C_4H_9NO_2$ Ref. 67

Resolution of α-methyl-β-phthalimidopropionic acid. A solution of (±)-α-methyl-β-phthalimidopropionic acid (Ia, 23·3 g, 0·1 mole) in 96% ethanol (400 cc) was added to a solution of brucine (46·6 g, 0·1 mole) in 96% ethanol (250 cc). The ethanol was removed under reduced pressure, and the yellow brucine salt of α-methyl-β-phthalimidopropionic acid was obtained, which crystallised on standing, yield 67 g, and showed $[\alpha]_D^{18}$ −26·6° (*c*, 1·96 in chloroform).

The crude brucine salt (67 g) was dissolved in warm ethyl acetate (1600 cc), filtered, petroleum ether added (b.p. 40–60°, 260 cc) and the solution left at 0° overnight. Yellow crystals separated (51 g) showing $[\alpha]_D^{18}$ − 28·4° (*c*, 1·5 in chloroform). Repeated recry tallisation of this salt from ethyl acetate-petroleum ether afforded the brucine salt of low solubility, $[\alpha]_D$ −44·2° (Found: C, 66·68; H, 6·03. Calc. for $C_{35}H_{37}O_8N_3$: C, 66·97; H, 5·94%).

A more soluble diastereomer could be isolated from the filtrates by fractional crystallisation.

(−)-α-*Methyl-β-phthalimidopropionic acid.* To a suspension of the brucine salt (11·4 g, $[\alpha]_D$ −44°) in water (500 cc) 4 N HCl (250 cc) was added and complete solution occurred; after standing for 1 hr at room temperature, crystals of (−)-α-methyl-β-phthalimidopropionic acid were filtered off, thoroughly washed with water, and dried, yield 3·5 g (83%), $[\alpha]_D^{16}$ −11·3° (*c*, 1·85 in chloroform). The filtrate was extracted with benzene (3 × 100 cc) and after removing the solvent under reduced pressure 0·46 g of (−)-α-methyl-β-phthalimidopropionic acid was obtained (total yield 92%), $[\alpha]_D^{18}$ −24°.

Fractional crystallisation of the acid (3·5 g, $[\alpha]_D$ −11°) from aqueous ethanol (1 : 1) was carried out; 0·9 g of racemic acid was isolated. The optically pure (−)-α-methyl-β-phthalimidopropionic acid was obtained from the filtrates (0·9 g). A small sample showing $[\alpha]_D^{18}$ −20·1° was sublimed for analysis at 110–115°/0·001 mm and showed $[\alpha]_D^{17}$ −24·4° (*c*, 0·98 in chloroform) and the m.p. 145–146° (Found: C, 62·02; H, 4·99. Calc. for $C_{12}H_{11}O_4N$: C, 61·80; H, 4·76%.

(−)-α-*Methyl-β-alanine.* A solution of (−)-α-methyl-β-phthalimidopropionic acid (0·93 g, 0·004 mole) in glacial acetic acid (16 cc) and 47% HI (4 cc) was refluxed 8 hr; the glacial acetic and hydriodic acids were removed under reduced pressure. Water was added to the residue, the separated phthalic acid filtered off, washed with water, and the filtrate extracted with ether. The aqueous layer was evaporated to dryness. Addition of water and evaporation was repeated till no trace of free hydriodic acid remained.

The pale yellow (−)-α-methyl-β-alanine hydriodide was dissolved in water (400 cc) and passed through a column containing IR-4B Amberlite ion-exchange resin (10 g, 21 cc). The filtrate (1000 cc) was evaporated to dryness under reduced pressure, dissolved in water, and filtered (with a small quantity of charcoal). After evaporating the water, crystals of (−)-α-methyl-β-alanine were obtained (0·4 g, 99%), m.p. 169–174° the $[\alpha]_D^{16}$ −9·4° (*c*, 1·37 in water). Sublimation at 110°/0·001 mm gave the pure acid with the m.p. 173–175° and $[\alpha]_D^{17}$ −14·2° (*c*, 0·42 in water) (Found: C, 46·63; H, 8·44. Calc. for $C_4H_9O_2N$: C, 46·59; H, 8·80%).

$C_4H_9NO_2$ Ref. 67a

N-Acetyl-(−)-AIB—Cinchonidine (30 g, 0.102 mole) and acetyl-DL-AIB (14.76 g, 0.102 mole) were dissolved in 300 ml of hot isopropyl alcohol, and the solution was chilled overnight: 20.7 g of crystals were collected; m.p. 181–184°. Recrystallization of this crop from 100 ml of isopropyl alcohol yielded 17.1 g; m.p. 184–187°; $[\alpha]_D^{25}$ −100.6° (c, 1, H_2O). Another recrystallization from 100 ml of isopropyl alcohol yielded 15.10 g; m.p. 186–188°; $[\alpha]_D^{25}$ −101.6° (c, 1, H_2O). Subsequent recrystallizations did not lead to any further change in physical properties.

The cinchonidine salt was dissolved in 50 ml of water, 10 ml of concentrated NH_4OH were added, the resulting mixture was cooled, and the precipitated cinchonidine was filtered and washed with three 20 ml portions of cold water. The filtrate and washings were combined, washed with two 100-ml portions of chloroform, acidified to pH 1 by the addition of concentrated HCl, and saturated with NaCl. Acetyl-(−)-AIB was extracted into six 200-ml portions of ethyl acetate; the combined extracts were dried over anhydrous sodium sulfate and concentrated to dryness under reduced pressure. The crystalline residue was recrystallized from 80 ml of dichloroethane to yield 4.14 g of product; m.p. 80–82°; $[\alpha]_D^{27}$ − 29.8° (c, 1, absolute ethanol). Recrystallization did not change these properties.

$C_6H_{11}NO_3$

Calculated: N 9.58
Found: N 9.40

(−)-AIB—*N*-Acetyl-(−)-AIB (13.0 g) was refluxed in 200 ml of 2 N HCl for 3 hours and the acid was removed under reduced pressure. The crystalline residue was taken up into 20 ml of water and passed through a column (2 × 22 cm) of Amberlite CG-120 (H^+ form) (100 to 200 mesh). The resin was washed with 100 ml of water and the AIB was eluted with 3 N NH_4OH. The first 75 ml of eluate were discarded and the following 50 ml were collected and concentrated to dryness. The crystalline AIB was recrystallized from 70 ml of 95% ethanol to yield 6.55 g of long rectangular plates; m.p. 194–196°; $[\alpha]_D^{27}$ −15.4° (c, 1, H_2O).

$C_4H_9NO_2$

Calculated: N 13.6
Found: N 13.3

An additional 1.54 g, m.p. 192–195°, was obtained from the mother liquor; total yield, 88%.

(+)-AIB—The mother liquor from the first crop of the cinchonidine salt of acetyl-(−)-AIB was concentrated to 60 ml and 2.28 g of material, which crystallized when the solution was cooled, were collected and discarded. The filtrate was concentrated to dryness and the oily residue was dissolved in 100 ml of water. Free acetyl-(+)-AIB was obtained from the salt as described above; yield, 5.73 g; m.p. 76–86°; $[\alpha]_D^{25}$ +24.5° (c, 1, absolute ethanol). Attempts to obtain an optically pure specimen of the salts of acetyl-(+)-AIB with brucine, quinine, quinidine, and cinchonine were unsuccessful. Impure acetyl-(+)-AIB (15.6 g) was hydrolyzed and the hydrochloride was converted to the free amino acid as described above for the isolated (−)-AIB. After recrystallization from 100 ml of 98% ethanol, 9.75 g (90%) were recovered; m.p. 183–192°; $[\alpha]_D^{25}$ +12.1° (c, 1, H_2O). From the mother liquor, two additional small crops of somewhat better optical purity were obtained. Recrystallization did not effect a significant improvement in the product, so all of the crops and dried mother liquors were combined, dissolved in 9.5 ml of concentrated HCl, and concentrated to dryness. The residue was dissolved in 100 ml of absolute ethanol, 150 ml of ether were added, and the solution was allowed to stand at room temperature for 6 hours; 9.0 g of shining plates were obtained; m.p. 131–140°, $[\alpha]_D^{25}$ +10.1° (c, 1, H_2O). This material was recrystallized by dissolving it in 100 ml of absolute ethanol and adding 100 ml ether; yield 7.34 g (50% based on the impure acetyl-(+)-AIB; m.p. 133–137°; $[\alpha]_D^{25}$ +10.8° (c, 1, H_2O).

$C_4H_9NO_2HCl$

Calculated: N 10.0
Found: N 9.84

For comparison, (−)-AIB hydrochloride was prepared in the same manner from 100 mg of pure (−)-AIB; m.p. 134–138°; $[\alpha]_D^{24}$ −10.7° (c, 1, H_2O).

Of the pure (+)-AIB hydrochloride, 7.1 g were dissolved in 30 ml of water and passed through a column (2 × 17 cm) of Amberlite CG-120 (H^+ form). The column was washed with 100 ml of water and the amino acid was eluted with 120 ml of 3 N NH_4OH. The last 50 ml of the eluate were collected and evaporated under reduced pressure. The residue was recrystallized from 150 ml of 99% ethanol to yield 4.2 g of long rectangular plates; m.p. 192–194°; $[\alpha]_D^{23}$ + 15.4° (c, 1, H_2O).

$C_4H_9NO_2$

Calculated: N 13.6
Found: N 13.4

Another crop of 0.62 g, m.p. 190–192°, was obtained from the mother liquor.

$HSCH_2CH_2-CH(NH_2)-COOH$

$C_4H_9NO_2S$ Ref. 68

All melting points are uncorrected

S-Benzyl-DL-β-N-formyl-homocysteine (I)

Acetic formic anhydride was prepared according to Huffman[17] from acetic acid anhydride (12.2 ml.) and formic acid (98%, 5.2 ml.). *S*-Benzyl-DL-β-homocysteine[15] (13.5 g., 60 mMoles) was gradually added for half an hour to the stirred mixture of anhydride, maintained at 50—60°, and the solution was kept under the same conditions for additional five hours. Water (10 ml.) was added and the solution evaporated to dryness *in vacuo*. The residue was dissolved in hot ethylacetate, treated with charcoal, filtered and cooled to 0°. 12.3 g. (81%) of I separated as white crystals m. p. 70—72°. Additional 1.5 g. was obtained from mother liquor on addition of petroleum-ether (total yield 91%).

The analytical sample, recrystallized twice from ethylacetate, showed m. p. 83—84.5°.

Anal. 6.92 mg. subst.: 14.40 mg. CO_2, 3.65 mg. H_2O
5.69 mg. subst: 0.275 ml. N_2 (20°, 745 mm)
$C_{12}H_{15}NO_3S$ (253.32) calc'd.: C 56.89; H 5.97; N 5.53%
found : C 56.83; H 6.03; N 5.52%

The resolution of S-benzyl-DL-β-N-formyl-homocysteine with brucine

12.6 g. (50 mMoles) of I and 21.0 g. (53.5 mMoles) of anhydrous brucine were dissolved in hot 40% acetone (100 ml.) and the solution cooled in the refrigerator for 24 hours. 16.9 g. of brucine salt showing $[\alpha]_D^{24}$ —18.2° ± 1 (c, 1.92 in ethanol) was collected. The salt was recrystallized from water (600 ml.) yielding 11.9 g. of white crystals with $[\alpha]_D^{24}$ —16.50 ± 1 (c, 1,97 in ethanol). The second crystallization from water (150 ml.) resulted in a product (11.1 g., 65.5%) m. p. 65—70° with no change in rotation: $[\alpha]_D^{25}$ —16.3° ± 1 (c, 2.00 in ethanol). From water mother liquors, after evaporation to the half of volume, an additional crop of 1.5 g. (total yield: 74.5%) having the same rotation was obtained. Analysis showed that the salt crystallized with one and half molecule of water.

Anal. 5.73 mg. subst.: 13.10 mg. CO_2, 3.41 mg. H_2O
4.31 mg. subst.: 0.233 ml. N_2 (20°, 748.5 mm)
$C_{35}H_{41}N_3O_7S \cdot 1.5\ H_2O$ (674.79)
calc'd.: C 62.29; H 6.57; N 6.23%
found : C 62.42; H 6.67; N 6.21%

S-Benzyl-L-β-homocysteine (IIa)

The brucine salt (11.1 g., $[\alpha]_D^{24}$ — 16.3°, in ethanol) was shaken with chloroform (25 ml.) and 1 *N* ammonium hydroxide (50 ml.) in a separatory funnel. The aqueous layer was extracted with three portions of chloroform (10 ml.) and then concentrated *in vacuo* (to about 10 ml.). Concentrated hydrochloric acid was added to make the solution approximately 1*N*. To hydrolyze the formyl derivative, the solution was refluxed for one hour, cooled, and evaporated to dryness. The residue was dissolved in hot water (20 ml.) and neutralized with conc. ammonia (*p*H 6—7). After standing overnight in the refrigerator 3.15 g. (85% calc'd. on brucine salt, 55.9% calc'd. on I) of IIa were collected showing $[\alpha]_D^{25}$ — 59.0° ± 1 (c, 1.12 in *N* HCl). One recrystallization from ethanol: water (4:1) raised the rotation to $[\alpha]_D^{25}$ — 64.0 ± 1 (c, 1.16 in *N* HCl), m. p. 171—174° (decomp.).

S-Benzyl-D-β-homocysteine (IIb)

The acetone mother liquor from the brucine salt of I was evaporated to dryness *in vacuo*. The remaining oil (20.8 g.) was decomposed as described for the L-isomer. 5.65 g. of crude product was obtained showing $[\alpha]_D^{24}$ + 24.8° ± 2 (c, 1.11 in *N* HCl). After two recrystallizations from absolute ethanol 2.05 g. (36.5% calc'd. on I) of IIb with $[\alpha]_D^{24}$ + 60.5° ± 1 (c, 1.17 in *N* HCl) was obtained, m. p. 173—175° (decomp.).

Anal. 6.14 mg. subst.: 13.15 mg. CO_2, 3.78 mg. H_2O
$C_{11}H_{15}NO_2S$ (225.30) calc'd.: C 58.64; H 6.71%
found : C 58.42; H 6.88%

```
          NH2
          |
 CH3-CH-CH-COOH
      |
      OH
```

$C_4H_9NO_3$ Ref. 69

N-Phtaloyl-DL-threonin wurde im wesentlichen nach den Angaben von SHEEHAN, GOODMAN & HESS[7]) synthetisiert. 100 g (0,84 Mol) fein pulverisiertes DL-Threonin (Biochemica ROCHE) und 150 g (1,01 Mol) Phtalsäureanhydrid in 450 ml Dioxan wurden 16 Std. unter Rühren und Rückfluss erhitzt und im Wasserstrahlvakuum bis zu einem zähflüssigen Sirup eingedampft. Dieser Sirup wurde in 650 ml Wasser aufgenommen und zur Kristallisation 2 Std. bei 0° belassen. Darauf wurde das ausgefallene N-Phtaloyl-DL-threonin abgesaugt, portionenweise mit total 400 ml Wasser gewaschen und im Wasserstrahlvakuum bei 80° bis zur Gewichtskonstanz getrocknet. Ausbeute: 178 g (85%), Smp. 119–121°. Zur Analyse wurde dreimal aus Essigester/Petroläther umkristallisiert und 16 Std. bei 25° im Hochvakuum über Phosphorpentoxyd ge trocknet. Smp. 122–123°.

$C_{12}H_{11}O_5N$ (249,16) Ber. N 5,62% Gef. N 5,61%

Spaltungsversuch mit der berechneten Menge Brucin. – Brucinsalz von N-Phtaloyl-L-threonin. 197 g (0,5 Mol) Brucin wurden in 250 ml Methylcellosolve bei einer Temperatur von 80° gelöst und unter mechanischem Rühren mit 124,5 g N-Phtaloyl-DL-threonin versetzt. Der sofort ent-

[9]) W. J. POPE & ST. J. PEACHY, J. chem. Soc. **75**, 1066 (1899).
[10]) Die angegebenen Smp. sind nicht korrigiert.

[7]) J. C. SHEEHAN, M. GOODMAN & G. P. HESS, J. Amer. chem. Soc. **78**, 1367 (1956).

stehende, gelb gefärbte Kristallbrei wurde zunächst noch 10 Min. bei 80° weitergerührt und dann 24 Std. bei 20° sich selbst überlassen[11]). Dann wurde das L-Salz abgesaugt, fünfmal mit je 80 ml Methanol gewaschen und bei 90° im Wasserstrahlvakuum getrocknet. Ausbeute: 159 g (99%), Smp. 220–223° (Zers.), $[\alpha]_D^{20} = -37{,}3° \pm 1°$ (c = 1,5; Methylcellosolve). Zur Analyse wurde zweimal aus Methylcellosolve umkristallisiert und 16 Std. bei 110° über Phosphorpentoxyd im Hochvakuum getrocknet. Smp. 222–223° (Zers.), $[\alpha]_D^{20} = -37{,}5° \pm 1°$ (c = 1,5; Methylcellosolve).

$C_{35}H_{37}O_9$ (643,67) Ber. C 65,31 H 5,79 N 6,53% Gef. C 65,15 H 5,82 N 6,64%

Brucinsalz von N-Phtaloyl-D-threonin. Die Methylcellosolve/Methanol-Mutterlauge, welche das leichtlösliche D-Salz enthält, wurde im Wasserstrahlvakuum eingedampft und bei 80° bis zur Gewichtskonstanz getrocknet. Die spröde, glasartige Substanz konnte bisher nicht kristallisiert werden. Ausbeute: 150 g (93%); $[\alpha]_D^{20} = +3{,}3° \pm 1°$ (c = 1,5; Methylcellosolve).

Spaltungsversuch mit der halben Menge Brucin. 35,5 g (0,09 Mol) Brucin, 70 ml Methylcellosolve und 44,8 g (0,18 Mol) N-Phtaloyl-DL-threonin wurden wie oben beschrieben angesetzt und zum schwerlöslichen Brucinsalz von N-Phtaloyl-L-threonin aufgearbeitet. Ausbeute: 48,2 g (83%), Smp. 220–223° (Zers.), $[\alpha]_D^{20} = -37{,}2° \pm 1°$ (c = 1,5; Methylcellosolve).

L-Threonin. 155 g (0,24 Mol) Brucinsalz von N-Phtaloyl-L-threonin wurden 30 Min. in einer Mischung von 362 ml 2-n. Natronlauge und 360 ml Chloroform kräftig gerührt. Hierbei löst sich das Brucinsalz vollständig auf und wird zerlegt. Die abgetrennte Phase wurde noch zweimal mit je 150 ml Chloroform ausgeschüttelt, während alle Chloroformphasen noch zweimal mit je 75 ml 0,5-n. Natronlauge nachgewaschen wurden. Die vereinigten und filtrierten wässerigen Lösungen wurden mit konz. Salzsäure auf ein pH von 3 gestellt, im Wasserstrahlvakuum auf $^1/_3$ des ursprünglichen Volumens eingedampft und nach Zusatz von 140 ml konz. Salzsäure $2^1/_2$ Std. unter Rückfluss gekocht. Nun wurde im Eisbad abgekühlt, die ausgefallene Phtalsäure abgenutscht, zweimal mit je 40 ml Wasser nachgewaschen und das Filtrat unter Wasserstrahlvakuum zur Trockne eingedampft. Der Rückstand wurde in 100 ml Wasser suspendiert, von nicht gelöstem Natriumchlorid und einer Spur Phtalsäure abgenutscht, mit etwas Wasser nachgewaschen und das Filtrat unter Wasserstrahlvakuum wieder zur Trockne eingedampft. Das gebildete L-Threonin-hydrochlorid wurde mit 200 ml Methanol herausgelöst, vom Natriumchlorid abgenutscht, mit Methanol nachgewaschen und die Lösung im Wasserstrahlvakuum eingedampft. Der anfallende Rückstand wurde erneut in 200 ml Methanol aufgenommen, von einer Spur Natriumchlorid abfiltriert und das Filtrat mit Diäthylamin auf ein pH von 6 eingestellt. Nach 4 Std. wurde das ausgefallene L-Threonin abgenutscht, gründlich mit Methanol gewaschen und im Wasserstrahlvakuum bei 90° getrocknet. Ausbeute: 25,8 g (90%), $[\alpha]_D^{20} = -27{,}0° \pm 1°$ (c = 2; Wasser). Dieses Rohprodukt wurde in siedendem Wasser gelöst, filtriert, mit dem doppelten Volumen Alkohol versetzt, nach 2 Std. abgenutscht, mit Alkohol gewaschen und im Wasserstrahlvakuum bei 90° getrocknet. Ausbeute: 23,8 g (83%), $[\alpha]_D^{20} = -28{,}2° \pm 1°$ (c = 2; Wasser). In Butanol/Eisessig/Wasser 4:1:1 verhält sich die Substanz nach 24 Std. Laufzeit (absteigend) papierchromatographisch einheitlich.

D-Threonin. 150 g (0,234 Mol) glasartiges Brucinsalz von N-Phtaloyl-D-threonin wurden mit 350 ml 2-n. Natronlauge und 350 ml Chloroform übergossen und 1 Std. kräftig gerührt. Die Substanz wird hierbei allmählich gelöst. Die weitere Aufarbeitung erfolgte wie beim L-Threonin. Das rohe D-Threonin wurde aus siedendem Wasser ohne Zusatz von Alkohol umkristallisiert. Ausbeute nach Aufarbeitung der Mutterlauge 20 g (70%), $[\alpha]_D^{20} = +27{,}6° \pm 1°$ (c = 2; Wasser). Über die papierchromatographische Reinheitsprüfung gilt das für L-Threonin Gesagte.

[11]) Es gelingt bei der angegebenen Konzentration kaum, alle Substanz in Lösung zu bringen, vielmehr fängt das schwerlösliche Salz an auszufallen, ehe alles N-Phtaloyl-DL-threonin in fester Form zugesetzt ist. Dank der grossen Löslichkeitsdifferenz der beiden Salze beeinträchtigt dies die Spaltung keineswegs, vorausgesetzt, dass der Kristallbrei nach Vorschrift 10 Min. bei 80° mit dem Rührer homogenisiert wird.

$C_4H_9NO_3$ Ref. 69a

N-*p*-Nitrobenzoyl-DL-threonine.—To a solution of 32 g. (0.269 mole) of DL-threonine in 1200 cc. of water and 270 cc. of normal sodium hydroxide at 0°, was added with vigorous agitation over a period of one hour, a total of 50 g. (0.269 mole) of *p*-nitrobenzoyl chloride (freshly distilled, m. p. 72–74°) in equal portions at three-minute intervals. Simultaneously, over this period of time, 135 cc. of 2 *N* sodium hydroxide solution were added dropwise. The solution was stirred for an additional twenty minutes and then acidified to congo red with 40 cc. of concentrated hydrochloric acid. After cooling in an ice-bath for one hour, the crude *p*-nitrobenzoyl-DL-threonine was filtered, and without drying was extracted with 300 cc. of boiling water. The insoluble portion, *p*-nitrobenzoic acid, weighed 4.1 g. The product began to crystallize from the hot solution immediately, and after chilling at 0–5° for one hour was filtered, washed with two 75-cc. portions of ice-cold water and dried at 60°; m. p. 159–162°. For further purification, the 56.5 g. of crude threonine derivative was extracted twice with hot ether to remove additional quantities of *p*-nitrobenzoic acid. The yield of pure *p*-nitrobenzoyl-DL-threonine was 50.7 g. (70.5%); m. p. 166–167°.

Brucine Salt of *p*-Nitrobenzoyl-DL-threonine.—To a warm solution (50–55°) of 64 g. (0.149 mole) of brucine Merck dissolved in 160 cc. of methanol was added 40 g. (0.149 mole) of *p*-nitrobenzoyl-DL-threonine, and the mixture was heated until solution was complete. With rapid stirring and scratching, the solution was cooled in an ice-bath, whereupon crystallization took place within a short time. The cooling and stirring were continued for a total of five minutes, during which time a heavy, yellowish precipitate separated out, and the temperature dropped to 25°. The flask was reheated to 50° with stirring, and then cooled to 25° over a five-minute period as in the previous case. The brucine salt of *p*-nitrobenzoyl-L-threonine was filtered, and the small amount of product adhering to the flask was transferred with the mother liquors. After washing the cake thoroughly in a mortar with 60 cc. of cold methanol, the slurry was transferred back to the funnel and washed, again using approximately 60 cc. of methanol for this operation. As much of the wash liquor as possible was removed by suction, after which the product was finally washed with two 50-cc. portions of ether, and these washes were collected in a separate receiver. The dried L-threonine salt, which was almost white in color, weighed 46 g. (88% yield) and melted at 145–150°.[5]

The combined filtrate and methanol washes (volume about 260 cc.) were allowed to stand at room temperature overnight, and then cooled in an ice-bath for two hours. The yellow brucine salt of *p*-nitrobenzoyl-D-threonine was filtered and dried without washing; weight 41 g. (78.5% yield); m. p. 190–192°.

By concentrating the mother liquors to a volume of 90 cc. *in vacuo*, scratching to induce crystallization, and cooling in an ice-bath for one hour, an additional 2 g. of the brucine salt of *p*-nitrobenzoyl-L-threonine (m. p. 147–150°) was obtained.

The resulting mother liquors and washes were again concentrated to a volume of 25 cc., whereupon on standing in the refrigerator for two hours an orange mass separated. After filtering, washing with a small amount of cold methanol and drying, 5 g. of the brucine salt of *p*-nitrobenzoyl-D-threonine was obtained.

The combined total yield of the brucine salt of *p*-nitrobenzoyl-L-threonine was 48 g. or 92%. The combined total yield of the brucine salt of *p*-nitrobenzoyl-D-threonine was 46 g. or 88.5%.

The brucine salts of the L-isomer were found to crystallize with two molecules of water.

Anal. Calcd. for $C_{34}H_{38}O_{10}N_4 \cdot 2H_2O$: C, 58.49; H, 6.07; N, 8.05. Found: C, 58.11; H, 6.44; N, 8.24.

L-Threonine.—To a mixture of forty-eight grams of the brucine salt of *p*-nitrobenzoyl-L-threonine in 500 cc. of water at 40–50° was added slowly with vigorous stirring 80 cc. of *N* sodium hydroxide solution, and the solution was stirred for an additional hour under the same conditions. After chilling in an ice-bath for thirty minutes, the brucine was collected by filtration and washed thoroughly with two 100-cc. portions of ice water. The recovered dried brucine weighed 26.5 g. Traces of brucine remaining in the mother liquors and washes were removed by extracting the combined solutions with 150 cc. of chloroform and 150 cc. of ether. The aqueous solution containing the sodium salt of *p*-nitrobenzoyl-L-threonine was concentrated *in vacuo* to a volume of 80 cc., and then refluxed for four hours with 56 cc. of constant boiling hydrobromic acid. After storing the mixture in the refrigerator overnight, the *p*-nitrobenzoic acid was filtered, washed with two 50-cc. portions of cold water and dried. The calculated amount of acid (11.7 g.) was recovered. The mother liquors, combined with the washings, were concentrated to a sirup *in vacuo*, redissolved in 50 cc. of water, and again concentrated *in vacuo*. The resulting sirup was warmed on the steam-bath with 50 cc. of absolute ethanol until all particles of oil were dissolved, and only the white, insoluble sodium bromide remained. To insure as complete removal of water and hydrobromic acid as possible, the solution was again concentrated and the resultant oil was finally dissolved in 100 cc. of absolute ethanol, after which the solution was treated with 0.2 g. of Darco G-60 and filtered. The residue was washed with a total of 40 cc. of cold absolute ethanol. To the filtrate and washings, warmed to 50°, were added 20 cc. of concentrated ammonium hydroxide, whereupon the crude L-threonine precipitated immediately. After chilling for fifteen minutes in an ice-bath, the product was filtered and washed with ethanol. The crude material (weight 7.0 g.) gave a slight test for bromide ion. It was dissolved in 30 cc. of hot water and treated with 0.2 g. of charcoal, and after filtration the charcoal cake was washed with 5 cc. of hot water. To the warm filtrate (50–55°) a total of 140 cc. of absolute ethanol was added slowly with stirring. The solution was allowed to stand at room temperature for fifteen minutes and then placed in an ice-bath for the same length of time. After filtering, washing with ethanol, and drying, pure L-threonine was obtained as brilliantly white leaflets. The yield was 6.25 g. This represents an over-all yield of 49.5% based on the DL-threonine used: $[\alpha]^{26}D$ −27.9° (5% aqueous solution); purity by solubility data, 99.7 ± 0.1%.

D-Threonine.—Using the same procedure as described above, pure D-threonine was obtained in an over-all yield of 43.6%. From 46 g. of the brucine salt of *p*-nitrobenzoyl-D-threonine, 5.88 g. of crude product was obtained. The recrystallized material weighed 5.5 g., $[\alpha]^{26}D$ +27.8°; purity by solubility data, 99.7 ± 0.1%.

(5) Two factors, time and temperature, are very critical in this method of resolution. Advantage is taken of the difference in solubility in methanol of the brucine salts of L- and D-*p*-nitrobenzoyl derivatives. This difference in solubility becomes smaller as the soluble orange form of the *p*-nitrobenzoyl-D-threonine salt slowly changes over to an insoluble light yellow form. This change is affected by temperature and length of time of standing. The volume should be kept to a minimum so that the L-salt will precipitate rapidly. It may then be removed before a significant quantity of the soluble form of the D-salt has had an opportunity to be converted to the insoluble form.

$C_4H_9NO_3$

Ref. 69b

Experimenteller Teil.

Die Schmelzpunkte wurden auf dem *Kofler*-Block bestimmt und sind korrigiert; Fehlergrenze ca. ± 2°. Die Mikroanalysen wurden im Mikrolabor der Organ.-chem. Anstalt, Basel (Leitung *E. Thommen*) ausgeführt.

1. N-Tosyl-DL-threonin. Zu einer Lösung von 1,0 g DL-Threonin (8,4 mMol) und 0,34 g NaOH (8,5 mMol) in 7 ml Wasser liess man unter starkem Rühren innert einer Stunde (nicht rascher!) bei einer Temperatur von 65—68° in kleinen Portionen getrennt 0,51 g NaOH (12,75 mMol) in 10 ml Wasser und 2,0 g p-Toluolsulfochlorid (10,5 mMol) in 30 ml Äther zutropfen, wobei der verdampfende Äther mit einem schwachen Luftstrom durch einen kleinen Rückflusskühler abgesaugt wurde. Mitgerissenes Toluolsulfochlorid wurde so kondensiert und liess sich nach Beendigung des Zutropfens mit wenig Äther in die Lösung spülen. Aus der noch zehn Minuten gerührten und dann auf Zimmertemperatur abgekühlten, nur noch schwach alkalischen Lösung (pH 7—8) fiel beim Ansäuern mit 5 ml konz. Salzsäure (stark kongosauer) ein dicker Niederschlag aus. Dieser wurde direkt in 40 ml Essigester aufgenommen. Nach der Trennung im Scheidetrichter extrahierte man noch zweimal mit je 10 ml Essigester und wusch die Auszüge zweimal mit je 5 ml Wasser. Die vereinigten Essigesterauszüge wurden mit Na_2SO_4 getrocknet und bei Normaldruck unter Vermeidung von Überhitzung unbenetzter Kolbenwände entweder bis zur beginnenden Kristallisation oder aber auf ca. 5 ml eingeengt; im letzten Fall führte Animpfen zur sofortigen Kristallisation. Nach mehrstündigem Stehen bei −15° wurde filtriert und mit wenig kaltem Essigester gewaschen. Ausbeute 1,80—1,89 g (78—82% d. Th.); Smp. 181—182°.

Das Analysenpräparat, aus Essigester umkristallisiert und im Hochvakuum bei 100° 4 Stunden getrocknet, schmolz bei 182—183° mit Sublimation in Tröpfchen ab 175°.

$C_{11}H_{15}O_5NS$	Ber.	C 48,34	H 5,53	N 5,13%
(273,30)	Gef.	„ 48,22	„ 5,47	„ 4,93%

2. N-Tosyl-DL-threonin-brucinsalz (partielles Racemat). Eine Lösung von 11,41 g N-Tosyl-DL-threonin (41,75 mMol) in 50 ml warmem Methanol wurde mit einer Lösung der äquivalenten Menge Brucin (16,48 g wasserfreie Base) in 50 ml warmem Methanol versetzt. Nach etwa zehnminütigem Erwärmen auf dem Wasserbad begann sich das Salz auszuscheiden. Nach mehrstündigem Stehen bei −15° wurde abgesaugt und mit etwas kaltem Methanol gewaschen. Die Ausbeute an Rohprodukt mit einem Schmelzpunkt zwischen 205° und 210° betrug 26,70 g (95,7% d. Th.). Das Salz wurde so weiter verarbeitet.

Aus Wasser umkristallisiert (100 mg in 4 ml Wasser) schmolzen die prismatischen Kristalltafeln variabel[1]) zwischen 210° und 215° mit vorheriger Sublimation in Tröpfchen.

3. N-Tosyl-L-threonin-brucinsalz und N-Tosyl-D-threonin-brucinsalz. a) *Verfahren mit Methanol allein:* 2,33 g (3,49 mMol) N-Tosyl-DL-threonin-brucinsalz (partielles Racemat) wurden in 55 ml Methanol durch dreiviertelstündiges Kochen am Rückfluss gelöst. Die heisse Lösung impfte man mit wenig N-Tosyl-L-threonin-brucinsalz an und liess sie über Nacht bei 0° stehen. Das L-Salz schied sich als zusammenhängender Kristallkuchen aus. Die überstehende Lösung wurde schnell durch eine Nutsche, welche rasches Durchlaufen gewährleistete, abdekantiert, der noch im Kolben haftende Kristallkuchen mit 20 ml Methanol vorsichtig gewaschen (kein Spatel, kein Glasstab!), das Methanol wieder abdekantiert, dann nach Zugabe von 10 ml Methanol der Kristallkuchen zerdrückt, aufs Filter gespült und mit 5 ml Methanol nachgewaschen. Nach dem Trocknen wogen die Kristalle 1,1 g (94% d. Th. an L-Salz) und schmolzen von 135—140°.

Aus dem Filtrat schied sich sofort oder wenige Minuten nach der Filtration das D-Salz in Nadeln aus. Nach mehrstündigem Stehen bei −15° erhielt man durch Filtration 0,82 g (70% d. Th.) D-Salz mit Smp. 131—137°.

Aus 1,1 g rohem L-Salz erhielt man durch Umkristallisieren aus 50 ml Methanol (zum Lösen ist wieder längeres Kochen nötig) unter Animpfen 0,95 g feine verfilzte Nädelchen mit einem Smp. von 138—140°[1]).

Aus 0,82 g rohem D-Salz erhielt man durch Umkristallisieren aus 19 ml Methanol unter Animpfen 0,74 g Nadeln und Stäbchen mit einem Smp. von 136—138°[1]).

Ein Gemisch der beiden Salze schmolz zwischen 90° und 110°, erstarrte bei weiterem Erwärmen und schmolz ein zweites Mal bei 200—205°.

b) *Verfahren mit Methylcellosolve und Methanol:* 26,6 g (39,8 mMol) N-Tosyl-DL-threonin-brucinsalz (partielles Racemat) wurden in 53 ml Methylcellosolve durch Erwärmen auf dem Wasserbad gelöst. Durch den Rückflusskühler liess man nicht zu schnell 145 ml Methanol zufliessen, impfte mit L-Salz an und liess vier Stunden bei Zimmertemperatur stehen (nicht im Kühlschrank!). Das Abdekantieren und Filtrieren erfolgte wie unter a) beschrieben, wobei zum ersten Nachspülen 80 ml eines Methylcellosolve-Methanol-Gemisches gleicher Zusammensetzung, zu einem zusätzlichen zweiten Nachspülen 80 ml Methanol, zum Herausspülen nochmals 80 ml Methanol und zum Nachwaschen noch 50 ml Methanol verwendet wurden. Die Ausbeute an L-Salz, das noch durch etwas D-Salz verunreinigt war, betrug 14,1 g.

Die weitere Reinigung des rohen L-Salzes erfolgte durch genaue Wiederholung des obigen Trennungsprozesses mit den gleichen Lösungsmittelmengen. Man erhielt so 12,2 g reines L-Salz (92% d. Th.) mit Smp. 140—142°[1]).

Aus dem Filtrat fiel auch hier sofort das D-Salz aus. Die Filtration erfolgte nach mehrstündigem Stehen bei −15° und lieferte 10,78 g Substanz. Durch einmaliges Umkristallisieren aus Methanol, wie unter a) beschrieben, erhielt man reines D-Salz.

4. N-Tosyl-L- und N-Tosyl-D-threonin. Die Suspension von 17,35 g (26,1 mMol) N-Tosyl-L-threoninbrucinsalz in 52 ml Wasser wurde auf 0° gekühlt und unter Rühren auf einmal mit der theoretischen Menge 0,5-n. Natronlauge (53 ml) versetzt. Kurze Zeit nach der Bildung einer fast klaren Lösung begann sich das Brucin auszuscheiden. Man rührte noch eine Stunde bei 0° weiter, filtrierte sodann, wusch mit 50 ml Eiswasser und säuerte das Filtrat mit konzentrierter Salzsäure an (stark kongosauer!). Nach einigen Minuten begann eine rasche Kristallisation in Schuppen. Das nach mehrstündigem Stehen bei 0° abfiltrierte und mit wenig Eiswasser gewaschene Produkt enthielt ein Mol Kristallwasser und schmolz bei vorsichtigem Erhitzen zwischen 80 und 90°. Die Ausbeute betrug 6,88 g (91% d. Th.). Umkristallisieren aus der neunfachen Menge Wasser lieferte praktisch verlustlos ein Produkt, welches nach Vertreiben des Kristallwassers durch langsames Erhitzen auf 100° im Vakuum einen Schmelzpunkt von 135—137° zeigte.

---

[1]) Vgl. Bemerkung über die Schmelzpunkte der Brucinsalze in der Einleitung.

Ein Analysenpräparat, aus Wasser umkristallisiert und im Hochvakuum 6 Stunden bei 90° getrocknet, schmolz bei 136—137°. $[\alpha]_D^{18} = +14,8° \pm 0,8°$ (c = 2 in Methanol).

$C_{11}H_{15}O_5NS$	Ber.	C 48,34	H 5,53	N 5,13%
(273,30)	Gef.	,, 48,49	,, 5,55	,, 4,93%

Die aus dem N-Tosyl-D-threonin-brucinsalz ganz analog dargestellte D-Form besass den gleichen Schmelzpunkt. $[\alpha]_D^{18} = -14,7° \pm 0,5°$ (c = 2 in Methanol). Gef. N 4,98%.

5. L-Threonin und D-Threonin. 3,48 g (11,95 mMol) kristallwasserhaltiges rohes N-Tosyl-L-threonin wurden mit 40 ml konz. Salzsäure (d = 1,19) bei 90—100° 7 Stunden im Bombenrohr verseift. Durch Abkühlen der blassgelben Lösung auf −10° konnte der grösste Teil der entstandenen Sulfosäure ausgefällt werden. Nach Filtration durch Glaswolle und Nachwaschen mit kalter, konz. Salzsäure wurde das Filtrat im Vakuum zum Sirup eingeengt. Durch zweimaliges Eindampfen mit je 50 ml Wasser wurde überschüssige Salzsäure und durch anschliessendes Eindampfen mit 50 ml Alkohol das Wasser entfernt. Bei der Neutralisation des in 35 ml absolutem Alkohol gelösten Sirups mit Diäthylamin auf Lackmus[2]) fiel das L-Threonin feinsandig aus. Filtration und Waschen mit Alkohol und Äther ergaben 1,24 g (82% d. Th.) Rohprodukt.

Die weitere Reinigung erfolgte durch Lösen in Wasser, kurzes Aufkochen mit wenig gewaschener Tierkohle, Filtration und Versetzen der heissen, wässerigen Lösung (18 ml) mit 72 ml siedendem, absolutem Alkohol, in welchem wenige Kristalle bereits vorhandener Substanz oder einige Körnchen zurückbehaltenen Rohprodukts suspendiert waren. Ausbeute 1,09 g (72% d. Th.) optisch und chemisch reines L-Threonin in Form hexagonaler Blättchen. Zersetzungspunkt ca. 260°[3]).

Zu Analyse und Drehung wurde aus Wasser/Alkohol umkristallisiert und 4 Stunden bei 90° im Hochvakuum getrocknet. $[\alpha]_D^{18} = -29,2° \pm 1°$ (c = 2 in Wasser)[3]). Ber. N 11,76%, Gef. N 11,62%.

D-Threonin liess sich analog aus N-Tosyl-D-threonin darstellen. $[\alpha]_D^{18} = +29,3° \pm 1°$ (c = 2 in Wasser). Gef. N 11,68%.

---

[1]) Vgl. Bemerkung über die Schmelzpunkte der Brucinsalze in der Einleitung.

[2]) Rotes Lackmuspapier wurde gerade schwach gebläut; pH 7—7,5.

[3]) *D. F. Elliott*, loc. cit., gibt für L-Threonin einen Smp. 262—263° (Zers.) und eine Drehung $[\alpha]_D^{21} = -28,5°$ (c = 2,4 in Wasser) an.

$C_4H_9NO_3$

Ref. 69c

Resolution of Racemic N-Benzoylthreonine

2.23 g. of N-benzoyl-DL-threonine and 2.46 g. of L-threo-2-amino-1-(p-methylsulfonylphenyl)-1,3-propanediol were dissolved together in 35 ml. of methanol. The solution was seeded with a few crystals of the L-threo-2-amino-1-(p-methylsulfonylphenyl)-1,3-propanediol salt of N-benzoyl-L-threonine and allowed to stand for one hour at 25° C. The crystalline solid which had separated from solution was then collected on a filter, washed with a few ml. of methanol, and dried at 70° C. There was thus obtained 2.2 g. of the L-threo-2-amino-1-(p-methylsulfonylphenyl)-1,3-propanediol salt of N-benzoyl-L-threonine which melted at 187–189° C. The melting point of this product remained unchanged after it had been recrystallized from 95% ethanol. This salt was mixed in aqueous solution with hydrochloric acid to yield a precipitate of N-benzoyl-L-threonine which after collection on a filter and drying melted at 150–151° C. and had $[\alpha]_D^{25}$ +26.15° (c. 2% in water).

By evaporating the filtrate from the L-threo-amine-L-acylamino acid salt, there is obtained the L-threo-2-amino-1-(p-methylsulfonylphenyl)-1,3-propanediol salt of N-benzoyl-D-threonine. By mixing an aqueous solution of this salt with hydrochloric acid, there is obtained a precipitate of N-benzoyl-D-threonine.

The racemic forms of N-formylthreonine, N-acetylthreonine, N-propionylthreonine, and N-butyrylthreonine can be resolved in a manner similar to that described above for racemic N-benzoylthreonine.

$C_4H_9NO_3$

Ref. 69d

EXAMPLE 6

Resolving of N,N-dibenzyl-DL-threonine

(A) PREPARATION OF THE SALT OF L(+)-THREO-1-p-NITRO PHENYL-2-AMINO PROPANE-1,3-DIOL AND N,N-DIBENZYL-D(−)-THREONINE

15 g. of N,N-dibenzyl-DL-threonine and 11 g. of L(+)-threo-1-p-nitro phenyl-2-amino propane-1,3-diol are dissolved in 150 cc. of aqueous 50% methanol while keeping the temperature near the boiling point under reflux. The solution is kept at a temperature of 60° C. for one hour and is then allowed to cool to a temperature of 35° C. within two hours. The crystallized salt is filtered off, washed with a small quantity of water and recrystallized from 100 cc. of aqueous 50% methanol. 10.15 g. of the salt of L(+)-threo-1-p-nitro phenyl-2-amino propane-1,3-diol and N,N-dibenzyl-D(−)-threonine are obtained thereby. The resolving yield amounts to 80%. The compound is obtained in the form of yellow platelets which are insoluble in benzene and chloroform and only slightly soluble in water, ethanol, ether, and acetone. The compound melts at 187–188° C.; its rotatory power is $[\alpha]_D^{20} = -57° \pm 1°$ (concentration: 1% in methanol).

$C_4H_9NO_3$

(B) PREPARATION OF N,N-DIBENZYL-D(—)-THREONINE

10.15 g. of the salt of L(+)-threo-1-p-nitro phenyl-2-amino propane-1,3-diol and N,N-dibenzyl-D(—)-threonine obtained according to the preceding Example 6(A) are treated with 21 cc. of N sodium hydroxide solution for 15 minutes while stirring vigorously. By following the procedure as indicated in the preceding examples, 4 g. of the resolving agent corresponding to a yield of 48% of the initially employed resolving agent are recovered. In addition, 5.85 g. of N,N-dibenzyl-D(—)-threonine, corersponding to a yield of 98%, are obtained thereby. Said acid, on recrystallization from aqueous 50% methanol, has a melting point of 94–96° C. and a rotatory power of $[\alpha]_D^{20}=-111°\pm1°$ (concentration: 2% in methanol). The compound is obtained in colorless needles which are soluble in ethanol, acetone, chloroform, and ether, only slightly soluble in warm water, and very slightly soluble in cold water.

(C) TREATMENT OF THE MOTHER LIQUORS

The mother liquors obtained on preparing and recrystallizing the salt of L(+)-threo-1-p-nitro phenyl-2-amino propane-1,3-diol and N,N-dibenzyl-D(—)-threonine are combined and concentrated by evaporation to a volume of about 150 cc. 32 cc. of N sodium hydroxide solution are added thereto. On following the procedure as described in the preceding examples, 5.2 g. of the resolving agent are recovered. The alkaline filtrate is acidified by the addition of 3.5 cc. of acetic acid and extracted with chloroform. The chloroform solution is dried and evaporated to dryness in a vacuum. The resulting residue is treated with 5.7 g. of D(—)-threo-1-p-nitro phenyl-2-amino propane-1,3-diol in 75 cc. of aqueous 50% methanol. Thereby, 9.8 g. of the salt of D(—)-threo-1-p-nitro phenyl-2-amino propane-1,3-diol and N,N-dibenzyl-L(+)-threonine are obtained. The melting point of said salt is 187–188°; its rotatory power is $[\alpha]_D^{20}=+57°\pm1°$ (concentration: 1% in methanol). The yield amounts to 77% calculated for the N,N-dibenzyl-L(+)-threonine contained in the initial racemic mixture. The compound forms yellow platelets which are insoluble in benzene and chloroform, and only slightly soluble in water, ethanol, ether, and acetone.

(D) HYDROGENOLYSIS OF N,N-DIBENZYL-D(—)-THREONINE

2 g. of N,N-dibenzyl-D(—)-threonine dissolved in 30 cc. of 80% ethanol are hydrogenated in the presence of 2 g. of a palladium black catalyst containing 6% of palladium and with the addition of 1 cc. of concentrated hydrochloric acid at a temperature of 70° C. The catalyst is filtered off. The filtrate is concentrated to a volume of 10 cc. 1.5 cc. of pyridine are added thereto, and the precipitate is removed by filtration. Optically pure D(—)-threonine is obtained thereby in a yield of 72%. The rotatory power of said acid is $[\alpha]_D^{20}=-28°\pm1°$ (concentration: 2% in water).

$C_4H_9NO_3$

Ref. 69e

Au cours de nos recherches sur le dédoublement par entraînement (1, 2), nous avons déjà envisagé l'application de cette méthode à la DL-thréonine (2).

Nous précisons ici le mode opératoire et le bilan pratique d'une opération prise comme exemple. Sur une quantité plus importante de matière, la reproductibilité des résultats s'est montrée meilleure que dans nos essais précédents (2).

Dédoublement de la DL-thréonine par entraînement.

N° des cristallisations	DL-thréonine introduite, en g	D-thréonine introduite, en g	D-thréonine recueillie		L-thréonine recueillie	
			poids en g	$[\alpha]_D$ (eau)	poids en g	$[\alpha]_D$ (eau)
1	45	5	8,5	+ 26°		
2	8,5				8,2	— 25°
3	8,2		9,3	+ 26°		
4	9,3				9,4	— 26°
5	9,4		11,6	+ 20°		
6	11,6				8,8	— 25°
7	8,8		9,4	+ 25°		
8	9,4				8,7	— 25°
9	8,7		10,6	+ 17,5°		
10	10,6				10,8	— 22°
11			6,6	+ 22°		
Total............	129,5	5	56,0		45,9	

a) Après avoir dissous, à 80°, 45 g de DL-thréonine et 5 g de D(+)—thréonine dans 150 cm³ d'eau (3 vol), on refroidit à 30°. Lorsque la cristallisation débute, on refroidit à 20° puis abandonne pendant une heure à cette température, en agitant de temps en temps. On essore. Le poids de D-thréonine recueillie est sensiblement le double de celui de la D-thréonine introduite en début d'expérience.

On ajoute à la solution mère un poids de DL-thréonine égal au poids de la D-thréonine recoltée précédemment. On chauffe à 80° jusqu'à dissolution complète, puis refroidit et cristallise comme ci-dessus, ce qui fournit un poids de L-thréonine sensiblement égal à celui de l'isomère D, déjà recueilli.

b) Le même cycle d'opérations a été renouvelé dix fois. Après la dixième cristallisation, la solution mère abandonnée pendant deux jours à 10° environ a laissé déposer la D-thréonine utilisée comme amorce, avec un peu de racémique.

pur, $[\alpha]_D = +28°$ ou $-28°$ ($c = 2$ %, eau). Pour cela, on dissout l'isomère optique dans cinq volumes d'eau bouillante, provoque la cristallisation par addition de 15 volumes d'alcool bouillant puis laisse refroidir.

Les jets successifs de D- et L-thréonine sont voisins de la pureté optique. Une seule recristallisation fournit le produit

On récupère la DL-thréonine qui reste dans les solutions mères de dédoublement et de recristallisation par concentration sous vide et addition d'alcool.

***

Le tableau ci-dessus résume l'ensemble de l'opération. A partir de 129 g de racémique et 5 g d'isomère D, on a obtenu finalement, après recristallisation des formes D et L :

D-thréonine	46 g
L-thréonine	39 g
DL-thréonine récupérée	42 g

Le rendement en D(+)-ou L(-)-thréonine est donc voisin de 90 %.

Bibliographie.

(1) L. Velluz. G. Amiard et R. Joly, *Bull. Soc. chim.*, 1953, p. 342.
(2) L. Velluz et G. Amiard, *Bull. Soc. chim.*, 1953, p. 903.

$C_4H_9NO_3$ Ref. 69f

Le présent travail a été suggéré par les résultats obtenus, dans notre laboratoire, lors du dédoublement du DL-*thréo* 1-*p*-nitrophényl 2-amino propane 1,3-diol (*ibid.*, 1953 [5], 20, 342).

$$CH_3-CH(OH)-CH(NH_2)-CO_2H$$

Thréonine.

1. Une première série d'expériences permet de construire la courbe en trait plein du graphique annexé, sur lequel il est porté (en g pour 100 cm³) :

— en abscisses : le poids de thréonine D ou L mis en œuvre;

— en ordonnées : le poids d'amino-acide restant dissous.

La droite OZ, dont les ordonnées et les abscisses sont égales, représente les concentrations des solutions lorsqu'aucune cristallisation n'est encore survenue.

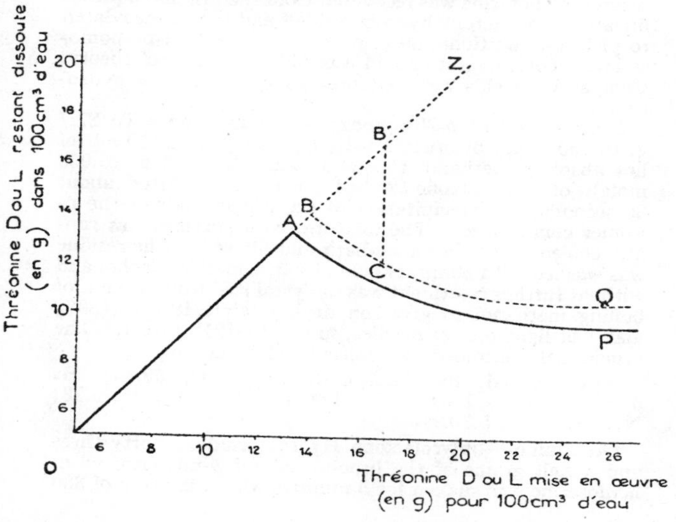

Après dissolution à 80°, puis refroidissement pendant une heure à 20°, les solutions contenant plus de 13,4 g d'amino-acide pour 100 cm³ d'eau (point A) cristallisent. La quantité restant dissoute dépend alors de la quantité initialement mise en œuvre. La représentation de cette fonction est fournie par AP. La courbe OAP correspond, en définitive, au poids d'amino-acide D ou L restant en solution à 20°, dans nos conditions expérimentales. L'existence du point de rebroussement A montre que le taux de sursaturation décroît sensiblement lorsque la quantité de produit précipité augmente.

2. Dans une seconde série d'expériences, on dissout des quantités connues de DL-thréonine, à 80°. Puis, après refroidissement à 20°, amorçage de la cristallisation et repos d'une heure, on pèse le produit resté en solution.

Ces nouvelles valeurs permettent de construire la courbe pointillée ABQ, qui traduit l'influence propre de l'amino-acide L, lorsque celui-ci est présent en quantités égales à celles de l'amino-acide D, sur le taux de sursaturation des liqueurs à 20°. L'existence du point de rebroussement B, dont les coordonnées sont supérieures à celles de A, démontre qu'en présence d'énantiomorphe L on observe le même phénomène ci-dessus décrit, à savoir la décroissance du taux de sursaturation lorsque le poids d'amino-acide précipité augmente.

3. Si, donc, on dissout à 80°, dans 100 cm³ d'eau, 28 g d'amino-acide racémique et 2,8 g d'amino-acide D, autrement dit 14 g d'amino-acide L et 16,8 g d'amino-acide D, on se trouve dans les conditions représentées respectivement, sur le graphique, par les points B et B'. En refroidissant à 20°, l'amino-acide L reste en sursaturation alors que l'amino-acide D précipite de telle façon que la quantité restant dissoute à 20° corresponde au point C (*). La quantité précipitée (4,7 g) est nettement supérieure à celle mise en œuvre (2,8 g). L'opération crée donc un apport en forme D. Après avoir séparé l'énantiomère, les liqueurs comportent un léger excès de l'antipode. Il suffit d'introduire dans ces liqueurs une nouvelle quantité de racémique et de procéder encore à une cristallisation par chaud et froid. On sépare l'énantiomère inverse et le cycle des opérations peut être poursuivi dans des conditions presque identiques aux précédentes.

Partie expérimentale.

Courbes de sursaturation.

Des quantités croissantes de D- ou L-thréonine sont dissoutes, à 80°, dans 2 cm³ d'eau. On refroidit à 20° et amorce la cristallisation. On abandonne ensuite à 20° pendant une heure, centrifuge rapidement, décante la solution limpide et prélève 1 cm³ en cristallisoir taré. On pèse, puis évapore à sec. Le résidu séché à poids constant représente le poids d'amino-acide dissous dans un volume d'eau correspondant à la différence entre le poids de la solution et le poids du résidu. On rapporte à 100 cm³ d'eau et obtient ainsi le poids de substance restée en solution dans 100 cm³ d'eau, après une heure à la température de 20°.

Dédoublement.

L(—) *thréonine.*

Après avoir dissous 9 g de DL-thréonine et 1 g de L (—)thréonine dans 30 cm³ d'eau à 80°, on refroidit la liqueur sous agitation dans un bain d'eau à 20°. Lorsque l'équilibre de température est obtenu, on amorce la cristallisation, abandonne pendant une heure à 20°, en agitant de temps en temps, essore à fond et sèche. On obtient

(*) En toute rigueur, les ordonnées en B et C devraient tenir compte du fait que la quantité de l'autre amino-acide en présence n'est pas absolument égale ; la correction est négligeable.

2,07 g de L(—) thréonine légèrement souillée de racémique,

$[\alpha]_D^{20} = -26° \pm 1$ ($c = 1$ %, eau) ; [théorie $[\alpha]_D^{20} = -28°$].

$C_4H_9NO_3$

D(+) thréonine.

Les eaux-mères provenant de l'opération précédente sont additionnées de 2,07 g de DL-thréonine. On chauffe à 80°, refroidit comme précédemment, amorce et abandonne pendant une heure à 20°. Après filtration et séchage, on obtient 2,95 g de D-thréonine impure, $[\alpha]_D^{20} = +17°$ (c = 1 %, eau). On dissout le produit dans 6 cm³ d'eau, chauffe à 80° en agitant puis laisse refroidir à 20°. On essore, lave à l'alcool absolu et sèche. Le rendement en D(+)thréonine, $[\alpha]_D^{20} = +28° \pm 1°$ (c = 1 %, eau), est de 1,7 g.

For experimental details of enzymatic methods of resolution of racemic threonine, see J.P. Greenstein and L. Levintow, J. Am. Chem. Soc., 72, 2814 (1950), S.M. Birnbaum, L. Levintow, R.B. Kingsley and J.P. Greenstein, J. Biol. Chem., 194, 455 (1952), and K.R. Rao, S.M. Birnbaum, R.B. Kingsley and J.P. Greenstein, J. Biol. Chem., 198, 507 (1952); also see M. Winitz, L. Bloch-Frankenthal, N. Izumiya, S.M. Birnbaum, C.G. Baker and J.P. Greenstein, J. Am. Chem. Soc., 78, 2423 (1956), R.L.M. Synge, Biochem. J., 33, 1931 (1939); M. Tanaka and K. Mineura, Japanese Patent 42-13447 (1967), M. Tanaka and K. Mineura, Japanese Patent 42-6325 (1967), K. Mineura and M. Tanaka, Nippon Nogeikagaku Kaishi, 42, 216 (1968), Y. Kameda, E. Toyoura, K. Matsui, Y. Kimura, Y. Kanaya, A. Nakatani, H. Saito and K. Kawase, Yakugaku Zasshi, 78, 769 (1958), Y. Kameda, E. Toyoura and K. Matsui, Chem. Pharm. Bull., 7, 702 (1959).

$$\text{HOOC-CH-CH-CH}_3 \\ \quad\;\;|\quad\;\;| \\ \quad\;\;\text{NH}_2\;\;\text{OH}$$

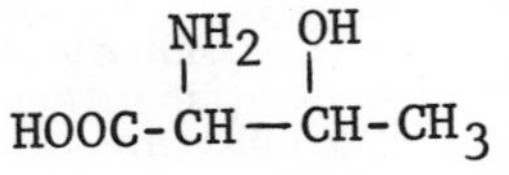

$C_4H_9NO_3$ Ref. 70

Beschreibung der Versuche

N-Phthaloyl-DL-allothreonin

Zu einer Lösung von 71,4 g DL-Allothreonin (0,6 Mol) und 172,8 g $Na_2CO_3 \cdot 10\ H_2O$ in 750 ml Wasser werden unter kräftigem Rühren 135 g feinpulverisiertes N-Carbäthoxyphthalimid gegeben und bei Zimmertemperatur so lange gerührt, bis alles in Lösung gegangen ist. Danach wird filtriert und unter Kühlung das pH der Lösung mit 6 n HCl auf 2 eingestellt. Es scheidet sich zunächst ein farbloses Öl ab, das bei weiterem kräftigem Rühren kristallin wird. Der Niederschlag wird abgesaugt, mit kaltem Wasser nachgewaschen und getrocknet. Ausbeute 112 g (70% d. Th.) fast reines N-Phthaloyl-DL-allothreonin. Schmp. 144—146°. Zur Analyse wurde noch zweimal aus Essigester umkristallisiert. Schmp. unverändert 144—146°.

$C_{12}H_{11}NO_5$ (249,22) ber.: C 57,83; H 4,45; N 5,62;
gef.: C 57,78; H 4,72; N 5,78.

N-Phthaloyl-D-allothreonin-brucinsalz

118,2 g Brucin werden in 150 ml Äthylenglykol-monomethyläther (Methylcellosolve) bei 80° gelöst und unter Rühren 74,4 g (0,3 Mol) N-Phthaloyl-DL-allothreonin zugegeben. Der sofort entstehende gelbe Kristallbrei wird noch 15 Minuten bei 80° gerührt und anschließend 24 Stunden bei Zimmertemperatur stehen gelassen. Danach wird das Salz angesaugt, 5mal mit je 50 ml Methanol gewaschen und getrocknet. Es werden 93,5 g Salz (97% d. Th.) vom Schmp. 223–225° (Zers.) erhalten. $[\alpha]_D^{20} + 13,1°$ (c = 1; Dimethylformamid). Zur Analyse wurde noch zweimal aus Äthylenglykol-monomethyläther umkristallisiert. Schmp. 227–228° (Zers.); $[\alpha]_D^{20} + 15,2°$ (c = 1; Dimethylformamid).

$C_{35}H_{37}N_3O_9$ (643,67) ber.: C 65,31; H 5,79; N 6,53;
gef.: C 65,34; H 5,49; N 6,79.

D-Allothreonin

60 g N-Phthaloyl-D-allothreonin-brucinsalz werden mit 145 ml 2 n NaOH und 200 ml Chloroform 1 Stunde kräftig gerührt. Das Brucinsalz löst sich dabei vollständig auf. Die wäßrige Phase wird noch zweimal mit je 100 ml Chloroform extrahiert, während die vereinigten Chloroform-Extrakte noch zweimal mit je 50 ml 0,5 n NaOH ausgeschüttelt werden. Die vereinigten wäßrigen Phasen werden mit HCl auf pH 3 eingestellt und im Vakuum auf etwa ⅓ des ursprünglichen Volumens eingedampft. Nach Zusatz von 56 ml konz. HCl wird 2½ Stunden unter Rückfluß gekocht und nach dem Abkühlen die ausgeschiedene Phthalsäure abgesaugt, zweimal mit je 20 ml Wasser nachgewaschen und das Filtrat im Vakuum zur Trockne eingedampft. Der Rückstand wird in 40 ml Wasser aufgenommen, von ungelöstem NaCl und etwas Phthalsäure durch Filtration befreit und die wäßrige Lösung erneut im Vakuum zur Trockne eingedampft. Das D-Allothreonin-hydrochlorid wird mit 80 ml Methanol herausgelöst, NaCl durch Filtration abgetrennt und das Filtrat mit Zugabe von Diäthylamin auf pH 6 gebracht. Nach mehrstündigem Stehen wird das D-Allothreonin abgesaugt, mit Methanol gründlich nachgewaschen und getrocknet. Ausbeute 9,5 g; $[\alpha]_D^{20} -31,7°$ (c = 1; 1 n HCl). Das Rohprodukt wird in siedendem Wasser gelöst und mit dem doppelten Volumen Alkohol ausgefällt, abgesaugt, mit Alkohol nachgewaschen und getrocknet. Ausbeute 8,9 g; $[\alpha]_D^{20} -32,4°$ (c = 1; 1 n HCl). Papierchromatographisch einheitlich bei 48 Stunden Laufzeit (aufsteigend) im System Butanol-Wasser-Ammoniak-(20proz.)-Aceton (100:69:18,5:12,5)[12].

L-Allothreonin

Das Filtrat vom N-Phthaloyl-D-allothreonin-brucinsalz wird im Vakuum zur Trockne eingedampft. Die Zersetzung des L-Salzes (etwa 92 g) erfolgt mit 245 ml 2 n NaOH in Gegenwart von 245 ml Chloroform. Die weitere Aufarbeitung erfolgt wie beim D-Allothreonin. Letzte Kristallisation aus Wasser allein. Ausbeute 13 g (67% d. Th.); $[\alpha]_D^{20} +32,8°$. (c = 1; 1 n HCl). Reinheitsprüfung wie oben angegeben.

[12] K. N. F. Shaw u. S. W. Fox, J. Amer. chem. Soc. 75, 3421 (1953).

$$\text{HOCH}_2\text{CH}_2\text{-CH-COOH} \\ \qquad\qquad\;\;| \\ \qquad\qquad\text{NH}_2$$

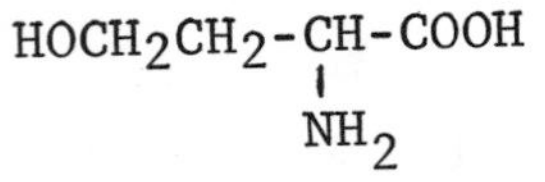

$C_4H_9NO_3$ Ref. 71

N-p-Nitrobenzoyl-DL-homoserine.—The preparation of this compound, with certain modifications, was carried out essentially by the procedure described for threonine.[11] One hundred and nineteen grams (1 mole) of DL-homoserine was dissolved in 1800 ml. of water and 574 ml. of 1.75 N sodium hydroxide and the solution cooled to 0°. With stirring, there were added simultaneously over a period of one hour 185.5 g. (1 mole) of p-nitrobenzoyl chloride and 574 ml. of 1.75 N sodium hydroxide. After being stirred for a total of one hour at 0° and then at room temperature, the solution, after cooling, was acidified with 265 ml. of concentrated hydrochloric acid and left overnight in the refrigerator. The precipitate was filtered, washed with ice-water, dried *in vacuo* over sulfuric acid and extracted 3 hours with ether in a Soxhlet. The white residue so obtained (201 g.), m.p. 142°, was of sufficient purity for the preparation of the brucine salt. Most of the unreacted homoserine was recovered from the original aqueous filtrate as the lactone hydrobromide[10] and then benzoylated to yield an additional 39 g. of N-p-nitrobenzoyl-DL-homoserine. Total weight of acid was 240 g. or 89% of theory. Comparable yields were obtained using the lactone hydrobromide.

Brucine Salt of p-Nitrobenzoyl-L-homoserine.—To 27.7 g. (0.059 mole) of brucine (4H₂O), dissolved in 150 ml. of hot absolute methanol (55–60°), was added 16 g. (0.059 mole) of N-p-nitrobenzoyl-DL-homoserine. After about 30 seconds, the precipitation of the brucine salt of the L-isomer commenced. The mixture, with stirring, was rapidly cooled to 30° in an ice-bath and filtered. The residue was washed with 50 ml. of cold absolute methyl alcohol and without further treatment was recrystallized from 600 ml. of boiling methanol to give, on drying (air), 18.7 g. (89% yield) of light orange needles, m.p. 119–121° (dec.). The brucine salt contained two molecules of water.

Anal. Calcd. for $C_{34}H_{38}O_{10}N_4\cdot 2H_2O$: C, 58.44; H, 6.06; N, 8.02; H₂O, 5.16. Found: C, 58.46; H, 6.23; N, 8.26; H₂O, 5.30.

L-α-Amino-γ-butyrolactone Hydrobromide.—Forty-three and a half grams of the brucine salt of p-nitrobenzoyl-L-homoserine was shaken for 5 minutes with a mixture of 360

---

(9) All melting points are uncorrected.
(10) J. E. Livak, E. C. Britton, J. C. VanderWeele and M. F. Murray, This Journal, **67**, 2218 (1945).
(11) A. J. Zambito, W. L. Peretz and E. E. Howe, *ibid.*, **71**, 2541 (1949).

ml. of 2 N ammonium hydroxide and 150 ml. of chloroform. The aqueous layer was extracted with two portions of 75 ml. of chloroform and the alkaline solution was evaporated to about 100 ml. *in vacuo*. On acidification of the cooled solution with hydrochloric acid, N-*p*-nitrobenzoyl-L-homoserine precipitated. The mixture was allowed to stand overnight in the refrigerator, filtered and the precipitate washed with ice-cold water. The filtrates and washings were combined, concentrated *in vacuo* to about 25 ml., whereby additional quantities of the product were obtained. Total weight on different runs was 15.6–16.4 g. (94–98% of theory).

Twelve grams of *p*-nitrobenzoyl-L-homoserine was refluxed with 240 ml. of 16% hydrobromic acid for 4 hours and the mixture left overnight in the refrigerator. The *p*-nitrobenzoic acid was filtered off, the filtrate was evaporated to dryness *in vacuo* and the residue was transferred to a funnel with the minimum amount of ice-cold absolute methanol. The light yellow solid was washed 3 times with 3 ml. portions of cold methyl alcohol and the white crystalline product dried over P_2O_5 *in vacuo* at 60°; yield 6.6 g., m.p. 242° dec., $[\alpha]^{24}_D$ −21.3 (1% in water) (lit.[8] −21.0).

Anal. Calcd. for $C_4H_8O_2NBr$: N, 7.69. Found: N, 7.82.

Brucine Salt of *p*-Nitrobenzoyl-D-homoserine.—The filtrates from which the L-brucine salt had been precipitated plus the washings were placed on the refrigerator overnight. The light yellow solid was filtered and recrystallized from 40 ml. of boiling methyl alcohol to give 14.0 g. of yellow needles, m.p. 119–121° (dec.). By concentrating the filtrate further and cooling, additional 2 g. of the salt was obtained. Total weight was 16 g. or 77% of theory. The brucine salt crystallized with two molecules of water.

Anal. Calcd. for $C_{34}H_{38}O_{10}N_4 \cdot 2H_2O$: C, 58.44; H, 6.06; N, 8.02; H_2O, 5.16. Found: C, 58.85; H, 6.58; N, 8.27; H_2O, 5.12.

D-α-Amino-γ-butyrolactone Hydrobromide.—Forty-three and a half grams of the brucine salt was decomposed in the same way as described previously, yielding 16.0 g. of N-*p*-nitrobenzoyl-D-homoserine. From 60 g. of this compound there was obtained 33.5 g. of D-α-amino-γ-butyrolactone hydrobromide; m.p. 241° dec., $[\alpha]^{24}_D$ +21.0 (1% in water) (lit.[8] +21.0).

Anal. Calcd. for $C_4H_8O_2NBr$: N, 7.69. Found: N, 7.68.

The over-all yield of the L-isomer was 60% while that of the D-isomer was 53%.

(8) M. D. Armstrong, This Journal, **70**, 1758 (1948).

C₄H₉NO₃   Ref. 71a

Resolution of N-Formyl-O-phenyl-DL-homoserine.—To a dry mixture of 112 g. (0.5 mole) of N-formyl-O-phenyl-homoserine and 170 g. (0.5 mole) of powdered strychnine was added 5 liters of hot 50% methanol and the suspension was swirled and heated in a water-bath until almost all of the solids had dissolved. The solution was filtered while hot and was allowed to stand overnight at room temperature. The crystalline strychnine salt of N-formyl-O-phenyl-D-homoserine was collected on a filter, washed with a small amount of cold water, and dried; yield, 152 g.; $[\alpha]^{27}_D$ −28° (1% in HOAc). One recrystallization from 4 liters of hot 50% methanol yielded 128 g.; $[\alpha]^{26}_D$ −27° (1% in HOAc). Four more recrystallizations of the salt from aqueous methanol produced a pure strychnine salt, $[\alpha]^{27}_D$ −25° (1% in HOAc), but in practice the best method of obtaining the pure isomer proved to be recrystallization of the free O-phenyl-D-homoserine resulting from the decomposition of the once recrystallized strychnine salt.

O-Phenyl-L-homoserine.—The original mother liquors from the crystalline strychnine salt were concentrated to a volume of approximately 2.5 liters, made alkaline by the addition of 20 ml. of concd. ammonia, cooled and filtered. The strychnine may be dried and reused. The filtrate was concentrated to a volume of about 1500 ml., made 1 N in hydrochloric acid by the addition of the proper amount of concd. hydrochloric acid and refluxed for two hours. The solution was then concentrated to dryness under reduced pressure and the residue was dissolved in 200 ml. of hot water; the hot solution was made neutral to congo red by the careful addition of concd. sodium hydroxide solution, the suspension was cooled and filtered. The residue was recrystallized from 500 ml. of hot water; yield, 35 g., m. p. 210–211° dec., $[\alpha]^{25}_D$ +21.5° (1% in 1 N HCl). The combined mother liquors were concentrated to a volume of approximately 500 ml., cooled and filtered, yielding an additional 13 g. of impure product; m. p. 196–204° dec.; $[\alpha]^{25}_D$ +5° (1% in 1 N HCl).

Two more recrystallizations of the pure derivative from 400-ml. portions of hot water yielded 22.0 g. of pure O-phenyl-L-homoserine; m. p. 241–242° dec., $[\alpha]^{26}_D$ +23.5° (1% in 1 N HCl).

Anal. Calcd. for $C_{10}H_{13}O_3N$: N, 7.17. Found: N, 7.21.

O-Phenyl-D-homoserine.—A solution of 125 g. of the strychnine salt of N-formyl-O-phenyl-D-homoserine in 4 liters of hot 50% methanol was made alkaline by the addition of 20 ml. of concd. ammonia. The solution was cooled overnight in a refrigerator, the strychnine was removed by filtration, and the crude O-phenyl-D-homoserine was prepared in the same manner as previously described for the L-isomer; yield 38.5 g., m. p. 216–219° dec., $[\alpha]^{24}_D$ −19° (1% in 1 N HCl). By reworking the mother liquors 6 g. of impure compound was obtained; m. p. 214–216° dec., $[\alpha]^{26}_D$ −10° (1% in 1 N HCl).

Two recrystallizations of the first crop from 300 ml. portions of hot water yielded 26.5 g., m. p. 218–220° dec., $[\alpha]^{26}_D$ −22.0° (1% in 1 N HCl). One more recrystallization from 500 ml. of hot water yielded 22 g., m. p. 241° dec., $[\alpha]^{24}_D$ −23.5° (1% in 1 N HCl).

Anal. Calcd. for $C_{10}H_{13}O_3N$: N, 7.17. Found: N, 7.32.

In spite of its low solubility the D isomer possesses a definitely sweet taste, whereas the L isomer is tasteless or nearly so.

C₄H₉NO₃

For experimental details of an enzymatic method of resolution of racemic homoserine, see S.M. Birnbaum and J.P. Greenstein, Arch. Biochem. Biophys., **42**, 212 (1953).

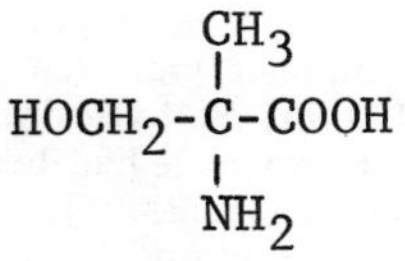

$$\begin{array}{c} CH_3 \\ | \\ HOCH_2-C-COOH \\ | \\ NH_2 \end{array}$$

C₄H₉NO₃   Ref. 72

EXAMPLE 4

Resolution of Racemic N-Benzoyl-Alpha-Methylserine

A 10% solution of the L-threo-2-amino-1-(p-methylsulfonylphenyl)-1,3-propanediol salt of N-benzoyl-DL-alpha-methylserine in 95% ethanol was seeded with a few crystals of the L-threo-2-amino-1-(p-methylsulfonylphenyl)-1,3-propanediol salt of N-benzoyl-L-alpha-methylserine and the solution was allowed to stand for one hour at 25° C. The crop of L-threo-2-amino-1-(p-methylsulfonylphenyl)-1,3-propanediol salt of L-alpha-methylserine, M.P. 168–170° C., which crystallized from solution was collected on a filter. An aqueous solution of this salt was mixed with a slight excess of hydrochloric acid to yield a precipitate of N-benzoyl-L-alpha-methylserine, M.P. 128–130° C. By heating the N-benzoyl-L-alpha-methylserine with hydrobromic acid, there is obtained L-alpha-methylserine, M.P. 284–285° C. (dec.), $[\alpha]_D^{25}$ −5.4° (c. 1% in water).

The L-threo-2-amino-1-(p-methylsulfonylphenyl)-1,3-propanediol salt of N-benzoyl-D-alpha-methylserine, which is recovered by evaporating the ethanolic mother liquor to dryness, is converted by mixing with an excess of hydrochloric acid to N-benzoyl-D-alpha-methylserine, M.P. 127–129° C.

$C_4H_9NO_3$ Ref. 72a

EXAMPLE 12

Resolution of N-Benzoyl-DL-alpha-Methylserine

22.3 g. of N-benzoyl-DL-alpha-methylserine and 21.3 g. of L-threo-2-amino-1-(p-methylmercaptophenyl)-1,3-propanediol were dissolved together in 100 ml. of methanol. Crystallization of solid from this solution was initiated by scratching the inside wall of the container below the surface of the solution. After the solution had stood for several hours at 25° C., a very heavy separation of solid from the solution had taken place. This solid was collected on a filter and washed with a few ml. of cold 95% ethanol and dried at 90° C. There was thus obtained 17.6 g. of the L-threo-2-amino-1-(p-methylmercaptophenyl)-1,3-propanediol salt of N-benzoyl-L-alpha-methylserine which melted at 175–177° C. The melting point of the product remained unchanged after recrystallization from 150 ml. of 95% ethanol.

The filtrate was evaporated under reduced pressure, and the residue thus obtained was dissolved in 100 ml. of water containing 5 ml. of concentrated hydrochloric acid. To the clear solution there was added 10 ml. of 35% aqueous sodium hydroxide solution, thus causing the precipitation of L-threo-2-amino-1-(p-methylmercaptophenyl)-1,3-propanediol. The mixture was cooled to 5° C. and the precipitated L-threo-2-amino-1-(p-methylmercaptophenyl)-1,3-propanediol was collected on a filter. The filtrate was acidified by mixing with 7 ml. of concentrated hydrochloric acid and the acidified solution was seeded with a few crystals of N-benzoyl-DL-alpha-methylserine. The seeded solution was allowed to stand at 5–10° C. for several hours and then the solid which had separated from solution was collected on a filter. There was thus obtained 4 g. of N-benzoyl-DL-alpha-methylserine which melted at 158–160° C. The filtrate was saturated with sodium chloride and allowed to stand overnight at 5° C. The solution was then filtered to collect 7.5 g. of N-benzoyl-D-alpha-methylserine which had separated from the solution. This product melted at 127–129° C.

N-Benzoyl-L-Alpha-Methylserine

475 g. of the L-threo-2-amino-1-(p-methylmercaptophenyl)-1,3-propanediol salt of N-benzoyl-L-alpha-methylserine (obtained, in the manner indicated above, from 637 g. of N-benzoyl-DL-alpha-methylserine) was dissolved in one liter of water to which had been added 90 ml. of concentrated hydrochloric acid. This solution was cooled and made alkaline by addition of 180 ml. of 35% sodium hydroxide solution. There was an essentially quantitative precipitation of L-threo-2-amino-1-(p-methylmercaptophenyl)-1,3-propanediol, and this product was collected on a filter. The filtrate was acidified by the addition of 92 ml. of concentrated hydrochloric acid and the acidified solution was saturated with sodium chloride and allowed to stand for several hours at 0° C. The crystalline solid, N-benzoyl-L-alpha-methylserine, which had separated from solution was collected on a filter. The filtrate was extracted several times with ethyl acetate to recover a further quantity of N-benzoyl-L-alpha-methylserine. The initial crop of crystalline product was combined with the material obtained by the extraction procedure, and the product was recrystallized from ethyl acetate. There was thus obtained 220 g. of N-benzoyl-L-alpha-methylserine which melted at 128–130° C.

N-Benzoyl-D-Alpha-Methylserine

This product was obtained from the L-threo-2-amino-1-(p-methylmercaptophenyl)-1,3-propanediol salt of N-benzoyl-D-alpha-methylserine in a manner similar to that described above for the preparation of N-benzoyl-L-alpha-methylserine.

When N-benzoyl-L-alpha-methylserine was refluxed with 34% hydrobromic acid, there was obtained L-alpha-methylserine which melted at 284–285° C. (dec.) and had $[\alpha]_D^{25}$ −5.4° (c. 1% in water). By hydrolysis of N-benzoyl-D-alpha-methylserine in similar fashion, there is obtained D-alpha-methylserine.

$$H_2NCH_2\underset{\displaystyle OH}{\underset{|}{C}}HCH_2COOH$$

$C_4H_9NO_3$ Ref. 73

(+)-4-Benzamido-3-hydroxybutyric acid

89.2 g (0.4 mole) of (I) was dissolved in 200 ml of 1 *N* sodium hydroxide and 625 ml of boiling water. 66.8 g of strychnine (0.2 mole) were added in portions to the boiling solution. On cooling, the strychnine salt crystallized in long needles. After two days at room temp. 98.8 g, m.p. 102°, were filtered off. Refrigeration of the mother liquor gave 7.5 g, m.p. 102°. Recrystallization of a sample from ethanol gave needles of m.p. 98° unchanged on further recrystallization.

80.0 g of the strychnine salt were suspended in 500 ml of water and adjusted to pH 10 with 6 *N* sodium hydroxide. The precipitated strychnine was filtered off, the solution extracted twice with chloroform and acidified to pH 1 with hydrochloric acid. Leaving overnight at room temperature gave 15.9 g, m.p. 160-163°. ($[\alpha]_{589}^{23°} = +4.6°$; $l = 2$; $c = 2$ in 1 *N* NaOH.) The mother liquor was concentrated and left at 0° overnight which gave an additional 9.32 g, m.p. 105-106°, $[\alpha]_{589}^{23°} = +20,6°$; $l = 0,75$; $c = 2$ in 1 *N* NaOH. A sample of this fraction which had been recrystallized from ethanol/ether and dried at 78° and 0.1 mm Hg had m.p. 100° (lit. [7] 116°).

$$[\alpha]_{589}^{23°} = +20.2°;\ l = 2;\ c = 2 \text{ in } 1\ N\ \text{NaOH}$$

Anal. Calc. for $C_{11}H_{13}O_4N$ (223.32): C 59.16; H 5.87; N 6.27.
Found: C 58.81; H 5.98; N 6.31.

(+)-4-Amino-3-hydroxybutyric acid

2.0 g of (+)-4-benzamido-3-hydroxybutyric acid were refluxed with 10 mm of 4 *N* hydrochloric acid for 90 min. Benzoic acid was filtered off after cooling, the solution extracted twice with ether and evaporated to dryness *in vacuo*. The residue was dissolved in the minimum amount of 50 per cent ethanol, adjusted to pH 6–7 with ammonium hydroxide and left at room temperature. 0.1 g of large prisms were obtained, m.p. 210–212° (decomp.).

$[\alpha]_{589}^{23°} = +20.4°$; $l = 1$; $c = 1.48$ in water (lit. [7] m.p. 214°; $[\alpha]_D^{20°} = +18.3°$ in water).

Anal. Calc. for $C_4H_9O_3N$ (119.12): C 40.33; H 7.62; N 11.76.
Found: C 40.28; H 7.69; N 11.61.

On refrigeration the mother liquor gave 0.44 g, m.p. 210–212° (decomp.); $[\alpha]_{589}^{23°} = +19.7°$; $l = 1$; $c = 0.8$ in water.

The mother liquor from the strychnine salt was concentrated to about 150 ml, made alkaline and extracted twice with chloroform. On acidification and cooling to 0° 45.9 g of crystals, m.p. 158–162°, $[\alpha]_{589}^{23°} = -7°$; $l = 0.75$; $c = 2$ in 1 *N* NaOH, where obtained. They were dissolved with 81.0 g of brucine in 250 ml of boiling water. After two days in the refrigerator 84.7 g of crystalline brucine salt, m.p. 97–98° (lit. [7] 87°) were obtained. The brucine salt was dissolved in 500 ml water, the solution made alkaline, extracted twice with chloroform and acidified to pH 1. The solution was kept for 6 days in the refrigerator when 7.09 g were filtered off, m.p. 159–160°, $[\alpha]_{589}^{25°} = -7,3°$; $l = 1$; $c = 2$ in 1 *N* NaOH. Concentration of the mother liquor to 200 ml and refrigeration gave 4.85 g, m.p. 108–110°, $[\alpha]_{589}^{23°} = -18.6°$; $l = 2$; $c = 2$ in 1 *N* NaOH. Further concentration to 75 ml gave 3.72 g, m.p. 100–101°, $[\alpha]_{589}^{23°} = -19.3°$; $l = 2$; $c = 2$ in 1 *N* NaOH. A sample of the fraction with m.p. 108–110° was recrystallized from ethanol/ether and gave needles of m.p. 97°.

$$[\alpha]_{589}^{23°} = -21.45°;\ l = 2;\ c = 2 \text{ in } 1\ N\ \text{NaOH}$$

after drying at 78° and 0.1 mm Hg (lit. m.p. 114° [7], $[\alpha]^D = 21.5–22.5$ in 0.5 *N* NaOH [16]).

Anal. Calc. for $C_{11}H_{13}O_4N$: C 59.16; H 5.87; N 6.27.
Found: C 58.27; H 5,94; N 6.32.

The mother liquor of the brucine salt on acidification gave 14.3 g, $[\alpha]_{589}^{23°} = +4.0°$, of benzamido acid.

(−)-4-Amino-3-hydroxybutyric acid

0.5 g of (−)-4-benzamido-3-hydroxybutyric acid was hydrolyzed as described for the (+)-form. Yield 0.15 g, m.p. 210–212° (decomp.)

$$[\alpha]_{589}^{27°} = -21.4°;\ l = 2;\ c = 1.18 \text{ in water}$$

(lit. [7] m.p. 212°, $[\alpha]_D^{20} = -21.06°$ in water).

Anal. Calc. for $C_4H_9O_3N$ (119.12): C 40.33; H 7.60; N 11.76.
Found: C 40.49; H 7.68; N 11.88.

C4H9NO3 Ref. 73a

The racemic acid was resolved using brucine as resolving agent in ethanol. The physical constants of the active acids obtained were: m.p. 214° decomp., $[\alpha]_D^{20}$ +18.30°; m.p. 212° decomp., $[\alpha]_D^{20}$ -20.98°.

```
 OH NH2
  |  |
HOCH2-CH-CH-COOH
```

C4H9NO4 Ref. 74

```
         OH
         |
HOCH2-CH-CH-COOH
            |
            NH2
```

Resolution of Racemic I_a. *N*-Acetyl-I_a was prepared in an 88% yield by the usual Schotten-Baumann method, mp 151—153°C, as CHA*[3] salt. The salt obtained (7.33 g, 0.02 mol) was dissolved in 2N sodium hydroxide, and the liberated CHA was extracted with ether. After the pH of the aqueous layer had been adjusted to 6.8, the solution was incubated with Takadiastase at 37°C for 4 days. The precipitate was then filtered off, and the filtrate was concentrated under reduced pressure until crystals appeared. The crystals of I_{a-1} were obtained in a 67% yield and were recrystallized from hot water; mp 194—195°C; $[\alpha]_D^{23}$ +21.9° (*c* 5.7, 1N HCl).

Found: C, 58.62; H, 6.75; N, 6.20%. Calcd for $C_{11}H_{15}O_4N$: C, 58.65; H, 6.71; N, 6.22%.

The mother liquor of I_{a-1} was acidified to pH 2.0 and was extracted with ethyl acetate. *N*-Acetyl-I_{a-2} was obtained from the concentrated extract as CHA salt in an 84% yield. Recrystallization from methanol-ethyl acetate gave pure crystals; mp 145—146°C; $[\alpha]_D^{23}$ −12.5° (*c* 2.5, EtOH). The partial hydrolysis of the CHA salt of *N*-acetyl-I_{a-2} with 1N hydrochloric acid gave free amino acid I_{a-2} in a 82% yield; mp 194°C; $[\alpha]_D^{23}$ −21.9° (*c* 5.5, 1N HCl).

Found: C, 58.78; H, 6.67; N, 6.25%. Calcd for $C_{11}H_{15}O_4N$: C, 58.65; H, 6.71; N, 6.22%.

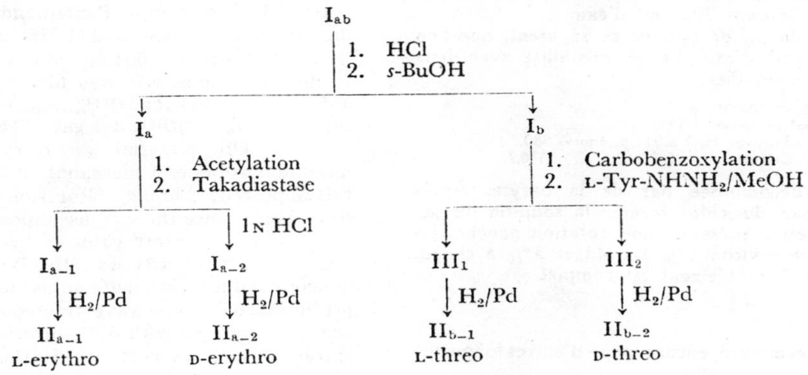

Fig. 1.

***N*-Carbobenzoxy-I_b (III).** *N*-Carbobenzoxy-I_b was prepared from I_b and carbobenzoxychloride by the usual method; it was obtained as CHA salt in a 79% yield; mp 158—159°C.

L-Tyrosine Hydrazide Salt of III. The free acid (III) was obtained from CHA salt (13.7 g, 0.03 mol) by the use of 3N hydrochloric acid and by ethyl acetate extraction. L-Tyrosine hydrazide (5.86 g, 0.03 mol) was added to a solution of III in methanol (50 m*l*), and the solution was warmed at 65°C. After the removal of insolubles, the solution was concentrated; the subsequent addition of ethanol gave L-tyrosine hydrazide salt of III in a quantitative yield (16.6 g), $[\alpha]_D^{23}$ +36° (*c* 1.0, water).

Four recrystallizations of this salt (5.96 g) from methanol gave a small amount (310 mg) of optically-pure L-tyrosine hydrazide salt (V_1), mp 158.5—159°C; $[\alpha]_D^{23}$ +30.6° (*c* 1.0, water). V_1 was used as the seed of the following partial crystallization.

Resolution of Racemic III. A solution of L-tyrosine hydrazide salt of III in methanol (50 m*l*) was similarly prepared from the CHA salt of III (6.88 g, 0.015 mol). After the insolubles had been filtered off, a small amount of V_1 was seeded to the filtrate; then the mixture was stirred for 1.5 hr and kept at 25°C for 22.5 hr. The precipitated first crop (V_1) was then collected (2.12g); mp 158.5°C; $[\alpha]_D^{23}$ +30.5° (*c* 1.0, water). The mother liquor was concentrated, and the resulting crystals were dissolved in methanol (50 m*l*). After the solution had been stirred at 25°C for 4 hr and then kept at 25—17°C for 20 hr, the second crop (V_2) was obtained (2.66 g, 64%); mp 163—165°C; $[\alpha]_D^{23}$ +39.3° (*c* 1.0, water). From the filtrate of V_2, an additional crop of V_1 was obtained; the total yield of V_A was 73%.

III_1 was obtained from V_1 by treatment with 3N hydrochloric acid; it was purified as CHA salt; mp 138.0°C; $[\alpha]_D^{23}$ +10.9° (*c* 2.0, EtOH).

Found: C, 65.61; H, 7.60; N, 6.18%. Calcd for $C_{25}H_{34}O_6N_2$: C, 65.48; H, 7.47; N, 6.11%.

The CHA salt of III_2 was obtained from V_2 in the same way; mp 137—138°C; $[\alpha]_D^{23}$ −10.3° (*c* 2.0, EtOH)

Found: C, 65.62; H, 7.57; N, 6.14%.

***α*-Amino-*β*,*γ*-dihydroxybutyric Acids (II).** A mixture of I_{a-1} (675 mg, 3 mmol), water (10 m*l*), palladium charcoal (160 mg), and N hydrochloric acid (3 m*l*) was stirred for 5 hr at room temperature under the bubbling of hydrogen gas. After the removal of the catalyst, the solution was neutralized and subsequently concentrated under reduced pressure. The resulting crystals were recrystallized twice from water-methanol to give II_{a-1} (305 mg, 75%); mp 194—195°C; $[\alpha]_D^{23}$ −11.3° (*c* 7.2, water).

Found: C, 35.56; H, 6.77; N, 10.30%. Calcd for $C_4H_9O_4N$: C, 35.55; H, 6.71; N, 10.37%.

II_{a-2}, II_{b-1}, and II_{b-2} were obtained from I_{a-2},

III_1, and III_2 by the catalytic hydrogenation described above.

II_{a-2}: mp 194°C; $[\alpha]^{23}_D$ +11.3° (*c* 7.0, water). Found: C, 35.99; H, 6.91; N, 10.63%.

II_{b-1}: mp 214°C; $[\alpha]^{23}_D$ −13.5° (*c* 2.0, water). Found: C, 35.32; H, 6.72; N, 10.21%.

II_{b-2}: mp 214—215°C; $[\alpha]^{23}_D$ +13.6° (*c* 4.8, water). Found: C, 35.61; H, 6.74; N, 10.98%.

*3 CHA=cyclohexylamine

TABLE 1. MELTING POINTS AND SPECIFIC ROTATIONS OF THE FOUR ISOMERS OF α-AMINO-β,γ-DIHYDROXYBUTYRIC ACID (II)

		II_{a-1}	II_{a-2}	II_{b-1}	II_{b-2}
Author	Mp (°C)	194—195	194	214	214—215
	$[\alpha]^{23}_D$ (°)	−11.3	+11.3	−13.5	+13.6
		L-erythro	D-erythro	L-threo	D-threo
Hamel	Mp (°C)	—	193—194	214—215	214—215
	$[\alpha]^{23}_D$ (°)		+15.3	−13.6	+13.1
Niemann	Mp (°C)	—	192—194	215	—
	$[\alpha]^{24}_D$ (°)		+16.0	−13.7	

CH_3-CH(Cl)-$CH_2CH_2AsO_3H_2$

$C_4H_{10}AsClO_3$ Ref. 75

1. *Acide 3-chlorebutane-1-arsonique.*

On ajoute 8 g (0.02 mol.) de quinine à une solution bouillante de 4.3 g (0.02 mol.) de l'acide dans 300 cm³ d'eau.

Par refroidissement 6 g du *sel de quinine* se séparent, que l'on fait recristalliser dans 200 cm³ d'eau. Le sel cristallise avec deux molécules d'eau en rosettes d'aiguilles.

Substance 0.1144 g: perte à 90° 0.0069 g.
" 0.4108 g: 5.82 cm³ de baryte 0.1217 n.
Trouvé: H_2O 6.03: p. équiv. 580.
$C_4H_{10}O_3AsCl . C_{20}H_{24}O_2N_2 . 2 H_2O$. Calculé: " 6.25: " " 576.7.

Une partie du sel est décomposée par de la baryte. Après extraction de la quinine par du chloroforme, la solution du sel, examinée dans un polarimètre, présente une rotation gauche. La solubilité du sel de quinine est environ 1 % à froid et 3 % à chaud. Après cinq cristallisations le dédoublement est complet.

Sel de baryum: $[M]_D$ = −21.4°.
Acide: " = −28°.

Le pouvoir rotatoire a été mesuré encore pour d'autres longueurs d'ondes:

λ (μμ)	656.3	589.3	546.3	486.1	435.8
[M] sel	16.9°	21.4°	24.5°	30.6°	33.7°
" acide	23.3°	28.0°	32.7°	42.0°	48.2°

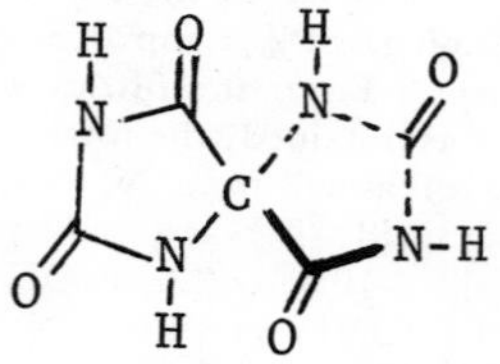

$C_5H_4N_4O_4$ Ref. 76

On fractional crystallisation of the brucine salt of the synthetic material from water, *l*-brucine-*l*-*spiro*-5 : 5-dihydantoin is readily obtained in clusters of thin colourless needles, melting at 260° ; from this the lævo-*spiro*dihydantoin is separated in the usual way.

The specific rotatory power of *l*-*spiro*-5 : 5-dihydantoin for the mercury green line is, in ethyl alcohol solution, [α] = −113° and, in aqueous solution, [α] = −115°. In dilute aqueous ammonia, however, the specific rotation changes widely with the concentration of the alkali; thus a 1·75% solution of 12% aqueous ammonia solution gave a specific rotatory power of [α] = + 8·7°.

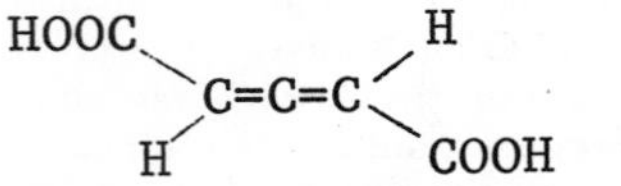

$C_5H_4O_4$ Ref. 77

Partial Resolution of Pentadiendioic Acid.—A solution of 200 mg. of pentadiendioic acid (V) in 32 ml. of ether was added dropwise to a solution of 504 mg. of quinine in 100 ml. of ether. The white amorphous salt was filtered off and dried *in vacuo;* it amounted to 644 mg. (91%), $[\alpha]_D$ −141°. Trituration of the salt (611 mg.) with water gave 315 mg. of cream crystals, $[\alpha]_D$ −167°. This material was recrystallized by dissolving in a minimum volume of methanol and adding water until tan crystals appeared; 230 mg. Rotation on this material was difficult to measure, since the salt decomposes somewhat on recrystallization; an approximate value of $[\alpha]_D$ −187 ± 7° was obtained. This 230 mg. of salt was dissolved in aqueous methanol and aqueous sodium carbonate added to pH 9. The resulting milky precipitate of quinine was extracted with chloroform. Solution pH was adjusted to 2 with 5 *M* sulfuric acid and pentadiendioic acid extracted with six portions of ether. The dried ether solution yielded 25 mg. of crude crystalline material, shown to be pentadiendioic acid by solid state infrared spectrum, $[\alpha]_D$ −104°.

This resolution was repeated on a larger scale using 1.48 g. of pentadiendioic acid and 4.17 g. of quinine in 200 and 800 ml. of ether, respectively. The 5.14 g. of crude amorphous salt was crystallized from water and decomposed directly to alkaloid and acid. There was thereby obtained 348 mg. of crude allene, $[\alpha]_D$ −53°, which was combined with the above 25 mg., $[\alpha]_D$ −104°, for purification. Recrystallization from dichloromethane–ether gave 271 mg. of pentadiendioic acid in two crops, $[\alpha]_D$ −45.6 ± 0.3°, infrared spectrum in potassium bromide virtually superposable on that of authentic racemic material.

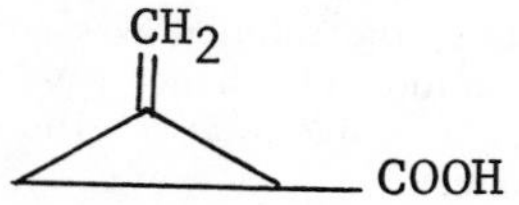

$C_5H_6O_2$ Ref. 78

EXAMPLE 1

Formation of (−)-cinchonidine-(−)-2-methylene-cyclopropane carboxylate

40.0 grams of *dl* 2-methylenecyclopropane-1-carboxylic acid is dissolved in 8.0 liters of water at 80° C., followed by the slow addition (over 3 minutes to prevent lumping) of mortar ground cinchonidine. The slurry is heated with stirring to reflux where it is maintained for 0.5 hour and almost total solution is present. This is then filtered hot (95–100° C.), seeded with (−)-cinchonidine(−)-2-

methylenecyclopropane carboxylate, and allowed to cool with stirring to ambient temperature. The melting point is 166–167° C.

The resulting slurry is filtered, the solid cake slurry washed with 400 ml. water and sucked on a funnel for an hour to afford a white crystalline solid which, for best results, is sent through the crystallization procedure described in the next step without further drying. Physical constants, determined on a small portion of this salt, vacuum dried at 100° C. for 3 hours, indicates a resolution yield of 82.3 grams, which is 102.5% based on the dried sample. The melting point is 159–161° C., $[\alpha]_D^{24}$ −75.8° (c.=1, $CHCl_3$).

EXAMPLE 2

First recrystallization of (−)-cinchonidine(−)-2-methylenecyclopropane

To 4.2 liters of water at 70° C., 81.7 grams of crude cinchonidine salt is added, the slurry heated with agitation to reflux, stirred until almost total solution (5 minutes), and filtered at the boil. The filtrate is seeded with (−)-cinchonidine(−)-2-methylenecyclopropane carboxylic acid, M.P. 166–167° C., and allowed to cool with stirring over 3-4 hours to ambient temperature. The solid is filtered, slurry washed with 400 ml. of water and sucked dry on a funnel, and for best results, used in the final recrystallization without further drying. Physical constants determined on a small portion of this salt, vacuum dried at 100° C. for 3 hours, indicates a recovery yield of 58.7 grams, which is 71.3% based on the dried sample. The melting point is 162-163° C., $[\alpha]_D^{24}$ −77.4° (c.=1, $CHCl_3$).

4

EXAMPLE 3

Second recrystallization of (−)-cinchonidine(−)-2-methylenecyclopropane carboxylate

58.0 grams of the partially purified cinchonidine salt is added to 2.95 liters of water at 70° C. The stirred slurry is heated to and maintained at reflux for 0.5 hour, filtered at the boil, and the hot filtrate seeded as in the prior crystallization.

The filtrate, which crystallizes at 95° C., is allowed to cool to ambient temperature, filtered, the crystalline precipitate slurry washed with 300 ml. water, and air-dried 0.5 hour. Physical constants determined on a fractional portion of this salt, vacuum dried at 100° C. for 3 hours, indicates a recovery yield of 43.8 grams, which is 75.4% based on the dried sample. The melting point is 164–165° C., $[\alpha]_D^{24}$ −78.7° (c.=1, $CHCl_3$).

EXAMPLE 4

Liberation of (−)-2-methylenecyclopropane carboxylic acid

36.0 grams of the purified cinchonidine salt is acidified with 470 ml. of 0.4 N sulfuric acid, stirred until solution is complete, and then extracted with six 100 ml. portions of chloroform. The combined chloroform extract is dried over $MgSO_4$, concentrated in vacuo at less than 40° C. to a thin syrup, and then, after de-gassing, distilled in vacuo.

Fraction	Pressure (mm. Hg)	Temp. (° C.)	Weight (g.)
1	0.4	81	0.15
2	0.4	82-84	7.19

Physical constants are determined on the major fraction, the weight is 7.19 g., which is 79.9% recovery.

$$[\alpha]_D^{24}-35.2°$$

(neat), $[\alpha]_D^{24}$ −38.3° (c.=1, $CHCl_3$).

Elemental analysis for $C_5H_6O_2$: Calculated (percent): C, 61.22; H, 6.16; N, 0.00. Found (percent): C, 61.12; H, 6.14; N, 0.00.

EXAMPLE 5

Resolution of (+)-2-methylenecyclopropane carboxylic acid

8.0 liters of filtrate from the formation of (−)-cinchonidine(−)-2-methylenecyclopropane carboxylate (crude) is concentrated under reduced pressure at 40° C. to .8 liter, filtered, and the filtrate further concentrated to dryness, keeping the temperature at 40° C. The residual solid weighs 74 grams and is recrystallized twice from the minimal amount of hot acetone:benzene (2:1) to give 48.0 grams of purified cinchonidine salt. The melting point is 163–165° C.

36.0 grams of the salt is liberated via the procedure employed in liberation of the *l*-acid to give 7.2 grams of (+)-2-methylenecyclopropane carboxylic acid,

$$[\alpha]_D^{24}+34.8°$$

(neat), $[\alpha]_D^{24}$ −38.1° (c.=1, $CHCl_3$).

$C_5H_6O_3$

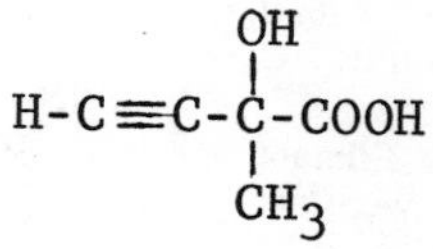

Ref. 79

(2) *Acides (+) et (−)-hydroxy-2 méthyl-2 butyne-3 oïques.* Dans un ballon de 1 litre surmonté d'un réfrigérant, on place 22·8 g (0·200 mole) d'acide (±)**1**, 66 g (0·204 mole) de quinine et 500 ml de méthanol anhydre. On chauffe à reflux pendant 2 heures et laisse reposer un jour entier. La solution dépose alors une certaine quantité de sel solide I, que l'on filtre et lave à l'alcool méthylique. On en obtient ainsi 38 g (43 %), F = 192°. Après évaporation à mi-volume de la liqueur mère et repos de 24 heures, une nouvelle fraction de sel précipite. On filtre et recueille 7 g de sel II (8 %), F = 190°. Le reste du solvant est alors évaporé complètement et le résidu constitue une dernière fraction de sel III: 38·5 g (44 %), F = 187°. Les acides sont libérés en agitant les sels de quinine en suspension dans l'éther avec HCl à 25 % pendant 7 heures. Après extraction au perforateur, séchage des extraits et évaporation du solvant, on obtient trois fractions d'acide cristallisé: Acide I: 8·6 g (88 %), $\{\alpha_D^{24}\} = -33°$; Acide II: 1·5 g (80 %), $\{\alpha_D^{24}\} = -23°$; Acide III: 8·3 g (84 %), $\{\alpha_D^{24}\} = +32°$.

A la suite de cinq opérations analogues à la précédente, les trois fractions d'acide sont soumises à une série de recristallisations dans le benzène. On obtient finalement trois échantillons de chacun des acides droit et gauche (Tableau 2).

TABLEAU 2

$\{\alpha^{24}_{D}\}$	Masse (g)	Rdt global (%)
−41°	23·8	21
−19°	1·6	1·5
−9°	9·7	8·5
+43°	21·5	19
+24°	0·8	0·7
+9°	8·2	7

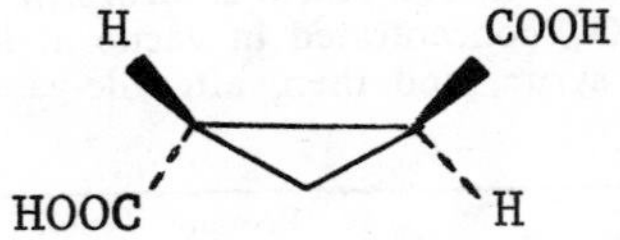

$C_5H_6O_4$ Ref. 80

Resolution of *trans*-Cyclopropane-1,2-dicarboxylic Acid (*trans*-I). **(a) Brucine.** A partial resolution has been reported.[15] *trans*-I (3.9 g, 30 mmol) in 100 ml of water was added to a solution of brucine (23.64 g, 60 mmol) in 100 ml of 95% ethanol, diluted with 110 ml of water, and stored for 5 hr at room temperature. As no solid separated during this period, most of the ethanol was removed under vacuum leaving an aqueous solution from which 14.5 g of colorless crystals, $[\alpha]^{27}{}_{578}$ −27° (*c* 0.868, EtOH), deposited on standing at room temperature overnight. Five recrystallizations from water gave 3.4 g of brucine salt, $[\alpha]^{27}{}_{578}$ +1° (*c* 0.830, EtOH), which was hydrolyzed with 10% aqueous NaOH, brucine was removed by filtration, and the aqueous solution was extracted with 20 ml of methylene chloride. The aqueous solution was then acidified with cold 10% dilute HCl, continuously extracted with ether for 12 hr, and the dried ethereal extract was concentrated to give 0.28 g of (+)-*trans*-I: mp 171–172.5°; $[\alpha]^{27}{}_{D}$ +218° (*c* 0.161, EtOH) [lit. $[\alpha]_D$ +84.5°[15] (*c* 2.02, H_2O), 227°[16] (*c* 2.34, EtOH)]. Crystallization of the diacid from ethyl acetate–petroleum ether did not change the rotation significantly.

(b) Quinine, One Equivalent. A solution of *trans*-I (53.8 g, 0.414 mol) and quinine (134 g, 0.414 mol) in 1:3 ethanol–ethyl acetate (1400 ml) was stored at 5° for 3 days to yield 37 g of crystalline salt. A sample of *trans*-I, recovered from 0.55 g of this salt by suspension in water, acidification with cold 20% HCl, extraction six times with 20-ml portions of ether, and concentration to a residue of 75 mg, had $[\alpha]^{27}{}_{578}$ −188° (*c* 0.5825, EtOH). The mother liquor on keeping at 5° for an additional 48 hr yielded a second crop of crystals (38 g) from which diacid, $[\alpha]^{27}{}_{578}$ +127° (*c* 0.837, EtOH), was obtained. The filtrate when stored at 20° for 30 hr deposited a third crop (30 g); diacid: $[\alpha]^{27}{}_{578}$ −208° (*c* 0.555, EtOH). Two recrystallizations of the first and the third crops furnished a total of 45 g of salt which gave 6 g of (−)-*trans*-I, $[\alpha]^{27}{}_{578}$ −230.2° (*c* 0.470, EtOH). A sample of this salt, crystallized from absolute MeOH and dried at 110° (0.05 mm) for 48 hr, had mp 160–163°. *Anal.* Calcd for $C_{45}H_{54}N_6O_8$: C, 69.4; H, 7.0; N, 7.2. Found: C, 69.0; H, 7.0; N, 7.3.

(c) Quinine, Two Equivalents, from Ethyl Acetate–Ethanol. *trans*-1 (2.6 g, 20 mmol) and quinine (13.96 g, 40 mmol) were dissolved in 50 ml of absolute ethanol, diluted with 150 ml of ethyl acetate, and kept at room temperature for 3 hr, whereupon a colorless solid had crystallized (6 g). A portion was converted to *trans*-I, $[\alpha]^{27}{}_{578}$ −166.4°. The remainder of the salt was crystallized twice from ethanol–ethyl acetate (3:5) to give material of mp 145–147° from which *trans*-I, $[\alpha]^{27}{}_{578}$ −231.5° (*c* 0.326, EtOH), was obtained.

(d) Quinine, One Equivalent, from Ethanol–Water. *trans*-I (3.9 g, 30 mmol) in 15 ml of water was added to a solution of quinine (9.72 g, 30 mmol) in 30 ml of ethanol and freed of most of the ethanol by distillation. The residue was then crystallized from 60 ml of water–ethanol (4:1) at room temperature. Recrystallization from the same solvent followed by a second recrystallization from absolute ethanol gave 0.55 g of needle-like crystals, mp 170–172°; $[\alpha]^{27}{}_{578}$ −135.7° (*c* 0.696, EtOH). Fractional crystallization of the filtrates from water–ethanol (4:1) gave another 1.7 g of material, mp 170–172°, $[\alpha]^{27}{}_{578}$ −137.1° (*c* 0.554, EtOH); regenerated *trans*-I (0.43 g from the combined samples) had $[\alpha]^{27}{}_{578}$ +198.7°, $[\alpha]^{27}{}_{589}$ +188.5° (*c* 0.382, EtOH).

$C_5H_6O_4$ Ref. 80a

Resolution of *trans*-1,2-Cyclopropanediacarboxylic Acid. The racemic *trans*-acid, mp 174—175°C. (Found:C, 46.25; H, 4.59%. Calcd for $C_5H_6O_4$: C, 46.16; H, 4.65%). (60 g; 0.46 mol) was dissolved in boiling water (3500 m*l*) and cinchonidine (136 g; 0.46 mol) was added and the mixture was allowed to stand still overnight at room temperature, when the cinchonidine salt crystallized out. After fractional triangular crystallization of the salt from 20% ethanol, the salt with a constant rotation was decomposed with dilute sulfuric acid to afford the optically active (−)-acid (6.9 g). Further recrystallization of the free acid from acetonitrile gave the optically pure acid, mp 171—173°C. (Found: C, 46.21; H, 4.53%. Calcd for $C_5H_6O_4$: C, 46.16; H, 4.65%). $[\alpha]^{25}_{D}$ −235° (*c* 2.0, water).

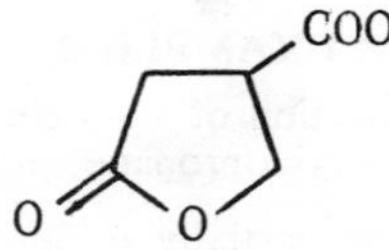

$C_5H_6O_4$ Ref. 81

Dédoublement de l'acide paraconique.

45 g d'acide paraconique, en solution dans 250 cm³ de méthanol absolu, sont additionnés de 42 g (quantité équimoléculaire) de de (—) α-phényl-éthylamine ($[\alpha]_D^{15}$ — 40,7°, en substance). A la solution obtenue on ajoute 500 cm³ d'éther anhydre, qui provoque la cristallisation du sel sous forme d'aiguilles brillantes (28 g). Après 5 recristallisations dans le même solvant on obtient 1,6 g de (+) paraconate d'α-phényl-éthylamine $[\alpha]_D^{22}$ + 25° (*c* = 1, méthanol absolu), dont le pouvoir rotatoire ne varie plus après une nouvelle cristallisation. Ce sel, filtré lentement sur une colonne contenant environ 20 cm³ d'Amberlite IR-120, fournit 0,830 g (rendement quantitatif) d'acide (+) paraconique, F = 48°, $[\alpha]_D^{20}$ + 49° (± 0,5°, *c* = 1,1, méthanol absolu).

L'estérification par le diazométhane fournit le paraconate de méthyle, $[\alpha]_D$ + 32° (± 0,5°; *c* = 2,44, méthanol absolu).

Par recristallisation fractionnée du paraconate de (—)-α-phényl-éthylamine et traitement par l'Amberlite IR-120, 3 g d'acide (+)-paraconique pur ont été obtenus, ainsi que de l'acide paraconique dextrogyre impur (mélangé à du racémique) qui a été utilisé tel quel pour divers essais.

L'acide paraconique enrichi en isomère lévogyre, récupéré après traitement du racémique par la (—)-α-phényl-éthylamine, a été traité par la (+)-α-phényl-éthylamine et a conduit à l'acide (—)-paraconique d'une manière analogue à celle qui a fourni l'acide (+)-paraconique.

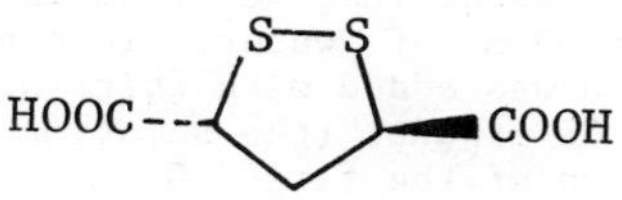

$C_5H_6O_4S_2$ Ref. 82

Resolution of racem-1,2-dithiolane-3,5-dicarboxylic acid.—8.6 g (0.044 mole) of the racemic acid and 20.4 g (0.044 mole) of brucine (4 H_2O) were dissolved in a hot mixture of 220 ml of water and 135 ml of ethanol. The solution was left for crystallization at room temperature. After one day the precipitate was filtered off, dried and weighed. About 0.3 g was taken for preliminary measurement of the optical activity (on the liberated acid) and the rest recrystallized from mixtures of ethanol and water. The results are given below.

Crystallization	1	2	3	4
ml of ethanol	135	150	100	80
ml of water	220	150	100	80
Salt obtained, in g	13.1	9.8	7.7	5.8
$[\alpha]_D^{25}$ of acid (abs. ethanol)	—	+518°	+542°	+538°

10.1 g (0.022 mole) of brucine (4 H_2O) was added to the above first mother liquor (with heating to 50°–60°) and the solution was placed over-night in a refrigerator. The salt obtained yielded almost pure active acid but was recrystallized twice.

Crystallization	1	2	3
ml of ethanol	135	100	75
ml of water	220	100	75
Salt obtained, in g	19.8	14.4	11.6
$[\alpha]_D^{25}$ of acid (abs. ethanol)	−542°	−545°	−546°

It may be noted that quinine yields acid and neutral salts (in alcohol) containing excess of the levorotatory acid but the resolution is much less than that obtained with brucine ($[\alpha]_D^{25}$ = −128° and −13° resp.).

(+)-1,2-Dithiolane-3,5-dicarboxylic acid

The acid brucine salt was decomposed with 4*N* sulphuric acid and the liberated acid was extracted with ether. After evaporation of the solvent (previous drying) the remaining acid was recrystallized from water (must be rapidly performed without heating above 80°) or from a mixture of ethyl acetate and ligroin. M.p. 185°–187°.

The melting point depends somewhat on the rate of heating but to a lesser extent than for the racemic acid. All melting points have been taken in the same manner in order to allow comparison.

Anal. Subst., 27.35 mg: 8.38 ml of 0.0334*N* NaOH. Equ. wt. Calc. 97.1; Found 97.7.

The solubility of the optically active acid is greater than that of the racemic in the above mentioned solvents.

The optical activity of the dextrorotatory acid was measured in different solvents (the solutions made up to 10.00 ml). The results are given below.

Solvent	mg of acid	$2\alpha_D^{25}$	$[\alpha]_D^{25}$	$[M]_D^{25}$
2*N* Hydrochloric acid	29.33	+4.25°	+725°	+1407°
Water	24.11	+3.38°	+701°	+1361°
Water (acid salt)	24.00	+2.75°	+573°	+1113°
Water (neutral salt)	25.56	+1.81°	+354°	+ 688°
Glacial acetic acid	30.46	+4.61°	+757°	+1470°
Abs. ethanol	45.88	+5.16°	+562°	+1092°
Ethyl acetate	23.07	+2.60°	+564°	+1095°
Acetone	30.61	+3.53°	+577°	+1120°

(−)-1,2-Dithiolane-3,5-dicarboxylic acid

The acid was liberated from the neutral brucine salt as described above for the enanthiomorph. The levorotatory acid resembled its antipode in all respects. M.p. 185°–187°.

Anal. Subst., 30.53 mg: 9.40 ml of 0.0334*N* NaOH. Equ. wt. Calc. 97.1; Found 97.2.
33.11 mg of the acid in 10.00 ml of abs. ethanol: $2\alpha_D^{25}$ = −3.71°; $[\alpha]_D^{25}$ = −560°; $[M]_D^{25}$ = −1088°.

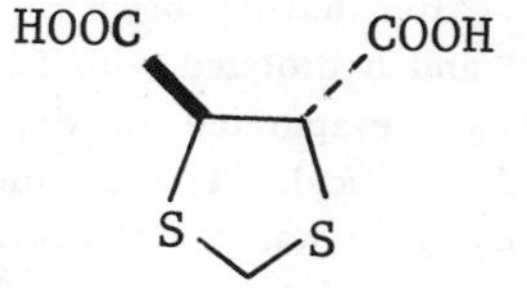

$C_5H_6O_4S_2$ Ref. 83

Preliminary experiments on resolution were carried out as usual. Salts of quinine and brucine gave (−)-acid of fairly high activity. Dextrorotatory acid was obtained via the salts of cinchonidine and strychnine. Other salts gave an acid of low activity or failed to crystallize. Quinine and cinchonidine were selected for the final resolution.

(−)-*Acid.* (±)-Acid (10 g, 0.051 mol) and quinine (39 g, 0.102 mol) were dissolved in 250 ml boiling methanol. The salt was collected after 24 hours at 0° and recrystallized from the same solvent. After each recrystallization, a small sample of the salt was decomposed and the rotatory power of the acid determined in abs. ethanol.

Crystallization	1	2	3	4	5
ml methanol	250	200	200	150	150
g salt obtained	22	18	12	8.4	6.2
$[\alpha]_D^{25}$ of the acid	−125°	−154°	−170°	−191°	−193°

The acid was liberated by means of dil. sulphuric acid, extracted with ether, dried and recrystallized from ethyl acetate-petroleum ether in the usual way. The salt yielded 1.5 g (15%) pure acid with m.p. 162–164° (175°), $[\alpha]_D^{25}$ = −193°(abs. ethanol).

$C_5H_6O_4S_2$ (194.2) calc. S 33.01
found S 32.89

(+)-*Acid* ((+)-*III*). From the mother liquors in the preceding paragraph, 5 g (26 mmol) (+)-acid of low activity was obtained. This was dissolved in 650 ml hot water and mixed with a solution of cinchonidine (15.0 g, 52 mmol) in 650 ml boiling ethanol. The salt was collected after 24 hours at 0° and recrystallized from dil. ethanol.

Crystallization	1	2	3	4	5
ml methanol	650	500	300	150	150
ml water	650	500	300	150	150
g salt obtained	15.1	9.2	6.3	4.8	3.7
$[\alpha]_D^{25}$ of the acid	+116°	+145°	+184°	+190°	+194°

The acid was liberated and recrystallized as above. Yield 0.6 g (12%), m.p. 162–164°, (175°), $[\alpha]_D^{25}$ = +194° (abs. ethanol).

$C_5H_6O_4S_2$ (194.2) calc. S 33.01
found S 32.80

Table 2. Optical activity of (+)-1,3-dithiolane-4,5-dicarboxylic acid ((+)-III).
g = mg acid in 10.00 ml solution.

Solvent	*g*	α_D^{25} (1 dm)	$[\alpha]_D^{25}$	$[M]_D^{25}$
Ethanol	41.56	+0.805	+194°	+377°
Methanol	22.33	+0.231	+103°	+200°
Acetone	19.94	+0.256	+128°	+248°
Water	14.15	+0.200	+141°	+274°

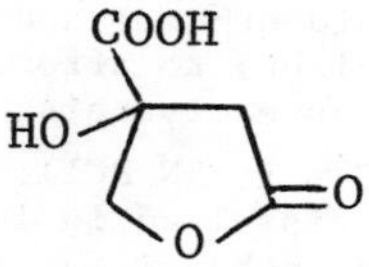

$C_5H_6O_5$ Ref. 84

To 40 g. of IV in 100 ml. of H_2O was added 109 g. of (−)-brucine in 40 ml. of hot MeOH. After 24 hr. at room temperature, the salt was harvested, washed with cold (4°) MeOH-H_2O (4:1), and dried. After 10 recrystallizations, the salt (3 g.), $[\alpha]_D^{25}$ (H_2O) −12.5 ± 0.2° (c 1.00), was dissolved in H_2O and passed through an acid ion-exchange resin[4] column. Evaporating the effluent and drying the residue gave 0.57 g. of (+)-IV: m.p. 87–89°; $[\alpha]_D^{23}$ (H_2O) +45.2 ± 0.2° (c 3.1); lit. (13) m.p. 86–90°, $[\alpha]_D^{15}$ (H_2O) −45.6° (c 2.43); the IR and NMR spectra were identical to those of IV obtained from III.

Anal.—Calcd. for $C_{28}H_{32}N_2O_9$: C, 62.21; H, 5.96; N, 5.18. Found: C, 61.9; H, 5.64; N, 5.00.

(4) H. Patel and G. Hite, *J. Org. Chem.*, **30**, 4336(1965).
(13) C. Katsuta and N. Sugiyama, *Bull. Chem. Soc. Japan*, **35**, 1194(1962).

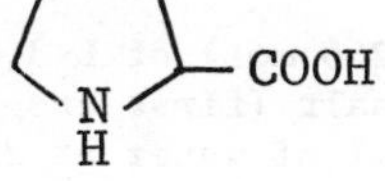

$C_5H_7NO_2$ Ref. 85

EXAMPLE 1

L-3,4-Dehydroproline (+) Tartaric Acid Salt

A 400 ml beaker was charged with 56.6 g (0.50 mol) of D,L-3,4-dehydroproline, 75.0 g (0.50 mol) of (+) tartaric acid (Baker) and 110 ml of water (distilled water is used through these examples). The mixture was stirred and heated at 80°C until all starting material was dissolved. The solution was stirred and allowed to cool to room temperature; at

48° the product began to crystallize. After the mixture had been stirred for three hours at room temperature and two hours in an ice bath, it was refrigerated overnight. The precipitate was collected on a glass sinter funnel (pre-cooled in the refrigerator), washed well with 2 x 50 ml = 100 ml of ice cold ethanol-water 1:1, 75 ml of ice cold ethanol-water 2:1, and 75 ml of ice cold ethanol and dried at room temperature (const. weight) to afford 52.3 g (40%) of product as off-white crystals, mp 173° (dec); $[\alpha]_D^{25}$ = -114.7° (c = 1, 5N HCl); e.e. 99.5%+.

The washings of the above described filtration were collected in a separate suction flask and concentrated on a rotavap (water bath 40°; aspirator vacuum). The residual gum (24 g) was taken up in 150 ml of warm water and combined with the mother liquor. The brown solution was heated at reflux (oil bath 120°) under nitrogen for 9 hours. 20 g of charcoal (Norite SG-SV) was added, and the mixture was heated at reflux for 20 minutes and filtered through a bed of Celite which was washed with 50 ml of water. The deep yellow filtrate was concentrated on a rotavap (water bath 45°; aspirator vacuum). The residual orange-brown paste (92 g) was dissolved in 50 ml of water at 70° and transferred into a beaker using an additional 12 ml of water to wash the flask. The warm solution was stirred for two hours at room temperature (when the temperature reached 40° the solution was seeded) and for two hours in an ice bath. The mixture was then refrigerated overnight. The precipitate was collected on a glass sinter funnel, washed with 2 x 25 ml = 50 ml of ice cold ethanol-water 1:1, 40 ml of ice cold ethanol-water 2:1, and 40 ml of ice cold ethanol and dried at room temperature (const. weight) to afford 25.5 g (19%) of product as light orange crystals, mp 173° (dec); $[\alpha]_D^{25}$ = -114.0° (C = 1, 5N HCl); e.e. = 99.5%+. The two crops were dissolved in 100 ml of water at 67°. Then 200 ml of ethanol was added with stirring (soon thereafter crystallization of the product began). The mixture was stirred in a 20° water bath for one hour and was then refrigerated overnight. The precipitate was collected by filtration, washed with 120 ml of ice cold ethanol and dried at room temperature (const. weight) to afford 71.5 g (54%) of product as off-white crystals, mp 173° (dec); $[\alpha]_D^{25}$ = -115.0° (c = 1, 5N HCl); e.e. = 99.8%+.

A second crop was obtained by concentration of the mother liquor and recrystallization from water-ethanol (1:2) as described above: 3.8 g (3%) of product as white crystals, mp 172° (dec); $[\alpha]_D^{25}$ = -114.5° (C = 1, 5N HCl).

EXAMPLE 2

L-3,4-Dehydroproline

70.0 g (0.266 mol) of L-3,4-dehydroproline (+) tartaric acid salt (first crop material) was dissolved in 320 ml of water at 25° and poured onto a 6.4 cm (diameter) X 17.5 cm (length) column of Dowex 50W - X4 cation exchange resin, 50-100 mesh (Biorad). The column was washed with a total of 1550 ml of water. The product was then eluted with 2.4 l of 0.5M pyridine acetate which was collected in eight 300-ml fractions. Each fraction was spotted on Silica Gel, sprayed with ninhydrin and developed at elevated temperature. The fractions, 5, 6 and 7 which contained the product showed a brown-orange spot. These three fractions were concentrated on a rotavap (water bath 25°, p = 1 mm) and the residue was dissolved twice in 100 ml of water and concentrated under the same conditions. The crystalline, wet residue (41.7 g) was dissolved in 30 ml of water. To the clear colorless solution was added with stirring over 30 minutes 240 ml of ethanol (the solution was seeded after addition of the first 60 ml). The mixture was refrigerated overnight. The precipitate was collected by filtration, washed with 50 ml of ice cold ethanol and 100 ml of ether and dried at room temperature/0.1 mm to afford 25.3 g. (84%) of product as colorless needles, mp 244° (dec); $[\alpha]_D^{25}$ = 278.7° (c = 1, 5N HCl); $[\alpha]_D^{25}$ = -403.1° (c = 1, H_2O); e.e. 99.8%+. The mother liquor was concentrated on a rotavap (water bath 25°, aspirator vacuum) and the residue evaporated from 50 ml of ethanol. The crystalline residue (4.6 g) was dissolved in 9 ml of water and 40 ml of ethanol was slowly added. The mixture was seeded, stirred at room temperature for one hour and refrigerated overnight. The precipitate was collected by filtration, washed with 10 ml of ice cold ethanol, and 10 ml of ether and dried at room temperature/0.1 mm to afford an additional 2.7 g (9%) of product as colorless crystals, mp 244° (dec); $[\alpha]_D^{25}$ = -278.8° (c = 1, 5N HCl); $[\alpha]_D^{25}$ = -404.3° (c = 1, H_2O); e.e. 99.8%+.

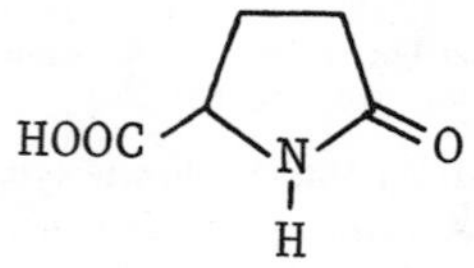

$C_5H_7NO_3$ Ref. 86

Resolution of DL-Pyro-glutamic Acid.

After DL-pyro-glutamic acid (12.9 g, 0.1 mol) and L-leucinamide (14.3 g, 0.11 mol) were dissolved in 400 ml of 99% ethanol on a water bath at 50°C, the solution was let to stand for 5 hrs at room temperature. A thick mass of fine needles produced was filtered and washed with a small amount of cold ethanol in which it was sparingly soluble (the yield; 10.4 g). The recrystallization from 300 ml of hot ethanol gave 8.4 g of L-pyro-glutamic acid salt. Yield, 60% based on L-form; m.p. 148~149°C; $[\alpha]_D^{21}$ −7.78° (c, 10: water). *Anal.* Found: N, 17.80. Calcd. for $C_{11}H_{21}N_3O_3\cdot\frac{1}{2}H_2O$: N, 17.77%.

At the second step the solution of the salt in 200 ml of water was passed through a column of Diaion SA 200 (OH-form), and then the resin was washed with water. The absorbed L-pyro-glutamic acid was recovered from the column by eluting with 2 N hydrochloric acid. After having been concentrated to a small volume, and hydrolyzed with 2 N HCl for 2 hrs the solution was evaporated to dryness to remove excess hydrochloric acid. The residue was dissolved in a small amount of water, and neutralized with 2 N sodium hydroxide solution to pH 3.2. Then the precipitated. L-glutamic acid was filtered off, washed with cold water and dried in vacuo. Yield, 96%; $[\alpha]_D^{20}$+31.50° (c, 5: 2 N HCl).

The effluent containing L-leucinamide was evaporated to dryness in vacuo and the residual crops were recrystallized from boiling benzene. L-Leucinamide suitable for re-use was recovered in 95% yied; m.p. 101~2°C; $[\alpha]_D^{20}$+7.82° (c, 5: water).

$$\begin{array}{c} CD_3NH_2 \\ || \\ CH_3-CH-\!\!-CHCOOH \end{array}$$

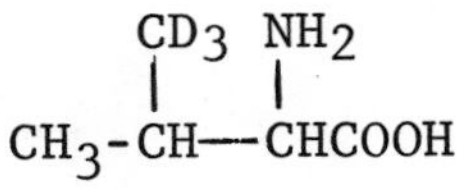

$C_5H_8D_3NO_2$ Ref. 87

Resolution of (2RS,3S)-[4,4,4-$^2H_3$]*Valine* (8c).—(2RS,3S)-[4,4,4-$^2H_3$]Valine (530 mg) was treated with glacial acetic acid (20 ml) and acetic anhydride (10 ml) at reflux for 30 min. Water (100 ml) was added and the solution evaporated to dryness. The product was dissolved in water (30 ml) and the pH was adjusted to 7·2 with dilute NH_4OH. Hog kidney acylase I (15 mg) was added and the solution was incubated at 37° for 44 h. Water (25 ml) was added and the solution was filtered, then applied to a 2·5 × 10 cm column of Rexyn 101 (H^+) cation exchange resin and eluted with water (350 ml). The eluate was freeze-dried, yielding (2R,3S)-N-acetyl[4,4,4-$^2H_3$]valine (16) (303 mg), $[\alpha]_D^{20}$ −7° (c 2, HOAc). A portion of (16) was methylated with diazomethane in ether; the n.m.r. spectrum of the methyl ester is discussed in the text.

The ion exchange column was then eluted with 1M-NH_4OH (200 ml) and the eluate freeze-dried, yielding (2S,3S)-[4,4,4-$^2H_3$]valine (15) (170 mg), $[\alpha]_D$ +70° (c 1, HOAc), δ (D_2O + ND_2OD, external Me_4Si) 1·39 (3H, d, J 7 Hz), 2·50 (1H, m), and 3·78 (1H, d, J 5 Hz).

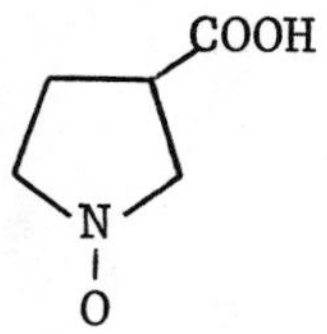

$C_5H_8NO_3$ Ref. 88

Resolution of (±)-4. Racemic acid (±)-4 (4.4 g) and 0.45 equiv (1.3 g) of (+)-α-methylbenzylamine ((+)-MBA) were mixed and dissolved in 70 mL of hot acetone. After ca. 1.5 h at 20 °C a first crop of 1.72 g of A1 crystals was collected. Evaporation to dryness of the mother liquor furnished mixture B1 (containing the impure more soluble salt and the unreacted acid) which was decomposed with a calculated amount of 0.5 N sodium hydroxide. Extraction with methylene chloride afforded the amine; acidification of the aqueous solution (slight excess of 1.2 N HCl) followed by extraction with ether yielded 3.4 g of partially resolved (+)-4. This compound was combined with 1.3 g of (−)-MBA in 70 mL of acetone to give, after 2 h, 1.90 g of crystals C1 and the corresponding D1 from the mother liquor.

Mixture D1 was worked up like B1 and furnished 2.25 g of partially resolved (−)-4, which was treated with 0.8 g of (+)-MBA in 40 mL of acetone to give 1.22 g of crystals A2 and mixture B2.

Similar work up of B2 afforded 1.47 g of impure (+)-4, which was combined with 0.7 g of (−)-MBA to give 0.85 g of crystals C2.

The corresponding crops A1 + A2 were recrystallized from 80 mL of acetone to yield A3, 1.96 g.

Similar treatment of C1 + C2 afforded C3, 1.7 g.

The decomposition of A3 ($NaOH/CH_2Cl_2$ followed by HCl/ether) gave 1.13 g of (−)-4: mp 204 °C (decomp); $[\alpha]_{578}^{30}$ −85°, $[\alpha]_{546}^{30}$ −117°, $[\alpha]_{364}^{30}$ +384° (ethanol, c = 0.37). In the same way, decomposition of C3 yielded 0.96 g of (+)-4: $[\alpha]_{578}^{25}$ +83° (ethanol, c = 0.38).

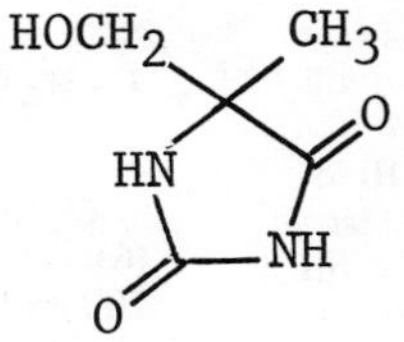

$C_5H_8N_2O_3$ Ref. 89

Resolution of DL-5-Hydroxymethyl-5-methylhydantoin (DL-VIII) into Its Optically Active Components by Brucine——A mixture of DL-VIII (2.9 g., 0.02 mole) and brucine (7.9 g., 0.02 mole) in EtOH (60 ml.) was warmed until solution occurred, and was kept in a refrigerator, the brucine salt of (+)-VIII (6.1 g.) crystallized out as prisms, m.p. 166.5∼189°, $[\alpha]_D^{24}$ −65.9° (C=1.096, EtOH), to which was added water (30 ml.) and NaOH (1 g.) dissolved in H_2O (5 ml.). The brucine separated was filtered and washed with water. The combined filtrate and the washings, were concentrated and poured onto a column (Amberlite IR-120, H^+-form, 50 ml.). The column was eluted with H_2O (200 ml.). The eluate was evaporated to dryness to give crude (+)-VIII as a pale yellow solid (0.9 g.), m.p. 165∼188°, $[\alpha]_D^{19}$ +14.1° (C=1.080, EtOH).

The mother liquor from which the first brucine salt of (+)-VIII was obtained was evaporated to dryness to give a yellow oil, which was treated as same as above to give (−)-VIII (1.41 g.), m.p. 171∼190°, $[\alpha]_D^{21}$ −143° (C=0.996, EtOH). One recrystallization from EtOH afforded white crystals (0.90 g.), m.p. 191°, $[\alpha]_D^{22}$ −4.4° (C=0.856, EtOH). The mother liquor of this recrystallization from EtOH was evaporated to dryness to give a solid, which was purified on column chromatography using silica gel (50 g., solvent : acetone-ethyl acetate). The fractions containing (−)-VIII only were evaporated to dryness to give white crystals (0.20 g.), m.p. 166.5∼168°. $[\alpha]_D^{15}$ −46.3° (C=0.704, EtOH). Further twice recrystallizations from EtOH-hexane afforded pure (−)-VIII, m.p. 168∼171°, $[\alpha]_D^{15}$ −45.9° (C=0.606, EtOH). *Anal.* Calcd. for $C_5H_8O_3N_2$: C, 41.66; H, 5.59; N, 19.44. Found: C, 41.88; H, 5.78; N, 19.55. IRν_{max}^{KBr} cm^{-1}: 3260, 3048, 1744, 1710, 1061. This IR spectrum was not identical with that of DL-VIII. Optical rotatory dispersion measurement: $[\alpha]^{24}$ (C=0.606, EtOH) (mμ): −1716 (250), −354 (300), −186 (350), −115 (400), −78.4 (450), −57.7 (500), −41.7 (589).

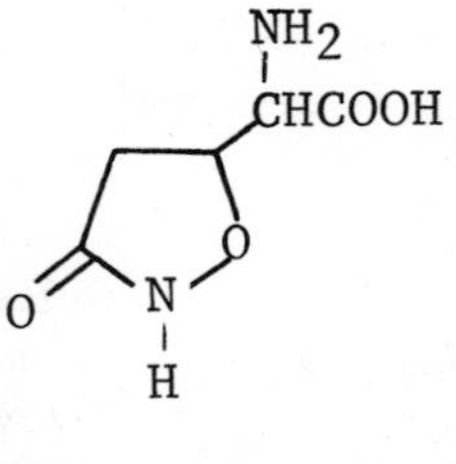

Ref. 90

L- and D-*erythro*-α-Amino-3-oxo-5-isoxazolidineacetic Acid (L- and D-*erythro*-I)——To solution of 590 mg of XVII and 1.5 g of quinine in 10 ml of MeOH was added H_2O until the solution had become opaque, and the mixture was left standing in a refrigerator overnight to give 1.4 g of quinine salt (XVIII). *Anal.* Calcd. for $C_{47}H_{59}O_{11}N_6F_3$ (DL-*erythro*-α-trifluoroacetamido-3-oxo-5-isoxazolidineacetic acid with 2 moles of quinine and 2 moles of H_2O): C, 59.99; H, 6.32; N, 8.93. Found: C, 60.25; H, 6.40; N, 8.55. XVIII was subjected to the fractional crystallization from ethanol to separate slightly-soluble crystal (crystal-1) from more-soluble ones (crystal-2). Crystal-1: 532 mg, $[\alpha]_D^{24}$ $-158°$ ($c=0.5$, $l=1$, MeOH), crystal-2: 468 mg, $-108°$ ($c=0.5$, $l=1$, MeOH). Each of both crystals was dissolved in 5 ml of MeOH and treated with a mixture of 20 ml of $CHCl_3$ and 10 ml of 14% aq. NH_3. The mixture was shaken well and the $CHCl_3$ layer was extracted three times with 14% aq. NH_3 (30 ml). The combined 14% NH_3 solution was washed twice with $CHCl_3$ and left standing for 5 hr at room temperature. The reaction mixture was evaporated *in vacuo* and the residue was dissolved in 1 ml of H_2O, then left standing in a refrigerator after acidifying with AcOH to give crystals. From the crystal-1 was obtained D-tricholomic acid (D-*erythro*-α-amino-3-oxo-5-isoxazolidineacetic acid), 51 mg colorless plates, mp 202—204° (decomp.), $[\alpha]_D^{22}$ $-104°$ ($c=0.2$, $l=1$, H_2O), while from the crystal-2 was obtained L-tricholomic acid (L-*erythro*-α-amino-3-oxo-5-isoxazolidineacetic acid), 60 mg, colorless plates, mp 201—202° (decomp.) with no depression on admixture with tricholomic acid, $[\alpha]_D^{22}$ $+105.5°$ ($c=0.2$, $l=1$, H_2O),[16] $[\alpha]_D^{21}$ $+185.5°$ ($c=0.2$, $l=1$, 0.1N HCl), $+34.5°$ ($c=0.2$, $l=1$, 0.1N NaOH). ORD ($c=1.94$ mg/2 ml in 0.1N HCl): positive Cotton effect.

$CH_2=CH-CH-COOH$
      $|$
      CH_3

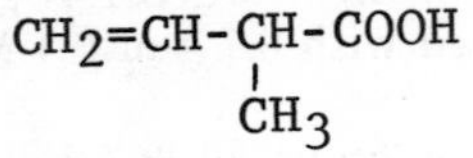

$C_5H_8O_2$

Ref. 91

Synthesis of Optically Active Multistriatin. Resolution of 5 with (+)- and (−)-α-Methylbenzylamine. (+)-α-Methylbenzylamine (13.0 g, 0.1 mol) and **5** (11.0 g, 0.1 mol) were dissolved in 80 ml of boiling acetone and allowed to cool to 20 °C and to stand for 14 h. Crystals were filtered from the acetone, washed with a small volume of ethyl ether, and vacuum dried. The amine salt was recrystallized five times from a minimum volume of acetone. The rate of cooling was controlled by placing the flask containing the hot acetone solution in a Dewar flask which contained water at 55 °C. The yield of the partially resolved amine salt was 2.1 g. This resolution process was repeated with (−)-α-methylamine as the resolving reagent, and 4.5 of the amine salt was obtained.

The free acids were recovered by decomposing the salts with 10 ml of 1 M hydrochloric acid for the (+) amine and 20 ml for the (−) isomer. The aqueous solutions were extracted four times with 15-ml portions of ethyl ether, and the extracts were dried with 4 Å sieves. Evaporation of the solvent gave 1.0 g of S-(+)-**5**, $[\alpha]^{20}D$ +28.8° (c 0.10, hexane), 70% optical purity; the (−) amine extract gave 2.1 g of R-(−)-**5**, $[\alpha]^{20}D$ 24.8° (c 0.42, hexane), 60% optical purity.

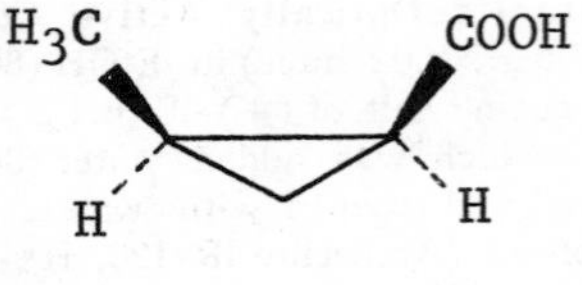

$C_5H_8O_2$

Ref. 92

III. Optically Active Series. Optical Resolution of *cis*-2-Methylcyclopropanecarboxylic Acid (8). A sample of 72 g of racemic acid **8** was dissolved in 300 ml of acetone at reflux and 23 g of quinine added. A homogeneous solution formed initially, but the salt began to crystallize soon after, so acetone was added until the boiling mixture was again homogeneous. The mixture was allowed to cool to room temperature, and after standing 2 days 140 g of crystals were collected, mp 142–148°. The material collected was recrystallized twice more from fresh acetone and the crystals obtained (90 g, mp 151–154°) were shaken with a mixture of 17% hydrochloric acid and ether. The phases were separated and the aqueous solution was salted heavily and extracted twice more with ether. After washing the combined organic phases with 5% hydrochloric acid and brine, they were dried over sodium sulfate and concentrated *in vacuo*. The greenish residue was distilled (bp 98°, 20 mm) to give 18.79 g of acid **8**, whose ir spectrum was identical with that of racemic material. A sample of this material showed only one symmetrical peak by vpc on column B and had $[\alpha]^{25}D$ +17.9° (95% ethanol).

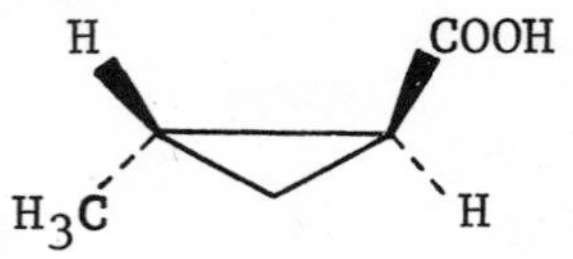

$C_5H_8O_2$

Ref. 93

Optical Resolution of *trans*-2-Methylcyclopropanecarboxylic Acid (6). The salt of the *trans*-acid **6** was prepared from 62 g of acid and 200 g of quinine as described for the *cis* isomer. Four recrystallizations from acetone, followed by regeneration of the acid in 17% hydrochloric acid and vpc purification, gave material whose ir was identical with racemic **6**. Its rotation was $[\alpha]^{25}D$ −46.4° (95% ethanol).

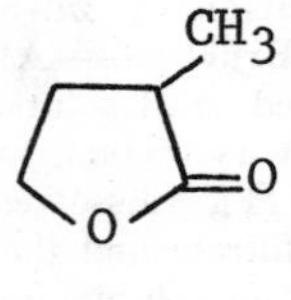

$C_5H_8O_2$

Ref. 94

Optical Resolution of DL-α-Methyl-γ-butyrolactone (VII).——To a solution of 15 g. of barium hydroxide octahydrate in 100 ml. of water, 8.4 g. of DL-α-methyl-γ-butyrolactone was added. After the solution had been heated at 90°C for 2 hr., it was neutralized by the addition of 10% sulfuric acid; the barium sulfate precipitated was removed by filtration. To this filtrate, 35 g. of quinine was added; it was then heated at 90°C until dissolution was complete. After standing overnight, the reaction mixture was evaporated under reduced pressure, and the residue was dissolved in 85 ml. of acetone. After the solution had stood overnight in a refrigerator, the less soluble quinine salt precipitated was collected and recrystallized 3 times from acetone-methanol (9 : 1). (−)-α-Methyl-γ-butyrolactone (yield, 1.5 g.; b. p., 92.5°C/20 mmHg, $[\alpha]_D^{15}$ −21.5°C (c 5.5, absolute ethanol)) was obtained from this purified, less soluble quinine salt by the usual manner.

Found: C, 59.69; H, 7.98. Calcd. for $C_5H_8O_2$: C, 59.98; H, 8.05%.

From the mother liquor, after removing less soluble salt, (+)-α-methyl-γ-butyrolactone (yield, 1.2 g.; b. p., 93∼95°C/22 mmHg, $[\alpha]_D^{15}$ +11.9° (c 6.9, absolute ethanol)) was obtained.

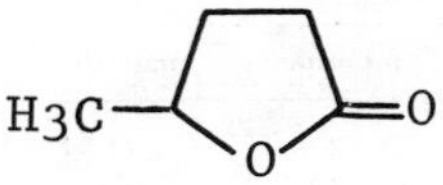

$C_5H_8O_2$ Ref. 95

Note:
Resolution of racemic lactone via cinchonidine (the cinchonidine salt of the (-)-lactone was recrystallized 20 times from acetone-methanol) gave (-)-lactone, $[\alpha]_D^{23}$ -17.2° ± 1° (c 3.911, dioxane) and (+)-lactone $[\alpha]_D^{21.5}$ +4.2° ± 2° (c 1.557, dioxane).

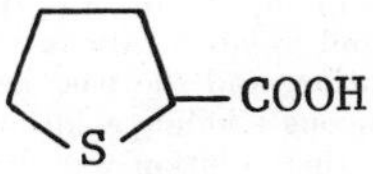

$C_5H_8O_2S$ Ref. 96

Preliminary tests on resolution of thiophane-2-carboxylic acid

0.001 mole of racemic acid and 0.001 mole of optically active base were dissolved hot ethanol (96 %). From the crystals formed, the acid was liberated and the op-al activity measured in 96 % ethanol. The results are given in Table 1.

Optical resolution of thiophane-2-carboxylic acid

13.2 g (0.1 mole) of racemic acid and 46.7 g (0.1 mole) of brucine were dissolved 370 ml of hot 96 % ethanol. The solution was left to crystallize at room tempera-re, and then in a refrigerator over night. The salt was filtered off, the acid isolated om a sample (about 0.2 g), and the optical activity measured in 96 % ethanol. ne course of the resolution is shown in Table 2.

The mother liquor from the first crystallization was evaporated *in vacuo*. The ucine salt was decomposed with 2 *M* sodium hydroxide and the brucine was tracted with 4 portions of chloroform. The water layer was acidified with 2 *M* lphuric acid, saturated with ammonium sulphate and extracted five times with her. Evaporation of the ether gave 4.0 g of acid with $[\alpha]_D^{25}$ = +98° (96 % ethanol). is acid and 8.2 g of cinchonidine were dissolved together in hot 96 % ethanol. e solution was left to crystallize at room temperature and then in a refrigerator er night. The salt (8.2 g) was filtered off, but the optical activity was unchanged in ite of the good result of the primary resolution (Table 1). Recrystallization in 80 % hanol gave $[\alpha]_D^{25}$ = +137°. With regard to this value, it seemed probable that maxi-um optical rotation should be obtained at the earliest after four crystallizations.

Table 1

Base	$[\alpha]_D^{25}$
Brucine	−94°
Cinchonidine	+34°
Quinine	±0°
(+)-α-Phenylethylamine	oil
Strychnine	—
Quinidine	oil
(+)-Phenylisopropylamine	oil
(−)-α-(2-Naphthyl)-ethylamine	−2°

In order to save valuable acid no measurements were made of the three following crystallizations in 80% ethanol. The 5th gave $[\alpha]_D^{25}$ = +151°, the 6th no measurement, the 7th $[\alpha]_D^{25}$ = +155° and the 8th $[\alpha]_D^{25}$ = +155.7°.

Table 2

Crystallization	1	2	3	4	5
ml 96 % ethanol	370	125	80	75	65
g salt obtained	21.5	14.4	12.5	11.1	9.8
$[\alpha]_D^{25}$ of the acid	−126°	−145°	−152°	−155°	−154.9°

(−)-Thiophane-2-carboxylic acid

The brucine salt (9.8 g) was decomposed as above. Evaporation of ether and recrystallization from ligroin gave 2.0 g of colourless needles, m.p. 36–37°.

Anal. Subst. 59.64 mg: 9.43 ml of 0.04964 *M* NaOH.
$C_5H_8O_2S$ (132.18) Calc. S 24.26 %; Equiv. wt. 132.18.
Found S 24.38 %; Equiv. wt. 132.4.
0.04713 g acid dissolved in 96 % ethanol to 10.00 ml: α_D^{25} = −1.46° (2 dm); $[\alpha]_D^{25}$ = −154.9°; $[M]_D^{25}$ = −204.7°.

(+)-Thiophane-2-carboxylic acid

The acid was liberated from 0.58 g of the cinchonidine salt with 2 *M* sulphuric acid and extracted five times with ether. Recrystallization from ligroin gave 0.12 g of colourless, staff-shaped crystals with m.p. 36–38°.

Anal. Subst. 33.89 mg: 5.15 ml of 0.04964 *M* NaOH.
$C_5H_8O_2S$ (132.18) Calc. S 24.26 %; Equiv. wt. 132.18.
Found S 24.46 %; Equiv. wt. 132.6.
0.04271 g acid dissolved in 96 % ethanol to 10.00 ml: α_D^{25} = +1.33° (2 dm); $[\alpha]_D^{25}$ = +155.7°; $[M]_D^{25}$ = +205.8°.

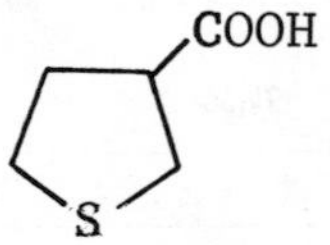

$C_5H_8O_2S$ Ref. 97

Preliminary tests on resolution

0.001 mole of the racemic acid and 0.001 mole of the optically active base were dissolved in hot solvent. The solutions were left at room temperature, and if no crystals formed water was added gradually (in the case of water-soluble solvents) until crystals or oils were formed. The crystals were filtered off, and the acid liberated with dilute sulphuric acid and extracted with ether. The ether was removed, the residue dissolved in ethanol and the optical activity measured in a 2 dm tube. The concentration of the acid was determined volumetrically with sodium hydroxide. The results are given in Table 1.

Optical resolution of thiophane-3-carboxylic acid

11.0 g (0.0834 mole) of racemic acid and 27.0 g (0.0834 mole) of quinine were dissolved in 90 ml of 96 % ethanol. The solution was left to crystallize at room temperature and then in a refrigerator. The salt was filtered off, dried and weighed. About 0.2 g of the salt was made alkaline with 2 *N* sodium hydroxide and the base extracted with chloroform. The acid was liberated from the aqueous alkaline solution with sulphuric acid and extracted three times with ether. The solvent was evaporated

Table 1

	$[\alpha]_D^{25}$ of acid, measured in ethanol					
	Crystallization medium					
Base	99.5% ethanol	96% ethanol	Methanol	Acetone	Ethyl-acetate	Carbon tet.-ligroin
Strychnine		− 3°				
Quinine		+ 30°	+ 24°	+ 18°		
Quinidine		Oil	Oil	Oil	Oil	
Brucine		+ 7°	± 0°			
Morphine	− 7°	Oil	Oil	Oil	Oil	
Cinchonidine		+ 26°				
(−)-α-Phenylethylamine		± 0°				
(+)-α-(2-Naphtyl)ethylamine		± 0°	± 0°	± 0°	± 0°	
(−)-Phenyl-*iso*-propylamine		Oil				± 0°
(+)-α-Phenylethylthiuronium-chloride		+ 4°				
(+)-Dehydroabietylamine		± 0°				
Ephedrine		Oil				
Cinchonine		± 0°	± 0°			

Table 2

Crystallization . . .	1	2	3	4	5	6	7
ml 96% ethanol	90	55	40	38	35	33	32
g salt obtained	16.8	12.4	10.7	9.7	8.9	8.0	7.2
$[\alpha]_D^{25}$ of the acid	+ 29°	+ 49°	+ 55°	+ 58°	+ 61°	+ 61°	+ 60°

and the acid dissolved in about 11 ml of ethanol and the optical activity measured in a 2 dm tube. The concentration of the solution was determined volumetrically with sodium hydroxide. The main part of the precipitate was recrystallized from ethanol until the liberated acid showed constant optical activity. The progress of the resolution is seen in Table 2.

The mother liquor from the first crystallization of the quinine salt was evaporated to dryness at room temperature and the acid isolated. A small part of the acid was neutralized with sodium hydroxide, and an equivalent amount of (−)-α-phenylethylthiuroniumchloride [25] was added. The salt obtained was recrystallized from ethanol, and the optical activity of the liberated acid (from 0.1 g samples) determined in the same manner as described for the (+)-acid. The results are shown in Table 3.

The specific rotation values shown in Tables 2 and 3 (the concentration of the crude acid in these samples was determined by titration with sodium hydroxide) are higher than those obtained from recrystallized and weighed samples. On recrystallization of the liberated acid from petroleum ether, some of the crude acid does not dissolve.

(+)-*Thiophane-3-carboxylic acid*

Water was added to 7.0 g quinine salt, and then an excess of 2 *N* sodium hydroxide solution. The quinine was extracted three times with chloroform, the aqueous solution treated with diluted sulphuric acid, and the liberated organic acid extracted five times with ether. The ether was removed, and the remaining 2.0 g was recrystallized three times from petroleum ether. The acid formed colourless needles with m.p. 53–55°.

Anal. $C_5H_8O_2S$ (132.2)
Calc. S 24.26 % Equiv. wt. 132.2
Found S 24.29 % Equiv. wt. 133.6

0.3473 g acid dissolved in 96 % ethanol to 10.00 ml: $\alpha = +3.87°$ (2 dm); $[\alpha]_D^{25} = +55.7°$; $[M]_D^{25} = +73.5°$.

Table 3

Crystallization . . .	1	2	3	4
ml 96% ethanol	20	10	5	2
mg salt obtained	750	450	150	64
$[\alpha]_D^{25}$ of the acid	−51°	−54°	—	−53°[a]

[a] This value was obtained from a recrystallized and weighed sample.

(−)-***Thiophane-3-carboxylic acid***

The acid was isolated from the (−)-α-phenylethylthiuronium-salt described above and the acid was recrystallized from ligroin. 0.01883 g acid dissolved in 96 % ethanol to 10.00 ml: $\alpha = -0.30°$ (3 dm); $[\alpha]_D^{25} = -53°$.

Authors' comments:

The dextrorotatory (55.7°) enantiomer of thiophane-3-carboxylic acid was obtained by fractional crystallization of the quinine salt from ethanol. In a preliminary resolution, the levorotatory form, obtained by recrystallization of its (−)-α-phenylethylthiuronium salt, had nearly the same optical rotation as the dextrorotatory form. However, attempts to repeat the latter resolution failed. Despite a large number of recrystallizations we could not raise the rotation of the levo-form above −37°. Other solvents were also tried, namely ethyl acetate, methanol and ethanol – acetone, but from all of them recrystallizations gave salt-compositions in which the acid had the same maximum rotation (−37°).

When the partly resolved acid (−37°) was recrystallized as a salt with morphine (in 99.5 % ethanol) and with strychnine (in ethanol-water), only acid with the initial value of rotation was recovered. We also added quinine corresponding to the amount of (+)-acid present in the −37° acid. However, the quinine salt, precipitated from 96 % ethanol, contained acid with the same rotation as the starting material. An analogous experiment with cinchonidine also gave unchanged rotation of the thiophanecarboxylic acid.

Recrystallizations of acid with $[\alpha]_D = -37°$ from formic acid or petroleum ether did not change the optical rotation. When the acid was shaken for a few seconds with these solvents, the dissolved part of the acid had the same optical rotation as the undissolved crystals. A careful sublimation also left the rotation unchanged. Finally, it was observed that crystallization of the partly resolved acid from water gave crystals with slightly lower rotation. The aqueous mother liquor thus contained acid enriched with the (−)-form. By repeated utilization of the difference in solubility of racemic and active acid in water, we were able to isolate acid with $[\alpha]_D^{22} = -45°$. (Limitation of available material prevented further purification.)

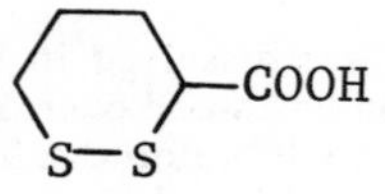

$C_5H_8O_2S_2$ Ref. 9

Preliminary tests on resolution

1,2-Dithiane-3-carboxylic acid. **0.001 mole of racemic acid and 0.001 mole of ticaly active base were dissolved in hot ethanol (96 %). The solutions were left room temperature, and if no crystals formed, the solutions were put in a refrigera until crystals or oils were formed. The crystals were filtered, and the acid libera and extracted with ether. The ether was removed and the optical activity measu in 96 % ethanol. The results are given in Table 1.**

Table 1

Base	ml ethanol	mg salt	$[\alpha]_D^{25}$ of the acid
Cinchonine	4.0	oil	—
Cinchonidine	4.0	105	−19°
Morphine	3.5	oil	—
Brucine	2.0	41	± 0°
Strychnine	10.0	161	−68°
Quinine	2.0	184	−31°
Quinidine	3.0	oil	—
Ephedrine	8.0	133	+22°

1,2-Dithiane-4-carboxylic acid. The preliminary tests on resolution have be published earlier [15].

Optical resolution

(−)-*1,2-Dithiane-3-carboxylic acid.* Racemic acid, 30.1 g (0.183 mole), and 61.3 (0.183 mole) of strychnine were dissolved in 3000 ml of hot 96 % ethanol. The sol tion was left to crystallize at room temperature, and then in a refrigerator overnigh The salt was filtered off, dried, and weighed. About 0.2 g of the salt was made alk line with sodium hydroxide solution, and the base extracted with chloroform. Th acid was liberated from the aqueous alkaline solution with hydrochloric acid, ar then extracted with ether. The ether solution was dried over magnesium sulphat the acid isolated, and the optical activity measured in 96 % ethanol. The ma part of the precipitate was recrystallized from ethanol until the liberated acid show constant optical activity (Table 2). The acid seemed to racemize and decompose easi under alkaline conditions. The heating time, necessary for dissolving the strychnir salt, was therefore kept as short as possible.

The acid from 3.90 g of the strychnine salt was isolated as described above, an 1.2 g of a pale yellow oil was obtained. The oil crystallized after a few minutes i refrigerator. Recrystallizations from petroleum ether (b.p. 30–60°) gave white crysta with m.p. 37.8–39.5°, $[\alpha]_D^{25} = -172°$ ($c = 1.0$, ethanol).

Anal. ($C_5H_8O_2S_2$)
Calc. Equiv. wt. 164.3
Found Equiv. wt. 164.0

(+)-*1,2-Dithiane-3-carboxylic acid.* The mother liquor from the first crystallizatic of the strychnine salt was evaporated to dryness at room temperature, and the aci isolated as described above. Recrystallization from petroleum ether gave 7.8 g

Table 2

Crystallization . . .	1	2	3	4	5	6	7	8
ml 96 % ethanol	3000	1700	1400	900	900	750	700	650
g salt obtained	54.8	38.0	33.1	26.0	25.0	21.0	19.3	18
$[\alpha]_D^{25}$ of acid	−78°	−121°	−139°	−152°	−160°	−166°	−168°	−168

Table 3

Crystallization . . .	1	2	3	4	5
ml 96 % ethanol	100	80	70	60	50
g salt obtained	8.7	7.3	5.64	4.60	4.15
$[\alpha]_D^{25}$ of the acid	140°	150°	162°	168°	167°

acid with $[\alpha]_D^{25} = +100°$ (96 % ethanol). This acid and 7.8 g of ephedrine were di solved together in hot ethanol. Recrystallizations and measurements of the optic activity were performed as described for the strychnine salt. The results are show in Table 3.

The acid was isolated from 3.9 g of ephedrine salt, and two recrystallizations fro petroleum ether (b.p. 30–60°) gave 1.4 g of white crystals with m.p. 39–39.5 $[\alpha]_D^{25} = +171°$ ($c = 1.1$, ethanol).

Anal. ($C_5H_8O_2S_2$)
Calc. Equiv. wt. 164.3
Found Equiv. wt. 164.9

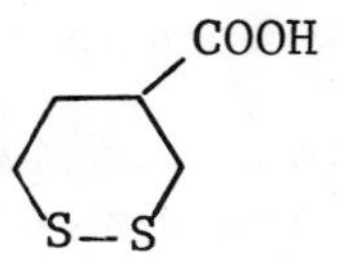

$C_5H_8O_2S_2$ **Ref. 99**

(−)-*1,2-Dithiane-4-carboxylic acid.* Racemic acid, 15.4 g (0.094 mole), and 30.5 g (0.094 mole) of quinine were dissolved in 1000 ml of boiling 96 % ethanol. After the solution stood 24 hours in a refrigerator, 30.0 g salt had separated; this was recrystallized four times. The change in optical activity of the acid was measured as described for 1,2-dithiane-3-carboxylic acid.

To 10.1 g of the quinine salt from the last recrystallization was added a slight excess of 2 *N* sodium hydroxide solution. The quinine was extracted with chloroform, and to the remaining aqueous solution hydrochloric acid was added carefully. At first, a small amount of a brown, oily precipitate formed. This was filtered and discarded. The colourless aqueous solution was cooled in an ice-bath and more hydrochloric acid added under stirring. The formed colourless crystals were filtered and dried. One crystallization from benzene–petroleum ether gave 2.1 g of acid with m.p. 103–104°, $[\alpha]_D^{25} = -167°$ ($c = 1.0$, ethanol). The acidic aqueous filtrate was extracted with ether and a further 0.4 g of acid was obtained. After two recrystallizations this gave the same melting point as the bulk of the acid.

Anal. ($C_5H_8O_2S_2$)
Calc. Equiv. wt. 164.3
Found Equiv. wt. 165.2

(+)-*1,2-Dithiane-4-carboxylic acid.* From the mother liquors of the first two quinine salt crystallizations, the quinine was removed and the acid obtained in an

Table 4

Crystallization...	1	2	3	4	5
ml 96 % ethanol	1000	1000	750	650	575
g salt obtained	30.0	17.1	15.0	12.4	10.3
$[\alpha]_D^{25}$ of the acid	−49°	−129°	−142°	−164°	−163°

alkaline, aqueous solution as described for the (−)-form. Acidification with hydrochloric acid gave 7.7 g of an acid with m.p. 102–113° and $[\alpha]_D^{25} = +65°$. Ether extraction of the acidic aqueous solution gave a further 1.2 g of acid with m.p. 106–109° and $[\alpha]_D^{25} = +117°$. To 7.6 g (0.046 mole) of the acid with $[\alpha]_D^{25} = +65°$ was added 13.6 g of cinchonidine. The mixture was dissolved in 125 ml of boiling ethanol. After the solution stood 48 hours in a refrigerator, the salt thus obtained was filtered. The acid was liberated from a small sample and its optical rotation measured; the main part of the salt was recrystallized five times. The optical activity of the liberated acid had then reached a constant value, and, therefore, the bulk of the salt was decomposed and the acid was recrystallized from benzene–petroleum ether. The melting point was 102–103° and $[\alpha]_D^{25} = +166°$ ($c = 1.1$, ethanol).

Anal. ($C_5H_8O_2S_2$)
Calc. Equiv. wt. 164.3
Found Equiv. wt. 163.6

$CH_2{=}CH{-}CH(OH){-}CH_2COOH$

$C_5H_8O_3$ **Ref. 100**

Note:
The racemic acid was treated with a half equivalent of quinine. The salt formed was decomposed to yield the active acid, $[\alpha]_D^{25}$ -9.8° (c 0.305, H_2O).

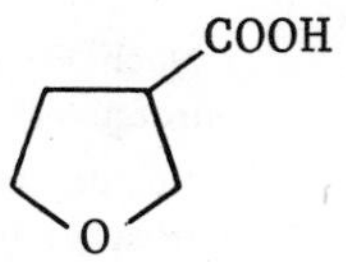

$C_5H_8O_3$ **Ref. 101**

Resolution of 3-tetrahydrofuroic acid. In about 550 ml. of hot acetone were dissolved 57.5 g. of 3-tetrahydrofuroic acid[5] and 160 g. of quinine. The fine needles which separated on cooling were recrystallized six more times from acetone, yielding finally 100 g. of the quinine salt, m.p. 134–135°, $[\alpha]_D^{24}$ −124.2° in ethanol. A solution of the salt in 500 ml. of 5% hydrochloric acid was continuously extracted with ether for 34 hr. The ether extracts were dried over magnesium sulfate and distilled to afford 23.0 g. of the dextrorotatory acid, b.p. 117.5–119° (6 mm.), $[\alpha]_D^{24}$ + 3.77° (*c*, 0.0623 in ethanol); +4.59° (*c*, 0.0746 in acetone).

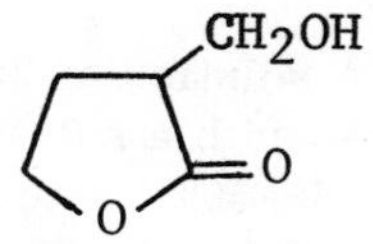

$C_5H_8O_3$ **Ref. 102**

Resolution of α-Hydroxymethyl-γ-butyrolactone (III).—To a solution of 18.0 g. of barium hydroxide octahydrate in 100 ml. of water, 11.6 g. of DL-α-hydroxymethyl-γ-butyrolactone (III) was added, and the solution was heated to 90°C over a 2-hr. period. After the solution had cooled, 10% sulfuric acid was added, and the precipitated barium sulfate was removed by filtration. To the filtrate, 36.0 g. of quinine was added, and the mixture was heated at 90°C to make a clear solution. After standing over night, the solution was concentrated under reduced pressure, and the residue was recrystallized from acetone-methanol six to ten times. From a purified, less soluble quinine salt, optically impure α-hydroxymethyl-γ-butyrolactone (yield 0.5 g., b. p. 105~107°C/1 mmHg, $[\alpha]_D^{14}$ −4.1° (*c* 12.2, ethanol)) was obtained, while the maximum value of its specific rotation was +10.4°.

(+)-α-Hydroxymethyl-γ-butyrolactone (+III).—It was obtained from (+)-alloisocitric lactone (+IV), as has been previously described[6].

Yield, 19%; b. p., 103~107°C/0.5~1 mmHg, $[\alpha]_D$ +10.4° (*c* 5.6, ethanol). Found: C, 51.83; H, 6.94. Calcd. for $C_5H_8C_3$: C, 51.72; H, 6.94%.

6) T. Kaneko, H. Katsura, H. Asano and K. Wakabayashi, *Chem. &. Ind.*, **1960**, 1187; H. Katsura, *J. Chem. Soc. Japan, Pure Chem. Sec.* (*Nippon Kagaku Zasshi*), **82**, 91 (1961).

$CH_3{-}CH(OCOCH_3){-}COOH$

$C_5H_8O_4$ **Ref. 103**

b) *Racematspaltung von DL-O-Acetyl-milchsäure mittels a-Phenyl-äthylamin:* 23.0 g (0.174 Mol) *DL-O-Acetyl-milchsäure*[11] in 100 ccm Methylenchlorid wurden mit 21.1 g (0.174 Mol) *L-a-Phenyl-äthylamin*[10] in 50 ccm Methylenchlorid versetzt. Nach Zugabe von 300 ccm Petroläther und Reiben bis zur beginnenden Kristallisation wurde 2 Tage bei Raumtemperatur aufbewahrt. Es konnten 15.3 g (70%) *L-a-Phenyläthylammoniumsalz der D-O-Acetyl-milchsäure* in farblosen, verfilzten Kristallen isoliert werden, $[\alpha]_D^{20}$: +13.3° (c = 3; Äthanol); Schmp. 147–149°.

11) *R. Anschütz* und *W. Bertram*, Ber. dtsch. chem. Ges. **37**, 3972 (1904).

Nach Freisetzen der *D-O-Acetyl-milchsäure* mit 1 *n* HCl in Äther, wie oben beschrieben, wurde i. Vak. destilliert: Sdp. 82°/0.1 Torr; $[\alpha]_D^{20}$: +44.7° (c = 2; Benzol).

Aus der Mutterlauge wurde mit 1 *n* HCl in Äther die angereicherte *L-O-Acetyl-milchsäure* freigesetzt. Ihre Umsetzung mit *D-α-Phenyl-äthylamin* ergab in 50proz. Ausb. (bez. auf das ursprünglich eingesetzte Racemat) das *D-α-Phenyläthylammoniumsalz:* $[\alpha]_D^{20}$: −13.0° (c = 6; Äthanol); Schmp. 147−149°.

Die aus diesem Salz freigesetzte *L-O-Acetyl-milchsäure* hatte eine spezif. Drehung von $[\alpha]_D^{20}$: −44.0° (c = 2; Benzol)[8].

```
HOOC-CH-S-CH2COOH
     |
     CH3
```

$C_5H_8O_4S$ Ref. 104

Salt (A). A solution of 20 g (0.121 mole) of 2-methyl-2, 2'-thiodiacetic acid in 70 ml of absolute alcohol was mixed with a solution of 40 g (0.242 mole) of ephedrine in 70 ml of absolute alcohol; the temperature rose to 45-50°. After 18-20 h the precipitate was filtered off, washed with two 10-ml portions of alcohol, and dried at 60-70°. The colorless crystals were repeatedly crystallized from 100 ml of absolute alcohol. Yield 26.5 g (88.4%); $[\alpha]_D$ −13.5°, m.p. 140-141° [salt (A)]. Found %: C 60.73; H 8.11; N 5.62; S 6.39. $C_{25}H_{38}N_2O_6S$. Calculated %: C 60.72; H 7.69; N 5.66; S 6.47.

Salt (C). The alcoholic mother solution from the salt (A) was evaporated down to one-third of the initial volume, and an equal volume of dry ether was added. The precipitate that formed was filtered off, and we obtained 10.6 g of a salt of $[\alpha]_D$ −36.0° [salt (B)]. The mother solution was again evaporated, and the residue was rubbed out with ether; we obtained 5.8 g of a salt of $[\alpha]_D$ −44°, m.p. 80-105° [salt (C)]. After the recrystallization of salt (B) from 15 ml of absolute alcohol we isolated 3.4 g of a salt of $[\alpha]_D$ −27.8° [salt (D)], and from the evaporated mother solution we obtained 6.9 g of a salt with $[\alpha]_D$ −42.7°, which was added to salt (C). After the further evaporation of the alcoholic mother solutions, the recrystallization of the less readily soluble salt, and treatment of the readily soluble residues with ether we isolated also: 3.28 g of salt (A), $[\alpha]_D$ −13°, 4.8 g of salt (C), $[\alpha]_D$ −43°, and 24 g of salt (D), $[\alpha]_D$ −28°. The total yields were 17.5 g (58.3%) of salt (C) and 5.8 g of the mixture of salts (D). Found %: C 60.57; H 8.0; N 5.47; S 7.06. $C_{25}H_{38}N_2O_6S$. Calculated %: C 60.72; H 7.69; N 5.66; S 6.47.

Optically Active Forms of the Acid. 26.5 g of the salt (A) was added to 35 ml of water. 100 ml of ether was added to the resulting suspension, and with vigorous stirring the mixture was acidified with 11 ml of concentrated hydrochloric acid so as to give pH 2.0. The aqueous layer was extracted with five 100-ml portions of ether, and a crystalline precipitate of ephedrine hydrochloride separated. The combined ether extracts were dried with sodium sulfate, ether was evaporated, and the residue crystallized on standing. We obtained dextrorotatory 2-methyl-2, 2'-thiodiacetic acid, $[\alpha]_D$ +134.2° (c 0.5; alcohol), m.p. 72-73° (chloroform, 1 : 2). Yield 8.2 g (93.75% on the ephedrine salt and 82.5% on the original racemate).

17.5 g of the salt (C) was treated analogously, and we obtained levorotatory 2-methyl-2, 2'-thiodiacetic acid, $[\alpha]_S$ −134.5° (c 0.5; alcohol), m.p. 69-71° (chloroform, 1 : 2). Yield 5.78 g (99.6% on the ephedrine salt and 57.8% on the original racemate).

A mixture of the levo and dextro forms of the acid melted at 82-84°.

Isolation of Ephedrine Base. 70 ml of ether was added to the combined aqueous acid mother solutions from the isolation of the dextro and levo isomers of 2-methyl-2, 2'-thiodiacetic acid, and the mixture was cooled to 0°. The mixture was stirred while 70 ml of 40% sodium hydroxide solution was added and then extracted with three 50-ml portions of ether; the ether extracts were combined and dried with sodium sulfate; ether was evaporated; m.p. 40-41°. Yield 25 g (77.1% on the salt and 62.5% on the ephedrine).

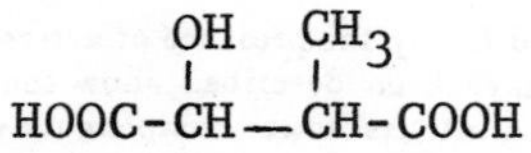

$C_5H_8O_5$ Ref. 105

The Optical Resolution of I.—Thirty grams of I and 86 g. of brucine dihydrate were dissolved in 800 ml. of boiling water, and the resulting solution was allowed to stand overnight in an ice box. The crystals of the brucine salt which precipitated were collected by filtration; seven recrystallizations of this salt from ten parts of boiling water gave 25 g. of the pure brucine salt of I. This salt was suspended in 200 ml. of cold water and decomposed with 50 ml. of a N sodium hydroxide solution. The precipitated brucine was removed by filtration, and the alkaline filtrate was washed twice with 30 ml. portions of chloroform, neutralized with concentrated hydrochloric acid, and then evaporated to a syrup. To this a small amount of water was added, and the resulting solution was decolorized with charcoal and evaporated to dryness in order to remove the free hydrochloric acid. The residue was extracted several times with 20 ml. of acetone. The acetone extracts were combined and again evaporated to a syrup. The syrupy product was dried over sulfuric acid in a vacuum desiccator. In this way, white crystals of (+)-I were obtained. Recrystallization from 10 ml. of ethyl acetate - petroleum ether (1 : 1) gave 3.6 g. of optically-pure (+)-I. m. p. 107°C, $[\alpha]^{20}_D$ +9.1° (*c* 3.7, water).

Found: C, 40.36; H, 5.51. Calcd. for $C_5H_8O_5$: C, 40.54; H, 5.44%.

The mother liquor of the brucine salt of (+)-I was concentrated to a paste. After it had been stored for a week in an ice box, the brucine salt of (−)-I was collected. Optically-pure (−)-I was obtained from this salt by the procedure described above. Yield, 1.3 g. M. p. 106—107°C, $[\alpha]^D_{20}$ −8.9° (*c* 3.7, water). Found: C, 40.40; H, 5.24%.

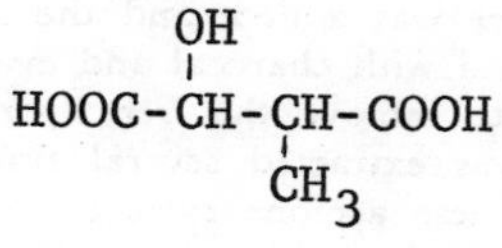

$C_5H_8O_5$ Ref. 106

The Optical Resolution of II.—Twenty grams of II and 58 g. of brucine dihydrate were dissolved in 900 ml. of boiling water, and then the resulting solution was treated as described for the resolution of I. Thus, 2.7 g. of optically pure (−)-II crystallized out. M. p. 111°C; $[\alpha]^D_{20}$ −5.3° (*c* 3.2, water).

Found: C, 40.32; H, 5.50. Calcd. for $C_5H_8O_5$: C, 40.54; H, 5.44%.

From the mother liquor of the brucine salt of (−)-II, 1.1 g. of optically pure (+)-II were obtained. M. p. 110—111°C, $[\alpha]^{20}_D$ +5.2° (*c* 3.2 water). Found: C, 40.54; H, 5.39%.

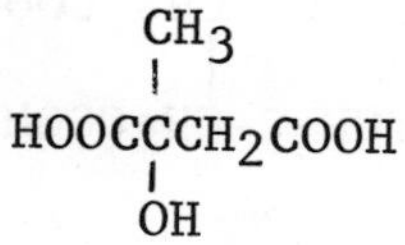

$C_5H_8O_5$ Ref. 107

38.4 g Brucinhydrat wurden in einer kochenden Lösung von 25.6 g *racem.* Citramalsäure in 256 ccm Wasser gelöst. Nach 15-stdg. Stehenlassen bei gewöhnlicher Temperatur hatten sich hexagonale Tafeln des sauren Brucinsalzes abgeschieden. Nach dem Trocknen an der Luft bei gewöhnlicher Temperatur betrug die Ausbeute 30 g. Das wasserfreie Salz schmolz unt. Zers. bei 228°. Das Salz wurde in 400 ccm Wasser bei 60° gelöst und das Brucin durch Ammoniak gefällt. Das Brucin wurde abfiltriert, das Filtrat mit Salzsäure angesäuert und zur Trockne eingedampft. Die (+)-Citramalsäure wurde mit Äther extrahiert und aus einem Gemisch von Äthylacetat und Petroläther (Sdp. 80—100°) umkrystallisiert. Ausbeute 4 g. Diese Säure (A) schmolz bei 109—110°; $[\alpha]^{14}_D$: +23° in Wasser (c = 6.316).

Eine weitere Spaltung von 30 g *racem.* Säure lieferte ebenfalls eine Säure (B) mit dem Schmp. 108—109° und $[\alpha]^{15.8}_D$: +22.7° in Wasser (c = 6.223). Ausbeute 4.5 g.

A und B wurden vereinigt und aus demselben Lösungsmittel umkrystallisiert. Ausbeute 6.2 g.

(+)-Citramalsäure krystallisiert in farblosen Prismen vom Schmp. 108—109°, während die *racem.* Säure bei 119° schmilzt.

36.6 mg Sbst.: 54.0 mg CO_2, 17.0 mg H_2O.
$C_5H_8O_5$. Ber. C 40.5, H 5.4. Gef. C 40.2, H 5.2.

Die optische Drehung wurde in Wasser bestimmt: l = 1; c = 6.245. α^{14}_D: +1.45°, α^{14}_{5461}: +1.73°; $[\alpha]^{14}_D$: +23.2°, $[\alpha]^{14}_{5461}$: +27.7°. l = 2; c = 1.56125. α^{20}_D: +0.74°, α^{20}_{5461}: +0.88°; $[\alpha]^{20}_D$: +23.7°, $[\alpha]^{20}_{5461}$: +28.2°.

Diese Werte änderten sich nach nochmaligem Umkrystallisieren der Säure nicht.

Marckwald und Axelrod erhielten die (+)-Säure durch Zersetzung ihres Bleisalzes mit Schwefelwasserstoff. Sie beschreiben ihre Säure als höchst zerfließlich und bei 95° schmelzend; sie wurde nicht umkrystallisiert, eine Analyse wurde nicht angegeben. Die Rotationsdispersion wurde in wäßr. Lösung[10]) bei 14° für verschiedene Konzentrationen mit Hilfe des Landoltschen Farbenfilters gemessen. Für p = 7.87 und d^{14}: 1.0295 werden angegeben: $[\alpha]^{14}_D$: +25.25° und $[\alpha]^{14}_{grün}$: +31.33°.

Später haben Buraczewski und Marchlewski[11]), ohne sich auf die frühere Arbeit von Marckwald und Axelrod zu beziehen, *racem.* Citramalsäure mit Strychnin gespalten. Nach den etwas unvollständigen Angaben dieser Autoren über ihre Versuche wurde die (+)-Säure erhalten; allerdings wird keine Analyse mitgeteilt. Der Schmelzpunkt wird zu 108—109°, das Drehungsvermögen ohne Angabe des Lösungsmittels zu $[\alpha]^{20}_D$: +22.83 (c = 1.5) angegeben.

[9]) Journ. prakt. Chem. [2] **46**, 285 [1892]. [10]) B. **32**, 712 [1899].

[11]) J. Buraczewski and L. Marchlewski, *Z. physiol. Chem.*, **43**, 410 (1905).

$C_5H_8O_5$ Ref. 107a

Note:

The procedure can be summarized as follows: 1) the acid is resolved with brucine. The less soluble brucine salt obtained has a m.p. 222° (dec.) 2) Acidify to obtain (+)citramalic acid, which after recrystallization has a m.p. of 112.2°-112.8° and a $[\alpha]^{22}_D$ +23.6° (3% solution in H_2O).

J. Buraczewski and L. Marchlewski, Z. Physiol. Chem. 43, 410 (1905) obtained the same enantiomer using strychnine as resolving agent.

$C_5H_8O_5$

```
      CH2-COOH
       |
  CH3O-CH-COOH
```

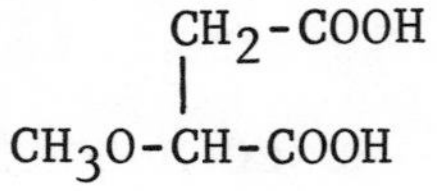

$C_5H_8O_5$ Ref. 108

With the view of deciding on the best method of effecting the resolution, the normal and acid strychnine salts of both active acids were prepared, the acids already obtained by the cinchonine method being used for the purpose. The following observations were made. *Normal dextro-salt*—long thin needles; crystallises readily. *Acid dextro-salt*—bundles of opaque radiating needles, much more soluble than any of the other salts. *Normal lævo-salt*—glassy square plates, more insoluble than either of the preceding. *Acid lævo-salt*—plates like the normal lævo-salt, but apparently more soluble. It seemed, therefore, best to remove the lævogyrate acid first as acid salt, and then the dextrogyrate acid as normal salt.

Twenty-three grams of strychnine were accordingly dissolved in an aqueous solution containing 10·4 grams of inactive methoxysuccinic acid, the quantities used being in molecular proportions. The solution, having been sown with acid lævo-salt, deposited a crop of plates, weighing 8 grams, in the course of 24 hours, and the mother liquor deposited, on evaporating, a further crop of the same salt, weighing 5·5 grams. The 13·5 grams gave on recrystallisation 10·8 grams of the dry salt, which was dissolved in water and decomposed by adding ammonia. The strychnine, which was separated by filtration and dried, weighed 6·77 grams, the calculated quantity for the acid salt being 7·48 grams. The ammoniacal solution having been evaporated to a small bulk, a little more strychnine separated, which was filtered off; the filtrate, being made up to 50 c.c., gave the rotation −1·74° in a 200 mm. tube. The mother liquor, from which the acid lævo-salt had been removed, was neutralised by warming slightly with excess of strychnine; the filtered liquid, having been sown with a nucleus of dextro-normal salt, gave a crop of long, thin needles, which, after being recrystallised twice from water and dried, weighed 15·2 grams. The ammoniacal solution, obtained as before from this salt, gave the rotation +59' in a 200 mm. tube.

The acid, having been evidently resolved into its optically active components by the process described, the experiment was carried out on a larger scale, 73 grams of acid being used and 167 grams of strychnine. The acid lævo-salt obtained was recrystallised twice from water, and then weighed in the dry state 73 grams. It gave on decomposition with ammonia 47 grams of strychnine, the calculated quantity for the acid salt being 50·6 grams. The calcium salt was obtained from the solution of the ammonium salt by treatment with calcium hydroxide, removal of the excess of the latter by means of carbonic anhydride, and evaporation of the filtered liquid. To insure the removal of traces of strychnine, the dry calcium salt was digested repeatedly with alcohol, and, to separate the active salt from any of the less soluble racemoid form with which it might be contaminated, it was finally kept in agitation with a quantity of cold water insufficient for complete solution. The acid was obtained by adding the calculated quantity of sulphuric acid to the solution of the calcium salt, and by extracting the residue left on evaporating the filtered liquid with alcoholic ether. The acid obtained weighed 15 grams; being slightly coloured, it was converted into the acid potassium salt, from which the colouring matter was removed by digestion with alcohol. An estimation of potassium in the salt dried at 100° gave 21·14 per cent., the calculated percentage being 21·01. The salt being recrystallised from water, two successive crops of crystals showed the same specific rotation, namely, −23·12° ($c = 8$, $t = 15°$), that of the purest salt obtained by the cinchonine method being −23·18° under similar conditions.

The normal strychnine salt of the dextrogyrate acid obtained by neutralising with strychnine the mother liquor, from which the acid salt of the lævogyrate acid had been separated, weighed 125 grams after being recrystallised twice from water, and gave 100 grams of dry strychnine when decomposed with ammonia, the calculated weight of strychnine for the normal salt being 102·3 grams. The dextro-acid, weighing 10 grams, was obtained by the same process as the lævo-acid, and was similarly converted into the acid potassium salt; an estimation of potassium in the latter gave 20·94 per cent., the calculated percentage being 21·01, and the specific rotation of the salt was +23·05° under the conditions mentioned above; a second fraction gave the number +22·75°; but as it was obtained by evaporating the solution to a very small bulk, the

$C_5H_8O_6$

lower activity may be accounted for by the presence of active normal salt. The experiments which have been described show that methoxysuccinic acid can be resolved into its active components with the aid of strychnine in two ways, namely, as normal salt, in which case the salt of the lævo-acid separates first from the solution, or as acid salt when the salt of the lævo-acid crystallises first, the dextro-acid being afterwards obtained as normal salt.

Solvent.	Dextro-acid.				Lævo-acid.			
	$t.°$	c	a	$[a]_D$	$t.°$	c	a	$[a]_D$
Water	15	24·6520	16·07°	+32·59°	15	22·0353	14·41	−32·70
„	14	16·6805	10·94	+32·79	—	—	—	—
„	15	8·7620	5·73	+32·70	15	7·9266	5·14	−32·42
Acetone	11	24·9640	28·51	+57·10	11	25·5833	28·78	−56·25
„	11	18·7693	21·88	+58·29	13	15·6140	18·17	−58·18
„	14	10·2960	11·95	+58·03	—	—	—	—
„	14	4·1184	4·90	+59·49	—	—	—	—
„	14	1·6474	1·98	+60·09	—	—	—	—
Ethylic acetate	11	20·5424	26·08	+63·48	11	25·5510	31·63	−61·90
„	12	15·8720	20·46	+64·45	13	19·0770	24·01	−62·93
„	12	8·9193	11·53	+64·04	—	—	—	—

```
           OH    CH3
            |    |
      HOOC-CH — C-COOH
                |
                OH
```

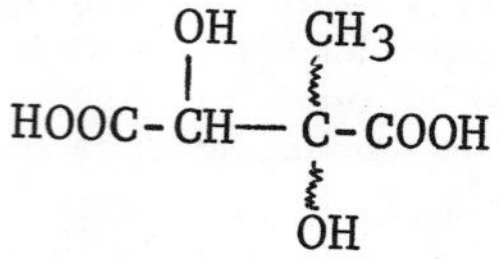

$C_5H_8O_6$ Ref. 109

The Optical Resolution of Ia.—Forty-nine grams of Ia and 118 g. of brucine were dissolved in 1000 ml. of boiling water; the resulting solution was allowed to stand overnight in a refrigerator. The needle-like crystals of the brucine salt which precipitated were collected by filtration; six recrystallizations of this salt from ten parts of boiling water gave 61 g. of the pure brucine salt of Ia. M. p. 236—237°C, $[a]_D^{20}$ −21.9° (c 0.4, in water). This salt was then suspended in 100 ml. of cold water and decomposed with 225 ml. of a N sodium hydroxide solution. The precipitated brucine was removed by filtration, and the alkaline filtrate was washed twice with 50 ml. of chloroform, neutralized with 20 ml. of concentrated hydrochloric acid, and then evaporated to a syrup. To the residual syrup a small amount of water was added, and the resulting solution was decolorized with charcoal and evaporated to dryness in order to remove the free hydrochloric acid. The residue was extracted several times with 20 ml. of acetone. These acetone extracts were then combined and evaporated to a syrup. The syrupy product was dried over sulfuric acid in a vacuum desiccator. In this way, white, hygroscopic crystals of (−)-Ia were obtained. Recrystallization from dry ethyl acetate gave 14.9 g. of optically pure (−)-Ia. M. p. 115—116°C, $[a]_D^{20}$ −8.94° (c 3.7, water).

Found: C, 36.53; H, 5.02. Calcd. for $C_5H_8O_6$: C, 36.59; H, 4.91%.

The mother liquor of the brucine salt of (−)-Ia was concentrated to a paste. After it had been stood for a week in a refrigerator, the brucine salt of (+)-Ia was collected. Optically pure (+)-Ia was obtained from this salt by the procedure described above. Yield, 8.0 g. M. p. 115°C, $[a]_D^{20}$ +8.90° (c 3.7, water).

Found: C, 36.57; H, 4.94. Calcd. for $C_5H_8O_6$: C, 36.59; H, 4.91%.

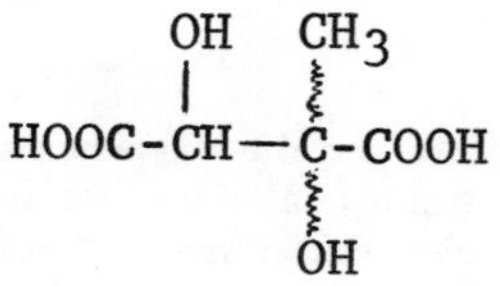

$C_5H_8O_6$ **Ref. 110**

The Optical Resolution of Ib.—Thirty-eight grams of Ib and 91.5 g. of brucine were dissolved in 1000 ml. of boiling water, and the resulting solution was treated as in the resolution of Ia. Forty grams of the pure brucine salt of (+)-Ib were, thus obtained. M. p. 234—235°C, $[\alpha]_D^{20}$ −16.5° (*c* 0.4, water). When the brucine salt was decomposed with alkali, 8.2 g. of optically pure (+)-Ib were obtained. M. p. 130—131°C, $[\alpha]_D^{20}$ +5.88° (*c* 3.2, water).

Found: C, 36.53; H, 4.88. Calcd. for $C_5H_8O_6$: C, 36.59; H, 4.91%.

From the mother liquor of the brucine salt of (+)-Ib, 5.1 g. of optically acitive (−)-Ib were obtained by the procedures described above. M. p. 129°C, $[\alpha]_D^{20}$ −5.60° (*c* 3.2, water).

Found: C, 36.63; H, 4.79. Calcd. for $C_5H_8O_6$: C, 36.59; H, 4.91%.

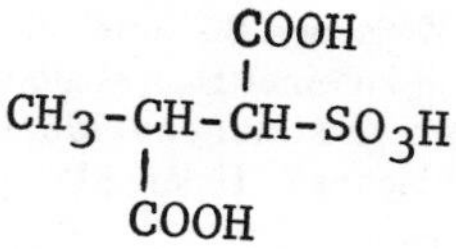

$C_5H_8O_7S$ **Ref. 111**

Le *dédoublement optique* de l'acide peut être réalisé à l'aide de strychnine et de quinine, qui tous les deux fournissent l'énantiomorphe dextrogyre.

Le sel secondaire de strychnine cristallise dans l'eau en petites aiguilles courtes, s'agglomérant en boules massives. Le sel peut cristalliser avec 4 molécules d'eau, qu'il perd à 105° ou dans le vide en présence d'anhydride phosphorique. Parfois le sel cristallise avec plus de molécules d'eau.

Substance 0.2021 g.: perte à 105° 0.0152 g.
Subst. 0.5007 g.: 14.35 cm³. de NaOH de 0.1076 n.

	Trouvé:	H_2O 7.55;	poids équiv. 325.
$C_5H_8O_7S \cdot 2\,C_{21}H_{22}O_2N_2 \cdot 4\,H_2O$ (970.59).	Calculé:	,, 7.42;	,, ,, 323.5.

On peut préparer ce sel de strychnine, en dissolvant deux molécules de strychnine à chaud dans une solution de l'acide.

Lorsqu'on ne connaît pas la tendance de l'acide actif à la racémisation, il vaut mieux le préparer par la „cristallisation froide".

Parmi plusieurs essais de dédoublement que nous avons entrepris nous ne citerons qu'une seule expérience, qui a fourni l'acide actif à l'état optiquement pur.

220 cm³. d'une solution de l'acide 1.48 n., neutralisés par une lessive de soude, ont été mélangés à la température ordinaire à 36 g. de strychnine, dissous dans l'eau renfermant 10 cm³. d'acide acétique. Volume total 750 cm³. Le sel a cristallisé très lentement. Après 2 jours on en a recueilli 11 g.

Cette quantité a été décomposée par 25 cm³. d'eau et 30.6 cm³. d'une lessive de soude caustique 1.076 n. et neutralisée par 0.4 cm³. d'acide sulfurique de 1.91 n. Concentration calculée d'après ce titrage, 0.0191 mol. g. dans 100 cm³. $\alpha_D = +\,0°.75$ (2 dm.). Donc $[M]_D = 19°.6$.

La solution du sel sodique actif, additionnée d'une solution acétique de 7 g. de strychnine, a donné au bout de 24 heures, dans un volume de 80 cm³., un dépôt de 7 g. du sel de strychnine.

Cette quantité, décomposée à l'aide de 21.2 cm³. d'une lessive de soude (1.076 n.) et de 0.4 cm³. d'acide sulfurique (1.91 n.), a accusé une rotation de + 1°.22 (2 dm.). D'après le titrage la concentration est de 0.0340 mol. g. dans 100 cm³. $[M]_D = 18°$. La rotation n'a plus augmenté.

En prenant le moyen des deux valeurs, on trouve pour la rotation moléculaire du sel sodique 19°. Les solutions du sel sodique, additionnées d'acide sulfurique, subissent une augmentation graduelle de leur pouvoir rotatoire. Une solution du sel de la rotation + 0°.21 (2 dm.), acidulée par un excès d'acide sulfurique, montra une rotation, corrigée pour la dilution, de + 0°.38, soit $[M]_D = +\,35°$.

Donc, rotation moléculaire de *l'acide libre*, ion monovalent

$CH_3 . CH(CO_2H) . CH(CO_2H) . SO_3'$: $[M]_D = +\,35°$.

Rotation moléculaire des *sels neutres*, ion trivalent

$CH_3 . CH(CO_2') . CH(CO_2') . SO_3'$: $[M]_D = +\,19°$.

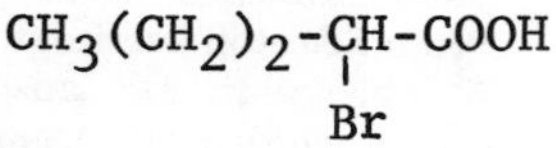

$C_5H_9BrO_2$ **Ref. 112**

Derivatives of n-Valeric Acid.

Resolution of α-Bromo-n-Valeric Acid.—In preliminary experiments on the resolution of α-bromo-*n*-valeric acid by means of brucine, cinchonidine, strychnine, and quinine, the best results were obtained with quinine. A warm solution of 100 gm. of the bromo acid in 600 cc. of acetone was treated with 179 gm. of quinine. On cooling, the salt separated in the form of white needle-like crystals. It was recrystallized several times from acetone. To decompose the quinine salt it was dissolved with cooling in a slight excess of dilute hydrochloric acid and the solution was extracted with ether. The ethereal extract was washed with water and dried over sodium sulfate. The ether was then removed and the residue fractionated under reduced pressure. The substance distilled at 123–124°C. (p = 15 mm.), and showed an optical activity of

$$[\alpha]_D^{20} = \frac{+\,2.22° \times 100}{1 \times 7.16} = +\,31.0° \text{ (in ether).}$$

$$[\alpha]_D^{20} = \frac{+\,0.56° \times 100}{2 \times 1.38} = +\,20.3°. \quad [M]_D^{20} = +\,36.7° \text{ (in water).}$$

30 gm. of the active acid were obtained.

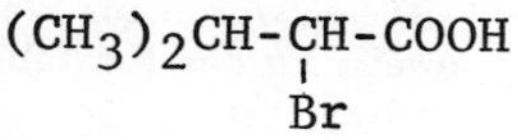

$C_5H_9BrO_2$ **Ref. 113**

Resolution of α-Bromoisovaleric Acid.—Berlingzzi and his coworker[16] have recently resolved this acid by means of brucine but by this method it is somewhat difficult to prepare the acid in quantity. Cinchonidine was found more convenient for this purpose.

50 gm. of the inactive bromo acid were dissolved in 350 cc. of acetone and 82 gm. of pure cinchonidine were then added. After cooling, the salt was filtered and recrystallized from acetone nine times. It was decomposed with a slight excess of dilute hydrochloric acid and extracted with ether. The ethereal extract was washed with water and dried over sodium sulfate. On removal of the ether, the bromo acid was obtained in crystalline form. It was purified by fractional distillation under reduced pressure. The fraction distilling at 119-120°C. (p = 14 mm.) showed an optical rotation of

$$[\alpha]_D^{20} = \frac{-\,0.50° \times 100}{1 \times 6.46} = -\,7.7° \text{ (in ether).}$$

The mother liquor of the above salt was allowed to stand overnight at 0°C. and another crop of crystals was obtained. The filtrate from these crystals was concentrated under reduced pressure and the residue decomposed as usual. The bromo acid thus obtained showed a rotation of

$$[\alpha]_D^{20} = \frac{+\,0.73° \times 100}{1 \times 9.82} = +\,7.43° \text{ (in ether).}$$

Another sample which had the optical activity of $[\alpha]_D^{20} = \dfrac{+0.54° \times 100}{1 \times 7.12} = +7.6°$ in ether, showed the following rotation in water.

$$[\alpha]_D^{20} = \frac{+\,0.14° \times 100}{2 \times 1.55} = +\,4.5°. \quad [M]_D^{20} = +\,8.1°.$$

[16] Berlingzzi, S., and Furia, M., *Gazz. chim. ital.*, 1926, lvi, 828.

$C_5H_9BrO_2$ Ref. 113a

(—)-α-Bromisovaleriansäure

98,4 g Cinchonidin wurden in 500 ml warmem Aceton suspendiert und die Suspension unter kräftigem Rühren mit einer Lösung von 60,0 g rac. α-Bromisovaleriansäure in 900 ml Aceton vereinigt. Es wurde rasch zur Lösung erwärmt, filtriert und allmählich auf 5° abgekühlt. Die ausgefallenen Nadeln des (—)-α-Bromisovaleriansäure-Cinchonidin-Salzes wurden abgetrennt, mit kaltem Aceton gewaschen und getrocknet. — Einengung des Filtrats i. Vak. auf die Hälfte erbrachte eine zweite Kristallfraktion. Das Filtrat wurde zur Trockne gebracht; der Rückstand (84,0 g) wurde auf (+)-α-Bromisovaleriansäure weiterverarbeitet. Das (—)-Säure-Cinchonidinsalz wurde aus Aceton weiter umkristallisiert. Zur Messung der Drehung wurden jeweils 2 g des Kristallisates zerlegt und die Säure mit Äther extrahiert. Ausbeuten und Drehwerte:

Kristallisation	Eingesetzt g	Ausbeute g Salz	$[\alpha]^{20}_{546}$ der Säure
1.	60,0 Säure + 98,4 Base	62,0	— 6,9°
2.	60,0 Salz	47,0	—11,7°
3.	45,0 „	34,0	—15,3°
4.	32,0 „	25,0	—19,8°
5.	23,5 „	18,5	—23,2°
6.	17,0 „	13,0	—26,4°
7.	11,0 „	7,0	—27,6°
8.	5,5 „	3,8	—27,6°

(+)-α-Bromisovaleriansäure

Der Rückstand (84,0 g) der zweiten Kristallisation des (—)-Säure-Cinchonidin-Salzes wurde zerlegt und die Säurefraktion bei 85° und 1 Torr destilliert. Zwischen 85° und 90° gingen 27,3 g über. Sie wurden in 200 ml Äther gelöst und die Lösung zu einer Suspension von 58,9 g Brucin in 4300 ml Wasser von 30° gegeben. Unter Rühren entstand eine klare Lösung, die bei 30°/15 Torr auf die Hälfte eingeengt wurde. Im Kühlschrank fielen 54,0 g eines feinkristallinen Niederschlages des (+)-Säure-Brucin-Salzes aus.

Dieses wurde aus Wasser wiederholt umkristallisiert. Zur Messung der Drehung wurden jeweils 2 g des Kristallisates zerlegt und die Säure mit Äther extrahiert. Ausbeuten und Drehwerte:

Kristallisation	Eingesetzt g	Ausbeute g Salz	$[\alpha]^{20}_{546}$ der Säure
1.	27,3 Säure + 58,9 Base	54,0	+ 8,5°
2.	52,0 Salz	39,0	+ 11,5°
3.	37,0 „	32,7	+ 17,4°
4.	30,7 „	27,7	+ 22,3°
5.	26,2 „	23,0	+ 24,8°
6.	21,5 „	19,7	+ 25,9°
7.	18,2 „	14,0	+ 26,6°
8.	12,5 „	10,0	+ 27,6°
9.	8,5 „	5,0	+ 27,6°

$C_5H_9BrO_2$ Ref. 113b

Resolution of DL-2-Bromo-3-methylbutyric Acid into its Optical Isomers. a) A solution of 50.3 g of (-)-α-methylbenzylamine in 500 ml of dry ether was added to a boiling solution of 75 g of 2-bromo-3-methylbutyric acid in 1 liter of dry ether. The mixture was cooled, and after one day the precipitated crystals were filtered off and dried in a vacuum desiccator. We obtained 65.5 g of a salt of m.p. 113-115°; after three crystallizations from an equal amount of dry benzene we isolated 42 g (67%) of the salt of D-2-bromo-3-methylbutyric acid with (-)-α-methylbenzylamine (salt I), m.p. 121-123° and $[\alpha]_D^{20}$ -18.5° (c 1.0 in acetone). Found: C 52.02; H 6.82; N 4.58; Br 26.49%. $C_{13}H_{20}O_2NBr$. Calculated: C 51.66; H 6.67; Br 26.44%.

b) 100 ml of ether and 10 ml of concentrated hydrochloric acid were added to a suspension of 28 g of salt I in 50 ml of water. The mixture was shaken until solution was complete, the ether layer was separated, and the combined ethereal solution was dried with magnesium sulfate. Solvent was distilled off, and the residue was vacuum-fractionated. We obtained 15.9 g (95%) of D-2-bromo-3-methylbutyric acid; b.p. 110-112° (12 mm); m.p. 43-44°; $[\alpha]_D^{20}$ +23.0° (c 1.0 in benzene) [6].

c) The mother solution remaining after the separation of salt I was washed with two 100-ml portions of 2 N HCl and with water; it was dried with anhydrous magnesium sulfate and evaporated. The residue was vacuum-fractionated, and we obtained L-2-bromo-3-methylbutyric acid (with some D-isomer impurity); $[\alpha]_D^{20}$ -16.0° (c 1.0 in benzene). A solution of 23.4 g of (+)-α-methylbenzylamine in 200 ml of dry ether was added to a boiling solution of the bromo acid in 600 ml of dry ether. After one day the precipitated salt (54.4 g, m.p. 116-119°) was filtered off and crystallized three times from an equal amount of dry benzene. We obtained 38.8 g (62%) of the salt of L-2-bromo-3-methylbutyric acid with (+)-α-methylbenzylamine (salt II); m.p. 121-123°; $[\alpha]_D^{20}$ +18.5° (c 1.0 in acetone).

d) By the method of Expt., (b) from 18 g of the salt II we obtained 10.2 g (94%) of L-2-bromo-3-methylbutyric acid; b.p. 110-112° (12 mm); m.p. 43-44°; $[\alpha]_D^{20}$ -23.0° (c 1.0 in benzene) [6].

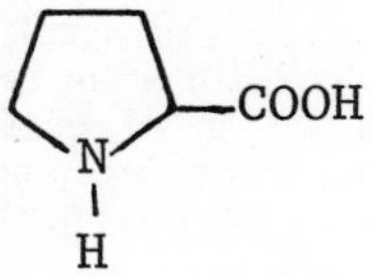

$C_5H_9NO_2$ Ref. 114

DL-prolin (I), prepared from diethyl malonate and acryl nitrile according to ALBERTSON and FILLMAN [5], m. p. 210—211°C.

3,5-Dinitrobenzoyl-DL-prolin, prepared from 3,5-dinitrobenzoyl chloride and DL-prolin, according to reference [2], in 92% yield, m. p. 221—222°C [2].

3,5-Dinitrobenzoyl-D-prolin. The filtered solution of 15,84 g of 3,5-dinitrobenzoyl-DL-prolin in 100 ml of acetone was treated with the filtered solution of 24,0 g of brucin.$4H_2O$ in 100 ml of acetone. On removing the solvent by distillation, the residual reddish yellow crystalline substance was dissolved in 450 ml of hot water, filtered and allowed to stand at room temperature, then at +3°C overnight. Recrystallization from water afforded (from 18,37 g of yellow platelets) 17,54 g of 3,5-dinitrobenzoyl-D-prolin brucinate, m. p. 111—112°C. $[\alpha]_D^{22} = +31{,}5°$ (c: 1,018; waterfree ethanol). Analysis of substance dried for 3 hours at 80°C under a pressure of 30 mm Hg. Calculated from formula $C_{36}H_{37}O_{11}N_5$, N 9,95; found 9,94%.

The solution of 17,54 g of 3,6-dinitrobenzoyl-D-prolin brucinate in 486 ml of hot water was made slightly alkaline with a 1,0 N solution of sodium hydroxyde, the precipitated brucin base (8,4 g) filtered on standing for a few hours at +3°C and the residual brucin base removed by extracting the aqueous solution with 3 × 200 ml portions of ether. On removing ether, the aqueous solution was adjusted with concentrated hydrochloric acid to pH 6, allowed to stand for 24 hours at +3°C, the yellow crystals (4,1 g; m. p. 174—176°C) filtered and the mother liquor concentrated to 75 ml (yield: further 1,95 g of substance). Recrystallization from water afforded 4,9 g (62%) of 3,5-dinitrobenzoyl-D-prolin of adequate purity, m. p. 179°C [2]. $[\alpha]_D^{30} = +94{,}8°$ (c: 1,292; 50% ethanol).

D-prolin (I), prepared by hydrolysis with 5 N hydrochloric acid and liberated with Amberlite IR—4B ion exchange resin, by the VELLUZ method [2], in 82% yield, m. p. 215—220°C (decomp.). $[\alpha]_D^{20} = +81°$ (c: 0,500, H_2O) [1], [2].

3,5-Dinitrobenzoyl-L-prolin. On concentrating the mother liquor obtained in the previous operations, containing mainly 3,5-dinitrobenzoyl-L-prolin to a volume of 200 ml, the precipitated crystalline substance (0,42 g; m. p. 135—150°C) was filtered, the filtrate made alkaline with 1,0 N sodium hydroxyde and allowed to stand for a few hours at +3°C. The precipitated brucin base (11.06 g) was filtered, the aqueous phase extracted with 3 × 100 ml portions of ether, adjusted with concentrated hydrochloric acid to pH 6, allowed to stand a few hours at +3°C, the crystalline substance filtered (4,26 g; m. p. 172—174°C) and the mother liquor concentrated to one third of original volume, thus affording further 2,37 g of substance. On repeated recrystallisation from water, yield 4,46 g (59%) of 3,5-dinitrobenzoyl-L-prolin of satisfactory purity, m. p. 178—179°C [2]. $[\alpha]_D^{24} = -94{\cdot}7°$ (c: 1,015; 50% ethanol).

$C_5H_9NO_2$ Ref. 114a

Preparation of N-3,5-dinitro benzoyl-DL-proline

2.7 g. of DL-proline are dissolved in 52 cc. of an N aqueous sodium hydroxide solution. The solution is cooled to 0° C. 6.25 g. of 3,5-dinitro benzoylchloride are added to said solution. The reaction mixture is removed from the cooling bath and is agitated at room temperature for 15 minutes. After filtering off sodium chloride, 5 cc. of concentrated hydrochloric acid are added. 6.8 g. of N-3,5-dinitro benzoyl-DL-proline melting at 221–222° C. are obtained thereby. The yield amounts to 95% of the theoretical yield.

EXAMPLE 2

Resolving N-3,5-dinitro benzoyl-DL-proline by means of L(+)-threo-1-(p-nitro phenyl)-2-amino propanediol-(1,3) and separation of D- and L-proline

(*a*) FORMATION OF THE SALT OF N-3,5-DINITRO BENZOYL-D-PROLINE WITH L(+)-THREO-1-(P-NITRO PHENYL)-2-AMINO PROPANEDIOL-(1,3)

5 g. of N-3,5-dinitro benzoyl-DL-proline are dissolved at about 70° C. in 25 cc. of water. A.75 g. of L(+)-threo-1-(p-nitro phenyl)-2-amino propanediol-(1,3) are added to said solution. The mixture is cooled to 40° C. within 30 minutes while stirring. The precipitated salt is filtered off and washed with a small amount of water. On recrystallization from water and drying in a drying oven at 60° C. the hydrated salt is obtained. Said salt melts at about 100° C. and has a rotatory power $[\alpha]_D^{20} = +73° \pm 1°$ (concentration: 0.5% in 50% ethanol). The yield is 3.5 g. corresponding to 85% of the theoretical yield.

(*b*) PREPARATION OF N-3,5-DINITRO BENZOYL-D-PROLINE

3.5 g. of said salt of N-3,5-dinitro dibenzoyl-D-proline with L(+)-threo-1-(p-nitro phenyl)-2-amino propanediol-(1,3) are treated at 40° C. with 7 cc. of N sodium hydroxide solution. The salt first dissolves and thereafter L(+)-threo-1-(p-nitro phenyl) - 2 - amino propanediol-(1,3) starts to crystallize. The crystals are filtered off and washed with a small amount of water. In this manner 1.4 g. of the resolving agent, corresponding to 37% of the theoretical yield, are recovered.

The resulting filtrate is acidified with 0.7 cc. of concentrated hydrochloric acid. N-3,5-dinitro benzoyl-D-proline crystallizes and is filtered off, washed with a small amount of water, and dried. In this manner 2.01 g. of crystalline N-3,5-dinitro benzoyl-D-proline are obtained. Its melting point is 179–180° C., its rotatory power $[\alpha]_D^{20} = +92° \pm 1°$ (concentration: 0.5% in 50% ethanol). The yield amounts to 80% calculated for N-3,5-dinitro benzoyl-DL-proline used as starting material.

(*c*) ISOLATION OF D-PROLINE

A solution of 1 g. of N-3,5-dinitro benzoyl-D-proline is heated under reflux in 10 cc. of 5 N hydrochloric acid during 30 minutes. The resulting solution is then filtered, evaporated to dryness in a vacuum, and the resulting hydrochloride of D-proline is recrystallized from acetone. The product is dissolved in 10 cc. of water and is treated with 2 g. of Amberlite IR4B for one hour. The mixture is filtered and the filtrate is evaporated to dryness in a vacuum. 300 mg. of crystalline D-proline, corresponding to a yield of 80% of the theoretical yield, and having a rotary power of $[\alpha]_D^{20} = +83.5° \pm 2°$ (concentration: 0.5% in water) are obtained thereby.

(*d*) ISOLATION OF L-PROLINE

The aqueous filtrate, obtained after removing the crystallized salt of N-3,5-dinitro benzoyl-D-proline with L(+)-threo-1-(p-nitro phenyl) - 2 - amino propanediol-(1,3), is treated with 1 cc. of sodium hydroxide solution. Thereby 2.1 g. of the resolving agent, corresponding to a yield of 56%, are recovered as described hereinabove under (*b*). 1 cc. of concentrated hydrochloric acid is added to the filtrate, thereby yielding 2.7 g. of impure

N-3,5-dinitro benzoyl-L-proline. Said compound can be purified by recrystallization according to conventional methods or by means of its ability of forming a difficultly soluble salt with D(—)-threo-1-(p-nitro phenyl)-2-amino propanediol-(1,3) as this will be described hereinafter in Example 3.

The product which can be recovered from the mother liquors of the recrystallization of N-3,5-dinitro benzoyl-L-proline can be added to a new charge of racemic mixture to be resolved.

L-proline is obtained from its N-acyl compound in the same manner as described hereinabove under (*c*) for D-proline, by hydrolyzing N-3,5-dinitro benzoyl-L-proline by means of hydrochloric acid.

EXAMPLE 3

Purification of N-3,5-dinitro benzoyl-L-proline by means of D(—)-threo-1-(p-nitro-phenyl)-2-amino propanediol-(1,3)

2.8 g. of impure N-3,5-dinitro benzoyl-L-proline obtained according to Example 2 (*d*), are treated with 2 g. of D(—)-threo-1-(p-nitro phenyl)-2-amino propanediol-(1,3) in 14 cc. of water at a temperature of 40° C. Thereby, 3.9 g. of the hydrated salt of N-3,5-dinitro benzoyl-L-proline with D(—)-threo-1-(p-nitro phenyl)-2-amino propanediol-(1,3) are obtained. Said salt has a melting point of 100° C. and a rotatory power $[\alpha]_D^{20}=-73°\pm1°$ (concentration: 0.5% in 50% ethanol). After treating said salt with alkali hydroxide, separating the resolving agent, and acidifying the filtrate as described hereinabove in Example 2 under (*b*), there are obtained 2.1 g. of pure N-3,5-dinitro benzoyl-L-proline having a melting point of 179–180° C. and a rotatory power $[\alpha]_D^{20}=-92°\pm1°$ (concentration: 0.5% in 50% ethanol).

$C_5H_9NO_2$ Ref. 114b

Experimenteller Teil [2])

A) Optische Spaltung von Z-DL-Prolin. – 1. *Z-DL-Prolin* wurde nach der Vorschrift für Z-L-Prolin [9] hergestellt und aus Essigester-Petroläther kristallisiert; Smp. 74–75°.

$C_{13}H_{15}O_4N$ (249,26) Ber. C 62,64 H 6,07 N 5,62% Gef. C 62,42 H 6,07 N 5,63%

2. *L-Tyrosinhydrazidsalz von Z-D-Prolin.* – a) *Mit der berechneten Menge L-Tyrosinhydrazid.* 97,5 g (0,5 Mol) L-Tyrosinhydrazid [5] und 124,5 g (0,5 Mol) Z-DL-Prolin wurden in 2,2 l siedendem Methanol gelöst, die Lösung sofort filtriert, mit dem L-Tyrosinhydrazidsalz von Z-D-Prolin angeimpft und 20 Std. bei 25° gerührt. Die Kristalle wurden abgenutscht, mit insgesamt 200 ml Methanol gewaschen und getrocknet. Ausbeute 103,2 g (93%) L-Tyrosinhydrazidsalz von Z-D-Prolin, Smp. 187–189°, $[\alpha]_D^{23} = +74,2°$ ($c = 1$, in Wasser).

$C_{22}H_{28}O_6N_4$ (444,48) Ber. C 59,44 H 6,35 N 12,61% Gef. C 59,50 H 6,55 N 12,55%

Die Mutterlauge, die das optisch noch nicht reine L-Tyrosinhydrazidsalz von Z-L-Prolin enthielt, wurde gemäss **A***5a*) verarbeitet.

b) *Mit der halben Menge L-Tyrosinhydrazid.* 48,75 g (0,25 Mol) L-Tyrosinhydrazid und 124,5 g (0,50 Mol) Z-DL-Prolin wurden in 750 ml siedendem Methanol gelöst. Die Lösung wurde heiss filtriert, mit dem L-Tyrosinhydrazidsalz von Z-D-Prolin angeimpft und 20 Std. bei 25° gerührt. Die Kristalle wurden abgenutscht, mit 100 ml Methanol gewaschen und getrocknet. Ausbeute 97,8 g (88%) L-Tyrosinhydrazidsalz von Z-D-Prolin, Smp. 187–189°, $[\alpha]_D^{22} = +73,9°$ ($c = 1$, in Wasser).

Die Mutterlauge wurde im Vakuum eingedampft, der Rückstand mit 300 ml Äther vermischt, 2 Std. bei 0° belassen, abgenutscht, mit 200 ml Äther gewaschen und getrocknet; Ausbeute 5,5 g Zwischenfraktion, Smp. 164–168°, $[\alpha]_D^{23} = +23,5°$ ($c = 1$, in Wasser).

Filtrat und Waschlösung enthielten zur Hauptsache optisch noch nicht reines Z-L-Prolin; sie wurden gemäss **A***5b*) weiterverarbeitet.

3. *Z-D-Prolin.* Eine Suspension von 222 g (0,5 Mol) L-Tyrosinhydrazidsalz von Z-D-Prolin in 200 ml Wasser und 100 ml konz. Salzsäure wurde im Scheidetrichter mit insgesamt 1 l Äther extrahiert. Die Ätherlösungen wurden mit Wasser gewaschen, über $MgSO_4$ getrocknet, im Vakuum eingedampft, der Rückstand aus Äther-Petroläther kristallisiert, abgenutscht und im Wasserstrahl- und Hochvakuum bei 60° getrocknet. Ausbeute 119,5 g (96%), Smp. 76–77°, $[\alpha]_D^{23} = +61,2°$ ($c = 5,3$, in Eisessig). Literaturwert für Z-L-Prolin [9]: $[\alpha]_D^{20} = -61,7°$ ($c = 5,3$ in Eisessig).

$C_{13}H_{15}O_4N$ (249,26) Ber. C 62,64 H 6,07 N 5,62% Gef. C 62,94 H 5,94 N 5,60%

Das L-Tyrosinhydrazid wurde aus den salzsauren Lösungen analog **B***2*) regeneriert.

4. *D-Prolin.* 25 g (0,1 Mol) Z-D-Prolin wurden in 150 ml 80-proz. Essigsäure mit 3 g 5-proz. Pd-Kohle in üblicher Weise hydriert, vom Katalysator abgetrennt, im Vakuum eingedampft und mehrmals mit Benzol abgedampft. Der Rückstand wurde in möglichst wenig siedendem Alkohol gelöst, filtriert, mit Äther ausgefällt, abgenutscht und aus Alkohol-Äther umkristallisiert. Ausbeute 10,2 g (89%), $[\alpha]_D^{23} = +85,2°$ ($c = 3,5$, in Wasser). Literaturwerte für L-Prolin: $[\alpha]_D = -84,9°$ [10], $-86,5°$ [11], $-85,6°$ [12].

$C_5H_9O_2N$ (115,13) Ber. C 52,16 H 7,88 N 12,17% Gef. C 52,19 H 8,00 N 12,28%

5. *Isolierung des optisch unreinen Z-L-Prolins.* – a) Die Mutterlauge aus der Spaltung **A***2a*) wurde im Vakuum eingedampft, der Rückstand in 100 ml Wasser und 50 ml konz. Salzsäure sus-

[2]) Die Smp. wurden auf dem Schmelzpunktsbestimmungsapparat nach TOTTOLI der Firma BÜCHI, Flawil (Schweiz), bestimmt; sie sind korrigiert. Fehlergrenze der Werte der spezifischen Drehungen: ± 2°. Die Analysenmuster wurden 18 Std. über P_2O_5 bei 0,01 Torr und bei 60° getrocknet.

pendiert und mehrmals mit Äther extrahiert. Die Ätherlösungen wurden mit Wasser gewaschen, über $MgSO_4$ getrocknet, im Vakuum eingedampft, aus Äther-Petroläther kristallisiert, abgenutscht und getrocknet. Ausbeute 63 g optisch unreines Z-L-Prolin, $[\alpha]_D^{23} = -55{,}4°$ ($c = 5$, in Eisessig); es wurde gemäss *A6b*) über das Dicyclohexylaminsalz optisch gereinigt. Das L-Tyrosinhydrazid wurde aus den salzsauren Lösungen analog **B**2) regeneriert.

b) Die Äthermutterlauge und der Waschäther aus der Spaltung **A**2*b*) wurden im Scheidetrichter mehrmals mit 1N Salzsäure und Wasser gewaschen, über $MgSO_4$ getrocknet, im Vakuum eingedampft, aus Äther-Petroläther kristallisiert und getrocknet. Ausbeute 62 g optisch unreines Z-L-Prolin, $[\alpha]_D^{23} = -53{,}5°$ ($c = 5$, in Eisessig); es wurde gemäss **A**6*b*) über das Dicyclohexylaminsalz optisch gereinigt.

6. *Optische Reinigung von Z-L-Prolin über das Dicyclohexylaminsalz.* – a) *Modellversuch.* Das Dicyclohexylaminsalz von Z-DL-Prolin wurde nach der Vorschrift für das entsprechende Z-L-Prolinsalz [13] hergestellt; Smp. 160°.

$C_{25}H_{38}O_4N_2$ (430,57) Ber. C 69,73 H 8,89 N 6,51% Gef. C 69,73 H 8,78 N 6,89%

Es ist etwas leichter löslich als das Dicyclohexylaminsalz von Z-L-Prolin; der Löslichkeitsunterschied ist ausreichend für eine optische Reinigung. Je 1 g Z-L-Prolin und Z-DL-Prolin wurden in 40 ml Essigester bei 50° gelöst, mit 1,45 g Dicyclohexylamin versetzt, nach 45 Min. abgenutscht, mit 40° warmem Essigester gewaschen und getrocknet. Ausbeute 1,6 g (95%) Dicyclohexylaminsalz von Z-L-Prolin, Smp. 178–180°, $[\alpha]_D^{23} = -25°$ ($c = 2$, in Methanol). Literaturwert [13]: Smp. 179–180°, $[\alpha]_D = -25{,}5°$ ($c = 2$, in Methanol).

Die Mutterlauge wurde im Vakuum stark eingeengt, mit Äther ausgefällt, abgenutscht, und getrocknet; Smp. 160–162°, $[\alpha]_D^{23} = -2°$ ($c = 2$, in Methanol) (Racemat).

b) *Anwendungsbeispiel.* 60 g optisch unreines Z-L-Prolin aus **A**5*a*) bzw. **A**5*b*), $[\alpha]_D^{23} = -54°$ ($c = 5$, in Eisessig), wurden in 600 ml Essigester auf 60° erwärmt, mit 51,6 g Dicyclohexylamin versetzt und sofort mit dem Dicyclohexylaminsalz von Z-L-Prolin angeimpft. Man liess zur Kristallisation langsam abkühlen, nutschte nach 45 Min. ab, wusch portionenweise mit insgesamt 250 ml 40° warmem Essigester nach und trocknete im Vakuum bei 60°. Ausbeute 91,4 g (83,5%) optisch reines Dicyclohexylaminsalz von Z-L-Prolin, Smp. 178°, $[\alpha]_D^{23} = -25{,}3°$ ($c = 2$, in Methanol). Literaturwert [13]: Smp. 179–180°, $[\alpha]_D = -25{,}5°$ ($c = 2$, in Methanol).

Aus der Mutterlauge konnten wir 13,2 g *rac.*-Salz isolieren, Smp. 161–164° $[\alpha]_D^{23} = -2{,}7°$ ($c = 2$, in Methanol).

7. *Z-L-Prolin.* 343 g optisch reines Dicyclohexylaminsalz von Z-L-Prolin wurden in 300 ml Wasser, 400 ml Essigester und 80 ml konz. Salzsäure 3 Std. stark gerührt. Das Dicyclohexylaminhydrochlorid wurde abgenutscht und mit 300 ml Essigester nachgewaschen. Das Filtrat wurde im Scheidetrichter mehrmals mit 1N Salzsäure und Wasser gewaschen, die Essigesterphase über $MgSO_4$ getrocknet, im Vakuum eingedampft und bei 60° getrocknet. Der dicke Sirup wurde aus Äther-Petroläther kristallisiert, abgenutscht, mit Petroläther gewaschen, im Wasserstrahl- und Hochvakuum bei 60° getrocknet. Ausbeute 179 g (90%), Smp. 76–77°, $[\alpha]_D^{22} = -61{,}7°$ ($c = 5{,}3$, in Eisessig). Literaturwert: $[\alpha]_D = -61{,}7°$ ($c = 5{,}3$, in Eisessig) [9]. Zur Kristallisation von Z-L-Prolin vgl. [14].

$C_5H_9NO_2$

Ref. 114c

d-m-Nitrobenzoyl-prolin.

27 g ganz reines *dl*-Nitrobenzoylprolin werden mit 30 g Cinchonin in 120 ccm Alkohol heiß gelöst, die filtrierte Flüssigkeit unter vermindertem Druck verdampft und der amorphe Rückstand in 1.5 l heißem Wasser gelöst. Beim Abkühlen scheidet die rötlich gefärbte Lösung zuerst eine kleine Menge eines dunkel gefärbten zähen Öls ab, das durch Dekantieren entfernt wird. Bei längerem Stehen der Flüssigkeit im Eisschrank erfolgt die Ausscheidung von farblosen, sehr feinen Nadeln. Ihre Menge betrug nach 12 Stunden 22.5 g. Die Mutterlauge gab noch 1 g. Die Gesamtausbeute entspricht 85 % der Theorie. Nach zweimaligem Umkrystallisieren aus der 40-fachen Menge heißen Wassers betrug die Menge des Salzes noch 12.5 g. Dieses Präparat war rein und schmolz gegen 150° zu einer braunen Flüssigkeit.

Zur Gewinnung des *d*-Nitrobenzoylprolins werden 12 g des Cinchoninsalzes in 500 ccm Wasser heiß gelöst, und nach Zusatz von 30 ccm *n*-Natronlauge rasch abgekühlt. Sofort beginnt die Ausscheidung des Cinchonins. Es wird nach ½-stündigem Stehen bei 0° abgesaugt, und das Filtrat nach dem Ansäuern mit 6 ccm 5-*n*. Schwefelsäure unter geringem Druck auf etwa 50 ccm eingedampft. Jetzt fügt man noch 20 ccm 5-*n*. Schwefelsäure hinzu und extrahiert das *d*-Nitrobenzoylprolin 5-mal mit je 120 ccm Äther. Beim Verdampfen des Äthers bleibt ein hellgelb gefärbtes Öl, das nach einigen Tagen krystallinisch erstarrt. Die Ausbeute war 5.3 g oder 91 % der Theorie. Zur Reinigung genügt einmaliges Umkrystallisieren aus heißem Wasser. Die Substanz fällt daraus beim Abkühlen zuerst ölig aus, erstarrt aber sehr bald krystallinisch.

Für die Analyse und die optischen Bestimmungen wurde bei 100° über Phosphorpentoxyd und unter 15 mm Druck getrocknet.

0.1629 g Sbst.: 0.3273 g CO_2, 0.0654 g H_2O. — 0.1674 g Sbst.: 0.3355 g CO_2, 0.0713 g H_2O.

$C_{12}H_{12}O_5N_2$ (264.12). Ber. C 54.52, H 4.58.
Gef. » 54.80, 54.66, » 4.46, 4.73.

Für die optischen Bestimmungen diente eine Lösung in *n*-Natronlauge.

0.0630 g Sbst. Gesamtgewicht der Lösung 2.7882 g. Prozentgehalt 2.26. Spez. Gewicht 1.037. Drehung bei 20° und Natriumlicht im 1-dm-Rohr 2.80° (± 0.02°) nach rechts.

Mithin $[\alpha]_D^{20} = +119.5°$ (± 0.8°).

0.1156 g Sbst. Gesamtgewicht der Lösung 2.9266 g. Prozentgehalt 3.95. Spez. Gewicht 1.04. Drehung bei 20° und Natriumlicht im 1-dm-Rohr 4.93° (± 0.02°) nach rechts.

Mithin $[\alpha]_D^{20} = +120.0°$ (± 0.5°).

Das *d-m*-Nitrobenzoyl-prolin schmilzt nicht ganz konstant bei 137—140°, also erheblich höher als die inaktive Verbindung. Es ist auch in Äther und Wasser schwerer löslich als diese. Aus verdünnter wäßriger Lösung krystallisiert es in mikroskopischen Prismen, die meist sternförmig angeordnet sind.

d-Prolin.

4.5 g krystallisiertes *d-m*-Nitrobenzoylprolin wurden mit 250 ccm 10-proz. Salzsäure am Rückflußkühler gekocht. Nach etwa 1 Stunde trat völlige Lösung ein, und nach 6 Stunden wurde die Hydrolyse unterbrochen. Die durch ausgeschiedene Nitrobenzoesäure getrübte Lösung wurde unter vermindertem Druck auf 50 ccm eingeengt, die Nitrobenzoesäure abfiltriert, der Rest ausgeäthert und die Lösung unter geringem Druck völlig verdampft. Dabei blieb das salzsaure Prolin als Sirup. Wir haben daraus in der üblichen Weise durch

Silbersulfat und Baryt das freie Prolin bereitet und dieses sofort ins Kupfersalz verwandelt.

Durch das lange Kochen mit Salzsäure bei der Hydrolyse war ein Teil des Prolins racemisiert, wie es auch stets bei der Spaltung der Proteine der Fall ist. Infolgedessen mußten wir das Gemisch der Kupfersalze durch Alkohol trennen. Zu dem Zweck wurde die wäßrige Lösung der Salze unter stark vermindertem Druck verdampft und der Rückstand mit 30 ccm Alkohol ausgekocht. Zur völligen Abscheidung des racemischen Salzes wurde die alkoholische Lösung unter geringem Druck verdampft und der Rückstand wieder mit 30 ccm heißem Alkohol ausgelaugt. Beim Verdampfen des alkoholischen Auszugs blieb das *d*-Prolinkupfer als völlig krystallinische Masse zurück, die ganz das Aussehen des schon bekannten krystallisierten *l*-Prolinkupfers zeigte. Die Ausbeute an reinem alkohollöslichem Kupfersalz war 1.6 g und an racemischem Salz 0.5 g. Die Gesamtmenge betrug also 84 % der Theorie, berechnet auf das angewandte Nitrobenzoylprolin.

Zur Bereitung des *d*-Prolins haben wir das reine Kupfersalz in wäßriger Lösung mit Schwefelwasserstoff zersetzt. Beim Verdampfen des farblosen Filtrats unter vermindertem Druck blieb die Aminosäure als rasch erstarrender Sirup. Sie wurde in 20 ccm warmem Alkohol gelöst und Äther bis zur beginnenden Trübung zugefügt. Das *d*-Prolin schied sich dann rasch in kleinen Prismen ab, die für die Analyse und optische Bestimmung bei 60° und unter 15 mm Druck über Phosphorpentoxyd getrocknet wurden. Da die trockne Substanz recht hygroskopisch ist, so müssen die quantitativen Bestimmungen mit großer Vorsicht ausgeführt werden.

Die Aminosäure hat keinen konstanten Schmelzpunkt, weil sie sich gleichzeitig zersetzt. Wir fanden, daß beim raschen Erhitzen im Capillarrohr die Schmelzung und Zersetzung zwischen 215—220° stattfindet.

0.1534 g Sbst.: 0.2911 g CO_2, 0.1112 g H_2O.

$C_5H_9O_2N$ (115.08). Ber. C 52.14, H 7.88.
Gef. » 51.75, » 8.05.

Für die optische Untersuchung diente die wäßrige Lösung.

0.1026 g Sbst. Gesamtgewicht der Lösung 2.6544 g. Prozentgehalt 3.865. Spez. Gewicht 1.01. Drehung bei 20° und Natriumlicht im 1-dm-Rohr 3.18° (± 0.02°) nach rechts.

Mithin $[\alpha]_D^{20} = +81.5°$ (± 0.5°).

0.1608 g Sbst. Gesamtgewicht der Lösung 3.1228 g. Prozentgehalt 5.15. Spez. Gewicht 1.015. Drehung bei 20° und Natriumlicht im 1-dm-Rohr 4.28° (± 0.02°) nach rechts.

Mithin $[\alpha]_D^{20} = +81.9°$ (± 0.4°).

0.1251 g Sbst. Gesamtgewicht der Lösung 2.7805 g. Prozentgehalt 4.50. Spez. Gewicht 1.012. Drehung bei 20° und Natriumlicht im 1-dm-Rohr 3.69° (± 0.01°) nach rechts.

Mithin $[\alpha]_D^{20} = +81.0°$ (± 0.2°).

l-Prolin.

Für seine Gewinnung haben wir die wäßrige Mutterlauge des Cinchoninsalzes benutzt, aus der das Salz des *d*-Nitrobenzoylprolins auskrystallisiert war. Zunächst wurde das Cinchonin in der gleichen Art, wie oben beschrieben, entfernt und das rohe Gemisch der Nitrobenzoylproline, in welchem die *l*-Verbindung überwiegt, und welches ein schlecht krystallisierendes Öl ist, mit Salzsäure ebenfalls in der oben angeführten Weise hydrolysiert. Auch die Isolierung des Prolins und die Trennung durch die Kupfersalze in Racemkörper und *l*-Verbindung geschahen in der gleichen Art. Auf 3 g aktives Kupfersalz traf 1.7 g Racemsalz, und diese Mengen entsprachen 25 g ursprünglichem *dl*-Nitrobenzoylprolin.

Vermischt man die alkoholischen Lösungen von *d*-Prolinkupfer, und *l*-Prolinkupfer, so fällt sofort das racemische Salz in Krystallen aus.

Das aus dem Kupfersalz isolierte *l*-Prolin schmolz ebenfalls, im Capillarrohr rasch erhitzt, unter Zersetzung zwischen 215—220° und zeigte auch die gleiche Form der Krystalle. Für die Analyse und optische Bestimmung wurde auch hier unter 15 mm Druck bei 60° getrocknet.

0.1571 g Sbst.: 0.2969 g CO_2, 0.1115 g H_2O.

$C_5H_9O_2N$ (115.08). Ber. C 52.14, H 7.88.
Gef. » 51.54, » 7.88.

Für die beiden ersten Bestimungen diente die wäßrige Lösung.

I. 0.1580 g Sbst., Gesamtgewicht der Lösung 2.4443 g, Prozentgehalt 6.46, spez. Gewicht 1.018, Drehung bei 20° und Natriumlicht im 1-dm-Rohr 5.32° (± 0.03°) nach links.

Mithin $[\alpha]_D^{20} = -80.9°$ (± 0.5°).

II. 0.1474 g Sbst., Gesamtgewicht der Lösung 3.3376 g, Prozentgehalt 4.42, spez. Gewicht 1.012, Drehung in 1-dm-Rohr 3.57° (± 0.02°) nach links.

Mithin $[\alpha]_D^{20} = -79.8°$ (± 0.5°).

Die dritte Bestimmung geschah in alkalischer Lösung.

0.0462 g *l*-Prolin gelöst in einem Gemisch von 3 Tln. *n*-Kalilauge und 2 Tln. Wasser, Gesamtgewicht der Lösung 1.9654 g, Prozentgehalt 2.35, spez. Gewicht 1.031, Drehung bei 20° und Natriumlicht im 1-dm-Rohr 2.25° (± 0.01°) nach links.

Mithin $[\alpha]_D^{20} = -93.0°$ (± 0.4°).

Für das Drehungsvermögen des *l*-Prolins aus Casein sind früher von E. Fischer[1]) ziemlich verschiedene Zahlen gefunden worden. Der höchste Wert war $[\alpha]_D^{20} = -77.40°$, aber wie er dazu bemerkte, war die Möglichkeit nicht ausgeschlossen, daß auch dieser Wert noch zu niedrig sei.

Die Untersuchung des synthetischen Präparates hat diese Vermutung in der Tat bestätigt, denn der höchste Wert beträgt hier für die wäßrige Lösung $[\alpha]_D^{20} = -80.9°$, und für das *d*-Prolin, das aus der reinen *m*-Nitrobenzoylverbindung bereitet war, wurde als höchste Zahl $[\alpha]_D^{20} = +81.9°$ gefunden.

Diese Schwankungen sind bei den synthetischen Präparaten wohl größtenteils durch die unangenehmen Eigenschaften der Aminosäure, d. h. ihre Hygroskopizität und ihre verhältnismäßig schwierige Krystallisation, bedingt. Bei den Präparaten, die aus Proteinen gewonnen werden, kommt dazu noch die Möglichkeit einer Verunreinigung des Prolins durch kleine Mengen anderer Aminosäuren.

$C_5H_9NO_2$ Ref. 114d

Solid complex of L*-proline with* L*-tartaric acid.* A mixture of L-proline (5.76 g) and L-tartaric acid (7.50 g) was dissolved in water (20 ml). To the solution, ethanol (300 ml) was gradually added and then the solution was allowed to stand overnight at room temperature. The precipitated crystals were collected, washed with ethanol, and dried to give 11.10 g (83.7%) of the solid complex of L-proline with L-tartaric acid (L-L complex). The products were almost pure, mp 153.5∼154.0°C, $[\alpha]_D^{25}$ −24.2° (c=1, water). Further purification was performed by dissolving the complex (11.00 g) in water (20 ml) and by precipitating it with the addition of ethanol (300 ml) to the solution. The yield was 9.50 g (86.4%). The specific rotation of the pure L-L complex was $[\alpha]_D^{25}$ −24.2° (c=1, water). The infrared absorption spectrum and the physical properties of L-L complex are shown respectively in Fig. 1 and Table I.

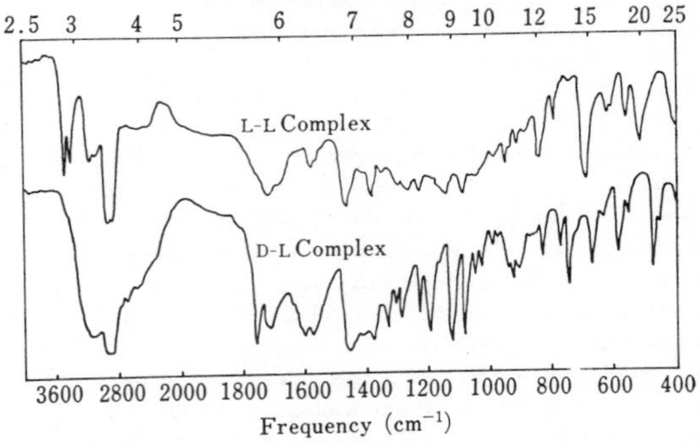

FIG. 1. Infrared Absorption Spectra of L-L and D-L Complex.

Solid complex of D*-proline with* L*-tartaric acid.* Since the solid complex of D-proline with L-tartaric acid (D-L complex) was more soluble than L-L complex, the former was prepared by the following manner. A mixture of D-proline (5.76 g) and L-tartaric acid (7.50 g) was dissolved in water (5 ml). Ethanol (35 ml) was gradually added to the warm solution and then cooled in an ice bath. The precipitated crystals were collected, washed with ethanol, and dried to give 8.17 g (61.6%) of D-L complex, mp 138.5~139.0°C, $[\alpha]_D^{25}$ +44.4° (*c*=1, water). This complex (8.00 g) was recrystallized by dissolving it in water (5 ml) and by precipitating it with the addition of ethanol (35 ml) to the solution. The yield was 4.70 g (58.8%). The specific rotation of the pure D-L complex was $[\alpha]_D^{25}$ +44.4° (*c*=1, water). The infrared absorption spectrum and the physical properties of D-L complex are shown respectively in Fig. 1 and Table I.

Optical resolution of DL*-proline by forming a new diastereoisomeric solid complex*

Procedure A. DL-Proline was resolved by using an equimolar amount of L-tartaric acid. A mixture of DL-proline (11.51 g) and L-tartaric acid (15.01 g) was dissolved in water (20 ml). Ethanol (70 ml) was gradually added to the solution under stirring and then the solution was seeded with a small amount of L-L complex. After the beginning of crystallization, ethanol (230 ml) was further added, and the mixture was stirred for 2 hr at room temperature. The precipitated crystals were collected, washed with ethanol, and dried to give 9.91 g (74.7%) of L-L complex, $[\alpha]_D^{25}$ −22.8° (*c*=1, water), optical purity 95.9%. The resolved L-L complex (9.80 g) was recrystallized by dissolving it in water (15 ml) and by precipitating it with addition of ethanol (230 ml) to the solution to give 8.31 g (84.8%) of pure L-L complex, $[\alpha]_D^{25}$ −24.2° (*c*=1, water). After the separation of less soluble L-L complex in the above resolution process, the mother liquor was evaporated *in vacuo* to dryness. The resulting residue was dissolved in water (5 ml). Ethanol (120 ml) was gradually added to the warm solution and the solution was stirred overnight at room temperature. The precipitated crystals were collected, washed with ethanol, and dried to give 10.70 g (80.7%) of optically impure D-L complex, $[\alpha]_D^{25}$ +31.4° (*c*=1, water), optical purity 62.1%.

Procedure B. The effect of the amount of resolving agent was investigated by changing its ratio to DL-proline. As a result, the use of 0.5 equimolar amount of the resolving agent was found to be suitable for practical resolution. A mixture of DL-proline (11.51 g) and L-tartaric acid (7.50 g) was dissolved in water (5 ml). Ethanol (30 ml) was gradually added to the solution under stirring. The solution was seeded with a small amount of L-L complex. After the beginning of crystallization, ethanol (45 ml) was further added and the mixture was allowed to stand overnight at room temperature. The precipitated crystals were collected, washed with ethanol, and dried to give 9.40 g (70.9%) of L-L complex, $[\alpha]_D^{25}$ −23.3° (*c*=1, water), optical purity 97.4%. The resolved L-L complex (9.00 g) was recrystallized by dissolving it in water (13 ml) and by precipitating it with addition of ethanol (200 ml) to the solution to give 7.70 g (85.6%) of pure L-L complex, $[\alpha]_D^{25}$ −24.2° (*c*=1, water).

Preparation of L*-proline.* Optically pure L-L complex (7.00 g) obtained by the above procedures was dissolved in water (30 ml). The solution was passed through a column of Amberlite IR-120 (40 ml, H^+ form). The column was washed with water and L-proline was eluted from the column with 5% NH_4OH (70 ml). The eluates were concentrated, treated with charcoal, and concentrated again to dryness. The residual crystals were dissolved in methanol (8 ml) and acetone (50 ml) was added to the solution at 5°C. The precipitated crystals were collected, washed with acetone, and dried to give 2.80 g (92.1%) of L-proline, $[\alpha]_D^{25}$ −85.5° (*c*=1, water), optical purity 100.0%.[9] *Anal.* Calcd for $C_5H_9NO_2$: C, 52.16; H, 7.88; N, 12.17. Found: C, 51.95; H, 7.90; N, 12.06.

Recovery of optically impure D*-proline from mother liquor.* After the separation of less soluble complex in procedure B of the above resolution process, the mother liquor was evaporated *in vacuo* to remove ethanol. The resulting residue was dissolved in water (50 ml) and passed through a column of Amberlite IR-120 (100 ml, H^+ form). The column was washed with water and eluted with 5% NH_4OH (200 ml). The eluates were concentrated, treated with charcoal, and concentrated again to dryness to give 7.00 g of optically impure D-proline, $[\alpha]_D^{25}$ +45.7° (*c*=1, water), optical purity 53.5%. *Anal.* Found: C, 51.96; H, 7.93; N, 12.05.

TABLE I. PROPERTIES OF L-L AND D-L COMPLEX

Solid complex (elemental composition)	Elemental anal., % Calcd.	Found	Mp, °C	$[\alpha]_D^{25}$, deg (*c*=1, water)	Solubility, g/100 ml[a] 20°C	30°C
L-L Complex ($C_9H_{15}NO_8$)	C 40.75 H 5.70 N 5.28	40.61 5.80 5.33	154.0~154.5	−24.2	0.6	0.9
D-L Complex ($C_9H_{15}NO_8$)	C 40.75 H 5.75 N 5.28	40.86 5.84 5.23	138.5~139.0	+44.4	1.7	2.4

[a] Solubility was determined in aqueous ethanol (water: ethanol; 1: 15, v/v).

Racemization of optically impure D*-proline.* Optically impure D-proline (6.00 g, optical purity 53.5%) obtained by the above procedure was dissolved in water (120 ml) containing an equimolar amount of sodium hydroxide (2.09 g). The mixture was heated in an autoclave at 170°C for 4 hr. The reaction mixture was passed through a column of Amberlite IRC-50 (50 ml, H^+ form). The column was washed with water. The effluent was concentrated, treated with charcoal, and concentrated again to dryness to give 5.82 g of DL-proline, $[\alpha]_D^{25}$ +0.2° (*c*=1, water). *Anal.* Found: C, 51.79; H, 7.90; N, 12.08.

9) The optical purity was calculated with the assumption that the specific rotation of the pure sample is $[\alpha]_D^{25}$ +85.5° (*c*=1, water).[2] [Lit. for D-proline: $[\alpha]_D^{23}$ +85.2° (*c*=3.5, water).[5] $[\alpha]_D^{23}$ +86.2° (*c*=1, water)[10]].

Authors' comments:

During the investigation of the behavior of DL-amino acids in the solutions containing a chiral compound or in the solvents having chirality, we unexpectedly found that DL-proline itself is capable of forming new diastereoisomeric solid complexes with L-tartaric acid in a molar ratio of 1 : 1 on crystallization from an aqueous ethanol solution. The complex of L-proline was less soluble than that of D-proline, and the difference of solubility between the two diastereoisomeric complexes was sufficient to perform the chemical resolution. In fact, DL-proline could be resolved in a good yield by separating the less soluble complex from the mixture of DL-proline and an equimolar amount of L-tartaric acid. The resolution was also achieved by the use of 0.5 equimolar amount of the resolving agent and this was advantageous for the practical production of L-proline. The optically impure D-proline recovered from mother liquor was racemized by heating and was reused for optical resolution.

$C_5H_9NO_2$ Ref. 114e

Preparation of Amidase—Fresh, frozen hog kidney was thawed and defatted and then ground in a Waring blendor with 3 times its weight of cold distilled water. The extract, after centrifugation at 2500 r.p.m., hydrolyzed a 0.05 M solution of DL-prolinamide at pH 8.0 at a rate of 0.5 μM per hour per mg. of N. The extract at pH 6.8 was brought to −7° and alcohol concentration of 15 per cent and centrifuged in the Sharples centrifuge. The precipitate was discarded. The supernatant fluid was brought to pH 5.7 and again centrifuged, the precipitate being discarded. The fluid was then adjusted to −15° and an alcohol concentration of 30 per cent and centrifuged, the precipitate again discarded. The pH of the supernatant fluid was lowered to 4.6, and the sediment, which is the active enzyme, was centrifuged and suspended in water at pH 6.8 ready for use. The activity of this preparation toward DL-prolinamide was about 2 μM per hour per mg. of N, or roughly a 4-fold concentration in activity over the crude extract. However, it was noted that the rate of hydrolysis with this preparation markedly increased with increase in concentration of the substrate up to 0.25 M, at which the rate was approximately 19 μM per hour per mg. of N (Fig. 1). The optimum pH for the hydrolysis of the amide was at about 8.0, as shown in Fig. 2. Finally, tests on the prolonged incubation of DL-prolinamide with an excess of this kidney enzyme preparation showed that the reaction went completely to 50 per cent hydrolysis of the racemate, corresponding to complete hydrolysis of the L-amide and no action on the D-amide. It was therefore considered that this enzyme acting upon a 0.25 M solution of DL-prolinamide at pH 8 would be a suitable agent for the resolution.

Resolution of DL*-Prolinamide*—28.5 gm. of DL-prolinamide were dissolved in water and brought to pH 8.0 by addition of acetic acid. 300 cc. of enzyme preparation (about 3.0 mg. of N per cc.), derived from 3.5 kilos of hog kidneys, were added; the mixture was made up to 1 liter with water and incubated at 38°, the progress of the hydrolysis being followed on aliquots at intervals (*a*) by determination of the ammonia produced and (*b*) by determination of carboxyl nitrogen liberated by the Van Slyke ninhydrin-CO_2 procedure. It was found that the hydrolysis was nearly complete in 24 to 30 hours of incubation. At this stage, 50 to 100 cc. more of the enzyme preparation were added, and the digestion was allowed to continue for about another 30 hours. The course of a typical hydrolysis is shown in Fig. 3. The ninhydrin estimations generally read slightly higher than 50 per cent because of increase in the blank value of the enzyme. The results for ammonia generally are a little low, owing to loss of this base from the solution. There was, however, no detectable change in either of these values in the last 24 hours of incubation, and the reaction

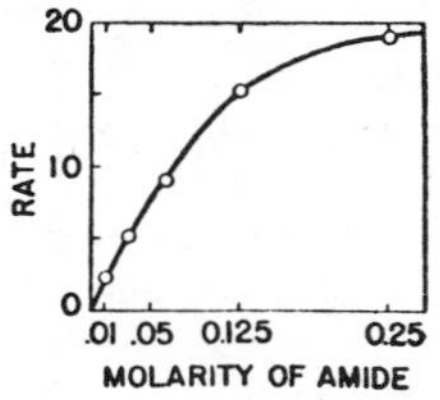

FIG. 1

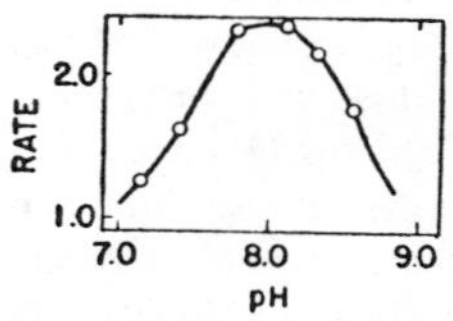

FIG. 2

FIG. 1. Effect of increasing substrate concentration on the rate of hydrolysis of DL-prolinamide by a hog kidney amidase preparation at pH 8.0.

FIG. 2. Effect of pH on the hydrolysis of DL-prolinamide by a hog kidney amidase preparation.

could be considered complete after 50 to 60 hours. The incubation mixture was brought to pH 5 with acetic acid and deproteinized with norit. The clear filtrate contains L-proline, ammonia, and D-prolinamide.

Separation of Isomers—The deproteinized solution was evaporated *in vacuo* to 300 cc. and brought to pH 10.5 by addition of saturated potassium carbonate solution. A vigorous stream of nitrogen was blown through the solution for several hours until all of the ammonia was removed. The solution was then chilled in ice and shaken with successive amounts of 60 to 70 gm. of fresh carbobenzoxy chloride. This treatment took 40 minutes, during which period a white, oily product separated. This was the carbobenzoxy-D-prolinamide. Addition of a small amount of ether generally brought the product to crystallization.[4] It was filtered with suction and washed with cold water. The filtrate and washings were combined and washed with ether, the ether discarded, and the aqueous layer acidified to pH 1.5 with concentrated HCl. A white, oily emulsion of carbobenzoxy-L-proline appeared. This was extracted into ether, and the compound was obtained as a viscous liquid after drying the solution and evaporating the solvent. The dried oil weighed about 10 to 12 gm. and, if pure, would correspond to 35 per cent of theory. The product has thus far resisted every effort to bring it to crystallization.

The precipitate of carbobenzoxy-D-prolinamide was dried and recrystallized from ethyl acetate as needles. The yield was 6 to 8 gm. or about 20 to 25 per cent of theory. In three successive resolutions, the melting point of this product was 93°, 94°, and 94°; $[\alpha]_D$ at 23° for 2 per cent solutions in ethanol, +31.0°, +33.6°, and +33.9°, respectively; N found, 11.2 per cent for all three. The lower value of the first product may have been

[4] When an unusually large excess of carbobenzoxy chloride was used, the addition of much ether at this point caused the carbobenzoxy-D-prolinamide to pass into solution. Evaporation of the ether solution resulted in crystallization of the pure compound.

C5H9NO2

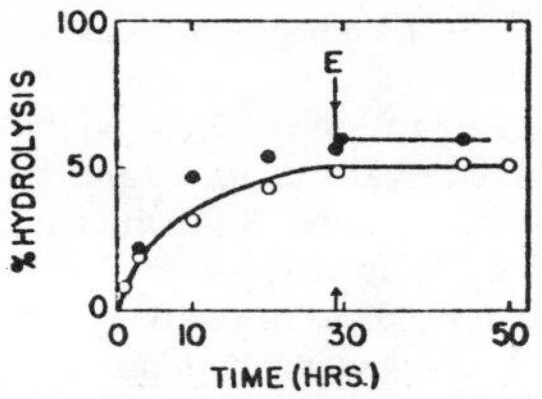

FIG. 3. Time-course of the hydrolysis at pH 8.0 of DL-prolinamide by a hog kidney amidase preparation. Substrate concentration 0.25 M. ●, van Slyke determinations; ○, ammonia determinations; E, point at which additional enzyme added.

due to incomplete removal of ammonia. It nevertheless yielded a pure specimen of D-proline.

The carbobenzoxy-D-prolinamide samples were dissolved in 10 times their weight of methanol, about 2 equivalents of 1 N HCl were added, and the solution was treated with hydrogen in the presence of palladium black. When the reaction was over, the solution was filtered and evaporated to dryness *in vacuo*. The residue was dissolved in 10 times its weight of 1 N HCl, refluxed for 2 hours, and evaporated to dryness *in vacuo*. The residue was dissolved in 50 times its weight of distilled water, and the solution treated carefully and shaken with solid silver carbonate in slight excess. When there was no further reaction, the silver salts were filtered and washed with cold water. The combined filtrate and washings were saturated at 25° with H₂S, and the silver sulfide was removed by filtration with the aid of norit. The filtrate was evaporated to dryness *in vacuo* and the solid residue dissolved in boiling alcohol. After filtration and cooling, the alcoholic solution deposited crystalline D-proline in a yield of nearly 50 per cent of theory, based on the carbobenzoxamide. Two preparations made in this way yielded $[\alpha]_D$ for 2 per cent solutions in water at 23° = +86.1° and +86.2°. N calculated 12.2, found 12.2, 12.2.

The oily carbobenzoxy-L-proline was dissolved in methanol, treated with a few drops of glacial acetic acid, and catalytically hydrogenated with palladium black as above. The solution was evaporated to dryness *in vacuo*, and the residue crystallized twice from alcohol. Two preparations made in this way yielded $[\alpha]_D$ for 2 per cent solutions in water at 23° = −85.8° and −85.9°. The yields were about 30 per cent of theory, based upon DL-prolinamide taken. N found 12.2, 12.2.[5]

When the ammonia was not completely removed prior to the carbobenzoxylation, 16 to 22 gm. of carbobenzoxyprolinamide were isolated (50 to 75 per cent of theory), $[\alpha]_D$ ranging from almost 0° to +25°, and N in every case close to the theoretical of 11.3 per cent. When one of these products with nearly zero optical rotation and melting point of 115° was catalytically hydrogenated and worked up as described, proline was readily obtained and possessed an optical rotation close to zero; N found 12.2. The carbobenzoxyprolinamide taken was unquestionably racemic. It was also observed in several instances that the yield of carbobenzoxy-L-proline was generally lower the lower the optical rotation of the carbobenzoxyprolinamide product. However, the carbobenzoxyproline obtained in these cases invariably yielded practically pure L-proline on catalytic hydrogenation. With two resolutions in which completely racemic carbobenzoxyprolinamide was obtained, no carbobenzoxy-L-proline could be isolated. It would appear that the optically inactive carbobenzoxyprolinamide was formed in part at the expense of the L-proline in the medium rather than by a direct racemization of the D-prolinamide by carbobenzoxy chloride. The yields of racemic carbobenzoxyprolinamide should thereby be higher than those of pure carbobenzoxy-D-prolinamide and this was indeed so; *e.g.*, 50 to 75 per cent for the former and 20 to 25 per cent for the latter. It seems likely, therefore, whatever the explanation may be, that the critical step for the preparation of the optically pure D enantiomorph, and of adequate amounts of the L enantiomorph, is the complete removal of ammonia prior to carbobenzoxylation.

[5] A sample of the L-proline obtained by the resolution was subjected to the action of D-amino acid oxidase by the method described (18). A maximum of 2 μM of the thousand employed in the test was oxidized by the preparation, corresponding to an amount of D-proline contamination no greater than 0.2 per cent. All of the commercially available specimens of L-proline (Nutritional Biochemicals Corporation) which formed the starting material for the resolution contained this same proportion of material oxidizable by D-amino acid oxidase. Whether this material, whatever its nature, was carried through the resolution procedure and remained with the L enantiomorph, or whether the fact that carbobenzoxy-L-proline could only be isolated from the resolution mixture as a non-crystallizable oil, and hence might not have been completely separated from traces of the enantiomorph, cannot be decided at this time. The analogous test for the presence of possible L-proline contamination in the D-proline preparations could not be performed because of the inability of snake venom to oxidize L-proline at a suitable rate.

For additional enzymatic methods for the resolution of proline, see V. E. Price, L. Levintow, L. Greenstein, J. P. Greenstein and R. B. Kingsley, Arch. Biochem., <u>26</u>, 92 (1950), and D. Hamer and J. P. Greenstein, J. Biol. Chem., <u>193</u>, 81 (1951).

$$\text{CH}_2=\text{CH}-\underset{\underset{\text{NH}_2}{|}}{\overset{\overset{\text{CH}_3}{|}}{\text{C}}}-\text{COOH}$$

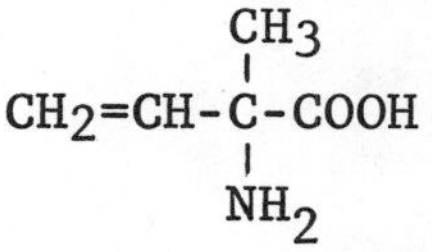

C5H9NO2   Ref. 115

(−)-1-Menthyl 2-Acetamido-2-methyl-3-butenoate ((−)-XIV)——To a suspension of Na-powder (6.6 g., 0.287 mole) in anhyd. benzene (400 ml.) was added l-menthol (35.8 g., 0.229 mole). The reaction mixture was kept standing overnight avoiding moisture and then refluxed for 2 hr. Unreacted Na-powder was decanted off, and washed with benzene (50 ml.). To the combined solution of supernatant and the washings was added a solution of DL-XIII (26.6 g., 0.191 mole) in anhyd. benzene (50 ml.). The whole was stirred for 9 hr. at room temperature and then kept standing overnight to give a clear orange-yellow solution, which was washed successively with 10% AcOH solution (300 ml. × 2), H₂O (300 ml.), 2.5% Na₂CO₃ solution (300 ml.) and H₂O (300 ml.), and then dried over Na₂SO₄. An evaporation of the solvent gave a white solid, which was dissolved in benzene-hexane on warming. The solution was seeded and kept standing to crystallize out crude (−)-XIV (32.5 g.), m.p. 112~135°, $[\alpha]_D^{25}$ −55.8° (C=0.850, EtOH). Seven recrystallizations from benzene-hexane, until specific rotation remained unchanged, afforded pure (−)-XIV as colorless small needles, m.p. 152~152.5°, $[\alpha]_D^{25}$ −69.0° (C=0.664, EtOH). *Anal.* Calcd. for C₁₇H₂₉O₃N : C, 69.11; H, 9.90; N, 4.74. Found : C, 69.27; H, 9.79; N, 4.85. IRν_{max}^{KBr} cm⁻¹ 3260, 1740, 1640, 1553, 1125, 987, 916.

(+)-2-Amino-2-methyl-3-butenoic Acid ((+)-XI) and its Hydantoin Derivative (−)-V——A mixture of (−)-XIV (6.0 g., m.p. 151.5~152.5, $[\alpha]_D^{25}$ −70.0° (C=0.980 in EtOH)) and conc. HCl (45 ml.) in H₂O (45 ml.) was refluxed for 9 hr. After addition of H₂O (45 ml.), the solution was extracted with benzene (50 ml. × 2) to remove l-menthol. Aqueous layer was evaporated to dryness to give a yellowish white solid. This solid was dissolved in H₂O (20 ml.) and poured through the ion exchanger column (Amberlite IR-120, H⁺-form, 100 ml.). The column was washed with H₂O until eluates became neutral, and then eluted with dil. NH₄OH until ninhydrin-test became negative. The eluates were combined and evaporated to dryness to give crude (+)-XI as a pale brown solid (2.0 g.), m.p. 231~232° (decomp. with sublimation). A part of this (+)-XI was recrystallized three times from H₂O-EtOH to give colorless crystals, which was dried completely at 40~50° *in vacuo* to give an analytical sample, m.p. 264° (decomp. with sublimation), $[\alpha]_D^{25}$ +33.0° (C=0.612, H₂O).

This sample was hygroscopic. *Anal.* Calcd. for $C_5H_9O_2N$: C, 52.16; H, 7.88; N, 12.17. Found : C, 51.41; H, 8.02; N, 11.94. IRν^{KBr}_{max} cm^{-1} : 3045, 1664, 1610, 1537, 1360, 998, 939. This IR spectrum was different from that of DL-XI.

A solution of another part of crude (+)-XI (1.2 g.) and potassium cyanate (4.0 g.) in H_2O (30 ml.) were refluxed for 2 hr. The cooled solution was acidified with conc. HCl and refluxed again 30 min., and then extracted with ethyl acetate (40 ml. × 3). The combined extracts were washed with satd. NaCl (40 ml.) and dried over Na_2SO_4. An evaporation of the solvent gave (−)-V as a white solid (0.33 g.), m.p. 162~164°, $[\alpha]^{23}_D$ −35.2° (C=0.762, EtOH). Twice recrystallizations from ethyl acetate afforded colorless needles, m.p. 165~166°. $[\alpha]^{27}_D$ −37.3° (C=0.782, EtOH). The mixed melting point with the authentic (−)-V showed no depression. *Anal.* Calcd. for $C_6H_8O_3N_2$: C, 51.42; H, 5.75; N, 19.99. Found : C, 51.77; H, 5.78; N, 20.17. IRν^{KBr}_{max} cm^{-1} : 3220, 1772, 1740, 1713, 1638, 1427, 994, 926. This IR spectrum was superimposable with that of the authentic (−)-V.

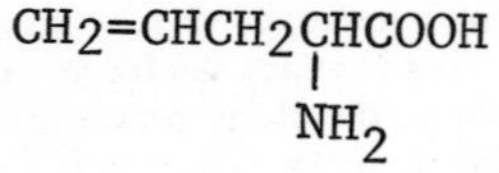

$C_5H_9NO_2$ Ref. 116

Note:

An enzymatic resolution of acetyl-DL-allylglycine which yielded acetyl-D-allylglycine with $[\alpha]_D$ -42.0° (ethanol) as the compound isolated may be found in S. Black and N.G. Wright, J. Biol. Chem., 213, 39 (1955).

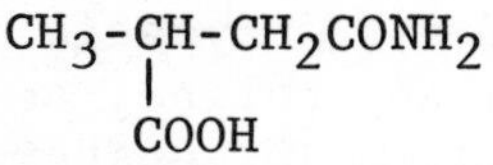

$C_5H_9NO_3$ Ref. 117

Resolution of β-Carboxy-*n*-butyramide.—A solution of 17.4 g. of β-carboxy-*n*-butyramide and 52.2 g. of anhydrous brucine in 700 ml. of methanol was permitted to stand for 24 hours at 15°. The crystalline precipitate was separated by filtration and was recrystallized from methanol (10 ml. of methanol per 1 g. of the salt). After five crystallizations, 13.5 g. of shining prismatic crystals resulted, m.p. 181–182°; rotation, 0.2315 g. made up to 5 ml. with methanol at 20° gave α_D −2.01°, *l* 2; $[\alpha]^{20}_D$ −21.7°.

No attempt was made to isolate the more soluble brucine salt from the mother liquors.

A solution of 13.5 g. of less soluble salt in 50 ml. of chloroform was treated with 40 ml. of a saturated aqueous solution of sodium bicarbonate. The aqueous layer was extracted with three 30-ml. portions of chloroform and neutralized with 10% hydrochloric acid until slightly acidic to congo. The water was removed under diminished pressure and the residue extracted with three 20-ml. portions of absolute ethanol. Evaporation of the ethanol gave 2.8 g. of crude product which was crystallized from 10 ml. of absolute ethanol (carbon). A yield of 1.4 g. resulted. Two more crystallizations from ethanol gave pure amide, m.p. 132–133.5°; rotation, 0.1044 g. made up to 5 ml. with absolute ethanol at 25° gave α_D −0.89°, *l* 2; $[\alpha]^{25}_D$ −21.4 ± 0.5°.

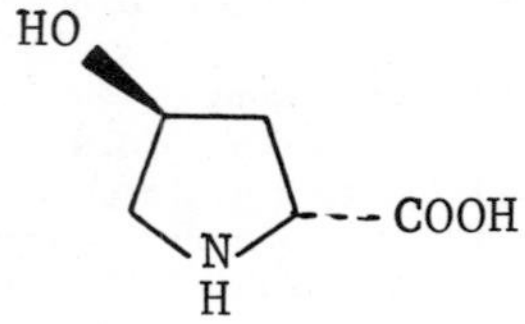

$C_5H_9NO_3$ Ref. 118

EXAMPLE 2.

Resolution and preparation of D- and L-hydroxyprolines.

a) Formation of the salt of N-3,5-dinitrobenzoyl-L-hydroxyproline with L-(+)-*threo*-1-*p*-nitrophenyl-2-aminopropane-1,3-diol.

20 g. of N-3,5-dinitrobenzoyl-D,L-hydroxyproline are dissolved in 160 cc. of dioxane containing 5% of water at 80° C. and 13.5 g. of L-(+)-*threo*-1-*p*-nitrophenyl-2-aminopropane-1,3-diol are added at this temperature. Cooling to 25° C. is effected and the resulting salt is filtered with suction while stirring; the salt is then washed twice with 10 cc. of dioxane containing 5% of water. The hydrated salt melts at 140—141° C, $[\alpha]_D^{20} = -78° \pm 2$ (1% concentrate in water). The salt of N-3,5-dinitrobenzoyl-L-hydroxyproline and L-(+)-*threo*-1-*p*-nitrophenyl-2-aminopropane-1, 3-diol has not yet been described. The yield of the dried product is 14.7 g (89%).

b) Preparation of N-3,5-dinitrobenzoyl-L-hydroxyproline.

14.7 g. of the above salt are treated with 32 cc. of normal sodium hydroxide solution. Dissolution of the product takes place and then crystallisation of L-(+)-*threo*-1-*p*-nitrophenyl-2-aminopropane-1,3-diol which is filtered with suction and washed with a little water. In this way 4.8 g. (35.5%) of the resolution reagent are recovered. The filtrate obtained as above is acidified by means of 3.5 cc. of concentrated hydrochloric acid, saturated with sodium chloride, and the resulting material is extracted four times with 15, 10, and 5 cc respectively of ethyl acetate. The organic solution is washed with a little water, drying is effected over magnesium sulphate and then removal of solvent by evaporation. 8.3 g. of N-3,5-dinitrobenzoyl-L-hydroxyproline (83%) (based on the N-3,5-dinitrobenzoyl-DL-hydroxyproline used as a starting material) are obtained. $[\alpha]_D^{20} = -147° \pm 2$ (1% concentration in 50% ethanol). This product is new.

c) Separation of L-hydroxyproline.

A solution of 2.5 g. of N-3,5-dinitrobenzoyl-L-hydroxyproline is heated for 40 minutes at reflux in 25 cc. of 5N hydrochloric acid. Filtration is effected, evaporation to dryness in a vacuum and crystallisation from acetone of the resulting hydrochloride of L-hydroxyproline. This material is taken up in 1 volume of water, 0.7 cc. of aniline are added and then 12 cc of boiling absolute alcohol. In this way, after cooling, 0.87 g (87%) of crystalline L-hydroxyproline are obtained, melting point 273—274° C, $[\alpha]_D^{20} = -77° \pm 1$ (1% concentration in water).

d) Separation of D-hydroxyproline.

There is evaporated to dryness the solution from which the salt of N-3,5-dinitrobenzoyl-L-hydroxyproline with L-(+)-*threo*-1-*p*-nitrophenyl-2-aminopropane-1,3-diol has crystallised and the resulting material is taken up in a little water. By treating this concentrated aqueous solution with 40 cc. of normal sodium hydroxide solution there are recovered as indicated at b) above 7.2 g. (53%) of the resolution reagent. The filtrate is saturated with sodium chloride and 5 cc. of cencentrated hydrochloric acid are added and extraction with ethyl acetate is effected as at b) above. The organic solution is concentrated to 20 cc. and by cooling 1.5 g (7.5%) of N-3,5-dinitrobenzoyl-DL-proline are recovered which may be used for a fresh resolution. Filtration is effected, then evaporation to dryness and 9.5 g. (95%) of N-3,5-dinitrobenzoyl-D-hydroxyproline are obtained which are contaminated with a small quantity of racemate; $[\alpha]_D = +115° \pm 2$ (1% concentration in 50% ethanol). The corresponding D-hydroxyproline is obtained by hydrolysis with hydrochloric acid effected in the manner indicated at c) for L-hydroxyproline.

$C_5H_9NO_3$ Ref. 118a

3,5-Dinitrobenzoyl-D-hydroxy-prolin. 3,5-dinitrobenzoyl-DL-hydroxy-prolin (14 g) and 20,02 g of brucin.4H_2O was dissolved in 250 ml of hot water, filtered, allowed to cool, a seed crystal added and allowed to stand for a day at room temperature. The mother liquor decanted carefully from the settled crystals forming yellow platelets, suspended in 40 ml of ethanol and filtered (16,2 g). On recrystallization from 130 ml of hot water afforded 13 g (84 %) of 3,5-dinitrobenzoyl-D-hydroxy-prolin brucinate, m. p. 147—150° C. $[\alpha]_D^{20} = +54°$ (c:0,500; abs. ethanol).

The obtained amount (13 g) of D-brucinate was dissolved in 55 ml of 1,0 N sodium hydroxyde, cooled, the precipitated brucin base filtered and the brucin extracted with 2 × 10 ml of 1,0 N sodium hydroxide. On acidifying the aqueous phase with 12 ml of concentrated hydrochloric acid, extracting the acid solution with 6 × 100 ml portions of ethylacetate, drying the organic phase and removing the solvent by distillation, 3,5-dinitrobenzoyl-D-hydroxy-prolin was obtained in form of a solid foam. Yield: 5,16 g (89 %). $[\alpha]_D^{20} = +147°$ (c:1,004; 50 % ethanol).

D-Hydroxy-prolin (II). 3,5-dinitrobenzoyl-D-hydroxy-prolin was refluxed with 90 ml of 5 N hydrochloric acid, allowed to stand for 4 hours, the separated 3,5-dinitro-benzoic acid collected and the filtrate extracted with 2×120 ml portions of ether. Having evaporated the decolorized aqueous phase, under reduced pressure to dryness, the residue was dissolved in 10 ml hot water, then decolorized again and to the clear hot solution the mixture of 50 ml of abs. ethanol and 3 ml of fresh distilled anilin was added. After standing for a day at 0°C the separated crystals filtered, washed with waterfree ethanol and ether and dried, yield 2,67 g, m. p. 265—266°C (decomp.). The obtained substance was dissolved in 9 ml of hot water, filtered and crystallized by adding of 40 ml of ethanol, allowed to stand for 3 hours at 0°C, filtered and washed with waterfree ethanol and ether, yield 2,26 g of pure D-hydroxy-prolin (65 %), m. p. 274—275°C (decomp.). $[\alpha]_D^{20} = +77°$ (c:1,001; H_2O) [2], [3].

3,5-Dinitrobenzoyl-L-hydroxy-prolin. Mother liquor obtained from resolvation combined with the washing alcohol of D-brucinate was inoculated with 3,5-dinitrobenzoyl-L-hydroxy-prolin brucinate and the solution was kept for 8 hours at 0°C. The separated crystals were filtered without washing and dried, yield 15,27 g. Recrystallization of the filtered product from 120 ml of water after cooling for 8 hours gave 12,4 g (80 %) pure, yellow colored 3,5-dinitrobenzoyl-L-hydroxy-prolin, m.p. 95—108°C. $[\alpha]_D^{20} = -50°$ (c:1,020; abs. ethanol).

The obtained amount (12,4 g) of L-brucinate was treated further as described under the heading of the D-modification, to afford 5,14 g (93%) of solid foam 3,5-dinitrobenzoyl-L-hydroxy-prolin. $[\alpha]_D^{20} = -145°$ (c:1,112; 50 % ethanol).

L-Hydroxy-prolin (II). Hydrolysis with 5 N hydrochloric acid of the product obtained in the previous step was fulfilled as described under the heading of the D-modification, to afford 1,72 g crude L-hydroxy-prolin, m. p. 264—265°C (decomp.). The substance was dissolved in 5 ml of hot water, filtered and crystallised by adding of 30 ml of ethanol, allowed to stand for 8 hours at 0°C, filtered and washed with waterfree ethanol and ether, yield 1,34 g of pure L-hydroxy-prolin (61 %), m. p. 273—274°C (decomp.). $[\alpha]_D^{20} = -76°$ (c:1,001; H_2O) [1], [4].

[1] *Fischer, E., G. Zemplén:* Ber. **42**, 2989 (1909).
[2] *Velluz, L., G. Amiard, R. Heynés:* Bull. Soc. Chim. France 1015 (1954).
[3] *Leuchs, H., J. F. Brewer:* Ber. **46**, 986 (1913).
[4] *Leuchs, H., K. Bormann:* Ber. **52**, 2086 (1919).
[5] *Albertson, M. F., J. L. Fillman:* J. Amer. Chem. Soc. **71**, 2818 (1949).
[6] *Gaudry, R., C. Godin:* J. Amer. Chem. Soc. **76**, 139 (1954).

$C_5H_9NO_4$ HOOC-CH(NH_2)-CH_2CH_2COOH Ref. 119

500 g feines entfettetes Bucheckernmehl aus geschälten Bucheckern werden mit 1500 ccm konz. Salzsäure (37-proz.) übergossen und im Verlauf 1 Stde. häufig geschüttelt, bis die Hauptmenge in Lösung gegangen ist. Nach 20-stdg. Kochen am Rückflußkühler wird die Lösung auf ca. 2500 ccm verdünnt und über Aktivkohle filtriert. Das hellbraune Filtrat wird im Vakuum bis zu einem dünnflüssigen Sirup eingeengt und bis zur Sättigung unter Eiskühlung HCl eingeleitet. Nach 3-tägigem Stehenlassen bei 0° ist die Hauptmenge des Glutaminsäure-hydrochlorids auskristallisiert. Nach dem Abnutschen und Umkristallisieren aus 20-proz. Salzsäure werden 35 g reines Glutaminsäure-hydrochlorid gewonnen. Die Darstellung der freien Glutaminsäure erfolgt nach der Vorschrift von Anslow u. King[12] durch Behandlung des Hydrochlorides mit Anilin und Äthylalkohol.

See also Ref. 1484

[12] W. K. Anslow u. H. King, Biochem. J. **21**, 1172 [1927].

12,9 g *d, l*-Pyrrolidoncarbonsäure vom Schmp. 163° (durch Hitzeracemisation von *l*(+)-Glutaminsäure nach Arnow-Opsahl erhalten) werden in 120 ccm Wasser unter portionsweiser Zugabe von 32,4 g wasserfreien Chinins, Schmp. 175°, in der Hitze gelöst. Nach 12 stdg. Stehenlassen unter Eiskühlung wird das noch nicht ganz optisch reine Chininsalz der *d*-Pyrrolidoncarbonsäure (20 g, Schmp. 211°, α_D^{20} — 114°) abgenutscht und mit wenig Aceton gewaschen. Das abfiltrierte Chininsalz wird in 50 ccm Wasser unter Erwärmen gelöst und 5 Stdn. unter Eiskühlung aufbewahrt. Das ausgeschiedene reine *d*-pyrrolidoncarbonsaure Chininsalz (Schmp. 115—116°, α_D^{20} — 118,8°, 15,4 g) wird abgenutscht, in 200 ccm Wasser unter Zusatz von 1 ccm konz. Salzsäure gelöst und mit einer in der Kälte gesättigten Lösung von 31 g Pikrinsäure in Methylpropylketon ausgeschüttelt. Etwas ausgefallenes Chininpikrat wird abfiltriert, ausgewaschen, die wässrige Lösung bis zur fast vollständigen Entfärbung mit Methylpropylketon ausgeschüttelt und im Vakuum unter Zusatz von 20 ccm konz. Salzsäure zur Trockne eingeengt. Der Rückstand wird in wenig eiskalter konz. Salzsäure aufgenommen, das reine *d*(—)-Glutaminsäure-hydrochlorid abfiltriert und auf dem Filter zur Entfernung überschüssiger Pikrinsäure mit trockenem Aceton gewaschen. Ausb. 6,0 g (65% d. Th.).

$$\alpha_D^{20} = \frac{100 \cdot 2,25}{1 \cdot 7} = -31,6° \text{ in 9-proz. HCl.}$$

Die Mutterlauge der ersten Fraktion wird auf 400 ccm mit Wasser unter Zusatz von 2 ccm konz. Salzsäure verdünnt und mit einer gesättigten Lösung von 52 g Pikrinsäure in Methylpropylketon ausgeschüttelt. Wie oben wird das ausgefallene Chininpikrat abfiltriert, gewaschen und die wässrige Lösung unter Zusatz von 40 ccm konz. Salzsäure im Vakuum eingeengt. Der Rückstand wird 3-mal mit einem Materialverlust von ca. 20% aus konz. Salzsäure umkristallisiert und zur Entfernung von Pikrinsäure mit Aceton auf dem Filter gewaschen. Ausb. 3,4 g reines *l*(+)-Glutaminsäure-hydrochlorid (37% d. Th.).

$$\alpha_D^{20} = \frac{100 \cdot 2,04}{1 \cdot 6,5} = +31,4° \text{ in 9-proz. HCl.}$$

$C_5H_9NO_4$ Ref. 119a

Spaltung der racemischen Benzoylglutaminsäure in die optischen Componenten.

Obschon die beiden Brucinsalze sehr schön krystallisiren, so sind sie doch für den vorliegenden Zweck wenig geeignet. Bessere Dienste leistet das neutrale Strychninsalz.

Zur Bereitung desselben werden 35 g wasserhaltige *r*-Benzoylglutaminsäure mit 87 g (2 Mol.) gepulvertem Strychnin in 450 ccm heissem Wasser gelöst. Nach dem Erkalten beginnt langsam die Krystallisation. Nach 20-stündigem Stehen im Eisschrank wird die Masse abgesaugt und wieder in 500 ccm heissem Wasser gelöst. Die Krystallisation erfolgt jetzt in der Kälte rascher, sodass nach 6-stündigem Stehen im Eisschrank filtrirt werden kann. Man wiederholt die Krystallisation aus der gleichen Menge Wasser noch zweimal, wobei es vortheilhaft ist, beim Auflösen einige Gramm gepulvertes Strychnin der Flüssigkeit zuzusetzen und dann zu filtriren. Bei jeder Operation werden die Krystalle des Salzes schöner, zur völligen Reinigung ist es aber nöthig, dieselben noch zweimal umzukrystallisiren, wozu wegen der geringeren Menge 400 ccm Wasser ausreichen. Das Salz bildet schliesslich feine, farblose, lange, schmale Blättchen, welche vielfach kugelförmig verwachsen sind. Die Ausbeute betrug zum Schluss 21 g, während theoretisch 60 g trocknes, actives, neutrales, benzoylglutaminsaures Strychnin der Menge der racemischen Verbindung entsprachen.

Die dem Salze zu Grunde liegende Benzoylverbindung entspricht der *l*-Glutaminsäure und ich bezeichne sie deshalb als

Benzoyl-*l*-glutaminsäure.

Zur Darstellung der freien Säure werden 20 g des Salzes in 400 ccm heissem Wasser gelöst, mit 48 ccm Normal-Kalilauge versetzt und das abgeschiedene Strychnin nach guter Abkühlung filtrirt. Die Mutterlauge wird mit 52 ccm Normal-Salzsäure angesäuert und im Vacuum stark eingedampft. Beim Abkühlen der concentrirten Lösung scheidet sich die active Benzoylglutaminsäure zunächst als zähes Oel ab, welches aber bei 0° nach einiger Zeit krystallinisch erstarrt. Die Ausbeute beträgt 4.6 g. Zur Reinigung wird sie aus heissem Wasser, wovon weniger, als zwei Theile zur Lösung genügen, umkrystallisirt. Sie scheidet sich beim Erkalten langsam in meistens dreieckig geformten Blättchen oder compacten Aggregaten aus, die keine scharfe Umgrenzung zeigen. An der Luft getrocknet, enthält die Säure kein Krystallwasser.

0.17 g Sbst.: 0.3584 g CO_2, 0.0851 g H_2O.

$C_{12}H_{13}NO_5$. Ber. C 57.37, H 5.18.
Gef. » 57.49, » 5.56.

Sie schmilzt bei 128—130° (corr. 130—132°), mithin 25° niedriger, als die racemische Verbindung. Von dieser unterscheidet sie sich auch durch die viel grössere Löslichkeit in Wasser. Denn sie löst sich bei 20° schon in 21 Theilen und beim Kochen in weniger, als 2 Theilen. Die wässrige Lösung dreht rechts.

Eine Lösung von 10.6982 g, die 0.5132 g Substanz enthielt und das spec. Gewicht 1.0114 besass, drehte bei 20° im 2 dm-Rohr das Natriumlicht 1.34° nach rechts. Mithin $[\alpha]_D^{20} = +13.81°$. In alkalischer Lösung dreht sie dagegen nach links.

1.0863 g, mit 2 Mol.-Gew. KOH gelöst, sodass das Gewicht der Lösung 11.4223 g und das spec. Gewicht 1.0588 betrug, drehten im 2 dm-Rohr das Natriumlicht 3.77° nach links. Daraus berechnet sich $[\alpha]_D^{20}$ in alkalischer Lösung zu — 18.7°. Eine stärkere Drehung wurde nicht erhalten, wenn auch das Strychninsalz noch öfters, als oben angegeben, umkrystallisirt war, woraus man schliessen darf, dass die untersuchte Säure rein war.

Verwandlung der Benzoylverbindung in die *l*-Glutaminsäure: 1.5 g wurden mit 12 ccm 10-procentiger Salzsäure 3½ Stdn. auf 100° erhitzt, die in Freiheit gesetzte Benzoësäure ausgeäthert und die salzsaure Lösung im Vacuum verdampft. Eine kleine Menge von Benzoylglutaminsäure war noch unzersetzt und mit der Benzoësäure in den Aether gegangen. In Folge dessen betrug die Ausbeute an salzsaurer Benzoylglutaminsäure nur 0.9 g, während 1.1 g berechnet sind. Das Chlorhydrat wurde mit der berechneten Menge Normal-Kalilauge zerlegt und die abgeschiedene Säure aus heissem Wasser unter Zusatz von Thierkohle umkrystallisirt. Sie schied sich beim Erkalten in feinen, schillernden Blättchen ab und wurde für die Analyse bei 100° getrocknet.

0.2063 g Sbst.: 0.3074 g CO_2, 0.1129 g H_2O.

$C_5H_9NO_4$. Ber. C 40.82, H 6.12.
Gef. » 40.64, » 6.08.

Die Säure schmolz beim raschen Erhitzen gleichzeitig mit einem Präparat von Glutaminsäure aus Caseïn bei 208° (corr. 213°) unter Zersetzung. Die Abweichung der Angaben von E. Schulze[1]) erklärt sich durch die Zersetzung. Für die optische Untersuchung diente die salzsaure Lösung, welche bekanntlich viel stärker dreht, als die freie Glutaminsäure.

0.2863 g *l*-Glutaminsäure wurden mit der äquimolekularen Menge

Salzsäure in Wasser gelöst. Die Flüssigkeit, welche 5.4078 g wog, das spec. Gewicht 1.0233 hatte und 5.3011 pCt. Glutaminsäure enthielt, drehte im 1 dm-Rohr das Natriumlicht 1.63° nach links, woraus sich in äquimolekularer salzsaurer Lösung $[\alpha]_D^{20}$ — 30.05° berechnet.

Zum Vergleich diente eine *d*-Glutaminsäure aus Caseïn. Eine Lösung von 12.1308 g, welche 0.65 g oder 5.3583 pCt. *d*-Glutaminsäure mit der äquivalenten Menge Salzsäure enthielt und das spec. Gewicht 1.0237 hatte, drehte im 2 dm-Rohr das Natriumlicht 3.34° nach rechts. Mithin $[\alpha]_D^{20}$ in äquimolekularer salzsaurer Lösung + 30.45°.

Schulze und Bosshard[1]) geben einen etwas höheren Werth 31.1° an, aber sie haben eine erheblich grössere Menge von Salzsäure angewandt, wodurch die kleine Differenz wahrscheinlich ihre Erklärung findet. Stark abweichend ist dagegen die Beobachtung von Scheibler[2]) $[\alpha]_D = + 25.5°$.

Benzoyl-*d*-glutaminsäure.

In den Mutterlaugen, welche von der Gewinnung des benzoyl-*l*-glutaminsauren Strychnins herrühren, befindet sich alle Benzoyl-*d*-glutaminsäure, selbstverständlich neben einer geringeren Menge des optischen Isomeren. Wird daraus die Säure in Freiheit gesetzt, so scheidet sich zuerst die schwerer lösliche racemische Verbindung ab, und aus der Mutterlauge lässt sich dann die Benzoyl-*d*-glutaminsäure durch blosse Krystallisation in fast reinem Zustand gewinnen. Dem entspricht folgende Vorschrift:

Die erwähnten Mutterlaugen, welche noch ungefähr 100 g Strychninsalz enthielten, wurden auf ca. 1 L im Vacuum eingedampft, mit 230 ccm Normal-Kalilauge versetzt, das ausgefällte Strychnin nach mehrstündigem Stehen der Flüssigkeit bei 0° abfiltrirt, das Filtrat mit 240 ccm Normal-Salzsäure übersättigt und im Vacuum auf ca. 150 ccm eingedampft. Beim Erkalten schied sich die racemische Säure ab, und nach gutem Kühlen in Eiswasser betrug ihre Menge 20 g. Als die Mutterlauge im Vacuum weiter stark eingedampft war, schied sich beim Stehen in der Kälte die active Säure als dickes Oel und der Rest des Racemkörpers in Krystallen ab. Das Oel konnte grösstentheils von den letzteren mit wenig Wasser abgeschlämmt werden; der Rest, welcher an den Krystallen haftete, wurde sammt den letzteren in wenig heissem Wasser gelöst, dann der Racemkörper durch Abkühlen wieder ausgeschieden und die Mutterlauge von Neuem zur Gewinnung des activen Productes eingedampft. Das Oel verwandelte sich im Laufe von 24 Stunden oder rascher beim Einimpfen von Kryställchen des activen Körpers in eine krystallinische Masse. Ihre Menge betrug 7.2 g. Das Präparat war gelb gefärbt, und da Thierkohle die Farbe nicht wegnimmt, so wurde es zur Reinigung in heissem Wasser gelöst und mit einem Ueberschuss von zweifach-basischem Bleiacetat versetzt. Das hierbei ausfallende Bleisalz ist zunächst undeutlich krystallinisch, verwandelt sich aber beim längeren Stehen mit der Mutterlauge in schöne Nädelchen. Es wurde filtrirt, dann in heissem Wasser suspendirt, durch Schwefelwasserstoff zersetzt und das farblose Filtrat im Vacuum bis auf einige Cubikcentimeter eingedampft. Beim längeren Stehen in der Kälte schieden sich jetzt feine, seidenglänzende Nädelchen aus, welche meist zu concentrischen Aggregaten vereinigt waren und die Flüssigkeit breiartig erfüllten. Das Präparat war noch nicht ganz frei von Racemkörpern; nach der optischen Bestimmung enthielt es von letzterem etwa 10 pCt. Eine Lösung von 11.5133 g, die 1.0715 g Substanz mit 2 Mol.-Gew. KOH enthielt und das spec. Gewicht 1.0571 hatte, drehte bei 20° im 2 dm-Rohr das Natriumlicht 3.38° nach rechts. Mithin $[\alpha]_D^{20}$ = + 17.18°, während bei der reinen Benzoyl-*l*-glutaminsäure —18.7° gefunden wurden.

Die Verunreinigung verrieth sich auch im Schmelzpunkt, denn die Substanz sinterte zwar stark bei 128°, war aber erst bei 137—139° vollständig geschmolzen.

Da eine völlige Entfernung des Racemkörpers durch blosse Krystallisation nicht zu erwarten war und auch kein anderes Salz mit einer activen Base gefunden wurde, welches die völlige Reinigung dieser Benzoyl-*d*-glutaminsäure ermöglicht hätte, so wurde das Präparat zur Darstellung der *d*-Glutaminsäure selbst benutzt, welche sich leicht als schwer lösliches Hydrochlorat von der kleinen Menge des Racemkörpers trennen lässt.

[1]) Diese Berichte 16, 314.

[2]) Zeitschr. f. physiol. Chemie 10, 143.

[3]) Diese Berichte 17, 1728.

$C_5H_9NO_4$ Ref. 119b

Resolution of Glutamic Acid.—In a typical run DL-glutamic acid[33] (73.5 g., 0.50 mole) and a 92% aqueous solution of (−)-1-hydroxy-2-aminobutane (49.5 g., 0.50 mole) were stirred vigorously, and 20 ml. of water was added in small portions. The reaction was noticeably exothermic. The sirup was heated with stirring on a water-bath at 60° until nearly all of the DL-glutamic acid dissolved. The sirup was filtered through No. 1 filter paper in a jacketed büchner funnel heated with CCl_4 vapors. It was necessary to use 40 ml. of hot water to wash the sirup through the filter paper. Enough ethanol was added to the solution to bring the volume to one l. After 20 hr. in the refrigerator 2.0 g. was collected as a first crop, m.p. 139–142°; $[\alpha]^{22}D$ −1.1° ± < 0.1° (H_2O, *c* 5).

The second crop was collected one day later, yielding 40 g., 68%, of (−)-1-hydroxybutane-2-ammonium hydrogen D-glutamate, m.p. 146–147°; $[\alpha]^{22}D$ −3.3° ± <0.1° (H_2O, *c* 5).

Anal. Calcd. for $C_9H_{20}O_5N_2$: N, 11.85. Found: N, 11.9.

The third crop was very small, while the fourth, fifth and sixth crops were (−)-1-hydroxybutane-2-ammonium hydrogen L-glutamate, yield 21 g., 36%. Later crops were optically impure.

(−)-1-Hydroxybutane-2-ammonium hydrogen D-glutamate and (−)-1-hydroxybutane-2-ammonium hydrogen L-glutamate were prepared from their optically pure components, and the constants checked with the salts obtained from the resolution. For the L-glutamate, m.p. 115–117° was found; $[\alpha]^{22}D$ −9.7° ± 0.2° (H_2O, *c* 5).

Anal. Calcd. for $C_9H_{20}O_5N_2$: N, 11.85. Found: N, 11.8.

In other runs, the yield of (−)-aminium D-glutamate was 64, 79, 70, 72 and 51%. The corresponding yields of L-glutamate in the first three cases were 32, 20 and 23%; the others were not worked up.

In two instances, the original precipitate of D-glutamate did not separate within 48 hr., but did so promptly on s eding. In these two instances, the proportions of water in the alcoholic solvent were 7 and 8%. Difficulties in precipitation were not encountered at 6% water or less. A few attempts to interfere with the normal succession of precipitation by seeding with the diastereomer were unsuccessful; the pattern of a major separation of the D-glutamate followed by a minor separation of L-glutamate was quite rigid. In those few instances in which the products failed to exhibit the requisite m.p. or rotation it proved to be easy to purify the material to the desired constants by washing with successive small portions of absolute ethanol, because of the solubility behavior which is noted in a later section.

Recovery of optically impure glutamic acid and of 1-hydroxy-2-aminobutane could be effected from the mother liquors of the D-glutamate. Concentration, adjustment of *p*H to 3.2, dilution with ethanol, precipitation and filtration gave the glutamic acid. In some cases relatively pure hydrochloride of the amine could be obtained by concentration of the mother liquors, in others the *p*H was raised to *ca.* 10.6 with alkali, and the amine recovered by distillation.

D-Glutamic Acid.—(−)-1-Hydroxybutane-2-ammonium hydrogen D-glutamate (25.0 g., 0.106 mole) was dissolved in 100 ml. of water and concd. hydrochloric acid was added to bring the *p*H to 3.16 ± 0.04. Absolute ethanol (200 ml.) was added to the suspension and it was filtered within ten min. The D-glutamic acid was washed with 200 ml. of absolute ethanol and the washings and mother liquor were allowed to stand at room temperature for the precipitation of further crops. A total of 15.0 g., 96%, was collected; $[\alpha]^{33}D$ −31.0° ± 0.0° (6 *N* HCl, *c* 5).

L-Glutamic Acid.—A 1.0-g. sample of (−)-1-hydroxybutane-2-ammonium hydrogen L-glutamate yielded 0.60 g., 97%, of L-glutamic acid; $[\alpha]^{23}D$ +31.0° ± 0.1° (6 *N* HCl, *c* 2).

(33) L. E. Arnow and J. C. Opsahl, *J. Biol. Chem.*, **134**, 649 (1940).

$C_5H_9NO_4$ Ref. 119c

Example

DL-glutamic acid in the amount of about 73.56 parts and L-tyrosinhydrazide in the amount of about 97.61 parts were dissolved with agitation in 425 parts of water by heating the mixture to about 45° C. An additional 4.9 parts of L-tyrosinhydrazide was added to the solution to adjust the pH from 5.7 to about 6.0. Five grams of activated carbon were added, and the solution was stirred for 30 minutes and filtered. The filtrate cake was washed with about 200 parts of water and the filter transferred to a tared vessel. Additional water was added to give a total solution amounting to 685 parts and to this solution was added with agitation about 1,284 parts of commercial grade isopropanol. Crystallization of L-tyrosinhydrazide D-glutamate salt was effected by seeding the solution with 0.2 part of L-tyrosinhydrazide D-glutamate crystals and agitating for 24 hours at 25° C. Solids which crystallized during the agitation period were removed by vacuum filtration and washed twice with about 22 parts of commercial grade isopropanol. After drying the crystals to constant weight in an 80° C. oven, it was found that 76.9 parts of L-tyrosinhydrazide D-glutamate crystals had been obtained. The crystal crop was found to be 86.7% pure with respect to L-tyrosinhydrazide D-glutamate salt. The L-tyrosinhydrazide D-glutamate salt had a melting point of 193° C. and optical rotation $[\alpha]_D^{25°}=28.28$ at a concentration of 4.00% in a 5% hydrochloric acid solution. The compound had a hydrazine content of 9.7% as compared with 9.4% for the theoretical. D-tyrosinhydrazide L-glutamate salt was prepared similarly and identified by means of melting point, optical rotation, and hydrazine content.

L-tyrosinhydrazide was recovered from the crystal crop comprising L-tyrosinhydrazide D-glutamate salt by dissolving 75.5 parts of the latter crystals in 200 parts water and adjusting the pH of the resulting solution to about 8.4 by the addition of 24.9 parts of 50% aqueous sodium hydroxide solution. Precipitation of L-tyrosinhydrazide from this solution was effected by stirring the solution at 25° C. for about 2 hours and then cooling to about 0–5° C. for about 12 hours. L-tyrosinhydrazide crystals were separated from the aqueous phase by filtration and washed with 25 parts of water. After again washing the crystals with an additional 5 parts of water, the crystals were dried to constant weight in an 80° C. oven. Recovery of L-tyrosinhydrazide in this manner was determined to be about 96% of the theoretical.

Mother liquor remaining following separation of L-tyrosinhydrazide was concentrated and the pH adjusted to about 3.2 with concentrated hydrochloric acid. The solution was cooled to 0–5° C. for about 4.5 hours and filtered. Solid glutamic acid crystals, which precipitated, were removed by filtration, washed with water and dried to constant weight. The yield of glutamic acid values amounted to 95.8% of the theoretical. These glutamic acid values contained about 66% of the D-glutamic acid present in the starting DL-glutamic acid.

Mother liquor remaining following separation of L-tyrosinhydrazide D-glutamate, which mother liquor contained L-tyrosinhydrazide, DL-glutamic acid and L-glutamic acid, was concentrated to remove excess water and isopropanol and adjusted to pH 8.4 by addition of 23.8 parts of 50% aqueous sodium hydroxide solution. The adjusted solution was agitated at 25° C. for about 2 hours and cooled to 0–5° C. for a period of about 12 hours. L-tyrosinhydrazide crystals which precipitated were removed by filtration, washed with two portions of water and dried to constant weight in an 80° C. oven. Recovery of L-tyrosinhydrazide amounted to 93.1% of the theoretical.

The filtrate remaining following separation of L-tyrosinhydrazide crystals was adjusted to a 4% solution with respect to the DL-glutamic acid present (18.9 parts) and the resulting dilute solution had a total weight of about 493 parts. Concentrated hydrochloric acid was added to the solution to adjust the pH from about 8.4 to about 3.2 and the adjusted solution was seeded with a few crystals of L-glutamic acid and agitated for two hours. L-glutamic acid preferentially crystallized under these conditions and the precipitated crystals were removed by filtration, washed with 5 parts of water and dried to constant weight in an 80° C. oven. The L-glutamic acid crystals obtained had an optical purity of 97.8% and were removed in the amount of 24.5 parts corresponding to a net yield of L-glutamic acid of 67%, based on the weight of L-glutamic acid in the starting material.

$C_5H_9NO_4$ Ref. 119d

5. Resolution of DL-Glutamic Acid with L-Leucinamide.

In a typical experiment, 660 g (4 mol) of DL-glutamic acid monohydrate and 572 g (4.4 mol) of L-leucinamide were mixed in 800 ml of water, and 3 l of absolute methanol was added.

The mixture was warmed on a water bath at 50°C with stirring to dissolve the crystals, then a small amount of the pure L-leucinamide L-glutamate was seeded at 42～43°C.

Crystalline L-glutamate grew readily to colorless long needle, during the solution was maintained at 42°C for three hours with constant stirring. After the solution was cooled to 20°C the depositted crystals were filtered, washed with 700 ml of 95% methanol, and dried under reduced pressure; yield 468 g (84.3% based on L-form); m.p. 187～8°C; $[\alpha]_D^{18}+3.23°$ (c, 5: water).

5.1 L-Glutamic Acid.

Two liter of aqueous solution containing 468 g L-leucinamide L-glutamate (1.69 mol) was passed through a column of Diaion SK #1 cation exchange resin (ammonium form), and washed with water. The effluent was freed of ammonia by adding 67 g of sodium hydroxide and concentrated under reduced pressure to 300 ml. The solution was brought to pH 3.2 by drop-wise addition of 6 N hydrochloric acid to precipitate L-glutamic acid. The precipitates were filtered, washed with 50 ml of water and dried under reduced pressure; yield 231 g (94%); $[\alpha]_D^{20}+31.65°$ (c, 5: 2 N HCl). Recrystallization from hot water gave pure crystals; $[\alpha]_D^{20}+31.84°$ (c, 5: 2 N HCl).

The resolving agent, L-leucinaimde was recovered from the resin bed by elution with 3 N aqueous ammonia, and the elute was freed of ammonia by evaporation. A recovery of of L-leucinamide was 98% of L-leucinamide L-glutamate used.

5.2 D-Glutamic Acid.

The mother liquor of L-leucinamide L-glutamate was concentrated under reduced pressure to recover the solvent used, the residual syrup was dissolved in water to make 2 l of solution. After the solution containing 468 g L-leucinamide D-glutamate and 86 g L-leucinamide L-glutamate was passed through a column of Diaion SK #1 cation exchange resin (ammonium form), suitable amount of water was passed for washing.

The effluent and washing solutions were collected, made to alkaline wtih 93 g of sodium hydroxide, and concentrated to 500 ml. After the solution was cooled to 10°C, 139 g of crude sodium DL-glutamate was crystallized, then filtered.

The pH of the filtrate was adjusted to 3.2 by dropwise addition of 6 N hydrochloric acid, and the precipitates were collected on filter, washed with 50 ml of water and dried under the reduced pressure; yield 224 g; $[\alpha]_D^{20}-31.58°$ (c, 5: 2 N HCl). Recrystallization from water gave pure D-glutamic acid; $[\alpha]_D^{20}-31.79°$ (c, 5: 2 N HCl).

TABLE II. THE EFFECT OF THE COMPOSITION (METHANOL-WATER RATIO) ON YIELD AND OPTICAL PURITY OF THE DIASTEREOISMERIC SALT

DL-glutamic acid/L-leucinamide Molar ratio	MeOH (ml)	Water (ml)	MeOH (%)	Gross yield based on the theoretical amount of L-glutamate (%)	Optical purity of product (%)
0.1/0.1	50	20	67	73	90
0.1/0.1	100	20	80	94	87
0.11/0.11	75	20	75	86	99
0.1/0.12	75	20	75	81	93
0.1/0.11	45	13	73	91	80
0.1/0.11	63	15	77	87	75
0.1/0.11	100	20	80	101	12
0.1/0.11	127	15	86	126	8.6

6. Resolution of DL-Glutamic Acid with L-Tyrosinamide.

6.1 Resolution in Aqueous Solution.

The preliminary experiments to search the effective condition for resolution of racemate were carried out in aqueous solution. In a typical experiment, 1080 g (6 mol) of L-tyrosinamide was dissolved in 1090 ml of hot water, the solution was mixed with 990 g (6 mol) of DL-glutamic acid monohydrate and then the mixture was heated on a water-bath until nearly all of the salts dissolved. The temperature of the warm solution was maintained at 72°±3°C, during the seeding with L-tyrosinamide L-glutamate. A gradient cooling to 52°±3°C gave the precipitates of L-tyrosinamide L-glutamate which were filtered, and washed with a small amount of water. The yield was 715 g (78% based on L-form). $[\alpha]_D^{22}+11.10°$ (c, 2: water) (The rotation value of the regenerated active L-glutamic acid: $[\alpha]_D^{22}+30.50°$ (c, 5: 2 N HCl)).

The hot mother liquor of L-tyrosinamide L-glutamate was seeded with L-tyrosinamide D-glutamate and allowed to stand in refrigerator for two days with occasional stirring. The crystallized L-tyrosinamide D-glutamate was collected on filter. The yield was 515 g (52.5% based on D-form); $[\alpha]_D^{22}+13.84°$ (c, 2: water); the rotation value of the regenerated D-glutamic acid: $[\alpha]_D^{21}+25.10°$ (c, 5: 2 N HCl).

6.2 Resolution in Aqueous Methanolic Solution.

Resolution of DL-glutamic acid with L-tyrosinamide was carried out in the various concentrations of aqueous methanolic solution. The results are given in Table III. The most useful solution system has been found to be methanol-water (1 : 4 by weight). In a typical run, 115.5 g (1.7 mol) of DL-glutamic acid monohydrate was mixed with 126 g (0.7 mol) of L-tyrosinamide in 160 ml of hot water, and the mixed solution was heated on boiling water bath to dissolve all of the crystals. After gradual cooling to 70~75°C, 40 g of methanol and then 0.1 g of pure L-tyrosinamide L-glutamate was added. The temperature of the seeded solution was maintained at 60~65°C for two hours, and then at 35°C for 45 minutes.

The precipitates were filtered, washed twice with 20 ml of 20% methanol, and dried under the reduced pressure. The yield was 107.3 g (89%); $[\alpha]_D^{22}+11.50°$ (c, 2: water). The rotation values of L-tyrosinamide L-glutamate after recrystallization from hot water and regenerated L-glutamic acid are $[\alpha]_D^{22}+11.19°$ (c, 2: water), $[\alpha]_D^{21}+29.40°$ (c, 5: 2 N HCl) respectively.

A recovery of optically active glutamic acid from L-tyrosinamide L-glutamate was carried out as follows:

Pure L-tyrosinamide L-glutamate was suspended in 100 ml of water, 29 g of ammonium chloride was added to the solution and the mixture was heated with vigorous stirring on a water bath at 40°C to dissolve the precipitates. L-Tyrosinamide hydrochloride was rapidly crystallized as narrow prism. The solution was chilled over night and filtered; yield, 18.3 g (95%). The pH of the filtrate was brought to 3.2 (the isoelectric point of glutamic acid) by the drop-wise addition of concentrated hydrochloric acid. To complete the precipitation 50 ml of ethanol was added finally. The total crop was 24 g; yield 94%; $[\alpha]_D^{23}-31.29°$ (c, 2: 2 N HCl).

TABLE I. THE EFFECT OF THE COMPOSITION (ETHANOL-WATER RATIO) ON YIELD AND OPTICAL PURITY OF THE DIASTEREOISOMERIC SALT

DL-glutamic acid/L-leucinamide Molar ratio	EtOH (ml)	Water (ml)	EtOH (%)	Gross yield based on the theoretical amount of L-glutamate (%)	Optical purity of product (%)
0.1/0.11	50	20	66	40	94
0.1/0.11	75	20	75	69	93
0.1/0.11	50	20	66	75	95
0.1/0.11	75	20	75	81	99

TABLE III. THE EFFECT ON THE COMPOSITION (METHANOL-WATER RATIO) ON YIELD AND OPTICAL PURITY OF THE DIASTEREOISOMERIC SALT

DL-glutamic acid/L-tyrosinamide Molar ratio	MeOH (ml)	Water (ml)	MeOH (%)	Gross yield based on the theoretical amount of L-glutamate. (%)	Optical purity of product (%)
0.7/0.7	30	216	10	84	82
0.7/0.7	53	170	20	92	86
0.7/0.7	250	180	50	85	71

L-Tyrosinamide hydrochloride (15 g) was suspended in 10 ml of water, a calculated amount of 6 N aqueous sodium hydroxide was added, and the solution was chilled overnight. The depositted L-tyrosinamide was filtered. Yield, 11.9 g (95.5%); m.p. 154°C; $[\alpha]_D^{20}$ +19.47° (c, 2: water).

7. Resolution of DL-Glutamic Acid with L-Phenylalaninamide.

DL-Glutamic acid monohydrate 9.8 g (0.059 mol) and L-phenylalaninamide 10.2 g (0.062 mol) were dissolved in 27 ml of hot water. After cooling, L-phenylalaninamide D-glutamate was crystallized slowly as fine prism from the seeded solution.

The crop was 7.5 g (81.4%); m.p. 186~7°C. Recrystallization from water gave 3.1 g (30.8%); m.p. 187°C; $[\alpha]_D^{23}$+12.35° (c, 5: water).

The recrystallized pure D-glutamate was decomposed with anion exchanger (OH form) in the same way as described in the case of L-leucinamide L-glutamate. D-Glutamic acid was eluted with N/2 aqueous acetic acid and the elute was concentrated under reduced pressure, until the crystallization occurred, and finally a small amount of ethanol was added. After being kept in a refrigerator overnight, the crystals of D-glutamic acid were collected, washed with ethanol, and dried under reduced pressure. Yield, 1.3 g; $[\alpha]_D^{23}$−31.82° (c, 2: 2 N HCl).

$C_5H_9NO_4$ Ref. 119e

Example 9

1.89 g. of *dl*-N-acetylglutamic acid and *l*-amphetamine are dissolved in a mixture of 45 ml. of isopropanol and 5 ml. of water with heating and the solution is allowed to stand for 16 hours at 23–25° C. The crop of *l*-amphetamine salt of *l*-N-acetylglutamic acid which forms is filtered off and dried in vacuo; $[\alpha]_D^{26}$=−2.7° (1.8% in water). An aqueous solution of the crystalline material is passed through a sulfonic acid ion exchanger (Dowex 50, acid form) and the effluent is concentrated to dryness to obtain *l*-N-acetylglutamic acid.

The alcoholic filtrate is concentrated to about 25 ml. with heating to obtain crystalline *l*-amphetamine salt of *d*-N-acetylglutamic acid. The crystals are separated and dissolved in water and the solution is made alkaline with dilute sodium hydroxide solution. The amphetamine is extracted with toluene and the residual solution is concentrated, cooled and acidified to obtain the opposite isomer, *d*-N-acetylglutamic acid.

$C_5H_9NO_4$ Ref. 119f

Resolution of Phthaloyl-DL-glutamic Acid.—A suspension of phthaloyl-DL-glutamic acid (5·54 g) prepared according to the direction of Kidd and King,[7] in water (15 ml), was exactly neutralized with $NaHCO_3$ (3·36 g) and the resulting solution added to a solution of *d-cis*-dinitrobis(ethylenediamine)cobalt(III) acetate (prepared from *d-cis*-dinitrobis(ethylenediamine)-cobalt(III) iodide (9 g) and silver acetate (3·73 g) in water (25 ml)).[10] After filtration from a small insoluble residue, ethanol (130 ml) was added to the filtrate, and the diastereoisomeric salt *d-cis*-dinitrobis(ethylenediamine)cobalt(III)-phthaloyl-L-glutamate [A] was collected, dried, and recrystallized by dissolving in warm water (40 ml) and reprecipitating with ethanol (130 ml) [6·5 g, 80%] α_D^{20} +26°, 1% in H_2O. (Found: C, 30·1; H, 5·5; N, 21·5%. Calc. for $C_{21}H_{41}O_{14}N_{13}Co_2.H_2O$ requires C, 30·2; H, 5·2; N, 21·8%).

The filtrates from the above were combined [B] and set aside for recovery of partly resolved phthaloyl-D-glutamic acid. The diastereoisomer [A] (6·5 g) was added in small quantities to a solution of KI (5 g) in water (20 ml), the suspension was stirred at room temperature for 15 min and the *d-cis*-dinitrobis(ethylenediamine)cobalt(III) iodide (4 g) was collected. The filtrate was acidified with 6N HCl and the phthaloyl-L-glutamic acid extracted several times with ethyl acetate. Evaporation of the ethyl acetate extracts followed by recrystallization from a little water yielded phthaloyl-L-glutamic acid, m.p. 158–159°C, α_D^{20} −45·4°, 1% in ethanol. (Tipson[6] reports −42·6° 1% in alcohol.)

The combined filtrates (B) were evaporated to dryness *in vacuo*, redissolved in water (30 ml) and KI (5 g) added. After removal of the precipitated *d*-iodide the filtrate was acidified with 6N HCl and the partially resolved phthaloyl-D-glutamic acid was extracted into ethyl acetate. Evaporation of the extract yielded 2·7 g. The product was again resolved with *l-cis*-dinitrobis(ethylenediamine)cobalt(III) acetate using the same procedure as described above. After recrystallization from water optically pure phthaloyl-D-glutamic acid (2 g), α_D^{20} +45·4°, 1% in ethanol, was obtained.

Recovery of D-Glutamic Acid from Phthaloyl-D-glutamic Acid.—The phthaloyl derivative (1 g) was suspended in water (5 ml) and the pH of the mixture adjusted to pH 6·5 by the addition of Na_2CO_3 (anhyd.) (0·4 g). Hydrazine hydrate (1 g; 80%) was then added and the solution left at room temperature for 2 days. The reaction mixture was then acidified with HI to a pH of 3·5, the precipitated phthalhydrazide was filtered off, the filtrate was conc. *in vacuo* and the D-glutamic acid was precipitated by the addition of alcohol. After recrystallization from aqueous ethanol optically pure D-glutamic acid was obtained; (0·4 g) α_D^{2} −17·2°, 4% in water. (Greenstein[14] reports α_D +17·7° in water for L-glutamic acid.)

[6] Tipson, R. S. (1956).—*J. Org. Chem.* **21**: 1353.

[7] King, F. E., and Kidd, D. A. A. (1949).—*J. Chem. Soc.* **1949**: 3315.

[10] Dwyer, F. P., and Garvan, F. L. (1960).—*Inorg. Synth.* **6**: 195.

[14] Greenstein, J. P., and Winitz, M. (1961).—"The Chemistry of the Amino Acids." Vol. 3. p. 1929. (Wiley: New York.)

Note: A summary of resolution procedures for glutamic acid which covers the literature up to the early 1960's is given in C. W. Huffman and W.G. Skelly, Chem. Rev. 63, 625 (1963).

```
HOCH2COCH2-CH-COOH
           |
           NH2
```

$C_5H_9NO_4$ Ref. 120

Brucine Salt of N-Acetyl-δ-hydroxy-γ-oxo-L-norvaline Dihydrate (XVIII)—A solution of 4 g. of the above oil (XVII) in 10 cc. of EtOH and 10 g. of brucine base in 30 cc. of EtOH were mixed and heated. To the clear solution, 300 cc. of Et_2O was added, the precipitate that appeared was collected by decantation, and recrystallized from EtOH to 600 mg. of colorless plates, m.p. 118~119° (decomp.); $[\alpha]_D^{20}$ −5° (c=1, EtOH). Yield, 8.5%. *Anal.* Calcd. for $C_{30}H_{41}O_{11}N_3$: C, 58.15; H, 6.67; N, 6.78. Found: C, 59.06; H, 6.58; N, 6.84.

δ-Hydroxy-γ-oxo-L-norvaline (L-HON)—A solution of 600 mg. of the above brucine salt (XVIII) dissolved in 20 cc. of H_2O was passed through a column of 20 cc. of Amberlite IR-120 (H). The resin was washed with H_2O, the effluent and the washing were combined and concentrated to 20 cc. *in vacuo.* The concentrate and 1 cc. of conc. HCl were heated for 1 hr. on a boiling water bath. After cool, the solution was concentrated *in vacuo* to 3 cc., treated with activated charcoal, and neutralized with pyridine. Addition of 15 cc. of EtOH gave colorless needles. Recrystallization from Me_2CO-H_2O gave 30 mg. of needles, $[\alpha]_D^{20}$ −6° (c=1, H_2O). Yield, 10%. This crystal showed no definite m.p. Ultraviolet and infrared absorptions, Rf values in paper partition chromatography, and the behavior in paper electrophoresis of this compound were identical with those of naturally occurring L-HON.

```
    H3C  NH2
     |    |
HOOC-CH-CH-COOH
```

$C_5H_9NO_4$ Ref. 121

For a resolution procedure using ephedrine in the direct resolution of the racemic N-benzyloxycarbonyl derivative of the above compound, see Ref. 20h.

```
     CH3
     |
HOOC-CH-CH-COOH
        |
        NH2
```

$C_5H_9NO_4$ Ref. 122

For a resolution procedure using ephedrine in the direct resolution of the racemic N-benzyloxycarbonyl derivative of the above compound, see Ref. 20h.

```
HOOC-CH-CH2COOH
     |
     NHCH3
```

$C_5H_9NO_4$ Ref. 123

DL-*N-Formyl-N-Methyl-asparaginsäure* (*X*). Die Lösung von 50 g (0,34 Mol) DL-N-Methyl-asparaginsäure (IX) in 100 ml wasserfreier Ameisensäure hielt man 3 Std. unter Rückfluss am Sieden. Dann wurden die leichtflüchtigen Anteile des Gemisches im Rotationsverdampfer entfernt und der Rückstand erneut in 100 ml Ameisensäure 2 Std. am Sieden gehalten. Nach erneutem Absaugen der Ameisensäure wurde der Rückstand wiederholt aus Methanol/Aceton umkristallisiert. Man erhielt so insgesamt 33,1 g (55%) drusige Kristalle vom Smp. 153–155° nach dem Trocknen im Hochvakuum (100°).

$C_6H_9O_5N$ Ber. C 41,1 H 5,2 O 45,7 N 8,0% Gef. C 41,1 H 5,2 O 46,0 N 7,6%

Brucinsalz der D-*N-Methyl-N-formyl-asparaginsäure* (*XI*). 32,5 g (0,1855 Mol) DL-N-Methyl-N-formyl-asparaginsäure (X) wurden in 250 ml Methanol heiss gelöst und zu einer heissen Lösung von 146,5 g (0,371 Mol) Brucin in 250 ml Methanol gegeben, wobei sofort Kristallisation erfolgte. Man liess auf 20° abkühlen und nutschte ab. Der Niederschlag wurde in 3,5 l Methanol heiss gelöst und die Lösung langsam auf Zimmertemperatur abgekühlt: 104 g Kristalle, welche noch zweimal aus je 3 l Methanol umkristallisiert wurden. Man erhielt nach Trocknen im Hochvakuum 70 g rechteckige, stengelige Kristalle, Smp. 149–154°, $[\alpha]_D^{20} = -13{,}4°$ (c = 3 in Wasser).

D-(−)-*N-Methylasparaginsäure* (*XII*). 69,5 g Brucinsalz der D-N-Methyl-N-formyl-asparaginsäure (XI) wurden mit 300 ml 1N Natronlauge versetzt und zur Entfernung des in Freiheit gesetzten Brucins dreimal mit Methylenchlorid extrahiert. Die wässerig/alkalische Lösung wurde dann, um das Natriumsalz der D-N-Methyl-N-formyl-asparaginsäure in die freie Säure XII zu überführen, durch eine mit 400 g (feucht gewogen) beschickte Ionenaustauscher-Säule von DOWEX-50W-X2 (H-Form) geschickt. Durch Eluation dieser Säule mit Wasser erhielt man nach dem Abdampfen des Wassers im Rotationsverdampfer 12,1 g D-N-Methyl-N-formyl-asparaginsäure (XII) als zähes, schlecht kristallisierendes Öl, welches direkt zur D-(−)-N-Methylasparaginsäure (XIII) oxydiert wurde.

Dazu wurde das Öl in 23,6 ml 30-proz. H_2O_2 und 23,6 ml Wasser gelöst und 2 Std. auf 60° erhitzt. Dann wurden die flüchtigen Bestandteile des Reaktionsgemisches am Vakuum schonend abdestilliert und der Rückstand aus wenig Wasser und Äthanol zweimal umkristallisiert. Man erhielt nach Trocknen im Hochvakuum bei 100° 3,8 g reine D-(−)-N-Methylasparaginsäure (XIII), Smp. 192–194°, $[\alpha]_D^{20} = -14{,}0°$ (*c* = 2 in Wasser).

$C_5H_9O_4N$ Ber. C 40,8 H 6,2 O 43,5 N 9,5% Gef. C 40,6 H 7,1 O 42,1 N 9,6%

$C_5H_9NO_4$ $CH_3CH(NO_2)\text{-}CH_2CH_2COOH$ Ref. 124

Optische Spaltung der racem. γ-Nitrovaleriansäure

Darstellung des Chininsalzes. 2,57 g Chinin und 1,17 g VII werden in trockenem Aceton unter leichtem Erwärmen gelöst. Beim Einengen der Lösung i. V. auf etwa 50 ccm kristallisieren 1,8 g (48% d. Th.) farblose Substanz vom Schmp. 133—133,5° u. Z. aus. Umkristallisieren aus Aceton ändert an dem Schmelzpunkt nichts. Wird die ursprüngliche Lösung stärker eingeengt, so läßt sich zwar die Ausbeute auf 60—80% erhöhen, das erhaltene Produkt ist aber mit einem Schmp. von rund 124° u. Z. unrein und muß mehrere Male aus Aceton umkristallisiert werden. Kleine Substanzmengen lassen sich gut reinigen, wenn man zu ihrer Aufschlämmung in Aceton unter Erwärmung tropfenweise Alkohol bis zur Lösung zugibt. Beim Einengen i. V. kristallisiert das reine Salz aus. Für größere Mengen ist diese Art der Reinigung nicht vorteilhaft, da durch zu langes Erwärmen merkliche Zersetzung eintritt.

$C_{25}H_{33}O_6N_3$ (471,5)	Ber.	C 63,68	H 7,05	N 8,91
	Gef.	» 63,71	» 7,01	» 8,73

$[\alpha]_D = -146{,}8°$ (c=0,7, Methanol) $[\alpha]_D = -106{,}7°$ (c=1,2, Aceton).

37,41 mg VII + 82,49 mg Chinin = 119,90 mg Salz in 10 ccm Aceton, 2 dm-Rohr: $\alpha_D = 2{,}27°$, $[\alpha]_D = -94{,}7°$.

Der Drehwert dieser Lösungen blieb innerhalb von 24 Stunden konstant.

Die Mutterlauge des Chininsalzes wurde zuerst durch Abdampfen i. V., zuletzt im Exsikkator vom Lösungsmittel befreit. Es blieb ein braunes, ziemlich zähes Öl zurück, das auch nach längerem Stehen nicht kristallisierte. $[\alpha]_D = -69{,}0°$ (c=1,9, Aceton).

(—)-γ-*Nitrovaleriansäure*

Eine Lösung von 3,26 g des Chininsalzes von VII in wenig Methanol wird unter Eiskühlung mit einem Äquivalent 20-proc. Schwefelsäure versetzt, wobei sich sofort die Hauptmenge an Chininsulfat ausscheidet. Man äthert aus, wäscht die ätherische Lösung, um sie völlig von Chininsulfat zu befreien, so lange mit wenig verd. Schwefelsäure, bis die blaue Fluoreszenz verschwunden ist, und trocknet sie mit Natriumsulfat. Nach dem Abdampfen des Äthers läßt sich die optisch aktive Säure i. V. destillieren und zeigt den gleichen Siedepunkt wie die Racemform. Ausbeute 0,70 g (69% d. Th.) farblose Säure vom Sdp. 118,5—120°/0,7.

$[\alpha]_D = -16{,}2°$ (c=0,5, Methanol).

Das Produkt einer anderen Spaltung ergab $[\alpha]_D = -16{,}5°$ (c=0,8, Methanol).

Nach Zusatz von 3 mol. Na-methylat zu einer methylalkoholischen Lösung verschwindet die optische Aktivität so schnell, daß bis zum Beginn der Messung keine Drehung mehr vorhanden ist.

Die optische Aktivität der Methanollösung der Säure bleibt über Wochen konstant, auch der Drehwert einer wäßrigen Lösung ändert sich nach Wochen nicht.

Ein Versuch, aus der Mutterlauge des Chininsalzes die (+)-Säure zu isolieren, mißlang. Dazu wurde das Lösungsmittel i. V. entfernt und der Rückstand mit der berechneten Menge verd. Schwefelsäure versetzt, wobei das Chininsulfat ausfiel. Nach dem Abfiltrieren des Alkaloidsalzes wurde die Hauptmenge des Alkohols i. V. entfernt, der Rückstand ausgeäthert und die ätherische Lösung über Chlorcalcium getrocknet. Der Ätherrückstand war in Sodalösung teilweise löslich, zeigte aber keine optische Aktivität.

$$\begin{array}{c} NH_2 \\ | \\ CH_3-C-CH_2COOH \\ | \\ COOH \end{array}$$

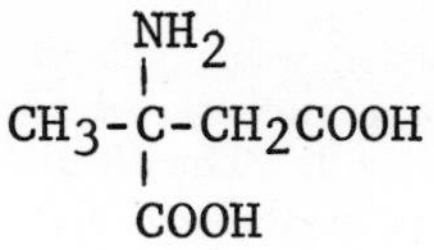

$C_5H_9NO_4$ Ref. 125

5. Aktivierung der Homoasparaginsäure

Man löst 3 g Homoasparaginsäure und 6,7 g Strychnin in der Siedehitze in 70 ccm Wasser. Beim Erkalten krystallisiert das Monostrychninsalz der rechtsdrehenden Säure in Form langer, seidiger Nadeln aus. (Ausbeute 5 g). Zur Analyse krystallisiert man mehrfach aus Wasser um. Der Schmelzpunkt ist unscharf. Von 210° ab verfärbt sich die Substanz, bei etwa 302° tritt weitgehende Zersetzung ein.

Hydrat. 3,612 mg Subst. verloren beim Erhitzen 0,246 g H_2O.

$C_5H_9O_4N$, $C_{21}H_{22}O_2N_2$, $2H_2O$ Ber. H_2O 6,96   Gef. H_2O 6,81

Wasserfreies Salz. 3,361 mg Subst.: 8,007 mg CO_2, 2,010 mg H_2O. — 2,236 mg Subst: 0,172 ccm N (22°, 758 mm).

$C_5H_9O_4N$, $C_{21}H_{22}O_2N_2$ Ber. C 64,86   H 6,44   N 8,73
                                   Gef. „ 64,97   „ 6,69   „ 8,88

Das Strychninsalz zeigt die Drehung: $[\alpha]_D^{24°} = -28{,}67°$. Man löst es in 250 ccm Wasser, kühlt mit Eis ab und gibt 37 ccm n-NaOH hinzu. Dann filtriert man nach einigem Stehen das ausgeschiedene Strychnin ab, neutralisiert das alkalische Filtrat mit 97 ccm n-HCl, dampft auf dem Wasserbad zur Trockne ein und extrahiert den Rückstand im Soxhlet erschöpfend mit 200 ccm absolutem Alkohol. Beim Einengen der alkoholischen Lösung scheidet sich die rechtsdrehende Homoasparaginsäure krystallinisch aus. Sie bildet, mehrfach aus verdünntem Aceton umkrystallisiert, farblose Nadeln vom Schmp. 240°.

3,452 mg Subst. (getr. bei 100°): 0,289 ccm N (23°, 763 mm).
$C_5H_9O_4N$ Ber. N 9,52   Gef. N 9,69

193,7 mg getr. Subst., gelöst in 1 ccm Wasser, zeigten im 1 dcm-Rohr bei 23° im Na-Licht eine Drehung von $+0{,}67°$; $[\alpha]_D^{23°} = +3{,}46°$.

337,6 mg getr. Subst., gelöst in 10 ccm Wasser, zeigten im 1 dcm-Rohr bei 20° im Na-Licht eine Drehung von $+0{,}12°$; $[\alpha]_D^{20°} = +3{,}55°$.

Das Filtrat des Strychninsalzes der linksdrehenden Aminosäure gibt bei entsprechender Aufarbeitung die linksdrehende Säure (farblose Nadeln).

4,327 mg Subst. (getr. bei 100°): 0,362 ccm N (24°, 756 mm).
$C_5H_9O_4N$ Ber. N 9,52   Gef. N 9,57

158,4 mg getr. Subst., gelöst in 1 ccm Wasser, zeigten im 1 dcm-Rohr bei 24° im Na-Licht eine Drehung von $-0{,}58°$; $[\alpha]_D^{24°} = -3{,}66°$.

431,8 mg getr. Subst., gelöst in 10 ccm Wasser, zeigten im 1 dcm-Rohr bei 20° im Na-Licht eine Drehung von $-0{,}15°$; $[\alpha]_D^{20°} = -3{,}47°$.

Die fast gleich großen absoluten Werte der Drehungen der d- und l-Form der Homoasparaginsäure zeigen, daß sie weitgehend optisch rein sind.

$$\begin{array}{c} NH_2 \quad\quad OH \\ | \quad\quad\quad\quad | \\ HOOC-CH-CH_2-CH-COOH \end{array}$$

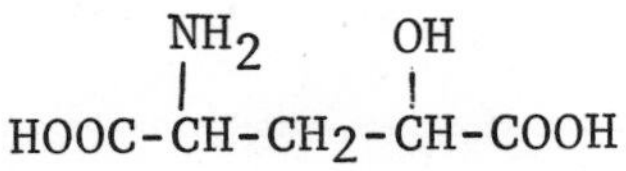

$C_5H_9NO_5$ Ref. 126

Resolution of N-*Benzoyl Derivative of One Diastereoisomer B of γ-Hydroxy-*DL-*glutamic Acid.* To an aqueous solution (300 ml) of N-benzoyl derivative (18 g) of one racemic diastereoisomer B of γ-hydroxyglutamic acid with mp 166 °C (decomp.),[7] strychnine (47 g) was added on warming. Insoluble material was filtered off while hot and the filtrate was evaporated *in vacuo*. To the residue, ethanol (30 ml) was added, and evaporation was repeated twice. To the residual syrup, 95% ethanol (200 ml) was added and the mixture was kept in a refrigerator overnight to yield crystals of crude salt (29 g). This was recrystallized from 95% ethanol (100 ml) three times to give pure salt (22 g). The strychnine salt was dissolved in water (600 ml) and then decomposed with 1 M sodium hydroxide. Strychnine deposited was filtered off and the filtrate was acidified with 20% hydrochloric acid to Congo red. It was evaporated *in vacuo* to give N-benzoyl-γ-hydroxy-D(+)-glutamic acid lactone, yield, 6 g (66%), mp 224—225 °C, $[\alpha]^{15}$ +70.8° (c 1.3, 99% ethanol).

Found: C, 57.95; H, 4.61; N, 5.55%. Calcd for $C_{12}H_{11}O_5N$: C, 57.83; H, 4.45; N, 5.62%.

The mother liquor from the above crystalline strychnine salt was evaporated to remove the salt of D(+)-lactone deposited. The residual syrup was treated as above to give crude crystals of N-benzoyl-L(−)-lactone. They were recrystallized from 1 M hydrochloric acid, mp 224—225 °C, $[\alpha]_D^{10}$ −71.3° (c 1.5, 99% ethanol).

Found: C, 57.79; H, 4.38; N, 5.54%.

*γ-Hydroxy-*D(−)*-glutamic Acid.* N-Benzoyl-D(+)-lactone (2.3 g) obtained above was dissolved in 20% hydrochloric acid (25 ml) and refluxed for 4 hr. The solution was evaporated *in vacuo*. Evaporation after addition of water was repeated. The residue thus obtained was dissolved in a little water and treated with pyridine and ethanol to give crude crystals of γ-hydroxy-D(−)-glutamic acid. Recrystallization from water afforded pure crystals, yield, 1 g, mp 166 °C (decomp.), $[\alpha]_D^{15}$ −19.5° (c 2.0, water), $[\alpha]_D^{15}$ −37.3° (c 1.5, 20% hydrochloric acid).

Found: C, 36.70; H, 5.60; N, 8.58%. Calcd for $C_5H_9O_5N$: C, 36.81; H, 5.56; N, 8.59%.

*γ-Hydroxy-*L(+)*-glutamic Acid.* In a similar procedure to that of D(−)-amino acid, γ-hydroxy-L(+)-glutamic acid was obtained from N-benzoyl-L(−)-lactone, mp 166 °C (decomp.), $[\alpha]_D^{15}$ +20.0° (c 2.0, water), $[\alpha]_D^{15}$ +38.0° (c 1.5, 20% hydrochloric acid).

Found: C, 36.59; H, 5.70; N, 8.66%.

7) T. Kaneko, Y. K. Lee, and T. Hanafusa, This Bulletin, 35, 875 (1962).

$$\begin{array}{c} NH_2 \\ | \\ HOOC-CH-CH_2-CH-COOH \\ | \\ OH \end{array}$$

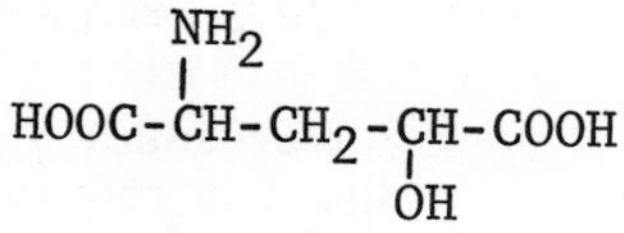

$C_5H_9NO_5$ Ref. 127

Resolution of N-*Benzyloxycarbonyl Derivative of One Diastereoisomer A of γ-Hydroxy-*DL-*glutamic Acid.* To a solution of N-benzyloxycarbonyl derivative (25 g) of a racemic diastereoisomer A hydrate of γ-hydroxyglutamic acid with mp 172—173 °C (decomp.)[7] in 95% ethanol (350 ml), brucine (67 g) was added. The mixture was heated for 10 min to obtain the complete dissolution and then evaporated *in vacuo*. Water (30 ml) was added to the residue and evaporation was repeated twice to remove ethanol completely. Residual syrup was dissolved in water (600 ml) and kept in a refrigerator overnight to give crude crystals (45 g). The crystals were filtered and recrystallized three times from water (200 ml each) to give pure brucine salt (38 g). The salt was dissolved in water (600 ml) and decomposed by addition of 1 M sodium hydroxide to phenolphthalein alkaline on cooling. Crystals of brucine deposited were removed by filtration. The filtrate was neutralized with 6 M hydrochloric acid and concentrated under reduced pressure to 200 ml.

The solution was acidified to Congo red with 6 M hydrochloric acid and extracted with ethyl acetate repeatedly. The combined extract was dried over anhydrous sodium sulfate and concentrated to afford crystals of N-benzyloxycarbonyl-γ-hydroxy-L(−)-glutamic acid. This was recrystallized from ethyl acetate, yield, 8 g (66%), mp 124—125 °C (decomp.), $[\alpha]_D^{20}$ −18.0° (c 3.3, 99% ethanol).

Found: C, 52.62; H, 5.36; N, 4.65%. Calcd for $C_{13}H_{15}O_7N$: C, 52.52; H, 5.09; N, 4.71%.

The mother liquor from crystals of the crude brucine salt mentioned above was concentrated to about 50 ml. Crystals deposited were removed by filtration. The filtrate was further concentrated and kept at room temperature for a few days. After removal of a small amount of crystals by filtration, the mother liquor was decomposed with aqueous sodium hydroxide as mentioned above to give N-benzyloxylcarbonyl-γ-hydroxy-D(+)-glutamic acid, yield, 6 g (50%), mp 124—125 °C (decomp.), $[\alpha]_D^{15}$ +18.0° (c 3.0, 99% ethanol).

Found: C, 52.61; H, 5.21; N, 4.68%.

γ-Hydroxy-L(−)-glutamic acid. N-Benzyloxycarbonyl-γ-hydroxy-L(−)-glutamic acid (2.5 g) obtained above was refluxed with 20% hydrochloric acid (25 ml) for 4 hr. The solution was evaporated *in vacuo*. Water was added to the residue and evaporation was repeated. Finally, the residue was dissolved in water and treated with silver carbonate to remove hydrochloric acid. The filtrate was again evaporated and the residue was crystallized from water and ethanol. Recrystallization from water gave pure γ-hydroxy-L(−)-glutamic acid, yield, 1 g, mp 172—173 °C (decomp.), $[\alpha]_D^{10}$ −12.5° (c 2.0, water), $[\alpha]_D^{10}$ +3.0° (c 5.0, 20% hydrochloric acid).

Found: C, 36.96; H, 5.52; N, 8.53%. Calcd for $C_5H_9O_5N$: C, 36.81; H, 5.56; N, 8.59%.

γ-Hydroxy-D(+)-glutamic Acid. In a similar manner to the above experiment, γ-hydroxy-D(+)-glutamic acid was obtained from its N-benzyloxycarbonyl derivative, mp 172—173 °C (decomp.), $[\alpha]_D^{10}$ +12.7° (c 3.0, water), $[\alpha]_D^{10}$ −3.2° (c 5.0, 20% hydrochloric acid).

Found: C, 36.96; H, 5.58; N, 8.56%.

$$C_5H_9O_4Se \qquad \begin{array}{c} CH_3(CH_2)_2\text{-CH-COOH} \\ | \\ SeO_2 \end{array}$$

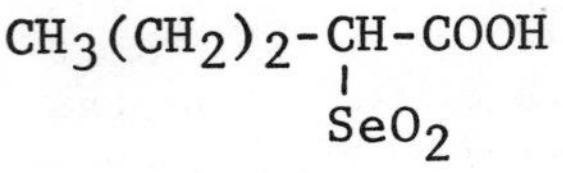

Ref. 128

Acide α-séléninevalérique.

Le dédoublement à l'aide de quinine donne l'énantiomorphe *lévogyre*.

Une solution de 0.15 mol.g de l'α-séléninevalérate de sodium dans ½ litre d'eau a été mélangée à une solution neutre et chaude de 50 g de quinine anhydre (48.6 g = 0.15 mol.g) dans ½ litre d'acide chlorhydrique dilué. En 24 heures 45 g du sel neutre de quinine se sont déposés. Après cinq recristallisations dans l'alcool dilué de la manière décrite le sel fut obtenu à l'état pur. Comme le pouvoir rotatoire est plus faible que dans le cas précédent, on a pris chaque fois au moins 0.8 g du sel de quinine pour examiner sa rotation à l'état de sel barytique, dans un volume de 25 cm³.

Cristallis. No.	1	2	3	4	5	6
Quantité	45 g	35	25	10	4.5	2
−[M]$_D$ Sel neutre	—	9.5°	10	12	12.5	12.5
−[M]$_D$ Acide	—	8°	—	11	11.5	11.5

Une solution de 3.8 mol. mg du sel de baryum dans 100 cm³ a donné la rotation α_D = −0.19° (4 dm); donc $[M]_D$ = −12.5°.

20 cm³ de cette solution, mélangés de 5 cm³ d'acide chlorhydrique 5 n et renfermant ainsi 3.04 mol.mg de l'acide non-dissocié dans 100 cm³, ont donné la valeur α_D = −0.14° (4 dm). $[M]_D$ = −11.5°.

La faible différence entre les valeurs de l'acide et des sels est frappante.

La dispersion rotatoire présente les valeurs suivantes.

λ (A.E.)	6265	5893	5605	5365	5160
[M] Sels	10.3°	12.5	14.6	17	19
[M] Acide	9 °	11.5	14	16.5	19

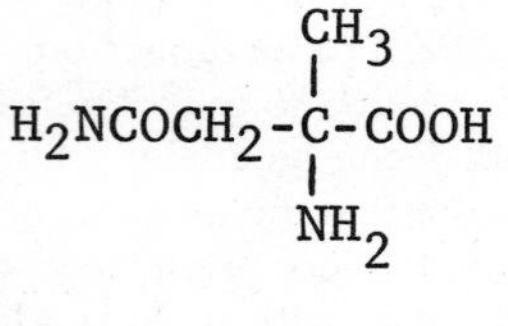

$$C_5H_{10}N_2O_3 \qquad \begin{array}{c} CH_3 \\ | \\ H_2NCOCH_2\text{-C-COOH} \\ | \\ NH_2 \end{array}$$

Ref. 129

Carbobenzossi-D.L.C-metil-asparagina: — Si pongono in un pallone, munito di agitatore, 18 g di D.L.metil-asparagina, 15 g di MgO, 145 cm³ di acqua e 70 cm³ di etere. Sotto forte agitazione e alla temperatura di 0° si aggiungono nel termine di un'ora, 27 g di clorocarbonato di benzile. Alla fine della reazione si aggiungono ancora 55 cm³ di acqua e si continua l'agitazione a temperatura ambiente fino alla scomparsa dell'odore del clorocarbonato. La soluzione viene allora acidificata, sotto raffreddamento, con HCl fino a pH 1-2. Si separa in primo tempo un olio che lentamente solidifica; si raccoglie il precipitato ancora untuoso e si macina in mortaio prima con molto etere e poi con acqua. Si purifica poi per cristallizzazione da acqua evitando di bollire la soluzione. In tal modo si ottiene come una sostanza incolora, in aghi a p.f. 119-120°.

Analisi:

trov.%: C 55,52; H 5,90; N 9,81;
per $C_{13}H_{16}O_5N_2$ calc. : 55,70; 5,75; 9,99.

Salificazione della carbobenzossi-D.L.metilasparagina con cinconina — Grammi 11,6 vengono sciolti, a caldo, in 880 cm³ di acqua e alla soluzione si aggiungono 12,22 g di cinconina sciolti in 530 cm³ di alcool bollente.

$C_5H_{10}N_2O_3$

La soluzione idroalcoolica viene evaporata lentamente a b.m. fino ad un volume di 900 cm³ e quindi lasciata in riposo per una notte; si ottiene un deposito microcristallino, del peso di 2,8 g, che mostra p.f. 183-184,5° e $[\alpha]_D = +111,5°$. Dopo altre dodici ore di riposo, si ottengono altri 4,5 g di precipitato avente le stesse caratteristiche del precedente. Queste due frazioni vengono riunite e cristallizzate da alcool fino a punto di fusione e potere rotatorio costanti; i valori ottenuti sono rispettivamente p.f. 186-186,5° e $[\alpha]_D = +102,8°$. I valori analitici concordano con quelli calcolati per il sale di cinconina della carbobenzossi-metilasparagina. Analisi:

trov.%: C 66,80; H 6,75; N 9,91;
per $C_{32}H_{38}O_6N_4$ calc.: 67,11; 6,69; 9,78.

Per ulteriore concentrazione della soluzione si separa un olio che si deposita sul fondo; si evapora allora completamente il solvente in essiccatore su ac. solforico e il residuo oleoso si riprende con poco alcool; questo lascia indisciolta una sostanza bianca pulverulenta che, filtrata e seccata, pesa 3 g. All'analisi non ha fornito i valori richiesti per un sale di cinconina della carbobenzossi-metilasparagina; il maggiore contenuto in cinconina fa pensare che sia presente anche ac. carbobenzossi-metilaspartico.

Evaporando a secco, sotto vuoto, la soluzione alcoolica abbiamo ottenuto un olio che poi solidifica in una massa pulverulenta a punto di fusione non netto, non cristallizzabile dai comuni solventi. Una determinazione del potere rotatorio specifico eseguito a scopo orientativo ha fornito il seguente risultato: $[\alpha]_D = +105,2°$. Questa frazione viene utilizzata tal quale per ottenere la carbobenzossi-metilasparagina negativa.

Carbobenzossi-(+)metilasparagina. — Si ottiene dal sale di cinconina a $[\alpha]_D = +102,8°$ e p.f. 186-186,5° per spostamento dell'alcaloide con idrato di sodio e acidificazione della soluzione alcalina e concentrazione della soluzione nel vuoto su H_2SO_4. Cristallizzata da acqua ha p.f. 135,5-136,5° e $[\alpha]_D = +5,2°$ (c = 0,7 in alcool). Analisi:

trov.%: C 55,48; H 5,89; N 10,05;
per $C_{13}H_{16}O_5N_2$ calc.: 55,70; 5,75; 9,99.

(—)metil-asparagina. — Si ottiene per idrogenazione catalitica del corrispondente carbobenzossi-derivato, nelle stesse condizioni descritte per l'α-asparagina. La soluzione filtrata dal catalizzatore viene portata a secco sotto vuoto e il residuo cristallizzato da alcool metilico contenente poca acqua. Si ottengono così dei bei cristalli che fondono a 240-241° (con decomp.) e mostrano $[\alpha]_D = -45,8°$ (c = 0,4 in ac. cloridrico; 1 mole di base: 30 di acido). Analisi:

trov.%: N 19,00;
per $C_5H_{10}O_3N_2$ calc.: 19,01.

Carbobenzossi-(—)metil-asparagina. — Si ottiene per spostamento dell'alcaloide, nelle condizioni descritte per l'antipodo, dal sale più solubile della metil-asparagina non purificato per le ragioni su esposte. Dopo cristallizzazione da acqua la sostanza fonde a 135-136° ed ha $[\alpha]_D = -5,3°$ (c = 0,35 in alcool). Analisi:

trov.%: C 55,73; H 5,82; N 10,07;
per $C_{13}H_{16}O_5N_2$ calc.: 55,70; 5,75; 9,99.

(+)metil-asparagina. — Si prepara per riduzione catalitica, nelle note condizioni, dal derivato carbobenzossilico. Cristallizzata da alcool metilico acquoso ha p.f. 240-241° e $[\alpha]_D = +46,3°$ (c = 0,3 in ac. cloridrico; 1 mole di base per 30 di acido). Analisi:

trov.%: N 18,90;
per $C_5H_{10}O_3N_2$ calc.: 19,01.

Firenze. — Istituto di Chimica organica dell'Università - 16 marzo 1959.

$C_5H_{10}O_2$

$$H_2N-\overset{O}{\overset{\|}{C}}-CH_2-CH_2-\underset{NH_2}{\overset{}{CH}}-COOH$$

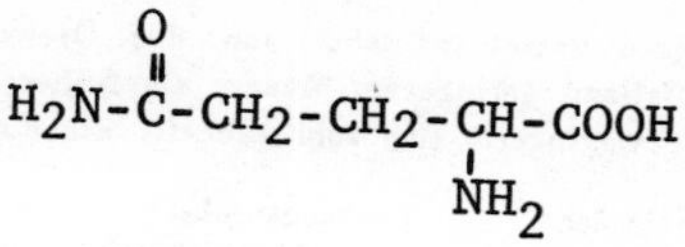

$C_5H_{10}N_2O_3$ Ref. 130

DL-Glutamine, 10·0 g, was dissolved in a mixture of 10·0 g of ammonium formate and 40 ml. of water at 80° C. The hot solution was filtered and cooled to 40° C. To this supersaturated solution, 0·50 g of L-glutamine was seeded and stirred. After 20 min of undisturbed standing, the precipitated crystals were filtered and washed with a small amount of cold water and ethanol. A weight of 2·05 g of L-glutamine was obtained. This was recrystallized from 16 ml. of water and 31 ml. of ethanol; 1·72 g of purified L-glutamine was obtained, $[\alpha]_D^{25} = +32·8°$ (c. 2·22, 1 N hydrochloric acid). (Found: C, 41·06; H, 7·00; N, 18·67. Calcd. for $C_5H_{10}O_3N_2$: C, 41·09; H, 6·90; N, 19·17 per cent.) To the filtrate, 0·50 g of D-glutamine was seeded. After 30 min of undisturbed standing, the precipitated crystals were filtered and washed with cold water and ethanol. D-Glutamine, 2·32 g, was obtained. This was recrystallized from water and ethanol; 1·97 g $[\alpha]_D^{25} = -31·4°$ (c. 2·31, 1 N hydrochloric acid). (Found: C, 41·48; H, 6·99; N, 19·21 per cent.)

DL-Glutamic acid and DL-asparagine were also resolved in essentially the same manner as described in the resolution of DL-aspartic acid and DL-glutamine. Summarized results are shown in Table 1. For further optical resolutions, the mother liquor can be used repeatedly by adding the same amount of DL-amino-acid which crystallized out from the solution.

Table 1. OPTICAL ACTIVITIES OF RESOLVED AMINO-ACIDS*

Amino-acid to be resolved	Configuration	$[\alpha]_D^{25}$ solvent (hydrochloric acid)	Optical purity (%)
DL-Aspartic acid	L	+24·6 6 N	99
	D	−23·6 6 N	95
DL-Glutamic acid	L	+28·5 6 N	91
	D	−27·5 6 N	88
DL-Asparagine	L	+28·6 3·6 N	93
	D	−29·7 3·6 N	97
DL-Glutamine	L	+32·8° 1 N	100
	D	−31·4° 1 N	96

* All elemental analyses of resolved amino-acids agree with the theoretical value after one recrystallization.

$$CH_3CH_2\underset{CH_3}{\overset{}{CH}}COOH$$

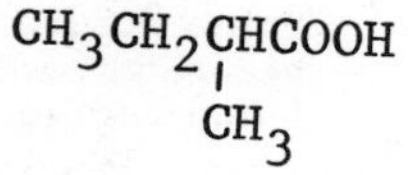

$C_5H_{10}O_2$ Ref. 131

Wir gingen von einem Rohbrucin aus, von welchem 990 g mit 200 g 2-Methylbutansäure, also etwas mehr als der berechneten Menge, und 1.8 L Wasser auf dem Wasserbade bis zur völligen Lösung erwärmt wurden. Beim Erkalten und nach zwölfstündigem Stehen schieden sich ca. 800 g Salz in grossen derben Krystallen aus. Die darin enthaltene Säure zeigte im 1 dm-Rohr den Drehungswinkel $\alpha_D = -2.5°$. Nach einmaligem Umkrystallisiren aus der anderthalbfachen Menge Wassers schieden sich 610 g Salz aus, welche eine Säure vom Drehungswinkel $\alpha_D = -3.7°$ lieferten. Das Drehungsvermögen hatte also um $-1.2°$ zugenommen. In gleichem Maasse nahm das Drehungsvermögen der Säure bei wiederholtem Umkrystallisiren des Salzes aus der anderthalbfachen Menge Wassers stetig zu. Gleichzeitig verringerte sich die Löslichkeit des Salzes in kaltem Wasser beträchtlich. Nach zwölfmaliger Krystallisation zeigte die Säure den Drehungswinkel $-14°$. Dabei war die Salzmenge auf 35 g zusammengeschmolzen. Man musste daher durch Aufarbeiten der Mutterlaugen hinreichend oft und rationell durchgeführtes Umkrystallisiren des daraus gewonnenen Salzes aus den besseren Laugen und schliesslich aus reinem Wasser grössere Mengen von jenem Salze beschaffen, dessen Säure ein Drehungsvermögen von ca. $-14°$ zeigte. Wir gelangten so zu 160 g von diesem Salze. Bei weiterem

Umkrystallisiren desselben nahm nun das Drehungsvermögen der Säure in auffallend geringerem Maasse als früher zu, und der Grad der Zunahme verringerte sich sehr schnell, wie die folgende Tabelle zeigt.

Anzahl der Krystallisationen	Drehungswinkel [1] α_D für 100 mm-Rohr.	Zunahme der Drehung.
2	−15.2°	1.2°
2	−15.9°	0.7°
2	−16.2°	0.3°
2	−16.3°	0.1°
1	−16.3°	0.0°

Nachdem so die Gewinnung der reinen *l*-2-Methylbutansäure gelungen war, wurde aus den Mutterlaugen des Brucinsalzes das Salz der Rechtssäure abzuscheiden unternommen. Aus den ersten Laugen schied sich zunächst ein Salz aus, das eine nahezu inactive Säure lieferte. Bei weiterem Einengen erhielt man Krystallisationen, die stark rechtsdrehende Säure enthielten. Die aus der Lauge abgeschiedene Säure aber zeigte, je mehr Salz auskrystallisirte, um so stärkere Rechtsdrehung. Das erreichte jedoch sein Ende, als die in der Lauge enthaltene Säure einen Drehungswinkel von + 10° im 100 mm-Rohr zeigte. Aus dieser Lauge schied sich fernerhin eine Krystallisation aus, welche eine Säure von gleichem Drehungsvermögen enthielt, wie das in der Lauge verbleibende Salz. Demnach war eine weitere Anreicherung der Rechtssäure auf diesem Wege nicht zu erzielen.

[1]) Diese Bestimmungen wurden ohne Messung der Temperatur im warmen Sommer ausgeführt.

$C_5H_{10}O_2$ Ref. 131a

(−)-2D-*Methylbutanoic acid.* — 82 g DL-2-methylbutanoic acid (b.p. 173°/760 mm, prepared from diethyl methylmalonate and ethyl iodide using the standard procedure for malonic ester synthesis) was dissolved in 400 ml of a mixture of light petroleum (b.p. 40–60°) and ether (1 : 1 v/v). A solution of 60 g (−)-α-phenethylamine in 400 ml of the same mixed solvent was cautiously added to the acid solution (heat is evolved during the formation of the salt). Crystallization was allowed to take place overnight at a temperature of −20°. The crystals formed were filtered off with suction in the cold, washed on the filter with a small volume of cold light petroleum, dissolved in a minimum volume of boiling light petroleum-ether mixture (1 : 3 v/v) and set for crystallization. The crystallization was now allowed to begin at room temperature and, after 6 hours, continued at a temperature of +10°.

The mother liquors from the crystallizations were evaporated to dryness and a 10 % solution of sodium hydroxide in water was added to the residue in order to liberate (−)-α-phenethylamine. The mixture was extracted three times with benzene, and the remaining alkaline solution acidified with hydrochloric acid. The 2-methylbutanoic acid was taken up in methylene chloride by extracting the water solution three times. After the combined methylene chloride layers had been washed with water, the organic solvent was distilled off and the acid distilled at reduced pressure.

The salt was recrystallized from successively smaller volumes of light petroleum-ether (1 : 3 v/v) in the manner described. The course of the resolution was followed be measuring the rotation of the acid in the mother liquor. The procedure of dissolving the salt in a minimum volume of boiling solvent, crystallizing first in room temperature and then at +10° gave the highest yield of the pure enantiomer. Twelve crystallizations were needed before the rotation of the acid in the mother liquor reached a constant value $\alpha_D = -3.61°$ (undiluted; *l*, 0.2). The salt crystallized in the form of long thin needles. Decomposition of the salt gave 4.7 g of (−)-2D-methylbutanoic acid of b.p. 67–68°, 12 mm. $d_4^{21.0} = 0.9338$.

Optical rotation: $\alpha_D^{21.0} = -3.63°$ (undiluted; *l*, 0.2)

$$[\alpha]_D^{21.0} = -19.34 \text{ degr. cm}^3\text{dm}^{-1}\text{g}^{-1};\ [M]_D^{21.0} = -19.70 \text{ degr. cm}^3\text{dm}^{-1}\text{g}^{-1}.$$

(+)-2L-*Methylbutanoic acid* was prepared in an analogous way using (+)-α-phenethylamine.

Optical rotation: $\alpha_D^{21.2} = +3.62°$ (undiluted; *l*, 0.2)

$$[\alpha]_D^{21.2} = +19.30 \text{ degr. cm}^3\text{dm}^{-1}\text{g}^{-1};\ [M]_D^{21.2} = +19.68 \text{ degr. cm}^3\text{dm}^{-1}\text{g}^{-1}.$$

```
CH3CH2-CH-COOH
       |
       CH3
```

$C_5H_{10}O_2$ Ref. 131b

(—)-2-*Methylbutansäure*

Die Spaltung der rac. Säure wurde von O. Schütz und W. Marckwald[23,24]) beschrieben. Die Autoren benutzten Brucin und kamen mit $[\alpha]_D^{20} = -17{,}85°$ zu etwa 90% der Enddrehung. Da die Arbeit mit Brucin sehr mühsam ist, wurde eine Reihe anderer Alkaloide in verschiedenen Lösungsmitteln versucht. Ephedrin, Strychnin, Morphin, Chinin, Cinchonin und Phenyläthylamin erwiesen sich gleichfalls als wenig geeignet. Das Chinidinsalz der (+)-Säure ist in Essigester schwerer löslich als das der (—)-Säure und gibt eine gute Trennung. Es kristallisiert aber so langsam, daß die Spaltung zu zeitraubend ist. Cinchonidin liefert Salze, die gut aus wäßrigem Methanol kristallisieren; es wurde zur Spaltung verwendet; z. B. wurden 140 g sehr reines Cinchonidin und 52 g rac. 2-Methylbutansäure in 400 ccm heißem Methanol gelöst und 700 ccm kaltes Wasser zugesetzt. Binnen 24 Stunden hatten sich 116 g Salz einer Säure von $[\alpha]_D^{25} = -3{,}2°$ ausgeschieden. Durch weiteres Umkristallisieren des Salzes aus 30-proc. Methanol unter Vereinigung passender Fraktionen erhielten wir schließlich eine Säure von $[\alpha]_D^{25} = -19{,}8°$ $n_D^{25} = 1{,}4049$ $D_4^{25} = 0{,}9313$; weiteres Umkristallisieren des Cinchonidinsalzes änderte die Drehung nicht mehr.

[23]) Ber. dtsch. chem. Ges. **29**, 53 (1896).
[24]) Ber. dtsch. chem. Ges. **32**, 1092 (1899).

```
(CH3)2CH-CH-COOH
         |
         SH
```

$C_5H_{10}O_2S$ Ref. 132

Resolution of α-Thiolisovaleric Acid.—This acid was resolved into its optical antipodes because the resolution of the corresponding bromo acid was more difficult. 25 gm. of the inactive thiol acid, which was prepared by heating the bromo acid and potassium hydrogen sulfide for 3 hours on the steam bath (not allowed to stand at 0°C.), were dissolved in 250 cc. of acetone. 55 gm. of cinchonidine were then added. The salt thus obtained was extracted with acetone (filtered at room temperature) eight times. It was decomposed with an excess of dilute hydrochloric acid and extracted with ether. The ethereal extract was washed with water and dried over sodium sulfate. The residue from the ether was fractionated under reduced pressure. The optical rotations were as follows:

$$[\alpha]_D^{25} = \frac{+0.44° \times 100}{1 \times 3.21} = +13.7° \text{ (in ether).}$$

$$[\alpha]_D^{25} = \frac{+0.50° \times 100}{1 \times 7.31} = +6.83°. \quad [M]_D^{25} = +9.2° \text{ (in 40 per cent alcohol).}$$

$(CH_3)_2CH-CH(OH)-COOH$

$C_5H_{10}O_3$ **Ref. 133**

Preparation of D-2-Hydroxy-3-methylbutyric Acid (D-I) (Expt. 2). a) The salt (D-I · L-II) (68 g) was dissolved in water (800 ml), and the resulting solution was passed through a column (diameter 4 cm, height 80 cm), filled with KU-2 resin. The fraction containing the acid (D-I) began to collect when the eluant turned acid. When the passage of the solution was complete, the column was immediately washed with water; in all, 5-6 liters of aqueous solution was collected, and this was vacuum-evaporated to dryness at 40-45°.* The residue was dried in a vacuum desiccator over P_2O_5 and recrystallized from dichloroethane. We obtained 19.6 g (81%) of the acid (D-I); m.p. 63-65°; $[\alpha]^{18}D$ −19.6 ± 2° (c = 1 in chloroform) [8].

b) The salt (D-I · L-II) was dissolved in 1 N HCl (300 ml), and the solution was extracted with ether (eight 60-ml portions). The ethereal solutions were combined and dried with magnesium sulfate; ether was vacuum-distilled off. After recrystallization from dichloroethane we obtained 20 g (83%) of the acid (D-I); m.p. 63-65° $[\alpha]^{18}D$ −19.4° (c = 1 in chloroform [8]).

Preparation of the D-threo-2-Amino-1-p-nitrophenyl-1,3-propanediol Salt of L-2-Hydroxy-3-methylbutyric Acid (L-I · D-II) (Expt. 4). The acetone mother liquor obtained in Expt. 1 was vacuum-evaporated to dryness. The oily residue, which contained the salt (L-I · D-II) with a little (D-I · L-II) impurity, was dissolved in six times its amount of water. The solution was filtered and passed through a column filled with KU-2 resin; the acid fraction was collected in the way described for Expt. 2a. The resulting aqueous solution was vacuum-evaporated to dryness. We obtained 55-60 g of a mixture of the acids (L-I) and (D-I).

This mixture of acids was dissolved in 50 ml of acetone and added to a hot suspension of 106 g of the amine (D-II) in 650 ml of acetone. The mixture was boiled for 5 min and then cooled and left in a refrigerator at 3-5° for 16-20 hr. The precipitate of the salt (L-I · D-II) was filtered off and washed three times with cooled acetone. After recrystallization of this salt, first from ten times the amount, and then from five times the amount of alcohol, we obtained 124.6 g (75.5%) of the salt (L-I · D-II); m.p. 161-162.5°; $[\alpha]^{18}D$ − 29.6° (c = 1 in alcohol). Found: C 50.97; H 6.63; N 8.30%. $C_{14}H_{22}O_7N_2$. Calculated: C 51.91; H 6.70; N 8.48%.

Preparation of L-2-Hydroxy-3-methylbutyric Acid (L-I) (Expt. 5). The preparation of the acid (L-I) was carried out as described for Expt. 2a or 2b for the preparation of the acid (D-I), one-half of the salt (L-I · D-II) obtained in Expt. 4 being taken for the decomposition. The yield of the acid (L-I) was 18.3 g (82.3%); m.p. 63-65°; $[\alpha]^{20}D$ +19.1° (c = 1 in chloroform) [8].

$CH_3-CH(COOH)-CH(OH)-CH_3$

$C_5H_{10}O_3$ **Ref. 134**

Optical Resolution of IIe. When a solution of cyclohexylammonium salt of IIe (150 g) in 1500 ml of water was passed through a column packed with 500 ml of Amberlite IR-120 (H+ type), 2000 ml of an acidic eluate were collected. To the eluate 220 g of quinine was added, after which the mixture was stirred until the added base was completely dissolved. After the decoloration of the solution with activated charcoal, it was evaporated to dryness under reduced pressure to give crude quinine salt. Ethanol was added to the crude salt in portions at the boiling temperature until the salt was completely dissolved (*ca.* 180 ml of ethanol was required); on cooling, crystals of the quinine salt were precipitated. Recrystallization from ethanol was carried out in the manner described above. The progress of resolution was followed by the measurement of the optical rotation of the sodium salt derived from a small portion of quinine salt by treatment with aqueous sodium hydroxide. After three recrystallizations, the rotation reached a steady value, are unchanged by two further recrystallizations. The quinine salt of IIe obtained amounted to 50 g; mp 140—141 °C, $[\alpha]_D^{20}$ −134 °C (*c* 1, H_2O). This was dissolved in 100 ml of water and treated with 5% aqueous sodium hydroxide with stirring until the pH of the mixture became 10. The removal of the liberated quinine by filtration and the evaporation of the filtrate gave a dense sirup, which was then dissolved in 50 ml of ethanol. After removing the insoluble matter by filtration, the filtrate was mixed with 250 ml of acetone to give 15 g of (−)-sodium salt of IIe as crystals, $[\alpha]_D^{20}$ −7.20 ° (*c* 10, H_2O), +13.61 ° (*c* 5, 3 M HCl).

The sodium salt (98 g), $[\alpha]_D^{20}$ +2.0 (*c* 10, H_2O), recovered from the combined mother liquids of the first and second crystallizations was dissolved in 1500 ml of water, after which the solution was passed through a column packed with 500 ml of Amberlite IR-120 (H+ type). Into the acidic eluate (1700 ml), a 280-g portion of quinidine was added, and the mixture was stirred until all the base had dissolved. After removing the insoluble matter, the solution was evaporated to dryness to give crude quinidine salt. Four successive recrystallizations of the salt from acetone gave 110 g of quinidine salt of optically pure IIe; $[\alpha]_D^{20}$ 172 ° (*c* 10, H_2O), mp 100 °C. This was then converted to (+)-sodium salt of IIe in a yield of 30 g; $[\alpha]_D^{20}$ +7.53 ° (*c* 10, H_2O), −13.74 ° (*c* 10, 3M HCl).

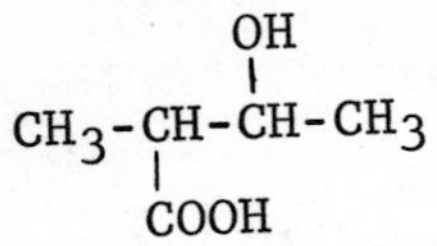

$C_5H_{10}O_3$ **Ref. 135**

Optical Resolution of IIt. A solution of sodium salt of IIt (98 g) in 1000 ml of water was passed through a column packed with 500 ml of Amberlite IR-120 (H^+ type). The acidic eluate (1700 ml) was mixed with 230 g of quinine and stirred until it was completely dissolved. The solution was then concentrated to 200 ml and allowed to stand overnight at 5 °C to give 137 g of quinine salt as crystals. After recrystallizing three times from water, the resulting 30-g portion of the salt (mp 154 °C, $[\alpha]_D^{20}$ −163 °(c 2, 50% ethanol)) was converted into (+)-sodium salt of IIt in a yield of 23 g, $[\alpha]_D^{20}$ +10.6 °(c 10, H_2O), +26.5 ° (c 5, 3M HCl).

The sodium salt (70 g, $[\alpha]_D^{20}$ −4.5 ° (c 10, H_2O)) recovered from the mother liquid was dissolved in 800 ml of water and treated with Amberlite IR-120 (H^+ type). The acid thus liberated was subjected to further resolution by the recrystallization of its cinchonidine salt from a mixture of acetone and 2-propanol (4 to 1). After recrystallizing five times, the salt was converted into (−)-sodium salt of IIt in a yield of 10 g, $[\alpha]_D^{20}$ −10.20 ° (c 10, H_2O), −26.30 ° (c 10, 3M HCl).

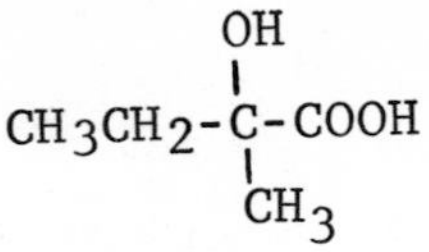

$C_5H_{10}O_3$ **Ref. 136**

Resolution of 2-Hydroxy-2-methylbutanoic Acid.—(±)-2-Hydroxy-2-methylbutanoic acid was resolved[10] with brucine to give a less soluble salt, m.p. 196—199°, $[\alpha]_D^{24}$ −18.4° (c 2.0 in EtOH), and a more soluble salt, m.p. 178—180°, $[\alpha]_D^{24}$ −17.4° (c 2.0 in EtOH). The free acids, recovered from acidified aqueous solutions of the salts by continuous extraction with ether, had, respectively, m.p. 75—77° $[\alpha]_{400}^{20}$ −15.4° (c 4.0 in EtOH) (lit.,[10] $[\alpha]_D$ −14.45°), and m.p. 72—75°, $[\alpha]_{400}^{20}$ + 7.5° (c 4.0 in EtOH), (lit.,[10] $[\alpha]_D$ +3.75°).

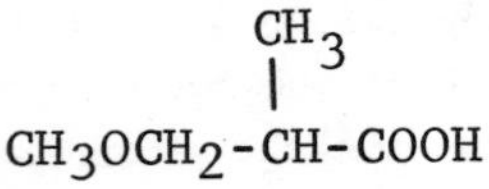

$C_5H_{10}O_3$ **Ref. 137**

Resolution of β-Methoxy-α-methylpropionic Acid.—The (±)-acid (60 g., 1 mol.) and quinine (164·6 g., 1 mol.) were dissolved in hot acetone (1·5 l.) and the mixture set aside. The quinine salt which crystallised overnight gave, after three recrystallisations from acetone, on decomposition with 2N-hydrochloric acid followed by ether-extraction and distillation, an acid with α_D^{22} +12·4°. After three further recrystallisations of the quinine salt the regained acid had α_D^{22} +14·64°; a further recrystallisation of the quinine salt gave acid with the same rotation. In further experiments six batches (in all 367 g.) of the racemic acid were resolved by this procedure. The (+)-*acid* so obtained had $\alpha_D^{17\cdot5}$ +15·02° and α_D^{19} +14·98° (for two different specimens), d_{19}^{19} 1·04, $[\alpha]_D^{19}$ +14·4° (Found: C, 50·8; H, 8·4; OMe, 25·8. $C_5H_{10}O_3$ requires C, 50·8; H, 8·4; OMe, 26·3%).

The mother-liquors from the first crystallisation of the quinine salt yielded acid having α_D^{21} −8·14°.

In other experiments the partially resolved (+)-acid was obtained by resolution with cinchonidine. After eight recrystallisations of the cinchonidine salt from acetone the regained acid had α_D^{24} +13·08°; the mother-liquors of the first crystallisation of the cinchonidine salt (from acetone) gave acid having α_D^{20} −4·74°.

α refers to homogeneous liquids (l = 1).

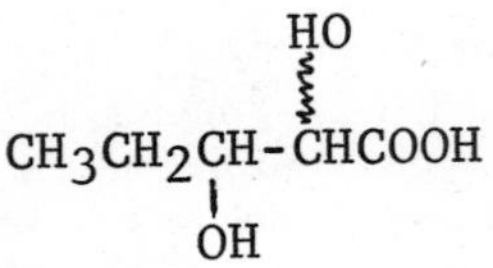

$C_5H_{10}O_4$ **Ref. 138**

450 mg der Säure wurden in wenig Alkohol gelöst und mit 1350 mg (etwa 1 Mol) Brucin in Chloroform versetzt. Da auch nach längerem Stehen des Gemisches in der Kälte keine Fällung erfolgte, wurde auf zwei Drittel des Volumens eingeengt und mit etwas Äther angespritzt. Über Nacht kleinkrystalline Fällung, 140 mg.

$$[\alpha]_D^{20} \text{ in Chloroform: } -\frac{1{,}44^\circ . 100}{0{,}5 . 1{,}03} = -279{,}2^\circ$$

Das Filtrat wurde mit dem gleichen Volumen Äther versetzt. Krystalline Ausscheidung, 88 mg.

$$[\alpha]_D^{20} \text{ in Chloroform: } -\frac{0{,}82^\circ . 100}{0{,}5 . 0{,}65} = -252{,}4^\circ.$$

Beide Fällungen wurden zusammen in Alkohol gelöst und in der Wärme vorsichtig bis zur Trübung mit Äther angespritzt. Krystalline Fällung, 170 mg.

$$[\alpha]_D^{20} \text{ in Chloroform: } -\frac{0{,}80^\circ . 100}{0{,}5 . 0{,}613} = -261{,}5^\circ.$$

Nochmals in derselben Weise umgefällt.

$$[\alpha]_D^{20} \text{ in Chloroform: } -\frac{0{,}84^\circ . 100}{0{,}5 . 0{,}644} = -260{,}8^\circ.$$

Das rechtsdrehende Salz ist sehr leicht löslich und konnte durch Anspritzen der Mutterlauge mit Petroläther nur in amorpher, mit Racemsalz verunreinigter Form gewonnen werden.

$$[\alpha]_D^{20} \text{ in Chloroform: } +\frac{0{,}24^\circ . 100}{0{,}5 . 0{,}634} = +75{,}7^\circ.$$

Umkrystallisations- und Umfällungsversuche führten stets nur zu amorphen Produkten von stark schwankenden, aber stets positiven α-Werten.

Das linksdrehende Brucinsalz wurde in möglichst wenig Wasser gelöst und tropfenweise mit 20-proc. Ammoniak versetzt. Nach Abfiltrieren des ausgefallenen Brucins wurde die Lösung mit Schwefelsäure schwach angesäuert und im Vakuum zur Trockne gedampft. Durch zweimaliges Ausziehen mit Alkohol und nachfolgendes Eindampfen ließ sich die organische Substanz vom Ammoniumsulfat trennen. Aus der konzentrierten alkoholischen Lösung schieden sich nach Anreiben mit wenig Äther und Aufbewahren im Exsiccator allmählich Krystalle aus. Schmelzp. 72°.

$$[\alpha]_D^{0°} \text{ in Wasser: } -\frac{0{,}92°.100}{0{,}5.0{,}972} = -189{,}2°.$$

Ein Teil dieser Säure wurde mit ω-Brom-p-phenylacetophenon verestert. Schmelzp. 207°.

$$[\alpha]_D^{0°} \text{ in Alkohol: } -\frac{0{,}29°\cdot 100}{0{,}5.1{,}675} = -34{,}6°.$$

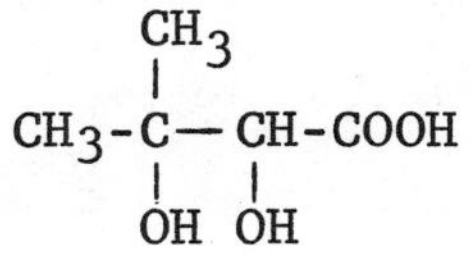

$C_5H_{10}O_4$ Ref. 139

Quinine Salt of Synthetic α,β-Dihydroxy-β-methylvaleric Acid.—DL-α,β-Dihydroxy-β-methylvaleric acid (4.0 g., 0.027 mole) was dissolved in ethanol and treated with anhydrous quinine (8.75 g., 0.027 mole) in ethanol. The solution (about 50 ml.) was seeded with the quinine salt of the natural acid from *Neurospora* and kept at 3° for three days. The sticky precipitate was recrystallized from ethanol to give a product (2.15 g.) melting at 188–191° dec. After seven more recrystallizations from ethanol, the compound (0.5 g.) melted at 203° dec. For analytical data, see Table II.

Quinine Salt of Natural α,β-Dihydroxy-β-methylvaleric Acid.—The calcium salt was first prepared, followed by regeneration of the acid and treatment with quinine according to the procedure described for the quinine salt of natural α,β-dihydroxyisovaleric acid. From 0.2 g. of calcium salt, 0.27 g. (50%) of quinine salt melting at 200–202° dec. was obtained. One recrystallization from ethanol gave material melting at 203–204°. For analytical data, see Table II.

Optically Active α,β-Dihydroxy-β-methylvaleric Acid.—The optically active acid was regenerated in 70% yield from the quinine salt according to the procedure described for α,β-dihydroxyisovaleric acid.

TABLE I

α,β-DIHYDROXYISOVALERIC ACID

	Quinine salt of synthetic α,β-dihydroxyisovaleric acid	Quinine salt of natural α,β-dihydroxyisovaleric acid
M.p., °C.	208–209 dec.	209–210 dec.
Mixed m.p., °C.	209–210 dec.	
Anal. Calcd. for $C_{23}H_{34}N_2O_6$:	Found:	Found:
C, 65.47	65.36	65.13
H, 7.47	7.21	7.23
N, 6.11	5.91	6.00
$[\alpha]^{23}D$ in MeOH	−142° (*c* 1)	−141° (*c* 1)

Infrared spectra: The spectra of the two salts as a solid mull in petrolatum were identical from 2–15 μ.

	Synthetic α,β-dihydroxyisovaleric acid	Natural α,β-dihydroxyisovaleric acid
$[\alpha]^{23}D$	−12.5° (*c* 2 in 0.1 *N* HCl)	−12.4° (*c* 2 in dilute HCl, *p*H 1)
	+ 9.5° (*c* 2 in water, *p*H 5.5–6.5)	+10° (*c* 2 in water, *p*H 5.5–6.5)

TABLE II

α,β-DIHYDROXY-β-METHYLVALERIC ACID

	Quinine salt of synthetic α,β-dihydroxy-β-methylvaleric acid	Quinine salt of natural α,β-dihydroxy-β-methylvaleric acid
M.p., °C.	203 dec.	204 dec.
Mixed m.p., °C.	203 dec.	
Anal. Calcd. for $C_{26}H_{36}N_2O_6$:	Found:	Found:
C, 66.09	66.47	66.16
H, 7.68	7.39	7.48
N, 5.93	5.98	5.81
$[\alpha]^{23}D$ in MeOH	−144° (*c* 1)	−144° (*c* 1)

Infrared spectra: The two substances gave identical spectra from 2–15 μ as solid mulls in petrolatum.

	Synthetic α,β-dihydroxy-β-methylvaleric acid	Natural α,β-dihydroxy-β-methylvaleric acid
$[\alpha]^{23}D$	+ 3° (*c* 2.3 in H_2O containing 1 eq. of $Ca(OH)_2$)	+3° (*c* 2.3 in H_2O containing 1 eq. of $Ca(OH)_2$)
	−15° (*c* 2.3 in dilute HCl, *p*H 1)	−16.7° (*c* 2.3 in dilute HCl, *p*H 1)

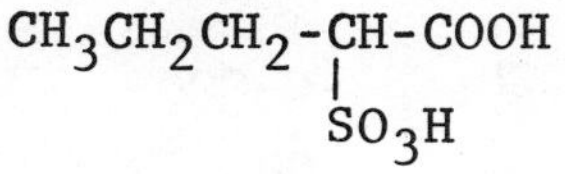

$C_5H_{10}O_5S$ Ref. 140

II. **Les composants actifs de l'acide α-sulfovalérique racémique.**

Séparation de l'acide lévogyre au moyen du sel acide de strychnine.

On fait recristalliser à plusieurs reprises le sulfovalérate acide de strychnine dans deux fois son poids d'eau chaude. Chaque fois on décompose une partie de l'eau-mère par addition de baryte et on élimine la strychnine par filtrage et par extraction au chloroforme. L'examen polarimétrique du sel barytique obtenu fournit un moyen convenable pour contrôler le degré de la séparation optique.

Les eaux-mères de plusieurs cristallisations successives ont fourni des sels barytiques, présentant les valeurs de rotation moléculaire: $[M]_D$ = +14, +8, −9, −24, −35, −36, −37, −37°.

La composition de l'eau-mère n'a donc plus changé après la sixième cristallisation et aussi le sel de strychnine enfin obtenu a donné un sel barytique de la même rotation. Ainsi la séparation était complète.

Séparation de l'acide dextrogyre au moyen du sel acide de brucine.

L'acide-d donne un sel acide de brucine qui est moins soluble que le sel de l'acide-l, mais les solubilités ne diffèrent pas beaucoup.

Après huit cristallisations successives du sel acide de brucine dans dix fois son poids d'eau, le sel obtenu était pur.

Le sel cristallisé, aussi bien que l'eau-mère, ont donné un sel barytique ayant une rotation moléculaire de + 37°.

Séparation de l'acide lévogyre au moyen du sel neutre de brucine.

Contrairement au sel acide de brucine, le sel neutre de brucine est le moins soluble pour l'acide-l.

On peut se servir des eaux-mères de la séparation précédente, renfermant un excès de l'acide-l, auxquelles on ajoute une molécule de brucine pour faire cristalliser le sel neutre.

Après huit cristallisations de ce sel dans six fois son poids d'eau, la séparation était complète. Le sel barytique a présenté la rotation −37°. Cependant, cette séparation est assez pénible, le sel neutre de brucine n'étant pas très stable.

L'acide l-α-sulfovalérique s'obtient en décomposant le sel barytique par la quantité théorique d'acide sulfurique.

L'examen polarimétrique a été effectué dans des tubes de 2 dm., à l'aide d'un polarimètre de Lippich, à la température ordinaire (environ 18°). Nous exprimons la concentration en grammes-équivalents par litre, la longueur d'onde en millimicrons, la rotation observée α et la rotation moléculaire [M] en degrés.

Acide l-α-sulfovalérique. Concentration 0.421 n., soit 0.02105 mol. gr. dans 100 cm³.

λ (μμ)	625	589	533	494
α (degrés)	—0.07	—0.08	—0.09	—0.10
[M] (degrés)	—1.7	—1.9	—2.1	—2.4

Acide d-α-sulfovalérique. Conc. 0.599 n.

λ	625	589	533	494
α	0.10	0.12	0.13	0.13
[M]	1.7	2.0	2.2	2.2

$$\underset{\underset{SO_3H}{|}}{CH_3CH}-\underset{\underset{CH_3}{|}}{CH}-COOH$$

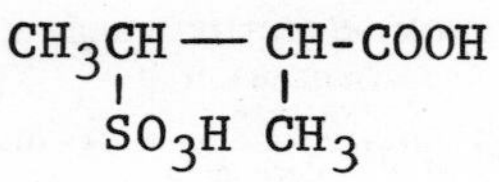

$C_5H_{10}O_5S$ Ref. 141

Le *sel neutre de quinine* cristallise en petites aiguilles tendres. Après séchage à l'air la teneur en eau varie pour plusieurs préparations de 7 à 9 %, correspondant à 4 mol. en moyenne (7.98 %).

Sel séché 1.1413 g: 23.05 cm³ de NaOH 0.1195 n
"   "   0.8546 g: 20.46   "   0.1003 n
"   "   0.4676 g: BaSO₄ 0.1461 g
                                Trouvé: p. équiv. 414.3. 416.4; S 4.29
$C_5H_{10}O_5S \cdot 2 C_{20}H_{24}O_2N_2$ (830.54). Calculé: " " 415.27; " 3.86

Ce sel de quinine donne par recristallisation l'acide lévogyre. On trouve pour le sel barytique $[M]_D = -40.5°$ et pour l'acide $[M]_D = -26°$. $[M]_D$ sel : $[M]_D$ acide $= +1.56$.

Dispersion rotatoire.

λ (mμ)	656.3	589.3	546.3	486.1
[M] acide A	20	26	32	43
[M] sel "	31	40.5	49	65

$$\underset{\underset{CH_3}{|}}{\overset{\overset{SO_3H}{|}}{CH_3CH_2-C-COOH}}$$

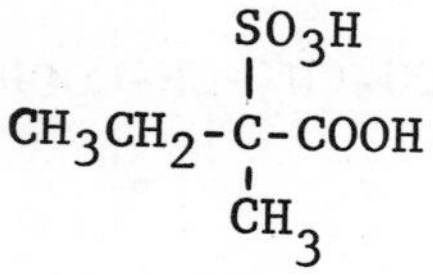

$C_5H_{10}O_5$ Ref. 142

3. *Dédoublement optique.* En faisant recristalliser à plusieurs reprises le sel acide de brucine et en le décomposant ensuite à la baryte, on obtient le sel barytique lévogyre. $[M]_D = -27°$. Le pouvoir rotatoire n'augmente plus par des cristallisations ultérieures du sel de brucine.

$C_5H_{11}NO_2$

Le sel neutre de quinine donne également le sel barytique lévogyre. En faisant recristalliser le sel neutre de strychnine, on obtient, à l'aide de baryte, le sel barytique dextrogyre.

En acidifiant la solution du sel barytique, on observe une augmentation de la rotation sans que le signe change. La rotation moléculaire de l'acide est $[M]_D = 36°$. $[M]_D$ sel : $[M]_D$ acide $= +0.75$.

Dispersion rotatoire.

λ (mμ)	659.3	589.3	546.3	486.1
[M] acide	27.9	36	44.3	60.8°
[M] sel	21	27	32	42°

$$\underset{\underset{AsO_3H_2}{|}}{CH_3CH_2CH_2-CH-COOH}$$

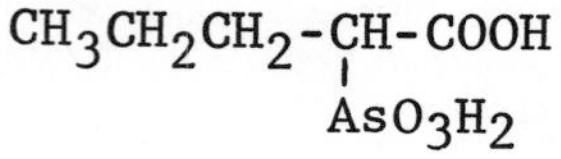

$C_5H_{11}AsO_5$ Ref. 143

3. *Acide α-arsonovalérique.* Une solution chaude renfermant 27 g d'acide α-arsonovalérique neutralisé et la quantité nécessaire (2 mol.) du chlorhydrure de quinine (volume total 1500 cm³) ont donné 100 g de sel.

Substance 0.3995 g; perte à 80° 0.0307 g
" 0.2481 g; Mg₂As₂O₇ 0.0402 g
                         Trouvé: H₂O 7.68; As 7.82
$C_5H_{11}O_5As \cdot 2C_{20}H_{24}O_2N_2 \cdot 4H_2O$ (946.53). Calculé: " 7.61; " 7.92

Après six recristallisations du sel le dédoublement était complet. La rotation n'augmente plus par des recristallisations ultérieures.
On a trouvé: $[M]_D$ sel $= -15.7°$. $[M]_D$ acide $= +19.3°$.

Dispersion rotatoire.

λ (μμ)	674.0	626.5	589.3	560.5	536.5	516.0
[M] sel	−12.2°	−13.7°	−15.7°	−17.7°	−20.0°	−22.0°
" acide	—	+16.7°	+19.3°	+22.1°	+24.8°	+28.3°

$$\underset{\underset{NH_2}{|}}{CH_3CH_2CH_2-CH-COOH}$$

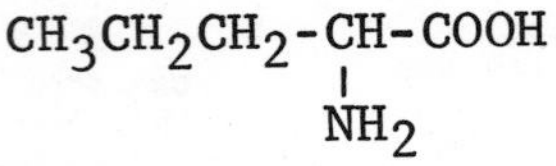

$C_5H_{11}NO_2$ Ref. 144

The melting points are uncorrected.

N-Acetyl-DL-norvaline

The substance was prepared in 65—70% yield according to the procedure given for the preparation of N-acetyl-DL-leucine by DeWitt and Ingersoll[9], and melted at 118—119°.

N-Acetyl-L-norvaline — (+)-α-Phenylethylamine Salt

N-Acetyl-DL-norvaline (5.20 g., 0.032 mole) and (+)-α-phenylethylamine (3.97 g., 0.032 mole) were dissolved in a boiling mixture of acetone (500 ml.) and methanol (30 ml.) and the solution allowed to stand at 0° overnight. The separated salt was collected (4.9 g., m. p. 165—168°) and recrystallized from a mixture of acetone (400 ml.) and methanol (30 ml.). By cooling the solution colourless crystals were obtained (2.42 g., m. p. 184—186°). The second crop (0.34 g., m. p. 184—186°) crystallized after evaporation of the mother liquid to half of its volume. The total yield on the less soluble diastereomeric salt was 2.76 g. (60.2%). A sample for analysis was recrystallized from acetone; colourless leaflets, m. p. 185—186°, $[\alpha]_D^{20}$ −15.3° (c 2.6, in water).

       Anal. 5.490 mg. subst.: 0.470 ml. N₂ (26.5°, 745 mm)
                 $C_{15}H_{24}O_3N_2$ (280.36) calc'd.: N 9.99%
                                  found: N 9.57%

L-(+)-Norvaline

The less soluble salt was dissolved in water (100 ml.) and shaken with concentrated hydrochloric acid (25 ml.). In order to remove (+)-α-phenylethylamine hydrochloride the solution was extracted with chloroform (400 ml.). The aqueous layer was evaporated *in vacuo* to dryness. The crude, oily residue — which could not be crystallized from the usual solvents — was dissolved in 12% hydrobromic acid (100 ml.) and refluxed for 3 hrs. After evaporation to dryness, the residue was dissolved in methanol (100 ml.) and the solution made alkaline (to pH 8) with concentrated ammonia. Thereby a crystalline precipitate of L-norvaline was obtained (3.0 g., 85.7%), $[\alpha]_D^{20}$ +20.5° (c 2.53, in 20% hydrochloric acid). One crystallization from a mixture methanol-water gave a product with $[\alpha]_D^{20}$ +23.4 (c 2.40, in water).

9. H. D. DeWitt and A. W. Ingersoll, *J. Am. Chem. Soc.* **73** (1951) 3359.

N-Acetyl-D-norvaline-(-)-α-Phenylethylamine Salt

From 5.20 g. (0.032 mole) of *N*-acetyl-DL-norvaline and 3.97 g. (0.032 mole) of (-)-α-phyenylethylamine following exactly the same procedure as described above 5.0 g. of crystalline product (m. p. 158—160°) were obtained. Recrystallization from a mixture acetone-methanol gave 1.52 g. (33.2%) of the pure salt, m. p. 182—184°. $[\alpha]_D^{20}$ +12.6° (c 2.52, in water)

Anal. 6.135 mg. subst.: 0.547 ml. N_2 (26.5°, 745 mm.)
$C_{15}H_{24}O_3N_2$ (280.36) calc'd.: N 9.99%
found: N 9.96%

D-(-)-*Norvaline*

The substance prepared from the other enantiomeric salt in the same manner showed $[\alpha]_D^{20}$ —25.1° (c 2.98, in 20% hydrochloric acid).

$C_5H_{11}NO_2$ Ref. 144a

EXAMPLE 3

Resolving of N,N-dibenzyl-DL-norvaline

(A) PREPARATION OF THE SALT OF L(+)-THREO-1-p-NITRO PHENYL-2-AMINO PROPANE-1,3-DIOL AND N,N-DIBENZYL-L(−)-NORVALINE

8.91 g. of N,N-dibenzyl-DL-norvaline and 6.45 g. of L(+)-threo-1-p-nitro phenyl-2-amino propane-1,3-diol are dissolved in 180 cc. of aqueous 50% ethanol at a temperature of 75–80° C. The solution is cooled to 30° C. within 1½ hours. The crystallized salt is filtered off, washed in the cold with a small quantity of 50% ethanol, and recrystallized from 35 cc. of 50% ethanol. After drying the filter residue, 6.2 g. of the salt of L(+)-threo-1-p-nitro phenyl-2-amino propane-1,3-diol and N,N-dibenzyl-L(−)-norvaline are obtained. The resolving yield corresponds to 81% of the theoretical yield. The resulting product is obtained in the form of yellow leaflets which are insoluble in water, only slightly soluble in ethanol, ether, and benzene, and soluble in acetone. The compound melts at 153–156° C.; its rotatory power is $[\alpha]_D^{20}=-24°\pm1°$ (concentration: 1% in methanol).

(B) PREPARATION OF THE SALT OF D(−)-THREO-1-p-NITRO PHENYL-2-AMINO PROPANE-1,3-DIOL AND N,N-DIBENZYL-D(+)-NORVALINE

The mother liquors obtained on preparing and recrystallizing the salt of L(+)-threo-1-p-nitro phenyl-2-amino propane-1,3-diol and N,N-dibenzyl-L(−)-norvaline are combined. They yield, on standing overnight at room temperature, 0.5 g. of a salt rich in the levorotatory diastereomeric compound. Said salt is separated from the mother liquor and can be added to a new charge of starting material to be resolved. The resulting filtrate is concentrated by evaporation in a vacuum to a volume of 100 cc. and is rendered alkaline by the addition of 19 cc. of N sodium hydroxide solution. Thereby L(+)-threo-1-p-nitro phenyl-2-amino propane-1,3-diol crystallizes. It is separated from the alkaline filtrate whereby 2.25 g., corresponding to 35% of the resolving agent employed, are recovered. The filtrate is acidified by the addition of 2 cc. of acetic acid. The precipitate is separated and is treated in the same manner as described hereinabove in Example 1 under (B). Thereby, the optically pure salt of D(−)-threo-1-p-nitro phenyl-2-amino propane-1,3-diol and N,N-dibenzyl-D(+)-norvaline is obtained. This compound forms yellow leaflets which are insoluble in water, only slightly soluble in ethanol, ether, and benzene and soluble in acetone. The salt melts at 153–156° C.; its rotatory power is $[\alpha]_D^{20}=+24°\pm1°$ (concentration: 1% in methanol).

Said salt, on decomposition according to the procedure described hereinabove in Example 2 under (C), yields N,N-dibenzyl-D(+)-norvaline. This compound forms white needles which are insoluble in water and soluble in ethanol, ether, acetone, benzene, and chloroform. Its melting point is 118–119° C.; its rotatory power is

$$[\alpha]_D^{20}=+71°\pm1°$$

(concentration: 1% in methanol).

(C) PREPARATION OF N,N-DIBENZYL-L(−)-NORVALINE

6.2 g. of the salt of L(+)-threo-1-p-nitro phenyl-2-amino propane-1,3-diol and N,N-dibenzyl-L(−)-norvaline obtained as described hereinabove under (A) is split up in the same manner as described hereinabove in Example 1 under (C) or Example 2 under (C). Thereby 3.3 g. of optically pure N,N-dibenzyl-L(−)-norvaline, corresponding to 92% of the theoretical yield, are obtained. Said compound forms white needles which are insoluble in water, and soluble in ethanol, ether, acetone, benzene, and chloroform. Its melting point is 118–119° C.; its rotatory power is $[\alpha]_D^{20}=-71°\pm1°$ (concentration: 1% in methanol).

In the course of said operation, 2.4 g. of L(+)-threo-1-p-nitro phenyl-2-amino propane-1,3-diol are recovered. The yield amounts to 37% of the resolving agent initially employed. Said recovered resolving agent can again be used for resolution.

(D) PREPARATION OF L(+)-NORVALINE

2 g. of N,N-dibenzyl-L(−)-norvaline are hydrogenated in 20 cc. of 80% ethanol in the presence of 1 g. of a palladium black catalyst containing 6% of palladium and with the addition of 0.7 cc. of concentrated hydrochloric acid. Hydrogenation and working up of the reaction mixture is carried out in the same manner as described hereinabove in Example 2 under (D). 630 mg. of optically pure L(+)-norvaline, corresponding to 79% of the theoretical yield, are obtained thereby. Its rotatory power is $[\alpha]_D^{20}=+23.3°\pm1°$ (concentration: 2% in 5 N hydrochloric acid).

$C_5H_{11}NO_2$ Ref. 144b

Note:

Using (+)-carvomenthoxyacetyl chloride as resolving agent, obtained an acid $[\alpha]_D^{20}$ = +21.8° (c, 3.7, 20% HCl) from the more insoluble salt with m.p. 84-85°C., $[\alpha]_D^{20}$ = +51.8° (c, 3, ethanol).

For an earlier resolution procedure using brucine as resolving agent acting upon the racemic N-formyl derivatives of norvaline which afforded both enantiomers, see E. Abderhalden and H. Kurten, Fermentforschung 4, 327 (1921).

Ref. 144c

$C_5H_{11}NO_2$

A. *Norvalin*

DL-Norvalin-benzylester-hydrochlorid: Durch zweimaliges Verestern der Aminosäure mit absol. Benzylalkohol/Chlorwasserstoff bei 100°. Ausb. 95 % d. Th. Nach dem Umkristallisieren aus Alkohol/Äther: Schmp. 124°.

$C_{12}H_{17}NO_2 \cdot HCl$ (243.7) Ber. C 59.00 H 7.78 N 5.74 Gef. C 58.57 H 6.61 N 5.88

DL-Norvalin-benzylester: Aus dem Hydrochlorid durch Behandeln mit ammoniakal. Äther[3]. Ausb. 92.8 % d. Th.

Racematspaltung: 11.9 g DL-Norvalin-benzylester werden in 70 ccm einer Mischung aus gleichen Teilen absol. Methanol und Äthanol gelöst und mit einer Lösung von 8.6 g D-Weinsäure in 195 ccm der Methanol/Äthanol-Mischung bei Zimmertemperatur vereinigt. Die Kristallisation ist nach 20 Stdn. beendet. Ausbeute an *L-Norvalin-benzylester-D-hydrogentartrat* 8.6 g; Schmp. 134–136°, $[\alpha]_D^{20}$: +5.43° (c = 1.56, in Wasser).

$C_{16}H_{23}NO_8$ (358.1) Ber. C 53.70 H 6.71 N 3.91 Gef. C 54.30 H 6.49 N 3.89

Aus der Mutterlauge des Spaltansatzes wird durch Einengen i. Vak. mit Äther das *D-Norvalin-benzylester-D-hydrogentartrat* isoliert. Ausb. 11.2 g; Schmp. 105–110°. $[\alpha]_D^{21}$: +15.6° (c = 5.0, in Wasser).

Gef. N 4.01

L- und D-Ester-hydrochlorid: Die diastereomeren Rohsalze werden in der 20fachen Menge absol. Äthers unter Zugabe von 2 % Methanol als Lösungsvermittler suspendiert und bei 0° Ammoniak bis zur Sättigung eingeleitet. Nach Abtrennen des Ammoniumtartrates und Absaugen des Äthers hinterbleibt der freie Ester, der mit absol. äther. Chlorwasserstoff in sein Hydrochlorid übergeführt wird (Äther/Ammoniak-Verfahren). Ausb. 90 % d. Th., bez. auf Tartrat.

L-Norvalin-benzylester-hydrochlorid: Schmp. 116–117°, $[\alpha]_D^{20}$: −9.0° (c = 5.2, in Wasser).

D-Norvalin-benzylester-hydrochlorid: Schmp. 108–110°, $[\alpha]_D^{20}$: +7.5° (c = 6.7, in Wasser).

Wie die Verseifung zur Aminosäure ergab, wurden damit die Esterhydrochloride in 86- bzw. 72-proz. optischer Reinheit gewonnen. Nach dem Umkristallisieren aus Alkohol/Äther gewinnt man in 50-proz. Ausbeute, bezogen auf das Rohprodukt, die optisch reinen Esterhydrochloride.

L-Ester-hydrochlorid: Schmp. 118.5°, $[\alpha]_D^{20}$: −9.84° (c = 4.8, in Wasser).

D-Ester-hydrochlorid: Schmp. 118.5°, $[\alpha]_D^{20}$: +9.87° (c = 5.7, in Wasser).

$C_{12}H_{17}NO_2 \cdot HCl$ (243.7)	Ber.	C 59.00	H 7.78	N 5.74
L-Verb.	Gef.	C 58.67	H 7.67	N 5.98
D-Verb.	Gef.	C 58.91	H 7.70	N 6.04

Die Benzylester-hydrochloride werden mit 20-proz. Salzsäure 30 Min. auf 100° erhitzt, anschließend wird der Benzylalkohol ausgeäthert und die wäßr. Lösung i. Vak. eingedampft. Ausb. an reinen Aminosäureantipoden 90 % d. Th.

L-Norvalin-hydrochlorid: Schmp. 223–224°, $[\alpha]_D^{20}$: +23.7° (c = 5.3, in 5n HCl).

D-Norvalin-hydrochlorid: Schmp. 223–224°, $[\alpha]_D^{20}$: −23.9° (c = 4.9, in 5n HCl).

$C_5H_{11}NO_2 \cdot HCl$ (157.7)	Ber.	C 39.20	H 7.84	N 9.14
L-Verb.	Gef.	C 39.10	H 7.97	N 9.07
D-Verb.	Gef.	C 39.06	H 7.99	N 9.09

Der Literaturwert[4] für Norvalin in 5n HCl beträgt $[\alpha]_D$: ±25.0°.

[3] S. M. McElvain und J. F. Vozza, J. Amer. chem. Soc. 71, 896 [1949].

[4] **J. P. Greenstein und Mitarbb., J. biol. Chemistry 182, 451 [1950].**

The optical purity of the isolated ester hydrogen tartrates was: L- 80-90%, and D- 70%.

$C_5H_{11}NO_2$ Ref. 144d

Papain.—In all of the experiments described, papain from a single batch from the American Ferment Company, Inc., was used. It had stood for about a year at room temperature prior to use. In a few experiments using an experimental papain prepared from frozen fresh papaya latex, yields slightly higher than those recorded here were obtained in the same period of time.

Buffer solutions were prepared from disodium phosphate and citric acid according to the procedure of McIlvaine[13] except only one-tenth as much water was used.

Anilides were prepared by the procedure of Fruton[3] with minor modifications as indicated in the following representative example.

Benzoyl-L-norvaline Anilide.—In an erlenmeyer flask was placed 12.8 g. of benzoyl-DL-norvaline, 58 ml. of normal sodium hydroxide, 10 ml. of aniline, 0.7 g. of cysteine hydrochloride in 20 ml. of water, 70 ml. of buffer (*p*H 4.23) and 300 ml. of distilled water. The enzyme solution, prepared by extracting 3.6 g. of papain with 50 ml. of water, was then added. The *p*H was brought to 5.04 by the addition of about 1.5 g. of citric acid and the flask was placed in an oven at 38° for 19 hours. The anilide was filtered, washed with water and recrystallized from methanol with practically no loss. (See Table II for data.) From the original filtrate, 5.2 g. of crude benzoyl derivative was obtained (81% recovery). This material had $[\alpha]^{32}D$ −25° (1% in one equivalent of sodium hydroxide).

The above stoichiometric ratios were maintained for other acylamino acids. Recovery of the crude benzoyl derivatives usually approached quantitative yields if the concentrated filtrates were extracted with ethyl acetate after acidification.

(3) (a) C. Dekker and J. Fruton, *ibid.*, **173**, 471 (1948); (b) the referee called our attention to the resolution of 3-fluorotyrosine; C. Niemann and M. Rapport, THIS JOURNAL, **68**, 1671 (1946).

(13) T. McIlvaine, *J. Biol. Chem.*, **49**, 183 (1921).

For the experimental details of enzymatic methods for the resolution of racemic norvaline, see K. R. Rao, S. M. Birnbaum, R.B. Kingsley and J.P. Greenstein, J. Biol. Chem., <u>198</u>, 507 (1952), and S. M. Birnbaum, L. Levintow, R. B. Kingsley, and J. P. Greenstein, J. Biol. Chem., <u>194</u>, 455 (1952).

$CH_3CH(NH_2)CH_2CH_2COOH$

$C_5H_{11}NO_2$ Ref. 145

20 g inaktive γ-Benzoylaminovaleriansäure werden mit 34 g Chinin (mit $3H_2O$) in 140 ccm heißem Alkohol gelöst und 400 ccm heißes Wasser zugefügt. Beim Erkalten trübt sich die Mischung.

Fügt man jetzt einige Impfkrystalle zu und läßt 15 Stunden im Eisschrank stehen, so scheidet sich das Chininsalz in feinen Nadeln aus. Es wird aus 70 ccm Alkohol und 200 ccm Wasser in der gleichen Weise umkrystallisiert, und ist dann so rein, daß weiteres Umlösen keinen Zweck hat. Die Ausbeute betrug in der Regel 20 g.

Die Bereitung der Impfkrystalle ist etwas mühsamer. Man muß dazu die oben erwähnte Lösung, die in der Kälte das Chininsalz als zähes Öl abscheidet, längere Zeit unter häufigem Reiben der Gefäßwände stehen lassen. Wie in vielen ähnlichen Fällen können bis zum Eintritt der Krystallisation Wochen vergehen. Sie hängt von Zufälligkeiten ab, die man nicht beherrscht.

Um aus dem Chininsalz die Benzoylverbindung zu gewinnen, wurden 20 g in 70 ccm warmem Alkohol gelöst, mit 400 ccm warmem Wasser versetzt, rasch abgekühlt und 34 ccm n-Natronlauge zugefügt. Das Chinin schied sich dabei milchig ab, ließ sich aber nach dem Abkühlen mit Eis und Schütteln mit Tierkohle filtrieren. Das Filtrat wurde mit 10 ccm 5n-HCl übersättigt, worauf bald die Krystallisation der aktiven Benzoyl-γ-aminovaleriansäure begann. Nach längerem Stehen in Eis betrug die Menge 4,5 g. Die unter stark vermindertem Druck eingeengte Mutterlauge gab noch 2,5 g, so daß die Gesamtausbeute 7 g oder 70 Proz. der Theorie betrug.

Beide Krystallisationen zeigten das gleiche Drehungsvermögen, das sich auch bei weiterem Umkrystallisieren aus heißem Wasser nicht mehr änderte.

Für die Analyse war im Vakuumexsiccator getrocknet.

0,1718 g gaben 0,4091 CO_2 und 0,1052 H_2O.
0,1500 g „ 8,3 ccm Stickgas über 33 prozentiger Kalilauge bei 23° und 767 mm Druck.

	Ber. für $C_{12}H_{15}O_3N$ (221,1)	Gef.
C	65,13	64,94
H	6,83	6,85
N	6,34	6,33

Die Substanz schmolz nach vorhergehendem Erweichen bei 131° (133° korr.), also fast bei der gleichen Temperatur wie der Racemkörper. Sie ist aber in Wasser etwas schwerer löslich und krystallisiert daraus in zentimeterlangen Nadeln. Aus einer Lösung von 1,62 g aktiver Substanz in 100 ccm heißem Wasser war nach 15 stündigem Stehen 1,12 g wieder ausgefallen.

Die Spaltung der inaktiven Benzoylverbindung kann auch mit dem Chinidinsalz ausgeführt werden, wobei ebenfalls die Verbindung der linksdrehenden Komponente auskrystallisiert. Da das Verfahren sonst keinen Vorteil vor der Chininmethode bietet, so verzichten wir auf die ausführliche Beschreibung.

0,1756 g Benzoylverbindung, gelöst in Alkohol. Gesamtgewicht der Lösung 1,7569 g. d_4^{20} = 0,8191. Drehung im 1 dm-Rohr bei 20° und Natriumlicht 1,79° (± 0,02°) nach links. Mithin $[\alpha]_D^{20} = -21,9°$ (± 0,2°) in Alkohol.

Drei weitere Bestimmungen an Substanz, die zum Teil aus dreimal umkrystallisiertem Chininsalz hergestellt und selbst mehrmals aus Wasser umkrystallisiert war, gaben innerhalb der Fehlergrenzen denselben Wert.

Die linksdrehende Benzoylverbindung entspricht der rechtsdrehenden und deshalb als d-Verbindung zu bezeichnenden γ-Aminovaleriansäure.

d-γ-Aminovaleriansäure.

Da die Benzoylverbindung schwer hydrolysiert wird, so haben wir 4,5 g mit 90 ccm 20 prozentiger Salzsäure 15 Stunden in einem Quarzkolben am Rückflußkühler gekocht. Nach dem Erkalten wurde die ausgeschiedene Benzoesäure abfiltriert und die Lösung unter vermindertem Druck zur Trockne verdampft, wobei das Hydrochlorid der Aminosäure krystallinisch zurückblieb. Seine wäßrige Lösung wurde zuerst ausgeäthert, um noch kleine Mengen von Benzoesäure und unveränderter Benzoylverbindung zu entfernen, dann mit Silberoxyd geschüttelt und in dem Filtrat das gelöste Silber durch Salzsäure genau ausgefällt. Beim Verdampfen des Filtrats unter vermindertem Druck blieb die freie Aminosäure krystallinisch zurück. Als ihre Lösung in wenig Wasser mit der achtfachen Menge Alkohol versetzt war, begann bald die Krystallisation und nach 10stündigem Stehen war der größte Teil (1,34 g) ausgefallen. Die zweite Krystallisation betrug 0,25 g und Zusatz von Äther gab noch 0,46 g. Gesamtausbeute 2,05 g oder 86 Proz. der Theorie.

Alle drei Krystallisationen zeigten nahezu die gleiche Drehung. Für die Analyse diente die erste Krystallisation nach dem Trocknen im Vakuumexsiccator.

0,1676 g gaben 0,3148 CO_2 und 0,1438 H_2O.

0,1557 g „ 16,4 ccm Stickgas über 33 prozentiger Kalilauge bei 22° und 751 mm Druck.

	Ber. für $C_5H_{11}O_2N$ (117,1)	Gef.
C	51,24	51,23
H	9,48	9,60
N	11,96	11,85

Das Präparat schmolz beim raschen Erhitzen unter Gasentwickelung gegen 209° (214° korr.). Ebenso verhält sich übrigens der Racemkörper und die Angabe von Tafel über den Schmelzp. 193° (unkorr.) ist offenbar auf andere Art des Erhitzens und Zersetzung der Substanz zurückzuführen.

Für die optische Untersuchung diente die analysierte Substanz.

0,1354 g Substanz, gelöst in Wasser. Gesamtgewicht der Lösung 1,3562 g. $d_4^{20} = 1,0237$. Drehung im 1 dm-Rohr bei 20° und Natriumlicht 1,23° (± 0,02°) nach rechts. Mithin

$$[\alpha]_D^{20} = + 12,0° (\pm 0,2°).$$

Die oben erwähnten späteren Krystallisationen der Aminosäure zeigten innerhalb der Fehlergrenzen dieselbe Drehung. Da bei dem langen Kochen mit Salzsäure eine partielle Racemisierung erfolgt sein konnte, so haben wir die Aminosäure in die Benzoylverbindung zurückverwandelt und diese optisch geprüft. Gefunden:

$$[\alpha]_D^{20} = - 21,7° (\pm 0,4°).$$

Wir glauben, aus diesem Resultet den Schluß ziehen zu dürfen, daß auch die freie Aminosäure optisch ziemlich rein gewesen ist.

l-γ-Aminovaleriansäure.

Ihre Benzoylverbindung befindet sich in der ersten Mutterlauge, die nach dem Auskrystallisieren des Chininsalzes ihres optischen Antipoden bleibt, und wird daraus in der gleichen Weise wie der Antipode isoliert. Das Rohprodukt enthielt ungefähr 30 Proz. Racemkörper. Durch häufiges Umkrystallisieren aus Wasser läßt sich dieser zum größten Teil entfernen. Aber dieses Reinigungsverfahren ist mit so großen Verlusten verbunden, daß es sich für die praktische Darstellung nicht eignet. Leider haben wir auch kein krystallisiertes Alkaloidsalz gefunden, das für diesen Zweck paßt. Wir mußten uns deshalb damit begnügen, in größerer Menge durch Umlösen aus Wasser ein Präparat von

$$[\alpha]_D^{20} = + 16,5°$$

darzustellen, mit dem auch die Analyse ausgeführt wurde.

0,1615 g gaben 0,3850 CO_2 und 0,1004 H_2O.

0,1504 g „ 8,5 ccm Stickgas über 33 prozentiger Kalilauge bei 19° und 758 mm Druck.

	Ber. für $C_{12}H_{15}O_3N$ (221,1)	Gef.
C	65,13	65,02
H	6,84	6,96
N	6,34	6,50

0,1158 g Substanz, gelöst in absolutem Alkohol. Gesamtgewicht der Lösung 1,1530 g. $d_4^{20} = 0,8191$. Drehung im 1 dm-Rohr bei 20° und Natriumlicht 1,36° (± 0,02°) nach rechts. Mithin

$$[\alpha]_D^{20} = + 16,5° (\pm 0,2°).$$

Aus einer solchen Benzoylverbindung wurde die freie linksdrehende γ-Aminovaleriansäure bereitet. Sie war natürlich auch durch Racemkörper verunreinigt, der sich in den ersten Krystallisationen anhäufte.

Für die Analyse diente ein Präparat von

$$[\alpha]_D^{20} = - 5,3° (\pm 0,2°).$$

0,1647 g gaben 0,3100 CO_2 und 0,1391 H_2O.

0,1501 g „ 15,8 ccm Stickgas über 33 prozentiger Kalilauge bei 23° und 761 mm Druck.

	Ber. für $C_5H_{11}O_2N$ (117,1)	Gef.
C	51,24	51,33
H	9,47	9,45
N	11,96	11,95

Aus den Mutterlaugen haben wir zwei weitere Krystallisationen von

$$[\alpha]_D^{20} = - 8,6° (\pm 0,2°)$$

und

$$[\alpha]_D^{20} = - 10,7° (\pm 0,2°)$$

erhalten.

$C_5H_{11}NO_2$ Ref. 145a

Resolution of N-carbobenzyloxy-4-aminopentanoic acid. Resolution was accomplished via multiple recrystallization of the quinine salt from EtOH–ether. The *d*-enantiomer was obtained from the more soluble salt: m.p. 75.5–77°, $[\alpha]_D^{20}$ 17.4° (24.5 mg/ml in EtOAc). The *l*-form had m.p. 77–78° (Found: C, 62.15; H, 6.80; N, 5.45. $C_{13}H_{17}NO_4$ requires: C, 62.14; H, 6.82; N, 5.57%).

A sample of *l*-4-aminopentanoic acid was obtained on cleavage of the *l*-carbobenzyloxy derivative by HBr in trifluoroacetic acid and recovery by ammonia neutralization of a 2-propanol soln of the hydrobromide: m.p. 196–202°, $[\alpha]_D^{20}$ −11.1° (27.4 mg/ml in water), reported, −12°.[14]

$H_2NCH_2CH_2$-CH-COOH
with CH_3 on the CH

$C_5H_{11}NO_2$ Ref. 146

Resolution of 2-Methyl-4-phthalimidobutyric Acid.—A solution of 24.7 g. of (±)-phthalimido acid and 32.4 g. of quinine in 1150 ml. of ethyl acetate was left overnight at a temperature of 15°. The crystalline product was separated by filtration and recrystallized from ethyl acetate (20 ml. of solvent per 1 g. of salt). The procedure was repeated until 28 g. (50%) of the crystalline salt was obtained (three to four crystallizations). This salt (fraction A) melted at 143–145° and had a rotation $[\alpha]^{26}D$ −97° (*c* 2.398% in ethanol). The ethyl acetate mother liquors were combined and the solvent evaporated to yield 28 g. of crystalline product (fraction B).

Fraction A.—In a separatory funnel 28 g. of salt was treated with 70 ml. of 10% hydrochloric acid and 250 ml. of benzene. The aqueous layer was removed and extracted with 20 ml. of benzene, the combined benzene solutions were washed with three 20-ml. portions of water, the washings were extracted with 20 ml. of benzene and combined benzene extracts washed once more with 20 ml. of water. After drying, the solvent was evaporated. The crystalline residue which weighed 11.9 g., $[\alpha]^{23}D$ −5.1° (*c* 2.558 % in benzene), was dissolved in 84 ml. of benzene, the solution cooled to room temperature, inoculated with racemic 2-methyl-4-phthalimidobutyric acid and allowed to stand overnight at room temperature (18°). The crystalline precipitate which weighed 8.1 g. was separated by filtration. It proved to be racemic 2-methyl-4-phthalimidobutyric acid ($[\alpha]^{24}D$ −0.14 ± 0.1° (*c* 3.46% in benzene)). On evaporation of the mother liquor 3.6 g. of crystalline product separated, $[\alpha]^{24}D$ −16.3° (*c* 4.84% in benzene). This product was dissolved in 24 ml. of benzene, inoculated with racemic phthalimido acid and kept at room temperature until a sample of mother liquor after evaporation of solvent gave a product with specific rotation higher than 18°. The crystalline portion, weighing 0.7 g., $[\alpha]^{24}D$ −3° (*c* 2.506% in benzene), was removed by filtration, and the filtrate evaporated to afford 2.9 g. of crystalline phthalimido acid. This product was dissolved in 10 ml. of benzene, and 10 ml. of petroleum ether (b.p. 40–60°) was added slowly until the solution became slightly turbid. 2-Methyl-4-phthalimidobutyric acid crystallized in fine needles; yield 2.8 g. (22.6% or 65% if calculated on the basis of recovered ±-acid), $[\alpha]^{24}D$ −18.4° (*c* 2.28% in benzene). After two recrystallizations from a mixture of benzene–petroleum ether (b.p. 40–60°) (1:1) the product had a m.p. 103–104° (softening at 102°);

rotation, 0.0875 g. made up to 5 ml. with benzene at 24° gave α_D −0.75°, *l* 2; $[\alpha]^{24}_D$ −21.5 ± 0.4°.

Anal. Found: C, 63.12; H, 5.01; N, 5.80.

Fraction B.—This fraction of salt was treated with 10% hydrochloric acid and benzene as described for the fraction A and 12 g. of crude (+)-2-methyl-4-phthalimidobutyric acid was obtained upon evaporation of the benzene. The crude acid was dissolved in 84 ml. of benzene, inoculated with racemic acid and crystallized overnight at room temperature without shaking. The 6.5 g. of crystalline portion was racemic, while evaporation of the benzene gave 5.07 g. of crystalline product, $[\alpha]^{24}_D$ +11° (*c* 3.73% in benzene). This product was dissolved in 40 ml. of benzene, inoculated with racemic acid and kept at room temperature as long as the rotation of a sample separated from the benzene solution reached a rotation of $[\alpha]^{24}_D$ +17.3° (*c* 3.692% in benzene); the yield was 2.9 g. of slightly yellow crystalline product. The crude acid was dissolved in 50 ml. of a saturated aqueous sodium bicarbonate solution, treated with charcoal and immediately acidified with 10% hydrochloric acid. The crystalline precipitate was extracted with 70 ml. of benzene, followed by three 15-ml. portions of benzene, the combined benzene solutions were dried and the solvent evaporated. A yield of 2.4 g. of acid was obtained which after crystallization from 20 ml. of benzene and 20 ml. of petroleum ether (b.p. 40–60°) gave 2.2 g. of pure (+)-2-methyl-4-phthalimidobutyric acid in the form of shiny white needles $[\alpha]^{25}_D$ +19.3° (*c* 1.662% in benzene) (17.9% or 37.5% on the basis of recovered (±)-acid). One more crystallization from a mixture of benzene–petroleum ether (b.p. 40–60°) (1:1) gave a product, m.p. 102–103.5°, rotation, 0.1144 g. made up to 5 ml. with benzene at 23° gave α_D +0.93°, *l* 2; $[\alpha]^{23}_D$ +20.3 ± 0.4°.

Anal. Found: C, 63.33; H, 5.18; N, 5.61.

(−)-2-Methyl-4-aminobutyric Acid.—A mixture of 2.8 g. of (−)-2-methyl-4-phthalimidobutyric acid ($[\alpha]_D$ −19.4°) and 14.6 ml. of M-hydrazine hydrate solution in ethanol was refluxed for one hour. The residue after evaporation of the ethanol was treated with 50 ml. of water, adjusted to *p*H 5 with acetic acid and allowed to stand 10 min. at 50° and one hour at room temperature. The phthalyl hydrazide weighing 1.6 g. (88%), was filtered and washed with four 5-ml. portions of water. The combined water and washings were evaporated *in vacuo*, the traces of water removed by repeated evaporation with absolute ethanol, 15 ml. of ethanol and 5 ml. of ether were added and left overnight in a refrigerator. The crystalline precipitate after separation by filtration and washing with ethanol, weighed 1.21 g. (91%), m.p. 192–193°, $[\alpha]^{24}_D$ −6.5° (*c* 5.1% in water). On purification by two recrystallizations from a mixture of water and ethanol (1:30), the m.p. was 196–197° dec.; rotation, 0.1385 g. made up to 5 ml. with water at 24° gave α_D −0.37°, *l* 2; $[\alpha]^{24}_D$ −6.7 ±0.3°.

Anal. Calcd. for $C_5H_{11}NO_2$: C, 51.26; H, 9.46; N, 11.96. Found: C, 51.50; H, 9.59; N, 12.13.

A sample of 141 mg. of (−)-2-methyl-4-aminobutyric acid was heated for 20 minutes at 135–140° with 205 mg. of phthalic anhydride. The reaction mixture was worked up as described for the inactive product. The yield was 150 mg. of (−)-2-methyl-4-phthalimidobutyric acid, m.p. 103–104°, rotation, 0.1419 g. made up to 5 ml. with benzene at 25° gave α_D −1.02°, *l* 2; $[\alpha]^{25}_D$ −18°.

```
CH3CH2-CHCOOH
       |
       CH2NH2
```

$C_5H_{11}NO_2$ Ref. 147

dl-Formyl-α-aminomethyläthylessigsäure.

100 g trockene Aminosäure werden mit 200 ccm möglichst wasserfreier Ameisensäure (99 Proz.) 3 Stunden am Rückflußkühler gekocht und dann die Säure unter 15—20 mm Druck möglichst vollständig abdestilliert. Diese Operation wird noch dreimal wiederholt, und der schließlich zurückbleibende Krystallbrei mit 300 ccm eiskaltem Wasser sorgfältig verrührt. Die scharf abgesaugte Masse (44 g) läßt sich dann aus etwa der neunfachen Menge heißem Wasser umkrystallisieren. Ausbeute an reinem Präparat 36,5 g = etwa 30 Proz. der Theorie.

Sämtliche Mutterlaugen können im Vakuum eingedampft und von neuem formyliert werden, wodurch die Ausbeute erheblich steigt.

Zur Analyse war nochmals aus warmem Wasser umkrystallisiert und bei 100° im Vakuum über Phosphorpentoxyd getrocknet.

0,1525 g gaben 0,2785 CO_2 und 0,1051 H_2O.
0,8951 g „ 33,7 ccm Stickgas bei 20° und 757 mm Druck über 33 prozentiger KOH.

Ber. für $C_6H_{11}O_3N$ (145,1)		Gef.
C	49,62	49,81
H	7,64	7,71
N	9,66	9,76

Die Substanz schmilzt bei raschem Erhitzen gegen 175,5—176° (korr.) unter Gasentwicklung. Sie löst sich ziemlich leicht in kaltem Alkohol, schwerer in Aceton und Essigester, sehr schwer in Äther und Benzol. Dagegen wird sie von Alkalien und Ammoniak leicht aufgenommen. Aus warmem Wasser krystallisiert sie in flachen, meist sechsseitigen Formen.

Zerlegung der Formylverbindung in die optisch-aktiven Komponenten.

70 g Formylverbindung und 190 g Brucin (wasserfrei) werden in 2060 ccm Alkohol von 85 Proz. unter Erwärmen gelöst. Bleibt diese Flüssigkeit über Nacht im Eisschrank, so findet reichliche Krystallisation statt. Zum Schluß wird noch eine Stunde auf 0° abgekühlt und der Krystallbrei abgesaugt. Ausbeute ungefähr 110 g. Da das Salz noch große Mengen des Antipoden enthält, so muß es 3—4 mal in der gleichen Weise aus 85 prozentigem Alkohol (auf 1 g ungefähr 80 ccm) umkrystallisiert werden, bis die in Freiheit gesetzte Säure, zu ungefähr 10 Proz. in ³/₄ n-Kalilauge gelöst, eine spezifische Drehung von ungefähr + 6,5° zeigt. Dabei bleibt ungefähr die Hälfte des Salzes in den Mutterlaugen, die natürlich durch rationelle Krystallisation wieder verwertet werden können.

Für die Gewinnung der freien Säure werden 12 g Brucinsalz in 60 ccm Wasser suspendiert, in Eis gekühlt, nach Zusatz von 24 ccm n-Kalilauge 10 Minuten stark geschüttelt, das gefällte Brucin abgesaugt und mit wenig kaltem Wasser gewaschen. Um den Rest des Brucins aus der Lösung zu entfernen, schüttelt man sie erst mit Chloroform, dann mit Äther. Schließlich wird sie mit 4 ccm 5n-Salzsäure versetzt und bei 10—15 mm Druck eingeengt, bis der größte Teil der Formylverbindung auskrystallisiert ist. Zum Filtrat fügt man noch 1,5 ccm 5n-Salzsäure zu, um das Kali ganz zu neutralisieren, und läßt durch Verdunsten den Rest der Formylverbindung auskrystallisieren. Ausbeute sehr gut.

Für die Analyse war nochmals aus warmem Wasser umkrystallisiert und unter 15—20 mm bei 100° getrocknet.

0,1475 g gaben 0,2686 CO_2 und 0,1000 H_2O.
0,1728 g „ 14,55 ccm Stickgas bei 18° und 755 mm Druck über 33 prozentiger KOH.

Ber. für $C_6H_{11}O_3N$ (145,1)		Gef.
C	49,62	49,66
H	7,64	7,59
N	9,66	9,68

Für die optische Untersuchung diente die alkalische Lösung.

0,1346 g Substanz; gelöst in ³/₄ n-KOH (1,45 Mol); Gesamtgewicht der alkalischen Lösung 2,0058 g; d_4^{22} = 1,044. Drehung im 1 dm-Rohr 0,50° nach rechts bei 22° und Natriumlicht; mithin $[\alpha]_D^{22}$ = + 7,14°.

Die Säure gleicht der inaktiven Verbindung in der

Form der Krystalle und in der Löslichkeit. Der Zersetzungspunkt lag einige Grade höher.

Der optische Antipode der *d*-Säure findet sich als Brucinsalz in den Mutterlaugen. Wir haben darauf verzichtet, sie in optisch reinem Zustand darzustellen, aber ein Präparat von $[\alpha]_D = -2{,}2^0$ für die Umwandlung in die Aminosäure und deren Zersetzung mit Nitrosylbromid benutzt.

d-α-Aminomethyläthylessigsäure.

2 g Formylverbindung vom höchsten Drehungsvermögen wurden mit 20 ccm 10 prozentigem Bromwasserstoff $1^1/_4$ Stunden im Wasserbad erhitzt, dann unter 15—20 mm Druck zur Trockne verdampft. Um den Bromwasserstoff zu entfernen, lösten wir in 25 ccm Wasser, kochten mit 6 g Bleioxyd 15—20 Minuten, fällten aus dem Fitrat das gelöste Blei mit Schwefelwasserstoff und verdampften die abermals filtrierte Flüssigkeit unter geringem Druck. Als die Lösung des Rückstandes in sehr wenig Wasser mit viel heißem Alkohol versetzt wurde, schied sich die Aminosäure in farblosen, glänzenden, sehr kleinen Nadeln aus. Im lufttrocknen Zustand enthielt sie ebenso wie der Racemkörper 1 Mol. Wasser, das beim einstündigen Erhitzen auf 100° unter 15—20 mm Druck über Phosphorpentoxyd völlig entwich.

0,1242 g verloren 0,0168 H_2O.
0,1621 g „ 0,0217 H_2O.

Ber. für $C_5H_{11}O_2N + H_2O$ (135,12)		Gef.	
H_2O	13,33	13,53	13,39

0,1217 g (wasserhaltig) gaben 11,2 ccm Stickgas bei 20° und 757 mm Druck über 33 prozentiger KOH.

Ber. für $C_5H_{11}O_2N + H_2O$		Gef.
N	10,37	10,53

0,1404 g trockne Säure gaben 0,2650 CO_2 und 0,1162 H_2O.

Ber. für $C_5H_{11}O_2N$ (117,1)		Gef.
C	51,24	51,48
H	9,47	9,26

0,1257 g Substanz; Gesamtgewicht der wäßrigen Lösung 1,3039 g; $d_4^{16} = 1{,}025$. Drehung im 1 dm-Rohr bei 16° und Natriumlicht 0,95° nach rechts; mithin $[\alpha]_D^{16} = +9{,}61$.

Das Präparat war aber noch nicht optisch rein, denn durch weiteres Umkrystallisieren der freien Aminosäure aus Wasser und Alkohol konnte das Drehungsvermögen bis auf 11,0° gesteigert werden.

0,1189 g Substanz; Gesamtgewicht der wäßrigen Lösung 1,3265 g; $d_4^{19} = 1{,}018$. Drehung im 1 dm-Rohr bei 19° und Natriumlicht 0,99° nach rechts; mithin $[\alpha]_D^{19} = +10{,}9^0$ (± 0,3).

0,1189 g Substanz; Gesamtgewicht der wäßrigen Lösung 1,3928 g; $d_4^{19} = 1{,}019$. Drehung im 1 dm-Rohr bei 19° und Natriumlicht 0,96° nach rechts; mithin $[\alpha]_D^{19} = +11{,}0^0$ (± 0,2).

Ob damit der Endwert wirklich erreicht ist, läßt sich allerdings nicht sagen.

Entsprechend der Beobachtung von Ehrlich mit der l-Säure dreht die d-Verbindung in Salzsäure ebenfalls nach rechts.

0,01724 g Substanz, gelöst in 20 prozentiger Salzsäure. Gesamtgewicht der Lösung 0,17280 g; $d_4^{21} = 1{,}105$. Drehung im $^1/_2$ dm-Rohr bei 21° und Natriumlicht 0,40° nach rechts; mithin $[\alpha]_D^{21} = +7{,}26^0$ (± 0,4°).

Die aktive Aminosäure ist dem Racemkörper sehr ähnlich, insbesondere sublimiert sie leicht, ohne zu schmelzen und bildet dann eine sehr lockere Masse.

Wie schon erwähnt, behält die Aminosäure sowohl in wäßriger wie in alkalischer Lösung auch bei längerem Erhitzen auf 100° ihre Aktivität, wie folgende Versuche zeigen.

a) 0,1373 g Substanz, gelöst in Wasser. Gesamtgewicht der Lösung 1,0235 g. Drehung im 1 dm-Rohr sofort + 1,57°, nach 10 stündigem Erhitzen auf 100° + 1,53°.

b) 0,1318 g Substanz, gelöst in 2n-Natronlauge. Gesamtgewicht der alkalischen Lösung 1,5555 g. Drehung im 1 dm-Rohr sofort + 0,69°; nach 4 stündigem Erhitzen auf 100° war die Drehung + 0,68°, nach weiteren 12 Stunden + 0,69°.

c) 0,0477 g Substanz, gelöst in 2n-Natronlauge. Gesamtgewicht der alkalischen Lösung 0,6975 g. Drehung im 1 dm-Rohr sofort + 0,57°, nach 16 stündigem Erhitzen im Wasserbade + 0,59°.

$$(CH_3)_2CHCHCOOH$$
$$\qquad\qquad |$$
$$\qquad\qquad NH_2$$

$C_5H_{11}NO_2$ Ref. 148

Formyl-*dl*-Valin.

Die Verbindung wird genau so wie Formylleucin [2]) durch Erhitzen der Aminosäure mit der $1^1/_2$-fachen Menge käuflicher wasserfreier Ameisensäure dargestellt. Die Erscheinungen sind ungefähr dieselben. Die Aminosäure geht allmählich in Lösung, und beim späteren Verdampfen der Ameisensäure bleibt ein krystallinisches Product zurück. Zur Erzielung einer guten Ausbeute ist aber eine zweimalige Wiederholung der Operation nöthig. Auch die Isolierung der Formylverbindung und die Rückgewinnung des unveränderten Valins geschah genau wie beim Leucin. Die Ausbeute betrug gleichfalls an Rohproduct ungefähr 80 pCt. und an Reinproduct ca. 70 pCt. der Theorie.

0.1647 g Sbst.: 0.3000 g CO_2, 0.1143 g H_2O. — 0.2333 g Sbst.: 19.6 ccm N (16.5°, 761 mm).

$C_6H_{11}O_3N$. Ber. C 49.66, H 7.59, N 9.66.
Gef. » 49.68, » 7.76, » 9.80.

Die Verbindung schmilzt ebenso wie das Formylleucin nicht ganz constant. Das analysirte Präparat begann bei 137° zu sintern und schmolz zwischen 139 und 144° (corr. 140—145°). Sie löst sich leicht in heissem Wasser, Alkohol, Aceton, dann successive schwerer in Essigäther, Aether, Benzol und fast garnicht in Petroläther. Aus Wasser krystallisirt sie in grossen, rhombenähnlichen Tafeln. Sie schmeckt sauer und löst sich leicht in Alkalien und Ammoniak.

Spaltung des Formyl-*dl*-Valins mit Brucin.

20 g des racemischen Formylkörpers werden mit 54.5 g Brucin in 600 ccm heissem Methylalkohol gelöst. Beim Abkühlen scheidet sich das Brucinsalz des Formyl-*l*-Valins in feinen Nädelchen ab. Nach zweistündigem Stehen bei 0° wird der Niederschlag abgesaugt und mit kaltem Methylalkohol gewaschen. Die Menge des Salzes beträgt nach dem Trocknen ungefähr 36 g. Es wird aus heissem Methylalkohol, wovon etwa 900 ccm nöthig sind, in derselben Weise umkrystallisirt, wobei der Verlust nur ca. 4 g beträgt. Absolut nöthig ist das Umkrystallisiren übrigens nicht, wie später auseinander gesetzt werden wird.

Zur Gewinnung des freien Formyl-*l*-Valins löst man 30 g Brucinsalz in 180 ccm Wasser, kühlt sorgfältig auf 0° ab und setzt einen geringen Ueberschuss von Alkali (60 ccm Normalnatronlauge) hinzu. Nachdem die Masse noch etwa 15 Minuten bei 0° gestanden hat, wird das Brucin scharf abgesaugt und mit eiskaltem Wasser nachgewaschen. Um den Rest des Brucins aus der Lösung zu entfernen, extrahirt man je einmal mit Chloroform und Aether. Dann wird die Flüssigkeit sofort mit 40 ccm Normalsalzsäure versetzt, wobei nahezu Neutralität eintritt und beim Druck von 10—15 mm aus einem Bade, dessen Temperatur nicht über 40° gehen darf, bis zur starken Krystallisation eingedampft. Jetzt kühlt man auf gewöhnliche Temperatur ab und fügt zur völligen Bindung des Alkalis noch 20 ccm Normalsalzsäure zu. Man lässt zur Vervollständigung der Krystallisation etwa eine Stunde bei 0° stehen, saugt sodann ab und wäscht mit wenig eiskaltem Wasser, bis das Kochsalz entfernt ist. Die Menge der Krystalle beträgt etwa 6 g. Die Mutterlauge wird im Vacuum zur Trockne verdampft, und der Rückstand mit 50 ccm lauwarmem Alkohol ausgelaugt. Beim Verdampfen bleibt der Rest des Formylvalins krystallinisch zurück. Bei gut verlaufender Operation beträgt die Ausbeute ungefähr 7 g. Sie kann aber erheblich sinken, wenn die angegebenen Bedingungen nicht innegehalten werden, weil das Formylvalin wie alle diese Formylkörper verhältnismässig leicht hydrolysirt wird. Das Product wird zur Reinigung in der 3—4fachen Menge heissen Wassers rasch gelöst und durch Abkühlen wieder abgeschieden, wobei eine erhebliche Menge in Lösung bleibt. Das so gewonnene Präparat scheint ganz rein zu sein, denn durch wiederholtes Umkrystallisiren des rohen Brucinsalzes und durch weiteres Umkrystallisiren des freien Formylkörpers konnte das Drehungsvermögen nicht gesteigert werden. Im Vacuum über Phosphorpentoxyd getrocknet, verlor die Substanz bei 70° nicht an Gewicht.

[2]) Diese Berichte 38, 3998 [1905].

0.1981 g Sbst.: 0.3605 g CO_2, 0.1358 g H_2O. — 0.1665 g Sbst.: 14.2 ccm N (17.5°, 760 mm).

$C_6H_{11}O_3N$. Ber. C 49.66, H 7.59, N 9.66.
Gef. » 49.63, » 7.67, » 10.05.

Die Substanz zeigt ähnliche Löslichkeitsverhältnisse wie der Racemkörper. Sie krystallisirt aus heissem Wasser beim Abkühlen in kleinen Prismen, die vielfach concentrisch verwachsen sind. Aus der verdünnten wässrigen Lösung scheidet sie sich nach längerem Stehen in ziemlich grossen Prismen ab. Sie hat ebensowenig wie die activen Formylleucine einen scharfen Schmelzpunkt. Im Capillarrohr beginnt sie gegen 150° zu sintern und ist bis 153° (corr. 156°) völlig geschmolzen. In alkoholischer Lösung dreht sie nach links und in wässriger Lösung nach rechts.

1. Eine Lösung in absolutem Alkohol vom Gesammtgewicht 5.2588 g, die 0.5891 g Substanz enthielt und das specifische Gewicht 0.8213 hatte, drehte bei 20° und Natriumlicht im 1 dm-Rohr 1.19° nach links.

2. Eine Lösung in absolutem Alkohol vom Gesammtgewicht 5.3726 g, die 0.6003 g Substanz enthielt und das specifische Gewicht 0.8214 hatte, drehte bei 20° und Natriumlicht im 1 dm-Rohr 1.20° nach links. Mithin $[\alpha]_D^{20}$ in alkoholischer Lösung

1. — 12.93° (± 0.2), 2. — 13.07° (± 0.2).

Eine Lösung in Wasser vom Gesammtgewicht 22.579 g, die 1.075 g Substanz enthielt und das specifische Gewicht 1.0068 hatte, drehte im 2 dm-Rohr bei 20° und Natriumlicht 1.62° nach rechts. Mithin in wässriger Lösung $[\alpha]_D^{20}$ = + 16.9° (± 0.2).

Formyl-*d*-Valin.

Sein Brucinsalz bleibt in der oben erwähnten methylalkoholischen Mutterlauge. Diese wird unter vermindertem Druck zur Trockne verdampft, der Rückstand in 200 ccm Wasser gelöst und daraus die Formylverbindung in derselben Weise isolirt, wie es zuvor für die *l*-Verbindung beschrieben ist. Die Ausbeute an Reinproduct ist ungefähr dieselbe, weil die Trennung der Brucinsalze durch die Krystallisation aus Alkohol ziemlich vollständig ist. Die kleinen Reste des optischen Isomeren werden, wie es scheint, beim Umkrystallisiren des Formylkörpers aus Wasser entfernt; denn das Drehungsvermögen des letzteren ist nach dem Umlösen aus Wasser ebenso stark wie im vorhergehenden Falle, aber natürlich im umgekehrten Sinne. Der Schmelzpunkt und die sonstigen Eigenschaften sind dieselben wie beim optischen Antipoden.

0.2478 g Sbst.: 0.4517 g CO_2, 0.1706 g H_2O. — 0.2052 g Sbst.: 16.8 ccm N (17.5°, 767 mm).

Ber. C 49.66, H 7.59, N 9.66.
Gef. » 49.71, » 7.70, » 9.59.

1. Eine Lösung in absolutem Alkohol vom Gesammtgewicht 6.2986 g, die 0.7194 g Substanz enthielt und das specifische Gewicht 0.8213 besass, drehte im 1 dm-Rohr bei 20° und Natriumlicht 1.20° nach rechts.

2. Eine Lösung in absolutem Alkohol vom Gesammtgewicht 5.8867 g, die 0.6591 g Substanz enthielt und das specifische Gewicht 0.8214 besass, drehte im 1 dm-Rohr 1.22° bei 20° und Natriumlicht nach rechts.

3. Eine Lösung in absolutem Alkohol vom Gesammtgewicht 5.3759 g, die 0.6057 g Substanz enthielt und das specifische Gewicht 0.8214 besass, drehte im 1 dm-Rohr 1.21° bei 20° und Natriumlicht nach rechts. Mithin $[\alpha]_D^{20}$ in alkoholischer Lösung 1. + 12.8°, 2. + 13.27°, 3. + 13.07° (± 0.2°).

Die 3 zuvor beschriebenen Formylkörper scheiden sich beim Verdunsten der wässrigen Lösung in ziemlich grossen Krystallen ab, deren Untersuchung Hr. Dr. F. von Wolff gütigst übernommen hat. Er wird darüber später berichten.

Hydrolyse der Formylverbindungen.

Die Abspaltung der Formylgruppe erfolgt beim Kochen mit verdünnten Mineralsäuren rasch. Die Isolirung der Aminosäure wird am bequemsten bei Anwendung von Bromwasserstoffsäure.

Man kocht die Formylverbindung mit der 10-fachen Menge Bromwasserstoffsäure von 10 pCt. 1 Stunde am Rückflusskühler, verdampft dann unter stark vermindertem Druck zur Trockne, bis das Bromhydrat krystallisirt, löst dies in kaltem Alkohol und fällt die Aminosäure mit einem kleinen Ueberschuss von concentrirter, wässriger Ammoniaklösung. Sie wird abfiltrirt und bis zur völligen Entfernung des Bromammons mit heissem Alkohol gewaschen. Die Ausbeute beträgt etwa 90 pCt. der Theorie. Zur völligen Reinigung löst man die Aminosäure in der 10-fachen Menge heissen Wassers und fällt mit viel absolutem Alkohol, wobei nur ein geringer Verlust eintritt.

In dieser Weise gewonnen, bilden die Aminosäuren sehr feine, wie Silber glänzende, mikroskopische Blättchen, die meist sechseckig ausgebildet sind. Die beiden Isomeren unterscheiden sich nicht allein durch ihr optisches Verhalten, sondern auch durch den Geschmack.

d-Valin.

Für die optischen Bestimmungen diente ein Präparat, dessen Reinheit durch die Analyse controllirt war.

0.1787 g Sbst.: 0.3363 g CO_2, 0.1515 g H_2O. — 0.1617 g Sbst.: 16.6 ccm N (18°, 760 mm).

$C_5H_{11}O_2N$. Ber. C 51.28, H 9.40, N 11.97.
Gef. » 51.33, » 9.50, » 11.88.

Das Präparat schmolz im geschlossenen Capillarrohr bei 306° (corr. 315°), also etwas höher wie der Racemkörper. Im offenen Gefäss sublimirte es sehr stark beim Erhitzen und zersetzte sich theilweise unter Anhydridbildung.

0.2181 g *d*-Valin, gelöst in 20-procentiger Salzsäure. Gesammtgewicht der Lösung 6.743 g. $d = 1.1$. Drehung im 1 dm-Rohr 1.02° bei 20° und Natriumlicht nach rechts. Mithin $[\alpha]_D^{20}$ = + 28.7° (± 0.4°).

0.4022 g *d*-Valin, gelöst in 20 procentiger Salzsäure. Gesammtgewicht der Lösung 13.200 g. $d = 1.1$. Drehung im 2 dm-Rohr bei 20° und Natriumlicht 1.93° nach rechts. Mithin $[\alpha]_D^{20}$ = + 28.8° (± 0.2°).

Erheblich kleiner ist das Drehungsvermögen in wässriger Lösung.

0.3877 g *d*-Valin gelöst in Wasser. Gesammtgewicht 8.0946 g. $d = 1.008$. Drehung im 2 dm-Rohr bei 20° und Natriumlicht 0.62° nach rechts. Mithin $[\alpha]_D^{20}$ + 6.42° (± 0.2°).

0.2507 g *d*-Valin gelöst in Wasser. Gesammtgewicht 7.0495 g. $d = 1.007$. Drehung im 2 dm-Rohr bei 20° und Natriumlicht 0.46° nach rechts. Mithin $[\alpha]_D^{20}$ = + 6.42° (± 0.3°).

Die in salzsaurer Lösung gefundene Drehung weicht von dem Werth, den E. Schulze und Winterstein im Mittel bei 16° für die natürliche Aminovaleriansäure gefunden haben (+ 27.9°), nur wenig ab, zudem haben diese Autoren schon die Möglichkeit angedeutet, dass das Drehungsvermögen ihres Präparates bei weiterer Reinigung noch etwas zunehme. Jedenfalls liegt kein Grund vor, eine Verschiedenheit des natürlichen und künstlichen Productes anzunehmen, zumal da die Gründe, die E. Fischer und Dörpinghaus für die Identität der racemisirten Aminovaleriansäure aus Horn mit dem synthetischen Product anführen, zum gleichen Schluss für die activen Körper führen. Eine weitere Unterlage für den Vergleich bieten die nachfolgenden Beobachtungen über die Löslichkeit des künstlichen *d*- und *l*-Valins und endlich über die Eigenschaften ihrer Verbindungen mit Phenylisocyanat.

l-Valin[1]).

Das für die nachfolgenden Bestimmungen dienende Präparat war ebenfalls analysirt.

0.1915 g Sbst.: 0.3606 g CO_2, 0.1653 g H_2O. — 0.1630 g Sbst.: 16.9 ccm N (18°, 760 mm).

$C_5H_{11}O_2N$. Ber. C 51.28, H 9.40, N 11.97.
Gef. » 51.36, » 9.66, » 12.00.

Löslichkeit in Wasser. Die fein gepulverte Aminosäure wurde mit Wasser 8 Stunden bei 25° geschüttelt.

4.2802 g Lösung enthielten 0.2365 g *l*-Valin. 1 Theil Valin löst sich also bei 25° in 17.1 Theilen Wasser.

Eine andere Bestimmung ergab 17.2 Theile Wasser.

Eine unter den gleichen Bedingungen ausgeführte Bestimmung ergab für *dl*-Valin die Löslichkeit 1:14.1.

Optische Bestimmungen.

0.3875 g *l*-Valin gelöst in 20-procentiger Salzsäure. Gesammtgewicht der Lösung 13.3123 g, spec. Gew. 1.1. Drehung im 2 dm-Rohr 1.86° bei 20° und Natriumlicht nach links. Mithin $[\alpha]_D^{20}$ = —29.04°.

0.3196 g *l*-Valin gelöst in 20-procentiger Salzsäure. Gesammtgewicht der Lösung 10.4143 g. $d = 1.1$. Drehung in 1 dm-Rohr bei 20° und Natriumlicht 0.96° nach links. Mithin $[\alpha]_D^{20}$ = —28.4° (± 0.4).

Um die spätere Identificirung der beiden activen Valine zu erleichtern, habe ich gerade so wie beim Racemkörper die Verbindung

[1]) Vor wenigen Tagen ist eine Mittheilung von F. Ehrlich über die Bildung dieses Valins bei der partiellen Vergährung des Racemkörpers durch Hefe erschienen. (Biochemische Zeitschrift 1, 28). Sein Präparat scheint aber nicht ganz rein gewesen zu sein, da das Drehungsvermögen in 20-procentiger Salzsäure etwas zu klein (27.36) gefunden wurde.

$C_5H_{11}NO_2$ Ref. 148a

Spaltung des *d,l*-Valin-äthylesters mit Dibenzoyl-*d*-weinsäure: Zu 7 g *d,l*-**Valin-äthylester** wird bei 21° eine Lösung von 9.1 g **Dibenzoyl-*d*-weinsäure** in 100 ccm absol. Äthanol gegeben. Es scheiden sich sehr langsam farblose Säulen ab. Nach 24 Stdn. wird abgesaugt und getrocknet: ***l*-Valin-äthylester-dibenzoyl-*d*-bi-tartrat**, Schmp. 180–181°; $[\alpha]_D^{17}$: –75.2° (*c* = 0.246, in Äthanol). Die Hydrolyse, wie oben beschrieben, ergibt *l*-Valin, $[\alpha]_D^{20}$: +12.2° (*c* = 0.9041, in Wasser). S. Fränkel und K. Gallia [6]) geben für das optisch reine *l*-Valin den Wert $[\alpha]_D^{26}$: +13.9° (*c* = 0.9, in Wasser).

Zur Mutterlauge wird so viel absol. Äther zugesetzt, bis eine Trübung auftritt. Nach der Kristallisation wird abgesaugt und getrocknet: Schmp. 180–181°, $[\alpha]_D^{18}$: –78.1° (*c* = 0.2496, in Äthanol). Die Hydrolyse, wie oben beschrieben, liefert ***l*-Valin**, $[\alpha]_D^{21}$: +12.1° (*c* = 0.913, in Wasser).

Von der Mutterlauge wird das Lösungsmittel i. Vak. entfernt. Beim Behandeln des Rückstandes mit Äther, Abtrennen und Verdampfen desselben hinterbleiben 3.2 g des überschüss. Esters. Der krist. Rückstand schmilzt zwischen 148 und 150°, $[\alpha]_D^{17}$: –57.6° (*c* = 0.243, in Äthanol). Hydrolyse wie oben beschrieben: ***d*-Valin**, $[\alpha]_D^{21}$: –12.2° (*c* = 0.9398, in Wasser).

In vier Ansätzen wurde auf Reproduzierbarkeit geprüft:

		Schmp.	$[\alpha]_D$	*l*-Valin
1. und 2. Fraktion:	1.	186°	–70.9°	+10.5°
		182–183°	–77.5°	
	2.	184–193°	–77.2°	+11.9°
		173–179°	–71.2°	
	3.	182–184°	–75.7°	+13.3°
		179–181°	–76.6°	
	4.	183–185°	–77.9°	+ 6.3°
		179–181°	–72.4°	

		Schmp.	$[\alpha]_D$	*d*-Valin
3. Fraktion:	1.	155–161°	–49.9°	– 6.9°
	2.	154–159°	–53.0°	– 6.8°
	3.	153–157°	–53.2°	–11.6°
	4.	146–152°	–50.9°	opt. inakt.

[5]) Liebigs Ann. Chem. 357, 9 [1907]. [6]) Biochem. Z. 184, 315 [1923].

$C_5H_{11}NO_2$ Ref. 148b

B. Racematspaltung des Valinhydrazides

Die Lösung von 4,0 g D,L-Valinhydrazid in 40 cm³ abs. Methanol wurde bei Zimmertemperatur mit einer Lösung von 11,5 g Dibenzoyl-D-weinsäure in 60 cm³ abs. Methanol vereinigt. Nach einem Tag hatte sich das **L-Valinhydrazid-dibenzoyl-D-hydrogentartrat** fast vollständig ausgeschieden.

Ausbeute: 6,7 g, entspr. 90% d. Th.; Schmp. 198–200°.
$[\alpha]_D^{21}$: –78,2° (c = 0,44 in abs. Methanol).

$C_{23}H_{27}N_3O_9$ (489,5)	ber.	C 56,43	H 5,56	N 8,58
	gef.	56,48	5,83	8,17.

Eine optisch unreine Zwischenfraktion von etwa 1 g wurde beim Einengen der Mutterlauge im Vakuum gewonnen. Durch völliges Eindampfen und Anreiben mit abs. Äther ließ sich das D-Valinhydrazid-dibenzoyl-D-hydrogentartrat gewinnen.

Ausbeute: 6,5 g, entspr. 87% d. Th., Schmp. 168°.
$[\alpha]_D^{21}$: –94,7° (c = 0,40 in abs. Methanol).

$C_{23}H_{27}N_3O_9$ (489,5) ber. N 8,58 gef. N 8,74.

Gewinnung der antipodischen Valinhydrazid-dihydrochloride

Das L- sowie D-Hydrazidtartrat aus dem Spaltansatz wurde zur Verbesserung der optischen Reinheit zweimal aus Methanol—Äther umkristallisiert. Aus den so gereinigten Tartraten ließen sich die antipodischen Hydrazid-dihydrochloride und die freien Aminosäuren in optisch reiner Form darstellen.

L-Valinhydrazid-dibenzoyl-D-hydrogentartrat:
Ausbeute: 65% d. Th., bezogen auf Rohsalz; Schmp. 212–213°.
$[\alpha]_D^{21}$: –75,7° (c = 0,45 in abs. Methanol).

D-Valinhydrazid-dibenzoyl-D-hydrogentartrat:
Ausbeute: 60% d. Th., bezogen auf Rohsalz; Schmp. 174–175°.
$[\alpha]_D^{21}$: –98,8° (c = 0,40 in abs. Methanol).

$C_{23}H_{27}N_3O_9$ (489,5)	ber.	C 56,43	H 5,56	N 8,58
L-Verb.	gef.	56,36	5,85	8,80
D-Verb.	gef.	56,44	5,82	8,30.

Zur Herstellung der antipodischen Valinhydrazid-dihydrochloride wurden die reinen diastereomeren Tartrate mit 3proz. wäßriger Salzsäure erwärmt, bis sich die Dibenzoylweinsäure als Öl absetzte und sich im Kutscher-Steudel-Apparat extrahieren ließ. Durch Eindampfen der salzsauren Lösung im Vakuum und mehrmaliges Waschen mit Äther wurden die Hydrazid-dihydrochloride gewonnen.

L-Verbindung: Ausbeute: 80% d. Th., bezogen auf Tartrat; Schmp. 200–202°.
$[\alpha]_D^{21}$: +47,5° (c = 0,65 in Wasser).

D-Verbindung: Ausbeute: 80% d. Th., bezogen auf Tartrat; Schmp. 200–201°.
$[\alpha]_D^{20}$: –47,8° (c = 0,71 in Wasser).

$C_5H_{13}O_3N \cdot 2HCl$ (204,1)	ber.	C 29,42	H 7,41	N 20,58
L-Verb.	gef.	29,51	7,53	20,50
D-Verb.	gef.	29,58	7,30	20,68.

Durch Umkristallisieren ließ sich der Drehwert der Hydraziddihydrochloride nicht weiter steigern. Ihre Verseifung führte zu den optisch reinen Aminosäureantipoden. Demgemäß lagen optisch reine Stoffe vor.

Die aus dem Spaltansatz isolierbaren diastereomeren Rohsalze lieferten ein L-Valinhydrazid-dihydrochlorid mit einer spezifischen Drehung $[\alpha]_D^{21}$: +39,5°, d. h. 83proz. optischer Reinheit und ein D-Valinhydrazid-dihydrochlorid mit einem Drehwert von $[\alpha]_D^{21}$: –31,2°, also 65proz. optischer Reinheit.

L-Verb.	gef.	C 29,02	H 7,53	N 20,26
D-Verb.	gef.	29,50	7,41	20,80.

Die Verseifung der reinen antipodischen Valinhydrazid-dihydrochloride wurde durch 1–2stündiges Erhitzen mit 20proz. Salzsäure ausgeführt. Nach Abdestillieren der Salzsäure im Vakuum wurde der Rückstand mit abs. Isopropylalkohol behandelt, wobei der größte Teil des Hydrazin-hydrochlorides ungelöst zurückblieb. Das Filtrat wurde im Vakuum eingedampft, der Rückstand in wenig Wasser gelöst und mit 2proz. Ammoniaklösung neutralisiert. Ein Teil der Aminosäure fiel aus, die Hauptmenge wurde durch Zugabe von Aceton gefällt.

Nach Umfällen aus Wasser—Aceton lieferten die so gewonnenen Valinantipoden folgende Daten:

L-Valin: Ausbeute: 60% d. Th.; Schmp. 270–275°.
$[\alpha]_D^{21}$: +13,8° (c = 1,70 in Wasser).

D-Valin: Ausbeute: 60% d. Th.; Schmp. 280°.
$[\alpha]_D^{20}$: –13,95° (c = 1,85 in Wasser).

$C_5H_{11}O_2N$ (117,2)	ber.	C 51,26	H 9,46	N 11,95
L-Verb.	gef.	51,22	9,52	11,62
D-Verb.	gef.	50,97	9,42	11,66.

$C_5H_{11}NO_2$ Ref. 148c

2. Spaltung des D,L-Valin-thiophenylesters

14,4 g frisch dargestellter D,L-Valin-thiophenylester werden mit 150 cm³ abs. Methanol verdünnt. Nach Zugeben einer filtrierten Lösung von 26 g Dibenzoyl-D-weinsäure in 74 cm³ abs. Methanol setzt die Kristallisation des D-Valin-thiophenylester-dibenzoyl-D-hydrogentartrates nach etwa 10 Minuten ein. Zur völligen Auskristallisation läßt man die Lösung 8 Tage erschütterungsfrei stehen. Nach Absaugen des D-Ester-dibenzoyl-D-hydrogentartrates dampft die Lösung im Vakuum bei Zimmertemperatur ein und erhält durch Ätherzusatz das diastereomere L-Ester-dibenzoyl-D-hydrogentartrat.

D-Valin-thiophenylester-dibenzoyl-D-hydrogentartrat

Ausbeute: 15,91 g (78,76% d. Th.) Schmp. (korr.) 173–174°.

$[\alpha]_D^{20}$: –73,74° (c = 0,956 in Methanol).

Das optisch reine Tartrat wurde durch zweimaliges Umkristallisieren aus Methanol/Äther erhalten: $[\alpha]_D^{20}$: –65,93° (c = 0,644 in Methanol) Schmp. (korr.) 180°.

$C_{29}H_{29}O_9NS$ (567,6)

ber. C 61,37; H 5,15; N 2,47;
gef. C 61,13; H 5,40; N 2,86.

L-Valin-thiophenylester-dibenzoyl-D-hydrogentartrat

Ausbeute: 15,06 g (74,06% d. Th.) Schmp. (korr.) 168—169°.

$[\alpha]_D^{20}$: −94,19° (c = 1,136 in Methanol).

Das optisch reine Tartrat wurde durch zweimaliges Umkristallisieren aus Äthanol/Äther erhalten: $[\alpha]_D^{20}$: −83,85° (c = 0,628 in Methanol) Schmp. (korr.) 171—172°.

$C_{29}H_{29}O_9NS$ (567,6)

ber.: C 61,37; H 5,15; N 2,47;
gef.: C 60,92; H 5,44; N 2,74.

Die Gesamtausbeute an Tartrat, bezogen auf eingesetztes Hydrobromid, beträgt 30,97 g (76,66% d. Th.).

3. Gewinnung der optisch aktiven Valin-thiophenylester-hydrochloride

D-Valin-thiophenylester-hydrochlorid

2 g rohes D-Estertartrat werden in 5 cm³ abs. Äthanol suspendiert. Unter Eiskühlung wird bis zur völligen Lösung trockener Chlorwasserstoff eingeleitet. Nach Zusatz von viel abs. Äther fallen 0,76 g (87,7% d. Th.) D-Esterhydrochlorid aus. Dampft man das Äthanol im Vakuum ab, so wird das Hydrochlorid als hygroskopisches Öl erhalten, das nach einiger Zeit mit viel abs. Äther kristallisiert.

$[\alpha]_D^{20}$: +55,11° (c = 0,998 in Wasser), Schmp. (korr.) 174°.

Nach zweimaligem Umkristallisieren aus i-Propanol/Äther wurden folgende Fraktionen erhalten:

$[\alpha]_D^{20}$: +106,80° (c = 0,632 in Wasser) Schmp. (korr.) 180—181°.
$[\alpha]_D^{20}$: +113,30° (c = 0,812 in Wasser) Schmp. (korr.) 180—181°.

Die spezifische Drehung der letzten Fraktion ließ sich durch Umkristallisieren nicht weiter steigern.

$C_{11}H_{16}ONSCl$ (245,78)

ber. C 53,75; H 6,56; N 5,70;
gef. C 53,47; H 6,63; N 5,82.

L-Valin-thiophenylester-hydrochlorid

Zur Darstellung wird das optisch reine L-Ester-D-hydrogentartrat in gleicher Weise wie beim Antipoden beschrieben in das L-Valin-thiophenylester-hydrochlorid überführt. Ausbeute: 89% Schmp. (korr,) 179—180°. $[\alpha]_D^{20}$: −103,52 (c = 0,966 in Wasser).

Nach Umkristallisieren aus i-Propanol/Äther: $[\alpha]_D^{20}$: −110,05° (c = 0,78 in Wasser) Schmp. (korr.) 180°.

$C_{11}H_{16}ONSCl$ (245,78)

ber. C 53,75; H 6,56; N 5,70;
gef. C 53,38; H 6,60; N 5,95.

$C_5H_{11}NO_2$ Ref. 148d

Resolution of N-Acetyl-DL-valine.

N-Acetyl-DL-valine (79.6 g, 0.5 mol) and L-leucinamide (71.5 g, 0.55 mol) were dissolved in 1 l of 99% ethanol on a water bath at 50°C. After being cooled at room temperature the L-leucinamide salt of N-acetyl-L-valine was crystallized as fine long lath. Yield, 36.5 g, 50.5% based on L-form; m.p. 178~180°C; $[\alpha]_D^{20}$+0.96° (c, 4: water). After twice recrystallization from 250 ml of hot ethanol, 25 g of pure N-acetyl-L-valine salt was obtained. Yield, 25% based on L-form; m.p. 181~182°C; $[\alpha]_D^{20}$+0.25° (c, 4: water).

The pure L-leucinamide salt (21.5 g) obtained as described above was dissolved in 50 ml of water. The solution was passed through a column of Diaion SK#1 (H-form) to separate N-acetyl-L-valine. The effluent was concentrated to 20 ml in vacuo. After being kept in a refrigerator overnight, 12 g of N-acetyl-L-valine was obtained; $[\alpha]_D^{20}$-0.4° (c, 2: methanol).

At the further step, N-acetyl-L-valine was hydrolyzed with 20% hydrochloric acid. The pure L-valine was obtained as same as described in the previous instance (Yield, 8.8 g), and the recrystallization from 75 ml of water and 100 ml of ethanol gave 8 g of pure crystal; $[\alpha]_D^{19}$+27.2° (c, 4: 6 N HCl). The value of +27.4° (c, 3: 6 N HCl) was reported by V. E. Price et al.[22)]

Resolution of N-Acetyl-DL-valine.

N-Acetyl-DL-valine 7.8 g (0.049 mol) and L-tyrosin-

22) V.E. Price, J.B. Gilbert and J.P. Greenstein, *J. Biol. Chem.*, **179**, 503 (1949).

amide 9 g (0.05 mol) were dissolved in 50 ml of methanol on a water bath at 50°C and the solution was let to stand at room temperature. Crystallization occurred slowly, giving 6.6 g of slightly impure salt of N-acetyl-L-valine. Twice recrystallization from methanol to remove a trace of D-salt gave pure L-tyrosinamide N-acetyl L-valinate. Yield, 3.7 g (44% based on L-form); m.p. 171°C.

L-Valine.

The solution of L-tyrosinamide salt 3 g in 200 ml of water was treated with a column of Diaion SK#1 (H-form) as described previously. After recrystallization from water, N-acetyl-L-valine was obtained. Yield, 1.35 g (90%); m.p. 164°C; $[\alpha]_D^{19}+0.38°$ (c, 4: MeOH). The hydrolysis of N-acetyl-L-valine in 20% hydrochloric acid gave pure L-valine. Yield, 88%; $[\alpha]_D^{20}+28.17°$ (c, 4: 6 N HCl).

Resolution of DL-Valinamide Attempted with N-Acetyl-L-leucine.

DL-Valinamide (17.4 g, 0.15 mol) and N-acetyl-L-leucine (28.5 g, 0.165 mol) were dissolved in 220 ml of ethanol on a water bath at 50°C and the solution was let to stand for 5 hrs, at room temperature. The deposited fine needle crystals were filtered off, washed with a little amount of cold ethanol and dried in vacuo. Yield, 10.2 g (44.8% based on L-form); m.p. 176°C; $[\alpha]_D^{19.5}-3.23°$ (c, 5: H_2O).

The salt of L-valinamide was dissolved in 200 ml of water and the solution was passed through a column of diaion SK#1 (H-form). The adsorbed L-valinamide was eluted with N-hydrochloric acid and the effluent was boiled under reflexing for one hour.

After having been evaporated to dryness in vacuo, the residue was dissolved in 200 ml of water, passed through a column of Diaion SK#1 (H-form). The absorbed L-valine was recovered by eluting with 2 N aqueous ammonia, and the eluate was concentrated in vacuo. When crystallization began, a small amount of ethanol was added. After having been kept in a refrigerator overnight, the crystals of L-valine were collected on a filter, washed with ethanol, and dried in vacuo. Yield, 3.77 g (93%); $[\alpha]_D^{20.5}+27.19°$ (c, 4: 6 N HCl). *Anal.* Found: N, 11.84. Calcd. for $C_5H_{11}O_2N$: N, 11.96%.

$C_5H_{11}NO_2$ Ref. 148e

EXAMPLE 2

Resolving of N,N-dibenzyl-DL-valine

(A) PREPARATION OF THE SALT OF L(+)-THREO-1-p-NITRO PHENYL-2-AMINO PROPANE-1,3-DIOL AND N,N-DIBENZYL-L(−)-VALINE

11.9 g. of N,N-dibenzyl-DL-valine and 8.9 g. of L(+)-threo-1-p-nitro phenyl - 2 - amino propane-1,3 - diol are heated under reflux in 150 cc. of methanol. 150 cc. of warm water are gradually added thereto. The mixture is allowed to cool slowly to 25° C. The precipitated salt is filtered off, washed with a small quantity of 50% methanol and recrystallized from 110 cc. of 50% methanol. The crystals are filtered off at 25° C., whereby 8.9 g. of the salt of L(+)-threo-1-p-nitro phenyl-2-amino propane-1,3-diol and N,N-dibenzyl-L(−)-valine are obtained. The yield amounts to 44% corresponding to a resolving yield of 88%. The salt is obtained in the form of yellow leaflets which are insoluble in water, rather soluble in ethanol, slightly soluble in acetone, benzene, and chloroform, and insoluble in ether. The salt melts at 158–160° C., its rotatory power is

$$[\alpha]_D^{20}=-41.5°\pm1°$$

(concentration: 1% in methanol).

(B) PREPARATION OF THE SALT OF D(−)-THREO-1-p-NITRO PHENYL-2-AMINO PROPANE-1,3-DIOL AND N,N-DIBENZYL-D(+)-VALINE

The mother liquors obtained on preparing and recrystallizing the salt of L(+)-threo-1-p-nitro phenyl-2-amino propane-1,3-diol and N,N-dibenzyl-L(−)-valine obtained as described hereinabove under (A), are collected and concentrated by evaporation in a vacuum to a volume of about 130 cc. 24 cc. of N sodium hydroxide solution are added thereto. The mixture is cooled and the resulting crystals are filtered off. 3 g. of the resolving agent, corresponding to 34% of the amount initially used for resolving, are recovered. The filtrate is acidified by the addition of 3 cc. of acetic acid. The oily precipitate is separated, dissolved in 55 cc. of warm methanol, and treated as indicated hereinabove under (A) with 4.7 g. of D(−)-threo-1-p-nitro phenyl-2-amino propane-1,3-diol, thereby yielding the corresponding salt of said D(−)-threo-1-p-nitro phenyl-2-amino propane-1,3-diol and N,N-dibenzyl-D(+)-valine. The yield amounts to 79% calculated for the initially employed racemic mixture. The compound is obtained in the form of yellow leaflets which are insoluble in water, rather soluble in ethanol, slightly soluble in acetone, benzene, and chloroform, and insoluble in ether. The salt melts at 158–160° C.; its rotatory power is $[\alpha]_D^{20}=+41°\pm1°$ (concentration: 1% in methanol).

(C) PREPARATION OF N,N-DIBENZYL-L(−)-VALINE

8.9 g. of the salt of L(+)-threo-1-p-nitro phenyl-2-amino propane-1,3-diol and N,N-dibenzyl-L(−)-valine obtained as described hereinabove under (A), are vigorously agitated with 19 cc. of N sodium hydroxide solution for 15 minutes. L(+)-threo-1-p-nitro phenyl-2-amino propane-1,3-diol crystallizes and is separated from the solution. 3.35 g. thereof corresponding to 38% of the initially employed resolving agent are recovered thereby. The filtrate is acidified by the addition of 2.5 cc. of acetic acid. The acidified mixture is extracted three times with 20 cc., 10 cc., and 10 cc. of chloroform. The combined chloroform solutions are washed with 15 cc. of water, dried, and evaporated to dryness in a vacuum. The resulting oily residue solidifies on trituration with petroleum ether. It is dissolved in dilute sodium hydroxide solution and is caused to crystallize therefrom by acidifying the solution with acetic acid. 4.75 g. of N,N-dibenzyl-L(−)-valine are obtained thereby in hydrated form. The yield amounts to 92% of the theoretical yield. The melting point of the anhydrous product is 75° C. and its rotatory power $[\alpha]_D^{20}=-119.5°\pm1°$ (concentration: 2% in methanol). The compound is obtained in the form of white needles which are insoluble in water and soluble in ethanol, ether, acetone, benzene, and chloroform.

(D) PREPARATION OF L(+)-VALINE

2 g. of N,N-dibenzyl-L(−)-valine are subjected to hydrogenolysis in 20 cc. of 80% ethanol in the presence of 1 g. of a palladium black catalyst containing 6% of palladium and with the addition of 0.7 cc. of concentrated hydrochloric acid at a temperature of 70° C. After hydrogen absorption ceases, the catalyst is filtered off and washer with ethanol. The filtrate is concentrated by evaporation to a small volume. 1.3 cc. of pyridine are added thereto. The crystallized L(+)-valine is filtered off and washed with ethanol. 660 mg. of optically pure L(+)-valine corresponding to a yield of 80% of the theoretical yield are obtained thereby. Its rotatory power is $[\alpha]_D^{20}=+27.5$ g.±1° (concentration: 2% in 5 N hydrochloric acid.

$C_5H_{11}NO_2$ Ref. 148f

Resolution of N-Acetyl-DL-Valine

A mixture of 2700 g. of N-acetyl-DL-valine, 3600 g. of L-threo-2-amino-1-(p-methylmercaptophenyl)-1,3-propanediol, and 40 liters of 95% ethanol was warmed to 50° C. in order to effect complete solution of the solid ingredients. The solution was then cooled to 40° C., and it was seeded with a few crystals of the L-threo-2-amino-1-(p-methylmercaptophenyl)-1,3-propanediol salt of N-acetyl-L-valine. The solution was allowed to stand for twenty-four hours at 25° C. with occasional stirring. There was very heavy crystallization of solid from the solution. The mixture was placed on a filter, sucked dry, washed with two liters of cold 95% ethanol, and dried at 50–70° C. There was thus obtained 1900 g. of the L - threo - 2 - amino - 1 - (p - methylmercaptophenyl)-1,3-propanediol salt of N-acetyl-L-valine which melted at 182–184° C.

The filtrate and wash liquid were combined, cooled to 5° C. seeded with a few crystals of the L-threo-2-amino - 1 - (p - methylmercaptophenyl) - 1,3-propanediol salt of N-acetyl-D-valine, and the solution was allowed to stand for twenty-four hours at 5° C. The solid which had separated from solution was then collected on a filter and washed with cold anhydrous ethanol. There was thus obtained 1300 g. of L-threo-2-amino-1-(p-methylmercaptophenyl)-1,3-propanediol salt of N-acetyl-D-valine which melted at 155–158° C.

Additional crops of the L-threo-amine-L-acid and L-threo-amine-D-acid salts were obtained as follows. The filtrate from the crop of L-threo-amine-D-acid salt was concentrated to a volume of fifteen liters by evaporation, seeded with a few crystals of the L-threo-amine-L-acid salt, and the solution was allowed to stand for twenty-four hours at 25° C. The solid which had separated from solution was collected on a filter and dried. There was thus obtained a second crop of 590 g. of the L-threo - 2 - amino - 1 - (p - methylmercaptophenyl) - 1,3-propanediol salt of N-acetyl-L-valine which melted at 183–185° C. The filtrate was cooled to 5° C. for twenty-four hours, and the solid which had separated from solution was collected on a filter. There was thus obtained a second crop of 500 g. of crude L-threo-2-amino-1-(p-methylmercaptophenyl)-1,3-propanediol salt of N-acetyl-D-valine which melted at 154–157° C.

N-Acetyl-L-Valine

2700 g. of the L-threo-2-amino-1-(p-methylmercaptophenyl)1,3-propanediol salt of N-acetyl-L-valine was dissolved in four liters of hot water, the solution was cooled until crystallization of solid began, and 600 ml. of 35% sodium hydroxide solution was added. The solution was then cooled to 5° C. The solid which separated from solution was collected on a filter, washed well with water and then with benzene, and dried at 50° C. There was thus obtained 1525 g. of L-threo-2-amino-1-(p-methylmercaptophenyl)-1,3-propanediol. The filtrate and the aqueous wash were combined and acidified by addition of 600 ml. of concentrated hydrochloric acid, and the solution was saturated with sodium chloride and cooled to 5° C. The solid which separated from solution was collected on a filter, washed with a few ml. of cold water and dried at 70° C. There was thus obtained 1115 g. of N-acetyl-L-valine which melted at 168–169° C. By extraction of the mother liquor with ethyl acetate there was recovered an additional 50 g. of crude N-acetyl-L-valine which melted at 150–158° C.

Recrystallization of the N-acetyl-L-valine from water yielded a purified product melting at 170–171° C. and having $[\alpha]_D^{25}$ −31.2° (c. 4% in water).

N-Acetyl-D-Valine

1950 g. of the L-threo-2-amino-1-(p-methylmercaptophenyl)-1,3-propanediol salt of N-acetyl-D-valine was dissolved in four liters of water, and the solution was mixed with 440 ml. of 35% aqueous sodium hydroxide solution. The solution was then cooled to 10° C. and the solid which separated from solution was collected on a filter. There was thus obtained 1100 g. of L-threo-2-amino - 1 - (p-methylmercaptophenyl)-1,3-propanediol. The filtrate was acidified with 440 ml. of concentrated hydrochloric acid, saturated with sodium chloride, and cooled to 5° C. The solid which separated from solution was collected on a filter, thus yielding 730 g. of N-acetyl-D-valine which melted at 167–169° C.; and a second crop of the same product which weighed 53 g. and melted at 150–160° C. was obtained by making the filtrate alkaline by addition of sodium hydroxide solution, removing the precipitated L-threo-2-amino-1-(p-methylmercaptophenyl)-1,3-propanediol by filtration, and reacidifying the filtrate with hydrochloric acid.

Recrystallization of the N-acetyl-D-valine from water yielded a purified product melting at 170–171° C. and having $[\alpha]_D^{25}$ +31.4° (c. 4% in water).

The N-acetyl-L-valine and N-acetyl-D-valine were hydrolyzed by heating with hydrochloric acid to yield L-valine and D-valine respectively.

$C_5H_{11}NO_2$ Ref. 148g

For a resolution procedure using ephedrine in the direct resolution of the racemic N-benzyloxycarbonyl derivative of the above compound, see Ref. 20h.

$C_5H_{11}NO_2$ Ref. 148h

Example 2

1.57 g. of *dl*-N-acetylvaline and 1.35 g. of *l*-amphetamine are dissolved with heating in 20 ml. of water. The solution is allowed to stand for 16 to 20 hours at room temperature and the resulting crystalline precipitate consisting of the *l*-amphetamine salt of *l*-N-acetylvaline is removed by filtration; $[\alpha]_D^{25} = -14°$ (1.37% in water). The crystalline product is dissolved in water and the solution is passed through an ion exchange column containing a sulfonate resin in the acid form (Dowex 50). The effluent is concentrated until crystals begin to form and is then cooled. The resulting crystalline product consisting of *l*-N-acetylvaline is filtered off and dried; $[\alpha]_D^{25} = -4°$ (2% in ethanol).

The filtrate first obtained above is made alkaline with dilute aqueous sodium hydroxide solution and extracted with several portions of ether. The residual solution is then concentrated and acidified to obtain the crystalline product, *d*-N-acetylvaline; $[\alpha]_D^{25} = +4°$ (2% in ethanol) after recrystallization from water.

$C_5H_{11}NO_2$ Ref. 148i

Resolution of N-Acetyl-DL-valine.—Crystalline salts were formed with α-phenylethylamine and with ephedrine, but no resolution was effected by crystallization from various solvents. Resolution was easily effected with (−)-α-fenchylamine. In a typical experiment the amine (153 g., 1 mole) and acetylvaline (160 g., 1 mole) were dissolved by heating in 1800 cc. of water. Slow cooling gave massive hexagonal prisms of the N-acetyl-D-valine salt and additional crops of this salt (total 156 g.) were obtained by concentrating the filtrate in stages to 200 cc. The viscous solution then gradually solidified but the more soluble salt was not obtained pure. The less soluble salt was already substantially pure; recrystallization from aqueous methanol gave 146 g. (93%) of the pure salt, m.p. 216°, $[\alpha]^{25}D$ −4.22° (*c* 8, methanol). The solubilities are: water, 4.5; methanol, 8.5. The salt can be recrystallized from a large volume of water but some volatilization of amine attends the dissolution of the salt and subsequent evaporations of the liquors. Loss from this cause may be avoided by distilling the original solution and recrystallization liquors under reduced pressure. Alternatively the salt may be dissolved in about 3 parts of hot methanol and the solution diluted with an equal volume of hot water. The salt is recovered by successive cooling and distilling of solvent without serious loss of amine. The less soluble salt can be crystallized from methanol but the original salt mixture is not readily resolved in this solvent.

N-Acetyl-D-valine was recovered from its salt (146 g.) by the general procedure already described. About 80% separated on addition of an exact equivalent of hydrochloric acid. The remainder was obtained by evaporating the filtrate to dryness and extracting the residue with acetone. The yield after recrystallization was 70.5 g. (88% based on the racemic form originally taken). The compound was purified by crystallization from water (laminated rhomboidal plates) or acetone (massive rhombs) and then had m.p. 164–165°; neut. equiv., 160; $[\alpha]^{25}D$ −0.5° (*c* 12, methanol); −3.4° (*c* 12, ethanol); −9.36° (*c* 4, glac. acetic acid); +20.05° (*c* 4, water). Synge[19] reports m.p. 164°; $[\alpha]^{25}D$ +4.0° (*c* 2, ethanol) for N-acetyl-L-valine. The solubilities are: water, 6.94; acetone, 5.54; ethyl acetate, 0.73; chloroform, 0.24. It may be noted that although the active form melts distinctly higher than the DL-form it is nevertheless more soluble in most solvents.

N-Acetyl-L-valine.—The crude substance (82 g.) obtained from mother liquors of the resolution had $[\alpha]^{25}D$ +8.27° (*c* 8, glac. acetic acid) and hence contained about 93% of the L-form. It could not be purified by crystallization from common solvents except that the less soluble DL-form was partially removed in the head fractions. It was purified by two methods. (a) A portion (50 g., 0.313 mole) was combined with 48 g. (0.313 mole) of (+)-α-fenchylamine in 300 cc. of methanol and 400 cc. of water and several crops of salt totalling 79 g. (89.5%) were obtained as previously described. The salt had $[\alpha]^{25}D$ +4.18° (*c* 8, methanol) and otherwise closely resembled the enantiomorphous form described above. Decomposition gave 38.6 g. of pure N-acetyl-L-valine, m.p. 164–165°; $[\alpha]^{25}D$ −20.08° (*c* 4, water) and other properties substantially identical with those of the D-form. The salts in the mother liquors gave 9.4 g. of crude acetylvaline containing excess D-form.

(b) A composite sample of crude N-acetyl-L-valine (54.4 g., 0.34 mole) calculated to contain 80% of the L-form and 20% of the DL-form was combined in aqueous methanol with 93 g. (0.61 mole) of DL-α-fenchylamine and 0.27 mole of hydrochloric acid. The proportions were calculated to produce 84.5 g. of N-acetyl-L-valine-(+)-fenchylamine salt, 21.3 g. of N-acetyl-DL-valine-DL-fenchylamine salt and 51.1 g. of (−)-fenchylamine hydrochloride. Fractionation in the usual manner gave 72 g. (85.2%) of the pure N-acetyl-L-valine-(+)-amine salt, $[\alpha]^{25}D$ +4.17° (*c* 8, methanol). Decomposition of this gave 34 g. of pure N-acetyl-L-valine and 33 g. of pure (+)-fenchylamine. Intermediate fractions in the resolution, rich in DL-acid-DL-amine salt, could not be purified. This salt was prepared from the pure components in a separate experiment and had m.p. 182–185°. The solubilities are: water 13.4; methanol, 25.3. The combined intermediate fractions and mother liquors of the resolution gave on decomposition 18.0 g. of mixed acetylvalines and 57 g. of mixed fenchylamines suitable for further processing.

D-Valine.—Pure N-acetyl-D-valine (8.0 g.) was hydrolyzed and the amino acid recovered as described for D-phenylalanine, except that two equivalents of hydrochloric acid were used. The initial product (5.9 g.) had $[\alpha]^{25}D$ −5.93° (*c* 4, water). Recrystallization from water gave 3.0 g., $[\alpha]^{25}D$ −6.1° (*c* 4, water); −23.6° (*c* 4.2, *N* HCl); −27.4° (*c* 4.2, 6 *N* HCl). A sample from another resolution had $[\alpha]^{25}D$ −27.7° (*c* 4, 6 *N* HCl). The value −27.1° (*c* 1, 6 *N* HCl) has been reported.[22a]

L-Valine.—N-Acetyl-L-valine (6 g.) similarly gave L-valine (4.3 g.) which, without recrystallization had $[\alpha]^{25}D$ +6.3° (*c* 4, water); +23.4° (*c* 4, *N* HCl); +27.4° (*c* 4, 6 *N* HCl). These values were not substantially changed by recrystallization.

$C_5H_{11}NO_2$ Ref. 148j

A resolution procedure for the above compound using menthoxyacetyl chloride as resolving agentmay be found in Ref. 18i.

Note:

A resolution procedure for racemic valine using acetylmandelyl chloride as resolving agent may be found in: S. Berlingozzi, G. Adembri and G. Bucci, Sperimentale, Sez. Chim. Biol., 6, 1 (1955).

$C_5H_{11}NO_2$ Ref. 148k

Experimental. Isobutyl ester of DL-*valine.* — 500 g of DL-valine are suspended in 2 500 ml of anhydrous isobutyl alcohol containing 10 per cent of anhydrous hydrochloric acid. The mixture is refluxed for three hours, after which a clear solution is obtained. The solution is evaporated *in vacuo* to an oil. This is mixed with 2 500 ml of anhydrous isobutyl ester containing 10 per cent of hydrochloric acid. After evaporation, the treatment with isobutyl alcohol and hydrochloric acid is repeated. The oil thus obtained is then mixed with 1 000 ml of ether and 500 ml of water and the mixture cooled to − 10° C. Concentrated ammonia is added with stirring until the reaction is faintly alkaline to phenolphthalein. The ether layer is separated and the aqueous solution extracted twice more, each time with 1 000 ml of ether. The ether extracts are dried over sodium sulfate. The sodium sulfate is removed and washed with 100 ml of anhydrous ether. The ether extracts are distilled *in vacuo* and the fraction distilling between 73—77° C at 5 mm contains the isobutyl ester of DL-valine; yield 610—650 g (83 to 88 per cent of theory). $C_9H_{19}O_2N$ Calculated, N 8.09%; found, N 8.12%

Enzyme preparation. — The enzymes used in the present investigation were produced as follows. — 180 g of pancreas dry powder according to Willstätter and Waldschmidt-Leitz [12] are shaken for two hours with 1 800 ml of glycerol. The mixture is filtered after the addition of 1 800 ml of water and to the clear solution two volumes of acetone are added. The precipitate is filtered and the precipitate is washed with 200 ml of acetone and 200 ml of ether. Yield 30—40 g.

The enzyme preparation is dissolved by shaking for 30 minutes in distilled water immediately before use. The rate of hydrolysis is shown in Fig. 1. It is evident from the figure that the D-ester is stable against the pancreatic enzymes for at least 72 hours and that, with the amount of enzymes used, practically all the L-ester is hydrolyzed after 24 hours. The fact that hydrolysis with enzymes is asymmetrical is shown by the method for the preparation of D-valine described in the following. The investigations showed that, under similar experimental conditions, the isobutyl ester of DL-valine is also hydrolyzed to some extent with water.

C₅H₁₁NO₂

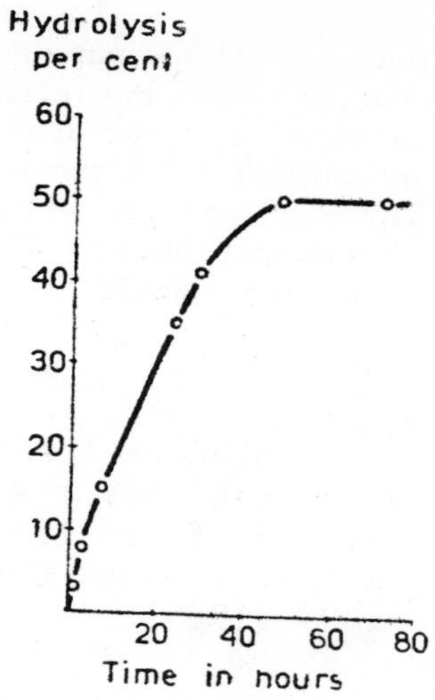

Fig. 1. Asymmetric hydrolysis of the isobutyl ester of DL-valine with a pancreatic enzyme preparation.

The experiments were made with 1 ml of the isobutyl ester of DL-valine and 5 ml of a 4 per cent pancreatic enzyme solution. The reaction was carried out at room temperature (22° C). Hydrolysis was followed by formol titration. The values given in the curve were obtained after the hydrolysis which occurs when 1 ml of the isopropyl ester is mixed with 5 ml of distilled water had been substracted from the experimental values. Under these conditions, hydrolysis of the isobutyl ester of DL-valine with water amounted to 4.0 per cent after 24 hours, 7.1 per cent after 48 hours and 9.1 per cent after 72 hours.

Preparation of D-valine. — 100 g of the isobutyl ester of DL-valine are shaken for 48 hours with 400 ml of a 5 per cent solution of pancreatic enzymes at room temperature. 400 ml of ether and 20 ml of concentrated ammonia are added to the mixture. The ether layer is separated and the aqueous solution extracted twice more, each time with 100 ml of ether. The ether extracts are dried over sodium sulfate and the ether evaporated after the sodium sulfate has been removed. The remaining oil is shaken for 48 hours at room temperature with 100 ml of 5 per cent pancreatic enzyme solution. While stirring 100 ml of ether and 5 ml of concentrated ammonia are added to the digestion mixture. The ether layer is separated and the aqueous solution extracted twice more, each time with 100 ml of ether. The ether extracts are dried over sodium sulfate. The sodium sulfate is removed and the ether is evaporated. The oil of the isobutyl ester of D-valine thus obtained is refluxed for 12 hours with 350 ml of 20 per cent hydrochloric acid. The solution is evaporated *in vacuo* to dryness and for a second time after adding 100 ml of water. The hydrochloride of D-valine thus obtained is dissolved in 200 ml of 96 per cent alcohol. 50 ml of pyridine are added and the mixture is placed overnight in the refrigerator. The precipitate is filtered off and washed with 50 ml of alcohol and 50 ml of ether. The impure D-valine is recrystallized by dissolving in hot water (11 ml per g), decolorized with 0.5 g of norit and the clear solution treated with an equal volume of 95 per cent alcohol. The precipitate is washed with absolute alcohol and ether as described. To obtain an analytically pure sample, the D-valine is again recrystallized and washed as outlined above. The product thus obtained weighs 14—16 g (41 to 47 per cent of theory). Analysis of the D-valine obtained with this method gave the following results:

C₅H₁₁O₂N Calculated, N 11.96; found, N 11.95
$\alpha_D^{23} = -28.2$ to $-28.6°$ ($c = 2$, in 6 N hydrochloric acid).

The figures given in the literature are $-29.0°$ (Fischer [1]), $-27.1°$ (Price, Gilbert and Greenstein [5]) and $-27.6°$ (Ehrlich and Wendel [6]) respectively.

1. Fischer, E. *Ber.* **39** (1906) 2320.
5. Price, V. E., Gilbert, J. B., and Greenstein, J. P. *J. Biol. Chem.* **179** (1949) 1169.
6. Ehrlich, F., and Wendel, A. *Biochem. Z.* **8** (1908) 399.
12. Willstätter, R., and Waldschmidt-Leitz, E. *Z. physiol. Chem.* **125** (1923) 150.

For the experimental details of other enzymatic methods for the resolution of racemic valine, see Ref. 18n and the following: R. Duchinsky and J. Jeannerat, Comp. Rend. <u>208</u>, 1359 (1939); F. Ehrlich, Biochem. Z., <u>1</u>, 8 (1906); V. E. Price, J. B. Gilbert and J. P. Greenstein, J. Biol. Chem., <u>179</u>, 1169 (1949); and S. M. Birnbaum, L. Levintow, R. B. Kingsley and J. P. Greenstein, J. Biol. Chem., <u>194</u>, 455 (1952).

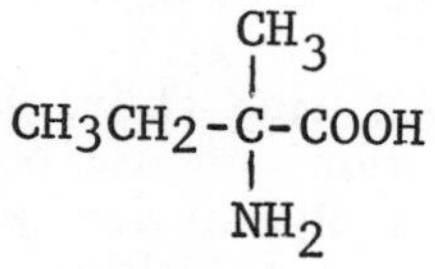

$$\begin{array}{c} CH_3 \\ | \\ CH_3CH_2-C-COOH \\ | \\ NH_2 \end{array}$$

C₅H₁₁NO₂

Ref. 149

Exemple 3. — 8,7 parties en poids de N-benzoylisovaline sont chauffées à reflux pendant 6 heures avec 80 parties en volume d'anhydride acétique. Après avoir retiré l'excès d'anhydride acétique et d'acide acétique glacial par distillation, on a obtenu 6,54 parties en poids de 4-éthyl-4-méthyl-2-phényloxazolone fondant à 46-47 °C par distillation de l'huile résiduaire brune sous pression réduite.

A une suspension de 1,01 partie en poids de sodium métallique en poudre dans 30 parties en volume de benzène, on a ajouté 6,86 parties en poids de 1-menthol ([α]$_D^{12}$ = —51,3°, C = 3,06 CH₃OH) et le mélange est chauffé à reflux pendant 2 heures. Le sodium métallique n'ayant pas réagi est retiré du mélange réactionnel et on ajoute 20 parties en volume de benzène et 6,54 parties en poids de 4-éthyl-4-méthyl-2-phényloxazolone.

Le mélange est agité pendant 4 heures à la température ambiante et on ajoute 50 parties en volume de benzène; la solution de benzène est lavée avec 50 parties en volume d'acide acétique dilué et d'eau, puis elle est séchée sur du sulfate de sodium anhydre. Après avoir retiré le benzène du mélange réactionnel par distillation, on obtient un résidu vitreux brun. La distillation de ce résidu sous pression réduite donne 9,16 parties en poids d'une fraction ayant un poids d'ébullition de 173-190 °C sous 0,08 mm Hg.

Lorsque cette fraction est dissoute dans 10 parties en volume de n-hexane et qu'on la laisse reposer pendant environ 12 heures, on obtient 5,4 parties en poids de cristaux blancs (appelés ci-après « cristaux A_3 »). Une autre concentration de la liqueur mère donne 1,9 partie en poids de cristaux blancs (désignés ci-après par « cristaux B_3 »).

Lesdits « cristaux A_3 » et lesdits « cristaux B_3 » bruts sont recristallisés à partir du n-hexane. Les premiers donnent des « cristaux A'_3 » ayant un point de fusion de 110-111,5 °C, $[\alpha]_D^{18} = -54{,}7°$ (C = 1,28 CH_3OH) et les derniers donnent des « cristaux B'_3 » ayant un point de fusion de 72,5-73,5 °C $[\alpha]_D^{12} = -41{,}9°$ (C = 1,48 CH_3OH).

0,5 partie en poids de « cristaux A'_3 » est dissoute dans 7 parties en volume de solution d'éthanol à 50 % contenant 0,78 partie en poids de potasse caustique et elle est chauffée à reflux pendant 6 heures. A la fin de la réaction, l'éthanol est enlevé par distillation : on ajoute 10 parties en volume d'eau et le 1-menthol est alors retiré en agitant avec du benzène.

Par addition d'acide chlorhydrique concentré à cette solution aqueuse pour la régler à un pH de 5, on obtient 0,28 partie en poids de cristaux en forme de plaques; la (—)-N-benzoylisovaline ayant un point de fusion de 176-178 °C, $[\alpha]_{\alpha} = -10{,}4°$ (C = 1,84 CH_3OH).

Le même traitement appliqué aux » cristaux B'_3 » donne 0,145 partie en poids de cristaux blancs en aiguilles, la (+)-benzoylisovaline fondant à 176-178°C, $[\alpha]_D^{12} = +10{,}1°$ (C = 1,44 CH_3OH).

0,5 partie en poids de (—)-N-benzoylisovaline est chauffée à reflux avec 15 parties en volume d'acide bromhydrique à 48 % et 15 parties en volume d'eau pendant 6 heures. Le mélange est refroidi et l'acide benzoïque est séparé par filtration; puis, l'acide benzoïque résiduaire est extrait avec du benzène.

La solution mère est évaporée à sec; le résidu est dissous dans 9 parties en volume d'eau et la solution aqueuse traverse une colonne d'amberlite dite IR 120 (forme H) et puis, après avoir lavé la colonne avec l'eau, l'isovaline est éluée avec une solution d'ammoniaque à 10 %.

L'ammoniaque est enlevée par distillation et le résidu est évaporé à sec. Par recristallisation de l'isovaline brute à partir d'une solution alcoolique, on obtient 0,195 partie en poids d'hydrate d'isovaline ($[\alpha]_D = -11°H_2O$).

Le même traitement de la (—)-N-benzoylisovaline donne 0,15 partie en poids de (—)-isovaline.

A resolution procedure for the above compound using the same resolving agent may also be found in S. Terashima, K. Achiwa and S. Yamada, Chem. Pharm. Bull. 12, 1399 (1965).

$C_5H_{11}NO_2$ Ref. 149a

Experimental Part

N-Chloroacetyl-DL-isovaline.—DL-Isovaline[8] was treated with chloroacetyl chloride and chilled NaOH in the usual manner. On acidification with concd. HCl to *p*H 1.7, N-chloroacetyl-DL-isovaline crystallized in 82% yield. It was recrystallized from water; m.p. 161.5–163.0° (cor.). A m.p. of 162° has been reported for this compound.[9]

Anal.[10] Calcd. for $C_7H_{12}O_3NCl$: C, 43.4; H, 6.3; N, 7.2; Cl, 18.3. Found: C, 43.4; H, 6.4; N, 7.2; Cl, 18.2.

Enzymatic Resolution of Chloroacetyl-DL-isovaline.—Fifty-three grams of N-chloroacetyl-DL-isovaline was dissolved in 2 liters of water and the solution brought to *p*H 7.5 with 2 *N* LiOH. Three grams of acylase I powder[4] was dissolved in the solution, and water added to bring the concentration of the racemic compound to 0.1 *M*. The enzymatic hydrolysis of the substrate could not be followed by the usual manometric ninhydrin procedure, because the liberated amino acid does not yield quantitative amounts of carbon dioxide, and therefore the nitrous acid method was employed. The rate of hydrolysis of the susceptible L-isomer of N-chloroacetyl-DL-isovaline by acylase I is 38 micromoles per hour per mg. protein N. The digest was treated with a few drops of toluene, and allowed to incubate at 38° for 24 hours. Analyses on an aliquot of the digest revealed that the hydrolysis of the compound had proceeded to 50%. Another gram of the enzyme was added, and the digest allowed to stand for 12 hours longer. Analysis again revealed 50% hydrolysis. Acetic acid was added to *p*H 5, and the protein filtered off with the aid of Norit. The filtrate was evaporated at 40° *in vacuo*, and the small amount of protein which flocculated was again removed by filtration. The filtrate contained L-isovaline, chloroacetic acid and chloroacetyl-D-isovaline. Treatment with excess ethanol in the usual manner[2-7] failed to bring about the separation of the highly soluble L-isovaline. Treatment with concd. HCl to *p*H 1.7 led to the separation in 50% yield of chloroacetyl-D-isovaline. After recrystallization from acetone–ether, the m.p. was 158° (cor.), and $[\alpha]^{25}D$ −9.0° for a 2% solution in absolute ethanol. In 2% aqueous solution, the rotation of the compound was imperceptible.

Anal. Calcd. for $C_7H_{12}O_3NCl$: N, 7.2. Found: N, 7.1.

Chromatographic Separation of the Enzymatic Products.—A general chromatographic procedure for the separation of the amino acid products obtained by the enzymatic resolution method has been developed in this Laboratory and will be described more fully in a subsequent publication.[11] A brief description of the procedure as it applied to the present problem is as follows. A 100-ml. aliquot of a deproteinized and concentrated isovaline resolution mixture (corresponding to 25.8 g. of chloroacetyl-DL-isovaline in the original digest) was poured onto the top of a column 87 cm. high and 6.5 cm. in diameter composed of 20 to 50 mesh Dowex 50 resin in the acid phase.[12]

Elution with water was carried out at a flow rate of 40 to 60 ml. per hour. Chloroacetyl-D-isovaline appeared in the effluent after approximately 250 ml. of water had passed through the column, as indicated by a fall in *p*H from about 7 to about 3. Aliquots taken from the hour-long fractions were hydrolyzed in 2 *N* HCl for 2 hours and tested for color development with ninhydrin. By this means it was demonstrated that the N-acyl derivative was eluted in approximately 3 liters of effluent. No free isovaline was present in the fractions collected during this interval since ninhydrin tests on unhydrolyzed aliquots were all negative. After further washing of the column with an additional 1.5 liters of water, elution was begun with 2.5 *N* HCl. L-Isovaline began to appear after about 4 liters of solution had passed through the column, as shown by positive ninhydrin tests. The entire L-isovaline was eluted after an additional 3800 ml. of solution had passed through the column.

All the fractions containing chloroacetyl-D-isovaline were combined and evaporated to dryness *in vacuo*, and the residue was taken up in absolute ethanol to remove sodium chloride[13] and any residual protein. The ethanol was evaporated and the residue taken up in acetone and filtered to ensure further the absence of any L-isovaline or sodium chloride. The chloroacetyl-D-isovaline was then isolated by evaporation of the acetone and crystallization from acetone–ether; m.p. 158° (cor.); yield 55% of theory, based on the original amount of chloroacetyl-DL-isovaline; $[\alpha]^{25}D$ $-9.0°$ for a 2% solution in absolute ethanol.

Anal. Calcd. for $C_7H_{12}O_3NCl$: N, 7.2. Found: N, 7.2.

Thus the chloroacetyl-D-isovaline isolated from the column was identical in properties with that obtained by acidification of the resolution mixture. Five grams of chloroacetyl-D-isovaline was refluxed for 2 hours with 100 cc. of 2 *N* HCl. The solution was decolorized with Norit, and the filtrate evaporated *in vacuo* to dryness. The residue was dissolved in 100 cc. water and the solution treated with a slight excess of silver carbonate. The silver chloride was filtered off, and the filtrate saturated with hydrogen sulfide gas. The final filtrate was evaporated to dryness *in vacuo* and the residual D-isovaline taken up in a little water, the solution filtered, and acetone added in excess to the clear filtrate. The D-isovaline crystallized as long needles in nearly quantitative yield, $[\alpha]^{25}D$ $-11.28°$ for a 5% solution in water.

Anal. Calcd. for $C_5H_{11}O_2N$: C, 51.2; H, 9.4; N, 12.0. Found: C, 51.0; H, 9.5; N, 12.0.

The combined fractions containing the L-isovaline were evaporated to dryness *in vacuo*, and the residue taken up in absolute ethanol and filtered to remove sodium chloride.[13] The ethanol was evaporated, and the residue treated successively with silver carbonate and hydrogen sulfide as described for the D-enantiomorph. The yield after crystallization from water with excess acetone was 77% of the theoretical, based on the original amount of chloroacetyl-DL-isovaline; $[\alpha]^{25}D$ $+11.13°$ for a 5% solution in water.

Anal. Calcd. for $C_5H_{11}O_2N$: C, 51.2; H, 9.4; N, 12.0. Found: C, 51.0; H, 9.5; N, 12.2.

(2) J. P. Greenstein, L. Levintow, C. G. Baker and J. White, *J. Biol. Chem.*, **188**, 647 (1951).

(3) L. Levintow and J. P. Greenstein, *ibid.*, **188**, 643 (1951).

(4) S. M. Birnbaum, L. Levintow, R. B. Kingsley and J. P. Greenstein, *ibid.*, **194**, 455 (1952).

(5) D. Rudman, A. Meister and J. P. Greenstein, THIS JOURNAL, **74**, 551 (1952).

(6) D. Hamer and J. P. Greenstein, *J. Biol. Chem.*, **193**, 81 (1951).

(7) S. M. Birnbaum and J. P. Greenstein, *Archiv. Biochem. Biophys.*, **39**, 108 (1952).

(8) P. A. Levene and R. Steiger, *J. Biol. Chem.*, **76**, 299 (1928).

(9) K. W. Rosenmund, *Ber.*, **42**, 4473 (1909).

(10) Analyses by R. J. Koegel and staff of this Laboratory.

(11) C. G. Baker and H. A. Sober, in preparation.

(12) Cationic exchange resin from the Dow Chemical Company. The resin was regenerated by two cycles of washing with 5 *N* HCl, water, 1 *N* NaOH and water, followed by a final 5 *N* HCl and water wash.

(13) Large volumes of water were used for the final wash of the resin during its regeneration. However, even after the effluent was neutral to phenolphthalein, additional sodium chloride was obtained. The coarse mesh resin employed probably requires a longer equilibration period than does the resin of a finer mesh.

A resolution procedure for the above compound (as the formyl derivative) using brucine as resolving agent may be found in E. Fischer and von Gravenitz, Ann., 406, 5 (1914).

$C_5H_{11}NO_2S$

$CH_3SCH_2CH_2CH(NH_2)COOH$

Ref. 150

***dl*-Formylmethionine.**—A mixture of 25 g. of *dl*-methionine and 38 g. of absolute formic acid was heated for three hours on a steam-bath in a small flask fitted with a short reflux condenser and protected from the air by a calcium chloride tube. After this heating period, the excess formic acid and water were removed by distillation under reduced pressure. The residue was again heated for three hours with 25 g. of absolute formic acid. The solution was then evaporated under diminished pressure and the treatment with a fresh 25-g. portion of absolute formic acid again repeated. After removing the excess formic acid and water, the hot residue was dissolved in about 400 cc. of hot ethyl acetate and the solution was filtered. If the filtrate was not completely clear, it was heated and again filtered. On cooling in an ice-bath, the formylmethionine crystallized. The crystals were collected on a filter, the filtrate was concentrated to about 100 cc. and a second fraction was obtained. The yield at this point was 21.5 g. of a product which melted at 99–100°. By careful evaporation of the mother liquor and recrystallization of the material thus obtained, another 2 g. of pure formyl derivative was isolated. The total yield of pure product reached 23.5 g. (79% of the theoretical amount).

Anal. (Parr bomb). Subs., 0.4168: $BaSO_4$, 0.5503. Calcd. for $C_6H_{11}O_3NS$; S, 18.10. Found: S, 18.13. Subs., 0.1875: 10.26 cc. of 0.1015 *N* NaOH. Calcd. for $C_6H_{11}O_3NS$: neutral equivalent, 177. Found: 180.

Ethyl acetate did not dissolve the unreacted amino acid and from the material insoluble in this solvent about 1.5 g. of pure *dl*-methionine was recovered by dissolving the crude product in water, adding a little (2–3 cc.) pyridine and precipitating the amino acid by the addition of two volumes of alcohol.

Formylmethionine was also prepared by refluxing the solution of methionine in absolute formic acid. However, the product was less pure and the yield was lower. The scheme of formylating by distilling a mixture of toluene, formic acid and methionine[7] was also tried but was not as satisfactory as the method already described.

Resolution of *dl*-Formylmethionine.—To a solution of 30 g. of formyl derivative in 5000 cc. of hot absolute alcohol was added 48 g. of recrystallized anhydrous brucine (slightly more than one-half the theoretical amount to form the neutral salt). The solution was allowed to stand at about 0° for four days and then filtered. The first crop of crystals weighed 51.5 g. (Fraction 1) and melted at 144–146°.

Rotation. 0.2000 g. made up to 25 cc. with water at 20°, 2-dm. tube: α_D −0.34° (±0.01°); $[\alpha]^{25}_D$ −21.25° (±0.5°).

Fraction 1 was recrystallized from 1500 cc. of absolute alcohol. On boiling, the salt did not go completely into solution. On cooling the solution for one day at 0° and filtering there was obtained 47 g. of salt, m. p. 144–145°.

Rotation. 0.2001 g. made up to 25 cc. with water at 20°, 2-dm. tube: α_D −0.38° (±0.01°); $[\alpha]^{25}_D$ −23.74° (±0.5°).

Anal. (Parr bomb). Subs., 0.4011: $BaSO_4$, 0.1601. Calcd. for $C_{29}H_{37}O_7N_3S$: S, 5.60. Found: S, 5.48.

The filtrate from this recrystallization was added to the original filtrate from Fraction 1 and 50 g. of brucine was added. After standing at 0° for about twenty-four hours, the solution was filtered and 10.5 g. (Fraction 2) of salt was obtained. This salt softened at about 140° and melted with decomposition at 185–190°. It was not used in further work.

The filtrate from Fraction 2 was evaporated to a volume of about 3 liters and allowed to stand at 0° for about twenty-four hours. The precipitate was collected on a filter and after air-drying weighed 29.5 g. (Fraction 3). It softened at 192° and melted at 194–196°.

The filtrate from Fraction 3 was evaporated to about 300 cc. and on cooling to 0° for about twenty-four hours an additional 7 g. (Fraction 4) of salt was obtained. This fraction softened slightly at 155° and melted at 185–187°. Fractions 3 and 4 were combined.

Rotation. 0.2000 g. made up to 25 cc. with water at 20°, 2-dm. tube: α_D −0.32° (±0.01°); $[\alpha]^{25}_D$ −20.00 (±0.5°).

The strychnine salt was prepared from formylmethionine in absolute alcohol solution. No fractionation was obtained from this solvent because the salt was too soluble. From ethyl acetate a crystalline product melting at 138–140° with a rotation of −20.57° was obtained. The more soluble fraction was obtained as a solid melting at 60–63° with a rotation of −16.25°. These products were analyzed for sulfur and found to be the desired strychnine salts. However the separation was not satisfactory and further work was abandoned in favor of the brucine salt.

The *d*-α-phenylethylamine salt of formylmethionine was formed in ethyl acetate solution and a product melting at 89–90° which analyzed correctly was isolated. However, the rotations were extremely low and no fractionation was obtained.

Isolation of *d*- and *l*-Formylmethionine.—To a solution of 47 g. of the least soluble brucine salt in 100 cc. of water was added 500 cc. of a saturated solution of barium hydroxide. The solution was chilled for thirty minutes in an ice–salt mixture and filtered. The filtrate was extracted with three 150-cc. portions of chloroform and one 150-cc. portion of ether. The solution was then exactly neutralized with 1 *N* sulfuric acid so that it gave no test for barium or sulfate ions. After removal of the barium sulfate by filtration, the water was evaporated under reduced pressure until the volume was about 40 cc. The residue was extracted with two 100-cc. portions of hot ethyl acetate. The ethyl acetate solution was filtered and concentrated to about 25 cc. on a hot-plate. It was then further concentrated by drawing a current of air over the solution at ordinary temperatures. Crystals separated and were collected on a filter. The first fraction amounted to 5.8 g. of a product which melted at 99–100°. Two grams of less pure material was obtained by evaporation of the mother liquors.

Titration. Subs., 0.1499: 7.79 cc. of 0.1070 *N* NaOH. Calcd. for $C_6H_4O_3NS$. Neutral equivalent, 177. Found: 179.8.

Anal. Subs., 0.1013: $BaSO_4$, 0.1334. Calcd. for $C_6H_{11}O_3NS$: S, 18.18. Found: S, 18.13.

Rotation. 0.2002 g. made up to 25 cc. with water at 20°, 2-dm. tube: α_D +0.17 (±0.01°); $[\alpha]^{25}_D$ +10.62° (±0.5°).

l-Formylmethionine was obtained by approximately the same procedure. From 36 g. of the more soluble brucine salt, 4.2 g. of product which melted at 99–100° was obtained. In order to obtain the *l*-isomer in the crystalline form, the ethyl acetate solution was evaporated to about 25 cc. and then ether was added to the hot solution until it became turbid. On cooling the formyl derivative separated.

Rotation. 0.2000 g. made up to 25 cc. with water 20°, 2-dm. tube: α_D −0.16 (±0.01°); $[\alpha]^{25}_D$ −10.00 (±0.5°).

Some of the *l*-formyl derivative remained in the water solution and was not extracted by the ethyl acetate. This water solution was boiled with a little hydrochloric acid and the derivative hydrolyzed as described below to give a sample of *l*-methionine of the highest rotatory power which was any sample of this resolution.

***d*- and *l*-Methionine.**—A solution of 1 g. of *d*-formylmethionine in 10 cc. of 10% hydrochloric acid was heated on the steam cone for one hour. The solution was concentrated to one-half of its original volume under reduced pressure, and to it was added 5 cc. of pyridine. The methionine was precipitated by adding three volumes of alcohol. After thoroughly chilling the solution, the *d*-methionine was collected on a filter. The yield was 0.450 g.

Rotation. 0.1997 g. made up to 25 cc. with water, 2-dm. tube: α_D +0.14 (±0.01°); $[\alpha]_D^{25}$ +8.76° (±0.5°).

This sample was recrystallized by the same procedure and gave 0.4 g. of product.

Rotation. 0.2000 g. made up to 25 cc. with water, 2-dm. tube: α_D +0.13° (±0.01°); $[\alpha]_D^{25}$ +8.12° (±0.5°). 0.2004 g. made up to 25 cc. with 5% sodium bicarbonate at 20°, 2-dm. tube: α_D −0.12° (±0.01°); $[\alpha]_D^{25}$ −7.47° (±0.5°). 0.2006 g. made up to 25 cc. with 0.2001 *N* hydrochloric acid at 20°, 2-dm. tube: α_D −0.34° (±0.01°); $[\alpha]_D^{25}$ −21.18° (±0.5°).

Anal. Subs., 0.1086: $BaSO_4$, 0.1705. Subs., 3.069 mg.: 0.263 cc. of N_2 at 27° and 750 mm.[8] Calcd. for $C_5H_{11}O_2NS$: S, 21.50; N, 9.40. Found: S, 21.53; N, 9.54.

l-Methionine was obtained by the same procedure. From 1 g. of *l*-formylmethionine there was obtained 0.5 g. of product.

Rotation. 0.2000 g. made up to 25 cc. with water at 20°, 2-dm. tube: α_D −0.12° (±0.01°); $[\alpha]_D^{25}$ −7.5° (±0.5°).

Anal. Subs., 3.130 mg.: 0.273 cc. of N_2 at 28° and 748 mm. Calcd. for $C_5H_{11}O_2$-NS: N, 9.40. Found: N, 9.73.

The water solution of *l*-formylmethionine left after the extraction with ethyl acetate was hydrolyzed as mentioned above and the *l*-methionine was isolated by the usual procedure. This yielded 0.3 g. of *l*-methionine.

Rotation. 0.2004 g. made up to 25 cc. with water at 20°, 2-dm. tube: α_D −0.13° (±0.01°); $[\alpha]_D^{25}$ −8.11 (±0.5°).

---

[7] Steiger, *J. Biol. Chem.*, **86**, 695 (1930).

$C_5H_{11}NO_2S$ Ref. 150a

Darstellung und fraktionierte Kristallisation des Brucinsalzes des *N*-Carbäthoxy-sulfanilyl-methionins

34 g (0.09 Mol) *N*-Carbäthoxy-sulfanilyl-methionin wurden in 800 ccm und 50 g Brucin in 2700 ccm frisch dest. absol. Äthanol gelöst und die warmen Lösungen vereinigt. Nach Zugabe einiger Impfkristalle, die aus Vorversuchen erhalten wurden, begann das Salz sofort auszukristallisieren.

Die systematischen Kristallisationen zur Zerlegung des Salzes wurden nach dem bekannten Dreieckschema durchgeführt. Hierzu wurde jeweils die erste Salzfraktion jeder Kristallisationsreihe in der rund 50fachen Menge heißem Alkohol gelöst. Die Lösung wurde nach Abkühlung auf 50° geimpft und in einem Wasserbad großer Wärmecapazität 2–4 Stdn. bei 45–50° gehalten. In der nach Abtrennung des Kristallisats verbliebenen Mutterlauge wurde die nächste Fraktion der Ausgangskristallisations-Reihe gelöst und die Probe nach Impfen erneut getempert usw.; aus den Mutterlaugen der letzten Kristallisation jeder Reihe wurden die bei Zimmertemperatur kristallisierenden Anteile vor dem Eindampfen abgetrennt. Nach zahlreichen Routine-Kristallisationen bis zur 10. Kristallisations-Reihe wurden Spitzen- und Endfraktionen mit unveränderlichen Drehwerten erhalten (vergl. Tafel 1).

Spitzenfraktion:

Salz, α: −1.73°; $[\alpha]_D^{20}$: −34.6° (c = 2.49, in Chloroform)
Säure, α: −0.94°; $[\alpha]_D^{20}$: −18.8° (c = 2.505, in Alkohol)

Endfraktion:

Salz, α: −0.32°; $[\alpha]_D^{20}$: − 3.1° (c = 5.01, in Chloroform)
Säure, α: +0.95°; $[\alpha]_D^{20}$: +18.9° (c = 2.50, in Alkohol)

Abhängigkeit von der OH-Ionen-Konzentration (Tafel 3)

100.0 mg reine (−)-Säure in 4 ccm 0.1 *n* KOH (c = 2.50)
99.2 ,, ,, (−) ,, ,, 4 ccm 1 *n* KOH (c = 2.48)
101.0 ,, ,, (−) ,, ,, 4 ccm 10-proz. Kalilauge (c = 2.525)
100.1 ,, ,, (+) ,, ,, 4 ccm 10-proz. Kalilauge (c = 2.50)

Cinchonidinsalz des linksdrehenden *N*-Carbäthoxy-sulfanilyl-methionins

2.2 g linksdrehende Säure ($[\alpha]_D$: −19.0°; Schmp. 145–147°) wurden mit 3 g Cinchonidin ($[\alpha]_D$: −113.6°, c = 1.13, in Alkohol) in etwa 100 ccm heißem absol. Alkohol gelöst. Aus der über Nacht im Eisschrank gekühlten Lösung schieden sich etwa 4 g säulenförmige, farblose Kristalle aus. Diese wurden zweimal aus 70–80 ccm absol. Alkohol umkristallisiert. Die Kristalle wurden mit viel Alkohol gewaschen und im heizbaren Vak.-Exsiccator über Calciumchlorid und Blaugel bis zur Gewichtskonstanz getrocknet; Ausb. etwa 2 g, Schmp. 195°.

$C_{33}H_{42}O_7N_4S_2$ (670.8) Ber. S 9.55 Gef. S 9.15

Aus den Mutterlaugen konnten 1.7—1.8 g einer 2. Fraktion isoliert werden. Aus den beiden Fraktionen konnte die freie Säure zurückgewonnen werden. Das Salz ging jedoch im Gegensatz zum Brucinsalz bei der Alkali-Behandlung nicht vollständig in Lösung.

Die Ausbeuten betrugen:

1. Fraktion 0.7—0.8 g vom Schmp. 146—147°,
2. Fraktion 0.5—0.6 g vom Schmp. 145—146°.

Die Ergebnisse der optischen Untersuchungen (für beide Fraktionen c ≈ 2.5 in Alkohol) sind in der Tafel 2 enthalten.

Chininsalz des rechtsdrehenden *N*-Carbäthoxy-sulfanilyl-methionins

Die Darstellung des Salzes erfolgte aus 2.3 g (+)-*N*-Carbäthoxy-sulfanilyl-methionin und 3 g Chinin in 80 ccm absol. Äthanol entsprechend der Darstellung des vorstehenden Salzes.

1. Fraktion: 1.8 g kleiner Nadeln und Säulen; Schmp. 164—167°,
2. Fraktion: 1.5 g.

$C_{34}H_{44}O_8N_4S_2$ (700.8) Ber. S 9.15 Gef. S 9.16

Durch die übliche Aufarbeitung wurden aus beiden Fraktionen je 0.6—0.7 g Säure (Schmp. 145—147°) isoliert. Die Ergebnisse der optischen Untersuchungen sind in der Tafel 2 enthalten.

Tafel 1. Spez. Drehungen des (±)-*N*-Carbäthoxy-sulfanilyl-methionins und seiner Brucinsalze

	$[\alpha]_D$ Salz c ≈ 2.5	$[\alpha]_D$ Säure c ≈ 2.5
8. Krist.-Reihe	−34.4°	−19.0°
9. „ „	−34.9°	—
10. „ „	−34.6°	−18.8°
10. „ „	−3.1°	+18.9°

Tafel 2. Rotationsdispersionen

Säure aus	6563 A (C)	5893 A (D)	5460 A (Hg I)	4861 A (F)	4358 A (Hg II)
1) schwererlösl. Brucinsalz ...	−14.7°	−19.0°	−22.8°	−32.8°	(−46°)
2) schwererlösl. Cinchonidinsalz	−14.6°	−18.7°	−22.7°	−31.8°	—
3) Mutterlauge nach 2	—	−18.6°	—	—	—
4) leichterlösl. Brucinsalz	+14.6°	+18.9°	+23.0°	+33.0°	(+47°)
5) schwererlösl. Chininsalz ...	+14.8°	+19.1°	+23.3°	+33.8°	(+48°)
6) Mutterlauge nach 5	—	+18.9°	—	—	—
$\frac{M \cdot [\alpha]}{100} = [M] =$ .... (Mol.-Gew. = M = 376, 436)	±55.7°	±71.5°	±86.7°	±127°	(±180°)

Tafel 3. Optische Drehung der Säure in alkal.-wäßr. Lösung

Lösungsmittel:	6563 A (C)	5893 A (D)	5460 A (Hg I)	4861 A (F)	4358 A (Hg II)
Alkohol	−14.7°	−18.7°	−22.7°	−31.8°	(−46°)
0.1 *n* KOH	+10.1°	+12.8°	+15.3°	+20.8°	(+30°)
1.0 *n* KOH	+34.6°	+44.2°	+53.9°	+74.7°	(+110°)
10-proz. Kalilauge	+38.0°	+48.1°	+59.1°	+82.2°	(+118°)
10-proz. Kalilauge	−38.5°	−48.7°	−58.8°	−80.5°	(−115°)
Alkohol	+14.8°	+19.1°	+23.2°	+33.4°	(+48°)

$C_5H_{11}NO_2S$ Ref. 150b

For a resolution procedure using ephedrine in the direct resolution of the racemic N-benzyloxycarbonyl derivative of the above compound, see Ref. 20h.

$C_5H_{11}NO_2S$ Ref. 150c

Example 5

19.1 g. of *dl*-N-acetylmethionine and 13.5 g. of *l*-amphetamine are dissolved with heating in 125 ml. of 95% ethanol. The solution is allowed to cool slowly for 16 hours, and the crystalline product consisting of the *l*-amphetamine salt of *d*-N-acetylmethionine is filtered off and dried; $[\alpha]_D^{25} = -17.3°$ (2% in water). The crystalline salt is dissolved in water and the solution is passed through a sulfonic acid ion exchanger (Dowex 50). The effluent is concentrated in vacuo and cooled to obtain the product, *d*-N-acetylmethionine; $[\alpha]_D^{25} = +20.1°$ (2% in water) after recrystallization from water.

l-N-acetylmethionine is obtained from the alcoholic filtrate described above by removing the alcohol and then treating the resulting residual product, *l*-amphetamine salt of *l*-N-acetylmethionine, in the same manner as described above for the recovery of the corresponding salt of the *d*-isomer.

$C_5H_{11}NO_2S$ Ref. 150d

EXAMPLE 2

Resolution of Racemic N-Acetylmethionine

19.1 g. of N-acetyl-DL-methionine and 24.5 g. of L-threo-2-amino - 1 - (p - methylsulfonylphenyl)-1,3-propanediol were dissolved together in 400 ml. of methanol and the solution was allowed to stand for twelve hours at 0–5° C. The crystalline needles which had separated from solution were then collected on a filter, washed with a few ml. of methanol and ether and dried at 70° C. There was thus obtained 19.8 g. of the L-threo-2-amino-1-(p-methylsulfonylphenyl)-1,3-propanediol salt of N-acetyl-D-methionine which melted at 182–184° C. The melting point of this product remained unchanged after it had been recrystallized from methanol. 15.4 g. of this salt was mixed with 4 ml. of concentrated hydrochloric acid and 10 ml. of water. When the solution thus obtained was saturated with sodium chloride and allowed to stand at 5° C., there separated from solution almost the theoretical quantity of L-threo-2-amino-1-(p-methylsulfonylphenyl) - 1,3 - propanediol hydrochloride. This product was collected on a filter, and the filtrate was extracted with ethyl acetate. The ethyl acetate solution was evaporated and the residue thus obtained was recrystallized from 20 ml. of ethyl acetate, thus yielding 5.0 g. of N-acetyl-D-methionine which melted at 102–104° C. and had $[\alpha]_D^{25}$ +20.5° (c. 4% in water). By evaporating the methanolic resolution liquors to dryness, dissolving the residue in water and mixing with a slight excess of hydrochloric acid, there was obtained N-acetyl-L-methionine, M.P. 104° C., $[\alpha]_D^{25}$ −20.3° (c. 4% in water).

$C_5H_{11}NO_2S$ Ref. 150e

Resolution of N-Acetyl-DL-methionine.

N-Acetyl-DL-methionine (recrystallized from water: 191 g, 1.0 mol) and L-leucinamide (143 g, 1.1 mol) were dissolved in 1.5 l of 99% ethanol on a water bath at 50°C. The L-leucinamide salt of N-acetyl-L-methionine crystallized slowly from the seeded solution in fine needles (120 g). Yield, 79.8% based on L-form; m.p. 143°C; $[\alpha]_D^{20}$+6.97° (c, 5: water). The recrystallization from 99% ethanol gave 110 g of fine crystal. Yield, 73.6%, based on L-form; $[\alpha]_D^{20}$+6.90° (c, 5: water); m.p. 144°C. *Anal.* Found. N, 11.85. Calcd. for $C_{13}H_{27}N_3O_4S$: N, 12.1%.

The pure L-leucinamide salt (50 g) was dissolved in 1.5 l of water. The solution was passed through a column of Diaion SK 1 (H-form) to let N-acetyl-L-methionine go into the effluent.

The effluent was concentrated under reduced pressure to 40 ml, then kept in a refrigerator overnight. The long prisms of N-acetyl-L-methionine deposited were collected, filtered, washed with a small amount of cold water, and dried in vacuo. Yield, 29 g (97%); m.p. 102~103°C; $[\alpha]_D^{20}$−20.8° (c, 4: methanol), this value was in close agreement with reported values[20]. *Anal.* Found: N, 7.50. Calcd. for $C_7H_{13}O_3NS$: N, 7.42%.

The absorbed L-leucinamide was recovered from the resin by eluting with 2 N aqueous ammonia. The effluent was collected until the fraction showed no further biuret reaction, evaporated to dryness in vacuo. The residual crops were recrystallized from boiling benzene, and thus L-leucinamide was recovered in about 95% yield. M.p. 101~2°C; $[\alpha]_D^{19}$+7.80° (c, 5: water).

N-Acetyl-L-methionine (37.6 g) which was obtained by the resolution using L-leucinamide as resolving agent was hydrolyzed with 120 ml of 20% hydrochloric acid for 2 hours and the subsequent solution was evaporated nearly to dryness. After the residue was extracted with 300 ml of absolute ethanol, the extracts were evaporated to dryness in vacuo. The residue was dissolved in 200 ml of water and the solution was passed through a column of Amberlite IR 4B (OH-form). The resin was washed with water until the effluent became ninhydrin negative. The aqueous effluent containing free L-methionine was concentrated in vacuo until the crystallization began and then a small amount of ethanol was added to the solution.

After having been kept in a refrigerator overnight, the crystals of L-methionine were collected, washed with ethanol, and dried in vacuo.

Twenty-nine gram of L-methionine was obtained, and the yield was 98.7% based on N-acetyl-L-methionine. $[\alpha]_D^{20}$+23.88° (c, 5: 6 N HCl); $[\alpha]_D$ value was in close agreement with the previously reported value[21].

20) S. Akabori and S. Mizushima, "Chemistry of Protein" Vol. I, Kyoritsu Shuppan Co., p. 154 (1954). S.M. Birnbaum, L. Levintow, R.B. Kingsley and J.P. Greenstein, *J. Biol. Chem.* **199**, 455 (1952).

21) A.W. Ingersoll and G.P. Wheeler, *J. Am. Chem. Soc.*, **73**, 4605 (1951).

$C_5H_{11}NO_2S$ Ref. 150f

N-Acetyl-DL-methionine.—Acetylation with acetic anhydride and aqueous alkali as previously described[3] gave yields of 78–84% but the considerable solubility of the product required prolonged extraction with chloroform to secure maximum yields. Acetylation as by Knoop and Blanco[19] gave crude yields of about 80% but the discolored product was difficult to purify. The best method was a modification of that of Kolb and Toennies.[20] Acetylation was effected in three hours at room temperature in 10 parts of acetic acid containing acetic anhydride (1.6 moles). The sirupy product remaining after evaporation of acetic acid *in vacuo* was crystallized directly from 1.5 parts of water. Overnight 75–80% of the calculated amount of substantially pure product (m.p. 114–115°) separated. Additional crops increased the yield to 85%. The later crops were discolored and contained some free methionine but recrystallization from ethyl acetate or acetone gave satisfactory material. Pure N-acetyl-DL-methionine forms large prisms from water, m.p. 114–115°; solubilities in water, 9.12; acetone, 10.0; ethyl acetate, 2.29; chloroform, 1.33; neut. equiv., 191.3 (calcd., 191.1). du Vigneaud and Meyer[21] reported m.p. 114–115°.

Resolution of N-Acetyl-DL-methionine.—The acetamino acid and (−)-α-fenchylamine (0.5 mole) were dissolved in 200 cc. of warm water; the seeded solution slowly deposited the salt of N-acetyl-DL-methionine as small prisms. Successive crops having $[\alpha]^{25}D$ −7.5° to −6.8° (*c* 4, water) were systematically recrystallized from about 1 part of water and eventually gave 70.6 g. (82%). The pure salt has m.p. 172–173°; solubility in water, 15.6; $[\alpha]^{25}D$ −7.5° (*c* 4, water); −22.6° (*c* 4, methanol). It may be recrystallized from 2-propanol but resolution in this or other common solvents was not satisfactory. The more soluble salt forms sirupy solutions and was not purified.

N-Acetyl-D-methionine.—The pure salt was decomposed with alkali and the amine extracted repeatedly with benzene as previously described.[3] The solution was carefully neutralized, evaporated to small volume and treated with the calculated amount of concentrated hydrochloric acid. The rather soluble N-acetyl-D-methionine partially crystallized. Additional crops contained similarly soluble sodium chloride but were purified by extraction with hot ethyl acetate or chloroform and crystallization from these solvents. The yield of well-crystallized compound was 66% but additional less distinctive fractions (16%) gave pure D-methionine on subsequent hydrolysis. The pure compound forms triangular aggregates of plates from water or ethyl acetate, m.p. 104–105°; neut. equiv., 191.5; solubilities in water, 30.7; acetone, 29.6; ethyl acetate, 7.04; chloroform, 6.43; $[\alpha]^{25}D$ +20.3° (*c* 4, water); $[\alpha]^{30}D$ +18.2° (same solution). The effect of temperature on the rotation values is noteworthy. N-Acetyl-D-methionine has not been previously described but a specimen of the antipode prepared from natural methionine was reported[21] to have m.p. 111–111.5°; $[\alpha]^{25}D$ −16.1° (water). Since our values were checked repeatedly on several samples, it seems likely that the sample from natural sources was slightly impure.

D-Methionine.—The pure acetyl derivative (5 g.) was hydrolyzed with 3 *N* hydrobromic acid and the amino acid isolated (88%) as described in other instances.[3,4] After washing with methanol it had $[\alpha]^{25}D$ +8.2° (*c* 0.8, water); −23.5° (*c* 3, 1 *N* HCl). A sample recrystallized from 67% ethanol had $[\alpha]^{27}D$ −23.3° (*c* 3, 1 *N* HCl); $[\alpha]^{25}D$ −24.0° (*c* 3, 5.99 *N* HCl) in close agreement with reported values.[5,17]

In later resolutions the procedure was abbreviated by omitting isolation of N-acetyl-D-methionine. The solution remaining after careful extraction of the amine was acidified with 3 equivalents of 48% hydrobromic acid, boiled for three hours and evaporated to dryness under reduced pressure. The hydrobromide was taken up in methanol and D-methionine precipitated and washed repeatedly with hot methanol. The main crop (71%) had $[\alpha]^{25}D$ −23.0° (*c* 3, 1 *N* HCl).

L-Methionine.—The mother liquors of the resolution gave 48 g. of mixed acetylmethionines ($[\alpha]^{29}D$ −14.2°, hence *ca.* 81% of L-form). Since the L-form is more soluble than the accompanying DL-form it could not be purified by crystallization. Free methionine, presumably formed by slight hydrolysis during the prolonged earlier operations, was also detected. The crude material accordingly was combined with (+)-α-fenchylamine and fractionation was conducted as for the antipode. Rotation values for the less soluble salt, however, were slightly low and free methionine was detected in some fractions. The salt (35 g.) was decomposed and L-methionine obtained by the abbreviated procedure. The rotation $[\alpha]^{25}D$ +22.5° (*c* 3, 1 *N* HCl) indicates about 98% purity. The procedure is not satisfactory but L-methionine presumably could be prepared as readily as the D-form by initial resolution with (+)-α-fenchylamine.

Attempted Resolution with Other Bases.—N-Acetyl-DL-methionine formed needles with (−)-α-phenylethylamine in water but only slight resolution occurred in common solvents. The quinine salt crystallized readily from acetone with partial resolution; acetylmethionine from head crops had $[\alpha]^{25}D$ −4.1° (*c* 1.8 water); m.p. 103–112.5°. The cinchonine salt was not crystalline. The brucine salt formed a crystalline mass when a sirupy solution in acetone was kept several weeks. Stirring with cold acetone, decantation and recrystallization from acetone gave 70% of a fairly uniform salt as coarse granules, $[\alpha]^{25}D$ −7.6° to −8.2° (*c* 4, methanol). Acetyl-L-methionine from this salt after crystallization from chloroform had m.p. 102–104°; $[\alpha]^{25}D$ −19.6° (*c* 4, water) but the method is rather tedious.

(3) L. R. Overby and A. W. Ingersoll, This Journal, **73**, 3363 (1951).
(4) W. A. H. Huffman and A. W. Ingersoll, *ibid.*, **73**, 3366 (1951).
(5) W. Windus and C. S. Marvel, *ibid.*, **53**, 3490 (1931).

(17) K. Vogler and F. Hunzicker, *Helv. Chim. Acta*, **30**, 2013 (1947).
(18) A. W. Ingersoll and S. H. Babcock, This Journal, **55**, 341 (1933).
(19) F. Knoop and J. G. Blanco, *Z. physiol. Chem.*, **146**, 267 (1925).
(20) J. J. Kolb and G. Toennies, *J. Biol. Chem.*, **144**, 193 (1942).
(21) V. du Vigneaud and C. E. Meyer, *ibid.*, **98**, 245 (1932).

$C_5H_{11}NO_2S$ Ref. 150g

Direct Resolution of DL-Methionine (a).—The amino acid (14.9 g., 0.1 mole) and ammonium (+)-α-bromocamphor-π-sulfonate[22] (32.8 g., 0.1 mole) were dissolved in 100 cc. of warm 1 *N* hydrochloric acid. The solution soon deposited the essentially pure L-methionine salt (21.1 g., 85.5%) as narrow transparent prisms with $[\alpha]^{25}D$ +61.7° (*c* 4, water), not substantially changed (+61.5°) after one or more crystallizations from 2.5 parts of water with small loss; solubility in water 5.4. The salt is a hydrate, stable at ordinary temperatures in air or over calcium chloride. Analytical data accord closely with a 7:4 hydration ratio.

Anal. Calcd. for $(C_{15}H_{26}O_6NS_2Br)_4 \cdot 7H_2O$: C, 36.53; H, 6.04; H_2O, 6.42. Found: C, 36.68; H, 6.04; H_2O, 6.47, 6.44, 6.44.

The anhydrous salt has $[\alpha]^{25}D$ +65.8°.

Additional crops (14.8 g.) from the original solution had $[\alpha]^{25}D$ +82° to +84° and consisted principally of ammonium (+)-α-bromocamphor-π-sulfonate dihydrate, $[\alpha]^{25}D$ +84.3° (*c* 4, water). This salt is apparently less soluble than the corresponding D-methionine salt, leaving D-methionine hydrochloride in the final liquors.

L-Methionine.—Pure L-methionine salt (12.4 g.) was dissolved in 2 parts of hot water and the amino acid was precipitated by addition of ammonia to *p*H 6, followed by 4 volumes of methanol. Recrystallization from 80% methanol removed traces of ammonium bromocamphorsulfonate and gave L-methionine (78%) having $[\alpha]^{25}D$ +23.4° (*c* 3, 1 *N* HCl). Evaporation of the methanol from filtrates gave a solution of ammonium bromocamphorsulfonate suitable for re-use.

D-Methionine (6.6 g.) having $[\alpha]^{25}D$ −17.5° (about 84% D-form) was obtained similarly from the foot liquors of the resolution.

(b).—DL-Methionine (119.3 g., 0.8 mole) and ammonium (+)-α-bromocamphor-π-sulfonate (131.2 g., 0.4 mole) were dissolved in 600 cc. of water containing 0.8 mole of hydrochloric acid and the warm solution was seeded. The initial crop of L-methionine salt (145 g.) and further crops obtained by successive concentrations to 200 cc. totaled 182 g. One crystallization series from 2.5 parts of water gave 163 g. (83%) of pure L-methionine salt (hydrate) as large prisms, $[\alpha]^{25}D$ +61.5°. Decomposition of this salt with ammonia gave 44.2 g. (74% based on the DL-form) of L-methionine, $[\alpha]^{25}D$ +23.4° (*c* 3, 1 *N* HCl). The combined mother liquors from the resolution and recrystallization

(18) A. W. Ingersoll and S. H. Babcock, This Journal, **55**, 341 (1933).

(22) W. J. Pope and J. Read, *J. Chem. Soc.*, **97**, 2199 (1910).

gave 60.5 g. of D-methionine, $[\alpha]^{25}D$ $-19.2°$.

(c).—D-Methionine (0.05 mole) from (b) was purified by use of ammonium (−)-α-bromocamphor-π-sulfonate.[18] Except for the sign of rotation the constants of the salt and amino acid were substantially identical to those given for the antipodes.

$C_5H_{11}NO_2S$

C. *Methionin* Ref. 150h

DL-Methionin: Durch Addition von Methylmercaptan an Acrolein über β-Methylmercapto-propionaldehyd und anschließende Strecker-Synthese[7].

DL-Methionin-methylester-hydrochlorid: Durch Veresterung der Aminosäure mit absol. Methanol/Thionylchlorid gewonnen[8]. Ausb. 95 % d. Th., Schmp. 116–118°.

DL-Methionin-methylester: Aus dem Hydrochlorid mit ammoniakalischem Äther[3], Ausb. 90 % d. Th.

Racematspaltung des DL-Methionin-methylesters: 7.2 g DL-Methylester und 13.3 g D-Weinsäure (Mol.-Verhältnis 1:2) werden in insgesamt 80 ccm absol. Methanol bei 20° angesetzt. Beginnt die Kristallisation nicht innerhalb einer Stunde freiwillig, so impft man die Lösung mit DL-Methionin-methylester-D-hydrogentartrat an. Die Impfkristalle gewinnt man durch Fällung einer dem Spaltansatz entnommenen Probe mit Äther. Nach 12 Stdn. hat sich *L-Methionin-methylester-D-hydrogentartrat* fast quantitativ abgeschieden. Ausb. 5.5 g; Schmp. 137–140°, $[\alpha]_D^{20}$: +26.5° (c = 3.17, in Wasser).

$C_{10}H_{19}NO_8S$ (313.3) Ber. C 38.40 H 6.08 N 4.48 Gef. C 38.03 H 6.26 N 4.74

Durch Zugabe von 20 ccm Äther zur Mutterlauge werden weitere 1–1.5 g Salz geringerer optischer Reinheit gewonnen, die man verwirft.

Die durch Ätherzugabe im Anschluß an die Zwischenfraktion ausfallenden Anteile sind *D-Methionin-methylester-D-hydrogentartrat.* Ausb. 5.8 g, Schmp. 130–135°, $[\alpha]_D^{20}$: +3.3° (c = 3.18, in Wasser).

Gef. C 38.43 H 6.29 N 4.70

Die Mutterlauge des D-Estersalzes enthält die überschüss. D-Weinsäure.

L- und D-Ester-hydrochlorid: Aus den so erhaltenen diastereomeren Rohtartraten gewinnt man nach dem Äther/Ammoniak-Verfahren in 90-proz. Ausbeute die rohen Esterhydrochloride:

L-Methionin-methylester-hydrochlorid: $[\alpha]_D^{20}$: +20.5° (c = 3.9, in Wasser).

D-Methionin-methylester-hydrochlorid: $[\alpha]_D^{20}$: −19.3° (c = 2.9, in Wasser).

Da der Literaturwert[8] für die optisch reinen Methionin-methylester-hydrochloride ±26.3° beträgt, wurden diese in einer optischen Reinheit von 70–80 % isoliert. Einmaliges Umkristallisieren aus Alkohol/Äther liefert in 40-proz. Ausbeute, bezogen auf Rohprodukt, die optisch reinen Esterhydrochloride.

L-Ester-hydrochlorid: Schmp. 151–155°, $[\alpha]_D^{20}$: +26.3° (c = 2.99, in Wasser).

D-Ester-hydrochlorid: Schmp. 152–154°, $[\alpha]_D^{20}$: −25.8° (c = 3.19, in Wasser).

$C_6H_{12}NO_2S \cdot HCl$ (199.7)	Ber.	C 36.10	H 7.02	N 7.02
L-Verb.	Gef.	C 36.48	H 7.37	N 7.18
D-Verb.	Gef.	C 36.05	H 7.06	N 7.29

Zur Verseifung werden die reinen Methylester-hydrochloride 30 Min. mit der 15fachen Menge 20-proz. Salzsäure unter Rückfluß erhitzt. Anschließend dampft man i. Vak. zur Trockne ein und behandelt den Rückstand mit absol. Äther. Ausbeute praktisch quantitativ.

L-Methionin-hydrochlorid: Schmp. 233–236° (Umwandlungspunkt 168–171°), $[\alpha]_D^{20}$: +21.8° (c = 1.68, in 0.2n HCl).

D-Methionin-hydrochlorid: Schmp. 232–235° (Umwandlungspunkt 168–172°), $[\alpha]_D^{20}$: −22.3° (c = 1.96, in 0.2n HCl).

$C_5H_{11}NO_2S \cdot HCl$ (185.7)	Ber.	C 32.34	H 6.47	N 7.55
L-Verb.	Gef.	C 32.59	H 6.71	N 7.30
D-Verb.	Gef.	C 32.81	H 6.71	N 7.69

M. BRENNER und V. KOCHER[9] geben für optisch reines Methionin-hydrochlorid eine spezif. Drehung von $[\alpha]_D^{18}$: ±21.0 ± 0.5° (c = 1.2, in 0.2n HCl) an.

---

7) E. PIERSON, M. GIELLA und M. TISHLER, J. Amer. chem. Soc. 70, 1450 [1948].

8) M. BRENNER und R. W. PFISTER, Helv. chim. Acta 33, 568 [1950]; 34, 2085 [1951].

$C_5H_{11}NO_2S$

1. Darstellung der 4′-Methyl-4-nitro-diphenylamin-2-sulfosäure.

Die Sulfosäure kann nach der Methode von *Ullmann*[3]) befriedigend dargestellt werden.

Das erhaltene Natriumsalz wurde in heisser wässriger Lösung mit der berechneten Menge $BaCl_2$-Lösung versetzt und das ausgefallene schwerlösliche Bariumsalz abfiltriert und getrocknet.

Aus dem Bariumsalz lässt sich auf einfache Art mit der berechneten Menge Schwefelsäure unter Rühren bei 80° eine Stammlösung der freien Sulfosäure mit bekannter Konzentration herstellen. Für die Fraktionierungsversuche wurde stets eine 0,5-n. Lösung der 4′-Methyl-4-nitro-diphenylamin-2-sulfosäure verwendet.

2. Darstellung des *d,l*-Methioninsalzes der 4′-Methyl-4-nitro-diphenylamin-2-sulfosäure.

1 g *d,l*-Methionin wurde in 20 cm³ heissem Wasser gelöst und mit 13 cm³ 0,5-n. 4′-Methyl-4-nitro-diphenylamin-2-sulfosäure versetzt. Diese Lösung wurde langsam auf Zimmertemperatur abgekühlt, wobei prachtvolle rotbraune Prismen entstanden. Nach eintägigem Stehen wurden die Krystalle abfiltriert, Rohausbeute 2,4 g. Das Salz wurde zweimal aus je 25 cm³ heissem dest. Wasser umkrystallisiert. Nach der Filtration blieben 1,8 g braunrote Prismen zurück, welche mit wenig Äther gewaschen und nachher im Hochvakuum bei Zimmertemperatur getrocknet wurden. Smp. 153—155° (korr., Sintern bei ca. 100°). Sie sind leicht löslich in Alkohol, Aceton und Dioxan, schwer löslich in Wasser, Äther und Chloroform und enthalten zwei Mol Krystallwasser, welches durch 9-stündiges Erhitzen bei 80° über P_2O_5 entfernt werden kann.

$C_{18}H_{23}O_7N_3S_2 \cdot 2\ H_2O$ (493,76) Ber. H_2O 7,28 Gef. H_2O 7,43%

Zur Elementaranalyse wurde das Salz 24 Stunden bei 110° im Hochvakuum getrocknet, wobei es gelb geworden und vollkommen entwässert war.

3,720 mg Subst. gaben 6,412 mg CO_2 und 1,763 mg H_2O
3,778 mg Subst. gaben 0,313 cm³ N_2 (21°, 728 mm)
4,541 mg Subst. verbrauchten 1,961 cm³ 0,02-n. KJO_3

$C_{18}H_{23}O_7N_3S_2$	Ber. C 47,20	H 5,03	N 9,24	S 13,95%
(457,73)	Gef. „ 47,04	„ 5,30	„ 9,21	„ 13,85%

3. Darstellung des *l*-Methioninsalzes der 4′-Methyl-4-nitro-diphenylamin-2-sulfosäure.

Die warme filtrierte Lösung von 1 g *l*-Methionin ($[\alpha]_D^{22} = +24,3° \pm 1°$ in 1-n. HCl) in 20 cm³ Wasser wurde mit 13,5 cm³ 0,5-n. 4′-Methyl-4-nitro-diphenylamin-2-sulfosäure-Lösung versetzt. Das Gemisch wurde unter vermindertem Druck auf ca. 10 cm³ eingedampft und über Nacht in den Eisschrank gestellt. Am andern Tag wurde das Salz abgesaugt, in 10 cm³ heissem Wasser gelöst, filtriert und eine halbe Stunde mit Eiswasser gekühlt. Nach dem Absaugen wurde erneut in 6 cm³ heissem Wasser gelöst und langsam krystallisiert. Es wurden 0,8 g gelbe Nadeln vom Smp. 198—199° (korr.) erhalten.

Das Salz ist schwer löslich in Äther, Aceton, Dioxan und Chloroform und schwerer löslich in Alkohol als das *d,l*-Isomere. Es enthält kein Krystallwasser. Zur Analyse wurde 24 Stunden im Hochvakuum bei 110° getrocknet.

3,180 mg Subst. gaben 6,560 mg CO_2 und 1,739 mg H_2O
4,201 mg Subst. gaben 0,350 cm³ N_2 (17°, 738 mm)
2,720 mg Subst. verbrauchten 1,98 cm³ 0,02-n. KJO_3

$C_{18}H_{23}O_7N_3S_2$	Ber. C 47,20	H 5,03	N 9,24	S 13,95%
(457,73)	Gef. „ 47,06	„ 5,02	„ 9,52	„ 14,12%

4. Löslichkeitsbestimmung des *l*- und *d,l*-Salzes von Methionin mit 4′-Methyl-4-nitro-diphenylamin-2-sulfosäure.

Je 200 mg Substanz des *d,l*- bzw. *l*-Salzes wurden in 100 cm³ Erlenmeyer-Kölbchen eingewogen und mit 40 bzw. 20 cm³ dest. Wasser überschichtet. Um das Löslichkeitsgleichgewicht zu erreichen, wurde über einem Quecksilberverschluss, dessen Ansatz die Flasche gut verschloss, 8 Stunden in Eiswasser gerührt. Dann wurde vorsichtig und rasch in eine gekühlte Vorlage abgesaugt und aus diesen Filtraten mit aliquoten Mengen der Gehalt an Trockensubstanz durch Eindampfen bestimmt. Diese approximative Löslichkeitsbestimmung ergab folgendes Resultat:

Vom *d,l*-Salz lösten sich 3,57 mg pro cm³ Wasser bei 0° C, vom *l*-Salz lösten sich 9,37 mg pro cm³ Wasser bei 0° C. Das Racemat besitzt somit eine Löslichkeit, welche weit kleiner ist als das Doppelte der Löslichkeit des *l*-Salzes, was auf ein echtes Racemat schliessen lässt.

[3]) *Ullmann*, B. **41**, 3751.

5. Fraktionierungsbeispiel Nr. 1 (Krystallisation des *l*-Salzes).

2,8 l einer wässrigen Lösung, welche 40 g Methionin von $[\alpha]_D^{22} = +18^0 \pm 1^0$ (1-n. HCl) enthielt, wurde auf 70° erwärmt und mit 700 cm³ (1,3 Mol) 0,5-n. 4′-Methyl-4-nitro-diphenylamin-2-sulfosäure-Lösung versetzt. Dann wurde die Reaktionsmischung langsam auf Zimmertemperatur abgekühlt, mit einem Impfkrystall des *d,l*-Methioninsalzes der 4′-Methyl-4-nitro-diphenylamin-2-sulfosäure versetzt und zwei Tage bei Zimmertemperatur belassen. Das während dieser Zeit in prachtvollen Prismen auskrystallisierte *d,l*-Salz wurde abfiltriert und bei 40° C am Vakuum getrocknet. Ausbeute 30,8 g Smp. 153–155° (unkorr., Sintern bei 98–100°).

Zersetzung des *d,l*-Salzes: die 30,8 g *d,l*-Sulfosalz wurden in 400 cm³ Wasser bei 40° gelöst und in einem geräumigen Becherglas gerührt. Dann wurde eine frisch filtrierte Lösung von 10 g (ber. 9,75 g) frisch umkrystallisierten Bariumhydroxyds in 200 cm³ Wasser ziemlich rasch zugetropft. Das Bariumsalz der Sulfosäure fiel aus und wurde nach 4-stündigem Stehen bei Zimmertemperatur abgesaugt, mit Wasser gut gewaschen und getrocknet. Es kann zur Gewinnung der Sulfosäure wieder verwendet werden. Das Waschwasser und das Filtrat wurden noch mit 2-n. H_2SO_4 ausbalanciert und das ausgefallene $BaSO_4$ abfiltriert. Dann wurde die Lösung im Vakuum beinahe zur Trockne verdampft. Das gelb gefärbte Methinonin wurde nun mit 100 cm³ Alkohol aus dem Kolben in ein Becherglas gespült und ca. 4 Stunden im Eisschrank stehen gelassen. Das nach dem Absaugen beinahe farblose Methionin wurde in der 10-fachen Menge Wasser bei ca. 90° gelöst, heiss filtriert und mit dem doppelten Volumen warmem Alkohol versetzt. Nach Krystallisation über Nacht wurde abfiltriert und am Vakuum bei 60° getrocknet. Ausbeute 5,6 g.

$[\alpha]_D^{25} = -0{,}06^0 \pm 1^0$ (c = 2,961; 6,30-n. HCl)

Ber. N 9,39 Gef. N 9,20% (*Kjeldahl*)

Die ursprüngliche Lösung (Volumen 3,5 l), aus der das *d,l*-Salz durch Krystallisation entfernt worden war, und welche noch das *l*-Salz enthält, wurde im Vakuum auf 500 cm³ eingeengt und 2½ Stunden im Kühlschrank stehen gelassen. Es fiel ein gelbes Salz aus, welches abfiltriert und getrocknet wurde. Ausbeute 35 g. Smp. 191–193° (unkorr., bei 170° leichtes Sintern). Durch weiteres Eindampfen wurden noch 9,2 g gelbes *l*-Salz gewonnen vom Smp. 190–192° (unkorr.). Die Mutterlauge enthielt noch 7,2 g eines orangegelben Niederschlages vom Zersetzungspunkt 224–227° (unkorr.), offenbar die überschüssige freie Sulfosäure.

Zersetzung des *l*-Salzes: die 1. und 2. Fraktion von 44,2 g des *l*-Salzes wurden unter Rühren in 800 cm³ warmem Wasser gelöst und mit einer Lösung von 15 g (ber. 14,3 g) Bariumhydroxyd in 300 cm³ Wasser versetzt. Dann wurde genau so verfahren, wie es bei der Zersetzung des *d,l*-Salzes beschrieben ist. Das durch Eindampfen erhaltene gelb gefärbte *l*-Methionin wurde mit 200 cm³ Alkohol in ein Becherglas gespült, nach 4-stündigem Stehen im Kühlschrank abgesaugt und mit 100 cm³ Alkohol gewaschen. Dieses Rohprodukt wurde wie oben aus der 10-fachen Menge heissem Wasser und dem doppelten Volumen Alkohol unkrystallisiert.

1. Fraktion 12,5 g. Aus den vereinigten Mutterlaugen konnte noch eine zweite Fraktion gewonnen werden. 2. Fraktion 2,6 g.

1. Fraktion $[\alpha]_D^{22} = +23{,}70^0 \pm 1^0$ (c = 3,041; 6,30-n. HCl)

Ber. N 9,39 Gef. N 9,28% (*Kjeldahl*)

2. Fraktion $[\alpha]_D^{22} = +24{,}32^0 \pm 1^0$ (c = 2,990; 6,30-n. HCl)

Ber. N 9,39 Gef. N 9,16% (*Kjeldahl*)

6. Fraktionierungsbeispiel Nr. 2 (ohne Krystallisation des *l*-Salzes).

(Dieses Beispiel entspricht den serienmässigen Ansätzen, wie sie sich durch die Erfahrung ergaben. Es war dabei nicht nötig, das *l*-Salz als solches zu isolieren, sondern man zersetzte die Mutterlauge, aus der das *d,l*-Salz entfernt worden war, direkt mit Bariumhydroxyd.)

50 g Methionin von $[\alpha]_D^{22} = +18^0 \pm 1^0$ (1-n. HCl) wurden in 2,8 l heissem Wasser gelöst und mit 720 cm³ 0,5-n. 4′-Methyl-4-nitro-diphenylamin-2-sulfosäure-Lösung versetzt (geringer Überschuss). Nachdem langsam auf Zimmertemperatur abgekühlt war, wurde mit dem *d,l*-Salz angeimpft und 48 Stunden stehen gelassen. Während dieser Zeit krystallisierten 48 g *d,l*-Sulfosalz, welches abfiltriert und getrocknet wurde. Dieses Salz wurde mit der berechneten Menge Bariumhydroxyd (16,2 g) gemäss Beispiel Nr. 1 zersetzt und daraus 8,2 g *d,l*-Methionin isoliert.

8,2 g, $[\alpha]_D^{23} = 0{,}03^0 \pm 1^0$ (c = 8,50; 1-n. HCl)

Die Mutterlauge des *d,l*-Salzes wurde direkt mit 40,8 g Bariumhydroxyd zersetzt und mit 2-n. H_2SO_4 ausbalanciert. Aus dem durch Eindampfen verbliebenen Rückstand wurden, wie oben beschrieben, zwei Fraktionen *l*-Methionin isoliert.

1. Fraktion 30,5 g.

$[\alpha]_D^{22} = +23{,}20^0 \pm 1^0$ (c = 9,54; 1-n. HCl)

2. Fraktion 3,2 g.

$[\alpha]_D^{22} = +23{,}78^0 \pm 1^0$ (c = 8,72; 1-n. HCl)

In einigen Fällen wurde bei der Zersetzung des *d,l*-Salzes eine kleine zweite oder dritte Fraktion Methionin isoliert, welche nicht mehr ganz inaktiv, sondern etwa halb racemisch war. Diese Fraktion wurde dann von neuem in den Prozess zurückgeführt. Das ursprüngliche Methionin enthält gemäss der Drehung (+ 18°) ca. 13 g *d,l*- und ca. 37 g *l*-Form. Davon wurden isoliert 8,2 g *d,l*-Methionin oder 63% und 33,7 g *l*-Methionin oder 91%.

7. Bemerkung zu den Drehungen.

Aus der nachstehenden Tabelle geht hervor, dass die spezifische Drehung von *l*-Methionin innerhalb von 1—10-proz. Lösungen von der Konzentration unabhängig ist. Zudem ist der Grenzwert der Drehung bereits in 1-n. HCl bis auf ca. 0,5° erreicht.

l-Methionin; N = 9,28% (*Kjeldahl*); 1. Fraktion, Beispiel 1	
HCl 6,30-n.	HCl 1,05-n.
C = 3,041, $[\alpha]_D^{24} = +23,70° \pm 1°$	C = 2,976, $[\alpha]_D^{22} = +23,78° \pm 1°$
C = 6,018, $[\alpha]_D^{20} = +23,68° \pm 0,5°$	C = 6,039, $[\alpha]_D^{20} = +23,50° \pm 0,5°$
C = 10,006, $[\alpha]_D^{20} = +23,97° \pm 0,3°$	C = 9,989, $[\alpha]_D^{20} = +23,25° \pm 0,3°$

Als Vergleichsmaterial diente ein *d*(+)-Methionin, welches aus synthetischem *d,l*-Methionin nach *Windus* und *Marvel*[1]) mit Brucin gespalten wurde.

d(+) Methionin, von uns gespalten:

$[\alpha]_D^{15,5} = -24,1° \pm 0,5°$ (c = 2,304; 6,30-n. HCl)

$[\alpha]_D^{16} = +\ 7,99° \pm 0,5°$ (c = 2,272; Wasser)

$[\alpha]_D^{20} = -21,45° \pm 0,7°$ (c = 3,220; 0,20-n. HCl)

d(+) Methionin nach *Windus* und *Marvel*:

$[\alpha]_D^{25} = +\ 8,12° \pm 0,5°$ (c = 0,80; Wasser)

$[\alpha]_D^{25} = -21,18° \pm 0,5°$ (c = 0,80; 0,2001-n. HCl)

Die Analysen verdanken wir teils Hrn. *W. Manser*, Mikroanalytisches Laboratorium der E.T.H., teils Hrn. Dr. *G. Frey* von der *Aligena A.G.*, Basel.

A resolution procedure for the above compound using 2-pyrrolidone-5-carboxylic acid as resolving agent may be found in J. H. Sbarklop, Japanese Patent 45-32250 (1970). Another resolution procedure for the above compound as the amide derivative using D-tartaric acid as resolving agent may be found in S. Tatsuoka and M. Honjo, Yakugaku Zasshi, 73, 357 (1953).

$C_5H_{11}NO_2S$ Ref. 150j

I. DL-Methionin-isopropylester.

Man sättigt ein Gemisch von 100 g DL-Methionin und 500 cm³ absolutem Isopropylalkohol mit trockenem HCl-Gas, kocht 30 Minuten am Rückfluss und entfernt den Alkohol im Vakuum (Bad 50—60°). Der Prozess wird nach Zusatz von 500 cm³ frischem Isopropylalkohol wiederholt. Der Rückstand wird mit einem Spatel aufgelockert, mit 200 cm³ Wasser und 500 cm³ Äther übergossen, das Gemisch auf − 10° abgekühlt und unter kräftigem Rühren und ständiger Kühlung mit 25-proz. wässerigem Ammoniak eben schwach phenolphtaleinalkalisch gestellt. Eine entstandene Fällung wird abfiltriert, mit Äther gewaschen und verworfen. Man trennt das Filtrat im Scheidetrichter, extrahiert die wässerige Phase nach Sättigung mit NaCl mehrmals mit frischem Äther, trocknet die vereinigten Ätherlösungen und destilliert. Ausbeute 112 g Ester (87%), Kp.$_{11 mm}$ 127—129°. Eine Probe wurde ein zweites Mal destilliert und die Mittelfraktion analysiert:

4,142 mg Subst. gaben 7,690 mg CO_2 und 3,378 mg H_2O

$C_8H_{17}O_2NS$ (191,28) Ber. C 50,23 H 8,96% Gef. C 50,66 H 9,13%

Dichte: $\varrho_{18°} = 1,0289$

II. Charakterisierung der Methionin-Nasylate.

Die Salze krystallisieren aus der heissen Lösung von 1 mMol Methionin und 1,2 mMol Naphtalin-β-sulfosäure in 1,2 cm³ 1-n. Salzsäure (kratzen!). Reinigung durch Umkrystallisieren aus der drei- bis vier-fachen Menge Wasser. Zur Analyse wird im Hochvakuum bei 60—80° getrocknet.

DL-Salz: Schmelzpunkt von intakten Krystallen: 187—190°
Schmelzpunkt von zerriebenen Krystallen: 176—185°

3,779 mg Subst. gaben 6,953 mg CO_2 und 1,852 mg H_2O
$C_{15}H_{19}O_5NS_2$ (357,43) Ber. C 50,40 H 5,36% Gef. C 50,21 H 5,48%

D-Salz: Schmelzpunkt von intakten Krystallen: 203—205°
Schmelzpunkt von zerriebenen Krystallen: 181—191°
Mischprobe mit DL-Salz (zerrieben): 150—176°

4,014 mg Subst. gaben 7,369 mg CO_2 und 1,920 mg H_2O
$C_{15}H_{19}O_5NS_2$ (357,43) Ber. C 50,40 H 5,36% Gef. C 50,10 H 5,35%
$[\alpha]_D^{19,5} = -15,6^0 \pm 0,2^0$ (c = 5 in trockenem Methylcellosolve)

L-Salz: Es zeigt dieselben Schmelzpunkte wie das D-Salz und denselben Drehwert mit umgekehrtem Vorzeichen.

III. Herstellung von D- und L-Methionin.

Das erwähnte technische Pankreasenzym der Firma *J. R. Geigy AG.*, mit dem die vorliegenden Versuche durchgeführt worden sind, kann direkt, d. h. in Substanz oder in Form eines wässerigen Auszuges, verwendet werden. Eine Reinigung durch Umfällen mit Ammonsulfat ist jedoch empfehlenswert. 30 g „Enzymhochkonzentrat" werden während einiger Stunden mit 150 cm³ kaltem Wasser digeriert und der Extrakt durch Filtration oder auf der Zentrifuge geklärt. Man versetzt die Lösung (140 cm³) portionenweise unter Rühren mit insgesamt 95 g Ammonsulfat, dekantiert die resultierende Fermentsuspension von ungelöstem Ammonsulfat ab und filtriert durch eine grosse Nutsche. Das zurückbleibende Ferment wird zweimal mit gesättigter Ammonsulfatlösung gewaschen, scharf abgesaugt, im Vakuum über Schwefelsäure getrocknet und gemahlen. Ausbeute 4 g. Das gereinigte Produkt wird im folgenden kurz als „Ferment" bezeichnet. Es ist monatelang haltbar.

100 g DL-Methionin-isopropylester werden mit der Lösung von 2 g „Ferment" in 100 cm³ Wasser versetzt und in einem mit Natronkalk-Rohr versehenen Gefäss unter langsamem Rühren bei 20° sich selbst überlassen. Das Reaktionsgemisch erstarrt im Verlauf einiger Stunden (Rührer abstellen). Nach insgesamt 48 Stunden — ein Kontrollversuch ohne Ferment bleibt während dieser Zeit unverändert — werden 260 cm³ Äther-Alkohol (1 : 1) zugesetzt und während 3—4 Stunden mit der Reaktionsmasse verrührt. Man saugt durch eine geräumige Nutsche vorsichtig ab, deckt den Rückstand in mehreren Portionen mit insgesamt 130 cm³ Äther-Alkohol (1 : 1) und schliesslich mit 70 cm³ Äther. Das Produkt wird im Vakuum vom Äther befreit, gemahlen und im *Soxleth*-Apparat nochmals über Nacht mit Äther extrahiert. Trockengewicht 36 g. Ausbeute an rohem L-Methionin nach Abzug von 2 g Fermentsubstanz: 34 g (87% der Theorie). Sämtliche Alkohol-Äther- und Ätherlösungen werden vereinigt und im Vakuum eingeengt. Der ölige Rückstand enthält D-Ester, etwas Wasser und geringe Mengen noch nicht näher untersuchter Umwandlungsprodukte des L-Esters. Das Öl wird deshalb vor der Destillation nochmals in 250 cm³ Äther aufgenommen, von nunmehr ausfallenden Reaktionsprodukten abfiltriert, von abgeschiedenem Wasser abgetrennt und über Natriumsulfat getrocknet. Durch Destillation erhält man 40 g D-Ester, Kp.$_{11\,mm}$ 126—128°, $[\alpha]_D^{18} = +7,64^0$ (unverdünnt)[1]).

D-Methionin.

60 g D-Methionin-isopropylester werden zur Verseifung mit 420 cm³ 2-n. Salzsäure 2 Stunden am Rückfluss gekocht. Zur heissen Lösung fügt man 90 g Naphtalin-β-sulfosäure-trihydrat, löst die Sulfosäure durch kurzes Aufkochen, impft möglichst heiss mit etwas reinem D-Methionin-nasylat an und lässt vor rascher Abkühlung geschützt über Nacht erkalten. Der resultierende Krystallkuchen wird auf einer Nutsche ausgepresst, so dass die Mutterlauge sofort abfliessen kann, und schliesslich mit Eiswasser gewaschen. Eine nochmalige Krystallisation aus der 3-fachen Menge Wasser (heiss impfen) liefert praktisch reines, konstant drehendes D-Methionin-nasylat. Ausbeute lufttrocken 73 g (65% bezogen auf verseiften D-Ester). Zur Gewinnung des freien Methionins werden 70 g Nasylat in einer Lösung von 26,5 cm³ Diäthylamin in 700 cm³ 96-proz. Alkohol suspendiert. Man rührt 3½ Stunden bei Zimmertemperatur und filtriert die freigesetzte, in Alkohol unlösliche Aminosäure ab. Sie wird auf der Nutsche mit Alkohol gewaschen und hierauf nochmals in 250 cm³ 96-proz. Alkohol suspendiert. Nach 3-stündigem Rühren wird wiederum abgesaugt, mit Alkohol gewaschen und getrocknet. Ausbeute fast quantitativ. Zur Analyse und Drehung wird das praktisch schon reine Produkt aus 66-proz. Alkohol umkrystallisiert und eine Probe während 2 Stunden bei 70° im Hochvakuum getrocknet. Ausbeute 22 g D-Methionin.

---

[1]) Die Drehung des Esters kann um mehr als 1° sinken, wenn er Spuren von Feuchtigkeit enthält. Zum Trocknen wird er in der 4-fachen Menge absolutem Äther gelöst, mit 25% gebranntem Kalk über Nacht auf der Maschine geschüttelt und destilliert.

3,239 mg Subst. gaben 4,80 mg CO_2 und 2,20 mg H_2O
3,124 mg Subst. gaben 0,271 cm^3 N_2 (28°, 743 mm)

$C_5H_{11}O_2NS$	Ber. C 40,23	H 7,43	N 9,39%
(149,15)	Gef. „ 40,44	„ 7,60	„ 9,62%

$[\alpha]_D^{19,5} = -23,4° \pm 0,4°$ (c = 5 in 6-n. HCl)

$[\alpha]_D^{18} = -20,8° \pm 0,5°$ (c = 1,2 in 0,20-n. HCl)

Die höchste von uns gemessene spezifische Drehung beträgt

$[\alpha]_D^{17,5} = -23,8° \pm 0,4°$ (c = 5 in 6-n. HCl)

Das betreffende, aus 66-proz. Alkohol krystallisierte und im Hochvakuum bei 80° getrocknete, analysenreine Präparat war aus einem über mehrere Krystallisationen hinweg konstant drehenden Nasylat bereitet worden. *K. Vogler* und *F. Hunziker*[2]) erhielten bei der Spaltung des DL-Methionins mit Brucin nach *Windus* und *Marvel*[3]) D-Methionin mit der spezifischen Drehung

$[\alpha]_D^{15,5} = -24,1° \pm 0,5°$ (c = 2,304 in 6,3-n. HCl)

Windus und *Marvel* geben an[3])

$[\alpha]_D^{25} = -21,18° \pm 0,5°$ (c = 0,80 in 0,2001-n. HCl).

L-Methionin.

Man behandelt 36 g rohes L-Methionin mit 540 cm^3 kochendem Wasser, filtriert heiss von unlöslichen Anteilen ab (3,6 g), versetzt das Filtrat mit dem gleichen Volumen konz. Salzsäure (d = 1,19), kocht zur Hydrolyse von vorhandenen Peptiden über Nacht am Rückfluss und verdampft im Vakuum zur Trockene. Der Rückstand wird in 100 cm^3 Wasser gelöst, die dunkelgefärbte Lösung mit 7 g säurebehandelter Kohle während 10 Minuten kochend gerührt und heiss filtriert. Man verdünnt das nur noch schwach gefärbte Filtrat (+ Waschwasser) mit Wasser auf ein Volumen von 290 cm^3, versetzt es heiss mit 58 g Naphtalin-β-sulfosäure-trihydrat, löst kochend und lässt nach erfolgtem Animpfen wie beim D-Methionin angegeben erkalten[1]). Ausbeute 50 g Nasylat (58% bezogen auf rohes L-Methionin). Zur weiteren Reinigung werden insgesamt 76 g Nasylat (aus 2 Ansätzen) ein zweites Mal aus der 3-fachen Menge Wasser umkrystallisiert (heiss impfen). Ausbeute 71 g (93%) praktisch reines L-Salz. Die Zerlegung des Salzes (vgl. oben) liefert 28 g L-Methionin. Eine Probe wird zur Analyse und Drehung aus 66-proz. Alkohol umkrystallisiert und bei 80° im Hochvakuum getrocknet.

4,397 mg Subst. gaben 6,520 mg CO_2 und 2,870 mg H_2O
3,175 mg Subst. gaben 0,274 cm^3 N_2 (27°, 743 mm)

$C_5H_{11}O_2NS$	Ber. C 40,23	H 7,43	N 9,39%
(149,15)	Gef. „ 40,46	„ 7,31	„ 9,60%

$[\alpha]_D^{18} = +23,4° \pm 0,4°$ (c = 5 in 6-n. HCl)

$[\alpha]_D^{18,5} = +21,0° \pm 0,5°$ (c = 1,2 in 0,20-n. HCl)

[1]) Das folgende, einfachere Vorgehen führt oft zum selben Ziel: Rohmethionin mit 7-facher Menge 1-n. HCl aufkochen, erkaltete Suspension filtrieren, Filtrat heiss mit Naphtalin-β-sulfosäure versetzen und impfen.

[2]) Helv. **30**, 2013 (1947). [3]) Am. Soc. **53**, 3490 (1931).

Authors' comments:

Das verwendete Fermentpräparat ist ein technisches Pankreas-Enzym der Firma *J. R. Geigy AG.*, Basel[2]). Es kann durch „*Difco*-Trypsin 1 : 250“[3]), „*Difco*-Pangestin 1 : 75“[3]), „Pankreatin Hochkonzentrat, roh“ (*Schweiz. Ferment AG.*, Basel) und andere Handelsprodukte ersetzt werden, welche sich in ihrer Wirkung auf die Verseifung des L-Esters beschränken. Die günstigsten Reaktionsbedingungen sind jeweilen durch Vorversuche zu ermitteln.

An Stelle des Isopropylesters können auch andere Ester, z. B. der n-Butyl-, i-Butyl-, n-Propyl- oder Äthylester der fermentativen Verseifung unterworfen werden. Wir haben den Isopropylester gewählt, weil er weniger zu spontanen Veränderungen neigt als die Ester der primären niedermolekularen Alkohole[4]) und eine hervorragende Affinität zum Ferment besitzt.

[1]) *M. Brenner, E. Sailer* und *V. Kocher*, Helv. **31**, 1908 (1948).
[2]) „Enzymhochkonzentrat“ mit einer Aktivität von ca. 35000 „*Geigy*-Einheiten“ pro kg. Das Produkt ist uns von der Firma *Geigg* in freundlicher Weise zur Verfügung gestellt worden.
[3]) *Difco Laboratories*, Inc., Detroit, Michigan.
[4]) Versuche der Herren *K. Menzi* und *W. Thommen* im hiesigen Laboratorium; vgl. ferner *Abderhalden* und *Suzuki*, Z. physiol. Ch. **176**, 101 (1928).

For the experimental details of additional biochemical methods for the resolution of racemic methionine, see the following: R. Duschinsky and J. Jeannerat, Compt. Rend., 208, 1359 (1939); P. K. Stumpf and D. E. Green, J. Biol. Chem., 153, 387 (1944); C. A. Dekker and J. S. Fruton, J. Biol. Chem., 173, 471 (1948); V. E. Price, J. B. Gilber and J. P. Greenstein, J. Biol. Chem., 179, 1169 (1949); V. Kocher and K. Vogler, Helv. Chim. Acta, 31, 352 (1948); K. A. J. Wretlind, Acta physiol. Scand., 20, 1 (1950); K. A. J. Wretlind and W. C. Rose, J. Biol. Chem., 187, 697 (1950); S. M. Birmbaum, L. Levintow, R. B. Kingsley and J. P. Greenstein, J. Biol. Chem., 194, 455 (1952); K. Michi, and H. Nonaka, Bull. Agr. Chem. Soc. Japan, 19, 153 (1955); C. Neuberg and J. Mandl, Enzymologia, 14, 28 (1950); S. Yamada, I. Chibata and S. Yamada, Yakugaku Zasshi, 75, 113 (1955); I. Chibata, T. Ishikawa and S. Yamada, Bull. Agr. Chem. Soc. Japan, 21, 304 (1957); S. Kameda, Japanese Patent 37-11662 (1962); Y. Kimura, Japanese Patent 47-671 (1972); T. Tosa, T. Mori, N. Fuse and I. Chibata, Biotechnol. Bioeng., 9, 603 (1967) and Ref. 18n.

$CH_3Se(CH_2)_2CH(NH_2)COOH$

$C_5H_{11}NO_2Se$ **Ref. 151**

Enzymatic resolution of γ-benzylselenohomocysteine into the optically active isomers

8 g DL-N-acetyl-γ-benzylselenohomocysteine were dissolved in 30 ml 1 *N* NaOH 122 ml 0.2 *M* citrate buffer of pH 5, and were mixed with 5.7 g freshly distilled in, shaken, and slightly warmed until a clear solution had been obtained. 2.4 g ain (Merck No. 7147, digestive power 1:350) were extracted with 21 ml water filtered. 0.45 g L-cysteine hydrochloride was dissolved in 16 ml citrate buffer. three solutions were mixed, and 93 ml citrate buffer were added. The flask was d with nitrogen, and the mixture was incubated for 48 hours at 40°C. The result- L-N-acetyl-γ-benzylselenohomocysteine anilide was collected, washed with some er, and recrystallized from ethanol-water. Yield 4.15 g, melting point 156–157°. filtrate, minus the washing water, was incubated for another 48 hours in order change all L-N-acetyl-γ-benzylselenohomocysteine into anilide. The product ch now resulted and which was of lesser purity was sucked off after the mixture been stored cool over night. This fraction was recrystallized as above. Yield g, melting point 155–156.5°. A further recrystallization raised the melting at to 156–157°. Total yield 4.95 g (90%). 0.1082 g dissolved in acetic acid to 0 ml: $\alpha_D^{25} = -0.160°$ $[\alpha]_D^{25} = -14.8°$.

rom the mother lye after the papain resolution the D-N-acetyl-γ-benzylseleno- ocysteine was obtained by acidification with concentrated hydrochloric acid to 2 and cooling in the refrigerator over night. The precipitated product was care- y collected and dissolved in acetone. The solution was boiled with charcoal and red, whereupon the acetone was evaporated. The remainder was recrystallized n methanol and some water. Yield 4.1 g (93%). Melting point 137–138°.

1009 g dissolved in acetic acid to 10.00 ml: $\alpha_D^{25} = -0.146°$. $[\alpha]_D^{25} = -14.5°$.

Benzylselenohomocysteine

.2 g (5.7 mmole) of L-N-acetyl-γ-benzylselenohomocysteine anilide were hydro- d with 150 ml 2 *N* HCl under stirring and reflux for 4½ hours. The hydrochloric was distilled off in vacuum upon a 60° water bath until the greatest part of hydrochloride of the amino acid had crystallized. After the addition of 25 ml er the mixture was refluxed for another 1½ hour, whereupon it was neutralized armth with NaOH. The product was collected and washed with water, methanol acetone. The amino acid was again dissolved in hot 2 *N* HCl, filtered, and pre- tated with LiOH. The precipitate was collected and washed with some water then with methanol. Yield 1.35 g (88%).

1025 g dissolved in 1 *N* HCl to 10.00 ml: $\alpha_D^{25} = +0.159°$. $[\alpha]_D^{25} = +15.5°$.

elenomethionine

.36 g L-γ-benzylselenohomocysteine was suspended in *c.* 30 ml of NH_3 liq., sodium vings being added until the blue colour persisted for about 10 minutes. A few ligrammes of ammonium chloride made the blue colour disappear, and 1 ml thyl iodide was added drop by drop. After 30 minutes at −70° the carbondioxide ohol cooling bath was removed, and the NH_3 was permitted to evaporate through rying tube filled with cotton. The last traces of ammonia were removed by coup- the reaction vessel to an evacuated desiccator containing a dish with concentrat- sulphuric acid. The remainder was neutralized with 2 *N* HCl, the resulting solu- n extracted once with ether and the water phase concentrated in vacuum to 18 ml. s solution was loaded on top of a 1.8×41 cm ion-exchange column of Bio Rad A8 G in the self-absorbed form, and the amino acid was eluted with distilled water. The amino acid front was detected with ninhydrin tests on paper, and the eluate collected until a test with silver nitrate revealed the presence of halide ions in the eluate. A second portion of eluate containing a mixture of amino acid and inorganic salts was collected, concentrated and, and once again desalted under the same conditions.

The salt-free amino acid solution was charcoaled, evaporated in vacuum to a small volume, and left to crystallize in the refrigerator. The thin flaky crystals were collected and washed with little methanol-acetone and dried. Yield 0.63 (65%).

Rf 0.60 using circular filter paper technique upon Munktell's No. 302 with *n*-butanol, acetic acid, H_2O 4:1:5 (upper phase). For IR-spectrum see Fig. 1.

Anal.

Found	Se	39.94%
Calc.	Se	40.26%

0.0244 g dissolved in 2.01 ml 1 *N* HCl: $\alpha_D^{25} = +0.201°$. $[\alpha]_D^{25} = +18.1°$.

D-*Selenomethionine*

D-Selenomethionine prepared in the same way from 0.80 g D-γ-benzylseleno-homocysteine yielded 0.42 g (74%) D-Selenomethionine. Rf-value and IR-spectrum identical as expected.

Anal.

Found	Se	40.00%
Calc.	Se	40.26%

0.0199 g dissolved in 2.01 ml 1 *N* HCl: $\alpha_D^{25} = -0.181°$, $[\alpha]_D^{25} = -18,3°$.

$CH_3-C(CH_3)(OH)-CH(NH_2)-COOH$

$C_5H_{11}NO_3$ **Ref. 152**

(—)-α-Methyl Benzylamine Salt of *N*-Benzoyl-β-hydroxyvaline (V). — DL-*N*-Benzoyl-β-hydroxyvaline (IV), 94.8 g. (0.4 mol.), and (—)-α-methylbenzylamine, 48.4 g. (0.4 mol.), ($[\alpha]_D^{25}=42.3°$, benzene), were dissolved in 50 ml. methanol. The solvent was evaporated in vacuo. The residual syrup was dissolved in hot acetone (150 ml.). After cooling, crystallization of (—)-α-methylbenzylamine salt (V) began by seeding with an authentic specimen which was obtained in another crystallization experiment. After standing overnight in a refrigerator, the crystals were filtered and washed with cold acetone. The crystals, 51.8 g. (72%), were purified from a hot mixture of methanol and acetone (5 : 95). Yield, 47.3 g. (66%), m. p. 155°C, $[\alpha]_D^{25}= -22.9°$ (*c* 2.32, water).

Found: C, 67.12; H, 7.43; N, 7.64. Calcd. for $C_{20}H_{26}N_2O_4$: C, 67.02; H, 7.31; N, 7.82%.

(—)-*N*-Benzoyl-β-hydroxyvaline (VI). — (—)-α-Methylbenzylamine salt (V), 36.0 g. (0.1 mol.), was dissolved in 100 ml. of water. To this solution, 110 ml.

of 2 N sodium hydroxide was added. The alkaline solution was extracted with ether twice (100 ml. × 2) to remove (—)-α-methylbenzylamine. Concentrated hydrochloric acid was added to the aqueous solution to bring the pH to 3.0. (—)-*N*-Benzoyl-β-hydroxyvaline (VI) was extracted with ethyl acetate (100 ml. × 2). The ethyl acetate solution was washed with water and dried with anhydrous sodium sulfate. After evaporation of ethyl acetate in vacuo, 20.7 g. (87%) of crude VI was obtained. This was recrystallized from methanol and ether. Yield, 19.7 g. (83%), m. p. 112°C, $[\alpha]_D^{25} = -30.5°$ (*c* 2.01, ethanol). Infrared spectrum showed bands at 1745 cm^{-1} (-COOH); 1645, 1550 cm^{-1} (-CONH-); 1150 cm^{-1} (>C-OH). The optically active *N*-benzoyl-β-hydroxyvaline also showed polymorphism which was detected by infrared spectrographically as was reported for DL-*N*-benzyl-β-hydroxyvaline.[15)]

Found: C, 60.85; H, 6.32; N, 5.70. Calcd. for $C_{12}H_{15}NO_4$: C, 60.75; H, 6.37; N, 5.90%.

(—)-β-Hydroxyvaline (VII).—A mixture of (—)-*N*-benzoyl-β-hydroxyvaline (VI), 2.37 g. (0.01 mol.), and 5 N hydrochloric acid, 50 ml., was refluxed for 5 hr. After cooling, the precipitated benzoic acid was separated by filtration. The filtrate was extracted with chloroform (50 ml. × 2) to remove the residual benzoic acid. After evaporation of the hydrochloric acid solution under reduced pressure, the residual syrup was dissolved in 50 ml. of water. The solution was treated with silver carbonate to remove hydrogen chloride. This mixture was filtered and washed with water. Hydrogen sulfide gas was bubbled through the solution. Silver sulfide was removed by filtration. The filtrate was concentrated in vacuo. A yield of 1.18 g. of (89%) of crude VII was obtained. This was recrystallized from water and ethanol, 0.997 g. (75%), m. p. 208°C (decomp.), $[\alpha]_D^{25} = -13.5°$ (*c* 4.82, 5 N HCl). Infrared spectrum showed bands at 1660, 1600, 1555, 1418 cm^{-1} (amino acid), 1145 cm^{-1} (>C-OH). R_f value 0.44, BuOH : AcOH : H_2O = 4 : 1 : 2. The phenomenon of polymorphism of the compound was also observed by infrared spectra. Infrared absorption curve varied depending on the ratio of water and ethanol in the recrystallization.

Found: C, 45.08; H, 8.31; N, 10.69. Calcd. for $C_5H_{11}NO_3$: C, 45.10; H, 8.33; N, 10.52%.

$C_5H_{11}NO_3$ C. Dihydro-lysergyl-dimethylserin-peptide. Ref. 152a

1. DL-Formyl-O-methyl-dimethylserin. 50 g DL-O-Methyl-dimethylserin[1)] löste man in 250 cm^3 wasserfreier Ameisensäure, versetzte unter Eiskühlung mit 50 cm^3 Acetanhydrid, liess ½ Std. im Eiswasser und 3 Std. bei Raumtemperatur stehen, verdünnte die Lösung mit 50 cm^3 Wasser, liess weitere 3 Std. bei Zimmertemperatur stehen, verdampfte im Vakuum zur Trockne und kristallisierte den Rückstand aus 100 cm^3 Wasser, woraus 50 g Kristallisat vom Smp. 170—171° erhalten wurden. Durch Konzentration der Mutterlaugen konnten weitere 2 g etwas weniger reiner Substanz gewonnen werden, so dass die Ausbeute 87% d.Th. betrug.

Für die Analyse wurde ein Präparat zweimal aus Wasser umkristallisiert, wodurch sich der Smp. auf 175° erhöhte.

$C_7H_{13}O_4N$	Ber. C 47,99	H 7,48	N 8,00%
(175,18)	Gef. „ 48,48	„ 7,53	„ 8,05%

2. Brucinsalz von L-Formyl-O-methyl-dimethylserin. 1,75 g DL-Formyl-O-methyl-dimethylserin und 4,0 g im Hochvakuum getrocknetes Brucin löste man zusammen in 60 cm^3 Alkohol. Durch Reiben mit dem Glasstab konnte aus der alkoholischen Lösung das L-formyl-O-methyl-dimethylserin-saure Brucin zur Kristallisation gebracht werden (2,92 g). Einmaliges Umkristallisieren aus der zehnfachen Menge Alkohol lieferte 2,66 g (93% d.Th.) reines Salz vom Smp. 150—151°.

3. Brucinsalz von D-Formyl-O-methyl-dimethylserin. Die Mutterlaugen des unter 2. beschriebenen L-formyl-O-methyl-dimethylserin-sauren Brucins dampfte man im Vakuum zur Trockne ein und kristallisierte den Rückstand aus der zehnfachen Menge Isopropylalkohol. Man erhielt so 2,52 g (88% d. Th.) reines D-Brucinsalz vom Smp. 194—196°.

Die Trennung des DL-formyl-O-methyl-dimethylserin-sauren Brucins konnte mit ähnlich guten Resultaten auch in der umgekehrten Reihenfolge durchgeführt werden, indem zuerst aus isopropylalkoholischer Lösung die D-Verbindung abgeschieden, und hierauf aus der zur Trockne gebrachten Mutterlauge das L-Brucinsalz aus Alkohol kristallisiert wurde.

4. L-Formyl-O-methyl-dimethylserin. 4,9 g des unter 2. beschriebenen L-formyl-O-methyl-dimethylserin-sauren Brucins wurden in Chloroform gelöst und die Lösung mit 12 cm^3 n. Natronlauge und etwas Wasser geschüttelt. Der gesammelte wässerige Auszug wurde mit der berechneten Menge n. Salzsäure versetzt und bis zur beginnenden Kristallisation eingeengt, worauf beim Abkühlen 1,08 g L-Formyl-O-methyl-dimethylserin vom Smp. 163° erhalten wurden.

Zweimaliges Umkristallisieren lieferte 0,82 g viereckige Blättchen vom Smp. 163 bis 165° und der spez. Drehung $[\alpha]_D^{20} = -5°$ (c = 1,0 in Wasser).

5. D-Formyl-O-methyl-dimethylserin. Aus 4,2 g D-formyl-O-methyl-dimethylserin-saurem Brucin wurde, wie unter 4. beschrieben, die Säure in Freiheit gesetzt und zweimal aus Wasser umkristallisiert, wobei viereckige Blättchen vom Smp. 164—165° und der spez. Drehung $[\alpha]_D^{20} = +5°$ (c = 1,0 in Wasser) erhalten wurden.

---

1) *W. Schrauth & H. Geller*, B. **55**, 2783 (1922).

6. L-Dimethylserin. 0,71 g L-Formyl-O-methyl-dimethylserin kochte man mit 4 cm³ 40-proz. Bromwasserstoffsäure 2 Std. am Rückfluss, verdampfte die Lösung im Vakuum zur Trockne, löste den Rückstand in absolutem Alkohol und neutralisierte vorsichtig mit Ammoniak, worauf L-Dimethylserin in feinen, sechseckigen Blättchen vom Smp. 205° auskristallisierte.

Für die Analyse wurde ein Präparat noch zweimal aus Wasser/Alkohol umkristallisiert, ohne dass sich sein Smp. veränderte.

$C_5H_{11}O_3N$	Ber. C 45,11	H 8,26	N 10,52%
(133,09)	Gef. ,, 44,84	,, 8,05	,, 11,01%

$[\alpha]_D^{20} = +4,7°$ (c = 1,0 in Wasser)

7. D-Dimethylserin. 0,4 g D-Formyl-O-methyl-dimethylserin wurden, wie unter 6. beschrieben, mit konzentrierter Bromwasserstoffsäure behandelt und aufgearbeitet. Man erhielt 0,23 g D-Dimethylserin in feinen, sechseckigen Blättchen vom Smp. 205°.

$C_5H_{11}O_3N$	Ber. C 45,11	H 8,26	N 10,52%
(133,09)	Gef. ,, 44,89	,, 8,28	,, 10,46%

$[\alpha]_D^{20} = -4,6°$ (c = 1,0 in Wasser)

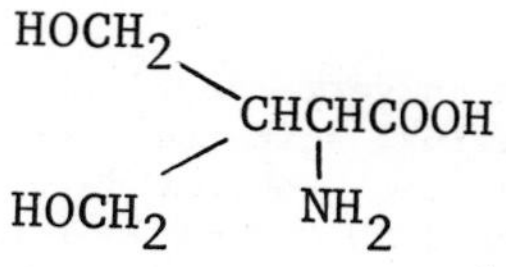

$C_5H_{11}NO_4$ Ref. 153

Resolution of DL-γ,γ'-Dihydroxyvaline.—A solution of 3.035 g. (18.1 mmoles) of DL-γ,γ'-dihydroxyvaline lactone hydrochloride (VI. X = OH) in 25 ml. of water was placed in a Beckman Automatic Titrator (Model K), in which an efficient vibration mixer replaced the stirrer. Upon setting the pH dial to 8.5, the titrator added 5.30 ml. of 3.56 *N* sodium hydroxide solution (18.8 mmoles, NaOH). The solution was then cooled and, under strong vibration, 6.20 g. (36.2 mmoles) of chloroacetic anhydride was added in ten equal portions over a period of 1 hr. Simultaneously, the titrator added 15.25 ml. of 3.56 *N* sodium hydroxide solution (54.3 mmoles, NaOH). After one more hour of vibration at room temperature, the solution was acidified to pH 1.5 by the addition of 4.0 ml. of concentrated hydrochloric acid and extracted with ethyl acetate (twelve 10-ml. portions). Evaporation of the dried ethyl acetate solution yielded an oil, from which most of the chloroacetic acid was eliminated by successive extractions with hot hexane (five 10-ml. portions) and high vacuum drying over solid potassium hydroxide. The residue consisted of 2.55 g. (50%) of crude **α-chloroacetamino-β-chloroacetoxymethyl-γ-butyrolactone**, which remained a viscous oil. λ_{max}^{neat} 2.92, 5.63–5.70, 5.99, 6.48, 6.52 and 8.45 μ.

Anal. Calcd. for $C_9H_{11}NO_5Cl_2$: C, 38.04; H, 3.90; Cl, 24.95; N, 4.93. Found: C, 38.64; H, 4.66; Cl, 22.93; N, 4.64.

A 2.37-g. sample (8.4 mmoles) of the above oil was vibrated in 50 ml. of water and by setting the pH dial of the autotitrator to 9.0, an 1.835 *N* lithium hydroxide solution was automatically added. The base uptake virtually stopped after addition of 8.30 ml. (15.3 mmoles LiOH; 180 min. at room temperature) and a clear solution resulted. After dilution to 180 ml. and adjusting the pH to 8.0, 60 mg. of hog kidney acylase powder (Nutrition Biochemical Corporation, Cleveland, Ohio) was added and the mixture kept at 38° ± 1° for 90 hr. The enzyme was then eliminated by stirring the solution (pH 6) with 1.0 g. of "Darco" for 1 hr. at room temperature and the filtered solution was evaporated *in vacuo* to yield 2.36 g. of an oil. The latter was redissolved in 25 ml. of water and passed through a column of Amberlite IR 120 cation exchange resin (H^+ form, 1.9 meq./ml., 20 ml.). The column was washed with 250 ml. of water and the combined solutions were evaporated *in vacuo* to yield 1.417 g. of an oil, from which the D-γ,γ'-**dihydroxyvaline** was obtained after hydrolysis with boiling 5 *N* hydrochloric acid (25 ml., 20 hr.) followed by adsorption on Amberlite IR 120 resin (20 ml. H^+ form) and elution with 1.5 *N* aqueous ammonia. Evaporation of the ammonia eluate gave 240 mg. (39% based on the dichloroacetyl lactone) of an oil, which soon solidified. After one recrystallization from water–acetone 1:3 v./v., colorless needles of m.p. 168–172° were obtained.

The L-γ,γ'-**dihydroxyvaline** was eluted from the first column with 1.5 *N* aqueous ammonia (250 ml.). The 454 mg. of oily product, obtained on evaporation, solidified on treatment with acetone. Recrystallization from water–acetone 1:3 yielded colorless needles, m.p. 174.5–175°.

The specific rotations of the D and L acids, measured on freshly prepared solutions in 0.1 *N* potassium hydrogen carbonate, were $[\alpha]_D$ +13.7° and $[\alpha]_D$ −12.2°, respectively (*c* was 1.90 and 1.49 respectively).

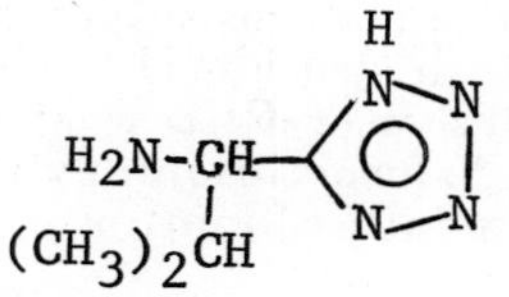

$C_5H_{11}N_5$ Ref. 154

For the experimental details of the resolution of this compound, see Ref. 21.

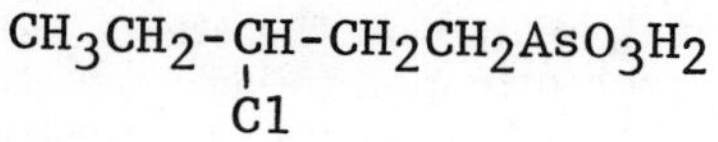

$C_5H_{12}AsClO_3$ Ref. 155

2. ***Acide 3-chlorepentane-1-arsonique.***

En dissolvant 20 g (0.087 mol.) de l'acide et 28 g (0.087 mol.) de quinine dans 1400 cm³ d'eau bouillante, nous avons obtenu par refroidissement de la solution filtrée 33 g du *sel de quinine*, qu'on fait recristalliser dans un litre d'eau. Le sel se dépose en rosettes d'aiguilles avec 2 mol. d'eau.

Substance 0 1613 g: perte à 90° 0.0099 g.
,, 0.4067 g: 5.70 cm³ de baryte 0.1217 n.
Trouvé: H_2O 6 14: p. équiv. 586.
$C_5H_{12}O_3AsCl \cdot C_{20}H_{24}O_2N_2 \cdot 2\,H_2O$ Calculé: ,, 6.10: ,, ,, 590.7.

Après cinq cristallisations le dédoublement du sel lévogyre est fini. On trouve: Sel de baryum: $[M]_D$ = —5.7°. Acide libre: $[M]_D$ = —12.4°.

Dispersion de la rotation:

λ (μμ)	656.3	589.3	546.3	481.1	435 8
[M] sel	4.7°	5.7°	7.5°	8.4°	9.4°
,, acide	10.5°	12.4°	14.3°	17 1°	20.0°

```
SH NH2
 |  |
(CH3)2CH-CH-COOH
```

$C_5H_{12}NO_2S$ Ref. 156

EXAMPLE.

*dl-iso*propylidenepenicillamine (13.6 g.) was dissolved in 98% formic acid (102 ccs.) and, with continuous stirring, acetic anhydride (41 ccs.) slowly run in over a period of 4 hours, the reaction vessel being cooled in a bath of cold water. When addition was complete, stirring was continued for a further 3 hours, water (41 ccs.) then slowly added, and the mixture allowed to stand for one hour. The product was concentrated *in vacuo* and the resulting resin crystallised from a mixture of benzene and petroleum ether (b.p. 80—100° C.), when colourless prisms of *dl*-N-formyl-*iso*-propylidenepenicillamine were obtained m.p. 140—1° C. (Yield 14 g.).

The *dl*-N-formyl compound (14.6 g.) together with brucine (33.4 g.) was dissolved in 700 ccs. hot water and allowed to stand for 48 hours to crystallise. The solid was filtered off and the filtrate concentrated to one-third of its bulk to obtain a further crop of crystals. The combined solids on treatment in aqueous suspension with concentrated hydrochloric acid (approximately 30% w/w HCl) went into solution and *d* - N-formyl-*iso*propylidenepenicillamine crystallised out, m.p. 177—8° C. $[\alpha]^{21}_{5461} = +61°$ (c = 0.5 in alc.). The aqueous filtrate from the resolution on acidification with concentrated hydrochloric acid (approximately 30% w/w HCl) yielded *l*-N-formyl-*iso*propylidenepenicillamine m.p. 177—8° $[\alpha]^{21}_{5461}$ − 61° (c = 0.8 in alc.).

d-N-formyl-isopropylidenepenicillamine (1 g.) was heated on the steam bath with 2N hydrochloric acid (50 ccs.), a stream of carbon dioxide being passed continuously through the reaction mixture. The formyl derivative rapidly passed into solution. After one hour the product was cooled and extracted three times with ether and concentrated *in vacuo*, when long colourless needles of *d*-penicillamine hydrochloride were obtained (0.8 g.).

The penicillamine hydrochloride had an optical rotation $[\alpha]^{23}_{5461}$ + 1.0° (c = 1.0 in water).

The identity of the product (*d*-penicillamine hydrochloride) was confirmed by conversion to its isopropylidene derivative, through which the free *d*-penicillamine may be conveniently prepared.

d-penicillamine hydrochloride (300 mg.) was heated with dry acetone (50 ccs.) for 1 hour. The solution was shaken with a little charcoal, filtered and concentrated *in vacuo*. The resulting *iso*propylidenepenicillamine hydrochloride, which melted with decomposition at 196—200° C. showed $[\alpha]^{23}_{5461}$ + 115° (c = 0.9 in water). The *iso*propylidenepenicillamine hydrochloride was dissolved in 10 ccs. water and treated with pyridine until neutral to congo red. The solution was concentrated *in vacuo* to half volume when *iso*propylidenepenicillamine crystallised. It was filtered off and recrystallised from acetone. The material had m.p. 196—198° C. and showed an optical rotation of $[\alpha]^{23}_{5461}$ + 131° (c = 0.7 in water).

$C_5H_{12}NO_2S$ Ref. 156a

Dans un réacteur en verre de 50 litres muni de réfrigérant à reflux, agitateur, ampoule à brome, tube adducteur de gaz, robinet de vidange, on dissout en chauffant a 50°C dans 20 litres d'ester acétique, 3,26 dg (15moles) de N-formyle-isopropylidène-D,L-pénicillamine, obtenus par réaction de 3 kg de D,L-pénicillamine-chlorhydrate (16 moles) avec 2,8 dg d'acétone (50 moles) suivie d'une formylation avec un mélange d'acide formique et d'anhydride acétique en neutralisant parallélement avec acetate de sodium. On poursuit le chauffage et l'on ajoute sóus agitation 3,28 kg (1,05 x 15 moles) d 1-noréphedrine dissous dans 7 litres d'ester acetique. On observe une augmentation de la température d'environ 5°C. Le produit d'addition (I) de N-formyle-isopropylidène-D-pénicillamine et de 1-noréphédrine précipite après quelques minutes. Sous chauffage sous reflux on agite pendant 30 autres minutes. Après refroidissement on essore et on lave avec environ 3 litres d'ester acétique, puis on sèche à environ 50°C sous vide. On obtient 2,75 kg = 98% du produit d'addition (I) P.F. 200-204°C $[\alpha]^{20}_D = +33°$.

Après évaporation à sec, on obtient de l'eau mère un produit d'addition brut (II) de N-formyle-isopropylidène-L-pénicillamine et 1-noréphédrine à partir duquel le produit d'addition pur, $[\alpha]^{20}_D$ = -74,6°; P.F. 116°, peut être préparé par recrystallisation avec isopropanol.

A 2,75 dg du produit d'addition (I) on ajoute à 25°C successivement 10 litres d'eau distillée et 1 litre d'acide chlorhydrique concentré. Après une heur d'agitation on essore, on lave avec 2 litres d'eau distillée et on sèche le résidu à 50°C sous vide. On obtient 1,49 kg = 92% de N-formyle-isopropylidène-D-pénicillamine de P.F. 183-184°C et $[\alpha]^{20}_D$ = +53°. De l'eau mère résultent après évaporation à sec et recristallisation avec isopropanol 1,19 kg de 1-noréphédrine. HCl P.F. 172-174°C.

1,49 kg de N-formyle-isopropylidène-D-pénicillamine sont ajoutés a 9,0 litres d'acide chlorhydrique à 15%, chauffé à 70°C. En distillant l'acétone libéré, on chauffe encore pendant 2 heures a la même température. Après évaporation à sec dans un évaporateur rotatif de 50 litres on obtient 1,08 kg de D-pénicillamine. HCl brute.

1,08 kg de D-pénicillamine. HCl sont dissous dans 8,7 litres d'alcool a 96%; on ajoute 0,59 kg (5,82 moles) de triéthylamine, précipitant la D-pénicillamine. Après essorage, lavage avec alcool à 96% et séchage sous vide, à 50°C, on obtient 0,78 kg de D-pénicillamine P.F. 212-214°C et $[\alpha]_D^{20}$ = -62,8°.

La finition du produit d'addition brut (II) se fait de la même façon que celle de l'adduction(I). On obtient 1,3 kg de N-formyle-isopropylidène-l-pénicillamine P.F. 182-184°C et $[\alpha]_D^{20}$ = -53°.

$C_5H_{12}N_2O_2$

$H_2NCH_2CH_2CH_2CH(NH_2)COOH$

Ref. 157

DL-Ornithin-dibenzoyl-D-hydrogentartrat: Die Lösung von 5.0 g *DL-Ornithin-monohydrochlorid* in 64 ccm Wasser wurde mit der äquivalenten Menge ca. 8*n* NaOH versetzt und eine warme Lösung von 22.4 g *Dibenzoyl-D-weinsäure* in 82 ccm Isopropylalkohol hinzugefügt. Nach 12stdg. Stehenlassen wurde i. Vak. zur Trockne eingedampft und der Rückstand zur Entfernung von Natriumchlorid und Dibenzoylweinsäureresten mit Wasser und Äther behandelt. Das so gewonnene Salz wurde zur Racematspaltung eingesetzt. Ausbeute quantitativ. Schmp. 154—155° (korr.).

$[\alpha]_D^{20}$: −93.1° (*c* = 0.66, in Wasser).

$C_{41}H_{40}N_2O_{18} \cdot 3\,H_2O$ (902.8) Ber. C 54.54 H 5.13 N 3.10
Gef. C 54.13 H 5.07 N 3.21

Racematspaltung des DL-Ornithin-dibenzoyl-D-hydrogentartrates: 10.0 g des DL-Salzes wurden in einem Gemisch von 40 ccm Wasser und 18 ccm Isopropylalkohol in der Wärme gelöst und filtriert. Nach kurzer Zeit setzte die Kristallisation von *L-Ornithin-Dibenzoyl-D-hydrogentartrat* ein, die nach 12 Stdn. vollständig war. Es wurde zur vollständigen Reinigung nochmals aus 58 ccm wäßr. Isopropylalkohol gleicher Zusammensetzung wie die Spaltlösung umkristallisiert. Ausb. 5.0 g. Schmp. 155—157° (korr.).

$[\alpha]_D^{20}$: −90.9° (*c* = 0.46, in Wasser).

Gef. C 54.23 H 5.34 N 3.17

Die D-Verbindung wurde aus der Mutterlauge des L-Aminosäuretartrates durch Einengen i. Vak. bis zur Trockne isoliert. Sie wurde durch fraktionierte Kristallisation aus Isopropylalkohol/Wasser (Verhältnis 1:2) in der Weise gereinigt, daß 1—2 g der schwerer löslichen Anteile abgetrennt wurden. Die leichter löslichen Anteile bestanden aus optisch reinem *D-Ornithin-dibenzoyl-D-hydrogentartrat.* Ausb. 3—4 g, Schmp. 142—147° (korr.).

$[\alpha]_D^{20}$: −97.1° (*c* = 0.47, in Wasser).

Gef. C 54.49 H 5.31 N 3.04

Gewinnung von L- und D-Ornithin-dihydrochlorid: 2.5 g jedes der gereinigten diastereomeren Ornithin-dibenzoyltartrate wurden mit 20 ccm 4-proz. Salzsäure versetzt und die Dibenzoylweinsäure im Kutscher-Steudel-Apparat quantitativ ausgeäthert. Nach Eindampfen der wäßr. Phase wurde der Rückstand in wenig Methanol aufgenommen und mit Aceton gefällt. Ausb. 90% d. Th., bezogen auf Tartrat.

L-Ornithin-dihydrochlorid: Schmp. 190—192° (korr.).

$[\alpha]_D^{20}$: +16.4° (*c* = 2.59, in Wasser).

$C_5H_{12}N_2O_2 \cdot 2\,HCl$ (205.1) Ber. C 29.28 H 6.88 N 13.66 Gef. C 29.43 H 6.58 N 13.68

D-Ornithin-dihydrochlorid: Schmp. 194—195° (korr.).

$[\alpha]_D^{20}$: −16.3° (*c* = 2.80, in Wasser).

Gef. C 28.92 H 7.06 N 13.72

LEVINTOW und GREENSTEIN[1] geben für die antipodischen Ornithin-dihydrochloride spezif. Drehungen von +16.5° bzw. −16.4° (4-proz. wäßrige Lösung) an.

[1] L. LEVINTOW und J. P. GREENSTEIN, J. biol. Chemistry **188**, 643 [1951].

For the experimental details of enzymatic methods for the resolution of this compound, see Ref. 18n and L. Levintow and J. P. Greenstein, J. Biol. Chem., 188, 643 (1951) and S. M. Birnbaum, L. Levintow, R. B. Kingsley and J. P. Greenstein, J. Biol. Chem., 194, 455 (1952).

$C_5H_{12}N_2O_3$

$$H_2NCH_2CH_2OCH_2-\underset{\displaystyle NH_2}{\underset{|}{CH}}-COOH$$

Ref. 158

Brucine salts of N_α,N_ε-diphthaloyl-DL-4-oxalysine

Method A (mixture of diastereoisomeric salts).

4 g of N_α, N_ε-diphthaloyl-DL-4-oxalysine were dissolved by heating in 15 ml of methanol and a solution of 4 g of brucine in 8 ml of methanol was added. The mixture was diluted with 10 ml of isoamyl alcohol, left in the refrigerator for 24 hours and filtered. The yellow residue (fraction A) was dried *in vacuo*. Yield: 7.7 g (95 %).

Method B (partially separated mixture).

4 g of N_α, N_ε-diphthaloyl-DL-4-oxalysine were dissolved in 8 ml of methyl cellosolve, previously distilled from cuprous chloride to remove peroxides (b.p. 126°). Under vigorous stirring a solution of 4 g of brucine in 8 ml of methyl cellosolve was added. An intense yellow color appeared. The solution was diluted with 16 ml of warm isoamyl alcohol and left at room temperature until crystallization had begun. After 16 hours at 1-4° the nearly white crystals were filtered, yielding 4.2 g of a product (fraction B1) with melting point 145-149°. The filtrate was diluted once more with 32 ml of methanol and cooled to 1-4°. After 48 hours a second yellow fraction (B2) had separated, which was filtered and dried, giving 3.3 g of crystals with m.p. 168°.

Separation and purification

The mixture of diastereoisomeric salts (fraction A, see above) was dissolved in a small amount of dichloromethane with gentle warming. Crystallization was induced by addition of 12 ml of isoamyl alcohol and a first white fraction was collected by filtration. A second, more soluble, yellow fraction was obtained by addition of an excess of methanol. The procedure was repeated with both fractions until constant melting points and specific rotations were obtained. Further purification of the fractions B1 and B2 (see above) could be performed in the same way.

The least soluble fraction (pure white crystals after three crystallizations) appeared to be solvated with isoamyl alcohol, which could be smelt during the subsequent decomposition with hydrochloric acid. M.p. 152°; $[\alpha]_D^{24°} = +12.5°$ (c = 1 in DMF). On crystallization from methyl cellosolve and thorough drying, an unsolvated product was obtained with m.p. 133°, which could be converted again into the solvated form by crystallization from methyl cellosolve-isoamyl alcohol. The more soluble yellow fraction melted at 178°; $[\alpha]_D^{18°} = -60.1°$ (c = 1 in DMF).

(+)-and (—)-N_α,N_ε-Diphthaloyl-4-oxalysine

1 g of each of the brucine salts was suspended in 2 ml of water and 2 ml of *N* hydrochloric acid were added. By rubbing the mixture with a glass rod a yellow viscous oil was formed which solidified on prolonged stirring (5-20 min). A pale yellow mother liquor, containing the brucine hydrochloride, remained. After filtration, washing with water and drying over phosphorus pentoxide, 450 mg of a crystalline product were obtained, which were crystallized from isoamyl alcohol. M.p. 174° (in both cases), $[\alpha]_D^{25°} = +65.5°$ (c = 1 in DMF) and $[\alpha]_D^{22°} = -65.1°$ (c = 1 in DMF) respectively. The dextrorotatory compound was obtained from the white brucine salt; the enantiomer from the yellow one.

Found	: C 61.7 ;	H 4.1 ;	N 6.9
Calc. for $C_{21}H_{16}N_2O_7$ (408.36):	„ 61.76;	„ 3.92;	„ 6.86.

Dephthaloylation of (+)-N_α,N_ε-diphthaloyl-D-4-oxalysine

a) *With alkali and hydrochloric acid*

0.50 g of (+)-N_α, N_ε-diphthaloyl-4-oxalysine were dissolved in 7.20 ml (3 equivalents) of 0.5 *N* sodium hydroxide. The solution was left for 2 hours in the refrigerator and acidified to pH 2 by dropwise addition of 6 *N* hydrochloric acid. After evaporation to dryness, a glassy foam was obtained, which was dissolved in absolute alcohol and filtered from insoluble sodium chloride. The solvent was removed by distillation and the residue boiled for 10 minutes with 10 ml of *N* hydrochloric acid. After cooling, the phthalic acid which separated was removed by filtration, the filtrate concentrated, poured into 90 % ethanol and partially neutralized to pH 5 by addition of diethylamine. The separating oil solidified after a short time but was optically inactive. Yield: 55 %. The racemic compound was identified as DL-4-oxalysine monohydrochloride by comparison with an authentic sample.

b) *By acid hydrolysis*

4.1 g of N_α, N_ε-diphthaloyl-D-4-oxalysine were heated at 105° for 6 hours with 40 ml of 6 *N* hydrochloric acid in a sealed tube, which was shaken now and then. The reaction mixture was cooled, filtered and evaporated to dryness. The residual oil was taken up in 15 ml of dry ethanol and cooled, when 1.32 g (60 %) of D-4-oxalysine dihydrochloride separated as beautiful colorless needles. Purification was effected by dissolving 400 mg in 0.2 ml of water and adding 10 ml of dry ethanol. M.p. 162°; $[\alpha]_D^{22°} = -5°$ (c = 2 in water).

$(CH_3)_3$-CH-COOH
      |
     NH_2

$C_5H_{13}NO_2$

Ref. 159

Optical Resolution of V and Isolation of L-*t*-Leucine (IIIa). Sixty-five grams of V and 110 g of cinchonidine were dissolved in 200 m*l* of hot alcohol, after which the solution was allowed to stand overnight in a refrigerator. The crystalline precipitates were then collected. Six recrystallizations of the salt from alcohol gave 12 g of pure cinchonidine salt of IIIa. Twelve grams of the salt were decomposed with 1 N sodium hydroxide, and the liberated cinchonidine was filtered off. The filtrate was shaken with chloroform, acidified with 2 N hydrochloric acid, concentrated to 100 m*l* and extracted with ethyl acetate. The extract was evaporated to dryness, and the residue was recrystallized from a mixture of water and alcohol. 1.2 g of (−)-*N*-acetyl-*t*-leucine (VI) were obtained; mp 228—229°C, $[\alpha]_D^{20}$ −40° (*c* 1, water).

1.2 g of VI were hydrolyzed with hydrochloric acid, and then the hydrolysate was concentrated and the excess hydrochloric acid was removed by azeotropic distillation with water. The resulting crystals were dissolved in 20 m*l* of water and neutralized with diethylamine, and then 2 m*l* of alcohol were added. After the solution had stood overnight, IIIa was obtained. Yield 820 mg; mp 280°C (sublimed), $[\alpha]_D^{20}$ −9.5° (*c* 2, water).

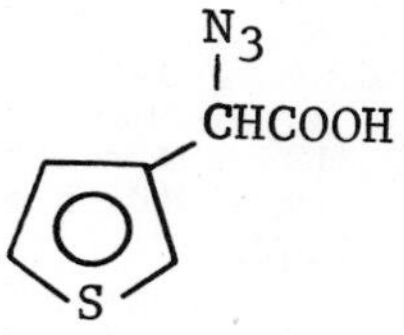

$C_6H_5N_3O_2S$

Ref. 160

D(−)-*α-azido-3-thienylacetic acid*

16.0 g (0.0874 moles) of α-azido-3-thienylacetic acid and 25.0 g (0.0849 moles) of cinchonine were dissolved by heating in 580 ml of boiling ethyl acetate and the solution left to crystallize in a refrigerator overnight. The precipitate was filtered off and dried. About 0.3 g of salt was set apart, the acid was isolated and its optical activity measured in acetone. The remaining salt was crystallized from ethyl acetate; the result is seen in Table 4.

It is evident that the best resolution was obtained after one crystallization and that continued recrystallization caused the activity to fall, probably due to partial racemization. The acid was therefore liberated by treatment with dilute hydrochloric acid followed by extraction with ether and evaporation of the ether. 2.0 g ($[\alpha]_D^{25}$ = −95.5°) was treated with an equivalent amount of (−)-α-phenylethyl amine in ethyl acetate. The salt which separated was recrystallized once more, yielding 1.5 g of salt. The acid was isolated in the usual manner and after one recrystallization from ligroin-petrol-ether 0.67 g of D(−)-α-azido-3-thienylacetic acid, m.p. 54–56°, $[\alpha]_D^{25}$ = −108° (acetone) was obtained. IR-spectrum in Fig. 9.

Anal. $C_6H_5N_3O_2S$ (183.2)
Calc. C 39.33 H 2.75 N 22.94
Found C 39.45 H 2.96 N 22.85

Table 4

Crystallization	1	2	3	4
ml ethyl acetate	580	1175	900	780
g salt	14.0	11.1	9.3	8.3
$[\alpha]_D^{25}$ of acid	−103°	−95°	−94°	−96°

L(+)-*α-Azido-3-thienylacetic acid*

The mother liquor from the first crystallization of the cinchonine salt was evaporated to dryness *in vacuo* and the acid liberated. 8.0 g of acid with $[\alpha]_D^{25}$ = +78 was obtained. The acid was dissolved, together with 8.0 g (0.043 moles) of (−)-ephedrine in 200 ml of boiling ethyl acetate and the solution left overnight in a refrigerator. Maximal resolution was apparently obtained at once and continued recrystallization of the ephedrine salt caused the activity to diminish slightly as can be seen in Table 5.

The acid was therefore liberated in the usual manner and 3.4 g with $[\alpha]_D^{25}$ = +105° was obtained. One recrystallization from ligroin-petroleum ether yielded 2.9 g of L(+)-α-azido-3-thienylacetic acid, m.p. 54–56° with the same IR-spectrum as the (−)-form. The rotations in different solvents are given in Table 6.

Anal. $C_6H_5N_3O_2S$ (183.2)
Calc. C 39.33 H 2.75 S 17.50
Found C 39.23 H 2.82 S 17.51

D(−)-*α-Azidophenylacetic acid*

m.p. 55–56°, has been described by Sjöberg *et al.* [15].

Reduction of (+)-*α-azido-3-thienylacetic acid*

200 mg of a palladium catalyst (5 % Pd on barium sulphate) was added to 20 ml of ethanol and hydrogenated. A solution of 500 mg of (+)-α-azido-3-thienylacetic

Table 5

Crystallization	1	2	3
ml ethyl acetate	200	300	225
g salt	11.0	9,3	8.0
$[\alpha]_D^{25}$ of acid	+107°	+105°	+103°

Table 6

Solvent	Conc./g acid/dl	$2\alpha_D^{25}$	$[\alpha]_D^{25}$	$[M]_D^{25}$
Benzene	0.547	+1.64°	+150°	+274°
Water (ion)	0.442	+1.23°	+139°	+255°
Ethanol	0.642	+1.70°	+132°	+241°
Methanol	0.968	+2.48°	+128°	+234°
Acetic acid	0.636	+1.68°	+128°	+234°
Acetone	0.877	+1.85°	+107°	+208°
Dioxane	0.607	+1.09°	+90°	+165°

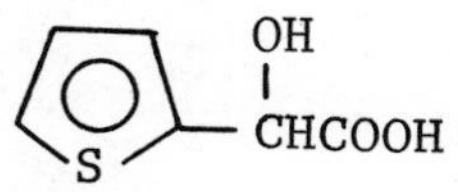

$C_6H_6O_3S$

Ref. 161

Experimental

Preliminary tests on resolution were made with 0.005 mole of optically active acid and 0.005 moles of base in dilute alcohol. Most alkaloids, namely quinine, quinidine, cinchonine, brucine and strychnine gave soluble salts. Gummy precipitates were obtained with yohimbine and morphine. A crystalline salt formed with (+)-benzedrine but the acid recovered was inactive.

With (−)-phenetylamine in ethyl acetate, however, an acid with $[\alpha]_D^{25}$ = +21.8°, and with cinchonidine an acid $[\alpha]_D^{25}$ = −56.5° were obtained in 20 % ethanol. These bases were therefore used for the resolution.

Optical resolution of 2-thienylglycolic acid

29.7 g (0.188 mole) of 2-thienylglycolic acid and 55.4 g (0.188 mole) of cinchonidine were dissolved by heating in 1400 ml of water and 300 ml of alcohol and the solution left in a refrigerator for 24 hours. The precipitate was filtered off, washed with cold water, and dried. About 0.4 g of the salt was set apart, the acid isolated and the optical activity measured in water. The remaining salt was recrystallized from dilute ethanol and the process repeated until the optical activity of the acid remained constant. The course of the resolution is seen below.

Crystallization	1	2	3	4	5
g salt	42	33	28	23.5	19.5
ml water	1400	1080	930	800	700
ml ethanol	300	230	180	150	125
$[\alpha]_D^{25}$ of acid	−73°	−96°	−95°	−95°	−96°

The mother liquor from the first crystallization with cinchonidine was evaporated in vacuum under nitrogen, and 43 g of salt was obtained. This salt yielded 13 g (0.082 mole) of acid having $[\alpha]_D^{25}$ = +72°. This acid was dissolved in 680 ml of hot ethyl acetate, and 9.9 g (0.082 mole) of (−)-α-phenylethylamine was added. The salt deposited almost instantly and was filtered off when the solution had cooled. The progress of the resolution is seen below.

Crystallization	1	2	3	4
g salt	19.5	15.5	13.2	9.2
ml ethylacetate	680	600	550	600
ml methanol	—	50	40	50
$[\alpha]_D^{25}$ of acid	—	+94°	+96°	+97°

(−)-2-Thienylglycolic acid

The cinchonidine salt (19 g) was decomposed with 1 *N* sulphuric acid and the active acid extracted 20 times with ether. The ether was removed in vacuo under nitrogen and the acid recrystallized from benzene. The yield of pure acid was 4.2 g and the m.p. 83–84°.

0.07107 g acid: 4.47 ml 0.1002 *N* NaOH.
$C_6H_6O_3S$ (158.2). Equiv. wt. calc. 158.2, found 158.7.
0.2037 g acid made up to 20.00 ml with water:
$2\alpha_D^{25} = -2.003$, $[\alpha]_D^{25} = -98.3°$, $[M]_D^{25} = -155.5°$.

(+)-2-Thienylglycolic acid

8.0 g of the phenylethylamine salt was decomposed with 1 *N* sulphuric acid and worked up in the same manner as above. After one recrystallization from benzene, 2.8 g of pure acid m.p. 83–84° was obtained.

0.11326 g acid: 7.11 ml 0.1002 *N* NaOH
$C_6H_6O_3S$ (158.2). Equiv. wt. calc. 158.2, found 159.0.

Weighed amounts of acid were made up to 20 ml with different solvents and the optical activity measured (Table 2).

Table 2.

Solvent	mg of acid	$2\alpha_D^{25}$	$[\alpha]_D^{25}$	$[M]_D^{25}$
Acetic acid	100.67	+1.094°	+108.7°	+172.0°
Water	202.02	+2.012°	+ 99.6°	+157.6°
Ethanol	111.26	+0.980°	+ 88.1°	+139.4°
Acetone	106.50	+0.908°	+ 85.3°	+134.9°
Ethyl acetate	167.60	+1.427°	+ 85.1°	+134.6°
Ether	111.67	+0.792°	+ 70.9°	+112.2°
Water (ion)	141.73	+0.823°	+ 51.0°	+ 91.9°

The optical rotation of 0.1246 g of (+)-2-thienylglycolic acid in 20 ml of water was measured at different wavelengths. The results are given in Table 3.

Table 3.

Wavelength Å	$2\alpha_D^{25}$	$[\alpha]_D^{25}$
6438	+0.981°	+ 78.7°
5893	+1.213°	+ 97.3°
5461	+1.429°	+114.6°
5086	+1.690°	+135.6°

Raney-nickel desulphurization

Raney-nickel catalyst was prepared as described by Mozingo (18) from 50.5 g of sodium hydroxide in 200 ml of water and 40 g Raney-nickel alloy, except that the washings with alcohol were omitted. The catalyst was added to a solution of 1.0 g (0.00632 mole) of 2-thienylglycolic acid and 0.54 g (0.00642 mole) sodium bicarbonate in 200 ml of water, and the mixture was stirred at room temperature for 18 hours. The nickel was filtered off and washed with water and 0.1 *N* sodium hydroxide solution. The combined acidified water phases were continuously extracted with ether. The ether solution was dried, and the ether was distilled off *in vacuo*, leaving 0.63 g of a crystalline residue m.p. 55–58°. After one recrystallization from 20 ml of petroleum-ether (b.p. 45–50°) and 3 ml of ether 0.36 g of α-hydroxycaproic acid m.p. 58–60° was obtained. This acid gave no m.p. depression with an authentic sample of α-hydroxycaproic acid (24).

0.10140 g acid: 7.68 ml 0.1002 *N* NaOH.
$C_6H_{12}O_3$ (132.1). Equiv. wt. calc. 132.1, found 131.7.
0.02287 g made up to 10.00 ml with water and the UV-absorption measured at 235 mμ gave D = 0.277, ε = 16. (2-Thienylglycolic acid gave ε = 7800 at this wavelength.)

When 1.0 g of (+)-2-thienylglycolic acid ($[\alpha]_D^{25} = +98°$ in water, $[\alpha]_D^{25} = +51°$ for the sodium salt in water) was treated in exactly the same manner, 0.32 g acid m.p. 59–60° was obtained after one recrystallization from petroleum-ether (25 ml) and ether (3 ml).

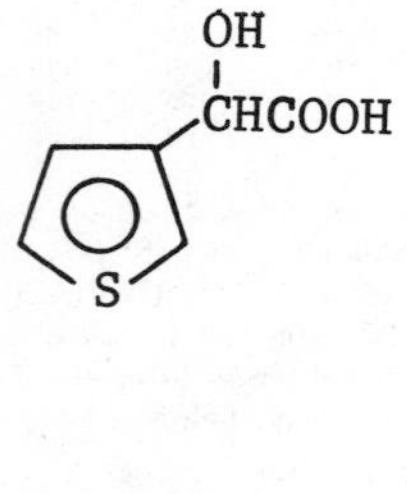

$C_6H_6O_3S$ Ref. 162

Preliminary tests on resolution were made with 0.005 moles of acid and 0.005 moles of base in dilute alcohol or ethyl acetate. The results are given below.

Base	$[\alpha]_D^{25}$ of acid in ethanol: From water	From ethyl acetate
Brucine	s	
Strychnine	+36°	
Cinchonine	+ 0°	
Cinchonidine	−25°	
Quinine	+89°	
Quinidine	oil	
(−)-α-Phenylethylamine	± 0°	+72°
(+)-Benzedrine	± 0°	± 0°
(+)-α-Naphthyletylamine	± 0°	± 0°

Optical resolution of 3-thienylglycolic acid

20.0 g (0.126 moles) of 3-thienylglycolic acid and 15.3 g (0.126 moles) (+)-α-phenylethylamine were dissolved in a hot mixture of 500 ml ethyl acetate and 225 ml of methanol. The crystals which separated on cooling were filtered off. About 0.3 g of the salt was set apart, the acid isolated and the optical activity measured in ethanol. The remaining salt was recrystallized from ethyl acetate-methanol mixtures, and the process repeated until the optical activity of the acid remained constant. The course of the resolution is seen below.

Crystallization	1	2	3	4	5	6
g salt	12.5	8.0	4.5	2.9	2.0	1.4
ml ethyl acetate	500	400	160	100	75	50
ml methanol	225	160	20	20	15	10
$[\alpha]_D^{25}$ of acid	−10°	−41°	−77°	−100°	−101°	−101°

The mother liquor from the first crystallization with (+)-phenylethylamine was evaporated in vacuo and 22.5 g of salt was obtained. This salt yielded 12.1 g (0.0764 moles) of 3-thienylglycolic acid having $[\alpha]_D^{25} = +7°$. This acid was dissolved together with 25.5 g (0.0764 moles) of strychnine in 600 ml water. The salt crystallized over night. The progress of resolution is seen below.

Crystallization	1	2	3	4	5	6
g salt	21	12.5	7.2	5.3	4.4	3.4
ml water	600	400	200	90	70	60
$[\alpha]_D^{25}$ of acid	+40°	+71°	+95°	+100°	+101°	+101°

(−)-3-Thienylglycolic acid

The (+)-phenylethylamine salt (1.2 g) was decomposed with 1-N sulphuric acid and the active acid extracted 20 times with ether. The ether was removed in vacuo and the acid recrystallized from 75 ml of benzene, yielding 0.62 g of pure acid, m. p. 123.5–124°, crystallizing in white flakes.

0.08536 g acid: 5.36 ml 0.1014-N NaOH
$C_6H_6O_3S$ (158.2). Equiv. wt. calc. 158.2, found 157.0
0.10724 g made up to 20.00 ml with ethanol: $2\alpha_D^{25} = -1.095°$;
$[\alpha]_D^{25} = -102.1°$, $[M]_D^{25} = -161.5°$.

(+)-3-Thienylglycolic acid

3.0 g of the strychnine salt was treated with 1-N sodium hydroxide and the strychnine rapidly extracted four times with chloroform. The aqueous phase was at once acidified with 1-N sulphuric acid and worked up in the same manner as above. After one recrystallization from benzene, 0.88 g of pure acid m. p. 123.5–124° was obtained.

0.09752 g acid: 6.07 ml 0.1014-N NaOH
$C_6H_6O_3S$ (158.2). Equiv. wt. calc. 158.2, found 158.4
Weighed amounts of acid were made up to 20.00 ml with different solvents and the optical activity measured.

Table 2. Optical activity of (+)-3-thienylglycolic acid in different solvents.

Solvent	mg of acid	$2\alpha_D^{25}$	$[\alpha]_D^{25}$	$[M]_D^{25}$
Acetic acid	83.80	+1.049°	+125.2°	+198.1°
Water	107.50	+1.306°	+121.5°	+192.2°
Ethanol	78.43	+0.804°	+102.5°	+162.2°
Acetone	93.11	+0.946°	+101.6°	+160.7°
Ethyl acetate	112.65	+1.121°	+ 99.5°	+157.4°
Ether	99.32	+0.839°	+ 84.5°	+133.7°
Water (ion)	118.07	+1.134°	+ 84.3°	+151.9°

The optical rotation of 0.10750 g of (+)-3-thienylglycolic acid in 20 ml of water was measured at different wavelengths. The results are given in Table 3.

Racemization experiments

To weighed amounts of the acids the calculated amounts of 1.000-N sodium hydroxide solution was added and then diluted to 25.0 or 50.0 ml with water.

Table 3. Rotation dispersion of (+)-3-thienylglycolic acid.

Wavelength, Å	$2\alpha^{25}$	$[\alpha]^{25}$
6438	+1.084°	+100.8°
5893	+1.306°	+121.5°
5461	+1.575°	+146.5°
5086	+1.890°	+175.8°

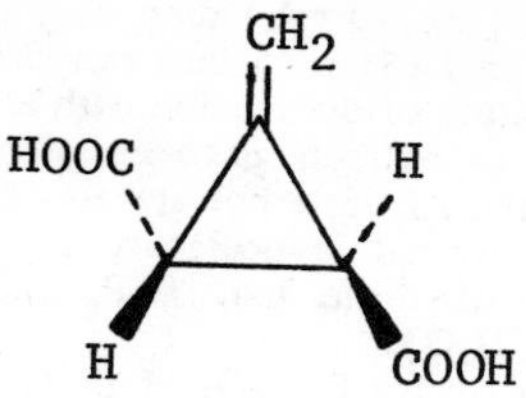

$C_6H_6O_4$ Ref. 163

Resolution of Feist's Acid. To 32 g of Feist's acid dissolved in a minimum volume of boiling ethanol was added 89 g of brucine dissolved in a minimum volume of boiling ethanol; then enough water was added to make a 3:1 ethanol–water mixture. Upon cooling, 75 g of a light brown powder was obtained which was recrystallized once from a 3:1 ethanol–water mixture. Obtained were a head crop of crystals, 46 g, and a second crop, 13 g. Regeneration of the acid from each portion of these by treatment with acid and extraction with ether followed by optical measurements gave $[\alpha]^{22}D$ −149.6° (*c* 1.275, EtOH) for the head crop and $[\alpha]^{22}D$ −136° for the second crop. A second recrystallization of the head crop resulted in Feist's acid with $[\alpha]^{22}D$ −147.0° (EtOH). Combination of head crops of subsequent recrystallizations of the second crop with the initial head crop gave 45 g of the salt which was treated with dilute hydrochloric acid in the presence of ether. Extensive extraction of the aqueous layer after saturation with sodium chloride followed by drying and evaporation of the solvent gave 5.0 g of Feist's acid: $[\alpha]^{22}D$ −149.4° (*c* 1.45, EtOH).[20]

(20) The low recovery here suggests that the head crop was actually the bisbrucine salt of Feist's acid.

$C_6H_6O_4$ Ref. 163a

Optische Spaltung. Eine siedende Lösung der Säure, nach und nach mit 1 Äq. Chinin versetzt, scheidet beim Abkühlen das *saure Chininsalz* der d-Säure ab, das nach Reinigung aus Wasser mit Tierkohle in seidigglänzenden Nadeln (Schmelzpunkt 157°) krystallisiert.

0,1533 g Salz: 0,3756 g CO_2, 0,0881 g H_2O.

$C_{26}H_{30}O_6N_2$	Ber. C 66,95	H 6,44
	Gef. „ 66,84	„ 6,43.

Die Krystallfraktionen aus der Mutterlauge wurden fraktioniert krystallisiert. Die letzten Anteile sind braun und sirupös. Die aus allen Salzfraktionen isolierten aktiven Säuremengen schmolzen sämtlich bei 200°. Bei verschiedenen Spaltversuchen ergab sich die spezifische Drehung der Säure

1. aus der schwerstlöslichen Fraktion zu $(\alpha)_D = +265,66°$,
   aus der letzten Mutterlauge zu $(\alpha)_D = -103,11°$,
2. aus der schwerstlöslichen Fraktion zu $(\alpha)_D = +215°$,
   aus der letzten Mutterlauge zu $(\alpha)_D = -141°$,
3. aus der schwerstlöslichen Fraktion zu $(\alpha)_D = +215°$,
   aus der letzten Mutterlauge zu $(\alpha)_D = -156°$.

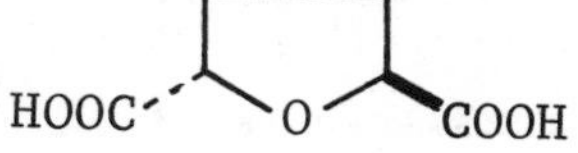

$C_6H_6O_5$ Ref. 164

Das Cinchoninsalz der *l*-Säure ist nur etwa halb so löslich wie das gleiche Salz der *d*-Säure, sodass ein fünfmaliges Umkrystallisiren der Verbindung aus Wasser zur Gewinnung von reinem Salz der *l*-Säure genügte. Die Hauptmenge der *d*-Säure verblieb hierbei in der Mutterlauge der ersten Krystallisation.

Die aus dieser Flüssigkeit isolirte Säure wurde in ihr Strychninsalz umgewandelt. Da sich das *d*-Salz etwas schwerer löslich zeigte als das *l*-Salz, so gelang es, Ersteres durch wiederholtes Umkrystallisiren zu reinigen. In jedem einzelnen Fall wurde das Umkrystallisiren solange fortgesetzt, bis das Drehungsvermögen der Säure constant geworden war. Dass die Säuren nunmehr rein waren, liess sich durch die Thatsache erhärten, dass das specifische Drehungsvermögen alsdann zwar von entgegengesetztem Vorzeichen, aber von gleicher Grösse war.

d-Säure				*l*-Säure			
c	α	l	$[\alpha]_D$	c	α	l	$[\alpha]_D$
10.15	+ 48.78°	1 dcm	+ 480.7°	10.04	− 48.04°	1 dcm	− 478.7°
5.08	+ 24.67°	1 »	+ 485.4°	5.01	− 24.22°	1 »	− 483.4°
2.525	+ 12.35°	1 »	+ 489.2°	2.517	− 12.29°	1 »	− 488.3°
0.629	+ 6.31°	2 »	+ 502°	0.627	− 6.35°	2 »	− 507°
0.157	+ 1.63°	2 »	+ 519°	0.158	− 1.67°	2 »	− 538°

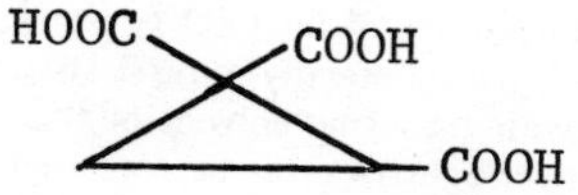

$C_6H_6O_6$ Ref. 165

Spaltung der *r*-1.1.2-Cyclopropantricarbonsäure.

Isolirung der dextrogyren Form mittels Brucin. Man benutzt das saure Salz, welches auf 1 Molekül Säure 2 Moleküle Base enthält. Die daraus gewonnene Säure zeigte den Zersetzungspunkt 187°; die specifische Drehung änderte sich bei nochmaliger Reinigung durch das Brucinsalz nicht mehr.

0.2022 g Säure, gelöst in Wasser zu 10 ccm. 26°, Natriumlicht. Drehung im 1 dcm Rohr: + 1.71°. Spec. Drehung: $[\alpha]_D^{26°} = +84.57°$.

Die Analyse des Brucinsalzes, welches lufttrocken 4 Moleküle Krystallwasser enthält und bei 130° wasserfrei wird, ergab:

0.3422 g Sbst. (auf 130° erhitzt): 0.0233 g H_2O. — 0.4539 g Sbst.: 0.0310 g H_2O.

$C_{52}H_{58}O_{14}N_4 + 4H_2O$. Ber. H_2O 6.97. Gef. H_2O 6.81, 6.83.

0.1350 g Sbst. (bei 130° getrocknet): 0.3214 g CO_2, 0.0747 g H_2O. — 0.1321 g Sbst.: 6.9 ccm N (17°, 757 mm).

$C_{52}H_{58}O_{14}N_4$. Ber. C 64.82, H 6.09, N 5.83.
Gef. » 64.92, » 6.20, » 6.05.

Mittels Chinin. Das zur Isolirung verwendete Salz enthielt auf 1 Molekül Tricarbonsäure 2 Moleküle Base. Erst nach vielmaligem Umkrystallisiren und erneuter Behandlung der Säure mit Chinin gelang es schliesslich, den reinen *d*-Antipoden zu isoliren.

0.2027 g Säure, in Wasser zu 10 ccm gelöst. 25°, Natriumlicht. Drehung im 1 dcm-Rohr: + 1.72°. Spec. Drehung: $[\alpha]_D^{25°} = +84.84°$.

Das Chininsalz enthält lufttrocken 2 Moleküle Krystallwasser, die bei 130° entweichen.

0.4626 g Sbst. (auf 130° erhitzt): 0.0196 g H_2O. — 0.5851 g Sbst.: 0.0240 g H_2O.

$C_{46}H_{54}O_{10}N_4 + 2H_2O$. Ber. H_2O 4.20. Gef. H_2O 4.24, 4.10.

0.1270 g Sbst. (bei 130° getrocknet): 0.3122 g CO_2, 0.0757 g H_2O. — 0.1420 g Sbst.: 8.4 ccm N (21°, 758 mm).

$C_{46}H_{54}O_{10}N_4$. Ber. C 67.08, H 6.63, N 6.83.
Gef. » 67.05, » 6.68, » 6.75.

Die Isolirung der *l*-1.1.2-Cyclopropantricarbonsäure gelang mit Hülfe des Cinchonidinsalzes, das auf 1 Molekül Säure 1 Molekül Base enthält, nach vielem Umkrystallisiren, wobei die Ausscheidung des Salzes manchmal erst nach Impfen mit einigen Krystallen erfolgte.

0.2020 g Säure, gelöst in Wasser zu 10 ccm. 26°, Natriumlicht. Drehung im 1 dcm-Rohr: − 1.71°. Spec. Drehung: $[\alpha]_D^{26°} = -84.65°$.

Analyse des bei 130° getrockneten Cinchonidinsalzes:

0.1285 g Sbst.: 0.3013 g CO_2, 0.0712 g H_2O. — 0.1446 g Sbst.: 7.7 ccm N (18°, 758 mm).

$C_{25}H_{28}O_7N_2$. Ber. C 64.04, H 6.04, N 6.00.
Gef. » 63.95, » 6.21, » 6.15.

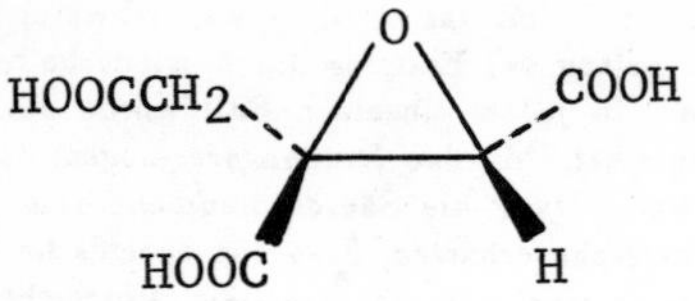

$C_6H_6O_7$ Ref. 166

***R*(+)-α-Methyl-*p*-nitrobenzylamine Salt of (+)-*threo*-Epoxyaconitic Acid:**
Racemic *threo*-epoxyaconitic acid[6] (**2**, 38.85 g, 200 mmol) dissolved in methanol/water (250 ml; 98:2 v/v) is treated with **1a** (59.8 g, 360 mmol) in the same solvent (150 ml). The warm solution begins to deposit solid immediately and is left at room temperature overnight. The mixture sets to a solid mass, which is mashed and filtered. The resulting solid (50.5 g, $[\alpha]_{436}^{25} = +13.07°$ in water, approximately 87% optically pure) is stirred and refluxed gently as a slurry with the same solvent (400 ml) for 2 h, then cooled and stirred overnight to give the hydrated bis-*R*(+)-α-methyl-*p*-nitrobenzylamine salt of (+)-*threo*-epoxyaconitic acid; yield: 44.8 g (83%); $[\alpha]_{436}^{25} = +14.83°$ ($c = 2\%$ in water); 99.1% optically pure by comparison with an authentic sample prepared from optically pure (+)-*threo*-epoxyaconitic acid[7] and *R*(+)-α-methyl-*p*-nitrobenzylamine. An analytical sample is recrystallized from ethanol/water; m.p. 171–172°; $[\alpha]_{436}^{25} = +14.97°$ ($c = 2\%$ in water).

$C_{22}H_{26}N_4O_{11} \cdot H_2O$	calc.	C 48.89	H 5.22	N 10.37	H_2O 3.34
(540.5)	found	48.87	5.30	10.37	3.19

[6] R. W. Guthrie, J. G. Hamilton, R. Kierstead, O. Neal Miller, A. C. Sullivan, *U. S. Patent* 3966772 (1976).

[7] Furnished by Dr. R. Guthrie, Chemical Research Department, Hoffmann-La Roche Inc., Nutley, N. J.

$C_6H_6O_7$ Ref. 166a

EXAMPLE 2

Resolution of (±)-threo-1,2-epoxy-1,2,3-propanetricarboxylic acid [(±)-threo-epoxyaconitic acid]

A solution of (±)-*threo*-epoxyaconitic acid (3.0 g.) in methanol (120 ml.) was heated to reflux. To the boiling solution cinchonidine (9.0 g.) was added followed by ethyl acetate (~250 ml.) and heating was continued until crystallization of the salt had started. The solution was allowed to cool and the cinchonidine salt was collected by filtration and washed with ethyl acetate (20 ml.) (Crop A). The filtrate was concentrated and two additional crops of crystalline material were obtained (Crops B and C). The mother liquors were evaporated to dryness *in vacuo* to give a white solid (Crop D). Crops A and D were processed as below. Crops B and C were combined for recycling.

(i) (+)-Threo-epoxyaconitic acid.—Crop A was dispersed in chloroform and extracted with 1 N sodium hydroxide solution (1× 25 ml.; 1× 10 ml.). The combined basic extracts were washed with chloroform (2× 10 ml.) then were acidified with 1 N hydrochloric acid (36 ml.) and concentrated *in vacuo*. The resulting residue was extracted with hot ethyl acetate and the residual sodium chloride was removed by filtration. The filtrate was concentrated under reduced pressure and crystallization of the product from ethyl acetate-carbon tetrachloride gave the epoxide as a carbon tetrachloride solvate. Air drying furnished the dextrarotatory *threo*-epoxide as the monohydrate, M.P. 108–112°; $[\alpha]_D^{25}$ +63.1° (c., 1.0, H_2O).

Analysis.—Calcd. for $C_6H_6O_7 \cdot H_2O$: C, 34.63; H, 3.87. Found: C, 34.91; H, 3.81.

(ii) (−)-Threo-epoxyaconitic acid.—Crop D was dispersed in chloroform (40 ml.) and extracted with two portions (35 ml. and 5 ml.) of 1 N sodium hydroxide solution. The combined aqueous extracts were washed with chloroform (2× 10 ml.) then were acidified with 1 N hydrochloric acid (51 ml.) and evaporated to dryness *in vacuo*. Trituration of the residue with ethyl acetate and evaporation of the extracts gave an oil which on fractional crystallization from ethyl acetate-carbon tetrachloride gave the solvated levorotatory epoxide. After air drying, the monohydrate had M.P. 108–112°; $[\alpha]_D^{25}$ −62.5° (c., 1.0, H_2O).

Analysis.—Calcd. for $C_6H_6O_7 \cdot H_2O$: C, 34.63; H, 3.89. Found: C, 34.91; H, 3.74.

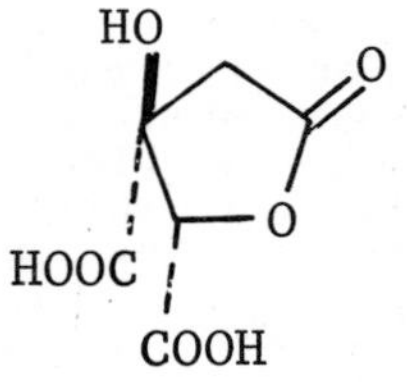

$C_6H_6O_7$ Ref. 167

Resolution of 3:
Racemic *threo*-hydroxycitric acid γ-lactone[8] (**3**, 9.12 g, 48 mmol) is dissolved in ethanol (80 ml) and treated with a solution of **1a** (10 g, 60 mmol) in methanol (20 ml). The solution is stirred overnight and the resulting precipitate is filtered and washed with methanol to give crude product; yield 9.3 g; $[\alpha]_D^{25} = +30.7°$ ($c = 1\%$ in methanol). This crude salt is refluxed with methanol (100 ml) for 3 h, cooled, and filtered to give the pure bis- *R*(+)-α-methyl-*p*-nitrobenzylamine salt; yield: 7.56 g; $[\alpha]_D^{25} = +38.2°$ ($c = 1\%$ in methanol). This salt (9.7 g, 18.6 mmol) is suspended in diethyl ether (75 ml) and treated with 1 normal ethereal hydrogen chloride (50 ml). After stirring at room temperature for 20 min, the mixture is filtered to afford the hydrochloride of **1a**; yield: 7.7 g; m.p. 245–247°. Concentration of the filtrates gives the lactone (3.35 g) which is crystallized from ethyl acetate/carbon tetrachloride to give the γ-lactone of (−)-*threo*-hydroxycitric acid; yield: 2.6 g (57%); $[\alpha]_D^{25} = +106.1°$ ($c = 1\%$ in water); m.p. 179–180.5°. A second crop, 0.33 g, m.p. 178–180°, $[\alpha]_D^{25} = +106.4°$, is also obtained; total yield: 2.93 g (64%).

[8] C. Martius, R. Mame, *Z. Physiol. Chem.* **269**, 33 (1941).

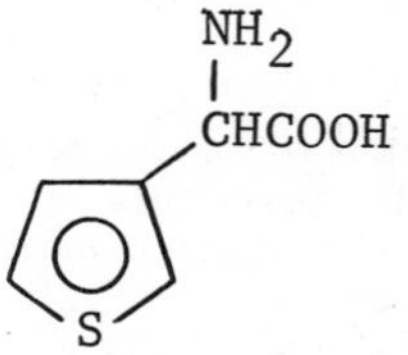

$C_6H_7NO_2S$ Ref. 168

α-Bromo-d-camphor-π-sulphonic acid

42 g of ammonium α-bromo-*d*-camphor-π-sulphonate was dissolved in 200 ml of water and passed through the hydrogen form of a Dowex 50 ion exchange column, which was eluated with water until a neutral reaction was produced. The eluate was evaporated *in vacuo* and dried over conc. sulphuric acid, yielding 39 g (98%) of α-bromo-*d*-camphor-π-sulphonic acid, m.p. 194–196°, $[\alpha]_D^{25} = +88$ (water).

Optical resolution of α-amino-3-thienylacetic acid

39.6 g (0.127 moles) of α-bromo-*d*-camphor-π-sulphonic acid and 20.0 g (0.127 moles) of α-amino-3-thienylacetic acid were dissolved in 250 ml of boiling ethyl acetate and left over-night in a refrigerator. 14.0 g of salt crystallized out, which was filtered off and dried. The amino acid was liberated from 0.5 g of salt and the optical activity measured in water. The remaining salt which was partially soluble in ethyl acetate was recrystallized from a mixture of ethyl acetate and ethanol which was then evaporated to about half its volume and allowed to crystallize in the refrigerator. This process was repeated until the optical activity of the acid remained constant. The course of resolution is seen in Table 1.

The mother liquor from the first recrystallization was evaporated and the amino acid liberated. 5.0 g (0.032 moles) of acid having $[\alpha]_D^{25} = -32°$ were dissolved together with 6.9 g (0.032 moles) of *d*-camphor-10-sulphonic acid in 80 ml of isobutyl alcohol and the salt allowed to crystallize at 0°. The progress of resolution is seen in Table 2.

L(+)-α-*Amino-3-thienylacetic acid*

2.5 g of the α-bromo-*d*-camphor-π-sulphonic acid salt was dissolved in 15 ml of water and an equivalent amount of 1.00 *N* sodium hydroxide (5.35 ml) was added. The amino acid separated in the form of colourless shiny crystals. It was washed with small amounts of water and dried over phosphoric anhydride. Yield 0.82 g, m.p. 218–221°, (IR-spectrum, Fig. 6.) $[\alpha]_D^{25} = +100°$, c; 0.250 in water.

Anal. $C_6H_7NO_2S$ (157.2)
Calc. C 45.84 H 4.49 N 8.91
Found C 45.53 H 4.49 N 8.78

Table 1

Crystalization	1	2	3	4
g salt disolved	—	10.5	5.6	3.8
g salt recovered	14.0	6.1	4.3	3.0
ml ethyl acetate	250	300	175	100
ml ethanol	—	100	50	25
Volume of solution after evaporation		150	80	60
$[\alpha]_D^{25}$ of acid	+ 85°	+ 97°	+ 100°	+ 100°

D(−)-α-*Amino-3-thienylacetic acid*

1.3 g of the *d*-camphor-10-sulphonic acid salt was dissolved in 10 ml of hot water and an equivalent amount of 1.00 *N* sodium hydroxide added. The amino acid was filtered off and dried. 0.50 g of (−)-α-amino-3-thienylacetic acid, m.p. 218–221°, and with the same IR-spectrum as the (+)-acid was obtained. The rotations under some different conditions are given in Table 3.

Anal. $C_6H_7NO_2S$ (157.2)
Calc. C 45.84 H 4.49 S 20.39
Found C 46.11 H 4.70 S 20.22

Table 3

Solvent	Conc./g acid dl	$2\alpha_D^{25}$	$[\alpha]_D^{25}$
Water	0.270	− 0.54°	− 100°
0.1–N NaOH	0.270	− 0.48°	− 89°
0.46–N HCl	0.230	− 0.47°	− 102°

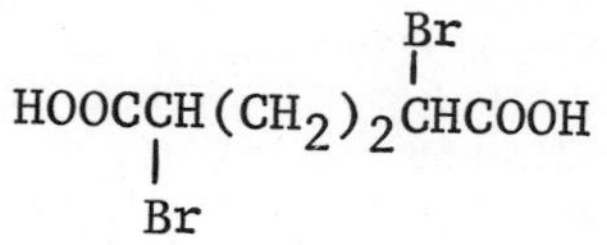

$C_6H_8Br_2O_4$ Ref. 169

Andererseits wurden 8.5 g der Dibrom-säure vom Schmp. 138—139° mit 3.4 g *l*-Phenäthylamin und 80 ccm Wasser verrührt, wobei teils Lösung, teils eine teigige Masse entstand. Bei mechanischer Bearbeitung wurde die Masse allmählich krystallinisch, und gleichzeitig schied sich eine voluminöse Masse von mikroskopischen, zu Bündeln und Rosetten vereinigten Nadeln oder flachen Prismen aus. Nach dem Absaugen und Waschen mit wenig Wasser wurden zusammen 5.1 g Salz erhalten, das bei 85—86° schmolz und bei Behandlung mit Salzsäure und Äther und freiwilliger Abdunstung des Äthers 3.3 g einer Säure ergab, die bei etwa 140° schmolz[12]) und für welche $[\alpha]_D = -15.9°$ in absol. Alkohol bei c = 5 gefunden wurde[13]).

Das Filtrat des eben beschriebenen Salzes wurde mit Schwefelsäure versetzt und mit Äther extrahiert, wonach beim freiwilligen Abdunsten des Äthers 4.3 g einer Säure erhalten wurde, welche bei 134—139° schmolz und $[\alpha]_D = +11.2°$ in Alkohol wie oben zeigte. Von dieser Säure wurden 4.1 g in 50 ccm gelinde erwärmtem Wasser gelöst und mit 1.6 g *d*-Phenäthylamin versetzt, wobei eine klare Lösung entstand, aus der wie bei der hochschmelzenden Säure oben nach dem Absaugen (Mutterlauge A) 2.0 g Salz als glasklare, flächenreiche Prismen oder Täfelchen gewonnen wurden. Dieses Salz schmolz nach langsamem Erhitzen bei 101—103°, beim schnellen bei 108—110° und gab bei der Analyse in lufttrockener Form:

0.2480 g Sbst.: 7.9 ccm N (17°, 754 mm). — $C_{14}H_{19}O_4NBr_2$. Ber. N 3.3. Gef. N 3.6.

Von diesem Salz wurden 1.3 g mit 10 ccm 2-*n*. Salzsäure behandelt, wobei neue Krystalle ausgeschieden wurden, ehe alles Salz in Lösung gegangen war. Das Gemisch wurde daher gelinde erwärmt, wobei klare Lösung entstand, aus welcher beim Erkalten 0.65 g *d*-α, α'-Dibrom-adipinsäure vom Schmp. 151—153° auskrystallisierten.

0.1563 g Sbst. verbrauchten zur Neutralisation 9.10 ccm 0.1131-*n*. Baryt und ergaben dann nach Erhitzen mit überschüssiger Lauge, Sauermachen mit Salpetersäure und Fällen mit Silbernitrat 0.1930 g AgBr.

$C_6H_8O_4Br_2$. Ber. Äquiv.-Gew. 152.0, Br 52.61. Gef. Äquiv.-Gew. 151.0, Br 52.55.

0.5100 g Sbst. in absol. Alkohol zu 10.04 ccm: $\alpha_D = +3.37°$; $[\alpha]_D = +66.3°$ und $[M]_D = +202°$.

Aus der Mutterlauge dieser Säure wurden mit Äther 0.35 g einer Säure gewonnen, welche den Schmp. 149—151° und $[\alpha]_D = +52{,}0°$ für 0.32 g in 10 ccm absol. Alkohol zeigte.

Die Mutterlauge A des *d*-Phenäthylamin-Salzes oben gab bei der Behandlung mit Salzsäure und Äther 2.0 g einer Säure vom Schmp. 125—130° und $[\alpha]_D = -18.4°$.

Von der Säure mit $[\alpha]_D = -15.9°$ wurden 3.15 g mit 30 ccm Wasser und 1.3 g *l*-Phenäthylamin wie oben behandelt, wobei kleine, körnige Aggregate und einige größere, dem *d*, *d*-Salz ähnelnden Krystalle abgeschieden wurden. Ein Paar dieser Krystalle wurden mechanisch herausgelesen und zeigten den Schmp. 104—105°, während die Hauptportion des ausgeschiedenen Salzes, die 1.3 g wog, bei 95—96° schmolz. Daraus wurden in gewöhnlicher Weise 0.8 g einer Säure isoliert, welche bei 144—145° schmolz und $[\alpha]_D = -31.2°$ zeigte. Aus der Mutterlauge dieses Salzes endlich wurde ein stark klebriges Säure-Gemisch gewonnen, das nicht weiter verarbeitet wurde.

Von der Säure mit $[\alpha]_D = -18.4°$ wurden 1.75 g genommen und zusammen mit 0.75 g von der anderen mit $[\alpha]_D = -31.2°$ in 25 ccm Wasser mit 1.0 g *l*-Phenäthylamin in Salz übergeführt, wobei 1.2 g Prismen oder Täfelchen vom Schmp. 108—109° erhalten wurden. Wie bei dem *d*, *d*-Salz oben wurden daraus direkt 0.5 g *l*-α, α'-Dibrom-adipinsäure vom Schmp. 151—153° und Äquiv.-Gew. 152.5 gewonnen.

0.4440 g Sbst. in absol. Alkohol zu 10.04 ccm: $\alpha_D = -2.88°$; $[\alpha]_D = -65.1°$ und $[M]_D = -198°$.

Die Mutterlauge der *l*-Säure gab bei der Extraktion mit Äther 0.25 g einer Säure vom Schmp. 147—149° und $[\alpha]_D = -55.7°$ für 0.24 g in 10 ccm absol. Alkohol, und aus der Mutterlauge des letzten *l*, *l*-Salzes wurde eine klebrige Masse von freien Säuren erhalten, welche nach dem Trocknen auf unglaciertem Porzellan bei etwa 135° schmolz des Äquiv.-Gew. 140 ergab und $[\alpha]_D = +5.2°$ zeigte.

Von der Dibrom-adipinsäure mit $[\alpha]_D = -65.1°$ wurden 0.1818 g zusammen mit 0.1785 g der Säure mit $[\alpha]_D = +66.3°$ in wenig Äther aufgelöst, wonach bei freiwilliger Abdunstung des Äthers eine *racem* α, α'-Dibrom-adipinsäure als weiße, grobkrystallinische Masse vom Schmp. 143—144° erhalten wurde.

0.1707 g Sbst. bei der Klason-Verbrennung: 0.2106 g AgBr.
$C_6H_8O_4Br_2$. Ber. Br 52.61. Gef. Br 52.50.

Stockholm, Organ.-chem. Laborat. d. Techn. Hochschule, im Juni 1925.

[12]) Wo nichts anderes gesagt wird, beziehen sich die Schmelzpunkts-Angaben auf bei schnellem Erhitzen gemachte Beobachtungen. Bei langsamem Erhitzen traten oft verschiedene Komplikationen ein.

[13]) Alle Drehungs-Bestimmungen wurden bei Zimmertemperatur, etwa 18°, ausgeführt.

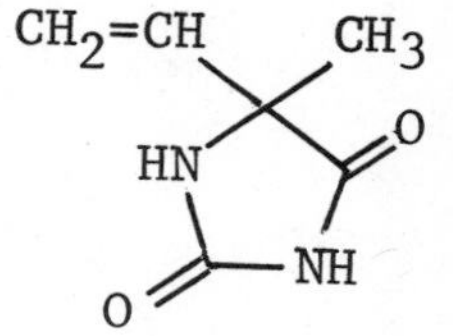

$C_6H_8N_2O_2$ Ref. 170

Resolution of DL-5-Methyl-5-vinylhydantoin (DL-V) into Its Optically Active Components by Brucine——A mixture of DL-V (14.0 g., 0.1 mole) and brucine (35.5 g., 0.09 mole) in EtOH (250 ml.) was warmed to become homogeneous. The EtOH solution was seeded*6 and kept standing in a refrigerator to crystallize out the brucine salt of (+)-V as prisms (27.8 g.), m.p. 157~160°, $[\alpha]^{16}_D$ −65.2° (C=0.840, EtOH), which was dissolved in 10% aq. HCl (50 ml.). The acidic aqueous solution was extracted with ethyl acetate (50 ml. × 4) and the combined extracts were washed with satd. NaCl, dried over Na_2SO_4, and evaporated to dryness *in vacuo* to give (+)-V as pale yellow needles (4.7 g.), m.p. 125~133°, $[\alpha]^{23}_D$ +4.1° (C=1.650, EtOH). Crude (+)-V obtained above was recrystallized several times from ethyl acetate to give pure (+)-V as colorless needles, m.p. 165~166°, $[\alpha]^{18}_D$ +37.8° (C=0.650, EtOH). *Anal.* Calcd. for $C_6H_8O_2N_2$: C, 51.42; H, 5.75; N, 19.99. Found: C, 51.50; H, 5.64; N, 19.67. IRν^{KBr}_{max} cm^{-1}: 3210, 1772, 1739, 1715, 1637, 993, 926.

The mother liquor from which the brucine salt of (+)-V crystallized out was evaporated to dryness to give a yellow oil, which was dissolved in 10% aq. HCl (50 ml.). The acidic aqueous solution was treated similarly to the above to give (−)-V (5.4 g.), m.p. 125~133°, $[\alpha]^{18}_D$ −5.7°(C=1.236, EtOH). Several recrystallizations from ethyl acetate afforded pure (−)-V as colorless needles, m.p. 165~166°, $[\alpha]^{16}_D$ −38.1° (C=0.536, EtOH). *Anal.* Calcd. for $C_6H_8O_2N_2$: C, 51.42; H, 5.75; N, 19.99. Found: C, 51.71; H, 5.85; N, 19.77. IRν^{KBr}_{max} cm^{-1}: 3215, 1773, 1740, 1715, 1638, 995, 927. This IR spectrum was superimposable with those of the authentic DL-V and (+)-V.

*6 The seeds were prepared by the following procedure. Another homogeneous ethanol solution was evaporated to dryness and the viscous oil obtained was triturated in an ice bath to give white crystals, which was recrystallized several times from ethanol to give a partially resolved brucine salt of (+)-V as prisms.

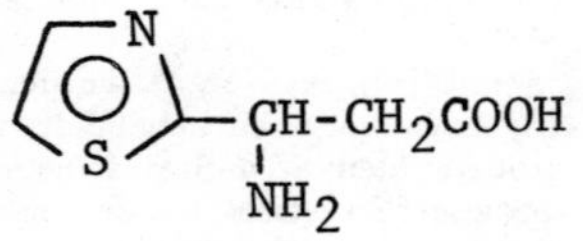

$C_6H_8N_2O_2S$ **Ref. 171**

An Optical Resolution of DL-*Phthalyl-β-(2-thiazolyl)-β-alanine.* To a solution of 10.7 g of DL-phthalyl-β-(2-thiazolyl)-β-alanine and 14.0 g of brucine in 100 ml of chloroform, we added 100 ml of acetone; after the mixture had been allowed to stand at room temperature overnight, the separated crystals were collected; yield, 5.4 g; $[\alpha]^{22}_D$ −3.2° (*c* 1, chloroform). After repeated recrystallizations from chloroform–acetone, the optical rotation settled upon $[\alpha]^{22}_D$ +10.6° (*c* 1, chloroform); yield, 4 g. From the mother liquor, the more soluble salt of $[\alpha]^{22}_D$ −39° (*c* 1, chloroform) was obtained; yield, 7 g.

To remove the brucine, the salt was dissolved in hot water; the solution was ice-cooled quickly and then after a small excess of aqueous sodium hydroxide had been added, the freed base was extracted with chloroform. The aqueous layer was acidified to pH 2 with 6 M hydrochloric acid and then concentrated. After ice-cooling, the separated phthalyl derivative was collected. No change in the optical rotation was observed after recrystallization from methanol–water. The properties of the phthalyl-β-(2-thiazolyl)-β-alanine thus obtained were as follows:

From the less soluble brucine salt; $[\alpha]^{22}_D$ +33.4° (*c* 1, methanol), mp 138—140 °C (solidified immediately and melted again at 162—163 °C).

From the easily soluble brucine salt; $[\alpha]^{22}_D$ −35.4° (*c* 1, methanol); the melting point is completely identical with that of the above sample.

(+)-β-(2-Thiazolyl)-β-alanine. ((+)-III). To a solution of the phthalyl derivative with $[\alpha]^{22}_D$ +33.4° in a small amount of methanol, we added 3 equivalents of hydrazine hydrate and then heated the mixture to reflux for 30 min. After the solvent had been distilled off, the residue was dissolved in water, acidified with acetic acid, and refluxed for 1 hr. The ice-cooled mixture was filtered, the filtrate was evaporated to dryness, and the remaining hydrazine hydrate was thoroughly removed under a high vacuum. By recrystallizing the residue from alcohol, (+)-β-(2-thiazolyl)-β-alanine was obtained as crystals; yield, 55%, mp 203.5—204.5 °C, $[\alpha]^{17}_D$ +22.4° (*c* 1, water). Further recrystallization from water–ethanol had no influence on the value of the specific rotatory power. Circular paper chromatographies; R_f 0.30 (Solvent A) and 0.29 (Solvent B).

Found: C, 41.62; H, 4.69; N, 16.26%. Calcd for $C_6H_8N_2O_2S$: C, 41.85; H, 4.68; N, 16.27%.

Lit,[7] mp 200—201 °C, $[\alpha]^{18}_D$ +9° (*c* 1, water).

((+)-III) in 6 M hydrochloric acid showed no perceptible change in optical rotation in the range of 270—700 nm after 24 hr at room temperature, but after 50 hr a little change was observed. When the solution was refluxed, however, for 1, 2.5, and 8 hr, the optical activity decreased to about 50%, less than 10%, and 0% respectively.

TABLE 1. SALT FORMATION OF ACYL DERIVATIVES OF β-(2-THIAZOLYL)-β-ALANINE WITH ALKALOIDS IN SOME SOLVENTS

Acyl-group	Quinine		Brucine		Ephedrine
	Acetone	Ethanol	Acetone	Ethanol	Ethanol
Z-	—	—	—	oil	—
Bz-	±	—	—	+	—
Pht-	—	±	+	oil	—
Tos-	—	+	—	—	

Z=benzyloxycarbonyl, Bz=benzoyl, Pht=phthalyl, and Tos=tosyl. +, crystalline salt; ±, amorphous solid; —, no precipitate

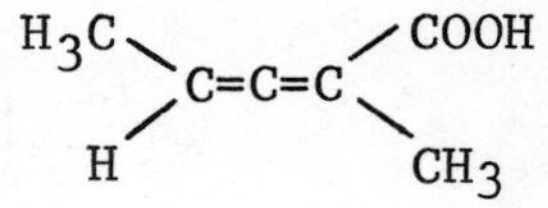

$C_6H_8O_2$ Ref. 172

*(+)- und (−)-***1e**: Nach den allgemeinen Vorschriften lieferten 2.609 g racem. **1e** und 3.501 g Cinchonidin nach vier Umkristallisationen aus Aceton 2.852 g (59%) farblose Nadeln mit Schmp. 154−155°C (Rotfärbung). $[\alpha]_{589}^{25} = -22.0°$, $[\alpha]_{350}^{25} = -106°$ ($c' = 0.012408$, C_2H_5OH).

$C_{25}H_{30}N_2O_3$ (406.6) Ber. C 73.85 H 7.44 N 6.89 Gef. C 73.31 H 7.58 N 7.11

Aus dem Filtrat der ersten Kristallisationsfraktion wurden 0.927 g (36%) optisch unreines (−)-**1e** gewonnen; Sdp. 75−78°C/0.06 Torr (Kugelrohr). $[\alpha]_{589}^{25} = -15.5°$ ($c' = 0.030202$, C_2H_5OH). − Aus 2.670 g Cinchonidinsalz wurden 0.430 g [51%] (+)-**1e** erhalten; Sdp. 75−77°C/0.06 Torr (Kugelrohr), Schmp. 55−58°C. $[\alpha]_{589}^{20} = +56.6°$ ($c' = 0.008391$, C_2H_5OH). − CD (C_2H_5OH): λ_{max} ($\Delta\varepsilon$) = 284 (−0.0038), 269 (+0.047), 250 (+0.10), 217 nm (+1.4).

$C_6H_8O_2$ (112.2) Ber. C 64.23 H 7.19 Gef. 63.73 H 6.97

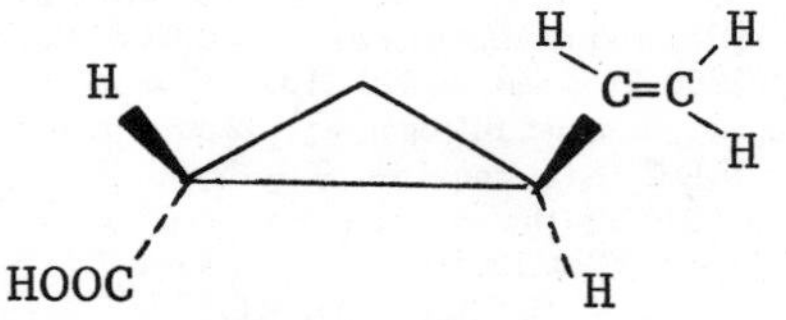

$C_6H_8O_2$ Ref. 173

Note:

Partially resolved using quinine as resolving agent by recrystallization from water. The acid obtained has b.p. 82-83°/2-3 Torr, $[\alpha]_D^{25}$ = -60.0° (c, 1.49, acetone).

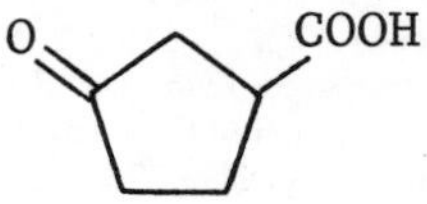

$C_6H_8O_3$ Ref. 174

Resolution of *dl*-Cyclopentanone-3-carboxylic Acid. — (+)-*Cyclopentanone-3-carboxylic Acid* (Ia). — A mixture of *dl*-cyclopentanone-3-carboxylic acid (48 g.), brucine tetrahydrate (178.5 g.) and 90% methanol (380 cc.) was heated until solution was effected. After cooling and storing at 11~12°C for 2 hr., the first crystalline crop was separated, and washed with a small quantity of 90% methanol. The weight of the dried product 103 g. On decomposing a small portion of the crop in the same way as described below, the free acid of $[\alpha]_D^{10}+12.0°$ (c 4, methanol) was obtained. Recrystallization of the crude salt from 90% methanol (120 cc.) gave 80 g. of the purified salt, which, on decomposition, gave the free acid of $[\alpha]_D^{9}+18.5°$ (c 4, methanol).

A second crop (41 g.) was obtained by evaporation of the mother liquor and recrystallized three times from 90% methanol to give 16.7 g. of the salt [free acid: $[\alpha]_D^{10}+19.0$ (c 4, methanol)]. The crops were collected and recrystallized from 90% methanol (1.2 cc./g.) to give 82.5 g. of the nearly pure salt [free acid: $[\alpha]_D^{9}+23.0°$ (c 4, methanol)]. The product was further recrystallized twice from the same solvent and subjected to decomposition.

To a solution of the recrystallized brucine salt (66 g.) of (+)-acid in boiling water (160 cc.) was added dropwise 14% sodium carbonate solution (142 cc.) with stirring. After being cooled in a refrigerator overnight, the brucine was filtered off. The filtrate was acidified with hydrochloric acid to pH 2.4, saturated with ammonium sulfate and extracted with ether repeatedly. After drying the extract, evaporation of ether gave 10.3 g. (43% from *dl*-acid) of (+)-cyclopentanone-3-carboxylic acid (Ia), $[\alpha]_D^{7.5}+27.2°$ (c 2.0, methanol)*. Recrystallization from ether gave colorless plates, m. p. 66~67°C.

Anal. Found: C, 56.49; H, 6.28. Calcd. for $C_6H_8O_3$: C, 56.24; H, 6.29%.

(−)-*Cyclopentanone-3-carboxylic Acid* (Ib).—By concentration of the mother liquor of the above-mentioned second crop, 54.6 g. of the third crop was obtained. The corresponding free acid obtained by decomposition showed $[\alpha]_D^{8}-17.5°$ (c 4, methanol). Recrystallization from 85% ethanol (55 cc.) gave 34.5 g. of crystals. The corresponding free acid showed $[\alpha]_D^{9.5}-15.7°$ (c 3.94, methanol). The recrystallized salt was mixed with the crystalline residue obtained by evaporation of the recrystallization mother liquors of the first and the second crops described above and dissolved in hot 85% ethanol. The solution was left at 14°C for 1.5 hr. and the less soluble salt there separated was filtered off. Evaporation of the mother liquor gave 8 g. (crystal A) of the more soluble salt. Evaporation of the mother liquor separated from the third crop and the recrystallization mother liquor of the third crop gave 14 g. (crystal B) and 13 g. (crystal C) of crystalline residues, respectively. Crystals A, B and C correspond to the brucine salt of (−)-cyclopentanone-3-carboxylic acid which is more soluble than the salt of the (+)-acid described above.

For decomposition 14% sodium carbonate solution (76 cc.) was added dropwise to a hot solution of the brucine salt of (−)-acid (35 g., collected crystals A, B and C) and the resulting mixture was processed, in the same manner as described

4) K. Toki, This Bulletin, 30, 450 (1957); ibid., 31, 333 (1958).

* Rotation reported by Toki[4] is $[\alpha]_D^{21}+22.1°$ (c 1.9, methanol).

in the preceding paragraph, to yield the crystals of (−)-cyclopentanone-3-carboxylic acid; yield 4.6 g. (19.2% from *dl*-acid), $[\alpha]_D^{10}$ −23.4° (c 2.0, methanol)**. A sample for analysis was prepared by recrystallization from ether, m. p. 65~67°C.

Anal. Found: C, 56.63; H, 6.20. Calcd. for $C_6H_8O_5$: C, 56.24; H, 6.29%.

** Rotation reported by Toki[4] is $[\alpha]_D^{21}$ −22.2° (c 2.0, methanol).

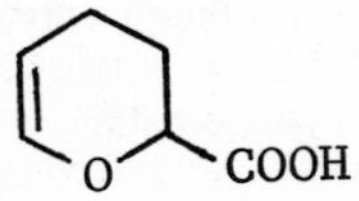

$C_6H_8O_3$ Ref. 175

(+)-(1*R*,5*R*)-6,8-Dioxabicyclo[3.2.1]octan-7-one (**7**) was successfully prepared through the optical resolution of racemic 3,4-dihydro-2*H*-pyran-2-carboxylic acid by using dehydroabietylamine, a base which has lately been used with success in optical resolution of several carboxylic acids.[4,5] An aqueous solution of sodium 3,4-dihydro-2*H*-pyran-2-carboxylate (**5**) was slightly acidified with 6 N hydrochloric acid, and the liberated carboxylic acid was extracted several times with ethyl ether. The ethyl ether extract was then added to an ice-cooled ethyl ether solution of dehydroabietylamine with occasional shaking. Immediately, a white mass was formed, which was separated and recrystallized repeatedly from methanol solution to yield pure (+)-dehydroabietylammonium 3,4-dihydro-2*H*-pyran-2-carboxylate (**6**) as white needles: $[\alpha]^{24}{}_D$ +11.7° (*c* 1 g/dL, ethanol); mp 168–171 °C. Anal. ($C_{26}H_{39}NO_3$) C, H, N. The diastereomeric ammonium salt was converted to the sodium salt on treatment with aqueous sodium carbonate, and the liberated dehydroabietylamine was removed by extraction with ethyl ether. The sodium salt was subsequently transformed to the free acid followed by immediate distillation under reduced pressure to afford an optically active monomer **7** with $[\alpha]^{24}{}_D$ +128° (*c* 1.0 g/dL, ethanol); bp 61 °C (5 mm).

(4) H. Budziekiewicz, C. Djerassi, and D. H. Williams, "Structure Elucidation of Natural Products by Mass Spectrometry", Vol. 1, Holden-Day Inc., San Francisco, 1964.
(5) G. Barth, R. E. Linder, E. Bunnenberg, and C. Djerassi, *Helv. Chim. Acta*, **55**, 2168–2178 (1972).

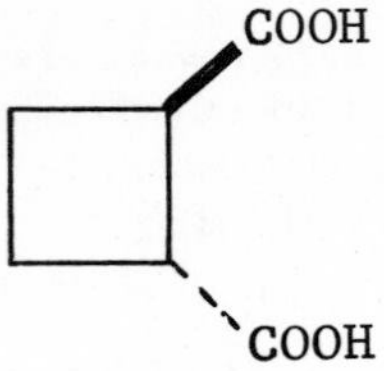

$C_6H_8O_4$ Ref. 176

Resolution by means of Quinine.—Brucine effected a partial resolution only, but the acid was completely resolved by quinine. Quinine, 21 gms. (1 mol.) in a little alcohol, was added to *trans-cyclo*butane-1 : 2-dicarboxylic acid, 9 gms. (1 mol.) in 300 c.c. of water, the alcohol was evaporated, and the quinine salt dissolved by addition of boiling water. On cooling, the filtered solution deposited needles of the quinine salt of the *l*-acid ($[\alpha]_D$ in alcohol −193·6° after one crystallisation, and −192·9° after five crystallisations, from water and drying at 120°). 0·1508 of the salt dried at 120° gave 0·3868 CO_2 and 0·1020 H_2O; C = 70·0; H = 7·5. 0·2697 gave 18 c.c. of moist nitrogen at 762 mm. and 28°; N = 7·3. $(C_{20}H_{24}O_2N_2)_2,C_6H_8O_4$ requires C = 69·7; H = 7·1; N = 7·1 per cent. 0·0910 of the crystals, air-dried at the ordinary temperature, lost 0·0140 at 120°; H_2O = 15·4.

$(C_{20}H_{24}O_2N_2)_2,C_6H_8O_4,8H_2O$

requires H_2O = 15·4 per cent.

The l-*acid*, magnificent, colourless needles, m. p. 105°, from hydrochloric acid, was prepared by decomposing the hot aqueous solution of the quinine salt by addition of ammonia, the quinine removed by filtration, and the acidified filtrate evaporated to dryness; the acid was extracted from the residue with ether. Rotation: α −2·10° in water (*c* = 0·8452, *l* = 2), whence $[\alpha]_D^{30°}$ −124·3°.

The d-*acid*, similarly isolated from the mother-liquor from the first crystallisation of the *l*-salt, separated from hydrochloric acid in crystals, m. p. 105°. 0·1094 neutralised 0·0611 NaOH [calc. for $C_4H_6(CO_2H)_2$, 0·0608]. Rotation: α + 3·01° in water (*c* = 1·2208, *l* = 2), whence $[\alpha]_D^{30°}$ +123·3°. The *ethyl* ester, b. p. 236°/761 mm., was prepared from the *d*-acid and boiling 10 per cent. alcoholic sulphuric acid (7 parts). α + 2·70° in acetone (*c* = 1·732, *l* = 2), whence $[\alpha]_D^{30°}$ +77·9°.

$C_6H_8O_4$ Ref. 176a

(−)-trans-cyclo*Butane*-1 : 2-*dicarboxylic Acid*.—Goldsworthy's resolution[5] was repeated. The diquinine salt of the (−)-acid, recrystallised from water and dried *in vacuo* at 120°, gave $[\alpha]_D^{17·5}$ −174°, $[\alpha]_D^{29·5}$ −168° (*c* 0·5), $[\alpha]_D^{29}$ −171° (*c* 1 in EtOH); it decomposed before melting (Found: C, 69·4; H, 7·3; N, 6·95. Calc. for $C_{46}H_{56}O_8N_4$: C, 69·7; H, 7·1; N, 7·1%). These values were not appreciably changed by further recrystallisation or by liberating the acid and reconverting it into the salt, and they differ considerably from Goldsworthy's, $[\alpha]_D$ −192·9° (concentration and temperature not stated). The discrepancy between our result and Goldsworthy's clearly cannot be accounted for by the differences of temperature and concentration, and it seems that it may be due mainly to the presence of water in the alcohol used by Goldsworthy, as shown:

Ethanol	Anhydrous	+10% water	+19% water
$[\alpha]_D^{19·5}$	−171°	−188°	−198°
$[\alpha]_D^{29}$	−167	−182	−192

Excess of ammonia was added to a solution of the salt in hot water, the precipitated quinine removed by filtration, and the filtrate acidified with hydrochloric acid and evaporated under reduced pressure; after thorough drying, the residual solid was extracted with anhydrous ether, and the extract evaporated to give the (−)-acid, recrystallising from benzene [in which it is more soluble than the (±)-acid] as prisms, m. p. 116—117°, $[\alpha]_D^{18·5}$ −158°, $[\alpha]_D^{29}$ −157°

[5] L.J. Goldsworthy, J. Chem. Soc., 124, 2012 (1924).

(c 0·75 in H_2O) (Found: C, 50·1; H, 5·7. Calc. for $C_6H_8O_4$: C, 50·0; H, 5·6%). Goldsworthy reported m. p. 105°, $[\alpha]_D^{30}$ −124·3° (c 0·85 in H_2O); the possibility that his specimen had been partially racemised by recrystallisation from concentrated hydrochloric acid was eliminated, since (−)-acid recovered after 2 hr. in boiling concentrated hydrochloric acid had $[\alpha]_D^{22}$ −151°. Changes of concentration and temperature had virtually no effect on the rotation of the (−)-acid.

The mother-liquor from the original crystallisation of the diquinine salt yielded acid having $[\alpha]_D^{29}$ +86° (from benzene): treatment of this with the theoretical amount of quinine for (−)-acid present, with care that the salt separated as crystals and not as an oil, yielded acid, after removal of the salt, having $[\alpha]_D^{22}$ +107°, which was then fractionally crystallised from a large volume of benzene. Ultimately a small quantity of material, $[\alpha]_D^{21}$ +150°, was obtained; thus, the (+)-acid was not isolated in a pure state. One of the crops from benzene had m. p. 105—108°, $[\alpha]_D^{21}$ +123·8° {cf. Goldsworthy's values for the (+)-acid, m. p. 105°, $[\alpha]_D^{30}$ +123·3°}.

$C_6H_8O_4$ Ref. 176b

(−)-(*R*)-*trans*-1,2-Cyclobutanedicarboxylic Acid. Racemic *trans*-1,2-cyclobutanedicarboxylic acid (commercially available from Aldrich Chemical Co.) was resolved by a modification of Goldsworthy's method.[14,15] To improve the result, 95% ethanol, instead of water, was used as a solvent in fractional recrystallization of the diquinine salt. The pure (−)-(*R*)-diacid salt which was obtained by four successive recrystallizations was dried at room temperature *in vacuo:* mp ~185° dec, $[\alpha]^{20}D$ −180° (EtOH, c 1.0 g/dl).

Anal. Calcd for $C_6H_8O_4\cdot(C_{20}H_{24}N_2O_2)_2\cdot H_2O$: C, 68.12; H, 7.21; N, 6.91. Found: C, 68.24; H, 7.10; N, 6.76.

The salt was further dried at 110° *in vacuo:* mp ~185° dec, $[\alpha]^{20}D$ −187° (EtOH, c 1.0 g/dl).

Anal. Calcd for $C_6H_8O_4\cdot(C_{20}H_{24}N_2O_2)_2$: C, 69.69; H, 7.12; N, 7.07. Found: C, 69.59; H, 7.10; N, 6.98.

The (−)-(*R*)-diacid was obtained from the salt as mentioned above, followed by recrystallization from benzene: mp 116–117°, $[\alpha]^{20}D$ −165° (H_2O, c 0.75 g/dl) (lit. mp 116–117°, $[\alpha]^{18.5}D$ −158° (H_2O, c 0.75 g/dl)[15]).

(14) L. J. Goldsworthy, *J. Chem. Soc.*, 2012 (1924).
(15) F. B. Kipping and J. J. Wren, *ibid.*, 3246 (1957).

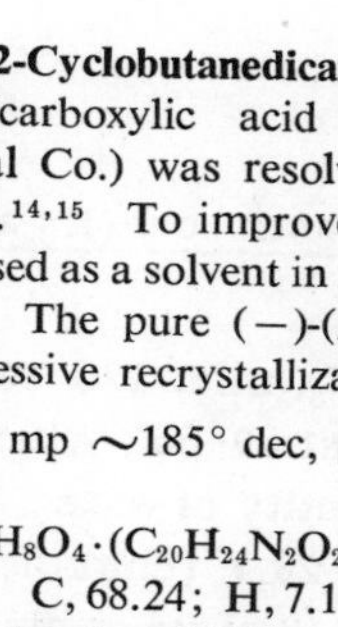

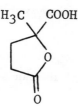

$C_6H_8O_4$ Ref. 177

Optical resolution of (±)-*2-hydroxy-2-methylpentane-1,5-dioic acid 5→2 lactone* (**2**+**2′**)

(a) *With quinine.* The (±)-acid (117 g) and quinine (267 g) were dissolved in hot 95% EtOH (2000 ml). The soln was left to stand overnight in a refrigerator to give less soluble crystals A-1 (249 g). This was recrystallized from 95% EtOH to give crystals A-2(135 g). The mother liquor after removal of A-2 was concentrated *in vacuo* and the residue was recrystallized from 95% EtOH to give A-3 (60 g). Recrystallization of A-3 from 95% EtOH gave A-4 (17 g). Since the crystals A-2 and A-4 exhibited the same IR and $[\alpha]_D$ values, they were combined to give pure salt A (152 g, 38%). *The salt A* was less soluble in EtOH than the salt B described below and had the following properties: needles from EtOH, m.p. 222 ~ 224° (dec); $[\alpha]_D^{23}$ − 138° (c = 0·74%, EtOH) [lit.[8] m.p. 199·5 ~ 201·3°; $[\alpha]_D^{28}$ − 134·41° (c = 0·5%, EtOH)]; ν_{max} (nujol) ~ 3100 (m), 1765 (s), 1615 (m), 1590 (m), 1510 (m), 1405 (m), 1375 (m), 1305 (w), 1280 (m), 1260 (s), 1220 (m), 1195 (w), 1170 (m), 1150 (w), 1095 (s), 1080 (m), 1040 (w), 1000 (w), 980 (w), 940 (m), 925 (w), 915 (m), 900 (w), 885 (w), 870 (w), 850 (w), 805 (m), 785 (w), 760 (w), 715 (m) cm^{-1}. (Found: C, 66·78; H, 6·77; N, 5·82. $C_{26}H_{32}O_6N_2$ requires: C, 66·65; H, 6·88; N, 5·98%). The mother liquors obtained after removals of A-1, A-3 and A-4 were combined and concentrated *in vacuo*. The residue was recrystallized from 95% EtOH to give crystals B-1 (87 g). The mother liquor gave further crystals (B-2, 33·5 g) after concentration *in vacuo*. Since B-1 and B-2 showed the same IR and $[\alpha]_D$ values, they were combined (salt B, 120·5 g, 30%). An analytical sample was obtained by further recrystallization from EtOH. *The salt B* crystallized from EtOH as prisms, m.p. 224 ~ 225° (dec); $[\alpha]_D^{23}$ − 133·6° (c = 0·68%, EtOH); ν_{max} (nujol) ~ 3200 − ~ 3000 (m), 1775 (s), 1615 (m), 1590 (s), 1510 (m), 1400 (m), 1380 (m), 1305 (w), 1280 (w), 1240 (s), 1220 (m), 1145 (m), 1095 (s), 1060 (w), 1030 (m), 1000 (w), 980 (w), 945 (m), 915 (m), 855 (m), 830 (m), 800 (w), 720 (m) cm^{-1}. (Found: C, 66·83; H, 6·83; N, 5·95. $C_{26}H_{32}O_6N_2$ requires: C, 66·65; H, 6·88; N, 5·98%).

(b) *With cinchonine.* The (±)-acid (50 g) and cinchonine (95 g) were dissolved in hot 95% EtOH (600 ml). The soln was left to stand overnight in a refrigerator to give crystalline salt (74 g). This was recrystallized from 95% EtOH to give 50 g (41%) of the salt. This crystallized from EtOH as prisms, m.p. 228 ~ 230° (dec); $[\alpha]_D^{23}$ + 147° (c = 0·81%, EtOH); ν_{max} (nujol) ~ 3160 (m), 1785 (s), ~ 1570 (m), 1400 (s), 1305 (m), 1250 (m), 1215 (s), 1140 (m), 1120 (m), 1110 (m), 1040 (w), 1010 (w), 995 (w), 950 (m), 930 (m), 920 (m), 880 (w), 860 (w), 840 (w), 800 (m), 790 (m), 780 (m), 750 (m), 720 (w) cm^{-1}. (Found: C, 68·54; H, 6·83; N, 6·22. $C_{25}H_{30}O_5N_2$ requires: C, 68·47; H, 6·90; N, 6·39%).

(R)-(+)-*2-Hydroxy-2-methylpentane-1,5-dioic acid 5→2 lactone* (**2**)

A soln of the quinine salt A (151 g) in 10% HCl (2500 ml) was extracted continuously with ether for 3 days. The ether soln was concentrated *in vacuo*, The residue was mixed with C_6H_6 and concentrated *in vacuo* to remove trace of H_2O. The residue was recrystallized from C_6H_6-CCl_4 (1:1) to give crude acid (39 g, 83%). This was recrystallized from C_6H_6-CCl_4 (1:1). Further recrystallization from EtOAc-C_6H_6 gave pure **2** (27·2 g, 58%). The acid **2** crystallized from EtOAc-light petroleum as elongated prisms, m.p. 87·5 ~ 88·5° (lit.[8] 87·8 ~ 89·8°); $[\alpha]_D^{24}$ + 16·2° (c = 3·15%, H_2O) [lit.[8] $[\alpha]_D^{28}$ + 15·39° (c = 2·5%, H_2O)]; ν_{max} (nujol) ~ 3200 (m), ~ 2600 (m), 1760 (s), 1710 (s), 1290 (m), 1280 (s), 1250 (s), 1220 (m), 1180 (s), 1160 (m), 1110 (s), 1080 (m), 1000 (w), 980 (w), 950 (w), 850 (w), 800 (w), 760 (w), 715 (w) cm^{-1}; CD (c = 0·302%, EtOH): $[\theta]_{255}$ 0; $[\theta]_{215}$ − 5770 (min); $[\theta]_{200}$ − 3100. (Found: C, 49·96; H, 5·45. $C_6H_8O_4$ requires: C, 50·00; H, 5·60%).

[8] R. Adams and F. B. Hauserman, *J. Am. Chem. Soc.* **74**, 694 (1952)

$C_6H_8O_4$ Ref. 177a

Resolution of DL-γ-Carboxy-γ-valerolactone.—A solution of 13.5 g. of the DL-acid and 30 g. of anhydrous quinine in 1000 ml. of hot absolute ethanol was heated to boiling and filtered. The straw-colored filtrate was cooled at 0° for 20 hours. The white crystals which separated were isolated by filtration; m.p. 198–200° (cor.) (dec.); yield 16.2 g. (37.2%). A small sample was recrystallized from ethanol for analysis; m.p. 199.5–201.3° (cor.) (dec.).

Anal. Calcd. for $C_{26}H_{32}N_2O_6$: C, 66.65; H, 6.89. Found: C, 66.85; H, 7.12.

Rotation.—0.1049 g. made up to 20 ml. with ethanol at 28° gave αD −1.41°; *l*, 2; $[\alpha]^{28}_D$ −134.41° (±0.95°).

The straw-colored solution obtained on dissolving 16 g. of the quinine salt in 250 g. of 10% hydrochloric acid was continuously extracted with ether for 34 hours. After removing the ether, the residual oil crystallized from a 1:1 mixture of benzene and carbon tetrachloride, yielding 3.86 g. (78.4%) of product. The pure enantiomorph was obtained on recrystallization from the same solvents; white crystals, m.p. 87.8–89.8° (cor.); yield 3.38 g. (69%).

Anal. Calcd. for $C_6H_8O_4$: C, 50.00; H, 5.60. Found: C, 50.20; H, 5.83.

Rotation.—0.5003 g. made up to 20 ml. with water at 28° gave αD +0.77°; *l*, 2; $[\alpha]^{28}_D$ +15.39° (±0.20°). 0.5088 g. made up to 20 ml. with absolute ethanol at 26° gave αD −0.05°; *l*, 2; $[\alpha]^{26}_D$ −0.99° (±0.20°).

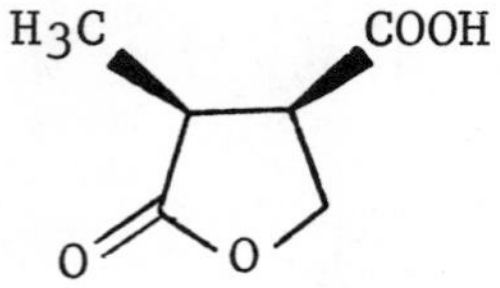

$C_6H_8O_4$ Ref. 178

II. Cleavage of labile racemic α-alkylparaconic acids (II) into the optical antipodes (IIa and IIb). Brucine salts of labile dextro and levo α-methylparaconic acids.

To a solution of 2.2 g of labile racemic α-methylparaconic acid (II, R = CH_3, m.p. 125-126°) in 12 ml of water was added 6 g of brucine, and the reaction mixture was heated for an hour at 50-55°. After cooling, the precipitate was separated and recrystallized three times from three times the quantity of water. Yield of brucine salt of labile dextro α-methylparaconic acid 2.7 g (65.7%). M.p. 118-123°, $[\alpha]^{20}_D$-20.7° (chloroform, c 2.548). The filtrate from the brucine salt of the dextro acid was evaporated in vacuo to 1/3 of its original volume, the precipitate was separated and the filtrate from the latter evaporated to dryness at 35-40° and 10 mm. The dry residue was washed with acetone twice (10 ml each time). Yield of brucine salt of labile levo α-methylparaconic acid 3 g (72.9%). M.p. 137-140°, $[\alpha]^{20}_D$-31.8° (chloroform, c 10.944).

Labile dextrorotatory and levorotatory α-methylparaconic acids (IIa and IIb, R = CH_3). The labile optically active α-methylparaconic acids were obtained from their brucine salts by the procedure described for isolation of stable levo α-isopropylparaconic acid. From 2.4 g of brucine salt of labile dextro α-methylparaconic acid was obtained the dextro acid in a yield of 0.45 g (70.1%). M.p. 123-124° (from benzene), $[\alpha]^{20}_D$ + 19.7° (water, c 9.908).

Found %: C 50.03; H 5.63. $C_6H_8O_4$. Calculated %: C 50.00; H 5.59.

The levo acid was isolated from 2.5 g of brucine salt of labile levo α-methylparaconic acid. Yield 0.5 g (64.7%). M.p. 122.5-124°, $[\alpha]^{20}_D$ - 19.15° (water, c 10.188).

Found %: C 49.90; H 5.38. $C_6H_8O_4$. Calculated %: 50.00; H 5.59.

A mixture of 0.02 g each of labile dextrorotatory and levorotatory α-methylparaconic acids melted at 129-130°.

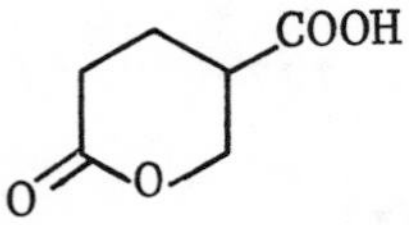

$C_6H_8O_4$

(*R*)-γ-Valerolactone-γ-carboxylic Acid (*VII*, R = H) Ref. 179

Racemic γ-valerolactone-γ-carboxylic acid, 15·4 g b.p. 165—175°C/0·1 mm Hg, prepared from levulinic acid[8], was mixed with 34·7 g of (−)-quinine in 500 ml of ethanol. After 24 h, 35 g of a salt separated, m.p. 203—205°C, which, after recrystallization from ethanol, had m.p. 208·5—210·5°C. After decomposing the salt with 300 ml of 10% hydrochloric acid and extraction with diethyl ether, 5·3 g of the acid was obtained, b.p. 162°C/0·3 mm Hg. Recrystallization from a mixture of benzene and carbon tetrachloride (1 : 1) afforded the acid having m.p. 82—84°C, $[\alpha]^{20}_D$ +15·3° (*c* 3·8; water). Lit.[9], m.p. 87·8—89·8°C, $[\alpha]^{20}_D$ +15·39° (water), −0·5° (methanol). Processing the mother liquor after crystallization of the quinine salt in the way described above afforded the other, optically impure, enantiomer of the acid. Crystallization from a mixture of benzene and carbon tetrachloride gave 7·4 g of the acid, m.p. 62—66°C, $[\alpha]^{20}_D$ −7·6° (*c* 5·1; water).

8. Kawai Y.: Nippon Kagaku Zasshi *80*, 647 (1959); Chem. Abstr. *55*, 3426 (1961).
9. Adams R., Hauserman F. B.: J. Am. Chem. Soc. *74*, 694 (1952).

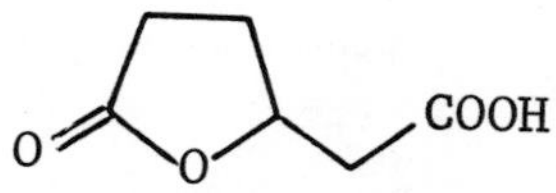

$C_6H_8O_4$

Ref. 180

(R)-(-)-γ-carboxymethylbutanolide (3)

A solution of crude (±)-γ-carboxymethylbutanolide (2) (3.0 g, 20.8 mmole) and (S)-(-)-α-phenylethylamine (2.52 g, 20.8 mmole) in 10 ml of 1:1 mixture of isopropyl alcohol and isopropyl ether was left at room temperature over night. The resulted crystal (3.91 g) was further recrystallized three times from 1:1 mixture of isopropyl alcohol-isopropyl ether, twice from 4:4:1 mixture of isopropyl alcohol-isopropylether-ethanol, and once from 3:1 mixture of isopropyl alcohol-ethanol to give 0.399 g of crystalline salt [m.p. 118-128°, $[\alpha]^{20}_{D}$-19.1° (C=1, EtOH)]. The purified salt (380 mg) was dissolved in 5.0 ml of 2N-Na_2CO_3, extracted with methylene chloride, and acidified with 6N-HCl to pH=2. The solution was evaporation to dryness to leave a solid which was then washed with 25 ml of 9:1 mixture of chloroform and methanol. Evaporation of the solvent left 246 mg (8.4%) of (R)-(-)-γ-carboxymethylbutanolide (3); [$[\alpha]^{25}_{D}$-28.2° (C=1, ethanol)]. Further purification by TLC (silica gel, ethyl acetate-acetic acid=9:1) gave an oil of 3 [222 mg, 7.62% from crude 3, $[\alpha]^{25}_{D}$-33.6° (C=1, EtOH), CD $[\theta]_{210\ nm}$+1159 (MeOH)].

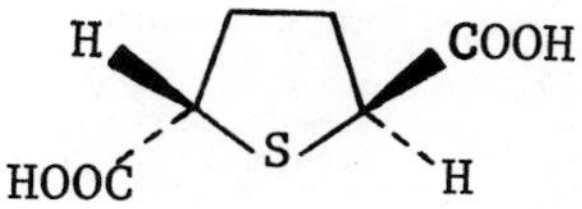

$C_6H_8O_4S$

Ref. 181

Spaltung der racem-Form in optische Antipoden

Wie bei den analogen Selenverbindungen erhält man die (−)-Form leicht über das primäre Brucinsalz und die (+)-Form über das primäre Chininsalz. Beide Salze gleichen in ihrem Aussehen und den Löslichkeitsverhältnissen den entsprechenden selenhaltigen Salzen; es dürfte daher schon die Spaltungsmethode als solche eine gute Stütze für die Annahme darstellen, daß Schwefel- und Selenverbindung mit gleichem Drehungssinn auch gleiche Konfiguration besitzen. Die Bildung des aktiven Racemats aus (+)-Schwefel- und (−)-Selenverbindung dürfte als entscheidender Beweis dafür zu betrachten sein.

17,7 g (0,1 Mol) racem-Säure und 46,6 g krystallwasserhaltiges Brucin wurden in 350 ccm heißem Wasser gelöst. Das beim Erkalten erhaltene Salz wurde aus dem etwa 8-fachen Gewicht Wasser auf konstante Drehung umkrystallisiert; da die Drehung der Säure bedeutend größer ist als die von Brucin, läßt sich der Gang der Spaltung durch Drehungsmessungen an salzsauren Lösungen des Salzes verfolgen. Das Salz bildet kompakte, kurzprismatische, ziemlich flächenreiche Krystalle ohne Krystallwasser.

Krystallisation	1	2	3	4	5	6	7
Ausbeute in g	43,3	25,1	19,0	16,8	14,5	13,1	11,0
$[\alpha]^{25}_{D}$ . . .	—	−73,6°	−85,7°	−89,3°	−91,2°	−91,4°	−91,4

Die Mutterlaugen von den beiden ersten Krystallisationen wurden bei Raumtemperatur zur Trockne eingedampft und aus dem Rückstand einige größere Krystalle der für das Salz der (−)-Form charakteristischen Gestalt ausgesucht. Aus dem Rest, der aus blättrigen oder tafelförmigen Krystallen bestand, wurden 7,5 g etwa 95%-ige (+)-Säure erhalten. Diese wurden

in vollständig neutralisierter wäßriger Lösung: 2 α = + 5,105°; + 6,11°. — 0,1112 g in Eisessig: 2 α = + 6,50°; + 7,72°. — 0,1126 g in Essigester: 2 α = + 5,37°; + 6,38°. — 0,1242 g in absolutem Alkohol: 2 α = + 6,42°; + 7,62°.

	5893 Å.-E.		5461 Å.-E.	
	$[\alpha]^{25}_{D}$	$[M]^{25}_{D}$	$[\alpha]^{25}_{D}$	$[M]^{25}_{D}$
0,4 n-Salzsäure	+225,9°	+398°	+268,1°	+472°
halb neutralisierte wäßrige Lösung	+226,1	+398	+269,7	+475
vollständig neutralis. „ „	+227,7	+401	+272,5	+480
Eisessig	+292,3	+515	+347,1	+611
Essigester	+238,5	+420	+283,3	+499
Alkohol (absolut)	+258,5	+455	+306,8	+540

Bemerkenswert ist der Umstand, daß die Drehung in wäßriger Lösung durch den Ionisationszustand nur so wenig beeinflußt wird.

Das primäre Chininsalz der (+)-Säure bildet weiße, asbestähnliche Nadeln. Es wurde nicht analysiert.

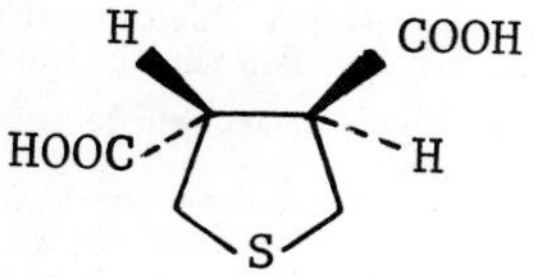

$C_6H_8O_4S$

Ref. 182

Levorotatory acid. 10.4 g (0.06 mole) racemic acid in 125 ml boiling ethanol was mixed with 9.8 g (0.03 mole) strychnine in 250 ml boiling ethanol. After cooling, 13.8 g salt was obtained which was recrystallised five times from 50 % aqueous ethanol (50 ml/g salt); the activity remained constant after three recrystallisations. The acid was liberated and recrystallised three times from acetone-benzene giving plates or flat needles with m.p. 137.5−138°. (Equiv. wt. found 88.4; calc. 88.1. $[\alpha]_D^{25}$ = −151.4° in absolute ethanol). The racemic acid melted at 133−136°, resolidifies and remelts at 140−141°.

Dextrorotatory acid. The first mother liquor of the strychnine salt yielded, after decomposition, 3.15 g acid having $[\alpha]_D^{25}$ = + 114°. This acid in 50 ml boiling ethanol was mixed with 5.8 g quinine in 125 ml boiling ethanol. The salt obtained after cooling was recrystallised ten times from 60 % ethanol (10 ml/g salt). The acid was liberated and recrystallised twice (acetone-benzene) giving plates or flat needles with m.p. 137.5−138°. (Equiv. wt. found 88.6; calc. 88.1. $[\alpha]_D^{25}$ = +151.8° in absolute ethanol).

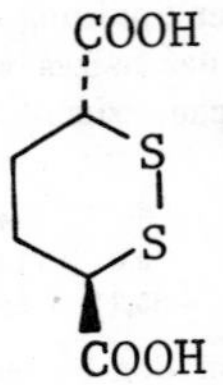

$C_6H_8O_4S_2$ **Ref. 183**

Spaltung der *racem*-Form in die optischen Antipoden. Wie schon oben erwähnt, bereitete es beträchtliche Schwierigkeiten, eine bequeme Spaltungsmethode aufzufinden. Die primären und sekundären Salze von *Strychnin* und *Brucin* sowie die primären Salze von *Chinidin* und *Cinchonin*, die alle gut kristallisierten, gaben nur unbedeutende Spaltung. Die Salze von *Morphin* und *Phenäthylamin* zeigten geringe Kristallisationsfähigkeit. Wenn die Säure mit *Chinin* im Molekularverhältnis 1 : 1 zusammengebracht wurde, erhielt man kein primäres Salz, sondern ein sekundäres mit grossem Überschuss an (—)-Säure; durch Umkristallisation wurde allmählich maximale Drehung aber schlechte Ausbeute an aktiver Säure erhalten. Schliesslich ergab sich, dass die Umkristallisation des sekundären *Chinidinsalzes* aus Aceton eine sehr bequeme Methode zur Reindarstellung der (—)-Säure war; das Salz scheint fest gebundenes Kristallaceton zu enthalten und kann aus keinen anderen Lösungsmitteln umkristallisiert werden.

Aus der Mutterlauge nach der ersten Kristallisation des Chinidinsalzes erhält man beim Eindampfen ein nur träge kristallisierendes Salz, das etwa 80%ige (+)-Säure enthält. Man kann nun entweder dieses Salz aus verdünntem Alkohol auf konstante Drehung umkristallisieren oder auch die Säure in das sekundäre *Cinchoninsalz* überführen, in welchem Falle man die maximale Drehung etwas rascher erreicht; hinsichtlich des Arbeitsaufwandes scheinen beide Methoden gleichwertig zu sein, möglicherweise verdient die letztere den Vorzug. Da auch die ungünstigen Spaltungsmethoden ein gewisses Interesse beanspruchen können als Kontrolle, dass tatsächlich maximale Drehung erreicht wurde, seien im folgenden auch diese Versuchsreihen beschrieben.

Von den einzelnen Salzfraktionen wurde aus kleinen Mengen (etwa 0,3 g) die Säure mit Schwefelsäure in Freiheit gesetzt und mit Äther extrahiert. Die nach Verdampfen des Äthers verbleibende Säure wurde pulverisiert, im Exsikkator getrocknet, eingewogen und in verdünnter Sodalösung zu einem bestimmten Volumen gelöst. Die dabei erhaltenen Werte der maximalen Drehung sind wie gewöhnlich 1 bis 2 % niedriger als die, welche man an reiner umkristallisierter Säure erhält.

1. *Versuche mit Chinin.* 21 g Säure und 36 g Chinin (mit Kristallwasser) wurden zusammen in 1500 ml 50 %igem Alkohol gelöst; das erhaltene Salz wurde aus dem 80- bis 100-fachen seines Gewichts desselben Lösungsmittels auf konstante Drehung umkristallisiert.

Kristallisation	1	2	3	4	5	6	7	8
Salzmenge in g	30,0	22,5	16,4	12,8	10,4	9,0	7,4	6,2
$[\alpha]_D^{25}$ der Säure	−211°	−258°	−287°	−306°	−318°	−326°	−331	−329°

2. *Versuche mit Chinidin.* 10,4 g Säure und 32,4 g Chinidin wurden einzeln in je 150 ml siedendem, mit 10 % Wasser verdünntem Aceton gelöst (in wasserfreiem Aceton ist die Löslichkeit bedeutend geringer) und die Lösungen vereinigt. Da das erhaltene Salz fast vollständig unlöslich in Aceton zu sein schien, wurde es zur Umkristallisation in etwa seinem eigenen Gewicht siedendem Alkohol gelöst und durch Zusatz des 8- bis 10-fachen Volumens warmes Aceton gefällt. Dabei kristallisiert das Salz sehr schön in glänzenden Schuppen.

Kristallisation	1	2	3	4
Salzmenge in g	19,7	18,5	17,6	17,2
$[\alpha]_D^{25}$ der Säure	−305°	−324°	−331°	−332°

Die Mutterlauge nach der ersten Fraktion wurde zur Trockne eingeengt und das Salz durch Lösen in siedendem Alkohol und sehr vorsichtiges Fällen mit warmem Wasser umkristallisiert.

Kristallisation	1	2	3	4	5	6	7	8
Salzmenge in g	14,8	12,3	10,5	9,5	8,3	7,4	6,2	5,5
$[\alpha]_D^{25}$ der Säure	+274°	+300°	+313°	—	+323°	+329°	+329°	+331°

3. *Versuche mit Cinchonin.* 9,5 g aus Mutterlaugen erhaltene Säure mit einer Drehung von +270° wurden mit 17,5 g Cinchonin in etwa 250 ml 40 %igem Alkohol zusammengebracht. Das Salz wurde aus demselben Lösungsmittel umkristallisiert.

Kristallisation	1	2	3	4	5
Salzmenge in g	22,9	17,2	15,8	14,0	11,8
$[\alpha]_D^{25}$ der Säure	+290°	—	—	+332°	+332°

Die aktiven Säuren wurden aus den Salzen mit Schwefelsäure in Freiheit gesetzt, mit Äther ausgeschüttelt und zweimal aus Wasser umkristallisiert.

(+)-1, 2-Dithian-3, 6-dikarbonsäure. Aus 11,0 g Cinchoninsalz wurden nach dem Umkristallisieren 2,25 g Säure erhalten. Sie kristallisiert in kleinen stengeligen Prismen oder Nadeln, bisweilen der *racem*-Form nicht unähnlich. Schmelzpunkt etwa 257° (die Substanz bei 250° in das Bad eingebracht). Unmittelbar nach dem Schmelzen tritt Zersetzung ein.

0,1899 g Subst.: 15,79 ml 0,1161-n NaOH.
$C_6H_8O_4S_2$ (208,2) Äqu.-Gew. ber. 104,1 gef. 104,0
0,1212 g Subst. gelöst in verd. Sodalösung zu 10,00 ml: $2\alpha = +8{,}165°$
$[\alpha]_D^{25} = +336{,}6°$ $[M]_D^{25} = +701°$

Die aktive Säure ist ausnahmslos leichter Löslich als die *racem*-Form; in Wasser ist sie etwa zwölfmal löslicher.

25,00 ml bei 25,00° gesättigte wässrige Lösung: 12,87 ml 0,1161-n NaOH.
Löslichkeit bei 25°: 0,0299 Mol je Liter = 6,22 g je Liter.

Das sekundäre Cinchoninsalz der (+)-Säure, das im Laufe der Spaltung erhalten wurde, bildet stengelige Nadeln oder Prismen mit einem Molekül Kristallwasser.

0,6016 g Subst.: Gewichtsverlust über P_2O_5 0,0140 g.
1 H_2O ber. 2,21 gef. 2,33
0,1888 g kristallwasserfreie Subst.: 0,1092 g $BaSO_4$.
$C_6H_8O_4S_2$, 2 $C_{19}H_{22}N_2O$ (796,6) S ber. 8,05 gef. 7,95

Das sekundäre Chinidinsalz der (+)-Säure bildet kleine vierseitige Tafeln oder Schuppen. Es wurde nicht analysiert.

(—)-1, 2-Dithian-3, 6-dikarbonsäure. Aus 16,9 g Chinidinsalz wurden 3,15 g zweimal umkristallisierte Säure erhalten. Hinsichtlich Aussehen und Schmelzpunkt stimmt sie mit der Antipode vollständig überein.

0,1818 g Subst.: 15,08 ml 0,1161-n NaOH. — 0,1300, 0,2440 g Subst. in 14,40 g Aceton: $\Delta = 0{,}068°$, $0{,}135°$.
$C_6H_8O_4S_2$ (208,2) ber. Äqu.-Gew. 104,1 Molgew. 208,2
gef. » » 104,3 » 227, 215.

Die Drehung wurde in verschiedenen Medien bei den Wellenlängen 5893 ÅE und 5461 ÅE gemessen. Dabei wurde in derselben Weise verfahren, wie schon oben S. 9 bei der (—)-Dimerkapto-adipinsäure beschrieben. Die aus den Messungen berechneten Werte sind in Tab. 3 wiedergegeben. Bemerkenswert ist der Umstand, dass die Säure in saurer wässriger Lösung nach rechts dreht; da sie aber in allen übrigen untersuchten Medien nach links dreht, dürfte sie am zweckmässigsten als (—)-Form zu bezeichnen sein.

5,02 ml 0,0299-molare (gesättigte) wässrige Lösung wurde mit 1 ml 4-n Salzsäure versetzt und auf 10,00 ml aufgefüllt: $2\alpha = +0{,}79°$; $+0{,}95°$. — 0,1206 g in verd. Sodalösung: $2\alpha = -8{,}10°$; $-9{,}78°$. — 0,0764 g in Eisessig: $2\alpha = -0{,}70°$; $-0{,}835°$. — 0,1028 g in Essigester: $2\alpha = -3{,}50°$; $-4{,}245°$. — 0,1137 g in abs. Alkohol: $2\alpha = -4{,}43°$; $-5{,}37°$.

Tabelle 3.

	5893 ÅE		5461 ÅE	
	$[\alpha]_D^{25}$	$[M]_D^{25}$	$[\alpha]_D^{25}$	$[M]_D^{25}$
0,4-n Salzsäure	+126	+263	+152	+316
Vollständig neutralisierte wässr. Lösung	−335,8	−699	−405,5	−844
Eisessig	− 46	− 95	− 54,5	−114
Essigester	−170,2	−354,5	−206,5	−430
Alkohol (abs.)	−194,8	−405,5	−236,1	−491,5

Bei der Löslichkeitsbestimmung wurde derselbe Wert wie für die Antipode erhalten:
10,01 ml bei 25,00° gesättigte wässrige Lösung: 5,16 ml 0,1161-n NaOH.
Löslichkeit bei 25°: 0,0299 Mol je Liter = 6,23 g je Liter.

Das sekundäre Chininsalz der (—)-Säure bildet kleine glänzende Nadeln mit zwei Molekülen Kristallwasser.

0,8116 g Subst.: Gewichtsverlust über P_2O_5 0,0329 g.
2 H_2O ber. 4,04 gef. 4,05.
0,2543 g kristallwasserfreie Subst.: 0,1378 g $BaSO_4$.
$C_6H_8O_4S_2$, 2 $C_{20}H_{24}N_2O_2$ (856,6) S ber. 7,49 gef, 7,45.

Das sekundäre Chinidinsalz der (—)-Säure kristallisiert sehr schön in glänzenden Schuppen. Über Phosphorpentoxyd verliert es nur ganz unbedeutend an Gewicht; im Trockenschrank bei 110° nimmt das Gewicht sehr langsam ab, ohne innerhalb gewöhnlicher Zeit einen konstanten Wert anzunehmen. Da das Salz aus Aceton glatt kristallisiert, aus anderen Lösungsmitteln aber als Öl erhalten wird, enthält es wahrscheinlich fest gebundenes Kristallaceton. Die Schwefelbestimmungen lassen erkennen, dass ein Solvat vorliegt; die Analysenergebnisse stimmen am besten mit 1 ½ Mol Aceton überein.

35,13 mg Subst.: 81,87 mg CO_2 und 21,83 mg H_2O. — 0,2672, 0,1944 g Subst.: 0,1313, 0,0965 g $BaSO_4$.
$C_6H_8O_4S_2$, 2 $C_{20}H_{24}N_2O_2$, 1 ½ C_3H_6O (943,6)

ber.	C 64,22	H 6,94	S 6,80
gef.	» 63,56	» 6,95	» 6,75 6,82.

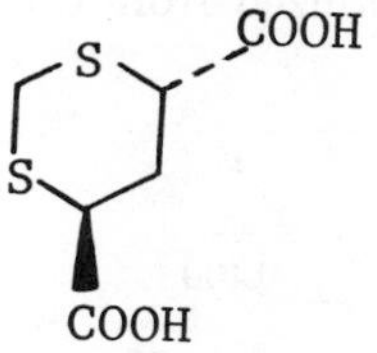

$C_6H_8O_4S_2$ Ref. 184

Preliminary experiments on resolution were performed with the common alkaloids and dehydroabietylamine as usual. The (−)-acid was best isolated as the neutral quinine salt while the (+)-acid could preferably be obtained from the acid strychnine salt.

(−)-*Acid.* (±)-Acid (15.0 g, 72 mmol) and quinine (54.5 g, 144 mmol) were dissolved with warming in 95% ethanol (450 ml). After 24 hours at 0° the precipitate was collected and recrystallized from dilute ethanol. After each crystallization, a small sample of the salt was decomposed and the rotatory power of the acid determined in abs. ethanol.

Cryst. no.	1	2	3	4
Ethanol (ml)	450	300	200	125
Water (ml)	—	150	125	125
Yield of salt (g)	37	22	18	14
$[\alpha]_D^{25}$ of acid	−90°	−148°	−158°	−159°

The acid was liberated with 2 M sulphuric acid and taken up in ether, which was dried and evaporated. The residue was recrystallized from ethyl acetate-petroleum ether yielding, 2,6 g (35%) m.p. 147–148°, $[\alpha]_D^{25}$159° (conc. 5 g/l, abs. ethanol). IR spectrum; 2920s (very broad), 1700s, 1435s, 1415s, 1265s, 1195s, 965m, 745m.

$C_6H_8O_4S_2$ (208.3)	calc.	C 34.6	H 3.8
	found	C 34.5	H 3.8

(+)-*Acid.* Acid (8 g, 38 mmol, $[\alpha]_D^{25}$ +80°) isolated from the mother liquors mentioned in the preceding paragraph, and strychnine (12.7 g, 38 mmol) were dissolved in 80% ethanol (175 ml). After 24 hours at 0°, the salt was filtered off and recrystallized from the same solvent mixture. After each recrystallization a small sample of the salt was decomposed and the rotatory power of the acid determined in abs. ethanol.

Cryst. no.	1	2	3
Solvent (ml)	175	120	100
Yield of salt (g)	12,5	9,0	7,2
$[\alpha]_D^{25}$ of acid	+138°	+157°	+159°

The acid was isolated and recrystallized as the enantiomer, yield 1.6 g (21%), m.p. 147–148°. The IR spectrum was identical with that of the enantiomer.

Solvent	Abs. ethanol	Methanol	Aceton	Water
Conc. (g/l)	4	3	5	6
$[\alpha]_D^{25}$	+159°	+158°	+146°	+148°

$C_6H_8O_4S_2$ (208.3)	calc.	C 34.6	H 3.8
	found	C 34.6	H 3.9

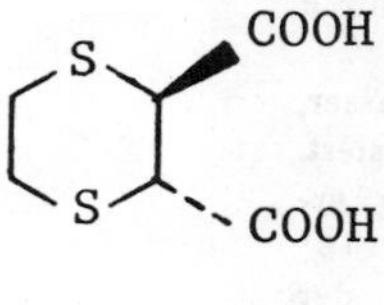

$C_6H_8O_4S_2$ Ref. 185

Preliminary experiments on resolution were carried out with several alkaloids. Dehydroabietylamine in methanol and cinchonidine in dil. ethanol gave (+)-acid of rather high activity. Further experiments revealed that the latter was preferable. (−)-Acid was isolated from salts of quinine and brucine in dil. methanol. Among these the latter gave the highest rotation.

(−)-*Acid.* Brucine (15.1 g, 38.5 mmol) and (±)-acid (8.0 g, 38.5 mmol) were each dissolved in 200 ml boiling methanol and the solutions mixed immediately. After 24 hours at room temperature, the deposited salt (17.5 g) was collected and recrystallized from dil. methanol or water. After each recrystallization, a small sample of the acid was liberated and the rotatory power determined in abs. ethanol.

Crystallization	1	2	3	4
ml methanol	400	700	800	—
ml water	—	500	350	1000
g salt obtained	15	11	7.2	5.8
$[\alpha]_D^{25}$ of the acid	−42°	−52°	−60°	−62.3°

The salt was treated with 2-M sulphuric acid and the acid extracted with ether. The extract was dried and evaporated to dryness without heating. The resulting crystals were dissolved in ethyl acetate and petroleum ether added. The turbid solution was left at 0° for 3 hours and the long slender needles of pure acid (1.9 g, 24%) collected, m.p. 173–175° (192°), $[\alpha]_D^{25} = -62.3°$ (abs. ethanol).

$C_6H_8O_4S_2$ (208.3) calc. S 30.79
found S 30.83

(+)-*Acid.* About 2 g (+)-acid of low activity was isolated from the mother liquors in the preceding paragraph. This acid (2 g) combined with another 6 g of pure (±)-acid (38.5 mmol) and cinchonidin (11.3 g, 38.5 mmol) were each dissolved in boiling ethanol (100 ml and 200 ml respectivily) and the hot solutions mixed immediately. After 24 hours, the salt was collected and recrystallized from dil. ethanol and water.

Crystallization	1	2	3	4	5
ml methanol	300	650	—	—	—
ml water	—	650	2200	1400	1200
g salt obtained	16	11	7.1	5.8	4.8
$[\alpha]_D^{25}$ of the acid	+27°	+54°	+60°	+62.1°	+62.3°

The acid was isolated and recrystallized as the enantiomer, yielding 2.0 g (25%), m.p. 173–175° (192°), $[\alpha]_D^{25} = +62.3°$.

$C_6H_8O_4S_2$ (208.3) calc. S 30.79
found S 30.70

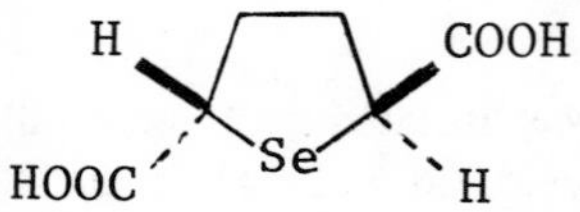

$C_6H_8O_4Se$ Ref. 186

(−)-Tetrahydroselenophen-α,α'-dicarbonsäure

10 g *racem*-Verbindung wurden in etwa 900 ccm heißem Wasser gelöst und mit einer Lösung von 17 g wasserfreiem Brucin in 75 ccm Alkohol versetzt. Beim Erkalten scheidet sich das primäre Brucinsalz der (−)-Säure in kurzen, derben Prismen aus. Das Salz ist wohlcharakterisiert und wasserfrei; da die Säure zudem eine bedeutend höhere Aktivität als das Brucin besitzt, kann man den Gang der Spaltung durch Drehungsmessungen an dem Salze verfolgen.

Das Salz wird aus der 30-fachen Menge heißem Wasser, das zweckmäßig mit etwas Alkohol versetzt wird, umkrystallisiert. Nach 2—3 Umkrystallisierungen ist maximale Drehung erreicht: das Salz zeigt dann in salzsaurer Lösung bei 16° [M] = − 773°. Ausbeute etwa 9 g: durch Aufarbeiten der Mutterlaugen nach den letzten Fraktionen lassen sich noch 1—2 g erhalten.

0,4089 g Subst.: 0,0524 g Se.

$C_6H_8O_4Se, C_{23}H_{26}O_4N_2$ (617,5) Ber. Se 12,83 Gef. Se 12,81

0,1742 g in 0,4 n-Salzsäure zu 10,02 ccm gelöst, zeigten in 2 dm-Rohr bei 16° die Drehung − 4,355°: [α] = − 124,4°, [M] = − 773°.

Das Brucinsalz wurde in verdünnter Schwefelsäure gelöst und 5 mal mit reinem Äther ausgeschüttelt. Nach freiwilligem Verdunsten des Äthers wurde der schwach bräunliche Rückstand unter Zusatz von Tierkohle aus wenig Wasser umkrystallisiert; durch nochmaliges Umkrystallisieren wurde die Drehung nicht verändert. Die aktive Säure ist im Gegensatz zur *racem*-Verbindung in Wasser leicht löslich; die Mutterlaugen enthalten deshalb nicht unbeträchtliche Mengen Säure, die man durch Eindampfen und wiederholtes Umkrystallisieren zurückgewinnen kann.

Die Säure bildet farblose, oft zentimeterlange Prismen, die gern etwas Mutterlauge einschließen. Die Messungen sind an fein gepulvertem, über P_2O_5 getrocknetem Material ausgeführt. Schmp. 173°.

0,1429 g Subst.: 11,64 ccm 0,1102 n-NaOH (Phenolphthalein).

$C_6H_8O_4Se$ (223,3) Ber. Äq.-Gew. 111,6 Gef. Äq.-Gew. 111,4

Die Drehungsmessungen sind an etwa 0,05 molare Lösungen in 2 dm-Rohr ausgeführt.

0,1143 g in 0,4 n-Salzsäure zu 10,02 ccm gelöst: α = − 6,63°.
0,1112 g mit NaOH zur Hälfte neutralisiert und zu 10,12 ccm verdünnt: α = − 6,67°.
0,1124 g mit NaOH neutralisiert und zu 10,02 ccm verdünnt: α = − 7,11°.

	[α]	[M]
Freie Säure	− 290,5°	− 649°
Primäres Salz . . .	− 303,5°	− 678°
Sekundäres Salz . .	− 317°	− 708°

Das saure Salz steht also bezüglich seiner Drehung genau in der Mitte zwischen der Säure und dem neutralen Salz. Die Aktivität wurde auch in einigen organischen Lösungsmitteln bestimmt:

0,1102 g in absolutem Alkohol zu 10,02 ccm gelöst: α = − 6,55°.
0,1114 g in Essigester zu 10,02 ccm gelöst: α = − 6,01°.
0,1113 g in Aceton zu 10,02 ccm gelöst: α = − 5,77°.

	[α]	[M]
In Alkohol . . .	− 298°	− 665°
„ Essigester . . .	− 270,5°	− 603,5°
„ Aceton	− 260°	− 580°

Die Drehungen der sauren und neutralen wäßrigen Lösungen waren nach Verlauf einer Woche unverändert.

(+)-Tetrahydroselenophen-α,α'-dicarbonsäure

Die Mutterlaugen nach den zwei ersten Fraktionen des Brucinsalzes werden, nach Ausfällen des Brucins mit NH_3, auf etwa 100 ccm eingedampft, mit Schwefelsäure angesäuert und mit Äther ausgeschüttelt. In dieser Weise erhält man etwa 3 g einer Säure, die 80—90% aktive Verbindung enthält. Sie wird in saures Chininsalz überführt; man nimmt dann etwas weniger als die berechnete Menge Alkaloid, weil sich sonst leicht etwas sekundäres Salz ausscheidet, was für die Spaltung ungünstig ist. Aus demselben Grunde ist es zweckmäßig, bei den folgenden Umkrystallisierungen einen Tropfen verdünnte Schwefelsäure zuzusetzen.

Für das Umkrystallisieren wird das Salz in 50prozent. Alkohol gelöst und die Lösung mit Wasser auf etwa 400 ccm verdünnt. Die Aktivität des Salzes ist ziemlich gering, da die Säure und das Chinin in entgegengesetzten Richtungen drehen; auch verliert es leicht einen Teil seines Krystallwassers. Um den Gang der Spaltung zu verfolgen, habe ich deshalb 0,25 g jeder Fraktion mit verdünnter Natronlauge zersetzt, das Chinin durch Ausschütteln mit Chloroform entfernt und die Aktivität nach Ansäuern mit HCl gemessen. Die Konzentration der Tetrahydroselenophen-dicarbonsäure in der Lösung wurde sodann durch Titrieren mit Jod bestimmt.

Nach 1—2 maligem Umkrystallisieren des Chininsalzes ist die molekulare Drehung der Säure auf 650° ± 7° gestiegen und bleibt dann bei weiterem Umkrystallisieren innerhalb der Fehlergrenzen unverändert. Das Salz bildet feine, glänzende Nadeln mit $1^1/_2$ Mol Wasser, das über P_2O_5 oder im Trockenschrank bei 90° völlig entweicht.

0,7872 g Subst.: 0,0377 g H_2O. — 0,8520 g Subst.: 0,0405 g H_2O.
$C_6H_8O_4Se$, $C_{20}H_{24}O_2N_2$, $1^1/_2H_2O$ (574,5)
Ber. H_2O 4,70 Gef. H_2O 4,79, 4,75

0,2769 g getrockneter Subst.: 0,0400 g Se.
$C_6H_8O_4Se$, $C_{20}H_{24}O_2N_2$ (547,5) Ber. Se 14,47 Gef. Se 14,45

Die Säure wurde, wie die (—)-Verbindung in Freiheit gesetzt, mit Tierkohle entfärbt und zweimal aus Wasser umkrystallisiert. Sie ist dem optischen Antipode äußerlich völlig ähnlich. Schmp. 173°.

0,1663 g Subst.: 13,51 ccm 0,1102 n-NaOH (Phenolphthalein).
$C_6H_8O_4Se$ (223,3) Ber. Äqu.-Gew. 111,6 Gef. Äqu.-Gew. 111,7
0,1118 g in 0,4 n-Salzsäure zu 10,02 ccm gelöst, zeigten in 2 dm-Rohr $\alpha = +6{,}50°$: $[\alpha] = +291{,}5°$, $[M] = +650{,}5°$.

Dieselbe Lösung wurde auch bei 15° gemessen; von der Dichteänderung der Lösung kann abgesehen werden.

$\alpha = +6{,}55°$: $[\alpha]^{15} = +293{,}5°$, $[M]^{15} = +655{,}5°$.

Mit Silbernitrat gibt die neutrale Lösung der Säure einen flockigen, sehr voluminösen Niederschlag, der in siedendem Wasser spurenweise löslich ist. Kupfersalze erzeugen eine grüne Färbung, die auf Komplexbildung deutet; die Drehung wird jedoch von der Gegenwart von Kupfersalzen nicht merklich beeinflußt. Vielleicht wäre ein Effekt in anderen Spektralgebieten zu suchen.

Beim Zusammenbringen gleicher Mengen rechts- und linksdrehender Säure in wäßriger Lösung krystallisierte eine Säure von dem Schmp. 194° aus. Sie war der früher beschriebenen *racem*-Form (Schmp. 195°) völlig ähnlich.

Während der vorläufigen Spaltungsversuche wurden noch folgende Salze hergestellt und analysiert.

Sekundäres Chininsalz der *racem*-Säure. Längliche Prismen oder Nadeln mit 2 Mol Wasser, die bei 90° entweichen. Das Salz enthielt einen kleinen Überschuß an (—)-Säure.

0,5874 g Subst.: 0,0231 g H_2O.
$C_6H_8O_4Se$, $2(C_{20}H_{24}O_2N_2)$, $2H_2O$ (907,6) Ber. H_2O 3,97 Gef. H_2O 3,93
0,5556 g getrocknete Subst.: 0,0505 g Se.
$C_6H_8O_4Se$, $2(C_{20}H_{24}O_2N_2)$ (871,6) Ber. Se 9,09 Gef. Se 9,09

Sekundäres Brucinsalz der (—)-Säure. Längliche, dünne Tafeln oder Schuppen mit gerader Auslöschung. Der Wassergehalt liegt zwischen $4^1/_2$ und 5 Mol; trotz vieler Versuche gelang es nicht, ein stöchiometrisch wohldefiniertes Hydrat zu erhalten. Das Salz wurde in wasserfreiem Zustande analysiert.

0,4907 g Subst.: 0,0884 g Se. — 0,5026 g Subst.: 0,0398 g Se.
$C_6H_8O_4Se$, $2(C_{23}H_{26}O_4N_2)$ (1011,7) Ber. Se 7,83 Gef Se 7,83, 7,80

Upsala, April 1931.

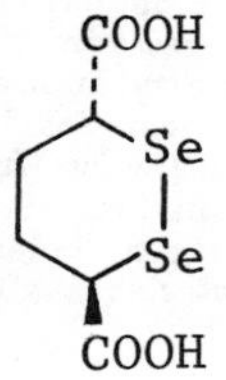

$C_6H_8O_4Se_2$ Ref. 187

Die optischen Antipoden lassen sich am besten über das sekundäre Chininsalz bzw. das primäre Strychninsalz darstellen. Spaltungsversuche mit anderen Alkaloiden (Brucin, Cinchonin, Chinidin, Morphin und Spartein) gaben weniger zufriedenstellende Resultate. Grösse und Vorzeichen der Drehung sind vom Lösungsmittel stark abhängig; im folgenden wird als (+)-Säure diejenige bezeichnet, die in den meisten Lösungsmitteln nach rechts dreht. Die Drehungsmessungen wurden mit der Wellenlänge des Na-Lichts bei 25,0° ausgeführt.

(—)-Zyklo-tetramethylen-diselenid-dicarbonsäure. $^1/_{20}$ Mol Racemsäure wird mit NaOH neutralisiert und mit $^1/_{20}$ Mol Chininchlorhydrat versetzt; das vereinigte Volumen der Lösungen soll etwa 600 ccm betragen. Etwa 45 % der Säure kristallisieren als sekundäres Chininsalz; die im Salz gebundene Säure ist linksdrehend und ist zu etwa 50 % aktiv. Die Mutterlauge wird auf die zweite Antipode verarbeitet. Das Chininsalz ist äusserst schwerlöslich in Wasser und in den meisten organischen Lösungsmitteln; es kann aus Methylalkohol umkristallisiert werden, wobei jedoch Anreicherung an rechtsdrehender Säure stattfindet. Es wurde daher beim Umkristallisieren folgenderweise verfahren. Das Salz wurde in heissem Wasser suspendiert und durch Zusatz von verdünnter HCl aus einer Bürette vollständig in Lösung gebracht. Dann wurde tropfenweise unter lebhaftem Umrühren Ammoniak zugesetzt in einer Menge entsprechend 75 % der zugesetzten Salzsäure. Dabei kristallisieren 60—70 % der Salzes; Bildung von primärem Chininsalz wurde nicht beobachtet. Nach jedesmaligem Umkristallisieren wurde eine kleine Menge Säure in Freiheit gesetzt durch Ansäuern mit Schwefelsäure und Ausschütteln mit Äther. Nach Eindampfen des Äthers wurde die Säure aus möglichst wenig Wasser umkristallisiert, in verdünnter Sodalösung gelöst und im Polarimeter untersucht. Der Verlauf der Spaltung ergibt sich aus folgenden Angaben.

Kristallisation:	1	2	3	4	5	6	7
[α]	−179°	−249°	−284°	−317°	−344°	−351°	−351°

Die Ausbeute ist verhältnismässig schlecht; aus insgesamt 29 g wurde nur etwa 1 g reine aktive Säure erhalten. Sie kristallisiert aus Wasser in stengeligen Nadeln mit undeutlich ausgebildeten Flächen. Aus Äther erhält man kompakte Kristalle von kurzprismatischem Habitus. Die Farbe ist rein gelb; im Polarisationsmikroskop beobachtet man Dichroismus (gelb bis fast farblos). Die aktive Säure ist leichter löslich als die Racemform; sie zersetzt sich ohne definierten Schmelzpunkt beim Erhitzen auf etwa 200°.

0,1986 g Subst.: 11,60 ccm 0,1107-n NaOH.
$C_6H_8O_4Se_2$ (302,0) Ber. Äqu.-Gew. 151,0 Gef. Äqu.-Gew. 150,8
0,0980 g in verdünnter Sodalösung zu 10,03 ccm gelöst gaben im 2 dm Rohr $\alpha = -6{,}585°$; $[\alpha] = -351°$; $[M] = -1059°$

Das *sekundäre Chininsalz* bildet feine Nadeln von blassgelber Farbe; es enthält zwei Mol Wasser, die bei 100° abgegeben werden.

0,6123 g Subst.: 0,0226 g H_2O
$C_6H_8O_4Se_2$, 2 $(C_{20}H_{24}N_2O_2)$, $2H_2O$ (986,6) Ber. H_2O 3,65 %; Gef. H_2O 3,63 %
0,3983 g getrocknete Substanz: 0,0655 g Se.
$C_6H_8O_4Se_2$, 2 $(C_{20}H_{24}N_2O_2)$ (950,6) Ber. Se 16,62 % Gef. Se 16,65 %

(+) Zyklo-tetramethylen-diselenid-dicarbonsäure. Die aus der Mutterlauge von der ersten Kristallisation des Chininsalzes in Freiheit gesetzte Säure ([α] in Sodalösung etwa +150°) wird im hundertfachen Gewicht heisses Wasser gelöst und mit 70 % der zur Bildung des primären Salzes nötigen Menge Strychnin versetzt. Es ist nicht angezeigt direkt von der Racemform

auszugehen, da in diesem Falle das kristallisierende Salz durch freie Säure von entgegengesetzter Drehungsrichtung verunreinigt ist. Das primäre Strychninsalz wird aus Wasser umkristallisiert; ein Teil Salz löst sich in etwa 100 Teilen siedendem Wasser. Nach jedesmaligem Umkristallisieren wurde aus 0,2—0,3 g Salz die Säure in Freiheit gesetzt und wie oben bei der (—)-Säure beschrieben im Polarimeter untersucht. Der Gang der Spaltung ergibt sich aus den folgenden Drehungswerten.

Kristallisation: 1   2   3   4   5   6
$[\alpha]$ +301° +334° +347,5° — +351,5° +351°

Aus der oben genannten Menge Racemsäure (29 g) wurden dieserart 6 g reine (+)-Säure erhalten. Die in den Mutterlaugen der Chinin- und Strychninsalze verbleibende unvollständig gespaltete Säure konnte zum grössten Teil zurückgewonnen und für Rückbildung der Racemform oder weitere Spaltung verwertet werden.

Die (+)-Zyklo-tetramethylen-diselenid-dicarbonsäure wurde aus möglichst wenig Wasser umkristallisiert. Sie gleicht äusserlich vollständig ihrer optischen Antipode.

0,2018 g Subst.: 12,07 ccm 0,1107-n NaOH. — 0,2006, 0,3175 g Subst. in 13,30 g Aceton: $\Delta = 0,087°$, 0,139°
$C_6H_8O_4Se_2$ (302,0) Ber. Äqu.-Gew. 151,0 Molgew. 302,0
   Gef.   »   151,0   »   296; 294
50,0 ccm der bei 25° gesättigten wässrigen Lösung verbrauchten 7,08 ccm 0,1107-n NaOH. Löslichkeit bei 25°: 0,00784 Mol/l entspr. 2,37 g/l.

Die Drehung wurde in verschiedenen Lösungsmitteln untersucht. Die freie Säure besitzt in wässriger Lösung Linksdrehung, in allen anderen untersuchten Fällen wurde Rechtsdrehung gefunden. Alle Messungen wurden im 2 dm-Rohr ausgeführt.

0,1472 g Säure in verdünnter Sodalösung zu 10,03 ccm gelöst: $\alpha = +10,30°$. — Von der bei 25° gesättigten wässrigen Lösung (siehe oben) wurden 10,0 ccm mit 2 ccm 1,5-n HCl versetzt und auf 15,00 ccm verdünnt: $\alpha = -0,77°$. — 0,1120 g Säure in Alkohol zu 10,02 ccm gelöst: $\alpha = +3,99°$. — 0,1135 g Säure in Äther zu 10,02 ccm gelöst: $\alpha = +5,13°$. — 0,1711 g Säure in Aceton zu 15,00 ccm gelöst: $\alpha = +3,895°$.

Lösungsmittel	$[\alpha]$	$[M]$
Verdünnte Sodalösung	+351°	+1060°
0,2-n Salzsäure	—244°	—737°
Alkohol	+178,5°	+539°
Äther	+226,5°	+684°
Aceton	+170,5°	+515°

$$CF_3-CH(CH_3)-OCH_2CH_2COOH$$

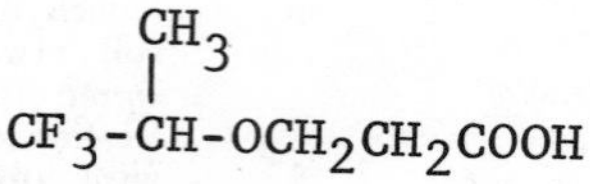

Ref. 188

Resolution of (±)-β-(1-*Trifluoromethylethoxy*)*propionic Acid.*—*Preparation of quinine salt.* (±)-β-(1-Trifluoromethylethoxy)propionic acid (558 g., 3 moles), dissolved in carbon tetrachloride (6 l.), was heated to 60°, and dehydrated quinine (924 g., 2·85 moles) added, with stirring at such a rate as to keep the liquid boiling. The resulting solution was cooled quickly in water, and the *quinine salt* that crystallised was separated in a basket centrifuge and air-dried (1090 g., 75%) (Found: N, 5·4. $C_{29}H_{33}F_3N_2O_5$ requires N, 5·5%).

Separation of isomeric quinine salts. The air-dried salt was treated with chloroform–carbon tetrachloride (1 : 3; 2·7 l.), and the whole equilibrated by being tumbled slowly for 3 days at room temperature. The residual salt was filtered in the centrifuge, and washed in the basket with carbon tetrachloride (300 ml.); the air-dried salt weighed 766 g. The series of fractional solutions was continued by using *ca.* 2·5 ml. mixed solvent, with 0·3 ml. carbon tetrachloride for washing, per g. of salt treated. Progress of the solution was followed from measurements of the specific rotation, in solution in n-butyl acetate, of the alkoxy-acid liberated from samples of residual salt and the corresponding extraction liquid. The specific rotations converged after seven extractions. The final residue, the *quinine salt* of (+)-β-(1-*trifluoromethylethoxy*)*propionic acid* (205 g., 27%), had $[\alpha]_D^{25}$ —129·8° (c 5, in MeOH) (Found: N, 5·4%).

(+)-β-(1-*Trifluoromethylethoxy*)*propionic Acid.*—The (+)-quinine salt (200 g., 0·39 mole) was stirred with 8% hydrochloric acid (500 ml., 1·1 mole) until it had decomposed. The alkoxy-acid was extracted with ether, the extract freed from quinine by washing it with acid (2N) and water, dried (Na_2SO_4), the solvent removed, and the residue fractionated. (+)-β-(1-*Trifluoromethylethoxy*)*propionic acid* (64·4 g., 89%) distilled steadily at 111·2°/11 mm. (corr.). It had n_D^{25} 1·3727, d_4^{25} 1·2705, $[\alpha]_D^{25}$ +1·70° (homogeneous), +1·85° (c 10, in n-butyl acetate). $\lambda_{max.}$ (cyclohexane) 213 mμ (ε 40) (Found: Equiv., 186. $C_6H_9F_3O_3$ requires Equiv., 186·1).

$$CH_3CH_2-C(CH_3)(CN)-COOH$$

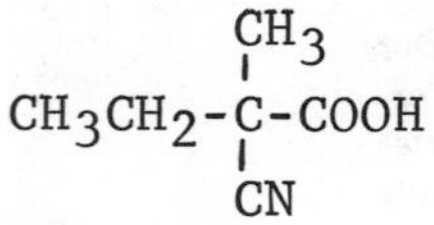

Ref. 189

Resolution. (i) After three recrystallisations, the brucine salt of the acid, prepared in the usual way, formed large prisms, m. p. 114—116°, from which only the (±)-acid could be obtained. (ii) A similarly recrystallised quinine salt (prisms) of m. p. 189—190° gave the (±)-acid on decomposition. (iii) The cinchonidine salt (needles) of m. p. 203—205° also afforded no resolution. (iv) No crystalline salt could be obtained with quinidine. (v) To the (±)-acid (4 g.) dissolved in aqueous ethanol (30 c.c. of 50%) was added strychnine (10·5 g.), and the whole warmed until dissolution was complete. After several weeks at room temperature, crystals (2·8 g.) of ill-defined crystalline form, m. p. 120—127°, were obtained.

To the (±)-acid (31 g.) in acetone (100 c.c.) was added strychnine (81·5 g.). The solution was concentrated to about half its volume, and seeded with the strychnine salt obtained earlier. After several weeks, when the solvent had almost completely evaporated, the needle-like crystals (15 g.) of m. p. 127—130° were removed from the viscous mother-liquor by decantation and quickly washed with small quantities of acetone. This salt was twice recrystallised from minimum quantities of acetone. Decomposition with dilute ammonia gave (+)-*ethylmethylcyanoacetic* (α-*cyano-α-methylbutyric*) *acid*, $[\alpha]_D^{20}$ +3·85° (l, 2; c, 5·99 in ethanol) (Found: equiv., 130·3. $C_6H_9O_2N$ requires equiv., 127·1).

$C_6H_9NO_3$

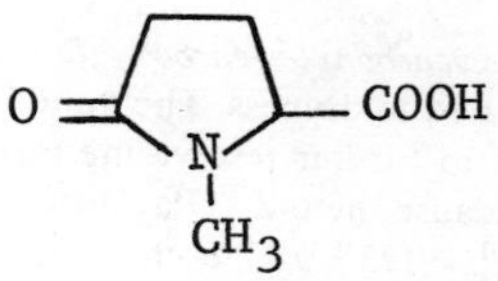

Ref. 190

(+)-N-Methyl-pyrrolidon-(5)-carbonsäure-(2) (XI) aus DL-N-Methyl-pyrrolidon-(5)-carbonsäure-(2) (VIa). 4,0 g DL-N-Methyl-pyrrolidoncarbonsäure (VIa) und 9,08 g Chinin wurden in 30 cm³ Alkohol gelöst und 150 cm³ Äther zugefügt. Aus der auf −10° gekühlten Lösung kristallisierten nach einiger Zeit 7,9 g Chininsalz vom Smp. 184—185° aus. Das mehrmals aus Alkohol-Äther umkristallisierte Präparat schmolz bei 185°. Das Analysenpräparat wurde 20 Std. bei 50° im Hochvakuum getrocknet. $[\alpha]_D = -124°$ (c = 0,52 in Alkohol).

$C_{26}H_{33}O_5N_3$, ½ H_2O Ber. C 65,53 H 7,19% Gef. C 65,49 H 7,45%

7,9 g (16,9 Millimol) Chininsalz wurden mit 25 cm³ 1-n. Natronlauge und 50 cm³ Chloroform geschüttelt. Das Chloroform wurde abgetrennt und die wässerige Lösung zur vollständigen Entfernung des Chinins noch dreimal mit Chloroform ausgeschüttelt. Die wässerige Lösung wurde hierauf durch 30 cm³ Wofatif KS laufengelassen. Nach dem Abdampfen des Wassers blieben 2,6 g kristallisierte rohe Säure XI zurück, die aus Alkohol oder Aceton umkristallisiert werden konnte. Das Analysenpräparat von XI, vom Smp. 158°, wurde 24 Std. bei 60° im Hochvakuum getrocknet. $[\alpha]_D = +7,6°$ (c = 2,6 in Wasser).

$C_6H_9O_3N$ Ber. C 50,34 H 6,34% Gef. C 50,29 H 6,36%

$C_6H_9NO_6$

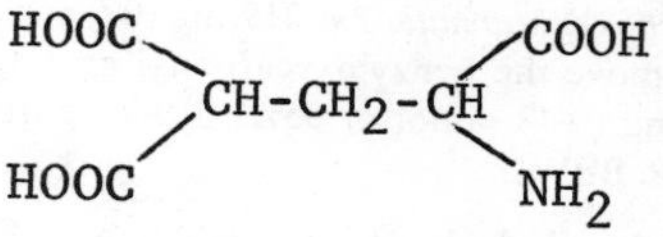

Ref. 191

General procedures and materials. See *Märki & Schwyzer* [8]. Additional solvents (volume parts) for thin-layer chromatography (TLC.) are: I: 2-propanol/water/pyridine 7:6:6; CM: chloroform/methanol 1:1; EMA: acetic acid/methanol/acetone 7:2:1.

Diastereomeric salts of Gla derivatives with optically active bases were decomposed as follows: The sulfonic acid resin Amberlyst 15 (*Rohm and Haas Co.*, Philadelphia) in its acid cycle was washed repeatedly with methanol. The alcaloid salts were dissolved in methanol and chromatographed over columns containing 10 ml of the resin per mmol salt. The columns were washed with 5 times their volume of methanol. This eluate was evaporated. The viscous residue consisted of the desired amino acid derivative.

The absolute configuration and the optical purity of the Gla derivatives were determined as follows: 0.1 mmol of the product was suspended in 1 ml of 6N HCl containing minute amounts of phenol as antioxidant and the ampoule sealed in a high vacuum. Hydrolysis was carried out at 110° for 3 h and the solution evaporated over P_2O_5 and KOH at about 0.001 Torr. The residue was not further purified and its optical rotation measured. Part of the solution was subjected to thin-layer electrophoresis to make sure that it contained pure glutamic acid. The specific rotations were calculated on the basis of the weights of the non-hydrolysed educts. The optical purity in per cent is defined as: ($[\alpha]_D$ found · 100)/($[\alpha]_D$calc.) wherein the calculated value is derived either from the literature or from measurements with the pure compounds, or both.

The amino acid analyses were carried out with a *Beckman* Model 121 Amino Acid Analyser using a column of *Beckman* AA 15 resin, 0.6 × 52 cm.

γ,γ'-Di-t-butyl DL-N-*benzyloxycarbonyl-γ-carboxyglutamate* **(1)**. This compound was prepared in crystalline form according to [8]. It was further characterized by its cyclohexylamine salt: colourless needles from ethyl acetate, m.p. 119–121°, $[\alpha]_D^{20} = 0°$ (c = 1.1, $CHCl_3$).

Quinine salt of γ,γ'-di-t-butyl D(−)-N-*benzyloxycarbonyl-γ-carboxyglutamate.* 2.66 g (6.08 mmol) of DL-Z · Gla(OtBu)$_2$ · OH **(1)** and 1.97 g (6.08 mmol) of quinine (m.p. 175°; $[\alpha]_{546}^{20} = -154 \pm 3°$, c = 1.5, $CHCl_3$) were dissolved in 50 ml of hot ethyl acetate and kept at RT. for 12 h. Crystallization was completed by cooling to 4° for 15 h. The colourless needles were gathered by filtration and washed with ice-cold ethyl acetate: 1.82 g (2.4 mmol, 40% of the theoretical yield), m.p. 132–133°; $[\alpha]_D^{20} = -76.8°$ and $[\alpha]_{546}^{20} = -94.1°$ (both c = 1, $CHCl_3$).

The filtrate and washings were evaporated to dryness, the residue dissolved in CCl_4 and treated with pentane. Another 204 mg of the salt (4.4% yield) were obtained.

γ,γ'-Di-t-butyl D(−)-N-*benzyloxycarbonyl-γ-carboxyglutamate* **(1a)**. The quinine was removed according to the general procedure. The viscous residue was crystallized from CCl_4/pentane in the form of dense aggregates of needles. 1.63 g (2.1 mmol) of the quinine salt yielded 781 mg (1.79 mmol, 83%) **1a**; m.p. 86–88°; Rf 0.68 CM, 0.37 CME. – NMR.: 10.05 (*broad signal*, 1H); 7.4 (*s*, 5H); 5.6–5.3 (*broad signal*, 1H); 5.15 (*s*, 2H); 4.7–4.3 (*m*, 1H); 3.35 (*t*, 1H); 2.6–2.2 (*many signals*, 2H); 1.45 (2*s*, 18H).

$C_{22}H_{31}NO_8$ (437.5) Calc. C 60.40 H 7.14 N 3.20% Found C 60.32 H 7.19 N 3.28%

[8] *W. Märki & R. Schwyzer*, Helv. *58*, 1471 (1975); *W. Märki, M. Oppliger & R. Schwyzer*, Helv. *59*, 901 (1976).

γ,γ'-Di-t-butyl L(+)-*N-benzyloxycarbonyl-γ-carboxyglutamate* (**1b**). The filtrate and washings of the crystalline quinine salt were evaporated to dryness. The quinine was removed from the residue according to the general procedure. Attempts to further resolve the mixture of L- and DL-acids *via* the quinidine salts were only partly successful, because the L-Z · Gla(O*t*Bu)$_2$ · OH salt was only obtained as a gel (m.p. 83-93°, $[\alpha]_D^{20} = +100 \pm 0.5°$ and $[\alpha]_{546}^{20} = 123 \pm 0.5°$, both $c = 1$, $CHCl_3$).

Therefore the acid mixture was simply recrystallized. The L-isomer appeared from CCl_4/pentane mixtures at 4° as dense aggregates of needles and from diisopropyl ether/pentane at 4° usually as colourless needles, m.p. 87-89°, in seldom cases as hard, thick crystals. The TLC. and NMR. results were identical with those of the D-isomer, **1a**. The (−)-ephedrine salt of **1b** (below) yielded an identical product in 62% over-all yield from **1**.

$C_{22}H_{31}NO_8$ (437.5) Calc. C 60.40 H 7.14 N 3.20% Found C 60.23 H 7.11 N 3.13%

The cyclohexylamine salt was obtained as colourless needles from ethyl acetate, m.p. 122-126°, $[\alpha]_D^{20} = +11.7°$, $[\alpha]_{546}^{20} = +14.7°$ (both $c = 1.1$, $CHCl_3$).

(−)-*Ephedrine salt of γ,γ'-di-t-butyl* L(+)-*N-benzyloxycarbonyl-γ-carboxyglutamate*. 437 mg (6.08 mmol) **1** were dissolved in 0.3 ml ethyl acetate and combined with a solution of 83 mg (0.5 mmol) (−)-ephedrine in 0.2 ml ethyl acetate. The clear mixture was diluted with much pentane and kept at 4° for 2 days. The solid precipitate was recrystallized from ethyl acetate/pentane: 173 mg (58%) large, colourless needles, m.p. 122-124°. $[\alpha]_{546}^{20} = -7.7°$, $[\alpha]_D^{20} = -6.7°$ (both $c = 1.2$, $CHCl_3$); $[\alpha]_{546}^{20} = -17.7°$, $[\alpha]_D^{20} = -14.5°$ (both $C = 1.0$, MeOH).

$C_{32}H_{46}N_2O_9$ (602.7) Calc. C 63.77 H 7.69 N 4.65 C 62.03 H 7.73 N 4.52%
with 2.73% EtOAc Found C 62.03 H 7.71 N 4.60%

The product was decomposed to **1b** with Amberlyst 15, yield 85%. Treatment of the mother liquors with (−)-ephedrine resulted in another 13% of pure salt, or 11% of **1b**. Total yield of **1b** from **1** = 62%.

γ,γ'-Di-t-butyl D(−)-*γ-carboxyglutamate*, **2a**. 219 mg (0.5 mmol) **1a** were catalytically hydrogenated according to [8] in order to remove the benzyloxycarbonyl group. The crude solid product was washed with diethyl ether, yield 144 mg (0.48 mmol or 95%). It was pure without recrystallization from water, m.p. 165.5-167.5 (dec.); Rf 0.46 BEWI.

$C_{14}H_{25}NO_6$ (303.36) Calc. C 55.43 H 8.31 N 4.62% Found C 55.42 H 8.43 N 4.76%

γ,γ'-Di-t-butyl L(+)-*γ-carboxyglutamate*, **2b**. This isomer was prepared in exactly the same manner as its antipode, **2a**. Yield 96% (precipitated from methanol with diethyl ether), pure solid, m.p. 166-167° (dec.). Crystallization from water with a small amount of methanol resulted in colourless needles, yield 70%; m.p. 170-171° (dec.).

$C_{14}H_{25}NO_6$ (303.36) Calc. C 55.43 H 8.31 N 4.62% Found C 55.23 H 8.30 N 4.55%

D(−) *γ-Carboxyglutamic acid monoammonium salt* (**3a**). 94 mg (0.31 mmol) of D-H · Gla(O*t*Bu)$_2$ · OH (**2a**) were dissolved in 1 ml of cold, concentrated hydrochloric acid and kept at 0° for 15 min. The solution was evaporated at 0.001 Torr over P_2O_5 and KOH. The hygroscopic residue was quickly dissolved in 1 ml of glacial acetic acid, treated with a slight excess of ammonium acetate (35 mg= 0.42 mmol), and lyophilized. The residue was repeatedly triturated with ethanol: yield 53 mg (0.25 mmol, 82%) colourless solid; m.p. 157-159° (decomposition with evolution of CO_2). At pH 6.4 and 40 V/cm, the compound migrates about 50 mm towards the anode in 50 min. This corresponds to about 1.4 times the distance travelled by aspartic acid under the same conditions. The electrophoretic and TLC. aspects are those of a pure compound, Rf 0.23 I. - ^{1}H-NMR. confirmed the expected structure and the analytical purity (see *Fig. 1*).

$C_6H_{12}N_2O_6$ (208.1) Calc. C 34.62 H 5.81 N 13.46% Found C 34.63 H 6.00 N 13.66%

L(+)-*γ-Carboxyglutamic acid* (**3b**). This isomer was obtained by dissolving **2b** in trifluoroacetic acid/water, 9:1 (*v/v*), evaporating the solution after 90 min at 20°, azeotropic removal of the reagent with toluene, triturating with ether, and recrystallizing from water/ethanol, 1:1 (*v/v*). Colourless, very small crystals, yield 75%, m.p. 167-167.5° (dec.).

$C_6H_9NO_6$ (191.14) Calc. C 37.70 H 4.75 N 7.32% Found C 37.59 H 4.85 N 7.29%

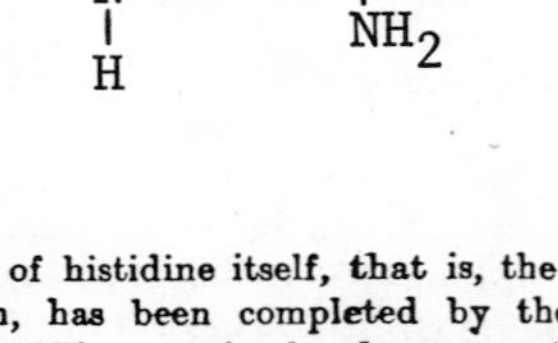

$C_6H_9N_3O_2$ Ref. 192

The synthesis of histidine itself, that is, the naturally occurring lævo-modification, has been completed by the resolution of the racemic variety. When equimolecular amounts of *r*-histidine and *d*-tartaric acid were crystallised from water, there separated first d-*histidine* d-*hydrogen tartrate* (melting point 234° (corr.); $[\alpha]_D$ +13·3°). This salt is sparingly soluble in water, and is obtained in a yield amounting to about 90 per cent. of the theoretical. The hitherto unknown d-*histidine* was regenerated from it, and found to melt at 287—288° (corr.), and to have $[\alpha]_D$ +39·3°. The mother liquors from the *d*-base-*d*-acid then deposited the easily soluble but magnificently crystalline l-*histidine* d-*hydrogen tartrate* (melting point 172—173° (corr.); $[\alpha]_D$ +17·4°) in a yield amounting to nearly 80 per cent. of the theoretical. The *l*-histidine regenerated from this was found to have $[\alpha]_D$ −36·6°, and was therefore further purified by conversion into the sparingly soluble l-*histidine* l-*hydrogen tartrate* (melting point 234° (corr.); $[\alpha]_D$ −12·1°). After regeneration from this salt, *l*-histidine melted at 287—288° (corr.), and had $[\alpha]_D$ −38·1°.

The specific rotatory power is thus substantially in agreement with that found for natural histidine, $[\alpha]_D$ −39·7° by Kossel and Kutscher (*Zeitsch. physiol. Chem.*, 1899, **28**, 382).

A resolution procedure for the above compound using the same resolving agent may also be found in: R. M. Conrad and C.P. Berg, J. Biol. Chem., 117, 351 (1937). These authors report that the optical rotation of d-histidine monohydrochloride determined on a 5 per cent solution in water containing 3 equivalents of HCl, was $[\alpha]_D^{20} = -8.29°$.

For the experimental details of enzymatic methods for the resolution of this compound, see L. Levintow, V. E. Price and J. P. Greenstein, J. Biol. Chem., 184, 55 (1950); S. M. Birnbaum and J. P. Greenstein, Arch. Biochem. Biophys., 39, 108 (1952); and Ref. 18n.

$C_6H_9N_3O_2$ Ref. 192a

In carefully working up hydrolyzed hæmoglobin not only were *l*- and *dl*-histidine isolated, but the author has also found *d*-histidine. It will be demonstrated below that the formation of *d*-histidine is due to a spontaneous resolution of the racemic compound formed during the hydrolysis. Thus by fractional crystallizations of a mixture of pure *l*- and *dl*-histidine monohydrochlorides from 1·5 parts of water, the *d*-isomeride was produced as follows:

(*a*) On rapidly cooling the hot solution to 20°, then rapidly filtering, crystals of more or less optically pure *l*-histidine monohydrochloride were separated.

(*b*) By adding alcohol and ether to the filtrate and further cooling to 0°, almost pure *dl*-salt was obtained.

(*c*) On concentrating the mother-liquor and treating as in (*a*) almost optically pure *d*-histidine monohydrochloride was deposited.

(*d*) The filtrate when treated as in (*b*) again furnished the inactive salt.

By recrystallizing each of these fractions, optically pure substances were easily obtained.

On comparing the quantitative results of the different recrystallizations it was found that the amount of racemic compound had diminished and was partly resolved into the optically isomeric salts.

In treating a batch of 1000 g. consisting of a mixture of 660 g. of *dl*-histidine hydrochloride and 340 g. of *l*-histidine hydrochloride, the following yields were finally obtained: 400 g. of *l*-histidine monohydrochloride, $[\alpha]_D^{20}$ $-38°$ to $-39{\cdot}5°$,[2] 540 g. of *dl*-histidine monohydrochloride, $[\alpha]_D^{20}$ 0°, 46 g. of *d*-histidine monohydrochloride, $[\alpha]_D^{20}$ $+38°$ to $+39{\cdot}5°$.

On repeating this procedure successively it was possible to prepare large quantities of the *d*-isomeride which has only been produced hitherto by biological methods or by combining with optically active substances.

It is interesting to note that the original *dl*-hydrochloride, as well as that recovered, corresponds to the formula $C_6H_9O_2N_3,HCl,2H_2O$, whilst the optically active salts fix only one molecule of water. Consequently the inactive compound seems to be a racemate and not an externally compensated mixture.

[1] A summary of the cases hitherto observed is given by L. Anderson and D. W. Hill, J.C.S., 1928, 993.

[2] After addition of one molecule of NaOH, $c = 3183$ (as free base).

HOOC-CH-(CH2)3COOH
       |
      N3

$C_6H_9N_3O_4$ Ref. 193

Separation of the enantiomers of **DL-2-*azidoadipic acid***

1. D-isomer

a) DL-2-azidoadipic acid (153.1 g, 0.817 mole) was dissolved in 1.5 l of hot acetone. After addition of 270 g (1.63 mole) of (−) ephedrine dissolved in 1 l of acetone, a precipitate was formed. The precipitate was dissolved by heating, and the solution was concentrated to a volume of 1.8 l. The solution was cooled slowly to room temperature, and 152 g of crystalline product was obtained. The product was recrystallized in acetone until a constant $[\alpha]_D$ was obtained. The rotation was determined on the acid prepared from 0.5 g of salt.

Number of crystallizations	1	2	3	4	5
g of salt	152	89.5	42.5	17	10
$[\alpha]_D^{25}$ (*c* 2, acetone) of the acid	+19.3	+32.3	+46.3	+53.3	+53.3

The ephedrine salt obtained after the last crystallization was dissolved in water and the acid was extracted with ether, after acidification with hydrochloric acid. The D(+) 2-azidoadipic acid was crystallized in benzene, 3.1 g mp 71-72.5°, $[\alpha]^{25}$ +53.3 (*c* 2, acetone).

Anal. Calcd. for $C_6H_9N_3O_4$: C, 38.50; H, 4.85.
Found: C, 38.40; H, 4.72.

b) The pure optical isomer could also be obtained by fractional crystalization of partially resolved 2-azidoadipic acid. 2-Azidoadipic acid (9.8 g) with $[\alpha]_D^{25}$ +35 was triturated with 80 ml of benzene. The nondissolved material (5.8 g) had $[\alpha]_D^{25}$ +24; the filtrate was concentrated to 10 ml, and 2.8 g D(+) 2-azidoadipic acid mp 70-72°, $[\alpha]_D^{25}$ +54 was obtained.

c) To a solution of 360.8 g (1.26 mole) of dehydroabiethylamine [4] with $[\alpha]_D^{25}$ +30 (*c* 5, acetone) in 8.5 l of hot ethanol was added to 117.8 g (0.63 mole) of DL-2-azidoadipic acid. The solution was cooled slowly and kept for 1 day at room temperature. The crystalline precipitate was then filtered off. The product was recrystallized in methanol until a constant $[\alpha]_D$ was obtained.

Number of crystallizations	1	2	3	4
g of salt	340	176	105	45
$[\alpha]_D^{25}$ (*c*, acetone) of the acid	+12	+21	+34	+50

The dehydroabiethylamine salt was dissolved in an ethanol-water mixture, acidified with HCl, and the acid extracted with ether. After evaporation of the organic solvent, the acid was recrystallized in benzene 7.2 g mp 71-72.5°, $[\alpha]_D^{25}$ +54 (*c* 3, acetone).

2. L-isomer.

Partially resolved L-2-azidoadipic acid was obtained from the preparation of the D-isomer. This acid was dissolved in ethanol and 2 moles of (+) ephedrine added. By fractional crystallization, L-2-azidoadipic acid mp 70-72°, $[\alpha]_D^{25}$ −53 (*c* 1, acetone), was obtained.

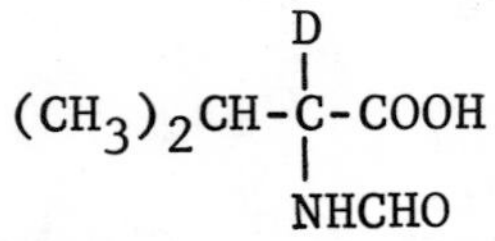

$$(CH_3)_2CH-\underset{NHCHO}{\overset{D}{C}}-COOH$$

Ref. 194

16.6 g so erhaltenes *Formyl-2-deutero-DL-valin* ließen sich nach E. Fischer[14] mit 43.4 g *Brucin* in die Enantiomeren spalten. Dabei kristallisierten aus ca. 800 ccm Methanol (+4°) insgesamt 30.2 g 4a-Brucin-Salz, aus dessen Lösung in 155 ccm Wasser das Brucin mit 29 ccm 2n NaOH bei 0° wieder abgeschieden wurde. Das mit Chloroform ausgeschüttelte Filtrat versetzte man mit 19.3 ccm 2n HCl, dampfte i. Vak. ein und kristallisierte den Rückstand bei 4° aus 19.4 ccm n HCl und wenig Wasser. Ausb. 6.6 g 4a, das aus wenig Wasser nochmals umkristallisiert und zur Analyse 6 Stdn. bei 80° i. Hochvak. getrocknet wurde. Schmp. 151—153°; $[\alpha]_D^{20}$: $-13.7 \pm 0.2°$ (c = 1.2 in absol. Äthanol); $-8.4 \pm 0.2°$ (c = 1.2 in Methanol).

NMR[42] (Aceton-d_6): CHO d δ 8.23 (J = 2.0 Hz); NH 7.50; 2-H (undeuterierter Anteil) dd 4.53 (9.0/5.0); 3-H m 2.21 (7.0); $(CH_3)_2$ dd 0.97/0.95 (7.0). Deuterierungsgrad: 0.85 ± 0.03.

$C_6H_{10}DNO_3$ (146.2)   Ber. C 49.30 H 8.27 N 9.58   Gef. C 49.18 H 7.85[43] N 9.45

Formyl-2-deutero-L-valin wurde auf gleiche Weise aus der methanol. Mutterlauge isoliert: $[\alpha]_D^{20}$: $+11.8 \pm 0.2°$ (c = 1.1 in absol. Äthanol).

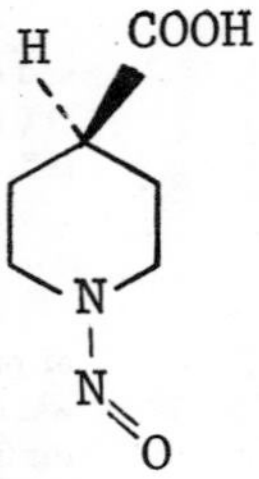

$C_6H_{10}N_2O_3$

Ref. 195

A mixture of equimolar solutions of quinine (in methanol) and (±)-1 (in ethyl acetate) was evaporated under reduced pressure in a rotary evaporator (bath temperature ca. 40°C) until crystals appeared. The crystallization was then allowed to proceed for ca. 1 h at ambient temperature, and the salt isolated (91 %) [9]. The acid was recovered by distribution of the salt between ethyl acetate and 1 N hydrochloric acid at 0°C, drying the organic solution and concentrating it in vacuum at 0 - 5°C (rot. evap., 1 torr). The resultant crystalline (-)-1 was chemically pure according to elemental analysis and spectral data [10] mp 137 - 139°C; $[\alpha]_{436}^{22}$ -80° (c = 2.0, methanol). Similar resolving experiments with other amines (1-phenylethylamine, ephedrine [11]) gave acids with smaller optical rotations.

[9] Since the acid had an enantiomeric purity of ca. 25 %, the yield of 91 % indicates that 2nd order asymmetric transformation had occurred.

$$CH_2=CH-CH_2-\underset{CH_3}{\overset{}{CH}}COOH$$

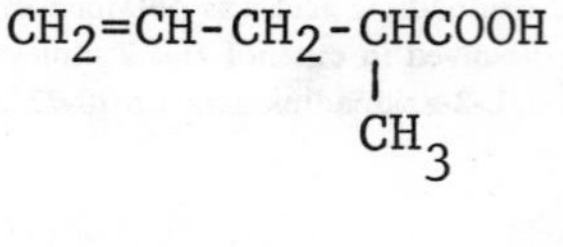

$C_6H_{10}O_2$

Ref. 196

Resolution of dl-methylallylacetic acid. 69 g of quinine was dissolved in warm acetone (400 ml), the solution filtered hot and 20 g of *dl*-methylallylacetic acid added. On cooling to room temperature the solution deposited a felted mass of very thin needles. The crystals were filtered off and recrystallized four times from hot acetone. The quinine salt was dissolved in dilute hydrochloric acid and the solution extracted with ether five times. The combined extracts were washed twice with a small amount of water. As methylallylacetic acid is appreciably soluble in water the washings were also extracted with ether and this ether solution added to the main portion. The combined ether extracts were dried by means of sodium sulphate and the ether evaporated. The slightly yellow residue weighed 5.8 g and had α_D^{18} + 7.35° (homogeneous, *l*, 1). To the material described was added the residue from another run (7.1 g, α_D^{18} + 6.95° (homogeneous, *l*, 1)). The quinine salt was prepared and recrystallized twice from hot acetone. After decomposition of the salt there was obtained a residue having α_D^{18} + 7.27°. Distillation of the liquid residue gave 7.15 g of (+)-*methylallylacetic acid*, b. p. 112.0—112.7°, 29 mm. D_4^{18} 0.956.

Optical rotation α_D^{18} + 7.62° (homogeneous, *l*, 1)

$[\alpha]_D^{18}$ + 7.97° $[M]_D^{18}$ + 9.10°

Anal.: Calcd. for $C_6H_{10}O_2$ (114.2) C, 63.13; H, 8.83 %

Found   C, 63.16; H, 8.80 %

$CH_2=CH-CH_2-CHCOOH$ (with CH_3 on the CH)

$C_6H_{10}O_2$ Ref. 196a

Experimental. (−)-2D-*Methyl-Δ*$^{4:5}$-*pentenoic acid.* Partially resolved laevorotatory acid (43 g, α_D^{25} −5.3° (undiluted, *l* 1)), obtained from the mother liquor of the first crystallization of the quinine salt of the dextrorotatory enantiomer, was dissolved in ether (700 ml) and a solution of (+)-phenylethylamine (44.5 g) in ether (800 ml) was added cautiously (heat is evolved during the formation of the salt). Crystallization was allowed to take place overnight at a temperature of −15°. The salt crystallized in the form of long thin needles. Acid isolated from the mother liquor of this crystallization had an optical rotation of α_D^{25} −2.2° (undiluted, *l* 1). The salt was recrystallized from successively smaller volumes of ether and the course of the resolution followed by measuring the rotation of acid isolated from the mother liquors. Nine crystallizations were needed before the rotation attained the constant value α_D^{25} −7.80° (undiluted, *l* 1). Decomposition of the salt gave 6.6 g of (−)-methylallylacetic acid of b.p. 87−88° at 12 mm; n_D^{25} 1.4275, n_D^{18} 1.4305; d_4^{25} 0.946; R_D calcd. 31.02, found 31.01.

Optical rotation: α_D^{25} −7.80° (undiluted, *l* 1); $[\alpha]_D^{25}$ −8.25°; $[M]_D^{25}$ −9.41°.

The specimen of the dextrorotatory enantiomer, prepared in the course of the present work had $[\alpha]_D^{25}$ +8.25°. Ställberg-Stenhagen [4] gives α_D^{20} +7.88° (undiluted, *l* 1), which corresponds to $[\alpha]_D^{20}$ +8.28°. Fray and Polgar [5] give $[\alpha]_D^{19}$ +8.24°.

4. Ställberg-Stenhagen, S. *Arkiv Kemi* **1** (1949) 153.
5. Fray, G. I. and Polgar, N. *J. Chem. Soc.* 1956 2036.

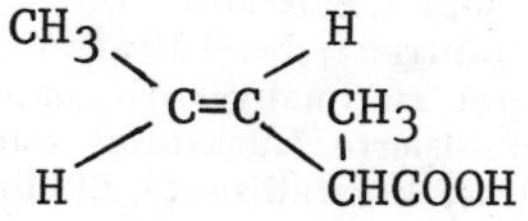

$C_6H_{10}O_2$ Ref. 197

Note:

The racemic acid formed a crystalline salt (from ether) with (+) or (-)-α-methylbenzylamine. Recrystallization of the salt from the (+)-amine afforded the (-)-acid. Similarly the (+)-acid with $[\alpha]_D^{23}$ = +63.6° (ether) was obtained using the (-)-amine.

CH_3CH-CH_2COOH (with $COCH_3$ on the CH)

$C_6H_{10}O_3$ Ref. 198

Resolution of α-Methyllevulinic Acid.—α-Methyllevulinic acid (1.11 g.), cinchonidine (2.52 g.) and water (15 ml.) were heated until a solution was obtained, the water was then completely removed under vacuum, and the residue recrystallized from acetone. Colorless needles radiating from a center were obtained (2.0 g.), m.p. 146–147°, $[\alpha]^{21.5}D$ −88° (ethanol, *c* 1.40). Another recrystallization from acetone did not change the rotation.

Anal. Calcd. for $C_{22}H_{32}O_4N_2$: C, 70.7; H, 7.6; N, 6.6. Found: C, 71.1; H, 7.6; N, 6.6.

The α-methyllevulinic acid (0.33 g.) was recovered by treatment with dilute 1:1 hydrochloric acid (20 ml.) and continuous extraction with ether. It had $[\alpha]^{12}D$ +22.6° (water, *c* 151). **The semicarbazone** of this optically active methyllevulinic acid had m.p. 178°, $[\alpha]^{21}D$ + 66.3° (ethanol, *c* 0.935). Recrystallization did not raise the melting point, and it was depressed when admixed with the semicarbazone from the natural product. **The *p*-bromophenacyl** derivative prepared in the usual way was obtained as colorless needles from alcohol, m.p. 74–75°.

Anal. Calcd. for $C_{14}H_{15}O_4Br$: C, 51.4; H, 4.6; Br, 24.4. Found: C, 51.4; H, 4.6; Br, 24.4.

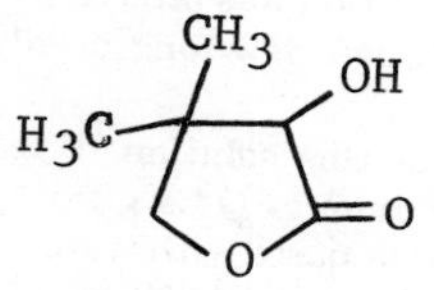

$C_6H_{10}O_3$ Ref. 199

The Quinine Salt of (+)α,γ-Dihydroxy-β,β-dimethylbutyric Acid.—A solution of 21 g. (0.1615 mole) of pure racemic lactone in 48.5 cc. of water was treated with 46 cc. of 3.872 *N* sodium hydroxide (10% in excess of theoretical) and heated to 80–90° when hydrolysis of the lactone was complete. After cooling, the excess alkali was neutralized with 2.5 *N* hydrochloric acid and the solution diluted to 400 cc. and again heated to 80–90° on a steam-bath. Thirty-two grams (0.0805 mole) of quinine hydrochloride was added to the hot solution with stirring. Separation of the crystalline quinine salt of (+)α,γ-dihydroxy-β,β-dimethylbutyric acid commenced after a small portion of the quinine hydrochloride had been added The solution and crystals were allowed to stand overnight in a refrigerator and then separated by filtration. After washing with water, the crystals were dried to constant weight in an oven at 60°. The yield of the first crop was 32.4 g. An additional 1.4 g. of pure material was obtained from the mother liquor on concentration. The total yield was 33.8 g. (86.5%); m. p. 189°; $[\alpha]^{25}D$ −130.5° in methanol; $C = 1\%$.

Anal. Calcd. for $C_{26}H_{36}O_6N_2$: C, 66.08; H, 7.67; N, 5.93. Found: C, 66.00; H, 7.69; N, 5.94.

(−)α-Hydroxy-β,β-dimethyl-γ-butyrolactone.—A solution of 20.1 g. (0.0425 mole) of the quinine salt $[\alpha]^{25}D$ −130°) in 50 cc. of 2.5 *N* hydrochloric acid (3 equivalents) was heated on the steam-bath for twenty minutes and then continuously extracted for eleven hours with ethyl ether. The ether solution was evaporated to dryness and the residue dried by distilling with alcohol and benzene. The crystalline residue was recrystallized from a little benzene and petroleum ether (b. p. 30–40°). The yield of (−)α-hydroxy-β,β-dimethyl-γ-butyrolactone was 3.95 g. (0.0304 mole) 71.5%); m. p. 89–90°; $[\alpha]^{25}_D$ −50.7° in water, C = 2.05%. A once recrystallized sample had a melting point and mixed melting point of 90–91° with a sample of lactone (m. p. 91–92°) isolated from natural pantothenic acid.

Anal. Calcd. for $C_6H_{10}O_3$: C, 55.37; H, 7.75. Found: C, 55.32; H, 7.80.

p-Nitrobenzoate of (−)α-Hydroxy-β,β-dimethyl-γ-butyrolactone.—One gram of the synthetic (—)lactone was treated in pyridine with 1.57 g. of p-nitrobenzoyl chloride. After heating on the steam-bath, it was poured onto several volumes of cracked ice. The solid which separated was recrystallized three times from 95% alcohol. The yield of the pure p-nitrobenzoate was 0.7 g.; m. p 112°. When mixed with the p-nitrobenzoate of the lactone from natural sources there was no depression in melting point.

Anal. Calcd. for $C_{13}H_{13}O_6N$: C, 55.91; H, 4.69; N, 5.02. Found: C, 56.10; H, 4.45; N, 4.94.

Quinine Salt of (−)α,γ-Dihydroxy-β,β-dimethylbutyric Acid.—The barium salt of α,γ-dihydroxy-β,β-dimethylbutyric acid was made by dissolving 65 g. (0.5 mole) of the d,l-lactone in 200 cc. of water and heating on the steam-bath with a 20% excess of barium hydroxide. On cooling crystallization took place, so the solution was made up to 500 cc., heated to 80° and neutralized with a stream of carbon dioxide. The solution was filtered from the barium carbonate and on standing a few long needle-like crystals separated.

An aliquot portion of this solution (0.1825 mole) was added to an equivalent amount (71.4 g.) of quinine sulfate which had been made to a paste with water. The quinine salt was brought into solution in 750 cc. of boiling water and quickly centrifuged from precipitated barium sulfate. Crystallization started almost immediately. These crystals proved to be the quinine salt of the dextrorotatory dihydroxy acid mixed with uncombined quinine sulfate. The aqueous filtrate contained the quinine salt of the levo rotatory dihydroxy acid. This was recovered by concentration to a small volume under reduced pressure when crystallization took place. The specific rotations of three successive crops were −143, −146, and −147°, respectively. The latter two crops were combined and recrystallized from 95% ethanol. It crystallized in fine needles in feather like designs which formed a much bulkier mass than did the quinine salt of the (+)acid; m. p. 176–178°; $[\alpha]^{25}_D$ −146° in methanol; C = 1%. Mixed melting point with the isomer melting at 188–189° showed a depression of only four degrees.

Anal. Calcd. for $C_{26}H_{36}O_6N_2$: C, 66.08; H, 7.67; N, 5.93. Found: C, 66.23; H, 7.72; N, 5.80.

(+)α - Hydroxy - β,β - dimethyl - γ - butyrolactone.—The quinine salt of the (−)acid (3.8 g.) ($[\alpha]^{25}_D$ −146°) was treated with 10 cc. of 2.5 N hydrochloric acid on a steam-bath for one hour and then extracted ten times with ether. The (+)lactone was recovered and purified in the same manner as described for the (−)lactone. The crude yield was 0.7 g. (67%). After recrystallization from benzene, the m. p. was 91°; $[\alpha]^{25}_D$ +50.1° in water; C = 2%.

Anal. Calcd. for $C_6H_{10}O_3$: C, 55.37; H, 7.75. Found: C, 55.19; H, 7.51.

$C_6H_{10}O_3$ Ref. 199a

Experimenteller Teil.

Herstellung der reinen $d(-)$- und $l(+)$-α-Oxy-β,β-dimethyl-butyrolactone (II) und (III).

20 g d,l-α-Oxy-β,β-dimethyl-butyrolacton wurden in 50 cm³ Methanol gelöst und ½ Stunde mit der Lösung von 27 g krystallisiertem Bariumhydroxyd in 500 cm³ Methanol unter Rückfluss gekocht. Dann wurde mit Kohlendioxyd neutralisiert und vom Bariumcarbonat abfiltriert. Das Filtrat wurde mit der heissen Lösung von Chininsulfat in Methanol genau ausgefällt, wozu etwa 60 g Chininsulfat in 800 cm³ Methanol nötig waren. Das Bariumsulfat wurde durch Zentrifugieren entfernt. Die Lösung gab beim Einengen zunächst 30 g rohes, schwerlösliches Chininsalz (A). Aus den Mutterlaugen wurden, wie früher beschrieben, 28 g rohes Chininsalz (B) erhalten. Aus den ersteren wurden durch Umkrystallisieren aus Methanol 26 g reines A-Salz gewonnen. Das rohe B-Salz wurde aus Methanol-Äther umkrystallisiert und gab 23 g gereinigtes B-Salz.

Zur Spaltung wurden 26 g A-Salz in 500 cm³ Methanol gelöst, mit der Lösung von 11 g Bariumhydroxyd in heissem Wasser versetzt und im Vakuum vom Methanol befreit. Dem Rückstand wurde durch Ausschütteln mit Chloroform das Chinin entzogen und Chloroformreste durch Ausschütteln mit Äther entfernt. Dann wurde mit Kohlendioxyd neutralisiert, vom Bariumcarbonat abfiltriert und die klare Lösung eingedampft. Der Rückstand wurde aus wenig Wasser durch Zusatz von Aceton umkrystallisiert und gab 10,5 g reines Bariumsalz (A) vom Smp. 198—200° (Zers.). Es zeigte eine spez. Drehung von $[\alpha]^{19°}_D = +5{,}5° \pm 1°$ (c = 2,8 in Wasser).

Die 23 g Chininsalz (B) wurden analog mit Bariumhydroxyd gespalten. Das rohe Bariumsalz (10 g) wurde in 10 cm³ Wasser gelöst und mit 30 cm³ Aceton versetzt. Es fielen dabei rasch Krystallnadeln aus, die abgenutscht und mit Methanol gewaschen wurden. Sie wogen 2,4 g und schmolzen nach einmaligem Umkrystallisieren aus Wasser mit Alkohol bei 220° korr. (Zers.). Es handelt sich um das racemische Bariumsalz. Dasselbe Salz wird nämlich erhalten, wenn gleiche Gewichtsmengen des (+)- und (−)-Salzes in Methanol gelöst zusammengegeben und mit einer Spur Wasser versetzt werden. Zur Analyse wurde an der Luft getrocknet.

1,518 mg Subst. gaben 0,828 mg $BaSO_4$

$C_{12}H_{22}O_8Ba$ Ber. Ba 31,82 Gef. Ba 32,10%

Die Mutterlauge der genannten Krystalle wurde mit Aceton bis fast zur Trübung versetzt und angeimpft, wobei sofort Krystallisation einsetzte, die durch längeres Stehen bei 0° und vorsichtigen Zusatz von Aceton möglichst vervollständigt wurde. Erhalten wurden 7,4 g Bariumsalz (B), das nach einmaligem Umkrystallisieren aus Wasser-Aceton bei 198—200° korr. (Zers.) schmolz. Die spez. Drehung betrug $[\alpha]^{20°}_D = -6{,}5° \pm 1{,}5°$ (c = 1,545 in Wasser).

Zur Gewinnung des freien Lactons wurden 10 g Bariumsalz (A) in absolutem Alkohol gelöst und mit etwas mehr als der berechneten Menge alkoholischer Salzsäure versetzt. Das ausfallende Bariumchlorid wurde abfiltriert, das Filtrat im Vakuum eingedampft und der Rückstand durch zweimaliges Abdampfen mit etwas Benzol getrocknet. Der krystallisierte Rückstand wurde im Hochvakuum sublimiert. Das farblose, krystallisierte Sublimat schmolz roh bei 87—89° und zeigte eine spez. Drehung von $[\alpha]^{19°}_D = -14{,}6° \pm 0{,}5°$ (c = 5,25 in Aceton). Einmaliges Umkrystallisieren aus Benzol-Petroläther brachte den Smp. auf 89—90°. Die spez. Drehung betrug $[\alpha]^{17°}_D = -17{,}4° \pm 0{,}5°$ (c = 3,843 in Aceton), bzw. $[\alpha]^{17.5}_D = -49° \pm 0{,}5°$ (c = 4,012 in Wasser).

Analog wurde das Bariumsalz (B) ins freie $l(+)$-Lacton (III) übergeführt. Das farblose Sublimat des Rohproduktes zeigte einen Smp. von 87—89° und eine spez. Drehung von $[\alpha]^{19°}_D = +14{,}1° \pm 1°$ (c = 1,98 in Aceton). Einmaliges Umkrystallisieren aus Benzol-Petroläther ergab farblose Nadeln vom Smp. 89—90°. Die spez. Drehung betrug $[\alpha]^{18°}_D = +14{,}5° \pm 0{,}5°$ (c = 4,751 in Aceton) bzw. $[\alpha]^{17°}_D = +51{,}5° \pm 0{,}5°$ (c = 4,331 in Wasser).

$C_6H_{10}O_3$

Ref. 199b

D-(-)-α-Hydroxy-β,β-dimethyl-γ-butyrolactone (VI). a. To a solution of 6.3 g sodium hydroxide in 40 ml water was added 20 g racemic α-hydroxy-β,β-dimethyl-γ-butyrolactone (I, mp 72-73°). The reaction mass was heated to 85° and after 30 min was cooled to 45°(pH 7.5). L-(+)-threo-1-(p-Nitrophenyl)-2-aminopropane-1,3-diol sulfate [III, X = OSO_3H, mp 226° (decomposition), $[\alpha]_D^{20}$ + 24.2° (c 5, water)] (20.15 g) was added to the solution of substance obtained. The mixture was stirred at this temperature until complete solution of L-(+)-threo-amine sulfate, then the solution was gradually cooled to 18-20°, stirred 2 h further after which it was kept 18 h at 3-5°. The salt of L-(+)-threo-1-(p-nitrophenyl)-2-amino-propane-1,3-diol and (-)-α,γ-dihydroxy-β,β-dimethylbutyric acid (IV) which separated was filtered off, washed, with 15 ml ice water and dried. Yield 27.7 g (98.7%) mp 84-85°, $[\alpha]_D^{20}$ + 11.82° (c 2, water).

After separation of the salt (IV), the filtrate was treated with conc. sulphuric acid (pH 1) for 2 h at 80-85°. D-(-)-Pantolactone (VI) was extracted with methylene chloride (4 · 30 ml). After evaporation of the solvent and recrystallization from benzene or methylene chloride 4.85 g (49%) D-(-)-pantolactone was obtained of mp 89.5°-90.5°, $[\alpha]_D^{20}$ -49.95° (c 2, water). Found %: C 55.18; H 7.66. $C_7H_{10}O_3$. Calculated %: C 55.37; H 7.72.

b. L-(+)-Threo-1-p-nitrophenyl)-2-aminopropane-1,3-diol (16.25 g) of mp 159-161°, $[\alpha]_D^{20}$ + 31° (c 5 1 N HCl), was added to a solution of 6.5 ml hydrochloric acid (d 1.17 in 30 ml water. The mixture was stirred at 50° until complete solution of the amine pH 6.5) (solution A).

To a solution of 6.3 g sodium hydroxide in 25 ml water, 20 g α-hydroxy-β,β-dimethyl-γ-butyrolactone was added. The reaction mass was heated for 30 min at 85°, then cooled to 45° (pH 7.5) (solution B).

Solution B was added at 45° to solution A, then the mixture gradually cooled to 18-20°, stirred for 2 h, after which it was kept at 3-5° for 18 h. The salt of L-(+)-threo-amine and (-)-α,γ-dihydroxy-β,β-dimethyl-butyric acid which precipitated was separated, washed with 15 ml ice water and dried. Yield 25.11 g (92%), mp 83.5-84.5°, $[\alpha]_D^{20}$ + 11.81° (c 2, water).

After separation of salt (IV), the filtrate was treated with hydrochloric acid (d 1.179) at 85° for 2 h (pH 1). D-(-)-Pantolactone (VI) was extracted with methylene chloride (4 · 30 ml). The dried extract was concentrated until viscous. D-(-)-Pantollactone crystallized in a yield of 4.39 g (44.3%), mp 88.5-89.5°, $[\alpha]_D^{20}$ - 49.83° (c 2, water).

L-(+)-α-Hydroxy-β,β-dimethyl-γ-butyrolactone, L-(+)-Pantolactone (VII). To a solution of 6 ml sulfuric acid (d 1.83) in 35 ml water was added 24.4 g salt of (-)-α,γ-dihydroxy-β,β-dimethyl-butyric acid and L-(+)-threo-amine. The mixture was stirred 2 h at 85° (pH 1). L-(+)-Pantolactone was extracted with methylene chloride. After distillation of the solvent the yield was 6.9 g (78.5%), $[\alpha]_D^{20}$ + 34.2° (c 2 water). After recrystallization from methylene chloride a further 5.62 g, $[\alpha]_D^{20}$ + 49.32° (c 2, water) was obtained.

$C_6H_{10}O_3$

Ref. 199c

(20) All melting points are uncorrected. Specific rotations were taken in water at 23°, at a concentration of 1-4% in a 2-decimeter tube.

The Resolution of Pantolactone with D(−)-Galactamine. (a) Typical Resolution Involving "α"-(−)-Dulcitylpantamide (IVb). The Isolation of "β"-(−)-Dulcitylpantamide (IVc).—Pantolactone (17.4 g., 0.134 mole), 11.2 g. (0.062 mole) of D(−)-galactamine and 200 ml. of absolute ethanol were heated to reflux with vigorous stirring in a nitrogen atmosphere for 7 hr. The cloudy reaction mixture was filtered yielding a clear pale yellow filtrate (*p*H 8). The ethanol was removed under reduced pressure and the residual sirup was dissolved in 40 ml. of hot water. The excess pantolactone was removed by extraction with five 40-ml. portions of methylene chloride, and the aqueous solution was concentrated under reduced pressure to a viscous pale yellow sirup which was dried by repeated distillation of absolute ethanol. The dried residue was dissolved in 65 ml. of hot ethanol, seeded with "α"-(−)-dulcitylpantamide and stored at 3°. Five crops of crystals were removed over a period of two weeks. The pertinent data for these crops are summarized in Table I. After the removal of the fifth crop, the mother liquor was concentrated to a sirup which was dissolved in 25 ml. of water and extracted well with methylene chloride. The aqueous solution was evacuated at 15 mm. to remove residual methylene chloride and was filtered through a column of Darco–Celite (1:2) (5 cm. × 1 cm.). The golden yellow solution was decolorized effectively by this treatment. The filtrate was concentrated under reduced pressure to a viscous sirup which was dissolved in 18 ml. of hot ethanol. The solution was seeded with (+)-dulcitylpantamide and stored at 3°. After two days, a white crystalline solid, crop 6, was removed by filtration, 2.0 g. (11%), m.p. 149–152°, $[\alpha]_D$ −25°. This material melted about 30° higher than any crop previously isolated in the resolution of pantolactone with galactamine. Crop 6 was recrystallized three times from ethanol–water to yield an analytical sample of "β"-(−)-dulcitylpantamide, m.p. 162.5–165°.

Anal. Calcd. for $C_{12}H_{25}NO_8$: C, 46.29; H, 8.09; N, 4.50. Found: C, 46.46; H, 8.32; N, 4.37.

(b) Resolution Involving "β"-(−)-Dulcitylpantamide.—Pantolactone (100 g., 0.769 mole, 99% excess) was dissolved in 500 ml. of absolute ethanol, and about 100 ml. of the ethanol was removed by distillation to dry the lactone. The solution was cooled below the boiling point, D(−)-galactamine (70 g., 0.387 mole) was added and the suspension was stirred vigorously at the reflux temperature in a nitrogen atmosphere for 17.5 hr. The reaction mixture (*p*H 7) was cooled to room temperature, and the solid which was present

TABLE I

SUMMARY OF CROPS REMOVED FROM THE RESOLUTION IN WHICH "β" (−)-DULCITYLPANTAMIDE WAS ISOLATED

Crop	1	2	3	4	5	6
Weight, g.	6.7	1.4	2.1	0.5	1.0	2.0
Theor. amides, %	35	7	11	3	5	11
M.p., °C.	107–112	100–110	111–117	115–122	92–102	149–152
[α]D (°)	−2	−20	−28	−29	0	−25
Major component,[a] %	71% D	71% "α"-L	89% "α"-L	91% "α"-L	75% D	82% "β"-L

[a] (+)-Dulcitylpantamide is designated D and (−)-dulcitylpantamide L.

was removed by filtration, 47.2 g. (29% of the theoretical amides), m.p. 161.5–163.5°, [α]D −30°. The filtrate was concentrated under reduced pressure to a sirup which was taken up in 200 ml. of warm water. The aqueous solution was extracted six times with 125 ml. portions of methylene chloride, the combined extracts were dried over anhydrous sodium sulfate and concentrated under reduced pressure to a semi-solid (50 g., 100%) recovered pantolactone. The aqueous solution was concentrated under reduced pressure to a burnt orange colored sirup which was dried by distillation of 100 ml. of absolute ethanol and dissolved in 200 ml. of hot alcohol. It was seeded with "β"-(−)-dulcitylpantamide and stored at 30° for 4 hr. The white crystalline solid which separated was removed by filtration, washed with absolute ethanol and dried in a vacuum oven at 50°, (34.1 g., 28%) of the theoretical amides, m.p. 112–115°, [α]D +9°. It is surprising that practically pure (+)-dulcitylpantamide was obtained in this crop in spite of the fact that the solution was seeded with the L(−)-isomer. After removal of two intermediate crops totaling 12.5 g. (10%), an additional crop of pure D(+)-diastereoisomer was isolated, 7.5 g. (6%), m.p. 103–111°, [α]D +9°. A summary of the crystalline crops isolated in this resolution is presented in Table II.

Hydrolysis of N-D-Dulcityl-2,4-dihydroxy-3,3-dimethylbutyramide. Proof of Structure of the "β"-(−)-Enriched Form.—"β"-(−)-N-D-Dulcityl-L-2,4-dihydroxy-3,3-dimethylbutyramide (6.6 g., 0.0212 mole, m.p. 160–162.5°, [α]D −30°, 93% (−)-diastereoisomer) was heated on a steam-bath with 40 ml. of 6% by volume hydrochloric acid for 3 hr. The resulting solution was extracted with five 25-ml. portions of methylene chloride and the combined extracts were dried over anhydrous sodium sulfate. Removal of the solvent under reduced pressure afforded 2.3 g. (83%) of pantolactone which was purified for analysis by sublimation ([α]D +43°, equivalent to 93% of the L(+)-enantiomorph). From another experiment in which the starting material to be hydrolyzed melted at 148–151°, the pantolactone which was obtained melted at 86.6–90°, [α]D +16° [66% L(+)].

Anal. Calcd. for $C_6H_{10}O_3$: C, 55.37; H, 7.75. Found: C, 55.45; H, 7.61.

The aqueous phase was concentrated under reduced pressure to a white crystalline solid which was further dried by distillation of absolute ethanol, reslurried in ethanol and filtered to yield galactamine hydrochloride in 97% yield, m.p. 158–161°.

Anal. Calcd. for $C_6H_{16}ClNO_5$: Cl, 16.29. Found: Cl, 16.66.

A mixed melting point with an authentic sample of galactamine hydrochloride was not depressed and the infrared spectra of the two samples were identical.

(b) D(+)-Form.—(+)-N-D-Dulcityl-D-2,4-dihydroxy-3,3-dimethylbutyramide (31.0 g., 99.7 millimoles) was hydrolyzed by the above procedure to yield 12.2 g. (94% yield) of D-(−)-pantolactone ([α]D −48°, 98% D(−)-enantiomorph). The galactamine hydrochloride was dissolved in 500 ml. of deionized water and the resulting solution was passed through a 3/4" × 18" column of Rohm and Haas IR-410 anion exchange resin in the hydroxide cycle at the rate of 5 ml. per minute. The column was washed with water until the *p*H of the effluent was 7.5. The aqueous solution of galactamine was then concentrated to dryness under reduced pressure in a nitrogen atmosphere. The white crystalline residue was dried by repeated distillation of absolute ethanol and finally by evacuation with an oil-pump to yield 17.3 g. of galactamine (96%), m.p. 142–146°.

Anal. Calcd. for $C_6H_{15}NO_5$: neut. equiv., 181.2. Found: neut. equiv., 188.

This material when treated with D(−)-pantolactone yielded (+)-dulcitylpantamide in 90% yield in one crop of crystals. The mother liquor was not worked up further for additional crops.

TABLE II

RESOLUTION OF PANTOLACTONE WITH D(−)-GALACTAMINE INVOLVING "β"-(−)-DULCITYLPANTAMIDE

Crop							
Weight % of theor. amides	39	28	6	5	6	1	12
Cum. weight, %	39	67	73	78	84	85	97
M.p., °C.	161–163	112–115	120–135	149–153	103–111	162–164	Sirup
[α]D (°)	−30	+9	−7	−23	+9	−32	......
Major component, %	93% L	96% D	59% D[a]	77% L[a]	96% D	98% L	*ca.* 63% D[a]

[a] Can be recycled.

$C_6H_{10}O_3$ Ref. 199d

(R)-Pantolacton (R-1) durch Racematspaltung mit (+)-3-Aminomethylpinan [(+)-5]

1. Eine Lösung von 260 g (2 mol) *RS*-**1** in 230 ml Wasser wird mit 168 g (2.1 mol) 50 proz. Natronlauge versetzt, 20 min auf 80°C erwärmt und durch tropfenweise Zugabe von 10 proz. Salzsäure auf pH 8.8 eingestellt. Man läßt eine Lösung von 203 g (1 mol) (+)-**5**-Hydrochlorid mit $[\alpha]_D^{25}$ = 44.5° (*c* = 4, in Methanol)[13] in 800 ml Wasser und 600 ml Methanol zulaufen, kühlt zur Kristallisation in 3 h von 60°C auf 5°C ab und saugt (+)-3-Pinanylmethylammonium-(*R*)-pantoat (**9**) auf der Nutsche ab. Das feuchte **9** wird in 250 ml Wasser aufgerührt (10 min, 20°C), und nochmals abgesaugt. Man suspendiert es in 320 ml Wasser, setzt 50 proz. Natronlauge bis pH 12 zu und erwärmt 0.5 h auf 70°C. (+)-**5** scheidet sich dabei als Oberphase ab und wird durch 2 malige Extraktion mit je 300 ml Heptan isoliert. Man säuert die wäßrige Lösung mit konz. Schwefelsäure bis pH 1 an, erhitzt 10 min auf ca. 100°C und extrahiert 5 mal mit Chloroform (2

[13] Für optisch reines (+)-**5**-Hydrochlorid nehmen wir den Drehwert $[\alpha]_D^{25}$ = 44.5° (*c* = 4, in Methanol) an.

× 250 ml, 3 × 100 ml). Aus der Chloroformlösung lassen sich durch Eindampfen 117 g (90%) *R*-**1** mit $[\alpha]_D^{25}$ = −48.5° (*c* = 5, in Wasser)[14] gewinnen.

2. *Rückgewinnung von* (+)-**5**: Zur Mutterlauge von **9** gibt man tropfenweise 50proz. Natronlauge bis pH 12.5, extrahiert (+)-**5** 2mal mit je 300 ml Heptan und vereinigt diese Heptanlösung mit der, die bei der Herstellung von *R*-**1** aus **9** angefallen ist. Die gaschromatographische Gehaltsbestimmung mit *N,N*-Dimethylanilin als innerem Standard zeigt, daß 96% (160 g) des ursprünglich eingesetzten (+)-**5** zurückgewonnen wurden[15]. Die Heptanlösung wird mit 800 ml Wasser und 600 ml Methanol versetzt und unter Kühlung im Eisbad mit Chlorwasserstoff begast bis die wäßrige Phase pH 2 erreicht. Man trennt, setzt der Unterphase 4 g (+)-**5**-Hydrochlorid zu, um Verluste auszugleichen und kann diese wäßrig-methanolische Lösung ohne weitere Reinigung mit dem gleichen Ergebnis für die nächste Spaltung von *RS*-**1** verwenden.

3. *Rückgewinnung und Racemisierung von S*-**1**: Das in der wäßrigen Lösung von Natrium-(*S*)-pantoat enthaltene Methanol wird über eine 30-cm-Füllkörperkolonne abdestilliert. Man säuert mit konz. Schwefelsäure bis pH 1 an, erhitzt 10 min auf etwa 100 °C und extrahiert 5mal mit Chloroform (2 × 250 ml, 3 × 100 ml). Nach dem Einengen werden 145 g *S*-**1** mit $[\alpha]_D^{25}$ = 42.2° (*c* = 5, in Wasser) erhalten[14]. – Zur Racemisierung erhitzt man *S*-**1** nach Zugabe von 8 g (0.1 mol) 50proz. Natronlauge 1 h auf 150 °C. Die Schmelze kann in der äquivalenten Menge Natronlauge aufgelöst (20 min, 80 °C) und als Natrium-(*RS*)-pantoat direkt wieder in die Racematspaltung eingesetzt werden.

[14] Eine optisch reine Probe mit dem Drehwert $[\alpha]_D^{25}$ = −52° erhält man durch Umkristallisieren aus Isopropylalkohol.

[15] Eine Probe wird unter Kühlung im Eisbad mit Chlorwasserstoff gesattigt. Der kristalline Niederschlag von (+)-**5**-Hydrochlorid hat nach Abfiltrieren und Trocknen (50 °C/ca. 10 Torr) den Drehwert $[\alpha]_D^{25}$ = 44.3° (*c* = 4, in Methanol).

$C_6H_{10}O_3$

Ref. 199e

*(R)-Pantolacton (R-*1*) durch Racematspaltung mit (−)-6,6-Dimethylbicyclo[3.1.1.]hept-2-ylamin[(−)-Nopinylamin,* (−)-**8**]

1. Eine Lösung von 80 g (0.6 mol) *RS*-**1** in 150 ml Wasser wird mit 24.8 g (0.6 mol) 97proz. Natriumhydroxid versetzt, 20 min bei 80 °C gehalten und dann bei 50 °C auf pH 8.8 gestellt. Man gibt eine Lösung von 52.4 g (0.3 mol) (−)-**8**-Hydrochlorid mit $[\alpha]_D^{25}$ = 20.7° (*c* = 4, in Wasser)[16] in 130 ml Wasser zu und rührt 2 h unter langsamem Abkühlenlassen auf 25 °C. Die Kristallisation setzt bei ca. 40 °C (nach etwa 5 min) ein. Man rührt noch 15 min bei 10 °C und filtriert das kristalline **11** ab. – Zur Mutterlauge von **11** gibt man tropfenweise 50proz. Natronlauge bis pH 12.5, extrahiert 2mal mit je 60 ml Ether, säuert die Wasserphase mit konz. Schwefelsäure bis pH 1 an und engt auf etwa die Hälfte des Volumens ein. Extraktion mit 5 × 40 ml Chloroform liefert nach dem Einengen 40.0 g *R*-**1** mit $[\alpha]_D^{25}$ = −30.0° (*c* = 4, in Wasser)[14]; GC-Gehaltsbestimmung mit Adipinsäure-diethylester als innerem Standard zeigt 92% an.

2. *Rückgewinnung von S*-**1** *und* (−)-**8**: Man suspendiert **11** in einer Mischung von 100 ml Wasser und 100 ml Ether, setzt 50proz. Natronlauge bis pH 12.5 zu, trennt die Etherphase ab und wäscht die Wasserphase mit 100 ml Ether. Die Wasserphase wird mit konz. Schwefelsäure auf pH 1 angesäuert, bis zur Trübung eingeengt und mit 5mal 40 ml Chloroform extrahiert. Nach Einengen erhält man 36.5 g (93.5%) *S*-**1** mit $[\alpha]_D^{25}$ = 37.8° (*c* = 4, in Wasser)[14]. – Die Etherphasen werden vereinigt und unter Kühlung im Eisbad mit gasförmigem Chlorwasserstoff gesättigt. Der kristalline Niederschlag von (−)-**8**-Hydrochlorid wird abgesaugt und bei 50 °C/ca. 10 Torr getrocknet; Ausb. 48.7 g (93%) mit $[\alpha]_D^{25}$ = 20.8° (*c* = 4, in Wasser) und Schmp. 237 – 239 °C.

[14] Eine optisch reine Probe mit dem Drehwert $[\alpha]_D^{25}$ = −52° erhält man durch Umkristallisieren aus Isopropylalkohol.

[16] Für optisch reines (−)-**8**-Hydrochlorid nehmen wir den Drehwert $[\alpha]_D^{25}$ = −21.5° (*c* = 4, in Wasser) an.

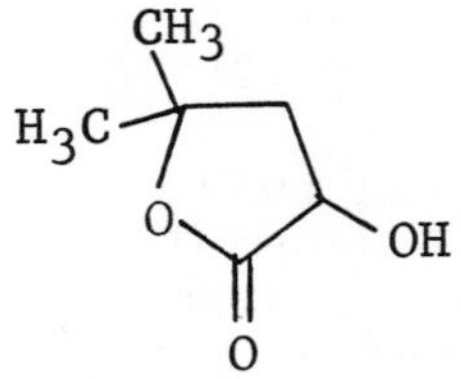

$C_6H_{10}O_3$

Ref. 200

(R)-(+)-4-*Methyl-2,4-dihydroxypentanoic acid* 1→4 *lactone* **20'**. The racemic lactone **20** (54 g) was mixed with 4N NaOH (130 ml) and the mixture was stirred and heated under reflux for 1.5 hr to effect hydrolysis. After cooling, the mixture was diluted with water (100 ml) and neutralized with N HCl (*ca.* 90 ml) to pH 7. A neutral soln (pH 7) of (+)-α-phenyl-β-(p-tolyl) ethylamine hydrochloride was prepared from the amine (78.0 g) and N HCl (*ca.* 350 ml). The carboxylate soln and the amine salt soln were combined and stirred and heated for 2 hr at 100°. On the next day, another batch of the amine hydrochloride soln (prepared from 5.2 g of the amine) was added to the mixture and it was stirred and heated at 100° for 2 hr. The soln was concentrated *in vacuo*. The residue was diluted with 99% EtOH (*ca.* 500 ml), warmed at 40° and filtered to remove NaCl. The filtrate was concentrated *in vacuo*. The residual crystalline mass was triturated with ether and filtered. The collected crystals were washed with ether to yield 91.6 g (65%) of crude **21**. This was twice recrystallized from acetone to give 13.10 g (9%) of pure **21**, m.p. 138.7°; $[\alpha]_D^{21}$ + 85.2° (*c* = 0.51, MeOH); ν_{max} (Nujol) 3400 (br), 3200 (br), 2600 (br), 1640 (m), 1620 (m), 1560 (s), 1520 (s), 1150 (s), 810 (s), 760 (s) cm^{-1}. The salt **21** (22.8 g) was mixed with 2.5 N HCl (78 ml). The mixture was stirred and heated at 60–65° for 30 min, cooled, neutralized with $NaHCO_3$ aq soln to pH 5, saturated with NaCl and filtered. The solid on the filter was thoroughly washed with ether and the filtrate was extracted with ether. The combined ether soln was dried ($MgSO_4$) and concentrated *in vacuo* to give 7.0 g (84% recovery from **21**) of **20'**. This was recrystallized from EtOAc–light petroleum to give 4.81 g of **20'** as needles, m.p. < 25°; $[\alpha]_D^{21}$ + 23.9° (*c* = 0.564, MeOH); ν_{max} 3400 (s), 2970 (m), 2940 (m), 2880 (m), 1770 (vs), 1380 (m), 1315 (s), 1280 (m), 1205 (s), 1160 (s), 1110 (s), 1035 (w), 1000 (m), 950 (m), 920 (m), 800 (m), 700 (m) cm^{-1}; δ 1.42 (3H, s), 1.54 (3H, s), 2.08 (1H, q, J_1 = 10 Hz, J_2 = 14 Hz), 2.56 (1H, q, J_1 = 10 Hz, J_2 = 14 Hz), 3.82 (1H, br), 4.72 (1H, t, J = 10 Hz); MS: *m/e* 130 (M^+); ORD: (*c* = 0.106%, MeOH), $[\phi]_{290}$ 0° $[\phi]_{250}$ − 1350° *cf* ORD of (*S*)-(+)-pantolactone (*c* = 0.062%, MeOH): $[\phi]_{300}$ + 420°, $[\phi]_{250}$ + 3570°.

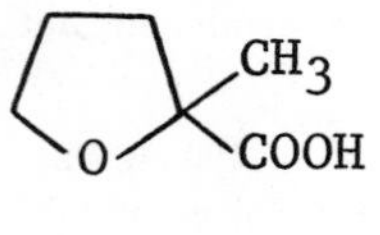

$C_6H_{10}O_3$

Ref. 201

(R)-2-Methyl-tetrahydrofuran-2-carbonsäure

130 g Racemische 2-Methyl-tetrahydrofuran-2-carbonsäure, gelöst in 135 ml Aethanol und 1 500 ml Aceton, werden mit 121 g (−)α-Phenyläthylamin versetzt. Nach 12 Stunden kristallisieren

ⓐ 152 g Salz, F. 145-147°, $[\alpha]_{578}$ − 5,1 (1 % in. CH_3OH), das sukzessive weiter umgelöst wird ⓑ − ⓓ :

ⓑ 152 g Salz a. + 1 300 ml Aceton + 250 ml Alkohol liefern 92 g, $[\alpha]_{578}$ − 4,9 (1 % in CH_3OH)

ⓒ 92 g Salz b + 250 ml Aethanol liefern 58 g $[\alpha]_{578}$ − 5,9 (1 % in CH_3OH)

ⓓ 58 g Salz c + 175 ml Aethanol liefern 42,5 g, F. 155 , $[\alpha]_{578}$ − 7,0 (1 % in CH_3OH).

Nach Zersetzung des Salzes mit n-H_2SO_4 und Extraktion mit Aether im Perforator erhält man 22 g der (R)-Säure, Kp_{15} 110°, $[\alpha]_{578}$ + 11,9 (3 % in H_2O); $[\alpha]_{578}$ + 28,9 (1,1 % in ln NaOH).

(S)-2-Methyl-tetrahydrofuran-2-carbonsäre

Aus den Mutterlaugen von ⓐ, ⓑ und ⓒ wird durch Umlösen aus Aceton/Alkohol nach Abfiltrieren der kristallinen Fraktion eine Mutterlauge gewonnen, die nach Eindampfen und Zersetzen mit verdünnter Schwefelsäure 28 g 2-Methyl-tetrahydrofuran-2-carbonsäure, $[\alpha]_{578}$ − 5,8 (3 % in H_2O), ergeben. Diese 28 g werden mit einer äquivalenten Menge (+)-α-Phenyläthylamin in 200 ml Aethanol versetzt und nach der oben beschriebenen Methode gereinigt. Man erhält 10,2 g reine (S)-2-Methyltetrahydrofuran-2-carbonsäure, Kp_{15} 110°, $[\alpha]_{578}$ − 11,1 (3 % in H_2O).

$C_6H_{10}O_3$

Ref. 201a

(S)-2-Methyl-tetrahydrofuran-2-carbonsäure :

Zu 7,8 g (0,06 Mol) racem. 2-Methyl-tetrahydrofuran-carbonsäure, gelöst in 30 ml Tetrahydrofuran, fügt man 10,3 g (0,05 Mol) Dicyclohexylcarbodiimid. Nach einer Stunde tropft man 6,75 g (0,05 Mol) (R)-N-Methyl-2-phenäthylamin ($[\alpha]_{578}$ + 61,27 [1,9 % in CH_3OH]), gelöst in 30 ml Tetrahydrofuran, ein und rührt 20 Stunden. Man verrührt mit 500 ml Aether, filtriert den ausgefallenen Dicyclohexylharnstoff ab und extrahiert die nicht umgesetzte Säure mit 80 ml 10 % iger Natronlauge. Zur Reinigung wird die alkalische Lösung mehrere Male ausgeäthert und abschließend mit verdünnter Schwefelsäure angesäuert. Die wäßrige Phase wird nun mit NaCl gesättigt und mit Aether im Perforator extrahiert. Nach dem Destillieren der ätherischen Lösung erhält man 1,2 g (S)-2-Methyltetrahydrofuran-2-carbonsäure, Kp_{10} 110°, mit einer Drehung von $[\alpha]_{578}$ − 0,6925, $[\alpha]_{365}$ 1,6860 (3 % in H_2O), $[\alpha]_{578}$ − 3,444 (2,8 % in ln NaOH). Der Methylester wird daraus durch Veresterung mit Diazomethan erhalten, Kp_{10} 60°, 0,7 g $[\alpha]_{578}$ + 0,971, $[\alpha]_{365}$ + 3,56 (3 % in CH_3OH), Monomethylamid, Kp_{10} 92° $[\alpha]_{578}$ − 1,435, $[\alpha]_{365}$ − 4,07 (2 % in CH_3OH).

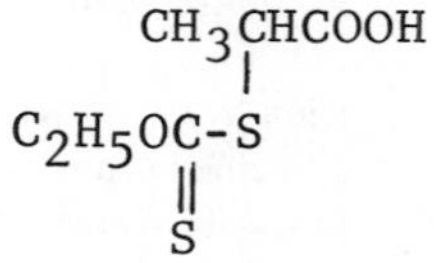

$C_6H_{10}O_3S_2$

Ref. 202

Spaltung der racem-Säure. Die (−)-Säure lässt sich am besten über das *Cinchonidinsalz* erhalten; die rechtsdrehende Säure aus der Mutterlauge wurde durch Umkristallisieren ihres *Strychninsalzes* auf maximale Drehung gebracht. Das Brucinsalz gibt eine schwach linksdrehende Säure. Spaltungsversuche mit Cinchonin, Chinidin, Chinin und α-Phenäthylamin hatten keinen Erfolg.

40 g (0,2 Mol) α-Xantogen-propionsäure wurden nebst 60 g (0,2 Mol) Cinchonidin in 1000 ml heissem Alkohol gelöst und mit 1500 ml heissem Wasser versetzt. Das nach 15-stündigem Stehen bei 10—15° ausgeschiedene Salz wurde mehrmals aus 40-%-igem Alkohol umkristallisiert (Auflösen in heissem Alkohol und nachherige Verdünnung mit Wasser). Von jeder Kristallisation wurden 0,5 g Salz entnommen, die Säure isoliert und ihre Drehung in Essigester bestimmt.

Kristallisation:	1	2	3	4	5	6	7
ml Alkohol . .	1000	500	340	290	225	200	200
ml Wasser . .	1500	750	510	430	340	300	300
g Salz	48	34	29	25	23	20	18
$[\alpha]_D^{25}$	−57,5°	−82,5°	−84,5°	−89°	−90,5°	−92°	−91,5°

Aus der Mutterlauge der ersten Kristallisation des Cinchonidinsalzes wurden 17,5 g Säure von $[\alpha]_D^{25}$ = + 36° (in Essigester) erhalten. Sie wurde in 435 ml heissem Aceton gelöst und mit 30 g Strychnin versetzt. Dann wurden allmählich 435 ml heisses Wasser zugefügt, wobei das Strychnin in Lösung ging. Das nach 15-stündigem Stehen bei 5—10° ausgeschiedene Salz wurde aus 50-%-igem Aceton umkristallisiert und der Verlauf der Spaltung wie oben verfolgt.

Kristallisation:	1	2	3	4	5	6
ml Aceton	435	200	150	150	125	120
ml Wasser	435	200	150	150	125	120
g Salz	28	23	20	17	14	12
$[\alpha]_D^{25}$	+83,5°	+87,5°	+87,5°	+90°	+91°	+90°

(—)-*Xantogen-propionsäure.* Beim Auflösen des Cinchonidinsalzes in verdünnter Schwefelsäure wurde die (—)-Xantogen-propionsäure als gelbliches Öl ausgeschieden, das in Äther aufgenommen wurde. Beim freiwilligen Eindampfen der mit $CaCl_2$ entwässerten Ätherlösung kristallisierte die Säure glatt. 18 g Salz gaben 5,7 g Säure; nach Umkristallisieren aus Petroleumäther verblieben 5,0 g kleine glänzende Prismen vom Schmp. 70—71°. Die Säure lässt sich auch bei vorsichtigem Arbeiten aus viel Wasser umkristallisieren und wird dann als dünne glänzende Schuppen erhalten.

14,17 mg Sbst.: 34,08 mg $BaSO_4$.

$C_6H_{10}O_3S_2$ (194,2) S Ber. 33,02 Gef. 32,99.

15,00 ml der bei 25,00° gesättigten wässrigen Lösung: 2,47 ml 0,1183-n NaOH. Löslichkeit bei 25°: 0,0195 Mol je Liter = 3,78 je Liter.

Dieselbe Lösung zeigte in 2-dm Rohr $\alpha = -0,50°$; $[\alpha]_D^{25} = -66°$; $[M]_D^{25} = -128°$.

Die Drehungsmessungen in anderen Lösungsmitteln sind in Tab. 1 zusammengestellt. Die eingewogenen Säuremengen wurden auf 10,00 ml gelöst und die Drehung (2 α) in 2-dm-Rohr gemessen.

Tabelle 1.

Lösungsmittel	g Säure	$2\alpha_D^{25}$	$[\alpha]_D^{25}$	$[M]_D^{25}$
Essigester	0,1582	−2,885°	−91,1°	−177,0°
Alkohol (abs.)	0,1186	−1,94	−81,8	−158,8
Äther	0,2176	−3,87	−88,9	−172,6
Aceton	0,1244	−2,21	−88,8	−172,5
Chloroform	0,1217	−2,445	−100,5	−195,0
Benzol	0,1169	−3,18	−136,0	−264,1
Verdünnte Sodalösung	0,1218	−0,60	−24,0	−46,5

Die Drehung in Essigester wurde auch bei höherer Konzentration gemessen; dabei wurde ein etwas niedrigerer Wert gefunden.

0,3057 g Sbst. in Essigester auf 10,00 ml gelöst: $2\alpha = -5,50°$. $[\alpha]_D^{25} = -90,0°$; $[M]_D^{25} = -174,8°$.

Das Cinchonidinsalz der (—)-Xantogen-propionsäure, das bei der Spaltung direkt erhalten wurde, bildet feine Nadeln. Der Wassergehalt des lufttrockenen Salzes stimmt am besten auf ein Mol Kristallwasser, doch scheint es zweifelhaft ob ein gut definiertes Hydrat vorliegt. Das Salz wurde deshalb in völlig entwässertem Zustande analysiert.

32,88 mg Sbst.: 31,10 mg $BaSO_4$.

$C_{19}H_{22}ON_2 + C_6H_{10}O_3S_2$ (488,4) S Ber. 13,13 Gef. 13,19

(+)-*Xantogen-propionsäure.* Das Strychninsalz wurde in Sodalösung suspendiert und das Strychnin mit Chloroform ausgeschüttelt. Dann wurde mit verdünnter Schwefelsäure angesäuert, die Xantogen-propionsäure in Äther aufgenommen und wie die Antipode umkristallisiert. 12 g Salz gaben 3,0 g Säure; nach dem Umkristallisieren verblieben 2,1 g vom Schmp. 70—70,5°.

16,70 mg Sbst.: 40,15 mg $BaSO_4$.

$C_6H_{10}O_3S_2$ (194,2) S Ber. 33,02 Gef. 33,02.

0,1058 g Sbst. in Essigester auf 10,00 ml gelöst: $2\alpha = +1,945°$. $[\alpha]_D^{25} = +92,0°$; $[M]_D^{25} = +178,5°$.

Das Strychninsalz der (+)-Xantogen-propionsäure, das bei der Spaltung dargestellt wurde, bildet dünne Nadeln, die nach der Schwefelanalyse ein Mol Kristallwasser enthalten. Das Salz lässt sich jedoch im Vakuumexsiccator ohne Gewichtsverlust aufbewahren.

35,86 mg Sbst.: 30,87 mg $BaSO_4$. — 36,80 mg Sbst.: 31,45 mg $BaSO_4$. — 38,66 mg Sbst.: 33,12 mg $BaSO_4$.

$C_{21}H_{22}O_2N_2 + C_6H_{10}O_3S_2 + H_2O$ (546,4) S Ber. 11,74
» Gef. 11,82, 11,74, 11,77

$$C_2H_5S-\underset{\underset{S}{\|}}{C}-O-\underset{\underset{CH_3}{|}}{CH}-COOH$$

$C_6H_{10}O_3S_2$ Ref. 203

(-)-Äthyl-carbothiolon-milchsäure. 35 g racem-Säure wurden nebst 22g (-)-Phenäthylamin unter Turbinieren und mässigem Erwärmen in 750 ml Wasser gelöst. Eine geringe Ölmenge blieb ungelöst und wurde abfiltriert. Die Lösung wurde mit einigen Impfkristallen (bei einem Vorversuch erhalten) versetzt. Nach einem Tage in Kälteschrank hatten sich 15,1g gut kristallisiertes Salz ausgeschieden. Es wurde wie oben in 250 ml Wasser gelöst. Nach Entfarben mit Kohlenpulver und Filtrieren kristallisierte im Kälteschrank allmählich 7,7g Salz in schön glänzenden, zentimeterlangen Prismen. Nach Zerlegen des Salzes mit Schwefelsäure wurde die Säure in alkoholfreiem Äther aufgenommen, und die Lösung mit $CaCl_2$ getrocknet. Beim freiwilligen Eindamfen im Exsickator kristallisierte die Säure nach enigem Stehen spontan. Sie ist in den meisten organischen Lösungsmitteln (auch Benzin) sehr leicht löslich, lässt sich jedoch aus Petroläther (Spkt 30-50°) umkristallisieren, und wird dann in grossen, würfelähnlichen, an käufliches Jodkalium erinnernden Kristallen erhalten Ausbeute 4,2g vom Schmp. 63-64°.

0, 2058g in Aceton auf 10,00 ml gelöst; $2\alpha = -0,40°$: $[\alpha]_D^{25} = -9,7°$; $[M]_D^{25} = -18,9°$. 0,2042g in Essigester auf 10,00 ml gelost: $2\alpha = -0,25°$; $[\alpha]_D^{25} = -6,1°$: $[M]_D^{25} = -11,9°$. 0,2081 g in abs. Alkohol auf 10,00 ml gelöst: $2\alpha = -0,185°$; $[\alpha]_D^{25} = -4,4°$; $[M]_D^{25} = -8,6°$; 0,2038 g in Eisessig auf 10,00 ml gelost: $2\alpha = -0,385°$; $[\alpha]_D^{25} = -9,4°$; $[M]_D^{25} = -18,3°$. 0,2071g in Chloroform auf 10,00 ml gelöst: $2\alpha = -0,59°$; $[\alpha]_D^{25} = -14,2°$; $[M]_D^{25} = -27,7°$. 0,2071g in Benzol auf 10,00 ml gelöst: $2\alpha = -1,985°$: $[\alpha]_D^{25} = -47,9°$: $[M]_D^{25} = -93,1°$.

Die Drehungswerte stimmen mit den von Holmberg (5) mitgeteilten, in konzentrierteren Lösungen (ca 1,0g Säure in 10 ml Losung) erhaltenen sehlecht überein. Er fand z. B. in Aceton $[\alpha]D = -12,6°$. Bei einer Messung bei etwa derselben Konzentration fand ich aber gute Übereinstimmung: 0,5162g in Aceton auf 5,00 ml gelöst: $\alpha = 1,345°$; $[\alpha]_D^{25} = -13,0°$.

Die spezifische Drehung zeigt also bei dieser Säure eine auffallend starke Konzentrationsabhängigkeit.

Die Drehungen sind in Tab. 1 zusammengestellt; die Werte für saure und neutralisierte wässrige Lösung sind der Arbeit von Holmberg entnommen.

(+)-Äthyl-carbothiolon-milchsäure. Aus der Mutterlauge der ersten Kristallisation des (-)-Phenäthylaminsalzes liess sich eine rechtsdrehende Säure isolieren die nebst 16,5g (+)-Phenäthylamin in 550 ml Wasser wie oben gelöst wurde. Die auskristallisierende Salzmenge (15,7g) wurde zweimal aus 200 bzw. 130 ml Wasser umkristallisiert; dabei verblieben 6,1 g. Die Säure wurde wie die Antipode isoliert und umkristallisiert. Ausbeute 3,6g vom Schmp. 63-64°.

Holmberg hat auch eine bei 36-37° schmelzende, instabile Modifikation der Säure beschrieben. Dies wurde bei meinen Arbeiten nie beobachtet.

$$C_2H_5SSCO\underset{\underset{CH_3}{|}}{CH}COOH$$

$C_6H_{10}O_3S_2$ Ref. 204

Bei einem Vorversuch wurden dabei 7.7 g der inaktiven Säure und 4.8 g (—)-Phenäthylamin mit 100 ccm Wasser gut durchgearbeitet; hierbei ging eine zuerst entstandene, klebrige Masse allmählich in Lösung, während sich an ihrer Stelle bald kleine Mengen eines schwach gelblichen Öles ausschieden. Nach einigem Stehen erstarrte das Öl von selbst; aus der Lösung schieden sich außerdem kleine, farblose Prismen ab, so daß zusammen 4.4 g Salz vom Schmp. ca. 125—129° gewonnen wurden. Das Salz wurde mit 15 ccm 2-*n.* Salzsäure und 25 ccm Äther geschüttelt, wonach die ätherische Schicht bei dem freiwilligen Eindunsten ein Öl ergab, welches zum Schluß von selbst in 2.5 g einer weißen Krystallmasse vom Schmp. ca. 40°, Äquiv.-Gew. 193.4 und $[\alpha]_D = -5.66°$

(für 1.01 g in 10 ccm absol. Alkohol) überging. Durch Behandlung mit Dimethyl-amin in der unten näher beschriebenen Weise wurde aus dieser Säure eine Dimethyl-amidocarbothion-milchsäure vom $[\alpha]_D = -49.2^0$ (für 0.67 g in 10 ccm absol. Alkohol) erhalten. Die Mutterlauge des Phenäthylamin-Salzes gab mit 25 ccm 2-*n*. Salzsäure eine Emulsion, welche nach Impfen mit inaktiver Säure 3.2 g einer Säure vom Schmp. ca. 65—70° und $\alpha_D = +0.11^0$ (für 1.08 g in 10 ccm absol. Alkohol) ergab; aus dem Filtrat von dieser Säure wurde durch Extraktion mit Äther eine ölige Äthyl-carbothiolon-milchsäure erhalten, aus welcher 1.0 g einer Dimethyl-amidocarbothion-milchsäure vom Schmp. ca. 118—123° (unter Aufschäumen), Äquiv.-Gew. 177.1 (ber. 177.2) und $[\alpha]_D = +59.2^0$ (für 0.60 g in 10 ccm absol. Alkohol) dargestellt wurde.

Bei einer Spaltungsreihe in größerem Maßstabe wurden dann 58 g einer Äthyl-carbothiolon-milchsäure, die 7 g bereits schwach linksdrehender, aus Vorversuchen stammender Säure enthielt, $^1/_2$ Stde. in gelinder Wärme mit 36 g (—)-Phenäthylamin und 1200 ccm Wasser turbiniert. Nach dem Abfiltrieren einer Spur ungelösten Öls emulgierte sich die Lösung beim Erkalten und schied dann nach Impfen mit Krystallen von einem Vorversuch bis zum nächsten Tage 25 g glasglänzender Prismen aus. Das Filtrat gab mit Schwefelsäure ein Öl, welches allmählich teilweise erstarrte, so daß nach 2 Tagen 25 g fast inaktiver Säure abgenutscht werden konnten, während mit der wäßrigen Lösung 8.8 g Öl durch das Filter gingen. Dieses Öl gab mit Dimethylamin-Lösung 5.3 g einer Dimethyl-amidocarbothion-milchsäure vom $[\alpha]_D = +58.1^0$ (für 0.5 g in 10 ccm absol. Alkohol), und der wäßrigen Lösung wurden mit Äther 7 g einer Äthyl-carbothiolon-milchsäure entzogen, welche 5.3 g einer Dimethyl-amidocarbothion-milchsäure vom $[\alpha]_D = +55.2^0$ ergab.

Das Phenäthylamin-Salz schmolz bei 132—134° unter Gasentwicklung und gab bei der Analyse[15]):

0.3665 g Sbst.: 14.80 ccm N (20°, 753 mm).
$C_{14}H_{21}O_3NS_2$ (315.3). Ber. N 4.45. Gef. N 4.55.

Aus 24 g des Salzes wurde mit Salzsäure ein Öl erhalten, welches mit Äther aufgenommen wurde und beim freiwilligen Verdunsten des Äthers zwar wieder als Öl zurückblieb, allmählich aber spontan zu einer kompakten Krystallmasse von 15 g Gewicht, Schmp. 60—62° und $[\alpha]_D = -8.8^0$ (für 1.015 g in 10 ccm absol. Alkohol) erstarrte. Von dieser Säure wurden 14.6 g mit 9 g (—)-Phenäthylamin und 400 ccm Wasser wie oben behandelt, wonach 12.2 g Salz erhalten wurden, während aus der Mutterlauge nach Zusatz von Schwefelsäure mit Äther 7 g einer Säure gewonnen wurden, welche in 5.5 g Dimethyl-amidocarbothion-milchsäure von $[\alpha]_D = -57.6^0$ (für 0.5 g in 10 ccm absol. Alkohol) übergeführt werden konnte. Das Salz schmolz bei 137—138° unter Aufschäumen und gab nach Behandlung mit Salzsäure und Äther beim Verdunsten des Äthers ein Öl, welches erst nach Impfen mit der unten beschriebenen *l*(+)-Äthyl-carbothiolon-milchsäure vom Schmp. 36—37°[16]) zu einer festen Masse vom gleichen Schmelzpunkt und dem $[\alpha]_D = -9.8^0$ (für 1.04 g in 10 ccm absol. Alkohol) erstarrte. Von dieser Säure wurden zum Schluß 6.8 g mit 4.2 g (—)-Phenäthylamin und 200 ccm Wasser wie oben behandelt, wobei 5.2 g Salz erhalten wurden. Die Mutterlauge ergab 3.2 g einer Säure, aus welcher eine reine *d*(—)-Dimethyl-amidocarbothion-milchsäure von $[\alpha]_D = -70.0^0$ (für 0.5 g in 10 ccm absol. Alkohol) dargestellt wurden, während aus dem Salz, welches bei 137—138° unter Aufschäumen schmolz, in der gewöhnlichen Weise 3.1 g reine *d*(—)-Äthyl-carbothiolon-milchsäure als eine strahlig-krystallinische, farblose Masse vom Schmp. 36—37° und unverändertem Drehungsvermögen isoliert wurden.

0.2319 g Sbst.: 10.62 ccm 0.1131-*n*. Baryt, also Äquiv.-Gew. gef. 193.1, ber. 194.2. Die titrierte Lösung gab beim freiwilligen Eindunsten einen Sirup, der zum Schluß in unansehnlichen Efflorescenzen und eine zum Teil strahlig-krystallinische, zum Teil glasige Masse überging.

---

[16]) Leider wurde keine Probe von der sehr rasch krystallisiert erhaltenen *d*(—)-Säure vom Schmp. 60—62° aufbewahrt, und es dauerte dann lange, bis die hochschmelzende Form gegen Ende der Untersuchung wieder auftauchte.

---

[15]) Sämtliche Elementaranalysen verdanke ich Hrn. Privatdozent Dr. S. Kallenberg.

1.028 g Sbst., in absol. Alkohol zu 10.04 ccm gelöst: $\alpha_D = -0.99^0$; $[\alpha]_D = -9.7^0$ und $[M]_D = -18.9^0$.

1.0258 g Sbst., in Aceton zu 10.04 ccm gelöst: $\alpha_D = -1.30^0$; $[\alpha]_D = -12.7^0$ und $[M]_D = -24.7^0$.

1.0063 g Sbst., in Essigester zu 10.04 ccm gelöst: $\alpha_D = -0.82^0$; $[\alpha]_D = -8.2^0$ und $[M]_D = -15.9^0$.

Beim Schütteln der pulverisierten Säure mit Wasser bei 20° blieb sie fest; 10.00 ccm der gesättigten Lösung verbrauchten zum Neutralisieren 2.90 ccm 0.1131-*n*. Baryt, was einer Löslichkeit von 0.0328 Mol. oder 6.37 g im Liter entspricht. Die gesättigte Lösung zeigte $2\,\alpha_D = +0.11^0$, $[\alpha]_D = +8.6^0$ und $[M]_D = +16.8^0$. Nach Abfiltrieren der ungelöst gebliebenen Säure wurde ihr Schmelzpunkt zu 63—64° gefunden; sie hatte sich also während des Aufbewahrens in die höher schmelzende Form verwandelt.

1.0245 g mit 9.32 ccm 0.5659-*n*. Natronlauge neutralisierter und mit Wasser zu 20.0 ccm verdünnter Säure zeigten $2\,\alpha_D = +0.55^0$, $[\alpha]_D = +5.4^0$ und $[M]_D = +10.4^0$. Die polarisierte Lösung wurde mit Salzsäure versetzt und mit Äther extrahiert, wonach der Äther beim freiwilligen Verdunsten ein Öl ergab, das nach dem Impfen mit der oben erwähnten *l*(+)-Säure zu einer strahlig-krystallinischen Masse von dem früheren Schmelzpunkt, 36—37°, erstarrte.

Von der nach der Löslichkeits-Bestimmung zurückgewonnenen Säure vom Schmelzpunkt 63—64° wurden 2.3 g mit Dimethylamin behandelt, wobei 1.8 g Dimethyl-amidocarbothion-milchsäure vom Schmp. 123—124° (unter Gasentwicklung) und $[\alpha]_D = -70.1^0$ (für 0.5 g in 10 ccm absol. Alkohol) gewonnen wurden.

c) *l*(+)-Äthyl-carbothiolon-milchsäure.

Zur Darstellung der rechtsdrehenden Äthyl-carbothiolon-milchsäure wurde die wahrscheinlich nur wegen anhaftender, öliger (+)-Säure schon schwach aktive, aus der Mutterlauge des ersten (—)-Phenäthylamin-Salzes (s. oben) erhaltene, krystallisierte Säure mit (+)-Phenäthylamin in der bei der Spaltung mit der (—)-Base beschriebenen Weise behandelt und dabei zum Schluß ein Salz vom Schmp. 136—137.5° (unter Aufschäumen) gewonnen, aus welchem eine reine *l*(+)-Säure vom Schmp. 36—37° und Äquiv.-Gew. 194.0 isoliert wurde.

1.0449 g Sbst., in absol. Alkohol zu 10.04 ccm gelöst: $\alpha_D = +1.02^0$; $[\alpha]_D = +9.80^0$ und $[M]_D = +19.0^0$.

1.0410 g Sbst., in Aceton zu 10.04 ccm gelöst: $\alpha_D = +1.31^0$; $[\alpha]_D = +12.6^0$ und $[M]_D = +24.5^0$.

Von der bei 20° gesättigten wäßrigen Lösung dieser Säure, welche beim Zusammenbringen mit dem Wasser sogleich zu einem schwach gelblichen Öl zerfloß, verbrauchten 10.00 ccm 3.65 ccm 0.1131-*n*. Baryt, der Löslichkeit von 0.0412 Mol. oder 8.02 g im Liter entsprechend. Im 2-dm-Rohr zeigte die Lösung $2\,\alpha_D = -0.14^0$, $[\alpha]_D = -8.7^0$ und $[M]_D = -17^0$. Beim freiwilligen Verdunsten des Wassers gab die polarisierte Lösung ein Öl, welches nach dem Impfen mit der oben erwähnten *d*(—)-Säure vom Schmp. 63—64° zu einer strahlig-krystallinischen Masse von demselben Schmelzpunkt erstarrte; als dann das Gemisch von öliger Säure und gesättigter wäßriger Lösung mit einer Probe dieser Form geimpft wurde, erstarrte das Öl, neben welchem sich auch eine Spur kleiner Prismen ausschied. Von der dann nach zwei Tagen abfiltrierten Lösung verbrauchten 10.00 ccm 2.95 ccm 0.1131-*n*. Baryt, was also fast genau derselben Löslichkeit, nämlich 0.0333 Mol. oder 6.48 g im Liter, entspricht, welche für die entsprechende Form der *d*(—)-Säure gefunden wurde. Die erstarrte Säure schmolz auch bei 63—64° und zeigte das unveränderte Drehungsvermögen $[\alpha]_D = +12.3^0$ für 1.0 g in 10 ccm Aceton, und die aus ihr gewonnene Dimethyl-amidocarbothion-milchsäure bestand aus reiner *l*(+)-Form vom Schmp. 123—124° (unter Aufschäumen) und $[\alpha]_D = +70.2^0$ für 0.5 g in 10 ccm absol. Alkohol.

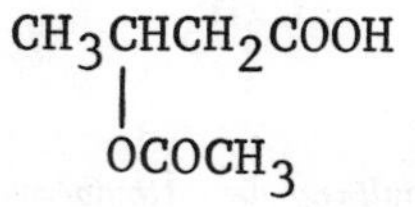

$C_6H_{10}O_4$ Ref. 205

(*R*)-(−)- and (*S*)-(+)-3-Acetoxybutyric Acid (1 and 6).—(±)-3-Acetoxybutyric acid (31.89 g, 0.218 mol) was mixed in ethyl acetate (*ca.* 100 ml) with 70.8 g (0.218 mol) of anhydrous quinine, and the mixture was heated until most of the quinine had dissolved. The quinine residue was removed by filtration through a glass wool plug. The filtrate was reheated to near the boiling point and petroleum ether (bp 60–110°) was added until cloudiness persisted. The solution was again rewarmed to produce homogeneity, seeded, and cooled to room temperature and then to 0°. After 3 days at 0°, the white crystalline salt was filtered and washed twice with ethyl acetate–petroleum ether (1:1). Fractional crystallization of this solid from ethyl acetate–petroleum ether gave 20 g of quinine salt, mp 101.5–104°.

(*S*)-(+)-3-Acetoxybutyric acid (6) was recovered quantitatively by treatment of this quinine salt with 5% hydrochloric acid, saturation of the solution with sodium chloride, extraction with ether, and concentration of the combined organic extracts. Molecular distillation at 70° (0.01 mm) gave a colorless oil, $[\alpha]^{24}D$ +3.6 ± 0.3° (*c* 10.175, C_2H_5OH) [lit.[6b] $[\alpha]^{23}D$ +2.78° (homogeneous, *l* 1)].

(*R*) (−)-3-Acetoxybutyric acid (1) was obtained similarly from the mother liquor residues from several resolutions by treatment with 5% hydrochloric acid and extraction in the above manner. The concentrated oil was distilled at 95–99° (0.05 mm) and there was obtained 25.4 g of 1. This acid was reduced directly to 2 for measurement of optical rotation.

(6) For the absolute configurational assignment, see (a) K. Serck-Hanssen, *Ark. Kemi*, 8, 401 (1955); (b) K. Serck-Hanssen, S. Stallberg-Stenhagen, and E. Stenhagen, *ibid.*, 5, 203 (1953).

$C_6H_{10}O_4$ Ref. 205a

(−)-3D-Acetoxybutanoic acid (II)

Resolution of the racemic acid (I) is remarkably easy, as the morphine salt of the levorotatory acid may be obtained in optically pure form without recrystallization [3, 4]. The yield of free acid has been raised from about 45 % [4] to 73 % by a diminution of the losses caused by the water solubility of the acid:

The morphine salt (10 g) prepared as already described [4], was dissolved in water (100 ml) and the alkaloid precipitated by addition of a slight excess of dilute ammonia water. Immediate filtration into an excess of dilute hydrochloric acid was at once followed by extraction with six portions of ethyl ether. After washing the ether solution with two small portions of water saturated with sodium sulphate, filtration and evaporation, the liquid acid was dried at about 15 mm Hg over sulphuric acid at 0° overnight. The yield (3.25 g) of crude (−)-3D-acetoxybutanoic acid (*M* 146.1) corresponds to the theoretical amount present in a 1:1 morphine salt (10 g) containing one molecule of water. The optical activity of undiluted acid changes slowly at room temperature [3], and the rotations recorded below were therefore measured on the freshly isolated sample.

Solvent	*c*	*l*	α_D^{21}	$[\alpha]_D^{21}$	$[M]_D^{21}$
None	–	1/2	− 1.33° ± 0.05°	–	–
CCl_4	4.8	1	− 0.12° ± 0.02°	− 2.5° ± 0.4°	− 3.7° ± 0.6°
$CHCl_3$	4.9	1	− 0.25° ± 0.02°	− 5.1° ± 0.4°	− 7.5° ± 0.6°
CH_3OH	5.7	1	− 0.44° ± 0.02°	− 7.7° ± 0.4°	− 11.3° ± 0.6°

Optical activity formerly observed [3, 4] for undistilled and undiluted acid: α_D^{24} − 2.86° (*l* 1).

3. Serck-Hanssen, K., Ställberg-Stenhagen, S., and Stenhagen, E., Arkiv Kemi *5*, 203 (1953).

4. Serck-Hanssen, K., Arkiv Kemi *8*, 401 (1955).

```
HOOC-CH-CH2CH2-COOH
     |
     CH3
```

$C_6H_{10}O_4$ Ref. 206

Optische Spaltung der α-Methylglutarsäure.

Die Spaltung der α-Methylglutarsäure haben wir vor einigen Jahren in einer vorläufigen Mitteilung[3]) beschrieben. Sie wurde am besten mittels Strychnin in der folgenden Weise durchgeführt: In eine warme Lösung von 10 g Säure in 100 ccm Wasser wurden 22,9 g Strychnin (1 Mol Alkaloid auf 1 Mol Säure) in kleinen Anteilen eingetragen und die Erhitzung fortgesetzt bis alles gelöst war. Schon nach ½ Stunde bei gewöhnlicher Temperatur fing das Salz der rechts-Säure in rosettenähnlichen Aggregaten zu krystallisieren an. Nach 2-tägigem Stehen im Kühlschrank wurde das Salz abfiltriert und zwischen Filtrierpapier getrocknet. 65 g Säure gaben 107 g Salz, das 3-mal aus Wasser umkrystallisiert wurde. Aus dem so erhaltenen Salz (50 g) isolierten wir 14,5 g einer Säure mit der spez. Drehung + 20,2° (Wasser, *p* = 4,8). Die Mutterlauge lieferte beim Eindampfen etwas mehr Salz, das nach Umkrystallisieren 7,5 g Säure von + 20,4° (*p* = 3,8) ergab. Die beiden Portionen der d-Säure wurden zusammen aufs neue mit Strychnin behandelt, wobei eine Säure mit der Drehung + 20,4° (*p* = 4,3) erhalten wurde. Eine dritte Behandlung mit Strychnin änderte diesen Wert nicht (+ 20,5° bei *p* = 3,8), und wir sehen die gewonnene Säure als die reine d-Komponente an. Schmelzp. 81°.

Aus den gesammelten Mutterlaugen von dem Strychninsalz isolierten wir 28,5 g l-Säure mit spez. Drehung ÷ 11° und 13,5 g mit ÷ 8°. Weitere Versuche zur Reinigung dieser Komponente wurden aber nicht unternommen.

Die Bestimmungen des Drehungsvermögens der d-Säure in Wasser und absolutem Alkohol bei verschiedenen Konzentrationen sind in Tab. XIV angegeben.

Tabelle XIV.

Spez. Drehungsvermögen der d-α-Methylglutarsäure in wäßriger und absolut alkoholischer Lösung.

In Wasser			In absolutem Alkohol		
p	$d_4^{t°}$	$[\alpha]_D^{t°}$ in °	*p*	$d_4^{t°}$	$[\alpha]_D^{t°}$ in °
7,280	1,0139	+ 20,04	5,268	0,8078	+ 21,74
14,112	1,0296	19,23	24,350	0,8703	21,38
25,759	1,0561	18,34	44,013	0,9441	21,22
35,065	1,0766	17,88	53,697	0,9837	21,25

[3]) Kgl. Norske Videnskabers Selskab. Forhandl. 7 Nr. 35 (1934).

$C_6H_{10}O_4$ Ref. 206a

(+)-*α-Methylglutaric acid* was isolated over the acid strychnine salt. The rotatory power was found to be in close agreement with the statements by Berner and Leonardsen (3), the melting point a little higher (82.5—83°).

0.5200 g made up to 10.00 ml with water: 2 α_D^{25} = + 2.18°.
$[\alpha]_D^{25}$ = + 20.5°; $[M]_D^{25}$ = + 29.9°.

The same solution was measured at 20.0°: 2 α_D^{20} = + 2.175°.
$[\alpha]_D^{20}$ = + 20.9°; $[M]_D^{20}$ = + 30.6°.

(—)-*α-Methylglutaric acid.* According to a preliminary statement by Berner and Leonardsen (16), the brucine salt of the levogyric acid is less soluble than that of the antipode. On the other side, a moderately levogyric acid could be isolated from the mother liquors from the first crystallisation of the strychnine salt. 44 g of this acid, showing $[\alpha]_D$ = — 9°, and 280 g of brucine were dissolved in 400 ml of hot water. The salt deposited on cooling amounted to 171 g. Contrary to expectation, it was found to contain a practically inactive acid. The mother liquor from the brucine salt, however, yielded 20 g of an acid showing $[\alpha]_D$ = —17,1° (after crystallisation from benzene).

19.3 g of this acid were finely powdered and mixed with 50 ml acetyl chloride. After refluxing for 20 minutes, acetic acid and excess of acetyl chloride were distilled off, first at ordinary pressure, finally in vacuum, the temperature of the bath not exceeding 80°. On cooling, the anhydride crystallised readily. It was dissolved in hot benzene; on careful addition of petroleum (b. p. 60—70°) it was deposited as short prismatic crystals. After three crystallisations the yield was 13.0 g. The anhydride had the same melting point as the antipode (3).

13.0 g of the anhydride were dissolved in 15 ml of hot water. The acid deposited on cooling was filtered off and recrystallised from benzene. Yield 9.2 g, m. p. 82.5—83°. A second crop amounting to 4.6 g (after two recrystallisations from benzene) was obtained from the mother liquors.

0.1309 g: 14.17 ml 0.1268-N sodium hydroxide.
$C_6H_{10}O_4$ E. W. calc. 73.07 found 73.1

The activity data for the acid are found in table 1, those for the (—)-αα′-dimethylglutaric acid in table 2.

Table 1.

(—)-α-Methylglutaric acid. *v* = 10.00 ml.

	g acid	2 α	$[\alpha]_D^{25}$	$[M]_D^{25}$
Water	0.5299	− 2.165°	− 20.4°	− 29.9°
Alcohol (abs.) . .	0.2060	− 0.895°	− 21.7°	− 31.7°
Ethyl acetate . .	0.2055	− 1.015°	− 24.7°	− 36.1°
Chloroform . . .	0.2243	− 1.125°	− 25.1°	− 36.7°
Acetic acid . . .	0.2153	− 1.02°	− 23.7°	− 34.6°

(3) **Berner** and **Leonardsen,** Ann. *538*, 1 (1939).
(16) **Berner** and **Leonardsen,** Kgl. Norske Videnskab. Selskabs Forh. *7*, 125 (1934). —

```
HOOC-CHCH2COOH
     |
     C2H5
```

$C_6H_{10}O_4$ Ref. 207

Optische Spaltung der Äthylbernsteinsäure.

Die Durchführung der optischen Spaltung von Äthylbernsteinsäure ist erstmalig von E. Berner und J. Molland[1]) in einer vorläufigen Mitteilung veröffentlicht. Sie verwendeten das saure Cinchoninsalz und erhielten eine rechtsdrehende Säure mit dem $[\alpha]_D$ = + 12,58° in 11,3-proc. wäßriger Lösung. Es hat sich aber gezeigt, daß diese Säure nicht die reine d-Komponente war. Sie bestand vielmehr zu ⅓ aus r-Säure und zu ⅔ aus d-Säure, für die wir jetzt den richtigen Wert + 18,5° bei derselben Konzentration bestimmt haben.

Auch H. Wren und F. Crawford[2]) spalteten die Äthylbernsteinsäure, für die sie ein Drehungsvermögen von + 20,65° und ÷ 20,8° der beiden Komponenten in 4-proc. Acetonlösung fanden. Ihre Säure hatte aber nicht die maximale Drehung, die nach unseren Bestimmungen 26,0° bei derselben Konzentration in Aceton ist. Die optisch aktiven

Säuren von Wren und Crawford zeigten auch einen niedrigeren und unscharfen Schmelzpunkt von 83—85°, gegen den von uns gefundenen scharfen Schmelzpunkt von 96° für die reinen Antipoden.

Wir haben die Spaltung der Äthylbernsteinsäure mit Strychnin durchgeführt. Zuerst verfuhren wir in der Weise, daß eine 2-proc. Lösung der Säure in Wasser kochend mit Strychnin gesättigt und das ausgeschiedene Salz mehrmals aus Wasser umkrystallisiert wurde. Die freigemachte Säure wurde noch einmal mit Strychnin in derselben Weise behandelt, und lieferte dann eine d-Komponente von maximaler Drehung. Später fanden wir, daß die Trennung der rac. Äthylbernsteinsäure in die beiden optisch reinen Komponenten sich nach dem folgenden Verfahren erreichen ließ. Die rac. Säure wurde zuerst mittels Strychnin in optisch aktive Komponenten gespalten, die eine spez. Drehung von mindestens 13° besaßen. Durch Einwirkung von Acetylchlorid oder Thionylchlorid wurden die entsprechenden Anhydride dargestellt und durch Destillation i. V. (Siedep. 129—130° bei 17 mm) gereinigt. Nach kurzem Stehen in einem Kühlschrank krystallisierten die reinen Komponenten des Anhydrids aus und konnten durch Abpressen in einem Filtertiegel von dem flüssigen inaktiven Anhydrid befreit werden. Eine weitere Reinigung geschah durch Lösen in Äther und Ausfällen mit Petroläther. Das hierdurch zuerst ausgeschiedene Anhydrid war flüssig, krystallisierte aber sofort. Durch Lösen der krystallisierten Anhydride in Wasser wurden die entsprechenden reinen aktiven Säuren erhalten. Das aus einer unreinen linksdrehenden Säure hergestellte Anhydrid hatte Schmelzp. 32,5° und gab eine Säure mit $[\alpha]_D^{20} = \div 18,5°$ in 9,6-proc. wäßriger Lösung. Das Anhydrid der rechtsdrehenden Säure hatte denselben Schmelzpunkt von 32,5° und gab beim Lösen in Wasser eine Säure mit $[\alpha]_D^{20} = +18,4°$ ($p = 9$).

Für die Ermittelung des spez. Drehungsvermögens bei verschiedenen Konzentrationen sowohl in Wasser wie in Alkohol müßten wir das spez. Gewicht der Lösungen kennen. Um Zeit zu ersparen bestimmten wir durch einige Versuche die Abhängigkeit des spez. Gewichtes (d_4^{20}) von der Konzentration (p = g/100 g) und erhielten die beiden folgenden Ausdrücke.

Für eine wäßrige Lösung:

(6) $$d_4^{20} = 0,9983 + 0,002229 . p$$

und für eine Lösung in absolutem Alkohol:

(7) $$d_4^{20} = 0,7906 + 0,002853 . p + 0,0000139 . p^2.$$

Die Resultate der Bestimmungen des Drehungsvermögens sind in Tab. VII zusammengestellt.

Tabelle VII.

Spez. Drehungsvermögen von d-Äthylbernsteinsäure in Wasser und in absolutem Alkohol. D-Licht bei 20°.

In Wasser			In absolutem Alkohol		
p	Gef. in °	Ber. in °	*p*	Gef. in °	Ber. in °
5,390	+ 18,03	+ 18,03	5,149	+ 20,78	+ 20,63
9,702	18,43	18,44	10,771	20,32	20,48
18,475	18,99	18,99	15,914	20,25	20,34
27,696	19,29	19,35	21,396	20,15	20,20
32,341	19,45	19,48	31,517	19,95	19,93
46,678	19,78	19,78	37,823	19,87	19,77
49,625	19,83	19,83	51,213	19,39	19,41
71,172	19,95	20,07	60,569	19,13	19,16
			73,059	18,82	18,84

Die Werte für die wäßrigen Lösungen lassen sich von 5—50 Proc. zufriedenstellend durch die Gleichung:

(8) $$[\alpha]_D^{20} = A + \frac{B\,q}{C + q}$$

[1]) Kgl. Norske Videnskabers Selskab. Forhandl. 9, Nr. 6 (1936).
[2]) Soc. 1937, 230.

$C_6H_{10}O_4$

Ref. 207a

r-Ethylsuccinic acid (26·95 g.) and quinine (59·9 g.) were dissolved in boiling alcohol (400 c.c.). The solution, on cooling, deposited 49·4 g. of crystals which were further purified by repeated crystallisation from alcohol. The course of the resolution was followed by observation of the specific rotation of the acids obtained from the successive filtrates, the following values of $[\alpha]_D$ in acetone being obtained: − 4·4° − 4·4°, − 1·8°, + 2·7°, + 7·7°, + 8·43°, +8·38°. The remaining crop (13·1 g.) on decomposition with hydrochloric acid and extraction with ether gave 3·8 g. of crude *d*-ethylsuccinic acid, m. p. 69—74°, $[\alpha]_D$ + 19·4°, in acetone. Repeated crystallisation of this product from benzene–light petroleum gave pure d-*ethylsuccinic acid*, m. p. 83·5—85°, $[\alpha]_D^{16·8°}$ + 20·65° in acetone ($l = 2$, $c = 3·7284$) (Found: C, 49·2; H, 6·9. $C_6H_{10}O_4$ requires C, 49·3; H, 6·9%).

The lævorotatory by-products from the resolution described above were converted into the brucine salts, which were crystallised repeatedly from absolute alcohol. The acid obtained from the filtrates had successively $[\alpha]_D$ + 2·85°, + 5·71°, + 4·3°, − 2·7°, − 5·6°, − 8·6° and − 13·4° in acetone. The residual crop yielded a crude *l*-acid with $[\alpha]_D$ − 20·9° in acetone. This, when crystallised repeatedly from benzene–light petroleum, yielded homogeneous l-*ethylsuccinic acid*, m. p. 83—85°, $[\alpha]_D^{23·8}$ − 20·80° in acetone ($l = 2$, $c = 4·5944$) (Found: C, 49·5; H, 6·8%).

$C_6H_{10}O_4$

Ref. 207b

S-(-)-2-Ethylsuccinic Acid (*S*-Ia).

Compound *RS*-Ia, prepared according to the method reported by Cloetzel (4), was resolved by fractional crystallization of its D-(-)-*threo*-1-*p*-nitrophenyl-2-amino-1,3-propanediol salt from ethanol. After four crystallizations *S*-Ia was obtained; $[\alpha]_D$ -24° (c = 3%, acetone) [lit. (5a) $[\alpha]_D$ +20.65°, (acetone) for the antipode].

Anal. Calcd. for $C_6H_{10}O_4$: C, 49.31; H, 6.90. Found: C, 49.49; H, 7.01.

(5a) H. Wren and J. Crawford, *J. Chem. Soc.*, 230 (1937);

```
      CH3
      |
HOOC-CH-CH-COOH
         |
         CH3
```

$C_6H_{10}O_4$

Ref. 208

I. The Active Dimethylbutanediols and Derivatives

D(2)A-(+)-2,3-Dimethylsuccinic Acid.—Racemic diacid[18,19] of m.p. 128–130° (24 g., 0.165 mole) and 99 g. (0.25 mole) of anhydrous *l*-brucine ($[\alpha]_D$ −127°, chloroform) were dissolved in 1300 ml. of boiling methanol. The filtered solution on cooling deposited 81.9 g. of crystals, m.p. 123–126°, $[\alpha]^{21}_D$ −40.0° (chloroform, *c* 2). After repeated recrystallization from methanol, 50 g. (65%) of the di-(−)-brucine (+)-dimethylsuccinate of constant rotation $[\alpha]^{21}_D$ −37.5° was obtained, m.p. 125–130°. The neutral brucine salt $(1B)_2dA$ must have been obtained, since the yield calculated for the acid salt (1B)HdA would be 113%.

A solution of 49 g. of the brucine salt in 500 ml. of chloroform was extracted thrice with 1 *M* sodium carbonate and once with water. The combined, filtered aqueous phases were strongly acidified with 5 *M* hydrochloric acid, saturated with sodium chloride and continuously extracted with ether for 20 hr. The (not dried) ether extract was evaporated to near-dryness, benzene added, and the mixture distilled to a small volume and filtered, giving 7.2 g. of cotton-like crystals, m.p. 128–133°. The product was recrystallized thrice from benzene, giving 5.7 g. (75% based on brucine salt) of pure (+)-diacid, m.p. 134–135° (preheat to 128°), $[\alpha]^{21}D$ +8.42° (water, *c* 4); reported[19b] m.p. 134–135°, $[\alpha]^{20}D$ +8.02° (water, *c* 4).

L(2)A-(−)-2,3-Dimethylsuccinic Acid.—The original methanolic mother liquor and first recrystallization filtrate from the above resolution were evaporated and the oily residue regenerated by the above procedure. This gave 12.5 g. (52%) of crude (−)-diacid, m.p. 122–130°, $[\alpha]^{23}D$ −4.24° (abs. ethanol, *c* 8); estimated aqueous rotation $[\alpha]^{23}D$ −7.50° (*c* 4). This crude product appears to contain about 11% of DL-impurity but would probably be satisfactory for preparation of the pure (+)-diol (see below).

Berner and Leonardsen reported[19b] that DL-impurity can be eliminated by a strychnine treatment, because the strychnine salt of the DL-diacid is relatively insoluble, but gave few details. Since the reported strychnine/diacid molar ratio of 1:2 failed to give a precipitate, we changed it to 3:2; actually this represents a large excess of strychnine with respect to the DL-diacid present. To 28 g. of the above crude (−)-diacid in 1100 ml. of hot *n*-butyl alcohol was added 96 g. of (−)-strychnine. On cooling, a crystalline precipitate with $[\alpha]^{23}D$ −138° (chloroform, *c* 2) was obtained, but appeared to consist largely of strychnine itself ($[\alpha]^{18}D$ −139°, chloroform, *c* 1). The yield including a second crop was 37 g.

The combined filtrates were evaporated, leaving an oil which was regenerated by the above brucine salt procedure, giving 14 g. of (−)-diacid with m.p. 122–125° and $[\alpha]^{23}D$ −4.31° (abs. ethanol, *c* 9) or −7.62° (water, *c* 4).

The strychnine treatment thus appeared nearly useless; however, when a filtered solution of this purified product in 28 ml. of warm water was treated with 14 ml. of 12 *M* hydrochloric acid and the mixture cooled, 8.7 g. of product melting at 132–134° (softens 126°) and having $[\alpha]^{25}D$ −8.06° (water, *c* 4) was obtained.

(18) (a) W. Bone and C. Sprankling, *ibid.*, **75**, 839 (1899); (b) most reported preparations of the racemic dimethylsuccinic acid have had a low m.p. We find that the pure compound is best obtained by brief hydrolysis of the anhydride (m.p. 91°) in warm water (2 parts). On addition at 0° of 12 *M* hydrochloric acid (1 part), crystals of m.p. 128–129.5° (preheat to 124°) were obtained; reported m.p. 129° (B. and S.).

(19) (a) A. Werner and M. Basyrin, *Ber.*, **46**, 3229 (1913); see also E. Ott, *ibid.*, **61**, 2134 (1928). The lower-melting (130°) dimethylsuccinic acid was incorrectly designated *meso* in the 1909 edition of Beilstein. Although the error was corrected in the 1919 supplement on the basis of W. and B.'s resolution, it has persisted in much subsequent literature. For example, on page 384 of the 1951 edition of F. Whitmore's "Organic Chemistry" it is incorrectly stated that the higher-melting acid (210°) is racemic and that "All attempts to separate either acid into optically active forms have failed." (b) E. Berner and R. Leonardsen, *Ann.*, **538**, 1 (1939).

$C_6H_{10}O_4$ Ref. 208a

2. Darstellung der optisch-aktiven Triäthylendiamin-kobaltibromide.

Das racemische Triäthylendiamin-kobaltibromid wurde aus *trans*-Dichloro-diäthylendiamin-kobaltisalzen dargestellt. Nach der von A. Werner[1]) beschriebenen Methode wurde Praseochlorid und aus den Mutterlaugen Praseonitrat dargestellt. Jedes dieser Salze wurde mit der berechneten Menge (1 Mol) 10-prozentiger Äthylendiaminlösung übergossen und ca. eine Stunde auf dem Wasserbad erwärmt. Hierauf wurde das Triäthylendiamin-kobaltibromid durch Bromnatrium ausgesalzen und das ausgeschiedene Salz umkrystallisiert. Die Spaltung der Triäthylendiamin-kobalti-Reihe erfolgte nach der schon beschriebenen Methode[2]) mit Hilfe des Bromid-tartrats.

3. Spaltungsversuche mit den Dimethyl-bernsteinsäuren.

Da in der Weinsäuregruppe von den beiden inaktiven Säuren (Traubensäure und Mesoweinsäure), die höher schmelzende spaltbar ist, wurde zunächst versucht, die höher schmelzende Dimethyl-bernsteinsäure zu zerlegen. Sowohl von *d*- als auch *l*-Triäthylendiamin-kobalt wurde das Dimethylsuccinat-bromid dargestellt, welche gut krystallisieren. Bei der Isolierung der Dimethyl-bernsteinsäure aus diesen Verbindungen konnte jedoch nur die inaktive Säure zurückgewonnen werden. Dieselben Versuche wurden deshalb mit der niedrigschmelzenden Säure durchgeführt, wobei sich zeigte, daß in diesem Falle eine Spaltung in aktive Formen eintritt. Die Spaltung wurde deshalb in größerem Maßstabe durchgeführt.

a) **Darstellung des Silbersalzes.** 56 g Dimethyl-bernsteinsäure vom Schmp. 127° (bei mehrmaligem Umkrystallisieren aus Wasser zeigte die Säure den richtigen Schmp. 128—129°) wurden in ca. 150 ccm heißem Wasser aufgelöst, mit Ammoniak neutralisiert und mit der berechneten Menge einer konzentrierten heißen Silbernitratlösung versetzt. Nach Abkühlen wurde das Silbersalz abfiltriert und mit heißem Wasser, Alkohol und Äther gewaschen und im Trockenschrank getrocknet.

b) **Darstellung der Triäthylendiamin-kobaltibromid-dimethylsuccinate.** 200 g aktives $[Co\ en_3]Br_3 + 2H_2O$ wurden pulverisiert, getrocknet und dann mit dem trocknen Silbersalz der Dimethyl-bernsteinsäure innig gemischt. Das Gemisch, mit heißem Wasser übergossen und auf dem Wasserbade einige Zeit verrieben, ergab eine gelbe Lösung, welche vom Niederschlag abfiltriert wurde. Der feste Rückstand wurde so lange mit heißem Wasser ausgezogen, bis die Waschwasser nur noch schwach gelb waren. Sämtliche Lösungen wurden vereinigt und dann soweit eingedampft, bis sich eine reichliche Menge von Krystallen ausgeschieden hatte. Das Eindampfen muß soweit erfolgen, weil der Löslichkeitsunterschied dieses Salzes in heißem und in kaltem Wasser nicht groß ist.

Die ausgeschiedenen Krystalle bestehen aus *d*-Triäthylendiamin-kobaltibromid-*l*-dimethylsuccinat. Zur vollständigen Reinigung werden sie noch drei- bis viermal aus heißem Wasser umkrystallisiert. Beim weiteren Eindampfen der Mutterlauge gewinnt man noch mehr Krystalle, die aber erst nach öfterem Umkrystallisieren rein erhalten werden. Zum Schluß bildet die Mutterlauge einen dicken Sirup, der bei sehr langem Stehen manchmal Krystalle ausscheidet. Dieser Sirup enthält das Salz der Rechtssäure.

c) **Gewinnung der aktiven Säuren.**

Zur Darstellung der *d*-Säure wird die Mutterlauge mit dem vierfachen Volumen heißem Wasser verdünnt und mit einer heißen konzentrierten Silbernitratlösung vermischt. Der gebildete Niederschlag, aus dem Silbersalz der *d*-Dimethyl-bernsteinsäure und Bromsilber bestehend, wird abgesaugt und mit heißem Wasser so lange ausgewaschen, bis das Waschwasser vollständig farblos ist. Dann wäscht man mit Alkohol und Äther nach und trocknet das Salzgemisch (1 Mol. Silberdimethylsuccinat + 1 Mol. Silberbromid). Hierauf wird es mit der zur Zerlegung des Silber-dimethylsuccinats nötigen Menge Salzsäure (1:10) während einer Viertelstunde auf dem Wasserbad erwärmt und die entstandene Lösung der aktiven Dimethyl-bernsteinsäure vom Halogensilber abfiltriert. Beim Eindampfen der Lösung, bis sie dickflüssig geworden ist, scheidet sich die aktive Säure in schönen Krystallen aus. Die Darstellung der *l*-Säure aus dem krystallisierten Bromid-dimethylsuccinat erfolgte in ganz gleicher Weise.

d) **Eigenschaften der aktiven Dimethyl-bernsteinsäuren.**

Die nach dem soeben beschriebenen Verfahren aus der Mutterlauge gewonnene Säure zeigte in 5-prozentiger Lösung ein spezifisches Drehungsvermögen von $[\alpha]_D = +6°$. Beim mehrmaligen Umkrystallisieren aus Wasser und dann aus Alkohol erhöhte sich die Drehung auf $[\alpha]_D = 7.8°$. Der Schmelzpunkt der Säure stieg dabei von 132° auf 135°. Bei der Schmelzpunktbestimmung darf man nicht langsam erhitzen, sondern man muß ein vorgewärmtes Heizbad anwenden, weil sonst der Schmelzpunkt zu tief gefunden wird. Eine Säure, welche während einer Stunde auf 125° erhitzt worden war, sinterte und schmolz dann bei dieser Temperatur zusammen. Beim Trocknen der Säure darf man nicht über 80° gehen, weil auch dadurch Schmelzpunkterniederung eintritt.

Die in gleicher Weise aus den Krystallen gewonnene *l*-Säure zeigte den Schmp. 135° und ein spezifisches Drehungsvermögen von $[\alpha]_D = -8°$.

I. Racemische Säure: 0.1586 g Sbst.: 0.2857 g CO_2, 0.0979 g H_2O.
II. *d*-Säure: 0.1463 g Sbst.: 0.2650 g CO_2, 0.0910 g H_2O.
III. *l*-Säure: 0.1371 g Sbst.: 0.2474 g CO_2, 0.0848 g H_2O.

$C_6H_{10}O_4$. Ber. C 49.32, H 6.85.
Gef. I. » 49.13, » 6.91.
» II. » 49.40, » 6.96.
» III. » 49.21, » 6.92.

Zürich, Universitätslaboratorium, Oktober 1913.

¹) B. 21, 3166 [1888]. ²) B. 34, 1733 [1901]. ³) B. 45, 121 [1912].

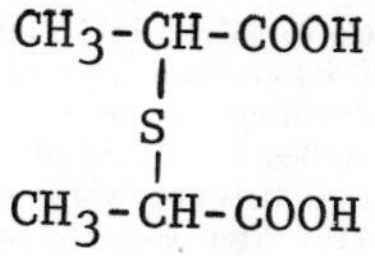

$C_6H_{10}O_4S$ **Ref. 209**

21,5 gr (0,12 Mol) Razem-Thiodilaktylsäure und 77 gr (0,24 Mol) wasserfreies Chinin wurden in 300 ccm warmem Methylalkohol gelöst. Der Becher wurde mit einem Uhrglas bedeckt und um das Verdampfen des Lösungsmittels zu vermeiden unter einer Glashaube verwahrt. Am nächsten Tage wurde von der auskrystallisierten Salzmenge (Fraktion I) filtriert; die Fraktion wog 26,9 gr; $[\alpha] = +55°$. Die Mutterlauge wurde etwas eingeengt und einige Tage stehen gelassen, wodurch 23,3 gr Salz (Fraktion II) erhalten wurden; $[\alpha] = -151°$. Durch weitere Einengung wurden 5,7 gr Salz (Fraktion III) erhalten; $[\alpha] = +161°$. Die jetzt stark sirupöse Mutterlauge wurde im Vakuum über Schwefelsäure zur Trockne eingedampft, dann noch weiter über P_2O_5 getrocknet und schliesslich in möglichst wenig siedendem Methylalkohol gelöst. Aus dieser Lösung krystallisierten 14,9 gr Salz (Fraktion IV); $[\alpha] = +121{,}5°$. Diese Fraktion wurde umkrystallisiert und gab dabei 9 gr Salz (Fraktion IV a); $[\alpha] = +175{,}5°$. Die oben genannte Fraktion I wurde ebenfalls umkrystallisiert, wobei 15,1 gr Salz erhalten wurden (Fraktion I a); $[\alpha] = +170°$. Die Mutterlaugen von I a und IV a, welche bedeutenden Überschuss an linksdrehender Säure enthalten mussten; wurden schliesslich vereinigt und zur Trockne eingedampft. Der Rückstand wurde umkrystallisiert und ergab 17 gr Salz (Fraktion V); $[\alpha] = -135{,}5°$.

Die drei rechtsdrehenden Fraktionen I a, III und IV a (insgesamt 29 gr) wurden nun vereinigt und auf konstante Drehung umkrystallisiert. Die Fortschritte der Anreicherung ergeben sich aus folgender Übersicht:

Krystallisation:	1	2	3
$[\alpha]$	+ 193,5°	+ 197°	+ 197,5°
Ausbeute an Salz	24,2 gr	19,7 gr	17,4 gr

Auch die beiden stark linksdrehenden Fraktionen II und V (insgesamt 40 gr) wurden vereinigt. Zum Umkrystallisieren wurde als Lösungsmittel Methylalkohol benützt, der mit einem Drittel seines Volumens Wasser versetzt worden war. Da die Löslichkeit des Salzes in dieser wässrig-methylalkoholischen Lösung bedeutend geringer ist als in reinem Methylalkohol, sind auch die Verluste bei den einzelnen Krystallisationen geringer; da es aber dabei andererseits nötig war öfter umzukrystallisieren, dürfte die Verwendung des wässrig-methylalkoholischen Lösungsmittels keinen wirklichen Substanzgewinn zur Folge gehabt haben. Der Verlauf der Anreicherungen ergibt sich aus folgender Übersicht:

Krystallisation:	1	2	3	4	5	6
$[\alpha]$	− 170°	− 184°	− 191°	− 195,5°	− 199°	− 198,5°
Ausbeute an Salz	34,6 gr	32,4 gr	30,1 gr	28,1 gr	26,4 gr	24,5 gr

Da die zur Spaltung angewandte Menge Razemsäure ungefähr 100 gr Chininsalz entspricht, beträgt die Ausbeute an reiner (+)- und (—)-Säure als Chininsalz gebunden 35 bzw. 49 % der Theorie. Einige orientierende Versuche mit den Mutterlaugen ergaben, dass man aus ihnen noch weitere nicht unbeträchtliche Mengen der beiden Antipoden gewinnen können dürfte.

Die freien Säuren wurden durch Ausschütteln der schwefelsauren Lösungen der Salze mit Äther isoliert. Sie wurden zweimal aus Benzol umkrystallisiert, worin sie bei Raumtemperatur nur unbedeutend löslich sind. Man erhält in diesem Falle die beiden Antipoden als dünne glänzende Tafeln oder Schuppen, deren Umriss die Gestalt eines verhältnismässig spitzwinkligen Rhombus hat: die Auslöschung ist diagonal. *Lovén* erhielt aus wässriger Lösung ziemlich grosse glasklare Krystalle, welche er als rhombisch, Kombination eines Prismas und eines Pinakoids bezeichnet.

Mit beiden Antipoden wurde der Schmelzpunkt 117—118° gefunden.

(+)-Thiodilaktylsäure.

0,1716 gr verbrauchten 17,61 ccm 0,1095-n NaOH (Phenolphtalein).
$C_6H_{10}O_4S$ (178,2) Äqu.-Gew. ber. 89,1 gef. 89,0
0,2134 gr in Wasser zu 10,03 ccm gelöst gaben im 2 dm-Rohr $\alpha = +8{,}53°$
$[\alpha]^{25}_{D} = +200{,}5°$; $[M]^{25}_{D} = +357{,}2°$.

Das Chininsalz der rechtsdrehenden Säure erhält man aus Wasser oder mit Wasser stark verdünnten Methylalkohol als lange, gerade, glänzende Nadeln, die $1^1/_4$ Mol Krystallwasser enthalten, das im Vakuum über P_2O_5 leicht abgegeben wird. Da ein derartiger Wassergehalt als wenig wahrscheinlich betrachtet werden muss, wurde die Bestimmung des Krystallwassers an drei verschiedenen Präparaten ausgeführt, die teilweise aus Wasser und teilweise aus verdünnten Methylalkohol umkrystallisiert und an der Luft bei Raumtemperatur auf Gewichtskonstanz getrocknet worden waren.

0,5788 gr Substanz: 0,0154 gr Gewichtsverlust über P_2O_5
0,5588 gr » : 0,0146 gr » » »
0,7172 gr » : 0,0192 gr » » »
$C_6H_{10}O_4S$, 2 ($C_{20}H_{24}N_2O_2$), $1^1/_4$ H_2O (849,1) H_2O ber. 2,65 % gef. 2,66; 2,61; 2,68 %
0,3071 gr wasserfreie Substanz: 18,30 ccm N_2 (18°, 733,5 mm)
$C_6H_{10}O_4S$, 2 ($C_{20}H_{24}N_2O_2$) (826,5) N ber. 6,78 % gef. 6,75 %.

(—)-Thiodilaktylsäure.

0,1520 gr verbrauchten 15,58 ccm 0,1095-n NaOH (Phenolphtalein).
$C_6H_{10}O_4S$ (178,2) Äqu.-Gew. ber. 89,1 gef. 89,1
0,2110 gr (einmal umkrystallisiert) in Wasser zu 10,02 ccm gelöst gaben im 2 dm-Rohr $\alpha = -8{,}46°$; $[\alpha] = {}^{25}_{D} = -200{,}9°$; $[M]^{25}_{D} = -358{,}0°$.
0,2057 gr (zweimal umkrystallisiert) in Wasser zu 10,03 ccm gelöst gaben im 2 dm-Rohr $\alpha = -8{,}24°$; $[\alpha]^{25}_{D} = -200{,}9°$; $[M]^{25}_{D} = -358{,}0°$.

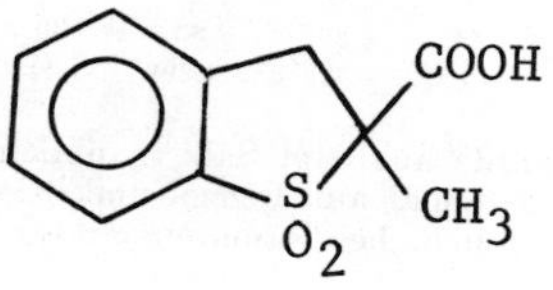

$C_6H_{10}O_4S$ **Ref. 210**

(+)-2-Methyltetrahydrothiophene-2-carboxylic Acid 1,1-Dioxide ((+)-III). A solution of 75.3 g of III and 169.3 g of brucine in 1600 ml of hot water was prepared. The solution was allowed to cool very slowly to 7°, and the salt (111.2 g) separated as large plates. Recrystallization of the salt from hot water yielded 93.0 g (38%) of material. A small portion of the salt was converted to the free acid which, after sublimation, exhibited $[\alpha]^{25}_{546}$ +20.4°. Additional crystallizations of the salt provided no change in rotation of the derived acid. The optically pure salt (88 g, 36%) was dissolved in 800 ml of warm water, and a solution of 21 g of potassium carbonate in 50 ml of water was added with rapid stirring. Precipitation of brucine was complete within 15 min. The slurry was filtered; the filtrate was extracted twice with 300-ml portions of chloroform and reduced in volume to 225 ml under vacuum. Sulfuric acid, 45 ml, was added with cooling, and the acidic solution was saturated with sodium chloride. Continuous extraction of the solution with chloroform yielded 22.8 g of crude (+)-III. Recrystallization of the acid from 50 ml of ether gave 16.6 g of (+)-III, mp 161.5–162.5°, $[\alpha]^{28}_{546}$ +20.4° (*c* 4.7, water).

(−)-2-Methyltetrahydrothiophene-2-carboxylic Acid 1,1-Dioxide ((−)-III). To the mother liquors from the first crystallization of the brucine salt of III was added a solution of 25 g of potassium carbonate in 50 ml of water with rapid stirring, and the acid was recovered as before. The crude, partially optically active acid was placed in 55 ml of ether and warmed; the warm solution was filtered. The residue from filtration was found to be racemic, $[\alpha]^{28}_{546}$ 0.0° (*c* 4.7, water). Cooling of the filtrate afforded 14.04 g of (−)-III, mp 159–160°, $[\alpha]^{28}_{546}$ −20.2° (*c* 4.7, water).

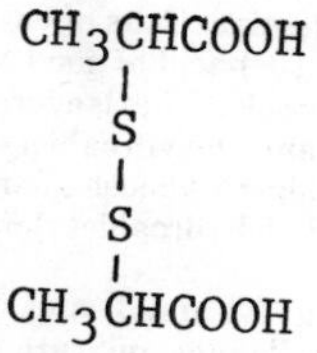

$C_6H_{10}O_4S_2$ **Ref. 211**

Spaltungsversuche mit Strychnin und Brucin.

Die *racem*-Form kann bekanntlich mit Hilfe von aktivem Phenäthylamin in ihre optischen Antipoden gespalten werden (3, 5). Da mir keine grösseren Mengen dieser Base zur Verfügung standen, führte ich einige Spaltungsversuche mit handelsüblichen Alkaloiden aus. Strychnin gab schon in der ersten Kristallisation eine zu etwa 75 % aktive, linksdrehende Säure. Durch zweimalige Umkristallisation des schwerlöslichen Strychninsalzes aus 40-prozentigem Aceton wurde eine Aktivität von etwa 90 % erhalten; bei weiteren Umkristallisationen stieg die Drehung nur sehr langsam, wodurch die Gegenwart einer schwerlöslichen Verunreinigung angedeutet wird. Es ist denkbar, dass beim wiederholten Umkristallisieren etwas *meso*-Form gebildet worden war, die nicht entfernt werden konnte; das Strychninsalz der aktiven Säure ähnelt nämlich in seinen Löslichkeitsverhältnissen dem der *meso*-Form. Andere Erklärungen wie beispielsweise das Auftreten eines Zersetzungs- oder Umlagerungsprodukts von Strychnin sind jedoch nicht ausgeschlossen.

Durch Überführung der Säure in das Phenäthylaminsalz wurde leicht maximale Linksdrehung erhalten.

Aus der Mutterlauge von der ersten Kristallisation des Strychninsalzes wurde rechtsdrehende Säure von etwa 50 % Aktivität erhalten, die in das Brucinsalz übergeführt wurde. Durch Umkristallisation des Brucinsalzes aus 50-prozentigem Methylalkohol wurde ziemlich leicht maximale Drehung erhalten:

Kristallisation	1	2	3	4	5	6
Salzausbeute in g	44	39	33,5	30,5	29,5	27,5
$[\alpha]_D^{25}$ der Säure	+320°	+374,5°	+399°	+413°	—	+412,5°

Die Säure wurde aus dem Salz in üblicher Weise in Freiheit gesetzt und zweimal aus Benzol umkristallisiert, worin sie bei Siedehitze reichlich, bei Raumtemperatur aber nur spärlich löslich ist.

0,2019 g Subst.: 16,72 ml 0,1146-n NaOH.
Äquiv.-Gew. ber. 105,1 gef. 105,3
0,2112 g Subst. gelöst in 0,1-n HCl zu 10,03 ml: $2\ \alpha = +18{,}185°$;
0,1022 g Subst. mit NaOH vollständig neutralisiert und zu 10,03 ml aufgefüllt: $2\ \alpha = +3{,}02°$.
Daraus folgt: für die freie Säure $[\alpha]_D^{25} = +431{,}8°$; $[M]_D^{25} = +907{,}5°$
für das Salz $[\alpha]_D^{25} = +148{,}2°$; $[M]_D^{25} = +311{,}5°$.

Die gefundene Drehung der freien Säure stimmt gut mit dem Werte $[\alpha]_D^{25} = 430{,}2°$ von Bernton (5) überein; unsere Werte für die Aktivität des Salzes weichen jedoch um etwa 3 % voneinander ab, was vielleicht darauf beruht, dass die Messungen bei verschiedenen Konzentrationen ausgeführt wurden.

Das *Brucinsalz* der aktiven Säure, das im Verlaufe der Spaltung erhalten wurde, kristallisierte in kleinen Nadeln, die 6 ½ Moleküle Kristallwasser enthielten.

0,3011 g wasserfreie Subst.: 15,32 ml N_2 (24°, 729,5 mm Hg).
$C_6H_{10}O_4S_2$, 2 $C_{23}H_{26}O_4N_2$ (998,6) N ber. 5,61 % gef. 5,61 %
0,6686, 0,6135, 1,0271 g Subst.: 0,0700, 0,0644, 0,1073 g H_2O;
6 ½ H_2O ber. 10,50 % gef. 10,47, 10,50, 10,45 %.

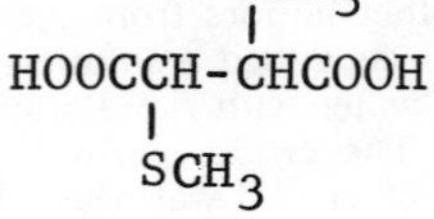

$C_6H_{10}O_4S_2$ Ref. 212

Preliminary experiments on resolution. The acid was tested against the common alkaloids and dehydroabietylamine. Salts of cinchonidine in dil. acetone and brucine in methanol or ethanol gave (+)-acid of good activity. Further experiments showed that cinchonidine was preferable. The laevorotatory acid could be isolated from salts of morphine, quinine and dehydroabietylamine. Among these the latter was superior to the others in giving an almost quantitative yield of (−)-acid in methanol. Thus, cinchonidine and dehydroabietylamine were chosen for the (+)- and (−)-acids, respectively.

(−)-*Acid.* (±)-Acid (10 g, 48 mmol) and dehydroabietylamine acetate (32.8 g, 95 mmol) were dissolved in a boiling mixture of 250 ml ethanol and 100 ml water. The salt was collected after 24 hours in a refrigerator and recrystallized once from ethanol. Further recrystallization did not raise the activity of the liberated acid.

Crystallization	1	2
ml ethanol	250	310
ml water	160	—
g salt obtained	19.2	15.2
$[\alpha]_D^{25}$ of the acid	−187°	−188°

The salt was carefully powdered and suspended in 100 ml 2-M sulphuric acid. The mixture was extracted with ether which was dried and evaporated to dryness. The remaining crystals were dissolved in ethyl acetate and petroleum ether added. The turbid solution was left at 0° for 3 hours and the glistening needles of pure acid (3.8 g, 38%) collected, m.p. (173°), $[\alpha]_D^{25} = -188°$ (abs. ethanol).

$C_6H_{10}O_4S_2$ (210.3) calc. S 30.50
found S 30.40

(+)-*Acid.* The mother liquor from the isolation of the (−)-isomer was partly evaporated, diluted with 50 ml 2-M sodium hydroxide and the dehydroabietylamine taken up in chloroform. The aqueous solution was immediately acidified and carefully extracted with ether, leaving 4 g acid on evaporation; $[\alpha]_D^{25} = +151°$ (abs. ethanol). This acid (19 mmol) was dissolved in 15 ml water and mixed with cinchonidine (11.4 g, 38 mmol) in 75 ml acetone. The mixture was refluxed until the salt dissolved when another 225 ml boiling acetone was added. The salt was filtered off after 24 hours at 0° and recrystallized from dil. acetone. It was very important that the ratio 1 : 20 was maintained between water and acetone as an excess of water lowered the yield considerably. After each recrystallization a small sample of the salt was decomposed and the activity of the acid measured in abs. ethanol.

Recrystallizations	1	2	3
ml acetone	300	220	160
ml water	15	11	8
g salt obtained	11	8	5.6
$[\alpha]_D^{25}$ of the acid	+180°	+186°	+188°

The salt was decomposed with 2-M sulphuric acid. The acid was extracted with ether and recrystallized as the enantiomer, yielding 1.0 g (25%), m.p. (173°), $[\alpha]_D^{25} = +188°$ (abs. ethanol).

$C_6H_{10}O_4S_2$ (210.3) calc. 30.50
found 30.28

Table 1. Optical activity of (−)-2,3-bis(methylthio)succinic acid (II).

g = mg acid in 10.00 ml solution.

Solvent	g	α_D^{25} (1 dm)	$[\alpha]_D^{25}$	$[M]_D^{25}$
Ethanol	27.29	−0.514	−188°	−395°
Methanol	25.19	−0.468	−186°	−391°
Acetone	21.69	−0.462	−213°	−447°
Water	29.24	−0.466	−159°	−334°

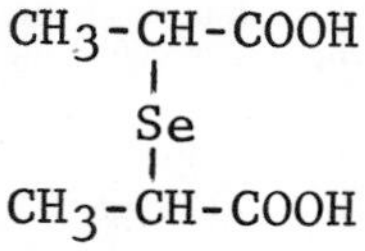

$C_6H_{10}O_4Se$ Ref. 213

Wie schon früher (43) in Kürze beschrieben wurde, kann die *racem*-Form durch Umkristallisation des Chininsalzes aus Methylalkohol leicht gespalten werden. Einige Versuche mit Umkristallisation aus Wasser waren erfolglos; die erhaltene Drehung war ganz geringfügig und änderte nach einigen Umkristallisationen ganz unerwartet das Vorzeichen. Die Chininsalze der beiden Antipoden scheinen also in Wasser angenähert gleiche Löslichkeit zu besitzen. Durch Umkristallisation aus Methylalkohol wurde jedoch das Salz der (+)-Form unmittelbar rein erhalten.

87.5 g (0.1 Mol) Chininsalz der *racem*-Form (dargestellt in wässriger Lösung) wurden aus 400 ml Methylalkohol umkristallisiert. Das erhaltene Produkt wurde noch dreimal aus möglichst wenig Methylalkohol umkristallisiert. Von jeder Fraktion wurde eine kleine Menge abgetrennt, aus der die Säure nach Ansäuern durch Ausschütteln mit Äther isoliert wurde. Das erhaltene Produkt wurde in Wasser gelöst, die Drehung bestimmt und der Gehalt der Lösung durch Titrieren mit Natronlauge ermittelt. Die spezifische Drehung muss sich derart mit einer Genauigkeit von ± 1.5 % bestimmen lassen. Der Verlauf der Spaltung ergibt sich aus folgender Zusammenfassung:

Kristallisation	1	2	3	4
Ausbeute in g	23.6	17.1	13.4	9.5
$[\alpha]_D$	+234.5°	+234°	+232°	+234°

Die Mutterlauge von der ersten Fraktion wurde zur Trockne eingedampft, wobei 53 g Salz erhalten wurden, das ziemlich stark linksdrehende Säure enthielt. Zur Umkristallisation dieses Salzes wurde nach einigen vorbereitenden Versuchen ein Gemisch von gleichen Teilen Aceton und Wasser gewählt. 100 ml dieses Gemisches lösen beim Siedepunkt etwa 5 g Salz. Der Gang der Spaltung wurde wie oben beim Salz der rechtsdrehenden Säure überprüft.

Kristallisation	1	2	3	4	5	6	7
Ausbeute in g	38.4	28.3	22.1	18.1	14.3	10.5	8.1
$[\alpha]_D$	−161°	−207°	−223°	−231.5°	—	−235°	−234.5°

Es ist auffallend, dass das Salz der (+)-Säure so leicht zu erhalten war. Es ist nicht ausgeschlossen, dass dabei Übersättigungserscheinungen in der verhältnismässig viskosen methylalkoholischen Lösung mitspielten. Die Beobachtungen sprechen jedoch dafür, dass das Salz der (—)-Säure aus reinem Methylalkohol nur schwierig oder vielleicht überhaupt nicht kristallisieren kann. Wie sich aus den unten angeführten Analysen ergibt, kristallisiert das Chininsalz der (+)-Säure wasserfrei, das der (—)-Säure hingegen mit einem Molekül Kristallwasser, das sehr fest gebunden ist und auch nach mehrmonatlicher Trocknung im Vakuum über P_2O_5 nicht vollständig abgegeben wird. Es liegt daher nahe anzunehmen, dass die geringe Neigung des Salzes der (—)-Säure aus Methylalkohol zu kristallisieren darauf zurückzuführen ist, dass das für die Kristallbildung erforderliche Wasser dem Methylalkohol nicht oder nur mit Schwierigkeit entzogen werden kann. Im Laufe der Umkristallisationen enthielt natürlich wenigstens die erste Lösung die stöchiometrisch erforderliche Wassermenge, da ja das Salz ursprünglich in wässriger Lösung dargestellt worden war.

Es kann hier darauf hingewiesen werden, dass die beiden aktiven Thio-di-laktylsäuren nach einer ähnlichen Spaltungsmethode erhalten werden können (28); bei ihnen kann die Reindarstellung der *beiden* Antipoden durch Umkristallisation der Chininsalze aus Methylalkohol erfolgen. In diesem Falle dürfte es sicher sein, dass beim Zustandekommen der ersten Spaltung Übersättigungserscheinungen eine bedeutende Rolle spielen. Die aktiven Thio-di-laktylsäuren gleichen in ihrem Aussehen vollständig den entsprechenden Selenverbindungen, sind aber etwas leichter löslich als diese.

Die aktiven Selen-di-laktylsäuren wurden in gewöhnlicher Weise durch Zersetzung der Chininsalze mit Schwefelsäure und anschliessendes Ausschütteln mit Äther isoliert. Beim Verdampfen des Äthers kristallisieren sie spontan. Sie wurden zweimal aus reinem Benzol umkristallisiert, worin sie bei Siedetemperatur reichlich, bei Raumtemperatur aber nur ganz unbedeutend löslich sind.

(+)-Selen-di-laktylsäure.

Die Säure wird aus Benzol in dünnen glänzenden Tafeln oder Schuppen erhalten, deren Umriss die Form eines ziemlich spitzwinkligen Rhombus, bisweilen auch eines Sechsecks hat. Beim langsamen Verdampfen von Lösungen in anderen Lösungsmitteln erhält man leicht verhältnismässig grosse, sechsseitige Tafeln. Die Säure ist in den gewöhnlichen Lösungsmittel viel leichter löslich als die *racem*-Form. Schmelzpunkt 123.5—124.5°.

0.2165 g Subst. : 17.44 ml 0.1100-n NaOH.
$C_6H_{10}O_4Se$ (225.1) Äqu.-Gew. ber. 112.5 gef. 112.9.
0.2206 g Subst. (einmal umkristallisiert) in 0.4-n HCl auf 10.02 ml gelöst: $2\alpha = +10.47°$; 0.2266 g Subst. (zweimal umkristallisiert) in 0.4-n HCl auf 10.03 ml gelöst: $2\alpha = +10.74°$.

$[\alpha]_D^{25} = +237.8°; +237.7°. [M]_D^{25} = +535°; +535°.$

Das *Chininsalz*, dessen Darstellung oben beschrieben wurde, bildet büschelige Nadeln. In lufttrockenem Zustand ist es frei von Kristallwasser.

0.4845 g Subst. : 0.0441 g Se.
$C_6H_{10}O_4Se, 2(C_{20}H_{24}O_2N_2)$ (873.5) Se ber. 9.04 % gef. 9.10 %.

(—)-Selen-di-laktylsäure.

Im Aussehen und ihren Eigenschaften gleicht die Säure vollständig ihrer optischen Antipode. Schmelzpunkt 123.5—124.5°.

0.2110 g Subst. : 16.98 ml 0.1100-n NaOH.
$C_6H_{10}O_4Se$ (225.1) Äqu.-Gew. ber. 112.5 gef. 112.9.
0.2314 g Subst. in 0.4-n HCl auf 10.02 ml gelöst: $2\alpha = -11.03°$.

$[\alpha]_D^{25} = -238.6°. [M]_D^{25} = -537°.$

Das *Chininsalz*, dessen Darstellung oben beschrieben wurde, kristallisiert in geraden glänzenden Nadeln, die ein Molekül Wasser enthalten. Dieses Wasser kann durch Erwärmen nicht ohne Zersetzung entfernt werden; auch nach mehrmonatlichem Aufbewahren in Vakuum über P_2O_5 ist das Salz nicht vollständig wasserfrei.

0.2921 g Subst. : 15.90 ml N_2 (18°, 746.5 mm);
0.5685 g, 0.5103 g Subst. : 0.0504 g, 0.0453 g Se.
$C_6H_{10}O_4Se, 2(C_{20}H_{24}O_2N_2), H_2O$ (891.5) ber. N 6.28 % Se 8.86 %
gef. » 6.27 » » 8.87, 8.88 %.

Beim Zusammenbringen gleicher Mengen der beiden Antipoden in Benzollösung wurde die bei 147—148° schmelzende *racem*-Form erhalten.

$$\begin{array}{c} CH_3CHCOOH \\ | \\ Se \\ | \\ CH_2CH_2COOH \end{array}$$

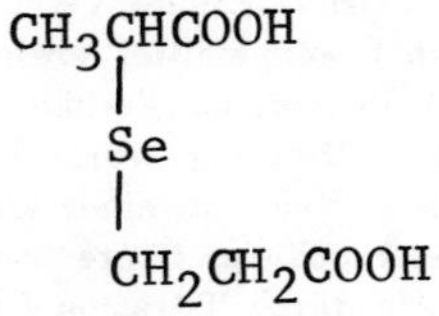

$C_6H_{10}O_4Se$ Ref. 214

(—)-*Säure*. Bei vorläufigen Spaltungsversuchen gaben *Strychnin*, *Brucin*, *Chinin*, *Chinidin* und *Cinchonidin* kristallisierende Salze. Die darin enthaltene Säure war in sämmtlichen Fällen linksdrehend. *Chinin* gab die beste Spaltung und wurde deshalb für die Reindarstellung der (—)-Säure gewählt. Auf die Isolierung der rechtsdrehenden Antipode wurde verzichtet.

42 g (0,187 Mol) Säure wurden nebst 140 g (0,37 Mol) kristallwasserhaltiges Chinin in 200 ml Aceton + 400 ml Wasser gelöst. Das nach eintägigem Stehen ausgeschiedene Salz, das auffällig schwerlöslich ist, wurde dreimal aus einer Mischung von etwa gleichen Mengen Wasser, Alkohol und Aceton umkristallisiert; aus jeder Kistallisation wurde eine kleine Quantität Säure isoliert und auf ihre Drehung in Essigester untersucht:

Kristallisation:	1	2	3	4
g Salz	84,5	71,5	67	61,5
$[\alpha]_D^{25}$ der Säure	—102°	—117,5°	—122°	—121,5°

Das Chininsalz wurde mit Schwefelsäure zerlegt, die aktive Säure in reinem Äther aufgenommen und nach Verdampfen des Äthers scharf getrocknet. Sie ist in den meisten Lösungsmitteln sehr leicht löslich, lässt sich jedoch aus einer kleinen Menge Toluol umkristallisieren. Um die allerdings recht erheblichen Verluste zu vermindern kann man während des Auskristallisierens allmählich kleine Mengen Benzin zusetzen. Ausbeute nach zwei Kristallisationen 7,8 g vom Schmp. 53,5—54,5°. Eine kleine Probe wurde nochmals umkristallisiert, wobei Drehung und Schmelzpunkt nicht verändert wurden. Nach längerem Stehen im gepulverten Zustande ging die Säure in eine andere Form vom Schmp. 61,5—62,5° über, die offenbar die stabile Modifikation darstellt. Durch Schmelzen und Impfen im Schmelzpunktsrohr liess sich die eine Form in die andere überführen.

0,1824 g Sbst.: 13,72 ml 0,1188-n NaOH.
$C_6H_{10}O_4Se$ (225,04) Äqu.-gew. Ber. 112,5 Gef. 112,4.
0,2251 g Subst. (zweimal umkristallisiert) in Essigester auf 10,00 ml gelöst: $2\alpha_D^{25} = -5,58°. [\alpha]_D^{25} = -123,9°; [M]_D^{25} = -278,9°.$

Drehungsmessungen an dreimal umkristallisierter Säure finden sich in Tabelle 4.

Tabelle 4.

	g Säure	ml Lösung	$2\alpha_D^{25}$	$[\alpha]_D^{25}$	$[M]_D^{25}$
Essigester	0,2220	10,00	—5,52°	—124,2°	—279,5°
Eisessig	0,1281	10,00	—2,98	—121,0	—272,4
Alkohol (absol.)	0,1340	9,99	—3,25	—121,4	—273,2
Aceton	0,1276	10,00	—3,045	—119,8	—268,5
Chloroform	0,1245	9,99	—2,75	—110,5	—248,7
0,4-n Salzsäure	0,1259	10,00	—2,76	—109,6	—246,6
neutr. Wasserlösung	0,1239	10,00	—1,295	—52,3	—117,5

$$\begin{array}{c}\text{CH}_3\text{-CH-COOH}\\|\\\text{Se}\\|\\\text{Se}\\|\\\text{CH}_3\text{-CH-COOH}\end{array}$$

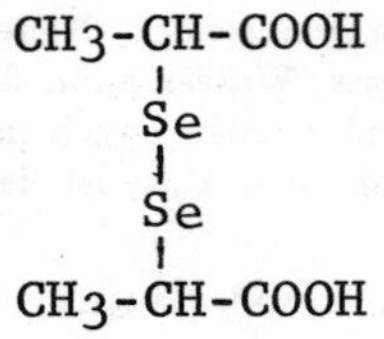

$C_6H_{10}O_4Se_2$ Ref. 215

30.4 g (0.1 Mol) *racem*-Diselen-di-laktylsäure wurden mit Ammoniak neutralisiert, wonach das Volumen 350 ml betrug. Die Lösung wurde mit 0.15 Mol Brucinhydrochlorid, gelöst in 350 ml Methylalkohol, versetzt. Das Salz kristallisiert ziemlich langsam; nach zwei Tagen wurde abgesaugt und 38 g Salz erhalten. Das Salz wurde aus 50-prozentigem Methylalkohol auf konstante Drehung umkristallisiert; von diesem Lösungsmittel wurden 5 ml je g Salz berechnet. Orientierende Versuche schienen zu zeigen, dass andere denkbare Lösungsmittel sowie Gemische von Methylalkohol und Wasser in anderen Verhältnissen weniger günstige Resultate gaben. Die Drehung wurde bei den einzelnen Fraktionen in gewöhnlicher Weise an einer kleinen Probe ätherextrahierter und in Wasser gelöster Säure bestimmt; die Konzentration der Lösung wurde durch Titration bestimmt. Die Drehung war nach der fünften Kristallisation konstant.

Kristallisation	1	2	3	4	5	6
Ausbeute in g	38.0	31.1	28.0	25.3	23.4	22.0
$[\alpha]_D^{25}$	+188.5°	+217.5°	+238.5°	+243°	+246°	+246°

Um die Reproduzierbarkeit der Methode zu prüfen und um mehr aktive Säure darzustellen, führte ich einen Parallelversuch mit 20 g (0.07 Mol) Säure aus. Wie sich aus der folgenden Übersicht ergibt, stimmen die beiden Versuchsreihen sehr gut miteinander überein.

Kristallisation	1	2	3	4	5	6	7
Ausbeute in g	28.0	21.1	19,4	17.3	15.6	14.3	12.6
$[\alpha]_D^{25}$	+190°	+224°	+239.5°	+243.5°	+247°	+245°	+247°

Das Salz besteht aus hellgelben Prismen oder Nadeln, die bei weiterem Fortschritt der Spaltung immer feiner werden und hellere Farbe annehmen. Es sei erwähnt, dass das Salz bei mässiger Erwärmung, z. B. beim Trocknen über einem Heizkörper, stark gelbe Farbe annimmt, die beim Erkalten wieder verblasst.

Aus der Mutterlauge der ersten Fraktion des Brucinsalzes erhält man eine Säure, deren spezifische Drehung bei etwa −85° liegt. Beim Einengen der Mutterlauge erhält man ein in ziemlich groben gelben Prismen kristallisierendes Salz, dessen Säure etwa dieselbe Drehung besitzt wie die des gelöst bleibenden Salzes. Es ist also unmöglich, die Linksform mit Hilfe des Brucinsalzes noch weiter anzureichern.

Die linksdrehende Säure wurde aus dem Brucinsalz und der Mutterlauge in Freiheit gesetzt und in Strychninsalz übergeführt, das aus mit einem Drittel Aceton versetztem Wasser umkristallisiert wurde (Methyl- und Äthylalkohol gaben bei Vorversuchen weniger gute Resultate und in Wasser allein ist die Löslichkeit des Strychninsalzes zu gering). Für 1 g Strychninsalz benötigt man etwa 20 ml Lösungsmittel. Wie sich aus der folgenden Zusammenstellung der Messungen an Proben der einzelnen Fraktionen ergibt, steigt die Drehung ziemlich langsam.

Kristallisation	1	3	5	7	9	10
Ausbeute in g	46.3	28.9	18.7	14.5	11.7	10.5
$[\alpha]_D^{25}$	−116°	−152°	−195°	−223°	−236°	−240°

Kristallisation	11	12	13	14	15
Ausbeute in g	9.3	8.4	7.4	6.5	5.7
$[\alpha]_D^{25}$	−243°	−245°	−247.5°	−249°	−248°

Bei einer Wiederholung des Versuches arbeitete ich mit etwas grösseren Flüssigkeitsmengen, wobei die Spaltung etwas rascher erfolgte.

Kristallisation	1	3	6	8	9	10
Ausbeute in g	24.8	18.5	7.9	4.9	4.2	3.4
$[\alpha]_D^{25}$	−118°	−182°	−231°	−241.5°	−247.5°	−247°

Diese Methode der Darstellung linksdrehender Säure kann nicht als leicht oder bequem bezeichnet werden. Es ist möglich, dass man durch systematische Untersuchung der leichter zugänglichen Alkaloide ein sich besser eignendes finden könnte.

Bei den Umkristallisationen kann man zwei Arten von Kristallen mit deutlich verschiedenem Aussehen unterscheiden. Die Kristalle des einen Typus (Typus 1) waren kurzprismatisch und bestanden gewöhnlich aus abgestumpften Pyramiden; bisweilen waren sie fast würfelförmig. Diese Art von Kristallen verschwand allmählich im Laufe der fortschreitenden Spaltung. Der andere Typus (Typus 2), der schliesslich allein übrig blieb und offenbar das Strychninsalz der linksdrehenden Säure darstellte, bildete dünne Blättchen oder Schuppen. Durch Aussuchen konnten geringe Mengen von Kristallen des Typus 1 isoliert werden; wenn diese neuerlich umkristallisiert wurden, wurde ein Gemisch von Kristallen des Typus 1 und 2 erhalten. Es ist daher unmöglich, den Typus 1 auf diesem Wege vollständig rein zu erhalten; durch Aussuchen von Kristallen allein kann man ja kaum vollständig reines Produkt bekommen, wenn es sich nicht um besonders grosse und gut ausgebildete Kristalle handelt. Aus einer Probe von ausgesuchten Kristallen vom Typus 1 wurde die Säure in gewöhnlicher Weise in Freiheit gesetzt und aus Kohlenstofftetrachlorid umkristallisiert; sie besass Schmelzpunkt und allgemeine Eigenschaften der *meso*-Form. Im Polarimeter zeigte sie schwache Aktivität, welche die Gegenwart einiger Prozente aktiver Form andeutete. Aus Analysen ergab sich, dass die beiden Salze vom Typus 1 und 2 verschiedenen Kristallwassergehalt besitzen.

Alle Beobachtungen scheinen dafür zu sprechen, dass Typus 1 das Strychninsalz der *meso*-Form ist. In der verdünnten Acetonlösung, die ja in gewissem Ausmasse mit einer wässrigen Lösung vergleichbar ist, wird ein gewisser Anteil *meso*-Form gebildet; die Umwandlung muss beim Lösen des Salzes bei Siedetemperatur sehr rasch erfolgen. Beim Erkalten kristallisiert dann teils das Strychninsalz der *meso*-Form, teils das der (−)-Form; beide Salze scheinen ziemlich schwerlöslich zu sein und die Entfernung der *meso*-Form erfordert daher viele Umkristallisationen. Das Strychninsalz der (+)-Säure scheint demgegenüber verhältnismässig grosse Löslichkeit und geringe Kristallisationsfähigkeit zu besitzen.

Beim Zusammenbringen von *racem*-Form einerseits und von *meso*-Form der Diselen-di-laktylsäure andererseits mit Strychnin in siedender acetonhaltiger wässriger Lösung erhält man, wie einige orientierende Versuche gezeigt haben, in beiden Fällen Salzfraktionen von etwa gleichem Aussehen, die sowohl Salz vom Typus 1 als auch solches vom Typus 2 enthalten. Auch die optische Aktivität der Säure des Salzes ist in beiden Fällen ungefähr gleich.

Das Salz der linksdrehenden Säure enthielt eine Verunreinigung, die nach allen Umständen zu beurteilen nicht sauer ist; beim Ausschütteln der Säure mit Äther folgt diese Verunreinigung mit und erschwert die Kristallisation, ohne aber den Schmelzpunkt nennenswert zu beeinflussen. Sie gab sich auch dadurch zu erkennen, dass sie der Säure eine bedeutend tiefer gelbe Farbe erteilte als die Antipode besass. Da das racemische Ausgangsmaterial vollständig rein zu sein schien (diesem Umstande wurde beim Kontrollversuche über die Spaltung der Säure besondere Aufmerksamkeit geschenkt), liegt hier wahrscheinlich ein Zersetzungsprodukt von Strychnin vor, das sich bei den wiederholten Umkristallisationen aus siedender Lösung gebildet hat. Die Verunreinigung konnte dadurch entfernt werden, dass die aus dem Strychninsalz in Freiheit gesetzte Säure in das Brucinsalz übergeführt wurde, das dann aus möglichst wenig Wasser umkristallisiert wurde. Die auf diesem Wege erhaltene Säure zeigte dann genau dasselbe Aussehen wie die optische Antipode.

Die aktiven Säuren wurden in gewöhnlicher Weise durch Ausschütteln mit Äther isoliert. Beim Verjagen des Äthers kristallisierten sie ohne grössere Schwierigkeit. Sie liessen sich vorteilhaft aus Kohlenstofftetrachlorid umkristallisieren, worin sie bei Siedehitze ziemlich leichtlöslich, bei Raumtemperatur aber fast unlöslich sind. Die Säuren wurden in hellgelben Nadeln von gleichem Aussehen wie die instabile *racem*-Form erhalten. Die Auslöschung ist gerade, die Kristallflächen sind schlecht ausgebildet und meist gewölbt.

(+)-Diselen-di-laktylsäure.

Schmelzpunkt 86.5—87.5° · Spez. Gew. bei 25° : 2,024.

0.2366 g, 0.2302 g Subst. : 14.17 ml, 13.77 ml 0.1098-*n* NaOH.

$C_6H_{10}O_4Se_2$ (304.0) Äqu.-Gew. ber. 152.0 gef. 152.3, 152.1.

0.1535 g Subst. (erste Spaltung), in Wasser auf 10.03 ml gelöst : 2 α = + 7.775°.

0.1520 g Subst. (zweite Spaltung), in Wasser auf 10.03 ml gelöst : 2 α = + 7.69°.

$[\alpha]_D^{25} = +254.0°$, $+253.7°$; $[M]_D^{25} = +772.5°$, $+771.5°$.

Wie schon früher (2) mitgeteilt wurde, erhielt ich aus reiner aktiver Selencyan-propionsäure eine rechtsdrehende Diselen-di-laktylsäure, welche geringe Neigung zu kristallisieren zeigte; beim Impfen mit (+)-Dithio-di-laktylsäure erstarrte sie zu sehr langen feinen Kristallnadeln, die bei 43—45° schmolzen und in den meisten organischen Lösungsmitteln leichtlöslich waren. Die molekulare Drehung betrug bei 25° und bei einer mit der in den obigen Messungen vergleichbaren Konzentration + 765°; das Äquivalentgewicht der Säure war richtig. Die molekulare Drehung war also nur weniger als 1 Prozent niedriger als bei dem neuen Präparat, während im Schmelzpunkt eine Differenz von reichlich 40° vorliegt. Bei Berücksichtigung der oben beschriebenen Verhältnisse kann dieser Unterschied der Schmelzpunkte kaum auf Verunreinigungen in dem früher beschriebenen Präparat zurückgeführt werden. Es ist daher am naheliegendsten anzunehmen, dass dieses Präparat eine instabile, niedrigschmelzende Modifikation der Säure darstellte. Es ist mir nicht gelungen, diese Modifikation neuerlich darzustellen.

Das *Brucinsalz*. Die Darstellung und die Eigenschaften des Salzes sind schon oben im Zusammenhang mit der Spaltung beschrieben worden. Das Salz enthält 6 Moleküle Kristallwasser, welche über P_2O_5 leicht abgegeben werden.

0.4532 g Subst. : 0.0406 g H_2O und 0.0602 g Se.

$C_6H_{10}O_4Se_2$, 2 ($C_{23}H_{26}O_4N_2$), 6 H_2O (1200.6) ber. H_2O 9.00 % Se 13.16 %
gef. » 8.96 » » 13.28 ».

(—)-Diselen-di-laktylsäure.

Schmelzpunkt 86—87.5°.

0.2309 g Subst. : 13.80 ml 0.1098-*n* NaOH.

$C_6H_{10}O_4Se_2$ (304.0) Äqu.-Gew. ber. 152.0 gef. 152.4.

0.1508 g Subst. in Wasser auf 10.03 ml gelöst : 2 α = − 7.625°

$[\alpha]_D^{25} = -253.6°$; $[M]_D^{25} = -771°$.

Das *Strychninsalz* wurde in der oben beschriebenen Weise als kleine glänzende Tafeln oder Schuppen erhalten. Über P_2O_5 verliert es ziemlich rasch 2 Moleküle Wasser, worauf das Gewicht langsam abzunehmen fortsetzt ohne auch nach mehreren Monaten konstanten Wert erreicht zu haben. Bei Beurteilung nach der Selenbestimmung muss der Wassergehalt 3 Molekülen entsprechen; auch der Gewichtsverlust über P_2O_5 scheint einem diesem Wassergehalt entsprechenden Werte zuzustreben.

0.5775 g Subst. : 0.0889 g Se.

$C_6H_{10}O_4Se_2$, 2 ($C_{21}H_{22}O_2N_2$), 3 H_2O (1026.4) Se ber. 15.39 % gef. 15.39 %

Wenn gleiche Mengen (+)- und (—)-Säure zusammen in Kohlenstofftetrachlorid gelöst werden, kristallisiert zunächst die hellgelbe, niedrigschmelzende Modifikation der *racem*-Form. Die Kristalle wurden in Berührung mit der Lösung stehen gelassen, wobei allmählich in gewöhnlicher Weise die Umwandlung in die stabile, tiefer gefärbte Modifikation stattfand.

```
              OH
              |
HOOC-CH-CH2-CH2-CH-COOH
                  |
                  OH
```

$C_6H_{10}O_6$ **Ref. 216**

l-(+)-Dioxy-adipinsäure.

Die Vorschrift von H. R. Le Sueur [1]) wurde in einigen Punkten abgeändert. Adipinsäure wird mit Thionylchlorid (statt Phosphorpentachlorid) in das Chlorid verwandelt, und dieses über die Dibrom-adipinsäure in die Dioxy-adipinsäure übergeführt. Diese wird aus der wäßrigen Lösung durch Einengen i. V. gewonnen. Der Trockenrückstand enthält Mesosäure und deren Monolacton, sowie Racemsäure und deren Dilacton. Unter Verzicht auf die Mesosäure wird das Rohprodukt unmittelbar auf Dilacton verarbeitet.

30 g des Gemisches werden in kleinen Portionen im Apparat für „Molekulardestillation" [2]) i. V. langsam auf 250° erhitzt. Die gesammelten Sublimate werden mehrmals mit Methylalkohol durchfeuchtet und rasch abgesaugt. Ausbeute 10—11 g; Schmelzp. 136°.

10 g Dilacton werden in 200 ccm Wasser 3 Stunden gekocht und mit 20 g Cinchonidin versetzt. Das entstehende Salz wird 5-mal aus Wasser umkrystallisiert und mit Ammoniak zerlegt. Die ammoniakalische Lösung wird mit Salzsäure eben angesäuert, i. V. auf ein kleines Volumen eingeengt, übersäuert und 8 Tage lang ausgeäthert. Die Säure wird aus Aceton durch Zusatz von Chloroform umkrystallisiert. Ausbeute 0,5—1 g.

$[\alpha]^{I}_{578}$, $^{II}_{546}$ in Wasser = + 0,12° (I); 0,16° (II) × 2,360/0,0672 × 1,009
= + 4,2° (I); 5,6° (II).

In Übereinstimmung hiermit findet Le Sueur + 3,8° für Natriumlicht.

$[\alpha]_{578}$ in neutraler (NaOH)-Lösung = − 1,02° × 1,913°/0,0629 × 1,025
= − 30,3°.

[1]) Soc. 93, 718 (1908).

```
                   OH
                   |
HOOC-CH2CH-CHCH2COOH
          |
          OH
```

$C_6H_{10}O_6$ **Ref. 217**

Résolution en antipodes optiques. Des essais ont été effectués au moyen de sels de divers alcaloïdes (strychnine, cinchonine, cinchonidine, quinine, brucine). Les résultats les plus favorables ont été obtenus au moyen des sels de strychnine.

43 g de DL-dihydroxy-3,4-adipate de baryum brut finement pulvérisé sont suspendus dans 500 cm³ d'eau. On ajoute une solution chaude de 110 g de sulfate neutre de strychnine dans 1 l d'eau et agite 10 h à la machine. On dilue à 2 l, chauffe à 70° et essore à chaud, sur charbon, le sulfate de baryum qu'on lave avec 200 cm³ d'eau chaude.

Après repos d'une nuit à 16°, il s'est séparé 31,6 g de sel cristallin qu'on recristallise à 3 reprises dans 17—20 parties d'eau. L'enrichissement en antipodes optiques est déterminé chaque fois sur une partie aliquote du produit qu'on transforme par l'intermédiaire du sel de baryum, de la manière indiquée plus loin, en acide libre dont on mesure le pouvoir rotatoire. Obtenu finalement 15,4 g de (+)-dihydroxy-3,4-adipate neutre de strychnine dont la rotation (acide libre) n'est plus modifiée par recristallisations ultérieures du sel d'alcaloïde.

$C_{48}H_{54}N_4O_{10}$ Calculé N 6,62% Trouvé N 6,59%

Le sel de strychnine suspendu dans l'eau tiède est additionné, avec agitation, d'une solution concentrée chaude d'hydroxyde de baryum jusqu'à début d'alcalinité à la phénolphtaléine. La strychnine est éliminée par plusieurs extractions au chloroforme. Le sel de baryum est précipité de la solution aqueuse par addition d'alcool. Après reprécipitation, il est desséché dans le vide sulfurique.

$C_6H_8O_6Ba$, 0,5 H_2O Calculé Ba 42,59% Trouvé Ba 42,48%

Le pouvoir rotatoire est mesuré avec une solution fraîchement préparée d'acide libre obtenue par décomposition du sel de baryum au moyen de la quantité strictement nécessaire d'acide sulfurique. Nous indiquons les chiffres fournis par deux échantillons différents.

$$\alpha_D^{20} = +0{,}29^0 \pm 0{,}02^0;\ c = 1{,}51\ (\text{eau});\ l = 1\ d;\ [\alpha]_D^{20} = +19{,}3^0 \pm 1{,}3^0$$

$$\alpha_D^{19} = +0{,}35^0 \pm 0{,}02^0;\ c = 1{,}75\ (\text{eau});\ l = 1\ d;\ [\alpha]_D^{19} = +20{,}0^0 \pm 1{,}1^0$$

Par évaporation à sec à basse température de la solution de l'acide libre, on obtient un résidu cristallisé consistant toujours en un mélange de γ,γ'-dilactone et d'acide libre. Ce dernier n'a pu être obtenu à l'état pur en raison de sa tendance à la lactonisation et de sa solubilité qui est notablement plus élevée que celle du racémique.

Les eaux-mères du sel de strychnine de l'acide dextrogyre sont traitées par un léger excès d'hydroxyde de baryum. Après extraction de la strychnine au chloroforme, la solution du sel de baryum est traitée par la quantité strictement nécessaire (calculée après dosage de Ba^{++}) de sulfate de cinchonine. On essore sur charbon le sulfate de baryum. La solubilité des sels de cinchonine des deux acides antipodes n'est pas très différente. Par amorçage, un des composants peut toutefois cristalliser en quantité prépondérante. Nous avons ainsi obtenu fortuitement, lors d'une expérience, une fraction qui nous a fourni un acide dihydroxy-3,4-adipique lévogyre.

$$\alpha_D^{21} = -0{,}16^0 \pm 0{,}02^0;\ c = 0{,}89\ (\text{eau});\ l = 1\ d;\ [\alpha]_D^{21} = -18{,}0^0 \pm 2{,}2^0$$

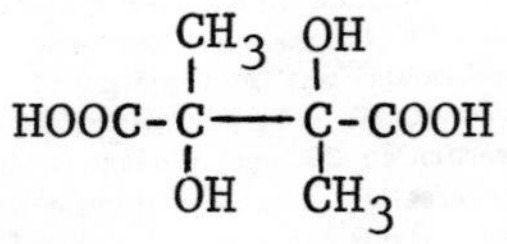

$C_6H_{10}O_6$ Ref. 218

Resolution of *racemic* 2,3-Dimethyltartaric Acid. The racemic acid (7 g) and (−)-ephedrine (6.5 g) were dissolved in hot ethyl acetate (300 m*l*) and the solution was allowed to stand overnight at room temperature, when the ephedrine salt (10.5 g) crystallized out. After triangular fractional recrystallization from acetone, the salt was decomposed with ammonia to afford, after acidification and usual work-up, the optically active acid of the constant rotation (0.7 g). Mp 161—162°C, $[\alpha]_D^{25}$ 13.6° (*c* 0.99, water).

The active acid was converted by the standard method with diazomethane into the dimethyl ester. bp 81—83°C/0.1 mmHg, n_D^{25} 1.4485, $[\alpha]_D^{25}$ 19.7° (*c* 2.65, methanol).

$C_6H_{10}O_6$ Ref. 218a

Note:

Resolved using brucine as resolving agent. Melting points: 158°C, $[\alpha]_D^{20}$ +13.4° (c 4.0, H_2O) and 160° $[\alpha]_D^{20}$ -13.4° (c 4.0 H_2O).

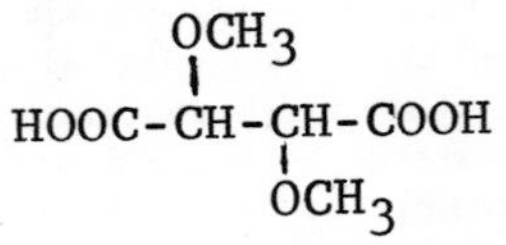

$C_6H_{10}O_6$ Ref. 219

Resolution of r-*Dimethoxysuccinic Acid by* l-*Menthol*.—*r*-Dimethoxysuccinic acid is conveniently prepared by the hydrolysis of ethyl *r*-dimethoxysuccinate with a small excess of boiling, aqueous barium hydroxide solution, precipitation of excess of alkali by carbon dioxide, filtration, and evaporation of the filtrate to dryness. The residual barium salt is dried at 105°, dissolved in water, and decomposed with the requisite quantity of sulphuric acid.

r-*Dimethoxysuccinic acid* separates from water, in which it is more sparingly soluble than the corresponding optically active acids, in colourless rhombs containing $2H_2O$. The anhydrous acid melts somewhat indefinitely at 168—171° (Found: H_2O = 16·6. $C_6H_{10}O_6,2H_2O$ requires H_2O = 16·8 per cent. Found: C = 40·6; H = 5·1. $C_6H_{10}O_6$ requires C = 40·4; H = 5·0 per cent.).

Di-*l*-menthyl *r*-dimethoxysuccinate was obtained by heating dehydrated *r*-dimethoxysuccinic acid with three times its weight of *l*-menthol for 20 hours at 120° in an intermittent current of dry hydrogen chloride or, more conveniently, by the action of *l*-menthol on ethyl *r*-dimethoxysuccinate under similar conditions during 30 hours. The product, dissolved in ether, was shaken with water and sodium carbonate solution to remove unchanged acid and hydrogen ester, and the menthol was distilled in steam. The normal ester, which was isolated in the usual manner and solidified readily ($[\alpha]_D^{21°}$ − 68·8° in chloroform), was repeatedly crystallised from ethyl alcohol, whereby ultimately *di*-l-*menthyl* l-*dimethoxysuccinate* was isolated. It crystallised in long, radiating needles, m. p. 88·5—89·5° (Found: C = 68·4; H = 9·8. $C_{26}H_{46}O_6$ requires C = 68·7; H = 9·9 per cent.). In chloroform: $l = 2$, $c = 2·138$, $\alpha_D^{16·5°}$ − 4·95°, $[\alpha]_D^{16·5°}$ − 115·1°; in benzene: $l = 2$, $c = 2·1456$, $\alpha_D^{16·4°}$ − 5·21°, $[\alpha]_D^{16·4°}$ − 121·4°; in ethyl alcohol: $l = 2$, $c = 2·0908$, $\alpha_D^{16·5°}$ − 4·95°, $[\alpha]_D^{16·5°}$ − 118·4°. The identity of the ester was established by direct comparison with the product of the action of *l*-menthol on ethyl *l*-dimethoxysuccinate and also by its hydrolysis to *barium* l-*dimethoxysuccinate*; the salt crystallises from water with approximately $5H_2O$ (Found : for the anhydrous salt, Ba = 43·4. Calc., Ba = 43·7 per cent.). In water : $l = 2, c = 0{\cdot}950, \alpha_D - 0{\cdot}48°$, $[\alpha]_D - 25{\cdot}2°$. For barium *d*-dimethoxysuccinate Purdie and Irvine (J., 1901, **79**, 959) record the value, $[\alpha]_D^{20°} + 27{\cdot}22°$ for $c = 1{\cdot}3775$.

The ester remaining in solution after the first crystallisation of the di-*l*-menthyl *dl*-dimethoxysuccinate was hydrolysed with barium hydroxide and the mixture of barium salts was converted into the corresponding acids, which were found to be strongly dextrorotatory in aqueous solution ; the isolation of the homogeneous *d*-acid could not, however, be effected.

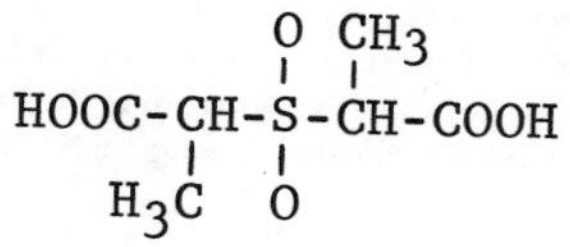

$C_6H_{10}O_6S$ Ref. 220

Résolution de l'acide α-sulfonyldipropionique.

On neutralise 42 g. (0.2 mol.) d'acide sulfonyldipropionique par un peu plus que 400 cm³. d'ammoniaque normale (réaction alcaline vis-à-vis de phénolphtaléine) et on refroidit la solution jusqu'à 0°. On introduit 117.7 g. (0.4 mol.) de cinchonine dans 400 cm³. d'acide chlorhydrique normal et par addition d'eau glacée on dilue la solution jusqu'à 4 litres.

A 0° on verse le sulfonyldipropionate d'ammonium dans la solution de l'alcaloïde en agitant énergiquement.

Le *sulfonyldipropionate neutre de cinchonine* se sépare. Le sel, essoré après dix minutes, pesa 46 g., ce qui correspond à 27 % de la quantité totale. Le sel renferme 3 mol. d'eau de cristallisation ; à l'état anhydre il est hygroscopique.

22.486 mg. de substance, séchés sous une pression de 12 mm. à 110° au-dessus d'anhydride phosphorique, ont perdu 1.460 mg.
Substance 9.020 mg., 0.523 cm³. de N_2 (22°, 76 mm.).

$C_6H_{10}O_6S . 2 C_{19}H_{22}ON_2 . 3 H_2O$ (852.53). Trouvé : H_2O 6.49 ; N 6.72.
Calculé : „ 6.34 ; „ 6.57.

1.7 g. de ce sel de cinchonine (= 2 mol.mg.), décomposés prudemment par une solution normale d'ammoniac (phénolphtaléine), libérés de la cinchonine par extraction au chloroforme et dilués d'eau jusqu'à 20 cm³., ont montré une rotation de 0°.09 pour la ligne D, dans un tube de 2 dm. La concentration du sel ammoniacal étant 0.01 mol.g. dans 100 cm³., on trouve $[M]_D = 4°.5$.

2me cristallisation. On a décomposé 42.6 g. du sel de cinchonine (0.05 mol.) à 0° par addition d'ammoniaque normale. Après filtrage de la cinchonine on a dilué d'eau glacée la solution du sel ammoniacal jusqu'à 200 cm³.

On a dissous la cinchonine dans 100 cm³. d'acide chlorhydrique normal et 800 cm³. d'eau glacée. A 0° on a ajouté le sel ammoniacal à la solution de l'alcaloïde.

Le sulfonyldipropionate de cinchonine, essoré après 5 minutes, pesa 15.5 g., soit 36 % de la quantité employée.

1.7 g. du sel de la seconde cristallisation ont donné un sel ammoniacal, offrant une rotation de 0°.14 dans 20 cm³.

3me cristallisation. On a fait recristalliser 15.5 g. du sel de la seconde cristallisation à 0° dans une volume de 300 cm³., selon la méthode décrite ci-dessus. Il s'est déposé 10.25 g. de sel de cinchonine, soit 66 % du sel mis en oeuvre.

1.7 g. de ce produit ont donné un sel ammoniacal d'une rotation de 0°.17 dans 20 cm³.

4me cristallisation. 10.25 g. du sel de cinchonine de la cristallisation précédente, soumis à la „cristallisation froide" dans 285 cm³. ont déposé 7.9 g. de sel, soit 77 %.

1.7 g. de ce sel, décomposés à l'ammoniaque, ont donné une rotation de 0°.16 dans 20 cm³.

On peut donc conclure que la rotation spécifique du sel ammoniacal ne change plus.

En utilisant la rotation, mesurée pour la dernière cristallisation (0°.16 pour 0.01 mol.g. dans 100 cm³., tube de 2 dm.), on trouve les valeurs suivantes pour les rotations moléculaire et spécifique du *sulfonyldipropionate d'ammonium*.

$SO_2 \{CH(CH_3) . CO_2NH_4\}_2$. P. mol. = 244.21 ; $[M]_D = 8°$; $[\alpha]_D = 3°.3$.

Il va sans dire que les liqueurs-mères de toutes les cristallisations ont servi à récupérer l'acide et l'alcaloïde.

L'acide α-sulfonyldipropionique actif.

On a décomposé 3.4 g. (= 4 mol.mg. du sel de cinchonine de la dernière cristallisation) très prudemment à 0°, par 12 cm³. (excès de 50 %) d'acide sulfurique normal et puis on a extrait à l'éther l'acide organique libéré.

La solution éthérée (50 cm³.), bien séchée sur du sulfate sodique fondu, a donné une rotation de 0°.43 pour 2 dm. Après 4 heures la rotation n'avait pas changé.

Comme d'après le titrage la solution éthérée renfermait 0.000946 mol.g. de l'acide sulfonyldipropionique dans 100 cm³., on trouve les constantes suivantes.

$SO_2 \{CH(CH_3)CO_2H\}_2$. Sol. éthérée. P. mol. = 210.15 ;
$[M]_D = 227°$; $[\alpha]_D = 108°$.

On a fait évaporer rapidement dans le vide la solution éthérée et on a repris le résidu dans 30 cm³. d'eau.

La solution aqueuse a donné dans un tube de 2 dm. une rotation de 0°.47.

Le titrage de la solution aqueuse a accusé la présence de 0.001577 mol.g. dans 100 cm³.

$SO_2 \{CH(CH_3)CO_2H\}_2$. Sol. aqueuse. P. mol. = 210.15 ;
$[M]_D = 149°$; $[\alpha]_D = 71°$.

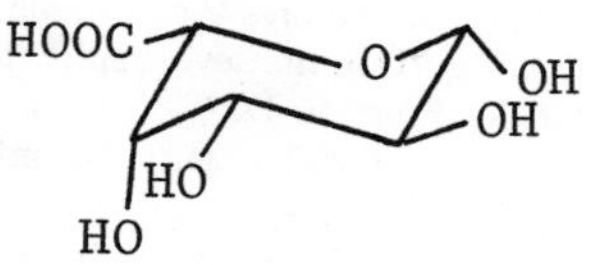

$C_6H_{10}O_7$ Ref. 221

Isolation of d-Galacturonic Acid—10.0 gm. of *dl*-galacturonic acid monohydrate were dissolved in 50 cc. of 90 per cent ethyl alcohol and added to a solution of 20.0 gm. of brucine dihydrate in 50 cc. of the same solvent. After standing overnight in the ice chest the crystalline precipitate was collected (about 28 gm.) and recrystallized from 150 cc. of a 70 per cent acetone water mixture. The product was again recrystallized from 200 cc. of 75 per cent acetone and after washing freely with methyl alcohol and chloroform was dried at room temperature over phosphorus pentoxide. The preparation (10 gm.) melted at 180–181° with decomposition.

12.0 gm. of the brucine salt were dissolved in 100 cc. of water and 100 cc. of saturated barium hydroxide added to the cold solution. The brucine was removed by filtration and extraction with chloroform. The excess alkali was then neutralized with carbon dioxide, the solution heated at 60° for 30 minutes, the barium carbonate removed, and the barium salt of the uronic acid isolated from the filtrate in the usual manner (14). The yield was about 5 gm. and the salt when thoroughly dried at 78° under 1 mm. pressure had an $[\alpha]_D^{25}$ of +25.8°.

4.0 gm. of the above barium salt were converted into the free acid (4, 14) and isolated as the monohydrate (1.9 gm.).

Melting Point—The acid sintered at 110° and melted with decomposition at 157–159°.

Rotation—$[\alpha]_D^{25}$ = +50.0° ± 2.0° (in water, c = 1.36 per cent).

Analysis

$C_6H_{10}O_7 \cdot H_2O$. Calculated. Neutralization equivalent 47.20 cc. 0.1 N alkali
Found. " " 48.50 " 0.1 " "

Isolation of l-Galacturonic Acid—The mother liquors resulting from the isolation and recrystallization of brucine *d*-galacturonate (above) were combined and concentrated, whereupon a brucine salt (7.0 gm.) of melting point 162–163° was isolated. 12.0 gm. of this salt upon treatment with barium hydroxide yielded 5.0 gm. of a barium salt which had an $[\alpha]_D^{25}$ of −7.0°. 4.26 gm. of this barium salt (35 per cent *d*- and 65 per cent *l*-) were treated with 14.20 cc. of N sulfuric acid. The uronic acid was isolated from the reaction mixture in the usual way (14) and fractionated according to the following scheme. All crystallizations were carried out at 25°.

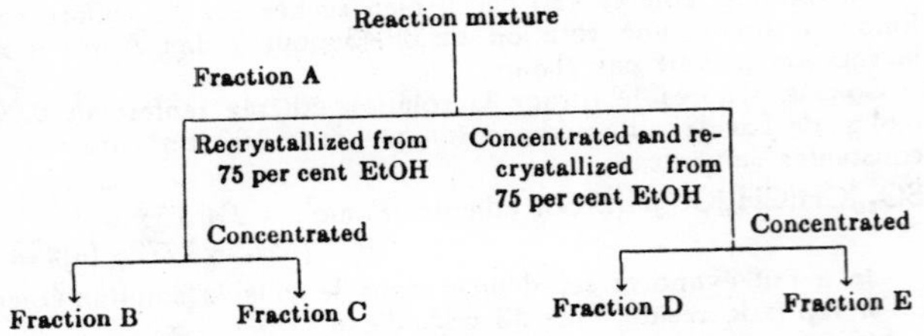

The yields and rotations of the various fractions are given in the accompanying tabulation.

Fraction	Yield	Rotation $[\alpha]_D^{23}$ in water
	mg.	*degrees*
B	1400	−5.0 ± 1.0
C	70	−40.0 ± 2.0
D	30	−52.0 ± 2.0
E	15	−43.0 ± 2.0

From the tabulation it is evident that Fraction D is an optically pure specimen of *l*-galacturonic acid. Fractions C and E can be combined and recrystallized to yield a further quantity of the *l* acid. A theoretical yield requires the isolation of 2.1 gm. of the racemic acid and 0.9 gm. of the levo acid. The following are constants exhibited by Fraction D.

Melting Point—The acid melts with decomposition between 156-158°.

Rotation—$[\alpha]_D^{23} = -52.0° \pm 2.0°$ (in water, c = 0.18 per cent).

Analysis

$C_6H_{10}O_7$. Calculated. Neutralization equivalent 51.52 cc. 0.1 N alkali
Found. " " 51.30 " 0.1 " "

$CH_3(CH_2)_3$-CH-COOH
Br

$C_6H_{11}BrO_2$ Ref. 222

Derivatives of n-Caproic Acid.

Resolution of α-Bromo-n-Caproic Acid.—60 gm. of inactive bromo acid were dissolved in 500 cc. of hot acetone and 102 gm. of pure strychnine were then added, whereupon the strychnine salt crystallized immediately. The salt was extracted with hot acetone seven times. It was dissolved in well cooled, concentrated hydrochloric acid and extracted with ether. The ethereal extract was washed with ice water until the washings contained no halogen ion and dried with sodium sulfate. The ether was distilled off and the residue fractionated under reduced pressure. The bromo acid boiled at 129–130°C. (p = 14 mm.). It had optical activities as follows:

$$[\alpha]_D^{20} = \frac{-1.27° \times 100}{1 \times 4.71} = -27.0° \text{ (in ether).}$$

$$[\alpha]_D^{20} = \frac{-0.57° \times 100}{2 \times 1.26} = -22.6°. \quad [M]_D^{20} = -44.1° \text{ (in 30 per cent alcohol).}$$

$(CH_3)_2CHCH_2$CHCOOH
Br

$C_6H_{11}BrO_2$ Ref. 223

l-α-Brom-isocapronsäure.

60 g inactiver α-Bromisocapronsäure werden in 800 ccm Wasser von 35—40° mit 145 g Brucin zusammen durch kräftiges Schütteln gelöst, von einem kleinen Rückstand abfiltrirt und die Flüssigkeit bei möglichst geringem Druck in einem Bade, dessen Temperatur nicht über 30° steigen darf, auf etwa ein Drittel eingedunstet. Dabei fällt eine grosse Menge des Brucinsalzes krystallinisch aus, das nach 15-stündigem Stehen bei 10 − 15° abfiltrirt wird. Seine Menge beträgt ungefähr 110 g. Man löst das Salz in etwa 3½ L Wasser von 30° und engt die Lösung ebenso wie das erste Mal auf etwa 250 ccm ein. Jetzt beträgt die Ausbeute an Brucinsalz etwa 60 g. Auf die gleiche Art muss das Salz im ganzen fünfmal umkrystallisirt werden, mit dem einzigen Unterschiede, dass man das dritte, vierte und fünfte Mal stärker einengt, sodass stets die Hauptmenge des Salzes zurückgewonnen wird. Zur Orientirung folgen hier die Ausbeuten: beim dritten Mal 50 g, beim vierten Mal 48 g und beim fünften Mal 45 g. Durch das Umlösen werden die Krystalle allmählich grösser und in Wasser viel schwerer löslich. Man kann deshalb zur Erleichterung auch in wenig Alkohol lösen und die mit Wasser verdünnte Flüssigkeit bei geringem Druek eindampfen. Zum Schluss resultirten feine Nadeln. Zur Umwandelung in die freie Säure löst man sie in Wasser von etwa 30°, fügt etwas mehr als die berechnete Menge verdünnter Salzsäure zu und extrahirt die *l-α*-Bromisocapronsäure mit Aether. Man trocknet diesen Auszug mit Natriumsulfat, verdampft den Aether am besten im Vacuum und destillirt den öligen Rückstand bei möglichst niedrigem Druck. Die Säure siedet unter 0.2—0.4 mm Druck bei ungefähr 94° (corr.).

Ein solches Präparat drehte im 1 dm-Rohr bei 20° Natriumlicht 66.81° nach links und hatte die Dichte 1.358, mithin

$$[\alpha]_D^{20} = -49.20°.$$

Bei Anwendung von 60 g racemischer Säure betrug die Ausbeute an Brucinsalz (5. Krystallisation) 45 g und an activer Säure von obigem Drehungsvermögen 8.64 g, mithin 14.4 pCt. der inactiven Säure und fast 29 pCt. der Theorie.

0.2122 g active Säure: 0.2040 g AgBr.

$C_6H_{11}O_2Br$ (Mol.-Gew. 195.07). Ber. Br 40.99. Gef. Br 40.91.

Zur Prüfung der Reinheit in optischer Beziehung wurde diese Säure noch einmal mit Hülfe des Brucinsalzes gereinigt. Sie drehte dann unter den gleichen Bedingungen wie oben 67.13° nach links. Mithin

$$[\alpha]_D^{20} = -49.43°.$$

Die Differenz von 0.23° ist so gering, dass sie auch ein Versuchsfehler sein kann; denn Temperaturschwankungen von 1° bewirken schon eine Veränderung der Drehung um 0.5°, und wir konnten bei unseren Beobachtungen die Temperatur nicht genauer als innerhalb ½ Grades halten.

d-α-Brom-isocapronsäure.

Als die Mutterlauge vom Brucinsalz der isomeren Säure ohne weitere Concentration eine Woche lang bei Zimmertemperatur stehen blieb, schieden sich auf's neue Krystalle ab, die man schon äusserlich an der viel dickeren Form als verschieden von dem ersten Brucinsalz erkennen konnte. Die Ausbeute daran betrug 35 g auf 60 g der inactiven Säure. Wurden diese 35 g nochmals in der dreifachen Menge Wasser gelöst, die filtrirte Lösung bei 10 mm vorsichtig bis auf ungefähr ⅓ Volumen eingedampft und dann bei 0° geimpft, so schieden sich beim längeren Stehen ebenfalls bei 0° 22 g derselben derben Krystalle aus. Aus ihnen wurde die Säure wie beim optischen Isomeren isolirt und gereinigt.

Die Ausbeute betrug 4 g, das sind 13.3 pCt. der Theorie, berechnet auf die racemische Säure. Die Untersuchung des Drehungsvermögens gab für Natriumlicht im 1 dcm-Rohr bei 20° $\alpha = +66.53°$ und Dichte = 1.358, mithin

$$[\alpha]_D^{20} = +48.99°.$$

Diese Zahl entspräche, wenn der für die linksdrehende Säure gefundene höchste Werth als richtig angenommen wird, einem Gehalt an Racemkörper von nicht ganz 1 pCt.

0.1912 g desselben Präparates gaben nach Carius 0.1848 g AgBr.

$C_6H_{11}O_2Br$ (Mol.-Gew. 195.07). Ber. Br 40.99. Gef. Br 41.13.

Eine Säure aus dem nicht umkrystallisirten Brucinsalz, dessen Menge 35 g betrug, hatte ein geringeres Drehungsvermögen und zwar

$$[\alpha]_D^{20} = +46.85°.$$

$(CH_3)_2CHCH_2\text{-}CH(Br)\text{-}COOH$

$C_6H_{11}BrO_2$

Ref. 223a

Resolution of α-Bromoisocaproic Acid.—α-Bromoisocaproic acid was resolved into its optical enantiomorphs by Fischer[18] by means of brucine but in our work quinine was used for this resolution. To 50 gm. of inactive bromo acid dissolved in 400 cc. of acetone, 95 gm. of pure quinine were added. After seven recrystallizations from acetone the salt was decomposed under cooling with a slight excess of dilute hydrochloric acid and the solution was extracted with ether. The ethereal extract was washed with water and dried over sodium sulfate. The ether was removed and the residue fractionated under reduced pressure. It distilled at 131–131.5° (p = 16 mm.). The optical activity was

$$[\alpha]^{20}_D = \frac{+1.79° \times 100}{1 \times 6.00} = +29.8° \text{ (in ether)}.$$

$$[\alpha]^{20}_D = \frac{+0.45° \times 100}{2 \times 0.84} = +26.8°. \quad [M]^{20}_D = +52.2° \text{ (in 20 per cent alcohol)}.$$

[18] Fischer, E., and Carl, H., *Ber. chem. Ges.*, 1906, xxxix, 3996.

$(CH_3)_3C\text{-}CH(Cl)\text{-}COOH$

$C_6H_{11}ClO_2$

Ref. 224

Note:

Resolved by fractional crystallization of its cinchonidine salts. It has m.p. 67-70°, b.p. 83-84°/1.5-2.0 mm., $[\alpha]^{27}_D$ -9.9° (c 0.323, methanol).

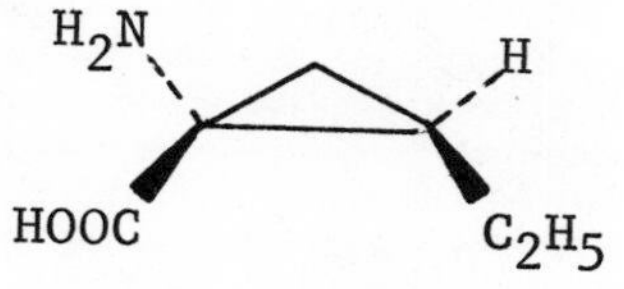

$C_6H_{11}NO_2$

Ref. 225

Optical resolution of 2 was carried out with quinine salts of N-formyl coronamic acid and after several fractional recrystallization, resultant two crystalline materials were hydrolized to yield optically active 2a, $[\alpha]^{20}_D$ +14.7 (c 1.67, H_2O) and 2b, $[\alpha]^{23}_D$ -14.2 (c 1.67, H_2O) respectively. Enzymatic resolution of dl-N-acetylcoronamic acid using L-acylase also gave directly 2a, and recovered 2b acetate was hydrolysed to 2b.

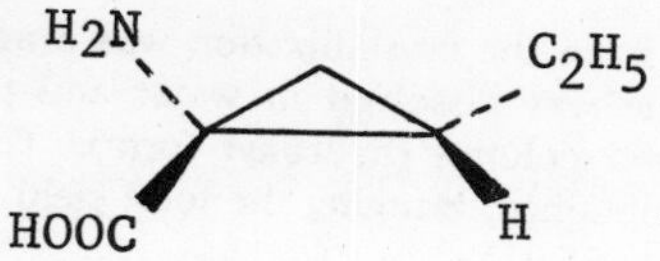

$C_6H_{11}NO_2$

Ref. 226

The synthesized (±)-allocoronamic acid was resolved through the quinine salt. The salts of N-formyl derivatives of **2** were recrystallized fractionally from hot ethyl acetate-ethanol. The less soluble salt was hydrolyzed to yield (+)-allocoronamic acid (**2a**); $[\alpha]^{23}_D$+65.0° (c=1.83, H_2O). The salt recovered from the mother liquor was recrystallized from acetone to yield the antipode (**2b**); $[\alpha]^{21}_D$−68.4° (c=1.15, H_2O).

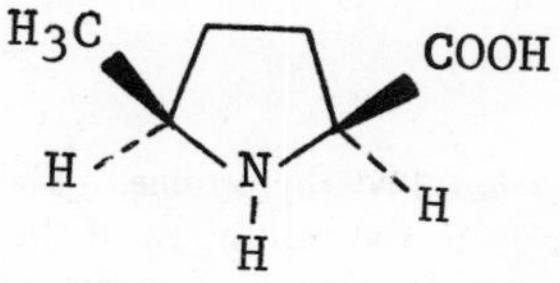

$C_6H_{11}NO_2$

Ref. 227

Resolution of *cis*-5-Methylproline. Finely powdered XIa (211.5 g, 1.18 mol) was suspended in 1 l. of absolute ether. Triethylamine (119 g, 1.18 mol) was added over 1 hr with vigorous stirring. After 2 hr, the solution was cooled in an ice bath and filtered from triethylamine hydrochloride. The triethylamine hydrochloride was suspended in 1 l. of ether, refluxed for 30 min, cooled, and filtered. The combined filtrates were concentrated at reduced pressure at a bath temperature not exceeding 30°. Liquid *cis*-5-methylproline methyl ester (XIIa) (158.8 g, 1.11 mol, 94%) remained. To XIIa, dissolved in 1 l. of absolute methanol, was added 166.5 g (1.11 mol) of tartaric acid in 1 l. of absolute methanol. The solution was heated almost to boiling. Upon cooling, 152 g of a dextrorotary salt with [α]D +30° crystallized. After three recrystallizations from absolute methanol, 105.5 g of (XIII), mp 159–161° dec (65% of the theoretical yield for one enantiomer), with [α]D +32.2° remained.

Anal. Calcd for $C_{11}H_{19}NO_8$ (mol wt 293.28): C, 45.05; H, 6.53; N, 4.78. Found: C, 45.14; H, 6.59; N, 4.57.

The filtrate from the first crystallization was evaporated to dryness. The oily residue was stirred with a refluxing mixture of 50 ml of methanol and 500 ml of acetone. The oil crystallized on cooling. An additional 250-ml increment of acetone was added. A salt (161.5 g) with [α]D +3.14° was collected after cooling for several hours. Five recrystallizations from a mixture of acetone–water (9:1) gave 72 g of a levorotary salt (XIV), mp 72–73°, with [α]D −2.8°. This is 44% of the theoretical yield for one enantiomer.

(+)-*cis*-5-Methyl-D-proline Hydrochloride (XV) and (+)-*cis*-5-Methyl-D-proline (XVI). Potassium chloride (27 g, 0.36 mol) in 100 ml of hot water was added to XIII (105.3 g, 0.36 mol) in 200 ml of hot water. Acetone (300 ml) was added and the solution was cooled overnight. The potassium tartrate was filtered and the filtrate was evaporated to dryness. The residue was redissolved in 300 ml of water and was refluxed for 1 hr to hydrolyze the methyl ester completely. The residue was crystallized from methanol–ether after evaporation to dryness to yield 42.3 g of XV (71.2%), mp 190.5–192°, [α]D +34.1°.

The filtrate from the crystallization was evaporated to dryness and the residue was dissolved in water and passed through an Amberlite IR-45 column (hydroxyl form). Crude XVI, 11.5 g (24.7%), was obtained, bringing the total yield of amino acid and hydrochloride to 95.9%.

Two recrystallizations gave the free amino acid with mp 202–203°, $[\alpha]_D$ +69.5°.

Anal. Calcd for $C_6H_{11}NO_2$ (mol wt 129.16): C, 55.79; H, 8.59; N, 10.85. Found: C, 55.95; H, 8.56; N, 10.90.

(−)-*cis*-5-Methyl-L-proline Hydrochloride (XVII) and (−)-*cis*-5-Methyl-L-proline (XVIII). The liberation of the amino acid from the levorotary salt was performed in the same way as described for the dextrorotary salt: from 69.5 g of XIV was obtained 21 g of XVII, mp 190–191°, $[\alpha]_D$ −33.6°, and 9 g of XVIII, mp 200–200.5° dec, $[\alpha]_D$ −68°. The total yield was 83%.

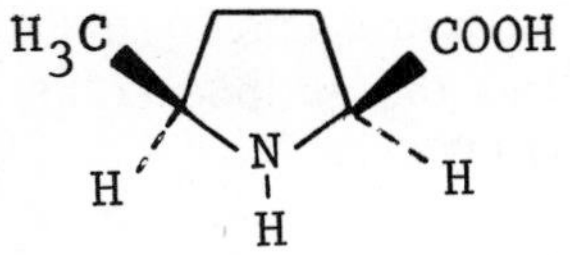

$C_6H_{11}NO_2$ Ref. 228

Resolution of *trans*-5-Methylproline. The resolution was performed analogously to the resolution of the cis isomer. The salt from 43.5 g of *trans*-5-methylproline methyl ester and 45.5 g of tartic acid was formed in 120 ml of acetone containing 15% methanol. A salt, 43 g, with $[\alpha]_D$ +3.35°, crystallized. A constant $[\alpha]_D$ −11.7° was reached after six recrystallizations from acetone–methanol. The yield of levorotary salt was 9.0 g, *i.e.*, 20% of the theoretical yield for one enantiomer, mp 132–136.5°.

The dextrorotary salt could not be obtained in crystalline form. Partially resolved *trans*-5-methylproline, 10 g, with $[\alpha]_D$ +50°, was recovered from the first filtrate.

(−)-*trans*-5-Methyl-L-proline Hydrochloride (XIX) and (−)-*trans*-5-Methyl-L-proline (XX). The free amino acid and its hydrochloride were recovered from the tartaric acid salt in the same way as described for the cis isomer: 9 g of the levorotary salt gave 4.2 g of XIX (86%), mp 180°, $[\alpha]_D$ −57.1°. XX had mp 227° and $[\alpha]_D$ −86.2°.

Anal. Calcd for $C_6H_{11}NO_2$ (XX, mol wt 129.16): C, 55.79; H, 8.59; N, 10.85. Found: C, 55.95; H, 8.64; N, 10.78.

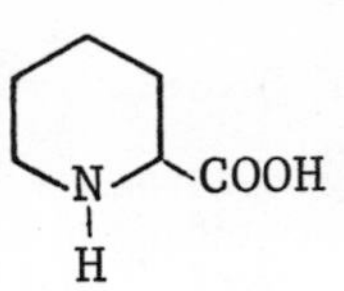

$C_6H_{11}NO_2$ Ref. 229

(+)-*Pipecolic acid.* Pipecolic acid (52 g) and (+)-tartaric acid (57 g) were dissolved in ethanol (1 litre), the solution boiled for 30 min and allowed to stand for 18 hr, whereupon the salt crystallized (57 g). The solid was crystallized three times from aqueous acetone to yield (+)-pipecolic acid hydrogen (+)-tartrate (19·5 g), $[\alpha]_D^{25}$ + 21·1° (*c*, 4·0 in H_2O). This salt, whose rotation was unchanged on further recrystallization, was dissolved in distilled water (250 ml), the solution passed through a column of Amberlite IR4b (OH^- form) resin (300 g), and the column washed with distilled water until the eluate was no longer alkaline. The total eluate was treated with charcoal (0·2 g) and filtered to remove the charcoal. The filtrate was evaporated under reduced pressure and the residue crystallized from aqueous acetone to yield (+)-pipecolic acid (7·0 g) as colourless needles, m.p. 280°, $[\alpha]_D^{23}$ + 24·6° (*c*, 3·35 in H_2O). [Heilbron & Bunbury (1946) quote m.p. 270°, $[\alpha]_D$ + 24·5°.]

Note:

H. C. Beyerman, Recl. Trav. Chim. Pays-Bas, <u>78</u>, 134 (1959) reported a m.p. of 192-193° and $[\alpha]_D^{20}$ +22.3° (±) 1° (c 6.4, H_2O) for the pipecolic acid bi-d-tartrate, and $[\alpha]_D^{19}$ +26.2° (±) 1° (c 3.106, H_2O) for the D-(+)-pipecolic acid.

$CH_2{=}CH{-}CH_2SCH_2{-}CH(NH_2){-}COOH$

$C_6H_{11}NO_2$ Ref. 230

b) Brucinsalz des S-Allyl-DL-N-formylcysteins. Unter vorsichtigem Erwärmen auf 75° werden 110 g S-Allyl-DL-N-formylcystein und 220 g wasserfreies Brucin in 1100 cm³ Butanol gelöst. Aus der filtrierten, leicht gelb gefärbten Lösung kristallisiert beim Erkalten das Brucinsalz des S-Allyl-D-N-formylcysteins in grossen, viereckigen Tafeln, die nach einigem Stehen bei 3° abfiltriert und mit wenig eiskaltem Butanol gewaschen werden. Die noch feuchten Kristalle werden erneut aus 800 cm³ Butanol kristallisiert: 173 g Rohkristalle, $[\alpha]_D^{20} = -26{,}3°$ (in Wasser). Nach zwei weiteren Umkristallisationen aus je 800 cm³ Butanol erhält man 155 g reines Brucinsalz des S-Allyl-D-N-formylcysteins.

Zur Analyse wurde eine Probe zweimal aus der 10fachen Menge Methanol umkristallisiert: Grosse, viereckige Tafeln vom Smp. ca. 150° (unter Zersetzung; Sintern bei 109—113°). $[\alpha]_D^{20} = -21{,}6°$ (in Wasser).

$C_{29}H_{37}O_6N_3S$	Ber. C 62,72	H 6,65	N 7,56%
(555,3)	Gef. „ 62,59	„ 6,85	„ 7,64%

c) Zerlegen des Brucinsalzes und Isolierung des S-Allyl-D-cysteins. 150 g Brucinsalz des S-Allyl-D-N-formylcysteins werden in 500 cm³ Wasser gelöst und nach Zugabe von 500 cm³ 2-n. Ammoniak zur Entfernung des Brucins zuerst mit 300 cm³, dann zweimal mit je 100 cm³ Chloroform ausgeschüttelt. Dann wird die wässerige Lösung im Vakuum solange (auf ca. 650 cm³) eingedampft, bis sie auf Lackmuspapier sauer anzeigt. Zu dieser Lösung fügt man 60 cm³ konz. Salzsäure hinzu, erhitzt zur Abspaltung des Formylrestes während 90 Minuten zum Sieden, entfärbt die klare, leicht gelb gefärbte Lösung durch Aufkochen mit etwa 5 g Noritkohle und dampft im Vakuum zur Trockne ein. Dem kristallinen Rückstand wird mit warmem absolutem Alkohol das S-Allyl-D-cystein-hydrochlorid entzogen und die Lösung im Vakuum zur Trockne eingedampft. Der kristalline Rückstand wird nun in 150 cm³ warmem Wasser gelöst, die Lösung mit konz. Ammoniak auf ein pH von 4,5 gestellt und dann mit 200 cm³ Aceton versetzt. Beim Erkalten kristallisieren 25,3 g S-Allyl-D-cystein aus. Smp. 235—236° (unter Zersetzung, von 213° an Braunfärbung). $[\alpha]_D^{20} = +15,8°$ (in Wasser). Die Aufarbeitung der Mutterlauge ergab noch 9 g rohes S-Allyl-D-cystein, das nach einmaligem Umkristallisieren aus verdünntem Aceton rein war.

Zur Analyse wurde eine Probe nochmals aus der 10-fachen Menge 60-proz. Aceton umkristallisiert.

$C_6H_{11}O_2NS$	Ber.	C 44,74	H 6,83	N 8,69%
(161,1)	Gef.	„ 44,73	„ 7,14	„ 8,58%

$$(CH_3)_2NCS_2\underset{\displaystyle CH_3}{\underset{|}{C}}HCOOH$$

$C_6H_{11}NO_2S_2$ Ref. 231

Optical resolutions. Preliminary experiments were carried out with the following bases:

Strychnine	Quinidine	Cinchonidine
Brucine	Morphine	(+)-α-Phenethylamine
Quinine	Cinchonine	(−)-β-Phenylisopropylamine [10]

Ethyl acetate alone or with a small quantity of methanol was found to be the most effective solvent. The salt of the acid to be resolved with the optically active base was recrystallized several times. After each recrystallization a small sample of the salt was decomposed with hydrochloric acid, the acid isolated and the rotatory power determined.

Table 5. Recrystallization of the strychnine salt of N,N-dimethyldithiocarbamyllactic acid. Initial quantity: 0.075 mole.

Recryst. No.	Solvent (ml)		Yield of salt (%)	$[\alpha]_D$ of the acid
	EtAc	MeOH		
1	300	20	54	−54.4°
2	200	15	40.5	
3	175	15	25.5	−70.1°
4	150	15	21.5	−76.8°
5	125	15	12	−77.2

Table 6. Recrystallization of the (+)-β-phenyl*iso*propylamine salt of N,N-dimethyldithiocarbamyllactic acid recovered from cryst. No. 1 above. ($[\alpha]_D = +15°$).

Recryst. No.	Solvent (ml)		Yield of salt (%)	$[\alpha]_D$ of the acid
	EtAc	MeOH		
1	160	40	44	50°
2	100	20	35	59.1°
3	75	20	30	71.7°
4	50	20	25	76.2°
5	50	20	22	75.0°
6	30	15	19.5	76.3°

10. Matell, M. *Acta Chem. Scand.* 7 (1953) 698.

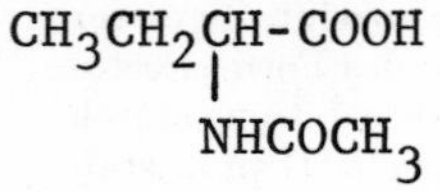

$C_6H_{11}NO_3$

$$CH_3CH_2\underset{\underset{NHCOCH_3}{|}}{CH}-COOH$$

$C_6H_{11}NO_3$ Ref. 232

Parte sperimentale (*)

Sale della (S)(—)-feniletilamina con l'acido (R)(+)-2-acetilaminobutirrico

7,25 g (0,05 mole) di acido (R)(S)-2-acetilaminobutirrico e 6,05 g (0,05 mole) di (S)(—)-feniletilamina sono stati sciolti in 100 cc di alcool etilico assoluto e la soluzione è stata mantenuta a ricadere per qualche minuto. Dopo raffreddamento si sono introdotti alcuni germi preventivamente preparati e si è lasciato a sè per due giorni. E' così cristallizzato il sale dell'acido (R)(+)-2-acetilaminobutirrico: 4,2 g, p.f. 180°, $[\alpha]_D^{27°}$ = +7,03 (c = 2 acqua). Il prodotto è stato cristallizzato da 25 cc di etanolo ass. ottenendo 3,35 g a p.f. 180°, $[\alpha]_D^{20°}$ = +7,70 (c = 2 acqua). Resa 50%.

Ulteriori cristallizzazioni non hanno portato a variazione di p.f. e potere rotatorio.

	trov. %	:	C 62,98;	H 8,27;	N 10,53
per $C_{18}H_{22}N_2O_3$	calc.	:	63,14;	8,32;	10,52

Una soluzione in acqua c = 2 ha dato i seguenti valori polarimetrici:

$[\alpha]_D^{20°}$	67,36	34,4	18,5	9,48	8,21	7,70
$\lambda_{m\mu}$	313	365	436	546	577	589,3

Per valori di $[\alpha]_D^t$ nell'intervallo di 15° e 35° si ha $[\alpha]_D^t$ = +7,70 + [(20 — t)·0,0963].

Sale della (R)(+)-feniletilamina con l'acido (S)(—)-2-acetilaminobutirrico

Le acque madri della scissione sono state evaporate a secco, il residuo è stato sciolto in 50 cc di acqua e la soluzione è stata resa basica con NaOH al 10%. La soluzione è stata estratta con 3 x 30 cc di etere, la soluzione acquosa è stata portata a pH = 3 con acido cloridrico ed è stata evaporata a secco in pallone rotante senza superare i 25°.

Il residuo è stato estratto quattro volte con 50 cc di acetone bollente e dopo essiccamento su cloruro di calcio, la soluzione acetonica è stata evaporata a secco.

Il residuo vetroso, che pesava 5 g, è stato sciolto in 60 cc di alcool contenente 4 g di (R)(+)-feniletilamina, ottenendo una soluzione limpida, che è stata posta a ricadere per pochi minuti. Dopo raffreddamento si è lasciato a temperatura ambiente per una notte. Il sale della (R)(+)-feniletilamina con l'acido (S)(—)-2-acetilaminobutirrico ottenuto pesava 4,1 g con p.f. 179°, $[\alpha]_D^{27°}$ = —6,71 (c = 2 acqua).

Per cristallizzazione da 25 cc di etanolo assoluto si sono ottenuti 3,45 g, p.f. 180°, $[\alpha]_D^{20°}$ = —7,72 (c = 2 acqua). Resa 51,5%.

	trov. %	:	C 63,22;	H 8,50;	N 10,48
per $C_{18}H_{22}N_2O_3$	calc.	:	63,14;	8,32;	10,52

$[\alpha]_D^{20°}$	—67,17	—34,3	—18,4	—9,47	—8,07	—7,72
$\lambda_{m\mu}$	313	365	436	546	577	589,3

Acido (R)(+)-2-acetilaminobutirrico

4 g (0,15 mmole) di sale dell'acido (R)(+)-2-acetilaminobutirrico con (S)(—)-feniletilamina sono stati sciolti in 30 cc di acqua e la soluzione fatta passare su Amberlite IR 120 (0,04 eq). La frazione contenente l'acido è stata raccolta sino a pH 3,5 evaporata a secco sotto vuoto ottenendo 2,1 g, p.f. 135°, $[\alpha]_D^{25°}$ = +41,96 (c = 2 acqua).

Il prodotto è stato cristallizzato da 14 cc di etanolo-acetato d'etile (1:4). Si sono ottenuti 1,95 g a P.f. 135°, $[\alpha]_D^{20°}$ = +43,6 (c = 2 acqua).

Questi valori non subiscono alcuna variazione con ulteriore cristallizzazione. Resa 90%.

	trov. %	:	C 49,43;	H 7,71;	N 9,63
per $C_6H_{11}NO_3$	calc.	:	49,65;	7,63;	9,65

$[\alpha]_D^{20°}$	228,4	145,2	90,62	52,08	45,8	43,6
$\lambda_{m\mu}$	313	365	436	546	577	589,3

Per temperature comprese tra 15° e 35°

$[\alpha]_D^t$ = +43,5 + [(20 — t)·0,155°]

(*) I poteri rotatori sono stati misurati con polarimetro Perkin-Elmer modello 141.

Acido (S)(—)-2-acetilaminobutirrico

Operando su 4 g di sale di (R)(+)-feniletilamina con l'acido (S)(—)-2-acetilaminobutirrico come nel caso precedente si sono avuti 2 g di prodotto grezzo a p.f. 135°, $[\alpha]_D^{27°}$ = —41,08 (c = 2 acqua).

Dopo cristallizzazione si sono ottenuti 1,9 g a p.f. 135° e $[\alpha]_D^{20°}$ = —43,5 (c = 2 acqua). Resa 87%.

$[\alpha]_D^{20°}$	—227,1	—144,5	—90,02	—52,10	—45,7	—43,5
$\lambda_{m\mu}$	313	365	436	546	577	589,5

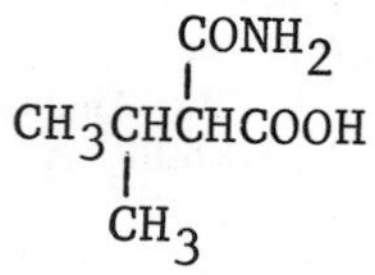

$C_6H_{11}NO_3$ Ref. 233

Darstellung der rechtsdrehenden Isopropyl-malonaminsäure.

Für die Spaltung der Racemverbindung diente das Chininsalz. Dabei ist es zweckmäßig, sich erst im Kleinen Krystalle eines möglichst reinen Salzes der *d*-Säure zu verschaffen, indem man 2 g Racemverbindung mit 5.22 g Chinin in 40 ccm heißem Wasser löst und die in der Kälte ausgeschiedenen Krystalle 3—4-mal aus heißem Wasser umlöst. Für die große Operation werden 134 g *d,l*-Verbindung mit 350 g wasserhaltigem Chinin in 2750 ccm heißem Wasser gelöst, dann etwas abgekühlt und Impfkrystalle eingetragen. Bald beginnt die Krystallisation, und wenn man schließlich über Nacht im Eisschrank stehen läßt, so ist die Flüssigkeit von einem dicken Brei feiner, meist kugelförmig angeordneter Nadeln erfüllt. Ausbeute: 180 g (Theorie 216.7 g). Die Mutterlauge enthält das Salz der *l*-Säure, auf deren Isolierung wir verzichtet haben. Man kann sie aber zur Gewinnung von neuen Mengen der *d*-Verbindung benutzen. Schon beim wochenlangen Stehen der Mutterlauge erfolgt wieder neue Krystallisation des Chininsalzes der *d*-Säure, die offenbar allmählich unter diesen Bedingungen entsteht. Rascher geht es beim längeren Kochen der Flüssigkeit. Endlich haben wir die Umwandlung durch Alkali benutzt, um aus der Mutterlauge Racemkörper zurückzugewinnen. Zu dem Zweck wurde die Mutterlauge mit 10-*n*. Natronlauge versetzt, bis alles Chinin gefällt war, dann abgesaugt, nochmals die gleiche Menge Natronlauge zugefügt und 3 Tage bei gewöhnlicher Temperatur aufbewahrt. Dann wurde die Flüssigkeit mit Salzsäure schwach angesäuert, unter vermindertem Druck stark eingeengt und ein Überschuß von Salzsäure zugefügt, worauf die racemische Isopropyl-malonaminsäure rasch krystallisierte. Abgesehen von unvermeidlichen Verlusten ist die Ausbeute an diesem regenerierten Produkt sehr befriedigend.

Um aus dem oben erwähnten krystallisierten Chininsalz die freie *d*-Isopropyl-malonaminsäure zu bereiten, ist einige Vorsicht geboten. 20 g Salz werden mit 200 ccm Chloroform und 300 ccm Wasser übergossen, das Gemisch auf 0° abgekühlt, nun mit 85 ccm $^n/_1$-Natronlauge versetzt und kräftig durchgeschüttelt, wobei das in Freiheit gesetzte Chinin von dem Chloroform gelöst wird. Die wäßrige Schicht wird sofort von dem Chloroform getrennt, nochmals mit 50 ccm Chloroform durchgeschüttelt, wieder getrennt, das suspendierte Chloroform durch Ausschütteln mit Äther entfernt und nach Abheben des Äthers sofort mit 45 ccm *n*.-Salzsäure versetzt. Alle diese Operationen sollen rasch und bei niederer Temperatur von statten gehen, um möglichst Racemisierung der aktiven Säure durch das Alkali zu vermeiden. Schließlich wird die saure Flüssigkeit bei etwa 10—15 mm Druck stark eingeengt. Dabei scheidet sich die aktive Säure in dünnen, farblosen Prismen ab. Ausbeute etwa 28% vom angewandten Chininsalz.

Zur Analyse wurden 2 g aus 15 ccm heißem Wasser rasch umkrystallisiert und im Vakuumexsiccator über Phosphorpentoxyd getrocknet.

0.1687 g Sbst.: 0.3062 g CO_2, 0.1161 g H_2O. — 0.1529 g Sbst.: 13.1 ccm N (22.5°, 757 mm).

$C_6H_{11}O_3N$ (145.1). Ber. C 49.62, H 7.64, N 9.66.
Gef. » 49.50, » 7.70, » 9.69.

Die Säure schmilzt ebenso wie die Racemverbindung bei 158° (korr.) unter Zersetzung, sie ist in Wasser etwas schwerer löslich und zeigt auch andre Krystallform, d. h. Prismen; dagegen ist sie sonst dem Racemkörper sehr ähnlich. Sie dreht sowohl in Wasser als auch in Alkohol nach rechts. Genauer bestimmt wurde die Drehung in alkoholischer Lösung, wobei sich ergab, daß dieselbe von der Konzentration abhängig ist.

0.1195 g Sbst. Gesamtgewicht der Lösung 2.7752 g. Mithin Prozentgehalt 4.31; $d_4^{18} = 0.8112$. Drehung im 1-dm-Rohr bei 18° und Na-Licht 1.74° nach rechts. Mithin $[\alpha]_D^{18} = +49.81°$.

0.1178 g Sbst. Gesamtgewicht der Lösung 2.8779 g. Mithin Prozentgehalt 4.09; $d_4^{18} = 0.8111$. Drehung im 1-dm-Rohr bei 18° und Na-Licht 1.64° nach rechts. Mithin $[\alpha]_D^{18} = +49.40°$.

0.7961 g der vorhergehenden Lösung, enthaltend 0.0326 g Säure, wurden auf das Gesamtgewicht 1.6171 g verdünnt. Mithin Prozentgehalt 2.02; $d_4^{18} = 0.8064$. Drehung im 1-dm-Rohr bei 18° und Na-Licht 0.76° nach rechts. Mithin $[\alpha]_D^{18} = +46.52°$.

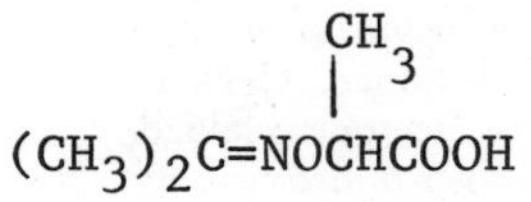

$$(CH_3)_2C=NOCH(CH_3)COOH$$

Ref. 234

(−)-Ephedrine (+)-α-(Isopropylidenaminooxy)propionic Acid.—The dl acid (29 g., 0.2 mole) and 35 g. (0.2 mole) of hydrated l-ephedrine were dissolved in 800 ml. of a solution made by diluting 48 ml. of commercial absolute alcohol to 800 ml. with ethyl acetate. The solution was cooled, seed crystals were added if available, and crystallization was allowed to proceed for 8–16 hr. in the refrigerator.

The mass of white crystals was filtered with suction and recrystallized from 10 vol. of ethyl acetate without addition of ethanol, using Eastman ethyl acetate (99%). A yield of 22–25 g. (70–80%) melting at 116–119° should be obtained. However, since the yield and its concomitant purity may vary with the impurities present in different brands or batches of solvent, another crystallization or even a change in the amount of ethyl acetate (containing ethanol) may be needed to bring the yield to within the indicated range. In this range both diastereoisomers were isolated with the desired purity.

(−)-Ephedrine (−)-α-(Isopropylidenaminooxy)propionic Acid.—The original ethyl acetate filtrate was combined with the filtrate from recrystallization and an amount of pentane equal to the total was added; the solution was kept in the cold for 8–16 hr. The crystals which formed were filtered and air dried to give 26 g. (80%) of the (−)-base (−)-acid monohydrate melting at 88–90°. It was recrystallized from 10 vol. of pure ethyl acetate. On standing over phosphorus pentoxide, the anhydrous salt resulted, m.p. 109–110°, $[\alpha]^{26}D -19.1 \pm 0.3°$.

Anal. Calcd. for $C_{16}H_{28}N_2O_5$: N, 8.53; H_2O, 5.49. Found: N, 8.79, 8.63; H_2O, 5.515.

(+)- and (−)-(Isopropylidenaminooxy)propionic Acid.—The (−)-base (+)-acid (20 g., 0.064 mole) was dissolved in 60 ml. of water and 14 ml. (0.07 mole) of 5 N hydrochloric acid was added. The solution was filtered from a small residue and extracted with ether. The ether solution was dried and the ether was evaporated. Petroleum ether (75 ml.) was added, and the solution was refrigerated overnight. The colorless crystals of the free acid weighed 7.5 g., m.p. 75–81°. The crude product was recrystallized by dissolution in 0.5 vol. of hot acetone and addition of 5 vol. of hexane. After refrigeration, 6.5 g. (70%) of the (+)-acid resulted, m.p. 83–85°.

In a similar manner the (−)-acid resulted from the (−)-base (−)-acid. In this case a melting point of 83–85° was obtained without recrystallization.

Author's comments:

In 1955 Newman and Lutz[1] introduced optically active α-(2,4,5,7-tetranitro-9-fluorenylidenaminooxy)-propionic acid, subsequently abbreviated TAPA,[2] for the resolution of polycyclic aromatic hydrocarbons which do not possess a functional group capable of salt formation with an optically active acid or base. This compound consists of the complexing agent, tetranitrofluorenone (TENF) to which is attached, by a very apt ketone exchange, an optically active aminooxypropionic acid side chain.

The present work deals with an alternative synthesis and resolution of the acid used to introduce the side chain. Advantage is taken of the greater ease of purification of the ethyl ester of the acid. The yield obtained more than compensates for the losses in the added step required to hydrolyze the ester. The proposed resolution permits the isolation of the (−)-ephedrine (−)-acid in addition to the previously characterized diastereoisomer, (−)-ephedrine (+)-acid, in equal yield and in a somewhat purer state.

(1) M. S. Newman and W. B. Lutz, *J. Am. Chem. Soc.*, **78**, 2469 (1956).
(2) M. S. Newman and D. Lednicer, *ibid.*, **78**, 4765 (1956).

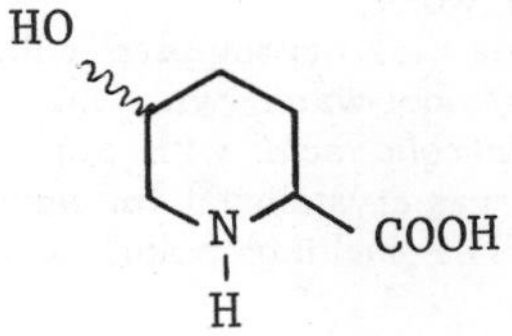

Ref. 235

(−)-5-Hydroxy-L-pipecolic acid.

A solution of DL-5-hydroxypipecolic acid hydrochloride (3.04 g, 16.8 mmoles) in 80 ml of water was neutralized by the addition of silver carbonate (3.2 g). Upon removal of the precipitate of silver chloride, a stream of hydrogen sulphide was introduced into the filtrate. Removal of silver sulphide, evaporation of the filtrate to dryness *in vacuo*, and drying over sulphuric acid in a vacuum desiccator yielded DL-5-*hydroxypipecolic acid* in a nearly theoretical yield as a vitreous product which could not be induced to crystallize.

The DL-5-hydroxypipecolic acid was dissolved in 20 ml of 96% alcohol at about 60° and this solution was added to a solution of (+)-tartaric acid (2.56 g, 17.1 mmoles) in 20 ml of 96% alcohol also brought to about 60°. This addition produced a precipitate, which could be dissolved by the addition of a few drops of water and heating. Slow cooling in the refrigerator (−8°) gave a precipitate of impurities, which was removed by filtration. Concentration of the filtrate, addition of a few drops of water and of some acetone, and cooling to —8° gave 1.9 g (77 %) of *5-hydroxy-L-pipecolic acid hydrogen-(+)-tartrate,* m.p. 196° (with dec.); the mother liquor was used for the preparation of the optical antipode, as described later.

After one crystallization the melting point increased to 202–203°; $[\alpha]_D$ 0 (± 2°), c = 3 in water. A sample was again crystallized from alcohol-water with the addition of some acetone and dried for six hours at 56°/0.2 mm over P_2O_5. The melting point was about 206–207° (with decomposition and dependent on the time of heating). This material was analysed.

Found	: C 40.9 ;	H 5.9 ;	N 4.8
Calc. for $C_6H_{11}NO_3.C_4H_6O_6$ (295.24)	: „ 40.68;	„ 5.80;	„ 4.74.

A portion of the 5-hydroxy-L-pipecolic acid hydrogen-(+)-tartrate (1.45 g, 4.9 mmoles, m.p. 202–203°) was dissolved in water (8 ml) and added to a solution of lead acetate.3 H_2O (8.56 g, 24.5 mmoles) in water (20 ml). This solution was left in the ice-box overnight, the precipitate of lead salts was removed by filtration, and hydrogen sulphide was introduced into the filtrate. By the usual procedure approximately the theoretical amount of a vitreous product was obtained. Crystallization from an alcohol-water mixture with the addition of some acetone yielded, after drying for four hours at 100°/2 mm, 650 mg (93 %) of *(—)-5-hydroxy-L-pipecolic acid,* m.p. about 257° with previous sintering at about 240°.

$[\alpha]_D^{23}$ – 22.2 (± 2°), c = 0.99 in water.

A sample was again crystallized and dried as before; the melting point, which was difficult to determine, remained unchanged.

Found	: C 49.6 ;	H 7.7. ;	N 9.8
Calc. for $C_6H_{11}NO_3$ (145.16)	: „ 49.64;	„ 7.64;	„ 9.65.

$[\alpha]_D^{22}$ – 22.7 (± 2°), c = 1.1 in water.

Infra-red spectrum (In KBr):

3280 strong, 3185 w, 2925 medium, 2703 m, 1957 very weak, 1605 ss, 1439 s, 1381 s, 1339 s, 1318 s, 1295 s, 1239 s, 1193 w, 1182 w, 1125 s, 1110 m, 1057 s, 1036 s, 1028 w, 980 s, 898 s, 873 m, 804 m, 788 s, 712 s, 687 m, cm^{-1}.

These wave-numbers correspond essentially with the major bands of the infra-red absorption spectrum (nujol mull) of (—)-5-hydroxy-L-pipecolic acid from dates [6].

The optical resolution was repeated with another batch of DL-5-hydroxypipecolic acid hydrochloride. The rotation of the L-(—)-acid was, within the limits of error, the same.

$[\alpha]_D^{21}$ – 22.8 (± 2°), c = 1.02, in water.

(+)-5-Hydroxy-D-pipecolic acid.

The mother liquor of the preparation of 5-hydroxy-L-pipecolic acid hydrogen-(+)-tartrate mentioned above was decomposed as usual with lead acetate and the lead salts were removed with hydrogen sulphide.

(—)-Tartaric acid was reacted with the residue analogously to the procedure described in the foregoing. The yield of *5-hydroxy-D-pipecolic acid hydrogen-(—)-tartrate,* after drying at about 70°/0.2 mm over P_2O_5, amounted to 1.4 g (74 %), m.p. 194–200° (with decomposition). Recrystallization and drying as before gave 1.14 g with m.p. 204–205° (with dec.).

$[\alpha]_D^{22}$ 0° (± 2°), c = 1, in water.

The hydrogen-(—)-tartrate was decomposed with lead acetate as usual and the residue crystallized from alcohol-water with the addition of some acetone. The yield of *(+)-hydroxy-D-pipecolic acid* with m.p. 248° (with dec.) amounted to 480 mg (86 %). A sample was crystallized for analysis and dried for six hours at 100°/0.1 mm over P_2O_5. The melting point, which was difficult to determine, remained about the same.

Found	: C 49.4 ;	H 7.6 ;	N 9.6
Calc. for $C_6H_{11}NO_3$ (145.16)	: „ 49.64;	„ 7.64;	„ 9.65.

$[\alpha]_D^{23}$ + 23.1° (± 2°), c = 1.08 in water.

The *infra-red spectrum* (in KBr) was identical with the spectrum of the optical antipode given above.

The optical resolution was repeated as described in the foregoing with another batch of DL-5-hydroxypipecolic acid hydrochloride. The rotation of the D-(+)-acid was, within the limits of error, the same.

$[\alpha]_D^{23}$ + 22.1 (± 2°), c = 0.95 in water.

$$(CH_3)_2\underset{\underset{S}{\|}}{N}CO\underset{\underset{CH_3}{|}}{C}HCOOH$$

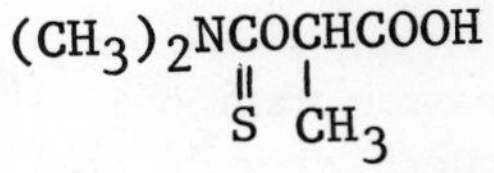

$C_6H_{11}NO_3S$ Ref. 236

Optical resolutions. Preliminary experiments were carried out with the following bases:

Strychnine	Quinidine	Cinchonidine
Brucine	Morphine	(+)-α-Phenethylamine
Quinine	Cinchonine	(−)-β-Phenylisopropylamine [10]

Ethyl acetate alone or with a small quantity of methanol was found to be the most effective solvent. The salt of the acid to be resolved with the optically active base was recrystallized several times. After each recrystallization a small sample of the salt was decomposed with hydrochloric acid, the acid isolated and the rotatory power determined.

Table 2. Recrystallization of the (−)-β-phenyl*iso*propylamine salt of N,N-dimethylthiocarbamyllactic acid. Initial quantity: 0.07 mole.

Recryst. No.	Solvent (ml)		Yield of salt (%)	$[\alpha]_D$ of the acid
	EtAc	MeOH		
1	250	100	42.5	56.7°
2	125	75	32	65.0°
3	100	50	27	75.6°
4	80	40		74.7°
5	50	30	20	(71.4°)
6	50	30	18	74.1°

Table 3. Recrystallization of the cinchonidine salt of N,N-dimetylthiocarbamyllactic acid recovered from cryst. No. 1 and 2 above ($[\alpha]_D = -36°$). Initial quantity: 0.042 mole.

Recryst. No.	Solvent (ml)		Yield of salt (%)	$[\alpha]_D$ of the acid
	EtAc	MeOH		
1	150	15	67	−57.6°
2	150	20	49	−73.1°
3	100	18	35.5	(−70.9°)
4	75	15	22	−76.1°

10. Matell, M. *Acta Chem. Scand.* **7** (1953) 698.

Note: A resolution procedure for the above compound using α-phenylethylamine as resolving agent may be found in B. Holmberg, Chem. Ber., **59**, 1558 (1926).

$$HOOC(CH_2)_3\underset{\underset{NH_2}{|}}{C}HCOOH$$

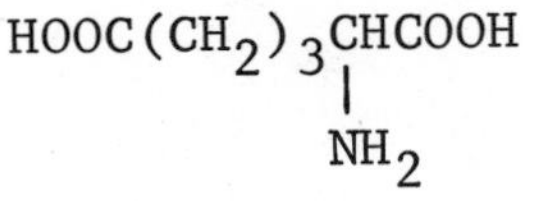

$C_6H_{11}NO_4$ Ref. 237

1. L-Isomer

DL-2-Carbobenzyloxyaminoadipic acid (80 g, 0.27 mole) and D(−) threo-1-*p*-nitrophenyl-2-aminopropane-1,3-diol were dissolved in 750 ml of hot absolute ethanol. The solution was slowly cooled, and a crystalline precipitate (58 g) was formed. By keeping the solution at room temperature for one day, another 44 g was obtained. The first crop of salt was recrystallized in hot absolute ethanol according to the scheme given below, and the progress of the resolution of the isomers was examined by determination of the rotation of the acid obtained from 0.5 g of salt.

Number of crystallizations	1	2	3	4
g of salt	102	74.8	56	49
$[\alpha]_D^{25}$ (c 2, in EtOH-2 N NaOH, 9:1) of the acid	+10	+13	+16	+18

The salt obtained in the last recrystallization was dissolved in 180 ml of water, and the solution was extracted four times with 100 ml of ethyl acetate after acidification with 5 N HCl to pH 2.0. The organic layer was washed with 2.5 N HCl and with water, and dried on Na_2SO_4. The residue obtained by evaporation of the organic solvent, was recrystallized in an acetone-cyclohexane mixture and 15.8 g of L-carbobenzyloxyaminoadipic acid mp 136-137° and $[\alpha]_D^{25}$ +17±1 (c 2, in EtOH-2 N NaOH, 9:1) was obtained.

2. D-Isomer

2-Carbobenzyloxyaminoadipic acid (42 g, 0.145 mole), obtained from the mother liquors of the preceding operation, and L(+) threo-1-*p*-nitrophenyl-2-aminopropane-1,3-diol were dissolved in 450 ml of hot absolute ethanol. The solution was slowly cooled and a first crop of 66.5 g of salt was obtained. After the filtrate was kept for one day at room temperature an additional 16.5 g of product was isolated. The amine salt was recrystallized in absolute ethanol according to the following scheme:

Number of crystallizations	1	2	3	4
g of salt	83	64	49	37
$[\alpha]_D^{25}$ (c 2, in EtOH-2 N NaOH, 9:1) of the acid	−7	−10	−13	−15

The acid was isolated from the salt obtained in the last crystallization, according to the method described for the L-isomer. An amount of 9.6 g of D-carbobenzyloxyaminoadipic acid mp 135-136° and $[\alpha]_D^{25}$ −15 (c 2, in EtOH-2 N NaOH, 9:1) was obtained.

$C_6H_{11}NO_4$ Ref. 237a

N-Benzoyl-DL-α-aminoadipic acid. Sixteen grams of DL-α-aminoadipic acid was added to a suspension of 90 g of sodium bicarbonate in 300 ml of water. When the brisk evolution of carbon dioxide had ceased, 42 g of benzoyl chloride was added gradually in 1.5 hours with vigorous stirring. The stirring was continued for another 4 hours, after which the undissolved excess sodium bicarbonate was filtered, and the filtrate was acidified with concentrated HCl to pH 2.0. The mixture of benzoic and N-benzoyl-α-aminoadipic acids was suction-filtered, dried, and extracted with ether in an extractor. The insoluble N-benzoyl-α-aminoadipic acid (19.5 g) was recrystallized from water (1:10). Yield 18 g (68%), m. p. 183-184°.

Found %: C 58.66, 58.50; H 5.81, 5.87; N 5.56, 5.31. $C_{13}H_{15}O_5N$. Calculated %: C 58.8; H 5.6; N 5.28.

Salt of N-benzoyl-DL-α-aminoadipic acid with brucine. A mixture of 16.8 g of N-benzoyl-DL-α-aminoadipic acid and 50.4 g of brucine was dissolved with heating in 300 ml acetone + 30 ml water. After cooling, the deposited crystalline salt was suction-filtered and dried at 50-60° to give 48.7 g of salt A_1. Recrystallization from 200 ml acetone + 20 ml water gave 39.6 g of salt A_2. A second recrystallization from 160 ml of acetone and 16 ml of water gave 34.4 g of salt A_3; further recrystallization from 140 ml of acetone and 14 ml of water gave 31.4 g of salt A_4.

Found %: C 64.10, 64.12; H 6.73, 6.67; N 6.34, 6.23. $C_{59}H_{67}O_{13}N_5 \cdot 3H_2O$. Calculated %: C 63.95; H 6.59; N 6.32.

N-Benzoyl-L-α-aminoadipic acid. Salt A_4 (31.4 g) was dissolved with heating in 160 ml of water. The cooled solution was then treated with 1 N NaOH solution to pH~9.0 . The deposited brucine was suction-filtered, and the filtrate was shaken with chloroform, after which it was acidified with 10% HCl to pH approximately 2.0, where 6.7 g of N-benzoyl-L-α-aminoadipic acid crystallized out, m. p. 179-180°, $[\alpha]_D$ + 17.3° (c 4.1; 1 N NaOH).

N-Benzoyl-D-α-aminoadipic acid. The aqueous-acetone mother liquor from salt A_1 was evaporated in vacuo; the residue was 21.3 g of semicrystalline salt B_1, which was dissolved by refluxing with 85 ml of acetone and 9 ml of water. Cooling of the solution gave about 1 g of crystalline salt A_1; evaporation of the filtrate gave 19.3 g of glassy salt B_2. Decomposition of salt B_2 with alkali, followed by removal of the brucine, and then acidification of the filtrate led to the isolation of 3.65 g of N-benzoyl-D-α-aminoadipic acid with m. p. 178-179°, $[\alpha]_D$ − 16.0° (c 3.3; 1 N NaOH).

L- and D-α-Aminoadipic acids. The obtained N-benzoyl derivatives, $[\alpha]_D$ + 17.3° and $[\alpha]_D$ − 16°, were hydrolyzed by heating with 10% HCl (1:15) for 15 hours at 110-115°. The deposited benzoic acid was removed by extraction with ether, and the acid solution was then evaporated in vacuo. The optically active α-aminoadipic acids were then isolated by adding excess aniline to water solutions of the obtained hydrochlorides, after which they were purified by dissolving in 1 N NaOH, treating the solution with activated carbon, and precipitation with 10% HCl at pH 3.5-4. From 6.6 g of N-benzoyl-L-α-aminoadipic acid we obtained 2.8 g of L-α-aminoadipic acid with m. p. 184-185°; $[\alpha]_D$ + 25.5° (c 1.3; 6 N HCl). From 3.6 g of the D-N-benzoyl derivative we obtained 1,8 g of D-α-aminoadipic acid, m. p. 183-184°, and $[\alpha]_D$ − 25° (c 1.3; 6 N HCl).

$C_6H_{11}NO_4$

Ref. 237b

Resolution of DL-α-aminoadipic acid with the mold Penicillium chrysogenum. The mold mycelium, washed free of culture liquid, was added to a water solution containing Na and K phosphates and glucose, and then the DL-α-aminoadipic acid was added (3% on the solution volume). After 6 days of culturing, the mycelium was separated and the filtrate (40 ml) was acidified with 6 N HCl to pH approximately 2.0, followed by evaporation in vacuo to dryness. The residue was dissolved in 8-10 ml of water, aniline was added, and the obtained precipitate of α-aminoadipic acid was separated and dissolved in several milliliters of 1 N NaOH. The alkaline solution was washed with ether to remove traces of aniline, and then it was acidified carefully with 15% HCl to pH 3.5-4. We obtained 0.3 g of D-α-aminoadipic acid, $[\alpha]_D$ − 25.9° (c 1.5; 6 N HCl).

$C_6H_{11}NO_4$ Ref. 237c

L-α-*Aminoadipic Acid (28)*

Racemic *N*-chloroacetylaminoadipic acid (6.02 g, 25 mmol) and potassium dihydrogen phosphate (0.68 g, 5 mmol) were suspended in 100 mL of water, and brought into solution by the addition of 3 *N* sodium hydroxide. The pH was adjusted to 7.0, 250 mg of hog kidney acylase I was added, and the solution was stirred for 22 h. The enzyme was removed by careful heating of the solution to boiling for 5 min, followed by filtration. The solution was concentrated to 50 mL, the pH adjusted to 2.9 with hydrochloric acid, and 150 mL of ethanol was then added. After refrigeration for 24 h, the crude L-α-aminoadipic acid was collected, dried, and recrystallized from water–ethanol: 1.65 g (81%); $[\alpha]_D$ + 23.5° (*c* 2, 5 *N* HCl) (lit. (28) +25°). The material was positive to ninhydrin.

The D-α-chloroacetylaminoadipic acid was recovered by removal of ethanol from the original filtrate, acidification to pH 1, and extraction with ethyl acetate. Workup in the usual manner and recrystallization from ethyl acetate – hexane gave 2.33 g (77%), mp 99–100°C.

28. J. P. Greenstein, S. M. Birnbaum, and M. C. Otey. J. Am. Chem. Soc. **75**, 1994 (1953).

Note:

For the experimental details of another enzymatic method for the resolution of the above compound, see H. Borsook, C. L. Deasy, A. J. Haagen-Smit, et al., J. Biol. Chem. 176, 1388 (1948).

$(CH_3)_2CCH_2$-CH-COOH
|
SO_3H

$C_6H_{11}O_5S$ Ref. 238

3. *Dédoublement optique et racémisation de l'acide α-sulfoisovalérique.* Le sel acide de brucine, recristallisé trois fois dans l'eau et décomposé à la baryte, donne le sel barytique dextrogyre. $[M]_D = +15°$.

Le sel neutre de quinine, recristallisé à plusieurs reprises, a donné un sel barytique présentant à peu près la même rotation moléculaire, +14.7°. Une solution de l'acide en présence d'un excès d'acide chlorhydrique donne une rotation gauche $[M]_D = -3°$.

$[M]_D$ sel : $[M]_D$ acide = —5.

Dispersion rotatoire du sel barytique.

λ (mμ)	656.3	589.3	546.3	486.1
[M]	11.8	15	18.8	29.4°

CH_3CHCONHCHCOOH
| |
NH_2 CH_3

$C_6H_{12}N_2O_3$ Ref. 239

Racematspaltung des DL-Alanyl-β-alanin-äthylesters: 2,0 g Ester werden mit 10 cm³ absol. Äthanol verdünnt, filtriert und mit einer filtrierten Lösung von 4,0 g Dibenzoyl-D-weinsäure in 20 cm³ absol. Äthanol versetzt. Nach 2 Stdn. beginnt die langsame Ausscheidung von L-Alanyl-β-alanin-äthylester-dibenzoyl-D-hydrogentartrat, die nach 48 Stdn. beendet ist. Ausb. 2,1 g, Schmp. 168—169°. $[\alpha]_D^{20}$: —79,6° (c = 1,04, in Äthanol).

$C_{26}H_{30}N_2O_{11}$ (546,6) Ber. C 57,20 H 5,54 N 5,14
Gef. C 56,98 H 5,58 N 4,87

Wie durch Abbau zum Peptid bestimmt wurde, beträgt die optische Reinheit dieser Fraktion 60—65%. Durch Einengen der Mutterlauge im Vak. auf zwei Drittel des Volumens werden weitere 0,6 g Salz geringerer optischer Reinheit isoliert, die man verwirft. Insgesamt hat sich damit das L-Estersalz quantitativ abgeschieden.

Die zuerst auskristallisierte Fraktion wird aus Methanol-Äther umkristallisiert, bis folgende Meßwerte erreicht sind: Schmp. 173—174°; $[\alpha]_D^{20}$: —77,7° (c = 0,68, in Äthanol).

Die Ausbeute an optisch reinem Salz beträgt 70%, bezogen auf das eingesetzte rohe Tartrat.

Gef. C 57,66 H 5,82 N 5,32

Völliges Eindampfen der Mutterlauge im Vak. und Anreiben des halbfesten Rückstandes mit absol. Äther führt zum rohen D-Alanyl-β-alanin-äthylester-dibenzoyl-D-hydrogentratrat. Ausb. 2,2 g; Schmp. 135—136°. $[\alpha]_D^{20}$: —76,0° (c = 0,79, in Äthanol).

$C_{26}H_{30}N_2O_{11}$ (546,6) Ber. C 57,20 C 5,54 N 5,14
Gef. C 56,63 H 5,61 C 4,94

Die optische Reinheit des D-Estertartrates beträgt 40%, wie durch Überführung in das Peptid bewiesen wurde.

Das rohe D-Salz wird in wenig absol. Methanol gelöst und dazu langsam die 3fache Menge absol. Äther gegeben. Das ausfallende Salz wird abfiltriert, die Mutterlauge im Vak. eingedampft und der Rückstand mit Äther angerieben. Die zweite Fraktion stellt die optisch reine dar. Schmp. 128—130°. $[\alpha]_D^{20}$: —78,1° (c = 1,0, in Äthanol).

Gef. N 5,52

Zu den folgenden Schritten wurde von den optisch reinen Estertartraten ausgegangen.

Gewinnung der optisch aktiven Alanyl-β-alanin-äthylester-hydrochloride: In die äther. Suspension von L-Alanyl-β-alanin-äthylester-dibenzoyl-D-hydrogentartrat wird bei 0° bis zur Sättigung Ammoniakgas eingeleitet. Dann saugt man ab, befreit die äther. Lösung im Vak. vom überschüss. Ammoniak und fällt das Hydrochlorid mittels absol. ätherischer Salzsäure. Das L-Alanyl-β-alanin-äthylester-hydrochlorid ist hygroskopisch. Ausb. 90% d. Th. $[\alpha]_D^{20}$: +1,95° (c = 1,7, in Methanol).

Die Gewinnung des D-Alanyl-β-alanin-äthylester-hydrochlorids aus dem Tartrat erfolgt wie die des Antipoden. Ausb. 90% d. Th.; $[\alpha]_D^{20}$: —1,98° (c = 2,3, in Methanol).

Gewinnung der optischen Antipoden des Alanyl-β-alanins: 1,5 g L-Alanyl-β-alanin-äthylester-dibenzoyl-D-hydrogentartrat werden 5 bis 10 Min. mit 25 cm³ 0,37n $Ba(OH)_2$ geschüttelt, dann 32 cm³ 0,37n H_2SO_4 hinzugegeben, abgesaugt und die Lösung 4 Stdn. im Perforator mit Äther extrahiert. Dann wird die Lösung mit 0,37n $Ba(OH)_2$ neutralisiert, abfiltriert, im Vak. eingeengt und das Peptid mit Aceton ausgefällt. Das L(+)-Alanyl-β-alanin wird aus Wasser-Aceton umkristallisiert. Ausb. 72% d. Th., Schmp. 224—225°. $[\alpha]_D^{20}$: +29,4° (c = 0,78, in Wasser).

$C_6H_{12}N_2O_3$ (160,2)	Ber. C 44,99	H 7,54	N 17,49
	Gef. C 44,60	H 7,56	N 17,20

Die Gewinnung des D(—)-Alanyl-β-alanins erfolgt wie die des Antipoden aus dem Tartrat. Ausb. 70% d. Th., Schmp. 224°. $[\alpha]_D^{20}$: —30,0° (c = 0,91, in Wasser).

Gef. C 45,01 H 7,66 N 16,93

```
     NH2
     |
HOOC-CH-CH2S2CH2-CH-COOH
                 |
                 NH2
```

$C_6H_{12}N_2O_4S_2$ **Ref. 240**

Brucine Salt of Diacetyl-l-Cystine

In a 1500 cc. beaker packed in ice and equipped with a mechanical stirrer, 25 gm. of cystine were dissolved in 105 cc. of 2 N NaOH. From burettes 500 cc. of 2 N NaOH and 50 cc. of acetic anhydride were added in portions over a period of about 30 minutes. The solution was allowed to stand at room temperature for about 30 minutes longer and an amount of 6 N sulfuric acid added corresponding exactly to the total amount of sodium hydroxide that had been used. The solution was then evaporated to dryness under diminished pressure.

Although acetylcystine is practically insoluble in absolute acetone it is quite soluble in watery acetone. This latter solvent was used for the extraction of acetylcystine from the evaporated reaction mixture. If too much water is present, some sodium sulfate will be dissolved. In this case absolute acetone can be added to throw out the sodium sulfate. The acetone method of extraction in spite of requiring a little more care is preferable to ethyl alcohol extraction because the latter usually results in some esterification. This will be referred to subsequently in greater detail.

The acetone extract of diacetylcystine was then evaporated under diminished pressure. Absolute acetone was added and distilled, the process being repeated a number of times to drive off the water. Towards the end of the treatment the thick viscous syrup fluffs up in the flask, finally solidifying, and can be broken up and removed from the flask. Macroscopically the material appears crystalline but under the microscope no definite crystalline structure has been observed. Since the material is not definitely crystalline, it seems preferable not to introduce the melting point and rotation into the literature. The crystalline brucine salt and the crystalline ethyl ester served to characterize the compound. The free diacetylcystine is exceedingly soluble in water and alcohol, but insoluble in acetone, chloroform, ether, and ethyl acetate. The yield of the product was about 90 per cent of the theoretical amount.

To 4.5 gm. of diacetyl-l-cystine in 15 cc. of water, powdered brucine was added in small portions until the solution was neutral to litmus. During the addition the solution was slightly warmed. After the solution was cooled and allowed to stand for a few hours, 9 gm. of crystals were obtained having a rotation of −62° and melting at 147–150°. Upon recrystallization from water large rectangular prisms were obtained. The salt, crystallized to constant rotation and dried at 78° and 18 mm. pressure, had a melting point of 148–150° (corrected) and a rotation of $[\alpha]_D^{27}$ = −66° for a 1 per cent solution in water.

Analysis

0.1715 gm. substance: 0.0724 $BaSO_4$.
0.2107 " " : 0.0893 "

$C_{56}H_{64}O_8N_4S_2$.	Calculated.	S	5.94
	Found.	"	5.80
		"	5.82

In such a typical resolution 100 gm. of inactive diacetylcystine were dissolved in about 300 cc. of water and brucine was added gradually with stirring until the solution was neutral to litmus. About 270 gm. of the hydrated brucine were required. The solution was then allowed to remain in the ice box for 24 hours. A thick precipitate of crystals, constituting the first crop, formed. This sample consisting of 46 gm. was recrystallized three times from methyl alcohol and four times from water. 6 gm. were finally obtained, having a rotation of $[\alpha]_D^{27}$ = +20°.

The mother liquor from the first crop of 46 gm. was evaporated and a second crop of 102 gm. was obtained. To this was added the evaporated mother liquor of the first recrystallization of the 46 gm. sample and the combined product weighing 113 gm. was recrystallized six times from methyl alcohol. At each succeeding recrystallization the mother liquor of the corresponding fraction from the recrystallization of the first crop was added to it. 34 gm. of crystals having a slight positive rotation were obtained. To this were added 11 gm. of a product having approximately the same rotation which had been obtained from the sixth recrystallization from methyl alcohol of the third crop of 80 gm. of crystals from the original mother liquor. In the recrystallization of the third crop the same procedure as indicated above was utilized by adding to succeeding fractions the evaporated mother liquors resulting from the corresponding recrystallizations of the second crop. The combined 11 and 34 gm. samples were then recrystallized six more times from water, yielding 8 more gm. of the pure dextro isomer.

The mother liquor of the third crop was quite levo and did not give further crops of crystals on evaporation. The differential solubility of the isomers in the alcohol series was then brought into play. The mother liquor was evaporated to dryness and recrystallized from butyl alcohol, 65 gm. of crystals being obtained. This was recrystallized twice from butyl alcohol and three times from ethyl alcohol, yielding 16.5 gm. of the pure levo isomer.

The mother liquors from the first six recrystallizations of the third crop were evaporated and likewise recrystallized from butyl alcohol and finally from ethyl alcohol. The uniting of mother liquors with corresponding fractions was of course resorted to as in the recrystallizations on the dextro side. The general attack of utilizing first methyl alcohol and water to remove dextro fractions and then butyl and ethyl alcohol to get out the levo fractions was employed on the remaining fractions intermediate in rotation. In this way the material was worked back and forth until the mother liquors would finally yield no further crystalline fractions. This attack not only made possible the isolation of the pure levo isomer but also actually increased the yield of the dextro because residues that would no longer yield dextro fractions would do so after they had been put through the butyl or ethyl alcohol treatment to remove levo fractions. Throughout the work rotations were taken in order to follow the course of the separation. In the first resolution rotations were taken at each step but in later resolutions rotations were determined less frequently.

From this resolution 30.5 gm. of the brucine salt of diacetyl-*d*-cystine having a specific rotation of $[\alpha]_D^{27} = +20°$ and 46 gm. of diacetyl-*l*-cystine salt with a rotation of $[\alpha]_D^{26} = -64°$ were finally obtained. Besides this, 12 gm. ($[\alpha]_D^{28} = +16°$), 19.5 gm. ($[\alpha]_D^{28} = -8°$), and 26 gm. ($[\alpha]_D^{28} = -53°$) of material were obtained. The original mother liquor and other mother liquors that no longer would yield crystalline fractions represented about 160 gm.

Other runs gave yields approximately the same so that we feel that we have subjected the material to as exhaustive a resolution as possible. The possible nature of the material remaining in the mother liquors will be discussed later.

Hydrolysis of the Brucine Salt of Diacetyl-d-Cystine

For the recovery of cystine from the brucine salt of diacetyl-*d*-cystine the brucine was filtered after the addition of alkali to the solution at a low temperature, and the remaining traces extracted with chloroform. The neutralized solution of the free diacetyl-*d*-cystine was then evaporated to a convenient volume and hydrochloric acid added to make 8 to 10 volumes of 2 N hydrochloric acid. After refluxing for 2 hours, the solution was evaporated almost to dryness, water was added, and the solution neutralized with sodium acetate. The yield of cystine varied from 65 to 80 per cent of the theoretical. The cystine can also be isolated in almost as good a yield without removing the brucine. The brucine salt itself was hydrolyzed with 10 volumes of 2 N hydrochloric acid for 2 hours and the cooled solution made just alkaline to Congo red with sodium acetate. The precipitated cystine was filtered and washed freely with water to remove any brucine acetate.

The rotation of the cystine samples so obtained varied from + 200° to +210° at 26°. Some racemization no doubt occurred. Under less drastic conditions for hydrolysis in which yield was sacrificed to get as high a rotating product as possible, the acetyl-*d*-cystine was refluxed with 1 N hydrochloric acid for ½ hour. Cystine having a rotation of $[\alpha]_D^{26} = +212°$ was obtained.

Analysis

2.759 mg. substance: 0.283 cc. N at 23° and 744 mm.
4.972 " " : 9.63 mg. $BaSO_4$.

$C_6H_{12}O_4N_2S_2$.	Calculated.	N 11.67, S 26.66
	Found.	" 11.58, " 26.60

Note:

For the experimental details of enzymatic methods for the resolution of this compound, see "Chemistry of the Amino Acids", Vol. 3, J.P. Greenstein and M. Winitz, Eds., John Wiley and Sons, Inc. (1961), pgs. 1914-1923.

$CH_3CH_2CH_2$-CH-COOH
(with CH_3 on the CH)

$C_6H_{12}O_2$ Ref. 241

Levo-Methyl-n-Propylacetic Acid—The inactive acid was dissolved in boiling acetone and an equivalent weight of quinine was added. After eighteen recrystallizations, the rotation of the free acid reached a constant value. B.p. 96° at 15 mm. $D_4^{25} = 0.920$. $n_D^{25} = 1.4117$.

$$[\alpha]_D^{25} = \frac{-16.90°}{1 \times 0.920} = -18.4°; [M]_D^{25} = -21.4° \text{ (homogeneous)}$$

4.025 mg. substance: 9.200 mg. CO_2 and 3.835 mg. H_2O

$C_6H_{12}O_2$.	Calculated.	C 62.0, H 10.4
116.1	Found.	" 62.3, " 10.7

0.063 gm. substance: 5.45 cc. 0.1 N NaOH (titration, phenolphthalein). Calculated, 5.43 cc.

$CH_3CH_2CHCH_2COOH$
(with CH_3 on the CH)

$C_6H_{12}O_2$ Ref. 242

Spaltung der racem. 3-Methyl-pentansäure.[1]) 25 g der racem. 3-Methyl-pentansäure wurden mit 100 g wasserfreiem Brucin in 500 ccm 70%igem Äthylalkohol unter Erwärmen gelöst, und die Lösung bei etwa 40° auf 350 ccm eingedunstet. Läßt man dann langsam auf 0° abkühlen, so krystallisiert allmählich ein Teil des Brucinsalzes der (−)3-Methyl-pentansäure aus. Nach etwa 2 Wochen werden die Krystalle abfiltriert, die Mutterlauge läßt man bei 40° auf etwa 200 ccm eindunsten und danach wiederum langsam auf 0° abkühlen. Nach weiteren 2—3 Wochen hat sich dann eine zweite Krystallisation abgeschieden, die ebenfalls abfiltriert wird. Aus der Mutterlauge krystallisiert nach längerem Stehenlassen (3—4 Wochen) bei 15—18° nochmals ein Teil des Brucinsalzes aus. Die 3 Krystallisate wiegen zusammen 70—72 g, also etwas mehr als der zu erwartenden Menge an Salz der linksdrehenden Säure entspricht (= 62 g). Nach 3maligem Umkrystallisieren aus Wasser konnte daraus die (−)3-Methyl-pentansäure vom Siedep. 196—197° und der Drehung[2]) $[\alpha]_D^{20} = -8,91°$ (in Alkohol) freigemacht werden.

Die für unsere Versuche erforderliche (+)-Säure konnte aus der öligen Mutterlauge des Brucinsalzes der linksdrehenden Säure gewonnen werden. Das Öl wurde in warmem Wasser gelöst, mit HCl angesäuert und die Säure dem Gemisch mit Äther entzogen. Siedep. nach 2maliger Rektifikation 198—199°. Spezifische Drehung[3]): $[\alpha]_D^{20} = +8,52°$. Ausbeute: 6—7 g.

[1]) Vgl. C. Neuberg u. B. Rewald, Biochem. Z. 9, 409 (1908).
[2]) Alle Drehungen wurden in einem 1 dm-Mikrorohr bestimmt.
[3]) v. Romburgh, Rec. Trav. chim. Pays-Bas 5, 222 (1886), gibt für die entsprechende Säure, die er durch Oxydation von römischem Kamillenöl gewonnen hat: $[\alpha]_D^{20} = +8,92°$ an; W. Marckwald u. E. Nolda, Ber. 42, 1583 (1909) erhielten die Säure durch Verseifung des optisch aktiven Nitrils und fanden: $[\alpha]_D^{20} = +8,37°$; C. Neuberg u. B. Rewald gelang es, durch Spaltung der racem. Säure eine d-Säure von $[\alpha]_D^{20} = +7,62°$ zu erhalten.

$C_6H_{12}O_2$ Ref. 242a

Zerlegung der 3-Methylpentansäure.

20 g der inaktiven Capronsäure wurden in einer Mischung von 250 g Alkohol und 250 g Wasser gelöst und mit 80 g Brucin, d. h. etwas mehr als der berechneten Menge, versetzt. Beim Umrühren und gelinden Erwärmen auf dem Wasserbade ging das Alkaloid in Lösung. Da, wie bereits erwähnt, beim schnellen Verdampfen nur ein Öl erhalten wird, läßt man die Mischung in einem Becherglase langsam verdunsten. Nach einiger Zeit scheiden sich an den Gefäßwandungen harte, große, durchsichtige Krystalle ab, und nach zwei bis drei Monaten findet man eine reichliche Menge auskrystallisierten Salzes, während ein zähes, gelbbraunes Öl, das ebenfalls von Krystallen durchsetzt ist, am Boden haftet. Der gesamte Inhalt wird sodann auf der Nutsche abgesaugt, das Filtrat wird der weiteren freiwilligen Verdunstung überlassen. Die auf der Nutsche zurückbleidenden Krystalle werden mit 50%igem Alkohol gewaschen. Die Waschflüssigkeit wird nicht mit der Mutterlauge vereinigt, sondern für sich der langsamen Verdunstung überlassen. Hieraus wie aus der Mutterlauge erhält man nach einiger Zeit aufs neue eine spärliche Krystallisation, die mit der Hauptmenge vereinigt wurde. Die Umkrystallisation dieses schneeweißen Salzes gelingt leicht aus Wasser, dem etwas Alkohol zugesetzt ist. Mit verschiedenen Proben wurde festgestellt, daß nach der dritten Krystallisation durch Destillation befreit werden kann. Die Hauptmenge siedet scharf zwischen 195 bis 196°. Es hinterbleiben nur eine kleine Quantität eines dunkel gefärbten höher siedenden Rückstandes. Die so gereinigte Säure zeigt im 2-Dezimeterrohr in 21,4%iger Lösung in (an sich selbstverständlich inaktivem) Ligroin eine Linksdrehung von 2°45', was einem spezifischen Drehungsvermögen von

$$[\alpha_{D_u}] = -8,98°$$

$$(\alpha = -2,75°, \quad d = 0,716, \quad c = 21,374, \quad l = 2)$$

entspricht. Um noch ein weiteres krystallisiertes Salz außer der Brucinverbindung zu erhalten, wurde das

das Brucinsalz von anfangs anhaftendem Salz der inaktiven Säure völlig befreit ist und eine konstant drehende

l-Capronsäure

ergibt. Die Hauptmenge des Brucinsalzes wurde nunmehr in wenig Wasser gelöst und mit verdünnter Schwefelsäure im Überschuß versetzt. Dann wird gepulvertes Ammonsulfat bis zur Sättigung eingetragen und 5- bis 6mal mit Äther extrahiert. Die vereinigten Ätherauszüge wurden mit geglühtem Natriumsulfat getrocknet und abdestilliert. Man gewinnt so eine schwach gelb gefärbte Säure vom typischen Capronsäuregeruch. Sie enthält noch kleine Mengen Verunreinigungen, von denen sie

d-Capronsäure

kann aus der ersten für sich belassenen Mutterlauge des krystallisierten Brucinsalzes der l-Säure gewonnen werden. Da geeignete Alkaloidverbindungen — wie eingangs erwähnt — nicht im krystallisierten Zustande erhalten worden sind, so wurde direkt das ölige Brucinsalz auf die freie Säure verarbeitet. Zu diesem Zweck wurde das Öl im warmen Wasser gelöst, mit Schwefelsäure angesäuert und in gleicher Weise mit Äther extrahiert, wie für den Antipoden angegeben ist. Die durch Destillation gereinigte Säure hatte auch ihren Siedepunkt scharf bei 195 bis 196°. Ihr spezifisches Drehungsvermögen war (ebenfalls in Ligroinlösung):

$$[\alpha_D] = +7{,}93°$$

$$(\alpha = +2{,}67°,\quad c = 22{,}24,\quad d = 0{,}7544,\quad l = 2),$$

was einem Gehalt von ca. 90% an aktiver d-Säure entspricht.

$(CH_3)_2CH\text{-}CH(CH_3)\text{-}COOH$

$C_6H_{12}O_2$ Ref. 243

Inactive methylisopropylacetic acid, prepared through the malonic ester synthesis, was dissolved in hot acetone; 1 equivalent of quinine was added and the solution allowed to cool to room temperature. The quinine salt precipitated in a hard cake from which the mother liquor could be decanted. After ten recrystallizations the rotation reached a maximum value. B.p. 90° at 16 mm.

$$[\alpha]_D^{25} = \frac{-17.54°}{1 \times 0.928} = -18.9°;\ [M]_D^{25} = -21.9° \text{ (homogeneous)}$$

$(CH_3)_2CHCH_2\text{-}CH(SH)\text{-}COOH$

$C_6H_{12}O_2S$ Ref. 244

Resolution of α-Thiolisocaproic Acid.—The inactive thiol acid was prepared in the same way as the active substance, but in this case the mixture of bromo acid and potassium hydrogen sulfide was heated for 3 hours on the steam bath without standing at 0°C.

A warm solution of 46 gm. of the inactive thiol acid in 300 cc. of acetone was treated with 117 gm. of pure quinine. The salt thus obtained was recrystallized four times from acetone and then decomposed as usual with dilute hydrochloric acid. The ethereal extract was evaporated and the residue fractionated under reduced pressure (p = 15 mm.). The thiol acid boiled at 127–128°C. and showed the optical rotations of

$$[\alpha]_D^{20} = \frac{+1.93° \times 100}{1 \times 9.95} = +19.4° \text{ (in ether).}$$

$$[\alpha]_D^{20} = \frac{+1.42° \times 100}{2 \times 4.71} = +15.1°.\quad [M]_D^{20} = +22.3° \text{ (in 40 per cent alcohol).}$$

$CH_3(CH_2)_3\text{-}CH(OH)\text{-}COOH$

$C_6H_{12}O_3$ Ref. 245

Resolution of α-Hydroxy-n-Caproic Acid.—65 gm. of α-hydroxy-n-caproic acid were added to a warm solution of 145 gm. of cinchonidine in 700 cc. of chloroform. The mixture was evaporated and the salt obtained in crystalline form. After twelve recrystallizations from water, the salt was dissolved in chloroform, cooled with ice, and decomposed with ammonium hydroxide. The aqueous layer was washed with chloroform to remove cinchonidine and evaporated to small volume. The residue was acidified with hydrochloric acid and extracted with ether. The ethereal extract was dried over sodium sulfate. On removal of the ether the residue was almost entirely crystalline. The crystals were filtered off from the syrup and recrystallized, first from dry ether and then from ether and petrolic ether. The pure hydroxy acid which crystallized in the form of plates was somewhat hygroscopic. It melted at 60–61°C. The optical rotation in water was

$$[\alpha]_D^{20} = \frac{-3.35° \times 100}{2 \times 44.6} = -3.75°.\quad [M]_D^{20} = -4.95°.$$

To the above solution 1 equivalent of sodium hydroxide was added and the solution was made up to 10 cc. For the sodium salt,

$$[\alpha]_D^{20} = \frac{+7.73° \times 100}{2 \times 26.17} = +14.77°.\quad [M]_D^{20} = +22.75°.$$

The substance analyzed by the micro method as follows:

3.920 mg. substance: 7.803 mg. CO_2 and 3.262 mg. H_2O.

$C_6H_{12}O_3$.	Calculated.	C 54.50, H 9.16.
	Found.	" 54.28, " 9.31.

$(CH_3)_2CH\text{-}CH(OH)CH_2COOH$

$C_6H_{12}O_3$ Ref. 246

(−) *Acide hydroxy-3 méthyl-4 valérique lévogyre* **15**

Nous avons utilisé la méthode que nous avons récemment décrite pour dédoubler rapidement l'acide α-phénylbutyrique.[19] A 14.1 g (0.106 mole) d'acide (±)**15** dissous dans 140 cm³ d'éther on ajoute, en agitant, à 0° et goutte à goutte, 6.41 g (0.053 mole) d'α-méthylbenzylamine lévogyre optiquement pure. On essore le précipité formé et le dissout dans une solution normale d'HCl. L'extraction à l'éther permet d'isoler 3.7 g d'acide dont le pouvoir rotatoire est $[\alpha]_D^{20} = +6.44°$ (c = 2.37, benzène). Le filtrat traité de la même façon donne 8 g d'acide de pouvoir rotatoire $[\alpha]_D^{20} = -2.77°$ (c = 2.88, benzène).

Afin de déterminer la pureté optique de ces échantillons et le pouvoir rotatoire spécifique maximum de l'acide **15** une deuxième fraction d'acide racémique a été partiellement dédoublée par cristallisation fractionnée de son sel d'α-phényléthylamine dextrogyre dans l'éthanol absolu. Après trois cristallisations, nous avons isolé un échantillon de pouvoir rotatoire $[\alpha]_D^{20} = -18.11°$ (c = 1.96, benzène). Son ester méthylique est obtenu par traitement au diazométhane. Eb = 110°/1 torr. $[\alpha]_D^{20} = -21.49°$ (c = 1.94, benzène). La pureté optique de cet ester déterminée par RMN 100 MHz en présence de complexes chiraux de terres rares est de 45%. Le pouvoir rotatoire spécifique maximum de l'acide hydroxy-3 méthyl-4 valérique est donc: $[\alpha]_D^{20} = (18.11 \times 100)/45 = 40.2°$ (c = 1.96, benzène).

[19] J. P. Vigneron et V. Bloy, *Ibid.* 649 (1976).

$(CH_3)_2CH\text{-}CHCH_2COOH$
|
OH

$C_6H_{12}O_3$ Ref. 246a

Optical Resolution of (±)-*3-Hydroxy-4-methylpentanoic Acid.*—(±)-Acid (5·44 g.) in ethyl acetate (50 ml.) was added to a warm solution of quinine in ethyl acetate (200 ml.). Concentration gave a quinine salt (8·0 g.), m. p. 147·5—151·5°, $[\alpha]_D$ −143·5° ± 0·5° (in EtOH). Recrystallisation from ethyl acetate gave quinine salt (5·7 g.), m. p. 151·5—153°, $[\alpha]_D^{25}$ −145° ± 2° (*c* 1·03 in EtOH). The acid (2·9 g.), $[\alpha]_D^{28}$ −24·7° ± 0·6° (*c* 0·98 in $CHCl_3$), was isolated from this salt by treatment with 10% sulphuric acid and extraction with ether. Its infrared spectrum was identical with that of the specimen of enantiomeric (+)-acid (above). The 4-*bromophenacyl ester* had m. p. 74—75°, $[\alpha]_D^{25}$ −14·4° (*c* 2 in $CHCl_3$), and the same infrared spectrum as the (+)-ester (Found: C, 51·4; H, 5·4%).

$C_6H_{12}O_3$ Ref. 246b

Spaltung der *dl*-Leucinsäure in die optisch-aktiven Komponenten.

Die Spaltung gelang mit dem Brucin-, Chinin- und Chinidinsalz. In allen 3 Fällen krystallisiert zuerst das Salz der *l*-Leucinsäure. Am bequemsten ist die Operation bei Anwendung von Chinidin.

30 g unkrystallisierter *dl*-Leucinsäure wurden in 500 ccm Wasser gelöst und mit einer Lösung von 84 g (1 Mol.) Chinidin (krystallalkoholhaltig) in 200 ccm Alkohol versetzt. Die Mischung wurde auf etwa 500 ccm eingeengt und bei Zimmertemperatur nach Zugabe einiger Impfkrystalle 15 Stunden aufbewahrt. Die Ausbeute an Chinidinsalz betrug etwa 45 g, das aber noch etwa zur Hälfte racemisch war. Durch zweimaliges Umkrystallisieren aus wenig Alkohol und viel Wasser wurde daraus 30 g Chinidinsalz oder etwa die Hälfte der theoretischen Menge erhalten, die einer Leucinsäure von $[\alpha]_D^{20} = -26.9°$ (in alkalischer Lösung) entsprachen. Dieses Präparat war annähernd rein und wurde zur Weiterverarbeitung verwandt. Durch mehrmaliges verlustreiches Umkrystallisieren kommt man aber schließlich zu einer Oxysäure von $[\alpha]_D^{20} = -27.8°$. Da erneute Krystallisation des Chinidinsalzes aus 1 Tl. Alkohol und 19 Tln. Wasser, wobei die Hälfte in der Mutterlauge blieb, keine höher drehende Säure lieferte, so scheint damit das Ende der Spaltung erreicht zu sein.

l-Leucinsäure. Das zweimal umkrystallisierte, fein gepulverte Chinidinsalz wurde mit einem Überschuß von *n*-Natronlauge 15 Minuten geschüttelt, die Flüssigkeit vom abgeschiedenen Chinidin abfiltriert, mit Schwefelsäure übersättigt und mehrmals mit Äther extrahiert. Die durch Verdampfen des Äthers gewonnene Säure läßt sich ebenso wie der Racemkörper aus Äther + Petroläther umkrystallisieren. Sie scheidet sich hieraus in dünnen Prismen ab, die bei langsamer Krystallisation bedeutende Länge erreichen. Zur Analyse und optischen Bestimmung wurde im Vakuumexsiccator über Phosphorpentoxyd getrocknet.

0.1609 g Sbst.: 0.3214 g CO_2, 0.1810 g H_2O.

$C_6H_{12}O_3$ (132.10). Ber. C 54.50, H 9.16.
Gef. » 54.48, » 9.11.

Die *l*-Leucinsäure sintert etwa von 78° an und schmilzt bei 81—82°. Die Löslichkeitsverhältnisse sind ähnlich wie beim Racemkörper.

In wäßriger Lösung zeigt die Oxysäure eine nicht besonders starke Linksdrehung. Stärker ist das Drehungsvermögen des Natriumsalzes; daher eignet sich für die optische Bestimmung am besten die Lösung in überschüssiger *n*-Natronlauge.

0.1732 g Sbst., gelöst in *n*-Natronlauge. Gesamtgewicht der Lösung 1.7622 g. $d^{20} = 1.044$. Drehung im 1-dm-Rohr bei 20° und Natriumlicht 2.85° (± 0.02°) nach links.

Mithin

$$[\alpha]_D^{20} = -27.8° (\pm 0.2°).$$

Nochmaliges Umkrystallisieren des entsprechenden Chinidinsalzes lieferte eine Oxysäure von gleichem Drehungsvermögen:

0.1321 g Sbst., gelöst in *n*-Natronlauge. Gesamtgewicht der Lösung 1.3307 g. $d^{20} = 1.044$. Drehung im 1-dm-Rohr bei 20° und Natriumlicht 2.88° (± 0.02°) nach links.

Mithin

$$[\alpha]_D^{20} = -27.8° (\pm 0.2°).$$

Von demselben Präparat wurde eine optische Bestimmung in wäßriger Lösung ausgeführt.

0.1219 g Sbst., gelöst in Wasser. Gesamtgewicht der Lösung 1.2366 g. $d^{20} = 1.010$. Drehung im 1-dm-Rohr bei 20° und Natriumlicht 1.04° (± 0.02°) nach links.

Mithin

$$[\alpha]_D^{20} = -10.4° (\pm 0.2°).$$

d-Leucinsäure. Sie wurde aus den Mutterlaugen vor der Bereitung des *l*-leucinsauren Chinidins gewonnen, aber nur mit einem Drehungsvermögen von $[\alpha]_D = +11.9°$ (in alkalischer Lösung). Die Säure bildet zwar mit Cinchonidin und anderen Alkaloiden krystallisierende Salze, durch welche sich aber das Drehungsvermögen nicht wesentlich steigern ließ. Wir haben deshalb auf die Reindarstellung verzichten müssen.

OH
|
$(CH_3)_2CH\text{-}C\text{-}COOH$
|
CH_3

$C_6H_{12}O_3$ Ref. 247

Resolution of (±)-α-*Hydroxy*-αβ-*dimethylbutyric Acid.*—The acid (3·3 g.) (Perkin, *J.*, 1896, **69**, 1457) was added to (−)-brucine (9 g.) dissolved in dry ethyl acetate (200 ml.). On being kept, this deposited (−)-*brucine* (±)-α-*hydroxy*-αβ-*dimethylbutyrate* in prisms (10·7 g.), m. p. 194—197° [Found: N, 5·3; OMe, 12·3. $C_{28}H_{32}O_5N_2(OMe)_2$ requires N, 5·2; OMe, 11·6%]. Fractional crystallisation of this salt from ethyl acetate gave as the less soluble component (−)-*brucine* (+)-α-*hydroxy*-αβ-*dimethylbutyrate*, m. p. 208° (decomp.), $[\alpha]_D^{20}$ −29·65° (*c*, 5 in $CHCl_3$). Liberated from the salt with dilute sulphuric acid and isolated with ether, the (+)-*acid* was distilled [130° (bath)/20 mm.] and then crystallised from light petroleum (b. p. 60—80°), forming plates, m. p. 69—71°, $[\alpha]_D^{20}$ 4·57° (*c*, 5 in $CHCl_3$) (Found: C, 55·0; H, 9·4%; equiv., 128. $C_6H_{12}O_3$ requires C, 54·5; H, 9·2%; equiv., 132). The more soluble salt could not be obtained optically pure but on decomposition furnished the (−)-*acid* which separated from light petroleum (b. p. 60—80°) in plates, m. p. 67—69°, $[\alpha]_D^{20}$ −3·34° (*c*, 5 in $CHCl_3$) (Found: C, 54·5; H, 9·3%; equiv., 132).

```
(CH3)3C-CH-COOH
        |
        OH
```

$C_6H_{12}O_3$ Ref. 248

Optical Resolution of I into L-(+)-3,3-Dimethyl-2-hydroxybutyric Acid (Ia). Seventy-five grams of I and 224 g of brucine were dissolved in 1 *l* of water, and then the solution was concentrated to 500 m*l*. After the solution had been allowed to stand overnight in a refrigerator, the brucine salts were collected. Six recrystallizations of the salt from water yielded the pure brucine salt of Ia. Yield 22 g; mp 223—224°C (decomp.), $[\alpha]_D^{20}$ −31° (*c* 1, alcohol).

Found: N, 5.32 Calcd for $C_{29}H_{38}N_2O_7$: N, 5.33%.

Twenty grams of the salts were treated with 5 N sodium hydroxide, and the liberated brucine was removed by filtration. The filtrate was treated with sulfonic resin to remove a trace of brucine and sodium ions, and then concentrated to dryness under reduced pressure. The crystallization of the crude material from benzene-petroleum benzin gave optically-pure Ia. Yield 3.9 g; mp 51—52°C, $[\alpha]_D^{20}$ +4.5° (*c* 4, water). $[\alpha]_D^{20}$−63° (*c* 1, water, 48.5 mg ammonium molybdate in 10 m*l*).

Found: C, 54.75; H, 9.30%.

Optical Resolution of I into D-(−)-3,3-Dimethyl-2-hydroxybutyric Acid (Ib). Thirty-five grams of cinchonidine and 15.8 g of I were dissolved in 100 m*l* of alcohol, and then the solution was allowed to stand overnight in a refrigerator to give 33 g of the cinchonidine salt. Five recrystallizations of the salt from alcohol gave pure cinchonidine salt of Ib. Yield 9 g; Mp 220—221°C, $[\alpha]_D^{20}$ −10.2° (*c* 1, alcohol).

Found: C, 69.46; H, 8.11; N, 6.24%. Calcd for $C_{25}H_{34}N_2O_4$: C, 70.09; H, 8.03; N, 6.57%.

Nine grams of the salt obtained were suspended in 100 m*l* of hot water, and then decomposed with 5 m*l* of 5 N sodium hydroxide under vigorous stirring for a day. After the removal of the cinchonidine by filtration, the filtrate yielded pure Ib when treated by the procedure described above. Yield 2.3 g; mp 50—51°C, $-[\alpha]_D^{20}$ −4.3° (*c* 4, water), $[\alpha]_D^{20}$ +65° (*c* 1, water, 51 mg of ammonium molybdate in 10 m*l*).

Found: C, 54.36; H, 9.38%.

```
       CH3
       |
HOCH-C-COOH
  |    |
 H3C  CH3
```

$C_6H_{12}O_3$ Ref. 249

(b) (+)-(S)- *and* (−)-(R)-*Enantiomers by resolution of racemic material.* A solution of brucine (62·2 g) in acetone (270 ml) and water (30 ml) was added to a hot, stirred solution of the (±)-3,5-dinitrobenzoyloxy-acid (4a) (51·4 g) in a solvent mixture (300 ml) of the same composition. The solution became deep red and crystals began to form immediately. The solution was left overnight at 0°, and the crystals were filtered off and dried to give a mixture of the diastereoisomeric brucine salts (72·6 g) as red needles, $[\alpha]_D$ −70·5° (*c* 1·75 in pyridine). Three recrystallisations from acetone–water (9 : 1; 1500 ml reduced to 1000 ml by distillation under reduced pressure after complete dissolution) gave the *brucine salt* of the (−)-(R)-*acid* (26·6 g) as deep red needles, m.p. 208—209°, $[\alpha]_D$ −82° (*c* 1·29 in pyridine) (Found: C, 60·3; H, 5·7; N, 7·6. $C_{36}H_{40}N_4O_{12}$ requires C, 60·0; H, 5·6; N, 7·75%), $\lambda_{max.}$ 214 and 301 nm (log ε 4·5 and 3·92); $\nu_{max.}$ (Nujol) 720, 730, 1640, and 1715 cm^{-1}.

The mother liquor (600 ml) from the initial crystallisation was evaporated under reduced pressure to *ca.* 300 ml, and was kept overnight at 0° to give the brucine salt of the (+)-(*S*)-acid (33·8 g) as red needles, $[\alpha]_D$ −55° (*c* 2·30 in pyridine).

The brucine salt of $[\alpha]_D$ −82° (26·5 g) was suspended in acetone (100 ml) and stirred while dilute hydrochloric acid (50 ml) was added. The deep red colour of the suspension was discharged immediately and a homogeneous solution was obtained. Addition of more dilute hydrochloric acid (150 ml), with stirring for a further 15 min gave a precipitate which was filtered off, washed successively with dilute hydrochloric acid and water, and recrystallised from aqueous ethanol to give the (−)-(R)-*acid* (4a) (10·68 g, 89%; 41% overall) as needles, m.p. 195—197°, $[\alpha]_D$ −35° (*c* 4·2 in acetone) (Found: C, 48·15; H, 4·45; N, 8·75. $C_{13}H_{14}N_2O_8$ requires C, 47·85; H, 4·35; N, 8·6%), $\lambda_{max.}$ 210 nm (log ε 4·42); $\nu_{max.}$ (Nujol) 725, 735, 1550, 1630, 1700, and 1720 cm^{-1}, n.m.r. and u.v. spectra identical with those of the racemic acid [i.r. spectra (Nujol mulls) differed considerably in the 700—1200 cm^{-1} region].

In the same way, the brucine salt of $[\alpha]_D$ −55° (33·8 g) was suspended in acetone (100 ml) and stirred with dilute hydrochloric acid (200 ml) to give, after work-up as described above and recrystallisation from aqueous ethanol, the (+)-(S)-*acid* (4a) (13·59 g, 88%; 53% overall) as needles, m.p. 189—194°, $[\alpha]_D$ +28° (*c* 3·78 in acetone) (Found: C, 47·95; H, 4·35; N, 8·55. $C_{13}H_{14}N_2O_8$ requires C, 47·85; H, 4·35; N, 8·6%), i.r. and u.v. spectra identical with those of the (−)-(*R*)-acid.

(b) (−)-(R)-*Enantiomer.* Aqueous potassium hydroxide (2·22N; 53 ml) was added to a stirred suspension of the (−)-acid (4a) (10·5 g), $[\alpha]_D$ −35°, in 95% ethanol (53 ml) to give a deep red solution, the colour of which slowly faded. The mixture was stirred for 2 h, and the precipitate of potassium 3,5-dinitrobenzoate was filtered off and washed with 95% ethanol. The combined filtrate and washings were acidified with dilute hydrochloric acid and extracted with ethyl acetate. The extract was washed once with brine. T.l.c. indicated that in addition to the desired product the solution contained some 3,5-dinitrobenzoic acid. The latter was removed by hydrogenation of the nitro-groups over 10% palladised charcoal (500 mg), followed by removal of the catalyst by filtration and successive washing of the filtrate with dilute hydrochloric acid and brine. The ethyl acetate solution thus obtained was dried and evaporated *in vacuo* to give a yellow liquid from which, by short-path distillation at 130° (1 mmHg), was obtained the (−)-(R)-*acid* (4b) (1·61 g, 38%) as a liquid which slowly crystallised. One recrystallisation from light petroleum (b.p. 40—60°) gave material, m.p. 42—44°, $[\alpha]_D$ −9·1° (*c* 2·38) (Found: C, 54·55; H, 9·15. $C_6H_{12}O_3$ requires C, 54·55; H, 9·15%), i.r. and n.m.r. spectra identical with those of the (±)-hydroxy-acid.

(c) (+)-(S)-*Enantiomer.* A suspension of the (+)-(*S*)-acid (4a) (13·4 g), $[\alpha]_D$ +28°, in 95% ethanol (65 ml) was stirred with aqueous potassium hydroxide (2·28N; 65 ml) for 1 h. The yellow liquid obtained by work-up as described in (b) was distilled under reduced pressure to give the (+)-(S)-*acid* (4b) (1·8 g, 33%) as a liquid, b.p. 96—100° (0·3 mmHg), which slowly crystallised. Two recrystallisations from light petroleum (b.p. 40—60°) gave crystals, m.p. 36—43°, $[\alpha]_D$ +8·0° (*c* 19·6), n.m.r. spectrum and t.l.c. behaviour identical with those of the (−)-(*R*)-enantiomer.

$C_6H_{12}O_3$

$$CH_3OCH_2CH_2\underset{}{\overset{CH_3}{\overset{|}{C}}}HCOOH$$

Ref. 250

Resolution of γ-Methoxy-α-methylbutyric Acid.—An aqueous solution of cinchonidine methohydroxide was prepared by the following alternative procedures. (i) Methyl iodide (23·6 g., 1 mol.) was added to cinchonidine (49 g., 1 mol.) and ethanol (50 c.c.), and the mixture set aside overnight; crystallisation from hot water gave pale yellow needles of cinchonidine methiodide (55·8 g.). This was shaken with freshly precipitated silver oxide (from 70 g. of silver nitrate) and water (150 c.c.) at 0° for 15 min., then at room temperature for a further 3 hr. The aqueous solution was then decanted and filtered, and the filtrate centrifuged to remove colloidal silver oxide.

(ii) Cinchonidine methiodide (30 g.; prepared as above) was shaken with freshly precipitated silver chloride (from 10 g. of silver nitrate) and water (175 c.c.) for 2 hr.; the mixture was then heated and filtered while hot. Saturation of the cold filtrate with sodium chloride precipitated crude cinchonidine methochloride which was reprecipitated from filtered solution in dry ethanol by the addition of ether. After repetition of this purification process, the resulting methochloride was shaken with silver oxide (from 12 g. of silver nitrate) and water (20 c.c.) as above; the mixture was then filtered and the filtrate set aside overnight. A further filtration through Whatman No. 2 paper gave a clear solution of the methohydroxide.

The resulting solutions of cinchonidine methohydroxide were standardised by titration with 0·088N-sulphuric acid, phenolphthalein being used as indicator.

(±)-γ-Methoxy-α-methylbutyric acid (6·45 g., 1 mol.; obtained by Prelog and Zalan's method [6] from 2-methoxyethyl bromide [10] and ethyl methylmalonate) was added to aqueous cinchonidine methohydroxide (0·32N; 153 c.c., 1 mol.), the solution filtered and then evaporated, and the residual salt dried *in vacuo* (P_2O_5). After four crystallisations from dioxan the cinchonidine methosalt (3·5 g.) was obtained as white needles, $[\alpha]_D^{16·5}$ −100° (*c*, 3·06 in EtOH); the rotatory power was unaltered on further recrystallisation. The salt was decomposed with 2N-hydrochloric acid and the liberated acid isolated by extraction with ether. On removal of the solvent the (+)-*acid* (0·8 g.) was obtained with $[\alpha]_D^{16·5}$ +12·2° (*c*, 15·28 in Et_2O).

Concentration of the mother-liquors from the above crystallisations, followed by the addition of dry ether, gave a further fraction of the cinchonidine methosalt which on decomposition as before afforded the partially resolved (−)-enantiomer (0·9 g.), $[\alpha]_D^{16·5}$ −6·9° (*c*, 17·30 in Et_2O).

In small-scale experiments the partially resolved (+)-enantiomer was obtained by resolution with quinine or quinine methohydroxide.[7] The quinine salt, prepared in acetone solution, crystallised from acetone–light petroleum (b. p. 40—60°), and the quinine methosalt (obtained as described above for the cinchonidine methosalt) from alcohol–ether.

[6] Prelog and Zalan, *Helv. Chim. Acta*, 1944, **27**, 531. [10] Palomaa and Kenetti, *Ber.*, 1931, **64**, 797.
[7] Major and Finkelstein, *J. Amer. Chem. Soc.*, 1941, **63**, 1368.

$C_6H_{12}O_4$

$$C_2H_5-\underset{OH}{\underset{|}{CH}}-\underset{OH}{\underset{|}{\overset{CH_3}{\overset{|}{C}}}}-COOH$$

Ref. 251

Resolution of (±)-erythro-2,3-Dihydroxy-2-methylvaleric Acid (I) into its Antipodes. 4.0 g of L-(+)-threo-2-amino-1-p-nitrophenyl-1,2-propanediol was dissolved in 30 ml of boiling ethanol, and the resulting solution was stirred while 3.0 g of carefully ground (±)-erythro-2,3-dihydroxy-2-methylvaleric acid (I) was added in one portion. The solution was left overnight at room temperature, and the precipitated crystals were filtered off. After four crystallizations from alcohol we obtained 2.3 g of the salt formed by erythro-2,3-dihydroxy-2-methylvaleric acid and L-(+)-threo-2-amino-1-p-nitrophenyl-1,3-propanediol; m.p. 158-159°; $[\alpha]_D^{20}$ +17.8° (c 0.5 in water; *l* = 2). Found: C 50.05; H 6.72; N 7.65%. $C_{15}H_{24}N_2O_8$. Calculated: C 50.00; H 6.71; N 7.77%.

The resulting salt was dissolved at 40-50° in 25 ml of water, and the solution was cooled to room temperature and passed through a column (300× 20 mm) filled with Cationite KU-2. Water was vacuum-distilled off, and three crystallizations from ethyl acetate gave 0.5 g of (+)-erythro-2,3-dihydroxy-2-methylvaleric acid; m.p. 153-154°; $[\alpha]_D^{20}$ + 13.8° (c 1 in water; *l* = 2). Found: C 48.70; H 8.14%. $C_6H_{12}O_4$. Calculated: C 48.64; H 8.17%.

The mother solution obtained after the separation of the salt of the L-(+)-threo amine with the (+) dihydroxy acid (I) was vacuum-evaporated to dryness, and the residue was dissolved in 10 ml of water at 40-50°. The resulting solution was cooled to room temperature and passed through a column (200× 15 mm) containing Cationite KU-2. The eluate was vacuum-evaporated to dryness, and 1.8 g of finely ground D-(-)-threo-2-amino-1-p-nitrophenyl-1,3-propanediol was added in one portion to a boiling solution of the residue in 5 ml of alcohol. When the solution was cooled to room temperature, 1.25 g of the salt of the (-)-erythro acid (I) with the D-(-)-threo amine was precipitated. After three crystallizations from alcohol we obtained 1 g of the salt; m.p. 158-159°; $[\alpha]_D^{20}$ -18° (c 1 in water; *l* = 2). Found: C 49.93; H 6.58; N 7.73%. $C_{15}H_{24}N_2O_8$. Calculated: C 50.00; H 6.71; N 7.77%.

By decomposition on Cationite KU-2 under the conditions for the isolation of the (+)-erythro acid, from 1 g of this salt we obtained 0.4 g of (-)-erythro-2,3-dihydroxy-2-methylvaleric acid; $[\alpha]_D^{20}$ -13.3° (c 1 in water; l = 1). Found: C 48.79; C 8.12%. $C_6H_{12}O_4$. Calculated: C 48.64; H 8.17%.

$C_6H_{12}O_4$ Ref. 251a

(iii) *Resolution*. The dihydroxy-acid (14 g) in ethanol (100 cm³) was treated with quinine (36 g) in ethanol (100 cm³). The mixture was set aside overnight at 3 °C to give the *quinine salt* (10 g). The mother liquor was kept at −10 °C for 5 days to give more (8.0 g) of the quinine salt. The fractions were combined and recrystallised (ethanol). After eight crystallisations the m.p. had reached a constant value of 200—202° (decomp.); $[\alpha]_D^{23}$ −140° (*c* 0.83 in MeOH) (Found: C, 65.95; H, 8.05; N, 5.95. $C_{26}H_{36}N_2O_6$ requires C, 66.1; H, 7.85; N, 5.95%). The free acid was recovered from the quinine salt as described above as a gum, $[\alpha]_D^{23}$ −15.6° [*c* 2.3 in HCl (0.1 mol dm⁻³)], τ (100 MHz) 4.34br (1 H, s, OH), 5.84 [1 H, s, C*H*(OH)], 8.30 (2 H, m, MeCH_2), 8.73 [3 H, s, *Me*C(OH)], and 9.04 (3 H, t, *J* 7.0 Hz, *Me*CH_2).

$C_6H_{12}O_4$ Ref. 251b

2R,3R- and 2S,3S-Isomers of 2,3-dihydroxy-3-methylpentanoic acid. Erythro-2,3-Dihydroxy-3-methylpentanoic acid was prepared and resolved with quinine, as previously described [14,15], to yield the 2R,3R-isomer. From the mother liquors of the crystallization of the quinine salts, the free acid (enriched in the 2S,3S-isomer) was recovered and converted into the dicyclohexylamine salt. This salt (2.66 g) was applied in 10 ml water to a column that contained 50 g Dowex 50W-X8 (H⁺) ion-exchange resin, and the free acid was eluted with water. The acidic eluate was evaporated under reduced pressure to yield the acid as a gum (1.2 g) which was dissolved in acetone and then treated with 1.6 g (+)-2-*p*-bromophenylethylamine. The solvent was removed, the residue was dissolved in 5 ml acetone and the solution was treated with 20 ml ether. After 15 h, the crystalline salt (1.59 g, m.p. 124—125°C) was filtered and recrystallized five times from acetone/ether to a constant m.p. of 128—129°C, $[\alpha]_D^{23}$ = +6.9 ± 0.7° (*c* 4.3 in methanol). (Found: C, 48.1; H, 6.3; N, 3.9; $C_{14}H_{22}BrNO_4$ requires C, 48.4; H, 6.1; N, 4.0%). The free acid was isolated by passage through Dowex 50W-X8(H⁺), as described above $[\alpha]_D^{23}$ = +18.9° (*c* 8.5 in H_2O). Since this rotation value indicated an optical purity of only 91% by comparison with that of the pure 2R,3R-isomer (−23.1°) [14], the acid was again converted into the (+)-2-*p*-bromophenylethylamine salt which was recrystallized. However, no increase in the optical rotation of the free acid was obtained. Accordingly, the acid was treated with an equimolar amount of dicyclohexylamine in acetone. The acetone was removed and the residual salt was recrystallized twice from acetone/ether to give the dicyclohexylamine salt of the 2S,3S-isomer, m.p. 187—188°C. (Found: C, 65.6; H, 10.9; N, 4.2; $C_{18}H_{35}NO_4$ requires C, 65.6; H, 10.7; N, 4.3%). The free acid was isolated as described above $[\alpha]_D^{23}$ = +23.2 ± 0.75° (*c* 3.84 in H_2O).

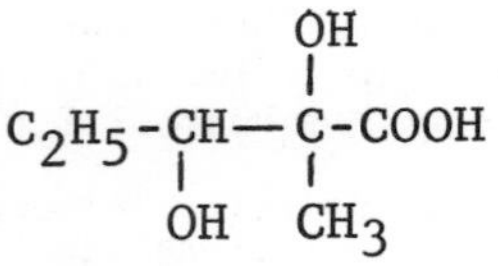

$C_6H_{12}O_4$ Ref. 252

Resolution of threo-2,3-Dihydroxy-2-methylvaleric Acid (II) into its Antipodes. 300 mg of (±)-threo-2,3-dihydroxy-2-methylvaleric acid was added to a hot solution of 630 mg of strychnine in 40 ml of ethyl acetate; the mixture was heated for five minutes and then set aside at room temperature for 20 hours. The precipitated crystals were filtered off, and after two crystallizations from ethyl acetate we obtained 220 mg of the salt formed by strychnine with (+)-threo-2,3-dihydroxy-2-methylvaleric acid; m.p. 104-106°; $[\alpha]_D^{20}$ -14° (c 5 in water; l = 0.5). A solution of 150 mg of the resulting salt in water was passed through a column (100× 10 mm) filled with Cationite KU-2. Water was distilled off in a vacuum, and two crystallizations from ethyl acetate gave 44 mg of (+)-threo-2,3-dihydroxy-2-methylvaleric acid; m.p. 121-122°; $[\alpha]_D^{20}$ + 28° (c 3 in water; l = 0.5). Found: C 48.73; H 8.17%. $C_6H_{12}O_4$. Calculated: C 48.64; H 8.17%.

The mother solution obtained after the separation of the salt of strychnine and (+)-threo-2,3-dihydroxy-2-methylvaleric acid was vacuum-evaporated to dryness; the residue was dissolved in 2 ml of water and passed through a column containing Cationite KU-2 (100× 10 mm). The eluate was evaporated to dryness, and we obtained 180 mg of an acid, $[\alpha]_D^{15}$ -11.1° (c 9 in water; l = 0.5). 170 mg of this acid was dissolved in 2 ml of ethyl acetate, and 285 g of morphine in 50 ml of ethyl acetate was added to the hot solution. The mixture was heated for five minutes and then left overnight at room temperature. The crystals that were precipitated were filtered off; two crystallizations from ethyl acetate gave 180 mg of the morphine salt of the (-)-threo dihydroxy acid (II); decomp. temp. 185-190°; $[\alpha]_D^{20}$ -87.1° (c 5 in water; l = 0.5). The salt was dissolved in 2 ml of water and passed through a column containing Cationite KU-2 (100× 10 mm). The eluate was vacuum-evaporated, and two crystallizations from ethyl acetate gave 55 mg of (-)-threo-2,3-dihydroxy-2-methylvaleric acid; m.p. 121-122°; $[\alpha]_D^{21}$ -27.1° (c 5 in water; l = 0.5). Found: C 48.40; H 8.18%. $C_6H_{12}O_4$. Calculated: C 48.64; H 8.17%

$CH_3(CH_2)_3CHCOOH$
$\quad\quad\quad\quad\quad\quad |$
$\quad\quad\quad\quad\quad NH_2$

$C_6H_{13}NO_2$ **Ref. 253**

l-Benzoyl-α-Amino-*n*-capronsäure.

5 Theile racemische Benzoyl-α-Amino-capronsäure[3]) werden mit 6.25 Theilen krystallisirtem käuflichem Cinchonin in 750 Theilen kochendem Wasser gelöst. Bei längerem Stehen der erkalteten Flüssigkeit und öfterem Reiben der Glaswandung krystallisirt das Salz der linksdrehenden Benzoylverbindung. Durch Einimpfen einiger Krystalle kann diese Operation sehr beschleunigt werden. 15—20 Stunden nach Eintritt der Krystallisation war nach öfterem Umrühren die Abscheidung des Salzes beendet. Es wurde filtrirt und aus der 70-fachen Menge Wasser umkrystallisirt. Die Menge des reinen Salzes betrug in der Regel 3 Theile oder 54 pCt. der Theorie. Auf die Verarbeitung der Mutterlaugen kommen wir gleich zurück. Zur Gewinnung der freien Benzoylverbindung werden 10 g fein zerriebenes Cinchoninsalz mit 250 ccm Wasser und 40 ccm Normalkalilauge etwa ½ Stunde auf dem Wasserbade digerirt, nach dem Erkalten vom Cinchonin abfiltrirt und die Lösung mit 50 ccm Normalsalzsäure versetzt. Nach kurzer Zeit beginnt die Krystallisation der activen Benzoylverbindung. Die Mutterlaugen liefern nach dem Einengen unter vermindertem Druck eine zweite Krystallisation.

Die Ausbeute ist nahezu quantitativ. Durch Umkrystallisiren aus heissem Wasser wird der Benzoylkörper in schönen, langen, farblosen Nadeln erhalten. Dieselben enthalten im lufttrockenen Zustande ½ Mol. Krystallwasser, welches schon im Vacuumexsiccator über Schwefelsäure zum Theil entweicht, wobei die Masse klebrig wird. Für die Bestimmung des Krystallwassers wurde bei 100° getrocknet.

0.1829 g verloren 0.0062 g, wobei die Masse völlig zusammensinterte.

½ H_2O. Ber. 3.55, Gef. 3.39.

Die krystallwasserhaltige Substanz schmilzt bei 53° (corr.). Das getrocknete Präparat gab folgende Zahlen:

0.1767 g Sbst.: 0.4305 g CO_2, 0.1180 g H_2O.

$C_{13}H_{17}NO_3$. Ber. C 66.39, H 7.23.
Gef. » 66.44, » 7.42.

Für die optische Bestimmung wurden 1.0672 g krystallwasserhaltige Substanz in 5 ccm Normalkalilauge (etwas mehr als die molekulare Menge) und ca. 6 ccm Wasser gelöst.

Gesammtgewicht der Flüssigkeit 11.2018 g, mithin Procentgehalt 8.88; spec. Gewicht 1.027; Drehung bei 20° im 1 dm-Rohr und Natriumlicht 2° nach links. Mithin für die krystallwasserhaltige Verbindung in alkalischer Lösung $[\alpha]_D^{20°} = -21.9°$.

Da auch nach öfterem Umkrystallisiren des Cinchoninsalzes dieser Werth constant blieb, so glauben wir, dass die active Benzoylverbindung ganz frei von Racemkörper war. Sie ist in Alkohol zerfliesslich, in Aether auch ziemlich leicht, in Ligroïn dagegen äusserst schwer löslich; beim Erwärmen mit Wasser schmilzt sie und löst sich beim Kochen in ungefähr 80 Theilen.

l-α-Amino-*n*-Capronsäure.

Wird die Benzoylverbindung mit der 150-fachen Menge zehnprocentiger Salzsäure am Rückflusskühler gekocht, so ist die Zersetzung nach 6 Stunden beendet. Nach dem Erkalten wird von der ausgeschiedenen Benzoësäure filtrirt, die Lösung unter vermindertem Druck stark eingeengt, mehrmals ausgeäthert, dann zur Trockne verdampft und das zurückbleibende Hydrochlorat in der üblichen Weise durch Kochen mit gelbem Bleioxyd zerlegt. Die so gewonnene freie Aminosäure krystallisirt aus heissem Wasser in farblosen, glänzenden, dem Leucin ähnlichen Blättchen, die im verschlossenen Capillarrohr beim raschen Erhitzen gegen 296° schmelzen, während im offenen Capillarrohr zwischen 260° und 270° Zersetzung ohne deutliche Schmelzung erfolgt.

Zur Analyse wurde das Präparat bei 120° getrocknet:
0.1793 g Sbst.: 0.3599 g CO_2, 0.1606 g H_2O.

$C_6H_{13}NO_2$. Ber. C 54.96, H 9.92.
Gef. » 54.74, » 9.95.

Für die optische Bestimmung diente die Lösung in 20-procentiger Salzsäure. Gewicht der Substanz 0.4890 g, Gewicht der Lösung 11.4703 g, mithin Procentgehalt 4.26; spec. Gewicht 1.10; Drehung im 2 dm-Rohr bei 20° für Natriumlicht 2.10° nach links. Daraus würde sich berechnen $[\alpha]_D^{20} - 22.4°$. Wie schon erwähnt, haben Schulze und Likiernik durch Vergährung der α-Amino-*n*-capronsäure mit Penicillium glaucum eine active Säure gewonnen, die unter den gleichen Bedingungen $[\alpha]_D - 26.5°$ zeigte. Nimmt man an, dass ihr Präparat reine active Säure war, so würde sich aus diesen Zahlen für unsere Substanz ein Gehalt von 16 pCt. Racemkörper ergeben.

d-Benzoyl-α-Amino-*n*-capronsäure.

Sie findet sich in den Mutterlaugen, aus welchen das Cinchoninsalz des optischen Antipoden auskrystallisirt ist.

Dieselben werden zunächst mit überschüssigem Alkali versetzt, das ausgeschiedene Cinchonin abfiltrirt und die Mutterlauge mit soviel Salzsäure versetzt, das alles Alkali dadurch gebunden wird. Lässt man jetzt die Flüssigkeit 1—2 Tage bei 0° stehen, so krystallisirt racemische Benzoyl-α-amino-*n*-capronsäure und zwar ist die Abscheidung so vollständig, dass fast gar keine *l*-Benzoylverbindung mehr in Lösung bleibt.

Verdampft man nun das Filtrat unter stark vermindertem Druck auf ¼ seines Volumens, so krystallisirt nach dem Erkalten allmählich die rechtsdrehende Benzoylverbindung in denselben Formen wie das optische Isomere. Die Mutterlaugen geben bei weiterer Concentration noch neue Mengen. Die Ausbeute entspricht ungefähr der für die *l*-Verbindung angegebenen Menge, und nach einmaligem Umkrystallisiren aus warmem Wasser ist das Präparat ganz frei von Racemkörper. Dieses Resultat, welches bei blosser Krystallisation von Gemischen activer und racemischer Substanzen sehr selten erreicht wird, ist um so erfreulicher, als es uns bisher nicht gelungen ist, ein krystallisirtes Salz der *d*-Benzoylaminocapronsäure mit einem activen Alkaloïd zu gewinnen.

Die Krystalle der *d*-Verbindung schmelzen gerade so wie das optische Isomere bei 53° (corr.) und enthalten Krystallwasser, welches bei 100° völlig entweicht, wobei eine ganz amorphe Masse zurückbleibt. Die Analyse der Letzteren gab folgende Zahlen:

0.1951 g Sbst.: 0.4735 g CO_2, 0.1293 g H_2O.

$C_{13}H_{17}NO_3$. Ber. C 66.39, H 7.23.
Gef. » 66.18, » 7.36.

Für die optische Bestimmung diente das krystallwasserhaltige Präparat. 0.9696 g wurden in 5 ccm Normalkalilauge und 6.1 ccm Wasser gelöst. Gesammtgewicht der Lösung 11.806 g, mithin Procentgehalt 8.21. Spec. Gewicht 1.026; Drehung im 1 dm-Rohr bei 20° im Natriumlicht 1.8° nach rechts. Mithin in alkalischer Lösung $[\alpha]_D^{20°} + 21.4°$, während für die *l*-Verbindung $- 21.9°$ gefunden wurde.

d-α-Amino-*n*-capronsäure.

Sie wurde ebenso dargestellt wie die *l*-Verbindung und zeigte mit derselben äusserlich die grösste Aehnlichkeit. Die Analyse ergab:

0.1701 g Sbst.: 0.3415 g CO_2, 0.1529 g H_2O.

$C_6H_{13}NO_2$. Ber. C 54.96, H 9.92.
Gef. » 54.75, » 9.98.

Für die optische Bestimmung diente wieder die Lösung in 20-procentiger Salzsäure.

Gewicht der Substanz 0.4411 g, Gewicht der Lösung 11.4090 g, mithin Procentgehalt 3.86; spec. Gewicht 1.10; Drehung bei 20° im 2 dm-Rohr für Natriumlicht 1.81° nach rechts, woraus sich $[\alpha]_D^{20} = + 21.3°$ berechnet. Das Präparat würde demnach ungefähr 20 pCt. Racemkörper enthalten haben.

Um diesen Schluss zu controlliren, haben wir die Säure durch Behandlung mit Natriumbicarbonat und Benzoylchlorid in die Benzoylverbindung zurückverwandelt und durch deren optische Untersuchung einen Gehalt von 17.6 pCt. Racemkörper gefunden.

---

[1]) Diese Berichte 32, 2451, 3638 [1899] und 33, 2370, 2383 [1900].
[2]) Zeitschr. f. physiol. Chem. 17, 523 [1893].
[3]) Diese Berichte 33, 2382 [1900].

$C_6H_{13}NO_2$ Ref. 253a

Darstellung der Formyl-d,l-α-aminocapronsäure.

159 g d, l-α-Aminocapronsäure wurden mit der 1½fachen Menge wasserfreier käuflicher Ameisensäure 3 Stunden lang auf dem Wasserbade erhitzt. Nun wurde das Lösungsmittel unter vermindertem Druck möglichst vollständig verdampft, der Rückstand wieder in Ameisensäure aufgenommen, 3 Stunden auf 100° erhitzt, dann eingedampft und diese Operation noch einmal wiederholt. Der feste Rückstand, der durch Absprengen des Kolbens gewonnen werden konnte, wurde zerkleinert, dann im Vakuumexsikkator über Kalihydrat von noch anhaftender Ameisensäure befreit und schließlich fein pulverisiert. Nun wurde, um die unveränderte Capronsäure zu entfernen, in kleinen Portionen mit der 1½fachen Menge eiskalter n-Salzsäure rasch zerrieben, scharf abgenutscht, und mit wenig eiskaltem Wasser bis zur Chlorfreiheit gewaschen. Das Rohprodukt wurde unter Anwendung von Tierkohle aus möglichst wenig Wasser umkrystallisiert. Ausbeute 90 g.

Die Formyl-d-l-α-aminocapronsäure krystallisiert in glänzenden Nadeln. Beim Erhitzen im offenen Kapillarrohr erweicht die Substanz bei 110—111° und schmilzt bei 114° (unkorr.).

0,2178 g Substanz verbrauchten bei der N-Bestimmung nach Kjeldahl 13,39 ccm n/10-n-Schwefelsäure.

Gefunden: N = 8,62%. Berechnet: 8,80%.

Spaltung der Formyl-dl-α-aminocapronsäure mit Brucin.

50 g Formyl-dl-α-aminocapronsäure wurden in 2 Liter absoluten Alkohols eingetragen und am Wasserbade erwärmt. Ein Teil der Substanz ging nicht in Lösung. Von diesem wurde abfiltriert. Es stellte sich bei der Prüfung heraus, daß der ungelöste Rückstand, der etwa 4,8 g wog, reine α-Aminocapronsäure war. Die Formylgruppe war also teilweise abgespalten worden.

112,5 g wasserfreies Brucin wurden in 1600 ccm absoluten Alkohols unter Erwärmen gelöst. Diese Lösung gaben wir zu der alkoholischen Lösung der Formyl-dl-α-aminocapronsäure. Nach 12 stündigem Stehen in der Kälte begann die Abscheidung von Brucinsalz. Sie nahm nach energischem Reiben plötzlich stark zu. Nach weiteren 12 Stunden wurde die Krystallmasse scharf abgesaugt und mit kaltem absolutem Alkohol gewaschen. Die Krystallmasse wog ca. 95 g. Sie bestand aus dem Brucinsalz der Formyl-d-α-aminocapronsäure.

Die Substanz wurde in 530 ccm Wasser gelöst, auf 0° abgekühlt und mit 158 ccm n-Natronlauge versetzt. Es fiel das Brucin aus. Nach einer Viertelstunde wurde abgesaugt und der Rückstand mit wenig kaltem Wasser gewaschen. Die letzten Reste von Brucin wurden durch Ausschütteln mit Chloroform vollständig entfernt. Nun wurden 20 ccm 5 fach Normalsalzsäure zugegeben. Dann engten wir die Lösung unter vermindertem Druck ein, übersättigten mit weiteren 14 ccm 5 fach n-Salzsäure und stellten das Gemisch in Eiswasser. Nach einer halben Stunde wurde die Krystallmasse abgenutscht und mit eiskaltem Wasser chlorfrei gewaschen. Ausbeute ca. 15 g.[1])

Die alkoholische Mutterlauge, welche das Brucinsalz der Formyl-l-α-aminocapronsäure enthielt, wurde unter vermindertem Druck eingedampft, der Rückstand in Wasser aufgenommen und weiter so behandelt, wie die d-Verbindung. Ausbeute ca. 15 g.

Beide optisch-aktiven Formylkörper wurden aus wenig heißem Wasser umkrystallisiert, abgesaugt, und mit wenig kaltem Wasser gewaschen. Um die Abspaltung der Formylgruppe möglichst zu beseitigen, wurden die Substanzen in einem Exsikkator über Schwefelsäure getrocknet, welches in einem Gefäß wasserfreie Ameisensäure enthielt.

Die beiden optisch-aktiven Antipoden haben bis auf die Drehung völlig übereinstimmende Eigenschaften. Sie krystallisieren in feinen seidenglänzenden, kürzeren oder längeren Nadeln, welche zu büschelförmigen Gruppen angeordnet sind. Sie lösen sich sehr leicht in Wasser, in Äthyl- und Methylalkohol und in Äther.

Beim Erhitzen im offenen Kapillarrohr erweichen beide bei 111°. Bei 114° findet ein richtiges Schmelzen statt.

Bei der optischen Untersuchung wurden folgende Werte erhalten:

I. 0,3769 g Substanz in 9,7547 g wässeriger Lösung drehen im 2 dm-Rohr — 0,90, $[\alpha]_D^{20} = -15{,}85$.

II. 0,2121 g Substanz in 10,0124 g wässeriger Lösung drehen im 1 dm-Rohr + 0,33, $[\alpha]_D^{20} = +15{,}53$.

Darstellung der optisch-aktiven α-Aminocapronsäuren.

Die optisch-aktiven α-Aminocapronsäuren wurden durch Hydrolyse mit der 10fachen Menge 10%iger Salzsäure aus den betreffenden Formylverbindungen dargestellt. Es wurde von 15 g der Formylkörper ausgegangen. Nach 1½stündigem Kochen am Rückflußkühler wurde die Lösung im Vakuum eingedampft, der Rückstand in Wasser aufgenommen und wieder eingedampft. Nun wurde der Rückstand in Wasser gelöst, in einen 500 ccm-Meßkolben gebracht, dieser bis zur Marke aufgefüllt und in 1 ccm der Lösung der Chlorgehalt nach Volhard bestimmt. Dann wurde die berechnete Menge einer n-Lithiumhydroxydlösung zugegeben und das Gemisch in einer Porzellanschale am Wasserbade eingeengt. Es erfolgte Krystallisation, welche durch Zugabe von Alkohol sich vervollständigte. Sie wurde abgesaugt, mit kaltem Wasser und kaltem Alkohol chlorfrei gewaschen und mit wenig Tierkohle aus heißem Wasser umkrystallisiert. Ausbeute: 9 bezw. 8 g.

Die beiden optisch-aktiven α-Aminocapronsäuren krystallisieren in glänzenden schuppenförmigen Blättchen. Sie sind sehr schwer löslich in Wasser und Äthylalkohol. Beim Erhitzen im offenen Kapillarrohr sintert die Substanz bei 275 bis 280° sehr stark und sublimiert teilweise weg. Bei 301° beobachtet man deutliches Schmelzen, doch ist bis zum Erreichen dieser Temperatur der weitaus größte Teil der Substanz sublimiert.

Die Bestimmung des optischen Drehungsvermögens ergab die folgenden Werte. Es sei bemerkt, daß aus der d-Formyl-α-aminocapronsäure die l-α-Aminocapronsäure und von der l-Formylverbindung die d-α-Aminocapronsäure sich gebildet. Die optisch-aktiven α-Aminocapronsäuren drehen sowohl in wässeriger Lösung, wie in 20%iger Salzsäurelösung in derselben Richtung.

Spaltung I.

I. 0,1079 g Substanz in 17,8865 g wässeriger Lösung drehen im 2 dm-Rohr — 0,05°[1])

$$[\alpha]_D^{20^\circ} = -4{,}14^\circ.$$

II. 0,0842 g Substanz in 11,1026 g wässeriger Lösung drehen im 2 dm-Rohr + 0,095°[1])

$$[\alpha]_D^{20^\circ} = +6{,}26^\circ.$$

I. 0,1080 g Substanz in 10,0693 g 20%iger Salzsäurelösung (spez. Gew. = 1,11) drehen im 2 dm-Rohr — 0,52

$$[\alpha]_D^{20^\circ} = -21{,}17^\circ;$$

auf HCl-freie Substanz umgerechnet:

$$[\alpha]_D^{20^\circ} = -17{,}00^\circ.$$

II. 0,0720 g Substanz in 10,1526 g 20%iger HCl-Lösung (spez. Gew. = 1,11) drehen im 2 dm-Rohr + 0,28°

$$[\alpha]_D^{20^\circ} = +18{,}56^\circ;$$

auf HCl-freie Substanz umgerechnet:

$$[\alpha]_D^{20^\circ} = +15{,}6^\circ.$$

[1]) Die Ausbeute läßt sich ohne Zweifel noch erheblich steigern.

[1]) Mittelwerte aus mehreren Ablesungen.

[1]) Hüfner, Journal f. prakt. Chemie, II. Folge, Bd. 1, S. 6 (1870).

Spaltung II.

I. 0,1268 g Substanz in 13,7940 g wässeriger Lösung drehen im 2 dm-Rohr + 0,095°[1])

$[\alpha]_D^{20°} = + 5,16°$.

II. 0,1597 g Substanz in 16,4970 g wässeriger Lösung drehen im 2 dm-Rohr — 0,087°[1])

$[\alpha]_D^{20°} = - 4,49°$.

I. 0,1120 g Substanz in 11,3896 g 20%iger HCl-Lösung (spez. Gew. = 1,11) drehen im 2 dm-Rohr + 0,45°

$[\alpha]_D^{20°} = + 20,44°$;

auf die HCl-freie Substanz umgerechnet:

$[\alpha]_D^{20°} = + 16,02°$.

II. 0,1360 g Substanz im 11,7498 g 20%iger HCl-Lösung (spez. Gew. = 1,11) drehen im 2 dm-Rohr — 0,54°

$[\alpha]_D^{20°} = - 20,82°$;

auf die HCl-freie Substanz umgerechnet:

$[\alpha]_D^{20°} = - 16,07°$.

Die d-α-Aminocapronsäure schmeckt fad süß, die l-α-Aminocapronsäure hat einen bitteren Geschmack.

Emil Fischer fand bei der Spaltung von dl-Aminocapronsäure mittels der Benzoylverbindungen die folgenden Werte für $[\alpha]_{20°}^{D}$:

$[\alpha]_D^{20} = - 22,4°$ und $[\alpha]_D^{20} = + 21,3°$.

$C_6H_{13}NO_2$ Ref. 253b

Resolution of N-Acetyl-DL-norleucine.—(a) The acid and (−)-α-fenchylamine (0.15 mole) were dissolved in 200 cc. of water and the salts crystallized in several fractions. Three systematic recrystallizations gave 15.7 g. (67%) of coarse needles $[\alpha]^{25}D$ −12.7° (*c* 4, methanol) and more soluble fractions of fine needles or liquors with lower levo or dextro rotations.

N-Acetyl-D-norleucine was recovered from the less soluble salt as usual (61% yield) and crystallized from 3–4 parts of water as long needles, m.p. 114–115.5°; $[\alpha]^{25}D$ +5.0° (*c* 4, water); −0.8° (*c* 2, ethanol); +19.6° (*c* 4, methanol); solubilities in water, 7.84; ethyl acetate, 13.4. The solubility in acetone was too great to permit crystallization. Synge[20] reported m.p. 112–114°, $[\alpha]^{23}D$ −0.2° (*c* 2.4, ethanol).

N-Acetyl-L-norleucine was recovered similarly in impure form from the more soluble fractions. Crystallization removed the less soluble DL-form (8.2 g.) as coarse masses and gave the L-form (5.5 g.) as long needles, m.p. 114–115.5°; $[\alpha]^{25}D$ −19.4° (*c* 4, methanol).

(b) A resolution was effected with (−)-α-phenylethylamine by essentially the same procedure as described in (a) above. The less soluble salt (55% yield) has $[\alpha]^{25}D$ −14.8° (*c* 4, methanol) and gave N-acetyl-D-norleucine $[\alpha]^{25}D$ +19.4° (*c* 4, methanol). The antipode obtained from the more soluble salts has m.p. 114–115°, $[\alpha]^{25}D$ −19.3°.

The Amino Acids.—Each of the racemic and active acetamino acids (5 g.) was boiled two hours under reflux with 11.6 cc. (1.2 equivs.) of 3 *N* hydrobromic acid and the solution then evaporated to dryness under reduced pressure in a boiling water-bath. The residue was taken up in 25 cc. of methanol and the amino acid obtained by dropwise addition of 27% aqueous ammonia to about *p*H 6, filtration and washing with warm methanol. Additional small crops were obtained by reworking the filtrates: yields 90–98%. The amino acids were then purified by one or more crystallizations from water or aqueous methanol.

The racemic amino acids were partially characterized by solubility determinations in water at 25°; DL-isoleucine, 2.02; DL-alloisoleucine, 5.35; DL-leucine, 1.02; DL-norleucine, 1.13. The active amino acids were characterized by the specific rotation values recorded in Table I for several solvents.

TABLE I

SPECIFIC ROTATIONS OF AMINO ACIDS, $[\alpha]^{25}D$

	Water	1.007 *N* HCl (*c* 4)	5.990 *N* HCl (*c* 4)	Ref.[d]
D-Isoleucine	−12.2[a]	−36.6	−40.7	5, 8, 14
L-Isoleucine	+12.2[a]	+36.7	+40.7	5, 8
D-Alloisoleucine	−15.6[b]	−34.8	−38.4	8
L-Alloisoleucine	+15.7[b]	+34.9	+38.5	8
D-Leucine	+10.6[b]	−14.9	−15.2	20
L-Leucine	−10.7[b]	+14.6	+15.2	21, 22
D-Norleucine	− 6.0[c]	−22.1	−23.7	20
L-Norleucine	+ 6.1[c]	+21.8	+23.5	20

[a] *c* = 3.2. [b] *c* = 2. [c] *c* = 1. [d] References to comparable literature values.

$C_6H_{13}NO_2$ Ref. 253c

Note:

Using (+)-carvomenthoxyacetyl chloride as resolving agent, obtained an acid $[\alpha]_D^{21}$ = +24.0° (c, 3.3, 20% HCl) from the more insoluble salt with m.p. 105.5-107°C and $[\alpha]_D^{20}$ = +73.4° (c, 2, ethanol).

$C_6H_{13}NO_2$ Ref. 253d

B. *Norleucin*

DL-Norleucin-benzylester-hydrochlorid wird entsprechend wie DL-Norvalin-benzylester-hydrochlorid gewonnen. Ausb. 76% d. Th. Nach dem Umkristallisieren Schmp. 128°.

$C_{13}H_{19}NO_2 \cdot HCl$ (257.7) Ber. C 60.58 H 7.78 N 5.44 Gef. C 60.35 H 7.86 N 5.86

DL-Norleucin-benzylester: Aus dem Hydrochlorid nach l.c.[3]). Ausb. 93% d. Th.

$C_{13}H_{19}NO_2$ (221.3) Ber. N 6.34 Gef. N 6.64

D-Weinsäure-monomethyläther: Durch Methylieren von D-Weinsäure mittels Dimethylsulfats nach W. N. HAWORTH[5]). Das rohe Methylierungsprodukt wird durch Extrahieren mit Äther im Soxhlet gereinigt. Schmp. 172°, $[\alpha]_D^{20}$: +43.0° (*c* = 3.0, in Wasser).

a) *Spaltung des DL-Norleucin-benzylesters mit D-Weinsäure*

5.0 g DL-Norleucin-benzylester werden mit 3.4 g D-Weinsäure in 200 ccm absol. Äthanol bei 20° vereinigt. Nach 40 Stdn. ist die Abscheidung von *L-Norleucin-benzylester-D-hydrogentartrat* beendet. Ausb. 4.0 g; Schmp. 153°, $[\alpha]_D^{20}$: +1.77° (*c* = 2.0, in Wasser).

$C_{17}H_{25}NO_8$ (371.4) Ber. N 3.77 Gef. N 3.79

[3]) S. M. MCELVAIN und J. F. VOZZA, J. Amer. chem. Soc. **71**, 896 [1949].

[4]) J. P. GREENSTEIN und Mitarbb., J. biol. Chemistry **182**, 451 [1950].

[5]) J. chem. Soc. [London] **107**, 8 [1915].

Nach dem Umkristallisieren aus Alkohol gewinnt man das Tartrat optisch rein. Schmp. 157–158°, $[\alpha]_D^{20}$: 0° (*c* = 1.9, in Wasser).

Ber. N 3.77 Gef. N 3.78

Aus der Mutterlauge des Spaltansatzes isoliert man durch Einengen i. Vak. und Fällung mit Äther das D-Ester-hydrogentartrat. Ausb. 3.5 g, Schmp. 145°. $[\alpha]_D^{20}$: +19.7° (*c* = 1.5, in Wasser).

Gef. N 3.88

Das durch Umkristallisieren erhaltene optisch reine D-Estersalz liefert folgende Daten: Schmp. 149°, $[\alpha]_D^{20}$: +20.3° (*c* = 1.1, in Wasser).

Gef. N 3.73

L- und D-Ester-hydrochloride: Die Verbindungen werden, ausgehend von den reinen Tartraten, wie beim Norvalin nach dem Äther/Ammoniak-Verfahren gewonnen. Ausb. 95 % d. Th.

L-Norleucin-benzylester-hydrochlorid: Schmp. 120°, $[\alpha]_D^{20}$: −13.6° (*c* = 1.2, in Äthanol).

D-Norleucin-benzylester-hydrochlorid: Schmp. 118°, $[\alpha]_D^{20}$: +13.3° (*c* = 1.8, in Äthanol).

$C_{13}H_{19}NO_2 \cdot HCl$ (257.7)	Ber.	C 60.58	H 7.78	N 5.44
L-Verb.	Gef.	C 60.40	H 7.67	N 5.54
D-Verb.	Gef.	C 60.09	H 7.91	N 5.43

Die Drehwerte der Ester-hydrochloride ließen sich durch Umkristallisieren nicht weiter steigern. Ihre Verseifung lieferte optisch reines L- und D-Norleucin-hydrochlorid. Die Ester-hydrochloride waren damit optisch rein.

Der aus ihnen mit ammoniakalischem Äther[3] freigesetzte reine L-Benzylester (D-Benzylester) besaß die spezif. Drehung −6.9° (*c* = 3.0, in Äthanol) [+7.0° (*c* = 3.1, in Äthanol)]. Die Überführung der diastereomeren Rohtartrate in die Esterhydrochloride lieferte Verbindungen mit den Werten:

L-Ester-hydrochlorid: Schmp. 117°, $[\alpha]_D^{20}$: −10.5°.

Gef. N 5.68

D-Ester-hydrochlorid: Schmp. 114°, $[\alpha]_D^{20}$: +7.3°.

Gef. N 5.78

Sie besaßen damit optische Reinheitsgrade von 76 bzw. 55 %.

L- und D-Norleucin-hydrochlorid: Die Verseifung wird, ausgehend von reinen Esterhydrochloriden, analog wie beim Norvalin ausgeführt. Ausb. 98 % d. Th.

L-Norleucin-hydrochlorid: Sublimiert bei 275–280°, $[\alpha]_D^{20}$: +23.9° (*c* = 1.2, in 10-proz. Salzsäure).

D-Norleucin-hydrochlorid: Sublimiert bei 270°, $[\alpha]_D^{20}$: −23.3° (*c* = 1.5, in 10-proz. Salzsäure).

$C_6H_{13}NO_2 \cdot HCl$ (167.6)	Ber.	C 42.98	H 8.35	N 8.35
L-Verb.	Gef.	C 43.08	H 8.61	N 8.45
D-Verb.	Gef.	C 43.38	H 8.41	N 8.48

Der Literaturwert[6] für optisch reines Norleucin-hydrochlorid beträgt ±23.9°.

[6] W. A. H. HUFFMAN und A. W. INGERSOLL, J. Amer. chem. Soc 73, 3366 [1951].

$C_6H_{13}NO_2$ Ref. 253e

b) *Spaltung des DL-Norleucin-benzylesters mit D-Weinsäure-monomethyläther*

2.3 g DL-Norleucin-benzylester und 1.7 g Monomethyl-weinsäure werden gesondert in insgesamt 80 ccm absol. Isopropylalkohol gelöst und die Lösungen bei Zimmertemperatur vereint. Nach 2 Tagen wird der Niederschlag von *L-Norleucin-benzylester-monomethyl-D-hydrogentartrat* abgesaugt. Ausb. 2.0 g, Schmp. 150–152°, $[\alpha]_D^{20}$: +3.44° (*c* = 1.5, in Wasser).

$C_{18}H_{27}NO_8$ (385.4) Ber. C 56.10 H 7.01 N 3.63 Gef. C 55.80 H 7.07 N 4.02

Aus der Mutterlauge des Spaltansatzes lassen sich 1.5 g *D-Norleucin-benzylester-monomethyl-D-hydrogentartrat* gewinnen. Schmp. 85–90°, $[\alpha]_D^{20}$: +23.4° (*c* = 1.2, in Wasser).

Gef. N 4.24

L- und D-Ester-hydrochlorid: Die diastereomeren Benzylester-monomethyl-D-hydrogentartrate werden ohne weitere Reinigung nach dem Äther/Ammoniak-Verfahren in die Esterhydrochloride übergeführt.

L-Ester-hydrochlorid: Schmp. 115°, $[\alpha]_D^{20}$: −8.0° (c = 2.4, in Äthanol).
D-Ester-hydrochlorid: Schmp. 112°, $[\alpha]_D^{20}$: +4.1° (c = 4.2, in Äthanol).

$C_{13}H_{19}NO_2 \cdot HCl$ (257.7)	Ber. C 60.58	H 7.78	N 5.44
L-Verb.	Gef. C 60.56	H 8.06	N 5.88
D-Verb.	Gef. C 60.32	H 8.01	N 5.89

Die Esterhydrochloride wurden also in 60-proz. bzw. 30-proz. optischer Reinheit gewonnen. Die optisch reinen Antipoden gewinnt man auch hier durch Umkristallisieren der Tartrate oder der Esterhydrochloride und Weiterverarbeitung, wie bei der Weinsäurespaltung des Norleucin-benzylesters beschrieben.

$(CH_3)_2CHCH_2CH(NH_2)COOH$

$C_6H_{13}NO_2$ Ref. 254

Spaltung des Formyl-*dl*-leucins mit Brucin.

Zu einer Lösung von 50 g Formyl-*dl*-leucin in 4 Liter absolutem Alkohol setzt man 124 g wasserfreies Brucin (1 Mol.) und erwärmt unter Umschütteln, bis völlige Lösung eingetreten ist. Beim Abkühlen erfolgt sofort die Krystallisation vom Brucinsalz des Formyl-*d*-leucins. Man lässt unter zeitweisem Schütteln 12 Stunden im Eisschrank stehen, saugt die Krystallmasse scharf ab und wäscht sie sorgfältig mit etwa 500 ccm kaltem Alkohol. Die Menge der Krystallmasse beträgt ungefähr 93 g. Umlösen des Salzes hat keinen Zweck, da dadurch kein reineres actives Formyl-leucin erhalten wird. Die alkoholische Mutterlauge, die das Brucinsalz des Formyl-*l*-leucins enthält, wird unter vermindertem Druck verdampft und der Rückstand in 450 ccm Wasser gelöst, dann die Flüssigkeit auf 0° abgekühlt und mit 175 ccm Normalnatronlauge versetzt; dadurch wird das Brucin gefällt. Man lässt noch etwa 10 Minuten in Eis stehen, saugt dann ab und wäscht mit wenig kaltem Wasser. Um den sehr kleinen Rest des Brucins zu entfernen, kann man das Filtrat erst mit Chloroform und dann nochmals mit Aether ausschütteln, bis die wässrige Lösung mit Salpetersäure keine Brucinreaction mehr giebt.

Um den grösseren Theil des Alkalis, das bei längerer Einwirkung eine partielle Hydrolyse des Formylkörpers bewirken kann, zu neutralisiren, fügt man möglichst bald 23 ccm 5-fach Normalsalzsäure zu, verdampft unter geringem Druck auf etwa 100 ccm und übersättigt jetzt durch weitere Zugabe von 15 ccm derselben Salzsäure. Dadurch wird die Abscheidung des Formyl-*d*-leucins, die schon während des Einengens begonnen hat, vervollständigt. Man lässt noch eine viertel Stunde in Eiswasser stehen, saugt dann ab und wäscht mit kaltem Wasser.

Die Ausbeute an rohem Formyl-*l*-leucin betrug 22 g und nach dem Umkrystallisiren aus der 5-fachen Menge Wasser 19 g.

Das Formyl-*d*-leucin wird aus dem krystallisirten Brucinsalz ganz in der gleichen Weise gewonnen. Seine Menge betrug nach dem Umkrystallisiren 18 g.

Die activen Formyl-leucine zeigen ähnliche Löslichkeitsverhältnisse wie das inactive Product, aber einen erheblich höheren Schmelzpunkt. Dieser ist leider auch nicht ganz constant; er lag bei den analysirten Präparaten zwischen 139—142° (141—144° corr.), nachdem bei 137° schon Erweichen eingetreten war. Sie krystallisiren aus warmem Wasser in häufig centimeterlangen Formen, die unter dem Mikroskop als lang gestreckte, schmale Prismen erscheinen und makroskopisch wie dicke Nadeln aussehen. Für die Analyse wurden sie im Vacuum über Schwefelsäure getrocknet. Für die Bestimmung des optischen Drehungsvermögens diente die Lösung in absolutem Alkohol, in der keine merkbare Veresterung im Laufe von einigen Stunden erfolgt.

Formyl-*d*-leucin.

Für die Analyse diente das einmal aus Wasser umkrystallisirte und im Vacuum über Schwefelsäure getrocknete Präparat.

0.1880 g Sbst.: 0.3638 g CO_2, 0.1405 g H_2O. — 0.1933 g Sbst.: 14.4 ccm N (15°, 763 mm) über 33-proc. KOH.

$C_7H_{13}O_3N$. Ber. C 52.77, H 8.23, N 8.82.
Gef. » 52.78, » 8.36, » 8.77.

Drehungsvermögen des analysirten Präparates.

1.3347 g Sbst.; dazu 13.3 ccm absoluter Alkohol; Gesammtgewicht der Lösung 11.8965 g; d_4^{20} = 0.8169. Drehung im 1 dcm-Rohr bei 20° und Na-Licht: +1.76°; mithin $[\alpha]_D^{20°}$ = +19.2°, wobei zu bemerken ist, dass der Beobachtungsfehler für $[\alpha]_D$ 0.2° betragen kann.

Das Präparat, welches die spec. Drehung +19.2° zeigte, wurde nochmals aus der 5-fachen Menge Wasser umkrystallisirt und ergab dann eine etwas geringere Drehung.

1.3446 g Sbst.; dazu 13.4 ccm absoluter Alkohol; Gesammtgewicht der Lösung 12.0247 g; d_4^{20} = 0.8169. Drehung im 1 dcm-Rohr bei 20° und Na-Licht: +1.72°. $[\alpha]_D^{20}$ = +18.8° (± 0.2°).

Formyl-*l*-leucin.

Für die Analyse und erste optische Bestimmung diente wieder das einmal aus Wasser umkrystallisirte Präparat.

0.1702 g Sbst.: 0.3293 g CO_2, 0.1283 g H_2O. — 0.1952 g Sbst.: 14.3 ccm N (12°, 753 mm) über 33-proc. KOH.

$C_7H_{13}O_3N$. Ber. C 52.78, H 8.23, N 8.82.
Gef. » 52.76, » 8.43, » 8.63.

1.3313 g Sbst.; dazu 13.3 ccm absoluter Alkohol; Gesammtgewicht der Lösung 11.8375 g; d_4^{20} = 0.8203. Drehung im 1 dcm-Rohr bei 20° und Na-Licht: −1.70°. $[\alpha]_D^{20}$ = −18.4° (± 0.2°).

Dieses Präparat wurde nochmals aus der 5-fachen Menge Wasser umkrystallisirt.

1.3412 g Sbst.; dazu 13.4 ccm absoluter Alkohol; Gesammtgewicht der Lösung 11.9639 g; d_4^{20} = 0.8168. Drehung im 1 dcm-Rohr bei 20° und Na-Licht: −1.69°. $[\alpha]_D^{20°}$ = −18.5° (± 0.2°).

In alkalischer Lösung gaben die Formyl-leucine eine viel stärkere Drehung.

Eine 5-procentige Lösung der *d*-Verbindung in Normal-Natronlauge zeigte bei 20° ungefähr die specifische Drehung +40°, aber die genaue Bestimmung des Werthes ist nicht möglich, weil schon bei der niederen Temperatur eine langsame Abspaltung der Formylgruppe erfolgt.

Hydrolyse der Formylverbindung.

Die Abspaltung der Formylgruppe lässt sich sowohl mit Säuren, wie mit Alkalien leicht bewerkstelligen. Für die Gewinnung der activen Form ist aber die saure Hydrolyse vorzuziehen, weil die Gefahr der Racemisirung geringer ist. Dementsprechend wurde für die Darstellung des *l*-Leucins die Formylverbindung mit der 10-fachen Menge 10-procentiger Salzsäure 1—1½ Stunden am Rückflusskühler gekocht und dann die Flüssigkeit unter stark vermindertem Druck möglichst stark eingedampft. Den Rückstand löst man in Wasser und verdampft jetzt auf dem Wasserbade, um den Rest der überschüssigen Salzsäure und geringe Mengen von Ameisensäure zum allergrössten Theil zu entfernen. Zum Schluss löst man wieder in Wasser, verdünnt auf ein bestimmtes Volumen und bestimmt in einem aliquoten Theil dieser Lösung titrimetrisch das Chlor. Dann fügt man zu dem Hauptheil der Flüssigkeit die nach dem Chlorgehalt berechnete Menge einer Normal-Lösung von Lithiumhydroxyd, verdampft auf dem Wasserbade auf ein geringes Volumen, bis der grösste Theil des Leucins auskrystallisirt ist, und fällt den Rest durch absoluten Alkohol, wobei das Lithiumchlorid in Lösung bleibt.

Die Ausbeute beträgt etwa 90 pCt. der Theorie, berechnet auf das active Formylleucin; sie ist nicht unerheblich grösser, als bei der Abscheidung der Aminosäure mittels Ammoniak, die beim Racemkörper so gute Dienste leistet. Das hängt zusammen mit der grösseren Löslichkeit des activen Leucins in Wasser und auch in einer wässrigen Lösung von Ammoniumchlorid.

Zur völligen Reinigung wird schliesslich das Präparat in heissem Wasser gelöst und durch starkes Eindampfen grösstentheils wieder zur Abscheidung gebracht.

Da das active Leucin in wässriger Lösung zu schwach dreht, so benutzt man nach dem Vorgange von E. Schulze gewöhnlich eine Lösung in 20-procentiger Salzsäure mit einer Concentration von 4—5 pCt.

Aber die einzelnen Beobachtungen zeigen ziemlich starke Abweichungen. Während Schulze für natürliches Leucin Werthe zwischen + 17.3° bis 17.8° fand, erhielt der Eine von uns für die beiden activen Leucine, die aus der synthetischen Benzoylverbindung dargestellt waren, Werthe von 15.6° bis 16° für $[\alpha]_D^{20°}$[1]). Da die letzteren Bestimmungen meist im 1 dcm-Rohr ausgeführt waren, so betrug der Beobachtungsfehler etwa 0.4° für $[\alpha]_D$.

Hr. Prof. E. Schulze hat uns jetzt auf unsere Bitte ein Präparat von *l*-Leucin übersandt, das aus Eiweisskörpern gewonnen und mit besonderer Sorgfalt gereinigt war. Nach der Beobachtung des Hrn. Schulze, die wir bestätigt fanden und mit Genehmigung des Autors mittheilen, zeigte es bei den gleichen Concentrationsverhältnissen den Werth $[\alpha]_D^{20°}$ + 16.9°. Wir stellen dem nun die Zahlen gegenüber, die wir für verschiedene Proben von Leucin erhielten, das aus der Formylverbindung durch Hydrolyse mit Salzsäure oder auch mit Bromwasserstoff nach obigem Verfahren gewonnen war.

d-Leucin. Erhalten aus der Formylverbindung durch 1½-stündiges Kochen mit der 10-fachen Menge Salzsäure am Rückflusskühler und isolirt aus dem Chlorhydrat mit Lithiumhydroxyd. Einmal aus heissem Wasser umkrystallisirt.

1.1781 g Leucin in 20-procentiger Salzsäure gelöst; Gesammtgewicht der Lösung 32.5207 g; d = 1.1. Drehung im 2 dm-Rohr bei 20° und Natriumlicht: —1.22° (± 0.02°). Mithin $[\alpha]_D^{20°}$ = —15.3° (± 0.2°).

Der mögliche Beobachtungsfehler bei der optischen Bestimmung ist stets den gefundenen Werthen beigefügt.

Das vorige Präparat wurde nochmals in wenig überschüssiger Salzsäure gelöst und durch Ammoniak gefällt, wobei fast die Hälfte in der Mutterlauge blieb. (Gef. N 10.5. Ber. N 10.7).

0.6588 g Leucin gelöst in 20-procentiger Salzsäure. Gesammtgewicht der Lösung 14.8296 g. *d* = 1.1. Drehung im 1 dcm-Rohr bei 20° und Natrium-Licht: — 0.76° (± 0.02°). Mithin $[\alpha]_D^{20}$ = — 15.6° (± 0.4°).

[1]) Diese Berichte 33, 2377 [1900].

$C_6H_{13}NO_2$ Ref. 254a

4. *L- und D-Leucinäthylester · HCl*

a) Die Bildung der diastereomeren Esterdibenzoyl-D-hydrogentartrate: 11.8 g *DL-Leucinäthylester*, Sdp.$_{12}$ 84 – 86°, wurden bei 21° mit einer Lösung von 13.9 g *Dibenzoyl-D-weinsäure* in 145 ccm absol. Äthanol versetzt. Nach kurzer Zeit begann die Kristallisation von *L-Leucinäthylester-dibenzoyl-D-hydrogentartrat*. Die abgeschiedenen Kristalle wurden nach 24 Stdn. abgesaugt, 5 ccm Waschalkohol der Mutterlauge zugegeben. Ausb. 3.9 g, Schmp. 188°, $[\alpha]_D^{20}$: –72.9°, $c = 0.192$ in Äthanol.

$C_{26}H_{31}O_{10}N$ (517.5) Ber. C 60.34 H 6.04 N 2.71 Gef. C 60.40 H 6.08 N 2.99

Weitere 9.0 g *L-Leucinäthylester-dibenzoyl-D-hydrogentartrat* kristallisierten aus der Mutterlauge nach Zusatz von absol. Äther. Schmp. 182°. Nach Abtrennen dieser zweiten Fraktion wurde die Mutterlauge bis zum Auftreten eines sirupartigen Rückstandes eingeengt (Wasserstrahl-Vak., 30° Badtemp.) und dieser durch Zusatz von absol. Äther zur Kristallisation gebracht. Ausb. 5.2 g *D-Leucinäthylester-dibenzoyl-D-hydrogentartrat*, Schmp. 151°, $[\alpha]_D^{20}$: –53.2°, $c = 0.460$ in Äthanol.

Aus der Mutterlauge dieser D-Fraktion wurden weiterhin 4.0 g Leucinäthylester, Sdp.$_{12}$ 84°, zurückgewonnen, der mit $[\alpha]_D^{22}$: –1.33° einem *D-Leucinäthylester* von ca. 12-proz. optischer Reinheit entsprach.

b) Die Gewinnung von D- und L-Leucinäthylester · HCl

L-Leucinäthylester · HCl: 3.5 g *L-Leucinäthylester-dibenzoyl-D-hydrogentartrat* wurden durch Einwirkung äthanol. Salzsäure in *L-Leucinäthylester · HCl* übergeführt (s. Phenylalaninäthylester). Ausb. 1.17 g (89% d. Th.); Schmp. 133°, $[\alpha]_D^{21}$: +18.2°, $c = 1.538$ in Äthanol. F. Röhmann[9]) gibt für L-Leucinäthylester · HCl $[\alpha]_D$: + 18.4°, $c = 5\%$ in Äthanol, an. Das erhaltene Produkt war demnach nahezu optisch rein.

$C_8H_{17}O_2N \cdot HCl$ (195.7) Ber. C 49.10 H 9.27 N 7.16 Gef. C 49.36 H 9.04 N 7.27

Die 9.0 g *L-Leucinäthylester-dibenzoyl-D-hydrogentartrat* der Fraktion II ergaben 2.96 g (87% d. Th.) *L-Leucinäthylester · HCl* mit $[\alpha]_D^{19}$: + 10.52°, $c = 1.520$ in Äthanol. Das entspricht einem optischen Reinheitsgrad von 57%.

D-Leucinäthylester · HCl: 5 g Estertartrat der D-Fraktion wurden mit äthanolischer Salzsäure in das Esterhydrochlorid übergeführt. Ausb. 1.68 g (89% d. Th.), $[\alpha]_D^{20}$: 15.33 , $c = 1.924$ in Äthanol. Optischer Reinheitsgrad: 83.5%.

$C_8H_{17}O_2N \cdot HCl$ (195.7) Ber. C 49.10 H 9.27 N 7.16 Gef. C 49.45 H 9.18 N 7.20

9) Ber. dtsch. chem. Ges. **30**, 1980 [1897].

$C_6H_{13}NO_2$ Ref. 254b

B. Racematspaltung des Leucinamides

5,0 g D,L-Leucinamid wurden in 25 cm³ absolutem Methanol gelöst und mit einer filtrierten Lösung von 14,5 g Dibenzoyl-D-weinsäure in 50 cm³ absolutem Methanol versetzt. Die Kristallisation setzte nach wenigen Minuten ein. Nach 24 Stunden wurde das L-Leucinamid-dibenzoyl-D-hydrogentartrat abgesaugt.

Ausbeute: 8,5 g (92% d. Th.); Schmp. 189 – 191°.
$[\alpha]_D^{20}$: –86,1° (c = 1,28 in Methanol).

$C_{24}H_{28}O_9N_2$ (488,5) ber. N 5,74 gef. N 6,03.

Nach dem Umkristallisieren aus absolutem Methanol wurden 4,8 g (57% d. Th.) optisch reines Salz erhalten.

Schmp. 192°. $[\alpha]_D^{20}$: –85,4° (c = 1,05 in Methanol).

$C_{24}H_{28}O_9N_2$ (488,5)	ber.	C 59,01	H 5,78	N 5,74
	gef.	58,83	6,08	6,10.

Durch Zufügen von Äther zur Mutterlauge des Spaltansatzes wurden noch 1–2 g Salz geringerer optischer Reinheit isoliert. Das Filtrat wurde im Vakuum zum Sirup eingedampft und das D-Leucinamid-dibenzoyl-D-hydrogentartrat mit viel Äther ausgefällt.

Ausbeute: 8,0 g entspr. 87% d. Th.; Schmp. 170 – 176°.
$[\alpha]_D^{20}$: –92,7° (c = 1,00 in Methanol).

Gef. N 6,04.

Nach mehrfachem Umkristallisieren aus Alkohol—Äther wurden 2,6 g (30% d. Th.) der optisch reinen Verbindung erhalten.

Schmp. 176 – 178°. $[\alpha]_D^{20}$: –92,2 (c = 1,42 in Methanol).

Gef. C 59,16 H 5,95 N 5,70.

Gewinnung der optisch aktiven Leucinamid-hydrochloride

L-Leucinamid-hydrochlorid: Reines L-Leucinamid-dibenzoyl-D-hydrogentartrat wurde in wenig absolutem Alkohol suspendiert und bis zur Sättigung trockener Chlorwasserstoff eingeleitet, wobei das Salz in Lösung ging. Mit absolutem Äther wurde das L-Leucinamidhydrochlorid ausgefällt.

Ausbeute: 95% d. Th., bezogen auf eingesetztes Tartrat.

Schmp. 244°. $[\alpha]_D^{20}$: +10,08° (c = 2,06 in Wasser),
$[\alpha]_D^{20}$: +10,96° (c = 2,64 in n HCl).

E. L. Smith und D. H. Spackmann[8]) geben für L-Leucinamidhydrochlorid die spezifische Drehung von $[\alpha]_D^{20}$: +10,0° (c = 1,0 in Wasser) an. D. S. Robinson[9]) fand für L-Leucinamidhydrochlorid den Drehwert von $[\alpha]_D^{20}$: +10,8° (c = 3,0 in n HCl. Danach lag die optisch reine Verbindung vor.

$C_6H_{14}ON_2 \cdot HCl$ (166,7)	ber.	C 43,24	H 9,07	N 16,81
	gef.	43,34	9,24	17,05.

D-Leucinamid-hydrochlorid: In gleicher Weise wie beim L-Antipoden wurde aus dem reinen D-Leucinamid-dibenzoyl-D-hydrogentartrat das D-Leucinamid-hydrochlorid hergestellt. Ausbeute 94% d. Th., bezogen auf eingesetztes Tartrat. Schmp. 235 bis 237°.

$[\alpha]_D^{20}$: −10,02 (c = 1,49 in Wasser),
$[\alpha]_D^{20}$: −10,92 (c = 1,96 in n HCl).

Gef. C 43,21 H 9,20 N 16,67.

D. S. Robinson[9]) gibt für D-Leucinamid-hydrochlorid den Drehwert von $[\alpha]_D^{20}$: −10,9° (c = 3,0 in n HCl) an. Der D-Antipode wurde also ebenfalls rein erhalten.

Die Überführung der aus dem Spaltansatz gewonnenen diastereomeren Rohtartrate in die antipodischen Amid-hydrochloride lieferte Verbindungen mit den Drehwerten $[\alpha]_D^{20}$: +6,80° und $[\alpha]_D^{20}$: −6,20°. Dies entspricht einer optischen Reinheit von 68 bzw. 62%.

Gewinnung der optisch aktiven Leucin-hydrochloride

L-Leucin-hydrochlorid: L-Leucinamid-hydrochlorid wurde mit der hundertfachen Menge 20proz. Salzsäure drei Stunden unter Rückfluß erhitzt und dann die Säure im Vakuum entfernt. Der vollständig getrocknete Rückstand wurde mit absolutem Isopropylalkohol behandelt, vom ungelösten Ammoniumchlorid abfiltriert und das L-Leucinhydrochlorid mit abs. Äther ausgefällt.

Ausbeute: 90% d. Th., bezogen auf das Amidhydrochlorid.

Schmp. 236–238°. $[\alpha]_D^{20}$: +15,8° (c = 1,82 in 20proz. Salzsäure).

E. Fischer und O. Warburg[10]) geben für L-Leucin die Drehung $[\alpha]_D^{20}$: +15,9° (in 20proz. Salzsäure) an. Das Produkt war demnach optisch rein.

$C_6H_{13}O_2N \cdot HCl$ (167,5)	ber.	C 43,01	H 8,36	N 8,36
	gef.	43,31	8,77	8,18.

[8]) E. L. Smith u. D. H. Spackmann, J. biol. Chem. 212, 290 (1955).
[9]) D. S. Robinson, J. biol. Chem. 202, 2 (1953).
[10]) E. Fischer u. O. Warburg, Ber. dtsch. chem. Ges. 38, 4002 (1905).

$C_6H_{13}NO_2$

Ref. 254c

EXAMPLE 5

Resolving of N,N-dibenzyl-DL-leucine

(A) PREPARATION OF THE SALT OF L(+)-THREO-1-p-NITRO PHENYL-2-AMINO PROPANE-1,3-DIOL AND N,N-DIBENZYL-L(−)-LEUCINE

18.7 g. of N,N-dibenzyl-DL-leucine and 13.2 g. of L(+)-threo-1-p-nitro phenyl-2-amino propane-1,3-diol are dissolved at a temperature of 50° C. in a mixture of 250 cc. of methanol and 175 cc. of water. The solution is cooled to a temperature of 15° C. within two hours. The crystallized salt is filtered off, washed with a small quantity of 50% methanol, and recrystallized from 220 cc. of 60% methanol. The salt is dried at 80° C. in a drying oven. Thereby, 12 g. of the salt of L(+)-threo-1-p-nitro phenyl-2-amino propane-1,3-diol and N,N-dibenzyl-L(−)-leucine are obtained. The resolving yield is 75%. The compound forms yellow leaflets which are insoluble in water, ether, and benzene, and soluble in ethanol and acetone. The compound melts at 140–141° C.; its rotatory power is $[\alpha]_D^{20}$=−23°±1° (concentration: 2% in methanol).

(B) PREPARATION OF N,N-DIBENZYL-L(−)-LEUCINE

8.3 g. of the salt of L(+)-threo-1-p-nitro phenyl-2-amino propane-1,3-diol and N,N-dibenzyl-L(−)-leucine obtained according to the above given Example 5(A) are treated with 18 cc. of N sodium hydroxide solution for 15 minutes while stirring vigorously. L(+)-threo-1-p-nitro phenyl-2-amino propane-1,3-diol crystallizes and is removed from the solution. In this manner 3.1 g. of the resolving agent, corresponding to a yield of 94%, are recovered. The filtrate is acidified by the addition of 1.5 cc. of acetic acid. The mixture is extracted with chloroform. The chloroform solution is washed with water, dried over magnesium sulfate, and evaporated to dryness in a vacuum. The oily residue is crystallized in aqueous acetic acid and yields N,N-dibenzyl-L(−)-leucine in hydrated form. The yield is quantitative. Said compound, in the anhydrous form, melts at 104–106° C.; its rotatory power is $[\alpha]_D^{20}$=−68°±1° (concentration: 2% in methanol).

(C) TREATMENT OF THE MOTHER LIQUORS

The mother liquors obtained on preparing and recrystallizing the salt of L(+)-threo-1-nitro phenyl-2-amino propane-1,3-diol and N,N-dibenzyl-L(−)-leucine are concentrated by evaporation to a volume of about 200 cc. 46 cc. of N sodium hydroxide solution are added thereto while heating. Thereby, 6.5 g. of the resolving agent are recovered in the same manner as described in the preceding examples. The resulting filtrate is acidified by the addition of 3 cc. of acetic acid and extracted by means of chloroform. The chloroform solution is evaporated to dryness and the residue is taken up, while heating, in 20 cc. of petroleum ether from which 7.7 g. of racemic N,N-di-benzyl leucine, corresponding to 41% of the starting racemic mixture, crystallize. The mother liquors of said crystallization are evaporated to dryness and the residue is treated with 4 g. of D(−)-threo-1-p-nitro phenyl-2-amino propane-1,3-diol in 130 cc. of aqueous 50% methanol. By following the procedure as described hereinabove in Examples 1(B), 2(B), or 3(B), there are obtained 6.2 g. of the salt of D(−)-threo-1-p-nitro phenyl-2-amino propane-1,3-diol and N,N-dibenzyl-D(+)-leucine. Said compound forms yellow leaflets which are insoluble in water, ether and benzene, and soluble in ethanol and acetone. Its melting point is 140–141° C.; its rotatory power is $[\alpha]_D^{20}$=+23°±1° (concentration: 2% in methanol).

(D) PREPARATION OF L(−)-LEUCINE

4.95 g. of L(−)-N,N-dibenzyl leucine prepared according to the preceding example 5(B) are dissolved in 50 cc. of 80% ethanol. 1.5 cc. of concentrated hydrochloric acid and 2 g. of palladium black catalyst containing 6% of palladium are added thereto and the mixture is hydrogenated at 70° C. After hydrogenolysis is completed, the catalyst is filtered off and washed with alcohol. The filtrate is concentrated by evaporation to a volume of about 5 cc. 15 cc. of absolute ethanol and then 2 cc. of pyridine are added thereto. The resulting crystals are filtered off, washed with absolute ethanol and dried in a drying oven at 100° C. 1.7 g. of optically pure I(−)-leucine, corresponding to a yield of 85%, are obtained thereby. The rotatory power of said acid is $[\alpha]_D^{20}$=+14°±1° (concentration: 3% in 5 N hydrochloric acid).

$C_6H_{13}NO_2$

Ref. 254d

For a resolution procedure using ephedrine in the direct resolution of the racemic N-benzyloxycarbonyl derivative of the above compound, see Ref. 20h.

$C_6H_{13}NO_2$ Ref. 254e

Resolution of N-Acetyl-DL-leucine.—(a) The racemic acid (0.2 mole) and (−)-α-fenchylamine (0.2 mole) were combined in 250 cc. of hot water and the salt was crystallized in several fractions. No satisfactory resolution occurred, even though the reciprocal resolution of the racemic amine with the L-acetamino acid[16] is easily effected in water. However, fractionation in methanol (2–3 cc./g.) after four series gave 23.5 g. (72%) of pure N-acetyl-D-leucine salt, $[\alpha]^{25}D$ +7.7–7.9° (*c* 4, methanol) identical in other properties with its antipode.[16] The more soluble salt was not purified. In later experiments it was found more expeditious to form the salt in methanol and, after brief fractionation, to take for decomposition somewhat impure salt (rotation +7.0° or higher).

N-Acetyl-D-leucine.—Decomposition of pure or somewhat impure salt in the usual manner and crystallization of the acetamino acid from 15 parts of water gave the pure D-form; the more soluble DL-form, if present, was recovered from the later fractions. The substance forms long needles from water or thin rhombic plates from acetone, m.p. 185–186°; $[\alpha]^{25}D$ + 24.2° (*c* 4, methanol); solubilities in water, 0.81; acetone, 1.73; ethyl acetate, 0.34, in agreement with reported values.[20]

N-Acetyl-L-leucine was obtained similarly in 50–70% yield from foot fractions of the resolution. It has m.p. 185–186°, $[\alpha]D$ −24.1° and other properties in close agreement with reported values for its antipode and for samples prepared from natural leucine.[21,22]

(b) The racemic acid (69.3 g.) was combined in 450 cc. of hot water with an equivalent amount of (−)-α-phenylethylamine and the salts were crystallized in several fractions ranging in rotations erratically from +8.3 to −9.1° (*c* 4, methanol). There was little difference in appearance and solubility of the various fractions upon recrystallization from water or the lower alcohols. Combination of fractions of similar rotations and recrystallization from 3–4 parts of water soon gave 31 g. (52%) of nearly pure dextrorotatory salt, $[\alpha]^{25}D$ +9.9° (*c* 4, methanol); +15.3° (*c* 4, water); solubility in water, 4.7. Decomposition and purification as previously described gave 17 g. (49%) of pure N-acetyl-D-leucine. The antipode was obtained similarly in smaller yield from levorotatory fractions.

(20) P L. M. Synge, *ibid.*, **33**, 1913 (1939).

(21) H. D. DeWitt and A. W. Ingersoll, THIS JOURNAL, **73**, 3359 (1951).
(22) A. J. P. Martin and R. L. M. Synge, *Biochem. J.*, **35**, 91 (1941).

Note:
For the method of hydrolysis to obtain the free acid, see Ref. 253b.

$C_6H_{13}NO_2$

Note:
A resolution procedure for the above compound using acetylmandelyl chloride as resolving agent may be found in S. Berlingozzi, G. Adembri and G. Bucci, Gazz. Chim. Ital. 84, 383 (1954).

$C_6H_{13}NO_2$ Ref. 254f

Note:
Using (+)-carvomenthoxyacetyl chloride as resolving agent, obtained an acid $[\alpha]_D^{20}$ = +9.4° (c, 4, H_2O) and $[\alpha]_D^{22}$ = -15.0° (c, 4, 20% HCl) from the more insoluble salt with m.p. 145-147°C. $[\alpha]_D^{24}$ = +43.02° (c, 2, ethanol).

$C_6H_{13}NO_2$ Ref. 254g

Darstellung von *d*(—)-Leucin: 5 g *racem.* Leucin, 18 g rohe Cholestenonsulfonsäure und 30 ccm absol. Alkohol werden in der angegebenen Weise vereinigt. Nach wenigen Min. erstarrt die Mischung zu einem Krystallbrei, der abgesaugt, in Alkohol gelöst und filtriert wird. Man engt im Vak. auf 20 ccm ein, läßt über Nacht krystallisieren, saugt ab und wäscht mit absol. Äther nach.

Die alkoholischen Mutterlaugen wurden zur Weiterverarbeitung auf *l*(+)-Leucin aufbewahrt.

Das Salz des *d*(—)-Leucins mit der Cholestenonsulfonsäure stellt weiße Nadeln dar. Schmp. 192—193° unter Gasentwicklung.

0.00759 g Sbst.: 0.158 ccm N (19.1°, 757 mm).
$C_{33}H_{57}O_6NS$. Ber. N 2.45. Gef. N 2.42.

Das Leucin wurde aus dem cholestenonsulfonsauren Salz, wie oben beschrieben, isoliert und gereinigt.

$[\alpha]_D^{21}$ = —0.74° × 2.1698/0.0850 × 1 × 1.10 = —17.17° (in 20-proz. HCl).

Das *d*(—)-Leucin wurde nach dieser Methode also optisch reiner erhalten, als es nach der Methode von E. Fischer möglich ist, wahrscheinlich weil in unserem Arbeitsgang keine Racemisierungsgefahr besteht.

Das aus den alkohol. Mutterlaugen isolierte Leucin zeigte:
$[\alpha]_D^{20}$ = +0.41° × 2.3128/0.1138 × 1 × 1.1 = +7.6° (in 20-proz. HCl).

For the experimental details of enzymatic methods for the resolution of this compound, see S. M. Birmbaum, L. Levintow, R. B. Kingsley and J. P. Greenstein, J. Biol. Chem., 194, 455 (1952); K. Michi and H. Tsuda, Bull. Agr. Chem. Soc. Japan, 21, 235 (1957); T. Tosa, T. Mori, N. Fuse and I. Chibata, Enzymologia 32, 153 (1967).

$C_6H_{13}NO_2$

$$CH_3CH_2-CH-CH-COOH$$
$$\qquad\quad | \quad\; |$$
$$\qquad\; H_3C \;\; NH_2$$

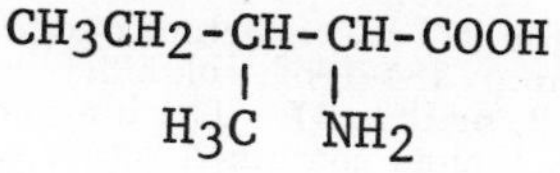

$C_6H_{13}NO_2$ Ref. 255

Resolution of N-Acetyl-DL-isoleucine.—Carefully purified material, m.p. 118–119° (0.68 mole) was dissolved in 1300 cc. of water with (−)-α-fenchylamine. Overnight needles separated and were filtered and washed with a little water. Stepwise concentration of filtrate and washings to about 500 cc. gave similar crops; the liquor was set aside. The moist salt was boiled briefly with 1.5 parts of water (insufficient for complete solution) and filtered after keeping overnight. The dry salt (96.8 g., 87.6%) had $[\alpha]^{25}_D$ −7.5° (c 4, methanol) and was the nearly pure salt of N-acetyl-D-isoleucine. The digestion procedure represents a simplification designed to expedite removal of most of the more soluble salt. Preliminary experiments had shown that the less soluble salt could be fully purified in about 70% yield ($[\alpha]^{25}_D$ −7.7°) by two or more further crystallizations from water or 2-propanol, but considerable evaporation and other manipulation was required. On the other hand the slight impurity of the salt obtained by the abbreviated procedure was of little consequence, since the correspondingly impure N-acetyl-D-isoleucine obtained by decomposition is less soluble than the DL-form and is readily purified.

N-Acetyl-D-isoleucine.—The slightly impure salt (95.8 g.) was decomposed with aqueous alkali and the amine extracted with benzene in the usual manner.[3] The crude acetamino acid was liberated with one equivalent of hydrochloric acid and crystallized in several crops. Brief systematic recrystallization from water gave 41.8 g. (71% based on the DL-form originally taken) of pure N-acetyl-D-isoleucine. This crystallizes in long flattened needles from water or rectangular plates from acetone, m.p. 150–151°; $[\alpha]^{25}_D$ +8.2° (c 2, water); −11.3° (c 4, methanol); −16.3° (c 2, abs. ethanol); −25.2° (c 2, acetone); solubilities in water, 4.15; acetone, 11.76; ethyl acetate, 2.80. For a sample prepared by resolution of isoleucine[6] and acetylation of the D-form Synge[20] reports m.p. 150–151°; $[\alpha]^{20}_D$ −15.6° (c 2.3, ethanol). The mother liquors of the recrystallization gave 12.2 g. of nearly pure N-acetyl-DL-isoleucine, m.p. 120°.

N-Acetyl-L-isoleucine.—The mother liquors of the resolution gave by similar procedures 36.0 g. (54.2%) of pure N-acetyl-L-isoleucine and 27.3 g. of slightly impure DL-form. The active form had m.p. 150–151°; $[\alpha]^{25}_D$ −8.2° (c 2, water); +11.4° (c 4, methanol) and solubilities substantially identical with those of the antipode.

The resolution of N-acetyl-DL-isoleucine was attempted with (−)-α-phenylethylamine. The salt formed moderately soluble fine needles in water, methanol and ethanol but was not resolved. It was nearly insoluble in acetone and ethyl acetate.

For the method of hydrolysis to obtain the free acid, see Ref. 253b.

$C_6H_{13}NO_2$ Ref. 255a

D) Optische Spaltung von Z-DL-Isoleucin. – 1. L-*Tyrosinhydrazidsalz von Z-D-Isoleucin*. 53 g (0,2 Mol) Z-DL-Isoleucin und 39 g (0,2 Mol) L-Tyrosinhydrazid wurden in 800 ml Methanol unter Erwärmen gelöst, filtriert, im Vakuum eingedampft und getrocknet. Das glasige Harz wurde in 300 ml siedendem Alkohol gelöst, auf 50° abgekühlt und mit authentischem L-Tyrosinhydrazidsalz von Z-D-Isoleucin angeimpft. Man liess 1 Std. bei 25° und 24 Std. bei 0° stehen, nutschte das Salz ab, wusch es sorgfältig mit 200 ml Isopropylalkohol und 150 ml Äther und trocknete im Vakuum bei 90°. Ausbeute 42,4 g, Smp. 154–156°, $[\alpha]^{23}_D = +46,5°$ ($c = 2$, in H_2O).

$C_{23}H_{32}O_6N_4$ (460,52) Ber. C 59,98 H 7,00 N 12,17% Gef. C 60,25 H 7,11 N 12,23%

Aus der Mutterlauge kristallisierten 3,8 g einer Zwischenfraktion, $[\alpha]^{23}_D = +66,8°$ ($c = 2$, in H_2O). – Die Mutterlauge wurde im Vakuum eingedampft und der Rückstand gemäss **D3**) zum Dicyclohexylaminsalz von Z-L-Isoleucin weiterverarbeitet.

2. *Dicyclohexylaminsalz von Z-D-Isoleucin*. 42 g L-Tyrosinhydrazidsalz von Z-D-Isoleucin wurden in 40 ml Wasser, 20 ml konz. Salzsäure und 80 ml Essigester suspendiert und im Scheidetrichter mit Essigester extrahiert. Die Essigesterphasen wurden mit Wasser gewaschen, über $MgSO_4$ getrocknet und im Vakuum eingedampft. Man erhielt 23,2 g (96%) öliges Z-D-Isoleucin, das in 45 ml Alkohol gelöst, mit 15,9 g Dicyclohexylamin und zur Vervollständigung der Kristallisation mit 45 ml Essigester versetzt wurde. Man liess 2 Std. bei 0° stehen, nutschte ab, wusch mit Essigester und Äther und trocknete im Vakuum bei 90°. Ausbeute 34,8 g (89%) Dicyclohexylaminsalz von Z-D-Isoleucin, Smp. 156–158°, $[\alpha]^{23}_D = -4,3°$ ($c = 2$, in Alkohol).

$C_{26}H_{42}O_4N_2$ (446,61) Ber. C 69,92 H 9,48 N 6,27% Gef. C 69,99 H 9,42 N 6,40%

3. *Dicyclohexylaminsalz von Z-L-Isoleucin*. 44 g Trockenrückstand der Spaltungsmutterlauge von **D1**) wurden analog **D2**) zersetzt. Man erhielt 22,8 g (90%) optisch noch nicht reines Z-L-Isoleucin. Es wurde in 45 ml Alkohol und 45 ml Essigester gelöst, mit 15,6 g Dicyclohexylamin versetzt, 1 Std. bei 25° aufbewahrt, abgenutscht, mit 40 ml Essigester gewaschen und getrocknet. Ausbeute 32,6 g (85%) Dicyclohexylaminsalz von Z-L-Isoleucin, Smp. 153–154°, $[\alpha]^{23}_D = +4,1°$ ($c = 2$, in Alkohol). Literaturwerte [13]: Smp. 153–154°, $[\alpha]_D = +4,1°$ ($c = 1$, in Alkohol).

4. *Dicyclohexylaminsalz von Z-DL-Isoleucin*. Herstellung nach [13] aus Z-DL-Isoleucin, Smp. 143–145°. Das Salz von Z-DL-Isoleucin war etwas leichter löslich als das entsprechende L- bzw. D-Salz. Der Löslichkeitsunterschied konnte zur optischen Reinigung von partiell racemischem Z-L-Isoleucin bzw. Z-D-Isoleucin ausgenützt werden.

$C_{26}H_{42}O_4N_2$ (446,61) Ber. C 69,92 H 9,48 N 6,27% Gef. C 69,86 H 9,30 N 6,26%

$C_6H_{13}NO_2$ Ref. 255b

For a resolution procedure using ephedrine in the direct resolution of the racemic N-benzyloxycarbonyl derivative of the above compound, see Ref. 20h.

$C_6H_{13}NO_2$ Ref. 255c

EXAMPLE 14

Resolution of Racemic N-Acetylisoleucine

183 g. of N-acetyl-DL-isoleucine and 225 g. of L-threo-2-amino-1-(p-methylmercaptophenyl)-1,3-propanediol were dissolved together in three liters of boiling 95% ethanol and the solution was allowed to stand overnight at 25° C. The solid which had separated from solution was then collected on a filter. There was thus obtained 165 g. of the L-threo-2-amino-1-(p-methylmercaptophenyl)-1,3-propanediol salt of N-acetyl-L-isoleucine which melted at 196–198° C. From this salt there was regenerated N-acetyl-L-isoleucine which melted at 148–150° and had $[\alpha]_D^{25}$ −8.0° (c. 2% in water).

From the filtrate, by evaporation, dissolving the residue in water and seeding, there is obtained the L-threo-2-amino-1-(p-methylmercaptophenyl)-1,3-propanediol salt of N-acetyl-D-isoleucine; from this salt N-acetyl-D-isoleucine, M.P. 148–150° C., is readily obtained.

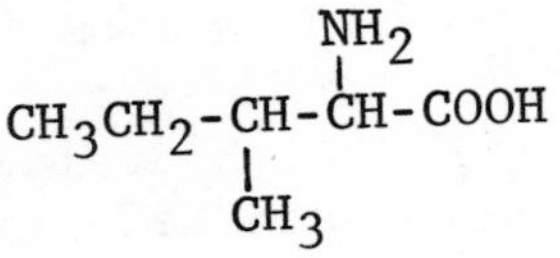

$C_6H_{13}NO_2$ Ref. 256

Resolution of N-Acetyl-DL-alloisoleucine.—Powdered pure material (52 g., 0.3 mole) was suspended in 500 cc. of gently boiling acetone and 113.5 g. (0.3 mole) of quinine trihydrate was added in small portions. The quinine dissolved gradually and the boiling solution became filled with fine needles consisting mainly of the N-acetyl-L-alloisoleucine salt. After boiling for an hour the suspension was allowed to cool and the salt (84 g.) was collected by filtration. A further crop (25 g.) was obtained after concentration to about 100 cc. The pale brown, sirupy liquor was set aside and the solid crops were digested separately in 2 parts of acetone. The head crop resisted solution and gave back 71.4 g., m.p. 181–183°. A sample crystallized readily from water, but the liquor and crystals (m.p. 185–187°) were discolored. A sample (2 g.) dissolved completely on prolonged boiling in 225 cc. of acetone and separated as well formed needles, m.p. 184–186°, solubility in acetone, 0.46; $[\alpha]^{25}D$ +126° (*c* 2, methanol). The intermediate crop (25 g.) and digestion liquors contained rather soluble material which could not be effectively fractionated.

N-Acetyl-L-alloisoleucine.—A part of the sparingly soluble salt (53 g.) was decomposed with alkali and the quinine extracted with chloroform. The recovered acetamino acid (18.8 g.) was crystallized from 3 parts of water in several crops ranging in rotation between +11° and +17° (*c* 4, methanol) and consisting mainly of rosettes of long pointed needles. Systematic recrystallization, assisted by combination of similar fractions and selective seeding, soon gave a head fraction (1.7 g.) of almost pure N-acetyl-DL-alloisoleucine, intermediate fractions (5.5 g.) of mixed forms and 9.8 g. (53%) of pure N-acetyl-L-alloisoleucine, m.p. 155–155.5°; $[\alpha]^{25}D$ −5.3° (*c* 2, water); +17.3° (*c* 4, methanol), values not changed by further crystallization from water or acetone. It forms small rectangular plates from acetone or ethyl acetate; solubilities in water, 4.37; acetone, 6.14; ethyl acetate, 1.20.

N-Acetyl-D-alloisoleucine.—The intermediate fractions and mother liquors of the resolution were freed from acetone and separately decomposed. The acetamino acid fractions recovered from both parts were found to be similar and were combined for fractionation. The crude material (29.6 g.) on fractionation gave 10.3 g. of the DL-form, 5.4 g. of mixed forms and 7.8 g. of pure N-acetyl-D-alloisoleucine, m.p. 154–155.5°; $[\alpha]^{25}D$ −17.2° (*c* 4, methanol); solubilities in water, 4.44; acetone, 5.90; ethyl acetate, 1.22.

Attempted resolutions with other bases were unsuccessful. Brucine, cinchonine and (−)-desoxyephedrine formed non-crystalline salts. (−)-α-Fenchylamine, (−)-α-phenylethylamine and (−)-ephedrine formed well-crystallized salts in water but no resolution occurred in this or other common solvents.

For the experimental details of enzymatic methods for the resolution of isoleucine and alloisoleucine, see "Chemistry of the Amino Acids," Vol. 3, J.P. Greenstein and M. Winitz, Eds., pgs. 2064-2085, John Wiley and Sons, Inc. (1961).

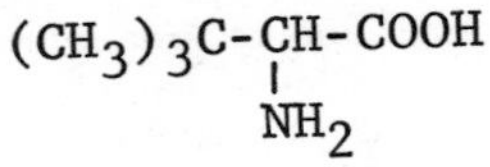

$C_6H_{13}NO_2$ Ref. 257

Formyl-d,l-pseudoleucin. 90 g d,l-Pseudoleucin wurden mit 140 g wasserfreier Ameisensäure 6 Stunden unter Rückfluß auf dem Wasserbade erhitzt. Es entstand eine vollkommen klare Lösung. Der nach Abkühlung auskrystallisierte Anteil wurde abgesaugt, mit Ameisensäure gewaschen und im Vakuum-Exsiccator über Alkali getrocknet. Das Filtrat wurde im Vakuum bis zur Trockne eingedampft. Beim Eindampfen in offener Schale auf dem Wasserbade treten Verluste durch Verflüchtigung ein. Die Formylierung wurde mit dem Eindampfungsrückstand wiederholt. Der ganze Prozeß wurde 3 mal durchgeführt. Aus der schließlich verbleibenden Mutterlauge wurden die letzten Mengen an Formylverbindung durch teilweises Einengen im Vakuum und Kühlen fraktionsweise isoliert. Ausbeute 101 g vom Schmelzp. 210°, nach vorherigem Sintern bei 206°. Die letzten Fraktionen (3 g) schmolzen unscharf bei 200—205°. Umkrystallisiert wurde aus Alkohol: längliche, säulenartige, rechteckige Täfelchen, schwer löslich in kaltem Wasser, leichter in Alkohol.

0,0670 g verbr. 4,2 ccm n/10-H_2SO_4.

Ber. für $C_7H_{13}O_3N$ (159,11) 8,81 % N, gef. 8,78 % N.

Gewinnung von Formyl-l(+)— und d(−)-pseudoleucin. 104 g Formyl-d,l-pseudoleucin wurden in 2,5 Liter kochendem absolutem Alkohol gelöst, dann Zusatz einer Lösung von 260 g reinem Brucin (wasserfrei) in 5 Liter heißem, absolutem Alkohol; das nach 48 stündigem Stehen im Eisschrank ausgeschiedene krystalline Brucinsalz (sternartige Büschel, schwer löslich in absolutem Alkohol, leicht löslich in Wasser) wurde abgenutscht, mit kaltem Alkohol gewaschen und getrocknet: 180 g, Schmelzp. 195° (Brucinsalz des Formyl-l(+)-pseudoleucins). Die Mutterlauge wurde im Vakuum eingedampft, der Rückstand (Brucinsalz des Formyl-d-(−)-pseudoleucins) in 1750 ccm Wasser gelöst, mit 400 ccm n-Natronlauge versetzt und das Filtrat nach Ausschütteln mit etwas Chloroform mit 50 ccm 5n-HCl versetzt. Das nach Abkühlen auskrystallisierte Formyl-d(−)-pseudoleucin wurde abgesaugt, das Filtrat nach Einengen mit weiteren 25 ccm 5 n-HCl versetzt und so eine zweite Krystallfraktion gewonnen. Gesamtausbeute an Formyl-d(—)-pseudoleucin nach Waschen mit kaltem Wasser und Trocknen 45 g. Schwer löslich in Wasser (etwa 1 : 500, bei 20° lösen sich in 20 ccm Wasser 0,0405 g), unlöslich in Aceton, wenig löslich in Alkohol. Krystallisiert aus Wasser in Prismen und auch Blättchen, spaltet beim Kochen mit Wasser nur schwer den Formylrest ab. In entsprechender Weise wurde das zuerst auskrystallisierte Brucinsalz zerlegt. Ausbeute 45 g Formyl-l(+)-pseudoleucin.

Wegen der geringen Löslichkeit der beiden Formylverbindungen in Wasser war eine genaue Bestimmung des Drehungsvermögens in diesem nicht möglich. Formyl-l(+)-pseudoleucin dreht nach rechts und Formyl-d(−)-pseudoleucin nach links. Formyl-pseudoleucin wurde durch Zufügung einer stöchiometrischen Menge n-NaOH in das Na-Salz überführt. Vgl. die Analysen- und Drehwerte in der folgenden Tabelle.

l(+) und d(−)-Pseudoleucin. Die beiden Formylverbindungen wurden mit der 10 fachen Menge 10%iger HBr 1½ Stunden unter Rückfluß gekocht und die Lösung zur Trockne verdampft. Das Hydrobromid der Aminosäure wurde in absolutem Alkohol gelöst und Ammoniak bis zum Neutralpunkt eingeleitet. Dabei schieden sich die beiden optisch-aktiven Aminosäuren in gallertiger, gequollener Form aus. Es erstarrte schließlich der ganze Kolbeninhalt in dieser Form. Nach einigem Stehen, besonders beim Erwärmen, erfolgte Krystallisation. Die Aminosäure ließ sich nunmehr

$C_7H_{13}O_3N$ 159,11	Analyse				Spez. Drehung 10 ccm-Kolben						
	Subst. g	ccm n/10-H_2SO_4 verbr.	N ber. %	N gef. %	Subst. g	ccm n-NaOH	g Na-Salz	Na-Salz, Konz. c	t °	α_D^t °	$[\alpha]_D^t$ °
Formyl-l(+)	0,1415	8,8	8,8	8,7	0,5242	3,30	0,597	5,97	21	+1,90	+31,8
Formyl-d(−)	0,0634	4,0	8,8	8,8	0,5273	3,32	0,600	6,00	22	−1,92	−32,0

filtrieren, was zuvor nicht möglich war. Der Filterrückstand wurde mit absolutem Alkohol bis zur Halogenfreiheit ausgewaschen.

Drehungsvermögen der wäßrigen Lösung von l(+)-Pseudoleucin:

0,0779 g Subst., aufgefüllt auf 1,5 ccm, $c = 5{,}19$, $\alpha_D^{20^\circ} = +0{,}52^\circ$ im dm-Rohr. $[\alpha]_D^{20^\circ} = +10{,}01^\circ$.

Von d(—)-Pseudoleucin:

0,0926 g Subst. zu 2 ccm gelöst, $c = 4{,}63$, $\alpha_D^{28,5^\circ} = -0{,}47^\circ$, $[\alpha]_D^{28,5^\circ} = -10{,}15^\circ$.

Drehungsvermögen in 20%iger Salzsäure: l(+)-Pseudoleucin: 0,0605 g Subst. zu 1,5 ccm gelöst, $c = 4{,}03$, $\alpha_D^{29^\circ} = -0{,}33^\circ$, $[\alpha]_D^{29^\circ} = -8{,}19^\circ$.

d(−)-Pseudoleucin: 0,0957 g Subst., zu 2 ccm gelöst, $c = 4{,}78$, $\alpha_D^{28,5^\circ} = +0{,}40^\circ$, $[\alpha]_D^{28,5^\circ} = +8{,}36^\circ$.

$C_6H_{13}NO_2$ Ref. 257a

***N*-*p*-Toluenesulfonyl-*tert*-leucine.** Under dry conditions, 1.45 g (7.61 mmol) of *p*-toluenesulfonyl chloride was added during 30 min to a stirred solution of 0.65 g (5.0 mmol) of *tert*-leucine, 2.2 mL of triethylamine, 4.0 mL of THF, and 8.0 mL of H_2O at 25 °C. The reaction mixture then was stirred at 25 °C for 1 h and rotary evaporated. The residue was added to 10 mL of H_2O, and the resulting solution was extracted twice with 10-mL portions of ether and acidified with concentrated HCl to precipitate 1.02 g (72%) of crude sulfonamide which was recrystallized from 4:1 (v/v) C_2H_5OH–H_2O to give *N*-*p*-toluenesulfonyl-*tert*-leucine; mp 230–231 °C (lit.[46] 226 °C).

Anal. Calcd for $C_{13}H_{19}NO_4S$: C, 54.72; H, 6.71. Found: C, 54.63; H, 6.53.

Resolution of *N*-*p*-Toluenesulfonyl-*tert*-leucine. A solution of 5.00 g (1.27 mmol) of brucine alkaloid, mp 176–177 °C, $[\alpha]_D^{25}$ −127° ($CHCl_3$), and 3.62 g (1.27 mmol) of (±)-*N*-*p*-toluenesulfonyl-*tert*-leucine in 375 mL of acetone was concentrated at its boiling point until salt began to precipitate at a volume of ca. 30 mL. Then 20 mL of acetone was added, and the solution was allowed to cool to 25 °C. A crop of 5.1 g of poorly defined but hard crystals, mp 191–198 °C, $[\alpha]_{546}^{25}$ −35.7° (*c* 0.387, acetone), was collected. This material was recrystallized from 75 mL of 4:1 (v/v) acetone–C_2H_5OH to give 2.5 g of very hard crystals, mp 201–203 °C, $[\alpha]_{546}^{25}$−36.1° (*c* 0.274, acetone), which was further recrystallized from 60 mL of 5:1 (v/v) acetone–C_2H_5OH to give 1.1 g of crystals, mp 202–203 °C, $[\alpha]_{546}^{25}$ −32.7° (*c* 0.416, acetone). A mixture of 420 mg of this last material and 20 mL of 9% aqueous NaOH was heated and then cooled and filtered to remove resultant brucine. The filtrate was extracted twice with 30-mL portions of CH_2Cl_2, acidified with concentrated HCl, and filtered to give crude sulfonamide, mp 241–242 °C, $[\alpha]_{546}^{25}$ +51.8° (*c* 1.00, absolute C_2H_5OH), which gave a single spot by TLC analysis on silica gel. Recrystallization of this material from aqueous C_2H_5OH gave (+)-*N*-*p*-toluenesulfonyl-*tert*-leucine, mp 242–243 °C, $[\alpha]_{546}^{25}$ +52.7° (*c* 1.01, absolute C_2H_5OH).

Anal. Calcd for $C_{13}H_{19}NO_4S$: C, 54.72; H, 6.71. Found: C, 54.82; H, 6.79.

$C_6H_{13}NO_2$ **Ref. 257b**

Racematspaltung von II: 16.4 g (0.113 Mol) des *racem. Esters II* wurden mit einer Lösung von 17.7 g (0.047 Mol) *Dibenzoyl-(R)-weinsäure-hydrat* in 190 ccm absol. Äthanol vermischt und 24 Stdn. bei Raumtemperatur zur Kristallisation stehengelassen. Nach Absaugen des Kristallisats wurde die Mutterlauge mit 45 ccm absol. Äther verdünnt, erneut 48 Stdn. stehengelassen und die resultierende Nachfraktion nach einmaligem Umkristallisieren aus absol. Äthanol mit der Hauptfraktion vereinigt. Nochmaliges Umkristallisieren des gesammelten *(S)-Ester-dibenzoyl-(R)-hydrogentartrates* lieferte 6.85 g (13.6 mMol) reines Salz, Schmp. ca. 185–190° (Zers.), $[\alpha]_D^{23}$: −83.7° (*c* = 1, in CH_3OH). Dieses wurde gemeinsam mit 3.15 g (6.26 mMol) einer sterisch etwas weniger reinen Nachfraktion nach l. c.[27] mit HCl in CH_3OH/Äther in das Hydrochlorid (3.3 g, $[\alpha]_D^{22}$: +14.0°) übergeführt, das nach dreimaligem Umkristallisieren aus Methanol/Äther rein war. Ausb.: 2.35 g (23% d. Th.) *(S)-tert.-Leucin-methylester-hydrochlorid*, Schmp. 175–185° (Einschlußküvette), $[\alpha]_D^{26}$: +17.6° (*c* = 1, in CH_3OH).

$C_7H_{16}NO_2]Cl$ (181.7) Ber. C 46.27 H 8.87 N 7.71 Gef. C 46.62 H 8.72 N 7.65

Aus der Mutterlauge der Erstkristallisationen des Dibenzoyl-*(R)*-tartrates wurde mit viel Äther eine weitere Nachfraktion gefällt und die Endmutterlauge durch Schütteln mit 30-proz. K_2CO_3-Lösung von Dibenzoylweinsäureresten befreit. Die Destillation der über K_2CO_3 getrockneten Ätherlösung lieferte 9.41 g (65 mMol) sterisch unreines *(−)-II*, das wie oben beschrieben mit 14.5 g (38.5 mMol) *Dibenzoyl-(S)-weinsäure* umgesetzt wurde. Aus dem umkristallisierten Dibenzoyltartrat erhielt man 2.6 g (25% d. Th.) reines *(R)-tert.-Leucin-methylester-hydrochlorid*, $[\alpha]_D^{21}$: −18.0° (*c* = 1, in CH_3OH). Die freien Aminoester (aus den Hydrochloriden mit K_2CO_3 in 90-proz. Ausb.) zeigten folgende Daten:

(S)-II: Sdp.$_{16}$ 69°, $[\alpha]_D^{23}$: +64.5° (*c* = 1, in absol. Äthanol), *(R)-II:* Sdp.$_{19}$ 71–73°, $[\alpha]_D^{25}$: −66.0° (*c* = 1, in absol. Äthanol), n_D^{20} 1.4324.

$C_7H_{15}NO_2$ (145.2) Ber. N 9.65 Gef. N 9.61

(R)- und (S)-tert.-Leucin (I): Der *aktive Ester II* wurde 20 Stdn. mit überschüss. 20-proz. *Salzsäure* gekocht, i. Vak. eingedampft und das in Wasser gelöste Hydrochlorid über einen Anionenaustauscher (Wofatit L 150) filtriert. Zur Entfernung nicht verseifter Aminosäure-esteranteile schickte man die Lösung danach über einen schwach sauren Kationenaustauscher (Wofatit CP 300) und dampfte sie i. Vak. zur Trockne ein. Aus 2.2 g *(R)*(−)-II (90% sterisch rein) erhielt man 1.6 g (80% d. Th.) *(R)(+)-I*, $[\alpha]_D^{19}$: + 9.2° (*c* = 1, in H_2O). Analog lieferte 93% reines *(S)*(+)-II die *(S)(−)-Aminosäure I* mit $[\alpha]_D^{21}$: −9.4° (*c* = 1, in H_2O).

27) G. Losse und H. Jeschkeit, Chem. Ber. 90, 1275 [1957].

$C_6H_{13}NO_2$ **Ref. 257c**

*Optical Resolution of N-Z-*DL*-t-Leucine(III).* Thirty-eight mmol of III[8] and quinine (or quinidine) were dissolved in hot ethanol (90 ml), and the precipitated white crystals recrystallized from ethanol four times. The purified salt was suspended in ethyl acetate, and shaken with 2 M HCl to remove the base, and subjected to the usual work-up. The optically active *t*-leucine was obtained by hydrogenation of the corresponding *N*-Z-derivative.

N-Z-D-*t*-Leucine quinine salt: mp 166—167 °C, $[\alpha]_D^{20}$ −106.5° (*c* 1.0, MeOH). Found: C, 69.46; H, 7.47; N, 7.12%. Calcd for $C_{34}H_{43}N_3O_6$: C, 69.24; H, 7.35; N, 7.13%.

N-Z-D-*t*-Leucine: syrup, $[\alpha]_D^{20}$ +5.9° (*c* 1.0, MeOH).

N-Z-D-*t*-Leucine dicyclohexylamine (DCHA) salt: mp 165.5—166 °C (from EtOH–Et_2O), $[\alpha]_D^{20}$ +8.5° (*c* 1.0, MeOH). Found: C, 69.90; H, 9.56; N, 6.04%. Calcd for $C_{26}H_{42}N_2O_4$: C, 69.92; H, 9.48; N, 6.27%.

D-*t*-Leucine: mp 250—252 °C (sublime), $[\alpha]_D^{21}$ +10.5° (*c* 1.0, H_2O), $[\alpha]_D^{26}$ −29.1° (*c* 1.0, AcOH) (lit,[9] $[\alpha]_D^{20}$ +10.01° (*c* 5.19, H_2O) (*cf.* Refs. 3 and 10)). Found: C, 54.74; H, 9.87; N, 10.35%.

N-Z-L-*t*-Leucine quinidine salt: mp 134—136 °C (*ca.* 98 °C sinter), $[\alpha]_D^{20}$ +127.7° (*c* 1.0, MeOH).

N-Z-L-*t*-Leucine: syrup, $[\alpha]_D^{20}$ −6.0° (*c* 1.0, MeOH).

N-Z-L-*t*-Leucine DCHA salt: mp 166—166.5 °C (from EtOH–Et_2O), $[\alpha]_D^{20}$ −8.7° (*c* 1.0, MeOH)(lit,[11] mp 165—168 °C, $[\alpha]_D$ −8.4° (*c* 0.59, MeOH)). Found: C, 69.79; H, 9.35; N, 5.98%.

L-*t*-Leucine: mp 250—252 °C (sublime), $[\alpha]_D^{21}$ −10.9° (*c* 1.0, H_2O), $[\alpha]_D^{26}$ +30.0° (*c* 1.0, AcOH) (lit, mp 250 °C (sublime),[12] $[\alpha]_D^{28.5}$ −10.15° (*c* 4.63, H_2O)[9] (*cf.* Refs. 3, 10, and 13), $[\alpha]_D^{25}$ +36.0° (*c* 2, AcOH)[3]). Found: C, 54.77; H, 9.91; N, 10.35%.

3) J. P. Greenstein and M. Winitz, "Chemistry of the Amino Acids," John Wiley & Sons, Inc., New York (1961), Vol. 3, p. 2580.

8) N. Izumiya, S.-C. J. Fu, S. M. Birnbaum, and J. P. Greenstein, *J. Biol. Chem.*, **205**, 221 (1953).

9) E. Abderhalden, W. Faust, and E. Haase, *Hoppe-Seyler's Z. Physiol. Chem.*, **228**, 187 (1934).

10) H. Pracejus and S. Winter, *Chem. Ber.*, **97**, 3173 (1964).

11) J. Pospíšek and K. Bláha, *Collect. Czech. Chem. Commun.*, **42**, 1069 (1977).

12) W. Steglich, E. Frauendorfer, and F. Weygand, *Chem. Ber.*, **104**, 687 (1971).

13) T. Tanabe, S. Yajima, and M. Imaida, *Bull. Chem. Soc. Jpn.*, **41**, 2178 (1968).

Note:

Compound III is the N-benzyloxy-carbonyl derivative of DL-t-leucine.

$C_6H_{13}NO_2$

$$(CH_3)_2CH\text{-}CH(NHCH_3)\text{-}COOH$$

Ref. 258

DL-N-3,5-Dinitrobenzoyl-N-methylvaline. 45 g of 3,5-dinitrobenzoyl chloride and 180 ml of 1 N NaOH were added in small portions with cooling to a solution of 20 g of DL-N-methylvaline in 160 ml of 1 N NaOH. The mixture was shaken for 1 hour at 20° and filtered. The filtrate was acidified to Congo Red with concentrated hydrochloric acid. The precipitated oil (48 g) completely crystallized out on standing. After three crystallizations from absolute alcohol we obtained 29 g (58%) of DL-N-3,5-dinitrobenzoyl-N-methylvaline, m.p. 200-202°. Found: C 47.99; H 4.61; N 12.98%. $C_{13}H_{15}O_7N_3$. Calculated: C 48.00; H 4.65; N 12.92%.

Resolution of DL-N-3,5-Dinitrobenzoyl-N-methylvaline into its Optical Isomers. a) 25 g of DL-N-3,5-dinitrobenzoyl-N-methylvaline and 16.3 g of the L-amine were dissolved at the boil in 200 ml of a mixture of ethanol and ethyl acetate (1:9); after one day the precipitated crystals (21 g) were filtered off and crystallized twice from a 1:1 mixture of alcohol and benzene. We obtained 15.15 g (73%) of the salt of D-N-3,5-dinitrobenzoyl-N-methylvaline with the L-amine (salt III), m.p. 146-147°; $[\alpha]_D^{20}$ +31.0° (c 1.0 in alcohol). Found: C 49.34; H 5.32; N 12.82%. $C_{22}N_{27}O_{11}N_5$. Calculated: C 49.16; H 5.06; N 13.03%.

b) 2 ml of concentrated hydrochloric acid was added to a suspension of 10 g of the salt III in 30 ml of water; the precipitated oil crystallized when rubbed out. The crystals were filtered off, washed with water, and dried in a vacuum over P_2O_5. We obtained 5.9 g (97%) of D-N-3,5-dinitrobenzoyl-N-methylvaline, m.p. 134-135° (from 50% alcohol); $[\alpha]_D^{20}$ +55.0° (c 1.0 in methanol). Found: C 48.06; H 4.64; N 12.81%. $C_{13}H_{15}O_7N_3$. Calculated: C 48.00; H 4.65; N 12.92%.

c) The mother solution remaining after the separation of the salt III was vacuum-evaporated, and 30 ml of water and 6 ml of concentrated hydrochloric acid were added to the residue; the precipitate formed was filtered off and vacuum-dried. The resulting crystals (11.8 g) and 7.9 g of the D-amine were dissolved in 70 ml of a hot 1:9 mixture of ethanol and ethyl acetate; the salt that precipitated with cooling was filtered off and crystallized twice from a 1:1 mixture of ethanol and benzene. We obtained 14.9 g (72%) of the salt formed by L-N-3,5-dinitrobenzoyl-N-methylvaline with the D-amine (salt IV), m.p. 146-147°; $[\alpha]_D^{20}$ -31.0° (c 1.0 in alcohol).

d) By the method of Expt. (b) from 8 g of the salt IV we prepared 4.66 g (96%) of L-N-3,5-dinitrobenzoyl-N-methylvaline, m.p. 134-135° (from 50% alcohol); $[\alpha]_D^{20}$ −55.0° (c 1.0 in methanol). Found: C 47.71; H 4.59; N 12.90%. $C_{13}H_{15}O_7N_3$. Calculated: C 48.00; H 4.65; N 12.92%.

Hydrolysis of the 3,5-Dinitrobenzoyl Derivatives of D- and L-N-Methylvalines. A mixture of 4.2 g of D-N-3,5-dinitrobenzoyl-N-methylvaline and 40 ml of 20% hydrochloric acid was boiled for one hour. The mixture was cooled, and the precipitate of 3,5-dinitrobenzoic acid (2.75 g; 100%) was filtered off; the filtrate was vacuum-evaporated to dryness. The residue was crystallized from a mixture of alcohol and ether, and we obtained 2.22 g (100%) of D-N-methylvaline hydrochloride, m.p. 171-173°; $[\alpha]_D^{20}$ -26.5° (c 1.0 in alcohol). Found: C 43.00; H 8.54; N 8.25; Cl 21.18%. $C_6H_{14}O_2NCl$. Calculated: C 42.99; H 8.42; N 8.36; Cl 21.15%.

Analogously, from 3 g of L-N-3,5-dinitrobenzoyl-N-methylvaline we obtained 1.55 g (100%) of L-N-methylvaline hydrochloride, m.p. 171-173° (from a mixture of alcohol and ether); $[\alpha]_D^{20}$ +26.5° (c 1.0 in alcohol).

$C_6H_{13}NO_2$

Ref. 258a

DL-N-(Benzyloxycarbonyl)-N-methylvaline. 62 ml of 4 N NaOH and a solution of 44 g of benzyl chloroformate in 100 ml of tetrahydrofuran were added in approximately equal portions with cooling and vigorous shaking to a solution of 32.8 g of DL-N-methylvaline in 125 ml of 2 N NaOH. The mixture was shaken further for 30 minutes and left overnight at 20°; tetrahydrofuran was vacuum-distilled off. The residue was acidified to Congo Red with concentrated hydrochloric acid, and the oil liberated was extracted with ether; the ethereal solution was washed with water and extracted with saturated sodium bicarbonate solution. The bicarbonate extract was acidified with concentrated hydrochloric acid, the product was extracted with ether, and the combined ethereal solution was washed with water, dried with magnesium sulfate, and evaporated. We obtained 64.5 g (97%) of DL-N-(benzyloxycarbonyl)-N-methylvaline as a colorless oil.

Resolution of DL-N-(Benzyloxycarbonyl)-N-methylvaline into its Optical Isomers. a) 101 g of DL-N-(benzyloxycarbonyl)-N-methylvaline and 72 g of the L-amine were dissolved in 400 ml of hot absolute alcohol, and the solution was left at 0° for 70 hours. The crystals precipitated (73.8 g) were filtered off and crystallized twice from double the amount of absolute alcohol. We obtained 59.2 g (65%) of the salt formed by L-N-(benzyloxycarbonyl)-N-methylvaline with the L-amine (salt V), m.p. 147-149; $[\alpha]_D^{20}$ -34.0° (c 1.0 in alcohol). Found: C 57.78; H 6.53; N 8.78%. $C_{23}H_{31}O_8N_3$. Calculated: C 57.85; H 6.54; N 8.80%.

b) 100 ml of ether and 10 ml of concentrated hydrochloric acid were added to a suspension of 39.5 g of the salt V in 50 ml of water, the mixture was shaken until solution was complete, the ether layer was separated, and the aqueous layer was extracted with ether; the combined ethereal solution was dried with magnesium sulfate and evaporated. After some time the residual oil crystallized out completely. After crystallization from a mixture of benzene and hexane we obtained 21.9 g (99%) of L-N-(benzyloxycarbonyl)-N-methylvaline, m.p. 68.5-69°; $[\alpha]_D^{20}$ -90.6° (c 0.5 in benzene)*. Found: C 63.32; H 7.10; N 5.50%. $C_{14}H_{19}O_4N$. Calculated: C 63.38; H 7.22; N 5.28%.

c) The mother solution remaining after the separation of the salt V was vacuum-evaporated, and 200 ml of water, 25 ml of concentrated hydrochloric acid, and 200 ml of ether were added to the residue. The mixture was shaken vigorously, the ether layer was separated, and the aqueous layer was extracted with ether; the combined ethereal solution was dried with magnesium sulfate and evaporated. We obtained 58 g of oil, which consisted of D-N-(benzyloxycarbonyl)-N-methylvaline with its L-isomer as impurity. The oil was dissolved in 250 ml of absolute alcohol, and 37.8 g of the D-amine was added. The mixture was heated until solution was complete, and the resulting solution was left at 0° for 70 hours. The precipitated crystals (64.5 g) were filtered off and crystallized twice from absolute alcohol. We obtained 55 g (61%) of the salt formed by D-N-(benzyloxycarbonyl)-N-methylvaline with the D-amine (salt VI), m.p. 147-149° and $[\alpha]_D^{20}$ +33.8° (c 1.0 in alcohol).

d) By the method of Expt. (b) from 49.1 g of the salt VI we obtained 27 g (99%) of D-N-(benzyloxycarbonyl)-N-methylvaline, m.p. 68.5-69° (from a mixture of benzene and hexane); $[\alpha]_D^{20}$ +90.0° (c 0.5 in alcohol). Found: C 63.66; H 7.27; N 5.57%. $C_{14}H_{19}O_4N$. Calculated: C 63.38; H 7.22; N 5.28%.

Hydrogenolysis of Benzyloxycarbonyl Derivatives of L- and D-N-Methylvalines. 2.05 g of L-N-(benzyloxycarbonyl)-N-methylvaline was hydrogenated in 30 ml of alcohol with passage of hydrogen and stirring in presence of 10 mg of palladium oxide at 20° for three hours. 30 ml of water was added, catalyst was filtered off, and the filtrate was vacuum-evaporated to dryness. The solid residue was dissolved in the least possible amount of water, and the amino acid was precipitated with acetone. We obtained 0.91 g (90%) of L-N-methylvaline, m.p. 278° (sublimes); $[\alpha]_D^{20}$ +20.4° (c 1.0 in water).

Analogously, from 1.65 g of D-N-(benzyloxycarbonyl)-N-methylvaline we obtained 0.73 g (90%) of D-N-methylvaline, m.p. 278° (sublimes); $[\alpha]_D^{20}$ -19.8° (c 0.6 in water).

* L-N-(benzyloxycarbonyl)-N-methylvaline was prepared by Brockmann in the course of the synthesis of actinomycin C_3; however, the author gives neither the constants of this compound, nor the method by which it was prepared [20].

$C_6H_{13}NO_2$ Ref. 258b

DL-N-(p-Nitrobenzyloxycarbonyl)-N-methylvaline. From 13.1 g of DL-N-methylvaline in 30 ml of 4 N NaOH, 27 g of p-nitrobenzyl chloroformate in 50 ml of tetrahydrofuran, and 30 ml of 4 N NaOH, by the procedure of the experiment on the preparation of DL-N-(benzyloxycarbonyl)-N-methylvaline, we obtained 30.8 g (99%) of DL-N-(p-nitrobenzyloxycarbonyl)-N-methylvaline, m.p. 110-112° (from a mixture of benzene and hexane). Found: C 54.26; H 5.62; N 8.97%. $C_{14}H_{18}O_6N_2$. Calculated: C 54.19; H 5.85; N 9.03%.

Resolution of DL-N-(p-Nitrobenzyloxycarbonyl)-N-methylvaline into its Optical Isomers. a) 16.35 g of DL-N-(p-benzyloxycarbonyl)-N-methylvaline and 10.6 g of the L-amine were dissolved with heating in 200 ml of absolute alcohol, and the solution was left at 0° for seven days. The crystals precipitated (11.7 g) were filtered off and crystallized twice from three times the amount of absolute alcohol. We obtained 8.4 g (64%) of the salt formed by L-N-(p-nitrobenzyloxycarbonyl)-N-methylvaline with the L-amine (salt VII), m.p. 147-149; $[\alpha]_D^{20}$ −31.0° (c 1.0 in methanol). Found: C 52.63; H 5.74; N 10.40%. $C_{23}H_{30}O_{10}N_4$. Calculated: C 52.87; H 5.79; N 10.72%.

b) 50 ml of ether and 1.5 ml of concentrated hydrochloric acid were added to a suspension of 6.95 g of the salt VII in 30 ml of water; the mixture was shaken until solution was complete, the ether layer was separated, and the aqueous layer was extracted with ether; the combined ethereal solution was dried with magnesium sulfate and evaporated. We obtained 4.09 g (99%) of L-N-(p-nitrobenzyloxycarbonyl)-N-methylvaline, m.p. 83.5-84° (from a mixture of benzene and hexane); $[\alpha]_D^{20}$ −85° (c 1.0 in benzene). Found: C 54.26; H 5.97; N 9.13%. $C_{14}H_{18}O_6N_2$. Calculated: C 54.19; H 5.85; N 9.03%.

c) The mother solution remaining after the separation of the salt VII was vacuum-evaporated, and to the oily residue we added 20 ml of water, 3 ml of concentrated hydrochloric acid, and 50 ml of ether. The mixture was shaken vigorously, the ether layer was separated, and the water layer was extracted with ether; the combined ethereal solution was dried with magnesium sulfate and evaporated. The residue (8.8 g), which consisted of D-N-(p-nitrobenzyloxycarbonyl)-N-methylvaline with some L-isomer impurity, was dissolved in 70 ml of absolute alcohol, and 5.3 g of the D-amine was added. The mixture was heated until solution was complete, and the solution was left at 0° for seven days. The precipitated crystals (10.1 g) were filtered off and crystallized twice from absolute alcohol. We obtained 8.2 g (63%) of the salt formed by D-N-(p-nitrobenzyloxycarbonyl)-N-methylvaline with the D-amine (salt VIII), m.p. 147-149°; $[\alpha]_D^{20}$ +30.5° (c 1.0 in methanol).

d) Under the conditions of Expt. (b) from 6.65 g of the salt VIII we obtained 3.9 g (99%) of D-N-(p-nitrobenzyloxycarbonyl)-N-methylvaline, mp. 83.5-84° (from a mixture of benzene and hexane): $[\alpha]_D^{20}$ +84° (c 1.0 in benzene). Found: C 54.49; H 5.99; N 9.17%. $C_{14}H_{18}O_6N_2$. Calculated: C 54.19; H 5.85; N 9.03%.

Hydrogenolysis of the p-Nitrobenzyloxycarbonyl Derivatives of D- and L-N-Methylvalines. 1.2 g of L-N-(p-nitrobenzyloxycarbonyl)-N-methylvaline was hydrogenated in 30 ml of a 1:1 mixture of alcohol and acetic acid with passage of hydrogen and stirring with a magnetic stirrer in presence of 20 mg of palladium oxide at 20° for three hours. Catalyst was filtered off, and the filtrate was vacuum-evaporated to dryness; the solid residue was vacuum-dried over P_2O_5 and KOH and dissolved in the least possible amount of water; the amino acid was precipitated with acetone. We obtained 0.45 g (88%) of L-N-methylvaline, m.p. 280° (sublimes); $[\alpha]_D^{20}$ +19.1° (c 1.0 in water).

Analogously, from 0.9 g of D-N-(p-nitrobenzyloxycarbonyl)-N-methylvaline we obtained 0.33 g (87%) of D-N-methylvaline, m.p. 278° (sublimes); $[\alpha]_D^{20}$ -19.3° (c 0.6 in water).

$HO(CH_2)_4CH(NH_2)COOH$

$C_6H_{13}NO_3$ Ref. 259

The preparation of optically active 5-substituted hydantoins (5-hydantoins) can be readily accomplished when the appropriate amino acid precursors are available in optically pure form.[1] In this reaction cyanate is added to the α-amino carboxylic acid, producing a carbamate intermediate. The carbamate is then cyclized to the hydantoin by heating in acid solutions. Suzuki *et al.* employed this reaction in the syntheses of L-hydantoins from 17 naturally occurring L-amino acids.[4] However, when the optically pure precursors are not available, the racemic 5-hydantoins must be resolved by conventional means. Such resolution is complicated by the fact that monosubstituted 5-hydantoins undergo spontaneous racemization in mildly alkaline solutions.[2]

We have developed a four step process which produces either the D or L stereoisomer directly from a racemic 5-hydantoin. This process takes advantage of the fact that 5-hydantoins readily yield racemic amino acids upon alkaline hydrolysis.[3] These racemic amino acids may then be stereospecifically oxidized by the application of commercially available amino acid oxidase enzymes. D-Amino acid oxidase (I.U.B. #1.4.3.3) reacts with most D-amino acids, but leaves the L-isomer unchanged, while L-amino acid oxidase (I.U.B. #1.4.3.2) acts upon L-amino acids without affecting the D-isomer. Both enzymes have a very high degree of optical specificity, but are relatively insensitive to the nature of the R-group in the amino acid.[5]

After one stereoisomer of the amino acid has been digested by the appropriate enzyme (D-amino acid oxidase for the synthesis of L-hydantoins; L-amino acid oxidase for the synthesis of D-hydantoins) the unreacted amino acid is recovered by ion exchange chromatography. Finally, the optically pure hydantoin is resynthesized by the method of Dakin.[1] The entire process is outlined diagramatically in Fig. 1.

This process is especially useful because of its ability to form both the D- and L-hydantoin for such purposes as comparing their biological activities. Furthermore, the broad substrate specificities of the stereospecific amino acid oxidase enzymes mean that this procedure should be applicable to the preparation of a variety of monosubstituted 5-hydantoins. The application of this process to the synthesis of L-5-(4-hydroxybutyl)hydantoin (L-HBH) is reported in this paper.

* To whom reprint requests should be sent.

B. *Oxidation of* D-*AHCA*. Stereospecific D-amino acid oxidase (DAO) was used to destroy the D-AHCA present in the racemic mixture, leaving optically pure L-AHCA in solution with α-keto-ε-hydroxy caproic acid.

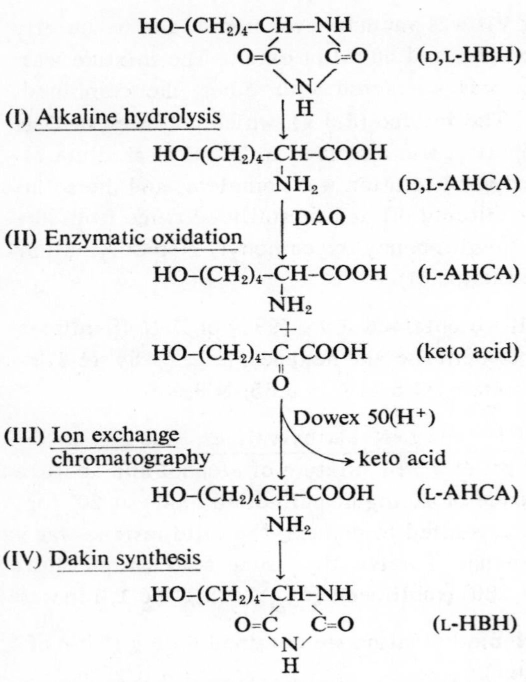
HO-(CH2)4-CH—NH
O=C C=O (D,L-HBH)
N
H
(I) Alkaline hydrolysis
HO-(CH2)4-CH-COOH
NH2 (D,L-AHCA)
DAO
(II) Enzymatic oxidation
HO-(CH2)4-CH-COOH (L-AHCA)
NH2
+
HO-(CH2)4-C-COOH (keto acid)
O
Dowex 50(H+)
(III) Ion exchange chromatography
keto acid
HO-(CH2)4CH-COOH (L-AHCA)
NH2
(IV) Dakin synthesis
HO-(CH2)4-CH-NH
O=C C=O (L-HBH)
N
H

FIG. 1. A Process for Preparation of L-5-(4-Hydroxybutyl) hydantoin (L-HBH) from D,L-HBH.

D,L-AHCA (10 g) and NaN_3 (0.1 g) were weighed into each of five flasks and dissolved in 800 ml H_2O. The solutions were sterilized by filtration and transferred to sterile 4000 ml shake flasks. Crude DAO (9.8 g) and catalase (5 ml) were dissolved in 200 ml saturated pyrophosphate buffer (pH 8.3) in each of five 500 ml flasks. The microbial content of the enzyme solutions was reduced to near zero by exposing them to 80Krads of radiation (27 min, 180 Krad/hr) at a Cesium γ-source.

Each of the substrate solutions in the large flasks received 200 ml of enzyme prep (total volume = 1 liter). The incubations were run with shaking at 28°C for 13 hr.

Nessler's assays for ammonia released during the incubation showed that 105 ~ 120% of the theoretical yield was obtained, indicating that the reaction had gone to completion. All of the solutions were acidified with 100 ml of concentrated HCl. Precipitated proteins were removed by centrifugation, and the supernatant fluid was clarified by vacuum filtration through Whatman No. 3 paper. The solution was stored at 4°C in four equal volumes.

C. Separation of keto acid from L-AHCA. Dowex 50W × 8 resin was cycled between the H^+ and Na^+ forms several times by washing with normal HCl and NaOH (2 ~ 3 volumes) alternated with H_2O rinses. Four 200 ml columns (43 × 2.5 cm) were poured from a thick slurry of the resin in water. The bottoms of the columns were sintered glass, below which a teflon stopcock served to control the elution rate. The columns were washed with water and then equilibrated with 1 N HCl before use.

Each of the four portions of supernatant liquid from *B* was applied to one of the ion exchange columns. Elution was done at a very slow rate to assure complete binding of the L-AHCA to the column. Each column was then washed with approximately 3 liters of H_2O. The eluents from these two steps, containing keto acid and protein, were discarded. L-AHCA was next released from the columns by elution with 500 ml concentrated NH_4OH. Heat was released as the solvent fronts moved down the columns, and in order to preserve the integrity of the columns, auxiliary cooling was necessary. The first 500 ml of eluent from each column was collected and evaporated to dryness, leaving 18.8 g of a yellow solid (75% yield).

Thin-layer chromatography using *n*-butanol: acetic acid: H_2O (4: 2: 1) showed a major ninhydrin-positive spot which co-chromatographed with AHCA and three minor contaminants which taken together comprised approximately 5% of the preparation. Recrystallization of the L-AHCA from ethanol/H_2O gave slight improvement in purity, but also gave significant loss of material. Therefore, the L-AHCA was used directly for L-HBH synthesis, with further purifications reserved for the final product (L-HBH).

When the L-AHCA recovered by chromatography was incubated with fresh D-amino acid oxidase, no ammonia was released, indicating that the preparation contained no residual D-AHCA.

The resynthesis of 5-hydantoins from the corresponding amino acids proceeds with full retention of stereoconfiguration.[1,4] Consequently the optical purity of the L-HBH preparation is estimated to be equal to the optical purity of the L-AHCA produced in steps 1 ~ 3 of this process. Since the L-AHCA preparation was characterized as containing more than 98% of the L-configuration, the optical purity of the L-HBH preparation is also estimated to be greater than 98%.

1) H. D. Dakin, *Biochem. J.*, **13**, 398 (1919).
2) H. D. Dakin, *J. Chem. Soc.*, **44**, 48 (1910).
3) R. Gaudry, *Can. J. Res.*, **26**, 387 (1948).
4) T. Suzuki, K. Igarashi, K. Hase and K. Tuzimura, *Agric. Biol. Chem.*, **37**, 411 (1973).
5) A. Meister, "Biochemistry of the Amino Acids," Academic Press (London), 1965, p. 296.

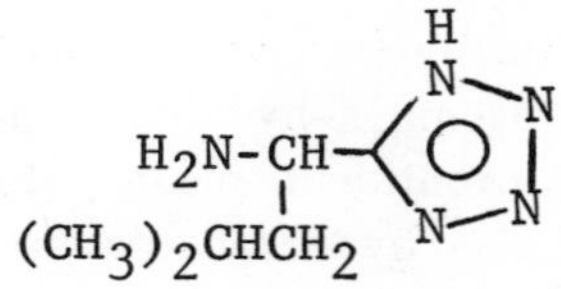
H
N
N
H2N-CH-
N
N
(CH3)2CHCH2 N

$C_6H_{13}N_5$

Ref. 260

Note:

For resolution procedure and physical constants, see Ref. 21.

$CH_3(CH_2)_2$-CH(Cl)-$CH_2CH_2AsO_3H_2$

$C_6H_{14}AsClO_3$ **Ref. 261**

3. *Acide 3-chlorehexane-1-arsonique.*

Une solution bouillante de 5.5 g (0.023 mol.) de cet acide avec 9 g (0.025 mol.) de quinine dans 350 cm³ d'eau, donne après filtrage, 7 g du sel de quinine. En le faisant recristalliser dans 230 cm³ d'eau, on obtient des rosettes d'aiguilles avec 2 molécules d'eau.

Substance 0 2224 g: perte à 90° 0.0135 g.
0.4415 g: 6.02 cm³ de baryte 0.1217 n.
Trouvé: H_2O 6.07; p. équiv. 603
$C_6H_{14}O_3AsCl . C_{20}H_{24}O_2N_2 . 2 H_2O$. Calculé: „ 5 96; „ „ 604.7.

La rotation ne change plus après six cristallisations du sel de quinine. Sel de baryum: $[M]_D = -4.3°$. Acide libre: $[M]_D = -6.8°$.

Dispersion rotatoire:

λ (μμ)	656.3	589 3	546.3	481.1	435 8
[M] sel	3.4°	4.3°	4.8°	6.2°	7.2°
„ acide	4.9°	6.8°	7.8°	9.7°	10.7°

$H_2N(CH_2)_4$-CH(NH_2)-COOH

$C_6H_{14}N_2O_2$ **Ref. 262**

Resolution of Lysine with Dibenzoyltartaric Acid.—Numerous trials with solutions of the acid salt in various concentrations of aqueous lower alcohols showed that the isomeric salts are not greatly different in solubility. However, upon undisturbed cooling of warm solutions the first deposit is usually nearly pure D-lysine salt; this crystallizes slowly as a hard crust of hexagonal plates. The somewhat more soluble L-lysine salt may later appear spontaneously or by seeding or disturbance, and then crystallizes rapidly as fine, fragile needles. When both salts crystallize together the deposit is best redissolved and another trial undertaken. Purification of each salt is effected by suitable mechanical separation and one or more recrystallizations.

The D-lysine salt crystallizes as a fairly stable dihydrate. When air-dry it has m.p. 136–140° (dec.), $[\alpha]^{25}D$ −86 ± 2° (*c* 4, water).

Anal. Calcd. for $C_{24}H_{28}O_{10}N_2 \cdot 2H_2O$: H_2O, 6.67; C, 53.30; H, 5.97. Found: H_2O, 6.60; C, 52.86; H, 5.87.

The L-lysine salt crystallizes as a dihydrate which rather rapidly loses water in air, forming a hemihydrate with m.p. 168–170°, $[\alpha]^{25}D$ −78 ± 2° (*c* 4, water).

Anal. Calcd. for $(C_{24}H_{28}O_{10}N_2)_2.H_2O$: H_2O, 1.75. Found: H_2O, 1.80.

The constants of the salts assist in guiding the selection of fractions but are not precisely characteristic. Estimation of purity is based mainly on experienced observation of crystalline appearance, with final confirmation through rotation values of regenerated lysine hydrochlorides.

In a typical experiment 29.2 g. (0.2 mole) of DL-lysine in 170 g. of water solution was mixed with 75.2 g. (0.2 mole) of dibenzoyltartaric acid monohydrate in 280 cc. of warm 2-propanol and the mixture was heated in a covered flask to dissolve all traces of salts. Undisturbed cooling overnight gave 40 g. of crude D-lysine salt. The mother liquor on decantation promptly gave 53 g. of crude L-lysine salt, which was collected by suction after an hour. Chilling the filtrate overnight at 5° gave a third crop (6 g.) rich in D-lysine salt. The first crop was dissolved in 2 parts of hot water and the solution diluted with twice its volume of 2-propanol. Substantially pure D-lysine salt then slowly crystallized at room temperature. The decanted liquor was used to recrystallize the third crop. In this manner 41.5 g. (82%) of D-lysine salt was obtained. The crude L-lysine salt was recrystallized similarly but more rapidly; a second crystallization from fresh solvent was required to attain substantial purity; the yield was 36.7 g. (76%). Minor crops from chilling filtrates were retained for crystallization with later runs. In early experiments combined liquors were evaporated to dryness *in vacuo* and the waxy residues, after washing with 2-propanol, were retained for reworking. Because of decomposition losses on evaporation, however, liquors are best re-used as solvent in later runs.

Active Lysine Hydrochlorides.—D-Lysine dibenzoyltartrate (10 g.) in 100 cc. of water was decomposed with 7 cc. of 6 *N* hydrochloric acid and the dibenzoyltartaric acid was extracted with 50 cc. of ether. The aqueous layer was evaporated to a sirup; this was dissolved in 10 cc. of warm 95% ethanol and **D-lysine dihydrochloride** was precipitated by 40 cc. of acetone. The salt was recovered by decantation, washed with acetone and dried *in vacuo;* yield 3.75 g. (87%), $[\alpha]^{25}D$ −15.3 ± 0.2° (*c* 4, water). The slightly hygroscopic material was triturated with a little absolute ethanol, washed with ether and redried. It then had $[\alpha]^{25}D$ −15.8° (*c* 4, water). Berg[2] reported $[\alpha]^{20}D$ −15.63° (*c* 3, water). The sample was converted to **D-lysine monohydrochloride** with the precautions indicated by Rice.[11] The carefully dried product had $[\alpha]^{25}D$ −10.0 ± 0.2° (*c* 4, water).

L-Lysine dihydrochloride and L-lysine monohydrochloride were prepared similarly from L-lysine dibenzoyltartrate and had, respectively, $[\alpha]^{25}D$ +16.0° and +10.1° (*c* 4, water).

(2) C.P. Berg, J.Biol. Chem., 115, 9 (1936)

(11) E. E. Rice, *Biochem. Preparations*, **1**, 63 (1949).

$C_6H_{14}N_2O_2$ **Ref. 262a**

Resolutions with Acetyl-3,5-dibromo-L-tyrosine (A) Direct Procedure.—The salts (0.2 mole) were formed by dissolving the acid (78.0 g.) in the calculated amount (170 g.) of hot 17% DL-lysine solution. The L-lysine salt separated as a thick mass of fine needles and was filtered off and washed with a little cold methanol in which it is sparingly soluble. Two further small crops were obtained by concentration; filtration was aided by dilution of the sirupy liquors with a little methanol. The much more soluble D-lysine salt remained in the liquors. The successive crystalline crops were recrystallized systematically from 1.5 parts of water and an equal volume of methanol. Further recrystallization in this way or from water alone caused no change in properties.

The pure L-lysine salt (monohydrate) had m.p. 166–168° (dec.), $[\alpha]^{25}D$ +36.8 ± 0.3° (*c* 4, water); yield 45.5 g. (84%).

Anal. Calcd. for $C_{17}H_{25}O_6N_3Br_2.H_2O$: C, 37.43; H, 4.99. Found: C, 37.49; H, 4.98.

L-Lysine Monohydrochloride.—A sample of the pure L-lysine salt (31.5 g.) was dissolved in 10 parts of hot water and 12 cc. of concentrated hydrochloric acid was added. The resolving agent (hemihydrate) then crystallized on cooling (97%) in condition suitable for re-use. The filtrate was evaporated to dryness *in vacuo*, the residue was taken up in 80 cc. of 95% ethanol and L-lysine monohydrochloride was precipitated by adding 5.9 g. of pyridine and keeping overnight at 5°. The salt was taken up in 10 cc. of water and reprecipitated with 40 cc. of ethanol.[11] After washing with absolute ethanol and ether and drying it weighed 8.75 g. (80%) and had $[\alpha]^{25}D$ +10.1° (*c* 4, water); +18.7° (*c* 4, 0.6 *N* HCl); +19.9° (*c* 4, 6 *N* HCl). Doherty and Popenoe[4] report $[\alpha]^{24}D$ +20.5° (*c* 3, 6 *N* HCl). The combined rotation samples were recovered as the dihydrochloride which had $[\alpha]^{25}D$ +16.0° (*c* 4, water).

The resolving agent and crude D-lysine hydrochlorides were recovered from the combined resolution liquors as for the L-form. The di- and monohydrochlorides had, respectively, $[\alpha]^{25}D$ −12.9° and −8.1° (*c* 4, water) and accordingly contained about 90% of the D-form.

(B) **Half-equivalent Salt Procedure.**—Acetyldibromo-L-tyrosine (0.10 mole) was dissolved in 100 cc. of hot water containing 0.12 mole of ammonia. DL-Lysine monohydrochloride (0.20 mole) was added and the L-lysine salt (monohydrate) obtained as before (three crops). After systematic recrystallization from aqueous methanol it weighed 42.5 g. (78%) and was identical with the salt previously described. In a further simplification L-lysine monohydrochloride was obtained directly from this salt by decomposition with exactly one equivalent of hot dilute hydrochloric acid; the resolving agent was filtered off after cooling, the filtrate was evaporated to small volume and the hydrochloride precipitated with ethanol. The salt (13.2 g.) had $[\alpha]^{25}D$ +20.2° (*c* 4, 6 *N* HCl). Crude D-lysine hydrochloride (20.3 g.) isolated similarly from the resolution liquors had $[\alpha]^{25}D$ −7.2° (*c* 4, water).

Supplementary Experiments (a).—Equivalent amounts (0.01 mole) of N-acetyl-3,5-dibromo-D-tyrosine[8] (ammonium salt) and the *ca.* 90% D-lysine monohydrochloride from (A) above gave the expected pure D-lysine salt. Except for sign of rotation the salt was identical with its antipode; m.p. 166–168° (dec.), $[\alpha]^{25}D$ −37.0° (*c* 4, water).

(b).—Attempted resolution of N-acetyl-3,5-dibromo-DL-tyrosine[8] (ammonium salt) with D-lysine (monohydrochloride) failed. The corresponding partially racemic double salt crystallized from water as the dihydrate in transparent thin plates, rapidly converted in air to the anhydrous form, m.p. 265–270° (charring). The salt was not resolved in methanol.

(c).—Attempts to obtain pure active lysine derivatives by fractional crystallization of partially active mixtures were not successful with the hydrochlorides, picrates and salicylidene derivatives. In each instance the DL-form was considerably less soluble; the active form was enriched in foot fractions but not completely purified.

Authors' comments:

Attempts to resolve diacetyl-DL-lysine with α-fenchylamine, α-phenylethylamine, brucine, cinchonine, quinine, ephedrine and desoxyephedrine were not successful.

$C_6H_{14}N_2O_2$ Ref. 262b

EXAMPLE I

There was added to 75 ml. of water contained in a 500 ml. round bottom flask 5.3 grams (0.0435 mole) of D-(+)-alphamethylbenzylamine, reagent grade. The resulting mixture was not a completely homogeneous solution. To this mixture was added at room temperature 10.0 grams (0.0435 mole) of DL-diacetyllysine. There was obtained a homogeneous solution immediately. The water solvent was removed by vacuum evaporation and 15.2 grams of a free flowing solid was obtained consisting of the salt of DL-diacetyllysine and the alphamethylbenzylamine. Since the theoretical yield of the salt is 15.26 grams the actual yield was substantially 100 percent, i.e. the reaction was quantitative.

The salt thus obtained consisted of the DD salt and the DL salt. Since these salts have different solubilities they can be separated by fractional crystallization. Therefore it was necessary to employ seed crystals of the salt of L-diacetyllysine and the D-(+)-alphamethylbenzylamine. The following example shows the preparation of the L-diacetyllysine.

EXAMPLE II

A mixture of 9.1 grams (0.05 mole) of reagent grade L-(+)-lysine·HCl and 16.8 grams (0.2 mole) sodium bicarbonate were dissolved in 150 ml. of water. To this solution was added 18.1 grams (0.1 mole) of p-nitrophenylacetate and 20 ml. of ethyl acetate. The mixture was stirred vigorously for about 16 hours at room temperature. The purpose of adding the ethyl acetate was to aid in the solubilization of the p-nitrophenylacetate. The solution was extracted three times with diethyl ether (100 ml. per treat) and the remaining aqueous solution was treated a second time with 18.1 grams of p-nitrophenylacetate and 8.4 grams of sodium bicarbonate. The aqueous solution containing the L-diacetyllysine sodium salt was again ether extracted in the same manner to remove the p-nitrophenol byproduct and unreacted reagents. Finally, the water solvent was removed by vacuum evaporation.

The residue after the water evaporation was heated to reflux in 200 ml. of absolute alcohol and filtered to remove the insoluble sodium bicarbonate. The filtrate containing the sodium salt of L-diacetyllysine was treated with gaseous hydrogen chloride at 5° C. to convert the sodium salt to L-diacetyllysine.

The acidified filtrate was vacuum evaporated to remove the solvent and the crude L-diacetyllysine was isolated as a viscous liquid.

The following example is provided to show the preparation of the salt of D-(+)-alphamethylbenzylamine and L-diacetyllysine to be used for seeding purposes.

EXAMPLE III

In 100 ml. of water 100 grams (0.0435 mole) of the crude L-diacetyllysine produced in Example II and 10 grams (0.082 mole) of D-(+)-alphamethylbenzylamine were heated to reflux for one hour, cooled to room temperature and extracted with diethyl ether to remove unreacted D-(+)-alphamethylbenzylamine. The aqueous solution was evaporated and 11.2 grams of the crude salt obtained. This crude salt was dissolved in 400 ml. of acetone and recrystallized by allowing the solution to stand for about 16 hours at about 5° C. Upon filtration 5.0 grams of the salt of D-(+)-alphamethylbenzylamine and L-diacetyllysine was obtained. This salt had a specific rotation of (+)9.24 in water solvent using a 1.0 decimeter tube in the polarimeter. This salt had not been reported in the literature and therefore the specific rotation of the pure salt was not known. In order to determine the purity of the salt thus prepared it was converted to L-lysine·HCl which has a known specific rotation. The method employed is set forth in the following example.

EXAMPLE IV

The conversion was carried out by dissolving 5.0 grams of the salt prepared as described in Example III in 100 ml. of water and the solution was passed downwardly through a bed (80 ml.) of a cation exchange resin (trade name, "Amberlyst-15") in the acid (hydrogen) cycle. This procedure exchanged the optically active base onto the resin so that the free L-diacetyllysine was recovered in the effluent. The resin was water-washed and the total volume of effluent was 300 ml. This effluent containing the L-diacetyllysine was refluxed 2 hours with 100 ml. of concentrated hydrochloric acid to hydrolyze the compound to L-lysine·2HCl. Removal of the solvent by vacuum evaporation gave 4.0 grams of crude L-lysine·2HCl. This 4.0 grams of crude L-lysine·2HCl was taken up in 90 ml. of 95 percent alcohol, and heated to reflux while in contact with about 8 ml. of pyridine. This procedure converted the L-lysine·2HCl to L-lysine·HCl. The solution was cooled to about 5° C. and the L-lysine·HCl was recovered by filtration (2.0 grams). The specific rotation of the L-lysine·HCl obtained was found to be (+)19.3, when the solvent consisted of 2 N HCl and a 1 decimeter tube was employed in the polarimeter.

The L-lysine·HCl starting material had a specific rotation of (+)21.03 determined in the same manner thus showing that the L-lysine·HCl recovered as described had a purity of 92 percent (19.3 divided by 21.03 times 100) and, accordingly, the purity of the salt of D-(+)-alphamethylbenzylamine and L-diacetyllysine was also at least 92 percent. This salt had a specific rotation of (+)9.21 with water as the solvent and a 1 decimeter tube.

Another run was made as described in Example III to prepare seed crystals of L-diacetyllysine and D-(+)-alphamethylbenzylamine. The crystals made in this second run had a specific rotation of (+)9.2 and thus also had a purity of at least 92 percent.

The following example shows the fractional crystallization of the salt of DL-diacetyllysine and D-(+)-alphamethylbenzylamine.

EXAMPLE V

An 8.2 gram sample of the salt of DL-diacetyllysine and D-(+)-alphamethylbenzylamine prepared as described in Example I was dissolved in a solvent consisting of 25 ml. of ethanol and 60 ml. of acetone. Seed crystals of the salt of L-diacetyllysine and D-(+)-alphamethylbenzylamine prepared as in the second run described above were added to this solution and the mixture stirred at room temperature for about 3 days but only a small amount of solids (0.5 gram) were produced. The mixture was then cooled to 5° C., stirred for one hour and filtered. There was obtained 2.4 grams of crystals which had a specific rotation of (+)8.25 in water solvent with a 1.0 decimeter tube indicating a purity of 83 percent. This salt was converted to L-lysine·HCl by dissolving 2 grams of the salt in 50 ml. of water and passing the resulting homogeneous solution down through a bed of cation exchange resin in the acid cycle (50 ml.). This resulted in the optically active base D-(+)-alphamethylbenzylamine being exchanged onto the resin while the effluent contained the L-diacetyllysine, excess water being used to wash the resin free of any L-diacetyllysine. It should be noted that the amount of water used to dissolve the salt is not critical except that sufficient water be used to completely dissolve the salt and likewise the amount of resin employed is not critical except that an excess of the resin be used so that its capacity for exchanging the amine is not exceeded. The resin, of course, can be regenerated in the usual manner by washing with a mineral acid solution, preferably hydrochloric acid.

The effluent containing the L-diacetyllysine from the ion exchange step amounting to about 400 ml. of solution was hydrolyzed for about 16 hours with 100 ml. of concentrated hydrochloric acid in order to convert the L-diacetyllysine to L-lysine·2HCl. The amount of concentrated hydrochloric acid is not critical however, it is preferred to use an excess in order to hydrolyze the L-diacetyllysine completely. The L-lysine·2HCl resulting from the hydrolysis step was dissolved in 50 ml. of hot 95 percent alcohol and then refluxed in the presence of 2 ml. of pyridine in order to convert the L-lysine·2HCl to L-lysine·HCl. The solution was cooled to about 5° C. and 0.8 grams of L-lysine·HCl was recovered by filtration. The specific rotation of this solid was (+)17.66 when using 2 N HCl as the solvent and a 1.0 decimeter tube in the polarimeter as compared to a specific rotation of (+)21.03 determined in the same manner for the commercially pure L-lysine·HCl indicating a purity of about 84 percent. In this example as well as in all of the preceding examples a specific rotation was determined as indicated using a Zeiss polarimeter, the D-line of sodium, and a temperature of 25° C.

The solubility of the racemic salt of DL-diacetyllysine and D-(+)-alphamethylbenzylamine in water, ethyl alcohol and isopropyl alcohol is greater than 60 weight percent (at reflux temperature) which prevents these solvents individually from being used for fractional crystallization to separate the DD salt from the DL salt as in Example V.

$C_6H_{14}N_2O_2$ Ref. 262c

Example 1

A water solution (50 cc.) containing 0.124 mole of free dl-lysine was mixed with l(+)-glutamic acid (18.23 g., 0.124 mole) and diluted with water to a total weight of 118 g. The mixture was warmed to dissolve the glutamic acid, then cooled to room temperature and mixed with 272 cc. of methanol. The solution was seeded with l(+)-lysine l(+)-glutamate monohydrate and stirred forty hours at room temperature. The solid which precipitated during the stirring was filtered off, washed with 100 cc. of methanol and dried. Yield 16.67 g. (86% of theoretical l(+)-lysine l(+)-glutamate monohydrate). The specific rotation of the product in water at 12% concentration and room temperature was +3.6°. Kjeldahl analysis of material prepared in the above manner showed 13.6% N; calc. for lysine glutamate monohydrate: 13.5% N.

The product remaining after the removal of analytical samples (12.96 g.) was heated in an oven at 80–85° C. for 39 hours. The loss in weight (0.75 g.) was equivalent to one mole of water per mole of lysine glutamate. The dehydrated product was dissolved in water to make 468 g. of solution and was passed through a 200 cc. bed of "Dowex 50" cation-exchange resin in ammonium form, which was then washed with water. Ammonium glutamate was recovered from the effluent. The resin bed was extracted with 10% aqueous ammonia and the effluent freed of ammonia by evaporation. The resulting lysine solution was shown by titration to be equivalent to the lysine glutamate fed to the resin bed. It was treated with the calculated quantity of HCl for conversion to lysine monohydrochloride and the latter recovered by partial evaporation, followed by the addition of ethyl alcohol to precipitate the product. The yield of l(+)-lysine monohydrochloride was 95%; its specific rotation (+9.0°) indicated 95% optical purity and its chloride ion content (19.1%) corresponded to 98.5% chemical purity.

$C_6H_{14}N_2O_2$ Ref. 262d

EXAMPLE 1

To a solution of 0.288 part of DL-lysineamide in 1 part by volume of methanol was added 0.32 part by volume of 6 N methanolic hydrochloric acid and 0.248 part of L-pyrrolidone carboxylic acid. A clear solution was obtained after refluxing one minute. Seed crystals of L-lysineamide - L - pyrrolidone carboxylate hydrochloride (0.005 part) were added and the solution allowed to stand for five hours at 25° C. with occasional stirring. The mixture was filtered, the crystalline product washed with 2 parts by volume of methanol and dried.

L-lysineamide-L-pyrrolidone carboxylate hydrochloride was obtained in 49% yield (0.150 part) having a melting point of 147–152° C.

The structure of the product was confirmed by potentiometric titration with silver nitrate which showed an argentimetric equivalent of the product of 310 (theoretical 311). The product had an optical rotation $[\alpha]_D^{20}$ of +2.84, which corresponds to a mixture containing 65.6% of the L-lysineamide-L-pyrrolidone carboxylate hydrochloride, the remainder being DL-lysineamide-L-pyrrolidone carboxylate hydrochloride.

EXAMPLE 2

A solution of 1.82 parts DL-lysineamide hydrochloride and 1.29 parts of L-pyrrolidone carboxylic acid in 10 parts by volume of refluxing methanol was prepared and cooled to 60° C. Seed crystals of L-lysineamide-L-pyrrolidone carboxylate hydrochloride were added and the mixture allowed to cool slowly to 25° C. After standing for 15 hours, the crystals were collected by filtration. A 60% yield (0.933 part) of the product was obtained having a melting point of 147–153° C.

$[\alpha]_D^{20}$ was +4.30 which corresponds to a mixture containing 78% of L-lysineamide-L-pyrrolidone carboxylate hydrochloride.

EXAMPLE 3

6.21 parts of L-lysineamide-L-pyrrolidone carboxylate hydrochloride was stirred in 27 parts by volume of a methanolic solution of 1.3 N hydrochloric acid. The temperature, which had risen to 35° C., was reduced to 25° C. and the mixture let stand several hours. The mixture was filtered and the crystalline product washed with methanol and dried.

An 85.3% yield (3.72 parts) of L-lysineamide dihydrochloride was obtained, as determined by potentiometric titration with silver nitrate.

$[\alpha]_D^{20}$ was +16.7, which corresponds to 100% pure product.

$C_6H_{14}N_2O_2$ Ref. 262e

EXPERIMENTAL

We have resolved *dl*-lysine by fractionally crystallizing its camphorates from 50 per cent methyl alcohol. The *l*(+)-lysine compound with *d*-camphoric acid is the less soluble and hence the more readily purified; with *l*-camphoric acid the reverse is true. The lysine isomers are readily isolated from the camphorates as the dihydrochlorides.

Preparation of d- and l-Camphoric Acids—The *d*-camphoric acid was obtained by oxidizing refined *d*-camphor with nitric acid, as directed by Noyes (13). Unchanged camphor was removed and the crude camphoric acid was purified through the anhydride, essentially as outlined by Aschan (1, *a*, *b*). From 100 gm. of *d*-camphor, 70 to 85 gm. of crude *d*-camphoric acid (53 to 64 per cent of the theoretical) and 50 to 65 gm. of the purified product (38 to 49 per cent) were readily obtained. Physical constants appear in Table I.

The *l*-camphoric acid was prepared from *l*-borneol,[2] a source recommended by Campbell (5) as being more satisfactory than *l*-camphor. The method used was essentially identical with that outlined above for the preparation of *d*-camphoric acid from *d*-camphor. From 100 gm. of *l*-borneol (used as purchased, without purification) 73 gm. of crude *l*-camphoric acid (56.2 per cent) and 42 gm. of the highly purified product (32.3 per cent) were readily obtained. Physical constants of the borneol and the camphoric acid are recorded in Table I.

TABLE I

Physical Constants of d-Camphor, l-Borneol, and d- and l-Camphoric Acids

Compound	M.p. (corrected)		$[\alpha]_D^{20}$ in absolute alcohol solution*	
	Observed	Reference	Observed	Reference
	°C.	°C.	*degrees*	*degrees*
d-Camphor†	178	178.6 (14)	+44.64	+44.20 (14)‡
l-Borneol†	206–207	208–209 (9)	−35.85	−37.74 (2)
d-Camphoric acid	188–189	188.2 (14)	+48.45	+47.75 (14)§
		187 (1, *c*)		+49.7 (1, *c*)‖
l-Camphoric "	188–189	187.5 (14)	−48.47	−48.12 (14)‡

The figures in parentheses denote bibliographic reference numbers.
* The concentration was that used in the reference cited. All readings by the author were at 20°.
† The camphor and borneol used by us were as purchased; the references are to purified preparations.
‡ Read at 16°.
§ Read at 17°.
‖ Read at 16.5°; white light used.

Preparation of dl-Lysine Dihydrochloride—The synthesis of Eck and Marvel (7) was used with only minor changes. It seemed more convenient to prepare the cyclohexanoneoxime according to the method of Bousquet (4), suggested by Eck and Marvel as an alternative procedure. Acetone was found more satisfactory than ether for the precipitation of lysine dihydrochloride from the alcohol solution of its syrup. When ether is used, the lysine dihydrochloride becomes gummy unless the syrup is sufficiently concentrated. Acetone prevents this because it also dehydrates. Physical properties and analyses may be found in Table III. The over-all yield and, with but one exception, also the yields at each step of the synthesis were as good as those claimed by the authors of the method.

Preparation of Lysine Camphorates—To 76.7 gm. of *dl*-lysine dihydrochloride (0.35 mole) dissolved in approximately 200 cc. of water was added slightly over 0.35 mole of freshly prepared silver oxide. The mixture was shaken to facilitate the precipitation of the chloride, the latter was filtered off, and excess silver was removed from the filtrate as the sulfide. The filtrate was partially concentrated *in vacuo*, 35.0 gm. of *d*-camphoric acid (0.175 mole) dissolved in methyl alcohol were added, and the concentration was carried to dryness. The residue showed $[\alpha]_D^{20} = +7.94°$.[3]

The mixture of *l*(+)- and *d*(−)-lysine *d*-camphorates was recrystallized from a minimal volume of 1:1 methyl alcohol-water solution. The mother liquors were concentrated to incipient crystallization, whereupon an equal volume of methyl alcohol was added. This process was repeated as long as fractions of appreciable size were obtained. Usually each crystallization was allowed to continue overnight in the cold. Each of the more insoluble fractions was similarly refractionated. Such a procedure must necessarily be followed closely by optical measurements, since practically unavoidable variations in technique influence the course of the fractionation. The following is a brief outline of the progress of one specific separation. Fraction I, 8.6 gm., showed $[\alpha]_D^{20} = +15.33°$; Fraction II, 28.5 gm., $[\alpha]_D^{20} = +13.93°$; Fraction III, 48.7 gm., $[\alpha]_D^{20} = +3.2°$. On recrystallization from 50 per cent methyl alcohol, Fraction I yielded 8.1 gm., $[\alpha]_D^{20} = +16.14°$, and Fraction II, 20.4 gm., $[\alpha]_D^{20} = +16.35°$. These recrystallized fractions were combined and again recrystallized, yielding 22.53 gm. of optically pure, $[\alpha]_D^{20} = +16.41°$, *l*(+)-lysine *d*-camphorate, or 52.3 per cent of the theoretical amount expected from 76.7 gm. of *dl*-lysine dihydrochloride. Further recrystallization did not change the rotation. Physical and chemical constants of this product are recorded in Table II.

Further fractionation of mixtures showing rotations of +4.5° or less proved too laborious to be practical. Consequently Fraction III and the dry residues from the various mother liquors were dissolved in water and made acid to Congo red with sulfuric acid. The *d*-camphoric acid was extracted with ether and barium hydroxide was cautiously added to effect the exact removal of the sulfate ion. The calculated amount of *l*-camphoric acid needed to replace the *d*-camphoric acid removed was next added and the solution was concentrated to dryness. Fractionation of the *l*-camphorate mixture, essentially as outlined for the *d*-camphorate, yielded 27.79 gm. of *d*(−)-lysine *l*-camphorate (64.5 per cent of the theoretical amount expected from the *dl*-lysine dihydrochloride used) which showed $[\alpha]_D^{20} = -16.39°$. The physical and chemical constants determined appear in Table II.

TABLE II

Melting Points and Analyses of the Lysine Camphorates

Compound	M.p. (corrected)	Nitrogen found*	$[\alpha]_D^{20}$, solution in water; c = 2.00
	°C.	*per cent*	*degrees*
l(+)-Lysine *d*-camphorate	245–246	11.25	+16.41
" *l*-camphorate	239–240	11.35	+ 0.35
d(−)-Lysine "	245–246	11.30	−16.39
" *d*-camphorate	239–240	11.24	−0.38

* Calculated for lysine camphorate, $(C_6H_{14}O_2N_2)_2 \cdot C_8H_{14}(COOH)_2$, 11.38 per cent.

On removing *l*-camphoric acid from the lower rotating *l*-camphorate fractions and substituting *d*-camphoric acid, 11.62 gm. more of optically pure *l*(+)-lysine *d*-camphorate were obtained, bringing the total yield to 34.15 gm. (79.3 per cent). A second substitution of *l*-camphoric acid instead of *d*- in this second set of low rotating *d*-camphorate fractions made possible the isolation of 7.00 gm. more of optically pure *d*(−)-lysine *l*-camphorate, increasing the yield to 34.79 gm. (80.8 per cent). Further fractionation in this way, though impractical, would undoubtedly increase these yields.

Of value for reference are the properties of *l*(+)-lysine *l*-camphorate and *d*(−)-lysine *d*-camphorate prepared from the free lysines and the purified camphoric acids. These are given in Table II.

Isolation of l(+)- and d(−)-Lysine As the Dihydrochlorides—To 19.7 gm. of *l*(+)-lysine *d*-camphorate (0.04 mole) dissolved in 100 to 200 cc. of water were added 20 cc. of concentrated hydrochloric acid. The bulk of the camphoric acid liberated was filtered off and the residue was extracted with ether. The hydrochloric acid-lysine solution was concentrated *in vacuo* to a thin

[2] Obtained from The British Drug Houses, Ltd., London.

[3] Aqueous solutions of the camphorates (*c* = 2.00 gm. per 100 cc.) were always employed.

TABLE III

Melting Points and Analyses of the dl-, l (+)-, and d(−)-Lysine Dihydrochlorides

Lysine dihydrochloride	M.p. (corrected)	Nitrogen found*	$[\alpha]_D^{20}$, solution in water
	°C.	per cent	degrees
dl	188–190†	12.76	0.0
l(+)	201–202‡	12.78	+15.63 (c = 3.00)§
			+16.55 (" = 16.00)‖
d(−)	201–202	12.73	−15.65 (" = 3.00)

* Calculated for $C_6H_{14}O_2N_2 \cdot 2HCl$, 12.79 per cent.
† 188–190°, Eck and Marvel (7).
‡ 200–201°, Cox, King, and Berg (6).
§ +15.57°, in water, c = 2.86, Lawrow (11); +15.4°, in water, c = 4.38, Lutz and Jirgensons (12).
‖ +16.44°, in water, c = 16.27, Bergmann and Zervas (3); +16.36° and 16.68°, in water, for c = 11.43 and c = 19.48 respectively, Lawrow (11).

syrup which was warmed on the steam bath and dissolved in a minimum amount of alcohol. 3 to 5 volumes of acetone were added to precipitate the lysine dihydrochloride. This was redissolved in alcohol and again precipitated with acetone. 17.1 gm. of *l*(+)-lysine dihydrochloride (97.6 per cent of the calculated amount) were thus isolated. The constants found appear in Table III.

19.7 gm. of *d*(−)-lysine *l*-camphorate, similarly treated, yielded 17.16 gm. (97.9 per cent of the calculated amount) of *d*(−)-lysine dihydrochloride. The properties of this product are also recorded in Table III.

The yields of pure camphorate and pure lysine dihydrochloride indicate that 77.4 per cent of the *l*(+)-lysine dihydrochloride and 79.1 per cent of the *d*(−)-lysine dihydrochloride can be separated from the racemic mixture by this procedure. More intensive fractionation would increase this yield.

DISCUSSION

The method of resolution described is simple and direct. The camphoric acids are easily prepared from relatively cheap sources. Their purification through the anhydrides is not difficult, but does cause considerable loss in yield. This is probably not essential, since good separations of *l*(+)-lysine have been secured with *d*-camphoric acid having a specific rotation of +44.6°. *d*-Camphoric acid of this grade can be obtained on the market.

Even when molecularly equivalent amounts of *dl*-lysine and *d*-camphoric acid were used, the compound yielded by recrystallization contained 1 molecule of camphoric acid to 2 of lysine.

Several modifications in fractionation procedure are possible. Thus when only the natural *l*(+)-lysine dihydrochloride is required, its yield may be increased by racemizing the lysine in the more soluble *d*-camphorate fractions to provide more *dl*-lysine. To secure *l*(+)-lysine alone, the use of *d*-camphoric acid only is necessary; for the isolation of *d*(−)-lysine, only *l*-camphoric acid is required. If *l*-camphoric acid is not available, a fair yield of *d*(−)-lysine should be obtainable from the more soluble *d*-camphorate fraction with *dl*-camphoric acid.

1. Aschan, O., (*a*) *Ber. chem. Ges.*, **27**, 2003 (1894); (*b*) *Chem. Zentr.*, **2**, 970 (1895); (*c*) *Ann. Chem.*, **316**, 210 (1901).
2. Beckmann, E., *Ann. Chem.*, **250**, 353 (1888).
3. Bergmann, M., and Zervas, L., *Z. physiol. Chem.*, **152**, 282 (1926).
4. Bousquet, E. W., in Marvel, C. S., *et al.*, Organic syntheses, New York, **11**, 54 (1931).
5. Campbell, A. N., *J. Am. Chem. Soc.*, **53**, 1661 (1931).
6. Cox, G. J., King, H., and Berg, C. P., *J. Biol. Chem.*, **81**, 755 (1929).
7. Eck, J. C., and Marvel, C. S., *J. Biol. Chem.*, **106**, 387 (1934).
8. Freudenberg, K., and Rhino, F., *Ber. chem. Ges.*, **57**, 1547 (1924).
9. Haller, A., *Ann. chim. et physiq.*, **27**, 392 (1892).
10. Karrer, P., Escher, K., and Widmer, R., *Helv. chim. acta*, **9**, 301 (1926).
11. Lawrow, D., *Z. physiol. Chem.*, **28**, 388 (1899).
12. Lutz, O., and Jirgensons, B., *Ber. chem. Ges.*, **64**, 1221 (1931).
13. Noyes, W. A., *Am. Chem. J.*, **16**, 501 (1894).
14. Ross, J. D. M., and Somerville, I. C., *J. Chem. Soc.*, 2770 (1926).

$C_6H_{14}N_2O_2$ Ref. 262f

Preparation of Mono-1-menthyl Phosphate (I)

Pyrophosphoric acid (40 g), 1-menthol (31.2 g, $[\alpha]_D^{20}$ −49.2 in methanol) and benzene (200 g) were placed in a 500 ml flask. After stirring

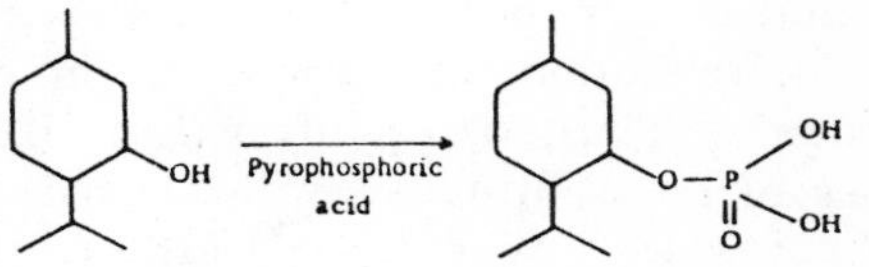

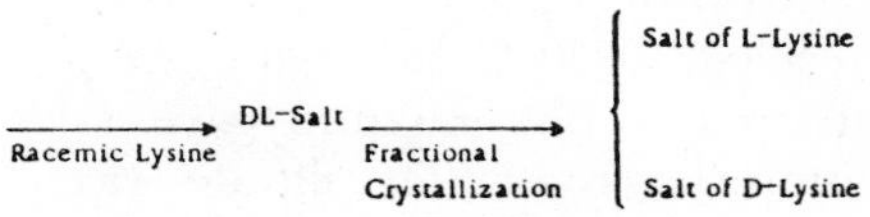

for 25—30 hr, the flask was allowed to stand for one day. After diethyl ether (40 ml) was added, the reaction mixture was washed with water to remove inorganic acid. The benzene-ether layer was washed with a warm solution, prepared from 40 g of sodium hydroxide and 250 g of water, to extract mono-1-menthyl phosphate (I) as disodium salt. The aqueous layer was adjusted to pH 1.0 with concentrated hydrochloric acid, and mono-1-menthyl phosphate (I) precipitated immediately as white crystals. They were dissolved in diethyl ether and the ether solution was washed twice with water and dried over anhydrous sodium sulfate. After evaporation of ether, 8 g of crude I was obtained; it was recrystallized twice from n-hexane to give 5.4 g of pure I. The purity of the product was ascertained by titration and thin layer chromatography using silica gel with a mixture of benzene (70%) and chloroform (30%) as solvent. Only one spot was obtained (Diagram). The physical properties were as follows: m.p. 113—114°, $[\alpha]_D^{20}$ − 40.9 (c = 2, in methanol), ν_{max}: 1180 cm^{-1} (> P=O, s) 1020 cm^{-1} and 1050 cm^{-1} (P-O-C, s), 2740 cm^{-1} (P-OH, m).
Anal. Calcd. for $C_{10}H_{21}O_4P$: C, 50.84; H, 8.90%.
Found: C, 50.45; H, 8.76%.

The benzene-ether layer was washed with sodium hydroxide solution and water, dried over sodium sulfate and fractionally distilled to give an oily product. It contained 9 g of unreacted menthol and 10 g of various terpenic hydrocarbons.

Attempts to prepare the mono-phosphates of borneol, geraniol and terpineol in similar fashion were unsuccessful. In all cases, only terpenic hydrocarbons were obtained.

Optical Resolution of DL-Lysine (II)

DL-Lysine (1.46 g, 0.01 mole) was dissolved in 13 cc of water and the solution mixed with an ethanolic solution (21 cc) of I (2.36 g). The mixture was warmed at 70—80°C during thirty minutes and cooled slowly to room temperature. Crystallization occurred slowly, giving 0.9 g of the nearly pure salt (III_A) of L-lysine (m.p. 131—133°C). Further crystallization occurred in the filtrate overnight, giving 1.2 g of the nearly pure salt (III_B) of D-lysine (m.p. 131—133°C).

Anal. Calcd. for $C_{16}H_{35}O_6N_2P$: C, 50.25; H, 9.22%.
Found for III_A: C, 50.01; H, 9.13%.
Found for III_B: C, 50.03; H, 9.18%.

The pure salt (III_A) (1 g) was suspended in 10 cc of 1 N hydrochloric acid, and the mixture warmed at 70–80°C during three minutes. I generated as white crystals, which was filtered off. The hydrochloric acid solution of L-lysine had $[\alpha]_D^{20}$ + 10.2 (in 0.6 N HCl) (reported [2] $[\alpha]_D^{25}$ + 14.6 (C = 2 in 0.6 N HCl)). Similarly, the hydrochloric acid solution of D-lysine ($[\alpha]_D^{20}$ – 13.8, in 0.6 N HCl) was obtained from the salt of (III_B).

$C_6H_{14}N_2O_2$ Ref. 262g

Here we report the resolution of DL-lysine-3,5-dinitrobenzoate (DL-Lys. B.A.) by the preferential crystallization procedure. Optically active and DL-Lys. B.A. were prepared at first in water as needle crystals by mixing an equimolar amount of 3,5-dinitrobenzoic acid with optically active and DL-lysine, respectively. Properties of L- and DL-Lys. B.A. are shown in Table I. DL-Lys. B.A. is approximately twice as soluble as L-Lys. B.A. L-Lys. B. A. is insoluble in a saturated solution of DL-Lys. B.A. IR spectrum and powder x-ray diffractionpattern were identical between L- and DL-Lys. B.A.

The resolution was carried out in a batch system. Thus, 67.2 g of DL-Lys. B.A. and 5.4 g of L-Lys. B.A. were dissolved in 200 ml of hot water. When cooled to 46°C, the solution was seeded with 2.2 g of L-Lys. B.A., and further cooled to 30°C in 50 min with stirring. Filtration of the mixture gave L-Lys. B.A. crystals (1st crop). The filtrate was heated after the addition of DL-Lys. B.A., the amount of which was equal to the 1st crop. The resulting solution was seeded at 46°C with 2.2 g of D-Lys. B.A. The mixture was cooled to 30°C in 50 min with stirring, and filtered to give D-Lys. B.A. crystals (2nd crop).

On repeating alternately the above procedures, additional crops of L- and D-Lys. B.A. (3rd and 4th crops) were obtained. The yield and optical purity of 1st to 4th crops are shown in Table II.

TABLE II. RESULTS OF THE RESOLUTION IN THE BATCH SYSTEM

	Crystals	Yield (g)	Optical purity (%)
1st	L-Lys. B.A.	14.5	91.3
2nd	D-Lys. B.A.	12.5	95.3
3rd	L-Lys. B.A.	8.5	94.7
4th	D-Lys. B.A.	12.5	85.8

The optical purity was determined by specific rotation of L- and D-lysine derived from L- and D-Lys. B.A. An aliquot of aqueous solution of the separated crystals of L- or D-Lys. B.A. was made acidic with hydrochloric acid to liberate 3,5-dinitrobenzoic aicd, which was removed by extraction with ether. The acidic solution containing L- or D-lysine was evaporated to dryness. The residue was dissolved in 6 N hydrochloric acid, and optical rotation was measured. The content of lysine in the solution used for optical rotation measurement was determined by Kjeldahl-N analysis.

The resolution was also carried out continuously in a fluidized bed system[11] consisting of two columns connected in parallel. The column, 8 cm long and 2 cm diameter, was

TABLE I. PROPERTIES OF L- AND DL-Lys. B.A.

	Mp, °C (dec.)	*Anal.* for: $C_{13}H_{18}O_8N_4$		$[\alpha]_D^{25}$	Solubility Temp.	g/100 g·H_2O
		Calcd:	C, 41.49; H, 5.36; N, 14.89			
L	210°	Found:	C, 41.80; H, 5.38; N, 14.96	+5.2° (c=4, H_2O)	30°	7.48
					50°	15.61
					65°	31.58
DL	213°	Found:	C, 41.47; H, 5.20; N, 15.08		30°	15.17
					50°	36.11
					65°	77.55

TABLE III. RESULTS OF THE RESOLUTION IN THE FLUIDIZED BED SYSTEM
Seed crystals, 1.0 g, 60~115 mesh

No.	Concentration of DL-Lys. B.A (%)	Temperature (°C)	Time (min)	Yield (g) L	Yield (g) D	Optical purity (%) L	Optical purity (%) D
1	23.0±0.2	35	45	2.2	2.4	78.5	72.2
2	23.0±0.2	35	85	2.8	2.8	71.2	68.3
3	23.0±0.2	35	160	3.8	3.4	58.1	59.5
4	36.0±0.2	55	125	2.8	2.8	95.1	95.1
5	36.0±0.2	55	200	3.2	3.2	92.2	93.1
6	36.0±0.2	55	300	4.4	4.2	89.7	92.5

11) K. Itoh, T. Akashi and S. Tatzumi, *Japanese Pat. Publication*, 36-17710 (1961).

separable in the middle, and fitted with sintered glass discs at both ends. Seed crystals (1.0 g, 60~115 mesh) of L- and D-Lys. B.A. were added separately to each column. A supersaturated aqueous solutions of DL-Lys. B.A. was circulated through the columns at a constant temperature and flow rate (250 ml per min). At the end of a given time, the selectively grown crystals in each column were separated by filtration, weighed and the optical purity was determined. The results are summarized in Table III.

The temperature of the circulated solution of DL-Lys. B.A. was 35°C and 55°C. It is apparent that the resolution is affected significantly by the temperature of the system. Higher concentration of DL-Lys. B.A. is required at higher temperature to get enough crystal growth. The optical purity of the grown crystals decreases considerably at 35°C, but slightly at 55°C as increasing the growth rate of seed crystals. The result in the case of No. 6 in Table III is useful for practical purpose.

It was also found that DL-lysine-1-chloronaphthalene-4-sulfonate and DL-lysine-anthraquinone-β-sulfonate are resolved by the preferential crystallization procedure. Details will be described elsewhere.

$C_6H_{14}N_2O_2$ Ref. 262h

EXAMPLE 1

Racemic lysine sulphanilate (30 g) was dissolved, with sufficient heating, in an aqueous solution (45 g) containing DL-lysine (21.8% by weight), whereupon the clear solution was supersaturated by cooling to 25°C. To the supersaturated solution solid L-lysine sulphanilate (2.5 g, crystal size smaller than 0.05 mm) was added, following which the suspension obtained was stirred at 25°C for 15 minutes. Next, the L-lysine sulphanilate which crystallized out was separated by filtration from the mother liquor, washed on the filter with methanol (approximately 20 ml), and dried. 8.7 g of L-lysine sulphanilate were obtained.

In order to determine the optical purity of the L-lysine sulphanilate so obtained, the salt was converted into L-lysine monohydrochloride. To this end, the 8.7 g of L-lysine sulphanilate obtained were dissolved in water (35 ml) and the resulting solution was passed over a column filled with about 150 ml of a strongly acid ion exchanger (Dowex 50) in the NH_4^+form. The column was reflushed with demineralized water until ammonium sulphanilate was no longer present in the eluate. The lysine bound to the ion exchanger was washed out with 3.5 N ammonia water, whereupon the eluate obtained was concentrated by evaporation at a reduced pressure in order to remove the ammonia. The lysine solution so obtained was neutralized with the required quantity of hydrochloric acid (1 mole per mole of lysine) and was then evaporated until fully dry.

L-lysine monohydrochloride (5 g) with a specific rotation of: $|\alpha|_D^{20} = 27.2$ (lysine concentration = 10; 6 N HCL) were obtained. From this follows an optical purity for the L-lysine sulphanilate obtained of 100%. The resolution efficiency (defined as

$$\frac{a \times \frac{b}{100} - c}{d} \times 100\%,$$

in which a, c and d respectively represent the quantity of L-lysine sulphanilate obtained, the quantity of seed material and the quantity of L-lysine sulphanilate in the starting material, expressed in grams, and b the optical purity, expressed in percents) amounts to 41.4%.

For the experimental details of enzymatic methods for the resolution of this compound, see J. P. Greenstein, J. B. Gilbert and P. J. Fodor, J. Biol. Chem., 182, 451 (1950); S. M. Birnbaum, L. Levintow, R. B. Kingsley and J. P. Greenstein, J. Biol. Chem., 194, 455 (1952); S. Utzino and T. Yoneya, J. Biochem. (Tokyo), 38, 343 (1951); S. Utzino and T. Yoneya, Chem. Ber., 85, 860 (1952).

```
           CH3
           |
H2N(CH2)3-C-COOH
           |
           NH2
```

$C_6H_{14}N_2O_2$ Ref. 263

(+)-α-Methylornithine Hydrochloride (4). To a solution of **18** (3.07 g, 0.0236 mol) in ethanol (50 mL) maintained at 50 °C was added a boiling solution of (–)-binaphthylphosphoric acid (8.2 g, 0.0236 mol) in ethanol (250 mL). The reaction mixture was maintained at 60 °C until crystallization started. Thereafter, it was allowed to cool slowly to room temperature and then maintained at 4 °C for 1.5 h. The crystals were collected by filtration and washed with ethanol (50 mL) and ether (2 × 100 mL) to give 5.47 g of the (–)-binaphthylphosphoric salt (95%) [mp 336 °C dec; $[\alpha]^{20}_D$ –432° (c 0.9, MeOH). Anal. ($C_{26}H_{25}N_2O_5P$) C, H, N], which was then heated at reflux temperature in 6 M HCl (40 mL) for 2 h. The solid which separated was filtered and washed with water (3 × 10 mL), and the filtrate was concentrated in vacuo. The crystalline residue was dissolved in water (~10 mL) and the pH of the solution was adjusted to 3.5 by adding triethylamine. Concentration under reduced pressure yielded a solid which was treated with hot $CHCl_3$ (3 × 10 mL). The insoluble residue was then crystallized (H_2O–EtOH) to yield 1.7 g of **4** (80%): mp 218 °C; $[\alpha]^{20}_D$ +8.4° (c 0.5, H_2O). Anal. ($C_6H_{15}N_2O_2Cl$) C, H, N.

(–)-α-Methylornithine hydrochloride (5) was obtained in a manner similar to that described for **4**. (+)-Binaphthylphosphoric acid was used in the place of the (–) enantiomer: mp 214 °C; $[\alpha]^{20}_D$ –10° (c 1, H_2O). Anal. ($C_6H_{15}N_2O_2Cl$) C, H, N.

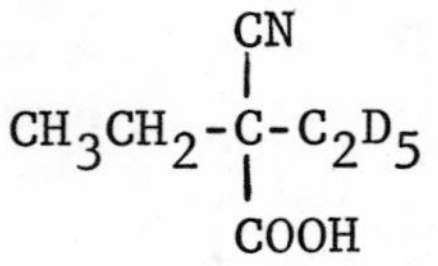

$$CH_3CH_2-\underset{\underset{COOH}{|}}{\overset{\overset{CN}{|}}{C}}-C_2D_5$$

Ref. 264

Optisch aktive Pentadeuteroethyl-propyl-cyanessigsäuren

53 g (0,32 mol) rac. Pentadeuteroethyl-propyl-cyanessigsäure wurden mit 56 g (0,33 mol) (−)-Threobase unter Erwärmen in 600 ml Ethanol gelöst. Nach 24stdg. Stehen bei 8° wurden 43,5 g Kristalle erhalten, Schmp. 155°. 2malige Umkristallisation aus 100 ml Ethanol ergab 30 g Diastereomerensalz, Schmp. 160°. Weitere 2malige Umkristallisation ergab 26 g Diastereomerensalz, Schmp. 160−161°. 26 g Diastereomerensalz wurden mit 10proz. HCl zerlegt. Ausb.: 11 g (42 % d. Th.), Schmp. 28−32°, Sdp.$_{0,5}$ 130°. $[\alpha]_D^{20} = + 3,4°$. $C_8H_8D_5NO_2$ (160,2) Ber.: C 60,0 H 8,18 N 8,7; Gef.: C 59,1 H 8,37 N 8,7.

24 g (0,15 mol) angereicherte (−)-Pentadeuteroethyl-propyl-cyanessigsäure wurden mit 25 g (0,15 mol) (+)-Threobase in 300 ml Ethanol gelöst. Nach 24stdg. Stehen bei 6°wurden 34 g Diastereomerensalz erhalten, Schmp. 154−157°. Weitere 2malige Umkristallisation ergab 30 g Diastereomerensalz, Schmp. 160−161°. 24 g Diastereomerensalz wurden mit 10proz. HCl zerlegt. Ausb.: 10,3 g (43 % d. Th.), Sdp.$_{0,4}$ 130°, $n_D^{20} = 1,4400$, $[\alpha]_D^{20} = − 3,5°$. $C_8H_8D_5NO_2$ (160,2) Ber.: C 60,0 H 8,18 N 8,7; Gef.: C 59,6 H 8,45 N 8,7.
Die IR-Spektren der deuterierten Cyanessigester und -säuren zeigen drei gegenüber den entsprechenden C-H-Schwingungen um etwa 800 cm^{-1} verschobene C-D-Valenzschwingungen (vgl.[7]).
C-H = 2955 (s), 2930 (m), 2870 (w) cm^{-1}
C-D = 2220 (m), 2107 (w), 2075 (w) cm^{-1}.

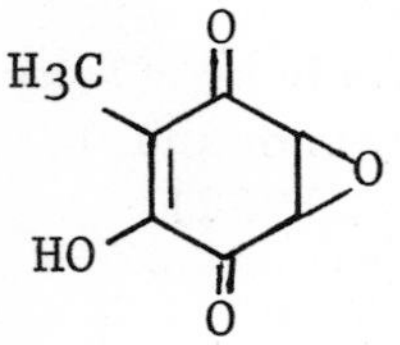

$C_7H_6O_4$

Ref. 265

Resolution of (±)-Terreic Acid (4) with (+)- and (−)-Ephedrine. (±)-Terreic acid (0.40 g, 2.94 mmol) was mixed with *l*-ephedrine (0.47 g, 2.85 mmol) in anhydrous ethyl ether (410 ml) and cooled in an acetone–Dry Ice bath. The deposited solid was redissolved in excess anhydrous ethyl ether. Partial evaporation of the solvent to the saturation point was carried out at reduced pressure. The saturated solution was cooled in the cold bath again. The salt was deposited in this manner ten times. Hydrochloric acid (1 N) was added to recover the resolved terreic acid from the last crop of deposited salt. The acidic solution was extracted with benzene and the extract was dried over anhydrous magnesium sulfate. Evaporation of the solvent gave crude terreic acid which was then purified by sublimation to give 67 mg (17%) of pale yellow crystals. The terreic acid recovered from the salt formed with *l*-ephedrine was the (+) enantiomer: $[\alpha]^{25}$D +26.5° (c 0.5, CHCl$_3$); mp 124.5–125°. The (−)-terreic acid could be obtained from the salt formed with *d*-ephedrine by the same method: $[\alpha]^{25}$D −24.9° (c 1.8, CHCl$_3$); mp 124.5–125°. The authentic terreic acid had $[\alpha]^{25}$D −26.1° (c 10.4, CHCl$_3$); mp 127–127.5°.

$$OC-COCH_2\underset{\underset{COOH}{|}}{\overset{\overset{\quad O\quad}{\frown}}{C}}-CH_2-COOH$$

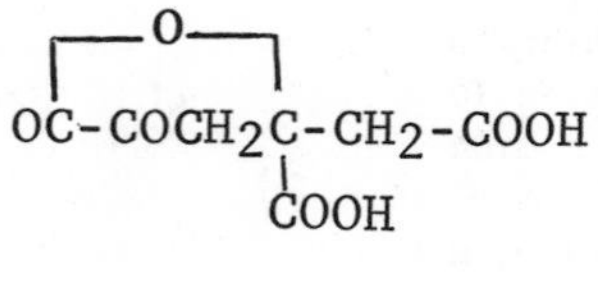

$C_7H_6O_7$

Ref. 266

Spaltung der d,l-Oxalcitramalsäure durch Brucin

8,3 g O. C. S. (= Oxalcitramalsäure) (Schmp. 177° Zers.) wurden in 50 ccm heißem Wasser gelöst und mit einer heißen Lösung von 9,7 g Brucin in 200 ccm Wasser versetzt. Nach 8-stündigem Stehen bei 0° wurde das ausgefallene saure Brucinsalz abfiltriert, in 5 ccm Wasser aufgeschlämmt und mit Soda auf p$_H$ 8,0 bis 8,4 gebracht. Das ausgefallene Brucin wurde abgesaugt, das Filtrat dreimal mit Chloroform durchgeschüttelt, mit 2n-H$_2$SO$_4$ auf p$_H$ 1,4 gebracht und erschöpfend mit Äther (peroxydfrei) extrahiert. Der Abdampfrückstand kristallisierte nach kurzem Stehen i. V.

Ausbeute: 1,9 g = 46%. $[\alpha]_{546}^{20} = — 167°$. Fünfmaliges Umkristallisieren aus Essigester/Petroläther gab farblose Kristalle. Schmp. 162—163° (Zers.).

$[\alpha]_{546}^{20} = — 215°$. Ausbeute 1,2 g = 29%.

Das Filtrat des sauren Brucinsalzes wurde i. V. auf 50 ccm eingeengt und acht Tage bei 0° stehen gelassen. Nach Abfiltrieren eines Niederschlages wurde wie oben aufgearbeitet. Ausbeute 2,2 g = 53% d. Th.

$[\alpha]_{546}^{20} = +158°$ in Wasser. Nach sechsmaligem Umkristallisieren wie oben. Schmp. 161,5—162,5° (Zers.). Ausbeute 1,5 g. (36% d. Th.). $[\alpha]_{546}^{20} = +213°$.

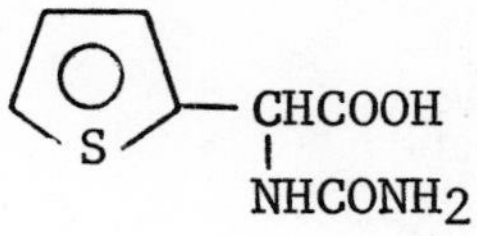

$C_7H_8N_2O_3S$

Ref. 267

Note:

Using N-benzyl-2-(hydroxymethyl)-cyclohexylamine as resolving agent, the authors obtained a less soluble diastereomeric salt after 2 recrystallizations from acetone ethanol (1:19) with $[\alpha]^{25}_{589}$ +30.6° (c 0.5, 95% ethanol) (melting point of authentic salt was 172-173°). The authentic more soluble salt had a m.p. 169-170° and $[\alpha]^{25}_{589}$ +29.6° (c 0.5 ethanol).

$C_7H_8O_3S$

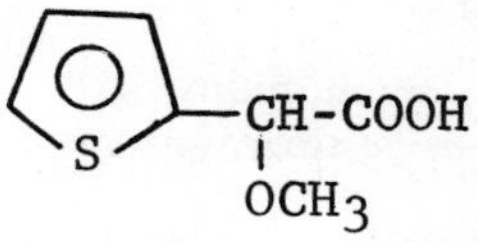

Ref. 268

Resolution of α-methoxy-2-thienylacetic acid

Preliminary experiments were carried out with 0.0015 mole of inactive acid and 0.0015 mole of various alcaloids in different solvent. In most cases, oily products or crystals with low activity only were obtained. Quinine in methanol and (−)-α-phenylethylamine in ethyl acetate yielded a dextrorotatory acid of high activity. Cinchonidine in 1:10 acetone–water solution gave levorotatory acid of moderate activity. Cinchonidine and quinine were therefore selected for the final resolution of the α-methoxy-2-thienylacetic acid.

14.7 g (0.0854 mole) of the inactive acid and 27.7 (0.0854 mole) of quinine were dissolved in 160 ml methanol. The salt was allowed to crystallize first at room temperature and finally in a refrigerator for three days. The salt was filtered off and recrystallized from methanol. The progress of the resolution was determined on small samples of the salt which were decomposed, the acid was isolated and the rotatory power determined in acetone solution.

Crystallization	1	2	3	4
ml methanol	160	75	20	15
g salt obtained	13.8	8.4	6.9	5.5
$[\alpha]^{25}_D$ of the acid	+73°	+81°	+83°	+83°

The mother liquor from the first crystallization was evaporated to a volume of 70 ml. The crystals (5.2 g) were filtered off and the filtrate evaporated to dryness. From the residue, 7.6 g of acid with $[\alpha]^{25}_D = -47°$ were isolated. This acid was dissolved together with 12.9 g (0.044 mole) of cinchonidine in 310 ml of 10 % aqueous acetone. The salt obtained was recrystallized from dilute acetone until the optical activity of the acid remained constant.

Crystallization	1	2	3	4	5	6	7
ml acetone–water	310	190	165	130	170	120	88
g salt obtained	15.6	12.8	10.4	8.7	5.9	4.2	2.9
$[\alpha]^{25}_D$ of the acid	−58°	−66°	−70°	−77°	−79°	−80°	−80°

(+)-α-Methoxy-2-thienylacetic acid

The acid was liberated from 5.2 g of the quinine salt with dilute hydrochloric acid and extracted with ether. 1.4 g of crude acid was obtained, which after recrystallization from benzene-petroleum-ether, gave 1.1 g, m.p. 85–86°.

0.1389 g acid: 8.00 ml 0.1003 *N* NaOH
$C_7H_8O_3S$ calc. Eq. weight 172.2
found Eq. weight 173.1
0.1003 g of acid dissolved in acetone to 10.00 ml: $2\alpha^{25}_D = +1.67°$; $[\alpha]^{25}_D = +83.2°$.

Weighed amounts of acid were made up to 5.00 ml with different solvents and the optical activity measured. The results are given below.

Solvent	g of acid	α_D^{23}(1 dm)	$[\alpha]_D^{23}$	$[M]_D^{23}$
Water	0.1005	+2.66°	+132°	+227°
Methanol	0.1006	+2.56°	+127°	+219°
Ethanol	0.1014	+2.28°	+112°	+193°
Chloroform	0.1005	+2.23°	+111°	+191°
Acetic acid	0.1010	+2.23°	+110°	+190°
Acetone	0.1021	+1.72°	+84.2°	+145°
Benzene	0.1008	+1.60°	+79.4°	+137°
Ethyl acetate	0.1015	+1.38°	+68.0°	+117°
Dioxane	0.1001	+1.23°	+61.3°	+106°
Ether	0.1009	+0.75°	+37°	+64°

(−)-*α-Methoxy-2-thienylacetic acid*

The acid was isolated from the cinchonidine salt in the usual way. 0.62 of crude acid was obtained, which after recrystallization gave 0.23 g pure acid, m.p. 85–86°.

0.1152 g acid: 6.64 ml 0.1003 *N* NaOH
$C_7H_8O_3S$ calc. Eq. weight 172.2
found Eq. weight 173.0
0.1010 g of acid dissolved in 10.00 ml acetone: $2\alpha_D^{25} = -1.68°$; $[\alpha]_D^{25} = -83.1°$.

$C_7H_8O_3S$ Ref. 268a

(+)-α-Methoxyphenylacetic acid was prepared by the methylation of (+)-mandelic acid with diazomethane using aluminium-*t*-butoxide as a catalyst. The resulting methyl ester was hydrolysed with dilute aqueous sodium hydroxide solution, the sodium salt precipitated with sodium chloride solution and the acid was recrystallised several times from a mixture of petrol-ether and benzene [6], m.p. 65—66°, $[\alpha]_D^{25} = +148.8°$ ($c = 1.028$, ethanol). Literature value [10,11]: m.p. 65—66°, $[\alpha]_D^{25} = -148{,}4°$ ($c = 1.025$, ethanol) resp. m.p. 63—64°, $[\alpha]_D^{25} = -150.6°$ ($c = 0.5412$, ethanol).

Spontaneous resolution of α-methoxy-2-thienylacetic acid. The inactive acid was dissolved in a 1:1 ether petrol-ether mixture and the solvent was allowed to evaporate slowly (during one month) at room temperature. After two weeks crystal formation was observed. It was possible to pick out individual prismatic crystals from the crystalline mass which ultimately formed. Each crystal selected in this way was split into three parts: one was dissolved in ether and the m.p. was determined after evaporating the solvent, a second was mixed with an authentic levorotatory acid and the m.p. of the mixture was determined after evaporating the ether from its ethereal solution, and the third was given similar treatment with a dextrorotatory acid. The m.p. of each of the unmixed samples was 85° while for the mixed samples the melting point was unchanged in one case and depressed in the other, showing that each crystal was a pure enantiomorph. Of ten selected crystals examined in this way, three gave undepressed m.p. with levorotatory acid and seven gave undepressed m.p. with dextrorotatory acid. The largest crystal selected (0.0223 g) was dissolved in 5 ml acetone and the rotation determined: its value was + 0.38° giving $[\alpha]_D^{25} = +86°$.

10. Bonner, W. A. *J. Am. Chem. Soc.* **73** (1951) 3126.
11. Mc Kenzie, A. *J. Chem. Soc.* **75** (1899) 753.

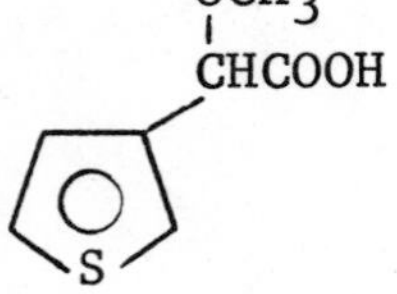

$C_7H_8O_3S$ Ref. 269

Resolution of α-methoxy-3-thienylacetic acid

Preliminary experiments were carried out with 0.0015 mole of the inactive acid and 0.0015 mole of various alcaloids in different solvents. Quinine in methanol yielded a dextrorotatory acid of high activity. Cinchonine in methanol-ethylacetate solution and strychnin in a 1:1 mixture of aceton-water yielded levorotatory acid. Quinine and strychnine were therefore used for the resolution of inactive acid.

14.6 g (0.0848 mole) and 27.6 g (0.0850 mole) of quinine were dissolved in 66 ml hot methanol and left to crystallize first at room temperature and than in a refrigerator. After three days, the salt was filtered off, which was then repeatedly recrystallized to maximum activity. The optical activities were measured in acetone as described above.

Crystallization	1	2	3	4	5	6
ml methanol	66	33	27	17	15	12
g salt obtained	15.4	11.2	8.3	6.9	5.6	4.1
$[\alpha]_D^{25}$ of the acid	+67°	+88°	+92°	+95°	+94°	+95°

From the mother liquor 6.1 g of acid with $[\alpha]_D^{25} = -50°$ (in acetone) was obtained. This acid was dissolved together with 12.0 g strychnine in 100 ml 1:1 aceton–water. The salt obtained was repeatedly recrystallized to maximum activity.

Crystallization	1	2	3
ml water–aceton 1:1	100	90	80
g salt obtained	12.3	10.2	8.8
$[\alpha]_D^{25}$ of the acid	−80°	−96°	−95°

(+)-*α-Methoxy-3-thienylacetic acid*

4.15 g of the quinine salt was decomposed with the dilute hydrochloric acid, and the acid thus liberated recrystallized from benzene–petroleum–ether, yielding 0.85 g, m.p. 86–87°.

0.1191 g acid: 6.68 ml 0.1003 − *N* NaOH
$C_7H_8O_3S$ calc. Eq. weight 172.2
found Eq. weight 172.6
0.1033 g of acid dissolved in acetone to 10.00 ml: $2\alpha_D^{25} = +1.955°$; $[\alpha]_D^{25} = +94.6°$.

The data for the rotatory power in various solvents are given below. Weighed amounts of acid were made up to 5.00 ml.

Solvent	g of acid	$\alpha_D^{24.5}$ (dm)	$[\alpha]_D^{24.5}$	$[M]_D^{24.5}$
Water	0.1008	+2.95°	+146°	+251°
Acetic acid	0.1001	+2.50°	+125°	+215°
Ethanol	0.1016	+2.44°	+120°	+207°
Acetone	0.1016	+1.92°	+95.5°	+164°
Ethylacetate	0.1016	+1.64°	+81.7°	+141°
Ether	0.1002	+0.91°	+45.5°	+78°

(−)-*α-Methoxy-3-thienylacetic acid*

The strychnine salt was decomposed with 1 *N* NaOH and the solution extracted twice with chloroform. The water phase was acidified and extracted with ether yielding 1.7 g of crude acid which after recrystallization gave 1.0 g pure acid m.p. 86–87°.

0.1822 g acid: 10.58 ml 0.1003 *N* NaOH
$C_7H_8O_3S$ calc. Eq. weight 172.2
found Eq. weight 171.7
0.1032 g of acid dissolved in acetone to 10.00 ml: $2\alpha_D^{25} = -1.96°$; $[\alpha]_D^{25} = -95.0°$.

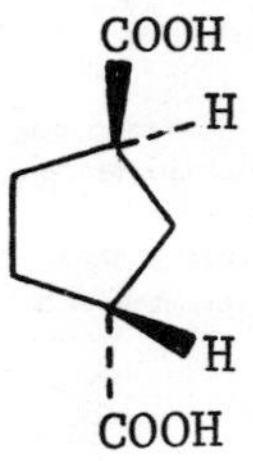

$C_7H_8O_4$ Ref. 270

Racemic *trans*-1,3-cyclopentanedicarboxylic acid was previously resolved by Birch and Dean.[18] However, the optical purities of both (+) and (−) isomers obtained were low, and an improved method was needed. It was found that the fractional recrystallization of the monostrychnine salt of the trans diacid from 95% ethanol gave either the (+)-diacid-rich salt or the (−)-diacid-rich salt as a precipitate, depending on the conditions. In a trial run, following Birch's method,[18] the partially resolved (+)-diacid {$[\alpha]^{20}D$ +12.4° (H_2O, *c* 8.0 g/dl)} and (−)-diacid ($[\alpha]^{20}D$ − 7.7°) were obtained. Starting with these diacids, the complete resolution could be carried out *via* the monostrychnine salt.

The partially resolved (−)-diacid, 3.16 g (2.00 × 10^{-2} mol), and strychnine, 6.69 g (2.00 × 10^{-2} mol), were dissolved on heating in 40 ml of 95% ethanol. The salt which precipitated at −15° was recrystallized four times from 95% ethanol (concentration ~10% w/v) to yield 1.0 g of the pure (−)-diacid strychnine (1:1) salt; after drying at room temperature *in vacuo*, mp 213–214° (becomes colored), $[\alpha]^{20}D$ −32.5 (50% EtOH–50% H_2O v/v, *c* 2.0 g/dl).

Anal. Calcd for $C_7H_{10}O_4 \cdot C_{21}H_{22}N_2O_2$: C, 68.28; H, 6.55; N, 5.69. Found: C, 68.35; H, 6.44; N, 5.73.

Pure (−)-*trans*-1,3-cyclopentanedicarboxylic acid was obtained from the above salt, followed by recrystallization from benzene–carbon tetrachloride (1:1): mp 87.5–88.5°, $[\alpha]^{20}D$ −35.1° (H_2O, *c* 8.0 g/dl) (lit. mp 85.0–85.7°, $[\alpha]^{15}D$ −22.6° (H_2O, *c* 8.0 g/dl)[17]).

Anal. Calcd for $C_7H_{10}O_4$: C, 53.16; H, 6.37. Found: C, 53.12; H, 6.27.

Starting from the partially resolved (+)-diacid and following the exact procedure as above, interestingly the pure (+)-diacid strychnine (1:1) salt could be obtained; after drying at room temperature *in vacuo*, the product had mp 202–203° (becomes colored), $[\alpha]^{20}D$ −10.7 (50% EtOH–50% H_2O v/v, *c* 2.0 g/dl).

Anal. Found: C, 68.05; H, 6.57; N, 5.45.

Pure (+)-*trans*-1,3-cyclopentanedicarboxylic acid was obtained from the above salt, followed by recrystallization from benzene–carbon tetrachloride (1:1): mp 87.5–88.5°, $[\alpha]^{20}D$ + 35.6° (H_2O, *c* 8.0 g/dl) (lit. mp 79.5–80.5°, $[\alpha]^{15}D$ +20.09° (H_2O, C, 8.0 g/dl)[18]).

Anal. Found: C, 53.24; H, 6.38.

When a solution of the equimolar (±)-diacid and strychnine in 95% ethanol (concentration 10–30% w/v) was allowed to stand at −15°, no salt was precipitated after standing for a few days. However, on addition of either the pure (+)-diacid salt or the pure (−)-diacid salt as a seed, a corresponding isomer-rich salt was precipitated. Thus, alternately using the pure (+)-diacid salt and the pure (−)-diacid salt as a seed, while the solvent was partially evaporated so that the concentration stayed at the range of 10–15% w/v, the salts of $[\alpha]^{20}D$ −12 to −13° and the salt of $[\alpha]^{20}D$ −27 to −29° were collected. The salts were then further recrystallized from 95% ethanol to afford the pure isomers.

(17) S. C. Temin and M. E. Baum, *Can. J. Chem.*, 43, 705 (1965).
(18) S. F. Birch and R. A. Dean, *J. Chem. Soc.*, 2477 (1953).

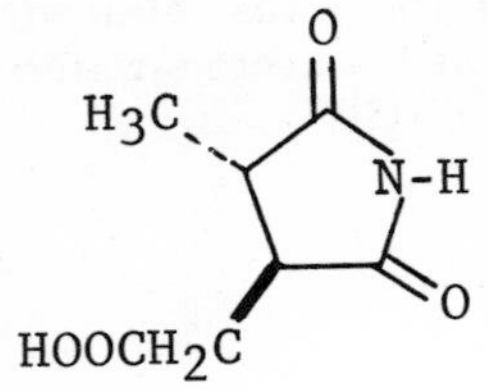

Ref. 271

$C_7H_9NO_4$

Resolution of trans-*Dihydrohaematinimide* (1) *and* (2).—*trans*-Dihydrohaematinimide [9] (11·5 g.) in ethyl acetate (100 ml.) was stirred while (−)-α-phenylethylamine [10] (7·6 g.) was added. Quantitative precipitation of the mixture of diastereoisomeric salts occurred instantly. This was recrystallised ten times from ethanol to a constant rotation to give (+)-trans-*dihydrohaematinimide* (−)-α-*phenylethylamine salt* (2·0 g.) as needles, m.p. 178—180°, $[\alpha]_D^{20}$ +39·5° (*c* 1·3 in 1 : 5 ethanol–water) (Found: C, 62·5; H, 7·4, N, 9·2. $C_{16}H_{22}N_2O_4$ requires C, 62·7; H, 7·2; N, 9·1%). The mother-liquors were concentrated and a second crop was recrystallised to constant rotation; the same procedure was followed with the new mother-liquors to give a third crop. (Concentration of the mother-liquors then caused the precipitation of the crude diastereoisomeric salt.) Total yield of fully resolved salt, 4·7 g.

The salt (4·2 g.) in water (100 ml.) was run slowly down a column of Zeocarb 225, (H^+), and washed through with water (200 ml.). Evaporation under reduced pressure gave the (+)-imide, an oil which could not be crystallised, $[\alpha]_D^{20}$ +67° (*c* 4·9 in MeOH); benzylamine salt, m.p. 166—169° (lit.,[4] 164—165°).

The mother-liquors from the resolution were evaporated to dryness and the imide was regenerated in a similar fashion. Repetition of the resolution technique on this imide, but with use of (+)-phenylethylamine,[10] gave (−)-trans-*dihydrohaematinimide* (+)-α-*phenylethylamine salt* (3·0 g.) as needles, m.p. 168—170°, $[\alpha]_D^{20}$ −38° (*c* 5·8 in 1 : 5 ethanol–water) (Found: C, 62·9; H, 7·1; N, 9·0%).

[4] G. E. Ficken, R. B. Johns, and R. P. Linstead, *J. Chem. Soc.*, 1956, 2272.

[9] G. E. Ficken, R. B. Johns, and R. P. Linstead, *J. Chem. Soc.*, 1956, 2280.

[10] A. Ault, *J. Chem. Educ.*, 1965, **42**, 269, based on W. Theilacker and H. G. Winkler, *Chem. Ber.*, 1954, **87**, 690.

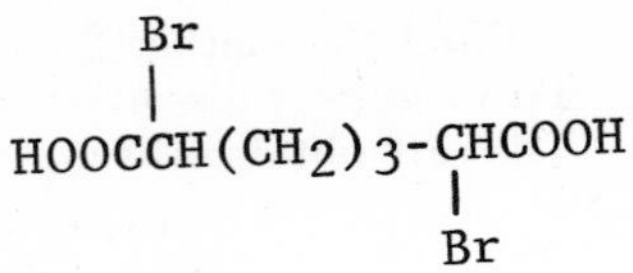

$C_7H_{10}Br_2O_4$ Ref. 272

Resolution of racem-α,α'-dibromopimelic acid.—9.7 g (0.0305 mole) of racemic acid and 8.9 g (0.030 mole) of cinchonidine were dissolved in 200 ml of hot acetone with addition of 20 ml of water. The solution was allowed to cool and then placed in the refrigerator. After one week the precipitate was filtered off, dried and weighed. About 0.3 g was taken for preliminary measurement of the optical activity (on the liberated acid) and the rest recrystallized from acetone, to which was added small amounts of water. The results are given below, where the optical activity of the acid is somewhat approximate, as the liberated acid was not purified before measurement.

Crystallization	1	2	3
ml of acetone	200	100	70
ml of water	20	5	3
Salt obtained, in g	4.2	2.9	1.9
$[\alpha]_D^{25}$ of acid (abs. ethanol)	−45°	−54°	−56°

The acetone was removed from the above first mother liquor and the remaining acid liberated. It was dissolved together with 10.5 g of brucine (4 H_2O) (0.023 mole) in a hot mixture of 85 ml of acetone and 85 ml of water. The solution was left at room temperature and filtered after one day. The salt obtained was then recrystallized until the optical activity of the liberated acid remained roughly constant.

Crystallization	1	2	3	4
ml of acetone	85	75	50	35
ml of water	85	75	50	35
Salt obtained, in g	10	6.0	3.9	2.4
$[\alpha]_D^{25}$ of acid (abs. ethanol)	—	+46°	+59°	+57°

It may be noted here that no salts were obtained in alcohol or acetone with quinine, strychnine and cinchonine.

(+)-α,α'-*Dibromopimelic acid.*—The acid brucine salt was decomposed with 4*N* sulphuric acid and the liberated acid was extracted with ether. After evaporation of the solvent an oily residue was obtained, which crystallized *in vacuo* after scratching the sides of the walls with a glass-rod. The optically active acid was recrystallized from a mixture of ether and light petroleum (b.p. 40°–60°) and obtained as rectangular transparent prisms with m.p. 60°–61°.

Due to the low melting point and the solubility conditions of the acid appreciable difficulties were encountered in the recrystallizations; hence only small amounts of the pure components were obtained.

Anal. Subst., 66.19 mg: 5.74 ml of 0.0721*N* NaOH. Equ. wt. Calc. 159.1; Found 159.9.

The optical activity of the dextrorotatory acid was measured in three different solvents (10.00 ml). The results are given below.

Solvent	mg of acid	$2\alpha_D^{25}$	$[\alpha]_D^{25}$	$[M]_D^{25}$
Abs. ethanol	66.2	+0.75°	+57°	+180°
Glacial acetic acid	34.2	+0.46°	+67°	+214°
Ethyl acetate	54.2	+1.06°	+98°	+311°

(−)-α,α'-*Dibromopimelic acid.*—The acid was liberated from the acid cinchonidine salt as described above for the enanthiomorph. The levorotatory acid resembled its antipode. M.p. 59°–61°.

Anal. Subst., 38.51 mg: 3.34 ml of 0.0721*N* NaOH. Equ. wt. Calc. 159.1; Found 159.9.
38.5 mg of the acid in 10.00 ml of abs. ethanol: $2\alpha_D^{25}$ = −0.43°; $[\alpha]_D^{25}$ = −56°; $[M]_D^{25}$ = −178°.

$C_7H_{10}NO_3P$

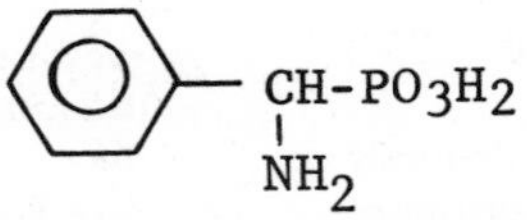

Ref. 273

1. (++) and (−−) α-phenylethylammonium salts of alkyl monoesters of N-carbobenzoxy-1-amino-1-phenylmethylphosphonic acid

To a solution of 10 mmole of (methyl or ethyl) alkyl monoesters of (+−) N-carbobenzoxy-1-amino-1-phenylmethylphosphonic acid in methyl alcohol 10 mmole of *R*(+) or *S*(−) α-phenylethylamine was added and the solvents were evaporated under reduced pressure. The mixture of diastereoisomeric (++) and (−−), or (−−) and (+−) salts obtained were crystallized from a suitable solvent. In this way enantiomeric (++) and (−−) α-phenylethylammonium salts of alkyl monoesters of N-carbobenzoxy-1-amino-1-phenylmethylphosphonic acid were obtained. The yields and properties of these salts are listed in Table 1.

2. Alkyl monoesters of (+) and (−) N-carbobenzoxy-1-amino-1-phenylmethylphosphonic acids

Two mmole of (++) or (−−) salts of (methyl or ethyl) alkyl monoester of N-carbobenzoxy-1-amino-1-phenylmethylphosphonic acid was dissolved in 5 ccm of 1 N NaOH. The reaction mixture was then extracted twice with 5 ccm of ethyl ether. The water layer was acidified with concentrated HCl to pH 1. The precipitated alkyl monoesters of (+) and (−) N-carbobenzoxy-1-amino-1-phenylmethylphosphonic acid were filtered off, dried and recrystallized from a suitable solvent. The yields and properties of these compounds are listed in Table 1.

Table 1

Compound		Recrystallization solvent	M.p. °C	$[\alpha]_D^{20}$	Yield %	Formula Molecular mass	Analysis, % N Calcd.	Analysis, % N Found
C_6H_5—CH(NHZ)—P(O)$(OCH_3)(O^-\overset{+}{A})$	(− −)	1. MeOH — ether (1 : 2)	196—198	−17.6 c = 2, MeOH	73	$C_{24}H_{29}N_2O_5P$ 456.5	6.1	6.2
	(+ +)	2. isopropanol		+17.5 c = 2, MeOH	70			
C_6H_5—CH(NHZ)—P(O)$(OCH_3)(OH)$	(−)	$CHCl_3$	155—156	−19.0 c = 2, 1 N NaOH	91	$C_{16}H_{18}NO_5P$ 335.3	4.1	4.0
	(+)	*n*-hexane		+18.9 c = 2, 1 N NaOH	92			
C_6H_5—CH(NHZ)—P(O)$(OCH_3)_2$	(−)	AcOEt	129—130	−14.0 c = 2, MeOH	95	$C_{17}H_{20}NO_5P$ 349.3	4.0	4.1
	(+)	*n*-hexane		+14.0 c = 2, MeOH	96			
C_6H_5—CH(NHZ)—P(O)$(OC_2H_5)(O^-\overset{+}{A})$	(− −)	MeOH — ether (1 : 5)	186—188	−18.1 c = 2, MeOH	71	$C_{25}H_{31}N_2O_5P$ 470.5	5.9	5.9
	(+ +)	twice		+18.0 c = 2, MeOH	70			
C_6H_5—CH(NHZ)—P(O)$(OC_2H_5)(OH)$	(−)	$CHCl_3$	172—173	−18.8 c = 2, 1 N NaOH	90	$C_{17}H_{20}NO_5P$ 349.3	4.0	3.9
	(+)	*n*-hexane		+19.0 c = 2, 1 N NaOH	91			
C_6H_5—CH(NHZ)—P(O)$(OC_2H_5)_2$	(−)	AcOEt	138—140	−14.3 c = 2, 1 N NaOH	85	$C_{19}H_{24}NO_5P$ 377.3	3.7	3.8
	(+)	*n*-hexane		+14.0 c = 2, 1 N NaOH	90			

Z — carbobenzoxy group, A—H_2N—CH(CH_3)—CH_3.

3. Methyl diesters of (+) and (−) N-carbobenzoxy-1-amino-1-phenylmethylphosphonic acids

To an ice-cooled solution of 10 mmole of methyl monoester of (+) or (−) N-carbobenzoxy-1-amino-1-phenylmethylphosphonic acid in methanol an etheral solution of diazomethane was added portionwise until the solution became light yellow. Volatile components of the reaction mixture were distilled off under reduced pressure and the remainder was crystallized from *n*-hexane and ethyl acetate mixture. The yields and properties of the compounds obtained are shown in Table 1.

4. Ethyl diesters of (+) and (−) N-carbobenzoxy-1-amino-1-phenylmethylphosphonic acids

The title compounds were prepared as described for methyl esters, using a hexane solution of diazoethane. The yields and properties of these esters are listed in Table 1.

5. Hydrolysis of alkyl monoesters and diesters of (+) and (−) N-carbobenzoxy-1-amino-1-phenylmethylphosphonic acids and isolation of enantiomers of 1-amino-1-phenylmethylphosphonic acid

Three mmole of (methyl or ethyl) alkyl mono or diester of (+) or (−) N-carbobenzoxy-1-amino-1-phenylmethylphosphonic acid was dissolved in a 1:1 mixture of 40 ccm of acetic acid and hydrochloric acid and heated to boiling for 10 hrs. The reaction mixture was evaporated to dryness; 50 ccm of water was then added and evaporated again; this was repeated five times. To the remainder 40 ccm of ethanol and 2 ccm of propylene oxide was added. The crystalline product was filtered off and dried. Its specific rotation was determined in 1 N NaOH. The yields and specific rotations of enantiomers of 1-amino-1-phenylmethylphosphonic acid obtained are shown in Table 2.

Table 2

Hydrolyzed compound		$[\alpha]_D^{20}$ (c = 2, 1 N NaOH) of isolated 1-amino-1-phenylmethylphosphonic acid	Yield %
C_6H_5–CH(NHZ)–P(=O)(OCH_3)(OH)	(−)	−19.4	88
	(+)	+19.2	85
C_6H_5–CH(NHZ)–P(=O)(OCH_3)$_2$	(−)	−19.3	91
	(+)	+19.4	90
C_6H_5–CH(NHZ)–P(=O)(OC_2H_5)(OH)	(−)	−19.3	90
	(+)	+19.4	88
C_6H_5–CH(NHZ)–P(=O)(OC_2H_5)$_2$	(−)	−19.4	92
	(+)	+19.3	89

Z – carbobenzoxy group.

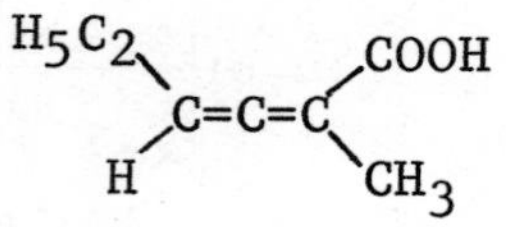

$C_7H_{10}O_2$ Ref. 274

*(+)- und (−)-***1b**: Nach den allgemeinen Vorschriften erhielt man aus 2.653 g racem. **1b** und 3.104 g Cinchonidin nach vier Umkristallisationen aus Aceton 2.315 g (52%) Salz als lange, farblose Nadeln mit Schmp. 148−149°C (Rotfärbung). $[\alpha]_{589}^{25} = -35.1°$, $[\alpha]_{350}^{25} = -162°$ (c' = 0.006166, C_2H_5OH).

$C_{26}H_{32}N_2O_3$ (420.6) Ber. C 74.25 H 7.69 N 6.66 Gef. C 74.31 H 7.69 N 6.96

Aus dem Filtrat der ersten Kristallisationsfraktion wurden 0.710 g (26.6%) optisch unreines (−)-**1b** gewonnen; Sdp. 65−69°C/0.04 Torr (Kugelrohr). $[\alpha]_{589}^{25} = -39.2°$ (c' = 0.04678, C_2H_5OH). — Aus 2.164 g Cinchonidinsalz wurden 0.508 g [78%] (+)-**1b** gewonnen; Sdp. 98−100°C/0.07 Torr (Kugelrohr). $[\alpha]_{589}^{25} = +79.4°$ (c' = 0.009109, C_2H_5OH). — CD (C_2H_5OH): λ_{max} ($\Delta\varepsilon$) = 284 (−0.0125), 250 (+0.13), 215 nm (+1.6).

$C_7H_{10}O_2$ (126.2) Ber. C 66.62 H 7.99 Gef. C 65.96 H 8.13

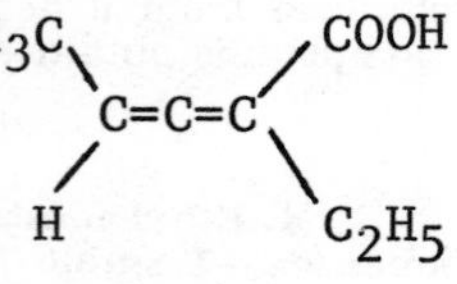

$C_7H_{10}O_2$ Ref. 275

*(+)- und (−)-***1c**. Nach den allgemeinen Vorschriften ergaben 1.476 g racem. **1c** und 1.738 g Cinchonidin nach vier Umkristallisationen aus Aceton 1.452 g (58%) lange, farblose Nadeln mit Schmp. 161−162°C (Rotfärbung). $[\alpha]_{589}^{25} = -55.0°$, $[\alpha]_{350}^{25} = -187°$ (c' = 0.005326, C_2H_5OH).

$C_{26}H_{32}N_2O_3$ (420.6) Ber. C 74.25 H 7.69 N 6.66 Gef. C 73.77 H 7.52 N 6.83

Aus dem Filtrat der ersten Kristallisationsfraktion wurden 0.438 g (29%) optisch unreines (−)-**1c** gewonnen; Sdp. 86−90°C/0.06 Torr (Kugelrohr). $[\alpha]_{589}^{25} = -11.6°$ (c' = 0.023226, C_2H_5OH). — Aus 2.250 g Cinchonidinsalz wurden 0.493 g [73%] (+)-**1c** erhalten; Sdp. 86−89°C/0.06 Torr (Kugelrohr), Schmp. 49−51°C. Umkristallisation dieses Produktes aus Petroläther ergab nur 0.322 g (+)-**1c** als kleine Stäbchen mit Schmp. 47−49°C und unverändertem Drehwert. Die GC zeigte, daß dieses Produkt einheitlich war. $[\alpha]_{589}^{25} = +27.3°$ (c' = 0.011907, C_2H_5OH). — CD (C_2H_5OH): λ_{max} ($\Delta\varepsilon$) = 275 (−0.0087). 252 (+0.059), 244 (+0.10), 215 nm (+1.0).

$C_7H_{10}O_2$ (126.2) Ber. C 66.62 H 7.99 Gef. C 66.67 H 7.68

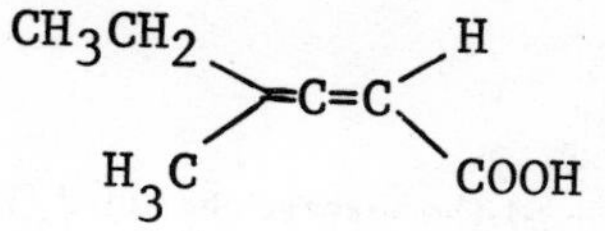

$C_7H_{10}O_2$

Ref. 276

Resolution of 4-Methylhexa-2,3-dienoic Acid.—Quinine (15 g) in hot ethyl acetate (300 ml) was added to the acid (10 g) in hot ethyl acetate (100 ml). The product which crystallised was recrystallised to constant rotation; m.p. 101—102°, $[\alpha]_D^{25}$ $-113{\cdot}5°$ ($\pm0{\cdot}1°$) (*c* 0·7 in $CHCl_3$). Decomposition (dil. HCl) gave (−)-*4-methylhexa-2,3-dienoic acid* (2·7 g, 54%), m.p. 44—45°, $[\alpha]_D^{24}$ $-36{\cdot}1°$ ($\pm0{\cdot}1°$) (*c* 0·34 in $CHCl_3$) (Found: C, 66·9; H, 7·9. $C_7H_{10}O_2$ requires C, 66·6; H, 8·0%). The mother liquor from the resolution gave, on work-up, the partially resolved (+)-acid, m.p. 34—35°, $[\alpha]_D^{24}$ $+21{\cdot}7°$ ($\pm0{\cdot}1°$) (*c* 1·36 in $CHCl_3$) (Found: C, 66·9; H, 7·95%).

The (−)-*methyl ester* (CH_2N_2) had b.p. 70—72° at 11 mmHg, n_D^{18} 1·4704, $[\alpha]_D^{24}$ $-29{\cdot}4°$ ($\pm0{\cdot}1°$) (*c* 0·48 in $CHCl_3$), $[\alpha]_D^{24}$ $-52{\cdot}6°$ ($\pm0{\cdot}1°$) (neat liquid) (Found: C, 68·7; H, 8·7. $C_8H_{12}O_2$ requires C, 68·5; H, 8·6%).

$C_7H_{10}O_2$

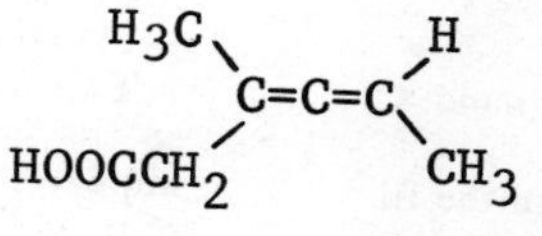

Ref. 277

2. Enantiomerentrennung von (*RS*)-2-Methyl-2,3-pentadiensäure (13) mit Cinchonidin sowie Überführung in die optisch aktiven Methylester (*R*)-3 und (*S*)-3. – Die Vorschrift zur Antipodenspaltung von *Runge et al.* [20a] wurde wie folgt modifiziert: Eine feinverteilte, feste Mischung von 11,21 g (100 mmol) (*RS*)-**13** und 15,33 g (52 mmol) Cinchonidin wurde langsam auf 60° erhitzt und dann mit soviel Aceton versetzt (*ca.* 800 ml), dass eine klare Lösung vorlag. Nach dem Abkühlen wurde über Nacht bei 0° auskristallisieren gelassen. Sowohl die abfiltrierte und im RV. eingedampfte Mutterlauge, wie auch ein Aliquot des gebildeten Salzes[26]) (*ca.* 4 g) wurden zur Freisetzung der Säure aufgehoben. Der Rest des Salzes wurde in obiger Art 9mal aus siedendem Aceton umkristallisiert, wobei weitere Aliquote entnommen wurden (vgl. *Tab. 4*). Zur Rückgewinnung der freien Säuren (*R*)-**13** bzw. (*S*)-**13** wurden die einzelnen Fraktionen mit verd. Schwefelsäure [20b] aufgearbeitet und mindestens 1mal aus siedendem Pentan umkristallisiert.

[20] a) *W. Runge, G. Kresze & E. Ruch*, Liebigs Ann. Chem. *1975*, 1361; b) *iidem*, ibid. *756*, 112 (1972).

[26]) $C_{25}H_{30}N_2O_3$ (406,53) Ber. C 73,86 H 7,44 N 6,89% Gef. C 73,60 H 7,49 N 6,87%.

Tabelle 4. *Physikalische Eigenschaften der optisch aktiven 2-Methyl-2,3-pentadiensäuren*

Messreihe Nr.	Säure	Umkr.[a]) (Aceton)	Umkr.[b]) (Pentan)	Smp. [°]	c[c]) mg/ml	$[\alpha]_\lambda$[d]) 365	436	546	578	589	e[e])	p[f])
1	(−)-(*R*)-13	1	1	54-55	9,1 und 20,5	−47,95	−27,46	−14,98	−13,02	−12,44	0,15 ± 0,02	0,17
2	(*RS*)-13	-	3	55	-						0,00	0,00
3	(+)-(*S*)-13	1	1	55	13,9	153,57	86,33	46,55	40,65	38,85	0,54 ± 0,02	0,53
4	(+)-(*S*)-13	2	1	56	13,5	184,52	104,22	56,30	48,59	46,67	0,62 ± 0,02	0,64
5	(+)-(*S*)-13	7	1	58	17,6	226,02	127,61	68,92	59,72	57,39	0,84 ± 0,02	0,80
6	(+)-(*S*)-13	10	2	64	7,7 und 15,2	279,87	157,40	85,06	74,01	70,39	0,92 ± 0,02	0,95

a) Anzahl Umkristallisationen des Cinchonidin-Adduktes (1. Kristallisation = 1).
b) Anzahl Umkristallisationen der freien Säure.
c) In mg Säure/1 ml Äthanol («Uvasol»).
d) Messtemp. 20°, λ in nm; Drehwerte in Grad, kein Vorzeichen bedeutet +.
e) Enantiomerenreinheiten e von den entsprechenden Estern 3 übernommen (vgl. *Tab. 5*).
f) Optische Reinheit p (vgl. Fussnote 2) berechnet aus den über alle Wellenlängen gemittelten Werten (rel. Fehler ≤ 2.5%).

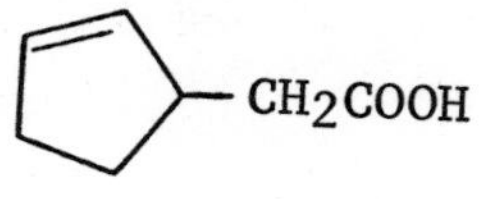

$C_7H_{10}O_2$

Ref. 278

Resolution of 2-Cyclopentene-1-acetic Acid.—Brucine, of a number of alkaloids tested,[32] proved the most practicable resolving agent. In a typical resolution, brucine, 55.0 g., was added to a hot solution of 14.9 g. of 2-cyclopentene-1-acetic acid in 100 ml. of acetone to which had been added 3.5 ml. of water.[33] From the head fraction of 36.2 g., $[\alpha]^{28}D$ $-18.84°$ (*c* 5.4, water), after eight recrystallizations from acetone–water, there was obtained 2.0 g. of product, $[\alpha]^{28}D$ $+0.99°$ (*c* 5.0, water), whose rotation remained substantially unchanged on further recrystallization. The resolutions were reproducible; the pure diastereomeric salt always had $[\alpha]^{28}D$ $+1.2 \pm 0.3°$ (water). One sample had $[\alpha]^{28}D$ $+1.54°$ (*c* 7.1, water), m.p. 147.5–148.5°.

Anal. Calcd. for $C_{30}H_{36}N_2O_6$: C, 69.2; H, 6.97; N, 5.38. Found: C, 69.0; H, 7.26; N, 5.44.

The diastereomeric salt was dissolved in warm water, and the solution was made basic with concentrated aqueous ammonia. The precipitated brucine was filtered and the filtrate was acidified with dilute hydrochloric acid and extracted with ether. The ether extracts were washed with saturated sodium chloride solution, dried (sodium sulfate) and distilled. The desired acid, b.p. 105° (8 mm.), had $n^{25}D$ 1.4673, $[\alpha]^{30}D$ $+109.2°$ (*c* 5.9, chloroform).

Anal. Calcd. for $C_7H_{10}O_2$: C, 66.6; H, 7.99. Found: C, 66.7; H, 7.76.

(32) The use of morphine, quinine, cinchonine, quinidine and strychnine proved less successful in our hands. Cinchonidine precipitated the (−)-acid, but recrystallization from acetone, ethanol or solvent mixtures (*e.g.*, acetone–water) did not yield a pure diastereomer.

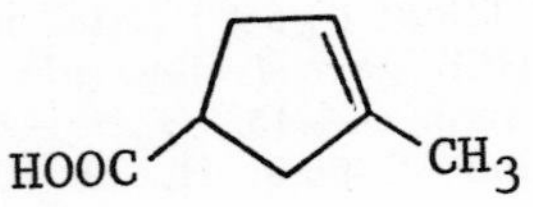

$C_7H_{10}O_2$ Ref. 279

Resolution of 1-Methylcyclopentene-4-carboxylic Acid. A suspension of quinine (4.2 g, 12.7 mmol) in 25 ml of ethyl acetate was combined with a solution of 1-methylcyclopentene-4-carboxylic acid (1.6 g, 12.7 mmol) in 15 ml of the same solvent When stored at room temperature for 20 hr, the solution deposited 1.5 g of the quinine salt, two recrystallizations of which from ethyl acetate furnished 1.01 g of colorless crystals: mp 135° after being dried over P_2O_5 in vacuo; $[\alpha]^{27}D$ −135.5° (*c* 0.5329, MeOH).

Anal. Calcd for $C_{27}H_{34}N_2O_4$: C, 72.0; H, 7.6; N, 6.2. Found: C, 72.0; H, 7.8; N, 6.3.

Fractional crystallization of the material obtained from the filtrates afforded 0.45 g of the salt of the same optical purity. The combined samples were suspended in water, acidified with ice-cold 3 *N* HCl to pH 2 and extracted with three 30-ml portions of ether. The extract was dried over $MgSO_4$ and concentrated by distillation to give 0.4 g of the acid: $[\alpha]^{27}D$ +21.2°, $[\alpha]_{365}{}^{27}$ +65.2° (0.530°, *c* 0.8130, absolute EtOH). The corresponding methyl ester, 4-carbomethoxy-1-methylcyclopentene, obtained by treatment with ethereal diazomethane, was purified by GLC on column B: $[\alpha]^{27}D$ +22.9; $[\alpha]_{365}{}^{27}$ +71.1° (+0.606°, *c* 0.8526, MeOH); NMR identical with that of the racemic sample described above.

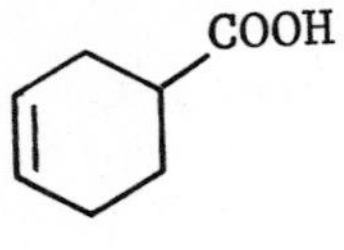

$C_7H_{10}O_2$ Ref. 280

Resolution of 3-Cyclohexene-1-carboxylic Acid. (a) (*R*)-(+)-3-Cyclohexene-1-carboxylic Acid. Ethyl 3-cyclohexene-1-carboxylate[55] (bp 90° (20 mm), $n^{23}D$ 1.4548) was saponified to the racemic acid (bp 118° (8 mm), $n^{22}D$ 1.4804).[56] Although partial resolution and assignment of absolute configuration have been accomplished,[33] resolution is better effected by crystallization of the brucine salt from acetone [124 g of acid, 388 g of brucine, 1980 ml of acetone; seeding; 12-hr standing]. The progress of the resolution is followed by recovery of small quantities of acid which was purified by evaporative distillation prior to polarimetry: (I) 275 g of brucine salt, recrystallized from 1375 ml of acetone; (II) 180 g, 1020 ml, $[\alpha]^{26}D$ +33.1° (neat, 0.5-dm tube); (III) 124 g, 1100 ml; (IV) 68 g, 800 ml, +53.8°; (V) 26.9 g, 400 ml; (VI) 12.2 g, 113 ml, +58.5°; (VII) 8.2 g, 113 ml, +59.2°; (VIII) 2.83 g, . . . , +58.6°.

The resolved salt (87.0 g) was dissolved in 400 ml of H_2O, acidified cautiously with 22 ml of concentrated HCl, and extracted continuously for 30 hr with 300 ml of ether. The residue from the dried (anhydrous Na_2SO_4) ethereal solution was distilled yielding 19.2 g of (*R*)-(+)-4-carboxycyclohexene, bp 114–115° (7 mm), $[\alpha]^{30}D$ +119.9°, $[\alpha]^{30}{}_{578}$ +123.5°, $[\alpha]^{30}{}_{546}$ +140.7° (neat, 1-dm tube).

Catalytic hydrogenation (Pt in ethyl acetate) afforded cyclohexanecarboxylic acid, mp 28–29°, $[\alpha]^{27}D$ 0.0°.

(b) (*s*)-(−)-3-Cyclohexene-1-carboxylic Acid. From the mother liquor from the first crystallization above (272 g in 1375 ml, very slow crystallization), the brucine salt of the enantiomeric acid was obtained: (I) 200 g, 1000 ml, $[\alpha]^{27}{}_{578}$ −36.2° (*c* 1.54, CCl_4) of the recovered acid; (II) 152 g, . . . , −53.8°; (III) 110 g, 775 ml, −69.4°; (IV) 83 g, 650 ml, −82.4°; (V) 50.6 g, 520 ml, −91.7°; (VI) 32.9 g, 350 ml, −101.3°; (VII) 23.7 g, . . . , −108.3°. (This value compares with $[\alpha]^{27}{}_{578}$ +109.9° (*c* 1.01, CCl_4) for the dextrorotatory enantiomer.)

(56) A. A. Petrov and N. P. Sopov, *J. Gen. Chem. USSR*, **17**, 2228 (1947).

$C_7H_{10}O_2$ Ref. 280a

R-(+)- and *S*-(−)-**4-Carboxycyclohexenes (IIIa and IIIb).**—*rac*-4-Carboxycyclohexene[14] (27 g., 0.21 mole) was dissolved in 100 ml. of acetone containing quinine (82 g., 0.25 mole) by gently agitating the warm mixture during 2 hr. After the solution was allowed to cool to room temperature, it was kept in the refrigerator overnight while it slowly deposited the first crop of crystalline salt. Collection and air drying of the salt provided 28 g. of levorotatory material, m.p. 140–142°, $[\alpha]^{26}D$ −142 ± 0.3° (*c* 3.01, methanol).

This material was dissolved in 150 ml. of fresh acetone and allowed to recrystallize slowly in the refrigerator. This first recrystallization yielded 22 g. of salt, m.p. 142.5–144°, $[\alpha]^{26}D$ −144 ± 0.4° (*c* 2.57, methanol).

Two more recrystallizations, carried out in the same manner as the first, left 18 g. of levorotatory salt with no significant improvement in either melting point or rotational magnitude, m.p. 142–144°, $[\alpha]^{26}D$ −146 ± 0.6° (*c* 1.70, methanol).

After treatment of the entire quantity (18 g.) of this salt with 6 *N* hydrochloric acid, the mixture was extracted with four 100-ml. portions of ether, and the combined ethereal extracts yielded, after drying over anhydrous sodium sulfate and evaporation, crude (brown-yellow oil) IIIb (3.8 g., 21% yield), $[\alpha]^{28}D$ −1.4° (*c* 7.88, methanol), which was directly converted to the corresponding levorotatory carbinol Xb according to the experimental account given below.

The mother liquors obtained from the three recrystallizations of the levorotatory diastereomeric salt described above were combined and evaporated. The residue was taken up in 6 *N* hydrochloric acid, and the resulting mixture was extracted with ether. Examination of the residue (brown-yellow oil), obtained after the dried ethereal solution was filtered and evaporated, indicated crude IIIa, $[\alpha]^{28}D$ +1.2° (*c* 6.92, methanol), which was reduced to the dextrototatory carbinol Xa without further purification.

(14) Prepared *via* silver oxide oxidation of 4-formylcyclohexene which was purchased from K and K Laboratories, Inc.

$C_7H_{10}O_2$ Ref. 280b

Note:

Obtained an acid, b.p. 66°C. (0.15 mm), $[\alpha]_D^{25}$ = -95.9° (c, 1, $CHCl_3$), using (+)-α-phenylethylamine as resolving agent. Using (-)-α-phenylethylamine as resolving agent obtained an acid b.p. 66°C (0.15 mm.), $[\alpha]_D^{25}$= +95.3°C. (c 1, $CHCl_3$).

$C_7H_{10}O_3$

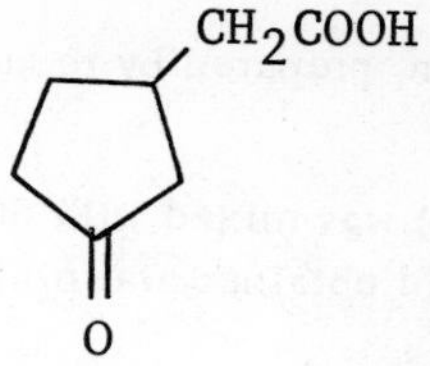

Ref. 281

Resolution of 3-oxocyclopentaneacetic acid (VI). A mixture of 3-oxocyclopentaneacetic acid[5] (5·68 g) and brucine (20 g) was dissolved in 100 ml acetone heated under reflux. The solution deposited crystals of the salt, m.p. 129–136°, on cooling and standing for several hr. Two crystallizations from acetone yielded colourless crystals (6·8 g), m.p. 169–170°, $[\alpha]_D^{21}$ −57·6° (*c* = 5 in $CHCl_3$).

A total of 217·8 g of the recrystallized brucine salt was dissolved in 1200 ml hot water and made alkaline to litmus with 1:1 NH_4OH. The brucine was filtered off and the filtrate acidified with 1:1 HClaq, then extracted exhaustively with ether. The ether solution was washed with brine, dried over $MgSO_4$ and concentrated, leaving 31·5 g of acid VI, $[\alpha]_D^{21}$ −24·6° (*c* = 5 in $CHCl_3$).

[5] E. Demole and M. Stoll, *Helv. Chem. Acta* 45, 692 (1962).

$C_7H_{10}O_3$

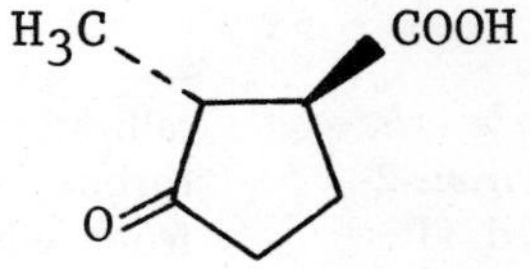

Ref. 282

Resolution of 2-Methylcyclopentanone-3-carboxylic Acid into Optical Antipodes

a) Preparation of quinine salts. A solution of 5 g of the racemic 2-methylcyclopentanone-3-carboxylic acid in 15 ml of anhydrous alcohol was mixed with a hot solution of 12.15 g of quinine. The mixture was heated to boiling and left for 2 hours at room temperature. The precipitate was filtered off and washed with cold alcohol (5 ml). The weight of the residue was 9.2 g, m.p. 162-163°, $[\alpha]_D^{20}$ − 127.4° (c 2, in water). The salt obtained (9.2 g) was recrystallized from 30 ml of anhydrous alcohol; we obtained 5.85 g of a salt, m.p. 169-170° $[\alpha]_D^{20}$ − 153.2° (c 2, in water). Repeated crystallization did not change the melting point or the specific rotation of the salt.

Found %: C 69.63, 69.66; H 7.47, 7.51; N 6.10, 6.29. $C_{27}H_{34}O_5N_2$. Calculated %: C 69.52; H 7.29; N 6.01.

From the mother liquor remaining after the isolation of the first salt a second salt precipitated after 24 hours, which was filtered off and washed with alcohol. The weight of the precipitate was 3.8 g, m.p. 167-168°, $[\alpha]_D^{20}$ − 93.8° (c 3, in water). After recrystallization from 8 ml of anhydrous alcohol, we obtained 3.3 g of a salt, m.p. 168-169°, $[\alpha]_D^{20}$ − 93.5° (c 3, in water).

Found %: C 69.65, 69.58; H 7.50, 7.45; N 5.94, 5.87. $C_{27}H_{34}O_5N_2$. Calculated %: C 69.52; H 7.29; N 6.01.

b) Isolation of the optically active acids. A 10% ammonia solution was added to a solution of 5.3 g of the first salt ($[\alpha]_D^{20}$ −153.2°) in 235 ml of water until the quinine was completely precipitated. The precipitated quinine was filtered off, weight 3.4 g (calculated 3.56 g) and the aqueous solution twice extracted with ether, concentrated in vacuum, acidified with sulfuric acid solution and saturated with magnesium sulfate. The acid separated as an oil, which quickly crystallized. The acid was extracted with ether, the ether solution dried with magnesium sulfate and the ether evaporated off. The residue (1.4 g) was recrystallized from 4 ml of benzene; we obtained 0.85 g of the (−)-acid, m.p. 98-98.5°, $[\alpha]_D^{20}$ −64.5° (c 2.5, in water).

Found %: C 59.40, 59.21; H 7.05, 7.18. $C_7H_{10}O_3$. Calculated %: C 59.15; H 7.09.

Similarly, from 3 g of the second salt ($[\alpha]_D^{20}$ −93.5°) we isolated 1 g of an acid, which was recrystallized from benzene and washed with petroleum ether (b.p. 30-60°); we obtained 0.55 g of the (+)-acid, m.p. 97.5-98.0°, $[\alpha]_D^{20}$ + 63.8° (c 2.5, in water).

Found %: C 59.30, 59.40; H 7.20, 7.21. $C_7H_{10}O_3$. Calculated %: C 59.15; H 7.09.

Literature data [1] for dihydrosarcomycin, prepared by reduction of the natural antibiotic: m.p. 99-99.5°, $[\alpha]_D^{20}$ +66.7° (c 1, in water).

50 mg of the (−)-acid (m.p. 98.0-98.5°) was mixed with 50 mg of the (+)-acid (m.p. 97.5-98.0°) and the mixture recrystallized from benzene. The acid obtained was optically inactive and had m.p. 94-95°.

[1] J.R. Hooper, L.C. Cheney, M.J. Cron, O.B. Fardig, D.A. Johnson, D.L. Johnson, M. Palermiti, H. Schmitz and W.B. Wheatley, Antibiotics and Chemotherapy 5, 585 (1955).

$C_7H_{10}O_3$ Ref. 283

Optical Isomers of 2-Hydroxy-3-cyclohexenecarboxylic Acid. The procedure for resolution of racemic *cis*- and *trans*-2-hydroxy-3-cyclohexenecarboxylic acids was patterned after that for the saturated compounds, described by Real and Pascual (1953); 0.025 mole of brucine (10 g) and 0.025 mole of 2-hydroxy-3-cyclohexenecarboxylic acid (3.5 g) were refluxed in acetone or ethyl acetate for 1–3 hr. The brucine salts were carefully and repeatedly recrystallized until an essentially constant rotation was achieved. To obtain the free hydroxycarboxylic acids, the salts were hydrolyzed with ten volumes of 2 N NaOH and the brucine which separated was removed by filtration. The filtrate was acidified with 2 N H_2SO_4, saturated with ammonium sulfate, and extracted with ether. The ether extract was evaporated, yielding a crystalline product. Recrystallization of the optically active 2-hydroxy-3-cyclohexenecarboxylic acids from ethyl acetate–ligroin (1:5, v/v) and/or from ether was performed until constant rotation was achieved. The final yield of each isomer of 2-hydroxy-3-cyclohexenecarboxylic acid varied from 0.5 to 0.75 g. Optical rotations were determined on a Durrum-JASCO optical rotatory dispersion spectrophotometer at a wavelength of 589 mμ and a temperature of 23 ± 1°. The properties of the brucine salts and of the free hydroxycarboxylic acids obtained are summarized in Table I.

Real, J. S., and Pascual, J. (1953), *Anales Real. Soc. Españ. Fis. Quim.* (*Madrid*) *49B*, 445.

TABLE I: Properties of the 2-Hydroxy-3-cyclohexenecarboxylic Acids.

	Optical Isomers			
Properties	(−)-*cis*	(+)-*cis*	(−)-*trans*	(+)-*trans*
Brucine salts				
$[\alpha]_{589}^{23}$ (deg) (*c* 0.60, H_2O)	−95	+35	−56	−5
Free acids				
$[\alpha]_{589}^{23}$ (deg) (*c* 0.40, H_2O)	−305	+312	−70	+77
Mp (°C)	62–64	62–64	78–80	79–81
Anal. Calcd for $C_7H_{10}O_3$				
C, 59.14	C, 59.17	C, 59.94	C, 58.93	C, 59.36
H, 7.09	H, 7.05	H, 6.91	H, 6.99	H, 7.05

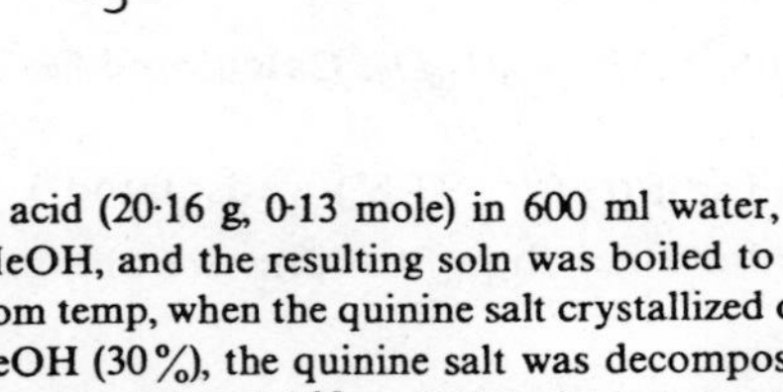

$C_7H_{10}O_4$ Ref. 284

To a hot soln of the racemic acid (20·16 g, 0·13 mole) in 600 ml water, was added a soln of quinine (42·17 g, 0·13 mole) in 200 ml MeOH, and the resulting soln was boiled to remove excess of MeOH and allowed to stand overnight at room temp, when the quinine salt crystallized out. After triangular fractional crystallization from aqueous MeOH (30%), the quinine salt was decomposed with dil HCl to afford the optically active acid of the constant rotation $[\alpha]_D^{25}$ −193° (*c*, 0·87, 1 dm, EtOH), m.p. 154°, yield 1·1 g; dimethylester b.p. 102°/23 mm, n_D^{25} 1·4438; $[\alpha]_D^{25}$ −175° (*c*, 0·63, 1 dm, EtOH).

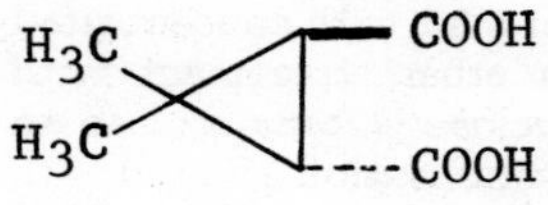

$C_7H_{10}O_4$ **Ref. 285**

Preliminary experiments on resolution.—The common optically active bases were tested; as the acid is dibasic, two series of salts could be expected. The data are given in Tables 1 and 2; the best results were obtained if acid and base were brought together in the mole ratio 1:1. The rotatory power of the acid obtained from the salts was measured in ethanol.

Table 1.

Acid/base = 1 : 1

Base	Solvent	$[\alpha]_D^{25}$
(−)-Ephedrine	acetone	−15.1°
Strychnine	methanol	+26.5°
Brucine	methanol	+ 3.2°
Quinine	methanol	+27.1°

Quinidine, cinchonidine, α-phenyl ethylamine and β-phenyl-*iso*propylamine gave no crystalline salts.

Table 2.

Acid/base = 1 : 2

Base	Solvent	$[\alpha]_D^{25}$
(−)-Ephedrine	acetone	− 9.7°
Strychnine	methanol	+14.8°
Brucine	methanol	+ 2.9°
Quinine	methanol	+ 1.6°

Quinidine and cinchonidine gave no crystalline salts.

The final resolution was carried out using (−)-ephedrine and quinine. 30.0 g (0.19 mole) *racem-trans*-Caronic acid and 35 g (0.19 mole) (−)-ephedrine were dissolved separately in 275 ml acetone + 10 ml water, the two solutions were heated and mixed. After standing for some time at room temperature, the solution was left in a refrigerator for 48 hours. The salt obtained (32 g) was recrystallised several times from 96-% ethanol. The course of the resolution was tested on samples from the different crystallisations, from which the acid was liberated and examined in absolute ethanol (Table 3).

Table 3.

Crystallisation	1	2	3	4			7	8
ml solvent	570	160	150	110			61	47
g salt obtained	32.0	21.3	15.0	12.4			6.4	5.0
$[\alpha]_{5461}^{25}$	−23.6°	−35.3°	−35.7°	−37.5°			−37.4°	−37.6°

The mother liquor from the first crystallisation of the ephedrine salt was evaporated to dryness and the caronic acid liberated. 17.2 g having $[\alpha]_{5461}^{25} = +14.1°$ (absolute ethanol) were obtained. This acid (0.109 mole) was dissolved with 35.3 g (0.109 mole) of quinine in 215 ml of boiling 96 % ethanol. The salt obtained after 48 hours in a refrigerator was recrystallised from the same solvent. As the salt was very sparingly soluble, it was dissolved by refluxing with a large volume of alcohol (given in brackets in the table) after which the solution was evaporated to a smaller volume and left to crystallise. The course of the resolution was tested as described above (Table 4).

Table 4.

Crystallisation	1	2	3			6
ml solvent	215	(600) 250	(510) 200			(450) 120
g salt obtained	31.3	24.9	21.5			13.3
$[\alpha]_{5461}^{25}$	+35.7°	—	+37.7°			+37.7°

As the active acids were isolated with the aid of different alkaloids and the maximum activity remained constant during three or four recrystallisations af the salts, there can hardly be any doubt that they represent the pure enanthiomorphs. The alkaloid salts were not analysed, but the yields of liberated acid indicate that they contained acid and base in the ratio 1:1.

(−)-trans-*Caronic acid.*—The (−)-ephedrine salt (5.0 g) was decomposed with dilute sulphuric acid and the caronic acid was isolated by extraction with ether. After evaporation of the ether, the residue was recrystallised from water. The yield of pure acid was 1.5 g. It forms prisms or needles with m.p. 214.5° (corr.).

0.1213 g acid: 15.20 ml 0.1006-*N* NaOH.
$C_7H_{10}O_4$ (158.15) Equiv. wt. calc. 79.1, found 79.3.
0.4592 g acid dissolved in *absolute ethanol* to 10.00 ml: $2\alpha_{5461}^{25} = -3.44°$. $[\alpha]_{5461}^{25} = -37.5°$; $[M]_{5461}^{25} = -59.2°$.

(+)-trans-*Caronic acid.*—The quinine salt (13.3 g) was decomposed as described above and the caronic acid recrystallised from water. The yield of pure acids was 4.1 g. In melting point and general appearance it was quite similar to the (−)-acid.

0.1164 g acid: 14.61 ml 0.1006-*N* NaOH.
$C_7H_{10}O_4$ (158.15) Equiv. wt. calc. 79.1, found 79.2.

The activity data are given in Table 5.

Table 5.

g = gram acid 10.00 ml solution

	g	$2\alpha_D^{25}$	$[\alpha]_D^{25}$	$[M]_D^{25}$	$2\alpha_{5461}^{25}$	$[\alpha]_{5461}^{25}$	$[M]_{5461}^{25}$
Ethanol (abs.)	0.3520	+2.23°	+31.7°	+50.1°	+2.65°	+37.6°	+59.5°
Ethanol (96 %)	0.3878	+2.27°	+29.3°	+46.3°	+2.68°	+34.6°	+54.6°
Methanol	0.3022	+2.15°	+35.6°	+56.3°	+2.55°	+42.2°	+66.7°
Acetone	0.3678	+2.55°	+34.7°	+54.8°	+3.03°	+41.2°	+65.1°
Dioxan	0.2762	+1.52°	+27.5°	+43.5°	+1.79°	+32.4°	+51.2°
Water	0.1551	+0.58°	+18.7°	+29.7°	+0.70°	+22.6°	+35.7°

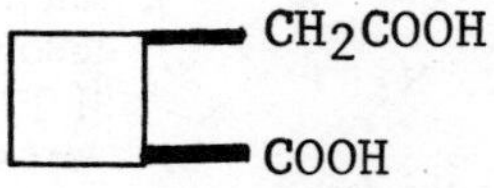

$C_7H_{10}O_4$ **Ref. 286**

Resolution of 2-*carboxy*cyclo*butylacetic acid*. Quinine methohydroxide solution (0·39N; 97·4 ml.) (Major and Finkelstein, *loc. cit.*) was exactly neutralised by the addition of the *cis*-(±)-acid (3 g.). The salt, recovered by evaporation under reduced pressure, was dissolved in absolute ethanol (20 ml.) and the resulting solution diluted with pure dioxan (200 ml.). The salt which separated rapidly in fluffy needles (8 g.), m. p. 220—224° (decomp.), $[\alpha]_D^{20}$ −140° (c = 2·105 in EtOH), was recrystallised six times by a similar procedure. The final product (3·5 g.), which had m. p. 225—226°, $[\alpha]_D^{20}$ −154° (c = 1·024 in EtOH), was decomposed with 10% w/w sulphuric acid; extraction with ether (15 × 100 ml.) yielded the crude *cis*-(−)-acid (350 mg.), m. p. 105—106°, $[\alpha]_D^{20}$ −34° (c = 1·216 in EtOH). Fractional crystallisation from benzene afforded cis-(−)-2-*carboxy*cyclo*butylacetic acid* (190 mg.) in clusters of long needles, m. p. 110—111°, $[\alpha]_D^{20}$ −45° (c = 1·254 in EtOH) [Found: C, 53·1; H, 6·4%; Equiv. (by titration), 79]. Fractional crystallisation of the acid recovered from the more soluble quinine metho-salt yielded cis-(+)-2-*carboxy*cyclo*butylacetic acid* (140 mg.) in rosettes of thin needles, m. p. 110—111°, $[\alpha]_D^{20}$ +43° (c = 2·15 in EtOH) [Found: C, 53·3; H, 6·2%; Equiv. (by titration), 79].

The pure *cis*-(+)-acid (500 mg.) was heated in a sealed tube at 180° for 5 hr. with concentrated hydrochloric acid (5 ml.). Evaporation, followed by dissolution in ether, treatment with charcoal (300 mg.), and re-evaporation afforded trans-(−)-2-*carboxy*cyclo*butylacetic acid* as an oil, $[\alpha]_D^{20}$ −71° (c = 0·92 in EtOH), which resisted all attempts at crystallisation [Found : C, 52·9; H, 6·1%; Equiv. (by titration), 78]. Similar treatment of the pure *cis*-(−)-acid yielded trans-(+)-2-*carboxy*cyclo*butylacetic acid* as an uncrystallisable oil, $[\alpha]_D^{20}$ +73° (c = 0·84 in EtOH) [Found : C, 52·9; H, 6·4%; Equiv. (by titration), 79·5].

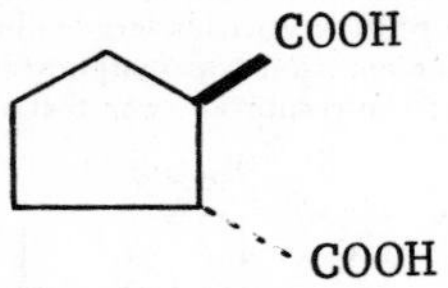

$C_7H_{10}O_4$ **Ref. 287**

d-trans-cyclo*Pentane*-1 : 2-*dicarboxylic Acid*.

The *dl-trans-cyclo*pentane-1 : 2-dicarboxylic acid employed in these experiments was prepared by the method described by Perkin (T., 1887, **51**, 240; compare T., 1894, **75**, 586). The pure acid, in quantities of 15 grams, dissolved in hot water, was mechanically stirred, and brucine (90 grams) gradually added, when the alkaloid readily dissolved. The excess of brucine was filtered off, well washed with hot water, and the filtrate and washings were concentrated on the water-bath until crystals just commenced to form on the surface. When the liquid was cooled and vigorously stirred, a copious crystallisation took place, and the whole became semi-solid; the crystals were then collected and repeatedly recrystallised from hot water. During this operation, the progress of the separation of the brucine salt of the *d*-acid from that of the *l*-modification was followed with the polarimeter, and the table given below shows that the separation is nearly complete after six crystallisations, since the difference between the rotation of this crop and of that obtained as the result of the twelfth crystallisation is very small.

No. of crystallisation.	Weight of substance. Gram.	Observed rotation.	Specific rotation.
1st crop	0·5250	−1·68°	−32·1°
3rd ,,	0·4566	−1·14°	−25·0°
6th ,,	0·5050	−1·15°	−22·8°
12th ,,	0·2612	−0·51°	−19·9°

A specimen of the pure brucine salt was subsequently prepared by adding excess of brucine to the hot dilute aqueous solution of the pure *d*-acid (see below), and, after filtering, the solution was allowed to crystallise slowly over sulphuric acid, when large, brilliant, tabular crystals separated. As these crystals effloresce in a vacuum desiccator, they were dried by exposure to air, and then analysed:

0·1314 gave 0·2782 CO_2 and 0·0859 H_2O. C=57·7; H=7·3.
0·5079 ,, 22·6 c.c. N_2 at 19·8° and 753 mm. N=5·1.
$2C_{23}H_{26}O_4N_2,C_7H_{10}O_4,9H_2O$ requires C=57·4; H=7·2; N=5·1 per cent.

That the salt has this composition was confirmed by the fact that 0·2316 gram, heated for one hour at 125°, lost 0·0336 H_2O, whereas the calculated loss for $9H_2O$ is 0·0339 gram.

In order to obtain the pure *d*-acid, the brucine salt from the twelfth crystallisation was dissolved in hot water, the brucine precipitated by ammonia and, after filtering and washing, the solution of the ammonium salt was concentrated and acidified with hydrochloric acid when, on cooling, the *d*-acid separated in plates, and melted at 178—180°. After completely decolorising with the aid of animal charcoal, and twice crystallising from water, the acid melted at 181°, and 0·1752 dissolved in water (20 c.c.) gave, in a 2-dcm. tube, a rotation of +1·535°, whence α_D +87·6°. On titration, 0·1778 required 0·0898 NaOH for neutralisation, whereas this amount of an acid, $C_5H_8(CO_2H)_2$, should neutralise 0·0900 NaOH

l-trans-cyclo*Pentane*-1 : 2-*dicarboxylic Acid*.

In order to obtain this acid, the mother liquors from the first six crystallisations of the brucine salt of the *dl*-acid (p. 2641) were concentrated until crystals began to appear on the surface of the brown liquid. The salt, which separated in quantity on cooling, was dissolved in hot water, the solution decolorised with animal charcoal, and the crude *l*-acid isolated in the manner described in the case of the *d*-acid

This acid is readily obtained pure simply by recrystallising four times from water, it then melted sharply at 180—181°, and 0·2363, dissolved in water (20 c.c.), gave, in a 2-dcm. tube, a rotation of −2·03°, whence α_D −85·9°. On titration, 0·3110 required 0·1568 NaOH for neutralisation, whereas this amount of an acid, $C_5H_8(CO_2H)_2$, should neutralise 0·1575 NaOH.

The l-*ethyl ester*, $C_5H_8(CO_2Et)_2$, obtained in the manner described in detail in the case of the ester of the *d*-acid, distilled at 170°/100 mm., and 0·3326, dissolved in acetone (20 c.c.), gave, in a 2-dcm. tube, a rotation of −2·32°, whence α_D −69·76°.

$C_7H_{10}O_4$ **Ref. 287a**

Acide cyclopentanedicarboxylique-1,2 trans (—) ($C_7H_{10}O_4$).

A partir de 5 g de diacide racémique et 6,7 g de (*p*-nitrophényl)-1 amino-2 propanediol-1,3 *thréo* ($[\alpha]_J^{22}$ = + 21,5°, méthanol) dans 130 cm³ d'éthanol, on obtient 7,1 g de sel qu'on recristallise 2 fois dans, respectivement, 220 et 150 cm³ d'éthanol. La décomposition du sel (4,87 g) fournit 1,05 g d'acide $[\alpha]_J^{22}$ = — 84,1° (c = 1, eau), F=183-185°, qu'on recristallise dans 8 cm³ d'eau : $[\alpha]_D^{22}$=—85,1°; $[\alpha]_J^{22}$ = — 88,3°; F = 184° [litt. (21) : $[\alpha]_D$ = 87,6°; F = 181°].

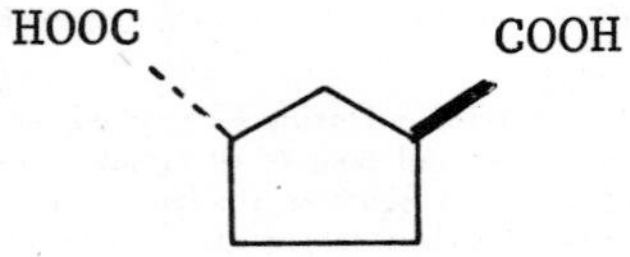

$C_7H_{10}O_4$ **Ref. 288**

Racemic *trans*-1,3-cyclopentanedicarboxylic acid was previously resolved by Birch and Dean.[18] However, the optical purities of both (+) and (−) isomers obtained were low, and an improved method was needed. It was found that the fractional recrystallization of the monostrychnine salt of the trans diacid from 95% ethanol gave either the (+)-diacid-rich salt or the (−)-diacid-rich salt as a precipitate, depending on the conditions. In a trial run, following Birch's method,[18] the partially resolved (+)-diacid {$[\alpha]^{20}$D +12.4° (H_2O, c 8.0 g/dl)} and (−)-diacid ($[\alpha]^{20}$D − 7.7°) were obtained. Starting with these diacids, the complete resolution could be carried out *via* the monostrychnine salt.

The partially resolved (−)-diacid, 3.16 g (2.00 × 10^{-2} mol), and strychnine, 6.69 g (2.00 × 10^{-2} mol), were dissolved on heating in 40 ml of 95% ethanol. The salt which precipitated at −15° was recrystallized four times from 95% ethanol (concentration ~10% w/v) to yield 1.0 g of the pure (−)-diacid strychnine (1:1) salt; after drying at room temperature *in vacuo*, mp 213–214° (becomes colored), $[\alpha]^{20}$D −32.5 (50% EtOH–50% H_2O v/v, c 2.0 g/dl).

Anal. Calcd for $C_7H_{10}O_4 \cdot C_{21}H_{22}N_2O_2$: C, 68.28; H, 6.55; N, 5.69. Found: C, 68.35; H, 6.44; N, 5.73.

Pure (−)-*trans*-1,3-cyclopentanedicarboxylic acid was obtained from the above salt, followed by recrystallization from benzene–carbon tetrachloride (1:1): mp 87.5–88.5°, $[\alpha]^{20}$D −35.1° (H_2O, c 8.0 g/dl) (lit. mp 85.0–85.7°, $[\alpha]^{15}$D −22.6° (H_2O, c 8.0 g/dl)[17]).

Anal. Calcd for $C_7H_{10}O_4$: C, 53.16; H, 6.37. Found: C, 53.12; H, 6.27.

(18) S. F. Birch and R. A. Dean, *J. Chem. Soc.*, 2477 (1953).

Starting from the partially resolved (+)-diacid and following the exact procedure as above, interestingly the pure (+)-diacid strychnine (1:1) salt could be obtained; after drying at room temperature *in vacuo*, the product had mp 202–203° (becomes colored), $[\alpha]^{20}D$ −10.7 (50% EtOH–50% H_2O v/v, *c* 2.0 g/dl).

Anal. Found: C, 68.05; H, 6.57; N, 5.45.

Pure (+)-*trans*-1,3-cyclopentanedicarboxylic acid was obtained from the above salt, followed by recrystallization from benzene–carbon tetrachloride (1:1): mp 87.5–88.5°, $[\alpha]^{20}D$ + 35.6° (H_2O, *c* 8.0 g/dl) (Lit. mp 79.5–80.5°, $[\alpha]^{15}D$ +20.09° (H_2O, C, 8.0 g/dl)[18]).

Anal. Found: C, 53.24; H, 6.38.

When a solution of the equimolar (±)-diacid and strychnine in 95% ethanol (concentration 10–30% w/v) was allowed to stand at −15°, no salt was precipitated after standing for a few days. However, on addition of either the pure (+)-diacid salt or the pure (−)-diacid salt as a seed, a corresponding isomer-rich salt was precipitated. Thus, alternately using the pure (+)-diacid salt and the pure (−)-diacid salt as a seed, while the solvent was partially evaporated so that the concentration stayed at the range of 10–15% w/v, the salts of $[\alpha]^{20}D$ −12 to −13° and the salt of $[\alpha]^{20}D$ −27 to −29° were collected. The salts were then further recrystallized from 95% ethanol to afford the pure isomers.

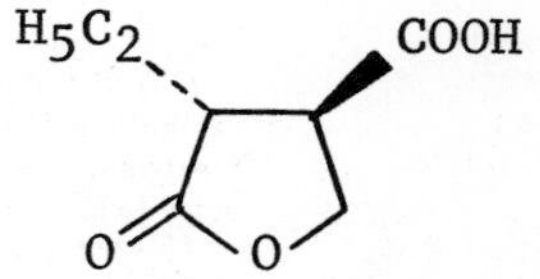

$C_7H_{10}O_4$ Ref. 289

Strychnine salts of stable levo and dextro α-ethylparaconic acids. Reaction of 26.9 g of stable racemic α-ethylparaconic acid (I, R = C_2H_5, m.p. 87.5-88°) with 53.46 g of strychnine in 55 ml of water by the procedure, described for the strychnine salts of optically active stable α-isopropylparaconic acids gave 39.5 g of the salt of stable levo α-ethylparaconic acid which was recrystallized twice from 20 and 18 ml of water respectively. Yield 32.8 g (78.3%).

From the filtrate was isolated the strychnine salt of the dextro acid which was recrystallized from anhydrous alcohol. Yield 26.9 g (64.2%).

Stable levorotatory and dextrorotatory α-ethylparaconic acids (Ia and Ib, R = C_2H_5). Levo α-ethylparaconic acid was prepared from 32.8 g of strychnine salt of stable levo α-ethylparaconic acid dissolved in 636 ml of water. Yield 8.5 g (80.5%). Colorless crystals, readily soluble in alcohol, ether, chloroform and hot water, less soluble in cold water. M.p. 105-105.5° (from water), $[\alpha]_D^{20}$ -58.3° (water, c 3.8835).

Found %: C 53.20; H 6.47. $C_7H_{10}O_4$. Calculated %: C 53.16; H 6.37.

The dextro acid was isolated from 26.9 g of strychnine salt of stable dextro α-ethylparaconic acid dissolved in 80 ml of water. Yield 7.8 g (90.3%). Colorless, silky plates. M.p. 104.5-105.5° (from alcohol), $[\alpha]_D^{20}$ + 58.55° (water, c 3.9126).

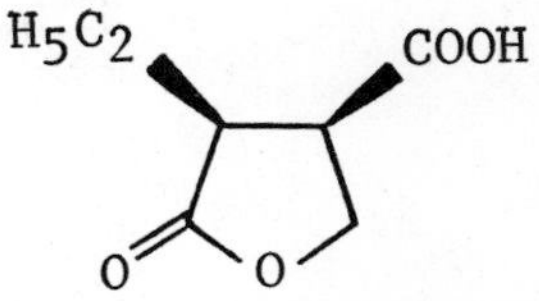

$C_7H_{10}O_4$ Ref. 290

Brucine salts of dextro and levo α-ethylparaconic acids. Reaction of 30 g of labile racemic α-ethylparaconic acid (II, R = C_2H_5, m.p. 89-90°), dissolved in 150 ml of water, with 73.5 g of brucine gave 50.8 g of the brucine salt of the dextro acid. Yield 96.9%. M.p. 99-102°. Colorless, small needles, poorly soluble in cold water, readily in hot water and acetone.

The filtrate was evaporated in vacuo and the dry residue washed with acetone (two lots of 50 ml each). The brucine salt of the levo acid was obtained with m.p. 105-107°. Acicular crystals, readily soluble in water, poorly in acetone. Yield 39.2 g (74.8%).

Labile dextrorotatory and levorotatory α-ethylparaconic acid (IIa and IIb, R = C_2H_5). From 50.8 g of brucine salt of labile dextro α-ethylparaconic acid, dissolved in 660 ml of water, was isolated the dextro acid in a yield of 10 g (68.7%). Colorless crystals, readily soluble in water, hot ether and benzene, poorly soluble in cold benzene. M.p. 121-122°, $[\alpha]_D^{20}$ + 53.55° (water, c 4.426).

Found %: C 53.32; H 6.43. $C_7H_{10}O_4$. Calculated %: C 53.16; H 6.37.

From 39.2 g of brucine salt of labile levo α-ethylparaconic acid was obtained 8.1 g of levo acid. Yield 72.2%. Colorless needles, readily soluble in hot ether and benzene, less soluble in cold benzene and water. M.p. 120-120.5°, $[\alpha]_D^{20}$ -53.75° (water, c 3.915).

Found %: C 53.02; H 6.39. $C_7H_{10}O_4$. Calculated %: C 53.16; H 6.37.

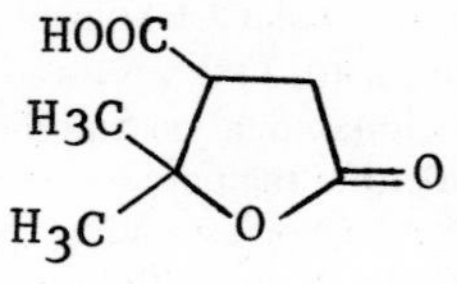

$C_7H_{10}O_4$ **Ref. 291**

b) **Sel de cinchonine de l'acide (+)-térébinique.**

Six grammes et demi d'acide (±)-térébinique (F. 175°) dans 50 cc. d'alcool sont mêlés avec 12.5 g. de cinchonine dans 50 cc. d'alcool. On laisse reposer la solution résultante une nuit. Les cristaux séparés sont recristallisés dans 80 cc. d'alcool bouillant. Le produit (5 g.) fond à 222°.

$C_{26}H_{32}O_5N_2$ calculé C% 68.35 H% 6.64 N% 6.68
trouvé 68.96 6.74 6.45

$[\alpha]_D^{25°} = +151°$ (c: 1.0 dans l'alcool à 12%).

c) **L'acide (+)-térébinique.**

Neuf grammes de cristaux de sel en haut sont dissous dans 160 cc. d'eau, on ajoute 10 cc. d'ammoniaque à 25%. Après refroidissement, la cinchonine qui est précipitée, est séparée et lavée à 100 cc. d'eau. La solution est évaporée à 25 cc. et acidulée en ajoutant 15 cc. d'acide hydrochlorique à 35%. Les cristaux se précipitent sous la forme de fragments carrés très jolis. Après recristallisation dans 30 cc. d'eau, l'acide (2 g.) fond à 201-204°.

$C_7H_{10}O_4$ calculé C% 53.16 H% 6.33
trouvé 52.84 6.00

$[\alpha]_D^{25°} = +9.0°$ (c: 4.0 dans l'alcool).

d) **L'acide (−)-térébinique.**

On laisse reposer 3 jours les liqueurs-mères du sel de cinchonine de l'acide (+)-térébinique et on débarrasse tous les cristaux séparés. Après l'évaporation d'alcool à sec, le résidu est repris dans 120 cc. d'eau et on ajoute 40 cc. d'ammoniaque à 25%. La cinchonine précipitée est débarrassée par filtration et la solution filtrée est évaporée à 40 cc. sur le bain-marie et on y ajoute 30 cc. d'acide chlorhydrique concentré. L'acide (−)-térébinique se sépare premièrement sous la forme de fragments carrés très jolis qui fondent à 198°. Après recristallisation dans 50 cc. d'eau, on obtient 1.5 g. du produit fondant à 200°.

$C_7H_{10}O_4$ calculé C% 53.16 H% 6.33
trouvé 52.60 6.54

$[\alpha_D]^{26.5°} = -7.95°$ (C; 3,333 dans l'alcool).

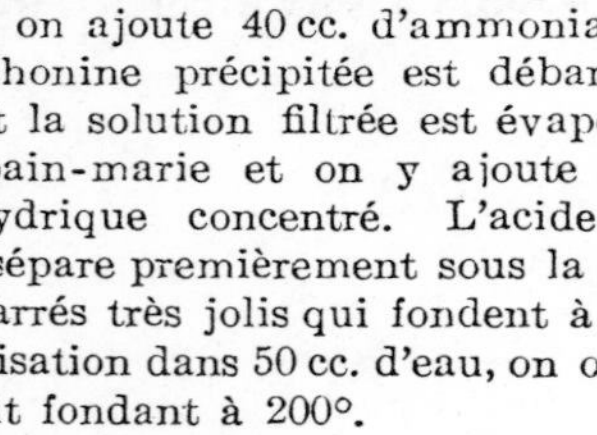

$C_7H_{10}O_4S$ **Ref. 292**

Preliminary experiments on resolution were made on a small scale (0.001 mole of the acid), in ethanol solution. The (+)-acid liberated from the acid and the neutral salts from quinine had $[\alpha]_D^{25} = +61.5°$ and $+57°$ respectively. The (+)-form liberated from the crystals obtained from 0.001 mole acid, 0.0006 mole $NaHCO_3$, and 0.0014 mole quinine had $[\alpha]_D^{25} = +40°$.

The (−)- form was obtained from the cinchonidine salt, and had $[\alpha]_D^{25} = -20°$. Schotte reported −62.3° as specific rotation while using strychnine. No crystals were obtained with cinchonine and quinidine.

It was thus decided to use strychnine and quinine for the main resolution.

(+)-*Tetrahydrothiapyran-2,6-dicarboxylic acid.* 20 g of the *trans*-isomer of I and 34 g quinine were dissolved in hot ethanol. The solutions were mixed while still hot, the total volume of ethanol amounting to 200 ml. The solution was left at 0° for 24 hours, but there was no sign of crystallization. It was then transferred to a chill-room at −10°, and after one week, 17 g of crystals separated, were filtered and dried. The quinine salt was then recrystallized several times, with some difficulty, from a mixture of ethanol and water. After each recrystallization, a small amount of the acid was liberated and its specific rotation measured in ethanol solution.

After the third recrystallization, the rotatory power of the acid decreased, and it was decided to free the whole amount of the acid from the quinine salt left.

Crystallization	1	2	3	4
ml ethanol	200	200	130	85
ml water	—	20	20	15
g salt obtained	17	12.6	7.5	4.0
$[\alpha]_D^{25}$ of acid	+61.5°	+78°	+91°	+81°

(−)-*Tetrahydrothiapyran-2,6-dicarboxylic acid.* The mother liquor from the first crystallization of the acid salt of quinine, was evaporated and the acid liberated by treatment with dilute sulphuric acid, and extracted with ethyl acetate. 9.4 g of acid, $[\alpha]_D^{25} = -60°$, were obtained on evaporation of the ethyl acetate.

The 9.4 g of (−)-form of I were dissolved in 50 ml hot ethanol and mixed while still hot to a hot solution of 15.6 g strychnine in 50 ml ethanol. The resulting solution was left to stand at room temperature, and crystals separated almost immediately.

The salt was recrystallized from dilute ethanol, and the rotatory power in ethanol measured for small amounts of acid liberated after each recrystallization.

Crystallization	1	2	3	4	5
ml ethanol	100	45	40	25	12
ml water	—	5	5	3	3
g salt obtained	13.6	9.8	8.6	6.0	4.0
$[\alpha]_D^{25}$ of acid	−73.8°	−87.6°	−104°	−103°	−103°

The acid was then liberated from the whole amount of strychnine salt left and extracted with ethyl acetate. After evaporation of the ethyl acetate, 1.4 g of the (−)-acid were left. Recrystallization from small amounts of formic acid yielded 1.1 g acid, $[\alpha]_D^{25} = -115.5°$ in ethanol. The melting point of the pure antipode was 148–149°.

0.0300 g acid: 6.10 ml 0.0516 *N* NaOH.
$C_7H_{10}O_4S$ Equiv. wt. calc. 95.1 found 95.2.

The rotatory power in various solvents is given in Table 1.

Table 1

g = g acid dissolved in 5.0 ml.

Solvent	g	α_D^{25} (1 dm)	$[\alpha]_D^{25}$	$[M]_D^{25}$
Ethanol	0.0375	−0.866°	−115.5°	−219°
Acetone	0.0100	−0.234°	−117°	−222°
Water	0.0075	−0.150°	−100°	−190°
Chloroform	0.0050	−0.096°	−96°	−182°

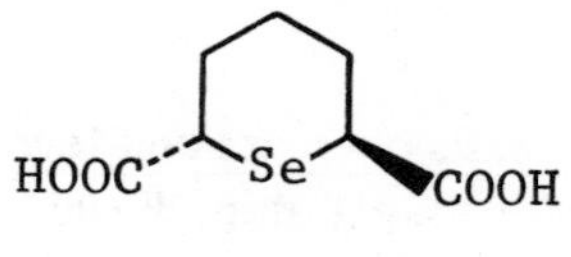

$C_7H_{10}O_4Se$ **Ref. 293**

Preliminary experiments on resolution were performed on a small scale (about 0.001 mole of acid). The secondary quinine salt obtained from dilute alcohol gave (+)-acid of high rotation. The secondary salts of brucine and strychnine, crystallising from water, yielded (−)-acid, the former alkaloid giving much better resolution. Cinchonine and morphine gave oily products, but the former salt crystallised partially after long standing. Attempts to prepare primary salts gave in some cases crystalline products but poor resolution. The acid obtained from the cinchonidine salt was quite inactive. Quinine and brucine were thus selected for the final resolution.

(+)-*Tetrahydroselenapyrane-2,6-dicarboxylic acid.*—A solution consisting of 16.6 g (0.07 mole) of racemic acid, and 3.3 g (0.04 mole) of sodium hydrogen carbonate in 160 ml of hot water was prepared and mixed with 32.4 g (0.10 mole) of quinine dissolved in 160 ml of 96% ethanol. The quinine salt started to separate almost immediately. The following day it was filtered off with suction and dried. It was then recrystallised several times from dilute ethanol; after each crystallisation a small amount of acid was liberated and examined in acetone solution.

Crystallisation	1	2	3	4	5
ml water	160	200	200	150	205
ml ethanol (96%)	160	250	240	220	225
g salt obtained	33.0	30.5	28.2	26.1	24.3
$[\alpha]_D^{25}$ of the acid	+105°	+110°	+111°	+114°	+112°

The quinine salt (24.3 g) was decomposed with strong sulphuric acid and the solution extracted with ether. After spontaneous evaporation of the ether, 6.0 g of active acid were obtained. It was recrystallised twice from a mixture of water and concentrated hydrochloric acid (the first time with charcoal). The yield of pure acid was 5.0 g. It forms small prisms melting at 146.5–148.5° and is very soluble in water, acetone, ethanol and formic acid; less so in hydrochloric acid.

0.0463 g acid: 7.82 ml 0.04986-*N* NaOH.
$C_7H_{10}O_4Se$ (237.1) Eq. weight calc. 118.6 found 118.7.

The rotatory power in various solvents is given in Table 1. Two measurements in acetone at different concentrations indicated that the variation with the concentration is very small.

Table 1. g = g acid in 10.00 ml solution.

	g	α_D^{25} (2 dm)	$[\alpha]_D^{25}$	$[M]_D^{25}$
Acetone	0.1730	+ 3.885°	+ 112.3°	+ 266.2°
,,	0.3531	+ 7.90°	+ 111.9°	+ 265.2°
Ethyl acetate	0.1527	+ 3.47°	+ 113.6°	+ 269.4°
Ethanol	0.1607	+ 4.475°	+ 139.2°	+ 330.1°
Chloroform	0.1601	+ 4.63°	+ 144.6°	+ 342.8°
Acetic acid	0.1560	+ 4.815°	+ 154.3°	+ 365.9°
0.4-*N* Hydrochloric acid	0.1664	+ 4.05°	+ 121.7°	+ 288.5°
Water (neutralised)	0.1528	+ 4.965°	+ 162.5°	+ 385.2°

(−)-*Tetrahydroselenapyrane-2,6-dicarboxylic acid.*—The mother liquor from the first crystallisation of the brucine salt was evaporated and the acid liberated; 7.1 g with $[\alpha]_D^{25} = -97°$ were obtained. This acid (0.03 mole) was dissolved with 23.5 g (0.06 mole) of anhydrous brucine in 50 ml of hot water. The salt was recrystallised from water and the course of the resolution followed as described for the (+)-acid.

Crystallisation	1	2	3	4	5
ml water	50	65	60	50	40
g salt obtained	27.9	24.2	20.5	18.0	15.2
$[\alpha]_D^{25}$ of the acid	− 104°	− 110°	− 112°	− 112°	− 112°

The acid was liberated from the salt (15.2 g) as described above and recrystallised three times (twice with charcoal). The yield of crude acid was 3.1 g; after the recrystallisations there remained 1.9 g. M.p. 146.5–148.5°.

0.0518 g acid: 8.76 ml 0.04986-*N* NaOH.
$C_7H_{10}O_4Se$ (237.1) Eq. weight calc. 118.6 found 118.6.
0.1689 g acid dissolved in acetone to 10.00 ml: $\alpha_D^{25} = -3.785°$ (2 dm). $[\alpha]_D^{25} = -112.0°$; $[M]_D^{25} = -265.7°$.

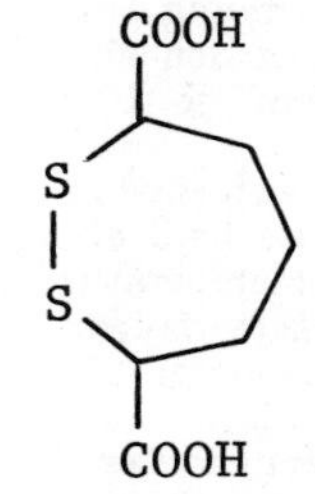

$C_7H_{10}O_4S_2$

Ref. 294

Preliminary experiments on resolution.—0.22 g (0.0010 mole) of *racem*-1,2-dithiepane-3,7-dicarboxylic acid was dissolved in 8 ml of hot acetone. 0.47 g (0.0010 mole) of brucine (4 H_2O) was added and water dropped into the mixture, until the solution was clear (about 1 ml). Crystallization over-night in a refrigerator yielded 0.35 g of salt, from which the acid was liberated in the usual manner. The rotatory power was measured in alcohol. M.p. 162°–166°. $[\alpha]_D^{25} = +220°$.

To the mother liquor from above (heated to about 50°) 0.23 g (0.0005 mole) of brucine (4 H_2O) was added together with 3 ml of water. On cooling down the solution, 0.30 g of salt deposited. The measurement of the rotatory power was performed as described above. M.p. 150°–160°. $[\alpha]_D^{25} = -382°$.

Anal. Subst. ((+)-form), 38.01 mg: 4.70 ml of 0.0721*N* NaOH. Equ. wt. Calc. 111.1; Found 112.2. — Subst. ((−)-form), 22.77 mg: 2.82 ml of 0.0721*N* NaOH. Equ. wt. Calc. 111.1; Found 112.0.

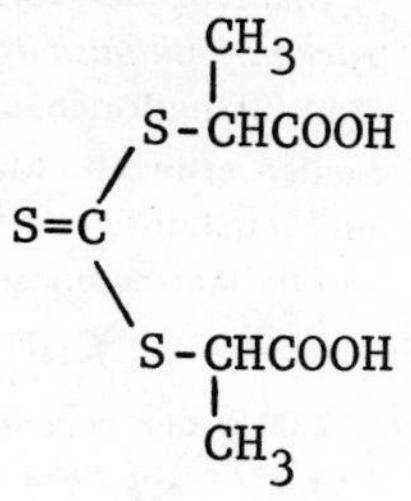

$C_7H_{10}O_4S_3$

Ref. 295

Resolution of the acid. Racemic acid (25.4 g, 0.1 mole) and cinchonine (58.8 g, 0.2 mole) were dissolved in 600 ml of hot ethanol and 600 ml of water were added. On cooling, the salt crystallised spontaneously. It was recrystallised from dilute ethanol; after each crystallisation, a small sample of the acid was liberated and the rotatory power determined in acetone solution.

Crystallisation	1	2	3	4
ml ethanol	600	140	60	45
ml water	600	50	20	20
g salt obtained	33.9	25.8	24.2	22.7
$[\alpha]_D^{25}$ of the acid	+ 136°	+ 170°	+ 176°	+ 175°

The mother liquor from the first crystallisation of the cinchonine salt yielded, after evaporation of the ethanol and acidification, 14.3 g acid having $[\alpha]_D = -102°$ (acetone). 14.0 g of this acid and 36.8 g of strychnine were dissolved in 280 ml of hot ethanol and 375 ml of water were added. The salt obtained was recrystallised and the course of the resolution followed as described above. It was found that the salt can be conveniently dissolved in 90 % ethanol; it then separates in a satisfactory yield without further addition of water.

Crystallisation	1	2	3	4
ml ethanol	280	625	260	275 (90 %)
ml water	375	800	220	—
g salt obtained	35.0	24.9	20.5	18.4
$[\alpha]_D^{25}$ of the acid	− 158°	− 170°	− 174°	− 174°

(+)-*Trithiocarbon-dipropionic acid.* The acid was liberated from the cinchonine salt and extracted with ether in the conventional way. 22.4 g salt yielded 6.5 g of the crude acid, which was recrystallised from 25 ml ethyl acetate + 35 ml benzene giving 5.9 g pure acid as yellow, glistening needles.

0.07361 g acid dissolved in *acetone* to 10.00 ml: $\alpha_D^{25} = +2.594°$ (2 dm). $[\alpha]_D^{25} = +176.2°$; $[M]_D^{25} = +448°$. Holmberg, who measured the activity at higher concentration, found $[\alpha]_D^{19} = +168.0°$.

(−)-*Trithiocarbon-dipropionic acid.* The acid was liberated as described above. 17.0 g salt yielded 4.3 g crude acid, which was recrystallised from 15 ml ethyl acetate + 30 ml benzene, giving 3.7 g of yellow glistening needles with m.p. 135.0–136.5°.

0.1903 g acid, dissolved in *acetone* to 10.00 ml: $\alpha_D^{25} = -6.574°$ (2 dm). $[\alpha]_D^{25} = -172.7°$; $[M]_D^{25} = -440°$. Holmberg found at a higher concentration $[\alpha]_D^{20} = -167.8°$.

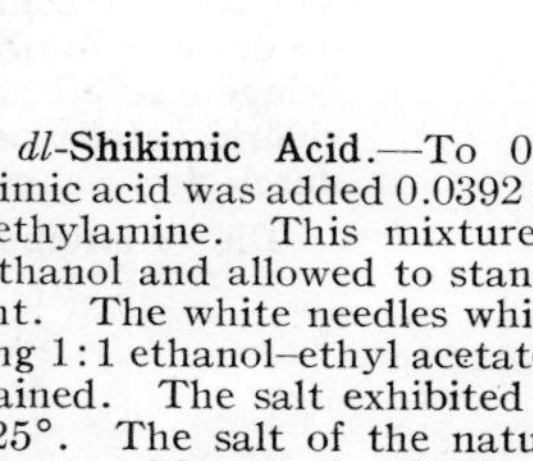

$C_7H_{10}O_5$

Ref. 296

Resolution of *dl*-Shikimic Acid.—To 0.0568 g. (0.33 mmole) of *dl*-shikimic acid was added 0.0392 g. (0.32 mmole) of (−)-α-phenylethylamine. This mixture was dissolved in hot absolute ethanol and allowed to stand at room temperature overnight. The white needles which formed were recrystallized using 1:1 ethanol–ethyl acetate until constant rotation was obtained. The salt exhibited a specific rotation $[\alpha]^{28.5}{}_D$ −125°. The salt of the naturally occurring shikimic acid had a specific rotation $[\alpha]^{29.0}{}_D$ −123°. The same procedure was utilized using (+)-α-phenylethylamine and the enantiomorph obtained.

Ref. 296a

$C_7H_{10}O_5$

Spaltung der racem. Shikimisäure in die optischen Antipoden: Eine Mischung aus 1.8 g *racem. Shikimisäure* (**3a**), 10 ccm Dimethylformamid, 10 ccm *Cyclohexanon,* 100 ccm Benzol und 30 mg Ionenaustauscher (Dowex 50 × 4) wird 2 Stdn. an einem Wasserabscheider zum Sieden erhitzt[1]. Man filtriert vom Austauscher ab, destilliert das Lösungsmittel i. Vak. ab und kristallisiert aus Essigester um. Man erhält 2.1 g (82%) *racem. 4.5-O-Cyclohexyliden-shikimisäure* vom Schmp. 171°.

$C_{13}H_{18}O_5$ (254.3) Ber. C 61.40 H 7.14 Gef. C 61.34 H 7.20

2.00 g der *racem. Cyclohexyliden-Verbindung* werden in 50 ccm Aceton gelöst und mit 0.96 g *(−)-α-Phenyl-äthylamin* versetzt. Nach dreimaligem Umkristallisieren aus Aceton ist der Drehwert des Salzes konstant. Man erhält 0.74 g Substanz (50%) vom Schmp. 191°; $[\alpha]_D^{20}$: −67° (*c* = 1, Methanol).

$C_8H_{12}N]C_{13}H_{17}O_5$ (375.5) Ber. C 67.18 H 7.79 N 3.73 Gef. C 67.45 H 7.74 N 3.90

Man löst das reine Salz in einem Wasser/Methanol-Gemisch (1 : 1), gibt die Lösung über einen Kationenaustauscher (Amberlite IR 120), fügt zum Eluat 2 ccm 2*n* HCl, destilliert das Lösungsmittel ab — dabei wird gleichzeitig das abgespaltene Cyclohexanon entfernt — und kristallisiert aus Essigester um. Man erhält 0.31 g (90%) *(−)-Shikimisäure* vom Schmp. 182°; $[\alpha]_D^{20}$: −159° (*c* = 1, Methanol). Lit.[16]: Schmp. 184°; Lit.[17]: $[\alpha]_D^{20}$: −157° (*c* = 0.94, Methanol) bzw. $[\alpha]_D^{20}$: −161° (*c* = 0.57, Methanol).

Die in der Aceton-Mutterlauge zurückgebliebenen Restsalze werden in analoger Weise zur Rohsäure aufgearbeitet und diese, wie oben beschrieben, erneut mit Cyclohexanon kondensiert. Die erhaltene rohe Cyclohexylidenverbindung wird ohne Reinigung in 50 ccm Aceton gelöst und mit 0.67 g *(+)-α-Phenyl-äthylamin* versetzt. Nach mehrmaligem Umkristallisieren aus Aceton erhält man 0.91 g (61%) Salz vom Schmp. 190°, $[\alpha]_D^{20}$: +66° (*c* = 1, Methanol).

Wenn man das Salz, wie oben bei der enantiomeren Form beschrieben, zerlegt, erhält man in 90-proz. Ausb. *(+)-Shikimisäure* vom Schmp. 182°; $[\alpha]_D^{20}$: +159° (*c* = 1, Methanol).

---

16) *R. Grewe* und *W. Lorenzen,* Chem. Ber. **86**, 928 (1953).

$HOOCCH_2CH_2CH-CH_2COOH$
with COOH on the CH:
$$HOOCCH_2CH_2\underset{\displaystyle |\atop \displaystyle COOH}{CH}-CH_2COOH$$

Ref. 297

$C_7H_{10}O_6$

Teilweise aktive D(+)β-Carboxyadipinsäure

In 8 Ansätzen wurden insgesamt 221 g reine β-Carboxyadipinsäure vom Schmp. 122,5—123,5° und 433 g wasserhaltiges l-Ephedrin[22]) in insgesamt 930 ccm abs. Methanol in der Wärme gelöst. Nach Abkühlung auf Zimmertemperatur wurde mit soviel abs. Aceton versetzt, daß eine erste Trübung auftrat. Hierzu waren insgesamt etwa 7500 ccm Aceton erforderlich. In den einzelnen Ansätzen wurden 30 g β-Carboxyadipinsäure eingesetzt. Auf je 10 g β-Carboxyadipinsäure kamen 42,0 ccm Methanol und 340 ccm Aceton.

Sobald sich die Trübungen verstärkten, wurden die Lösungen kurz aufgekocht. Hierbei schied sich eine erste Fraktion von Di-Ephedrinsalz ab, die heiß abgesaugt wurde. Die klaren Mutterlaugen ließ man langsam auf Zimmertemperatur abkühlen. Es dauerte in der Regel einige Stunden bis sich aus den Mutterlaugen genügend Salz abgeschieden hatte, um eine weitere Fraktion zu isolieren. In Abständen von einigen Stunden bis zu Tagen wurden so 4—5 Fraktionen abgefangen. Hatte sich nach längerem Stehen bei Zimmertemperatur kein weiteres Salz mehr ausgeschieden, so wurde durch Einstellen in den Eisschrank eine weitere, meist kleine Fraktion des Salzes erhalten. Aus den Mutterlaugen wurden nach Abdampfen der Lösungsmittel, zähe gelb gefärbte Sirupe gewonnen die, mit einem beträchtlichen Volumen abs. Aceton versetzt (pro Ansatz zwischen 400 und 600 ccm Aceton), nach 1—2 Tagen im Eisschrank durchkristallisierten. Je dunkler die Sirupe anfielen, um so schwerer waren sie zur Kristallisation zu bringen. Alle Fraktionen wurden bei 14 mm mit Phosphorpentoxyd und Calciumchlorid getrocknet. Die Trennwirkung der fraktionierten Kristallisationen wurde durch den Schmp. (unkorr.) der Ephedrinsalze kontrolliert. (IV. Abh.)

Die erhaltenen Fraktionen wurden in folgende Gruppen zusammengefaßt:

1) Fraktionen über 147°
2) Fraktionen von 145—147°
3) Fraktionen von 141—145°
4) Fraktionen unter 141°

Die Fraktionen mit einem Schmp. unter 141° wurden nicht weiter verarbeitet. Die Mittelfraktionen und die Spitzenfraktion wurden getrennt der erneuten fraktionierten Kristallisation unterworfen. Je 10 g Di-Ephedrinsalz wurden dabei in 14,5 ccm abs. Methanol gelöst. Nach 3—4-maliger fraktionierter Kristallisation erhielt man Spitzenfraktionen, die zwischen 150° und 152° schmolzen.

---

22) Der Chemischen Fabrik Knoll AG. in Ludwigshafen danken wir für die Überlassung des Ephedrins.

Die Fraktionen 142—150° wurden in Gruppen mit Schmelzpunktsintervallen von 2—3° zusammengefaßt und so lange fraktioniert, bis ihre Schmp. 150° erreichten.

Schließlich wurden erhalten

60,1 g Schmp. 150,5—152°
67,0 g Schmp. 149,5—151°

Beide Fraktionen wurden in der Kälte mit überschüssiger 2n-Schwefelsäure zerlegt und die gelben Lösungen einer kontinuierlichen Ätherextraktion unterworfen. Nach Vertreiben des Äthers wurden Sirupe erhalten, die im Wasserstrahlvakuum von Resten von Äther und mitgerissenem Wasser befreit wurden. Die Sirupe erstarrten beim Abkühlen alsbald zu steinharten Massen, die nach Zerkleinerung bei 14 mm über Phosphorpentoxyd getrocknet wurden.

Es wurden erhalten:

aus 60,1 g Salz 18,9 g rohe β-Carboxy-adipinsäure = 92,6% d. Th.;
aus 67,0 g Salz 20,2 g rohe β-Carboxy-adipinsäure = 89,4% d. Th.

Nach Umkristallisation aus gereinigtem abs. Essigester verblieben:

1. von 18,9 g Rohsäure 16,8 g = 88,8% des Rohproduktes
2. von 20,2 g Rohsäure 18,0 g = 89,1% des Rohproduktes

$$1.\ [\alpha]_D^{20} \text{ in Aceton } \frac{+0{,}49 \cdot 5}{0{,}1495 \cdot 1} = +16{,}39^\circ$$

$$2.\ [\alpha]_D^{20} \text{ in Aceton } \frac{+1{,}26 \cdot 1}{0{,}081 \cdot 1} = +15{,}55^\circ$$

$NC-CH_2-CH(C_2H_5)CH_2COOH$

$C_7H_{11}NO_2$ Ref. 298

γ-Cyano-β-ethylbutyric acid was resolved in the following way.[21] Quinine methochloride (14 g.) and silver oxide (15 g.) in water (70 ml.) were shaken for 3 hr. The mixture was filtered through Celite and the clear solution titrated against γ-cyano-β-ethylbutyric acid (6·0 g.) in ethanol (60 ml.) to pH 9—10. The solvent was removed *in vacuo*, and the residue dried azeotropically with benzene, taken up in acetonitrile (70 ml.), filtered through Celite, and left at room temperature. The crystalline precipitate (6·0 g.), m. p. 162—166°, $[\alpha]_D$ −157° (*c* 1·2 in water), was recrystallised from acetonitrile three times (4·0 g.), m. p. 166—168°, $[\alpha]_D$ −154° (*c* 1·4 in water). This salt in water (8 ml.) was acidified with dilute sulphuric acid (1·0N) and extracted with ether, to furnish (+)-γ-cyano-β-ethylbutyric acid (1·1 g.), b. p. 95—100°/10^{-4} mm., $[\alpha]_D$ +28° (*c* 5·2), $\nu_{max.}$ (film) 1705, 2280 $cm.^{-1}$.

[21] R. T. Major and J. Finkelstein, *J. Amer. Chem. Soc.*, **1941**, **63**, 1368.

$CH_3-C(CH_2CH_3)(CN)-CH_2COOH$

$C_7H_{11}NO_2$ Ref. 299

2.) Racematspaltung der 3-Cyanopropionsäuren **1a – 1d**

Äquimol. Mengen Säure und Threobase (vgl. Tab. 2) wurden in siedendem Methanol gelöst und die erkaltete Lösung bis zur Trübung mit Äther versetzt. Das im Kühlschrank über Nacht auskristallisierte Salz wurde abgetrennt und 6 mal aus Methanol/Äther umkristallisiert. Die Lösung dieses Salzes in verd. HCl wurde mit Essigester ausgeschüttelt und die erhaltene Säure aus Essigester/Petroläther umkristallisiert.

Tabelle 2: 3,3-Disubstituierte 3-Cyanopropionsäuren

Verbdg.	Schmp	$[\alpha]_D^{20}$*)	Spaltbasen	Summenformel		Ber.: Gef.:			Ausb. % d. Th.
1d	38–39°	---	---	$C_7H_{11}NO_2$		(141,2)			
				C	59,6	H	7,85	N 9,9	
					59,8		8,06	9,7	60
(+) 1d	45°	+ 6,6°	(–)Threobase II		59,2		7,92	9,9	16
(–)1d	46°	– 6,8°	(+)Threobase II		59,6		8,06	9,6	19

* Ber. aus den bei 578 nm und 546 nm gemessenen Drehwerten (Methanol)

Threobase II: threo-l-(4'-Nitrophenyl)-2-amino-propandiol-(1,3)

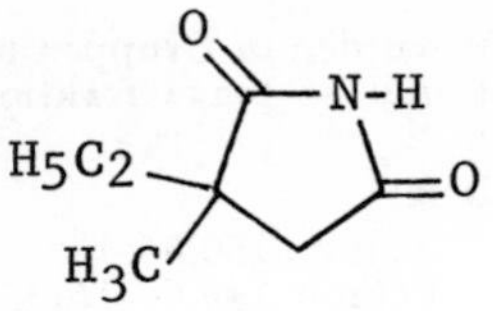

$C_7H_{11}NO_2$ Ref. 300

Resolution of emo. Since the EMO isolated from urine proved to be optically active, it was desirable to know the specific rotation of the optical isomers of EMO. The resolution was accomplished through formation of the brucine salt in the following manner. In 100 ml. of absolute ethanol is dissolved 0.2 mole of EMO and to this is added 0.2 mole of brucine in 300 ml. of absolute ethanol. The mixture is allowed to stand at room temperature for several hours. The precipitate that forms consists largely of the brucine salt of the dextrorotatory form of EMO. This salt is recrystallized from absolute ethanol until there is no further change in optical composition (usually three times). The salt is then taken up in water and the solution acidified with H_2SO_4. The EMO is extracted from this solution with ether. The ether is washed with dilute H_2SO_4 and dried over Na_2SO_4. The ether is removed by distillation and the residue of EMO distilled under reduced pressure. The highest rotation obtained was $[\alpha]_D^{29}$ + 38° (20 mgm. per ml. in ethanol).

The mother liquors from the initial crystallization and the first recrystallization of the brucine salt of *d*-EMO are evaporated to dryness. The solid residue is dissolved in water. The solution is acidified and the EMO extracted into ether as described above. After drying and subsequent removal of the ether, the residue of EMO is weighed and redissolved in absolute ether. One equivalent of piperidine is added. The piperidine salt of EMO precipitates. This salt is recrystallized by suspending in boiling absolute ether, adding sufficient chloroform to effect solution, and cooling to about − 20° C. After three recrystallizations the piperidine salt is decomposed and the EMO isolated in the same manner as described above for the brucine salt. The highest rotation obtained for *l*-EMO was $[\alpha]_D^{32}$ − 35° (16 mgm. per ml. in ethanol).

The rotations of the piperidine salts of the purest samples of *d*-EMO and *l*-EMO were, respectively, $[\alpha]_D^{23}$ + 28° and $[\alpha]_D^{25}$ − 25° (50 mgm. per ml. in ethanol). The piperidine salts of both isomers melt at 105–106° C, corr. Mixtures of the salts in any proportions all melt at the same point (indicative that the salts of the two isomers form solid solutions).

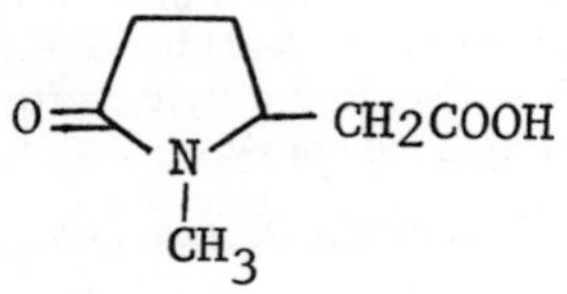

$C_7H_{11}NO_3$ Ref. 301

L-Ecgoninsäure (VIII) aus DL-Ecgoninsäure (VIIa). 25,8 g (164 Millimol) DL-Ecgoninsäure (VIIa) und 76 g (164 Millimol) kristallwasserhaltiges Brucin wurden in 200 cm³ Aceton gelöst. Beim Abkühlen auf − 10° kristallisierte Brucinsalz in nahezu quantitativer Ausbeute. Es wurde aus Methanol-Äther und aus Methanol-Aceton umkristallisiert und wies dann die praktisch konstante spez. Drehung von − 27,4° (c = 0,52 in Alkohol) auf. Das Präparat schmolz unscharf bei 93—100°. Das Analysenpräparat wurde 20 Std. bei 60° im Hochvakuum getrocknet.

$C_{30}H_{37}O_7N_3$, 3½ H_2O	Ber. C 58,61	H 7,22	N 6,84%
	Gef. „ 58,71	„ 7,46	„ 6,68%

26,0 g (42 Millimol) Brucinsalz der L-Ecgoninsäure wurden mit 60 cm³ 1-n. Natronlauge und 100 cm³ Chloroform geschüttelt. Das Chloroform wurde abgetrennt und die wässerige Lösung noch viermal mit je 60 cm³ Chloroform ausgeschüttelt. Die wässerige Lösung wurde durch 80 cm³ Wofatit KS laufen gelassen. Nach dem Abdampfen des Wassers blieben 8,9 g kristallisierte Substanz (ber. 6,6 g), aus welcher durch einmaliges Umkristallisieren aus Aceton-Äther 6,1 g L-Ecgoninsäure (VIII) vom Smp. 120° und $[\alpha]_D$ = − 38,8° (c = 0,85 in Wasser) erhalten wurden. Das dreimal aus Aceton umkristallisierte Analysenpräparat vom Smp. 120° wurde 24 Std. bei Zimmertemperatur im Hochvakuum getrocknet. $[\alpha]_D$ = − 43,7° (c = 0,94 in Wasser).

$C_7H_{11}O_3N$ Ber. C 53,49 H 7,04% Gef. C 53,38 H 6,93%

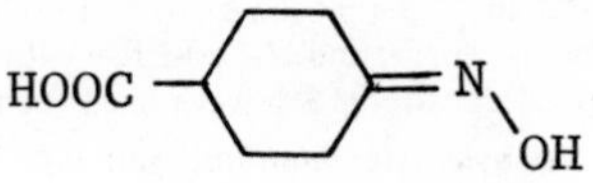

$C_7H_{11}NO_3$

A resolution procedure for the above compound may be found in W. H. Mills and A. M. Bain, J. Chem. Soc., 97, 1866 (1910).

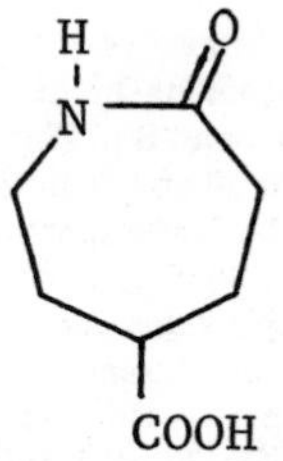

$C_7H_{11}NO_3$ Ref. 302

Resolution of (±)-Hexahydro-5-carboxy-2*H*-azepin-2-one

(*S*)-Hexahydro-5-carboxy-2*H*-azepin-2-one (XVII). An intimate mixture of (±)-hexahydro-5-carboxy-2*H*-azepin-2-one (7.85 g, 0.05 mole) and brucine (19.9 g, 0.0505 mole) was slowly added with stirring to 1.5 liters of isoamyl acetate.* After stirring for 2 hr, an additional 2.35 liters of isoamyl acetate was added, and the solution was heated to 60°C and filtered. The filtrate was then heated to 85°C and allowed to stir at that temperature for 3 days. A fine precipitate formed during this time (seeding with previously obtained brucine salt of the (−)-isomer accelerates this crystallization, although some runs required up to 7 days). This solid was filtered and washed with isoamyl acetate three times. A sandlike powder, 10 g (mp dec 217°C) was obtained by air drying on a Büchner funnel. The filtrate was then concentrated on a rotary evaporator with the pressure adjusted so that the solution remained above 65°. As the solution volume was reduced to 1.5 liters additional solid precipitated. The solid was initially flocculent, but by maintaining the temperature above 65°C and continuing the agitation, it became a fine granular powder. The hot solution was filtered to isolate this salt, 2.5 g (mp dec 215°C). The combined precipitates, 12.5 g (90.1% of the theoretical amount of one diastereoisomeric salt), were recrystallized from 190 ml *n*-propanol, to give 10 g of brucine salt; mp 220°C (dec), $[\alpha]_D^{25}$ +4.4 (*C* = 1.25, methanol). The brucine salt was dissolved in 40 ml water and passed through a Bio-Rex 70 ion-exchange column (column dimensions 70 cm × 5 cm, cationic weak carboxylic acid type minus 400 mesh previously converted from its sodium form by treating with 4*N* hydrochloric acid followed by washing with deionized water until free of excess mineral acid). The free acid lactam was eluted with 2 liters of water. Removal of water under reduced pressure left a colorless residue, 2.8 g, which solidified on standing. Recrystallization three times from *p*-dioxane gave a white solid, 2.5 g (63%), mp 168–169°C; $[\alpha]_D^{23}$ −31.9° (*C* = 3.0, methanol); infrared and NMR identical to (±)-isomer.

Anal. Calcd for $C_7H_{11}NO_3$: C, 53.49%; H, 7.05%; N, 8.91%. Found: C, 53.55%; H, 7.10%; N, 8.78%.

(*R*)(+)-Hexahydro-5-carboxy-2*H*-azepin-2-one (XXVIIa). The filtrate, obtained after isolating the above brucine salt of the (−)-isomer of hexahydro-5-carboxy-2*H*-azepin-2-one (12.5 g), was brought to dryness on a rotary evaporator leaving a dark residue which was extremely soluble in *n*-propanol, as opposed to the low solubility of the other diastereoisomeric salt. This salt was also very hygroscopic. The residue was dissolved in water and extracted four times with ether. The excess ether was removed on a rotary evaporator, and the aqueous solution was passed through a Bio-Rex 70 ion-exchange column (column dimensions 70 cm × 5 cm) (acid form) with 2.5 liters of water. Removal of water under reduced pressure left a viscous liquid which solidified on standing. Recrystallization from *p*-dioxane six times gave a white solid, 1.1 g (28%); mp 168–169°C; $[\alpha]_D^{25}$ +31.7 (*C* = 2.5 methanol); infrared and NMR identical to (±)-isomer.

Anal. Calcd for $C_7H_{11}NO_3$: C, 53.49%; H, 7.05%; N, 8.91%. Found: C, 53.58%; H, 7.09%; N, 8.98%.

(*S*)(−)-Hexahydro-5-carbomethoxy-2*H*-azepin-2-one (XVIII)

A solution of diazomethane in ether was prepared from 41 g (0.40 mole) of *N*-nitroso-*N*-methylurea. This solution was added to a slowly stirred slurry of finely powdered (−)-hexahydro-5-carboxy-2*H*-azepin-2-one 11 g, 0.07 mole $[\alpha]_D^{25}$ −31.9° at 0°C. After stirring for 3 hr the solution was allowed to come to room temperature and stir for an additional 12 hr. The ether and excess diazomethane were then removed under reduced pressure in a hood leaving a white solid. Recrystallization from 1500 ml ether resulted in long needles 11.1 g (93%), mp 93–94°C, $[\alpha]_D^{25}$ −32.1°C which did not increase in specific rotation on further recrystallization; infrared (nujol) ν_{max} 3340, 1720, 1655, 1620, 1210, 1200, 1188 cm⁻¹; infrared (CCl_4) ν_{max} 3400, 3280, 3200, 3075, 1730, 1660, 1172 cm⁻¹; NMR identical to (±)-isomer.

Anal. Calcd for $C_8H_{13}NO_3$: C, 56.13%; H, 7.65%; N, 8.18%. Found: C, 56.14%; H, 7.82%; N, 8.29%

(*R*)(+)-Hexahydro-5-carbomethoxy-2*H*-azepin-2-one (XVIIIa)

To a fine powder of (+)-hexahydro-5-carboxy-2*H*-azepin-2-one 3.4 g (0.02 mole) $[\alpha]_D^{25}$ +31.7° suspended in 400 ml of ether was added a ethereal diazomethane solution prepared from 13.6 g (0.13 mole) of *N*-nitroso-*N*-methylurea. The mixture was stirred at 0°C for 3 hr and at room temperature overnight. The ether and excess diazomethane were then evaporated into a hood by slight warming. The residue was dissolved in 300 ml of ether which, upon cooling, deposited the ester. Two recrystallizations from ether gave long colorless needles 2.7 g, mp 93°, $[\alpha]_D^{25}$ +31.9° (*C* = 3.0 methanol); infrared and NMR identical to (*S*)(−)-isomer.

Anal. Calcd for $C_8H_{13}NO_3$: C, 56.13%; H, 7.65%; N, 8.18%. Found: C, 56.13%; H, 7.68%; N, 8.10%.

* Fisher Certified Reagent #A719, isoamyl acetate, bp 139.3–141.9°C; specific gravity 0.878; assay 99.2%.

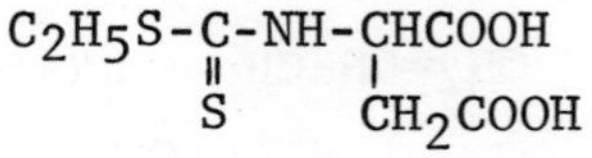

$C_7H_{11}NO_4S_2$ Ref. 303

(—)-*Äthyl-dithiocarbamino-bernsteinsäure.* 9,0 g *racem*-Säure wurden nebst 4,5 g (—)-Phenäthylamin in 150 ml Methanol gelöst. Beim Erkalten kristallisierte 6,2 g Salz; nach zweimaligem Umkristallisieren wie oben für das Antipodensalz beschrieben verblieben 5,5 g. Aus dem Salz wurden 3,1 g einmal aus Wasser umkristallisierte Säure erhalten.

0,2330 g Sbst.: 17,66 ml 0,1111-n NaOH; Äq.-gew. ber. 118,6 gef. 118,8.

0,2393 g in *Aceton* auf 9,99 ml gelöst: 2 α = —2,42°; $[\alpha]_D^{25}$ = —50,6°; $[M]_D^{25}$ = = —120,0°.

$$\begin{array}{c} CH_3 \\ C=N \\ CH_3 \end{array} \begin{array}{c} COOH \\ | \\ O-CH-CH_2COOH \end{array}$$

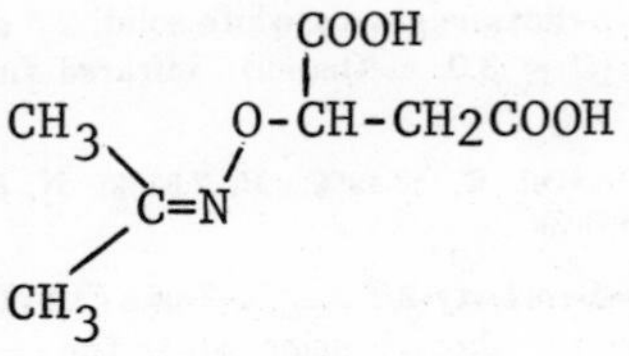

$C_7H_{11}NO_5$ Ref. 304

Dimethyl (−)-Isopropylideneaminooxysuccinate (3).—A suspension of 262 g of purified dehydroabietylamine acetate[6,12] in 1 l. of water was treated with a solution of 65 g of sodium hydroxide in 100 ml of water in portions. The cooled reaction mixture was extracted with ether. After drying over potassium hydroxide pellets, the ether solution was added to a solution of 75.6 g of dl-4 in 1 l. of ether–benzene (1:1). After standing at room temperature for 4 hr, the colorless solid was collected by filtration. Successive crops were obtained from the filtrate until a total of 130.8 g had been obtained. After four recrystallizations from methanol–chloroform (1:1), 76 g of the (−)(−) salt of 4, mp 165–168°, was obtained. Attempts to obtain the (+)(−) salt from the mother liquors were unsuccessful. A solution of 75.9 g of the (−)(−) salt in 200 ml of dry methanol was added to 800 ml of methanol saturated at 0° with dry HCl. After standing overnight at room temperature 400 ml of methanol was distilled under reduced pressure. The mixture was diluted with water and extracted with ether. After the usual work-up distillation afforded 12.3 g (56%) of (−)-3, bp 145° (11 mm), $[\alpha]^{24}_D$ 23.01° (c 3, CCl_4).

(6) B. Sjöberg and S. Sjöberg, *Ark. Kemi*, **22**, 447 (1964).

(12) L. F. Fieser and M. Fieser, "Reagents for Organic Synthesis," Wiley, New York, N. Y., 1967, p 183. We thank the Hercules Powder Co., Wilmington, Del., for a generous gift of Amine D.

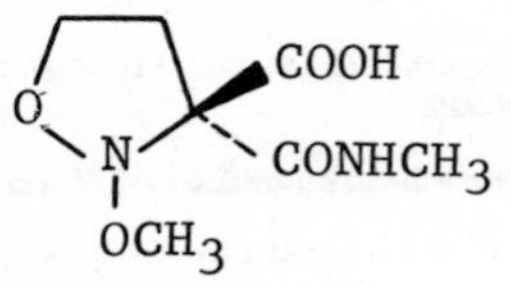

$C_7H_{12}N_2O_5$ Ref. 305

Separation of diastereomeric 2-methoxyisoxazolidine-3,3-dicarboxylic acid trans-methylamide α-phenylethylammonium salts. Separation of α-phenylethylamine into antipodes was performed as described.[25] S-(−)-PEA with $[\alpha]^{20}_{546}$ −46.4° (pure liquid), optical purity 96.3%, and R-(+)-PEA with $[\alpha]^{20}_{546}$ + 47.58° (pure liquid), optical density 98.8% were used for separation.

(A) *S-(−)-α-Phenylethylammonium salt of 2-methoxyisoxazolidine-3,3-dicarboxylic acid* trans-*methylamide* **(8)**. A soln of **6** (3.59 g; 0.013 mole) in 30 ml distilled water was passed through an ion exchange column (Dowex 50WX12, 10-fold molar excess, in the form of $PhCH(Me)NH_3^+$). Fractions output was controlled by UV spectrum. The eluate (120 ml) was evaporated *in vacuo* on rotor evaporator with further liophile drying yield **8** (3.79 g) (Table 2). Found: C, 55.23; H, 7.24; N, 12.80. Calc. for $C_{15}H_{23}N_3O_5$: C, 55.37; H, 7.13; N, 12.92%. The NMR spectrum (80 MHz, CD_3OD, δ ppm from HMDS, J Hz): 7.34 (Ph), 3.51 (MeO), 2.66 (MeN), 1.54 (Me, J=7), 4.46–3.51 (the region of –CH and cyclic proton signals), 2.89–2.66 (cyclic proton signals).

(B) *Diastereomerically pure S-(−)-α-phenylethylammonium salt of 2-methoxyisoxazolidine-3,3-dicarboxylic acid* trans-*methylamide* (+ **10**). Compound **8** (3.79 g; 0.0116 mole) was successively crystallized from 50 and 30 ml MeOH and +**10** (0.6 g) was obtained (Table 2). The NMR spectrum is similar to that of **8**.

(C) *R-(+)-α-Phenylethylammonium salt of 2-methoxyisoxazolidine-3,3-dicarboxylic acid* trans-*methylamide* **(9)**. Crystallization liquids from a previous run were evaporated *in vacuo*, the residue was dissolved in 30 ml distilled water and the soln was passed through the ion exchange column (Dowex 50WX12, 10-fold molar excess, in K^+ form). The eluate (100 ml) was partly evaporated *in vacuo* on rotor evaporator to ca 20 ml and passed through the ion exchange column (Dowex 50WX12, 10-fold molar excess, in $R-(+)-PhCH(Me)NH_3^+$ form). The fraction output was controlled by UV spectrum. The eluate (100 ml) was evaporated *in vacuo* on rotor evaporator. After liophile drying 1.95 g of **9** was obtained (Table 2). Found: C, 55.31; H, 7.02; N, 12.78. Cal. for $C_{15}H_{23}N_3O_5$: C, 55.37; H, 7.13; N, 12.92%. The NMR spectrum was similar to that of **8**.

D. *Diastereomerically pure R-(+)-α-phenylethylammonium salt of 2-methoxyisoxazolidine-3,3-dicarboxylic acid* trans-*methylamide* (−**10**). Compound **9** (1.95 g) was successively crystallized from 20 and 17 ml of MeOH and −**10** (0.41 g) was obtained (Table 2). The NMR spectrum is similar to that of **8**.

[25] W. Theilacker and H. G. Winkler, *Chem. Ber.* **87**, 690 (1954).

Table 2. Optically active derivatives 2-methoxyisoxasolidine-3,3-dicarboxylic acid

Compound	Yield,%	M.p.,°C	$[\alpha]^{20}_D$ (deg.)	$[\alpha]^{20}_{546}$ (deg.)	$[\theta]^{20}_{max} \cdot 10^{-2}$ (λ nm)	Concentration,vol.%, (solvent)
(**8**)	89.0	129–145	−	−1.7	−	7.63(H_2O)
(**9**)	61.0	131–146	−	−36.3	−	5.50(H_2O)
(+**10**)	31.7[a]	155–156	170.4	205.8	−	0.41(H_2O)
(−**10**)	22.8[a]	149	−	−210.4	−	0.48(H_2O)

a) per diastereomericalli pure form starting from (**8**).

$$\begin{array}{c} CH_2=CH-CH_2-CHCOOH \\ | \\ C_2H_5 \end{array}$$

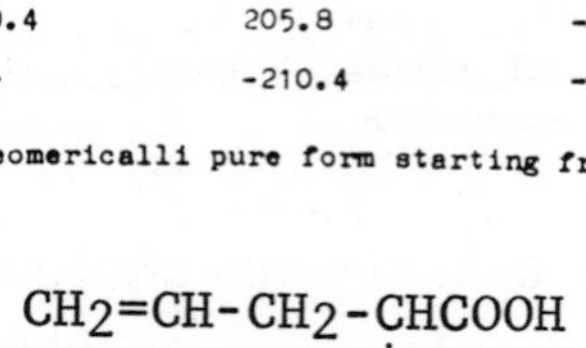

$C_7H_{12}O_2$ Ref. 306

(−)-*2L-Ethylpent-4-enoic acid.* The salt from 2-ethylpent-4-enoic acid (75 g) and quinine (190 g) was recrystallised eight times from successively smaller amounts of acetone (0.8–0.2 l). The quinine salt was decomposed with dilute hydrochloric acid giving (−)-2L-ethylpent-4-enoic acid (12 g, 16 %), b.p. 90.5–93° (0.5 mm), $[\alpha]^{20}_D$ −2.40° (undiluted, l 0.5), $[\alpha]^{20}_D$ −2.8° ($CHCl_3$, c 6.20), d^{20}_4 0.963. The configuration was established by degra-

(+)-*2D-Ethylpent-4-enoic acid.* The mother liquiors from the preparation of the quinine salt of (−)-2L-ethylpentenoic acid were concentrated and subjected to further crystallisation. Evaporation of the solvent from the most soluble part and subsequent decomposition of the quinine salt with hydrochloric acid yielded (+)-2D-ethylpent-4-enoic acid (15.5 g, 21 %), b.p. 110–112°/13 mm, $[\alpha]^{20}_D$ + 2.0° (unidluted, l 0.5), d^{20}_4 0.962.

$C_7H_{12}O_2$

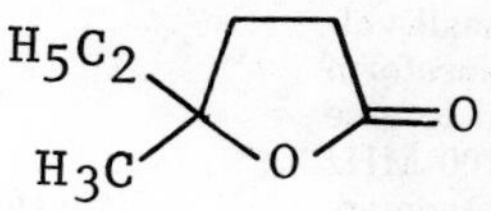

Ref. 307

Preparation of optically active γ-methyl-γ-caprolactone. 25.6 g (0.20 mole) of γ-methyl-γ-caprolactone was dissolved by warming in 30 ml of 6.7 N sodium hydroxide. The solution was neutralized to phenolphthalein with dilute hydrochloric acid and added to a boiling solution of 23.8 g (0.05 mole) of the neutral sulphate of (+)-1-(2-naphthyl)-ethylamine in 200 ml of water. The resulting clear solution was cooled down with swirling in ice-water for a few minutes and then left at room temperature for some hours and finally in the refrigerator over-night. Yield 24.2 g (76 %) of air-dried salt, m.p. 129—132°. Three recrystallizations from acetone gave 10.4 g of salt, m.p. 139—141°. The salt was treated with dilute sodium hydroxide, and the liberated amine extracted with ether. The aqueous phase was acidified with dilute hydrochloric acid, and the hydroxy acid extracted repeatedly with ether. The acid was lactonized by repeated distillation with benzene. Yield 2.65 g of lactone, b.p. 99°/10 mm, n_D^{25} 1.4402, α_D^{25} + 17.99° (2 dm). Kenyon and Symons [11] reported α_D^{25} + 7.15° (0.5 dm).

To the mother liquor from the first crystallization, previously warmed to 50°, was added a hot solution of 21.7 g (0.10 mole) of (−)-1-phenylethylthiuronium chloride in 25 ml of water. The salt began to separate almost immediately. After cooling to room temperature the mixture was left over night in the refrigerator. Yield 22.8 g (70 %) of air-dried salt. Unfortunately the salt tended to decompose when recrystallized from water, and two recrystallizations from acetone plus a little methanol did not seem to improve the quality of the product. The remaining salt (10.4 g, m.p. 128—132°) was shaken between dilute hydrochloric acid and ether and the aqueous phase extracted a few times more with ether, giving 2.4 g of lactone, b.p. 98°/9 mm, n_D^{25} 1.4402, α_D^{25} −11.78° (2 dm).

11. Kenyon, J. and Symons, M. C. R. *J. Chem. Soc.* **1953** 3583.

$C_7H_{12}O_2$ **Ref. 307a**

(−)-4-Hydroxy-4-methylhexanoic lactone was prepared from the (−)-1-phenylethylamine salt of the hydroxy-acid. Since no reaction occurred when 1-phenylethylamine was warmed with the lactone, and since the hydroxy-acid is very readily lactonised, the following procedure was adopted. A vigorously stirred, ice-cold mixture of ether and an aqueous solution of the sodium salt of the hydroxy-acid was acidified with dilute hydrochloric acid, and the ethereal solution of the liberated hydroxy-acid immediately added to an ethereal solution of an equivalent quantity of (−)-1-phenylethylamine, α_D^{20} −44·7° (*l*, 1·0). The precipitated salt, m. p. 112°, was recrystallised five times from acetone, the resulting needles, m. p. 122·5°, were dissolved in water, the solution was cooled and acidified, and the liberated hydroxy-acid extracted with ether. Repeated distillation with benzene to ensure complete lactonisation gave (−)-4-hydroxy-4-methylhexanoic lactone, b. p. 64°/0·5 mm., α_D^{20} −7·2° (*l*, 0·5). Two further crystallisations altered neither the m. p. of the salt nor the rotation of the lactone.

(+)-4-Hydroxy-4-methylhexanoic lactone was prepared from the brucine salt of the hydroxy-acid. This salt, prepared as described for the 1-phenylethylamine salt, had m. p. 91°, raised to 97·5° after ten recrystallisations from acetone. Decomposition of the resulting needles gave (+)-4-hydroxy-4-methylhexanoic lactone, b. p. 64°/0·1 mm., α_D^{20} +7·15° (*l*, 0·5). The m. p. of the salt and the rotation of the lactone were not altered by two further recrystallisations.

$CH_3COCH_2CH(CH_3)CH_2COOH$

$C_7H_{12}O_3$ **Ref. 308**

Resolution of 3-Methyl-5-oxohexanoic Acid (5). To a solution of 100 g (0.83 mol) of *l*-(−)-α-methylbenzylamine in 4700 ml of ethyl ether was added 118.5 g (0.82 mol) of **5** in 200 ml of ethyl ether. The solid which separated after standing at 10° for 3 days was isolated by filtration, and the filtrate was retained for further examination. The salt obtained was recrystallized five more times from ethyl ether to give 21.1 g of *l*-(−)-α-methylbenzylamine (−)-3-methyl-5-oxohexanoate as a hygroscopic salt.[40]

Anal Calcd for $C_{15}H_{23}NO_3$: C, 67.89; H, 8.74; N, 5.28. Found: C, 68.03; H, 8.70; N, 5.35.

To a solution of the salt in 200 ml of water was added 10 ml of concentrated hydrochloric acid, and the solution was extracted with chloroform. The chloroform extracts were dried (Na_2SO_4) and concentrated on a rotatory evaporator. The resulting liquid was distilled to give 7.6 g of **(−)-5:** bp 110° (0.05 mm); $[\alpha]^{29.5}D$ −2.3°; $[\alpha]_{365}^{29.5}$ −33.2° (*c* 0.519, C_2H_5OH). The ir and nmr spectral properties were identical with those of **(±)-5.**

The filtrates retained from the preparation of **(−)-5** were concentrated *in vacuo* and the free acid regenerated. The liquid obtained was distilled to give 75.4 g (0.52 mol) of partially resolved **5.** A solution of the acid in 200 ml of ethyl ether was added to 63.3 g (0.52 mol) of *d*-(+)-α-methylbenzylamine in 2000 ml of ethyl ether. The solid which separated on standing at 10° for 3 days was recrystallized three more times from ethyl ether to give 18.5 g of *d*-(+)-α-methylbenzylamine (+)-3-methyl-5-oxohexanoate as a hygroscopic salt.[40]

Anal. Calcd for $C_{15}H_{23}NO_3$: C, 67.89; H, 8.74; N, 5.28. Found: C, 68.11; H, 8.92; N, 5.45.

Using the same procedure described for the preparation of **(−)-5,** the salt above gave 8.2 g of **(+)-5:** bp 110° (0.05 mm); $[\alpha]^{26}D$ +2.8°; $[\alpha]_{365}^{26}$ +35° (*c* 0.50, C_2H_5OH).

In a separate experiment conducted in the same manner as above, **(+)-5** having $[\alpha]^{23}D$ +2.64°, $[\alpha]_{365}^{23}$ +33.6° (*c* 0.568, C_2H_5OH), was obtained.

Determination of the Optical Purity of (+)- and (−)-5. The salt (0.200 g) obtained from (±)- or (+)-**5** with *l*-(+)-2-methylbenzylamine was dissolved in 10 ml of tetrahydrofuran containing 0.202 g of 1-ethoxycarbonyl-2-ethoxy-1,2-dihydroquinoline (EEDQ), and the mixture was heated at 50° for 16 hr. The reaction mixture was concentrated on a rotary evaporator, and the remaining residue was dissolved in benzene. The benzene extracts were washed with 5% hydrochloric acid solution and water, and dried

(Na_2SO_4). The benzene solution was concentrated to a small volume and chromatographed on alumina using benzene–chloroform (3:1) as the eluent. The product fractions were combined to give 0.150–0.175 g of the amides from (±)- and (+)-**5**. The 100-MHz nmr spectrum ($CDCl_3$) of 0.026 g of the mixture of diastereomers obtained from (±)-**5** in the presence of 0.120 g of Eu(fod)$_3$ exhibited two doublets at δ 4.82 and 5.08 and two singlets at 5.85 and 6.02 ppm for the NCHCH_3 and CH_3CO resonances, respectively. The 100-MHz nmr spectrum ($CDCl_3$) of the amide (**6b**) from (+)-**5** ($[\alpha]_{365}$ +35°) showed one doublet at 5.08 and one singlet at 6.02 ppm for the NCHCH_3 and CH_3CO resonances indicating that this compound is optically pure. The calculated optical purities of (−)-**5** ($[\alpha]_{365}$ −33.2°) and (+)-**5** ($[\alpha]_{365}$ +33.6°) are 95 and 96%, respectively. These values were substantiated by nmr analyses of their respective α-methylbenzylamine amides as described for the analysis of the racemic amide of (+)-**5**.

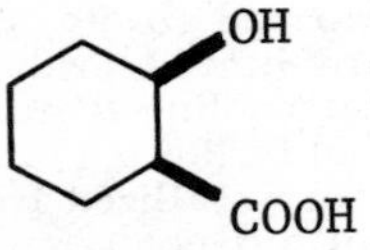

$C_7H_{12}O_3$ **Ref. 309**

Resolution of (±)-cis-2-*Hydroxycyclohexanecarboxylic Acid*.—(±)-*cis*-2-Hydroxycyclohexanecarboxylic acid (31 g.) was resolved by the previously reported method [15] by use of (−)-brucine (85 g.) in acetone (1050 ml.). The crystalline product (60·1 g., 51·8%), $[\alpha]_D^{20}$ −31·6 (*c* 5 in H_2O) (lit.,[15] $[\alpha]_D^{20}$ −32·5°, solvent not stated) showed no significant changes in rotation on further recrystallisation. The mother liquors were evaporated to dryness to give a white crystalline solid containing predominantly the other diastereoisomeric salt.

Each diastereoisomeric salt was separately dissolved in water and the solution was made alkaline with sodium hydroxide (5N) and extracted with chloroform. The aqueous solution was acidified (5N-HCl), saturated with ammonium sulphate, and extracted with ether. The combined extracts were dried ($MgSO_4$) and filtered, and the ether was distilled off. The residual oil was then fractionally distilled. The enantiomers had the following properties: (−)-*cis*-2-hydroxycyclohexanecarboxylic acid (10·96 g.) [from (−)-diastereoisomeric salt], b.p. 142°/1 mm., $[\alpha]_D^{20}$ −23·0° (*c* 3 in H_2O), $[M]_D^{20}$ 33·1°, $[\alpha]_D^{20}$ −17° (*c* 3 in $CHCl_3$) (lit.,[15] $[\alpha]_D^{20}$ −34·7°, solvent and concentration not stated); (+)-*cis*-2-hydroxycyclohexanecarboxylic acid (8·2 g.) (from soluble diastereosiomeric salt), b.p. 142°/1 mm., $[\alpha]_D^{20}$ +25·5° (*c* 3 in H_2O), $[M]_D^{20}$ 36·7°, $[\alpha]_D^{20}$ +18·0° (*c* in $CHCl_3$) (lit.,[15] $[\alpha]_D^{20}$ +30·7°, concentration and solvent not stated).

[15] D. S. Noyce and D. B. Denny, *J. Amer. Chem. Soc.*, 1954, 76, 768.

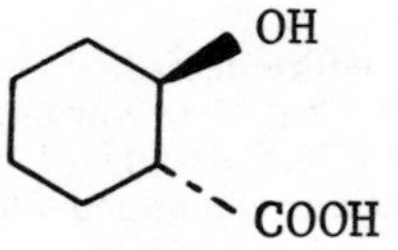

$C_7H_{12}O_3$ **Ref. 310**

Note:

Resolved via brucine in acetone. Gave a brucine salt m.p. 166-167°, $[\alpha]_D^{15}$ = -40.3° from which there was obtained the acid, m.p. 107-109° $[\alpha]_D^{16.5}$ = 47.3°.

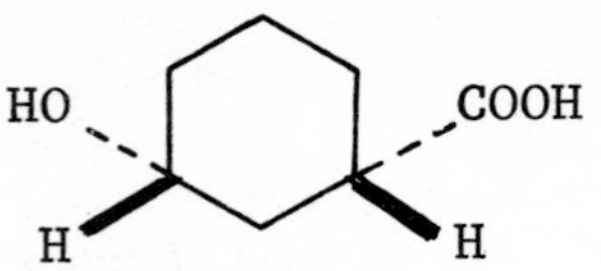

$C_7H_{12}O_3$ **Ref. 311**

The carboxylic acid 8 (55.5 g, 0.386 mol) and cinchonidine (113 g, 0.385 mol) were heated under reflux for 1 h in ethanol (400 mL). Standing overnight at room temperature deposited a solid, which was recrystallized three times from ethanol to give 81.5 g of a salt; $[\alpha]^{25}_D$ −89.0° (*c* 1.61, EtOH). A mixture of the salt (81.5 g) and NaOH aqueous solution (NaOH 15.0 g, water 200 mL) was stirred for 12 h at room temperature and cinchonidine was filtered off. The filtrate was made acidic with 18 N sulfuric acid and extracted with ether. Evaporation of the ether gave 25.5 g of 8: mp 124–127 °C; $[\alpha]^{24}_D$ −6.2° (*c* 1.61, EtOH).

Anal. Calcd for $C_7H_{12}O_3$: C, 58.31; H, 8.39. Found: C, 58.28; H, 8.35.

Concentration of the mother liquor of the cinchonidine salt gave 19.0 g of a semisolid, which was treated with aqueous NaOH (NaOH 2.0 g, water 70 mL). The same procedure described above afforded 3.85 g of 8: mp 123–127.5 °C; $[\alpha]^{28}_D$ +5.6° (*c* 0.930, EtOH).

Anal. Calcd for $C_7H_{12}O_3$: C, 58.31; H, 8.39. Found: C, 58.61; H, 8.41.

$C_7H_{12}O_3$ **Ref. 311a**

Resolution of ***cis*****-3-Hydroxycyclohexanecarboxylic Acid (III).**—Quinine trihydrate, 500 g. (1.30 moles), was dissolved in 1000 ml. of warm methanol; to this solution, a solution of 188.3 g. (1.30 moles) of III in 400 ml. of methanol was added in one portion. The combined solutions were placed in a large water-bath at 50° and allowed to cool slowly to room temperature. The solution was seeded and placed in an ice-bath for three hours. The resulting mixture was filtered to give 526 g. of crystals and mother liquor A. Recrystallization from 3 liters of methanol afforded 156 g. of salt, m.p. 193.4–194.2°. Further concentration of the mother liquor gave 128 g. of salt and mother liquor B. The mother liquors A and B were evaporated to dryness to yield 325 g. of the soluble salt. Fractional crystallization from methanol of the two head fractions afforded 162 g. (53%) of the quinine salt of (+) III, m.p. 197.1–198.1° (dec.), $[\alpha]^{25.6}$D −135° (*c* 1, in methanol).

Anal. Calcd. for $C_{27}H_{36}N_2O_5$. C, 69.20; H, 7.75; N, 5.97. Found: C, 69.08; H, 7.58; N, 5.94.

Decomposition of the Soluble Quinine Salt.—The soluble quinine salt, 309 g. (0.694 mole) was suspended in a stirred solution of 40 g. (1.0 mole) of sodium hydroxide in 750 ml. of water. The mixture was stirred and heated on a steam-bath for two hours at which time 100 ml. of water was added as the mixture became very thick and heavy. After 30 minutes more, 150 ml. of water was added, and the stirring and heating were continued for an additional 1.5 hours, and then the mixture was allowed to stand at room temperature for 12 hours. The suspension was filtered, and the aqueous solution was acidified to *p*H 3 with ice-cold 18 *N* sulfuric acid. The acidic solution was continuously extracted with ether for 36 hours. The ether extract was not dried as this caused precipitation of the acid. The ethereal solution was evaporated to dryness to afford 78 g. of partially resolved (−)-III. The recovered quinine was dissolved in chloroform and extracted with 100 ml. of 1.5 *N* sodium hydroxide solution. The basic extract was acidified with ice-cold 18 *N* sulfuric acid and continuously extracted with ether for six hours. Evaporation of the ether extract yielded 17 g. of partially resolved (−)-III. The total yield of crude (−)-III was 95 g. (100%), m.p. 122–128°, $[\alpha]^{18.5}$D −6.2° (*c* 2, in methanol).

Resolution of Partially Resolved (−)-III with Cinchonidine.—Crude (−)-III, 106 g. (0.736 mole), m.p. 123–129°, $[\alpha]^{25°}$D −6.2° (*c* 2, in methanol), was dissolved in 250 ml. of 95% ethanol. The above solution was added to a warm solution of 217 g. (0.736 mole) of cinchonidine in 1500 ml. of 95% ethanol. After the solution had cooled to room temperature, it was seeded and placed in an ice-bath for 12

$C_7H_{12}O_3$ $C_7H_{12}O_3S_2$

hours. The resultant mixture was filtered and the crystals were dried. The yield of the cinchonidine salt was 217 g., m.p. 175–181°. Fractional crystallization of the first crop and material from its mother liquor afforded 170 g. of the cinchonidine salt of (−)-cis-3-hydroxycyclohexanecarboxylic acid (III). The yield was 57%, and the salt had the following properties: m.p. 179.5–180.3° (dec.), $[\alpha]^{25}_D$ −91.5° (c 1, in 95% ethanol).

Anal. Calcd. for $C_{26}H_{34}N_2O_4$: C, 71.20; H, 7.81; N, 6.39. Found: C, 71.00; H, 7.89; N, 6.50.

Decomposition of the Quinine Salt of (+) III.—In a 2000-ml. round-bottomed three-necked flask equipped with a stirrer and a reflux condenser, there were placed 20 g. (0.50 mole) of sodium hydroxide and 750 ml. of water. The solution was stirred and warmed on a steam-bath, and to it was added 161.4 g. (0.344 mole) of the quinine salt of (+) III. The solution was stirred and warmed on the steam-bath for one hour. After stirring the mixture at room temperature for four hours, it was filtered and the quinine was dissolved in 300 ml. of chloroform. The chloroform solution was extracted with 80 ml. of 1.25 N sodium hydroxide solution. The combined basic extracts were acidified with ice-cold 18 N sulfuric acid, and the acidic solution was continuously extracted with ether for 36 hours. The ether extract was evaporated to yield 49.4 g. (100%) of (+)-cis-3-hydroxycyclohexanecarboxylic acid. One crystallization from ethyl acetate afforded 45 g. (91%) of (+)-III, m.p. 131.0–131.5°, $[\alpha]^{23.6}_D$ +9.82° (c 4, in methanol), $[\alpha]^{23.6}_D$ +9.75° (c 4, in 95% ethanol).

Decomposition of the Cinchonidine Salt of (−)-III.—The cinchonidine salt of (−)-III, 168 g. (0.383 mole), was suspended in a solution of 22 g. (0.55 mole) of sodium hydroxide in 750 ml. of water. The suspension was stirred and heated on the steam-bath for one hour and then stirred at room temperature for four hours. The suspension was filtered, and the cinchonidine was dissolved in 800 ml. of 1.25 N sodium hydroxide solution. The combined basic extracts were acidified with ice-cold 18 N sulfuric acid, and the acidic solution was continuously extracted with ether for 36 hours. The ether extract was evaporated to give 53.1 g. of (−) III. One crystallization from ethyl acetate afforded 48.7 g. (88%) of pure (−) III, m.p. 130.9–131.3°, $[\alpha]^{19.6}_D$ −9.93° (c 4, in methanol), $[\alpha]^{19.7}_D$ −9.84° (c 4, in 95% ethanol).

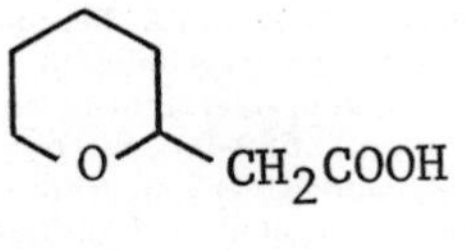

$C_7H_{12}O_3$ Ref. 312

Resolution of dl-2-Tetrahydropyranylacetic Acid.—To 52.6 g. of quinine in 450 cc. of hot benzene there was added 23.3 g. of II in 15 cc. of the same solvent. The hot solution was filtered, allowed to crystallize at 20°, and cooled to 5° for two days. The crystals were washed with three 10-cc. portions of cold benzene to yield 59.0 g. (79%) of mixed quinine 2-tetrahydropyranylacetates, m.p. 140–150°, $[\alpha]^{26}_D$ −133.2° (ethanol, c 0.4). A second crop of 9.9 g., m.p. 130–135°, $[\alpha]^{26}_D$ −123.7° (ethanol, c 0.4) was not further purified. After eight recrystallizations from benzene, the physical properties were essentially constant, m.p. 162–163°, $[\alpha]^{27}_D$ −136.3° (ethanol, c 0.7). The yield of quinine l-2-tetrahydropyranylacetate was 10.1 g. (28%).

Anal. Calcd. for $C_{27}H_{36}O_5N_2$: C, 69.23; H, 7.69; N, 5.90. Found: C, 69.00; H, 7.72; N, 6.10.

The salt was hydrolyzed by dissolving 10.0 g. in 50 cc. of chloroform and shaking with 60 cc. of 2 N aqueous sodium hydroxide. The chloroform layer was washed with two 20-cc. portions of water. The combined aqueous phase was then extracted with chloroform in a liquid-liquid extractor for four hours. The basic solution was finally neutralized to methyl orange indicator with 1.5 N hydrochloric acid and extracted with fresh chloroform for six hours. Distillation of the chloroform solution gave 3.4 g. (74%) of l-2-tetrahydropyranylacetic acid (IIa), b.p. 120–125° (4 mm.). One recrystallization from petrolcum ether, b.p. 60–90°, gave a product melting at 37–38°, $[\alpha]^{27}_D$ −5.67° (ethanol, c 15).

D-Desoxyephedrine was somewhat less satisfactory as a resolving agent. After preparation in, and five recrystallizations from butanone, the physical properties of the salt were essentially constant. At the end of seven, the yield of D-desoxyephedrine l-2-tetrahydropyranylacetate was 30%, m.p. 103–104°, $[\alpha]^{25}_D$ +2.00° (ethanol, c 6.5).

Anal. Calcd. for $C_{17}H_{27}O_3N$: C, 69.62; H, 9.21; N, 4.78. Found: C, 69.63; H, 9.28; N, 4.68.

Hydrolysis of 4.5 g. of salt gave 0.13 g. (3%) of impure IIa, m.p. 36–37°, $[\alpha]^{25}_D$ −4.54° (ethanol, c 6).

$$CH_3CH_2\underset{\underset{\displaystyle C_2H_4O-C-S}{|}}{CHCOOH}$$
$$\parallel$$
$$S$$

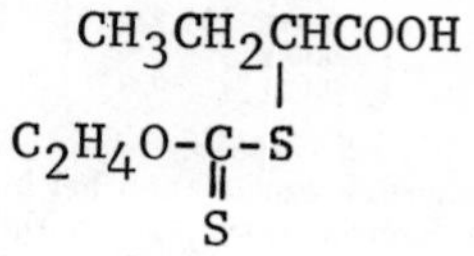

$C_7H_{12}O_3S_2$ Ref. 313

Spaltung der racem-Säure in die optisch aktiven Antipoden. Die (−)-Säure lässt sich über das *Cinchonidinsalz*, die (+)-Verbindung über das *Strychninsalz* darstellen. Kristallisierte Salze wurden auch von Brucin, Chinin und Cinchonin erhalten, aber die aus diesen Salzen regenerierten Säurepräparate zeigten keine messbare Drehung.

40 g racem-Säure wurden nebst 59 g Cinchonidin in 2000 ml siedendem Alkohol gelöst und mit 4000 ml heissem Wasser versetzt. Das nach 15-stündigem Stehen bei etwa 10° ausgeschiedene Salz wurde mehrmals aus verdünntem Alkohol umkristallisiert. Von jeder Kristallisation wurden etwa 0,5 g entnommen, die Säure in Freiheit gesetzt und ihre Drehung in Essigester gemessen. Die Genauigkeit dieser Messungen an nicht umkristallisierter Säure beträgt etwa ± 2%.

Kristallisation	1	2	3	4	5	6
ml Alkohol	2000	900	375	275	75	75
ml Wasser	4000	600	200	125	35	35
g Salz	60	30	26	22	20	18
$[\alpha]^{25}_D$	−45°	−84°	−92,5°	—	−93,5°	−94°

Aus den Versuchsdaten scheint hervorzugehen, dass man vorteilhaft mit höheren Alkoholkonzentrationen und kleineren Lösungsmengen arbeiten kann. Es wurde deshalb ein neuer Spaltungsversuch angestellt mit denselben Substanzmengen; dabei wurden noch 23 g Cinchonidinsalz aus den späteren Mutterlaugen der vorigen Spaltungsreihe zugefügt. Wie ersichtlich wird der Grenzwert auch hier in der dritten Kristallisation erreicht.

Kristallisation	1	2	3	4
ml Alkohol	1400	500	100	75
ml Wasser	800	250	75	60
g Salz	39	29	26,5	22,5
$[\alpha]^{25}_D$	−77,5°	−88,5°	−93°	−92°

Aus der Mutterlauge der ersten Kristallisation des Cinchonidinsalzes wurden 12,5 g Säure von $[\alpha]^{25}_D$ = +45° (in Essigester) gewonnen. Sie wurde nebst 20 g Strychnin in 150 ml siedendem Alkohol gelöst und die Lösung mit 900 ml heissem Wasser versetzt. Das Salz wurde aus verdünntem Alkohol umkristallisiert und die Spaltung wie oben verfolgt.

Kristallisation	1	2	3	4
ml Alkohol	150	115	70	45
ml Wasser	900	700	400	270
g Salz	25,5	18	13	10
$[\alpha]^{25}_D$	+87°	+89,5°	+92°	+92,5°

(−)-*Xantogen-buttersäure.* Das Cinchonidinsalz (18 g) wurde mit verdünnter Schwefelsäure zersetzt und die freigemachte Säure in reinem Äther aufgenommen. Nach Trocknen mit Chlorcalcium und freiwilligem Eindampfen bei Raumtemperatur verblieb ein gelbliches Öl, das beim Impfen mit kristallisierter Säure aus den Vorversuchen vollständig erstarrte. Die Säure ist in Wasser sehr wenig, in den gewöhnlichen organischen Lösungsmitteln aber leicht löslich. Sie kann aus Petroleumäther umkristallisiert werden, wenn man sie in einer kleinen Menge löst und im verschlossenen Gefäss im Eisschrank auskristallisieren lässt. Ausbeute an Rohprodukt 6,3 g; nach zweimaligem Umkristallisieren verblieben 3,2 g vom Schmp. 31–32°. Die Säure bildet ziemlich grosse, flächenreiche Kristalle von kurzprismatischem Habitus.

16,55 mg Sbst.: 37,13 mg $BaSO_4$.

$C_7H_{12}O_3S_2$ (208,2) S, Ber. 30,80 Gef. 30,81

Die Drehungsmessungen an der reinen Säure sind in Tab. 1 zusammengestellt. Die eingewogenen Säuremengen wurden auf 10,00 ml gelöst und die Drehung (2α) in 2 dm Rohr gemessen.

Tabelle 1.

Lösungsmittel	g Säure	$2\alpha_D^{25}$	$[\alpha]_D^{25}$	$[M]_D^{25}$
Essigester	0,0896	−1,665°	−92,9°	−193,5°
Alkohol (abs.)	0,1014	−1,71	−84,8	−175,5
Aceton	0,1099	−2,01	−91,5	−190,5
Chloroform	0,1096	−2,24	−102,0	−212,5
Benzol	0,1032	−2,90	−140,5	−292,5
Verdünnte Sodalösung	0,1025	−0,64	−31,0	−64,5

Die Drehung in Essigester wurde auch bei höherer Konzentration gemessen; eine Konzentrationsabhängigkeit der spezifischen Drehung liess sich dabei nicht feststellen.

0,8019 g Säure, in Essigester auf 10,00 ml gelöst: $2\alpha_D^{25} = -5{,}605°$. $[\alpha]_D^{25} = -92{,}8°$; $[M]_D^{25} = -193{,}8°$.

Das Cinchonidinsalz der (—)-Xantogen-buttersäure, das bei der Spaltung erhalten wurde, bildet feine Nadeln. Der Wassergehalt stimmt am besten auf ein Mol Kristallwasser, schwankt jedoch von Tag zu Tag je nach den Witterungsverhältnissen. Die Analysen wurden deshalb an völlig entwässertem Salz ausgeführt.

37,54, 35,68 mg Sbst.: 34,65, 33,27 mg $BaSO_4$.

$C_{19}H_{22}ON_2 + C_7H_{12}O_3S_2$ (502,4) S Ber. 12,76 Gef. 12,68, 12,81.

(+)-Xantogen-buttersäure. Die Säure wurde wie die Antipode aus dem Alkaloidsalz isoliert und umkristallisiert. 10 g Strychninsalz gaben 2,8 Säure; nach zweimaligem Umkristallisieren verblieben 1,0 g vom Schmp. 31—32°. Sie war der (—)-Säure äusserlich völlig ähnlich.

16,97 mg Sbst.: 38,12 mg $BaSO_4$.

$C_7H_{12}O_3S_2$ (208,2) S Ber. 30,80 Gef. 30,85.

0,1005 g Sbst. in Essigester auf 10,00 ml gelöst: $2\alpha_D^{25} = +1{,}865°$. $[\alpha]_D^{25} = +92{,}8°$; $[M]_D^{25} = +193{,}2°$.

Das Strychninsalz der (+)-Xantogen-buttersäure, das bei der Spaltung erhalten wurde, bildet dünne Nadeln oder längliche Blättchen, die zwei Mol Kristallwasser enthalten.

0,5866 g Sbst.: Gewichtsverlust über P_2O_5 0,0384 g. — 42,17 mg Sbst.: 34,82 mg $BaSO_4$.

$C_{21}H_{22}N_2O_2 + C_7H_{12}O_3S_2 + 2H_2O$ (578,4) Ber. H_2O 6,22 S 11,09
Gef. » 6,23 » 11,18

HOOC-CHCH$_2$-CH$_2$-COOH
 |
 C$_2$H$_5$

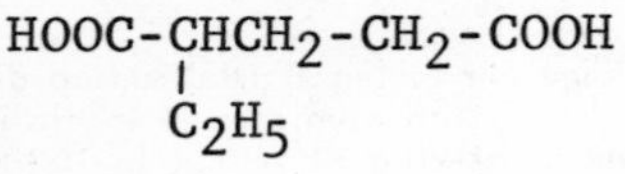

$C_7H_{12}O_4$ Ref. 314

Optische Spaltung der α-Äthylglutarsäure.

Die Spaltung gelang uns bei der Anwendung von Strychnin in wäßriger Lösung. Auf 1 Mol Säure wurde 1 Mol Strychnin verwendet, z. B. 10 g Säure + 20,9 g Strychnin in 250 ccm Wasser. Beim Stehen krystallisierte das neutrale Salz der rechtsdrehenden Säure. Es wurde 2-mal aus Wasser umkrystallisiert, dann in Wasser gelöst, das Strychnin mit Ammoniak gefällt und die aktive Äthylglutarsäure nach Zusatz von Schwefelsäure mit Äther extrahiert. Aus mehreren Portionen à 10 g erhielten wir eine Säure mit spez. Drehung +8,2 bis +8,4° (p = 7—8). Diese Säure wurde noch einmal mit Strychnin behandelt, wobei die Drehung auf etwa + 8,8° stieg, ein Wert, der durch einen neuen Spaltungsversuch nicht erhöht werden konnte. Die reine d-Säure hatte den Schmelzp. 42°.

Die Messungen des Drehungsvermögens in wäßriger und absolut alkoholischer Lösung bei verschiedenen Konzentrationen sind in Tab. XVI zusammengestellt.

Tabelle XVI.

Spez. Drehungsvermögen der d-α-Äthylglutarsäure in Wasser und absolutem Alkohol.

In Wasser			In absolutem Alkohol		
p	d_4^{20}	$[\alpha]_D^{20}$ in °	p	d_4^{20}	$[\alpha]_D^{20}$ in °
3,094	1,0045	+ 9,17	2,841	0,799	+ 14,32
5,909	1,0094	8,97	5,864	0,809	14,12
11,506	1,0191	8,68	11,057	0,824	14,00
15,624	1,0262	8,48	21,469	0,855	13,81
21,450	1,0363	8,32	43,855	0,929	13,90
31,106	1,0531	8,20	60,384	0,988	14,24
48,395	1,0843	8,39	75,459	1,047	14,84
64,847	1,1117	8,88			
74,674	1,1275	9,53			

Das Drehungsvermögen der geschmolzenen reinen Säure konnte direkt bei Zimmertemperatur gemessen werden, da sie erst nach ½—¾ Stunde zu krystallisieren anfing.

Abgelesener Winkel beim „Polarimeterrohr" von $l = 0{,}1757$, $\alpha_D^{20} = +3{,}40°$. Das spez. Gewicht der flüssigen Säure erhielten wir aus der Gleichung:

$$d_4^{20} = 0{,}9991 + 0{,}001735 \cdot p,$$

Tabelle XVII.

Drehungsvermögen der d-α-Äthylglutarsäure in absolutem Alkohol; beobachtete und nach Gleichung (16) berechnete Werte.

q	0	24,54	39,62	56,15	78,53	88,94	94,14	97,16	100
$[\alpha]_D^{20}$ ber.	16,51	14,88	14,24	13,84	13,83	14,02	14,16	14,27	14,37
$[\alpha]_D^{20}$ beob.	16,51	14,84	14,24	13,90	13,81	14,00	14,12	14,32	—

HOOC-CH$_2$-CH-COOH
 |
 (CH$_2$)$_2$CH$_3$

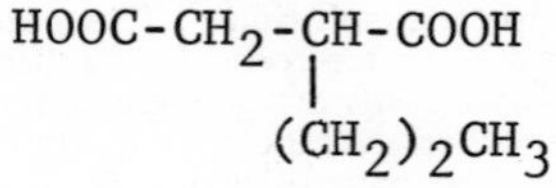

$C_7H_{12}O_4$ Ref. 315

Resolution of synthetic dl-n-*propylsuccinic acid.* The resolution was effected by strychnine using the method of Braun & Reinhard [1929]. The acid (7·5 g.) was dissolved in water (450 ml.) and strychnine (31·2 g.) was added, the solution warmed on the water-bath and kept overnight in the cold room. The precipitate was dissolved in water (1500 ml.) at 80–90°, filtered from dissociated strychnine NH_4OH (150 ml.: sp. gr. 0·88) added. The strychnine separating was filtered and the filtrate was evaporated to a small volume, acidified and extracted continuously with ether. 3·55 g. of acid were obtained, M.P. 95–97°, unchanged by recrystallization from light petroleum (B.P. 80–90°), $[\alpha]_{5461}^{18°} + 20{\cdot}6°$ ($c = 1{\cdot}204$ in H_2O). The resolution was repeated. The whole of the material was dissolved in warm water (280 ml.), strychnine (14·8 g.) was added and the propylsuccinic acid isolated from the precipitate as before. After recrystallization from light petroleum it melted at 103° and did not depress the M.P. of the natural product, M.P. 103°, $[\alpha]_{5461}^{18°} + 26{\cdot}8°$ ($c = 1{\cdot}12$ in water). (Found (Weiler): C, 52·67, 52·60; H, 7·51, 7·41%. $C_7H_{12}O_4$ requires C, 52·48; H, 7·55%.) 0·0691 g. required 8·65 ml. N/10 NaOH for neutralization to phenolphthalein giving equiv. wt. 79·9 ($C_7H_{12}O_4$ titrating as a dibasic acid requires 80).

$C_7H_{12}O_4$ Ref. 315a

S-(-)-2-*n*-Propylsuccinic Acid (*S*-Ib).

Compound *RS*-Ib, prepared by the known method (4), was resolved by four crystallizations of its salt with D-(-)-*threo*-1-*p*-nitrophenyl-2-amino-1,3-propanediol in ethanol. Upon hydrolysis of the salt with 2*N* hydrochloric acid, *S*-Ib was obtained, m.p. 88°; $[\alpha]_D$ -23° (c = 2%, water) [lit. (5b) m.p. 94° $[\alpha]_D$ +22.5° (water) for the antipode].

Anal. Calcd. for $C_7H_{12}O_4$: C, 52.49; H, 7.55. Found: C, 52.52; H, 7.80.

(b) P. W. Clutterbuck, H. Raistrik and F. Reuter, *Biochem. J.*, 31, 987 (1937).

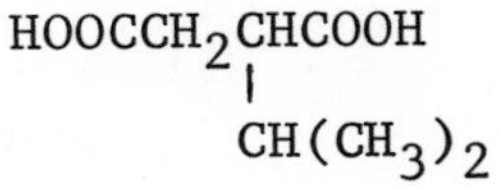

$C_7H_{12}O_4$ Ref. 316

1 Mol. Säure (A) und die 13-fache Menge Wasser wurde auf dem Wasserbade angewärmt und 2 Mol. Strychnin eingerührt Nachdem alles in Lösung gegangen war, wurde langsam erkalten und 24 Stdn. bei Zimmer-Temperatur stehen gelassen: das Strychninsalz der *rechts*-Säure (B) scheidet sich unter diesen Bedingungen in gut ausgebildeten, farblosen, bei 129—131° schmelzenden Krystallen, jedoch nur in geringer Ausbeute (etwas über 15% des Ausgangsmaterials), ab. Nach dem Lösen in Wasser, Ausfällen des Strychnins mit Soda, Ansäuern mit HCl und Eindampfen liefert es eine Säure (C), die von 82—88° schmilzt und in etwa 10-proz. wäßriger Lösung eine spezif. Drehung von ungefähr + 18° zeigt.

Das Filtrat von B liefert, wenn man es von Strychnin befreit und auf Isopropyl-bernsteinsäure verarbeitet, in mehr als 80% Ausbeute ein bei 75—100° schmelzendes, schwächer linksdrehendes Produkt (D), das durch Umkrystallisieren aus möglichst wenig Wasser etwas mehr als $^2/_3$ als reine inaktive Isopropyl-bernsteinsäure (A') zu isolieren gestattet. Aus dem Filtrat kann eine bei 75—90° schmelzende linksdrehende Säure (F) gefaßt werden, deren spezif. Drehung etwa halb so groß wie die Rechtsdrehung von C ist. Dieselbe Trennung, auf A' angewandt, führt mit annähernd derselben prozentualen Ausbeute zur *rechts*-Säure C', *d,l*-Säure A'' und *links*-Säure F' und kann beliebig oft weiter wiederholt werden.

Um aus A' + A'' + A''' usw. die rechtsdrehende Säure möglichst rein und einheitlich zu fassen, wurden die Krystall-Fraktionen, die schwach gefärbt und von sehr ähnlichem Drehungsvermögen waren, vereinigt, in wäßriger Lösung mit Tierkohle entfärbt und von neuem mit Strychnin, ähnlich wie A, angesetzt. Das nunmehr in 30% Ausbeute gewonnene Salz zeigte den Schmp. 130—132° und lieferte eine Säure, die bei 86—88° schmolz und in 10-proz. H_2O-Lösung eine spezif. Drehung von + 23.8° zeigte. Ein nochmaliger Ansatz mit Strychnin führte zu einem Produkt, das nur noch wenig verschieden war und seine Eigenschaften beim abermaligen Binden an Strychnin bzw. Umkrystallisieren aus Wasser nicht mehr änderte. Der Schmp. der reinen (+)-Isopropyl-bernsteinsäure liegt bei 87—88°.

0.1134 g Sbst.: 0.2166 g CO_2, 0.0795 g H_2O.

$C_7H_{12}O_4$. Ber. C 52.47, H 7.55. Gef. C 52.10, H 7.84.

Die spezif. Drehung in 10.16-proz. wäßriger Lösung beträgt + 24.01°, eine bei 20° gesättigte wäßrige Lösung enthält 47.89 g Säure: die Verhältnisse liegen hier also genau umgekehrt wie in der β-Isopropyl-adipinsäure-Reihe, wo die aktiven Formen, die übrigens nur wenig tiefer als die *d,l*-Form schmelzen, in Wasser schwerer löslich sind.

Zur Charakterisierung der Säure stellten wir ihr Anilid dar, indem wir auf sie erst unter Kühlung, dann unter schwachem Erwärmen 2 Mol. PCl_5 und dann Anilin im Überschuß einwirken ließen; das in der üblichen Weise isolierte Anilid zeigte nach dem Umkrystallisieren aus Holzgeist den Schmp. 200° und erwies sich stark linksdrehend.

0.1294 g Sbst.: 10.3 ccm N (24°, 758 mm).

$C_{19}H_{22}O_2N_2$. Ber. N 9.03. Gef. N 9.13.

$[\alpha]_D^{20} = (-0.49° \times 100) : (0.5 \times 0.8197 \times 3.275) = -36.5°$ (in Alkohol).

Zur Darstellung der Linksform der Isopropyl-bernsteinsäure wurden die Fraktionen F, F', F'' usw., deren Linksdrehung viel geringer als 24° war, vereinigt, mit Tierkohle gereinigt und mehrmals aus Wasser umkrystallisiert, wobei der sich zuerst abscheidende, schwerer lösliche Teil jedesmal entfernt und die Mutterlauge eingedampft wurde. Wir konnten so nach 3-maliger Wiederholung der Operation zu einer Säure kommen, deren Schmp. bei 70—75° und deren spezif. Drehung zwischen — 18° und — 19° lag. Da ein weiteres Umkrystallisieren keine Änderung herbeiführte und es nahe lag anzunehmen, daß eine kleine, durch die vielen Operationen erzeugte Verunreinigung störend wirkte, wurde eine Reinigung durch den Ester hindurch angewandt.

Der Äthylester der (—)-Isopropyl-bernsteinsäure ließ sich leicht in der üblichen Weise gewinnen und gestattete, den Hauptteil als einheitlich bei 119—120° siedende, analysenreine Flüssigkeit zu fassen.

0.1293 g Sbst.: 0.2894 g CO_2, 0.1101 g H_2O.

$C_{11}H_{20}O_4$. Ber. C 61.11, H 9.26. Gef. C 61.04, H 9.53.

$[\alpha]_D^{20} = -14.9°/0.9896 = -15.05°$ (unverdünnt).

Seine Verseifung lieferte eine Säure, die sich von der +-Säure nur noch wenig verschieden erwies: sie schmolz bei 85—90° und zeigte in 10-proz. wäßriger Lösung eine spezif. Drehung, die etwas größer als 21° war.

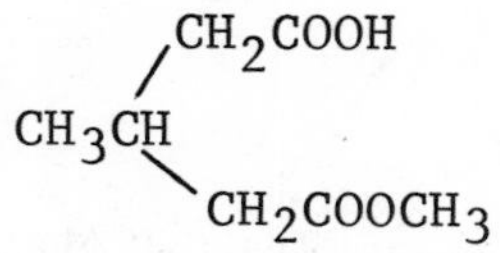

$C_7H_{12}O_4$ Ref. 317

The resolution of methyl hydrogen β-methylglutarate. The resolution of the *dl*-compound was attempted under different conditions with a number of different alkaloids, including cinchonidine, cinchonine, quinine, strychnine and brucine. The amino-acid arginine (cf ref. 15) was also tried, but the salts were too soluble. It was found that the cinchonidine salt of the dextrorotatory enantiomorph crystallized readily from water or from water containing 10 % acetone. The brucine salt of the same enantiomorph could also be obtained from water but was rather too soluble to be useful. A quinine salt of the laevorotatory enantiomorph crystallized from water (originally containing 10 % acetone) after standing for two months in the refrigerator at a temperature of + 4° C. On account of the slow crystallization of the quinine salt prepared from the *dl*-compound it was found convenient to start the preparation of the laevorotatory enantiomorph using the laevorotatory mono-ester from the mother liquors obtained from the crystallization of the cinchonidine salt.

(+)-Methyl hydrogen β-methylglutarate. To 1100 ml of water was added 100 ml of acetone, and the mixture heated to boiling. *dl*-methyl hydrogen β-methylglutarate (129 g, 0.806 mole) was dissolved in the hot solution, and cinchonidine (230 g, 0.781 mole) added in portions by stirring. The hot solution was filtered through glass wool, allowed to cool, seeded and left in the refrigerator (+ 4° C) overnight. The cinchonidine salt separated in the form of fine needles arranged in rosettes. The crystals were filtered off and washed on the filter with ice-water. 15 g of wet crystals (total weight about 230 g) were taken off, and the salt decomposed by means of dilute hydrochloric acid. The mixture was extracted with ether, the ether solution carefully washed with water, dried with sodium sulphate, the ether removed and the residue distilled, giving methyl hydrogen β-methylglutarate, b. p. 101°, 0.5 mm, $\alpha_D^{23} + 0.49°$ (homogeneous, *l*, 1). Mono-ester obtained from the mother liquor had $\alpha_D^{23} - 0.28°$ (undistilled). The second crystallization of the cinchonidine salt gave mono-ester of $\alpha_D^{22} + 0.55°$, and mono-ester recovered from the mother liquor had $\alpha_D^{23} - 0.24°$. After three crystallizations the rotations were $\alpha_D^{22} + 0.65°$ and $\alpha_D^{22} - 0.10°$ respectively. The cinchonidine salt now crystallized in the form of large rosettes of fine, silky needles. After the fourth crystallization, the distilled mono-ester (13.3 g) had $\alpha_D^{25} + 0.64°$. d_4^{22} 1.113. n_D^{22} 1.4389.

The yield, corrected for material taken off in the crystallizations, was about 29 % of the theoretically possible.

Optical rotation. $\alpha_D^{22} + 0.65°$ (homogeneous, *l*, 1)
$[\alpha]_D^{22} + 0.58°$ $[M]_D^{22} + 0.94°$.

Anal.: Calcd. for $C_7H_{12}O_4$ (160.2) C, 52.49; H, 7.55 %
Found C, 52.69; H, 7.58 %

(—)-Methyl hydrogen β-methylglutarate. 73 G (0.456 mole) of partly resolved mono-ester, $\alpha_D^{22} - 0.26°$, was dissolved in a hot mixture of water (1000 ml) and acetone (100 ml), and quinine (153 g, 0.405 mole) was added in portions by stirring. The hot solution was filtered through glass wool, allowed to cool to room temperature, seeded, and kept in the refrigerator at + 2° for 4 days. The quinine salt separated like the cinchonidine salt in the form of rosettes of fine silky needles. The salt was filtered off and washed with ice-water in the usual manner. The methyl hydrogen β-methylglutarate obtained after decomposition of the salt had, after distillation, b. p. 98°, 0.4 mm, $\alpha_D^{22} - 0.64°$, (homogeneous, *l*, 1). After two crystallizations the rotation was $\alpha_D^{22} - 0.61°$, after three $\alpha_D^{22} - 0.65$, and after four $\alpha_D^{22} - 0.64°$. Corrected yield about the same as for the dextrorotatory enantiomorph. Mono-ester recovered from the mother liquor after the fourth crystallization (about 10% was lost in this crystallization) had after distillation $\alpha_D^{22} - 0.63°$. d_4^{22} 1.112. n_D^{22} 1.4387.

Optical rotation. $\alpha_D^{22} - 2.57°$ (homogeneous, *l*, 4)
$[\alpha]_D^{22} - 0.58$ $[M]_D^{22} - 0.93°$.

Anal.: Calcd. for $C_7H_{12}O_4$ (160.2) C, 52.49; H, 7.55 %
Found C, 52.60; H, 7.55 %

$C_7H_{12}O_4$

```
(CH3)2CHCHCOOH
         |
       OCOCH3
```

Ref. 318

30.8 g (0.19 Mol) *DL-O-Acetyl-α-hydroxy-isovaleriansäure* wurden in 100 ccm Methylenchlorid bei 20° portionsweise mit 23.3 g (0.19 Mol) *D-α-Phenyl-äthylamin* in 60 ccm Methylenchlorid versetzt. Nach Zugabe von 320 ccm Petroläther (Sdp. 30—50°) wurde durch Reiben die Kristallisation eingeleitet. Nach 2tägigem Aufbewahren bei Raumtemperatur konnten 10.9 g (40%) *D-α-Phenyläthylammoniumsalz der L-O-Acetyl-α-hydroxy-isovaleriansäure* isoliert werden, $[\alpha]_D^{20}$: −11.5° (c = 0.9; Äthanol); Schmp. 143—145°.

Das Salz wurde in wenig Methylenchlorid unter Rühren mit der äquivalenten Menge 1*n HCl* in absol. Äther versetzt. Vom ausgefallenen Phenyläthylammoniumchlorid wurde abgesaugt, das Filtrat i. Vak. eingedampft und der Rückstand bei 0.05 Torr fraktioniert. *L-O-Acetyl-α-hydroxy-isovaleriansäure:* Sdp. 90°/0.05 Torr; $[\alpha]_D^{23}$: −25.4° (c = 6; Äthanol)[6].

Analog wurde mit *L-α-Phenyl-äthylamin* das *Salz der D-O-Acetyl-α-hydroxy-isovaleriansäure* kristallisiert: $[\alpha]_D^{20}$: +12.0° (c = 1; Äthanol); Schmp. 142—143°.

Die aus diesem Salz freigesetzte *D-O-Acetyl-α-hydroxy-isovaleriansäure* zeigte Sdp. 90°/0.05 Torr; $[\alpha]_D^{23}$: +24.6° (c = 6; Äthanol)[6].

```
HOOC-CH——CH-COOCH3
     |    |
    CH3  CH3
```

$C_7H_{12}O_4$

Ref. 319

The Resolution of the Monomethylester.

Attempts were made to resolve the monoester by means of strychnine, brucine, cinchonine and morphine, trying different solvents. The only alcaloid with which a crystalline salt could be obtained was strychnine, and it was necessary to proceed in the following way.

1 g of the monoester was dissolved in 5 ml hot water, and 2,09 g of strychnine added in portions. Spontaneous crystallization did not occur, even if the solution was kept in a frigigating box and the walls of the beaker were rubbed from time to time. A small part of the solution was therefore evaporated in a vacuum desiccator, when crystals were formed, which could be used for inocculating the main solution. Keeping the inocculated solution in the frigigating box over night, the salt of the dextrorotatory monoester had separated in very small, lustrous crystals. The salt was filtered on a Büchner funnel, pressed between filter paper and recrystallized two times from twice its weight of water. The strychnine salt was decomposed by means of hydrochloric acid, and the monoester extracted with ether.

The rotatory power was measured in absolute alcoholic solution, using a capillary tube. The density of the solution was determined by means of a small pycnometer.

Several (mostly greater) portions of the monoester were resolved in the same way as described above, giving active products with a specific rotation varying from + 8,68° to + 8,75° at a concentration p of about 14,7 per cent by weight. A second treatment of this active monoester with strychnine was now carried out, resulting in an ester with the following rotation:

0,4023 g of monoester dissolved in alcohol to 2,7489 g, $p = 14{,}63$, $d_4^{20} = 0{,}827$, $\alpha_D^{20} = +2{,}16°$, $l = 2$ dm.

$$[\alpha]_D^{20} = +8{,}93°.$$

As the rotation by several experiments varied only between + 8,92° and + 8,94° it seemed likely that the optically pure dextro-ester was obtained. In order to test this a sample of the ester was treated with strychnine for a third time, using a more diluted solution (20 ml water to 1 g ester). The rotation was then found to be:

0,3988 g of monoester dissolved in alcohol to 2,7258 g, $p = 14{,}62$, $d_4^{20} = 0{,}827$, $\alpha_D^{20} = +2{,}16°$, $l = 2$ dm.

$$[\alpha]_D^{20} = +8{,}93°.$$

After the active monoester had been recrystallized from light petroleum it melted at 50°.

The monoester, which was isolated from the mother liquid from the strychnine salt, showed specific rotations from − 3,8° to − 4,3°. When the mother liquid was kept for a long time at about 0°, some great, waterclear crystals were formed, which could not obviously be the salt of the dextro-ester. After recrystallizing three times from the equal weight of water, an ester was isolated, which had the specific rotation − 6,72° at p 14,72, corresponding to a mixture of 75,3 per cent laevo-ester and 24,7 per cent racemic ester. Recrystallized from light petroleum it melted at 50°. This melting point, which is the same as for the racemic monoester, was also found for any mixture of the two active components. We have not offered more work in obtaining the pure laevo-ester.

True Specific Rotation of the Dextro-ester.

In order to find the true specific rotation [6] of the dextro-ester, the rotation was measured at different concentrations in absolute ethyl alcohol and in dioxan, giving the necessary data for an extrapolation to the pure substance. The results of the measurements are given in the tables I and II.

Table I. *Rotations in absolute ethyl alcohol.*

Sol. No.	p g/100 g	d_4^{20}	l	α_D^{18}	$[\alpha]_D^{18}$
1	10,158	0,8163	2	+ 1,51°	+ 9,11°
2	14,693	0,8283	2	2,17°	8,92°
3	29,958	0,8707	2	4,47°	8,57°
4	50,181	0,9306	1	3,83°	8,20°
5	68,229	0,9918	1	5,30°	7,83°

Table II. *Rotations in dioxan.*

Sol. No.	p g/100 g	d_4^{20}	l	α_D^{20}	$[\alpha]_D^{20}$
1	10,180	1,0473	2	+ 1,76°	+ 8,25°
2	35,294	1,0670	2	6,16°	8,18°
3	58,950	1,0853	1	5,23°	8,18°

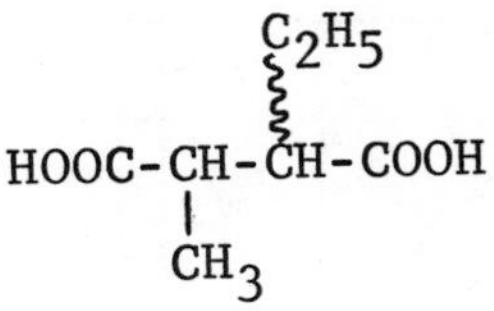

$C_7H_{12}O_4$ Ref. 320

Optische Spaltung der hochschmelzenden α-Methyl-α'-äthylbernsteinsäure.

Die Spaltung der Säure, Schmelzp. 180°, gelang leicht mit Hilfe von Strychnin. Es wurde das saure Salz bereitet, z. B. aus 1 g Säure, 2,09 g Strychnin und 20 ccm Wasser, und das schwer lösliche Salz der d-Säure schied sich dann sofort aus. Das Salz wurde 3-mal aus der 3-fachen Menge Wasser umkrystallisiert, und lieferte dann in einer Ausbeute von 80 Proc. d. Th. eine rechtsdrehende Säure, deren Drehungsvermögen sich weder durch Umkrystallisieren noch durch erneute Behandlung mit Strychnin erhöhen ließ. Der Schmelzpunkt war 180°.

Durch Titrieren einer abgewogenen Menge wurde festgestellt, daß die Säure kein Krystallwasser enthielt: 0,2278 g Säure verbrauchten 28,43 ccm $^n/_{10}$-NaOH, gegen berechnet für $C_7H_{12}O_4$ 28,44 ccm.

Da die Säure schwer löslich in Wasser ist, wurde das Drehungsvermögen nur bei einer einzigen Konzentration bestimmt:

$$\alpha_D^{20} = +0{,}42°, \quad p = 2{,}053, \quad d_4^{20} = 1{,}0019, \quad l = 4.$$
$$[\alpha]_D^{20} = +5{,}11°.$$

Bei einem Parallelversuch wurde + 5,10° gefunden.

Die Messungen des Drehungsvermögens in absolutem Alkohol und in Dioxan sind in Tab. XII zusammengestellt

Die Abhängigkeit des Drehungsvermögens von der Konzentration läßt sich sowohl in Alkohol wie in Dioxan durch lineare Funktionen ausdrücken:

In absolutem Alkohol:

$$[\alpha]_D^{20} = \div 0{,}05188\, p + 8{,}414 \quad \text{oder} \quad [\alpha]_D^{20} = +0{,}05188\, q + 3{,}226,$$

und in Dioxan:

$$[\alpha]_D^{20} = \div 0{,}03925\, p + 6{,}907 \quad \text{oder} \quad [\alpha]_D^{20} = +0{,}03925\, q + 2{,}982.$$

Tabelle XII.

Spez. Drehungsvermögen der d-α-Methyl-α'-äthylbernsteinsäure vom Schmelzp. 180° in absolutem Alkohol und in Dioxan.

In Alkohol			In Dioxan		
p	d_4^{20}	$[\alpha]_D^{20}$ in °	p	d_4^{20}	$[\alpha]_D^{20}$ in °
7,645	0,8120	+7,97	8,584	1,0474	+6,56
11,883	0,8214	7,79	17,185	1,0573	6,25
23,758	0,8573	7,19	27,524	1,0646	5,82
37,177	0,9007	6,47			

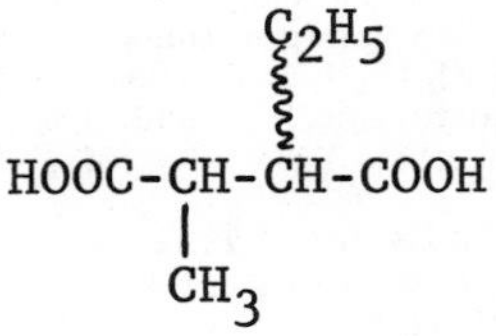

$C_7H_{12}O_4$ Ref. 321

Optische Spaltung der niedrigschmelzenden α-Methyl-α'-äthylbernsteinsäure.

Im Gegensatz zu dem Verhältnis bei der stereoisomeren Säure war die optische Spaltung der niedrigschmelzenden Säure mit Schwierigkeiten verbunden. Nach orientierenden Versuchen mit einer Reihe von Alkaloiden in verschiedenen Lösungsmitteln fanden wir, daß Brucin in Methanol und Strychnin in Wasser die besten Resultate gaben. In beiden Fällen krystallisierte das Salz der rechtsdrehenden Säure aus. Da das Arbeiten mit Strychnin bequemer ist, zogen wir dies vor. Es wurde das saure Salz bereitet, z. B. durch Lösen von 1 g Säure und 2,09 g Strychnin in 20 ccm Wasser. Krystallisation trat erst nach anhaltendem Reiben der Gefäßwand ein und schritt sehr langsam fort. Nachdem aber das Salz durch mehrere Umkrystallisationen gereinigt war, verlief die Krystallisation, wenn sie durch Reiben eingeleitet war, fast momentan. Die aus dem mehrmals umkrystallisierten Salz isolierte d-Säure besaß maximale Drehung, denn diese wurde durch erneute Spaltungsversuche nicht verändert. Die optisch aktive Säure war sehr leicht löslich in Wasser und hatte den Schmelzp. 81°.

0,2679 g Säure: 33,46 ccm $^n/_{10}$-NaOH.
$C_7H_{12}O_4$ Ber. 33,45 ccm $^n/_{10}$-NaOH.

Das Drehungsvermögen der reinen d-Säure wurde in Wasser und absolutem Alkohol bei verschiedenen Konzentrationen bestimmt. Die Resultate findet man in Tab. XIII.

Tabelle XIII.

Spez. Drehungsvermögen der d-α-Methyl-α'-äthylbernsteinsäure vom Schmelzp. 81° in Wasser und absolutem Alkohol.

In Wasser			In absolutem Alkohol		
p	d_4^{20}	$[\alpha]_D^{20}$ in °	p	d_4^{20}	$[\alpha]_D^{20}$ in °
4,832	1,0070	+25,69	5,263	0,8063	+13,55
9,313	1,0152	25,28	10,330	0,8213	13,79
17,119	1,0295	24,97	18,177	0,8448	14,72

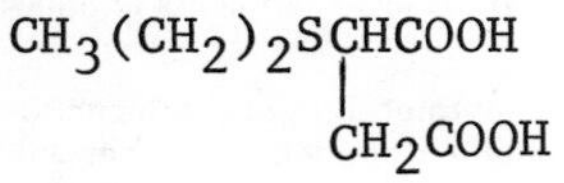

$C_7H_{12}O_4S$ Ref. 322

Resolution of S-propylmercaptosuccinic acid. Preliminary experiments on resolution showed that the (+)-acid could be obtained from the neutral quinine salt and the acid cinchonidine salt, but the neutral strychnine salt crystallized from dilute alcohol gave the best result. The (−)-acid could be obtained from the neutral strychnine and brucine salts, the former crystallized from water gave the best result.

30 g (0.16 mol) *racem*-acid was dissolved in 300 ml water and 350 ml alcohol. The solution was brought to the boiling point and 100 g (0.30 mol) strychnine was added. A small quantity of strychnine did not dissolve and was removed by filtering. The solution was allowed to stand for 6 hours in a refrigerator giving 61 g salt which was recrystallized as is shown by the scheme below; the optical activity *of the acid* was measured in acetone.

Crystallization	1	2	3	4	5	6
{ml alcohol	350	165	100	80	40	30
{ml water	300	135	100	80	40	30
g salt	61	40	34.5	28.5	25.5	24
$[\alpha]_D^{25}$	+84°	+118°	+132°	+133°	—	+133°

The mother liquor from crystallization 1 was boiled to remove the alcohol, diluted with water to 800 ml and heated to boiling. The solution was placed in a refrigerator for one day giving 40 g salt which was recrystallized several times from hot water, the optical activity *of the acid* being measured in acetone. The following series of crystallizations was obtained.

Crystallization	1	2	3	4	5	6	7	8	9
ml water	800	600	300	200	100	75	75	60	60
g salt	40	37	18	13.5	12	11	10.5	8	7
$[\alpha]_D^{25}$	−76°5	−83°	−118°5	−122°	—	−128°	—	−129°5	

Owing to shortage of substance the series was broken before maximum rotatory power was attained.

The salts were treated with sodium carbonate, the strychnine removed and the acids liberated with dilute sulphuric acid and extracted with ether.

(+)-S-Propylmercaptosuccinic acid. The crude acid (4.0 g) was recrystallized from ethyl acetate + petroleum ether. Yield 2.2 g. M. p. 118.5—119°5.

Analysis: 69.08 mg acid: 7.375 ml 0.0977-N NaOH
31.62 » » : 50.89 mg CO_2 and 18.11 mg H_2O
$C_7H_{12}O_4S$ calc. E. W. 96.1; C 43.73; H 6.29
found » 95.9; » 43.92; » 6.41
0.0878 g acid in *acetone*, made up to 10.00 ml:
$\alpha_D^{18} = +1°265$; $[\alpha]_D^{18} = +144°1$; $[M]_D^{18} = +277°0$
0.0932 g acid in *ethyl acetate*, made up to 10.00 ml:
$\alpha_D^{18} = +1°385$; $[\alpha]_D^{18} = +148°6$; $[M]_D^{18} = +285°7$

0.1011 g acid in *chloroform*, made up to 10.00 ml:
$\alpha_D^{20} = + 1^{\circ}365$; $[\alpha]_D^{20} = + 135^{\circ}0$; $[M]_D^{20} = + 259^{\circ}5$
0.0982 g acid in *alcohol*, made up to 10.00 ml:
$\alpha_D^{20} = + 1^{\circ}347$; $[\alpha]_D^{20} = + 137^{\circ}2$; $[M]_D^{20} = + 263^{\circ}7$

Strychnine salt of (+)-S-propylmercaptosuccinic acid. Crystallization from dilute alcohol yielded colourless plates with three moles of water of crystallization.

Analysis: 0.5600 g salt: loss in weight over P_2O_5 0.0330 g
$C_7H_{12}O_4S$, $2C_{21}H_{22}O_2N_2 \cdot 3H_2O$ calc. H_2O 5.91
found » 5.89
0.1497 g dry salt: 8.385 ml N_2 at 19°7 and 764.9 mm
$C_7H_{12}O_4S$, $2C_{21}H_{22}O_2N_2$ calc. N 6.51
found » 6.58

(—)-S-Propylmercaptosuccinic acid. The crude acid (1.9 g) was recrystallized from ethyl acetate + petroleum ether. Yield 1.4 g. M. p. 115—117°5.

Analysis: 60.64 mg acid: 6.500 ml 0.0977-N NaOH
31.26 » » : 49.63 mg CO_2 and 17.40 mg H_2O
$C_7H_{12}O_4S$ calc. E. W. 96.1; C 43.73; H 6.29
found » 95.5; » 43.33; » 6.23
0.1006 g acid in *acetone*, made up to 10.00 ml:
$3\,\alpha_D^{25} = -3^{\circ}968$; $[\alpha]_D^{25} = -131^{\circ}5$; $[M]_D^{25} = -252^{\circ}8$

The rotatory power shows that the product was not quite pure but contained about 4 % of the other antipode. For experiments on quasi-racemate formation this has little significance.

Strychnine salt of (—)-Propylmercaptosuccinic acid. Crystallization from water yielded a white, crystalline powder with four moles of water of crystallization.

Analysis: 0.4113 g salt: loss in weight over P_2O_5 0.0311 g
$C_7H_{12}O_4S$, $2C_{21}H_{22}O_2N_2 \cdot 4H_2O$ calc. H_2O 7.72
found » 7.56
0.1770 g dry salt: 9.975 ml N_2 at 21°4 and 767.6 mm
$C_7H_{12}O_4S$, $2C_{21}H_{22}O_2N_2$ calc. N 6.51
found » 6.60

S-Methylmercaptoethylsuccinic acid

Ethyl butane-4-methylsulfide-1.1.2-tricarboxylate (III). 11.5 g (0.5 gram-atom) sodium was dissolved in 225 ml absolute alcohol. 123 g (0.5 mol) ethyl ethane-1.1.2-tricarboxylate (9) was added while stirring and after a few minutes 55 g (0.5 mol) β-chloroethylmethylsulfide (10) was slowly added. The reaction was initiated by gentle heating with a small flame. After the reaction ceased the mixture was refluxed for one hour. The alcohol was distilled off, the residue shaken with equal volume water and the oily layer separated, washed with water and dried with anhydrous sodium sulfate. The ester was fractionated *in vacuo* through a WIDMER column. The main fraction (96 g) passed over at 187—188°/11 mm. It had a faint yellow colour and a slight, very disagreeable smell. $n_D^{17} = 1.4637$.

***rac.*-S-Methylmercaptoethylsuccinic acid.** 90 g ethyl butane-4-methylsulfide-1.1.2-tricarboxylate was heated on a water bath with excess of alcoholic potash. In a short time the potassium salt separated and was filtered off. The salt was treated with dilute sulphuric acid and the acid extracted with ether. On evaporation 54 g was obtained. This product was decarboxylated by heating at 130° as long as carbon dioxide was evolved. When cooling the product crystallized and was recrystallized with charcoal from benzene. After recrystallization from ethyl acetate + petroleum ether 26 g pure acid was collected. M. p. 92—93°. The compound is soluble in water, alcohol, ethyl acetate, acetone, ether and chloroform, less soluble in benzene and insoluble in petroleum ether.

Analysis: 0.2063 g acid: 21.98 ml 0.0977-N NaOH
31.57 mg » : 50.44 mg CO_2 and 17.49 mg H_2O
51.16 » » : 7.980 ml 0.0662-N NaOH (combustion according to *Grote-Krekeler*)
$C_7H_{12}O_4S$ calc. E. W. 96.1; C 43.73; H 6.29; S 16.68
found » 96.1; » 43.59; » 6.20; » 16.57

Resolution of S-methylmercaptoethylsuccinic acid. Preliminary experiments on resolution showed that the (+)-acid could be obtained from the neutral strychnine, cinchonidine and quinidine salts, strychnine giving the best result. The (—)-acid could be obtained from the neutral quinidine salt.

18.4 g (0.1 mol) *racem*-acid was dissolved in 800 ml water. 64 g (0.2 mol) strychnine was added to the boiling solution, which was filtered to remove some undissolved strychnine. After one day the salt (30 g) was removed and recrystallized from hot water, the optical activity of samples *of the acid* being measured in acetone. The following series of crystallizations was obtained.

Crystallization	1	2	3	4	5
ml water	800	200	175	150	15
g salt	30	24	21.5	15	1
$[\alpha]_D^{25}$	+27°	+34°5	+35°5	+36°5	+3

The mother liquor from crystallization 1 was treated with sodium ca nate, the strychnine removed and the acid liberated with dilute sulphuric The acid was extracted with ether, which after evaporation left a residu 9 g partially active acid. $[\alpha]_D^{25} = -16°$ (in acetone). This acid (0.05 mol) dissolved in 125 ml acetone and 125 ml water, 35.5 g (0.1 mol) quinidine added during boiling and the solution placed in a refrigerator over night. salt (32 g) was recrystallized from dilute acetone. The following series of tallizations was obtained; the rotatory power *of the acid* was measured in ace

Crystallization	1	2	3	4	5	6	7	8
ml acetone	125	75	50	50	50	50	40	40
ml water	125	75	50	50	50	50	40	40
g salt	32	21	19.5	17	—	15	—	12
$[\alpha]_D^{25}$	—26°5	—29°	—	—32°	—	—34°	—	—35°5

The (+)-acid was liberated from its salt after removing the strych The (—)-acid was obtained directly from its salt by acidifying with d sulphuric acid and extraction with ether in the usual manner.

(+)-S-Methylmercaptoethylsuccinic acid. The crude acid was recrystal from ethyl acetate + petroleum ether. Yield 1.3 g. M. p. 84.5—85.°5.

Analysis: 74.27 mg acid: 7.870 ml 0.0977-N NaOH
28.27 » » : 45.36 mg CO_2 and 16.45 mg H_2O
$C_7H_{12}O_4S$ calc. E. W. 96.1; C 43.73; H 6.29
found » 96.6; » 43.78; » 6.51
0.0996 g acid in *acetone*, made up to 10.00 ml:
$3\,\alpha_D^{25} = + 1^{\circ}16$; $[\alpha]_D^{25} = + 38^{\circ}8$; $[M]_D^{25} = + 74^{\circ}6$

Strychnine salt of (+)-S-methylmercaptoethylsuccinic acid. Crystal tion from water yielded colourless plates with varying content of water.

Analysis: 0.0963 g dry salt: 5.96 ml N_2 at 18°5 and 737.2 mm
0.1013 » » » : 6.14 » » » 18°1 » 747.5 »
0.1533 » » » : 9.45 » » » 20°2 » 745.3 »
$C_7H_{12}O_4S$, $2C_{21}H_{22}O_2N_2$ calc. N 6.51
$C_7H_{12}O_4S$, $3C_{21}H_{22}O_2N_2$ calc. N 7.03
found » 7.03; 6.99; 7.04

From the analysis data it is evident that the substance obtained after crystallization 5 of the mixture of S-methylmercaptoethylsuccinic acid strychnine was not the pure neutral salt. The compound $C_7H_{12}O_4S$, $3C_{21}H_{22}$(on the other hand seems less probable in spite of the fact that it is in g agreement with the analysis. In the resolution of n-butylsuccinic acid and butylsuccinic acid FREDGA and SAHLBERG noted that the strychnine salts w partially hydrolyzed on recrystallization (11). It is possible that the s phenomenon is responsible to the abovementioned complication in which the consistency between the nitrogen content found and that calculated the compound $C_7H_{12}O_4S$, $3C_{21}H_{22}O_2N_2$ is a mere chance.

(—)-S-Methylmercaptoethylsuccinic acid. The acid was recrystallized f ethyl acetate + petroleum ether. Yield 1.6 g. M. p. 84.5—85°5.

Analysis: 64.95 mg acid: 6.855 ml 0.0977-N NaOH
30.00 » » : 48.22 mg CO_2 and 17.26 mg H_2O
$C_7H_{12}O_4S$ calc. E. W. 96.1; C 43.73; H 6.29
found » 97.0; » 43.86; » 6.44

0.0939 g acid in *acetone*, made up to 10.00 ml:
$3\,\alpha_D^{25} = -1^{\circ}088$; $[\alpha]_D^{25} = -38^{\circ}6$; $[M]_D^{25} = -74^{\circ}2$
0.0971 g acid in *ethyl acetate*, made up to 10.00 ml:
$\alpha_D^{20} = -0^{\circ}397$; $[\alpha]_D^{20} = -40^{\circ}9$; $[M]_D^{20} = -78^{\circ}6$
0.1007 g acid in *alcohol*, made up to 10.00 ml:
$\alpha_D^{20} = -0^{\circ}333$; $[\alpha]_D^{20} = -33^{\circ}1$; $[M]_D^{20} = -63^{\circ}6$
0.1088 g acid in *chloroform*, made up to 10.00 ml:
$\alpha_D^{20} = -0^{\circ}261$; $[\alpha]_D^{20} = -24^{\circ}0$; $[M]_D^{20} = -46^{\circ}1$

Quinidine salt of (—)-S-methylmercaptoethylsuccinic acid. The salt c stallized from dilute acetone in colourless needles containing two moles of w of crystallization.

Analysis: 0.4187 g salt: loss in weight over P_2O_5 0.0174 g
$C_7H_{12}O_4S$, $2C_{20}H_{24}O_2N_2 \cdot 2H_2O$ calc. H_2O 4.11
found » 4.16
0.0968 g dry salt: 5.590 ml N_2 at 18°2 and 746.9 mm
$C_7H_{12}O_4S$, $2C_{20}H_{24}O_2N_2$ calc. N 6.66
found » 6.65

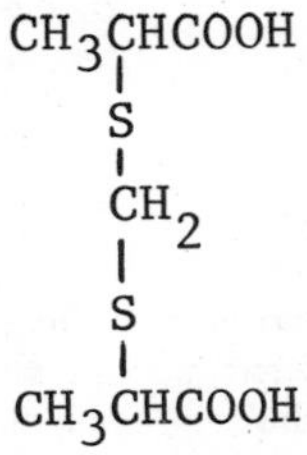

$C_7H_{12}O_4S_2$ Ref. 323

Spaltung der *racem*-Form.

Spaltungsversuche wurden mit Strychnin und Chinin ausgeführt. Strychnin gab ein in kompakten Tafeln kristallisierendes Salz, das stark rechtsdrehende Säure enthielt; aus der Mutterlauge schieden sich beim Stehen allmählich Kristalle von ganz anderem Aussehen (feine Nadeln) ab.

Die Vorversuche mit Chinin gaben noch bessere Resultate, weshalb dieses Alkaloid zur schliesslichen Spaltung angewandt wurde. Aus verdünntem Aceton wurde dabei zuerst ein Salz mit hochaktiver (wie sich später zeigte etwa 85 %-iger) Linkssäure erhalten. Aus der Mutterlauge kristallisierte dann ein Salz von etwa gleichem Aussehen, das aber Rechtssäure von noch höherer Aktivität als die der zuerst erhaltenen Linkssäure enthielt. In beiden Fällen konnten die Salze leicht auf maximale Drehung umkristallisiert werden.

Es liegen Anzeichen dafür vor, dass bei der Spaltung Übersättigungserscheinungen mitspielen können, weshalb es denkbar ist, dass sich das Verfahren nicht immer mit gleichem Erfolge wiederholen lässt.

22,5 g (0,1 Mol) razemische Säure wurden mit 60 g wasserfreiem Chinin in 800 ml warmem 45 %-igen Aceton gelöst. Das im Laufe von 24 Stunden abgeschiedene Salz wurde abgesaugt, mit Wasser gewaschen und aus der zehnfachen Menge verdünntem Aceton umkristallisiert, bis die im Salze enthaltene Säure konstante Drehung zeigte. Der Verlauf der Spaltung ergibt sich aus der folgenden Zusammenstellung; $[\alpha]_D$ bezieht sich auf die Drehung der aus dem Salz in Freiheit gesetzten Säure in Wasser bei 25°.

Kristallisation	1	2	3	4	5
Gewicht des Salzes in g	34	27	26	24	22
$[\alpha]_D$	−310°	−354°	−363,5°	−360°	−362,5°

Die Mutterlauge von der ersten Kristallisation sowie das mit ihr vereinigte Waschwasser (zusammen etwa 1 Liter) wurde zwecks vollständiger Lösung des beim Waschen abgeschiedenen Salzes erwärmt und dann an einem kühlen Ort 24 Stunden stehen gelassen. Das dabei ausgeschiedene Salz wurde in der oben bei der Antipode beschriebenen Weise umkristallisiert; wie aus der folgenden Zusammenstellung ersichtlich, erfolgt die Spaltung hier etwas langsamer.

Kristallisation	1	2	3	4	5	6	7
Gewicht des Salzes in g	22,4	17,2	15.6	13,3	12,3	10,5	9,6
$[\alpha]_D$	+314°	+343°	+353,5°	+357°	+360°	+365°	+360°

Die Salze wurden in Sodalösung suspendiert, das Chinin mit Chloroform extrahiert, die Lösung mit Schwefelsäure angesäuert und mit Äther ausgeschüttelt. Beim Verdampfen der mit Chlorcalcium getrockneten ätherischen Lösung im Exsikkator kristallisierte die aktive Säure spontan. Kohlenstofftetrachlorid eignete sich sehr gut zur Umkristallisation.

(−)-Methylen-bis-thiomilchsäure.

Aus 21,5 g Chininsalz wurden 5,2 g Säure erhalten; nach zwei Umkristallisationen verblieb ein Rückstand von 4,8 g. Die Säure bildet kurzprismatische, ziemlich flächenreiche Kristalle; bei rascher Kristallisation erhält man neben komplizierteren Formen vierseitige Tafeln mit den Seitenwinkeln 85° und 95° und diagonaler Auslöschung. Schmelzpunkt 82,5—83,5°.

0,2023 g Subst.: 15,68 ml 0,1152-n NaOH.
$C_7H_{12}O_4S_2$ (224,2) Äquiv.-Gew. ber. 112,1 gef. 112,0.

Die Ergebnisse von Drehungsmessungen in verschiedenen Lösungsmitteln sind in Tab. 2 zusammengestellt. In der Tabelle bezeichnet g die Menge eingewogene Substanz und v das Volumen der Lösung in ml.

Das Chininsalz der Linkssäure, das im Gange der Spaltung erhalten wurde, bildet feine Nadeln. Bei herrschender hoher Luftfeuchtigkeit kann das Salz an der Luft auf konstantes Gewicht getrocknet werden und enthält dann vier Moleküle

Tabelle 2.

	g	v	2 α	$[\alpha]_D^{25}$	$[M]_D^{25}$
0,5-n Salzsäure	0,1179	10,02	−8,855	−376,3	−843,5
Zur Hälfte neutralisierte wässrige Lösung	0,1123	10,03	−7,485	−335,2	−751,5
Völlig neutralisierte wässrige Lösung	0,1089	10,03	−6,48	−296,1	−664
Alkohol	0,1190	10,09	−13,125	−556,5	−1248
Essigester	0,1122	10,02	−11,95	−533,6	−1196
Eisessig	0,1136	10,02	−12,465	−549,7	−1233
Chloroform	0,1015	10,02	−8,285	−408,9	−917

Kristallwasser. Bei trockenerem Wetter erhält man niedrigere Werte und das Gewicht des Salzes ändert sich von dem einen Tage zum anderen.

0,2085 g wasserfreie Subst.: 0,1103 g $BaSO_4$.
$C_7H_{12}O_4S_2$, 2 $C_{20}H_{24}O_2N_2$ (872,7) S ber. 7,35 % gef. 7,27 %.
0,5915 g kristallwasserhaltige Subst.: 0,0450 g H_2O.
4 H_2O ber. 7,68 % gef. 7,61 %.

(+)-Methylen-bis-thiomilchsäure.

Die Säure wurde in derselben Weise wie die Antipode isoliert; sie glich ihr auch vollständig im Schmelzpunkt und den äusseren Eigenschaften.

0,2017 g Subst.: 15,60 ml 0,1152-n NaOH.
$C_7H_{12}O_4S_2$ (224,2) Äquiv.-Gew. ber. 112,1 gef. 112,3.
0,1084 g Subst. in 0,5-n HCl gelöst zu 10,03 ml: 2 α = + 8,11°.
$[\alpha]_D^{25}$ = + 375,2°; $[M]_D^{25}$ = + 841,5°.

Das Chininsalz der Rechtssäure, das bei der Spaltung erhalten wurde, ist dem der Antipode ähnlich, enthält aber im lufttrockenen Zustande nur ein Molekül Kristallwasser.

0,2463 g wasserfreie Subst.: 0,1322 g $BaSO_4$.
$C_7H_{12}O_4S_2$, 2 $C_{20}H_{24}O_2N_2$ (872,7) S ber. 7,35 % gef. 7,37 %.
0,5082 g kristallwasserhaltige Subst.: 0,0103 g H_2O.
1 H_2O ber. 2,02 % gef. 2,05 %.

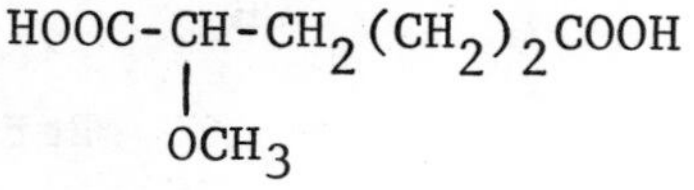

$C_7H_{12}O_5$ Ref. 324

Resolution of III.—Cinchonidine, when mixed with III in the molar proportions of two-to-one gave a salt which crystallized readily from water, m.p. 181–182° dec.

Anal. Calcd. for $C_{45}H_{56}O_7N_4$: C, 70.65; H, 7.38; N, 7.32. Found: C, 70.42; H, 7.27; H, 7.50.

After several crystallizations, the α-methoxyadipic acid regenerated from the head fractions had m.p. 80–85°, $[\alpha]^{23}D$ −6.85° (*c* 2 in ethyl acetate). Crystallization from methanol–ethyl acetate was also feasible, and using this solvent pair, resolution proceeded somewhat further. From 200 g. of salt there was obtained after fractional crystallization 4.5 g. of (−)α-methoxyadipic acid from the head fraction of the salt, m.p. 90–97°, which when further purified by partition chromatography on silicic acid[6] had m.p. 94.5–97°, $[\alpha]^{22.5}D$ −50.6° (*c* 2 in ethyl acetate), neut. equiv., 88.0. This material showed an infrared spectrum identical with that obtained by Noyce and Denney.[6] From the second fraction there was obtained 2.2 g. of (−)III, m.p. 84–95°, $[\alpha]^{20.5}D$ −47.8° (*c* 2 in ethyl acetate).

From the appropriate mother liquors there was obtained 16.7 g. of partially resolved (+)-III, m.p. 76–81°, $[\alpha]^{20}D$ +4.75°.

(6) D. S. Noyce and D. B. Denney, *ibid.*, 76, 768 (1954).

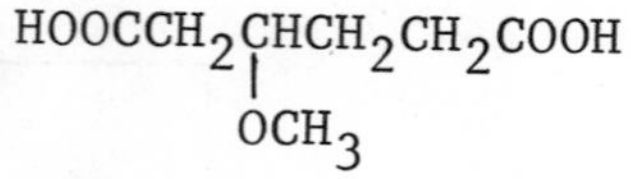

HOOCCH$_2$CHCH$_2$CH$_2$COOH
|
OCH$_3$

Ref. 325

Rechtsdrehende β-Methoxy-adipinsäure.

Die Mutterlauge des Strychninsalzes aus der alle krystallisierbaren Anteile nach starkem Einengen und mehrstündigem Stehen bei 0° möglichst entfernt waren, wurde genau wie bei der linksdrehenden Säure beschrieben gespalten. Die erhaltene rohe Säure wurde in Äther gelöst, von einer geringen Menge ätherunlöslichen Verunreinigungen durch Filtration befreit und das klare Filtrat eingedampft. Der Rückstand wurde im Molekularkolben bei 140—160° destilliert. Das Destillat (450 mg) gab aus Äther-Petroläther zwei Fraktionen, die unscharf bei 80—85° schmolzen. Dieses Material zeigte eine Drehung von $[\alpha]_D^{17} = +9.7°$ in Chloroform. Die Mutterlauge lieferte weitere Fraktionen, die zwischen 60 und 70° schmolzen. Aus diesen konnten nach mehrmaligem Umkrystallisieren aus Äther-Petroläther 90 mg farblose Nadeln vom Smp. 74—75° gewonnen werden, deren Schmelzpunkt sich bei weiterem Umkrystallisieren nicht mehr änderte. $[\alpha]_D^{17} = +13.1° \pm 1°$ (c = 2,203 in Chloroform).

22,237 mg Subst. zu 1,0094 cm³; $l = 1$ dm; $\alpha_D^{17} = +0.29° \pm 0.02°$

Zur Analyse wurde 2 Tage über P_2O_5 ohne Vakuum bei 20° getrocknet (Schweinchen).

3,884 mg Subst. gaben 6,821 mg CO_2 und 2,443 mg H_2O
3,144 mg Subst. verbr. 5,42 cm³ 0,02-n. $Na_2S_2O_3$ (*Zeisel-Vieböck*)

$C_7H_{12}O_5$ Ber. C 47,73 H 6,86 —OCH_3 17,62%
(176,16) Gef. ,, 47,92 ,, 7,04 ,, 17,76%

Die Mischprobe mit möglichst gleicher Menge linksdrehender Säure schmolz zur Hauptsache bei 82—85°, nach starkem Sintern von 74° an.

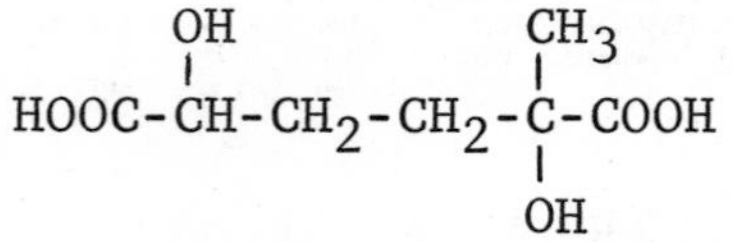

$C_7H_{12}O_6$

OH   CH$_3$
| |
HOOC-CH-CH$_2$-CH$_2$-C-COOH
              |
              OH

Ref. 326

Rechtsdrehende racemoide 2-Methyl-2,5-dioxy-adipinsäure.

26 g der aus dem Dilacton stammenden Säure wurden in 175 ccm Wasser gelöst und bei 50—60° mit 119 g reinem Brucin versetzt. Aus der klaren Lösung krystallisierten bei 4° über Nacht 92 g eines Salzes, das aus 92 ccm Wasser umkrystallisiert wurde (12 Stunden, 4°). Die Krystallmasse wurde abgesaugt, feucht abgeschleudert und mit 10 ccm kaltem Wasser abgedeckt. Die erhaltenen 38 g wurden in ihrem Gewicht Wasser umkrystallisiert; dieser Prozeß[2]) wurde noch 2-mal wiederholt, bis 22 g übrig blieben.

Das Salz wurde mit dem halben Gewicht wasserhaltigen Kaliumbisulfats und wenig Wasser zum Brei verrieben[3]), nach ½ Stunde mit der 5-fachen Menge wasserfreiem Natriumsulfat verrieben und zur Bindung des Wassers einige Stunden beiseite gestellt. Dann wurde mit Sand verrieben und im Apparat ausgeäthert. Die Säure (3 g) krystallisiert aus dem Äther aus. Sie schmilzt bei 173—174° (also höher als das Racemat) unter Gasentwicklung.

3,960 mg Subst.: 6,350 mg CO_2, 2,215 mg H_2O.
$C_7H_{12}O_6$ (192) Ber. C 43,75 H 6,30 Gef. C 43,73 H 6,26.

$[\alpha]^{I\ II}_{578,546}$ in Wasser = + 1,86°(I), 2,13°(II) × 2,071/1,0 × 1,05 × 0,266
= + 13,8° ± 0,1°(I); 15,8° ± 0,1°(II).

Die Eigenschaften der Säure änderten sich nicht bei weiteren Krystallisationsversuchen am Salz (z. B. Butylalkohol) und der Säure selbst (sehr wenig Wasser). Auch nach der Umwandlung in das Dilacton und Regenerierung aus diesem bleiben die Werte dieselben.

In der zur genauen Neutralisation benötigten Menge 10-proc. Kalilauge hatte die Säure folgende Drehung:

$[\alpha]^{I\ II}_{578,546}$ = — 0,44°(I); — 0,52°(II) × 1,33/0,0719 × 1,04 × 1,0
= — 7,8° (I); — 9,2°(II).

Mit mehr Alkali nahm die spezifische Drehung noch ein wenig zu.

²) Hierfür diente eine metallene Zentrifugenhülse, in die mit Korkringen zu unterst ein Glasbecher, darauf ein konischer Gummiring und auf diesem ein Goochtiegel befestigt war. Der konische Gummiring umfaßte den unteren Teil des Goochtiegels und reichte mit seiner engeren Öffnung in den Becher.
³) B. **47**, 2035 (1914).

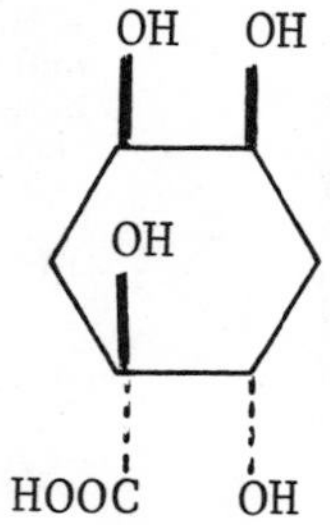

$C_7H_{12}O_6$

Ref. 327

Das Brucinsalz der inaktiven Chinasäure liefert aus Methanol eine schwerlösliche Fraktion vom Schmp. 200° (Zers.); $[\alpha]_D^{21}$: —7°, aus welchem nach dem Zerlegen durch den Austauscher IR 120 die (+) Säure vom Schmp. 168°, $[\alpha]_D^{23}$: +43° (in Wasser, c = 2) gewonnen wird. Wird dieses Produkt mit der gleichen Menge (—)-Chinasäure vom Schmp. 168°, $[\alpha]_D^{22}$: —43° (in Wasser, c = 2) gemischt, so entsteht wieder die inaktive Chinasäure vom Schmp. 149°; $[\alpha]_D^{22}$: 0°.

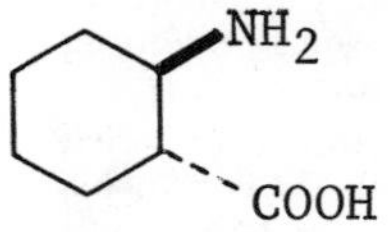

$C_7H_{13}NO_2$ **Ref. 328**

(−)-*N*-Benzoyl-*trans*-2-aminocyclohexanecarboxylic Acid [(−)-2]. To a mixture of 3.71 g (0.015 mol) of (±)-**2** and 5.18 g (0.016 mol) of quinine was added 130 m*l* of 70% methanol (methanol 70 and water 30 in volume). The mixture was briefly heated on a steam bath to dissolve all solid, then allowed to stand at room temperature. After 10 hr, the resulting fluffy needles were collected by filtration to give 6.0—6.5 g of salt, mp 125—130°C. Two or three recrystallizations of this material using 13—15 m*l* of 70% methanol per gram of salt gave 1.2—1.8 g of material whose properties did not change upon further recrystallization, mp 143—145°C, $[\alpha]_D^{20}$ −111.6° (*c* 2.5, ethanol).

With 3N hydrochloric acid the salt was converted into crude (−)-**2**, which was purified by recrystallization from 70% methanol to give 0.22—0.37 g (12—20%) of (−)-**2**, mp 257—258°C (in a capillary tube), subliming at near 240°C on an open plate, $[\alpha]_D^{20}$ −44.5° (*c* 0.67, ethanol).

Found: C, 68.26; H, 7.11; N, 5.71%. Calcd for $C_{14}H_{17}NO_3$: C, 67.99; H, 6.93; N, 5.66%.

l-Ephedrine salt of (−)-**2** melts at 186—188°C, $[\alpha]_D^{18}$ −39.7 (*c* 2.5, ethanol).

(+)-*N*-Benzoyl-*trans*-2-aminocyclohexanecarboxylic Acid [(+)-2]. To a mixture of 3.71 g (0.015 mol) of (±)-**2** and 2.75 g (0.015 mol) of *l*-ephedrine monohydrate was added 20 m*l* of 70% methanol. The mixture was heated on a steam bath to dissolve all solid. The solvent was then evaporated to dryness. The residue was recrystallized from 10 m*l* of 30% methanol to give 4.0—4.3 g of salt, melting at 160—175°C. Two or three recrystallizations of this material using 1.0—1.5 m*l* of 30% methanol per gram of salt gave 0.9—1.3 g of a material melting at above 190°C. With 3N hydrochloric acid the salt was converted into crude (+)-**2**, which was purified by recrystallization with 70% methanol, yielding 0.20—0.30 g (11—16%) of (+)-**2**, mp 257—258°C (in a capillary tube), subliming at near 240°C on an open plate, $[\alpha]_D^{20}$ +44.5° (*c* 0.67, ethanol).

Found: C, 68.13; H, 7.08; N, 5.75%. Calcd for $C_{14}H_{17}NO_3$: C, 67.99; H, 6.93; N, 5.66%.

The pure salt of *l*-ephedrine and (+)-**2** melts at 195—197°C, $[\alpha]_D^{18}$ +10.0° (*c* 2.5, ethanol). Quine salt of (+)-**2** melts at 122—123°C, $[\alpha]_D^{16}$ −72.3° (*c* 2.5, ethanol).

$(CH_3)_2CHCH(NH_2)CONHCH_2COOH$

$C_7H_{14}N_2O_3$ **Ref. 329**

1. Racematspaltung des Valyl-glycin-propylesters

a) Spaltung des Esters mit Dibenzoylweinsäure

DL-Valyl-glycin: α-Brom-isovalerylbromid[4] wurde mit Glycin gekuppelt und aminiert[5]. Schmp. 245° (Zers.).

DL-Valyl-glycin-propylester: Das durch zweimalige Veresterung mit absol. Propanol-Chlorwasserstoff gewonnene hygroskop. Propylester-hydrochlorid wurde mit ammoniakal. Äther[6] unter Zugabe von etwas Natriumsulfat in den freien Ester übergeführt. Ausb., bezogen auf Peptid, 86% d. Th.

Rohester:

Spaltung des Esters: 26,0 g DL-Valyl-glycin-propylester werden mit 50 cm³ absol. Methanol verdünnt und mit einer Lösung von 46,0 g Dibenzoyl-D-weinsäure in 370 cm³ absol. Methanol bei 20° vereinigt. Nach 24 Stdn. haben sich 23 g rohes L-Valyl-glycin-propylester-dibenzoyl-D-hydrogentartrat vom Schmp. 184° und etwa 50proz. optischer Reinheit (durch Abbau zum Peptid ermittelt) abgeschieden. Durch Umkristallisieren aus Methanol erhält man die optisch reine Verbindung. Schmp. 193°, $[\alpha]_D^{20}$: —54,3° (c = 1,2, in Methanol).

$C_{28}H_{34}N_2O_{11}$ (574,6)	Ber.	C 58,50	H 5,90	N 4,88
	Gef.	C 58,35	H 6,10	N 5,14

Durch Eindampfen der Mutterlauge des Spaltansatzes im Vak. auf die Hälfte isoliert man weitere 6 g L-Estertartrat geringerer optischer Reinheit, die verworfen werden.

Die resultierende Lösung enthält das rohe D-Estertartrat, welches durch völliges Eindampfen der Lösung und Fällung mit Äther isoliert wird. Sein Abbau zum Peptid zeigte, daß es eine optische Reinheit von 50% besaß. Ausb. 29,1 g, Schmp. 155°.

Durch Umkristallisieren aus Methanol gewinnt man das optisch reine D-Propylester-dibenzoyl-D-hydrogentartrat. Schmp. 145—147°, $[\alpha]_D^{20}$: —87,6° (c = 1,40, in Äthanol).

Gef. C 58,83 H 5,93 N 5,27

Die gereinigten diastereomeren Estersalze werden in äther. Suspension bei 0° mit NH_3 gesättigt, vom Ammonium-dibenzoyltartrat abfiltriert, der Äther abgesaugt und die so gewonnenen freien Ester mit absol. äther. HCl in ihre Hydrochloride übergeführt. Ausb. 92% d. Th., bezogen auf Tartrat.

L-Valyl-glycin-propylester-hydrochlorid: $[\alpha]_D^{20}$: +18,0° (c = 1,90, in Methanol).

D-Valyl-glycin-propylester-hydrochlorid: $[\alpha]_D^{20}$: —18,0° (c = 1,70, in Methanol).

[4] B. Schleicher, Liebigs Ann. Chem. **267**, 115 [1892].
[5] E. Fischer u. J. Schenkel, Liebigs Ann. Chem. **354**, 12 [1907].
[6] S. M. McElvain u. J. F. Vozza, J. Amer. chem. Soc. **71**, 896 [1949].

Die freien Peptidantipoden werden aus den reinen Estertartraten entweder nach dem Ammoniak-Äther-Verfahren[6] über den Ester und dessen Verseifung mit 0,37*n* $Ba(OH)_2$ (10 Min., 20⁰) oder auf folgendem Wege gewonnen:

Diastereomeres Valyl-glycin-propylester-dibenzoyl-D-hydrogentartrat wird 10 Min. mit überschüss. 0,37*n* $Ba(OH)_2$ bei 20⁰ geschüttelt, dann 0,37*n* H_2SO_4 bis zur Bildung des sauren Peptidfiltrates hinzugefügt, filtriert, Reste der Dibenzoylweinsäure im Kutscher-Steudel-Apparat mit Äther extrahiert, mit 0,37*n* $Ba(OH)_2$ auf p_H7 gebracht, filtriert und das optisch reine Peptid nach dem Einengen der wäßr. Lösung mit Aceton gefällt. Ausb. 70—80% d. Th., bezogen auf Tartrat.

L-Valyl-glycin: $[\alpha]_D^{20}$: +93,7⁰ (in Wasser) Schmp. 272⁰ (Zers.).

D-Valyl-glycin: $[\alpha]_D^{20}$: —93,5⁰ (in Wasser) Schmp. 272⁰ (Zers.).

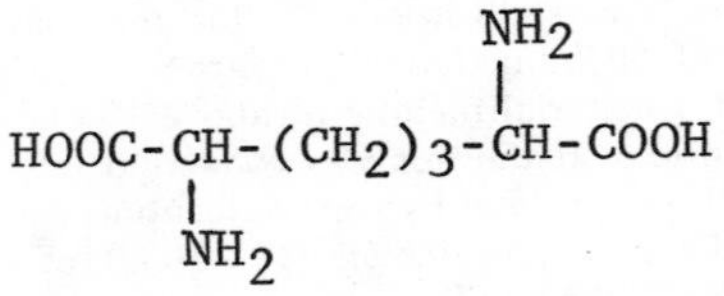

$C_7H_{14}N_2O_4$ **Ref. 330**

Cristallisation fractionnée
du N,N'-bis-dibenzoyl-D-tartrate acide du diamide du DL-DAP.

On dissout 60,3 g (0,32 mole) de diamide du DL-DAP dans 1,5 l de méthanol absolu puis, à la solution maintenue sous très forte agitation, on ajoute une solution d'acide dibenzoyl-D-tartrique (242 g; 0,64 mole; Fluka) dissous dans 1 l de méthanol absolu. En 5 à 10 mn de fins cristaux commencent à précipiter. On laisse la cristallisation se poursuivre une nuit à température ambiante. Après filtration, on lave abondamment les cristaux avec du méthanol chaud puis avec de l'éther. On obtient ainsi une première fraction (164,5 g, Rdt = 54,6 %) fondant à 215,5° avec brunissement à 214-215°. En concentrant la solution-mère on obtient une deuxième fraction (83 g, Rdt = 27,6 %) fondant entre 165 et 168°.

Obtention de L,L-DAP stéréochimiquement pur.

a) *DAP riche en forme L,L.*

Le lot de N,N'-*bis*-dibenzoyl-D-tartrate acide du diamide de DAP (164,5 g) fondant à 215,5° est mis en suspension dans l'éthanol absolu et la solution est saturée à température ambiante avec du gaz chlorhydrique sec. Le produit se dissout totalement puis le dichlorhydrate du diamide de DAP précipite partiellement; la précipitation complète est obtenue par l'addition d'éther. Après filtration et dessiccation, le produit obtenu (48,6 g) est mis à reflux 6 h dans l'acide chlorhydrique 4 N. Après évaporation et dessiccation sur soude, le résidu est dissous dans un minimum d'eau et le pH de la solution est ajusté à 6 avec de la lithine. Par addition d'éthanol absolu, on obtient 33 g de DAP pur ayant un pouvoir rotatoire de + 36,5° (*c* = 1; HCl 5 N). Litt. (14) :

$$[\alpha]_D^{24} = + 45{,}1°$$

(*c* = 2,6; HCl 5 N). La teneur en forme L,L est donc de 90,5 %.

b) *Sel de dicyclohexylamine du N,N'-bis-benzyloxycarbonyl-L,L-DAP.*

A partir du lot (33 g) de DAP riche en forme L,L on prépare le dérivé N,N'-benzyloxycarbonylé suivant le mode opératoire déjà décrit. Le produit obtenu est dissous dans un minimum d'éthanol absolu et de la dicyclohexylamine (69 ml, 0,35 mole) est ajoutée à la solution. Après une nuit à température ambiante, on recueille 40 g de cristaux (F = 179-181°). En recristallisant le produit entre chaud et froid dans un minimum d'éthanol absolu, on obtient un produit stéréochimiquement pur (26,7 g; F = 184-185°) avec un rendement de 20 % par rapport au diamide du DL-DAP. L'analyse est effectuée sur le produit séché 16 h à 61°.

Analyse $C_{47}H_{72}N_4O_8$ = 821,08 :

Calc. % :	C 68,75	H 8,84	N 6,82
Tr. :	68,61	8,77	6,89.

$[\alpha]_D$ = + 9,5° (*c* = 1; éthanol).

c) *N,N'-bis-benzyloxycarbonyl-L,L-DAP.*

Le composé précédent (26,7 g) est mis en suspension dans l'acétate d'éthyle et le mélange est extrait plusieurs fois par une solution à 10 % d'acide citrique. Après dessiccation sur sulfate de sodium puis concentration à sec, on obtient presque quantitativement le produit (14,2 g; Rdt = 95 %). Pour l'analyse une partie aliquote de ce résidu est cristallisé dans un mélange acétate d'éthyl-éther de pétrole (Rdt = 95 %, F = 153-155°) et le produit obtenu est séché 16 h à 80°.

Analyse $C_{23}H_{26}N_2O_8$ = 458,45 :

Calc. % :	C 60,25	H 5,72	N 6,11
Tr. :	60,01	5,58	6,45.

$[\alpha]_D$ = — 4,4° (*c* = 2; éthanol).

d) *L,L-DAP.*

Le L,L-DAP chromatographiquement pur (22) est obtenu quantitativement (5,8 g) par hydrogénation catalytique du dérivé N,N'-*bis*-benzyloxycarbonylé selon la technique déjà décrite. $[\alpha]_D^{25}$ = + 44,5° (*c* = 1; HCl 5 N). Litt. (14) : $[\alpha]_D^{24}$ = + 45,1° (*c* = 2,6; HCl 5 N) et $[\alpha]_D^{24}$ = + 45,0 (*c* = 1; HCl N).

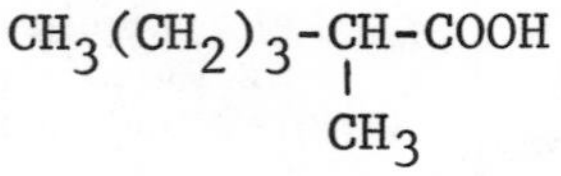

$C_7H_{14}O_2$ **Ref. 331**

Levo-Methyl-n-Butylacetic Acid—The inactive acid was dissolved in hot 66 per cent acetone and an equivalent weight of cinchonidine added. After eight recrystallizations from 66 per cent acetone the rotation of the free acid reached a constant value. B.p. 105° at 5 mm. $D\,\frac{25}{4}$ = 0.909. n_D^{25} = 1.4189.

$$[\alpha]_D^{25} = \frac{-17.0°}{1 \times 0.909} = -18.7°; [M]_D^{25} = -24.3° \text{ (homogeneous)}$$

3.980 mg. substance: 9.490 mg. CO_2 and 3.920 mg. H_2O

$C_7H_{14}O_2$.	Calculated.	C 64.6, H 10.8
130.1	Found.	" 65.0, " 11.0

0.1030 gm. substance: 7.86 cc. 0.1 N NaOH (titration, phenolphthalein). Calculated, 7.92 cc.

$C_7H_{14}O_2$ **Ref. 331a**

Resolution of 2-n-Butylpropionic Acid.—To a hot solution of 65 gm. of the acid in 500 cc. of acetone was added one equivalent (189 gm.) of quinine and the solution was then cooled to crystallization. The salt was subjected to five recrystallizations from acetone. An ethereal solution of the free acid obtained from the last crop of crystals had a rotation of $[\alpha]_D^{22}$ = + 19.7° (No. 228). After three more recrystallizations the value was

No. 244. $$[\alpha]_D^{22} = \frac{+1.08° \times 100}{1 \times 5.50} = +19.6°.$$

$(CH_3)_2CHCH_2-\underset{\underset{CH_3}{|}}{CH}-COOH$

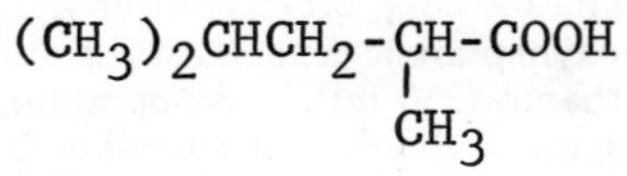

$C_7H_{14}O_2$　　　　　　　　　　　　　　　　　Ref. 332

Risoluzione dell'acido 2,4-dimetilpentanoico. — Grammi 334.4 (0,848 moli) di brucina anidra p.f. 172°-173°C, $[\alpha]_D^{17}-78.4°$ (c=5.348, etanolo 95%) e g 111.6 (0,857 moli) di acido 2,4-dimetilpentanoico (p.eb. 114°-115°C a 31 mm Hg, n_D^{20} 1.4182) preparato secondo quanto descritto in letteratura ($^{3, 12}$), sono stati introdotti in un pallone a due colli da 1500 ml, munito di refrigerante a ricadere, e agitatore meccanico e contenente 560 cc di acetone accuratamente purificato.

La sospensione ottenuta è stata riscaldata per due ore: dopo circa 15 minuti è stata osservata la dissoluzione pressoché completa della brucina.

Dopo raffreddamento la soluzione è stata decantata in una beuta in modo da allontanare le impurezze solide non reagite, quindi sono stati aggiunti 560 cc di acqua raffreddando poi con miscela frigorifera di ghiaccio e sale; dopo aver mantenuto la miscela di reazione a bassa temperatura per circa 30 minuti si è filtrato rapidamente seccando poi il prodotto cristallino su $CaCl_2$ in essiccatore da vuoto.

Si sono ottenuti g 201 (0,380 moli) di sale di brucina con p.f. 72°-76°C, $[\alpha]_D^{22}-27,1°$ (c=8,201, etanolo 95%).

Il sale di brucina così ottenuto è stato decomposto secondo quanto descritto da Levene (^{3}) con acido solforico diluito.

Dopo distillazione si sono avuti g 37 di acido (+)(S)-2,4-dimetilpentanoico avente: p.eb. 110°C a 20 mm Hg, n_D^{20} 1,4183, d_4^{20} 0,9101, $[\alpha]_D^{20}+13,08°$ (omogeneo) $[\alpha]_D^{22}+13,54°$ (etere etilico; c = 5,23).

La letteratura riporta per l'acido (+)(S)-2,4-dimetilpentanoico: $[\alpha]_D^{22}+19,4°$ (etere etilico; c = 5,23) (^{3}).

Dalle acque madri della cristallizzazione, dopo aver eliminato l'acetone a pressione ridotta, sono stati ottenuti, sempre seguendo il procedimento descritto da Levene (^{3}), g 69,1 di acido (−)(R)-2,4-dimetilpentanoico avente: p.eb. 110°C a 20 mm Hg, n_D^{20} 1,4184, $[\alpha]_D^{20}-7,39°$ (omogeneo).

(12) W. H. BENTLEY e M. W. BURROWS, *J. Chem. Soc.*, 67, 511 (1895); *Chem. Zent.*, I, 827 (1895).

(3) P. A. LEVENE e L. W. BASS, *J. Biol. Chem.*, 70, 211 (1926).

$C_7H_{14}O_2$　　　　　　　　　　　　　　　　　Ref. 332a

The inactive acid, prepared by means of the malonic ester synthesis, was dissolved in boiling acetone and 1 equivalent of cinchonidine added. The salt crystallized on cooling to −10° in the refrigerator. After ten recrystallizations a maximum specific rotation was reached. The acid was liberated from the cinchonidine salt and distilled. B.p. 117° at 17 mm. D 31/4 = 0.899.

$$[\alpha]_D^{25} = \frac{-12.4°}{1 \times 0.899} = -13.8° \text{ (homogeneous)}$$

$CH_3(CH_2)_2\underset{\underset{CH_3}{|}}{CH}CH_2COOH$

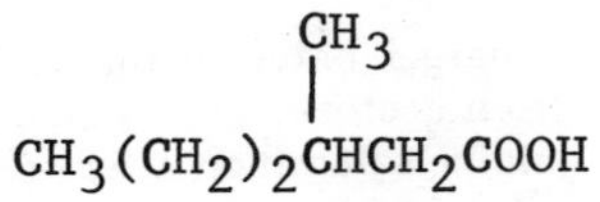

$C_7H_{14}O_2$　　　　　　　　　　　　　　　　　Ref. 333

(S)-(−)-3-Methylhexanoic Acid (III).—The acid III was obtained by six recrystallizations of the cinchonidine salt from an aqueous ethanol solution according to the procedure of Levene and Marker.[6] From 415 g of dl-3-methylhexanoic acid, 90 g of III was obtained: bp 109–110° (13 mm), n_D^{25} 1.4205, $[\alpha]_D^{24}$ −2.63° (neat); lit.[6] bp 113° (17 mm), n_D^{25} 1.4214, $[\alpha]_D^{27}$ −2.52° (neat).

(6) P. A. Levene and R. E. Marker, *J. Biol. Chem.*, 91, 77 (1931).

$CH_3(CH_2)_2-\underset{\underset{C_2H_5}{|}}{CH}-COOH$

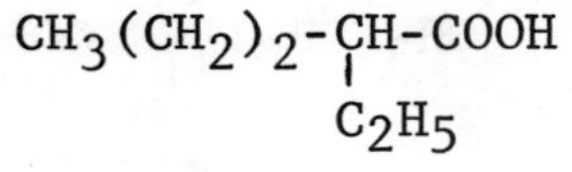

$C_7H_{14}O_2$　　　　　　　　　　　　　　　　　Ref. 334

The ethylpropylmalonic acid was recrystallized until it showed a constant melting point of 117–118°. The ethylpropylacetic acid was resolved by recrystallizing its quinine salt from acetone many times. In another experiment the resolution with cinchonidine proceeded at a faster rate.

The levo acid was recovered from the less soluble cinchonidine salt, and the dextro acid from the quinine salt. B.p. 95°, $p = 8$ mm.; $n_D^{25} = 1.4178$; $d_4^{25} = 0.9098$ (*in vacuo*).

$$[\alpha]_D^{25} = \frac{-1.55°}{1 \times 0.910} = -1.70°; [M]_D^{25} = -2.21°$$

Least maximum $[M]_D^{25} = -6.53°$ (homogeneous)
0.1067 gm. substance required 8.169 cc. 0.1 N NaOH. Mol wt. 130
5.110 mg. substance: 12.090 mg. CO_2 and 4.840 mg. H_2O
　　　　$C_7H_{14}O_2$. Calculated. C 64.56, H 10.84
　　　　130.1　　Found. 　　" 64.57, " 10.60

$(CH_3)_3C\underset{\underset{CH_3}{|}}{CH}COOH$

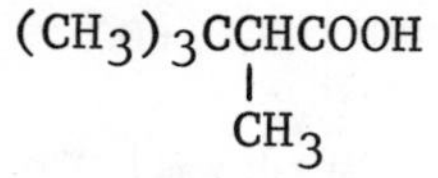

$C_7H_{14}O_2$　　　　　　　　　　　　　　　　　Ref. 335

(−) Acido 2,3,3-trimetilbutirrico [II].

Ad una soluzione di 126,5 g (0,443 moli) di deidroabietilammina (ottenuta secondo il metodo recentemente descritto ($^{13, 14}$) dall'*Ammine D* della «Hercules Powder Co.» rappresentata da «Eigenmann e Veronelli», Milano) in 910 ml di etere si aggiungono 57 g di acido (+) [II] sciolto in etere. Si lascia in agitazione per 2 ore a 15 °C, ottenendo un precipitato cristallino che viene filtrato e lavato con etere (175 g, $[\alpha]_D^{25} = +27$, p.f. 176 °C). Dopo 6 cristallizzazioni da una miscela di $CHCl_3$-C_2H_5OH 1:2 si ottengono 30 g di sale puro, avente $[\alpha]_D^{25} = 15,0$ e p.f. 183,5 °C. Dal sale si isola l'acido [II] otticamente attivo mediante trattamento con *NaOH*, estrazione dell'ammina con etere, acidificazione della soluzione acquosa ed estrazione con etere. Dopo essiccamento su $MgSO_4$ ed eliminazione del solvente si ottengono 9,6 g di acido (−) [II], avente $[\alpha]_D^{25} = -41,0$ (in etanolo assoluto al 2%) e p.f. 53 °C (vedi tab. 1).

TABELLA 1

Potere rotatorio specifico a 25 °C del 2,3,3-trimetilbutanolo-1 [I] e dell'acido 2,3,3-trimetilbutirrico [II] in etanolo assoluto.

λ (mμ)	589	578	546	436	365
$[\alpha]$ [I]	−41,4	−43,0	−48,8	−82,5	−126,8
$[\alpha]$ [II]	−41,0	−42,8	−49,0	−85,3	−137,2

Notevoli quantità dell'antipodo (+) vengono ottenute dai solventi di cristallizzazione: da un sale avente $[\alpha]_D^{25} = +33$ si ottiene l'acido [II] (+) con $[\alpha]_D^{25} = +22$.

(13) B. SJÖBERG, S. SJÖBERG, *Ark. Kemi* 22, 447 (1964).
(14) W. J. GOTTSTEIN, L. C. CHENEY, *J. Org. Chem.* 30, 2072 (1965).

$CH_3(CH_2)_4CHCOOH$
OH

$C_7H_{14}O_3$ Ref. 336

D-(−)-2-Hydroxyheptanoic Acid. Four separate but identical resolutions were performed simultaneously. Anhydrous quinine (55.75 g, 0.172 mol) and 2-hydroxyheptanoic acid (25.0 g, 0.171 mol) were combined in methanol (400 ml) and heated to boiling. Hot (70°) water (1 l.) was added with stirring, and the solution became cloudy. Enough hot methanol was added to clarify the solution. After cooling, the salt which had crystallized from solution was collected, washed (2:5 methanol–water), and dried. The collected salt weighed 38 g (47%). This and the remaining recrystallizations are shown in Table I.

Table I

Crystallization	Methanol, ml	Water, ml	Weight, g	% yield
1	400	1000	38	47
2	150	700	35	44.5
3	175	700	36	43.7
4	200	600	26	33.2

The final rotation of the salt was $[\alpha]^{27}D$ +122.53 ± 0.15 (*c* 4.0, MeOH), and the liberated acid had an average rotation of $[\alpha]^{27}D$ −5.55 ± 0.05° (*c* 5.8, $CHCl_3$), after extraction into base, washing with ether, reacidification, extraction, and recrystallization of the collected acids from hexane.

L-(+)-2-Hydroxyheptanoic Acid. The filtrates from the first crystallization were treated with acid and extracted into ether, dried, and evaporated to give 49.26 g of L-(+)-2-hydroxyheptanoic acid.

Dehydroabeitylamine (50.38 g, 0.175 mol) and the partially resolved L-(+)-2-hydroxyheptanoic acid (23.45 g, 0.159 mol) were dissolved in hot methanol (250 ml), and hot water (100 ml) was added. The solution became cloudy and a small additional amount of methanol was added to clarify the solution. This and the remaining crystallization are shown in Table II.

Table II

Crystallization	Methanol, ml	Water, ml	Weight, g	% yield
1	250	100	63.4	91.6
2	450	200	62.3	89.9

The final rotation of the dehydroabeitylamine salt was $[\alpha]^{26}D$ +14.47 ± 0.06 (*c* 4.0, MeOH) while the liberated acid had a rotation of $[\alpha]^{26}D$ +5.53 ± 0.05° (*c* 5.8, $CHCl_3$).

Optical Purity Analysis. The methyl esters of individual acid samples were prepared by reaction with diazomethane in ether. One drop of the neat ester in five drops of pyridine was treated with four drops of L-menthol chlorocarbonate solution[3] (an excess). Glc analysis was performed at 180° on a 6-ft 1% QF, 1% OV-17 column with a 30-cm^3/min flow rate. Flame ionization was used for detection. The retention times for the derivatized methyl esters follow: (+), 6.57 min; (−), 7.20 min. The major components of the glc trace were identified by glc–mass spectrometry and by comparison of the individual components with authentic samples.

(3) J. Westly and B. Halpern, *J. Org. Chem.*, **33**, 3978 (1968).

Authors' comments:

The method of Westly and Halpern,[3] which involves the glc analysis of the *l*-menthol carbonate ester derivative of the methyl ester of (±)-4 proved to be a facile check on the completeness of the resolution. When we applied this method to (±)-4, we found that the glc of (±)-4 showed two peaks (one for each diastereoisomer) for the *l*-menthol ester derivative, while the glc of the individually resolved and derivatized acids shows only one peak indicating that each acid was >98% optically pure.

CH_3
$CH_3CH_2CH_2-C-COOH$
OCH_3

$C_7H_{14}O_3$ Ref. 337

Resolution of 2-*Methoxy*-2-*methylpentanoic Acid.*—In a preliminary experiment a 0·95-g. sample of the crude acid, resulting from two methylations of methyl 2-hydroxy-2-methylpentanoate and subsequent hydrolysis as described in the preceding section, was dissolved in hot acetone (10 c.c.), and quinine (2·2 g.) was gradually added; the resulting solution was filtered and left overnight. The quinine salt which separated as fine needles was recrystallised from acetone, then decomposed with dilute hydrochloric acid: the liberated acid was isolated by ether-extraction and distillation. Its infrared spectrum showed a strong methoxyl band, and no absorption due to the presence of hydroxyl.

The above procedure was used to prepare the quinine salt from a larger batch of the acid (160 g.). After seven recrystallisations of the quinine salt (261 g.) from acetone, the regained acid (12·2 g.) had α_D^{14} +12·09°. This was again converted into the quinine salt which after one recrystallisation from acetone gave acid (7·5 g.) having α_D^{14} +12·26°, $[\alpha]_D^{14}$ +24·04°. The mother-liquors of the last recrystallisation of the quinine salt yielded 3·8 g. of acid having α_D^{14} +11·3°. This was converted into the quinine salt which after two recrystallisations gave acid (0·92 g.) with $\alpha_D^{9.5}$ +12·80°, $\alpha_D^{11.5}$ +12·64°. A further formation and recrystallisation of the quinine salt gave on decomposition the (+)-acid (0·42 g.), $\alpha_D^{11.5}$ +12·65°, α_D^{15} +12·39°, $[\alpha]_D^{15}$ +24·30°.

$$\begin{array}{c} \text{H}\text{CH(CH}_3)_2 \\ || \\ \text{CH}_3\text{-C}\!-\!\!-\!\text{C-COOH} \\ || \\ \text{OH}\text{OH} \end{array}$$

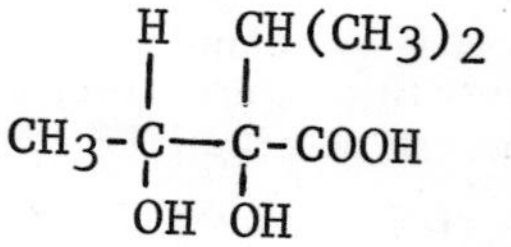

Ref. 338

Resolution of $(\pm)$-*viridifloric acid* VII

$(+)$-*Viridifloric acid* $(+)$-VII. To a solution of acid (VII) (7·77 g) in 10 ml of dry EtOH, 6·22 g $(-)$-α-phenylethylamine in 2 ml of dry EtOH was added and the mixture boiled under reflux for 10 min. On cooling, 9·0 g of the salt of $(+)$-acid VII deposited, m.p. 137–141°, $[\alpha]_D^{20}$ $-9\cdot88°$ (c, 1 in EtOH). Recrystallization from EtOH afforded 2·92 g of the salt. m.p. 158–159°, $[\alpha]_D^{20}$ $-3\cdot5°$ (c, 1 in EtOH). Found: C, 63·48; H, 8·88; N, 4·99%. $C_7H_{14}O_4$. $C_8H_{11}N$ requires: C, 63·57; H, 8·89; N, 4·94%. The salt was dissolved in water, acidified with HCl and continuously extracted with ether. 1·54 g of $(+)$-VII was obtained, m.p. 126–127°, $[\alpha]_D^{21}$ $+1\cdot97°$ (c, 1 in water). Found: C, 52·21; H, 8·75. $C_7H_{14}O_4$ requires: C, 51·85; H, 8·64%.

$(-)$-*Viridifloric acid* $(-)$-VII. The first mother liquor of the above experiment was evaporated, 3·24 g of acid isolated from residue as described above, and resolved using $(+)$-α-phenylethylamine, as described above. Recrystallization from EtOH afforded 2·15 g of salt, m.p. 158–159°, $[\alpha]_D^{21}$ $+8\cdot5°$ (c, 1 in EtOH). Found: C, 63·87; H, 8·87. $C_7H_{14}O_4$. $C_8H_{11}N$ requires: C, 63·57; H, 8·89%. 1 g of $(-)$-VII was isolated from the salt obtained, m.p. 126–126·5°, $[\alpha]_D^{21}$ $-2\cdot0°$ (c, 1 in H$_2$O). Found: C, 52·13; H, 8·81. $C_7H_{14}O_4$ requires: C, 51·85; H, 8·64%. Ref. 4: $(+)$-acid, m.p. 127·5, $[\alpha]_D^{26}$ $+1\cdot8°$ (c, 1 in H$_2$O); $(-)$-acid, m.p. 127·5°, $[\alpha]_D^{26}$ $-1\cdot6°$ (c, 1 in H$_2$O).

Natural viridifloric acid VII has been obtained by hydrolysis of viridiflorine II (0·8 g) with 10% ethanolic KOH (8 ml)[11] for 3 hr under reflux. Yield 100%, m.p. 122–123°, $[\alpha]_D^{20}$ $-1\cdot7°$ (c, 1·6 in H$_2$O). The mixture of equal amounts of natural VII and of $(+)$-VIII had m.p. 146–148°, of natural VII and $(-)$-VII-123–126°.

$$\begin{array}{c} \text{OH}\text{OH} \\ || \\ \text{H}\!-\!\text{C}\!-\!\!-\!\text{C-COOH} \\ || \\ \text{CH}_3\text{CH(CH}_3)_2 \end{array}$$

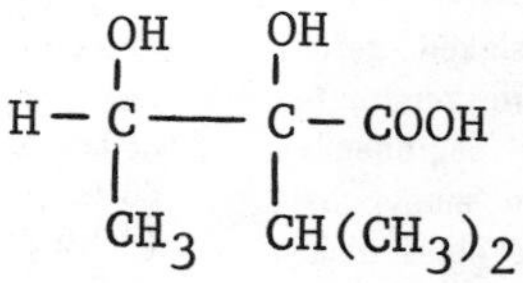

Ref. 339

Resolution of $(\pm)$-*trachelanthic acid*

$(+)$-*Trachelanthic acid* $[(+)$-V$]$. The solution of $(\pm)$-acid (5 g) in dry EtOH (3 ml) was treated with $(-)$-α-phenylethylamine (4 g) in 2 ml of EtOH, the mixture boiled under reflux for 30 min and left overnight. EtOH was evaporated and 20 ml of dry ether were added to the residue. On storage at 5°, the mixture crystallized to afford 5·65 g of the salt. Recrystallization from ethanol afforded 2·34 g of the salt, m.p. 156–158°, $[\alpha]_D^{22}$ $-9\cdot4°$ (c, 1 in EtOH). Found: C, 63·60; H, 8·80; N, 4·97. $C_{17}H_{14}O_4$. $C_8H_{11}N$ requires: C, 63·57; H, 8·83; N, 4·94%.

This salt was dissolved in minimum amount of water, acidified with HCl and continuously extracted with ether. Evaporation of ethereal extract afforded 1·33 g (52%) of the acid $(+)$-V, m.p. 80°. After recrystallization from light petroleum–benzene, m.p. 89–90°, $[\alpha]_D^{24}$ $+3\cdot8°$ (c, 1 in water). Lit.[32]: m.p. 89–90°, $[\alpha]_D^{20}$ $+2\cdot2°$ (in EtOH);[4] m.p. 89°, $[\alpha]_D^{25}$ $+2\cdot9°$ (c, 2·5 in H$_2$O).

$(-)$-*Trachelanthic acid* $[(-)$-V$]$. In an analogous manner, 0·55 g (50%) of acid $(-)$-V was obtained from 2·2 g of $(\pm)$-acid and 1·77 g $(+)$-α-phenylethylamine; m.p. 90–91°, $[\alpha]_D^{22}$ $-2\cdot4°$ (c, 1 in H$_2$O). Lit.[4]: m.p. 89°, $[\alpha]_D^{25}$ $-3\cdot4°$ (c, 2·5 in H$_2$O).

$$\begin{array}{c} \text{CH}_3(\text{CH}_2)_4\text{-CH-COOH} \\ | \\ \text{NH}_2 \end{array}$$

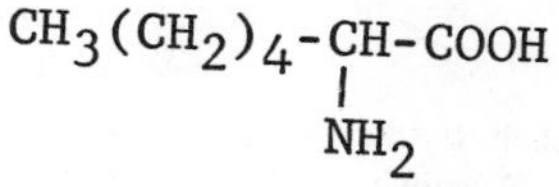

Ref. 340

Dédoublement de l'α-aminoheptanoïque. On chauffe 3 h au bain d'huile 45 g de $(\pm)$-α-amino-heptanoïque avec 65 g d'acide formique *puriss.* 100% et refroidit à 0°. On filtre les cristaux, les dissout dans l'acétate de méthyle et filtre l'acide aminé non transformé. On ajoute de l'éther de pétrole (Eb. 40–60°) à la solution d'acétate de méthyle dès que le dérivé formylé cristallise. F. 90–100°. L'acide aminé récupéré est traité à nouveau avec de l'acide formique comme précédemment. Rendement: 25 g de dérivé formylé; ramollissement 110°; F. 116–117°.

$C_8H_{15}O_3N$ Calc. N 8,09%   Tr. N 8,19%

On ajoute une solution de 60 g de brucine dans 150 ml d'éthanol absolu à 25 g du dérivé formylé dissous dans 400 ml d'éthanol absolu. Après plusieurs jours à 0°, on filtre les cristaux de sel de brucine, les lave bien à l'éthanol froid et les recristallise dans l'éthanol. Rendement: 37,5 g; ramollissement 125°, F. 133–135°; $[\alpha]_D^{20} = -12,8°$ ($c = 1,250$; méthanol; $l = 1$). Ces valeurs ne subirent aucune modification, même après plusieurs recristallisations.

Un excès d'ammoniaque concentrée est ajouté au sel de brucine dissous dans 350 ml d'eau refroidis à la glace. Après 1 h à 0°, on filtre le précipité de brucine. Le filtrat est rendu tout juste acide au congo par HCl 6N, puis laissé 20 h au repos à 0°. Il se forme des aiguilles blanches, brillantes du dérivé formylé. Rendement: 8,84 g; F. 108–110°; $[\alpha]_D^{18} = -14,86°$ ($c = 1,380$; méthanol; $l = 1$).

On hydrolyse le dérivé formylé en le chauffant 1 h 30 au bain-marie, en suspension dans 200 ml d'une solution aqueuse de HCl 10%. L'acide aminé se sépare après refroidissement et neutralisation par NH_4OH. Rendement: 5,60 g; $[\alpha]_D^{18} = -24{,}45°$ (c = 1,325; HCl 20%; l = 1).

$C_7H_{15}O_2N$ Calc. N 9,65% Tr. N 9,74%

Le sel de brucine du dérivé formylé dextrogyre est plus soluble. C'est un sirop qui contient 3,83 g d'acide aminé; $[\alpha]_D^{18} = +20{,}12°$ (c = 1,640; HCl 20%; l = 1).

$C_7H_{15}NO_2$ Ref. 340a

Darstellung der l-Aminoheptansäure aus razemischer α-Aminoheptansäure durch lebende Hefe.

Eine Lösung von 400 g Rohrzucker in 3 l gewöhnlichem Wasser wird zum Sieden erhitzt und in diese 10 g fein gepulverte Aminoheptansäure in kleinen Portionen eingetragen. Nach einigem Kochen löst sich die Säure in der Zuckerlösung und bleibt auch in Lösung beim Erkalten der Flüssigkeit. Hierauf werden 200 g frische Hefe mit einem Teil der so bereiteten Flüssigkeit in einer Reibschale angerührt, in eine 10 l Flasche gespült, dazu die Hauptlösung getan und die Flasche mit einem Ventil abgeschlossen. Nach einigen Minuten setzt eine kräftige Gärung ein. Diese ist nach 5 Tagen beendet. Während dieser Zeit wird die Flasche öfters umgeschüttelt. Sobald die Fehlingsche Reaktion negativ ist, wird die Flüssigkeit mit Kieselgur gut durchgeschüttelt, auf eine Nutsche gebracht und abgenutscht. Man erhält sofort ein klares Filtrat, nur ist darauf zu achten, daß die Nutsche bis zuletzt mit Flüssigkeit gefüllt bleibt. Der Rückstand wird mit Wasser gut gewaschen. Die vereinigten Filtrate und Waschwasser werden im Vakuum bis zur beginnenden Kristallisation eingedampft. Dann wird die Flüssigkeit in eine Schale übergeführt und auf dem Wasserbad bis zum Entstehen eines dicken Kristallbreies eingedampft. Nach einigem Stehen werden die Kristalle auf eine Nutsche gebracht, abgesaugt, mit eiskaltem Wasser gewaschen und auf einem Tonteller gut abgepreßt. Das so erhaltene Produkt wird in Salzsäure gelöst, mit wenig Tierkohle aufgekocht, filtriert und die Tierkohle mit heißem Wasser gut ausgewaschen. Das Filtrat wird im Vakuum bis zur Trockne eingeengt, der Rückstand mit Wasser aufgenommen und abermals abgedampft, um die Salzsäure möglichst zu entfernen. Nun löst man den Rückstand in wenig Wasser, filtriert die etwas trübe Lösung und prüft diese auf ihr Drehungsvermögen. Sie dreht nach links. Die freie l-Aminoheptansäure kann leicht aus dem salzsauren Salz gewonnen werden, wenn man in seine Lösung einen langsamen Strom von Ammoniak bis zur schwach alkalischen Reaktion einleitet. Die Aminoheptansäure fällt sofort in schneeweißen Nadeln aus. Nach zweistündigem Stehen im Eisschrank werden diese abgesaugt, mit wenig eiskaltem Wasser gewaschen und im Vakuumexsikkator über Schwefelsäure getrocknet. Ausbeute 4 g.

0,2000 g Subst.: 0,0190 g N.
Ber.: 9,63 Proz. N,
Gef.: 9,50 „ „

Die l-Aminoheptansäure ist in Wasser etwas leichter löslich als die razemische Form, in Chloroform wieder schwerer löslich als dl-Aminoheptansäure. Sehr leicht löst sie sich in Xylol und Benzol.

Im Kapillarrohr erhitzt, ist bei 240° eine leichte Braunfärbung zu sehen, bei 274° zersetzt sich die Substanz.

0,8072 g Aminoheptansäure in 10 ccm 20 proz. Salzsäure gelöst, drehen im 1 mm-Rohr 0,34° nach links.

$$\alpha_D^{20°} = -4{,}21°.$$

$H_2N(CH_2)_4CH(CH_3)COOH$

$C_7H_{15}NO_2$ Ref. 341

(+)-6 Phthalimido-2-methylhexanoic Acid (VI)

Racemic 6-phthalimido-2-methylhexanoic acid (IV), 200 g. (0.73 mole), was dissolved in 6 liters of boiling ethyl acetate, and to this solution was added 236 g. (0.73 mole) of quinine. After the dissolving was complete, the solution was cooled and allowed to stand in a refrigerator for 2 days. The crystalline precipitate weighed 384 g. (88%); m.p., 141 to 145°. Twenty-eight recrystallizations from ethyl acetate gave 65.1 g. (33.8%) of the resolved diastereoisomer (V); m.p., 152 to 153°.

The quinine salt (V), 65.1 g. (0.108 mole), was dissolved in 870 ml. of boiling benzene and shaken with 775 ml. of 12% aqueous hydrochloric acid in a separatory funnel. The aqueous layer was separated and twice extracted with 150 ml. portions of benzene. The combined benzene extracts were evaporated to dryness, and the residue was dissolved in 210 ml. of hot benzene and filtered. Addition of a large excess of hexane (b.p., 66 to 69°) to the filtered solution yielded a white, flocculant precipitate, (+)-6-pthalimido-2-methylhexanoic acid. The yield after filtration was 28.96 g. (96.5%); m.p., 83 to 84°; $[\alpha]_D^{25}$ = 9.77° (c 4.50, chloroform) [51%; m.p., 89 to 90°; $[\alpha]_D^{25}$ = +5.1° (c 3.21, benzene)].[2]

The infrared spectrum of the resolved product (VI) is approximately equivalent to that of the racemic 6-phthalimido-2-methylhexanoic acid (2).

$(CH_3)_2CH-CH(CH_3)-CH(NH_2)-COOH$

$C_7H_{15}NO_2$ Ref. 342

***N*-Benzyloxycarbonyl-*erythro*-β-methyl-L-leucine·(−)-ephedrine (XVIL) and *N*-Benzyloxycarbonyl-*erythro*-β-methyl-D-leucine·(+)-ephedrine (XVID).** To a solution of 5.58 g (0.02 mol) of XV in 6.0 m*l* of ether, there was added a solution of 1.91 g (0.011 mol) of (−)-ephedrine hemihydrate in 4.0 m*l* of ether. The reaction mixture was then allowed to stand overnight in a refrigerator to form a gelatinous material. The product was collected by filtration after it had been crushed to pieces with a glass rod and a small amount of ether had been added. A somewhat crude (−)-salt (1.6 g) (mp 120—124°C) was thus obtained. The ethereal mother liquor was washed wih 2N hydrochloric acid and then with water, and dried over anhydrous sodium sulfate. The ethereal solution was evaporated to dryness under reduced pressure; then the residue thus obtained was redissolved in 6.0 m*l* of ether. To the solution, there was added a solution of 1.91 g of (+)-ephedrine hemihydrate in 4.0 m*l* of ether. Then (+)-salt (1.4 g) (mp 124—125°C) was obtained by a way similar to that described above. The same treatment was possible again reciprocally with (−)- and (+)-ephedrine hemihydrate. Optically-active (−)-salt consisting of (−)-ephedrine was combined and recrystallized from ethyl acetate to give 2.0 g (45%) of a pure material with a melting point of 124—125°C; $[\alpha]_D^{25} = -17.6°$ (c 0.74 in ethyl alcohol). Found: C, 67.50; H, 8.09; N, 6.28%. Calcd for $C_{25}H_{36}O_5N_2$: C, 67.54; H, 8.16; N, 6.30%.

In the same way, optically-pure (+)-salt with a melting point of 124—125°C was obtained in a 49.5% (2.2 g) yield; $[\alpha]_D^{25}=+17.7°$ (*c* 0.95 in ethyl alcohol). Found: C, 67.60; H, 8.14; N, 6.32%. The mixed-melting-point of minus and plus salt was depressed to 101—104°C.

***erythro*-β-Methyl-L-leucine (XIIL) and the D-Isomer (XIID).** To a suspension of 1.9 g (0.0034 mol) of (−)-salt in 10 m*l* of ethyl acetate, there were added 3.4 m*l* of 2N hydrochloric acid; then the mixture was shaken in a separatory funnel. The organic layer was treated as usual to give 1.3 g of an oily residue. The oil was dissolved in 3.4 g of 30% hydrogen bromide in glacial acetic acid, after which the mixture was allowed to stand for 1 hr at room temperature and then for 24 more hour in a refrigerator after the addition of 100 m*l* of ether. The deposited crystals were collected by filtration on a glass-fritted funnel and washed with ether. Recrystallization from ethyl alcohol and ether afforded a pure material with a melting point of 195—196°C in a 69% (0.66 g) yield. Similarly, from 2.18 g of (+)-salt D-isomer hydrobromide was obtained in a 73% (0.8 g) yield; mp 203—205°C. A solution of 0.6g of hydrobromide of the L-isomer in ethyl alcohol was neutralized with triethylamine. The crystals thus obtained were collected and recrystallized from water and dioxane to give 0.28 g (73% from hydrobromide) of the amino acid XVIIL; $[\alpha]_D^{24}=+38.9°$ (*c* 0.485, in 5N hydrochloric acid), $[\alpha]_D^{25}=+29.7°$ (*c* 0.37, in water). Found: C, 57.92; H, 10.38; N, 9.70%. From 0.65 g of the hydrobromide of the D-isomer, free amino acid XVIID was obtained in a 67% (0.3 g) yield; $[\alpha]_D^{25}=-39.4°$ (*c* 0.38, in 5N hydrochloric acid). Found: C, 57.82; H, 10.34; N, 9.62%.

***erythro*-N,β-Dimethyl-L-leucine (XVIIIL) and the D-Isomer (XVIIID).** The methylation of the amino groups of XVIIL and XVIID was accomplished, step by step, by *N*-benzylation, *N*-methylation, and subsequent catalytic hydrogenolysis, in a way similar to that used for the corresponding racemic *threo*-isomer. *N*-Benzyl-L-isomer: mp 218—219°C, 74% yield. Found: C, 71.62; H, 8.95; N, 5.82%. *N*-Benzyl-D-isomer: mp 218—219°C, 43% yield. Found: C, 71.70; H, 8.93; N, 6.03%. The mixed-melting-point of the two isomers was depressed to 208—210°C. *N*-Benzyl-*N*-methyl-L-isomer: mp 102—105°C, 68% yield. Found: C, 72.41; H, 9.08; N, 5.60%. *N*-Benzyl-*N*-methyl-D-isomer: mp 102—105°C, 74% yield. Found: C, 72.20; H, 9.40; N, 5.55%. The mixed-melting-point of these two isomers was depressed to 76—80°C. *erythro*-*N*,β-Dimethyl-L-leucine: mp>270°C, 51% yield; $[\alpha]_D^{25}=+38.0°$ (*c* 0.50, in 5N hydrochloric acid). Found: C, 60.52; H, 10.48; N, 8.91%. *erythro*-*N*,β-Dimethyl-D-leucine: mp >270°C, 52% yield; $[\alpha]_D^{25}=-38.3°$ (*c* 0.60, in 5N hydrochloric acid). Found: C, 60.18; H, 10.51; N, 8.76%.

$C_7H_{15}NO_2$ Ref. 342a

Resolution of the Amino Acid (Ia, Ib, IIa, IIb). IIa·HCl was acetylated with 2 N sodium hydroxide and acetic anhydride by the ordinary method. Fifty-six grams of the acetyl derivative of IIa were dissolved in 1.5 *l* of water, and the pH was adjusted to 6.8 with 2 N sodium hydroxide. To this solution, 3 g of prozyme acylase were then added, after which the mixture was allowed to stand for 50 hr at 42°C. After 50 hr, the pH of the solution was again adjusted to 6.8 with 2 N NaOH, and the hydrolysis was continued. Then, the solution was evaporated to 100 m*l* *in vacuo*; the crystals of the L-amino acid were subsequently collected by filtration and washed with a small amount of water. The crude L-amino acid was purified by recrystallization from the water by treatment with charcoal. Yield, 12 g (55.3%). $[\alpha]_D^{13}+40.6$ (*c* 1, 6N HCl).

The mother liquor was acidified with 6 N hydrochloric acid. The *N*-acetyl-D-amino acid thus precipitated was collected and washed with dilute hydrochloric acid and then water. Yield, 24 g (85.7%). $[\alpha]_D^{25}$ −2.3 (*c* 8.3 EtOH).

One portion of the *N*-acetyl-D-amino acid was refluxed with 10% hydrochloric acid; the solution was then neutralized with triethylamine to afford, quantitatively, D-amino acid. $[\alpha]_D^{25}$ −39.9 (*c* 1.2, 6 N HCl).

```
              NH2
              |
(CH3)2CH-CH-CH-COOH
         |
         CH3
```

$C_7H_{15}NO_2$ Ref. 343

***N*-Benzyloxycarbonyl-*threo*-*N*,β-dimethyl-L-leucine·(−)-ephedrine Salt (XIIL) and *N*-Benzyloxycarbonyl-*threo*-*N*,β-dimethyl-D-leucine (XIIID).** XI (8.3 g, 0.028 mol) and (−)-ephedrine hemihydrate (2.7 g, 0.0155 mol) were dissolved in 28 m*l* of warming ethyl acetate. The mixture was then allowed to stand for 24 hr at room temperature. The deposited crystals were filtered and washed with ethyl acetate to give 6.1 g (94%) of XIIL; mp 157°C. Recrystallization from an ethyl alcohol-ethyl acetate mixture (1:1) afforded 5.7 g of the pure material with a melting point of 160—161°C, $[\alpha]_D^{24}=-75.0°$ (*c* 0.40, in ethyl alcohol). Found: C, 68.26; H, 8.24; N, 6.27%. Calcd for $C_{26}H_{38}O_5N_2$: C, 68.09; H, 8.35; N, 6.11%.

The mother liquor obtained after the removal of XIIL was washed twice with a small amount of 2N hydrochloric acid and water. The organic layer was dried over anhydrous sodium sulfate. The filtrate was concentrated to dryness to give a crystalline residue in a 94% (3.9 g) yield; mp 90—91°C. Recrystallization from benzene-petroleum ether afforded 3.4 g of almost optically-pure XIIID; mp 94—95°C, $[\alpha]_D^{24}=+72°$ (*c* 1.00, in ethyl alcohol).

***N*-Benzyloxycarbonyl-*threo*-*N*,β-dimethyl-D-leucine·(+)-ephedrine Salt (XIID).** To a solution of 0.70 g (0.0024 mol) of XIIID ($[\alpha]_D^{24}=+72°$) in 1.5 m*l* of ethyl acetate, there was added a solution of 0.42 g (0.0024 mol) of (+)-ephedrine hemihydrate in 1.0 m*l* of ethyl acetate. The mixture was set aside for 24 hr at room temperature, and then the crystals deposited were filtered, washed with ethyl acetate, and dried to afford XIID (mp 155—157°C) in an 89% (0.975 g) yield. Recrystallization from an ethyl alcohol-ethyl acetate mixture (1:1) gave crystals with a melting point of 159—160°C and $[\alpha]_D^{25}=+73.5°$ (*c* 0.30, in ethyl alcohol). Found: C, 67.97; H, 8.31; N, 6.16%.

***N*-Benzyloxycarbonyl-*threo*-*N*,β-dimethyl-L-leucine (XIIIL).** To a suspension of 4.8 g (0.0105 mol) of the XIIL salt in 15 m*l* of ethyl acetate, there were added 7.8 m*l* (0.0156 mol) of 2N hydrochloric acid (7.8 m*l*, 0.0156 mol), and the mixture was shaken effectively in a separatory funnel. After washing further with water, the ethyl acetate layer was dried over anhydrous sodium sulfate. The ethyl acetate was evaporated to dryness to afford 3 g (98%) of crystals; mp 94—96°C. Recrystallization from benzene-petroleum ether gave 2.8 g of XIIIL with a melting point of 98—99°C, $[\alpha]_D^{26}=-75.9°$ (*c* 1.10, in ethyl alcohol). Found: C, 65.25; H, 7.99; N, 4.89%. Calcd for $C_{16}H_{23}O_4N$: C, 65.51; H, 7.90; N, 4.78%.

***N*-Benzyloxycarbonyl-*threo*-*N*,β-dimethyl-D-leucine (XIIID).** Following the method used for the preparation of XIIIL, from 3.2 g (0.007 mol) of XIID, 1.9 g (93%) of XIIID with a melting point of 98—99°C was obtained. $[\alpha]_D^{25}=+75.3°$ (*c* 0.7, in ethyl alcohol). Found: C, 65.72; H, 8.04; N, 4.79%.

threo-N,β-**Dimethyl-L-leucine** (**XIV**L). A) To 1.0 g (0.0034 mol) of XIIIL, 3.5 g of 30% hydrogen bromide in glacial acetic acid was added. The reaction mixture was then set aside in a tight-stoppered vessel for 1 hr at room temperature. There after 30 ml of ether were added to complete the precipitation of the product, which was collected on a glass-fritted funnel and washed with ether. The yield of the hydrobromide was 0.75 g (92%), and its melting point was 209—210°C. Free amino acid XIVL was obtained by treatment with triethylamine in a solution of 4 ml of ethyl alcohol, followed by recrystallization from a water-ethyl alcohol mixture (1:1). The yield was 0.43 g (86% from hydrobromide), and it had a melting point of over 270°C; $[\alpha]_D^{25}=+38.3°$ (c 0.90, in 5N hydrochloric acid), $[\alpha]_D^{23}=+31.5°$ (c 0.50, in water). Found: C, 60.39; H, 10.57; N, 8.93%. Calcd for $C_8H_{17}O_2N$: C, 60.34; H, 10.74; N, 8.80%.

B) A solution of 1.0 g (0.0034 mol) of XIIIL in 10 ml of glacial acetic acid was hydrogenated using 200 mg of palladium black as a catalyst under mechanical shaking for 5 hr. After the catalyst had then been removed by filtration, the filtrate was evaporated to dryness to give a crystalline residue and dried in a vacuum desiccator over concentrated sulfuric acid and solid sodium hydroxide. The residue was recrystallized from water and dioxane to give 0.47 g (87%) of XIVL with a melting point of over 270°C; $[\alpha]_D^{25}=+37.3°$ (c 0.70, in 5N hydrochloric acid), $[\alpha]_D^{23}=+29.1°$ (c 0.90, in water).

threo-N,β-**Dimethyl-D-leucine** (**XIV**D). A way similar to the method A) used in the preparation of XIVL was used. From the reaction of 1.0 g (0.0034 mol) of XIIID and 3.5 g of 30% hydrogen bromide in glacial acetic acid, followed by neutralization with amine, XIVD with a melting point of over 270°C was obtained in almost the same yield. $[\alpha]_D^{25}=-39.3°$ (c 0.60, in 5N hydrochloric acid), $[\alpha]_D^{25}=-31.8°$ (c 0.40, in water). Found: C, 60.56; H, 10.63; N, 9.01%.

N-Benzyloxycarbonyl-*erythro*-β-methyl-DL-leucine (**XV**). By the usual method of Schotten-Baumann, a pure product with a melting point of 62—64°C was obtained from VII in an 84% yield. The recrystallization was performed from benzene-petroleum ether. Found: C, 64.37; H, 7.42; N, 5.18%. Calcd for $C_{15}H_{21}O_4N$: C, 64.49; H, 7.58; N, 5.01%.

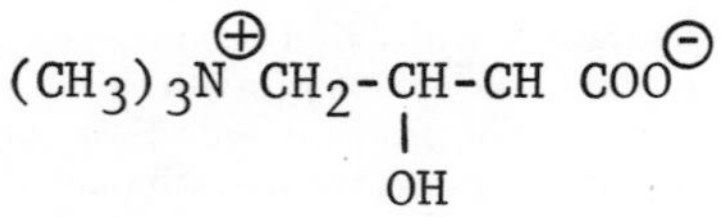

$$(CH_3)_3\overset{\oplus}{N}-CH_2-CH-CH-COO^{\ominus}$$
$$\phantom{(CH_3)_3\overset{\oplus}{N}-CH_2-CH-}|$$
$$\phantom{(CH_3)_3\overset{\oplus}{N}-CH_2-CH}OH$$

Ref. 344

1 Mol (=197,7 g) DL-Carnitin-hydrochlorid wird als 10proz., wäßrige Lösung über 600 ml eines schwach basischen Anionenaustauschers geschickt. Das Eluat wird im Vak. zur Trockne gebracht. Die so gewonnenen 161,2 g (= 1 Mol) DL-Carnitin werden mit 200,3 g (=1 Mol) D(+)-Camphersäure in 500 ml Methanol heiß gelöst. Die sich abkühlende Lösung wird mit D(×)-Carnitin-D(×)-camphoratkristallen angeimpft. Bei 20°C werden die abgeschiedenen Kristalle abfiltriert, mit wenig Äthanol nachgewaschen und aus 400 ml Methanol umkristallisiert. Erhalten werden 108 g D(+)-Carnitin-D(+)-camphorat (=60% d. Th.). Die Mutterlaugen werden vereinigt, zur Trockne eingeengt und mit Wasser und Anionenaustauscher in 140 g D(+)-Camphersäure und 113 g DL-Carnitin (80,5 g L(−)-Carnitin und 32,5 g D(+)-Carnitin enthaltend) wie unten beschrieben zerlegt.

Die so erhaltenen 113 g DL-Carnitin werden in 350 ml Methanol zusammen mit 140 g L(−)-Camphersäure heiß gelöst. Die sich abkühlende Lösung wird mit L(−)-Carnitin-L(−)-camphoratkristallen angeimpft. Bei 20°C werden die abgeschiedenen Kristalle abfiltriert und aus 210 ml Methanol umkristallisiert. Erhalten werden 154 g L(−)-Carnitin-L(−)-camphorat (=85% d. Th.). Aus den Mutterlaugen können durch fraktionierte Kristallisation noch 22 g L(−)-Carnitin-L(−)-camphorat (=12% d. Th.) und 36 g D(+)-Carnitin-L(−)-camphorat (=20% d. Th.) erhalten werden. (Wird das in den Mutterlaugen verbliebene DL-Carnitin-L(−)-camphorat in L(−)-Camphersäure und DL-Carnitin zerlegt, und dieses bei weiteren Racematspaltungen eingesetzt, ist die quantitative Auftrennung des eingesetzten DL-Carnitins in die optisch aktiven Carnitine besonders rationell.)

Ausbeute:

108 g D(+)-Carnitin-D(+)-camphorat (=0,3 Mol)

176 g L(−)-Carnitin-L(−)-camphorat (=0,49 Mol)

36 g D(+)-Carnitin-L(−)-camphorat (=0,1 Mol)

Diese Salze werden in die optisch aktiven Carnitine und die Camphersäuren zerlegt. Es werden 79 g L(−)-Carnitin (=98% d. Th.) und 64,3 g D(+)-Carnitin (=80% d. Th.) erhalten.

Die zurückgewonnenen Camphersäuren sind rein und können erneut zur Racematspaltung eingesetzt werden. Die in diesem Beispiel verwendete L(−)-Camphersäure kann durch Dibenzoyl-D(−)-weinsäure oder L(+)-Weinsäure ersetzt werden.

Zerlegung der Carnitinsalze

1. Durch Wasser: 0,1 Mol (=36,1 g) Carnitincamphorat wird in 100 ml heißem Wasser gelöst, dann auf 4°C abgekühlt, abfiltriert und mit wenig Wasser nachgewaschen. Erhalten: 18 g Camphersäure (=90% d. Th.). Aus Filtrat und Waschwasser wird die restliche Camphersäure (∼2 g) durch Anionenaustauscher entfernt. Das neutrale Eluat wird eingeengt. Es werden 16,1 g (= 0,1 Mol) reines Carnitin erhalten. Vom Anionenaustauscher wird die Camphersäure mit NaOH zurückgewonnen.

2. Durch Ätherextraktion: 0,1 Mol (=36,1 g) Carnitincamphorat wird in 50 ml Wasser bzw. 0,1 Mol (=51,9 g) Carnitindibenzoyltartrat in 100 ml Wasser suspendiert und die Suspension 16 bis 24 h im Extraktor mit Äther extrahiert. In der eingeengten ätherischen Phase werden 19 g Camphersäure bzw. 33 g Dibenzoylweinsäure rein erhalten. Aus der wäßrigen Phase werden die restlichen Säurebestandteile durch Anionenaustauscher entfernt. Die neutralen Eluate werden im Vak. zur Trockne eingeengt. Es werden je 16,1 g (=0,1 Mol) reines Carnitin erhalten.

3. Durch Salzsäure: 0,1 Mol (=36,1 g) Carnitincamphorat wird mit 0,1 Mol HCl in 30 ml Wasser bzw. 0,1 Mol (=51,9 g) Carnitindibenzoyltartrat mit 0,1 Mol HCl in 100 ml Wasser suspendiert. Die ausgefallenen Carbonsäuren werden verrieben, abfiltriert und mit wenig Wasser nachgewaschen. Filtrat und Waschwasser werden zusammen im Vak. eingeengt und der Rückstand aus Alkoholen, vorteilhaft aus 2-Propanol, umkristallisiert. Es werden 90—98% der eingesetzten Carnitinsalze als Carnitin-hydrochlorid erhalten. Das restliche Carnitin-hydrochlorid sowie die restlichen Carbonsäuren können aus der alkoholischen Mutterlauge ebenfalls gewonnen werden.

4. Durch Anionenaustauscher: 0,1 Mol eines Carnitinsalzes der optisch aktiven Säuren wird in wäßriger Lösung bzw. als Suspension mit 0,12 bzw. 0,25 Äquiv.

der OH-Form eines schwach basischen Anionenaustauschers (Wofatit AK 40, Wofatit AD 41 oder Merck II) ~5 min geschüttelt. Der Anionenaustauscher wird abfiltriert und mit Wasser nachgewaschen. Aus dem Filtrat werden die restlichen Säurebestandteile über eine mit 20–30 m*l* Wofatit SBK (OH-Form) gefüllte Anionenaustauschersäule entfernt. Das neutrale Eluat wird im Vak. zur Trockene eingeengt. Als Rückstand verbleiben 16,1 g (=0,1 Mol) reines Carnitin.

Eigenschaften der Salze der optisch aktiven Carnitine mit einigen optisch aktiven Säuren

Zur Schmelzpunktbestimmung legten wir die Carnitinsalze 20°C unterhalb des erwarteten Schmelzpunktes in die Boethius-Apparatur ein und erwärm-

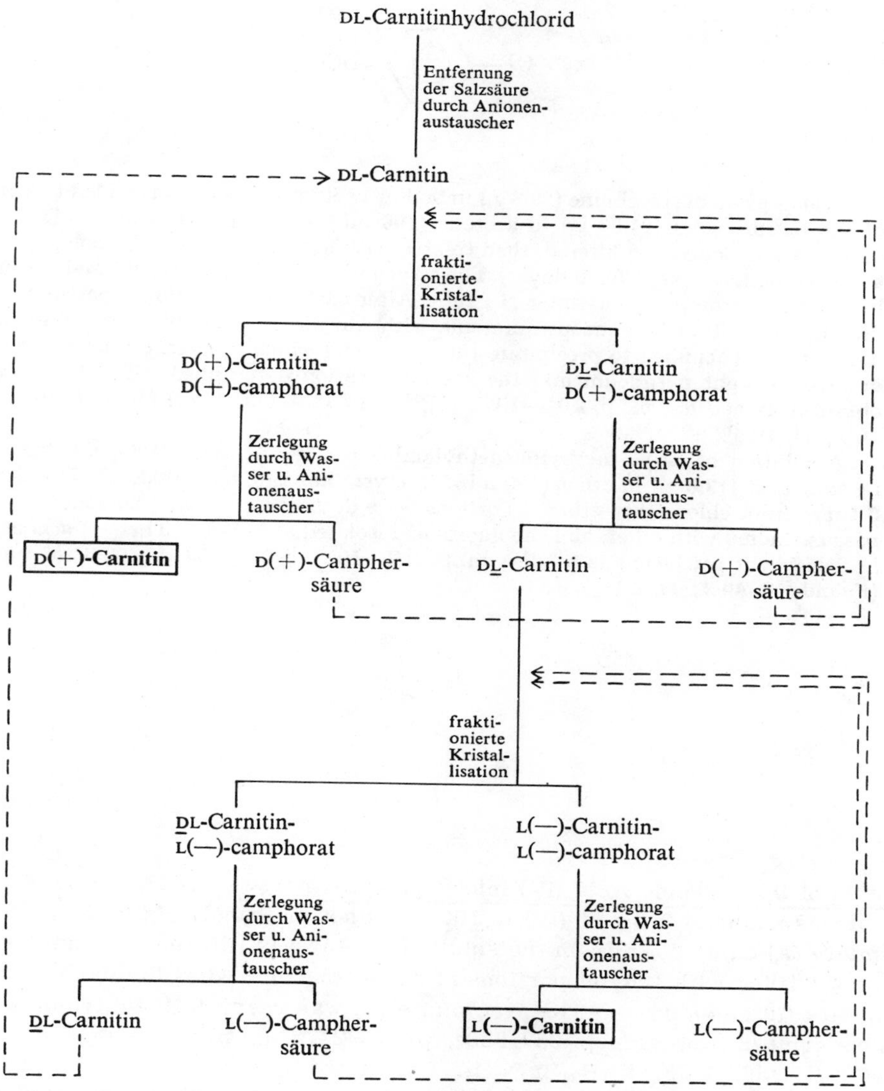

Schema. Verfahrensprinzip der zweistufigen Racematspaltung von DL-Carnitin in die optisch aktiven Carnitien (bei optisch aktiven Mischfraktionen ist die angereicherte Komponente unterstrichen).

	Schmp. [°C]	$[\alpha]_D^{22}$
D(+)-Carnitin-D(+)-camphorat[a]	183	+37,3
L(−)-Carnitin-L(−)-camphorat[a]	183	−37,3
D(+)-Carnitin-L(−)-camphorat[b]	168	−10,4
L(−)-Carnitin-D(+)-camphorat[b]	168	+10,4
L(−)-Carnitin(dibenzoyl-D(−)-tartrat)[a]	154	−95,2
D(+)-Carnitin(dibenzoyl-D(−)-tartrat)[b]	138	−73,3
Di-L(−)-carnitin-L(+)-tartrat[a,c]	176/8	−10,0
Di-D(+)-carnitin-L(+)-tartrat[b,c]	157/8	+29,9

a) In Methanol löslich, in Äthanol und 2-Propanol wenig löslich, in Äther nicht löslich.
b) In Methanol leicht löslich, in Äthanol löslich, in 2-Propanol wenig löslich, in Äther nicht löslich.
c) In Wasser s. leicht löslich.

ten um 4°C/min. Der Säuregehalt der Salze und damit die Molekulargewichte wurden durch Titration mit 0,1N NaOH gegen Phenolphthalein bestimmt. Der spezifische Drehwert der optisch aktiven Carnitinsalze wurde in 5 bzw. 10proz. Lösung in Methanol, bei den L(+)-Tartraten in Wasser, bestimmt. Die Genauigkeit der angegebenen Drehungswerte beträgt ±0,3°.
Die Carnitincamphorate hydrolysieren in Wasser vollständig, die Carnitindibenzoyltartrate teilweise. Die Camphorate und Tartrate kristallisieren aus Äthanol in gedrungenen Prismen, die Dibenzoyltartrate in nadelförmigen Kristallen.

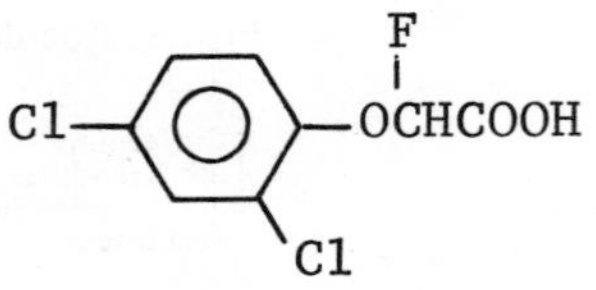

$C_8H_5Cl_2FO_3$ Ref. 345

A suspension of strychnine (29·8 g.) in boiling acetone (250 ml.) was added to a warm solution of the racemic acid (21·35 g.) in acetone (100 ml.) and water (150 ml.). The clear solution obtained was boiled and filtered, then treated with hot water (400 ml.), cooled to room temperature, seeded, and kept for 2 days. The strychnine salt (58 g.) was filtered off and recrystallised 8 times from acetone–water (1 : 10). After each crystallisation a portion of the salt was decomposed with 7·5N-aqueous ammonia with warming. Strychnine was filtered off and the cooled filtrate acidified to precipitate the fluoro-acid which was recrystallised from *cyclo*hexane or benzene–light petroleum and the specific rotation determined. The (+)-*acid* was finally obtained as needles, m. p. 104—105°, $[\alpha]_D^{22.5}$ +57·1° ± 0·3° * (*c* 1·505 in $CHCl_3$) (Found: C, 40·1; H, 1·95%).

A solution of (+)-α-methylphenethylamine (7·6 g.) in ether (100 ml.) was added to the racemic acid (13·8 g.) in ether (150 ml.); a crystalline salt separated. This was recrystallised 6 times from chloroform–ether. Portions were decomposed with 1·5N-sodium hydroxide, the base extracted with ether, and the fluoro-acid isolated as before. The (−)-*acid* separated from benzene–light petroleum as needles, m. p. 103—105°, $[\alpha]_D^{21.5}$ −57·1° ± 0·3° (*c* 1·504 in $CHCl_3$) (Found: C, 40·2; H, 2·1%).

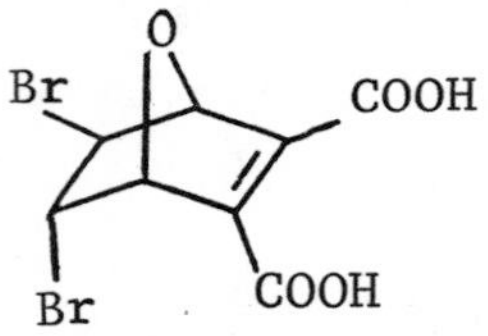

$C_8H_6Br_2O_5$ Ref. 346

Resolution of the Racemic Acid (IV) into Its Enantiomers. A hot solution of 11.1 g of brucine in 100 ml of methanol was added to 9.6 g of (IV) in 100 ml of hot methanol. After 5 min at the boil a white crystalline precipitate (A) came down from the solution, and it was filtered off from the hot solution and washed three times on the filter with 100-ml portions of hot methanol. When the mother solution was cooled a second precipitate (B) came down. The precipitate (A) was recrystallized from water until the specific rotation was constant (6 times); m.p. 257-258°, $[\alpha]_D$ + 24.3° (in nitromethane, c 0.99, *l* 1 dm). Found %: C 50.59, 50.70; H 4.56, 4.58. $C_{31}H_{32}Br_2N_2O_5$. Calculated %: C 50.56; H 4.38. The precipitate (B) was recrystallized from methanol until the specific rotation was constant (6 times); m.p. 252-253°, $[\alpha]_D$ −11.4° (in nitromethane, c 1, *l* 1 dm). Found %: C 50.84, 51.00; H 4.68, 4.48. $C_{31}H_{32}Br_2N_2O_5$. Calculated %: C 50.56; H 4.38.

The brucine salt was suspended in ether, and concentrated hydrochloric acid was added in small portions until the salt had dissolved completely. The ether layer was separated, and the aqueous layer was extracted with ether; the combined ether extracts were dried with calcium chloride and after the removal of ether we isolated the optically active acid (IV).

From the salt (A) we obtained the acid (IV) of m.p. 203-204°; $[\alpha]_{578}$ −32.5°, $[\alpha]_{546}$ −30.8°, $[\alpha]_{436}$ −2.9°, $[\alpha]_{405}$ +28.8°, $[\alpha]_{365}$ +121.4° (in methanol, c 1, *l* 1 dm). Found %: C 27.65, 27.78; H 2.10, 2.19. $C_8H_6Br_2O_5$. Calculated %: C28.07; H 1.85.

From the salt (B) we obtained the acid (IV) of m.p. 198-199°; $[\alpha]_{589}$ +15.1°, $[\alpha]_{578}$ +14.3°, $[\alpha]_{546}$ +13.4°, $[\alpha]_{436}$ +35.6°, $[\alpha]_{405}$ −43.4°, $[\alpha]_{365}$ −131.7° (in methanol, c 1, *l* 1 dm). Found %: C 28.24, 28.22; H 1.90, 2.11. $C_8H_6Br_2O_5$. Calculated %: C 28.07; H 1.85.

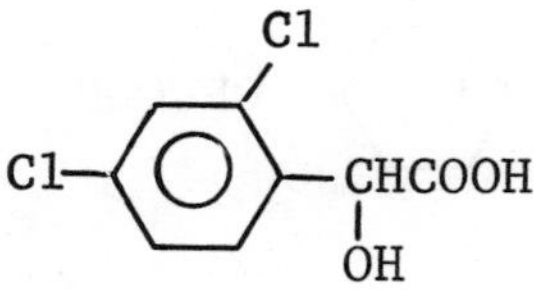

$C_8H_6Cl_2O_3$ Ref. 347

The authors reported that the above compound was custom prepared by Niels Clauson-Kaas Chemical Research Laboratory, Rugmarken 28, Farum, Denmark. The reported m.p. of the resolved acid was 121-123°C. The $[\alpha]_D^{25}$ = -141°C. (c, 1.5, ethanol).

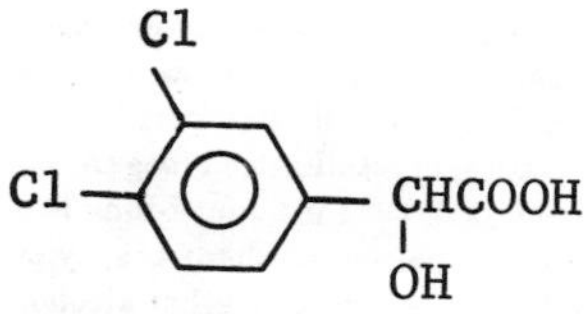

$C_8H_6Cl_2O_3$ Ref. 348

Resolution was accomplished by serial crystallization of the salt of the desired acid with (+)-α-methylbenzylamine. The m.p. of the resolved acid was 117-118°C. The $[\alpha]_D^{25}$ = -113°C (c, 0.3, H_2O).

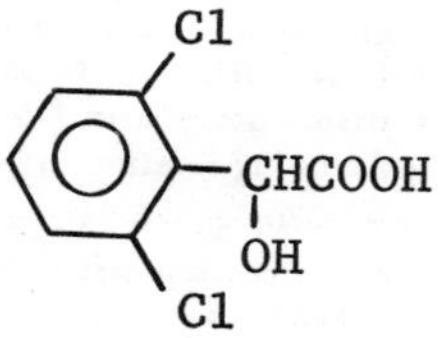

$C_8H_6Cl_2O_3$ Ref. 349

Resolution was accomplished by serial crystallization of the salt of the desired acid with (-)-α-methylbenzylamine. The m.p. of the resolved acid was 145-147°C. The $[\alpha]_D^{25}$ = -90°C. (c, 0.3, ethanol).

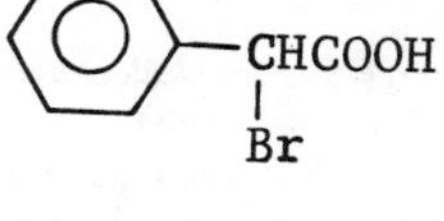

$C_8H_7BrO_2$ Ref. 350

α-Bromophenylacetic Acid (Resolution) ($C_8H_8O_3$)

Twenty grams of α-bromophenylacetic acid and fifteen grams of brucine were added to 100 ml. methanol. The mixture was stirred vigorously to dissolve the solutes. After standing for two hours, needle-like crystals started to separate from the solution. Additional standing resulted in more precipitation. The crystals were collected overnight and acidified with cold, dilute sulfuric acid. The L-isomer was obtained in 21% yield after repeated resolution. M.p. 87-8°C, $(\alpha)_D^{20}$ +149° (c=0.150, water). The D-isomer was obtained from the mother liquor, m.p. 87-9°C, $(\alpha)_D^{20}$ -143° (c=0.131, water).

Reference: P. Walden, Ber. 28, 1295 (1895).

$C_8H_7BrO_2$ Ref. 350a

Resolution of r-Phenylbromoacetic Acid.

Sixty-three grams of morphine (½ mol.) were dissolved by heating with 900 c.c. of methyl alcohol, and, after the solution had been cooled to 20°, 90 grams of *r*-phenylbromoacetic acid (1 mol.) were added. The acid dissolved readily, and glassy prisms soon began to separate. When the crystallisation had proceeded for one and a-half hours, the mixture was stirred, and, four and a-half hours later, the salt was removed and washed with methyl alcohol. The resulting 76 grams were decomposed with cold dilute sulphuric acid, and extracted six times with ether as quickly as possible. The ethereal solution was dried over sodium sulphate, and then allowed to evaporate spontaneously at the ordinary temperature. The acid obtained from the first two extractions with ether had $[\alpha]_D$ −119·8° in benzene solution, and amounted to 26·3 grams; it was somewhat purer than that (2·3 grams) obtained from the last four extractions, which had $[\alpha]_D$ −113°, and presumably contained more mandelic acid than did the former product. After four crystallisations from light petroleum (b. p. 60—70°), the acid was pure. Yield, 15 grams.

When the *r*-acid and morphine were employed in molecular quantities with methyl alcohol as solvent, the acid, obtained from the morphine salt, was less active than when the above conditions were adopted; thus, in one experiment, the acid had $[\alpha]_D$ −69°. In another experiment, where the *r*-acid and morphine were taken in molecular quantities with ethyl alcohol as solvent, a dextrorotatory acid with $[\alpha]_D$ +26° was obtained from the morphine salt.

The morphine salt, which separates in the resolution as described above, is sparingly soluble in the commoner organic solvents, and it is not practical to crystallise it from water, since morphine hydrobromide is formed in considerable quantity when the temperature is raised. This behaviour had already been encountered during the resolution of *iso*dibromosuccinic acid (T., 1912, **101**, 1201). The ease with which bromine is eliminated also renders it advisable not to allow the crystallisation of the morphine salt to be continued longer than about six hours.

l-*Phenylbromoacetic acid* separates from light petroleum in large, glistening leaflets. It melts at 87—88°, whereas the *r*-acid melts at 83—84°:

0·2109 gave 0·3442 CO_2 and 0·0620 H_2O. C=44·5; H=3·3.
$C_8H_7O_2Br$ requires C=44·7; H=3·3 per cent.

The acid is sparingly soluble in cold water, and readily so in ether, acetone, benzene, ethyl alcohol, methyl alcohol, or chloroform.

The following polarimetric values were obtained in benzene solution:

l=2, c=2·0065, $\alpha_D^{13\cdot5}$ −5·90°, $\alpha_{5461}^{13\cdot5}$ −7·15°; whence $[\alpha]_D^{13\cdot5}$ −147·0° and $[\alpha]_{5461}$ −178·2°.

$C_8H_7BrO_2$ Ref. 350b

Resolution of 1-*Menthyl* dl-*Phenylbromoacetate into its Diastereoisomerides.*—The following is a description of a typical preparation.

r-Phenylbromoacetic acid (50 gms.) was esterified by heating at about 100° for 9 hours with *l*-menthol (150 gms.) in presence of hydrogen chloride. The ethereal solution was washed with a dilute solution of sodium hydrogen carbonate, the ether and the excess of menthol were then removed, the latter by steam-distillation.

The residual oil gradually solidified when kept in the ice-chest. After 6 crystallisations from rectified spirit, the crystals showed a constant specific rotatory power. In this resolution, the ester of the *d*-halogen acid is the more sparingly soluble of the diastereoisomerides, as is the case also in the resolution of *l*-menthyl *dl*-phenylchloroacetate (*loc. cit.*). Yield = 16 grams.

l-*Menthyl* d-*phenylbromoacetate*, $C_6H_5\cdot CHBr\cdot CO_2\cdot C_{10}H_{19}$, m. p. 78—79°, fine needles from ethyl alcohol (Found: Br = 22·7. $C_{18}H_{25}O_2Br$ requires Br = 22·6 per cent.), is readily soluble in acetone, chloroform, light petroleum, benzene, toluene, xylene, trichloroethylene, carbon tetrachloride and ethyl acetate, but somewhat sparingly soluble in ethyl alcohol. Its specific rotation was determined in ethyl-alcoholic solution: l = 2, c = 2·7744, $\alpha_D^{15°}$ + 0·52°, $[\alpha]_D^{15°}$ + 9·4°; $\alpha_{5461}^{15°}$ + 0·73°, $[\alpha]_{5461}^{15°}$ + 13·2°.

Shimomura and Cohen (J., 1921, **119**, 1816) combined *dl*-phenylbromoacetyl bromide with *l*-menthol and crystallised the product repeatedly from alcohol. They obtained an ester melting at 84°

and showing $[\alpha]_D^{18°} + 2{\cdot}2°$ ($c = 0{\cdot}68$, $l = 2$) in alcoholic solution, the suggestion being made that this small dextrorotation is probably due to partial resolution of the acid radical. We have never succeeded in obtaining the melting point of the ester so high as 84°. The polarimetric data given by Shimomura and Cohen are not such as to enable one to judge if the resolution had been carried to a completion; the concentration selected by them is small, the observed angle of rotation when calculated from their figures being $\alpha_D + 0{\cdot}03°$.

l-*Menthyl* l-*phenylbromoacetate*, $C_6H_5{\cdot}CHBr{\cdot}CO_2{\cdot}C_{10}H_{19}$, was obtained by prolonged fractional crystallisation of the mother-liquors from the preparation of the *d*-ester. It is readily soluble in ethyl alcohol and crystallises from aqueous ethyl alcohol in fine needles, m. p. 40—41° (Found: C = 61·0; H = 7·3. $C_{18}H_{25}O_2Br$ requires C = 61·2; H = 7·1 per cent.). In ethyl alcohol: $l = 2$, $c = 3{\cdot}7205$, $\alpha_D^{18°} - 10{\cdot}87°$, $[\alpha]_D^{18°} - 146{\cdot}1°$; $\alpha_{5461}^{18°} - 12{\cdot}86°$, $[\alpha]_{5461}^{18°} - 172{\cdot}8°$. In benzene: $l = 2$, $c = 3{\cdot}2835$, $\alpha_D^{17°} - 9{\cdot}18°$, $[\alpha]_D^{17°} - 139{\cdot}8°$; $\alpha_{5461}^{17°} - 10{\cdot}96°$, $[\alpha]_{5461}^{17°} - 166{\cdot}9°$.

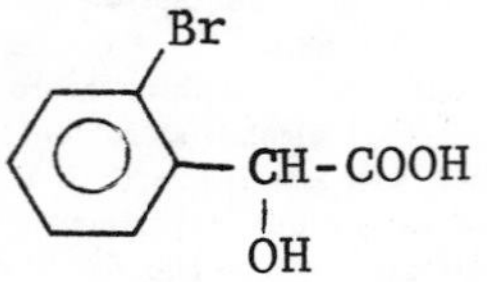

$C_8H_7BrO_3$ **Ref. 351**

Acides (+) *et* (—) *o-bromomandéliques.*

4,6 g d'acide (±) *o*-bromomandélique (F = 89°) et 3,25 g de (—) éphédrine dans 25 cm³ de benzène donnent 2,1 g de cristaux **A**, que l'on décompose directement en 1,2 g d'acide $[\alpha]_{578}^{25} = -127{,}5°$ (acétone, $c = 1$); l'acide lévogyre optiquement pur est obtenu par recristallisation dans 80 cm³ de benzène: 0,55 g, F = 122°, $[\alpha]_{578}^{25} = -143°$ (acétone, $c = 1{,}1$).

Après évaporation à sec, les eaux-mères de **A** sont décomposées en 2,7 g d'acide, $[\alpha]_{578}^{25} = +48°$ (acétone); trois recristallisations dans le benzène (25, 50 et 50 cm³) permettent d'obtenir 0,75 g d'acide dextrogyre, F = 122°, $[\alpha]_{578}^{25} = +141°$ (acétone, $c = 1$). Litt.: F = 121-123° (13).

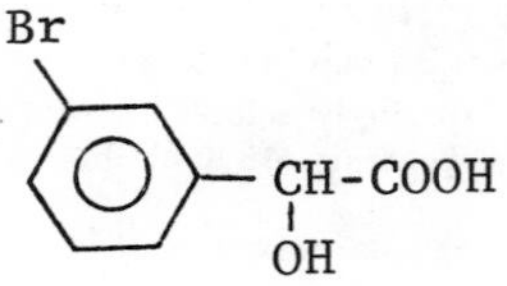

$C_8H_7BrO_3$ **Ref. 352**

Acide (+) *m-bromomandélique.*

2,31 g d'acide (±) *m*-bromomandélique (F = 121°) et 1,62 g de (—) éphédrine dans 5 cm³ d'éthanol fournissent 1,4 g de sel. Celui-ci est recristallisé dans 4 cm³ d'éthanol (Rdt = 0,65 g) et décomposé par HCl en 0,34 g d'acide dextrogyre, $[\alpha]_{578}^{25} = +94°$. Recristallisation dans 10 cm³ de benzène, 0,26 g, F = 105°, $[\alpha]_{578}^{25} = +99°$ (acétone, $c = 0{,}5$), optiquement pur.

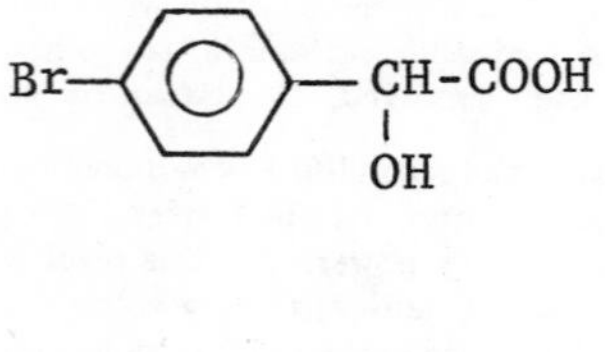

$C_8H_7BrO_3$ **Ref. 353**

Acide (—) *p-bromomandélique.*

A partir de 2 g d'acide (±) *p*-bromomandélique (F = 119°) et 1,4 g de (—) éphédrine dans 5 cm³ d'éthanol on obtient 2,25 g de sel, que l'on recristallise dans 10 cm³ d'éthanol, Rdt = 1 g. Ce sel est décomposé, l'acide lévogyre obtenu (0,5 g) recristallisé dans 7 cm³ de chloroforme, 0,27 g, F = 131°, $[\alpha]_{578}^{25} = -115°$ (acétone, $c = 0{,}6$), proche de la pureté optique.

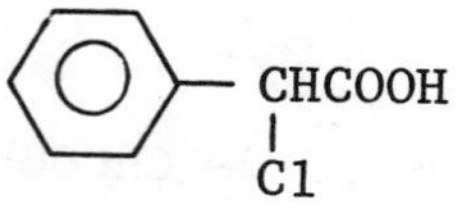

$C_8H_7ClO_2$ **Ref. 354**

Of the various alkaloids tried for the resolution of the *r*-acid, morphine was found to be the most suitable. Water is not a convenient solvent, since the acid is very sparingly soluble in it in the cold, and the halogen is displaced with great readiness if the aqueous solution is warmed. The best results were obtained by the use of methyl alcohol as solvent, and the following experiment may be quoted as typical.

To a solution of the *r*-acid (40 grams) in 200 c.c. of cold methyl alcohol, 71 grams of morphine were added, this being the calculated amount for the formation of the normal salt. The mixture was heated with constant shaking, and, as soon as the boiling point was reached, the morphine had all dissolved. During cooling to the laboratory temperature, crystallisation began, and, after the mixture had been maintained at 8—9° for twenty-four hours, the crystals were drained off and dried on porous earthenware; yield, 76 grams. These were dissolved in 300 c.c. of warm methyl alcohol, and the solution set aside at 8—10° for twenty-four hours. The resulting 56 grams were dissolved in 300 c.c. of warm methyl alcohol, and, after twenty-four hours at about 6°, 45 grams were obtained. This product was not yet, however, a homogeneous morphine salt, but it was found convenient at this point to decompose it and then crystallise the resulting acid mixture, instead of proceeding with the crystallisation from methyl alcohol. *N*-Hydrochloric acid (120 c.c.) was accordingly added, and the chloro-acid extracted four times with ether. (In subsequent experiments, the powdered morphine salt was shaken with aqueous sodium carbonate solution, the morphine removed, the sodium salt decomposed by mineral acid, and the chloro-acid extracted with ether.) The ethereal solution, which had been dried by sodium sulphate, yielded a lævorotatory acid (12 grams), which melted at 58—62° and had the following rotation in benzene solution:

$$l = 1, \ c = 3{\cdot}070, \ \alpha_D^{16°} - 5{\cdot}54°, \ [\alpha]_D^{16°} - 180{\cdot}5°.$$

This product, after one crystallisation from light petroleum, gave 7 grams of the pure *l*-acid. It is important that the crystallisation at this point is effected with low-boiling light petroleum. The petroleum, boiling at 60—80°, which was used for the purification of the *r*-acid is not suitable.

l-*Phenylchloroacetic acid*, $C_6H_5{\cdot}CHCl{\cdot}CO_2H$, separates from light petroleum in long, glassy needles, or, if the crystallisation is conducted slowly, in hexagonal, glassy plates. It melts at 60—61°. On analysis:

0·1502 gave 0·3102 CO_2 and 0·0562 H_2O. C = 56·3; H = 4·2.
$C_8H_7O_2Cl$ requires C = 56·3; H = 4·1 per cent.

$C_8H_7ClO_2$ **Ref. 354a**

Resolution of l-*Menthyl* dl-*Phenylchloroacetate into its Diastereoisomerides.*

A current of dry hydrogen chloride was passed for fifteen minutes into a molten mixture of *r*-phenylchloroacetic acid (10 grams) and *l*-menthol (30 grams), and heating on a water-bath was continued for four hours. After further treatment of the mixture with hydrogen chloride and subsequent heating for one and a half hours, the ethereal solution of the oil was washed first with water, and then with a dilute solution of sodium hydrogen carbonate. The ether and menthol were removed, the latter by distillation with steam, and the residual oil solidified after twenty-four hours in the ice-chest. Yield 17·3 grams; m. p. 24—29°. After crystallisation from rectified spirit at − 5°, the crystals (11·7 grams) melted over a larger range, namely 28·5—51·5°, and gave $[\alpha]_D - 63{\cdot}5°$ for $c = 4{\cdot}3808$ in ethyl-alcoholic solution. The resolution proceeded slowly when the crystallisation was conducted at temperatures below 0°, since a further crystallisation at − 7° gave a product (10·4 grams) which was only slightly less optically active than before, the value being $[\alpha]_D - 60{\cdot}9°$ for $c = 4{\cdot}2184$ in ethyl-alcoholic solution. The next crystallisation from rectified spirit took place at the ordinary temperature, and the resulting crystals (3·8 grams) were now slightly dextro-, giving $[\alpha]_D + 1{\cdot}5°$ for $c = 4{\cdot}3264$ in ethyl-alcoholic solution, whilst the crystals (1·3 grams) which separated from the filtrate were strongly lævo-, giving $[\alpha]_D$ — 141·1° for $c = 4·0736$ in ethyl-alcoholic solution. The dextro-product with $[\alpha]_D$ + 1·5° gave the pure ester of the *d*-acid after one further crystallisation; yield = 3 grams.

l-*Menthyl* d-*phenylchloroacetate*, $C_6H_5·CHCl·CO_2·C_{10}H_{19}$, separates from aqueous ethyl alcohol in fine needles, and melts at 56—57° (Found: Cl = 11·5. $C_{18}H_{25}O_2Cl$ requires Cl = 11·5 per cent.). In ethyl alcohol: $l = 2$, $c = 4·1732$, $\alpha_D^{15·8°}$ + 0·47°, $[\alpha]_D^{15·8°}$ + 5·6°; $\alpha_{5461}^{15·8°}$ + 0·66°, $[\alpha]_{5461}^{15·8°}$ + 7·9°.* This ester was also prepared directly from *d*-phenylchloroacetic acid, which was obtained by resolving the *r*-acid with morphine in ethyl-alcoholic solution (McKenzie and Clough, T., 1909, **95**, 777). The acid was esterified by *l*-menthol by means of hydrogen chloride, and the resulting ester after crystallisation from rectified spirit was found to be identical with the above; m. p. 56—57° (Found: Cl = 11·6; calc., 11·5 per cent.). In ethyl alcohol: $l = 2$, $c = 4·3672$, $\alpha_D^{14·8°}$ + 0·50°, $[\alpha]_D^{14·8°}$ + 5·7°; $\alpha_{5461}^{14·8°}$ + 0·70°, $[\alpha]_{5461}^{14·8°}$ + 8·0°. By fractional crystallisation of he products from the preceding resolution, the diastereoisomeride was also isolated.

l-*Menthyl* l-*phenylchloroacetate*, $C_6H_5·CHCl·CO_2·C_{10}H_{19}$, separates from aqueous ethyl alcohol in fine needles, and melts at 44·5—45·5° (Found: Cl = 11·7. $C_{18}H_{25}O_2Cl$ requires Cl = 11·5 per cent.). In ethyl alcohol: $l = 2$, $c = 3·9912$, $\alpha_D^{15°}$ — 11·96°, $[\alpha]_D^{15°}$ — 149·8°.

The homogeneity of this ester was controlled by its preparation from *l*-phenylchloroacetic acid, which was obtained by resolving the *r*-acid with morphine in methyl-alcoholic solution (McKenzie and Clough, T., 1908, **93**, 811). On esterification with *l*-menthol in the presence of hydrogen chloride, this acid gave an ester which, after crystallisation from rectified spirit, melted at 44·5—45·5°. In ethyl alcohol: $l = 2$, $c = 4·0748$, $\alpha_D^{15°}$ — 12·23°, $[\alpha]_D^{15°}$ — 150·1°; $\alpha_{5461}^{15°}$ — 14·65°, $[\alpha]_{5461}^{15°}$ — 179·8°.

* The determinations of optical rotatory power of solid substances made in this and in all other cases recorded in the present paper were conducted with products which were dried over sulphuric acid at the ordinary temperature and under diminished pressure until the weight was constant.

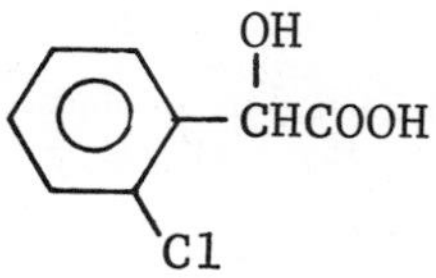

$C_8H_7ClO_3$

Ref. 355

Acides (+) *et* (—) *o-chloromandéliques.*

4 g d'acide (±) *o*-chloromandélique et 3,4 g de (—) éphédrine dans 10 cm³ d'éthanol donnent en 2 h, 2,4 g de sel **A**. Ce dernier est décomposé en 1,1 g d'acide $[\alpha]_{578}^{25}$ = — 120° (acétone, c = 1). Deux recristallisations dans 20 et 30 cm³ de benzène donnent 0,66 g d'acide lévogyre optiquement pur, F=119,5°, $[\alpha]_{578}^{25}$=—160° (acétone, c = 0,6).

Les eaux-mères de **A** sont évaporées à sec et décomposées en 2,1 g d'acide dextrogyre, $[\alpha]_{578}^{25}$ = + 57° (acétone). Trois recristallisations dans le benzène donnent 0,53 g d'acide optiquement pur, F = 119°, $[\alpha]_{5}{}^{4}{}_{8}$ = + 159° (acétone, c = 0,4). Litt.: F=116-118° (13).

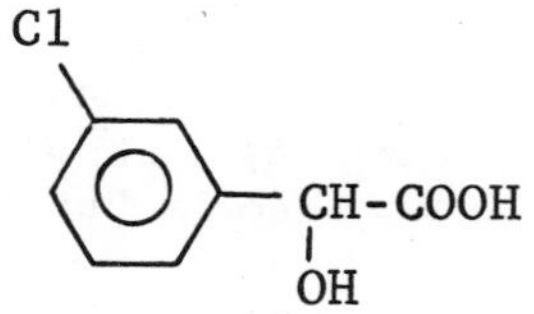

$C_8H_7ClO_3$

Ref. 356

Acide (+) *m-chloromandélique.*

3,73 g d'acide (±) *m*-chloromandélique (F = 115°) et 3,24 g de (—) éphédrine dans 20 cm³ d'éthanol donnent, à — 20°, 2 g de sel cristallisé. Celui-ci est recristallisé dans 5 cm³ d'éthanol (Rdt = 1,45 g) puis décomposé par HCl en 0,82 g d'acide dextrogyre, $[\alpha]_{578}^{5}$ = + 111° (acétone, c = 0,7). Recristallisation dans 15 cm³ de benzène, Rdt = 0,55 g, F = 107°, $[\alpha]_{578}^{25}$ = +122° (acétone; c = 0,9), optiquement pur.

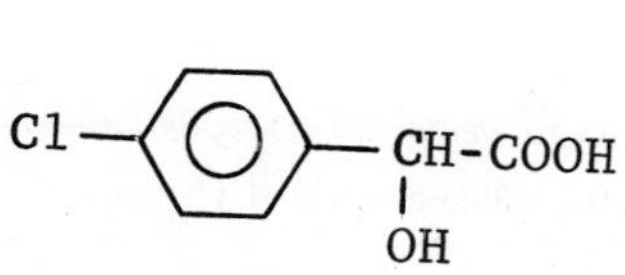

$C_8H_7ClO_3$

Ref. 357

Acide (—) *p-chloromandélique.*

4 g d'acide (±) *p*-chloromandélique (F = 121°) et 3,4 g de (—) éphédrine dans 10 cm³ d'éthanol déposent après une nuit 5 g de sel. Ce dernier est recristallisé deux fois dans 15 cm³ d'éthanol (Rdt = 2,2 g), puis décomposé en acide $[\alpha]_{578}^{25}$ = — 132° (acétone, c = 0,7). Recristallisation dans le benzène ou le chloroforme, F = 121°, $[\alpha]_{578}^{25}$ = — 136° (acétone, c = 0,7), optiquement pur.

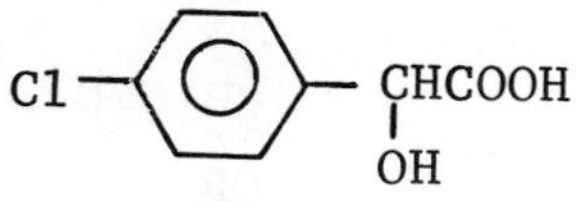

$C_8H_7ClO_3$

Ref. 357a

Resolution was accomplished by serial crystallization of the salt of the desired acid with (+)-α-methylbenzylamine. The m.p. of the resolved acid was 119-121°C. The $[\alpha]_D^{25}$ = -134° (c, 0.3, ethanol).

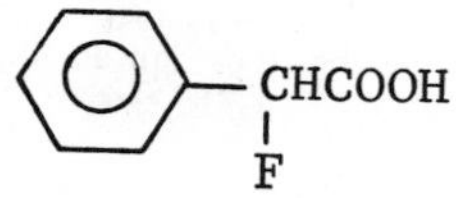

$C_8H_7FO_2$

Ref. 358

Acido (l)-fenilfluoroacetico

Una soluzione di 155 g di acido (d,l)-fenilfluoroacetico in 2 l di etere viene aggiunta sotto agitazione ad una soluzione di 165 g di (l)-efedrina in 2 l di etere. Dopo una notte a riposo, si raccolgono per filtrazione 299 g di sale a p.f. 163°; $[\alpha]_D^{20}$ = —59,6 (c = 1% CH_3OH). Si cristallizza il sale per tre volte consecutive usando ogni volta una miscela di 1500 cc di etanolo e 150 cc di metanolo. Resa 139 g. P.f. 182°; $[\alpha]_D^{20}$ = —85 (c = 1,2881, CH_3OH). Analisi ($C_{18}H_{22}FNO_3$) C, H, N.

50 g di sale di efedrina sono sciolti in HCl 5%. La soluzione è estratta ripetutamente con etere e gli estratti eterei, riuniti, sono filtrati ed evaporati. Il residuo si cristallizza da cicloesano. Resa 18 g. P.f. 105-106°; $[\alpha]_D^{20}$ = = —135,2 (c = 1,0388, $CHCl_3$). Analisi ($C_8H_7FO_2$) C, H, N, F.

Spettro I.R.: 3100-2200 cm^{-1} (ν OH), 1750 e 1700 cm^{-1} (ν C=O) (in soluzione in $CHCl_3$, si ha una banda unica a 1710 cm^{-1}), 1170 cm^{-1} (ν C—O), γ CH aromatici 695 e 725 cm^{-1}.

Spettro R.M.N. ($CDCl_3$): 4,15 (d, 1H, J_{HF} = 47,5 Hz, CHF), 2,52 (s, 5H, H aromatici), 1,43 τ (s, 1H, —COOH).

Acido (d)-fenilfluoroacetico

I tre filtrati alcolici della cristallizzazione del sale di efedrina della precedente preparazione, sono riuniti, concentrati a metà volume e trattati con sodio solfato. Dopo sei ore si filtra il solfato di sodio contenente del prodotto cristallino. Il filtrato, che lascia un residuo oleoso non cristallizzabile, viene scartato. Il solido viene scaldato a ricadere in un litro di acetone. Si filtra e si concentra nel pieno sino a 350 cc. Si raccoglie dopo una notte il precipitato e lo si ricristallizza da acetone per due volte. Resa 25 g. P.f. 122-124°; $[\alpha]_D^{20}$ = +23,6 (c = 1,12%, CH_3OH). Analisi ($C_{18}H_{22}FNO_3$) C, H, N.

Da questo sale di efedrina si ottiene l'enantiomero destrogiro nello stesso modo della preparazione precedente. Resa 8 g. P.f. 105-106°; $[\alpha]_D^{20}$ = +133,2 (c = 1,038%, $CHCl_3$). Analisi ($C_8H_7FO_2$) C, H, F.

Gli spettri I.R. e R.M.N. risultano uguali a quelli dell'enantiomero levogiro.

Cloruro dell'acido (l)-fenilfluoroacetico

27 g di acido (l)-fenilfluoroacetico si sciolgono in 110 cc di cloruro di tionile. Si scalda a ricadere per due ore, si allontana sotto vuoto il cloruro di tionile e si distilla il residuo Resa 26 g. P.e. 60° a 0,6 mm Hg; $[\alpha]_D^{20} = -158$ (c = 2,25%, $CHCl_3$).

Spettro I.R.: 1785 cm^{-1} (ν C=O), 695 e 750 cm^{-1} (γ CH aromatici).

Cloruro dell'acido (d)-fenilfluoroacetico

Il prodotto si ottiene dall'acido (d)-fenilfluoroacetico nello stesso modo e con le stesse rese della precedente preparazione. P.e. 60° a 0,6 mm Hg; $[\alpha]_D^{20} = +154{,}2$ (c = 1,33%, $CHCl_3$).

Spettro I.R.: risulta uguale a quello dell'enantiomero levogiro.

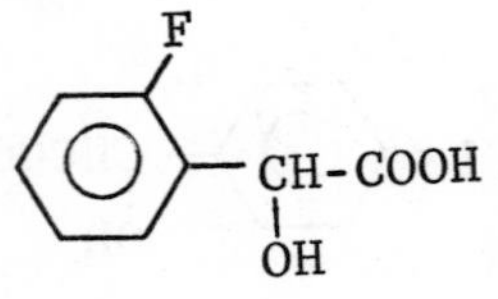

$C_8H_7FO_3$ **Ref. 359**

Acide (—) o-fluoro mandélique.

4 g d'acide (±) *o*-fluoromandélique (F = 118°) et 3,8 g de (—) éphédrine dans 10 cm³ d'éthanol donnent, après 12 h, 3,65 g de sel recristallisés à nouveau dans 15 cm³ d'éthanol. On recueille ainsi 3 g de sel que l'on décompose par HCl dilué; l'acide obtenu (1,1 g), $[\alpha]_{578}^{25} = -140°$ (acétone, $c = 1$) est optiquement pur après recristallisation dans le benzène ou le chloroforme: F = 90°, $[\alpha]_{578} = -145°$ (acétone, $c = 1{,}2$).

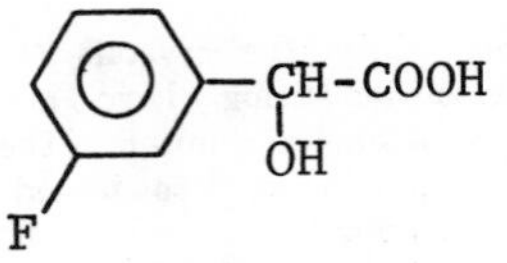

$C_8H_7FO_3$ **Ref. 360**

Acides (+) et (—) m-fluoro mandéliques.

A partir de 4 g d'acide (±) *m*-fluoromandélique (F = 97°) et 3,8 g de (—) éphédrine dans 25 cm³ d'éthanol, on obtient par grattage, et après 1 h à 0°, 2,45 g de sel **A**. Ce sel, recristallisé dans 10 cm³ d'éther (Rdt = 1,3 g) et décomposé par HCl dilué, donne 0,58 g d'acide $[\alpha]_{578}^{25} = -122°$ (acétone, $c = 1{,}3$). Par recristallisation dans le benzène on obtient 0,3 g d'acide lévogyre optiquement pur, F = 121°, $[\alpha]_{578}^{25} = -129°$ (acétone, $c = 1$).

On évapore à sec les eaux-mères de **A**, et les reprend par 15 cm³ d'un mélange $CCl_4/CHCl_3$; après 36 h, on essore 2,7 g de cristaux **B** dont la décomposition donne un acide faiblement dextrogyre. Les eaux-mères de **B** sont évaporées à sec et décomposées par HCl. On obtient 1,9 g d'acide $[\alpha]_{578}^{25} = +89°$ (acétone, $c = 1{,}5$). Deux recristallisations, dans 50, puis 20 cm³ de benzène donnent 0,45 g d'acide optiquement pur, F = 121°, $[\alpha]_{578}^{25} = +129°$ (acétone, $c = 1{,}8$).

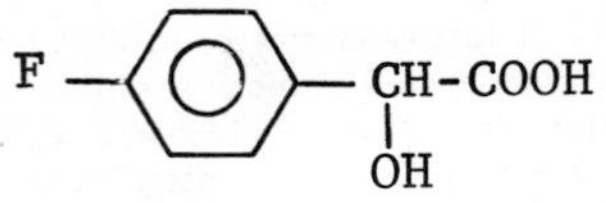

$C_8H_7FO_3$ **Ref. 361**

Acides (+) et (—) p-fluoromandéliques.

A partir de 2 g d'acide (±) *p*-fluoromandélique (F = 130°) et 1,9 de (—) éphédrine dans 10 cm³ d'éthanol on obtient 1,4 g de sel **A**; une recristallisation du sel dans le minimum d'éthanol et décomposition par HCl donnent 0,45 g d'acide lévogyre, $[\alpha]_{578}^{25} = -129°$ (acétone $c = 1$). L'acide (—) est obtenu optiquement pur par recristallisation dans 100 cm³ de toluène: 0,32 g, F = 153°, $[\alpha]_{578}^{25} = -140°$ (acétone, $c = 1$).

Les eaux-mères de **A** sont évaporées à sec et décomposées en acide dextrogyre, 0,8 g, $[\alpha]_{578}^{25} = +97°$ (acétone). Par recristallisation dans l'éther/hexane, puis le toluène, on obtient l'acide (+) optiquement pur, F = 153°, $[\alpha]_{578}^{25} = +140°$ (acétone, $c = 1$).

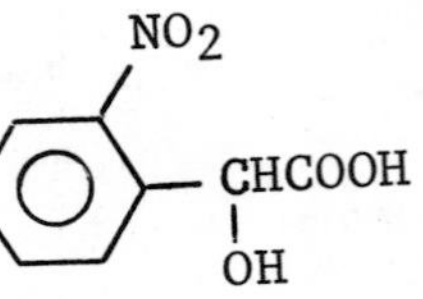

$C_8H_7NO_5$ **Ref. 362**

Resolution of r-o-*Nitromandelic Acid*.—100 G. of brucine (1 mol.) were added with stirring to a hot solution of 50 g. of the acid (1 mol.) in water (2 l.). After 16 hours, crystals (A, 68 g.) had separated, and from the filtrate crystals (B, 60 g.) were deposited after 36 hours longer. On crystallising A thrice from water, the homogeneous brucine (—)*o*-nitromandelate (46 g.) crystallised with $4H_2O$ as hexagonal plates (m. p. 91—92°), which after heating under diminished pressure at 80° had m. p. 137° (decomp.) and then gave in acetone ($l = 2$, $c = 2{\cdot}4305$) $\alpha_{5461}^{20{\cdot}5°} - 16{\cdot}63°$, $[\alpha]_{5461}^{20{\cdot}5°} - 342°$. The brucine was removed by precipitating the aqueous solution of this salt with ammonia, the filtrate being then acidified by hydrochloric acid. The resulting acid, obtained by extraction with ether, was crystallised once from chloroform, yielding 13 g. of (—)o-*nitromandelic acid* in colourless hexagonal plates, m. p. 100—101° (Found: C, 48·5; H, 3·7. $C_8H_7O_5N$ requires C, 48·7; H, 3·6%). Its rotatory power was determined in acetone ($l = 2$, $c = 1{\cdot}5785$, $t = 20°$):

λ	6708	6563	5893	5461	4861	4358
α	− 10·21°	− 10·89°	− 14·79°	− 18·74°	− 28·90°	− 48·35°
[α]	− 323°	− 345°	− 468°	− 594°	− 915°	−1532°

In acetone ($l = 2$, $c = 1{\cdot}8955$), $\alpha_{5791}^{20°} - 18{\cdot}56°$, $[\alpha]_{5791}^{20°} - 490°$.

The three successive filtrates from which A had been crystallised gave acids with $[\alpha]_{5791}$ − 268°, − 473°, and − 483° respectively in acetone ($c = 2$) after removal of the brucine.

Crystallisation of B thrice from water gave the homogeneous brucine (+)*o*-nitromandelate, needles (35 g.) which when dried in air contained $2H_2O$. When dried under diminished pressure at 80°, it had m. p. 137° (decomp.). In acetone ($l = 2$, $c = 2{\cdot}336$), $\alpha_{5461}^{20°} + 13{\cdot}61°$, $[\alpha]_{5461}^{20°} + 291°$.

(+)o-*Nitromandelic acid* from the brucine salt crystallised from chloroform in colourless hexagonal plates, m. p. 100—101° (Found: C, 48·9; H, 3·8. $C_8H_7O_5N$ requires C, 48·7; H, 3·6%). Yield, 10 g. The rotation in acetone was in agreement with that for its antimeride (l = 2, c = 1·896): $\alpha^{20°}_{5791}$ + 18·59°, $[\alpha]^{20°}_{5461}$ + 490°. The rotatory power decreases with rise of temperature: in acetone (l = 2, c = 1·8965),

λ.	$\alpha^{1·5°}$.	$[\alpha]^{1·5°}$.	$\alpha^{20°}$.	$[\alpha]^{20°}$.	$\alpha^{40°}$.	$[\alpha]^{40°}$.
5461	+23·92°	+ 631°	+22·54°	+ 594°	+21·14°	+ 557°
4358	+62·67	+1652	+58·54	+1543	+54·86	+1446

The three successive filtrates from which B had been crystallised gave acids with $[\alpha]_{5791}$ + 256°, + 444°, and + 477° respectively in acetone (c = 2).

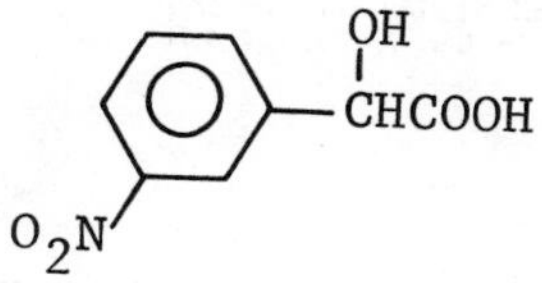

$C_8H_7NO_5$

Ref. 363

Spaltung der *racem*-m-Nitro-mandelsäure.

Die *racem*-m-Nitro-mandelsäure wurde nach Brode und Adams (2) aus m-Nitro-benzaldehyd dargestellt. Bei Vorversuchen mit einer kleinen Substanzmenge (0,01 Mol) wurden die Brucinsalze der beiden Antipoden durch fraktionierte Kristallisation rein erhalten; das Salz der linksdrehenden Säure hat die grössere Kristallisationsfähigkeit. Die Trennung kann in Wasserlösung ausgeführt werden, scheint aber in Alkohol etwas schneller vor sich zu gehen. Bei der Bearbeitung der Hauptmenge konnte indessen das Drehungsvermögen der rechts-Säure durch Umkristallisation des Brucinsalzes nicht bis zum Maximalwert gebracht werden. Die Möglichkeit, dies beim rascheren Arbeiten mit kleinen Substanzmengen auszuführen, ist wohl als ein Übersättigungsphänomen zu deuten. Die Säure wurde deshalb in das Cinchoninsalz übergeführt, das bei Umkristallisation aus Wasser schnell rein erhalten wurde. Von anderen Alkaloiden, die in Wasserlösung geprüft wurden, sei erwähnt, dass Strychnin und Chinin gut kristallisierende Salze gaben, die schwach linksdrehende Säure enthielten; die Morphin- und Chinidinsalze wurden nicht kristallisiert erhalten.

Zur Kontrolle der Spaltung wurde 0,1—0,2 g Alkaloidsalz in verdünnter Schwefelsäure gelöst und die Nitro-mandelsäure durch Ausschütteln mit Äther isoliert. Das erhaltene Produkt wurde in Wasser gelöst, die Drehung bestimmt und der Gehalt der Lösung durch Titrieren mit Natronlauge ermittelt.

19,7 g (0,1 Mol) *racem*-Säure und 30 g (0,065 Mol) Brucin wurden in 450 ml siedendem Alkohol gelöst. Die ausgeschiedenen Kristalle wurden nach 5 Stunden abgesaugt. Aus der Mutterlauge wurde nach Zugabe von 16,6 g (0,035 Mol) Brucin eine zweite Kristallfraktion erhalten. Der Verlauf der Spaltung der ersten Fraktion ergibt sich aus folgender Zusammenfassung:

Kristallisation	1	2	3	4	5	6
Ausbeute in g	33,2	25,7	24,6	20,7	19,9	19,1
$[\alpha]_D$	−68°	−100°	−104,5°	−112,5°	−116,5°	−117,5°

Aus der späteren Fraktion, die rechtsdrehende Säure, $[\alpha]_D$ = + 74°, enthielt, konnte das Salz der rechts-Säure durch Umkristallisieren nicht rein erhalten werden. Die Säure wurde deshalb aus dem Salz isoliert; ihr Gewicht betrug 10,3 g. Sie wurde nebst 14,5 g Cinchonin in 900 ml siedendem Wasser gelöst. Das Cinchoninsalz kristallisierte unmittelbar beim Erkalten aus; es wurde abgesaugt, während die Lösung noch lauwarm war, und dreimal aus Wasser umkristallisiert.

Kristallisation	1	2	3	4
Ausbeute in g	16,3	13,3	11,3	9,3
$[\alpha]_D$	+111,5°	+117°	+118,5°	+118,5°

(—)-m-Nitro-mandelsäure.

Aus 18,5 g Brucinsalz wurden 4,5 g einmal aus Toluol umkristallisierte Säure gewonnen; Schmp. 133—134°. Die Säure bildet schwach gelblich gefärbte, mikroskopisch kleine Prismen, die zu baumähnlichen Aggregaten zusammengewachsen sind.

0,2486 g Subst.: 12,37 ml 0,1019-n NaOH.
$C_8H_7O_5N$, Äqu.-Gew. ber. 197,1 gef. 197,2.

Die Drehungsmessungen wurden im 3 dm-Polarimeterrohr ausgeführt.

0,1034 g Subst. in Wasser auf 6,01 ml gelöst:
3 α_D = − 6,82°; $[\alpha]^{25}_D$ = − 122,4°; $[M]^{25}_D$ = − 241,3°.
3 α_{5461} = −7,48°; $[\alpha]^{25}_{5461}$ = −144,9°; $[M]^{25}_{5461}$ = −285,6°.

Das Brucinsalz bildet stark gelb gefärbte Nadeln, die unter Zersetzung bei 148—150° schmelzen. Im Vakuum über P_2O_5 verliert es bei Zimmertemperatur sehr langsam Wasser in einem Betrag, der ungefähr ½ Molekül Kristallwasser entspricht.

0,3915 g Subst.: 0,0063 g H_2O.
0,10833 g Subst.: 6,50 ml N_2 (762,3 mm Hg, 18,3 °C).
$C_8H_7O_5N$, $C_{23}H_{26}O_4N_2$, ½H_2O; H_2O ber. 1,50 %, gef. 1,61 %.
N » 7,00 » » 7,05 ».

(+)-m-Nitro-mandelsäure.

Aus 8,7 g Cinchoninsalz wurden 2,9 g einmal aus Toluol umkristallisierte Säure vom Schmelzp. 133—134° gewonnen. Die Säure bildet schwach gelblich gefärbte Kristalle, die ähnlich denen der links-Säure zusammengehäuft sind.

0,2888 g Subst.: 14,41 ml 0,1019-n NaOH.
$C_8H_7O_5N$, Äqu.-Gew. ber. 197,1, gef. 196,7.

Die Ergebnisse der Drehungsmessungen sind in folgender Tabelle zusammengestellt:

Lösungsmittel	g Subst. in 6,01 ml Lsg.	λ Å.-E.	3 α_λ	$[\alpha]^{25}_\lambda$	$[M]^{25}_\lambda$
Wasser	0,1011	5893	+ 6,185°	+122,6°	+241,5°
»	»	5461	+ 7,30°	+144,6°	+285,1°
0,1-n NaOH . .	0,1016	5893	+ 4,97°	+ 98,0°	+193,1°
1-n HCl . . .	0,1058	»	+ 6,69°	+126,7°	+249,6°
Aceton	0,1005	»	+ 5,53°	+110,2°	+217,2°
Eisessig . . .	0,1010	»	+ 6,635°	+131,6°	+259,8°
Essigester . .	0,1027	»	+ 6,11°	+119,2°	+234.9°

Das Brucinsalz, das aus Wasser oder Alkohol in kleinen Nadeln kristallisiert, ist im Unterschied zum stark gefärbten Brucinsalz der (—)-Säure nur schwach gelblich.

Das Cinchoninsalz bildet farblose Nadeln mit einem Stich ins gelblich Grüne. Über P_2O_5 verliert es schnell ½ Molekül Kristallwasser.

0,3760 g Subst.: 0,0066 g H_2O.
$C_8H_7O_5N$, $C_{19}H_{22}ON_2$, ½H_2O; H_2O ber. 1,75 %, gef. 1,76 %.

Die Stickstoffbestimmung wurde an wasserfreiem Material ausgeführt.

0,11819 g Subst.: 8,81 ml N_2 (735,8 mm Hg, 23,6 °C).
$C_8H_7O_5N$, $C_{19}H_{22}ON_2$; N ber. 8,29 %, gef. 8,81 %.

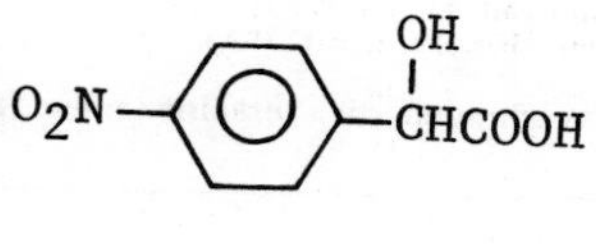

$C_8H_7NO_5$

Ref. 364

Spaltung der *racem*-p-Nitro-mandelsäure.

20 g (0,1 Mol) *racem*-Säure wurden nebst 15 g Chinidin in einer siedenden Mischung aus 1250 ml Wasser und 325 ml Alkohol gelöst, und der Lösung dann noch 12,5 g Chinidin (insgesamt 0,075 Mol), in 300 ml Alkohol gelöst, zugefügt. Da das Chinidinsalz eine ausgeprägte Neigung zeigt, sich als Öl auszuscheiden, wurde im voraus eine kleine Menge Salz dargestellt, das nach zwei Umkristallisationen genügend rein war, um als Impfmaterial dienen zu können. Nachdem die Temperatur der filtrierten Lösung bis auf 55° gesunken war, wurde die Lösung mit Chinidinsalz geimpft. Trotzdem schied sich ein Teil des Salzes als Emulsion aus. Beim Erwärmen auf 50—55° löste sich indessen die Emulsion und das Salz kam

beim Erkalten kristallisiert heraus. Das Chinidinsalz wurde aus 30-prozentigem Alkohol umkristallisiert. Der Verlauf der Spaltung war der folgende:

Kristallisation	1	2	3	4
Ausbeute in g	25,8	17,0	12,8	10,9
$[\alpha]_D$ der Säure	+81°	+117,5°	+123,5°	+125,5°

Die Mutterlaugen der zwei ersten Kristallisationen wurden mit 100 ml 2-m Schwefelsäure versetzt und bei Zimmertemperatur zu einem Volumen von etwa 300 ml eingeengt. Durch Ätherextraktion wurden daraus 11,7 g Säure isoliert. Diese wurde nebst 18 g Strychnin in 700 ml siedendem 30-prozentigem Alkohol gelöst. Beim Erkalten kristallisierte das Strychninsalz in glänzenden Schuppen.

Kristallisation	1	2	3	4	5	6	7
Ausbeute in g	24,3	20,8	18,9	16,7	15,4	12,8	11,2
$[\alpha]_D$	−85,5°	−108°	−116,5°	−124,5°	−126,5°	−130°	−129,5°

Aus den Mutterlaugen der zwei ersten Strychninsalz- und der zwei letzten Chinidinsalzkristallisationen wurden in oben angegebener Weise 4,3 g Säure isoliert. Als diese nebst 7,5 Chinidin in 375 ml 30-prozentigem Alkohol gelöst wurde, kristallisierten 7,5 g Chinidinsalz aus, das Säure vom $[\alpha]_D = +126°$ enthielt. Es wurde zusammen mit der Hauptmenge des Chinidinsalzes noch zweimal umkristallisiert:

Kristallisation	5	6
Ausbeute in g	13,0	10,9
$[\alpha]_D$	+130°	+130°

(+)-p-Nitro-mandelsäure.

Das Chinidinsalz der Säure (10,5 g) wurde in verdünnter Schwefelsäure gelöst und die filtrierte Lösung mit Äther ausgeschüttelt. Die Ätherlösung wurde mit Natriumsulfat getrocknet, und nachdem der Äther, gegen Ende im Vakuum über Schwefelsäure, abdestilliert worden war, erstarrte der Rückstand, 3 g, zu einer blassgelben Masse, die bei 91—93° schmolz. Die Säure ist in Wasser, Dioxan und sauerstoffhaltigen aliphatischen Lösungsmitteln spielend leicht löslich und kristallisiert aus solchen Lösungsmitteln nicht. In Benzol ist sie bei gewöhnlicher Temperatur wenig löslich, löst sich indessen darin bei Siedetemperatur etwas besser und kann aus grossen Mengen des Lösungsmittels umkristallisiert werden. Doch konnte sie dadurch nicht rein erhalten werden. Umkristallisation aus Chlorbenzol führte auch nicht zum Ziel. Für die Reinigung wurde zuletzt das folgende Verfahren gewählt. Die Säure wurde in 0,1-m Bariumhydroxyd zur genauen Neutralisation gelöst, die Lösung auf etwa 50° erwärmt und ein Überschuss warmer Bleiacetatlösung zugefügt. Dabei schied sich das Bleisalz der Säure als farbloser, grobkristalliner Niederschlag aus. Nach zwei Stunden wurde das Salz abgesaugt und auf dem Filter mehrmals mit kaltem Wasser gewaschen. Es wurde in heissem Wasser aufgeschlämmt und ein Überschuss verdünnter Schwefelsäure zugefügt. Nach ein paar Stunden wurde filtriert und die Säure durch Ätherextraktion isoliert. Nachdem der Äther bei Zimmertemperatur verjagt worden war, kristallisierte die Säure. Sie wurde feinpulverisiert einen Tag im Vakuum über Kaliumhydroxyd aufbewahrt, um etwa mitgerissene Spuren von Essigsäure zu entfernen. Das so erhaltene Präparat war farblos und schmolz bei 93—94°. Eine Wiederholung des Verfahrens veränderte den Schmelzpunkt und das Drehungsvermögen nicht.

0,2163 g Subst.: 10,74 ml 0,1021-n NaOH.
$C_8H_7O_5N$, Äqu.-Gew. ber. 197,1, gef. 197,8.

Das Drehungsvermögen in verschiedenen Lösungsmitteln beträgt:

Lösungsmittel	g Subst. in 6,01 ml Lsg	λ Å.-E.	$3\ \alpha_\lambda$	$[\alpha]_\lambda^{25}$	$[M]_\lambda^{25}$
Wasser	0,0999	5893	+6,48°	+128,9°	+254,1°
»	»	5461	+7,57°	+151,8°	+299,1°
0,1-n NaOH	0,1001	5893	+4,975°	+ 99,6°	+196,2°
1-n HCl	0,1013	»	+6,76°	+133,7°	+263,4°
Aceton	0,1008	»	+6,635°	+131,9°	+259,9°
Eisessig	0,1008	»	+7,815°	+145,4°	+286,5°
Essigester	0,0995	»	+6,54°	+131,7°	+259,5°
Dioxan	0,1012	»	+6,11°	+121,0°	+238,3°

Das Chinidinsalz kristallisiert aus 30-prozentigem Alkohol wasserfrei in schwach gelb gefärbten Nadeln.

0,18218 g. Subst.: 12,87 ml N_2 (745 mm Hg, 19,1° C).
$C_8H_7O_5N$, $C_{20}H_{24}O_2N_2$, N ber. 8,06 %, gef. 8,10 %.

(−)-p-Nitro-mandelsäure.

Das Strychninsalz der Säure wurde in warmer, verdünnter Schwefelsäure gelöst und die Säure dann, wie dies für den optischen Antipoden beschrieben worden ist, isoliert und gereinigt. Farblose Kristallmasse vom Schmelzpunkt 93—94°.

0,2050 g Subst.: 10,18 ml 0,1021-n NaOH.
$C_8H_7O_5N$, Äqu.-Gew. ber. 197,1, gef. 197,2.

0,1005 g Subst. in Wasser auf 6,01 ml gelöst:
$3\ \alpha_D = -6{,}48°$, $[\alpha]_D^{25} = -129{,}2°$, $[M]_D^{25} = -254{,}5°$.
$3\ \alpha_{5461} = -7{,}63°$, $[\alpha]_{5461}^{25} = -152{,}1°$, $[M]_{5461}^{25} = -299{,}7°$.

Das Strychninsalz kristallisiert aus 30-prozentigem Alkohol in Form farbloser Schuppen mit 2 Molekülen Kristallwasser.

0,4171 g Subst.: 0,0268 g H_2O.
$C_8H_7O_5N$, $C_{21}H_{22}O_2N_2$, 2 H_2O, H_2O ber. 6,35 %, gef. 6,48 %.
0,12598 g Subst.: 8,66 ml N_2 (750 mm Hg, 18,4° C).
$C_8H_7O_5N$, $C_{21}H_{22}O_2N_2$, N ber. 7,91 %, gef. 7,95 %.

$C_8H_7NO_5$ Ref. 364a

Resolution was accomplished by serial crystallization of the salt of the desired acid with (+)-α-methylbenzylamine. The m.p. of the resolved acid was 93-95°C. The $[\alpha]_D^{25}$ = 129°C. (c 1, ethanol).

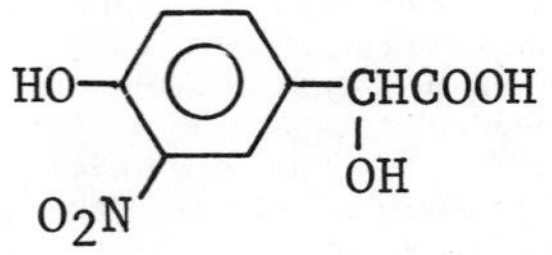

$C_8H_7NO_6$ Ref. 365

Scissione ottica dell'acido DL 3.nitro-4.ossimandelico

Sale di brucina dell'acido (+)3.nitro-4.ossimandelico. — 21,3 g (0,1 moli) di acido DL 3.nitro-4.ossimandelico vengono sciolti in 590 cc di alcool etilico all'80% bollente; si aggiungono alla soluzione g 39,45 (0.1 moli) di brucina. Si lascia raffreddare la soluzione e si insemenza con cristalli del sale di brucina ottenuti in una prova in piccolo.

Dopo riposo di una notte a 0°, si ottengono g 38.5 di precipitato cristallino (frazione A). Il sale di brucina così preparato viene ripetutamente cristallizzato da etanolo all'80% finchè l'acido ottenuto dal sale mediante trattamento con la quantità calcolata di acido solforico 1:10, estrazione con acetato di etile, ed evaporazione del solvente, non presenta un potere rotatorio costante.

L'andamento della scissione è rappresentato nella Tab. I in cui sono riportati le rese del sale, il potere rotatorio dell'acido, il p.f. dell'acido ed il p.f. del sale.

Tabella I

Cristallizzazione		1	2	3	4	5	6
$[\alpha]_D$ dell'acido (c = 0,5 — 0,6 in etanolo	+28.7°	+49,3°	+62,1°	+65,2°	+71,5°	+78,0°	+78,5°
P.F. dell'acido	140-145°	144-145°	145-146°	149-150°	149-150°	151-152°	152°
P.F. del sale	99-101°	140-142°	145-146°	147°	147-148°	147-148°	147-148°
Rese in g di sale	38,5	17,5	12,4	10,7	8,9	7	5,6

Il potere rotatorio del sale non è indicativo per la scissione, data la sua scarsa solubilità in acqua ed in solventi neutri ed il suo piccolo potere rotatorio.

Analisi del sale di brucina:

	trov. % :	C 61,00; H 5,70; N 7,08
per $C_8H_7O_6N \cdot C_{23}H_{26}O_4N_2$ (607,6)	calc. :	61,27; 5,47; 6,92

Dalle acque madri della scissione per ulteriore riposo precipitano 1,1 g di sale (frazione B) dal quale si ottiene un acido a p.f. 140-143° ed $[\alpha]_D^{25} = +25°{,}3$ (c = 0,61 in alcool).

Le acque madri provenienti dalla scissione vengono concentrate sotto vuoto a bassa temperatura fino a secchezza.

Si ottengono g 21,0 (frazione C) di sale di brucina da cui si libera l'acido come sopra indicato. Si ottengono in tal modo g 6,92 di acido 3.nitro-4.ossimandelico levogiro avente $[\alpha]_D^{25} = -47°{,}5$ (c = 0,55 in alcool).

Sale di chinina dell'acido (—)3.nitro-4.ossimandelico

G 6,6 di acido (—)3.nitro-4.ossimandelico ad $[\alpha]^{25}_{D}$ = — 47°,5 (c = 0,55 in alcool) proveniente dal sale di brucina (frazione C) (0,03 moli) vengono sciolti in 80 cc di etanolo 80% bollente. Si aggiungono g 10,05 (0,03 moli) di chinina sciolti in 15 cc di etanolo 80% bollente. Dopo riposo di una notte si filtrano g 7,70 di sale di chinina dell'acido (—)3.nitro-4.ossimandelico.

Il sale di chinina viene cristallizzato da alcool 80% finchè l'acido da esso ottenuto presenta potere rotatorio costante. L'andamento della purificazione è rappresentato nella Tab. II dove sono riportati le rese del sale, il p.f. ed il potere rotatorio dell'acido e del sale.

TABELLA II

Cristallizzazione		1	2
$[\alpha]_D$ dell'acido (c = 0,5 — 0,6 in etanolo)	— 72,2°	— 73,1°	— 75,7°
P.F. dell'acido	148-150°	150-152°	152°
$[\alpha]_D$ del sale (c=0,5 — 0,55 in etanolo)	— 139,2°	— 138,5°	— 139,3°
P. dec. del sale	152-154°	154-156°	155-156°
Rese in g di sale	7,70	5,40	4,10

Analisi del sale di chinina:

	trov. % :	C 62,40; H 5,97; N 7,99
per $C_8H_7O_6N \cdot C_{20}H_{24}N_2O_2$ (537,5)	calc. :	62,55; 5,81; 7,82

Acido (+) 3.nitro-4.ossimandelico

G 5 di sale di brucina proveniente dalla sesta cristallizzazione della frazione A, vengono sciolti in 7,5 cc di H_2SO_4 10%, ed estratti con acetato di etile. Si lavano gli estratti con qualche cc di acqua e si seccano su solfato di sodio anidro. Per evaporazione del solvente sotto vuoto rimangono g 1,6 di acido che presenta p.f. 152°; $[\alpha]^{25}_{D}$ = + 78°,5 (c = 0,480 in alcool). Resa 96%.

	trov. % :	C 44,91; H 3,42; N 6,72
per $C_8H_7O_6N$ (213,1)	calc. :	45,08; 3,31; 6,57

Acido (—) 3.nitro-4.ossimandelico

G 4 di sale di chinina proveniente dalla seconda cristallizzazione precedentemente descritta, vengono sciolti in 7,5 cc di acido solforico 10% ed estratti ripetutamente con acetato di etile. Gli estratti lavati con acqua, seccati su solfato di sodio anidro ed evaporati sotto vuoto lasciano g 1,50 di acido (—)3.nitro-4.ossimandelico che possiede p.f. 152°; $[\alpha]^{25}_{D}$ = — 75°,7 (c = 0,538 in alcool). Resa 95%.

	trov. % :	C 44,85; H 3,50; N 6,80
per $C_8H_7O_6N$ (213,1)	calc. :	45,08; 3,31; 6,57

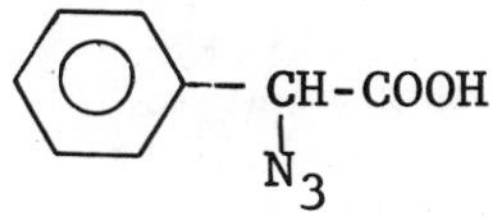

$C_8H_7N_3O_2$ Ref. 366

Beispiel 1:

d (-)-α-Azidophenylessigsäure. Es wird eine Lösung von 212,5 g der gemäss C erhaltenen Säure, gelöst in 900 ml Aceton, mit 254,6 g 1(+)-threo-1-p-Nitrophenyl-2-amino-1,3-propandiol neutralisiert. Die anfallende Lösung wird durch Abdestillieren des Lösungsmittels unter vermindertem Druck auf ein kleines Volumen eingeengt. Dann wird das anfallende Öl, das noch eine gewisse Menge Aceton enthält, unter Rühren und leichtem Erwärmen in 3 Litern Methylenchlorid gelöst. Nahezu sofort beginnt die Abtrennung des Diastereomeren der d(-)-Säure und 1(+)-Base. Nach 2 Stunden wird dieses abfiltriert, mit mehr Methylenchlorid gewaschen und getrocknet, um 233 g. eines weissen Feststoffes zu ergeben. Dann wird er bei leichtem Sieden in 24 Litern 99%-igem Äthanol gelöst und rekristallisiert, wenn man auf Raumtemperatur abkühlen lässt. Es ergeben sich 196 g reines Produkt mit einem Smp. von 151-3°C. Der Rest wird aus den Äthanollaugen nach Eindampfen rekristallisiert, um einen weiteren Teil zu liefern.

Die 196 g. des Diastereomeren werden in 500 ml Wasser und 500 ml Methylenchlorid suspendiert und unter Rühren werden 100 ml konzentrierte Chlorwasserstoffsäure zugesetzt. Die untere organische Phase wird, nachdem sie gewaschen ist, dekantiert, getrocknet und das Lösungsmittel wird bei vermindertem Druck verdampft, um 88 g d(-)-α-Azidophenylessigsäure mit praktisch quantitativer Ausbeute zu ergeben, Smp. 50-57°C (d) und $[\alpha]^{22}_{D}$ = -146,5 (DMSO). Weitere 10,5 g werden von dem wiedergewonnenen, rekristallisierten Diastereomeren isoliert, so dass sich insgesamt 98,5 g ergeben.

1(+)-α-Azidophenylessigsäure. Die Mutterlaugen von der Herstellung des Diastereomeren d(-)-Saure/1(+)-threo-Base werden mit 500 ml Wasser, das 100 ml konzentrierte Chlorwasserstoffsäure enthält, extrahiert. Die organische Phase wird getrocknet und bei vermindertem Druck destilliert, um ein Öl zu liefern, das sich schliesslich verfestigt, um 105 g 1(+)-α-Azidophenylessigsäure zu ergeben. Durch Umkristallisieren in Benzol wird zuerst ein Teil mit geringem Drehvermögen erhalten, dann wird mit Zusatz von n-Heptan eine optisch reine Säure mit Smp. 52-57°C und $[\alpha]^{22}_{D}$ = +146,4 (DMSO) erhalten.

$C_8H_7O_2N_3$ Ref. 366a

Note:
Treatment of the racemic acid in 2 propanol with (-)-α-phenylethylamine in 2-propanol afforded the less soluble salt, m.p. 152° and $[\alpha]^{20}_{D}$ -75.5° (c 1, CH_3OH). This salt yielded active acid, m.p. 62°, $[\alpha]^{20}_{D}$ -142.6° (c 5, CH_3OH).

Treatment of the racemic acid in 2-propanol with (+)-α-phenylethylamine afforded the less soluble salt, m.p. 139-140°, $[\alpha]^{20}_{D}$ +78.5° (c 1, CH_3OH). This salt yielded active acid m.p. 59°, $[\alpha]^{20}_{D}$ +142.0° (c 5, CH_3OH).

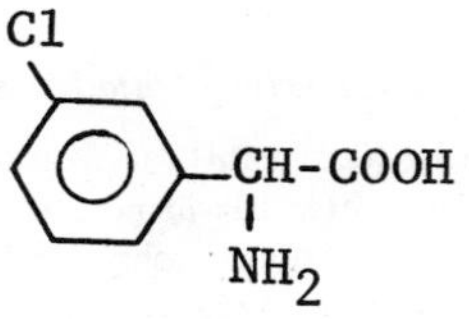

$C_8H_8ClNO_2$ Ref. 367

D-(—)-α-amino-3-chlorophenylacetic acid.—α-Formamido-3-chlorophenylacetic acid (721 g.) and one kg. of dehydroabietylamine were combined in 4 l. of methanol. After storing in the cold for two hours the crystalline salt was collected by filtration. The product was recrystallized from methanol-water; yield 598 g., $[\alpha]_D^{25°}$=—225° (C.=0.4 methanol). The salt was slurred in 2 l. of methanol and 2 liters of saturated sodium bicarbonate solution. The mixture was diluted with 2 l. of water, layered with methyl isobutyl ketone, and stirred vigorously. The aqueous phase was separated and acidified to pH 2 with conc. hydrochloric acid. The acid was collected by filtration and dried. The dried product was combined with 2 l. of 6 N hydrochloric acid and 750 ml. of methanol, the mixture heated for two hours, and filtered. The solution was adjusted to pH 5 with ammonium hydroxide. The solid was collected by filtration and washed with water and acetone giving 112 g. of D-(—)-α-amino-3-chlorophenylacetic acid; $[\alpha]_D^{23°}$=—125° (C.=0.4 1 N HCl).

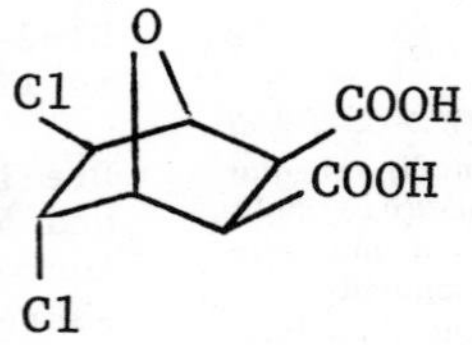

$C_8H_8Cl_2O_5$ Ref. 368

Resolution of Dichloroacid (Vc) into Antipodes. A solution of 14 g of dichloroacid (m.p. 147-148°) (Vc) in 60 ml hot methanol was quickly stirred with a hot solution of 21.6 g brucine in 60 ml methanol. The mixture was boiled for about an hour; the precipitate (A) was filtered hot and washed on the filter with 300 ml hot methanol. Crystals A were boiled with 1000 ml methanol, filtered hot, and crystallized 8 times from nitromethane to constant specific rotation of the brucine salt $[\alpha]_D$ −23.3°, m.p. 156-157°, yield 10 g.

The methanolic mother liquor (B), remaining after separation of precipitate A, was evaporated to dryness in vacuo, and after 10 crystallizations from methanol the brucine salt was obtained with $[\alpha]_D$ + 20°, m.p. 147-148°, yield 9.5 g.

Salt A (2 g) was suspended in 50 ml ether in a separating funnel, and 20 ml concentrated hydrochloric acid cautiously added portionwise with shaking. The ethereal layer was separated, the acid extracted with 100 ml ethyl acetate, and the combined extracts dried with magnesium sulfate. After removal of the solvent there was obtained 1.1 g of one of the antipodes of (Vc) with m.p. 142-143° (from water): $[\alpha]_{589}$ −61°, $[\alpha]_{578}$ −79.85°, $[\alpha]_{546}$ −88.75°, $[\alpha]_{436}$ −146.1°, $[\alpha]_{405}$ −189.8°, $[\alpha]_{365}$ −224.8°, $[\alpha]_{313}$ −364.05°, $[\alpha]_{302}$ −378.8°, $[\alpha]_{297}$ −424.5°, $[\alpha]_{289}$ −488.8°. Found %: C 37.69, 37.60; H 3.52, 3.49. $C_8H_8Cl_2O_5$. Calculated %: C 37.67; H 3.16.

Similarly from 2 g of salt B was isolated 1 g of the other antipode of dichloroacid (Vc) with m.p. 154-155°; $[\alpha]_{592.5}$ + 66°, $[\alpha]_{590}$ + 67.33°, $[\alpha]_{582}$ + 76.17°, $[\alpha]_{538}$ + 128.5°, $[\alpha]_{518}$ + 152.67°, $[\alpha]_{484}$ + 198.5°, $[\alpha]_{450}$ + 248°, $[\alpha]_{420}$ + 273.29°, $[\alpha]_{400}$ + 299.5°, $[\alpha]_{390}$ + 315.01°, $[\alpha]_{373}$ + 349.01°. Found %: C 37.90, 38.00; H 3.20, 3.28, $C_8H_8Cl_2O_5$. Calculated %: C 37.67; H 3.16.

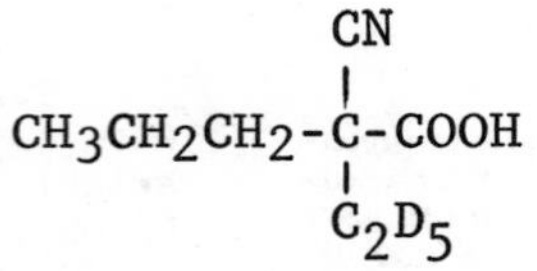

$C_8H_8D_5NO_2$ Ref. 369

Optisch aktive Pentadeuteroethyl-propyl-cyanessigsäuren

53 g (0,32 mol) rac. Pentadeuteroethyl-propyl-cyanessigsäure wurden mit 56 g (0,33 mol) (−)-Threobase unter Erwärmen in 600 ml Ethanol gelöst. Nach 24stdg. Stehen bei 8° wurden 43,5 g Kristalle erhalten, Schmp. 155°. 2malige Umkristallisation aus 100 ml Ethanol ergab 30 g Diastereomerensalz, Schmp. 160°. Weitere 2malige Umkristallisation ergab 26 g Diastereomerensalz, Schmp. 160–161°. 26 g Diastereomerensalz wurden mit 10proz. HCl zerlegt. Ausb.: 11 g (42 % d. Th.), Schmp. 28–32°, Sdp.$_{0,5}$ 130°. $[\alpha]_D^{20}$ = + 3,4°. $C_8H_8D_5NO_2$ (160,2) Ber.: C 60,0 H 8,18 N 8,7; Gef.: C 59,1 H 8,37 N 8,7.

24 g (0,15 mol) angereicherte (−)-Pentadeuteroethyl-propyl-cyanessigsäure wurden mit 25 g (0,15 mol) (+)-Threobase in 300 ml Ethanol gelöst. Nach 24stdg. Stehen bei 6° wurden 34 g Diastereomerensalz erhalten, Schmp. 154–157°. Weitere 2malige Umkristallisation ergab 30 g Diastereomerensalz, Schmp. 160–161°. 24 g Diastereomerensalz wurden mit 10proz. HCl zerlegt. Ausb.: 10,3 g (43 % d. Th.), Sdp.$_{0,4}$ 130°, n_D^{20} = 1,4400, $[\alpha]_D^{20}$ = − 3,5°. $C_8H_8D_5NO_2$ (160,2) Ber.: C 60,0 H 8,18 N 8,7; Gef.: C 59,6 H 8,45 N 8,7.

Die IR-Spektren der deuterierten Cyanessigester und -säuren zeigen drei gegenüber den entsprechenden C-H-Schwingungen um etwa 800 cm^{-1} verschobene C-D-Valenzschwingungen (vgl.[7]).
C-H = 2955 (s), 2930 (m), 2870 (w) cm^{-1}
C-D = 2220 (m), 2107 (w), 2075 (w) cm^{-1}.

7 J.F. Thomson, Biological Effects of Deuterium, S. 23, Pergamon Press, London 1963.

Note: (-)-Threobase = (-)1R/2R-2-Amino-1-phenyl-propandiol(1,3)

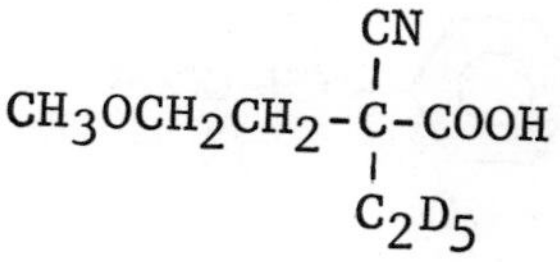

$C_8H_8D_5NO_3$

$$CH_3OCH_2CH_2-\underset{\underset{C_2D_5}{|}}{\overset{\overset{CN}{|}}{C}}-COOH$$

Ref. 370

Die Racematspaltung erfolgte mit den Enantiomeren der Threobase. Ein Teil rac. 7 wurde mit 0,9 Teilen Spaltbase in wenig MeOH gelöst, bis zur ersten bleibenden Trübung mit Ether versetzt und 2 d bei 14° der Kristallisation überlassen. Dieser Vorgang wurde bis zur Schmelzpunktkonstanz des isolierten Salzes wiederholt. Die isolierten Diastereomerensalze hatten einen Schmp. von 121–123°. 26 g Diastereomerensalz wurden mit 10proz. HCl zerlegt. Ausb. 13,5 g (45 % d. Th.) (+)-7. $C_8H_8D_5NO_3$ (176,2) $[\alpha]_D^{20} = +10,9°$ (c = 10,5).

19 g Diastereomerensalz wurden mit 10proz. HCl zerlegt. Ausb. 9,6 g (32 % d. Th.) (−)-7. $C_8H_8D_5NO_3$ (176,2) $[\alpha]_D^{20} = -10,4°$ (c = 3,34).

Note: Threobase = (−)1R/2R-2-Amino-1-phenyl-propandiol (1,3)

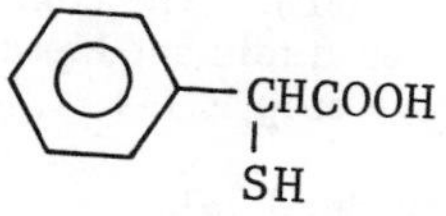

$C_8H_8O_2S$

Ref. 371

Resolution of α-Mercaptophenylacetic Acid.—The above racemic α-mercapto acid (13.35 g) was dissolved in acetone, as was 23.30 g of cinchonidine. The solutions were mixed and diluted to 150 ml with acetone, seeded with material from a small-scale resolution, and allowed to stand 2 hr. The salt obtained (a), 11.38 g, had mp 162.5° and $[\alpha]^{25}D$ −91.0° (c 1.3, 95% ethanol). Only a small amount of additional material (0.22 g, mp 158–161°) resulted on placing the mother liquors at 0° overnight. The mother liquors were accordingly evaporated to dryness and the residue was redissolved in 45 ml acetone, seeded, and chilled. This produced 5.10 g of additional salt (b): mp 147–153° (softening from 70°); $[\alpha]^{25}D$ −81.7° (c 1.0, 95% ethanol). The first product (a) was recrystallized at 0° from a mixture of ethanol (25 ml) and acetone (150 ml) to yield 4.64 g of solid (c): mp 164–165.5°; $[\alpha]^{25}D$ −99.6° (c 0.7, 95% ethanol). Product b was recrystallized from acetone (25 ml) to produce 2.40 g of d: mp 154.5–156.5°; $[\alpha]^{25}D$ −92.9° (c 0.7, 95% ethanol). The mother liquors from c were evaporated to dryness, the residue was combined with d, and the mixture was recrystallized from ethanol (6 ml) and acetone (50 ml), affording 5.04 g of e: mp 159.5–162°; $[\alpha]^{25}D$ −90.0° (c 1.5, 95% ethanol). The latter was again recrystallized from ethanol (4 ml) and acetone (25 ml) to give 2.33 g of f: mp 168°; $[\alpha]^{25}D$ −98.1° (c 1.0, 95% ethanol). Crops c and f were combined and recrystallized from ethanol (8 ml) and acetone (15 ml) to obtain 4.80 g of g: mp 171°; $[\alpha]^{25}D$ −101° (c 0.7, 95% ethanol). In view of the little change in rotation during the last recrystallization, it was (erroneously) concluded that the final cinchonidine salt (g) was practially optically pure.

Anal. Calcd for $C_{27}H_{30}N_2O_3S$: C, 70.11; H, 6.54; N, 6.06; S, 6.92. Found: C, 70.25; H, 6.64; N, 6.10; S, 6.91.

The salt g (4.4 g) was dissolved in a little warm ethanol and the solution was added to an excess of 10% sulfuric acid. The mixture was extracted thrice with ether, and the extract was washed with water, dried, and stripped of solvent. There resulted 1.64 g (102%) of thin syrup which soon crystallized as long needles: mp 76–83°; $[\alpha]^{25}D$ −81.0° (c 1.3, 95% ethanol). These were recrystallized from a mixture of benzene (2 ml) and hexane (6 ml) to produce 1.00 g of product having mp 87–87.5° and $[\alpha]^{25}D$ −104.8° (c 0.5, 95% ethanol). Another recrystallization from benzene (1.3 ml) and hexane (5 ml) gave 0.90 g of (−)-α-mercaptophenylacetic acid having mp 88–88.5 and $[\alpha]^{25}D$ −106.2° (c 0.5, 95% ethanol), −93.6° (c 0.9, ether).

Anal. Calcd for $C_8H_8O_2S$: C, 57.12; H, 4.79; S, 19.06. Found: C, 57.46; H, 4.86; S, 19.17.

While the above results indicate that the original cinchonidine salt (g) was not optically pure, the reasonably similar melting point and rotation before and after the final recrystallization of the recovered acid led us, again erroneously, to assume that optical purity had been substantially achieved. The infrared spectrum (CHCl₃ solution) of the final (−)-α-mercaptophenylacetic acid and its racemic oily precursor were identical in all respects. The structure of the resolved acid was confirmed by its integrated nmr spectrum (CDCl₃, tetramethylsilane internal standard, Varian A60 nmr spectrometer). The single thiol proton appeared as a doublet centered at δ 2.60 (J = 8 Hz); the lone α proton appeared as a doublet centered at 4.68 (J = 8 Hz); the five aromatic protons appeared as a complex multiplet centered around 7.37; and the single carboxyl proton appeared as a singlet at 11.47.

The original cinchonidine salt mother liquors from the above resolution were evaporated to dryness and the residue (19.95 g) was decomposed with dilute sulfuric acid as above. Similar processing yielded 6.72 g (92.6%) of a semisolid paste. This was slurried with benzene and filtered. The insoluble residue (2.03 g) had mp 217–218° and $[\alpha]^{25}D$ 90.8° (c 1.2; acetone). The residue from evaporation of the filtrate was 4.51 g of amber oil which gradually crystallized, $[\alpha]^{25}D$ +8.6° (c 4.4; 95% ethanol). The insoluble residue proved to be the dextrorotatory form of the disulfide VIII, of unknown optical purity. Its recrystallization from acetone–benzene did not lead to optical enrichment but yielded 0.59 g: mp 220–221° dec; $[\alpha]^{25}D$ +75.3° (c 0.76, acetone).

Anal. Calcd for $C_{16}H_{14}O_4S_2$: C, 57.46; H, 4.22; S, 19.18. Found: C, 57.67; H, 4.34; S, 19.15.

The decomposition with acid of several other intermediate fractions of cinchonidine salts from the above resolution similarly led to residues from which high melting (209–215°) optically active products, presumably also optically impure analogous disulfides, were isolated.

At the outset, resolution of the above racemic α-mercaptophenylacetic acid was attempted with both α-phenylethylamine and with (+)-α-(1-naphthyl)ethylamine. In each case crystalline salts were readily obtained, but their recrystallization led to little apparent optical enrichment. Acid decomposition of the thrice recrystallized salt from the latter amine gave an optically inactive sample of recovered acid which again contained about 10% of the disulfide VIII.

Note:

Resolution was also attempted with (−)-α-phenylethylamine and (+)-α-(1-naphthyl)ethylamine but were unsuccessful.

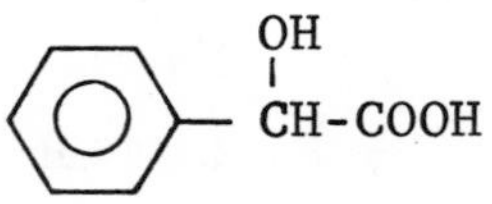

$C_8H_8O_3$ Ref. 372

The Resolution of r-*Mandelic Acid with* (—)*Ephedrine*. By ROBERT ROGER.

r-MANDELIC acid was resolved by Manske and Johnson (*J. Amer. Chem. Soc.*, 1929, **51**, 1909) and Skita, Keil, and Meiner (*Ber.*, 1933, **66**, 974) by means of both optically active forms of ephedrine. By the following process the resolution has been achieved with (—)ephedrine alone.

r-Mandelic acid (50 g. in 60 c.c. of absolute alcohol) was added slowly to (—)ephedrine (50 g. in 90 c.c. of absolute alcohol). The solution was gently warmed for 2 hours and then cooled in the ice-chest. The crystalline complex was collected, washed with fresh alcohol (30 c.c.), boiled with alcohol (75 c.c.), collected when cold, and decomposed with hydrochloric acid; the mandelic acid was then extracted (crop A, 22 g., $[\alpha]_{5461}$ — 177° in acetone). (—)-Ephedrine (40 g.) was dissolved in the original resolution liquor and to it was added *r*-mandelic acid (40 g.) dissolved in the alcoholic washings from the previous stage. After gentle warming for an hour, the complex was separated and treated as before (crop B, 20 g., $[\alpha]_{5461}$ — 147·4°). This process was repeated with a further 40 g. of (—)ephedrine and 40 g. of *r*-mandelic acid, giving crop C (24 g., $[\alpha]_{5461}$ — 78·3°). The final liquors from this series of operations were concentrated under diminished pressure to half the volume; on cooling, a solid separated, which was collected, washed with alcohol, and decomposed (crop D, 12 g., $[\alpha]_{5461}$ + 90°). The combined filtrate and washings were finally completely denuded of alcohol under diminished pressure at as low a temperature as possible; decomposition of the residue gave crop E (44 g., $[\alpha]_{5461}$ + 160°).

Crop A, recrystallised from acetone–benzene, gave 15 g. of (—)mandelic acid, $[\alpha]_{5461}$ — 189·8°. Similarly, crop E after recrystallisation gave 30 g. of (+)mandelic acid, $[\alpha]_{5461}$ + 189·1°. Crops B and C were united and quickly combined with fresh (—)ephedrine; a crop of lævo-acid (30 g.) was then obtained, $[\alpha]_{5461}$ — 182° (the rotatory power of pure lævorotatory mandelic acid in acetone is $[\alpha]^{20°}_{5461}$ — 189·9°; Roger, J., 1932, 2168).

130 G. of *r*-mandelic acid thus gave 45 g. of the lævorotatory acid and 30 g. of the dextrorotatory acid; 124 g. of ephedrine were recovered.

$C_8H_8O_3$ Ref. 372a

Preparation of Cinchonine Mandelates.—To (±) mandelic acid (50 g.; 1 mol.) dissolved in acetone (100 ml.) added a solution of cinchonine (125 g.; 1·25 mol.) in chloroform (150 ml.) and warmed. The clear solution on standing for one hour started depositing the less soluble cinchonine mandelate, and the whole of it separated after keeping overnight (74·5 g.; m.p. 173–75°). After three crystallisations from chloroform-acetone (1 : 2) it had m.p. 179–80° (71·25 g.; 95%).

The filtrate did not deposit any more salt even after a week. The mother liquors were combined with the filtrate and concentrated to 100 ml. and mixed with acetone (50 ml.). After keeping overnight it started depositing the more soluble cinchonine mandelate and continued depositing the salt for four days until a viscous solution was left behind which did not give any crystals even after keeping for a month. The more soluble cinchonine mandelate (73·6 g.; m.p. 158–61°) after three crystallisations from acetone had m.p. 165–66° (59·8 g.; 79·8%).

Wood, Chrisman and Nicholas[4] : m.p. 176–77°; 165·66°. McKenzie[2] : m.p. 79–80°; 165–66°.

Liberation of Optically Active Mandelic Acids.—The less soluble cinchonine mandelate (70 g.) was made into paste with acetone (15 ml.) and decomposed with ice-cold hydrochloric acid (1 : 1) until the resulting solution was just acidic to congo red. Diluted the solution to one litre and extracted twice with ether (750 ml.). Removed cinchonine from aqueous solution with ammonia and extracted the aqueous filtrate again with ether (750 ml.) after acidifying with dilute hydrochloric acid.

The combined ether extracts were washed with water, dried and the solvent distilled off. (—) Mandelic acid (22·4 g), m.p. 124–29°, after three crystallisations from acetone-benzene had m.p. 133–34° (16·3 g.; 69·9%).

The mother liquors on evaporation left behind (±) mandelic acid (7·32 g.), m.p. 118° C.

Similarly the more soluble cinchonine mandelate (58·1 g.) yielded (+) mandelic acid (12·8 g.; 65·3%), m.p. 132–23° $(\alpha)^{alc}_{D}$ + 156° (1 : 1) and (±) mandelic acid (6·2 g.; m.p. 118° $(\alpha)^{alc}_{D}$ ± 0·00.

McKenzie[2] $(\alpha)^{H_2O}_{D}$ — 159·7° Ward Chrisman and Nicholas[4] $(\alpha)^{H_2O}_{D}$ — 157·2° (*C* 2·01, *l* 2); — 143·3°.

1. Lewkowitsch, *Ber.*, 1883, **16**, 1569.
2. McKenzie, *Jour. Chem. Soc.*, 1904, **85**, 380.
3. Rimbach, *Ber.*, 1889, **32**, 2385.
4. Wood, Chrisman and Nicholas, *Jour. Chem. Soc.*, 1928, **131**, 2180.
5. Levene and Co-workers, *J. Biol. Chem.*, 1932, **97**, 379; 1935, **112**, 195; 1937, **118**, 315.
6. Richard and Johnson, *Jour. Amer. Chem. Soc.*, **1929**, **51**, 1906, 1909.
7. Skita, Keil add Meiner, *Ber.*, 1933, **66**, 974.

See also L. Smith, J. Prakt. Chem., 84, 743 (1911).

Note:

The above references contain resolutions of mandelic acid using various resolving agents.

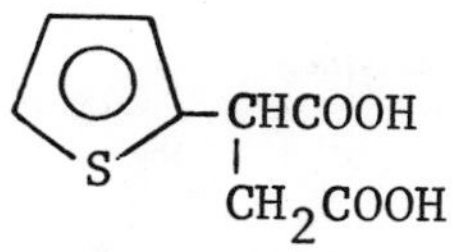

$C_8H_8O_4S$

Ref. 373

Preliminary tests on resolution: 0.5 g (0.0025 moles) of optically inactive acid and 0.005 ›les of the bases were dissolved in 15 ml of *ethanol*. The solutions were allowed to stand, d some water gradually added until crystals or oils were formed. The crystals were collected, e acid isolated and the optical activity measured in absolute ethanol.

Brucine in dilute ethanol: $[\alpha]_D^{25} = -34°$. The activity increases by about 10° with each re- ystallisation.

Cinchonidine in dilute ethanol: $[\alpha]_D^{25} = +72°$.

Strychnine in dilute ethanol: $[\alpha]_D^{25} = +23°$.

Quinine in dilute ethanol gave a practically inactive acid.

Quinidine in dilute ethanol yielded a precipitate, which was shown to be the free alkaloid.

l-Ephedrine in dilute ethanol gave an oil.

0.2 g (0.001 moles) of the optically inactive acid and 0.002 moles of the bases were dissolved 100 ml of *ethyl acetate* and a little *methanol*. The solutions were allowed to cool, and the ›cipitates were handled as described above.

(+)-α-*Phenylethylamine* yielded an acid with $[\alpha]_D^{25} = +48°$.

(+)-*Benzedrine* (21) yielded an acid with $[\alpha]_D^{25} = +37°$.

Resolution of 2-thienylsuccinic acid.

20 g (0.10 moles) of 2-thienylsuccinic acid and 60 g (0.20 moles) of cinchonidine re dissolved in 400 ml of hot ethanol. 400 ml of water was added and the solution t at room temperature for 24 hours. The precipitate was filtered off, washed with d dilute ethanol, and dried. About 1 g of the salt was set apart, the acid isolated d the optical activity measured in absolute ethanol. The remaining salt was rystallised from dilute alcohol, and the process repeated until the optical activity the acid remained constant. The course of the resolution is seen below:

Crystallisation	1.	2.	3.
Ml of solvent	800	500	400
g salt obtained	41	30	25
$[\alpha]_D^{25}$ of acid	+80°	+100°	+100°

Cinchonidine (+)-2-thienylsuccinate crystallises in threadlike needles, which form ge, hemispherical crystal aggregates. In air-dry condition it is free from crystal vent.

165.2 mg of the product gave 10.10 ml of nitrogen of 750.0 mm Hg at a temperature of 20.8°.
$C_{46}H_{52}O_6N_4S$ N calc. 7.10 % found 7.06 %
73.21 mg in 10.00 ml of ethanol. $2\ \alpha_D^{25} = -1.083$, $[\alpha]_D^{25} = -73.9°$.

The mother liquor from the first crystallisation with cinchonidine was evaporated *vacuo* at room temperature. The acid was liberated and 5.0 g was used for the isola- n of the (−)-antipode. The acid, which showed an optical activity of about 30°, and 6.0 g of (−)-α-phenylethylamine were dissolved in a mixture of 300 ml ethyl acetate and 50 ml of methanol, and left at room temperature for 24 hours. e crystals formed were filtered off and washed with cold ethyl acetate. The progress the resolution was as described above.

stallisation	1.	2.	3.	4.	5.
of ethyl acetate	300	200	130	90	60
of ethanol	50	50	50	30	25
salt obtained	6.5	4.2	3.0	2.0	1.2
$_D^{25}$ of acid	−78°	−91°	−96°	−99°	−98°

−)-α-*Phenylethylamine (−)-2-thienylsuccinate* crystallises in small colourless nee- s. In air-dry condition it is free from crystal solvent.

114.2 mg of the product gave 6.26 ml of nitrogen of 750.0 mm Hg at a temperature of 20.2°.
$C_{24}H_{30}O_4N_2S$ N calc. 6.33 % found 6.35 %
16.01 mg in 10.00 ml of water $2\ \alpha_D^{25} = -0.628°$, $[\alpha]_D^{25} = -19.6°$.

+)-*2-Thienylsuccinic acid:* 24 g of the cinchonidine salt were decomposed with sulphuric acid. The solution was extracted several times with ether, the ether ers then combined, and the ether removed. 5.5 g of crude acid was obtained and rystallised several times from water. Sometimes the acid crystallised in transparent s and sometimes as a colourless, granular powder. The acid was dried in air and desiccator over phosphorous pentoxide.

transparent rod of the acid was placed in a capillary tube and heated. Between °–130° it lost its transparency without changing its shape. This clearly indicates ransformation to another modification, which has also been verified by X-ray vder analysis. The acid melts at 158°–162° (dec.). The comparatively large melting- nt interval is not due to the presence of impurities, but to the thermal instability he acid or to the existence of polymorphic modifications.

25.75 mg of acid (rods), 7.95 ml of 0.03215-N NaOH
44.95 mg of acid (powder), 6.46 ml of 0.06880-N NaOH
$C_8H_8O_4S$. Equiv. wt. calc. 100.0, found rods 100.7, powder 99.9
129.9 mg in 10.00 ml of ethanol. $2\ \alpha_D^{25} = +2.690°$, $[\alpha]_D^{25} = +103.5°$ $[M]_D^{25} = +207.0°$.

(−)-*2-Thienylsuccinic acid:* 1.2 g of (−)-α-phenylethylamine (−)-2-thienylsuccinate was decomposed with 2-N sulphuric acid. The solution was extracted several times with ether, the ether removed, and the acid recrystallised and dried as described for the (+)-acid, which it resembles in all respects.

26.11 mg of acid 8.12 ml of 0.03215-N NaOH
$C_8H_8O_4S$. Equiv. wt. calc. 100.0, found 100.0
100.3 mg in 10.00 ml of absolute ethanol. $2\ \alpha_D^{25} = -2.095°$
$[\alpha]_D^{25} = -103.9°$. $[M]_D^{25} = -207.8°$.

Author's comments:

The primary alkaloid salts of the acid are apparently too soluble in the usual solvents to be employed in resolving the acid into the optical antipodes. Several crystallised secondary salts have been obtained, however, some of which possess strong resolving power. Cinchonidine in dilute ethanol and (−)-α-phenylethylamine in ethyl acetate give satisfactory yields of the (+)- and (−)-2-thienylsuccinic acids respectively.

(+)-*2-Thienylsuccinic acid* is obtained in a state of purity by recrystallisation from water of the crude acid liberated from the cinchonidine salt. It crystallises in colourless, transparent needles or rods, which may attain a length of several centimeters if the solvent is slowly evaporated. If a hot solution of the acid is allowed to cool rapidly, the acid is obtained as a granular, colourless powder. A similar behaviour of optically active phenylsuccinic acid is reported by Wren and Williams (12), who also found an instable modification, changing to the stable modification on standing in water for a night in a refrigerator. Investigations in a hot stage microscope lead Fredga and Matell to assume optically active phenylsuccinic acid to be dimorphic (11). Since it was possible, that (+)-2-thienylsuccinic acid might likewise exist in two modifications, the two above-mentioned preparations were examined with X-rays. The X-ray powder photographs were identical, however. Now, if this modification is heated above 125°, it changes to another modification, without melting or racemisation (13), indicating the acid to be at least dimorphic. Later it will also be shown that the "low temperature" modifications of the optically active 2-thienylsuccinic and phenylsuccinic acids are isomorphic (13).

(+)-Thienylsuccinic acid melts at 158°–162°, with decomposition. As in the case of the optically inactive acid a gas is evolved, indicating anhydride formation, but in addition racemisation occurs.

(−)-*Thienylsuccinic acid* is obtained in a state of purity by recrystallisation from water of the crude acid liberated from the (−)-α-phenylethylamine salt. It resembles its optical antipode in all respects. Melting point 158°–162° (dec).

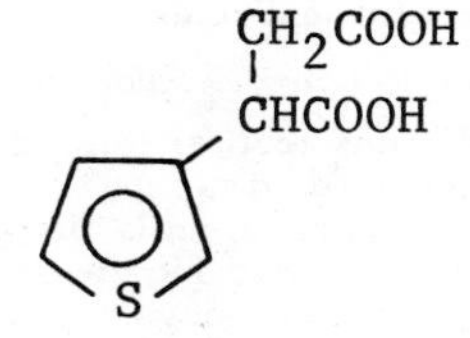

$C_8H_8O_4S$

Ref. 374

Preliminary test on resolution.—0.0025 moles of optically inactive acid and 0.005 moles of the base were dissolved in 25 ml of dilute ethanol. The solutions were allowed to stand in the refrigerator and some water gradually added until crystals were formed. Experiments with (+)-phenylethylamine were made in ethyl acetate. The results are given below.

Base	$[\alpha]_D^{25}$ of acid in acetone
Brucine	+10.5°
Strychnine	+17.4°
Cinchonidine	+82.7°
Quinine	+6.5°
(+)-α-phenylethylamine	+14.0°

Optical resolution of 3-thienylsuccinic acid.—20 g (0.1 moles) of 3-thienylsuccinic acid (8) and 60 g (0.2 moles) cinchonidine were dissolved in 800 ml 50% ethanol and the solution left in a refrigerator for 24 hours. The precipitate was filtered off, washed with cold dilute ethanol and dried. About 0.3 g of the salt was set apart, the acid isolated and the optical activity measured in ethanol. The remaining salt was recrystallized from dilute ethanol and the process repeated until the optical activity of the acid remained constant. The course of the resolution is seen below.

Crystallization	1	2	3	4	5
g salt	42	26	19	14.5	8
ml 50% ethanol	800	700	400	350	300
$[\alpha]_D^{25}$ of acid	+65°	+92°	+102°	+103°	+103°

The cinchonidine salt obtained in the course of resolution as needle bundles, was not further examined.

The mother liquor from the first crystallization with cinchonidine was evaporated at room temperature and 38 g salt was obtained. This salt yielded 9 g acid, having $[\alpha]_D^{25} = -62°$ in ethanol. This acid was dissolved in a mixture of 540 ml ethyl acetate and 90 ml methanol and 10.9 g of (−)-α-phenylethylamine was added. The progress of the resolution is seen below.

Crystallization	1	2	3	4	5	6
ml, ethylacetate	540	550	440	220	150	105
ml, methanol	90	340	270	135	90	65
g salt	17	11.5	7.2	4.9	3.4	2.4
$[\alpha]_D^{25}$ of acid	−78°	−92°	−99°	−101°	−102°	−102°

The phenylethylamine salt obtained in the course of the resolution was not further examined.

(+)-*3-Thienylsuccinic acid.*—The cinchonidine salt (7.7 g) was decomposed with 2 *N* sulphuric acid and the active acid extracted with ether. The ether was evaporated at room temperature and the acid was recrystallized two times from water. The yield of pure acid was 1.5 g and the m.p. 170–173°. The acid crystallized in colourless plates.

0.06562 g : 8.90 ml 0.0734 *N* NaOH. $C_8H_8O_4S$ (200.0). Equiv. wt. calc. 100.0, found 100.5.

Weighed amounts of acid were made up to 10.00 ml with different solvents and the optical activity measured. In the case of water the acid was neutralized with 0.1 *N* NaOH and made up to 10.00 ml.

The results are given below.

Solvent	mg of acid	$2[\alpha]_D^{25}$	$[\alpha]_D^{25}$	$[M]_D^{25}$
Ethanol	97.00	+ 1.985°	+ 102.3°	+ 204.6°
Ethyl acetate	92.45	+ 2.38°	+ 128.8°	+ 257.6°
Acetone	79.88	+ 1.98°	+ 123.7°	+ 247.4°
Acetic acid	87.25	+ 1.965°	+ 112.6°	+ 225.2°
Water (ion) ;	67.60	+ 0.545°	+ 40.3°	+ 80.6°

(−)-*3-Thienylsuccinic acid.*—2.1 g of the (−)-phenylethylamine salt was decomposed with 2 *N* sulphuric acid and the active acid extracted with ether. The ether was evaporated and the acid recrystallized in the same way as the antipode which it resembles in appearance and properties. The yield was 0.85 g having m.p. 170–173°.

65.29 mg: 8.84 ml 0.0734 *N* NaOH, $C_8H_8O_4S$ (200.0). Equiv. wt. calc. 100.0, found 100.6. 0.10363 g made up to 10.00 ml with ethanol. $2[\alpha]_D^{25} = -2.125$, $[\alpha]_D^{25} = -102.5°$.

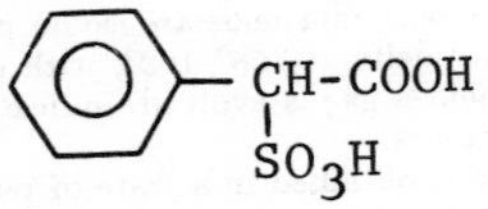

$C_8H_8O_5S$ **Ref. 375**

V. **Les composants actifs de l'acide phénylsulfoacétique.**

Séparation du composant lévogyre à l'aide de brucine.

On fait cristalliser à plusieurs reprises le phénylsulfoacétate acide de brucine dans dix fois son poids d'eau.

Chaque fois on examine l'eau-mère, en la décomposant à l'ammoniaque, en éliminant la brucine par filtrage et par extraction au chloroforme et en mesurant enfin le pouvoir rotatoire de la solution du sel ammoniacal. La concentration de l'eau-mère est fixée par un titrage à la baryte en présence de chloroforme. Dix-huit cristallisations ont été nécessaires pour que l'eau-mère et les cristaux présentassent la même rotation moléculaire. Il s'en suit que la différence entre les solubilités des sels de brucine des deux énantiomorfes est très faible.

Voici les rotations moléculaires des eaux-mères successives, mesurées à l'état de sel ammoniacal:

+90°, +73, +68, +66, +63, + 54, +50, 0, −29.6, −30, −63, −79, −97, −120, −145, −151, −150, −151°.

Il va sans dire qu'on ne doit pas s'attendre à une série régulière, puisque la durée de la cristallisation n'est pas toujours la même.

On obtient l'acide lévogyre libre en décomposant le sel de brucine prudemment à froid avec une solution de baryte, en extrayant la brucine au chloroforme et en précipitant le baryum par de l'acide sulfurique.

La solution donne par évaporation dans le vide sulfurique l'*acide phénylsulfoacétique lévogyre* en fines aiguilles, se décomposant vers 225°.

Lorsqu'on fait réagir à froid en solution étendue le phénylsulfoacétate de sodium et l'acétate de brucine, il se dépose d'abord la combinaison de brucine avec l'acide sulfonique dextrogyre. Dans une expérience le sel ammoniacal, obtenu par décomposition de ce sel de brucine, a montré la rotation moléculaire de +26°.

Les tableaux suivants donnent pour l'acide et quelques sels les valeurs de la rotation observée et de la rotation moléculaire. Les mesures ont été faites à la température ordinaire (environ 18°), à l'aide de tubes de 2 dm. de longueur.

L'acide l-phénylsulfoacétique.

Concentration: a. 0.1628 n., b. 0.0814 n., c. 0.0407 n.; l = 2 dm. température 18° C.

	λ (U.A°.)	6420	6210	5890	5500	5335	5050
a.	α	− 0°.67	− 0.74	− 0.84	− 0.97	− 1.03	− 1.20
,,	[M]	−41°.2	−45.5	−51.6	−59.6	−63.3	−73.7
b.	α	− 0°.34	− 0.37	− 0.42	− 0.49	− 0.53	− 0.61
,,	[M]	−41°.8	−45.5	−51.6	−60.2	−65.1	−74.9
c.	α	− 0°.17	− 0.19	− 0.21	− 0.24	− 0.26	− 0.29
,,	[M]	−41°.8	−46.7	−51.6	−59.0	−63.9	−71.3

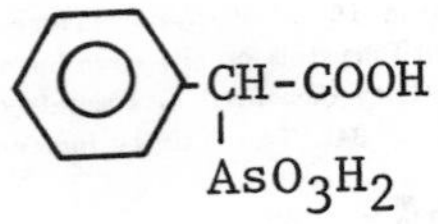

$C_8H_9AsO_5$ **Ref. 376**

8. *Acide phénylarsono-acétique.* Le phénylarsono-acétate de potassium et le chlorhydrure de quinine donnent le sel de quinine en cristaux feutrés renfermant 6 mol. d'eau.

Substance 0.3742 g; perte à 90° 0.0405 g
,, 0.2893 g; $Mg_2As_2O_7$ 0.0462 g

$C_8H_9O_5As \cdot 2C_{20}H_{24}O_2N_2 \cdot 6H_2O$ (1016.54). Trouvé: H_2O 10.82; As 7.71
Calculé: ,, 10.63; ,, 7.37

On a fait recristalliser le sel quatre fois dans l'alcool dilué; la quantité a diminué de 42 à 12 g. En décomposant le sel de quinine par de la soude caustique et en éliminant la quinine par extraction au chloroforme, on a obtenu le sel sodique secondaire.

$[M]_D$ sel = +4.2°. $[M]_D$ acide = +3.1°.

Dispersion rotatoire du sel sodique secondaire.

λ (μμ)	626.5	589.3	560.5	536.5	516.0
[M]	+ 3.7°	+ 4.2°	+ 4.8°	+ 5.3°	+ 5.9°

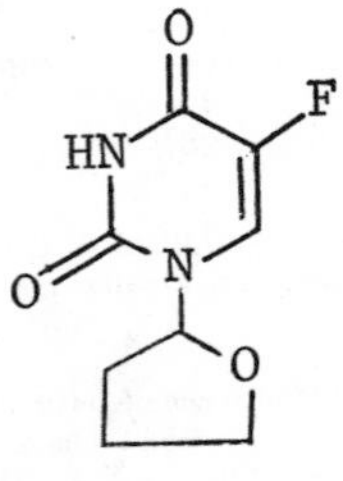

$C_8H_9FN_2O_3$ **Ref. 377**

Optical Resolution of Thf-FU (3). (1) Formation of Diastereoisomers with Brucine. A saturated solution of the Thf-FU (10 g, 0.05 mol) in EtOH at 70 °C and a saturated solution of brucine dihydrate (21.5 g, 0.05 mol) in EtOH at 70 °C were mixed and the resulting solution stood at room temperature until crystallization was complete. The solid was recrystallized from EtOH at 70 °C to give 12.5 g (42.1%) of a pure diastereoisomer (4): mp 187–188 °C; $[\alpha]^{23}{}_D$ −46.0° (*c* 0.5, $CHCl_3$). Anal. ($C_{31}H_{35}FN_4O_7$) C, H, N. Concentration of the mother liquor gave a solid, which was recrystallized from EtOH at 70 °C to give 10.7 g (36.0%) of another pure diastereoisomer (5): mp 162.5–164 °C; $[\alpha]^{23}{}_D$ −89.6° (*c* 0.5, $CHCl_3$). Anal. ($C_{31}H_{35}FN_4O_7$) C, H, N.

(2) Recovery of *R* and *S* Isomers from the Diastereoisomers. (a) Treatment with Silica Gel. The crystals (4, 11.9 g) were dissolved in $CHCl_3$–EtOH (5:1, v/v) and applied to a column of silica gel (500 g) equilibrated with the same solvent. Elution with the same solvent gave 3.8 g (95%) of the *R*-(+) isomer of Thf-FU (3a): mp 174–175.5 °C; $[\alpha]^{23}{}_D$ +70.0°, $[\alpha]^{23}{}_{436}$ +182.0° (*c* 0.5, $CHCl_3$); UV $\lambda_{max}{}^{pH2}$ 271 nm (ε 9100); UV $\lambda_{max}{}^{pH7}$ 270 nm (ε 8800); UV $\lambda_{max}{}^{pH12}$ 270 nm (ε 7000); ^1H NMR (pyridine-d_5) δ 3.73 and 4.10 [m(2), 2, $C_{4'H}$], 6.13 (qd, 1, $C_{1'H}$), 7.72 (d, 1, C_{6H}, J_{H-F} = 7.0 Hz). Anal. ($C_8H_9FN_2O_3$) C, H, N. The same treatment of crystals of 5 gave 3.7 g (92.5%) of the *S*-(−) isomer of Thf-FU (3b): mp 175–177 °C; $[\alpha]^{23}{}_D$ −70.0°, $[\alpha]^{23}{}_{436}$ −187.0° (*c* 0.5, $CHCl_3$); UV $\lambda_{max}{}^{pH2}$ 271 nm (ε 9000); UV $\lambda_{max}{}^{pH7}$ 270 nm (ε 8700); UV $\lambda_{max}{}^{pH12}$ 270 nm (ε 6900); ^1H NMR (pyridine-d_5) δ 3.73 and 4.10 [m(2), 2, $C_{4'H}$], 6.12 (qd, 1, $C_{1'H}$), 7.72 (d, 1, C_{6H}, J_{H-F} = 7.0 Hz). Anal. ($C_8H_9FN_2O_3$) C, H, N.

(b) Treatment with Dilute HCl. A solution of the crystals

(4, 5.9 g) in EtOH (150 mL) was mixed with dilute HCl keeping the pH at 3.5–4.5 at room temperature. The mixture was stirred for 1 h, then the solvent was removed in vacuo, and the residue was treated with water (30 mL)–$CHCl_3$ (90 mL). The $CHCl_3$ layer was separated, washed with water, and evaporated to dryness. Recrystallization of the residue from EtOH gave 1.8 g (90.1%) of the *R* isomer **3a**. The same treatment of crystals of **5** gave 1.7 g (85.6%) of the *S* isomer **3b**.

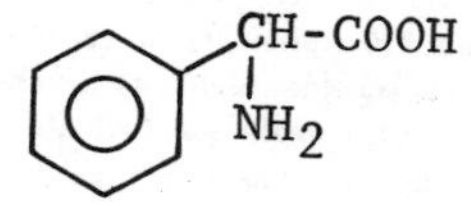

$C_8H_9NO_2$ Ref. 378

Aktive Formyl-Phenyl-aminoessigsäure.

70 g Racemverbindung werden mit 115 g Cinchonin (1 Mol.), beide fein gepulvert, in 2 l kochendem Wasser gelöst, und die Flüssigkeit wenn nötig filtriert. Beim Erkalten tritt bald Krystallisation ein, besonders wenn man mit einigen Kryställchen von einer früheren Darstellung impft; solche Impfkrystalle sind verhältnismäßig leicht zu beschaffen durch Verdunsten eines Teiles der wäßrigen Lösung im Vakuumexsiccator unter öfterem Reiben. Läßt man nach Beginn der Krystallisation die Flüssigkeit noch 12 Stunden unter zeitweisem Umschütteln stehen, so scheidet sich das Cinchoninsalz der *l*-Formyl-phenyl-aminoessigsäure zum größeren Teil ab. Die Ausbeute beträgt etwa 95 g. Das Produkt wird aus der 9-fachen Menge heißem Wasser umkrystallisiert. Zur völligen Reinigung genügt in der Regel nochmaliges Umkrystallisieren aus der 9-fachen Menge heißem Wasser. Die Ausbeute geht dabei erheblich herunter und beträgt schließlich 50—60 % der Theorie. Das Cinchoninsalz bildet schöne, farblose, manchmal zentimeterlange Krystalle, die prismatisch ausgebildet sind.

Zur Bereitung der freien

l-Formyl-Phenyl-aminoessigsäure

löst man 37 g Cinchoninsalz in 500 ccm heißem Wasser, versetzt noch warm mit 25 ccm *n*-Natronlauge, wobei schon ein Teil des Cinchonins ausfällt, kühlt jetzt rasch ab und gibt noch 57.6 ccm *n*-Natronlauge zu, so daß die Gesamtmenge des Alkalis wenig mehr als 1 Mol. beträgt. Nachdem das Gemisch in Eiswasser sorgfältig gekühlt ist, wird das Cinchonin scharf abgesaugt und mehrmals mit wenig eiskaltem Wasser gewaschen. Zum Filtrat fügt man 8 ccm *n*-Salzsäure, um den Überschuß an Alkali zu neutralisieren, verdampft dann die Lösung unter 10—20 mm Druck auf etwa 60 ccm und versetzt schließlich mit 37.3 ccm doppeltnormaler Salzsäure. Dabei fällt die aktive Formylverbindung sofort krystallinisch aus; sie wird nach 1-stündigem Stehen bei 0° abgesaugt und mit eiskaltem Wasser gewaschen. Die Ausbeute betrug 12.5 g oder 88% der Theorie. Zur völligen Reinigung wird aus etwa 10 Teilen kochendem Wasser umkrystallisiert. Zur Analyse wurde im Vakuumexsiccator getrocknet.

0.1238 g Sbst.: 0.2736 g CO_2, 0.0553 g H_2O. — 0.1120 g Sbst.: 7.41 ccm N (14.5°, 764 mm).

$C_9H_9O_3N$ (179.1). Ber. C 60.31, H 5.07, N 7.82.
Gef. » 60.27, » 5.00, » 7.75.

Die Verbindung schmilzt etwas höher als der Racemkörper, aber wegen der Zersetzung auch nicht konstant. Beim raschen Erhitzen beginnt sie gegen 187° zu sintern und schmilzt gegen 190° (korr.) unter Gasentwicklung und Gelbfärbung.

In siedendem Wasser ist sie leicht löslich, doch etwas schwerer als der Racemkörper, und krystallisiert daraus in mikroskopisch sehr dünnen, langgestreckten Platten, die vielfach verwachsen sind und manchmal wie Nadeln aussehen. Leicht wird sie ferner von Alkohol und Aceton aufgenommen und krystallisiert aus Alkohol in sehr feinen Nadeln, die oft zu Büscheln vereinigt sind.

Für die optische Bestimmung diente die alkoholische Lösung.

0.2555 g Sbst., Gesamtgewicht der Lösung 6.3777 g. Mithin Prozentgehalt 4.006. $d_4^{20} = 0.8048$. Drehung im 2-dm-Rohr bei 20° und Natriumlicht 16.76° nach links. Mithin $[\alpha]_D^{20} = -259.9°$.

Das für die folgende Bestimmung dienende Präparat war aus dem gleichen Cinchoninsalz bereitet, nachdem es noch zweimal aus Wasser umkrystallisiert war.

0.2566 g Sbst., Gesamtgewicht der Lösung 6.3321 g. Mithin Prozentgehalt 4.052. $d_4^{20} = 0.8053$. Drehung im 2-dm-Rohr bei 20° im Natriumlicht 16.95° nach links. Mithin $[\alpha]_D^{20} = -259.7°$.

Da das Drehungsvermögen hierbei nicht mehr erhöht wurde, so ist das öftere Umkrystallisieren des Cinchoninsalzes überflüssig. Ferner glauben wir, aus dem Resultat den Schluß ziehen zu dürfen, daß die Formylverbindung sehr wahrscheinlich optisch rein war.

d-Formyl-Phenyl-aminoessigsäure.

Für ihre Bereitung dienten die ersten beiden wäßrigen Mutterlaugen, die bei der Gewinnung des Cinchoninsalzes des optischen Antipoden resultieren. Man versetzt sie mit soviel *n*-Natronlauge, daß stark alkalische Reaktion eintritt, filtriert vom abgeschiedenen Cinchonin auf der Pumpe und fügt zum Filtrat eine dem Alkali entsprechende Menge *n*-Salzsäure. Zu dieser Flüssigkeit, die ungefähr 4 l beträgt, fügt man etwa 95 g gepulvertes Chinin und kocht, wobei Lösung eintritt. Bei mehrtägigem Stehen scheiden sich reichliche Mengen von Krystallen zugleich mit einem Öl aus. Man filtriert, preßt die Krystalle zwischen Fließpapier und krystallisiert sie aus nicht zu viel heißem Alkohol. Zur völligen Reinigung haben wir das Chininsalz aus etwa der 40-fachen Menge heißem Wasser umkrystallisiert, wobei ungefähr ¹/₆ verloren geht. Es bildet dann seideglänzende, farblose Nadeln.

Für die Rückverwandlung in den Formylkörper übergießt man 30 g Chininsalz mit 60 ccm kalter *n*-Natronlauge und fügt Äther hinzu. Beim kräftigen Umschütteln wird das ölig ausgeschiedene Chinin vom Äther leicht aufgenommen. Nachdem die ätherische Schicht abgetrennt und der Rest des Chinins durch abermaliges Ausäthern entfernt ist, versetzt man die alkalische Lösung mit 60 ccm *n*-Salzsäure. Die abgeschiedene Formylverbindung wird nach dem Abkühlen auf 0° abgesaugt und mit wenig kaltem Wasser gewaschen. Diese Isolierung aus dem Chininsalz ist nur mit geringem Verlust verbunden. Berechnet man die Ausbeute an Formyl-*d*-Verbindung auf den ursprünglich angewandten Racemkörper, so beträgt sie 35—40% der Theorie.

Für die Analyse und optische Bestimmung war nochmals aus heißem Wasser umgelöst und im Vakuumexsiccator getrocknet worden.

0.1324 g Sbst.: 0.2931 g CO_2, 0.0605 g H_2O. — 0.1153 g Sbst.: 8 ccm N (20°, 759.5 mm).

$C_9H_9O_3N$ (179.1). Ber. C 60.31, H 5.07, N 7.82
Gef. » 60.38, » 5.11, » 7.96.

Für die optische Bestimmung diente die alkoholische Lösung.

0.1642 g Sbst., Gesamtgewicht der Lösung 4.2658 g. Mithin Prozentgehalt 3.849. $d_4^{20} = 0.8045$. Drehung im 1-dm-Rohr bei 20° im Natriumlicht 8.03° nach rechts (± 0.015°). Mithin $[\alpha]_D^{20} = +259.3°$ (± 0.5°).

Für die zweite Bestimmung wurde ein Präparat benutzt, das aus demselben Chininsalz, aber erst nach abermaliger Krystallisation aus Wasser dargestellt war.

0.2331 g Sbst., Gesamtgewicht der Lösung 6.0223 g. Mithin Prozentgehalt 3.871. $d_4^{20} = 0.8044$. Drehung im 2-dm-Rohr bei 20° im Natriumlicht 16.16° nach rechts (± 0.04°). Mithin $[\alpha]_D^{20} = +259.45°$ (± 0.6°).

Wie man sieht, stimmen diese Werte sowohl unter einander, als auch mit dem Drehungsvermögen des Antipoden überein. Dasselbe gilt für Schmelzpunkt, Aussehen der Krystalle und Löslichkeit.

Aktive Phenyl-aminoessigsäuren.

Die Hydrolyse der Formylkörper geht sehr leicht von statten; es genügt, mit der zehnfachen Menge 10-prozentiger Bromwasserstoffsäure ¹/₂ Stunde am Rückflußkühler zu kochen. Die Lösung wird dann unter 10—15 mm Druck zur Trockne verdampft, wobei das Bromhydrat in schönen weißen Nadeln zurückbleibt. Man löst es in etwa 25 Teilen 50-prozentigen Alkohols und fügt Ammoniak in geringem Überschuß hinzu. Dann scheidet sich die Aminosäure in glänzenden Nadeln als dicker Brei ab. Sie wird abgesaugt und mit verdünntem Alkohol gewaschen. Die Ausbeute ist fast quantitativ.

Für die Analyse und optische Bestimmung diente ein Präparat, das aus der 70-fachen Menge kochendem Wasser umkrystallisiert und im Vakuumexsiccator über Schwefelsäure getrocknet war.

l-Phenyl-aminoessigsäure.

0.1735 g Sbst.: 0.4030 g CO_2, 0.0940 g H_2O. — 0.1110 g Sbst.: 9.1 ccm N (23°, 763 mm).

$C_8H_9O_2N$ (Mol.-Gew. 151.1). Ber. C 63.54, H 6.00, N 9.27.
Gef. » 63.35, » 6.06, » 9.33.

Die Aminosäure zeigt wie die meisten anderen Aminosäuren keinen konstanten Schmelzpunkt; im geschlossenen Capillarrohr fanden wir ihn je nach der Schnelligkeit des Erhitzens bei 305—310° (korr.), was mit der Angabe von Ehrlich, 303—305°, im wesentlichen übereinstimmt.

Sie ist in heißem Wasser erheblich leichter löslich als der Racemkörper, verlangt davon aber noch ungefähr 70 Teile. Für die Bestimmung der Löslichkeit in kaltem Wasser wurde die gepulverte Substanz bei 25° im Thermostaten 20 Stunden im Silberrohr mit Wasser geschüttelt: 1 Teil Aminosäure verlangt unter diesen Umständen 207.6 Teile Wasser von 25°.

In absolutem Alkohol ist sie so gut wie unlöslich, löst sich dagegen in kochendem 50-prozentigem Alkohol und krystallisiert daraus in kurzen, glänzenden Nadeln. Aus Wasser krystallisiert sie in langen, ganz schmalen und sehr dünnen Platten, die manchmal wie Nadeln aussehen und auch häufig zu ausgezackten Aggregaten vereinigt sind. In Alkali ist sie leicht löslich, nicht so leicht wird sie von verdünnter Salzsäure aufgenommen, schwer löslich in starker Salzsäure.

Für die Bestimmung des Drehungsvermögens diente eine Lösung in wenig mehr als der berechneten Menge Salzsäure von ungefähr 2.5 %.

0.7438 g Sbst. wurden gelöst in 6.7 ccm Normalsalzsäure und 3 ccm Wasser, Gesamtgewicht der Lösung 10.3400 g. Mithin Prozentgehalt 7.193. $d_4^{20} = 1.0286$. Drehung im 2-dm-Rohr bei 20° im Natriumlicht 23.35° nach links. Mithin $[\alpha]_D^{20} = -157.78°$ (± 0.5°).

Nach nochmaligem Umlösen der Aminosäure aus heißem Wasser wurde fast der gleiche Wert gefunden:

0.5387 g Sbst. wurden in 4.85 ccm Normalsalzsäure und 2.17 ccm Wasser gelöst; Gesamtgewicht der Lösung 7.6671 g. Mithin Prozentgehalt 7.026. $d_4^{20} = 1.0277$; Drehung im 2-dm-Rohr bei 20° im Natriumlicht 22.80° nach links. Mithin $[\alpha]_D^{20} = -157.87°$ (± 0.5°).

Um einen Vergleich mit dem von F. Ehrlich bei seinem besten Präparat gefundenen Wert zu haben, haben wir mit obiger Substanz einen dritten Versuch ausgeführt, wobei 10-prozentige Salzsäure zur Anwendung kam.

0.3032 g Sbst., Gesamtgewicht der Lösung 8.2277 g. Mithin Prozentgehalt 3.686. $d_4^{20} = 1.0586$. Drehung im 2-dm-Rohr bei 20° im Natriumlicht 12.91° nach links (± 0.03°). Mithin $[\alpha]_D^{20} = -165.43°$ (± 0.4°).

Man ersieht daraus, daß das Präparat von F. Ehrlich, der $[\alpha]_D^{20}$: − 144.83° fand, optisch noch nicht rein war, wie er schon selbst vermutete, sondern daß es mindestens noch 6.2 % des Antipoden enthielt. Ferner ist zu beachten, daß die spezifische Drehung durch die starke Salzsäure etwas erhöht wird.

Dem schon erwähnten, schön krystallisierten Bromhydrat ist das Chlorhydrat sehr ähnlich. Es läßt sich aus warmer, ganz verdünnter Salzsäure leicht umkrystallisieren und bildet lange, farblose Nadeln. Es löst sich leicht in warmem Wasser und Alkohol; aus letzterem wird es durch Äther gefällt. Im Capillarrohr schmilzt es nicht konstant gegen 246° (korr.) unter Gasentwicklung. Für die Analyse wurde im Vakuumexsiccator getrocknet:

d-Phenyl-aminoessigsäure.

Da die Eigenschaften dieselben sind wie beim Antipoden, so begnügen wir uns damit, die Resultate der Analyse und optischen Bestimmung anzuführen.

0.1686 g Sbst.: 0.3928 g CO_2, 0.0909 g H_2O. — 0.1129 g Sbst.: 9.2 ccm N (20.5°, 760 mm).

$C_8H_9O_2N$ (151.1). Ber. C 63.54, H 6.00, N 9.27.
Gef. » 63.54, » 6.03, » 9.34.

Optische Bestimmung: 0.5410 g Sbst. wurden in 4.85 ccm Normalsalzsäure und 2.17 ccm Wasser gelöst; Gesamtgewicht der Lösung 7.6217 g. Mithin Prozentgehalt 7.098. $d_4^{20} = 1.0281$. Drehung im 2-dm-Rohr bei 20° im Natriumlicht 23.04° nach rechts. Mithin $[\alpha]_D^{20} = +157.86°$ (± 0.5°).

Nach nochmaligem Umkrystallisieren der Aminosäure aus Wasser war die Drehung nicht geändert.

0.5427 g Sbst. wurden in 4.88 ccm Normalsalzsäure und 2.18 ccm Wasser gelöst; Gesamtgewicht der Lösung 7.6831 g. Mithin Prozentgehalt 7.064. $d_4^{20} = 1.0281$. Drehung im 2-dm-Rohr bei 20° und Natriumlicht 22.96° nach rechts. Mithin $[\alpha]_D^{20} = +158.09°$ (± 0.5°).

Endlich haben wir noch das Drehungsvermögen der *d*-Phenyl-aminoessigsäure in wäßriger Lösung untersucht und dafür die bei 25° gesättigte Lösung benutzt.

Prozentgehalt 0.4795. $d_4^{20} = 1.0002$. Drehung im 2-dm-Rohr bei 20° und Natriumlicht 1.08° nach rechts (± 0.025°). Mithin $[\alpha]_D^{20} = +112.6°$ (± 3°).

Beide Aminosäuren sind nahezu geschmacklos. Von dem charakteristischen Unterschied im Geschmack, der wiederholt bei den optischen Antipoden der Aminosäuren beobachtet wurde, ist hier also nichts zu merken.

$C_8H_9NO_2$ Ref. 378a

Aktive Formyl-Phenyl-aminoessigsäure.

70 g Racemverbindung werden mit 115 g Cinchonin (1 Mol.), beide fein gepulvert, in 2 l kochendem Wasser gelöst, und die Flüssigkeit wenn nötig filtriert. Beim Erkalten tritt bald Krystallisation ein, besonders wenn man mit einigen Kryställchen von einer früheren Darstellung impft; solche Impfkrystalle sind verhältnismäßig leicht zu beschaffen durch Verdunsten eines Teiles der wäßrigen Lösung im Vakuumexsiccator unter öfterem Reiben. Läßt man nach Beginn der Krystallisation die Flüssigkeit noch 12 Stunden unter zeitweisem Umschütteln stehen, so scheidet sich das Cinchoninsalz der *l*-Formyl-phenyl-aminoessigsäure zum größeren Teil ab. Die Ausbeute beträgt etwa 95 g. Das Produkt wird aus der 9-fachen Menge heißem Wasser umkrystallisiert. Zur völligen Reinigung genügt in der Regel nochmaliges Umkrystallisieren aus der 9-fachen Menge heißem Wasser. Die Ausbeute geht dabei erheblich herunter und beträgt schließlich 50—60 % der Theorie. Das Cinchoninsalz bildet schöne, farblose, manchmal zentimeterlange Krystalle, die prismatisch ausgebildet sind.

Zur Bereitung der freien

l-Formyl-Phenyl-aminoessigsäure

löst man 37 g Cinchoninsalz in 500 ccm heißem Wasser, versetzt noch warm mit 25 ccm *n*-Natronlauge, wobei schon ein Teil des Cinchonins ausfällt, kühlt jetzt rasch ab und gibt noch 57.6 ccm *n*-Natronlauge zu, so daß die Gesamtmenge des Alkalis wenig mehr als 1 Mol. beträgt. Nachdem das Gemisch in Eiswasser sorgfältig gekühlt ist, wird das Cinchonin scharf abgesaugt und mehrmals mit wenig eiskaltem Wasser gewaschen. Zum Filtrat fügt man 8 ccm *n*-Salzsäure, um den Überschuß an Alkali zu neutralisieren, verdampft dann die Lösung unter 10—20 mm Druck auf etwa 60 ccm und versetzt schließlich mit 37.3 ccm doppeltnormaler Salzsäure. Dabei fällt die aktive Formylverbindung sofort krystallinisch aus; sie wird nach 1-stündigem Stehen bei 0° abgesaugt und mit eiskaltem Wasser gewaschen. Die Ausbeute betrug 12.5 g oder 88 % der Theorie. Zur völligen Reinigung wird aus etwa 10 Teilen kochendem Wasser umkrystallisiert. Zur Analyse wurde im Vakuumexsiccator getrocknet.

0.1238 g Sbst.: 0.2736 g CO_2, 0.0553 g H_2O. — 0.1120 g Sbst.: 7.41 ccm N (14.5°, 764 mm).

$C_9H_9O_3N$ (179.1). Ber. C 60.31, H 5.07, N 7.82.
Gef. » 60.27, » 5.00, » 7.75.

Die Verbindung schmilzt etwas höher als der Racemkörper, aber wegen der Zersetzung auch nicht konstant. Beim raschen Erhitzen beginnt sie gegen 187° zu sintern und schmilzt gegen 190° (korr.) unter Gasentwicklung und Gelbfärbung.

In siedendem Wasser ist sie leicht löslich, doch etwas schwerer als der Racemkörper, und krystallisiert daraus in mikroskopisch sehr dünnen, langgestreckten Platten, die vielfach verwachsen sind und manchmal wie Nadeln aussehen. Leicht wird sie ferner von Alkohol und Aceton aufgenommen und krystallisiert aus Alkohol in sehr feinen Nadeln, die oft zu Büscheln vereinigt sind.

Für die optische Bestimmung diente die alkoholische Lösung.

0.2555 g Sbst., Gesamtgewicht der Lösung 6.3777 g. Mithin Prozentgehalt 4.006. $d_4^{20} = 0.8048$. Drehung im 2-dm-Rohr bei 20° und Natriumlicht 16.76° nach links. Mithin $[\alpha]_D^{20} = -259.9°$.

Das für die folgende Bestimmung dienende Präparat war aus dem gleichen Cinchoninsalz bereitet, nachdem es noch zweimal aus Wasser umkrystallisiert war.

0.2566 g Sbst., Gesamtgewicht der Lösung 6.3321 g. Mithin Prozentgehalt 4.052. $d_4^{20} = 0.8053$. Drehung im 2-dm-Rohr bei 20° im Natriumlicht 16.95° nach links. Mithin $[\alpha]_D^{20} = -259.7°$.

Da das Drehungsvermögen hierbei nicht mehr erhöht wurde, so ist das öftere Umkrystallisieren des Cinchoninsalzes überflüssig. Ferner glauben wir, aus dem Resultat den Schluß ziehen zu dürfen, daß die Formylverbindung sehr wahrscheinlich optisch rein war.

d-Formyl-Phenyl-aminoessigsäure.

Für ihre Bereitung dienten die ersten beiden wäßrigen Mutterlaugen, die bei der Gewinnung des Cinchoninsalzes des optischen Antipoden resultieren. Man versetzt sie mit soviel *n*-Natronlauge, daß stark alkalische Reaktion eintritt, filtriert vom abgeschiedenen Cinchonin auf der Pumpe und fügt zum Filtrat eine dem Alkali entsprechende Menge *n*-Salzsäure. Zu dieser Flüssigkeit, die ungefähr 4 l beträgt, fügt man etwa 95 g gepulvertes Chinin und kocht.

wobei Lösung eintritt. Bei mehrtägigem Stehen scheiden sich reichliche Mengen von Krystallen zugleich mit einem Öl aus. Man filtriert, preßt die Krystalle zwischen Fließpapier und krystallisiert sie aus nicht zu viel heißem Alkohol. Zur völligen Reinigung haben wir das Chininsalz aus etwa der 40-fachen Menge heißem Wasser umkrystallisiert, wobei ungefähr 1/6 verloren geht. Es bildet dann seideglänzende, farblose Nadeln.

Für die Rückverwandlung in den Formylkörper übergießt man 30 g Chininsalz mit 60 ccm kalter *n*-Natronlauge und fügt Äther hinzu. Beim kräftigen Umschütteln wird das ölig ausgeschiedene Chinin vom Äther leicht aufgenommen. Nachdem die ätherische Schicht abgetrennt und der Rest des Chinins durch abermaliges Ausäthern entfernt ist, versetzt man die alkalische Lösung mit 60 ccm *n*-Salzsäure. Die abgeschiedene Formylverbindung wird nach dem Abkühlen auf 0° abgesaugt und mit wenig kaltem Wasser gewaschen. Diese Isolierung aus dem Chininsalz ist nur mit geringem Verlust verbunden. Berechnet man die Ausbeute an Formyl-*d*-Verbindung auf den ursprünglich angewandten Racemkörper, so beträgt sie 35—40% der Theorie.

Für die Analyse und optische Bestimmung war nochmals aus heißem Wasser umgelöst und im Vakuumexsiccator getrocknet worden.

0.1324 g Sbst.: 0.2931 g CO_2, 0.0605 g H_2O. — 0.1153 g Sbst.: 8 ccm N (20°, 759.5 mm).

$C_9H_9O_3N$ (179.1). Ber. C 60.31, H 5.07, N 7.82.
Gef. » 60.38, » 5.11, » 7.96.

Für die optische Bestimmung diente die alkoholische Lösung.

0.1642 g Sbst., Gesamtgewicht der Lösung 4.2658 g. Mithin Prozentgehalt 3.849. $d_4^{20} = 0.8045$. Drehung im 1-dm-Rohr bei 20° im Natriumlicht 8.08° nach rechts (± 0.015°). Mithin $[\alpha]_D^{20} = +259.3°$ (± 0.5°).

Für die zweite Bestimmung wurde ein Präparat benutzt, das aus demselben Chininsalz, aber erst nach abermaliger Krystallisation aus Wasser dargestellt war.

0.2331 g Sbst., Gesamtgewicht der Lösung 6.0223 g. Mithin Prozentgehalt 3.871. $d_4^{20} = 0.8044$. Drehung im 2-dm-Rohr bei 20° im Natriumlicht 16.16° nach rechts (± 0.04°). Mithin $[\alpha]_D^{20} = +259.45°$ (± 0.6°).

Wie man sieht, stimmen diese Werte sowohl unter einander, als auch mit dem Drehungsvermögen des Antipoden überein. Dasselbe gilt für Schmelzpunkt, Aussehen der Krystalle und Löslichkeit.

Aktive Phenyl-aminoessigsäuren.

Die Hydrolyse der Formylkörper geht sehr leicht von statten; es genügt, mit der zehnfachen Menge 10-prozentiger Bromwasserstoffsäure ½ Stunde am Rückflußkühler zu kochen. Die Lösung wird dann unter 10—15 mm Druck zur Trockne verdampft, wobei das Bromhydrat in schönen weißen Nadeln zurückbleibt. Man löst es in etwa 25 Teilen 50-prozentigen Alkohols und fügt Ammoniak in geringem Überschuß hinzu. Dann scheidet sich die Aminosäure in glänzenden Nadeln als dicker Brei ab. Sie wird abgesaugt und mit verdünntem Alkohol gewaschen. Die Ausbeute ist fast quantitativ.

Für die Analyse und optische Bestimmung diente ein Präparat, das aus der 70-fachen Menge kochendem Wasser umkrystallisiert und im Vakuumexsiccator über Schwefelsäure getrocknet war.

l-Phenyl-aminoessigsäure.

0.1735 g Sbst.: 0.4030 g CO_2, 0.0940 g H_2O. — 0.1110 g Sbst.: 9.1 ccm N (23°, 763 mm).

$C_8H_9O_2N$ (Mol.-Gew. 151.1). Ber. C 63.54, H 6.00, N 9.27.
Gef. » 63.35, » 6.06, » 9.33.

Die Aminosäure zeigt wie die meisten anderen Aminosäuren keinen konstanten Schmelzpunkt; im geschlossenen Capillarrohr fanden wir ihn je nach der Schnelligkeit des Erhitzens bei 305—310° (korr.), was mit der Angabe von Ehrlich, 303—305°, im wesentlichen übereinstimmt.

Sie ist in heißem Wasser erheblich leichter löslich als der Racemkörper, verlangt davon aber noch ungefähr 70 Teile. Für die Bestimmung der Löslichkeit in kaltem Wasser wurde die gepulverte Substanz bei 25° im Thermostaten 20 Stunden im Silberrohr mit Wasser geschüttelt: 1 Teil Aminosäure verlangt unter diesen Umständen 207.6 Teile Wasser von 25°.

In absolutem Alkohol ist sie so gut wie unlöslich, löst sich dagegen in kochendem 50-prozentigem Alkohol und krystallisiert daraus in kurzen, glänzenden Nadeln. Aus Wasser krystallisiert sie in langen, ganz schmalen und sehr dünnen Platten, die manchmal wie Nadeln aussehen und auch häufig zu ausgezackten Aggregaten vereinigt sind. In Alkali ist sie leicht löslich, nicht so leicht wird sie von verdünnter Salzsäure aufgenommen, schwer löslich in starker Salzsäure.

Für die Bestimmung des Drehungsvermögens diente eine Lösung in wenig mehr als der berechneten Menge Salzsäure von ungefähr 2.5%.

0.7438 g Sbst. wurden gelöst in 6.7 ccm Normalsalzsäure und 8 ccm Wasser, Gesamtgewicht der Lösung 10.3400 g. Mithin Prozentgehalt 7.193. $d_4^{20} = 1.0286$. Drehung im 2-dm-Rohr bei 20° im Natriumlicht 23.35° nach links. Mithin $[\alpha]_D^{20} = -157.78°$ (± 0.5°).

Nach nochmaligem Umlösen der Aminosäure aus heißem Wasser wurde fast der gleiche Wert gefunden:

0.5387 g Sbst. wurden in 4.85 ccm Normalsalzsäure und 2.17 ccm Wasser gelöst; Gesamtgewicht der Lösung 7.6671 g. Mithin Prozentgehalt 7.026. $d_4^{20} = 1.0277$; Drehung im 2-dm-Rohr bei 20° im Natriumlicht 22.80° nach links. Mithin $[\alpha]_D^{20} = -157.87°$ (± 0.5°).

Um einen Vergleich mit dem von F. Ehrlich bei seinem besten Präparat gefundenen Wert zu haben, haben wir mit obiger Substanz einen dritten Versuch ausgeführt, wobei 10-prozentige Salzsäure zur Anwendung kam.

0.3032 g Sbst., Gesamtgewicht der Lösung 8.2277 g. Mithin Prozentgehalt 3.686. $d_4^{20} = 1.0586$. Drehung im 2-dm-Rohr bei 20° im Natriumlicht 12.91° nach links (± 0.03°). Mithin $[\alpha]_D^{20} = -165.43°$ (± 0.4°).

Man ersieht daraus, daß das Präparat von F. Ehrlich, der $[\alpha]_D^{20}$: — 144.83° fand, optisch noch nicht rein war, wie er schon selbst vermutete, sondern daß es mindestens noch 6.2% des Antipoden enthielt. Ferner ist zu beachten, daß die spezifische Drehung durch die starke Salzsäure etwas erhöht wird.

Dem schon erwähnten, schön krystallisierten Bromhydrat ist das Chlorhydrat sehr ähnlich. Es läßt sich aus warmer, ganz verdünnter Salzsäure leicht umkrystallisieren und bildet lange, farblose Nadeln. Es löst sich leicht in warmem Wasser und Alkohol; aus letzterem wird es durch Äther gefällt. Im Capillarrohr schmilzt es nicht konstant gegen 246° (korr.) unter Gasentwicklung. Für die Analyse wurde im Vakuumexsiccator getrocknet:

0.1168 g Sbst.: 0.089 g AgCl.

$C_8H_9O_2N.HCl$ (187.5) Ber. Cl 18.90. Gef. Cl 18.84.

l-Phenyl-aminoessigsäure-äthylester. Suspendiert man 5 g Aminosäure in 50 ccm absolutem Alkohol und leitet trockne Salzsäure ohne Kühlung bis zur Sättigung ein, so findet in der Regel klare Lösung statt. Ist dies nicht der Fall, so muß schließlich bis zur völligen Lösung erhitzt werden. Man verdampft dann unter stark vermindertem Druck, löst den Rückstand in Alkohol und fällt mit Äther.

Für die Analyse wurde im Vakuumexsiccator über Schwefelsäure getrocknet:

0.1906 g Sbst.: 0.1286 g AgCl.

$C_{10}H_{13}O_2N.HCl$ (215.6). Ber. Cl 16.44. Gef. Cl 16.68.

Das Salz ist in Wasser leicht löslich. Es schmilzt unter Gasentwicklung bei ungefähr 203° (korr.).

0.2103 g Sbst. wurden in 4 ccm Wasser gelöst, Gesamtgewicht der Lösung 4.1880 g. Mithin Prozentgehalt 5.021. $d_4^{18} = 1.0097$. Drehung im 1-dm-Rohr bei 20° im Natriumlicht 4.51° nach rechts (± 0.02°). Mithin $[\alpha]_D^{20} = +88.95°$ (± 0.4°).

Versetzt man die nicht zu verdünnte Lösung des Salzes mit Alkalicarbonat, so scheidet sich der freie Ester als Öl ab und kann leicht ausgeäthert werden.

d-Phenyl-aminoessigsäure.

Da die Eigenschaften dieselben sind wie beim Antipoden, so begnügen wir uns damit, die Resultate der Analyse und optischen Bestimmung anzuführen.

0.1686 g Sbst.: 0.3928 g CO_2, 0.0909 g H_2O. — 0.1129 g Sbst.: 9.2 ccm N (20.5°, 760 mm).

$C_8H_9O_2N$ (151.1). Ber. C 63.54, H 6.00, N 9.27.
Gef. » 63.54, » 6.03, » 9.34.

Optische Bestimmung: 0.5410 g Sbst. wurden in 4.85 ccm Normalsalzsäure und 2.17 ccm Wasser gelöst; Gesamtgewicht der Lösung 7.6217 g. Mithin Prozentgehalt 7.098. $d_4^{20} = 1.0281$. Drehung im 2-dm-Rohr bei 20° im Natriumlicht 23.04° nach rechts. Mithin $[\alpha]_D^{20} = +157.86°$ (± 0.5°).

Nach nochmaligem Umkrystallisieren der Aminosäure aus Wasser war die Drehung nicht geändert.

0.5427 g Sbst. wurden in 4.88 ccm Normalsalzsäure und 2.18 ccm Wasser gelöst; Gesamtgewicht der Lösung 7.6831 g. Mithin Prozentgehalt 7.064. $d_4^{20} = 1.0281$. Drehung im 2-dm-Rohr bei 20° und Natriumlicht 22.96° nach rechts. Mithin $[\alpha]_D^{20} = +158.09°$ (± 0.5°).

Endlich haben wir noch das Drehungsvermögen der *d*-Phenylaminoessigsäure in wäßriger Lösung untersucht und dafür die bei 25° gesättigte Lösung benutzt.

Prozentgehalt 0.4795. $d_4^{20} = 1.0002$. Drehung im 2-dm-Rohr bei 20° und Natriumlicht 1.08° nach rechts (± 0.025°). Mithin $[\alpha]_D^{20} = +112.6°$ (± 3°).

Beide Aminosäuren sind nahezu geschmacklos. Von dem charakteristischen Unterschied im Geschmack, der wiederholt bei den optischen Antipoden der Aminosäuren beobachtet wurde, ist hier also nichts zu merken.

$C_8H_9NO_2$ **Ref. 378b**

The levorotatory form was obtained by following the procedure of Ingersoll and Adams,[6] which is essentially that of Betti and Mayer,[7] of fractional crystallization of the *d*-camphorsulfonate, the *l*-phenylaminoacetic acid *d*-camphorsulfonate being less soluble. Since the *d*-camphorsulfonate hydrolyzes readily in hot water, it was found best to use an excess of *d*-camphorsulfonic acid in this resolution. The separation of the forms was carried out as follows: 110 g. of *dl*-phenylaminoacetic acid, 206 g. of *d*-camphorsulfonic acid and 500 cc. of water were heated together, giving a clear solution which on cooling deposited crystals of the *l*-phenylaminoacetic acid *d*-camphorsulfonate. These were recrystallized from hot water using about 10% excess of *d*-camphorsulfonic acid to prevent hydrolysis and about 2 g. of Norite to decolorize the solution. The *l*-phenylaminoacetic acid was precipitated from a cold, saturated, aqueous solution of the *d*-camphorsulfonate by adding dilute ammonium hydroxide until the solution was faintly alkaline. The crystals, after washing with water and drying, gave a rotation $[\alpha]^{20}_D$ $-152°$ in 0.133 *N* hydrochloric acid; $[\alpha]^{20}_D$ $-162.8°$ in 10% hydrochloric acid; $[\alpha]^{20}_D$ $-111°$ in water, this last value checking that of $-111.02°$ given by Betti and Mayer.[7] It was found that on slow evaporation of a saturated aqueous solution of the *l*-phenylaminoacetic acid, long needle-like crystals formed. One of the most perfect of these was used in the rotating crystal studies; m. p., subl. 245–248°, uncorr.; $[\alpha]^{20}_D$ (in 0.140 *N* hydrochloric acid) $-149°$. It may be noted here that the rotations observed varied widely with the concentration of the hydrochloric acid used as a solvent. No consistent melting point was obtained, which is in harmony with the results previously published. For example, St. Minovici[8] reports m. p. 227° (subl.); Fischer and Weichhold[9] report m. p. 305–310°; Betti and Mayer[7] report m. p. above 305°.

The *d*-phenylaminoacetic acid was obtained by extraction of the partially resolved (80% *d*-, 20% *dl*-phenylaminoacetic acid) acid precipitated from the mother liquor of the *d*-camphorsulfonate crystallization with ammonium hydroxide. The active form of phenylaminoacetic acid is more readily soluble in hydrochloric acid than the racemic, and is thus extracted from the racemic.[10] The *d*-phenylaminoacetic acid was precipitated by adding dilute ammonium hydroxide. The mother liquor from this precipitation, on standing for several days, yielded needle crystals, one of the most perfect of which was used in the x-ray work; m. p. 242–244°, uncorr.; $[\alpha]^{20}_D$ (in 0.105 *N* hydrochloric acid) $+147°$.

[6] Ingersoll and Adams, This Journal, **44**, 2930 (1922).
[7] Betti and Mayer, *Ber.*, **41**, 2071 (1908).
[8] St. Minovici, *Bull. Soc. Chim. Romania*, **2**, 8 (1910).
[9] Fischer and Weichhold, *Ber.* **41**, 1286 (1908).
[10] Unpublished work from this Laboratory.

$C_8H_9NO_2$ **Ref. 378c**

Cloruro di D.(—).acetilmandelile. — Per le nostre ricerche abbiamo preferito questo acicloruro perchè durante il suo impiego non subisce racemizzazione. In primo tempo si era pensato di usare, quale mezzo acilante dissimmetrico, il cloruro dell'acido α.clorofenilacetico, direttamente ottenibile dall'acido mandelico per azione del cloruro di tionile; ma dovemmo constatare che esso è facilmente racemizzabile, e quindi non adatto allo scopo.

La preparazione del cloruro di D.(—).acetilmandelile è stata condotta modificando opportunamente le condizioni descritte nella letteratura pel prodotto racemico ([18]). Siamo partiti dall'acido D.(—).mandelico chimicamente ed otticamente puro; ed abbiamo impiegati anche tutti gli altri materiali allo stato di purezza, onde giungere ad un buon prodotto tecnico finale da potersi usare successivamente senza sottoporlo a rettificazione.

Grammi 105 di acido D.(—).mandelico (0,69 Mol.) sono trattati, in pallone codato, con g 151 (1,92 Mol.) di cloruro di acetile a temperatura ordinaria. Dopo lieve agitazione e qualche ora di riposo si ottiene soluzione completa; allora si scalda cautamente (non oltre i 30°) e si attacca il vuoto per eliminare l'eccesso di cloruro di acetile. A questo prodotto acetilato si aggiungono g 250 (2,1 Mol.) di cloruro di tionile e si scalda a ricadere per quattro ore alla temperatura di 55°. Finalmente si allontana l'eccesso di cloruro di tionile sotto vuoto alla temperatura di 30°. Il cloruro di D.(—).acetilmandelile così ottenuto contiene circa il 90 % di prodotto reale (come abbiamo potuto controllare chimicamente e otticamente)e può essere senz'altro usato tenendo presente, naturalmente, tale contenuto medio.

([18]) Cfr. Org. Syntheses, *I*, 12 (1948).

Acidi D.(—).N.mandelil-α.aminofenilacetici (form. I e II, $R = C_6H_5$)
Grammi 3 di acido D.L.α. aminofenilacetico (0,020 Mol.) sono sci
in un leggero eccesso (21 cm³) di NaOH.N/$_1$. La soluzione viene tratt
sotto forte agitazione ed a temperatura non oltre i 15°, alternativame
ed a piccole porzioni, con g 4,3 di cloruro di D.(—).acetilmande
(0,022 Mol.) sciolti in poco etere, e con soluzione di NaOH.N/$_1$ in m
che il liquido resti sempre leggermente alcalino. Il trattamento richi
un'ora circa. Si sospende quindi il raffreddamento esterno, si aggiu
ancora NaOH.N/$_1$ fino ad un totale di cm³ 45 e si continua l'agitazi
per altri 40'. (Complessivamente, la NaOH.N/$_1$ impiegata per la salif
zione dell'aminoacido, per l'acilazione e per la disacetilazione dei p
dotti formatisi, assomma dunque a cm³ 66). A questo punto si aggiu
acido cloridrico fino a debole acidità al tornasole, si estrae con et
per eliminare alcune impurezze, e finalmente si aggiunge ancora ac
cloridrico fino a netta acidità al rosso-congo. Dal liquido si separa
tal modo un prodotto microcristallino bianco che dopo alcune ore
raccoglie per filtrazione alla pompa e si asciuga all'aria. Pesa circa
e fonde con decomposizione verso i 190°. E' poco solubile in acqua,
in alcool, pochissimo in etere. Per la osservazione polarimetrica si a
pera la soluzione acquosa del sale sodico ottenuta sciogliendo il prod
nella quantità calcolata di NaOH.N/$_1$ e portando a volume con acq
Si trova così $[\alpha]^{20}_D = +26°4$. La purificazione chimica ed ottica si
giunge diluendo con acqua questa soluzione e riacidificando con ac
cloridrico: si ottiene allora la separazione di piccoli aggregati cristal
a rosetta che raccolti ed essiccati pesano g 1,5 e fondono a 200°-
(decomp.). Il potere rotatorio (sale sodico) risulta $[\alpha]^{20}_D = +91°$. Qu
caratteri non variano per ulteriori operazioni di frazionamento.
l'analisi:

trov.% N 5,15;
per $C_{16}H_{15}O_4N$ calc. 4,91.

Le acque madri acide dalle quali si era separato il prodotto grezzo vengono sottoposte a lenta evaporazione a temperatura ordinaria nel vuoto. I prodotti che si separano progressivamente mostrano punto di fusione più basso e potere rotatorio quasi nullo, dapprima; poi presentano potere rotatorio negativo che va aumentando in valore. Finalmente si raccoglie una sostanza (circa g 1,5) che, purificata dall'alcool acquoso, si presenta in piccoli aggregati cristallini bianchi, con p. f. 149°-151° (decomp.). discretamente solubili in acqua e in alcool, poco in etere. La soluzione del sale sodico mostra $[\alpha]_D^{20} = -83°$. All'analisi:

trov.% N 5,19;
per $C_{16}H_{15}O_4N$ calc. 4,91.

Acido L.(+).α.aminofenilacetico. — Il composto precedentemente descritto (p. f. 200°-202°, $[\alpha]_D^{20} = +91°$) viene idrolizzato nelle seguenti condizioni: g 1 di sostanza sono trattati con 40 cm³ di acido cloridrico al 20%. Si scalda prima a bagnomaria, poi all'ebullizione per circa mezz'ora ottenendosi soluzione completa. Dopo raffreddamento si estrae ripetutamente con etere per asportare l'acido D.(—).mandelico che si ricupera per evaporazione del solvente. Il liquido acquoso viene concentrato nel vuoto su calce sodata, a temperatura ordinaria: si ottiene così la separazione di cristalli aghiformi bianchi che corrispondono, per tutti i loro caratteri, al cloridrato dell'acido L.(+).α.aminofenilacetico. Da essi può essere liberato l'aminoacido per neutralizzazione con alcali della soluzione acquosa concentrata. L'aminoacido si presenta in rosette od anche in scagliette madreperlacee poco solubili in acqua, fusibili con decomposizione oltre i 300°. Mostra in soluzione acquosa (0,75 in 100 cm³) $[\alpha]_D = +112°$. Questi dati corrispondono esattamente a quelli noti in letteratura ([19]). Ciò dimostra che il mandelilderivato fusibile a 200°-202° ($[\alpha]_D = +91°$) è da considerarsi acido D.(—).N.mandelil-L.(+).α.aminofenilacetico (form. I).

Acido D.(—).α.aminofenilacetico. D'altronde, il mandelilderivato p. f. 149°-151° e $[\alpha]_D = -83°$, idrolizzato nelle stesse condizioni, fornisce (accanto ad acido D.(—).mandelico) l'acido D.(—).α.aminofenilacetico in buona purezza ottica. Al mandelilderivato fusibile a 149°-151° corrisponde dunque la costituzione di acido D.(—).N.mandelil-D.(—).α.aminofenilacetico (form. II).

$C_8H_9NO_2$ Ref. 378d

Example 1

The adduct of dehydroabietylamine and carbon dioxide (82g) was dissolved in boiling methylated spirits (250ml) and added to a suspension of DL-N-acetylphenylglycine (110 g) in hot methylated spirits (750ml). After precipitation was complete, the product was filtered off and washed with methylated spirits (2×90ml) and dried to yield the salt of L-N-acetylphenylglycine and dehydroabietylamine (85g) of specific rotation $[\alpha]_{546}$ +106°.

To the resulting filtrate was then added a solution of the adduct of dehydroabietylamine and carbon dioxide (82g) in boiling methylated spirits (200ml) and the mixture cooled to room temperature. The crystalline precipitate was filtered to yield the salt of D-N-acetylphenylglycine and dehydroabietylamine (95g) of specific rotation $[\alpha]_{546} -65°$.

The foregoing pure salts (48.2g) were heated with sodium hydroxide (5g) in water (50ml) at 50°C for 30 minutes and the liberated dehydroabietylamine extracted with ether (50ml). Acidification of the aqueous phase with concentrated hydrochloric acid to pH 1 precipitated the requisite L- or D-isomer of N-acetylphenylglycine having a specific rotation $[\alpha]_{546}$ of +240° or −242° respectively.

The foregoing pure enantiomorphs of N-acetylphenylglycine (19.3g) were heated under reflux in 6N hydrochloric acid (200ml) for 4 hours and the solution evaporated to dryness. The resulting solid was dissolved in water (50ml) and neutralised by the addition of 4N sodium hydroxide. The resulting precipitate was washed with water and dried to yield L- or D-phenylglycine having a specific rotation $[\alpha]_{546}$ of +132° or −132° respectively.

$C_8H_9NO_2$ Ref. 378e

Note: For a resolution procedure using l-menthoxyacetyl chloride as resolving agent, see Ref. 18i

$C_8H_9NO_2$ Ref. 378f

Note:

Using trans-1,2-cyclohexane dicarboxylic acid anhydride as resolving agent, obtained a less soluble diastereomeric amide from dioxane with m.p. 206-206.5° and $[\alpha]_{435}^{25}$ -252.4° (c 1, ethanol). The acid hydrochloride obtained from the amide had a $[\alpha]_D^{25}$ -95.3° (c 1, 2N HCl).

$C_8H_9NO_2$ Ref. 378g

Resolution of (±)-Phenylglycine by Acid of 8b. The hydroxy sulfonate sodium salt **8b** (0.092 mol) in 90% ethanol was passed through an ion-exchange column in the acid form (ANGC-242). Evaporation of the solvent yielded the oily sulfonic acid, which was dissolved in water (90 mL), and to this was added (±)-phenylglycine (0.066 mol). The reaction mixture was heated and filtered and the filtrate was allowed to stand at room temperature overnight. Filtration gave 5.6 g of a crystalline adduct: mp 201–203 °C; $[\alpha]_D$ −48.2 ° (c 8.5, H_2O). Anal. Calcd for $C_{18}H_{29}NSO_6$: C, 55.79; H, 7.54; N, 3.62; S, 8.28. Found: C, 56.06; H, 7.49; N, 3.74; S, 8.03. To the adduct (2.0 g) in water (50 mL) was carefully added ammonium hydroxide to pH 8. Filtration of the resulting white plates gave (+)-phenylglycine (0.37 g): $[\alpha]_D$ +154° (1.0 N HCl) (lit.,[18] 158.6 ± 0.8°); sublimes 249–254 °C (lit.[18] 245–250 °C).

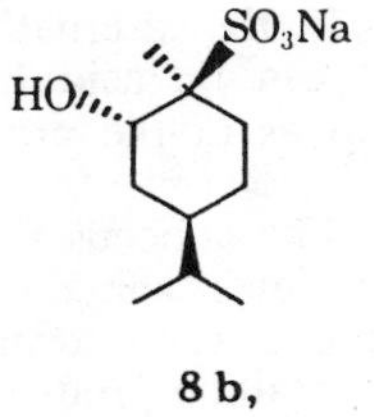

8 b,

(18) D(−)-Phenylglycine is required for the synthesis of antibiotics such as Ampicillin: see J. C. Clark, G. H. Phillipps, M. R. Steer, and L. Stephenson. *J. Chem. Soc., Perkin Trans. 1*, 471 (1976).

Note: The authors give the following procedure for the preparation of the resolving agent 8b.

General Conditions for Preparation of β-Hydroxy Sulfonates for Epoxides. A three-neck, 3-L Morton flask was charged with epoxide (0.86 mol), water (800 mL), and sodium sulfite (0.91 mol). The reaction was refluxed with high-speed mechanical stirring until the oil layer had disappeared. The crystalline hydrated hydroxy sulfonate precipitated on cooling. Filtration of the crystalline product was followed by ether washing the crystals to remove diol side products. Recrystallization from 90% ethanol yielded the pure crystalline hydrated β-hydroxy sulfonates.

(14) The *cis*-epoxide **7b** required 111 h of reflux compared to 40 h for the *trans*-epoxide **4b**. Alternatively, **7a** was completely reacted after 18 h at 150 °C under pressure in the presence of a phase-transfer catalyst.

For the experimental details of an enzymatic method for the resolution of this compound, see D. Rudman, A. Meister and J. P. Greenstein, J. Am. Chem. Soc., 74, 551 (1952).

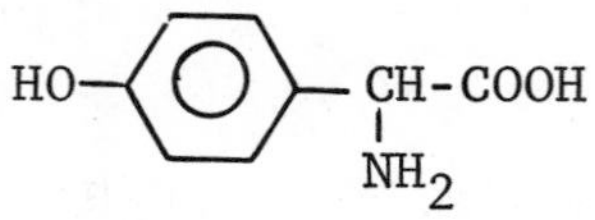

$C_8H_9NO_3$ Ref. 379

Resolution of DL-N-Chloroacetyl-2-(*p*-hydroxyphenyl)-glycine (II) with Dehydroabietylamine

A. Salt of D-N-chloroacetyl-2-(*p*-hydroxyphenyl)-glycine with dehydroabietylamine (III)

A mixture of 5.0 g. (0.0205 mole) of racemic N-chloroacetyl-2-(*p*-hydroxyphenyl)-glycine (II), 3.52 g. (0.01025 mole) of dehydroabietylamine acetate, 240 ml. of 2-propanol and 170 ml. of water was heated on a steam bath. The resultant hot solution was filtered and the filtrate was stored in an insulated container and permitted to cool slowly. Crystallization was allowed to proceed for 16 hrs. Long needle-like crystals (III) were obtained. The crystals (III) were collected by filtration, washed sparingly with 2:1 2-propanol-water and dried at 40°C. 2.6 g. of product (corresponding to a yield of 48%) having a melting point of 206—212°C (decomp.) were obtained.

The filtrate containing L-N-chloroacetyl-2-(*p*-hydroxyphenyl)-glycine was rendered alkaline in pH and heated (e.g., with added NaOH followed by cautious acidification) to racemize the L isomer to the original DL ractmate used as the starting material.

B. D-N-Chloroacetyl-2-(p-hydroxyphenyl) glycine (IV)

A vigorously stirred suspension of 2.6 g. of the dehydroabietylamine salt (III) in 20 ml. of water and 50 ml. of methylene chloride was adjusted to pH 10.7 by the slow addition of 20% sodium hydroxide whereby solution of the solids was effected. The aqueous phase was separated, washed 3 times with methylene chloride, filtered and acidified to pH 2 with 6N HCl. Scratching induced crystallization of the product as large clumps of fluffy crystals. After 1 hr. in the refrigerator the product (IV) was collected by filtration, washed sparingly with water and dried at 40°C. 0.73 g. of product (IV) (corresponding to a yield of 61.2%), having a melting point of 193—196°C (decomp.) and $[\alpha]_D^{24°} = -205.1°$ (C 1 95% EtOH) were obtained. The IR and NMR spectra were fully consistent for the desired product.

Anal. Found: C, 49.24; H, 4.19; N, 5.56. Calc'd for $C_{10}H_{10}ClNO_4$: C, 49.29; H, 4.14; N, 5.75.

The dehydroabietylamine can be recovered from the solvent phase by evaporation of the solvent or by salt formation.

In the alternative, the solvent-extracted aqueous phase containing the compound (IV) can be acid hydrolyzed to the compound (V) (to remove the N-chloro-acetyl group as in the procedure below) directly without isolation of the compound (IV) as a solid.

The aquecus filtrate was extracted 3 times with ethyl acetate. The combined and dried (over anhydrous Na_2SO_4) extracts were concentrated to dryness yielding, after drying in vacuo over phosphorus pentoxide, 0.24 g. (corresponding to a yield of 20%) of crystalline product (IV) having a melting point of 189—190°C (decomp.) $[\alpha]_D^{24°} = -201.1°$ (C 1 95% EtOH).

D-2-(*p*-Hydroxyphenyl) glycine (V)

A mixture of 1.19 g. of D-N-chloroacetyl-2-(*p*-hydroxyphenyl) glycine (IV), $[\alpha]_D^{24°} = -206.6°$, and 12 ml. of 2N HCl was heated at reflux for 1.5 hrs. The mixture was concentrated to dryness. Water (5 ml.) was added to the residue and the resultant solution was again concentrated to dryness leaving as residue a crystalline hydrochloride salt of D-2-(p-hydroxyphenyl) glycine having a melting point of 215—215°C (decomp.).

A solution of the residue in 10 ml. of water was adjusted to pH 4.5 with 20% sodium hydroxide. A gelatinous mass formed. The mixture was warmed to about 50°C on a steam bath whereupon another crystalline form (granular) started to separate. The mixture was concentrated to approximately one-half of its initial volume, permitted to stand for one hour at room temperature, then stored in the refrigerator overnight. The product was collected by filtration, washed sparingly dropwise with ice water, and dried *in vacuo* at 65°C over phosphorus pentoxide. 61 g. of product (V) (corresponding to a yield of 81.3%) having a melting point of 223—226°C (decomp.), $[\alpha]_D^{24°} = -159.1°$ (C 1 1N HCl), $[\alpha]_D^{24°} = -108.5°$ (C 1 H_2O) was obtained. The IR and MNR spectra were fully consistent.

Anal. Found: C, 57.56; H, 5.63; N, 8.38. Calc'd for $C_8H_9NO_3$: C, 57.48; H, 5.44; N, 8.38.

$C_8H_9NO_3$ Ref. 379a

Beispiel 1

10.0 g DL-N,O-Diacetyl-p-hydroxyphenylglycin, 5,28 g d-(+)-α-Phenyläthylamin und 100 ml Wasser wurden miteinander gemischt und die Mischung wurde so lange auf etwa 80°C erhitzt, bis sie sich vollstandig in Wasser gelöst hatte. Dann wurde die erhaltene Lösung 1 Stunde lang bei 40°C stehen gelassen, um das d-(+)-α-Phenyläthylamin-D-N-O-diacetyl-p-hydroxyphenylglycinat, F. 186 bis 188°C, $[\alpha]_D^{25} = -98{,}0°$ (c - 1, MeOH), zu kristallisieren. Die erhaltenen Kristalle wurden durch Filtrieren abgetrennt und dann unter Erwärmen in 60 ml Wasser gelöst, dann wurde die erhaltene Lösung 1 Stunde lang bei 40°C stehengelassen, wobei 4,2 g gereinigte Kristalle des D-Antipoden, F. 190,5 bis 191,5°C, $[\alpha]_D^{24} = -102{,}4°$ (c = 0,3, MeOH), erhalten wurden.

Elementaranalyse fur $C_{20}H_{24}N_2O_5$:

	C	H	N
ber.:	64,33	6,47	7,50
gef.:	64,53	6,48	7,74 %

Eine 10%ige wässrige Natriumcarbonatlösung wurde zu 4,2 g der Kristalle zugegeben, um den pH-Wert auf etwa 9 einzustellen. Das freie d-(+)-α-Phenyläthylamin wurde mit Äther extrahiert. Der pH-Wert der wässrigen Schicht wurde mit Chlorwasserstoffsäure auf 1 bis 2 eingestellt, wobei 1,5 g D-N,O-Diacetyl-p-hydroxyphenylglycin, F. 213°C. $[\alpha]_D^{20} = -213{,}3°$ (c = 1, MeOH), erhalten wurden, das dann mit 2nChlorwasserstoffsäure 2 Studen lang unter Ruckfluss hydrolysiert und anschliessend mit einer 10%igen wässrigen Natriumcarbonatlösung behandelt wurde, um den pH-Wert der erhaltenen Lösung auf 6-7 zu bringen, wobei 0,9g D-p-Hydroxyphenylglycin, F. 240 bis 242°C (Zers), $[\alpha]_D^{24} = -160{,}0°$ (c = 1, n -HCl), erhalten wurden.

Note:

A resolution procedure for the above compound using the same resolving agent may also be found in G. Bison and W. Wolfes, German Offen. 2322412 (1974).

$C_8H_9NO_3$ Ref. 379b

EXAMPLE 1

Resolution of α-Benzyloxycarbonylamino-p-hydroxyphenylacetic Acid.

A solution of the title acid (224 g.) and quinine trihydrate (285 g.) in boiling ethanol (2.5 l.) was allowed to cool. The crystals which separated were collected and recrystallized twice from ethanol to give a 76 percent yield of the quinine salt of the laevorotatory acid, $[\alpha]_D^{20}$ -158.5° (C, 1 in MeOH). Treatment of the quinine salt (68 g.) with dilute sodium hydroxide, removal of quinine by ether-extraction, and acidification of the aqueous solution gave the laevorotatory acid. This was crystallized twice from aqueous ethanol to give 28 g. (85 percent) of (-)-α-benzyloxycarbonylamino-p-hydroxyphenyl-acetic acid, m.p. 159°-161°, $[\alpha]_D 13$ -120.0° (C, 1 in MeOH). (Found C, 64.1; H, 5.2; N, 4.7. $C_{24}H_{21}NO_7$ requires C, 63.9; H, 4.9; N, 4.7%).

The mother liquor from the crude quinine salt was evaporated to dryness in vacuo to leave a syrup, which was treated with aqueous sodium hydroxide and the quinine removed by ether-extraction. Acidification of the aqueous layer gave the crude dextrorotatory acid (108 g.) which was collected, dried, and treated with ephedrine (63 g.) in boiling ethanol (450 ml.). On cooling the solution the ephedrine salt of the dextrorotatory acid separated, and was collected and recrystallized from ethanol. Yield 111 g., $[\alpha]_D^{21}$ +46.8° (C, 1 in H_2O). The acid was recovered from this salt in the usual way to give (+)-α-benzyloxycarbonylamino-p-hydroxphenylacetic acid (96 percent) which, after re-crystallization from 50 percent aqueous ethanol, had m.p. 158°-161°, $[\alpha]_D 21$ +120.2° (C, 1 in MeOH). (Found: C,64.1; H, 5.3; N, 4.8. $C_{24}H_{21}NO_7$ requires C, 63.9; H, 4.9; N, 4.7%).

$C_8H_9NO_3$ Ref. 379c

Materials. DL-*p*-HPG manufactured by our company, Tanabe Seiyaku Co., Ltd., was used. *d*-BCS was prepared from commercial *d*-camphor ($[\alpha]_D^{20}$ +44.0° *c*=7.5, in EtOH) by the method of Kipping and Pope.[6,7] Namely, bromine (320 g) was added dropwise to *d*-camphor (304 g) at 80°C over a period of 3 hr under stirring and the liquified reaction mixture was kept at the same conditions for 3 hr. After hydrogen bromide was releasing by bubbling, the reaction mixture was poured into iced water (3 liters) and the resulting precipitate was recrystallized from ethanol (230 ml) to give *d*-3-bromocamphor (302 g), mp 76°C, $[\alpha]_D^{20}$ +134.6° (*c*=10, EtOH). *d*-3-Bromocamphor (231 g) was dissolved in chloroform (400 ml) and chlorosulfonic acid (233 g) was added dropwise to this solution over 1 hr at 50°C. The reaction mixture was refluxed for 12 hr and poured into iced water (1 liter). The water layer and washings were neutralized with $Ca(OH)_2$ (120 g) and the precipitated $CaSO_4$ was filtered off. To the filtrate $(NH_4)_2CO_3$ (128 g) was added and the precipitated $CaCO_3$ was removed. The filtrate was concentrated and crystallized crude ammonium *d*-BCS (152 g) was recrystallized from water (270 ml) to give ammonium *d*-BCS (102 g), mp 270~272°C (dec.), $[\alpha]_D^{20}$ +85.3° (*c*=2, water), [lit, $[\alpha]_D^{22}$ +85.3°

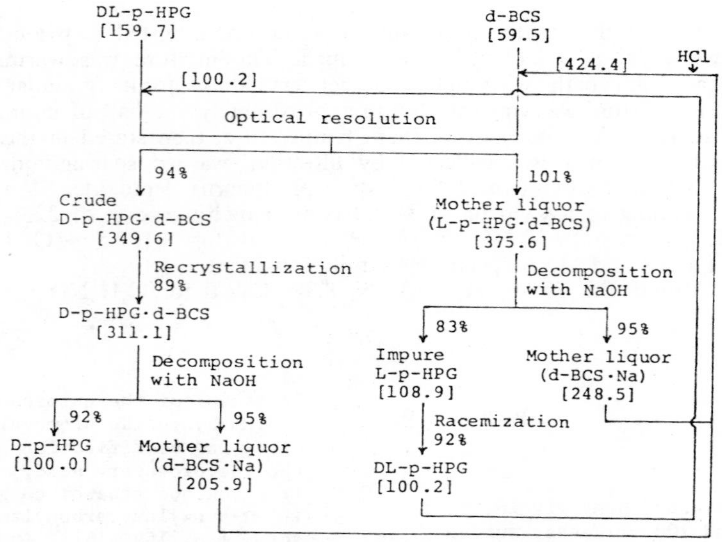

SCHEME 1. Flowsheet for Preparation of D-*p*-Hydroxyphenylglycin.
Figures in [ ] are weight. Substances in ( ) were used in the next process without separation.

(c=4, water),[8] $[\alpha]_D^{20}$+84.5° (c=1.6, water)[9]]. *Anal.* Found: C, 36.55; H, 5.60; N, 4.25. Calcd. for $NH_4\cdot C_{10}H_{14}O_4SBr$: C, 36.59; H, 5.53; N, 4.27%. Ammonium *d*-BCS obtained above was passed through Amberlite IR-120, and the effluent was concentrated to dryness and used as free *d*-BCS·monohydrate.

Optical resolution of DL-*p-hydroxyphenylglycine with d-3-bromocamphor-8-sulfonic acid.* A mixture of DL-*p*-HPG (30.0 g) and *d*-BCS·monohydrate (59.1 g) was dissolved in water (290 ml) at 95°C and was stirred at 25°C for 2 hr. The precipitated crystals were filtered, washed with a small amount of cold water, and dried to give crude D-*p*-HPG·*d*-BCS (40.2 g), $[\alpha]_D^{25}$+4.9° (c=1, 1N HCl). The crude salt (40.0 g) was recrystallized from 0.5% *d*-BCS aqeous solution (300 ml) to give D-*p*-HPG·*d*-BCS (35.5 g), $[\alpha]_D^{25}$+2.9° (c=1, 1N HCl), mp 243~245°C (dec.). *Anal.* Found: C, 45.17; H, 5.11; N, 2.93; S, 6.94. Calcd. for $C_{18}H_{24}O_7NSBr$: C, 45.20; H, 5.06; N, 2.93; S, 6.70%. The product was optically and chemically pure. The specific rotation of a mixture of DL-*p*-HPG and equivalent amount of *d*-BCS was $[\alpha]_D^{25}$+54.7° (c=1, 1N HCl), and that of an authentic D-*p*-HPG·*d*-BCS was $[\alpha]_D^{25}$ +2.9° (c=1, 1N HCl).

Preparation of D-*p-hydroxyphenylglycine.* The pure D-*p*-HPG·*d*-BCS (30.0 g) obtained above was dissolved in water (250 ml) at 95°C. The solution was adjusted at pH 6 with 2N NaOH (*ca.* 31 ml), concentrated to about 70 g, and stirred at 5°C for 2 hr. The precipitated crystals were filtered, washed with water, and dried to give D-*p*-HPG (9.6 g), $[\alpha]_D^{25}$−158.3°C (c=1, 1N HCl). *Anal.* Found: C, 57.70; H, 5.41; N, 8.33. Calcd. for $C_8H_9NO_3$: C, 57.48; H, 5.43; N, 8.38%.

Recovery of optically impure L-*p-hydroxyphenylglycine.* After the separation of less soluble D-*p*-HPG·*d*-BCS in the above resolution process, the mother liquor was adjusted at pH 6 with 2N NaOH, concentrated to about 130 g, and stirred at 5°C for 2 hr. The precipitated crystals were filtered, washed with water, and dried to give optically impure L-*p*-HPG (12.6 g). $[\alpha]_D^{25}$+129.3° (c=1, 1N HCl).

Racemization of optically impure L-*p-hydroxyphenylglycine.* Optically impure L-*p*-HPG (10.0 g) obtained by the above procedure was dissolved in 2N HCl (30 ml). The mixture was heated in an autoclave at 140°C for 12 hr. After the reaction, the mixture was adjusted at pH 6 with 2N NaOH and was stirred at 5°C for 2 hr. The precipitated crystals were filtered, washed with water, and dried to give DL-*p*-HPG (9.2 g), $[\alpha]_D^{25}$±0.0 (c=1, 1N HCl). The racemized *p*-HPG could be reused for resolution.

Reuse of d-3-bromocamphor-8-sulfonic acid. Sodium salt of *d*-BCS contained in the mother liquors after the separations of D- and L-*p*-HPG could be reused as resolving agent by addition of equivalent amount of hydrochloric acid. In the mother liquor after the separation of D-*p*-HPG (9.6 g) in the preceding procedure, DL-*p*-HPG (9.1 g) and 2N HCl (31 ml) were added. The mixture was heated at 95°C for dissolution and was stirred at 25°C for 2 hr. The precipitated crystals were filtered, washed with a small amount of cold water, and dried to give crude D-*p*-HPG·*d*-BCS (14.7 g), $[\alpha]_D^{25}$+3.9° (c=1, 1N HCl).

REFERENCES

1) L. B. Crast, Jr., U. S. Patent, 3489750 (1970) [*C. A.*, **72**, 100725v (1970)].
2) L. B. Crast, Jr., U. S. Patent, 3517023 (1970) [*C. A.*, **73**, 66567d (1970)].
3) A. A. W. Long, J. H. C. Nayler, H. Smith, T. Taylor and N. Ward, *J. Chem. Soc. C*, **1971**, 1920.
4) S. Yamada, C. Hongo and I. Chibata, *Agric. Biol. Chem.*, **42**, 1521 (1978).
5) K. Kariyone, H. Yagi and K. Yoshida, Ger. Offen. 2540735 (1976) [*C. A.*, **85**, 33392n (1976)].
6) F. S. Kipping and W. J. Pope, *J. Chem. Soc.*, **63**, 548 (1893).
7) F. S. Kipping and W. J. Pope, *ibid.*, **67**, 354 (1895).
8) W. J. Pope and J. Read, *ibid.*, **97**, 2199 (1910).
9) F. Hein and K.-H. Vogt, *Ber.*, **98**, 1691 (1965).

$C_8H_9NO_3$ Ref. 379d

Optically active D-*p*-hydroxyphenylglycine is useful as a starting material for the preparation of semisynthetic penicillins and cephalosporins.[1~3] Most of the known procedures for preparing optically active *p*-hydroxyphenylglycine have employed the optical resolution of the racemic N-acyl derivatives by selected chiral resolving agents[3~8] or stereospecific enzyme actions.[9] Although DL-*p*-hydroxyphenylglycine can be easily synthesized in a good yield by the known methods,[1~3,10] the optical resolution by the above conventional methods are laborious and practically disadvantageous since they require two additional processes, *i.e.* conversion to N-acyl derivatives and then release of the corresponding free amino acids. From the practical view point, it is most desirable that the optical resolution is performed in the form of free amino acids or simple salts by the preferential crystallization procedures, which enable the desired optically active isomer to crystallize preferentially from a supersaturated solution of the racemic modification.

In our previous reports, optical resolution of DL-amino acids in the form of aromatic sulfonates by a preferential crystallization procedure has been developed and many amino acids have been resolved successfully by this method.[11~14] Namely, through the use of many different sulfonic acids, the salts crystallized as the racemic mixture resolvable by the preferential crystallization procedure can easily be selected. Succeeding to these works, we attempted to resolve DL-*p*-hydroxyphenylglycine by the same method for industrial production of optically active D-*p*-hydroxyphenylglycine. The advantages of this simple resolution method and the screening method for resolvable salts were described in our previous reports.[12,13]

In this study, we prepared a wide variety of aromatic sulfonates of DL-*p*-hydroxyphenylglycine, screened the salts, and selected the ones suitable for this resolution method, because DL-*p*-hydroxyphenylglycine itself, as well as many other common amino acids, formed a racemic compound unsuitable for the preferential crystallization procedure. As a result, the benzenesulfonate (DL-*p*-HPG·BS), the *o*-toluenesulfonate (DL-*p*-HPG·*o*-TS), the *p*-toluenesulfonate (DL-*p*-HPG·*p*-TS), the *p*-ethylbenzenesulfonate (DL-*p*-HPG·*p*-EBS), the sulfosalicylate monohydrate (DL-*p*-HPG·SS·H_2O) and the 2-naphthol-6-sulfonate (DL-*p*-HPG·NS) of DL-*p*-hydroxyphenylglycine (DL-*p*-HPG) were found to be crystallized from water or a dilute aqueous solution of the respective sulfonic acids as racemic mixtures suitable for this resolution method.

Of the sulfonates selected here, DL-*p*-HPG·*p*-TS has been previously reported to be resolvable.[15] However, this sulfonate could not always be crystallized from water as the typical salt which is composed of an equimolar amount of DL-*p*-HPG and *p*-toluenesulfonic acid (*p*-TS). Sometimes, a solid complex composed of 3 moles of DL-*p*-HPG and 2 moles of *p*-TS was deposited from water and the optical resolution in water was unsuccessful. Consequently, recrystallization and optical resolution were carried out in a dilute solution of *p*-TS. In both cases for DL-*p*-HPG·BS and DL-*p*-HPG·*p*-EBS, small amounts of the corresponding free amino acids were partly liberated from the dilute aqueous solutions of the respective sulfonates. Therefore, recrystallization and optical resolution of both sulfonates were also performed in an aqueous solution containing the respective sulfonic acids.

The elemental analyses and the physical properties of the selected sulfonates are shown in Table I and Table II. These racemic sulfonates were actually resolved into each of the antipodes in good yield by the preferential crystallization procedure described in the Experimental Section. The conditions for the resolution in each case and the analyses for separated crystals are summarized in Table III. These successful examples indicate that the preferential crystallization procedure by using aromatic sulfonates is generally applicable for the resolution of synthetic amino acids.

Among these sulfonates, DL-*p*-HPG·*o*-TS was most suitable for practical resolution, primarily because of its adequate solubility in water and characteristics suitable for easy handling. Therefore, detailed studies were made with DL-*p*-HPG·*o*-TS and the results of the successive resolution of DL-*p*-HPG·*o*-TS are shown in Table IV.

The optically active sulfonates of *p*-HPG obtained above were almost optically pure. If the optical purity is not satisfactory and further purification is required, the crude products can be purified by recrystallization without loss of the optically active isomer, because the optically active isomer no longer dissolves in the saturated solution of the racemic modifi-

TABLE II. PHYSICAL PROPERTIES OF AROMATIC SULFONATES OF *p*-HYDROXYPHENYLGLYCINE

Aromatic sulfonate	Isomer	mp, °C	$[\alpha]_D^{25}$, deg (c=1, water)	Solubility, g/100 ml 20°C	30°C	40°C	Solvent
p-HPG·BS	DL	221～223		15.4	24.1	35.0	0.5 M BS
	D	228～230	−68.9	7.7	11.0	16.7	
p-HPG·*o*-TS	DL	213～215 (dec.)		12.3	16.0	21.0	Water
	D	222～224 (dec.)	−66.6	7.7	9.6	12.2	
p-HPG·*p*-TS	DL	220～221 (dec.)		13.8	20.9	31.5	0.5 M *p*-TS
	D	226～227 (dec.)	−66.5	6.2	9.0	12.7	
p-HPG·*p*-EBS	DL	192～194		19.3	33.2	52.1	3 M *p*-EBS
	D	210～211	−62.7	10.3	14.2	21.8	
p-HPG·SS·H_2O	DL	217～218 (dec.)		18.3	26.6	37.6	Water
	D	219～221 (dec.)	−56.6	10.0	14.0	20.3	
p-HPG·NS	DL	218～219 (dec.)		9.0	14.5	17.5	Water
	D	227～228 (dec.)	−55.3	5.1	7.1	9.0	

TABLE III. OPTICAL RESOLUTIONS OF AROMATIC SULFONATES OF DL-*p*-HYDROXYPHENYLGLYCINE[a]

Aromatic sulfonate	Run[b] No.	Composition of solution: DL form, B, g	Active form, A, g	Solvent, ml	Crystallization: Seed crystals, S, g	Time, min	Temp., °C	Crystals separated: Yield, W, g	Optical purity P, %	Degree[c] of resolution, R, %
DL-*p*-HPG·BS	1 (D)	29.60	2.80	100	0.10	40	25	5.97	98.6	19.8
	2 (L)	29.60	2.93	(0.5 M BS)	0.10	40	25	6.03	99.9	20.9
DL-*p*-HPG·*o*-TS	1 (D)	24.00	2.50	100	0.10	70	30	5.30	97.0	21.2
	2 (L)	24.00	2.54	(water)	0.10	70	30	6.11	95.1	26.4
DL-*p*-HPG·*p*-TS	1 (D)	25.08	0	100	0.10	20	30	2.03	92.5	14.2
	2 (L)	25.08	1.78	(0.5 M *p*-TS)	0.10	10	30	3.78	94.7	13.6
DL-*p*-HPG·*p*-EBS	1 (D)	32.00	0	100	0.40	50	25	2.43	95.0	11.9
	2 (L)	32.00	1.91	(3 M *p*-EBS)	0.20	50	25	4.81	94.9	15.3
DL-*p*-HPG·SS·H_2O	1 (D)	25.50	0.50	100	0.10	120	25	1.63	67.4	3.9
	2 (L)	25.50	0.50	(water)	0.10	60	25	1.47	79.5	4.5
DL-*p*-HPG·NS	1 (D)	16.00	0	100	0.30	70	25	1.42	96.4	13.4
	2 (L)	16.00	1.07	(water)	0.10	80	25	2.31	96.9	13.4

[a] The procedure for the optical resolution of DL-*p*-HPG·*o*-TS is described in the text and other sulfonates were also resolved in the same manner.

[b] (D) or (L) means that D-isomer or L-isomer was preferentially crystallized by seeding D-isomer or L-isomer, respectively.

[c] $R=100\times(W\times P/100-A-S)\times 2/B$.

TABLE IV. SUCCESSIVE RESOLUTION OF *p*-HYDROXYPHENYLGLYCINE·*o*-TOLUENESULFONATE[a]

Run No.	Added DL form, g	Composition of solution: DL form, g	Active form, g	Crystallization: Time, min	Temp., °C	Crystals separated: Yield, g	Optical purity %
1 (D)		24.00	2.50	70	30	5.30	97.0
2 (L)	5.38[b]	24.00	2.54[c]	70	30	5.91	95.1
3 (D)	6.25[b]	24.00	2.98[c]	80	30	6.52	95.8
4 (L)	7.03[b]	24.00	3.17[c]	80	30	6.74	95.5
5 (D)	7.50[b]	24.00	3.17[c]	70	30	6.40	98.5
6 (L)	12.38[b]	29.40	3.03[c]	80	37	7.45	92.2
7 (D)	8.51[b]	29.40	3.74[c]	80	37	7.60	93.0
8 (L)	7.59[b]	29.40	3.23[c]	70	37	7.53	93.8
9 (D)	13.62[b]	35.00	3.73[c]	60	45	7.77	95.3
10 (L)	8.07[b]	35.00	3.57[c]	80	45	8.43	95.9
11 (D)	8.52[b]	35.00	4.41[c]	70	45	9.23	93.1

[a] Resolutions were carried out on a 100 ml scale using 0.10 g of seed crystals.

[b] The amount of DL-*p*-HPG·*o*-TS which was added to the mother liquor of the previous operation to adjust the concentration of the DL-form to that of the next column.

[c] Values calculated theoretically from the analysis of the crystals separated previously.

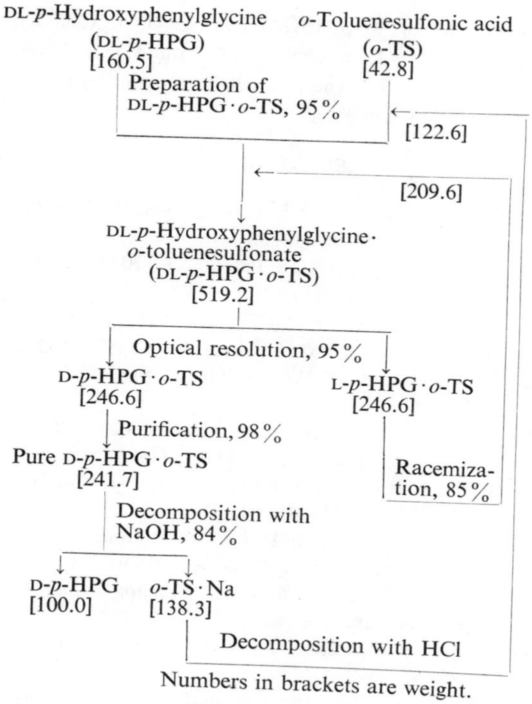

Numbers in brackets are weight.

SCHEME 1.

cation. An example for *p*-HPG·*o*-TS was shown in the Experimental Section. Thus obtained optically pure sulfonates were easily decomposed to optically pure amino acids by neutralization with alkali or by use of ion exchange resins.

On the other hand, undesired optically active L-*p*-HPG·*o*-TS could be completely racemized by dissolving it in water and heating at 140°C. Our proposed process for preparing D-*p*-HPG is shown in Scheme 1.

In the present work, we could not reveal a relationship between the kind of aromatic sulfonic acid and applicability for the preferential crystallization procedure. However, the practical optical resolution of DL-*p*-hydroxyphenylglycine could be accomplished by converting it to various aromatic sulfonates. Industrial production of D-*p*-HPG by the present simple method is considered to be very promising if combined with a proper synthetic method for DL-*p*-HPG.

Preparation of aromatic sulfonates of p-HPG

Aromatic sulfonates of *p*-HPG were easily prepared from *p*-HPG and an equimolar amount or a slightly excess of the corresponding aromatic sulfonic acids in aqueous solution. In the case of DL-*p*-HPG·*o*-TS, a mixture of DL-*p*-HPG (200.0 g) and 1.05 equimolar amount of *o*-TS·$2H_2O$ (261.6 g) was dissolved in water (800 ml) by heating, treated with charcoal, and cooled in a refrigerator. The resulting precipitates and the second crop obtained by concentration of the mother liquor to about a half volume were collected, washed with cold water and dried. The total yield of DL-*p*-HPG·*o*-TS was 393.5 g (96.9%). The products were almost pure and could be used for optical resolution without further purification. D- and L-*p*-HPG·*o*-TS were prepared in the same way. The racemic modifications and the optically active isomers of *p*-HPG·BS, *p*-HPG·*p*-TS, *p*-HPG·*p*-EBS, *p*-HPG·SS·H_2O, and *p*-HPG·NS were similarly prepared as above in a high yield (85~95%). For elemental analysis and determination of properties, *p*-HPG·*o*-TS, *p*-HPG·SS·H_2O, and *p*-HPG·NS were recrystallized from water. *P*-HPG·BS, *p*-HPG·*p*-TS, and *p*-HPG·*p*-EBS were recrystallized from aqueous solutions of 0.5 M BS, 0.5 M *p*-TS, and 3 M *p*-EBS, respectively. Results of the elemental analyses are shown in Table I.

The physical properties of the aromatic sulfonates thus obtained are shown in Table II.

Optical resolution

Optical resolution of the aromatic sulfonates of DL-*p*-HPG by the preferential crystallization procedure was carried out in the usual manner. In the case of DL-*p*-HPG·*o*-TS, DL-*p*-HPG·*o*-TS (24.00 g) and D-*p*-HPG·*o*-TS (2.50 g) were dissolved in water (100 ml) at an elevated temperature. The solution was cooled to 30°C, seeded with D-*p*-HPG·*o*-TS (0.10 g), and stirred at the same temperature. By refractometric and polarimetric measurements of the liquid phase, it was observed that preferential crystallization of D-*p*-HPG·*o*-TS occured, while L-*p*-HPG·*o*-TS remained in the solution. After 70 min, the precipitated crystals were collected by filtration, washed with a small amount of cold water, and dried to give D-*p*-HPG·*o*-TS (5.30 g), $[\alpha]_D^{25}$ −64.6° (c=1, water), optical purity 97.0%. In order to adjust the concentration of DL-*p*-HPG·*o*-TS in the mother liquor to the same as that in the previous operation, DL-*p*-HPG·*o*-TS (5.38 g) and a small amount of water were added to the mother liquor after the separation of D-*p*-HPG·*o*-TS. Amounts of the added DL-*p*-HPG were calculated from the analysis of the solution. Thus, the composition of the solution was the same as that in the initial state except that the solution contained the L-isomer in excess. By seeding this supersaturated solution with L-*p*-HPG·*o*-TS (0.10 g), preferential crystallization of L-*p*-HPG·*o*-TS was carried out in the same manner as described above. By repeating these procedures, D- and L-isomers were successively obtained as shown in Table IV.

Other sulfonates of DL-*p*-HPG could also be resolved in the same manner as described above. Conditions for resolution and the analyses for separated crystals are summarized in Table III.

Purification of optically impure D-p-HPG·o-TS

The optical isomers obtained by the above procedure were practically pure. However, if further optical purification is required, it can be performed as follows. Optically crude D-*p*-HPG·*o*-TS (20.00 g, optical purity 82.3%) was mixed with water (28.8 ml) for 20 hr at 20°C. The residual crystals were collected by filtration, washed with a small amount of cold water, and dried to give optically pure D-*p*-HPG·*o*-TS (16.20 g), the yield being 98.4% based on D-isomer in the original optically crude D-*p*-HPG·*o*-TS.

Preparation of optically active D-*p-HPG*

From the optically pure sulfonates of *p*-HPG obtained above, optically pure *p*-HPG was easily obtained by neutralization with alkali or by use of ion exchange resins. In the case of *p*-HPG·*o*-TS, optically pure D-*p*-HPG·*o*-TS (14.00 g) was dissolved in water (40 ml) at an elevated temperature and treated with charcoal. The solution was adjusted to pH 6 with 5 N sodium hydroxide and allowed to stand in a refrigerator overnight. The resulting precipitate was collected, washed with water, and dried to give D-*p*-HPG (5.82 g), $[\alpha]_D^{25}$ −158.4° (*c*=1, N–HCl). *Anal.* Calcd for $C_8H_9NO_3$: C, 57.48; H, 5.43; N, 8.38. Found: C, 57.63; H, 5.63; N, 8.29.

Preparation of DL-*p-HPG·o-TS by using recovered o-TS*

To the mother liquor obtained after the separation of D-*p*-HPG in the above experiment, DL-*p*-HPG (5.82 g) was added with 12 N HCl (3.5 ml) and dissolved by heating. The solution was concentrated to about 40 g, and cooled in a refrigerator. The resulting precipitates and the second crop obtained by concentration to about 13 g were collected, washed with cold water, and dried. The total yield of DL-*p*-HPG·*o*-TS was 12.10 g, $[\alpha]_D^{25}$ −11.3° (*c*=1, water). The product could be reused for resolution.

Racemization of optically active p-HPG·o-TS

A mixture of L-*p*-HPG·*o*-TS (4.00 g) and water (4 ml) was heated at 140°C for 12 hr in a sealed tube. The rate of racemization was as follows: 4 hr, 74.3%; 8 hr, 93.0%; 10 hr, 95.3%; 12 hr, 96.3%. The reaction mixture was cooled to 5°C. The resulting precipitates and the second crop obtained by concentration were collected, washed with cold water, and dried. The total yield of DL-*p*-HPG·*o*-TS was 3.4 g (85.0%), $[\alpha]_D^{25}$ +1.8° (*c*=1, water). *Anal.* Found: C, 53.06; H, 5.08; N, 4.11. The product itself could be reused for resolution step.

REFERENCES

1) L. B. Crast, Jr., U. S. Patent, 3489750 (1970) [*C. A.*, **72**, 100725_v (1970)].
2) L. B. Crast, Jr., U. S. Patent, 3517023 (1970) [*C. A.*, **73**, 66567_d (1970)].
3) A. A. W. Long, J. H. C. Nayler, H. Smith, T. Taylor and N. Ward, *J. Chem. Soc. C*, **1971**, 1920.
4) G. Bison and W. Wolfes, Ger. Offen. 2322412 (1974) [*C. A.*, **82**, 58112_x (1975)].
5) T. Shirai, Y. Tashiro and S. Aoki, Ger. Offen. 2449492 (1975) [*C. A.*, **83**, 43749_g (1975)].
6) C. T. Holdrege, U. S. Patent, 3796748 (1974) [*C. A.*, **80**, 121327_x (1974)].
7) J. M. H. Gubbels and J. P. M. Houbiers, Ger. Offen. 2355785 (1974) [*C. A.*, **81**, 105018_n (1974)].
8) S. Horiuchi, A. Takami, H. Yamada and S. Takenaka, Japan. Kokai 52–83345 (1977).
9) S. Horiuchi, A. Takami and U. Nakamura, Japan. Kokai 52–76483 (1977).
10) M. J. Elton, J. W. Herrison and A. Jackson, Ger. Offen. 2134251 (1972) [*C. A.*, **76**, 113526_p (1972)].
11) I. Chibata, S. Yamada, M. Yamamoto and M. Wada, *Experientia*, **24**, 638 (1968).
12) S. Yamada, M. Yamamoto and I. Chibata, *J. Agric. Food Chem.*, **21**, 889 (1973).
13) S. Yamada, M. Yamamoto and I. Chibata, *J. Org. Chem.*, **38**, 4408 (1973).
14) S. Yamada, M. Yamamoto, C. Hongo and I. Chibata, *J. Agric. Food Chem.*, **23**, 653 (1975).
15) T. Shirai, Y. Tashiro and S. Aoki, Ger. Offen. 2501957 (1975) [*C. A.*, **83**, 179614_s (1975)].

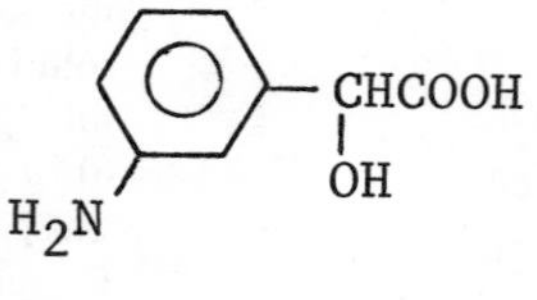

$C_8H_9NO_3$ Ref. 380

Resolution of *m*-Amino-mandelic Acid.—A suspension of 20 g. of crystalline cinchonine and 11 g. of *m*-aminomandelic acid in 500 cc. of water was heated on the water-bath for an hour, and shaken frequently. After cooling, the solution was filtered and allowed to stand in an open vessel at room temperature. The first crystals appeared after 4 days. Later, however, seeding decreased the time to a few hours. The crystals were roset-shaped and on drying fell to a powder. They were recrystallized from hot water (25 cc. of water for each g. of the dry salt); m. p., 156.5°. The pure crystals were dissolved again in hot water and the solution was cooled to room temperature, treated with ammonium hydroxide until all the cinchonine was precipitated and the solution gave the odor of ammonia. The cinchonine was filtered off and the filtrate was freed from ammonia by evaporation at room temperature over sulfuric acid in a vacuum desiccator. It was then acidified with hydrochloric acid and the evaporation was continued at room temperature. The dextrorotatory amino acid crystallized in a few days. It was recrystallized twice from water; m. p. 130° (decomp.); $[\alpha]_D$ = +33.80°. After several crops of mixed crystals had been removed from the mother liquor the levo product was obtained nearly pure by fractional recrystallization at ordinary temperatures. The yield was poor; m. p., 130° (decomp.); $[\alpha]_D$ = —33.65°.

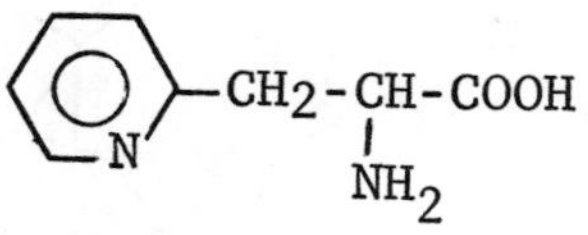

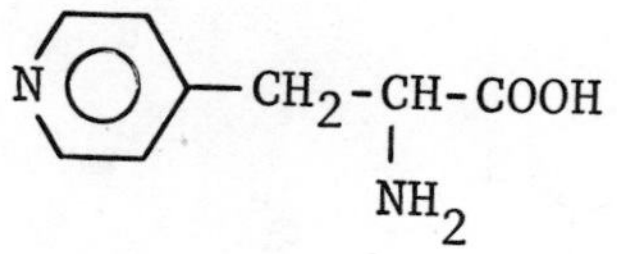

$C_8H_{10}N_2O_2$

Ref. 381

Salt of (−)-3-(4-Pyridyl)alanine with L-(+)-Tartaric Acid [(−)-I · II]. Warm solutions of 3.4 g of DL-3-(4-pyridyl)alanine (I) in 85 ml of water and 3 g of L-(+)-tartaric acid (II) in 7 ml of water were mixed, and the mixture was left at room temperature for 2.5 h. The precipitated salt was filtered off and dried at 60°. Yield 3.6 g. The salt was crystallized from water.

Salt of (+)-3-(4-Pyridyl)alanine with L-(+)-Tartaric Acid [(+)-I · II]. The filtrate from the isolation of the salt [(−)-I · II] was evaporated, and the residue was ground with absolute alcohol and dried. Yield 2.9 g. The salt was crystallized from 70% alcohol.

(−)-3-(4-Pyridyl)alanine [(−)-I]. A hot solution of 2.8 g of the salt [(−)-I · II] in 100 ml of water was passed through a column of Amberlite IR-4B (85 cc) at a rate of 25-30 drops per min. The amino acid was eluted with warm water until the ninhydrin reaction was negative. The aqueous solution was vacuum-evaporated, and the residue obtained was ground with absolute alcohol and dried. We obtained 1.4 g.

In an analogous way, from 4.4 g of the salt [(+)-I · II] we obtained 2.1 g of (+)-3-(4-pyridyl)alanine [(+)-I].

Salt of (+)-3-(2-Pyridyl)alanine with L-(+)-Tartaric Acid [(+)-III · II]. Hot solutions of 3.5 g of (III) in 14 ml of water and 3.2 g of L-(+)-tartaric acid in 3 ml of water were mixed, the mixture was left to cool to room temperature, 17 ml of alcohol was added, and the mixture was left for 2 h. The precipitated salt was filtered off and dried at 60°. Yield 2.9 g. For analysis the salt was crystallized from 80% alcohol.

Salt of (−)-3-(2-Pyridyl)alanine with L-(+)-Tartaric Acid [(−)-III · II]. The filtrate from the isolation of the salt [(+)-III · II] was vacuum-evaporated; and the residue was ground with absolute alcohol, filtered off, and dried at 60°. Yield 3.3 g. For analysis the salt was crystallized from absolute alcohol and vacuum-dried at 40°.

By the procedure described for [(−)-I], from 2 g of the salt [(+)-III · II] we obtained 0.9 g of (+)-3-(2-pyridyl)alanine [(+)-III] (for analysis it was crystallized from 80% alcohol), and from 2 g of the salt [(−)-III · II] we isolated 0.91 g of (−)-3-(2-pyridyl)alanine [(−)-III].

TABLE 1

No.	Compound	Yield, %	mp	$[\alpha]_D^{18}$
1	Salt (−)-I · L-(II)	87.5	224–228°	+6°±1 [c 1, $(C_2H_5)_3N$ in H_2O
2	Salt (+)-I · L-(II)	91	205–207	+27 (c 1, H_2O)
3	Salt (+)-III · L-(II)	87	184–185	+38 (c 1, H_2O)
4	Salt (−)-III · L-(II)	98.5	98–100	− 5 (c 1, H_2O)
5	(−)-I	95	248–250	−35 (c 1, 1 N HCl)
6	(+)-I	91	248–250	+35 (c 1, 1 N HCl)
7	(+)-III	82	199–200	+36 (c 1, 1 N HCl)
8	(−)-III	86	199–200	−35 (c 1, 1 N HCl)

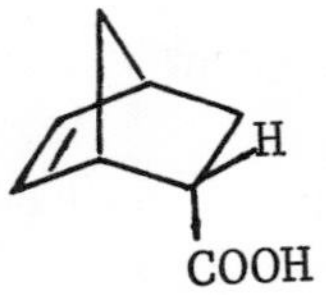

$C_8H_{10}O_2$ Ref. 382

(−)-endo-5-Carboxybicyclo[2.2.1]hept-2-ene (**7a**). To a boiling mixture of 50.0 g of cinchonidine in 3 l of acetone was added 24.0 g of pure *endo*-5-carboxybicyclo[2.2.1]hept-2-ene (**7a**). Cooling of the clear solution produced a salt $[\alpha]^{20}_{D}$ −84.8° (*c* 1.1, 99% ethanol), (47.0 g). Recrystallization six times from acetone gave the salt $[\alpha]^{19}_{D}$ −122° (*c* 1.2, 99% ethanol), (11.7 g). The salt was treated with 10% aqueous sodium hydroxide at room temperature. After acidification, the solution was extracted with ether and the extract was washed with saturated aqueous sodium chloride and dried over magnesium sulfate. After filtration and evaporation of the solvent, the residue was distilled to give 2.64 g of (−)-*endo*-5-carboxybicyclo[2.2.1]hept-2-ene (**7a**), bp 139—140°C/20 mmHg, $[\alpha]^{20}_{D}$ −119° (*c* 1.4, 95% ethanol), (lit, $[\alpha]^{26}_{D}$ −70.4°, *c* 1.4, 95% ethanol[10])).

Found: C, 69.45; H, 7.37%. Calcd for $C_8H_{10}O_2$: C, 69.54; H, 7.30%.

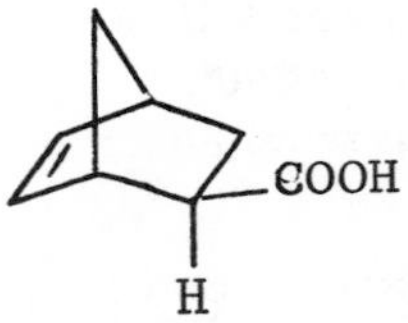

$C_8H_{10}O_2$ Ref. 383

Resolution of 2-*exo*-Norbornene-5-carboxylic Acid (34).—A solution of 142 g. of pure racemic acid[39a] and 320 g. of cinchonidine in 1800 ml. of hot absolute ethanol was allowed to cool to room temperature and stored overnight in the refrigerator. The solution was concentrated and triturated until crystallization began. The first crop of salt (260 g.), from which acid of $[\alpha]_D$ +6.64° (ethanol) could be regenerated, was recrystallized once from absolute ethanol to give 127 g. of salt from which acid of $[\alpha]_D$+10.0° (ethanol) was regenerated. The mother liquors from the salt recrystallizations were subjected to systematic triangular fractionation. From the tail fraction there was obtained salt (140 g.) which gave acid of $[\alpha]_D$ −11.7° (ethanol) (68% optically pure[10b]), m.p. 37–39°, after bulb-to-bulb distillation. The infrared spectrum was identical with that of a sample of the racemate. We have observed that samples of the *exo*-methyl ester, prepared from either optically active or racemic acid and diazomethane, sometimes contain a few per cent. of an impurity which is not the *endo*-ester and is not readily removed by distillation. Pure ester can be obtained by preparative gas chromatography. In this way, *exo*-acid of $[\alpha]_D$+13.6° (ethanol) gave ester of $[\alpha]_D$ +22.8° (ethanol). This requires a minor revision of the maximum rotation of the ester previously[10b] reported from 26.5° to 28.7°.

J. Am. Chem. Soc.,

(10) (a) J. A. Berson and D. A. Ben-Efraim, *ibid.*, **81**, 4094 (1959); (b) **81**, 4083 (1959).

(39) (a) C. D. ver Nooy and C. S. Rondestvedt, *J. Am. Chem. Soc.*, **77**, 3583 (1955); (b) K. Alder and G. Stein, *Ann.*, **525**, 247 (1936).

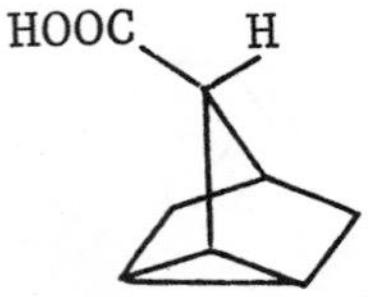

$C_8H_{10}O_2$ Ref. 384

Scheme I. Optical Activation of Nortricyclene-3-carboxylic Acid (7). The cinchonidine salt was prepared in methanol, 95% ethanol, and acetone, but crystals were obtained only from the latter solvent. A solution of 79.0 g of the acid[8] in 250 ml of acetone was heated to boiling on a steam bath, and 189 g of solid cinchonidine was added. After having been heated at reflux for 0.5 hr, the mixture was filtered, concentrated on the steam bath, and cooled to room temperature. The solution was allowed to stand for 8 days at room temperature, and the crystals were collected on a Büchner funnel. Four recrystallizations from boiling acetone gave a total of 110 g (wet) of salt in the head crop. A 50.0-g portion of this material was reconverted to carboxylic acid in the manner described for the ephedrine salt of 3-*endo*-methyl-2-*exo*-norbornyl acid phthalate,[4] yielding 14.6 g of optically active material as a greenish oil. Reaction with diazomethane in ether solution and evaporation of the solvent gave **methyl 3-nortricyclenecarboxylate (8)** which distilled as a clear oil shown to be 98.5% pure by capillary vpc. The nmr and infrared spectra were identical with those of the racemic compound. A sample was chromatographed on column E at 175° to give material of greater than 99.9% purity which was redistilled bulb to bulb. This material had $[\alpha]^{24.1}_{D}$ +15.3° (95% ethanol).

(4) Paper V of this series: J. A. Berson, R. G. Bergman, J. H. Hammons, and A. W. McRowe, *ibid.*, **89**, 2581 (1967).

(8) Racemic material was prepared according to J. D. Roberts, E. R. Trumbull, Jr., W. Bennett, and R. Armstrong, *J. Am. Chem. Soc.*, **72**, 3116 (1950).

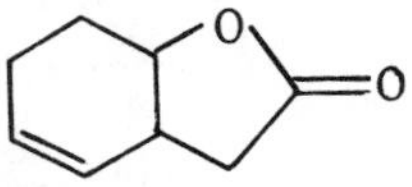

$C_8H_{10}O_2$ Ref. 385

Resolution of Lactone 1. The lactone 1[1] (1.10 g, 8 mmol) was added to a solution of lithium hydroxide (250 mg, 10 mmol) in 5 ml of water and 2 ml of dimethoxyethane. The solution was stirred for 10 hr at 25° and then evaporated to dryness. The white, crystalline residue was taken up in 2 ml of water and 5 ml of ethyl acetate and cooled to 0°. Oxalic acid solution (10% in water) was added at 0° until pH 3.5 was reached. Saturated sodium chloride solution (10 ml) was added, and the reaction mixture was extracted with four portions of ethyl acetate which were dried and evaporated at 10°. The residual hydroxy acid was taken up in 10 ml of ethyl acetate, and (−)-1-(1-naphthyl)ethylamine[3] (1.54 g, 9.0 mmol) was added. After standing for 24 hr at 25° and 12 hr at 0°, white crystals had formed. The crystals were filtered, dried, and recrystallized from ethyl acetate–methanol, giving a salt with $[\alpha]^{27}_{D}$ +59° (*c* 1.4, CH_3OH). An additional recrystallization gave a product with $[\alpha]^{27}_{D}$ +62° (*c* 0.8, CH_3OH). A final recrystallization gave 700 mg (53%) of a salt with $[\alpha]^{27}_{D}$ +63° (*c* 1.2, CH_3OH), indicating completion of the resolution.

Relactonization of the resolved (−)-1-(1-naphthyl)ethylammonium salt was effected by standard procedures. The salt (700 mg, 2.14 mmol) was dissolved in 1.2 equiv of aqueous sodium hydroxide solution. This solution was extracted with two portions of ether to remove the amine. The aqueous layer was acidified with 10 *N* hydrochloric acid and was extracted with five portions of methylene chloride. Tlc analysis of this solution indicated that relactonization had occurred spontaneously. The combined organic extracts were dried and evaporated, giving the lactone 1 (250 mg, 1.82 mmol, 85%) with $[\alpha]^{27}_{D}$ +28° (*c* 0.6, CH_3OH).

(1) E. J. Corey and T. Ravindranathan, *Tetrahedron Lett.*, **4753** (1971).
(2) E. J. Corey and B. B. Snider, *Tetrahedron Lett.*, **3091** (1973).
(3) Available from Norse Laboratories, Inc., Santa Barbara, Calif.

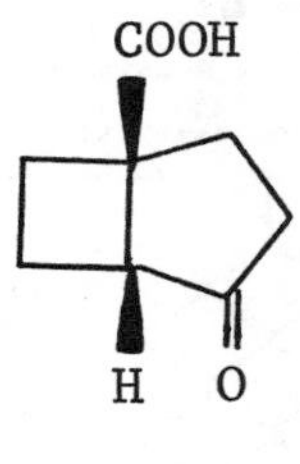

$C_8H_{10}O_3$ Ref. 386

Optical resolution of (±)-2

(a) *With quinine.* Quinine (17 g) and (±)-2 (8.0 g) were dissolved in hot 95% EtOH (70 ml). After cooling, the separated crystals were collected and recrystallized 4 times from EtOH to give 11.0 g of the quinine salt as prisms, m.p. 214–215°; $[\alpha]_D^{20}$ − 155° (c = 0.79, EtOH); ν_{max} ~ 3480 (w), ~2700 (m), ~2350 (m) 1740 (s), 1650 (w), 1625 (m), 1600 (m), 1520 (s), 1460 (s), 1440 (m), 1410 (s), 1255 (s), 1240 (s), 1040 (m), 860 (m) cm^{-1}. (Found: C, 70.12; H, 7.08; N, 5.82. $C_{28}H_{34}O_5N_2$ requires: C, 70.27; H, 7.16; N, 5.85%). This salt (11 g) was dissolved in N-HCl (50 ml). The soln was saturated with NaCl and thoroughly extracted with ether. The ether soln was washed with NaCl soln, dried ($MgSO_4$) and concentrated *in vacuo* to give 3.0 g of crystalline acid (2), $[\alpha]_D^{20}$ − 54.2° (c = 0.929, EtOH). The resolution was repeated to give 7.0 g of the partially resolved acid. This was recrystallized from ether–pet. ether. The racemate was less soluble and separated first. The ppts were removed and the mother liquor was concentrated *in vacuo*. The residue was again dissolved in ether–pet. ether. This operation was repeated for 4 times to give 2.0 g of low melting acid (2), m.p. 20–22°, $[\alpha]_D^{20}$ − 136° (c = 0.66, EtOH). The IR spectrum was identical with that of the racemate. However, polymorphism was observed in some cases resulting in a different IR spectrum, ν_{max} ~ 3400 (m), ~ 3200 (m), ~2600 (m), 1740 (vs), 1700 (vs), 1270 (m), 1250 (m), 1240 (m), 1210 (m), 1180 (s), 1105 (m), 1005 (w), 940 (w) cm^{-1}.

(b) *With cinchonine.* Cinchonine (12 g) and (±)-2 (6.0 g) were dissolved in hot 95% EtOH (75 ml). After cooling, the separated crystals were collected and recrystallized 4 times from EtOH to give 6.1 g of the cinchonine salt as prisms, m.p. 189–191°, $[\alpha]_D^{20}$ + 159° (c = 0.74, EtOH); ν_{max} ~ 3150 (m), 1720 (vs), 1580 (m), 1560 (m), 1500 (m), 1380 (s), 1310 (m), 1280 (m), 1160 (m), 1100 (m), 800 (m), 780 (m), 765 (m), 750 (m) cm^{-1}. (Found: C, 71.99; H, 7.23; N, 6.28. $C_{27}H_{32}O_4N_2$ requires: C, 72.29; H, 7.19; N, 6.25%). This salt (6.0 g) was dissolved in N-HCl (50 ml). The soln was thoroughly extracted with ether. The ether soln was washed with NaCl soln, dried ($MgSO_4$) and concentrated *in vacuo* to give 2.0 g of crystalline acid 2′, $[\alpha]_D^{20}$ + 37.1° (c = 1.02, EtOH).

(c) *With brucine.* Brucine (15 g) and (±)-2 (7.5 g) were dissolved in hot 99% EtOH (60 ml). After cooling, the separated crystals were collected and recrystallized 3 times from EtOH to give 4.0 g of the brucine salt as prisms, m.p. 115–118°, $[\alpha]_D^{20}$ − 6.8° (c = 0.887, EtOH); ν_{max} 3550 (m), ~3400 (m), 1740 (s), 1650 (vs), 1620 (m), 1510 (vs), 1425 (m), 1380 (m), 1300 (m), 1215 (m), 1115 (m), 850 (m) cm^{-1}. (Found: C, 67.82; H, 6.65; N, 4.93. $C_{31}H_{36}O_7N_2$ requires: C, 67.86; H, 6.61; N, 5.11%). This was dissolved in N-HCl (40 ml). The soln was thoroughly extracted with ether. The ether soln was washed with NaCl soln, dried ($MgSO_4$) and concentrated *in vacuo* to give 1.0 g of oily acid 2′, $[\alpha]_D^{22}$ + 83° (c = 0.762, EtOH). This crude 2′ was purified by fractional crystallization in the manner described for 2. Thus 3.1 g of crude 2′ gave 1.9 g of purified 2′, m.p. 20–23, $[\alpha]_D^{22}$ + 116° (c = 0.384, EtOH), The IR spectrum was identical with that of 2.

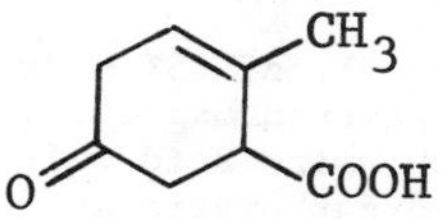

$C_8H_{10}O_3$ Ref. 387

Resolution of the Keto-acid rac-(21).—To a cooled (15°) solution of the keto-acid (14·81 g.) in acetone (150 c.c.) (−)-quinine (31·2 g.) was added gradually with shaking. The homogeneous solution was kept for 18 hr. at −7°. The salt was collected, washed with ice-cold acetone (300 c.c.), and dried (wt. 20·7 g.); it had $[\alpha]_D^{22}$ −232° (in $CHCl_3$). To its solution in chloroform (400 c.c.) were added water (18 c.c.) and aqueous 2N-sodium hydroxide (21·7 c.c.); the mixture was shaken, the phases were separated, and the chloroform phase was re-extracted, first with water (39 c.c.) and aqueous 2N-sodium hydroxide (1·1 c.c.) and then with water (20 c.c.). The combined alkaline phases were extracted with chloroform (40 c.c.); the extract was discarded. The alkaline phase was then acidified at 0° with 2·3N-hydrochloric acid (18·9 c.c.), and the product was extracted with chloroform (4 × 200 c.c.). The extract was dried and evaporated under reduced pressure giving an oil (5·75 g.). Crystallisation from ether gave material (4·34 g.), m.p. 98—100°, $[\alpha]_D^{22}$ −375° (in $CHCl_3$), and material (0·14 g.), m.p. 95—100°, $[\alpha]_D^{22}$ −355° (in $CHCl_3$). Optically pure (−)-*keto-acid* (21) had m.p. 101°, $[\alpha]_D^{22}$ −386° (in $CHCl_3$) (Found: C, 62·5; H, 6·4%). The (−)-quinine salt had $[\alpha]_D$ −270° (in $CHCl_3$).

The mother liquor and washings from the separation of the crude (−)-acid (−)-quinine salt were evaporated under reduced pressure, and the residue (25·3 g.) was dissolved in chloroform (400 c.c.); the crude (+)-acid was regenerated in a manner similar to that used for the enantiomer. This gave material (7·66 g.) which on crystallisation from ether yielded material (4·14 g.), $[\alpha]_D^{20}$ +373° (in $CHCl_3$) and material (0·41 g.), $[\alpha]_D^{20}$ +380° (in $CHCl_3$). Pure (+)-*keto-acid ent*-(21) had m.p. 100°, $[\alpha]_D^{22}$ +385° (in $CHCl_3$) (Found: C, 62·35; H, 6·35%); the (−)-quinine salt had $[\alpha]_D$ +20° (in $CHCl_3$).

From the resolution was recovered racemic keto-acid (1·18 g.), m.p. 67—69°, suitable for recycling.

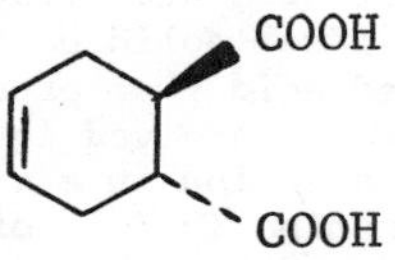

$C_8H_{10}O_4$ Ref. 388

b. **One equivalent quinine**

Using the method of Walborsky (38), 190 g. racemic and 138 g. of quinine in 1100 ml. methanol yielded a first crop of 29 g. of enriched (+) acid salt, m.p. 200-202°, which on decomposition gave 6.3 g. (+) acid. $[\alpha]_D^{30}$ +42.6°, m.p. 142-146°. Concentration of the mother liquor and changing to ethyl acetate solvent gave an amorphous precipitate of salt, which on recrystallization and decomposition with dilute acid and extraction, yielded from benzene 5.04 g. of racemic acid, m.p. 170-171°. Concentration of the filtrates above gave only viscous organic gels which were decomposed with 30% hydrochloric acid. The product which precipitated was slightly yellow racemic acid, 25 g., m.p. 167-168°. Extraction of the hydrochloric acid solution with ether gave, after drying, evaporation in vacuo, and precipitation from benzene, two further crops of racemic acid, 7.0 g., m.p. 167-168°, and 5.0 g., m.p. 166-167°.

The viscous ethyl acetate mother liquor was hydrolyzed in 25% hydrochloric acid, and the aqueous solution (1.2 l.) was extracted with six 300 ml. portions of ether. The yellow organic solution was decolorized with activated charcoal, dried over anhydrous magnesium sulfate, filtered, and evaporated in a stream of nitrogen. The residual solution and solid was swirled with 250 ml. of benzene and filtered by suction. The product collected weighed 34.5 g., m.p. 161-162°, and was enriched in (-)-acid.

The benzene-ether filtrate above was heated on a steam bath to remove excess ether. On cooling, several consecutive crops were collected from the benzene solution: 13.5 g., m.p. 154-155°; 11.5 g., m.p. 161-162°; and 7.0 g., m.p. 130-132°, enriched in (-) acid. The fractions enriched in (-) acid were combined for further resolution with cinchonine.

c. Two salt simultaneous resolution

The general scheme of the resolution is charted in Figure XII, p. 91.

A solution of 73.5 g. of partially enriched (+) acid from previous resolutions, in 500 ml. of absolute ethanol was brought to a boil. To this was carefully added a solution of 53.2 g. of (-)-$\underline{\alpha}$-methylbenzylamine (89) in 200 ml. absolute ethanol and the clear solution concentrated to 500 ml. and allowed to cool overnight. The precipitated solid was collected and immediately recrystallized from ethanol. The original mother liquor was concentrated to one-half volume. After standing 24 hours, the head fraction yielded 43.2 g. of salt, m.p. 140-159°. The filtrate was again concentrated to one-half volume. An additional 4.94 g., m.p. 149-151°, of solid precipitated. After three recrystallizations of the respective fractions, three crops of solid were obtained, m.p. 173-174°, 33 g. (total), which were combined and held for final purification.

The mother liquors, enriched in (-)-acid, from the recrystallizations of the head fractions were combined and hydrolyzed with 30% hydrochloric acid. The acidic solution was extracted with four 300 ml. portions of ether. These were decolorized, dried over magnesium sulfate, and evaporated in vacuo to dryness. The residual solid obtained, together with other (-) enriched acid from procedure b. above (80.7 g. total) was dissolved in 1200 ml. ethyl acetate. To the solution in a 4 l. beaker, a slurry of 85 g. of cinchonine in 2 l. of boiling ethyl alcohol was added. (38) The mixture became clear almost immediately and no solid was visible. The solution was boiled vigorously on a steam bath to remove the ethanol as an azeotrope. (144) Additional ethyl acetate was added occasionally. When the volume had been concentrated to about 800 ml., crystal formation began on the surface of the glass. The solution was set aside to cool overnight after seeding with crystals from a preliminary trial cinchonine resolution. A crystalline product, 26.8 g., was collected. The salt was recrystallized from acetone and the mother liquor concentrated successively as indicated by the generalized resolution scheme indicated in Figure XII, p. 91.

After four recrystallizations, crops of m.p. 173-176° were combined and recrystallized from acetone. All cinchonine salts turned pink about one degree below the melting point. The solid obtained had m.p. 176-177°, $[\alpha]_D^{29}$ +10.6° (c 0.7, methanol). Decomposition of a fraction of the salt with 25% hydrochloric acid and isolation gave (-) acid, m.p. 145.5-147.5°, $[\alpha]_D^{29}$ -122°, 72.8% optically pure. (38) Cinchonine salt fractions of rotations +11°, +10°, +8.4° and +9.8° were combined and again recrystallized from acetone. The crystalline product obtained had an m.p. 175-176° (pink at 174°), $[\alpha]_D^{28}$ +8.9 ± 0.6°, lit. (38) m.p. 175-177°, $[\alpha]_D^{26}$ +8.8 ± 1.6°.

Decomposition of the salt with 25% hydrochloric acid, extraction with ether and workup gave, on recrystallization from benzene-petroleum ether, (-)-trans-4-cyclohexene-1,2-dicarboxylic acid, 23r, m.p. 147-148°, $[\alpha]_D^{28}$ -165.9 ± 0.6° (c 0.9, methanol), lit. (38) m.p. 144-146°, $[\alpha]_D^{26}$ -160.0 ± 1.3° (c 2.7, 95% ethanol).

Figure XII

General Two Salt Resolution

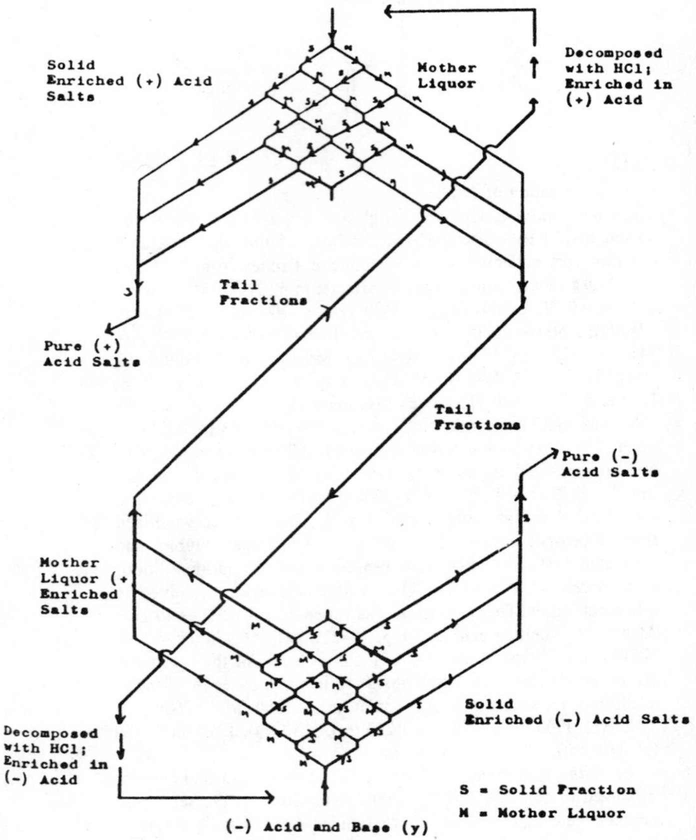

91

The combined tail fractions from the cinchonine resolution were decomposed with 25% hydrochloric acid. The acidic solution was extracted into ether, which was dried, filtered, and evaporated in vacuo, yielding 35.5 g. of partly enriched (+) acid which was returned for further resolution with (-)-$\underline{\alpha}$-methylbenzylamine.

All fractions of (-)-$\underline{\alpha}$-methylbenzylamine sale, 22d, of m.p. 171-176° were combined. The salts were recrystallized from 1.2 l. (5:1) acetone-ethanol. The product, m.p. 174-176°, had a rotation $[\alpha]_D^{28}$ +87.1°. Three additional recrystallizations from (5:1) acetone-ethanol gave colorless crystals, m.p. 174.5-176°, $[\alpha]_D^{27}$ +90.2 ± 0.6° (c 0.8, methanol).

Anal. Calculated for $C_{16}H_{21}NO_4$: C, 65.96; H, 7.27; O, 21.97. Found: C, 65.95; H, 7.26; O, 21.88.

Evaporation of the filtrate from the above fractions to dryness yielded a salt, m.p. 175.5-176.5°, $[\alpha]_D^{27}$ +90.5° (c 0.6, methanol).

A solution of 300 ml. of 25% hydrochloric acid in a 1 l. beaker was stirred magnetically and 53.0 g. of (-)-$\underline{\alpha}$-methylbenzylamine acid salt of (+)-trans-4-cyclohexene-1,2-dicarboxylic acid was slowly added. All of the salt readily dissolved, and no exotherm was noted. The colorless acidic solution (pH 1, test paper) was extracted with five 200 ml. portions of ether. The ethereal solution was dried overnight over 12 g. anhydrous magnesium sulfate, then filtered and evaporated in vacuo slowly. The residual

(38) H.M. Walborsky, L. Barash and T.C. Davis, Tetrahedron 19, 2333 (1963).

colorless solid was dissolved in minimum hot benzene, transferred to a 500 ml. Erlenmeyer flask quantitatively and saturated with ligroin (60-75°). After cooling two hours, the crystalline product was collected by suction and air dried. (+)-(1S:2S)-trans-4-cyclohexene-1,2-dicarboxylic acid, 23s, weighed 29.3 g., m.p. 145-146°, $[\alpha]^{28}_D$ +167.1 ± 1.1° (c 0.3, methanol), lit. (38) m.p. 144-146°, $[\alpha]^{26}_D$ -160.0 ± 1.3° (c 2.7, 95% ethanol, enantiomer). Concentration of the filtrate gave an additional 1.03 g. of acid, m.p. 146.5-147.5°.

$C_8H_{10}O_4$ Ref. 388a

A. Preparation of the Quinidine Salt. In a 1-l. flask were refluxed until homogeneous anhydrous quinidine (Brocades, The Hague, The Netherlands, or Lamers & Indeman, Bois-le-Duc, The Netherlands) (81 g), 14 (21.2 g) (2 mol of base to 1 mol of acid), and alcohol (500 ml). In case the mixture was not homogeneous after 30 min, it was filtered. The alcohol was removed with a rotatory evaporator, and to the residue were added water (500 ml) and alcohol (150 ml); on heating a clear solution was obtained which was left to cool. Every time it became turbid, the turbidity was removed by the addition of some alcohol (about 60 ml of alcohol was necessary). Seeding is recommended. After standing for 2 days at room temperature the crystals were removed by filtration and recrystallized. We used water (6.5 l.) and alcohol (2.5 l.) for the recrystallization of quinidine salt from 14 (292 g) and quinidine (1120 g), seeded the mixture and left it to crystallize for 2 days, and obtained 486.25 g of salt.

B. Regeneration of the Quinidine Salt. Recrystallized quinidine salt (108 g) was stirred with $CHCl_3$ (340 ml) and a solution of NaOH (20 g) in water (120 ml) for 1.5 hr. Then the $CHCl_3$ layer containing the alkaloid was separated from the alkaline layer containing the sodium salt of (1*R*,2*R*)-**14**. The latter solution was acidified with concentrated HCl, and the dicarboxylic acid was isolated by ether extraction. From quinidine salt (486.25 g from 292 g of racemic **14**) was thus obtained (1*R*,2*R*)-**14** (98.0 g, 67.1% of one antipode), $[\alpha]_D$ −161° (*c* 0.5, absolute EtOH) [lit.[5] $[\alpha]_D$ −161° (*c* 2.7, EtOH).

The mother liquor of the resolution of **14** (292 g) with anhydrous quinidine (1120 g) was evaporated to dryness to give an oil, which was treated with NaOH solution and $CHCl_3$, etc., as indicated above to give optically impure (1*S*,2*S*)-**14** (166.5 g), $[\alpha]_D$ +90° (*c* 0.5, absolute EtOH).

[5] H.M. Walborsky, L. Barash and T.C. Davis, Tetrahedron 19, 2333 (1963).

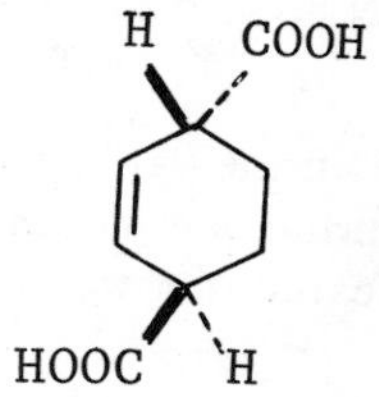

$C_8H_{10}O_4$ Ref. 389

l-trans-Δ^2-*Tetrahydroterephthalic Acid.*—30 G. of the *trans*-Δ^2-acid and 141 g. of brucine were dissolved in 1800 c.c. of hot water, and the salt allowed to crystallise (colourless prisms). The product was crystallised four times from water and the rotation of the brucine salt, dried in a vacuum at 100°, was tested after each crystallisation: $[\alpha]^{16°}_{5461}$ (for 0·4 g. in 50 c.c. of chloroform) − 92·5°, − 112·0°, − 116·0°, − 117·0°, − 117·0°. The last fraction (8 g.) was pure and the brucine was removed by treatment with ammonia, filtration, and extraction with chloroform. The ammonium salt was acidified, and the *trans*-Δ^2-acid extracted with ether. It was crystallised from water; m. p. 222° (Found: C, 56·3; H, 5·8%). 0·2 G. of the acid in 50 c.c. of ethyl alcohol at 16° had a rotation of − 4·465° for the Hg green line in a 4 dm. tube. Hence $[\alpha]^{16°}_{5461}$ = − 279°. The brucine mother-liquors yielded a dextrorotatory acid, $[\alpha]^{16°}_{5461}$ = + 132°. Attempts to get the dextrorotatory acid pure failed. The strychnine salt can be crystallised to give the lævo-acid, but attempts with the acid and neutral salts of quinine, cinchonine, cinchonidine, morphine, and nor-ψ-ephedrine gave only slightly active acids.

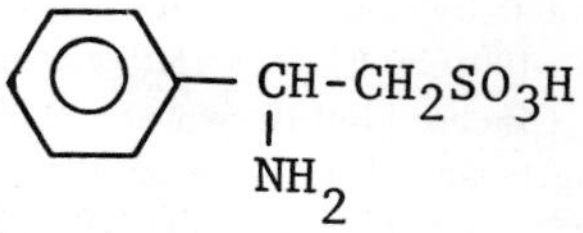

$C_8H_{11}NO_3S$ Ref. 390

Barium salt of N-formyl-β-phenyltaurine. The ammonium salt of N-formyl-β-phenyltaurine, which was prepared from acetophenone-ω-sulfonic acid after Leuckhart [1] was heated with freshly precipitated barium carbonate in a steam-distillation apparatus until the evolution of ammonia ceased. The excess carbonate was filtered off and the filtrate evaporated to dryness. The substance was recrystallized twice from water.

Found %: Ba 18.98, 18.64. $C_9H_{10}O_4NSBa_{1/2}$. Calculated %: Ba 19.06.

Resolution of β-phenyltaurine into antipodes. A solution of 38 g of the barium salt of N-formyltaurine and one of α-phenylethylamine sulfate, prepared by neutralizing 9.7 g of the free amine ($[\alpha]^{20}D$ −40°) with 2 N sulfuric acid to the neutral point (methyl red indicator), were mixed. The barium sulfate precipitate was filtered out and boiled with water. The combined filtrates were evaporated to the point of incipient crystallization, and the precipitate obtained was recrystallized from water until the melting point and value of specific rotation became constant.

Results of the experiments are given in the table.

No. of recrystallization	Weight of precipitate (in g)	Melting point	Weighed portion (in g)	Weight of solution (in g)	α	$[\alpha]_D^{20}$
0	16.2	110—112°, 135—138	0.9547	8.33	−2.72	−23.7
1	8.87	139—141	0.9492	8.23	−6.73	−58.2
2	4.91	140—141	0.9103	8.75	−6.69	−65.1
3	2.83	140—141	0.9100	7.91	−7.72	−67.5

On the fourth recrystallization $[\alpha]^{20}D$ did not change.

The salt (2.83 g) obtained after three recrystallizations was dissolved in water, and 18 ml of NaOH (0.4977 N) was added with cooling; the α-phenylethylamine which separated out was extracted with ether. The specific rotation of the resulting solution of the sodium salt of N-formyl-β-phenyltaurine amounted to −83.1°. A 23-ml quantity of concentrated hydrochloric acid was added to the solution, and the latter was evaporated in vacuo at 40-50°. The precipitate formed was filtered out and twice recrystallized from hot water. M.p. 317-320°, $[\alpha]^{20}D$ +29.1° (water, c 17). The substance gave a qualitative reaction for sulfur and nitrogen; according to the data of elementary analysis, it was β-phenyltaurine.

Found %: C 47.61, 47.54; H 5.72, 5.65; N 6.88, 6.81. $C_8H_{11}O_3NS$. Calculated %: C 47.72; H 5.50; N 6.96.

The value of specific rotation of β-phenyltaurine varies with the pH of the medium. In 10% NaOH solution $[\alpha]^{20}D$ +14.9°. Optically active β-phenyltaurine does not racemize on heating with alkalis. A solution of 0.3916 g of β-phenyltaurine in 10% alkali was heated in a water bath for ten hours, after which the rotation was measured: $[\alpha]^{20}D$ +14.9°.

[1] A.P. Terent'ev, V.M. Potapov and I.Z. Semion, J. Gen. Chem. 26, 2934 (1956).*

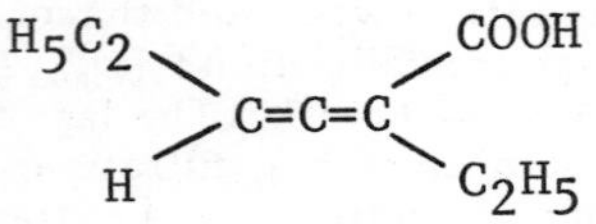

$C_8H_{12}O_2$ Ref. 391

*(+)- und (−)-***1d**: Nach den allgemeinen Vorschriften erhielt man aus 2.358 g racem. **1d** und 2.433 g Chinchonidin nach vier Umkristallisationen aus Aceton 1.131 g (32%) als farblose Nadeln mit Schmp. 149°C (Rotfärbung). $[\alpha]_{589}^{25} = -36.6°$, $[\alpha]_{350}^{25} = -162°$ ($c' =$ 0.005666, C_2H_5OH).

$C_{27}H_{34}N_2O_3$ (434.6) Ber. C 74.62 H 7.89 N 6.45 Gef. C 74.91 H 7.72 N 6.65

Aus dem Filtrat der ersten Kristallisationsfraktion wurden 1.158 g (49%) optisch unreines (−)-**1d** gewonnen; Sdp. 96−100°C/0.07 Torr (Kugelrohr). $[\alpha]_{589}^{25} = -12.4°$ ($c' = 0.039852$, C_2H_5OH). − Aus 3.552 g Cinchonidinsalz wurden 0.694 g [55%] (+)-**1d** erhalten; Sdp. 103−106°C/0.08 Torr (Kugelrohr). $[\alpha]_{589}^{25} = +53.4°$ ($c' = 0.016468$, C_2H_5OH). − CD (C_2H_5OH): λ_{max} ($\Delta\varepsilon$) = 285 (−0.0020), 267 (+0.020), 244 (+0.17), 215 nm (+1.3).

$C_8H_{12}O_2$ (140.2) Ber. C 68.54 H 8.63 Gef. C 68.17 H 8.68

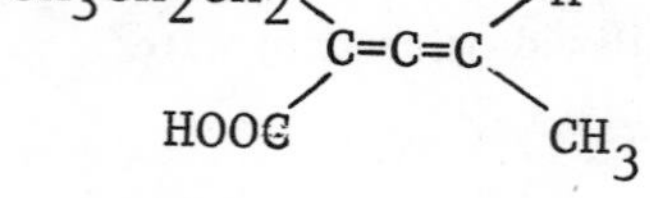

$C_8H_{12}O_2$ Ref. 392

The Resolution of 2,3-Octadien-4-carboxylic Acid into Its Optically Active Components.—All optical measurements were made on a Bausch and Lomb saccharimeter in chloroform solutions using a two decimeter tube. A chloroform solution of 2.1735 g. (0.0065 mole) of strychnine and 1.9225 g. (0.0125 mole) of III, after standing for 12 hours, was evaporated to a thick sirup, 5 cc. of ethyl acetate was added causing a white precipitate to form. Six recrystallizations from a mixture of chloroform and ethyl acetate gave a white solid, m.p. 126–127°,[14] $[\alpha]^{25}D$ −72.69°. *Anal.* Calcd. for $C_{30}H_{36}N_2O_4$: C, 73.8; H, 7.4; N, 5.9. Found: C, 73.8; H, 7.3. N, 5.6.

Decomposition of the Salt with Solvent.—A small portion of the salt was dissolved in chloroform and allowed slowly to evaporate until crystals were formed. The remaining solvent was then removed under vacuum. The residue was extracted with ether, and the ether soluble material crystal-

(14) After collapsing at 126–127°, there remains a mixture of liquid melt and solid up to the point where the solid begins to turn dark, usually about 140°.

lized from petroleum ether (30–60°) to a melting point of 46–47°.

Decomposition of the Salt with Acid.—The salt was stirred with a mixture of dilute hydrochloric acid and ether. The aqueous layer was washed with three portions of ether, all the ether layers were combined and the ether soluble material crystallized from petroleum ether (30–60°) to a maximum constant melting point of 49–50° and a rotation of $[\alpha]^{25}D$ −30.69°. *Anal.* Calcd. for $C_9H_{14}O_2$: C, 70.3; H, 9.1. Found: C, 70.2; H, 9.1.

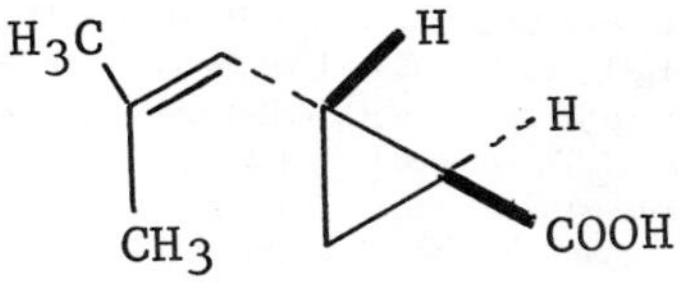

$C_8H_{12}O_2$ Ref. 393

Samples of *trans*-**4** from several isomerizations, 9.6 g (68 mmol), were allowed to react with 22.0 g (68 mmol) of quinine, and the resulting salt was partially resolved by recrystallization from 40:60 ethyl acetate–diethyl ether. After four recrystallizations, treatment of a less soluble fraction with dilute hydrochloric acid followed by extraction with diethyl ether yielded 1.54 g of optically active carboxylic acid, $[\alpha]^{25}D$ +72.5° (*c* 4.95, CH_3OH). From a more soluble fraction of the quinine salt, (−)-*trans*-**4**, $[\alpha]^{25}D$ −90.7° (*c* 4.84, CH_3OH), was obtained: nmr δ ($CDCl_3$) 0.7–2.3 (4, m, H at C_1, C_2, and C_3), 1.67 and 1.74 (6, two d, methyls at $C_{2'}$, $J \simeq$ 1.5 Hz), and 4.63 ppm (1, d of septets, H at $C_{1'}$, $J_{2,1'}$ = 9 Hz).

Anal. Calcd for $C_8H_{12}O_2$: C, 68.55; H, 8.63. Found: C, 68.80; H, 8.75.

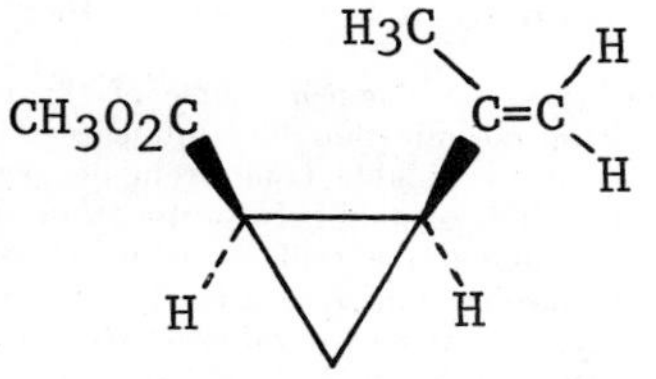

$C_8H_{12}O_2$ Ref. 394

Resolution of *cis*-II (Acid) with Ephedrine. A solution of *cis*-II (acid) (0.63 g) in ethyl acetate (2 ml) combined with a solution of ephedrine (0.83 g) in ethyl acetate (3 ml) was diluted with 15 ml of petroleum ether (45–60°). On keeping 16 hr at room temperature 0.3 g of the salt crystallized: mp 85–92°, $[\alpha]^{27}_{578}$ −46.5° (*c* 0.560, MeOH). A first recrystallization from 8 ml of ethyl acetate–petroleum ether (1:2) for 8 hr at 5° gave 0.2 g of salt, $[\alpha]^{27}_{578}$ −55.4° (*c* 0.470, MeOH); a second recrystallization gave 0.13 g, $[\alpha]^{27}_{578}$ −60.8 (*c* 0.360, MeOH). Fractional crystallization of the filtrates furnished an additional 65 mg of salt of this purity. The two samples were combined; acid was regenerated by acidifying with cold dilute HCl and extracting with ether. The extracts were treated with ethereal diazomethane, dried over $MgSO_4$, and concentrated by distillation to a residue which was isolated pure (80 mg) by preparative glpc (column C at 125°): $[\alpha]^{27}_{578}$−14.6°, $[\alpha]_{365}$ −59.4° (*c* 0.281 MeOH); nmr spectrum identical with that of (−)-*cis*-II obtained from the rearrangement of (−)-trans-II.

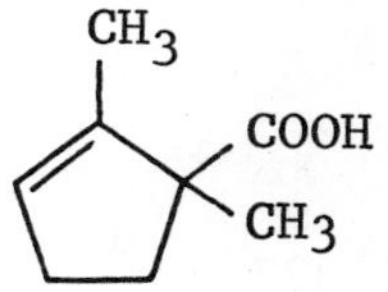

$C_8H_{12}O_2$ Ref. 395

Resolution of (±)-*1,2-Dimethyl-2-cyclopentenecarboxylic acid* [5][11]. Equimolar amounts of *rac*-acid and (+)-dehydroabietyl amine [prepared from the acetate (*Fluka AG*) of constant m.p. 144–145° by decomposition in aqueous 10% KOH-solution at 100°] were dissolved in CH_3OH. Evaporation of the solvent afforded a crystalline mass which by fractional crystallization in benzene gave the (−)-diastereoisomeric salt with a constant optical rotation, $[\alpha]^{20}_D = -22°$ (0.9). This salt was dissolved in CH_2Cl_2 and decomposed by shaking with aqueous satd. $NaHCO_3$ solution. The aqueous layer was brought to pH 2 with 2N H_2SO_4 and extracted with CH_2Cl_2. Distillation of the crude product at 75°/0.1 Torr gave (S)-(−)-*1,2-dimethyl-2-cyclopentenecarboxylic acid*; $[\alpha]^{20}_D = -152°$ (0.8).

[11]) For the analytical data of the racemic compounds see [5]. The spectral and chromatographic comparisons with the optically active substances were satisfactory in all cases.

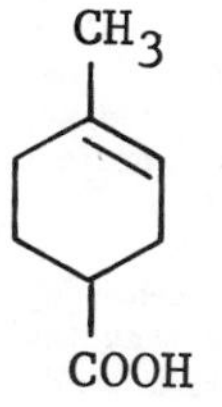

$C_8H_{12}O_2$ Ref. 396

Attempted resolution. The alkaloids brucine, cinchonidine, morphine, quinidine, uinine and strychnine, the synthetic bases (+)-α-phenyl-ethylamine and (+)-α-(2-naphthyl)-ethylamine and (−)-α-phenylethylthiuronium chloride [11] were all ested in the solvents acetone, methanol, ethanol and ethyl acetate. Significant esults were obtained only with brucine in methanol and ethanol ($[\alpha]^{25}_D$ of acid −27° and −20° respectively), with naphthylethylamin in all solvents ($[\alpha]^{25}_D$ ranging rom +15° to 25°) and with the thiuronium chloride in methanol (−20°).

50.0 g (0.357 mole) of the racemic acid and 141 g (0.357 mole) of brucine were issolved by warming in 1400 ml of methanol. After several hours at room temerature and one night in the refrigerator 111 g of salt were obtained. The acid rom this had $[\alpha]^{25}_D = -21.5°$. The salt was recrystallized from methanol eight imes, after which 53.5 g remained. The acid was liberated and recrystallized twice from light petrol, yield 8.3 g m.p. 96–99°. Equiv. wt. calc. 140.2, found 140.6 $[\alpha]^{25}_D = -58.2°$.

The mother liquor from the first crystallization was evaporated to dryness and the acid liberated, yielding 17.8 g with $[\alpha]^{25}_D = +29.0°$.

This acid was dissolved in ether and mixed with an ethereal solution of (+)-α-(2-naphthyl)-ethylamine obtained from 31 g of the neutral sulphate. The salt obtained was recrystallized from mixtures of acetone with methanol (5–10 % by volume) the latter being omitted for the last four recrystallizations. The course of the resolution is shown in Fig. 1. After the thirteenth recrystallization only 2.0 g of salt remained, and the acid from this was liberated and recrystallized from *n*-heptane. It then had $[\alpha]^{25}_D = +99.2°$ with m.p. 98–99°. All optical activities were measured in absolute alcohol.

Author's comments:

In spite of great efforts the present author could not find a satisfactory procedure for the resolution, and as the necessary amount of optically active acid was not obtained by the method finally adopted, an alternative approach was made to the original problem [1]. The resolving bases used in the present investigation were brucin in methanol ((−)-acid) and (+)-α-(2-naphthyl)-ethylamine in acetone ((+)-acid). The resolution was interrupted at values of $[\alpha]^{25}_D$ of −58° and +99° respectively (abs. alcohol); the latter value was reached after thirteen recrystallizations but it is doubtful if this is the maximum activity. Fisher and Perkin obviously believed they had obtained the pure antipodes with values of $[\alpha]_D$ of

$-58°$ and $+56°$ respectively. It is notable that they reported melting points "about 99°C".

It is not clear wheather the sluggish course of the resolution, illustrated by Fig. 1, is due to a partial racemization. The influence of the resolving base might cause a transformation of the double bond from one side of the methyl group to the other. Another possibility would be a protonation of the double bond during the isolation of the optically active acid. It was, however, found that when the acid obtained from the mother liquor of a recrystallization of the amine salt was tested for optical activity, the calculated value was obtained. Furthermore, it could be shown that the optically active acid did not racemize in 2-*N* sodium hydroxide solution, and as will be seen later, the hydration of the double bond is so slow that it is unlikely to be the operative process. Instead, the explanation may be that the difference between the solubilities of the diastereomeric salts is unusually small, since the enantiomers can be said to differ only in the position of a double bond. Consequently the separation difficulties may be further complicated by the formation of mixed crystals.

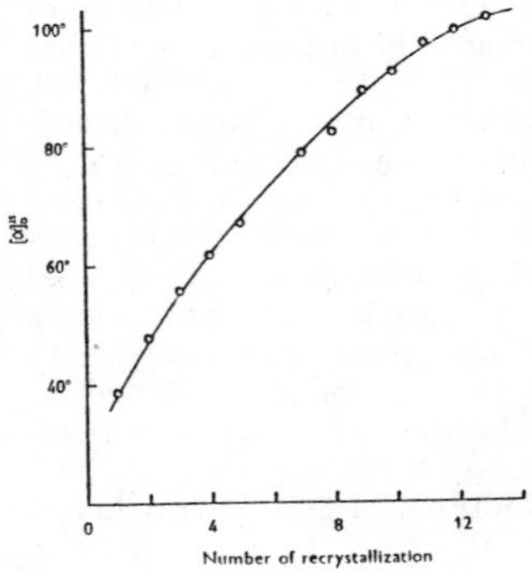

Fig. 1. $[\alpha]^{25}_D$ of (+)-acid against number of recrystallization of the salt with (+)-α-(naphthyl)-ethylamin.

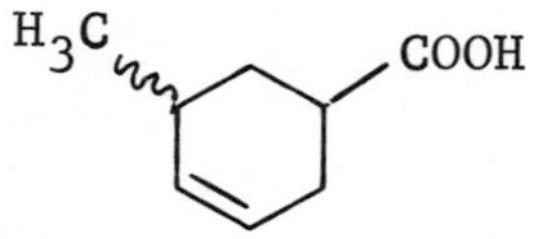

$C_8H_{12}O_2$ Ref. 397

The resolution of this acid into its active modifications may be accomplished with the aid either of *l*-menthylamine or of quinine.

I. *Experiments with* l-*Menthylamine*.—The acid available for this purpose weighed rather more than 160 grams, and was divided into two parts. The oil (80 grams) was dissolved in 570 c.c. of *N*/10-sodium carbonate, heated to boiling, and mixed with a solution of pure *l*-menthylamine hydrochloride (120 grams), when a viscid syrup separated which soon began to crystallise. After remaining overnight, the aqueous liquid was decanted * from the semi-solid cake; the latter was then washed, and left in contact with porous porcelain until quite hard and dry; it then weighed 105 grams. The crude salt was rubbed with a little pure acetone in a mortar, quickly filtered, and the colourless residue crystallised from acetone, from which it separated in long, slender needles:

0·9028, made up to 20 c.c. with alcohol, gave α_D $-0·48°$ in a 2-dcm. tube at 16°, whence $[\alpha]_D$ $-5·3°$.

After two more crystallisations, the salt had $[\alpha]_D$ $-2·8°$, and after two further crystallisations, $[\alpha]_D$ $-1·7°$, and it therefore consists of the *l*-menthylamine salt of *d*-1-methyl-Δ⁵-*cyclo*hexene-3-carboxylic acid:

0·1108 gave 5·1 c.c. N_2 at 18° and 760 mm. N=5·3.

$C_{18}H_{33}O_2N$ requires N=4·7 per cent.

By extracting the porous plates which had been employed in purifying the crude *l*-menthylamine salt as explained above, and carefully working up all mother liquors, about 120 grams of the pure *l*-menthylamine salt of rotation $[\alpha]_D$ $-1·7°$ were ultimately obtained, and this was decomposed by dilute sodium hydroxide, and after the *l*-menthylamine had been extracted with ether, the alkaline solution was acidified, the active acid extracted with ether, and distilled under diminished pressure:

0·1931 gave 0·4819 CO_2 and 0·1532 H_2O. C=68·1; H=8·8.

$C_8H_{12}O_2$ requires C=68·5; H=8·6 per cent.

d-1-*Methyl*-Δ^5-cyclo*hexene-3-carboxylic acid* boils at 145°/20 mm., and has the following rotation:

* These aqueous liquors yield, on acidifying and extracting with ether, nearly 10 grams of acid, which was used in another experiment.

1·0088, made up to 20 c.c. with ethyl acetate, gave α_D $+3·34°$ in a 2-dcm. tube at 16°, whence $[\alpha]_D$ $+33·1°$.

Ethyl d-1-*Methyl*-Δ^5-cyclo*hexene-3-carboxylate* was prepared by mixing the acid (40 grams) with alcohol (200 grams) and sulphuric acid (15 c.c.), and, after remaining at the ordinary temperature for a week, water was added, and the oily ester extracted with ether. The ethereal solution was well washed with sodium carbonate, dried, and evaporated. The ester distilled at 140—141°/100 mm. as a mobile liquid with a penetrating and very unpleasant odour:

0·1856 gave 0·4849 CO_2 and 0·1626 H_2O. C=71·2; H=9·8.

$C_{10}H_{16}O_2$ requires C=71·4; H=9·5 per cent.

0·9520, made up to 20 c.c. with ethyl acetate, gave α_D $+2·91°$ in a 2-dcm. tube at 16·5°, whence $[\alpha]_D$ $+30·5°$.

II. *Experiment with Quinine*.—This method of resolution, which appears to give good results, was carried out subsequent to the experiments with *l*-menthylamine, just described, and with a comparatively small quantity of the *dl*-acid. The acid (45 grams) was digested in ethyl acetate solution with quinine (125 grams), and the clear solution left in the ice-chest for eight days, when a considerable quantity of a crystalline crust had separated. This was collected, twice recrystallised from ethyl acetate, and then decomposed in the usual manner, when it yielded an acid which distilled at 142°/20 mm., which had the following high rotation:

1·0052, made up to 20 c.c. with ethyl acetate, gave α_D $+3·31°$ in a 2-dcm. tube at 16°, whence $[\alpha]_D$ $+32·9°$.

It would have been interesting again to have converted the active acid into the quinine salt in order to determine whether a higher rotation would have resulted than the $+33·1°$ obtained in the case of the resolution with *l*-menthylamine, but unfortunately the acid was accidentally lost.

d-Δ^5-m-*Menthenol*(8) *and* d-$\Delta^{5:8(9)}$-m-*Menthadiene*.

The conversion of ethyl *d*-1-methyl-Δ^5-*cyclo*hexene-3-carboxylate into *d*-Δ^5-*m*-menthenol(8) was brought about by adding the ester (15 grams) to an ethereal magnesium methyl iodide solution containing 6 grams of magnesium, the temperature being kept below 25° during mixing and subsequently. After twenty-four hours, the product was decomposed by water and hydrochloric acid in the usual manner, the ethereal solution washed well, dried, and evaporated. The residual oil distilled remarkably constantly at 115°/30 mm.:

0·1438 gave 0·4120 CO_2 and 0·1551 H_2O. C=78·1; H=12·0.

$C_{10}H_{18}O$ requires C=77·9; H=11·7 per cent.

0·8810, made up to 20 c.c. with alcohol, gave α_D $+3·25°$ in a 2-dcm. tube at 16°, whence $[\alpha]_D$ $+36·7°$.

d-Δ^5-m-*Menthenol*(8) is a colourless, rather viscid oil, possessing a strong odour of terpineol and menthol; even when kept for several months, it showed no signs of crystallising, and an attempt to prepare a crystalline phenylurethane was also unsuccessful.

l-1-*Methyl*-Δ^5-cyclo*hexene-3-carboxylic Acid*, l-Δ^5-m-*Menthenol*(8), *and* l-$\Delta^{5:8(9)}$-m-*Menthadiene*.

The mother liquors from the separation of the *l*-menthylamine and quinine salts of *d*-1-methyl*cyclo*hexene-3-carboxylic acid were decomposed in the usual manner, and the acid (60 grams), which distilled at 142—146° and had $[\alpha]_D$ $-21·4°$, was systematically treated with *l*-menthylamine and quinine, with the result that an acid was ultimately obtained which distilled at 142°/20 mm. and had the following rotation:

1·1009, made up to 20 c.c. in ethyl acetate, gave α_D $-3·41°$ in a 2-dcm. tube at 15°, whence $[\alpha]_D$ $-30·9°$.

It follows therefore that this acid is nearly pure l-1-*methyl*-Δ^5-cyclo*hexene-3-carboxylic acid*, since the rotation of the corresponding *d*-acid was found to be $+33·1°$. Unfortunately no suitable crystalline salt of the *l*-acid was discovered, although experiments were made with *d*-bornylamine, *d*-*iso*menthylamine, and most of the usual alkaloids, and therefore complete separation could not be carried out.

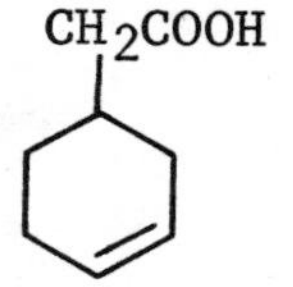

$C_8H_{12}O_2$ Ref. 398

Note:

Obtained an acid, b.p. 88°C (0.5 mm.), $[\alpha]_D^{25}$ = -71.0° (c, 1, $CHCl_3$) using (+)-α-phenylethylamine as resolving agent. Using (-)-α-phenylethylamine as resolving agent obtained an acid b.p. 88°C (0.5 mm.) $[\alpha]_D^{25}$ = +73.6° (c, 1, $CHCl_3$).

$C_8H_{12}O_2$ Ref. 399

Resolution of Ia.—Crystalline salts of Ia were obtained with quinine, brucine and cinchonidine. The quinine and brucine salts fractionated slowly, and attention was given to the cinchonidine salt. Fractionation occurred from benzene, ethyl acetate or ethanol. After a number of preliminary experiments, a large scale separation was run.

To a mixture of 1115 g. of cinchonidine in 2700 ml. of boiling absolute ethanol was added 530 g. of pure Ia. The resulting clear solution was allowed to cool and then was seeded with a few crystals of cinchonidine salt obtained from an earlier experiment; 580 g. of salt gradually precipitated. A small sample of this material was converted with 2 *N* sulfuric acid and a chloroform extraction to free acid of $[\alpha]_D$ −12.9°, m.p. 48–62°. The main body of the salt (crop 1), was recrystallized from absolute ethanol to give crop 1A and mother liquor 1A. The mother liquor from the original precipitation of the salt was concentrated to give crop 2 and mother liquor 2. Crop 2 was recrystallized from mother liquor 1A, while the head fraction (crop 1A) was recrystallized from fresh absolute ethanol. The tail fraction (mother liquor 2) was concentrated further to give mother liquor 3 and crops 3 and 4. Crops 2, 3 and 4 and the appropriate mother liquors were submitted to systematic triangular fractionation. The head fraction 1C, which gave acid of $[\alpha]_D$ −30.8°, was combined with two other fractions from a late stage of the triangle scheme, which gave acids of $[\alpha]_D$ −29.7°, −30.7° and −30.1°. Recrystallization of the last salt fraction from ethyl acetate and regeneration gave acid of $[\alpha]_D$ −30.1°. Further salt crops from the mother liquors gave acid of $[\alpha]_D$ −30.3°. An analytical sample of the completely resolved acid was prepared by distillation. It had m.p. 25.5–26.5°, infrared spectrum identical with that of authentic racemic material, $[\alpha]^{26}_D$ −30.6° (*c* 1.2, *l* 4). A total of 17 g. of optically pure levorotatory material was obtained.

Anal. Calcd. for $C_8H_{12}O_2$: C, 68.55; H, 8.63. Found: C, 68.42; H, 8.59.

The cinchonidine salt of optically pure material had m.p. 143–153°.

Anal. Calcd. for $C_{19}H_{22}ON_2 \cdot 2C_8H_{12}O_2 \cdot C_2H_6O$: C 71.58; H, 8.44; N, 4.51. Found: C, 71.86; H, 7.99; N, 4.83.

Mother liquor 3 from the triangle scheme above was treated with enough water to bring the alcohol concentration to 95%, whereupon a salt of dextrorotatory acid crystallized. Regeneration of a small portion gave material of $[\alpha]_D$ +17.7°. Triangular crystallization from 95% ethanol gave 3.15 g. of (+) −Ia, $[\alpha]^{26}_D$ +30.1° (*c* 1.3, *l* 4), m.p. 25.0–26.5°, infrared spectrum identical with that of the racemate.

Anal. Calcd. for $C_8H_{12}O_2$: C, 68.55; H, 8.63. Found: C, 68.44; H, 8.71.

The cinchonidine salt had m.p. 133–150°.

Anal. Calcd. for $C_{19}H_{22}ON_2 \cdot C_8H_{12}O_2 \cdot {}^1/_2C_2H_6O$: C, 73.49; H, 8.15; N, 6.12. Found: C, 73.68; H, 8.01; N, 5.68.

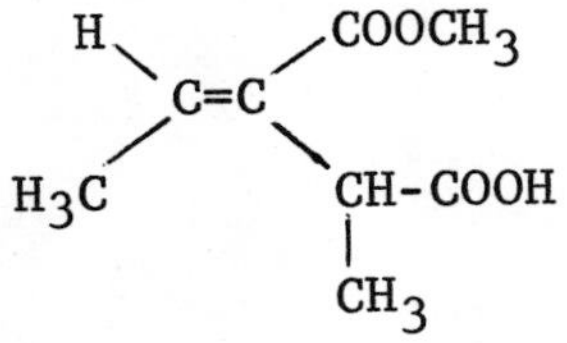

$C_8H_{12}O_4$ Ref. 400

The half-ester (VIII) in acetone was successfully resolved by means of brucine, and one of the diastereomeric salts (mp 157-159°C) was decomposed with dilute hydrochloric acid to give VIIIa, mp 46-47°C, $[\alpha]_D$ +117° ($CHCl_3$). The mother liquor, after the separation of the above brucine salt, gave another salt, mp 78° (sinter). This was also treated with dilute hydrochloric acid to give VIIIb, mp 47-48°C, $[\alpha]_D$ -114° ($CHCl_3$).

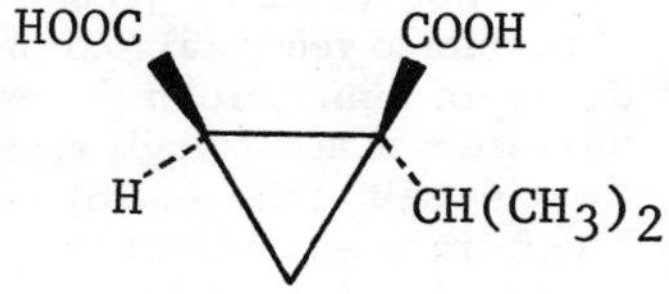

$C_8H_{12}O_4$ Ref. 401

A procedure for the resolution of the above compound with brucine may be found in the following reference: S. K. Ranganathan, J. Indian Chem. Soc., 13, 419 (1936). A more detailed procedure for this compound with the same resolving agent may be found on the following page.

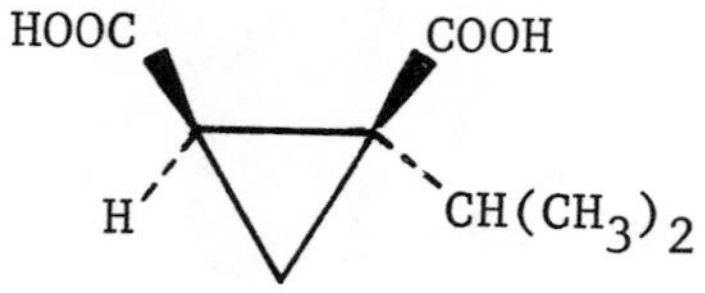

$C_8H_{12}O_4$

Ref. 402

Resolution of the cis-*Acid.*—6·6 G. of the monohydrate of the *dl-cis*-acid were dissolved in 200 c.c. of hot water, and 16·5 g. (1 mol.) of brucine stirred in. After 12 hours, a mixture (15·08 g.) of plates and needles was deposited. Six crystallisations from water afforded the pure acid brucine salt (5·61 g.) of the *d*-acid in transparent plates, m. p. 110—120° (decomp.), $[\alpha]_{5461}^{20°}$ + 6·2° (*l* = 2, *c* = 2·020 in acetone). On decomposition with ammonia in the usual manner, these yielded 1·49 g. of the *d-cis*-acid. d-cis-1-iso*Propyl*cyclo*propane*-1 : 2-*dicarboxylic acid monohydrate* crystallised from water in interlacing, stout, prismatic needles, m. p. 78—79° (Found : C, 50·4; H, 7·3; H_2O, 9·5%). The *anhydrous acid* crystallised from benzene–light petroleum (b. p. 60—80°) in rosettes of stout elongated prisms, m. p. 119—120°, $[\alpha]_D^{16°}$ + 86·9° (*l* = 2, *c* = 1·221 in chloroform) (Found : C, 55·6, 55·8; H, 7·2, 7·1%).

A long series of systematic fractional crystallisations of the mother-liquors failed to yield the *l*-acid in a state approaching purity. A similar lack of success attended conversion into the neutral brucine salt and attempts to use cinchonine, cinchonidine, and strychnine for the resolution. After many trials the pure *l*-acid was obtained by the following procedure. 13·0 G. of the monohydrate of the *dl-cis*-acid were dissolved in 500 c.c. of hot water, and 31·9 g. (1 mol.) of powdered recrystallised brucine (tetrahydrate) stirred in. The solution deposited over-night 32·6 g. of a mixture of plates and needles, the former much more rapidly than the latter. The mixture was accordingly crystallised from 600 c.c. of water and the resulting first crop of plates (brucine salt of the *d*-acid) was filtered off after 3 hours while the solution was still slightly warm. The filtrate was heated to boiling and kept over-night; the resulting crop of needles, after one recrystallisation, yielded the pure acid brucine salt of the *l*-acid (6·9 g.) in prismatic needles, $[\alpha]_{5461}^{16°}$ − 29·5° (*l* = 2, *c* = 1·982 in acetone), unchanged by further crystallisation. This was decomposed with ammonia in the usual manner; crystallisation of the product (2·0 g.) from water yielded l-cis-1-iso*propyl*cyclo*propane*-1 : 2-*dicarboxylic acid monohydrate* in interlacing, stout, prismatic needles, m. p. 78—79° (Found : C, 50·4; H, 7·4; H_2O, 9·8%). The *anhydrous acid* crystallised from benzene–light petroleum (b. p. 60—80°) in rosettes of prisms, m. p. 119—120°, $[\alpha]_D^{16°}$ − 88·8° (*l* = 2, *c* = 2·020 in chloroform) (Found : C, 56·2; H, 7·1%).

A mixture of approximately equal amounts of the anhydrous *d*- and *l-cis*-acids melted at 106°, and a similar mixture of the monohydrates at 99° after softening at 76°.

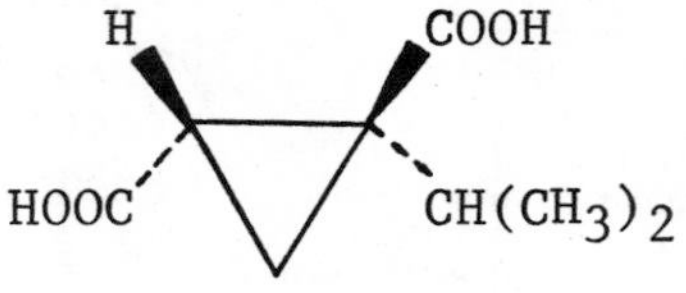

$C_8H_{12}O_4$

Ref. 403

Resolution of the trans-*Acid.*—8 G. of the acid were dissolved in 1700 c.c. of hot water, and 40 g. (2 mols.) of powdered brucine added in small portions, with stirring. After 12 hours, the neutral *brucine* salt of the *l*-acid (19·8 g.) was deposited in feathery aggregates of prisms, m. p. 149—151° (decomp.), $[\alpha]_{5461}^{25°}$ − 98·8° (*l* = 2, *c* = 2·060 in chloroform), unchanged on recrystallisation (Found : C, 60·9; H, 7·1. $C_{54}H_{64}O_{12}N_4,6H_2O$ requires C, 60·7; H, 7·1%). This salt was decomposed with ammonia, the brucine removed by filtration, and the residual solution acidified and extracted with ether. The dried extract, on evaporation, yielded 3·27 g. (41%) of crude *l*-acid; on crystallisation from water, this yielded interlacing clusters of fine needles consisting of l-trans-1-iso*propyl*cyclo*propane*-1 : 2-*dicarboxylic acid*, m. p. 155°, $[\alpha]_{5461}^{21°}$ − 236·2° (*l* = 2, *c* = 1·317 in acetone) (Found : C, 56·0; H, 6·9%).

On evaporation of the mother-liquor from the brucine salt of the *l*-acid, a crop of needles (13·7 g.) was obtained. On recrystallisation this afforded the neutral brucine salt of the *d*-acid in prismatic needles (12·3 g.), $[\alpha]_{5461}^{22°}$ − 15·39° (*l* = 2, *c* = 2·047 in chloroform), unchanged by further crystallisation. This was decomposed with ammonia in the usual manner; the recovered acid was crystallised from water, yielding d-trans-1-iso*propyl*cyclo*propane*-1 : 2-*dicarboxylic acid* in radiating clusters of fine needles, m. p. 155°, $[\alpha]_{5461}^{17°}$ + 232·1° (*l* = 2, *c* = 0·983 in acetone) (Found : C, 55·9; H, 7·15%). A mixture of this acid and the *l-trans*-acid melted at 194—195°.

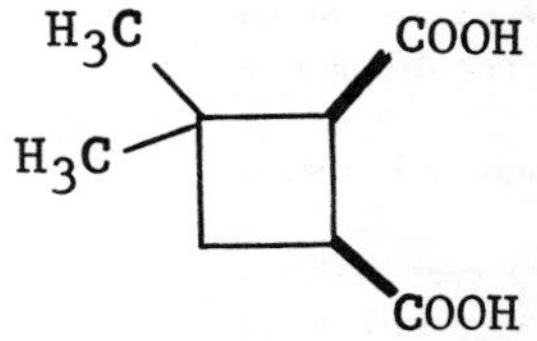

$C_8H_{12}O_4$ Ref. 404

Resolution of dl-cis-*Norcaryophyllenic Acid.*—5·0 G. of *dl-cis*-norcaryophyllenic acid, m. p. 150—151°, were dissolved in 375 c.c. of hot water, and 8·55 g. of cinchonidine stirred in. The filtered solution, on standing, deposited 4·29 g. of salt as flattened needles. Systematic fractional crystallisation from water yielded finally 3·49 g. of the neutral cinchonidine salt of the *d*-acid in transparent prisms, m. p. 215° (decomp.), $[\alpha]_{5461}^{20°}$ − 138·0° (l = 2; c = 0·8772 in ethyl alcohol). This was dissolved in 350 c.c. of hot water containing a little alcohol and decomposed with ammonia. The precipitated cinchonidine was filtered off, and the acidified filtrate extracted with ether. Evaporation of the dried extract yielded 1·07 g. of active acid, m. p. 160—162°. Crystallisation from water (in which the active acid was less soluble than the *dl*-acid) yielded d-cis-*norcaryophyllenic acid* in plates, m. p. 163—165°, $[\alpha]_{5461}^{20°}$ + 4·9° (l = 2; c = 1·8496 in chloroform) (Found: C, 55·6; H, 7·0. $C_8H_{12}O_4$ requires C, 55·8; H, 7·0%).

The more soluble cinchonidine salt crystallised in silky needles which were too highly coloured to allow of the determination of its rotation. Decomposition of the most soluble fraction from a long series of crystallisations yielded 1·15 g. of crude active acid, m. p. 165°. One crystallisation from water afforded l-cis-*norcaryophyllenic acid* as plates, m. p. 165°, $[\alpha]_{5461}^{20°}$ − 5·9° (l = 2; c = 1·0984 in chloroform) (Found: C, 55·9; H, 7·2%). An intimate mixture of approximately equal weights of the *d*- and the *l*-acid melted at 150°.

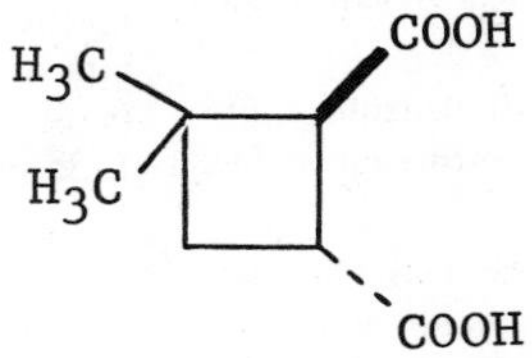

$C_8H_{12}O_4$ Ref. 405

Resolution of dl-trans-*Norcaryophyllenic Acid.*—8·22 G. of *dl-trans*-norcaryophyllenic acid, m. p. 148—149°, were dissolved in 300 c.c. of hot water and 37·64 g. of anhydrous brucine added with stirring. The filtered solution deposited, on standing, 25·44 g. of thin, hair-like needles. Two recrystallisations from water gave 13·76 g. (29·3%) of the neutral brucine salt of the *l*-acid in silky needles, $[\alpha]_{5461}^{16·5°}$ − 81·46° (l = 2; c = 1·6992 in acetone). These were dissolved in 150 c.c. of hot water and decomposed with ammonia. The filtrate from the precipitated brucine was acidified and extracted with ether. Evaporation of the dried extract left 2·32 g. of crude active acid; three crystallisations from water (in which the active acid is more soluble than the *dl*-acid) gave 0·93 g. of l-trans-*norcaryophyllenic acid* in transparent aggregates of prisms, m. p. 126°, $[\alpha]_{5461}^{18°}$ − 129·0° (l = 2; c = 2·0000 in chloroform) (Found: C, 56·2; H, 7·1. $C_8H_{12}O_4$ requires C, 55·8; H, 7·0%).

As with the *cis*-acid, the most soluble salts were too highly coloured to permit of the determination of rotation; decomposition of the most soluble fraction with ammonia yielded 3·54 g. of crude *l*-acid. Three crystallisations from water afforded d-trans-*norcaryophyllenic acid* in transparent aggregates of prisms, m. p. 123—125°, $[\alpha]_{5461}^{18°}$ + 122·3° (l = 2; c = 1·9960 in chloroform) (Found: C, 55·9; H, 6·9%).

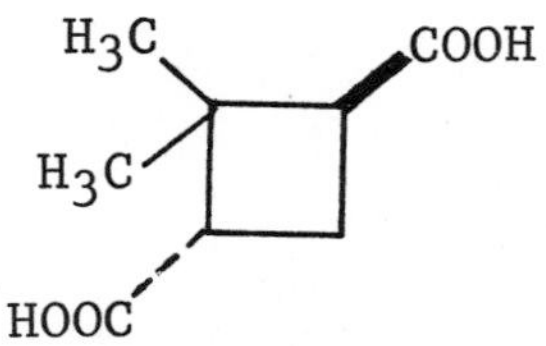

$C_8H_{12}O_4$ Ref. 406

Smistamento dell'acido d,l « trans » norpinico a p.f. 178-9° C negli antipodi ottici. - A) *operando con chinina.* — Ad una soluzione di g 7,7 (2 moli) di chinina anidra in 470 cm³ di acetato di etile, mantenuto all'ebollizione, vengono addizionati g 2 (1 mole) di acido d,l «trans» (¹) norpinico sciolti in 100 cm³ di acetato di etile. Si mantiene a ricadere la miscela per 2 ore, si filtra a caldo il sale insolubile, che viene lavato sul filtro con 50 cm³ di acetato di etile bollente; si ottengono 4 g di un sale che viene sciolto in acqua (~ 300 cm³) acidificato con HCl e la soluzione estratta a fondo con etere. Dall'evaporazione dell'etere, preventivamente essiccato su $CaCl_2$, si ottengono g 0,790 di un acido che presenta un potere rotatorio $[\alpha]_D^{20}$ = —22°.

Come si è già riferito nella parte generale trattando all'ebollizione il sale insolubile con acetato di etile e filtrando a caldo non si realizza nessun aumento nel valore del potere rotatorio dell'acido corrispondente.

Sciogliendo l'acido a potere rotatorio —22° in 40 cm³ di acetato di etile e ripetendo la salificazione con 2,92 g di chinina anidra in 200 cm³ di acetato di etile si ottengono dalla filtrazione a caldo 2,48 g di sale che forniscono per decomposizione g 0,450 di un acido il cui potere rotatorio risulta di $[\alpha]_D^{20}$ = —27°.

Una successiva salificazione con chinina non varia molto il valore del potere rotatorio, pur portando a perdite dell'ordine del 40%. Risulta più conveniente riprendere i 450 mg di acido a potere rotatorio —27° con 9 cm³ di benzolo, filtrando all'ebollizione l'indisciolto che risulta essere la frazione racema. Per raffreddamento cristallizza un acido il cui valore del potere rotatorio risulta di $[\alpha]_D^{20}$ = —30°,5/-31°,5.

Dall'acetato di etile proveniente dalla prima salificazione, lasciato a se una notte, cristallizzano g 2,41 di un sale il cui corrispondente acido è risultato praticamente inattivo. Concentrando il solvente ad un volume di circa 250 cm³ si ottengono 1,94 g di sale che fornisce, per decomposizione un acido a potere rotatorio $[\alpha]_D^{20} = +23°$.

Non risulta possibile, anche con successive risalificazioni con chinina ottenere un acido con potere rotatorio più elevato.

Dalla cristallizzazione da benzolo, operando come nel caso dell'isomero a potere rotatorio (—), si ottiene un acido che presenta un $[\alpha]_D^{20} = +31°$.

B) *operando con brucina.* — Ad una soluzione di g 4,6 (1 mole) di brucina in 400 cm³ di acetato di etile, mantenuta all'ebollizione, vengono addizionati in 15' g 2 (1 mole) di acido d,l « trans » norpinico sciolti in 100 cm³ di acetato di etile. Si mantiene la miscela a ricadere per 1 ora e si filtra poi rapidamente, a caldo, il sale insolubile formatosi; si ottengono g 3,15 di un sale che viene sciolto in acqua e decomposto con HCl. La soluzione acida viene ripetutamente estratta con etere, e l'etere, essiccato su $CaCl_2$, allontanato; residuano dall'evaporazione g 0,910 di un acido che presenta un potere rotatorio $[\alpha]_D^{20} = +23°$. Sciogliendo l'acido a potere rotatorio +23° in 50 cm³ di acetato di etile e ripetendo la salificazione con 2,1 g di brucina in 225 cm³ di acetato di etile all'ebollizione si ottengono, per filtrazione, g 2,2 di sale dal quale è possibile ottenere per acidificazione ed estrazione, g 0,620 di un acido che presenta un potere rotatorio di $[\alpha]_D^{20} = +31,5°$ e un p.f. 161-162°C.

Dalle acque madri della prima salificazione, lasciate a sè una notte, cristallizzano g 3 di un sale che sciolto in acqua, acidificato con HCl, ed estratto con etere, dà per evaporazione dell'etere g 0,860 di un acido che presenta un potere rotatorio $[\alpha]_D^{20} = -23°$.

Riprendendo questo acido all'ebollizione con 17 cm³ di benzolo e filtrando a caldo l'indisciolto, si ottiene, per raffreddamento, un acido che presenta un potere rotatorio di $[\alpha]_D^{20} = -30°$.

Si può anche salificare l'acido a potere rotatorio —23° con chinina, con le modalità viste nel caso precedente, ottenendosi un acido a potere rotatorio —27-28° e cristallizzare in definitiva quest'ultimo da benzolo.

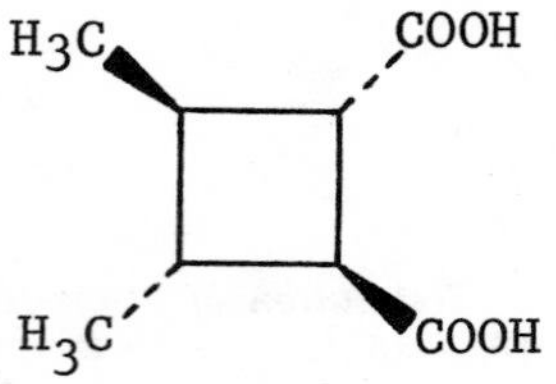

$C_8H_{12}O_4$ **Ref. 407**

A mixture of 9 g of the diacid above and 33 g of brucine was added to 160 ml of not absolute ethanol. After filtration to remove suspended materials, the solution was cooled to 5° to give 18 g of a solid. Three recyrstallizations from ethanol and combination with the head crop from another resolution gave 13.5 g of a brucine salt which, upon treatment with aqueous hydrocloric acid and extraction with diethyl ether followed by drying of the ether solution and evaporation, gave 3.5 g of a moist white solid. After heating with 350 ml of methanol and 3 ml of concentrated sulfuric acid followed by the usual work-up and distillation at 124–126° (25 Torr), 3.8 g of an optically active dimethyl *trans,trans,trans*-3,4-dimethyl-1,2-cyclobutanedicarboxylate was obtained: NMR (CCl_4) δ 1.15 (d, J = 6.54 Hz, 6 H), 2.9 (sym, m, 2 H), 2.67 (sym, m, 2 H), 3.64 (s, 6 H); ir (neat) broad 1735 cm⁻¹; $[\alpha]^{23}D$ −35.6° (c 1.52 MeOH); *m/e* 200.1031.

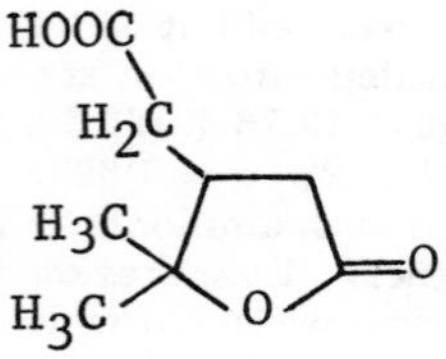

$C_8H_{12}O_4$ **Ref. 408**

The racemic terpenylic acid was prepared from ethyl β-acetoglutarate according to a method first given by Simonsen[11]. The preparation of the glutarate and some modifications of the following procedure will be described elsewhere.

Preliminary experiments on resolution. The common alkaloids were tested in water solution. The data for the bases giving crystalline salts are given in table 2; the rotatory power of the acid isolated from the salts was measured in water solution. Strychnine, brucine, cinchonine and quinidine gave only oily products. Brucine gave, however, a crystallising salts from acetone solution: the acid from this salt had $[\alpha]_D = +9°$.

Table 2.

	$[\alpha]_D^{25}$
Quinine	— 2.5°
Cinchonidine	+ 14.5°
(+)- α-(2-Naphthyl)-ethylamine	+ 54.5°

The final resolution was carried out using (+)- and (—)- α-(2-naphthyl)-ethylamine. 25.0 g (0.13 mole) of *racem*-terpenylic acid (hydrated, m.p. 55—57°) were dissolved in 700 ml water heated to 40°. A solution of (+)- α-(2-naphthyl)-ethylamine in ether, prepared from 31.5 g (0.066 mole) of tue neutral sulphate, was added in portions with stirring and the ether evaporated. After standing at room temperature for some hours, the solution was left in a refrigerator over-night. The salt obtained was recrystallised from water. The course of the resolution is shown in table 3. To avoid racemisation during the recrystallisations, the salt was dissolved in hot water and the solution cooled as rapidly as possible.

11. Simonsen, J. L. *J. Chem. Soc.* **91** (1907) 184.

Table 3.

Crystallisation	1	2	3	4	5	6
ml solvent	700	310	170	170	100	100
g salt obtained	22.5	17.3	15.0	12.8	11.3	9.4
$[\alpha]^{25}_D$ of the acid	+39°	+52°	+56°	+56°	+55.5°	+55.5°

The mother liquor from the first crystallisation was evaporated to dryness at room temperature and the terpenylic acid liberated; 10.3 g having $[\alpha]^{25}_D = -30.5°$ (in water) were obtained. This acid (0.06 mole) was dissolved in ether and mixed with an ethereal solution of (—)-α-(2-naphthyl)-ethylamine obtained from 14.0 g (0.029 mole) of the neutral sulphate. The salt thus obtained was filtered off and recrystallised from water as described above (see table 4).

Table 4.

Crystallisation	1	2	3	4
ml solvent	150	110	110	100
g salt obtained	14.7	12.6	11.1	9.8
$[\alpha]^{25}_D$ of the acid	—42°	—52.5°	—55°	—56°

(+)-*Terpenylic acid.* The amine salt (9.4 g) was decomposed with dilute sulphuric acid and the terpenylic acid isolated by repeated extractions with ethyl acetate. The solvent was evaporated at room temperature. Recrystallisation was effected by dissolving the crude acid (3.7 g) in ether, this solution being filtered to remove traces of amine sulphate, and then adding 10 ml of ethyl acetate and 40 ml of gasoline. As the ether was allowed to evaporate in the course of several days, the acid crystallised in big prisms. After two crystallisations, there remained 2.5 g of pure acid, m.p. 92—94°.

0.0758 g acid: 8.61 ml 0.05110-N NaOH.

$C_7H_{10}O_4$. Equiv. wt. calc. 172.2 found 172.3.

0.0931 g acid dissolved in water to 9.98 ml: $\alpha^{25}_D = +1.05°$ (2 dm).

$[\alpha]^{25}_D = +56.3°$; $[M]^{25}_D = +96.9°$.

(—)-*Terpenylic acid.* The acid was liberated from the amine salt (9.8 g) as described above and recrystallised in the same way as the (+)-acid. The yield of pure acid with m.p. 92—94° was 2.7 g.

0.0719 g acid: 8.20 ml 0.05110-N NaOH.

$C_7H_{10}O_4$. Equiv. wt. calc. 172.2 found 171.6.

Activity data in different solvents are given in table 5.

Table 5.

g = gram acid in 9.98 ml solution

-	g	α^{25}_D(2 dm)	$[\alpha]^{25}_D$	$[M]^{25}_D$
Water	0.1099	—1.245°	—56.5°	—97.3°
Ethanol (abs.)	0.1064	—1.195°	—56.1°	—96.6°
Ethyl acetate	0.0982	—1.22°	—62.0°	—106.7°
Acetone	0.1148	—1.49°	—64.8°	—111.5°

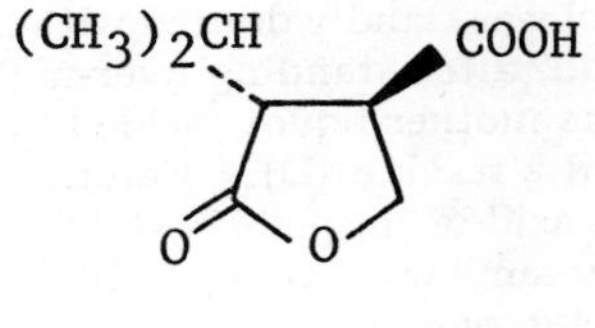

$C_8H_{12}O_4$ Ref. 409

I. Resolution of stable racemic α-alkylparaconic acids (I) into the optical antipodes (Ia and Ib). a) Strychnine salts of stable levo and dextro α-isopropylparaconic acids.

To a solution of 5 g of stable, racemic α-isopropylparaconic acid (I, R=iso-C_3H_7, m.p. 94-95°) in 12 ml of water at 80° was added 9.7 g of strychnine. After 2-hours' heating at 80°, the solution was left overnight in a refrigerator. The deposited crystals were separated and recrystallized from 15 ml of water. Yield of strychnine salt of the levo acid 5.7 g (77.5%).

The filtrate from the strychnine salt of the levo form of α-isopropylparaconic acid was concentrated at 30-40° and 10 mm to 1/3 of its original volume. After cooling, the crystals were separated and the filtrate evaporated to dryness in vacuo. The residue was dried by addition of anhydrous alcohol and subsequent distillation of the latter in vacuo. To the resulting glassy mass was added 5 ml of anhydrous alcohol. The resulting crystals were collected and recrystallized three times from alcohol. Yield of strychnine salt of dextro acid 5.2 g (70.7%).

Stable levorotatory and dextrorotatory α-isopropylparaconic acids (Ia and Ib, R= iso-C_3H_7). To a cooled saturated solution of 5.7 g of strychnine salt of stable levo α-isopropylparaconic acid in 114 ml of water was added (with cooling) the calculated quantity of 0.2 N potassium hydroxide solution. The precipitated strychnine was separated and the strychnine residues in the filtrate extracted with ether. The solution was then neutralized with 5% hydrochloric acid and evaporated to dryness in vacuo. The levo α-isopropylparaconic acid was extracted from the dry residue with ether. The solvent was distilled off and the acid twice recrystallized from 2 ml of water. Yield 1.62 g (83.6%). M.p. 89-90°, $[\alpha]_D^{18}$ - 48.5° (water, c 3.650).

Found %: C 55.85; H 6.73. $C_8H_{12}O_4$. Calculated %: C 55.81; H 7.02

The dextro acid was isolated by the same method from 10 g of the strychnine salt of the stable dextro form of α-isopropylparaconic acid. Yield 3.17 g (93.2%). M.p. 89-89.7° (from benzene), $[\alpha]_D^{19}$+ 46.3° (water, c 3.00).

Found %: C 55.78; H 7.07. $C_8H_{12}O_4$. Calculated %: C 55.81; H 7.02.

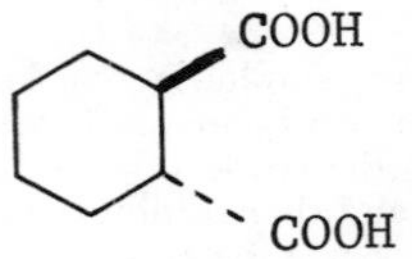

$C_8H_{12}O_4$ Ref. 410

Resolution of (±)-*trans*-1,2-Cyclohexanedicarboxylic Acid.—A solution of 43 g. (0.25 mole) of (±)-*trans*-1,2-cyclohexanedicarboxylic acid in 370 ml. of 95% ethanol was heated to boiling and then added slowly (heat was evolved) to a hot solution of 162 g. (0.50 mole) of quinine in 950 ml. of 95% ethanol. The combined solution was heated to boiling and then filtered. The filter was washed with 195 ml. of hot 95% ethanol. The solution was cooled overnight and the solid was then removed by filtration to give 117 g. (95%) of a white crystalline material. After three more recrystallizations from 95% ethanol, the diastereomer was sufficiently pure to give (+)-*trans*-1,2-cyclohexanedicarboxylic acid with high optical purity. The yield of pure quinine salt was 55 g. (45% based upon isolation of one diastereomer). Fifty grams (0.089 mole) of the quinine salt was added in portions to 500 ml. of a sulfuric acid–water solution (1:4) in an ice bath. The solution was then continuously extracted with ether for 2 days. The ether was removed on a steam bath and 10.3 g. (100%) of (+)-*trans*-1,2-cyclohexanedicarboxylic acid was obtained, m.p. 183.5–185.0° (lit.,[4] m.p. 179–183°); $[\alpha]^{30}D$ in acetone +22.3° (c = 5.3) (lit.,[4] $[\alpha]^{20}D$ +18.2°).

(4) A. Werner and H. E. Conrad, *Ber.*, **32**, 3046 (1899).

$C_8H_{12}O_4$ Ref. 410a

Optical Resolution of Cyclohexane-trans-1,2-*dicarboxylic Acid.*—A warm solution of cinchonidine (47 g., 2 mol.) in ethanol (400 ml.) was added to a solution of the diacid [9] (13·8 g., 1 mol.) in 95% ethanol (100 ml.). The mixture was concentrated to 200 ml. and kept at room temperature for 3 days. Colourless needles (17·3 g.), m.p. 199—200° (decomp.), were filtered off and the filtrate was diluted with water until cloudy. The prisms deposited (32·5 g.), m.p. 180—181° (decomp.), gave the *dicinchonidine salt* (22·9 g.), m.p. 190—191° (decomp.) [from aqueous ethanol (80%)], $[\alpha]_{578}$ —89·6° (c 0·48 in EtOH) (Found: C, 72·4; H, 7·65; N, 7·1. $C_8H_{12}O_4,2C_{19}H_{22}N_2O$ requires C, 72·6; H, 7·4; N, 7·4%). This salt was decomposed with cold 7·5N-sulphuric acid (50 ml.) and extracted with ether (5 × 50 ml.). The extracts gave (1S,2S)-cyclohexane-1,2-dicarboxylic acid (4·17 g., 60% based on half of racemate), m.p. 179—180°, $[\alpha]_{578}$ +22·4° (c 0·214 in Me_2CO) (lit.,[9] m.p. 183·5—185°, $[\alpha]_D$ +22·3°; lit.,[8] m.p. 179—183°, $[\alpha]_D$ +18·2°; both obtained by resolution of the quinine salt).

[8] A. Werner and H. E. Conrad, *Ber.*, 1899, **32**, 3046.

[9] D. E. Applequist and N. D. Werner, *J. Org. Chem.*, 1963, **28**, 48.

$C_8H_{12}O_4$ Ref. 410b

A solution of 15·7 g. of the acid and 47·2 g. of cinchonidine in 160 c.c. of boiling ethyl alcohol was rapidly diluted with 2400 c.c. of boiling water. White crystalline balls rapidly appeared and, after standing over-night, were filtered off and washed thoroughly with cold water (A). The mother-liquor yielded further crops, (B) and (C), by successive evaporations to half bulk, and a residue (D). Fraction (A) (30 g.) after two crystallisations from boiling water yielded an acid of m. p. 149—150°, fern-like tufts of opaque needles, $[\alpha]^{18\cdot5°}$ + 41·9° (c = 2·564), probably similar to that of Hückel and Friedrich. 3·5 G. of this acid were resolved as before with 10·4 g. of cinchonidine. The first crop of salt was crystallised twice from water, decomposed with ammonia, and filtered. The acid was isolated on acidification, partly as a precipitate and partly by extraction with ether, processes which gave identical material. d-trans-cyclo*Hexane*-1 : 2-*diacetic acid* formed opaque needles from water, m. p. 152°, $[\alpha]^{18°}$ + 54·9° (c = 2·535) (Found : C, 60·2; H, 8·0. $C_{10}H_{16}O_4$ requires C, 60·0; H, 8·0%). The process of resolution was repeated, but the cinchonidine salt after three crystallisations yielded an acid with unchanged m. p. and specific rotation.

Fraction (B) (6 g.) gave mainly *dl*-acid (m. p. 163—165°) and fraction (C) (20 g.) gave an impure *l*-acid. Fraction (D) (5 g.) after three crystallisations from water gave small needles of the neutral *cinchonidine* salt of the *l*-acid (Found : C, 69·4; H, 7·8. $C_{48}H_{60}O_6N_4,2H_2O$ requires C, 69·8; H, 7·8%). This yielded a l-*acid*, not quite optically pure, m. p. 148—149°, $[\alpha]^{19°}$ − 47·9° (c = 1·832) (Found : C, 60·1; H, 7·9. $C_{10}H_{16}O_4$ requires C, 60·0; H, 8·0%). A mixture obtained by grinding the *d*- and the *l*-acid together melted at 164—166°; crystallisation of an equal mixture from water yielded an acid of m. p. 166—167°.

$C_8H_{12}O_4$ Ref. 410c

Resolution of *cis*-hexahydrohomophthalic acid

To a solution of the *cis*-acid (9 g) in warm water (200 ml) 45 g of brucine (2 mole alkaloid to 1 mole acid) were added in portions. It was necessary to keep the solution warm and to stir continuously in order to dissolve the alkaloid. One should, however, be careful not to heat too long as the solution then would darken and the subsequent crystallization of the salt be impeded. Generally the crystallization of the brucine salt took place over night especially on keeping the solution at a cool place. The salt was recrystallized 3—4 times from water, the purified salt dissolved in hydrochloric acid and the *cis*-acid extracted with ether. From the dried ethereal solution a residue was obtained which partly crystallized and obviously consisted of a mixture of active and racemic acid. Great difficulties were encountered in obtaining the active acid in the crystalline state. After decolorizing the crude acid with charcoal crystallizations from water gave fractions of practically inactive acid, the active acid being enriched in the mother liquor. From the latter the active acid was obtained as a thick syrup the crystallization of which in the end was achieved in the following way. The syrup was placed at the bottom of a low glass dish and covered with a layer of water. Placing the dish at zero degree for several days big clear crystals of the pure dextrorotatory *cis*-acid separated which after washing and drying had m.p. 82°.

$C_9H_{14}O_4$ (186.2)	Calc.	C 58.05	H 7.58
	Found	» 57.76	» 7.53

The levorotatory *cis*-acid was isolated from the easily soluble brucine salt in the same way as described for the dextrorotatory acid, even greater difficulties being met with, however, in obtaining the pure (—) -acid which also had m.p. 82°.

The specific rotations of the optical components at various concentrations in dioxan are given in Table 1. The concentration c is g acid in 100 ml solution; all measurements taken in a 1 dm tube. The specific gravity at various concentrations was determined on solutions of the racemic acid and could be expressed sufficient accurately by means of the linear equation.

$$d_4^{20} = 1.035 + c \cdot 0.00168.$$

Table 1. Specific rotations of the cis-acids in dioxan.

(+)-*cis*-acid				(—)-*cis*-acid			
c	d_4^{20}	α_D^{20}	$[\alpha]_D^{20}$	c	d_4^{20}	α_D^{20}	$[\alpha]_D^{20}$
4.80	1.043	+0.60°	+12.0°	3.42	1.041	—0.39°	—11.0
9.49	1.051	1.25	12.5	5.48	1.044	0.67	11.6
12.33	1.056	1.62	12.5	10.29	1.052	1.31	12.1
14.35	1.059	1.95	12.8	14.57	1.059	1.93	12.5
19.89	1.068	2.65	12.5	23.41	1.074	3.28	13.0
26.70	1.080	3.84	13.3				

The melting point curve obtained for mixtures of the inactive *cis*-acid with its levorotatory component (see Fig. 5) shows that the inactive acid is a true racemic form.

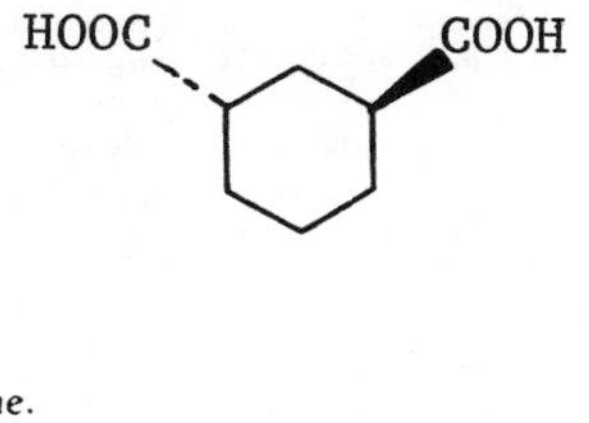

$C_8H_{12}O_4$ Ref. 411

Avec la strychnine.

200 mgr. de l'acide trans ont été dissous avec 194 mgr. de strychnine dans 2 cm³. d'alcool; il se forme alors de beaux cristaux limpides. Cette expérience fut répétée avec 5 gr. de l'acide trans sur 4.85 gr. de strychnine (1 mol. de l'acide: ½ mol. du base). Les cristaux qui s'étaient formés lentement, furent recristallisés trois fois avec de l'alcool.

2.921 mgr. donnent 0.150 cM³. N_2 à 15° et 768 mM. = 5.86 %.
8.211 .. .. 0.490 .. N_2 à 17° .. 761 .. = 6.83 %.
Calculé pour $C_8H_{12}O_4 . C_{21}H_{22}O_2N_2$ = 5.53 %.
.. .. $C_8H_{12}O_4[C_{21}H_{22}O_2N_2]_2$ = 6.67 %.

Il est donc probable que nous avons affaire au sel neutre, chaque groupe carboxyle se combinant avec une molécule du base.

Ce sel fut bouilli avec de l'ammoniaque et séparé par filtrage de la strychnine; la solution fut extraite au chloroforme jusqu'à ce qu'elle ne donnait plus de coloration pourpre avec du $K_2Cr_2O_7$ et de l'acide sulfurique concentré. La solution montre une rotation droite au polarimètre; elle fut concentrée et acidulée à l'acide chlorhydrique concentré; il se déposa des carreaux totalement différents de l'acide racémique. Ces cristaux furent recristallisés dans de l'eau. P. d. f. = 134°.

L'eau-mère les cristaux donna une deuxième cristallisation, mais celle-ci était constituée pour la plus grande partie d'acide racémique.

0,4328 gr. de l'acide *d* furent dissous dans 10 cm³. d'eau; la rotation $\alpha^{26°}$ était —1°.04 dans une tube d'un décimètre; $[\alpha]_D^{26°}$ = 24° (moyenne de 5 observations et après correction). L'eau-mère du sel de strychnine fut évaporée; le résidu ne cristallisa plus, la strychnine fut enlevée de la manière décrite. La solution tournait maintenant à gauche, mais elle contenait encore beaucoup d'acide racémique, de sorte que nous n'avons pu isoler l'acide gauche. D'abord nous avons tâché de précipiter directement l'acide gauche avec un alcaloïde qui fait tourner le plan de polarisation à droite, p. ex. la cinchonine, mais en vain.

L'analyse du sel de strychnine de l'acide droit ayant donné une indication qu'il se forme le sel neutre, et considérant que les eaux-mères contenaient encore beaucoup de strychnine et d'acide racémique, il est évident que la strychnine se combine aussi bien avec l'acide gauche qu'avec l'acide droit. Pour précipiter autant que possible ce dernier il est donc utile d'employer un excès de strychnine.

Mais ayant trouvé que l'eau-mère du sel de strychnine de l'acide droit, même après trois cristallisations, contient encore de l'acide racémique, nous avons ajouté cet excès en deux parties.

A 9.152 gr. de l'acide racémique dissous dans 200 cm³. d'alcool bouillant on ajoute 8.950 gr. de strychnine; après évaporation jusqu'à 150 cm³. environ on laisse refroidir. Après quelque temps le sel de strychnine est filtré et recristallisé par de l'alcool. On prend soin de ne rien perdre, ni du sel ni des solutions.

Le sel est décomposé par de l'ammoniaque et la strychnine est recueillie de la manière indiquée = *4.56 gr.* Les eaux-mères furent évaporées; elles ne donnèrent plus de cristillisation du sel d. l', même en évaporant tout le dissolvant et en refroidissant à —15°.

La quantité d'acide droit obtenue était de + 2 gr., c'est-à-dire beaucoup moins que la moitié du sel racémique initial; ce résidu devait donc contenir encore du sel de strychnine *d. l'*, qui ne voulait pas cristalliser. Nous avons dilué ce résidu et ajouté une nouvelle quantité de strychnine (8.95 gr.). Celle ne se dissolvait pas totalement dans le liquide bouillant; il resta 1.5 gr.; quantité dissoute 8.95—1.5 = 7.45 gr.

Après refroidissement et évaporation partielle, il se déposa une nouvelle quantité du sel *d. l'*, qui fut décomposée par de l'ammoniaque; la quantité de strychnine récupérée était de 2.86 gr. Nous avons donc dissous 8.95 + 7.45 = 16.4 gr. de strychnine, dont 4.56 + 2.86 = 7.42 gr. ou presque la moitié furent regagnés.

En admettant que le sel *d. l'* a la constitution $C_8H_{12}O_4$. [2 strychnine] la scission serait encore très incomplète. De l'eau mère du deuxième dépôt du sel d. l' il cristallisa un sel qui en différait distinctement.

Nous l'avons décomposé; 3.14 gr. de strychnine furent ainsi enlevés et la solution ammoniacale fit tourner le plan de polarisation à *gauche*: mais pas très fort.

C'est pourquoi nous n'avons pas employé cette solution pour préparer l'acide et nous l'avons isolé de l'eau-mère. Celle-ci fut privée de strychnine, et la solution restante fut sursaturée par de l'acide chlorhydrique et extraite par de l'éther. L'éther chassé, le résidu fut recristallisé par de l'eau, jusqu'à ce que l'acide gauche avait le p. d. f. = 134° et le même aspect que l'antipode droit.

Pouvoir rotatoire.

0.2158 gr. dissous dans 10 cm³. H_2O dans un tube de 10 cm. à 22° donnaient α = —0°.50.

$$[\alpha]^{22°}_D = \frac{100 \times -0.50}{2.158} = -23°.10'.$$

Une deuxième détermination de l'acide droit donna dans les mêmes conditions = +23°46':

0.2230 gr. dans 10 cm³. H_2O, tube de 1 dm. à 22°. α = +0.53

$$[\alpha]^{22°}_D = \frac{100 \times +0.53}{2.230} = +23°.46'.$$

Analyse micro-élémentaire de l'acide droit.

10.160 mgr. donnèrent 21.064 mgr. CO_2 et 5.863 mgr. H_2O.
11.560 .. 23.182 .. .. 8.319 .. ..
Trouvé: 56.54 % et 54.69 % C.; 6.41 et 7.90 % H_2O.
Calculé: 55.82 %: 7.37 % p. $C_8H_{12}O_4$.

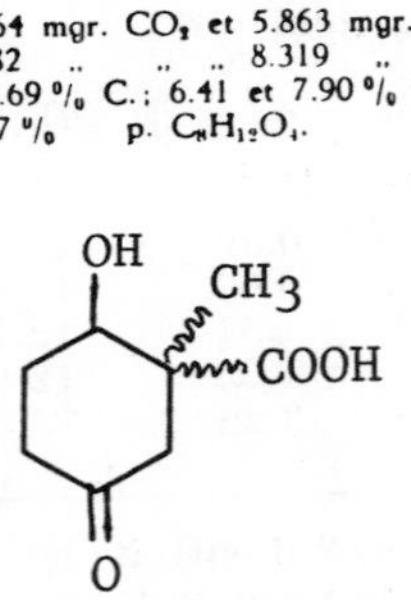

$C_8H_{12}O_4$ **Ref. 412**

Optical Resolution of X.—A dry mixture of 2.2 g. of X and 4.9 g. of anhydrous brucine (m.p. 176–178°) was dissolved in 60 cc. of boiling acetone. Upon standing at room temperature the solution began to deposit the salt of the (−)-acid in rosettes of fine needles. When deposition of the rosettes of needles had ceased, thick prisms of the other diastereomeric salt began to separate. The supernatant liquid was immediately decanted into another vessel, where deposition of the prisms continued. After three recrystallizations from acetone the brucine salts of (+)- and (−)-X had the following constants: salt of (+)-acid, m.p. 188–190°; $[\alpha]^{25}D$ −16 ± 1° (water, *c* 1.8); salt of (−)-acid, m.p. 146–148°; $[\alpha]^{23}D$ −29 ± 1° (water, *c* 1.7).

(+)- and (−)-1-Methyl-2-hydroxy-5-ketocyclohexanecarboxylic Acid (X).—To a solution of 6.5 g. of the brucine salt of the (+)-acid (X) in 50 cc. of water was added 10 cc. of 25% aqueous potassium hydroxide. This mixture was shaken well, and the liberated brucine was extracted with three 15-cc. portions of chloroform. The resulting homogeneous aqueous solution was brought to *p*H 2 with 96% sulfuric acid, and was extracted continuously with chloroform for 16 hours. The chloroform extract was then evaporated, and the oily residue was crystallized from acetone–petroleum ether, giving 1.8 g. (91%) of (+)-1-methyl-2-hydroxy-5-ketocyclohexanecarboxylic acid (X), m.p. 138–140°, $[\alpha]^{25}D$ +48 ± 1° (water, *c* 1.3). A similar treatment of 6.0 g. of the brucine salt of the (−)-acid gave 1.6 g. (88%) of (−)-1-methyl-2-hydroxy-5-ketocyclohexanecarboxylic acid (X), m.p. 137–139°, $[\alpha]^{25}D$ −48 ± 1° (water, *c* 1.3). A mixture of equal amounts (23 mg.) of the (+)- and (−)-acids upon recrystallization from acetone–petroleum ether gave 35 mg. of the *dl*-acid (X), m.p. and mixed m.p. 155–157°.

HOOC-CH$_2$-CH-CH$_2$-COOCH$_3$
OCOCH$_3$

$C_8H_{12}O_6$ **Ref. 413**

Resolution of methyl hydrogen β-acetoxyglutarate

Several alkaloids were tested for their ability to give readily crystallizing salts with the undistilled racemic half-ester. Aqueous solutions of the brucine, cinchonine, morphine, quinidine, and quinine salts gave on evaporation at room-temperature only glassy residues. Cinchonidine and strychnine, however, gave crystalline salts which were readily soluble in water and many of the common organic solvents, but relatively insoluble in ether.

The cinchonidine salt crystallized by addition of ether to its acetone, chloroform, or ethyl acetate, solutions. The half-ester isolated from the filtrates was laevorotatory in each case, and was most strongly so from ethyl acetate. This solvent was therefore adopted.

The strychnine salt crystallized by addition of ether to its chloroform solution. The half-ester isolated from the filtrate was laevorotatory in one case and dextrorotatory in another. When, however, partly resolved laevorotatory half-ester from filtrates of the cinchonidine salt was used to make the strychnine salt, the resolution proceeded smoothly in the desired direction.

(+)-Methyl hydrogen β-acetoxyglutarate (XXV)
(+)-*Methyl 3D-acetoxy-4-carboxybutanoate*

41 g (0.2 mole) of distilled racemic half-ester was dissolved with 59 g (0.2 mole) of cinchonidine in 300 ml of warm ethyl acetate, and 300 ml of ether added to the solution. The salt crystallizing at 0° was recrystallized from three parts of ethyl acetate to which was added decreasing amounts of ether. The salt became less soluble in ethyl acetate as its purity increased, and no addition of ether was necessary in the final crystallizations. To follow the resolution, each filtrate was extracted with three portions of water, and the half-ester isolated from the acidified aqueous phases by extraction with three portions of ethyl acetate. The solvent was removed by suction, and the residue dried for 10 minutes at 100°C and 10 mm. Because of the high viscosity of the half-ester, the rotatory power of the resulting samples was measured in solution. About four recrystallizations were needed to give an undistilled half-ester of constant rotation[1] ($[\alpha]^{25}_D$ +5.8° (chloroform, *c* 20, *l* 1), n^{25}_D 1.445). The yield of nicely crystalline salt (soft needles in rosettes) was approximately 18 g (36%); m.p. 89°, $[\alpha]^{22}_D$ −62.5°±0.4° (chloroform, *c* 5, *l* 1); partly resolved salt had a higher negative rotation. The melting-point was diffuse in a capillary tube, but fairly sharp and reproducible on a Kofler bench (48). Isolation and determination of the rotatory power of the half-ester should be unnecessary until the above constants for the salt are reached.

18.0 g of optically pure salt was dissolved in excess dilute hydrochloric acid saturated with sodium sulphate, the solution extracted with six portions of ether, and the extract washed once with a saturated aqueous solution of Na_2SO_4. Rapid vacuum distillation of the extract residue (6.6 g) from a small Claisen flask gave 6.0 g of viscous (+)-*methyl hydrogen β-acetoxyglutarate* (29% yield from the racemic half-ester); b.p. 140–150°/0.3 mm, n^{20}_D 1.4474, Δn_D 0.00034, d^{20}_4 1.236, R_D 44.17 (44.15).

$[\alpha]^{25}_D$	$[M]^{25}_D$	Solvent	c	l
+6.1° ±0.1°	+12.5° ±0.2°	chloroform	20	1
+5.28°±0.03°	+10.78°±0.06°	none	123	0.5
+3.7° ±0.1°	+ 7.6° ±0.2°	dry ether	20	1

Analysis:	47.0 % C	6.0 % H
$C_8H_{12}O_6$ (204.2):	47.06 % C	5.92 % H

Rapid re-distillation of 3.0 g gave a product with $[\alpha]^{23}_D$ +5.20°±0.03° (undiluted). The cold trap contained 0.02 g of a condensate smelling like acetic acid.

(−)-Methyl hydrogen β-acetoxyglutarate
(−)-*Methyl 3L-acetoxy-4-carboxybutanoate*

43 g (0.21 mole) of distilled, partly resolved half-ester ($[\alpha]^{20}_D$ −1.0° (chloroform, *c* 20, *l* 1), thus containing 25 g (0.12 mole) of the laevorotatory antipode) was dissolved with 57 g (0.17 mole) of finely-powdered strychnine in 300 ml of warm chloroform, and 300 ml of ether added to the solution. The salt crystallizing at room-temperature was recrystallized from chloroform by adding an equal volume of ether. To follow the resolution, the strychnine was precipitated from the mother liquours with ammonia–water, and the half-esters were isolated from the acidified filtrates as described in the case of the cinchonidine salt. As tests had shown that the rotatory power of the half-ester may diminish in ammonia solution, precautions were taken to avoid using a large excess, and to filter off the alkaloid and acidify the filtrate as quickly as possible. About five recrystallizations were needed to give an undistilled half-ester of constant rotation ($[\alpha]^{20}_D$ −5.9° (chloroform, *c* 20, *l* 1); n^{25}_D 1.445). The

[1] Optical rotations were measured with polarimeters of the Lippich triple-field type. The angles could usually be read to ±0.01°.

yield of crystals (which looked much like those of the cinchonidine salt) was approximately 23 g (35 %); m.p. 70° (unsharp); $[\alpha]_D^{22}$ −31.3°±0.4° (chloroform, *c* 5, *l* 1); partly resolved salt had a higher negative rotation.

Ether-extracted half-ester (6.9 g) from 23.0 g of optically pure strychnine salt gave, on rapid distillation at 0.2 mm, 6.2 g of (−)-*methyl hydrogen β-acetoxyglutarate* (25 % of the amount present in the partly resolved half-ester used for making the strychnine salt); n_D^{25} 1.4450, Δn_D −0.00033; d_4^{25} 1.231; R_D 44.14 (44.15).

$[\alpha]_D^{21}$	$[M]_D^{21}$	Solvent	*c*	*l*
−6.2° ±0.1°	−12.7° ±0.2°	chloroform	20	1
−5.27° ±0.05°	−10.76° ±0.1°	none	123	0.5
−3.7° ±0.1°	− 7.6° ±0.2°	dry ether	20	1

Analysis:	47.1 % C	6.2 % H
$C_8H_{12}O_6$ (204.2):	47.06 % C	5.92 % H

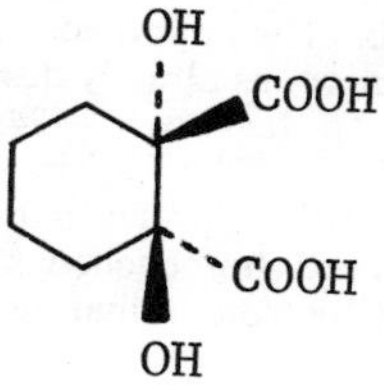

$C_8H_{12}O_6$

Ref. 414

The Optical Resolution of II.—Twenty grams of II and 39 g. of brucine were dissolved in 600 ml. of boiling water, and the resulting solution was allowed to stand overnight in an ice box. The crystals of the brucine salt which precipitated were collected by suction; six recrystallizations of this salt from ten parts of boiling water gave 16 g. of the pure brucine salt of II. This salt was suspended in 150 ml. of water and decomposed with 35 ml. of N sodium hydroxide. The precipitated brucine was removed by filtration, and the alkaline filtrate was washed with 20 ml. of chloroform, neutralized with concentrated hydrochloric acid, and then evaporated to a syrup. To this a small amount of water was added, and the resulting solution was decolorized with charcoal and evaporated to dryness. The residue was extracted several times with 10 ml. volumes of acetone. The acetone extracts were combined and again evaporated to a syrup. The syrupy product was dried over sulfuric acid in a vacuum desiccator. In this way, 4.4 g. of (+) II was obtained. Recrystallization from ethyl acetate - petroleum ether (3 : 1) gave 3.8 g. of optically pure (+)-II. $[\alpha]_D^{20}$ +31.0° (*c* 2, water).

Found: C, 47.11; H, 5.81. Calcd. for $C_8H_{12}O_6$: C, 47.06; H, 5.92%.

The mother liquor of the brucine salt of (+)-II was concentrated to dryness under reduced pressure. The brucine salt of (−)-II was collected and treated as has been described above. 2.0 g. of optically pure (−)-II were thus obtained. $[\alpha]_D^{20}$ −30.5° (*c* 2, water),

Found: C, 47.14; H, 5.94%.

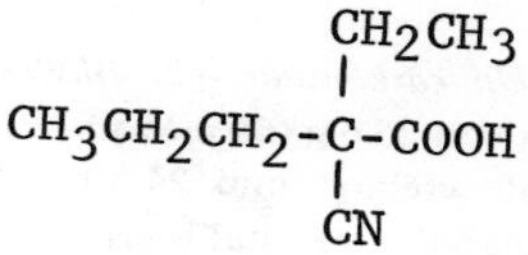

$C_8H_{13}NO_2$

Ref. 415

Razematspaltung

Razemische bzw. mit einem Enantiomer angereicherte Säure wurde mit der äquimolaren Menge optisch aktivem threo-1-Phenyl-2-aminopropandiol-(1,3) (Threobase) in Wasser zum Diastereoisomerensalz umgesetzt. Das Kristallisat wurde 2mal aus Wasser umkristallisiert (Schmp. 160°), danach in 10 proz. HCl gelöst, die Säure ausgeäthert und destilliert. Aus 50 g (0,32 Mol) **1** wurden mit (-)-Threobase 10,4 g (+)-**1** (41,5 % d. Th.); Schmp. 30–37°; $[\alpha]_D^{24}$ = + 3,34° (c = 4,34) und anschließend mit (+)-Threobase 8,4 g (-)-**1** (~ 33 % d. Th.); Schmp. 29–38°; $[\alpha]_D^{22}$ = -3,66° (c = 3,15) erhalten.

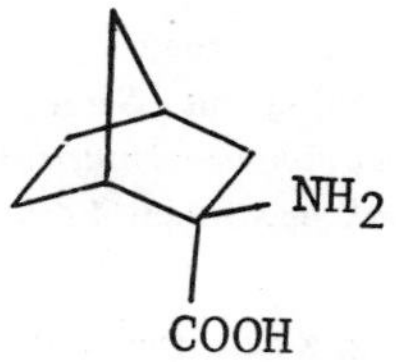

$C_8H_{13}NO_2$

Ref. 416

Preparation of *N*-Formyl Amino Acids. The *N*-formyl a and b isomers were prepared by the method of Sheehan and Yang.[20] The yields from the first crop of crystals were 70–75%. *N*-Formyl-(±)-a-BCH melted at 236–237° and *N*-formyl-(±)-b-BCH, at 209–211°.

Optical Resolution of (±)-a-BCH. *N*-Formyl-(±)-a-BCH (5.490 g, 0.03 mol) and 11.835 g (0.03 mol) of anhydrous *l*-brucine were dissolved in 50 ml of hot water. The solution was allowed to stand at room temperature for 4 hr and then at 2° for 4 hr. The crystals were collected and washed with 10 ml of ice-cold water, the mother liquor being set aside. The product was recrystallized from 50 ml of water, and this mother liquor was discarded. The yield of the less soluble *N*-formyl-a-BCH brucine salt was 8.40 g (48.6% of the total weighted components), melting at 136–137°; $[\alpha]^{25}$D was −68.1°, 1% in methanol, and did not change upon recrystallization.

The salt (7.0 g) was dissolved in 50 ml of warm water and the solution was brought to pH 10 with 2 *N* NaOH with vigorous stirring. The resulting thick slurry was cooled to 2° and filtered. The residue was washed with three 10-ml portions of ice-cold water, and the brucine precipitate was discarded. The filtrate and washings were combined and brought to dryness, and the residue was refluxed in 6 *N* HCl for 2 hr, the hydrolysate then being brought to dryness. The product was dissolved in a minimum amount of water and further treated as described in **Preparation of (±)-a-BCH** to obtain the pure amino acid. The yield was 1.40 g (75% of theoretical for one optical isomer) melting at 309–311° with decomposition; $[\alpha]^{25}$D was −61.4°, 1% in water and −47.9°, 0.5% in 5 *N* HCl. The CD in water (*c* 0.062%) had $[\theta]_{275}$ 0, $[\theta]_{225}$ + 1115, $[\theta]_{223}$ + 1056, $[\theta]_{216}$ 0, $[\theta]_{206}$ − 4837, and $[\theta]_{201}$ − 4818. The crystals of (−)-a-BCH·HBr used in the X-ray crystallographic studies reported separately[8] were obtained by allowing diethyl ether to diffuse slowly into an ethanol solution of the amino acid salt.

The mother liquor set aside from the first crystallization of the *N*-formyl amino acid brucine salt was treated as above to remove the brucine and obtain the pure amino acid. The yield was 1.45 g (77% of theoretical), melting at 309–311°; $[\alpha]^{25}$D was +61.2°, 1% in water and +48.5°, 0.5% in 5 *N* HCl. The CD in water (*c* 0.062%) gave values for $[\theta]$ equal and opposite in sign to those reported for (−)-a-BCH.

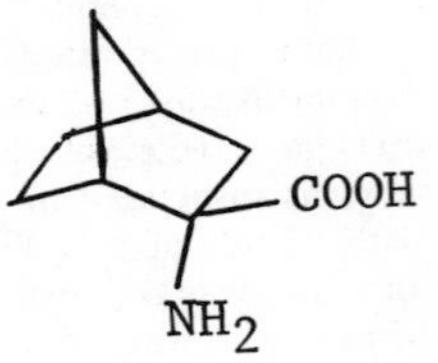

$C_8H_{13}NO_2$ Ref. 417

Optical Resolution of (±)-b-BCH. For the resolution, we used our best preparation of BCH synthesized by the Bucherer method which contained approximately 97% of the b isomer. The amino acid was formylated as described.

The *N*-formyl amino acid preparation (20.90 g, 0.114 mol) and 45.05 g (0.114 mol) of anhydrous brucine were dissolved in 1000 ml of warm water, and the solution was allowed to stand at room temperature for 24 hr. The crystals were collected, washed, and recrystallized twice from 100-ml portions of water, the original mother liquor being set aside. The yield of the less soluble *N*-formyl-b-BCH brucine salt was 29.2 g (44% of the total weighed components), melting at 133–134°; $[\alpha]^{25}$D was −2.7°, 1% in methanol, and did not change on recrystallization.

The amino acid was isolated as described under **Optical Resolution of (±)-a-BCH.** The yield was 6.7 g (78% of theoretical for one optical isomer). In both this and the following case determination of the melting point was prevented by rapid sublimation; $[\alpha]^{25}$D was +24.4°, 1% in water and +21.2°, 0.5% in 5 *N* HCl. The CD in water (*c* 0.062%) had $[\theta]_{275}$ 0, $[\theta]_{206}$ − 1973, and $[\theta]_{203}$ − 1700.

The mother liquor from the first crystallization of the less soluble salt was concentrated to approximately 600 ml and allowed to stand at room temperature for 24 hr. The 2.3 g of crystals which appeared were discarded. The amino acid was isolated on further concentration of the solution, as already described. The yield was 7.1 g (84% of theoretical); $[\alpha]^{25}$ was −24.7°, 1% in water and −20.3°, 0.5% in 5 *N* HCl. The CD in water (*c* 0.062%) gave values for $[\theta]$ equal and opposite in sign to those reported for (+)-b-BCH. Both isomers were homogeneous on the amino acid analyzer, indicating that the contaminating a isomer had been removed.

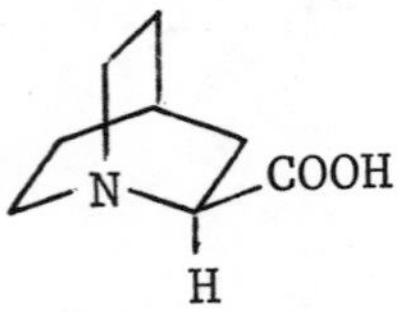

$C_8H_{13}NO_2$ Ref. 418

Racemat-Trennung von Chinuclidin-carbonsäure-(2)-äthylester (**17**). — 12.5 g (68 mMol) (R, S)-**17** wurden mit einer Lösung von 25.6 g (68 mMol) *(R)(−)-Dibenzoylweinsäure-monohydrat* in 136 ccm absol. Äthanol vereinigt und 24 Stdn. bei Raumtemperatur belassen. Dann wurde 3mal aus absol. Äthanol umkristallisiert. Es resultierten 8.67 g (16 m Mol) *(S)*-**17**-*(R)-Dibenzoyl-hydrogentartrat*, Schmp. 147° (Zers.), $[\alpha]_D^{19.5}$ = −104.1° (*c* = 1; absol. Äthanol).

$C_{28}H_{31}NO_{10}$ (541.6) Ber. N 2.58 Gef. N 2.86

Spaltung: Dieses Salz wurde zunächst mit HCl in Alkohol + Äther (vgl. Lit.[29]) in das rohe *Hydrochlorid* übergeführt und dieses mit gesätt. K_2CO_3-Lösung + Äther in der Kälte gespalten. Der destillativ isolierte *Ester* (S)(−)-**17** zeigte folgende Daten: Sdp.$_3$ 89°, $[\alpha]_D^{22}$ = −88.4° (*c* = 1; absol. Äthanol), n_D^{20} = 1.4779; Ausbeute 2.15 g (35%).

$C_{10}H_{17}NO_2$ (183.2) Ber. C 65.54 H 9.35 Gef. C 65.45 H 9.42

Die vereinigten Mutterlaugen des obigen (R)-Dibenzoyl-hydrogentartrates wurden i. Vak. eingedampft und der daraus freigesetzte Ester (8.8 g, $[\alpha]_D^{21}$ = +23.4°) analog mit *(S)(+)-Dibenzoyl-weinsäure* umgesetzt. Aus dem umkristallisierten *(S)-Dibenzoyl-hydrogentartrat* (Schmp. 138°, Zers., $[\alpha]_D^{21}$ = +105°, 18 g) wurden 3.22 g (52%) (R)(+)-**17** gewonnen, Sdp.$_3$ 89°, $[\alpha]_D^{22}$ = +85.5° (*c* = 1; absol. Äthanol).

(R)- und (S)-Chinuclidin-carbonsäure-(2) (**17a**) erhielten wir durch 4stdg. Kochen von (R)- bzw. (S)-**17** mit 5*n HCl* als *Hydrochloride. (R)-Form:* Schmp. 283−288°, $[\alpha]_D^{24}$ = +93.3° (*c* = 1; H_2O); *(S)-Form:* Schmp. 284−288°, $[\alpha]_D^{21}$ = −96.2° (*c* = 1; H_2O). — *ORD* (in 0.2*n* HCl; *c* = 0.13): $[\Theta]_{400}$ = −333°, $[\Theta]_{250}$ = −1360°, $[\Theta]_{240}$ = −1520°, $[\Theta]_{233}$ = −1490°, $[\Theta]_{210}$ = −3520°. Filtration der wäßr. Lösungen dieser Salze über eine Anionen-Austauscher-Säule (Wofatit L 150) lieferte die *freien Säuren*, die aus dem noch feuchten Eindampf-Rückstand durch Digerieren mit Aceton farblos und chloridfrei erhalten wurden. (R)-**17a**: Schmp. 268−271°, $[\alpha]_D^{23}$ = +120.6° (*c* = 1; H_2O); (S)-**17a**: Schmp. 268−272°, $[\alpha]_D^{21}$ = −123.6°.

---

[29] *G. Losse* und *H. Jeschkeit*, Chem. Ber. **90**, 1275 (1957).

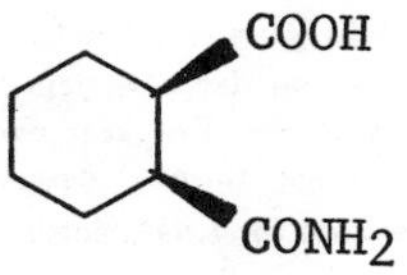

$C_8H_{13}NO_3$ Ref. 419

Dédoublement. - Nous avons employé également la quinine. Dans 500 cc. d'alcool chaud on dissout 60 gr. de quinine et 32 gr. de l-amide. Par refroidissement, il se forme une masse gélatineuse assez difficile à essorer. Le dédoublement se fait cependant bien; on facilite les cristallisations et les essorages en ajoutant une assez grande quantité d'éther ordinaire.

Après 1 cristallisation on a F. 150-152° (déc.).
$[\alpha]_{578}$ = -131° (alcool c = 0,02).

Après 4 cristallisations on a F. 149-150° (déc.).
$[\alpha]_{578}$ = -128° (alcool c = 0,02).

On ne peut guère suivre la purification au point de fusion ou au pouvoir rotatoire. Après 4 cristallisations, nous avons isolé l'amide comme il est dit plus loin. Son pouvoir rotatoire dans l'alcool (c = 0,02) était $[\alpha]_{578}$ = -15°,4.

Deux autres cristallisations faites sur une portion non détruite du sel de quinine n'ont presque pas modifié ce pouvoir rotatoire de l'amide $[\alpha]_{578}$ = -16°,7 Quatre a six cristallisations semblent donc suffire pour obtenir l'isomere gauche sensiblement pur.

Pour libérer l'amide active, on met le sel en solution dans un peu d'alcool chaud. On ajoute de la soude aqueuse diluée jusqu'à reaction alcaline a la phtaléine. On précipite la quinine par l'eau et on extrait quatre fois au chloroforme, puis deux fois à l'éther. On concentre la solution alcaline dans le vide et on acidule par HCl. L'amide cristallise; on essore et on lave à l'eau froide. Elle fond vers 165° avec un commencement de décomposition.

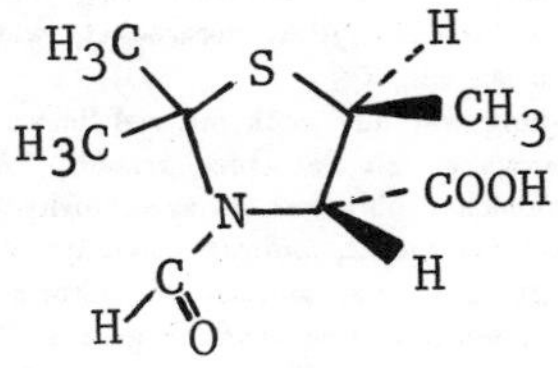

$C_8H_{13}NO_3S$ Ref. 420

Resolution of DL-*threo*-*N*-**Formyl-2,2,5-trimethyl-4-carboxythiazolidine.** DL-4a (65.6 g, 321.5 mmol) and 142 g (330 mmol) of brucine dihydrate were dissolved by gentle heating in 800 ml of water. The solution was filtered warm and kept for 48 hr at room temperature. The needle-like crystals (32 g) of the brucine salt of D-4a were collected. Evaporation of the filtrate yielded three other fractions (43.7 g) of the D enantiomer. The mother liquid of the preceding operation, containing the L enantiomer, was evaporated to an oil which was taken up in 500 ml of water. Crystallization at room temperature afforded 76.5 g of the crude brucine salt of L-4a. Concentration of the filtrate yielded a second fraction (41.7 g) of the L enantiomer. Both D- and L-brucine salts were purified by fractional crystallization from water, yielding 73.6 g of the D and 65 g of the L enantiomer.

Recovery of the Enantiomers of ***N*-Formyl-2,2,5-trimethyl-4-carboxythiazolidine. A. Threo Isomers (D- and L-4a).** The brucine salt of D-4a (73 g) was dissolved in 750 ml of $CHCl_3$. The solution was extracted three times with 70 ml of aqueous NH_4OH (0.5 *N*). The aqueous layer was covered with 150 ml of EtOAc and acidified to pH 1.5 with 6 *N* HCl. The EtOAc layer was dried and evaporated. The residue was taken up in 50 ml of anhydrous benzene. The solution was evaporated to dryness and triturated with petroleum ether (bp 40–60°), yielding 22.23 g of amorphous **D-4a**, $[\alpha]^{25}_D$ +81° (*c* 1, EtOH). Conversion of the brucine salt of **L-4a** (65 g) into the free acid was performed in a similar way, yielding 19.8 g of pure **L-4a**, $[\alpha]^{25}_D$ −78.5° (*c* 1, EtOH).

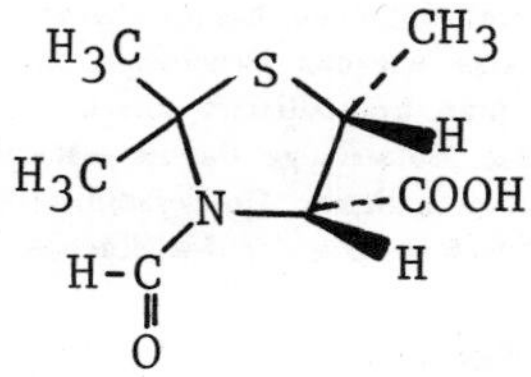

$C_8H_{13}NO_3S$ Ref. 421

Resolution of DL-*erythro*-*N*-**Formyl-2,2,5-trimethyl-4-carboxythiazolidine.** To a stirred solution of 25.37 g (125 mmol) of **DL-4b** and 35.62 g (125 mmol) of dehydroabietylamine in 300 ml of methanol was added 250 ml of water. The mixture was kept overnight in a cool place and the crude dehydroabietylamine salt of **D-4b** (27.9 g) was isolated. The residue obtained on evaporation of the mother liquid was crystallized from benzene–petroleum ether (bp 30–60°) (1:20) yielding 23.4 g of the pure dehydroabietylamine salt of **L-4b**. The crude salt of the D isomer was recrystallized from methanol–water (1:1), yielding 25.1 g.

B. Erythro Isomers (D- and L-4b). The dehydroabietylamine salt of **D-4b** (25.1 g) was added to 100 ml of 1 *N* NH_4OH and extracted twice with ether. Acidification of the aqueous layer to pH 1 with 6 *N* hydrochloric acid afforded on cooling a precipitate, which was collected, washed with water, and dried, yielding 9.6 g of **D-4b**, mp 223–224° dec, $[\alpha]^{25}_D$ +89° (*c* 1, EtOH). Similar treatment of the salt of the L isomer (23.4 g) gave **L-4b**, which was recrystallized from boiling water, yielding 6.6 g of **L-4b**, mp 220.5–221.5° dec, $[\alpha]^{25}_D$ −88.5° (*c* l, EtOH).

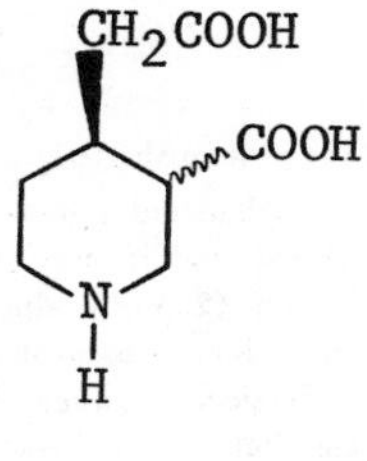

$C_8H_{13}NO_4$ Ref. 422

8 g reinste *rac.-α*-Cincholoiponsäure (1 Mol.) wurden in 35 ccm Wasser gelöst und in die heiße Lösung 8 g Brucin (1/2 Mol.) eingetragen. Im Laufe von 6 Wochen krystallisierten 7.2 g Brucinsalz aus. (Schön ausgebildete Prismen vom Schmp. 105°.) Trotz des einheitlichen Aussehens der Krystalle war die Spaltung nur unvollkommen verlaufen, und es lag ein Gemenge von *rac.* mit dem Salz der *l*-Säure vor, das durch weiteres Umkrystallisieren nicht getrennt werden konnte, da die Salze dabei immer teilweise in Brucin und freie Säure zerfallen. Die Reindarstellung der aktiven Form gelang aber durch Überführung in die freien Säuren. 7 g des einmal aus Alkohol umkrystallisierten Brucinsalzes wurden zur Entfernung des Brucins in wenig Wasser gelöst und mit einer Lösung von 6 g Barythydrat versetzt, von Brucin abfiltriert und das Barium mit fünffach-normaler Schwefelsäure genau ausgefällt. Das brucinfreie Filtrat wurde mit Tierkohle etwas gekocht und auf dem Wasserbade auf ein kleines Volum eingedunstet.

Es krystallisierten daraus zunächst 2.2 g *rac.-α*-Cincholoiponsäure in Form feiner Krystalle; die Mutterlauge erwies sich als stark aktiv, und im Laufe einer Woche schieden sich daraus große tafelige Krystalle (1.5 g) der *l-α*-Cincholoiponsäure ab. Die einmal umkrystallisierte Säure wurde im Lippichschen Polarisationsapparat untersucht.

Eine 4.42-prozentige Lösung vom spez. Gewicht 1.0137 drehte im 1-dm-Rohr — 1.57°. Daraus berechnet sich

$$[\alpha]_D^{20} = -35.00°.$$

Bei weiterem Umkrystallisieren blieb die Drehung unverändert. Die *l*-*α*-Cincholoiponsäure krystallisiert in dicken, gut ausgebildeten Prismen, die kein Krystallwasser enthalten und bei 253° (korr.) schmelzen. Sie ist leicht löslich in Wasser, unlöslich in Alkohol und Äther.

d-*α*-Cincholoiponsäure.

Das sirupöse Filtrat des oben beschriebenen Brucinsalzes wurde in genau gleicher Weise wie das Brucinsalz von Brucin befreit und aufgearbeitet. Auch hier krystallisiert zuerst *rac.* Verbindung aus (ca. 2 g). Die sirupöse Mutterlauge davon erstarrte nach 6 Wochen krystallinisch. Nach einmaligem Umkrystallisieren wurden daraus reine, groß ausgebildete Krystalle der *d*-*α*-Cincholoiponsäure gewonnen (Schmp. 253°).

Eine 4.54-prozentige Lösung dieser Krystalle vom spez. Gewicht 1.0151 drehte im 1-dm-Rohr um + 1.61°, woraus sich berechnet

$$[\alpha]_D^{20} = +34.90°.$$

Wie der Habitus der Krystalle und die folgende Zusammenstellung zeigt, sind die aktiven Komponenten der *α*-Säure von der Cincholoiponsäure aus Chinin sicher verschieden.

	Cincholoiponsäure aus Chinin	*d*- bezw. *l*-Cincholoiponsäure
Zusammensetzung	Säure + 1 aq.	krystallwasserfrei
Schmelzpunkt	125° (ohne Krystallwasser 221—222°) [1])	253°
Drehungsvermögen	+ 30.1° [2])	± 35°

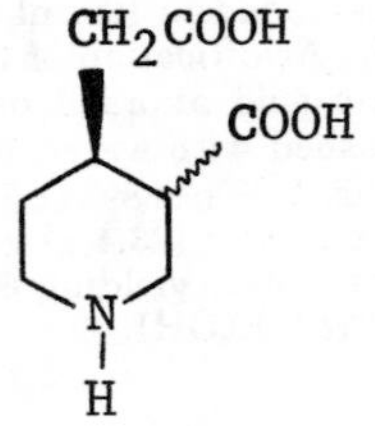

$C_8H_{13}NO_4$ Ref. 423

d-*β*-Acetyl-cincholoiponsäure.

5 g reinste *rac.*-*β*-Acetylcincholoiponsäure (1 Mol.) wurden in 35 ccm heißem Wasser gelöst und in die heiße, auf dem Wasserbad stehende Lösung 17 g Brucin (2 Mol.) eingetragen, nach beendigter Lösung noch ½ Stde. auf dem Wasserbad belassen und die dünn sirupöse Lösung in den Exsiccator über Ätznatron gestellt. Nach 12 Stdn. waren 2.5 g ziemlich dicker Nadeln ausgeschieden. Eine zweite Fraktion ergab 2.7 g, eine dritte Fraktion 1 g. Dann erschienen neben den Nadeln augenscheinlich andere, weiße, mikrokrystallinische Produkte. Die Krystalle wurden jeweils gut mit Wasser gedeckt und, insgesamt 7.2 g, aus 6 ccm Wasser umkrystallisiert. Das so erhaltene Brucinsalz schmolz unscharf bei 107°, war aber zur Reinabeidung der aktiven Säure genügend einheitlich; es wurden demgemäß zur Entfernung des Brucins 6 g Salz in 20 ccm Wasser gelöst, mit einer heißen konzentrierten Lösung von 2 g Barythydrat (das Doppelte der berechneten Menge) versetzt, abfiltriert, der Rückstand nochmals mit wenig Wasser ausgekocht, wiederum filtriert und die vereinigten Filtrate zur Entfernung von Spuren Brucin viermal mit Chloroform und dreimal mit Äther ausgeschüttelt. Aus der vollkommen brucinfreien Lösung wurde das Barium in der Hitze mit 2.4 ccm fünffach-normaler Schwefelsäure genau gefällt und das Filtrat (ca. 50 ccm) auf dem Wasserbad auf ein kleines Volumen verdampft. Nach 24 Stdn. begann der über Schwefelsäure gestellte Sirup zu krystallisieren und erstarrte bald zu einer strahligen Krystallmasse, die reine *d*-Acetylcincholoiponsäure darstellend. Ausbeute 1.2 g; scharfer Schmp. 167—168°, übereinstimmend mit der Acetylcincholoiponsäure von Skraup. Für das so erhaltene Produkt wurde im Lippichschen Polarisationsapparat die Drehung bestimmt. Eine 8.36-prozentige Lösung vom spez. Gewicht 1.0228 drehte im 1-dm-Rohr (ca. 1 ccm Flüssigkeit fassend) + 1.70°, entsprechend:

$$[\alpha]_D^{20} = +19.86°.$$

Da Skraup die Drehung der Acetylcincholoiponsäure aus Chinin nicht angegeben hat, wurde das Präparat nach seiner Vorschrift dargestellt; eine 5.53-prozentige Lösung desselben vom spez. Gewicht 1.0142 drehte im 1-dm-Rohr + 1.09°, entsprechend $[\alpha]_D^{20} = +19.43°$.

d-*β*-Cincholoiponsäure-chlorhydrat.

0.6 g *d*-*β*-Acetylcincholoiponsäure wurden mit der 20-fachen Menge 20-prozentiger Salzsäure während 5 Stdn. zum Sieden am Rückflußkühler erhitzt, auf dem Wasserbad eingedunstet und die filtrierte Lösung mit einem Krystallsplitter von Cincholoiponsäure-chlorhydrat aus Chinin geimpft. Über Nacht hatten sich 0.3 g abgeschieden, dicke Prismen, drusenförmig angeordnet, die abgesaugt und mit Alkohol gewaschen wurden. Schmp. 192—194°, in genauer Übereinstimmung mit der Angabe von Skraup und dem Befunde an der nach seiner Vorschrift hergestellten Verbindung; Misch-Schmp. 192—194°.

Eine 4.07-prozentige Lösung vom spez. Gewicht 1.0122 gab im 1-dm-Rohr + 1.57°, entsprechend

$$\alpha_D^{20} = +38.04°.$$

Skraup fand fur die Drehung seines Cincholoiponsäure-chlorhydrats zuerst + 34.4° [1]), aus einer anderen Beobachtung von ihm [2]) berechnet sich + 38.32°, Skraups Mitarbeiter gaben einmal [3]) + 40.2°, ein andermal + 39.6° [4]) an.

Mit Rücksicht auf diese schwankenden Befunde haben wir mit der durch Aufbau erhaltenen Säure ein Originalpräparat Skraups, für dessen gütige Überlassung wir Hrn. Hofrat Prof. Skraup auch hier bestens danken, und die von uns nach seiner Vorschrift aus Cinchonin hergestellte Säure unmittelbar verglichen.

Eine 3.88-prozentige Lösung des einmal umkrystallisierten Präparats von Skraup (spez. Gewicht 1.0118) drehte im 1-dm-Rohr + 1.49°, entsprechend

$$[\alpha]_D^{20} = +37.87°.$$

Eine 4.24-prozentige Lösung (spez. Gewicht 1.0133) des aus Cinchonin gewonnenen Cincholoiponsäure-chlorhydrats zeigte im 1-dm-Rohr + 1.63°, entsprechend

$$[\alpha]_D^{20} = +37.96°.$$

Es wurde also in allen drei Fällen innerhalb üblicher Fehlergrenzen der gleiche Wert gefunden.

Endlich verdanken wir dem Assistenten am Mineralogisch-geologischen Institut der hiesigen Hochschule, Hrn. Dr. Lehmann, eine eingehende krystallographische Vergleichung der drei Präparate, über die derselbe unten berichtet, und die weiter jeden Zweifel an der Übereinstimmung ausschließt.

l-*β*-Cincholoiponsäure-chlorhydrat.

Das Filtrat von dem Brucinsalz der *d*-Acetylcincholoiponsäure stellt einen dicken Sirup dar, der wieder krystallisierte. Nach Abscheidung weiterer 1.5 g Brucinsalz wurde abgesaugt, das Filtrat von Brucin befreit und auf gleiche Weise auf *l*-Cincholoiponsäurechlorhydrat verarbeitet, wie es bei der *d*-Verbindung beschrieben worden ist.

Die *l*-Acetylverbindung war hier noch mit viel Racemat vermengt, wurde aber ohne weitere Reinigung mit Salzsäure verseift. Aus der salzsauren Lösung krystallisierte zunächst 1.3 g *rac.*-*β*-Säurechlorhydrat, dann ca. 1.2 g ziemlich reines *l*-*β*-Säurechlorhydrat, zuletzt herrschte wieder Racemat vor. Die gute Ausbildung der Krystalle, sowohl der *l*-Form (Prismen) als auch des Racemats (Polyeder), erlaubte eine ziemlich genaue Trennung der in der Löslichkeit nicht stärk verschiedenen Chlorhydrate durch Auslesen.

Nach einmaligem Umkrystallisieren des sorgfältig vom Racemat gereinigten *l*-Säurechlorhydrats zeigte eine 2.20-prozentige Lösung desselben vom spez. Gewicht 1.0050 eine Drehung von — 0.75°, entsprechend $[\alpha]_D^{20} = -33.89°$. Nach dreimaligem Umkrystallisieren konnte die Drehung beinahe auf den Wert des *d*-Säurechlorhydrats hinaufgebracht werden, eine 3.79-prozentige Lösung vom spez. Gewicht 1.0112 gab nun — 1.40°, entsprechend

$$[\alpha]_D^{20} = -36.51° \text{ (statt 38.04°).}$$

Der Schmelzpunkt des reinen *l*-Cincholoiponsäure-chlorhydrats liegt wie der der *d*-Form bei 192—194°.

[1]) Wiener Monatsh. 9, 799.
[2]) Wiener Monatsh. 10, 46.
[3]) Wiener Monatsh. 10, 59.
[4]) Wiener Monatsh. 10, 70.

$C_8H_{13}NO_5$

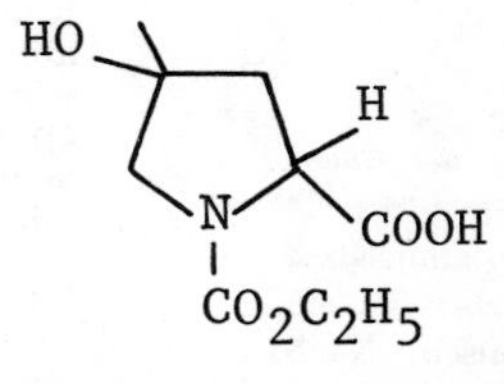

Ref. 424

Abtrennung des L-Antipoden über das kristallisierte Brucinsalz: 136 g *N-Carbäthoxy-allo-hydroxy-DL-prolin* (III) wurden in 3 gleichen Portionen mit je 90 g feinstgepulvertem Brucin in 2 *l* Butanon auf dem Wasserbad unter kräftigem Umrühren zum Sieden erwärmt. Nach 10 Min. erschienen an Stelle der ursprünglichen Komponenten die Kristalle des Salzes, die man nach 2tägigem Stehenlassen bei 0° absaugte. Die Rohkristallisate der 3 Ansätze (etwa 200 g) wurden mit 5 *l* trockenem Butanon unter Rückfluß 15 Min. auf dem Wasserbad ausgekocht, die Suspension rasch filtriert und das Filtrat bei 0° zur Kristallisation aufbewahrt. Es schieden sich Kristalle vom $[\alpha]_D^{20}$: $-22°$ ($c = 2$, in Wasser) ab*). Mit der Mutterlauge dieser Kristallisation wurde der beim ersten Auskochen ungelöst gebliebene Salzrückstand ein zweites Mal 15 Min. unter Rückfluß aufgekocht. Aus der filtrierten Lösung schied sich beim Aufbewahren in der Kälte eine Kristallisation vom $[\alpha]_D^{20}$: $-25.5°$ ab, deren Mutterlauge, mit frischem Butanon auf 5 *l* ergänzt, zur Heißextraktion des immer noch ungelösten Salzanteils verwendet wurde. Hierbei ging alles bis auf einen unbedeutenden Rückstand in Lösung. Nach dem Klarfiltrieren kristallisierte eine weitere Menge Brucinsalz vom $[\alpha]_D^{22}$: $-27°$ aus. Die Mutterlauge wurde i. Vak. auf etwa 1/4 eingeengt und neuerlich in der Kälte zur Kristallisation aufgestellt. Es kristallisierte ein Salz vom $[\alpha]_D^{22}$: $-27.5°$, nach völligem Abdampfen der Mutterlauge i. Vak. blieb ein Rückstand vom $[\alpha]_D^{22}$: $-20°$ zurück.

Zur völligen Reingewinnung des L-Salzes wurde weiter so verfahren, daß alle Fraktionen so lange aus frischem Butanon umkristallisiert wurden, bis ihre spezifische Drehung den Endwert von $-34°$ erreichte. Die jeweiligen Mutterlaugen wurden i. Vak. konzentriert und zur neuerlichen Kristallisation kalt gestellt, die Endmutterlaugen schließlich völlig abgedampft. Kristallisate, deren spezifische Drehung innerhalb von 2 Graden übereinstimmte, wurden jeweils vereinigt. Es gelang so in ziemlich mühevoller Weise 85 g (21 %) Brucinsalz der reinen L-Verbindung III vom $[\alpha]_D^{22}$: $-34.0 \pm 0.5°$ ($c = 2$, in Wasser) zu isolieren.

Zur *Zerlegung des Salzes* wurden die erhaltenen 85 g in einem 2-*l*-Scheidetrichter in 500 ccm Wasser gelöst und mit verd. Natronlauge auf p_H 12 gebracht. Das hierbei in dicker Flockung ausfallende Brucin wurde durch zweimaliges Ausschütteln mit je 250 ccm Chloroform entfernt, die vereinigten Chloroformphasen wurden 4mal mit je 100 ccm Wasser re-extrahiert und alle wäßrigen Phasen mit 2*n* HCl auf p_H 7 gebracht. Die neutrale Lösung engte man i. Vak. auf etwa 250 ccm ein, säuerte mit 2*n* HCl auf p_H 2 an und extrahierte durch 5maliges Ausschütteln mit je 100 ccm Essigester den größten Teil der freien Säure. Den Rest gewann man, ebenfalls durch Ausschütteln mit Essigester, nachdem die wäßrige Lösung bis zur starken Natriumchlorid-Abscheidung weiter i. Vak. eingedampft worden war. Die vereinigten Extrakte wurden mit Na_2SO_4 getrocknet und i. Vak. völlig abgedampft. Es hinterblieben 31 g *Säure III*, die nach Aufbewahren über KOH im Exsikkator 3mal aus Essigester umkristallisiert wurden. Man erhielt 23 g der optisch reinen Säure vom Schmp. 97° und $[\alpha]_D^{20}$: $-56.5°$ ($c = 2$, in Wasser).

---

*) Die spezifische Drehung des Gemisches vom Brucinsalz gleicher Mengen D- und L-Form beträgt $-12°$, die vom Brucinsalz der reinen L-Form $-34.0°$.

6) E. Gross, in Th. Wieland und G. Pfleiderer, Angew. Chem. **69**, 200 [1957].

$(CH_3)_2CHCH_2-CH-COOH$
$\qquad\qquad\qquad\quad |$
$\qquad\qquad\quad NHCOCH_2Cl$

$C_8H_{14}ClNO_3$

Ref. 425

4,0 g Chloracetyl-d,l-leucin wurden in 40 ccm Wasser mit 3,2 g *a*-Phenyläthylamin-carbonat (= 1 Mol.) versetzt. Unter lebhafter Abspaltung von Kohlensäure begann zunächst Lösung einzutreten. Ehe diese vollkommen erreicht war, kristallisierte das gebildete Phenyläthylaminsalz aus, so daß die Flüssigkeit zu einem dicken Brei erstarrte. Unter häufigem Umschütteln und Aufrühren ließen wir das Reaktionsgemisch noch 24 Stunden stehen. Dann wurde scharf abgesaugt und der Filterrückstand mit Wasser gut ausgewaschen. Filtrat und Waschwasser wurden vereinigt und zur Gewinnung der d-Form benutzt.

Die Ausbeute an Phenyläthylaminsalz des Chloracetyl-l-leucins betrug 2,2 g (= 60 % der Theorie). In heißem Wasser gelöst, kristallierte es beim Erkalten in prachtvollen makroskopischen, sternförmig angeordneten Nadeln. Ihr Schmelzpunkt lag bei 178°. 6,985 mg gaben bei 757 mm und 28° 0,539 ccm N_2.

Berechnet für $C_{16}H_{25}O_3N_2Cl$ (328,7) 8,52 %, gefunden 8,71 % N.

1 g des Phenyläthylaminsalzes wurde in 10 ccm fünffach normaler Salzsäure gelöst. Dabei geht das Phenyläthylamin als Chlorhydrat in Lösung. Es kann durch Versetzen der Lösung mit Natronlauge und Ausäthern zurückgewonnen werden. Das Chloracetyl-l-leucin kristallisierte sofort aus. Nach mehrstündigem Stehen wurde es abfiltriert und kurz mit verdünnter Salzsäure und Wasser ausgewaschen. Die Ausbeute betrug 0,6 g = 75 % der Theorie. Das Chloracetyl-l-leucin kristallisierte aus Wasser in rechteckigen Prismen.

Die optische Drehung wurde in absoluter alkoholischer Lösung bestimmt. 0,133 g in 10 ccm Alkohol drehten das polarisierte Licht

0,18° nach links bei 20° C.

Hieraus ergibt sich $[\alpha]_D^{20°} = -13{,}4°$.

III. Gewinnung von Chloracetyl-d-leucin aus der Mutterlauge des Chloracetyl-l-leucin-a-phenyläthylaminsalzes mittels l-Phenyläthylamins.

Die Mutterlauge des Chloracetyl-l-leucin-a-phenyläthylaminsalzes wurde mit fünffach normaler Salzsäure angesäuert. Hierbei fiel ein Teil des Chloracetyl-leucins aus. Er erwies sich als racemisch. Nach starkem Einengen der Lösung konnte der übrige Teil des in der Lösung vorhandenen Chloracetyl-leucins gewonnen werden. Er zeigte nur eine schwache Linksdrehung.

Beide Kristallisationen des Chloracetyl-leucins wurden hierauf wieder vereinigt und in der gleichen Weise, wie oben beschrieben, mit l-Phenyläthylamin-carbonat behandelt. Auch hier fiel das Phenyläthylaminsalz sofort aus. Aus Wasser kristallisierte es in langen Nadeln. Sein Schmelzpunkt lag bei 169 bis 170°.

6,240 mg gaben bei 750 mm und 18° 0,465 ccm N_2.

Berechnet für $C_{16}H_{25}O_3N_2Cl$ (328,7) 8,52 %, gefunden 8,63 % N.

Die Zerlegung des Salzes erfolgte, wie oben mitgeteilt, durch Behandlung mit fünffach normaler Salzsäure. Das Chloracetyl-d-leucin schied sich erst nach mehrstündigem Stehen in der Kälte aus der Lösung aus. Aus Wasser kristallisierte es in rhombischen Tafeln. Die Ausbeute betrug etwa 70 % der Theorie.

0,165 g in 10 ccm absolutem Alkohol gelöst, drehten bei 20° das polarisierte Licht 0,21° nach rechts. Hieraus ergibt sich $[\alpha]_D^{20°} = +12{,}7°$.

$ClCH_2CH_2$-CH($CH_2)_4$COOH (with Cl on the CH)

$C_8H_{14}Cl_2O_2$ Ref. 426

(+)-6,8-Dichlorooctanoic Acid (−)-Ephedrine Salt.—A 12.06-g. (0.073 mole) portion of (−)-ephedrine was dissolved in a solution of 15.57 g. (0.073 mole) of DL-6,8-dichlorooctanoic acid in 100 ml. of ethyl acetate. The solution, cooled to −18°, soon crystallized to a continuous mass. The crystals were filtered cold and pressed dry. A sample was dried *in vacuo*, m.p. 81.5–94.5°, $[\alpha]^{24.5}D$ −19.1° (*c* 1.2, ethyl alcohol). The pressed-dry crystals were redissolved in the minimum amount of ethyl acetate at 35°. Crystals formed at −18°. The crystals were recrystallized in this manner a total of six times to yield 3.85 g. of (+)-6,8-dichlorooctanoic acid (−)-ephedrine salt, m.p. 106.5–108°, $[\alpha]^{24.2}D$ −11.1° (*c* 1, ethyl alcohol). *Anal.* Calcd. for $C_{18}H_{29}O_3NCl_2$: N, 3.70; Cl, 18.72. Found: N, 3.69; Cl, 18.87.

The filtrates from the six recrystallizations were combined and concentrated under reduced pressure at 35° almost to dryness. The crystals which formed were recrystallized six times from ethyl acetate as described above to yield an additional 2.6 g. of salt, m.p. 105–107°. The filtrates obtained were treated by a like procedure to yield another 1 g. of salt, m.p. 105–106°. The total yield was thus 7.45 g. (54%) of (+)-6,8-dichlorooctanoic acid (−)-ephedrine salt.

(+)-6,8-Dichlorooctanoic Acid (VII).—A 7.43-g. (0.0196 mole) portion of (+)-6,8-dichlorooctanoic acid (−)-ephedrine salt was treated with 60 ml. of ice-cold 5% hydrochloric acid. The aqueous layer was saturated with potassium chloride, and the mixture extracted four times with a total of 400 ml. of ether. The ether extract was washed four times with a total of 120 ml. of ice-cold saturated potassium chloride solution and dried over anhydrous sodium sulfate in the cold. After removal of drying agent and ether, the remaining liquid was distilled to yield 4.07 g. (97.4%) of (+)-6,8-dichlorooctanoic acid, b.p. 121.5–123° (0.2 mm.), $n^{25}D$ 1.4768–1.4778. A middle fraction, $n^{25}D$ 1.4776, $[\alpha]^{23.8}D$ +30.5° (*c* 2, benzene), was analyzed. *Anal.* Calcd. for $C_8H_{14}O_2Cl_2$: C, 45.10; H, 6.63; Cl, 33.22; neut. equiv., 213. Found: C, 45.54; H, 6.79; Cl, 33.17; neut. equiv., 208.

(−)-6,8-Dichlorooctanoic Acid (−)-Ephedrine Salt.—The filtrate remaining after filtration at −18° of the crude (+)-6,8-dichlorooctanoic acid (−)-ephedrine salt from 15.57 g. of DL-6,8-dichlorooctanoic acid, as described above, was concentrated under reduced pressure at 35° to 25 ml. and stored at −18°. White crystals formed, m.p. 68.5–78°, $[\alpha]^{23.6}D$ −30.8° (*c* 0.8, ethyl alcohol). The crystals were recrystallized five times by dissolving them in the minimum amount of ethyl acetate at 35° and cooling to −18°. There was obtained 0.36 g. of (−)-6,8-dichlorooctanoic acid (−)-ephedrine salt, m.p. 78.5–82°, $[\alpha]^{24.2}D$ −33.4° (*c* 1, ethyl alcohol). *Anal.* Calcd. for $C_{18}H_{29}O_3NCl_2$: N, 3.70; Cl, 18.72. Found: N, 3.78; Cl, 18.76.

The filtrates from the five recrystallizations were combined and concentrated to yield crystals which were recrystallized three times. The operation then was repeated on these filtrates. There was thus obtained two additional portions of (−)-6,8-dichlorooctanoic acid (−)-ephedrine salt amounting to 2.29 g., m.p. 77–80°, and 1.93 g., m.p. 77.5–81°. The total yield was thus 4.58 g. (33%).

(−)-6,8-Dichlorooctanoic Acid (VII).—A 4.58-g. (0.0121 mole) portion of (−)-6,8-dichlorooctanoic acid (−)-ephedrine salt was treated with cold 5% hydrochloric acid and the (−)-6,8-dichlorooctanoic acid isolated as described above for (+)-6,8-dichlorooctanoic acid. Distillation of the product yielded 2.5 g. (97%) of (−)-6,8-dichlorooctanoic acid, b.p. 122–126° (0.3 mm.), $n^{25}D$ 1.4753–1.4779. A middle fraction, $n^{25}D$ 1.4776, $[\alpha]^{23.8}D$ −29.3° (*c* 2, benzene), was analyzed. *Anal.* Calcd. for $C_8H_{14}O_2Cl_2$: C, 45.10; H, 6.63; Cl, 33.22; neut. equiv., 213. Found: C, 45.51; H, 6.78; Cl, 33.17; neut. equiv., 208.

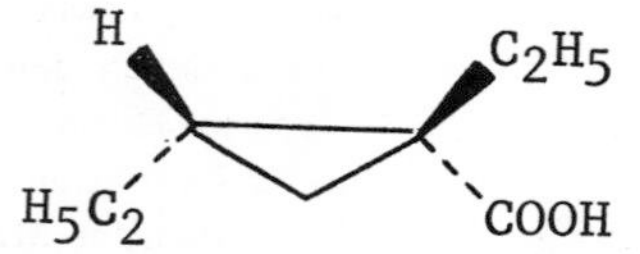

$C_8H_{14}O_2$ Ref. 427

Resolution of (±)-*trans*-2,3-Diethylcyclopropanecarboxylic Acid. To a solution of 9.85 g (30.3 mmol) of quinine in 100 ml of 95% ethanol, 8.0 g (56.2 mmol) of (±)-*trans*-2,3-diethylcyclopropanecarboxylic acid was added. The solution was heated to reflux for 10 min and then diluted with 250 ml of 20% ethanol. Slow cooling to room temperature followed by cooling in the refrigerator overnight yielded 12.0 g of the diastereomer as a white solid. Recrystallization of this material from hot 60% ethanol yielded 5.8 g (59%) of the diastereomer, mp 126–127°. A mixture of 5.8 g (12.4 mmol) of the diastereomer dissolved in 30 ml of methanol, 60 ml of a solution of 4% sodium hydroxide in methanol, and 60 ml of water was heated at 50° for 3 hr. The mixture was evaporated at reduced pressure to a volume of 50 ml, cooled to room temperature, and filtered to remove the precipitated quinine. The filtrate was washed with ether to remove the last traces of quinine, then acidified to pH 2 with 10% sulfuric acid. Extraction with ether, drying with magnesium sulfate, and removal of the solvent by use of a rotary evaporator yielded 2.05 g (25%) of the acid as a colorless liquid, $[\alpha]^{26}D$ −4.55° (*c* 0.0314, 95% ethanol). The filtrates from the first two crystallizations of the diastereomer were combined and saponified in an identical manner to give 5.19 g (65%) of the acid, $[\alpha]^{23}D$ +1.74° (*c* 0.110, 95% ethanol).

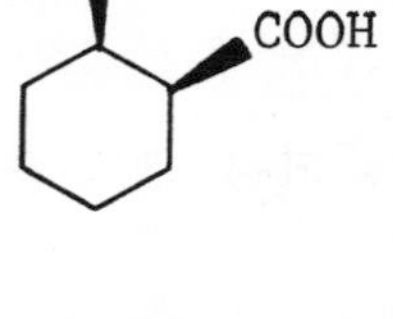

$C_8H_{14}O_2$ Ref. 428

39,76 g d'acide racémique ci-dessus sont traités par un équivalent de déhydroabiétylamine (venant de 92,68 g d'acétate) dans 1 700 cm³ d'éthanol; 53,70 g de sel cristallisent en longues aiguilles. L'acide est régénéré de son sel par addition de soude 0,5 N et d'éther; la phase éthérée est lavée 2 fois à la soude 0,5 N, et les phases aqueuses alcalines sont acidifiées par l'acide chlorhydrique concentré et extraites à l'éther: $[\alpha]_J^{20} = +1{,}6°$; $[\alpha]_{436}^{20} = +2{,}1°$ (éthanol; $c = 9{,}0$).

Après 13 *nouvelles cristallisations* de ce sel dans l'éthanol (environ 1 g dans 15 cm³ pour les premières cristallisations, 1 g dans 5 cm³ à la fin), on en obtient 410 mg qui fournit un acide $[\alpha]_D^{21,5} = +7{,}6°$; $[\alpha]_J^{21,5} = +7{,}9°$; $[\alpha]_{436}^{21,5} = +10{,}6°$ (éthanol $c = 5{,}95$) (voir note p. 2899 sur la pureté optique de cet échantillon).

Les eaux-mères de la première cristallisation sont évaporées à sec et le résidu, recristallisé dans 500 cm³ d'éthanol et 150 cm³ d'eau, fournit 38,7 g de produit qui est mélangé aux 26,9 g obtenus par évaporation des eaux-mères de la deuxième cristallisation.

Ces 65,6 g sont cristallisés dans 950 cm³ d'éthanol. Après trois nouvelles cristallisations dans l'éthanol, on obtient 10,0 g de sel qui fournit un acide brut $[\alpha]_J^{21} = +4{,}6°$, $[\alpha]_{436}^{21} = +6{,}2°$ (éthanol; $c = 7{,}5$).

L'acide venant des dernières cristallisations du dédoublement est distillé au tube à boules, sous vide, en recueillant la fraction qui distille entre 95 et 162° sous 21 mm. On obtient 3,9 g; $[\alpha]_D^{21,5} = +5{,}2°$ (éthanol; $c = 1{,}11$) (la pureté chimique de ce composé est sensiblement égale à celle de l'acide racémique de départ: environ 97 %).

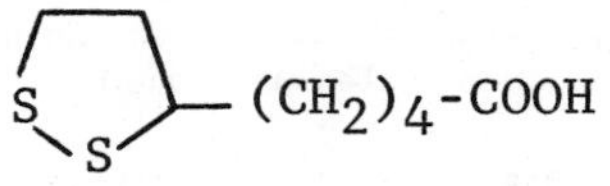

$C_8H_{14}O_2S_2$ **Ref. 429**

B. From DL-α-Lipoic Acid.—A 10.00-g. (0.04845 mole) portion of DL-α-lipoic acid was dissolved in 100 ml. of ethyl acetate at room temperature. To this solution was added a solution of 14.28 g. (0.04845 mole) of cinchonidine in 4100 ml. of ethyl acetate. As an aid to crystallization, 1 ml. (1.1 moles/mole of salt) of water was added to the solution. The container was then scratched with a glass rod and the whole stored at −18°. Crystals developed over a period of several days. The crystals were filtered cold, pressed dry, and then redissolved in the minimum amount of ethyl acetate by stirring at room temperature. Crystals formed during storage at −18°. The crystals were recrystallized in this manner a total of five times. After the fifth recrystallization, there was obtained 0.35 g. (there was some loss due to spillage) of (+)-α-lipoic acid (−)-cinchonidine salt as light yellow crystals which melted over the range 99.5–156°, $[\alpha]^{24.1}D$ −50° (*c* 0.9, ethyl alcohol). *Anal.* Calcd. for $C_{27}H_{36}O_3N_2S_2$: N, 5.60; S, 12.80. Found: N, 5.57; S, 13.20.

All of the above-described operations were carried out in the dark or in red light and under nitrogen to minimize side reactions of lipoic acid such as oxidation and polymerization. A 0.35-g. portion of (+)-α-lipoic acid (−)-cinchonidine salt was treated with 50 ml. of 5% hydrochloric acid, and the mixture extracted three times with ether. The ether extract was washed five times with water and dried over sodium sulfate in the cold. Removal of drying agent and ether left a yellow solid which was taken up in 30 ml. of *n*-pentane. The solution crystallized to yield 0.026 g. (0.5%) of (+)-α-lipoic acid as yellow leaflets, m.p. 44.5–46.5°, $[\alpha]^{23.8}D$ +87.4° (*c* 0.18, benzene). Its activity in the enzymatic POF assay[20] was 196% of that of DL-α-lipoic acid.

Upon repeating the above preparation, the least soluble salt fraction after four recrystallizations consisted of 0.9 g. of (+)-α-lipoic acid (−)-cinchonidine salt as light yellow crystals which melted over the range 99.5–152°, $[\alpha]^{24.6}D$ −54.6° (*c* 0.6, ethyl alcohol). Treatment of the 0.9 g. of salt with 5% hydrochloric acid resulted in a yellow solid which was recrystallized from *n*-pentane to yield 0.22 g. (4.4%) of (+)-α-lipoic acid as yellow leaflets, m.p. 46–48°, $[\alpha]^{24.5}D$ +115.9° (*c* 1, benzene). Its activity in the enzymatic POF assay was 188% of that of DL-α-lipoic acid.

B. From DL-α-Lipoic Acid.—In the preparation of (+)-α-lipoic acid (−)-cinchonidine salt described above, the filtrate remaining after removal of the crystalline salt from the original reaction mixture was reduced in volume by stages by distillation under reduced pressure at 35°. At each stage, additional crystalline material which formed by cooling the solution to −18° was removed by filtration. There finally remained 4.5 g. of a yellow sirup which did not crystallize, $[\alpha]^{24.7}D$ −79.6° (*c* 1, ethyl alcohol). The sirup was treated with 50 ml. of 5% hydrochloric acid, and the mixture extracted three times with ether. The ether extract was washed five times with water and dried over sodium sulfate in the cold. Removal of drying agent and ether left a yellow solid. Recrystallization of the solid from *n*-pentane yielded 2.29 g. of (−)-α-lipoic acid as yellow leaflets, m.p. 57–59°, $[\alpha]^{24.4}D$ −50° (*c* 1, benzene). Its activity in the POF assay[20] was 26% of that of DL-α-lipoic acid.[22]

(22) The optical rotation and POF assay values for this sample of (−)-α-lipoic acid were not in good agreement but indicated that the sample contained 13–22% of (+)-α-lipoic acid.

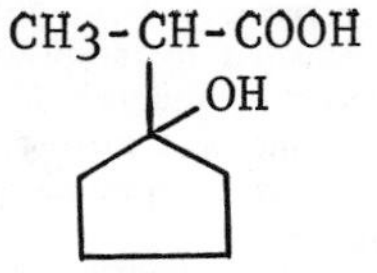

$C_8H_{14}O_3$ **Ref. 430**

The racemic β-hydroxy acid was resolved by combining a hot solution of 20.4 g (0.129 mol) of the acid **16** in 50 mL of absolute ethanol with a hot solution of 42.2 g (0.13 mol) of quinine in 300 mL of absolute ethanol and allowed to stand overnight. The profuse precipitate was collected on a Büchner funnel and dried at reduced pressure (ca. 35 °C, 10 mm), affording 30.0 g (50%) of crude quinine salt, mp 165–173 °C. This salt was recrystallized from ether–ethanol (1:3) until constant melting point (ca. ten times) to give 23 g (38%) of quinine salt, mp 176–176.5 °C. The free β-hydroxy acid was recovered by treating an 800 mL CH_2Cl_2 suspension of the quinine salt with 10% aqueous HCl. The CH_2Cl_2 layer was washed with 3 × 250 mL of 10% aqueous HCl and with 2 × 100 mL of H_2O and dried over $MgSO_4$. Rotoevaporation of the solvent (ca. 28 °C, 30 mm) gave 7.2 g (35%) of crude β-hydroxy acid **16**, mp 50–56 °C. The pure acid was obtained by two recrystallizations from ether–hexane (1:10): 6.5 g (31.8%, based on racemic acid), mp 62–62.5 °C, α^{20}_{589} +3.75°, α^{20}_{578} +3.95°, α^{20}_{546} +4.50°, α^{20}_{436} +7.71°, and α^{20}_{365} −11.8° (*c* 12.53, CCl_4). The spectral data follow: IR (CCl_4) 3500–2600 (associated OH), 2950 and 2875 (aliphatic CH), 1700 (acid C=O), and 1450 and 1375 cm^{-1} (aliphatic CH); NMR (60 MHz) δ (CCl_4, Me_4Si) 1.22 (3 H, d, *J* = 7 Hz, CH_3CH), 1.60 (8 H, s, $(CH_2)_4C$), 1.45 (1 H, q, *J* = 7 Hz, $CHCH_3$), and 7.15 (2 H, s, OH).

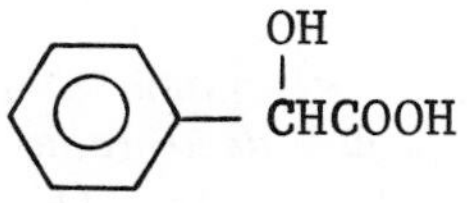

$C_8H_{14}O_3$ **Ref. 431**

Resolution of Hexahydromandelic Acid.—Complete resolution of the acid into the optically active enantiomorphs was effected by the use of quinine and morphine. A good criterion of optical purity was given by the fairly high value of the rotation of the methyl ester derived from both forms. Further evidence of completeness of resolution was obtained from the rotation of esters other than the methyl, of both forms of the acid, and of the dextro- and lævo-ammonium salts. The outline of the methods of resolution adopted is similar to that for the resolution of lactic acid by morphine (Wood, Such, and Scarf, *loc. cit.*).

Isolation of optically pure dextro acid. Hexahydromandelic acid (20 g.), exactly neutralised with *N*-potassium hydroxide, was added to 20 g. of the acid in 3000 c.c. of hot water containing 41 g. of quinine. The quantity of quinine added was sufficient to give the basic salt; the normal salt crystallises with difficulty. The whole was brought gradually to the boiling point; a quantity of the dark-coloured, syrupy quinine salt remained undissolved, which became solid on cooling. The solution was left over-night in ice, the mass of long, colourless needles separated, the liquid again brought to the boiling point in contact with the quinine salt, and a further crop of crystals obtained. This process was repeated eight times. Crystallisation took place below 45°, and was assisted by inoculation. The quinine salt was recrystallised twice, from 4000 c.c. and 3000 c.c., respectively, of water. To the hot solution was added excess of dilute ammonia, the precipitated quinine separated, and the solution reduced to 250 c.c. and rendered just alkaline with ammonia. Traces of quinine were separated, and the acid was obtained from the ammonium *d*-hexahydromandelate by acidification with dilute hydrochloric acid, filtration, and ether-extraction. The acid was purified by redissolving in warm water, leaving until cold, extracting the supersaturated solution with ether, and crystallising from ether. The resolution gave 58% of the theoretical quantity of *d*-acid.

Recovery of the partly resolved hexahydromandelic acid from the mother-liquors was effected by addition of excess of dilute ammonia, filtration, reduction in volume to 300 c.c., further addition of ammonia until just alkaline, and further procedure as detailed above. Resolution by means of quinine proved more difficult than by the use of morphine.

d-Hexahydromandelic acid crystallises from ether in prisms which appear to be monoclinic, m. p. 129·7° (corr.); $[\alpha]_D^{25} = + 13·51°$ ($\alpha = + 1·68°$ for 0·8570 g. of acid in 6·9 c.c. of solution in absolute alcohol).

Quinine *d*-hexahydromandelate crystallises from water in long needles, m. p. 45°; $[\alpha]_D^{25} = - 123·24°$ ($\alpha = - 5·21°$ for 0·4366 g. of salt in 10·3 c.c. of solution in absolute alcohol).

Ammonium *d*-hexahydromandelate is of little value as a criterion of the optical purity of the acid, owing to its small rotatory power: $[\alpha]_D^{25} = - 7·78°$ ($\alpha = - 0·56°$ for 0·5 g. of acid and 2·0 c.c. of ammonium hydroxide, d 0·88, made up to 7·7 c.c. of solution with distilled water). The rotations of the sodium and potassium salts also are of opposite sign to that of the acid.

Isolation of optically pure lævo-acid by means of morphine was carried out as follows: Half the available acid (27 g.) was neutralised by N-potassium hydroxide solution, and the other half by the equivalent quantity (51·8 g.) of morphine in 900 c.c. of boiling water. The solutions were mixed, cooled, and crystals of morphine *l*-hexahydromandelate were precipitated. On evaporation to half volume, an additional crop was obtained. The morphine salt was recrystallised twice from water, the rotation then being constant. To the boiling solution (900 c.c.) of the salt, ammonia was added, and the morphine filtered off. The solution, kept just alkaline with ammonia, was reduced in volume to 250 c.c. and left overnight. The small quantity of precipitated morphine was separated, and the *l*-hexahydromandelic acid isolated by acidification of the solution of its ammonium salt in a similar manner to that for the dextro-acid. The resolution gave 85% of the theoretical quantity of *l*-acid.

The mother-liquors were worked up to obtain partly resolved *d*-acid, for resolution with quinine.

l-Hexahydromandelic acid crystallises from ether in hexagonal clusters, m. p. 129·7° (identical with that of the dextro-enantiomorph); $[\alpha]_D^{25} = - 13·62°$ ($\alpha = - 1·04°$ for 0·5047 g. of the acid in 6·6 c.c. of solution in absolute alcohol).

Morphine *l*-hexahydromandelate crystallises from water in short needles, m. p. 127·6° (decomp.); $[\alpha]_D^{25} = - 63·73°$ ($\alpha = - 3·51°$ for 0·6392 g. in 11·6 c.c. of solution in absolute alcohol). The rotation of the salt was constant after the second recrystallisation.

Ammonium *l*-hexahydromandelate gave $[\alpha]_D^{25} = + 7·51°$ ($\alpha = + 0·54°$; 0·50 g. of acid, 2 c.c. ammonium hydroxide, d 0·88, made up to 7·7 c.c. of solution with water).

Resolution by strychnine and brucine. Both alkaloids gave as the less soluble salts in water and acetone, those derived from the dextro-acid. The partly resolved strychnine salt is precipitated from water as a syrup. The brucine salt is similar, but it may with difficulty be crystallised from acetone as extremely deliquescent feathery needles. After three recrystallisations from acetone, the acid derived from this salt was not completely resolved.

$$A_1 \quad \begin{array}{c} CH_3 \quad COOH \\ | \quad\quad | \\ CH_3CH_2-CH—CH-CH_2COOH \end{array} \qquad \begin{array}{c} CH_3 \\ | \\ CH_3CH_2-CH-CH-CH_2COOH \\ | \\ COOH \end{array} \quad A_2$$

$C_8H_{14}O_4$ Ref. 432

Resolution of the diastereoisomeric racemic compounds A_1 and A_2 into the respective dextrorotatory enantiomers d_1 and d_2. — To a suspension of strychnine (42 g, 0.12 mole) in H_2O (1 500 ml), were added, with stirring, 13 g (0.075 mole) of racemic compound A_1 dissolved in 300 ml of H_2O.

The mixture was heated under reflux for 10 h, filtered to remove the excess of strychnine, and left overnight. The strychnine salt that precipitated was collected and recrystallized three times from water and finally hydrolyzed with a solution of NaOH 10%, to yield 0.033 g of acid d_1, having m.p. 101-2 °C and constant specific rotatory power $[\alpha]_{D\,max}^{25}$ +30.6 ± 0.5 (c = 0.660, $CHCl_3$). The racemic compound A_2 was resolved into the enantiomer d_2 in the same way: 0.047 g of acid d_2, m.p. 81-2 °C and $[\alpha]_{D\,max}^{25}$ +27.2 ± 0.3 (c = 0.934, $CHCl_3$) were obtained from 13 g (0.075 mole) of diastereoisomer A_2.

$$(CH_3)_2CH-CH-CH_2CH_2COOH$$
$$|$$
$$COOH$$

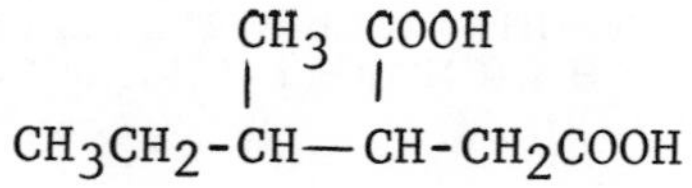

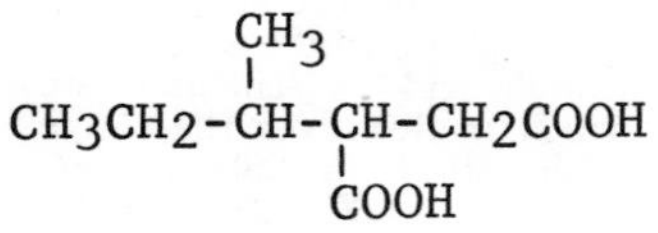

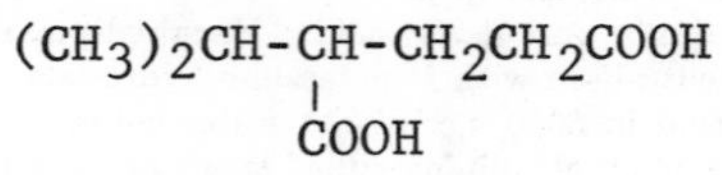

$C_8H_{14}O_4$ Ref. 433

(+)-*iso-Propylglutaric acid*. 30.5 g (0.175 mol) of racemic acid and 166 g of brucine (0.35 mol) were dissolved together in 350 ml hot water. After 24 hours the salt was collected, dried and recrystallized several times from its own weight of hot water. After each crystallization a small sample was set aside, the acid liberated, and its rotatory power determined.

Crystallization:	1	2	3	4
g salt:	87.2	69.4	65.4	57.7
$[\alpha]_D^{25}$ in alcohol:	+11.1°	+16.8°	+17.0°	+17.8°

After several crystallizations from water the liberated acid had $[\alpha]_D = + 18.2°$. The powdered acid was then mixed with double its weight of acetyl chloride. After refluxing for 15—20 minutes, acetic acid and excess of acetyl chloride were distilled off, first at ordinary pressure, finally in vacuum, the temperature of the bath not exceeding 80°. On cooling the residue crystallized readily. It was dissolved in benzene; on careful addition of petroleum (b. p. 60—70°) to the hot solution, the anhydride was deposited as a felted mass of long, hairlike needles.

10 g of the dextrogyric acid described above were converted into anhydride; after two recrystallizations the yield was 8 g. 1 g of this product was converted into acid which was recrystallized from water.

0.4774 g made up to 10.00 ml with *alcohol*: $2\alpha_D^{25} = + 1.815°$.
$[\alpha]_D^{25} = + 19.0°$; $[M]_D^{25} = + 33.1°$.

The remaining 7 g of anhydride gave after another two recrystallizations 6.1 g. 5.2 g of this product were converted into acid and recrystallized from water yielding 4.9 g. The acid forms short prisms, melting at 93.5—94.2°.

0.1625 g: 14.93 ml 0.1249-N sodium hydroxide.
$C_8H_{14}O_4$ E. W. calc. 87.06   found 87.1.

0.4936 g made up to 10.00 ml with *alcohol*: $2\alpha_D^{25} = + 1.89°$.
$[\alpha]_D^{25} = + 19.1°$; $[M]_D^{25} = + 33.3°$.

0.5181 g made up to 10.00 ml with *aceton*: $2\alpha_D^{25} = + 1.92°$.
$[\alpha]_D^{25} = + 18.5°$; $[M]_D^{25} = + 32.3°$.

0.5121 g made up to 10.00 ml with *ether*: $2\alpha_D^{25} = + 2.165°$.
$[\alpha]_D^{25} = + 21.1°$; $[M]_D^{25} = + 36.8°$.

0.2326 g made up to 10.00 ml with *water*: $2\alpha_D^{25} = +0.57°$.
$[\alpha]_D^{25} = +12.3°$; $[M]_D^{25} = +21.3°$.

0.2649 g carefully *neutralized* with NaOH and made up to 10.00 ml with *water*: $2\alpha_D^{25} = +0.555°$.
$[\alpha]_D^{25} = +10.5°$; $[M]_D^{25} = +18.2°$.

The anhydride of the (+)-*acid* crystallizes readily in long hairlike needles as described above. Strangely enough it has no well-defined m. p.; it softens at about 65° and is completely liquid at 75°. The substance probably has a transition point not far from the melting point. READ and REID have described an optically active anhydride with m. p. 55—56° (4).

The rotatory power varies considerably with the solvent used.

24.92 mg: 56.18 mg CO_2 and 17.68 mg H_2O.
$C_8H_{12}O_3$ (156.09) calc. C 61.50 H 7.75.
found » 61.48 » 7.94.

0.3108 g made up to 10.00 ml with *aceton*: $2\alpha_D^{25} = -0.215°$.
$[\alpha]_D^{25} = -3.5°$; $[M]_D^{25} = -5.4°$.

0.3024 g made up to 10.00 ml with *chloroform*: $2\alpha_D^{25} = +0.345°$.
$[\alpha]_D^{25} = +5.7$; $[M]_D^{25} = +8.9°$.

0.2346 g made up to 10.00 ml with *benzene*: $2\alpha_D^{25} = -0.86°$.
$[\alpha]_D^{25} = -18.3°$; $[M]_D^{25} = -28.6°$.

(—)-iso-*Propylglutaric acid.* A crude acid was isolated from the mother liquor from the first crystallization of the brucine salt; its rotatory power was $[\alpha]_D^{25} \sim -10°$. 14 g of this acid (collected from two experiments) were converted into anhydride as above. After three crystallizations from benzene + petroleum, the yield was 6.2 g. 1.0 g was converted into acid and recrystallized from water.

0.4272 g made up to 10.00 ml with *alcohol*: $2\alpha_D^{25} = -1.64°$.
$[\alpha]_D^{25} = -19.2°$; $[M]_D^{25} = -33.4°$.

The remaining 5.2 g were recrystallized once more. 4.2 g of the product were converted into acid and recrystallized as above, yielding 4.0 g, m. p. 93.5—94.2°.

0.1600 g: 14.71 ml 0.1249-N sodium hydroxide.
$C_8H_{14}O_4$ E. W. calc. 87.06 found 87.1.

0.5147 g made up to 10.00 ml with *alcohol*: $2\alpha_D^{25} = -1.965°$.
$[\alpha]_D^{25} = -19.1°$; $[M]_D^{25} = -33.2°$.

The anhydride of the (—)-*acid* had the same appearance and melting point as that of the antipode.

HOOC-CH(CH$_3$)-(CH$_2$)$_2$-CH(CH$_3$)-COOH

$C_8H_{14}O_4$

Ref. 434

Separation of the Racemic α,α'-Dimethyladipic Acid into its Optical Isomers.—Twenty-seven grams of the acid melting at 70° were dissolved in 450 cc. of water, and 32 grams of brucine were gradually added to the hot solution. 131 grams brucine are necessary to neutralize completely the 27 grams of the acid. We have used about one-fourth of the weight of brucine, theoretically required, with the object of forming the acid-salt of the less soluble optically active acid. The solution, while hot, was inoculated with a few crystals of the brucine salt from a former preparation, and the liquid left overnight. The crystals were collected on a filter, and washed four times with about 15 cc. of water each. They were then dissolved in 250 cc. of hot water and crystallized. One more crystallization from 150 cc. of water is all that is necessary.

The salt was dissolved in about 150 cc. water, decomposed with sodium hydroxide, cooled in ice, the solution of the sodium salt of the acid filtered off from the alkaloid, and the latter washed several times with cold water.

The aqueous solution was treated with hydrochloric acid, a little in excess of the amount required to neutralize the sodium hydroxide used. The acid was then extracted 5 times with ether, the latter evaporated off, and the acid crystallized once from a little water. The acid thus obtained gives a positive rotation of $(\alpha)_D^{28°} = +31.3°$ in 10 per cent. alcoholic solution. It begins to soften at 103.5° and melts completely at 105.5°. Yield, 7 grams A sample of the acid (0.1741 gram) titrated required 20.02 cc. 0.1 *N* potassium hydroxide. Theory, 20.00 cc. For the *l*-acid the mother liquor from the first crop of the brucine salt, in the preparation of the *d*-acid, was extracted with ether. The acid thus obtained melted at 70–95°.

Ten grams of this acid were dissolved in 150 cc. water and 11 grams of brucine added. The crystals that separated were recrystallized from diminishing amounts of water four times. The acid that was liberated and purified as in the case of the dextro acid, melted at 85–103° and gave a rotation of $(\alpha)_D^{28°} = -23.4$ in 10 per cent. alcoholic solution.

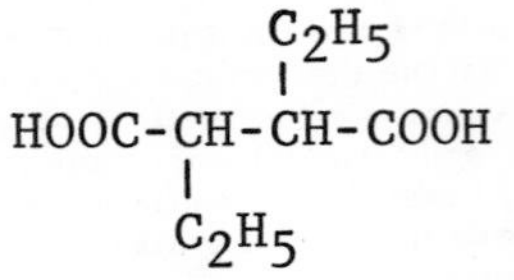

$C_8H_{14}O_4$ Ref. 435

Optische Spaltung der rac.-α,α'-Diäthylbernsteinsäure.

In der Absicht zu entscheiden, welche von den beiden stereoisomeren sym. Diäthylbernsteinsäuren die racemische war, versuchte Bischoff[1]) 1891 die Spaltung mittels verschiedener Alkaloide zu erzielen. Seine Versuche scheiterten aber. Wir haben nun die Spaltung der bei 133° schmelzenden Säure mit Hilfe von Strychnin durchgeführt und damit diese Säure als die *rac.* Form erkannt.

In einer Lösung von 1 g Säure und 1,92 g Strychnin (1 Mol Säure auf 1 Mol Alkaloid) in 10 ccm Wasser bildet sich anfangs selbst nach längerem Stehen in Kühlschrank kein krystallinisches Salz. Erst nach anhaltendem Reiben der Gefäßwand mit einem Glasstab fing das Salz der rechtsdrehenden Säure zu krystallisieren an. Wenn bei einem neuen größeren Versuch mit den erhaltenen Krystallen geimpft wurde, fiel im Laufe von 1 Tag 67 Proc. der gesamten Salzmenge aus. Nach 2-maligem Umkrystallisieren des Salzes wurde eine Säure mit einer spez. Drehung von + 39,9° (p = 4,9) isoliert. Ausbeute aus 21,5 g rac. Säure 6,2 g d-Säure. Diese Säure wurde noch 2-mal mit Strychnin behandelt und zeigte dann bei gleicher Konzentration eine Drehung von + 40,6° bzw. 41,5°. Eine erneute Behandlung mit dem Alkaloid veränderte den letzten Wert nicht. Die reine rechtsdrehende Säure hatte den Schmelzp. 126°.

Aus der Mutterlauge des Strychninsalzes erhielten wir eine linksdrehende Säure von ungefähr ÷ 25°. Die Aufarbeitung dieser auf reine links-Säure haben wir bis auf weiteres zurückgestellt.

Das spez. Drehungsvermögen der reinen d-Säure bei verschiedenen Konzentrationen in Wasser und absolutem Alkohol geht aus der Tab. XI hervor.

[1]) B. 24, 1068 (1891).

Tabelle XI.

Spez. Drehungsvermögen der d-α,α'-Diäthylbernsteinsäure in wäßriger und absolut alkoholischer Lösung.

In Wasser			In absolutem Alkohol		
p	d_4^{20}	$[\alpha]_D^{20}$ in °	p	d_4^{20}	$[\alpha]_D^{20}$ in °
2,623	1,0024	+42,04	3,742	0,8026	+26,14
5,140	1,0063	41,58	7,208	0,8117	26,23
			13,253	0,8291	26,85

Wegen der geringen Löslichkeit konnte nicht bei größeren Konzentrationen gemessen werden. Da eine Extrapolation auf p = 100 unmöglich ist, haben wir die Drehung einer unterkühlten, optisch nicht reinen links-Säure gemessen und in üblicher Weise auf reine Säure umgerechnet. Zur Anwendung kam eine Säure mit $[\alpha]_D^{20}$ = ÷ 23,24° (p = 5,26 in Wasser). Aus den Messungen an der d-Säure ergibt sich eine Drehung der reinen l-Säure bei derselben Konzentration zu − 41,56°. Der Drehungswinkel der geschmolzenen Säure in dem früher erwähnten „Rohr" von 0,1757 dm war bei etwa 20° ÷ 4,05°. Das spez. Gewicht berechneten wir zu $d_4^{20} \simeq$ 1,153. Das Drehungsvermögen der reinen, homogenen l-Säure ist folglich:

$$[\alpha]_D^{20} \simeq \div 35{,}8°, \quad [M]_D \simeq \div 62{,}3°.$$

Nachdem wir die Spaltung der sym. Diäthylbernsteinsäure vom Schmelzp. 133° durchgeführt hatten, erschien eine Abhandlung von H. Wren und J. W. E. Haller[1]), in der die Spaltung derselben Säure durch fraktionierte Krystallisation des Brucinsalzes beschrieben wird. Die Trennung war aber nach ihrer Beschreibung umständlich — das zuerst ausgeschiedene Salz mußte 13-mal umkrystallisiert werden, bevor eine reine Säure erhalten werden konnte — und die Ausbeute war sehr schlecht. Die von uns angegebene Anwendung von Strychnin ist daher vorzuziehen. Die von Wren und Haller dargestellte d-Säure hatte den Schmelzp. 125° und die folgende Drehung in Aceton:

$$[\alpha]_{5461}^{24} = + 34{,}5° \quad \text{und} \quad [\alpha]_{\text{Hg-gelb}}^{24} = + 30{,}9° \ (p = 5).$$

Zum Vergleich bestimmten wir das Drehungsvermögen unserer d-Säure in Aceton und fanden: c = 5,202 g/100 ccm, α_D^{20} = + 3,01°, l = 2.

$$[\alpha]_D^{20} = + 28{,}9°.$$

Wenn man Rücksicht auf die verschiedenen Temperaturen und Wellenlängen des verwendeten Lichtes nimmt, stimmen die Messungen an den beiden Säuren ganz gut überein.

$C_8H_{14}O_4$ Ref. 435a

4. Absolute Konfiguration der α,α'-Diäthylbernsteinsäure. – *4.1. Racematspaltung der rac.-α,α'-Diäthylbernsteinsäure:* 15 g der *rac.*-Säure und 28,8 g Strychnin wurden in 80 ml heissem Wasser gelöst. Die Lösung wurde abgekühlt und mit einigen Kristallen des Strychnin-monosalzes der (+)-α,α'-Diäthylbernsteinsäure, das in einem Vorversuch nach längeren Bemühungen kristallin erhalten worden war, angeimpft. Nach Stehen über Nacht wurde das Salz abgesaugt, mit wenig Wasser gewaschen (Filtrat A) und noch zweimal aus Wasser umkristallisiert. Man erhielt 19 g Salz mit $[\alpha]_D^{26} = -0{,}27°$ (c = 10,0; in Aceton). Dieses Salz wurde in 30 ml Wasser gelöst, mit 15 ml konz. Salzsäure versetzt und die freigesetzte Diäthylbernsteinsäure mit Äther extrahiert: 6,02 g Säure mit $[\alpha]_D^{25} = +26{,}3°$ (c = 10,0; Aceton). Aus 9,38 g einer solchen Säure hat man nochmals mit 18,01 g Strychnin das Salz hergestellt und dieses wieder aus Wasser umkristallisiert. Man erhielt daraus 8,3 g *(+)-α,α'-Diäthylbernsteinsäure* mit $[\alpha]_D^{25} = +40{,}8°$ (c = 5; Wasser). Die Drehung erfuhr nach dem Umlösen der Säure aus Äther keine signifikante Veränderung mehr. Smp. 127–128,5°. In der Literatur [11]) [17]) wird für diese Säure der Smp. 126° und $[\alpha]_D = +42{,}04°$ (c = 2,6; Wasser) angegeben.

Das Filtrat A der ersten Aufspaltung wurde nach Zusatz überschüssiger konz. Salzsäure mit Äther extrahiert; man erhielt 7,3 g einer Mischung der enantiomeren Diäthylbernsteinsäuren mit $[\alpha]_D^{26} = -20{,}5°$ (c = 10,0; Aceton). 10,9 g eines solchen Gemisches löste man in 50 ml Äther, impfte mit *rac.*-α,α'-Diäthylbernsteinsäure an und liess einige Std. bei 20° stehen, wobei sich 2,65 g (±)-Säure ausschieden. Die eingedampfte Mutterlauge (8,2 g) wurde abwechselnd aus Benzol und Äther umkristallisiert, wobei am Anfang hauptsächlich die racemische, später die (−)-Säure auskristallisierte. Man erhielt schliesslich aus Äther ungefähr 5 g reine *(−)-α,α'-Diäthylbernsteinsäure* vom Smp. 127–128° und $[\alpha]_D^{23} = -41{,}08°$ (c = 5,00; Wasser). Literatur-Smp. 125° [17]).

[11]) E. Berner & R. Leonardsen, Liebigs Ann. Chem. *538*, 27 (1939).

[17]) H. Wren & J. W. E. Haller, J. chem. Soc. *1937*, 230.

$(CH_3)_2CHCH_2CHCOOH$
|
CH_2COOH

$C_8H_{14}O_4$ Ref. 436

Aktive n-Butyl-bernsteinsäuren. Vorläufige Spaltungsversuche mit Säure und Alkaloid im Verhältnis 1:1 hatten wenig Erfolg; dagegen sind die sekundären Salze schön kristallisiert und geben in einigen Fällen gute Spaltung.

0,17 g Säure, nebst 0,67 g *Strychnin* in 10 ml Wasser + 10 ml Alkohol gelöst, gaben 0,46 g Salz. Die darin enthaltene Säure zeigte in Wasserlösung $[\alpha]_D^{25} = +17°$.

0,18 g Säure, nebst 0,62 g *Chinin* in 10 ml Wasser + 10 ml Alkohol gelöst, gaben 0,52 g Salz. Die Säure zeigte $[\alpha]_D^{25} = -11{,}5°$

Bei ähnlichen Versuchen gab *Cinchonidin* eine Säure von $[\alpha]_D^{25} = +5{,}2°$. *Chinidin* eine sehr schwach linksdrehende Säure und *Brucin* eine Säure ohne merkbare optische Aktivität. Für die Spaltung in grösserem Masstabe wurden folglich Strychnin und Chinin gewählt.

Bei Darstellung des Strychninsalzes wurde beobachtet, dass zum völligen Auflösen des Strychnins ein nicht unbeträchtlicher Überschuss an Säure erforderlich ist. Beim Umkristallisieren wurde das Salz zum Teil hydrolysiert; das dabei ausgeschiedene Strychnin wurde durch Zusatz der nötigen Menge Essigsäure (einige ml) wieder in Lösung gebracht.

In einem Versuch wurde von 133 g (0,4 Mol) Strychnin und 46 g (0,26 Mol) Säure ausgegangen; das Salz wurde dreimal aus verdünntem Alkohol umkristallisiert.

Kristallisation:	1	2	3	4
ml Alkohol	500	200	150	100
ml Wasser	700	250	200	125
g Salz	114,3	90,8	78,0	65,8
$[\alpha]_D^{25}$ der Säure	+16,5°	--	+21,5°	+21,5°

In einem anderen Versuch wurde mit 100 g Strychnin und der zum Lösen genau erforderlichen Menge Säure gearbeitet:

Kristallisation:	1	2	3	4
ml Alkohol	400	250	100	75
ml Wasser	600	250	100	75
g Salz	60,8	43,5	33,7	29,2
$[\alpha]_D^{25}$ der Säure	+19°	+21,5°	+21°	+21,5°

Wie ersichtlich ist die Säure schon in der zweiten Kristallisation praktisch rein.

Die Mutterlauge nach der ersten Kristallisation des erst-erwähnten Versuches wurde in der Hitze mit Strychnin gesättigt; beim Erkalten kristallisierten 16 g Salz, das ebenfalls stark rechtsdrehende Säure enthielt. Aus der Mutterlauge dieses Salzes wurden 20 g Säure von $[\alpha]_D = -20°$ erhalten. Diese wurde nebst der äquivalenten Menge (74 g) Chinin in verdünntem Alkohol gelöst; schon in der ersten Kristallisation wurde maximale Drehung erhalten.

Kristallisation:	1	2	3
ml Alkohol	160	100	100
ml Wasser	250	125	125
g Salz	89,5	86	82,5
$[\alpha]_D^{25}$ der Säure	−22,5°	−22°	−22,5°

(+)-n-Butyl-bernsteinsäure. Das Strychninsalz wurde in Wasser gelöst, die Base mit überschüssigem Ammoniak gefällt und die Säure durch Ansäuern mit Schwefelsäure und Ausschütteln mit Äther isoliert. Das Umkristallisieren bereitete gewisse Schwierigkeiten. In vielen organischen Lösungsmitteln ist die Säure sehr leicht löslich, aus anderen wird sie in sehr voluminöser Form ausgeschieden, so dass die ganze Lösung zu einem gelähnlichen Gebilde erstarrt. In Wasser löst sich die Säure bei Raumtemperatur sehr leicht, kristallisiert aber im Kälteschrank zum grössten Teil aus. Bei langsamem Kristallisieren erhält man harte Krusten, bei raschem einen Kristallbrei aus sehr kleinen Nadeln, der auch bei scharfem Absaugen ziemlich viel Wasser zurückhält. 65 g Salz gaben 11,9 g Säure; nach dreimaligem Umkristallisieren aus wenig Wasser (einmal mit Tierkohle) verblieben 9,5 g vom Schmp. 82,5—83,5°.

0,1240 g Sbst.: 11,50 ml 0,1240-n NaOH.

$C_8H_{14}O_4$ (174,1) Äq.-gew. Ber. 87,05 Gef. 87,0.

Die spezifische Drehung der Säure in wässriger Lösung nimmt mit steigender Konzentration etwas ab (Tab. 1). Die Drehung wurde auch in absolutem Alkohol gemessen.

0,5210 g Säure, in abs. *Alkohol* auf 10,00 ml gelöst: $2\alpha_D^{25} = +2{,}76°$.
$[\alpha]_D^{25} = +26{,}5°$; $[M]_D^{25} = +46{,}1°$.

Tabelle 1.

Drehung der (+)-n-Butyl-bernsteinsäure in wässr. Lösung. C = g Säure in 10 ml Lösung; 2α = Drehung in 2-dm-Rohr.

C	2α	$[\alpha]_D^{25}$	$[M]_D^{25}$
0,1521	+0,705°	+23,2°	+40,4°
0,5138	+2,82°	+22,6°	+39,8°
1,0081	+4,435°	+22,0°	+38,8°

Das Strychninsalz der (+)-Säure kristallisiert aus verdünntem Alkohol in glänzenden Schuppen; beim Absaugen und Waschen mit Wasser scheint eine Phasenumwandlung stattzufinden. Aus Wasser erhält man kleine, undeutliche Schuppen; der Wassergehalt des lufttrockenen Salzes stimmt am besten auf 3 ½ Mol Kristallwasser.

0,7870 g Sbst.: Gewichtsverlust über P_2O_5 0,0521 g.

$C_8H_{14}O_4$, 2 $C_{21}H_{22}O_2N_2$, 3 ½ H_2O (905,5) H_2O Ber. 6,96 Gef. 7,07.

0,2089 g entwässertes Salz: 12,18 ml N_2 (756,0 mm Hg, 18,0°).

$C_8H_{14}O_4$, 2 $C_{21}H_{22}O_2N_2$ (842,5) N Ber. 6,65 Gef. 6,65.

(—)-n-Butyl-bernsteinsäure. Das Chininsalz wurde mit Schwefelsäure zerlegt und die Säure in Äther aufgenommen. 82 g Salz gaben 16 g Säure, die viermal aus Wasser umkristallisiert wurde. Sie war der (+)-Verbindung völlig ähnlich. Schmp. 82,5—83,5°.

0,1256 g Sbst.: 11,64 ml 0,1240-n NaOH.

$C_8H_{14}O_4$ (174,1) Äq.-gew. Ber. 87,05 Gef. 87,0.

0,5145 g Säure, in *Wasser* auf 10,00 ml gelöst: $2\alpha_D^{25} = -2{,}32°$.

$[\alpha]_D^{25} = -22{,}5°$; $[M]_D^{25} = -39{,}3°$.

Das Chininsalz der (—)-Säure bildet kleine, glänzende Nadeln, die im lufttrockenen Zustande kein Kristallwasser enthalten.

0,2079 g Sbst.: 12,64 ml N_2 (721,9 mm Hg, 18,0°).

$C_8H_{14}O_4$, 2 $C_{20}H_{24}O_2N_2$ (822,5) N Ber. 6,81 Gef. 6,77.

0,1292 g Sbst.: 11,99 ml 0,1240-n NaOH.

$C_8H_{14}O_4$ (174,1) Äq.-gew. Ber. 87,05 Gef. 87,0.

Aktive iso-Butyl-bernsteinsäuren. Auch bei dieser Säure waren Vorversuche mit Säure und Alkaloid im Verhältnis 1:1 wenig erfolgreich, dagegen gaben die sekundären Salze in einigen Fällen gute Spaltung.

0,19 g Säure, nebst 0,72 g *Strychnin* in 10 ml Wasser + 10 ml Alkohol gelöst, gaben 0,55 g Salz. Die darin enthaltene Säure zeigte in wasserlösung $[\alpha]_D = +16{,}5°$.

0,18 g Säure, nebst 0,67 g *Chinidin* in 10 ml Wasser + 10 ml Alkohol gelöst, gaben 0,68 g Salz. Die Säure zeigte $[\alpha]_D = -7°$.

Gut kristallisierende Salze wurden auch mit *Brucin*, *Cinchonin* und *Chinin* erhalten. Die Säure aus dem Brucinsalz zeigte die spezifische Drehung — 6,5°, die aus dem Cinchoninsalz + 2°. Mit Chinin wurde in einem Versuch rechtsdrehende, in einem anderen linksdrehende Säure erhalten. Für die endgültige Spaltung wurden Strychnin und Chinidin gewählt.

Auch bei der *iso*-Butyl-bernsteinsäure war zum völligen Auflösen des Strychnins ein Überschuss an Säure nötig. Das beim Umkristallisieren ausgeschiedene Strychnin wurde hier nicht mit Essigsäure in Lösung gebracht sondern abfiltriert; diese Methode ist jedoch, insbesondere beim Arbeiten mit grösseren Mengen, ziemlich unbequem und deshalb kaum zu empfehlen. Im Hauptversuch wurden 133 g (0,4 Mol) Strychnin nebst 48 g (0,28 Mol) Säure in Alkohol mit etwas Wasser gelöst, die Lösung mit heissem Wasser verdünnt, und das nach 17-stündigem Stehen ausgeschiedene Salz viermal umkristallisiert.

Kristallisation:	1	2	3	4	5
ml Alkohol	570	500	250	100	50
ml Wasser	706	200	100	50	25
g Salz	98,2	69,9	50,9	33,5	22,5
$[\alpha]_D^{25}$ der Säure	+20°	+22,5°	+22,5°	+23,5°	+24,5°

Durch Aufarbeiten verschiedener Mutterlaugen wurden 13 g Salz erhalten, das Säure von $[\alpha]_D^{25} = +20°$ enthielt. Zwecks Kontrolle wurde dieses auf konstante Drehung der Säure umkristallisiert.

Kristallisation:	1	2	3	4
ml Alkohol	20	14	10	6
ml Wasser	10	7	5	3
g Salz	6,9	4,2	2,3	1,8
$[\alpha]_D^{25}$ der Säure	+23°	+22,5°	+24°	+23,5°

Da man bei diesen Messungen an kleinen, nicht umkristallisierten Säuremengen mit einer Fehlergrenze von etwa ± 1,5°

rechnen muss, scheint es sichergestellt, dass wirklich maximale Drehung erreicht ist.

Die Mutterlauge nach der ersten Kristallisation des Hauptversuches gab beim Eindampfen 16 g Salz, das Säure von $[\alpha]_D = +9{,}5°$ enthielt. Die neue Mutterlauge gab bei Sättigen mit Strychnin und weiterem Eindampfen 20,1 g Salz mit einer Säure von $[\alpha]_D = -11°$. Aus der Mutterlauge dieses Salzes wurde endlich 16 Säure isoliert, die $[\alpha]_D = -24°$ zeigte und also aus praktisch reiner (—)-Säure bestand. Sie wurde in Chinidinsalz überführt, und dieses zweimal aus verdünntem Alkohol umkristallisiert; die Drehung der Säure blieb dabei unverändert.

(+)-*iso*-Butyl-bernsteinsäure. Die Säure wurde aus dem Strychninsalz in üblicher Weise isoliert; 22,5 g Salz gaben 4,1 g Säure. Nach Umkristallisieren aus 5 ml Wasser verblieben 3,9 g. Die Säure bildet vierseitige, anscheinend rechtwinklige Tafeln mit gerader Auslöschung vom Schmp. 95—96°.

0,1261 g Sbst.: 11,69 ml 0,1240-n NaOH.

$C_8H_{14}O_4$ (174,1) Äq.-gew. Ber. 87,05 Gef. 87,0.

0,5008 g Säure, in abs. *Alkohol* auf 10,00 ml gelöst: $2\alpha_D^{25} = +2{,}68°$.
$[\alpha]_D^{25} = +26{,}8°$; $[M]_D^{25} = +46{,}6°$.

Das Strychninsalz der (+)-*Säure*, das bei der Spaltung direkt erhalten wurde, enthält nach Lufttrocknen in gelinder Wärme kein Kristallwasser.

0,1907 g Sbst.: 10,92 ml N_2 (764,1 mm Hg, 21°).

$C_8H_{14}O_4$, 2 $C_{21}H_{22}O_2N_2$ (842,5) N Ber. 6,65 Gef. 6,68.

(—)-*iso*-Butyl-bernsteinsäure. 28 g Chinidinsalz gaben 5,0 g Säure; nach Umkristallisieren aus 5 ml Wasser verblieben 4,8 g. Schmp. 95—96°.

0,1288 g Sbst.: 11,45 ml 0,1240-n NaOH.

$C_8H_{14}O_4$ (174,1) Äq.-gew. Ber. 87,05 Gef. 87,2.

Die spezifische Drehung ist nur wenig höher als die der n-Butyl-bernsteinsäure. Bei steigender Konzentration nimmt sie etwas ab (Tab. 2).

Tabelle 2.

Drehung der (—)-*iso*-Butyl-bernsteinsäure in wässr. Lösung. C = g Säure in 10 ml Lösung; 2 α = Drehung in 2-dm-Rohr.

C	2 α	$[\alpha]_D^{25}$	$[M]_D^{25}$
0,1008	—0,48°	—24,0°	—41,5°
0,5018	—2,825°	—28,2°	—40,8°
1,0020	—4,55°	—22,7°	—39,5°

Die Drehung in absolutem Alkohol stimmt mit der der (+)-Säure gut überein.

0,5009 g Säure, in abs. *Alkohol* auf 10,00 ml gelöst: $2\alpha_D^{25} = -2{,}70°$.
$[\alpha]_D^{25} = -27{,}0°$; $[M]_D^{25} = -46{,}9°$.

Das Chinidinsalz der (—)-*Säure* bildet Nadeln, die nach Lufttrocknen in gelinder Wärme kein Kristallwasser enthalten.

0,2084 g Sbst.: 12,15 ml N_2 (764,1 mm Hg, 20,5°).

$C_8H_{14}O_4$, 2 $C_{20}H_{24}O_2N_2$ (822,5) N Ber. 6,81 Gef. 6,82.

$$CH_3O_2C-\underset{CH_3}{\overset{CH_2CH_3}{C}}-CH_2COOH$$

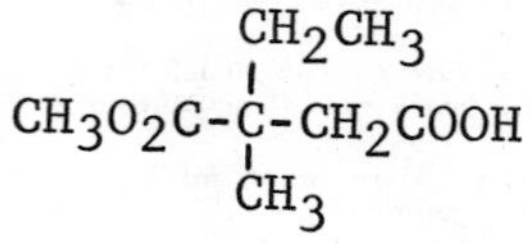

$C_8H_{14}O_4$ Ref. 437

(—)-*3-Methyl-3-ethyl-3-carbmethoxypropionic acid* (I). Quinine (217 g) was dissolved in hot acetone (750 ml) and the solution filtered hot. Low-melting DL-monomethyl ester (I) (100 g) was added, followed by water (1 700 ml), and the solution left at room temperature overnight. The crystalline material from two similar crystallizations was combined, and the wet crystals (299 g) dissolved in 2 000 ml of 30 % acetone. The resulting crystalline material was recrystallized from 1 100 ml of 30 % acetone. The salt now obtained weighed 141 g. Part of the salt (15 g) was decomposed and the half-ester distilled, giving 2.55 g, b.p. 88°, 0.5 mm, $\alpha_D^{23} = -11{.}44°$ (homogeneous; l, 1). The main portion was further crystallized from 900 ml of 30 % acetone. Recovered half-ester now had a rotation $\alpha_D^{23} = -12{.}12°$ (hom., l, 1). The rotation increased to $\alpha_D^{23} = -12{.}46°$ after five crystallizations, and finally attained a value of $\alpha_D^{23} = -12{.}55°$ after the sixth crystallization.

d_4^{23} 1.091.

Optical rotation $\alpha_D^{23} = -12{.}55°$ (homogeneous; l, 1)
                 $[\alpha]_D^{23} = -11{.}50°$ $[M]_D^{23} = -20{.}03°$

Anal.: Calcd for $C_8H_{14}O_4$ (174.2)   C, 55.16; H, 8.10 %
       Found   C, 55.20; H, 8.15 %

Ref. 437a

(b) A solution of (±)-3-ethyl-3-methyl-3-carboxymethoxypropionic acid (97 g.) in methanol (300 ml.) was treated with dehydroabietylamine (162 g.) in methanol (300 ml.). Next day the salt was collected and crystallised five times from methanol to give a salt (40 g.), m. p. 149—151°, whose decomposition yielded (+)-*3-ethyl-3-methyl-3-carbomethoxypropionic acid*, b. p. 135°/0·1 mm., $[\alpha]_D^{20}$ +8·2° (c, 8·5 in alcohol), $[\alpha]_D^{20}$ +8·92° (homogeneous), d_4^{23} 1·096 [Found: C, 54·4; H, 8·1; OMe, 16·7. $C_7H_{11}O_3$(OMe) requires C, 55·2; H, 8·1; OMe, 17·7%]. Hydrolysis of this ester (1·5 g.) with excess of 20% sodium hydroxide for 45 min. on the steam-bath gave (+)-*2-ethyl-2-methylsuccinic acid* (1 g.) which formed needles, m. p. 65—66°, from benzene-light petroleum (b. p. 60—80°), $[\alpha]_D^{20}$ +2·97° (c, 6·7 in $CHCl_3$) (Found: C, 53·0; H, 7·7. $C_7H_{12}O_4$ requires C, 52·5; H, 7·6%). Prepared quantitatively by the use of diazomethane the (+)-*dimethyl ester* had b. p. 100°/20 mm., $[\alpha]_D^{19}$ +6·5° (c, 0·7 in methanol) [Found: C, 56·9; H, 8·6; OMe, 31·5. $C_7H_{10}O_2$(OMe)$_2$ requires C, 57·4; H, 8·6; OMe, 32·9%].

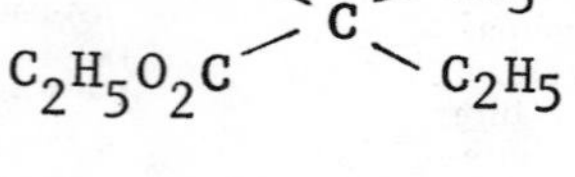

$$\underset{C_2H_5O_2C}{\overset{HO_2C}{\diagdown}}C\underset{C_2H_5}{\overset{CH_3}{\diagup}}$$

Ref. 437.1

Resolution. Quinine (72·5 g.) was added to a solution of (±)-ethyl hydrogen ethylmethylmalonate (38·5 g.) in aqueous ethanol (400 c.c. of 80%), and the whole warmed until dissolution was complete. After 2 days at 0°, the separated crystals were removed by filtration (43 g.; m. p. 103—110°). Two further recrystallisations from the same solvent gave a quinine salt (33 g.) as needles, m. p. 98—100°, which on decomposition with hydrochloric acid yielded (+)-*ethyl hydrogen ethylmethylmalonate* of constant rotatory power, $[\alpha]_D^{18}$ +3·38° (l, 2; c, 15·0 in chloroform).

A lævorotatory acid (22 g.) was obtained by decomposition of the more soluble fractions of the quinine salt. A portion of this acid (4 g.) was dissolved in acetone (30 c.c.), cinchonidine (6·8 g.) added, and the whole warmed until dissolution was complete. Three recrystallisations of the salt obtained from minimum quantities of acetone gave needles (5·6 g.) of m. p. 154—155°. The remainder of the acid (18 g.) was dissolved in acetone (250 c.c.), and cinchonidine (37·5 g.) was added. The solution was seeded with the recrystallised cinchonidine salt obtained earlier. Two recrystallisations from acetone gave needles (22 g.) of m. p. 155—156°, which on decomposition yielded the (—)-*ester* of $[\alpha]_D^{25}$ —3·47° (l, 2; c, 5·03 in chloroform); in one equivalent of aqueous sodium hydroxide, $[\alpha]_D^{18}$ was +3·64° (l, 2; c, 3·57).

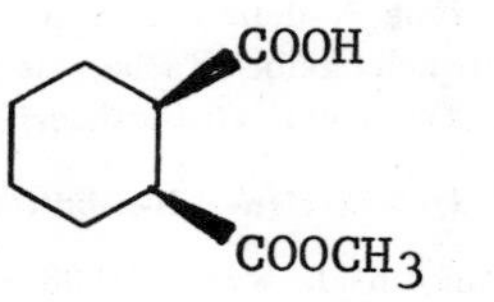

$C_8H_{14}O_4$ Ref. 438

Dedoublement de l'hexahydrophtalate de methyle.

Il s'obtient facilement à l'aide de la quinine. On mélange au sein de l'alcool 23 gr. d'éther-sel,, et 40 gr. de quinine anhydre. Par évaporation de l'alcool, on obtient un sel brut fondant à 127-132° et de pouvoir rotatoire $[\alpha]_{578}$ = -125° (alcool c = 0,1).

Par cristallisation fractionnée dans l'alcool a 50 %, le pouvoir rotatoire s'élève jusque - 136°,5 et le point de fusion se fixe à 141°,5-142° (n.c).

Quatre cristallisations suffisent et on recueille environ 15 gr. de sel de premier jet.

Nous n'avons pas cherché a isoler l'autre isomère.

Pour libérer l'éther actif, on met le sel en solution alcoolique concentrée et on ajoute un léger excès de soude en présence de phtaléine. La quinine mise en liberté est précipitée par addition d'eau et extraite trois fois au chloroforme. On extrait une fois encore a l'éther puis on acidule et extrait de nouveau. La solution éthérée est lavée a l'eau et évaporée.

L'éther actif cristallise à la longue F. 48-49°.

Pouvoir rotatoire dans l'alcool c = 0,1:

$[\alpha]_{578}$ = -6°,67

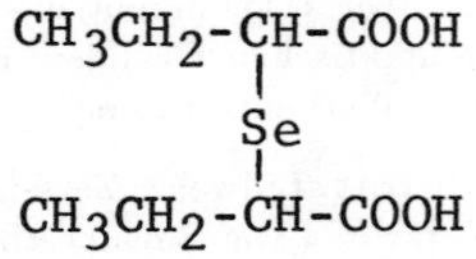

$C_8H_{14}O_4Se$ Ref. 439

Spaltung der *racem*-Form.

Die rechtsdrehende Antipode wurde sehr leicht durch Spaltung der *racem*-Form mit Brucin erhalten. Bei Versuchen mit Chinin wurde ebenfalls die rechtsdrehende Säure, aber mit geringerer Aktivität erhalten; das Strychninsalz hingegen gibt eine schwach linksdrehende Säure. Ich habe keine geeignete Methode finden können zur Reindarstellung der linksdrehenden Säure ohne allzu grossen Arbeitsaufwand. Zwecks Kontrolle der Spaltung wäre es wünschenswert gewesen, auch diese Säure in reiner Form zu besitzen. Da jedoch die (+)-Form leicht zugänglich ist und als gut kristallisierte Substanz mit definiertem Schmelzpunkt erhalten wurde, liegt kein Anlass vor, die Reinheit dieser Form in Zweifel zu ziehen.

Der Verlauf der Spaltung lässt sich geeigneterweise durch Drehungsmessungen an einer Lösung des von Kristallwasser befreiten Brucinsalzes in verdünnter Salzsäure verfolgen, Brucin ist bekanntlich linksdrehend und seine Aktivität von derselben Grössenordnung wie die der Säure; die spezifische Drehung des Salzes ist daher schwach. Die Genauigkeit der Messungen wird dadurch jedoch nicht beeinflusst, da eine Veränderung der Aktivität der Säure eine gleich grosse absolute und daher eine sehr beträchtliche relative Veränderung der Aktivität der Salzlösung verursacht. Das Salz ist verhältnismässig leichtlöslich und muss daher aus möglichst wenig Wasser umkristallisiert werden. Von jeder Fraktion wurde eine kleine Menge abgetrennt (0.2—0.3 g), die auf Gewichtskonstanz getrocknet, gewogen und in 0.4-*n* HCl auf ein bekanntes Volumen (etwa 10 ml) gelöst wurde. Die Drehung wurde im 2 dm-Rohr gemessen.

Kristallisation	1	2	3	4
$[\alpha]_D^{25}$	+17.2°	+18.8°	+18.8°	+18.9°
$[M]_D^{25}$	+179°	+196°	+196°	+197°

Aus 20 g *racem*-Säure wurden 14.5 g Brucinsalz der (+)-Säure erhalten, entsprechend 35 % der theoretischen Menge. Wie aus der Übersicht zu entnehmen, ist die Säure schon in der zweiten Kristallisation rein und da die Aktivität der Säure jetzt bekannt ist, muss sich die Spaltung mit bedeutend geringeren Verlusten durchführen lassen.

Die molekulare Drehung der Säure ist, wie unten gezeigt wird, +511.5°, was einer molekularen Drehung des Brucinsalzes von +196.5° entspricht. Das in der ersten Fraktion erhaltene Salz mit einer molekularen Drehung von +179° muss also Säure von [M]= etwa +494° enthalten haben, da man berechtigt sein dürfte mit Additivität der Aktivitäten von Säure und Brucin zu rechnen. Man erhält also schon in der ersten Kristallisation eine zu etwa 96 % aktive Säure, was für sehr grosse Löslichkeit oder geringe Kristallisationsfähigkeit des Brucinsalzes der (—)-Säure spricht.

Das Salz wurde in verdünnter Schwefelsäure gelöst und mit Äther ausgeschüttelt. Nach Verdunsten des Äthers im Vakuum wurde ein farbloser Sirup erhalten, der nach einigen Schwierigkeiten zur Kristallisation gebracht werden konnte. Das Rohprodukt (aus 14.5 g Salz) wog 3.1 g und schmolz bei 74—76°. Die Säure konnte nach scharfem Trocknen aus einer grösseren Menge Benzin (Siedeintervall 60—70°) umkristallisiert werden.

Nach der ersten Umkristallisation wurden 2.6 g Säure erhalten, die bei 76—77° schmolzen, nach der zweiten Umkristallisation 2.0 g von unverändertem Schmelzpunkt.

(+)-Selen-α-di-n-buttersäure.

Die Säure kristallisiert in gut ausgebildeten, oft rosettenförmig gruppierten Prismen mit gerader Auslöschung. Sie ist leichtlöslich in den gewöhnlichen Lösungsmitteln mit Ausnahme von Benzin und Petroleumäther. Schmelzpunkt 76—77°.

0.2014 g Subst. : 14.46 ml 0.1100-*n* NaOH.

$C_8H_{14}O_4Se$ (253.1) Äqu.-Gew. ber. 126.6 gef. 126.6.

0.1403 g Subst. (einmal umkristallisiert) in 0.4-*n* HCl gelöst auf 10.02 ml : 2 α = +5.65°; 0.1266 g Subst. (zweimal umkristallisiert) in 0.4-*n* HCl gelöst auf 10.03 ml : 2 α = +5.105°.

$[\alpha]_D^{25}$ = +201.8°; +202.0°. $[M]_D^{25}$ = +511°; 511.5°.

Das *Brucinsalz* erhält man, wie oben beschrieben, bei der Spaltung. Es bildet längliche Tafeln oder Schuppen, deren Kristallwasser schon an der Luft teilweise verloren geht. Eine Bestimmung des Wassergehalts scheint ohne besondere Massnahmen nicht möglich zu sein; aus einigen orientierenden Analysen ergab sich, dass wenigstens 6 Moleküle vorhanden sind. Die Selenbestimmung wurde an wasserfreiem Material ausgeführt.

0.3168 g Subst. : 0.0239 g Se.

$C_8H_{14}O_4Se$, 2 ($C_{23}H_{26}O_4N_2$) (1041.5) Se ber. 7.59 % gef. 7.54 %.

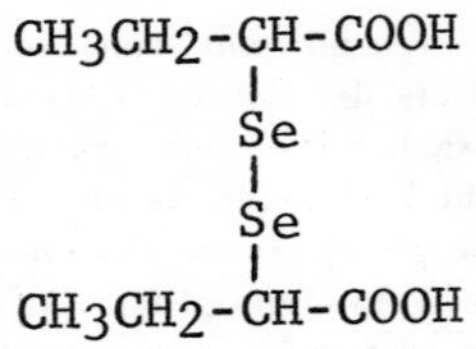

$C_8H_{14}O_4Se_2$ Ref. 440

33.2 g (0.1 Mol) Säure wurden in 1600 ml Wasser gelöst und mit 48.5 g (0.16 Mol) Morphin und 4 g (0.04 Mol) Kaliumbikarbonat versetzt. Aus der Lösung kristallisierten allmählich 20.6 g Salz : [α] der im Salz enthaltenen Säure −102°. Die Mutterlauge wurde auf halbes Volumen eingeengt, wobei 43.5 g Salz erhalten wurden : [α] der in dieser Salzfraktion enthaltenen Säure −56°. Dieses Salz wurde aus etwa 500 ml Wasser umkristallisiert, wobei 27.9 g erhalten wurden : [α] der darin enthaltenen Säure −108°. Diese Fraktion wurde mit der zuerst erhaltenen, deren Säure ja etwa gleiche Aktivität besass, vereinigt und das Produkt wurde wiederholten Umkristallisationen unterworfen. Nach dreimaliger Umkristallisation, wobei die Mutterlauge jedesmal in der oben angedeuteten Weise behandelt wurde, war die Drehung der Säure auf −210° gestiegen. Da das Morphin während dieser Operationen eine chemische Veränderung zu erleiden begonnen hatte (Oxydation oder Umlagerung?), die sich durch Braunfärbung und Abnahme der Kristallisationsfähigkeit des Salzes zu erkennen gab, wurde die Säure in Freiheit gesetzt und mit Tierkohle behandelt. Die Menge des erhaltenen Produkts betrug 10 g.

Diese Säure wurde nun in Wasser gelöst, neutralisiert und mit 22 g frischem Morphinhydrochlorid versetzt. Das Salz wurde dann aus Wasser umkristallisiert, bis konstante Drehung erhalten wurde. Ein Teil Salz wird in Siedehitze von etwa 25 Teilen Wasser gelöst; der Verlauf der Spaltung wurde in derselben Weise wie bei der Diselen-di-laktylsäure verfolgt.

Kristallisation	1	2	3	4	5	6	7
Ausbeute in g	23.3	19.1	15.8	13.2	10.9	9.5	7.9
$[\alpha]_D^{25}$	−238.5°	−255°	−266°	−273.5°	−279°	−281.5°	−280°

Das Morphinsalz der Säure mit niedriger Aktivität kristallisiert in kurzen, kompakten Kristallen, die an Oktaeder oder Würfel erinnern. Sobald die Drehung der im Salz enthaltenen Säure auf etwa −250° gestiegen ist, ändert das Salz seinen Habitus und tritt dann in flachen Nadeln oder länglichen Tafeln auf.

Die Mutterlaugen nach den ersteren Fraktionen des Morphinsalzes wurden eingeengt, bis sie verhältnismässig dickflussig waren; von dem dabei kristallisierenden Salze wurde filtriert. Aus dem Filtrat wurden 10 g Säure mit der spezifischen Drehung +217° erhalten. Diese Säure wurde in das Chinidinsalz übergeführt, das dann aus mit 15 bis 20 % Alkohol versetztem Wasser umkristallisiert wurde. Von dem Lösungsmittel berechnet man etwa 100 ml je g Salz. Wenn man mit stärker konzentrierter Lösung oder mit höherem Alkoholgehalt arbeitet, scheidet sich das Salz leicht als Öl ab.

Maximale Drehung der im Salz enthaltenen Säure wurde erst nach 7 Umkristallisationen erhalten.

Kristallisation	1	2	3	4	5	6	7	8
Ausbeute in g	19.7	16.2	11.7	8.7	6.8	5.0	3.4	2.3
$[\alpha]_D^{25}$	+231.5°	—	+256°	—	+267°	+272°	+279.5°	+280°

Die aktiven Säuren wurden in gewöhnlicher Weise durch Extraktion der schwefelsauren Lösungen der Alkaloidsalze mit reinem Äther erhalten. Beim Eindampfen der ätherischen Lösungen im Vakuum über Schwefelsäure kristallisierten die Säuren ziemlich leicht. Nach sorgfältiger Trocknung konnten sie aus Benzin (Siedeintervall 60—70°) umkristallisiert werden. In dieser Weise wurden sie als hervorragend schöne, zentimeterlange, scharf abgeschnittene Nadeln mit gerader Auslöschung erhalten. Sie besitzen schwach gelbe Farbe, die bedeutend heller ist als die der aus den Antipoden synthetisierten *racem*-Form.

(+)-Diselen-α-di-n-buttersäure.

Aus 2 g Chinidinsalz wurden 0.33 g einmal umkristallisierte Säure erhalten. Schmelzpunkt 50—52°.

0.2000 g Subst. : 10.95 ml 0.1098-*n* NaOH.
$C_8H_{14}O_4Se_2$ (332,1) Äqu.-Gew. ber. 166.0 gef. 166.4.
0.1026 g Subst. in Wasser auf 10.03 ml gelöst : 2 α = +5.985°.
$[\alpha]_D^{25} = +292.5°$; $[M]_D^{25} = +971.5°$.

Das *Chinidinsalz* wurde bei der Spaltung erhalten; es bildet längliche dünne Prismen von hellgelber Farbe. Das an der Luft auf Gewichtskonstanz getrocknete Salz scheint 4¼ Moleküle Kristallwasser zu enthalten. Die Selenbestimmung wurde an wasserfreier Substanz ausgeführt.

0.5173 g Subst. : 0.0838 g Se.
$C_8H_{14}O_4Se_2$, 2 $(C_{20}H_{24}O_2N_2$ (980.5) Se ber. 16.12 % gef. 16.20 %.
0.5579 g, 0.5195 g kristallwasserhaltige Subst. : 0.0406 g, 0.0379 g H_2O.
4¼ H_2O ber. 7.24 % gef. 7.28, 7.29 %.

(—)-Diselen-α-di-n-buttersäure.

Aus 7 g Morphinsalz wurde 1 g zweimal umkristallisierte Säure erhalten; Aufarbeitung der Mutterlaugen gab noch etwa 1 g. Die Säure ist schwerlöslich in Benzin und Petroleumäther, leichtlöslich in den übrigen gewöhnlichen organischen Lösungsmitteln. In Wasser ist sie mässig löslich; wenn sie mit einer kleineren Menge Wasser zusammengebracht wird, zerfliesst sie zu einem Öl. Schmelzpunkt 50.5—52°.

0.2930 g Subst. : 16.08 ml 0.1098-*n* NaOH.
$C_8H_{14}O_4Se_2$ (332.1) Äqu.-Gew. ber. 166.0 gef. 166.0.
0.1020 g Subst. in Wasser auf 10.02 ml gelöst : 2 α = −5.97°.
$[\alpha]_D^{25} = -293°$; $[M]_D^{25} = -973.5°$.

Das *Morphinsalz* wurde wie oben beschrieben bei der Spaltung in länglichen Tafeln oder flachen Nadeln von gelber Farbe erhalten. Im Vakuum über P_2O_5 nimmt das Gewicht sehr langsam ab ohne in aus praktischen Gründen abzuwartendem Zeitraum einen konstanten Wert anzunehmen. Im Trockenschrank bei etwas über 100° tritt teilweise Zersetzung ein, wobei sich das Präparat braun färbt. Die Selenbestimmung deutet einen Kristallwassergehalt von 2 Molekülen an; der beobachtete Gewichtsverlust über P_2O_5 steht damit nicht in Widerspruch.

0.4982 g Subst. : 0.0842 g Se.
$C_8H_{14}O_4Se_2$, 2 $(C_{17}H_{19}O_3N)$, 2 H_2O (938.4) Se ber. 16.84 % gef. 16.88 %.

Wie schon oben erwähnt, wurde bei Vereinigung von gleichen Teilen (+)- und (—)-Säure in Benzinlösung ein bei 84.5—85.5° schmelzendes Produkt erhalten, dessen Aussehen vollkommen dem des inaktiven Ausgangsmaterials vor der Spaltung glich.

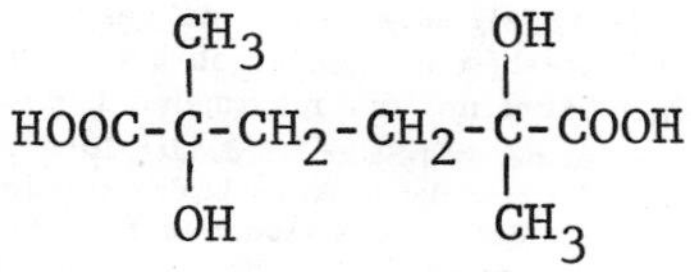

$C_8H_{14}O_6$

$$HOOC-\underset{\underset{OH}{|}}{\overset{\overset{CH_3}{|}}{C}}-CH_2-CH_2-\underset{\underset{CH_3}{|}}{\overset{\overset{OH}{|}}{C}}-COOH$$

$C_8H_{14}O_6$ Ref. 441

(—)-*Dimethyl-dioxy-adipinsäure.*

17 g Säure aus Dilacton werden mit 24 g Cinchonidin in 600 ccm heißem Wasser gelöst. Das in der Kälte auskrystallisierte Salz wird aus 1500 ccm Wasser, das Krystallisat aus 1200 und dann noch einmal aus 1000 ccm, 700, 600 ccm Wasser umkrystallisiert. Das Salz (8—10 g) wird mit Ammoniak zerlegt, das Filtrat vom Cinchonidin angesäuert und die auskrystallisierte Säure ausgewaschen. Die Mutterlaugen werden 14 Tage lang ausgeäthert. Ausbeute 3—4 g.

$[\alpha]_{578,\ 546}^{I\quad II}$ in Wasser
$= -0{,}31°(I);\ -0{,}34°(II) \times 2{,}557/0{,}0464 \times 1{,}005 \times 1{,}0$
$= -17{,}0°(I);\ -18{,}6°(II).$

Drehung der Säure als neutrales Kaliumsalz, bezogen auf Säure: $[\alpha]_{578} = -0{,}03° \times 1{,}968/0{,}125 \times 1{,}025 \times 1{,}0 = -0{,}5° \pm 0{,}2.$

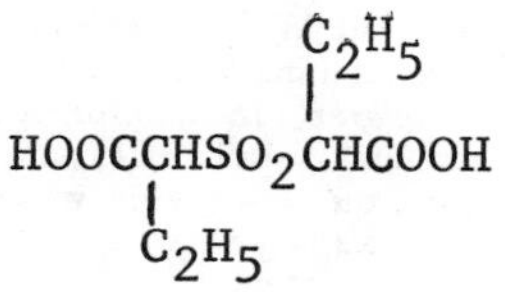

$C_8H_{14}O_6S$ Ref. 442

(+)-α-Sulfon-di-*n*-buttersäure.

Die Racemsäure kann mittels der Alkaloide Brucin, Cinchonin und Cinchonidin in die aktiven Komponenten zerlegt werden[2]). Sämtliche Alkaloid-Salze bestehen aus 2 Mol. Base und 1 Mol. Säure. Mit den beiden ersten Alkaloiden gibt die (+)-Säure das schwerstlösliche Salz, während mit Cinchonidin das Salz der (—)-Säure ausgeschieden wird. Es ist jedoch möglich, die (—)-Säure aus der Mutterlauge des ausgefällten Brucin-Salzes in reiner Form, und sogar mit besserer Ausbeute, zu erhalten.

Zerlegung über das Brucin-Salz: Die früher beschriebene Methode[1]) ist wie folgt abgeändert worden, wodurch Umlagerung während der Arbeit möglichst vermieden wird: 0.075 Mol. Brucin (35.0 g; 75% der für die vollständige Umsetzung nötigen Menge) wurden in der berechneten Menge *n*/₂-Chlorwasserstoffsäure gelöst und bei Zimmer-Temperatur mit einer höchstens 10-proz. Lösung von 0.05 Mol. Ba-Salz der *racem.* α-Sulfon-dibuttersäure (21.4 g bei Zimmer-Temperatur an der Luft getrocknetes Salz) vermischt. Unmittelbar darauf trat Krystallisation ein, und nach 10 Min. (unter zeitweiligem Umrühren) konnte die Mutterlauge von dem Krystallbrei abgesaugt und mit etwa 10 + 50 ccm Wasser gewaschen werden, von welchen die ersten 10 ccm der Mutterlauge zugefügt wurden. (Die Mutterlauge mit der (—)-Form wurde sogleich in der unten bei der „(—)-α-Sulfon-dibuttersäure" beschriebenen Weise weiter behandelt.) Die an der Luft getrocknete Krystallmasse wog 29—30 g. Wegen seiner Schwerlöslichkeit kann das Brucin-Salz nicht umkrystallisiert werden. In Aceton und Äthylalkohol ist es nicht merklich, in Methylalkohol wenig löslich[4]).

Aus dem ausgefallenen Brucin-Salz wurde die (+)-Säure mit Salzsäure freigemacht und mit Äther isoliert. Um schnellstmögliches Lösen des Salzes zu erreichen und einer Abscheidung von Brucin-Hydrochlorid nach Möglichkeit zu entgehen, wurden zu je 10 g zerkleinertem Brucin-Salz 50 ccm Wasser und danach Äther und 5 ccm konz. Salzsäure zugesetzt. Dann wurde rasch 2- oder 3-mal mit etwa dem 1½-fachen Volumen Äther extrahiert, der Äther auf dem Wasserbade größtenteils abdestilliert und der Rest schließlich im Vakuum vollständig verdunstet. Es erschien vorteilhaft, hiernach wieder eine kleine Menge getrockneten Äthers zuzusetzen und abermals im Vakuum eintrocknen zu lassen. Dann wurde in getrocknetem Äther vollständig gelöst, das Lösungsmittel aus einem weithalsigen Kolben größtenteils abdestilliert und der Rest zur Krystallisation beiseite gestellt. Die so erhaltene Säure wurde noch 2-mal aus trocknem Äther umkrystallisiert, wobei die (+)-Säure, mit einer maximalen Drehung von $[\alpha]_D^{14} = +107.6°$, in mehr als 80% der theoretischen Ausbeute erhalten wurde.

Beim Zerlegen der obigen 29.6 g Brucin-Salz wurden in 2 Extraktionen 450 ccm Äther-Lösung mit einer Drehung von $2\ \alpha_D^{14} = +2.38°$ erhalten. Von dieser Lösung verbrauchten 3.00 ccm 3.01 ccm 0.1056-*n*. Bariumhydroxyd-Lösung. Eine dritte Extraktion mit 400 ccm Äther, auf 25 ccm eingeengt, zeigte die Drehung $1\ \alpha_D = +0.60°$. Die ersten beiden Extraktionen gaben somit 5.75 g Säure mit einem $[\alpha]_D^{14} = +94.5°$ in wasserhaltigem Äther, während die dritte Extraktion nur 0.15 g aktive Säure enthielt. Die Lösungen wurden zusammengegeben und wie oben beschrieben weiter behandelt. Sie lieferten hierbei 5.3 g einmal umkrystallisierter Säure und nach zwei weiteren Umkrystallisationen 5.0 g reiner (+)-Säure.

0.1660 g 3-mal umkrystallisierte Säure, in getrocknetem Äther zu 7.00 ccm gelöst, zeigten bei 14° $1\ \alpha_D = +2.55°$; $[\alpha]_D^{14} = 107.5°$. — 0.1664 g 4-mal umkrystallisierte Säure, in getrocknetem Äther zu 7.00 ccm gelöst, ergaben $1\ \alpha_D^{14} = 2.56°$; $[\alpha]_D^{14} = 107.7°$, $[M]_D^{14} = 256.5°$. — 0.2037 g 3-mal umkrystallisierte Säure wurden in wasser-haltigem Äther zu 17.0 ccm gelöst: $2\ \alpha_D = 2.53°$; $[\alpha]_D^{14} = 106°$.

Die beiden ersten Extraktionen gaben also aus dem Brucin-Salz ein Produkt, das in Bezug auf die reine aktive Säure 89.5-proz. war (vergl. unten Fußnote[6])).

Die Zerlegung der Säure über das Cinchonin-Salz wurde folgendermaßen durchgeführt: Es wurden 1.19 g *racem.* Säure und 2.94 g Cinchonin in 8 ccm 65-proz. Alkohol aufgelöst und dann 12 ccm Wasser zugefügt. Unmittelbar darauf begann eine recht langsame Krystallisation, die etwa 2—2.2 g Salz ergab. Zum Umkrystallisieren wurde jedes Gramm Salz in 5 ccm 50-proz. Alkohol auf dem Wasserbade gelöst; dann wurden etwa 6—8 ccm Wasser zugesetzt. Nach 3-maligem Umlösen waren von den 2.0 g nur noch 1.15 g übrig.

0.2503 g luft-trockne Sbst.: 0.0712 g $BaSO_4$. — 0.5295 g Sbst. wurden über Schwefelsäure nicht leichter.

$2\ C_{19}H_{22}N_2O,\ C_8H_{14}O_6S$. Ber. S 3.88. Gef. S 3.91.

3.5 g 3-mal umkrystallisiertes Salz gaben bei 1-maliger Extraktion 0.95 g als Rückstand der Äther-Lösung; 0.1687 g dieser Säure, in trocknem Äther zu 7.00 ccm gelöst, zeigten bei 14° eine Drehung von $1\ \alpha = 2.49°$. Hieraus ergibt sich ein $[\alpha]_D^{14}$ von 104°, 0.1674 g, aus Äther 2-mal umkrystallisierter Säure, ebenfalls in Äther zu 7.00 ccm gelöst, zeigten bei 13° $1\ \alpha = 2.58°$, also $[\alpha]_D^{13} = 107.9°$. Weiteres Umkrystallisieren des Salzes und der Säure erhöhte die Aktivität nicht. Die Säure hatte den ungefähren Schmp. 156—160°[5]).

Über das Chinin-Salz ist nur eine partielle Zerlegung der *racem.* Säure erreicht worden; die auf diesem Wege erhaltene Sulfon-dibuttersäure hatte im Maximum ein $[\alpha]_D^{14} = 85°$ (in Äther-Lösung). Die Mutterlauge des zuerst auskrystallisierten Salzes änderte in 80 Stdn. bei 14° ihre Drehung nicht nachweisbar. Eine Analyse zeigte, daß eine 8-mal umkrystallisierte Probe des Chinin-Salzes etwa 20% durch Hydrolyse entstandenes saures Salz enthielt.

Die Reindarstellung der aktiven Formen der α-Sulfon-di-*n*-buttersäure bietet, wie schon oben erwähnt, vor allem wegen ihrer Unbeständigkeit gewisse Schwierigkeiten, ist aber dank anderen Faktoren gelungen, nämlich 1. der genügenden Schwerlöslichkeit der aktiven Säuren im Verhältnis zu den anderen Formen und 2. ihrer relativen Beständigkeit in trocknem Äther, denn beim Freimachen der aktiven Säuren aus ihren Alkaloid-Salzen tritt stets eine partielle Racemisation ein. So kann z. B. durch direktes Ausfällen ihres Brucin-Salzes eine (+)-Form vom $[\alpha]_D^{14} = 94.5°$ erhalten werden[6]), ferner durch Darstellen und Umkrystallisieren des Cinchonin-Salzes, aber ohne Umkrystallisieren der freigemachten Säure eine (+)-Form von der spez. Drehung 104°. Da nun das Freimachen und Ausziehen der aktiven Säure mit Äther mit einer Inaktivierung im Betrage von vielleicht 2—4% verbunden ist (vergl. unten bei dem Ba-Salz der (+)-Säure), so würde hieraus folgen, daß diese Säure ohne Umlagerung eine spez. Drehung

[1]) Journ. prakt. Chem. [2] **107**, 241 [1924]; vergl. auch B. **55**, 1279 [1922].

[2]) oft mit sichtbaren Zwischenräumen, die mit Mutterlauge gefüllt sind (vergl. unter „Gewichtsverlust beim Übergang des Salzes mit 8 H_2O in das mit 3 H_2O").

[3]) Mit Chinin ist eine partielle Zerlegung durchgeführt worden.

[4]) Analyse des Salzes: Journ. prakt. Chem. [2] **107**, 289 [1924].

[5]) Ein bestimmter Schmelzpunkt ist weder für die *racem.* Säure, noch für die aktiven Säuren erhalten worden; so lag z. B. bei einer *racem.* Säure, die mehrmals aus Wasser umkrystallisiert worden war, die Schmelztemperatur bei 152—156°, während durch 2-maliges Umkrystallisieren aus Äther der Schmp. auf 150—154° erniedrigt wurde. Sicherlich findet mithin bei dieser Temperatur eine schnelle Umlagerung statt, welche die Schmelztemperatur beeinflußt.

[6]) Aus der Brucinsalz-Mutterlauge wurde eine (—)-Säure von $[\alpha] = -99°$ erhalten; das ausgefallene Brucin-Salz betrug 108% der theoretischen Menge, war somit zu mehr als 8% mit den Barium- und Brucin-Salzen der Salzsäure und der (—)-Sulfon-dibuttersäure verunreinigt. (Das Brucin-Salz der Mesosäure ist leichtlöslich.)

von etwa 106—108° gezeigt haben müßte. Tatsächlich sind auch bei der 4- bzw. 2-mal aus Äther umkrystallisierten Säure aus dem Brucin- bzw. Cinchonin-Salz spez. Drehungen von $[\alpha]_D$ = 107.6° bzw. 107.9° bei 14 bzw. 13° festgestellt worden. Die Annahme, daß diese spez. Drehung von 107.8° etwa einer beständigen Mischung von (+)-Säure und z. B. *racem.* Säure zukäme, die vielleicht wegen eines bei dieser Zusammensetzung auftretenden Minimums der Löslichkeit beständig wäre, muß jedoch verworfen werden, weil dieselbe Drehung, soweit dies nachweisbar ist, der Säure in dem Cinchonin-Salz zukommt und wahrscheinlich auch der Säure aus einem gereinigten Brucin-Salz zukommen würde.

Die reine (+)-α-Sulfon-di-*n*-buttersäure wird am bequemsten aus dem Brucin-Salz erhalten. Diese Methode liefert nicht nur eine gute Ausbeute, sondern ermöglicht auch eine bequeme Gewinnung des Antipoden.

Die Drehung in Äther wurde bereits weiter oben angeführt[7]); außerdem sind noch die Drehungen in Aceton und absol. Alkohol bei 11° und 25° in 0.1-molarer Lösung (mit 4-mal umkrystallisierter Säure) bestimmt worden.

0.1668 g Säure, in Aceton zu 7.00 ccm gelöst: $1\alpha_D^{11}$ = 2.17°, $1\alpha_D^{25}$ = 2.11°; $[\alpha]_D^{11}$ = 91.1°, $[\alpha]_D^{25}$ = 88.6°; $[M]_D^{11}$ = 217°, $[M]_D^{25}$ = 211°.

0.1664 g Säure, in absol. Alkohol zu 7.00 ccm gelöst: $1\alpha_D^{11}$ = 2.12°, $1\alpha_D^{25}$ = 2.01°; $[\alpha]_D^{11}$ = 89.2°, $[\alpha]_D^{25}$ = 84.4°; $[M]_D^{11}$ = 212°, $[M]_D^{25}$ = 201°.

In Wasser ist mit 0.332 g Säure die ungefähre Drehung für 0.2-molare Lösungen bestimmt worden. Schon in der ersten Minute war beinahe alles aufgelöst. Nach vollständigem Lösen und Filtrieren bei 14° wurde im vorgewärmten Polarimeterrohr auf 25° erwärmt. Nach etwa 15 Min. bei 14° und etwa 9 Min. bei 25° wurde die Drehung 1α = 1.82° abgelesen, die ohne Racemisation einer Drehung von α_D^{25} = 1.95° entspricht[8]); $[\alpha]_D^{25}$ = 41°; $[M]_D^{25}$ = 98°. Bei niedriger Temperatur ist die Drehung größer ($[\alpha]_D^{12}$ ungefähr gleich 44°).

Löslichkeit der (+)-α-Sulfon-di-*n*-buttersäure in getrocknetem Äther: Bei 13° enthalten 100 ccm gesättigter Lösung 4.30 g und bei 25° 5.50 g Säure. In Benzin, Kohlenstofftetrachlorid und Benzol ist die Säure schwer, in Alkohol und Essigester leicht löslich.

Äquivalentgewicht: 0.0738 g Säure verbrauchten 5.86 ccm 0.1057-*n*. Bariumhydroxyd-Lösung: Äquiv.-Gew. gef. 119.2, ber. 119.1.

0.3000 g (+)-Säure vom spez. Drehungsvermögen $[\alpha]_D^{14}$ = 107.6° wurden in feinpulvrigem Zustande mit der zur vollständigen Neutralisation berechneten Menge Bariumhydroxyd-Lösung versetzt. Die auf 26.0 ccm verdünnte, kein überschüssiges Alkali enthaltende Lösung zeigte im 4-dm-Rohr bei 14° eine Drehung von 4α = −1.22° und bei 25° 4α = −1.08°. Hieraus berechnet sich das molare Drehungsvermögen des Bariumsalzes der reinen (+)-Säure in 0.0485-molarer Wasser-Lösung zu $[M]_D^{14}$ = 63° und $[M]_D^{25}$ = 55°. Nach 3 Tagen wurde bei 14° dieselbe Drehung 4α = −1.22° abgelesen, was die Beständigkeit des Salzes in neutraler Lösung beweist.

Die Säure wurde nun mit Salzsäure freigemacht, mit Äther extrahiert, der Äther abdestilliert, der Rückstand in trocknem Äther aufgelöst und die Drehung bei 14° gemessen. Es wurde gefunden 4α = 3.78°. 6.00 ccm dieser Lösung verbrauchten 4.32 ccm 0.1057-*n*. Bariumhydroxyd, woraus sich die Konzentration zu 0.0381 Mol im Liter oder 0.906 g in 100 ccm Lösung, das molare Drehungsvermögen zu $[M]_D^{14}$ = 248° und das spez. Drehungsvermögen $[\alpha]_D^{14}$ = 104° ergibt. Durch Neutralisieren, 3-tägiges Aufbewahren des Salzes in Wasser-Lösung, Ansäuern und Ausziehen mit Äther waren also nur etwa 3% der Aktivität verloren gegangen. Beim Ausziehen mit Äther usw. von größeren Mengen Säure ist natürlich diese Schnelligkeit beim Arbeiten nicht erreichbar, und dürften deshalb wohl größere Mengen umgelagert werden.

Beständigkeit der (+)-α-Sulfon-di-*n*-buttersäure in wasserhaltigem Äther: Eine Äther-Lösung mit einer Drehung von 4α = 2.78° zeigte nach 190 Tagen bei 12—16° und demselben Volumen 4α = 1.62°; es hatten sich also 42% racemisiert.

(—)-α-Sulfon-di-*n*-buttersäure.

Diese Säure läßt sich, wie oben erwähnt, durch Zerlegen der *racem.* Säure mittels des Cinchonidin-Salzes oder besser aus der Mutterlauge des ausgefällten Brucin-Salzes ihres Antipoden gewinnen. Diese Mutterlauge wurde mit 2 ccm konz. Salzsäure pro Gramm des angewandten Ba-Salzes versetzt und die Säure so rasch wie möglich mit Äther ausgezogen. In dem oben angeführten Fall, bei welchem von 0.05 Mol. Ba-Sulfondibutyrat ausgegangen wurde, ließen sich mit 3 × 250 ccm Äther 5.5 g Säure erhalten von der spez. Drehung $[\alpha]_D^{14}$ = −99°. 3-maliges Umkrystallisieren aus Äther, wie bei der (+)-Säure beschrieben, ergab 4.3 g oder 72% der berechneten Ausbeute an einer Säure mit der maximalen Drehung $[\alpha]_D^{14}$ = 106.5°.

Bei dem Ausziehen der (—)-Säure wurde der dritte Äther-Extrakt für sich verarbeitet. Die miteinander vereinigten, ersten beiden Auszüge zeigten die Drehung 2α = —2.43°, und 3.00 ccm des Extraktes verbrauchten 2.94 ccm 0.1057-*n*. Bariumhydroxyd-Lösung. Die Säure hatte also ein $[\alpha]_D^{14}$ = —99°. Die dritte Extraktion zeigte bei einem Volumen von 226 ccm 2α = —0.49°; sie enthielt, da 9.00 ccm davon 1.81 ccm 0.1057-*n*. Bariumhydroxyd verbrauchten, 0.57 g einer Säure vom $[\alpha]_D^{14}$ = —97°. 0.1661 g 2-mal umkrystallisierte Säure, in trocknem Äther zu 7.00 ccm gelöst, zeigten bei 14° 1α = 2.51°; $[\alpha]_D^{14}$ = 105.8°; 0.1669 g 3-mal umkrystallisierte Säure, in Äther zu 7.00 ccm gelöst: 1α = —2.54°, $[\alpha]_D^{14}$ = 106.5°. 0.1662 g 4-mal umkrystallisierte Säure, in Äther zu 7.00 ccm gelöst: 1α = —2.53°, $[\alpha]_D^{14}$ = 106.5°, $[M]^{14}$ = 254°.

In der sicher berechtigten Annahme, daß bei dem Ausziehen der (—)-Säure 2—4% der Säure sich racemisierten, braucht die Brucinsalz-Mutterlauge nur 2% Brucin-Salz des Antipoden zu enthalten, um die gefundene Drehung $[\alpha]_D^{14}$ = −99° in wasser-haltigem Äther zu erklären[9]). Nun ist aber hier als maximale Drehung ein Wert gefunden worden, der um 1% kleiner als die Drehung der (+)-Säure ist; hierbei dürfte es sich jedoch lediglich um einen Ablesungsfehler des Nullpunktes unseres Polarimeters usw. handeln. Übrigens wurde in einem anderen Falle, in welchem allerdings die Ausbeute nicht bestimmt wurde, ein Maximum von $[\alpha]_D^{14}$ = −108.2° für 3-mal umkrystallisierte Säure festgestellt.

Das Zerlegen mittels des Cinchonidin-Salzes: 11.9 g *racem.* Säure wurden in 80 ccm Alkohol gelöst, 50 ccm Wasser zugesetzt und danach unter schwachem Erwärmen 29.4 g Cinchonidin in der Mischung aufgelöst. Wurde dann die Salzlösung mit 120 ccm Wasser versetzt, so trat gewöhnlich sofort Krystallisation ein. Nach 12-stdg. Stehen an einer kühlen Stelle wurden so 22—23 g (luft-trocknes) Salz gewonnen. Letzteres wurde in der Weise 3-mal umkrystallisiert, daß es in der $1^1/_2$-fachen Menge 65-proz. Alkohols unter gelindem Erwärmen aufgenommen und die Lösung dann bis zum 3-fachen Volumen mit Wasser verdünnt wurde. So ergaben sich 13—15 g Salz, die nicht weiter umkrystallisiert wurden. 0.433 g des 2-mal umkrystallisierten Salzes zeigten, in 96-proz. Alkohol zu 15 ccm gelöst, 1α = −3.12°; 3- und 4-mal umkrystallisiertes Salz, in derselben Weise gelöst, zeigte 1α = −3.18° bzw. 1α = −3.20°, während 0.298 g Cinchonidin und 0.119 g *racem.* Säure (= 0.435 g Salz), zu 15.00 ccm in 96-proz. Alkohol gelöst, bei 14° 1α = −2.23° zeigten.

0.3359 g Sbst.: 0.0899 g $BaSO_4$. — 0.3107 g Sbst. verloren über Schwefelsäure in 1 bzw. 2 Tagen 0.0123 bzw. 0.0124 g.

2 $C_{19}H_{22}N_2O$, $C_8H_{14}O_6S$ + 2 H_2O. Ber. S 3.72, H_2O 4.18. Gef. S 3.68, H_2O 3.99.

Aus dem Salz wurde die Säure durch verd. Schwefelsäure freigemacht und 2-mal mit etwa 150 ccm Äther ausgeschüttelt, wobei man nicht darauf wartete, daß die Äther-Lösung klar durchsichtig erschien. Sie mußte deshalb filtriert werden; im übrigen wurde wie oben bei der (+)-Form verfahren. Die in trocknen Äther übergeführte Säure zeigte ein spez. Drehungsvermögen von $[\alpha]_D^{14}$ = 66°, ließ sich aber schon durch 2-maliges Umkrystallisieren aus Äther, das letzte Mal unter Benzin-Zusatz, zu einer Säure mit der früher gefundenen maximalen Drehung $[\alpha]_D^{14}$ = 106.7° reinigen. So wurden aber nur 1.3—1.4 g reine (—)-Säure oder 11—12% der angewandten *racem.* Säure erhalten. Wenn man mithin mittels des Cinchonidin-Salzes nur eine partielle Zerlegung erreichen kann, so genügt diese doch, um reine aktive Säure durch Umkrystallisieren zu gewinnen.

[7]) Der Äther wurde für sämtliche Drehungs-Bestimmungen mit Calciumchlorid getrocknet.

[8]) vergl. auch Tabelle 34 auf S. 825 der folgenden Abhandlung.

[9]) Da das Brucin-Salz der Mesoform (l. c., Fußnote 1) leichtlöslich ist, ist hierdurch die Reinheit des Barium-*racem.*-sulfondibutyrats bewiesen.

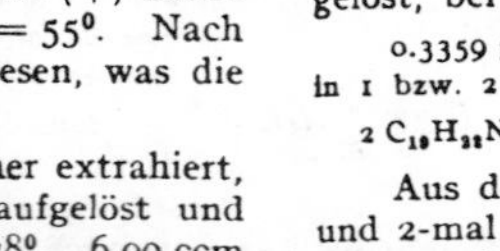

$C_8H_{15}NO_2$ Ref. 443

The inactive acid (6·8 g.) and anhydrous quinine (8·45 g.) were dissolved in boiling 3 : 2-aqueous alcohol (130 ml.). The filtered solution was cooled to opalescence and seeded with a crystal of quinine salt from an initial experiment. Separation of quinine salt as small prisms commenced immediately and was completed by cooling and storage overnight. The filtered salt was washed with 3 : 2 aqueous alcohol (50 ml.) and recrystallised twice from the same solvent. After drying *in vacuo* at 100° (P_2O_5) *quinine* D-*α-benzamido-α-*cyclo*hexylacetate* (5 g.) had m. p. 149° after softening at 141°, $[\alpha]_D^{23}$ −112° (*c* 0·4 in EtOH) (Found : C, 69·5; H, 7·5; N, 6·85. $C_{35}H_{43}O_5N_3,H_2O$ requires C, 69·6; H, 7·5; N, 7·0%).

The quinine salt (8 g.), suspended in water (150 ml.), was acidified with dilute hydrochloric acid to Congo-red, and the mixture extracted with ether. The residue after evaporation of the dried ether solution, when recrystallised from benzene, gave the D-*benzamido-acid* (3 g.), m. p. 171—172°, $[\alpha]_D^{23}$ −26·1° (c 0·803 in 0·5N-NaOH) (Found: C, 69·2; H, 7·3; N, 5·05%).

D-α-*Amino*-α-*cyclohexylacetic Acid*.—The D-benzamido-acid (1 g.) was refluxed with concentrated hydrochloric acid (40 ml.) and water (20 ml.) for 8 hr., then evaporated to dryness under reduced pressure and the residue freed from benzoic acid by several extractions with ether, redissolved in a little water, and made faintly alkaline with ammonia. The acid, which separated on ice-cooling, recrystallised from water as plates (sublimed above 260°), $[\alpha]_D^{25}$ −33·8° (c 0·4 in 5N-hydrochloric acid) (lit.,[13] $[\alpha]_D^{26}$ −35·0°) (Found: C, 61·0; H, 9·65; N, 8·95. Calc. for $C_8H_{15}O_3N$: C, 61·1; H, 9·6; N, 8·9%).

L-α-*Benzamido*-α-*cyclohexylacetic Acid*.—The mother-liquor separated from the quinine salt of the D-acid was evaporated under reduced pressure to remove most of the alcohol. The residue was acidified with dilute hydrochloric acid and extracted with ether. Evaporation of the dried ethereal extract gave 2·48 g. of crude acid which was extracted with boiling benzene (75 ml.) and filtered from some insoluble material. The solution, on cooling, gave crystals (1·27 g.) of *acid* which on recrystallising from benzene (80 ml.) had m. p. 172—173°, $[\alpha]_D^{23}$ +25·9° (c 1·082 in 100 ml. of water containing 4·4 ml. of N-sodium hydroxide) (Found: C, 69·25; H, 7·4; N, 4·95. $C_{15}H_{19}O_3N$ requires C, 68·95; H, 7·3; N, 5·4%). The benzene-insoluble residue above (0·95 g.) was optically inactive and had m. p. 198° undepressed on admixture with DL-α-benzamido-α-*cyclo*hexylacetic acid.

$C_8H_{15}NO_2$ Ref. 443a

Resolution Procedure.—The acylated derivatives were resolved by the enzymatic procedure previously described.[1] The yields were 60 to 80% for the L-isomers and 40 to 60% for the D-isomers. The nitrogen analyses and optical properties of the isomers are collected in Table I. The resolution of DL-α-amino-β-phenylpropionic acid has been described,[6] and the isomers of this amino acid are included in the table for comparative purposes. The unusually high rotation values of the α-aminophenylacetic acid enantiomorphs are noteworthy.

TABLE I
PROPERTIES OF RESOLVED ISOMERS

Isomers, acid	N, analyses, % Found	N, analyses, % Calcd.	$[\alpha]_D^a$ (water)	$[\alpha]_D^a$ (5 N HCl)
L-α-Aminophenylacetic[b]	9.2	9.3	+114.0[oc]	+168.0[oc]
L-α-Aminocyclohexylacetic	8.9	8.9	d	+ 35.5[oe]
L-α-Amino-β-phenylpropionic	8.5	8.5	− 34.6[oe]	− 4.5[oe]
L-α-Amino-β-cyclohexyl-propionic[f]	8.2	8.2	− 9.0[oe]	+ 15.0[oe]
D-α-Aminophenylacetic[b]	9.2	9.3	−114.0[oc]	−169.0[oc]
D-α-Aminocyclohexylacetic	8.9	8.9	d	− 35.0[oe]
D-α-Amino-β-phenylpropionic	8.5	8.5	+ 34.4[oe]	+ 4.5[oe]
D-α-Amino-β-cyclohexyl-propionic[f]	8.1	8.2	+ 8.5[oe]	− 15.0[oe]

[a] At 26°. [b] E. Fischer (*Ber.*, **41**, 1290 (1908)) reported $[\alpha]_D^{20}$ values of −157.87° (in 0.65 N hydrochloric acid) and −165.43° (in 10% hydrochloric acid) for the D-isomer, and +158.09° (in 0.65 N hydrochloric acid) and +112.6° (in water) for the L-isomer. [c] 1.000% solution; 2-dm. tube. [d] These isomers were too insoluble to permit accurate determinations. [e] 2.000% solution; 2-dm. tube. [f] E. Waser and E. Brauchli (*Helv. Chim. Acta*, **7**, 740 (1924)) reported $[\alpha]_D^{20}$ values (in 4% hydrochloric acid) of +13.30° and and −13.32°, respectively, for the L- and D-isomers of this amino acid.

Optical Purity of the Isomers.—The optical purity of the isomers was determined by the enzymatic technique previously described.[7] Hog kidney oxidase was used to detect the presence of D-isomer in the L-amino acids, and rattlesnake and cobra venom oxidases were employed, respectively, to test the D-enantiomorphs of the acetic and propionic acid derivatives.[8] No evidence of contamination was observed under conditions whereby one part in 1000 of the enantiomorph could be detected, indicating an optical purity of greater than 99.9%.

---

(1) For the literature see S. M. Birnbaum, L. Levintow, R. B. Kingsley and J. P. Greenstein, *J. Biol. Chem.*, **194**, 455 (1952).

(6) J. B. Gilbert, V. E. Price and J. P. Greenstein, *J. Biol. Chem.*, **180**, 473 (1949).
(7) A. Meister, L. Levintow, R. B. Kingsley and J. P. Greenstein, *ibid.*, **192**, 535 (1951).
(8) The susceptible isomers were rapidly oxidized by the respective enzymes under the conditions employed.

$$\begin{array}{c} CONH_2 \\ | \\ CH_3CH_2-C-COOH \\ | \\ CH(CH_3)_2 \end{array}$$

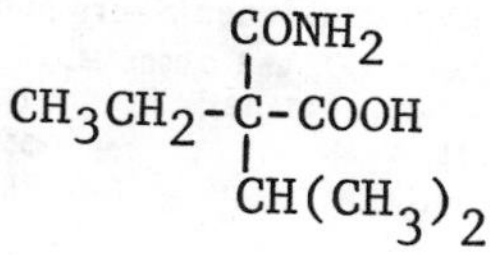

$C_8H_{15}NO_3$ Ref. 444

Spaltung der d,l-Äthylisopropyl-malonaminsäure in die optisch-aktiven Komponenten.

70 g d,l-Säure wurden in 90 ccm warmem Alkohol gelöst, mit einer Lösung von 153,5 g wasserhaltigem Chinin in 375 ccm Alkohol vermischt, mit 750 ccm Wasser verdünnt und die Flüssigkeit unter vermindertem Druck auf etwa 700 ccm eingeengt. Schon während der Operation schied sich das Chininsalz der d-Säure in flächenreichen Krystallen ab, deren Menge in 15 Stunden sehr zunahm (96 g). Diese wurden in 225 ccm heißem Alkohol gelöst und nach Zusatz von 300 ccm Wasser unter vermindertem Druck auf 400 ccm eingeengt. Dabei fand wieder Krystallisation statt. Nach ½ stündigem Stehen in Eis wurde filtriert. Erhalten 73 g. Diese Operation haben wir noch dreimal wiederholt, wobei das Drehungsvermögen der aus dem Salz isolierten freien Säure mikropolarimetrisch gefunden wurde: nach der zweiten Krystallisation +9,18°, nach der vierten +13,98°, nach der fünften +14,25°. Weiteres Umkrystallisieren des Salzes erhöhte nicht mehr das Drehungsvermögen der Säure. Ausbeute bei der fünften Krystallisation 32 g Chininsalz; selbstverständlich läßt sie sich durch Verarbeitung der Mutterlaugen erhöhen. Das Salz bildet feine, lange Nadeln. Es wurde fein gepulvert, mit verdünnter Salzsäure zerlegt und die d-*Äthylisopropyl-malonaminsäure* ausgeäthert. Aus der nicht zu verdünnten Ätherlösung fiel sie durch Petroläther in derben Nadeln, welche denselben Schmelz- bzw. Zersetzungspunkt (116°) wie die inaktive Säure zeigten.

0,1576 g (im Vakuumexsiccator getrocknet) gaben 0,3190 CO_2 und 0,1230 H_2O

	Ber. für $C_8H_{15}O_3N$ (173,13)	Gef.
C	55,45	55,2
H	8,73	8,73

Für die optische Bestimmung diente die alkoholische Lösung.

0,2206 g Substanz: 2,2187 g Lösung, mithin Prozentgehalt 10,06. d = 0,813. Drehung bei Natriumlicht und 22° im 1 dm-Rohr 1,18° nach rechts. Mithin

$$[\alpha]_D^{22} = + 14{,}59°.$$

0,2151 g Substanz: 2,1145 g Lösung, mithin Prozentgehalt 10,196. d = 0,8135. Drehung bei Natriumlicht und 20° im 1 dm-Rohr 1,20° nach rechts. Mithin

$$[\alpha]_D^{20} = + 14{,}5°.$$

Die *l-Äthylisopropyl-malonaminsäure* befindet sich in den Mutterlaugen des Chininsalzes. Sie wird daraus durch Verdampfen des Alkohols, Übersättigen mit Salzsäure und Ausäthern gewonnen. Das aus der ersten Hauptmutterlauge des Chininsalzes isolierte Präparat zeigte in Alkohol $[\alpha]_D^{20} = - 6{,}5°$, enthielt also noch viel von dem optischen Antipoden. Zur Reinigung diente das Morphinsalz.

26,4 g des Präparates wurden deshalb in 100 ccm Alkohol gelöst, mit einer Lösung von 46,25 g käuflichem Morphin in 450 ccm Alkohol vermischt, nach Zusatz von 450 ccm Wasser unter vermindertem Druck eingeengt, bis eine reichliche Krystallisation eintrat und nach dem Abkühlen auf 0° abgesaugt (50 g). Das Salz wurde wiederholt aus heißem Wasser oder verdünntem Alkohol umkrystallisiert, bis das Drehungsvermögen einer Probe der aus dem Salz abgeschiedenen Säure nicht mehr stieg.

0,1200 g gaben 0,2431 CO_2 und 0,0905 H_2O.

	Ber. für $C_8H_{16}O_3N$ (173,13)	Gef.
C	55,45	55,25
H	8,73	8,44

Für die optische Bestimmung diente die alkoholische Lösung.

0,1772 g Substanz: 1,7717 g Lösung, mithin Prozentgehalt 10,002. d = 0,8136. Drehung im 1 dm-Rohr bei Natriumlicht und 22° 1,19° nach links. Mithin

$$[\alpha]_D^{22} = - 14{,}62°.$$

0,1777 g Substanz: 1,7697 g Lösung, mithin Prozentgehalt 9,959. d = 0,8192. Drehung im 1 dm-Rohr bei Natriumlicht und 17° 1,19° nach links. Mithin

$$[\alpha]_D^{17} = - 14{,}47°.$$

Die Säure zeigt im übrigen ganz das Verhalten der d-Verbindung. Da die Reinigung der beiden optischen Antipoden sowohl über das Chininsalz wie über das Morphinsalz zum gleichen Endresultat führte, so ist es wahrscheinlich, daß die Präparate optisch rein waren.

```
HOOC-(CH2)5-CH-COOH
            |
            NH2
```

$C_8H_{15}NO_4$ Ref. 445

***N*-Carbobenzoxy-L-α-aminosuberic Acid (V).** D-Tyrosine hydrazide (29.3 g, 0.15 mol) and III (32.3 g, 0.1 mol) were dissolved in boiling methanol (600 m*l*), and a small amount of an insoluble material was filtered off. The resulting clear solution was concentrated to about 300 m*l*, and V was precipitated as the salt of D-tyrosine hydrazide by the careful addition of ethyl acetate (about 350 m*l*). When oily material appeared during the precipitation procedure, the addition of ethyl acetate was stopped and the oil was rubbed vigorously with a glass rod until it changed to crystals. After the solution had stood overnight at room temperature, the crystals were collected by filtration, washed with a mixture of methanol and ethyl acetat , and dried; yeild, 32.3 g (90%), mp 151—153°C. (The mother liquor was used later for the isolation of VI.). This material, which represented the salt of D-tyrosine hydrazide and V (2: 1 mol/mol), was recrystallized from ethanol (800 m*l*) for purification; yeild, 25.8 g (72%), mp 153—154°C, $[\alpha]_D^{24}$ −56.0° (*c* 1.8, water). A second crop of crystals was recovered from the mother liquor; weight, 2.7 g (7.4%), mp 153—154°C, $[\alpha]_D^{25}$ −56.1° (*c* 2, water).

Found: C, 57.20; H, 6.94; N, 13.66%. Calcd for $C_{34}H_{47}O_{10}N_7$: C, 57.21; H, 6.64; N, 13.73%. This salt (25.6 g, 0.036 mol) was suspended in ethyl acetate (120 m*l*), and the mixture was shaken with 4N hydrochloric acid (30 m*l*). The ethyl acetate layer was separated, washed with water, dried over sodium sulfate, and concentrated to a residue, which was then recrystallized from ethyl acetate and petroleum ether; yield of V, 10.8 g (93%), mp 119—121°C, $[\alpha]_D^{26}$ −3.3° (*c* 7.7, acetic acid), $[\alpha]_D^{25}$ −9.1° (*c* 4, dimethylformamide).

Found: C, 59.39; H, 6.56; N, 4.29%. Calcd for $C_{16}H_{21}O_6N$: C, 59.43; H, 6.55; N, 4.33%. The optical purity of this compound was confirmed after the removal of the protective group by catalytic hydrogenolysis; the compound IV thus obtained showed a mp of 246—248°C (decomp.), $[\alpha]_D^{28}$ +20.0° (*c* 2, 5N hydrochloric acid).

***N*-Carbobenzoxy-D-α-aminosuberic Acid (VI).** The mother liquor mentioned above was concentrated, and crude VI was extracted from the residue with ethyl acetate in the presence of hydrochloric acid. On the concentration of the extract, a residue of crystals was obtained; yield, 17.3 g, $[\alpha]_D^{25}$ +6.3° (*c* 4, dimethylformamide). A part of this material (3.23 g, 0.01 mol) was dissolved in boiling methanol, together with L-tyrosine hydrazide (3.90 g, 0.02 mol), and the mixture was treated as in the case of V. The yield of the salt of L-tyrosine hydrazide and VI was 4.86 g (73%); mp 153—154°C, $[\alpha]_D^{27}$ +56.1° (*c* 2, water). Compound VI was isolated from the salt as has been described previously; yield, 97%, mp 119—121°C, $[\alpha]_D^{28}$ +9.1° (*c* 4, dimethylformamide). D-α-Aminosuberic acid was obtained from VI by the removal of the protecting group; mp 246—248°C (decomp.), $[\alpha]_D^{28}$ −20.1° (*c* 2, 5N hydrochloric acid).

$C_8H_{15}NO_4$ Ref. 445a

Isolation of L-α-Aminosuberic Acid (IV) and of Its D-Isomer by Enzymic Digestion of II. A solution of Takadiastase (250 g) made by extraction with ice water (1.8 *l*) was added to a mixture of II (266 g, 1 mol) and cobaltous chloride (460 mg) in 4N sodium hydroxide; the pH of the solution was adjusted to 7.2 with acetic acid and with sodium hydroxide before and after the addition of the enzyme solution respectively. During the incubation at 37°C the solution was frequently adjusted to pH 7.2 with 4N sodium hydroxide. After 8 days the mixture was boiled for 10 min to precipitate proteins, which were then removed by filtration with Hyflo Super-cel. The unchanged chloroacetyl derivative (mainly the D-form) was extracted with ethyl acetate at pH 1.7, and then the water layer was adjusted to pH 3 and stored overnight in a refrigerator. The crude IV was collected by filtration and recrystallized from hot water after decolorization with active charcoal; yeild, 80 g (84%), mp 243—245°C (decomp.), $[\alpha]_D^{17}$ +20.1° (*c* 1.9, 5N hydrochloric acid), $[\alpha]_D^{26}$ +19.6° (*c* 1.4, 6N hydrochloric acid). Reported, $[\alpha]_D^{25}$ +20.2° (*c* 0.1, 5N hydrochloric acid).[2)]

Found: C, 50.73; H, 8.35; N, 7.64%. Calcd for $C_8H_{15}O_4N$: C, 50.78; H, 7.99; N, 7.40%.

The ethyl acetate extract, which was rich in the D-form, was dried over sodium sulfate and concentrated *in* vacuo. The oily residue was refluxed with concentrated hydrochloric acid for 5 hr, and then the solution was concentrated to dryness. The residue was dissolved in water, the pH of the solution was adjusted to 3, and the D-α-aminosuberic acid which precipitated was collected by filtration. The crude crystals were recrystallized from hot water, as in the case of the L-form; yield, 42.5 g (45%), mp 243—245°C (decomp.), $[\alpha]_D^{15}$ $-19.9°$ (c 2.1, 6N hydrochloric acid).

$$CH_3(CH_2)_4\text{-}CH\text{-}COOH$$
$$\phantom{CH_3(CH_2)_4\text{-}}\underset{CH_3}{|}$$

$C_8H_{16}O_2$ Ref. 446

Resolution of 2-Methylheptanoic Acid (Methyl Amyl Acetic Acid)—900 gm. of dl-2-methylheptanoic acid were neutralized with 1825 gm. of cinchonidine in acetone. The mixture was filtered, and the mother liquor was evaporated under reduced pressure and the acid recovered. B.p. 94–96°, 1 mm.; yield, 150 gm.; $n_D^{25} = 1.4235$.

$$[\alpha]_D^{25} = +8.9° \text{ (homogeneous)}$$

The crystals were recrystallized from acetone six times, and from methyl ethyl ketone three times more. The acid was extracted and distilled. Yield, 50 gm.; $d_4^{25} = 0.902$ (in vacuo); $n_D^{25} = 1.4233$.

$C_8H_{16}O_2$. Calculated, C 66.63, H 11.19; found, C 66.56, H 11.35

$[\alpha]_D^{25} = -15.6°$; $[M]_D^{25} = -22.5°$, i.e. 86.5 per cent of the maximum $[M]_D^{25} = -26.0°$ (homogeneous), calculated from the maximum rotation of 4-methylnonanoic acid given by Levene and Marker,[a] assuming no racemization in reactions (I) to (V).

[a] Levene, P. A., and Marker, R. E., *J. Biol. Chem.*, **103**, 299 (1933).

$$\underset{\phantom{CH_3(CH_2)_3\text{-}}}{\overset{CH_3}{|}}$$
$$CH_3(CH_2)_3\text{-}CH\text{-}CH_2COOH$$

$C_8H_{16}O_2$ Ref. 447

Resolution of 2-n-Butylbutyric Acid-4—This acid was resolved by crystallizing its quinine salt from acetone at $-15°$ until the rotation of the acid reached a constant value. This was after the eighth crystallization.

$$[\alpha]_D^{17} = \frac{-3.81°}{1 \times 0.905} = -4.21°; [M]_D^{17} = -6.06° \text{ (homogeneous)}$$

4.885 mg. substance: 11.945 mg. CO_2 and 4.950 mg. H_2O
$C_8H_{16}O_2$. Calculated. C 66.6, H 11.2
              Found.   " 66.7, " 11.3

$$CH_3(CH_2)_3\text{-}CH\text{-}COOH$$
$$\phantom{CH_3(CH_2)_3\text{-}}\underset{C_2H_5}{|}$$

$C_8H_{16}O_2$ Ref. 448

Ethylbutylacetic Acid (2-Ethylcaproic Acid (1))—The acid was made by the malonic ester synthesis and resolved by recrystallizing the quinine salt from 50 per cent acetone. After seven crystallizations, an acid was obtained, which had the following properties. B.p. 120°, $p = 13$ mm.; $n_D^{25} = 1.4229$; $d_4^{25} = 0.9031$ (in vacuo).

$$[\alpha]_D^{25} = \frac{-3.79°}{1 \times 0.903} = -4.20°; [M]_D^{25} = -6.05° \text{ (homogeneous)}$$

Maximum $[M]_D^{25} = -23.5°$ (homogeneous)
0.1046 gm. substance required 7.205 cc. 0.1 N NaOH. Mol. wt. 145
4.800 mg. substance: 11.695 mg. CO_2 and 4.810 mg. H_2O
$C_8H_{16}O_2$. Calculated.   C 66.61, H 11.19
144.1   Found.   " 66.44, " 11.21

The acid recovered from the mother liquors was converted into the cinchonidine salt in acetone. The salt which crystallized very slowly was recrystallized from acetone, then from 60 per cent acetone. An acid having a rotation of

$$[\alpha]_D^{25} = \frac{+5.18°}{1 \times 0.903} = +5.74° \text{ (homogeneous)}$$

was obtained from the crystals.

$$CH_3CH_2CH_2CH\text{-}CH_2COOH$$
$$\underset{C_2H_5}{|}$$

$C_8H_{16}O_2$ Ref. 449

Dextro-3-Propyl Valeric Acid (5)—The inactive acid was prepared from 3-bromohexane and ethyl malonate. 228 gm. of the inactive acid were dissolved in 1 liter of hot acetone and 650 gm. of quinine were added. The solution was filtered and allowed to crystallize in a refrigerator at 0°. It required 2 days for crystallization. After five crystallizations the salt was decomposed with 10 per cent hydrochloric acid and the organic acid extracted with ether. B.p. 106° at 5 mm.; $D_4^{20} = 0.911$; $n_D^{25} = 1.4287$.

$$[\alpha]_D^{25} = \frac{+1.69°}{1 \times 0.911} = +1.86°. \ [M]_D^{25} = +2.67° \text{ (homogeneous)}$$

4.375 mg. substance: 10.680 mg. CO_2 and 4.510 mg. H_2O.
$C_8H_{16}O_2$. Calculated.   C 66.62, H 11.19
              Found.   " 66.58, " 11.53

```
CH3CH2CH-COOH
       |
     (CH2)3CH3
```

$C_8H_{16}O_2$ **Ref. 449a**

(—)-n-*Heptane-γ-carboxylic Acid.*—Quinine (324 g.) was dissolved in a hot solution of the *dl*-acid (144 g.) in 50% aqueous acetone (1500 c.c.), and the salt which separated overnight was recrystallised six times from aqueous acetone. The resulting quinine (—)-*n*-heptane-γ-carboxylate (140 g.) had m. p. 64—65°; $[\alpha]_{5893}$ —112·0°, $[\alpha]_{5461}$ — 136·0°, $[\alpha]_{4358}$ — 270° (*c*, 5·084; *l* = 2, in ethyl alcohol); on decomposition it yielded the (—)acid, b. p. 218—220°, the rotatory powers of which are recorded in Tables I and II.

(+)-n-*Heptane-γ-carboxylic Acid.*—The dextrorotatory acid (80 g.), obtained by decomposition of the more soluble fractions of the quinine salt described above, was combined with cinchonidine (161 g.) in aqueous-acetone solution; the resulting salt after four recrystallisations from aqueous acetone yielded the cinchonidine salt of (+)-*n*-heptane-γ-carboxylic acid (135 g.), m. p. 67—69°; $[\alpha]_{5893}$ — 74·8°, $[\alpha]_{5461}$ — 89·1°, $[\alpha]_{4358}$ — 165·5° (*c*, 4·646; *l* = 2, in ethyl alcohol). The (+)acid obtained from this salt had b. p. 218—220°, $d_{4°}^{20°}$ 0·9064; its rotatory powers are included in Table I.

TABLE I.

Specific Rotatory Powers (l = 2) *and Densities of* (+)-n-*Heptane-γ-carboxylic Acid at Various Temperatures.*

t.	$d_{4°}^{t°}$.	$[\alpha]_{6438}^{t°}$.	$[\alpha]_{5893}^{t°}$.	$[\alpha]_{5780}^{t°}$.	$[\alpha]_{5461}^{t°}$.	$[\alpha]_{5086}^{t°}$.	$[\alpha]_{4358}^{t°}$.
14°	0·9112	+ 7·65°	+ 9·00°	+ 9·47°	+ 10·64°	+ 12·43°	+ 18·19°
20	0·9064	7·34	8·95	9·43	10·62	—	17·50
31	0·8975	7·38	8·89	9·37	10·56	12·25	17·55
48	0·8840	7·43	8·78	9·22	10·38	—	16·74
68	0·8678	7·14	8·71	9·17	10·28	12·26	14·81
90	0·8500	—	8·61	9·18	10·21	12·20	—
15*	0·9104	—	8·92	9·43	10·64	12·33	17·36

* Recooled.

TABLE II.

Specific Rotatory Powers of (—)-n-*Heptane-γ-carboxylic Acid* ($[\alpha]_{5461}^{16}$ — 10·32°) *in Various Solvents* (l = 2).

Solvent.	*t.*	*c.*	$[\alpha]_{6438}^{t°}$.	$[\alpha]_{5893}^{t°}$.	$[\alpha]_{5780}^{t°}$.	$[\alpha]_{5461}^{t°}$.	$[\alpha]_{5086}^{t°}$.	$[\alpha]_{4358}^{t°}$.
Et_2O	20°	5·008	— 8·59°	— 9·88°	— 10·30°	— 12·28°	— 14·87°	— 18·37°
CH_2Cl_2	20	4·990	6·91	8·62	9·12	9·41	11·42	16·43
$CHCl_3$	20	5·079	5·12	7·88	8·17	8·86	10·14	13·98
1 : 4-Dioxan	21	5·043	6·45	6·94	7·54	8·64	9·92	12·60
CCl_4	20	5·087	6·29	7·08	7·27	7·67	9·04	12·78
CS_2	19	5·075	—	6·60	6·80	7·39	8·28	11·22
C_6H_6	18	5·127	4·29	5·66	5·85	6·24	7·31	9·85
C_6H_5Me	19	4·994	4·01	5·01	5·41	6·11	7·0	10·0

```
                C2H5
                |
(CH3)2CHCH2CH-COOH
```

$C_8H_{16}O_2$ **Ref. 450**

Dextro-2-Isobutylbutyric Acid-4—The inactive acid was prepared from methylisobutylbromomethane and malonic ester.

1 mol of the inactive acid was dissolved in 400 cc. of boiling acetone and 1 mol of quinine was added. The solution was then filtered and 100 cc. of water added. The solution was allowed to stand in a refrigerator at −15° until crystallization took place. The salt was filtered and recrystallized three times from 80 per cent acetone. It is very soluble and crystallizes with difficulty. For these experiments the mother liquors from the first crystallization were used. The acetone was evaporated and the quinine salt decomposed with hydrochloric acid. The organic acid was extracted with ether, purified through its sodium salt, and then distilled. B.p. 124° at 20 mm. $D\frac{30}{4} = 0.899$.

$$[\alpha]_D^{30} = \frac{+1.48°}{1 \times 0.899} = +1.65°; [M]_D^{30} = +2.37° \text{ (homogeneous)}$$

3.881 mg. substance: 9.433 mg. CO_2 and 4.000 mg. H_2O

$C_8H_{16}O_2$. Calculated. C 66.6, H 11.2
Found. " 66.3, " 11.5

```
            C2H5
            |
CH3(CH2)4-C-COOH
            |
            CH3
```

$C_8H_{16}O_2$ **Ref. 451**

Resolution of 2-ethyl-2-methylpentanoic acid with brucine. A mixture of 21.6 g. of the *dl*-acid, 59.1 g. of brucine, and 130 ml. of absolute ethanol was allowed to crystallize producing 39 g. of salt, m.p. 81–82°. Six more crystallizations furnished 13.8 g. of salt: m.p. 83°; $[\alpha]_D^{25}$ −38.06° (95% ethanol; c, 8). Decomposition of this brucine salt in hydrochloric acid, extraction with ether, and distillation gave 2.1 g. (10%) of (+)-*2-ethyl-2-methylpentanoic acid;* b.p. 127° (20 mm.); α_D^{25} +0.970° (homogeneous, 1 dm.); $[\alpha]_D^{25}$ +1.05°.

The mother liquors were concentrated according to the usual diamond scheme to give 29.0 g. of yellowish crystals: m.p. 80–81°, $[\alpha]_D^{25}$ −40.62° (95% ethanol; c, 8). Decomposition of this salt gave 2.7 g. of (−)-*2-ethyl-2-methylpentanoic acid;* b.p. 127° (20–21 mm.); $[\alpha]_D^{25}$ −0.58° (homogeneous).

An attempted resolution with quinine produced a 24% yield of quinine salt after seven crystallizations from ethanol: m.p. 131.5°; [α] −131.4° (95% ethanol; c, 8). Decomposition of this salt gave 2.3 g. of the acid: b.p. 128° (21 mm.); $[\alpha]^{25}_D$ −0.14°. The tail fraction gave the (+)-acid, $[\alpha]^{25}_D$ +0.12°. Cinchonine, strychnine, and (−)-α-phenylethylamine failed as resolving agents.

Authors' comments:

Cinchonine, strychnine and (-)-α-phenylethylamine failed as resolving agents.

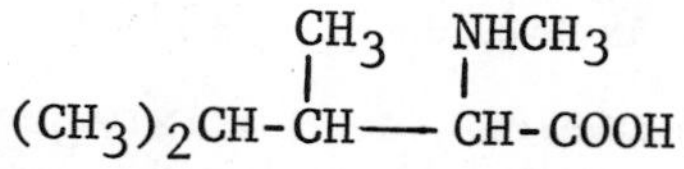

$C_8H_{17}NO_2$ Ref. 452

***N*-Benzyloxycarbonyl-*erythro*-β-methyl-DL-leucine (XV).** By the usual method of Schotten-Baumann, a pure product with a melting point of 62—64°C was obtained from VII in an 84% yield. The recrystallization was performed from benzene-petroleum ether. Found: C, 64.37; H, 7.42; N, 5.18%. Calcd for $C_{15}H_{21}O_4N$: C, 64.49; H, 7.58; N, 5.01%.

***N*-Benzyloxycarbonyl-*erythro*-β-methyl-L-leucine·(−)-ephedrine (XVIL) and *N*-Benzyloxycarbonyl-*erythro*-β-methyl-D-leucine·(+)-ephedrine (XVID).** To a solution of 5.58 g (0.02 mol) of XV in 6.0 m*l* of ether, there was added a solution of 1.91 g (0.011 mol) of (−)-ephedrine hemihydrate in 4.0 m*l* of ether. The reaction mixture was then allowed to stand overnight in a refrigerator to form a gelatinous material. The product was collected by filtration after it had been crushed to pieces with a glass rod and a small amount of ether had been added. A somewhat crude (−)-salt (1.6 g) (mp 120—124°C) was thus obtained. The ethereal mother liquor was washed wih 2N hydrochloric acid and then with water, and dried over anhydrous sodium sulfate. The ethereal solution was evaporated to dryness under reduced pressure; then the residue thus obtained was redissolved in 6.0 m*l* of ether. To the solution, there was added a solution of 1.91 g of (+)-ephedrine hemihydrate in 4.0 m*l* of ether. Then (+)-salt (1.4 g) (mp 124—125°C) was obtained by a way similar to that described above. The same treatment was possible again reciprocally with (−)- and (+)-ephedrine hemihydrate. Optically-active (−)-salt consisting of (−)-ephedrine was combined and recrystallized from ethyl acetate to give 2.0 g (45%) of a pure material with a melting point of 124—125°C; $[\alpha]^{25}_D$=−17.6° (*c* 0.74 in ethyl alcohol). Found: C, 67.50; H, 8.09; N, 6.28%. Calcd for $C_{25}H_{36}O_5N_2$: C, 67.54; H, 8.16; N, 6.30%.

In the same way, optically-pure (+)-salt with a melting point of 124—125°C was obtained in a 49.5% (2.2 g) yield; $[\alpha]^{25}_D$=+17.7° (*c* 0.95 in ethyl alcohol). Found: C, 67.60; H, 8.14; N, 6.32%. The mixed-melting-point of minus and plus salt was depressed to 101—104°C.

***erythro*-β-Methyl-L-leucine (XIIL) and the D-Isomer (XIID).** To a suspension of 1.9 g (0.0034 mol) of (−)-salt in 10 m*l* of ethyl acetate, there were added 3.4 m*l* of 2N hydrochloric acid; then the mixture was shaken in a separatory funnel. The organic layer was treated as usual to give 1.3 g of an oily residue. The oil was dissolved in 3.4 g of 30% hydrogen bromide in glacial acetic acid, after which the mixture was allowed to stand for 1 hr at room temperature and then for 24 more hour in a refrigerator after the addition of 100 m*l* of ether. The deposited crystals were collected by filtration on a glass-fritted funnel and washed with ether. Recrystallization from ethyl alcohol and ether afforded a pure material with a melting point of 195—196°C in a 69% (0.66 g) yield. Similarly, from 2.18 g of (+)-salt D-isomer hydrobromide was obtained in a 73% (0.8 g) yield; mp 203—205°C. A solution of 0.6g of hydrobromide of the L-isomer in ethyl alcohol was neutralized with triethylamine. The crystals thus obtained were collected and recrystallized from water and dioxane to give 0.28 g (73% from hydrobromide) of the amino acid XVIIL; $[\alpha]^{24}_D$=+38.9° (*c* 0.485, in 5N hydrochloric acid), $[\alpha]^{25}_D$=+29.7° (*c* 0.37, in water). Found: C, 57.92; H, 10.38; N, 9.70%. From 0.65 g of the hydrobromide of the D-isomer, free amino acid XVIID was obtained in a 67% (0.3 g) yield; $[\alpha]^{25}_D$= −39.4° (*c* 0.38, in 5N hydrochloric acid). Found: C, 57.82; H, 10.34; N, 9.62%.

***erythro*-*N*,β-Dimethyl-L-leucine (XVIIIL) and the D-Isomer (XVIIID).** The methylation of the amino groups of XVIIL and XVIID was accomplished, step by step, by *N*-benzylation, *N*-methylation, and subsequent catalytic hydrogenolysis, in a way similar to that used for the corresponding racemic *threo*-isomer. *N*-Benzyl-L-isomer: mp 218—219°C, 74% yield. Found: C, 71.62; H, 8.95; N, 5.82%. *N*-Benzyl-D-isomer: mp 218—219°C, 43% yield. Found: C, 71.70; H, 8.93; N, 6.03%. The mixed-melting-point of the two isomers was depressed to 208—210°C. *N*-Benzyl-*N*-methyl-L-isomer: mp 102—105°C, 68% yield. Found: C, 72.41; H, 9.08; N, 5.60%. *N*-Benzyl-*N*-methyl-D-isomer: mp 102—105°C, 74% yield. Found: C, 72.20; H, 9.40; N, 5.55%. The mixed-melting-point of these two isomers was depressed to 76—80°C. *erythro*-*N*,β-Dimethyl-L-leucine: mp>270°C, 51% yield; $[\alpha]^{25}_D$=+38.0° (*c* 0.50, in 5N hydrochloric acid). Found: C, 60.52; H, 10.48; N, 8.91%. *erythro*-*N*,β-Dimethyl-D-leucine: mp >270°C, 52% yield; $[\alpha]^{25}_D$=−38.3° (*c* 0.60, in 5N hydrochloric acid). Found: C, 60.18; H, 10.51; N, 8.76%.

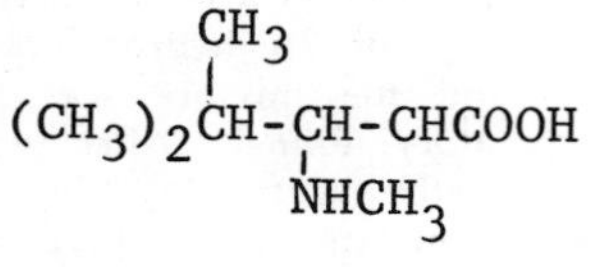

$C_8H_{17}NO_2$ Ref. 453

***N*-Benzyloxycarbonyl-*threo*-*N*,β-dimethyl-DL-leucine (XI).** To a solution of 8.8 g (0.55 mol) of X in 14 m*l* (0.055 mol) of 4N sodium hydroxide, there were added 16.7 m*l* (0.066 mol) of 4N sodium hydroxide and 10 g (0.059 mol) of benzyloxycarbonyl chloride according to the Schotten-Baumann procedure; the subsequent reaction proceeded well with concurrent stirring and cooling at 0—5°C in an ice bath. The reaction was continued for 1 hr at 0—5°C and then for an additional hour at room temperature. The reaction mixture was washed with 15 m*l* of ether, and then the aqueous layer was neutralized with 6N hydrochloric acid to pH 4. The oily product which was thus deposited was extracted twice with 30-m*l* portions of ethyl acetate, and the combined extract was washed with dilute hydrochloric acid and water. The ethyl acetate layer was dried over anhydrous sodium sulfate. The filtrate was evaporated to dryness to give 13.7 g of an oily residue. The oil was purified by redissolving it in an aqueous solution of sodium hydrogen carbonate and by washing it with ether, followed by a treatment similar to that described above. The oily product thus obtained was triturated with petroleum ether to afford a crystalline solid, which was then recrystallized from ether-petroleum ether to give 12 g (75%) of XI with a melting point of 79°C. Found: C, 65.78; H, 7.70; N, 4.94%. Calcd for $C_{16}H_{23}O_4N$: C, 65.51; H, 7.90; N, 4.78%.

***N*-Benzyloxycarbonyl-*threo-N,β*-dimethyl-L-leucine·(—)-ephedrine Salt (XIIL) and *N*-Benzyloxycarbonyl-*threo-N,β*-dimethyl-D-leucine (XIIID).** XI (8.3 g, 0.028 mol) and (−)-ephedrine hemihydrate (2.7 g, 0.0155 mol) were dissolved in 28 m*l* of warming ethyl acetate. The mixture was then allowed to stand for 24 hr at room temperature. The deposited crystals were filtered and washed with ethyl acetate to give 6.1 g (94%) of XIIL; mp 157°C. Recrystallization from an ethyl alcohol-ethyl acetate mixture (1 : 1) afforded 5.7 g of the pure material with a melting point of 160—161°C, $[\alpha]^{25}_{D}=-75.0°$ (*c* 0.40, in ethyl alcohol). Found: C, 68.26; H, 8.24; N, 6.27%. Calcd for $C_{26}H_{38}O_5N_2$: C, 68.09; H, 8.35; N, 6.11%.

The mother liquor obtained after the removal of XIIL was washed twice with a small amount of 2N hydrochloric acid and water. The organic layer was dried over anhydrous sodium sulfate. The filtrate was concentrated to dryness to give a crystalline residue in a 94% (3.9 g) yield; mp 90—91°C. Recrystallization from benzene-petroleum ether afforded 3.4 g of almost optically-pure XIIID; mp 94—95°C, $[\alpha]^{24}_{D}=+72°$ (*c* 1.00, in ethyl alcohol).

***N*-Benzyloxycarbonyl-*threo-N,β*-dimethyl-D-leucine·(+)-ephedrine Salt (XIID).** To a solution of 0.70 g (0.0024 mol) of XIIID ($[\alpha]^{24}_{D}=+72°$) in 1.5 m*l* of ethyl acetate, there was added a solution of 0.42 g (0.0024 mol) of (+)-ephedrine hemihydrate in 1.0 m*l* of ethyl acetate. The mixture was set aside for 24 hr at room temperature, and then the crystals deposited were filtered, washed with ethyl acetate, and dried to afford XIID (mp 155—157°C) in an 89% (0.975 g) yield. Recrystallization from an ethyl alcohol-ethyl acetate mixture (1 : 1) gave crystals with a melting point of 159—160°C and $[\alpha]^{25}_{D}=+73.5°$ (*c* 0.30, in ethyl alcohol). Found: C, 67.97; H, 8.31; N, 6.16%.

***N*-Benzyloxycarbonyl-*threo-N,β*-dimethyl-L-leucine (XIIIL).** To a suspension of 4.8 g (0.0105 mol) of the XIIL salt in 15 m*l* of ethyl acetate, there were added 7.8 m*l* (0.0156 mol) of 2N hydrochloric acid (7.8 m*l*, 0.0156 mol), and the mixture was shaken effectively in a separatory funnel. After washing further with water, the ethyl acetate layer was dried over anhydrous sodium sulfate. The ethyl acetate was evaporated to dryness to afford 3 g (98%) of crystals; mp 94—96°C. Recrystallization from benzene-petroleum ether gave 2.8 g of XIIIL with a melting point of 98—99°C, $[\alpha]^{18}_{D}=-75.9°$ (*c* 1.10, in ethyl alcohol). Found: C, 65.25; H, 7.99; N, 4.89%. Calcd for $C_{16}H_{23}O_4N$: C, 65.51; H, 7.90; N, 4.78%.

***N*-Benzyloxycarbonyl-*threo-N,β*-dimethyl-D-leucine (XIIID).** Following the method used for the preparation of XIIIL, from 3.2 g (0.007 mol) of XIID, 1.9 g (93%) of XIIID with a melting point of 98—99°C was obtained. $[\alpha]^{25}_{D}=+75.3°$ (*c* 0.7, in ethyl alcohol). Found: C, 65.72; H, 8.04; N, 4.79%.

***threo-N,β*-Dimethyl-L-leucine (XIVL).** A) To 1.0 g (0.0034 mol) of XIIIL, 3.5 g of 30% hydrogen bromide in glacial acetic acid was added. The reaction mixture was then set aside in a tight-stoppered vessel for 1 hr at room temperature. There after 30 m*l* of ether were added to complete the precipitation of the product, which was collected on a glass-fritted funnel and washedwith ether. The yield of the hydrobromide was 0.75 g (92%), and its melting point was 209—210°C. Free amino acid XIVL was obtained by treatment with triethylamine in a solution of 4 m*l* of ethyl alcohol, followed by recrystallization from a water-ethyl alcohol mixture (1 : 1). The yield was 0.43 g (86% from hydrobromide), and it had a melting point of over 270°C; $[\alpha]^{25}_{D}=+38.3°$ (*c* 0.90, in 5N hydrochloric acid), $[\alpha]^{23}_{D}=+31.5°$ (*c* 0.50, in water). Found: C, 60.39; H, 10.57; N, 8.93%. Calcd for $C_8H_{17}O_2N$: C, 60.34; H, 10.74; N, 8.80%.

B) A solution of 1.0 g (0.0034 mol) of XIIIL in 10 m*l* of glacial acetic acid was hydrogenated using 200 mg of palladium black as a catalyst under mechanical shaking for 5 hr. After the catalyst had then been removed by filtration, the filtrate was evaporated to dryness to give a crystalline residue and dried in a vacuum desiccator over concentrated sulfuric acid and solid sodium hydroxide. The residue was recrystallized from water and dioxane to give 0.47 g (87%) of XIVL with a melting point of over 270°C; $[\alpha]^{25}_{D}=+37.3°$ (*c* 0.70, in 5N hydrochloric acid), $[\alpha]^{23}_{D}=+29.1°$ (*c* 0.90, in water).

***threo-N,β*-Dimethyl-D-leucine (XIVD).** A way similar to the method A) used in the preparation of XIVL was used. From the reaction of 1.0 g (0.0034 mol) of XIIID and 3.5 g of 30% hydrogen bromide in glacial acetic acid, followed by neutralization with amine, XIVD with a melting point of over 270°C was obtained in almost the same yield. $[\alpha]^{25}_{D}=-39.3°$ (*c* 0.60, in 5N hydrochloric acid), $[\alpha]^{25}_{D}=-31.8°$ (*c* 0.40, in water). Found: C, 60.56; H, 10.63; N, 9.01%.

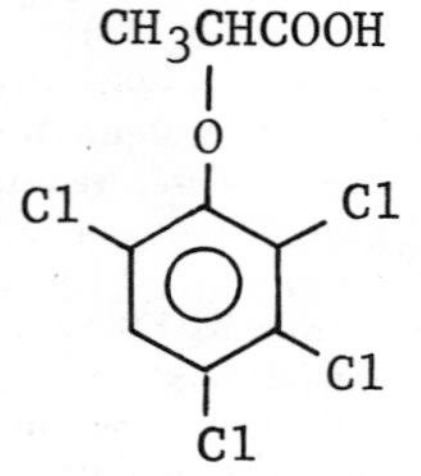

$C_9H_6Cl_4O_3$ Ref. 454

Resolution of the racemic acid

Preliminary experiments were carried out on a small scale. In dilute (50 %) *acetone*, cinchonine, cinchonidine and quinine gave laevoratatory acid of moderate or low activity, with the first-named base giving the best resolution. Brucine yielded dextrorotatory acid of low activity, and with strychnine no crystalline salt was obtained. Experiments with dilute *ethanol* as solvent met with no success. In *benzene* solution, cinchonidine yielded a crystalline salt containing dextrorotatory acid of rather high activity. Cinchonidine in benzene and cinchonine in dilute acetone were thus selected for the final resolution.

30.4 g (0.1 mole) of racemic acid and 29.4 g (0.1 mole) of cinchonidine were dissolved together in 700 ml of hot benzene. After four days, the salt was filtered off and washed with a small amount of cold benzene. It was recrystallised to maximum activity of the acid; after each recrystallisation a small sample of the salt was decomposed, the acid liberated and the rotatory power determined in acetone solution:

Crystallisation ...	1	2	3	4	5	6
ml benzene	750	640	300	300	300	300
g salt obtained	27.8	13.6	4.8	3.2	2.5	2.2
$[\alpha]_D$ of the acid	+ 18.5°	+ 25°	+ 26°	+ 26.5°	+ 26°	+ 26°

Further crops were obtained by systematic fractionation of the salt contained in the mother liquors; the total yield of pure (+)-salt was 8.8 g.

17.3 g (0.057 mole) of acid obtained from the first mother liquors and showing strong laevorotation was dissolved with 16.7 g (0.057 mole) of cinchonine in dilute acetone. The salt was filtered off after four days and recrystallised to constant activity of the acid:

Crystallisation ...	1	2	3
ml acetone	700	400	550
ml water	700	400	350
g salt obtained	20.6	15.1	4.5
$[\alpha]_D$ of the acid	− 25°	− 25.5°	− 25.5°

Further crops were obtained by working up the mother liquors; the total yield of pure (−)-salt amounted to 8.9 g.

(+)-2,3,4,6-Tetrachlorophenoxy-propionic acid

The cinchonidine salt was decomposed with dilute sulphuric acid and the organic acid was taken up in ether. 8.8 g salt yielded 4.0 g of crude acid; after one recrystallisation from 250 ml of ligroin, 3.1 g remained. The acid forms long, hairlike needles, which represent an unstable modification, melting at about 116°. The stable form, which crystallises from the melt, has the m.p. 123.5–124.5°.

0.1161 g acid: 7.86 ml 0.04855 *N* NaOH. 39.59 mg acid after combustion according to Grote-Krekeler: 10.24 ml 0.05069 *N* NaOH.

$C_9H_8O_3Cl_4$ (304.0) calc. Eq. weight 304.0 Cl 46.65
found ,, ,, 304.3 ,, 46.56

The data for the rotatory power in various solvents are given in Table 2.

Table 2. Rotatory power of (+)-2,3,4,6-tetrachlorophenoxy-propionic acid.
g = g acid in 10.00 ml solution.

	g	α_D^{25} (2 dm)	$[\alpha]_D^{25}$	$[M]_D^{25}$
Acetone	0.1478	+0.78°	+26.4°	+80.2°
Ethanol (abs.)	0.1449	+0.46°	+15.9°	+48°
Benzene	0.1443	+0.565°	+19.6°	+59.5°
Chloroform	0.1282	+0.12°	+4.7°	+14°
Water (neutr.)	0.1518	−0.065°	−2.1°	−6.5°

(−)-2,3,4,6-Tetrachlorophenoxy-propionic acid

The acid was isolated from the cinchonine salt as described above. 8.9 g salt yielded 4.1 g of crude acid; after recrystallisation from 250 ml ligroin, 3.1 g remained. M.p. (of the stable form) 123.5–124.5°

0.1069 g acid: 7.23 ml 0.04855 *N* NaOH. 26.36 mg acid after combustion according to Grote-Krekeler: 6.80 ml 0.05069 *N* NaOH.

$C_9H_8O_3Cl_4$ (104.0) calc. Eq. weight 304.0 Cl 46.65
found ,, ,, 303.9 ,, 46.44

0.1602 g acid dissolved in *acetone* to 10.00 ml: α_D^{25} = −0.85° (2 dm). $[\alpha]_D^{25}$ = −26.5°; $[M]_D^{25}$ −80.5°.

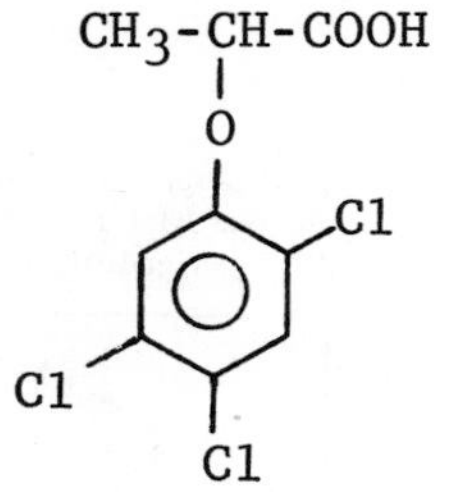

$C_9H_7Cl_3O_3$

Ref. 455

Resolution

This was achieved in two stages by forming (1) the salt of the (−)-acid with (−)-strychnine, and (2) the salt of the (+)-acid with (+)-yohimbine. In each case one equivalent of the racemic acid was treated with half an equivalent of the optically active base and half an equivalent of sodium hydroxide.

Preparation of salt of (−)-acid with (−)-strychnine

16·7 g. of finely powdered strychnine (m.p. 273–275·5° C.) was suspended in 2 l. boiling 20% ethanol. To the vigorously stirred suspension, 27 g. α-(2:4:5-trichlorophenoxy)propionic acid in 50 ml. N-NaOH and 50 ml. ethanol was added. The mixture was brought to boiling with continued stirring. The almost clear solution was filtered and, on cooling, the salt separated. Yield 26·9 g. It was recrystallized five times from 20% ethanol. Yield 7·8 g.

Preparation of salt of (+)-acid with (+)-yohimbine

19·5 g. of yohimbine hydrochloride was dissolved in 1 l. boiling water. The base was precipitated with excess ammonium hydroxide, filtered and washed with hot water. The moist base was suspended in 2 l. 25% ethanol, and a mixture of 27 g. α-(2:4:5-trichlorophenoxy)propionic acid in 50 ml. N-NaOH and 50 ml. ethanol added to the vigorously stirred suspension. The solution cleared completely, and in a few minutes the salt began to crystallize as a mass of very fine needles. Yield 28 g. Recrystallized three times from acetone. Yield 9·4 g.

Regeneration of acids

Each salt was treated with a slight excess of sodium hydroxide and the base removed by filtration. The acids were recovered by addition to an excess of ice-cold dilute hydrochloric acid and recrystallized from 30% ethanol.

Yield of (−)-acid 3 g. (from 7·8 g. strychnine salt).

Yield of (+)-acid 3·3 g. (from 9·4 g. yohimbine salt).

Specific rotations were as follows:

(−)-Acid (m.p. 143–144·5° C.). 550·1 mg. in 25 ml. ethanol; 2 dm.; α = −2·19°; $[\alpha]_D^{20.5}$ = −49·7°.

(+)-Acid (m.p. 143–144·5° C.). 589·9 mg. in 25 ml. ethanol; 2 dm.; α = +2·31°; $[\alpha]_D^{22}$ = +48·9°.

$C_9H_7Cl_3O_3$

Ref. 455a

Preliminary experiments were always carried out using *brucine, cinchonine, cinchonidine, quinine, quinidine, morphine, strychnine*, and (+)-*α-phenylethylamine*. In some cases *yohimbine* was tried. In most cases the salts were obtained as non-crystallizing oils while the crystalline salts often gave little or no resolution. These facts are illustrated in Table 3 where the results of some resolution experiments, carried out with dilute alcohol as solvent, have been summarized.

The strychnine salt was recrystallized 6 times from 25% alcohol.

After each recrystallization, $[\alpha]_D$ of the acid, liberated from a small sample and from the mother liquor, was measured in alcohol. The results are found in Table 4.

From each of the fractions 3–5 a somewhat larger sample than usual was taken away and the acid liberated. After recrystallization from 30% alcohol the rotatory power and biological activity were measured

TABLE 3. Resolution experiments

(+), +, + + and (−), −, − −, respectively, means slight, moderate, large excess of (+)- and (−)-form, respectively, in the first crystal fraction from dilute alcohol. 0 represents no resolution. For abbreviations see Table 20.

Acid	Brucine	Cinchonine	Cinchonidine	Quinine	Quinidine	Morphine	Strychnine	(+)-Phenethyl-amine	Configuration of (+)-acid
2.4 Cl_2POP	(+)	oil	(+)	+ +	oil	oil	0	oil	D
3.4 Cl_2POP	−	−	0	(−)	oil	oil	+ +	(+)	D
2 Me 4 ClPOP	oil	oil	oil	+	oil	oil	(−)	+	D
1 Cl 2 NOP	oil	(−)	+ +	− −	(−)	− −	oil	+	D
1 NMeP	oil	+	oil	(−)	oil	oil	oil	oil	L[1]
2 NMeP	(−)	oil	+	(−)	oil	oil	oil	−	D[1]
1 NSP	oil	oil	oil	(−)	oil	oil	oil	oil	
2 NSP	oil	(−)	oil	+	oil	oil	oil	+ +	D[1]
POB	oil	(−)	oil	(−)	oil	+	− −	oil	D
1 NOB	oil	+ +	+ +	oil	oil	oil	+	oil	L
2 NOB	oil	oil	(+)	+ +	oil	oil	oil	−	D
2.4 Cl_2POB	oil	oil	+	−	oil	oil	+ +	+ +	D
POC	oil	oil	(+)	(−)	oil	oil	oil	+	D[2]
2 NOC	oil	oil	+	(−)	oil	0	oil	(+)	D[1, 2]

[1] Based upon biological effects.
[2] „ „ optical activity data.

TABLE 4. Recrystallization of the strychnine salt of α-(2.4.5-trichlorophenoxy)-propionic acid

Cryst. no.		1	2	3	4	5	6
$[\alpha]_D$	acid from salt	−13.8°	−32.5°	−47.0°	−48.1°	−48.7°	−48.8°
	„ „ mother liquor	+39.3°	+30.0°	+24.8°	−33.1°	−46.6°	−48.7°

** − 537319

(see Table 19). The acid from fraction 6 was liberated and after recrystallization from 30 % alcohol it melted at 143.5–144.7°.

114.02 mg (−)-acid: 8.705 ml 0.04839 *N* NaOH
$C_9H_7O_3Cl_3$ calc. equiv. wt. 269.5
found „ „ 270.7
0.1901 g acid from fract. 3 dissolved in *abs. alcohol* to 20.06 ml:
$2\alpha_D^{25} = -0.912°$. $[\alpha]_D^{25} = -48.1°$.
0.1860 g acid from fract. 4 dissolved in *abs. alcohol* to 20.06 ml:
$2\alpha_D^{25} = -0.914°$. $[\alpha]_D^{25} = -49.3°$.
0.1815 g acid from fract. 5 dissolved in *abs. alcohol* to 20.06 ml:
$2\alpha_D^{25} = -0.890°$. $[\alpha]_D^{25} = -49.2°$.
0.1951 g acid from fract. 6 dissolved in *abs. alcohol* to 20.06 ml:
$2\alpha_D^{25} = -0.962°$. $[\alpha]_D^{25} = -49.5°$; $[M]_D^{25} = -133.3°$.

From the mother liquors 1–3 (*vide supra*) an acid with $[\alpha]_D = +38°$ (in alcohol) was recovered. The yohimbine salt of this product was recrystallized 4 times from acetone. The acid liberated from the pure salt was recrystallized from 30 % alcohol. M. p. 144–145.0°.

104.00 mg (+)-acid: 7.945 ml 0.04839 *N* NaOH
$C_9H_7O_3Cl_3$ calc. equiv. wt. 269.5
found ,, ,, 270.5
0.1847 g (+)-acid dissolved in *abs. alcohol* to 20.06 ml:
$2\alpha_D^{25} = +0.909°$. $[\alpha]_D^{25} = +49.4°$; $[M]_D^{25} = +133.0°$.
0.2022 g (+)-acid dissolved in *acetone* to 20.06 ml:
$2\alpha_D^{25} = +1.194°$. $[\alpha]_D^{25} = +59.2°$; $[M]_D^{25} = +159.6°$.
0.2076 g (+)-acid dissolved in *chloroform* to 20.06 ml:
$2\alpha_D^{25} = +0.714°$. $[\alpha]_D^{25} = +34.5°$; $[M]_D^{25} = +93.0°$.
0.1879 g (+)-acid dissolved in *benzene* to 20.06 ml:
$2\alpha_D^{25} = +0.342°$. $[\alpha]_D^{25} = +18.3°$; $[M]_D^{25} = +49.2°$.
0.0987 g (+)-acid, neutralized with *aqueous NaOH* and made up to 10.03 ml:
$\alpha_D^{25} = +0.122°$. $[\alpha]_D^{25} = +12.4°$; $[M]_D^{25} = +33.4°$.

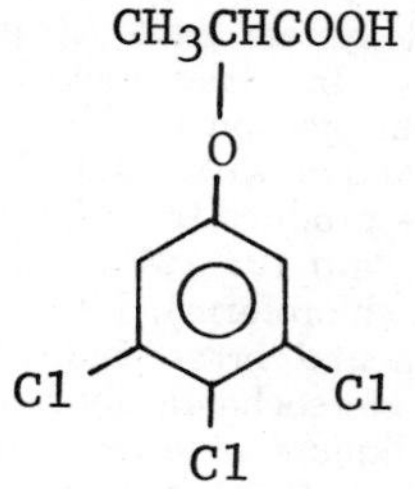

$C_9H_7Cl_3O_3$ **Ref. 456**

Preliminary experiments on resolution were carried out with the common alkaloids and dehydroabietylamine [4] in dilute ethanol. The *cinchonidine* salt gave a strongly laevorotatory acid, the *cinchonine* salt a (−)-acid of lower activity. The acid obtained from the *brucine* salt was racemic. No crystallising salts were obtained from *strychnine*, *quinine* and *quinidine*. *Dehydroabietylamine* gave a dextrorotatory acid of moderate activity. Cinchonidine and dehydroabietylamine were thus selected for the final resolution.

(−)-*3,4,5-Trichlorophenoxy-propionic acid.* 18.9 g (0.07 mole) of racemic acid and 20.6 g (0.07 mole) of cinchonidine were dissolved in 210 ml of hot ethanol + 70 ml of water. After 48 h at room temperature, 11.2 g salt had separated. It was recrystallised from dilute ethanol. After each recrystallisation, a small sample of the salt was decomposed, the acid isolated and its rotatory power determined in ethanol solution.

Crystallisation	1	2	3	4
ml ethanol	210	42	25	12
ml water	70	14	8.3	4
g salt obtained	11.2	9.0	8.1	7.2
$[\alpha]_D^{25}$ of the acid	−35°	−40°	−39.5°	−40.5°

The mother liquor from the first crystallisation was placed in a refrigerator for 24 h. A second crop of salt (13.0 g) was obtained. This salt was recrystallised from the successive mother liquors after the first series of recrystallisations:

Recrystallisation	1	2	3
g salt obtained	5.1	4.1	3.7
$[\alpha]_D^{25}$ of the acid	−28°	−39.5°	−40°

The salt from the first series of recrystallisations was decomposed with dilute sulphuric acid and the trichloro-phenoxy acid was isolated by extraction with ether. The yield of crude acid was 3.4 g. After recrystallisation from 80 ml of ligroin, there remained 3.3 g acid forming glistening needles with m.p. 128–128.5°.

$C_9H_7O_3Cl_3$ (269.5) calc. equiv.wt 269.5 Cl 39.47
found. equiv.wt 268.9 Cl 39.59, 39.43

Table 1. g = g acid dissolved to 10.00 ml.

Solvent	g	α_D^{25} (2 dm)	$[\alpha]_D^{25}$	$[M]_D^{25}$
Acetone	0.2032	−2.23°	−54.9°	−147.9°
Ethanol (abs.)	0.2053	−1.64°	−39.9°	−107.6°
Chloroform	0.1618	−0.835°	−25.8°	−69.5°
Water (neutr.)	0.2090	−0.715°	−17.1°	−46.1°
Benzene	0.2042	−0.57°	−14.0°	−37.6°

The rotatory power in various solvents is given in Table 1.

The salt from the second series of recrystallisations yielded 1.6 g of recrystallised acid with the same m.p.

0.2109 g acid dissolved in abs. ethanol to 10.00 ml:
$[\alpha]_D^{25} = -1.685°$. $[\alpha]_D^{25} = -39.9°$; $[M]_D^{25} = -107.7°$.

From the salt contained in the mother liquor after the second crop of the cinchonidine salt, 15.4 g acid was isolated. Its rotatory power was $[\alpha]_D^{25} = +17°$ (ethanol solution). This acid (0.057 mole) was dissolved with 19.7 g (0.057 mole) dehydroabietylamine acetate and 5.7 g (0.057 mole) potassium hydrogen carbonate in 350 ml hot methanol + 80 ml water. The yield of salt obtained on standing was 10.6 g. The mother liquor was diluted with 250 ml water and after 24 h in a refrigerator a second crop of salt (11.5 g) had separated.

The first crop of salt was recrystallised to maximum activity of the acid; the course of the resolution was followed as described above.

Crystallisation	1	2	3	4	5
ml ethanol	350	90	45	42	20
ml water	80	20	10	10	4.5
g salt obtained	10.6	7.3	6.1	4.9	4.2
$[\alpha]_D^{25}$ of the acid	+34.5°	—	+39.5°	+40°	+39.5°

The second crop of salt contained an acid with $[\alpha]_D^{25} = +10°$. This salt was recrystallised from the successive mother liquors after the recrystallisations of the first crop:

Recrystallisation	1	2	3	4	5
g salt obtained	7.8	7.4	6.5	6.0	5.1
$[\alpha]_D^{25}$ of the acid	+29.5°	+34.5°	+37°	+38.5°	+39.5°

The first crop of the salt yielded 2.0 g of crude acid. Recrystallisation from 40 ml of ligroin gave 1.9 g of glistening needles with m.p. 128–128.5°.

$C_9H_7O_3Cl_3$ (269.5) calc. equiv.wt 265.5 Cl 39.47
found equiv.wt 269.8 Cl 39.46, 39.42

The rotatory power was determined at two different concentrations:

0.2285 g acid dissolved in abs. ethanol to 10.00 ml: $\alpha_D^{25} = +1.815°$ (2 dm). $[\alpha]_D^{25} = +39.7°$; $[M]_D^{25} = +107.0°$. – 0.1162 g acid dissolved in abs. ethanol to 10.00 ml: $\alpha_D^{25} = +0.92°$ (2 dm). $[\alpha]_D^{25} = +39.6°$; $[M]_D^{25} = +106.7°$.

The second crop of dehydroabietylamine salt gave 2.1 g of recrystallised acid, having m.p. 127–127.5° and $[\alpha]_D^{25} = +39.6°$ (ethanol solution).

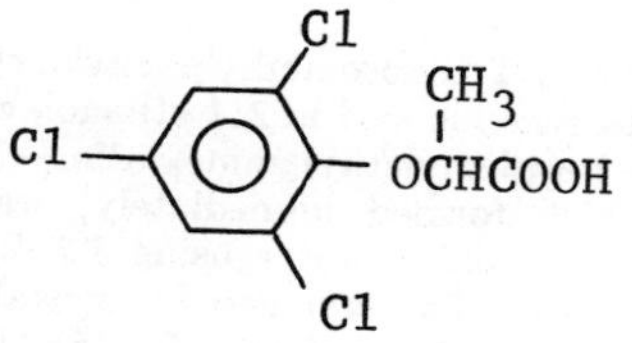

$C_9H_7Cl_3O_3$ **Ref. 457**

(+)-α-*(2.4.6-Trichlorophenoxy)propionic acid.*—To 40.4 g (0.15 mole) racemic acid dissolved in hot ethanol (750 ml) were added 50.2 g (0.15 mole) strychnine. After cooling to room temperature the solution was seeded with crystals from a fore run and was allowed to stand over night in a refrigerator. The crystalline salt was collected, the acid was isolated from a small sample and the rotatory power determined in abs. alcohol. The salt was recrystallized according to Table 2.

Table 2. Recrystallization of the strychnine salt of α-(2.4.6-trichlorophenoxy)-propionic acid.

Cryst. no.	Solvent (ml)		Yield of salt %	[α]D of the acid
	Water	Ethanol		
1	1200	750	69	+ 6.8°
2	800	400	55	+ 11.1°
3	600	300	45	+ 16.3°
4	500	250	40	+ 15.9°
5	450	225	33	+ 17.6°
6	400	200	31.5	+ 17.8°
7	400	200	28	—
8	400	200	25	+ 17.6°

The crude (+)-acid (6.3 g) obtained from the last fraction was recrystallized from petroleum yielding 5.7 g pure (+)-acid, m.p. 125–126°.

94.01 mg acid: 6.62 ml 0.0522_4 *N* NaOH
$C_9H_7O_3Cl_3$ calc. equiv. wt. 269.5
found ,, ,, 271.8

0.2371 g dissolved in abs. *ethanol* to 20.06 ml: $2\alpha_D^{25} = +0.428°$. $[\alpha]_D^{25} = +18.1°$; $[M]_D^{25} = +48.8°$.
0.2168 g dissolved in *acetone* to 20.06 ml: $2\alpha_D^{25} = +0.649°$. $[\alpha]_D^{25} = +30.0°$; $[M]_D^{25} = +80.9°$.
0.1932 g dissolved in *benzene* to 20.06 ml: $2\alpha_D^{25} = +0.264°$. $[\alpha]_D^{25} = +13.7°$; $[M]_D^{25} = +36.9°$.
0.2069 g dissolved in *chloroform* to 20.06 ml: $2\alpha_D^{25} = +0.093°$. $[\alpha]_D^{25} = +4.5°$; $[M]_D^{25} = +12.1°$.
0.2131 g neutralized with *aqueous NaOH* and made up to 20.06 ml: $2\alpha_D^{25} = +0.016°$. $[\alpha]_D^{25} = +0.8°$; $[M]_D^{25} = +2.0°$.

(−)-α-(2.4.6-Trichlorophenoxy)propionic acid.—The acid was regenerated from the mother liquours from the first three recrystallizations of the strychnine salt. This acid (21 g, $[\alpha]_D = -10°$) and an equivalent quantity of (−)-β-phenyl*iso*propylamine (10.6 g) were dissolved in 500 ml hot acetone. Water (40 ml) was added and the solution allowed to stand over night in a refrigerator. The salt was collected and recrystallized several times from successively smaller volumes of the same solvent. The mother liquors contained the more active products, which were again recrystallized from dilute acetone. By repeating the procedure a salt was finally obtained, which on decomposition yielded optically pure (−)-acid. After recrystallization from petroleum it melted at 125–126°.

93.45 mg acid: 6.63_5 ml 0.0522_4 *N* NaOH
$C_9H_7O_3Cl$ calc. equiv. wt. 269.5
found ,, ,, 269.6

0.1925 g dissolved in abs. *ethanol* to 20.06 ml: $2\alpha_D^{25} = -0.340°$. $[\alpha]_D^{25} = -17.7°$; $[M]_D^{25} = -47.7°$.

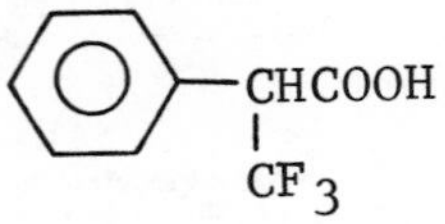

$C_9H_7F_3O_2$ Ref. 458

Resolution of α-Trifluoromethylphenylacetic Acid.—A solution of 100 g of the racemic acid in 2:1 ethanol–water was mixed with 59.4 g of (−)-α-phenylethylamine, $\alpha^{12}D$ −37.43 (l = 1, neat). The salt, which formed immediately, was crystallized three times from 2:1 ethanol–water using 1.5–2.0 ml of solvent for each gram of salt. This was twice recrystallized without changing its properties. A fraction (20.2 g), $[\alpha]^{24}D$ −27.4 ± 0.5 (ethanol, c 3.86), mp 164–165°, was decomposed in 2:1 water-concentrated hydrochloric acid to give, after isolation and recrystallization from hexane, 11.9 g of material, $[\alpha]^{24}D$ −71.4 ± 0.5 ($CHCl_3$, c 2.24), mp 93–94°. In this manner 22 g of material was obtained with specific rotation greater than 70°. The remaining salt fractions were decomposed in the above manner, and the isolated acid was treated with (+)-α-phenylethylamine, $\alpha^{24}D$ +37.56 (l = 1, neat). Recrystallization of this salt gave a fraction (40.0 g) with $[\alpha]^{21}D$ +25.8 ± 0.8 (ethanol, c 2.59). Decomposition of the salt gave 23.7 g of acid, $[\alpha]^{24}D$ +73.4 ± 0.5 ($CHCl_3$, c 4), $[\alpha]^{24}D$ +65.8 ± 1.7 (ethanol, c 1), mp 93–94°, after hexane crystallization

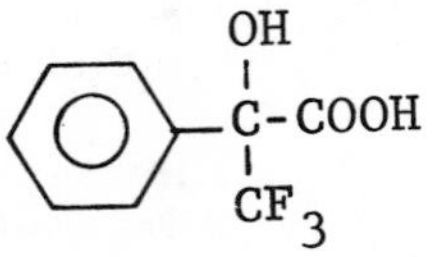

$C_9H_7F_3O_3$ Ref. 459

Resolution of α-Hydroxy-α-trifluoromethylphenylacetic Acid with (+)-α-(1-Naphthyl)ethylamine.—A mixture of 22.0 g (0.1 mol) of racemic α-hydroxy-α-trifluoromethylphenylactic acid and 17.2 g (0.1 mol) of (+)-α-(1-naphthyl)ethylamine (Aldrich Chemical Co.), $\alpha^{20}D$ 81.35° (neat, l = 1) in 80 ml of 6:1 benzene-ethanol was prepared. Heat was evolved and the salts precipitated immediately. Two recrystallizations of this material from the same solvent system gave 7.5 g of salt, $[\alpha]^{22}D$ 15.5 ±1.2° (c 1.68, ethanol). In a trial resolution, material with this rotation remained unchanged on further recrystallization. Decomposition of this fraction with dilute hydrochloric acid and recrystallization of the product from hexane–benzene gave, in three fractions, 4.09 g, mp 123–124°. All fractions had $[\alpha]^{20}D$ −22.5 ± 0.7° (c 2.70, chloroform), $[\alpha]^{20}D$ 31.1 ± 0.7° (c 2.76, water) within experimental error, indicating that the initial product was probably stereochemically homogeneous. Concentration of the mother liquors gave two additional salt fractions of 3.09 g, $[\alpha]^{23}D$ 14.3 ± 0.6° (c 2.93, ethanol), and 12.6 g, $[\alpha]^{22}D$ 0.0° (c 5.5, ethanol). Decomposition of these fractions gave, after recrystallization, 1.0 g, $[\alpha]^{20}D$ −23.2 ± 0.7° (c 2.93, chloroform), and 5.4 g, $[\alpha]^{22}D$ 8.8 ± 0.6° (c 3.62, chloroform). Decomposition of the remaining salt fractions gave, after recrystallization, 4.3 g, $[\alpha]^{19}D$ 9.3 ± 0.3° (c 3.56, chloroform).

A subsequent exploratory resolution with (−)-α-phenylethylamine in methylene chloride was also successful.

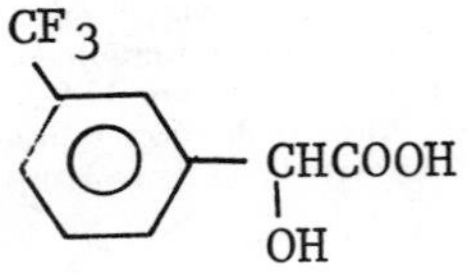

$C_9H_7F_3O_3$ Ref. 460

Resolution was accomplished by serial crystallization of the salt of the desired acid with (+)-α-methylbenzylamine. The m.p. of the resolved acid was 50-54°C. The $[\alpha]_D^{25}$ = 102°C. (c, 0.3, ethanol).

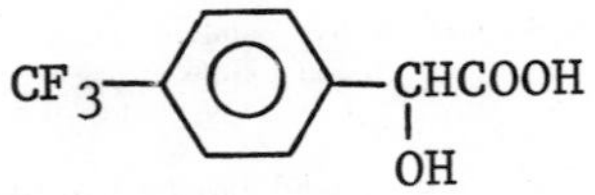

$C_9H_7F_3O_3$ Ref. 461

Resolution was accomplished by serial crystallization of the salt of the desired acid with (-)-ephedrine. The m.p. of the resolved acid was 146-150°C. The $[\alpha]_D^{25}$ = -105° (c, 0.3, ethanol).

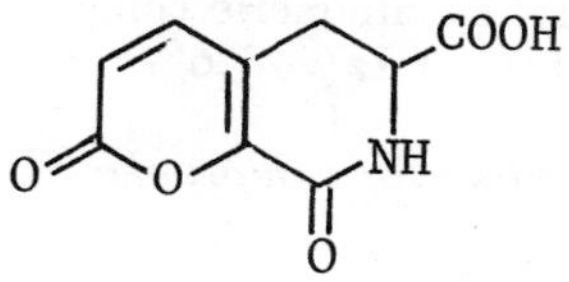

$C_9H_7O_5$

Ref. 462

Optical Resolution of DL-β-(6-Carboxy-α'-pyron-5-yl)-alanine Lactam (III B). *(−)- and (+)-β-(6-Carboxy-α'-pyron-5-yl)alanine Lactam.* A mixture of 9.32 g of brucine and 4.82 g of DL-β-(6-carboxy-α'-pyron-5-yl)alanine lactam monomethanolate was recrystallized three times from 70% methanol and once from 50% aqueous methanol. The crystals weighed 3.7 g; $[\alpha]_D^{20}$ −40° (*c* 1, 50% aqueous methanol).

The less soluble brucine salt obtained above was dissolved in water and decomposed by passing through a Dowex 50W×2 (H^+ form) column. The effluent was concentrated to dryness. Recrystallization from water gave 1.2 g of the (−)-lactam; $[\alpha]_D^{25}$ −30.0° (*c* 1, water).

Found: C, 51.27; H, 3.47; N, 6.72%. Calcd for $C_9H_7NO_5$: C, 51.68; H, 3.37; N, 6.70%.

The mother liquor of the brucine salt of (−)-lactam was evaporated to dryness and the residue was recrystallized three times from 90% aqueous methanol to yield 1.9 g of the (+)-brucine salt; $[\alpha]_D^{20}$ +12° (*c* 1, 50% aqueous methanol).

The decomposition of the (+)-salt and recrystallization of the product was carried out by the same procedure mentioned above, and 0.5 g of the (+)-lactam, $[\alpha]_D^{25}$ +28.5° (*c* 1, water), was obtained.

Found: C, 51.36; H, 3.56; N, 6.81%. Calcd for $C_9H_7NO_5$: C, 51.68; H, 3.37; N, 6.70%.

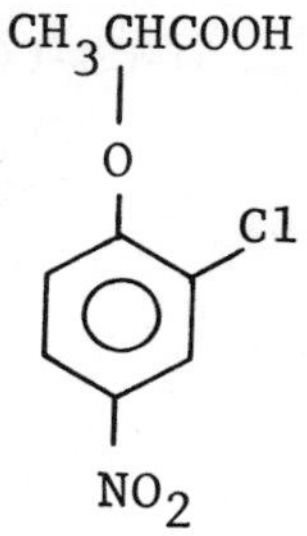

$C_9H_8ClNO_5$

Ref. 463

Preliminary experiments on resolution were carried out on a small scale using 0.001 mole of acid and base in dilute ethanol. The cinchonine salt gave a strongly laevorotatory acid. Strychnine, brucine, quinine, cinchonidine and dehydroabietyl amine gave crystallised salts but no resolution. The salts from quinidine, phenylethylamine and 2-naphthyl-ethylamine failed to crystallise.

(−)-2-Chloro-4-nitrophenoxy-propionic acid. Racemic acid (12.4 g, 0.05 mole) and cinchonine (14.9 g, 0.05 mole) were dissolved with heating in 150 ml ethanol and 100 ml water was added. The salt separated on cooling and was filtered off after some hours. It was recrystallised from dilute ethanol; a small sample from each crystallisation was decomposed and the activity of the acid determined in acetone solution.

Crystallisation	1	2	3	4
ml ethanol	150	120	80	52
ml water	100	80	55	28
g salt obtained	10.3	7.4	6.0	5.2
$[\alpha]_D^{25}$	−47.5°	−50°	−50°	−49.5°

The first mother liquor was evaporated to about half its volume. A mixture of oil and crystals separated. The liquor was decanted and the semi-solid residue recrystallised from 55 ml ethanol + 30 ml water. The crystalline salt obtained (3.7 g) was recrystallised from the successive mother liquors after the main resolution. In this way a further crop of 2.7 g pure salt was obtained.

The cinchonine salt from the main experiment (5.2 g) was decomposed with dilute sulphuric acid and the organic acid extracted with ether. Recrystallisation from dilute formic acid gave 1.8 g pure (−)-acid as small, colourless needles with m.p. 116–117°.

$C_9H_8O_5ClN$ (245.62)

Calc.	C 44.01	H 3.28	Cl 14.44	N 5.70
Found	C 43.91	H 3.27	Cl 14.47	N 5.64

Table 1

g = g acid in 10.00 ml solution

	g	α_D^{25} (1 dm)	$[\alpha]_D^{25}$	$[M]_D^{25}$
Acetone	0.1520	−0.755°	−49.7°	−122.0°
Ethanol (abs.)	0.1019	−0.295°	−29.0°	−71.1°
Chloroform	0.1128	−0.333°	−29.5°	−72.5°
Water (neutr.)	0.1090	−0.041°	−3.8°	−9.2°
Benzene	0.1150	+0.356°	+31.0°	+76.2°

The rotatory power in some solvents is given in Table 1 above.

(+)-2-Chloro-4-nitrophenoxy-propionic acid. From the decanted mother liquor after the cinchonine salt (see above), 3.3 g acid was regenerated. The rotatory power was +38.5°, corresponding to 77% of active acid. The acid was powdered, dried, shaken for a few seconds with a small volume af hot carbon tetrachloride and filtered. The procedure was repeated several times until the residue showed a melting-point near that of the racemic form. The three first extracts gave acid with practically the same melting-point, a little lower than the that of the active form. These fractions were combined and recrystallised twice from dilute formic acid giving 0.8 g (+)-acid with m.p. 115–116°, quite similar to the (−)-form.

$C_9H_8O_5ClN$ (245.62)

Calc.	C 44.01	H 3.28	Cl 14.44	N 5.70
Found	C 44.17	H 3.32	Cl 14.57	N 5.63

0.1623 g acid dissolved in *acetone* to 10.00 ml: α_D^{25} = +1.607° (2 dm). $[\alpha]_D^{25}$ = +49.5°; $[M]_D^{25}$ = +121.6°.

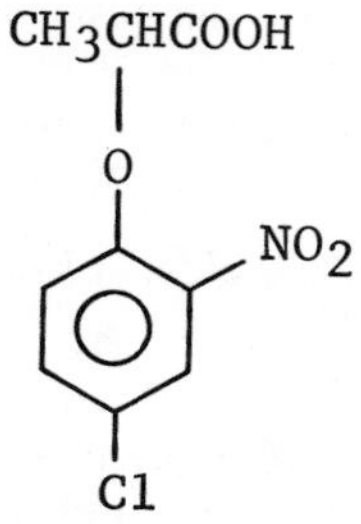

$C_9H_8ClNO_5$

Ref. 464

Preliminary experiments on resolution were carried out with small samples (0.001 mole) of the acid in dilute ethanol. Strychnine gave a strongly laevorotatory acid, while cinchonine yielded dextrorotatory acid of lower activity. The brucine salt gave (+)-acid of low activity and the acid from the cinchonidine salt was quite inactive. Quinine, quinidine and phenyl-isopropylamine gave no crystalline salts.

(−)-4-Chloro-2-nitrophenoxypropionic acid. Racemic acid (24.6 g, 0.10 mole) and strychnine (33.4 g, 0.10 mole) were dissolved in 300 ml ethanol + 100 ml water. The salt was filtered off after 24 h and recrystallised to constant activity of the acid as seen below.

Crystallisation	1	2	3	4
ml ethanol	300	100	50	50
ml water	100	70	40	40
g salt obtained	28.2	25.2	24.2	23.2
$[\alpha]_D^{25}$ of the acid	−85°	−97.5°	−97.5°	−97.5°

The salt (23.2 g) was decomposed with sodium carbonate solution and the liberated alkaloid extracted with chloroform. The acid was then liberated by excess sulphuric acid, extracted with ether and recrystallised four times from dilute formic acid (once with charcoal). It forms glistening needles with m.p. 91–92°; the yield of pure acid was 6.0 g.

$C_9H_8O_5ClN$ (245.62)

Calc.	C 44.01	H 3.28	Cl 14.44	N 5.70
Found	C 44.00	H 3.26	Cl 14.50	N 5.70

The rotatory power is given in Table 2.

Table 2

g = g acid in 10.00 ml solution

	g	α_D^{25} (1 dm)	$[\alpha]_D^{25}$	$[M]_D^{25}$
Acetone	0.1236	−1.245°	−100.7°	−247.4°
Ethanol (abs).	0.1108	−1.476°	−133.2°	−327.2°
Chloroform	0.1103	−0.732°	−66.4°	−163.0°
Water (neutr.)	0.1172	−1.953°	−166.6°	−409.3°
Benzene	0.1150	−0.655°	−52.7°	−129.4°

(+)-4-Chloro-2-nitrophenoxy-propionic acid. The mother liquor after the first crystallisation of the strychnine salt gave 11.1 g of a strongly dextrorotatory acid. It was converted to the cinchonine salt, which was recrystallised several times from dilute ethanol. After the first crystallisation, the rotatory power of the acid was +67.5° and the activity did not increase on further recrystallisation.

Various experiments on the solubility of active and racemic acid indicated that the racemic acid was more soluble in carbon tetrachloride. The acid liberated from the cinchonine salt was therefore extracted several times with small quantities of this solvent. After four extractions, there remained 2.9 g acid with $[\alpha]_D^{25} = +99°$. This product was recrystallised three times from dilute formic acid and twice from toluene + petrol ether, yielding 1.4 g of pure (+)-acid, quite similar to the antipode. M.p. 91–92°.

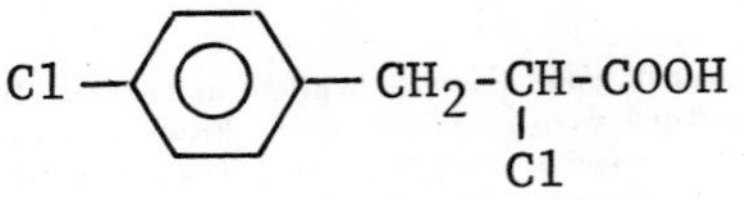

$C_9H_8Cl_2O_2$ Ref. 465

Brucine salt of 2-chloro-3-(4-chlorophenyl)-propionic acid

Water was added dropwise to a stirred solution of 4.38 g (0.02 mol) **1b** and 7.88 g (0.02 mol) brucine in 50 ml of absolute ethanol until the mixture started to become turbid. After being allowed to stand overnight, the precipitate was filtered off and dried at 50 °C in vacuum. Recrystallisation twice from ethanol/water gave 4.7 g (38.4%) salt, m.p. 107 °C.

d(+)2-Chloro-3-(4-chlorophenyl)-propionic acid (d(+) **1b**)

4.7 g (0.0077 mol) brucine salt of d(+) 2-chloro-3-(4-chlorophenyl)-propionic acid, 50 ml of ether, 50 ml of water and 2 ml of concentrated hydrochloric acid were placed in a separatory funnel and shaken until all solid had dissolved. The organic layer was separated, washed with 10 ml 2 N hydrochloric acid and twice with 10 ml of water and dried over sodium sulfate. After evaporation of solvent 1.6 g (95%) crystalline acid were obtained, m.p. 104 °C; $[\alpha]_D^{20} = +0.7°$ (10% in ethanol).

Determination of enantiomeric composition by NMR

50 mg of d(+) 2-chloro-3-(4-chlorophenyl)-propionic acid $[\alpha]_D^{20} = +0.7°$ were esterfied with diazomethane, dissolved in carbon tetrachloride and the enantiomeric composition determind from the NMR-spectrum using the optically active europium shift reagent tris-[3-(heptafluoro-*n*-propylhydroxymethylene)-d-camphorato]-europium (III) complex. The ester was found to consist of 68.6% of d(+) and 31.4% of l(−) isomer. From the $[\alpha]_D^{20}$ value of +0.7° for the original acid, the optical rotation of the pure enantiomeric 2-chloro-3-(4-chlorophenyl)-propionic acid was calculated as $[\alpha]_D^{20} = \pm 1.88°$.

d(+) Methyl-2-chloro-3-(4-chlorophenyl)-propionate (d(+) **1a**)

In a somewhat bigger scale the salt from 0.1 mol brucine and 0.1 mol 2-chloro-3-(4-chlorophenyl)-propionic acid (**1b**) was prepared as described above. After nine recrystallisations 15.2 g (24.8%) brucine salt of the d(+) acid **1b** were obtained, leading to 5.2 g (90.3%) of the d(+) ester **1a**, b.p. 97 – 100 °C/0.1 mmHg, $[\alpha]_D^{20} = +5.9987°$ (10% in ethanol), enantiomeric composition d(+) **1a** = 92.4% and l(−) **1a** = 7.6%.

l(−) Methyl-2-chloro-3-(4-chlorophenyl)-propionate (l(−) **1a**)

The mother liquors from the preparation of the brucine salt of d(+) 2-chloro-3-(4-chlorophenyl)-propionic acid were combined and the solvent evaporated at 50 °C. The residue was dissolved in absolute ethanol and water added dropwise very slowly over a period of several days to precipitate as much of the less soluble brucine salt as possible. After filtration the solution was evaporated, yielding 10.5 g (17.1%) of solid salt which led to 3.1 g (78.4%) of l(−) **1a**, b.p. 97 – 100 °C/0.1 mmHg $[\alpha]_D^{20} = -4.8876°$ (10% in ethanol), enantiomeric composition l(−) **1a** = 86.4% and d(+) **1a** = 15.6%.

From the two enantiomeric compositions and their corresponding optical rotations, the $[\alpha]_D^{20}$ was calculated for optically pure **1a**

$$\frac{5.9987}{92.4 - 7.6} \cdot 100 = 7.06,$$

$$\frac{4.8876}{84.6 - 15.4} \cdot 100 = 7.06,$$

leading to a value of $[\alpha]_D^{20} = \pm 7.06°$.

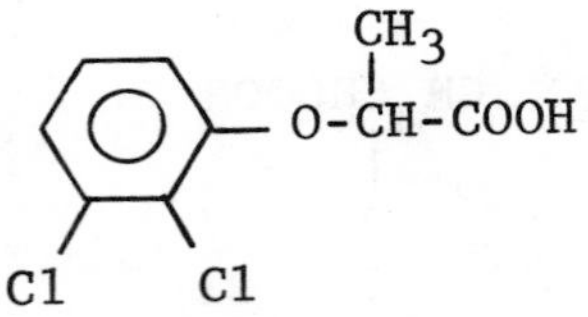

$C_9H_8Cl_2O_3$ Ref. 466

2,3-Dichlorphenoxy-propionsäure

Wird eine siedende Lösung der Säure in Essigester mit Strychnin gesättigt (wobei nur etwa 2/3 der äquivalenten Menge aufgenommen wird), so krystallisiert beim Erkalten ein Salz, das etwa 95% an aktiver (rechtsdrehender) Säure enthält. Nach Sättigen der Mutterlauge mit Strychnin wird eine zweite Fraktion erhalten. Maximale Drehung wird leicht durch Umkrystallisieren aus Essigester erreicht, wobei allerdings ein Teil des Strychnins als freie Base ausgeschieden wird, was die Ausbeute beeinträchtigt. Das durch Eindampfen der ersten Mutterlauge und Sättigen mit Strychnin erhaltene Salz enthält etwa 90% an linksdrehender Säure, die durch zweimaliges Umkrystallisieren des Salzes aus verdünntem Äthanol auf maximale Drehung gebracht wird.

Die (+)-2,3-*Dichlorphenoxy-propionsäure* bildet kleine Prismen (aus verdünnter Ameisensäure). Schmp. 139—140.5°. Da die Drehung in Aceton und Äthanol sehr niedrig ist, wurde sie in neutralisierter wässriger Lösung bestimmt, wo die Säure nach l i n k s dreht.

Anal. 31.31 mg Subst.: 5.28 ml 0.05040-*N* NaOH
$C_9H_8O_3Cl_2$ (235.07) Cl ber. 30.17 gef. 30.1

0.1834 g Subst in Wasser neutralisiert und auf 10.00 ml. verdünnt: $\alpha_D^{25} = -1.83°$ (2 dm). $[\alpha]_D^{25} = -49.9°$; $[M]_D^{25} = -117.5°$.

Die (—)-2,3-*Dichlorphenoxy-propionsäure* ist der (+)-Form äusserlich völlig ähnlich. Schmp. 139—140.5°.

Anal. 28.73 mg Subst.: 4.88 ml 0.05040-*N* NaOH
$C_9H_8O_3Cl_2$ (235.07) Cl ber. 30.17 gef. 30.4

0.1821 g Subst. in Wasser neutralisiert und auf 10.00 ml verdünnt: $\alpha_D^{25} = +1.81°$ (2 dm). $[\alpha]_D^{25} = +49.7°$; $[M]_D^{25} = +117°$.

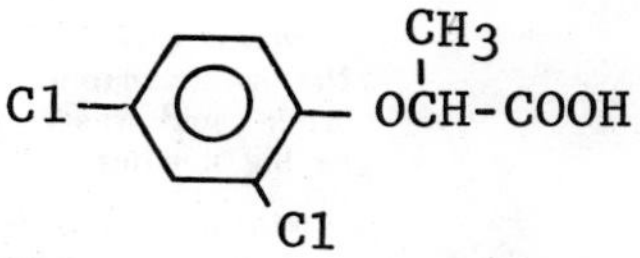

$C_9H_8Cl_2O_3$ Ref. 467

Pure α-(2 : 4-dichlorophenoxy)propionic acid (23·51 g., m.p. 117°) was dissolved in 60 ml. of acetone by shaking. To the solution 39·44 g. of anhydrous brucine was added and dissolved by warming. The solution was stood in a refrigerator for a long period until a deposit of crystals was obtained. The crop was filtered, washed with a little acetone, and dried *in vacuo* over sulphuric acid. Crystals (30·45 g.) $[\alpha]^{17}_{5461}$ − 15·0° (*c*, 2·3 in acetone) was obtained. An approximate determination of the rotation of the brucine salt remaining in solution gave the value $[\alpha]^{17}_{5461}$ − 42°, hence the less soluble isomer would yield the (+)-isomer of the acid. The solid was repeatedly crystallized from acetone until the specific rotation became constant at $[\alpha]^{17}_{5461}$ − 11° (*c*, 1·33 in acetone), the melting point being 100°.

To isolate the (+)-acid, 14·6 g. of the purified brucine salt was dissolved in the minimum volume of warm alcohol and the solution was then stirred with a mixture of 150 ml. of water and 10 ml. of hydrochloric acid. The oil that formed soon changed to crystals, which were powdered and filtered after standing overnight. Crystallization from benzene gave 3·4 g. of colourless needles, melting point 124°, $[\alpha]^{21}_{D}$ + 26·6° (*c*, 1·23 in ethanol). No racemization of an aqueous alcoholic solution of the acid was observed over a period of ten weeks.

Pure α-(2 : 4-dichlorophenoxy)propionic acid (11·75 g.) and 14·72 g. of cinchonine were shaken with 60 ml. of methanol until they dissolved. Water (85 ml.) was added and a crystalline salt was obtained after standing a few days. The crystals were washed with a little dilute methanol and dried *in vacuo* over sulphuric acid. The dry crystals, 13·2 g., had $[\alpha]^{19}_{D}$ + 113° (*c*, 2·0 in methanol) and the cinchonine salt remaining in solution was found to be more strongly dextrorotatory, indicating that the less soluble salt would yield the (−)-isomer of the acid. After repeated crystallizations from aqueous methanol a salt having a constant specific rotation $[\alpha]^{21}_{D}$ + 110° (*c*, 2·0 in methanol) and of indefinite melting point was obtained.

To isolate the (−)-acid 3·5 g. of the purified cinchonine salt was dissolved in 25 ml. of methanol and added to 200 ml. of stirred 1% sulphuric acid. When the oil formed had crystallized, it was collected, washed with water and dried. The dry solid, 1·38 g., was crystallized from 5 ml. of benzene, giving 0·9 g. of colourless needles, melting point 124°, $[\alpha]^{20}_{D}$ − 26·3° (*c*, 1·22 in ethanol).

$C_9H_8Cl_2O_3$ Ref. 467a

Preliminary experiments on resolution. Crystalline salts were obtained with cinchonidine, strychnine and quinine, but except for the latter, no or very slight resolution could be detected. From the quinine salt an acid containing excess of (+)-acid could be isolated.

(+)-α-(2.4-Dichlorophenoxy)-propionic acid. 23.5 g (0.1 mole) inactive acid and 32.4 g (0.1 mole) quinine were dissolved in a warm mixture of 450 ml acetone and 350 ml water. The solution was allowed to crystallise in a refrigerator over night. The salt was recrystallised according to the scheme below ($[\alpha]_D$ in alcohol):

Crystallisation	1	2	3	4	5
Ml acetone	450	225	180	180	180
» water	350	175	140	140	140
Weight of salt (g)	29.1	23.0	21.1	19.7	19.0
$[\alpha]_D$ of the acid	+16.7°	+25.4°	+25.1°	+26.8°	+26.8°

The pure salt yielded 7.65 g crude acid which was recrystallised twice from petroleum (b.p. 60−80°). 7.2 g pure acid were collected as a wadlike mass of thin needles with m.p. 122.2−123.2°.

72.48 mg acid: 6.46 ml 0.04792-*N* NaOH
$C_9H_8O_3Cl_2$ calc. equiv.wt. 235.1
found » 234.1

0.1004 g dissolved in abs. *alcohol* to 10.00 ml: $\alpha^{25}_{D} = +0.282°$. $[\alpha]^{25}_{D} = +28.1°$; $[M]^{25}_{D} = +66.0°$.
0.1049 g dissolved in *acetone* to 10.00 ml: $\alpha^{25}_{D} = +0.369°$. $[\alpha]^{25}_{D} = +35.2°$; $[M]^{25}_{D} = +82.7°$.
0.1049 g dissolved in *benzene* to 10.00 ml: $\alpha^{25}_{D} = -0.179°$. $[\alpha]^{25}_{D} = -17.1°$; $[M]^{25}_{D} = -40.1°$.
0.1031 g dissolved in *chloroform* to 10.00 ml: $\alpha^{25}_{D} = +0.216°$. $[\alpha]^{25}_{D} = +21.0°$; $[M]^{25}_{D} = +49.2°$.
0.0956 g neutralised with *aqueous sodium hydroxide* and made up to 10.00 ml: $\alpha^{25}_{D} = -0.135°$. $[\alpha]^{25}_{D} = -14.1°$; $[M]^{25}_{D} = -33.2°$.

(−)-α-(2.4-Dichlorophenoxy)-propionic acid. As no suitable base for the isolation of the (−)-form was known it was tried to concentrate the salt of the (−)-form in the mother liquor. For that reason the mother liquors from crystallisation 1 and 2 above were evaporated to about half the volume. The salt of the (+)-acid did not, however, separate but an oil was obtained. After some hours this crystallised. A sample showed that this fraction gave an acid with $[\alpha]_D = -26°$ in alcohol. An attempt to purify this salt further failed. Therefore the acid was liberated from the salt, the crude (−)-acid (3.4 g) was recrystallised twice from petroleum (b.p. 60−80°) and four times from cyclohexane. 1.9 g pure (−)-acid could be collected as thin needles. M.p. 122.5−123.2°.

72.34 mg acid: 6.41 ml 0.04792-*N* NaOH
$C_9H_8O_3Cl_2$ calc. equiv.wt. 235.1
found » 235.5

0.0978 g dissolved in abs. *alcohol* to 10.00 ml: $\alpha^{25}_{D} = -0.279°$. $[\alpha]^{25}_{D} = -28.5°$; $[M]^{25}_{D} = -67.1°$.

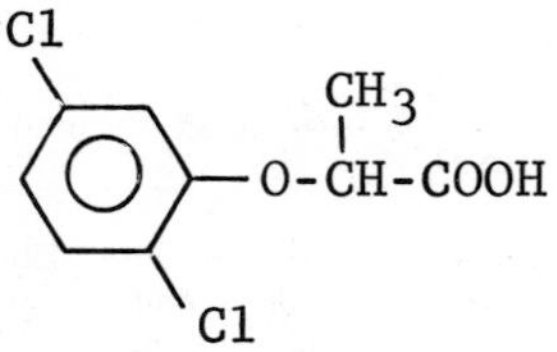

$C_9H_8Cl_2O_3$ Ref. 468

2,5-Dichlorphenoxy-propionsäure

Die (—)-Form wurde durch Krystallisieren des Strychninsalzes aus verdünntem (etwa 45-%-igem) Aceton erhalten. Die Mutterlauge nach der ersten Krystallisation gab eine etwa 93-%-ige (+)-Säure, die durch Umkrystalliseren aus verdünnter Ameisensäure leicht auf maximale Drehung gebracht wurde.

Die (+)-2,5-*Dichlorphenoxy-propionsäure* bildet glänzende Nadeln (aus verdünnter Ameisensäure). Schmp. 136.5—137.5°.

Anal. 29.50 mg Subst.: 4.98 ml 0.05040-*N* NaOH
$C_9H_8O_3Cl_2$ (235.07) Cl ber. 30.17 gef. 30.2

0.1298 g Subst. in Aceton auf 10.00 ml gelöst: $\alpha^{25}_{D} = +1.60°$ (2 dm). $[\alpha]^{25}_{D} = +61.6°$; $[M]^{25}_{D} = +145°$.

Die (—)-2,5-*Dichlorphenoxy-propionsäure* ist der (+)-Form äusserlich völlig ähnlich. Schmp. 136.5—137.5°.

Anal. 30.65 mg Subst.: 5.20 ml 0.05040-*N* NaOH
$C_9H_8O_3Cl_2$ (235.07) Cl ber. 30.17 gef. 30.3

0.2080 g Subst. in Aceton auf 10.00 ml gelöst: $\alpha^{25}_{D} = -2.55°$ (2 dm). $[\alpha]^{25}_{D} = -61.3°$; $[M]^{25}_{D} = -144°$.

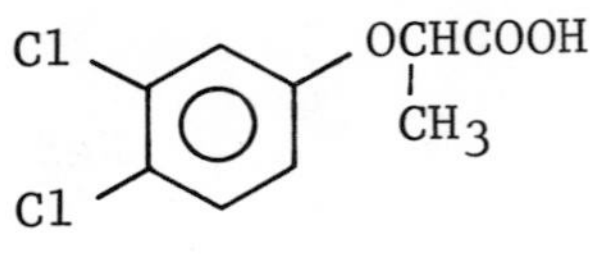

$C_9H_8Cl_2O_3$ **Ref. 469**

***rac.*-α-(3.4-Dichlorophenoxy)-propionic acid. (I).** 1.73 g (0.075 mole) sodium were dissolved in 40 ml abs. alcohol, 12.2 g (0.075 mole) 3.4-dichlorophenol and 13.6 g (0.075 mole) ethyl α-bromopropionate were added and the solution refluxed for two hours. 60 ml 2-*N* sodium hydroxide were added and the ester was hydrolysed by heating for one hour. The alcohol was distilled off and the residue acidified with hydrochloric acid when an oil separated which in a few minutes solidified. The crude acid was recrystallised twice from petroleum (b.p. 60 – 80°). 14.7 g pure acid were collected as white plates. M.p. 129.2 – 130.8°. Yield 83 %.

88.40 mg acid: 7.82 ml 0.04792-*N* NaOH
47.69 » » 8.40 » 0.04844-*N* » (combustion according to GROTE-KREKELER)
$C_9H_8O_3Cl_2$ calc. equiv.wt. 235.1; Cl 30.17
found » 235.9; » 30.25.

Preliminary experiments on resolution. Brucine, quinine, cinchonine and (–)-α-phenethylamine yielded crystalline salts containing excess of (+)-acid, brucine giving the best resolution. With strychnine and (+)-α-phenethylamine crystalline salts were obtained, which contained the (–)-form in excess. The latter base gave only slight resolution, the former being much more effective.

(—)-α-(3.4-Dichlorophenoxy)-propionic acid. 19.7 g (0.05 mole) brucine and 11.8 g (0.05 mole) inactive acid were dissolved in a hot mixture of 75 ml alcohol and 250 ml water. On cooling, the solution was seeded with crystals from a preliminary run and left over night in a refrigerator. The crystals were collected, the acid was liberated from a small sample and the rotatory power in abs. alcohol determined. The salt was recrystallised several times from dilute alcohol according to the scheme below:

Crystallisation	1	2	3	4	5
Ml alcohol	75	75	50	50	45
» water	250	250	200	200	175
Weight of salt (g)	23.4	16.3	14.2	12.9	11.8
$[\alpha]_D$ of the acid	– 12.6°	– 31.9°	– 36.8°	– 37.4°	– 37.3°

The acid liberated from the salt 5 weighed 4.3 g. It was recrystallised twice from cyclohexane. 3.65 g of pure acid were obtained as small glistening needles with m. p. 119.5 – 121.0°.

67.90 mg acid: 6.02 ml 0.04792-*N* NaOH
$C_9H_8O_3Cl_2$ calc. equiv.wt. 235.1
found » 235.4
0.0996 g dissolved in abs. *alcohol* to 10.00 ml: $\alpha_D^{25} = -0.391°$.
$[\alpha]_D^{25} = -39.3°$; $[M]_D^{25} = -92.3°$.

(+)-α-(3.4-Dichlorophenoxy)-propionic acid. The partially active acid from the mother liquors from crystallisation 1 and 2 above was liberated. 6.0 g were obtained with $[\alpha]_D = +31°$ in alcohol. This acid (0.025 mole) and 8.5 g (0.025 mole) strychnine were dissolved in 75 ml acetone and 200 ml water in the hot. The solution was seeded with crystals from a preliminary experiment and left over night in a refrigerator. The salt was treated as above, giving the following series of crystallisations ($[\alpha]_D$ in alcohol):

Crystallisation	1	2	3
Ml acetone	75	70	60
» water	200	180	160
Weight of salt (g)	12.4	11.5	11.0
$[\alpha]_D$ of the acid	+36.9°	+37.9°	+37.9°

The pure salt on decomposition yielded 4.3 g crude acid. After three recrystallisations from cyclohexane, 3.6 g pure acid were collected. It closely resembled the (–)-acid and melted at 119.8 – 121.2°.

80.75 mg acid: 7.19 ml 0.04792-*N* NaOH
$C_9H_8O_3Cl_2$ calc. equiv.wt. 235.1
found » 234.4

0.1063 g dissolved in abs. *alcohol* to 10.00 ml: $\alpha_D^{25} = +0.416°$. $[\alpha]_D^{25} = +39.1°$; $[M]_D^{25} = +92.0°$.
0.0966 g dissolved in *acetone* to 10.00 ml: $\alpha_D^{25} = +0.529°$. $[\alpha]_D^{25} = +54.8°$; $[M]_D^{25} = +128.7°$.
0.1075 g dissolved in *benzene* to 10.00 ml: $\alpha_D^{25} = +0.038°$. $[\alpha]_D^{25} = +3.5°$; $[M]_D^{25} = +8.3°$.
0.0984 g dissolved in *chloroform* to 10.00 ml: $\alpha_D^{25} = +0.235°$. $[\alpha]_D^{25} = +23.9°$; $[M]_D^{25} = +56.1°$.
0.1035 g neutralised with *aqueous sodium hydroxide* and made up to 10.00 ml: $\alpha_D^{25} = +0.151°$. $[\alpha]_D^{25} = +14.6°$; $[M]_D^{25} = +34.3°$.

***rac.*-α-(2.4-Dichlorophenoxy)-propionic acid (II).** This acid was prepared in the same manner as the 3.4-substituted isomer using 2.4-dichlorophenol. The crude acid was recrystallised from petroleum (b.p. 60 – 80°) and was obtained as big needles with m.p. 117 – 118°.

90.99 mg acid: 8.06 ml 0.04792-*N* NaOH
45.32 » » : 7.905 » 0.04844-*N* » (combustion according to GROTE-KREKELER)
$C_9H_8O_3Cl_2$ calc. equiv.wt 235.1; Cl 30.17
found » 235.6 » 29.96

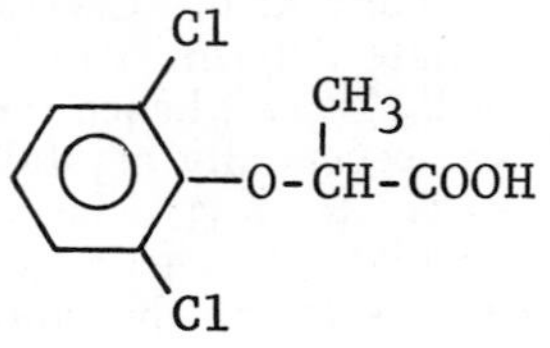

$C_9H_8Cl_2O_3$ **Ref. 470**

3,5-Dichlorphenoxy-propionsäure.

Die (+)-Säure wird am besten durch Umkristallisieren des Chininsalzes aus Essigester erhalten. Die stark linksdrehende Säure aus der Mutterlauge nach der ersten Krystallisation wurde in das Cinchonidinsalz übergeführt, das durch Umkristallisieren aus verdünntem Aceton auf maximale Drehung gebracht wurde.

Die (+)-3,5-*Dichlorphenoxy-propionsäure* bildet glänzende Nadeln (aus Cyclohexan oder verdünnter Ameisensäure). Schmp. 120.5—121.5°.

Anal. 35.05 mg Subst.: 5.85 ml 0.05089-*N* NaOH
$C_9H_8O_3Cl_2$ (235.07) Cl ber. 30.17 gef. 30.1

0.1800 g Subst. in Aceton auf 10.00 ml gelöst: $\alpha_D^{25} = +1.99°$ (2 dm). $[\alpha]_D^{25} = +55.3°$; $[M]_D^{25} = +130.0°$.

Die (—)-3,5-*Dichlorphenoxy-propionsäure* ist der (+)-Form äusserlich völlig ähnlich. Schmp. 120.5—121.5°.

Anal. 31.64 mg Subst.: 5.31 ml 0.05040-*N* NaOH
$C_9H_8O_3Cl_2$ (235.07) Cl ber. 30.17 gef. 30.0

0.1809 g Subst. in Aceton auf 10.00 ml gelöst: $\alpha_D^{25} = -2.005°$ (2 dm). $[\alpha]_D^{25} = -55.4°$; $[M]_D^{25} = -130.3°$.

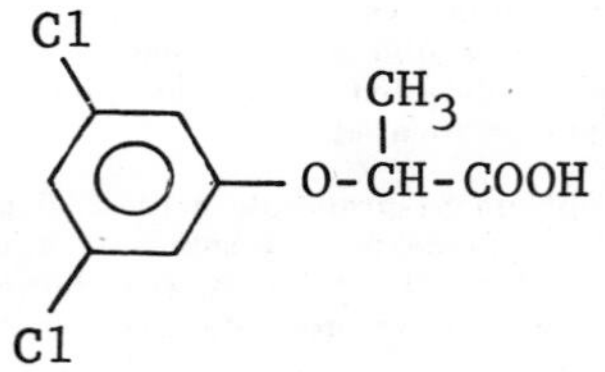

$C_9H_8Cl_2O_3$ **Ref. 471**

2,6-Dichlorphenoxy-propionsäure

Dr. M. S. Smith hat die (+)-Form mit Strychnin in verdünntem Äthanol da gestellt; mit Yohimbin erhielt sie eine (—)-Säure, die wahrscheinlich nicht ga rein war[10]. Der Verfasser erhielt die (+)-Form mit Strychnin in etwa 40-%-ige Aceton. Aus der Mutterlauge liess sicht bei zweckmässiger Arbeitsweise eine fa reine (—)-Säure isolieren, die durch Umkrystallisieren aus verdünnter Ameise säure leicht auf maximale Drehung gebracht wurde.

Die (+)-2,6-*Dichlorphenoxy-propionsäure* bildet lange, glänzende Nadeln (a verdünnter Ameisensäure). Schmp. 94.5—96°.

Anal. 36.54 mg Subst.: 6.18 ml 0.05040-*N* NaOH
$C_9H_8O_3Cl_2$ (235.07) Cl ber. 30.17 gef. 30.2

0.2518 g Subst. in Aceton auf 10.00 ml gelöst: $\alpha_D^{25} = +1.18°$ (2 dm). $[\alpha]_D^{25} = +23.$ $[M]_D^{25} = +55.1°$.

Die (—)-2,6-*Dichlorphenoxy-propionsäure* ist der (+)-Form äusserlich völ ähnlich. Schmp. 94.5—96°.

Anal. 37.52 mg Subst.: 6.30 ml 0.05040-*N* NaOH
$C_9H_8O_3Cl_2$ (235.07) Cl ber. 30.17 gef. 30.0

0.2572 g Subst. in Aceton auf 10.00 ml gelöst: $\alpha_D^{25} = -1.21°$ (2 dm). $[\alpha]_D^{25} = -23.5°$; $[M]_D^{25} = -55.3°$.

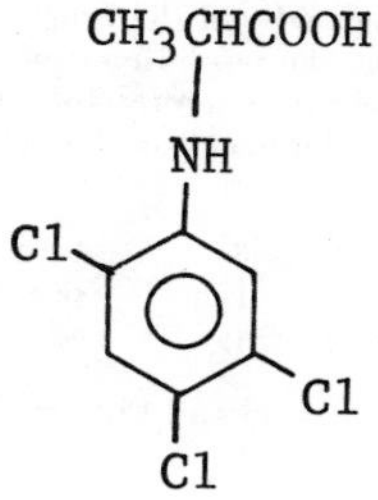

$C_9H_8Cl_3NO_2$

Ref. 472

Preliminary experiments on resolution. Optically active α-phenyl-ethylamine gave readily crystallising salts and very good resolution; the mode of rotation of the acid is opposite to that of the amine used. Levorotatory acid can also be isolated through the cinchonine salt. Cinchonidine and brucine yielded salt containing levorotatory acid of moderate activity. The acid regenerated from the quinine salt was quite inactive and the strychnine salt could not be brought to crystallisation.

Resolution of the racemic acid. 12.1 g acid (0.045 mole) and 5.4 g (−)-α-phenyl-ethylamine (0.045 mole) were dissolved in 100 ml of acetone and 100 ml of hot water was added. After standing in a refrigerator over-night, the salt was filtered off and recrystallised from dilute acetone. After each crystallisation, a small sample of the salt was decomposed, the acid liberated and the rotatory power determined in acetone solution.

Crystallisation	1	2	3
ml water	100	25	20
ml acetone	100	25	20
g salt obtained	6.8	6.1	5.4
$[\alpha]_D^{25}$ of the acid . . .	+34°	+35°	+34.5°

The mother liquor from the first crystallisation was evaporated and the acid isolated; 8.2 g having $[\alpha]_D^{25} = -22.8°$ (acetone solution) was obtained. This acid was dissolved with 3.7 g of (+)-α-phenyl-ethylamine in dilute acetone and the salt was recrystallised as described above.

Crystallisation	1	2	3
ml water	30	30	25
ml acetone	30	25	20
g salt obtained	7.3	6.3	5.7
$[\alpha]_D^{25}$ of the acid . . .	−33°	−34.5°	−34°

(+)-2.4.5-Trichloroanilino-propionic acid. The (−)-phenyl-ethylamine salt (5.4 g) was decomposed with sulphuric acid (acidification to pH = 3) and the organic acid was extracted with ether. It was recrystallised once from ethyl acetate + light petroleum and once from dilute (45%) ethanol. The yield of crude acid was 3.6 g; after the recrystallisations there remained 2.7 g. Small glistening needles with m.p. 166–167°.

29.79 mg substance: 6.66 ml 0.0942 *N* NaOH.
$C_9H_8O_2NCl_3$ (268.5) Cl calc. 39.61 found 39.48
0.2286 g substance dissolved in *acetone* to 10.00 ml: $2\alpha = +1.57°$. $[\alpha]_D^{25} = +34.3°$; $[M]_D^{25} = +92.1°$.

(−)-2.4.5-Trichloroanilino-propionic acid. The (+)-phenyl-ethylamine salt (5.5 g) was decomposed as described above and the acid was recrystallised once from benzene and once from dilute ethanol. Yield of crude acid 3.8 g; after recrystallisation 3.4 g. M.p. 166–167°.

28.65 mg substance: 6.40 ml 0.0942 *N* NaOH.
$C_9H_8O_2NCl_3$ (268.5) Cl calc. 39.61 found 39.59
0.2384 g substance dissolved in *acetone* to 10.00 ml: $2\alpha = -1.63°$. $[\alpha]_D^{25} = -34.2°$; $[M]_D^{25} = -91.8°$. — 0.2005 g substance dissolved in abs. *ethanol* to 10.00 ml: $2\alpha = -1.145°$. $[\alpha]_D^{25} = -28.6°$; $[M]_D^{25} = -76.7°$. — 0.1816 g substance dissolved in *chloroform* to 10.00 ml: $2\alpha = -1.095°$. $[\alpha]_D^{25} = -30.1°$; $[M]_D^{25} = -81.0°$.

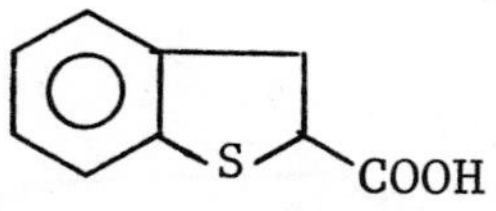

$C_9H_8O_2S$

Ref. 473

Dextrorotatory acid. Racemic acid was converted to brucine salt, which was recrystallised four times from 50 % aqueous methanol; the activity remained constant after two recrystallisations. The acid was liberated and recrystallised twice from ligroine (b. p. 85—110°). Plates or flat needles with m. p. 94.5—95°. (Equiv. wt. found 180.5; calc. 180.2. $[\alpha]_D^{25} = +365.4°$ in absolute ethanol.)

Levorotatory acid. The acid liberated from the first mother liquor of the brucine salt was converted to cinchonine salt, which was recrystallised four times from dilute ethanol; the activity remained constant after two recrystallisations. The acid was liberated and recrystallised as described above. Plates or flat needles with m. p. 94.5—95°. (Equiv. wt. found 180.3; calc. 180.2 $[\alpha]_D^{25} = -365.8°$ in absolute ethanol.)

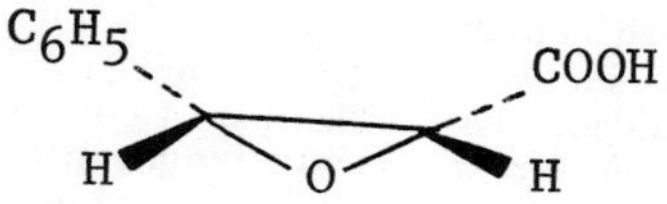

$C_9H_8O_3$

Ref. 474

Optical Resolution of cis-*β-Phenylglycidic Acid (III').* Potassium β-phenylglycidate (III) (8.3 g) was dissolved in 60 ml of water. Crushed ice and 40 ml of ether were added to the solution. To this mixture, 40 ml of 1 M hydrochloric acid was added and the liberated free glycidic acid was extracted with ether. Two additional ether extractions (40 ml × 2) were carried out. The combined ether solution was dried with anhydrous sodium sulfate. To the dried ether solution, 6.78 g of (−)-ephedrine in 20 ml of ether was added. The mixture was kept at room temperature overnight. The resulting salt seemed to be amorphous; however, crystallization took place. After evaporation of ether, the syrupy residue which contained crystalline salt was digested with 30 ml of acetone and filtered. White crystals (IV) were obtained, yield 3.45 g; mp 138—140 °C. This was recrystallized from ethanol. Yield, 3.0 g; mp 145–146 °C, $[\alpha]_D^{25}$ −27.2° (*c* 2.77, ethanol). Additional crystals (1.18 g) were obtained from the salt mixture.

The resolved (−)-salt (1.0 g) in 15 ml of water was acidified to pH 2 with 6 M hydrochloric acid under ice cooling. The liberated free glycidic acid was extracted with ether. The ether solution was dried with anhydrous sodium sulfate and the solvent was evaporated. The residual oil was dissolved in 5 ml of ethanol. To this, potassium hydroxide (0.17 g) in 1 ml of ethanol was added. The precipitated potassium salt of optically active *cis*-phenylglycidic acid (V) was collected by filtration. Yield, 150 mg; mp 203—204 °C (decomp.), $[\alpha]_D^{25}$ +4.4° (*c* 0.98, H_2O).

$C_9H_8O_3$

$C_9H_8O_3$

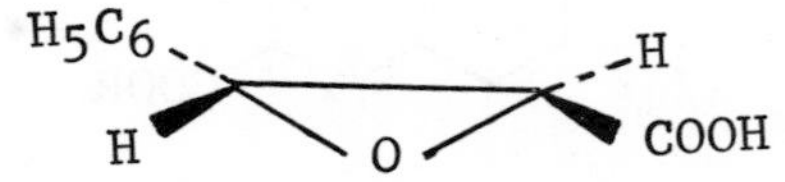

$C_9H_8O_3$ **Ref. 475**

Optical Resolution of β-Phenylglycidic Acid.—Potassium β-phenylglycidate (I, 16.2 g) was dissolved in 100 ml of water. Crushed ice and 80 ml of ether were added to the solution. To this mixture, 80 ml of 1 *N* HCl was added and the liberated free G acid was extracted with ether. Two additional ether extractions (two 50-ml portions) were carried out. The ether solutions were combined and dried with anhydrous sodium sulfate. To the dried ether solution, 9.70 g (0.08 mole) of (−)-amine, $[\alpha]^{27}_D$ −40.6° (benzene), was added. The precipitated oil was crystallized by rubbing with a glass rod. After 2 hr, the crystals were collected by filtration. Uncrystallized oily materials [diastereomeric salt (−)-amine-(+)-acid salt] passed through the filter paper. The crystals were washed with acetone. (−)-Amine-(−)-acid salt IIIa (8.5 g) was obtained, mp 157–158° dec. This was recrystallized by dissolving in 34 ml of ethanol and by precipitating with 60 ml of acetone: yield 6.80 g, mp 161–162° dec, $[\alpha]^{25}_D$ −125.4° (*c* 0.95, absolute EtOH).

Anal. Calcd for $C_{17}H_{19}NO_3$: C, 71.56; H, 6.71; N, 4.91. Found: C, 71.82; H, 6.69; N, 4.88.

To the mother liquor, 100 ml of ice–water and 40 ml of 1 *N* hydrochloric acid were added. Liberated G acid (II) was extracted with ether three times and the combined ether solution was dried with anhydrous sodium sulfate. To the dried ether solution, (+)-amine, 4.85 g (0.04 mole), $[\alpha]^{27}_D$ +39.3° (benzene), was added. Crystallization of (+)-amine-(+)-acid salt IIIb began after 1 min.: yield 8.90 g, mp 156–157° dec. This was recrystallized from ethanol and acetone: yield 6.90 g, mp 161–162° dec, $[\alpha]^{25}_D$ +125.5° (*c* 0.96, absolute EtOH).

Anal. Found: C, 71.67; H, 6.83; N, 4.93.

Ammonium (−)- and (+)-β-Phenylglycidate (IVa and b).—(−)-Amine-(−)-acid salt (2.0 g) was dissolved in 40 ml of water and cooled to 5°. To this, 20 ml of 1 *N* hydrochloric acid was added. Precipitated free (−)-G acid was extracted with ether and dried with anhydrous sodium sulfate. Dry ammonia gas was introduced to precipitate (−)-G acid ammonium salt (IVa): 1.05 g, $[\alpha]^{25}_D$ −154.0° (*c* 1.03, H_2O).

(+)-G acid ammonium salt (IVb) was prepared in the same way: $[\alpha]^{25}_D$ +153.5° (*c* 1.10, H_2O).

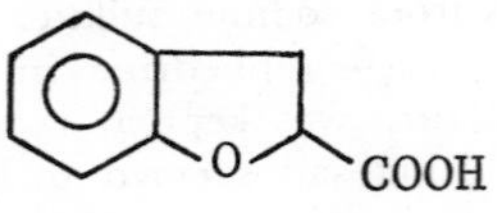

$C_9H_8O_3$ **Ref. 476**

Racemic hydrocoumarilic acid was prepared according to STOERMER and KÖNIG (11). After several recrystallisations, alternately from formic acid and benzene, it melted at 116–117°.

Preliminary experiments on resolution were carried out with the usual alkaloids in different solvents (water, ethyl acetate, dilute ethanol, dilute methanol, dilute acetone). The dextrorotatory acid is best isolated through the brucine salt. Optically active benzedrine (β-phenyl-*iso*-propylamine) recommended for such purposes by MATELL (12), gives fairly good resolution; the acid obtained has the same direction of rotation as the base. (−)-Benzedrine was therefore used for the isolation of the (−)-acid.

Quinine and cinchonine give levorotatory acid of low acitivity, and cinchonidine gives excess of the (+)-form. Strychnine gives (+)- or (−)-acid according to the solvent used, α-phenyl-ethylamine acid having the same direction of rotation as the amine. The quinidine salt could only be obtained as an oil. None of these bases was suitable for the isolation of the pure enanthiomorphs.

11. STOERMER, R. and KÖNIG, W., Ber. *39*, 493 (1906).

16.4 g (0.1 mole) acid and 46.6 g (0.1 mole) brucine tetrahydrate were dissolved in the necessary amount of hot water and the salt obtained was recrystallised several times from the same solvent. The acid and the brucine can be dissolved in a rather moderate volume of water, but as the resolution proceeds, the solubility of the salt decreases very markedly. The course of the resolution was tested on samples from each crystallisation, from which the acid was liberated and examined in ethanol solution.

Crystallisation	1	2	3	4	5	6	7
g salt obtained . .	27	21	19.5	16.2	12.6	9.1	6.3
$[\alpha]^{25}_D$ of the acid . .	+14.5°	+18°	+19°	+22°	+22.5°	+22°	+22°

In a second run, 49.2 g (0.3 mole) acid and 139.8 g (0.3 mole) brucine tetrahydrate were used.

Crystallisation	1	2	3	4	5	6
g salt obtained . .	81.5	57.9	49.4	45.7	41.9	34.5
$[\alpha]^{25}_D$ of the acid . .	+13°	+19.5°	+20.5°	+21°	+21.5°	+22°

The mother liquor from the first crystallisation of the second run was evaporated to dryness and the salt obtained decomposed with dilute sulphuric acid. A hydrocoumarilic acid having $[\alpha]^{25}_D = -11.7°$ (ethanol) was obtained. 10.4 g (0.063 mole) of this acid and 8.6 g (0.063 mole) of (−)-benzedrine (prepared by recrystallisation of the neutral tartrate to constant activity) were dissolved in hot water and the salt obtained recrystallised from the same solvent. The course of the resolution was tested as described above.

Crystallisation	1	2	3	4
g salt obtained	9.9	7.3	4.9	3.1
$[\alpha]^{25}_D$ of the acid	−15.5°	−20.5°	−21°	−20.5°

As seen from the data, the maximum activity is somewhat lower than that found for the (+)-acid. A closer examination revealed that this effect was due to a slight contamination with the amine.

(+)-*Hydrocoumarilic acid.* The brucine salt (34.1 g) was decomposed with dilute sulphuric acid and the hydrocoumarilic acid isolated by extraction with ether. The yield of crude acid was 9.1 g. It was recrystallised three times from carbon tetrachloride, yielding 6.1 g of glistening needles with m.p. 104–105°. The melting point did not change further recrystallisation.

0.2211 g acid: 12.47 ml 0.1075 *N* NaOH.

$C_9H_8O_3$ Equiv. wt. calc. 164.2 found 164.9.

0.2473 g dissolved in *benzene* to 10.00 ml: $2\alpha^{25}_D = +2.73°$. $[\alpha]^{25}_D = +55.2°$; $[M]^{25}_D = +90.6°$. — 0.2584 g dissolved in absolute *ethanol* to 10.00 ml: $2\alpha^{25}_D = +1.17°$. $[\alpha]^{25}_D = +22.6°$; $[M]^{25}_D = +37.2°$. — 0.2497 g dissolved in *chloroform* to 9.99 ml: $2\alpha^{25}_D = +2.695°$. $[\alpha]^{25}_D = +54.0°$; $[M]^{25}_D = +88.6°$. — 0.2452 g dissolved in glacial *acetic acid* to 10.00 ml: $2\alpha^{25}_D = +0.665°$. $[\alpha]^{25}_D = +13.6°$; $[M]^{25}_D = +22.3°$. — 0.2648 g dissolved in *acetone* to 10.00 ml: $2\alpha^{25}_D = +0.055°$. $[\alpha]^{25}_D = +1.0°$; $[M]^{25}_D = +1.6°$. — 0.2413 g *neutralised with aqueous NaOH* and made up with *water* to 9.99 ml: $2\alpha^{25}_D = +3.695°$. $[\alpha]^{25}_D = +76.6°$; $[M]^{25}_D = +125.7°$.

Preliminary melting point determinations on mixtures of active and racemic acid indicate that the latter is a true racemate. The crystal habit is also quite different from that of the active acid.

(−)-*Hydrocoumarilic acid.* The (−)-benzedrine salt (3.1 g) was decomposed with excess of dilute sulphuric acid in the usual way and the hydrocoumarilic acid was extracted with ether. The residue obtained after drying and evaporation of the ether was, however, still contaminated with some benzedrine. It was therefore dissolved in dilute sodium hydroxide solution, and the benzedrine removed by repeated extractions with ether. The alkaline solution was then acidified with dilute sulphuric acid and the hydrocoumarilic acid isolated as usual. After drying, it was recrystallised twice from carbon tetrachloride yielding 0.95 g of glistening needles with m.p. 104–105°.

0.1904 g acid: 10.71 ml 0.1075 *N* NaOH.

$C_9H_8O_3$ Equiv. wt. calc. 164.2 found 165.3.

0.2356 g dissolved in *benzene* to 9.98 ml: $2\alpha^{25}_D = -2.595°$. $[\alpha]^{25}_D = -55.0°$; $[M]^{25}_D = -90.2°$.

$C_9H_8O_3$ **Ref. 476a**

Resolution of Hydrocoumarilic Acid.—Our initial attempts to employ brucine as a resolving agent for this acid after the procedure of Fredga[12] were unsuccessful. In our experience racemic brucine hydrocoumarilate failed to crystallize, as did also the brucine salts of the resolved samples obtained below. We subsequently found that amphetamine served conveniently to resolve hydrocoumarilic acid and, since it is commercially available in both enantiomeric forms, had the additional advantage of permitting procurement of both enantiomers of this acid. Fredga has used amphetamine only on the mother liquors from his brucine resolution.[12]

(12) A. Fredga and C. Vazquez de Castro y Sarmiento, *Arkiv. Kemi*, **7**, 387 (1954); *C. A.*, **49**, 14746a (1955).

In a typical experiment hydrocoumarilic acid (4.30 g.) was treated with the amine liberated from (+) amphetamine sulfate (5.0 g.) and the mixture was dissolved in water (60 ml.) at 80°. Cooling deposited 3.75 g. of salt, m.p. 150–181°. Recrystallization of the salt from water (50 ml.) afforded 1.73 g. of salt, m.p. 185–187°, the acid recovered from which had m.p. 92–96°, $[\alpha]^{25}D$ +20.0° (*c*, 6.1, ethanol). A second recrystallization gave 0.54 g. of salt having m.p. 186–187°, the acid (0.22 g.) from which had m.p. 100–104° and $[\alpha]^{25}D$ 22.6° (*c*, 7.3, ethanol), in agreement with previously reported values.[12] By conversion of the free acid obtained from the mother liquors of this resolution into its (−) amphetamine salt, the enantiomeric (−) hydrocoumarilic acid could similarly be obtained.[12] Subsequently it was found that recrystallization of such salts from 95% ethanol permitted somewhat more rapid resolution. In larger scale resolutions full purification of the amphetamine salts was not undertaken, since complete resolution was attended by heavy losses and it was further found that the succeeding crystalline acetate XI could be rendered optically pure with less over-all loss by simple recrystallization.

(12) A. Fredga and C. Vazquez de Castro y Sarmiento, *Arkiv. Kemi*, **7**, 387 (1954); *C. A.*, **49**, 14746a (1955).

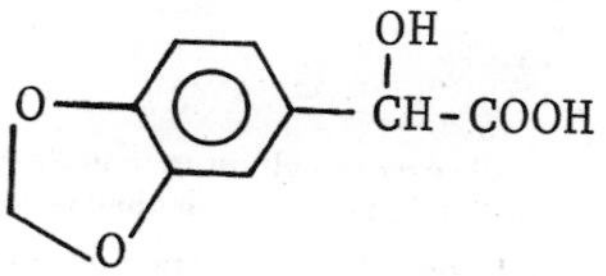

$C_9H_8O_5$ Ref. 477

(f) (±)-3,4-*Methylenedioxymandelic acid* (If). The (±)-acid (If) (59 g.), m.p. 158—160° (from aqueous ethanol) was added to (−)-ephedrine (51·5 g.) in ethanol–water (9 : 1; 300 ml.). The solution was boiled, then left at room temperature. The solid deposited was recrystallised twice more from the same solvent mixture. The complex, m.p. 152—154° was worked up in the usual way with sulphuric acid to give (−)-3,4-*methylenedioxymandelic acid* (If) (18 g.) m.p. 129—131° [from benzene–methanol (20 : 1)], [α] −128·5° (EtOH) (Found: C, 55·4; H, 4·0. $C_9H_8O_5$ requires C, 55·1; H, 4·1%). The (+)-acid was not obtained pure. At ambient temperatures, methanol and traces of sulphuric acid in benzene readily converted the (−)-acid into *methyl* (−)-3,4-*methylenedioxymandelate* (75%), [α] −130·3° (EtOH), m.p. 103—104° (Found: C, 56·8; H, 4·8. $C_{10}H_{10}O_5$ requires C, 57·1; H, 4·8%).

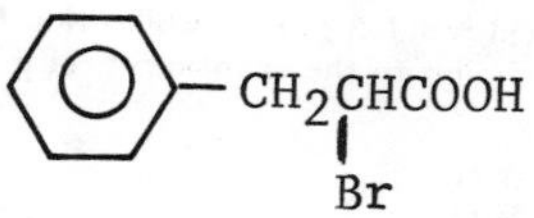

$C_9H_9BrO_2$ Ref. 478

l-α-Brom-hydrozimmtsäure.

Die Spaltung der racemischen Säure kann sowohl durch Brucin, wie durch Chinin bewirkt werden; wir beschreiben hier nur das erste Verfahren.

Wegen der Neigung der Bromhydrozimmtsäure, bei Gegenwart von Basen in Zimmtsäure überzugehen, müssen auch ihre Alkaloïdsalze mit grosser Vorsicht behandelt werden. Abdampfen der wässrigen Lösung selbst bei geringem Druck bewirkt schon theilweise Zersetzung. Für die Krystallisation des Brucinsalzes diente deshalb folgendes Verfahren.

60 g reine, inactive α-Bromhydrozimmtsäure wurden mit 105 g wasserfreiem Brucin in möglichst wenig Alkohol bei gewöhnlicher Temperatur gelöst und zur filtrirten Lösung die 6-fache Menge Wasser gegeben. Dabei schieden sich 110 g Brucinsalz krystallinisch aus. Es wurde fein gepulvert, wieder in möglichst wenig Alkohol von 25° gelöst, mit etwas mehr als der berechneten Menge verdünnter Salzsäure versetzt und die Flüssigkeit mit viel Wasser verdünnt.

Die hierdurch gefällte Bromhydrozimmtsäure wurde mehrmals ausgeäthert, die ätherische Lösung mit Natriumsulfat getrocknet, filtrirt und nach dem Verdampfen des Aethers bei 0.2—0.4 mm destillirt. Sie ging bei 138—143° über und zeigte dann im 1 dcm-Rohr bei 20° und Natriumlicht eine Drehung von 4.25° nach links.

Mit dieser linksdrehenden Säure wurde zum zweiten Mal das Brucinsalz wie oben dargestellt. Diesmal war die Ausbeute 49 g. Glänzende Nadeln. Die optische Untersuchung der daraus isolirten freien Säure ergab nun den Drehungswinkel $\alpha = -8.8°$.

Nach der dritten Krystallisation (36 g) stieg der Werth auf −11.68° und nach der vierten auf −11.96°. Die letzte Steigerung war so gering, dass die Fortsetzung des Verfahrens zwecklos erschien. Die Ausbeute an activer Säure betrug schliesslich nur 18 pCt. der Theorie. Die Verluste werden aber viel kleiner, wenn man die öftere Isolirung der freien Säure vermeidet und das Umkrystallisiren des Brucinsalzes durch einfaches Lösen in kaltem Alkohol und Fällen mit Wasser bewirkt. Die Ausbeute betrug dann nach zweimaligem Umlösen 44 pCt. der Theorie, und das Präparat hatte die Drehung $\alpha = -11.92°$.

Eine Probe der Säure vom Drehungswinkel $\alpha = -11.96°$ wurde analysirt. 0.1821 g Sbst.: 0.1493 g AgBr.

$C_9H_9O_2Br$ (Mol. 229.05). Ber. Br 34.91. Gef. Br 34.89.

Das Präparat hatte die Dichte 1.48, mithin

$$[\alpha]_D^{20°} = -8.08°.$$

Die *l*-α-Bromzimmtsäure konnte nicht krystallisirt erhalten werden. Dagegen krystallisirte aus einer Säure, die das polarisirte Licht nur 5° nach links drehte, bei mehrtägigem Stehen bei 0° der Racemkörper so vollständig aus, dass für das abgesaugte Oel im 1 dcm-Rohr eine Drehung von 12.28° nach links oder $[\alpha]_D^{20°} = -8.3°$ gefunden wurde. Das ist sogar 0.2° mehr als bei der Säure, die aus dem 4 Mal krystallisirten Brucinsalz gewonnen und destillirt war. Vielleicht rührt das von einer geringen Racemisirung bei der Destillation her. Selbstverständlich ist es vortheilhaft, die Krystallisation durch Impfen einzuleiten.

Da es sehr unwahrscheinlich ist, dass bei dem Ausfrieren aller oder nahezu aller Racemkörper entfernt wird, so können wir auch nicht annehmen, dass die Säure von $\alpha = -12.28°$ schon ganz rein war.

d-Phenyl-alanin aus *l*-α-Brom-hydrozimmtsäure.

2 g Säure von $\alpha = -11.96°$ und 10 ccm wässriges Ammoniak von 25 pCt. wurden getrennt in einer Kältemischung gekühlt und dann zusammengebracht. Nachdem die Lösung einen halben Tag auf niedriger Temperatur gehalten war, blieb sie im verschlossenen Gefässe 5 Tage bei gewöhnlicher Temperatur stehen. Dann wurde auf dem Wasserbade verdampft, der Rückstand mit Alkohol ausgekocht und abfiltrirt. Um die letzten Spuren Bromammonium zu entfernen, haben wir das Phenylalanin noch aus heissem Wasser umkrystallisirt. Die Ausbeute betrug 0.7 g oder 50 pCt. der Theorie.

Zur optischen Untersuchung diente die wässrige Lösung. 0.2134 g Substanz, Gewicht der Lösung 18.8354 g, Dichte 1.00. Drehung im 2 dcm-Rohr bei 20° und Natriumlicht 0.72° nach rechts. Mithin

$$[\alpha]_{20}^{D} = +31.78°.$$

Wenn man den von E. Fischer und Mouneyrat[1]) für *d*-Phenylalanin ermittelten Werth

$$[\alpha]_{20}^{D} = +35.08°$$

als richtig annimmt, so hätte obiges Phenylalanin 9.4 pCt. Racemkörper enthalten.

Ein Rückschluss auf die Reinheit der *l*-Bromhydrozimmtsäure lässt sich daraus aber nicht ziehen, da die Ausbeute an Phenylalanin nur 50 pCt. der Theorie betrug, und da man auch nicht weiss, wie stark die Racemisation bei der Umwandlung der Bromverbindung in die Aminosäure ist.

d-α-Brom-hydrozimmtsäure.

Sie ist in den Mutterlaugen enthalten, die beim Auskrystallisiren des *l*-α-bromhydrozimmtsauren Brucins resultiren, und lässt sich daraus durch Uebersättigen mit Salzsäure und Ausäthern gewinnen. War die Menge des auskrystallisirten Brucinsalzes etwa ²/₃ des Ganzen, so zeigte die gewonnene *d*-Bromhydrozimmtsäure einen Drehungs-

[1]) Diese Berichte 33, 2385 [1900].

winkel von ungefähr 7.5°. Durch Impfen mit dem Racemkörper und mehrtägiges Aufbewahren bei 0°, dann Absaugen des Oeles, konnte die Activität so gesteigert werden, dass bei 20° und Natriumlicht $\alpha = -11.1°$ oder $[\alpha]_D^{20°} = -7.9°$ war.

Wie schon erwähnt, lässt sich die Zerlegung der racemischen α-Bromhydrozimmtsäure auch mit Chinin bewerkstelligen. Zur Darstellung des Chininsalzes verfährt man folgendermaassen: Man löst äquimolekulare Mengen der Säure und der Base in der 2½-fachen Menge Alkohol von gewöhnlicher Temperatur. Nach kurzer Zeit tritt ohne Hinzufügen von Wasser eine reichliche Krystallisation des Chininsalzes ein. Sie enthält ebenfalls hauptsächlich die *l*-α-Bromhydrozimmtsäure. Aus der Mutterlauge lässt sich dann durch Uebersättigung mit Schwefelsäure und Ausäthern die isomere Säure, natürlich im unreinen Zustand, gewinnen. Als die Menge des auskrystallisirten Chininsalzes 65 pCt. der Gesammtmenge betrug, wurde für die aus der Mutterlauge isolirte *d*-α-Bromhydrozimmtsäure bei 20° und Natriumlicht der Drehungswinkel 7.6° gefunden, und konnte dann durch Ausfrieren, wie oben schon erwähnt, gesteigert werden. Will man sich auf die Gewinnung der *d*-α-Bromhydrozimmtsäure beschränken, so würde das Chininsalz den Vorzug verdienen.

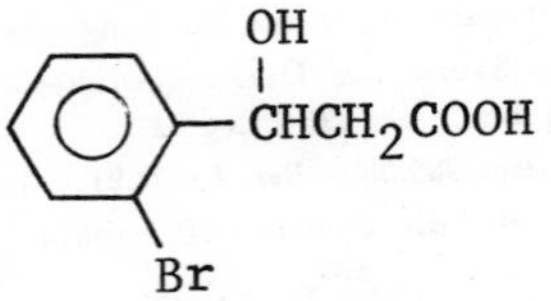

$C_9H_9BrO_3$ **Ref. 479**

Acide o-bromophényl hydracrylique R (+).

0,49 g d'acide, 0,79 g de brucine et 4 cm³ d'éthanol donnent 0,44 g de sel, recristallisé dans 3 cm³ d'éthanol, et décomposé en acide F = 87° (recristallisé dans l'éther/hexane), $[\alpha]_J^{25} = +81°$ ($CHCl_3$, c = 1,3).

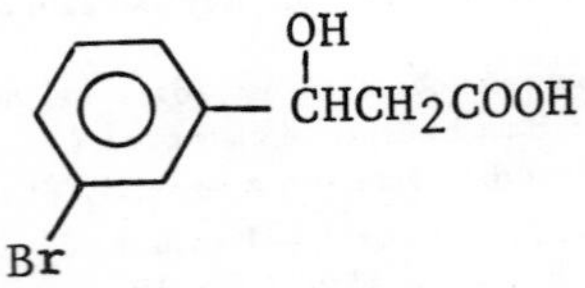

$C_9H_9BrO_3$ **Ref. 480**

Acide m-bromophényl hydracrylique R (+).

0,8 g d'acide et 1,2 g de brucine dans 4 cm³ d'éthanol; 1,1 g de sel sont obtenus après cristallisation au réfrigérateur. Une recristallisation de ce sel dans 4 cm³ d'éthanol donne 0,5 g, décomposé en 0,15 g d'acide $[\alpha]_J^{25} = +32°$. Celui-ci, recristallisé dans le chloroforme/hexane donne 60 mg d'acide optiquement pur, F = 77°, $[\alpha]_J^{25} = +39,6°$ ($CHCl_3$, c = 0,8).

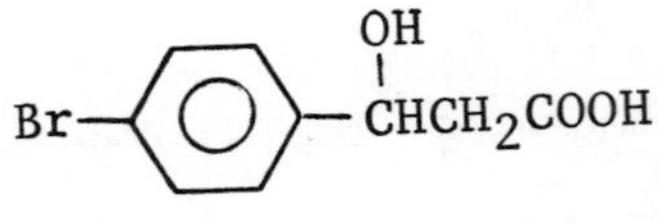

$C_9H_9BrO_3$ **Ref. 481**

Acide p-bromophényl hydracrylique R (+).

1,6 g de brucine et 1 g d'acide dans 10 cm³ d'éthanol donnent 1,3 g de sel, Celui-ci, recristallisé dans 10 cm³ d'éthanol et décomposé donne 0,21 g d'acide $[\alpha]_J^{25} = +35,5°$; recristallisation dans l'éther/hexane: F = 125°, $[\alpha]_J^{25} = +42,5°$ ($CHCl_3$, c = 0,5).

L'acide (—) optiquement pur est facilement obtenu en décomposant les eaux-mères du premier sel de brucine et recristallisant l'acide partiellement dédoublé obtenu.

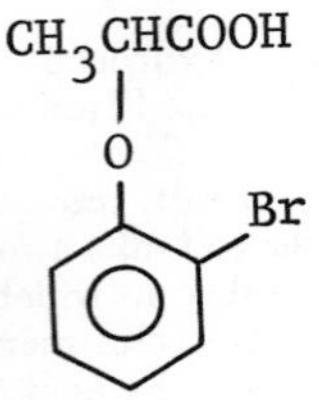

$C_9H_9BrO_3$ **Ref. 482**

Preliminary experiments on resolution were carried out with various bases in dilute ethanol. *Strychnine* gave (−)-acid of rather good activity while *cinchonine* gave only poor resolution. Dextrorotatory acid was obtained with *cinchonidine* and *dehydroabietylamine* [14], the latter base giving the higher activity. *Quinine* gave only salt of the racemic acid. The salts of *brucine*, *quinidine* and *β-phenyl-isopropylamine* failed to crystallise.

(−)-*2-Bromophenoxy-propionic acid*. Racemic acid (24.5 g, 0.1 mole) and strychnine (29.5 g, 0.088 mole) were dissolved in 200 ml ethanol and 200 ml water were added. The salt (21.4 g) was filtered off after 24 h and recrystallised from dilute ethanol as seen below. After each crystallisation a small sample of the salt was decomposed and the rotatory power of the liberated acid determined in acetone solution.

Crystallisation	1	2	3	4	5
ml ethanol	200	125	55	50	40
ml water	200	125	55	50	40
g salt obtained	21.4	19.3	16.4	15.3	14.4
$[\alpha]_D^{25}$ of the acid	−14°	−15.5°	−17°	−17°	−17°

The salt (14.4 g) was shaken with excess sodium carbonate solution and the liberated strychnine taken up in chloroform. The acid was then isolated from the aqueous layer by acidification with sulphuric acid and extraction with ether. The yield of crude acid was 5.3 g; recrystallisation from 50% formic acid gave 4.9 g pure acid as fine, glistening needles with m.p. 115–116°.

$C_9H_9O_3Br$ (245.08)

Calc.	C 44.10	H 3.69	Br 32.61
Found	C 44.27	H 3.77	Br 32.42

0.2520 g dissolved in *acetone* to 10.00 ml: $\alpha_D^{25} = -0.85°$ (2 dm). $[\alpha]_D^{25} = -16.9°$; $[M]_D^{25} = -41.3°$.
0.2515 g dissolved in absolute *ethanol* to 10.00 ml: $\alpha_D^{25} = -0.93°$ (2 dm). $[\alpha]_D^{25} = -18.5°$; $[M]_D^{25} = -45.3°$.

(+)-*2-Bromophenoxy-propionic acid*. The mother liquor from the first crystallisation of the strychnine salt was treated with sodium carbonate solution and the acid isolated as described above; 13.2 g having $[\alpha]_D^{25} = +7.5°$ were obtained. This acid (0.054 mole) was carefully neutralised with ammonia in 50 ml water. Dehydroabietylamine acetate (19.8 g, 0.054 mole), prepared according to Sjöberg and Sjöberg [14], was dissolved in 90 ml ethanol and the solutions were mixed. The readily crystallising salt was filtered off after 24 h and recrystallised to maximum activity of the acid; the course of the resolution was followed as described above.

Crystallisation	1	2	3	4
ml ethanol	90	110	180	150
ml water	50	50	30	75
g salt obtained	20.6	15.8	13.5	12.3
$[\alpha]_D^{25}$ of the acid	+14°	+15°	+17°	+17°

The salt was decomposed and the acid liberated as described above for the strychnine salt. The yield of crude acid was 5.3 g; recrystallisation from 50% formic acid gave 5.0 g of pure acid, quite similar to the enantiomer. M.p. 115–116°.

$C_9H_9O_3Br$ (245.08)

Calc.	C 44.10	H 3.69	Br 32.61
Found	C 44.23	H 3.69	Br 32.42

The rotatory power in various solvents is given in Table 1.

Table 1

Solvent	g[a]	α_D^{25} (2 dm)	$[\alpha]_D^{25}$	$[M]_D^{25}$
Acetone	0.2507	+0.85°	+17.0°	+41.5°
Ethanol (abs.)	0.2514	+0.92°	+19.3°	+44.8°
Chloroform	0.2195	+0.08°	+1.8°	+4.5°
Water (neutr.)	0.2067	−1.295°	−31.3°	−76.8°
Benzene	0.2099	−1.225°	−29.2°	−71.5°

[a] g = g acid dissolved to 10.00 ml.

2-Methylphenoxy-propionic acid

The racemic acid was prepared from commercial *o*-cresol and ethyl α-bromopropionate following conventional methods. The crude acid separated as an oil, which crystallised rather slowly. It was recrystallised twice from 50% formic acid and obtained as glistening scales or flakes with m.p. 92.5–94°.

$C_{10}H_{12}O_3$ (180.20)

Calc. equiv. wt.	180.2	C 66.65	H 6.70
Found equiv. wt.	179.3	C 66.68	H 6.71

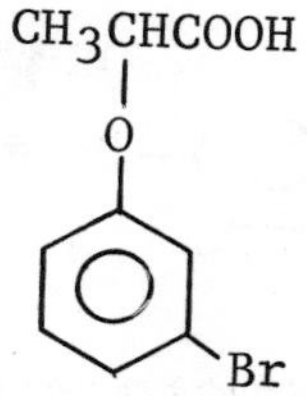

$C_9H_9BrO_3$

Ref. 483

3-Bromo-phenoxypropionic acid

The racemic acid was prepared in the usual way from 3-bromophenol and ethyl α-bromopropionate. It was recrystallised three times from benzene and once from 50-% formic acid. M.p. 110–111.5°.

Preliminary experiments on resolution. The bases found suitable for resolving the analogous 3-chloro acid [1] were first tested. Strychnine in dilute ethanol gave a strongly laevorotatory acid. (−)-*β*-Phenyl-isopropylamine (benzedrine) yielded

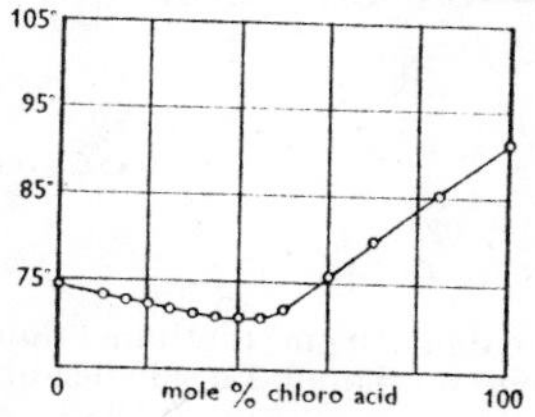

Fig. 5.

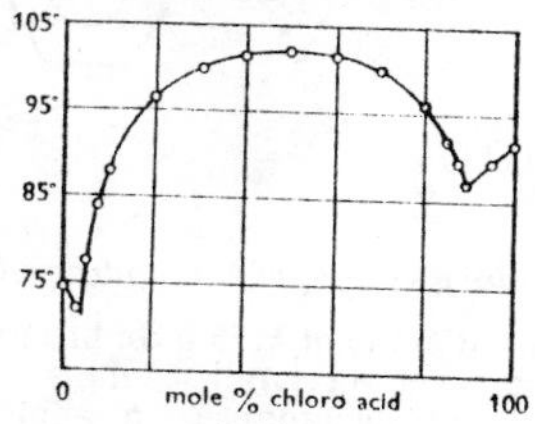

Fig. 6.

Fig. 5. (−)-3-Methyl-phenoxypropionic acid and (−)-3-chloro-phenoxypropionic acid.
Fig. 6. (+)-3-Methyl-phenoxypropionic acid and (−)-3-chloro-phenoxypropionic acid.

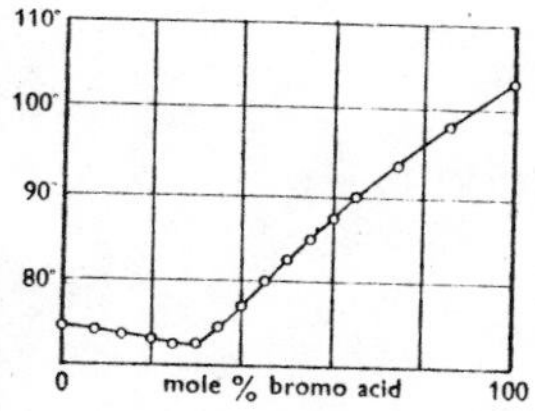

Fig. 7.

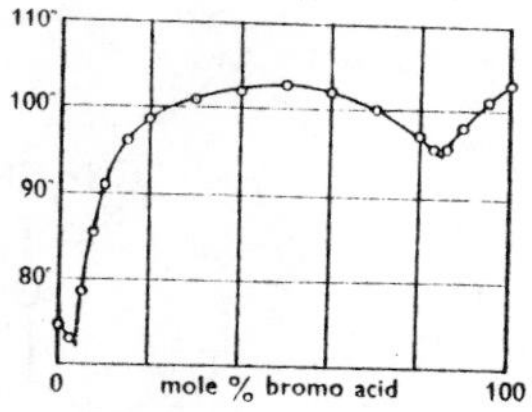

Fig. 8.

Fig. 7. (−)-3-Bromo-phenoxypropionic acid and (−)-3-methyl-phenoxypropionic acid.
Fig. 8. (−)-3-Bromo-phenoxypropionic acid and (+)-3-methyl-phenoxypropionic acid.

(+)-acid of rather low activity. The base could, however, be used for bringing to maximum activity the strongly dextrorotatory acid isolated from the mother liquor after the strychnine salt. Experiments with other bases were not encouraging.

(−)-3-Bromo-phenoxypropionic acid. 24.5 g (0.10 mole) racemic acid and 33.4 g (0.10 mole) strychnine were dissolved in a mixture of 220 ml ethanol and 250 ml water. After cooling and seeding with crystals from a preliminary experiment, the solution was left in a refrigerator for 24 hours. The salt obtained was recrystallised from dilute ethanol; after each crystallisation a small sample was decomposed, the acid isolated and its rotatory power determined in acetone solution.

Crystallisation	1	2	3	4	5
ml ethanol	220	55	45	45	45
ml water	250	70	55	55	55
g salt obtained	25.6	24.0	22.1	21.6	20.7
$[\alpha]_D^{25}$ of the acid	−44.5°	−45.5°	−46.5°	−47°	−46.5°

The acid is thus already optically pure in the third crystallisation. The strychnine salt was decomposed with sodium carbonate solution and the strychnine extracted with chloroform. The solution was then acidified with excess sulphuric acid and the active acid extracted with ether. After recrystallisation from ligroin, it was obtained as colourless needles with m.p. 105–106°. The yield was 8.0 g (65.3 %).

0.38345 g acid: 15.62 ml 0.1003 *N* NaOH.
$C_9H_9O_3Br$ Equiv. wt. calc. 245.1 found 244.8

The rotatory power in various solvents is given in Table 1. A sample was recrystallised from dilute formic acid; the rotatory power remained unchanged.

Table 1. g = g acid dissolved to 10.00 ml.

Solvent	g	α_D^{25} (2 dm)	$[\alpha]_D^{25}$	$[M]_D^{25}$
Acetone	0.2003	−1.858°	−46.4°	−113.7°
Ethanol (abs.)	0.2038	−1.319°	−32.4°	−79.4°
Chloroform	0.1987	−0.763°	−19.2°	−47.1°
Water (neutr.)	0.1957	−0.320°	−8.2°	−20.1°
Benzene	0.1985	±0.00°	0°	0°

(+)-Bromo-phenoxypropionic acid. On evaporation of the mother liquor from the first crystallisation of the strychnine salt, the salt of the impure (+)-acid separated as a syrup. After decomposition it yielded 11.9 g (0.048 mole) of strongly dextrorotatory acid. This acid was dissolved with 6.55 g (0.048 mole) of (−)-*β*-phenyl-isopropylamine in 150 ml water + 65 ml ethanol. The salt has a tendency to separate as an emulsion, but if the solution is cooled with efficient stirring it is obtained as aggregates of fine hairlike needles, not unlike cotton-wool. It was recrystallised in the same way from dilute ethanol to maximum activity of the acid. As the rotatory power increased rather slowly and several crystallisations were necessary, data are not given for every crystallisation.

Crystallisation	1	2	3	6	8	10	12
ml ethanol	65	30	15	10	8	10	12
ml water	150	75	50	45	40	6	7
g salt obtained	12.9	11.6	10.2	8.1	6.8	30	35
$[\alpha]_D^{25}$ of the acid	+39°	+42°	+42.5°	+44.5°	+45.5°	5.8	4.5
						+46°	+46.5°

The salt (4.5 g) was decomposed with dilute sulphuric acid and the active acid extracted with ether. After recrystallisation from 50-% formic acid it was obtained as colourless, glistening needles. M.p. 105–106°. The yield was 2.6 g.

0.1244 g acid: 5.05 ml 0.1003 *N* NaOH.
$C_9H_9O_3Br$ Equiv. wt calc. 245.1 found 245.6
0.1895 g dissolved in *acetone* to 10.00 ml: $\alpha_D^{25} = +1.758°$ (2 dm).
$[\alpha]_D^{25} = +46.4°$; $[M]_D^{25} = +113.7°$.

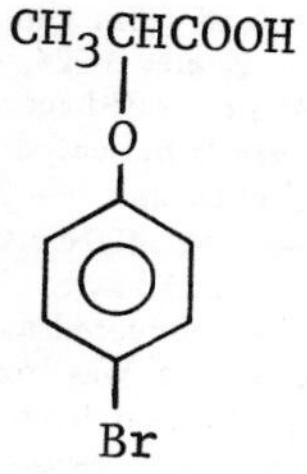

$C_9H_9BrO_3$

Ref. 484

(−)-4-Bromophenoxy-propionic acid. 27.0 g (0.11 mole) of racemic acid and 43.3 g (0.11 mole) of anhydrous brucine were dissolved in 200 ml of hot methanol and 225 ml of water were added. Crystallisation of the salt occurred spontaneously. After standing over-night in a refrigerator, the salt was filtered off, washed with a little water and dried. It was recrystallised several times from dilute ethanol; after each crystallisation a small sample of the acid was liberated and the rotatory power determined in acteone solution. As seen from the data, the (−)-acid was already pure after a single recrystallisation.

Crystallisation	1	2	3	4
ml ethanol	200	400	400	350
ml water	225	425	425	400
g salt obtained	37.5	34.7	32.8	30.6
$[\alpha]_D^{25}$ of the acid	−47°	−50°	−50°	−50°

The salt (30.6 g) was suspended in sodium carbonate solution and the brucine extracted with chloroform. The acid was precipitated from the aqueous solution (ca. 200 ml) by adding a slight excess of concentrated hydrochloric acid. It was twice recrystallised from dilute formic acid, yielding 9.9 g of small short-prismatic crystals. M.p. 108.5–109.5°.

$C_9H_9O_3Br$ (245.08) calc. equiv. wt 245.08 Br 32.61
found equiv. wt 245.5 Br 32.39, 32.47
0.2144 g acid dissolved in *acetone* to 10.00 ml: $\alpha_D^{25} = -2.165°$ (2 dm). $[\alpha]_D^{25} = -50.5°$; $[M]_D^{25} = -123.7°$.

(+)-4-Bromophenoxy-propionic acid. The brucine salt obtained in the first crystallisation (see above) contained a (−)-acid of 94 % activity. As this fraction was about 52 % of the total amount of brucine salt (calculated on racemic acid), it could be concluded that the mother liquor must contain practically pure (+)-acid.

The mother liquor was evaporated, the remaining salt decomposed with sodium hydroxide solution and the brucine extracted with chloroform. The remaining aqueous solution was diluted to 650 ml and acidified by adding concentrated hydrochloric acid in excess. The (+)-acid separated as a voluminous mass of felted needles, which on standing changed into a rapidly sedimenting precipitate of small, compact short-prismatic crystals. The needles obviously represent an unstable modification. The precipitate was filtered off and dried, yielding 12.8 g acid with $[\alpha]_D^{25} = -50.1°$ (acetone). After two recrystallisations from dilute formic acid 11.3 g remained. It was quite similar to the (−)-acid. M.p. 108.5–109.5°.

$C_9H_9O_2Br$ (245.08) calc. equiv. wt 245.08 Br 32.61
found equiv. wt 245.5 Br 32.41, 32.66

The rotatory power in various solvents is given in the table below.

Solvent	g	α_D^{25} (2 dm)	$[\alpha]_D^{25}$	$[M]_D^{25}$
Acetone	0.1930	+1.945°	+50.4°	+123.5°
Ethanol (abs.)	0.2006	+1.43°	+35.6°	+87.4°
Chloroform	0.1846	+0.76°	+20.6°	+50.4°
Benzene	0.1922	−0.26°	−6.8°	−16.6°
Water (neutr.)	0.2219	+0.75°	+16.9°	+41.4°

g = g acid dissolved to 10.00 ml

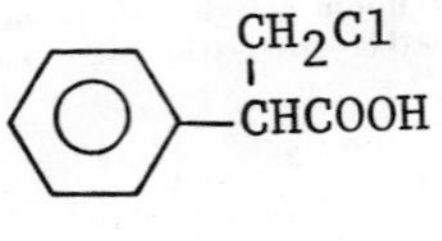

$C_9H_9ClO_2$ **Ref. 485**

Resolution of r-β-*Chlorohydratropic Acid*.—The solution obtained by boiling 49·5 g. of morphine (1 mol.) in 820 c.c. of methyl alcohol was cooled to 42°, and 30 g. of *r*-β-chlorohydratropic acid (1 mol.; prepared from atropic acid: McKenzie and Wood, *loc. cit.*) were added in one instalment with vigorous stirring. The acid dissolved quickly. When the temperature had fallen to 39°, glassy prisms started to separate. After 18 hours in the ice-chest, 56 g. of solid were deposited. After the evaporation of the methyl alcohol from the filtrate, the acid obtained from the latter gave the following rotation in benzene: $l = 2$, $c = 3{\cdot}924$, $\alpha_D^{16°} + 3{\cdot}71°$, $[\alpha]_D^{16°} + 47{\cdot}3°$, so that the resolution had proceeded markedly after one crystallisation. The further progress is indicated as follows: 56 g. crystallised from 1350 c.c. of methyl alcohol ⟶ 35 g., 850 c.c. of alcohol ⟶ 19 g., 550 c.c. of alcohol ⟶ 13 g., 400 c.c. of alcohol ⟶ 9 g., 360 c.c. of alcohol ⟶ 6 g. pure morphine *l*-salt. The solution of the crystals in methyl alcohol should be conducted as expeditiously as possible. The progress of the resolution was tested by the polarimetric examination of the acids obtained from the mother-liquors. The morphine salt, which separates in rectangular prisms, was decomposed by dilute sulphuric acid, the solution extracted with ether nine times, and the ethereal extract dried with anhydrous sodium sulphate. The resulting acid (2·1 g.), after drying in a vacuum until constant in weight, gave the following rotation in benzene: $l = 2$, $c = 3{\cdot}536$, $\alpha_D^{16°} - 8{\cdot}66°$, $[\alpha]_D^{16°} - 122{\cdot}4°$. The acid was crystallised from light petroleum (b. p. 60—80°), and gave a value for the rotatory power agreeing with the above within the limits of experimental error: $l = 2$, $c = 3{\cdot}536$, $\alpha_D^{16°} - 8{\cdot}67°$, $[\alpha]_D^{16°} - 122{\cdot}6°$.

l-β-*Chlorohydratropic acid*, m. p. 62·5—63·5°, separates from light petroleum in rosettes of glassy plates. It is sparingly soluble in light petroleum, and readily so in ether, carbon disulphide, benzene, and ethyl alcohol (Found: Cl = 19·39. $C_9H_9O_2Cl$ requires Cl = 19·24%).

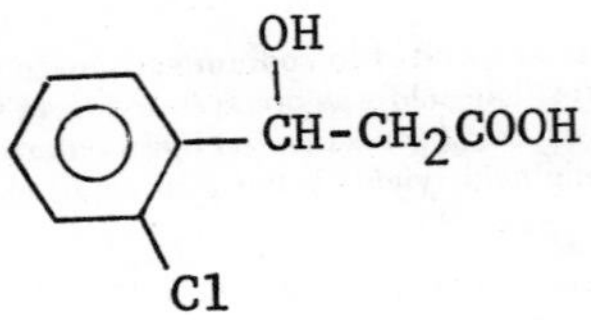

$C_9H_9ClO_3$ **Ref. 486**

Acide o-chlorophényl hydracrylique R (+).

1 g d'acide, 2 g de brucine et 10 cm³ d'éthanol donnent 1 g de sel, recristallisé 2 fois dans 10 cm³ d'éthanol et décomposé par HCl dilué. L'acide obtenu (0,2 g, $[\alpha]_J^{25} = +90{,}4°$) est recristallisé dans l'éther hexane: F = 84°, $[\alpha]_J^{25} = +94{,}8°$ ($CHCl_3$, $c = 1$).

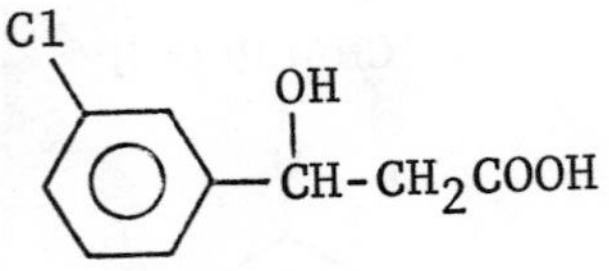

$C_9H_9ClO_3$ **Ref. 487**

Acide m-chlorophényl hydracrylique R (+).

4 g d'acide, 7,7 g de brucine et 15 cm³ d'éthanol, sont abandonnés à 5°. Une recristallisation du sel obtenu (4,4 g) dans 4,5 cm³ d'éthanol au réfrigérateur donne 2,8 g de sel, décomposé en acide $[\alpha]_J^{25} = +37°$. Ce dernier (0,7 g), recristallisé dans l'éther-hexane donne 0,3 g d'acide optiquement pur, F = 95°, $[\alpha]_J^{25} = +57{,}5°$ ($CHCl_3$, $c = 0{,}55$).

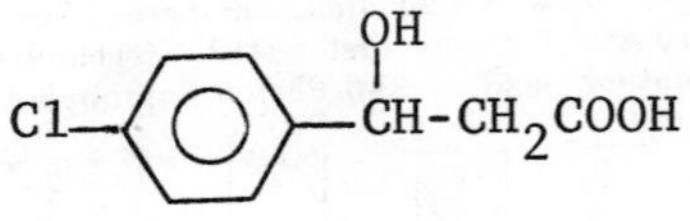

$C_9H_9ClO_3$ **Ref. 488**

Acide p-chlorophényl hydracrylique R (+).

0,9 g d'acide et 1,75 g de brucine dans 10 cm³ d'éthanol donnent 1,22 g de sel, recristallisés dans 8 cm³ d'éthanol. Le sel ainsi obtenu (0,7 g) est décomposé en acide $[\alpha]_J^{25} = +41°$ (0,18 g). Après recristallisation dans l'éther/hexane, 0,1 g, F = 112°, $[\alpha]_J^{25} = +50°$ ($CHCl_3$, $c = 0{,}65$).

L'acide (—) est obtenu optiquement pur sans difficulté après décomposition des eaux-mères du premier sel de brucine et recristallisation de l'acide.

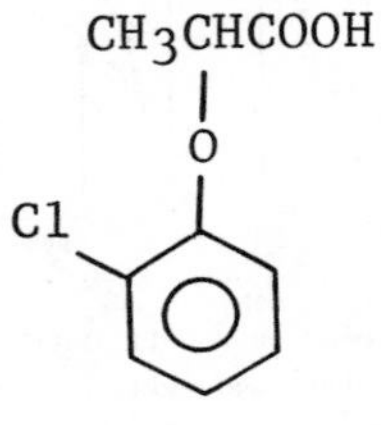

$C_9H_9ClO_3$ **Ref. 489**

24.0 g (0.12 mole) acid and 40 g (0.12 mole) strychnine were dissolved in a mixture of 120 ml 96 % ethanol and 200 ml water. The salt was filtered off after 20 hours and recrystallised three times from dilute ethanol. After each crystallisation, a small sample of the salt was decomposed and the activity of the acid determined in 96 % ethanol. The results are given in Table 3; the (−)-acid is practically pure already in the first crystallisation.

Table 3.

Crystallisation	1	2	3	4
ml ethanol	120	50	20	20
ml water	200	100	40	40
g salt obtained	21.0	18.8	16.5	15.5
$[\alpha]_D^{25}$ of the acid	−25.5°	−26°	−25.5°	−25°

(−)-2-Chlorophenoxy-propionic acid. 14.6 g of the strychnine salt was decomposed with 120 ml of 2-*M* sodium carbonate solution and the strychnine removed by repeated extractions with chloroform. After acidification with sulphuric acid in excess, the 2-chloro acid was extracted with ether. The yield of crude product was 4.95 g; after two recrystallisations from cyclohexane 4.2 g remained. The acid forms very thin glistening needles with m.p. 95–95.5°

$C_9H_9O_3Cl$ (200.62) Cl calc. 17.67; found 17.69.

0.2036 g acid dissolved in *acetone* to 10.00 ml: $\alpha_D^{25} = -1.225°$ (2 dm). $[\alpha]_D^{25} = -30.1°$; $[M]_D^{25} = -60.4°$. 0.2032 g acid dissolved in *abs. ethanol* to 10.00 ml: $\alpha_D^{25} = -1.23°$ (2 dm). $[\alpha]_D^{25} = -30.3°$; $[M]_D^{25} = -60.7°$. 0.2075 g acid dissolved in *96 % ethanol* to 10.00 ml: $\alpha_D^{25} = -1.09°$ (2 dm). $[\alpha]_D^{25} = -26.3°$; $]M]_D^{25} = -52.7°$. 0.2038 g acid dissolved in *chloroform to* 10.00 ml: $\alpha_D^{25} = -0.59°$ (2 dm). $[\alpha]_D^{25} = -14.5°$; $[M]_D^{25} = -29.0°$. 0.2027 g acid dissolved in *benzene* to 10.00 ml: $\alpha_D^{25} = +0.785°$ (2 dm). $[\alpha]_D^{25} = +19.4°$; $[M]_D^{25} = +38.8°$. 0.2045 g acid *neutralised* and dissolved in *water* to 10.00 ml: $\alpha_D^{25} = +0.995°$ (2 dm). $[\alpha]_D^{25} = +24.3°$; $[M]_D^{25} = +48.8°$.

(+)-2-Chlorophenoxy-propionic acid. The mother liquor from the first crystallisation of the strychnine salt yielded 10 g acid having $[\alpha]_D \sim +10°$ (abs. ethanol). This was powdered and digested with two 40 ml portions of cold carbon tetrachloride. The acid obtained from the extracts (3.8 g) was recrystallised twice from cyclohexane, yielding 3.2 g with $[\alpha]_D^{25} = +27.5°$ (acetone). Further recrystallisations from the solvent had no effect. 2.3 g of the acid were therefore recrystallised from dilute formic acid until maximum activity was obtained (Table 4).

Table 4.

Crystallisation	1	2	3	4
g acid	2.0	–	1.4	1.2
$[\alpha]_D^{25}$ in acetone	+28.3°	+29.7°	+30.2°	+30.1°

The acid forms very thin, glistening needles, and is quite similar to the (−)-form. M.p. 95–95.5°

$C_9H_9O_3Cl$ (200.62) Cl calc. 17.67; found 17.69.

0.2036 g acid (3rd crystallisation) dissolved in *acetone* to 10.00 ml: $\alpha_D^{25} = +1.23°$ (2 dm). $[\alpha]_D^{25} = +30.2°$; $[M]_D^{25} = +60.6°$. 0.2086 g acid (4th crystallisation) dissolved in *acetone* to 10.00 ml: $\alpha_D^{25} = +1.255°$ (2 dm). $[\alpha]_D^{25} = +30.1°$; $[M]_D^{25} = +60.3°$.

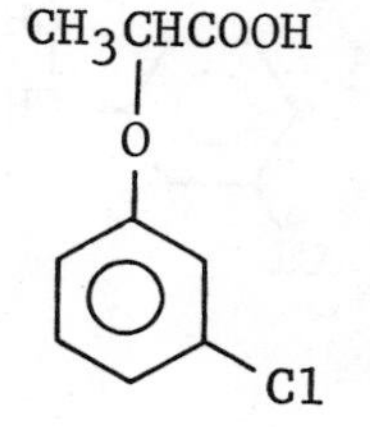

$C_9H_9ClO_3$ Ref. 490

Resolution of the 3-chlorophenoxy-propionic acid (Experiments by C. Grönwall). Preliminary experiments were carried out as described above. As seen from Table 5, strychnine gives by far the best resolution. (−)-Phenyl-isopropylamine (benzedrine) could be used for purification of the (+)-form.

Table 5.

Base	Solvent	Yield of salt, %	$[\alpha]_D^{25}$
Strychnine	Ethanol (50 %)	56	−35°
Strychnine	Ethanol (60 %)	30	−39°
Cinchonidine	Ethanol (50 %)	70	+2°
Quinine	Ethanol (50 %)	76	−10°
(−)-α-Phenyl-isopropylamine	Ethanol (15 %)	52	+9°

30 g (0.15 mole) acid and 47 g (0.15 mole) strychnine were dissolved together in 225 ml of hot 96 % ethanol, and 225 ml of hot water were added. The salt was filtered off after 8 hours and recrystallised until maximum activity. The course of the resolution was followed as described above (Table 6).

Table 6.

Crystallisation	1	2	3	4	5
ml ethanol	225	100	70	60	60
ml water	225	150	105	90	90
g salt obtained	24.1	18.7	16.4	13.8	12.3
$[\alpha]_D^{25}$ of the acid	−32°	−36.5°	−35.5°	−36°	−36°

The mother liquor from the first crystallisation was left in a refrigerator overnight. 9.8 g Salt separated and was recrystallised utilising the successive mother liquors from the previous experiment. In this way an additional crop of 9.4 g pure (−)-salt was obtained.

(−)-3-Chlorophenoxy-propionic acid. The total amount of pure strychnine salt (20.8 g) was decomposed and the acid liberated as described above for the 2-chloro acid. The yield of crude product was 6.6 g; after two recrystallisations from dilute formic acid 6.4 g remained. The acid forms glistening needles with m.p. 93.5–94.5°.

$C_9H_9O_3Cl$ (200.62) Cl calc. 17.67; found 17.54.

0.2050 g acid dissolved in *acetone* to 10.00 ml: $\alpha_D^{25} = -2.23°$ (2 dm). $[\alpha]_D^{25} = -54.4°$; $[M]_D^{25} = -109.1°$. 0.2038 g acid dissolved in *abs. ethanol* to 10.00 ml: $\alpha_D^{25} = -1.59°$ (2 dm). $[\alpha]_D^{25} = -39.0°$; $[M]_D^{25} = -78.3°$. 0.2093 g acid dissolved in *96 % ethanol* to 10.00 ml: $\alpha_D^{25} = -1.495°$ (2 dm). $[\alpha]_D^{25} = -35.7°$; $[M]_D^{25} = -71.7°$. 0.2044 g acid dissolved in *chloroform* to 10.00 ml: $\alpha_D^{25} = -0.995°$ (2 dm). $\alpha_D^{25} = -24.3°$; $[M]_D^{25} = -48.8°$. 0.2023 g acid dissolved in *benzene* to 10.00 ml: $\alpha_D^{25} = +0.055°$ (2 dm): $[\alpha]_D^{25} = +1.4°$; $[M]_D^{25} = +2.7°$. 0.2058 g acid *neutralised* and dissolved in *water* to 10.00 ml: $\alpha_D^{25} = -0.43°$ (2 dm). $[\alpha]_D^{25} = -10.4$; $[M]_D^{25} = -21.0°$.

(+)-3-Chlorophenoxy-propionic acid. On evaporation, the mother liquors from the first crystallisation of the strychnine salt gave several salt fractions, from which in all 13.5 g more or less dextrorotatory acid was isolated. Three extractions with 40 ml portions of cold carbon tetrachloride yielded 8.2 g acid having $[\alpha]_D^{25} \sim +50°$ (acetone). On recrystallisation from dilute formic acid, the rotatory power increased very slowly. In this respect there seems to be an obvious difference between the 2-chloro and the 3-chloro acids.

5.0 g (0.025 mole) of the strongly dextrorotatory 3-acid were therefore dissolved with 3.3 g (0.025 mole) of (−)-phenyl-isopropylamine in 40 ml water + 8 ml ethanol. The salt obtained (fine needles) was recrystallised from 30 ml water + 6 ml ethanol. On decomposition it yielded 3.4 g acid. After recrystallisation from dilute formic acid 3.2 g of glistening needles, quite similar to the (−)-form were obtained. M.p. 93.5–94.5°.

$C_9H_9O_3Cl$ (200.62) Cl calc. 17.67; found 17.57.

0.2061 g acid dissolved in *acetone* to 10.00 ml: $\alpha_D^{25} = +2.245°$ (2 dm). $[\alpha]_D^{25} = -54.5°$; $[M]_D^{25} = +109.3°$.

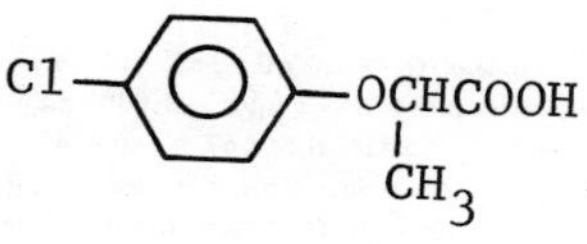

$C_9H_9ClO_3$ Ref. 491

α-*(4-Chlorophenoxy)propionic acid (I)*.—The racemic acid was prepared in the usual manner by refluxing equimolecular parts of sodium 4-chlorophenolate and ethyl α-bromopropionate in abs. ethanol, followed by hydrolysis with aqueous sodium hydroxide. The crude acid was recrystallised from 50 % formic acid. Yield of pure acid 70 %, m.p. 114.5–115.5° in good agreement with the literature (7).

80.31 mg acid: 8.06 ml 0.0495-*N* NaOH

$C_9H_9O_3Cl$ calc. equiv. wt. 200.6

found ,, ,, 201.1

Preliminary experiments on resolution.—Of the common alkaloids only brucine gave an appreciable resolution; from the salt a laevorotatory acid was isolated. Cinchonine, quinidine and strychnine yielded crystalline salts from which nearly quite inactive products were obtained. Good results were recorded with optically active α-phenethylamine and β-phenyl*iso*propylamine. The latter base and brucine were used for the final resolution.

(−)-α-*(4-Chlorophenoxy)propionic acid.*—Inactive acid (30.1 g, 0.15 mole) and an equivalent amount of brucine (38.6 g) were dissolved in a hot mixture of ethanol (200 ml) and water (300 ml). On standing at room temperature the salt crystallised and was collected after a day. The salt was recrystallised several times as seen from Table 2; the rotatory power of the acid was measured in abs. ethanol.

Table 2. Recrystallisation of the brucine salt of α-(4-chlorophenoxy)-propionic acid.

Cryst. No.	Solvent (ml)		Weight of salt (g)	$[\alpha]_D$ of the acid
	Water	Ethanol		
1	200	300	65.9	–
2	300	300	39.1	−35.7°
3	250	250	35.1	−38.9°
4	250	250	31.7	−38.3°
5	200	200	29.4	−39.4°

28.4 g of the pure salt was decomposed with dilute hydrochloric acid yielding 8.9 g crude acid. It was recrystallised first from 50 % formic acid and then from petroleum (b.p. 60–80°). The pure (−)-acid (7.2 g) melted at 103.5–104.5°.

76.58 mg (−)-acid: 8.18_5 ml 0.0464-*N* NaOH

$C_9H_9O_3Cl$ calc. equiv. wt. 200.6

found ,, ,, 201.6

0.2022 g dissolved in abs. *ethanol* to 20.06 ml: $2\alpha_D^{25} = -0.808°$. $[\alpha]_D^{25} = -40.1°$; $[M]_D^{25} = -80.4°$.

(+)-α-*(4-Chlorophenoxy)propionic acid.*—From the mother liquors 1 and 2 (Table 2) an acid with $[\alpha]_D = +28°$ was isolated. This acid (15.1 g, 0.075 mole) and an equivalent amount of (−)-β-phenyl*iso*propylamine (10.1 g) were dissolved in hot 50 % ethanol (150 ml). On adding water (175 ml) the salt crystallised and was collected after standing some hours in a refrigerator. The salt was recrystallised according to Table 3. The rotatory power of the acid was measured in abs. ethanol.

Table 3. Recrystallisation of the (−)-β-phenyl*iso*propylamine salt of α-(4-chlorophenoxy)propionic acid.

Cryst. No.	Solvent (ml)		Yield of salt (g)	$[\alpha]_D$ of the acid
	Water	Ethanol		
1	75	250	17.8	+39.1°
2	75	200	14.2	+39.1°
3	50	150	11.7	+39.8°

The acid was isolated from 11.2 g salt from the last fraction. The crude acid (6.9 g) was recrystallised first from 50 % formic acid and then from petroleum (b.p. 60–80°) yielding 4.5 g of the pure (+)-acid, m.p. 103.5–104.7°.

81.76 mg (+)-acid: 8.76_5 ml. 0.0464-*N* NaOH
$C_9H_9O_3Cl$ calc. equiv. wt. 200.6
found ,, ,, 201.0

0.1753 g dissolved in abs. *ethanol* to 20.06 ml: $2\alpha_D^{25} = 0.716°$. $[\alpha]_D^{25} = +41.0°$; $[M]_D^{25} = +82.2°$.
0.1816 g dissolved in *acetone* to 20.06 ml: $2\alpha_D^{25} = +1.039°$. $[\alpha]_D^{25} = +57.4°$; $[M]_D^{25} = +115.1°$.
0.2028 g dissolved in *chloroform* to 20.06 ml: $2\alpha_D^{25} = +0.474°$. $[\alpha]_D^{25} = +23.4°$; $[M]_D^{25} = +47.0°$.
0.2101 g dissolved in *benzene* to 20.06 ml: $2\alpha_D^{25} = -0.232°$. $[\alpha]_D^{25} = -11.1°$; $[M]_D^{25} = -22.2°$.
0.1779 g neutralized with *aqueous NaOH* and made up to 20.06 ml: $2\alpha_D^{25} = +0.318°$. $[\alpha]_D^{25} = +17.9°$; $[M]_D^{25} = +36.0°$.

$C_9H_9ClO_3$ Ref. 491a

L-(−)-α-(4-Chlorophenoxy)propionic Acid (7).—Finely powdered (−)-brucine (16.7 g, 0.042 mole) was suspended in 2 l. of boiling 20% EtOH in H_2O. To the stirred suspension was added 20 g (0.1 mole) of DL-7 in 50 ml of 1 *N* NaOH and 50 ml of EtOH. The mixture was stirred with boiling until a clear solution resulted. The solution was filtered and on cooling the salt crystallized. Five additional recrystallizations from 20% EtOH–H_2O afforded 10.6 g (44%) of brucine salt, $[\alpha]^{25}_D$ −25.4° (*c* 1.9720, MeOH). Utilizing twelve times these amounts 122 g of optically pure salt was obtained. The acid was liberated from 122 g of salt by acidification with 5% H_2SO_4 and extraction with ether. The ether was dried (Na_2SO_4), filtered, and removed under reduced pressure affording after recrystallization from petroleum ether (60–80°), 28.8 g (70%) of L-(−)-7: mp 104–105°, lit.[19a] mp 103.5–104.5°; $[\alpha]^{25}_D$ −34.95° (*c* 5.0078, MeOH), lit.[19a] $[\alpha]^{25}_D$ −40.1° (EtOH); RD (*c* 0.0242, MeOH) (24–25°), $[\phi]_{325}$ −567°, $[\phi]_{310}$ −803°, $[\phi]_{300}$ −1320°, $[\phi]_{290}$ −2650°, $[\phi]_{283}$ −1800 (sh), $[\phi]_{277}$ 0°, $[\phi]_{267}$ +1130°, $[\phi]_{264}$ +1180, $[\phi]_{260}$ +990, $[\phi]_{250}$ 0°, $[\phi]_{245}$ −710°, $[\phi]_{240}$ −1460°; CD (*c* 0.2420, MeOH) (24–25°), $[\theta]_{300}$ 0°, $[\theta]_{289}$ −2620°, $[\theta]_{285}$ −1750°, $[\theta]_{282}$ −2990°, $[\theta]_{273}$ −2310°, $[\theta]_{254}$ −500°, $[\theta]_{240}$ −2500°.

D-(+)-α-(4-Chlorophenoxy)propionic acid (7) was obtained from DL-7 by the method of Smith, *et al.*,[20] for resolving similar compounds. Resolution was accomplished utilizing (+)-yohimbine obtained from the HCl salt. Three recrystallizations of the yohimbine salt of D-(+)-7 from Me_2CO afforded white crystals, $[\alpha]^{25}$ +64.8° (*c* 2.3220, MeOH). The D-(+) acid was liberated from the salt as in L-(−)-7 above. Recrystallization from petroleum ether (60–80°) afforded white crystals: mp 104–105°, lit.[19a] mp 103.5–104.7°; $[\alpha]^{25}_D$ +34.1° (*c* 3.6720, MeOH), lit.[19a] $[\alpha]^{25}_D$ +39.8 (EtOH); RD (*c* 0.0261, MeOH) (24–25°), $[\phi]_{325}$ +610°, $[\phi]_{310}$ +830°, $[\phi]_{300}$ +1180°, $[\phi]_{290}$ +2580°, $[\phi]_{283}$ +1700° (sh), $[\phi]_{277}$ 0°, $[\phi]_{269}$ −660°, $[\phi]_{264}$ −880°, $[\phi]_{250}$ 0°, $[\phi]_{24}$ +790°, $[\phi]_{240}$ +1490°; CD (*c* 0.2610, MeOH) (24–25°), $[\theta]_{300}$ 0°, $[\theta]_{288}$ +1850°, $[\theta]_{285}$ +2600°, $[\theta]_{270}$ +1560°, $[\theta]_{268}$ +1450°, $[\theta]_{253}$ +460°, $[\theta]_{240}$ +2020°.

(19) (a) M. Matell, *Arkiv Kemi*, **7**, 437 (1954); (b) D. Osborne, R. L. Wain, and R. Walker, *J. Hort. Sci.*, **27**, 44 (1952).
(20) M. S. Smith, R. L. Wain, and F. Wightman, *Ann. Appl. Biol.*, **39**, 295 (1952).

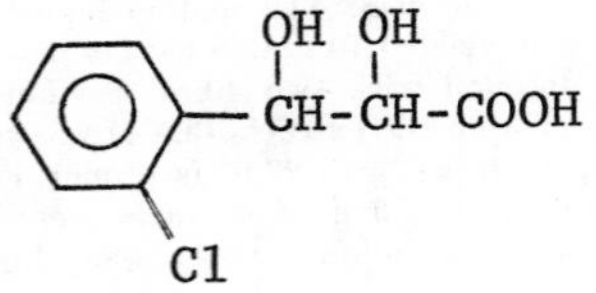

$C_9H_9ClO_4$ Ref. 492

Acides (+) *et* (—) érythro *o-chlorophénylglycériques.*

4,33 g d'acide (±) **4b** et 4,24 g de S (—) *thréo* *p*-nitrophényl-1 amino-2 propanediol-1,3 dans 20 ml d'éthanol donnent 2,42 g de sel, recristallisés dans 30 ml du même solvant; rdt = 1,90 g. La décomposition de ce sel par HCl et l'extraction de l'acide à l'acétate d'éthyle donnent (+) **4b**, 0,77 g, F = 145-146°, $[\alpha]^{25}_{578}$ = + 90,5° (acétone, *c* = 0,8). Recristallisation dans l'éther/hexane, F=147°, pouvoir rotatoire pratiquement inchangé. Les eaux-mères du premier sel sont décomposées et l'acide repris par la base dextrogyre. On obtient ainsi l'acide (—) **4b**, F = 147°, $[\alpha]^{25}_{578}$ = — 90,5°.

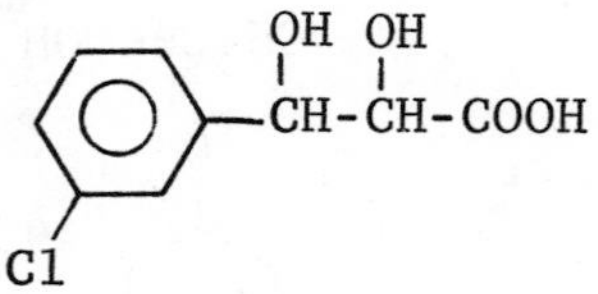

$C_9H_9ClO_4$ Ref. 493

Acides (+) *et* (—) érythro *m-chlorophénylglycériques.*

695 mg d'acide (±) **4c** et 795 mg de R (+) *thréo* diméthylamino-2 *p*-nitrophényl-1 propanediol-1,3 dans 4 ml d'éthanol donnent en 12 h 590 mg de sel, recristallisés dans 4 ml du même solvant, rdt = 407 mg. Après décomposition et extraction à l'acétate d'éthyle, on obtient 173 mg d'acide lévogyre, F = 124°. L'acide (—) **4c** optiquement pur est obtenu par recristallisation dans l'éther/pentane, F = 126°, $[\alpha]^{25}_{578}$ = — 25° (acétone, *c*=0,65).

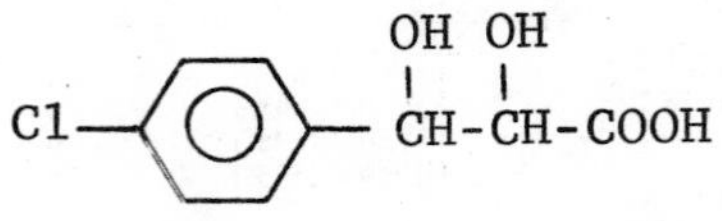

$C_9H_9ClO_4$ Ref. 494

Acides (+) *et* (—) érythro *p-chlorophénylglycériques.*

4,23 g d'acide (±) **4d** et 2,84 g de (+) désoxyéphédrine dans 20 ml d'éthanol donnent 2,1 g de sel, recristallisé successivement dans 15 puis 10 ml d'éthanol; rdt = 1,12 g, F = 151°. Ce sel est décomposé par HCl et l'acide extrait à l'acétate d'éthyle. On obtient (+) **4d**, 0,64 g, F = 123° [autre forme F = 102°], $[\alpha]^{25}_{578}$ = + 38° (acétone). Cet acide est suspendu dans 5 ml de chlorure de méthylène bouillant et isolé par décantation; F = 123°, $[\alpha]^{25}_{578}$ = + 39° (acétone, *c* = 1,4).

Les eaux-mères du sel de l'amine (+) sont décomposées et l'acide traité par la base lévogyre. On obtient ainsi l'acide (—) **4d**, F = 123°, $[\alpha]^{25}_{578}$ = — 40° (acétone, *c* = 0,7).

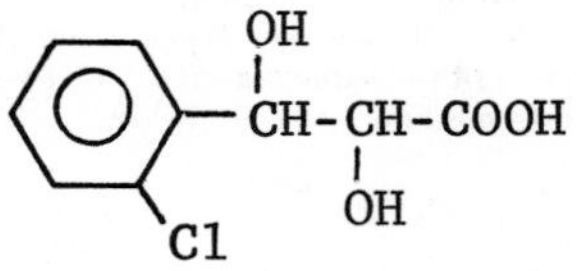

$C_9H_9ClO_4$ **Ref. 495**

Acides (+) *et* (—) thréo *o-chlorophénylglycériques.*

1 g d'acide **3***b* racémique et 1,36 g de cinchonidine dans 70 ml d'éthanol donnent au bout d'1 h 15 0,96 g de sel. Ce sel est suspendu dans 80 ml d'éthanol bouillant, essoré après refroidissement, et à nouveau repris par 10 ml du même solvant. On obtient ainsi 0,5 g de cristaux. Leur décomposition par HCl et l'extraction de l'acide à l'acétate d'éthyle donnent (+) **3***b*, 0,18 g, $[\alpha]^{25}_{578} = +46°$ (acétone, $c = 0,8$). Recristallisation dans 3 ml d'acétate d'éthyle, 90 mg, F = 189°, $[\alpha]^{25}_{578} = +49°$ (acétone, $c = 0,5$).

La décomposition des eaux-mères du sel de cinchonidine donne 0,5 g d'acide (—) **3***b*, $[\alpha]^{25}_{578} = -25°$ (acétone). Ce dernier est repris par 0,65 g de cinchonine dans 10 ml d'acétate d'éthyle. On obtient 0,45 g de sel dont la décomposition donne 135 mg d'acide F = 186°, $[\alpha]^{25}_{578} = -48°$ (acétone). Recristallisation dans l'acétate d'éthyle, 40 mg, F = 189°, $[\alpha]^{25}_{578} = -49°$ (acétone, $c = 0,3$).

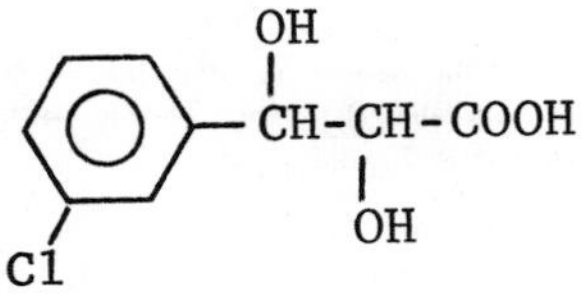

$C_9H_9ClO_4$ **Ref. 496**

Acide (+) thréo *m-chlorophénylglycérique.*

2,16 g d'acide (±) **3***c* et 2,94 g de cinchonidine dans 10 ml d'éthanol donnent 1,76 g de sel, recristallisé successivement dans 16 puis 15 ml d'éthanol : rdt = 1,15 g. Ce sel est décomposé par HCl et l'acide (+) **3***c* extrait à l'acétate d'éthyle : 395 mg, $[\alpha]^{25}_{578} = +35°$ (acétone). Deux recristallisations dans un mélange acétate d'éthyle/cyclohexane donnent 160 mg, F=143°, $[\alpha]^{25}_{578} = +39°$.

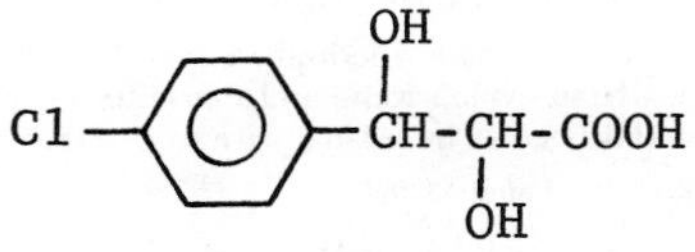

$C_9H_9ClO_4$ **Ref. 497**

Acide (—) thréo *p-chlorophénylglycérique.*

0,8 g d'acide (±) **3***d* et 0,45 g de (+) α-phényléthylamine dans 8 ml d'éthanol donnent au bout de 2 h 0,37 g de sel que l'on recristallise dans 5 ml d'éthanol puis décompose par HCl. L'extraction à l'acétate d'éthyle donne 126 mg d'acide (—) **3***d*, $[\alpha]^{25}_{578} = -38°$ (acétone). Recristallisation dans 2 ml d'acétate d'éthyle, F=177,5°, $[\alpha]^{25}_{578} = -40,5°$ (acétone, $c = 0,4$).

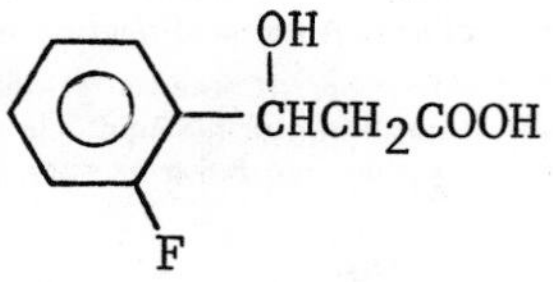

$C_9H_9FO_3$ **Ref. 498**

Acide o-fluorophényl hydracrylique R (+).

0,7 g d'acide, 1,4 g de brucine, 5 cm³ d'éthanol. Une recristallisation du sel obtenu (0,9 g) dans 4 cm³ d'éthanol, donne 0,65 g de sel, décomposé en acide (0,2 g). Recristallisation dans le benzène, F = 75°, $[\alpha]_J^{25} = +67°$ ($CHCl_3$, $c = 0,8$).

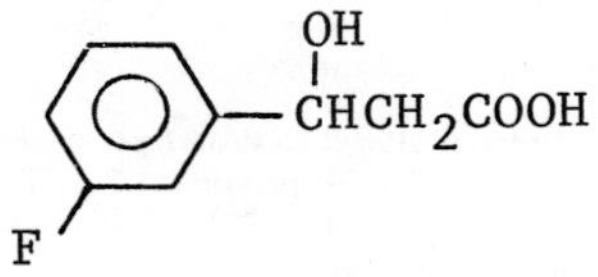

$C_9H_9FO_3$ **Ref. 499**

Acide m-fluorophényl hydracrylique R (+).

1,9 g d'acide et 4 g de brucine dans 20 cm³ d'éthanol donnent, par cristallisation au réfrigérateur 1,1 g de sel. Une recristallisation dans 6 cm³ d'éthanol et décomposition du sel donne l'acide $[\alpha]_J^{25} = +48°$ ($CHCl_3$, $c = 0,6$), F = 38°, proche de la pureté optique.

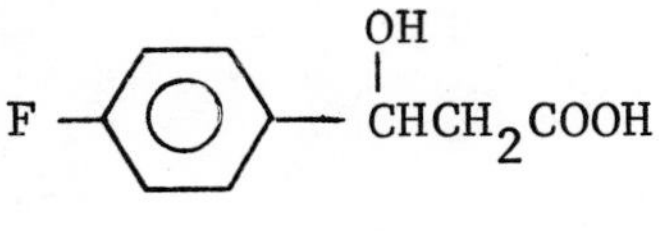

$C_9H_9FO_3$ **Ref. 500**

Acide p-fluorophényl hydracrylique R (+).

1,84 g d'acide et 1,65 g de (—) éphédrine dans 10 cm³ d'éthanol donnent 1,7 g de sel, recristallisés dans 10 cm³ d'éthanol; la décomposition du sel obtenu (0,55 g) conduit à un acide $[\alpha]_J^{25} = +44°$. Celui-ci (0,24 g), recristallisé dans l'éther/hexane donne 0,16 g d'acide optiquement pur, F = 108°, $[\alpha]_J^{25} = +52,7°$ ($CHCl_3$, $c = 0,7$).

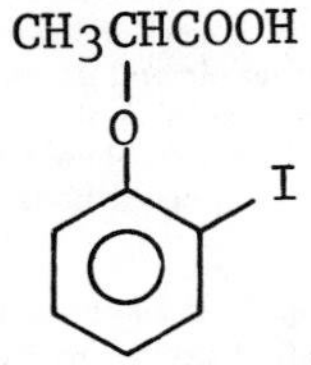

$C_9H_9IO_3$ **Ref. 501**

Preliminary experiments on resolution were carried out with several alkaloids. *Strychnine* gave (+)-acid of high activity; after standing for several days the mother liquor deposited crystals of the salt of the (−)-acid. No alkaloid gave (−)-acid in the first crystallisation. *Brucine* and *cinchonidine* gave dextrorotatory acid of lower activity than strychnine. The latter base was thus chosen for the final resolution.

(+)-*2-Iodophenoxy-propionic acid.* Racemic acid (26.4 g, 0.090 mole) and strychnine (30.0 g, 0.090 mole) were dissolved together in 200 ml ethanol and 200 ml water were added. The salt was filtered off after 24 h at room temperature and recrystallised from dilute ethanol; after each crystallisation, a small sample of acid was liberated and the rotatory power determined in dilute ammonia (the rotation in acetone and ethanol is very low).

Crystallisation	1	2	3	4
ml ethanol	200	100	75	60
ml water	200	100	75	60
g salt obtained	26.9	25.0	23.9	22.8
$[\alpha]_D^{25}$ of the acid	+ 39.5°	+ 40.5°	+ 41°	+ 40.5°

The salt obtained was decomposed with sodium carbonate solution and the strychnine extracted with chloroform. After acidification of the water phase, the acid was isolated by extraction with ether. It was recrystallised from dilute formic acid, chloroform and once more from dilute formic acid. The yield of crude acid was 10.8 g; after recrystallisation there remained 8.8 g as thin, glistening needles with m.p. 132–132.5°.

$C_9H_9O_3I$ (292.07) I calc. 43.45 found 43.20

0.2280 g dissolved in *dilute ammonia* to 10.00 ml: $\alpha_D^{25} = +1.86°$ (2 dm). $[\alpha]_D^{25} = +40.8°$; $[M]_D^{25} = +119.1°$.

(–)-*2-Iodophenoxy-propionic acid.* The mother liquor from the first crystallisation of the strychnine salt was diluted with water to opalescence and seeded with the salt of the (–)-acid from a preliminary experiment. On standing, 26.4 g salt were obtained; the acid liberated from a small sample showed $[\alpha]_D^{25} = -38.5°$. It was recrystallised to maximum activity as seen below.

Recrystallisation	1	2	3
ml ethanol	150	100	70
ml water	250	190	120
g salt obtained	22.2	20.0	18.5
$[\alpha]_D^{25}$ of the acid	– 40.5°	– 41°	– 41°

The acid was isolated and recrystallised as described above for the (+)-form. The yield of crude acid was 8.2 g and of pure product 7.2 g. The acid forms glistening needles, quite similar to the (+)-form. M.p. 132–132.5°.

$C_9H_9O_3I$ (292.07) I calc. 43.45 found 43.32

The rotatory power in various solvents is given in Table 1.

Table 1

g = g acid dissolved to 10.00 ml.

Solvent	g	α_D^{25} (2 dm)	$[\alpha]_D^{25}$	$[M]_D^{25}$
Acetone	0.2154	– 0.08°	– 1.9°	– 5.4°
Ethanol (abs.)	0.2168	±0.00°	±0°	±0°
Chloroform	0.2392	– 0.685°	– 14.3°	– 41.8°
Water (neutr.)	0.2175	– 1.785°	– 41.0°	– 119.9°
Benzene	0.2226	– 1.91°	– 42.9°	– 125.3°

(+)-*3-Iodophenoxy-propionic acid.* On standing, the mother liquor from the fi crystallisation of the cinchonidine salt deposited 8.0 g salt; a small sample of ac from this salt had $[\alpha]_D^{25} = +37°$. The mother liquor from the second crystallisati gave 2.9 g salt, containing acid of the same activity. The two portions were combin and recrystallised from dilute methanol as seen below.

Crystallisation	1	2	3
ml methanol	50	30	30
ml water	50	25	26
g salt obtained	9.5	8.8	7.9
$[\alpha]_D^{25}$ of the acid	+ 39°	+ 39.5°	+ 39°

The acid was isolated as described above. The crude product (3.5 g) was recrys lised twice from dilute formic acid, yielding 3.1 g of small glistening needles with 92.5–94°

$C_9H_9O_3I$ (292.07) I calc. 43.45 found 43.27

The rotatory power in various solvents is given in Table 2.

Table 2

g = g acid dissolved to 10.00 ml.

Solvent	g	α_D^{25} (2 dm)	$[\alpha]_D^{25}$	$[M]_D^{25}$
Acetone	0.2001	+ 1.595°	+ 39.9°	+ 116.4°
Ethanol (abs.)	0.2307	+ 1.26°	+ 27.3°	+ 79.8°
Chloroform	0.2247	+ 0.65°	+ 14.5°	+ 42.2°
Water (neutr.)	0.2550	+ 0.39°	+ 7.6°	+ 22.3°
Benzene	0.2252	– 0.08°	– 1.8°	– 5.2°

Reduction to (+)-phenoxy-propionic acid. 0.30 g acid were reduced with zinc and ammonia as described above for the 2-iodo acid. The yield of recrystallised duct was 0.12 g.

0.1071 g acid dissolved in *acetone* to 10.00 ml: $\alpha_D^{25} = +0.855°$ (2 dm). $[\alpha]_D^{25} = +39.9°$. product showed no melting-point depression with an authentic specimen of (+)-phenoxy-pionic acid having $[\alpha]_D^{25} = +39.3°$.

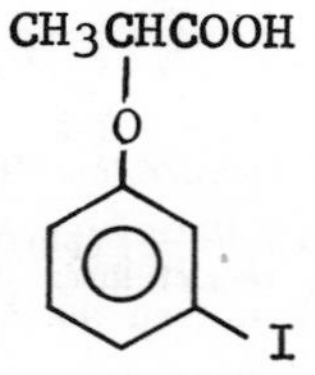

$C_9H_9IO_3$ Ref. 502

Preliminary experiments on resolution were performed with the common alkaloids. *Cinchonidine* and *strychnine* in dilute methanol gave (–)-acid of rather good activity. *Brucine* in dilute ethanol gave (–)-acid of lower rotation. The *quinine* salt gave racemic acid and the salts of *cinchonine* and *quinidine* failed to crystallise. Optically active α-(2-naphthyl)-ethylamine was also tested; it gave an acid with the same direction of rotation as the base. An experiment with cinchonidine on a somewhat larger scale showed that the (+)-acid could be isolated from the salt in the mother liquors.

(–)-*3-Iodophenoxy-propionic acid.* Racemic acid (23.4 g, 0.080 mole) and cinchonidine (23.5 g, 0.080 mole) were dissolved with gentle heating in 400 ml methanol + 100 ml water. A further 300 ml water were added. The salt, which separated on standing, was recrystallised from dilute methanol. The course of the resolution was followed as described for the 2-acid. The rotatory power of the samples was measured in acetone solution.

Crystallisation	1	2	3	4	5	6
ml methanol	400	250	200	115	55	40
ml water	400	250	200	60	50	25
g salt obtained	26.3	18.0	12.2	10.8	9.7	8.7
$[\alpha]_D^{25}$ of the acid	– 10.5°	– 23.5°	– 36.5°	– 37.5°	– 38°	– 38.5°

The acid was liberated and extracted with ether in the usual way. The crude product (3.7 g) was recrystallised from cyclohexane and twice from dilute formic acid, yielding 3.35 g of pure acid. It forms small needles with m.p. 92.5–94°

$C_9H_9O_3I$ (292.07) I calc. 43.45 found 43.18

0.2069 g acid dissolved in *acetone* to 10.00 ml: $\alpha_D^{25} = -1.648°$ (2 dm). $[\alpha]_D^{25} = -39.8°$; $[M]_D^{25} = -116.3°$. 0.1937 g acid dissolved in absolute *ethanol* to 10.00 ml: $\alpha_D^{25} = -1.062°$ (2 dm). $[\alpha]_D^{25} = -27.4°$; $[M]_D^{25} = -80.1°$.

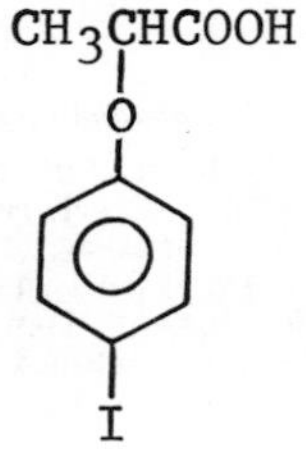

$C_9H_9IO_3$ Ref. 50

4-Iodophenoxy-propionic acid

The *racemic acid* was prepared from 4-iodophenol as described above for the is mers. It was recrystallised from cyclohexane and dilute formic acid, twice from ea solvent, and obtained as flat needles or prisms with m.p. 118.5–119°

$C_9H_9O_3I$ (292.07) I calc. 43.45 found 43.28

Preliminary experiments on resolution with *brucine* in acetone seemed very pr mising, yielding (–)-acid of high activity. For work on a larger scale, howev acetone was found less satisfactory. The (+)-acid could be isolated via the (– β-phenyl-isopropylamine salt.

(–)-*4-Iodophenoxy-propionic acid.* Racemic acid (16.1 g, 0.055 mole) was dissolv in 150 ml acetone. Brucine tetrahydrate (25.6 g, 0.055 mole) was dissolved in 250 acetone + 5 ml water and the solutions were mixed. Crystallisation took place readi but the salt obtained (34.5 g) represented over 80% of the total amount and t resolution was rather poor. The salt was therefore recrystallised from dilute ethar as seen below; obviously it would have been better to use this solvent from the begi ning. The rotatory power of the samples was determined in acetone solution.

Recrystallisation	1	2	3	4
ml ethanol	45	190	180	170
ml water	115	200	200	190
g salt obtained	18.6	17.2	16.6	15.3
$[\alpha]_D^{25}$ of the acid	– 44°	– 45°	– 46°	– 46°

The acid was isolated in the usual way and recrystallised from cyclohexane twice from dilute formic acid. The yield of crude product was 6.0 g and of pure a 5.1 g. It forms fine, glistening needles with m.p. 120.5–121.5°

$C_9H_9O_3I$ (292.07) calc. equiv. wt. 292.07 I 43.45
found equiv. wt. 292.3 I 43.33

0.1921 g acid dissolved in *acetone* to 10.00 ml: $\alpha_D^{25} = -1.782°$ (2 dm). $[\alpha]_D^{25} = -46.4°$; $[M]_D^{25}$ – 135.5°.

(+)-*4-Iodophenoxy-propionic acid.* The first two mother liquors from the brucine salt were evaporated to dryness and the acid liberated; 8.2 g having $[\alpha]_D^{25} = +41°$ were obtained. This acid (0.028) mole was dissolved with 3.8 g (0.028 mole) (−)-*β*-phenyl-isopropylamine in 40 ml ethanol and 40 ml water were added. The salt obtained was recrystallised as seen below.

Crystallisation	1	2	3
ml ethanol	40	35	30
ml water	40	35	30
g salt obtained	10.7	10.0	9.6
$[\alpha]_D^{25}$ of the acid	+46°	+46°	+46°

The acid was isolated and recrystallised as described for the (−)-form. The yield of crude product was 6.4 g and of pure acid 5.7 g. It forms fine, glistening needles, quite similar to the enantiomer. M.p. 120.5–121.5°.

$C_9H_9O_3I$ (292.07)	calc. equiv. wt. 292.07	I 43.45
	found equiv. wt. 292.2	I 43.29

The rotatory power in various solvents is given in Table 3.

Table 3

g = g acid dissolved to 10.00 ml.

Solvent	g	α_D^{25} (2 dm)	$[\alpha]_D^{25}$	$[M]_D^{25}$
Acetone	0.1842	+1.72°	+46.7°	+136.4°
Ethanol (abs.)	0.1874	+1.24°	+33.1°	+96.6°
Chloroform	0.1829	+0.78°	+21.3°	+62.3°
Water (neutr.)	0.2424	+0.875°	+18.0°	+52.7°
Benzene	0.1894	−0.105°	−2.8°	−8.1°

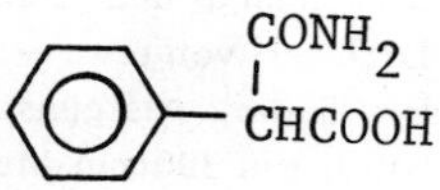

$C_9H_9NO_3$ Ref. 504

Benzylmalonamic acid gave similar results.

Cinchonidinyl D-*phenylmalonamate.* (±)-Phenylmalonamic acid (2·25 g) and cinchonidine (3·65 g) were dissolved in acetone (125 ml). The solution was concentrated, under vacuum, to *ca.* 50 ml. Crystallisation of the salt was completed by the addition of water. The salt (5·8 g, 97%), m.p. 115—117°, $[\alpha]_D^{20}$ −58·0° (*c* 2), gave, on decomposition, the active *acid*, m.p. 121°, $[\alpha]_D^{20}$ +45·0° (*c* 2) (Found: C, 60·4; H, 5·1; N, 7·7. $C_9H_9O_3N$ requires C, 60·3; H, 5·1; N, 7·8%).

Similar experiments using ethanol (100 ml) or chloroform (200 ml) and acetone (25 ml) yield salts with $[\alpha]_D^{20}$ −78·0 and −66·0° respectively, which on decomposition yielded acid with m.p. 120—121°, $[\alpha]_D^{20}$ +5·0 and +32·0° (*c* 2), respectively.

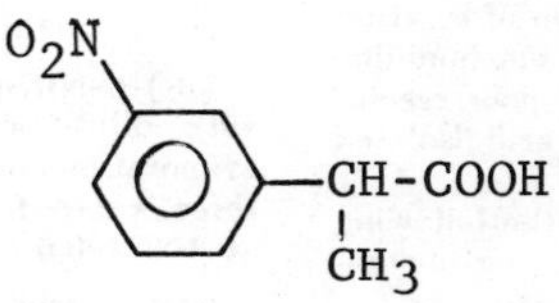

$C_9H_9NO_4$ Ref. 505

Spaltung der m-Nitro-hydratropasäure mit (+)-Bornylamin: m-Nitro-hydratropasäure[4] ist wenig löslich in Petroläther und Schwefelkohlenstoff, leicht löslich in Methanol, Äthanol, Aceton, Äther, Dioxan und Chloroform. Sie läßt sich umkristallisieren aus Benzol, Tetrachlorkohlenstoff, Benzin, Methanol/Wasser und aus etwa 35-proz. Essigsäure. Das letzte Lösungsmittel ist am bequemsten, jedoch bleibt die Säure dabei stets etwas gefärbt.

28.0 g (+)-Bornylammoniumchlorid, $[\alpha]_D^{22.5} = +21$ bis 21.5° (Wasser), haben wir in 100 ccm Wasser mit 50 ccm 20-proz. Natronlauge zersetzt und die Lösung 3 mal mit je 50 ccm Äther ausgeschüttelt. Die äther. Lösung wurde nach kurzem Trocknen mit Na_2SO_4 zur Lösung von 25.0 g *m*-Nitro-hydratropasäure in 250 ccm Äther gegeben. Insgesamt wurden 42.8 g (96% d. Th.) Salz gewonnen. Dieses Salz wurde aus 450 bis 250 ccm Benzol 14 mal umkristallisiert. Der bei etwa +11° liegende spezif. Drehwert des racemischen Salzes (in Alkohol) war nach 4 maligem Umkristallisieren auf +1° gesunken, beim weiteren Umkristallisieren veränderte er sich nur noch langsam und blieb nach dem 12. Mal mit −2 bis −3° konstant. Die zuletzt erhaltenen 1.1 g Salz, $[\alpha]_D^{22} = -2.5°$, wurden mit 5 ccm Salzsäure (1:1) zersetzt, die rohe Säure wurde aus 2.5 ccm 30-proz. Essigsäure umkristallisiert: 0.3 g schwach gelbliche Kristalle vom Schmp. 85.5−86.5°, $[M]_D^{25} = +91°$ (Methanol). Die aus den benzolischen Mutterlaugen zurückgewonnenen Salzmengen mußten noch mehrmals umkristallisiert werden; erschwerend war, daß sich Salze mit einem Drehwert von etwa +12° nur wenig differenzierten oder aus den Lösungen bevorzugt auskristallisierten. Offenbar handelte es sich um ein partielles Racemat, dessen Löslichkeit nur wenig größer als die des d-Salzes ist. Aus der höherdrehenden Fraktion entstanden beim Umkristallisieren 4.6 g Salz mit $[\alpha]_D^{25} = +25.5°$, aus der die (−)-Säure vom Schmp. 85.5−86.5°, $[M]_D^{25} = -88.5°$ (Methanol), gewonnen wurde. Die Schmpp. der diastereomeren Salze, die beide feine, weiße Nadeln bildeten, waren nicht sehr charakteristisch und nur wenig voneinander verschieden; der des d-Salzes lag bei 157−158°, der des l-Salzes bei 153.5−154°.

Spaltung der m-Nitro-hydratropasäure mit (−)-α-Phenyl-äthylamin: 40.0 g Säure wurden in 1750 ccm Benzol gelöst und 25.9 g Amin zugegeben. Es kristallisierten bei 25° Badtemperatur 31.7 g Salz in feinen, weißen Nadeln vom Schmp. 157.5−159°, $[\alpha]_D^{24} = +5.0°$, aus, beim völligen Abkühlen weitere 31.1 g (zusammen 97% d. Th.) vom Schmp. 110−115°. Zur Gewinnung

[4] F. Nerdel und G. Kresze, Chem. Ber. 85, 168 (1952).

der reinen d- und l-Salze mußte mehrfach aus Benzol umkristallisiert werden. Salze mit einem Drehwert von etwa $-12°$ differenzierten sich meistens nicht nennenswert [Salz mit etwa 50% (+)-Säure]. Als günstiger erwies sich die Darstellung des Salzes in *Methanol:* 58.9 g Säure wurden in 300 ccm Methanol gelöst und 36.5 g Amin in der Wärme zugegeben. Bei 29° Badtemperatur kristallisierten 15.6 g kleine, plättchenförmige Kristalle vom Schmp. 160.5–162°, $[\alpha]_D^{24} = +8.1°$. Beim Kühlen im Eisschrank weitere 14.3 g mit $[\alpha]_D^{24} = +6.9°$, nach dem Abdestillieren von 100 ccm Methanol 12.0 g mit $[\alpha]_D^{24} = +3.9°$. Nach dem Versetzen mit 400 ccm Benzol 11.4 g, $[\alpha]_D^{24} = +0.7°$, im Filtrat entstanden dann 39 g feine Nadeln mit $[\alpha]_D^{24} = -13.6°$. Die erste und die letzte Fraktion lassen sich direkt zur Gewinnung der (−)- bzw. (+)-Säure verwenden. l- bzw. d-Salze aus Methanol umkristallisiert: Schmp. 163–164.5°, $[\alpha]_D^{24} = +10.5°$ bzw. Schmp. 125°, $[\alpha]_D^{24} = -20°$. Das Phenyläthylaminsalz mit $[\alpha]_D^{24} = +10.5°$ liefert die reine (−)-Säure vom Schmp. 86–86.5°, $[M]_D^{24} = -92°$ (aus wäßr. Essigsäure). Die reine (+)-Säure wurde aus Phenyläthylaminsalzen mit $[\alpha]_D$-Werten zwischen -11 und $-15°$ gewonnen durch anschließende fraktionierte Kristallisation aus wäßr. Essigsäure: Schmp. 85.5–86.5°, $[M]_D^{24} = +91°$.

$C_9H_9NO_4$ (195.2) Ber. N 7.18 Gef. N 7.21

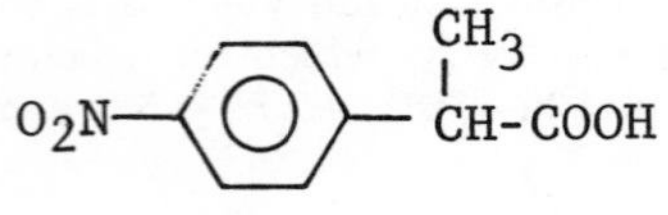

$C_9H_9NO_4$ Ref. 506

Resolution of the acid into the optical antipodes. Good resolution was obtained with α-phenyl-ethylamine, giving acid of the same direction of rotation as the base. The (+)-acid could also be isolated through the cinchonidine salt. Strychnine and (+)-benzedrine gave dextrorotatory acid but poor resolution, quinine and brucine levorotatory acid of low activity. The acid isolated from the cinchonine salt was quite inactive.

In order to economize the rather expensive α-phenyl-ethylamine, the following procedure was worked out. 58.5 g (0.3 mole) acid, 24.2 g (0.2 mole) (−)-α-phenylethylamine and 5.6 g (0.1 mole) potassium hydroxide were dissolved in a hot mixture of 100 ml ethanol and 50 ml water. 450 ml of hot water were carefully added. The salt was filtered off after 24 hours and recrystallised several times from dilute ethanol; the progress of the resolution was tested on samples from each crystallisation (about 0.3 g), from which the acid was liberated and examined in absolute ethanol.

Crystallisation	1	2	3	4	5	6
ml water	500	270	200	150	130	115
ml ethanol	100	70	70	60	50	40
g salt obtained	46.8	36.8	30.7	26.2	23.3	21.9
$[\alpha]_D^{25}$ of the acid	−31.4°	−44.9°	−49.6°	−53.2°	−54.6°	−54.6°

The mother liquor from the first crystallisation on standing deposited a second crop of salt which was filtered off and added to a second run. The mother liquor was then evaporated to dryness at room temperature and acid was liberated. 22.0 g (0.113 mole) having $[\alpha]_D^{25} = +28.3°$ (ethanol) w obtained. The acid was dissolved with 33.2 g (0.113 mole) of cinchonidine hot dilute ethanol and the resulting salt recrystallised until the acid had reac maximum activity.

Crystallisation	1	2	3	4
ml water	700	400	350	325
ml ethanol	100	100	75	65
g salt obtained	40.9	34.4	30.0	26.8
$[\alpha]_D^{25}$ of the acid	+47.9°	+53.6°	+55.0°	+54.8°

(+)-*p*-Nitrohydratropic acid. The cinchonidine salt (26.8 g) was decompo with dilute sulphuric acid and the organic acid extracted with ether. Af evaporation of the ether, the crude acid (10.2 g) was dried and recrystalli three times from carbon tetrachloride. 9.0 g of faintly yellow needles melt at 100–100.5° were obtained.

0.2891 g acid, 13.79 ml 0.1074-*N* NaOH.

$C_9H_9O_4N$ Equiv. wt. calc. 195.2 found 195.3

0.3914 g dissolved in absolute *ethanol* to 10.00 ml: $2\alpha_D^{25} = +4.32°$. $[\alpha]_D^{25} = +55.2°$; [M] $= +107.7°$.

(−)-Nitrohydratropic acid. The α-phenyl-ethylamine salt (21.9 g) was composed with dilute sulphuric acid and the organic acid was extracted w ether as described above. The crude acid (13.1 g) was recrystallised from carb tetrachloride, from ether + petroleum and once more from carbon tetrachlori 10.3 g of pure acid, melting at 100–100.5°, were obtained.

0.2981 g acid: 14.21 ml 0.1074-*N* NaOH.

$C_9H_9O_4N$ Equiv. wt. calc. 195.2 found 195.3

0.4015 g dissolved in absolute *ethanol* to 10.00 ml: $2\alpha_D^{25} = -4.44°$. $[\alpha]_D^{25} = -55.3°$; [M] $= -107.9°$ — 0.3559 g dissolved in *methanol* to 9.99 ml: $2\alpha_D^{25} = -4.005°$. $[\alpha]_D^{25} = -56.$ $[M]_D^{25} = -109.8°$.

$C_9H_9NO_4$ Ref. 506a

Spaltung der p-Nitro-hydratropasäure in die optischen Antipoden: Zu einer siedenden Lösung von 97.5 g *p-Nitro-hydratropasäure* in 750 ccm 30-proz. wäßrigem Äthanol ließ man 30.3 g *(+)-α-Phenyl-äthylamin* zutropfen. Innerhalb von 24 Stdn. kristallisierten 58.5 g Salz aus. Die aktive Säure ließ sich daraus durch Hinzufügen der doppelten molaren Menge an konz. Salzsäure bei 0° und anschließendes Verdünnen auf das 5fache Volumen gewinnen. $[M]_D^{20} = +107.8°$ (Äthanol). Ausbeute 28.8 g (29.5% d. Th.), Schmp. 100–100.5°.

Die Mutterlaugen der Salzkristallisation wurden i. Vak. zur Trockne gedampft. Nach Abtrennen des Amins blieben 61.3 g Säure zurück von $[M]_D^{20} = -46.2°$ und Schmp. 92°. Diese wurden mit 27.3 g (−)-α-Phenyl-äthylamin in der gleichen Weise gespalten. $[M]_D^{20} = -107.4°$ (Äthanol). Ausbeute 27 g (28.7% d. Th.), Schmp. 100–100.5°.

Tabelle 1. Rotationsdispersionen von *p*-substituierten Hydratropasäuren

Substituent	Lösungsmittel	Konz. in g/25 ccm	$[M]^{20}$ bei Wellenlänge (mμ) 656.3	589.3	546.0	506.0	486.1
p-NO_2	Benzol	0.5259	88°	113°	136°	166°	183°
	Dioxan	0.4427	103	129	156	190	213
	Aceton	0.5621	119	152	183	222	246
	Eisessig	0.5148	92	117	140	169	187
	Äthanol	0.5120	84	108	127	156	173
	Methanol	0.4537	88	110	131	159	176
	Chloroform	0.4845	78	97	116	141	158
	Na-Salz in H_2O	0.5791	−29	−39	−52	−65	—

$C_9H_9NO_4$

Ref. 506a

Spaltung der p-Nitrohydratropasäure.

3 g reine p-Nitrohydratropasäure werden in 100 ccm Methylalkohol gelöst, dann gibt man eine Lösung von 3 g Chininhydrat in 40 ccm Methylalkohol zu und läßt 3 Tage im Eisschrank stehen. Von dem Niederschlag wird abgetrennt und 5-mal aus Methylalkohol umkrystallisiert. Man bekommt so die Fraktion A_1, welche die reine (+)-Säure als Salz enthält. Die Mutterlaugen werden mit dem Filtrat vereinigt, das i. V. bei Zimmertemperatur auf das ursprüngliche Volumen eingedampft wird. Hierauf versetzt man es mit einer Lösung von 3 g Chininhydrat in 20 ccm Methylalkohol, läßt 6—8 Stunden bei Zimmertemperatur stehen und filtriert ab. Der Niederschlag wird 7-mal aus Methylalkohol umkrystallisiert und man gewinnt so die Fraktion A_2, die ebenfalls die reine (+)-Säure enthält. Die Mutterlaugen werden zum Filtrat gegeben und wieder auf das ursprüngliche Volumen eingeengt. Der Methylalkohol wird dabei stets i. V. bei Zimmertemperatur weggenommen. Nach 8—10-stündigem Stehen ist die Fraktion A_3 ausgefallen, sie wird filtriert und das Filtrat auf die Hälfte des vorhandenen Volumens eingedampft, über Nacht im Eisschrank stehen lassen und so die vierte Fraktion gewonnen. Diese Operation wird wiederholt bis zur Fraktion A_7. A_4—A_6 enthalten ein Gemisch der optisch aktiven Nitrosäure. Aus ihnen wird die freie Nitrosäure zurückgewonnen, die den Trennungsprozeß von neuem durchläuft. A_7 dekantiert man einige Male mit kaltem Aceton und krystallisiert 2-mal aus Methylalkohol um. Diese Fraktion enthält das Chininsalz der (—)-Nitrosäure. Die Mutterlaugen werden mit dem Filtrat vereinigt, das zur Trockne eingedampft wird. Der Rückstand besteht aus einer braunen, harzigen Schmiere, aus der sich durch Behandeln mit kaltem Aceton noch etwas farbloses Chininsalz gewinnen läßt. Man krystallisiert dieses 2-mal aus Methylalkohol um und erhält A_8, das ebenfalls die reine (—)-Nitrosäure als Chininsalz enthält. Ausbeute: A_1 1,8 g, A_2 3,6 g, A_3 1,5 g, A_4 0,3 g, A_5 0,2 g, A_6 0,6 g, A_7 0,08 g, A_8 0,08 g.

Zur *Darstellung der freien (+)-p-Nitrohydratropasäure* wurden 5 g A_1A_2 unter Eiskühlung mit 5 ccm konz. Salzsäure versetzt. Nach einiger Zeit gibt man 50 ccm eiskaltes Wasser in kleinen Portionen zu, knetet die Masse gut durch und läßt über Nacht im Eisschrank stehen. Dann wird abfiltriert, mit kaltem Wasser gut ausgewaschen und aus 85-proc. Essigsäure bei 40° umkrystallisiert. Schmelzp. 88,5°. Ausbeute 1,7 g = 93 Proc. d. Th.

4,976 mg Subst.: 10,140 mg CO_2, 2,140 mg H_2O.

$C_9H_9O_4N$ Ber. C 55,38 H 4,62 Gef. C 55,4 H 4,8.

Drehung: 0,092 g in 25 ccm CH_3OH, 2 dm-Rohr.

$[\alpha]_D^{22} = +10,9°$; $[M]_D^{22} = +21,3°$.

(—)-p-Nitrohydratropasäure. 1 g A_7A_8 wird in der gleichen Weise mit 1 ccm konz. Salzsäure und 10 ccm Wasser versetzt, wie es oben beschrieben wurde. Es wurde ebenfalls aus 85-proc. Essigsäure umkrystallisiert. Schmelzp. 86,5°. Ausbeute 0,33 g = 90 Proc. d. Th.

5,109 mg Subst.: 10,445 mg CO_2, 2,200 mg H_2O.

$C_9H_9O_4N$ Ber. C 55,38 H 4,62 Gef. C 55,7 H 4,8.

Drehung: 0,089 g in 25 ccm CH_3OH, 2 dm-Rohr.

$[\alpha]_D^{22} = -11,2°$; $[M]_D^{22} = -21,8°$.

$C_9H_9NO_4$

Ref. 506b

(+)-Bornylammoniumsalz der *p*-Nitro-hydratropasäure: 18 g (16% Überschuß) (+)-Bornylamin-Hydrochlorid wurden mit verd. Kalilauge in wäßr. Lösung zersetzt und mit Äther ausgeschüttelt. Zu der getrockneten äther. Lösung wurde eine Lösung von 15 g *p*-Nitro-hydratropasäure in Äther gegeben. Es fiel sofort ein dicker Niederschlag aus, der mit Äther gewaschen wurde. Ausb. 26.1 g (97.4% d.Th.); Schmp. 152–153° C (Zers.), $[\alpha]_D^{21}$: +13.8° (0.1443 g in 25 ccm Methanol, 2-dm-Rohr).

Die Trennung erfolgte durch Umkristallisation des Salzes aus Chloroform und wurde durch Bestimmung der spezif. Drehung verfolgt. Als Beispiel sei die Umkristallisation eines Teilansatzes beschrieben. 13.5 g Salz wurden in 150 ccm Chloroform unter Rückfluß bis zur vollständigen Lösung erhitzt, filtriert und der Kristallisation bei Zimmertemp. überlassen. Ausb. 9.2 g (68%); $[\alpha]_D^{20}$: +5.5° (Methanol).

Nochmaliges Umkristallisieren der 9.2 g aus 110 ccm Chloroform ergaben: 5.7 g (42% von 13.5 g); $[\alpha]_D^{20}$: +3.8° (Methanol).

Im allgemeinen wurde die Umkristallisation des Salzes, aus dem die (+)-*p*-Nitrohydratropasäure für die weiteren Reaktionen gewonnen werden sollte, bei einer spezif. Drehung zwischen 3 und 4° abgebrochen. Zur Bestimmung des Enddrehwertes wurden 7.8 g (+)-Bornylammoniumsalz bis zur Konstanz der spezif. Drehung umkristallisiert, was folgende Übersicht zeigen soll:

Einwaage (in g)	Lösungsmittel-menge	Ausbeute	$[\alpha]_D^{20}$ °	*c* in Methanol
7.8	110	5.6	+9.7	0.5676
5.5	100	3.9	+5.1	0.6856
3.7	70	2.5	+4.2	0.6568
2.3	55	1.7	+3.2	0.4232
1.6	50	1.1	0	2.080
1.0	35	0.74	0	1.522

(+)-*p*-Nitro-hydratropasäure: 0.367 g (+)-Bornylammoniumsalz mit $[\alpha]_D^{18}$: –0° wurden mit 1 ccm konz. Salzsäure bei 0° durchgeknetet, nach und nach mit 5 ccm Wasser versetzt und ca. 3 Stdn. stehen gelassen. Dann wurde abgesaugt und aus 2 ccm 35-proz. Essigsäure umkristallisiert. Ausb. 0.22 g (59%); Schmp. 95.5–96.5°.

$[\alpha]_D^{19}$: +49.4° (0.2176 g in 25 ccm Methanol, 2-dm-Rohr)

$C_9H_9O_4N$ (195.1) Ber. C 55.38 H 4.65 N 7.18 Gef. C 55.48 H 4.68 N 7.25

Die Hauptmenge der (+)-*p*-Nitro-hydratropasäure wurde aus einer (+)-Bornylammoniumsalzfraktion mit $[\alpha]_D^{22}$: +3 bis +4° (Methanol) durch Zersetzung unter den beschriebenen Bedingungen dargestellt. 7.8 g Salz ergaben 3.63 g (83%) (+)-*p*-Nitro-hydratropasäure, wobei diesmal die salzsaure Lösung noch ausgeäthert wurde. Schmp. 90–91.5°.

$[\alpha]_D^{22}$: +37° (0.1212 g in 25 ccm Methanol, 2-dm-Rohr).

Diese (+)-*p*-Nitro-hydratropasäure besaß demnach eine optische Reinheit von 75%, was bei der Berechnung der wahren Drehwerte der weiter unten beschriebenen Hydratropasäure-Derivate berücksichtigt wurde.

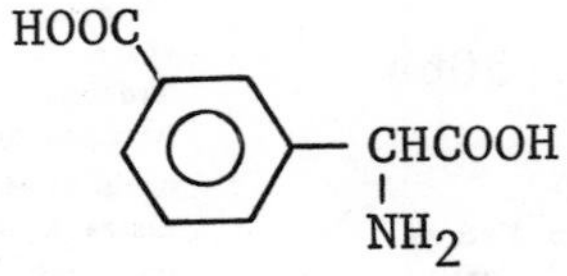

$C_9H_9NO_4$ **Ref. 507**

(−)-*N-(Menthoxyacetyl)*-m-*carboxyphenylglycine*. *m*-Carboxy-DL-phenylglycine was treated under Schotten-Baumann conditions with (−)-menthoxyacetyl chloride to give a crude product, which was recrystallized from 50 % ethanol several times, yet without any noticeable fractionation into diastereoisomerides. An analytically pure specimen, m.p. 202°, of unknown steric composition, was readily obtained, $[\alpha]_D^{23}$ −53.5° (Found: C 64.65; H 7.60; N 3.57. Calc. for $C_{21}H_{29}NO_6$: C 64.45; H 7.46; N 3.57). Further attempts to utilize this derivative for the resolution were abandoned.

Resolution with L*-arginine*. *m*-Carboxy-DL-phenylglycine (1.32 g) and L-arginine monohydrochloride (1.18 g) were dissolved in hot water (20 ml). Ethanol (4 ml) was slowly added to the 60° hot solution, which was then allowed to cool and set aside in the icebox overnight. A crop of crystals (1.20 g), $[\alpha]_D^{25}$ + 34° (*c* 0.9, H_2O), was filtered off and a little ethanol was added to the mother liquor. After being kept at 0° overnight, the solution had deposited a second crop of crystals $[\alpha]_D^{23}$ −10.5° (*c* 0.9, H_2O). Both crystalline fractions were subjected to repeated fractional recrystallizations, yet without preparations of sterical homogeneity were obtained.

Therefore, the mother liquor from the second recrystallization was diluted with more ethanol, and the resulting crystalline material (380 mg), $[\alpha]_D^{22}$ −29° (*c* 1.2, H_2O), was subjected to two additional recrystallizations from aqueous methanol resulting in separation of the pure *salt of* L*-arginine and* (−)-m*-carboxyphenylglycine* (34 mg), m.p. 203° (decomp.), $[\alpha]_D^{26}$ −40.4° (*c* 0.8, H_2O) (Found: C 45.44; H 6.56; N 17.52. Calc. for $C_{15}H_{21}N_5O_5$, 2 1/2 H_2O: C 45.45; H 6.61; N 17.67). Attempts to remove the water of crystallization at 100° over phosphorus pentoxide led to gradual decomposition of the salt.

The aromatic amino acid was liberated from the salt by passage of the latter in aqueous solution through a strongly basic ion exchange resin (Dowex 1−8, 0.8 × 5 cm) in the acetate form. Arginine passed through the column, and the strongly acid, aromatic amino acid was then eluted with 1 N acetic acid. After two recrystallizations from hot water, pure (−)-*m*-carboxyphenylglycine was obtained as colourless crystals (28 mg), $[\alpha]_D^{25}$ −89° (*c* 0.6, H_2O). An analytical specimen was dried at room temperature over calcium chloride (Found: C 53.20; H 5.08; N 7.03. Calc. for $C_9H_9NO_4$, 0.5 H_2O: C 52.93; H 4.94; N 6.86). This sample, recrystallized from pure water, possessed a solid phase infra-red spectrum significantly different from that of the synthetic preparation described above. The latter was recrystallized from aqueous ethanol, and the spectroscopic differences are obviously attributable to the different degrees of solvation, apparent also from the analytical compositions. The solid phase infra-red spectra of the hydrochlorides and triethylammonium salts of the two preparations were identical in all details.

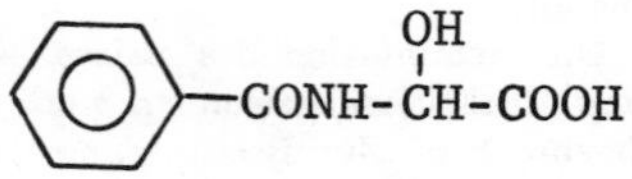

$C_9H_9NO_4$ Ref. 508

The Resolution of N-Benzoyl-α-hydroxyglycine Into Its Optical Antipodes. After dissolving 25 g of (I) and 27.2 g of L-amine in 180 ml of boiling 60% aqueous acetone, the solution was allowed to stand for a day at 18-20°. The crystals (19.2 g) which formed were filtered off and recrystallized three times from 60% acetone, yielding 13.6 g (52.2%) of the L-amine salt of N-benzoyl-*l*-α-hydroxyglycine ("*l*¹/₂ (I) salt"), mp 127-129°, $[\alpha]_D^{20}$ +6.1° (c 2, water). Found %; C 59.92; H 5.00; N 10.35. $C_{18}H_{21}N_3O_8$. Calculated %: C 53.08; H 5.19; N 10.17.

A solution of 4 g of *l*-(I) salt in 100 ml of water was stirred with 2 g of KU-2 resin, filtered, and the filtrate evaporated to a volume of 5-7 ml, after which it was allowed to stand for 12 h at 5°. 1.4 g (73%) of *l*-(I), mp 158-160°, $[\alpha]_D^{20}$ −48.1° (c 1, water) was filtered off. Found %: C 55.36; H 4.61; N 7.41. $C_9H_9NO_4$. Calculated %: C 55.39; H 4.64; N 7.17.

After removal of the *l*-(I) salt, the mother liquor was evaporated to half volume and kept at 4° for 12 h. The precipitate (25.4 g) was filtered off and recrystallized three times from 60% acetone, yielding 16.3 g (62.5%) of the salt of d-(I), mp 130-131°, $[\alpha]_D^{20}$ +27.2° (c 1, water). Found %: C 53.08; H 5.01; N 10.30.

By procedures analogous to those for the *l*-isomer, 4 g of the salt of d-(I) yielded 1.4 g (73%) of d-(I), mp 148-150°, $[\alpha]_D^{20}$ +48.0° (c 1, water). Found %: C 55.62; H 4.56; N 7.10.

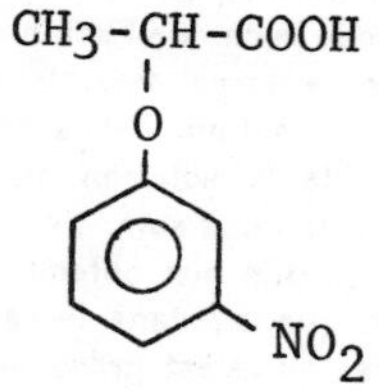

$C_9H_9NO_5$

Ref. 509

Preliminary experiments with conventional bases gave negative results, but the acid could be resolved with optically active β-phenyl-isopropylamine (benzedrine). The first crystallisation of the salt from dilute ethanol gives an acid containing about 45 % of active form with the same direction of rotation as the amine used. Maximum activity is easily obtained by recrystallisation of the salt. For practical purposes it may, however, be most convenient to isolate the (−)-form *via* the strychnine salt as described by Fourneau and Sandulesco.[1]

Experimental. Laevorotatory acid was prepared *via* the strychnine salt.[1] From the mother liquors after this salt, an acid having $[\alpha]_D^{25}$ about +20° was isolated. This acid (24.7 g, 0.117 mole) was dissolved with 15.8 g (0.117 mole) (+)-β-phenyl-isopropylamine in 45 ml ethanol + 100 ml water. The salt was recrystallised to constant activity of the acid (measured in absolute ethanol on samples isolated from the successive salt fractions; see Table 1).

The acid was isolated in the conventional way and recrystallised three times from carbon tetrachloride (once with charcoal). It formed pale yellow needles with m.p. 100−101°. 0.2210 g in 10.00 ml abs. ethanol: $\alpha_D^{25} = +2.283°$ (2 dm). $[\alpha]_D^{25} = +51.65°$. The rotation was practically unchanged when the same solution was investigated at 20°C.

The (−)-form isolated according to Ref. 1 was quite similar to the (+)-form. M.p. 100−101°. 0.2202 g in 10.00 ml abs. ethanol: $\alpha_D^{25} = -2.276°$ (2 dm). $[\alpha]_D^{25} = -51.68°$.

1. Fourneau, E. and Sandulesco, G. *Bull. Soc. Chim. France* [4] **31** (1922) 988; **33** (1923) 459.
2. Fredga, A. and Åberg, B. *Ann. Rev. Plant. Physiol.* **16** (1965) **53**.

Table 1.

Crystallisation	1	2	3	4	5
ml ethanol	45	25	25	20	15
ml water	100	60	50	50	40
g salt obtained	26.5	24.5	22.5	20.7	19.7
$[\alpha]_D^{25}$ of the acid	+45.5°	+48.5°	+50°	+50°	+50°

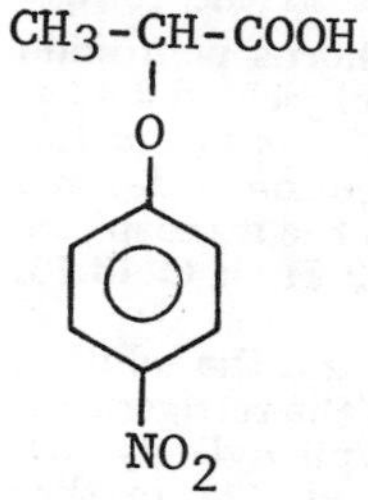

$C_9H_9NO_5$

Ref. 510

Dédoublement de l'acide para-nitrophénoxypropionique.

1° *Par la quinidine.* — On dissout $4^{gr},220$ d'acide para-nitrophénoxypropionique et $7^{gr},380$ de quinidine dans 400 cc. d'alcool à 20°, à la température du bain-marie bouillant. On filtre. On laisse refroidir. Il se forme de très beaux cristaux brillants qui peu à peu se laissent englober par un précipité gommeux, surtout abondant quand on porte le mélange à la glacière. Au bout de 3-4 jours, on décante le liquide. On triture la masse cristalline avec de l'alcool dilué à 50 0/0 qui dissout peu à peu la masse gommeuse. On fait, enfin, recristalliser le sel de quinidine; dans ce but, on en prend $5^{gr},800$ qu'on mélange à 70 cc. d'alcool à 96 étendu de 200 cc. d'eau. On porte à l'ébullition et on laisse refroidir à la température ordinaire. On obtient $4^{gr},800$ d'un produit absolument pur, car une deuxième cristallisation ne modifie pas le pouvoir rotatoire. Nous avons donc un produit brut et les cristaux de deux cristallisations successives.

Les observations polarimétriques sont faites, toujours de la même façon, en dissolvant $0^{gr},20$ de sel dans 25 cc. d'alcool absolu et en observant le pouvoir rotatoire dans un tube de 20 cm. à la T de 21°. Ces conditions seront maintenues dans toutes nos expériences :

Produit brut..........	$\alpha = +2°42'$	$[\alpha]_D = +168°7$
1re cristallisation......	$\alpha = +2,46$	$[\alpha]_D = +172,5$
2e cristallisation	$\alpha = +2,46$	$[\alpha]_D = +172,5'$

On met l'acide en liberté par la soude normale. On essore la quinidine. On extrait la liqueur alcaline avec de l'éther pour enlever les dernières traces d'alcaloïde, puis on y ajoute de l'acide chlorhydrique étendu en léger excès : l'acide nitrophénoxypropionique droit se précipite à l'état huileux. On l'extrait à l'éther; après évaporation de ce solvant, il reste sous la forme d'une huile qui se solidifie très rapidement par un léger frottement. Le pouvoir rotatoire est pris dans les mêmes conditions que pour le sel :

On trouve.............. $\alpha = 0°,52'$ $[\alpha]_D = 53°,7$

On peut faire recristalliser l'acide dans un mélange, à parties égales, de benzène sec et d'éther de pétrole; il se précipite sous la forme d'aiguilles blanches sol. dans l'alcool, l'acétone, peu sol. dans l'eau. P. F. = 89-90°.

2° *Par la yohimbine.* — On prend : acide $10^{gr},550$, yohimbine $17^{gr},700$. On dissout ces deux substances dans un mélange de 200 cc. d'alcool à 96° et de 1000 cc. d'eau. On porte à l'ébul. On filtre et on laisse reposer. Après 24 h. de repos à la *température ordinaire*, il se sépare $12^{gr},95$ d'un sel bien cristallisé. Après une seule cristallisation, ce sel est tout à fait pur, car une deuxième n'a pas d'influence sur le pouvoir rotatoire.

Pour le produit brut, on observe :

$\alpha = +0°,26'$ $[\alpha]_D = +26°,8$

La recristallisation du sel diminue le pouvoir rotatoire et cela tient à ce que la yohimbine est lévogyre. Cette cristallisation se fait en dissolvant 12 gr. du sel dans 650 cc. d'alcool à 15°. On porte au bain-marie bouillant; quand tout est dissous, on laisse déposer dans la glacière. On obtient $9^{gr},15$ de sel pour lequel on observe la rotation suivante (en se plaçant toujours dans les conditions déjà définies au début) :

$\alpha = +0°,20'$ $[\alpha]_D = +20°,6$

Une 2e cristallisation donne des résultats exactement semblables.
L'acide isolé dans les mêmes conditions que pour le sel de quinidine est lévogyre. Il possède le même pouvoir rotatoire que l'acide droit en valeur absolue :

$\alpha = -0°,52'$ $[\alpha]_D = -53°,7$

Le filtrat restant après la séparation de 12gr,95, *placé dans une glacière*, laisse déposer, après un repos de 2 jours, 10gr,5 d'un sel tout aussi bien cristallisé que le premier.

L'acide séparé à partir d'un échantillon de cette substance s'est montré doué d'un pouvoir rotatoire droit assez élevé : $\alpha = +0°42'$. Mais malgré toutes les cristallisations que nous avons fait subir, le pouvoir rotatoire ne dépasse pas un certain chiffre, qui est légèrement inférieur à celui qu'on doit observer :

$\alpha = +0°,50'$ $[\alpha]_D = +51°,8$

Quant à l'acide isolé du sel, il a le même pouvoir rotatoire en valeur absolue soit : 51,°8, très légèrement inférieur, par conséquent, à celui qu'on a obtenu avec la quinidine. Comme son isomère, il cristallise bien dans un mélange de benzine et d'éther de pétrole.

Les eaux-mères restantes, après la séparation des deux fractions, évaporées à siccité dans le vide, ont fourni un acide inactif.

Voilà donc un exemple intéressant et qui nous paraît assez instructif. Il montre que deux cristallisations successives, donnant chacune de bons rendements et des produits presque purs, se sont séparées dans le même liquide et sans qu'on ait été obligé de le concentrer, l'une à la température ordinaire, l'autre à 0°. Il est évident que si nous avions mis d'emblée notre liquide à la température de 0° nous n'aurions pas pu réaliser la séparation des isomères. Du reste, dans le cas de l'acide méta, où le dédoublement est fait avec de la strychnine, tandis qu'il faut trois cristallisations pour obtenir le sel pur de l'acide gauche quand on n'a pas amorcé au préalable la solution, la pureté du sel est presque parfaite d'emblée si on a soin d'ensemencer le liquide par un cristal du sel de l'acide pur obtenu dans une opération précédente. Il est certain que si, dans le cas où le sel qui se sépare est un mélange, cas qui s'est présenté souvent dans nos expériences, on amorçait la cristallisation avec un sel pur d'un acide actif, on obtiendrait de tout autres résultats.

Comme on l'a vu au début de ce mémoire nous avons essayé d'autres alcaloïdes pour effectuer le dédoublement.

3° *Avec la cinchonine* il se forme d'abord un louche qui laisse déposer un abondant précipité gommeux à la température ordinaire. Après un repos de 24 h. la solution fournit, à la glacière, un petit dépôt cristallin qui vient recouvrir la couche gommeuse (c'est tout le contraire de ce qui se passe avec la quinidine). On n'obtient que 2gr,3 de sel cristallisé en partant de 25gr,25 de sel de cinchonine. L'acide contenu dans les cristaux est lévogyre.

4° *Avec la cinchonidine* on obtient les mêmes résultats qu'avec la cinchonine.

5° *Avec la strychnine*, la cristallisation est belle, mais il n'y a pas de dédoublement.

6° *La brucine* donne les mêmes résultats que la strychnine.

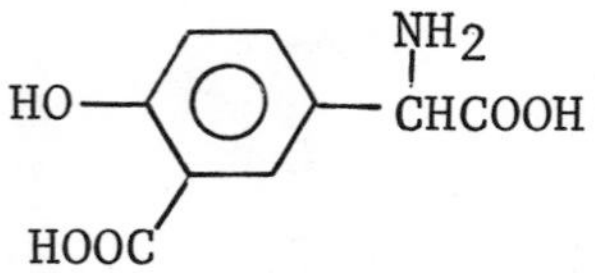

$C_9H_9NO_5$

Ref. 511

Resolution of synthetic (3-carboxy-4-hydroxyphenyl)-glycine. **The racemic amino acid dihydrate (12.3 g) and L-arginine (8.62 g) were dissolved in hot water (300 ml). Ethanol (220 ml) was added to the hot solution, which was left overnight in the refrigerator. A crystalline salt was isolated and dried at 0.01 mm over phosphorus pentoxide (4.36 g), $[\alpha]_D^{21}$ −41.7° (*c* 1.3, water). Recrystallization from aqueous ethanol yielded 3.42 g, $[\alpha]_D^{24}$ −42.9° (*c* 1.2, water), an additional recrystallization 2.32 g, $[\alpha]_D^{26}$ −43.3° (*c* 1.3, water). This preparation, believed to represent a homogeneous diastereoisomeride, was very hygroscopic and took up water to a sesquihydrate when exposed to the atmosphere (Found: C 44.08; H 6.45; N 16.68; H_2O 6.04. Calc. for $C_{15}H_{23}O_7N_5$, 1 1/2 H_2O: C 43.70; H 6.32; N 16.99; H_2O: 6.56).**

The salt (670 mg of the sesquihydrate)was dissolved in water (25 ml) and the solution adjusted to pH 2.4 with hydrochloric acid (1.66 ml, 1 N). After a night in the refrigerator, a small precipitate (47 mg) of partly racemic D-(3-carboxy-4-hydroxyphenyl)-glycine, H_2O ($[\alpha]_D^{23}$ −85.6°, (*c* 0.86, phosphate buffer pH 7, 0.2 M)) was removed. The mother liquor yielded a second crop consisting of pure D-*(3-carboxy-4-hydroxyphenyl)-glycine*, *H_2O* (Found: C 47.27; H 4.97; N 6.09. Calc. for $C_9H_9O_5N$, H_2O: C 47.17; H 4.84; N 6.11), $[\alpha]_D^{24}$ −90.1° (*c* 0.7, phosphate buffer pH 7, 0.2 M), $[\alpha]_D^{22}$ −121° (*c* 0.75, 1 N HCl), $[\alpha]_D^{25}$ −105° (*c* 0.7, 1 N formic acid), $[\alpha]_D^{25}$ −90.3° (*c* 0.88, 1 N NH_3, aq), $[\alpha]_D^{25}$ −90.3° (*c* 0.68, 1 N NaOH), $[\alpha]_D^{22}$ −95° (*c* 0.71, water, supersaturated). The water of crystallization could not be removed at 50° and 0.01 mm over phosphorus pentoxide. The rotatory dispersion curve was determined in a 0.04 % solution in 1 N HCl as shown in Fig. 3.

To the first mother liquor, more ethanol was added. A crystalline precipitate (11.4 g after drying over phosphorus pentoxide, $[\alpha]_D^{23}$ −2.1° (*c* 1.3 water)) was removed by filtration. The mother liquor was evaporated to dryness *in vacuo*, and the residue dissolved in water (100 ml). The solution was adjusted to pH 2.4 with hydrochloric acid (4.22 ml, 4 N). After cooling, a precipitate of partly racemic material was obtained (1.34 g, $[\alpha]_D^{25}$ + 15.1° (*c* 0.75, phosphate buffer pH 7, 0.2 M). The mother liquor was concentrated *in vacuo*. After cooling overnight in the refrigerator, a crystalline product, (2.76 g, $[\alpha]_D^{25}$ + 88.5° (*c* 0.73, phosphate buffer pH 7, 0.2 M)) was isolated by filtration. Recrystallization from water afforded an analytical specimen of L-*(3-carboxy-4-hydroxyphenyl)-glycine*, *H_2O* (Found: C 47.28; H 4.84; N 6.18), $[\alpha]_D^{24}$ + 85.6° (*c* 0.72, phosphate buffer pH 7, 0.2 M). The infra-red absorption spectrum was indistinguishable from that of the enantiomer.

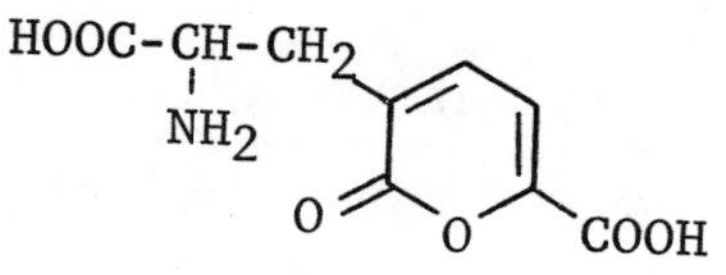

$C_9H_9NO_6$ Ref. 512

Optical Resolution of DL-Stizolobinic Acid (I). Method A: *(+)- and (−)-Stizolobinic Acid.* Free L-lysine was freshly prepared from the hydrochloride by Dowex 50W×2 (H^+ form) column chromatography. A mixture of 2.27 g of DL-stizolobinic acid and 1.5 g of L-lysine was dissolved in 10 *ml* of water and filtered. After the addition of the same volume of ethanol, the solution was allowed to stand overnight in a refrigerator. The precipitate was collected, recrystallized three times from 50% aqueous ethanol, and then three times from 55% aqueous methanol. Yield of the less soluble lysine salt was 0.33 g; $[\alpha]_D^{20}$ +16.5° (*c* 1, water).

The aqueous solution of this salt was acidified with hydrochloric acid, the crystalline precipitate formed was collected and recrystallized from 3 N hydrochloric acid to give 106 mg of (+)-stizolobinic acid hydrochloride. The hydrochloride was converted easily to the free acid by treatment with water; $[\alpha]_D^{25}$ +19.5° (*c* 1, 3 N HCl).

Found: C, 47.73; H, 4.07; N, 6.31%. Calcd for $C_9H_9NO_6$: C, 47.58; H, 3.99; N, 6.17%.

All the filtrates from the less soluble lysine salt were combined, and evaporated to dryness *in vacuo*. The residual material was dissolved in 55% aqueous methanol, seeded with (+)-stizolobinic acid - L-lysine salt, kept in a refrigerator overnight and filtered. The filtrate was evaporated to dryness, dissolved in water and acidified with hydrochloric acid until crystalline precipitate had formed. The precipitate was collected and recrystallized four times from 3 N hydrochloric acid to give 89 mg of (−)-stizolobinic acid hydrochloride. Treatment with water produced the free acid; $[\alpha]_D^{25}$ −20.5° (*c* 1, 3 N HCl).

Found: C, 47.80; H, 4.01; N, 6.20%. Calcd for $C_9H_9NO_6$: C, 47.58; H, 3.99; N, 6.17%.

(+)- and (−)-Stizolobinic acid are identical with natural stizolobinic acid with regard to infrared and ultraviolet spectra.

$C_9H_9NO_6$ Ref. 512a

Method B: *N-Acetyl-DL-stizolobinic Acid.* To a mixture of 1 g of DL-stizolobinic acid and 3 g of sodium bicarbonate in 25 *ml* of water was added dropwise 2.5 *ml* of acetic anhydride under vigorous agitation in an ice bath. Stirring was continued for 1 hr at 0°C and allowed to stand overnight at room temperature. The reaction mixture was acidified with 3 *ml* of concentrated hydrochloric acid and chilled. The precipitate was collected and recrystallized from hot water, yielding 0.84 g of colorless crystals, mp 251—252°C (decomp.).

Found: C, 49.32; H, 4.11; N, 5.21%. Calcd for $C_{11}H_{11}NO_7$: C, 49.07; H, 4.12; N, 5.20%.

(+)-Acetylstizolobinic Acid. A mixture of 1.12 g of strychnine and 0.9 g of *N*-acetyl-DL-stizolobinic acid was dissolved in 50 *ml* of hot ethanol and filtered. The filtrate was chilled and the bulky precipitate formed was collected. Three recrystallization from ethanol gave 370 mg of colorless crystals, mp 255—257°C (decomp.); $[\alpha]_D^{20}$ +11.6° (*c* 1, water).

Found: C, 63.41; H, 5.78; N, 6.98%. Calcd for $C_{32}H_{33}N_3O_9$: C, 63.67; H, 5.51; N, 6.96%.

The strychnine salt (340 mg) was decomposed by dissolving in 20 *ml* of aqueous sodium bicarbonate solution (100 mg). The aqueous solution was extracted with chloroform and acidified with hydrochloric acid to yield the crystalline precipitate. The precipitate was collected and recrystallized from hot water to give (+)-acetylstizolobinic acid, mp 230—232°C; $[\alpha]_D^{20}$ +98.5° (*c* 1, methanol).

Found: C, 49.24; H, 4.20; N, 5.18%. Calcd for $C_{11}H_{11}NO_7$: C, 49.07; H, 4.12; N, 5.20%.

(+)-Stizolobinic Acid. (+)-Acetylstizolobinic acid (100 mg) in 5 *ml* of concentrated hydrochloric acid was refluxed for 3 hr. Then the solution was cooled and the precipitate was collected. Recrystallization from hot water gave 40 mg of free (+)-stizolobinic acid; $[\alpha]_D^{25}$ +18.5° (*c* 1, 3 N HCl).

Found: C, 47.71; H, 3.96; N, 5.87%. Calcd for $C_9H_9NO_6$: C, 47.58; H, 3.99; N, 6.17%.

$C_9H_9NO_6$ Ref. 512b

Method C. *N-Carbobenzyloxy-DL-stizolobinic Acid.* To a mixture of 1 g of DL-stizolobinic acid and 3 g of sodium bicarbonate in 20 *ml* of water and 15 *ml* of ether was added dropwise 13.6 g of 30% carbobenzyloxychloride in toluene solution under vigorous agitation in an ice bath. Stirring was continued for 3 hr at 0°C and for 3 hr at room temperature. After separation from the organic layer, the aqueous layer was

Found: C, 51.27; H, 3.47; N, 6.72%. Calcd for $C_9H_7NO_5$: C, 51.68; H, 3.37; N, 6.70%.

The mother liquor of the brucine salt of (−)-lactam was evaporated to dryness and the residue was recrystallized three times from 90% aqueous methanol to yield 1.9 g of the (+)-brucine salt; $[\alpha]_D^{20}$ +12° (*c* 1, 50% aqueous methanol).

The decomposition of the (+)-salt and recrystallizaion of the product was carried out by the same procedure mentioned above, and 0.5 g of the (+)-lactam, $[\alpha]_D^{25}$ +28.5° (*c* 1, water), was obtained.

Found: C, 51.36; H, 3.56; N, 6.81%. Calcd for $C_9H_7NO_5$: C, 51.68; H, 3.37; N, 6.70%.

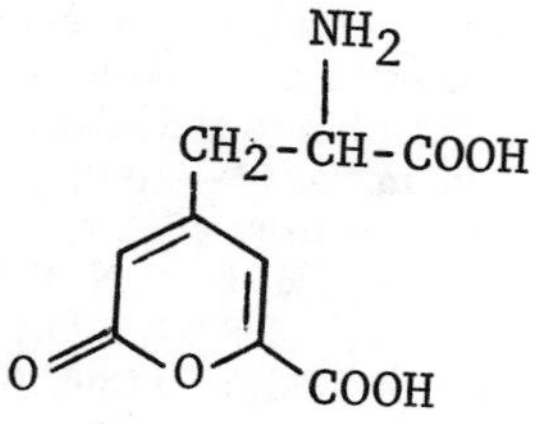

$C_9H_9NO_6$ Ref. 513

The Optical Resolution of DL-Stizolobic Acid (II). *N-Chloroacetyl-DL-stizolobic Acid.* *N*-Chloroacetyl-derivative was prepared from 1.5 g of DL-stizolobic acid, 4.1 g of chloroacetyl chloride and 4.5 g of sodium bicarbonate by the same procedure for the preparation of *N*-carbobenzyloxy-DL-stizolobinic acid. The product was recrystallized twice from acetone-benzene-petroleum ether and once from acetone, mp 175°C.

Found: C, 43.35; H, 3.78; N, 4.57%. Calcd for $C_{11}H_{10}NO_7Cl$: C, 43.51; H, 3.32; N, 4.61%.

Preparation of Aminoacylase.[6)] Three grams of "Takadiastase" was suspended in 30 *ml* of water, extracted one hour at 0°C under stirring, and filtered through a mat of Dicalite. The filtrate was used for the resolution.

Isolation of (+)-Stizolobic Acid and (−)-Chloroacetylstizolobic Acid. A reaction mixture consisting of 1 g of *N*-chloroacetyl-DL-stizolobic acid, 30 *ml* of aminoacylase solution, 0.3 *ml* of 37% formaldehyde solution,

100 ml of 0.1 M phosphate buffer (pH 6.8) and water to make a final volume of 170 ml, was incubated at 37°C for 3 weeks with the occassional addition of the enzyme solution. The progress of the reaction was examined by the spot test with ninhydrin.

After filtration, the reaction mixture was applied to a column of Dowex 50W×8 (H^+ form, 2×30 cm) and eluted with water. The eluate could be divided into two fractions by the measurement of the optical density at 303 mμ. The first fraction (A) was acidic and contained phosphate and the second (B) was neutral and free from phosphate. The fraction B was concentrated to dryness under reduced pressure, and the residue (0.24 g) was recrystallized from hot water. The crystalline material (0.16 g) was dissolved in 20 ml of water and applied to Amberlite IR 45 column (OH^- form, 2×15 cm). After washing with 200 ml of water, eluted with 0.01 N hydrochloric acid. The eluate was evaporated to dryness, and the residue was recrystallized from hot water to give 65 mg of (+)-stizolobic acid; $[\alpha]_D^{25}$ +16.0° (*c* 0.5, 1 N HCl).

Found: C, 47.50; H, 4.13; N, 6.07%. Calcd for $C_9H_9NO_6$: C, 47.58; H, 3.99; N, 6.17%.

(+)-Stizolobic acid was identical with respect to infrared spectra and mixed melting point with the natural acid.

The fraction A was extracted with ethyl acetate, and the extract was evaporated to dryness. The residual oil solidified partially by storage at room temperature and was separated by filtration. The solid product was dissolved in hot ethanol, filtered, and evaporated to dryness. This procedure was repeated several times. Weighed 0.3 g; $[\alpha]_D^{25}$ −10° (*c* 1, ethanol).

(−)-*Stizolobic Acid.* A solution of (−)-chloroacetylstizolobic acid (0.3 g) obtained above in glacial acetic acid saturated with dried hydrogen chloride was saponified on a steam bath for 48 hr. The precipitate was collected, washed with glacial acetic acid and recrystallized twice from hot water to yield 0.15 g of colorless crystals; $[\alpha]_D^{25}$ −4.5° (*c* 1, 1 N HCl).

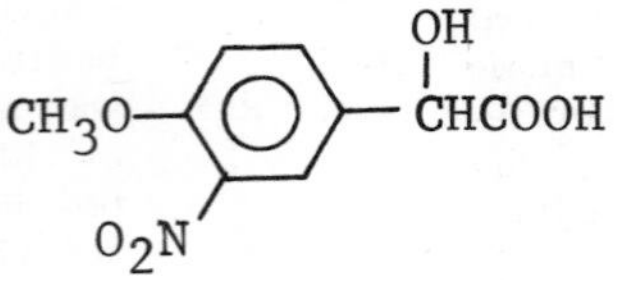

$C_9H_9NO_6$ Ref. 514

Resolution of DL-*4-Methoxy-3-nitromandelic Acid* (V).—Cinchonine (58·9 g., 0·2 mole) was added to a solution of 4-methoxy-3-nitromandelic acid (45·4 g., 0·2 mole) in boiling 1 : 4 (v/v) aqueous ethanol (560 ml.). The hot solution was treated with charcoal and filtered, and crystals (*A*) (55·3 g.) were collected 24 hr. later. Fraction A had m. p. 166—167°, $[\alpha]_D^{18}$ +156·0° (0·3—0·4% w/v in $CHCl_3$) and gave an acid with $[\alpha]_D^{18}$ +63·4° (0·2—0·3% w/v in EtOH). Four crystallisations from 80% ethanol gave *cinchonine* (+)-*4-methoxy-3-nitromandelate* (29·5 g.) of constant m. p. (190—191°) and rotation {$[\alpha]_D^{18}$ +175·2° (0·3—0·4% w/v in $CHCl_3$)} (Found: C, 62·0; H, 6·4; N, 7·8; H_2O, 3·4. $C_{19}H_{22}ON_2,C_9H_9O_6N,H_2O$ requires C, 62·3; H, 6·2; N, 7·8; H_2O, 3·3%).

This cinchonine salt (12·5 g.), dissolved in 0·1N-sulphuric acid (18 ml.) and extracted with ether, gave the (+)-*acid* (V) (4·9 g.), m. p. 96—97°, $[\alpha]_D^{17}$ +119·0° (0·26% w/v in EtOH) (Found: C, 47·3; H, 4·0; N, 6·1. $C_9H_9O_6N$ requires C, 47·6; H, 4·0; N, 6·2%).

The mother-liquor from fraction *A* was concentrated *in vacuo* below 40° to a syrup (*B*) from which crude (−)-4-methoxy-3-nitromandelic acid (18 g.), $[\alpha]_D^{17}$ −79·5° (0·44% w/v in EtOH), was obtained. This acid (17·45 g.) in boiling 80% ethanol (100 ml.) was treated with brucine (30·2 g.), and the solution cooled, seeded, and left overnight. The *brucine salt* obtained (31·3 g.) had m. p. 95—100°, $[\alpha]_D^{17}$ −58·1° (0·7% w/v in $CHCl_3$); crystallised to constant m. p. and rotation from 80% ethanol, it had m. p. 106—107°, $[\alpha]_D^{17}$ −60·3° (0·7% w/v in $CHCl_3$) (Found: C, 60·0; H, 6·1; N, 6·5; H_2O, 2·7. $C_{23}H_{26}O_4N_2,C_9H_9O_6N,H_2O$ requires C, 60·1; H, 5·8; N, 6·6; H_2O, 2·8%). The (−)-*acid* (V) (2·05 g.), obtained from the brucine salt (6·2 g.), had m. p. 96—97°, $[\alpha]_D^{18}$ −118·5° (0·25% w/v in EtOH) (Found: C, 47·5; H, 4·1; N, 6·3%).

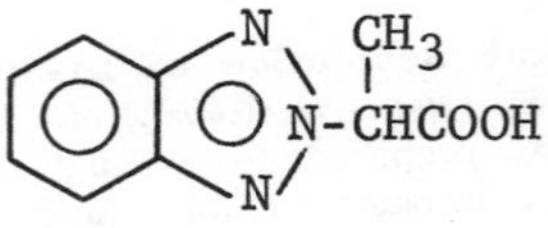

$C_9H_9N_3O_2$ Ref. 515

R-(+)-α-*(Benzotriazolyl-2) propionic acid (+II)*. 20.0 g (0.105 mol) α-(benzotriazolyl-2)-propionic acid and 30.8 g (0.105 mol) cinchonidine were dissolved in 320 ml boiling methanol-water (7:1). After one day in a refrigerator the salt obtained was filtered off. The acid was liberated from a small sample of the salt after each recrystallization and the rotary power was determined. The salt was recrystallized several times until the optical activity remained constant.

After nine recrystallizations the acid was liberated by addition of 2 M sulfuric acid. The acid obtained was extracted with ether and the ether evaporated *in vacuo*. 1.27 g (13 %) (+)-II was isolated, which was recrystallized from 170 ml boiling water and finally dried over concentrated sulfuric acid in a desiccator. M.p. 230–231 °C. $[\alpha]_D^{25}$ = +33.4°, $[\alpha]_{365}^{25}$ = +74.4° (*c*=0.548 g/100 ml *l*=1 dm, α_D^{25} = +0.183°, α_{365}^{25} = +0.408° in acetone). $[\theta]_{max}$ is +1945° at 280 nm in methanol solution.

S-(−)-α-*(Benzotriazolyl-2) propionic acid* [(−) *II*]. The mother liquor from the first crystallization was evaporated to dryness and the acid isolated as before. 7.5 g (0.04 mol) acid and 4.8 g (0.040 mol) (+)-α-methylbenzylamine was dissolved in 280 ml boiling ethyl acetate. The solution was cooled gently and then put in a refrigerator. 8.5 g salt was obtained, which was recrystallized until the optical activity remained constant.

The acid was isolated as described above. 2.33 g (11 %) acid was obtained. The acid was recrystallized from boiling water and dried over sulfuric acid in a desiccator. M.p. 230–231 °C. The rotatory power of the (–)-II acid in various solvents is given in Table 1.

Table 1. The optical activity of (–)-II in some different solvents. g = grams acid dissolved in the solvent to 10.0 ml. l = the tube length in dm.

Solvent	$[\alpha]_D^{25}$	l	g
Dimethylformamide	–42.6°	1	0.352
Acetone	–33.2°	1	0.750
Abs. ethanol	–7.7°	2	0.325
Water (neutral.[a])	+41.0°	2	0.305

[a] The acid was dissolved in 10.0 ml of 1.5 M ammonia solution.

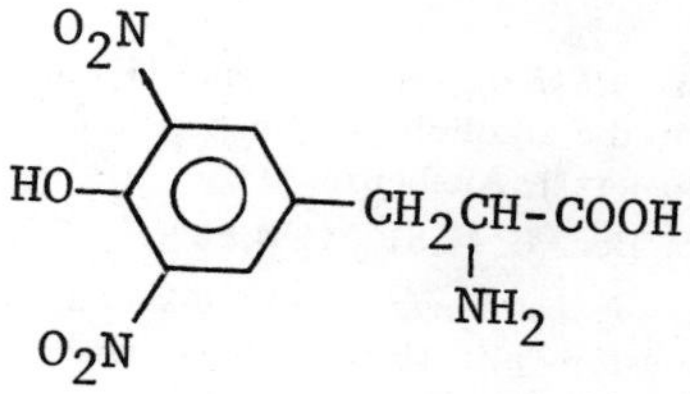

$C_9H_9N_3O_7$ Ref. 516

Salz aus (—)-α-Phenäthylamin und 3.5-Dinitro-*N*-acetyl-D-tyrosin: Zu einer Lösung von 12,1 g (0,1 Mol) (—)-α-Phenäthylamin in 430 m*l* absol. Äthanol werden 31,3 g (0,1 Mol) 3.5-Dinitro-DL-tyrosin gegeben und unter Rückfluß bis zur vollständigen Lösung erhitzt. Nach 6 Tagen trennt man das Kristallisat ab und wäscht den Rückstand mit 25 m*l* absol. Äthanol, anschließend gründlich mit Äther. Ausb. 16,7 g. Aus dem im Vak. auf 100 m*l* eingeengten Filtrat wurden nach Animpfen und 6tägigem Aufbewahren im Kühlschrank noch 2,6g erhalten, insgesamt 19,3 g (89% d. Th.). Zweimalige Umkristallisation aus Wasser führt zu 14,3 g (66% d. Th.) der reinen Verbindung, orange Nadeln vom Schmp. 208—209° (Zers.), $[\alpha]_D^{21}$: —8,8° (in Wasser, c = 0,5). Löslichkeit in Äthanol: 0,6 g in 100 g Lösungsmittel bei 22°.

$C_{19}H_{22}N_4O_8$ (434,4)	Ber.	C 52,53	H 5,11	N 12,90
	Gef.	C 52,88	H 5,42	N 12,66

Salz aus (+)-α-Phenäthylamin und 3.5-Dinitro-*N*-acetyl-L-tyrosin: Die Mutterlauge wird im Vak. zur Trockne eingedampft, der erhaltene Rückstand von 23,6 g in 240 m*l* Wasser gelöst und zur wäßrigen Lösung 50 m*l* konz. Salzsäure hinzugefügt. Die entstehende kristalline Fällung wird nach 30 Min. Stehenlassen im Eisbad abgesaugt, mit wenig schwach salzsaurem Wasser, mit Wasser, mit Äther/Äthanol 1:1 und schließlich gründlich mit Äther gewaschen. Ausb. 13,7 g (86% d. Th.). Die optische Reinheit des so erhaltenen 3.5-Dinitro-*N*-acetyl-L-tyrosins beträgt 62%, $[\alpha]_D^{21}$: +7,6° (in Dioxan, c = 1,0). Diese Verbindung wird zur Lösung von 5,3 g (+)-α-Phenäthylamin in 190 m*l* absol. Äthanol gegeben und unter Rückfluß bis zur vollständigen Lösung erhitzt. Nach 6tägigem Aufbewahren im Kühlschrank trennt man das Kristallisat ab. Ausb. 13,3 g vom Schmp. 203—204° (Zers.). Die Umkristallisation aus 200 m*l* Wasser führte sofort zu 11,3 g (52% d. Th.) reinem Aminsalz vom Schmp. 208—209° (Zers.) und $[\alpha]_D^{21}$: +8,8° (in Wasser, c = 0,5).

$C_{19}H_{22}N_4O_8$ (434,4)	Ber.	C 52,53	H 5,11	N 12,90
	Gef.	C 52,60	H 4,98	N 12,78

3.5-Dinitro-*N*-acetyl-D-tyrosin: 10 g (—)-α-Phenäthylaminsalz des 3.5-Dinitro-*N*-acetyl-D-tyrosins werden unter schwachem Erwärmen in 300 m*l* Wasser gelöst und mit 10 m*l* konz. Salzsäure versetzt. Die beim langsamen Abkühlen entstehende kristalline Fällung wird nach 30 Min. Stehenlassen im Eisbad abgesaugt, mit wenig schwach salzsaurem Wasser, mit Äther/Äthanol 1:1 und gründlich mit Äther gewaschen. Nach Umkristallisieren aus Wasser werden 5,9 g (82% d. Th.) optisch reine Verbindung, $[\alpha]_D^{21}$: —12,2°, vom Schmp. 189—190° erhalten.

3.5-Dinitro-*N*-acetyl-L-tyrosin: Die Verbindung wird analog wie die der entsprechenden D-Konfiguration hergestellt. Ausb. 5,8 g (80% d. Th.) vom Schmp. 189—190°, $[\alpha]_D^{21}$: +12,2°.

Salz aus (—)-α-Phenäthylamin und 3.5-Dinitro-*N*-acetyl-L-tyrosin: Dieses Salz wird aus einer alkohol. Lösung der optisch reinen Komponenten im äquimolaren Verhältnis durch Ausfällen mit Äther erhalten. Orange feine Nadeln vom Schmp. 173—174°, $[\alpha]_D^{21}$: +7,7° (in Wasser, c = 0,5). Löslichkeit in Äthanol: 9,9 g in 100 g Lösungsmittel bei 22°.

$C_{19}H_{22}N_4O_8$ (434,4)	Ber.	C 52,53	H 5,11	N 12,90
	Gef.	C 52,16	H 4,66	N 12,63

Salz aus (+)-α-Phenäthylamin und 3.5-Dinitro-*N*-acetyl-D-tyrosin: Dieses Salz wird analog der vorstehenden Verbindung hergestellt. Orange feine Nadeln vom Schmp. 173—174°, $[\alpha]_D^{21}$: —7,7° (in Wasser, c = 0,5).

$C_{19}H_{22}N_4O_8$ (434,4)	Ber.	C 52,53	H 5,11	N 12,90
	Gef.	C 52,62	H 5,01	N 12,75

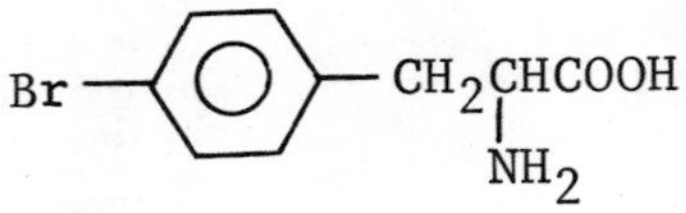

$C_9H_{10}BrNO_2$ **Ref. 517**

3) *D-p-Bromphenylalanin-äthylester-dibenzoyl-D-tartrat (III)*: 0,01 Mol DL-*p*-Bromphenyl-*alanin-äthylester* (II) wurden zu einer auf 21° gehaltenen Lösung von 0,005 Mol Dibenzoyl-D-weinsäure in 19 ml abs. Äthanol gegeben, worauf sich die Reaktionsmischung erwärmte. Nach 15 Min. setzte Kristallisation ein, die nach 2 Std. beendet war. Zweimaliges Umkristallisieren aus Äthanol (nicht aber aus Isopropanol[7]), war nötig, um einen maximalen Drehungswert zu erreichen. Ausbeute 70–75%. Wird anstelle von Dibenzoyl-D-weinsäure Dibenzoyl-L-weinsäure verwendet, so wird das enantiomere Salz V erhalten, welches genau gleich weiterverarbeitet werden kann.

4) *D-(+)-p-Bromphenylalanin-äthylester-hydrochlorid (VII)*: Das Dibenzoyl-D-tartrat III wurde durch Einleiten von HCl in die alkoholische Suspension zerlegt und das Hydrochlorid VII durch Zugabe von Äther kristallisiert[8]), Ausbeute 70%.

$C_{11}H_{15}O_2NBrCl$ Ber. C 42,81 H 4,89% Gef. 42,71 4,99%

5) *D-(+)-p-Bromphenylalanin-hydrochlorid (IX)*: Gleichzeitige Zerlegung des Salzes III und Hydrolyse des Aminosäure-esters mit 10-proz. wässeriger Salzsäure, wie am Beispiel des Phenylalanins ausgeführt[7]), führte wegen der Schwerlöslichkeit nicht zum Ziele, weshalb das Salz III mit der fünffachen Menge eines Gemisches von konz. HCl und Eisessig (1:1, v:v) kurze Zeit auf dem siedenden Wasserbad erhitzt wurde. Darauf wurde das Reaktionsgemisch mit Wasser verdünnt und mit Essigester von Dibenzoyl-D-weinsäure befreit. Beim Eindampfen der wässerigen Phase kristallisierte D-(+)-p-Bromphenylalanin-hydrochlorid (IX) in Nadeln aus, Ausbeute 90–95%.

6) *D-(+)-p-Bromphenylalanin (XI)*: Aus der mit 5-proz. Ammoniak neutralisierten, heiss gesättigten, wässerigen Lösung des Hydrochlorids kristallisierte in der Kälte die freie Aminosäure in Blättchen aus.

$C_9H_{10}O_2NBr$ Ber. C 44,25 H 4,13% Gef. 44,28 4,24%

7) *Drehungsverschiebung von D-(+)-p-Bromphenylalanin bei Änderung des pH:* Bei D-α-Aminosäuren verändert sich allgemein die spezifische Drehung, ausgehend vom isoelektrischen Punkt, mit zunehmender H-Ionenkonzentration nach negativeren Werten. Für das rechtsdrehende Isomere des p-Bromphenylalanins wurden folgende Werte für [α] gefunden: bei pH 11: +5°; pH 7: +15–20°; pH 2: +11°; pH 0: +7,5°. Die Abnahme der Drehung von pH 7 bis gegen pH 0 deutet auf das Vorliegen der D-Form hin.

7) W. Langenbeck & O. Herbst, Chem. Ber. *86*, 1524 (1953).
8) G. Losse & H. Jeschkeit, Chem. Ber. *90*, 1275 (1957).

Schmelzpunkte und opt. Drehungen

Verbindung	Schmelzpunkt	[α]D	Konz.	Lösungsmittel
III	135–142° (147° Zers.)	− 69,7° ± 1,1°	1,67	Äthanol
IIIa	133–144° (147° Zers.)	− 73,1° ± 0,6°	1,50	Äthanol
VII	193°	− 27° ± 4°	1,80	Äthanol
VIIa	210–211°	− 17,5° ± 0,1°	5,05	Methanol
IX	255–256°	+ 9,3° ± 0,8°	1,51	Wasser
XI	267°	+ 20° ± 2,6°	0,40	Wasser
		+ 6,4° ± 1,5°	0,78	Eisessig
V	135–142° (147° Zers.)	+ 67° ± 4°	0,8	Äthanol
Va	129–144° (147° Zers.)	+ 71,9° ± 0,5°	1,46	Äthanol
VIII	193°	+ 27° ± 4°	1,80	Äthanol
VIIIa	212–213°	+ 18,6° ± 0,1°	5,70	Methanol
X	256°	− 9,8° ± 1,2°	1,30	Wasser
XII	268,5°	− 20,5° ± 1,6°	0,39	Wasser
		− 5,9° ± 1,3°	0,35	Eisessig
		− 0,41° ± 0,35°	2,39	2 N NaOH

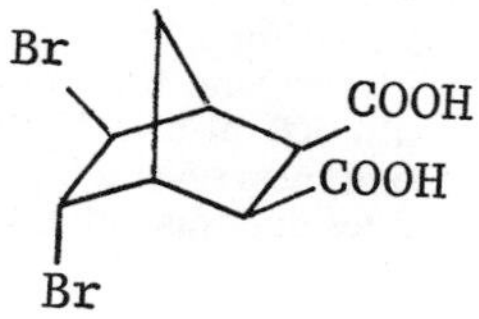

$C_9H_{10}Br_2O_4$

Ref. 518

Optical Resolution of the Dibasic Acid VIa.—The racemic acid was prepared (95% yield) by boiling the anhydride IVA in 50% acetonitrile for 2.5 hours. It melted sharply at 159–160° with evolution of gas, resolidified as the temperature was raised and remelted at 184–185°, reported[3] m.p. 160–161° dec. The over-all material balance in the resolution of this acid was poor, principally because of the pronounced solubility of the optically active acid in water and organic solvents. A hot solution of 3.94 g. of VIa in 20 cc. of acetone was treated with a hot, filtered solution of 4.15 g. of quinine trihydrate in 100 cc. of acetone. The salt (7.28 g.) precipitated immediately. The mother liquor A was preserved. The salt was recrystallized from 250 cc. of methanol and the mother liquor B was preserved. The precipitate, 3.0 g. of a microcrystalline powder, was treated with 10 cc. of ice-cold 15% hydrochloric acid, whereupon it dissolved. The organic acid failed to crystallize from this solution. The solution was made basic with ammonia and extracted successively with chloroform, methylene chloride and ether. When the aqueous phase was acidified with hydrochloric acid and chilled, 0.23 g. of a colorless solid separated. The mother liquor C was preserved. The solid contained inorganic salt which was removed during recrystallization from ethyl acetate–ligroin. The resulting acid D was preserved. The aqueous mother liquor C was evaporated to dryness on the steam-bath *in vacuo* and the residue leached successively with boiling ethyl acetate and chloroform. The salt residue E was preserved. The extract was combined with D and the whole evaporated to dryness. The residue was recrystallized from ethyl acetate–ligroin to give 0.105 g. of the dibasic acid, sintering at 155° and melting at 183–184°. This material showed $[\alpha]_D$ −23.1° (*c* 1.05 in acetone, *l* = 4). The organic mother liquor F was preserved.

An additional 0.04 g. of material was obtained by leaching the salt residue E with methylene chloride. This was combined with the residue obtained by evaporation of the solvent from F and the whole recrystallized from ethyl acetate–ligroin to give 0.085 g. of sharply defined triangular plates. This material melted at 146–147° to a turbid liquid which became clear at 160°; in acetone, $[\alpha]_D$ −56.6° (*c* 0.48, *l* = 4).

Anal. Calcd. for $C_9H_{10}O_4Br_2$: C, 31.61; H, 2.95. Found: C, 31.72; H, 2.79.

Dextrorotatory material was obtained from the tail fractions (A and B). These solutions were combined and concentrated by evaporation to a volume of 100 cc. The solution was allowed to stand in the refrigerator for three days. The precipitated quinine salt (3.94 g.) was filtered off, treated with 10 cc. of concentrated ammonia water and the mixture was washed with chloroform. The aqueous phase was acidified with concentrated hydrochloric acid and exhaustively extracted with ethyl acetate. The organic solution was dried with sodium sulfate, evaporated and the residue (2.1 g.) recrystallized from ethyl acetate–ligroin. The acid, 0.60 g., melted at 184–186° and was virtually optically inactive. The mother liquor was concentrated and the resulting crystalline product collected. It weighed 1.25 g., sintered at 138° and melted at 185–186°; in acetone, $[\alpha]_D$ +11.3° (*c* 2.13, *l* = 4). The greater solubility of the enantiomeric modification compared to that of the racemate parallels our experience with the previous members of the series.[2,5]

J. Amer. Chem. Soc.,

(2) J. A. Berson and R. Swidler, This Journal, **76**, 4060 (1954).
(3) D. Craig, *ibid.*, **73**, 4889 (1951).
(4) H. Kwart and L. Kaplan, *ibid.*, **75**, 3356 (1953).
(5) J. A. Berson, *ibid.*, **76**, 4069 (1954).

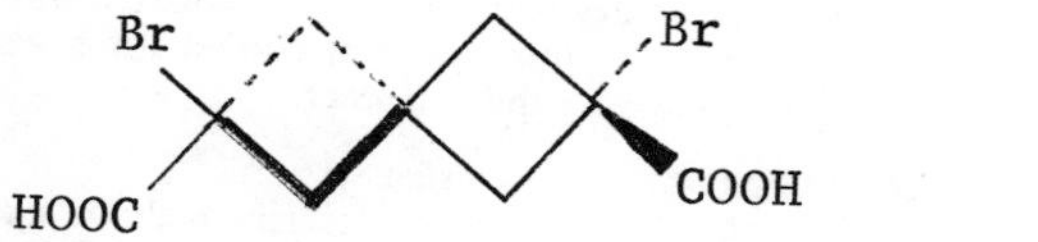

$C_9H_{10}Br_2O_4$

Ref. 519

Dédoublement optique.

Après plusieurs expériences préliminaires avec la strychnine et la brucine, le dédoublement optique a été effectué de la manière suivante.

On a mélangé 16 g de l'acide et 36.9 g de brucine, tous les deux dissous dans 280 cm³ d'alcool chaud. Le lendemain, le sel de brucine déposé a été recristallisé dans 700 cm³ d'alcool additionnés d'un peu d'eau. On a répété plusieurs fois la recristallisation dans la quantité nécessaire d'alcool. Après dix recristallisations, la rotation de l'acide qu'on obtient aux dépens du sel de brucine n'a plus augmenté.

2 g de sel de brucine ont été décomposés par 35 cm³ d'acide sulfurique 2 *n*; l'extraction à l'éther a donné une solution, qui, examinée dans un tube de 4 dm, a présenté les valeurs de rotation suivantes. La concentration de la solution éthérée (30 cm³) a été déterminée par évaporation; résidu, séché à 120°, 0.4490 g.

Acide dibromospiroheptanedicarboxylique.

λ	656.3	589.3	546.3	486.1 m
α	+ 0.305°	0.39°	0.465°	0.615°
$[\alpha]$	+ 5.10°	6.51°	7.77°	10.27°
[M]	+ 17.4°	22.3°	26.6°	35.1°

L'acide actif, dissous dans un peu d'alcool, neutralisé par de la soude caustique, dilué à 50 cm³ et examiné dans un tube de 4 dm, a présenté une rotation de +0.15°. Rotation moléculaire du sel sodique: $[M]_D = +\ 14.4°$.

Rotation de l'acide dissous dans l'alcool absolu $[M]_D = +\ 5.4°$.

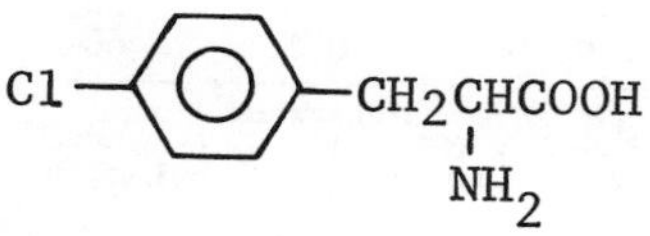

$C_9H_{10}ClNO_2$

Ref. 520

Resolution of p-Substituted DL-Phenylalanines[4]

Hydrogen chloride was bubbled through a suspension of 0.5 g of the DL-amino acid in 30 ml of absolute ethyl alcohol for 20 min. Next day, the solvent was evaporated off, and the whole operation was repeated. Excess HCl was then removed by repeating the evaporation three times after the addition of ethanol. The residue was dissolved in 15 ml of water, the pH of the solution was adjusted to 5.0 with 0.2 *M* LiOH, α-chymotrypsin[4] was added, and the mixture was incubated at room temperature for ½–1½ h,[4] the pH being kept constant by the automatic addition of 0.2 *M* LiOH from the titrator. After the digestion, the mixture was concentrated until crystals appeared, cooled for 1 h, filtered, and the precipitate was washed with ethanol. The L-isomer so obtained was recrystallized by dissolving in hot water containing 2 ml of *N* HCl, filtering the solution (Celite), and adding 2 ml of *N* LiOH.

The filtrate obtained after removal of the L-isomer was brought to pH 9.0 with 0.2 *M* LiOH and extracted with ethyl acetate (3 × 50 ml) which was then dried over $MgSO_4$ and filtered into ethyl acetate containing hydrogen chloride. The solvent was removed by evaporation, the residue was dissolved in 20 ml of 0.2 *M* LiOH (*i.e.* to pH 12.0), and the solution was kept at 45 °C for 1 h. The pH was then adjusted to 5.0 with *N* HCl, the solution was

[4] See Table 1.

evaporated to dryness, the residue was triturated in hot ethanol, and the mixture then cooled for several hours and filtered to give the D-isomer.

Resolution of DL-*m-Tyrosine*

This was carried out as described above except that the esterification step was not repeated, and after the enzymatic digestion, the mixture was evaporated to dryness and the residue was triturated with ethanol to obtain the L-isomer. The filtrate was then evaporated to dryness, the residue was dissolved in water, and the D-amino acid obtained as described above.

Resolution of DL-*o-Tyrosine*

This was carried out as for the *m*-tyrosine isomers except that the esterification step was repeated as for the *p*-substituted derivatives. In addition, after saponification of the D-amino acid ester, the L-isomer present due to the incomplete enzymatic reaction was destroyed by digestion with L-amino acid oxidase. For this purpose, 5 ml of *N* HCl and 3 ml of 10% NaH_2PO_4 were added to bring the pH to 8.0 and the mixture was incubated at 37 °C for 24 h, with oxygen being bubbled through, in the presence of 100 mg of *Crotalus adamanteus* L-amino acid oxidase. Dowex 50 (H^+) (~120 ml) was added, the suspension was stirred for 20 min and filtered, the resin was washed with water, and the amino acid then eluted from the resin with 500 ml of 3 *N* NH_4OH. The eluate was evaporated to dryness several times after the addition of water, and the residue was collected with the help of ethanol.

Resolution of DL-*Dopa*

This was carried out as described for the *p*-substituted phenylalanines except that the esterification step was not repeated, sodium hydroxide was used in all cases instead of lithium hydroxide, both isomers were crystallized from an aqueous solution at pH 5.5, and the ester was hydrolyzed by refluxing in *N* HCl for 1 h instead of being saponified.

Determination of Optical Purity of Isomers

The optical purity of the isomers was determined by chromatography of their L-alanine dipeptides formed by reaction with L-alanine *N*-carboxyanhydride according to the method of Manning and Moore (4) as used in our laboratory (5). After the coupling, the reaction mixture was left at pH 10.8 for 30 min to hydrolyze any phenolic esters.

TABLE 1. Data for the resolution of ring-substituted DL-phenylalanines

	Incubation conditions[a]			L-Isomer			D-Isomer		
Amino acid	Substrate (g)	Enzyme (mg)	Time (min)	Yield (%)	$[\alpha]_D^{25b}$	Optical purity (%)	Yield (%)	$[\alpha]_D^{25b}$	Optical purity (%)
o-Tyrosine	2.0	200	60	75	−26.8	100	50	+25.4	
m-Tyrosine	0.5	45	30	75	− 7.9	100	78	+ 7.9	>99.5
Tyrosine	0.5	20	30	80	−10.2	100	78	+10.0	99.8
Dopa	4.0	140	30	70	−11.7[c]	>99.8	60	+11.6[c]	99.5
					− 9.5[d]			+ 9.5[d]	
p-Chloro-	0.5	50	90	64	− 3.5	100	60	+ 3.3	99.5
p-Fluoro-	0.5	60	30	60	− 5.6	100	60	+ 5.6	>99.5

[a]pH 5.0, room temperature.
[b]$c = 2$, *N* HCl.
[c]20 °C; Literature 11.6 ° ($c = 3$, 4% HCl) (6).
[d]30 °C.

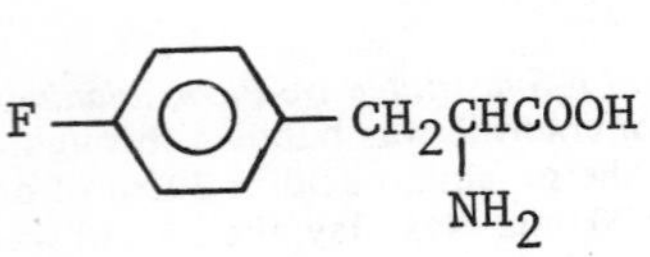

$C_9H_{10}FNO_2$ Ref. 521

For the resolution procedure and for physical constants of the resolved compound, see Ref. 520.

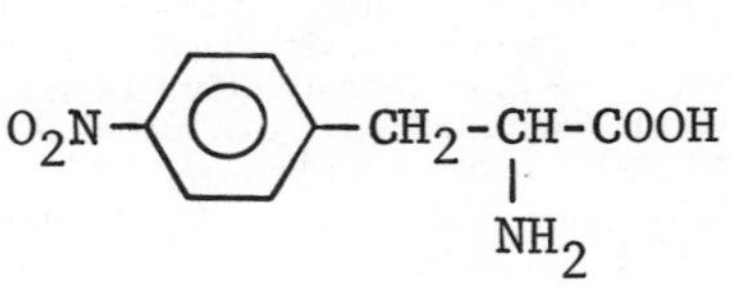

$C_9H_{10}N_2O_4$ Ref. 522

(1) Diethyl α-acetamido-α-p-nitrobenzylmalonate (50 g.) was refluxed for 24 hrs. with a solution of sodium carbonate (50 g.) in water (500 ml.)(cf. Albertson, J. Amer. Chem. Soc., 1950, 72, 1396). To the mechanically-stirred hot filtered solution was slowly added concentrated hydrochloric acid (110 ml.). Solid separated, the solution frothed vigorously and its temperature rose. The mixture was boiled with stirring for a few minutes, cooled and left in the ice-box overnight. The pale brown crystalline product (28 g.; 78%) was filtered off and had M.P. 205–209° C. Recrystallisation from water (charcoal) gave colourless needles of N-acetyl-p-nitro-DL-phenylalanine, M.P. 207–209° C. (Found: C, 52.7; H, 5.0; N, 10.7. $C_{11}H_{12}O_5N_2$ requires C, 52.4; H, 4.8; N, 11.1%.)

To a solution of the DL acid (23.55 g.) in hot ethanol (200 mil.) was added a solution of brucine (36.9 g.; 1.00 mol.) in hot ethanol (200 ml.), and the mixture left overnight at room temperature. Pale yellow prisms were deposited. The mixture was cooled in ice-water for an hour, and the product, M.P. 203–208° C. (25.35 g.; 84%) filtered off. Recrystallisation from ethanol gave pale yellow prisms of the brucine salt of N-acetyl-p-nitro-L-phenylalanine, M.P. 207–209.5° C., and $[\alpha]_D^{21}$ +19.2°±0.5° (c., 1.59 in 1:1 H_2O dioxan). Recrystallisation did not significantly affect the rotation or M.P. (Found: C, 61.1; H, 5.85; N, 8.6. $C_{34}H_{38}O_9N_4.H_2O$ requires C, 61.0; H, 6.0; N, 8.4%.) (Found (after drying at 100° C. and 1 mm. pressure for 4 hours): C, 62.7; H, 6.2; N, 8.6. $C_{34}H_{38}O_9N_4$ requires C, 63.1; H, 5.9; N, 8.7%.)

Treatment of an aqueous solution of the salt with ammonia or sodium hydroxide, removal of the brucine by filtration, acidification of the filtrate, and recrystallisation of the product from water yielded colourless prisms of N-acetyl-p-nitro-L-phenylalanine, initial M.P. 170–172° C., resolidification, final M.P. 206–209° C., $[\alpha]_D^{24}$ +49.7°±1° (c., 1.55 in EtOH). (Found: C, 52.5; H, 4.8; N, 11.0. $C_{11}H_{12}O_5N_2$ requires C, 52.4; H, 4.8; N, 11.1%.)

The ethanolic mother liquors from the brucine salt separation were evaporated to dryness under vacuum, and the residual gum was tak n up in hot water. Crystallisation set in on cooling slightly. After an hour at ice temperature, the product (36.95 g.) was collected and recrystallised from water, yielding yellow prisms of the brucine salt pentahydrate of N-acetyl-p-nitro-D-phenylalanine, M.P. 98–99° C., $[\alpha]_D^{21}$ −36.9°±0.5° (c., 1.63 in 1:1 H_2O dioxan). Recrystallisation raised the $[\alpha]_D^{21}$ to −37.6±0.5° (c., 1.60). (Found (on sample dried in a vacuum desiccated over H_2SO_4): C, 55.3; H, 6.6; N, 7.9. $C_{34}H_{38}O_9N_4.5H_2O$ requires C, 55.4; H, 6.6; N, 7.6%.) (Found (on sample dried to constant weight at 80° C. in a high vacuum): C, 62.8; H, 5.65; N, 9.0. $C_{34}H_{38}O_9N_4$ requires C, 63.1; H, 5.9; N, 8.7%.) Treatment of an aqueous solution of the salt in water with aqueous ammonia, removal of the precipitated brucine, acidification with hydrochloric acid and recrystallisation of the product from water gave colourless prisms of N-acetyl-p-nitro-D-phenylalanine, initial M.P. 170–172° C., resolidification, final M.P. 205–206° C., $[\alpha]_D^{23}$ −44°±0.5° (c., 1.45 in EtOH). (Found: C, 52.3; H, 4.8; N, 11.2.)

N-acetyl-p-nitro-L-phenylalanine (70 mg.) from the brucine separation was esterified in the cold with 2 N ethanolic hydrogen chloride (2 days at room temperature). Evaporation of the solvent gave a colourless gum which recrystallised from water in tiny colourless needles (90%) of the ethyl ester of N-acetyl-p-nitro-L-phenylalanine, M.P. 115–117° C., $[\alpha]_D^{19}$ +13.0°±0.5° (c., 1.33 in EtOH).

In order to show that this compound belonged to the L series p-nitro-L-phenylalanine ethyl ester hydrochloride (2.0 g.) was prepared from L-phenylalanine (Bergel and Stock, J.C.S., 1954, 2409) and acetylated in 75% yield by heating for 10 min. with acetic anhydride-potassium carbonate. Crystallisation of the product from water gave colourless needles of the L-acetyl compound, M.P. 115–117° C., $[\alpha]_D^{19}$ +13.5°±0.5° (c., 1.32 in EtOH). (Found: C, 55.9; H, 5.8; N, 10.2. $C_{13}H_{16}O_5N_2$ requires C, 55.7; H, 5.75; N, 10.0%.) When this reference compound was mixed with the compound obtained as described above the melting point was unchanged.

The D ester was obtained similarly in comparable yield from the D acid ($[\alpha]_D^{20}$ −43°±1° (c., 1.5 in EtOH)). Recrystallisation from water gave a product of M.P. 116–118° C., $[\alpha]_D^{22}$ −11.0°±1.0° (c., 1.58 in EtOH). (Found, C, 55.9; H, 5.5; N, 9.9.)

N-acetyl-p-nitro-L-phenylalanine (0.5 g.: $[\alpha]_D^{24}$ +50°) was refluxed for 2.5 hours with 6 N hydrochloric acid (10 ml.). The solution was evaporated to dryness (vacuum), the crystalline residue evaporated with ethanol (vacuum) and then refluxed 1.5 hours with 2 N ethanolic hydrogen chloride (8 ml.). Evaporation of the solvent and crystallisation of the residue from acetone-methanol gave the p-nitro-L-phenylalanine ethyl ester hydrochloride (0.440 g.), M.P. 203–205° C. (decomp.), unchanged on admixture with the L compound, M.P. 204–205° C. (decomp.), prepared from p-nitro-L-phenylalanine (Bergel and Stock, loc. cit.), $[\alpha]_D^{20}$ +11.7°±0.5° (c., 2.3 in H_2O).

This compound was then converted to p-bis-(2-chloroethyl)-amino-L-phenylalanine as described in Example 3 of United States application Ser. No. 415,964.

(2) Diethyl sodium phthalimidomalonate (Barger and Weichselbaum, Organic Syntheses, 1943, Coll. Vol. II, 384) (6.52 g.) was dissolved in boiling methyl ethyl ketone (80 ml.) and a solution of p-nitrobenzyl chloride (3.44 g.; 1.0 mol.) in the same solvent (20 ml.) was added. Sodium iodide (ca. 0.5 g.) dissolved in hot methyl ethyl ketone (10 ml.) was introduced, and produced an immediate precipitation. The mixture was refluxed for 1.5 hours, cooled, filtered, evaporated under vacuum and the residual gum crystallised from ethanol. The diethyl-p-nitrobenzyl-phthalimidomalonate formed colourless prisms (88%), M.P. 103–105° C., sharpening to 104–105° C. on recrystallising from ethanol. (Found: C, 59.8; H, 4.5; N, 6.4. $C_{22}H_{20}O_8N_2$ requires C, 60.0; H, 4.6; N, 6.4%.)

Diethyl-p-nitrobenzyl-phthalimidomalonate (70 g.) and sodium carbonate (70 g.) in water (700 ml.) were refluxed overnight with mechanical stirring (to avoid bumping). The clear brown solution was acidified with hydrochloric acid and refluxing and stirring were continued for a further 40 minutes. The mixture was cooled and the colourless precipitate (31 g.) collected. A second crop (18.5 g.) was obtained on evaporation of the mother liquors. Crystallisation from aqueous ethanol gave the compound N-carboxybenzoyl-p-nitro-DL-phenylalanine as small needles, M.P. 198–200° C. (Found (on sample dried at 100° C. in high vacuum): C, 56.8; H, 4.1; N, 7.8. $C_{17}H_{14}O_7N_2$ requires C, 57.0; H, 3.9; N, 7.8%.)

The N-carboxybenzoyl compound (2.7 g.) was refluxed for 30 minutes with acetic anhydride (10 ml.), the mixture taken to dryness (vacuum) and the residue heated with water. The cooled gummy product became granular on rubbing and crystallised from m thyl ethyl ketone-petrol or aqueous ethanol in almost colourless needles, M.P. 184–186° C., of p-nitro-N-phthaloyl-DL-phenylalanine.

A solution of p-nitro-N-phthaloyl-DL-phenylalanine (1.0 g.) in methanol (25 ml.) and a solution of cinchonidine (0.865 g.; 1.00 mol.) in methanol (30 ml.) were mixed. Crystallisation soon set in. The mixture was left overnight, and the colourless needles (0.97 g.), M.P. 209–210° C., collected. After two recrystallisations from methanol the cinchonidine salt of the D-acid had M.P. 211° C. and $[\alpha]_D^{21}$ +82°±1.0° (c., 0.84 in dioxan). (Found (after drying at 100° C. in high vacuum): C, 67.0; H, 5.8; N, 8.8. $C_{36}H_{34}O_7N_4.MeOH$ requires C, 66.7; H, 5.7; N, 8.4%.)

To the salt (2.9 g.) in warm ethanol (50 ml.) was added water (50 ml.) and a slight excess (ca. 10 ml.) of N aqueous sodium hydroxide. The mixture was diluted with water, cooled, filtered from the precipitated base and the filtrate acidified with hydrochloric acid. The tiny needles of p-nitro-N-phthaloyl-D-phenylalanine (1.05 g.) had, after recrystallisation from ethanol, M.P. 207–208° C., $[\alpha]_D^{20}$ +240°±2° (c., 1.01 in EtOH). (Found: C, 60.1; H, 3.7; N, 8.2. $C_{17}H_{12}O_6N_2$ requires C, 60.0; H, 3.55; N, 8.2%.) Refluxing with 2 N ethanolic hydrogen chloride yielded p-nitro-N-phthaloyl-D-phenylalanine ethyl ester, M.P. 82–83° C., $[\alpha]_D^{20}$ +206°±1.0°.

Evaporation of the mother liquors from the original cinchonidine experiment gave a gum which crystallised readily from aqueous ethanol in almost colourless needles (0.73 g.), M.P. 191–192.5° C. Two recrystallisations from aqueous ethanol gave the cinchonidine salt of the L-acid, M.P. 192.5–194° C., $[\alpha]_D^{20}$ −170°±1.0° (c., 1.32 in EtOH). (Found after drying at 100° C. in high vacuum): C, 66.2; H, 5.6; N, 8.6. $C_{36}H_{34}O_7N_4.H_2O$ requires C, 66.3; H, 5.5; N, 8.6%.)

The acid was isolated as for the D-isomer. The recrystallised p-nitro-N-phthaloyl-L-phenylalanine had M.P. 209–211° C., $[\alpha]_D^{21}$ −233°±2°. (Found, C, 60.3; H, 3.8; N, 8.4%.)

Conversion to the L ethyl ester gave, after two crystallisations, a product of M.P. 84–85° C. (unchanged on admixture with an authentic specimen of the same M.P.) and $[\alpha]_D^{20}$ −212°±2°.

$C_9H_{10}O_2$ 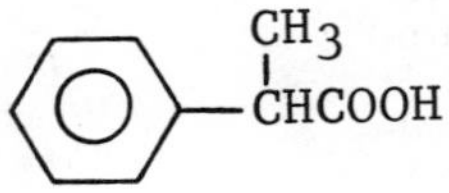 Ref. 523

(+)*Hydratropic Acid.*—A solution of strychnine (90 g.) and *dl*-hydratropic acid (40 g.) in warm 75% aqueous alcohol (400 c.c.), after standing overnight in the ice-chest, deposited 60 g. of strychnine salt in the form of glassy rhombs; concentration of the filtrate yielded a second crop (25 g.). After five recrystallisations, optical purity was reached, and the strychnine salt yielded *d*-hydratropic acid (8·5 g.), b. p. 143°/12 mm., which sets to a mass of transparent, flat rhombs, m. p. 29°: in view of its considerable alteration of rotatory power with change of temperature, the following values were determined (*l*, 0·5).

t.	a_{5893}.	a_{5780}.	a_{5461}.	a_{4358}.
13·6°	—	+55·04°	+63·08°	+112·5°
16·0	+51·81°	—	62·16	111·2
16·2	—	54·30	62·30	111·7
17·75	—	—	61·85	—
19·4	—	53·55	61·30	110

t.	a_{5780}.	a_{5461}.	a_{4358}.
23°	+53·03°	+60·73°	+109°
25	52·30	59·80	108
30	51·30	58·62	105·7

Specific rotatory powers of the acid in solution are given below.

Collected Specific Rotatory Powers.

Solute.	Solvent.	*c.*	*l.*	$[a]_{5893}$.	$[a]_{5461}$.	$[a]_{4358}$.
d-Hydratropic acid *	$CHCl_3$	3·060	2	+ 74·8°	+ 90·9°	+162°
,, ,, ...	C_6H_6	3·4825	2	92·5	110·3	197
,, ,, (Na salt)	H_2O	3·247	2	5·8	6·3	9·2
d-Hydratropamide	$CHCl_3$	2·733	2	58·3	71·5	130
d-a-Phenylethylacetamide	EtOH	2·340	2	−165	−195	−370

* *d*-Configuration is assigned arbitrarily to (+)hydratropic acid for the purpose of description.

(−)*Hydratropic Acid.*—From the first mother-liquor from the strychnine salt there was recovered hydratropic acid (12 g.) with $\alpha^{19°}_{5893}$ −32·28° (*l*, 0·5), which, on being kept in the ice-chest for several days, deposited a crop of crystals; the liquid acid drained from the crystals had $\alpha^{19°}_{5893}$ −26·6° (*l*, 0·5). The crystals were melted, allowed to recrystallise, and again drained from the liquid acid. When the process was repeated a third time there was obtained (−)hydratropic acid (4 g.) in hard, glassy, flat rhombs, m. p. 29°, and (supercooled) $\alpha^{17°}_{5461}$ −61·68° (*l*, 0·5).

In view of the readiness with which this separation takes place the—presumed—optically pure (+)hydratropic acid described above was melted, inoculated, and when crystallisation was about two-thirds complete, the liquid portion was drained away. The rotatory power of the re-melted acid was identical with that of the original acid, thus increasing very considerably the probability that optical purity had been reached by the fractional crystallisation of strychnine hydratropate.

This is the more desirable because Levene and Marker (*J. Biol. Chem.*, 1933, **100**, 692), who fractionally crystallised its quinine salt, give $[\alpha]^{25°}_{5893}$ −74·1° as the maximum value of *l*-hydratropic acid in the homogeneous condition, whereas Ott and Krämer (*Ber.*, 1935, **68**, 1657) state, without recording any experimental details, that optically pure *d*-hydratropic acid has $[\alpha]^{20°}_{5893}$ + 89·7°. Also, during the present investigation it was found that lævorotatory hydratropic acid with $\alpha^{19°}_{5893}$ − 27·1° (*l*, 0·5) gives a beautifully crystalline cinchonidine salt which was recrystallised successively from acetone, methyl acetate (twice), and aqueous alcohol without the rotatory power of the liberated hydratropic acid being increased beyond $\alpha^{19°}_{5893}$ − 36° (*l*, 0·5). On the other hand, Raper (*loc. cit.*), whose method we have used, gives for the *d*-acid $[\alpha]_{5893}$ + 76·2° in chloroform solution, but records no value for the homogeneous acid, owing to the small amount available.

Even in this most favourable case we found it advantageous to allow the crystallisation to proceed undisturbed and to decant the mother-liquor; when the containing vessel is scratched, a considerable amount of strychnine salt of low optical activity is brought out of solution.

An attempt was made to obtain *l*-hydratropic acid by the recrystallisation from ether of *l* + *dl*-hydratropamide (from hydratropic acid of $\alpha^{19°}_{5893}$ − 25·5°, *l* 0·5), but after three crystallisations the resulting amide had only $[\alpha]_{5461}$ − 43·9° in chloroform solution.

A crystal of *d*-hydratropic acid was at room temperature placed in contact with a similar crystal of *l*-hydratropic acid: they slowly melted and the resulting *dl*-acid remained liquid after being kept at − 5° for several days.

J. P. Engstrom and F. D. Greene, J. Org. Chem., 37, 968 (1972) have recommended the following modification in the above procedure.

S-(+)-2-Phenylpropionic Acid.—2-Phenylpropionic acid was resolved with strychnine according to the procedure of Arcus and Kenyon[24] with the modification that the salt was dissolved in an excess of 75% ethanol–water and then the excess solvent was removed under vacuum. This method was superior to heating the solvent to dissolve the salt, because heating led to slow decomposition of the salt.

(24) C. L. Arcus and J. Kenyon, *J. Chem. Soc.*, 916 (1939); H. I. Bernstein and F. C. Whitmore, *J. Amer. Chem. Soc.*, **61**, 1324 (1939).

$C_9H_{10}O_2$

Ref. 523a

Resolution of the hydratropic acid. The method followed was essentially that of Raper (10). The experiment started with 40 g racemic acid and 90 g strychnine in 300 ml ethanol + 100 ml water. The salt obtained (40 g) was recrystallised from 75 % ethanol and the course of the resolution was tested on samples of each fraction, from which the acid was liberated and examined in ethanol solution. After five recrystallisations, maximum activity was obtained. For the sake of control, the salt was recrystallised five times more and the acid was examined for every fraction. The activity remained quite constant. The yield of salt was 11.0 g and 3.2 g after five and ten recrystallisations respectively. The acid was now liberated from the whole portion and recrystallised from a small volume of petrol ether (b.p. 30–50°). The crystallisation must take place at about 0°. The acid was obtained as glassy prisms, melting at 30.2–30.9° (precision thermometer, divided in 0.1° C, efficient stirring of the heating bath, rise of temperature 1° in 4 minutes).

0.1829 g acid: 11.32 ml 0.1075-*N* NaOH.
$C_9H_{10}O_2$ Equiv. wt. calc. 150.2 found 150.3

0.1674 g dissolved in absolute *ethanol* to 10.00 ml: $2\alpha_D^{25} = +2.645°$. $[\alpha]_D^{25} = +79.0°$; $[M]_D^{25} = +118.6°$. The rotation was also determined at 20.00°: $2\alpha_D^{20} = +2.725°$. After correction for the change in volume this gives: $[\alpha]_D^{20} = +81.0°$; $[M]_D^{20} = +121.6°$.

Systematic fractionation of the strychnine salt from the mother liquors yielded in all 14.3 g salt containing acid of the same maximum activity.

16.5 g acid recovered from the first mother liquors of the strychnine salt and containing considerable excess of (−)-form was dissolved with 35.7 g quinine in 800 ml acetone; 25 g salt were obtained. It was recrystallised six times; the activity of the acid remained constant from the fourth recrystallisation. The yield was 9.3 g and 6.3 g after four and six crystallisations respectively. The acid was liberated and recrystallised as described above. The melting point was 30.3–31.0°.

0.2270 g acid: 14.08 ml 0.1075-*N* NaOH.
$C_9H_{10}O_2$ Equiv. wt. calc. 150.2 found 150.0

0.1567 g dissolved in absolute *ethanol* to 10.00 ml: $2\alpha_D^{25} = -2.48°$. $[\alpha]_D^{25} = -79.1°$; $[M]_D^{25} = -118.8°$. — 0.1544 g dissolved in *benzene* to 10.00 ml: $2\alpha_D^{25} = -2.875°$. $[\alpha]_D^{25} = -93.1°$; $[M]_D^{25} = -139.8°$. — 0.1482 g dissolved in *acetone* to 9.98 ml: $2\alpha_D^{25} = -2.835°$. $[\alpha]_D^{25} = -95.5°$; $[M]_D^{25} = -143.3°$. — 0.1801 g dissolved in glacial *acetic acid* to 9.99 ml: $2\alpha_D^{25} = -3.145°$. $[\alpha]_D^{25} = -87.3°$; $[M]_D^{25} = -131.1°$. — 0.1587 g dissolved in *chloroform* to 10.00 ml: $2\alpha_D^{25} = -2.39°$; $[\alpha]_D^{25} = -75.3°$; $[M]_D^{25} = -113.1$. The rotation was also determined at 20.00°: $2\alpha_D^{20} = -2.45°$. After correction for the change in volume this gives: $[\alpha]_D^{20} = -76.7°$; $[M]_D^{20} = -115.3°$.

10. Raper, H. S., J. Chem. Soc. *1923*, 2557.

$C_9H_{10}O_2$

Ref. 523b

Resolution of Racemic Hydratropic Acid.—Racemic hydratropic acid (24 g) and strychnine (44 g) were dissolved in 200 ml of 75% (v/v) aqueous ethanol, and the resulting solution was kept in the refrigerator (*ca.* 10°) during 3 days. The crystallized strychnine hydratropate was collected, redissolved in fresh 75% aqueous ethanol, and allowed once again to crystallize slowly in the refrigerator. This procedure was repeated through the sixth recrystallization since the fifth and sixth recrystallizations did not give rise to any improvement in the melting point (176–177°) of the white, crystalline salt. The constant-melting strychnine hydratropate was taken up in 6 *N* hydrochloric acid, and the liberated organic acid was extracted into ether. Distillation of the oily residue obtained from evaporation of the combined and dried ether extracts gave (+)-(*S*)-**hydratropic acid (VIIb)**: 4.5 g; bp 101–103° (0.4 mm); $[\alpha]^{23}_D$ +75.2 ± 0.7° (*c* 1.76, chloroform), lit.[9] $[\alpha]_D$ +76.3 ± 0.6° (*c* 1.613, chloroform).

The mother liquors and washings obtained from the six recrystallizations of the strychnine hydratropate described above were combined and acidified, and the liberated hydratropic acid was extracted into ether. The combined and dried ether extracts yielded an oily residue upon evaporation, which afforded partially resolved, levorotatory hydratropic acid {$[\alpha]^{25}_D$ −24.0 ± 0.1° (*c* 6.75, acetone)}, upon distillation: 8.5 g, bp 120–121° (4 mm). This sample of partially resolved acid was dissolved in hot acetone containing quinine (18.6 g), and the resulting solution was allowed to cool slowly to room temperature and remain there during several days to ensure complete crystallization of quinine hydratropate. The white, crystalline salt displayed a melting point of 176–177° and a specific rotation of −123°, both of which were improved after successive recrystallizations of the salt from hot acetone, giving, after the third and fourth recrystallizations, material which possessed a melting range of 178–179° and a rotation of $[\alpha]^{27}_D$ −99.2 ± 0.8 (*c* 0.870, chloroform).

This material was taken up in 6 *N* hydrochloric acid, and the freed organic acid was extracted into ether. After drying and evaporation of the combined ether extracts, distillation of the residue gave (−)-(*R*)-**hydratropic acid (VIIa)**: 3.0 g, $[\alpha]^{27}_D$ −77.0 ± 0.3° (*c* 2.380, chloroform), lit.[9] $[\alpha]^{25}_D$ −76.1 ± 0.6 (*c* 1.599, chloroform).

(9) S. P. Bakshi and E. E. Turner, *J. Chem. Soc.*, 171 (1961).

$C_9H_{10}O_2$

Ref. 523c

Optical resolution of hydratropic acid

(−)-α-*Phenylethylamine (−)-hydratropoate:* 28 g (0.19 moles) of racemic hydratropic acid and 22.6 g of (−)-α-phenylethylamine were dissolved in a hot mixture of 400 ml of benzene and 100 ml of ethanol and allowed to crystallize. The precipitate was filtered off, washed with a small amount of cold benzene, dried and weighed. 0.4 g of the salt was set apart, the acid liberated with dilute sulfuric acid, extracted with ether and isolated by evaporation of the solvent. The acid was dissolved in ethanol and the optical activity measured in a 2 dm tube. The principal part of the salt was recrystallized from a mixture of benzene and ethanol 4:1 until the optical activity remained constant. The pure (−)-phenethylamine salt of (−)-hydratropic acid was obtained as colourless needles. The progress of the resolution is seen in Table 5.

(+)-α-*Phenylethylamine salt of (+)-hydratropic acid:* 17.5 g (0.12 moles) of hydratropic acid ($[\alpha]_D^{25} = +22°$) and 14 g (0.12 moles) of (+)-α-phenethylamine were recrystallized from a 4:1 mixture of benzene and ethanol until the optical activity of the liberated acid remained constant. The results are collected in Table 5.

Table 5.

Crystallization number	ml of solvent	g of (−)-acid salt obtained	$[\alpha]_D^{25}$	ml of solvent	g of (+) acid salt obtained	$[\alpha]_D^{25}$
1	500	16	−65°	250	14	+75°
2	150	11.5	−78°	87	9.6	+79°
3	87	8.7	−78°	62	7.2	+80°
4	65	5.4	−78°	42	6.1	+80°

(−)-*Hydratropic acid:* The (−)-phenethylamine salt was decomposed with dilute sulfuric acid; the liberated acid was extracted with ether, and the solvent removed. The residue crystallized as transparent prisms, and was recrystallized from petroleum ether. M.p. 30–31°.

100.4 mg of the acid were dissolved in ethanol and made up to 10.00 ml.
$2\alpha_D^{25} = -1.577°$, $[\alpha]_D^{25} = -78.4°$ (lit. 79.1° (1)).

(+)-*Hydratropic acid:* From the (+)-phenethylamine salt as described for its optical antipode. Transparent prisms, m.p. 30–31°.

100.4 mg of the acid was dissolved in 10.00 ml of ethanol.
$2\alpha_D^{25} = +1.582°$, $[\alpha]_D^{25} = +78.9°$ (lit. +79.0° (1)).

Amide: Glistening leaves, m.p. 101.5–102°. Optical activities in different solvents are collected in Table 7.

Table 7.

Solvent	mg of amide	$2\alpha_D^{25}$	$[\alpha]_D^{25}$	$[M]_D^{25}$
Benzene	92.8	+1.700°	+91.6°	+136.7°
Ether	131.9	+2.065°	+78.3°	+116.8°
Acetone	89.1	+0.837°	+47.0°	+70.1°
Ethanol	97.8	+0.909°	+46.5°	+69.3°
Acetic acid	129.7	+0.369°	+14.2°	+21.2°

Ref. 523d

$C_9H_{10}O_2$

Resolution of (±)-*Hydratropic Acid with* (−)-*Ephedrine*.—(−)-Ephedrine (40 g.) and (±)-hydratropic acid (32 g.) were dissolved in warm aqueous alcohol (160 ml.; water : alcohol = 3 : 1 v/v). The salt which crystallised overnight at 0° was crystallised five times from 75% aqueous alcohol (120 ml.). The final crop was decomposed with 5N-sulphuric acid to give (−)-hydratropic acid (6 g.), b. p. 147°/11 mm., m. p. 29°, α_{5461}^{19} −60·30° (homogeneous, *l* 0·5), $[\alpha]_{5461}^{19}$ −113·1° (*c* 3·58 in benzene).

The following Table shows the specific rotations obtained by various workers for the optically active hydratropic acids.

Alkaloid used	[α] 5893	[α] 5461	Solvent	Temp.	Authors
Strychnine	+76·2°	—	$CHCl_3$	20°	Raper [1]
Quinine	−74·1	—	Homog.	25	Levene and Marker [3]
Strychnine	+94·2	—	Homog.	16	Arcus and Kenyon [2]
	—	+111·5°	Homog.	19·4	
,,	+74·8	—	$CHCl_3$	—	
,,	—	+110·3	C_6H_6	—	
Quinine	−79·1	—	EtOH	25	Fredga [4]
,,	−75·3	—	$CHCl_3$	25	
Strychnine	+79·0	—	EtOH	25	
Ephedrine	—	−109·6	Homog.	19	This paper
,,	—	−113·1	C_6H_6	19	

[1] Raper, *J.*, 1923, 2557.
[2] Arcus and Kenyon, *J.*, 1939, 916.
[3] Levene and Marker, *J. Biol. Chem.*, 1933, **100**, 685.
[4] Fredga, *Arkiv Kemi*, 1954, **7**, 241.

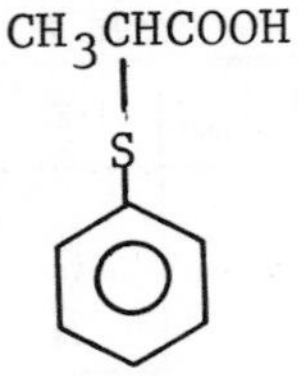

$C_9H_{10}O_2S$

Ref. 524

Bei vorbereitenden Versuchen mit mehreren optisch aktiven Basen ergab sich, dass es schwierig war, mit Brucin, Cinchonin und Morphin kristallisierende Salze zu erhalten; mit Chinidin, Strychnin und (+)-Phenäthylamin wurden zwar mehr oder weniger gut kristallisierende Salze erhalten, die aus den ersten Salzfraktionen isolierte Säure zeigte aber nur geringe Drehung. Mit Chinin wurden jedoch zufriedenstellende Ergebnisse erhalten. In allen untersuchten Fällen besass das Salz der rechtsdrehenden Säure geringste Löslichkeit.

Im folgenden werden Angaben über einige typische Spaltungsversuche mitgeteilt. Die angeführten Werte von $[\alpha]_D^{25}$ wurden folgenderweise erhalten. Eine geringe Menge Chininsalz wurde mit verdünnter Schwefelsäure versetzt und mit Äther ausgeschüttelt. Der Äther wurde im Luftstrom abgetrieben und die rückständige Säure in Wasser gelöst. Nach Bestimmung der Drehung dieser Lösung im 3-dm-Rohr wurde sie mit etwa n/40 Barytlauge titriert. Da die spezifische Drehung der Säure in wässriger Lösung ziemlich beträchtlich mit der Konzentration variiert, wurden die [α]-Werte mit Hilfe der Angaben in Tab. 1 auf die Konzentration 0,03 Mol je Liter umgerechnet.

Versuch I. 40,5 g durch wiederholtes Ausfrieren gereinigte inaktive α-Phenylmercapto-propionsäure und 72,8 g Chinin wurden in 400 ml 95 %igem Äthylalkohol gelöst. Die Lösung wurde erwärmt und mit etwa 500 ml Wasser von ungefähr 50 °C versetzt. Das beim Erkalten abgeschiedene Salz (Kristallisation Nr. 1 in der folgenden Zusammenstellung) wurde dreimal aus warmem, 30 bis 40 %igen Äthylalkohol nach folgendem Schema umkristallisiert.

Kristallisation Nr.	1	2	3	4
Lösungsmittel, ml	900	650	650	450
Auskristallisiertes Salz[1], g	83	72	57	48
$[\alpha]_D^{25}$ der im Salz enthaltenen Säure	+64°	+103°	+121,0°	+123,9°

Die Ausbeute an (+)-Säure betrug somit 85 % der möglichen Menge.

Bei erneuter Umkristallisation von Nr. 4 wurde Salz einer Säure von $[\alpha]_D^{25} = +122,4°$ erhalten.

Aus den Mutterlaugen von Nr. 1 und 2 wurde die Säure in Freiheit gesetzt und in (−)-Phenäthylaminsalz übergeführt. Daraus wurden nach 13 zweckmässig durchgeführten Kristallisationen 1,7 g Säure von $[\alpha]_D^{25} = -123°$ gewonnen.

Versuch II. Salz aus 42 g Säure und 75 g Chinin wurde fünfmal aus ungefähr 20 %igem Alkohol umkristallisiert (das Salz wurde in 95 %igem Alkohol gelöst und mit Wasser von ungefähr 50° C versetzt). Bei Darstellung des Salzes und bei der ersten Umkristallisation wurden etwa 2 Liter Lösungsmittel angewandt. Die durch Umkristallisation erhaltenen einzelnen Salzportionen enthielten Säure von $[\alpha]_D^{25} = +52°$ bezw. +83°, +118°, +123° und +123°. Die Ausbeute an Salz maximal drehender Säure betrug 44,5 g, entsprechend 76 % der möglichen Menge. Die in den Mutterlaugen von der Darstellung des Salzes und der ersten Umkristallisation enthaltene Säure zeigte bemerkenswert hohe Linksdrehung, nämlich $[\alpha]_D^{25} = -103°$ bezw. −110°. Durch Einengen dieser Mutterlaugen zur Trockne und sechsmalige Umkristallisation des Salzes aus Aceton + Wasser wurden schliesslich 2 g Säure von maximaler Linksdrehung erhalten.

Zur Isolierung der maximal rechtsdrehenden Säure aus ihrem Chininsalz wurde folgendermassen verfahren. Das Salz wurde unter Zusatz verdünnter Schwefelsäure mit frisch destilliertem Äther ausgeschüttelt. Die Ätherlösung wurde mit Wasser gewaschen, der Äther abgetrieben und die rückständige Säure durch Erwärmen auf 60 °C und Durchsaugen trockener Luft unter vermindertem Druck getrocknet.

Äquivalentgewichtsbestimmung an derart erhaltener Säure. 0,2996 g Subst.: 14,84 ml 0,1108-n NaOH.
Äquiv.-Gew. ber. 182,14 gef. 182,2.

Das eben beschriebene Verfahren musste deshalb gewählt werden, weil bei Vakuumdestillation der optisch aktiven Säure eine beträchtliche Razemisierung eintrat. Ein Präparat, das nach Vakuumdestillation $\alpha_{5461} = 47,67°$ im 2,5-cm-Rohr (ohne Lösungsmittel) zeigte, wurde nochmals bei einem Drucke von 0,4 mm Hg und 156° Badtemperatur destilliert. Das Destillat zeigte im 2,5-cm-Rohr $\alpha_{5461} = 46,71°$. Die Drehung hatte also durch die letzte Destillation um 2 % abgenommen.

Alle Versuche, die optisch aktiven Säuren zur Kristallisation zu bringen, sind bisher gescheitert.

In Tab. 1 ist die spezifische Drehung der maximal aktiven Säuren bei 25,0 °C in wässrigen Lösungen verschiedener Kon-

[1] Die Kristallisationen Nr. 1 bis 3 waren beim Wägen nicht ganz trocken.

zentration angegeben; $3\alpha_D$ ist der direkt bestimmte Drehungswinkel (Mittel aus in der Regel wenigstens 6 Ablesungen) der Lösungen im 3-dm-Rohr, *C* die Konzentration in Mol je Liter, bestimmt durch Titration (in einigen Fällen wurden die Lösungen durch genaue Verdünnung titrierter Lösungen hergestellt). Die Messungen wurden mit einem Halbschattenpolarimeter nach Lippich-Landolt mit zweiteiligem Polarisator (von Franz Schmidt & Haensch) ausgeführt; die Rohre befanden sich während der Messungen in einem Thermostaten von dem von

Tab. 1.

Rechtsdrehende Säure					
C	0,03787	0,02469	0,01188	0,00592	--
$3\alpha_D$ $[\alpha]_D^{25}$	$+2{,}55_8°$ $+123{,}4°$	$+1{,}65_8°$ $+122{,}6°$	$+0{,}77_8°$ $+120°$	$+0{,}38°$ $+118°$	— —
Linksdrehende Säure					
C	0,03645	0,02591	0,01481	0,00767	0,00435
$3\alpha_D$ $[\alpha]_D^{25}$	$-2{,}45_5°$ $-123{,}2°$	$-1{,}78_5°$ $-122{,}_5°$	$-0{,}98°$ $-121°$	$-0{,}50°$ $-119°$	$-0{,}28°$ $-117°$

Ramberg und Heuberger[8] beschriebenen Typus. Als Lichtquelle diente eine Scheinwerferlampe, deren Strahlung durch einen auf die mittlere Wellenlänge der D-Linien eingestellten Askania-Geradsicht-Monochromator nach Ewald und Schulz trat.

Die spezifische Drehung nimmt somit, wie aus Tab. 1 ersichtlich, mit fallender Konzentration ab. Aus den in der Tabelle angegebenen Werten erhält man graphisch für eine 0,03-m wässrige Lösung $[\alpha]_D^{25} = \pm 123{,}0°$; bei dieser Konzentration ist also das molekulare Drehvermögen $[M]_D^{25} = \pm 224{,}0°$. Der Temperaturkoeffizient der Drehung wässriger Lösungen ist verhältnismässig klein; mit einer 0,02469-m Lösung wurden folgende Werte erhalten.

Temperatur	°C	17,7	19,8	25,0.
$3\alpha_D$		1,68°	1,67°	$1{,}65_5°$.

Eine 0,0586-m Lösung des Natriumsalzes der Säure, die durch Neutralisation von 0,1600 g (+)-Säure mit kohlensäurefreier Natronlauge und Auffüllen auf 14,99 ml bereitet wurde, zeigte bei 25,0 °C $3\alpha_D = +2{,}45°$, somit $[\alpha]_D^{25} = +76{,}5°$. Das Anion der Säure besitzt somit bedeutend geringere spezifische Drehung als die nicht dissoziierte Säure; die Abhängigkeit der Drehung von der Konzentration, wie sie sich in den Werten der Tab. 1 ausdrückt, wird hauptsächlich durch diesen Umstand verursacht.

Ein Präparat der maximal rechtsdrehenden Form zeigte ohne Lösungsmittel im 5-cm-Rohr bei 25 °C $\alpha_D = +100{,}30°$. $D_4^{25} = 1{,}178$. Es ist somit $[\alpha]_D^{25} = +170{,}3°$ und $[M]_D^{25} = +310{,}2°$.

[8] Sv. Kem. Tidskr. *40*, 227 (1928).

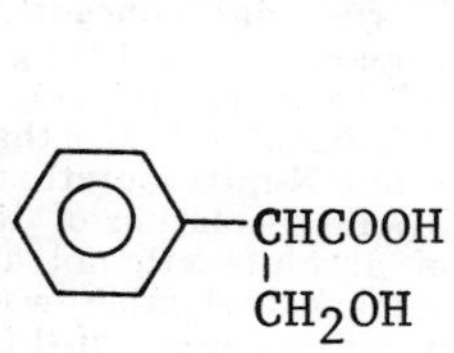

$C_9H_{10}O_3$ Ref. 525

Preparation of the Optically Active Tropic Acids.

85·4 Grams of hydrated quinine (1 mol.) were dissolved in 850 c.c. of absolute ethyl alcohol at 17°, and 37·5 grams of *r*-tropic acid (1 mol.) were added with stirring until the solution was homogeneous. Crystallisation began after five minutes, and proceeded rapidly. After nineteen hours at the ordinary temperature, the crystals were separated; after remaining on a porous plate for three days, they amounted to 71 grams. Crystallisation was carried out another eight times from ethyl alcohol. The alcohol was distilled off from the filtrate from each successive crystallisation, the residue decomposed by dilute sulphuric acid, and the tropic acid extracted with ether and dried with sodium sulphate. The progress of the resolution was indicated by determining the specific rotatory power of the successive acids in ethyl-alcoholic solution, the following values for $[\alpha]_D$ being obtained: −54·2°, −36·7°, −6·1°, +23·1°, +41·2°, +60·8°, +67·1°, +69·4°, +69·6°. The quinine *d*-salt amounted to 25·5 grams; it separated from ethyl alcohol in felted leaflets grouped in rosettes melting at 190—191°; 100 c.c. of its ethyl-alcoholic solution contain about 0·55 gram of the salt at 13°. The salt was decomposed by dilute sulphuric acid, and the tropic acid repeatedly extracted with ether, 7·2 grams of pure *d*-acid being obtained.

d-*Tropic acid* is sparingly soluble in benzene, from which it separates in lustrous needles melting at 128—129°. It crystallises from water in glassy needles and plates grouped in rosettes:

0·1858 gave 0·4421 CO_2 and 0·1043 H_2O. C = 64·9; H = 6·3.

$C_9H_{10}O_3$ requires C = 65·0; H = 6·1 per cent.

Its specific rotation was determined in the following solvents:

(*a*) Ethyl alcohol:

$c = 2·695$, $l = 2$; α_D^{16} +3·89°; $[\alpha]_D^{16}$ +72·2°.

(*b*) Water:

$c = 1·515$, $l = 2$; α_D^{16} +2·43°; $[\alpha]_D^{16}$ +80·2°.

(*c*) Acetone:

$c = 2·19$, $l = 2$; α_D^{13} +3·67°; $[\alpha]_D^{13}$ +83·8°.

For the preparation of the *l*-isomeride, 9 grams of acid with $[\alpha]_D$ −54·2° for $c = 2·65$ in ethyl-alcoholic solution, and 4 grams of acid with $[\alpha]_D$ −36·7° for $c = 2·574$ in ethyl-alcoholic solution (obtained from the first two filtrates in the resolution just described), were united and added to a warm solution of 23·8 grams of morphine in 350 c.c. of ethyl alcohol. On cooling, glassy crystals began to separate, and these were collected on the following day. The product, dried in air at the ordinary temperature, amounted to 32·5 grams. It was crystallised from ethyl alcohol six times. The tropic acid recovered from the mother liquors from these crystallisations gave the following values for $[\alpha]_D$ in ethyl-alcoholic solution: −13·5°, −28°, −56°, −59·4°, −60·7°, −63°. Although the morphine salt (14 grams) was not yet quite pure, it was deemed advisable at this stage to decompose it with dilute sulphuric acid. The resulting acid (4 grams) was nearly pure, giving the value $[\alpha]_D$ −71·3° in ethyl-alcoholic solution. The pure acid was obtained after one crystallisation from water.

l-*Tropic acid* melts at 128—129°. Equivalent: Found: 166·5. Calc.: 166·1. The values for its rotatory power were determined in the following solvents, and are in agreement with the corresponding values for its *d*-isomeride:

(*a*) Ethyl alcohol:

$c = 2·578$, $l = 2$; α_D^{15} −3·74°; $[\alpha]_D^{15}$ −72·5°.

(*b*) Water:

$c = 1·538$, $l = 2$; α_D^{15} −2·43°; $[\alpha]_D^{15}$ −79·0°.

(*c*) Acetone:

$c = 1·806$, $l = 2$; α_D^{15} −3·01°; $[\alpha]_D^{15}$ −83·3°.

The authors desire to acknowledge the able assistance rendered by the late Mr. Harold Halcro Johnston, B.Sc., in the earlier stages of this research. They are also indebted to Professors G. G. Henderson and F. S. Kipping for a supply of atrolactinic acid.

$C_9H_{10}O_3$ **Ref. 525a**

Preparation of *S*(−)tropic acid. 26,7 g (±)tropic acid and 34 g (1*R* : 2*R*)(−)1-p-nitrophenyl-2-aminopropane-1.3-diol were dissolved in 275 ml of water at 40—50° in an Erlenmayer flask and the solution kept at room temperature overnight. The salt of (−)tropic acid, 27 g, was crystallised, filtered by suction and washed with 15 ml ice-water. Yield 27 g (89%) (−)tropic acid salt, m. p. 107° C.

The salt was then dissolved in 50 ml hot water and the solution alkalinised with concentrated ammonium hydroxide. After standing for an hour in ice-water the aminodiol base, 15 g, was filtered off. The mother liquor was acidified with concentrated hydrochloric acid to pH = 1, and cooled by ice-water, whereby (−)tropic acid separated. The mother liquor was concentrated in vacuum giving a second crop of (−)tropic acid. 10 g *S*(−)tropic acid was obtained, overall yield 73%, m. p. 126—128°, $[\alpha]_D^{20} = -72°$ (c = 0.5; water).

Preparation of *R*(+)tropic acid. 1. The mother liquor of the salt of *S*(−)tropic acid with the aminodiol was alkalinised with concentrated ammonium hydroxide to pH 9, whereby the aminodiol crystallised. The solution was allowed to stand for an hour before filtration. The filtrate was then acidified to pH 1, the (+)acid filtered off and twice recrystallised each time from 30 ml hot water. Yield, 5 g *R*(+)tropic acid, 37%, m. p. 107°; $[\alpha]_D^{20} = +72°$ (c = 0.5; water).

2. By using the 1*S* : 2*S*(+) enantiomer of the aminodiol and proceeding in the same manner as described for the resolution of (±)tropic acid by (1*S* : 2*S*) aminodiol salt of R(+) tropic acid separated first, yielding after isolation 10 g optically pure *R*(+)tropic acid, m. p. 107—108°, $[\alpha]_D^{20} = +72°$.

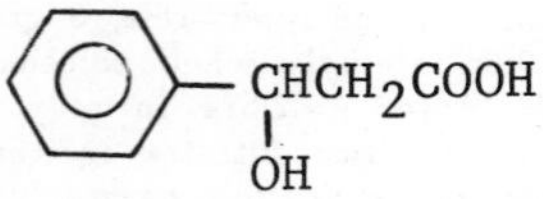

$C_9H_{10}O_3$ **Ref. 526**

Powdered morphine (61 grams) was added to a solution of 36 grams of the hydroxy-acid in 750 c.c. of boiling water. Crystallisation began after the solution was allowed to cool at the ordinary temperature for one hour; the solution was then stirred occasionally, and left overnight at the ordinary temperature. About half of the total morphine salt separated. The crystals, which melted and decomposed at about 206°, were suspended in 50 c.c. of water and the morphine precipitated by means of a slight excess of ammonia. The addition of an excess of hydrochloric acid to the filtrate, from which the morphine had been separated, caused the gradual separation, in the form of needles, of the *l*-acid, which is sparingly soluble in water. The acid was drained off and, after crystallisation from 300 c.c. of benzene, was pure. The yield amounted to 9 grams. Its melting point and its specific rotation did not alter after it had been recrystallised several times from benzene.

l-β-Hydroxy-β-phenylpropionic acid, $C_6H_5·CH(OH)·CH_2·CO_2H$, is sparingly soluble in water and in benzene. It separates in colourless needles and melts at 115—116°:

0·1925 gave 0·4579 CO_2 and 0·1032 H_2O. C = 64·9 ; H = 6·0.
$C_9H_{10}O_3$ requires C = 65·0 ; H = 6·1 per cent.

Its rotation was determined in ethyl-alcoholic solution:

$l = 2$, $c = 5·153$, α_D^{20} − 1·95°, $[\alpha]_D^{20}$ − 18·9°.

In order to obtain the enantiomorphously related isomeride, the mother liquor, from which the morphine *l*-salt had been separated, was concentrated by evaporation to 150 c.c., when no separation of salt took place. The dextro-acid was then separated in the manner described above and crystallised from benzene. The yield amounted to 10 grams.

d-β-Hydroxy-β-phenylpropionic acid melts at 115—116°, and resembles its *l*-isomeride in other particulars:

0·194 gave 0·4638 CO_2 and 0·1051 H_2O. C = 64·9 ; H = 6·0.
$C_9H_{10}O_3$ requires C = 65·0 ; H = 6·1 per cent.

A determination of its specific rotation in ethyl-alcoholic solution gave a value in agreement with that of the *l*-acid:

$l = 2$, $c = 5·194$, α_D^{18} + 1·99°, $[\alpha]_D^{18}$ + 19·2°.

Although the inactive acid had not been resolved previously, the active acids have been obtained by Barkow (*Inaug. Diss.*, Strasburg, 1906), working in Erlenmeyer's laboratory, in the course of an investigation dealing with the α-halogen-β-hydroxy-β-phenylpropionic acids. Barkow found that when *d*-α-bromo-β-hydroxy-β-phenylpropionic acid, $C_6H_5·CH(OH)·CHBr·CO_2H$, was reduced by sodium amalgam, it was converted into *d*-β-hydroxy-β-phenylpropionic acid with $[\alpha]_D$ + 19° in ethyl-alcoholic solution.

$C_9H_{10}O_3$ **Ref. 526a**

Brucine (16.2 g., 0.042 mole) was added to a hot solution of 16.5 g. (0.10 mole) of this compound in 120 ml. of ethyl acetate, and the solution was allowed to cool.[28,29] Crystals were collected and the filtrate was set aside. The salt was recrystallized six times from ethyl acetate, yielding 8.11 g. (0.014 mole, 33%), m.p. 99–108° dec., lit.[28] m.p. 89–108° dec. The brucine salt was treated with 100 ml. of *N* hydrochloric acid, the solution was taken to dryness, the residue was extracted with ether, and the ether was dried and concentrated, leading to L(+)-**β-phenyl-β-hydroxypropionic acid**, 1.70 g. (0.010 mole), 25% yield, m.p. 117–118° from chloroform, lit.[30] m.p. 116°, $[\alpha]^{22}_D$ +21.1°, 2.2% in ethanol; lit.[28] $[\alpha]^{21}_D$ +18.2°. The ethyl acetate filtrate which had been set aside was extracted with *N* hydrochloric acid, dried, and concentrated leading to D(−)-β-phenyl-β-hydroxypropionic acid, 1.60 g. (0.0096 mole), 23% yield, m.p. 118–119° from chloroform, lit.[29] m.p. 115–116, $[\alpha]^{22}_D$ −18.9, 2.3% in ethanol. Diazoethane was prepared from 3 g. (0.02 mole) of ethyl N-nitrosourethane in 100 ml. of ether and was added to a boiling solution of 15 g. of potassium hydroxide in 60 ml. of absolute ethanol and 100 ml. of ether. Distillation of diazoethane and ether was continued until the distillate was colorless, ether being added to the distillation flask. D(−)-β-Phenyl-β-hydroxypropionic acid (1.0 g., 0.0060 mole) was added to the diazoethane solution, magnesium sulfate was added, and the solution was allowed to stand overnight. The solution was filtered, concentrated, and distilled leading to D(−)-**ethyl β-phenyl-β-hydroxypropionate**, 0.34 g. (0.0018 mole), 30% yield, b.p. 75° (0.05 mm.), $[\alpha]^{22}_D$ −54.9°, 3.5% in chloroform. L(+)-**Ethyl β-phenyl-β-hydroxypropionate** was prepared similarly from 1 g. of the L(+) acid, yielding 0.40 g. (35%), b.p. 68–70° (0.03 mm.), lit.[29] b.p. 90–91° (<0.1 mm.), $[\alpha]^{22}_D$ +53.2°, 3.4% in chloroform; lit.[29] +19.17° (neat).

(28) D. S. Noyce and C. A. Lane, *J. Am. Chem. Soc.*, **84**, 1635 (1962).

(29) J. Kenyon, H. Phillips, and G. R. Shutt, *J. Chem. Soc.*, 1663 (1935).

(30) H. Wieland, W. Koschara, E. Dane, J. Renz, W. Schwarze, and W. Lind, *Ann.*, **540**, 103 (1939).

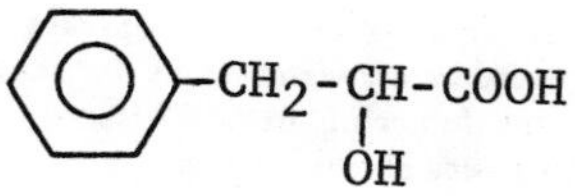

$C_9H_{10}O_3$ Ref. 527

Resolution of r-α-Hydroxy-β-phenylpropionic Acid by l-Menthol.

A. *By Crystallisation from Light Petroleum.*—*r*-α-Hydroxy-β-phenylpropionic acid (10 grams) was heated with *l*-menthol (23 grams) during sixteen hours at 120°, an intermittent current of dry hydrogen chloride being passed through the mixture. The product was dissolved in ether and the solution thoroughly agitated with aqueous sodium carbonate solution, which, however, did not remove any hydroxy-acid, thus showing that the esterification had been carried to completion. The crude ester was purified from excess of *l*-menthol in the usual manner and ultimately obtained as a mass of crystalline needles impregnated with oil. The whole product was dissolved in boiling light petroleum and the solution allowed to crystallise; the crop which separated weighed 6 grams, melted at 54—55°, and had $[\alpha]_D$ −39·6° when dissolved in chloroform. After several further crystallisations from the same solvent, *l*-menthyl *d*-α-hydroxy-β-phenylpropionate was isolated in the pure condition; it melted at 63—63·5° and had $[\alpha]_D^{15}$ −37·5° in chloroform, whereas McKenzie and Wren (*loc. cit.*) give m. p. 63—63·5°, $[\alpha]_D^{15}$ −36·9° for the synthetic ester (Found: C=74·6; H=9·6. Calc., C=74·9; H=9·3 per cent.).

For the isolation of the *d*-acid, the final crop of the *l*-menthyl *d*-ester was united with the residues obtained by the evaporation of the mother-liquors from all the crystallisations with the exception of the first, and hydrolysed with a slight excess of aqueous alcoholic potassium hydroxide solution. The crude acid was isolated from the solution in the usual manner and purified by crystallisation from water. The final crop melted at 123° and had $[\alpha]_D^{15}$ +18·2° in ethyl-alcoholic solution; these data are in harmony with those (m. p. 124—125°, $[\alpha]_D^{18}$ +18·5°) recorded by McKenzie and Wren (T., 1910, **97**, 1357) for the *d*-acid prepared by resolving the *r*-acid with morphine.

The residue obtained by the evaporation of the filtrate from the initial crystallisation was hydrolysed similarly and yielded a mixture of *r*- and *l*-α-hydroxy-β-phenylpropionic acids which had $[\alpha]_D$ −7·0° in ethyl-alcoholic solution. Successive crystallisations of this material from benzene and water gave the pure *l*-acid, which melted at 123—124° and had $[\alpha]_D^{17}$ −18·7° in ethyl alcohol.

B. *By Crystallisation from Alcohol* (96 per cent.).—The *l*-menthyl ester obtained from 17 grams of the *r*-acid by the method described above was crystallised from ethyl alcohol (96 per cent.) and the crop and residue obtained by evaporation of the filtrate were hydrolysed separately to the corresponding acids.

That obtained from the former melted at 97—110° and had $[\alpha]_D$ +13·7°; by successive crystallisations from water and benzene, it readily yielded the pure *d*-acid, m. p. 123—124°, $[\alpha]_D^{18}$ + 18·4° in ethyl-alcoholic solution. In a similar manner the latter gave the pure *l*-acid, which melted at 123° and had $[\alpha]_D^{17}$ −18·7° when dissolved in ethyl alcohol.

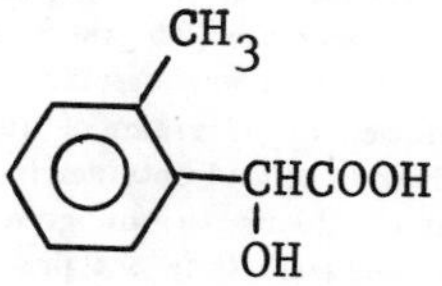

$C_9H_{10}O_3$ Ref. 528

The authors reported that the above compound was custom prepared by Niels Clauson-Kaas Chemical Research Laboratory, Rugmarken 28, Farum, Denmark. The reported m.p. of the resolved acid was 61-63°C. The $[\alpha]_D^{25}$ = -175°C (c, 2, ethanol).

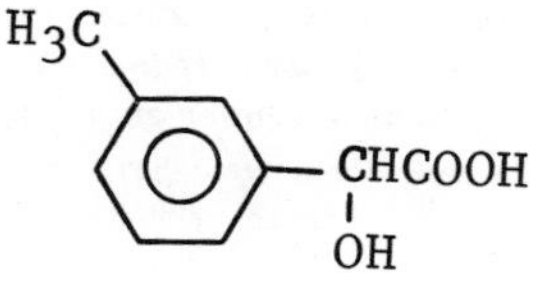

$C_9H_{10}O_3$ Ref. 529

Resolution was accomplished by serial crystallization of the salt of the desired acid with (=)-α-methyl-benzylamine. The m.p. of the resolved acid was 109-112°C. The $[\alpha]_D^{25}$ = -144°C. (c, 0.3, ethanol).

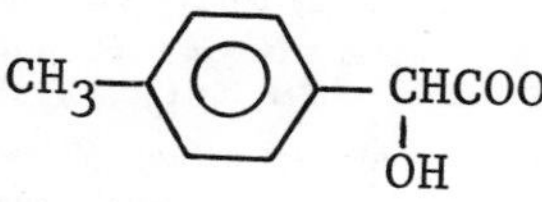

$C_9H_{10}O_3$ Ref. 530

Resolution was accomplished by serial crystallization of the salt of the desired acid with (+)-α-methyl-benzylamine. The m.p. of the resolved acid was 132-134°C. The $[\alpha]_D^{25}$ = -153°C. (c, 0.3, ethanol).

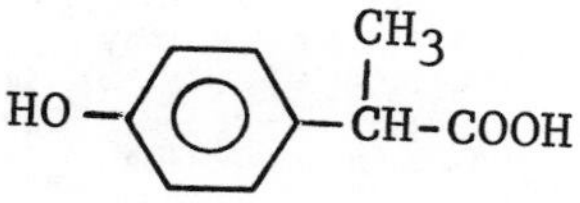

$C_9H_{10}O_3$ Ref. 531

Note:

The compound was resolved using morphine as resolving agent. The less soluble salt gave an acid with m.p. 130°, $[\alpha]_D$ -71° (c 2, H_2O).

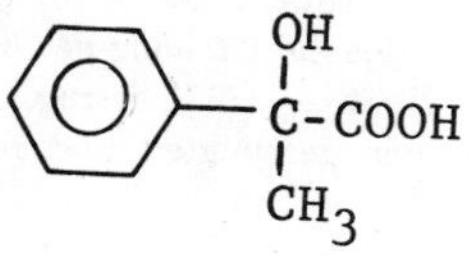

$C_9H_{10}O_3$ Ref. 532

Spaltung der Säure in ihre aktiven Komponenten.

17,5 g (1/10 Mol) der Säure wurden in 60 ccm Wasser gelöst und mit 12,1 g l-Phenäthylamin (aus dem sauren weinsauren Salze)[1]) versetzt. Die Lösung erstarrte nach einigen Augenblicken zu einem Brei von Krystallen, die nach einigen Stunden abgesaugt wurden. Ausbeute: 13,2 g. Nach dreimaligem Umkrystallisieren aus dem doppelten Gewichte Wasser zeigte die bei 20° gesättigte Lösung im 1 dm-Rohr eine Drehung von + 2° 15′, welche nach fortgesetztem Umkrystallisieren unverändert blieb.

Die beiden Salze scheinen gleiche Zusammensetzung zu haben. Von der reinen schwer löslichen Form konnte ich, wegen Ermangelung an Substanz, nur eine Stickstoffbestimmung

[1]) Lovén, Dies Journ. 72. 307.

machen. Vollständig analysiert wurde das aus der ersten Mutterlauge sich später abscheidende leicht lösliche Salz, das jedoch in reinem Zustande schwer zu erhalten ist.

Stickstoffbestimmung des schwerlöslichen Salzes ergab:

	Gef.:	Ber. für $C_{17}H_{21}O_3N$:
N	5,10	4,88%.

Analyse des leichtlöslichen Salzes:

	C	H	N
Gef.:	71,62	7,15	5,02%
Ber. nach Obenstehendem:	71,02	7,37	4,88 „ .

Aus dem schwer löslichen Salze setzte ich die Säure in Freiheit mit Schwefelsäure, extrahierte mit Äther und krystallisierte das erhaltene Produkt aus kaltem Wasser um, das im Vakuumexsiccator verdunstet wurde. Die Säure wurde dann als 4—5 cm lange seidenglänzende Stäbe erhalten. Bisweilen scheiden sich dicke Prismen ab, besonders beim Impfen der Lösung. Die Säure wird nicht leicht racemisiert.

0,1079 g Säure verbrauchten 6,26 ccm 0,1036 n-$Ba(OH)_2$.
Äqu.-Gew. Gef.: 166,5. Ber. für $C_9H_{10}O_3$: 166,1.

Die Säure schmilzt unscharf zufolge Wasserabgabe in der Nähe des Schmelzpunktes. Bei ziemlich raschem Erhitzen fand ich 116,5°—117,0°.

Mc. Kenzie und Clough bestimmten die Drehung in Alkohol, Aceton und Wasser. Ich beschränkte mich auf einige Bestimmungen in Wasser, hauptsächlich bei 20°. Der Temperaturkoeffizient der Drehung ist wahrscheinlich negativ, jedenfalls sehr klein und in diesem Falle beinahe von der Größe der Versuchsfehler.

c = g auf 100 ccm der Lösung. l = Länge des Rohres.

$c = 1{,}591$;	$l = 1$;	$T = 13°$;	$\alpha = +53{,}7'$;	$[\alpha_D]^{13} = +56{,}0°$;
„	„	$T = 20°$;	$\alpha = +51{,}2'$;	$[\alpha_D]^{20} = +53{,}5°$;
$c = 2{,}078$;	$l = 2{,}2$;	$T = 14{,}5°$;	$\alpha = +142{,}5'$;	$[\alpha_D]^{14{,}5} = +52{,}0°$;
„	„	$T = 20°$;	$\alpha = +142{,}5'$;	$[\alpha_D]^{20} = +52{,}0°$;
$c = 2{,}087$;	$l = 2{,}2$;	$T = 14°$;	$\alpha = +141{,}7'$;	$[\alpha_D]^{14} = +51{,}5°$;
„	„	$T = 20°$;	$\alpha = +140{,}5'$;	$[\alpha_D]^{20} = +51{,}0°$;
$c = 2{,}148$;	$l = 1$;	$T = 20°$;	$\alpha = +67{,}2'$;	$[\alpha_D]^{20} = +52{,}0°$;
$c = 6{,}02$;	$l = 1$;	$T = 20°$;	$\alpha = +180{,}5'$;	$[\alpha_D]^{20} = +50{,}0°$.

Eine Probe, die zweimal aus kaltem Wasser umkrystallisiert worden war:

$c = 2{,}068$;	$l = 2{,}2$;	$T = 13°$;	$\alpha = +140{,}5'$;	$[\alpha_D] = +51{,}5°$.
„	„	$T = 20°$;	$\alpha = +138{,}0'$;	$[\alpha_D] = +50{,}5°$.

Mc. Kenzie und G. Clough[1]) führen an:

$c = 1{,}5892$;	$l = 4$;	$[\alpha_D]^{13} = -52{,}5°$;	$T = 13°$.
$c = 2{,}160$;	$l = 2$;	$[\alpha_D]^{14{,}5} = -51{,}1°$;	$T = 14{,}5°$.

Die l-Form wurde auf die folgende Weise gewonnen. Aus der ersten Mutterlauge vom Salze von l-Phenäthylamin und d-Atrolactinsäure wurde durch Schwefelsäure und Äther die Säure in Freiheit gesetzt und dann mit d-Phenäthylamin (aus dem äpfelsauren Salze) behandelt. Die abgeschiedenen Krystalle wurden umkrystallisiert, bis die Drehung der bei 20° gesättigten Mutterlauge im 1-dm-Rohr — 2°17′ betrug.

Smp. der wie vorher gereinigten und im Exsiccator getrockneten Säure: 116°.

Drehung: $c = 2{,}001$; $l = 2{,}2$; $T = 20°$; $\alpha = -137{,}5'$;
$[\alpha_D]^{20} = -52{,}0°$.

Die Löslichkeit der reinen rechtsdrehenden Säure betrug bei 18,0°, 25,0° und 30,0° (im Mittel) 52,0 g, 64,3 g und 76,2 g pro Liter resp. Die zu diesen Bestimmungen benutzte Substanz wurde zuletzt regeneriert und auf ihr Drehungsvermögen geprüft: $c = 2{,}697$; $l = 1$; $T = 20°$; $[\alpha_D]^{20} = +50{,}0°$.

$C_9H_{10}O_3$ Ref. 532a

Resolution of r-Phenylmethylglycollic Acid into its Optically Active Components.

Thirty-seven grams of *r*-phenylmethylglycollic acid (1 mol.), prepared from acetophenone and hydrogen cyanide by modifying Spiegel's method (*Ber.*, 1881, **14**, 1353), were dissolved in 2300 c.c. of boiling water, and 64 grams of powdered morphine (1. mol.) were gradually added. The alkaloid dissolved with ease, and the boiling was continued for a few minutes until the solution was clear. On cooling, glassy prisms grouped in rosettes began to separate. The solution, when still warm, was placed in an ice-chest at 10°, and 44 grams of salt had separated on the following day. After five more crystallisations from water, 19 grams of the pure morphine salt of the *l*-acid were obtained. The progress of the resolution was tested by removing the morphine from successive filtrates, and polarimetrically examining the resulting acid in ethyl-alcoholic solution, this method being preferred to the customary one of examining successive crops of the alkaloidal salt. The last four filtrates gave acids with the values $[\alpha]_D$ −20·9°, −33°, −36·2°, and −37·9° respectively. When the 19 grams of salt were again crystallised, the filtrate gave an acid with $[\alpha]_D$ −37·9°. The *l*-acid was then obtained from the homogeneous morphine salt by acidification with sulphuric acid and extraction with ether.

It appeared from another experiment, where 44 grams of *r*-phenylmethylglycollic acid were employed, that the resolution is fairly complete, even after two crystallisations, under conditions similar to the above. The resulting acid was crystallised once from benzene, and then amounted to 14 grams, with $[\alpha]_D$ −35° in ethyl-alcoholic solution.

l-*Phenylmethylglycollic acid*, OH·CMePh·CO_2H, dissolves readily in ethyl alcohol or acetone. It is easily soluble in hot benzene, but sparingly soluble in the cold solvent, from which it separates in silky needles, m. p. 116—117°. It may also be crystallised from water, in which it is more soluble than is the *r*-isomeride. It differs from the *r*-acid (m. p. 93—94°) in being devoid of water of crystallisation:

0·1360 gave 0·3232 CO_2 and 0·0736 H_2O. C = 64·8; H = 6·1.
$C_9H_{10}O_3$ requires C = 65·0; H = 6·1 per cent.

Analysis of the *silver* salt gave:

Found: Ag = 39·4. $C_9H_9O_3Ag$ requires Ag = 39·5 per cent.

The purity of the acid was further controlled by finding that, when it is crystallised several times either from benzene or from water, the value for the specific rotation of the product is not altered.

The following polarimetric observations were made.

In ethyl-alcoholic solution:

$l = 2$, $c = 3{\cdot}354$, $\alpha_D^{14\cdot8}$ −2·53°, $[\alpha]_D^{13\cdot8}$ −37·7°.

In acetone solution:

$l = 2$, $c = 4{\cdot}142$, α_D^{13} −3·02°, $[\alpha]_D^{13}$ −36·5°.

In aqueous solution:

$l = 2$, $c = 2{\cdot}160$, $\alpha_D^{14\cdot5}$ −2·21°, $[\alpha]_D^{14\cdot5}$ −51·1°.
$l = 4$, $c = 1{\cdot}5892$, α_D^{13} −3·34°, $[\alpha]_D^{13}$ −52·5°.
$l = 2$, $c = 0{\cdot}688$, $\alpha_D^{14\cdot5}$ −0·74°, $[\alpha]_D^{14\cdot5}$ −53·8°.

The specific rotation of the acid in aqueous solution increases with dilution. An aqueous solution of the sodium salt should, therefore, give a value higher than any of the three preceding values, and this is the case. The following observations were made with an aqueous solution of sodium salt:

$l = 2$, $c = 1{\cdot}920$, $\alpha_D^{13\cdot6}$ −2·17°, $[\alpha]_D^{13\cdot6}$ −56·5°.
$l = 2$, $c = 4{\cdot}916$, $\alpha_D^{14\cdot7}$ −5·60°, $[\alpha]_D^{14\cdot7}$ −57·0°.

It is not unimportant to note that the polarimetric observations with the free acid were made with samples which had previously been dried under diminished pressure at the ordinary temperature over sulphuric acid. The active acid should not be dried at 100°, since water is slowly eliminated from it at this temperature with the consequent formation of atropic acid.

The first two mother liquors from the morphine resolution quoted gave a dextrorotatory acid mixture (22 grams) with $[\alpha]_D$ +19·4° in ethyl-alcoholic solution. This acid was dissolved in 550 c.c. of boiling ethyl alcohol, and neutralised by 47 grams of quinine. Forty grams of quinine salt separated, and the acid obtained from it had $[\alpha]_D$ +28·3° in ethyl-alcoholic solution. Owing to the greater solubility of the active as compared with the *r*-acid, it is not practical to separate the *d*-acid from this mixture by crystallisation. After two crystallisations from water, the acid was in fact less active than before, having $[\alpha]_D$ +26° in ethyl-alcoholic solution.

For the preparation of the *d*-acid, 64 grams of quinine (1 mol.) were dissolved in 700 c.c. of boiling ethyl alcohol, and 30 grams of *r*-phenylmethylglycollic acid (1 mol.) were added. After fifty minutes, crystallisation started, and the solution was then placed in an ice-chest at 10°. The resulting crystals, which contain a preponderance of the quinine salt of the *d*-acid, were withdrawn next day, and repeatedly crystallised from the minimum amount of boiling ethyl alcohol. The resolution proceeds rather more slowly than when morphine was used, the progress being tested by decomposing the mother liquors as before. After four crystallisations, the salt amounted to 19 grams, but was not quite pure, the acid obtained from a portion of it having $[\alpha]_D$ +34·6° in ethyl-alcoholic solution. After three additional crystallisations, the salt, which separated in glassy needles, grouped in rosettes, was pure, and amounted to 11 grams; it melts and decomposes at about 216°. The *d*-acid was obtained from it by acidification and extraction with ether.

d-*Phenylmethylglycollic acid* melts at 116—117°, and in other respects resembles the *l*-isomeride:

0·1558 gave 0·3712 CO_2 and 0·0865 H_2O. C=65·0; H=6·2.
$C_9H_{10}O_3$ requires C=65·0; H=6·1 per cent.

Its specific rotation was determined in ethyl-alcoholic solution:
l=4, c=3·500, $\alpha_D^{10·5}$ +5·27°, $[\alpha]_D^{10·5}$ +37·7°.

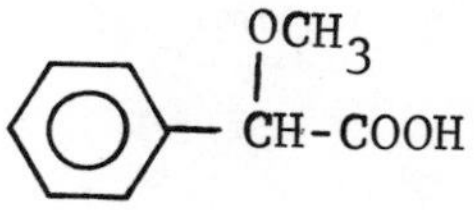

$C_9H_{10}O_3$ Ref. 533

105 G. of hydrated quinine (1 mol.) and 46 g. of *r*-phenylmethoxyacetic acid (1 mol.) were dissolved by warming in 552 c.c of a mixture of equal volumes of ethyl alcohol and water. The solution was cooled in the ice-chest for 5 hours, whereupon 90 g. of a salt crystallised in rosettes of needles. This salt, on being recrystallised from 300 c.c. of the alcohol-water mixture, gave 70 g. of salt (after drying in air on porous plate for 1 day) giving in ethyl alcohol (l = 1, c = 2·208) α_D − 1·92°, whence $[\alpha]_D$ − 87·0°. After two further crystallisations, the salt weighed 60 g. and gave in ethyl alcohol (l = 1, c = 2·12) α_D − 1·74°, whence $[\alpha]_D$ − 82·1°. The quinine salt was decomposed by dilute sulphuric acid, and the solution extracted with ether. 16 G. of dextrorotatory phenylmethoxyacetic acid (m. p. 64—65°) were obtained, having in ethyl alcohol (l = 1, c = 1·716) α_D + 2·6°, whence $[\alpha]_D$ + 151·5°. This value, which did not increase on crystallisation of the acid from petroleum (b. p. 80—100°) to which a little benzene had been added, is in agreement with that ($[\alpha]_D$ − 151·1°) of McKenzie (*loc. cit.*) for the lævorotatory acid.-

$C_9H_{10}O_3$ Ref. 533a

Ephedrine (40 g.) and this (±)-acid (40 g.) were dissolved in warm ethanol (150 ml.) and left for 24 hr. at room temperature. The ephedrine salt (40 g.) which was precipitated was filtered off and crystallised once from ethanol (200 ml.) to give a salt (28·8 g.), $[\alpha]_{5461}^{19}$ −87·8° (c 0·668 in methanol; l 2). One further crystallisation from ethanol (200 ml.) gave a salt of $[\alpha]_{5461}^{20}$ −91·2° (c 0·534 in methanol; l 2). Decomposition of this salt with 20% sulphuric acid gave (−)-acid (11 g.) which after one crystallisation from light petroleum (b. p. 80—100°)-benzene had $[\alpha]_{5461}^{20}$ −173·3° and $[\alpha]_D^{20}$ −150·7° (c 0·574 in ethanol; l 2). Evaporation and treatment of the reaction liquor in a similar way yielded (+)-acid (9·5 g.), $[\alpha]_{5461}^{20}$ +172·6°, $[\alpha]_D^{20}$ +150·0° (c 0·494 in ethanol; l 2). The acids had m. p. 65—66° and gave no depression on admixture with authentic active acids.

[1] Pirie and Smith, *J.*, 1932, 338.
[5] McKenzie, *J.*, 1899, **75**, 753.
[6] Weygand and Gölz, *Chem. Ber.*, 1954, **87**, 707.

The Table shows the specific rotations obtained by various workers.

$[\alpha]_D$ in EtOH	Temp.	Concn. (c) *	Ref.	Method of prep.
−150·6°	13°	0·541	5	From (−)-mandelic acid
−151·1	,,	2·706	,,	,, ,,
−150·7	20	0·574	This paper	Ephedrine resoln.
+150·0	20	0·494	,,	,, ,,
−150·9	20	0·685	6	From sugars
+151·5	—	1·716	1	Quinine resoln.

* The specific rotations of these acids increase with increasing concentration.[5]

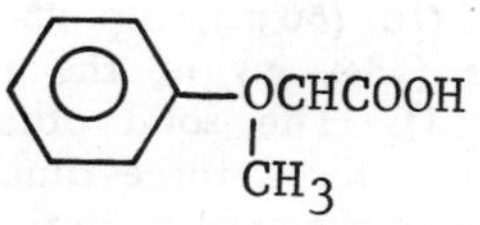

$C_9H_{10}O_3$ Ref. 534

Dehydroabietylammonium (+)-α-Phenoxypropionate.—To a solution of 914 g. (3.2 moles) of dehydroabietylamine dissolved in 7 l. of methanol was added 537 g. (3.2 moles) of racemic α-phenoxypropionic acid. The stirred solution was slowly diluted with 5.5 l. of water and stored at 10° for 5 hr. The crystals were collected and air dried to obtain 850 g., m.p. 168–170°. Recrystallization from a mixture of 8 l. of methanol and 3.5 l. of water during storage at 10° for 7 hr. gave 650 g. of salt, m.p. 173–177°. A final recrystallization was made from a mixture of 6 l. of methanol and 1.5 l. of water to yield 287 g. of colorless crystals, m.p. 188–189.5°, $[\alpha]^{23}D$ +27.7° (c 1, alcohol). Further recrystallization did not raise the melting point or change the rotation.

Anal. Calcd. for $C_{29}H_{41}NO_3$: C, 77.12; H, 9.15. Found: C, 77.03; H, 9.36.

(+)-α-Phenoxypropionic Acid.—To 1 l. of a saturated solution of sodium carbonate were added 287 g. (0.635 mole) of finely ground dehydroabietylammonium (+)-α-phenoxypropionate and 1 l. of ether. The mixture was shaken vigorously until all of the solid had dissolved. The ether layer was separated and the aqueous solution was washed twice with ether and acidified to pH 2 with concentrated hydrochloric acid. The mixture was cooled to 10° for 2 hr., and the white crystals were collected to obtain 76 g. of acid with m.p. 88–89°, $[\alpha]^{23}D$ +40.0° (c 1, absolute alcohol), lit.[8] $[\alpha]^{20}D$ +39.3°. A second crop of 22 g. was obtained from the mother liquor on further storage at 10° which had m.p. 88–89°, $[\alpha]^{23}D$ +39.1° (c 1, absolute alcohol).

(8) E. Fourneau and G. Sandulesco, *Bull. soc. chim. France*, [4] **31**, 988 (1922).

$C_9H_{10}O_3$ Ref. 534a

The *racemic acid was resolved* by the method of FOURNEAU and SANDULESCO (7) using yohimbine. The acid isolated from the alkaloid salt was recrystallised from carbon tetrachloride and cylohexane. M. p. 86°—87°.

0.2069 g acid: 11.65 ml 0.1066-*N* NaOH.

$C_9H_{10}O_3$ Equiv. wt. calc. 166.2, found 166.6.

0.2484 g acid dissolved in absolute ethanol to 10.00 ml: $2\,\alpha_D^{20} = +1.955°$. $[\alpha]_D^{20} = +39.3°$; $[M]_D^{20} = +65.4°$.

The rotatory power found is in excellent agreement with the values given by FOURNEAU and SANDULESCO (7).

[7] **E. Fourneau** and **G. Sandulesco**, Bull. soc. chim. France [4] *31*, 988 (1922). —

On evaporation of the mother liquor from the first crystallisation of the yohimbine salt at room temperature to about half its volume, a second crop of crystalline salt was obtained. The remaining mother liquor gave, on evaporation to dryness, a gum-like residue. The crystalline salt was combined with salt obtained from the mother liquors from the higher fractions of the yohimbine salt. After several recrystallisations a second portion of the salt of the pure (+)-acid was obtained. When the gum-like residue was treated with dilute sulphuric acid, it gave a phenoxy acid of $[\alpha]_D^{20} = -28°$. As the antipodes are far more soluble than the racemic acid, this fraction was rapidly extracted with hot chloroform and washed with a little cold carbon tetrachloride. The undissolved acid had a very low activity while the chloroform extract on evaporation yielded an acid having $[\alpha]_D^{20} = -36°$. Maximum activity was obtained by repeated crystallisations from cyclohexane and carbon tetrachloride. M. p. 86—87°.

0.1923 g acid: 10.82 ml 0.1066-*N* NaOH.

$C_9H_{10}O_3$ Equiv. wt. calc. 166.2, found 166.7.

0.2226 g acid dissolved in absolute ethanol to 10.00 ml: $2\alpha_D^{20} = -1.75°$. $[\alpha]_D^{20} = -39.3°$; $[M]_D^{20} = -65.3°$.

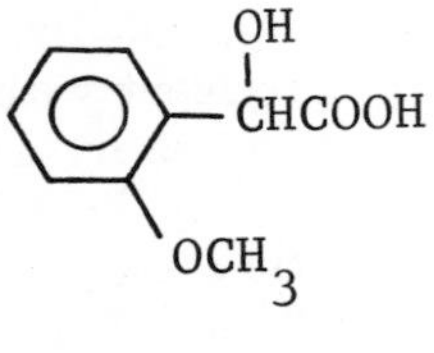

$C_9H_{10}O_4$ Ref. 535

The crude (±)-acid (30 g.) so obtained was heated with ephedrine (25 g.) in ethanol (60 ml.) and set aside to crystallise at 0°. The resultant solid was recrystallised from the minimum of ethanol and then decomposed with dilute sulphuric acid to give (+)-*o*-methoxymandelic acid which, after crystallisation from benzene–cyclohexane, had $[\alpha]_{5461}^{20}$ +175·9° (*c* 0·52 in acetone) and m. p. 87—88°. The original liquors yielded, in the same way, the (−)-isomer, $[\alpha]_{5461}^{20}$ −168·0° (*c* 0·86 in acetone) and with m. p. 80—83°.

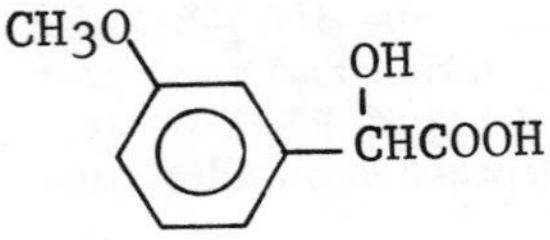

$C_9H_{10}O_4$ Ref. 536

(c) (±)-3-*Methoxymandelic acid* (Ic).—The (±)-acid (Ic) (60 g.), m.p. 56—58°, was treated with (−)-ephedrine (50 g.) in boiling chloroform–carbon tetrachloride (2 : 3; 300 ml.). The resultant solid was recrystallised three times from the same solvent mixture, giving a complex, m.p. 102—104° which on work-up in the usual way gave (−)-3-methoxymandelic acid (17 g.), m.p. 80—81°, [α] −117·3° (EtOH) (Found: C, 59·2; H, 5·5. $C_9H_{10}O_4$ requires C, 59·3; H, 5·5%). The original liquors yielded the (+)-acid (Ic), m.p. 79—81°, [α] +69·5° (EtOH).

$C_9H_{10}O_4$ Ref. 536a

Resolution was accomplished by serial crystallization of the salt of the desired acid with (+)-α-methylbenzylamine. The m.p. of the resolved acid was 78-79°C. The $[\alpha]_D^{25}$ = 117°C. (c, 0.3, ethanol).

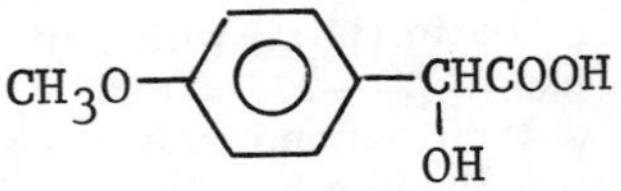

$C_9H_{10}O_4$ Ref. 537

Resolution was accomplished by serial crystallization of the salt of the desired acid with (+)-α-methylbenzylamine. The m.p. of the resolved acid was 102-104°C. The $[\alpha]_D^{25}$ = -141° (c 0.3, H_2O).

$C_9H_{10}O_4$ Ref. 537a

(d) (±)-4-*Methoxymandelic acid* (Id). The following procedure was an improvement on the published resolution.[3] The (±)-acid (Id) (56 g.), m.p. 95—96°, was treated with (−)-ephedrine (46·4 g.) in the minimum of hot methanol–water (9 : 1). The solid obtained at 0° was filtered off and recrystallised three times from the same solvent mixture to give a complex, m.p. 144—145°. This yielded, in the usual way, the (−)-acid (20 g.) m.p. 101—103°, $[\alpha]_D$ −141·4° (H_2O) {lit.,[3] $[\alpha]_D$ −145·2° (H_2O) m.p. 104—105°}.

[3] Y. Izumi and T. Ninomiya, *Mem. Inst. Protein Research, Osaka Univ.*, 1968, **10**, 70 (*Chem. Abs.*, 1968, **69**, 95,880).

$C_9H_{10}O_4$ Ref. 537b

Spaltung von *racem.* *p*-Methoxy-mandelsäure in die optisch aktiven Komponenten.

Die Methode von Knorr (l. c.) wurde folgendermaßen modifiziert: 81 g Cinchonin löste man in einer warmen wäßrigen Lösung (2500 ccm) von 50 g *racem.* Säure vom Schmp. 108—109°. Nach 1-stdg. Stehenlassen bei gewöhnlicher Temperatur wurden die Krystalle (41 g) abgetrennt und aus 800 ccm Wasser umgelöst, das erhaltene Salz (29 g) dann mit verdünnter Salzsäure zersetzt und die Säure mit Äther extrahiert. Ausbeute 9.1 g Säure von $[\alpha]_D$: +123° (l = 1, c = 2.86 in Wasser). Eine andere Portion von 50 g *racem.* Säure gaben bei der Spaltung unter ähnlichen Bedingungen 9.8 g Säure von $[\alpha]_D$: +136° (l = 1, c = 1.764), eine 3. Spaltung mit 30 g *racem.* Säure lieferte 4.6 g mit $[\alpha]_D$: +136° (l = 1, c = 1.85). Nun wurden die 3 Partien mit frischem Cinchonin (37 g) in 850 ccm heißem Wasser behandelt. Beim Abkühlen krystallisierten Nadeln von einheitlichem Cinchoninsalz (45 g) der (+)-Säure aus, die optisch reine (+)-Säure (16 g) vom Schmp. 104—105° ergaben.

Drehung in Wasser (c = 2.266):

λ ...	6563	6162	5893	5461	5106	4861	4358
$[\alpha]^{20}$	+115.0°	+132.0°	+146.1°	+176.2°	+208.2°	+237.1°	+315.2°.

Der von Knorr (l. c.) angegebene Wert ist $[\alpha]_D^{19}$: +146.14° in Wasser.

Drehung in Chloroform (c = 1.830):

λ ...	6708	6563	6162	5893	5461	5106	4861	4358
$[\alpha]^{20.3}$	+130.0°	+139.9°	+159.3°	+177.3°	+213.7°	+253.3°	+288.6°	+385.5°.

Die spezif. Drehungen von (—)-Mandelsäure[28]) und von (—)-α-Naphthyl-glykolsäure[29]) seien zum Vergleich angeführt; dabei soll bemerkt werden, daß die Drehwerte bei diesen Säuren in Chloroform höher sind als in Wasser:

	$[\alpha]_D$ Wasser	$[\alpha]_D$ Chloroform
(—)-Mandelsäure	—157.5°	—188°
(—)-α-Naphthyl-glykolsäure .	—156.5°	—204.5°.

Die spezif. Drehung von (+)-*p*-Methoxy-mandelsäure wird mit steigender Temperatur kleiner, was an das entsprechende Verhalten von (—)-Mandelsäure und (—)-α-Naphthyl-glykolsäure erinnert. So findet man in Chloroform (c = 1.83):

t	3.5°	20.3°	32.9°
$[\alpha]_{6563}$	+151.1°	+139.9°	+129°
$[\alpha]_{5893}$	+195.1°	+177.3°	+165.6°
$[\alpha]_{5461}$	+233.6°	+213.7°	+206.6°
$[\alpha]_{4861}$	+303°	+288.5°	+262.3°.

Das rohe Cinchoninsalz (99 g) der (—)-Säure, das durch Krystallisation aus jedem der 3 Filtrate der oben beschriebenen Spaltungen erhalten wurde, wurde zweimal aus Wasser umgelöst, dann zersetzt, und die Säure einmal aus Benzol umkrystallisiert. Ausbeute 13 g vom Schmp. 104—105°.

$[\alpha]_D$: —146.5° (l = 2, c = 2.375 in Wasser).

[28]) Roger, Journ. chem. Soc. London 1932, 2168.

[29]) McKenzie u. Gow, Journ. chem. Soc. London 1933, 32; Roger u. Gow, Journ. chem. Soc. London

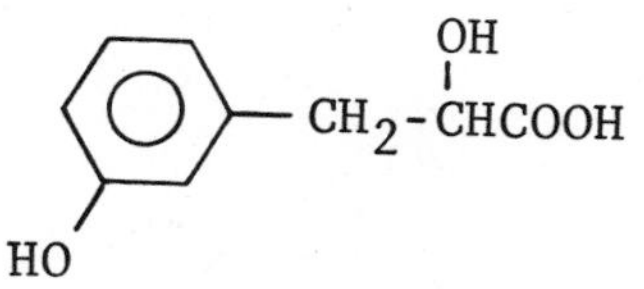

$C_9H_{10}O_4$ Ref. 538

Resolution of m-hydroxyphenyl-DL-lactic acid. *m*-Hydroxyphenyl-DL-lactic acid (6.38 g., 0.035 mole) and 11.35 g. (0.035 mole) of anhydrous quinine were dissolved in 200 ml. of boiling methanol. The solution was treated with 2.0 g. of charcoal, the filtrate was concentrated to dryness under reduced pressure and nitrogen, and the residue was dissolved in 75 ml. of boiling methanol. The crystals that had separated after the solution had stood overnight at room temperature were collected, washed with three 5-ml. portions of cold methanol, and desiccated over potassium hydroxide and phosphorus pentoxide at 10 mm.; crop A, 8.35 g., m.p. 194–202° (slight softening at 113°). Crop A was dissolved in 135 ml. of boiling methanol and the solution was stored overnight at 5°; 3.48 g. of crop B were recovered as colorless cuboids, m.p. 204–206°. The filtrate from this material was concentrated to dryness under reduced pressure and nitrogen; the residue was dissolved in 35 ml. of boiling methanol and the solution was allowed to stand at room temperature to yield 1.33 g. more of crop B, colorless cuboids, m.p. 204–206°; total crop B, 4.81 g. Crop B was recrystallized from 90 ml. of boiling methanol with charcoal treatment to give 3.32 g. of crop C as large colorless cuboids, m.p. 205–206° (initial bath temperature 200°, heating rate 2° per minute), $[\alpha]_D^{24}$ −161.6° (*c*, 1.0, acetic acid). A sample was recrystallized again from methanol, m.p. 205–206°, $[\alpha]_D^{24}$ −160.5° (*c*, 1.4, acetic acid).

Crop C was suspended in 75 ml. of water, treated with 13.2 ml. of 1 *N* sodium hydroxide, and extracted with four 75-ml. portions of dichloromethane to remove quinine. The aqueous phase was acidified to *p*H 1.6 with concentrated hydrochloric acid and extracted with six 75-ml. portions of ethyl acetate. The ethyl acetate extracts were dried over sodium sulfate and concentrated under nitrogen and reduced pressure to clear amber oil. This oil was dissolved in 25 ml. of ether, 50 ml. of cyclohexane was added in several portions, and the mixture was seeded with a rubbed aliquot when clouding occurred; rapid precipitation ensued. The mixture was allowed to digest for two days at room temperature, and then was stored overnight at 5°; 0.88 g. (74% yield from pure quinine salt) of *m*-hydroxyphenyl-D-lactic acid was obtained, m.p. 84–86°. After recrystallization from 1,2-dichloroethane with charcoal treatment, the D-acid was obtained as shimmering colorless flakes, 88% recovery, m.p. 85–86°, $[\alpha]_D^{21}$ +19.5° (*c*, 1.0, water), $[\alpha]_D^{22}$ +48.2° (*c*, 1.0, 1 *N* sodium hydroxide).

Anal. Calc'd for $C_9H_{10}O_4$: C, 59.34; H, 5.53. Found: C, 59.11; H, 5.66.

The filtrate from crop A was concentrated to dryness under nitrogen and reduced pressure, and the residue was crystallized from 30 ml. of boiling methanol with cooling to 5° to produce 6.39 g. of crop D, which softened to a froth at 106°, solidified again at 115°, and melted at 161–163°. Crop D was crystallized from 20 ml. of boiling methanol with cooling to 5° to give 3.81 g. (after desiccation) of crop E, m.p. 112–120°, $[\alpha]_D^{25}$ −173.0° (*c*, 1.2, acetic acid). After drying in air, crop E was in the form of colorless clusters of elongated rhombs, which decreased in weight by 0.25 g. and visibly decrepitated during desiccation. This behavior indicates the loss of one molecule of methanol of crystallization. This crop was recrystallized from 15 ml. of boiling methanol with charcoal treatment. A felted mass of long fine needles rapidly precipitated as the solution cooled. After eight hours at room temperature, numerous clusters of elongated rhombs had appeared; in 18 hours, transformation of the needles to rhombs was complete. The mixture was stored at 5° and 3.03 g. (after desiccation) of crop F was recovered, m.p. 114–120°, $[\alpha]_D^{25}$ −174.0° (*c*, 1.2, acetic acid). Crop F showed the same behavior on desiccation as crop E. It was converted to free acid in the manner described for the D-isomer; 0.77 g. (71% yield from pure quinine salt) of white powder was recovered, m.p. 84–86°. After recrystallization from 1,2-dichloroethane with charcoal treatment, *m*-hydroxyphenyl-L-lactic acid was obtained as shimmering colorless flakes, 80% recovery, m.p. 85–86°, $[\alpha]_D^{22}$ −19.1° (*c*, 1.0, water), $[\alpha]_D^{22}$ −45.0° (*c*, 1.0, 1 *N* sodium hydroxide).

Anal. Calc'd for $C_9H_{10}O_4$: C, 59.34; H, 5.53. Found: C, 59.12; H, 5.58.

The configuration of the *m*-hydroxyphenyllactic acid isomers is assigned on the basis of the observed effect of alkali on their rotations as compared to the change observed with phenyl-L-lactic acid, $[\alpha]_D^{24}$ −21.4° (*c*, 1, water),[34] $[\alpha]_D^{22}$ −31.3° (*c*, 1, 1 *N* sodium hydroxide). In addition, the magnitude and direction of rotation of the isomer assigned the L-configuration are similar to those of phenyl-L-lactic acid, and *p*-hydroxyphenyl-L-lactic acid [$[\alpha]_D$ −18.1° (*c*, 1.57, water)].[35]

(34) Armstrong, Shaw, and Robinson, *J. Biol. Chem.*, **213**, 797 (1955).

(35) Kotake, *Z. physiol. Chem.*, **65**, 397 (1910).

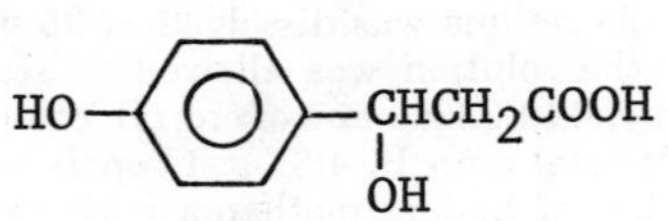

$C_9H_{10}O_4$ **Ref. 539**

Sale di chinidina dell'acido (+)4-benzilossifenilidracrilico

Grammi 40,8 (0,15 mole) di acido DL-4-benzilossifenilidracrilico vengono sciolti in 188 cc di metanolo bollente; si aggiungono g 48,6 (0,15 mole) di chinidina.

Dopo riposo di 18 ore a temperatura ambiente e di 6 ore in ghiacciaia, si ottengono g 41,5 di precipitato cristallino (frazione A), con p.f. 175-179° e $[\alpha]_D^{26}$ = +139,2° (c = 0,22% in etanolo assoluto).

Dalle acque madri per ulteriore riposo di diversi giorni, precipitano altri 4,36 g di sale (frazione B) a p.f. 174-179° e $[\alpha]_D^{26}$ = +147,4° (c = 0,23% in etanolo assoluto).

Le frazioni A e B riunite (g 45,8) vengono cristallizzate da metanolo.

Cristallizzazione		I	II
g. sale	45,8	32,3	26,6
p.f. sale	174-179°	181-183°	181-183°
$[\alpha]_D^{20}$ del sale (0.23% in etanolo ass.)	-	+152,6°	+152,8°
quantità di metanolo	1380 cc	900 cc	

Acido (+)4-benzilossifenilidracrilico

Grammi 26,6 di sale di chinidina vengono sospesi in 50 cc di acqua; la miscela viene acidificata con HCl conc. spappolando bene. Il sale di chinidina si scioglie e immediatamente precipita l'acido. Si filtra. Si ottengono g 9,3 di (+)acido a p.f. 161-164° ed $[\alpha]_D^{28}$ = +21,5° (c = 0,75% in acetone).

Per cristallizzazione da acetato di etile-etere di petrolio si ottengono g 8,2 di acido a p.f. 161-163° e $[\alpha]_D^{28}$ = +21,9° (c = 0,72% in acetone).

Questo acido viene di nuovo salificato con chinidina; si ottiene il sale di chinidina con p.f. 181-183° e $[\alpha]_D^{28}$ = +152,4° (c = 0,23% in etanolo assoluto); il p.f. e il potere rotatorio non variano dopo ulteriori cristallizzazioni da metanolo.

Da questo sale di chinidina per trattamento con acido cloridrico, si ottiene di nuovo acido (+)4-benzilossifenilidracrilico con p.f. 161-163° e $[\alpha]_D^{25}$ = +21,4° (c = 0,72% in acetone).

Acido (-)4-benzilossifenilidracrilico

Le acque madri della scissione dopo precipitazione del sale di chinidina dell'acido (+)4-benzilossifenilidracrilico (frazioni A e B), vengono concentrate a pressione ridotta fino a secco. Il sale di chinidina ottenuto (g 39,2; p.f. 82-85°; $[\alpha]_D^{28}$ = +114,2° (c = 0,24% in etanolo assoluto)) viene sospeso in poca acqua; si acidifica con acido cloridrico conc. spappolando bene e si filtra l'acido levogiro.

Si ottengono così g 17,7 di acido a p.f. 150-160° e $[\alpha]_D^{30}$ = -13,4° (c = 0,8% in acetone).

Dopo 6 cristallizzazioni da acetato di etile-etere di petrolio, se ottengono g 8,3 di acido a p.f. 161-163° e $[\alpha]_D^{25}$ = -21,0° (c = 0,75% in acetone).

Acido (-)-4-ossifenilidracrilico (XXXI)

Da g 1,0 di acido (-)4-benzilossifenilidracrilico idrogenati in presenza di Pd/C come indicato per il composto racemico, si ottengono g 0,3 (44,8% d.t.) di acido (-)4-ossifenilidracrilico a p.f. 93-96° e $[\alpha]_D^{20}$ = -26,3° (c = 2,15% in acetato d'etile).

Acido (+)-4-ossifenilidracrilico (XXXII)

Si ottiene in modo analogo all'antipodo levogiro e mostra p.f. 94,5-96° e $[\alpha]_D^{20}$ = +26,1° (c = 2,1% in acetato d'etile).

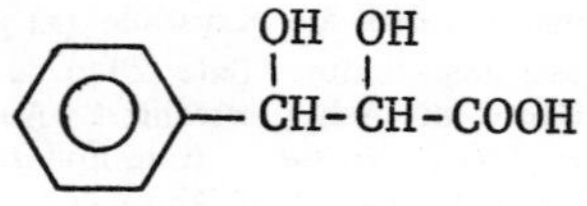

$C_9H_{10}O_4$ **Ref. 540**

Acide (+) érythro *phénylglycérique.*

[Voir également réf. (16)]. 4 g d'acide (±) **4a** et 7,34 g de strychnine dans 50 ml d'éthanol *absolu* donnent au bout d'une nuit 6,2 g de cristaux. Ces derniers sont recristallisés dans 350 ml du même solvant. Le sel obtenu (4,2 g) est décomposé par 20 ml de soude 2 N, la strychnine extraite au chloroforme. La solution aqueuse, acidifiée par 10 ml d'acide chlorhydrique 6 N et extraite à l'acétate d'éthyle (sel) donne 1,4 g d'acide (+) **4a**, F = 95-96°; $[\alpha]^{26}_{578}$ = + 28,6° (acétone). Recristallisation dans le chlorure de méthylène, F = 98,5°, $[\alpha]^{25}_{578}$ = + 30° (acétone, *c* = 0,5).

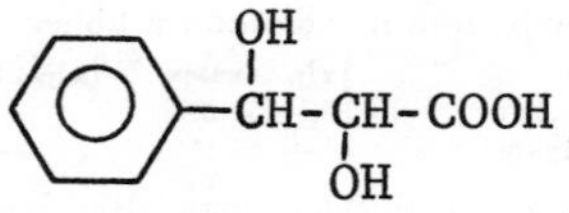

$C_9H_{10}O_4$ **Ref. 541**

Dédoublement des acides phénylglycériques.

Sauf autre indication, le solvant utilisé est l'éthanol à 95°.

Acides (+) *et* (—) thréo *phénylglycériques.*

On ajoute 5,5 g d'acide **3***a* racémique à une solution chaude de 3,68 g de (+) α-phényléthylamine dans 25 ml d'éthanol. On obtient après 12 h 3,1 g de sel que l'on décompose par la soude. L'amine est extraite à l'éther et la phase aqueuse, acidifiée, extraite à l'acétate d'éthyle, donne 1,55 g d'acide (—) **3***a*, $[\alpha]^{25}_{578}$ = — 30° (acétone, *c* = 1). Recristallisation dans 5 ml d'eau, 1,1 g, F = 166°, $[\alpha]^{25}_{578}$ = — 36° (acétone, *c* = 0,8). Les eaux-mères du sel d'α-phényléthylamine sont évaporées à sec et décomposées en 3,2 g d'acide (+) **3***a*, $[\alpha]_{578}$ = + 14° (acétone). 0,2 g de cet acide recristallisé dans 1,5 ml d'eau donne 40 mg, F = 165°, $[\alpha]^{25}_{578}$ = + 35° (acétone, *c* = 0,6). Un échantillon très pur d'acide (—) **3***a* est obtenu du chlorure de méthylène: F = 166,5°.

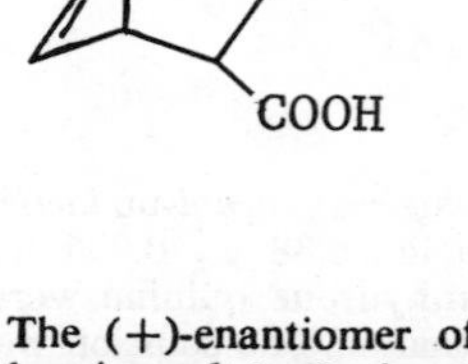

$C_9H_{10}O_4$ **Ref. 542**

Resolution of 1. The (+)-enantiomer of **1** was obtained by crystallization of its brucine salt.[8] Rather than recrystallizations from water, multiplet recrystallizations from 80% acetone–20% water by volume seemed more efficient and ultimately (after *ca.* 20 cycles) gave more highly resolved product. The acid was regenerated by treatment of the salt with 5 *M* NaOH, filtering off the brucine, and acidifying the filtrate. Recrystallization from water gave (in different preps) material of rotations $[\alpha]^{26}D$ +147 and +137° (in acetone), with mp 177–179 and 176–178° (lit.[8] $[\alpha]^{20}D$ +89°, mp 166–168°).

(8) A. T. Blomquist and E. C. Winslow, *J. Org. Chem.*, **10**, 149 (1945).

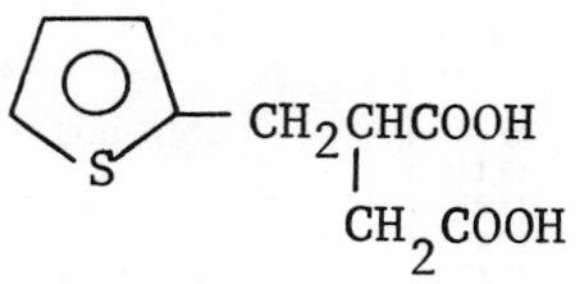

$C_9H_{10}O_4S$ Ref. 543

Resolution of the racemic acid. Preliminary experiments were made with several alkaloids and different solvents (water, acetone, ethyl acetate). The brucine salt crystallised well but gave no resolution. Quinine and cinchonidine gave likewise crystallising salts, but the acid showed very poor activity. The salts of quinidine and cinchonine were deposited as gums. The best results were obtained with α-phenylethylamine in ethyl acetate and strychnine in dilute alcohol.

26.8 g (0.125 mol) racemic acid was dissolved in 1500 ml of hot ethyl acetate and 30.3 g (0.25 mol) of (+)-α-phenylethylamine was added. The salt was deposited almost instantly and was filtered off after about 5 minutes. For recrystallisation it was dissolved in a small volume of hot methanol and precipitated by the addition of hot ethyl acetate. The progress of the resolution was tested on small samples (about 0.25 g) from each crystallisation, from which the acid was liberated and examined in acetone solution.

Crystallisation	1	2	3	4	5	6
g salt obtained	40.7	28.0	24.8	19.7	15.6	12.3
$[\alpha]_D^{25}$ of the acid	—4.5°	—11.9°	—12.8°	—14.4°	—14.6°	—14.7°

By evaporating the mother liquors from the two first crystallisations, 23.6 g salt was obtained. This salt yielded 10.7 g acid, showing $[\alpha]_D^{25} = + 11°$. The acid (0.05 mol) was dissolved together with 33.4 g (0.1 mol) of strychnine in hot dilute alcohol (34 % alcohol and 66 % water); the salt was recrystallised once.

Crystallisation	1	2
ml 34 % alcohol	1900	1170
g salt obtained	36.6	29.4
$[\alpha]_D^{25}$ of the acid	+ 14.5°	+ 14.7°

(—)-*α-Thenylsuccinic acid.* The phenylethylamine salt (12.7 g) was decomposed with dilute sulphuric acid and the active acid was extracted with ether. The ether was evaporated at room temperature. The acid was decolourised with charcoal in acetone solution and finally recrystallised twice from water. The yield of pure acid was 4.7 g and the m. p. 156.5°. The acid forms long, pointed prisms or needles with parallel extinction. It is very soluble in alcohol and acetone, but only moderately so in ethyl acetate, acetic acid and ether.

0.1522 g acid: 15.55 ml 0.0914-N NaOH.
$C_9H_{10}O_4S$ (214.2) Equiv.wt. calc. 107.1 found 107.4
0.0702 g made up to 15.00 ml with *water*: $3\alpha = -0.33°$. $[\alpha]_D^{25} = -23.3°$; $[M]_D^{25} = -50.5°$. — 0.1826 g *neutralised* with NaOH and made up to 10.00 ml with *water*: $2\alpha = -0.80°$. $[\alpha]_D^{25} = -21.9°$; $[M]_D^{25} = -46.9°$. — 0.1855 g made up to 10.00 ml with *acetone*: $2\alpha = -0.55°$. $[\alpha]_D^{25} = -14.8°$; $[M]_D^{25} = -32°$. — 0.1590 g made up to 10.00 ml with *ethyl acetate*: $2\alpha = -0.39°$. $[\alpha]_D^{25} = -12.3°$; $[M]_D^{25} = 26.5°$. — 0.1506 g made up to 10.00 ml with absolute *alcohol*: $2\alpha = -0.23°$; $[\alpha]_D^{25} = -7.6°$; $[M]_D^{25} = -16.5°$. — 0.2313 g made up to 10.00 ml with *ether*: $2\alpha = -0.325°$. $[\alpha]_D^{25} = -7.0°$; $[M]_D^{25} = -15.1°$. — 0.2079 g made up to 10.00 ml with glacial *acetic acid*: $2\alpha = -0.055°$. $[\alpha]_D^{25} = -1.3°$; $[M]_D^{25} = -2.8°$.

The (+)-α-phenylethylamine salt obtained in the course of the resolution formed microscopic needles. It was not further examined.

(+)-*α-Thenylsuccinic acid.* The strychnine salt (29.4 g) was decomposed with 2-N sodium hydroxide solution and the strychnine was extracted with chloroform. The aqueous solution was acidified with sulphuric acid in excess and the active acid was extracted with ether. It was decolourised and recrystallised in the same way as the (—)-isomeride, which it closely resembles in appearance and properties. The yield was 5.8 g having m. p. 156.5°.

0.1564 g acid: 15.98 ml 0.0914-N NaOH.
$C_9H_{10}O_4S$ (214.2) Equiv.wt. calc. 107.1 found 107.1
0.1871 g made up to 10.00 ml with acetone: $2\alpha = + 0.55°$. $[\alpha]_D^{25} = + 14.7°$; $[M]_D^{25} = + 31.5°$.

The strychnine salt obtained in the course of the resolution was not further examined.

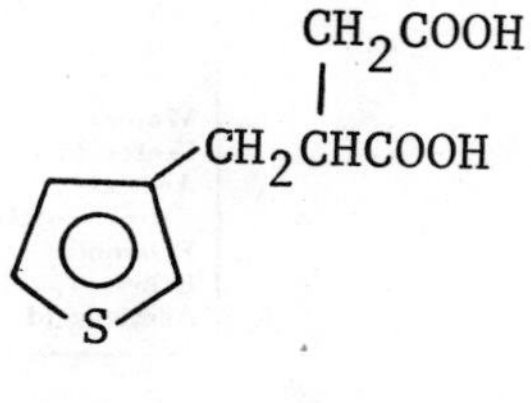

$C_9H_{10}O_4S$ Ref. 544

Preliminary tests on resolution.—0.0025 moles of optically inactive acid and 0.005 moles of the base were dissolved in 25 ml of dilute ethanol. The solutions were allowed to stand and some water gradually added until crystals were formed. In some cases the salts were too soluble to precipitate even when a large excess of water was added. Experiments with (+)-α-phenylethylamine and (+)-benzedrine were made in ethyl acetate. The results are given below.

Base	$[\alpha]_D^{25}$ of acid in acetone
Strychnine	+ 14.7°
Brucine	+ 3.5°
Quinine	± 0°
Cinchonine	s
Cinchonidine	± 0°
Chinidine	s
(+)-α-Phenylethylamine	− 9.4°
(+)-Benzedrine	+ 5.7°

Optical resolution of 3-thenylsuccinic acid.—24.65 g (0.115 moles) of 3-thenylsuccinic acid and 76.8 g (0.230 moles) of strychnine were dissolved in 3300 ml of 33 % ethanol and the solution left in a refrigerator for 24 hours. The precipitate was filtered off, washed with cold dilute ethanol, and dried. About 0.3 g of the salt was set apart, the acid isolated and the optical activity measured in acetone. The remaining salt was recrystallised from dilute ethanol and the process repeated until the optical activity of the acid remained constant. The course of the resolution is seen below.

Crystallisation	1.	2.	3.	4.	5.
g salt	52.0	39.0	27.0	21.0	17.2
ml 33 % ethanol	3400	3200	1800	900	800
$[\alpha]_D^{25}$ of acid	+ 13.9°	+ 16.8°	+ 18.0°	+ 17.9°	+ 17.9°

The strychnine salt obtained in the course of resolution, which crystallised in heavy plates, was not further examined.

The mother liquor from the first crystallisation with strychnine was evaporated at room temperature and 39.0 g salt was obtained. This salt yielded 8.76 g acid, having $[\alpha]_D^{25} = -14.1°$. This acid was dissolved in 450 ml of hot ethyl acetate and 9.92 g of (+)-α-phenylethylamine was added. The salt was deposited almost instantly and was filtered off when the solution had cooled. The progress of the resolution is seen below.

Crystallisation	1.	2.	3.
g, salt	17.8	14.5	11.6
ml ethyl acetate	450	450	450
$[\alpha]_D^{25}$ of acid	− 16.3°	− 18.0°	− 18.0°

The phenylethylamine salt (which crystallised in fine needles) obtained in the course of the resolution was not further examined.

(+)-*3-Thenylsuccinic acid.*—The strychnine salt (17.2 g) was decomposed with 2 *N* sodium hydroxide and the strychnine extracted four times with chloroform. The aqueous phase was acidified with hydrochloric acid and the active acid extracted six times with ether. The ether was evaporated at room temperature and the acid recrystallised three times from water and decolourised with charcoal. The yield of pure acid was 3.0 g and the m.p. 159.0°. The acid crystallised in rectangular plates.

0.07489 g acid: 9.57 ml 0.07336 *N* NaOH.
$C_9H_{10}O_4S$ (214.2). Equiv. wt. calc. 107.1, found 106.7.
0.2102 g acid made up to 10.00 ml with acetone:
$2\alpha_D^{25} = + 0.84°$, $[\alpha]_D^{25} = + 20.0°$, $[M]_D^{25} = + 42.8°$.

(−)-*3-Thenylsuccinic acid.*—11.6 g of the (+)-α-phenylethylamine salt was decomposed with 2 *N* sulphuric acid and the active acid extracted six times with ether. The ether was evaporated and the acid decolourised and recrystallised in the same way as the antipode which it resembles in appearance and properties. The yield was 4.8 g having m.p. 159.0°.

0.07056 g acid: 8.97 ml 0.07325 *N* NaOH.
$C_9H_{10}O_4S$ (214.2). Equiv. wt. calc. 107.1, found 107.4.

Weighed amounts of acid were made up to 10.00 ml with different solvents and the optical activity measured. The results are given below.

Solvent	mg of acid	$2\alpha_D^{25}$	$[\alpha]_D^{25}$	$[M]_D^{25}$
Water	41.43	−0.20°	−24.1°	−51.6°
Water (neutr.)	84.55	−0.34°	−20.1°	−43.1°
Acetone	182.5	−0.73°	−20.0°	−42.8°
Ethyl acetate	180.5	−0.71°	−19.7°	−42.2°
Ethanol	176.5	−0.53°	−15.0°	−32.1°
Ether	180.3	−0.50°	−13.9°	−29.8°
Acetic acid	175.4	−0.33°	−9.4°	−20.1°

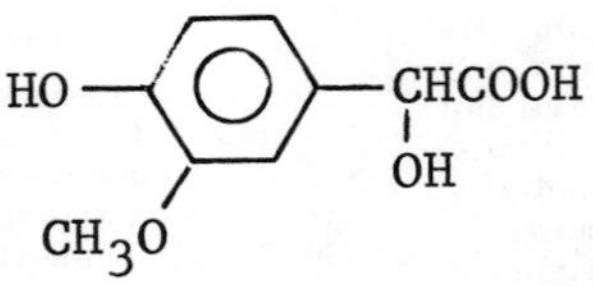

$C_9H_{10}O_5$ **Ref. 545**

Authentic DL-(I) was prepared by the method of GARDNER AND HIBBERT[4]; m.p. 129–130° dec.§. L- and D-(I) were prepared by fractional crystallization of the cinchonine salt of DL-(I). The first crop of less soluble salt was enriched in D-(I) and the mother liquor was enriched in L-(I). Impure L- and D-(I) were regenerated from their crude salts and recrystallized (acetonitrile) to constant rotation. L-(I), m.p. 151° dec., $[\alpha]_D^{25}$, +133° (1 % in water); D-(I), m.p. 152° dec., $[\alpha]_D^{23}$, −133° (1 % in water). The cinchonine salt of D-(I) was recrystallized (abs. ethanol) to constant rotation; $[\alpha]_D^{24}$, +89° (1 % in EtOH), m.p. 203–204° dec. D-(I) was regenerated from this salt; $[\alpha]_D^{23}$, −131° (1 % in water), m.p. 152° dec. L-(I), m.p. 152° dec., $[\alpha]_D^{22}$, +128° (0.7 % in water) was also prepared by enzymic hydrolysis of DL-(I)-amide with purified leucine aminopeptidase.

[4] J. A. F. GARDNER AND H. HIBBERT, *J. Am. Chem. Soc.*, 66 (1944) 607.

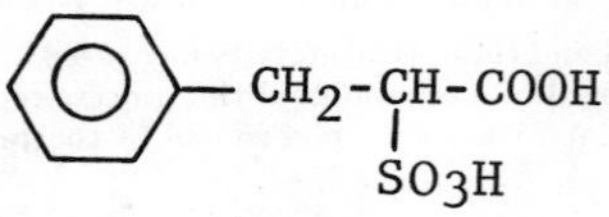

$C_9H_{10}O_5S$ **Ref. 546**

La majeure partie du sel potassique a été transformée en sel de quinine, afin de réaliser le dédoublement optique. Dans ce but 75 cm³ du mélange réactionnel sont dilués avec 1 litre d'eau bouillante et ajoutés à une solution chaude de 64.8 g de quinine (0.2 mol.), 200 cm³ d'alcool, 100 cm³ d'acide chlorhydrique 2 n et 1 litre d'eau. Le β-phényl-α-sulfopropionate de quinine cristallise aussitôt en petites aiguilles. Poids 80 g = 0.09 mol. Point de fusion 228°.

Sel de quinine. 0.5180 g; perte à 110° 0.0151 g. Subst. 0.3255 g: $BaSO_4$ 0.0711 g. Subst. 0.7342 g; 15.95 cm³ de NaOH 0.1027 n
Trouvé: p. équiv. 448.2; H_2O 2.91; S 3.00
Calculé: „ „ 452.8; „ 2.98; „ 3.54
$C_9H_{10}O_5S \cdot 2\,C_{20}H_{24}O_2N_2 \cdot 1^1/_2\,H_2O$ (905.57).

Recristallisée dans l'alcool à 50 %, la substance renferme un peu moins d'eau: 2.62 %. Le sel de quinine est recristallisé à plusieurs reprises dans l'alcool de 50 %. A partir de la quatrième cristallisation, une petite quantité est décomposée chaque fois par la quantité calculée de soude caustique; la solution du sel sodique est examinée dans un polarimètre. Les rotations moléculaires observées sont respectivement: +49, +61, +70, +80, +92, +97½, +104, +108.9, +109.1° (λ = 589.3 μμ). Les eaux-mères, évaporées dans le vide, ont donné chaque fois une seconde fraction du sel de quinine. Quelques-unes de ces fractions ont été examinées encore après leur transformation en sel sodique. Les rotations moléculaires observées sont respectivement: −103, −59, −25, +4, +52, +65, +89, +107° (λ = 589.3 μμ).

Après onze cristallisations le dédoublement est donc complet. La quantité du sel de quinine de la onzième cristallisation était encore 6 g. Une solution, renfermant 0.00513 mol. g dans 100 cm³ a servi à la détermination de la dispersion rotatoire. Tube de 4 dm.

Dispersion rotatoire du sel sodique à 20°.

λ(μμ)	626.5	589.3	560.5	536.5
α	1.96°	2.24	2.54	2.85
[M]	95.5°	109.1	123.8	138.9

Il nous a paru intéressant de savoir comment la rotation change quand on ajoute l'acide sulfurique dilué à une solution du sel sodique actif. La solution contient 0.00128 mol g dans 25 cm³. Il faut donc ajouter 2.47 cm³ d'acide sulfurique 1.038 n, pour mettre en liberté l'acide β-phényl-α-sulfopropionique. Le tableau et la figure montrent que l'addition du premier équivalent d'acide sulfurique (1.24 cm³) fait diminuer la rotation de 109° jusqu' à 78° (rotation du sel acide); donc une diminution de 31°, tandis que la diminution provoquée par le second équivalent n'est que 5°. Cela permet d'affirmer davantage encore que la dissociation du groupe carboxylique est la cause du changement de la rotation. On trouve pour la rotation de l'acide β-phényl-α-sulfopropionique: $[M]_D$ = 73°, $[M]_{acide} : [M]_{sel}$ = 0.67.

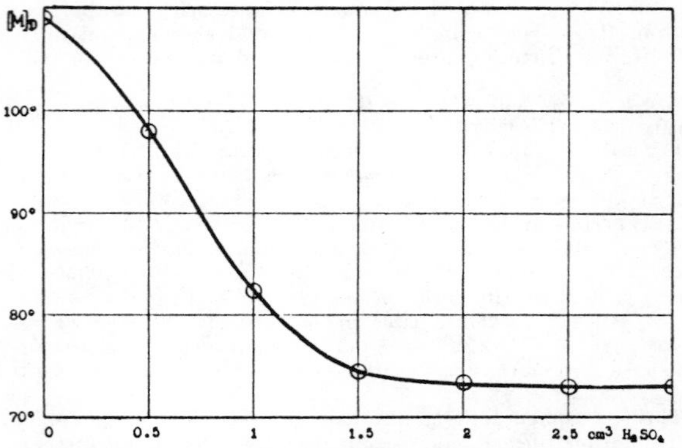

Changement de la rotation moléculaire du β-phényl-α-sulfopropionate de potassium, par l'addition d'acide sulfurique.

25 cm³ du sel potassique (c = 0.00513) avec de l'acide sulfurique 1.038 n. Tube de 4 dm.

H_2SO_4 cm³	0	½	1	1½	2	2½	3
α	2.235°	1.97	1.63	1.44	1.395	1.36	1.34
$[M]_D$	108.9°	97.9	82.7	74.4	73.5	73.0	73.1

Dispersion rotatoire de l'acide (c = 0.0045; tube de 2 dm).

λ(μμ)	626.5	589.3	560.5	536.5
α	0.555°	0.65	0.73	0.81
[M]	61.7°	72.2	81.1	90.0

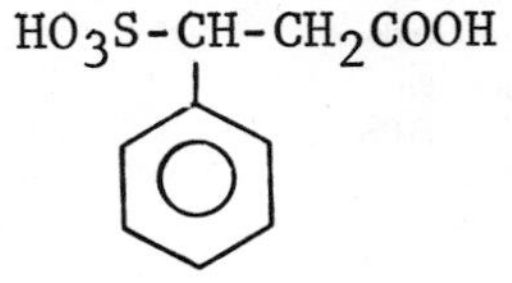

$C_9H_{10}O_5S$ **Ref. 547**

On peut préparer directement les sels de quinine, de strychnine et de brucine en ajoutant leurs solutions dans l'acide acétique dilué, à une solution neutralisée du sel potassique acide.

Les sels de ces trois alcaloïdes sont peu solubles dans l'eau froide et donnent par recristallisation répétée le sel barytique *lévogyre*.

Sel de quinine.

Substance 0.2048 g: perte à 115° 0.0209 g. Subst. 0.2268 g: $BaSO_4$ 0.0528 g.
Trouvé: H_2O 10.25: S 3.20.
$C_9H_{10}O_5S$. 2 $C_{20}H_{24}O_2N_2$. 5½ H_2O (977.65). Calculé: „ 10.14: „ 3.28.

Sel de strychnine.

Substance 0.1786 g: perte à 125° 0.0121 g. Subs. 0.5503 g: $BaSO_4$ 0.1314 g.
„ 0.4122 „: 8.51 cm³ de NaOH 0.1008 n.
Trouvé: p. équiv. 481: H_2O 6.77: S 3.28.
$C_9H_{10}O_5S$. 2 $C_{21}H_{22}O_2N_2$. 3½ H_2O (961.59). Calculé: p. équiv. 480.8; H_2O 6.56; S 3.33.

Sel de brucine.

Substance 0.4950 g: perte à 125° 0.0484 g. Subst. 0.3891 g:
6.87 cm³ de NaOH 0.1008 n.
Trouvé: p. équiv. 5.62: H_2O 9.78.
$C_9H_{10}O_5S$. 2 $C_{23}H_{26}O_4N_2$. 6 H_2O (1126.69). Calculé: „ 563.3: „ 9.60.

Le sel de quinine a servi au dédoublement optique. Nous l'avons fait recristalliser plusieurs fois dans l'alcool de 50 %, en examinant chaque fois le pouvoir rotatoire du sel barytique.

Le tableau suivant montre l'accroissement de la rotation.

Cristallisation No.	1	2	3	4	5	6	7
Poids du sel de quinine	85 g	46	28	20	15	12	10
Sel de baryum $[M]_D$		—3.5°	—7.7	—11.6	—13.2	—14.5	—14.8

Après sept cristallisations le dédoublement a été complet.

Une solution, renfermant 0.00549 mol. g dans 100 cm³ a servi à la détermination de la dispersion rotatoire. Tube de 4 dm.

Dispersion rotatoire du sel barytique.

λ	674.0	626.5	589.3	560.5	536.5	516.0
α	—0.24°	—0.28	—0.325	—0.375	—0.435	—0.49
[M]	—10.9°	—12.8	—14.8	—17.1	—19.8	—22.3

Afin de connaître le pouvoir rotatoire de l'acide libre nous avons décomposé une solution du sel barytique par la quantité nécessaire d'acide sulfurique.

La solution filtrée et concentrée, renfermait 0.01098 mol. g dans 100 cm³. Tube de 4 dm.

Dispersion rotatoire de l'acide β-sulfo-β-propionique.

λ	674.0	626.5	589.3	560.5	536.5
α	+0.115°	+0.135	+0.14	+0.14	+0.145
[M]	+2.6°	+3.1	+3.2	+3.2	+3.3

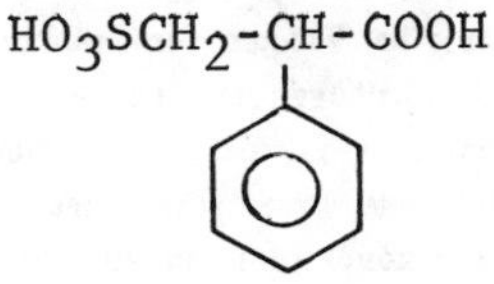

$C_9H_{10}O_5S$ **Ref. 548**

La préparation d'autres sels se fait le mieux par l'intermédiaire d'alcaloïdes qui donnent des sels cristallisant excellemment.

On les prépare en ajoutant les alcaloïdes, dissous dans l'acide acétique dilué, à une solution neutralisée du sel potassique acide que nous venons de décrire.

Le *sel de strychnine* cristallise avec 3 mol. d'eau.

Substance 0.1749 g: 3.65 cm³ de NaOH 0.1003 n.
„ 0.5663 „: perte à 110° 0.0327 g.
„ 0.1915 „: $BaSO_4$ 0.0489 g.
Trouvé: p. éq. 478; H_2O 5.77: S 3.51.
$C_9H_{10}O_5S$. 2 $C_{21}H_{22}O_2N_2$. 3 H_2O (952.58). Calculé: p. éq. 476.29; „ 5.67; „ 3.37.

Le *sel de brucine* cristallise avec 6 mol. d'eau.

Substance 0.4303 g: perte à 130° 0.0430 g: Subst. 0.1843 g: $BaSO_4$ 0.0404 g.
Trouvé: H_2O 9.99; S 3.01.
$C_9H_{10}O_5S$. 2 $C_{23}H_{26}O_4N_2$. 6 H_2O (1126.69). Calculé: „ 9.60: „ 2.85.

Les deux sels peuvent servir à préparer les énantiomorphes dextrogyre et lévogyre.

Le sel de strychnine est recristallisé dans l'eau à plusieurs reprises et chaque fois on en transforme une partie en sel barytique, dont on détermine le pouvoir rotatoire.

Après 10 cristallisations le sel de strychnine était optiquement pur; il renfermait alors 2 mol. d'eau de cristallisation.

A l'aide de baryte on obtient le *sel barytique dextrogyre*, cristallisant avec une molécule d'eau.

Substance 0.3056 g; perte à 130° 0.0149 g. Subst 0.1753 g: $BaSO_4$ 0.1080.
Trouvé: H_2O 4.87; S 8.46.
d-$C_9H_8O_5SBa$. H_2O (383.52). Calculé: „ 4.70; „ 8.36.

La rotation optique a été mesurée à 20° pour une solution renfermant 0.0043 mol. g dans 100 cm³. Tube de 2 dm. $[M]_D = +185°$.

$\lambda(\mu\mu)$	674.0	626.5	589.3	560.5	536.5	516.0
α(20°)	+1.19°	+1.39	+1 59	+1.80	+2.00	+2.21
[M]	+ 138°	+162	+185	+209	+233	+257

Le sel de brucine, recristallisé plusieurs fois, donne le sel barytique lévogyre. Après 8 cristallisations une solution renfermant 0.0047 mol. g dans 100 cm³ avait pour rotation α_D (2 dm) = —1.67°. Donc $[M]_D$ = —178°.

Les sels ont été examinés à plusieurs concentrations, afin de reconnaître l'influence de la dilution.

sel	mol. g dans 100 cm³	α_D (2 dm)	$[M]_D$
barytique-d	0.0049	+1.80°	+184°
„	0.0043	+1.59°	+185°
sodique-d	0.0023	+0.90°	+193°
barytique-l	0.0047	—1.67°	—178°
sodique-l	0.0022	—0.82°	—184°

La rotation augmente donc un peu par la dilution.

On observe une très faible *racémisation* en chauffant une solution alcaline renfermant des quantités équimoléculaires du sel sodique et de soude caustique. Après 14 heures de chauffe à 100° la rotation s'était abaissée de —0.79 jusqu'à —0.76°.

Le sel barytique racémique ne renferme qu'une demi-molécule d'eau.

Substance 0.4788 g: perte à 130° 0.0123 g. Subst. 0.1514 g: $BaSO_4$ 0.0952 g.
Trouvé: H_2O 2.57: Ba 37.00.
$C_9H_8O_5SBa$. ½ H_2O (374.51). Calculé: „ 2.41: „ 36.69

Une solution de l'acide dextrogyre, préparée par décomposition du sel barytique-d à l'aide de l'acide sulfurique, a présenté la rotation α_D (2 dm) = +0.95° pour une concentration de 0.0024 mol. g dans 100 cm³. Donc $[M]_D$ = +194°.

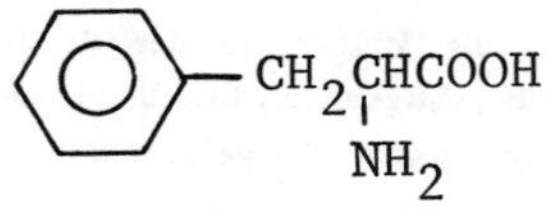

$C_9H_{11}NO_2$ **Ref. 549**

Formyl-d-phenylalanin.

50 g Formyl-dl-phenylalanin und 102 g wasserfreies Brucin (gleiche Mol.) werden beide gepulvert, gut durchgemengt, mit 600 ccm siedendem, trocknem Methylalkohol übergossen und das Lösen auf dem Wasserbade vervollständigt. Nach wenigen Minuten beginnt bereits in der Hitze das Brucinsalz des Formyl-d-phenylalanins krystallinisch auszufallen. Nach 12-stündigem Stehen im Eisschranke, wobei die Lösung gegen Feuchtigkeit zu schützen ist, sind etwa 75 g Brucinsalz auskrystallisirt.

Die Masse besteht zu etwa 95 pC. aus dem Salz des Formyl-d-phenylalanins. Zur Reinigung werden die 75 g Brucinsalz aus 1,5 Liter siedendem Methylalkohol umgelöst, wobei ungefähr 60 g in der Kälte ausfallen.

Weiteres Umkrystallisiren des Salzes hat keinen Zweck, da das Drehungsvermögen des Formylkörpers dadurch nicht mehr erhöht wird.

Um den letzteren aus dem Salz zu gewinnen, löst man 60 g in 500 ccm Wasser von 60°, kühlt dann rasch auf etwa 5° ab und giebt, bevor das Brucinsalz wieder auskrystallisirt, die berechnete Menge (104 ccm) abgekühlte Normalnatronlauge hinzu.

Beim Eingiessen der Natronlauge entsteht ein weisser Niederschlag, welcher aber sofort wieder verschwindet; nach etwa einer Minute beginnt dann die Ausscheidung des Brucins, meist in kugelförmigen Aggregaten; bei gutem Rühren ist in 20 Minten fast alles Brucin ausgefallen, wobei die Lösung zu einem dicken, weissen Brei erstarrt. Nach sorgfältigem Abnutschen wird mit wenig eiskaltem Wasser ausgewaschen und sofort mit 20 ccm fünffach Normalsalzsäure der grösste Theil des Formyl-d-phenylalanins in Freiheit gesetzt, welches sich in Form von Nadeln oder Blättchen abscheidet. Die nach einer Stunde bei 0° auskrystallisirte Menge betrug nach dem Trocknen im Vacuum 15,3 g oder 77 pC. der Theorie.

Die Mutterlauge wurde im Vacuum unter 20 mm Druck auf 40 ccm eingedampft, durch Ausschütteln mit Chloroform vom Rest des Brucins befreit und durch überschüssige Salzsäure (5 ccm fünffach normal) der Rest des Formylkörpers gefällt.

Die Ausbeute an rohem Formyl-d-phenylalanin betrug 19 g oder 76 pC., bezogen auf das racemische Ausgangsmaterial.

Zur Analyse wurde aus der fünffachen Menge heissem Wasser umkrystallisirt und im Vacuum über Schwefelsäure getrocknet.

0,1735 g gaben 0,3946 CO_2 und 0,0901 H_2O.
0,2369 g „ 15,1 ccm Stickgas über 33-procentigem KOH bei 20° und 759 mm Druck.

Berechnet für $C_{10}H_{11}NO_3$ (193,09)		Gefunden
C	62,15	62,02
H	5,74	5,81
N	7,26	7,30

Zur optischen Bestimmung diente die alkoholische Lösung.

0,2566 g Substanz. 6,1106 g Gewicht der Lösung. Drehung im 2 dm-Rohre bei 20° und Natriumlicht 5,09 nach links.

$$d_4^{20} = 0{,}8034;\ [\alpha]_D^{20} = -75{,}43° (\pm 0{,}2°).$$

Durch weiteres Umkrystallisiren wurde keine höhere Drehung erzielt.

Formyl-l-phenylalanin.

Lässt man die Mutterlauge des Brucinsalzes unter Abschluss der Luftfeuchtigkeit im Eisschranke stehen, so schiessen nach zwei Tagen warzenförmige Krystalle an, deren Menge nach acht Tagen etwa 50 pC. des noch im Methylalkohol enthaltenen Brucinsalzes beträgt; die Krystallisation tritt indessen nur ein, wenn der Methylalkohol wasserfrei ist.

Um den Rest des Brucinsalzes zu gewinnen, wird die methylalkoholische Lösung unter vermindertem Druck zur Trockne verdampft und der Rückstand in der achtfachen Menge Wasser von 60° gelöst. Beim 8—10-tägigen Stehen dieser Flüssigkeit im Eisschranke fällt noch eine reichliche Menge des Brucinsalzes aus; die Gesammtmenge desselben beträgt ungefähr 70 pC. der Theorie.

Zur Reinigung wurde das Brucinsalz wiederum in der achtfachen Menge warmem Wasser gelöst und die Lösung nach dem Abkühlen auf 0° und Einimpfen einiger Kryställchen mehrere Tage im Eisschranke aufbewahrt. Die Abscheidung erfolgt dann so vollständig, dass kaum 10 pC. des Salzes in der Mutterlauge bleiben.

Die Gewinnung des freien Formyl-l-phenylalanins geschah in der gleichen Weise wie bei dem Antipoden. Nach einmaligem Umkrystallisiren des Rohproductes aus der zehnfachen Menge heissem Wasser war die specifische Drehung noch zu gering (72°), nachdem aber die Krystallisation in der gleichen Weise wiederholt war, wurde fast derselbe Werth wie bei dem Antipoden gefunden.

Zur Analyse war ebenfalls im Vacuum über Schwefelsäure getrocknet.

0,1672 g gaben 0,3803 CO_2 und 0,0862 H_2O.
0,1909 g „ 12 ccm Stickgas über 33-procentigem KOH bei 17° und 770 mm Druck.

Berechnet für $C_{10}H_{11}NO_3$ (193,09)		Gefunden
C	62,15	62,03
H	5,74	5,77
N	7,26	7,40

Optische Bestimmung.

0,2540 g Substanz. Gesammtgewicht der alkoholischen Lösung 6,0896 g. Spec. Gew. 0,8030 g. Drehung bei 20° und Natriumlicht im 2 dm-Rohre 5,04° nach rechts.

$$\text{Mithin } [\alpha]_D^{20} = +75{,}2° (\pm 0{,}2°).$$

Der Werth ist nahezu identisch mit der spec. Drehung des optischen Antipoden; auch in den sonstigen physikalischen Eigenschaften zeigen beide gute Uebereinstimmung.

Sie erweichen gegen 163° (corrigirt) und schmelzen gegen 167° (corrigirt). Aus warmem Wasser krystallisiren sie in sehr schiefen, vierseitigen Täfelchen, die aber häufig durch Abstumpfung der spitzen Ecken sechsseitig erscheinen. Sie lösen sich im allgemeinen etwas leichter als der Racemkörper. Zum Vergleich führen wir einen Versuch über die Löslichkeit des Formyl-d-phenylalanins in Wasser von 27° an, wobei ungefähr 145 Theile Wasser zur Lösung erforderlich waren. Bemerkenswerth ist der seidenartige Glanz, den die Krystalle der activen Formen im optisch reinen Zustande haben, den sie aber durch die Beimengung einer verhältnissmässig geringen Menge des Racemkörpers verlieren.

Darstellung
von activem Phenylalanin aus der Formylverbindung.

Die Abspaltung der Formylgruppe erfolgt sehr leicht beim Kochen mit verdünnten Säuren; für die spätere Isolirung des Phenylalanins ist die Verwendung von Bromwasserstoffsäure zu empfehlen.

Dementsprechend werden 10 g active Formylverbindung mit 150 g Normalbromwasserstoffsäure eine Stunde am Rückflusskühler gekocht. Dann dampft man unter 10—15 mm Druck zur Trockne, wobei sich das Bromhydrat in prächtigen Nadeln abscheidet. Will man es reinigen, so nimmt man mit der doppelten Menge absolutem Alkohol auf und fällt mit viel Aether, wobei sich das Salz in seideglänzenden Nadeln abscheidet. Ausbeute 95 pC. der Theorie.

Aus dem Bromhydrat lässt sich die Aminosäure auf zwei Arten isoliren. Man nimmt entweder in der fünffachen Menge Wasser auf, versetzt mit wässrigem Ammoniak in geringem Ueberschuss, dampft im Vacuum zur Trockne und entfernt das Bromammon durch Auslaugen mit heissem absolutem Alkohol.

Oder bequemer ist es, das rohe beim Verdampfen der Lösung zurückbleibende Bromhydrat direct in der 30-fachen Menge absolutem Alkohol zu lösen und die Aminosäure durch tropfenweises Zugeben von starkem wässrigem Ammoniak in geringem Ueberschuss auszufällen. Das Bromammon bleibt im Alkohol gelöst und das Phenylalanin scheidet sich als dicker Brei ab, der aus mikroskopischen, verfilzten Nädelchen besteht; durch Auswaschen mit warmem Alkohol lässt es sich leicht von Spuren Bromammon befreien.

Aus 5 g activem Formylkörper wurden so 3,8 g Phenylalanin gewonnen, was 89 pC. der Theorie entspricht. Zur völligen Reinigung wird in etwa 30 Theilen heissem Wasser gelöst; beim Eindampfen scheidet es sich in schönen glänzenden Blättchen ab.

Zur Analyse wurde im Vacuum über Schwefelsäure getrocknet.

d-Phenylalanin.

0,1390 g Substanz. 6,8342 g Gewicht der Lösung. d_4^{20} = 1,0045, Drehung im 2 dm-Rohre bei 20° und Natriumlicht 1,43° nach rechts.

$$[\alpha]_D^{20} = +35,0° (\pm 0,5°).$$

Nach nochmaligem Umlösen wurde folgender Werth gefunden:

0,1259 g Substanz. 7,2187 g Gewicht der Lösung. d_4^{20} = 1,0035, Drehung im 2 dm-Rohre bei 20° und Natriumlicht 1,23° nach rechts.

$$[\alpha]_D^{20} = +35,14° (\pm 0,5°).$$

Die specifische Drehung war somit innerhalb der Fehlergrenze constant geblieben.

l-Phenylalanin.

0,1357 g Substanz. 7,0397 g Gewicht der Lösung. d_4^{20} = 1,0040, Drehung im 2 dm-Rohre bei 20° und Natriumlicht 1,36° nach links.

$$[\alpha]_D^{20} = -35,14° (\pm 0,5°).$$

Nach nochmaligem Umlösen wurde folgender Werth gefunden:

0,1295 g Substanz. 7,1892 g Gewicht der Lösung. d_4^{20} = 1,0038 Drehung im 2 dm-Rohre bei 20° und Natriumlicht 1,27° nach links.

$$[\alpha]_D^{20} = -35,09° (\pm 0,5°).$$

Die für die Drehung der beiden Phenylalanine gefundenen Werthe stimmen demnach innerhalb der Fehlergrenze überein:

l-Phenylalanin	d-Phenylalanin
− 35,14°	+ 35,00°
− 35,09°	+ 35,14°.

E. Fischer und Mouneyrat[3]) fanden für d-Phenylalanin aus der Benzoylverbindung

$$[\alpha]_D^{16} = +35,08° (\pm 0,5°).$$

[3]) Ber. d. deutsch. chem. Ges. **33**, 2383 (1900).

$C_9H_{11}NO_2$ Ref. 549a

Spaltung des *d,l*-Phenylalanin-äthylesters mit Dibenzoyl-*d*-weinsäure: 7 g *d,l*-Phenylalanin-äthylester werden bei 21° mit einer Lösung von 6.8 g Dibenzoyl-*d*-weinsäure in 50 ccm absol. Äthanol vereinigt. Nach etwa 30 Min. beginnen sich farblose, fadenförmige Kristalle abzuscheiden. Nach 2stdg. Stehenlassen wird abgesaugt und getrocknet: *d*-Phenylalanin-äthylester-dibenzoyl-d-tartrat; Schmp. 137—139°, $[\alpha]_D^{17}$: −71.4° (c = 0.2872, in Äthanol). Liegt der Drehwert unterhalb von −71.0° ± 1°, so wird aus absol. Isopropanol unter Verwendung von 15 ccm auf 1 g des Salzes umkristallisiert.

Zur Hydrolyse wird das Salz mit 50 ccm 10-proz. Salzsäure, u. U. unter Erwärmen, geschüttelt, von ausgeschiedener Dibenzoyl-*d*-weinsäure abfiltriert, deren noch in Lösung befindlicher Rest mit Äther extrahiert und das gesamte Wasser i. Vak. abgedampft. Aus dem zurückgebliebenen Hydrochlorid wird das *d*-Phenyl-alanin mit 15-proz. Ammoniak in Freiheit gesetzt; $[\alpha]_D^{20}$: +32.7° (*c* = 1.4816, in Wasser). E. Fischer und W. Schoeller[5]) ermittelten für das natürliche *l*-Phenyl-alanin den Wert $[\alpha]_D^{20}$: −35.1°.

Von der Mutterlauge wird möglichst sofort das Lösungsmittel i. Vak. abgedampft. Der getrocknete Rückstand schmilzt bei 143—145°, $[\alpha]_D^{22}$: −30.1° (*c* = 0.2828, in Äthanol). Hydrolyse wie oben beschrieben: *l*-Phenyl-alanin, $[\alpha]_D^{20}$: −25.6° (*c* = 0.528, in Wasser).

In vier Ansätzen wurde auf Reproduzierbarkeit geprüft:

		Schmp.	$[\alpha]_D$	*d*-Phenyl-alanin
1. Fraktion:	1.	136—139°	−62.6°	+32.6°
	2.	134—139°	−71.0°	+31.8°
	3.	139—144°	−64.2°	+33.9°
	4.	136—144°	−67.4°	+34.0°

		Schmp.	$[\alpha]_D$	*l*-Phenyl-alanin	Aufgearb. nach:
2. Fraktion:	1.	139—142°	−45.8°	−15.6°	48 Stdn.
	2.	149—154°	−48.6°	− 9.4°	96 Stdn.
	3.	145—149°	−47.1°	−21.0°	24 Stdn.
	4.	149—152°	−39.1°	−25.8°	4 Stdn.

$C_9H_{11}NO_2$ Ref. 549b

D. Racematspaltung des Phenylalanin-hydrazides

Die Lösung von 5,0 g D,L-Phenylalanin-hydrazid in 100 cm^3 abs. Äthanol wurde bei 20° mit einer Lösung von 10,5 g Dibenzoyl-D-weinsäure in 120 cm^3 abs. Äthanol versetzt. Nach ungefähr 10 Minuten bildeten sich die ersten Kristallkeime, nach 16-18 Stunden hatte sich das D-Phenylalanin-hydrazid-dibenzoyl-D-hydrogentartrat vollständig ausgeschieden.

Ausbeute: 7,6 g; Schmp. 158-159°
$(\alpha]_D^{23}$: -92,5° (c = 0,40 in abs. Methanol).

Aus der Mutterlauge des Spaltansatzes kristallisierte das L-Hydrazidsalz nach 4-5 tägigem Stehen aus.

Ausbeute: 5,0 g; Schmp. 190-191°.
$[\alpha]_D^{23}$: -58,5° (c = 0,35 in abs. Methanol).

Gewinnung der antipodischen Phenylalaninhydrazid-dihydrochloride

Zur Verbesserung der optischen Reinheit wurden die diastereomeren Hydrazidtartrate zweimal aus Alkohol-Ather umkristallisiert:

D-Phenylalanin-hydrazid-dibenzoyl-D-hydrogentartrat:
Ausbeute: 50% d. Th., bezogen auf Rohtartrat; Schmp. 163°.
$[\alpha]_D^{23}$: -94,7° (c = 0,41 in abs. Methanol).

Durch fünfstündiges Erhitzen des D-Hydrazid-dibenzoyl-D-hydrogentartrates in der Trockenpistole uber P_2O_5 auf 100° und im Vakuum liess sich das wasserfreie Salz erhalten.

L-Phenylalanin-hydrazid-dibenzoyl-D-hydrogentartrat:
Ausbeute: 50% d. Th., bezogen auf Rohtartrat; Schmp. 192-193°.
$[\alpha]_D^{23}$: -56,4° (c = 0,39 in abs. Methanol).

Aus den so gewonnenen optisch reinen diastereomeren Hydrazidtartraten wurden die Hydrazid-dihydrochloride entsprechend wie beim Valin und Leucin gewonnen.

D-Verbindung: Ausbeute: 85% d.Th.; Schmp. 153-155°.
$[\alpha]_D^{20}$: -59,8° (c = 0,78 in Wasser).
L-Verbindung: Ausbeute: 85% d.Th.; Schmp. 150-151°.
$[\alpha]_D^{20}$: +60,2° (c = 0,83 in Wasser).

Die kristallwasserfreien Phenylalanin hydrazid-dihydrochloride (gewonnen durch dreistündiges Erwärmen im Vakuum auf 140° über P_2O_5 in der Trockenpistole) lieferten golgend Daten:

D-Verbindung: $[\alpha]_D^{23}$: -64,10°[15]) (c = 0,72 in Wasser),
L-Verbindung: $[\alpha]_D^{23}$: +64,35° (c = 0,75 in Wasser).

Da die Verseifung dieser Hydrazid-dihydrochloride zu den freien Aminosäureantipoden optisch reine Stoffe lieferte, lagen auch die Hydrazidsalze in optisch reiner Form vor.

Phenylalanin-hydrazid-dihydrochloride, dargestellt aus den Rohtartraten des Spaltansatzes:

D-Verbindung: $[\alpha]_D^{23}$: -53,5° (c = 0,90 in Wasser); Schmp. 148-151°.
L-Verbindung: $[\alpha]_D^{23}$: +58,8° (c = 0,86 in Wasser); Schmp. 149-150°.

Die Rohprodukte besassen somit eine optische Reinheit von 90% bzw. uber 95%.

Die Verseifung der reinen antipodischen Phenylalaninhydrazid-dihydrochlorid-monohydrate zum D- und L-Phenylalanin erfolgte, wie beim Valin beschrieben.

D-Phenylalanin: Ausbeute: 70% d.Th.; Schmp. 260-265°.
$[\alpha]_D^{23}$: +34,9° (c = 0,95 in Wasser)[17]).
L-Phenylalanin: Ausbeute: 70% d.Th.; Schmp. 264-268°.
$[\alpha]_D^{20}$: -34,9° (c = 1,02 in Wasser).

[17]) E. Fischer u. W. Schoeller, Liebigs Ann. Chem. 357, 9 (1907).

$C_9H_{11}NO_2$ Ref. 549c

II. Racematspaltung des D,L-Phenylalanin-p-nitrothiophenylesters

1. Ausgangsstoffe

Cbo-D,L-phenylalanin

Nach M. BERGMANN und L. ZERVAS[10]) aus D,L-Phenylalanin und Chlorkohlensäurebenzylester. Schmp. (korr.) 102°, Ausbeute 80% d. Th.

Cbo-D,L-phenylalanin-p-nitrothiophenylester

a) Nach J. A. FARRINGTON, P. J. HEXTALL, G. W. KENNER und J. M. TURNER[4]) werden 3 g Cbo-Phenylalanin in 50 cm³ abs. Dimethylformamid gelöst und mit der äquivalenten Menge Lithium in abs. Methanol versetzt. Das Methanol wird im Vakuum abdestilliert und bei 25° C unter Rühren mit 5 g Tri-(p-nitrophenyl)-thiophosphit versetzt. Nach völliger Lösung wurde der Ansatz über Nacht stehen gelassen, das Dimethylformamid im Vakuum abdestilliert und der Rückstand in 100 cm³ Essigester gelöst. Nach Waschen mit $NaHCO_3$-Lösung und Wasser wird der Essigester im Vakuum abdestilliert. Aus Essigester/Petroläther wurden 3,8 g (87% d. Th.) Ester erhalten. Schmp. (korr.) 158 bis 159°.

b) 12 g Cbo-Phenylalanin und 5,4 cm³ abs. Triäthylamin werden in 100 cm³ abs. Tetrahydrofuran gelöst. Bei —20° C werden 3,84 cm³ Chlorameisensäureäthylester unter Rühren zugetropft. Hierauf rührt man noch 20 Minuten bei gleicher Temperatur und gibt in rascher Tropfenfolge eine filtrierte Lösung von 6,2 g p-Nitrothiophenol[11]) in 20 cm³ abs. Tetrahydrofuran zu. Man rührt noch 1/2 Stunde bei etwa —10 bis —20° C und 1 Stunde bei Zimmertemperatur. Nach Stehen über Nacht und Zusatz von Wasser wird das Tetrahydrofuran im Vakuum bei Zimmertemperatur abdestilliert.

Der kristallisierte Ester wird abgesaugt, mit Wasser neutral gewaschen und mit kleinen Äthermengen von nicht umgesetztem p-Nitrothiophenol befreit. Der Ester wird dann in Dichlormethan gelöst, mit $NaHCO_3$-Lösung, 1proz. HCl und Wasser gewaschen. Nach Abdestillieren des Dichlormethans im Vakuum erhält man aus Essigester/Petroläther 14,9 g Ester (85,2% d. Th.) in weißen Nadeln. Schmp. (korr.) 159°.

$C_{23}H_{20}O_5N_2S$ (436,5)

ber. C 63,28; H 4,62; N 6,42;
gef. C 62,99; H 4,49; N 6,55.

D,L-Phenylalanin-p-nitrothiophenylester-hydrobromid

7,66 g Cbo-ester werden nach R. SCHWYZER[6]) mit 14 cm³ 36proz. HBr-Eisessig-Lösung übergossen. Wenn die CO_2-Entwicklung beendet ist, wird das D,L-Phenylalanin-p-nitrothiophenylester-hydrobromid mit viel abs. Äther völlig ausgefällt. Zur Umkristallisation wird das Salz in wenig kaltem Methanol gelöst und mit viel abs. Äther wieder ausgefällt. In Wasser und siedendem Alkohol zersetzt sich das Hydrobromid. Ausbeute: 6,36 g (83% d. Th.) Schmp. (korr.) 179°.

$C_{15}H_{15}O_3N_2SBr$ (383,3)

ber.: C 47.00; H 3,95; N 7,31;
gef.: C 46,94; H 4,18; N 7,21.

D,L-Phenylalanin-p-nitrothiophenylester

In eine Suspension von 15 g D,L-Phenylalanin-p-nitrothiophenylester-hydrobromid in 125 cm³ abs. Dichlormethan wird unter ständigem Rühren bei —10° C trockenes Ammoniak eingeleitet. Die Gaszufuhr wird so geregelt, daß die Ammoniakblasen ganz langsam in die Suspension einperlen. Beim Auftreten einer leichten bleibenden Gelbfärbung wird das Einleiten sofort abgebrochen. Die Lösung besitzt jetzt einen p_H-Wert von 8—9. Nach etwa 10 Minuten wird das ausgefallene Ammoniumbromid schnell abgesaugt und die Dichlormethanlösung im Vakuum bei Zimmertemperatur eingedampft.

2. Spaltung des D,L-Phenylalanin-p-nitrothiophenylesters

15 g D,L-Phenylalanin-p-nitrothiophenylester-hydrobromid werden in den freien Ester überführt, der sofort in 30 cm³ abs. Dichlormethan gelöst wird. Die Lösung wird schnell filtriert, mit 255 cm³ abs. Äthanol gemischt und eine filtrierte Lösung von 40 g

4) J. A. FARRINGTON, P. J. HEXTALL, G. W. KENNER u. J. M. TURNER, J. chem. Soc. 1957, 1407.

6) R. SCHWYZER, Helv. chim. Acta 37, 647 (1954).

10) M. BERGMANN, L. ZERVAS, Chem. Ber. 65, 1192 (1932).

11) CH. C. PRICE, G. W. STACY, J. Amer. chem. Soc. 68, 498 (1946).

Dibenzoyl-D-weinsäure in 45 cm³ abs. Äthanol mit einem Mal zugesetzt. Die Lösung wird einmal schnell umgeschüttelt und sogleich wieder ruhig stehen gelassen. Nach etwa 1 Minute beginnt die Kristallisation des D-Ester-hydrogentartrates, das nach 1 Stunde abgesaugt und mit Äther gut gewaschen wird. Das L-Ester-hydrogentartrat wird nach Eindampfen des Filtrates im Vakuum bei Zimmertemperatur zunächst als gelber schmieriger Rückstand gewonnen, der nach gründlichem Durcharbeiten mit abs. Äther weiß und kristallin wird. In Alkohol zersetzen sich die Tartrate nach einigen Stunden schon bei Zimmertemperatur.

D-Phenylalanin-p-nitrothiophenylester-dibenzoyl-D-hydrogentartrat

Ausbeute: 7,87 g (51,44% d. Th.). Schmp. (korr.) 140°.

$[\alpha]_D^{20}$: —49,37° (c = 0,638 in Methanol).

Zur Reinigung wird das Tartrat durch Lösen in wenig kaltem abs. Äthanol und Zusatz von viel abs. Äther umgefällt.

$[\alpha]_D^{20}$: —45,17° (c = 0,642 in Methanol) Schmp. (korr.) 141—142°.

$C_{33}H_{28}O_{11}N_2S$ (660,7)

ber.: C 59,99; H 4,27; N 4,24;
gef.: C 60,31; H 4,77; N 4,30.

L-Phenylalanin-p-nitrothiophenylester-dibenzoyl-D-hydrogentartrat

Ausbeute: 8,10 g (52,94% d. Th.) Schmp. (korr.) 159—160°.

$[\alpha]_D^{20}$: —79,48° (c = 0,736 in Methanol).

Nach Lösen in wenig Äthanol und Ausfällen mit viel abs. Äther:

$[\alpha]_D^{20}$: —77,32° (c = 0,632 in Methanol) Schmp. (korr.) 162°.

$C_{33}H_{28}O_{11}N_2S$ (660,7)

ber.: C 59,99; H 4,27; N 4,24;
gef.: C 60,28; H 4,61; N 4,18.

Die Gesamtausbeute an Tartrat, bezogen auf eingesetztes Hydrobromid, beträgt 52,19% d. Th.

3. Gewinnung der optisch aktiven Phenylalanin-p-nitrothiophenylester-hydrochloride

D-Phenylalanin-p-nitrothiophenylester-hydrochlorid

2 g rohes D-Ester-hydrogentartrat in 20 cm³ abs. Dichlormethan werden, wie beim Hydrobromid beschrieben, mit Ammoniak in den freien Ester überführt. Nach Abfiltrieren des Ammoniumdibenzoyltartrates wird in die Dichlormethanlösung unter Eiskühlung trockener Chlorwasserstoff eingeleitet, bis die Lösung sauer reagiert und einen geringen Überschuß an HCl enthält. Nach Abdestillieren des Dichlormethans im Vakuum bei Zimmertemperatur erhält man das Hydrochlorid als kristallinen Rückstand.

Ausbeute: 0,78 g (76,8% d. Th.) Schmp. (korr.) 173—175°.

$[\alpha]_D^{20}$ + 69,66° (c = 0,646 in Methanol).

Das Hydrochlorid wird mit 10 cm³ abs. i-Propanol einige Minuten erwärmt und abgeaugt: $[\alpha]_D^{20}$: + 138,29° (c = 0,888 in Methanol). Nach Umkristallisieren aus Methanol-Äther:

$[\alpha]_D^{20}$: + 179,31° (c = 1,054 in Methanol) Schmp. (korr.) 181—182° (Zersetzungspunkt bei 264—270°).

$C_{15}H_{15}O_3N_2Cl$ (338,8)

ber.: C 53,17; H 4,47; N 8,27;
gef.: C 52,90; H 4,56; N 8,33.

L-Phenylalanin-p-nitrothiophenylester-hydrochlorid

Aus dem umkristallisierten L-Ester-hydrogentartrat wird in gleicher Weise wie beim D-Antipoden das L-Esterhydrochlorid in 74proz. Ausbeute gewonnen. Nach Umkristallisieren aus Methanol/Äther ergeben sich folgende Daten:

$[\alpha]_D^{20}$: —176,22° (c = 1,050 in Methanol) Schmp. (korr.) 179—181° (Zersetzungspunkt über 265°).

$C_{15}H_{15}O_3N_2SCl$ (338,8)

ber.: C 53,17; H 4,47; N 8,27;
gef.: C 53,21; H 4,67; N 8,55.

$C_9H_{11}NO_2$ Ref. 549d

Resolution of N-Carbobenzoxy-DL-phenylalanine. a. Separation of the Diastereoisomeric Salts with (−)-α-Phenylethylamine.—N-Carbobenzoxy-DL-phenylalanine (30 g.) and (−)-α-phenylethylamine (12.2 g.) were dissolved in 100 cc. of warm benzene. Crystallization at room temperature yielded 21 g., $[\alpha]^{25}D$ +17.75° (*c* 6.5, 95% ethanol). The theoretical amount of each diastereoisomeric salt was 21 g., but the salt did not consist solely of one form. Recrystallization from 95% ethanol gave 10.8 g., $[\alpha]^{25}D$ +27.50°, unchanged by further recrystallization. Systematic fractionation of the ethanol mother liquor gave an additional 2.2 g. of the pure salt of the L-acid, $[\alpha]^{25}D$ +27.91° (*c* 6.5, ethanol). Total recovery was 62% based on the amount taken in the racemic mixture.

b. N-Carbobenzoxy-L-phenylalanine and (−)-phenylethylamine were recovered from the less soluble salt by alkaline decomposition. The salt was dissolved in 25 cc. of warm 95% ethanol and poured into 25 cc. of water containing an exact equivalent of sodium hydroxide. The amine was extracted with benzene and the extracts dried with solid sodium hydroxide and distilled to recover the benzene and the amine. The aqueous ethanol layer containing the sodium salt of the carbobenzoxy derivative was evaporated to remove most of the ethanol, and the volume made to 50 cc. with water. A slight excess of hydrochloric acid was added with ice cooling. N-Carbobenzoxy-L-phenylalanine separated as an oil which solidified; yield 9.2 g. (100% based on the salt taken). It was recrystallized from freshly distilled xylene after filtering off traces of sodium chloride. The first crop of crystals was 7.5 g., m.p. 86–87°, $[\alpha]^{25}D$ nearly zero in acetone or methanol, +4.98° (*c* 4, glacial acetic acid), and −5.80° (*c* 4, *N* NaOH). These values were unchanged by recrystallizations from xylene. The recovery from xylene mother liquors was almost quantitative, but prolonged heating to remove solvent gave colored material. Bergmann and associates[21] prepared N-carbobenzoxy-L-phenylalanine from natural phenylalanine and reported $[\alpha]^{25}D$ +4.9° (glacial acetic acid) and m.p. 126–128°. Our recrystallized product showing m.p. 86–87° gave optically pure phenylalanine upon removal of the carbobenzoxy group.

c. **N-Carbobenzoxy-D-phenylalanine** mixed with some of the racemic compound and the (−)-amine were recovered from the more soluble salts present in the original benzene and ethanol liquors as described above. The partially resolved carbobenzoxy derivative (30 g.) and (+)-α-phenylethylamine (12.2 g.) were combined in 100 cc. of warm 95% ethanol. The estimated amounts of diastereoisomeric salts of the L-acid and D-acid were 29 and 13 g., respectively. Crystallization in three crops gave a total of 23 g. of solids and a sirupy residue. Recrystallization from 50 cc. of 95% ethanol gave 20.3 g., $[\alpha]^{25}D$ −27.63° (*c* 6.5, 95% ethanol). The salt was decomposed with alkali as above, giving N-carbobenzoxy-D-phenylalanine. After recrystallizing from xylene it had m.p. 86–87°, $[\alpha]^{25}D$ −4.90° (*c* 4, glacial acetic acid). The more soluble salts were decomposed and reserved for reworking with other similar material.

d. L-Phenylalanine and D-phenylalanine were prepared from the corresponding crude carbobenzoxy derivatives by hydrogenolysis as described above for alanine; $[\alpha]^{25}D$ −35.16 and +34.94° (*c* 1.7, water). The accepted value for L-phenylalanine is −35.1°.[18]

(18) M. S. Dunn and L. B. Rockland, *Advances in Protein Chem.*, **3**, 296 (1947).

(20) T. A. Henry, "The Plant Alkaloids," Fourth Edition, The Blakiston Co., Philadelphia, Pa., 1949, p. 636.

(21) M. Bergmann, L. Zervas, H. Rinke and H. Schleich, *Z. physiol. Chem.*, **224**, 33 (1934).

$C_9H_{11}NO_2$ Ref. 549e

Benzoyl-*d*-Phenylalanin.

Wie schon erwähnt, gelingt die Spaltung des racemischen Productes mit Cinchonin, welches mit dem Benzoyl-*d*-Phenylalanin das schwerer lösliche Salz bildet. Da die Krystallisation desselben einige Schwierigkeiten bietet, so ist es rathsam, den Versuch zuerst im Kleinen auszuführen. Man löst zu dem Zweck 1.4 g Cinchonin und 1.3 g Benzoylphenylalanin in 350 ccm kochendem Wasser. Beim Abkühlen trübt sich die Flüssigkeit und scheidet eine kleine Menge eines bräunlich gefärbten Syrups ab. Nach völligem Erkalten lässt man die abgegossene klare Lösung im Eisschrank mehrere Tage stehen und vertheilt die allmählich erscheinenden Kryställchen durch häufiges Umrühren in der Flüssigkeit. Ist man auf diese Weise in den Besitz von Impfmaterial gelangt, so wird der Versuch mit 14.8 g Cinchonin, 13.4 g Benzoylphenylalanin und 3½ L Wasser wiederholt und die abgekühlte decantirte Lösung bei Zimmertemperatur mit den Krystallen versetzt. Beschleunigt man dann durch häufiges Umrühren die Krystallisation, so erhält man nach 12 Stdn. ungefähr 9 g oder ⅔ der Theorie des krystallisirten Salzes. Eine weitere, aber nicht beträchtliche Krystallisation gewinnt man durch Eindampfen der Mutterlauge im Vacuum bis zur Abscheidung eines Syrups und längeres Stehenlassen nach Einimpfen einiger Kryställchen. Das zuerst ausgeschiedene Salz wird durch zweimalige Krystallisation aus der hundertfachen Menge heissen Wassers ganz rein erhalten, wobei der Verlust sehr gering ist, und bildet dann lange, farblose Nadeln vom Schmp. 180—181° (uncorrigirt).

Um das Salz zu zerlegen, löst man 15 g in 2 L kochendem Wasser, fügt 48 ccm Normalnatronlauge hinzu, kühlt auf 0° ab, filtrirt das gefällte Cinchonin und versetzt die Mutterlauge mit 60 ccm Normalsalzsäure. Dabei scheidet sich der grösste Theil des Benzoyl-*d*-phenylalanins als farblose, leichte, mikrokrystallinische Masse ab. Da aber ein nicht unbeträchtlicher Theil der Verbindung in den Mutterlaugen bleibt, so ist es nöthig, dieselben im Vacuum auf dem Wasserbade stark einzudampfen, wobei eine zweite Krystallisation erfolgt. Die Ausbeute ist fast quantitativ. Zur vollständigen Reinigung wird die Substanz aus der zweihundertfachen Menge kochendem Wasser unter Zusatz von etwas Thierkohle umkrystallisirt. Sie bildet dann schöne, farblose Nadeln, welche bei 142—143° (corr. 145—146°), mithin erheblich niedriger als der Racemkörper, schmelzen.

Für die Analyse wurde sie bei 100° getrocknet.

0.2066 g Sbst.: 9.6 ccm N (21°, 754 mm). — 0.2008 g Sbst.: 0.5236 g CO_2, 0.1020 g H_2O.

$C_{16}H_{15}O_3N$. Ber. C 71.4, H 5.61, N 5.20.
Gef. » 71.1, » 5.64, » 5.25.

Für die optische Bestimmung diente die alkalische Lösung, da die Substanz in Wasser zu wenig löslich ist.

Gewicht der Gesammtlösung 11.939 g, enthaltend 0.756 g Benzoyl-*l*-phenylalanin und 3 ccm Normalkalilauge; spec. Gewicht 1.022; Drehung bei 20° im 2-Decimeterrohr bei Natrium-Licht 2.21°.

Mithin $[\alpha]_D^{20°} = -17.1°$ (für die alkalische Lösung).

d-Phenylalanin.

Wird die Benzoylverbindung mit der 120-fachen Menge 10-procentiger Salzsäure auf dem Wasserbade unter öfterem Umschütteln erhitzt, so löst sie sich im Laufe von etwa 4 Stunden, und nach weiterem 6-stündigem Erhitzen ist die Zersetzung beendet. Nach dem Erkalten wird die Benzoësäure ausgeäthert, die salzsaure Lösung im Vacuum und zum Schluss in einer Schale auf dem Wasserbade zur Trockne verdampft, dann der Rückstand in der 4-fachen Menge Wasser gelöst und mit einer concentrirten Lösung von Natriumacetat das Phenylalanin ausgefällt. Zur völligen Reinigung löst man die filtrirte Krystallmasse in der 25-fachen Menge heissem Wasser, entfärbt mit Thierkohle und verdampft das Filtrat grösstentheils auf dem Wasserbade, wobei das Phenylalanin in schönen Blättchen krystallisirt.

Dieselben wurden für die Analyse bei 100° getrocknet.

0.1709 g Sbst.: 12.5 ccm N (19°, 763 mm). — 0.2018 g Sbst.: 0.4835 g CO_2, 0.1212 g H_2O.

$C_9H_{11}NO_2$. Ber. C 65.42, H 6.66, N 8.48.
Gef. » 65.34, » 6.66, » 8.44.

Die Verbindung schmilzt beim raschen Erhitzen im Capillarrohr unter starker Gasentwickelung bei 283—284° (corr.), mithin ungefähr bei derselben Temperatur wie der Racemkörper, dagegen ist sie in Wasser leichter löslich als dieser. 1 Th. *d*-Phenylalanin verlangt 35.3 Th. Wasser bei 16°.

Für die optische Untersuchung diente die wässrige Lösung.

Gewicht der Lösung 13.633 g. Gewicht der Substanz 0.2768 g. Spec. Gewicht 1.0043. Procentgehalt 2.03. Temperatur 16°. Drehung bei Natriumlicht im 2-Decimeterrohr + 1.43°.

Mithin $[\alpha]_D^{16°} = +35.08°$.

Schulze [1]) fand unter den gleichen Bedingungen für den optischen Antipoden das natürliche Phenylalanin, $[\alpha]_D^{16°} = -35.3°$. Die Differenz liegt innerhalb der Versuchsfehler.

[1]) Zeitschr. für physiol. Chem. 9, 85.

Viel geringer ist die Drehung einer Lösung des *d*-Phenylalanins in 18-procentiger Salzsäure.

Gewicht der Lösung 18.0239 g. Gewicht der Substanz 0.6311 g. Spec. Gewicht 1.0895. Procentgehalt 3.5 pCt. Temperatur 20°. Drehung bei Natriumlicht im 2-Decimeterrohr + 0.54°.

Mithin $[\alpha]_D^{20°} = +7.07°$.

Benzoyl-*l*-phenylalanin.

Dasselbe befindet sich in den Mutterlaugen, aus welchen das Cinchoninsalz des optischen Antipoden ausgefallen ist. Fällt man dieselben mit Alkali und übersättigt die vom Cinchonin abfiltrirte Lösung mit Salzsäure, so entsteht ein reichlicher Niederschlag von feinen Nadeln, welcher hauptsächlich aus racemischem Benzoylphenylalanin besteht, da der active Körper in kaltem Wasser leichter löslich ist. Werden die Mutterlaugen durch Eindampfen im Vacuum stark concentrirt, so resultirt eine neue reichliche Krystallisation, welche viel actives Benzoylphenylalanin, allerdings neben Racemkörper, enthält. Zur annähernden Trennung derselben kocht man das gepulverte Product mit der 35-fachen Menge Wasser, wobei der schwerer lösliche Racemkörper theilweise zurückbleibt, und überlässt das Filtrat der Krystallisation. Wir erhielten so farblose Nadeln, welche keinen scharfen Schmelzpunkt hatten und nach der optischen Bestimmung ein Gemisch von Benzoyl-*l*-phenylalanin mit ungefähr 20 pCt. Racemkörper waren. Es ist uns leider nicht gelungen, ein krystallisirtes Salz mit einer activen Base darzustellen und dadurch die völlige Reinigung zu erzielen. Durch Spaltung mit Salzsäure gewinnt man aus diesem Product natürlich ein noch unreines *l*-Phenylalanin.

$C_9H_{11}NO_2$ Ref. 549f

2. Racematspaltung des N-Acetyl-DL-phenylalanins

a) in Wasser:

2,625 kg L(+)-threo-1-[p-Nitrophenyl]-2-aminopropandiol-(1,3) und 2,590 kg N-Acetyl-DL-phenylalanin wurden in 12,52 l Wasser bei 90° gelöst. Nach zweitägigem Stehenlassen bei Zimmertemperatur wurde das auskristallisierte Salz des N-Acetyl-L-phenylalanins abgesaugt.

Ausbeute: 2,4 kg; Schmp. 186—188° C;
$[\alpha]_D^{25} = +40°$ (c = 2 in Wasser).

Das in der Mutterlauge gelöste Salz des N-Acetyl-D-phenylalanins wurde nicht isoliert, sondern durch Zugabe von Salzsäure gleich das N-Acetyl-D-phenylalanin ausgefällt.

Ausbeute: 1,07 kg N-Acetyl-D-phenylalanin; Schmp. 160°;
$[\alpha]_D^{25} = -38,3°$ (c = 2 in Alkohol).

Das Salz des N-Acetyl-L-phenylalanins trugen wir unter Rühren in 6,3 l n-Salzsäure ein und saugten nach einiger Zeit das N-Acetyl-L-phenylalanin ab.

Ausbeute: 1,06 kg N-Acetyl-L-phenylalanin; Schmp. 166°;
$[\alpha]_D^{25} = +40,0°$ (c = 2 in Alkohol).

b) in n-Propylalkohol:

In 850 ml n-Propylalkohol wurden in der Siedehitze 20,7 g N-Acetyl-DL-phenylalanin und 21,2 g L(+)-threo-1-[p-Nitrophenyl]-2-aminopropandiol-(1,3) gelöst. Nach zehnstündigem Stehenlassen bei Zimmertemperatur wurde das Salz des N-Acetyl-L-phenylalanins abgesaugt und getrocknet. Das Filtrat hinterließ nach dem Abdestillieren des Lösungsmittels das Salz des N-Acetyl-D-phenylalanins. Beide Salze wurden mit je 120 ml n-Salzsäure versetzt, das N-Acetyl-L- bzw. D-phenylalanin abgesaugt und mit Wasser gewaschen.

Ausbeuten: 8,1 g N-Acetyl-L-phenylalanin; Schmp. 160°;
$[\alpha]_D^{25} = +38,0°$ (c = 2 in Alkohol),
7,0 g N-Acetyl-D-phenylalanin; Schmp. 159°;
$[\alpha]_D^{52} = -39,0°$ (c = 2 in Alkohol).

c) in Äthylalkohol:

20,7 g N-Acetyl-DL-phenylalanin und 21,2 g L(+)-threo-1-[p-Nitrophenyl]-2-aminopropandiol-(1,2) wurden in 425 ml Äthanol in der Siedehitze gelöst. Nach zehnstündigem Stehenlassen bei Zimmertemperatur wird der Ansatz in analoger Weise aufgearbeitet, wie bei der Spaltung in n-Propylalkohol.

Ausbeuten: 9,8 g N-Acetyl-L-phenylalanin; Schmp. 154°;
$[\alpha]_D^{25} = +34,1°$ (c = 2 in Alkohol),
6 g N-Acetyl-D-phenylalanin; Schmp. 162°;
$[\alpha]_D^{25} = -42,3°$ (c = 2 in Alkohol).

3. D- bzw. L-Phenylalanin

N-Acetyl-D-phenylalanin wurde mit der 1,7fachen Menge Salzsäure (21proz.) eine Stunde unter Rückfluß gekocht, dann über Nacht mit Eis gekühlt, das auskristallisierte D-Phenylalaninhydrochlorid abgesaugt und nach Auflösen in der 10fachen Menge Methanol das D-Phenylalanin durch Zusatz von Ammoniak (oder Pyridin) ausgefällt. Nach Umkristallisation aus Wasser erhielten wir das reine D-Phenylalanin, Schmp. 260°; $[\alpha]_D^{25} = +32°$ (c = 2 in Wasser).

Das L-Phenylalanin wurde in analoger Weise gewonnen, Schmp. 260°; $[\alpha]_D^{25} = -32°$ (c = 2 in Wasser).

$C_9H_{11}NO_2$ Ref. 549g

Example 4

10.3 g. of *dl*-N-acetylphenylalanine is dissolved with heating in 250 ml. of 95% ethanol. 6.75 g. of *l*-amphetamine is added, and the solution is allowed to stand for 16 to 20 hours at room temperature. The resulting crystalline precipitate consisting of the *l*-amphetamine salt of *l*-N-acetylphenylalanine is filtered off and dried; $[\alpha]_D^{25}=-50°$ (2.5% in water). The crystalline material is dissolved in water and the aqueous solution is passed through a sulfonic acid ion exchanger (Dowex 50), and the effluent is evaporated to dryness. The product is *l*-N-acetylphenylalanine; $[\alpha]_D^{25}=-48.5°$ (1.5% in absolute alcohol).

The filtrate first obtained above is allowed to evaporate slowly at room temperature. The crystalline product which forms, consisting of the *l*-amphetamine salt of *d*-N-acetylphenylalanine, is filtered off and dried. The crystals are dissolved in water and the aqueous solution is passed through a sulfonic acid ion exchanger (Dowex 50). The effluent is evaporated to dryness to obtain the product, *d*-N-acetylphenylalanine; $[\alpha]_D^{25}=+51°$ (1.5% in absolute ethanol).

$C_9H_{11}NO_2$ Ref. 549h

For a resolution procedure using ephedrine in the direct resolution of the racemic N-benzyloxycarbonyl derivative of the above compound, see Ref. 20h.

$C_9H_{11}NO_2$ Ref. 549i

Resolution of Formylphenylalanine.—Formylphenylalanine (165 g., 0.84 mole) was dissolved with 75 g. (0.84 mole) of (−)-1-hydroxy-2-aminobutane in 600 ml. of 1-butanol and decolorized with Norit. To this was added 600 ml. each of benzene and of Skelly D. One crop of crystals was removed after 6 hr., another after overnight standing. Both crops were virtually pure (−)-1-hydroxybutane-2-ammonium formyl-D-phenylalaninate, m.p. 128–129°; 58 g. (50%). Recrystallization from 200 ml. of 1-butanol gave 40 g. of rosettes, m.p. 129–130°; $[\alpha]^{29}D$ −52.5° ± 0.2° (*c* 4.4 in 95% ethanol).

Anal. Calcd. for $C_{14}H_{22}O_4N_2$: N, 9.93. Found: N, 9.97.

To the filtrate was added another 1000 ml. of Skelly D, which deposited a small amount of oil, followed by 24 g. of (−)-1-hydroxybutane-2-ammonium formyl-L-phenylalaninate. Recrystallization from 70 ml. of 1-butanol and 70 ml. of benzene gave 20 g., m.p. 106–107°; $[\alpha]^{29}D$ +42.7° ± 0.6° (*c* 1.6 in 95% ethanol).

Anal. Calcd. for $C_{14}H_{22}O_4N_2$: N, 9.93. Found: N, 9.96.

Solubility determinations indicated closely similar values for the two diastereomers in the solvent system employed.

Recovery of D- and of L-Phenylalanine.—The formylphenylalanines were recovered from the salts by solution in four parts of water and addition of about an equal part of 5–6 *N* hydrochloric acid. After being washed with water, the formylphenylalanine precipitates were hydrolyzed to yield the phenylalanine isomers by the method of Fischer and Schoeller.[36] Rotations were $[\alpha]^{29}D$ +35.1° ± 1.0° and −35.0° ± 1.0° for 1% solutions in water.

Further crops of formylphenylalanine salt could be recovered from the mother liquors of the treatment of the formylphenylalanine with amine. Attempts to perfect the method were not continued inasmuch as it was found that unsubstituted phenylalanine could be resolved by use of methylcinchoninium hydroxide.[37]

$C_9H_{11}NO_2$ Ref. 549j

Preparation of N-Formyl-DL-phenylalanine.—The method was a simplification of that of Fischer and Warburg.[4b] DL-Phenylalanine (82.5 g.) was dissolved in 100 cc. of 90% formic acid and boiled gently under reflux for two hours. An additional 100 cc. of formic acid was added and the solution allowed to cool. Formylphenylalanine then crystallized in readily filtrable form and was essentially pure (m.p. 165–166°) after filtering, washing with a little formic acid and drying in a desiccator over solid sodium hydroxide. A second pure crop was obtained similarly after concentrating the filtrate by distillation to 75 cc. The yield was 83 g. (87%). Crops obtained by further concentration were slightly colored and contained some free phenylalanine or its formate as shown by failure to dissolve completely in hot acetone. This material was best combined with similar material from other runs and crystallized from fresh formic acid. Working in this manner, the yield is nearly quantitative. Crystallization from acetone left the melting point unchanged at 165–166° (uncor.)[28] in agreement with reported values.[9,24]

Resolution of N-Formyl-DL-phenylalanine (a) with (−)-α-Fenchylamine.—Formylphenylalanine (19.3 g., 0.1 mole) was dissolved in 35 cc. of water containing 0.1 mole of sodium hydroxide and the solution mixed with a nearly boiling solution of (−)-α-fenchylamine hydrochloride (19.0 g., 0.1 mole) in 60 cc. of water. Crystallization occurred promptly, giving 14.0 g. of the nearly pure salt of formyl-D-phenylalanine, m.p. 177–178°, $[\alpha]^{26}D$ −33.04° (*c* 4, water). Further crystallization occurred in the filtrate overnight, giving 9.6 g. of fairly pure salt of formyl-L-phenylalanine, $[\alpha]^{26}D$ +26.5°. Systematic recrystallization of these crops and further crops from the original solution rather readily gave 15.4 g. of salt with m.p. 177–178°, $[\alpha]^{26}D$ −33.05° and 16.4 g. with m.p. 164–165°, $[\alpha]^{26}D$ +27.53°; calculated for each salt, 17.4 g.

The salts were decomposed by a generally applicable procedure. The powdered salt was suspended in about twice its weight of water and an equal volume of sulfur-free benzene in a separatory funnel and a few drops of phenolphthalein solution was added. Concentrated sodium hydroxide solution was then gradually added with vigorous shaking and occasional cooling until the solid dissolved and a faint permanent pink color appeared in the aqueous layer. The benzene solution of the amine was removed and the aqueous layer was extracted five or six times with 25-cc. portions of benzene. The united benzene extracts were washed once with a little water, dried with solid sodium hydroxide and distilled through a moderately effective column for recovery of the benzene and amine. With care to exclude carbon dioxide, the recovery was nearly quantitative. The aqueous layer and washings were promptly acidified with the calculated amount of concentrated hydrochloric acid and the liberated acyl derivative was isolated.

In the present instance the aqueous layer from the levorotatory salt gave a nearly quantitative precipitate of crystalline N-formyl-D-phenylalanine. This was dried and recrystallized from acetone; yield 83.5%. However, the product was found to contain traces of (insoluble) free phenylalanine. This was readily removed, but further traces appeared in subsequent crystallizations or when samples were made up in ethanol for rotation determinations. The rotation, $[\alpha]^{26}D$ −72.7° (*c* 1, 95% ethanol) and m.p. 163–164°, indicate slight impurity since recorded values[9] are $[\alpha]^{26}D$ −75.2° and m.p. 167°. Samples warmed for some time in ethanol or methanol gave increased amounts of the free amino acid. Similar behavior was later observed with formyl-DL-norleucine and formyl-DL-leucine.

The dextrorotatory salt gave N-formyl-L-phenylalanine with $[\alpha]^{26}D$ +60.9° and m.p. 161–162°. This contained free amino acid, but the low values indicate that this salt was not completely purified in the resolution.

(b) With (−)-α-Phenylethylamine.—Formylphenylalanine (96.5 g., 0.5 mole) was suspended in 200 cc. of water and the amine (60.6 g., 0.5 mole) added. The salt dissolved readily on warming and on cooling 46.0 g. separated promptly as needles, $[\alpha]^{26}D$ +29.7° (*c* 1, water). One recrystallization gave 24.6 g. with $[\alpha]^{26}D$ +32.8° and m.p. 175–176°, values not further changed on recrystallization. Successive crops from the original solution were systematically recrystallized and about 45% of the calculated amount of dextrorotatory salt was obtained pure. The later crops were levorotatory. At this point free phenylalanine appeared in some of the solutions and the fractionation was discontinued. The dextrorotatory salt was decomposed as previously described. The resulting N-formyl-L-phenylalanine, $[\alpha]^{26}D$ +73.7 (*c* 1, 95% ethanol), was obtained in 95% yield, but repeated crystallization did not completely remove traces of free phenylalanine.

Hydrolysis and isolation of the amino acid in the usual manner,[24] followed by one crystallization from water gave pure L-phenylalanine, $[\alpha]^{27}D$ −35.2° (*c* 1.8, water); $[\alpha]^{27}D$ −7.32 (*c* 3.5, *N* HCl) in agreement with reported values.[24]

Preparation of N-Acetyl-DL-phenylalanine.—DL-Phenylalanine was acetylated in 1-mole lots by the procedure previously described for acetyl-L-leucine.[29] Most of the acetylphenylalanine separated during acidification. After an additional hour at 5–10° the solid was filtered sharply by suction and washed with two 100-cc. portions of ice-water by slurrying and filtering. After drying, the product was already essentially pure; m.p. 152°; yield 195 g. (94%). The filtrate is best discarded. However, it was extracted for 24 hours with chloroform in a continuous counter-current extractor. The extract was evaporated to dryness under reduced pressure to remove chloroform, water and acetic acid. Hot acetone then extracted an additional 8 g. of product from the brown residue.

Recrystallization of N-acetyl-DL-phenylalanine from water, acetone, ethyl acetate and chloroform gave an anhydrous product (neut. equiv., 207); m.p. 152.5–153°. Literature values[30,9] range from 143 to 151°. The substance forms glistening plates from water and characteristic long hexagonal tablets from dilute acetone solutions. The solubilities (expressed throughout this paper as g./100 cc. solution at 25 ± 0.5°) are: water, 0.73; acetone, 4.31; ethyl acetate, 0.79; chloroform, 0.34. It is conveniently recrystallized from acetone.

Resolution of N-Acetyl-DL-phenylalanine.—N-Acetyl-DL-phenylalanine (48.3 g., 0.233 mole) was suspended in 350 cc. of water, exactly neutralized to phenolphthalein with sodium hydroxid and the solution heated to boiling. A nearly boiling solution of (−)-α-fenchylamine hydrochloride (44.3 g., 0.223 mole) in about 150 cc. of water was added. Crystallization of coarse needles began at once. The solution was digested in a boiling water-bath for 30 minutes and filtered by suction while hot. The crystals were washed on the filter with two 50-cc. portions of boiling water and dried. The salt thus obtained (29.4 g.) is the nearly pure salt of acetyl-D-phenylalanine [(−)-B·D-A salt], $[\alpha]^{22}D$ −43.6° (*c* 6, methanol). The filtrate on cooling gave 12.9 g. of salt with rotation −1.8°. After evaporation to 200 cc. the filtrate slowly deposited 18.1 g. of irregular granules with rotation +21.2°. A final small crop from sirupy liquors had $[\alpha]^{22}D$ +38° but no attempt was made to purify completely the very soluble (−)-B·L-A salt. Digestion of the second crop with 100 cc. of boiling water left undissolved 4.2 g. with rotation −37.3° and systematic recrystallization of the third crop gave 4.7 g. of similar salt.

The combined levorotatory crops (38.3 g.) were crystallized from 250 cc. of methanol in several crops and thus gave 35.2 g. of the pure (−)-B·D-A salt (87% yield), m.p. 214–215°; $[\alpha]^{25}D$ −46.5° (*c* 4, methanol), values not changed by further crystallization. Methanol can be used as solvent in the resolution but is less convenient than water.

N-Acetyl-D-phenylalanine and (−)-α-fenchylamine were recovered from the salt (35.2 g.) by the general procedure already given, except that the salt was dissolved in minimum methanol (150 cc.) before alkalinization. The solution of the sodium salt was acidified with a few drops of acetic acid and evaporated to 150 cc. to remove methanol before addition of hydrochloric acid. The chilled solution deposited practically pure acetyl-D-phenylalanine. The substance was recrystallized from 175 cc. of acetone in several crops totalling 19.6 g. or 81% based on the racemic form taken. It forms coarse tablets or crusts from acetone, m.p. 171–172°; $[\alpha]^{25}D$ −47.5° (*c* 4, abs. ethanol); −40.3° (*c* 4, methanol); −38.1° (*c* 0.6, water). Slightly higher rotation values in absolute ethanol have been reported[9] but our values were repeatedly checked on samples crystallized from various solvents. The solubilities are: water, 0.85; acetone, 4.14; ethyl acetate, 0.93; chloroform, 0.16.

N-Acetyl-L-phenylalanine was recovered from the mother liquors of the resolution. The crude product (24 g., calcd., 27 g.) had $[\alpha]^{22}D$ +37.8° (*c* 4, ethanol) and hence contained about 80% of the L-form (dextrorotatory) and 20% of DL-form. Although the L-form is only slightly less soluble in acetone than the DL-form it crystallizes more rapidly, as also noted by Martin and Synge,[32] and is readily purified by two crystallizations. In this way there was obtained 18.5 g. of pure L-form (77% based on the racemic form taken), m.p. 171–172°; $[\alpha]^{25}D$ +47.5° (*c* 4, ethanol). Within experimental error the solubilities were identical with those given for the D-form. The sirupy, slightly colored acetone mother liquors slowly deposited a small amount of nearly inactive product.

D-Phenylalanine.—Pure N-acetyl-D-phenylalanine (5.0 g.) was boiled five hours under reflux with one and one-half equivalents of normal hydrochloric acid. The solution was evaporated to dryness under reduced pressure and the residue taken up in 40 cc. of 95% ethanol. The amino acid was precipitated by dropwise addition of concentrated aqueous ammonia to *p*H 6, filtered and washed freely with ethanol. The dry product (3.8 g., 98%) was already pure, $[\alpha]^{25}D$ +35.3° (*c* 1.6, water); +7.70° (*c* 4, *N* HCl); +3.48° (*c* 2, 6 *N* HCl), values not changed by recrystallization. In later experiments hydrolysis was effected in two hours with 1.2 equivalents of 3 *N* hydrobromic acid.

L-Phenylalanine.—The hydrolysis of N-acetyl-L-phenylalanine and isolation of the amino acid were carried out as just described. The initial product had $[\alpha]^{25}D$ −34.78°, changing after crystallization from water to $[\alpha]^{25}D$ −35.16° (*c* 1.6, water); −7.72° (*c* 4, *N* HCl); −3.48° (*c* 2, 6 *N* HCl). The reported value[24] in water solution is −35.1°.

4b) E. Fischer and O. Warburg, Ber., 38, 3997 (1905).

(9) V. du Vigneaud and C. E. Meyer, *ibid.*, **98**, 295 (1932); **99**, 143 (1932).

(29) H. D. DeWitt and A. W. Ingersoll, This Journal, **73**, 3359 (1951).

(30) A. H. Gordon, A. J. P. Martin and R. L. M. Synge, *Biochem. J.*, **35**, 1358 (1941).

$C_9H_{11}NO_2$ Ref. 549k

Almost all chemical procedures for the resolution of neutral α-amino-acids, have employed amino-acids acylated on the amino groups. This makes them essentially acids and hence able to form salts with optically active bases.[1] The resolution process, therefore, involves at least three distinct stages; preparation of the acyl derivative, resolution and finally hydrolysis of the resolved compound. To overcome this difficulty Bergmann and Zervas[2] suggested that the ease of Schiff's base formation and subsequent hydrolysis of amino-acid salts with aromatic aldehydes, might be exploited for the resolution of amino-acids; but their experiments using the conventional alkaloid bases were unsuccessful and the work was abandoned.

We have now found that the metal complex cation *cis*-dinitrobis(ethylenediamine)cobalt (III) ion[3] can be used successfully for the resolution of phenylalanine *via* the salicylidene derivative prepared and hydrolysed *in situ* during the resolution procedure.

The Schiff's base of the amino-acid salt was optically stable in neutral or weakly alkaline aqueous solutions and the parent aldehyde was conveniently removed by extracting the aqueous solution with chloroform containing a stronger primary amine than the amino-acid. This procedure preserved the optical purity of the amino-acid. It is believed that

References

[1] Greenstein, J. P. & Winitz, M., "The Chemistry of the Amino-acids," 1961, Vol. 1, (New York: John Wiley)

[2] Bergmann, M. & Zervas, L., *Z. physiol. Chem.*, 1926, **152**, 282

[3] Dwyer, F. P. & Halpern, B., *Nature, Lond.*, 1962, **196**, 270

the method could be adapted to the resolution of other amino-acids by a suitable choice of aldehyde and optically active complex ion.

D,L-Phenylalanine (3·3 g., 0·020 mol.) was dissolved in 1N sodium hydroxide (20·0 ml., 0·020 mol.). Salicylaldehyde (2·44 g., 0·020 mol.) was added with stirring until dissolved, followed by a solution of *d-cis*-dinitrobis(ethylenediamine)cobalt(III) acetate [I] (3·30 g., 0·010 mol.) in water (40 ml.). *d-cis*-Dinitrobis(ethylenediamine)cobalt(III) salicylidene - L - phenylalaninate [II] precipitated from solution and was recrystallised from formamide—ethanol (2·0 g.), $[\alpha]_D$ −73° (*c.*, 1·0 in $HCONH_2$) (Found: C, 44·6; H, 5·5; N, 18·3. $C_{20}H_{30}N_7O_7Co$ requires C, 44·5; H, 5·6; N, 18·2%).

The recovery of L-phenylalanine was effected by making a slurry of compound [II] in water and adding sodium iodide which precipitated the insoluble iodide of [I], leaving an aqueous solution of the sodium salt of salicylidene-L-phenylalanine. The salicylidene derivative was broken by several extractions with cyclohexylamine in chloroform. The aqueous solution was then passed through a Dowex 50W column, and the adsorbed L-phenylalanine eluted with 2·5% pyridine in water. After evaporation and recrystallisation from aqueous ethanol optically pure L-phenylalanine (0·45 g.), $[\alpha]_D$ −33° (*c.*, 1·5 in H_2O, 1 dm. tube) was obtained.

Partially resolved D-phenylalanine was recovered from the mother liquors from the salt (II). By repeating the resolution with the optical enantiomer of [I], optically pure D-phenylalanine, $[\alpha]_D$ +33° (*c.*, 1·5 in H_2O) was obtained.

$C_9H_{11}NO_2$ Ref. 549-1

Resolution of N-Acetyl-DL-phenylalanine.

N-Acetyl-DL-phenylalanine (207 g, 1 mol) and L-leucinamide (143 g, 1.1 mol) were dissolved in 3.5 l of ethanol on a water bath at 50°C, and the solution was let to stand at room temperature, after being seeded by 10 mg of pure L-leucinamide N-acetyl-L-phenylalanine. The deposited massive fine needles were filtered off, washed with ethanol and dried. Yield, 77 g (45.6% based on L-form); m.p. 187°C; $[\alpha]_D^{21}$+41.52° (c, 5: water).

The solution of L-leucinamide salt (77 g) in 300 ml of water was treated with a column of Diaion SK#1 (H-form) as described previously. Yield, 47 g (99.7%); m.p. 170~171°C; $[\alpha]_D^{24}$−40.2° (c, 2: methanol).

At the next step the hydrolysis of N-acetyl-L-phenylalanine with 20% hydrochloric acid gave 37 g of the L-phenylalanine similarly as described in the previous instances; $[\alpha]_D^{20}$−35.14° (c, 1.9: water). *Anal.* Found: N, 8.51. Calcd. for $C_9H_{13}O_2N$: 8.48%.

$C_9H_{11}NO_2$ Ref. 549m

The racemic tert-butyl ester of phenylalanine was resolved via carbobenzoxy-L-phenylalanine was resolved via carbobenzoxy-L-phenylalanine as resolving agent. Only the L-L salt deposits (nearly quantitatively) from a 0.15-0.17 M ethereal solution. From the resolved salts, the enantiomeric esters were isolated as the hydrochloride salts with $[\alpha]_D^{25}$ +43.8 (c 2, ethanol) and $[\alpha]_D^{25}$ -44.0 (c 2, ethanol).

$C_9H_{11}NO_2$ Ref. 549n

AZIONE DEL CLORURO DI D.(—).ACETILMANDELILE SULLE FENIL-ALANINE

D.(—).N.mandelil-fenilalanine. (Form. I e II, $R=C_6H_5—CH_2$). — Grammi 6,25 di D.L.fenilalanina (0,037 Mol.) sono sciolti in 38 cm³ di NaOH.N/$_1$. Sotto agitazione ed a temperatura di 12°-15°, la soluzione viene trattata, alternativamente ed a piccole porzioni, con g 8,1 di cloruro di D.(—).acetilmandelile (0,038 Mol.) sciolti in poco etere, e con NaOH.N/$_1$ cm³ 76. Il trattamento richiede circa due ore; alla fine si aggiungono altri 10 cm³ di soluzione alcalina e si continua ad agitare per mezz'ora a temperatura ambiente. Si filtra per eliminare poche sostanze resinose, si diluisce con 100 cm³ di acqua e si aggiunge acido cloridrico fino a netta acidità al rosso-congo. In tal modo si ottiene la separazione di un prodotto bianco polverulento che raccolto, dopo alcune ore, ed asciugato, pesa g 2,5 e fonde con decomposizione a 188°-191°. Il sale sodico in soluzione acquosa mostra $[\alpha]_D^{21}$ = —74°,9 (c=1,88). Il composto è già da considerarsi puro: esso non cambia sensibilmente i caratteri per ulteriore passaggio per soda e riprecipitazione con acido cloridrico. Separandosi da soluzioni acquose diluite prende l'aspetto di minutissime scagliette bianche. Analisi:

	trov.% N	4,56;
per $C_{17}H_{17}O_4N$	calc.	4,68.

Le acque madri acide dalle quali si era separato il prodotto precedente vengono concentrate nel vuoto a temperatura ordinaria. Si separa in primo tempo un po' di sostanza che si trascura; poi si ottiene un deposito cristallino bianco di aspetto diverso da quello del prodotto meno solubile. Raccolto ed essiccato, pesa g 2,2 e fonde con decomposizione a 124°-126°. La soluzione del sale sodico mostra $[\alpha]_D^{20}$ = —1°,58 (c=0,47). Analisi:

	trov.% N	4,52;
per $C_{17}H_{17}O_4N$	calc.	4,68.

IDROLISI DELLE D.(—).N.MANDELIL-FENILALANINE

D.(+).fenilalanina. — Grammi 2 del composto fusibile a 188°-191° ($[\alpha]_D$= —74°,9) sono trattati con cm³ 40 di acido cloridrico al 20% e disciolti a caldo (prima a bagnomaria, poi all'ebullizione). Dopo mezz'ora di riscaldamento si fa raffreddare, si estrae ripetutamente con etere che asporta l'acido mandelico, e finalmente si evapora il liquido acquoso nel vuoto su calce sodata. Resta il cloridrato sotto forma di aghetti bianchi che si sciolgono in poca acqua e si trattano con idrossido sodico fino a pH∼ 5,4. Si separa così una sostanza cristallina bianca in fogliette che possiede tutti i caratteri della D.(+).fenilalanina; in particolare p. f. 280°-284° e $[\alpha]_D$ = +34°,2. Pertanto, il mandelil-derivato fusibile a 188°-191° deve considerarsi D.(—).N.mandelil-D.(+).fenilalanina (form. II).

L.(—).fenilalanina. — In analoghe condizioni di idrolisi, il mandelil-derivato fusibile a 124°-126° fornisce L.(—).fenilalanina; e pertanto esso deve considerarsi D.(—).N.mandelil-L.(—).fenilalanina (form. I).

AZIONE DEL CLORURO DI D.(—).ACETILMANDELILE SULLE LEUCINE

D.(—).N.mandelil-leucine. (Form. I e II, $R=\frac{CH_3}{CH_3}>CH—CH_2$). — L'acilazione della D.L.leucina si effettua nelle seguenti condizioni: g 11 di aminoacido (0,084 Mol.) sono sciolti in 85 cm³ di NaOH.N/$_1$. La soluzione, mantenuta a 10°-15°, è trattata, alternativamente nel solito modo con g 18 (0,084 Mol.) di cloruro di D.(—).acetilmandelile sciolti in etere e con 168 cm³ di NaOH.N/$_1$. L'agitazione viene continuata a temperatura ordinaria ancora per 40' previa aggiunta di altri 45 cm³ di soluzione alcalina. Il liquido viene allora estratto con etere per allontanare alcune impurezze resinose, e poi acidificato nettamente al rosso-congo. Si separa in tal modo una massa oleosa che tende a solidificarsi. Dopo alcune ore si decanta il liquido sovrastante, e si scioglie in etere la massa pastosa. La soluzione eterea, per lenta evaporazione, lascia un deposito solido che viene disciolto in idrossido sodico diluito e riottenuto per acidificazione con acido cloridrico. Si ha così la lenta separazione di cristalli incolori che si distinguono in due specie: una in lunghi aghi e l'altra in minutissimi prismi tabulari. La loro separazione si raggiunge per cristallizzazione frazionata dall'acqua, essendo i cristalli aghiformi più solubili.

Questi ultimi fondono a 121°-122° e mostrano (sale sodico in acqua, c=0,82) $[\alpha]_D^{22}$ = —28°,1. Il prodotto in prismetti fonde a 144°-145° e mostra (sale sodico in acqua, c=0,77) $[\alpha]_D^{21}$ = —70°,1. All'analisi:

Composto fusibile a 121°-122°,	trov.% N	5,11;
» » 144°-145°,	»	5,14;
Calc. per $C_{14}H_{19}O_4N$	. . .	5,28.

$C_9H_{11}NO_2$ Ref. 549-o

EXAMPLE 1

D(−)-2-(2,5-Dimethylbenzylamino)-1-Butanol

2,5-Dimethylbenzyl chloride (78.4 grams; 0.50 mole) was added to D(−)-2-amino-1-butanol (160.0 grams; 1.79 moles) at 65°-72° C. over a period of about 30 minutes while stirring the reaction mixture. The reaction mixture was —stirred at 85°-90° C. for one hour, cooled to about 45° C., and poured into 500 mls of water. The resulting slurry was made alkaline to pH 11 by adding 50% caustic soda thereto and filtered. The filter cake was washed with water and recrystallized from 200 mls of acetone to obtain 83.5 grams (79.5% of theoretical) of D(−)-2-(2,5-dimethylbenzylamino)-1-butanol, m.p. 85°-88° C., [α]D25 = −28.0° (C, 2.5; methanol).

EXAMPLE 3

D(−)-2-(2,5-Dimethylbenzylamino)-1-Butanol Salt of N-Acetyl-L(+)-Phenylalanine

N-Acetyl-DL(±)-phenylalanine (30.0 grams; 0.145 mole)and D(—)-2-(2,5-dimethylbenzylamino)-1-butanol (30.0 grams; 0.145 mole) were dissolved in 300 mls of water at 80° C. and the solution was gradually cooled to 40° C. to form a heavy, crystalline slurry. The slurry was cooled to 20° C., and filtered; the filter cake was washed with 50 mls of ice cold (5° C.) water and dried to obtain 21.7 grams (72.3% of theoretical) of the D(−)-2-(2,5-dimethylbenzylamino)-1-butanol salt of N-acetyl-L(+)-phenylalanine.

Following the above procedure substituting equivalent amounts of N-benzoyl-DL(±)-phenylalanine, N-n-butyryl-DL(±)-phenylalanine, and N-n-caproyl-DL(±)-phenylalanine for the N-acetyl-DL(±)-phenylalanine the D(−)-2-(2,5-dimethylbenzylamino)-1-butanol salts of N-benzoyl-L(+)-phenylalanine, N-n-butyryl-L(+)-phenylalanine, and N-n-caproyl-L(+)-phenylalanine, respectively, are obtained.

EXAMPLE 4

N-Acetyl-L(+)-Phenylalanine

The D(−)-2-(2,5-dimethylbenzylamino)-1-butanol salt of N-acetyl-L(+)-phenylalanine (10.0 grams; 0.024 mole) from Example 3 was slurried in 40 mls of water at 30°-35° C. and acidified with acetic acid to obtain a weak blue spot on Congo Red indicator paper. The resulting slurry was cooled to 5° C. and filtered. The resulting filter cake was washed with ice cold water and dried to obtain 4.5 grams (90% of theoretical) of N-acetyl-L(+)-phenylalanine, m.p. 168.5°-169.8° C., [α]D25 = 50° (C = 1; ethanol).

In the manner described above utilizing equivalent amounts of the other products obtained in Example 3 N-benzoyl-L(+)-phenylalanine, N-n-butyryl-L(+)-phenylalanine, and N-n-caproyl-L(+)-phenylalanine are obtained.

EXAMPLE 6

Recovery of N-Acetyl-D(−)-Phenylalanine

The mother liquors obtained in Example 3 were treated with hydrochloric acid to produce a weak blue spot on Congo Red indicator paper (pH 4.5-5.0), cooled to 5° C. and allowed to stand for about 3 hours. The resulting precipitate was filtered, washed with ice cold water and dried to obtain 13.1 grams (87.3% of theoretical) of N-acetyl-D(−)-phenylalanine, m.p. 157°-161° C., [α]D25 = −49.8° (C, 1.0; ethanol).

EXAMPLE 7

Recovery of D(−)-2-(2,5-Dimethylbenzylamino)-1-Butanol

The mother liquor and wash liquor obtained in Example 4 were combined and made alkaline by adding 50% caustic soda thereto. After standing at ambient conditions for several hours, the slurry was filtered, washed with water and dried to obtain 4.9 grams (98% of theoretical) of D(−)-2-(2,5-dimethylbenzylamino)-1-butanol.

$C_9H_{11}NO_2$ Ref. 549p

Versuch 8: Spaltung von racem. Formylphenylalanin

2,2 g racem. N-Formylphenylalanin[8] und 2,8 g (—)-α-Phenyläthylthiuroniumacetat werden in 6 ccm Methanol gelöst und 40 ccm Wasser von 40 bis 50° zugesetzt. Der Methylalkohol wird bei dieser Temperatur im Vak. abdestilliert und die zurückbleibende, wäßr. Lösung noch warm filtriert. Man läßt nun den Ansatz (der bereits Kristallisationsansätze zeigt) 2 Stdn. bei 30° stehen, kühlt dann langsam auf 20° und läßt schließlich bei 0° 24 Stdn. stehen. (Bei zu schnellem Kühlen tritt ölige Fällung ein.) Nach dieser Zeit werden die ausgeschiedenen Kristalle isoliert (1,8 g, 83% d. Th., Schmp. 150°). Die Mutterlauge (A) wird zur Gewinnung des 2. Antipoden verwendet. Das Salz, bestehend aus (—)-Thiuroniumbase und d-(—)-Formylphenylalanin, wird zur Reinigung aus Methanol/Äther umgefällt (Ausbeute 1,2 g), Schmp. 154°. $[\alpha]_D^{20}$ = — 137,5°; c = 2 in Alkohol (96%), l = 2 dm, α_D^{20} = — 5,5°.

$C_{19}H_{23}O_3N_3S$ (373,46). Ber. N 11,25. Gef. N 11,11.

0,5 g dieses Salzes werden im Scheidetrichter mit 10 ccm 1 n HCl und 10 ccm Essigester geschüttelt. Die salzsaure Schicht wird nochmals mit etwas Essigester nachgeschüttelt. Die vereinigten Essigesterlösungen werden mit 5 ccm Wasser neutralgeschüttelt, getrocknet und im Vak. verdampft. Der Essigester hinterläßt 0,2 g d-(—)-Formylphenylalanin[8], Schmp. 164 bis 167°. $[\alpha]_D^{20} = -72°$; $c = 2,5$ in Alkohol, $l = 1$ dm, $\alpha_D^{20} = -1,8°$.

Die obengenannte, wäßr. Mutterlauge (A) wird mit konz. Salzsäure eben kongosauer gemacht und 3mal mit Essigester (je 10 ccm) ausgeschüttelt. Die vereinigten Essigesterlösungen hinterlassen nach Waschen mit etwas Wasser, Trocknen und Abdestillieren 1,1 g noch nicht optisch reines l-(+)-Formylphenylalanin. Zur weiteren Reinigung wird 1,1 g der rohen Verbindung, zusammen mit 1,4 g (+)-α-Phenyläthyl-thiuroniumacetat (Vers. 5) in 5 ccm Methanol gelöst, 15 ccm heißes Wasser zugesetzt und der Methylalkohol im Vak. abdestilliert. Beim langsamen Erkalten tritt meist ölige Fällung auf, die auf Anreiben kristallisiert. Man erhält so 1,5 g rohes Salz, bestehend aus (+)-α-Phenyläthyl-thiuroniumbase und l-(+)-Formylphenylalanin. Schmp. 150°. Aus Methylalkohol/Äther umgefällt, schmilzt das Salz bei 154°. $[\alpha]_D^{20} = +140°$; $c = 2$ in Alkohol (96%), $l = 1$ dm, $\alpha_D^{20} = +2,8°$.

0,3 g dieses Salzes werden in 7 ccm heißem Wasser gelöst und die Lösung mit einigen Tropfen konz. Salzsäure kongosauer gemacht. Beim Erkalten scheiden sich 0,11 g l-(+)-Formylphenylalanin[8], Schmp. 164 bis 167°, ab. $[\alpha]_D^{20} = +74°$; $c = 5$ in Alkohol, $l = 1$ dm, $\alpha_D^{20} = +3,7°$.

Versuch 9: Darstellung von akt. α-Phenyläthyl-thiuroniumacetat aus akt. α-Phenyläthylchlorid

3,5 g d-α-Phenyläthylchlorid, $[\alpha]_D^{20} = +48°$ (Substanz)[5] werden mit 1,9 g Thioharnstoff in 10 ccm 80% Methanol 4 Stdn. gelinde gekocht. Dann wird das Lösungsmittel im Vak. abdestilliert und der Rückstand in 20 ccm Wasser aufgenommen. Die entstandene, milchige Trübung wird durch Filtration mit Tierkohle entfernt und das Filtrat mit einer konz. Lösung von 3,6 g kristall. Natriumacetat in Wasser versetzt. Das ausgeschiedene Thiuroniumacetat wiegt nach der Trocknung 3,4 g. $[\alpha]_D^{20} = -33,3°$; $c = 7,5$ in Alkohol.

Desgleichen werden 4,6 g l-α-Phenyläthylchlorid, $[\alpha]_D^{20} = -47,5°$, mit 2,5 g Thioharnstoff in 10 ccm 80% Methanol 1 Std. gekocht. Die Aufarbeitung ergibt 4,6 g Thiuroniumacetat. $[\alpha]_D^{20} = +41,5°$; $c = 5$ in Alkohol.

[8] *E. Fischer* und *W. Schoeller*, Ann. Chem. **357**, 1 (1907).

$C_9H_{11}NO_2$ Ref. 549q

Reagents.—Acetyl-DL-phenylalanine, lustrous plates, m.p. 152–154°, was prepared in yields of 93–96% from DL-phenylalanine (Dow) by acetylation of 0.8 mole of the amino acid with 2.4 moles of acetic anhydride and 1.6 moles of sodium hydroxide at a temperature below 10°. Thirty grams of finely ground papain (Wallerstein) was stirred with 150 ml. of water for 3 hours at 5°, the suspension centrifuged for 15 minutes at 2000 r.p.m. and the supernatant solution reserved for use. *p*-Toluidine (Merck and Co., Inc.) was used without further purification.

Acetyl-L-phenylalanine-*p*-toludide (I).—To 2 l. of a 0.5 *M* acetic acid–0.5 *M* sodium acetate buffer containing 9 g. of L-cysteine hydrochloride was added 155.5 g. (0.75 mole) of acetyl-DL-phenylalanine, the suspension warmed to 50° to effect complete solution, 80 g. (0.75 mole) of *p*-toluidine and the above enzyme solution added, the total volume brought to 3 liters with the acetate buffer, and the clear solution, *p*H 4.6, incubated at 40° for 7 days. The reaction mixture was maintained at 5° for 2 hours prior to the collection of the precipitate which was washed with 1 l. of water and air-dried to give 102–106 g. (92–95%) of I, m.p. 215–217°. This product was dissolved in 1.25 l. of hot 96% ethanol, the hot solution filtered, the filtrate held at 5° overnight, the precipitate collected, washed with 500 ml. of cold 96% ethanol, and air-dried to give 86–89 g. of I, m.p. 219°; $[\alpha]^{25}D$ +35 ± 1° (*c*, 4 in pyridine). *Anal.* Calcd. for $C_{18}H_{20}O_2N_2$ (296): C, 73.0; H, 6.8; N, 9.5. Found: C, 72.8; H, 6.7; N, 9.4. An additional 9–10 g. of I, m.p. 219°, was recovered from the mother liquor to give a total yield of recrystallized I of 86–89%.

L-Phenylalanine (II).—A suspension of 86 g. of recrystallized I in 1100 ml. of 20% hydrochloric acid was heated under reflux for 16 hours, the clear solution evaporated to dryness *in vacuo*, the residue dissolved in 300 ml. of water, again evaporated to dryness, the residue dissolved in 250 ml. of water and 450 ml. of 28% aqueous ammonia cautiously added to this solution. After the reaction mixture was held at 5° for 2 hours the precipitated *p*-toluidine was collected, washed with cold water, the filtrate and washings combined, the solution extracted with two 300-ml. portions of chloroform, the volume of the aqueous phase reduced to *ca.* 500 ml. by boiling, the solution cooled to 5° (2 hours), the precipitate collected, washed successively with 100 ml. of water and 40 ml. of 96% ethanol and air-dried to give 28 g. of II, lustrous flat plates; $[\alpha]^{25}D$ −34 ± 1° (*c*, 2 in water). Concentration of the mother liquor gave two additional crops of II of 6 and 4 g., respectively, or a total yield of 38 g. (82%).

D-Phenylalanine (III).—The filtrate remaining after the collection of I was evaporated *in vacuo* below 50° to one-half of its original volume, acidified with 120 ml. of concentrated hydrochloric acid, stored at 5° overnight, the crystalline precipitate collected, washed with 200 ml. of cold water and recrystallized from 20% aqueous methanol to give 50–65 g. (77–84%) of acetyl-D-phenylalanine (IV), m.p. 162–164°; $[\alpha]^{25}D$ −32 ± 1° (*c*, 2 in methanol). A suspension of 50 g. of IV in 300 ml. of 20% hydrochloric acid was heated under reflux for 5 hours, the clear solution evaporated to dryness, the residue dissolved in 300 ml. of water, 100 ml. of 28% aqueous ammonia added, the solution boiled to remove excess ammonia, decolorized with 5 g. of Norite, the clear colorless filtrate stored at 5° overnight, the crystalline precipitate collected, washed successively with 100 ml. of water and 50 ml. of 96% ethanol, and dried to give 27 g. of III; $[\alpha]^{25}D$ +34 ± 1° (*c*, 1 in water). A second crop of 3 g. of III was obtained from the mother liquor to give a total yield of 30 g. (77%).

The average over-all yields of D and L-phenylalanine from DL-phenylalanine were 59 and 68%, respectively.

$C_9H_{11}NO_2$ Ref. 549r

Note:

This reference contains a procedure for the resolution of phenylalanine based upon the asymmetric hydrolysis of the isopropyl ester of racemic phenylalanine by an enzyme preparation derived from pancreas. The reported $[\alpha]$ is $[\alpha]_D^{23}$ -35° to -35.2° (c 2, H_2O) and $[\alpha]_D^{23}$ +35.0° to +35.2° (c 2, H_2O).

For the experimental details of other enzymatic methods for the resolution of this compound, see Ref. 18n, and S. M. Birnbaum, L. Levintow, R. B. Kingsley and J. P. Greenstein, J. Biol. Chem., 194, 455 (1952) and J. B. Gilbert, V. E. Price and J. P. Greenstein, J. Biol. Chem., 180, 473 (1949).

Also see T. Tosa, T. Mori, N. Fuse and I Chibata, Biotech. Bioeng., 9, 603 (1967).

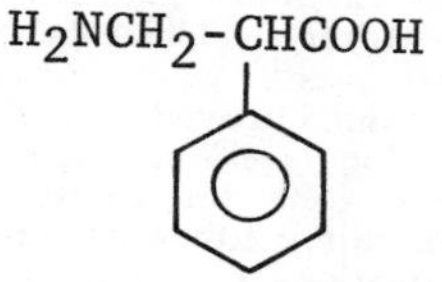

$C_9H_{11}NO_2$ Ref. 550

Acido D(—)-α-fenil-β-(N-carbobenzilossiamino)propionico (IIb)

Ad una soluzione bollente e limpida di 16,3 g (0,05 mole) di chinina base in 67 cc di etanolo si aggiunge una soluzione bollente di 14,6 g (0,048 mole) di acido DL-α-fenil-β-(N-carbobenzilossiamino)propionico in 56 cc di etanolo al 90%. Si raffredda per 4 ore in bagno di acqua e ghiaccio innescando la precipitazione del sale di chinina mediante sfregamento con una bacchetta di vetro. Il precipitato viene ricristallizzato da 55 cc di etanolo all'80% e lasciato una notte in frigorifero. Si filtra e si essicca sotto vuoto a 80°; si ottengono 11,4 g del sale di chinina dell'acido D(—)-α-fenil-β-(N-carbobenzilossiamino)propionico (IIb). P.f. 177-178°; $[\alpha]_D$ = —137,9° (C = 0,2%. etanolo).

Si sospende il sale di chinina, il cui potere rotatorio non si modifica per ulteriore ricristallizzazione, in 80 cc di acqua, si acidifica al Congo con HCl 10% e si estrae l'olio separatosi con etere solforico. L'estratto etereo viene lavato con acqua, anidrificato su sodio solfato, filtrato e concentrato sotto vuoto. L'olio residuo, ripreso con etere di petrolio, solidifica e si ricristallizza da alcool etilico acquoso. Resa 5,4 g (74,1% d.t.). P.f. 97-100° e $[\alpha]_D$ = —94° (C = 1%, etanolo).

	trov. % :	C 68,15;	H 5,57;	N 4,72
per $C_{17}H_{17}NO_4$	calc. :	68,21;	5,72;	4,68

Acido D(—)-α-fenil-β-aminopropionico (Ib)

30 g di acido D(—)-α-fenil-β-(N-carbobenzilossiamino)propionico vengono sciolti in 500 cc di alcool etilico ed idrogenati a pressione e temperatura ordinaria in presenza di 5 g di Pd al 5% su carbone. Il quantitativo teorico di idrogeno viene assorbito in ca. due ore. Si filtra e si raccoglie il catalizzatore ed il prodotto che è presente allo stato solido. Si tratta con acido cloridrico diluito separando così il prodotto dal catalizzatore. Le acque madri acide vengono neutralizzate a pH 6 mediante una soluzione diluita di idrato di sodio e concentrate a piccolo volume sotto vuoto. Il prodotto che si separa viene prima spappolato in poca acqua fredda poi purificato per riprecipitazione da una soluzione acquosa acida. Si essicca in stufa a vuoto. Resa 9,5 g a p.f. 225-226°; $[\alpha]_D$ = —83,5° (0,2%, H_2O).

	trov. % :	N 8,61
per $C_9H_{11}NO_2$	calc. :	8,48

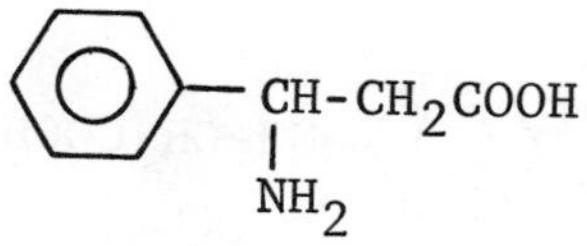

$C_9H_{11}NO_2$ **Ref. 551**

This compound, 0.21 mole, and 68 g. (0.21 mole) of quinidine were dissolved in 350 ml. of boiling methanol and allowed to crystallize. The salt was collected and recrystallized from methanol, leading to the **quinidine salt of L(+)-N-formyl-β-phenyl-β-aminopropionic acid**, 34 g. (0.065 mole), 62% yield, m.p. 192–193°, $[\alpha]^{22}D$ +181°, 5% in ethanol. A solution of 26 g. (0.050 mole) of this salt in 130 ml. of ethanol was treated with 52 ml. of *N* sodium hydroxide and concentrated to about 40 ml. *in vacuo*. The quinidine was filtered, and the filtrate was treated with 12 ml. of 5 *N* hydrochloric acid, leading to **L(+)-N-formyl-β-phenyl-β-aminopropionic acid**, 7.5 g. (0.039 mole), 36% yield, softens at 134°, m.p. 144–145°, lit.[32] m.p. 142–143°; $[\alpha]^{22}D$ +103°, 0.64% in ethanol; lit.[32] $[\alpha]D$ +115°. This compound (7.0 g., 0.036 mole) was boiled under reflux for 1 hr. in 125 ml. of 10% hydrochloric acid and concentrated to dryness *in vacuo*, and the residue was crystallized from methanol–ether leading to the hydrochloride of **L(+)-β-phenyl-β-aminopropionic acid**, 6.0 g. (0.030 mole), 83% yield. This was dissolved in water and treated with 30 ml. of *N* sodium hydroxide at 40°, leading to **L(+)-β-phenyl-β-aminopropionic acid**, 2.2 g. (0.013 mole), 44% yield, m.p. 232–233°, lit.[32] m.p. 234–235°; $[\alpha]^{22}D$ +10.6°, 0.65% in water; lit.[32] $[\alpha]^{20}D$ +6.9°; $[\alpha]^{22}D$ −10.2°, 1.5% in *N* sodium hydroxide; reported[32] $[\alpha]^{20}D$ −9.1°. L(+)-β-phenyl-β-aminopropionic acid, 1.0 g. (0.0060 mole), in 15 ml. of acetic anhydride was boiled under reflux for 3 hr. and concentrated *in vacuo*. The residue was taken up in chloroform and shaken with water, leading to **L(+)-β-phenyl-β-acetamidopropionic acid**, 0.64 g. (0.0031 mole), 51% yield, m.p. 198–199°, melting point of sample from Dr. B. Sjoberg, 200–202°; m.m.p. 199–200°, $[\alpha]^{22}D$ +103°, 1% in ethanol; $[\alpha]^{22}D$ of sample from Dr. Sjoberg, +95°.

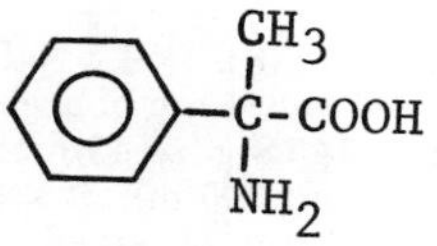

$C_9H_{11}NO_2$ **Ref. 552**

Resolution of 2-Formylamino-2-phenylpropanoic Acid.—Racemic starting material was prepared by formylation of 2-amino-2-phenylpropanoic acid as follows.[12] A solution of amino acid in 98% formic acid (100 g. in 228 ml.) was heated at 60° for 15 minutes. Acetic anhydride, 175 ml., was added dropwise at such a rate as to maintain a temperature of 60–62°.[13] Water, 36 ml., was then added cautiously. The resulting solution was cooled, the solid collected, washed with water and dried. The desired formamido acid amounted to 97 g. (83%), m.p. 176–177° (lit.[14] 178–179°). This material was resolved[14] through its quinine and cinchonine salts to give a 23% yield of (+)-isomer, $[\alpha]^{25}D$ +90.4° (*c* 3.3, ethanol) [lit.[14] $[\alpha]^{16}D$ +91.9° (*c* 3.2, ethanol)], m.p. 193–194° dec., and an 18% yield of (−)-isomer, $[\alpha]^{26}D$ −89.3° (*c* 3.4, ethanol) [lit.[14] $[\alpha]^{17}D$ −91.6° (*c* 3.7, ethanol)], m.p. 193–194° dec.

(+)- and (−)-2-Amino-2-phenylpropanoic Acid.—Hydrolysis of the two formamido acids with aqueous hydrobromic acid[14] gave a 64% yield of (+)-isomer, $[\alpha]^{26}D$ + 70.3° (*c* 2.0, water), m.p. 295° dec.; lit.[14] $[\alpha]^{18}D$ +70.0° (*c* 2.0, water), m.p. 295° dec. An 80% yield of (−)-isomer was obtained, $[\alpha]^{26}D$ −72.0° (*c* 2.0, water), lit.[14] $[\alpha]^{18}D$ − 69.5° (*c* 1.8, water).

(12) R. E. Sleiger, "Organic Syntheses," Coll. Vol. III, John Wiley and Sons, Inc., New York, N. Y., 1955, p. 88.

(13) V. du Vigneaud, R. Dorfmann and H. S. Loring, *J. Biol. Chem.*, **98**, 577 (1932).

(14) A. McKenzie and G. W. Clough, *J. Chem. Soc.*, **101**, 390 (1912).

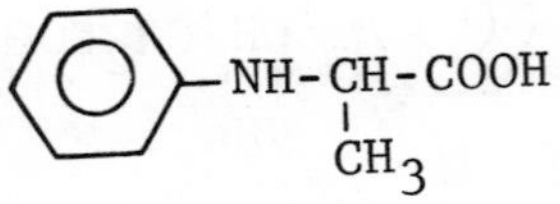

$C_9H_{11}NO_2$ Ref. 553

Experimental. Optically active α-anilino-propionic acids. The strychnine salt of the racemic acid (m.p. 161—161.5°) was recrystallised several times from dilute ethanol. The pure salt on decomposition yielded the *(—)-acid* which was recrystallised from dilute ethanol; m.p. 149.5—150.2°. An acid with $[\alpha]_D = +50°$ (in ethanol) was obtained from the mother liquor from the first recrystallisation of the strychnine salt. This product with (+)-α-phenylethylamine yielded a crystalline salt which was recrystallised several times from acetone. From the pure salt the *(+)-acid* was obtained. After recrystallisation from dilute ethanol it melted at 149.5—150.4°.

0.1004 g (—)-acid in abs. ethanol to 10.00 ml: $\alpha_D^{25} = -0.627°$. $[\alpha]_D^{25} = -62.5°$; $[M]_D^{25} = -103.2°$.

0.1031 g (+)-acid in abs. ethanol to 10.00 ml: $\alpha_D^{25} = +0.643°$. $[\alpha]_D^{25} = +62.4°$; $[M]_D^{25} = +103.0°$.

Found:

Equiv. wt.	165.0	Calc. for (—)-$C_9H_{11}O_2N$	165.2
» »	166.0	» » (+)- »	165.2

$C_9H_{11}NO_2$ Ref. 553a

(+)-N-Phenylalanine (IXa).—(±)-N-Phenylalanine (74.3 g., 0.45 mole) was dissolved in 675 ml. of acetone and combined with a solution consisting of 145.8 g. (0.45 mole) of quinine dissolved in 135 ml. of methanol and 315 ml. of acetone. After 24 hr. at room temperature, there was obtained 59.5 g. of quinine salt, m.p. 199–201°, $[\alpha]^{22}_D$ −126° (2% in ethanol), in the form of large prisms. Two subsequent recrystallizations did not alter the melting point or specific rotation.

Anal. Calcd. for $C_{29}H_{35}N_3O_2$: C, 71.20; H, 7.22; N, 8.58. Found: C, 71.11; H, 7.33; N, 8.90.

The quinine salt (50.0 g., 0.10 mole) was shaken in a separatory funnel with 150 ml. of *M* NaOH and 100 ml. of chloroform. The organic phase was drawn off and the aqueous phase was washed with a portion of chloroform. The aqueous phase was then acidified to pH 4, and the product was filtered, washed with water, and dried *in vacuo*. The yield of IXa, m.p. 149–150°, $[\alpha]^{23}_D$ +71° (2% in ethanol), was 13.1 g. The infrared spectrum (mull) was very similar to racemic IX.

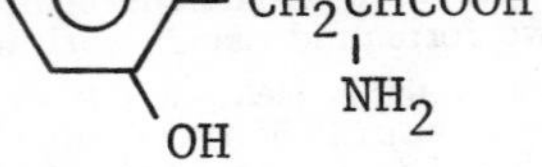

$C_9H_{11}NO_3$ Ref. 554

For the resolution procedure and physical constants, see Ref. 520.

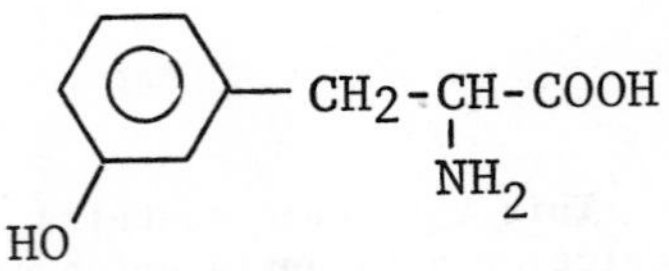

$C_9H_{11}NO_3$ Ref. 555

Resolution of Formyl-DL-*m*-tyrosine.—A mixture of 37 g. (0.176 mole) of the acylamino acid and an equimolar amount, 69.3 g., of anhydrous brucine was dissolved in 1900 ml. of hot 95% ethyl alcohol. The solution was filtered and allowed to cool slowly to room temperature during which time crystallization began. With frequent stirring the process was allowed to continue for four days in the cold. The 81.7 g. which was obtained exhibited a rotation of $[\alpha]^{26}_D$ −22.4° (1% solution in water). Three crystallizations from 20 volumes of alcohol in each case yielded 43.5 g. (82%) of salt with a rotation of $[\alpha]^{25}_D$ −37.3°. Additional recrystallizations did not increase the rotation observed, although with later resolutions on a larger scale an additional treatment was sometimes needed in order to achieve maximum rotation. All rotations were determined after drying over phosphorus pentoxide at 100°. The alcohol insoluble salt when air-dried contained one molecule of water of hydration.

In order to obtain the other diastereoisomer, all the alcoholic mother liquors were combined and concentrated to dryness *in vacuo*. The residue was dissolved in 3 volumes of water and treated with Norite. After a somewhat slow crystallization in the cold (it may be hastened by seeding) 37 g. of anhydrous salt was obtained. Two recrystallizations gave a maximum rotation of +0.73° (1% of anhydrous salt in water). The total yield calculated on the anhydrous basis was 27 g. corresponding to a 46% yield. The salt as obtained proved to be the tetrahydrate. In subsequent resolutions concentration of the aqueous mother liquors gave variable but significant increases in the total yield of this salt. Likewise, resolutions carried out with 0.5 mole quantities proved equally successful.

Formyl-D-*m*-tyrosine.—The alcohol insoluble salt obtained above was dissolved in 20 volumes of hot water and quickly cooled to 40°. The brucine was precipitated by making the solution alkaline to phenolphthalein with 2 *N* sodium hydroxide. After cooling overnight, the brucine was removed and the last traces extracted with six 50-ml. portions of chloroform. To the alkaline solution 6 *N* sulfuric acid was added in an amount exactly equivalent to the sodium hydroxide. After concentration to a thick sirup *in vacuo*, the compound was extracted with acetone. The acetone was removed *in vacuo* and the residue crystallized from 1 volume of water. A 40% yield of plate crystals melting at 146–148° was obtained. The mother liquor was reserved for isolation of the free amino acid after hydrolysis. This procedure is of particular advantage if the free amino acid is the primary requirement, for the formyl derivative is relatively soluble even in the cold. If desired, the isolation of the formyl derivative may be omitted entirely.

Anal. $[\alpha]^{27}_D$ −44.7°, 1% in water. Calcd. for $C_{10}H_{11}O_4N$: N, 6.69. Found: N, 6.51, 6.56.

Formyl-L-*m*-tyrosine.—The same procedure applied to the water insoluble brucine salt gave a 36.5% yield of compound melting at 148–149° and possessing a rotation of +45.7°. The mother liquor was again reserved for hydrolysis.

D-*m*-Tyrosine.—The formyl-D-*m*-tyrosine was refluxed for four hours with 10 volumes of 10% hydrochloric acid. The excess acid was removed *in vacuo* and the residue dissolved in 1 volume of water. A slight excess of concentrated ammonium hydroxide was added and immediately afterward sufficient glacial acetic acid to make the reaction slightly acid to litmus. The immediate crystallization was complete in twenty-four hours in the cold. The crystals were carefully washed with ice-cold water in small portions and finally with alcohol and ether. A 56% yield of plate crystals was obtained, which melted with decomposition at 275–276°. Recrystallization could be accomplished from either hydrochloric acid or hot water.

Anal. Calcd. for $C_9H_{11}O_3N$: N, 7.73. Found: N, 7.87, 7.68.

L-*m*-Tyrosine.—The identical procedure described above was used with the formyl-L-*m*-tyrosine. A quantitative yield of amino acid identical in all respects except with regard to direction of rotation was obtained.

$C_9H_{11}NO_3$ Ref. 555a

For the resolution procedure and physical constants, see Ref. 520.

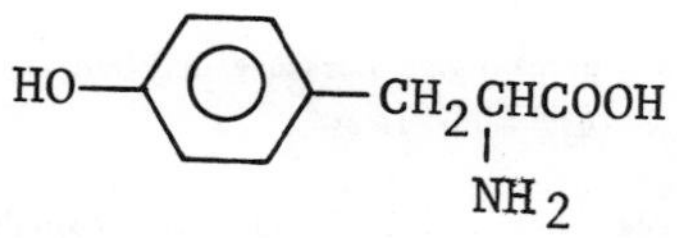

$C_9H_{11}NO_3$ Ref. 556

Spaltung des racemischen Benzoyltyrosins in die activen Componenten.

20 g fein gepulvertes, umkrystallisirtes *r*-Benzoyltyrosin werden mit 33 g Brucin in 2 L siedendem Wasser gelöst. Beim Erkalten trübt sich die Flüssigkeit rasch, und nach 15-stündigem Stehen ist ein gelblicher, zäher Syrup in reichlicher Menge abgesetzt. In der abgegossenen Mutterlauge beginnt dann nach einiger Zeit, zumal beim häufigen Reiben oder Impfen, die Krystallisation des Brucinsalzes und wird durch häufiges Umschütteln so beschleunigt, dass nach weiteren 12 Stunden der grösste Theil der einen Modification abgeschieden ist. Man filtrirt jetzt die Krystalle, löst den ersten Syrup durch Erhitzen mit der Mutterlauge auf dem Wasserbade und erhält beim längeren Stehen der erkalteten Flüssigkeit eine zweite Krystallisation. Die Wiederholung der Operation giebt dann noch eine dritte, aber recht geringe Krystallisation. Die Gesammtausbeute an krystallisirtem Brucinsalz betrug ebensoviel, wie die Menge des angewandten Benzoyltyrosins. Zur Reinigung wird das Salz in der 40-fachen Menge kochendem Wasser gelöst, wobei es zuerst schmilzt. Das beim Erkalten ausfallende Oel verwandelt sich nach kurzer Zeit in Krystalle, welche durch abermalige Krystallisation aus derselben Menge Wasser in glänzenden, vierkantigen Täfelchen, deren Ecken häufig schief abgeschnitten sind, erhalten werden. Seine Menge geht dabei auf $^2/_3$ des krystallisirten Rohproducts zurück. Beim weiteren Umkrystallisiren des Salzes ändert sich weder die Krystallform noch das Drehungsvermögen des daraus abgeschiedenen Benzoyltyrosins, sodass man das Salz schon nach 2-maliger Krystallisation aus Wasser als rein ansehen darf. Es wurde nicht analysirt.

Zur Umwandlung in Benzoyltyrosin löst man 10 g des reinen Salzes in 450 ccm kochendem Wasser, fügt 30 ccm Normalkalilauge zu, kühlt auf 0° ab, filtrirt nach 1-stündigem Stehen das auskrystallisirte Brucin ab und versetzt die Mutterlauge mit 45 ccm Normalsalzsäure.

Beim Verdampfen der Flüssigkeit unter stark vermindertem Druck auf etwa 15 ccm fällt das Benzoyltyrosin zuerst als harzige, farblose Masse aus, welche sich aber nach einiger Zeit schon während des Eindampfens in Krystalle verwandelt. Dieselben werden nach völligem Erkalten abfiltrirt, mit schwach salzsäurehaltigem Wasser gewaschen, in etwa 80 ccm siedendem Wasser gelöst, mit etwas Thierkohle behandelt und das Filtrat mit ein Paar Tropfen Salzsäure versetzt. Beim Abkühlen scheidet sich das active Benzoyltyrosin zuerst in Oeltropfen ab, welche aber beim längeren Stehen zu schönen glänzenden Blättern oder Tafeln erstarren. Da diese Verbindung dem natürlichen *l*-Tyrosin entspricht, so ist sie als

Benzoyl-*l*-tyrosin

zu bezeichnen. Die Ausbeute an reinem krystallisirtem Product betrug nach obigem Verfahren 22 pCt. des Racemkörpers, mithin 44 pCt. der Theorie.

Für die Analyse wurde bei 100° getrocknet.

0.2008 g Sbst.: 0.4952 g CO_2, 0.0959 g H_2O.

0.2061 g Sbst.: 8.9 ccm N (20°, 761 mm).

$C_{16}H_{15}NO_4$. Ber. C 67.37, H 5.26, N 4.90.

Gef. » 67.25, » 5.30, » 4.94.

Die Verbindung schmilzt bei 162—163° (corr. 165—166°), mithin 30° niedriger, als der Racemkörper. Sie ist auch in heissem Wasser erheblich leichter löslich als jener.

Für die Bestimmung des Drehungsvermögens diente eine wässrige Lösung, welche die für 1 Molekül berechnete Menge Kaliumhydroxyd enthielt. Das Drehungsvermögen ändert sich etwas mit der Concentration.

	I	II
Gesammtgewicht der Lösung	11.5477 g,	11.8693 g,
Gewicht des Benzoyltyrosins .	0.928 g,	0.607 g,
Gehalt an Alkali	3.4 ccm,	2.3 ccm Normalkalilauge,
specifisches Gewicht	1.0343,	1.0211,
Drehung bei 20° im 2 dm-Rohr	3.20°,	1.91°.

Mithin $[\alpha]_D^{20°} = +19.25°$ in 8-procentiger alkalischer Lösung und $[\alpha]_D^{20°} = +18.29°$ in 5-procentiger alkalischer Lösung.

Künstliches *l*-Tyrosin.

Beim vierstündigen Erhitzen des Benzoyl-*l*-tyrosins mit der 20-fachen Menge Salzsäure vom spec. Gewicht 1.19 auf 100° findet eine vollständige Abspaltung des Benzoyls statt, aber gleichzeitig wird ein erheblicher Theil, d. h. fast die Hälfte des Tyrosins, racemisirt. Auch bei Anwendung von 20-procentiger Salzsäure wurde eine solche Racemisirung, allerdings in viel schwächerem Maasse, beobachtet. Dagegen gab eine 10-procentige Salzsäure gute Resultate.

Dementsprechend wird 1 g des gepulverten Benzoyl-*l*-tyrosins mit 40 ccm 10-procentiger Salzsäure im verschlossenen Gefäss 8 Stunden auf 100° erhitzt. Im Anfang ist es nöthig, öfter umzuschütteln, um die Substanz zu lösen. Schliesslich wird die abgekühlte Flüssigkeit zur Entfernung der Benzoësäure ausgeäthert, dann unter vermindertem Druck verdampft, der Rückstand von salzsaurem Salz mit Natriumacetat völlig zersetzt, und das abfiltrirte Tyrosin aus heissem Wasser unter Zusatz von Thierkohle umkrystallisirt. Das so erhaltene Präparat zeigte genau die äussere Form des activen Tyrosins, d. h. es schied sich aus Wasser in ganz langen, seidenglänzenden, biegsamen Nadeln ab.

Für die Bestimmung des Drehungsvermögens diente die salzsaure Lösung, und zwar zunächst die Lösung in 21-procentiger Salzsäure, auf welche sich auch die meisten früheren Angaben über das optische Verhalten des natürlichen Tyrosins beziehen. Eine Lösung in Salzsäure von 21 pCt., welche 3.94 pCt. Base enthielt und das spec. Gewicht 1.116 hatte, drehte bei 20° das Natriumlicht im 2 dm-Rohr 0.76° nach links. Mithin

$$[\alpha]_D^{20} = -8.64.$$

Für natürliches Tyrosin wurde bei gleicher Concentration

von Mauthner[1]) $[\alpha]_D^{16} = -7.98°$,

von Landolt[2]) $[\alpha]_D^{20} = -8.07°$,

von Schulze und Bossbard[3]) $[\alpha]_D = -8.48°$

gefunden.

Wie man sieht, sind die Differenzen recht gering, und diejenigen zwischen den Bestimmungen von Schulze und Bosshard und den meinigen liegen sogar innerhalb der Beobachtungsfehler. Dass Mauthner und Landolt geringere Werthe gefunden haben, erklärt sich vielleicht durch die geringere Reinheit der von ihnen benutzten Basen. Ich halte es aber auch für möglich, dass dem Tyrosin, welches aus Eiweisskörpern durch Spaltung mit starker Salzsäure, die man in der Regel für diesen Zweck verwendet, gewonnen ist, eine wechselnde Menge racemischer Base beigemischt ist; denn ein Präparat, welches aus reinem Benzoyl-*l*-tyrosin mit 20-procentiger Salzsäure bei 100° dargestellt war, zeigte unter obigen Bedingungen

$$[\alpha]_D^{20} = -7.4°,$$

und obschon das Product sonst ganz dem activen Tyrosin glich, enthielt es also offenbar schon eine erhebliche Menge von Racemkörpern.

Bei geringerer Concentration der Salzsäure wächst das Drehungsvermögen des Tyrosins ziemlich stark, wie schon Schulze und Bosshard bei ihrem Präparat beobachteten; denn sie fanden für eine Lösung von 1 g Base in 20 ccm 4-procentiger Salzsäure

$$[\alpha]_D = -15.6°.$$

Das synthetische *l*-Tyrosin gab allerdings einen anderen Werth. Eine Lösung in 4-procentiger Salzsäure, welche 4.68 pCt. Tyrosin enthielt und das spec. Gewicht 1.034 hatte, drehte bei 20° im 2 dm-Rohr 1.28° nach links. Mithin

$$[\alpha]_D^{20} = -13.2°.$$

Ich habe deshalb zum Vergleich die Bestimmung mit natürlichem Tyrosin, welches aus Caseïn durch fünfstündiges Kochen mit 20-procentiger Salzsäure dargestellt war, wiederholt und unter den gleichen Bedingungen gefunden

$$[\alpha]_D^{20} = -12.56°.$$

[1]) Monatsh. f. Chemie 3 (1882), 345. [2]) Diese Berichte 17, 2838.

[3]) Zeitschr. f. physiolog. Chemie 9, 98.

Also auch hier ist das Drehungsvermögen des natürlichen Productes etwas geringer.

Die stärkere Abweichung der Zahl von Schulze und Bosshard kann ich mir deshalb nicht erklären.

Benzoyl-*d*-tyrosin.

Wie schon erwähnt, lässt sich die Verbindung aus dem Racemkörper durch das Cinchoninsalz gewinnen.

Löst man 1 g Benzoyltyrosin und 1 g Cinchonin in 150 ccm heissem Wasser, so scheidet sich nach dem Erkalten zunächst ein Syrup ab, der wenig Neigung zur Krystallisation hat. Aber beim mehrwöchentlichen Stehen erfolgt die Krystallisation, und wenn man einmal im Besitz der Krystalle ist, so wird die Gewinnung grösserer Quantitäten sehr einfach. Man löst dann 20 g des *r*-Benzoyltyrosins mit 20.6 g Cinchonin in 3 L kochendem Wasser und trägt in die nur wenig abgekühlte Lösung eine Probe des krystallisirten Salzes ein. Bei fortschreitender Abkühlung erfolgt dann sogleich die Abscheidung des Salzes in farblosen, ziemlich breiten Nadeln und ist nach 24 Stunden beendet. Zur völligen Reinigung wird dasselbe einmal aus 2 L, dann noch zweimal aus je 1½ L heissem Wasser umkrystallisirt und schliesslich in derselben Weise zerlegt, wie es oben für das Brucinsalz des Benzoyl-*l*-tyrosins beschrieben wurde. Die Ausbeute an Benzoyl-*d*-tyrosin ist eben so gross wie dort, sie betrug 42 pCt. der Theorie.

Die Substanz schmolz glatt bei 162° (corr. bei 165.5°). Die specifische Drehung wurde ebenfalls in alkalischer Lösung, unter ganz ähnlichen Bedingungen, wie bei der *l*-Verbindung, bestimmt.

Eine Lösung von 7.716 pCt. und dem spec. Gewicht 1.048 drehte bei 20° im 2 dm-Rohr 3.17° nach links. Mithin in alkalischer Lösung

$$[\alpha]_{20}^{D} = -19.59°,$$

während für die *l*-Verbindung + 19.25° gefunden wurde.

d-Tyrosin.

Die Darstellung der Base aus dem Benzoylderivat geschah in der gleichen Weise wie bei dem optischen Antipoden. Sie schmolz unter Zersetzung, gerade so wie das Isomere, und gab nach dem Trocknen bei 100° folgende Zahlen:

0.2007 g Sbst.: 0.4378 g CO_2, 0.1117 g H_2O.

$C_9H_{11}NO_3$. Ber. C 59.66, H 6.07.

Gef. » 59.49, » 6.18.

Für die optische Bestimmung diente wiederum eine Lösung in Salzsäure von 21 pCt

Sie enthielt 4.6071 pCt. Tyrosin, hatte das spec. Gewicht 1.1175 und drehte bei 20° im 2 dm-Rohr das Natriumlicht 0.89° nach links. Daraus berechnet sich

$$[\alpha]_D^{20} = +8.64°.$$

Ueber rechtsdrehendes Tyrosin liegt bisher nur eine recht unvollkommene Angabe vor. E. v. Lippmann fand in den bleichen Schösslingen der Zuckerrüben eine Base, welche er für Tyrosin erklärte, deren Lösung in 25-procentiger Salzsäure aber nur

$$[\alpha]_D = +6.85° \text{ zeigte.}$$

Ob das Product durch die Analyse, durch Schmelzpunkt und Löslichkeit mit dem gewöhnlichen Tyrosin verglichen wurde, ist nicht erwähnt. Es bleibt also zweifelhaft, ob von Lippmann wirklich den optischen Antipoden des *l*-Tyrosins unter Händen hatte.

Bevor die Spaltung mit Cinchonin aufgefunden war, wurde der Versuch gemacht, das *d*-Tyrosin aus den Mutterlaugen zu gewinnen, welche bei der Spaltung des *r*-Benzoyltyrosins resultirten. Dieselben wurden zunächst in der oben beschriebenen Weise von Brucin befreit und die Benzoylverbindungen durch Eindampfen der Lösung isolirt.

Da aus dem Gemisch das Benzoyl-*d*-tyrosin durch Krystallisation nicht in reinem Zustande zu isoliren war, so wurde das Rohproduct durch Kochen mit der 60-fachen Menge 10-procentiger Salzsäure zerlegt und das Tyrosin ebenfalls in der vorher beschriebenen Art isolirt. Zur Abscheidung des Racemkörpers diente dann das schwer lösliche Hydrochlorat. Zu dem Zweck wurde die rohe Base in der 20-fachen Menge Salzsäure vom spec. Gewicht 1.1 warm gelöst und die Flüssigkeit 12 Stunden bei 0° gehalten, wobei sich eine reichliche Menge des racemischen Tyrosinhydrochlorats in Nadeln ausschied. Die aus der Mutterlauge durch Verdampfen und Behandeln mit Natriumacetat isolirte Base wurde dann mehrmals aus heissem Wasser unter Zusatz von Thierkohle umkrystallisirt. Wenn man hierbei die Krystallisation durch rasches Abkühlen bewerkstelligt und schnell filtrirt, so lässt sich der Rest des Racemkörpers, wie es scheint, vollständig entfernen. Das so isolirte *d*-Tyrosin zeigt in der Krystallform, dem Schmelzpunkt und in der Zusammensetzung ($C_9H_{11}NO_3$. Ber. C 59.66, H 6.07. Gef. C 59.46, H 6.21) völlige Uebereinstimmung mit der reinen activen Base; aber das Drehungsvermögen war wesentlich grösser. Denn eine Lösung des Products in 21-procentiger Salzsäure zeigte ungefähr

$$[\alpha]_D^{20} = +11.6°,$$

während eine Lösung in 4-procentiger Salzsäure ungefähr

$$[\alpha]_D^{20} = +16.4°$$

ergab.

Ich bin nicht in der Lage, den Grund dieser auffälligen Erscheinung angeben zu können, halte es aber für möglich, dass die stärkere Drehung von einer sehr ähnlichen, isomeren Substanz herrührt, deren Benzoylverbindung schon dem angewandten *r*-Benzoyltyrosin beigemengt sein könnte.

$C_9H_{11}NO_3$ Ref. 556a

EXPERIMENTAL

Preparation of N-Acetyl-dl-tyrosine—100 gm. of *l*-tyrosine were dissolved in 1120 ml. of N sodium hydroxide and 400 ml. of redistilled acetic anhydride were added in ten portions at 10 minute intervals with continuous stirring. The reaction mixture was allowed to stand in a water bath at 60–70° for 6 hours, after which the calculated amount of 6 N sulfuric acid exactly to neutralize the sodium hydroxide was added. In order to remove the acetic acid, this solution was distilled to a thick syrup *in vacuo* and again concentrated with two portions of water. The material soluble in wet acetone, after removal of the acetone by distillation *in vacuo*, was dissolved in 100 ml. of water for purpose of hydrolysis of the acetoxy group. 400 ml. of 2 N sodium hydroxide were added to adjust the reaction to pH 10 to 11. After 30 minutes at room temperature the calculated amount of 6 N sulfuric acid was added, and the mixture concentrated to a thick syrup. The acetyltyrosine was separated from sodium sulfate with wet acetone, and then crystallized from 300 ml. of water, 116 gm. being obtained. After concentration of the mother liquor and the removal of a second crop of crystals, the combined material weighed 120 gm. Samples dried *in vacuo* over phosphorus pentoxide at 65° lost 7.60 and 7.68 per cent respectively, whereas 1 molecule of water of hydration is calculated as 7.47 per cent. A yield of 90.3 per cent was thus obtained.

TABLE I

Tyrosine and Acetyltyrosine Derivatives

All melting points are corrected. Optical activity was measured with the D line of sodium. Following the specific rotation are the items of concentration, solvent, and temperature respectively.

Derivative	Formula	Mol. wt.	Melting point	Optical rotation	Nitrogen analysis: Calculated	Nitrogen analysis: Found	Bibliographic reference No.
			°C.	*degrees*	*per cent*	*per cent*	
l-Tyrosine	$C_9H_{11}O_3N$	181		−10.6 (4%, N HCl, 22°)	7.73	7.74	
d-Tyrosine	"	181		+10.3 (4% " " 25°)	7.73	7.71	
N-Acetyl-*dl*- (monohydrate)	$C_{11}H_{13}O_4N \cdot H_2O$	241	94–95	0	5.81	5.85	
N-Acetyl-*l*-	$C_{11}H_{13}O_4N$	223	153–154	+47.3 (0.5%, H_2O, 26°)	6.28	6.33	(4, 7)
N-Acetyl-*d*-	"	223	153–154	−48.3 (0.5%, " 25°)	6.28	6.33	
Diacetyl-*dl*-	$C_{13}H_{15}O_5N$	265	168–170	0	5.28	5.22	(4)
Diacetyl-*l*-	"	265	171–172	+40.4 (0.5%, H_2O, 27°)	5.28	5.19	(6, 8)
Diacetyl-*d*-	"	265	171–172	−38.7 (0.5%, " 26°)	5.28	5.13	

d-TYROSINE

Resolution of Acetyl-dl-tyrosine—54.1 gm. of acetyl-*dl*-tyrosine monohydrate and an equimolar quantity (88.5 gm.) of anhydrous *l*-brucine were dissolved in 550 ml. of boiling absolute alcohol. The 75 gm. (54 per cent of the total amount) of impure brucine salt of acetyl-*d*-tyrosine which separated after remaining overnight or longer in the refrigerator were then recrystallized three or four times from 5 volumes of 95 per cent alcohol. From repeated resolutions in this fashion the pure salt was obtained in average yields of 75 per cent of the theoretical, calculated on the basis of anhydrous material. The air-dried material melting at 150–155° (shrinking at 130°) when further dried *in vacuo* over phosphorus pentoxide at 100° was found to decrease from 7.85 to 10.6 per cent in weight. Consequently, all samples for analytical purposes were dried to constant weight under the above conditions. A 0.5 per cent solution of the anhydrous salt in 95 per cent alcohol gave $[\alpha]_D^{26} = -42.3°$. Nitrogen analysis gave 6.76 per cent, as compared to the calculated value of 6.80 per cent.

For purposes of reference, the brucine salt of acetyl-*l*-tyrosine was prepared. The original mother liquor was evaporated to dryness *in vacuo*, the alcohol being completely removed by three distillations with small portions of water. The residue was then recrystallized from 5 volumes of water. After two recrystallizations, the air-dried material, m.p. 110–112° with shrinking at 94°, represented 74 per cent of the theoretical and contained 5.8 per cent moisture. A 1 per cent solution of the anhydrous salt in alcohol gave $[\alpha]_D^{25} = +10.7°$.

Ordinarily the mother liquors obtained from the recrystallization of the *d* salt are used only for the purpose of recovering the brucine and tyrosine. The brucine salt is decomposed and the acetyltyrosine solution treated with hydrochloric acid. The free tyrosine is again subjected to the racemization procedure.

Preparation of Acetyl-d-tyrosine—The brucine salt of acetyl-*d*-tyrosine was dissolved in 20 volumes of warm water and, after quickly cooling to approximately 40°, 2 N sodium hydroxide was added until the reaction was alkaline to phenolphthalein. The brucine was removed by filtration after 12 hours in the cold and thoroughly washed with cold water. The combined alkaline filtrates were extracted five times with chloroform to remove traces of brucine. After addition of the calculated amount of 6 N sulfuric acid and evaporation to dryness, the acetone-soluble material was crystallized from 100 ml. of water. From 88 gm. of brucine salt 22 gm. of acetyl-*d*-tyrosine representing a 69 per cent yield were obtained. The mother liquor may be concentrated to obtain additional material, although in practice it has been preferable to hydrolyze the residue to free *d*-tyrosine.

d-Tyrosine—The unnatural isomer has been obtained by suspending the crystalline acetyl derivative in 10 volumes of 5 N hydrochloric acid and gently refluxing for 2.5 hours. The excess acid was removed by concentrating *in vacuo*, and redistilling to dryness with three portions of water. The residue was dissolved in water, treated with norit, and free *d*-tyrosine precipitated by the addition of excess sodium hydroxide and sufficient acetic acid to make the solution just acid to litmus. By this method pure *d*-tyrosine is obtained in 93 per cent yield. The over-all yield from the original tyrosine is thus 62 to 66 per cent.

4. du Vigneaud, V., and Meyer, C. E., *J. Biol. Chem.*, **98**, 295 (1932).
5. Herriott, R. M., *J. Gen. Physiol.*, **19**, 283 (1935).
6. Bergmann, M., and Zervas, L., *Biochem. Z.*, **203**, 280 (1928).
7. Takenaka, Y., *Acta schol. med. univ. imp. Kioto*, **4**, 367 (1922).
8. Bergmann, M., and Stern, F., *Ber. chem. Ges.*, **63**, 437 (1930).

See Also R. R. Sealock, Biochem. Prep., 1, 71 (1949).

$C_9H_{11}NO_3$ Ref. 556b

D. Racematspaltung des Tyrosinamides

4,0 g D,L-Tyrosinamid wurden in 150 cm³ n-Propanol gelöst, filtriert und mit einer filtrierten Lösung von 16,7 g (2 Mole auf 1 Mol Amid) Dibenzoyl-D-weinsäure in 70 cm³ n Propanol versetzt. Nach 60 Stunden wurde das L-Tyrosinamid-dibenzoyl-D-hydrogentartrat abgesaugt.

Ausbeute: 5,2 g entspr. 87% d. Th.; Schmp. 172–175°.
$[\alpha]_D^{20}$: −74,3° (c = 1,28 in Methanol).

$C_{27}H_{26}O_{10}N_2$ (538,5)	ber. N 5,20	gef. N 5,36.	

Die Überführung dieses Rohsalzes in L-Tyrosinamid und L-Tyrosin zeigte, daß es in 65proz. optischer Reinheit isoliert wurde. Durch Umkristallisieren aus Alkohol–Äther wurden 3,1 g (60% d. Th.) optisch reines Salz erhalten. Schmp. 177–178°.
$[\alpha]_D^{20}$: −69,3° (c = 1,15 in Methanol).

		C	H	N
$C_{27}H_{26}O_{10}N_2$ (538,5)	ber.	60,20	4,83	5,20
	gef.	60,20	5,22	5,36.

Das Filtrat des Spaltansatzes wurde im Vakuum eingedampft und mit Äther das D-Tyrosinamid-dibenzoyl-D-hydrogentartrat ausgefällt.
Ausbeute: 5,9 g entspr. 98% d. Th.; Schmp. 165–168°.
$[\alpha]_D^{20}$: −79,7° (c = 0,62 in Methanol).

Gef. N 5,26.

Gewinnung der optisch aktiven Tyrosinamid-hydrochloride

L-Tyrosinamid-hydrochlorid-monohydrat: In die Suspension von reinem L-Tyrosinamid-dibenzoyl-D-hydrogentartrat in absolutem Alkohol wurde bis zur Sättigung trockener Chlorwasserstoff eingeleitet. Das Salz ging dabei in Lösung und L-Tyrosinamid-hydrochlorid-monohydrat wurde mit absolutem Äther ausgefällt.

Ausbeute: 80% d. Th.; Schmp. 238–239°.
$[\alpha]_D^{20}$: +28,8° (c = 0,74 in 80proz. Alkohol).

Die Verseifung zum L-Tyrosin zeigte, daß die Verbindung in optisch reiner Form vorlag.

$C_9H_{12}O_2N_2 \cdot HCl \cdot H_2O$ (234,7)	ber.	C 46,06	H 6,40	H 11,94
	gef.	46,11	6,84	12,00.

Zum Nachweis des Kristallwassers wurden 227,9 mg L-Tyrosinamid-hydrochlorid-monohydrat zwei Stunden über P_2O_5 in der Trockenpistole im Vakuum auf 100° erwärmt. Auswaage 209,9 mg. Verlust 18,0 mg (theoret. 17,5 mg).

D-Tyrosinamid-hydrochlorid-monohydrat: Die Verbindung wurde entsprechend wie der Antipode aus dem rohen diastereomeren D-Tartrat isoliert. Ausbeute: 80% d. Th.; Schmp. 235–237°.

$[\alpha]_D^{20}$: −14,12° (c = 0,66 in 80proz. Äthanol) entspr. einer 50proz. optischen Reinheit.

$C_9H_{12}O_2N_2 \cdot HCl \cdot H_2O$ (234,7)	ber.	C 46,06	H 6,40	N 11,94
	gef.	46,05	5,94	12,30.

Durch fraktioniertes Kristallisieren aus Alkohol—Äther wurde das optisch reine D-Tyrosinamid-hydrochlorid-monohydrat erhalten.

Ausbeute: 70% d. Th., bezogen auf rohes Amidhydrochlorid.
Schmp. 237–238°. $[\alpha]_D^{20}$: −27,9° (c = 0,79 in 80proz. Alkohol).

Die Verseifung zu D-Tyrosin zeigte, daß das D-Tyrosinamid-hydrochlorid-monohydrat in einer optischen Reinheit von 96,7% vorlag.

Gef. C 46,19 H 6,61 N 12,05.

Gewinnung der optisch aktiven Tyrosine

L-Tyrosin: L-Tyrosinamid-hydrochlorid-monohydrat wurde mit der 40fachen Menge 5 n HCl eine Stunde im Ölbad auf 120° erhitzt. Die Salzsäure wurde im Vakuum entfernt, der Rückstand in wenig Wasser gelöst und die 1,3fache der berechneten Menge Anilin, in Alkohol gelöst, zugefügt. Das ausgefallene L-Tyrosin wurde aus Wasser umkristallisiert.
Ausbeute: 80% d. Th. Schmp. 280–285°.
$[\alpha]_D^{20}$: −10.0° (c = 1,86 in n HCl).

D-Tyrosin. Seine Darstellung erfolgte entsprechend aus dem gereinigten D-Amid-hydrochlorid-monohydrat. Ausbeute: 81% d. Th. Schmp. 280–283°.
$[\alpha]_D^{20}$: +9,88° (c = 1,64 in n HCl),
$[\alpha]_D^{20}$: +8,63° (c = 1,05 in 21proz. HCl).

Nach den Literaturangaben[14] waren die so gewonnenen Tyrosinantipoden optisch rein.

11) E. ABDERHALDEN, Hoppe-Seylers Z. physiol. Ch. **131**, 280 (1924).

$C_9H_{11}NO_3$ Ref. 556c

Example 10

2.2 g. of *dl*-N-acetyltyrosine and 1.7 g. of l-amphetamine are dissolved in 50 ml. of absolute alcohol with heating. The solution is allowed to stand for 18 to 20 hours at room temperature and the crop of *l*-amphetamine salt of *l*-N-acetyltyrosine is removed by filtration; $[\alpha]_D^{26} = +40.25°$ (2% in water). The crystals are dissolved in water, the solution is made alkaline with dilute sodium hydroxide solution and is extracted with ether. The residual solution is passed through a sulfonic acid ion exchanger (Dowex 50, acid form) and the effluent is concentrated to dryness to obtain *l*-N-acetyltyrosine.

The alcoholic filtrate is taken to dryness in vacuo and the residue is dissolved in water. The aqueous solution is subjected to ion exchange and the effluent is concentrated in the same manner described for the recovery of the *l*-isomer. The product is *d*-N-acetyltyrosine.

$C_9H_{11}NO_3$ Ref. 556d

B. Spaltung des D,L-Äthylesters

6,6 g D,L-Tyrosinäthylester werden in 30 cm³ absol. Alkohol durch gelindes Erwärmen in Lösung gebracht und mit einer heißen Lösung von 9,5 g D-Weinsäure in 35 cm³ absol. Alkohol vereint. Nach zweitägigem Stehen hatten sich 5,2 g D-Tyrosinäthylester-D-hydrogentartrat ausgeschieden, das ist 91% d. Th.; Schmp.: 83–85° (korr.); $[\alpha]_D^{21}$: + 4,0° (c = 1,12 in Methanol). Nach dem Umkristallisieren aus absol. Alkohol: Schmp. 92–94° (korr.); $[\alpha]_D^{20}$: – 0,8° (c = 0,97 in Methanol).

$C_{15}H_{21}O_9N$ (359,2)

Ber.	C 50,14	H 5,85	N 3,90
Gef.	C 49,83	H 6,15	N 3,91

Durch portionsweisen Ätherzusatz (insgesamt 100–120 cm³) zur Mutterlauge des Spaltansatzes werden weitere 4–5 g Salz als L-Tyrosinäthylester-D-hydrogentartrat isoliert. Der Weinsäureüberschuß verbleibt in der Mutterlauge. Schmp.: 82–84° (korr.); $[\alpha]_D^{21}$: + 13,6° (c = 0,88 in Methanol)

Gef. N 4,03.

C. Gewinnung von optisch reinem D- und L-Tyrosinäthylesterhydrochlorid

Rohes D-Tyrosinäthylester-D-hydrogentartrat (das aus Alkohol umkristallisierte Salz besitzt keinen höheren optischen Reinheitsgrad) bzw. das L-Ester-D-hydrogentartrat wird, wie beim D,L-Esterhydrochlorid beschrieben, in Suspension mit der 30fachen Menge Chloroform bei 0° mit Ammoniak behandelt, wobei 2% absol. Methanol als Lösungsvermittler und etwas wasserfreies Natriumsulfat zugesetzt werden. An Stelle des mit Äther isolierten L-Estersalzes kann so auch das durch Eindampfen der Mutterlauge des Spaltansatzes gewonnene Gemisch von L-Ester-D-hydrogentartrat und D-Weinsäure zersetzt werden.

Nach dem Vertreiben des Chloroforms im Vakuum gewinnt man die rohen Antipoden des Esters in 70–80proz. Ausbeute.

D-Tyrosinäthylester: Schmp.: 105–106° (korr.); $[\alpha]_D^{20}$: – 12,5° (c = 3,96 in Alkohol).

L-Tyrosinäthylester: Schmp.: 103–106° (korr.); $[\alpha]_D^{21}$: + 11,7° (c = 3,47 in Alkohol).

E. FISCHER[13]) gibt für L-Tyrosinäthylester einen Schmp. von 108–109° und eine spez. Drehung von $[\alpha]_D$: + 20,4° (c = 4,9 in Alkohol) an, die Rohprodukte besitzen demnach optische Reinheitsgrade von etwa 60%.

Die Ester werden in wenig absol. Äthanol gelöst, Chlorwasserstoff bis zur sauren Reaktion eingeleitet und ihre Hydrochloride mit absol. Äther quantitativ gefällt.

D-Tyrosinäthylester-hydrochlorid: Schmp.: 164–165° (korr.); $[\alpha]_D^{20}$: – 18,9° (c = 2,11 in Alkohol).

L-Tyrosinäthylester-hydrochlorid: Schmp.: 165–167° (korr.); $[\alpha]_D^{21}$: + 17,6° (c = 1,86 in Alkohol).

Durch ein- bis zweimaliges Umkristallisieren aus absol. Alkohol-Äther gewinnt man die Hydrochloride optisch rein:

D-Tyrosinäthylester-hydrochlorid: Schmp.: 169–171° (korr.); $[\alpha]_D^{20}$: – 29,1° (c = 2,10 in Alkohol).

L-Tyrosinäthylester-hydrochlorid: Schmp.: 170–171° (korr.); $[\alpha]_D^{20}$: – 28,9° (c = 1,98 in Alkohol).

F. RÖHMANN[14]) gibt für L-Tyrosinäthylester-hydrochlorid einen Schmp. von 166° an.

$C_{11}H_{15}O_3N \cdot HCl$ (245,6)

	Ber.	C 53,60	H 6,50	N 5,69
D	Gef.	C 53,73	H 6,72	N 5,91
L	Gef.	C 53,75	H 6,86	N 6,01.

Aus den reinen Hydrochloriden lassen sich nun praktisch quantitativ entweder wie bei der D,L-Verbindung beschrieben, die reinen Ester in Freiheit setzen oder durch halbstündiges Kochen mit 20proz. Salzsäure die reinen Aminosäurehydrochloride gewinnen.

D-Tyrosinäthylester: Schmp.: 108–109° (korr.); $[\alpha]_D^{20}$: – 19,2° (c = 4,20 in Alkohol).

L-Tyrosinäthylester: Schmp. 108:–109° (korr.); $[\alpha]_D^{20}$: + 19,4° (c = 4,07 in Alkohol).

$C_{11}H_{15}O_3N$ (209,1)

	Ber.	C 63,16	H 7,17	N 6,69
D	Gef.	C 62,74	H 7,38	N 6,81
L	Gef.	C 62,80	H 7,06	N 6,41

D-Tyrosinhydrochlorid: $[\alpha]_D^{20}$: + 8,2° (c = 4,15 in 20proz. Salzsäure).

L-Tyrosinhydrochlorid[15]): $[\alpha]_D^{20}$: – 8,4° (c = 3,96 in 20proz. Salzsäure).

Die hieraus auf übliche Weise durch Neutralisation mit Ammoniak gegen Kongopapier gewonnenen freien Aminosäureantipoden lieferten folgende Daten:

D-Tyrosin: Schmp.: 294–300° (korr., Zersetzung)

L-Tyrosin: Schmp.: 294–300° (korr., Zersetzung)

$C_9H_{11}O_3N$ (181,1)

	Ber.	C 59,66	H 6,08	N 7,73
D	Gef.	C 59,59	H 6,55	N 7,88
L	Gef.	C 59,43	H 6,15	N 7,83.

[13]) E. FISCHER, Ber. dtsch. chem. Ges. **34**, 451 (1901).

[14]) F. RÖHMANN, Ber. dtsch. chem. Ges. **30**, 1978 (1897).

[15]) E. FISCHER, Ber. dtsch. chem. Ges. **32**, 3643 (1899).

$C_9H_{11}NO_3$ Ref. 556e

Die Darstellung von *l*(+)-Tyrosin geschah aus 5 g *racem.* Tyrosin, 13.3 g Cholestenonsulfonsäure und 20 ccm absol. Alkohol. Die Krystallisation trat nach wenigen Min. ein. Das Salz wurde umkrystallisiert. Auf eine Analyse wurde verzichtet.

Nach dem Ausfällen der Cholestenonsulfonsäure mit Bleioxyd wurde filtriert und das klare Filtrat mit Essigsäure angesäuert. Die Hauptmenge des in Wasser schwer löslichen Tyrosins fiel nach mehrstündigem Stehenlassen aus. Durch Einengen der Mutterlauge im Vak. konnte die Ausbeute etwas verbessert werden. $[\alpha]_D^{20} = +0.55° \times 3.4082/0.0980 \times 2 \times 1.112 = +8.6°$ (in 21-proz. HCl).

For the experimental details of enzymatic methods for the resolution of this compound, see S. M. Birnbaum, L. Levintow, R. B. Kingsley and J.P. Greenstein, J. Biol. Chem., <u>194</u>, 455 (1952), and J. B. Gilbert, V. E. Price and J.P. Greenstein, J. Biol. Chem., <u>180</u>, 473 (1949). and Ref. 520

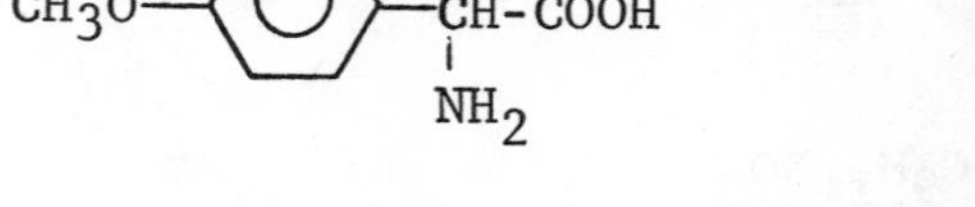

$C_9H_{11}NO_3$ Ref. 557

Example 1

a) **Preparation of DL - α - acetylamino - 4 - methoxyphenylacetic acid from DL - α - amino - 4 - methoxyphenylacetic acid**

To a solution containing 181 g (1.0 mole) of DL - α - amino - 4 - methoxyphenylacetic acid and 80 g (2.0 moles) of sodium hydroxide in 1 l of water was added 153 g (1.5 moles) of acetic anhydride at room temperature. This mixture was stirred for half an hour and then, adding conc. hydrochloric acid, brought to pH 1. The precipitate filtered off and dried, the yield was 208 g (=0.93 mole=93%) of DL - α - acetylamino - 4 - methoxyphenylacetic acid I.

b) Isolation of D(−) - α - acetylamino - 4 - methoxyphenylacetic acid from DL - α - acetylamino - 4 - methoxyphenylacetic acid I

While heating, 54 g (0.185 mole) of cinchonidine (grade purum, manufactured by Fluka AG) and 34 g (0.152 mole) of the acid I were dissolved in 600 ml of chloroform. Some crystal nuclei were introduced and the mixture was stirred at room temperature for approximately 12 hours. The cinchonidine salt was filtered off and washed twice with 100 ml of chloroform. After drying at reduced pressure the yield of salt was 35 g. In addition to the D(−) - α - acetylamino - 4 - methoxyphenylacetic acid II the salt thus obtained also contained cinchonidine and chloroform in a molar ratio of 1:1:1. Thus, these 35 g of salt corresponded to 0.055 mole of acid II (=a yield of 72.5%). The specific rotation of this salt was $[\alpha]_{546}=-142°$ and $[\alpha]_{578}=-123°$, respectively (c=1 in methanol at room temperature). The resulting $[\alpha]_D$ value proved to be −117°. To isolate the pure acid II, 24 g of the salt obtained above were shaken with 100 ml of water, 12 ml of conc. ammonium hydroxide and 70 ml of chloroform. The chloroform layer was separated and the extracting process was repeated, once with 70 ml of chloroform and once with 100 ml of ether. The aqueous layer was adjusted to pH 1 by the addition of conc. hydrochloric acid; the precipitated acid was filtered off and washed. After drying, the yield was 7.6 g, i.e. 74% pure D(−) - α - acetylamino - 4 - methoxyphenylacetic acid II. The specific rotation amounted to $[\alpha]_{546}=-255.3°$ and $[\alpha]_{578}=-221°$, respectively (c=1, methanol).

Instead of chloroform dioxane may be used. In that case, however, the resulting product has to be recrystallized more frequently, which generally leads to a lower yield.

c) Conversion of the D(−) - α - acetylamino - 4 - methoxyphenylacetic acid II into D(−) - α - amino - 4 - methoxyphenylacetic acid

A mixture consisting of 6.5 g of the acid II, 8 ml of conc. hydrochloric acid and 38 ml of water was refluxed for 3 hours. After cooling, the mixture was adjusted to pH 5.5 by the addition of ammonium hydroxide, the acid precipitating. The precipitate was filtered off, washed with a little water, and dried. The yield was 5.0 g=95% of D(−) - α - amino - 4 - methoxyphenylacetic acid III. The specific rotation amounted to $[\alpha]_{546}=-190°$ and $[\alpha]_{578}=-164°$, respectively (c=1, 1N HCl).

The resulting $[\alpha]_D$ proved to be −156°, whereas the highest value known in literature is the value mentioned in the British Patent Specification 1,240,687, viz. $[\alpha]_D^{25}=150.4°$ (c=1, 1N HCl).

$C_9H_{11}NO_3$

Ref. 557a

II. dl-2-(p-methoxyphenyl)-N-(chloroacetyl)glycine

To a stirred suspension of 36 g. (0.2 mole) of dl-2-(p-methoxyphenyl)-glycine in 500 ml. of H_2O was added 8 g. (0.2 mole) of NaOH pellets and when a clear solution was obtained the solution was cooled to 5° C. and with vigorous stirring 68.2 g. (0.4 mole) of chloroacetic anhydride (warm) was added all at once. Then a solution of 16 g. (0.4 mole) of NaOH in 100 ml. of H_2O was added over a 10 to 15 minute period. More 20% NaOH was added as needed to keep the pH at about 9 for a 1.5 hour period. Next, the pH was adjusted to 2 with 40% H_3PO_4. The product crystallized immediately and was filtered off, washed with water and recrystallized from ethanol-water to give 38 g. of product melting at 182°–183° C.

Analysis.—Calcd. for $C_{11}H_{12}ClNO_4$: C, 51.21; H, 4.69. Found: C, 51.49; H, 4.90.

III. D-(—)-2-(p-methoxyphenyl)-N-(chloroacetyl) glycine and L-(+)-2-(p-methoxyphenyl)-glycine

To 800 ml. of H_2O stirred at 37° C. was added 38 g. (0.148 mole) of dl-2-(p-methoxyphenyl)-N-(chloroacetyl)-glycine and NH_4OH added dropwise until pH 7.8 was obtained. To the resulting solution was added 2 g. of Hog Kidney Acylase (Sigma Chemical Company) and stirring continued at 37° C. (internal) for 21 hours. The solids containing crude L-(+)-2-(p-methoxyphenyl) - glycine were then filtered off and washed with 2×100 ml. H_2O and the pH of the combined filtrates adjusted to 4–5 with glacial acetic acid. This solution was heated on the steam bath for 30 min. with 5 g. of decolorizing carbon and then filtered. The carbon cake was washed with 50 ml. of warm water and the combined filtrates cooled and acidified to pH 2 with 40% H_3PO_4. After one hour cooling at 0° C. the crystalline product was filtered off and washed with cold water (3×) and air dried. The yield was 16 g. A second run using five times the previous amounts gave a yield of 83 g. (87% yield); M.P. 170–171° C.; $[\alpha]_D^{25°\,C.}$ −193 (C=1% ethanol).

Analysis.—Calcd. for $C_{11}H_{12}ClNO_4$: C, 51.21; H, 4.69. Found: C, 51.50; H, 4.99.

When the solids containing crude L-(+)-2-(p-methoxyphenyl)-N - chloroacetylglycine are treated with hot 3 NHCl (200 ml.) and carbon followed by filtration and pH adjustment to 5.5 there is obtained 6 g. (first run) of pure L-(+)-2-(p-methoxyphenyl)glycine.

$[\alpha]_D^{25°\,C.}$ +150.4° (C=1%, 1 NCHl)

IV. D-(—)-2-(p-methoxyphenyl)-glycine

The 16 g. of D-(—)-2-(p-methoxyphenyl)-N-chloroacetylglycine was refluxed 1.5 hours in 170 ml. of 2 NHCl. The resulting clear solution was filtered and cooled at 5° C. and the pH adjusted to 5.5 with NH_4OH. The product was then filtered off after cooling 30 min. and washed with 3×25 ml. of cold water. The dried material D-(—)-2-(p-methoxyphenyl)-glycine weighed 9.5 g. A second run gave 54 g. using the 83 g. of starting material from III.

$[\alpha]^D_{25°\,C.}$ −149.9° (C=1%, 1 NHCl) (first run)
$[\alpha]^D_{25°\,C.}$ −148.1° (C=1%, 1 NHCl) (second run)

Analysis.—Calcd. for $C_9H_{11}NO_3$: C, 59.67; H, 6.13; N, 7.74. Found: C, 59.38; H, 6.16; N, 8.00.

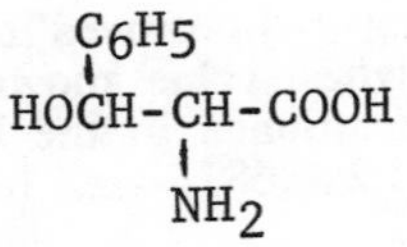

$C_9H_{11}NO_3$ Ref. 558

300 g. of the above DL-*threo*-β-phenyl-serine ethyl ester product are dissolved in 600 cc. of glacial acetic acid by heating on the steam bath. The solution is cooled to room temperature and reacted with 177 cc. of acetic anhydride (1.3 mol). The temperature is maintained below 40° C. by cooling with ice water. Upon completion of the reaction, the reaction mixture is cooled and the DL-*threo*-N-acetyl derivative crystallized out. It is filtered under suction and washed with water until the odour of acetic acid disappears. The filter cake is dried at 95° C. Upon recrystallization from alcohol, the product, principally DL-*threo*-N-acetyl-β-phenyl-serine ethyl ester, melts at 179—180° C. (uncorrected).

305 g. of the latter product are reacted with 487 cc. of 10 per cent. sodium hydroxide (1 mol), and heated on the steam bath for 30 minutes while shaking occasionally. The major part of the material goes into solution. A small residue is filtered off. The solution is then concentrated on the steam bath under diminished pressure until the sodium salt begins to crystallize. The precipitate is again dissolved by gentle warming, and the solution set congo-acid with concentrated hydrochloric acid. After cooling for 2 hours, the precipitated DL-*threo*-N-acetyl-β-phenyl-serine product is filtered off. M. p. 142—146° C. (uncorrected) with decomposition. Upon recrystallization from ethyl acetate-methanol, the m.p. rises to 148—150° C. (uncorrected) with decomposition.

112 g. of the above DL-*threo*-N-acetyl-β-phenyl-serine product are dissolved in 500 cc. of hot water and added to a suspension of 162 g. of quinine in 2 litres of water at 90° C. Solution is effected by vigorous shaking. A slight residue is filtered off and the solution is allowed to stand at room temperature. Thereupon, the quinine salt of L-*threo*-N-acetyl-β-phenyl-serine crystallizes in colourless prisms. After 48 hours, this is filtered off, the filter cake washed with a little cold water and dried M.p. 154—156° C. (uncorrected) with decomposition; $[\alpha]_D^{20} = -107.5°$ (c=2 in CH_3OH).

This less soluble quinine salt is dissolved in about 700 cc. of 50 per cent. alcohol and reacted with a solution of 9.4 g. of caustic soda in 100 cc. of water. The major portion of the alcohol is distilled off under reduced pressure, and the precipitated quinine is filtered off. The solution is then extracted twice with chloroform, 150 cc. in each portion, in order to remove the last traces of quinine, and then the solution is concentrated to a volume of about 100 cc. under reduced pressure. The optically active L-*threo*-N-acetyl-β-phenyl-serine is precipitated with concentrated hydrochloric acid. This intermediate has a m.p. of 163—164° C. (uncorrected) with decomposition, and a rotation of $[\alpha]_D^{20} = +16.3°$ (c=2 in CH_3OH). Upon heating for 2 hours with about 150 cc. of 3N HCl, this product is deacylated. The reaction mixture is boiled down *in vacuo,* taken up twice with 50 cc. each of water and again evaporated to dryness each time. The residue is dissolved in the smallest possible quantity of water, the solution mixed with approximately four times its volume of alcohol, and neutralised with concentrated ammonia. The amino acid crystallizes out. and is removed by filtration. This product, L-*threo*-β-phenyl-serine, has a rotation of $[\alpha]_D^{18} = -46.0°$ (c=2 in 6N HCl), and a decomposition point of 195—198° C. (uncorrected). The product is freed from ammonium chloride by recrystallization from water, whereupon the rotation rises to $[\alpha]_D^{18} = -49°$ (c=2 in 6N HCl).

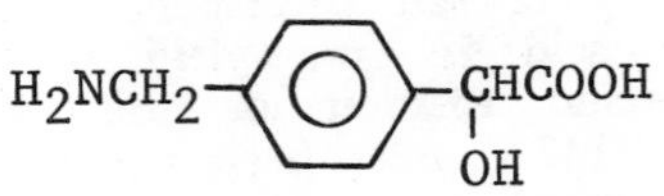

$C_9H_{11}NO_3$ Ref. 559

The authors reported that the above compound was custom prepared by Niels Clauson-Kaas Chemical Research Laboratory, Rugmarken 28, Farum, Denmark. The reported m.p. of the resolved acid was > 200°C. The $[\alpha]_D^{25}$= -124°C (c, 1.5, in HCl).

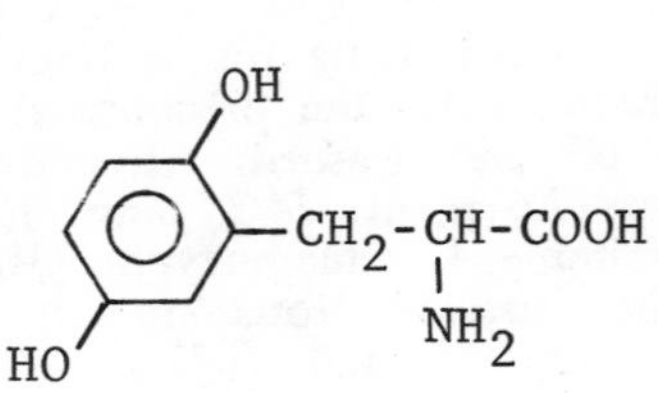

$C_9H_{11}NO_4$ Ref. 560

(–)-*Phenylethylamine salt of* L-*2-benzamido-3-(2′:5′-dimethoxyphenyl)-propionic acid.* The racemic acid (49 g.) and (–)-phenylethylamine (18·8 g.) were dissolved in methanol (300 ml.) and the solution evaporated to dryness. The crystalline residue was dissolved in boiling water (1500 ml.) and allowed to cool to 15°. The crystalline solid was filtered off, and again dissolved in water (1500 ml.) and cooled to 15°. A third recrystallization was done from 2 l. of water. Rotation and m.p. had by then become constant. Yield was 90% of the theoretical (31·5 g.). M.p. was 180°; $[\alpha]_D^{18°}$ –13·7° in methanol (c, 2·0). The salt crystallized as a *monohydrate.* (Found: C, 66·4; H, 6·8; N, 6·0. $C_{26}H_{30}O_5N_2.H_2O$ requires: C, 66·7; H, 6·8; N, 5·7%.)

(–)-*Phenylethylamine salt of* D-*2-benzamido-3-(2′:5-dimethoxyphenyl)-propionic acid.* The mother liquors amounting to 5 l. were combined and concentrated under reduced pressure (100 mm.) to about 1 l. Small amounts of crystalline material which separated out during the concentration, and which consisted of mixtures of diastereoisomers, were collected and preserved for a later resolution. The solution was then cooled to 0°, and the crystalline precipitate of m.p. 160° filtered off; a further crop was obtained by concentrating the solution to 300 ml. and cooling to 0°. The combined solids were again recrystallized first from water and then from ethanol. The yield was 24 g. The mother liquors were combined and concentrated. M.p. was 167–168°, $[\alpha]_D^{18°}$ +9·1° in methanol (*c*, 2·0). This salt also crystallized as *monohydrate*. (Found: C, 66·3; H, 6·5; N, 5·9. $C_{26}H_{30}O_5N_2.H_2O$ requires: C, 66·7; H, 6·8; N, 6·0%.)

The two salts of (+)-phenylethylamine were also prepared and had the expected properties: +D-*salt*: m.p. 180°; $[\alpha]_D^{19°}$ +13·8° in methanol (*c*, 2·0); +L-*salt*: m.p. 166°; $[\alpha]_D^{19°}$ −9·0° in methanol (*c*, 2·0). The −D- and +L-salts were probably optically not completely pure, but further recrystallization did not raise the rotation or m.p. any further.

D- *and* L-*2-Benzamido-3-(2′:5′-dimethoxyphenyl)-propionic acids.* The −L- and +D-salts respectively were decomposed as follows: 45 g. of the salt were dissolved in hot water (1 l.) and 5 N-NaOH (50 ml.) was added, whilst the solution was shaken. The phenylethylamine was removed by repeated extraction with chloroform and recovered in the usual manner. The aqueous solution was acidified with 5 N-HCl (55 ml.) and the precipitate filtered off and dried. The material was recrystallized first from chloroform-light petroleum (b.p. 120°) and then from 30% (v/v) acetic acid. M.p. of both compounds was 170°. The L-*acid* had $[\alpha]_D^{17°}$ −32·5° in ethanol (*c*, 2·0), whilst the D-*acid* had $[\alpha]_D^{16°}$ +32·8° in ethanol (*c*, 2·0). The L-acid only was analyzed. (Found: C, 65·6; H, 5·9; N, 4·1. $C_{18}H_{19}O_5N$ requires: C, 65·7; H, 5·9; N, 4·25%.) The +L- and −D-salts were decomposed in similar manner. The corresponding D- and L-benzamido acids had m.p. 167 and 168° respectively and the following rotations: D-acid: $[\alpha]_D^{17°}$ +30·5° in ethanol (*c*, 2·0); L-acid: $[\alpha]_D^{17°}$ −31·0° in ethanol (*c*, 2·0).

D- *and* L-*2:5-Dihydroxyphenylalanine.* 10 g. of the active benzoyl compound were hydrolyzed with 80 ml. of a mixture of equal parts of HI (sp.gr. 1·7) and glacial acetic acid which had previously been refluxed for 15 min. with red P (0·4 g.) and filtered. The mixture was refluxed for 15 min., methyl iodide removed by distillation, and the hydrolysis continued for another 1·25 hr. The amino-acids were isolated as described above. Yield of recrystallized amino-acid was 65–70%. The L-*amino-acid* had m.p. 265°, $[\alpha]_D^{17°}$ −8·1° in N-HCl (*c*, 2·0). (Found: C, 50·1; H, 6·2; N, 6·45; $C_9H_{11}O_4N.H_2O$ requires: C, 50·2; H, 6·1; N, 6·5%.) Loss of weight on drying at 110° (1 mm.) 8·1%, the *monohydrate* requires 8·4%. The D-*amino-acid* had m.p. 265°, $[\alpha]_D^{18°}$ +7·9° in N-HCl (*c*, 3·0), $[\alpha]_D^{16°}$ +40·0°±3·0 in water (*c*, 0·15).

Hydantoin. The L-amino-acid was converted into the L-*hydantoin*; 0·5 g. of the amino-acid and KCNO (0·5 g.) were dissolved in hot water (30 ml.). The solution was heated on the water bath for 40 min., whilst H_2 was passed through the solution. 5 N-HCl (7·5 ml.) was then added and heating continued for 2 hr. On cooling crystals appeared which were filtered off and, after treatment with charcoal, recrystallized from water. The L-*hydantoin* had m.p. 220–221°, $[\alpha]_D^{16°}$ −124·5° in 90% (v/v) ethanol (*c*, 0·92). (Found: N, 12·3. $C_{10}H_{10}O_4N_2$ requires N, 12·6%.)

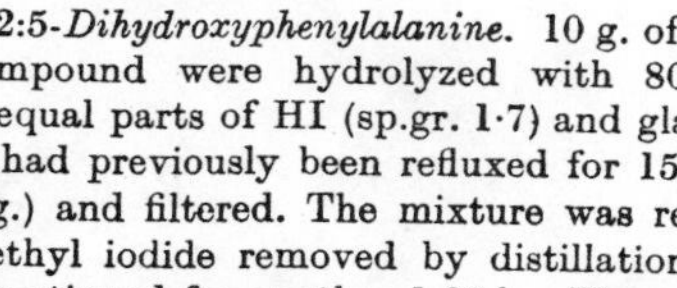

$C_9H_{11}NO_4$

Ref. 561

Beispiel 1

a) Herstellung des Cinchonin-Salzes von D-N-Benzoyl-DOPA

In einem Erlenmeyer-Kolben löst man 159,5 g rac. N-Benzoyl-DOPA-monohydrat (0,5 Mol) und 147 g Cinchonin (0,5 Mol) in etwa 750 ml Methanol und filtriert die Lösung in einen 4-l-Rundkolben und entfernt das Lösungsmittel unter Erwärmen im Wasserstrahlvakuum. Das zurückbleibende, sirupöse Salz löst man unter kräftigem Schütteln oder Rühren in einem siedenden Gemisch aus 2,5 l Aceton und 250 ml Wasser. Unmittelbar danach beginnt das Cinchonin-Salz des D-N-Benzoyl-DOPA auszukristallisieren. Zur Vervollständigung der Kristallisation läßt man die Mischung über Nacht bei Raumtemperatur oder im Kühlschrank stehen. Die Kristalle werden abgesaugt und mit gut gekühltem, wäßrigem Aceton (etwa 10 % Wasser) gewaschen. Ausbeute: 150,25 g (= 98 % der Theorie).

Für die weitere Verarbeitung zu D-N-Benzoyl-DOPA ist die Substanz rein genug und braucht daher nicht weiter gereinigt zu werden.

Für die analytischen Untersuchungen wurde ein Teil der Substanz aus dem Aceton-Wasser-Gemisch umkristallisiert. Sie zeigt dann folgende Werte:

Fp. = 168° C;
$[\alpha]_D^{20}$ = + 85° ± 2° (c = 1; Methanol).

b) Herstellung von D-N-Benzoyl-DOPA

Zur Isolierung von D-N-Benzoyl-DOPA versetzt man 150 g des unter 1, a) aus dem Lösungsmittelgemisch auskristallisierten Cinchonin-Salzes des D-N-Benzoyl-DOPA mit 100 ml halbkonzentrierter HCl und 400 ml Essigester und schüttelt die Mischung so lange, bis Lösung eingetreten ist. Die organische Schicht wird abgetrennt, die wäßrige Schicht wird noch fünfmal mit je 200 ml Essigester ausgezogen.

Die vereinigten Essigesterschichten werden mit Wasser gewaschen, mit Na_2SO_4 getrocknet und im Wasserstrahlvakuum eingeengt. Als Rückstand erhält man D-N-Benzoyl-DOPA, welches für die Weiterverarbeitung zu D-DOPA rein genug ist.

Ausbeute: 71 g (96,50 % der Theorie).

Für die analytischen Untersuchungen wurde ein Teil der Substanz aus Wasser umkristallisiert. Danach wurden folgende Werte erhalten.

Fp. = 170° C;
$[\alpha]_D^{20}$ = + 30° ± 2° (c = 1; Methanol).

c) Herstellung von D-Dihydroxy-phenylalanin

50 g D-N-Benzoyl-DOPA werden 2 Stunden lang mit etwa 255 ml 20%iger Salzsäure unter Stickstoff als Schutzgas zum Sieden erhitzt. Beim anschließenden Abkühlen kristallisiert ein Teil der freigewordenen Benzoesäure aus und wird abfiltriert. Der Rest wird aus der Lösung mit Äther extrahiert. Die wäßrige Lösung wird unter Schutzgas (Stickstoff) zur Trockene eingedampft.

Das D-3,4-Dihydroxy-phenylalanin-hydrochlorid wird in Wasser aufgenommen, die erhaltene Lösung mit einer Base, z. B. Triäthylamin, auf den isoelektrischen Punkt eingestellt. Nach längerem Stehen, am besten über Nacht, im Kühlschrank kristallisieren 31,5 g D-DOPA (96 % der Theorie).

Nach dem Umkristallisieren aus SO_2-haltigem Wasser erhält man 30,10 g reines D-DOPA (= 92 % der Theorie) vom Fp. 274° C (Zers.);

spezifische Drehung $[\alpha]_D^{20}$ = 11,3° ± 0,5° (c = 1; 1 n HCl).

d) Gewinnung des Cinchonin-Salzes von L-N-Benzoyl-DOPA

Dampft man die unter 1, a) nach dem Absaugen und Waschen der D-Salz-Kristalle erhaltene Mutterlauge im Wasserstrahlvakuum zur Trockene ein und wäscht den Rückstand mit Essigester, so erhält man das rohe Cinchonin-Salz des L-N-Benzoyl-DOPA in einer Ausbeute von 95 bis 99 % der Theorie. Für die Weiterverarbeitung zu L-N-Benzoyl-DOPA bzw. L-DOPA ist die Verbindung rein genug.

Für die analytischen Untersuchungen wurde ein Teil der Substanz aus Essigester umkristallisiert. Folgende Werte wurden erhalten:

Fp. = 154 bis 155° C;
$[\alpha]_D^{20} = +135° \pm 2°$ (c = 1; Methanol).

e) Herstellung von L-N-Benzoyl-DOPA

Man versetzt 150 g des unter 1, d) erhaltenen Cinchonin-Salzes des L-N-Benzoyl-DOPA mit 100 ml halbkonzentrierter Salzsäure und 400 ml Essigester und schüttelt die Mischung so lange, bis Lösung eingetreten ist. Die Essigester-Schicht wird abgetrennt, die wäßrige Schicht wird noch fünfmal mit je 200 ml Essigester ausgeschüttelt. Die vereinigten Essigesterextrakte werden mit Wasser gewaschen, mit Na_2SO_4 getrocknet und im Wasserstrahlvakuum eingeengt. Als Rückstand erhält man L-N-Benzoyl-DOPA, welches für die Weiterverarbeitung zu L-DOPA rein genug ist.

Ausbeute: 72 g (= 96 % der Theorie).

Ein Teil der Substanz wurde aus Wasser umkristallisiert zur Bestimmung der analytischen Daten.

Fp. = 172° C;
$[\alpha]_D^{20} = -30° \pm 2°$ (c = 1; Methanol).

f) Herstellung von L-DOPA

50 g L-N-Benzoyl-DOPA werden 2 Stunden lang mit etwa 225 ml 20%iger Salzsäure unter Stickstoff als Schutzgas zum Sieden erhitzt. Beim anschließenden Abkühlen kristallisiert ein Teil der freigewordenen Benzoesäure aus und wird durch Filtrieren entfernt. Der Rest wird aus der Lösung mit Äther extrahiert. Die wäßrige Lösung wird unter Schutzgas (Stickstoff) zur Trockene eingedampft. Das L-3,4-Dihydroxyphenylalaninhydrochlorid wird in Wasser aufgenommen, die erhaltene Lösung mit einer Base, z. B. Triäthylamin, auf den isoelektrischen Punkt eingestellt. Nach längerem Stehen (am besten über Nacht) im Kühlschrank kristallisieren 31 g L-DOPA (= 92 % der Theorie) aus. Nach dem Umkristallisieren aus SO_2-haltigem Wasser erhält man 30,0 g L-DOPA (= 89 % der Theorie) vom

Fp. 290° C (Zers.);
$[\alpha]_D^{20} = -11{,}3° \pm 0{,}5°$ (c = 1; 1 n HCl).

$C_9H_{11}NO_4$ Ref. 561a

2. *Spaltung des DL-DOPA-äthylesters*

4.6 g DL-DOPA-äthylester werden in 34.8 ccm absol. n-Propanol, gemeinsam mit 20.8 g Dibenzoyl-D-weinsäure gelöst und schnell mit 111 ccm absol. Äther versetzt. Nach 12 stdg. Stehenlassen bei 21° haben sich 6.4 g *D-DOPA-äthylester-dibenzoyl-D-hydrogentartrat* ausgeschieden (107 % d. Th.). Schmp. 153–157°, $[\alpha]_D^{22}$: −68.0° (*c* = 1.559, in Äthanol).

$C_{29}H_{29}NO_{12}$ (583.5) Ber. C 59.80 H 5.01 N 2.40 Gef. C 59.44 H 5.05 N 2.65

Durch Einengen der Mutterlauge i. Vak. bis zur Trockne und anschließendem Zusatz von 200 ccm absol. Äther kristallisieren 3.0 g *L-DOPA-äthylester-dibenzoyl-D-hydrogentartrat* (50.4 % d. Th.), Schmp. 138–141°, $[\alpha]_D^{22}$: −45.0° (*c* = 0.812, in Äthanol).

Gef. C 60.10 H 5.08 N 2.81

3. *Gewinnung von D- und L-DOPA-äthylester-hydrochlorid*

Die Darstellung von D-DOPA- und L-DOPA-äthylester-hydrochlorid aus den Dibenzoyl-D-hydrogentartraten erfolgt durch Ausfällung mit absol. Äther aus absol. äthanolischer Salzsäure[4].

D-DOPA-äthylester-hydrochlorid: Ausb. 94.5 % d. Th. Nach Umkristallisieren aus absol. Äthanol/Äther Schmp. 128–132°. $[\alpha]_D^{22}$: +9.59° (*c* = 1.332, in Wasser).

$C_{11}H_{15}NO_4 \cdot HCl$ (261.1) Ber. C 50.05 H 6.15 N 5.35 Gef. C 49.98 H 6.42 N 5.43

L-DOPA-äthylester-hydrochlorid: Ausb. 96.0 % d. Th. Nach Umkristallisieren aus absol. Äther/Äthanol Schmp. 126–129°, $[\alpha]_D^{22}$: −8.95° (*c* = 1.053, in Wasser).

Gef. C 50.00 H 6.50 N 5.30

4. *Darstellung der optisch reinen DOPA-Antipoden*

Die reinen *DOPA-äthylester-hydrochloride* lassen sich durch halbstündiges Kochen mit 20-proz. Salzsäure verseifen. Durch Neutralisation mit Ammoniak gegen Kongopapier werden hieraus auf übliche Weise die freien, optisch reinen Aminosäureantipoden gewonnen[17].

D-DOPA: Ausb. 78 % d. Th., Schmp. 282° (Zers.), $[\alpha]_D^{22}$: +12.05° (*c* = 2.012, in *n* HCl).

L-DOPA: Ausb. 50 % d. Th., Schmp. 280–282° (Zers.), $[\alpha]_D^{22}$: −12.5° (*c* = 0.640, in *n* HCl).

$C_9H_{11}NO_4$ (197.1) Ber. N 7.11 Gef. (D) N 7.26 (L) N 7.00

$C_9H_{11}NO_4$ Ref. 561b

EXAMPLE 1

31.5 g. of D,L-N-benzoyl-3-(4-hydroxy-3-methoxyphenyl)-alanine and 28.5 g. of dehydroabietylamine are placed in about 80 ml. of absolute methanol and heated. On boiling, all of the reactants dissolve. The resulting solution is then allowed to cool so that a precipitate is formed. After 22 hours at room temperature the precipitate is filtered off and rinsed portionwise with a total of 50 ml. of ice-cold methanol, then with diethyl ether, and dried at 60°C. under vacuum. For recrystallization, this salt is dissolved in 800–1,000 ml. of boiling methanol, filtered off hot, concentrated to about 100 ml. (whereby the substance begins to crystallize out), left to stand at room temperature for an hour, thereupon treated with 700 ml. of diethyl ether and left to stand overnight at 0°C. On the next morning, it is filtered, washed with ether and dried. 28.6 g. of dehydroabietylamine salt of the L-antipode are obtained in the form of white crystals with a melting point of 232°–233°C.; $[\alpha]_D^{24}=+55.0°$ ($c=1$ in methanol). A second recyrstallization does not raise the rotation.

EXAMPLE 2

27 g. of the dehydroabietylamine salt of the L-antipode prepared in Example 1 are treated in a shaking funnel with 800 ml. of cold methanol and 200 ml. of 2-N caustic soda. The resulting clear solution is extracted with two portions each of 600 ml. of petroleum ether. The methanol is evaporated off from the aqueous-methanolic layer. The residue is made congo-acidic with concentrated hydrochloric acid. To this acidified residue, ice is added and then the residue is extracted twice with 600 ml. of ethyl acetate each time. The ethyl acetate layers are washed twice with saturated aqueous sodium chloride solution, dried over anhydrous sodium sulfate and evaporated to an oil. This residue is sprinkled with isopropyl ether and allowed to stand at room temperature to form crystals. The residue is then treated with a little petroleum ether and allowed to stand overnight at 0°C. After filtering off and rinsing with petroleum ehter, for recrystallization this substance is dissolved in hot ethyl acetate, then left to cool to room temperature (whereupon recrystallization begins) and treated gradually with a little petroleum ether. After standing overnight at 0°C., the crystalline precipitate is filtered off and rinsed with petroleum ether. There are obtained 11.7 g. of pure L-N-benzoyl-3-(4-hydroxy-3-methoxyphenyl)-alanine with a melting point of 154°C. and a rotation of $[\alpha]_D^{24}=-32.7°$ ($c=1$ in methanol). Further recrystallization does not alter the rotation value.

EXAMPLE 3

5 g. of L-N-benzoyl-3-(4-hydroxy-3-methoxyphenyl)-alanine are treated under a nitrogen atmosphere with 40 ml. of colorless aqueous solution containing 48% by weight hydrobromic acid and heated for 2½ hours in a 140°C. bath. The dark-colored residue is cautiously evaporated on the rotary evaporator. After evaporation there resulted a mass permeated with crystals. This mass is added to about 80 ml. of distilled water and extracted with 2 portions of diethyl ether. In this manner, the benzoic acid formed is thereby extracted. The aqueous layer is evaporated to an oil, so that as much hydrobromic acid as possible is removed. A little distilled water is thereupon added and the mixture is once more vigorously evaporated. This operation is repeated.

The resulting dark-colored oil is thereupon dissolved in 20 ml. of distilled water and 10 ml. of acetonitrile to form a dark clear solution. This solution is cooled to room temperature and then treated with 5 ml. of propylene oxide. After about 15 minutes, the pH is tested, and if necessary adjusted to a value of 5.5–6 with further propylene oxide. At this pH almost colorless crystals precipitate. 100 ml. of acetonitrile are gradually added and the mixture is left to stand overnight at 0°C. The crystals are filtered and washed first with distilled water/acetonitrile 1:5 by volume and then with acetonitrile and finally then with diethyl ether. The crystals are recrystallized by dissolving them in distilled water in a sulfonation flask under constant stirring while saturated with SO_2 gas. The contents of the flask are warmed during this procedure. After all of the crystals have been dissolved, the heat source is thereupon removed and an inert nitrogen gas stream is led through the solution whereupon the crystallization begins. The mixture is left at 0°C. and filtered off on the next morning. The crystals are then rinsed with distilled water-/acetonitrile (1:1 by volume mixture) then with acetonitrile alone and finally with diethyl ether and dried. There are obtained 2.2 g. of L-3-(3,4-dihydroxyphenyl)-alanine with a melting point of 281°–282°C. $[\alpha]_D^{24}=-11.8°$ ($c=1$ in 1-N HCl).

$C_9H_{11}NO_4$ Ref. 561c

Among many types of resolution methods, a preferential crystallisation procedure[2] is considered to be one of the best. Although DOPA was reported to be resolvable by this simple method,[3] DOPA was sparingly soluble and the degree of resolution was so low, even in the acidic solution, that an attempt to resolve DL-DOPA seemed to be impractical. Now the salt of DL-DOPA with 2-naphthol-6-sulphonic acid has been found to be resolved in good yield by a preferential crystallisation procedure. Aromatic sulphonic acids were chosen because they vary greatly in properties and it seemed very likely that some of their salts form racemic mixtures suitable for use in this resolution method. Determinations of infrared spectra, mp, and solubility relationships were very useful to establish whether a given racemic modification forms a desirable racemic mixture or a racemic compound.

DOPA·NS was easily prepared from DOPA and an equivalent amount of 2-naphthol-6-sulphonic acid: DL-DOPA·NS·1·5H_2O, mp 152–154°C; L-DOPA·NS·1·5H_2O, mp 162–164°C, $[\alpha]_D^{25}-8{\cdot}55°$ (c, 2 in H_2O), $[\alpha]_{365}^{25}-19{\cdot}65°$ (c, 2 in H_2O). Typical resolution procedures are as follows: DL-DOPA·NS·1·5H_2O (16·0g) and L-DOPA·NS·1·5H_2O (3·0g) were dissolved in water (100cm³) at elevated temperature and cooled slowly to 50°C. The solution was nucleated with finely pulverised L-DOPA·NS·1·5H_2O (0·1g) and then stirred for 25min. at the same temperature. The precipitated crystals were filtered and L-DOPA·NS·1·5H_2O (6·44g) was obtained. The product was optically pure, $[\alpha]_{365}^{25}-19{\cdot}65°$ (c, 2 in H_2O). L-DOPA·NS·1·5H_2O (6·0g) was dissolved in hot water (60cm³) and decomposed at pH 6 with 1N-lithium hydroxide. The precipitated crystals were filtered and L-DOPA (2·53g) was obtained (95·9 per cent), mp 278–279°C (decomp.), $[\alpha]_D^{20}-11{\cdot}6°$ (c, 2 in N-HCl), $[\alpha]_D^{20}-12{\cdot}2°$ (c, 4 in N-HCl) (Found: C, 54·9; H, 5·6; N, 7·1. Calc. for $C_9H_{11}NO_4$: C, 54·8; H, 5·6; N, 7·1 per cent).

For successive optical resolution, the mother liquor can be used repeatedly to separate the other enantiomorph. After the separation of the L-isomer, 6·5g of DL-DOPA·NS·1·5H_2O and a small amount of water were added to the mother liquor according to refractometric measurement and weight of the solution. Thus, almost the same composition as in the previous resolution was obtained, except that the predominant enantiomorph was the D-isomer. This supersaturated solution was nucleated with D-isomer and crystallised as described above. The precipitated crystals were filtered to give D-DOPA·NS·1·5H_2O (6·5g) which was 99·7 per cent optically pure, $[\alpha]^{25}_{365}$ + 19·6° (*c*, 2 in H_2O). By repeating these procedures, L- and D-DOPA·NS·1·5H_2O were successively obtained. The mean of the results was 6·5g and 99 per cent in optical purity.

The optical resolution method now presented is very advantageous because it requires neither an optically active resolving agent nor conversion of DL-DOPA into complicated derivatives. Also, the yield per unit volume is very high and the operation is so simple that all processes are expected to be operated automatically in a sequence control system. Consequently, the application of the method for the industrial production of L-DOPA is considered to be very promising if combined with a proper synthetic method for DL-DOPA.

It is possible also to carry out the optical resolutions of DL-alanine·benzenesulphonate, DL-alanine·*p*-chlorobenzenesulphonate, DL-leucine·benzenesulphonate, DL-lysine·*p*-aminobenzenesulphonate, DL-serine·*m*-xylene-4-sulphonate, DL-tryptophan·benzenesulphonate and DL-3-(3,4-methylenedioxyphenyl)-2-methylalanine·*p*-phenolsulphonate. These results support the possibility that this simple method can be applied generally for the resolution of synthetic amino acids.

[2] Secor, R. M., *Chem. Rev.*, 1963, **63**, 297
[3] Krieger, K. H., Lago, C. J., Hill, M. & Wantuck, J. A., US Pat. No 3,405,159 (8 October 1968)

For an enzymatic resolution procedure for the above compound, see Ref. 520.

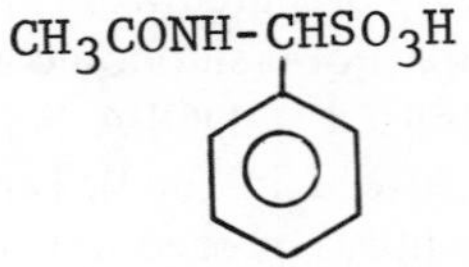

$C_9H_{11}NO_4S$ Ref. 562

Dédoublement optique.

On ajoute une solution chaude de l'acétate de strychnine à une solution concentrée de l'acétyl-1,1-phénylaminométhanesulfonate sodique (quantités équimoléculaires).

Après refroidissement le sel de strychnine cristallise en petites aiguilles blanches.

On a recristallisé le sel deux fois dans l'eau et ensuite encore quelques fois dans l'alcool de 50 %, parce que la solubilité dans l'eau avait diminué notablement.

On contrôle le dédoublement en décomposant, après chaque cristallisation, 1 g du sel de strychnine par l'eau de baryte et en éliminant la strychnine par extraction au chloroforme. Il faut prendre soin à ce que la solution du sel barytique ne colore que faiblement la phénolphtaléine. La solution filtrée est examinée dans un tube de polarisation de 4 dm; puis elle est analysée par dosage du baryum.

a. Sel de strychnine recristallisé deux fois.
Sel de baryum 10 cm³: $BaSO_4$ 0.1112 g.
α = —0 42° (4 dm). $[M]_D$ = —22°.

b. Sel de strychnine recristallisé cinq fois.
Sel de baryum 10 cm³: $BaSO_4$ 0 1080 g.
α = —0.44° (4 dm). $[M]_D$ = —23.8°.

c. Sel de strychnine recristallisé six fois.
Sel de baryum 10 cm³: $BaSO_4$ 0.1275 g.
α = —0 52° (4 dm). $[M]_D$ = —23.8°.

L'acétyl-1,1-aminophénylméthanesulfonate gauche de baryum présente donc une rotation moléculaire de —23.8°.

Analyse du sel de strychnine de l'acide gauche.
Substance 0.4467 g; $BaSO_4$ 0 1827 g.
Trouvé: S 5.62.
$C_{30}H_{33}O_6N_3S$ (563.34). Calculé: „ 5.69.

L'eau-mère de la première cristallisation, évaporée à l'air partiellement, a abandonné quelques petits cristaux représentant le sel de strychnine de *l'acide dextrogyre* à l'état pur.
Sel de baryum 10 cm³; $BaSO_4$ 0.0126 g.
α = +0.05° (4 dm). $[M]_D$ = +23°.

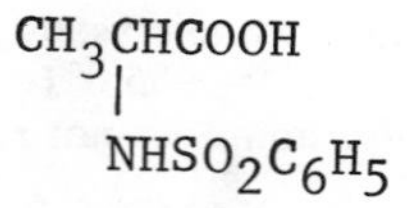

$C_9H_{11}NO_4S$ Ref. 563

EXPERIMENTAL.

Purification of Nor-d-*ψ-ephedrine Sulphate.*—The crude sulphate as supplied had $[\alpha]^{20°}_{5461}$ = + 38·9° * in aqueous solution. It was crystallised three times from water, after which its rotatory power was constant: *c* = 1·0608, α = + 2·07°, [α] = + 48·7°.

The free base was obtained by the addition of a slight excess of potassium hydroxide solution to an aqueous solution of the pure sulphate. A portion of the base which crystallised from the cooled solution was removed, and the filtrate extracted thoroughly with ether. The ethereal solution was washed, dried with potassium carbonate, and evaporated, and the residue was recrystallised from benzene. The base crystallised in colourless plates, m. p. 77°. In methyl alcohol, *c* = 0·6566, α = + 0·98°, [α] = + 37·5°. In water, *c* = 1·1992, α = + 1·19°, [α] = + 24·7°.†

Resolution of dl-*Benzenesulphonylalanine.*—The *dl*-benzenesulphonylalanine was prepared in the usual manner from *dl*-alanine, benzenesulphonyl chloride, and sodium hydroxide. It was recrystallised from water and obtained in colourless needles, m. p. 124—125°. Its properties agreed with those described by Hedin (*loc. cit.*).

To a warm solution of *dl*-benzenesulphonylalanine (4·58 g.) in sodium hydroxide (5·59 c.c., 1·79*N*), nor-*d*-ψ-ephedrine (1·51 g.) in water (10 c.c.) was added. The *salt* began to crystallise almost immediately and, after 24 hours, it was filtered off (3·2 g.) and recrystallised from hot water (10 c.c.). It was obtained in colourless prisms or fine needles and, after two crystallisations, its rotatory power was constant: In water, c = 0·3744, α = + 0·70°, [α] = + 46·6°. The salt had m. p. 147—148° (Found: N, 7·5. $C_{18}H_{24}O_5N_2S$ requires N, 7·4%).

* All rotatory powers in this work were carried out at 20°, 4 dm. tubes and the mercury green line (λ 5461) being used.

† These values for the rotatory powers of nor-*d*-ψ-ephedrine sulphate and for nor-*d*-ψ-ephedrine itself differ appreciably from those previously given (Smith, J., 1928, 51). We accordingly consulted with Dr. Sydney Smith, who confirmed our value for the sulphate and writes as follows:

"It has been found necessary to redetermine the rotatory powers of nor-*d*-ψ-ephedrine and its derivatives, as the previous measurements were made with the mercury yellow line (λ 5790) and not with the mercury green line (λ 5461) as recorded. The values at 20° for the line λ 5461 are as follows:

Nor-*d*-ψ-ephedrine sulphate, [α] = + 48·7° (*c* = 1·4 in water).
Nor-*d*-ψ-ephedrine, [α] = + 37·9° (*c* = 3·0 in methyl alcohol).
d-ψ-Methylephedrine methiodide, [α] = + 42·3° (*c* = 2·3 in water).
Nor-*d*-ψ-ephedrine hydrogen tartrate, [α] = + 49·5° (*c* = 2·3 in water).
Dibenzoylnor-*d*-ψ-ephedrine, [α] = + 32·8° (*c* = 2·2 in methyl alcohol).
N-Benzoylnor-*d*-ψ-ephedrine, [α] = + 67·2° (*c* = 1·2 in methyl alcohol).
O-Benzoylnor-*d*-ψ-ephedrine hydrochloride, [α] = − 37·6° (*c* = 1·0 in water).

"In the original paper (*loc. cit.*), '5461' should be corrected to '5790' throughout.—S. SMITH."

d-*Benzenesulphonylalanine* was obtained from an aqueous solution of the above nor-*d*-ψ-ephedrine salt by addition of an aqueous solution of ammonia. The mixture was thoroughly extracted with chloroform to remove the base and the aqueous solution was then acidified with concentrated hydrochloric acid. The acid separated as an oil which rapidly solidified. It was recrystallised from water and obtained in colourless prisms, m. p. 126—127°, which were found to be optically pure (Found: N, 6·6. $C_9H_{11}O_4NS$ requires N, 6·1%). In aqueous solution as the sodium salt: $c = 0·3994$, $\alpha = +0·53°$, $[\alpha] = +33·4°$.

l-*Benzenesulphonylalanine*.—After the above *dAdB* salt had been separated, the mother-liquor was treated with an excess of aqueous ammonia and extracted thoroughly with chloroform to remove the alkaloid. The aqueous solution was evaporated on the water-bath and then acidified with concentrated hydrochloric acid. The benzenesulphonylalanine so obtained had $[\alpha] = -9·8°$ as sodium salt in aqueous solution, and this rotatory power was unaffected on further recrystallisation from water. This acid (3·0 g.) was dissolved in sodium hydroxide solution (4·71 c.c., 1·79*N*—sufficient to neutralise the *l*-acid present) and nor-*d*-ψ-ephedrine (0·706 g.—sufficient to neutralise the *d*-acid present) and water (10 c.c.) were added. Complete solution was effected on warming. The crystalline salt which had separated after 24 hours was filtered off; on further standing, more crystalline salt separated. The mother-liquor from this salt was worked up in the usual way; the acid obtained (1·1 g.) was optically pure after one crystallisation from water. It had m. p. 126—127° and was superficially indistinguishable from its optical isomeride (Found: N, 6·3. $C_9H_{11}O_4NS$ requires N, 6·1%). As the sodium salt in water it had the following rotatory power: $c = 0·3326$, $\alpha = -0·44$, $[\alpha] = -33·1°$.

The same compound was prepared from *d*-alanine. The amino-acid (1·0 g.), dissolved in an equivalent quantity of sodium hydroxide solution (6·3 c.c., 1·79*N*), was treated with a further quantity of the alkali (18·8 c.c.) and the requisite amount of benzenesulphonyl chloride, added in small quantities, the mixture being vigorously shaken and warmed towards the end of the reaction. The solution was evaporated to about half its original volume and the acid was precipitated on acidification with hydrochloric acid. The product (1·6 g.; yield, 65%) was recrystallised from water and proved to be pure *l*-benzenesulphonylalanine identical in all respects with that described above. As sodium salt in water ($c = 0·4120$), it had $\alpha = -0·55°$, whence $[\alpha] = -33·6°$. In ethyl alcohol; $c = 0·4031$, $\alpha = -0·17°$, $[\alpha] = -10·5°$.

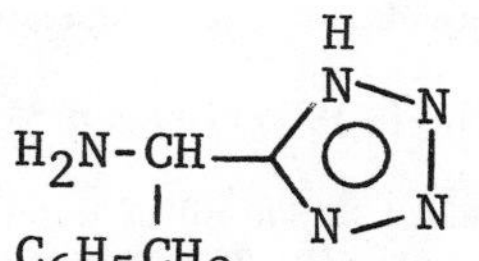

$C_9H_{11}N_5$

Ref. 564

For the experimental details of the resolution procedure for this compound, see Ref. 21.

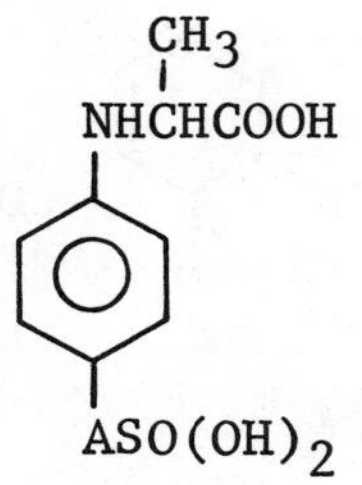

$C_9H_{12}AsNO_5$

Ref. 565

Resolution of dl-*Phenylalanineamide-4-arsinic Acid*.—The amide-acid was prepared by the method described by Jacobs and Heidelberger (*loc. cit.*; compare Fourneau and Nicolitch, *loc. cit.*). The amide-acid (36 g.) was dissolved in boiling water (2000 c.c.) and aqueous sodium hydroxide (0·516*N*; 121 c.c.), and quinine (22·3 g., allowing 10% excess on account of water content) added. The whole was brought into solution and allowed to cool during 20 hours. The salt crystallised in soft needles, and was separated from the mother-liquor. The salt was recrystallised from boiling water (2000 c.c.), and another recrystallisation from water (1500 c.c.) failed to affect the rotatory power. The salt was anhydrous (Found: N, 9·1; As, 12·2. Calc. for $C_9H_{13}O_4N_2As,C_{20}H_{24}O_2N_2$: N, 9·15; As, 12·3%). The rotatory power was determined in aqueous solution: $c = 0·2412$, $l = 4$, $\alpha = -1·20°$, whence $[\alpha] = -123·8°$.

l- *and* d-N-*Phenylalanineamide-4-arsinic Acids*.—The above quinine salt was decomposed by grinding with aqueous ammonia (*d* 0·880), and the liquid filtered. The separated quinine was re-extracted twice with further quantities of ammonia solution. The filtrate was acidified with concentrated hydrochloric acid (Congo red), and the precipitated amide-acid recrystallised from boiling water (decolorising charcoal); its rotatory power was then constant. The pure *l*-amide-acid crystallised in colourless needles, m. p. 247° (decomp.) (Found: N, 9·4. Calc. for $C_9H_{13}O_4N_2As$: N, 9·7%). Its rotatory power was determined in aqueous solution containing the exact quantity of sodium bicarbonate to form the sodium salt: $c = 0·5868$, $l = 4$, $\alpha = -0·42°$, whence $[\alpha] = -17·88°$.

The mother-liquor after separation of the above salt was evaporated to about half its original volume, and the quinine precipitated by addition of a slight excess of an aqueous solution of ammonia. After filtration, the liquid was acidified with concentrated hydrochloric acid (Congo red), and the amide-acid precipitated as before. This was recrystallised three times from boiling water; its rotatory power was then constant. In appearance and general behaviour the pure *d*-amide acid was similar to its enantiomorphous isomeride and it had the same melting-decomposition point (247°) (Found: N, 9·5%). Its rotatory power was determined as in the previous case: $c = 0·7770$, $l = 4$, $\alpha = +0·51°$, whence $[\alpha] = +16·5°$.

Before the effect of hydrolysis on these pure optically active amide-acids was examined, the rotatory power of another specimen of the pure *d*-amide-acid was determined under the same conditions as the above: $c = 0·7050$, $l = 4$, $\alpha = +0·46°$, whence $[\alpha] = +16·4°$.

This solution (40 c.c.) was treated in an exactly similar way to those in the previous experiments on the hydrolysis of the optically active amide-acids: the rotatory power of the final solution was $[\alpha] = -28·5°$.

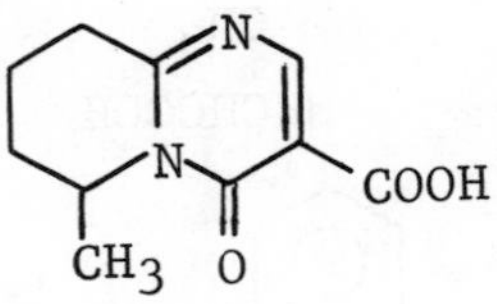

$C_9H_{12}N_2O_3$

Ref. 566

Examples

1. *Preparation of (—)—I—L—(+)—IV salt*

368 g (1.77 mol) of racemic-I is suspended in 900 ml of water, warmed to 40 to 50 °C and basified to pH 7 with cc. NH_4OH.

To this stirred solution is added 460.5 g (1.73 mol) of solid L—(+)—IV. HCl · H_2O ($[\alpha]_D^{20}$ = +30° (c : 2; water)).

The arising clear solution (pH 6.5) is cooled to room temperature, seeded by (—)—I—L—(+)—IV salt. The crystallization is beginning. The precipitated salt is filtered, washed with 2×50 ml of water.

Yield: 290 g, 78%

$[\alpha]_D^{20}$ = —45° (c : 2; MeOH)

mp: 138°

2. *Preparation of (—)—I*

a) The salt obtained as in 1 is suspended in 2 l of water and warmed to 60 °C. The solution contains 4% of (—)—I. The solution is acidified to pH 2 with cc. HCl, then cooled in ice bath for two hours. The precipitate is filtered.

Yield: 22.1 g

$[\alpha]_D^{20}$ = —35° (c : 2; MeOH)

mp: 132°

b) The filtrate is extracted with 4×200 ml of chloroform, the organic phase is separated, dried with Na_2SO_4, and concentrated in vacuo. The residue is the (—)—I, a good pulverisable material.

Yield: 108.7 g, 58.9%

$[\alpha]_D^{20}$ = —93° (c : 2; MeOH)

mp: 119°

optical purity (o. p.): 91%

3. *Preparation of (+)—I*

The mother liquor of 1 is further diluted with 1 l of water. (The solution contains 4% (+)—I.) The further treatment is analog to the ones described in item 2.

a) yield: 118 g

$[\alpha]_D^{20}$= —15° (c : 2; MeOH)

mp: 138 °C

b) yield: 107 g, 58.3%

$[\alpha]_D^{20}$= +90° (c : 2; MeOH)

mp: 119 °C

op: 90.5%

4. *Recovery of (±)—I*

The combined materials (I) obtained according to 2)a ($[\alpha]_D^{20}$= —35°) and 3)a ($[\alpha]_D^{20}$= —15°) are resolvable again. The loss of one step of the procedure is 3—4% calculated for the (±)—I-acid.

5. *Purification of (—)—I*

a) 108.7 g of (—)—I ($[\alpha]_D^{20}$= —93° (c : 2; MeOH)) is suspended in 500 ml of water and basified with cc. NH_4OH to pH 7, then acidified to pH 1.5 with cc. HCl, allowed to stand at room temperature, then the precipitated crystals are filtered.

Yield: 84.7 g, 78%

$[\alpha]_D^{20}$= —68° (c : 2; MeOH)

mp: 128 °C

b) The filtrate is extracted with ×100 ml of chloroform, the combined organic phases are dried, concentrated in vacuo.

Yield of residue: 22.1 g, 20.5%

$[\alpha]_D^{20}$= —115° (c : 2; MeOH)

mp: 114 °C

6. *Purification of (+)—I*

a) The 107 g ($[\alpha]_D^{20}$= +90° (c : 2; MeOH)) of (+)—I obtained as 3)b is treated as described for 5)a.

Yield: 82.1 g, 76%

$[\alpha]_D^{20}$= +65° (c : 2; MeOH)

mp: 130 °C

b) The filtrate obtained as in 6)a is treated as in 5)b.

Yield of residue: 20.4 g, 19%

$[\alpha]_D^{20}$= +115° (c : 2; MeOH)

mp: 113—5 °C

7. *Recovery of L—(+)—IV*

The aqueous phase obtained by chloroform extraction according to 2)b and 3)b are combined, and basified with cc. NH_4OH to pH 9. The solution is allowed to stand overnight at room temperature and the crystalline salt is filtered, dried.

Yield of obtained L—(+)—IV: 320 g, 87% (calculated to L—(+)—IV. HCl. H_2O).

The base is suspended in 300 ml of water and to this suspension cc. HCl is added drop by drop to pH 1. The solution is allowed to stand overnight at room temperature. The L—(+)—IV. HCl · H_2O is filtered and dried.

$[\alpha]_D^{20}$= +30° (c: 2; MeOH)

mp: 159—160 °C

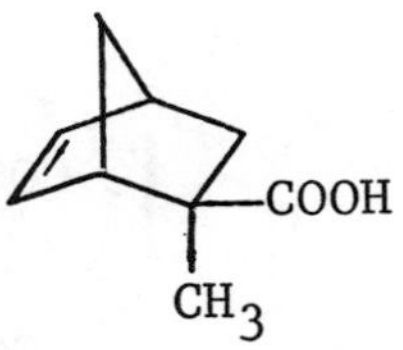

$C_9H_{12}O_2$ Ref. 567

Resolution of 5-Norbornene-2-*endo*-methyl-2-*exo*-carboxylic Acid (42).—To a solution of 76.2 g. of pure racemic acid, m.p. 81.5–83° (from aqueous acetic acid), reported 79–80°,[45] 83°,[46] in 4500 ml. of acetone was slowly added (from a Soxhlet extractor) 194.7 g. of quinine trihydrate. Crystals formed during the addition of the alkaloid. After all the quinine had been added, the solution was allowed to cool to room temperature. The crystals were filtered off and successively recrystallized from acetone six times, using a Soxhlet extractor. The course of the resolution was followed by regenerating small samples of acid at each stage with ether–10% sulfuric acid. The third, fourth, fifth and sixth recrystallizations of the salt gave acid of $[\alpha]_D$ +57.3°, +61.5°, +64.9° and +66.0°. The salt from a seventh recrystallization gave 6.0 g. of acid of $[\alpha]_D$ +67.3°, m.p. 47–48°. The infrared spectrum was identical with that of the pure racemic acid. This material was assumed to be optically pure, a point that is supported below. All rotations for this substance are in 95% ethanol.

Anal. Calcd. for $C_9H_{12}O_2$: C, 71.10; H, 7.90. Found: C, 71.18; H, 8.00.

From the early mother liquors, 10.3 g., $[\alpha]_D$ −45.4°, of partially resolved levo acid was obtained; the mother liquors from the middle of the fractionation gave 5.0 g. of the racemate, and those from the late mother liquors gave 8.0 g., $[\alpha]_D$ +32.1°, of partially resolved dextro isomer.

With ethereal diazomethane a sample of acid $[\alpha]_D$+62.5° gave methyl ester 44, $[\alpha]_D$+60.4°, vapor chromatographically homogeneous, infrared spectrum identical with that of the racemate.

(45) J. S. Meek and W. B. Trapp, *J. Am. Chem. Soc.*, **79**, 3909 (1957).

(46) S. Beckmann, R. Schaber and R. Bamberger, *Chem. Ber.*, **97**, 997 (1954).

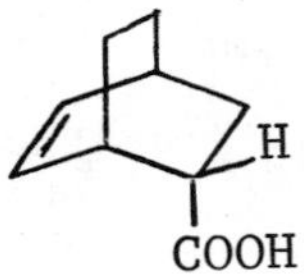

$C_9H_{12}O_2$ Ref. 568

Racemic *endo* acid (3 g) was mixed with (−)-1-(*p*-nitrophenyl)-2-amino-1,3-propanediol (4 g) in ether (20 ml). The obtained needles melted after three crystallizations from water at 62 to 66°C. The optically active acid, liberated from the salt, had m.p. 49—51°C (light petroleum), $[\alpha]_D^{20}$ +34·8° (*c* 4·2, ethanol). For $C_9H_{22}O_2$ (152·3) calculated: 71·10% C, 7·95% H; found: 71·10% C, 8·22% H.

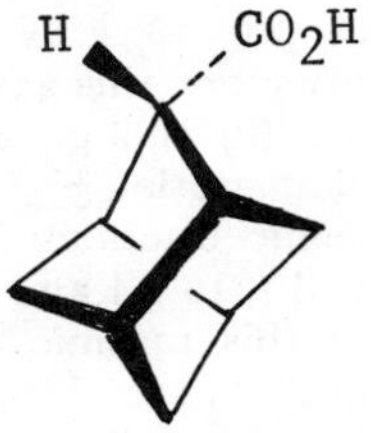

$C_9H_{12}O_2$ Ref. 569

Optical Resolution of Tricyclo[3.3.0.0^{3,7}]octane-2-carboxylic Acid (23). To a solution of (±)-**23** (34.9 g, 0.230 mol) in acetone (1.9 L) was added a solution of (+)-2-(1-aminoethyl)naphthalene (39.2 g, 0.230 mol) in acetone (300 mL) with stirring. The mixture was heated under reflux for 5 h, and 1 L of acetone was then distilled away. The salt solution was allowed to stand overnight at room temperature, and the deposited solid was collected by filtration (the filtrate was reserved for isolation of (+)-**23**) to give 57.2 g of dextrorotatory salt, $[\alpha]^{16}_D$ +12.4° (*c* 0.305, $CHCl_3$). Several fractional recrystallizations of the (+)-salt from acetone afforded 24.5 g of salt with $[\alpha]^{14}_D$ +5.3° (*c* 0.311, $CHCl_3$), mp 181–183 °C.

Anal. Calcd for $C_{21}H_{25}O_2N$: C, 77.98; H, 7.79; N, 4.33. Found: C, 77.69; H, 7.74; N, 4.19.

To this dextrorotatory salt (13.5 g, 0.0420 mol) was added 5% NaOH aqueous solution (160 mL), and the mixture was stirred for 3 h at room temperature. The reaction mixture was extracted with ether to remove the amine and then made acidic with HCl. The acidic solution was extracted with ether, and the extract was washed with water and dried over $MgSO_4$. Evaporation of the solvent gave 6.22 g of a white solid, $[\alpha]^{15}_D$ −16.2° (*c* 0.576, $CHCl_3$), which was recrystallized from *n*-hexane to yield 3.30 g of (−)-**23**: $[\alpha]^{15}_D$ −22.5° (*c* 0.800, $CHCl_3$); mp 85–86 °C (in a sealed tube).

Anal. Calcd for $C_9H_{12}O_2$: C, 71.02; H, 7.95. Found: C, 70.81; H, 7.95.

The filtrate that separated from the salt of (−)-**23** was concentrated to give a crystalline material which was recrystallized from acetone to afford 5.8 g of the dextrorotatory salt, $[\alpha]^{18}_D$ +24.8° (*c* 0.320, $CHCl_3$). This salt (5.50 g) was treated with 5% NaOH aqueous solution (70 mL), and the same workup described for (−)-**23** gave 2.30 g of (+)-**23**, $[\alpha]^{18}_D$ +19.1° (*c* 0.305, $CHCl_3$). Several recrystallizations of the (+)-carboxylic acid from *n*-hexane afforded 850 mg of (+)-**23**: $[\alpha]^{18}_D$ +22.1° (*c* 0.581, $CHCl_3$); mp 80–84 °C.

Anal. Calcd for $C_9H_{12}O_2$: C, 71.02; H, 7.95. Found: C, 70.81; H, 7.96.

(3) T. A. Elwood, and M. M. Bursey, *Org. Mass Spectrom.*, **1**, 537 (1968).

(7) A. K. Youssef and M. A. Ogliaruso, *J. Chem. Educ.*, **52**, 473 (1975).

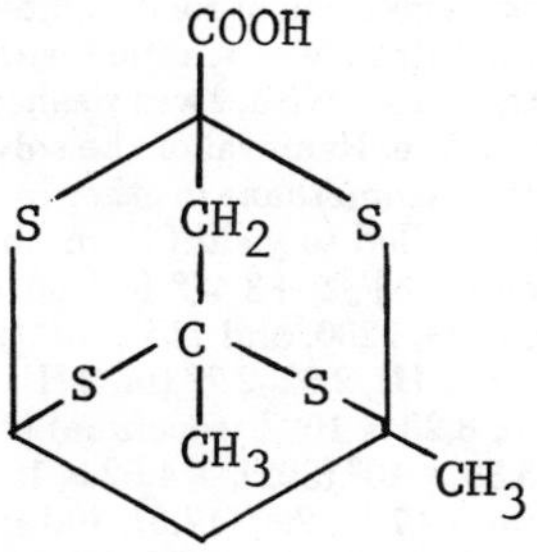

$C_9H_{12}O_2S_4$ Ref. 570

Preliminary experiments on resolution. The acid was tested against the common alkaloids and dehydroabietylamine in various solvents. The cinchonine and brucine salts in ethanol gave (+)-acid of the highest activity, and further experiments showed that the former base was preferable. The laevorotatory acid was isolated from the strychnine salt in ethanol, acetone and water. On a larger scale, dilute ethanol gave the best results.

(+)-3,5-Dimethyl-2,4,6,8-tetrathiaadamantane-1-carboxylic acid ((+)-*II*)

Cinchonine (16.1 g, 55 mmol) and (±)-acid (15.4 g, 55 mmol) were dissolved in boiling 95% ethanol (575 ml). After 24 hours at 0°, the precipitate was collected and recrystallized from the same solvent. After each crystallization, a small sample of the salt was decomposed and the rotatory power of the acid determined in acetone:

Cryst. no.	1	2	3	4	5
Ethanol (ml)	575	550	330	280	400
Yield of salt (g)	15.9	11.9	10.0	8.5	5.9
$[\alpha]^{25}_{365}$ of acid	+11°	+18°	+21°	+24°	+24.6°

(+)-II was liberated with 2-M sulphuric acid and taken up in ether, which was dried and evaporated. The residue was recrystallized from methanol, yielding white needles (2.2 g, 29%), m.p. 260–262°.

Solvent	abs. ethanol	methanol	acetone	chloroform
Conc. (g/l)	8	9	7	3
$[\alpha]^{25}_{365}$	+21.7°	+22.7°	+24.6°	+30.4°
$[\alpha]^{25}_{D}$	—	—	+3.1°	—

$C_9H_{12}O_2S_4$ (280.5) calc. C 38.5, H 4.3, S 45.7;
found C 38.5, H 4.3, S 45.7.

(−)-3,5-Dimethyl-2,4,6,8-tetrathiaadamantane-1-carboxylic acid ((−)-II)

Acid (6.5 g, 23 mmol, $[\alpha]_{365}$ −7.0°), isolated from the mother liquors mentioned in the preceding paragraph, and strychnine (7.8 g, 23 mmol) were dissolved in a boiling mixture of 95 % ethanol (210 ml) and water (210 ml). After 24 hours at 0°, the precipitate was collected and recrystallized from the same solvent. After each crystallization, a small sample of the salt was decomposed and the rotatory power of the acid determined in acetone:

Cryst. no.	1	2	3	4	5
Solvent (ml)	420	350	320	300	250
Yield of salt (g)	11.2	8.8	7.2	6.1	5.8
$[\alpha]^{25}_{365}$ of acid	−18°	−21°	−24°	−25°	−24.9°

The acid was isolated and recrystallized as the enantiomer, yield 2.0 g (26 %), m.p. 260–262°. The IR spectrum in KBr differed considerably from that of (±)-II. $[\alpha]^{25}_{365}$ −24.9° (conc. 9 g/l, acetone), $[\alpha]^{25}_{365}$ −31.0° (conc. 5 g/l, chloroform). ORD and CD (conc. 0.8 g/l, Fig. 6):

λ (nm)	287	278	257	249	241	227	221	215
[Φ]	−1 750°	0°	+11 800°	0°	−11 800°	0°	+3 080°	0°

λ (nm)	307	273	259	249	238	229	218	213
[Θ]	0	−4 740	0	+15 300	0	−6 240	0	+4 120

$C_9H_{12}O_2S_4$ (280.5) calc. C 38.5, H 4.3, S 45.7;
found C 38.5, H 4.3, S 45.6.

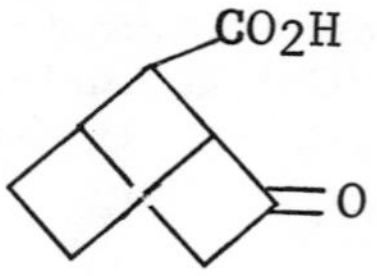

$C_9H_{12}O_3$

Ref. 571

Optical Resolution of 7-*syn*-Carboxybicyclo[2.2.1]heptan-2-one (15). Racemic 7-*syn*-carboxybicyclo[2.2.1]heptan-2-one (**15**) was prepared by the procedure reported previously.[12] A salt from the carboxylic acid (62.5 g, 0.405 mol) with cinchonidine (119 g, 0.405 mol) was systematically recrystallized from acetone. The levorotatory salt (50.2 g, yield 28%), mp 155–157° dec, $[\alpha]^{15}D$ −72.5° (*c* 0.614, ethanol), was obtained as a sparingly soluble crystal, which was treated with 5% sodium hydroxide solution at room temperature. After filtration of cinchonidine, the filtrate was acidified with hydrochloric acid and extracted with ether. The extract was washed with water and dried over magnesium sulfate. Removal of the solvent gave a solid which was esterified with diazomethane in ether by the usual manner. The methyl ester was distilled to yield (+) methyl ester **12** (16.8 g): bp 138–141 °C (13 mm); $[\alpha]^{13}D$ +3.47° (*c* 1.60, ethanol); ir (neat film) 1750, 1725, 1440, 1298, 1200, and 1145 cm^{-1}; NMR (CCl_4) δ 1.4–2.0 (m, 5 H), 2.1–2.4 (m, 1 H), 2.62–2.72 (m, 2 H), 2.75–2.88 (m, 1 H), and 3.64 (s, 3 H); CD (*c* 8.23 × 10^{-3}, isooctane) [θ] (nm) 0 (264), +4.01 × 10^2 sh (292.5), +5.89 × 10^2 (302), + 4.62 × 10^2 (313), 0 (332); uv max (isooctane) 284 nm (ε 17.0), 290 (17.5), 300 sh (16.0).

Anal. Calcd for $C_9H_{12}O_3$: C, 64.27; H, 7.19. Found: C, 64.09; H, 7.26.

Condensation of the mother liquor of the cinchonidine salt gave the levorotatory salt (21.1 g, yield 12%), mp 193–195 °C dec, $[\alpha]^{18}D$ −89.7° (*c* 0.836, ethanol). This was treated by the same manner described above to yield (−) methyl ester **16** (6.05 g), bp 109–111 °C (5 mm), $[\alpha]^{17}D$ −11.6° (*c* 1.08, ethanol).

Anal. Calcd for $C_9H_{12}O_3$: C, 64.27; H, 7.19. Found: C, 63.91; H, 7.29.

(12) S. Beckmann and H. Geiger, *Chem. Ber.*, **94**, 48 (1961).

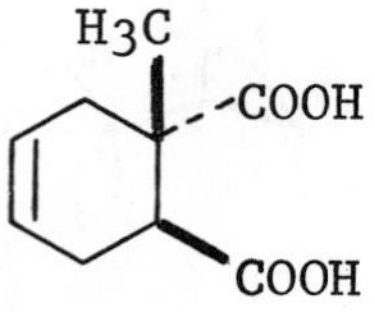

$C_9H_{12}O_4$

Ref. 572

Resolution of the Dibasic Acid rac-(**1**).—Solutions of the dibasic acid (215 g.) and quinine (189 g.), each in 50% aqueous ethanol (400 c.c.) at 70° were mixed and kept overnight at 18°. The precipitated salt was collected, washed with a little 50% ethanol, and dried. It had m.p. 210° (263 g.) and was used for regeneration of the laevorotatory enantiomer.

The mother liquor was evaporated to dryness at 25° under reduced pressure, and the residue was mixed with 2N-sulphuric acid (300 c.c.) and extracted with ether (4 × 200 c.c.). The ethereal extract was washed with 2N-sulphuric acid (10 c.c.) and with water (10 c.c.), and then dried and evaporated. The residue (108 g.) crystallised from hot water (1·35 l.) giving material (81·7 g.), $[\alpha]_D^{25}$ +4·4°. Evaporation of the mother liquor gave material (26·35 g.), $[\alpha]_D^{25}$ +62°. Recrystallisation of the first crop (81·7 g.) from water (450 c.c.) gave nearly racemic material (75·9 g.), $[\alpha]_D^{25}$ +1·1°; the mother liquor provided on evaporation partially resolved material (5·8 g.), $[\alpha]_D^{25}$ +46·6°.

The crystalline quinine salt (263 g.) was decomposed in the same manner as the mother liquor material, and the regenerated acid was processed by recrystallisation from water as for the enantiomer; this gave fractions as follows: (i) 16·8 g., $[\alpha]_D^{25}$ −65°; (ii) 10·9 g., $[\alpha]_D^{25}$ −59·5°; and (iii) 64·3 g., nearly racemic material, $[\alpha]_D^{25}$ −2°.

The combined racemic fractions (140·2 g.) were united with fresh racemic acid (210 g.) and the resolution process was repeated. From this racemic acid (227 g.) was recovered, and was subjected to resolution again; racemic acid recovered 153 g. The partially resolved dextrorotatory fractions from these experiments were united and purified by crystallisation from water, yielding the *dibasic acid* (**1**) (80·1 g.), m.p. 173°, $[\alpha]_D^{20}$ +74° (Found: C, 58·75; H, 6·5. $C_9H_{12}O_4$ requires C, 58·7; H, 6·6%), and nearly racemic material (42 g.), m.p. 182—184°.

The laevorotatory enantiomer *ent*-(**1**), $[\alpha]_D^{25}$ −74° (*ca.* 75 g.) was obtained from the appropriate fractions, together with racemic material (*ca.* 37 g.). Thus 425 g. of racemic acid provided *ca.* 232 g. of racemic acid, 80·1 g. of (+)-acid (≡42% of racemate used), and *ca.* 75 g. of (−)-acid.

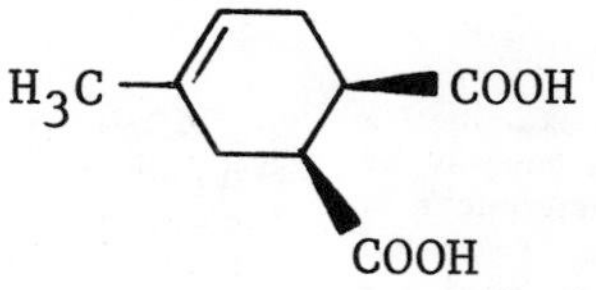

$C_9H_{12}O_4$

Ref. 573

Cis-4-methyl-4-cyclohexene-1,2-dicarboxylic acid was prepared according to Alder, m.p. 152–155°, lit [6], m.p. 155°.

Resolution. The acid (105 g) and brucine (255 g) were dissolved in ethanol. After standing over night, the precipitated salt was collected and recrystallised 6 times from ethanol, giving 68 g salt which was treated with hydrochloric acid to yield (+)-*cis*-4-methyl-4-cyclohexene-1,2-dicarboxylic acid (15.0 g) m.p. 136–138°, $[\alpha]_D^{25}$ +21.6° (c 6.30, ethanol-chloroform 1:1), lit m.p. 146–147°, $[\alpha]_D^{25}$ +16.5° [7].

6. Alder, K., and Dortmann, H. A., Ber. *85*, 556 (1952).
7. Böeseken, J., and de Rijck van der Gracht, W. J. F., Rec. Trav. Chim. *56*, 1203 (1937).

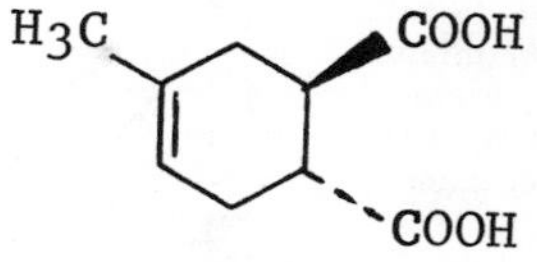

$C_9H_{12}O_4$

Ref. 574

Resolution of trans-4-*methyl*-4-*cyclohexene*-1,2-*dicarboxylic acid*

trans-4-Methyl-4-cyclohexene-1,2-dicarboxylic acid (78 g, 0·42 mole) was dissolved in 350 ml acetone and added to a refluxing solution of 67 g (0.21 mole) quinine in 1l. acetone. The solution was filtered, reduced in volume to 750 ml, and allowed to stand for 24 hr. The salt was recrystallized from abs ethanol to a constant rotation $[\alpha]_D^{26}$ −70 ± 1° (*c* 3, methanol), m.p. 193–195°.

The optically active salt was decomposed by addition of a 3:1 mixture of ice and HBr. The optically active acid was recovered by extraction with ether. The ether solution was dried (Na_2SO_4). The optically active *trans*-4-methyl-4-cyclohexene-1,2-dicarboxylic acid, m.p. 152–154°, $[\alpha]_D^{25}$ +160 ± 1° (*c* 3·0, methanol) was recovered by evaporation of the ether and recrystallization from benzene, yield 11 g. (Found: C, 58·65, H, 6·63; Calc. for $C_9H_{12}O_4$: C, 58·68; H, 6·57%).

$C_9H_{12}O_4$

Ref. 574a

Experimental. *(−)-4-Methyl-4-cyclohexene-(1R:2R)-dicarboxylic acid (1)* was obtained by repeated crystallizations from water of the brucine salt of *trans*-4-methyl-4-cyclohexene-1,2-dicarboxylic acid (compare Ref. 1). On acidification of an aqueous solution of the salt, followed by extraction with ethyl acetate and evaporation of the solvent, a low yield of the acid (*1*) was obtained (5 %), partial melting at 139—140°, followed by recrystallisation and melting at 146—149°, $[\alpha]_D^{25}$ −150 ± 5° (lit.[1] m.p. 152—164°, $[\alpha]_D^{25}$ + 160 ± 1° for the enantiomer).

1. Walborsky, H. M., Barash, L. and Davis, T. L. *Tetrahedron* 19 (1963) 2333.

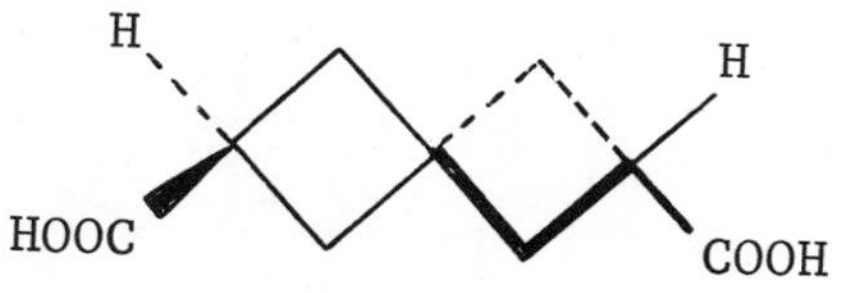

$C_9H_{12}O_4$

Ref. 575

Dédoublement optique de l'acide dicarboxylique.

Les alcaloïdes brucine, strychnine et quinine donnent des sels cristallins.

Une solution chaude de l'acide donne avec de la brucine le *sel neutre de brucine*, soit qu'on prenne deux molécules de l'alcaloïde, soit une. On obtient le même sel en mélangeant en milieu aqueux le sel d'ammonium et l'acétate de brucine (1 ou 2 mol.).

Le sel cristallise en prismes épais, qui peuvent atteindre la longueur d'un demi centimètre et qui renferment 6 molécules d'eau.

Substance 0.2336 g; perte à 80° 0.0241 g

	Trouvé:	H_2O	10.31.
$C_9H_{12}O_4 . 2 C_{23}H_{26}O_4N_2 . 6 H_2O$ (1080.7).	Calculé:	,,	10.03.

Le sel anhydre fond à 133°; on l'obtient aussi par „cristallisation froide" de l'acide avec deux molécules de brucine en milieu d'alcool à 96 %.

Le *sel neutre de strychnine* cristallise en aiguilles, fondant avec décomposition vers 238°.

Le dédoublement optique peut être effectué au moyen du sel de brucine. Il n'est pas possible de juger du progrès du dédoublement en examinant la rotation du sel de brucine. La solubilité dans l'eau froide est faible, inférieure à 1 %, et la rotation de l'alcaloïde l'emporte notablement sur celle de l'anion.

D'abord nous avons fait une série d'expériennces préliminaires, chaque fois sur deux grammes du sel de brucine pulvérisé.

On prépare une solution du sel d'ammonium en décomposant le sel de brucine par de l'ammoniaque diluée, en présence de phénolphtaléine et en éliminant l'alcaloïde par filtrage et extraction au chloroforme.

L'acide est trop peu soluble dans l'eau, pour permettre un examen polarimétrique de la solution aqueuse.

Mais on peut étudier l'acide en solution éthérée. On additionne le sel de brucine ou d'ammonium, dans un entonnoir à brome, d'acide chlorhydrique dilué en petit excès, et de 30 cm³ d'éther.

Une cristallisation de 20 g de l'acide dicarboxylique avec la quantité bimoléculaire de brucine dans un volume de 2 litres d'eau a donné le sel de brucine en cristaux assez grands avec un rendement de 75 %. Ce sel, tranformé en sel d'ammonium, présentait une faible rotation positive. La solution éthérée de l'acide avait à peu près la même rotation.

Ainsi on a prouvé la possibilité du dédoublement, car la brucine et ses sels sont lévogyres.

a. *Cristallisation d'une solution chaude.*

On a préparé le sel de brucine aux dépens de 40 g d'acide et de la quantité nécessaire de brucine dans 3 litres d'eau chaude, en laissant se refroidir très lentement la solution. Rendement 80 %. On a fait recristalliser prudemment le sel neuf fois dans l'eau chaude dans un volume tel que seuls 60—70 % se déposent. Nous avons examiné les cristallisations impaires (I, III, V, VII, IX). Chaque fois nous avons transformé 10 g du sel en sel d'ammonium. On peut déterminer la concentration en évaporant, après l'examen polarimétrique, la solution dans le vide, ou au bain-marie. Dans le dernier cas le sel perd une partie de l'ammoniac; on détermine cette perte par un titrage de l'acide formé. Cette correction ne s'est pas faite pour I et III. On a déterminé chaque concentration en double.

I. Sel d'ammonium, obtenu par décomposition de 10 g du sel de brucine. Concentration 0.144 g du sel d'ammonium dans 3.5 cm³. Tube de 2 dm. α = +0.03 à 0.04°. $[M]_D$ = +0.9°

III.	0.208 g dans 3.5 cm³	α = 0.11°	$[M]_D$ = +2.0°.
V.	0.232 g „ „ „	α = 0.13°	„ = +2.1°.
VII.	0.182 g „ „ „	α = 0.12°	„ = +2.1°.
IX.	0.215 g „ „ „	α = 0.13°	„ = +2.3°.

b. *Cristallisation froide.*

Puis nous avons préparé et purifié le sel neutre de brucine par „cristallisation froide"[13]), qui constitue une méthode commode pour les substances actives qui se racémisent à chaud. On prépare le sel de brucine en mélangeant les solutions froides du sel d'ammonium et de l'acétate de brucine en concentration appropriée. La recristallisation se fait en décomposant le sel de brucine par l'ammoniaque et en faisant réagir le sel ammoniacal de la même manière avec l'acétate de brucine.

On a examiné les cristallisations III, IV et V.

III.	0.204 g dans 3.5 cm³	α = 0.09°	$[M]_D$ = 1.7°.
IV.	0.233 g „ „ „	α = 0.13°	„ = 2.1°.
V.	0.208 g „ „ „	α = 0.24°	„ = 2.4°.

Les deux méthodes donnent ainsi le même résultat.

Nous pouvons admettre pour la rotation du sel d'ammonium $[M]_D$ = + 2.3°. La dispersion rotatoire est normale.

$\lambda(\mu\mu)$	656.3 (C)	589.5 (D)	546.3 (Hg)	486.1 (F)
α (2 dm)	+0.11°	0.13	0.15	0.19
[M]	+1.9°	2.3	2.6	3.4

La solution de l'acide libre dans l'éther présente la rotation $[M]_D$ = +1.9°.

$C_9H_{12}O_4$ Ref. 575a

d-**Spiro[3.3]heptane-2,6-dicarboxylic Acid** (1).—The resolution of the *dl* acid 1 was accomplished according to Backer and Schurink[16] using 171 g (0.43 mol) of brucine, 40 g (0.22 mol) of *dl*-spiro[3.3]-heptane-2,6-dicarboxylic acid, and 3 l. of distilled water. The boiling solution was filtered and placed in a haybox. After cooling to room temperature the supernatant liquid was decanted and the residue was dissolved again. After five crystallizations the brücine salt was dissolved in water and 50 ml of concentrated ammonia was added. Brucine separated and after one night it was filtrated over a Büchner funnel. The filtrate was washed with chloroform, evaporated to a volume of 50 ml, and acidified with concentrated hydrochloric acid, and the precipitated acid was recrystallized from water, yielding 8–12 g (40–60%) of the *d* acid 1: $[\phi]_{578}$ +8.7°, $[\phi]_{546}$ +9.9°, $[\phi]_{436}$ +15.9°, $[\phi]_{405}$ +18.8°, $[\phi]_{365}$ +23.9° (*c* 5.3, acetone); $[\phi]_{578}$ +2.7°, $[\phi]_{546}$ +3.3°, $[\phi]_{436}$ +4.5°, $[\phi]_{405}$ +5.1°, $[\phi]_{365}$ +5.4° (*c* 5.1, ethanol 96%); $[\phi]_{578}$ +3.1°, $[\phi]_{546}$ +3.5°, $[\phi]_{436}$ +5.6°, $[\phi]_{405}$ +6.3°, $[\phi]_{365}$ +7.7° (*c* 5.3, ammonia); $[\theta]_{203}$ +570 ± 60 in water (ref 16).

(16) H. J. Backer and H. B. J. Schurink, *Proc. Kon. Ned. Akad. Wetensch.*, **37**, 384 (1928); *Recl. Trav. Chim. Pays-Bas*, **50**, 921 (1931). The authors report the following values for the optical activity of the Fecht acid after resolution with brucine: $[\phi]_{589}$ +2.3°, $[\phi]_{546}$ +2.6°, $[\phi]_{486}$ +3.4° (c 6.0, water for the ammonium salt); and $[\phi]_{589}$ +1.9° (for the acid in ether solution).

l-**Spiro[3.3]heptane-2,6-dicarboxylic Acid** (1).—The mother liquor obtained after the first crystallization of the brucine salt of *dl*-spiro[3.3]heptane-2,6-dicarboxylic acid (1) was boiled in a beaker until the volume (originally 3 l.) was reduced to 2 l. A small amount of the salt separated on cooling. This was removed and the filtrate was concentrated again in the same way to 1 l. After separation and removal of a second crop of the salt, 40 ml of concentrated ammonia was added to the filtrate. The work-up, in the same way as described for the *d* acid 1 in the preceding section, gave 5–7 g (25–35%) of the *l* acid 1 with optical purity varying from 70 to 80% (related to the optical activity of the *d* acid 1).

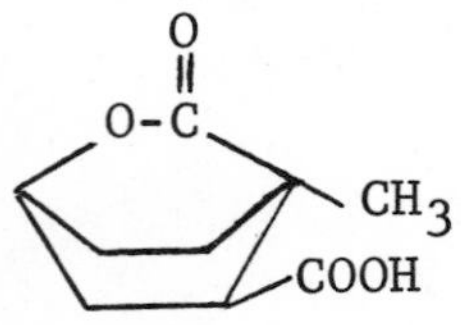

$C_9H_{12}O_4$ Ref. 576

Optical Resolution of the Racemic δ-Lactonic Acid r-(17).—Hot solutions of quinine (87·4 g.) in acetone (500 c.c.) and the δ-lactonic acid (49·55 g.) in acetone (75 c.c.) were mixed, and a part (200 c.c.) of the solvent was removed by distillation before the solution was allowed to cool. The crystalline material which separated and the mother-liquor material were subjected to a systematic fractional crystallisation from acetone–water, and provided, ultimately, two salts. The (−)-acid (−)-quinine salt, m. p. 171—172°, $[\alpha]_D^{25}$ −134° (*c* 1·0 in chloroform) was the least soluble; material of $[\alpha]_D$ −125°, suitable for regeneration of the optically pure acid, was obtained in about 70% yield. The more soluble (+)-acid (−)-quinine salt, plates, m. p. 146—147°, $[\alpha]_D^{21}$ −84·5° (*c* 1·0 in chloroform) was obtained in approximately 40% yield.

For regeneration of the δ-lactonic acid, quinine salt of $[\alpha]_D$ −125° (10 g.) was dissolved in 2N-hydrochloric acid, and the solution was continuously extracted with ether, which provided crude crystalline material (3·31 g., 91%). Recrystallisation from chloroform (30 c.c.) and benzene (100 c.c.) gave the (−)-*δ-lactonic acid* as prisms, m. p. 158—160°, $[\alpha]_D^{22}$ −79° (*c* 1·0 in chloroform) (Found: C, 58·75; H, 6·5. $C_9H_{12}O_4$ requires C, 58·7; H, 6·5%). The (+)-*δ-lactonic acid*, similarly prepared, had m. p. 158—159°, $[\alpha]_D^{21}$ +77·5° (*c* 1·0 in chloroform) (Found: C, 58·65; H, 6·45%).

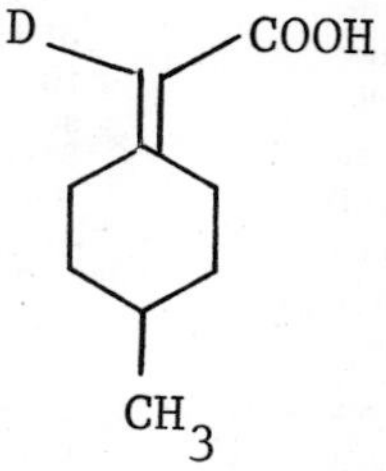

$C_9H_{13}DO_2$

Ref. 577

(+) *und* (−)-α-^{2}*H-4-Methylcyclohexylidenessigsäure* (*I*). Zu einer Lösung von 15,4 g (0,1 Mol) der racemischen Säure in 150 ml Methylacetat gab man 6 g (0,05 Mol) (−)-α-Phenyläthylamin. Das bei Raumtemperatur auskristallisierte Salz der rechtsdrehenden Säure nutschte man ab (9 g, Smp. 108–120°) und isolierte aus der Mutterlauge das Gemisch der freien Säuren. Diese wurden in Methylacetat gelöst und mit der halben molaren Menge (+)-α-Phenyläthylamin versetzt. Dabei fiel das Salz der linksdrehenden Säure aus. Nach dreimaliger Wiederholung dieser Prozedur wurden die rohen Salze des (−) und (+)-α-Phenyläthylamins mit (+) bzw. (−)-α-^{2}H-4-Methylcyclohexylidenessigsäure zweimal aus der zehnfachen Menge Methylacetat umkristallisiert. Der Smp. stieg dabei auf 124–125°. Durch Zersetzung der enantiomeren α-Phenyläthylammoniumsalze mit Alkali wurden 6,5 g rechtsdrehende und 5,3 g linksdrehende Säure, beide vom Smp. 52–54°, erhalten. Die spezifische Drehung der rechtsdrehenden Säure betrug $[\alpha]_{546} = +95°$ ($c = 1,2$, Feinsprit) in Übereinstimmung mit den Angaben der Literatur [26].

[26] W. H. Perkin & W. J. Pope, J. chem. Soc. *99*, 1511 (1911).

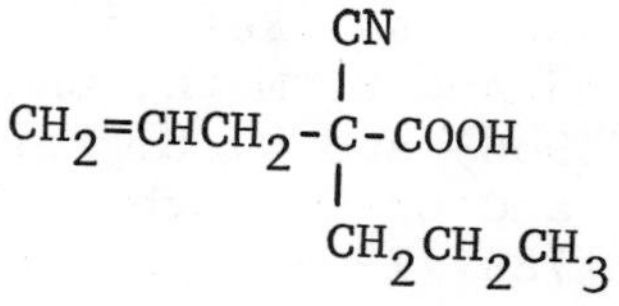

$C_9H_{13}NO_2$ Ref. 578

Spaltung der *d,l*-Allyl-propyl-cyan-essigsäure in die optischen Isomeren.

Sie gelingt am besten mit Brucin in wäßriger Lösung, wobei zuerst das Salz der *d*-Verbindung krystallisiert, während aus den Mutterlaugen das *l*-Salz gewonnen werden kann. Die völlige Reinigung durch Krystallisation ist aber recht schwierig. Wir haben deshalb noch das Morphinsalz zu Hilfe genommen.

Bei Anwendung von Cinchonin krystallisiert aus Wasser zuerst das Salz der *l*-Säure, aber auch in ziemlich unreinem Zustand.

Für die Versuche in größerem Maßstabe diente folgendes Verfahren. 150 g *d,l*-Säure und 420 g wasserfreies Brucin werden in 2250 ccm heißem Wasser gelöst. Beim langsamen Abkühlen krystallisieren lange, farblose Nadeln. Nach 24-stündigem Stehen im Eisschrank werden sie abgesaugt, mit eiskaltem Wasser gewaschen, abgepreßt und dann noch zweimal, zuerst aus 250, dann aus 200 ccm heißem Wasser umkrystallisiert. Ausbeute 130 g exsiccatortrocknes Salz.

Dieses Salz ist noch nicht rein, denn die daraus bereitete Säure zeigte, in *n*-Natronlauge gelöst, im besten Falle $[\alpha]_D^{16} = +11.9°$. Da erneutes Umkrystallisieren nur geringe Wirkung hatte, so haben wir die weitere Reinigung mit dem Morphinsalz ausgeführt.

Zu dem Zweck wird zunächst die freie Säure in folgender Weise dargestellt. Man suspendiert 120 g fein zerriebenes Brucinsalz in 1500 ccm kaltem Wasser und fügt 240 ccm 5*n*-Salzsäure und etwa 300 ccm Äther zu. Beim kräftigen Schütteln ist das Salz rasch zerlegt und die Fettsäure in den Äther übergegangen. Die ätherische Lösung wird noch zweimal etwa 15 Minuten mit 5*n*-Salzsäure durchgeschüttelt, um alles Brucin zu entfernen, dann mit Wasser gewaschen und der Äther verdampft. Die sämtlichen Mutterlaugen müssen nochmals ausgeäthert werden, um Verluste an organischer Säure zu vermeiden.

Zur Bereitung des Morphinsalzes haben wir 20 g der Allyl-propyl-cyan-essigsäure ($[\alpha]_D^{16} = +11.9°$ in *n*-Natronlauge) und 34.1 g Morphin mit 400 ccm Wasser übergossen, wobei eine lebhafte, strudelähnliche Bewegung zu beobachten ist, die von der Einwirkung der ölig werdenden Säure auf die Base herrührt. Beim Erwärmen tritt völlige Lösung ein, und beim Abkühlen scheidet sich das Morphinsalz rasch krystallinisch aus. Es wurde aus 300 ccm Wasser unter Zusatz von wenig Tierkohle umkrystallisiert. Es bildet farblose, derbe Prismen, welche etwas Wasser enthalten, das im Hochvakuum rasch entweicht. Ausbeute 30 g.

Die aus einer Probe des Salzes isolierte Säure zeigte, in *n*-Natronlauge gelöst, $[\alpha]_D^{16} = +16.52°$. Die Hauptmenge des Salzes wurde dann noch zweimal aus 150 bezw. 130 ccm Wasser umkrystallisiert, wobei die Ausbeute auf 20 g zurückging. Zur Umwandlung in die freie Säure diente dasselbe Verfahren wie beim Brucinsalz, nur wurde die Salzsäure durch Schwefelsäure ersetzt. Das Drehungsvermögen der freien Säure war dann nur auf $[\alpha]_D^{16} = +16.74$ gestiegen. Offenbar läßt sich also durch weiteres Umkrystallisieren des Morphinsalzes kaum noch eine reinere Säure gewinnen. Wir konnten auch wegen der großen Verluste den Versuch nicht weiter treiben und müssen es selbstverständlich unentschieden lassen, ob damit der Endwert der Drehung schon erreicht ist.

d-Allyl-propyl-cyan-essigsäure

bleibt beim Verdampfen der ätherischen Lösung als völlig krystallinische Masse zurück, die aus farblosen, makroskopischen Nadeln oder Prismen vom Schmp. 42° (korr.) besteht. Unter 0.5 mm sublimiert sie schon bei 36°, allerdings sehr langsam. Sie ist der *d,l*-Säure in den Lösungsverhältnissen sehr ähnlich und verflüssigt sich wie diese in Berührung mit Wasser. Die alkalische Lösung dreht ziemlich stark nach rechts. Überschuß des Alkalis sowie Verdünnung mit Wasser setzen das Drehungsvermögen herab.

Für die genaue Bestimmung diente die Lösung in *n*-Natronlauge:

$$[\alpha]_D^{16} = \frac{+1.55° \times 11.3638}{1 \times 1.0118 \times 1.040} = +16.74°.$$

In möglichst genau neutralisierter Lösung betrug die Drehung:

$$[\alpha]_D^{21} = \frac{+2.57° \times 2.7736}{1 \times 0.3769 \times 1.038} = +18.22°.$$

Eine zweite Bestimmung ergab:

$$[\alpha]_D^{21} = \frac{+2.65° \times 1.8302}{1 \times 0.2501 \times 1.036} = +18.72°.$$

0.8195 g dieser Lösung wurden mit Wasser verdünnt. Gesamtgewicht 1.7632 g; $\alpha_D = +1.12°$; d = 1.0145. Also $[\alpha]_D^{21} = +17.38°$.

In Eisessig zeigte dieselbe Säure die Drehung:

$$[\alpha]_D^{22} = \frac{+0.76° \times 2.3757}{1 \times 0.2161 \times 1.048} = +7.97°.$$

Für die Analyse war im Hochvakuum bei gewöhnlicher Temperatur getrocknet.

0.1486 g Sbst.: 0.3520 g CO_2, 0.1044 g H_2O. — 0.1182 g Sbst.: 0.2804 g CO_2, 0.0840 g H_2O. — 0.1648 g Sbst.: 11.8 ccm N (16°, 756 mm).

$C_9H_{13}O_2N$ (167.11). Ber.	C 64.63,	H 7.84,	N 8.38.
Gef.	» 64.60, 64.70,	» 7.86, 7.95,	» 8.32.

Zur Darstellung der Säure aus der *d,l*-Verbindung kann auch direkt das Morphinsalz dienen. Da aber die Verluste hier noch größer waren, haben wir die Kombination von Brucin- und Morphinsalz vorgezogen.

l-Allyl-propyl-cyan-essigsäure.

Wenn die wäßrigen Mutterlaugen, die bei der Darstellung des

Brucinsalzes der *d*-Säure aus dem Racemkörper resultieren, wochenlang stehen bleiben, so scheiden sich neben den Nadeln des *d*-Salzes ziemlich große, manchmal fächer- oder drusenförmig vereinigte, farblose Tafeln ab, die sich mit einiger Mühe mechanisch auslesen lassen. Bequemer ist es, das Gemisch der Salze mit der Mutterlauge kräftig zu schütteln, und die hierbei zerbrechenden Nadeln des *d*-Salzes durch Abschlämmen von den Tafeln des *l*-Salzes zu trennen. Aus dem so bereiteten Brucinsalz konnten wir direkt eine Säure gewinnen, die in *n*-Natronlauge $[\alpha]_D^{19} = \frac{-2.25^0 \times 2.9870}{1 \times 0.4081 \times 1.044} = -15.77^0$ zeigte. Sie war also optisch annähernd so rein, wie die beste, über das Morphinsalz gereinigte *d*-Säure. Der Schmelzpunkt war aber hier nicht so scharf und auch niedriger als bei der *d*-Säure.

Wie schon erwähnt, läßt sich die *l*-Säure auch direkt aus dem Racemkörper durch das Cinchoninsalz erhalten, aber die Reinigung erfordert dann viel häufigere Krystallisation.

Decomposition of the Salt with Solvent.—A small portion of the salt was dissolved in chloroform and allowed slowly to evaporate until crystals were formed. The remaining solvent was then removed under vacuum. The residue was extracted with ether, and the ether soluble material crystallized from petroleum ether (30–60°) to a melting point of 46–47°.

Decomposition of the Salt with Acid.—The salt was stirred with a mixture of dilute hydrochloric acid and ether. The aqueous layer was washed with three portions of ether, all the ether layers were combined and the ether soluble material crystallized from petroleum ether (30–60°) to a maximum constant melting point of 49–50° and a rotation of $[\alpha]^{25}D$ −30.69°. *Anal.* Calcd. for $C_9H_{14}O_2$: C, 70.3; H, 9.1. Found: C, 70.2; H, 9.1.

(14) After collapsing at 126–127°, there remains a mixture of liquid melt and solid up to the point where the solid begins to turn dark, usually about 140°.

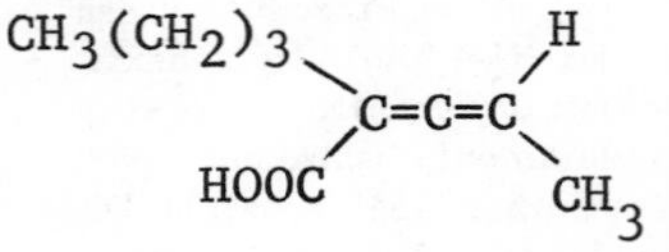

$C_9H_{14}O_2$ Ref. 579

The Resolution of 2,3-Octadien-4-carboxylic Acid into Its Optically Active Components.—All optical measurements were made on a Bausch and Lomb saccharimeter in chloroform solutions using a two decimeter tube. A chloroform solution of 2.1735 g. (0.0065 mole) of strychnine and 1.9225 g. (0.0125 mole) of III, after standing for 12 hours, was evaporated to a thick sirup, 5 cc. of ethyl acetate was added causing a white precipitate to form. Six recrystallizations from a mixture of chloroform and ethyl acetate gave a white solid, m.p. 126–127°,[14] $[\alpha]^{25}D$ −72.69°. *Anal.* Calcd. for $C_{30}H_{36}N_2O_4$: C, 73.8; H, 7.4; N, 5.9. Found: C, 73.8; H, 7.3. N, 5.6.

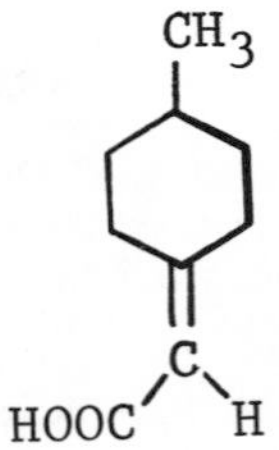

$C_9H_{14}O_2$

For the resolution procedure for the above compound, see Ref. 577 and also W. Marckwald and R. Meth, Chem. Ber., 39, 1175 (1906) and W. H. Perkin, W. J. Pope and O. Wallach, J. Chem. Soc., 91, 1789 (1909).

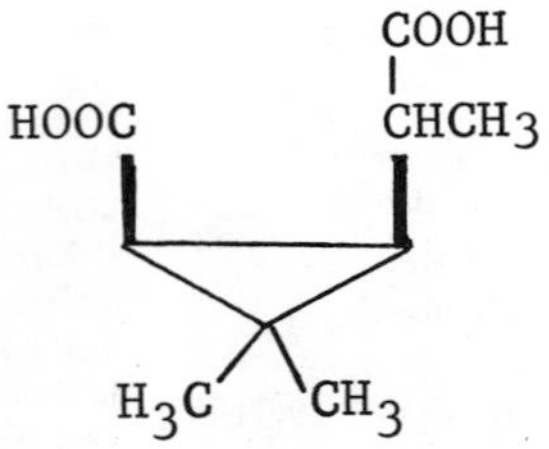

Ref. 580

$C_9H_{14}O_4$

Resolution of cis-dl-3-*Carboxy*-1 : 1-*dimethyl*cyclo*propane*-2-*propionic Acid.*—To a hot solution of the acid (8·9 g.) in water (430 c.c.) and sodium hydroxide solution (2·548*N*; 18 c.c.), morphine (14·5 g.) was added; the cooled solution deposited a crystalline solid (6·4 g.), m. p. 177—178°, and after concentration, two further crops of crystals (2·1 g., m. p. 178—180°; 0·25 g., m. p. 164—167°) were obtained. These were combined and recrystallised twice from dilute alcohol, the *dimorphine* salt of the acid being obtained in needles, m. p. 177—178° (sintering 168°) (Found: C, 66·0; H, 7·1. $C_{43}H_{52}O_{10}N_2,H_2O$ requires C, 66·7; H, 7·0%). The salt was decomposed in the usual manner, and *cis*-*d*-3-carboxy-1 : 1-dimethyl*cyclo*propane-2-propionic acid crystallised from water in needles, m. p. 104—105° (softening 102°), α_{5461} + 1·01°, $[\alpha]_{5461}$ + 39·0° (*c*, 2·59 in chloroform) (Found: C, 57·8; H, 7·8. Calc. for $C_9H_{14}O_4$: C, 58·0; H, 7·5%). On admixture with a specimen of acid prepared by the oxidation of *d*-Δ^4-carene, no depression of m. p was observed, and the acids appeared identical in all respects.

The acid ($[\alpha]_{5461}$ − 11·62°) recovered from the more soluble dimorphine salt contained approximately 30% of the *l*-acid; to a solution of this (2·9 g.) in water (120 c.c.), sodium hydroxide solution (2·548*N*; 8·8 c.c.) and strychnine (2·88 g.) were added, and the hot filtered solution concentrated to 70 c.c.; on cooling, a strychnine salt (2·5 g.) separated. The *distrychnine* salt (11·2 g.), which was very sparingly soluble in the usual solvents, was recrystallised twice from dilute alcohol, separating in needles, m. p. 189—190° (sintering 185°) [Found: H_2O, 6·9 (over P_2O_5). $C_{51}H_{58}O_8N_4,4H_2O$ requires $3\frac{1}{2}H_2O$, 6·8%. Found, in dried salt: C, 70·6; H, 7·4. $C_{51}H_{58}O_8N_4,\frac{1}{2}H_2O$ requires C, 70·9; H, 6·8%].

cis-l-3-*Carboxy*-1 : 1-*dimethyl*cyclo*propane*-2-*propionic acid*, recovered from the strychnine salt, crystallised from water in needles, m. p. 104—105° (softening 102°), α_{5461} − 1·21°, $[\alpha]_{5461}$ − 37·8° (*c*, 3·204 in chloroform) (Found: C, 57·8; H, 7·5%).

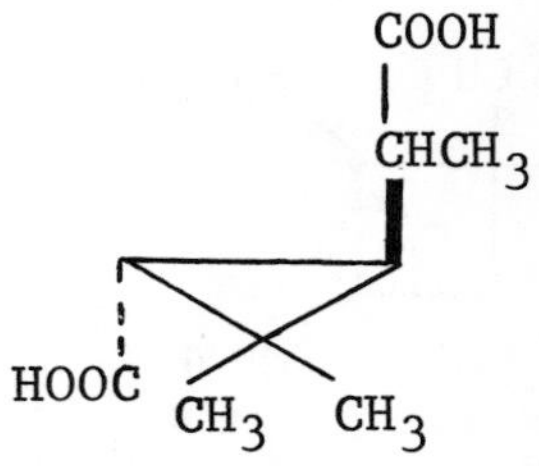

$C_9H_{14}O_4$ **Ref. 581**

Resolution of trans-dl-3-*Caroxy*-1 : 1-*dimethyl*cyclo*propane-2-propionic Acid.*—To a hot solution of nor-*d*-ψ-ephedrine (11·7 g.) in acetone (160 c.c.) the *trans*-acid (7·3 g.) was added, and the filtered solution kept for 48 hours; the needles which separated (4·8 g., m. p. 191—193°) were recrystallised twice from water (60 c.c.), and the *di-nor-*d*-ψ-ephedrine* salt of the acid was obtained in needles, which, after drying in a vacuum, had m. p. 192—193° (sintering 187°), α_{5461} + 0·535°, $[\alpha]_{5461}$ + 18·16° (*c*, 2·876 in methyl alcohol). The rotatory power was unaltered by further crystallisation (Found: C, 64·3; H, 8·1. $C_{27}H_{40}O_6N_2,H_2O$ requires C, 64·0; H, 8·3%).

trans-l-3-*Carboxy*-1 : 1-*dimethyl*cyclo*propane-2-propionic acid*, regenerated from the salt, crystallised from water in prisms, m. p. 112°, α_{5461} − 1·52°, $[\alpha]_{5461}$ − 37·1° (*c*, 4·1 in ethyl acetate) (Found: C, 57·9; H, 7·8%). The sodium salt of the acid was dextrorotatory in water, α_{5461} + 0·29°, $[\alpha]_{5461}$ + 11·6° (*c*, 2·496).

From the more soluble di-nor-*d*-ψ-ephedrine salt, the *trans*-acid ($[\alpha]_{5461}$ + 8·7°) was recovered and treated with nor-*l*-ψ-ephedrine under the conditions described above. The *di-nor-*l*-ψ-ephedrine* salt of the *trans-d*-acid separated very rapidly, and after crystallisation from water had m. p. 192—193°, α_{5461} − 0·45°, $[\alpha]_{5461}$ − 18·5° (*c*, 2·428 in methyl alcohol) (Found: C, 64·5; H, 8·2%). The trans-d-*acid*, after recrystallisation from water, had m. p. 112°, α_{5461} + 1·09°, $[\alpha]_{5461}$ + 37·4° (*c*, 2·912 in ethyl acetate) (Found: C, 58·1; H, 7·4%).

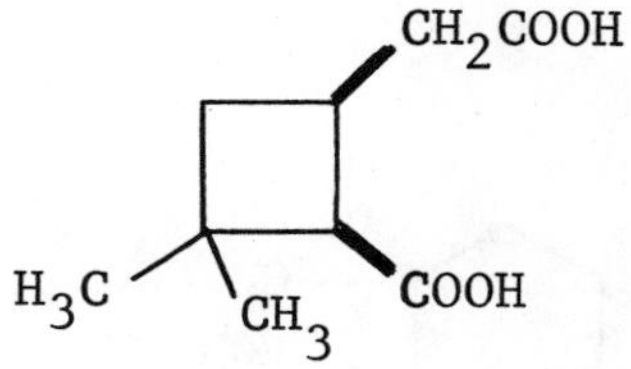

$C_9H_{14}O_4$ **Ref. 582**

Resolution of (±)-cis-2-*Carboxy*-3 : 3-*dimethyl*cyclo*butylacetic Acid.*—The (±)-*cis*-acid (1·000 g.) and anhydrous (−)-quinine (1·742 g.) were dissolved in ethanol (30 c.c.) and water (50 c.c.), and on being kept the filtered solution deposited the salt (1·48 g.) as needles. Two crystallisations from aqueous ethanol gave this (−)-acid-(−)-quinine salt (0·76 g.) in large clusters of fine needles, m. p. 202°. Decomposition of the salt with ammonia gave finally the (−)-cis-*acid* (0·25 g.) which crystallised from *cyclo*hexane in prisms, m. p. 157°, $[\alpha]_D^{19}$ −11·1° (in acetone; *c*, 3·650) (Found: C, 58·0; H, 7·3. $C_9H_{14}O_4$ requires C, 58·0; H, 7·6%). Treatment of the more soluble salt fractions in the manner described for the *trans*-acid, gave the (+)-cis-*acid* (0·35 g.) which crystallised from *cyclo*hexane in prisms, m. p. 157—158°, $[\alpha]_D^{20}$ +10·9° (in acetone; *c*, 3·960) (Found: equiv., 92·9. $C_9H_{14}O_4$ requires equiv., 93·0).

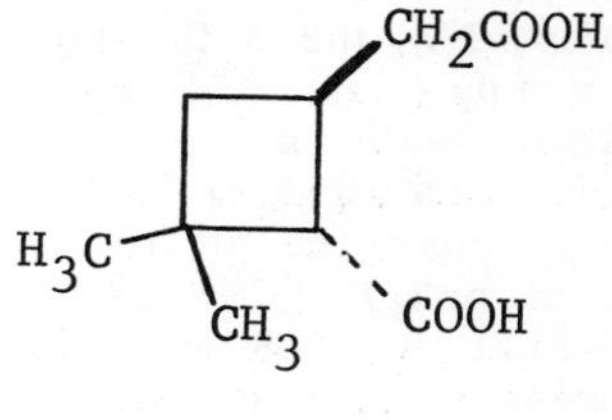

$C_9H_{14}O_4$ **Ref. 583**

Resolution of (±)-trans-2-*Carboxy*-3 : 3-*dimethyl*cyclo*butylacetic Acid.*—The (±)-*trans*-acid (1·000 g.) and anhydrous (−)-quinine (1·742 g.) were dissolved in ethanol (16 c.c.) and water (26 c.c.). The solution was filtered and on being kept deposited the salt (1·56 g.) as needles. Three crystallisations from aqueous ethanol (2 : 1 by vol.) gave the neutral (+)-acid-(−)-quinine salt in large rosettes of fine needles, m. p. 121°, $[\alpha]_D^{21}$ −114·4° (in ethanol; *c*, 4·036). The salt was dissolved in aqueous methanol and decomposed by addition of ammonia. The precipitated quinine was filtered off and the alcohol distilled from the filtrate which was then acidified and extracted with ether. Removal of the solvent gave the (+)-trans-*acid* (0·26 g.) which crystallised from *cyclo*hexane in prisms, m. p. 144—145°, $[\alpha]_D^{21}$ +24·2° (in acetone; *c*, 1·115) (Found: C, 57·8; H, 7·6. $C_9H_{14}O_4$ requires C, 58·0; H, 7·6%).

The acid recovered from the more soluble quinine salt fractions (0·71 g.), m. p. 123—125°, $[\alpha]_D^{21}$ −17·4°, was again treated with quinine (1·24 g.) in alcohol (5 c.c.) and water (10 c.c.). After one further crystallisation from alcohol the less soluble salt (0·77 g.) was collected, and the combined filtrates containing the more soluble salt were decomposed as above and gave the (−)-trans-*acid*, m. p. 142—143°, $[\alpha]_D^{21}$ −24·6° (in acetone; *c*, 1·084) (Found: C, 57·7; H, 7·7%).

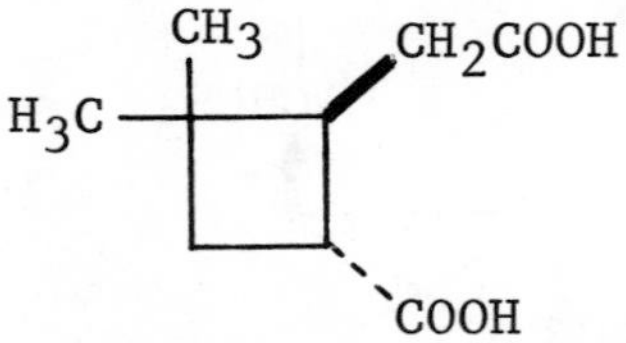

$C_9H_{14}O_4$ Ref. 584

Resolution of Caryophyllenic Acid.—Quinine methohydroxide solution (0·41N; 79 ml.) (Major and Finkelstein, *J. Amer. Chem. Soc.*, 1941, **63**, 1368) was exactly neutralised by the addition of *trans*-(±)-caryophyllenic acid (3 g.). The salt (16·9 g.), obtained by evaporating the solution to dryness, was dissolved in absolute ethanol (15 ml.) and the solution diluted with pure dioxan (150 ml.). The salt which crystallised out in small balls of fine needles (9·5 g.), $[\alpha]_D^{20}$ −134° (*c* = 1·121 in EtOH) was thrice recrystallised similarly, giving the pure quinine metho-salt of *trans*-(−)-caryophyllenic acid in rosettes of fine needles (3·5 g.), m. p. 202—203°, $[\alpha]_D^{20}$ −148° (*c* = 1·421 in EtOH). This salt was dissolved in 10% w/w sulphuric acid, and the organic acid removed by extraction with ether (10 × 100 ml.) as a yellow oil, which rapidly crystallised. Purification through the sparingly soluble potassium salt, followed by crystallisation from *cyclo*hexane, yielded trans-(−)-4-*carboxy*-2 : 2-*dimethyl*cyclo*butylacetic acid* [*trans*-(−)-caryophyllenic acid] as rosettes of long needles, m. p. 81—82°, $[\alpha]_D^{20}$ −28° (*c* = 2·125 in C_6H_6) (Found : C, 58·3; H, 7·3%; Equiv. (by titration), 93].

Recovery of the acid from the more soluble metho-salt of the *cis*-acid, contained in the filtrate from the original crude *trans*-salt, gave an oil (2·3 g.) which, on fractional crystallisation from *cyclo*hexane, yielded pure trans-(+)-4-*carboxy*-2 : 2-*dimethyl*cyclo*butylacetic acid* [*trans*-(+)-caryophyllenic acid] (0·15 g.) as long needles, m. p. 81—82°, $[\alpha]_D^{20}$ +28·5° (*c* = 1·452 in C_6H_6) [Found : C, 57·9; H, 7·4%; Equiv. (by titration), 93]. A 50% mixture of this acid with authentic *trans*-(+)-caryophyllenic acid, m. p. 81—82°, $[\alpha]_D^{20}$ +28° (*c* = 1·468 in C_6H_6), prepared from caryophyllene essentially by the method of Evans, Ramage, and Simonsen (*J.*, 1934, 1807), but omitting the preliminary permanganate oxidation, melted at 81—82°.

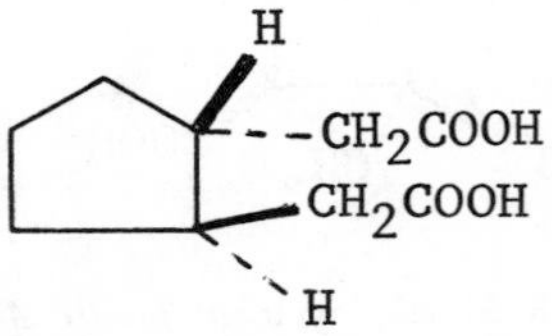

$C_9H_{14}O_4$ Ref. 585

cyclo*Pentane and* bicyclo*Octane Series.*—*dl-trans-cyclo*Pentane-1 : 2-diacetic acid could not be resolved by fractional crystallisation of its neutral or acid salts with cinchonine, cinchonidine, quinine, or strychnine, and the acid brucine salt was too soluble, but the neutral salt yielded the desired separation. A solution of 10 g. of the pure acid (m. p. 132—133°; Barrett and Linstead, *loc. cit.*) in 250 c.c. of hot water was treated with 49·4 g. (2 mols.) of brucine. The solution deposited large, flat, transparent prisms (A), the mother-liquor from which on concentration yielded four further crops (B) and a syrupy residue (C). Fraction (A) (20·2 g. after being dried in a vacuum desiccator) crystallised from water in large prisms, often aggregated in radiating clusters. The once-crystallised salt yielded an acid, isolated by means of ether, with m. p. 149°, $[\alpha]^{18°}$ − 52·4°. After 3 and 5 crystallisations of the salt from water, the liberated acid had m. p. 151°, $[\alpha]^{19°}$ − 63·3° (*c* = 2·340), and m. p. 151°, $[\alpha]^{18°}$ − 63·6° (*c* = 2·555), respectively. The neutral *brucine* salt of the *l*-acid melted at 105—108° (Found : N, 6·15. $C_{56}H_{66}O_{12}N_4$ requires N, 6·0%). 17·5 G. of the pure thrice-crystallised salt yielded 3·3 g. (calc., 3·4 g.) of 1-trans-cyclo*pentane*-1 : 2-*diacetic acid*, massive, flat, hexagonal prisms or small octahedra from water, m. p. 151°, $[\alpha]^{18°}$ − 63·5 (*c* = 2·630) (Found : C, 58·0; H, 7·6; equiv., 93·0. $C_9H_{14}O_4$, dibasic, requires C, 58·0; H, 7·6%; equiv., 93·05). An equal mixture with the *dl*-acid melted at 138—144°.

The intermediate crops (B) yielded mixtures of *dl*- and active acids even after thorough crystallisation. The syrupy residue (C) was decomposed to the nearly pure d-*acid*, which crystallised in flat prisms from water, 0·6 g., m. p. 150—151°, $[\alpha]^{19°}$ + 62·3° (*c* = 2·014) (Found : equiv., 93·2). An equal mixture of the *d*- and the *l*-acid melted at 132—136° when ground together, and after crystallisation from water, at 132—133°. The active acids appear to be less soluble than the racemic acid in water.

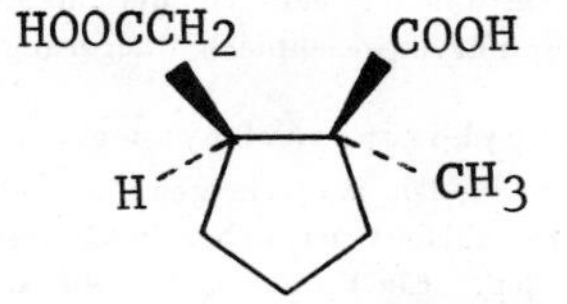

$C_9H_{14}O_4$ **Ref. 586**

Resolution of cis-2-methyl-2-carboxycyclopentane-1-acetic acid (I). The racemic acid[3] (m.p. 110–111°) (459.5 g.; 2.46 moles) was dissolved in a minimum amount of hot water and 1948 g. (2.46 moles) of brucine was added with enough hot water to dissolve it. The brucine salt separated when the solution was cooled; twenty-one such recrystallizations of the salt were carried out before the diacid regenerated by treatment of the salt with 10% hydrochloric acid and extraction with ether gave a constant rotation. The resolved diacid (70.5 g.) had the m.p. 114° and $[\alpha]^{21°}_D$ +37° (4% in chloroform). Racemic acid (186 g.) recovered from the mother liquor and put through a similar treatment with extensive recrystallization of the brucine salt gave an additional 18.5 g. of resolved acid, making a total of 89 g. Partially resolved (−) acid (109 g.) $[\alpha]^{25°}_D$ −6.4° was also obtained from the mother liquor.

(3) H. Conroy, *J. Am. Chem. Soc.*, **73**, 1889 (1951); **74**, 491 (1952); **74**, 3046 (1952).

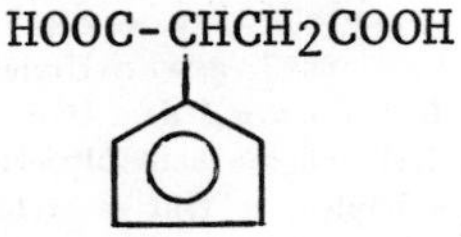

$C_9H_{14}O_4$ **Ref. 587**

*a-Cyclo*pentyl-succinic acid has been resolved with the help of brucine (neutral salt). The *d* and *l* acid had $[\alpha]^{26}_D$ + 17·81° (l = 1, c = 2·358 in acetone) and $[\alpha]^{26}_D$ − 16·94° (l = 1, c = 1·830 in acetone) respectively. The active forms melted at 135°.

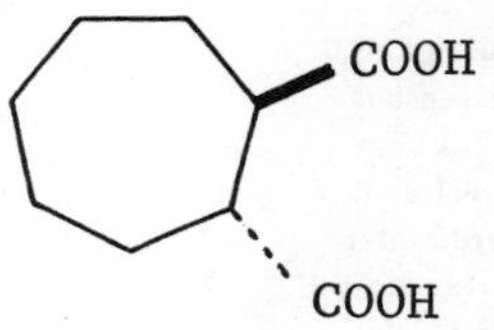

$C_9H_{14}O_4$ **Ref. 588**

Resolution of the Acid as the Brucine Salts.—*trans-cyclo*Heptane-1 : 2-dicarboxylic acid (1·97 g., 0·011 mole; m. p. 142—146°) and anhydrous brucine (8·3 g.; m. p. 167—171°) were dissolved in hot ethyl acetate (30 ml.). The brucine salt began to separate from the warm solution almost immediately; it was filtered off after 48 hours. The white solid product (8·25 g., 80% of total) was extracted with boiling ethyl acetate (30 ml.) for 45 min. and the mixture filtered. In 24 hr. the filtrate deposited a crystalline product (*salt A*; 3·8 g., 37%), m. p. 129—132°, α −0·44°, $[\alpha]^{19}_D$ −15·2° ± 0·7° (*c* 2·9 in EtOH) unchanged in rotation by further crystallisation (Found: C, 62·6; H, 6·7; N, 5·7. $C_{55}H_{66}O_{12}N_4,4H_2O$ requires C, 63·0; H, 7·1; N, 5·35%).

Evaporation of the original ethyl acetate liquor gave a thick oil (*salt B*; 1·1 g.) which solidified when rubbed in ether, and then had α −0·68°, $[\alpha]^{20}_D$ −39·5° ± 0·7° (*c* 1·7 in EtOH). Evaporation of the liquor from the first crystallisation of salt A gave more of salt B (3·7 g.), α −0·38°, $[\alpha]^{20}_D$ −38° ± 1·0° (*c* 2·0 in EtOH, *l* 0·5 dm.). This did not crystallise; purified by precipitation from alcohol by ether, it had m. p. 101—108°, $[\alpha]^{19}_D$ −46·0° ± 1·0° (Found: C, 62·6; H, 6·9; N, 5·5%). The salts probably became hydrated on exposure to the atmosphere. There was no weight loss at 76°/0·05 mm., and a loss of only 1% after 3 hr. at 110°/10^{-3} mm.

The brucine salts A and B (3·8 g.), $[\alpha]^{19}_D$ −15·2° and $[\alpha]^{20}_D$ −39·5°, were separately decomposed by stirring them with 2N-hydrochloric acid (25 ml.) for 4 min. at 0°. The acid solution was ether-extracted (3 × 20 ml.), and the combined extracts were washed with water (2 × 5 ml.), dried ($MgSO_4$), and evaporated below room temperature. Salt A gave an acid, m. p. 138—142°, α +0·41°, $[\alpha]^{19}_D$ +3·4° (±0·3°) (*c* 12·5 in EtOH). After one crystallisation from benzene this had m. p. 142—145°, $[\alpha]^{19}_D$ +1·6° (±0·4°). Partial racemisation also occurred during crystallisation from benzene–light petroleum (b. p. 60—80°). Salt B gave an acid, m. p. 138—141°, α −0·19°, $[\alpha]^{18}_D$ −4·7° (±0·5°) (*c* 8·2 in EtOH, *l* 0·5 dm.). The rotation of this sample was unchanged when the ethanolic solution was kept overnight at room temperature, but after crystallisation from benzene the material had $[\alpha]^{19}_D$ −1·0° (±0·2°). The time interval between the commencement of the acid treatment and the first reading of the polarimeter was 50 min. for the (+)-acid and 56 min. for the (−)-acid.

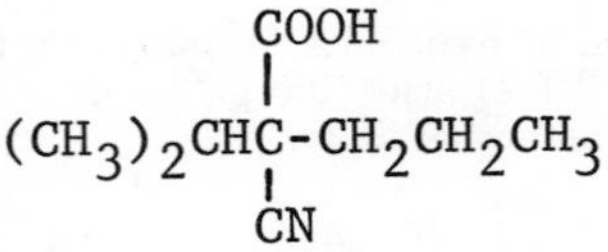

$C_9H_{15}NO_2$ **Ref. 589**

d-Propyl-isopropyl-cyan-essigsäure.

31 g reine inaktive Säure und 73 g wasserfreies Brucin werden in 450 ccm Wasser durch Erwärmen auf dem Dampfbade und schließlich über freier Flamme gelöst und die heiß filtrierte Flüssigkeit langsam abgekühlt. Kann man impfen, so tritt recht bald Krystallisation ein. Nach längerem Stehen im Eisschrank wird die Krystallmasse scharf abgesaugt, mit wenig eiskaltem Wasser ausgewaschen und schließlich gut abgepreßt (42 g). Die Krystalle werden zweimal aus je 150 ccm heißem Wasser unter Zusatz von sehr wenig Tierkohle umkrystallisiert. Die Lösung filtriert man zweckmäßig durch einen Heißwassertrichter, da das Salz schon bei geringer Abkühlung ausfällt. Nach dem Absaugen und Waschen mit Wasser wird jedesmal scharf abgepreßt. Schließlich wurde nochmals aus 100 ccm Wasser umkrystallisiert und im Vakuumexsiccator über Phosphorpentoxyd getrocknet. Die Ausbeute betrug dann 19.5 g. Das Brucinsalz bildet farblose, glänzende Krystalle, die man unter dem Mikroskop als schöne, fast rechtwinklige Tafeln erkennt.

Eine 10-prozentige Lösung des unter 15 mm Druck bei 75° getrockneten Brucinsalzes in Alkohol zeigte

$$\alpha = -0.29°.$$

Es schmilzt bei 120° (korr. 121°) zu einer zähen Flüssigkeit. Die getrocknete Substanz nimmt an der Luft sehr rasch Wasser auf.

Zur Gewinnung der freien Säure werden 19 g im Vakuumexsiccator getrocknetes Brucinsalz (entsprechend 18.5 g wasserfreiem Salz) in 300—400 ccm heißem Wasser gelöst, mit 30 ccm verdünnter Schwefelsäure versetzt und das abgeschiedene Öl ausgeäthert. Man schüttelt die ätherische Lösung nochmals mit verdünnter Schwefelsäure, wäscht mit Wasser und trocknet schließlich einige Stunden über Natriumsulfat. Beim Verdampfen der filtrierten Lösung bleibt ein farbloser Sirup zurück, der beim Reiben bald erstarrt. Die Krystalle wurden im Vakuumexsiccator über Phosphorpentoxyd getrocknet (5 g), dann auf der Presse scharf abgepreßt, in einer Reibschale mit etwas trocknem Petroläther verrieben, abgesaugt und mit Petroläther gewaschen. Die Ausbeute betrug 4.4 g oder 79 % der Theorie.

Die Säure schmilzt bei 94—95° (korr.), mithin erheblich höher als die inaktive Form, zu einer farblosen Flüssigkeit.

0.1308 g Sbst.: 0.3067 g CO_2, 0.1073 g H_2O. — 0.1226 g Sbst.: 0.2869 g CO_2, 0.1001 g H_2O. — 0.1670 g Sbst.: 12.2 ccm N (19°, 765 mm) über 33-prozentiger Kalilauge.

$C_9H_{15}O_2N$ (169.12). Ber. C 63.86, H 8.94, N 8.28.
Gef. » 63.95, 63.82, » 9.18, 9.13, » 8.48.

In Berührung mit Wasser verwandelt sich die Säure rasch in ein farbloses Öl. Sie ist in den gewöhnlichen organischen Solvenzien mit Ausnahme des Petroläthers leicht löslich und gleicht auch sonst sehr der inaktiven Verbindung.

Zu folgenden optischen Bestimmungen wurde über Phosphorpentoxyd getrocknetes und destilliertes Toluol als Lösungsmittel verwandt.

I. 0.3448 g Säure. Gesamtgewicht der Lösung 1.6560 g. $d^{20} = 0.896$. Drehung im 1-dm-Rohr bei 20° und Natriumlicht 2.15° (± 0.02°) nach rechts. Demnach

$$[\alpha]_D^{20} = +11.5° (\pm 0.1°).$$

II. 0.7014 g Sbst. Gesamtgewicht der Lösung 6.7186 g. $d^{20} = 0.881$. Drehung im 1-dm-Rohr bei 20° und Natriumlicht 1.04° (± 0.02°) nach rechts. Mithin

$$[\alpha]_D^{20} = +11.3° (\pm 0.2°).$$

III. 0.1968 g Säure. Gesamtgewicht der Lösung 2.0062 g. $d^{20} = 0.878$. Drehung im 1-dm-Rohr bei 20° und Natriumlicht 0.98° (± 0.02°) nach rechts. Demnach

$$[\alpha]_D^{20} = +11.4° (\pm 0.2°).$$

Mikropolarimetrisch haben wir noch für folgende Lösungsmittel die Drehung annähernd ermittelt. Die Werte gelten für 10-prozentige Lösungen bei 20° ohne Berücksichtigung des spezifischen Gewichts.

Reines, thiophenfreies Benzol	$\alpha = +0.72°$
Xylol	$\alpha = +1.21°$
Cymol	$\alpha = +1.01°$
Brombenzol	$\alpha = +1.52°$
Frisch destilliertes Nitrobenzol	$\alpha = +0.94°$
Alkohol	$\alpha = +0.28°$
Essigäther	$\alpha = +0.43°$
Chloroform	$\alpha = +0.96°$
n-Natronlauge	$\alpha = +0.49°$

Zusatz von Natriummolybdat zu der alkalischen Lösung verursachte keine oder nur unwesentliche Steigerung der Drehung.

l-Propyl-isopropyl-cyan-essigsäure.

Sie findet sich in den Mutterlaugen, aus denen das Brucinsalz der *d*-Säure auskrystallisiert ist, aber noch vermischt mit großen Mengen des Antipoden. Ein Präparat, das wir aus den ersten Mutterlaugen des Brucinsalzes auf ähnliche Art wie die *d*-Säure isolierten, drehte nach links, aber das Drehungsvermögen betrug kaum ein Drittel des berechneten Wertes. Zur Reinigung haben wir die Säure in das Morphinsalz übergeführt. Zu dem Zweck werden 4 g Säure und 7.3 g Morphin in 50 ccm siedendem Wasser gelöst, wenn nötig mit wenig Tierkohle entfärbt und heiß filtriert. Beim Abkühlen auf gewöhnliche Temperatur scheiden sich feine, meist büschelförmig gruppierte Nadeln ab. Sie wurden zweimal aus je 30 ccm und zum Schluß noch aus 20 ccm heißem Wasser umkrystallisiert, wobei allerdings erhebliche Verluste eintraten. Zur Gewinnung der freien Säure wurde das so gereinigte Salz in 50—60 ccm warmem Wasser gelöst, mit einem erheblichen Überschuß verdünnter Schwefelsäure versetzt und das gefällte Öl ausgeäthert. Beim Verdampfen der mit Natriumsulfat getrockneten ätherischen Lösung blieb die Säure als farblose Krystallmasse zurück, die abgepreßt und mit kaltem Petroläther gewaschen wurde. Die Ausbeute betrug 0.85 g.

Genau in derselben Weise läßt sich die linksdrehende Säure auch direkt aus dem Racemkörper durch das Morphinsalz gewinnen, nur ist dann die Ausbeute selbstverständlich geringer. Sie betrug nur 0.5 g aus 4 g Racemkörper, mithin 25 % der Theorie.

Für die Analyse und optische Bestimmung war im Vakuumexsiccator über Phosphorpentoxyd getrocknet.

0.1164 g Sbst.: 0.2718 g CO_2, 0.0934 g H_2O. — 0.1300 g Sbst.: 9.2 ccm N (22°, 761 mm).

$C_9H_{15}O_2N$ (169.12). Ber. C 63.86, H 8.94, N 8.28.
Gef. » 63.68, » 8.98, » 8.07.

Zur optischen Bestimmung wurde die Substanz in Toluol gelöst.

Präparat aus Racemkörper: 0.1530 g Sbst. gelöst in Toluol. Gesamtgewicht der Lösung 1.4707 g. Drehung im 1-dcm-Rohr bei 20° und Natriumlicht 0.96° (± 0.02°) nach links. $d^{20} = 0.882$.

$$\text{Mithin } [\alpha]_D^{20} = -10.5° (\pm 0.2°).$$

Verglichen mit der *d*-Verbindung, war die Säure also optisch noch nicht ganz rein. Sie enthielt ungefähr 4 % des Antipoden. Dem entspricht auch der niedrigere Schmelzpunkt, der bei 90° nach vorhergehender Sinterung beobachtet wurde.

Nahezu dasselbe Resultat gaben zwei mikropolarimetrische Bestimmungen, für welche ein Präparat direkt aus dem Racemkörper, das andere aus den Mutterlaugen des Brucinsalzes gewonnen war.

Die Werte gelten für 10-proz. Lösungen in Toluol bei 20° ohne Berücksichtigung des spezifischen Gewichts.

Präparat aus Racemkörper: $\alpha = -0.85°$,
Präparat aus den Mutterlaugen: $\alpha = -0.83°$,

während für die reine *d*-Säure unter denselben Bedingungen $\alpha = +0.94°$ beobachtet wurde.

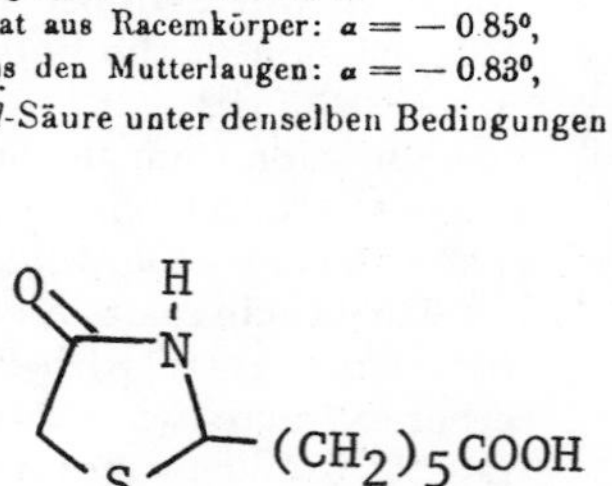

$C_9H_{15}NO_3S$ **Ref. 590**

Resolution of Synthetic I.—A mixture of 5 g. of racemic I and 9.05 g. of brucine was dissolved in 50 ml. of ethanol and crystallized at 5°. The resulting brucine salt, weighing 6.2 g. (m.p. 55–56°), was recrystallized from ethanol to give 1.9 g. of material, m.p. 60–61°, which was dissolved in 14 ml. of water and acidified with 6 *N* hydrochloric acid to give 0.47 g. of I, m.p. 136–138°. After being recrystallized from water, the product (0.39 g.) had m.p. 139–140° and $[\alpha]^{25}_D$ −51.4° (*c* 1, methanol); it was identical with I isolated from fermentation broth.

The original ethanolic filtrate of the brucine salts was concentrated *in vacuo* to a heavy sirup, which crystallized after being kept for 4 weeks; m.p. 85–90°. The crude product (weight 7.2 g.) was dissolved in 30 ml. of water and acidified with 6 *N* hydrochloric acid. The crude (+) I (weight 2.24 g., m.p. 114–124°) was dissolved in 90 ml. of hot water, the solution cooled to 45° and the resulting crystals filtered. The product (1.35 g.) had m.p. 128–132° and $[\alpha]^{25}D$ +31°. An additional recrystallization from water in the same manner yielded (+) I (0.62 g.) m.p. 138–139°, $[\alpha]^{25}D$ +57°. The microbiological activity of (+) I is one-fifth that of (−) I against *Mycobacterium tuberculosis*.

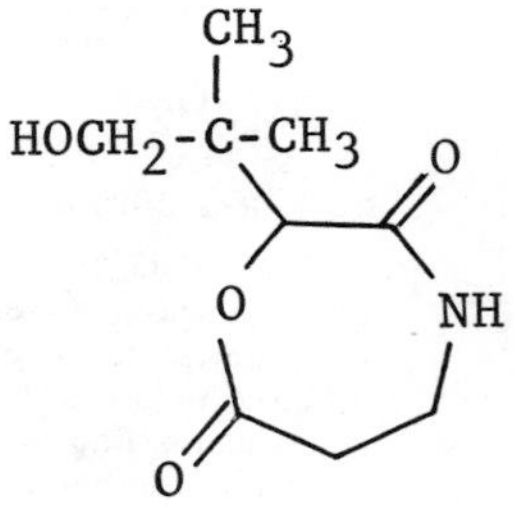

$C_9H_{15}NO_4$ Ref. 591

1.) **Strychnin-*d*(—)-pantolactonsulfat**: Eine Lösung von 52 g *d,l*-**Pantolacton** in 75 ccm Chloroform wird in einem Rundkolben unter Rühren und Eiskühlung nach und nach mit 28 ccm **Chlorsulfonsäure** versetzt. Alsbald beginnt eine Chlorwasserstoff-Entwicklung und hält ununterbrochen während der Zugabe der Säure an. Nachdem diese aufgebraucht ist, wird die Mischung plötzlich halbfest, indem sich das **saure Lactonsulfat** ausscheidet. Das Rühren wird noch 3 Stdn. unter Eiskühlung und dann bei Zimmertemperatur während weiterer 12 Stdn. fortgesetzt. Danach wird die Reaktionsmischung mit etwa 300 ccm Eiswasser geschüttelt, die gelbliche Chloroformschicht abgetrennt und diese nochmals mit je 200 und 150 ccm Eiswasser extrahiert. Die vereinigten wäßr. Lösungen und Auszüge werden zur Entfernung der letzten Spuren Chloroform mit 50 ccm Benzol ausgeschüttelt, von diesem abgetrennt und in einem Rundkolben unter kräftigem Rühren mit 45 g Natriumhydrogencarbonat in kleinen Anteilen versetzt. Die Lösung wird nun vorsichtig mit etwa 10 ccm 2*n* NaOH auf p_H 6 eingestellt und unter weiterem Rühren mit einer warmen Lösung von 79.5 g **Strychnin-hydrochlorid** in 750 ccm Wasser versetzt. Nach kurzer Zeit scheiden sich Kristalle des **Strychnin-*d*(—)-pantolactonsulfats** aus. Nachdem man noch 1—2 Tage die Mischung bei Zimmertemperatur stehengelassen hat, wird abfiltriert und mit etwa 300 ccm dest. Wasser gewaschen. Nach dem Trocknen i. Vak. bei 60—65° zeigt die Verbindung den Schmp. etwa 270° (unkorr., Zers.); Ausb. 70—80% d. Theorie. Sie kann durch Umkristallisieren aus Wasser in langen Nadeln erhalten werden; $[\alpha]_D^{20}$: —24.5 bis —25.5° (2% in 70-proz. Äthanol).

$C_6H_{10}O_6S \cdot C_{21}H_{22}O_2N_2 \cdot H_2O$ (562.6) Ber. C 57.6 H 6.1 N 5.0 S 5.7
Gef. C 57.5 H 5.9 N 4.7 S 5.4

Titration mit $n/_{10}$ NaOH (in zwei Stufen, die zweite nach Hydrolyse des Lactons durch Kochen):

Ber. Mol.-Gew. (Monohydrat) 562.6 Gef. Mol.-Gew. 563, 564
Ber. Mol.-Gew. (wasserfrei) 544.6 Gef. Mol.-Gew. 545, 546 (über P_2O_5 getr.)

2.) ***d*(—)-Pantolacton**: 81.7 g **Strychnin-*d*(—)-pantolactonsulfat** werden in eine Mischung von 4.5 ccm konz. Schwefelsäure und 320 ccm dest. Wasser in einem Rundkolben eingetragen. Der Inhalt des Kolbens wird unter Rückfluß erhitzt, und, nachdem das Sulfat unter Umschütteln in Lösung gegangen ist, wird weitere 5 Stdn. gekocht. Danach läßt man die Mischung auf 60° abkühlen und fügt noch 8.5 ccm konz. Schwefelsäure hinzu. Hierauf läßt man bei Zimmertemperatur oder im Kühlschrank etwa 12 Stdn. stehen und filtriert dann das in großen Kristallen ausgeschiedene saure Strychninsulfat ab. Der Filterrückstand wird zweimal mit je 20 ccm $2n\,H_2SO_4$ gewaschen. Das Filtrat und die damit vereinigten Waschflüssigkeiten werden nun in einem Flüssigkeitsextraktor 48 Stdn. mit Äther ausgezogen. Nach Entfernung des Äthers wird der Rückstand mit zweimal 5 ccm Benzol versetzt und i. Vak. erhitzt. Das **Rohlacton** bleibt als gelblich-wachsartige Masse zurück; Ausb. etwa 95% d. Theorie. $[\alpha]_D^{18}$: —50.6° (2% in Wasser), Schmp. 90.5—91.5° (unkorr.) (Lit. $[\alpha]_D^{20}$: —50 bis —51°, Schmp. 90—91°).

$HOOC-CH_2-CH(NHCOCH_3)-CH_2COOC_2H_5$

$C_9H_{15}NO_5$ Ref. 592

Resolution of *dl*-Ethyl-hydrogen-β-acetamidoglutarate.— Equimolar quantities of the *dl*-half acid and cinchonidine, 5.75 g. and 7.82 g., respectively, were dissolved in 110 ml. of hot ethyl acetate, and the solution was refrigerated overnight, leading to crystals of the partially resolved salt of the (+) enantiomorph. The crystals were recrystallized four times from ethyl acetate leading to the cinchonidine salt of constant m.p. and rotation, 1.42 g., 21% yield, m.p. and mixed m.p. with the sample above, 124°, $[\alpha]^{21}D$ − 83°, 0.375% in chloroform.

Anal. Calcd. for $C_{28}H_{37}N_3O_6$: C, 65.73; H, 7.29. Found: C, 65.3; H, 7.0. A portion of the salt, 0.320 g., was dissolved in 20 ml. of dilute hydrochloric acid saturated with sodium sulfate and extracted with ethyl acetate leading to (+)-ethyl hydrogen-β-acetamidoglutarate, 0.112 g., m.p. and mixed m.p. with a sample from the α-chymotrypsin hydrolysis, 92–93°, $\alpha_{obsd.}$ + 0.29°, 2.5% in acetone, $[\alpha]^{25}D$ + 5.8.

The original filtrate was taken down to dryness, leaving a solid residue, 8.5 g., melting 110–130°. This was decomposed as described above and the recovered half ester was crystallized from acetone–ether leading to partially resolved (−)-ethyl hydrogen-β-acetamidoglutarate, 2.65 g., melting 78–94°, $[\alpha]^{25}D$ −1.8°. An equimolar mixture of this ester and strychnine, 2.60 g. and 4.00 g., respectively, was crystallized five times from chloroform–ether to constant rotation, $[\alpha]^{25}D$ −51.8°, 0.375%, chloroform, 0.90 g., 13% yield, melting unsharply at 105°. This salt was decomposed as described above, leading to (−)-ethyl hydrogen-β-acetamidoglutarate, 0.275 g., m.p. 93°, mixed m.p. with the (+) enantiomorph, 77–78°, $\alpha_{obsd.}$ −0.295°, 2.5% in acetone, $[\alpha]^{25}D$ −5.9°.

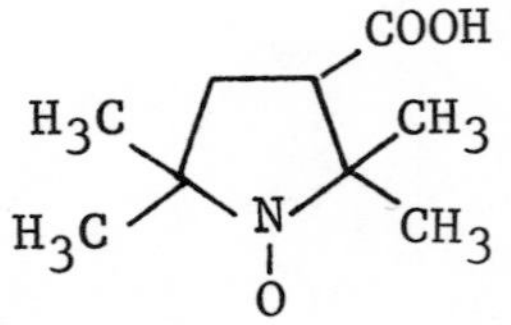

$C_9H_{16}NO_3$ Ref. 593

Resolution of 3-Carboxy-2,2,5,5-tetramethylpyrrolidinyl-1-oxy (IIb). Compound IIb was resolved by making a salt of (±)-IIb with an optically active amine and separating the two diastereomers of the salt formed by fractional crystallization (see Table II). Each salt was prepared from equimolar amounts of (±)-IIb and the resolving agent; the solvent used for the recrystallizations was acetone. Either of two different resolving agents, brucine or L(−)-α-methylbenzylamine, was used to prepare (+)-IIb. The (−) isomer of IIb was prepared using D(+)-α-methylbenzylamine. After two recrystallizations of the salt formed between IIb and any of the resolving agents used, further recrystallization did not result in any further change in the melting point or specific rotation of the acid extracted from the salt. Acid IIb was extracted from the recrystallized brucine or α-methylbenzylamine salt by adjusting an aqueous solution of the salt to pH 4 and then extracting with ether. The yield of (−)-IIb from a resolution starting with 2.0 g of (±)-IIb and an equimolar amount of D(+)-α-methylbenzylamine was 0.18 g [9% of the starting material or 18% of the theoretical yield of (−)-IIb].

Table II. Resolution of 3-Carboxy-2,2,5,5-tetramethylpyrrolidinyl-1-oxy

Isomer	Resolving agent	Mp, °C	[α]D (*c* 0.05–0.68, EtOH), deg
(+)	Brucine	206.5–208.5	78 ± 3
(+)	L(−)-α-Methylbenzylamine	202.5–205.0	79 ± 1
(−)	D(+)-α-Methylbenzylamine	205–207	−83 ± 2

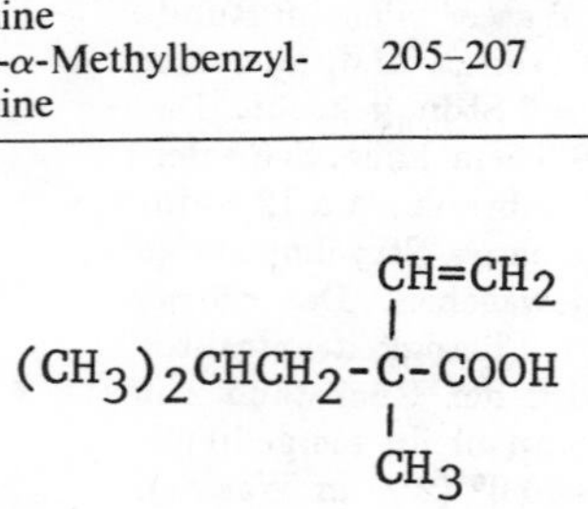

$C_9H_{16}O_2$ Ref. 594

Resolution of Methylisobutylvinylacetic Acid (VI).—The α-phenylethylamine ($[\alpha]^{25}D$ +39.3°, neat) salt was prepared from 31 g. of VI, and fractionally crystallized from ethyl acetate. After five recrystallizations there was obtained 16 g. of salt, m. p. 93–96°, which on acidification gave 8 g. of VI, b. p. 123–125° at 12 mm., $[\alpha]^{25}D$ +13.79° (neat, d^{25} 0.927), $[\alpha]^{25}D$ +8.36° (*c* = 10.17, 95% ethanol); *M*D calcd., 44.83; found, 44.94.

Anal. Calcd. for $C_{17}H_{27}NO_2$: C, 73.60; H, 9.81; N, 5.05. Found: C, 73.49; H, 9.75; N, 5.21.

A solution of 7 g. of (+)VI was dissolved in 20 ml. of ethanol and hydrogenated at room temperature and 3–4 atm. pressure using 1 g. of "W-5" Raney nickel.[7] On distillation, 6.5 g. of methylethylisobutylacetic acid (I) was obtained, b. p. 123–125° (12 mm.), $[\alpha]^{25}D$ +6.02° (neat), $n^{25}D$ 1.4326. If optically pure I is assumed to have a rotation of 21.6°, the rotation of VI is calculated to be 49.4°.

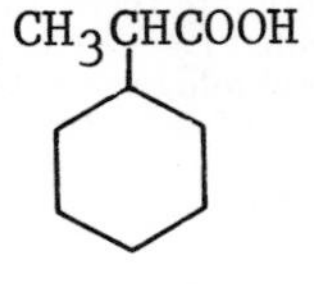

$C_9H_{16}O_2$ Ref. 595

Risoluzione dell'acido α-cicloesil-propionico

Grammi 15 di (−) α-fenil-etil-ammina ([11]) e g. 15,6 di acido α-cicloesil-propionico vengono sciolti a caldo in cm³ 100 di acetato di etile. La soluzione ancora tiepida viene innescata, se possibile, con alcuni cristalli di sale puro del (−) acido, e dopo circa un'ora si filtrano g. 7,2 di sale, p.f. 109-113° (I A). Il filtrato è riscaldato nuovamente fino a scomparsa dell'eventuale intorbidamento e innescato come sopra con sale puro. Dopo riposo di 12 ore si filtrano g. 10,6, p.f. 85-88° (I B). Il solvente filtrato viene evaporato a secchezza e il residuo (I C) utilizzato come descritto più avanti. La frazione (I B) viene ricristallizzata da acetato di etile e unita a (I A), e le due frazioni sono nuovamente cristallizzate fino a p.f. 118-119° costante. Il sale di (−) α-cicloesil-propionato della (−) ammina viene alcalinizzato con un eccesso di NaOH acquosa-10%. La (−) α-fenil-etil-ammina è estratta con etere e ricuperata, mentre le acque acidificate con acido solforico-30%, sono nuovamente estratte con etere: si ottengono g. 1,9 di acido (−) α-cicloesil-propionico, p.f. 41-43°. Si cristallizza da etere di petrolio ottenendosi un prodotto a p.f. 46-47°, α_D = −18,1 (etanolo).

Dalla frazione (I C), riunita alle frazioni residue di richistallizzazione della frazione (I B), si liberano g. 7,4 di (+) acido impuro, p.f. 39-50°. Questo viene sciolto a caldo in cm³ 50 di acetato di etile con g. 7,5 di (+) α-fenil-etil-ammina ([9]): per raffreddamento si separano g. 7,5 di sale, p.f. 110-113°, che ricristallizzati forniscono g. 5,3 di (+) α-cicloesil-propionato della (+) ammina, p.f. 118-119°. Da questo si liberano come sopra g. 2,5 di acido (+) α-cicloesil-propionico, p.f. 40-43°; e dopo cristallizzazione da etere di petrolio, p.f. 46-47°, α_D = +18,2 (etanolo).

Acido (+) α-cicloesil-perpropionico

Grammi 1,4 di carbonil-diimidazolo ([12]) vengono sciolti in cm³ 15 di tetraidrofurano anidro, in pallone a 3 colli munito di imbuto separatore, refrigerante e termometro. Nella soluzione riscaldata a 35-40° si sgocciola lentamente sotto agitazione una soluzione di g. 1,3 di acido (+) α-cicloesil-propionico in cm³ 10 di tetraidrofurano anidro. Si ha violento sviluppo di anidride carbonica; terminata l'aggiunta si continua a riscaldare per circa 20 minuti e quindi si evapora nel vuoto gran parte del solvente.

Grammi 0,90 di NaOH si sciolgono in cm³ 3,6 di acqua e cm³ 4,2 di etanolo. Si raffredda in ghiaccio e quindi si sgocciolano lentamente, sotto agitazione, cm³ 1,3 di acqua ossigenata-36%, mantenendo la temperatura sotto i 10°. Si ha separazione di un precipitato bianco, denso di sodio etil-perossido. Si raffredda esternamente con bagno di ghiaccio-sale e si sgocciola molto lentamente (45 minuti) sotto agitazione la soluzione concentrata dell'imidazolide nella sospensione del perossido. La temperatura viene mantenuta fra 0 e +5°. Si agita ancora per 5 minuti, si diluisce con poca acqua ghiacciata, e la soluzione alcalina viene estratta 3 volte con cloroformio raffreddato in ghiaccio. Alla fase acquosa, mantenuta fredda per aggiunta saltuaria di ghiaccio, si aggiungono ancora cm³ 3 di cloroformio ghiacciato, e sotto continua agitazione si acidifica cautamente fino a pH 1 con H_2SO_4 in modo da solubilizzare il peracido nel cloroformio non appena liberato dal sale. Si estrae ancora con 2 porzioni di cm³ 1,5 di cloroformio, gli estratti cloroformici riuniti sono lavati con soluzione acquosa di solfato ammonico al 10%, e essiccati su $MgSO_4$ anidro. La soluzione viene titolata con tiosolfato nelle usuali condizioni e direttamente impiegata per la sintesi asimmetrica. La conversione in peracido è del 39,2%.

Analoghi risultati si ottengono per l'*acido (−) α-cicloesil-perpropionico.*

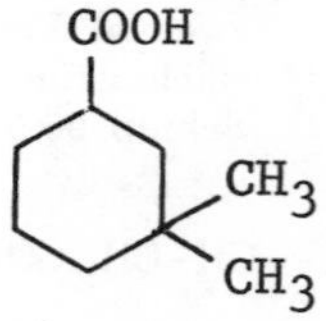

$C_9H_{16}O_2$ Ref. 596

B. Optically Active 3,3-Dimethylcyclohexanecarboxylic Acid.—*dl*-3,3-Dimethylcyclohexanecarboxylic acid was resolved by dissolving 31.2 g. (0.2 mole) of inactive acid and 65 g. (0.2 mole) of quinine in dry acetone, and recrystallizing the resultant salt twice from dry acetone. The salt was hydrolyzed with dilute hydrochloric acid to give (−)-3,3-dimethylcyclohexanecarboxylic acid, b.p. 99.5–100° (0.12 mm.), $[\alpha]^{26}D$ 1.14° (*l* 1 dm., neat). Reduction of 1 g. (0.0064 mole) of this active acid by the same procedure as in Part A gave 0.6 g. (66%) of (−)-3,3-dimethylcyclohexylcarbinol, b.p. 89° (10 mm.), $n^{25}D$ 1.4600, $[\alpha]^{24}D$ 2.07° (*l* 1 dm., neat).

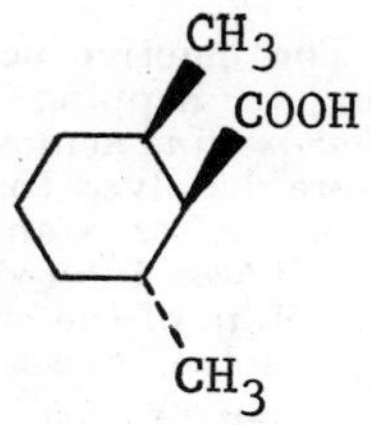

$C_9H_{16}O_2$

Ref. 597

Resolution of Iab.—Attempts at resolution with strychnine were unsuccessful because the salts hydrolyzed too readily. Accordingly a strong base, cinchoninium methohydroxide, was employed.[12] The cinchoninium metho salt of Iab was fractionally recrystallized from acetone and the less soluble fraction gave an acid, m.p. 82.3–83.0°; *a*, −1.586° (1 dm. tube); *c*, 9.8 g./100 ml. in acetone; $\alpha^{23.7}D$ −16.18°. The more soluble fraction finally gave an acid, m.p. 82.1–83.1°; *a*, 0.810° (1 dm. tube); *c*, 9.8 g./100 ml. in acetone; $\alpha^{24}D$ +8.26°. Mixed m.p.'s of these fractions with Iab gave intermediate m.p.'s, but they gave depressions with II or III.

(12) Major and Finkelstein, This Journal, **63**, 1368 (1941).

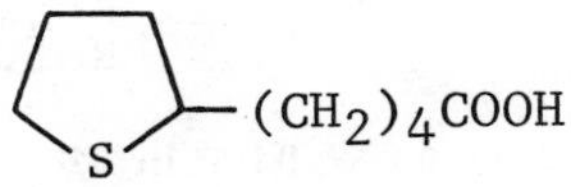

$C_9H_{16}O_2S$

Ref. 598

Optical resolution of thiophane-2-valeric acid

(+) *Thiophane-2-valeric acid.* 8.2 g (0.044 mole) of racemic acid and 12.8 g (0.044 mole) of cinchonidine were dissolved in 400 ml of refluxing ethyl acetate. The solution was allowed to cool slowly to room temperature, and was then left in a refrigerator for two days. The staff-formed crystals (6.4 g) were filtered and identified by means of m.p., mixed m.p. and IR spectrum as nearly pure cinchonidine. The filtrate was left at −10° overnight. The star-shaped clusters of needles were filtered off. About 0.2 g of the salt was made alkaline with a sodium hydroxide solution, and the base extracted with chloroform. The acid was liberated from the aqueous alkaline solution with sulphuric acid, and then extracted with ether. The ether solution was dried over magnesium sulphate, the acid isolated, and the optical activity measured in 96 % ethanol. The main part of the precipitate was recrystallized from ethyl acetate until the liberated acid showed constant optical activity (Table 1). An equivalent weight determination (with perchloric acid in pyridine) of a small sample of the salt from the fourth crystallization gave the value 245; calc. 241.2 for a molar ratio 1:1 of acid and alcaloid.

Table 1

Crystallization ...	1	2	3	4	5
ml ethyl acetate	400	300	200	100	50
g salt obtained	8.1	5.0	2.4	1.8	1.2
$[\alpha]_D$ of acid	+18°	+30°	+70°	+73°	+73°

The (+)-acid was liberated from the cinchonidine salt (1.0 g) as described above. Recrystallization from *n*-hexane gave colourless needles (0.32 g), m.p. 39–41°, $[\alpha]_D^{22} = +73.6°$ ($c = 0.8$, 96 % ethanol), equiv. wt. 189.0 (calc. 188.3).

(−)-*Thiophane-2-valeric acid.* The mother liquor from the first recrystallization of the cinchonidine salt was evaporated to dryness at room temperature, and the acid isolated as described above. The racemic acid was less soluble than the optically active acid. Thus, on crystallization, the acid which first precipitated was removed, and the filtrate reduced in volume, cooled and filtered again. This was repeated four times from water an three times from *n*-hexane. The acid remaining in the solution had then reached the same optical rotation as the (+)-form, and the rotation could not be raised further. The levorotatory (−)-thiophane-2-valeric acid formed long colourless needles, m.p. 38–40, $[\alpha]_D^{25} = -72.9$ ($c = 0.7$, 96 % ethanol), equiv. wt. 188.9 (calc. 188.3).

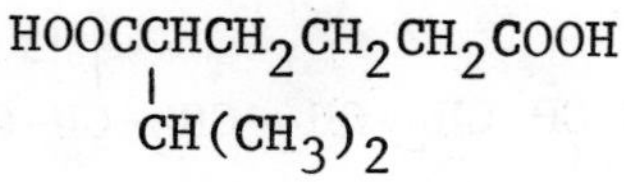

$C_9H_{16}O_4$

Ref. 599

Die optisch aktiven Formen der *d,l*-Isopropyl-adipinsäure.

Orientierenden Versuchen zufolge eignen sich für die Spaltung der inaktiven Isopropyl-adipinsäure weder Morphin, noch Cinchonin, noch Brucin. Gute Resultate liefert aber das Strychnin.

Übergießt man die Säure (25 g) mit 500 ccm H_2O und setzt 45 g (2 Mol.) Strychnin zu, so erhält man beim Erwärmen eine klare Lösung, aus der sich beim langsamen Erkalten 48 g eines bei 167—168° schmelzenden Strychnin-Salzes (A) abschieden. Die daraus durch Zerlegen mit Soda, Ansäuern der Sodalösung mit HCl, Eindampfen und Herausholen mit Äther isolierte Säure (14.5 g) schmolz bei 72—74° und zeigte nur geringe Drehung. Die Wiederholung des Versuches mit den 14.5 g und 26 g Strychnin ergab 30 g einer Abscheidung (A'), aus der 10 g einer bei 69—73° schmelzenden Säure isoliert wurden, deren Drehwert in Wasser wegen der verhältnismäßigen Schwerlöslichkeit auch nur sehr gering war. Es wurde daher hier und bei den folgenden Fraktionen die Drehung des sorgfältig dargestellten Na-Salzes gemessen, und zwar wurde für 1.1860 g Salz in 2.7566 g H_2O $[\alpha]_D = +2.013°$ gefunden.

Die 10 g, mit 18.8 g Strychnin umgesetzt, lieferten ca. 20 g eines Niederschlages (A'') vom Schmp. 176°, aus dem 7.5 g Säure mit dem Drehwert $[\alpha]_D = +3.76°$ für das Na-Salz (0.5725 g Salz in 1.3289 g H_2O) gewonnen wurden; als dieselbe Behandlung noch einmal wiederholt wurde, resultierten 12 g Strychnin-Salz vom Schmp. 180° und daraus eine Säure, die fast scharf bei 66° schmolz und als Na-Salz (0.6578 g Salz in 1.5674 g H_2O) $[\alpha] = +5.25°$ zeigte. Die weiteren Wiederholungen verschoben die Werte kaum: die Schmelzpunkte der Strychnin-Salze erhöhten sich auf 181° und 182°, die Drehwerte der Na-Salze auf +5.26° und +5.4°, der Schmelzpunkt der Säure wurde bei 66° ziemlich scharf.

Um die linksdrehende Säure zu fassen, wurden die Filtrate von A und A' mit Na_2CO_3 vom Strychnin befreit, mit HCl angesäuert, eingedampft und die Säure mit Äther herausgeholt. Sie zeigte eine Drehung (als Na-Salz) von etwa −3° und schmolz unscharf um 60°. Um die rechtsdrehende Säure, die mit Strychnin ein schwerer lösliches Salz bildet, herauszuschaffen, wurde mehrere Mal hintereinander in wechselnder Verdünnung mit Strychnin zur Salzbildung angesetzt und der sich beim Erkalten abscheidende Teil des Salzes entfernt; zum vollen Erfolg führten aber diese sehr mühsamen Versuche nicht, denn es gelang uns nur, zu einem Drehwert von −4.1° (0.6862 g Salz in 1.8121 g H_2O) zu kommen. Auch der Schmp. der freien Säure war noch nicht scharf.

Die rechtsdrehende β-Isopropyl-adipinsäure ist in Wasser weniger leicht löslich als die Racemform. Läßt man auf sie in der Kälte Thionylchlorid einwirken, so geht sie bald in Lösung. Nach dem Absaugen des überschüssigen $SOCl_2$ im Vakuum, destilliert das Säure-chlorid unter 15 mm bei 145—146° als farblose Flüssigkeit.

0.1458 g Sbst.: 12.7 ccm n/10-$AgNO_3$-Lösung.

$C_9H_{14}O_2Cl_2$. Ber. Cl 31.55. Gef. Cl 31.56.

$d_4^{20} = 1.1023$, $\alpha_D^{20} = +1.25°$ (1-dm-Rohr), $[\alpha]_D^{20} = +1.134°$.

Durch Eindampfen mit Wasser wurde das Chlorid zur Säure verseift; diese ergab als Na-Salz $[\alpha]_D^{20} = +5.2°$, so daß die Säurechlorid-Bildung ohne Racemisierung erfolgt. Mit wäßrigem konz. Ammoniak liefert das Säurechlorid eine Abscheidung des in Wasser schwer löslichen Säure-amids, das nach dem Umkrystallisieren aus Methanol bei 169.5° schmilzt.

0.1122 g Sbst.: 15.00 ccm N (22°, 750 mm).

$C_9H_{18}O_2N_2$. Ber. N 15.05. Gef. N 15.29.

$[\alpha]_D^{20} = +9.5°$ (in 2.22-proz. wäßriger Lösung).

Der mit HCl-Gas und Äthylalkohol dargestellte Äthylester zeigt den Sdp. des Esters der racemischen Säure (145—150° unter 13 mm).

0.1345 g Sbst.: 0.3161 g CO_2, 0.1214 g H_2O.

$C_{13}H_{24}O_4$. Ber. C 63.93, H 9.84. Gef. C 64.10, H 10.10.

$d_4^{20} = 0.9776$, $[\alpha]_D^{20} = -1.534°$ (unverdünnt).

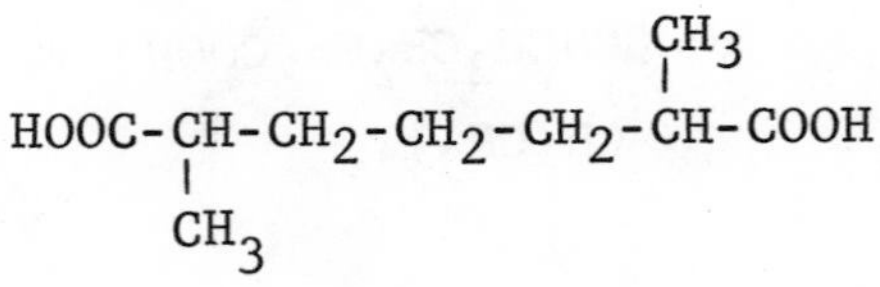

$C_9H_{16}O_4$ **Ref. 600**

Owing to lack of material, the optical antipodes have not yet been prepared in a state of purity. The acid from the first crystallisation of the strychnine salt had $[\alpha]_D^{25} = -20°$ (abs. ethanol). In the same solvent, dimethylglutaric and dimethyladipic acid show 36.5° and 30.3° [4], respectively. The maximum activity of the dimethylpimelic acid can be assumed to be somewhat lower, probably near 25°. The strychnine salt thus gives a fairly good resolution.

Experimental. The inactive acids were prepared according to Kipping [3].

0.250 g of the *low-melting* acid and 0.445 g of strychnine were dissolved together in hot dilute ethanol. After standing for several days, the salt was filtered off and the acid liberated. M. p. of the crude acid 67—68°. 0.0235 g dissolved in abs. ethanol to 10.00 ml: $2\alpha_D^{25} = -0.09°$. $[\alpha]_D^{25} = -20°$.

The mother liquor from the strychnine salt was evaporated to dryness and the acid liberated. 0.0293 g dissolved in abs. ethanol to 10.00 ml: $2\alpha_D^{25} = +0.105°$. $[\alpha]_D^{25} = +18°$.

3. Kipping, F. S. *J. Chem. Soc.* **67** (1895) 139.

4. Hedlund, I. *Arkiv Kemi* **8** (1955) 89.

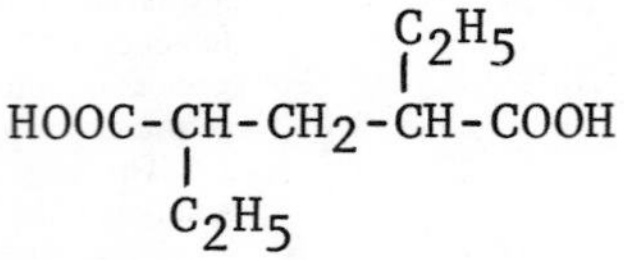

$C_9H_{16}O_4$ **Ref. 601**

In a warm solution of 10 g *a,a'*-diethylglutaric acid with the m.p. 93.5—94.5° in 240 ml water 21 g of brucine (a little less than 1 mole alkaloid to 1 mole acid) were dissolved in portions. Placing the solution in a refrigerator and inoculating with crystals from a preliminary experiment 10.4 g brucine salt of the (+)-acid crystallized in the course of two days. After recrystallizing once from water the salt was dissolved in hydrochloric acid and the active acid extracted with ether.

$$[\alpha]_D^{20} = +\ 28.74° \ (p = 4.02 \text{ in abs. ethanol})$$

On concentrating the mother liquor some more salt of the (+)-acid was obtained. From the rest of the solution a levorotatory acid (3 g) was obtained having a specific rotation of − 19.4°.

In another case, starting with 14 g of racemic acid, the least soluble brucine salt was recrystallized 6 times from water. The acid isolated from this salt had the specific rotation + 29.5°. A new treatment with brucine did not lead to an increase of the rotation.

In the mean time it had been found that the cinchonidine salt (1 mole alkaloid to 1 mole acid) crystallized well from acetone. Also in this case the salt of the (+)-acid crystallized first. In a series of experiments in which the cinchonidine salt had been recrystallized 1, 3 and 6 times from acetone the following specific rotations for the acid were found: + 29.03°, + 29.65° and + 30.75°. A final treatment of the acid + 30.75° with cinchonidine resulted in an acid with rotation +30.3°. This active acid on recrystallization from water was obtained in well-developed flat prisms with m.p. 75.5—76.5°.

$$[\alpha]_D^{20} = +\ 30.33° \ (p = 4.0 \text{ in abs. ethanol})$$

In order to determine the specific rotation at various concentrations it was necessary to measure the variation of the specific gravity with concentration. The results obtained, using the racemic acid, are given in Table 1 in which p denotes g acid in 100 g ethanolic solution.

Table 1. Specific gravity of solutions of racemic acid in ethanol.

p	d_4^{20}
0	0.791
3.94	0.801
10.21	0.816
13.52	0.824
21.17	0.843
39.15	0.889

According to these figures the relationship between specific gravity and percentage is practically linear.

The specific rotation of the (+)-acid at various concentrations will be found in Table 2.

Table 2. Specific rotation of (+)-a,a'-diethylglutaric acid in abs. ethanol.

p	d_4^{20}	l	α_D^{20}	$[\alpha]_D^{20}$
4.014	0.801	2	+ 1.95°	+ 30.33°
10.29	0.816	2	5.09	30.30
13.26	0.824	2	6.63	30.33
21.26	0.844	2	10.92	30.44
40.86	0.893	1	11.36	31.13
49.09	0.914	1	14.12	31.47
57.41	0.936	1	17.34	32.26

The isolation of pure levorotatory acid was successful only once. In that case 0.9 g of a levorotatory acid isolated from the light soluble cinchonidine salt was dissolved in 10 ml hot water and 1 ml 2 *N* hydrochloric acid added to the cooled solution. The next day a crystalline acid was filtered off consisting mostly of racemic acid (m.p. 89—91°). When the filtered solution had been left for another day some rather big rhombic crystals had separated (0.2 g). Recrystallized from water the m.p was 75.5—76.5°.

$$[\alpha]_D^{20} = -30.34° \ (p = 3.54 \text{ in abs. ethanol})$$

```
    CH3CH3
      |  
HOOCCH2CH-CH-CH2COOCH3
```

$C_9H_{16}O_4$

Ref. 602

Laevorotatory enantiomorph of the monomethyl ester (I) *of* meso-*3,4-dimethyladipic acid.*

It was found in preliminary experiments that the cinchonidine salt of the half-ester readily crystallized from 60 % acetone at 0°. No quinine salt was formed under similar conditions, and the experiments with quinine were not further pursued. DL-half-ester (85 g) was dissolved in 60 % acetone (800 ml) and cinchonidine (132 g) added. The mixture was warmed and stirred until the cinchonidine had dissolved. The solution was filtered while still hot and left overnight in the refrigerator at 0°. The cinchonidine salt crystallized in the form of large rosettes of long, thin needles. Part of the salt was decomposed with dilute hydrochloric acid, and the recovered half-ester showed after distillation on optical rotation α_D^{24} —0.98° (homogeneous; *l*, 1). The salt was recrystallized from 300 ml of 60 % acetone in the cold, and a specimen of recovered half-ester now had α_D^{23} —1.35°. After a further crystallization from 200 ml of 60 % acetone the recovered half-ester had α_D^{23} —1.49° (homogeneous; *l*, 1). Further recrystallization did not change the rotation of the half-ester. The yield of cinchonidine salt after three crystallizations was 30 g. The salt was decomposed by adding an excess of dilute hydrochloric acid. The half-ester was extracted with ether, and the ether solution washed carefully with water to remove all hydrochloric acid. The ether solution was dried with sodium sulphate and the solvent distilled off. The residue was distilled under reduced pressure, giving 9.9 g (23 % of the theoretical) of *laevorotatory monomethyl ester* of meso-*3,4-dimethyladipic acid*. The conditions for optimum yield are fairly critical, and the yields obtained in different runs differed considerably. Sometimes more than three crystallizations were needed to bring the rotation to a constant value, and in such cases the yields were only about 10 % of the theoretical. d_4^{20} 1.079. d_4^{25} 1.075.

Optical rotation α_D^{20} —1.51° (homogeneous; *l*, 1)
$[\alpha]_D^{20}$ —1.40° $[M]_D^{20}$ —2.64°
α_D^{25} —1.465° (homogeneous; *l*, 1)
$[\alpha]_D^{25}$ —1.36° $[M]_D^{25}$ —2.56°

Anal.: Calcd for $C_9H_{16}O_4$	(188.2)	C, 57.43; H, 8.57 %
Found		C, 57.40; H, 8.50 %

```
CH3(CH2)4CHCOOH
         |
         CH2COOH
```

$C_9H_{16}O_4$

Ref. 603

Resolution of the racemic acid. In preliminary experiments, strychnine and brucine re found to give the best results.

22.6 g (0.12 mole) of *rac.-n*-pentylsuccinic acid were dissolved in 450 ml of 50 % anol. The solution was heated to boiling, saturated with strychnine (c:a 75 g), filtered and placed in a refrigerator. After 24 hours, the salt was filtered off and recrystallised several times from dilute ethanol. On dissolving the salt, a small amount of strychnine separated owing to hydrolysis; it was brought into solution by adding a few drops of acetic acid. The progress of the resolution was tested on small samples from each crystallisation, from which the acid was liberated and examined in ethanolic solution.

Crystallisation	1	2	3	4
ml water	225	350	150	130
ml ethanol.................	225	150	70	60
g salt obtained	45.7	34.3	30.6	27.4
$[\alpha]_D$ of the acid	+ 21°	+ 26°	+ 26.5°	+ 26.5°

The mother liquor from the first crystallisation was evaporated to dryness and the residue recrystallised from the successive mother liquors obtained in the first series of crystallisations. The resulting salt was recrystallised from 60 ml water + 40 ml ethanol, yielding 15.6 g salt of pure (+)-acid. The total yield of strychnine salt was thus 42.0 g.

The mother liquors from the two first crystallisations of the second series were combined and the acid was liberated; 10.5 g having $[\alpha]_D$ = —22.5° were obtained. This acid (0.055 mole) was dissolved with 44.0 g (0.11 mole) of anhydrous brucine in 150 ml of water and the salt was recrystallised to maximum activity.

Crystallisation	1	2	3	4
ml water	150	100	100	100
g salt obtained	51.5	48.0	44.2	41.8
$[\alpha]_D$ of the acid	— 24°	— 25.5°	— 26°	— 26°

(+)-*n-Pentylsuccinic acid.* The strychnine salt (27.4 g) was decomposed with sodium carbonate solution and the strychnine extracted with chloroform. The acid was isolated from the sodium salt solution by acidification with sulphuric acid and extraction with ether. It was recrystallised from a little water and twice from ligroine (b.p. 60-70°). The yield of crude acid was 5.5 g; after the recrystallisations there remained 3.6 g. The acid forms microscopic needles with m.p. 84—86°.

0.1589 g acid: 15.84 ml 0.1065-*N* NaOH.
Equiv.wt. calc. 94.1 found 94.2
0.4773 g acid dissolved in absolute *ethanol* to 10.00 ml:
$2\,\alpha_D^{25} = +2.55°$. $[\alpha]_D^{25} = +26.7°$; $[M]_D^{25} = +50.3°$.

(—)-*n-Pentylsuccinic acid.* The brucine salt (41.3 g) was decomposed with dilute sulphuric acid and the organic acid was extracted with ether. It was recrystallised twice from ligroine and twice from water. The yield of crude acid was 6.7 g; after the recrystallisations there remained 5.2 g. M.p. 83-85°.

0.1509 g acid: 14.99 ml 0.1069-*N* NaOH.
Equiv.wt. calc. 94.1 found 94.2
0.4768 g acid dissolved in absolute *ethanol* to 10.00 ml:
$2\,\alpha_D^{25} = -2.505°$. $[\alpha]_D^{25} = -26.3°$; $[M]_D^{25} = -49.4°$.

0.5273 g acid dissolved in *water* to 10.00 ml:
$2\,\alpha_D^{25} = -2.16°$. $[\alpha]_D^{25} = -20.5°$; $[M]_D^{25} = -38.6°$.
0.2579 g acid dissolved in *water* to 10.00 ml:
$2\,\alpha_D^{25} = -1.12°$. $[\alpha]_D^{25} = -21.7°$; $[M]_D^{25} = -40.9°$.

$$\text{HOOCCH}_2-\underset{\underset{\text{CH}(\text{CH}_3)_2}{|}}{\overset{\overset{\text{CH}_3}{|}}{\text{C}}}-\text{COOCH}_3$$

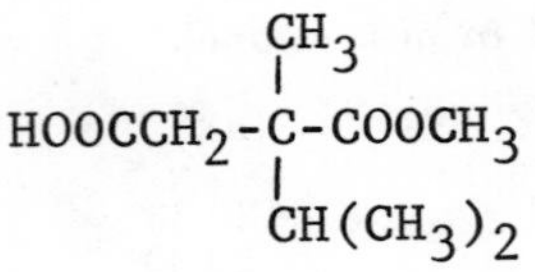

$C_9H_{16}O_4$ Ref. 604

(+) *and* (−)-1-*Methyl Hydrogen 2-Isopropyl-2-methylsuccinate*.—A solution of the (±)-half-ester (30.7 g) in ether (100 ml) was added to an ethereal solution (30 ml) of dehydroabietylamine (50 g). The crystalline salt (44 g) which separated during 24 h was collected and repeatedly recrystallised from light petroleum–methanol and light petroleum–chloroform to give the dehydroabietylamine *salt*, m.p. 154° of the (+)-half-ester in needles (26 g), $[\alpha]_D^{25}$ +26.1° (c 4.98 in MeOH) (Found: C, 72.8; H, 9.3; N, 3.1. $C_{29}H_{47}NO_4$ requires C, 73.5; H, 10.0; N, 3.0%). This salt (26 g) was shaken with an excess of ice-cold 2N-sodium hydroxide until dissolution was complete; liberated amine was removed by extraction with light petroleum and the residual aqueous solution after evaporation to small volume *in vacuo* was acidified (at 0°) with 10N-hydrochloric acid. After isolation with ether, the (+)-*half-ester* (8.9 g) separated from light petroleum in prisms, m.p. 45°, $[\alpha]_D^{24}$ +11.6° (c 5.88 in MeOH), $[\alpha]_D^{24}$ +9.4° (c 9.04 in CHCl$_3$) [Found: C, 57.2; H, 8.6; OMe, 16.4. $C_8H_{13}O_3$(OMe) requires C, 57.4; H, 8.6; OMe, 16.4%].

The mother liquors remaining from the purification of the salt of m.p. 154° were evaporated to dryness *in vacuo* and the residue was repeatedly crystallised from light petroleum–methanol and light petroleum–chloroform to yield the dehydroabietylamine *salt* (11 g) in silky needles, m.p. 136° of the (−)-acid, $[\alpha]_D^{25}$ +19.0° (c 5.23 in CHCl$_3$) (Found: C, 73.1; H, 10.2; N, 3.0. $C_{29}H_{47}NO_4$ requires C, 73.5; H, 10.0; N, 3.0%). Decomposition of this derivative as for the diastereoisomer gave the (−)-*half-ester* in prisms (from light petroleum), m.p. 47°, $[\alpha]_D^{24}$ −10.0° (c 12.2 in CHCl$_3$) [Found: C, 57.5; H, 8.6; OMe, 15.8. $C_8H_{13}O_3$(OMe) requires C, 57.4; H, 8.6; OMe, 16.4%].

cyclohexyl-CH$_2$-CH(NH$_2$)-COOH

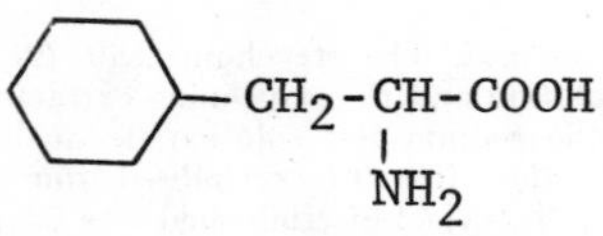

$C_9H_{17}NO_2$ Ref. 605

Note: For an enzymatic resolution procedure, see S.M. Birnbaum, L. Levintow, R.B. Kingsley and J.P. Greenstein, J. Biol. Chem., **194**, 455 (1952). These authors reported $[\alpha]_D^{26}$ = −9.0° (c 2.000%, H$_2$O), $[\alpha]_D^{26}$ = +15.0° (c 2.000%, 5N HCl), $[\alpha]_D^{26}$ = +8.5° (c 2.000%, H$_2$O) and $[\alpha]_D^{26}$ = −15.0° (c 2.000% 5N HCl) for the resolved acids of optical purity > 99.9%.

$$\text{HOCH}_2-\underset{\underset{\text{H}_3\text{C}}{|}}{\overset{\overset{\text{CH}_3}{|}}{\text{C}}}-\underset{\underset{\text{OH}}{|}}{\text{CH}}-\text{CONHCH}_2\text{CH}_2\text{COOH}$$

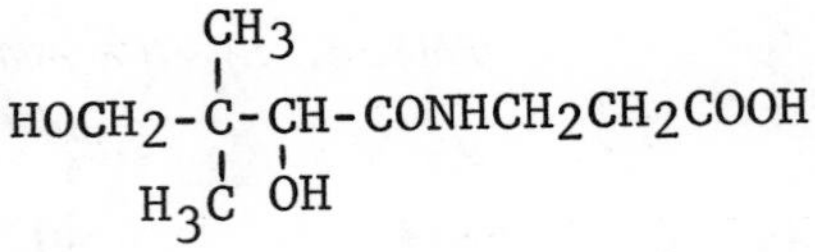

$C_9H_{17}NO_5$ Ref. 606

Resolution of d,l-Pantothenic Acid.—An aqueous solution of 32 g. of *d,l*-pantothenic acid in 50 cc. of water was cooled and neutralized to pH 7.5 with an aqueous solution of quinine methohydroxide, prepared by stirring quinine methochloride in aqueous suspension with silver oxide. After standing for some hours at 0° a small amount of crystalline material (0.27 g.; m. p. 196°) was obtained which had no effect on the growth stimulation of *Lactobacillus casei*. The aqueous solution was concentrated to half its volume *in vacuo* below 25°, and a further small quantity of bacteriologically inactive material was removed (0.7 g.). The solution was then evaporated to dryness below 25° and the residue was dried by dissolving in absolute alcohol (20 cc.), adding 75 cc. of benzene and removing the solvents by distillation *in vacuo*. The residue was finally dried in a desiccator over sulfuric acid.

The dry, partially crystalline residue was ground to a fine powder with 250 cc. of ethyl acetate in order to remove a small amount of gummy material which hindered crystallization of the product; yield, 65.6 g. This material had m. p. 177°; $(\alpha)^{27°}_D$ −132.1° (c, 0.71% in MeOH); bacterial assay 40%.

By fractional crystallization of the product from 100 cc. of alcohol by the successive addition of 100-cc. portions of ether, four crops of crystals were obtained as shown in Table I.

TABLE I

Fraction	M. p., °C.	$[\alpha]_D$	Weight, g.	Bacterial assay, %
A	182–184	−129°	22.6	46.4
B	177	−132.2°	3.7	51.0
C	177	−127.8°	0.5	36.6
D	163	−150.0°	6.3	5.4

After six recrystallizations of Fraction A from the alcohol–ether mixture (1:2) 4.8 g. clusters of small, fine, colorless needles was obtained which showed 100% activity on bacterial assay. It also showed $(\alpha)^{25}_D$ −122.0°, (c, 4.17% in MeOH); m. p. 196–197° with material prepared from (+)-pantothenic acid.

Anal. Calcd. for $C_{30}H_{43}O_7N_3$: C, 64.63; H, 7.72; N, 7.54. Found: C, 64.12; H, 7.90, 8.05; N, 7.48, 7.43, 7.49.

Fraction D was recrystallized five times from alcohol–ether (1:2) and was then obtained in clusters of fine colorless needles, m. p. 170°, mixed m. p. with a sample prepared from (−)-pantothenic acid, 170°. It showed $(\alpha)^{25}_D$ −156.5° (C, 0.96% in MeOH); and bacterial assay 0.1%; yield 4.1 g.

Anal. Calcd. for $C_{30}H_{43}O_7N_3$: C, 64.63; H, 7.72; N, 7.54. Found: C, 63.97; H, 7.97; N, 7.53.

Quinine Metho Salt of (+)-Pantothenic Acid.—A solution of 0.9814 g. of calcium (+)-pantothenate in 4 cc. of water was treated with an equivalent amount of N sulfuric acid with cooling. The calcium sulfate was removed by filtration and washed with a small quantity of ice-water. The combined filtrates were adjusted to pH 5.5 with pyridine and evaporated to dryness *in vacuo*, at 25°. The residue after drying in high vacuum was extracted with 20 cc. of acetone and the insoluble calcium sulfate removed by filtration. On removal of the acetone by evaporation *in* *vacuo*, the pure (+)-pantothenic acid was obtained as a colorless viscous oil; yield 0.8967 g.

The product was dissolved in 5 cc. of water and neutralized to *p*H 7.5 with an aqueous solution of quinine methohydroxide. After standing for some hours at 0° the solution was evaporated to dryness at 25° *in vacuo*, thoroughly dried in high vacuum, and then triturated with ethyl acetate. The crystalline product was recrystallized from alcohol–ether (1:2) and obtained as clusters of fine colorless needles; m. p. 196–197°; $(\alpha)_D$ −118.5° (*C*, 1.20% in MeOH); bacterial assay 101; 99.7%; yield 1.39 g.

Anal. Calcd. for $C_{30}H_{43}O_7N_3$: C, 64.63; H, 7.72; N, 7.54. Found: C, 64.01, 64.08, 64.10; H, 7.92, 7.72, 7.63; N, 7.41, 7.52.

Quinine Metho-(−)-pantothenate.—The salt was prepared as described above for the (+)-salt. The product was obtained as clusters of fine colorless needles and showed m. p. 170°; $(\alpha)^{25}_D$ −156.0° (*C*, 0.80% in MeOH); bacterial assay, negative.

$C_9H_{17}NO_5$ Ref. 606a

The Resolution of *d,l*-Pantothenic Acid by Means of the Cinchonidine Salt.—To a solution of 27 g. of *d,l*-pantothenic acid in 100 cc. of acetone, 36 g. of cinchonidine was added and the mixture boiled under reflux. Methyl alcohol was added until complete solution of the cinchonidine was obtained (140 cc.), and the refluxing continued for thirty minutes. On cooling, crystallization did not take place and the solvents were removed by distillation at 25° *in vacuo*. The residue was dissolved in 100 cc. of alcohol and 50 cc. of ether added. On standing at 0°, colorless fluffy needles were deposited (20 g.). After two recrystallizations from alcohol–ether, 13 g. of colorless needles was obtained, m. p. 176–177°, which showed no depression of the melting point when mixed with an authentic sample It showed $(\alpha)^{28.4}_D$ −60.6° (*C*, 1.35% in MeOH) and bacterial assays showed 101–102% activity.

Anal. Calcd. for $C_{28}H_{39}O_6N_3$: C, 65.49; H, 7.65; N, 8.18. Found: C, 65.55; H, 7.78; N, 8.11.

By the addition of further quantities of ether to the original mother liquor, and recrystallization of the product, a further 7 g. of the pure cinchonidine salt of (+)-pantothenic acid was obtained.

It was not found possible to isolate the cinchonidine salt of (−)-pantothenic acid. By fractional crystallization of the mother liquors, a fraction was obtained with some difficulty showing a bacterial activity of 19.6% (m. p. 143–145°). This salt was apparently somewhat unstable and crystallized with difficulty. During a subsequent recrystallization, it partially decomposed and the pantothenic acid was, therefore, recovered by the usual methods and showed a bacterial activity of 10.3%.

Cinchonidine Salt of (+)-Pantothenic Acid.—The cinchonidine salt of (+)-pantothenic acid was prepared from 0.5 g. (+)-pantothenic acid as described above. It was obtained as colorless needles, m. p. 177–178°. It had $(\alpha)^{25°}_D$ −61.3° (*C*, 0.67% in MeOH); bacterial assay, 101%.

Anal. Calcd. for $C_{28}H_{39}O_6N_3$: C, 65.49; H, 7.65; N, 8.18. Found: C, 65.31; H, 7.60; N, 8.28.

$C_9H_{17}NO_5$ Ref. 606b

5) Chininsalz der linksdrehenden Pantothensäure.

2 g *d,l*-Pantothensäure werden in wenig Wasser gelöst und mit Barytwasser auf p_H 8.5 gebracht. Zu dieser Lösung gibt man eine heiße wäßrige Lösung von neutralem Chininsulfat vorsichtig zu, bis alles Ba genau gefällt ist. Dann wird abzentrifugiert und die wäßrige Lösung im Vak. eingeengt. Der hinterbliebene Sirup fängt nach einiger Zeit an zu krystallisieren und geht beim Verreiben mit Aceton in ein weißes, feinkrystallines Pulver über, Dieses wird mehrmals aus Aceton-Methanol (1:1) umkrystallisiert, woraus es sich in farblosen, weichen, seidig glänzenden Nadeln vom Schmp. 165—167° (Berl) abscheidet.

Zur Analyse wurde 12 Stdn. bei 100° und 0.5 mm über P_2O_5 getrocknet.
3.785 mg Sbst.: 0.265 ccm N_2 (21°, 755 mm).
$C_{29}H_{41}O_7N_2$ (543.2). Ber. N 7.73. Gef. N 8.08.
$[\alpha]^{24}_D$: (1.15° × 100) : (1 × 1) = —115° (in Wasser).

Aus den Mutterlaugen der Umkrystallisationen läßt sich nach dem Konzentrieren ein Chininsalz vom Schmp. 148—152° gewinnen, dessen Zerlegung zu rechtsdrehender Pantothensäure führt.

6) Linksdrehende Pantothensäure.

1 g des Chininsalzes vom Schmp. 165—167° wurde in 10 ccm Wasser gelöst und mit Barytwasser phenolphthalein-alkalisch gemacht. Das dabei ausgefallene Chinin wurde durch je 3-maliges Ausschütteln mit Chloroform und Äther entfernt und die wäßrige Lösung mit Schwefelsäure genau vom Ba befreit. Nach dem Abzentrifugieren des Bariumsulfats wurde die wäßrige Lösung im Vak. abgedampft. Die freie (—)-Pantothensäure hinterblieb als sirupartige, in Wasser und den Alkoholen spielend lösliche Masse, die noch mit keinem Lösungsmittel zur Krystallisation zu bringen war.

8.1 ccm einer wäßrigen Lösung, die je ccm 3.50 mg N (Kjeldahl) (d. s. 54.7 mg (—)-Pantothensäure) enthalten, verbr. *2.65* ccm $n/_{100}$-NaOH (ber. auf 5.47 mg $C_9H_{17}O_5N$ (219.1): *2.49* ccm).
$[\alpha]^{21}_D$: (—1.46° × 100) : (5.47 × 1) = —26.7° (in Wasser).
$[\alpha]^{24}_D$: (—1.35° × 100) : (2.40 × 1) = —56.3° (in Methanol).

Das Bariumsalz der (—)-Pantothensäure, das durch Neutralisieren der wäßrigen Lösung mit Barytwasser und Eindunsten im Exsiccator dargestellt wurde, stellte eine bitter schmeckende, leicht in Wasser und Alkohol lösliche, glasartige Masse dar, die nicht krystallisierte.

$[\alpha]^{24}_D$: (—1.00° × 100) : (4.88 × 1) = —20.4° (in Wasser).
Daraus berechnet sich für das Pantothenat-Ion $[\alpha]^{21}_D$: —26.8°.

```
          CH3
          |
CH3(CH2)4-CH-CH2COOH
```

$C_9H_{18}O_2$ **Ref. 607**

Resolution of 2-n-Amylbutyric Acid-4—This acid was resolved by crystallizing its quinine salt from acetone at $-15°$ until the rotation of the free acid reached a constant value. This was reached on the eighth crystallization.

$$[\alpha]_D^{27} = \frac{-4.61°}{1 \times 0.896} = -5.14°; [M]_D^{27} = -8.12° \text{ (homogeneous)}$$

5.964 mg. substance: 15.025 mg. CO_2 and 6.145 mg. H_2O

$C_9H_{18}O_2$. Calculated. C 68.3, H 11.5

Found. " 68.7, " 11.5

```
                   C2H5
                   |
(CH3)2CHCH2CH2-CH-COOH
```

$C_9H_{18}O_2$ **Ref. 608**

Dextro-2-Isoamylbutyric Acid-4—The inactive acid was prepared from methylisoamylbromomethane and ethyl malonate.

1 mol of the acid was dissolved in 500 cc. of hot acetone and 1 mol of quinine added. When the quinine had dissolved completely, 500 cc. of water were added and the solution cooled at $-15°$ until crystallization set in. The resolution was very difficult, owing to the solubility of the quinine salt. A better result was obtained by making use of the filtrate from the first crystallization than by attempting the recrystallization of the levo form. The acetone was evaporated from the filtrate and the quinine salt decomposed with hydrochloric acid. The organic acid was extracted with ether, purified through its sodium salt, then distilled. B.p. 140° at 25 mm. $D\frac{26}{4} = 0.901$.

$$[\alpha]_D^{26} = \frac{+2.20°}{1 \times 0.901} = +2.44°; [M]_D^{26} = +3.86° \text{ (homogeneous)}$$

4.637 mg. substance: 11.675 mg. CO_2 and 4.665 mg. H_2O

$C_9H_{18}O_2$. Calculated. C 58.3, H 11.5

Found. " 68.6, " 11.3

```
(CH3)2CH(CH2)3-CH-COOH
                |
                CH3
```

$C_9H_{18}O_2$ **Ref. 609**

Spaltung der α,ε-Dimethyl-heptansäure. Für die Spaltung der Säure war das Cinchonidinsalz am besten geeignet. 10 g Säure wurden mit der ber. Menge 0.1-n NaOH neutralisiert und mit einer Lösung von 18.8 g Cinchonidin in der ber. Menge Salzsäure versetzt. Das ausgeschiedene Cinchonidinsalz (20 g) wurde aus Aceton umkrystallisiert. Es war nicht möglich, bei der Fraktionierung reine Cinchonidinsalze zu erhalten, da stets teilweiser Zerfall in die Komponenten auftrat. Um den Verlauf der Spaltung verfolgen zu können, musste die spez. Drehung der jeweils aus den Cinchonidinsalz-Fraktionen freigemachten Säure bestimmt werden. Hierbei wurde schliesslich eine Säure erhalten, deren spez. Drehung zu —17.3° gefunden wurde; nach nochmaliger Reinigung über das Cinchonidinsalz wurde für die Säure $(\alpha)_D^{20} = -17.5°$ gefunden. Dieses Präparat wurde in das Amid übergeführt. Schmelzp. des (—)α,ε-Dimethylheptansäure-amids 75—77°.

Spez. Drehung: C = 10.15, $\alpha = -0.92°$, $(\alpha)_D^{20} = -18.1°$ (in Alkohol).

```
(CH3)2CHCHCH2CH2COOH
         |
         C2H5
```

$C_9H_{18}O_2$ **Ref. 610**

Resolution of 4-Ethyl-5-methylhexanoic Acid.—The acid and cinchonidine (1 mol each) were dissolved in acetone; the mixture was heated until the solution became clear and then allowed to crystallize at room temperature. The crystals of the salt were filtered on a Büchner funnel as soon as they formed. After five recrystallizations from acetone the salt was treated with dilute hydrochloric acid and the acid was extracted with methylene chloride and distilled under reduced pressure. The optical rotation of the product was $[\alpha]_D -8°$.

Optical rotations were measured in chloroform.

```
CH3(CH2)2CHCOOH
         |
         CH2CH(CH3)2
```

$C_9H_{18}O_2$ **Ref. 611**

d-Propyl-isobutyl-essigsäure.

20 g Säure wurden mit 50 g trocknem **Brucin** in 180 ccm heißem Alkohol gelöst und in ein Gemisch von 1500 ccm Wasser und 200 ccm Alkohol, das auf 30° erwärmt war, unter Umschütteln eingegossen. Beim Abdampfen im Vakuum auf etwa 550 ccm fielen 38 g Salz krystallinisch aus; diese wurden abgepreßt, mit Wasser gewaschen, wieder mit Alkohol aufgenommen und in 600 ccm Wasser gegossen; die Lösung wurde dann auf ungefähr 100 ccm im Vakuum abgedampft, wobei wieder ein Teil des Salzes krystallinisch ausfiel. Dieser Prozeß wurde im ganzen 21-mal wiederholt. Das in mikroskopisch kleinen Prismen ausfallende Salz ist in Alkohol leicht, in Wasser aber schwer löslich. Zur Gewinnung der **freien Säure** wurde es mit 2-*n*.-Schwefelsäure zersetzt, sorgfältig ausgeäthert, die ätherische Lösung nochmals mit Schwefelsäure ausgeschüttelt, mit Wasser gewaschen und über Natriumsulfat getrocknet. Das nach dem Verdampfen des Äthers bleibende Öl wurde bei 0.5 mm Druck destilliert und ging bei ungefähr 100° über. Die so gewonnene Säure ist ein farbloses Öl vom spezifischen Gewicht 0.8876 bei 22° und zeigte $[\alpha]_D^{22} = +9.8°$. Das Drehungsvermögen war nach einem halben Jahre unverändert. Um den Gang der Spaltung durch das Brucinsalz zu verfolgen, haben wir auch einige Zwischenkrystallisationen und zwar die erste, dritte, neunte und zwölfte in die Säure verwandelt und diese optisch geprüft. Wir führen folgende Zahlen an, die aus der beobachteten Drehung im ½-dm-Rohr bei 22° mit der Dichte 0.8876 berechnet sind.

1. Krystallisation	$[\alpha]_D^{22} = +6.15°$,
3. »	$[\alpha]_D^{22} = +6.65°$,
9. »	$[\alpha]_D^{22} = +9.24°$,
12. »	$[\alpha]_D^{22} = +9.80°$,
21. »	$[\alpha]_D^{22} = +9.80°$.

Die aus der 12. Krystallisation erhaltene Säure wurde auch noch durch die Analyse auf ihre Reinheit geprüft.

0.1578 g Sbst.: 0.3950 g CO_2, 0.1619 g H_2O.

$C_9H_{18}O_2$ (158.14). Ber. C 68.29, H 11.47.

Gef. » 68.21, » 11.48.

Selbstverständlich ist das Verfahren sehr verlustreich. Nach der 16. Krystallisation betrug die Menge der Säure nur noch 3 g und nach der 21. Krystallisation nur 0.5 g oder 5 % der Theorie. Man sieht aus den Resultaten, daß mit dem Brucinsalz sich keine Säure von höherem Drehungsvermögen erhalten läßt. Ob damit aber der Endwert schon erreicht ist, müssen wir, wie in ähnlichen Fällen, unentschieden lassen.

$$CH_3-\underset{\underset{CH_3}{|}}{\overset{\overset{CH_3}{|}}{C}}-CH_2CH_2-\underset{\underset{CH_3}{|}}{CH}-COOH$$

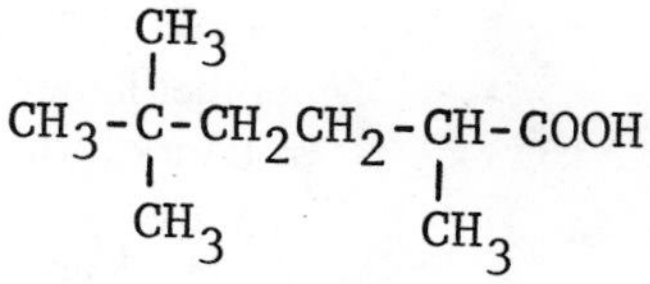

$C_9H_{18}O_2$ Ref. 612

(+)(s)-2,5,5-*Trimethylhexanoic acid* [*XVII*]. — (R)(s)-2,5,5-Trimethylhexanoic acid [XVI] (40 g) is treated with quinine base (82.3 g) in acetone solution (355 ml). The quinine salt is partially crystallized by cooling the solution at −25 °C.

After one more crystallization from acetone (177 ml) of the salt separated, a new portion of the salt is collected (32.6 g) which, by treatment with 10% H_2SO_4, gives (+)(s)-2,5,5-trimethylhexanoic acid (11.6 g), having b.p. 125 °C/16 torr, n_D^{25} 1.4265, d_4^{25} 0.8809 $[\alpha]_D^{25}$ +3.52, o p. 26.3%.

$$CH_3(CH_2)_3-\underset{\underset{CH_3}{|}}{\overset{\overset{C_2H_5}{|}}{C}}-COOH$$

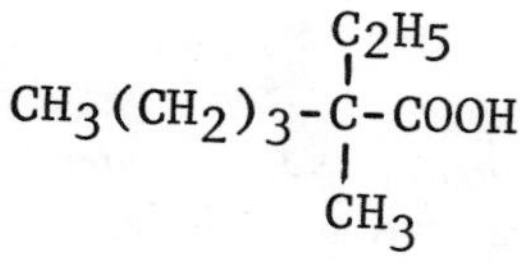

$C_9H_{18}O_2$ Ref. 613

(+)-*2-Ethyl-2-methylhexanoic acid* was resolved in 5.1% yield by nine crystallizations of the brucine salt in ethyl acetate: the salt, m.p. 89°; the acid, b.p. 137° (17 mm.); α_D^{25} +1.12° (homogeneous, 1 dm.); $[\alpha]_D^{25}$ +1.23°.

The (−)-acid from the tail fraction had the value, $[\alpha]_D^{25}$ −0.58° (homogeneous).

$$(CH_3)_2CHCH_2-\underset{\underset{C_2H_5}{|}}{\overset{\overset{CH_3}{|}}{C}}-COOH$$

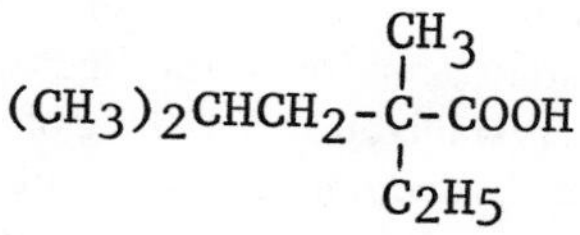

$C_9H_{18}O_2$ Ref. 614

Resolution of Methylethylisobutylacetic Acid (I).—The brucine salt prepared from 158 g. of I was fractionally crystallized from ethyl acetate. The crystallization was inconvenient due to the viscous nature of the concentrated solutions, but other solvents appeared to be less satisfactory. One crystallization using a seed of highly resolved material gave 190 g. of salt which melted to a cloudy liquid at 77–78° and became clear at 102°, and which gave I, $[\alpha]^{29}D$ + 8.7° (c = 8.6, 95% ethanol). After thirteen recrystallizations, head fractions of the brucine salt (70 g.), m. p. 105–108° with clearing at 115° was obtained which gave I with an average rotation $[\alpha]_D$ +15.3° (c = 9, 95% ethanol). Further fractional crystallization of 50 g. of this material from the same solvent gave 23 g. of the brucine salt, m. p. 114–115° with clearing at 117°, from which I, $[\alpha]^{24}D$ +21.55° (neat), $[\alpha]^{24}D$ +21.42° (c = 10.3, 95% ethanol), was obtained. This rotation could not be increased by further crystallization. (+) I has been shown to have the same infrared absorption curve as *dl*-I (Fig. 1) and has $n^{25}D$ 1.4322, and d^{25} 0.906; M_D calcd. 45.30; found, 45.29.

Anal. Calcd. for $C_9H_{18}O_2$: C, 68.31; H, 11.46; neut. equiv., 158.2. Found: C, 68.25; H, 11.46; neut. equiv., 158.3.

Anal. (brucine salt) Calcd. for $C_{32}H_{44}O_6N_2$: C, 69.54; H, 8.02; N, 5.07. Found: C, 68.95; H, 8.09; N, 5.20.

From the mother liquors and tail fractions there was obtained 36 g. of I, $[\alpha]^{25}D$ −5.30° (neat), and 55 g. of I, $[\alpha]^{23}D$ −2.89° (neat).

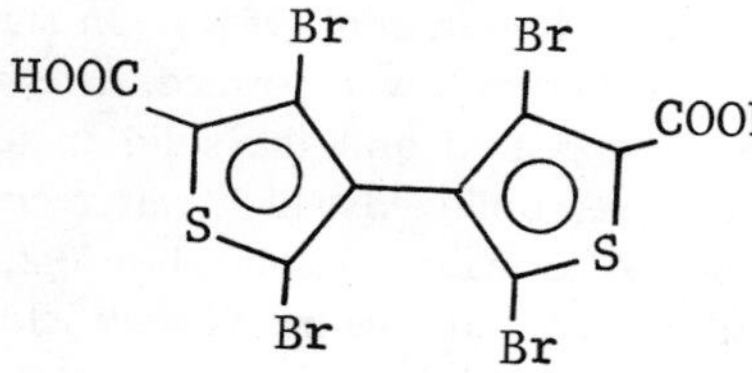

$C_{10}H_2Br_4O_4S_2$ Ref. 615

Optical resolution of 2,2',4,4'-tetrabromo-5,5'-dicarboxy-3,3'-bithienyl. Preliminary resolution experiments were made with 0.001 mole of inactive acid and 0.001 or 0.002 mole of optically active bases in 96% ethanol and in ethyl acetate. With dehydroabiethylamine, (−)-2-phenylethylamine, (+)-β-naphtylamine, strychnine and brucine, crystallizable salts were obtained in alcohol, but only the monosalt of brucine gave optically active acid. The salts of quinine and quinidine were very soluble in alcohol and gave oils in aqueous alcohol and in ethyl acetate.

10.0 g (0.018 mole) of 2,2',4,4'-tetrabromo-5,5'-dicarboxy-3,3'-bithienyl was dissolved in hot 96% ethanol and added to a hot alcoholic solution of 8.2 g (0.018 mole) of brucine tetrahydrate, giving a total volume of 550 ml. The solution was seeded with the salt of the levorotatory form of the acid. On standing for 36 hours, 11.5 g of this salt crystallized as a cake on the surface of the glass vessel. A small amount of the salt of the dextroratatory acid had appeared too, forming mobile needles which could be separated from the first crop of crystals with the mother liquor. After three more days, 3.1 g of the salt of the dextrorotatory form of 2,2',4,4'-tetrabromo-5,5'-dicarboxy-3,3'-bithienyl had crystallized. The two crops of crystals were individually recrystallized from ethanol. After each crystallization, the acid from 0.5–0.6 g of salt was liberated by treating it with 2-N sulphuric acid and extracting with ether. The ether phase was washed with 2-N sulphuric acid and with water, and dried over magnesium sulphate. Carbon tetrachloride was added and the ether distilled off until crystals of the acid appeared. The activity of the dried acid was measured in absolute ethanol. The progress of the resolution is given below:

Crystallization	1	2	3	4	5	6	7
ml, 96% ethanol	550	425	350	300	250	225	180
g, salt	11.5	8.8	7.3	6.2	5.2	4.3	3.4
$[\alpha]_D^{25}$ of acid (ethanol)	−5.8°	−7.4°	−7.7°	−7.9°	−7.8°	−7.9°	−7.9°
ml, 96% ethanol	550	140	80				
g, salt	3.1	1.7	1.1				
$[\alpha]_D^{25}$ of acid (ethanol)	+7.7°	+7.9°	+7.9°				

(+)-*2,2',4,4'-Tetrabromo-5,5'-dicarboxy-3,3'-bithienyl* was liberated from 1.1 g of its mono brucine salt as described above, giving 0.6 g of crystals, m.p. 247–250°C, containing one mole of carbon tetrachloride per two moles of acid.

Anal. $C_{10}H_2Br_4O_4S_2 \cdot \frac{1}{2} CCl_4$ (646.8)
Calc.   C 19.50   H 0.31
Found   C 19.64   H 0.28

(−)-*2,2',4,4'-Tetrabromo-5,5'-dicarboxy-3,3'-bithienyl.* From 3.4 g of its mono brucine salt, 1.9 g of the title compound, m.p. 248–250°C was obtained. Recrystallization from ether-carbon tetrachloride and drying at 95°C/10 mm Hg for five hours did not cause any change in weight or elementary analyses. G.l.c. analysis showed that an ethereal solution of the dried crystals contained a substance with the same retention time as carbon tetrachloride. On evaporation of the ethereal solution, a viscous residue was obtained which did not crystallize. 0.2006 g of crystals dissolved in 10 ml of absolute ethanol had $[\alpha]_D^{25}$ = −7.9° (α = −0.159). If the crystal solvent (12%) is taken into consideration, 0.2006 g of crystals contain 0.1756 g of acid. This corresponds to $[\alpha]_D^{25}$ = −9.0° for solvent free (−)-2,2',4,4'-tetrabromo-5,5'-dicarboxy-3,3'-bithienyl.

Anal. $C_{10}H_2Br_4O_4S_2 \cdot \frac{1}{2} CCl_4$ (646.8)
Calc.   C 19.50   H 0.31   S 9.91
Found   C 19.55   H 0.43   S 9.96

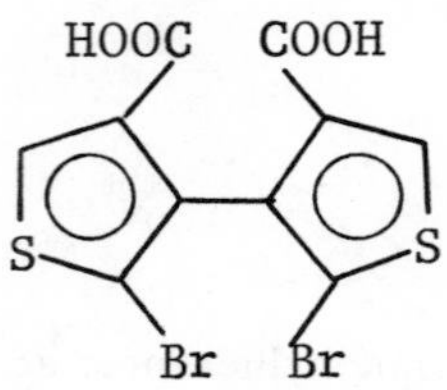

$C_{10}H_4Br_2O_4S_2$ **Ref. 616**

Resolution of 2,2'-dibromo-4,4'-dicarboxy-3,3'-bithienyl (IV). Preliminary tests on resolution showed that dicarboxylic acid IV with high optical activity could be obtained from the dibrucine salt.

8.5 g (21 mmol) of racemic IV and 16.2 g (41 mmol) of brucine were dissolved in 340 ml of hot ethanol (95%). On standing at room temperature overnight crystals were formed. After two days another type of crystals appeared and the solid material was collected. 11.1 g of salt was obtained which was recrystallized repeatedly from ethanol. After each crystallization 0.2 g of salt was set apart, the acid was liberated with 5 N hydrochloric acid and taken up in ether. After evaporation of the solvent the activity of IV was measured on ethanol solutions. The progress of the resolution is shown below.

Crystallization	1	2	3	4	5
g, salt	11.1	9.2	8.2	7.4	6.2
ml, ethanol	340	125	125	125	125
$[\alpha]_D^{25}$ of acid (ethanol)	−90°	−104°	−104°	−104°	−104°

The mother liquor from the first crystallization was evaporated to dryness, yielding 13.5 g of salt of which 0.2 g was set apart and treated as above. The activity of the liberated acid was $[\alpha]_D^{25} = +59°$ (ethanol). The main part of the salt (13.3 g) was dissolved in 250 ml of hot ethanol (95%) and 70 ml of ethyl acetate was added. 5.7 g of salt was obtained, which was recrystallized from the same solvent mixtures. The ethyl acetate was always added after the salt had been dissolved in hot ethanol. The progress of the resolution, followed as above, is shown below.

Crystallization	1	2	3
g, salt	5.7	4.1	2.8
ml, ethanol/ethyl acetate	250/70	200/100	150/75
$[\alpha]_D^{25}$ of acid (ethanol)	+104°	+104°	+104°

(R)-(−)-2,2'-Dibromo-4,4'-dicarboxy-3,3'-bithienyl (IV). The mother liquors of crystallizations 3–5 of the salt of the levorotatory form were combined and evaporated to dryness and the residual salt added to that left after crystallization 5 (a total of about 8.5 g). The acid was liberated as described above and recrystallized by dissolving it in 30 ml of ethanol (95%) and adding 60 ml of water at room temperature. (During a couple of days, crystals grew from the milky suspension.) 2.7 g (64%) of (R)-(−)-IV was obtained; m.p. 205–206°C (see the text), $[\alpha]_D^{25} = -104°$ (ethanol, $c = 1.5$ g/100 ml). [Found: C 29.2; H 1.31; Br 39.0; S 15.4 Calc. for $C_{10}H_4Br_2O_4S_2$ (412.1): C 29.15; H 0.98; Br 38.78; S 15.56.] IR: $\nu_{CO} = 1\,685$ cm^{-1}.

(S)-(+)-2,2'-Dibromo-4,4'-dicarboxy-3,3'-bithienyl (IV). The mother liquors of crystallizations 2 and 3 of the salt of the dextrorotatory form were evaporated and the residue added to the salt left after crystallization 3 (a total of about 5.0 g). The acid was liberated as above and recrystallized as described for (R)-(−)-IV, yielding 2.2 g (52%) of the title compound, m.p. 205–206°C (see the text), $[\alpha]_D^{25} = +104°$ (ethanol, $c = 1.2$).

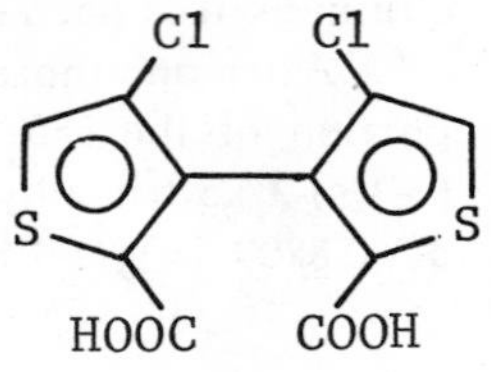

$C_{10}H_4Cl_2O_4S_2$ **Ref. 617**

Resolution of 2,2'-dicarboxy-4,4'-dichloro-3,3'-bithienyl (V). Preliminary tests on resolution were carried out with 1 mmol of racemic V and 1 or 2 mmol of an optically active base such as strychnine, brucine, quinine, quinidine, chinchonine, chinchonidine and dehydroabiethylamine in ethanol or ethanol-water mixtures. Brucine in ethanol-water gave crystalline salts with both enantiomers of V, which partly could be separated mechanically. However, a crystalline salt with only one form of V was obtained with dehydroabiethylamine in 95% ethanol. The resolution was performed by dissolving the carboxylic acid and the base in the hot solvent. After standing at room temperature for 24 h, the crystalline salt was collected and recrystallized repeatedly until constant activity of the liberated acid was obtained. After each recrystallization 0.1–0.2 g of salt was set apart and the acid liberated by decomposing the salt with 2 N sodium hydroxide. The dehydroabiethylamine was extracted out twice with benzene and once with ether. The remaining aqueous solution was acidified with 5 N hydrochloric acid, the precipitated acid filtered off and the activity of the acid was measured on an ethanol solution. The progress of the resolution is shown below.

Crystallization	1	2	3	4
ml, 95% ethanol	60	45	20	10
g, salt	1.3[a]	0.8	0.4	0.2
$[\alpha]_D^{25}$ of acid (ethanol)[b]	+48°	+50°	+50°	+50°

[a] From 2.0 g (6.2 mmol) of racemic V and 1.8 g (6.3 mmol) of dehydroabiethylamine.

[b] The levorotatory form: see (S)-(−) V below.

(R)-(+)-2,2'-Dicarboxy-4,4'-dichloro-3,3'-bithienyl (V). 1.8 g (5.6 mmol) of racemic V and 1.6 g (5.6 mmol) of dehydroabiethylamine were dissolved in 55 ml of hot ethanol (96%), and the solution was left at room temperature for 48 h. 1.5 g of salt was filtered off and the acid was liberated from the salt as described above. 0.5 g (56%) of (R)-(+)-V, $[\alpha]_D^{25} = +50°$C (ethanol, $c = 1.0$) was obtained, m.p. 280°C dec. Recrystallization from ethanol-water afforded crystals with the same $[\alpha]_D^{25}$ value as for the crude product. [Found: C 36.9; H 1.09; S 19.4. Calc. for $C_{10}H_4Cl_2O_4S_2$ (323.2): C 37.17; H 1.25; S 19.84.] IR: $\nu_{CO} = 1\,680$ cm^{-1}. The ^{1}H-NMR spectrum was identical to that of racemic V.

(S)-(−)-2,2'-Dicarboxy-4,4'-dichloro-3,3'-bithienyl (V). The mother liquor of the crystallization of the salt of the dextrorota-

tory form of V was evaporated to dryness, and the acid was liberated from the salt in the same way as described above. 0.65 g (2.0 mmol) of acid was obtained which was dissolved in 40 ml of ethanol-water 1 : 1 together with 1.6 g (4.0 mmol) of brucine. After standing at room temperature for 24 h, 1.2 g of salt was filtered off. The acid was liberated from the salt with 5 N hydrochloric acid and taken up in ether. The solution was dried over $MgSO_4$ and the solvent evaporated affording 0.25 g (28%) of (*R*)-(−)-V, $[\alpha]_D^{25} = -49°$ (ethanol, *c* = 1.0), m.p. 285°C dec. IR: $\nu_{CO} = 1665\ cm^{-1}$.

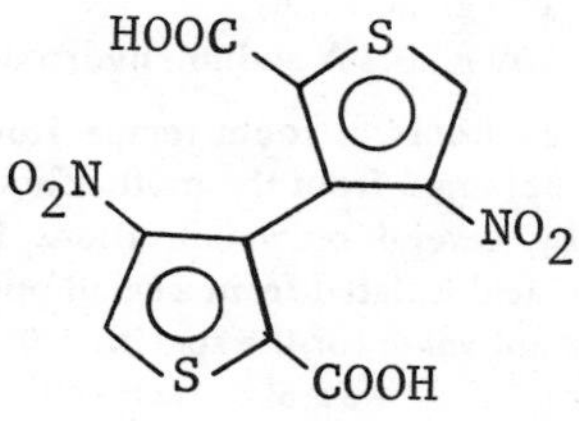

$C_{10}H_4N_2O_8S_2$ **Ref. 618**

Optical resolution of 2,2'-dicarboxy-4,4'-dinitro-3,3'-bithienyl

19.5 g (0.0566 moles) of 2,2'-dicarboxy-4,4'-dinitro-3,3'-bithienyl was dissolved in 0 ml of hot ethanol and added to a hot solution of 36.7 g (0.112 moles) of quinidine 140 ml of ethanol and left to crystallize in the refrigerator. The quinidine salt was tered off and dried. About 0.5 g of the acid was set apart; the acid was isolated d its activity measured in dioxane. The remaining salt was recrystallized from ueous ethanol. The progress of the resolution is seen below.

stallization	1	2	3	4	5
C_2H_5OH	420	215	225	225	225
H_2O	—	100	50	25	—
alt	24.5	21	19	17	10
^{25} of acid	−175°	−206°	−205°	−205°	−205°

It is obvious that resolution was complete after two recrystallizations. An addinal 7 g of pure salt was obtained by working-up the mother liquor from the last ee recrystallizations.

The mother liquor from the first recrystallization was evaporated in vacuo and residue decomposed with acid to give 12.6 g of acid, having $[\alpha]_D^{25} = +111°$. 5 g (0.0363 moles) of this acid and 21.3 g (0.0724 moles) of cinchonidine were disved in 700 ml of alcohol and the salt allowed to crystallize. The progress of the olution is seen below.

Crystallization	1	2	3
ml C_2H_5OH	700	1300	800
g salt	20.0	13.0	8.8
$[\alpha]_D^{25}$ of acid	+157°	+205°	+205°

R-(+)-2,2'-dicarboxy-4,4'-dinitro-3,3'-bithienyl

.0 g of the cinchonidine salt was decomposed with 2N hydrochloric acid and the eous phase extracted with ether, yielding after recrystallization from 100 ml of er, 2.51 g of (+)-acid in well-crystallized flakes, m. p. 285–287° dec. ($[\alpha]_D^{25}$ = 25°, c; 0.805, dioxane) R.D. in dioxane (c. 0.489–0.00489), temp. 26°. $[M]_{700}$ 70°, $[M]_{600}$ +743°, $[M]_{589}$ +774°, $[M]_{500}$ +1420°, $[M]_{400}$ +6090°, $[M]_{372}$ +8840°, $]_{337}$ ±0°, $[M]_{300}$ −6500°, $[M]_{266}$ −30600°, $[M]_{254}$ ±0°, $[M]_{244}$ +25600°, $[M]_{227}$ 00°, $[M]_{216}$ +3900°.

nal. $C_{10}H_4N_2O_8S_2 \cdot H_2O$ (362.3)
Calc. C 33.11 H 1.68 S 17.70
Found C 33.11 H 1.63 S 17.24

R(+)-2,2'-dicarbomethoxy-4,4'-dinitro-3,3'-bithienyl

This was prepared from the acid and diazomethane. M.p. 157.5°–159°, after recrylization from methanol. ($[\alpha]_D^{25}$ = +81°, c; 0.573, ethanol). R.D. in dioxane (c, 22–0.00422), temp. 26°. $[M]_{700}$ +366°, $[M]_{600}$ +591°, $[M]_{589}$ +618°, $[M]_{500}$ +1150°, $_{400}$ +6440°, $[M]_{372}$ +8150°, $[M]_{343}$ ±0.°, $[M]_{309}$ −13000°, $[M]_{300}$ −6000°.

nal. $C_{12}H_8N_2O_8S_2$ (372.3)
Calc. C 38.71 H 2.17
Found C 38.79 H 2.43

S(−)-2,2'-dicarboxy-4,4'-dinitro-3,3'-bithienyl

12.0 g of the quinidine salt was decomposed with 2-N hydrochloric acid and the acid isolated as described above. Recrystallization from water yielded 3.49 g of (−)-acid, melting with decomposition at 285–287°, the rotations in different solvents are given below.

Solvent	conc./g acid dl	$2[\alpha]_D^{25}$	$[\alpha]_D^{25}$	$[M]_D^{25}$
0.1-N NaOH	0.704	−4.18°	−297°	−760°
Dioxane	1.291	−5.70°	−221°	−1020°
Dimethyl sulphoxide	0.924	−3.30°	−179°	−616°
Acetone	1.544	−4.35°	−141°	−485°
Ethanol	0.764	−1.81°	−118°	−406°
Methanol	0.943	−2.25°	−119°	−409°
Water	0.278	−0.59°	−106°	−365°

Anal. $C_{10}H_4N_2O_8S_2 \cdot H_2O$ (362.3)
Calc. C 33.11 H 1.68 S 17.70
Found C 33.36 H 1.81 S 17.42

After drying at 120°, the anhydrous acid was obtained.

Anal. $C_{10}H_4N_2O_8S_2$ (344.2)
Calc. C 34.89 H 1.21 S 18.62
Found C 34.89 H 1.24 S 18.68

S(−)-2,2'-dicarbomethoxy-4,4'-dinitro-3,3'-bithienyl

This was prepared from the acid and diazomethane. M.p. 157.5–159° after recrystallization from methanol. ($[\alpha]_D^{25}$ = −81°, c; 0.574, ethanol).

Anal. $C_{12}H_8N_2O_8S_2$ (372.3)
Calc. C 38.71 H 2.17
Found C 38.89 H 2.38

S(−)-4,4'-dicarboxy-2,2'-dinitro-3,3'-bithienyl

Rotation dispersion in dioxane (c. 0.028, 0.413), temp. 27° $[M]_{550}$ −850°, $[M]_{500}$ −1150°, $[M]_{450}$ −1760°, $[M]_{400}$ −3390°, $[M]_{350}$ −7200°, $[M]_{327}$ −14000°, $[M]_{300}$ −7900°, $[M]_{287}$ ±0°, $[M]_{258}$ +28000°, $[M]_{250}$ +26000°, $[M]_{234}$ ±0°, $[M]_{227}$ −7200°, $[M]_{222}$ ±0°, $[M]_{210}$ +12000°.

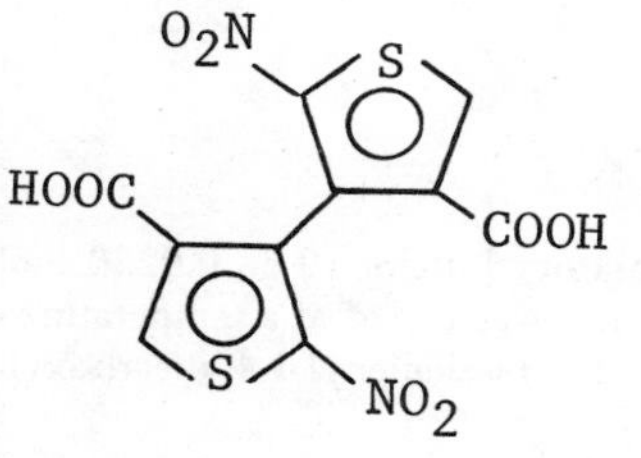

$C_{10}H_4N_2O_8S_2$ **Ref. 619**

Preliminary tests on resolution were made with about 0.005 moles of inactive acid and 0.01 moles of optically active bases in water, ethanol, and in some cases ethyl acetate. Most alkaloids such as strychnine, cinchonine, cinchonidine and quinidine give oils from water and excessively soluble salts in alcohol. The (−)-α-phenylethylamine salt was soluble in water and alcohol and gave a salt of the inactive acid from ethyl acetate. Acids having $[\alpha]_D^{25}$ = +98° and $[\alpha]_D^{25}$ = −50° were obtained from brucine and quinine respectively using 95% ethanol as solvent in each case. These bases were therefore used for the resolution.

Optical resolution of 2,2'-dinitro-4,4'-dicarboxy-3,3'-bithienyl. 20.7 g (0.060 moles) of 2,2'-dinitro-4,4'-dicarboxy-3,3'-bithienyl and 53.4 g (0.124 moles) of brucine were dissolved by heating in 2 l of ethanol and the solution left at room-temperature overnight. The precipitate was filtered off and dried. About 0.5 g of the salt was set apart, the acid isolated, and its optical activity measured in ethanol. The remaining salt was recrystallized from 95% ethanol and the process repeated until the optical activity of the acid remained constant. The results of the resolution are seen below.

Crystallization	1	2	3	4	5
g salt	26.5	17.1	14.5	12.0	10.1
ml ethanol	2000	1000	1000	750	725
$[\alpha]_D^{25}$ of acid	+74°	+97°	+96°	+96°	+96°

It is apparent that the optically pure salt was obtained after two recrystallizations. The mother liquor from the first crystallization with brucine (from which an additional 7.5 g of the salt had separated) was evaporated in vacuo, and the residue decomposed with acid to give 13.0 g of the acid having $[\alpha]_D^{25}$ = −43°. Crystallization of 12.7 g (0.0369 moles) of this acid with 28 g (0.074 moles) of quinine in 600 ml of hot alcohol, on standing overnight yielded 21.2 g of the quinine salt. This gave an acid having $[\alpha]_D^{25}$ = −78°. In order to further recrystallize 20.7 g of this quinine salt, 2 l of alcohol were needed (and even then a residue was left). From this solution only 3.5 g of the quinine salt were obtained giving an acid having $[\alpha]_D^{25}$ = −85°. By stepwise concentration of the mother-liquor *in vacuo*, an additional 8.1 g of the salt with the same activity were obtained. As further recrystallization did not raise the rotation, the quinine salt fractions, giving an acid having $[\alpha]_D^{25} = -85°$, were used directly for the preparation of the (−)-acid. In a repetition of this experiment no attempts were made to recrystallize the quinine salt, as the pure (−)-acid could be obtained by recrystallization of the partially resolved acid.

(R)−(+)-2,2′-dinitro-4,4′-dicarboxy-3,3′-bithienyl. 10 g of the brucine salt was decomposed with 2 *N* hydrochloric acid and the aqueous phase extracted with ether, yielding 2.5 g of (+)-acid, which was recrystallized twice from water (Norite), m.p. 289–293° with dec. R.D. in methanol (c. 0.114–0.228), temp. 26° $[M]_{700}$ +265°, $[M]_{600}$ +335°, $[M]_{589}$ +355°, $[M]_{500}$ +505°, $[M]_{400}$ +890°, $[M]_{392}$ +930°. The rotations in different solvents are given below.

Solvent	conc./g acid/dl	$2\alpha_D^{25}$	$[\alpha]_D^{25}$	$[M]_D^{25}$
0.1 *N* NaOH	0.225	+0.77°	+171°	+590°
Dimethyl sulphoxide	0.405	+1.15°	+142°	+488°
Chloroform	0.316	+0.72°	+115°	+396°
Acetone	0.387	+0.87°	+111°	+382°
Ethanol	0.462	+0.91°	+ 98°	+337°
Methanol	0.356	+0.62°	+ 87°	+303°
Water	0.255	+0.26°	+ 52°	+179°

Anal. $C_{10}H_4N_2O_8S_2$ (344.3)
Calc. N 8.14 S 18.62
Found N 8.14 S 18.55

(S)-(−)-2,2′-dinitro-4,4′-dicarboxy-3,3′-bithienyl. 16 g of quinine salt was decomposed with 2-*N* hydrochloric acid and the aqueous phase extracted with ether, yielding 5.0 g of acid ($[\alpha]_D^{25} = -84°$). Two recrystallizations from water (Norite) gave 3.5 g of the optically pure acid, m.p. 289–293° dec., having the same IR-spectrum (Fig. 2) as the (+)-acid. ($[\alpha]_D^{25} = -97°$c; 0.625, ethanol)

Anal. $C_{10}H_4N_2O_8S_2$ (344.3)
Calc. N 8.14 S 18.62
Found N 8.00 S 18.55

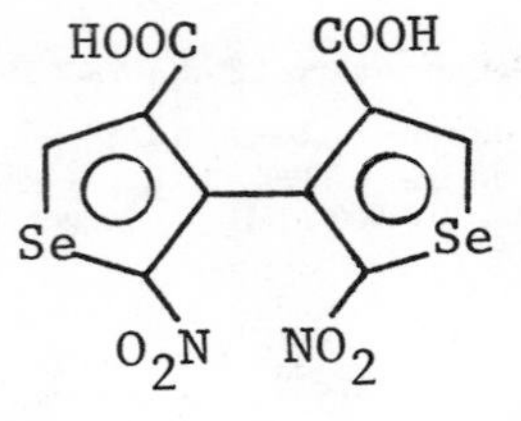

$C_{10}H_4N_2O_8Se_2$ Ref. 620

The levorotatory brucine (9 g., 0.0228 mole) dissolved in 95% ethanol (685 ml.) was added at a temperature of 40° to a solution of 2,2′-dinitro-3,3′-biselenienyl-4,4′-dicarboxylic acid (5 g., 0.0114 mole) in ethanol (540 ml.). The solution was left at room temperature for 20 hours. The brucine salt which precipitated (5.7 g.) was filtered off. This salt was recrystallized from ethanol until the samples of acid, isolated from a small portion of its brucine salt, showed constant specific rotation. The results are listed in Table I.

The dextrorotatory acid was recrystallized from water as colourless needles, m.p. 288°. The specific rotatory power for the dextrorotatory acid measured in some solvents were respectively (in bracket the weight of acid dissolved in 100 ml. of solvent):

$[\alpha]_D^{25°} = +37°$ (0.601 g., methanol)
$[\alpha]_D^{25°} = +44°$ (0.480 g., ethanol)
$[\alpha]_D^{25°} = +46°$ (0.176 g., acetone)
$[\alpha]_D^{25°} = +66°$ (0.169 g., 0.1 *N* sodium hydroxide)

After another 24 hours at room temperature another brucine salt (4.73 g.) was obtained from the mother liquor of brucine salt. It was purified by several recrystallizations from ethanol (see Table II) until the acid isolated from a small portion of its brucine salt showed a constant value for the specific rotation.

The levorotatory acid was also recrystallized from water as colourless needles, m.p. 288°.

(R)-(+)-2,2′-Dinitro-3,3′-biselenienyl-4,4′-dicarboxylic Acid Dimethyl Ester.

An excess of an ethereal solution of diazomethane was added to the (R)-(+)-2,2′-dinitro-3,3′-biselenienyl-4,4′-dicarboxylic acid (1 g.). After 6 hours at room temperature the excess diazomethane was removed. After being washed with 10% aqueous sodium bicarbonate, the solvent was distilled off from the residual ethereal solution. The remaining solid was crystallized from xylene; yellow prisms, m.p. 204-205°; $[\alpha]_D^{25°} = +63.5$ (0.45 g. in 100 ml. of acetone).

(S)-(−)-2,2′-Dinitro-3,3′-biselenienyl-4,4′-dicarboxylic Acid Dimethyl Ester.

Working in the above-mentioned way a product was obtained from the other enantiomer, which was crystallized from xylene as yellow prisms, m.p. 204-205°; $[\alpha]_D^{25°} = -63.5$ (0.30 g. in 100 ml. of acetone).

TABLE I

Crystallization	1	2	3	4	5
g. of salt	5.7	4.82	4.39	3.96	3.20
ml. of ethanol	1225	470	250	250	235
$[\alpha]_D^{25°}$ of acid in EtOH	+ 42°	+ 43°	+ 44°	+ 44°	+ 44°

TABLE II

Crystallization	1	2	3	4
g. of salt	4.73	4.10	3.50	3.00
ml. of ethanol	1225	180	140	100
$[\alpha]_D^{25°}$ of acid in ethanol	−42°	−43°	−44°	−44°

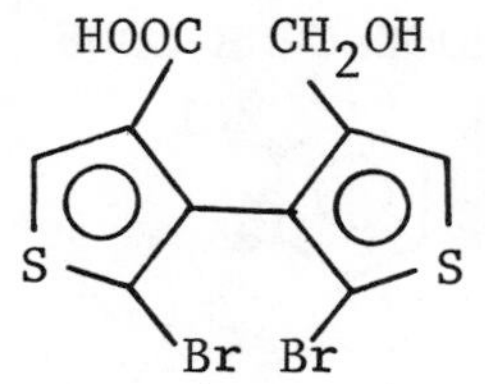

$C_{10}H_6Br_2O_3S_2$ **Ref. 621**

Resolution of 2,2′-dibromo-4-carboxy-4′-hydroxymethyl-3,3′-bithienyl (V). Preliminary tests on resolution showed that optically active V could be obtained with brucine as resolving agent in ethanol (95%).

5.0 g (13 mmol) of racemic hydroxy acid V and 5.0 g (13 mmol) of brucine were dissolved in 200 ml of hot 95% ethanol. After 24 h 5.7 g of salt was obtained, which was repeatedly recrystallized from ethanol. The progress of the resolution, followed as described for IV, is shown below. The variations of the rotational values may depend on difficulties in removing solvent from the oily products (see below and the text).

Crystallization	1	2	3	4	5	6	7
g, salt	5.7	3.8	3.1	2.5	2.2	1.9	1.6
ml, ethanol	200	200	150	125	100	75	60
$[\alpha]_D^{25}$ of acid (ethanol)	+32°	+60°	+59°	+64°	+63°	+64°	+65°

(S)-(+)-2,2′-Dibromo-4-carboxy-4′-hydroxymethyl-3,3′-bithienyl (V). From the salt (1.6 g) of the last crystallization in the resolution above the acid was liberated as previously described. The product solidified to a glassy sample which could not be crystallized; $[\alpha]_D^{25} = +65°$ (ethanol, $c = 1.5$). IR: $\nu_{CO} = 1\,690$ cm^{-1}.

(R)-(−)-2,2′-Dibromo-4-carboxy-4′-hydroxymethyl-3,3′-bithienyl (V) was obtained by treating the residue from the evaporation of the mother liquor from the first crystallization in the same manner as previously described. 2.0 g of (R)-(−)-V was obtained; $[\alpha]_D^{25} = -47°$, 72% or less optically pure (ethanol, $c = 1.3$).

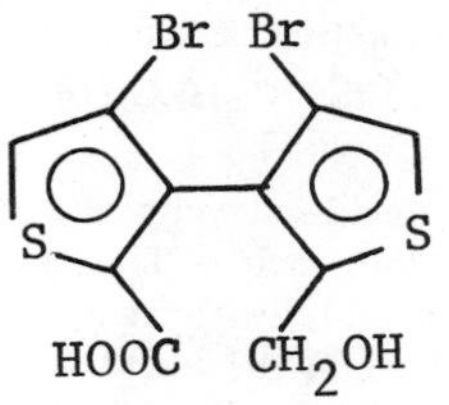

$C_{10}H_6Br_2O_3S_2$ **Ref. 622**

Resolution of 4,4′-dibromo-2-carboxy-2′-hydroxymethyl-3,3′-bithienyl (IV). Preliminary resolution experiments were carried out with 1 mmol of racemic IV and 1 mmol of an optically active base such as quinine, quinidine, cinchonine, cinchonidine, strychnine or brucine in ethanol and ethanol–water mixtures. Only brucine in 95% ethanol gave a crystalline salt.

7.96 g (20 mmol) of hydroxy acid IV and 7.88 g (20 mmol) of anhydrous brucine were dissolved in 1 000 ml of hot ethanol (95%). After standing at room temperature for 24 h, 7.9 g of salt was filtered off. (7.92 g is the theoretical amount of salt of one enantiomer.) The acid was liberated from 0.2 g of salt with 5 N hydrochloric acid and taken up in ether. The solution was dried over $MgSO_4$ and the solvent evaporated. The activity of the residual acid was measured on ethanol solutions. The main part of the salt was recrystallized from ethanol. Further crystallizations were continued with only 1.0 g of salt. 0.2 g of salt was set apart after each crystallization and treated as above. The progress of the resolution is seen below, and it is evident that the hydroxy acid was almost optically pure already after the first crystallization.

Crystallization	1	2	3	4	5
g, salt	7.9	7.6	0.9	0.65	0.40
ml, ethanol (95%)	1 000	750	75	50	50
$[\alpha]_D^{25}$ of acid (ethanol)	−9.5°	−9.7°	−9.7°	−9.6°	−9.7°

The mother liquor from the first crystallization was evaporated to dryness *in vacuo*. The residue was decomposed with 5 N hydrochloric acid, yielding optically pure (+)-hydroxy acid IV (see below).

(S)-(−)-4,4′-Dibromo-2-carboxy-2′-hydroxymethyl-3′3-bithienyl (IV). The mother liquors of crystallizations 3–5 of the salt of the levorotatory form of hydroxy acid IV were combined and evaporated. The residual salt was added to that remaining after crystallizations 2 and 5 (a total of 6.8 g), treated with 5 N hydrochloric acid, and the resulting carboxylic acid extracted out with ether. The ether solution was washed with water, dried ($MgSO_4$) and evaporated, and the residue was recrystallized from ethyl acetate–petroleum ether (40–60°). 3.2 g (80% of the theoretical 3.98 g) was obtained; m.p. 155–157°C. $[\alpha]_D^{25} = -9.8°$ (ethanol, $c = 2.8$ g/100 ml). [Found: C 30.2; H 1.69; Br 39.2; S 15.8. Calc. for $C_{10}H_6Br_2S_2O_3$ (398.1): C 30.17; H 1.52; Br 40.15; S 16.11.] IR: $\nu_{CO} = 1\,700$ cm^{-1} and 1 650 cm^{-1}.

(R)-(+)-4,4′-Dibromo-2-carboxy-2′-hydroxymethyl-3,3′-bithienyl (IV). The mother liquor from the first crystallization of the salt of the levorotatory form of IV was evaporated to dryness. The acid was liberated as above, yielding 3.5 g of the title compound (88% of the theoretical 3.98 g); m.p. 155–157°C. $[\alpha]_D^{25} = +9.8°$ (ethanol, $c = 2.9$).

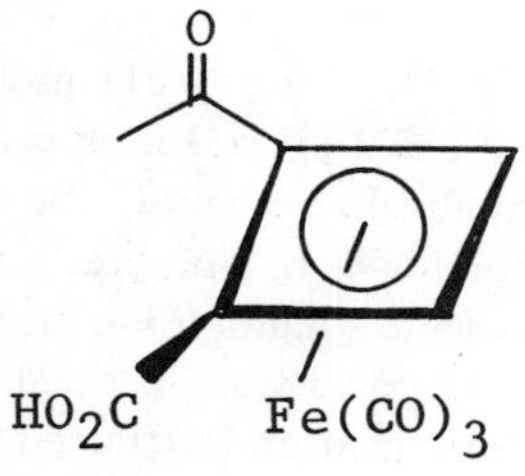

$C_{10}H_6FeO_6$ **Ref. 623**

Resolution of the acid by quinine afforded optically active (−)-(*2*), $[\alpha]_{578}^{24} = -102°$ (in ethanol).

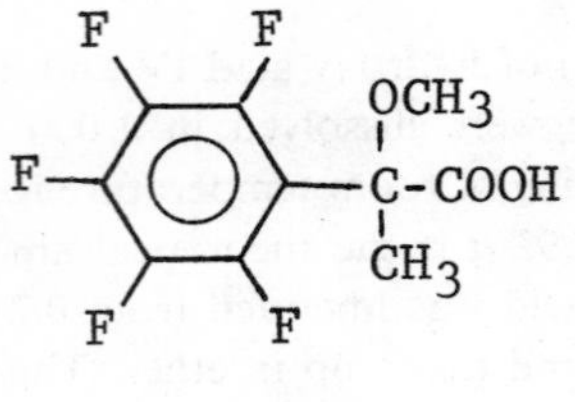

$C_{10}H_7F_5O_3$ **Ref. 624**

Resolution of *rac*-1. *rac*-1, 35 g (0.13 mol), quinine, 42 g (0.13 mol), and a 1:1 mixture of hexane and Me_2CO (200 ml) were mixed, and resulting salt was dissolved by boiling the mixture and recrystallizing by slowing cooling the hot solution to room temperature. The salt was filtered, washed with a minimum of cold hexane–acetone (1:1), and recrystallized several more times to give a white crystalline (needle) salt: mp 203–204°; 12.3 g; $[\alpha]^{22}D$ –118° (*c* 3.00, EtOH). A small amount of the salt was decomposed with dilute HCl, and the regenerated acid was extracted (ether), dried ($MgSO_4$), and vacuum evaporated to give a colorless, viscous oil, $[\alpha]^{22}D$ –34° (*c* 3.44, MeOH). During the reprocessing of the solids from the initial filtrates, another salt was collected: white powder; 12.9 g; mp 173–174°; $[\alpha]^{22}D$ –82° (*c* 3.00, EtOH). It was decomposed to give a colorless, viscous oil, $[\alpha]^{22}D$ +44° (*c* 3.27, MeOH). Both enantiomers gave the same nmr and ir spectra as the racemic acid.

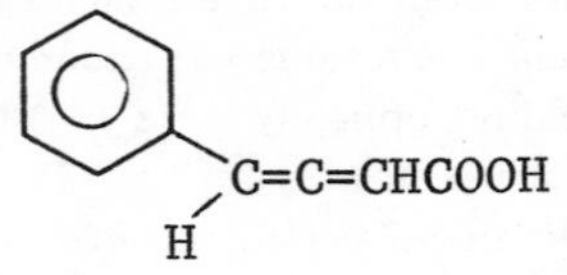

$C_{10}H_8O_2$ **Ref. 625**

Note:

The racemic acid was resolved via the cinchonidine salt. The melting point and $[\alpha]$ obtained for the active acid were: m.p. 109-110°C.; $[\alpha]_D$ = +420° (methanol).

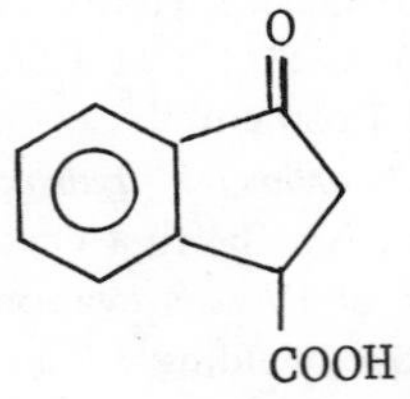

$C_{10}H_8O_3$ **Ref. 626**

The ketoacid ((±)-**15**), 90.0 g (0.511 mol), dried over P_2O_5 in vacuum for 48 h, and 222.0 g (0.511 mol) brucine·2 H_2O were dissolved in 7.8 l warm absolute ethanol. The solution was stored at –5° in a refrigerator for 48 h. The precipitated crystals were recrystallized from absolute ethanol using the same ketoacid/ethanol ratio as above. The ketoacid was liberated from the brucine salt with 2 M hydrochloric acid and extracted with ether. The ether solution was washed several times with hydrochloric acid and water. After evaporation of the ether the ketoacid was dried over P_2O_5 in vacuum for 48 h. The yield was 21.1 g with $[\alpha]^{25.2}_{578}$ = +25.1 (absolute ethanol).

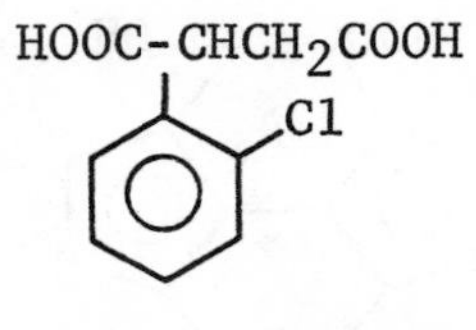

$C_{10}H_9ClO_4$ **Ref. 627**

d-*o*-**Chlorophenylsuccinic Acid.**—Attempts to fractionally crystallize the *d*- and *l*-acid-*l*-quinine salts from ethyl acetate and from ethanol were unsuccessful. The resolution was accomplished with strychnine in alcoholic solution. An equimolecular quantity (7.3 g.) of racemic *o*-chlorophenylsuccinic acid was added to 10.7 g. of strychnine dissolved in 500 cc. of hot 95% ethyl alcohol. The solution was refluxed two hours and then cooled to 20°. No deposit was formed after twelve hours, so the solution was evaporated to 130 cc. Since the sirupy residue showed no tendency to crystallize, 100 cc. of water was added and the solution stirred and then cooled overnight at 20°. A crop of 8.5 g. of transparent, monoclinic crystals which melted at 122–126° was deposited. The product was recrystallized from hot water to a constant melting point of 126–128°.

Anal. Calcd. for $C_{21}H_{22}O_2N_2{\cdot}C_{10}H_9ClO_4{\cdot}2H_2O$: N, 4.67. Found: N, 4.67.

The *d*-acid-*l*-strychnine salt was hydrolyzed with dilute hydrochloric acid and the *d*-*o*-chlorophenylsuccinic acid extracted with ether; yield, 2.5 g., m. p. 166–168°. The acid was recrystallized once from benzene and then from 350 cc. of chlorobenzene containing 5 cc. of 95% ethanol. The pure acid melted at 167–168° and gave the following optical rotations: $[\alpha]^{28}D$ +115.0° (*c* 0.997 in ethanol), $[\alpha]^{28}{}_{5770}$ +120.0° (*c* 0.997 in ethanol), $[\alpha]^{28}{}_{5461}$ +136.0° (*c* 0.997 in ethanol), $[\alpha]^{30}D$ +133.4° (*c* 1.388 in ethyl acetate) $[\alpha]^{31}D$ +110.6° (*c* 0.408 in chloroform), $[\alpha]^{31}D$ +112.2° (*c* 0.865 in acetone), $[\alpha]^{31}{}_{5461}$ +126.0° (*c* 0.865 in acetone).

Anal. Calcd. for $C_{10}H_9ClO_4$: C, 52.53; H, 3.97; Cl, 15.50. Found: C, 53.10; H, 4.21, Cl, 15.30.

l-*o*-**Chlorophenylsuccinic Acid.**—The mother liquor from which the *d*-acid-*l*-strychnine salt was obtained was diluted with 20 cc. of water and set aside to crystallize. A crop of 2.9 g. of transparent plates melting at 135–138° was deposited. Evaporation of the solution resulted in the formation of another crop with the same melting point. The combined crops were recrystallized from hot water and yielded 2.6 g. of *l*-*o*-chlorophenylsuccinic acid-*l*-strychnine salt melting at 138°.

Anal. Calcd. for $C_{21}H_{22}O_2N_2{\cdot}C_{10}H_9ClO_4$: N, 4.97. Found: N, 4.92.

The crop of *l*-acid-*l*-strychnine salt was hydrolyzed with dilute hydrochloric acid. The solution was cooled and extracted thoroughly with ether. After removal of the ether by distillation, the *l*-acid was washed with benzene and dried; m. p. 163–165°, yield, 2.2 g. The levo acid was further purified by recrystallization from hot chlorobenzene containing a small amount of 95% ethyl alcohol; m. p. 166–168°, $[\alpha]^{32}D$ –101.3° (*c* 1.005 in ethanol). The anhydride prepared from the above acid had the following rotation: $[\alpha]^{31}D$ –45.7° (*c* 0.835 in ethanol). This value is in good agreement with the rotation observed for pure *d*-*o*-chlorophenylsuccinic anhydride: $[\alpha]^{31}D$ +45.5° (*c* 0.907 in ethanol). The dextro anhydride was prepared from a sample of dextro acid having a specific rotation of $[\alpha]^{28}D$ +115.0°.

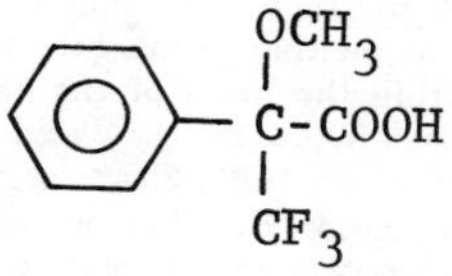

$C_{10}H_9F_3O_3$ **Ref. 628**

Resolution of α-Methoxy-α-trifluoromethylphenylacetic Acid. —Racemic MTPA (87.3 g), (+)-α-phenylethylamine (45.0 g, $\alpha^{25}{}_D$ 37.34°, neat, l = 1) and ethanol (300 ml) were mixed. The salt which formed immediately was dissolved by heating on the steam bath, and the solution was insulated and allowed to cool slowly without being disturbed for 48 hr. The salt was collected by filtration, washed with a minimum of cold ethanol, and recrystallized twice from ethanol to give 29.0 g, mp 195–198°, $[\alpha]^{26}{}_D$ 59.1 ± 1.1° (c 1.32, EtOH). Reprocessing solids from the filtrate gave an additional 14.5 g, $[\alpha]^{19}{}_D$ 62.5 ± 1.6° (c 1.23, EtOH). These combined crystals were decomposed with dilute hydrochloric acid and the regenerated acid extracted with ether. The extracts were dried ($MgSO_4$), the solvent was evaporated, and the residue distilled to give the (+) was acid, 28.6 g, bp 116–118° (1.5 mm), $[\alpha]^{25}{}_D$ 68.5 ± 1.3° (c 1.49, CH_3OH). The more soluble salt fractions were decomposed in the usual manner and the isolated acid treated with (−)-α-phenylethylamine (30 g, $[\alpha]^{25}{}_D$ −36.34°, neat l = 1) in ethanol (230 ml). Processing the salt as above gave a total of 55.4 g, $\alpha^{28}{}_D$ −60° ± 2° (c 1, EtOH) which was decomposed as above to give 34.7 g, bp 115–117° (1.5 mm), $[\alpha]^{24}{}_D$ −71.8 ± 0.6° (c 3.28, CH_3OH). The total recovery on the resolution was 72%.

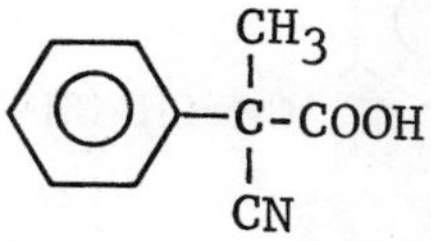

$C_{10}H_9NO_2$ **Ref. 629**

(+) **2d** *mit Brucin*

40 g racem. **2d** wurden in 150 ml absol. Äther gelöst und mit einer Lösung von 40 g Brucin in 150 ml absol. Methanol gemischt. Nach 24 Std. bei − 30° wurden 50 ml absol. Äther zugefügt, nach weiteren 24 Std. das ausgefallene Diastereomerensalz abfiltriert, aus Methanol/Äther umkristallisiert und mit Salzsäure zerlegt: 5,5 g (+) **2d** (27,5 % d. Th.); nach Umkristallisation aus Benzol/Petroläther Schmp. 66 – 68°.

$C_{10}H_9NO_2$ (175,2)	Ber.:	C 68,6	H 5,18	N 8,0
	Gef.:	C 68,3	H 5,13	N 8,0

Mit äther. Diazomethanlsg. entstand aus (+) **2d**, $[\alpha]^{20}_D$ = + 8,2°, quantitativ (+) **3d** vom Sdp. 102°/0,8 Torr.

$C_{11}H_{11}NO_2$ (189,2)	Ber.:	C 69,8	H 5,86	N 7,4
	Gef.:	C 69,7	H 5,74	N 7,4

$[\alpha]^{20}_{nm}$	(+) 2d		(−) 2d	
589	+	8,2°	−	8,7°
578	+	8,5°	−	9,2°
546	+	9,5°	−	10,0°
436	+	16,3°	−	16,3°
405	+	19,1°	−	19,6°
365	+	24,9°	−	25,9°

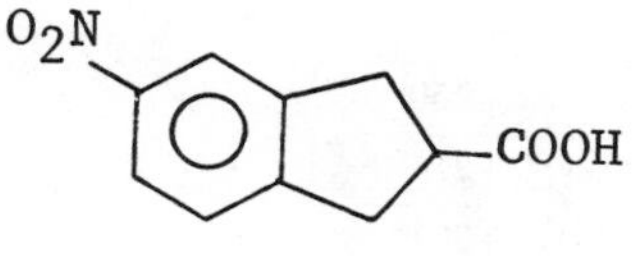

$C_{10}H_9NO_4$ **Ref. 630**

Resolution of 5-Nitrohydrindene-2-carboxylic Acid.

To an ethyl-alcoholic solution of nitrohydrindenecarboxylic acid (10·5 grams) and anhydrous quinine (16·4 grams) water was added in amount just insufficient to produce a permanent turbidity at the temperature of the room, the total volume of the solution being 400 c.c. Nuclei, obtained by cooling a small portion of the solution to 0° and rubbing with a glass rod, were then added, and the salt slowly crystallised. It was then recrystallised eight times from aqueous alcohol. The melting point of the successive crystallisations showed little variation, the first crystallisation having melted at 103—108°, and the eighth at 104—107°, but the magnitude of the specific rotation increased, and test portions of the acid isolated from the successive crops were increasingly lævorotatory, the specific rotation of the acid from the fourth, fifth, sixth, and eighth crystallisations having been found to be $[\alpha]^{20}_D$ −19·7°, −26·7°, −27·1°, −26·6° respectively. From the last three values it was concluded that the resolution had approximately reached its limit at the fifth crystallisation. The quinine salt contained two molecules of the l-acid combined with one of quinine. This was established by titration with N/10-alkali and phenolphthalein, for a nitrogen determination would not have discriminated between a mono- and a di-acid salt:

0·6814 required 18·6 c.c. N/10-sodium hydroxide. Acid = 56·5.
$C_{20}H_{24}O_2N_2,2C_{10}H_9O_4N$ requires acid = 56·1 per cent.

The following polarimetric observations were made on the seventh and eighth crystallisations of the quinine salt:

VII. $l=2$; $c=0·464$; α^{20}_D −0·945°; α^{20}_{5461} −1·135°; $[\alpha]^{20}_D$ −101·8°; $[\alpha]^{20}_{5461}$ −122·3°.
VIII. $l=4$; $c=0·24$; α^{20}_D −0·98°; α^{20}_{5461} 1·17°; $[\alpha]^{20}_D$ −102·1°; $[\alpha]^{20}_{5461}$ −121·9°.

The quinine salt obtained after the final recrystallisation was

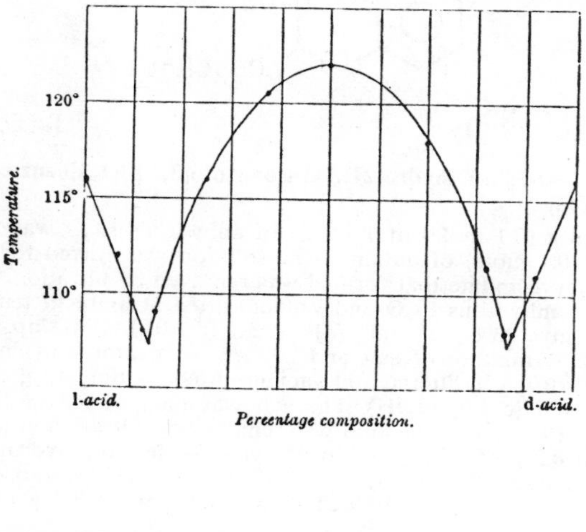

decomposed with sulphuric acid, and the *l*-acid was extracted with ethyl acetate, in which it is considerably more soluble than in ether. The solution was washed repeatedly with dilute sulphuric acid in order to remove the quinine. The acid was crystallised twice from water, and the two crops were polarimetrically examined.

I. $l=2$; $c=1{\cdot}752$; α_D^{17} $-1{\cdot}015°$; α_{5461}^{17} $-1{\cdot}28°$; $[\alpha]_D^{17}$ $-29{\cdot}0°$; $[\alpha]_{5461}^{17}$ $-36{\cdot}5°$.

II. $l=2$; $c=1{\cdot}022$; α_D^{16} $-0{\cdot}59°$; α_{5461}^{16} $0{\cdot}74°$; $[\alpha]_D^{16}$ $-28{\cdot}9°$; $[\alpha]_D^{16}$ $-36{\cdot}2°$.

The lævo-acid melts at 116°, 6·5° lower than the racemic compound.

To obtain the dextro-modification, the acid was extracted from the mother liquors of the quinine salt and crystallised from benzene. The crystals deposited directly had $[\alpha]_{5461}^{20}$ 18·5°, whilst the specific rotation of the second crop, which crystallised on the addition of light petroleum, was $[\alpha]_{5461}^{20}$ 32·5°. The *r*-acid consequently appears to be very much less soluble in benzene than the active acids.

The second crop, which, according to its specific rotation, must have contained nearly 90 per cent. of *d*-acid, was then crystallised twice from toluene. The melting point rose to 116°, and the specific rotation was, within the limits of the experimental error, equal and opposite to that found for the *l*-acid: $l=4$; $c=0{\cdot}405$; α_D^{20} 0·48°; α_{5461}^{20} 0·59°; $[\alpha]_D^{20}$ 29·6°; $[\alpha]_{5461}^{20}$ 36·4°.

The melting-point composition curve of mixtures of the *d*- and *l*-modifications was also investigated. It was found that the mixture of the enantiomerides in equal parts melted at 122°. The racemic acid was then melted together with different proportions of the active modifications, and the melting point of the powdered mixtures was determined in narrow glass tubes, the temperature at which the last portions of the solid were on the point of disappearing being taken as the melting point. The results are represented on the accompanying diagram, from which it may be seen that each of the two eutectics melts at 107·5°, and contains 11·7 per cent. of the one enantiomeride and 88·3 per cent. of the other.

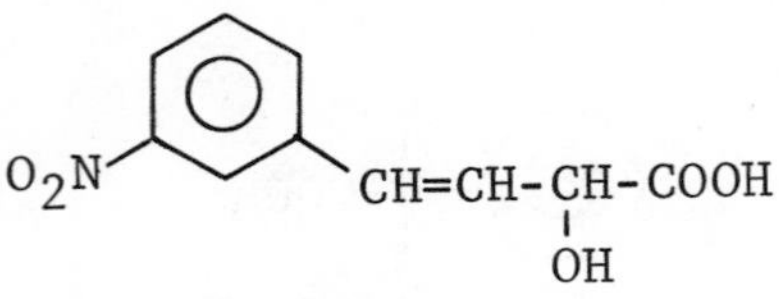

$C_{10}H_9NO_5$ Ref. 631

(+)-[3-Nitro-styryl]-glykolsäure: Zur Spaltung in die optischen Antipoden wurde die rohe Säure in möglichst wenig Methanol gelöst mit einer äther. Lösung von (+)-Bornylamin versetzt. Diese Lösung wurde auf −20° abgekühlt, wobei etwa 70% des Salzes mit einer Drehung von $[\alpha]_D^{20}$: +14.0° auskristallisierten. Die Mutterlauge mußte verworfen werden, weil sie die gesamten Verunreinigungen der Ausgangssäure enthielt. Das Salz wurde bei Zimmertemperatur in Methanol gelöst und durch Kühlen der Lösung auf −20° wieder zur Kristallisation gebracht. Es gelang so, die Drehung auf $[\alpha]_D^{20}$: −24.0° (in Methanol) zu steigern. Die Kristallisation des diasteromeren Salzes gelang wegen der starken Verharzung nicht.

Das Bornylaminsalz wurde mit Salzsäure zerlegt; hierbei schied sich die Säure als schwach rosa gefärbter Kristallbrei ab, der gründlich mit Eiswasser gewaschen wurde. Schmp. 112–117° (Zers.). $[\alpha]_D^{20}$: +71.0°; $[M]_D^{20}$: +158.5°.

$C_{10}H_9O_5N$ (223.2) Ber. N 6.28 Gef. N 6.30

Tafel 5. Rotationsdispersion [M] der (+)-[3-Nitro-styryl]-glykolsäure in verschiedenen Lösungsmitteln bei 20°

Lösgm.	$[M]_C$	$[M]_D$	$[M]_{HgI}$	$[M]_F$
Methanol (c = 0.5)	+128°	+158.5°	+208°	+279°
Äthanol (c = 0.5)	+114°	+139.5°	+174.5°	+250°
Dioxan (c = 0.5)	+ 95°	+115.5°	+143°	+214°
Eisessig (c = 0.5)	+139°	+176°	+214°	+301°
Aceton (c = 0.5)	+107°	+143°	+173°	+243°
NaOH (c = 0.5)	+ 82°	+100°	+122°	+176°

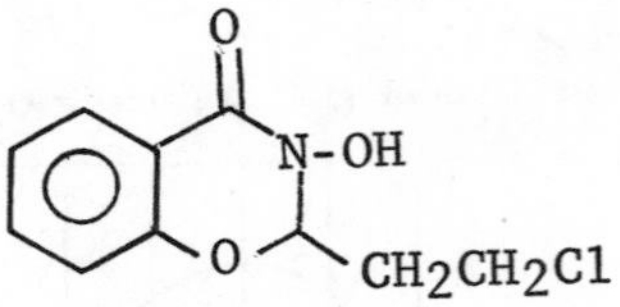

$C_{10}H_{10}ClNO_3$ Ref. 632

5.3.3. (—)-3,3a-Dihydro-2H,9H-isoxazolo[3,2-b][1,3]benzoxazin-9-one (45)

To 22.8 g (0.1 mole) of **1** in 1 l of anhydrous Et_2O was added 16.2 g (0.1 mole) of quinine. The solution was stirred for 1.5 h and the precipitate that formed was removed by filtration, washed with anhydrous Et_2O and vacuum-dried at ambient temperature to give 23.1 g of salt, $[\alpha]^{24}$—137° (c 1.0, EtOH), mp. 101—104° C. A mixture of salt and 200 ml of trichloroethylene was stirred for 0.5 h, filtered and vacuum-dried yielding 18 g of salt, $[\alpha]^{24}$ —45° (c 1.0, EtOH). The salt was then stirred in 100 ml of dilute HCl, and the solid was removed by filtration and dried to give 6.2 g of (—)-2-(2-chloroethyl)-2,3-dihydro-3-hydroxy-4H-1,3-benzoxazin-4-one, $[\alpha]^{24}$ —23.6° (c 5.0, $CHCl_3$), mp. 108—110° C. The solid was treated with a solution of 2.2 g of 50% NaOH in 50 ml of H_2O at room temperature for 0.5 h, and the mixture was filtered. The precipitate was washed with H_2O and air-dried. The crude product, $[\alpha]^{24}$ —332° (c 1.0, $CHCl_3$) weighed 4.5 g and after recrystallization from toluene (charcoal), the desired (—)-isomer weighed 2.5 g, $[\alpha]^{24}$ —414° (c 1.0, $CHCl_3$) and mp. 108—109.5° C.

5.3.4. (+)-3,3a-Dihydro-2H,9H-isoxazolo[3,2-b][1,3]benzoxazin-9-one (46)

A solution of 16 g of quinine in 150 ml of trichloroethylene was added to a warm solution of 11.5 g of **1** in 350 ml of Et_2O. On cooling a precipitate formed which was removed by filtration. The filtrate was evaporated to dryness in vacuo at room temperature and the residue was treated with dilute HCl. The mixture was stirred and filtered. The solid was washed with H_2O, air-dried and recrystallized twice from cyclohexane yielding 1.2 g of solid, $[\alpha]^{24}$ + 21.8° (c 5.0, $CHCl_3$), mp. 108—110° C. 0.7 g (0.003 mole) of the solid was added to an aqueous solution containing 0.12 g (0.003 mole) of NaOH and the mixture was stirred at room temperature for about 0.5 h. The solid was removed by filtration, washed with H_2O and air-dried to yield the (+)-isomer of **16** $[\alpha]^{24}$ + 417° (c 1.0, $CHCl_3$) and mp. 108—110° C.

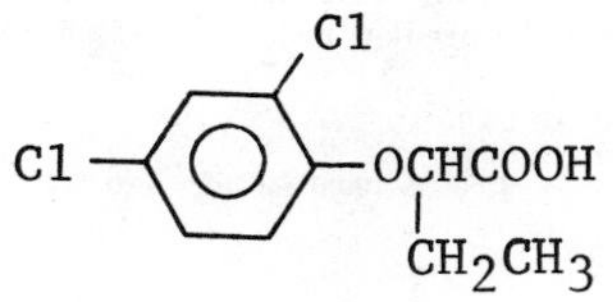

$C_{10}H_{10}Cl_2O_3$ **Ref. 633**

(+)-α-(2.4-Dichlorophenoxy)-*n*-butyric acid. Racemic acid (15.0 g 0.06 mole) and trychnine (20.1 g, 0.06 mole) were dissolved in hot alcohol (300 ml). Water (400 ml) vas added and the solution was allowed to crystallise in a cool place over night. The salt was recrystallised according to Table 4, where $[\alpha]_D$ refers to the rotatory ower of the acid in alcohol.

Table 4. Recrystallisation of the strychnine salt of α-(2.4-dichlorophenoxy)-*n*-butyric acid.

Cryst. No.	Solvent (ml)		Weight of salt (g)	$[\alpha]_D$ of the acid
	Ethanol	Water		
1	300	400	31.2	—
2	300	400	12.8	—
3	150	200	10.6	+ 40.2°
4	125	175	8.75	+ 41.6°
5	125	175	8.25	+ 41.4°

The last salt fraction was treated with hydrochloric acid and the organic acid taken ıp in ether, which was driven off. The crude acid (3.5 g) was recrystallised from etroleum (b.p. 60–80°). The pure (+)-acid (3.35 g) melted at 91–91.6°.

95.97 mg (+)-acid: 7.855 ml 0.04888-*N* NaOH
$C_{10}H_{10}O_3Cl_2$ calc. equiv. wt. 249.1
found » » 249.9

0.2052 g dissolved in abs. *alcohol* to 10.00 ml: $\alpha^{25}_D = +0.858°$. $[\alpha]^{25}_D = +41.8°$; $[M]^{25}_D = +104.2°$.
0.2279 g » » *acetone* to 20.06 ml: $2\alpha^{25}_D = +1.183°$. $[\alpha]^{25}_D = +52.1°$; $[M]^{25}_D = +129.7°$.
0.1846 g » » *chloroform* to 20.06 ml: $2\alpha^{25}_D = +0.650°$. $[\alpha]^{25}_D = +35.3°$; $[M]^{25}_D = +88.0°$.
0.2428 g » » *benzene* to 20.06 ml: $2\alpha^{25}_D = -0.179°$. $[\alpha]^{25}_D = -7.4°$; $[M]^{25}_D = -18.4°$.
0.2116 g neutralised with *aqueous NaOH* and made up to 20.06 ml: $2\alpha^{25}_D = +0.118°$. $[\alpha]^{25}_D = +5.6°$; $[M]^{25}_D = +13.9°$.

(–)-α-(2.4-Dichlorophenoxy)-*n*-butyric acid. From the mother liquor 1 above, a ævorotatory acid ($[\alpha]_D = -19°$ in alcohol) was recovered. This acid (7.2 g, 0.029 nole) and (–)-α-phenethylamine (3.5 g, 0.029 mole) were dissolved in hot, 30% lcohol. The solution was set aside in a refrigerator and the crystals were collected ext day. The purification of the salt was carried out as seen from Table 5. The otatory power of the acid isolated from small samples of the salt fractions was neasured in alcohol as usual.

Table 5. Recrystallisation of the (–)-α-phenethylamine salt of α-(2.4-dichlorophenoxy)-*n*-butyric acid.

Cryst. No.	Solvent (ml)		Weight of salt (g)	$[\alpha]_D$ of the acid
	Ethanol	Water		
1	45	110	6.2	– 29.9°
2	25	65	5.0	– 35.9°
3	25	65	4.3	– 36.1°
4	35	50	2.70	– 36.2°

The salt fraction 4 was decomposed and the acid liberated in the usual way. The rude acid (2.0 g) was recrystallised from petroleum (b.p. 60–80°). 1.1 g of the –)-acid was collected as very small, white needles, m.p. 90.5–91.5°.

105.85 mg (–)-acid: 8.660 ml 0.04888-*N* NaOH
$C_{10}H_{10}O_3Cl_2$ calc. equiv. wt. 249.1
found » » 250.1

0.1354 g dissolved in abs. *alcohol* to 10.02 ml: $\alpha^{25}_D = -0.559°$. $[\alpha]^{25}_D = -41.4°$; $[M]^{25}_D = -103.1°$.

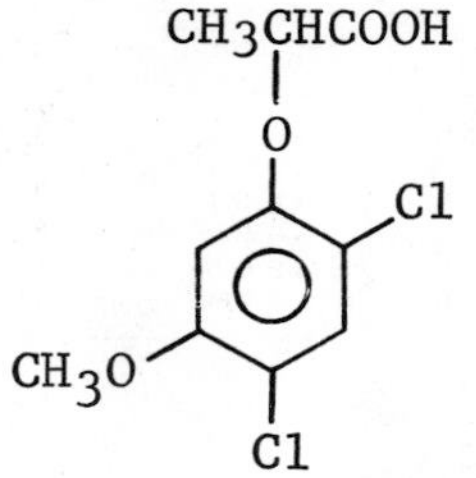

$C_{10}H_{10}Cl_2O_4$ **Ref. 634**

Preliminary experiments on resolution were performed in dilute ethanol (in several cases also in other solvents) with a number of alkaloids and synthetic bases. Crystallised salts were obtained from strychnine, cinchonidine, quinidine, quinine and α-phenylisopropylamine but the liberated acid was in all cases racemic. The salts from brucine and dehydroabietylamine failed to crystallise. Only cinchonine and optically active α-phenylethylamine gave salts containing optically active acids. The former salt had a quite strong tendency to separate as an oil and gave a dextrorotatory acid of a rather low activity. The (–)-α-phenylethylamine gave a laevorotatory acid of a somewhat higher activity (about – 17°).

Thus, (–)- and (+)-α-phenylethylamine were selected for the preparation of the optically active acids.

(–)-2,4-Dichloro-5-methoxyphenoxy-propionic acid

An experiment was started on a moderate scale with 0.05 mole of the racemic acid. This experiment showed that 96% ethanol as solvent gives better resolution than 50 or 75% ethanol. The acid used for the final resolution was a mixture of racemic acid and (–)-acid regenerated from the preliminary experiments.

This acid (23.0 g; 0.09 mole) and the (–)-α-phenylethylamine (10.5 g; 0.09 mole) were dissolved in 50 ml 96% ethanol solution. After standing over-night in the refrigerator, the salt was filtered off and recrystallised to constant activity of the acid. The progress of the resolution was tested on small samples of each crystallisation from which the acid was liberated and measured in acetone solution. The salt was decomposed with dilute sulphuric acid and the organic acid was extracted with ether. The course of the resolution was followed as described below.

Crystallisation	1	2	3	4	5
ml ethanol	50	60	30	30	30
g salt obtained	26.4	8.5	7.0	6.7	5.4
$[\alpha]^{25}_D$ of the acid	– 7°	– 46.5°	– 51.5°	– 52.5°	– 52°

The low activity in the first crystallisation compared with the high activity in the second is notable. Probably the resolution depends on different rates of crystallisation rather than on difference in solubility.

6.4 g salt gave 4.2 g (–)-acid, which was recrystallised three times from dilute formic acid (once with charcoal) and twice from toluene-petroleum ether. The yield of the pure acid was 3.0 g. It forms very small needles with m.p. 115.5–116.5°C.

$C_{10}H_{10}O_4Cl_2$ (265.09)
Calc. C 45.31 H 3.80 Cl 26.75
Found C 45.24 H 3.76 Cl 26.92
0.1125 g acid dissolved in acetone to 10.00 ml: $\alpha^{25}_D = -0.594°$ (1 dm); $[\alpha]^{25}_D$ – 52.8°; $[M]^{25}_D = -140.0°$.

(+)-2,4-Dichloro-5-methoxyphenoxy-propionic acid

The ethanolic mother liquors from the first and second crystallisations and from the first experiment were evaporated to dryness. From the salt obtained, 17.9 g of crude dextrorotatory acid of moderate activity was isolated. This acid (0.07 mole) was dissolved with 8.2 g (+)-α-phenylethylamine in 50 ml ethanol.

Crystallisation	1	2	3	4
ml ethanol	50	25	20	20
g salt obtained	13.9	11.1	8.5	7.2
$[\alpha]^{25}_D$ of the acid	+ 20.5°	+ 42.5°	+ 51°	+ 52°

7.2 g salt gave 4.9 crude acid. To get a better yield the mother liquor from the first crystallisation of the salt was evaporated to dryness, and the residue was recrystallised from the successive mother liquors after the original crystallisation. This gave a second crop of (+)-salt (3.7 g). In all 7.3 g (+)-acid was obtained, which was recrystallised three times, twice from dilute formic acid (once with charcoal) and once from toluene-petroleum ether. The yield, forming small glistening needles, was 6.3 g melting 115.5–116.5°C.

Table 1

g = g acid in 10.00 ml solution

Solvent	g	α_D^{25} (1 dm)	$[\alpha]_D^{25}$	$[M]_D^{25}$
Acetone	0.1065	+0.563°	+52.9°	+140.2°
Ethanol (abs.)	0.1214	+0.616°	+50.7°	+134.4°
Methanol (abs.)	0.1026	+0.464°	+45.2°	+119.9°
Chloroform	0.1102	+0.408°	+37.0°	+ 98.1°
Benzene	0.1030	+0.166°	+16.1°	+ 42.7°
Water (neutr.)	0,1015	+0.148°	+14.6°	+ 38.7°

$C_{10}H_{10}O_4Cl_2$ (265.09)
Calc. C 45.31 H 3.80 Cl 26.75
Found C 45.40 H 3.87 Cl 26.77

The values of the rotatory power in various solvents are given in Table 1.

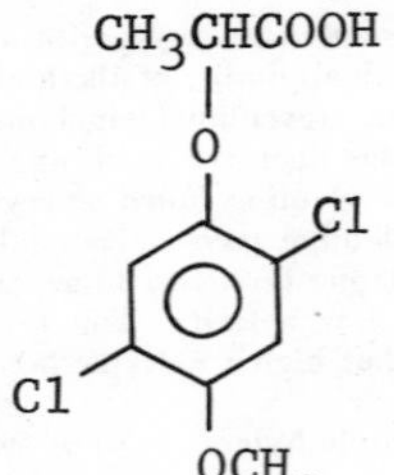

$C_{10}H_{10}Cl_2O_4$ Ref. 635

Preliminary experiments on resolution were performed in dilute ethanol with a number of alkaloids and synthetic bases. Cinchonine and quinidine gave laevorotatory acids of low activity. Strychnine, (+)-α-phenylethylamine and (+)-α-phenylisopropylamine gave laevorotatory acids of higher activity (about −48°) and fairly good yield (40−45%). Cinchonidine and quinine gave dextrorotatory acids of lower activity but (−)-α-phenylethylamine and (−)-α-(2-naphtyl)-ethylamine gave acids of high activity (about +45°) and the yield of the salts was in both cases nearly 50%.

For practical reasons the alkaloid strychnine and the synthetic base (−)-α-phenylethylamine were chosen for the resolution.

(−)-2,5-Dichloro-4-methoxyphenoxy-propionic acid

Racemic acid (26.5 g; 0.1 mole) and strychnine (33.4 g; 0.1 mole) were dissolved in a hot mixture of 300 ml ethanol and 100 ml water. The salt crystallised readily and was recrystallised as seen below.

Crystallisation	1	2	3	4
ml ethanol	300	120	185	130
ml water	100	40	—	—
g salt obtained	26.9	24.3	22.4	21.4
$[\alpha]_D^{25}$ of the acid	−51°	−55.5°	−55.5°	−55.5°

The strychnine salt was decomposed by excess sodium carbonate solution and the strychnine was taken up in chloroform. The aqueous solution was acidified with dilute sulphuric acid and the organic acid was taken up in ether. After evaporation of the ether, the residue, 9.0 g (−)-acid, was recrystallised three times from dilute formic acid (once with charcoal) and twice from toluene-petroleum ether. The yield of the pure (−)-acid was 7.8 g with a m.p. of 122.5–123.0°C, forming slender glistening needles.

$C_{10}H_{10}O_4Cl_2$ (265.09)
Calc. C 45.31 H 3.80 Cl 26.75
Found C 45.38 H 3.79 Cl 26.85

The rotatory power in various solvents is given in Table 2.

Table 2

g = g acid in 10.00 ml solution

Solvent	g	α_D^{25} (1 dm)	$[\alpha]_D^{25}$	$[M]_D^{25}$
Acetone	0.1118	−0.642°	−56.5°	−149.8°
Ethanol (abs.)	0.1010	−0.565°	−56.0°	−148.5°
Methanol (abs.)	0.1034	−0.465°	−45.0°	−119.2°
Chloroform	0.1056	−0.379°	−35.9°	− 95.2°
Benzene	0.1290	−0.174°	−13.5°	− 35.8°
Water (neutr.)	0.1085	−0.090°	− 8.3°	− 22.0°

(+)-2,5-Dichloro-4-methoxyphenoxy-propionic acid

13.2 g acid (0.05 mole) having $[\alpha]_D^{25}$ of about +35° was obtained by isolating the acid from several mother liquors in earlier experiments. This acid and 6.01 g (−)-α-phenylethylamine were dissolved in 150 ml 96% ethanol. On cooling and efficient stirring, the salt separated easily. It was filtered off after standing over-night in the refrigerator. The salt was recrystallised once more to ensure maximum activity of the acid.

Crystallisation	1	2
ml ethanol	150	150
g salt obtained	17.0	16.0
$[\alpha]_D^{25}$ of the acid	+55.5°	+55.5°

The salt was decomposed with dilute sulphuric acid and the organic acid isolat by extraction with ether. The yield of the crude acid was 10.9 g, which was recr tallised twice from dilute formic acid and once from toluene-petroleum ether. T acid formed small glistening needles, melting at 122.5–123.0°C. Yield 9.8 g.

$C_{10}H_{10}O_4Cl_2$ (265.09)
Calc. C 45.31 H 3.80 Cl 26.75
Found C 45.44 H 3.92 Cl 26.83
0.1037 g acid dissolved in acetone to 10.00 ml: α_D^{25} = +0.587° (1 dm); $[\alpha]_D^{25}$ = +56.6°; $[M]_D^{25}$ +150.0°

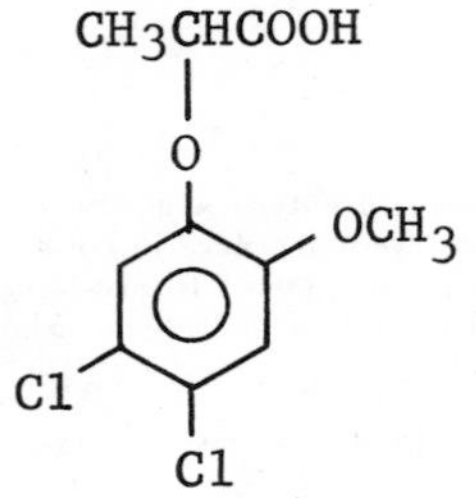

$C_{10}H_{10}Cl_2O_4$ Ref. 6

Resolution of the racemic acid

Preliminary experiments were carried out in the usual way. The optically act α-(2-naphthyl)-ethylamine and α-phenylethylamine gave crystalline salts and go resolution. The solubility of the naphthylethylamine salt was too low for conveni recrystallization, and the acid regenerated from the quinine, cinchonine and bruc salts had too low an activity.

A mixture of 11.2 g (0.040 mole) of racemic acid and 4.9 g (0.040 mole) of (−) phenylethylamine was dissolved in 400 ml of ethyl acetate. The salt was allowed crystallize at room temperature and then, after standing overnight in a refrigera it was filtered off, dried and weighed. From a small sample of the salt, the acid v liberated and its optical rotation determined in absolute ethanol. The salt was crystallized several times and the progress of the resolution is shown in Table 1

The combined mother liquors from the three first crystallizations were evapora to dryness and 7.7 g of acid isolated with $[\alpha]_D^{25}$ = −38° (absolute ethanol). This a was dissolved with 3.4 g (+)-α-phenylethylamine in ethyl acetate, and the s recrystallized as described above. The results are given in Table 2.

Table 1. Recrystallization of the (−)-α-phenylethylamine salt of α-(2-naphthy leno)-propionic acid.

Cryst. no.	Ethyl acetate (ml)	Weight of salt (g)	$[\alpha]_D^{25}$ of acid
1	400	8.8	+23°
2	300	6.0	+46°
3	300	3.8	+89°
4	200	3.3	+98°
5	200	2.9	+98°

Table 2. Recrystallization of the (+)-α-phenylethylamine salt of α-(2-naphthyl leno)-propionic acid.

Cryst. no.	Ethyl acetate (ml)	Weight of salt (g)	$[\alpha]_D^{25}$ of acid
1	400	5.5	−77°
2	300	4.6	−87°
3	300	3.9	−96°
4	200	3.6	−99°
5	200	3.2	−98°

(+)-α-(2-Naphthylseleno)-propionic acid

The acid was liberated from 2.8 g of the (−)-phenylethylamine salt using dilu sulphuric acid and extracted with ether. The yield of crude acid was 1.9 g. After tv recrystallizations from cyclohexane 1.5 g remained; m.p. 79–80°.

Anal. $C_{13}H_{12}O_2Se$: Equiv.wt. calc. 279.2; found 279.0.
0.1010 g dissolved in absolute ethanol to 10.00 ml: 2 α_D^{25} −1.98°; $[\alpha]_D^{25}$ −98.0°; $[M]_D^{25}$ −274°

(−)-α-(2-Naphthylseleno)-propionic acid

The (+)-phenylethylamine salt (3.0 g) was decomposed and the organic ac isolated and recrystallized as for the (+)-antipode. The yield of crude acid was 2.1 and after two recrystallizations 1.7 g remained; m.p. 79–80°.

Anal. $C_{13}H_{12}O_2Se$: Equiv.wt. calc. 279.2; found 280.5.

Optical activity in different solvents

Weighed amounts of the acid were dissolved in different solvents and made to 10.00 ml in each case. The rotatory power of these solutions was measured a the results are given in Table 3.

The solvent effect is similar to that of α-(2-naphthylthio)-propionic acid [3].

Table 3. Optical activity of (−)-α-(2-naphthylseleno)-propionic acid.

Solvent	mg of acid	$2\alpha_D^{25}$	$[\alpha]_D^{25}$	$[M]_D^{25}$
Benzene	102.3	−2.42°	−118°	−329°
Acetone	104.3	−2.40°	−115°	−321°
Chloroform	101.7	−2.04°	−100°	−279°
Ethanol	101.7	−2.00°	− 98.3°	−274°
Water (ion)	102.4	−0.93	− 45°	−126°

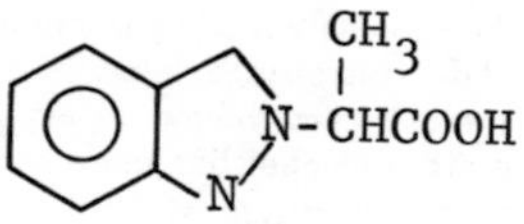

$C_{10}H_{10}N_2O_2$ Ref. 637

Preliminary tests on resolution

Preliminary experiments were carried out with the common alkaloids and some synthetic bases in different solvents. 0.001 mole acid and 0.001 mole base were dissolved in small amounts of solvent. Crystals formed were filtered off, the acid liberated with hydrochloric acid and extracted with ether. The ether was evaporated and the optical activity of the acid obtained was measured in dimethyl sulfoxide. Best results were obtained with (−)-α[2-naphthyl]ethylamine in ethanol and cinchonidine in ethyl acetate-ethanol (3:1) which both yielded $[\alpha]_D^{25}$ −13° and (+)-α-phenethylamine in ethyl acetate yielding $[\alpha]_D^{25}$ +7°.

According to this result the resolution was performed with cinchonidine in ethyl acetate-ethanol (3:1) and (−)-α-phenethylamine in ethyl acetate.

(+)-α[*Indazolyl-2*]*propionic acid.* 25 g (0.13 mole) acid and 15.9 g (0.13 mole) (+)-α-phenethylamine [8] were dissolved in 330 ml hot ethyl acetate. The salt was allowed to crystallize at room temperature and then in a refrigerator overnight. The salt was then filtered off, dried and recrystallized several times from ethyl acetate. After each recrystallization the acid was liberated from 0.12 g of the salt and the optical activity determined in dimethyl sulfoxide. After seven such recrystallizations the maximum activity of the acid was obtained. (Table 1.)

Table 1.

Crystallization	1	2	3	4	5	6	7	8
Ethyl acetate, ml	330	230	150	150	40	40	60	45
g salt obtained	19.0	9.7	7.2	3.8	3.1	2.3	1.7	1.2
$[\alpha]_D^{25}$ of the acid	+5.4°	+10.0°	+11.0°	+14.2°	+17.5°	+17.5°	+20.6°	+20.0°

Table 2.

g = g acid in 10.00 ml solution. The tube length is 1 dm.

Solvent	g	α_D^{25}	$[\alpha]_D^{25}$	$[M]_D^{25}$
DMF	0.0869	+0.343°	+39.5°	+75.1°
DMSO	0.0395	+0.084°	+21.3°	+40.5°
Glacial acetic acid	0.0488	+0.021°	+4.3°	+8.2°
Ethanol (Abs.)	0.0410	−0.014°	−3.4°	−6.5°

Table 3.

Crystallization	1	2	3	4	5	6	7
Ethyl acetate-ethanol (3:1), ml	100	93	80	75	45	25	15
g salt obtained	11.0	7.6	5.0	3.2	1.9	1.2	0.77
$[\alpha]_D^{25}$ of the acid	−14.7°	−15.2°	−17.9°	−19.4°	−20.0°	−20.4°	−20.6°

The acid was liberated from the salt by addition of dilute sodium hydroxide. (+)-α-Phenethylamine was removed by extraction with ether. The water phase was acidified to pH 3.5 with 2 *M* hydrochloric acid and the precipitated acid was filtered off, dried and recrystallized from ethanol-water (2:1). 0.60 g of pure acid forming white glistening plates with m.p. 220–220.5°C was obtained. The rotatory power of the acid in various solvents is given in Table 2.

$C_{10}H_{10}N_2O_2$
Calc. C 63.14 H 5.30 N 14.73
Found C 63.31 H 5.41 N 14.34

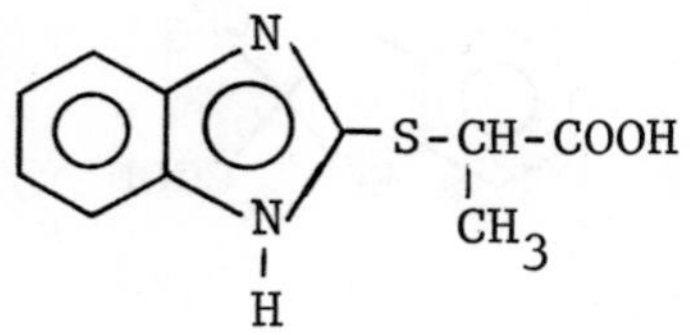

$C_{10}H_{10}N_2O_2S$ Ref. 638

Preliminary experiments on resolution. Salts of common alkaloids were prepared on a 0.001 mol scale in dilute ethanol. The quinine salt gave the acid with $[\alpha]_D^{25}$ −115° while the acid from the quinidine salt had $[\alpha]_D^{25}$ +107°. Strychnine and cinchonidine gave acid with very low rotation. The cinchonine salt was obtained in a very low yield and the brucine salt failed to crystallise.

(−)-α-(Benzimidazolyl-2-thio)-propionic acid. Racemic acid (26.6 g, 0.12 mol) was dissolved in 350 ml of 96 % ethanol and quinine (38.9 g, 0.12 mol) in 100 ml of the same solvent was added. After standing overnight, the salt (34.9 g) was filtered off: $[\alpha]_D^{25}$ of the acid −112°. Since the salt was only sparingly soluble, it was divided into two portions of 17 g each. The first portion was recrystallised as shown in Table 1.

The second portion of the salt was treated in the same way. Systematic fractionation of the salt from the mother liquors gave a third crop. The total yield of salt containing acid of maximum activity was 21.5 g. It was decomposed with excess aqueous ammonia, the quinine extracted with chloroform and the remaining solution acidified with hydrochloric acid to pH 3. The crude product (7.4 g) was recrystallised from 96 % ethanol, yielding 6.7 g of colourless needles with m.p. 176 − 177 °C: Equiv.wt. Found: 222.2. Calc. for $C_{10}H_{10}N_2O_2S$, 222.3. $[\alpha]_D^{25}$ −124°; $[M]_D^{25}$ −276° (*c* 0.522, 96 % ethanol).

(+)-α-(Benzimidazolyl-2-thio)-propionic acid. The various mother liquors from the quinine salt yielded 11.2 g of acid having $[\alpha]_D^{25}$ +119°. This acid (0.05 mol) was dissolved with quinidine (16.2 g, 0.05 mol) in 350 ml of ethanol and 50 ml of water were added. After evaporation to half the volume, the solution yielded 9.7 g of salt containing acid with $[\alpha]_D^{25}$ +123°. Recrystallisation from 80 ml of 75 % ethanol gave 6.9 g salt with acid of $[\alpha]_D^{25}$ +124°. Systematic fractionation of the salt from the mother liquors gave a further 13.8 g salt with acid of maximum activity. The acid was isolated and recrystallised as described for the enantiomer. The salt (19.0 g) gave 6.5 g of pure acid, m.p. 176 − 177 °C. Equiv.wt. Found: 222.1. Calc. for $C_{10}H_{10}N_2O_2S$: 222.3. $[\alpha]_D^{25}$ +124°; $[M]_D^{25}$ +276° (*c* 0.488, 96 % ethanol).

Table 1. Recrystallisation of the quinine salt from dilute ethanol to constant activity of the acid.

Re-cryst.	Ethanol/ml/	Water/ml	Salt obtained/g	$[\alpha]_D^{25}$/°
1	380	190	11.8	−119
2	330	170	8.7	−123
3	230	120	6.6	−124
4	190	90	5.2	−124

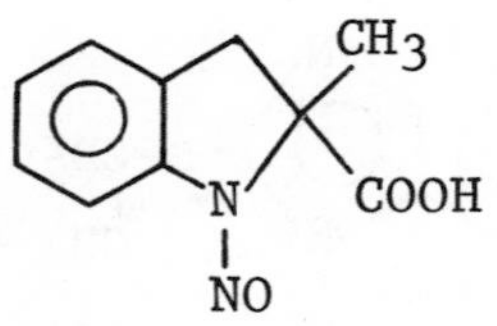

$C_{10}H_{10}N_2O_3$ Ref. 639

Resolution of (±)-1-Nitroso-2-methylindoline-2-carboxylic Acid (13) with (−)-Ephedrine. Isolation of (*S*)(+)-16. To a solution of 15 g (0.073 mol) of the nitroso acid **13** in ethyl acetate (100 ml) was added 12 g (0.073 mol) of (−)-ephedrine dissolved in 30 ml of the same solvent. In a few minutes a light yellow solid crystallized which was collected on filter and washed with ethyl acetate to give 11 g of the salt, mp 155–158°. One recrystallization from methanol–ethyl acetate mixture followed by another from methanol–benzene mixture furnished 8 g (60%) of light pale crystals, mp 165–166°, $[\alpha]^{24}D$ −44.7° (*c* 1.3, ethanol).

Anal. Calcd for $C_{20}H_{25}N_3O_4$: C, 64.47; H, 6.78; N, 11.31. Found: C, 65.05; H, 7.00; N, 11.22.

To obtain (*S*)(+)-**16**, the ephedrine salt (8 g) was dissolved in 200 ml of 1% sodium hydroxide solution and extracted with three 50-ml portions of ether to remove ephedrine. The aqueous solution was acidified with hydrochloric acid to pH 4 and extracted with four 50-ml portions of ether. The organic solution was washed with water (20 ml), dried (magnesium sulfate), and evaporated *in vacuo* to furnish 4 g (91%) of pale crystals, mp 115–116°, $[\alpha]^{24}D$ +12° (*c* 2, ethanol). The infrared spectrum was identical with that of the racemic sample.

Resolution of (±)-13 with Brucine. Isolation of (*R*)(−)-14. Following the above procedure, the mother liquor gave 7.5 g of **13** enriched in the (−)-antipode. To a solution of this in ethyl acetate (100 ml) was added brucine (18 g) dissolved in 50 ml of the same solvent, and the mixture was warmed for a few minutes. A crystalline solid separated which was filtered after cooling to give 13 g of the salt, mp 160–165°. Two crystallizations from methanol–ethyl acetate mixture furnished 9.5 g (45%) of the salt of (*R*)(−)-**14** as a glistening solid, mp 175–176°, $[\alpha]^{24}D$ −55.1° (*c* 1.3, ethanol).

Anal. Calcd for $C_{33}H_{36}N_4O_7$: C, 65.99; H, 6.04; N, 9.33. Found: C, 65.92; H, 6.23; N, 9.24.

To obtain the free acid (*R*)(−)-**14**, the above salt (9.5 g) was suspended in 200 ml of 1% sodium hydroxide and stirred well. The insoluble liberated base was filtered and washed with water. The alkaline filtrate was repeatedly extracted with ether to remove traces of the alkaloid. The basic solution was acidified with hydrochloric acid to pH 4–5, and the liberated acid was extracted into three 100-ml portions of ether. The washed and dried (magnesium sulfate) extract on evaporation *in vacuo* gave 3 g (88%) of the acid (*R*)(−)-**14**, mp 114–115°, $[\alpha]^{24}D$ −12° (*c* 2, ethanol). Its infrared spectrum was identical with that of the antipode (*S*)(+)-**16**.

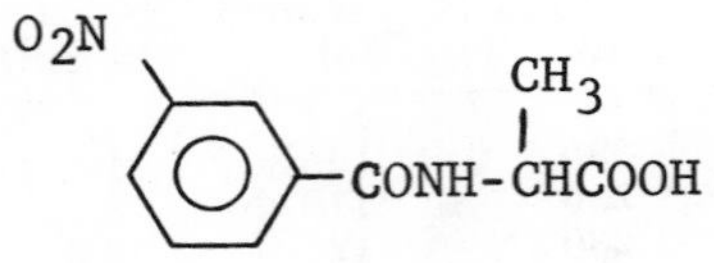

$C_{10}H_{10}N_2O_5$ Ref. 640

Resolution of dl-m-*Nitrobenzoylalanine*.—To a boiling solution containing the *dl*-acid (15 g.), sodium hydroxide (0·976*N*-solution, 32·4 c.c.) and water (1000 c.c.), quinine (11·92 g.) was added gradually and finally alcohol (10 c.c.). The clear solution was separated from a small quantity of insoluble gummy material and left to crystallise during 16 hours. The crystalline air-dried material (17 g.; theoretical wt., 19·5 g.), *lAlB* salt, was recrystallised three times from hot water (10 g. in 600 c.c.), but the rotatory power of the anhydrous salt in ethyl alcohol at 20° did not change after the first crystallisation: *c*, 1·829; *l*, 4; α_{5461} = —10·03°; $[\alpha]_{5461}$ = —137·1°.

The *quinine* salt of *l*-*m*-nitrobenzoylalanine crystallises from water in long soft needles containing $2H_2O$, which it loses in a vacuum over sulphuric acid in 5—6 days (Found in air-dried material: C, 60·6; H, 6·4; H_2O, 5·9, 6·0, 6·0. $C_{30}H_{34}O_7N_4,2H_2O$ requires C, 60·2; H, 6·35; H_2O, 6·0%). The anhydrous salt has no definite melting point, being converted into a glass-like mass at about 125°.

The mother-liquor after separation of the above salt was treated with excess of concentrated ammonia solution, and the alkaloid extracted by means of chloroform. The acid obtained by acidification of the evaporated aqueous solution was recrystallised from the minimum quantity of water; it had (as ammonium salt in water) $[\alpha]_{5461}$ = + 36·6° and contained 90·4% of the *d*-acid. This crude *d*-acid (23·4 g.) with brucine (38·3 g.) and sodium hydroxide (0·38 g.) was boiled with water (250 c.c.), complete solution being effected. The *dAlB* salt (58·2 g.) crystallised and was separated after some 16 hours. It was recrystallised three times from water (250 c.c.), its rotatory power being constant after the first crystallisation.

The *brucine* salt of *d*-*m*-nitrobenzoylalanine crystallises from water in colourless plates which turn yellow on exposure to light. It is sparingly soluble in ethyl alcohol but more soluble in acetone. In ethyl alcohol at 20° it has $[\alpha]_{5461}$ = − 9·1° (*c*, 0·5511; *l*, 4; α_{5461} − 0·20°). The air-dried salt appears to contain $3\frac{1}{2}H_2O$, which is lost in 7 days in a vacuum over sulphuric acid (Found in air-dried salt: C, 56·5; H, 6·1; H_2O, 8·95. $C_{33}H_{36}O_9N_4,3\frac{1}{2}H_2O$ requires C, 56·9; H, 6·2; H_2O, 9·1%). When the air-dried salt is heated at 110°, slight decomposition takes place, the loss in weight being 9·2%.

l-m-*Nitrobenzoylalanine*, obtained from the above quinine salt in the usual way, was recrystallised twice from water, in which it is more soluble than the *dl*-acid (Found: C, 50·8; H, 4·3. $C_{10}H_{10}O_5N_2$ requires C, 50·4; H, 4·2%). d-m-*Nitrobenzoylalanine*, similarly isolated from the above brucine salt, was also recrystallised from water (Found: N, 11·65. $C_{10}H_{10}O_5N_2$ requires N, 11·8%). These two acids both crystallise in long, almost colourless needles, m. p. 158°. The melting point of either is depressed by admixture with a small quantity of the other and since the *dl*-acid has m. p. 163—164°, the latter must be a racemic compound. The following determinations of rotatory powers were made at 20° in 4 dm. tubes.

l-*m*-Nitrobenzoylalanine,	*d*-*m*-Nitrobenzoylalanine,
as ammonium salt in water.	
c, 1·342; α_{5461}, − 2·40°; $[\alpha]_{5461}$, − 44·7°	*c*, 1·358; α_{5461}, + 2·43°; $[\alpha]_{5461}$, + 44·7°
in ethyl alcohol.	
c, 1·361; α_{5461}, + 0·32°; $[\alpha]_{5461}$, + 5·87°	*c*, 1·335; α_{5461}, − 0·30°; $[\alpha]_{5461}$, − 5·62°

The corresponding rotatory powers of the optically active *p*-nitrobenzoylalanines are $[\alpha]_{5461}$ = ± 51·65° and ± 15·8° respectively (Gibson and Colles, *loc. cit.*).

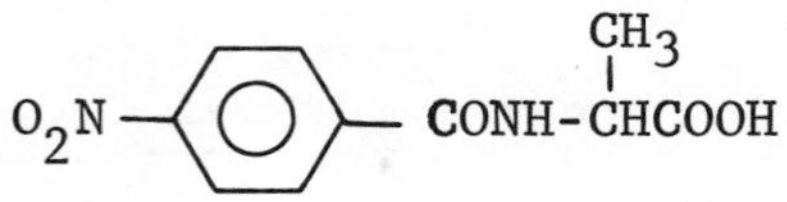

$C_{10}H_{10}N_2O_5$ Ref. 641

Resolution of dl-p-*Nitrobenzoylalanine*.—In a preliminary experiment with strychnine and *N*-sodium hydroxide (0·5 mol. of each) and the *dl*-acid (1 mol.), an acid, m. p. 167°, [α] + 13·52° * in alcoholic solution, was obtained from the salt that separated on cooling and had been once recrystallised from water. This salt, therefore, was the *dAlB* salt.

When brucine was used instead of strychnine, a readily soluble salt separated from the deep orange solution after considerable concentration at the ordinary temperature. The salt was recrystallised from hot water, and the acid liberated from a portion in the usual way. The acid had m. p. 194° and was optically inactive. The *brucine* salt of dl-p-*nitrobenzoylalanine* crystallises in thin, yellow plates which are readily soluble in water and in alcohol and appear to contain $4H_2O$ (Found: N, 8·3; H_2O, 10·8. $C_{33}H_{36}O_9N_4,4H_2O$ requires N, 7·95; H_2O, 10·2%).

By means of cinchonidine an almost insoluble salt (10·8 g. instead of 11·2 g.) was obtained which proved also to be the partial racemate. The *cinchonidine* salt of dl-p-*nitrobenzoylalanine* was recrystallised from water and obtained in colourless, fine needles containing $3H_2O$ (Found: N, 9·5; H_2O, 7·6. $C_{29}H_{32}O_6N_4,3H_2O$ requires N, 9·6; H_2O, 7·3%). The acids obtained from the recrystallised salt and from the mother-liquor were optically inactive.

* All rotatory powers were determined at 20° for the mercury-green (5461) line.

The salt obtained under similar conditions by means of cinchonine was also a partial racemate, and that obtained from quinine was an inseparable mixture of the two diastereoisomerides.

In the large-scale resolution, the *dl*-acid (100 g.; 1 mol.) and *N*-sodium hydroxide (210 c.c.; 0·5 mol.) were boiled with water (500 c.c.), strychnine (70·2 g.; 0·5 mol.) being added gradually, until complete solution was effected. The crystalline strychnine salt (123 g.) was filtered off after standing in the ice-chest for 16 hours. The mother-liquor (A) was worked up later.

The salt obtained was a mixture, a part of which was phototropic, turning deep yellow on exposure to daylight. The salt was recrystallised from 700 c.c. of boiling water and 111·4 g. crystallised. This in alcoholic solution had [α] — 11·19° (*c* = 1·4092). The rotatory power gradually diminished on recrystallisation and became constant after four recrystallisations under similar conditions. This constant value was confirmed after a fifth crystallisation; the rotatory power of the salt which crystallised was then identical with that of the salt obtained from the mother-liquor. Further crystallisation from absolute alcohol caused no alteration in the rotatory power. 47·7 G. of the pure strychnine salt of the *d*-acid were obtained.

The *strychnine* salt of d-p-*nitrobenzoylalanine*,

$CHMe(NH{\cdot}CO{\cdot}C_6H_4{\cdot}NO_2){\cdot}CO_2H,C_{21}H_{22}O_2N_2,1\frac{1}{2}H_2O$,

crystallises from hot water (solubility at 20°, 0·4; at 100°, 4) in colourless needles which are not phototropic (Found: H_2O, 4·1. Required for $1\frac{1}{2}$ H_2O: H_2O, 4·5%. Found in anhydrous material: N, 9·9. $C_{31}H_{32}O_7N_4$ requires N, 9·8%). Its rotatory power in ethyl alcohol was [α] — 1·90° (*c* = 1·5018, *l* = 4, α = — 0·114°). Other values obtained under similar conditions were [α] — 2·00° and — 1·83°. That the salt was pure was proved by the rotatory power of the acid obtained from it (see below). A solution of the anhydrous salt in absolute alcohol slowly deposits large, colourless prisms and the solubility of the salt in alcohol diminishes from 3·7 to 0·54 at the ordinary temperature. The prisms, kept in a vacuum desiccator or exposed to air, rapidly effloresce.

The acid liberated from the mother-liquor (A) in the usual way was recrystallised from 2400 c.c. of water. It then weighed 44·1 g. and had m. p. 189—190° (softening at 168°). In aqueous solution as the ammonium salt, it had [α] — 10·35° (*c* = 2·028, *l* = 4). It contained only 60% of the *l*-acid.

The mother-liquors from the first four crystallisations of the above strychnine salt were concentrated in two portions, and the strychnine salts separating on cooling collected. After repeated recrystallisation from boiling water, decolorisation with charcoal, and recrystallisation from absolute alcohol, 9·2 and 13·3 g., respectively, of samples of strychnine salt of identical and definitely constant rotatory power were obtained; [α] — 40·16°, — 40·71°, — 40·42° (*c* = 0·9130, 0·9828, 1·0222; *l* = 4 in ethyl alcohol). Although at the time it was believed that this was pure strychnine *l*-acid salt, subsequent work showed that it still contained *dAlB* salt.

In view of the difficulties indicated above, an approximate determination of the solubilities of the two diastereoisomeric salts in absolute alcohol was made. The *lAlB* salt was the less soluble in warm alcohol. At 50°, the solubilities in alcohol were: 2·68 for the *lAlB* and 3·64 for the *dAlB* salt. This difference in solubility, although small, indicated a method for the isolation of the pure salts.

The following conditions are typical of a series of experiments: To a solution of the *dl*-acid (10 g.; 1 mol.) in boiling alcohol (200 c.c.), strychnine (14·04 g.; 1 mol.) was added gradually. When the alkaloid had dissolved, the solution was placed in a thermostat at 50° and seeded with the purest strychnine *l*-acid salt available. A nodular deposit of radiating tufts of needles formed rapidly and the deposition was complete in 5 hours. The mother-liquor (B) was poured off and the deposit was washed with warm absolute alcohol and dried as rapidly as possible by suction (yield, 10—12 g.). The rotatory power in ethyl alcohol was [α] — 36·2° (*c* = 0·7916, *l* = 4). The mother-liquor (B) was treated as described later.

The purification of the above crude strychnine *l*-acid salt was very difficult. Successive fractional crystallisations from absolute alcohol at 50° was the only effective method; recrystallisation from water or from alcohol at the ordinary temperature was useless. The method was naturally wasteful of material. The acid liberated from a salt which had reached a rotatory power of [α] — 45·0° had [α] — 47·8° (as ammonium salt in water; *c* = 1·5040, *l* = 4) and contained 96·5% of the *l*-acid. A specimen of a salt which had a rotatory power of [α] — 46·6° gave an acid having [α] — 48·35° under similar conditions and containing 97·0% of *l*-acid. Finally, 42·5 g. of strychnine *l*-acid salt having [α] — 40·0°, after four fractional crystallisations at 50° from absolute alcohol, gave 9·0 g. of pure strychnine *l*-acid salt having a rotatory power identical with that of the salt remaining in the mother-liquor.

The *strychnine* salt of l-p-*nitrobenzoylalanine*,

$CHMe(NH{\cdot}CO{\cdot}C_6H_4{\cdot}NO_2){\cdot}CO_2H,C_{21}H_{22}O_2N_2,C_2H_5{\cdot}OH$,

crystallises from absolute alcohol in colourless needles which are not phototropic (Found: $C_2H_5{\cdot}OH$, 7·15. $C_{31}H_{32}O_7N_4,C_2H_5{\cdot}OH$ requires $C_2H_5{\cdot}OH$, 7·3%. Found in "anhydrous" substance: N, 9·95. $C_{31}H_{32}O_7N_4$ requires N, 9·8%). The "anhydrous" material in ethyl alcohol at 20° gave α — 1·433° (*c* = 0·7340, *l* = 4); whence [α] — 48·8°.

The resolution described above was frequently complicated by the separation, along with the nodular deposit, of large prisms (partial racemate) which were strongly phototropic. This invariably occurred when the solution was not seeded at once. When it did occur, the salt was redissolved and the experiment repeated. This explanation was proved correct by the following experiments: (a) 10 G. of the *dl*-acid were treated exactly as described above and the solution in the thermostat at 50° was seeded with the above-mentioned phototropic partial racemate. The crystalline deposit consisted of needles and prisms (no nodules), formed during 5 hours. The markedly phototropic crystals were separated as previously described. The rotatory power in ethyl alcohol (*c* = 0·8570, *l* = 4) was [α] — 26·6°. (b) 10 G. of the *dl*-acid were treated with the equivalent quantity of strychnine in double the quantity of alcohol and the solution was allowed to cool to the ordinary temperature. After 8 hours, 76% of the salt had crystallised and consisted of small, well-formed, phototropic, efflorescing prisms and needles. The rotatory power in ethyl alcohol (*c* = 1·2438, *l* = 4) was [α] — 24·06°. The deposit obtained from the mother-liquor on standing had [α] — 21·8° under the same conditions.

From the mother-liquor (B), allowed to cool to the ordinary temperature, a crystalline mixture of phototropic needles and non-phototropic, efflorescing prisms was deposited (8—10 g.). The rotatory power of an average specimen of this mixture was [α] — 15° (in alcohol). This salt therefore contained the greater proportion of the *dAlB* salt. Under the conditions of the first resolution, this would have required five successive recrystallisations from water for complete purification. The solubilities of the *dAlB* and the *lAlB* salt in alcohol at 20° being 0·69 and 0·91, respectively, this difference was used to obtain pure *dAlB* salt from the mixture. The mixture was dissolved in sufficient boiling absolute alcohol to make a 1·5% solution, and this was seeded with pure *dAlB* salt and kept for at least 24 hours. The crystalline deposit (45—50% yield) had an average rotatory power of [α] — 3° (in alcohol). One recrystallisation of this from boiling water was sufficient for complete purification. The mother-liquor was concentrated, and the crystalline salt which separated (30—35%) had an average rotatory power of [α] — 40° (in alcohol) and therefore contained an excess of the *lAlB* salt.

Optically Active p-*Nitrobenzoylalanines*.—d-p-*Nitrobenzoylalanine* was isolated from the pure strychnine *d*-acid salt by grinding it with a slight excess of dilute aqueous ammonia, filtering off the strychnine, and acidifying the filtrate, after thorough extraction with chloroform, with an excess of hydrochloric acid. The acid was recrystallised from boiling water (10 g. in 110 c.c.) and obtained in long, very pale yellow needles, m. p. 168·5—169° (Found: C, 50·8; H, 4·2; N, 11·8. $C_{10}H_{10}O_5N_2$ requires C, 50·4; H, 4·2; N, 11·8%). The rotatory power was (as ammonium salt) [α] + 51·40° in water (*c* = 1·7940, *l* = 4, α = + 3·688°) and [α] + 15·76° in ethyl alcohol (*c* = 1·9882, *l* = 4, α = + 1·254°).

The optical purity of this compound was proved by comparison with that of the *p*-nitrobenzoyl derivative of pure *d*-alanine (kindly supplied by Sir William Pope, F.R.S.) prepared by the method already described. This preparation (as ammonium salt) had [α] + 51·32° in water (*c* = 1·5964, *l* = 4, α = + 3·277°). The m. p. was 168·7—169° and was not affected by admixture with the above compound.

d-*p*-Nitrobenzoylalanine has a solubility in water at 15° of 0·26 and is easily soluble in alcohol. A mixture with a small quantity of pure *dl*-acid had m. p. 162—167°, and with a larger quantity 191·5°. A specimen having [α] + 15·15° (as ammonium salt in water) and therefore containing 64·7% of the pure *d*-acid softened at 165° and melted at 188·5—189·5°. Another specimen having [α] + 45·82° (as ammonium salt in water) and containing 94·6% of the pure *d*-acid had m. p. 167°.

l-p-*Nitrobenzoylalanine* was obtained in the same manner from the pure strychnine *l*-acid salt. It crystallised from water in long, pale yellow needles, m. p. 167·5—168° (Found: *M*, by titration,

238·1. $C_{10}H_{10}O_5N_2$ requires *M*, 238). As ammonium salt, it had [α] − 51·66° in water ($c = 1·6462$, $l = 4$, $\alpha = -3·402°$) and [α] − 15·81° in ethyl alcohol ($c = 0·9960$, $l = 4$, $\alpha = -0·628°$). A specimen of acid having [α] − 47·8° (as ammonium salt in water) and containing 96·5% of *l*-acid had m. p. 165·5—166·5°. Another specimen having [α] − 48·35° (as ammonium salt in water) and containing 97·03% of *l*-acid had m. p. 166—167°. A third specimen having [α] − 44·99° (as ammonium salt in water) and containing 93·7% of *l*-acid had m. p. 162·5—163·5° (? eutectic point).

Owing to the difficulty of obtaining the pure strychnine *l*-acid salt, attempts were made to purify the optically impure acid by various methods. On recrystallisation from water, the *l*-acid content only changed from 79·5% to 78·7%. Although the solubility of the optically active acid in alcohol is markedly greater than that of the racemic compound, recrystallisation from alcohol was not completely successful. 10·75 G. of a specimen of the acid containing 78·7% of the *l*-acid, dissolved in 50 c.c. of boiling alcohol, yielded after 48 hours 6 g. of prismatic needles which contained 69·7% of *l*-acid. The residue recovered from the mother-liquor, after recrystallisation from water, weighed 2·75 g. and contained 93·7% of *l*-acid (? eutectic).

The method of purifying optically impure *l*-α-naphthalenesulphonylalanine (Colles and Gibson, *loc. cit.*) was tried, cinchonidine being used (see above). 10 G. of acid containing 60% of *l*-acid were dissolved in water, and sufficient cinchonidine to combine with the racemic acid present and sufficient standard sodium hydroxide to combine with the excess of *l*-acid were added. The whole was brought into solution on boiling and the crystalline salt separated on cooling. The acid recovered from the filtrate was optically inactive, showing that the cinchonidine *l*-acid salt has the same solubility as the partial racemate (see above). A similar experiment was attempted, sufficient strychnine being used to combine with the *d*-acid salt, which is the less soluble in water. The acid recovered from the mother-liquor contained 77% of *l*-acid, whereas the acid used contained 60%.

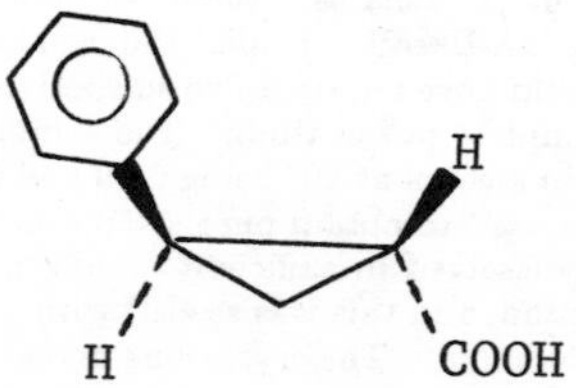

$C_{10}H_{10}O_2$ Ref. 642

Racemic *trans*-2-phenylcyclopropanecarboxylic acid, 125.2 g (0.771 mol), and quinine, 264.2 g (0.815 mol), were dissolved in 5 l. of ethyl acetate at reflux. The solution was filtered and allowed to stand at 20–25° for a few days. The salt which precipitated was recrystallized four times from ethyl acetate (concentration ~5% w/v) until the salt showed no change in rotation in two successive recrystallizations and the solute in the mother liquor gave the same rotation as the precipitate; yield 52.1 g. The recovered salt was recrystallized in the same way to yield an additional 41.1 g of the pure (+)-(*S*)-acid salt, total 93.2 g; after drying at room temperature *in vacuo*, mp 153–154°, $[\alpha]^{20}_D$ −10.9 (EtOH, *c* 1.0 g/dl).

Anal. Calcd for $C_{10}H_{10}O_2 \cdot C_{20}H_{24}N_2O_2$ (1:1 salt): C, 74.05; H, 7.04; N, 5.76. Found: C, 73.88; H, 7.08; N, 5.69.

The (+)-(*S*)-acid salt was added to 1 *N* aqueous HCl, and the free (+)-(*S*)-acid was extracted with diethyl ether by means of a continuous extractor,[13] followed by recrystallization from petroleum ether (30–60°): mp 48.5–49.5°, $[\alpha]^{20}_D$ +405° ($CHCl_3$, *c* 1.0 g/dl) (lit. mp 25–26°C, $[\alpha]^{14}_D$ +381° ($CHCl_3$, *c* 0.96 g/dl)[10]).

The (−)-(*R*)-isomer-rich acid, which was recovered from the mother liquor in the first fractionation of the quinine salt, 43.2 g (0.266 mol), and brucine, 110.4 g (0.279 mol), were dissolved in 850 ml of acetone at reflux. The solution was filtered and allowed to stand at 0° for a few days. After fractional recrystallization was repeated four more times, 21.2 g of the pure (−)-(*R*)-acid salt was obtained. The recovered salt was recrystallized in the same way to yield an additional 54.9 g of the pure (−)-(*R*)-acid salt, total 76.1 g; after drying at room temperature *in vacuo*, mp 106–108°, $[\alpha]^{20}_D$ −172° (EtOH, *c* 1.0 g/dl).

Anal. Calcd for $(C_{10}H_{10}O_2)_2 \cdot C_{23}H_{26}N_2O_4 \cdot H_2O$: C, 70.09; H, 6.57; N, 3.80. Found: C, 70.21; H, 6.57; N, 3.78.

Microanalysis showed that the composition of the salt was acid: brucine:H_2O = 2:1:1. On drying at 80° *in vacuo*, a part of the acid was lost to leave a mixture of the 1:1 salt and the 2:1 salt.

The (−)-(*R*)-acid was liberated quantitatively from the salt as described above, followed by recrystallization from petroleum ether: mp 48.5–49.5°, $[\alpha]^{20}_D$ −410° ($CHCl_3$, *c* 1.0 g/dl) (lit. mp 51–52°, $[\alpha]^{24}_D$ −368° ($CHCl_3$, *c* 0.931 g/dl)[12]).

(10) Y. Inouye, T. Sugita, and H. M. Walborsky, *Tetrahedron*, **20**, 1695 (1964).
(11) A. Burger and W. L. Yost, *J. Amer. Chem. Soc.*, **70**, 2198 (1948).
(12) H. M. Walborsky and L. Plonsker, *ibid.*, **83**, 2138 (1961).

(13) Y. Nishimura, Ph.D. Thesis, Polytechnic Institute of Brooklyn, Brooklyn, N. Y., 1966.

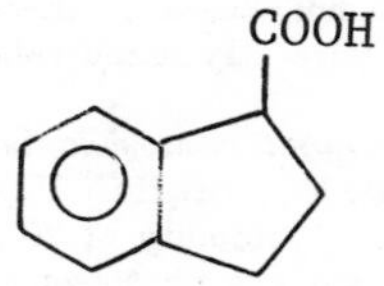

$C_{10}H_{10}O_2$ Ref. 643

Ausführung der Spaltung: 32.4 g (0.2 Mol) *racem.* Säure wurden nebst 93.2 g (0.2 Mol) kristallwasserhaltigem Brucin in einer Mischung von 200 ccm Aceton und 300 ccm Wasser gelöst. Nach 42stdg. Aufbewahren im Eisschrank wurden 35.1 g Salz erhalten (Frakt. I); die darin enthaltene Säure zeigte $[\alpha]_D$: +40.2° (Benzol). Die Mutterlauge wurde auf etwa 280 ccm eingedampft und lieferte nach 18stdg. Aufbewahren im Eisschrank noch 43.7 g Salz (Frakt. II) eines sehr schwach linksdrehenden Säuregemisches. Frakt. II wurde aus 50 ccm Aceton + 75 ccm Wasser umkristallisiert, wobei 14.4 g Salz (Frakt. IIa) erhalten wurden. Die vereinigten Fraktionen I und IIa (48.8 g) wurden dreimal aus 40 ccm Aceton + 100 ccm Wasser umkristallisiert. Nach jeder Kristallisation wurde etwa 0.3 g Salz entnommen, die Säure isoliert und ohne besondere Reinigung in Benzol gemessen:

Umkristallisation	1	2	3
g Salz	44.6	40.6	37.6
$[\alpha]_D$ der Säure	+40.8°	+41.1°	+41.1°

Aus den vereinigten Mutterlaugen der Fraktionen II und IIa wurden etwa 15 g Säure von $[\alpha]_D$: ~−35° gewonnen. 14.6 g (0.09 Mol) dieser Säure wurden nebst 26.7 g (0.09 Mol) Cinchonin in verd. Aceton gelöst und bis zur konstanten Drehung umkristallisiert; etwa infolge Hydrolyse ausgeschiedenes Cinchonin ließ sich dabei durch Zusatz einiger Tropfen Essigsäure in Lösung bringen.

Kristallisation	1	2	3
ccm Aceton	180	150	150
ccm Wasser	210	200	200
g Salz	33.4	25.6	19.5
$[\alpha]_D$ der Säure	−39°	−41.5°	−41°

(+)-Indan-carbonsäure-(1): Die Säure wurde durch Zerlegung des Brucinsalzes (37.2 g) mit verd. Schwefelsäure und Ausschütteln mit Äther isoliert. Ausb. an Rohprodukt 9.6 g; nach zweimaligem Umkristallisieren aus Petroläther (einmal mit Entfärbungspulver) verblieben 7.3 g farblose Nadeln vom Schmp. 45.7–46.6°.

$C_{10}H_{10}O_2$ (162.2) Äquiv.-Gew. Ber. 162.2 Gef. 162.2

Die Drehungsmessungen zeigt Tafel 1. Die Säureproben, die im Gang der Spaltung isoliert wurden, besaßen etwas niedrigere maximale Drehung; dies kommt z. Tl. davon, daß diese Säure nicht durch Umkristallisieren gereinigt wurde, z. Tl. liegt auch ein Konzentrationseffekt vor (siehe unten).

Tafel 1. Drehungswerte von (+)-Indan-carbonsäure-(1)

Lösungsmittel	g Säure	ccm Lösung	α^{25}_D (2 dm)	$[\alpha]^{25}_D$	$[M]^{25}_D$
Benzol	0.2588	9.99	+2.245°	+43.3°	+70.3°
2.2.4-Trimethyl-pentan	0.2250	9.98	+1.835°	+40.7°	+66.0°
Eisessig	0.2424	10.00	+1.40°	+28.9°	+46.8°
Wasser (neutral.)	0.2536	10.00	+1.26°	+24.8°	+40.3°
Chloroform	0.2420	9.98	+0.745°	+15.4°	+24.9°
Äthanol (absol.)	0.2635	10.00	+0.655°	+12.4°	+20.2°
Aceton	0.2522	9.99	+0.16°	+3.2°	+5.2°

In Benzol wurde bei höherer Konzentration eine etwas größere spezif. Drehung gefunden.

0.5090 g Säure in Benzol auf 10.00 ccm gelöst: α^{25}_D: +4.52° (2 dm). $[\alpha]^{25}_D$: +44.4°; $[M]^{25}_D$: +72.0°.

(−)-Indan-carbonsäure-(1): Die Säure wurde aus dem Cinchoninsalz, wie oben beschrieben, isoliert und dreimal aus Petroläther (einmal mit Entfärbungspulver) umkristallisiert. Ausb. an reiner Säure 3.6 g. Sie ist der (+)-Form äußerlich völlig ähnlich. Schmp. 45.7–46.5°.

$C_{10}H_{10}O_2$ (162.2) Äquiv.-Gew. Ber. 162.2 Gef. 162.0

0.2721 g Säure in Benzol auf 10.00 ccm gelöst: α^{25}_D: −2.36° (2 dm). $[\alpha]^{25}_D$: −43.4°; $[M]^{25}_D$: −70.3°.

$C_{10}H_{10}O_2$ Ref. 643a

Resolution of indan-1-carboxylic acid (55)

Racemic indan-1-carboxylic acid (55, 5.22 gm, 32.2 mmol) was added to a boiling solution of (+)-α-phenethylamine* (3.9 gm, 32 mmol) in acetone (50 mls). The crystals that formed on cooling to 0° were air dried and the optical rotation of the salt was measured. The salt was recrystallized to constant rotation from acetone (5 recrystallizations) to give material of $[\alpha]_D^{25}$ +40.9°. The salt was dissolved in 6M hydrochloric acid and the solution was extracted with hexane (2x50 mls).

The organic layer was dried ($MgSO_4$) and evaporated. The residue was recrystallized from hexane to yield as white needles (+) R indan-1-carboxylic acid (55 I) m.p. 44-45° (lit. (53) 45.7-46.6°) $[\alpha]_D^{25}$ +43.6° (c, 2.51 in benzene) (lit. (53) $[\alpha]_D^{25}$ +43.3° (c, 2.59 in benzene) with i.r. and p.m.r. spectra identical with the authentic material.

In a similar manner, using (-)-α-phenethylamine a salt of $[\alpha]_D^{25}$ -40.7° was obtained. The resulting acid (-) S indan-1-carboxylic acid (55 II) was obtained as white crystals. m.p. 45-47° (lit. (52) 45.7-46.5°).

$[\alpha]_D^{25}$ -43.8° (c, 2.50 in benzene) (lit. (52) -43.4° (c, 2.72 in benzene)) with i.r. and p.m.r. spectra identical with those of the authentic material.

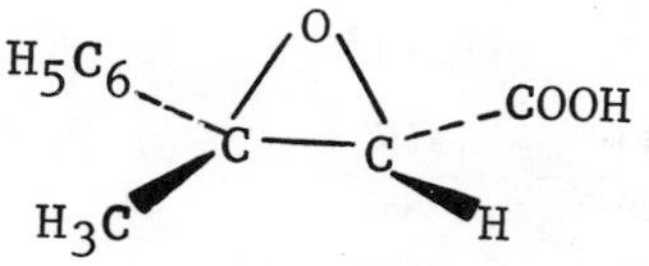

$C_{10}H_{10}O_3$ Ref. 644

Procedure of Optical Resolution

(−)-*cis-Ethyl β-methyl β-phenyl glycidate* (−)-(**1**). (±)-*cis*-Glycidic acid sodium salt (50 g) was dissolved in water (300 ml). Crushed ice was added to the solution. This mixture was acidified with 1 *N* HCl (congo red) and the liberated free glycidic acid was extracted with ether (100 ml). Additional ether extraction (100 ml) was carried out. To the combined ether solution, (−)-quinine (75 g) in MeOH (500 ml) was added. The mixture was kept at room temperature over night. The precipitate was filtered and recrystallized from MeOH. (−)-*cis*-Glycidic acid–(−)-quinine salt was obtained (12 g; mp 194.1–194.3°C).

The resolved (−)-salt (11.5 g) in water (300 ml) was acidified to pH 2 with 6 *N* HCl under ice cooling. The liberated free glycidic acid was extracted with ether. The ether solution was washed with water and dried with anhydrous sodium sulfate and the solvent was evaporated below 15°C. To the residual oil was added water (20 ml) and 2 *N* NaOH (11 ml). The sodium salt was treated with silver nitrate (4.5 g) in water at room temperature for 30 min. The precipitate was filtered off, dried in vacuo with heating and resulted in 3.7 g. The silver salt was suspended in ethyl iodide (50 ml) under cooling to control the reaction. Distillation at 123–135°C (12 mmHg) resulted in 2.0 g. $[\alpha]_D^{28}$ −6.1° (*C* 3.25 95% EtOH).

Table 3. Configuration and specific optical rotation of four optical isomers.

	(−)−*Cis*	(+)−*Cis*	(+)−*Trans*	(−)−*Trans*
	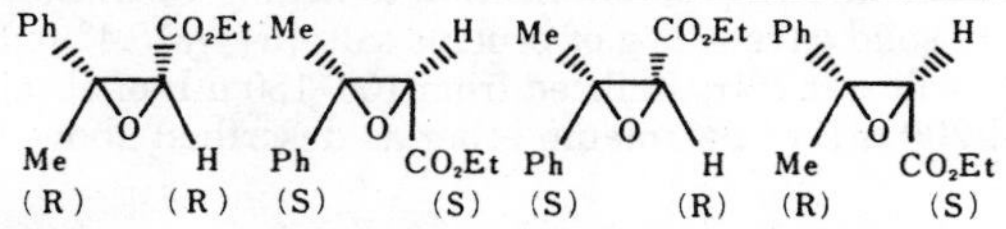			
$[\alpha]_D^{28}$ 95%EtOH	−6.1° (C 3.25)	+6.0° (C 3.19)	+140.6° (C 2.87)	−93.9° (C 2.33)
Diol $[\alpha]_D^{28}$ Benzene	+59.9° (C 3.25)	−55.9° (C 3.02)	−43.4° (C 2.99)	+26.4° (C 3.01)
B* salt m.p	Quinine 194.1-194.3℃	Brucine 187.0-188.9℃	Brucine 178.9-180.1℃	Brucine 134.2-137.1℃

Table 4. Characteristic odors.

Cis-Isomer	Round, soft, juicy, fruity-sweet
(−)-*cis*	Round, soft, sweet (molasses-like), fruity
(+)-*cis*	Faint, fruity, sweet
Trans-Isomer	Sharp, hard, fruity, sour-sweet
(−)-*trans*	Floral, fatty, light woody
(+)-*trans*	Faint, fruity, floral, fatty

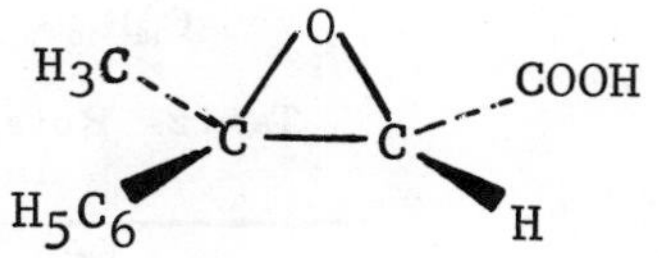

$C_{10}H_{10}O_3$ Ref. 645

Resolution of Sodium (*E*)-3-Methyl-3-phenylglycidate (4). In a modification of a reported procedure,[10] 12.0 g (60.0 mmol) of (*E*)-**4** was dissolved in 75 mL of water and 60 mL of ethyl ether in a two-phase system. The solution was cooled to −5 °C and 82.4 mL of 0.73 M hydrochloric acid, at 0 °C, was added in 2–3 min with rapid stirring. After the addition of acid was complete, the etheral layer was separated and stored at −10 °C. The water layer was again extracted with 60 mL of ether, and the ethereal fractions were combined, dried (Na_2SO_4), and kept at −35 °C. This solution of the glycidic acid **4a** must be used as soon as possible after drying.

To the solution of **4a**, after removal of the desiccant by rapid filtration, was added 28 g (60 mmol) of brucine tetrahydrate, dissolved in 70 mL of benzene–absolute ethanol (1:1), at 55 °C with rapid stirring. The oily residue initially formed slowly dissolved, and a white powder gradually appeared as the mixture was stirred an additional 6 h. Vacuum filtration afforded 12.3 g (77%) of the (+)-brucine–glycidic acid salt: mp 180–182 °C; $[\alpha]^{25}_D$ 3.12° (*c* 1.7, $HCCl_3$). The brucine complex was recrystallized from 225 mL of hot absolute ethanol. After 8 h, filtration gave 4.60 g of the brucine salt: mp 183–185 °C; $[\alpha]^{25}_D$ 15.0° (*c* 2.0, $HCCl_3$). A second recrystallization from ethanol yielded 2.34 g of salt (mp 185–189 °C; $[\alpha]^{25}_D$ 16.9° (*c* 1.1, $HCCl_3$)) and a third gave 1.75 g (mp 188–191 °C (lit.[10] mp 192–193 °C); $[\alpha]^{25}_D$ 19.5° (*c* 1.1, $HCCl_3$) (lit.[10] $[\alpha]^{25}_D$ 19.5°)). This method of resolution was not always reproducible, especially when performed on a large scale.

An alternate method of resolution involved the dissolution of

(10) Abidi, S. L. Ph.D. Thesis, University of Maine, 1969 (University Microfilm 70-12, 389).

12.3 g (21.3 mmol) of brucine salt in 200 mL of hot absolute ethanol with approximately 1 g of charcoal (Norit). This solution was boiled for 5 min and then gravity filtered, while hot, to yield upon cooling a clear, colorless solution. To this solution was added approximately 100 mL of petroleum ether (bp 30–45 °C) in small portions allowing crystallization to begin. Filtration of the fine, white solid gave 6.23 g of brucine salt; $[\alpha]^{25}_D$ 15.4° (*c* 1.9, $HCCl_3$). This salt was recrystallized from 100–150 mL of absolute ethanol and 200 mL of petroleum ether as described above to give 2.96 g of the brucine salt; $[\alpha]^{25}_D$ 19.6° (*c* 1.3, $HCCl_3$).

To 6.13 g (0.011 mol) of the brucine salt, $[\alpha]^{25}_D$ 15.4°, in 80 mL of chloroform was added, at 0 °C with rapid stirring, 10.6 mL of 1.0 M sodium hydroxide in 25 mL of water, and the mixture was stirred for 30 min. After the mixture had come to room temperature, the water layer was separated and the organic layer reextracted with 25 mL of water. The combined extracts were concentrated in vacuo at 75 °C, giving 2.04 g (96%) of (+)-(2*S*,3*R*)-4; $[\alpha]^{25}_D$ 58.7° (*c* 1.8, H_2O).

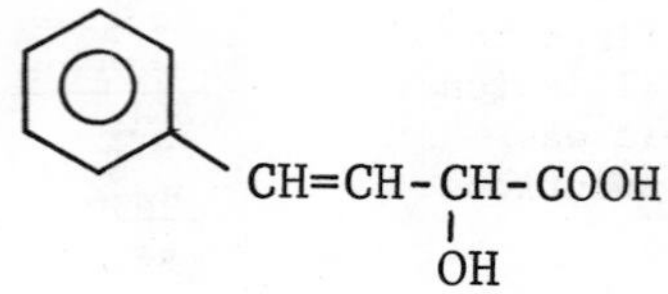

$C_{10}H_{10}O_3$ Ref. 646

Spaltung der Styrylglykolsäure in die optische Antipoden: Die Lösungen von 30 g (+)-Bornylamin in 200 ccm Äther und 30 g Säure in 200 ccm Methanol wurden vereinigt und langsam i. Vak. eingedampft; hierbei schied sich das Salz in gut ausgebildeten Kristallen ab. Nachdem etwa ²/₃ des Lösungsmittels verdampft waren, wurde abfiltriert und der Rest zur Trockne gedampft. Das erste Kristallisat hatte die Drehung $[\alpha]_D$: +18°, das zweite $[\alpha]_D$: +7°. Die Salze wurden in der Siedehitze in Isopropylalkohol gelöst und bei Zimmertemperatur der Kristallisation überlassen.

Nach 7- bis 8maliger Umkristallisation stieg der Drehwert der 1. Fraktion auf dem Endwert $[\alpha]_D$: +28.0° (in Methanol); Ausb. 20 g. Die leichter lösliche Fraktion ließ sich auch aus Wasser umkristallisieren, ihr Drehwert fiel auf den Endwert $[\alpha]_D$: −6.0° (in Methanol); Ausb. 10 g.

Die beiden Salze wurden mit Salzsäure zerlegt und die ausgeschiedene Säure gründlich mit Eiswasser gewaschen. Schmp. 139° (Zers.). Das rechtsdrehende Salz lieferte die Säure mit $[\alpha]_D$: +100.0°, $[M]_D$: +178.2°, das linksdrehende mit $[\alpha]_D$: −98.0°, $[M]_D$: −174.7°.

$C_{10}H_{10}O_3$ (178.2) Ber. C 67.40 H 5.66 Gef. C 67.27 H 5.71

Tafel 2. **Rotationsdispersion [M] der (+)-Styrylglykolsäure in verschiedenen Lösungsmitteln bei 20°**

Lösgm.	$[M]_c$	$[M]_D$	$[M]_{Hg_I}$	$[M]_F$	$[M]_{Hg_{II}}$[1])
Methanol (*c* = 1)	+139°	+178°	+215°	+304°	+415°
Äthanol (*c* = 1)	+126°	+167.5°	+201°	+288°	+388°
Dioxan (*c* = 1)	+ 92.5°	+121°	+151°	+221°	+301°
Aceton (*c* = 1)	+137°	+181.5°	+221°	+294°	+394°
Eisessig (*c* = 1)	+154°	+197°	+242°	+340°	+447°
Isopropyläther (*c* = 0.5)	+ 72°	+ 99°	+127°	+196°	+285°
NaOH (*c* = 1)	+102°	+131°	+156°	+220°	—

[1]) Hg_I und Hg_{II} sind die Quecksilberlinien λ = 5460 und 4358 A.

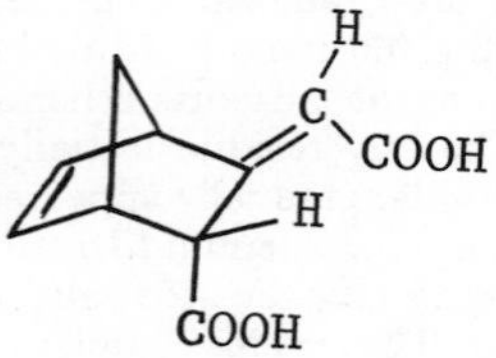

$C_{10}H_{10}O_4$ Ref. 647

Resolution of ***endo*****-3-Carboxy-5-norbornene-Δ²,α-acetic Acid (IX).**—Solutions of 1.00 g. of IX and 1.67 g. of quinine, each in 10 ml. of methanol, were mixed and taken to dryness *in vacuo.* The residual white foam was crystallized from ethyl acetate to give a first crop of 2.35 g. of fluffy white needles, $[\alpha]_D$ −53° ($CHCl_3$). This material was carried through four further recrystallizations, from ethyl acetate at first and later from ethyl acetate containing a small amount of methanol. This yielded 675 mg. of needles, $[\alpha]_D$ −87° ($CHCl_3$). To regenerate IX this salt was dissolved in a minimum volume of methanol, treated with 2.30 ml. of 5% aqueous sodium hydroxide solution (approximately 10% excess), and then diluted with water. Chloroform extraction removed the precipitated quinine, and the resulting clear solution was acidified with 5.83 ml. of 0.5 *M* hydrochloric acid. The water was removed at reduced pressure to leave 426 mg. of white crystals, a mixture of IX and sodium chloride. This residue was dissolved in 12 ml. of water and extracted continuously with ether for 7 hr. Removal of ether after drying left 257 mg. This was recrystallized from nitromethane to give 202 mg. (80%), m.p. 189–191°, $[\alpha]_D$ −122°. The solid state infrared spectrum of this active material was quite similar to, but not identical with, that of racemic IX. This was the most active sample of IX obtained.

From the mother liquors of the above partial resolution there was isolated by the same procedure 788 mg. of dextrorotatory diacid, m.p. 198–206°, $[\alpha]_D$ +39.4 ± 0.2°.

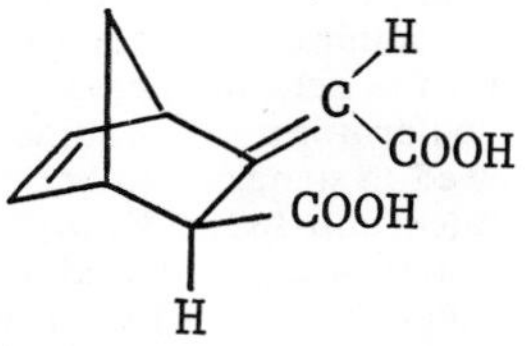

$C_{10}H_{10}O_4$ Ref. 648

Resolution of *exo*-3-Carboxy-5-norbornene-$\Delta^{2,\alpha}$-acetic Acid (X).—In methanol solution 500 mg. of adduct acid and 835 mg. of quinine were mixed, and the solution was then taken to dryness to leave a colorless foam, $[\alpha]D$ −113°. This material separated from solution in ethyl acetate as small needles, 858 mg. (64%). Recrystallization from ethyl acetate–methanol gave 608 mg. (45%), $[\alpha]D$ −160.0 ± 0.3°. This salt was decomposed to quinine and the dicarboxylic acid as detailed below for the isomeric adduct. There was recovered 105 mg. (44%), $[\alpha]D$ −67°. Recrystallization from aqueous methanol gave 78 mg., $[\alpha]D$ −81 ± 2°, m.p. 237–244°, infrared spectrum identical with authentic racemic material.

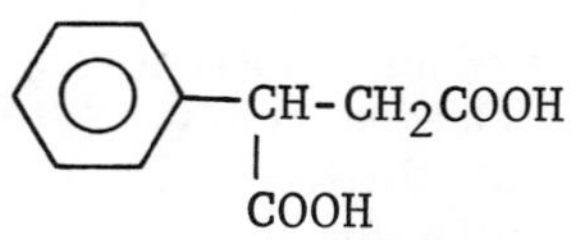

$C_{10}H_{10}O_4$ Ref. 649

Resolution of r-*Phenylsuccinic Acid by Brucine in Ethyl-alcoholic Solution.*

A mixture of *r*-phenylsuccinic acid (30 grams) and brucine (144 grams) was dissolved by gentle warming in rectified spirit (2400 c.c.). The filtered solution was allowed to cool to the temperature of the laboratory with frequent stirring, when a voluminous, white, crystalline precipitate separated. After remaining for three hours at 15° the crop was filtered and dried, when 106 grams of product were obtained which was purified by repeated crystallisation from rectified spirit, 20 c.c. of the solvent being used for each gram of salt. The course of the resolution was followed by determining the specific rotation of the acid recovered from the successive filtrates. The following values for $[\alpha]_D$ were thus obtained, the observations being made in acetone solution: −102·5°, −43·5°, +40·9°, +102·2°. The crop from which the last filtrate had been obtained, which consisted of indistinct crystals weighing 40·5 grams, was decomposed by treatment with dilute hydrochloric acid and extracted with ether, whereby 6·7 grams of *d*-phenylsuccinic acid, having $[\alpha]_D$ +168·0° in acetone solution, were isolated. After two crystallisations of this acid from water in the proportion of 50 c.c. of solvent to each gram of acid, this value had increased to +173·4° and then remained unchanged after a further crystallisation.

d-Phenylsuccinic acid separates from water in granular aggregates or well-defined prisms. It is readily soluble in the hot solvent, very sparingly so in the cold. It dissolves freely in cold acetone or ether, less readily in cold methyl and ethyl alcohols, very sparingly in boiling benzene. When rapidly heated in a capillary tube it melts at 173—174° after softening from 171°, and without suffering visible decomposition; if the molten mass is allowed to solidify, however, and the solid again heated, an indefinite melting point, 130—155°, is observed. Analyses and determinations of specific rotation were made with samples which had been dried in a vacuum over calcium chloride until constant in weight:

0·1566 gave 0·3562 CO_2 and 0·0736 H_2O. C=62·0; H=5·3.

$C_{10}H_{10}O_4$ requires C=61·8; H=5·2 per cent.

The following determinations of the specific rotation were made:

In acetone solution:

$l=2$, $c=1·8025$, $\alpha_D^{15·4}+6·25°$, $[\alpha]_D^{15·4}+173·4°$.
$l=2$, $c=1·8235$, $\alpha_D^{16}+6·32°$, $[\alpha]_D^{16}+173·3°$.

In ethyl-alcoholic solution:

$l=2$, $c=1·5340$, $\alpha_D^{16·5}+4·55°$, $[\alpha]_D^{16·5}+148·3°$.

In ethyl acetate solution:

$=2$, $c=1·5500$, $\alpha_D^{15·5}+5·40°$, $[\alpha]_D^{15·5}+174·2°$.

Preparation of l-*Phenylsuccinic Acid.*

The isolation of *l*-phenylsuccinic acid was effected by fractional crystallisation from water of the crude lævorotatory mixture of acids obtained from the filtrate from the first crop of brucine salt (see above). In one experiment, crude *l*-phenylsuccinic acid (9·7 grams, $[\alpha]_D-102·5°$ in acetone solution) was dissolved in boiling water (600 c.c.). A granular, crystalline crop separated on cooling which, after being dried, weighed 3·8 grams and had $[\alpha]_D$ −168·4° when dissolved in acetone. After two further crystallisations from water (60 c.c. of solvent being used for each gram of acid) this value had increased to −173·3°, which agrees well with the constant specific rotation determined for the *d*-acid.

l-*Phenylsuccinic acid* closely resembles its *d*-isomeride in appearance and properties. When heated in a capillary tube it melts at 173—174°, after marked softening at 171°; when the molten mass is allowed to solidify and again heated, an indefinite melting point is observed:

0·1548 gave 0·3510 CO_2 and 0·0725 H_2O. C=61·8; H=5·2.

$C_{10}H_{10}O_4$ requires C=61·8; H=5·2 per cent.

The specific rotation was determined in acetone and in methyl-alcoholic solution:

In acetone solution:

$l=2$, $c=1·4830$, $\alpha_D^{14·8}-5·14°$, $[\alpha]_D^{14·8}-173·3°$.

In methyl-alcoholic solution:

$l=2$, $c=1·9415$, $\alpha_D^{12·8}-5\ 71°$, $[\alpha]_D^{12·8}-147·1°$.

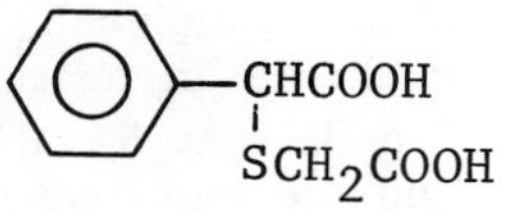

$C_{10}H_{10}O_4S$ Ref. 650

Note:

Resolved the racemic compound with α-phenylethylamine and reported m.p. 142-143.5° and $[\alpha]_D^{17}$ +242.7 and -243.7° (c ~ 6, ethyl acetate) for the resolved acids. A. Fredga, Arkiv Kemi, Mineral. Geol. 24B, No. 15 (1947) reported m.p. 143-144°, $[\alpha]_D^{17}$ +247.0° (c 6.3, ethyl acetate).

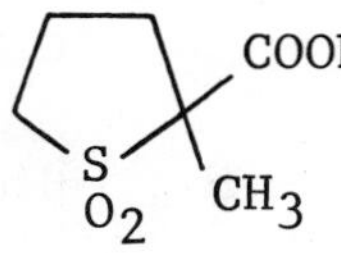

$C_{10}H_{10}O_4S$ Ref. 651

(−)-2-Methyl-2,3-dihydrobenzothiophene-2-carboxylic Acid 1-Dioxide ((−)-I). A solution of 169.7 g of I and 295.8 g of brucine in 1793 ml of refluxing 50:50 acetone–water was prepared. The flask containing the salt solution was sealed tightly and allowed to cool slowly to 25° in a large water bath, and the salt (137.7 g) separated as large plates. Recrystallization of the salt from 975 ml of the same solvent yielded 124.3 g (26.6%). A small portion of the salt was converted to the free acid which, after one crystallization from ether, exhibited $[\alpha]^{29}_{546}$ −83.9° (*c* 2.9 ethanol). Additional crystallizations of the salt provided no change in rotation of the derived acid after one crystallization from ether. The optically pure salt (132.3 g) was dissolved in 5000 ml of boiling water, and a solution of 66 g of potassium carbonate in 100 ml of water was added with rapid stirring. The solution was cooled to 0° for 1 hr to complete precipitation of brucine. The slurry was filtered, and the filtrate was reduced in volume to 400 ml under reduced pressure and filtered. Sulfuric acid, 130 ml, was added slowly to the concentrated filtrate with stirring and cooling, and crude (−)-I separated as an oil which was recovered by extraction with four 400-ml portions of chloroform. The combined extracts were dried and evaporated to dryness, yielding a white solid which was recrystallized from ether (26.8 g, 15.8% as large, colorless crystals), $[\alpha]^{29}_{546}$ −83.9° (*c* 2.9, ethanol), mp 169–171°; lit.[3e] $[\alpha]^{26}_{546}$ +59.0° (not given, ethanol), mp 161–162°; lit.[5] $[\alpha]^{27}_{546}$ −77.8° (*c* 1.8, 95% ethanol), mp 168–170.5°. *Anal.* Calcd for $C_{10}H_{10}SO_4$: C, 53.09; H, 4.45; S, 14.17. Found: C, 53.14; H, 4.50; S, 14.50.

(+)-2-Methyl-2,3-dihydrobenzothiophene-2-carboxylic Acid 1-Dioxide ((+)-I). To the mother liquor from the first crystallization of the brucine salt of I was added a solution of 60 g of potassium carbonate in 1000 ml of water with rapid stirring at 0°. The slurry was filtered, reduced in volume to 450 ml under reduced pressure, and filtered. Chloroform, 700 ml, was added to the basic solution; the solution was cooled in an ice bath and stirred rapidly, and 100 ml of 96% sulfuric acid was added slowly. The layers were separated, and the aqueous phase was extracted with three 430-ml portions of chloroform. The combined chloroform extract was dried and allowed to stand for 18 hr at 25°, whereupon 20.1 g of racemic I separated, mp 182–183°, $[\alpha]^{26}_{546}$ 0.0° (*c* 3.3, ethanol). The chloroform solution was evaporated to dryness under reduced pressure to give 57.0 g of crude (+)-I. This material was refluxed with 1000 ml of ether for 4 hr, and the solution was filtered, reduced in volume to 300 ml, cooled, and filtered again, thereby removing an additional 6.4 g of racemic I. Controlled evaporation of the filtrate yielded 28.1 g of crystalline (+)-I, mp 168–170.5°, $[\alpha]^{28}_{546}$ +80.5° (*c* 2.9, ethanol).

(e) E. J. Corey, H. König, and T. H. Lowry, *Tetrahedron Letters*, 515 (1962); (f) E. J. Corey and T. H. Lowry, *ibid.*, 793, 803 (1965).

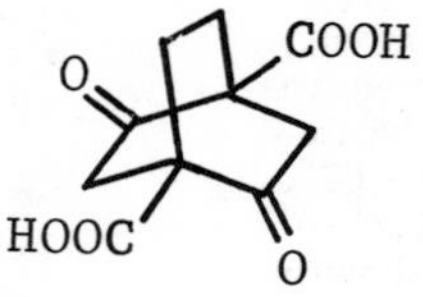

$C_{10}H_{10}O_6$ Ref. 652

Beschreibung der Versuche.

Das Brucinsalz: 8 g reine Brückensäure vom Schmp. 268° (1 Mol.) wurden mit 28 g wasserfreiem Brucin (2 Mol.) gemischt und in 1800 ccm siedendem Wasser gelöst. Aus der heiß filtrierten Lösung schieden sich sofort Krystalle ab. Nach etwa 30-stdg. Stehenlassen wurde filtriert. Die Ausbeute an lufttrocknem Salz betrug 20.2 g. Das Salz wurde aus siedendem Wasser 5-mal umkrystallisiert. Folgende Tafel zeigt die Wassermenge und die entsprechende Menge Salz, die bei den Krystallisationen benutzt wurden.

Wassergehalt in ccm	1500	1250	1000	750	650	—
Salzgewicht in g	20.2	25.7	12.6	10.3	8	6.3

Da das Brucinsalz in den üblichen Lösungsmitteln unlöslich war, wurde Pyridin für die Bestimmung der Drehung benutzt.

Für das 3-mal umkrystallisierte Salz ergab sich: $[\alpha]^{26.5}_{D}$: −70.66° (c = 2.25, *l* = 1),

für das 5-mal umkrystallisierte: $[\alpha]^{26.5}_{D}$: −70.87° (c = 2.354, *l* = 1).

Das Brucinsalz krystallisierte aus Wasser in langen derben Prismen.

$C_{56}H_{62}O_{14}N_4 + 3H_2O$. Ber. C 62.9, H 6.4, N 5.2, H_2O 5.1.
Gef. ,, 62.6, ,, 6.75, ,, 5.6, ,, 4.8.

Herstellung der aktiven Säure: *d*-Bicyclo-[2.2.2]-octan-dion-(2.5)-dicarbonsäure-(1.4): 5 g Brucinsalz wurden in siedendem Wasser gelöst und mit überschüss. Ammoniak behandelt. Nach dem Abkühlen wurde der voluminöse Niederschlag von Brucin abfiltriert, die klare Lösung 2-mal mit Chloroform extrahiert, um Brucin, das in Lösung geblieben war, zu entfernen, dann wurde die Lösung mit verd. Schwefelsäure angesäuert und mehrmals mit Äther extrahiert. Die ätherische Lösung wurde getrocknet, der Äther dann entfernt, wobei der Rückstand fest wurde. Er krystallisierte aus Alkohol in Tafeln, die in Wasser die Drehung $[\alpha]^{25}_{D}$: +23.85° (c = 2.138, *l* = 1) zeigten. Die aktive Säure war leicht löslich in Wasser, sehr wenig löslich in Benzol, Chloroform, Aceton, Äther, Alkohol, Petroläther und Äthylacetat. Schmp. 271°.

Behandlung der Mutterlauge zur Erlangung der linksdrehenden Form: Die Mutterlaugen vom Brucinsalz der *d*-Säure wurden auf die Hälfte ihres Volumens eingedampft; das Salz, das sich nach dem Stehenlassen ausgeschieden hatte, wurde filtriert und dieser Prozeß 3-mal wiederholt. Die restliche Lösung des Salzes der *l*-Säure wurde, wie beim Salz der *d*-Säure beschrieben, mit Ammoniak versetzt, um die *l*-Säure freizumachen. Die so erhaltene rohe aktive Säure krystallisierte aus Alkohol in Tafeln und zeigte in wäßriger Lösung die Drehung $[\alpha]^{28.5}_{D}$: −23.24° (c = 0.9, *l* = 1). Sie hatte denselben Schmelzpunkt wie die *d*-Säure und glich dieser in ihrer Löslichkeit in den verschiedenen Lösungsmitteln.

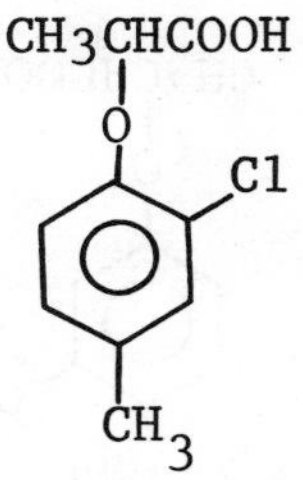

$C_{10}H_{11}ClO_3$ Ref. 653

Preliminary experiments on resolution were performed with small quantities of acid (about 0.3 g) in dilute ethanol.

The *cinchonidine* salt gave a (−)-acid with rather high activity. The *quinidine* salt, which crystallised after long standing, gave also a strongly laevorotatory acid. The acid from the *quinine* salt was practically inactive. *Dehydroabietylamine* gave (+)-acid of fairly good activity. The *strychnine* and *brucine* salts could not be brought to crystallisation. Cinchonidine and dehydroabietylamine were thus selected for the final resolution.

(−)-*2-Chloro-4-methylphenoxy-propionic acid.* Racemic acid (21.5 g, 0.1 mole) and cinchonidine (29.4 g, 0.1 mole) were dissolved in a mixture of 270 ml hot ethanol and 100 ml water. During the cooling, 170 ml water were added. Crystallisation started after about 1 h and after 24 h the salt was filtered off. After drying and weighing, it was recrystallised from dilute ethanol to maximum activity of the acid. After each crystallisation, the acid was liberated from a small sample of the salt and its rotatory power was determined in acetone solution.

Crystallisation	1	2	3	4	5	6
ml ethanol	270	200	150	140	135	110
ml water	270	200	140	135	135	110
g salt obtained	24.1	17.4	15.1	14.4	12.8	10.8
$[\alpha]_D^{25}$ of the acid	−31.5°	−35.5°	−37°	−37.5°	−37.5°	−37.5°

As seen from the data, the (−)-acid is practically pure after the third crystallisation.

The acid was liberated from the salt by acidification with dilute sulphuric acid and extraction with ether. It was recrystallised from dilute formic acid and obtained as glistening needles, more slender than those of the racemic acid. The yield from 10.8 g salt was 4.1 g (38 % calculated on 21.5 g racemic acid). M.p. 85–86°.

$C_{10}H_{11}ClO_3$ (214.65): calc. equiv.wt. 214.6 Cl 16.52
found equiv.wt. 214.4 Cl 16.38, 16.48

Data for the rotatory power in various solvents are given in Table 1.

Table 1

g = g acid dissolved to 10.00 ml.

Solvent	g	α_D^{25} (2 dm)	$[\alpha]_D^{25}$	$[M]_D^{25}$
Acetone	0.2077	−1.56°	−37.6°	−80.6°
Ethanol (abs.)	0.2140	−1.51°	−35.3°	−75.7°
Chloroform	0.1992	−0.70°	−17.6°	−37.7°
Water (neutr.)	0.2310	−0.58°	−12.6°	−26.9°
Benzene	0.2088	+0.10°	+2.4°	+5.1°

(+)-*2-Chloro-4-methylphenoxy-propionic acid.* The mother liquor from the first crystallisation of the cinchonidine salt was evaporated to dryness. From this salt, 11.3 g of strongly dextrorotatory acid were liberated. This acid (0.0525 mole) was dissolved in 225 ml water and carefully neutralised with ammonia. 18.1 g (0.0525 mole) of dehydroabietylamine acetate [8] were dissolved in 350 ml of hot ethanol and the solutions were mixed. On standing, the salt separated as small, glistening needles. It was filtered off after 24 h and recrystallised to maximum activity of the acid as described above.

Crystallisation	1	2	3	4
ml ethanol	350	285	250	200
ml water	225	305	265	225
g salt obtained	19.0	16.7	14.9	11.5
$[\alpha]_D^{25}$ of the acid	+36°	+37.5°	+37.5°	+37.5°

The salt was decomposed by shaking with a mixture of chloroform and sodium carbonate solution. The dehydroabietylamine was taken up by the chloroform and the acid stayed in the water phase as sodium salt. It was then liberated with dilute sulphuric acid and extracted with ether. After recrystallisation with dilute formic acid, it was obtained as glistening needles, quite similar to those of the (−)-form. The m.p. was 85–86° and the yield 4.2 g (39 % calculated on 21.5 g racemic acid.)

$C_{10}H_{11}ClO_3$ (214.65): calc. equiv.wt. 214.6 Cl 16.52
found equiv.wt. 214.2 Cl 16.50, 16.50

0.2053 g acid dissolved in acetone to 10.00 ml: α_D^{25} = +1.54° (2 dm). $[\alpha]_D^{25}$ = +37.5°; $[M]_D^{25}$ = +80.5°.

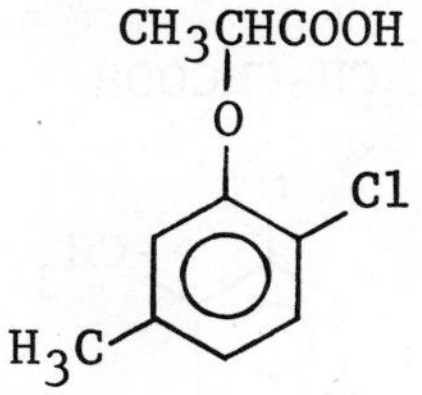

$C_{10}H_{11}ClO_3$ Ref. 654

Preliminary experiments on resolution were carried out with the common alkaloids and some synthetic bases in various solvents. The best results were obtained with *cinchonidine* in ethyl acetate (laevorotatory acid) and *strychnine* in dilute (55 %) methanol (dextrorotatory acid). The salts of *brucine*, *quinine*, *quinidine*, *cinchonine*, *morphine*, *dehydroabietylamine*, *α-phenylethylamine* and *β-phenyl-isopropylamine* failed to crystallise.

(−)-*2-Chloro-5-methylphenoxy-propionic acid.* Racemic acid (21.5 g, 0.1 mole) and cinchonidine (29.4 g, 0.1 mole) were dissolved in 270 ml of hot ethyl acetate. The salt crystallised rapidly. For recrystallisation it was dissolved in a small volume of hot methanol, after which a larger volume of hot ethyl acetate was added. After each crystallisation, a small quantity of acid was liberated and the rotatory power determined in acetone solution.

Crystallisation	1	2	3	4	5
ml methanol	–	30	40	30	25
ml ethyl acetate	270	180	150	120	100
g salt obtained	29.3	18.7	12.8	9.4	6.8
$[\alpha]_D^{25}$ of the acid	−13°	−27.5°	−40°	−44°	−44°

The acid was liberated with sulphuric acid and extracted with ether. The yield of crude product was 2.90 g. Two recrystallisations from ligroin and three from cyclohexane gave 2.02 of pure acid as fine needles with m.p. 112.5–113°. Fractionation of the salt from the mother liquors gave a second crop yielding 1.06 g of pure acid.

$C_{10}H_{11}O_3Cl$ (214.65) Calc. C 55.95 H 5.17 Cl 16.52
Found C 55.99 H 5.22 Cl 16.49

The rotatory power in various solvents is given in Table 1.

Table 1.

g = g acid in 10.00 ml solution.

Solvent	g	α_D^{25} (1 dm)	$[\alpha]_D^{25}$	$[M]_D^{25}$
Acetone	0.0932	−0.406°	−43.6°	−93.5°
Ethanol (abs.)	0.1099	−0.475	−43.2°	−92.8°
Chloroform	0.1057	−0.210°	−19.9°	−42.6°
Water (neutr.)	0.1044	−0.052°	+5.1°	+10.9°
Benzene	0.1013	−0.116°	−11.1°	−23.8°

(+)-*2-Chloro-5-methylphenoxy-propionic acid.* The mother liquor from crystallisations 1 and 2 of the cinchonidine salt were evaporated giving 21.2 g salt. The liberated acid (8.65 g) showed $[\alpha]_D^{25}$ = +24°. This acid (0.04 mole) was dissolved with 13.4 g (0.04 mole) strychnine in 100 ml methanol + 80 ml water. The salt was filtered off after 24 h and recrystallised as seen below.

Crystallisation	1	2	3
ml methanol	100	40	40
ml water	80	32	32
g salt obtained	10.1	8.3	7.3
$[\alpha]_D^{25}$ of the acid	+38.5°	+43°	44°

The salt was decomposed with excess sodium carbonate solution and the strychnine extracted with chloroform. The acid was then isolated in the usual way and recrystallised twice from cyclohexane. 2.34 g crude acid gave 2.06 g pure product as fine needles quite similar to the enantiomer. M.p. 112.5–113°.

$C_{10}H_{11}O_3Cl$ (214.65) Calc. C 55.95 H 5.17 Cl 16.52
Found C 55.96 H 5.16 Cl 16.46

0.0869 g dissolved in *acetone* to 10.00 ml: α_D^{25} = +0.377° (1 dm).
$[\alpha]_D^{25}$ = +43.4°; $[M]_D^{25}$ = +93.2°.

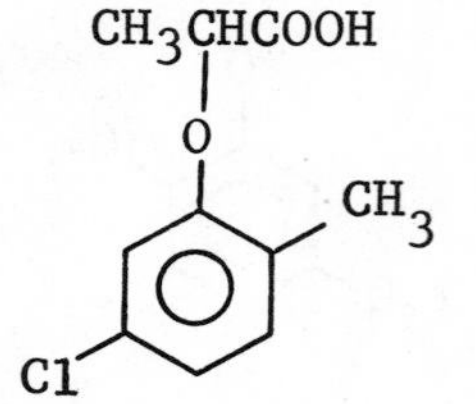

$C_{10}H_{11}ClO_3$ Ref. 655

Preliminary experiments on resolution were carried out as described for the isomer. *Ephedrine* and *pseudoephedrine* gave strongly dextrorotatory acid, while *strychnine* gave the (−)-form. *β-Phenyl-isopropylamine* also gave good resolution. The results obtained with *cinchonidine* were not reproducible. *Quinine* and *α (2-naphthyl)-ethylamine* gave very poor resolution. The salts of *α-phenylethylamine* and *dehydroabietylamine* gave completely inactive acid. No crystalline salts were obtained from *brucine, cinchonine* and *quinidine*. The commercial alkaloids ephedrine and strychnine were selected for the final resolution.

(+)-*5-Chloro-2-methylphenoxypropionic acid*. Racemic acid (21.5 g, 0.1 mole) and (−)-ephedrine (18.3 g, 0.1 mole) were dissolved in 200 ml ethanol + 300 ml water. The salt was filtered off after 24 h; the yield was 9.0 g and the rotatory power of the acid +63°. Recrystallisation from 50 ml ethanol + 75 ml water gave 7.5 g salt. The rotatory power of the acid was unchanged. Systematic fractionation of the salt in the mother liquors gave further 4.0 g salt with acid of maximum activity.

The salt (11 g) was decomposed in the usual way, yielding 6.5 g of crude acid. Two recrystallisations from ligroin (one with charcoal) gave 6.2 g of pure acid as small plates with m.p. 115–115.5°.

$C_{10}H_{11}O_3Cl$ (214.65) Calc. C 55.95 H 5.17 Cl 16.52
Found C 56.10 H 5.20 Cl 16.44

The rotatory power in various solvents is given in Table 2.

Table 2.

g = g acid in 10.00 ml solution.

Solvent	g	α_D^{25} (1 dm)	$[\alpha]_D^{25}$	$[M]_D^{25}$
Acetone	0.1899	+1.179°	+62.1°	+133.3°
Ethanol (abs.)	0.2055	+1.152°	+56.1°	+120.3°
Chloroform	0.1919	+0.849°	+44.2°	+ 95.0°
Water (neutr.)	0.0942	+0.282°	+29.9°	+ 64.3°
Benzene	0.2027	+0.321°	+15.9°	+ 34.0°

(−)-*5-Chloro-2-methylphenoxy-propionic acid*. The final mother liquors from the ephedrine salt gave 7.2 g acid with $[\alpha]_D^{25} = -38°$. 7.0 g (0.033 mole) of this acid and 11.5 g (0.033 mole) of strychnine were dissolved in 100 ml acetone + 50 ml water. The yield of salt was only 5.2 g. The mother liquor was therefore evaporated to about 65 % of its volume and the salt, which had separated during the evaporation was dissolved by gentle heating. On standing, a second crop of salt was obtained (9.6 g). Recrystallisation of the two crops from dilute acetone gave 9.6 g salt having acid with $[\alpha]_D^{25} = -62.5°$. A second recrystallisation from 40 ml acetone + 20 ml water gave 8.5 g salt; the activity of the acid was practically unchanged.

The salt was decomposed and the acid isolated as described for the strychnine salt of the isomer (p. 305). The yield of crude acid was 2.3 g. Two recrystallisations from ligroin (one with charcoal) gave 2.0 g of pure acid quite similar to the enantiomer. M.p. 115–115.5°

$C_{10}H_{11}O_3Cl$ (214.65) Calc. C 55.95 H 5.17 Cl 16.52
Found C 56.09 H 5.16 Cl 16.59

0.1096 g dissolved in *acetone* to 10.00 ml: $\alpha_D^{25} = -0.682°$ (1 dm). $[\alpha]_D^{25} = -62.2°$; $[M]_D^{25} = -133.6°$.

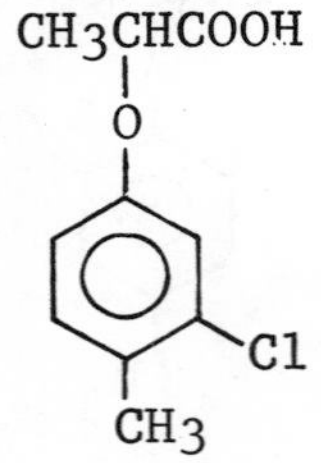

$C_{10}H_{11}ClO_3$ Ref. 656

Preliminary experiments on resolution. Laevorotatory acid of high activity was obtained with *brucine* and (−)-*α-phenyl-ethylamine*. Only *morphine* gave a strongly dextrorotatory acid. *Cinchonine* and *quinine* gave (−)-acid and *quinidine* (+)-acid, in all cases of low activity. The salts of *cinchonidine* and *α-(2-naphthyl)-ethyl-amine* gave racemic acid and the *strychnine* salt failed to crystallise.

(−)-*4-Chloro-3-methylphenoxy-propionic acid*. Racemic acid (17.1 g, 0.08 mole) and brucine tetrahydrate (37.2 g, 0.08 mole) were dissolved in 130 ml ethanol + 80 ml water. The crystallisation took place very slowly. After three days in a refrigerator, 4.8 g salt had separated. Standing for further 3 days at 0°C and 2 days at −10°C gave another 4.5 g. The two portions were combined and recrystallised as seen below. In the first two recrystallisations the solution was left to stand for 18 h at −10°C and in the third over-night at 0°C.

Crystallisation	1	2	3	4
ml ethanol	130	55	25	15
ml water	80	40	20	20
g salt obtained	9.3	7.3	5.5	4.5
$[\alpha]_D^{25}$ of the acid	—	−47°	−51°	−51°

The acid was liberated as described for the isomers. Recrystallisation from ligroin gave 1.3 g of pure acid as long glistening needles with m.p. 80.5–81°.

$C_{10}H_{11}O_3Cl$ (214.65) Calc. C 55.95 H 5.17 Cl 16.52
Found C 55.90 H 5.11 Cl 16.76

0.1128 g acid dissolved in *acetone* to 10.00 ml: $\alpha_D^{25} = -0.564°$ (1 dm). $[\alpha]_D^{25} = -50.0°$; $[M]_D^{25} = -107.3°$.

Systematic fractionation of mother liquors gave another crop of salt yielding 0.9 g of pure (−)-acid.

(+)-*4-Chloro-3-methylphenoxy-propionic acid*. The mother liquor from the first crystallisation of the brucine salt gave 12.2 g acid with $[\alpha]_D^{25} = +13°$. This acid (0.057 mole) was dissolved with 16.3 g (0.057 mole) morphine in 120 ml acetone + 60 ml water. After 24 h in a refrigerator, 12.9 salt g had separated. It was recrystallised from dilute acetone as seen below.

Crystallisation	1	2	3	4
ml acetone	120	15	7	7
ml water	60	20	18	18
g salt obtained	12.9	11.0	9.4	7.2
$[\alpha]_D^{25}$ of the acid	+48°	+46°	+48°	+50°

The acid was isolated and recrystallised from ligroine like the enantiomer. 7.2 g salt gave 2.6 pure acid as long glistening needles with m.p. 80–81°.

$C_{10}H_{11}O_3Cl$ (214.65) Calc. C 55.95 H 5.17 Cl 16.52
Found C 55.86 H 5.10 Cl 16.62

The rotatory power in various solvents is given in Table 2.

Table 2. g = g acid in 10.00 ml solution.

Solvent	g	α_D^{25} (2 dm)	$[\alpha]_D^{25}$	$[M]_D^{25}$
Acetone	0.1143	+1.152°	+50.4°	+108.2°
Ethanol (abs.)	0.1025	+0.72°	+35.1°	+75.4°
Chloroform	0.1024	+0.37°	+18.1°	+39.8°
Water (neutr.)	0.1046	+0.29°	+13.9°	+29.8°
Benzene	0.1004	−0.19°	−9.5°	−20.3°

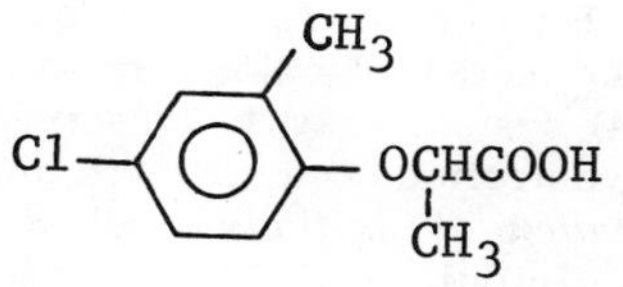

$C_{10}H_{11}ClO_3$ **Ref. 657**

Preliminary experiments on resolution. *α-(2-Methyl-4-chlorophenoxy)-propionic acid* gave crystalline salts with strychnine, quinine and α-phenethylamine. Only the latter base gave a satisfactory resolution. The (+)-amine was used for the isolation of the (+)-acid and the (−)-amine for the (−)-acid.

α-(2.4-Dichlorophenoxy)-n-butyric acid yielded crystalline salts with strychnine, quinine, cinchonidine and α-phenethylamine. From the quinine and (−)-phenethylamine salts a lævorotatory acid was obtained and the latter base was used for the isolation of the (−)-acid. The other bases yielded dextrorotatory products; strychnine gave the best resolution and was used for the preparation of the (+)-acid.

(+)-α-(2-Methyl-4-chlorophenoxy)-propionic acid. Equimolecular quantities of the racemic acid (28.2 g, 0.13 mole) and (+)-α-phenethylamine (16.0 g) were dissolved in a hot mixture of ethanol (100 ml) and water (500 ml). The solution was set aside to crystallise in a refrigerator over night. A small sample was treated with hydrochloric acid and the organic acid was taken up in ether, which was evaporated. The rotatory power of this product was measured in abs. alcohol. The salt was recrystallised until the optical activity did not increase any more (Table 2).

Table 2. Recrystallisation of the (+)-α-phenethylamine salt of α-(2-methyl-4-chlorophenoxy)-propionic acid.

Cryst. No.	Solvent (ml)		Weight of salt (g)	$[\alpha]_D$ of the acid
	Ethanol	Water		
1	100	500	29.8	+ 3.5°
2	100	500	17.5	+12.5°
3	75	375	12.2	+17.3°
4	50	250	9.35	+18.7°
5	50	250	7.35	+19.5°
6	40	200	5.65	+19.2°

The organic acid was isolated from the last salt fraction and the crude acid (3.4 g) was recrystallised from 75% formic acid. 2.77 g (+)-acid were collected as small, white needles. M.p. 95–96°.

86.39 g (+)-acid: 8.210 ml 0.04888-*N* NaOH
$C_{10}H_{11}O_3Cl$ calc. equiv. wt. 214.6
found » » 215.3

0.1836 g dissolved in abs. *alcohol* to 20.06 ml: $2\alpha_D^{25} = +0.349°$. $[\alpha]_D^{25} = +19.0°$; $[M]_D^{25} = +40.9°$.
0.1876 g » » *acetone* to 20.06 ml: $2\alpha_D^{25} = +0.548°$. $[\alpha]_D^{25} = +29.3°$; $[M]_D^{25} = +62.9°$.
0.2197 g » » *chloroform* to 20.06 ml: $2\alpha_D^{25} = +0.357°$. $[\alpha]_D^{25} = +16.3°$; $[M]_D^{25} = +35.0°$.
0.2002 g » » *benzene* to 20.06 ml: $2\alpha_D^{25} = +0.315°$. $[\alpha]_D^{25} = +15.8°$; $[M]_D^{25} = +33.9°$.
0.1873 g neutralised with *aqueous NaOH* and made up to 20.06 ml: $2\alpha_D^{25} = +0.099°$. $[\alpha]_D^{25} = +5.3°$; $[M]_D^{25} = +11.4°$.

(−)-α-(2-Methyl-4-chlorophenoxy)-propionic acid. The organic acid was recovered from the mother liquours 1 and 2 (Table 2). This acid (16.5 g, 0.077 mole, $[\alpha]_D =$ = −8° in alcohol) and an equimolecular quantity of (−)-α-phenethylamine (9.3 g) were dissolved in a hot mixture of alcohol (100 ml) and water (500 ml). The solution was set aside in a cool place and the crystalline product was collected next day. The rotatory power of a small sample of the acid was measured (in alcohol) in the usual way. The course of the purification is seen from Table 3.

Table 3. Recrystallisation of the (−)-α-phenethylamine salt of α-(2-methyl-4-chlorophenoxy)-propionic acid.

Cryst. No.	Solvent (ml)		Weight of salt (g)	$[\alpha]_D$ of the acid
	Ethanol	Water		
1	100	500	14.4	−16.8°
2	75	375	10.6	−18.0°
3	60	300	7.8	−17.8°
4	50	250	5.17	−17.9°

From the salt fraction 4, 3.23 g of the crude (−)-acid were isolated. It was recrystallised from 75% formic acid, yielding 2.35 g of the pure (−)-acid. M.p. 94.5–96°.

83.55 mg (−)-acid: 8.010 ml 0.04888-*N* NaOH
$C_{10}H_{11}O_3Cl$ calc. equiv. wt. 214.6
found » » 213.4

0.1740 g dissolved in abs. *alcohol* to 20.06 ml: $2\alpha_D^{25} = -0.326°$. $[\alpha]_D^{25} = -18.8°$; $[M]_D^{25} = -40.3°$.

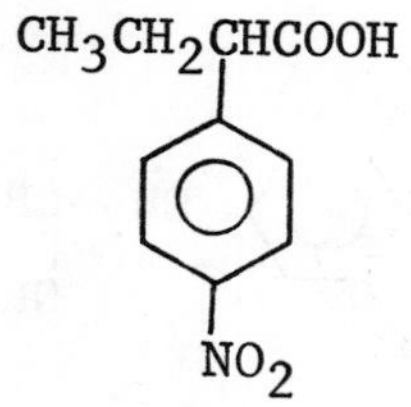

$C_{10}H_{11}NO_4$ **Ref. 658**

Dédoublement de l'acide para-nitrophényléthylacétique.

Les essais sont résumés dans le tableau suivant :

Cinchonine	Belles cristallisations. Pas de dédoublement.
Cinchonidine	— —
Quinine	— —
Quinidine	Très mauvaise cristallisation. Acide gauche.
Strychnine	Non cristallisable.
Brucine	Belle cristallisation. Pas de dédoublement.
Morphine..........	Belle cristallisation. Acide droit.

De tous les alcaloïdes, seule la morphine nous a permis d'avoir l'acide droit pur. L'acide gauche pur est très difficile à obtenir.

Une première opération effectuée sur 20 gr. de sel de morphine nous a conduit, après 3 cristallisations successives, à un acide de pouvoir rotatoire $(\alpha)_D = 445°,28$.

Le sel de morphine étant extrêmement peu soluble dans l'eau à froid et dans la plupart des solvants organiques, nous fûmes obligés de suivre la variation du pouvoir rotatoire de l'acide. D'ailleurs nous avons eu l'occasion, au cours de nos recherches, de constater que la variation du pouvoir rotatoire en valeur absolue du sel d'un alcaloïde peut provenir non seulement d'un enrichissement en l'un des deux acides énanthiomorphes, mais aussi d'une dissociation partielle, suffisante pour accuser des variations de pouvoir rotatoire notables.

Pour avoir donc une garantie absolue sur le degré de pureté de l'acide actif contenu dans le sel d'alcaloïde au cours de la purification, il faut suivre son pouvoir rotatoire et non pas celui du sel de l'alcaloïde.

La morphine que nous avons employée a été purifiée avec soin. Elle avait pour pouvoir rotatoire $(\alpha)_D = -130°,5$ (morphine pure $(\alpha)_D = -130°,9$) (4).

Le sel de morphine obtenu dans la première opération et contenant un acide de $(\alpha)_D = +15°,28$ nous a servi pour amorcer les cristallisations de la deuxième opération; ceci nous a permis d'accélérer la purification.

Dédoublement. **— Acide, 4gr,180 ; morphine, 6gr,060. Il vaut mieux pulvériser la morphine car la dissolution est lente. On dissout les quantités ci-dessus dans 700 cc. d'eau. Après un repos de 24 h. on obtient 6gr,500 à 6gr,700 de sel de morphine de l'acide droit.**

Nous avons réuni quatre opérations identiques ce qui nous a fourni 26gr,500 de sel. C'est une quantité minimum car la purification est longue. L'acide contenu dans cette fraction a un pouvoir rotatoire de : $(\alpha)_D = +25°,33$.

On fait recristalliser le sel de morphine dans 30 cc. d'eau environ. Les rendements de 83-86 0/0 à la T de 20°. Au bout de la quatrième cristallisation, le produit est pur, car le pouvoir rotatoire de l'acide qu'on en retire reste constant.

Isolement de l'acide. **— On dissout 1gr,5 de sel de morphine dans 70 cc. d'eau, à chaud. On filtre, on traite par 2cc,9 de soude normale, en amorçant avec une trace de morphine et on refroidit. On sépare la morphine et on précipite l'acide par l'acide chlorhydrique en refroidissant fortement la solution pour éviter la formation d'un précipité huileux. L'acide actif droit recristallisé dans l'alcool à 50 0/0, fond à 108-109° (acide racémique 123-124°) $(\alpha)_D = +53,33$ (solution dans l'alcool absolu contenant 1,50 0/0 d'acide).**

Traitement des eaux-mères. **— Les eaux-mères sont évaporées à sec dans le vide. Le sel recristallisé dans l'eau donne un acide à pouvoir rotatoire $(\alpha)_D = -38°,66$ (1gr,5 dans 100 cc. de solution alc. abs., tube 20 cm.).**

La troisième recristallisation nous a donné un acide $(\alpha)_D = -33°,3$ (mêmes conditions).

On peut déduire que le sel s'enrichit en acide droit.

Nous avons isolé l'acide et essayé de faire cristalliser le sel de brucine, mais il cristallise mal et, en somme, nous n'avons pas réussi jusqu'ici à obtenir l'acide gauche à l'état pur.

Nous poursuivons les essais de dédoublement en partant de l'acide para-acétylaminophényléthylacétique qui a l'avantage de fondre plus haut que l'acide nitré et dont les sels cristallisent très bien.

(4) *Chem. Soc.*, t. 77, p. 1037.

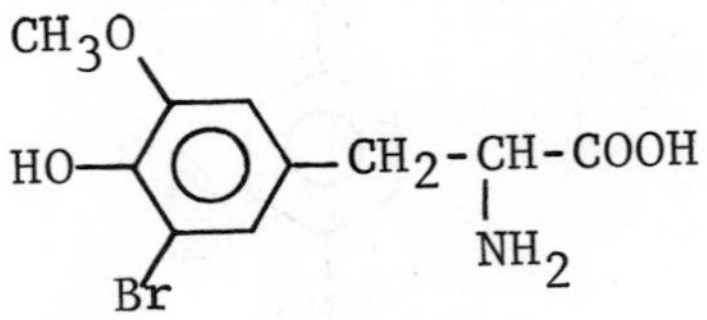

$C_{10}H_{12}BrNO_4$ **Ref. 659**

Resolution of (Ia) *into its Enantiomers by Formation of Diastereoisomers with Ephedrine.* (−)-3-*Bromo*-4-*hydroxy*-5-*methoxy*-N-*acetylphenylalanine.*—DL-(XI) (45 g.) is added to an ethanolic solution (31·6 ml.) of 4·3M-L-(−)-ephedrine, diluted with acetone (174·5 ml.). The solution obtained is kept for 5 days at room temperature. The crystals are then filtered off, washed with acetone, dried, powdered, and taken up in acetone (400 ml.). The suspension is stirred for 1 hr. and then filtered. The crystals collected are dissolved under stirring in hot anhydrous ethanol (80 ml.). By slow addition of ether (360 ml.) under stirring and with cooling, a precipitate is obtained which is allowed to stand for 4 hr. at room temperature. The crystals are filtered off, washed with ether, and dried to yield 25·4 g. (75·4%) of the L-(−)-ephedrine salt of (−)-(XI), m.p. 128—131°, $[\alpha]_D$ −66° (*c* 1 in ethanol). The above salt (21·5 g.) is dissolved in ethanol–water (75 : 25) (1075 ml.) and poured onto a 5-cm. diameter column of Amberlite IR 124 (H^+), the flow rate being adjusted to 5 ml./min. Elution is performed with the same solvent mixture. The eluate is concentrated to a small volume which is taken up with ethanol, treated with charcoal, and filtered. The solvent is evaporated under reduced pressure at 40° to yield (−)-(XI) (13·2 g., 92%), $[\alpha]_D$ −40° (*c* 1 in ethanol). Purity by phenolic group titration, 93·4%.

(+)-3-*Bromo*-4-*hydroxy*-5-*methoxyphenylalanine Hydrochloride.*—A suspension of (−)-(XI) (11·5 g.) in 1·2 N-hydrochloric acid (230 ml.) is saturated with sulphur dioxide. A solution is obtained after 30 min. heating under stirring at 110° (bath temperature), and heating and stirring are maintained for a further 3½ hr. to reach complete hydrolysis.

After this reaction time, the solution is discoloured with charcoal and filtered. The solution is concentrated to a small volume, taken up with water, and evaporated to dryness by azeotropic distillation with ethanol. The residue is taken up with anhydrous ether. After filtration and drying under reduced pressure at 40°, (+)-(Ia) (10·1 g.) (90·3%) is obtained with a 98·5% titration purity (NH_2), m.p. 207—210°, $[\alpha]_D$ +10° (*c* 1 in water).

(+)-3-*Bromo*-4-*hydroxy*-5-*methoxy*-N-*acetylphenylalanine.*—(a) *Isolation of the partially resolved* (+)-(XI). After crystallization and filtration of the L(−)-ephedrine salt of (−)-(XI), the mother-liquor is concentrated under a vacuum to yield impure (+)-(XI). The L(−)-ephedrine salt, $[\alpha]_D$ +16·5° (*c* 1 in ethanol) (43 g.), is dissolved in ethanol–water (80 : 20) (2·151), passed through a column of Amberlite IR/124 (H^+ form), and eluted with the same solvent mixture.

The eluate (9 l.) is concentrated under a vacuum until disappearance of the ethanol. A white powder is then obtained, which is filtered off, washed with water, and dried under a vacuum to yield partially resolved (+)-(XI) (9·1 g., 31·7%), $[\alpha]_D$ +4° (*c* 1 in ethanol).

The mother-liquor is discoloured with charcoal, concentrated to dryness, taken up with ethanol, again concentrated to dryness, and dried under a vacuum to yield impure (+)-(XI) (18 g., 62·75%), $[\alpha]_D$ +38·5° (*c* 1 in ethanol). Purity by titration: CO_2H, 92; OH (phenol), 97%. Total yield, 94·45%. By t.l.c. on silica gel, only one spot, R_F 0·61 ± 0·2, in the methanol–chloroform–ammonium hydroxide (20 : 20 : 4) system may be observed by detection with $FeCl_3$. No detection occurred with ninhydrin.

(b) *Salt formation of partially resolved* (+)-(XI) *with* D(+)-*ephedrine.* Impure (+)-(XI) (12·7 g.), $[\alpha]_D$ +38·5° (*c* 1 in ethanol), are dissolved in acetone (49 ml.) and treated with a 3·4M-ethanolic solution (11·25 ml.) of D(+)-ephedrine. After 2½ hr. at room temperature and the addition of acetone (150 ml.), the crystals obtained are filtered off, washed with acetone and ether, and dried under a vacuum. These crystals are dissolved in hot ethanol (40 ml.). The solution is cooled and ether (70 ml.) is added to yield, after washing with ether and drying, 11·15 g. (58·7%) of (+)-(XI), D(+)-ephedrine. $[\alpha]_D$ +60° (*c* 1 in ethanol).

(c) (+)-(XI). The above diastereoisomeric salt (10·1 g.) dissolved in ethanol–water (80 : 20) (610 ml.) is poured on a column of Amberlite IR-124 (H^+) and eluted with the same solvent. The acid eluate is collected and concentrated under a vacuum until free from ethanol. The aqueous solution is then diluted with water, treated with charcoal, filtered, concentrated, and dried under a vacuum. The residue is taken up with ethanol and concentrated to dryness to yield 6·4 g. (95·7%) of (+)-(XI). Purity by titration: CO_2H, 93%, $[\alpha]_D$ +40·5° (*c* 1 in ethanol). By t.l.c., one spot, R_F 0·605, was detected with $FeCl_3$, but not with ninhydrin. The solvent was methanol–chloroform–ammonium hydroxide (20 : 20 : 4).

(d) (−)-(Ia). (+)-(XI) (5·5 g.) is dissolved in 1·2N-hydrochloric acid (110 ml.) and hydrolysed at 110° during 4½ hr., the reaction being followed by t.l.c. The solution is then discoloured with charcoal, filtered, and concentrated to dryness. The residue is taken up with ethanol, concentrated to a small volume, and diluted with one volume of ether. (−)-(Ia) crystallized on cooling (5 g., 92·5%), m.p. 204—210°, $[\alpha]_D$ −9° (*c* 2 in water). Purity by titration: NH_2, 92%.

By paper chromatography on Whatman No. 1 (n-butanol saturated with 3N-hydrochloric acid) DL-(Ia) is resolved in its enantiomers, R_F 0·645 and 0·827. For the enantiomers of (Ia), we obtained the following values: (+)-(Ia), R_F 0·645 (single spot); (−)-(Ia), R_F 0·827 (single spot).

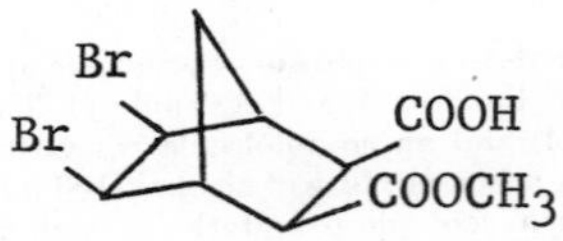

$C_{10}H_{12}Br_2O_4$ **Ref. 660**

Resolution of the Monomethyl Ester.—A hot solution of 5.40 g. of VIIIb in 50 cc. of acetone was treated with a hot solution of 5.75 g. of quinine trihydrate in 100 cc. of acetone. The solution was filtered while still hot and concentrated to a volume of 50 cc. After being allowed to cool to room temperature, the solution was placed in the refrigerator for two hours, after which time 1.50 g. (A) of a chalky, granular solid had separated. A second crop (B) of 8.0 g. and a third crop (C) of 1.55 g. were also obtained by successive concentration of the mother liquors. Crops A and B were combined and dissolved in 250 cc. of boiling acetone. The solution was filtered hot, concentrated to a volume of 75 cc. and allowed to cool to give 1.30 g. of quinine salt. This was triturated with 6 *N* hydrochloric acid, the resulting gummy precipitate was extracted with chloroform, the chloroform solution extracted with aqueous potassium carbonate, and the aqueous layer, after being washed with fresh chloroform, acidified with concentrated hydrochloric acid. The precipitated solid was recrystallized from dilute acetic acid and then from dilute acetone to give 0.25 g. of the (−)-monomethyl ester as tiny, glistening needles, melting at about 150°, evolving gas at 165–170°, resolidifying and then remelting at 251–254°. This material showed $[\alpha]_D$ −6.7° (*c* 1.85 in acetone, *l* 4).

Anal. Calcd. for $C_{10}H_{12}O_4Br_2$: C, 33.73; H, 3.40; neut. equiv., 356. Found: C, 33.44; H, 3.74; neut. equiv., 356.

The infrared spectrum (Nujol mull) was very similar to that of racemic VIIIb, although the slight differences to be expected between a racemate and an enantiomer[10] were noted.

Crop C from the original formation of the quinine salt was worked up by decomposition with hydrochloric acid, extraction with chloroform, extraction with carbonate and regeneration with hydrochloric acid to give 0.84 g. of (+)-VIIIb, $[\alpha]_D$ +1.8° (*c* 4.30 in acetone, *l* 4).

The acetone mother liquor from the recrystallization of crops A and B was concentrated by evaporation until the onset of crystallization, then cooled, the solid collected and recrystallized from acetone to give 1.7 g. of salt. This was decomposed in the usual manner to give 0.68 g. of monomethyl ester of very low optical activity. Further concentration of the mother liquors gave two further crops of quinine salt which yielded monomethyl ester of very low rotation. These samples of ester were combined (2.4 g.), dissolved in 15 cc. of hot acetone and the solution added to a hot solution of 2.2 g. of quinidine in 40 cc. of acetone. The mixture was concentrated to 20 cc. and allowed to cool to give 1.25 g. of quinidine salt. This was decomposed with 6 *N* hydrochloric acid, the resulting solid filtered, washed with water and dissolved in potassium carbonate solution. This solution was washed with ether, acidified with concentrated hydrochloric acid and the precipitated solid filtered, washed with water and dried to give 0.56 g. of (+)-monomethyl ester, $[\alpha]_D$ +3.6° (*c* 5.6 in acetone, *l* = 4). This substance melted at 145–150°, resolidified at about 170–175° and remelted at 251–253°.

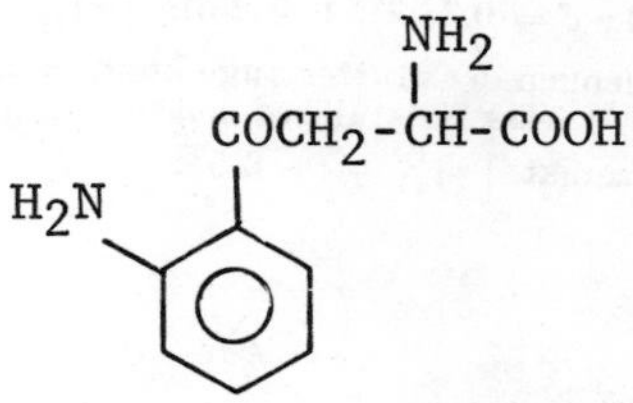

$C_{10}H_{12}N_2O_3$

Ref. 661

Spaltung von d, l-Kynurenin in seine optischen Antipoden über die l-Kynurenin-Rohrzuckerverbindung

A. Herstellung der Kynurenin-Zucker-Verbindung. Eine wäßrige Lösung von 4,5 g synthetischem d, l-Kynureninsulfat wurde mit Bariumhydroxyd neutralisiert. Das ausgefallene Bariumsulfat wurde abzentrifugiert, die klare Lösung im Vakuum im Stickstoffstrom auf ein geringes Volumen eingeengt, mit 18 g Rohrzucker versetzt und weiter eingeengt, bis freies Kynurenin sich auszuscheiden begann. Die Lösung wurde filtriert, das Filtrat (etwa 250 ccm) mit 1250 ccm Alkohol versetzt und 14 Tage bei etwa 0° stehen gelassen.

Die sich allmählich ausscheidenden nadelförmigen Kristalle der l-Kynurenin-Rohrzuckerverbindung sind von schwach gelber Farbe und rosettenförmig geordnet. Gewicht: 1,18 g. Sie wurden mehrfach aus wenig Wasser und der 10fachen Menge abs. Alkohol durch Stehenlassen bei — 12° umgelöst.

Schmelzpunkt: Zersetzung zwischen 142—155°.
Analyse: 3,434 mg Sbst.: 5,825 mg CO_2, 2,010 mg H_2O
2,791 mg Sbst.: 0,119 ccm N (24°, 775 mm)
$C_{22}H_{34}O_{14}N_2 \cdot H_2O$ Ber. C 46,47, H 6,38, N 4,93
Gef. C 46,28, H 6,55, N 5,00

Titration nach Graßmann und Heyde[7]) in 90proz. Alkohol mit Thymolphthalein: 5,4 mg Sbst. verbrauchen 0,44 ccm n/40 KOH
Ber. für 1 COOH 0,38 ccm n/40 KOH

(Der Umschlagspunkt ist schwer festzustellen, da die Lösung an und für sich gelb ist und daher zum Grün umschlägt.)

B. Gewinnung von l-Kynureninsulfat und l-Kynurenin. 400 mg der Kynurenin-Zuckerverbindung wurden in möglichst wenig warmem 90proz. Alkohol gelöst, mit 0,3 ccm konz. Schwefelsäure versetzt und zur Kristallisation bei 0° aufbewahrt. Es kristallisierten 240 mg l-Kynureninsulfat, die aus 90proz. Alkohol mehrmals umgelöst wurden.

Optische Drehung der letzten 3 Umkristallisationen des l-Kynureninsulfats:

$\alpha = +0{,}151$; $c = 0{,}9500$; $l = 2$ dm $[\alpha]_D^{19} = +7{,}9$

$\alpha = +0{,}146$; $c = 0{,}9750$; $l = 2$ dm $[\alpha]_D^{19} = +7{,}5$

$\alpha = +0{,}157$; $c = 1{,}005$; $l = 2$ dm $[\alpha]_D^{19} = +7{,}8$

Mittel $[\alpha]_D^{19} = +7{,}7°$

Das aus dem Kynureninsulfat dargestellte freie l-Kynurenin ergab folgende optische Drehung:

$\alpha = -0{,}200$; $c = 0{,}3175$; $l = 2$ dm $[\alpha]_D^{19} = -31{,}5°$

Vergleich mit der optischen Drehung des natürlichen l-Kynureninsulfats:

$\alpha = +0{,}077$; $c = 1{,}010$; $l = 1$ dm $[\alpha]_D^{19} = +7{,}6$

$\alpha = +0{,}070$; $c = 0{,}9850$; $l = 1$ dm $[\alpha]_D^{19} = +7{,}1$

$\alpha = +0{,}070$; $c = 0{,}9700$; $l = 1$ dm $[\alpha]_D^{19} = +7{,}2$

Mittel $[\alpha]_D^{19} = +7{,}3°$

7) H.-S. 183, 32 (1929)

und der optischen Drehung des natürlichen freien l-Kynurenins:

$$\alpha = -0{,}210;\ c = 0{,}3575;\ l = 1\ \text{dm}\ [\alpha]_D^{19} = -29{,}4°$$

C. Gewinnung von d-Kynureninsulfat und d-Kynurenin. Die Mutterlauge der l-Kynurenin-Rohrzucker-Verbindung wurde mit 0,1 ccm konz. Schwefelsäure versetzt. Das d-Kynureninsulfat begann langsam auszukristallisieren; die Lösung wurde auf etwa $^2/_3$ ihres Volumens eingeengt und das Sulfat abfiltriert. Gewicht: 1,0 g; es wurde aus 90proz. Alkohol mehrmals umgelöst.

Optische Drehung der 4 letzten Umkristallisationen des d-Kynureninsulfats:

$$\alpha = -0{,}086;\ c = 1{,}010;\ l = 1\ \text{dm}\ [\alpha]_D^{19} = -8{,}5$$
$$\alpha = -0{,}083;\ c = 1{,}015;\ l = 1\ \text{dm}\ [\alpha]_D^{19} = -8{,}2$$
$$\alpha = -0{,}088;\ c = 1{,}0175;\ l = 1\ \text{dm}\ [\alpha]_D^{19} = -8{,}6$$
$$\alpha = -0{,}083;\ c = 0{,}9725;\ l = 1\ \text{dm}\ [\alpha]_D^{19} = -8{,}5$$
$$\text{Mittel}\ [\alpha]_D^{19} = -8{,}5°$$

Aus dem d-Kynureninsulfat wurde freies d-Kynurenin dargestellt, dessen optische Drehung sich ergab zu:

$$\alpha = +0{,}231;\ c = 0{,}3850;\ l = 2\ \text{dm}\ [\alpha]_D^{19} = +30{,}0°$$

Durch weiteres Einengen der Mutterlauge konnten noch 130 mg Kynureninsulfat gewonnen werden. Dieses Kristallisat zeigte sich als Gemisch von l-Kynureninsulfat mit dem Razemat. $[\alpha]_D^{19} = +6{,}5°$.

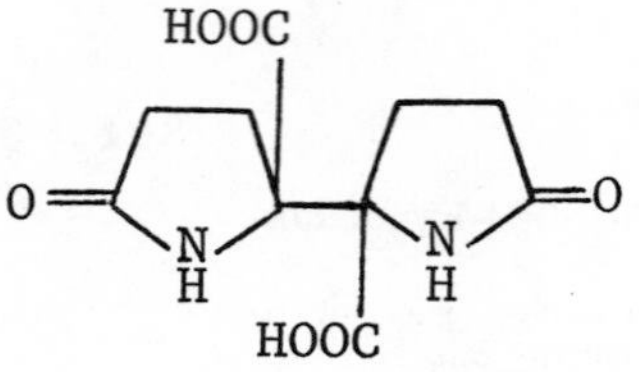

$C_{10}H_{12}N_2O_6$ Ref. 662

To 1.0 g of (±) 5,5'-di-pyroglutamic acid in 60 ml of hot water was added 3.36 g of brucine hydrochloride. The solution was filtrated and concentrated to half of its volume and allowed to stand at 4°c overnight. The resulting crystalline precipitate was filtrated and washed with cold water and recrystallized from water three times, affording 0.79 g of (+)-A-(-)-brucine complex as white needles, showing $[\alpha]_D^{30}$ -14.1° (c = 1, $CHCl_3$). This complex (0.22 g) was dissolved in 2 ml of 0.03N KOH and the alkaline solution was extracted with methylene chloride to remove the precipitated brucine. The alkaline solution was acidified to pH. 1 with conc hydrochloric acid or trifluoroacetic acid, affording white precipitate. The precipitate was collected and recrystallized from water three times, and 0.038 g of white crystals was obtained, showing $[\alpha]_D^{30}$ +5.63° (c = 1.6, 1.3% KOH soln.). The mother liquor mentioned above was concentrated to 15 ml and allowed to stand at 4°C overnight. The resulting crystalline precipitate was filtrated and washed with cold water and recrystallized from water, affording 0.90 g of (-)-A-(-)-brucine complex as white needles with $[\alpha]_D^{30}$ -20.5° (c = 1, $CHCl_3$). This complex (0.19 g) was also treated in the similar manner mentioned above and gave white precipitate. The precipitate was recrystallized from water, affording 0.035 g of granular crystals, showing $[\alpha]_D^{30}$ -6.50° (c = 1.6, 1.3% KOH soln.). No increase in optical rotation was observed on further crystallization.

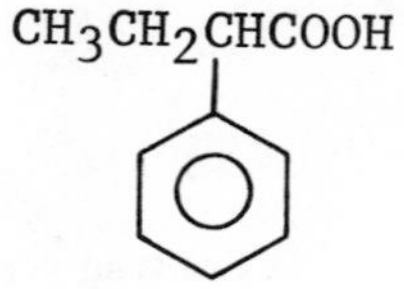

$C_{10}H_{12}O_2$ Ref. 663

Optical resolution of α-phenylbutyric acid

(+)-*Phenylethylamine* (−)-*α-phenylbutyrate:* 31 g (0.19 moles) of racemic α-phenylbutyric acid and 23 g (0.19 moles) of (+)-α-phenethylamine were dissolved in a hot mixture of 300 ml of benzene and 75 ml of ethanol and allowed to crystallize. The salt was filtered off and recrystallized from a mixture of benzene and ethanol 4:1 in the same manner as described for hydratropic acid. The pure (+)-phenethylamine salt of (−)-α-phenylbutyric acid was obtained as colourless needles. The progress of the resolution is seen in Table 8.

(−)-*α-Phenylethylamine* (+)-*α-phenylbutyrate:* 17.0 g of α-phenylbutyric acid ($[\alpha]_D^{25} = +50°$) and 12.5 g of (−)-α-phenethylamine were dissolved in a hot mixture of benzene and ethanol 4:1 and the resulting precipitate recrystallized until the optical activity of the liberated acid remained constant. The results are collected in Table 8.

Table 8.

Crystallization number	ml of solvent	g of (−)-acid salt obtained	$[\alpha]_D^{25}$	ml of solvent	g of (+)-acid salt obtained	$[\alpha]_D^{25}$
1	375	21	−60°	175	24	+72°
2	125	19	−69°	143	19.5	+77°
3	113	17	−77°	112	17.2	+77°
4	100	15	−77°			

(−)-*α-Phenylbutyric acid:* The (+)-phenethylamine salt was decomposed with dilute sulfuric acid and the organic acid was extracted with ether. After evaporation of the solvent the acid remained as a colourless oil, which was distilled *in vacuo*. B.p. 102–104°/0.4 mm Hg.

156.8 mg of the acid required 19.22 ml of 0.0495*N* NaOH

$C_{10}H_{12}O_2$: Equiv. wt. calc. 164.2, found 164.8

Optical acitivity in homogenous state: $0.5\alpha_D^{23} = -47.9°$, $[\alpha]_D^{23} = -95.8°$ $[M]_D^{23} =$ 157.2°

Optical activities in different solvents are collected in Table 9.

Table 9.

Solvent	mg of acid	$2\alpha_D^{25}$	$[\alpha]_D^{25}$	$[M]_D^{25}$
Acetone	100.9	−2.058°	−102.0°	−167.5°
Benzene	100.6	−1.918°	− 95.3°	−156.5°
Acetic acid	130.4	−2.191°	− 84.1°	−138.1°
Ether	100.0	−1.617°	− 80.9°	−132.8°
Ethanol	100.6	−1.580°	78.5°	−129.0°
Water (ion)	100.2	−0.129°	− 6.4°	− 10.6°

Amide: Colourless needles, m.p. 80.5–81.5°. After premelting it melted at 78.5°.

Anilide: Colourless needles, m.p. 79.5–80.5. Optical activities in different solvents are collected in Table 10.

Table 10.

Solvent	mg of anilide	$2\,\alpha_D^{25}$	$[\alpha]_D^{25}$	$[M]_D^{25}$
Ether	88.3	− 2.789°	− 158.1°	− 378.3°
Benzene	93.9	− 2.569°	− 136.8°	− 327.2°
Ethanol	98.1	− 2.073°	− 105.7°	− 252.8°
Acetic acid	92.7	− 1.698°	− 91.6°	− 219.2°
Acetone	91.8	− 1.508°	− 82.1°	− 196.6°

(+)-*α-Phenylbutyric acid:* From the (−)-α-phenethylamine salt as described for the (−)-acid. Colourless oil b.p. 98–100°/0.3 mm Hg.

157.5 mg of the acid consumed 19.28 ml 0.0495*N* NaOH.

$C_{10}H_{12}O_2$: Equiv. wt. calc. 164.2, found 165.0. Optical activity in the homogenous state: $0.5\alpha_D^{23} = +47.7°$, $[\alpha]_D^{23} = +95.4°$, $[M]_D^{23} = +156.7°$.

142.3 mg of the acid was dissolved in ethanol and made up to 10.00 ml.
$2\alpha_D^{25} = +2.236°$, $[\alpha]_D^{25} = +78.6°$, $[M]_D^{25} = +129.0°$.

$C_{10}H_{12}O_2$ Ref. 663a

Les F. ne sont pas corrigés.

1. Acide (+)-*α-phényl-α-éthyl-acétique* (*II*). La méthode de DELÉPINE a été quelque peu simplifiée. Dans 2 l d'alcool éthylique à 50%, on dissout à chaud 200 g d'acide phényléthylacétique racémique et 300 g de cinchonidine. Il cristallise 360 g de sel dont l'acide libre présente $[\alpha]_D^{20} = +58{,}2°$ (c = 2,6, benzène) soit 80% d'isomère (+). F. du sel brut 100°.

Recristallisés dans 1500 ml d'alcool à 50%, ces 360 g de sel fournissent 300 g d'un sel dont l'acide libre a $[\alpha]_D^{18} = +68{,}5°$ (c = 5,33, benzène). Cette cristallisation répétée encore 3 fois conduit au sel optiquement pur, de F. 127°. Acide libre: $[\alpha]_D^{16} = +96{,}6°$ (c = 1,5, benzène). Rendement 60% en acide (+).

2. Acide (−)-*α-phényl-α-éthyl-acétique* (*III*). Les liqueurs-mères des deux premières cristallisations ci-dessus, sont additionnées à chaud de 55 g de cinchonidine. Il cristallise au bout d'une nuit un sel d'acide lévogyre: acide libre $[\alpha]_D^{16} = -65°$ (c = 1,4, benzène), soit 84% d'isomère (−). D'une apparence différente du sel précédent, ce sel de cinchonidine est floconneux, léger, et cristallise plus lentement. Au cours des recristallisations soit dans l'alcool, à certaines concentrations, soit dans l'eau, nous avons constaté la synchristallisation observée par DELÉPINE et qui bloque le dédoublement pour un acide à $[\alpha]_D^{19} = -77°$ (c = 1,48, benzène). La barrière peut être le plus aisément franchie si l'on travaille dans l'alcool éthylique à 50%, ce qui permet d'obtenir le sel d'un acide de $[\alpha]_D^{15} = -85°$ (c = 1,4, benzène) puis, par deux recristallisations dans le même solvant, le sel dont l'acide optiquement pur a $[\alpha]_D^{19} = -96{,}0°$ (c = 1,36, benzène). Rendement 15% en acide (–).

Un autre procédé pour l'obtention de cet acide optiquement pur consiste à traiter les liqueurs-mères du sel brut de cinchonidine dextrogyre (selon 1) par l'acide chlorhydrique concentré de façon à libérer un acide déjà lévogyre. Celui-ci est extrait par le benzène, le solvant est éliminé et on purifie comme suit l'acide lévogyre obtenu: à la solution de 6,3 g de cet acide dans 120 ml d'éthanol bouillant, on ajoute 11,6 g de codéine. La cristallisation, qui est lente, donne un sel de codéine dont l'acide a $[\alpha]_D^{18} = -75°$ (c = 1,48, benzène). Par deux recristallisations dans l'eau, nous avons obtenu le sel de codéine optiquement pur, de F. 159°. Acide libre: $[\alpha]_D^{19} = -96{,}0°$ (c = 2,45, benzène). Rendement 20% en acide (−).

[3]) M. DELÉPINE & F. LARÈZE, Bull. Soc. chim. France **22**, 104 (1955).

$C_{10}H_{12}O_2$ Ref. 663b

Préparation d'acides α-phénylbutyriques optiquement actifs.

L'acide dextrogyre, optiquement pur, a été préparé par la méthode de M. DELÉPINE et F. LARÈZE (5).

100 g d'acide α-phénylbutyrique racémique sont dissous à chaud dans un mélange de 1 330 cm³ d'eau et 60 cm³ de lessive de soude à 40 %. A cette solution d'α-phénylbutyrate de sodium sont ajoutés 666 g d'éthanol et 113 g de sulfate de cinchonidine; tout se dissout à l'ébullition. Par refroidissement, il précipite un beau sel blanc; poids: 131 g (A). Après quatre recristallisations dans trois fois son poids d'éthanol à 96°, on obtient 59 g de sel, lequel est mis en suspension dans l'eau; on acidifie avec de l'acide chlorhydrique concentré et extrait deux fois l'acide avec du benzène. Après évaporation du solvant, on obtient 20,5 g de liquide jaunâtre qui est distillé. Il vient 17,15 g de liquide incolore: $Eb_{17} = 159\text{-}159{,}5°$.

$\alpha_{D\,liquide}^{21} = +97{,}3°$ ($l = 1$); $[\alpha]_D^{21} = +97{,}8°$ (benzène; $c=3$).

Le pouvoir rotatoire varie beaucoup avec la température, aussi bien quand on examine sa rotation à l'état liquide, qu'en solution dans le benzène: $\alpha_{D\,liquide}^{15} = +100{,}1°$ ($l = 1$) et pour les mêmes conditions à 17°: + 99,06°; à 19°: + 98,16°; à 21°: + 97,24°; à 23°: + 96,26° et à 25°: + 95,36°, soit une variation d'environ 1° de rotation pour deux degrés de température (la densité de ce liquide ne varie que de $d_4^{15} = 1{,}067$ à $d_4^{25} = 1{,}061$).

En solution à 2 % dans le benzène, on a les chiffres suivants: $[\alpha]_D^{15} = +101{,}9°$; $[\alpha]_D^{17} = +100{,}7°$; $[\alpha]_D^{19} = +99{,}2°$; $[\alpha]_D^{21} = +97{,}8°$; $[\alpha]_D^{23} = +96{,}5°$; $[\alpha]_D^{25} = +95{,}4°$.

5) M. Delepine and F. Lareze, Bull. Soc. chim. France 22, 104 (1955).

L'acide lévogyre, d'une pureté optique suffisante pour nos expériences, a été obtenu des deux manières suivantes:

1) *Par cristallisation du sel d'amphétamine.*

Les eaux-mères ayant fourni les 131 g de sel (A) précédent sont acidifiées et extraites au benzène et on obtient 34 g d'acide $\alpha_{D\,liquide}^{21} = -50°$ ($l = 1$); on ajoute 28,1 g (1 mol) d'amphétamine lévogyre $\alpha_{D\,liquide}^{21} = -14{,}92°$ ($l = 0{,}5$) et 220 cm³ de benzène; on dissout l'ensemble à chaud et recueille par refroidissement 46 g de sel incolore; ce sel est recristallisé trois fois dans de la méthyléthylcétone (5 cm³ pour 1 g) et les poids obtenus sont de 36 g, 31,5 g et 28 g; une dernière cristallisation dans le même solvant (10 cm³ pour 1 g) fournit 23 g de beau sel: $F_{(instantané)} = 122°$, $[\alpha]_D^{22} = -16{,}44°$ (chloroforme; $c = 4{,}4$);

Analyse $C_{19}H_{25}O_2N = 299{,}40$:

	C	O	H	N
Calc. % :	76,2	10,7	8,4	4,7
Tr. :	76,0	10,8	8,3	4,9.

Ce sel est mis en suspension dans l'eau, additionné de 25 cm³ d'acide chlorhydrique concentré et l'acide libéré est extrait plusieurs fois au benzène. L'évaporation du benzène fournit 12 g d'acide brut qui est distillé. On obtient 10,25 g de liquide incolore $Eb_{18} = 159°$; $\alpha_{D\,liquide}^{21} = -94°$ ($l = 1$).

2) *Par séparation du racémique.*

On dispose de 93 g d'acide provenant, comme précédemment, du dédoublement par le sulfate de cinchonidine $\alpha_{D\,liquide}^{20} = -54°$ ($l = 1$). On met les 93 g d'acide liquide à la glacière. Il y a prise en masse. On réchauffe légèrement à la température ordinaire, puis on essore. Il reste sur le buchner 30 g d'acide cristallisé (racémique en grande partie). Poids du filtrat: 63 g d'acide lévogyre liquide, de pouvoir rotatoire : $\alpha^{21}_{D\ liquide} = -70°$ ($l = 1$).

A 1 g de cet acide, on ajoute 5 cm³ de pentane. On abandonne à — 20° pendant une nuit après avoir amorcé avec de l'acide racémique; une cristallisation se produit. On décante rapidement la solution et on chasse le pentane. Il vient 711 mg d'huile. Pouvoir rotatoire : $[\alpha]_D^{21} = -87{,}2°$ (benzène).

Ce procédé rapide pour obtenir un produit de pureté optique suffisante est évidemment applicable au produit dextrogyre. On est cependant limité dans tous les cas par la solubilité du racémique dans le pentane qui est de 1,6 % à — 20°.

Préparation du chlorure d'acide α-phénylbutyrique lévogyre.

On met en suspension dans 30 cm³ de benzène anhydre, 4 g de sel de sodium d'acide α-phénylbutyrique lévogyre provenant d'un acide de pureté optique de 96,7 % ($\alpha_D = -93{,}4°$ pour $-96{,}5°$). La fiole étant placée dans un bain d'eau glacée, on ajoute goutte à goutte avec agitation magnétique 3,65 cm³ (soit deux fois la quantité théorique) de chlorure d'oxalyle en 5 mn. On observe un abondant dégagement gazeux. Petit à petit, le sel de sodium entre en réaction et la suspension devient de moins en moins épaisse. On continue l'agitation dans le bain glacé pendant 30 mn et on abandonne pendant une nuit à la température ordinaire, avec une garde de chlorure de calcium. Il y a un abondant dépôt de chlorure de sodium. On porte alors la fiole dans un bain d'huile à 60° pendant 30 mn, puis on chauffe (température maximum 110°), de façon à chasser l'excès de chlorure d'oxalyle et le benzène à la pression ordinaire. On ajoute deux fois 15 cm³ de benzène anhydre et on chasse dans les mêmes conditions. On filtre le mélange et on distille le benzène à fond sous vide. On obtient une huile jaune qui est mise au dessiccateur sur P_2O_5 jusqu'à poids constant; poids = 3,54 g (Rdt = 90 %). Pouvoir rotatoire : $\alpha^{20}_{D\ liquide} = -105°$ ($l = 1$). Ce chlorure d'acide brut ne se racémise pas facilement et au bout de quatre mois, on trouve $\alpha^{21}_{D\ liquide} = -100°$ ($l = 1$). On le distille alors et obtient un liquide incolore : $Eb_{21} = 117{,}5\text{-}118°$; $\alpha^{21}_{D\ liquide} = -97{,}5°$ ($l = 1$). La distillation ne produit qu'une racémisation insignifiante puisqu'un nouveau fractionnement ($Eb_{22} = 119°$) fournit un produit $\alpha^{21}_{D\ liquide} = -96{,}5°$ ($l = 1$); afin d'évaluer la pureté optique de ce chlorure d'acide, on en chauffe 0,598 g à reflux dans un mélange de 5 cm³ de dioxanne et 0,5 cm³ d'eau pendant 20 mn. On reprend largement par l'eau et extrait 0,487 g d'acide α-phénylbutyrique incolore, $[\alpha]_D^{21} = -83°$ (benzène; $c = 2{,}4$); la pureté optique du chlorure d'acide initial est donc (au minimum) de $\frac{83}{97{,}8} = 85$ %, ce qui porte le pouvoir rotatoire du produit optiquement pur à :

$$\alpha^{21}_{D\ liquide} = \frac{96{,}5 \times 100}{85} = -113° \ (l = 1).$$

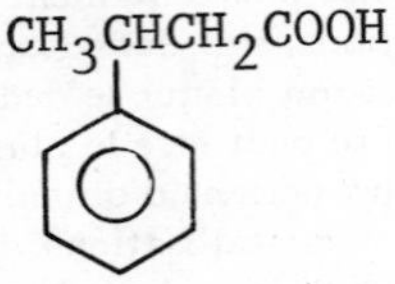

$C_{10}H_{12}O_2$ Ref. 664

(+)- *and* (−)-*β-Phenylbutyric acid.* 34.4 g (0.285 moles) of (−)-α-phenylethylamine were added to a solution of 46.6 g (0.285 moles) of rac. β-phenylbutyric acid in 127 ml of ethanol and 136 ml of water. After standing at room temperature overnight, the salt was collected and dried. It was recrystallized from ethanol and water and, after each recrystallization, the acid was liberated from a small amount of the salt and its rotation determined in benzene solution (Table 1).

Table 1.

Crystallization	Ethanol (ml)	Water (ml)	Salt obtained (g)	m.p. of salt	$[\alpha]_D^{25}$ of acid (in benzene)
1	127	136	32.5	135—140°	—
2	45	45	25.5	142—143.5°	+41.5°
3	25	35	19.7	144—145.5°	+48.5°
4	20	40	16.8	144.5—146°	+52.2°

From 16.8 g of α-phenylethylamine salt the acid was liberated. The resulting (+)-β-phenylbutyric acid (9.1 g 20 % $[\alpha]_D^{25} = +52.3°$) was used in the subsequent reactions without further purification. From the combined mother liquors, (−)-β-phenylbutyric acid (13.5 g, 0.082 moles, $[\alpha]_D^{25} = -30.8°$) was liberated. It was dissolved in 37 ml of ethanol and 39 ml of water. 9.6 g (0.082) moles) of (+)-α-phenylethylamine were added. The collected salt was recrystallized from ethanol and water (Table 2). Liberation of the acid gave 6.6 g (14 %) of (−)-β-phenylbutyric acid, $[\alpha]_D^{25} = -51.1°$.

Table 2.

Crystallization	Ethanol (ml)	Water (ml)	Salt obtained (g)	m.p. of salt	$[\alpha]_D^{25}$ of acid (in benzene)
1	37	39	15.1	142.5—144°	−48.2°
2	20	40	12.7	144—145.5°	−51.2°

$C_{10}H_{12}O_2$

Ref. 664a

(S)-β-Phenylbutyric Acid [(S)-2]. Resolution of (±)-**2** was accomplished using (S)-α-phenylethylamine, $[\alpha]^{25}_D$ −39° (neat). Five recrystallizations from ethanol–water gave a pure salt (44%): mp 145–147 °C; $[\alpha]^{26}_D$ +8.5° (c 1.6, CH_3OH) (lit.[19] mp 144.5–146 °C). Decomposition of this salt gave (S)-**2** (77%): bp 103–110 °C (0.3 mm); $[\alpha]^{26}_D$ +55° (c 1.5, benzene) [lit.[22] $[\alpha]^{25}_{589}$ −57.6° (c 2.7, benzene) for (R)-**2**].

(19) A.-M. Weidler and G. Bergson, *Acta Chem. Scand.*, **18**, 1483 (1964).
(20) E. L. Martin, *Org. React.*, **1**, 155 (1942).
(21) J. Grimshaw and P. G. Millar, *J. Chem. Soc. C*, 2324 (1970).
(22) J. Almy and D. J. Cram, *J. Am. Chem. Soc.*, **91**, 4459 (1969).

$C_{10}H_{12}O_2$

Ref. 664b

Note:

The racemic acid was resolved via its (-)-menthyl ester. The (-)-enantiomer had a b.p. 153-153.5°/10 mm. and $[\alpha]^{20}_D$ -57.0 (c 1.0, benzene).

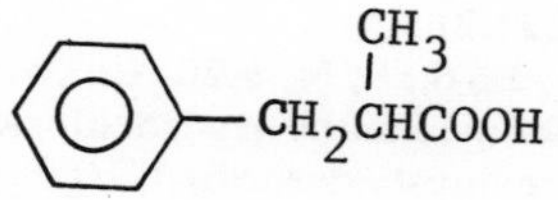

$C_{10}H_{12}O_2$

Ref. 665

L-(+)-*α-Methylhydrocinnamic acid.*[10] A cold solution of 41.0 g. of racemic I in 350 ml. of ethyl acetate was treated with 30.25 g. of (−)-α-methylbenzylamine (III),[30] heated to dissolve the precipitate, and allowed to cool slowly. The solid was recrystallized six times from ethyl acetate to yield 16.1 g. of the (+)-I-(−)-III salt as long colorless needles, m.p. 127–129.5°, $[\alpha]^{20}_D$ −1.48° (c 2.00, chloroform). Systematic recrystallization of the second crops gave another 1.8 g. of pure salt, bringing the total yield to 50%.

Anal. Calc'd for $C_{18}H_{23}NO_2$: C, 75.75; H, 8.12; N, 4.91. Found: C, 75.69; H, 8.08; N, 4.90.

The free acid was obtained in 95% yield by shaking the salt with ether and excess 2 N sulfuric acid and extracting the aqueous phase[31] with additional ether. The extracts were washed four times each with 2 N sulfuric acid and with sodium chloride solution, dried, and evaporated. Distillation gave (+)-I, b.p. 112°/0.25 mm., d^{20}_4 1.0654, $[\alpha]^{20}_D$ +23.51° (l 1, undiluted) (lit.[3] +22.65°).

D-(−)-*α-Methylhydrocinnamic acid.*[10] The impure (−)-I, recovered from the mother liquors of the (+)-I-(−)-III salt and treated with (+)-III,[30] yielded 51% of the (−)-I-(+)-III salt, m.p. 127.3–129.5°, $[\alpha]^{20}_D$ +1.36° (c 2.00, chloroform), which was also prepared from racemic I (after 10 recrystallizations: m.p. 128–130°, $[\alpha]^{20}_D$ +1.13°). The free acid had b.p. 111°/0.22 mm., n^{20}_D 1.5145, d^{22}_4 1.065, $[\alpha]^{22}_D$ −24.56° (l 1, undiluted).

(3) R. H. Pickard and J. Yates, *J. Chem. Soc.*, **95**, 1011 (1909).

(30) A. W. Ingersoll, *Org. Syntheses*, **Coll. Vol. 2**, 506 (1943); W. Theilacker and H. G. Winkler, *Ber.*, **87**, 690 (1954). Both methods appear equally satisfactory.

(31) Pure (−)-III is recovered from the aqueous solution by treatment with alkali, ether extraction, and distillation.

TABLE I

PHYSICAL CONSTANTS OF α-METHYLBENZYLAMINE SALTS

Salt	M.p., °C	$[a]_D$[a]	Solubility[b]
(+)-I-(+)-III[c]	120–121	+22.7°	2.64
(−)-I-(+)-III	127.3–129.5	+1.4°	1.00
(+)-I-(−)-III	127–129.5	−1.5°	—
(+)-IV-(+)-III[c]	107–109	+22.3°	1.39
(−)-IV-(+)-III[c]	113.5–114.5	−2.1°	1.47
(−)-IV-(−)-III	107–109	−22.0°	—
(+)-IV-(±)-III[c,d]	110.4–111.4	+12.8°	—

[a] 2% solution in chloroform. [b] Determined in ethyl acetate at 23° and expressed as grams/100 g. of solution. [c] Prepared from the two pure components and crystallized from ethyl acetate. [d] Racemic III was recovered from this salt.

$C_{10}H_{12}O_2$

Ref. 665a

Resolution of β-Phenylisobutyric Acid.[10]—To a solution of 500 g. of quinine in 650 ml. of hot ethanol was added 250 g. of *dl*-β-phenylisobutyric acid. The solution was cooled to 30° and seeded with the pure *d*-salt (obtained from a previous small scale run). The crystallization was allowed to proceed at room temperature for 24 hours and at 5° for another 24 hours. Filtration gave 265 g. of dry salt, $[\alpha]^{20}_D$ +110° (c 0.0203 g./cc., in ethanol). Three additional crops totaling 167 g. and having the same or slightly higher rotation were obtained by partial evaporation of the solvent and re-seeding. It is worth notice that the rotation of the quinine salt remained essentially constant even though the acid was not completely resolved.[10]

To 450 ml. of 10 N sulfuric acid was added 398 g. of the recrystallized *d*-salt. The liquid (+)β-phenylisobutyric acid was extracted with ether, washed, dried and distilled; yield 96%; α^{20}_D +13.73° (l = 1), $[\alpha]^{20}_D$ +11.5° (c 0.0327, ethanol, l = 2); reported α^{24}_D +21.6° (no solvent),[11] $[\alpha]^{21}_D$ + 17.87° (in ethanol[11]).

The (−)acid was obtained from the residue resulting from evaporation of the mother liquors to dryness, b.p. 123–124° at 1.5 mm.; $[\alpha]^{20}_D$ −11.2° (c 0.0245, ethanol, l = 2).[21]

The *p*-bromophenacyl ester of the (+)acid was obtained in 0.83-g. yield from 0.4 g. of acid and 0.71 g. of *p*-bromophenacyl bromide; m.p. 54.5–56°; $[\alpha]^{21}_D$ +23.8° (c 0.0136, ethanol, l = 2).

Anal. Calcd. for $C_{18}H_{17}BrO_3$: C, 59.8; H, 4.7; Br, 22.1. Found: C, 59.7, H, 5.0; Br, 22.0.

(10) (a) F. S. Kipping and A. E. Hunter, *J. Chem. Soc.*, **83**, 1005 (1903); (b) R. H. Pickard and J. Yates, *ibid.*, **95**, 1019 (1909).
(11) J. Kenyon, H. Phillips and V. P. Pittman, *ibid.*, 1072 (1935).

$C_{10}H_{12}O_2$

Ref. 665b

To a solution of 234 g. (1.058 moles) of DL-N-acetyl-α-methylphenylalanine in 438 ml. of absolute ethanol is added a solution of 312 g. (1.058 moles) of cinchonidine in 438 ml. of absolute ethanol. The mixture is cooled to room temperature, diluted with ether (1930 ml.), seeded with adduct and allowed to stand overnight at 25°. The precipitate is washed with three 75 ml. portions of 3:1 ether:ethanol. After drying in vacuo at 50°, the yield of adduct amounts to 222.5 g., m.p. 207°-210°. Recrystallization from ethanol-ether as before yields 163.0 g., m.p. 215°-216° $[\alpha]_{578}^{23}$ −49.2° (C 1, abs. ethanol). The analytical material is crystallized to constant values, m.p. 218°-219° dec., $[\alpha]_{578}^{23}$ = −47.0° (C 1, abs. ethanol) u/v max. (CH_3OH) 225 mμ (logε4.57), 285 (3.71), 302 (3.59), 315 (3.50).

Anal. Calcd. for $C_{31}H_{37}N_3O_4$: C, 72.20; H, 7.23; N, 8.13.

Found: C, 71.96; H, 7.25; N, 8.27.

L(−)-N-Acetyl-α-methylphenylalanine

To 1750 ml. of water and 131 ml. of 2.5 N hydrochloric acid is added 163.0 g. (0.316 mole) of acetylaminoacid cinchonidine salt and the mixture stirred at 0°–5° for 1 hour. An additional 172 ml. portion of 2.5 N hydrochloric acid is added and the stirring continued at 0°–5° for 1 hour. The mixture is filtered and the precipitate washed with three 75 ml. portions of cold water. After drying at 50° in vacuo the yield of L(−)-N-acetyl-α-methylphenylalanine, m.p. 203°–205°, $[\alpha]_{578}^{23}$ −42.1° (C 1, abs. ethanol), amounts to 64.4 g. The analytical material is recrystallized from acetone, m.p. 206°–207°, u/v max. (CH_3OH) 252 mμ (log ε2.17), 257.5 (2.24), 263 (2.09), sh 247.5 (2.00).

Anal. Calcd. for $C_{12}H_{15}NO_3$: C, 65.14; H, 6.83; N, 6.33; Eq. wt. 221.26.

Found : C, 65.17; H, 6.73; N, 6.59; Eq. wt. 222.

L(−)-α-Methylphenylalanine Hydrochloride

To 536 ml. of 6 N hydrochloric acid is added 64.4 g. (0.291 mole) of L(−)-N-acetyl-α-methylphenylalanine. After 6 hours at reflux, the mixture is cooled to room temperature spontaneously then at 15° for 30 minutes. The mixture is filtered and the precipitate is washed with three 25 ml. portions of 2 N hydrochloric acid at 0°–5°. After drying in vacuo at 50°, the yield of L(−)-α-methylphenylalanine hydrochloride amounts to 55.1 g., m.p. 217°–219°, $[\alpha]_{578}^{23}$ −2.10° (C 0.2, 1 N HCl). pH ½ 2.6, u/v max. (CH_3OH) 252 mμ (logε2.24), 258 (2.30), 263 (2.19), sh 247.5 (2.11).

Anal. Calcd. for $C_{10}H_{13}NO_2 \cdot HCl$: C, 55.69; H, 6.54; N, 6.49; Cl, 16.44; Eq. wt. 215.69.

Found: C, 55.99; H, 6.51; N, 6.48; Cl, 16.29; Eq. wt. 215.5.

L(−)-α-Methylphenylalanine

The L(−)-α-methylphenylalanine hydrochloride is dissolved in water and passed over resin, IR-120 (H_3^+O cycle). The column is eluted with water to neutrality and following that with 500 ml. of 1 N ammonium hydroxide. Concentration of the ammoniacal eluate yields crude product, m.p. 290°–291°, dec. sub. The crude L(−)-α-methylphenylalanine is dissolved in 178 ml. of refluxing methanol and upon concentration to about 100 ml., crystallization begins. The mixture is allowed to cool spontaneously and stand overnight at room temperature, chilled to 5°, filtered and the precipitate washed with methanol. After drying at 50° in vacuo, the yield of L(−)-α-methylphenylalanine, m.p. 307.5°, dec. sub., $[\alpha]_{578}^{23}$ −2.5° (C 1, 1 N HCl), u/v max. (CH_3OH) 237 mμ, (log ε2.02), 252.5 (2.23), 258 (2.30), 264 (2.18), sh 247 (2.13).

Anal. Calcd. for $C_{10}H_{13}NO_2$: C, 66.97; H, 7.31; N, 7.82.

Found: C, 66.86; H, 7.14; N, 7.60.

Enriched D(+)-N-Acetyl-α-methylphenylalanine

By the method previously described, 27.55 g. of enriched D(+)-N-acetyl-α-methylphenylalanine, m.p. 193°–195°, is obtained from the mother liquors of the L(−)-enantiomorph, cinchonidine salt.

D(+)-N-Acetyl-α-methylphenylalanine Cinchonine Salt Monohydrate

Equimolar amounts (0.020 mole) of enriched D(+)-N-acetyl-α-methylphenylalanine and cinchonine are dissolved in 40 ml. of warm ethanol. The mixture is cooled to room temperature, seeded and allowed to stand for 64 hours. After cooling the mixture to 0°, the precipitate is separated by filtration, washed and dried at room temperature. The pure salt monohydrate amounts to 3.25 g., m.p. 116°–118°, $[\alpha]_{578}^{27}$ +88.7° (C 2, abs. ethanol). The salt is recrystallized from absolute ethanol with no change in rotation or melting point, u/v max. (CH_3OH) 226 mμ (log ε4.54), 285 (3.68), 303 (3.56), 316 (3.49) IR spectrum consistent.

Anal. Calcd. for $C_{31}H_{37}N_3O_4 \cdot H_2O$: C, 69.80; H, 7.37; N, 7.87.

Found: C, 70.00; H, 7.72; N, 7.40.

D(+)-N-Acetyl-α-methylphenylalanine

By the method described for the L(−)-enantiomorph 0.42 g. of D(+)-N-acetyl-α-methylphenylalanine, m.p. 204°–207°, $[\alpha]_{578}^{26}$ +42.7° (C 1, abs. ethanol) is obtained. The material is recrystallized from 33 percent ethanol to yield an analytical sample, 0.30 g., m.p. 206°–207°, $[\alpha]_{578}^{24}$ +44.0° (C 1, abs. ethanol), u/v max. (CH_3OH) 247 mμ, (log ε2.03), 252.5 (2.19), 258 (2.26), 263 (6.03), IR consistent, pH ½ 4.95 (50 percent methanol).

Anal. Calcd. for $C_{12}H_{15}NO_3$: C, 65.14; H, 6.83; N, 6.33; Eq. wt. 221.26.

Found: C, 65.43; H, 6.69; N, 6.20; Eq. wt. 222.

D(+)-α-Methylphenylalanine Hydrochloride

By the method previously described, 10.0 g. of D(+)-α-methylphenylalanine hydrochloride, m.p. 217°–219° is obtained. The analytical material, 0.245 g. is obtained by recrystallization from acetone/n-hexane, m.p. 214°–215°, dec. $[\alpha]_{578}^{22}$ +2.1° (C 1.2, 1 N HCl), u/v max. (CH_3OH)252.5 (logε2.14), 258 (2.21), 264 (2.10).

Anal. Calcd. for $C_{10}H_{13}NO_2 \cdot HCl$: C, 55.69; H, 6.54; N, 6.49; Cl, 16.44.

Found: C, 55.76; H 6.43; N, 6.64; Cl, 16.15.

D(+)-α-Methylphenylalanine

By the method previously described, 7.4 g. of D-α-methylphenylalanine, m.p. 307°–308°, dec. sub., is obtained. This material is recrystallized as was the L(−)-enantiomorph to yield 3.55 g. of product, m.p. 315°, dec. sub., purity by phase solubility $[\alpha]_{578}^{23}$ +20.6°(C 1, H_2O), u/v max. (CH_3OH) 247 mμ (logε1.99), 252 (2.15), 258 (2.25), 264 (2.12), IR spectrum identical with that of L(−)-compound.

Anal. Calcd. for $C_{10}H_{13}NO_2$: C, 66.97; H, 7.31; N, 7.82.

Found: C, 66.98; H, 7.59; N, 7.84.

DL-α-Methylphenylalanine Hydrochloride

Equimolar amounts (0.88 g., 0.00408 mole) of D(+)- and L(−)-α-methylphenylalanine are dissolved in 15 ml. of concentrated hydrochloric acid and crystallized. The yield of racemic material is 0.75 g., m.p. 241°–244°, racemic at all wave lengths, u/v max. (CH_3OH) 252.5 mμ (logε2.15), 258 (2.28), 264 (2.18), sh 243 (1.88), sh 248 (2.02), sh 267.5 (1.82), IR spectrum identical with that obtained by Stein et al., J. Am. Chem. Soc. 77, 700 (1955).

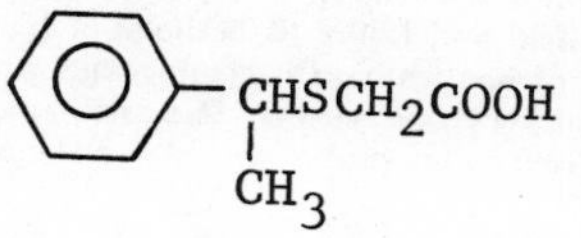

$C_{10}H_{12}O_2S$

Ref. 666

Beschreibung der Versuche.

1. Die Spaltung der inaktiven α-Phenäthylthioglykolsäure.

Zu einer neutralen Lösung von 98 g inaktiver α-Phenäthylthioglykolsäure und 20 g Natriumhydroxyd in 500 ccm Wasser wurde in mehreren Portionen eine Lösung von 61 g (+)-Phenäthylamin in 250 ccm 2-n. Salzsäure zugesetzt und, als bleibende Emulgierung eintrat, wurde das Gemisch mit bei einem Vorversuch[1] erhaltenen Kristallen geimpft. Am folgenden Tage waren dann 46 g Salz vom Schmp. 117°—120° und $[\alpha]_D = +139^\circ{,}6$[2] ausgeschieden. Die Mutterlauge ergab beim Ansäuern ein visköses Öl, welches zusammen mit einer durch Extraktion der sauren Lösung mit Äther gewonnenen, kleineren Portion Säure mit 358 ccm 1-n. Natronlauge neutralisiert wurde. Die hierdurch erhaltene Lösung wurde mit 42 g (—)-Phenäthylamin in 173,5 ccm 2-n. Salzsäure versetzt, wobei eine Emulsion entstand, welche nach Impfen mit (— —)-Kristallen von dem Vorversuch 67 g Salz vom Schmp. 112°—116° und $[\alpha]_D = -125^\circ{,}6$ lieferte. Die aus der hier entstandenen Mutterlauge zurückgewonnene Säure wurde durch 142 ccm 1-n. Natronlauge neutralisiert und ergab dann wie vorher mit 17,2 g (+)-Phenäthylamin in 71 ccm 2-n. Salzsäure 25 g Salz vom Schmp. 113°—117° und $[\alpha]_D = +124^\circ{,}8$. — Das erhaltene, linksdrehende Salz wurde in der 10-fachen Menge heissen Wassers gelöst und, als die Lösung beim Erkalten milchig wurde, wurde sie mit einer Spur von (— —)-Salz geimpft und dann umgerührt, bis sie breiartig erstarrt war. Das am folgenden Tage abgenutschte und mit Wasser gewaschene Salz wog 38 g, schmolz bei 121°—123° und zeigte $[\alpha]_D = -166^\circ{,}3$, wärend die aus der Mutterlauge regenerierte Säure nach Neutralisierung mit 82,5 ccm 1-n. Natronlauge $0{,}5\ \alpha_D = -6^\circ{,}72$ zeigte.[3] Beim Umkristallisieren aus 380 ccm Wasser ergab das Salz 24 g Kristalle vom Schmp. 122°—124° und $[\alpha]_D = -178^\circ{,}5$; nach Eindunstung der Mutterlauge auf dem Wasserbade auf ca. 100 ccm Volumen wurden noch 10 g Salz vom Schmp. 119°—122° und $[\alpha]_D = -168^\circ{,}9$ erhalten, während die aus der Mutterlauge isolierte Säure von 11,2 ccm 1-n. Natronlauge neutralisiert wurde und dabei eine Lösung von $0{,}5\ \alpha_D = -9^\circ{,}36$ lieferte. Das Salz von $[\alpha]_D = -178^\circ{,}5$ wurde noch zweimal in derselben Weise wie früher umkristallisiert, wobei sein Drehungsvermögen bei der letzten Umkristallisierung nicht verändert wurde und die aus den Mutterlaugen regenerierten Säuren in 1-n. Natronlauge bzw. $0{,}5\ \alpha_D = -16^\circ{,}2$ und $-18^\circ{,}4$ zeigten.

Das reine (— —)-Salz bestand aus farblosen, flachen, fast nadelförmigen Prismen vom Schmp. 124°—125° und enthielt 4,48, ber. 4,42, % N. Eine Lösung von 0,5147 g Salz in abs. Alkohol zu 10,04 ccm zeigte $\alpha_D = -9^\circ{,}27$, $[\alpha]_D = -180^\circ{,}8$ und $[M]_D = -573^\circ{,}7$.

Die beiden Portionen rechtsdrehendes Salz wurden wie oben zweimal aus der 10-fachen Menge heissem Wasser umkristallisiert, wobei zuerst aus 70 g Salz 44 g vom Schmp. 122°—124° und $[\alpha]_D = +172^\circ{,}1$ und dann 30 g reines (+ +)-Salz gewonnen wurden. Das zuletzt erhaltene Präparat schmolz bei 124°—125°, enthielt 10,01, ber. 10,10, % S und zeigte für 0,5101 g in abs. Alkohol zu 10,04 ccm gelöst $\alpha_D = +9^\circ{,}18$, $[\alpha]_D = +180^\circ{,}7$ und $[M]_D = +573^\circ{,}2$.

Bei Versuchen zur Darstellung von Salzen aus aktiven α-Phenäthylthioglykolsäuren und Phenäthylaminen der entgegengesetzten Drehungsrichtungen wurden nur visköse, nicht kristallisierbare Öle erhalten. Aus den oben mitgeteilten Versuchen geht jedoch hervor, dass auch kristallisierte Präparate in gewissem Grade aus Gemischen diastereomerer Salze bestehen können.

[1] Hierbei wurde die Lösung von dem abgeschiedenen, klebrigen Öl abdekantiert und freiwilliger Eindunstung überlassen, wobei Kristalle von rechtsdrehendem Salz abgeschieden wurden. Die aus der Mutterlauge zurückgewonnene, linksdrehende Säure lieferte in derselben Weise ein öliges Salz mit (—) Phenäthylamin, welches jedoch schon beim Anreiben kristallisierte.

[2] Wo nichts anderes gesagt wird, beziehen sich die Drehungsangaben auf ca. 5-proz. Lösungen in abs. Alkohol bei Raumtemperatur.

[3] Das maximale Drehungsvermögen beträgt unter diesen Verhältnissen (siehe S. 6) $0{,}5\ \alpha_D = -18^\circ{,}9$.

2. Optisch aktive α-Phenäthylthioglykolsäuren.

Aus 18 g des oben beschriebenen, reinen (— —)-Salzes wurde die Säure mit Salzsäure freigemacht und mit Äther aufgenommen. Nach freiwilliger Verdunstung des Äthers und Trocknen des Rückstandes im Vakuum über Schwefelsäure waren 11 g einer farblosen, dickflüssigen Flüssigkeit zurückgeblieben, von welcher 0,3360 g zur Neutralisation 15,89 ccm 0,1063-n. Natronlauge verbrauchten, was ein Äquiv.-Gewicht von 198,9, ber. 196,15, ergibt. Von diesem Präparat zeigten 0,5256 g, gemäss der Titrierung 0,5183 g reiner Säure entsprechend, in abs. Alkohol zu 10,04 ccm gelöst $\alpha_D = -17^\circ{,}20$, $[\alpha]_D = -333^\circ{,}2$, $[M]_D = -653^\circ{,}5$, und eine mit 19,43 ccm 0,1063-n. Natronlauge neutralisierte und dann auf 20,0 ccm Volumen verdünnte Lösung von 0,4108 g davon, 0,4505 g Na-Salz entsprechend, zeigte $\alpha_D = -4^\circ{,}66$, also für die Säure $[\alpha]_D = -230^\circ{,}1$, $[M]_D = -451^\circ{,}3$, und für das Salz $[\alpha]_D = -206^\circ{,}9$, $[M]_D = -451^\circ{,}3$.

Von einer durch Schütteln von überschüssiger Säure mit Wasser erhaltenen Lösung verbrauchten 10,00 ccm 2,06 ccm 0,1063-n. Natronlauge, was einer Konzentration von 0,0430 g Säure in 10 ccm Lösung entspricht. Die Lösung zeigte $2{,}2\ \alpha_D = -2^\circ{,}21$, $[\alpha]_D = -234^\circ$, $[M]_D = -458^\circ$, und nach 4-stündigem Erhitzen im Wasserbade ergab sie $2{,}2\ \alpha_D = -2^\circ{,}07$. Eine schwache Racemisierung scheint also unter diesen Verhältnissen eingetreten zu sein. Als andererseits 18 ccm der Lösung mit 2 ccm 2-n. Salzsäure versetzt wurden, entstand eine Lösung von $2{,}2\ \alpha_D = -1^\circ{,}77$, $[\alpha]_D = -208^\circ$, $[M]_D = -408^\circ$, und nach 4-stündigem Erhitzen von $2{,}2\ \alpha_D = -1^\circ{,}78$. Auch eine Lösung der Säure in der entsprechenden Menge 1-n. Natronlauge vertrug Erhitzen ohne merkbare Racemisierung, denn sie zeigte sowohl direkt als auch nach Erhitzen wie oben $0{,}5\ \alpha_D = -18^\circ{,}9$. Nach Verdünnen dieser Lösung mit dem gleichen Volumen 1-n. Lauge zeigte sie $0{,}5\ \alpha_D = -9^\circ{,}76$, und nach Erhitzen wie üblich $0{,}5\ \alpha_D = -9^\circ{,}69$, was auf eine sehr langsame Racemisierung oder auch auf eine geringfügige Zersetzung dank der Einwirkung des Hydroxylions deutet. Wurden andererseits je 3 g der Säure mit 50 ccm Wasser, 0,1-n. oder 1-n. Salzsäure 4 Stunden im Wasserbade turbiniert, so ergaben die danach zurückgewonnenen Präparate mit 1-n. Natronlauge Lösungen von maximalen Drehungsvermögen.

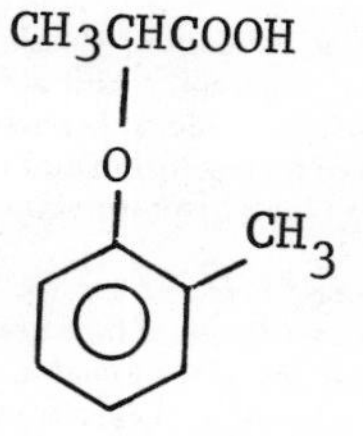

$C_{10}H_{12}O_3$

Ref. 667

Preliminary experiments on resolution were carried out in dilute ethanol. *Strychnine* and *cinchonidine* gave laevorotatory acid but the former base was far more efficient as resolving agent. The salts of *dehydroabietylamine*, *α-phenyl-ethylamine* and *β-phenyl-isopropylamine* yielded only racemic acid. The *brucine*, *quinine*, *quinidine* and *cinchonine* salts failed to crystallise. The (−)-acid was thus isolated via the strychnine salt and the (+)-acid was isolated from the mother liquor as described below.

(−)-2-Methylphenoxy-propionic acid. Racemic acid (25.2 g, 0.14 mole) and strychnine (46.8 g, 0.14 mole) was dissolved in 150 ml water + 150 ml ethanol. The salt was filtered off after 24 h and recrystallised to maximum activity of the acid as described for the 2-bromo acid:

Crystallisation	1	2	3	4	5
ml ethanol	150	50	35	35	25
ml water	150	50	35	35	25
g salt obtained	25.4	19.1	17.7	16.7	16.1
$[\alpha]_D^{25}$ of the acid	−15°	−20°	−20.5°	−21°	−21°

The acid was liberated from the strychnine salt as described above for the 2-bromo acid. The crude product (4.8 g) was recrystallised from dilute formic acid yielding 4.2 g of pure acid as hairlike, glistening needles with m.p. 70–71°.

$C_{10}H_{12}O_3$ (180.20)
Calc. C 66.65 H 6.71
Found C 66.67 H 6.69

The rotatory power in various solvents is given in Table 2. It should be noted that the acid is dextrorotatory in benzene and that the rotatory power in neutral water solution is to low to be measured. In the case of the 2-bromo acid, the direction of rotation in both benzene and water is opposite to that in the other solvents.

Table 2

Solvent	g[a]	α_D^{25} (2 dm)	$[\alpha]_D^{25}$	$[M]_D^{25}$
Acetone	0.2613	− 1.20°	− 23.0°	− 41.4°
Ethanol (abs.)	0.2508	− 1.02°	− 20.3°	− 36.7°
Chloroform	0.2049	− 0.614	− 15.0°	− 27.0°
Water (neutr.)	0.2098	0.00°	0.0°	0.0°
Benzene	0.2372	+ 1.835°	+ 38.7°	+ 69.7°

[a] g = g acid dissolved to 10.00 ml.

(+)-*2-Methylphenoxy-propionic acid*. The mother liquor from the first crystallisation of the strychnine salt yielded 11 g of an acid having $[\alpha]_D^{25} = +10°$. It was used for preliminary experiments on isolation of pure (+)-acid. The final isolation was then performed with acid of the same activity from a new resolution experiment on a larger scale.

The acid was shaken with ligroin (about 6 ml for 1 g acid) at room temperature for about 20 seconds and the ligroin was filtered off. The residue was dried and the extraction repeated three times. The residue was then practically pure racemic acid. The three first extracts yielded on evaporation acid of high activity and m.p. 66–69°. Recrystallisation from dilute formic acid did not rise the m.p. above 67–69°. The extraction with ligroin was thus repeated yielding an acid with m.p. 68.5–70°. This product was extracted once with ligroin and the acid obtained from the extract recrystallised from a small amount of ligroin. In this way, 2.8 g of pure (+)-acid with m.p. 70–71° were obtained. It formed glistening needles, quite similar to the (−)-form. Further quantities could be obtained from the residues after the extractions.

$C_{10}H_{12}O_3$ (180.20)
Calc. C 66.65 H 6.71
Found C 66.69 H 6.73

0.2544 g dissolved in acetone to 10.00 ml: $\alpha_D^{25} = +1.156°$ (2 dm). $[\alpha]_D^{25} = +22.7°$; $[M]_D^{25} = +40.9°$.

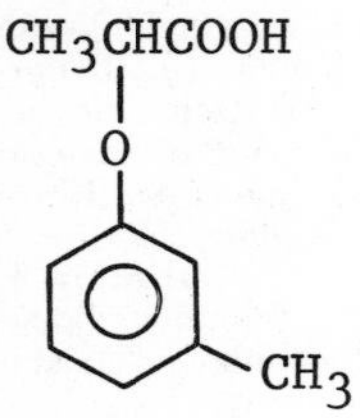

$C_{10}H_{12}O_3$ Ref. 668

Preliminary experiments on resolution. Small-scale experiments were performed with the common alkaloids. Cinchonine and strychnine gave salts containing (−)-acid of high activity, the latter alkaloid, however, in rather low yield. The salts of cinchonidine and quinine gave no resolution and the brucine salt could not be brought to crystallisation. Only (−)-phenyl-isopropylamine gave a salt containing excess of (+)-acid.

(−)-*Methyl-phenoxypropionic acid*. 27.0 g (0.15 mole) of racemic acid and 44.2 g (0.15 mole) of cinchonine were dissolved in a hot mixture of 350 ml ethanol and 200 ml water. After cooling, a further 200 ml water were added. Crystallisation of a salt (small cube-like crystals) started almost immediately. After 24 hours in a refrigerator the salt was filtered off and recrystallised from dilute ethanol to maximum activity of the acid; the course of the resolution was followed as described above for the bromo acid. The acid was already practically pure in the second crystallisation.

Crystallisation	1	2	3	4
ml ethanol	350	85	45	45
ml water	400	95	55	55
g salt obtained	27.1	24.7	23.5	21.4
$[\alpha]_D^{25}$ of the acid	− 45°	− 47°	− 48°	− 47.5°

The salt was decomposed with dilute sulphuric acid and the organic acid isolated by extraction with ether. After recrystallisation from ligroin, it formed long glistening needles with m.p. 74–75°. Yield of pure acid 7.9 g (58.5 %).

0.3033 g acid: 16.75 ml 0.1003 *N* NaOH.
$C_{10}H_{12}O_3$ Equiv. wt calc. 180.2 found 180.5

The rotatory power in various solvents is given in Table 2.

Table 2. g = g acid dissolved to 10.00 ml.

Solvent	g	α_D^{25} (2 dm)	$[\alpha]_D^{25}$	$[M]_D^{25}$
Acetone	0.1990	− 1.886°	− 47.4°	− 85.4°
Ethanol (abs.)	0.1980	− 1.398°	− 35.3°	− 63.6°
Chloroform	0.1966	− 0.680°	− 17.3°	− 31.2°
Water (neutr.)	0.1885	− 0.413°	− 11.0°	− 19.8°
Benzene	0.2011	+ 0.363°	+ 9.0°	+ 16.2°

(+)-*3-Methyl-phenoxypropionic acid*. From the mother liquor after the first cr
stallisation of the cinchonine salt, 16.3 g acid having $[\alpha]_D^{25} = +26.5°$ were isolat
16.0 g (0.09 mole) of this acid and 12.0 g (0.09 mole) of (−)-phenyl-isopropylami
were dissolved in 175 ml of hot water. On cooling and with efficient stirring, t
salt separated as a mass of very fine needles. It was filtered off after 24 hours in
refrigerator and recrystallised to maximum activity of the acid.

Crystallisation	1	2	3	4	5
ml water	175	110	85	60	60
g salt obtained	20.0	16.9	14.1	12.9	12.2
$[\alpha]_D^{25}$ of the acid	+ 41.5°	+ 46.5°	+ 47°	+ 47°	+ 47.5°

The acid was liberated as described above and recrystallised from ligroin.
formed long, glistening needles, quite similar to those of the (−)-form. M.p. 74–7
Yield 5.9 g.

0.2885 g acid: 15.93 ml 0.1003 *N* NaOH.
$C_{10}H_{12}O_3$ Equiv. wt calc. 180.2 found 180.6
0.1039 g dissolved in *acetone* to 10.00 ml; $\alpha_D^{25} = +0.986°$ (2 dm).
$[\alpha]_D^{25} = +47.3°$; $[M]_D^{25} = +85.2°$.

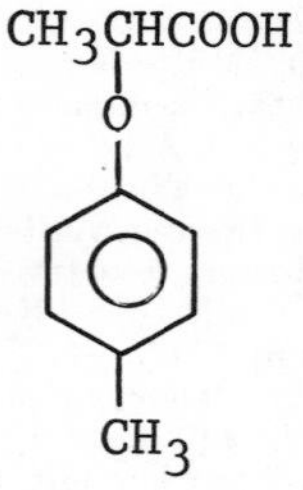

$C_{10}H_{12}O_3$ Ref. 66

(−)-*4-Methylphenoxy-propionic acid*. Racemic acid (27.0 g, 0.15 mole) and bruc
tetrahydrate (70 g, 0.15 mole) were dissolved in 150 ml ethanol. The salt, which c
stallised readily, was recrystallised to maximum activity of the acid. For this purp
it was dissolved in the necessary amount of hot methanol, and water (3.5 ml for e
ml of methanol) was added. From each crystallisation, a small sample of acid was li
rated and its rotatory power determined in absolute ethanol.

Crystallisation	1	2	3	4	
g salt obtained	47.5	37.0	34.5	32.5	2
$[\alpha]_D^{25}$ of the acid	− 29.5°	− 36.5°	− 40.5°	− 40.5°	− 4

The acid was liberated with sulphuric acid and extracted with ether. The yi
of crude acid from 29.7 g salt was 8.5 g. Recrystallisation from a small volum
ligroine gave 8.2 g of pure acid as colourless, glistening prisms or plates with m.p. 74–

$C_{10}H_{12}O_3$ (180.20) Calc. equiv. wt. 180.2 C 66.65 H 6.71
Found equiv. wt. 179.4 C 66.96 H 6.77

The rotatory power in various solvents is given in Table 1.

Table 1. g = g dissolved to 10.00 ml.

Solvent	g	α_D^{25} (2 dm)	$[\alpha]_D^{25}$	$[M]_D^{25}$
Acetone	0.2034	− 2.251°	− 55.3°	− 99.7°
Ethanol (abs.)	0.1192	− 0.96°	− 40.3°	− 72.5°
Chloroform	0.1243	− 0.52°	− 20.9°	− 37.5°
Water (neutr.)	0.1000	− 0.42°	− 21.0°	− 37.8°
Benzene	0.1170	+ 0.09°	+ 3.8°	+ 6.9°

(+)-*4-Methylphenoxy-propionic acid*. The mother liquors from the two
crystallisations of the brucine salt were combined. About 25 % of the solution
acidified with excess sulphuric acid and extracted with ether. Evaporation of
ether yielded 2.3 g 4-methylphenoxy-propionic acid having $[\alpha]_D = +30°$ (ethan
The acid was dissolved with an equivalent amount of quinine (4.2 g) in a mixtu
12 ml ethanol and 10 ml water. The salt obtained was recrystallised from
ethanol and the course of the resolution followed as described above.

Crystallisation	1	2	3	4
g salt obtained	5.2	4.0	3.5	2.9
$[\alpha]_D^{25}$ of the acid	+ 33.5°	+ 39°	+ 40°	+ 40°

The remaining 75 % of the combined mother liquors were evaporated to dry
and the resulting brucine salt dissolved in the necessary amount of boiling wa
On cooling, 3.5 g salt of the *laevorotatory* acid separated. The remaining solution
acidified and extracted with ether as described above, yielding 7.1 g 4-methylphen
propionic acid having $[\alpha]_D^{25} = +37.5°$. The rotatory power did not increase on

crystallisation of the free acid. The whole portion of acid was therefore dissolved with 14.8 g quinine in 40 ml ethanol + 30 ml water. The salt obtained was recrystallised as described above.

Crystallisation	1	2	3
g salt obtained	18.5	18.1	17.5
$[\alpha]_D^{25}$ of the acid	+39.5°	+40°	+40°

The two portions of the quinine salt were combined and the acid was isolated and recrystallised as the enantiomer. Glistening prisms or plates with m.p. 74–75°.

$C_{10}H_{12}O_3$ (180.20) Calc. equiv. wt. 180.2 C 66.65 H 6.71
Found equiv. wt. 180.0 C 66.74 H 6.68

0.2122 g acid dissolved in *acetone* to 10.00 ml; $\alpha_D^{25} = +2.341°$ (2 dm). $[\alpha]_D^{25} = +55.2°$; $[M]_D^{25} = +99.4°$. – 0.2029 g dissolved in abs. *ethanol* to 10.00 ml; $\alpha_D^{25} = +1.631°$ (2 dm). $[\alpha]_D^{25} = +40.2°$; $[M]_D^{25} = +72.4°$.

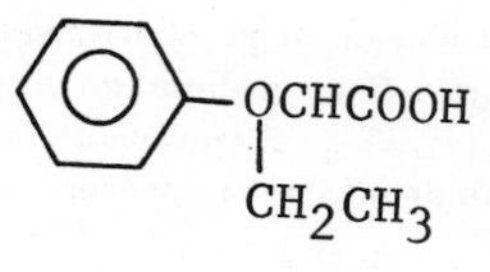

$C_{10}H_{12}O_3$ Ref. 670

Optically active α-phenoxy-n-butyric acid (I). In preliminary experiments crystalline salts were obtained with cinchonine, quinine and strychnine. These salts yielded levorotatory acids but only the salt with strychnine gave a product with an appreciable rotatory power. As no crystalline products yielding the (+)-form of the acid were obtained when using the racemic acid further attempts were made with a product containing about 85 % of the dextrorotatory acid. Under these conditions the morphine salt crystallised and on decomposition yielded almost pure (+)-acid.

Table 2. α-Aryloxy-*n*-butyric acids

α-Aryloxy-*n*-butyric acid	Recrystallised from	M.p. lit.	M.p. found	Equiv. wt. calc.	Equiv. wt. found
α-Phenoxy-	Petroleum (b.p. 60–80°)	82–83° (10)	81.5–83.3°	180.2	180.4
α-(1-Naphthoxy)-	Cyclohexane and petroleum (b.p. 60–80°)	113–114° (11)	113.5–115.3°	230.3	228.9
α-(2-Naphthoxy)-	Ethanol(50 %) and petroleum (b.p. 60–80°)	126.5° (11)	125.5–127°	230.3	231.0

α-Phenoxy-*n*-butyric acid (22.6 g, 0.125 mole) and strychnine (42.0 g, 0.125 mole) were dissolved in hot 25 % alcohol (400 ml). On cooling the salt crystallised and was collected after standing over night in a refrigerator. The salt was recrystallised as seen from Table 3; $[\alpha]_D$ refers to the rotatory power of the acid in abs. alcohol.

Table 3. Recrystallisation of the strychnine salt of α-phenoxy-*n*-butyric acid.

Cryst. No.	25 % ethanol (ml)	Weight of salt (g)	$[\alpha]_D$ of the acid
1	400	37.0	−29.9°
2	320	27.4	−43.4°
3	240	25.2	—
4	240	23.2	−46.5°
5	240	21.1	−46.1°
6	240	18.7	−48.3°
7	240	17.0	−49.0°
8	240	15.2	−49.1°

The pure strychnine salt was treated with aqueous ammonia, the alkaloid removed and the residue acidified when the organic acid was precipitated as colourless crystals. The crude acid (5.3 g), m.p. 79–80°, was recrystallised twice from petroleum (b.p. 60–80°). 3.7 g of the pure (−)-acid were collected as small colourless needles, m.p. 79.3–80.3°.

78.75 mg acid: 9.05 ml 0.04823-*N* NaOH
$C_{10}H_{12}O_3$ calc. equiv. wt. 180.2
found » » 180.4

0.0810 g dissolved in abs. *alcohol* to 10.00 ml: $\alpha_D^{25} = -0.416°$. $[\alpha]_D^{25} = -51.4°$; $[M]_D^{25} = -92.6°$.
0.1035 g dissolved in *acetone* to 10.00 ml: $\alpha_D^{25} = -0.678°$. $[\alpha]_D^{25} = -65.5°$; $[M]_D^{25} = -118.0°$.
0.1072 g dissolved in *chloroform* to 10.00 ml: $\alpha_D^{25} = -0.390°$. $[\alpha]_D^{25} = -36.4°$; $[M]_D^{25} = -65.6°$.
0.1087 g dissolved in *benzene* to 10.00 ml: $\alpha_D^{25} = -0.056°$. $[\alpha]_D^{25} = -5.2°$; $[M]_D^{25} = -9.3°$.
0.0844 g neutralised with *aqueous sodium hydroxide* and made up to 10.00 ml: $\alpha_D^{25} = -0.174°$. $[\alpha]_D^{25} = -20.6°$; $[M]_D^{25} = -37.2°$.

The mother liquids from the recrystallisations 1 and 2 (Table 3) gave an acid containing about 85 % of the (+)-form. This acid (10.8 g, 0.060 mole) and an equimolecular quantity of morphine (18.2 g) were dissolved in 150 ml of hot 20 % ethanol. After cooling to room temperature the solution was seeded with crystals from a preliminary run and allowed to stand in a refrigerator over night. The solid was collected, the acid liberated from a small sample and the rotatory power determined in abs. alcohol. As seen from Table 4 only three recrystallisations were necessary for the purification of the salt.

Table 4. Recrystallisation of the morphine salt of α-phenoxy-*n*-butyric acid.

Cryst. No.	Solvent (ml) Water	Solvent (ml) Ethanol	Weight of salt (g)	$[\alpha]_D$ of the acid
1	25	125	17.2	+48.6°
2	30	100	16.4	+49.2°
3	30	100	12.9	+49.7°

The organic acid was isolated from the pure morphine salt and the crude acid (3.5 g) recrystallised from petroleum (b.p. 60–80°) yielding 3.05 g of the pure (+)-acid. M.p. 79.2–80.2°.

64.70 mg acid: 7.42 ml 0.04823-*N* NaOH
$C_{12}H_{10}O_3$ calc. equiv. wt. 180.2
found » » 180.8

0.1077 g dissolved in abs. *alcohol* to 10.00 ml: $\alpha_D^{25} = +0.551°$. $[\alpha]_D^{25} = +51.2°$; $[M]_D^{25} = +92.2°$.

$C_{10}H_{12}O_3$ Ref. 670a

Preparation of D(+) - α - phenoxy - n - butyric acid

A mixture of (−)-ephedrine (165g., 1 mole) and (±) - α - phenoxy - n - butyric acid (180g., 1 mole) was dissolved in hot amyl acetate (337 ml.), filtered from any insoluble impurities, allowed to cool to room temperature and finally chilled for 2 hours. The resulting crystalline solid was filtered off, washed with fresh solvent and dried. Yield 169g. (98%); $[\alpha]_D^{20}$ −9.1° (C, 1 in ethanol).

The salt (165g.) was recrystallised from hot amyl acetate (297 ml.) and isolated in the same way to give 124g. (75% of a salt with a $[\alpha]_D^{20}$ −4.2° (C, 1 in ethanol).

A further recrystallisation using the same amount of solvent (1.8 ml./g.) gave 110g. (88%) of a salt with $[\alpha]_D^{20}$ −3.2°.

The salt (100 g.) was suspended in water (500 ml.) and ether (500 ml.), and acidified. The ether was separated and the aqueous phase extracted with ether (250 ml.).

The combined ether extracts were washed with water (60 ml.) and extracted with sodium bicarbonate solution. The alkaline extract was acidified and chilled for two hours. The precipitated acid was filtered off, washed with water and dried at 40°. Yield 47g., (90%) of acid with $[\alpha]_D^{20}$ +50.5° (C, 1 in ethanol).

Example 2

Preparation of L(−) - α - phenoxy - n - butyric acid

A mixture of (−)-cinchonidine (294 g., 1 mole) and (±) - α - phenoxy - n - butyric acid (180 g., 1 mole) was dissolved in hot acetone (215 l.) filtered from any insoluble material, allowed to cool and finally chilled for 2 hours. The salt was filtered off, washed with acetone and dried. Yield 208 g. with $[\alpha]_D^{20}$ −94.5° (C, 1 in ethanol).

This salt was twice recrystallised from acetone (19 ml./g.) to give the pure salt in 45% yield with $[\alpha]_D^{20} -105°$.

The L(−)- acid was obtained in the same manner as in the previous example. Yield 90%; $[\alpha]_D^{20} -50.1°$ (C, 1 in ethanol).

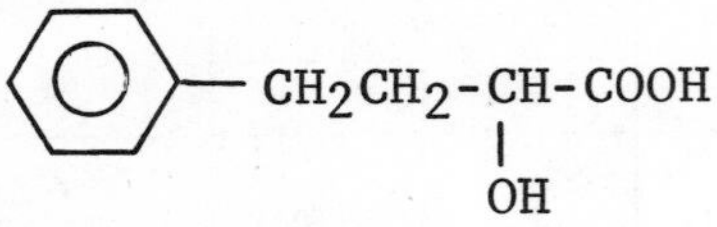

$C_{10}H_{12}O_3$ Ref. 671

130 g. d'acide benzyllactique racémique et 300 g. de menthol (2 fois 1/2 la quantité théorique) sont chauffés pendant 16 heures à 125° C. (²), un courant de ClH anhydre traversant le mélange. Le traitement du produit obtenu se fait exactement comme pour l'acide phényllactique. On sépare par distillation, sous pression réduite, le menthol en excès (P. Eb. : 104° C./20 mm.) et l'éther menthylique de l'acide benzyllactique (P. Eb. : 230-232° C./20 mm.).

Dans cette éthérification nous avons obtenu de 78 à 80 o/o de rendement.

Nous avons ainsi le mélange des deux éthers menthyliques de l'acide gauche et droit. On sépare ces deux éthers par des cristallisations fractionnées dans l'éther de pétrole. Après plusieurs cristallisations, on obtient l'éther menthylique de l'acide benzyllactique gauche à l'état optiquement pur, l'éther menthylique de l'acide énantiomorphe restant dans les liqueurs mères de cristallisation.

L'éther menthylique de l'acide benzyllactique gauche $C^6H^5.CH^2.CH^2.CHOH.CO^2C^{10}H^{19}$ ($C^{20}H^{30}O^3$), cristallise de l'éther de pétrole en de fines aiguilles blanches soyeuses ; il fond à 88° C. et, en solution dans le chloroforme, il présente pour la raie D du sodium un pouvoir rotatoire spécifique de :

$$[\alpha]_D = -65°4$$

($\alpha_D = -4°43$; $p =$ 0 g. 676 ; $v =$ 20 cm³ ; $l =$ 2 dm ; $t =$ 22° C.).

Analyse. — Subst. 0,1986 ; H^2O, 0,1686 ; CO^2, 0,5476. Trouvé o/o : H, 9,50 ; C, 75,20. Calculé pour $C^{20}H^{30}O^3$: H, 9,50 ; C, 75,41.

L'acide benzyllactique gauche est obtenu par saponification de son éther menthylique au moyen d'un excès d'une solution alcoolique de potasse à 15 o/o. Ici encore on opère comme pour la saponification du phényllactate de menthyle. La solution éthérée de l'acide benzyllactique gauche, par distillation du solvant, laisse l'acide à l'état solide. On recristallise cet acide dans le benzène puis dans un mélange d'éther et d'éther de pétrole.

L'acide benzyllactique gauche a ainsi été obtenu à l'état optiquement pur. C'est un solide blanc de P. F. : 114°-116° C. En solution dans l'alcool éthylique à 95°, il présente pour la raie D du sodium le pouvoir rotatoire spécifique :

$$[\alpha]_D = -9°9$$

($\alpha_D = -1°63$; $v =$ 10 cm³ ; $l =$ 2 dm ; $p =$ 0 g. 8175 ; $t =$ 27° C.).

Analyse. — Subst. 0,1522 ; H^2O, 0,0956 ; CO^2, 0,3703. Trouvé o/o : H, 7,03 ; C, 66,36. Calculé pour $C^{10}H^{12}O^3$: H, 6,72 ; C, 66,63.

L'acide benzyllactique gauche optiquement pur a été obtenu avec 30 à 35 o/o de rendement à partir de l'acide racémique.

(¹) Il est à supposer que les auteurs ont obtenu l'acide benzyllactique *lévogyre*, car l'acide gauche, ici étudié en solution dans la soude, faisait tourner à droite le plan de polarisation de la lumière.

(²) La temperature du bain d'huile était maintenue entre 125° et 130° C.

Ann. de Chim., 10e série, t. XX (Juillet-Août 1933).

Dans le cas de l'acide benzyllactique, il est à noter que la valeur du pouvoir rotatoire de l'acide actif impur n'est pas améliorée par des cristallisations ; c'est-à-dire que l'on ne peut pas séparer par des cristallisations fractionnées chacun des acides actifs de l'acide racémique. Il est donc nécessaire d'obtenir les éthers menthyliques à l'état pur par des cristallisations répétées dans l'éther de pétrole.

Acide benzyllactique droit. — L'éther menthylique de l'acide droit étant le plus soluble, il reste dans les liqueurs mères de cristallisation et on l'obtient forcément mélangé d'un peu d'éther de l'acide gauche. Donc par saponification du résidu de l'évaporation des liqueurs mères, on a un mélange d'acide droit et de racémique ; nous avons vainement essayé d'éliminer cet acide racémique par des cristallisations fractionnées au sein du benzène puis dans un mélange d'éther et d'éther de pétrole. L'échantillon le plus pur, d'acide benzyllactique droit obtenu, ne présentait, en solution dans C^2H^5OH et pour la raie D du sodium, qu'un pouvoir rotatoire spécifique de $[\alpha]_D = +7°$ tandis que l'isomère énantiomorphe pur aurait donné dans les mêmes conditions :

$$[\alpha]_D = -10°$$

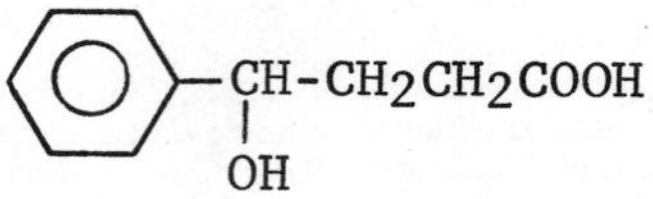

$C_{10}H_{12}O_3$ Ref. 672

Die aus 10 g kristallwasserfreiem Natriumsalz dargestellte Säure wurde in Alkohol gelöst, die alkoholische Lösung, welche nach dem Ergebnis der Titration eines aliquoten Teils 8,2 g freie Säure enthielt, auf 110 ccm gebracht und mit 220 ccm einer alkoholischen Brucinlösung, welche 9,1 g wasserfreies Brucin (etwas über ½ Mol.) enthielt, versetzt. Bei unter 0° kristallisierten 6,95 g Brucinsalz (Fraktion 1) aus, in einem zweiten mit denselben Mengen und in gleicher Weise ausgeführten Versuche 7,99 g (Fraktion 1). Aus den vereinigten Mutterlaugen schieden sich nach wiederholtem Einengen im Vakuum noch drei weitere Fraktionen aus in Mengen von 3,07 g (Fraktion 2), 4,07 g (Fraktion 3) und 1,12 g (Fraktion 4). Die Gesamtausbeute betrug also 23,20 g (statt der der Rechnung nach möglichen 28,4 g).

1. Kristallisationen. Die Fraktionen 1—3 wurden gemeinsam aus 150 ccm Alkohol umkristallisiert. Es scheiden sich 18,55 g wieder aus (a). Aus der mit der Fraktion 4 vereinigten Mutterlauge wurden 4 g erhalten (b). a lieferte ein Natriumsalz, das aus Alkohol umkristallisiert (5 g) die spezifische Drehung + 12,91° zeigte. Die aus ihm gewonnene freie Säure wurde unter Einhaltung derselben Konzentrationsverhältnisse wieder in alkoholischer Lösung mit einer alkoholischen Brucinlösung (½ Mol.) versetzt, das ausgeschiedene Brucinsalz in das Natriumsalz verwandelt und dieses polarisiert, dann das ganze Verfahren nochmals wiederholt. Die spezifische Drehung änderte sich nicht. Sie betrug nach der ersten Wiederholung der Behandlung mit Brucin + 12,72°, nach der zweiten + 12,87°.

0,2163 g Substanz. Gesamtgewicht der Lösung 7,6807 g. Spez. Gew. 1,0095. Prozentgehalt 2,816. Drehung bei 15° C. im 2-dm-Rohr bei Natriumlicht 0,73° nach rechts. Also $[\alpha]_D^{15°} = +12,87°$.

b gab ein Natriumsalz, das aus Alkohol umkristallisiert (0,97 g) die spezifische Drehung — 13,14° zeigte.

0,4842 g Substanz. Gesamtgewicht der Lösung 7,9580 g. Spez. Gew. 1,0188. Prozentgehalt 5,456. Drehung bei 14° im 2-dm-Rohr bei Auerlicht 1,46° nach links. Also $[\alpha]_D^{14°} = -13,14°$.

2. Mutterlauge. Die aus ihr zurückgewonnenen 10,5 g Natriumsalz (statt der berechneten 11,84 g) wurden in die Säure übergeführt. Die alkoholische Lösung, welche nach Titration eines aliquoten Teils 8,06 g freie Säure enthielt,

wurde auf 65 ccm gebracht und mit 212 ccm einer alkoholischen Brucinlösung, enthaltend 9 g Brucin (etwas über 1/2 Mol.), versetzt. Bei unter 0° kristallisierten 10,86 g Brucinsalz aus. Das aus ihm gewonnene Natriumsalz zeigte nach Umkristallisation aus Alkohol (2,7 g) die spezifische Drehung −12,77°.

0,3940 g Substanz. Gesamtgewicht der Lösung 7,4208 g. Spez. Gew. 1,0181. Prozentgehalt 5,309. Drehung bei 12° im 2-dm-Rohr bei Auerlicht 1,38° nach links. Also $[\alpha]_D^{14°} = -12,77°$.

Das aus der Mutterlauge der Brucinkristallisation wiedererhaltene Natriumsalz (5,92 g) erwies sich als ganz schwach linksdrehend.

Kristallwassergehalt der aus 96%igem Alkohol kristallisierten Natriumsalze. Das inaktive Salz verlor während 1 1/2 stündigen Trocknens bei 100° im Vakuum 15,36% seines Gewichts, das Linkssalz 15,14%, das Rechtssalz 15,28%. Berechnet für 2 Mol. Kristallwasser 15,12%.

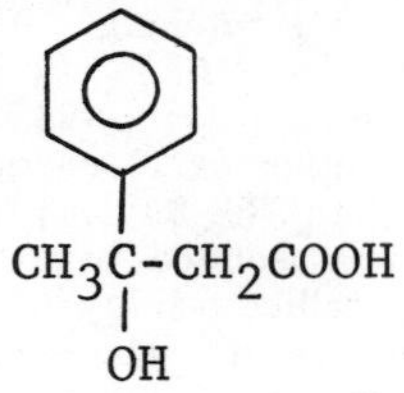

$C_{10}H_{12}O_3$ Ref. 673

Dédoublement de l'acide β-hydroxy β-phénylbutyrique par la phényléthylamine

50 g (0·301 mole) d'acide β-hydroxy β-phénylbutyrique racémique sont dissous à chaud dans 200 cm³ d'éthanol absolu. A cette solution on ajoute 36·5 g (0·301 mole) de (−) phényléthylamine dans 200 cm³ d'éthanol absolu et laisse cristalliser.

On obtient 21·5 g de sel, qui recristallisés dans l'éthanol absolu fournissent 3 fractions: (a) 10·7 g $[\alpha]_D^{22} = -20·4°$ (c = 0·99, EtOH); (b) 2·47 g $[\alpha]_D^{22} = -20·1°$ (c = 1, EtOH) et (c) 3·02 g $[\alpha]_D^{22} = -19·8°$ (c = 1, EtOH).

La fraction de sel $[\alpha]_D^{22} = -20·4°$ après traitement à l'acide chlorhydrique fournit 7·1 g d'acide β-hydroxy, β-phénylbutyrique $[\alpha]_D^{30} = -8·4°$ (c = 1·69, éthanol absolu).

Les eaux mères de cristallisation du premier jet de sel (−) sont évaporées à sec et liberent 58 g de sel qui par traitement à l'acide chlorhydrique fournissent 33·2 g (0·2 mole) d'hydroxyacide partiellement dédoublé. Cet hydroxyacide est mis en présence de 24·2 g (0·2 mole) de (+) phényléthylamine. On obtient 21·77 g de sel $[\alpha]_D^{22} = +19·3°$ (c = 0·99, éthanol). Ce sel recristallisé dans l'éthanol fournit 2 fractions: 12·77 g $[\alpha]_D^{22} = +20·2°$ et 4·05 g $[\alpha]_D^{22} = +20·0°$ (c = 1, éthanol).

La fraction de sel $[\alpha]_D^{22} = +20·2°$, après traitement à l'acide chlorhydrique fournit 7·4 g d'acide β-hydroxy β-phénylbutyrique $[\alpha]_D^{22} = +8·6°$ (c = 1·74, éthanol absolu). Litt.[29] optiquement pur $[\alpha]_D^{20} = 8·92°$ (c = 1·76, éthanol).

[29] M. H. Palmer, J. A. Reid, *J. Chem. Soc.* 931 (1960)

$C_{10}H_{12}O_3$ Ref. 673a

Experimental.—The (±)-acid (9·0 g., 1 mol.) and morphine monohydrate (15·1 g., 1 mol.) were dissolved in 300 c.c. of hot water. After some hours, 15·5 g. of salt separated with $[\alpha]_{5780}^{20}$ −81·3°. Five crystallisations of this salt from water gave a *morphine* salt, in rhombic prisms, $[\alpha]_{5780}^{20}$ −68·2° and $[\alpha]_{5461}^{20}$ −76·9° in water (*c*, 0·9820) (Found: C, 68·3; H, 6·9. $C_{17}H_{19}O_3N,C_{10}H_{12}O_3,0·5H_2O$ requires C, 68·3; H, 6·8%). The salt was shaken with chloroform and dilute alkali. The aqueous solution was extracted three times with chloroform and then once with ether. It was then acidified and extracted five times with ether. The combined ether layers were washed twice with water and evaporated. The residue became crystalline after some time and was recrystallised from light petroleum (b. p. 80—100°). It had m. p. 71—72°, $[\alpha]_{5780}^{20}$ +6·3° and $[\alpha]_{5461}^{20}$ +7·5°, in absolute ethyl alcohol (*c*, 1·660) (Found: C, 66·8; H, 7·0. Calc. for $C_{10}H_{12}O_3$: C, 66·6; H, 6·7%).

Concentration of the original morphine salt mother-liquor, followed by repeated crystallisation from water, gave a morphine salt, in spherical nodules, $[\alpha]_{5780}^{21}$ −87·8° and $[\alpha]_{5461}^{21}$ −99·7° in ethyl alcohol, but material with this approximate rotation was more readily obtained by carrying out a partial resolution in water, as described above, and then recrystallising, from absolute alcohol, crops with a specific rotation of about $[\alpha]_{5780}$ −75°. This led to what is apparently the *morphine* salt of the pure (−)-acid, having $[\alpha]_{5780}^{26}$ −90·3° and $[\alpha]_{5461}^{26}$ −103·4° in water (*c*, 1·000) (Found: C, 69·3; H, 6·8. $C_{17}H_{19}O_3N,C_{10}H_{12}O_3$ requires C, 69·6; H, 6·7%). The rotation of the salt was unchanged by recrystallisation.

Decomposition of the salt in the manner described above gave the (−)-acid, which after being crystallised from light petroleum (b. p. 80—100°) had m. p. 79—80°, $[\alpha]_{5780}^{23}$ −8·0°, $[\alpha]_{5461}^{23}$ −8·9° in absolute ethyl alcohol (*c*, 1·7544) (Found: C, 66·8; H, 7·0%). Recrystallisation from light petroleum did not alter the specific rotation.

An attempt to obtain one morphine salt preferentially by using racemic acid and morphine salt in the molecular ratio 3 : 1, in either aqueous or alcoholic solution, was unsuccessful. Brucine formed a well-crystallised salt from aqueous solution, but led to no appreciable resolution.

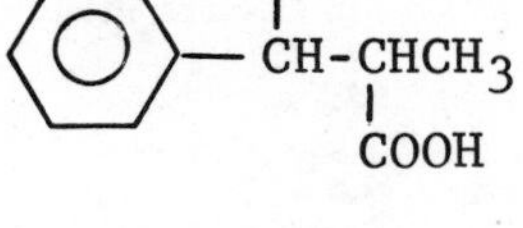

$C_{10}H_{12}O_3$ Ref. 674

Acide (—) hydroxy-3 S méthyl-2 R phényl-3 propionique (thréo).

A partir de 12 g d'acide *thréo* pur et 18 g de quinine dans 200 cm³ d'alcool, on récolte après une nuit, 8 g de sel qu'on recristallise 4 fois dans l'éthanol (1 g dans 15 cm³). Après 4 cristallisations, l'acide est régénéré de son sel par traitement à l'acide chlorhydrique en présence d'éther. Cet acide (4,2 g) $[\alpha]_D^{21} = -28,1°$ (*c* = 1, méthanol), $[\alpha]_J^{20} = -48,4°$ (*c* = 0,94, $CHCl_3$), est recristallisé dans l'eau F = 103-105°, $[\alpha]_J^{20} = -29,2°$ (*c* = 0,97, méthanol).

Analyse $C_{10}H_{12}O_3$ (180,20) :

Calc. % :	C 66,65	H 6,7
Tr. :	66,45	6,65.

7,15 g d'acide, $[\alpha]_J^{21} = +2,7°$ (méthanol, *c* = 0,6), régénéré du sel de quinine le plus soluble, sont traités par 11 g de déhydroabiétylamine dans 200 cm³ de méthanol. Après 3 cristallisations du sel, l'acide obtenu (2,8 g) présente un $[\alpha]_J^{20} = +10,6°$ (méthanol, *c* = 1). Dans une autre opération sensiblement identique, après 5 recristallisations de son sel de déhydroabiétylamine, l'acide obtenu a un $[\alpha]_J^{21} = +20,0°$ (méthanol, *c* = 0,96). Le dédoublement complet de l'acide dextrogyre n'a pas été achevé.

$C_{10}H_{12}O_3$ Ref. 674a

The Resolution of (±)-threo-*3-Hydroxy-2-methyl-3-phenylpropionic Acid (IIIb) with Cinchonidine.* A solution of (±)-IIIb (1.20 g), mp 97—98°C (lit,[7] mp 96.5—97.5°C) in ethyl acetate was added to a hot solution of cinchonidine (1.96 g) in ethyl acetate (100 m*l*), and then the solution was concentrated to *ca.* 40 m*l*. After the solution had stood at room temperature, the precipitates were collected and then recrystallized from ethyl acetate to give cinchonidine salt as colorless crystals; mp 170—172°C, $[\alpha]_D^{25}$ −87.6° (*c* 0.083, EtOH); yield, 1.22 g.

Found: C, 73.66; H, 7.28; N, 5.80%. Calcd for $C_{29}H_{34}O_4N_2$: C, 73.39; H, 7.22; N, 5.90%.

The cinchonidine salt (1.36 g) was decomposed with dilute sulfuric acid (5%: 20 m*l*) and extracted with ether, and the extract was washed with a saturated sodium chloride solution. After it had been dried over sodium sulfate and the ether had been removed, the residue (425 mg) was recrystallized from benzene to give (−)-IIIb as colorless crystals; mp 106—107°C; $[\alpha]_D^{25}$ −19.7° (*c* 0.094, EtOH).

Found: C, 66.87; H, 6.65%. Calcd for $C_{10}H_{12}O_3$: C, 66.65; H, 6.71%.

The acid, (−)-IIIb, was methylated with ethereal diazomethane to give (−)-IIb, $[\alpha]_D^{25}$ −57.1° (*c* 0.123, $CHCl_3$) or −41.6° (*c* 0.114, EtOH). The IR spectrum was identical with that of authentic (±)-IIb.

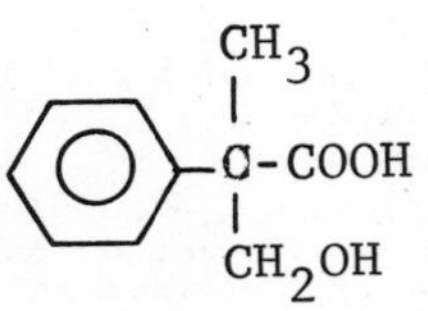

$C_{10}H_{12}O_3$ Ref. 675

Resolution of α-Methyltropic Acid (3). Using a modification of the literature procedure,[2a] 5.00 g (28.0 mmol) of α-methyltropic acid (3) and 9.77 g (28.5 mmol) of quinine was added to 22.5 mL of absolute ethanol and 22.5 mL of distilled water. The mixture was heated until all the solids dissolved. Crystallization was very slow, but seeding or scratching induced too rapid a rate of crystal formation. After 18 h, the crystals were removed by filtration and dried at 50 °C (0.1 mm) to give 3.40 g (46%) of the quinine salt, mp 169–172 °C. Recrystallization of this salt from 40 mL of ethanol and water (50:50) gave, after drying, 2.21 g of quinine salt, mp 178–179 °C.

This entire sequence was repeated twice on a 10-g scale with identical results. The combined batches of quinine salt, 14.08 g, were recrystallized from 400 mL of ethyl acetate–ethanol (9:1). Filtration of the solids and drying for 10 h (0.1 mm) gave 9.3 g of quinine-α-methyltropic salt, mp 183–185 °C (lit.[2a] mp 185–186 °C). Free (-)-(*S*)-α-methyltropic acid was obtained according to the literature procedure.[2a] Thus, 9.26 g of quinine salt afforded 2.56 g (78%) of (-)-(*S*)-α-methyltropic acid, $[\alpha]^{25}_D$ −29.9° (*c* 1.1, absolute EtOH) (lit.[2a], $[\alpha]^{25}_D$ −28°).

In a similar fashion, the quinine salt from the combined first mother liquors gave, after removal of the quinine, 10.83 g (43%) of (+)-(*R*)-3 $[\alpha]^{25}_D$ 6.14° (*c* 1.1, absolute ethanol).

(2) (a) Domagala, J. M.; Bach, R. D.; Wemple, J. *J. Am. Chem. Soc.*, **1976**, *98*, 1975;

$C_{10}H_{12}O_3$ Ref. 675a

Optical Resolution of α-Methyltropic Acid.—A solution of 70 g. of (±)-α-methyltropic acid in 315 ml. of acetone was added rapidly to a refluxing solution of 174 g. of quinine in 2200 ml. of acetone and the mixture was allowed to stay at room temperature for 4 hr. The crystals which separated were filtered and washed with 200 ml. of acetone. The filtrates (A) were collected and stored for recovery of the (+)-antipode. The solid material, after drying at 50° *in vacuo*, yielded 127 g. of crude quinine (−)-α-methyltropate (m.p. 180–181°), which was recrystallized from 4400 ml. of water to give 110 g. of quinine (−)-α-methyltropate (m.p. 183–184°), $[\alpha]^{20}D$ −120° (*c* = 2, ethanol). This salt was suspended in 125 ml. of water, acidified with 10% sulfuric acid to pH 3 with stirring and cooling, and the oil which separated was extracted three times with 300 ml. of ether. The extracts were collected, dried over sodium sulfate and evaporated to give an oil which solidified by scratching. After trituration with petroleum ether and filtration, 24.5 g. of (−)-α-methyltropic acid was obtained, m.p. 85–86°, $[\alpha]^{20}D$ −23.8°. Recrystallization from benzene–petroleum ether (1:1) yielded 20 g. (57%) of (−)-α-methyltropic acid, m.p. 88–89° $[\alpha]^{20}D$ −28° (*c* = 2, ethanol). The (+)-isomer was isolated by evaporating the filtrates A, suspending the crystalline residue in water and acidifying to pH 3 with 10% sulfuric acid. The separated oil was thoroughly extracted with ether, dried and the solvent evaporated to give 25.6 g. of crude (+)-α-methyltropic acid, $[\alpha]^{20}D$ +21° (*c* = 2, ethanol). The product was crystallized from benzene–petroleum ether to give 19.2 g. (55%) of (+)-α-methyltropic acid, m.p. 88–89°, $[\alpha]^{20}D$ +27° (*c* = 2%, ethanol).

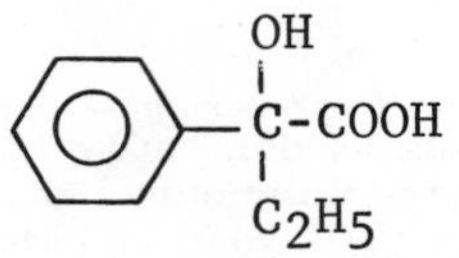

$C_{10}H_{12}O_3$ Ref. 676

56.7 g Chinin-trihydrat (1 Mol.) wurden in siedendem Äthylalkohol (325 ccm) gelöst und allmählich mit 27 g *racem.* Säure (1 Mol.) versetzt. Die Krystalle, die sich über Nacht bei 0° abgeschieden hatten, wurden 5-mal aus Äthylalkohol umkrystallisiert, wobei rechtwinklige Prismen (19.5 g) von reinem Chininsalz, Schmp. 205—206° (unt. Zers.) erhalten wurden. Das Fortschreiten der Spaltung wurde beobachtet, indem man die Filtrate aufeinanderfolgender Krystallisationen mit verd. Schwefelsäure zersetzte und die Drehung der so erhaltenen Säuren in Aceton bestimmte. Es wurde gefunden:

Fällung	Gewicht d. Säure	$[\alpha]_D$ der Säuren aus den Filtraten	c
I	39.8 g	—21.9°	4.069
II	30.5 g	— 5.4°	4.093
III	26.8 g	+17.9°	4.186
IV	24.4 g	+27.4°	4.194
V	21.5 g	+28.9°	4.046
VI	19.5 g	+28.8°	4.057

Gesamtausbeute an (+)-Säure 8 g.

(+)-Phenyl-äthyl-glykolsäure hat den Schmp. 128—129°. Ihre Drehung wurde in verschiedenen Lösungsmitteln gemessen.

Lösungsmittel	*l*	c	α_D^{20}	α_{5461}^{20}	$[\alpha]_D^{20}$	$[\alpha]_{5461}^{20}$
Wasser	2	0.696	+0.45°	+0.53°	+32.3°	+38.1°
Aceton	2	4.009	+2.30°	+2.86°	+28.7°	+35.7°
Äthylalkohol	2	3.996	+2.61°	+3.19°	+32.7°	+39.9°

Lennart Smith[25]) trennte die *racem.* Säure mittels (+)-α-Phenyl-äthyl-amins; er gibt den Wert $[\alpha]_D^{18}$: — 14° (c = 0.7032 in Wasser) an. Die von ihm beschriebene Spaltung ist unvollständig.

[24]) Journ. prakt. Chem. [2] **84**, 744 [1911]. [25]) l. c.

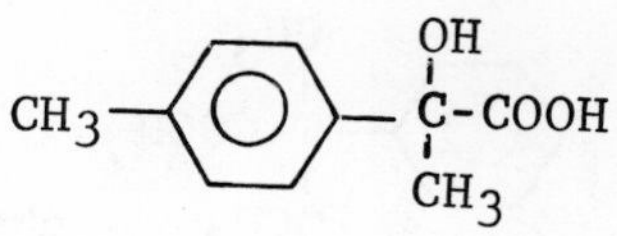

$C_{10}H_{12}O_3$

Ref. 677

Resolution of r-p-*Tolylmethylglycollic Acid.*—Hydrated quinine (28·8 g.) was dissolved in a boiling solution of the *r*-acid (15 g.) in ethyl alcohol (200 c.c.), the amount calculated for quinine trihydrate being 31·5 g. After 5 hours at the ordinary temperature, the crystals when dried under diminished pressure at 90° amounted to 28 g., and had m. p. 160—184°. These were crystallised repeatedly from ethyl alcohol: the diastereoisomeride of the (—)acid was the more sparingly soluble of the two, but the resolution was slow. The successive crops amounted to 28, 22, 15, 9·3, 5·7, 4·8, 4·3, 3·8, 3·2, and 2·3 g. respectively when dried in a vacuum at 80°, and the corresponding acids recovered from the mother-liquors gave for $[\alpha]_{5461}$ + 8·6°, + 1·1°, + 15·4°, + 5·6°, + 0·9°, — 15·3°, — 29·5°, —42·4°, — 47·8° and — 50° in ethyl alcohol. The final crystals (m. p. 202—204°) of the quinine salt of the (—)acid amounted to 2·3 g., and gave after acidification with dilute sulphuric acid and extraction with ether an acid (m. p. 140—142°) with $[\alpha]^{20^\circ}_{5790}$ — 43·4° and $[\alpha]^{20^\circ}_{5461}$ — 50·6° (c = 1·935) in ethyl alcohol. This acid was recrystallised from ethyl alcohol–light petroleum (b. p. 80—100°).

(—)p-*Tolylmethylglycollic acid* separated in long prismatic needles, m. p. 140—142° (Found: C, 66·6; H, 6·9. $C_{10}H_{12}O_3$ requires C, 66·7; H, 6·7%). In ethyl alcohol: l = 2, c = 1·5435, $\alpha^{20^\circ}_{5790}$ — 1·36°, $[\alpha]^{20^\circ}_{5790}$ — 44·0°; $\alpha^{20^\circ}_{5461}$ — 1·58°, $[\alpha]^{20^\circ}_{5461}$ — 51·2°.

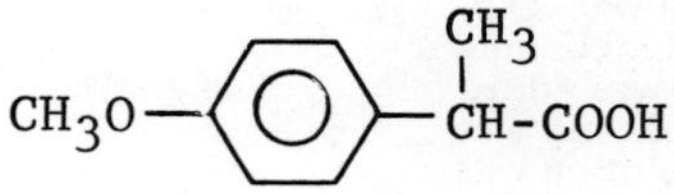

$C_{10}H_{12}O_3$

Ref. 678

(*p*) *Resolution of 2-(*p*-Methoxyphenyl)propionic Acid* (5*a*)

(i) Dextro *Isomer.*—A solution of 2-(*p*-methoxyphenyl)propionic acid[34] (36 g, m.p. 56–57°) and (—)-quinine (76 g) in ethanol (250 ml) and water (200 ml) was allowed to crystallize slowly. The product, m.p. 160–166°, was subjected to careful fractional crystallization from ethanol to give pure (—)-*quinine* (+)-*2-(*p*-methoxyphenyl)propionic acid salt* (24·2 g), m.p. 170–172°, $[\alpha]_D$ —101° (c, 0·66 in ethanol) (Found: C, 71·9; H, 7·3; N, 5·8. $C_{10}H_{12}O_3,C_{20}H_{24}N_2O_2$ requires C, 71·4; H, 7·2; N, 5·6%).

Treatment of the salt with 4N sulphuric acid and crystallization of the product from light petroleum yielded (+)-*2-(*p*-methoxyphenyl)propionic acid*, m.p. 77–79°, $[\alpha]_D$ +77·3° (c, 1·35 in dioxan) (Found: C, 66·7; H, 6·8. $C_{10}H_{12}O_3$ requires C, 66·7; H, 6·7%). The i.r. spectrum was identical with that of the racemic acid.

(ii) Laevo *Isomer.*—The *laevo*-enriched acid (1·8 g) recovered from (i) was treated with (—)-ephedrine (1·65 g) in water (20 ml) and crystallization of the salt induced by scratching. The crude salt (1·55 g) had m.p. 121–136°, which was raised to 130–137° upon crystallization from acetone. Three further crystallizations from acetone afforded the pure (—)-*ephedrine salt*, m.p. 139–141°, $[\alpha]_D$ —24·4° (c, 1·02 in ethanol) (Found: C, 69·7; H, 8·1; N, 3·9. $C_{10}H_{12}O_3,C_{10}H_{15}NO$ requires C, 69·5; H, 7·9; N, 4·1%).

Decomposition of the salt in the usual way, and crystallization from light petroleum yielded (—)-*2-(*p*-methoxyphenyl)propionic acid*, m.p. 77–79°, $[\alpha]_D$ —76·9° (c, 1·03 in dioxan), —64·0° (c, 1·06 in ethanol) [lit.[21] m.p. 73–74°, $[\alpha]_D$ —58° (ethanol)] (Found: C, 66·7; H, 6·9. $C_{10}H_{12}O_3$ requires C, 66·7; H, 6·7%).

[21] Ollis, W. D., Ramsay, M. V. J., and Sutherland, I. O., *Aust. J. Chem.*, 1965, **18**, 1787.

[34] Kenyon, W. G., Kaiser, E. M., and Hauser, C. R., *J. org. Chem.*, 1965, **30**, 2937.

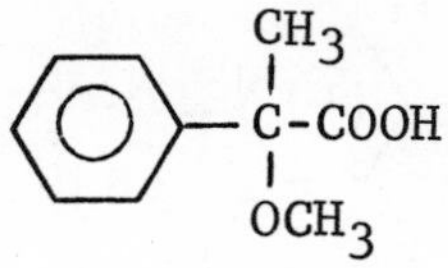

$C_{10}H_{12}O_3$

Ref. 679

Optical resolution of (±)-*2-methoxy-2-phenyl-propionic-acid* (VII) (*atrolactic-acid-methylether*)

Compound (±)-VII (20·45 g, 0·11 mol) and an equimolecular amount of brucine in acetone (100 ml) gave, after 1–2 hr, 54 g of the corresponding salt, which was then suspended in 100 ml boiling acetone and the warm mixture filtered. The solid was again treated as above with 70 ml acetone. The residual brucine salt (23 g) had m.p. 178–180°, $[\alpha]_D$ −11° (c = 1, MeOH). (Found: C, 69·1; H, 6·7; N, 4·95. $C_{33}H_{38}N_2O_7$ requires: C, 68·95; H, 6·65; N, 4·85%). Free *R*(−)-acid-VII[4]: $[\alpha]_D$ −32·5° (Neat, l = 0·1 dm), = −26° (c = 1, MeOH).

From the first two acetone mother liquors, 25 g of salt were obtained, m.p. 110–112°, $[\alpha]_D$ +3° (c = 1, MeOH). (Found: C, 68·0; H, 6·4; N, 4·7%). Free S(+)-acid VII[4]: $[\alpha]_D$ +31·5° (neat, l = 0·1 dm), = +25° (c = 1, MeOH).

[4] D. J. Cram and K. R. Kopecky, *J. Am. Chem. Soc.* **81**, 2748 (1959).

$C_{10}H_{12}O_3$

Ref. 679a

Resolution of 2-Methoxy-2-phenylpropionic Acid.—The above methyl 2-methoxy-2-phenylpropionate was stirred with excess 1 *N* sodium hydroxide solution until the mixture became clear. The basic solution was then washed with ether three times, acidified, and extracted with ether. The latter ether extract was dried and evaporated to an oil which was used directly in the resolution as follows. Quinine, 1152 g. or 4 moles, was dissolved in 2100 ml. of boiling ethanol, and 546 g. (3.3 moles) of the above oil was added followed by 1300 ml. of water. The resulting solution was cooled to 0°, and after 2 weeks the supernatant solution was decanted from the crystalline cake that separated. The crystals were washed twice with cold 50% aqueous ethanol, and were recrystallized twice from ethanol–water to give 260 g. of salt. After each crystallization 1.0 g. of salt was converted to the acid, and in ether was converted to the methyl ester with diazomethane. Rotations were taken on the methyl ester (neat, l = 1 dm.) as follows: 1st, $\alpha^{30}D$ −50.2°; 2nd, $\alpha^{28}D$ −52.4°; 3rd, $\alpha^{24}D$ −52.2°. A rotation for this compound has been reported, $\alpha^{20}{}_{578}$ −56.4° (neat, l = 1 dm.).

The entire crop of salt from the second recrystallization was shaken with a mixture of cold dilute sulfuric acid (excess) and ether. The water layer was washed with ether, and the combined ether extracts were washed with water, dried and evaporated. The residual oil amounted to 94 g. (35%), and was not further purified, $\alpha^{26}D$ +24° (l = 1 dm., neat).

The mother liquors from the original crystallization of the salt of the (+)-acid was diluted with 300 ml. of water. After standing for one week at 0°, 180 g. of salt was precipitated. Conversion of a sample of this material to the ester of the acid gave material, $\alpha^{25}D$ +50° (neat, l = 1 dm.).

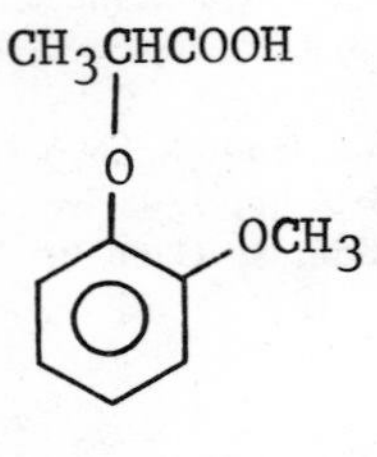

$C_{10}H_{12}O_3$

Ref. 680

Preliminary experiments on resolution were carried out in dilute ethanol with the common alkaloids and some synthetic bases. *Strychnine* gave a laevorotatory acid of high activity. *Cinchonidine* and *quinine* yielded (−)-acid of lower rotation. The acid liberated from the *brucine* and *dehydroabietylamine* salts was racemic. The salts of *cinchonine*, *quinidine*, *ephedrine*, *β-phenylisopropylamine* and *α-(2-naphthyl)-ethylamine* failed to crystallise. None of the bases tested gave extrorotatory acid.

(−)-*2-Methoxyphenoxy-propionic acid.* Strychnine (50.1 g, 0.15 mole) and racemic acid (29.4 g, 0.15 mole) were dissolved in 300 ml of hot ethanol and 300 ml of water were added. After 24 h in a refrigerator, 38.6 g salt had separated. It was recrystallised three times; measurements on small samples of acid, isolated from the different fractions and dissolved in ethanol, showed that the (−)-acid was practically pure after the second crystallisation:

Crystallisation	1	2	3	4
ml ethanol	300	200	125	100
ml water	300	200	125	100
g salt obtained	38.6	31.4	28.4	25.0
$[\alpha]^{25}_D$ of the acid	−37°	−43.5°	−43.5°	−43.5°

The salt was decomposed with excess sodium carbonate solution and the strychnine extracted with chloroform. The acid was then isolated by acidification with excess sulphuric acid and extraction with ether. The yield of crude acid was 9.0 g. After recrystallisation from dilute (80%) formic acid there remained 7.0 g with m.p. 82.5–83°.

$C_{10}H_{12}O_4$ (196.20) calc. equiv.wt 196.2 C 61.21 H 6.17
found equiv.wt 196.6 C 61.17 H 6.15

0.2226 g acid dissolved in *acetone* to 10.00 ml: α^{25}_D = −2.349° (2 dm).
$[\alpha]^{25}_D$ = −52.8°; $[M]^{25}_D$ = −103.5°.

(+)-*2-Methoxy-phenoxy-propionic acid.* About 50% of the salt contained in the mother liquor from the first crystallisation of the strychnine salt was decomposed and the acid liberated as described above; 7.7 g having $[\alpha]^{25}_D$ = +36.5° (ethanol) were obtained. The acid was recrystallised from dilute formic acid as seen below:

Crystallisation	1	2	3	4
ml formic acid	5.0	3.0	3.0	2.0
ml water	1.0	1.0	1.0	0.5
g acid obtained	6.5	5.2	4.7	4.2
$[\alpha]^{25}_D$ in ethanol	+45°	+45.5°	+47°	+48°

The rest of the salt from the mother liquor of the strychnine salt was treated in the same way, yielding 3.1 g acid with $[\alpha]^{25}_D$ = +48°. The total yield was thus 7.3 g having m.p. 82.5–83°.

$C_{10}H_{12}O_4$ (196.20) calc. equiv.wt 196.2 C 61.21 H 6.17
found equiv.wt 196.5 C 61.02 H 6.15

Data for the rotatory power in various solvents are given in Table 1.

Table 1. g = g acid in 10.00 ml solution.

Solvent	g	α^{25}_D (2 dm)	$[\alpha]^{25}_D$	$[M]^{25}_D$
Acetone	0.2067	+2.167°	+52.4°	+102.8°
Ethanol (abs.)	0.2293	+2.195°	+47.9°	+93.8°
Chloroform	0.2050	+0.205°	+5.0°	+9.8°
Water (neutr.)	0.2020	−0.528°	−13.1°	−25.6°
Benzene	0.2120	+0.278°	+6.6°	+12.0°

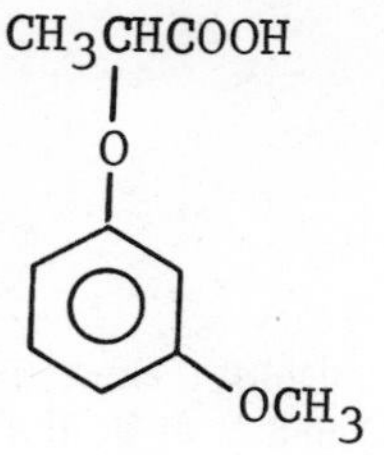

$C_{10}H_{12}O_4$ Ref. 681

Preliminary experiments on resolution were performed in dilute ethanol (in several cases also in other solvents) with a number of alkaloids and synthetic bases. Crystallised salts were obtained from *quinine, cinchonidine, ephedrine, (−)-α-phenyl-ethylamine, (+)-α-(2-naphthyl)ethylamine* and *dehydroabietylamine* but the acid liberated was in all cases racemic. The salts from *strychnine, brucine, cinchonine* and *quinidine* failed to crystallise. Only the *(−)-β-phenyl-isopropylamine* salt gave optically active (dextrorotatory) acid. The salt had, however, a certain tendency to separate as an oil, which gradually solidified more or less completely. This complicates the resolution on a preparative scale and makes it difficult to give an accurate and reproducible description of the procedure.

(+)-3-Methoxyphenoxy-propionic acid. Racemic acid (23.5 g, 0.12 mole) and (−)-β-phenyl-isopropylamine (16.2 g, 0.12 mole) were dissolved together in 40 ml ethanol +50 ml water. On standing, part of the salt separated as an emulsion. About 25 % of the solvent was evaporated, the solution was inoculated with crystals from the preliminary experiments and a further 50 ml of water were added. Now a mixture of crystals and oil was obtained. The crystals were filtered off by suction, washed free from oil and recrystallised five times from water. After each crystallisation, a small sample of the salt was decomposed and the activity of the acid determined in ethanol solution:

Crystallisation	1	2	3	4	5	6
$[\alpha]_D^{25}$ of the acid	+12.5°	+19°	+22.5°	+24°	+24.5°	+24°

The acid was liberated with dilute sulphuric acid and extracted with ether. It was recrystallised three times from carbon tetrachloride and twice from dilute formic acid. The yield of pure acid was 3.0 g, forming small needles with m.p. 67.5–69°

$C_{10}H_{12}O_4$ (196.20) calc. C 61.21 H 6.17
found C 60.81 H 6.22

The rotatory power in various solvents is given in Table 2.

Table 2. g = g acid in 10.00 ml solution.

Solvent	g	α_D^{25} (2 dm)	$[\alpha]_D^{25}$	$[M]_D^{25}$
Acetone	0.2063	+1.428°	+34.6°	+67.9°
Ethanol (abs.)	0.2151	+1.050°	+24.4°	+47.9°
Chloroform	0.2079	+0.257°	+6.2°	+12.1°
Water (neutr.)	0.2216	+0.235°	+5.3°	+10.4°
Benzene	0.1570	−0.618°	−19.7°	−38.6°

(−)-3-Methoxyphenoxy-propionic acid. The filtrates from the first two crystallisations of the (−)-β-phenyl-isopropylamine salt were evaporated and the acid liberated; 16.2 g having $[\alpha]_D^{25} = -9.1°$ were obtained. The product was powdered and shaken with carbon tetrachloride at room temperature. The residue had a somewhat higher activity (−11.6°). The extract contained mostly racemic acid and soluble impurities. The residue was extracted once more with carbon tetrachloride. This time active acid was dissolved, the residue being practically pure racemic acid. On spontaneous evaporation of the extract in a flat dish, the acid was obtained partly as very small crystals covering the bottom of the dish and partly as crusts on the walls. The small crystals were nearly pure racemic acid while the acid from the crusts had $[\alpha]_D^{25} = -23°$ (ethanol). This acid was recrystallised twice from formic acid and twice from carbon tetrachloride and the rotatory power measured after each recrystallisation:

Recrystallisation	1	2	3	4
$[\alpha]_D^{25}$ of the acid	−23.5°	−24°	−24.5°	−24.5°

The yield of pure acid, quite similar to the enantiomer, was 3.6 g. M.p. 67.5–69°.

$C_{10}H_{12}O_4$ (196.20) calc. C 61.21 H 6.17
found C 60.45 H 6.11

0.2144 g dissolved in *acetone* to 10.00 ml: $\alpha_D^{25} = -1.477°$ (2 dm).
$[\alpha]_D^{25} = -34.4°$; $[M]_D^{25} = -67.6°$.

The results of the extraction experiments may seem somewhat puzzling. However, we have found also in other cases that the first extraction of a crude acid isolated by ether extraction of an acidified mother liquor, may give unexpected results [11]., This is most probably due to the presence of soluble impurities (perhaps also moisture), which influence the solubility of the active and racemic acid.

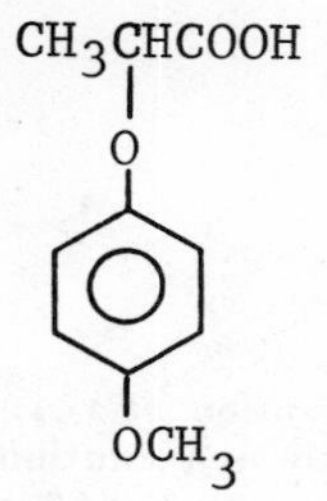

$C_{10}H_{12}O_4$ Ref. 682

Preliminary experiments on resolution were performed with the common bases in dilute ethanol or acetone. *Cinchonidine* gave a salt containing (+)-acid with $[\alpha]_D = +22°$. The acid isolated from the *quinine* and *quinidine* salts had very low activity. The *brucine* salt, which was at first somewhat difficult to bring to crystallisation, gave an acid with $[\alpha]_D = -32°$. The salt of *α(2-naphthyl)-ethylamine* yielded only racemic acid and the *strychnine* and *cinchonine* salts failed to crystallise. Brucine and cinchonidine were thus selected for the final resolution.

(−)-4-Methoxyphenoxy-propionic acid. Racemic acid (30 g, 0.15 mole) and anhydrous brucine (60 g, 0.15 mole) were dissolved in 300 ml hot 15 % ethanol. The salt, which crystallised very slowly after seeding, was filtered off after 4 days in a refrigerator. It was recrystallised from water to maximum activity of the acid; the progress of the resolution was tested on small samples of the salt, from which the acid was liberated and tested in ethanol.

Crystallisation	1	2	3	4
ml water	255	55	35	40
ml ethanol	45	—	—	—
g salt obtained	23.0	20.5	18.7	17.5
$[\alpha]_D^{25}$ of the acid	−34.5°	−43.5°	−43°	−44°

The mother liquor from the first crystallisation was evaporated to about half its volume and left in a refrigerator for one week. A second crop of salt (6.3 g) separated. It was recrystallised from the successive mother liquors from the first series of crystallisations yielding 6.7 g of pure salt.

The two salt fractions were combined and the acid was liberated by acidification with sulphuric acid and extraction with ether. Recrystallisation from 200 ml ligroin yielded 6.5 g of pure acid forming long, glistening needles with m.p. 65–66°.

$C_{10}H_{12}O_4$ (196.20) calc. equiv. wt. 196.2 C 61.21 H 6.17
found equiv. wt. 196.3 C 61.18, 61.26 H 6.09, 6.18

0.2167 g acid dissolved in absolute *ethanol* to 10.00 ml: $\alpha_D^{25} = -1.869°$ (2 dm). $[\alpha]_D^{25} = -43.1°$; $[M]_D^{25} = -84.6°$. — 0.2052 g acid dissolved in *acetone* to 10.00 ml: $\alpha_D^{25} = -2.416°$ (2 dm). $[\alpha]_D^{25} = -58.9°$; $[M]_D^{25} = -115.5°$.

(+)-Methoxyphenoxy-propionic acid. The mother liquors from the brucine salt yielded 18 g acid having $[\alpha]_D^{25} = +23.6°$ (ethanol). This acid (0.09 mole) was dissolved with 27 g (0.09 mole) cinchonidine in 175 ml ethanol and 100 ml water were added. The salt was recrystallised to maximum activity of the acid.

Crystallisation	1	2	3
ml water	100	265	210
ml ethanol	175	110	90
g salt obtained	25	20	18
$[\alpha]_D^{25}$ of the acid	+37.5°	+43°	+44°

A further crop of pure salt could be obtained by systematic fractionation of the salt in the mother liquors.

The acid was isolated in the usual way and recrystallised twice from 200 ml ligroin (the second time with charcoal). 23 g salt yielded 7.3 g pure acid, quite similar to the (−)-form. M.p. 65–66°.

$C_{10}H_{12}O_4$ (196.20) calc. equiv. wt. 196.2 C 61.21 H 6.17
found equiv. wt. 196.3 C 61.08, 60.98 H 6.18, 6.16

The rotatory power in various solvents is given in Table 1.

Table 1

g = g acid dissolved to 10.00 ml

Solvent	g	α_D^{25} (2 dm)	$[\alpha]_D^{25}$	$[M]_D^{25}$
Acetone	0.2032	+2.398°	+59.0°	+115.8°
Ethanol (abs.)	0.2024	+1.746°	+43.1°	+84.6°
Chloroform	0.2056	+1.011°	+24.6°	+48.2°
Water (neutr.)	0.2019	+0.918°	+22.7°	+44.6°
Benzene	0.2042	+0.114°	+2.8°	+5.5°

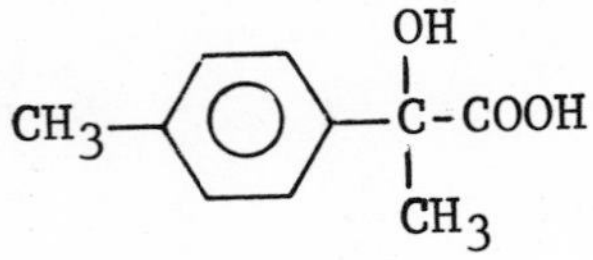

$C_{10}H_{12}O_4$ Ref. 683

Resolution of r-*Anisylmethylglycollic Acid.*—42·8 G. of morphine (1 mol.) were dissolved gradually in a solution of 24·9 g. of *r*-anisylmethylglycollic acid (1 mol.) in 1 l. of boiling water. The crystals (A, **33** g.) which had separated over-night in the ice-chest were collected and recrystallised seven times from boiling water, rosettes of prisms of the homogeneous morphine salt (14·9 g.) of the (—)acid being obtained. The acid (5·3 g.) got by decomposing this salt with dilute sulphuric acid and extraction with ether had m. p. 146—147° and gave in ethyl alcohol: $l = 1$, $c = 3{\cdot}997$, $\alpha_{5461}^{19\cdot5°}$ $-$ 2·47°, $[\alpha]_{5461}^{19\cdot5°}$ $-$ 61·8°, whereas McKenzie and P. D. Ritchie (*loc. cit.*) give m. p. 146—147° and $[\alpha]_{5461}^{25°}$ $-$ 61·7° ($c = 4{\cdot}0645$). The acid underwent partial dehydration when heated at 70° under diminished pressure owing to the formation of 4-methoxyatropic acid, which takes place more readily than does the formation of atropic acid from atrolactinic acid under similar conditions. In the determination of the specific rotatory power it is advisable to conduct the drying at the ordinary temperature under diminished pressure.

The filtrate, from which A had been removed, gradually deposited rectangular crystals (17 g.), which were crystallised twice from water. The resulting crystals (11 g.) were decomposed by dilute sulphuric acid and gave an acid, m. p. 145—146°; in ethyl alcohol $[\alpha]_{5461}$ + 60° ($c = 3{\cdot}946$) This consisted of the almost pure (+)acid, but some 4-methoxyatropic acid was also present, as was shown by the action on potassium permanganate and on bromine water.

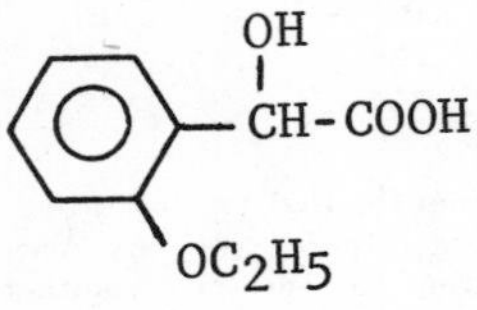

$C_{10}H_{12}O_4$ Ref. 684

7. Zerlegung der o-Äthoxymandelsäure in die Antipoden.

7 g o-Äthoxymandelsäure wurden in 500 ccm Wasser gelöst und in der Wärme mit 10,5 g Cinchonin versetzt. Nach 1-stündigem Erwärmen wurde von wenig ungelöstem Cinchonin abfiltriert, langsam erkalten und im Eisschrank stehen gelassen. Nach einigen Tagen schied sich das Cinchoninsalz der (—)-o-Äthoxymandelsäure in großen Krystallen ab, die nach 2-maligem Umkrystallisieren aus wäßrigem Alkohol als Impfkrystalle für die Spaltung einer größeren Menge (56 g) Säure dienten. Nach 3 Tagen war das Cinchoninsalz der (—)-o-Äthoxymandelsäure in einer Rohausbeute von 66 g auskrystallisiert.

$[\alpha]_D^{20}$ des ungereinigten Salzes war + 70°, gemessen in 1-proc. äthylalkoholischer Lösung. Nach 2-maligem Umkrystallisieren aus etwa 80-proc. Alkohol blieben 56 g reines Cinchoninsalz von der spezifischen Drehung $[\alpha]_D^{20}$ + 75° in 1-proc. äthylalkoholischer Lösung.

55 g dieses Cinchoninsalzes wurden in verdünnter Schwefelsäure gelöst, und die Lösung wurde nach dem Sättigen mit Ammonsulfat ausgeäthert. Die aus der ätherischen Lösung nach Trocknen mit Natriumsulfat und Abdestillieren des Äthers gewonnene (—)-o-Äthoxymandelsäure war sofort rein. Schmelzp. 125,5—126,5° (unkorr.), 126,5 bis 127,5 (korr.).

$[\alpha]_D^{20}$ in 1-proc. alkoholischer Lösung − 144,9°. Beide Daten blieben nach Umkrystallisieren aus Wasser unverändert. Ausbeute 22 g (78 Proc. d. Th.).

Die Mutterlauge des Cinchoninsalzes wurde i. V. auf $^1/_3$ ihres Volumens eingedampft. Nach dem Abkühlen schied sich das Cinchoninsalz der (+)-o-Äthoxymandelsäure als braunes Öl ab, das nicht zur Krystallisation gebracht wurde. Aus ihm wurde die (+)-o-Äthoxymandelsäure durch Ausschütteln der angesäuerten und mit Ammonsulfat gesättigten Lösung mit Äther isoliert. Schmelzpunkt nach Umkrystallisieren aus Wasser 125,5—126,5°.

$[\alpha]_D^{20}$ in 1-proc. alkoholischer Lösung + 145,5°.

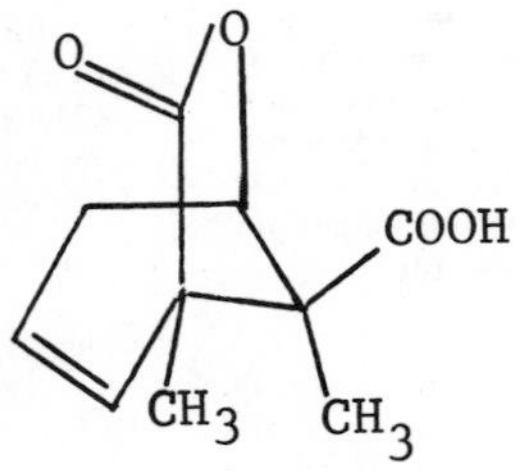

$C_{10}H_{12}O_4$ Ref. 685

1.3. ***Spaltung von (±)-Cantharsäure ((±)-2) in (+)-Cantharsäure ((+)-2)***[2] (vgl. [5c]). – 1,96 g (10,0 mmol) (±)-Cantharsäure und 3,95 g (10,0 mmol) Brucin löste man in heissem Wasser, fügte eine Spatelspitze Aktivkohle zu und kristallisierte nach dem Filtrieren 6mal aus Wasser um. Nach der dritten und den nachfolgenden Kristallisationen (vereinigte Mutterlaugen M-1) entnahm man eine Probe von etwa 50 mg, versetzte diese mit 30 ml 2N Natronlauge, extrahierte 3mal mit Chloroform, stellte die wässerige Phase mit 6N Salzsäure auf pH ≈ 2 und extrahierte die optisch aktive Cantharsäure durch 6maliges Ausschütteln mit Chloroform. Der Eindampfrückstand der vereinigten und über Natriumsulfat getrockneten Chloroformauszüge wurde bei 170–180°/0,01 Torr sublimiert. Anschliessend bestimmte man den Drehwert der Cantharsäure. Die Drehung erfuhr nach der dritten Kristallisation keine signifikante Änderung mehr.

Die Aufarbeitung des nach der sechsten Kristallisation erhaltenen Brucinsalzes ergab (+)-*Cantharsäure*, die einmal aus Wasser (Aktivkohle) umkristallisiert und anschliessend bei 0,01 Torr über Phosphorpentoxid getrocknet wurde. Ausbeute: 420 mg. Smp. 265° (Zers.). $[\alpha]_D^{22} = +89{,}4° \pm 2{,}0°$ ($c = 2{,}814$; Äthanol) (Literaturwert: $[\alpha]_D = +89{,}6°$ ($c = 1{,}6$; Äthanol [5c])). – CD.: siehe Tab. 2. – IR. (KBr): 3150 (Hydroxyl, geb.), 1765 (Fünfringlacton), 1730 (Carboxyl). – NMR. (100 MHz; Aceton-d_6): 10,3 (br. *s*; COOH), 5,86 ($d \times d \times t$; $J_{1,6} = 9$ Hz, $J_{6,5\alpha} \approx J_{6,5\beta} \approx 3$ Hz, $J_{6,4\alpha} = 2$ Hz; H-(6)), 5,52 ($d \times t$; $J_{1,6} = 9$ Hz, $J_{1,5\alpha} = J_{1,5\beta} = 2$ Hz; H-(1))[6]), 4,71 (*m*; H-(4)), 2,54–2,40 (*m*; H_2-(5)), 1,32 und 1,30 (zwei *s*; H_3-(8) und H_3-(9)).

Die vereinigten Mutterlaugen M-1 wurden eingeengt. Die aus dem auskristallisierten Brucinsalz (Mutterlauge M-2) freigesetzte Cantharsäure (581 mg) zeigte $[\alpha]_D^{22} = -42{,}4° \pm 2{,}0°$ ($c = 2{,}845$; Äthanol). – Die aus der Mutterlauge M-2 erhaltene (–)-*Cantharsäure* (446 mg) zeigte $[\alpha]_D^{22} = -79{,}1° \pm 2{,}0°$ ($c = 2{,}992$; Äthanol). Optische Reinheit: 88,5%.

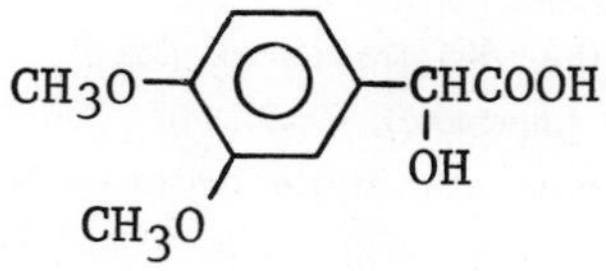

$C_{10}H_{12}O_5$ **Ref. 686**

Scissione ottica dell'acido DL-3,4-dimetossimandelico

Sale di cinconina dell'acido (-)3,4-dimetossimandelico. G 5.0 (0.023 moli) di acido DL-3,4-dimetossimandelico vengono sciolti all'ebollizione in 60 cc di alcool etilico 95% insieme a g 7,0 di cinconina. La soluzione viene, dopo raffreddamento, posta in ghiacciaia ed insemenzata con del sale ottenuto in una'prova in piccolo. Dopo 48 ore si filtra il precipitato bianco separatosi racogliendo g 4,0 di (-)3,4-dimetossimandelato di (+)cinconina a p.f. 190-193° ed $[\alpha]_D^{18}$ = 90.6° (c = 0.55% in acqua).

Il sale di cinconina cosi preparato viene ripetutamente cristallizzato da etanolo 95% finchè l'acido ottenuto dal sale mediante trattamento con la quantità calcolata di acido solforico dil., estrazione con acetato di etile et concentrazione del solvente, non presenta potere rotatorio costante.

L'andamento della scissione è riportato nella Tab. II in cui sono rappresentate le rese del sale, i poteri rotatori del sale e del rispettivo acido, i p.f. del sale e dell'acido.

Tabella II

Cristallizzazione		I.	II.
Rese in g di sale	4.0	3.4	2.9
p.f. del sale	190-3°	196-9°	203-5°
[α]D del sale (c= 0.6% in acqua)	+90,6°	+79,7°	+68,9°
p.f. dell'acido	95-6°	104-6°	104-6°
[α]D dell'acido (c= 0.65% in acqua)	-82,3°	-166,0°	-124,2°

Acido (-)3,4-dimetossimandelico. G. 2,0 di sale di cinconina provenienti dall'ultima cristallizzazione vengano sospesi in 30 cc di acqua e trattati con 4 cc di acido solforico 1:1. l'acido liberato viene estratto con acetato di etile (25x5). Si lavano gli estratti con qualche cc di acqua e si seccano su solfato di sodio anidro. Per evaporazione del solvente sotto vuoto rimangono g 0,8 di acido oleoso che, cristallizzato da benzolo, danno g 0,5 di acido (-)3,4-dimetossimandelico a p.f. 104-106° ed $[\alpha]_D^{18}$ = -124,2° (c = 0.6% in acqua).

Questo acido viene di nuovo salificato con cinconina; si ottiene il sale a p.f. 203-205° ed $[\alpha]_D^{18}$ = +69.2° (c = 0.6% in acqua). Il p.f. ed il potere rotatorio non variano dopo ulteriore cristallizzazione da etanolo 95%. Per trattamento di questo sale con acido solforico si ottiene di nuovo acido (-)3,4-dimetossimandelico a p.f. 104-106° ed $[\alpha]_D^{18}$= -124,6° (c = 0,58% in acqua).

Sale di cinconina dell'acido (+)3,4-dimetossimandelico. Le acque madri della scissione vengono concentrate sotto vuoto fino a secchezza. Il residuo (g 3,2) viene cristallizzato da 20 cc di etanolo 95%. Si ottengono g 2,3 di (=)3,4-dimetossimandelato di (+)cinconina a p.f. 172-175° ed $[\alpha]_D^{20}$ = +139,3° (c = 0.56% in acqua).

Acido (+)3,4-dimetossimandelico. G 2,3 di sale di cinconina dell'acido (+)3,4-dimetossimandelico vengono sospesi in acqua, acidificati con acido solforico ed estratti con acetato di etile come indicato per l'antipodo. Per concentrazione degli estratti si ottiene acido (+)3,4-dimetossimandelico a p.f. 97-101° ed $[\alpha]_D^{25}$ = +74,2° (c = 0,62% in acqua).

Tutti i tentatavi eseguiti per aumentare la purezza ottica di questo acido sia per cristallizzazione sia per salificazione con diverse basi in diversi solventi ed a differenti concentrazioni, sono risultati infruttuosi. Non si è mai riusciti ad ottenere acido (+)3,4-dimetossimandelico a potere rotatorio superiore a quello sopra riportato.

$C_{10}H_{12}O_5S$

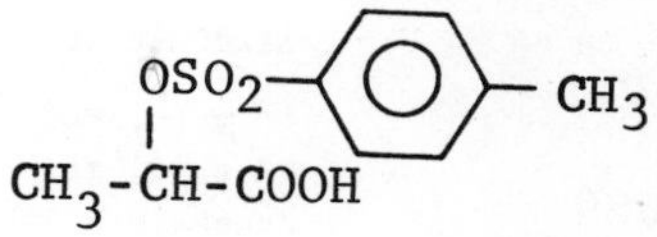

Ref. 687

Partial Resolution of α-p-*Toluenesulphonoxypropionic Acid.*—To an ice-cold solution of *dl*-α-*p*-toluenesulphonoxypropionic acid (15 g.) and quinine (20 g.) in acetone (25 c.c.), ethyl acetate (75 c.c.) was added, and the mixture cooled in ice. The quinine salt (18 g.) obtained proved too unstable to be recrystallised; when decomposed with dilute sulphuric acid, it gave (+)α-*p*-toluenesulphonoxypropionic acid, $[\alpha]_{5461}$ + 27·2° (*l*, 2; *c*, 5·00) in methyl alcohol. The filtrate, after decomposition, yielded the (—)acid with approximately the same rotatory power. The rotatory power of the *d* + *dl*-acid increases slowly when it is recrystallised from benzene and light petroleum, but the optically pure acid is more readily obtained from *d*-lactic acid by Kenyon, Phillips, and Turley's method (*loc. cit.*).

(Kenyon, Phillips, and Turley, J., 1925, **127**, 399)

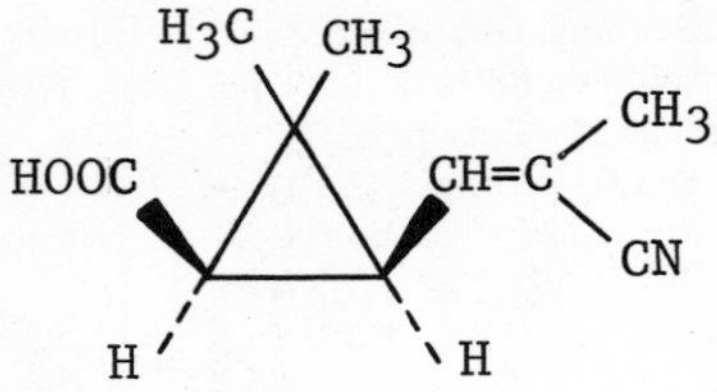

$C_{10}H_{13}NO_2$ **Ref. 688**

8) Optical Resolution of (±)-*cis*-2,2-Dimethyl-3-(2'-cyano-1'-propenyl)cyclopropanecarboxylic Acid.

Quinine (8.6 g) in methanol (20 ml) was added to the (±)-*cis*-acid (5 g) in methanol (15 ml). The mixture was kept in a refrigerator for four hours. It gave the quinine salt of the (+)-*cis*-acid (5.3 g), m.p. 146°C. Recrystallization twice from methanol gave the pure quinine salt of the (+)-*cis*-acid (3 g), m.p. 151°C, $[\alpha]_D^{17}$ −125.90° (methanol, C=1.189). *Anal.* Found: C, 70.21; H, 7.66; N, 8.01. Calcd. for $C_{30}H_{37}O_4N_3$: C, 70.53; H, 7.39; N, 8.23%.

The decomposition of this salt with dilute hydrochloric acid gave the (+)-*cis*-acid (1 g), m.p. 105°C, $[\alpha]_D^{17}$ +6.94° (methanol, C=10.160). After 20 hours, the second fraction was separated and the mother liquor was concentrated and decomposed with dilute hydrochloric acid.

Extraction with ether gave the crude (−)-*cis*-acid. (+)-α-Phenethylamine (2.3 g) in methanol was added to the (−)-*cis*-acid in methanol (10 ml). The first fraction was separated after 15 hours and recrystallized from methanol to give pure (+)-α-phenethylamine salt (10.8 g), m.p. 165°C, $[\alpha]_D^{17}$ +26.25° (methanol, C=1.683). The decomposition of this salt with dilute hydrochloric acid afforded the (−)-*cis*-acid (0.4 g), m.p. 105°C, $[\alpha]_D^{17}$ −6.45° (methanol, C=11.630).

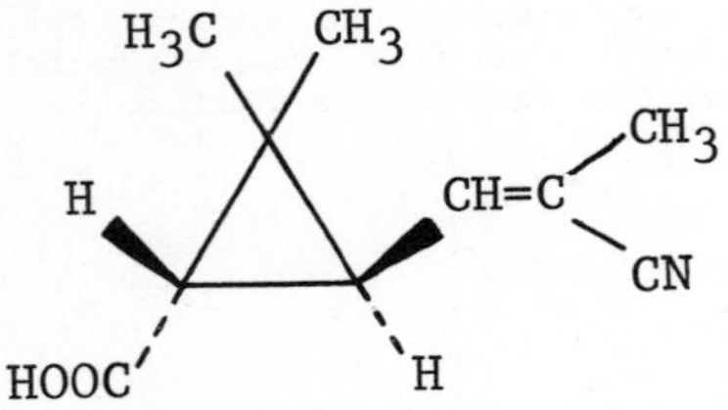

$C_{10}H_{13}NO_2$ **Ref. 689**

7) Optical Resolution of (±)-*trans*-2,2-dimethyl-3-(2'-cyano-1'-propenyl)cyclopropanecarboxylic Acid.

Strychnine (11.2 g) was added to the (±)-*trans*-acid (6 g) in methanol (45 ml). After 15 hours the strychnine salt (9.8 g) crystallized. This salt was recrystallized three times from methanol to give the optically pure strychnine salt of the (+)-*trans*-acid (5.2 g), $[\alpha]_D^{17}$ +16.89° (chloroform, C=2.960). *Anal.* Found: N, 8.24. Calcd. for $C_{31}H_{35}N_3O_4$: N, 8.1%.

The decomposition of this salt with dilute ammonium hydroxide gave (+)-*trans*-chrysanthemumdicarboxylic acid mononitrile (1.4 g). M.p. 98°C, $[\alpha]_D^{17}$ +95.39° (methanol, C=2.967).

The mother liquor was concentrated and the salt was decomposed with dilute ammonium hydroxide to give the crude (−)-*trans*-acid (2.9 g), m.p. 69~84°C, $[\alpha]_D^{17}$ −75.2° (methanol, C=2.683). (−)-α-Phenethylamine (1.9 g) in methanol (10 ml) was added to the crude (−)-*trans*-acid in methanol (10 ml). The mixture was kept overnight in a refrigerator. Then the (−)-α-phenethylamine salt of the (−)-*trans*-acid (1.8 g) was separated, m.p. 145~148°C, $[\alpha]_D^{17}$ −34.85° (methanol, C=3.300).

Recrystallization of the salt from methanol gave the pure salt of the (−)-*trans*-acid (1.2 g), m.p. 150°C, $[\alpha]_D^{16}$ −35.74° (methanol, C=3.133). The decomposition of this salt with dilute hydrochloric acid gave the (−)-*trans*-acid (0.6 g), m.p. 98°C, $[\alpha]_D^{18}$ −95.75° (methanol, C=3.200).

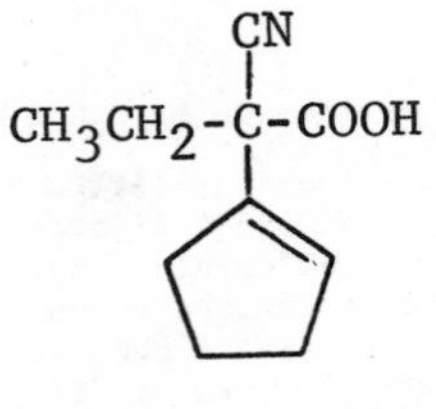

$C_{10}H_{13}NO_2$ **Ref. 690**

Note:

The acid was resolved with codeine using methanol/H_2O as solvent to obtain the (+)-enantiomer, m.p. 23°, $[\alpha]_D$ +278° and with quinine using methanol as solvent to obtain the (−)-enantiomer, m.p. 23°, $[\alpha]_D$ −25.6°.

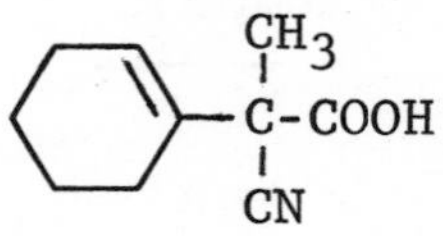

$C_{10}H_{13}NO_2$ **Ref. 691**

Methyl-(1'-cyclohexenyl-)-cyanessigsäure [3A]

(+)Antipode [(+)3A] mit Chinin

54 g (0,3Mol) razem. [3A] wurden unter gelindem Erwärmen mit 97 g (0,3Mol) Chinin in 250 ml abs. Methanol gelöst. Nach Zugabe von 200 ml abs. Äther wurde die Lösung langsam auf 4° abgekühlt und nach 6 Std. weitere 225 ml Äther zugefügt. Danach ließen wir etwa 36 Std. in der Tiefkühltruhe auskristallisieren und erhielten nach dem Trocknen bei 20° i.Vak. zwischen 65 g und 80 g = yg Diastereomerensalz in feinen filzigen Nadeln. Es wurde 3 x umkristallisiert aus

1.) 2y ml abs. Methanol / 2y ml abs. Äther
2.) und 3.) 2y ml abs. Methanol / y ml abs. Äther

Danach erhielten wir etwa 40% der eingesetzten yg in Form von derben Nadeln mit einem Schmp. von 117°–120° (Zers.). Ausbeute zwischen 34% und 43% d.Th.

54 g Diastereomerensalz aus 2 Ansätzen wurden mit H_2SO_4 zerlegt. Wir erhielten 18 g [(+)3A] als farbloses hygroskopisches Öl; Ausbeute 92,5% d.Th.

Spezifische Drehung

c = 2,91 , 1 dm Rohr , t = 20° , abs. Äthanol

$\alpha_{578} = + 1{,}000^\circ$ $[\alpha]^{20}_{578} = + 34{,}3^\circ$ $[\alpha]^{20}_{D} = + 32{,}7^\circ$

$\alpha_{546} = + 1{,}155^\circ$ $[\alpha]^{20}_{546} = + 36{,}7^\circ$

$\alpha_{436} = + 2{,}120^\circ$ $[\alpha]^{20}_{436} = + 72{,}8^\circ$

$\alpha_{405} = + 2{,}630^\circ$ $[\alpha]^{20}_{405} = + 90{,}3^\circ$

$\alpha_{365} = + 3{,}670^\circ$ $[\alpha]^{20}_{365} = +126{,}0^\circ$

Die Veresterung von [(+)3A] mit ätherischer Diazomethanlsg.

$\alpha_{365} = + 4{,}730^\circ$ $[\alpha]^{20}_{365} = + 176{,}6^\circ$

Beide Antipoden von [3A] mit Codein

(+)Antipode [(+)3A]

64 g (0,2Mol) Codein wurden in 400 ml abs. Methanol gelöst und mit einer Lösung von 36 g (0,2Mol) razem. [3A] in 200 ml abs. Äther gemischt. Nach 3 Std. bei 20° wurden abermals 200 ml abs. Äther zugefügt und die Mischung 24 Std. bei 4° und dann 6 bis 7 Std. bei -30° gehalten. Wir erhielten nach dem Trocknen i.Vak. 45 g feinkristallinen Niederschlages, der nach dreimaligem Umkristallisieren aus abs. Methanol / abs. Äther = 1:1 (Vol.) 30 g Diastereomerensalz in glänzenden Blättchen vom Schmp. 128°-130° ergab. Die Lösungen wurden dabei nicht über 50° erwärmt. Nach der Zerlegung dieses Salzes erhielten wir 9,5 g [(+)3A] als farbloses Öl. Ausbeute 53% d.Th. $[\alpha]^{20}_{D} = + 30{,}9^\circ$ (c = 4,19 , 1 dm Rohr , t = 20° , abs. Äthanol; $\alpha_{578} = + 1{,}360^\circ$; $\alpha_{546} = + 1{,}58o^\circ$.) Die Veresterung mit ätherischer Diazomethanlsg. ergab quantitativ [(+)3M] vom Siedep. 103° - 104° /1,5 Torr, $[\alpha]^{20}_{D} = + 45{,}7^\circ$ (c = 2,66 , 1 dm Rohr , t = 20° , abs. Äthanol; $\alpha_{578} = + 1{,}275^\circ$; $\alpha_{546} = + 1{,}460^\circ$.)

(-)Antipode [(-)3A]

Die Mutterlösung des Codeinansatzes wurde ohne Säurezusatz bei 40° am Rotationsverdampfer auf 1/3 ihres Vol. eingeengt und bei 20° 3 x mit je 25 ml abs. Äther versetzt, worauf in 24 Std. 22 g einer Mischfraktion aus beiden Diastereomerensalzen auskristallisierte. Diese wurden abgetrennt. Aus dem Filtrat kristallisierten nach 48 Std. bei 4° derbe, glasige Platten, welche nach zweimaligem Umkristallisieren aus 80 ml abs. Methanol/ 20 ml abs. Äther 15 g des zweiten Codein-Diastereomerensalzes vom Schmp. 87°-89° ergaben. Die Zerlegung dieses Salzes lieferte 5 g [(-)3A] als farbloses Öl, welches ungetrocknet bei 0° erstarrte und aus Benzol/Petroläther in Nadeln vom Schmp. 19°-22° kristallisierte. Nach dem Trocknen wurde [(-)3A] wieder als farbloses hygroskopisches Öl erhalten. Ausbeute 28% d.Th.

$C_{10}H_{13}NO_2$	(179,2)	Ber.:	C 67,02	H 7,31	N 7,82
		Gef.:	C 66,4	H 7,41	N 8,0

Spezifische Drehung

c = 4,34 , 1 dm Rohr , t = 20° , abs. Äthanol

$\alpha_{578} = - 1{,}49o^\circ$ $[\alpha]^{20}_{578} = - 34{,}3^\circ$ $[\alpha]^{20}_{D} = - 32{,}8^\circ$

$\alpha_{546} = - 1{,}750^\circ$ $[\alpha]^{20}_{546} = - 40{,}2^\circ$

$\alpha_{436} = - 2{,}080^\circ$ $[\alpha]^{20}_{436} = - 68{,}7^\circ$

$\alpha_{405} = - 3{,}820^\circ$ $[\alpha]^{20}_{405} = - 88{,}0^\circ$

$\alpha_{365} = - 5{,}335^\circ$ $[\alpha]^{20}_{365} = -122{,}9^\circ$

Die Veresterung von [(-)3A] mit ätherischer Diazomethanlsg. ergab quantitativ den **Methylester [(-)3M]** vom Siedep. 98°-100°/ 1 Torr.

$C_{11}H_{15}NO_2$	(193,2)	Ber.:	C 68,37	H 7,82	N 7,25
		Gef.:	C 68,00	H 7,75	N 7,2

Spezifische Drehung

c = 2,87 , 1 dm Rohr , t = 20° , abs. Äthanol

$\alpha_{578} = - 1{,}435^\circ$ $[\alpha]^{20}_{578} = - 50{,}7^\circ$ $[\alpha]^{20}_{D} = - 47{,}9^\circ$

$\alpha_{546} = - 1{,}650^\circ$ $[\alpha]^{20}_{546} = - 57{,}6^\circ$

$\alpha_{436} = - 3{,}005^\circ$ $[\alpha]^{20}_{436} = - 104{,}5^\circ$

$\alpha_{405} = - 3{,}735^\circ$ $[\alpha]^{20}_{405} = - 130{,}3^\circ$

$\alpha_{365} = - 5{,}245^\circ$ $[\alpha]^{20}_{365} = - 183{,}1^\circ$

Beide Antipoden von [3A] mit D(-) und L(+)Threobase *).

(+)Antipode mit D(-)Threobase.

24 g (0,13Mol) razem. [3A] wurden zusammen mit 29 g (0,13Mol) D(-)Threobase bei 45°-50° in 30 ml abs. Methanol und 250 ml abs. Aceton gelöst. Die über Kohle filtrierte Lösung wurde 48 Std. bei 4° gehalten. Wir erhielten nach Waschen mit Äther und Trocknen an der Luft 26 g = yg feine gelbliche Nadeln. Es wurde 3 x umkristallisiert aus

1.) 5y ml abs. Methanol / 4y ml abs. Äther,
2.) 6y ml abs. Methanol / 3y ml abs. Äther und aus
3.) 7y ml abs. Methanol / 3y ml abs. Äther

Danach wurden 12 g derbe, gelbliche Nadeln vom Schmp. 117°-119° (Zers.) erhalten, welche nach Zerlegen mit HCl 5,3 g [(+)3A] als farbloses Öl lieferten. Ausbeute 44% d.Th. $[\alpha]^{20}_{D} = + 30{,}8^\circ$ (c = 4,98 , 1 dm Rohr , t = 20° , abs. Äthanol ; $\alpha_{578} = + 1{,}610^\circ$; $\alpha_{546} = + 1{,}850^\circ$.)

(-)Antipode mit L(+)Threobase.

24 g razem. [3A] wurden auf die gleiche Weise mit 29 g D(-)Threobase umgesetzt und ergaben 12,6 g Diastereomerensalz vom Schmp. 117°-120° (Zers.), welches nach Zerlegen mit HCl 5,5 g [(-)3A] als farbloses Öl lieferte. Ausbeute 46% d.Th. $[\alpha]^{20}_{D} = - 31{,}o^\circ$ (c = 5,00 , 1 dm Rohr , t = 20° , abs. Äthanol ; $\alpha_{578} = - 1{,}630^\circ$; $\alpha_{546} = - 1{,}880^\circ$).

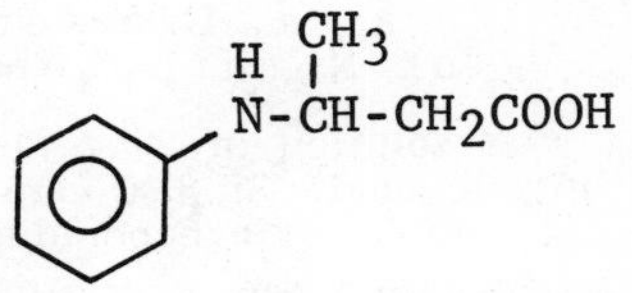

$C_{10}H_{13}NO_2$ Ref. 692

Equimolar amounts of (±)-amino-acid and quinine gave a *salt* which was crystallised four times from acetone to constant m.p. and specific rotation: m.p. 148—149°, $[\alpha]_D$ −130° (*c* 1.0 in MeOH) (Found: C, 71.5; H, 7.4; N, 8.3. $C_{30}H_{37}N_2O_4$ requires C, 71.55; H, 7.4; N, 8.35%).

Treatment of the salt with aqueous N-sodium hydroxide, followed by extraction with chloroform and acidification of the aqueous solution to pH 4, afforded the (−)-(*R*)-amino-acid (12) as an oil, $[\alpha]_D$ −24° (*c* 3.0 in MeOH), identical in i.r. spectrum with racemic β-anilinobutyric acid.

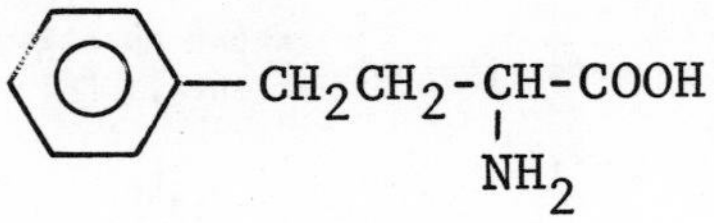

$$\text{C}_6\text{H}_5-\text{CH}_2\text{CH}_2-\underset{\underset{\text{NH}_2}{|}}{\text{CH}}-\text{COOH}$$

Ref. 693

Preparation of dl-γ-Phenyl-α-Formylaminobutyric Acid—71.5 gm. of pure dl-γ-phenyl-α-aminobutyric acid dissolved in 860 cc. of redistilled formic acid were treated gradually with 260 cc. of pure acetic anhydride. The temperature was held at 70–75° for 30 minutes after the addition of the acetic anhydride had been completed. 600 cc. of water were then added and the solution concentrated *in vacuo* with the addition of water from time to time, until the acetic acid had been removed. The residue was dissolved in the minimum amount of methyl acetate and crystallized by the addition of petroleum ether. 79 gm. of dl-γ-phenyl-α-formylaminobutyric acid melting at 130–131° (corrected) were obtained. This is 95 per cent of the theoretical amount. The compound gave the following analysis.

$C_{11}H_{13}O_3N$. Calculated, N 6.76; found, N 6.84

Resolution of dl-γ-Phenyl-α-Formylaminobutyric Acid and Isolation of the Isomers of the Free Amino Acid—57 gm. of the above formyl derivative were intimately mixed in a mortar with 109 gm. of brucine. The mixture was then dissolved in a minimum amount of boiling anhydrous methyl alcohol. After the solution was refluxed for 15 minutes, the brucine salt began to crystallize. The solution was allowed to cool and an equal volume of methyl acetate was added. The mixture was allowed to stand in the ice box for 48 hours and the product was filtered. The precipitate amounted to 55 gm. and had a rotation of $[\alpha]_D^{30} = -19.0°$ for a 1 per cent solution in methyl alcohol. The filtrate was concentrated to one-fifth its volume and was allowed to stand in the ice box overnight. A second crop of crystals amounting to 10.2 gm., with a rotation of $[\alpha]_D^{30} = -23.0°$, was obtained. Both fractions melted at 160–162° (corrected). Recrystallization of the salt from boiling absolute methyl alcohol gave a product with a rotation of $[\alpha]_D^{30} = -23.2°$ and a melting point of 160–162° (corrected).

To a cold solution of 44 gm. of this brucine salt in 1 liter of distilled water were added 65 cc. of cold concentrated NH₄OH. After the mixture had stood for 10 to 15 minutes in an ice bath, the brucine was filtered and washed with ice water. The filtrate was extracted with ten 60 cc. portions of chloroform followed by two extractions with ether. The solution was aerated for 1 hour to remove any ether and excess NH₃. After neutralization to litmus with HCl, sufficient concentrated HCl was now added to make the solution 1 N. The mixture was heated in a boiling water bath for 2 hours and then made faintly alkaline to litmus with NH₄OH. Crystallization of the l-γ-phenyl-α-aminobutyric acid began promptly. The yield was 9.2 gm., with a rotation of $[\alpha]_D^{30} = -47.0°$ for a 1 per cent solution in 1 N HCl, and a melting point of 323–325° (corrected). Recrystallization from water did not change the rotation. The compound gave the following analysis.

$C_{10}H_{13}O_2N$. Calculated, N 7.82; found, N 7.71

The original mother liquors from the crystalline brucine salt of l-γ-phenyl-α-formylaminobutyric acid were concentrated *in vacuo* to a heavy syrup. An equal volume of methyl acetate was added and the mixture allowed to stand for 48 hours in the ice box. A very small amount of the brucine salt of the l acid precipitated and was filtered. To the filtrate, 500 cc. of water were added, the solution was cooled to 0°, and then was decomposed as described above for the preparation of l-γ-phenyl-α-aminobutyric acid. 10 gm. of the d-phenylaminobutyric acid were thus obtained possessing a rotation of $[\alpha]_D^{30} = +48.8°$ for a 1 per cent solution in 1 N HCl. The product melted at 326–328° (corrected) and gave the following analysis.

$C_{10}H_{13}O_2N$. Calculated, N 7.82; found, N 7.74

$$\text{C}_6\text{H}_5-\underset{\underset{\text{CH}_2\text{NH}_2}{|}}{\text{CH}}-\text{CH}_2\text{COOH}$$

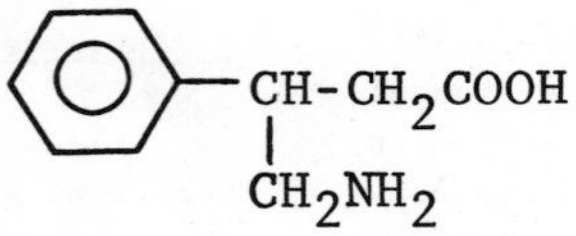

$C_{10}H_{13}NO_2$

Ref. 694

1. Resolution of racemic β-phenyl-γ-aminobutyric acid into its enantiomers

a. Diastereoisomeric cynchonidine salts of N-carbobenzoxy-β-phenyl-γ-aminobutyric acid

A solution of 20 g (0.063 mole) of racemic N-carbobenzoxy-β-phenyl-γ-aminobutyric acid [6] in 600 ccm of absolute ethanol was added to a solution of 18.3 g (0.062 mole) of cynchonidine in 500 ccm of absolute ethanol. Immediately, 21.7 g of a substance was precipitated which was designated as salt **1**. The filtrate which remained after separation of salt **1** was cooled in an ice bath to yield another portion of the precipitate (1.35 g), referred to as salt **2**. After separation of salt **2**, the filtrate was evaporated to 350 ccm under reduced pressure and cooled down to ambient temperature to afford 1.35 g of salt **3**. The filtrate was again evaporated to 150 ccm under reduced pressure to give 5.9 g of salt **4**. The remaining solution was evaporated to dryness under reduced pressure to give 3.9 g of salt **5**. Each of the salts was purified by crystallization from dimethylformamide or ethyl acetate. Physical properties of the fractions obtained have been listed in Table 1.

b. Enantiomers of N-carbobenzoxy-β-phenyl-γ-aminobutyric acid

Each of the cynchonidine N-carbobenzoxy-β-phenyl-γ-aminobutyrates obtained in point 1a was suspended in ethyl acetate followed by addition of 2 N HCl (for 1 g of a salt, 9 ccm of ethyl acetate and 3 ccm of 2 N HCl were consumed). The mixture

6. Kupryszewski G., Sobocińska M., Chaunina R., Pierekalin W., *Rozpr. Wydz. III Gdańskiego Tow. Nauk.*, 5, 249 (1968).

Table 1

Cynchonidine salts of N-carbobenzoxy-β-phenyl-γ-aminobutyric acid

No	Designation	Crude product		After crystallization			
		Yield %	M.p. (°C)	Solvent	Yield %	M.p. (°C)	$[\alpha]_D^{20}$
1	Salt 1	55	193—196 (decomp.)	DMFA	49	198—200 (decomp.)	−30.6° (*c* = 0.6, (DMFA)
2	Salt 2	3.5	193—195 (decomp.)	DMFA	2.2	198—200 (decomp.)	−36.1° (*c* = 0.6, DMFA)
3	Salt 3	3.5	189—191 (decomp.)	DMFA	3.5	195—198 (decomp.)	−27.7° (*c* = 0.7, DMFA)
4	Salt 4	15	132—135	AcOEt	14	132—135	−27.2° (*c* = 0.5, DMFA)
5	Salt 5	12	132—136	AcOEt	10.8	132—135	−35.6° (*c* = 0.4, DMFA)

Overall yield: a) of raw product 89%, b) after crystallization 79.5%. DMFA and AcOEt denote dimethylformamide and ethyl acetate, respectively.

Table 2

Enantiomers of N-carbobenzoxy-β-phenyl-γ-aminobutyric acid

No	Designation	Crude product		After crystallization		
		Yield %	M.p. (°C)	Yield %	M.p. (°C)	$[\alpha]_D^{20}$
1	Z-acid 1	80	108—111	77	109—111	−17.7° (*c* = 1, methanol)
2	Z-acid 2*	68	108—110	57	108—110	−13.5° (*c* = 1, methanol)
3	Z-acid 3*	31	104—106	31	105—107	−10.7° (*c* = 1, methanol)
4	Z-acid 4	76	110—113	74	111—113	+14.8° (*c* = 1, methanol)
5	Z-acid 5*	85	107—110	57	109—111	+19.6° (*c* = 1, methanol)

Z-acid denotes N-carbobenzoxy-β-phenyl-γ-aminobutyric acid
*** Obtained in too small quantities and not studied further.**

was then shaken for several minutes, the aqueous layer was discarded and washed with several portions of ethyl acetate. The combined organic solutions were repeatedly washed with water, dried with anhydrous magnesium sulfate, and the solvent was distilled off under reduced pressure. The products obtained were crystallized from dilute methanol. Physical properties of the enantiomers of N-carbobenzoxy-β-phenyl-γ-aminobutyric acid have been shown in Table 2.

c. Enantiomers of β-phenyl-γ-aminobutyric acid

Enantiomers of N-carbobenzoxy-β-phenyl-γ-aminobutyric acid obtained in point 1b were hydrogenolyzed on a Pd black as follows: the acid was dissolved in methanol and after addition of PdO (50 mg of PdO was taken for 1 g of N-carbobenzoxy-β-phenyl-γ-aminobutyric acid, and 60 ccm of abs. methanol) hydrogen was passed through the solution under normal pressure over 3—5 hrs. Then the catalyst was filtered off and the solvent was removed in vacuo. Physical properties of the enantiomers of β-phenyl-γ-aminobutyric acid have been summarized in Table 3.

Table 3

Enantiomers of β-phenyl-γ-aminobutyric acid

No	Designation	Yield	M.p. (°C)	$[\alpha]_D^{20}$	ν_{max}^{Nujol} cm^{-1}
1	Acid 1	83	154—156	+7.0° (*c* = 1, AcOH)	1550 s (—NH_2), 1620 w (phenyl), 1730 vs (COOH), 3300 w (NH H-bonded)
2	Acid 4	83	149—152	−5.3° (*c* = 1.1, AcOH)	1550 s (—NH_2), 1620 w (phenyl), 1730 vs (COOH), 3300 w (NH H-bonded)

"Acid" stands for β-phenyl-γ-aminobutyric acid. AcOH — acetic acid

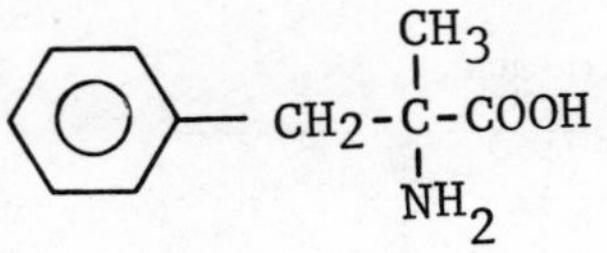

$C_{10}H_{13}NO_2$ **Ref. 695**

L-(−)-*N*-**Acetyl-α-methylphenylalanine, Cinchonidine Salt.**—To a soln of 234 g (1.058 moles) of *N*-acylamino acid in 438 ml of abs EtOH was added a soln of 312 g (1.058 moles) of cinchonidine in 438 ml of abs EtOH. The mixt was cooled to room temp, dild with Et_2O (1930 ml), seeded with crystals from a probe experiment, and allowed to stand overnight at 25°. The ppt was washed with three 75-ml portions of 3:1 Et_2O–EtOH. The yield of salt amounted to 222.5 g (82.3%), mp 207–210°. Recrystn from EtOH–Et_2O as before yielded 163.0 g (59.7%), mp 215–216°; $[\alpha]_{578}^{23}$ −49.2 ± 1.0° (*c* 1, abs EtOH) suitable for further processing. The analytical material was crystd to constant values: mp 218–219° dec; $[\alpha]_{578}^{23}$ −47.0 ± 1.0° (*c* 1, abs EtOH); uv max (MeOH) 225, 285, 302, and 315 mμ (log ε 4.57, 3.71, 3.59, and 3.50); ir spectrum consistent. *Anal.* ($C_{31}H_{37}N_3O_4$) C, H, N.

L-(−)-*N*-**Acetyl-α-methylphenylalanine.**—To 1750 ml of H_2O and 131 ml of 2.5 *N* HCl was added 163.0 g (0.316 mole) of the above cinchonidine salt and the mixt was stirred at 0–5° for 1 hr. An additional 172-ml portion of 2.5 *N* HCl was added and the stirring was continued at 0–5° for 1 hr. The mixt was then filtered and the ppt was washed with three 75-ml portions of cold H_2O. The yield of product amounted to 64.4 g (92.2%): mp 203–205°; $[\alpha]_{578}^{24}$ −42.1 ± 1.0° (*c* 1, abs EtOH). The anal. material was recrystd from Me_2CO: mp 206–207° (lit.[1] mp 196.5–200.5°); uv max (MeOH) 252, 257.5, 263, and sh 247.5 mμ (log ε 2.17, 2.24, 2.09, and 2.00); ir spectrum consistent. Further crystn did not change the melting point or rotation. *Anal.* ($C_{12}H_{15}NO_3$) C, H, N, equiv wt.

L-(−)-**α-Methylphenylalanine·HCl.**—To 536 ml of 6 *N* HCl was added 64.4 g (0.291 mole) of the acetylamino acid. After 6 hr at reflux the mixt was cooled to 25° spontaneously then at 15° for 30 min. The mixt was filtered and the ppt was washed with three 25-ml portions of 2 *N* HCl at 0–5°. The yield of L-(−)-α-methylphenylalanine·HCl amounted to 55.1 g (88.0%): mp 217–219°; $[\alpha]_{578}^{23}$ −2.1 ± 0.5°; $pH_{1/2}$ 2.6 (H_2O); Cu salt complex $[\alpha]_{546}^{25}$ +142° (*c* 0.2, $CuSO_4$ soln);[7] uv max (MeOH) 252, 258, 263, and sh 247.5 mμ (log ε 2.24, 2.30, 2.19, and 2.11); ir spectrum consistent. *Anal.* ($C_{10}H_{13}NO_2$·HCl) C, H, Cl, N, equiv wt.

L-(−)-**α-Methylphenylalanine.**—The amino acid·HCl was dissolved in H_2O and passed over a column of resin, IR-120 (H_3^+O cycle). The column was eluted with H_2O to neutrality and following that with 500 ml of 1 *N* NH_4OH. Concn of the ammoniacal eluate yielded a crude product: mp 290–291° dec. The crude amino acid was dissolved in 178 ml of refluxing MeOH and upon concn to about 100 ml crystn began. The mixt was allowed to cool spontaneously and stand overnight at 25°, chilled to 5°, and filtered and the ppt was washed with MeOH: yield, 1.85 g (41.5%); mp 306.5° dec; purity by phase solubility 98.5 ± 0.5%; (abs EtOH) $[\alpha]_{578}^{23}$ −2.5° (*c* 1, 1 *N* HCl); uv max (MeOH) 237, 252.5, 258, 264, and sh 247 (log ε 2.02, 2.23, 2.30, 264, and 2.13); Cu salt complex $[\alpha]_{546}^{23}$ +182.5° (*c* 0.2, $CuSO_4$ soln). This material proved difficult to dehydrate and showed a tendency to rehydrate. Anal. was obtained on a sample showing 2.3% loss on drying (100°, 1 mm) corrected to the anhyd basis. *Anal.* ($C_{10}H_{13}NO_2$) C, H, N.

Another sample with a purity by phase solubility of 97.4 ± 0.5% (abs EtOH) showed $[\alpha]_{578}^{24}$ −22.8° (*c* 1, H_2O).

Enriched D-(+)-*N*-**Acetyl-α-methylphenylalanine.**—From the mother liquors of the L-(−) enantiomorph, cinchonidine salt by the method previously described for the L-(−) isomer, 27.55 g (89.4%) of enriched D-(+)-*N*-acetyl-α-methylphenylalanine was obtained, mp 193–195°.

D(+)-*N*-**Acetyl-α-methylphenylalanine Cinchonine Salt Monohydrate.**—Equimolar amts (0.020 mole) of enriched D-(+)-*N*-acetyl-α-methylphenylalanine and cinchonine were dissolved in 40 ml of warm EtOH. The mixture was cooled to 25°, seeded with crystals from a probe experiment, and allowed to stand for 64 hr. After cooling the mixt to 0°, the ppt was sepd by filtration, washed, and dried at 25°. The pure salt monohydrate amounted to 3.25 g (25.4% based on the D-(+) content of enriched starting material): mp 116–118°; $[\alpha]_{578}^{27}$ +88.7° (*c* 2, abs EtOH). The salt was recrystd from abs EtOH in 46.2% yield with no change in rotation or melting point; uv max (MeOH) 226, 285, 303, and 316 mμ (log ε 4.54, 3.68, 3.56, and 3.49); ir spectrum consistent. *Anal.* ($C_{31}H_{37}N_3O_4 \cdot H_2O$) C, H; N: calcd, 7.87; found, 7.40.

D-(+)-*N*-**Acetyl-α-methylphenylalanine.**—By a method analogous to that described for the L-(−) isomer 0.42 g (84.3%) of D-(+)-*N*-acetyl-α-methylphenylalanine was obtained: mp 204–207°; $[\alpha]_{578}^{26}$ +42.7° (*c* 1, abs EtOH). The material was recryst from 33% EtOH to yield an anal. sample: 0.30 g (52.8%); mp 206–207° (lit.[1] mp 200.5–202.5°); $[\alpha]_{578}^{24}$ +44.0° (*c* 1, abs EtOH); uv max (MeOH) 247, 252.5, 258, and 263 mμ (log ε 2.03, 2.19, 2.26, and 6.03); ir consistent; $pH_{1/2}$ 4.95 (50% MeOH). *Anal.* ($C_{12}H_{15}NO_3$) C, H, N, equiv wt.

D-(+)-**α-Methylphenylalanine·HCl.**—By the method described for the L-(−) isomer 10.0 g (100%) of D-(+)-α-methylphenylalanine·HCl was obtained: mp 217–219°. The anal. material, 0.245 g (49.0%), was obtained by recrystn from Me_2CO–hexane: mp 214–215° (lit.[1] mp 210–214.5° dec); $[\alpha]_{578}^{22}$ +2.1° (*c* 1.2, 1 *N* HCl); uv max (MeOH) 252.5, 258, and 264 mμ (log ε 2.14, 2.21, and 2.10); ir spectrum (Nujol) similar to but differs from that of the L-(−) enantiomorph due to a polymorphic crystalline form. *Anal.* ($C_{10}H_{13}NO_2$·HCl) C, H, Cl, N.

D-(+)-**α-Methylphenylalanine.**—By the method described for the L-(−) isomer 7.4 g (94.8%) of D-(+)-α-methylphenylalanine was obtained: mp 307–308° dec. This material was recrystd as was the L-(−) enantiomorph to yield 3.55 g (48.0%) of product: mp 315° dec; purity by phase solubility 99.1 ± 0.9%, (abs EtOH), $[\alpha]_{578}^{23}$ +20.6° (*c* 1, H_2O); uv max (MeOH) 247, 252, 258, and 264 mμ (log ε 1.99, 2.15, 2.25, and 2.12); ir spectrum identical with that of L-(+) isomer, ord data in Table I. *Anal.* ($C_{10}H_{13}NO_2$) C, H, N.

TABLE I

OPTICAL ROTATORY DISPERSION

λ, mμ	L-(−)-AA·HCl φ,[a] degrees	D-(+)-AA·HCl φ,[b] degrees	L-(−)-AA φ,[a] degrees	D-(+)-AA φ,[a] degrees
400	0	−8.4	−10	+10
350	+15	−22.9	0	0
300	+60	−68.8	+40	−40
250	+320		+300	−300
220	+4000		+3100	−2800
C[c]	0.102	0.257	0.033	0.101

[a] The values of φ, the molecular rotation at 25°, are ±10% or ±20° whichever is larger. Spectra were obtained in 1 *M* HCl. [b] At 18.5° in 3 *M* HCl. [c] The concn is listed in per cent soln.

(1) S. Terashima, K. Achiwa, and S. Yamada, *Chem. Pharm. Bull.*, **14**, 1138 (1966).

$C_{10}H_{13}NO_2$ Ref. 695a

N-Acetyl-α-methylphenylalanine l-Menthyl Esters (VIa, VIb)——To the suspension of Na powder (5.8 g., 0.25 mole) in anhyd. benzene (200 ml.) was added l-menthol*6 (39.0 g., 0.25 mole). The reaction mixture was kept standing overnight at room temperature avoiding moisture and refluxed for 2 hr. Unreacted Na powder was decanted off and washed with anhyd. benzene (100 ml.). To the combined benzene solution of the supernatant and the washings was added a solution of DL-V (42.2 g., 0.208 mole) in anhyd. benzene (100 ml.). The reaction mixture was stirred at room temperature for 5 hr., and then kept standing overnight. The benzene solution was washed with 10% AcOH (300 ml. × 2), H_2O (300 ml. × 1), 2.5% Na_2CO_3 (300 ml. × 1), and H_2O (300 ml. × 2) successively, and then dried with anhyd. Na_2SO_4. Filtration and evaporation *in vacuo* of this benzene solution gave pale yellow oil, which solidified on standing. Recrystallization from iso-Pr_2O (700 ml.) afforded colorless plates (42.6 g.), m.p. 123~147°, $[\alpha]_D^{25}$ −28.8° (c=1.366, MeOH). Another twice recrystallizations from iso-Pr_2O gave crude VIa as colorless needles (18.6 g., 50%), m.p. 169.5~172°, $[\alpha]_D^{21}$ +35.0° (c=1.618, MeOH). Recrystallization of crude VIa from iso-Pr_2O-AcOEt (5:4) afforded pure VIa as colorless needles (15.1 g., 40%), m.p. 171~173°, $[\alpha]_D^{23}$ +37.5° (c=1.588, MeOH). Analytical sample obtained from further twice recrystallizations from the same solvent showed m.p. 171.5~172.5°, $[\alpha]_D^{22}$ +37.4° (c=1.112, MeOH). *Anal.* Calcd. for $C_{22}H_{33}O_3N$: C, 73.50; H, 9.25; N, 3.90. Found: C, 73.91; H, 9.16; N, 3.67. IR ν_{max}^{KBr} cm^{-1}: 3260, 1744, 1640, 1563, 1197, 735, 699. IR $\nu_{max}^{CHCl_3}$ cm^{-1}: 3440, 3400, 1730, 1677, 1505.

The combined mother liquor of the first two recrystallizations was evaporated *in vacuo* to give yellow solid, which was recrystallized from hexane (300 ml.) to afford pale yellow crystals (30.0 g.), m.p. 114~118°, $[\alpha]_D^{22}$ −71.5° (c=1.122, MeOH). Successive recrystallization from hexane, iso-Pr_2O, 70% aq. EtOH, and iso-Pr_2O (×2) gave crude partially resolved VIb*7 as white crystals (3.7 g., 9.9%), m.p. 121.5~123°, $[\alpha]_D^{21}$ −87.6° (c=1.242, MeOH). Another twice recrystallizations from hexane gave partially resolved VIb as white crystals, m.p. 121.5~123.5°, $[\alpha]_D^{19}$ −80.9° (c=1.318, MeOH). *Anal.* Calcd. for $C_{22}H_{33}O_3N$: C, 73.50; H, 9.25; N, 3.90. Found: C, 72.95; H, 9.04; N, 4.14. IR ν_{max}^{KBr} cm^{-1}: 3360, 1727, 1672, 1535, 1122, 749, 709. IR $\nu_{max}^{CHCl_3}$ cm^{-1}: 3440, 3400, 1727, 1677, 1504.

(+)-N-Acetyl-*R*-α-methylphenylalanine (*R*(+)-III)——A mixture of (+)-VIa (m.p. 171~172.5°, $[\alpha]_D^{22}$ +37.4° (c=1.168, MeOH)) (11.0 g., 0.0307 mole) and KOH (17.2 g., 0.307 mole) in 50% aq. EtOH (200 ml.) was refluxed for 5 hr., condenced to ca. half volume and extracted with benzene (50 ml. × 3). Aqueous layer was acidified with dil. HCl and kept in an ice bath for 2 hr. to crystallize out the crude *R*(+)-III as white powdery crystals (5.5 g., 81%), m.p. 200~201.5°, $[\alpha]_D^{21}$ +78.2° (c=1.334, MeOH). Recrystallization from 50% aq. EtOH gave pure *R*(+)-III as colorless needles (5.0 g., 74%), m.p. 200.5~202°, $[\alpha]_D^{19}$ +80.3° (c=1.052, MeOH). Analytical sample was prepared by the repeated recrystallization from the same solvent, m.p. 200.5~202.5°, $[\alpha]_D^{20}$ +79.3° (c=1.082, MeOH) (lit.,[5] m.p. not described. $[\alpha]_D$ +74.4° (c=1, MeOH)). *Anal.* Calcd. for $C_{12}H_{15}O_3N$: C, 65.14; H, 6.83; N, 6.33. Found: C, 65.14; H, 6.76; N, 6.36. IR ν_{max}^{KBr} cm^{-1}: 3340, 1722, 1633, 1560, 752, 706. This IR spectrum was different from that of DL-III in solid state. Another hydrolysis using 70% aq. EtOH as a solvent raised the yield of the crude *R*(+)-III up to 97%.

(−)-N-Acetyl-*S*-α-methylphenylalanine (*S*(−)-III)——A mixture of (−)-VIb (m.p. 120.5~122.5°, $[\alpha]_D^{25}$ −83.5° (c=1.424, MeOH)) (2.5 g., 0.00696 mole) and KOH (3.9 g., 0.0696 mole) in 50% aq. EtOH (45 ml.) was treated similarly to the case of *R*(+)-III to give crude partially resolved *S*(−)-III as white powdery crystals (1.5 g., 97%), m.p. 187~190°, $[\alpha]_D^{21}$ −46.8° (c=1.034, MeOH) (optical purity 59%).*8 Recrystallization from 50% aq. EtOH afforded colorless needles (1.1 g., 72%), m.p. 189.5~193.5°, $[\alpha]_D^{18}$ −55.6° (c=1.012, MeOH) (optical purity 70%).*8 Further recrystallization from the same solvent gave analytical sample as colorless needles, m.p. 196.5~200.5°, $[\alpha]_D^{19}$ −78.4° (c=1.036, MeOH) (optical purity 99%)*8 (lit.,[5] m.p. not described, $[\alpha]_D$ −74.3° (c=1, MeOH)). *Anal.* Calcd. for $C_{12}H_{15}O_3N$: C, 65.14; H, 6.83; N, 6.33. Found: C, 65.18; H, 6.99; N, 6.37. IR ν_{max}^{KBr} cm^{-1}: 3440, 1721, 1633, 1577, 752, 706. This IR spectrum was identical with that of *R*(+)-III in solid state.

***R*(+)-α-Methylphenylalanine Hydrochloride (*R*(+)-I-HCl)**——A mixture of *R*(+)-III (m.p. 200.5~202°, $[\alpha]_D^{19}$ +80.3° (c=1.052, CH_3OH)) (2.0 g., 0.00906 mole) and 10% HCl (40 ml.) was refluxed for 3 hr. and evaporated to dryness *in vacuo* to afford *R*(+)-I-HCl monohydrate as white solid (1.9 g., 90%), m.p. 210~215° (decomp.), $[\alpha]_D^{17}$ +6.8° (c=0.976, H_2O). Recrystallization from 95% aq. EtOH (6 ml.)-ether (50 ml.) gave white crystals (1.6 g.), m.p. 209~213.5° (decomp.), $[\alpha]_D^{17}$ +6.3° (c=1.280, H_2O). Another several recrystallizations from the same solvent gave pure *R*(+)-I-HCl monohydrate as white crystals, m.p. 210~213.5° (decomp.), $[\alpha]_D^{15}$ +6.6° (c=1.034, H_2O). *Anal.* Calcd. for $C_{10}H_{14}O_2NCl \cdot H_2O$: C, 51.40; H, 6.90; N, 5.99. Found: C, 51.31; H, 6.77; N, 6.46. IR ν_{max}^{KBr} cm^{-1}: 3385, 3310, 3025, 1743, 1603, 1590, 1523, 726, 695. Pure *R*(+)-I-HCl monohydrate was dried overnight *in vacuo* at ca. 60° to afford *R*(+)-I-HCl, m.p. 210~214.5° (decomp.), $[\alpha]_D^{16}$ +6.7° (c=0.920, H_2O). *Anal.* Calcd. for $C_{10}H_{14}O_2NCl$: C, 55.68; H, 6.54; N, 6.50. Found: C, 55.53; H, 6.39; N, 6.66. IR ν_{max}^{KBr} cm^{-1}: 3010, 1740, 1590, 1499, 729, 705, 696. This IR spectrum was different from those of DL-I-HCl and *R*(+)-I-HCl monohydrate in solid state.

Optical rotatory dispersion curve measurements*11 of *R*-I. $[M]^{18.5}$ (mμ) (c=0.257, 3*N* HCl)*12: 0.0° (700), 0.0° (589), 0.0° (500), −0.7° (450), −8.4° (400), −22.9° (350), −68.8° (300), −158° (270). $[M]^{18.5}$ (mμ) (c=0.175, H_2O)*13: +20.4° (700), +34.8° (589), +51.2° (500), +65.3° (450), +83.8° (400), +115° (350), +168° (300).

*6 $[\alpha]_D^{19}$ −51.1° (c=3.326, EtOH).

*7 Partial resolution of (−)-l-menthyl ester VIb was deduced from the fact that VIb showing $[\alpha]_D^{25}$ −83.5° (c=1.424, MeOH) gave *S*(−)-III whose optical purity was 59%.

*8 Optical purity was calculated based on the assumption that *R*(+)-III showing $[\alpha]_D^{20}$ +79.3° (MeOH) was optically pure.

*11 ORD curve measurements were carried on with a Spectrophotometer Model ORD/UV-5, Japan Spectroscopic Co., Ltd.

*12 15.5 mg. of *R*(+)-I-HCl was dissolved directly in 3*N* HCl and total volume was made up to 5 ml.

*13 10.5 mg. of *R*(+)-I-HCl was dissolved in 4.9 ml. of 0.01*N* NaOH solution and total volume was made up to 5 ml. with H_2O.

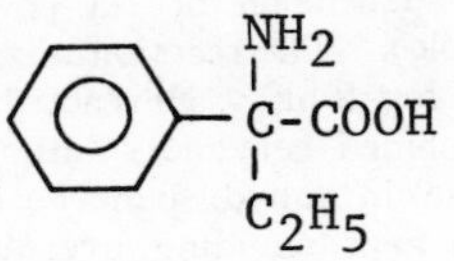

$C_{10}H_{13}NO_2$ **Ref. 696**

—The acid (II) was obtained (83%) from (±)-2-amino-2-phenylbutyric acid [7] by the method of Cram and his co-workers; [16] m.p. 197—198° (from water) (lit.,[8] 193°). The (±)-acid (20·3 g) and (−)-cinchonidine (28·8 g), $[\alpha]_D^{20}$ −127·5° (*c* 5·0 in EtOH) were dissolved in boiling water (900 ml) and kept in a refrigerator overnight. The precipitate was filtered off to give crude (−)-*2-formamido-2-phenylbutyrate cinchonidine salt* (23·1 g), $[\alpha]_D^{21}$ −57·3° (*c* 1·11 in EtOH). Two recrystallisations from water gave crystals (19·2 g, 76%) of m.p. 196—197°, $[\alpha]_D^{20}$ −65·2° (*c* 1·01 in EtOH) (Found: C, 71·8; H, 7·1. $C_{30}H_{35}N_3O_4$ requires C, 71·8; H, 7·0%). A suspension of the (−)-cinchonidine salt in aqueous sodium carbonate (15% w/v; 300 ml) was stirred at room temperature, and free cinchonidine was extracted into chloroform. The aqueous layer was acidified with dilute hydrochloric acid, evaporated to 100 ml, and kept at 0° for 24 h. The (+)-formamido-acid was filtered off (11·3 g); two recrystallisations from water produced white crystals (9·6 g, 73%), m.p. 202—203°, $[\alpha]_D^{26}$ +64·6°, $[M]_D$ +134° (*c* 1·07 in EtOH) {lit.,[8] m.p. 212°, $[\alpha]_D^{26}$ +126° (*c* 1·00 in aq. alkali)}. The i.r. spectrum was identical with that of the racemic form.

(+)-*2-Amino-2-phenylbutyric Acid* (I).—(+)-2-Formamido-2-phenylbutyric acid (9 g) was warmed with hydrobromic acid (N; 70 ml) until a clear solution resulted. Evaporation to dryness gave a pale yellow solid. The crude (+)-amino-acid hydrobromide was dissolved at 0° in dry ethanol (25 ml) saturated with dry ammonia gas. After a further 4 days at 0° the solution was filtered and solid was recrystallised from aqueous ethanol giving the (+)-amino-acid (I) (5 g, 64%), m.p. 255—265° (decomp.), $[\alpha]_D^{20}$ +34·4°, $[M]_D$ +63° (*c* 1·63 in H_2O), $[\alpha]_D^{27}$ +75·8°, $[M]_D$ +136° (*c* 0·99 in 6N-HCl) {lit.,[8] $[\alpha]_D$ +41° (*c* 1·01 in aq. alkali)}; the i.r. spectrum was identical with that of the racemic amino-acid.

[7] R. E. Steiger, *Org. Synth.*, 1955, Coll. Vol. III, p. 88.
[8] H. Sobotka, M. F. Holzman, and J. Kahn, *J. Amer. Chem. Soc.*, 1932, **54**, 4697.

[16] D. J. Cram, L. K. Gaston, and H. Jager, *J. Amer. Chem. Soc.*, 1961, **83**, 2183.

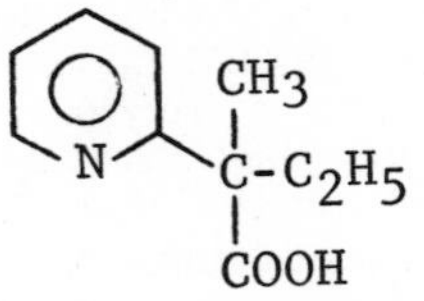

$C_{10}H_{13}NO_2$ **Ref. 697**

(c) **Resolution of I.**—The resolution was effected by fractional crystallization from absolute ethanol of the crystalline hydrogen dibenzoyl-*d*-tartrate of ethyl methylethyl-α-pyridylacetate, initial m. p. 121°.

Anal. Calcd. for $C_{30}H_{31}NO_{10}$: C, 63.70; H, 5.52. Found: C, 62.95; H, 5.87.

Regeneration of a sample of salt, m. p. 123–124° (9.7 g.), obtained after four crystallizations, gave 2.75 g. of distilled VIII, $[\alpha]^{28}$D −0.50° (neat). From the mother liquors, distilled ester of $[\alpha]^{27}$D + 0.25 to +0.28° was obtained.

(d) **Hydrolysis of VIII.**—A solution of 8.0 g. of VIII, $[\alpha]^{27}$D +0.27°, in concentrated hydrochloric acid was refluxed for eight hours. Removal of the solvent *in vacuo* gave crystalline material which was washed with acetone and dried; 4.80 g., m. p. 98–100°, $[\alpha]^{26}$D +0.32 (*c* = 46.7, hydrochloric acid).

Anal. Calcd. for $C_{10}H_{14}NO_4Cl$: C, 55.69; H, 6.54; N, 6.49. Found: C, 55.81; H, 6.68; N, 5.90.

Recrystallization from acetone containing 10% alcohol raised the m. p. to 103–104° and the $[\alpha]^{25}$D to +0.40° (*c* = 25.0, dil. hydrochloric acid).

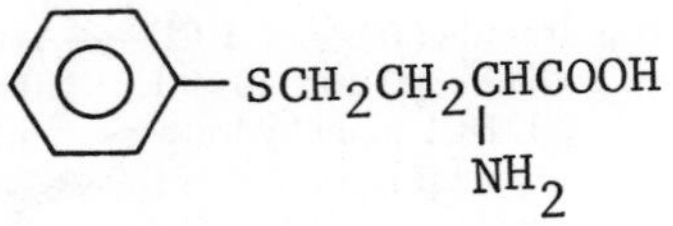

$C_{10}H_{13}NO_2S$ **Ref. 698**

*Resolution of N-acetyl-S-phenyl-*DL*-homocysteine.* A suspension of 28 g. of N-acetyl-S-phenylhomocysteine and 37.5 g. of strychnine in 1 l. of water was heated on a steam-bath until a clear solution resulted. The hot solution was allowed to cool slowly and a partially crystalline oil separated. The oil redissolved when the mixture was warmed, leaving some undissolved crystals, and when the solution was again allowed to cool slowly, more oil and crystalline material separated. This procedure was repeated until the solution could be cooled to room temperature without the separation of oil. The solution was then allowed to stand at room temperature overnight, was cooled in an ice-bath for two hours, and the crystalline precipitate was collected, washed once with cold water, and dried; 33.5 g. of crude salt was obtained; m.p. 206–212°.

*N-Acetyl-S-phenyl-*D*-homocysteine.*[4] The strychnine salt obtained above was dissolved in 900 ml. of hot water and the solution was treated with Norit and filtered. The hot filtrate was seeded with a crystal saved from the original crop and was allowed to cool slowly to room temperature. After it had stood at room temperature overnight, it was cooled in an

[3] All melting points were made on the micro hot stage and are corrected.
[4] The configuration was established, as shown later, by the formation of a phenylhydrazide from the N-acetyl derivative of the dextrorotatory isomer and phenylhydrazine by the action of papain-cysteine.

ice-bath for four hours, and the salt was collected, washed two times with cold water, and dried; 28.8 g. (89%) was obtained, m.p. 216–218°, $[\alpha]_D^{26.5}$ −30.7° (*c*, 1, H_2O). A small sample was recrystallized two more times from hot water, but showed no change in appearance, melting point, or rotation.

The pure strychnine salt (25.0 g.) was dissolved in 1 l. of hot water and the solution was brought to *p*H 8 by the addition of concentrated NH_4OH. The strychnine that separated was collected and the filtrate was further extracted with three 250-ml. portions of chloroform to complete the removal of strychnine. The clear aqueous layer was made acid to Congo Red with concentrated HCl, the solution was cooled in an ice-bath, and the N-acetyl-S-phenyl-D-homocysteine that separated was collected, washed with water, and dried. Yield, 10.4 g. (96%); m.p. 175–176°; $[\alpha]_D^{24}$ −10.1° (*c*, 1, 95% EtOH).

A sample was recrystallized two more times from aqueous ethanol for analysis; m.p., 178–179°; $[\alpha]_D^{22}$ −9.9° (*c*, 1, 95% EtOH).

Anal. Calc'd for $C_{12}H_{15}NO_3S$: N, 5.53; S, 12.66.
Found: N, 5.21; S. 12.10.

*N-Acetyl-S-phenyl-*L*-homocysteine.* The filtrate from the first crystallization of the strychnine salt of the D-isomer was treated for the removal of strychnine and the acetyl derivative of the L-isomer was prepared in the same manner as was the D; 12.7 g. was obtained; $[\alpha]_D^{25}$ +8.9° (*c*, 1, 95% EtOH). Two recrystallizations from aqueous ethanol yielded 10.5 g. (75%); m.p. 178–179°; $[\alpha]_D^{25}$ +9.6° (*c*, 1, 95% EtOH).

Anal. Calc'd for $C_{12}H_{15}NO_3S$: N, 5.53; S, 12.66.
Found: N, 5.29; S, 12.29.

*S-Phenyl-*D*-homocysteine.* N-Acetyl-S-phenyl-D-homocysteine (5.5 g.) was suspended in 400 ml. of 2 *N* HCl and the mixture was heated under reflux for three hours. The resulting solution was concentrated to dryness *in vacuo*. The residue was redissolved in a small amount of water, and the procedure was repeated. The residue was then dissolved in 50 ml. of hot water, the solution was adjusted to *p*H 4 by the addition of concentrated NH_4OH, cooled in an ice-bath, and the product was collected, washed with cold water, and dried. Yield, 4.3 g. (93%); $[\alpha]_D^{26}$ −28.7° (*c*, 1, *N* HCl). Two recrystallizations from water yielded material, $[\alpha]_D^{24}$ −30.2° (*c*, 1, *N* HCl); m.p. 233–236° d.

Anal. Calc'd for $C_{10}H_{13}NO_2S$: C, 56.85; H, 6.22; N, 6.63; S, 15.18.
Found: C, 56.94; H, 6.33; N, 6.46; S, 14.71.

*S-Phenyl-*L*-homocysteine.* N-Acetyl-L-homocysteine (4.5 g.) was hydrolyzed with 2 *N* HCl and worked up by the above procedure; yield, 3.6 g. (96%); $[\alpha]_D^{24}$ +30.2° (*c*, 1, *N* HCl). One recrystallization from water yielded material, $[\alpha]_D^{20}$ +30.3° (*c*, 1, *N* HCl); m.p. 234–236° d.

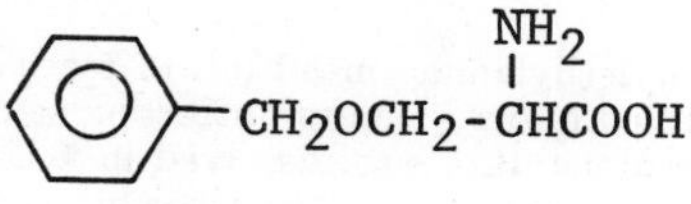

$C_{10}H_{13}NO_3$ Ref. 699

Sale della (S)-(-)-feniletilamina con O-benzil-N-formil-(R)-(-)-serina

17,84 g (0.08 mole) di O-benzil-N-formil-(R)-(S)-serina (3) e 9,7 g (0,08 mole) di (S)-(-)-feniletilamina sono stati portati in soluzione con 75 cc di isopropanolo e posti a blando ricadere per 15 minuti. La soluzione è stata lasciata a sè per una notte e quindi si è filtrato il sale della (S)-(-)-feniletilamina con la O-benzil-N-formil-(R)-(+)-serina separatosi. Si sono ottenuti 12,0 g a p.f. 140° $[\alpha]_D^{20}$ = -34,8° (etanolo ass. c = 1).

Dopo cristalizzazione da 67 cc di isopropanolo si sono avuti 10,65 g a p.f. 149-150° $[\alpha]_D^{20}$ = -40,5 (etanolo ass. c = 1). Resa 77%.

Ulteriori cristallizzazioni non hanno portato a variazioni delle caratteristiche.

Una soluzione in etanolo ass. c = 1 ha dato i seguenti valori polarimetrici:

$[\alpha]^{20}$	-218,3	-137,1	-85,0	-48,2	-42,4	-40,5
λ mμ	313	364	436	546	578	589

per valori di $[\alpha]_D^t$ nell'intervallo da 15 a 35° si ha una diminuzione di 0,14° per l'aumento di 1°.

Sale dell (R)-(+)-feniletilamina con O-benzil-N-formil-(S)-(+)-serina

Le acque madri della scissione sono state evaporate sotto vuoto a secco ed il residuo è stato sciolto in 30 cc di acqua cui si sono aggiunti 10 cc di NaOH 1:1. La soluzione è stata estratta con 2 x 30 cc di etere etilico e quindi portata a pH = 2 con HCl 1:1, ottenendo così la precipitazione della O-benzil-N-formil-(S)-(+)-serina grezza. Dopo due ore in frigorifero si è filtrato e lavato con 15 cc di acqua. Si sono ottenuti 6,8 g a p.f. 130° $[\alpha]_D^{20}$ = +27,3° (etanolo 80%, c = 3,5%). Tale prodotto e 3,7 cc di (R)-(+)-finiletilamina sono stati portati a blando ricadere con 63 cc di alcool isopropilico e si è lasciato a sè per una notte. Dopo filtrazione e lavaggio con 20 cc di etere si sono ottenuti 8,4 g di sale della (R)-(+)-feniletilamina con O-benzil-N-formil-(S)-(+)-serina a p.f. 149° $[\alpha]_D^{20}$ = +40,0°. Resa 80%.

O-benzil-N-formil-(R)-(-)-serina

10 g di sale di (S)-(-)-feniletilamina con O-benzil-N-formil-(R)-(-)-serina avente $[\alpha]_D^{20}$ = 40,5° (etanolo ass. c = 1) sono stati sciolti in 50 cc di acqua contenente 10 cc di NaOH 1:1. La soluzione è stata estratta con 2 x 30 cc de etere etilico e quindi acidificata a pH = 2 con HCl 1:1.

La O-benzil-N-formil-(R)-(-)-serina precipitata è stata filtrata, lavata con 2 x 25 cc di acqua e seccata. Si sono ottenuti 5,9 g a p.f. 132°; P.eq. 224 (calc. 223,2); $[\alpha]_D^{20}$ = -47,2° (etanolo 80% c = 3,5). Resa 91,5%.

Una soluzione in alcool 80% c = 3,5 ha dato i seguenti valori polarimetrici:

$[\alpha]^{20}$	-272,2	-166,3	-100,6	-55,35	-49,47	-47,22
λ mμ	313	364	436	546	578	589

Per valori di $[\alpha]_D^t$ nell'intervallo da 15 a 35° si ha una diminuzione di 0,15° per l'aumento di 1°.

O-benzil-N-formil-(S)-(+)-serina

Il composto è stato ottenuto da 10 g di sale della (R)-(+)-feniletilamina con la O-benzil-N-formil-(S)-(+)-serina. Si sono ottenuti 5,7 g a p.f. 132° $[\alpha]_D^{20}$ = +48,1° (etanolo 80% c = 3,5). P. eq. = 224. Resa 88%.

O-benzil-(R)-(-)-serina

5,0 (0,022 mole) di O-benzil-N-formil-(R)-(-)-serina sono stati sospesi in 26 cc di HCl 2N e, dopo aver ottenuto una soluzione per riscaldamento su bagno a vapore, si è mantenuto all'ebollizione per 2 ore e 30 minuti. Dopo raffreddamento si è neutralizzato con NH_4OH conc. ottenendo cosi la precipitazione del prodotto in scaglie lucenti.

La O-benzil-(R)-(-)-serina è stata poi filtrata e lavata con acqua. Si sono ottenuti 2,85 g a p.f. 211-213° dec. $[\alpha]_D^{28}$ = -21,8° (c = 2 in 80% CH_3COOH + 1 equiv. HCl) [lett. (3) p.f. 212-213° $[\alpha]_D^{20}$ = -22,65° ± 0,5].

O-benzil-(S)-(+)-serina

12,2 g (0,055 mole) di O-benzil-N-formil-(S)-(+)-serina sono stati ripresi con 62 cc di HBr N e posti su bagno di vapore per 2 ore e 30 minuti.

Si è quindi proceduto come nel caso precedente ottenendo 7,7 g di O-benzil-(S)-(+)-serina a p.f. 211-213° dec. (5) p.f. 210-213° (dec.) $[\alpha]_D^{27}$ = +21,60° (c = 2 in 80% CH_3COOH + 1 equiv. di HCl).

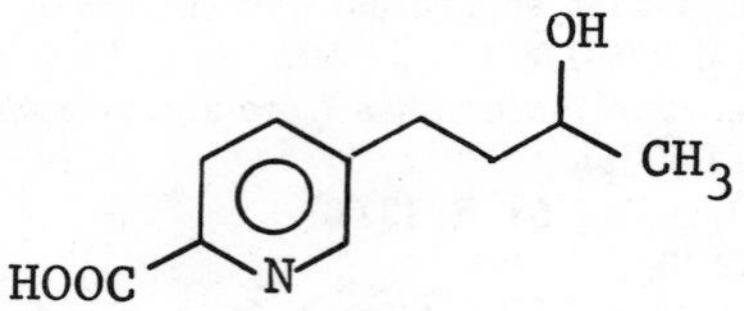

$C_{10}H_{13}NO_3$ Ref. 700

b) *Aus* (±)-*Fusarinolsäure* (**10a**) *über das* R(+)-*1-Phenyläthylammoniumsalz* **11b**. 700 mg (±)-Fusarinolsäure wurden in 10 ml Äthanol mit überschüssigem (+)-1-Phenyläthylamin kurz erwärmt, wobei sich die Lösung rotbraun färbte, und das Gemisch zur Trockene eingedampft. Viermal aus Äthanol/Äther gab 341 (60%) **11b**, weisse Nadeln, Smp. 146–148°; $[\alpha]_D = +12,2°$ (c = 2, Äthanol).

$C_{18}H_{24}N_2O_3$ (316) Ber. C 68,33 H 7,65 N 8,85% Gef. C 68,17 H 7,63 N 8,83%

632 mg **11b** wurden in 30 ml Wasser mit 8 ml ges. Bariumhydroxidlösung kurz erwärmt, mit Chloroform und mit Äther ausgeschüttelt, mit 2N H_2SO_4 auf pH 5,0 gestellt, vom Bariumsulfat abzentrifugiert und zur Trockene eingedampft. Chromatographie an *Whatman* Cellulose CF11 mit 2-Butanol/Ameisensäure/Wasser 75:15:10 gab 390 mg (100%) Öl. Aus Cyclohexan/Petroläther, Smp. 123–124°; $[\alpha]_D = +7,5°$ (c = 1, Äthanol).

$C_{10}H_{13}NO_3$ (195) Ber. C 61,52 H 6,71 N 7,18% Gef. C 60,91 H 6,79 N 7,12%

R(−)-*Fusarinolsäure* (**12**) *über das* S(−)-*1-Phenyläthylammoniumsalz* **12a**. Wie beim *R*(+)-Salz **11b**. $[\alpha]_D = -10,6°$ (c = 2,4, Äthanol).

$C_{18}H_{24}N_2O_3$ (316) Ber. C 68,33 H 7,65 N 8,85% Gef. C 68,17 H 7,56 N 8,78%

R(−)-*Säure* **12** *aus* **12a**. Wie **11** aus **11b**. $[\alpha]_D = 6,0°$ (c = 1, Äthanol).

$C_{10}H_{13}NO_3$ (195) Ber. C 61,52 H 6,71 N 7,18% Gef. C 61,52 H 6,76 N 7,12%

Die Analysen wurden in unserer mikroanalytischen Abteilung (Leitung *W. Manser*) ausgeführt.

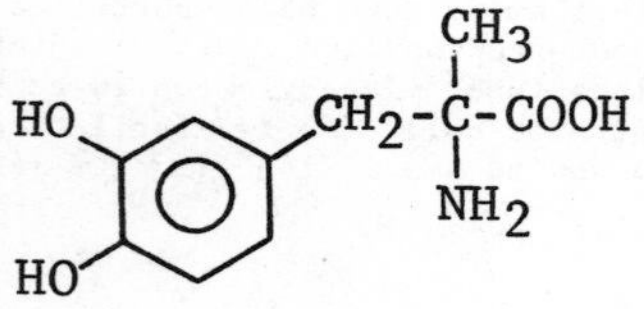

$C_{10}H_{13}NO_4$ Ref. 701

DL-**N-Acetyl-3-(3,4-dimethoxyphenyl)-2-methylalanine.**—A mixture of 69.0 g. (0.288 mole) of DL-3-(3,4-dimethoxyphenyl)-2-methylalanine in 207 ml. (2.56 moles) of pyridine and 276 ml. (2.92 moles) of acetic anhydride was heated at 90° with stirring for 3 hr. The solution was concentrated under vacuum to a thick sirup. The residue was poured into 300 ml. of ice–water and stirred for 10 min. The product crystallized on addition of 250 ml. of 2.5 *N* hydrochloric acid.[18] The mixture was aged at 5° for 1 hr. and filtered. After washing successively with cold water and 100 ml. of ethanol, and drying over phosphorus pentoxide, the N-acetyl derivative weighed 70.0 g. (86.% yield) and melted at 213–215°; λ_{max}^{Nujol} 3.1, 3.8, 5.87, 6.1, 6.16, and 6.6 μ.

Anal. Calcd. for $C_{14}H_{19}NO_5$: C, 59.78; H, 6.81. Found: C, 60.05; H, 6.60.

L-**N-Acetyl-3-(3,4-dimethoxyphenyl)-2-methylalanine (−)-1-Phenylethylamine Salt.**—To a slurry of 77 g. (0.274 mole) of DL-N-acetyl-3-(3,4-dimethoxyphenyl-2-methylalanine) in 200 ml. of methanol slowly was added a solution of 33.2 g. (0.274 mole) of (−)-1-phenylethylamine in 50 ml. of methanol. The methanol was distilled under vacuum until copious crystallization occurred. The precipitate was dissolved in 1 l. of water at 90°. The hot solution was filtered, cooled slowly to 25°, and aged at 8° for 40 hr. The collected salt was dried under vacuum at 55°. The yield was 54.5 g. (99%); $[\alpha]^{25}D$ +55° (c 1, methanol). Recrystallization from water gave 42.5 g. (77%), m.p. 212–215°, $[\alpha]^{25}D$ +69° (c 1, methanol). Titration with base gave an equivalent weight of 396 (theory, 402.5).

Anal. Calcd. for $C_{22}H_{30}N_2O_5$: C, 65.65; H, 7.51. Found: C, 65.55; H, 7.43.

L-**N-Acetyl-3-(3,4-dimethoxyphenyl)-2-methylalanine.**—The L-N-acetyl-3-(3,4-dimethoxyphenyl)-2-methylalanine (−)-1-phenylethylamine salt (25 g., 0.062 mole) was dissolved in 100 ml. of water and 27.5 ml. of 2.5 *N* sodium hydroxide. The solution was extracted with two 50-ml. and two 25-ml. portions of chloroform. The solution was heated to 70° and 30 ml. of 2.5 *N* hydrochloric acid was added. The N-acetyl acid, which crystallized immediately, was cooled to 10°, filtered, washed with cold water, and dried under vacuum at 60°. The product weighed 16.8 g. (96%); m.p. 192–194°; $[\alpha]^{25}D$ −55° (c 1, methanol); $\lambda_{max}^{CH_3OH}$ 230 mμ (ϵ 8950), 279 (2950).

Anal. Calcd. for $C_{14}H_{19}NO_5$: C, 59.78; H, 6.81. Found: C, 59.74: H, 6.77.

L-(−)-**3-(3,4-Dihydroxyphenyl)-2-methylalanine.**—A solution of L-N-acetyl-3-(3,4-dimethoxyphenyl)-2-methylalanine (10.0 g., 0.0356 mole) in 100 ml. of 48% hydrobromic acid was heated at reflux under a nitrogen atmosphere for 12 hr. The solution was concentrated to dryness and flushed successively with 50-ml. portions of water, *t*-butyl alcohol, and water. The partly crystalline residue was dissolved in 80 ml. of water by warming. The pH was adjusted to 6.4 with 6 *N* ammonium hydroxide under a nitrogen atmosphere. The hot solution was treated with 1.2 g. of decolorizing carbon and filtered. The amber-colored filtrate was concentrated under vacuum to a volume of 30 ml. The mixture was aged in an ice bath for 1 hr., filtered, and washed with a minimum amount of cold water. After drying under vacuum at 100°, the product weighed 5.38 g. (72%); m.p. 306–308°; $[\alpha]^{25}_D$ −4° (*c* 2, 0.1 *N* hydrochloric acid); λ^{Nujol}_{max} 2.79, 3.08, 4.2, 5.3, 6.17, 6.33, and 6.58 μ.

Anal. Calcd. for $C_{10}H_{13}NO_4$: C, 56.87; H, 6.20. Found: C, 57.06; H, 6.37.

(18) The acid is added to speed hydrolysis of the azlactone.

$C_{10}H_{13}NO_4$ Ref. 701a

DL-N-Acetyl-3-(3,4-diacetoxyphenyl)-2-methylalanine.—A slurry of 50 g. (0.237 mole) of α-methyldopa in 50 ml. (0.62 mole) of pyridine and 125 ml. (1.32 moles) of acetic anhydride was heated on the steam bath with stirring. The solid dissolved and the temperature rose to 118°. After 3 hr. on the steam bath (96°), the reddish solution was concentrated under vacuum. The residual oil was dissolved in 50 ml. of acetone and was diluted with 200 ml. of water and 50 ml. of 2.5 *N* hydrochloric acid.[18] After holding at 0–5° for 2 hr., the precipitated product was filtered, washed with water, and dried under vacuum at 50°. The product weighed 74 g. (92.6% yield); m.p. 197–199°; λ^{Nujol}_{max} 3.1, 3.8, 5.68, 5.86, 6.08, 6.4, and 6.61 μ; $\lambda^{CH_3OH}_{max}$ 265 mμ (ε 540), 271 (506).

Anal. Calcd. for $C_{16}H_{19}NO_7$: C, 56.97; H, 5.68. Found: C, 56.93; H, 6.00.

L-(−)-N-Acetyl-3-(3,4-diacetoxyphenyl)-2-methylalanine Quinine Salt.—Quinine (96.4 g., 0.297 mole) and 100 g. (0.297 mole) of D,L-N-acetyl-3-(3,4-diacetoxyphenyl)-2-methylalanine were placed in a 2-l. flask and 960 ml. of acetone was added. The solids dissolved upon stirring and product started to precipitate within 15 min. After stirring at 0–5° for 4 hr., the product was filtered, washed with acetone, and dried under vacuum at 40°. The yield was 91 g. (93%); m.p. 164–166°; $[\alpha]^{25}_D$ −72.7° (*c* 1, 96% ethanol). The quinine salt titrated with base gave an equivalent weight of 665 (theory, 661.8).

L-(−)-N-Acetyl-3-(3,4-diacetoxyphenyl)-2-methylalanine.—The quinine salt of L-(−)-N-acetyl-3-(3,4-diacetoxyphenyl)-2-methylalanine (17.7 g., 0.0268 mole) was dissolved at 0–5° in 11.0 ml. of 2.5 *N* hydrochloric acid and 60 ml. of water. To the clear solution was added 10.6 ml. of 2.5 *N* hydrochloric acid which caused product to precipitate. After holding overnight at 0–5°, the product was filtered, washed with cold water, and dried under vacuum at 40°. The yield was 7.49 g. (83%); m.p. 181–183°; $[\alpha]^{25}_D$ −74.5° (*c* 1, 96% ethanol). Base titration showed an equivalent weight of 336 (theory, 337.3).

L-(−)-3-(3,4-Dihydroxyphenyl)-2-methylalanine.—A solution of L-(−)-N-acetyl-3-(3,4-diacetoxyphenyl)-2-methylalanine (25.0 g., 0.074 mole) in 200 ml. of 6 *N* hydrochloric acid was refluxed for 2 hr. The solution was concentrated to dryness under vacuum and the residual yellow oil was concentrated to dryness three times with 50-ml. portions of *t*-butyl alcohol to remove hydrochloric acid. The gummy residue was dissolved in 45 ml. of water and the solution was filtered to remove a trace amount of insoluble material. The filtrate was adjusted to pH 7.0 with concentrated ammonia. After adding 1.0 g. of sulfur dioxide, the mixture was held at 0–5° overnight. The crystals were filtered, washed with cold water, and dried under vacuum at 50°. The product weighed 14.9 g., but contained 11.3% water by Karl Fischer titration (84.5% yield calculated for $C_{10}H_{13}NO_4 \cdot 1.5 H_2O$).[19]

The L-(−)-α-methyldopa, m.p. 295° dec., $[\alpha]^{25}_D$ −3° (*c* 2, 0.1 *N* hydrochloric acid) had an equivalent weight by base titration of 239 (theory, 238) and had an absorption maximum at 281 mμ (ε 2780). The dried product gave an infrared spectrum identical with material resolved through the 1-phenylethylamine salt.

(18) The acid is added to speed hydrolysis of the azlactone.

(19) X-Ray analysis reveals that L-(−)-α-methyldopa exists in three crystalline forms. Normally, when isolated from aqueous solutions, a sesquihydrate is obtained. Vigorous drying of the sesquihydrate (100°, under vacuum) gives an anhydrous form which, when exposed to air, absorbs water and is transformed back to the hydrate. A second, nonhygroscopic, anhydrous form has been isolated from isopropyl alcohol solutions.

$C_{10}H_{13}NO_4$ Ref. 701b

The importance of L-3-(3,4-dihydroxyphenyl)-2-methylalanine (L-α-methyl DOPA) as an antihypertensive agent is well recognized. DL-α-Methyl DOPA can be synthesized from 3,4-methylenedioxyphenylacetone (MDPA) via DL-3-(3,4-methylenedioxyphenyl)-2-methylalanine (DL-MDPMA).[1,2] In this case, it is most desirable that an intermediate in the chemical synthetic process, namely, DL-MDPMA, is resolved into the optical antipodes, because the undesired D-isomer can be easily degraded to MDPA and it can be reused for a starting material.[3] Therefore, the optical resolution of DL-MDPMA has been investigated by the preferential crystallization procedure which is considered one of the most practical methods for industrial application.

In our previous reports[4,5] it was shown that both the *p*-phenolsulfonic acid salt of DL-MDPMA and the hydrazine salt of N-acetyl-DL-MDPMA are resolved by this simple resolution method. In those days, the salt of DL-MDPMA with hydrochloric acid was also tested but could not be selected as a resolvable salt. As a result of further detailed experiments, however, the monohydrochloride of DL-MDPMA (DL-MDPMA·HCl) has now been found to crystallize as a racemic mixture from a dilute solution of hydrochloric acid and to be easily resolved into each of the antipodes by the preferential crystallization procedure. The failure to identify DL-MDPMA·HCl as a resolvable salt in the previous experiments seems to be due to the fact that DL-MDPMA forms the hemihydrochloride (MDPMA·1/2 HCl) or DL-MDPMA·HCl depending upon the acidity or the excess hydrochloric acid content of the solution.

In this report, properties of the optically active and racemic MDPMA·HCl, and the method of resolution are described. The optical resolution method now presented is more advantageous than those previously reported, because it requires neither a specific aromatic sulfonic acid nor conversion of MDPMA into a complicated derivative. Furthermore, L-MDPMA·HCl can be hydrolyzed by hydrochloric acid directly into L-α-methyl DOPA, and D-MDPMA·HCl can be degraded to the starting material, as shown in Scheme I.

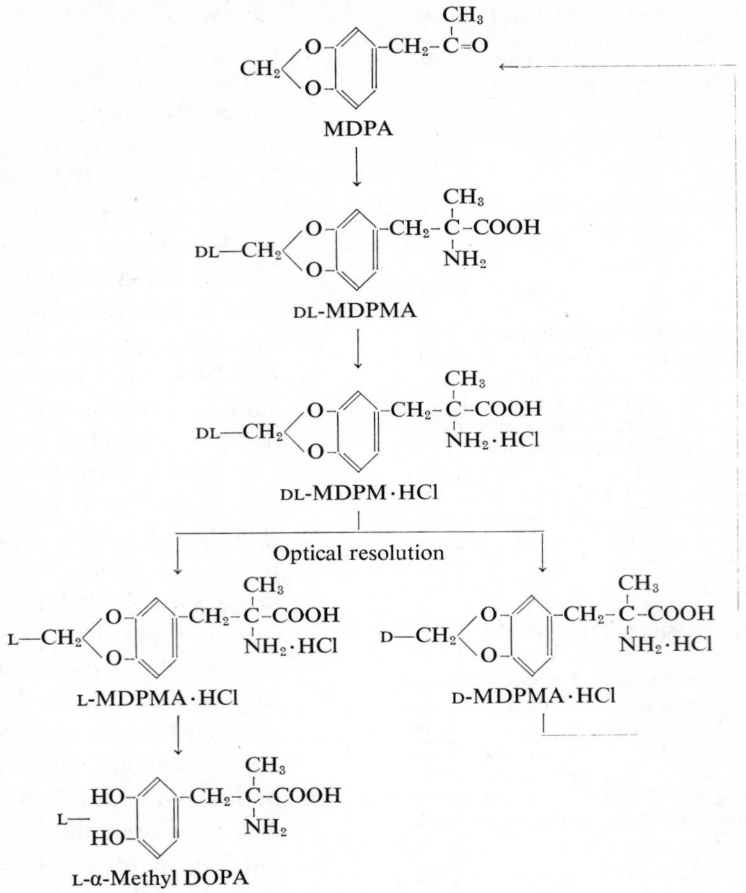

SCHEME I

TABLE I. SUCCESSIVE RESOLUTIONS OF DL-MDPMA·HCl[a]

Expt	Amount of addition		Composition of solution		Separated crystals	
	DL-Form (g)	Active form (g)	DL-Form (g)	Active form (g)	Yield (g)	Optical purity[b] (%)
1 (L)	28.00	2.00	28.00	2.00	4.66	93.1
2 (D)	4.75	—	28.00[c]	2.29[c]	4.67	94.5
3 (L)	4.78	—	28.00[c]	2.07[c]	4.58	92.6
4 (D)	4.70	—	28.00[c]	2.12[c]	4.62	93.5
mean	4.74	—	28.00[c]	2.11[c]	4.63	90.9

a) Resolutions were carried out at 25°C on a 50 ml scale. Crystallization time was 50 min in every case.
b) The optical purity was calculated based on the maximum rotation $[\alpha]_{365}^{25} \pm 21.8°$ (c=1, 1 N HCl).
c) Values calculated theoretically from analysis of separated crystals and mother liquors.

EXPERIMENTAL[6]

Preparation of L-, D-, and DL-MDPMA·HCl. Water (210 ml) and 12 N HCl (100 ml, 1.2 mole) were added to DL-MDPMA[7] (223.2 g, 1 mole). The mixture was heated, treated with charcoal, and allowed to stand in a refrigerator overnight. The resulting precipitate was collected, washed with cold water and dried *in vacuo*. The initial crop of DL-MDPMA·HCl (173.5 g) was obtained, mp 230~232°C (dec), and further crops were obtained by successive concentrations of the combined filtrates. The total yield was 233.2 g (89.8%). The products were almost pure and could be used for optical resolution without further purification. Crystallization from 0.25 N HCl gave a racemic mixture, colorless prisms, mp 233~234°C (dec). *Anal.* Calcd for $C_{11}H_{14}NO_4Cl$: C, 50.87; H, 5.43; N, 5.39; Cl, 13.65. Found: C, 50.99; H, 5.50; N, 5.30; Cl, 13.77. Solubility in 0.25 N HCl (g/100 ml): 26.7 (15°C), 28.9 (25°C), 43.9 (40°C).

The optically active L- and D-MDPMA·HCl were prepared from L- and D-MDPMA[8] respectively in the same way as described above. The L-isomer: colorless prisms, $[\alpha]_D^{25}$ +0.9° (c=1, 1 N HCl), $[\alpha]_{365}^{25}$ +21.8° (c=1, 1 N HCl); mp 244~245°C. *Anal.* Found: C, 50.98; H, 5.51; N, 5.30; Cl, 13.94. Solubility in 0.25 N

HCl (g/100 ml): 19.5 (15°C), 22.9 (25°C), 29.8 (40°C). The D-isomer: $[\alpha]_D^{25}$ −0.9° (c=1, 1 N HCl), $[\alpha]_{365}^{25}$ −21.8° (c=1, 1 N HCl); mp 244~245°C. The infrared spectra of L-, D-, and DL-MDPMA·HCl in KBr were identical. IR ν_{max}^{KBr} cm^{-1}: 3140, 2990, 2790, 2550, 1955, 1730, 1585, 1495, 1445, 1400, 1380, 1360, 1250, 1215, 1190, 1110, 1040, 935, 810, 760.

When DL-MDPMA·HCl (2.0 g) was dissolved in water (4 ml) at elevated temperature and the solution was cooled to room temperature, DL-MDPMA·1/2HCl (0.4 g) crystallized as colorless prisms, mp 245~246°C (dec). *Anal.* Calcd for $C_{11}H_{13}NO_4$·1/2HCl: C, 54.72; H, 5.64; N, 5.80; Cl, 7.34. Found: C, 54.68; H, 5.62; N, 5.53; Cl, 7.54. IR ν_{max}^{Nujol} cm^{-1}[9]: 3150, 2950~2850, 2550, 1720~1670, 1600, 1500, 1260, 1240, 1190, 1115, 1095, 1035, 930, 880, 810. On the other hand, recrystallization of optically active MDPMA·HCl from water gave the monohydrochloride, and the hemihydrochloride of optically active isomers was not obtained.

Optical resolution of DL-MDPMA·HCl. In a typical experiment, DL-MDPMA·HCl (28.00 g) and L-MDPMA·HCl (2.00 g) were dissolved in 0.25 N HCl (50 ml) at elevated temperature. The solution was cooled to 25°C, seeded with fine pulverized crystals of L-MDPMA·HCl (0.05 g) and stirred for 50 min at the same temperature. The precipitated crystals were collected by filtration, washed with a small amount of cold water (1 ml) and dried to give L-MDPMA·HCl (4.66 g), $[\alpha]_{365}^{25}$ +20.3° (c=1, 1 N HCl). Its optical purity was 93.1%, based on the maximum rotation $[\alpha]_{365}^{25}$ +21.8° (c=1, 1 N HCl). After the separation of the L-isomer, DL-MDPMA·HCl (4.75 g) and a small amount of 0.25 N HCl were added to the mother liquor in order to prepare the supersaturated solution of almost the same composition as in the previous resolution except that the D-isomer predominated. The solution thus obtained was cooled to 25°C, seeded with D-MDPMA·HCl (0.05 g) and stirred. After 50 min, the precipitated crystals were treated in the same manner as described above to yield D-MDPMA·HCl (4.67 g), $[\alpha]_{365}^{25}$ −20.6° (c=1, 1 N HCl), optical purity 94.5%. By repeating these procedures, L- and D-isomers were successively obtained. The examples of the several runs are shown in Table I.

Purification of optically impure MDPMA·HCl. The optically impure L-MDPMA·HCl (16.00 g, optical purity 87.5%) obtained by the above procedure was recrystallized from 2 N HCl (125 ml). The precipitated crystals were filtered off, washed with water and dried to give optically pure L-MDPMA·HCl (11.50 g), $[\alpha]_{365}^{25}$ +21.8° (c=1, 1 N HCl), mp 244~245°C (dec). Their mp and specific rotation did not change on further recrystallization. The L-MDPMA·HCl (3.00 g) obtained above was dissolved in hot water (10.5 ml) and adjusted to pH 6 with 5 N ammonium hydroxide. The precipitate was collected, washed with water and dried to give L-MDPMA (2.41 g). Recrystallization from water gave pure L-MDPMA, $[\alpha]_{365}^{25}$ +25.4° (c=1, 1 N HCl), mp 282~283°C, [lit.[4] $[\alpha]_{365}^{25}$ +25.4° (c=1, 1 N HCl)]. *Anal.* Calcd for $C_{11}H_{13}NO_4$: C, 59.19; H, 5.87; N, 6.28. Found: C, 58.88; H, 5.89; N, 6.39.

Preparation of L-α-methyl DOPA. The MDPMA·HCl (13.0 g) obtained above was added to a mixture of 20% HCl (200 ml) and phenol (10 ml). The mixture was refluxed for 17 hr under stirring. After filtration, the filtrate was treated with charcoal and concentrated to dryness to remove excess HCl. The resulting residue was dissolved in water (25 ml) and treated with charcoal. The solution was adjusted to pH 5.8 with 10% NaOH (*ca.* 20 ml) and then to pH 3.3 with 5% sodium bisulfite. After cooling in a refrigerator overnight, the precipitate was collected, washed with cold water and air-dried at 60°C to give L-α-methyl DOPA·3/2H_2O, 8.6 g (72.3%). Recrystallization from a diluted sulfurous acid solution (0.5%) gave a white powder of L-α-methyl DOPA·3/2H_2O, and drying of the sesquihydrate *in vacuo* at 100°C gave the anhydrous form $[\alpha]_D^{25}$ −5.2°, $[\alpha]_{578}^{25}$ −5.5° (c=2, 0.1 N HCl), mp 306~307°C (dec). [lit.[10] $[\alpha]_D^{25}$ −4° (c=2, 0.1 N HCl), and $[\alpha]_{578}^{25}$ +5.5° (c=2, 0.1 N HCl) for D-α-methyl DOPA]. *Anal.* Calcd for $C_{10}H_{13}NO_4$: C, 56.86; H, 6.20; N, 6.63. Found: C, 56.65; H, 6.23; N, 6.56.

REFERENCES AND NOTES

1) T. Kurano, M. Fukuda and M. Horiuchi, Japanese Patent 45–2733 (1970).
2) G. A. Stein, H. A. Bronner and K. Pfister, 3rd, *J. Amer. Chem. Soc.*, **77**, 700 (1955).
3) H. L. Slates, D. Taub, C. H. Kuo and N. L. Wendler, *J. Org. Chem.*, **29**, 1424 (1964).
4) S. Yamada, M. Yamamoto and I. Chibata, *ibid.*, **38**, 4408 (1973).
5) S. Yamada, M. Yamamoto and I. Chibata, *ibid.*, **40**, 3360 (1975).
6) All the experimental methods and the instruments were identical to those described in the previous report[4] unless otherwise noted.
7) The authors are grateful to Dr. M. Miyoshi and Dr. K. Matsumoto for a donation of this compound.
8) Resolved in the previous report.[4]
9) A good spectrum was not obtained in KBr.
10) E. W. Tristram, J. ten Broeke, D. F. Reinhold, M. Sletzinger and D. E. Williams, *J. Org. Chem.*, **29**, 2053 (1964).

$C_{10}H_{13}NO_4$

Ref. 702

For a resolution procedure for the above compound, see Ref. 561b.

$$CH_3-\underset{\displaystyle SO_3H}{CH}-CH_2CONHC_6H_5$$

$C_{10}H_{13}NO_4S$ Ref. 703

III. Dérivés fonctionnels des acides actifs.

Dédoublement de l'acide butyranilide-β-sulfonique racémique:
$CH_3 . CH(SO_3H) . CH_2 . CONHC_6H_5$.

Le sel de strychnine de cet acide se sépare sous forme d'une huile, lorsque sa solution est évaporée au bain-marie. L'évaporation à la température ordinaire donne le *butyranilide-β-sulfonate de strychnine* en cristaux bien développés, renfermant trois molécules d'eau.

Substance 0.1662 g.; 0.2709 m. équiv. de baryte. Subst. 0.1552 g.; perte de poids à 110° 0.0088 g.
Trouvé: H_2O 5.67; poids mol. 613.5
$C_{10}H_{13}O_4NS . C_{21}H_{22}O_2N_2 . 2 H_2O$. Calculé: „ 5.87; „ „ 613.4

Le sel de strychnine, décomposé à la baryte, a donné une solution inactive. Comme la recristallisation du sel est pénible en raison de sa tendance à se séparer à l'état huileux, nous avons recouru au sel de brucine, qui du premier coup a donné de bons résultats.

Le *butyranilide-β-sulfonate de brucine* cristallise dans l'eau sous forme de prismes. Le titrage indique la présence de dix molécules d'eau, dont neuf molécules s'échappent vers 100°.

Substance 0.4728 g.: perte de poids à 100° 0.0934 g.; 0.5767 m. équiv. de baryte.
Trouvé: H_2O 19.75; poids mol. 819.8
$C_{10}H_{13}O_4NS . C_{23}H_{26}O_4N_2 . 10 H_2O$. Calculé: 9 H_2O 19.83; „ „ 817.6

Le sel de brucine, recristallisé dans l'eau et décomposé à la baryte, a donné un sel barytique lévogyre.

L'examen systématique des eaux-mères des cristallisations successives a révélé que le composant gauche était pur après sept cristallisations.

La rotation spécifique du sel barytique préparé aux dépens de l'eau-mère n'a plus changé par une cristallisation ultérieure.

Elle a été de — 5°.3, ainsi que le produit obtenu par décomposition du sel de brucine cristallisé.

Le même composant gauche de l'acide butyranilide-β-sulfonique s'obtient par *la réaction de l'acide sulfobutyrique lévogyre avec l'aniline.*

Une solution de l'acide actif, évaporée avec une molécule d'aniline, a abandonné le 1-sulfobutyrate d'aniline, qui, chauffé à 190° pendant deux heures avec cinq fois son poids d'aniline, a donné le 1-butyranilidesulfonate d'aniline.

Le sel barytique, qu'on en a dérivé, a montré une rotation spécifique de — 4°.3. Comme le $[\alpha]_D$ du sel composant pur est de — 5°.3, on voit que le produit ne renferme que dix pour cent de l'antipode, c. à d. qu'un cinquième du produit s'est racémisé pendant la réaction.

L'acide l-butyranilide-β-sulfonique:
$CH_3 . CH(SO_3H) . CH_2 . CONHC_6H_5$.

En décomposant le sel barytique, décrit ci-dessous, à l'acide sulfurique (quantité théorique), on obtient, après filtration, une solution qui par évaporation spontanée dépose de petites aiguilles non hygroscopiques de l'acide 1-butyranilide-β-sulfonique. Elles fondent vers 165°—168°, mais pas nettement.

Substance 0.1847 g.; 0.7590 m. équiv. de baryte.
Trouvé: poids mol. 243.3
$C_{10}H_{13}O_4NS$. Calculé: „ „ 243.2

La rotation a été mesurée pour la solution aqueuse et aussi pour les solutions dans l'alcool absolu et dans l'acétone, qui ont montré une rotation en sens inverse.

Acide l-butyranilide-β-sulfonique. I. Solution aqueuse 0.0065 mol. gr. dans 100 cm³. II. Solution dans l'alcool absolu, 0.0044 mol. gr. dans 100 cm³. III. Solution dans l'acétone, 0.0044 mol. gr. dans 100 cm³.

	λ	642	589	550	505
I.	α	— 0°.16	— 0°.22	— 0°.26	— 0°.33
„	[M]	—12 .3	—17 .0	—20 .0	—25 .5
II.	α	+ 0 .14	+ 0 .16	+ 0 .17	+ 0 .20
„	[M]	+16 .0	+18 .2	+19 .3	+22 .8
III.	α	+ 0 .05	+ 0 .06	+ 0 .07	+ 0 .09
„	[M]	+ 5 .6	+ 6 .8	+ 8 .0	+10 .1

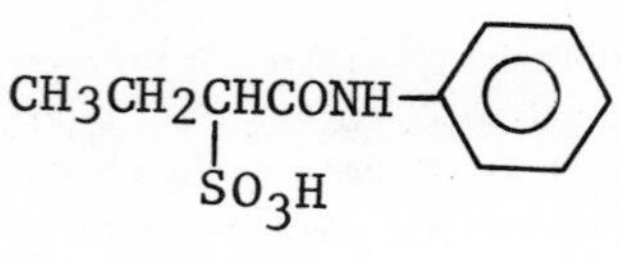

$C_{10}H_{13}NO_4S$ Ref. 704

III. **Les dérivés actifs de l'acide sulfobutyrique.**

Dédoublement de l'acide butyranilide-α-sulfonique racémique.
$C_2H_5 . CH(CONHC_6H_5)SO_3H$.

Une solution de 11.3 g. de l'acide butyranilide-α-sulfonique a été traitée à chaud par 15.6 g., soit la quantité équimoléculaire, de strychnine. Refroidie et triturée à l'aide d'une baguette de verre, elle a déposé 20 gr. du sel cristallin de strychnine.

Dans le but d'obtenir le composant actif le moins soluble à l'état pur, on a fait recristalliser ce sel à plusieurs reprises. Il se dépose dans l'eau sous forme de petits groupes concentriques d'aiguilles.

Afin de juger du degré de la séparation obtenue nous nous sommes servis de la „méthode des eaux-mères", décrite pour le dédoublement de l'acide sulfobutyrique racémique [9]).

L'eau-mère de chaque cristallisation est traitée par la quantité équimoléculaire de baryte. Filtrée et épuisée au chloroforme pour l'élimination de toute trace de strychnine, elle donne par évaporation le sel de baryum, dont on détermine l'activité optique.

Voici les valeurs de la rotation moléculaire, obtenues de la sorte, depuis la troisième jusqu'à la douzième cristallisation: + 9°.5, + 10°.5, + 8°.5, 0°, 0°, — 7°, — 14°.5, — 28°, — 32°, — 34°, — 34°.

La constance de la rotation semble indiquer la fin de la séparation.

En effet, le produit cristallisé de la dernière eau-mère a donné un sel de baryum présentant la même rotation moléculaire, soit — 34°.

Le *l-butyranilide-α-sulfonate de strychnine* cristallise en petites aiguilles renfermant une molécule d'eau.

Par suite de sa faible solubilité, une mesure exacte de sa rotation n'est pas réalisable.

Substance 0.7731 g.; mg. équiv. de baryte 1.296. Subst. 0.2069 g.; perte de poids à 110° 0.0063 g. Trouvé: poids mol. 596.5; H_2O 3.04.
$C_{10}H_{13}O_4NS . C_{21}H_{22}O_2N_2 . H_2O$. Calculé: „ „ 595.4; „ 3.02.

L'eau-mère de la première cristallisation a donné, par évaporation spontanée à l'air, un sel de strychnine renfermant le composant *d* en excès. Un échantillon, décomposé par la baryte, a fourni un sel dextrogyre.

Cependant, comme une cristallisation dans l'alcool absolu diminue la teneur en composant *d*, nous ne l'avons pas séparé à l'état pur.

Nous avons encore essayé si le dédoublement de l'acide racémique réussirait plus vite en appliquant la moitié de la quantité théorique de strychnine. Mais, cette méthode n'a pas donné de meilleurs résultats que celle que nous venons de décrire.

L'acide l-butyranilide-α-sulfonique, obtenu par l'intermédiaire de son sel barytique, a montré les valeurs de la rotation moléculaire [M], que voici:

Solution 0.048 n.; tubes de 2 dm.

λ =	621	589 (D)	530
α =	—0°.14	—0°.16	—0°.24
[M] =	—15°	—16°.5	—25°

Le l-butyranilide-α-sulfonate de baryum.

Le sel de strychnine, decomposé par de la baryte, donne le sel barytique, qu'on purifie en le faisant cristalliser dans l'eau.

Substance: 0.2529 g.; perte de poids à 150° 0.0234 g.; $BaSO_4$ 0.0862 g.
Subst. 0.2954 g.; perte de poids 0.0273 g., $BaSO_4$ 0 1003 g.
Trouvé: H_2O 9.25, 9.24; Ba 20.06, 19.98.
$C_{20}H_{24}O_8N_2S_2Ba . 3\frac{1}{2} H_2O$ (684.7). Calculé: „ 9.20; „ 20.07.
Substance 0.730 g. dans 20 cm³., tube 2 dm.

λ =	621	589	530
α =	—0°.31	—0°.36	—0°.54
[M] =	—29°	—34°	—51°

Le l-butyranilide-α-sulfonate de cuivre a été obtenu en décomposant le sel actif de baryum ($[M]_D$ = — 32°) par la quantité équimoléculaire du sulfate de cuivre.

La solution verte, évaporée, a abandonné un sel vert jaune très hygroscopique, qu'on a séché à 100° avant l'analyse.

Substance, séchée à 100°, 0.0895 g.; CuO 0.0131 g.
Trouvé; Cu 11.69.
$C_{20}H_{24}O_8N_2S_2Cu$ (548.0). Calculé: „ 11.60.

La rotation a été déterminée pour plusieurs longueurs d'onde, tant avec l'eau qu'avec l'alcool absolu comme dissolvants. Les tableaux suivants ne donnent pas les valeurs observées de α, mais se bornent à l'indication des rotations moléculaires qu'on en a déduites.

[9]) Ce Recueil 43, 309 (1924).

Solution de 0.8 % dans l'eau.

λ =	677	638	605	547	551	530	513	498	485
[M] =	−20°.5	−27°.5	−32°.5	−37°.5	−41°	−48°	−55°	−55°	−58°

Solution dans l'alcool absolu, contenant 0.7 g. dans 100 cm³.

λ =	638	605	589	574	551	530
[M] =	−23°.5	−27°.5	−31°.5	−35°	−41°	−48°

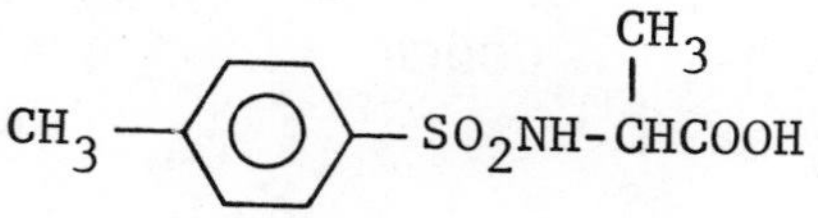

$C_{10}H_{13}NO_4S$ Ref. 705

First Resolution of dl-p-*Toluenesulphonylalanine.*

Using the equilibrium method, two equivalents of *dl-p*-toluenesulphonylalanine (25 grams), one equivalent of sodium hydroxide (2·06 grams) in normal solution, and one equivalent of brucine (24 grams) were dissolved in just sufficient boiling water, and, after keeping the mixture for some hours, an almost theoretical quantity of the *brucine* salt of *l-p*-toluenesulphonylalanine separated. The mother liquor from the separation of the brucine salt was evaporated, made alkaline, and the solution extracted thoroughly with chloroform until free from brucine. The solution was then acidified with concentrated hydrochloric acid, and the precipitated crude *d*-acid, which was first obtained as an oil, quickly crystallised. This acid, in the form of the sodium salt in aqueous solution, had the specific rotatory power $[\alpha]_{5461}$ +20·0°, and therefore contained 80 per cent. of the pure *d*-acid. The crude acid (13·1 grams) was accordingly mixed with 13·7 grams of strychnine and 0·5102 gram of sodium hydroxide in normal aqueous solution, and the whole brought into solution with sufficient boiling water. On cooling and keeping for some hours, an almost theoretical quantity of the *strychnine* salt of *d-p*-toluenesulphonylalanine was obtained. The resolution was repeated, with practically identical results.

Brucine l-p-*Toluenesulphonylalanine*,
$C_{23}H_{26}O_4N_2,C_{10}H_{13}O_4NS,3H_2O$.

The salt, separated as described, is very easily obtained pure on crystallisation from dilute aqueous alcohol. It was recrystallised four times, although it was found to be pure after two crystallisations. The salt crystallises in large, colourless, glistening plates, which are very brittle. It melts at 148—149°, losing water of crystallisation. It is very sparingly soluble in water, but more readily so in alcohol:

0·1244 gave 0·2601 CO_2 and 0·0729 H_2O. C=57·0; H=6·5.
1·020 lost 0·0738 at 130°. H_2O=7·2.
$C_{10}H_{13}O_4NS,C_{23}H_{26}O_4N_2,3H_2O$ requires C=57·3; H=6·5; $3H_2O$=7·8 per cent.

0·1928 made up to 30·0 c.c. with absolute alcohol and examined in a 4-dcm. tube at 29—30° gave α_{5461} −0·99°, whence $[\alpha]_{5461}$ −38·51°; $[M]_{5461}$ −266·2°.

Strychnine d-p-*Toluenesulphonylalanine*,
$C_{21}H_{22}O_2N_2,C_{10}H_{13}O_4NS,2H_2O$.

The strychnine salt is far more readily soluble in water than the above brucine salt, and is easily purified by recrystallisation from that solvent. The salt was recrystallised four times. On being allowed to crystallise slowly, it forms large, colourless, glistening prisms, melting at 188—189°; only one molecule of water is lost on heating at 125°:

0·1272 gave 0·2835 CO_2 and 0·0778 H_2O. C=60·8; H=6·8.
0·6954 lost 0·0245 at 125°. H_2O=3·5.
$C_{10}H_{13}O_4NS,C_{21}H_{22}O_2N_2,2H_2O$ requires C=60·7; H=6·4; $1H_2O$=3·2 per cent.

0·2076 made up to 30·0 c.c. with water and examined in a 4-dcm. tube at 29—30° gave α_{5461} −0·31°, whence $[\alpha]_{5461}$ −11·2°; $[M]_{5461}$ −68·7°.

Second Resolution of dl-p-*Toluenesulphonylalanine.*

This was carried out in an exactly similar manner to the first resolution, using first strychnine and then brucine. The resolution proceeds quite normally, and almost theoretical quantities of the strychnine *d-p*-toluenesulphonylalanine and brucine *l-p*-toluenesulphonylalanine are obtained in this order. The salts were purified as above, and had identical melting points and rotatory powers with those previously described; it was thus easily seen that they were identical. (Found, strychnine salt, m. p.=188—189°, C=60·8; H=6·2; $[\alpha]_{5461}^{29-30}$ −11·3° in water. Brucine salt, m. p.=148—149°, $[\alpha]_{5461}^{29-30}$ −38·76° in absolute alcohol.)

This method of resolution was repeated, and exactly similar results were obtained. It may be stated that, of the two methods of resolution employed, the second method is to be preferred. In the first method, after the precipitation of the brucine remaining in the solution after the separation of the brucine salt of *l-p*-toluenesulphonylalanine with alkali, unless the filtered liquid is thoroughly extracted with chloroform, the brucine remaining is likely to contaminate the strychnine salt obtained in the next process. This tedious extraction is avoided by first isolating the strychnine salt of *l-p*-toluenesulphonylalanine.

l-p-*Toluenesulphonylalanine*, $CH_3{\cdot}CH(CO_2H){\cdot}NH{\cdot}SO_2{\cdot}C_6H_4{\cdot}CH_3$.

This acid was liberated from the brucine salts obtained in the first and the second resolutions by boiling the salt with an excess of aqueous ammonia, filtering from the precipitated brucine, and extracting the filtrate at least six times with chloroform. The aqueous solution was evaporated to a small bulk on the water-bath, and acidified with concentrated hydrochloric acid. The precipitated acid was at first oily, but quickly solidified on stirring. It was recrystallised three times from very dilute aqueous alcohol. On allowing the solution to evaporate slowly, the acid was obtained in very long, colourless needles, melting at 131—132°. The substance is easily soluble in ethyl alcohol, methyl alcohol, or acetone, but much less readily so in water. It may be crystallised from boiling water:

0·1249 gave 0·2234 CO_2 and 0·0596 H_2O. C=48·8; H=5·3.
$C_{10}H_{13}O_4NS$ requires C=49·3; H=5·4 per cent.

As was anticipated, the acid from the brucine salt obtained by the first method of resolution (*a*) was identical with that from the brucine salt obtained in the second resolution (*b*):

(*a*) 0·2018 as sodium salt made up to 30·0 c.c. with water and examined in a 4-dcm. tube at 29—30° gave α_{5461} −0·91°, whence $[\alpha]_{5461}$ −33·8°; $[M]_{5461}$ −82·1°.
(*b*) 0·2020 as sodium salt made up to 30·0 c.c. in water and examined as above gave α_{5461} −0·91°, whence $[\alpha]_{5461}$ −33·8°; $[M]_{5461}$ −82·1°.

The following determinations of rotatory powers were made on the acid obtained from both resolutions:

In water.—0·5116 as sodium salt made up to 30·0 c.c. and examined in a 4-dcm. tube at 29—30° gave α_{5461} −2·41°, $[\alpha]_{5461}$ −35·33°; $[M]_{5461}$ −85·8°.
2·0044 as sodium salt, examined as above, gave α_{5461} −10·7°, $[\alpha]_{5461}$ −40·04°; $[M]_{5461}$ −97·3°.

It will be seen that, as was found in the case of optically active benzoylalanine, the rotatory power of the acid in the form of its salt increases with the concentration; in the present example this increase is very marked.

In absolute ethyl alcohol.—0·0063 made up to 30·0 c.c. and examined in a 4-dcm. tube at 29—30° gave α_{5461} −1·23°, $[\alpha]_{5461}$ −9·17°; $[M]_{5461}$ −22·2°.

In acetone.—1·0006 made up and examined as above gave α_{5461} −1·16°, $[\alpha]_{5461}$ −8·69°; $[M]_{5461}$ −21·1°.

In the case of optically active benzoylalanine, it was observed that the rotatory power in acetone was considerably greater than in alcohol. The above figures show that in the case of *p*-toluenesulphonylalanine the rotatory power in alcohol is almost the same as that in acetone.

d-p-*Toluenesulphonylalanine*, $CH_3{\cdot}CH(CO_2H){\cdot}NH{\cdot}SO_2{\cdot}C_6H_4{\cdot}CH_3$.

This acid was obtained from the strychnine salts obtained in both resolutions by boiling with concentrated aqueous ammonia, allowing to cool, and filtering from the precipitated strychnine. The filtrate was evaporated to a small bulk on the water-bath, and the acid precipitated by the addition of concentrated hydrochloric acid. The acid was then crystallised three times from dilute aqueous alcohol. It was obtained in colourless needles, melting at 131—132°, which often measured as much as 4 cm.:

0·1306 gave 0·2357 CO_2 and 0·0654 H_2O. C=49·2; H=5·6.
$C_{10}H_{13}O_4NS$ requires C=49·4; H=5·4 per cent.

The acid from the strychnine salts obtained in the first resolution (*a*) was proved to be identical with that obtained in the second resolution (*b*) by a comparison of the melting points and of the rotatory powers.

In water.—(*a*) 0·2009 as sodium salt made up to 30·0 c.c. with water and examined in a 4-dcm. tube at 29—30° gave α_{5461} +0·90°, $[\alpha]_{5461}$ +33·6°; $[M]_{5461}$ +81·6°.

(*b*) 0·2021 as sodium salt made up and examined as above gave α_{5461} +0·91°, $[\alpha]_{5461}$ +33·7°; $[M]_{5461}$ +81·8°.

The solubility of the *d*-acid was exactly similar to that of the *l*-acid, and the acids behave similarly in all respects.

It will be seen that the resolution of externally compensated *p*-toluenesulphonylalanine can be very readily accomplished in either of the above ways, and is not attended with the formation of mixed crystals. Owing to the conditions of working, it has, so far, been impossible to determine the rotatory dispersions. Further work on the resolution of analogous compounds is in progress.

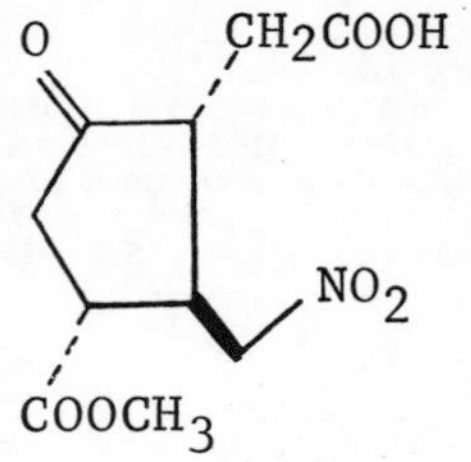

$C_{10}H_{13}NO_7$ Ref. 706

Authors' comments:

Preliminary experiments using (+) -and (−)-α-phenylethylamine afforded the (+) acid, mp 112–113°, $[\alpha]^{25}_D$ +84.28° (*c* 1.01, CH_3OH), and the (−) acid, mp 111–112°, $[\alpha]^{25}_D$ −79.9° (*c* 1.15, CH_3OH).

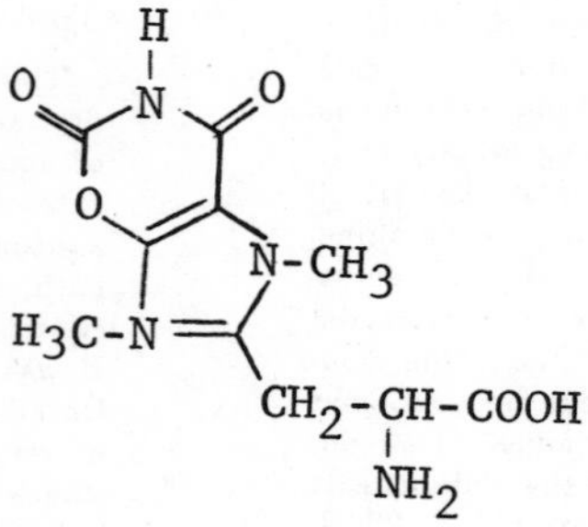

$C_{10}H_{13}N_5O_4$ Ref. 707

Salt of *L*-Threo-1-(p-nitrophenyl)-2-amino-1,3-propanediol and d-N-Carbobenzyloxy-β-(8-theobrominyl)-α-alanine (*L*-III · d-VIII). a) A hot solution of 3.3 g *dl*-N-carbobenzyloxy-β-(8-theobrominyl)-α-alanine in 130 ml 92% alcohol was mixed with a solution of 1.7 g *L*-threoamine (*L*-III) in 35 ml alcohol. After 2.5 h the precipitate of salt (*L*-III · d-VIII) was filtered off and dried in vacuum at 100°. Yield 2.5 g (92%), m.p. 186-189°; $[\alpha]_D^{20}$ +19° (c 1% in 45% tetrahydrofuran). Found %: C 53.18; H 5.34; N 16.19. $C_{27}H_{31}N_7O_{10}$. Calculated %: C 52.85; H 5.09; N 15.98. The alcoholic filtrate was utilized for the isolation of (*l*-II).

b) The aqueous alcoholic filtrate after removal of salt (D-III · *l*-VIII) was evaporated nearly to dryness in vacuum. The residue of salt (D-III · d-VIII) was treated with ethyl acetate, 5 N HCl was run into the suspension until solution of the precipitate, the aqueous layer separated and extracted twice with 40 ml ethyl acetate. The combined ethyl acetate layer was shaken with $MgSO_4$, filtered and evaporated to dryness. The residue was dissolved in 20 times the quantity of alcohol and an equivalent amount (*L*-III) added. The solution was heated to boiling and then left at room temperature. After 5-6 h the salt (*L*-III · d-VIII) which had precipitated was filtered off and washed with alcohol. M.p. 186-189°; $[\alpha]_D^{20}$ +19°, (c 1% in 45% tetrahydrofuran).

d-N-Carbobenzyloxy-β-(8-theobrominyl)-α-alanine (d-VIII). A suspension of 1.5 g salt (*L*-III · d-VIII) in 50 ml ethyl acetate was shaken with 25 ml 5% HCl. After dissolution of the solid the ethyl acetate layer was separated, and the aqueous layer was quickly extracted 4 times with 30 ml ethyl acetate. The combined ethyl acetate solution was shaken with $MgSO_4$, filtered and evaporated to dryness in vacuum. The residue was boiled with 25 ml 70% alcohol, filtered and 0.5 g (51%) isomer (d-VIII) was obtained. A sample was dried 2 h in vacuum at 100° for analysis, m.p. 230-231°. Found %: C 53.62; H 4.86; N 17.66. $C_{18}H_{19}N_5O_6$. Calculated %: C 53.86; H 4.78; N 17.46.

d-β-(8-Theobrominyl)-α-alanine (d-II). a) 1.6 g (d-VIII) was hydrogenated in 90 ml 60% methanol at 35° in the presence of Pd-black (from 1 g $PdCl_2$) for 15 h. The catalyst was filtered off and boiled with water until the absence of a ninhydrin reaction. The combined aqueous alcoholic filtrate was evaporated in vacuum to 60-70 ml volume and cooled. d-β-(8-Theobrominyl)-α-alanine (0.3 g) with m.p. 297-299° (decomposition) separated out. From the filtrate a further 0.1 g (d-II) was isolated. Overall yield 0.4 g (36%); $[\alpha]_D^{18}$ -22° (c 5% in 5 N HCl). Found %: C 42.11; H 5.40; N 24.33. $C_{19}H_{13}N_5O_4 \cdot H_2O$. Calculated %: C 42.10; H 5.35; N 24.54.

b) 2.8 g salt (*L*-III · d-VIII) was put into a flask fitted with a calcium chloride tube, 15-17 ml 33% HBr in glacial CH_3COOH run in and after 20 min (at the end of CO_2 evolution) 100 ml absolute ether was gradually added while shaking. After 30 min the ethereal layer was decanted from the precipitate, the solid washed with ether again and dissolved in 30 ml water. The aqueous solution was washed 2 times with 10 ml ether, neutralized to pH 6.5 with NH_4OH and 90 ml alcohol added. After 5 h crystals were filtered from the cooled mass, dissolved in 20 ml hot water, filtered and 100 ml added. 0.8 g (66%) d-β-(8-theobrominyl)-α-alanine with m.p. 297-299° (decomposition) was obtained: $[\alpha]_D^{20}$ -22° (c 0.5% in 5 N HCl).

c) Similarly (d-II) was obtained from the filtrate (after breakdown of the salt D-III · *l*-VIII) after evaporation of the alcoholic solution and treating the residue with HBr in CH_3COOH.

Salt of D-Threo-1-(p-nitrophenyl)-2-amino-1,3-propanediol and *l*-N-Carbobenzyloxy-β-(8-theobrominyl)-α-alanine (D-III · *l*-VIII). To a hot solution of 6.6 g d*l*-N-carbobenzyloxy-β-(8-theobrominyl)-α-alanine (VIII) in 300 ml 80% alcohol was added a hot solution of 3.5 g D-threoamine (D-III) in 35 ml alcohol. The mixture was boiled for 2-3 min then left at room temperature for 5 h. The salt (D-III · *l*-VIII) precipitated in the form of white grains. The yield was quantitative,* m.p. 187-189°; $[\alpha]_D^{20}$ -20° (c 1% in 50% tetrahydrofuran). Found %: C 51.34; H 5.16; N 15.82. $C_{27}H_{31}N_7O_{10} \cdot H_2O$. Calculated %: C 51.47; H 5.26; N 15.52. The filtrate was used either for the preparation of salt (*L*-III · d-VIII) or for the isolation of (d-II).

l-N-(Carbobenzyloxy-β-(8-theobrominyl)-α-alanine (*l*-VIII) was obtained similarly to the isomer (d-VIII). From 1.1 g salt (D-III · *l*-VIII) the yield of (*l*-VIII) was 0.3 g (41%), m.p. 227-229°; $[\alpha]_D^{25}$ -14° (c 1% in 2 N HCl).† Found %: C 53.97; H 4.94; N 17.35. $C_{18}H_{19}N_5O_6$. Calculated %: C 53.86; H 4.78; N 17.46.

l-β-(8-Theobrominyl)-α-alanine (*l*-II) was obtained analogously to the derivation of (d-II) by means of the hydrogenation of the N-carbobenzyloxy derivative (*l*-VIII) over palladium black (yield 18%) or in 68% yield by the action of a 33% solution of HBr in glacial CH_3COOH on the salt (D-III · *l*-VIII) or salt (*L*-III · *l*-VIII). The latter was obtained from the alcoholic filtrate from the resolution of (d*l*-VIII) with the aid of (*L*-III). M.p. 298-299° (decomposition); $[\alpha]_D^{18}$ +24° (c 0.5% in 5 N HCl). Found %: C 41.95; H 5.33; N 24.22. $C_{10}H_{13}N_5O_4 \cdot H_2O$. Calculated %: C 42.10; H 5.33; N 24.54.

Resolution of d*l*-N-Carbobenzyloxy-β-(8-theobrominyl)-α-alanine with Quinine. To a solution of 0.4 g carbobenzyloxy derivative (d*l*-VIII) in 12 ml 80% alcohol was added a solution of 0.38 g quinine in 5 ml absolute alcohol and the mixture left for 3 days. The finely divided crystalline precipitate was filtered off and 0.35 g salt of quinine with (*l*-VIII) was obtained. M.p. 194-197° (from water); $[\alpha]_D^{20}$ -98° (c 0.5% in alcohol). Found %: C 62.61; H 6.14; N 13.27. $C_{38}H_{43}N_7O_8$. Calculated % C 62.89; N 5.96; N 13.49. The alcohol was evaporated off from the alcoholic filtrate and the residue dried over P_2O_5. The salt of quinine with (d-VIII) was obtained. M.p. 140-144°; $[\alpha]_D^{20}$ -72° (c 0.5% in alcohol). By the treatment of both salts with HBr in CH_3COOH, (d-II) was obtained in 33% yield, m.p. 296-299° (decomposition); $[\alpha]_D^{20}$ -20° (c 0.5% in 5N HCl) and (*l*-II) in 50% yield, m.p. 294-297° (decomposition), $[\alpha]_D^{20}$ + 24° (c 0.5% in 5N HCl).

The authors express deep appreciation to E. S. Golovchinskaya for assistance rendered during the carrying out of this work.

* If the weight of the precipitate exceeded 5 g, it was purified by boiling with 20 times its volume of alcohol.
† As the isomers (d-VIII) and *l*-VIII) are very poorly soluble substances, a precise determination of their $[\alpha]_D$ was not always possible.

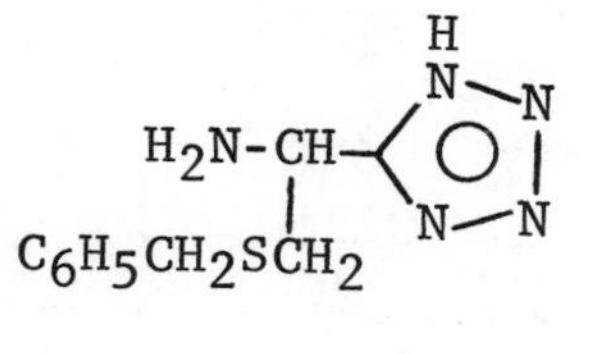

$C_{10}H_{13}N_5S$ Ref. 708

For a resolution procedure for the above compound, see Ref. 21.

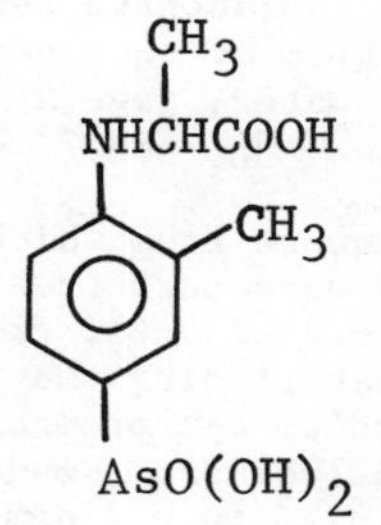

$C_{10}H_{14}AsNO_5$ Ref. 709

Resolution of dl-N-*Phenylalanineamide-4-arsinic Acid* (compare Gibson, Johnson, and Levin, *loc. cit.*).—The amide-acid (43·2 g.) was dissolved in sodium hydroxide solution (83·8 c.c., 1·79*N*) and water (167 c.c.). To the boiling solution, nor-*d*-ψ-ephedrine sulphate (15 g.) was added, a clear solution being obtained from which the *lAdB* salt crystallised in colourless soft needles after standing for 4 hours in the ice chest. The *salt* (24 g.) was recrystallised from water (225 c.c.) and obtained in colourless well-formed prisms, m. p.

217—223° (decomp.). The rotatory power was not increased by further recrystallisation (Found : N, 9·45; As, 17·4. $C_{18}H_{25}O_5N_3As$ requires N, 9·55; As, 17·1%). In aqueous solution, $c = 1{\cdot}0790$, $\alpha = +0{\cdot}39°$, $[\alpha] = +8{\cdot}97°$.

l-N-Phenylalanineamide-4-arsinic acid was liberated from the above *lAdB* salt in the usual manner, 2·4 g. being obtained from 4 g. of salt. It was optically pure after one crystallisation; as sodium salt in water, $[\alpha] = -18{\cdot}6°$ ($c = 0{\cdot}7804$). It had m. p. 247° (decomp.).

From the mother-liquor from the above *lAdB* salt the optically impure *d-N*-phenylalanineamide-4-arsinic acid was obtained in the manner described above. After one crystallisation from water, it had, as sodium salt in water, $[\alpha] = +9{\cdot}82°$, and further recrystallisation from water seemed to have little or no effect on the rotatory power.

The above optically impure *d-N*-phenylalanineamide-4-arsinic acid (6 g.) was dissolved in sodium hydroxide solution (8·71 c.c., 1·79*N*), and nor-*d*-ψ-ephedrine (0·793 g.—to neutralise the calculated amount of *l*-acid present) added. On warming, a homogeneous solution was obtained, which was kept for 20 hours at the ordinary temperature. The crystalline precipitate (pure *lAdB* salt) was filtered off, and the mother-liquor worked up in the usual manner. The crude acid, as sodium salt, had $[\alpha] = +13{\cdot}4°$. After four crystallisations from water this acid, as sodium salt in aqueous solution ($c = 0{\cdot}3271$), had $[\alpha] = +17{\cdot}12°$, approximating to the specific rotatory power of the sodium salt of pure *d-N*-phenylalanine-amide-4-arsinic acid.

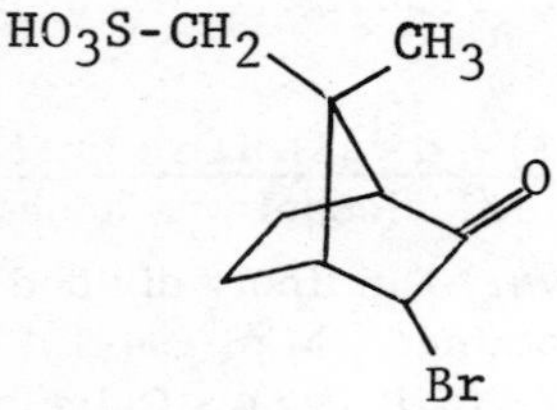

$C_{10}H_{14}BrO_4S$ Ref. 710

An aqueous solution was prepared containing d-pavine (one equivalent), ammonia (one equivalent), and dl-α-bromocamphor-π-sulphonic acid; a clear solution is obtained on boiling with addition of a little alcohol, and, on cooling, the greater part of the alkaloid separates as the crystalline d-pavine l-α-bromocamphor-π-sulphonate. After recrystallisation from water, the latter salt was treated with ammonia, and the resulting ammonium l-α-bromocamphor-π-sulphonate separated by crystallisation; 0.1080 gram of this ammonium salt, made up to 20 c.c. with water, gave α_D -0.915° in a 2-dcm. tube at 22°, whence $[\alpha]_D$ -84.7°. ...

The mother liquors from which the d-pavine salt had been separated were next treated with ammonia, filtered, and evaporated until crystallisation occurred; on recrystallising the residue from aqueous alcohol, pure ammonium d-α-bromocamphor-π-sulphonate was obtained. 0.1082 gram, made up to 20 c.c. with water, gave α_D +0.915° in a 2-dcm. tube at 22°, whence $[\alpha]_D$ +84.6°, which is the value ordinally assigned to pure ammonium d-α-bromocamphor-π-sulphonate.

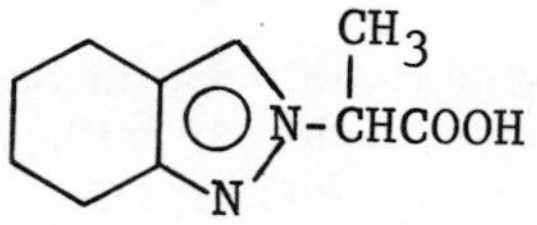

$C_{10}H_{14}N_2O_2$ Ref. 711

Preliminary tests on the resolution of α(4,5,6,7 tetrahydroindazolyl-2)propionic acid. Preliminary experiments were performed with the common alkaloids and some synthetic bases. Acid (0.001 mole) and base (0.001 mole) were dissolved in small amounts of solvent, and the salt obtained was filtered off and dried. The acid was then liberated and the rotatory power determined in absolute ethanol. In most cases, only oily products were obtained, but (−)α(2-naphthyl)-ethylamine and l-ephedrine yielded salts from ethyl acetate with $[\alpha]_D^{25} = -2°$ and $+2°$, respectively, in dimethyl-sulfoxide. The best results were obtained with cinchonidine in acetone-ethyl acetate yielding both enantiomers with $[\alpha]_D^{25} \approx \pm 10°$ (ethanol). The resolution was therefore performed with cinchonidine and with acetone-ethyl acetate as solvent.

(+)*α(4,5,6,7-Tetrahydroindazolyl-2)propionic acid.* 14.5 g (0.075 mole) acid and 22.2 g (0.075 mole) cinchonidine were dissolved in a mixture of 680 ml hot acetone and 340 ml ethyl acetate. The salt obtained was filtered off after 24 hours at room temperature and recrystallized several times from acetone-ethyl acetate (2:1) until the optical activity of the liberated acid remained constant. After each recrystallization the acid was liberated from 0.12 g of the salt and the activity of the acid was determined in absolute ethanol. The results are given in Table 1.

Table 1

Crystallization	1	2	3	4	5	6
Acetone-ethyl acetate (2:1), ml	1020	750	540	480	270	180
g salt obtained	16.8	10.0	7.1	4.4	3.1	2.0
$[\alpha]_D^{25}$ of the acid	+5.7°	+6.0°	+11.5°	+11.7°	+11.8°	+11.8°
$[\alpha]_{365}^{25}$ of the acid	+22.7°	+23.0°	+41.8°	+43.3°	+45.3°	+45.4°

Table 2

Recrystallization	1	2	3	4	5
Acetone-ethyl acetate (2:1), ml	360	300	240	180	120
g salt obtained	10.2	8.5	7.2	6.0	5.3
$[\alpha]_D^{25}$ of the acid	−10.3°	−10.9°	−11.8°	−11.8°	−11.6°
$[\alpha]_{365}^{25}$ of the acid	−39.4°	−43.0°	−45.3°	−45.8°	−46.0°

The acid was liberated from the cinchonidine salt by the addition of 5 ml 2 *M* sodium hydroxide and 25 ml water. The precipitated cinchonidine was filtered off and the alkaline solution was acidified to pH 3.5 with concentrated hydrochloric acid. After standing in a refrigerator over night the acid was filtered off and recrystallized from 24 ml ethanol-water (2:1). 0.65 g of pure acid, forming white glistening plates with m.p. 204°C was obtained.

0.1008 g acid dissolved in *chloroform* to 10.00 ml; tube length 1 dm: $\alpha_D^{25} = +0.367°$, $\alpha_{365}^{25} = +1.249°$, $[\alpha]_D^{25} = +36.4°$, $[\alpha]_{365}^{25} = +123.9°$, $[M]_D^{25} = +70.8°$, $[M]_{365}^{25} = +241°$.

(−)*α(4,5,6,7-Tetrahydroindazolyl-2)propionic acid.* The mother liquor from the first crystallization was evaporated to half of its volume. After 24 hours in a refrigerator, 12.2 g salt was isolated. A sample of the salt taken yielded acid having $[\alpha]_D^{25}$ −9.7° and $[\alpha]_{365}^{25}$ −37° in absolute ethanol. The salt was allowed to crystallize from acetone-ethyl acetate (2:1) several times; first at room temperature, then in a refrigerator over night and after that the salt was filtered off. After five such recrystallizations the maximum activity of the acid was obtained (Table 2). The course of the resolution was followed as described above.

Table 3

g = g acid dissolved in 10.00 ml solution; the tube length = 1 dm.

Solvent	g	α_D^{23}	α_{365}^{23}	$[\alpha]_D^{25}$	$[\alpha]_{365}^{25}$	$[M]_D^{23}$	$[M]_{365}^{25}$
DMF	0.1014	+0.167°	+0.526°	+16.5°	+51.9°	+32.0°	+108°
DMSO	0.1020	+0.055°	+0.116°	+5.4°	+11.4°	+10.5°	+22.1°
Glacial acetic acid	0.1003	+0.009°	+0.054°	+0.9°	+5.4°	+1.7°	+10.5°
Methanol	0.1036	−0.066°	−0.312°	−6.4°	−30.1°	−12.4°	−58.5°
Ethanol (Abs.)	0.1007	−0.121°	−0.475°	−12.0°	−47.2°	−23.3°	−91.6°
Chloroform	0.1004	−0.360°	−1.227°	−35.9°	−122°	−69.6°	−237°

The acid was isolated as above and recrystallized from ethanol-water (2:1). 1.35 g pure acid with m.p. 204° was obtained. The rotatory power of the (−) acid in various solvents is given in Table 3.

$C_{10}H_{14}N_2O_3S$

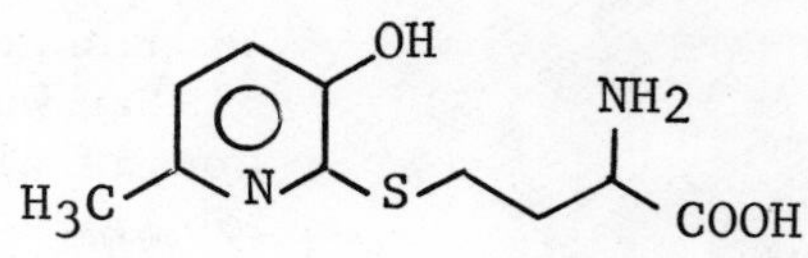

Ref. 712

2-Formylamino-4-(3-hydroxy-6-methyl-2-pyridylthio)butyric acid (XI). Acetic anhydride (21 ml) was added dropwise with stirring to an ice-cold solution of 2-amino-4-(3-hydroxy-6-methyl-2-pyridylthio) butyric acid (7.26 g, 0.3 mol) in formic acid (100 ml). The resultant solution was left at room temperature overnight, evaporated at reduced pressure, redissolved in boiling ethanol and treated with charcoal. A hygroscopic solid was obtained which was recrystallized from ethanol; yield 7.0 g (86 %), m.p. *ca.* 98°C. The molecular ion ($C_{11}H_{14}N_2O_4S$) on mass spectrometry corresponds to the title compound. The hygroscopic nature of the substance made elementary analyses difficult but chromatography shows a homogeneous product. NMR in TFA: 7.6 τ ($2H^3$); 7.2 τ (pyridyl-CH_3); 6.5 τ ($2H^4$); 4.9 τ (H^2); 2.0 and 2.3 τ (pyridyl-H^4, H^5).

Optical resolution of the 2-formylamino derivative (XI). XI (8.1 g, 0.03 mol) was dissolved in methanol (20 ml) and added to a solution of brucine. $4H_2O$ (14.0 g, 0.03 mol) in methanol (30 ml) containing water (3 ml). On cooling the warm solution, 14.3 g of salt was precipitated, $[\alpha]_D^{25} = -15.4°$. The specific sodium D-line rotation for the racemic salt was $-18°$. All the rotations were measured as 1 % solution in N HCl at 25°C. The salt was redissolved in methanol (50 ml) and water (9.5 ml) by heating. Two well defined crystal modifications were formed which were separated mechanically. The major part consisted of fine needles (7.3 g), $[\alpha]_D^{25} = -14.3°$. The cubic crystals (1.8 g) had $[\alpha]_D = -21.5°$. The major product was recrystallized from methanol (65 ml) and water (15 ml) which gave $[\alpha]_D^{25} = -12.8°$. Two more recrystallisations hardly changed the rotation which was $-12.7°$; final yield 4.4 g. This salt was dissolved in 1 N ammonia (130 ml), the brucine extracted into chloroform, and the solution evaporated to yield 2.0 g of the title compound shown to be homogeneous by chromatography; $[\alpha]_D^{25} = +11.5°$ ($c = 1$ in N HCl).

(R)-*2-Amino-4-(3-hydroxy-6-methyl-2-pyridylthio)butyric acid (X)*. The *N*-formyl derivative (XI) (2.0 g) was heated in 1 N HCl for 2 h and the pH of the cold solution adjusted to 5.5 with sodium bicarbonate. The title compound was slowly precipitated; yield 1.3 g (73 %), $[\alpha]_D^{25} = -18.0°$ ($c = 1$ in N HCl).

$C_{10}H_{14}O_2$

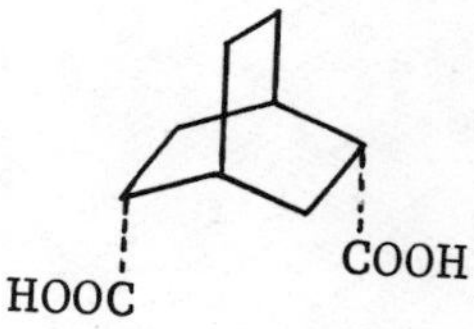

Ref. 713

(−)-*endo,endo*-Bicyclo[2,2,2]octane-2,5-dicarboxylic Acid

To a warm saturated ethanolic solution of the racemic *endo,endo*-bicyclo[2,2,2]octane-2,5-dicarboxylic acid (50 g) warm saturated ethanolic solution of brucine (100 g) was added. After cooling, the precipitated salt was filtered off, washed with cold ethanol and ether and dried. Two crystallisations from water (about 5 l) gave 41.5 g of the pure brucine salt, $[\alpha]_D^{25}$ −118·3° (0·5, dimethylformamide). The mother liquors afforded further 1·5 g of the pure salt, increasing thus the yield to 55%. For $C_{33}H_{40}N_2O_8.2\,H_2O$ (624·7) calculated: 63·04% C, 7·05% H, 4·46% N; found: 63·11% C, 7·27% H, 4·61% N. The salt (42·9 g) was decomposed with dilute hydrochloric acid, the product was extracted repeatedly with ether, the ethereal layer was washed several times with small amount of water and dried. Evaporation of the solvent afforded 13·0 g of the desired acid, m.p. 198—199°C (water), $[\alpha]_D^{25}$ −125° (*c* 0·50, methanol). For $C_{10}H_{14}O_4$ (198·2) calculated: 60·59% C, 7·12% H; found: 60·32% C, 7·07% H.

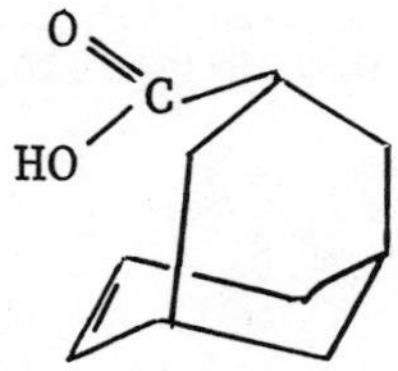

$C_{10}H_{14}O_2$

Ref. 714

Note:

Resolution of the endo-bicyclo [3.3.1]non-6-en-3-carboxylic acid, I, was achieved using (+)-dehydroabiethylamine. The diastereomeric salts could be separated by fractional crystallisation from 96% ethanol. Decomposition of the salt, $[\alpha]_{578}^{22} = +71.0°$, gave optically active acid (+) I, $[\alpha]_{578}^{22} = 115.9°$.

Analytical separation by high pressure liquid chromatography of the mixture of diastereomeric amides, prepared using the acylchloride of acid (+) I and (−)-α-phenylethylamine, showed that the enantiomeric excess of (+) I, $[\alpha]_{578}^{22} = +115.9°$ was 77 ± 3%.

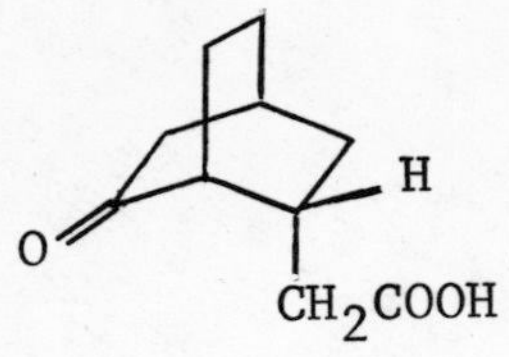

$C_{10}H_{14}O_3$ Ref. 715

Resolution via cinchonidine afforded the active acid with m.p. 81-81.5°, $[\alpha]_D^{25}$ -37.7° (c. 0.850 in EtOH).

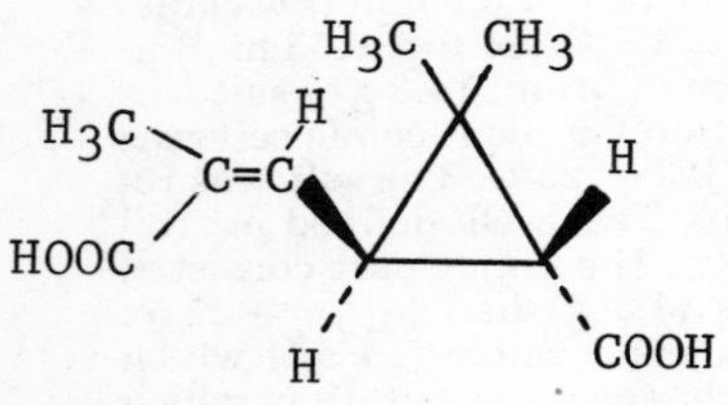

$C_{10}H_{14}O_4$ Ref. 716

The partial resolution of (±)-*trans*-3-(*trans*-2′-carboxypropenyl)-2,2-dimethylcyclopropane-1-carboxylic acid[1] by means of a synthetic optical active (−)-α-phenylethylamine to give the dextrorotatory acid, which was completely identical with the naturally derived acid, has already been descrived by the authors, and here, the complete resolution was achieved by obtaining the enantiomorphic laevorotatory acid with the results summarized below (rotations in ethanol).

	m.p.	$[\alpha]_D^{11}$
(+)-acid	163–4°	+70.9°
(±)-acid	206–8°	0
(−)-acid	163–4°	−70.5°

The combined filtrate from (−)-base (+)-acid salt[1] was evaporated, decomposed with dilute sulphuric acid and recovered by extraction. To a boiling solution of the recovered acid (3.4 g) in 80 ml of methanol, was added 2.1 g of (+)-α-phenylethylamine, having $[\alpha]_D^{11}$+39.4°, dissolved in 15 ml of methanol. After the solution was kept overnight the first crop (1.4 g) was obtained and the mother liquor concentrated, this procedure being repeated so that four crops were obtained, the rotations of which varied in a range from −15° to +3°. After successive recrystallizations and by combining crops with approximately the same rotation and crystallizing them as many times as before, there was finally obtained a pure (+)-α-phenylethylamine salt of the (−)-acid (0.6 g) m.p. 223–4°, $[\alpha]_D^{11}$ −28.0° (*c*, 1.1, methanol) (*Anal.* Found: C, 67.93; H, 8.08; N, 4.28, Calcd. for $C_{10}H_{14}O_4 \cdot C_8H_{11}N$: C, 67.69; H, 7.89; N, 4.39) Further recrystallization did not alter these values. The salt (0.5 g) was decomposed by pouring an aqueous solution into a dilute sulphuric acid, and extraction and recrystallization gave the pure (−)-acid (0.26 g) m.p. 163–4°, $[\alpha]_D^{11}$ −70.5° (*c*, 1.0, ethanol) (*Anal.* Found: C, 60.87; H, 7.28, Calcd. for $C_{10}H_{14}O_4$: C, 60.59; H, 7.12) *Di-p-phenylphenacylester* m.p. 151–2°, $[\alpha]_D^{11}$ −123.0° (*c*, 0.91, chloroform) (*Anal.* Found: C, 77.97; H, 6.05, Calcd. for $C_{38}H_{34}O_6$: C, 77.79; H, 5.84).

1) Y. Inouye and M. Ohno, Bull. Inst. Chem. Res. Kyoto Univ., 34, 90 (1956). cf. Y. Inouye et al. This Bulletin 19, 193 (1955), Botyu-Kagaku, 20, 102 (1955).

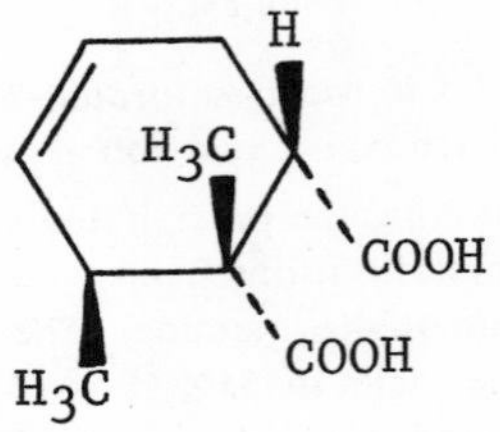

$C_{10}H_{14}O_4$ Ref. 717

Note:

Resolved with α-methylbenzylamine. An acid was obtained with $[\alpha]_D^{25}$ -110.1° (c 98.2, ethanol).

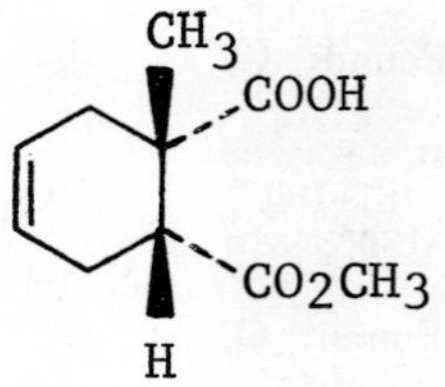

$C_{10}H_{14}O_4$

Ref. 718

Endo-bornyl amine hydrochloride (186.6 g, 0.995 mol) was dissolved in one liter of water and made basic by the addition of a 15% sodium hydroxide solution. The basic solution was extracted with ether (3 x 700 ml). The ether extract was washed with water, a saturated salt solution, and dried (Na_2SO_4). The solution was concentrated by distillation of most of the ether.

The half-acid-ester (197 g, 0.995 mol) was dissolved in the minimum amount of warm ether and added to the amine solution with stirring. The salt immediately precipitated out and was collected by suction filtration. The salt was recrystallized three times from decreasing amounts of warm absolute ethanol beginning with 2000 ml to yield 65.0 g (22%) of highly crystalling material, mp 159-161°, $[\alpha]_D$ = +3.4° (c, 10.8, abs. EtOH), whose melting point or optical rotation could not be improved by a further recrystallization.

The salt was treated with 2N hydrochloric acid and extracted with ether (5 x 200 ml); the extract was washed with a saturated salt solution, and dried ($CaSO_4$). The ether was removed in vacuo to yield 32.7 g of optically active half-acid-ester, mp 86-91°, $[\alpha]_D$ = -14.1° (c, 36.0, abs. EtOH).

That essentially complete resolution had been achieved was established by conversion of the half-acid-ester to its acid chloride (oxalyl chloride, C_6H_6, 0°) and subsequent reaction with d-(+)-α-methylbenzylamine. The resulting amide, $[\alpha]_D$ = +66.5° (c, 18.8, abs. EtOH) exhibited a single methyl ester absorption (δ 3.59) with no sign of the second signal which is present (ratio 1:1) when racemic half-acid-ester is similarly treated (δ 3.59 and 3.66).

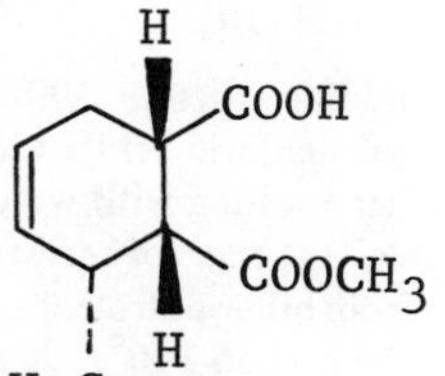

$C_{10}H_{14}O_4$

Ref. 719

Resolution of the half-ester 6. A soln of 6 (19·8 g; 0·1 mol) in 100 ml warm EtOAc was treated with 31·4 g (0·1 mol + 10%) of dehydroabietylamine (90 + %). When seeded, the desired salt separated in a pure condition almost immediately. After standing 18 h there was obtained 20·2 g of the DAA salt (*nat* series), m.p. 163–164·5° (80%); $[\alpha]_D^{MeOH}$ − 16·1°. (Found: C, 74·67; H, 9·23; N, 2·75. Calc. for $C_{30}H_{45}NO_4$: C, 74·53; H, 9·31; N, 2·90%). In another run 64 g of 6 in 250 ml acetone treated with 105 g dehydroabietylamine afforded 60 g, after crystallization from $CHCl_3$-acetone, m.p. 163–164° of the desired DAA salt. The mother liquors after due processing and repeated recrystallization from acetone gave 16 g (20%) of diastereomeric salt, m.p. 141–143°; $[\alpha]_D^{MeOH}$ + 65·32°. (Found: C, 74·83; H, 9·56; N, 2·70%).

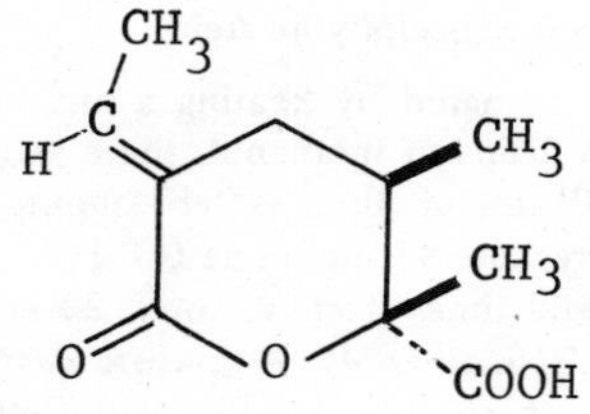

$C_{10}H_{14}O_4$

Ref. 720

Resolution of (±) VIIIb.—To a hot solution of 151.5 mg of brucine in 3.0 ml of ethyl acetate there was added 76 mg of (±) VIIIb. After cooling to room temperature, the solution was kept in a refrigerator for 24 hr. The colorless crystals, 93.9 mg, mp 194–197°, were filtered and washed with ethyl acetate. Recrystallization from ethyl acetate gave 91.8 mg of crystals, mp 198–200°, $[\alpha]_D$ +21.1° (*c* 1.40, ethanol).

Anal. Calcd for $C_{33}H_{40}N_2O_8$: C, 66.87; H, 6.80; N, 4.73. Found: C, 66.63; H, 6.88; N, 4.78.

The above filtrate was evaporated to dryness under vacuum and the residue was recrystallized first from ethyl acetate–ligroin (bp 60–90°) and then from acetone–ligroin to give 71.3 mg of colorless crystals, mp 187–189°, $[\alpha]_D$ −24.8° (*c* 1.98, ethanol). A mixture melting point determination with the above (+)-brucine salt was 175–180°.

Anal. Calcd for $C_{33}H_{40}N_2O_8$: C, 66.87; H, 6.80; N, 4.73. Found: C, 66.45; H, 6.88; N, 4.96.

(+) **VIIIb.**—To 5.0 ml of 5% sulfuric acid there was added 91.0 mg of the above (+)-brucine salt (mp 198–200°). The solution was extracted five times with ether and after washing with water and drying over sodium sulfate, the ether was evaporated under vacuum. The residue (27.9 mg) was recrystallized from ether–petroleum ether to give 27.1 mg of colorless crystals, mp 134–136°, $[\alpha]_D$ +10.8° (*c* 1.19 ethanol).

Anal. Calcd for $C_{10}H_{14}O_4$: C, 60.59; H, 7.12. Found: C, 60.58; H, 6.96.

(−) **VIIIb**.—This compound was prepared in the same way as (+) VIIIb from 37.9 mg of the (−)-brucine salt (mp 187–189°) and 7.5 mg of colorless crystals was obtained, mp 134–136°, $[\alpha]_D$ −10.0° (*c* 2.40, ethanol).

Anal. Calcd for $C_{10}H_{14}O_4$: C, 60.59; H, 7.12. Found: C, 60.41; H, 7.03.

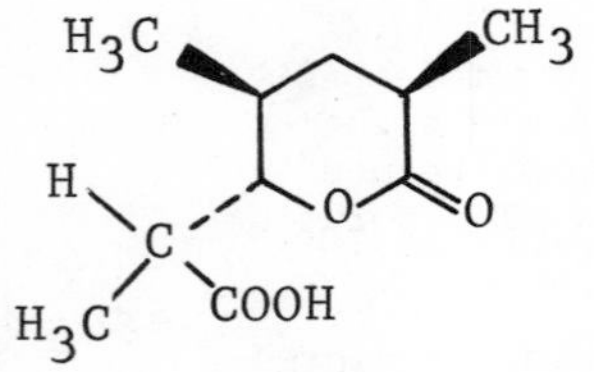

$C_{10}H_{14}O_4$ **Ref. 721**

Note:
Resolution of the racemic acid with (-)-D-threo-2-amino-1-(4-methylthiophenyl)-1,3-propanediol gave the active acid, m.p. 123-125° $[\alpha]_D^{25}$ +43.3° (c 2.40, $CHCl_3$).

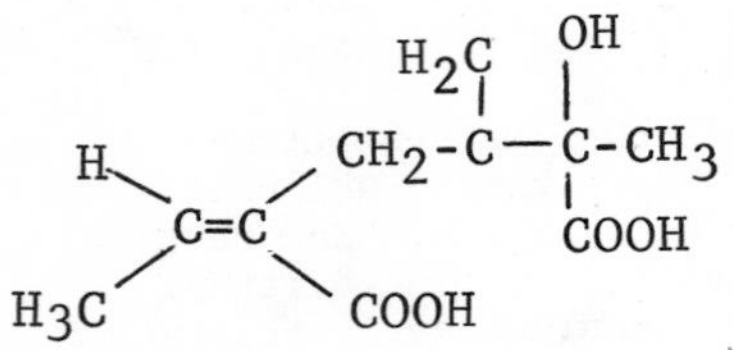

$C_{10}H_{14}O_5$ **Ref. 722**

Resolution of (±)-*cis*-Seneciphyllic Acid.

To a solution prepared by heating a mixture of 41.2 mg. of cinchonidine in 8 drops of methanol, there was added 0.7 ml. of ethyl acetate and 0.6 ml. of ether. After standing in the refrigerator for 6 days, the crystals, 43 mg. m.p. 160-165°, were filtered and recrystallized several times from acetone, 22 mg., m.p. 165-167°; $[\alpha]_D^{27°}$ -92° (c. 1.016, ethanol). A mixture melting point with the cinchonidine salt of *cis*-seneciphyllic acid (m.p. 165-167°) was undepressed.

Anal. Calcd. for $C_{29}H_{36}N_2O_6 \cdot CH_3COCH_3$: C, 67.82; H, 7.47; N, 4.94. Found: C, 68.06; H, 7.46; N, 5.01.

The above filtrate was evaporated to dryness and the residue dissolved in 20 ml. of ether by refluxing. This solution was condensed to 2 ml. and placed in a refrigerator to give 25 mg. of crystals. After three recrystallizations from water, the melting point was 151-153°. On standing *in vacuo* for 24 hours, the melting point was 184-187° and on recrystallization from acetone, 192-194°; $[\alpha]^{26°}$ -87° (c. 0.937, ethanol).

Anal. Calcd. for $C_{29}H_{36}N_2O_6$: N, 5.51. Found: N, 5.27.

cis-Seneciphyllic Acid (II).

(a) From 14 mg. of the cinchonidine salt, m.p. 165-167°, there was obtained in the same way as from the *trans*-cinchonidine salts, 5.2 mg. of crude *cis*-acid. Recrystallization from ether-petroleum ether, 30-60°, gave colorless crystals, m.p. 114-116°. A mixture melting point with *cis*-seneciphyllic acid (m.p. 114-115°) was undepressed. The infrared spectra (potassium bromide) were identical.

(b) From 20 mg. of the cinchonidine salt, m.p. 190-192°, there was in the same way obtained 4.2 mg. of the crude *cis*-acid. Recrystallization from ether-petroleum ether gave crystals, m.p. 112-114°. A 1:1 mixture of these two enantiomeric acids gave on crystallization the racemate, m.p. 94-96°.

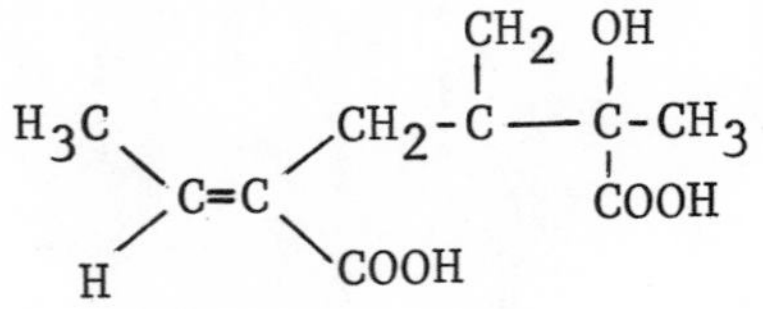

Ref. 723

$C_{10}H_{14}O_5$

Resolution of (±)-*trans*-Seneciphyllic Acid.

A mixture of 0.16 g. (0.75 mmole) of synthetic (±)-*trans*-seneciphyllic acid and 0.22 g. (0.75 mmole) of cinchonidine was dissolved in 25 drops of methanol and heated to the boiling point and 4.8 ml. of ethyl acetate were added. After thorough stirring and heating, the solution was placed in a refrigerator and after 2 days the crystals were filtered and washed with ethyl acetate, 230 mg., m.p. 181-184°. After 4 recrystallizations from methanol-ethyl acetate, the melting point was 199-200°, $[\alpha]_D^{27°}$, -80° (c. 1.067 ethanol). A mixture melting point with the cinchonidine salt (m.p. 196-198°) of *trans*-seneciphyllic acid was 198-199°. The infrared spectra (potassium bromide) of these two salts were identical.

Anal. Calcd. for $C_{29}H_{36}N_2O_6$: C, 68.48; H, 7.14; N, 5.51. Found: C, 68.53; H, 7.04; N, 5.23.

The filtrate from the above crystals (m.p. 181-184°) was evaporated to dryness under reduced pressure to give a residue of 140 mg. On recrystallization from acetone, 80 mg. of colorless crystals, m.p. 158-161°, were obtained and after three recrystallizations, the melting point was 161-162° (50 mg.); $[\alpha]_D^{27°}$ -68° (c. 0.801, ethanol).

Anal. Calcd. for $C_{29}H_{36}N_2O_6$: C, 68.48; H, 7.14; N, 5.51. Found: C, 68.49; H, 7.17; N, 5.35.

trans-Seneciphyllic Acid (III).

(a) The cinchonidine salt, m.p. 199-200° (40 mg.) was dissolved in 5 ml. of 5% hydrochloric acid and extracted three times with 4 ml. of ether. After washing with water and drying, the ether was evaporated to give 11 mg. of crystals, m.p. 144-146°. Two recrystallizations from ether-petroleum ether, 30-60°, gave 8 mg. of colorless crystals, m.p. 145-146°. A mixture melting point with seneciphyllic acid (m.p. 144-145°) was 145-146°. The infrared spectra (potassium bromide) were identical.

(b) In the same way, 20 mg. of the cinchonidine salt, m.p. 161-162°, gave 5 mg. of an acid, m.p. 144-145°; $[\alpha]_D^{25°}$ +9.0° (c. 1.021, ethanol). Equal amounts of this acid and the above enantiomer were dissolved in ether. After evaporation of the ether, the crystals had a m.p. 158-160°. A mixture melting point with the racemate (m.p. 161-162°) was 160-162°.

(±)-*cis*-Seneciphyllic Acid.

This racemate (45.9 mg.) was prepared from the (±)-*cis*-X (45 mg.) in the same way as the above *trans* racemate. After several recrystallizations from ether-petroleum ether, the colorless crystals had a m.p. 97-98°.

Anal. Calcd. for $C_{10}H_{14}O_5$: C, 56.07; H, 6.59. Found: C, 56.13; H, 6.60.

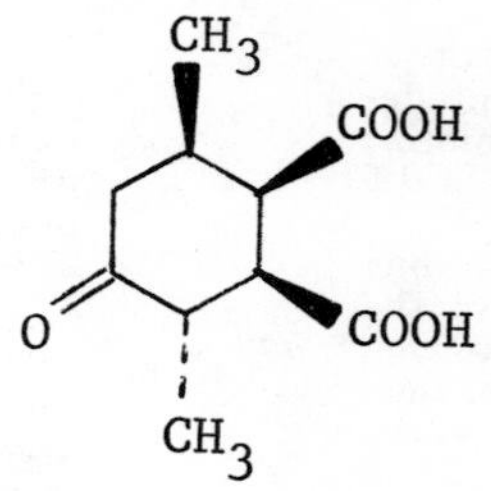

$C_{10}H_{14}O_5$

Ref. 724

Optical resolution of the acid VII. Preliminary experiments gave crystalline salt of the acid with strychnine when used in the molecular ratio 1:1. The acid (2.7 g) and strychnine (4.2 g) were dissolved in water (50 ml) at room temperature and the solution inoculated. The strychnine salt (2.1 g) which had separated during the night was decomposed by adding 10 ml N hydrochloric acid. After removing the strychnine hydrochloride by filtration the active acid was extracted with ether and was obtained as a crystalline mass (0.35 g) which melted at 195—199° and had $[\alpha]_D^{20} = -21.5°$ (water, *c* 3).

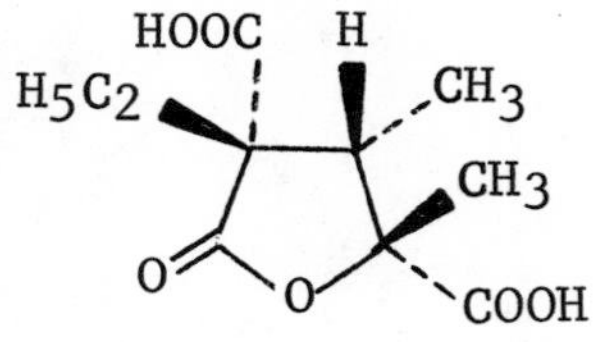

$C_{10}H_{14}O_6$

Ref. 725

Note:

Obtained an acid, m.p. 166-167°, $[\alpha]_D$ -45° using cinchonidine as resolving agent.

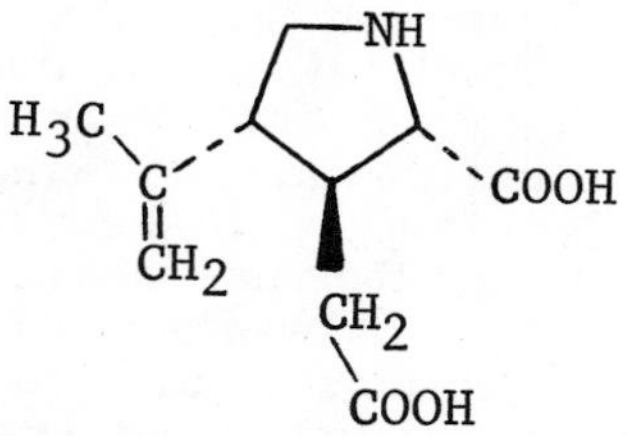

$C_{10}H_{15}NO_4$

Ref. 726

Note:

Resolved using (-)-ephedrine. Obtained acid with m.p. 238-242°C (decomp.) $[\alpha]_D^{20} = +7.7°C$ (c 1.3, H_2O).

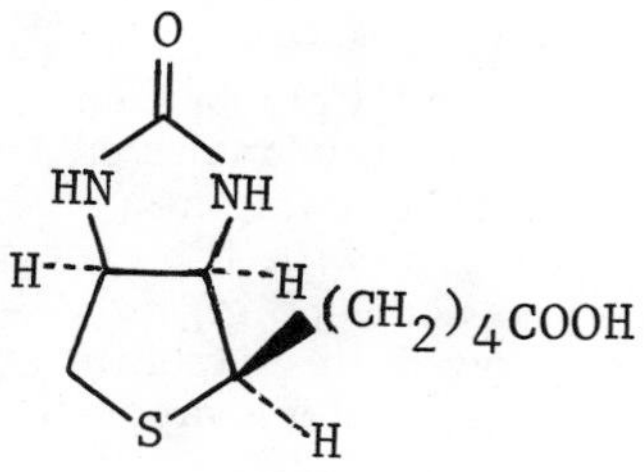

$C_{10}H_{16}N_2O_3S$

Ref. 727

Resolution of *d*(−)-Mandelic Acid Esters of *dl*-Biotin.—One and two-tenths grams of *dl*-biotin was treated with excess thionyl chloride at slightly below room temperature until all of it was in solution. The excess reagent was removed *in vacuo*. To the residual *dl*-biotin acid chloride, was added a hot solution of 0.9 g. of *d*(−)-mandelic acid in 16 ml. of chloroform. The mixture was agitated until all the residue had dissolved. After standing overnight, the solvent was removed from the ester, leaving it as a gum.

The crude *d*(−)-mandelic acid esters of *dl*-biotin were extracted five or six times with hot ethyl acetate. The ethyl acetate solution was concentrated to dryness *in vacuo* leaving a tan gum. This gum was extracted twice with hot water. The aqueous solution on standing deposited a partially crystalline precipitate. This was recrystallized by dissolving it in hot methanol and diluting the solution with water until it was slightly cloudy. Crystallization occurred giving an ester melting at 100–124°. Two further recrystallizations gave a product melting at 181–189° (micro-block). From 110 mg. of this crude ester, 67 mg. of crude biotin, m. p. 225–229° (micro-block), was obtained by hydrolysis with 5% aqueous sodium hydroxide at 70° for thirty minutes. After four recrystallizations from water, the biotin was pure and melted at 230–231° (micro-block); $[\alpha]^{25}D$ +90.7° (*c*, 2.04 in 0.1 *N* sodium hydroxide); yield 54 mg.

Anal. Calcd. for $C_{10}H_{16}N_2O_3S$: C, 49.16; H, 6.60; N, 11.46. Found: C, 49.12; H, 6.47; N, 11.23.

The *d*(−)-mandelic acid ester of biotin obtained from biotin of natural origin melted at 188–189° (micro-block).

Anal. Calcd. for $C_{18}H_{22}N_2O_5S$: N, 7.40. Found: N, 7.55.

The combined residual material from the resolution was hydrolyzed to the free acid and this in turn was converted to the *l*(+)-mandelic acid ester as described above. Fractional crystallization yielded an ester melting at 184–186° (micro-block). Hydrolysis of this ester gave *l*-biotin which after two recrystallizations from water melted at 229–230° (micro-block); $[\alpha]^{25}D$ −90.6° (*c*, 0.5 in 0.1 *N* sodium hydroxide).

Anal. Calcd. for $C_{10}H_{16}N_2O_3S$: C, 49.16; H, 6.60. Found: C, 49.52; H, 6.87.

$C_{10}H_{16}N_2O_3S$

Ref. 727a

Resolution of Quinidine Metho Salts of *dl*-Biotin.—One and seventeen-hundredths grams of *dl*-biotin was dissolved in 31.7 ml. of water containing one equivalent of quinidine methohydroxide. The solution was evaporated *in vacuo* and concentrated with methanol to remove residual water. The residue of quinidine metho salts was dissolved in dry methanol and diluted with dry acetone. A crystalline product, the quinidine metho salt of *l*-biotin, was obtained which was recrystallized from dry methanol–acetone; $[\alpha]^{25}D$ +122.7° (*c*, 1.94 in water); yield 0.92 g. The salt was dissolved in water and decomposed with 3 *N* hydrochloric acid to give *l*-biotin, m. p. 228–230° (micro-block); $[\alpha]^{25}D$ −90.5° (*c*, 1.78 in 0.1 *N* sodium hydroxide); yield 332 mg.

A second crop of crystalline quinidine metho salt of *l*-biotin was obtained which was recrystallized and decomposed as described above to yield 135 mg. of *l*-biotin; m. p. 228–231° (micro-block); $[\alpha]^{25}D$ −86° (*c*, 1.43 in 0.1 *N* sodium hydroxide).

The mother liquor from the second crop of quinidine metho salt when further diluted with acetone gave the quinidine metho salt of biotin as an oil. Decomposition of this product with aqueous hydrochloric acid gave crude biotin. Recrystallization from water purified the product, m. p. 229–231° (micro-block); $[\alpha]^{25}D$ +89.2° (*c*, 1.29 in 0.1 *N* sodium hydroxide); yield 112 mg.

The *l*-biotin was recrystallized from water and dried at 100° for analysis; m. p. 228–231°.

Anal. Calcd. for $C_{10}H_{16}N_2O_3S$: C, 49.16; H, 6.60; N, 11.46. Found: C, 49.21; H, 6.81; N, 11.25.

$C_{10}H_{16}N_2O_3S$ Ref. 727b

Resolution of *l*(+)-Arginine Salts of *dl*-Biotin.—A mixture of 1.44 g. of *dl*-biotin and 1.15 g. (10% excess) of *l*(+)-arginine was dissolved in 20 ml. of water and the solution was diluted with isopropyl alcohol. Crystallization was allowed to proceed in the refrigerator overnight. The crystals were collected on a filter and washed with acetone; m. p. 214–218° (micro-block); $[\alpha]^{25}D$ +49.09° (*c*, 1.039 in water); yield 1.13 g. (92%). The *l*(+)-arginine salt of biotin when pure had the following properties: m. p. 228–230° (micro-block); $[\alpha]^{25}D$ +59.9° (*c*, 1.37 in water).

Anal. Calcd. for $C_{16}H_{30}N_6O_5S$: C, 45.91; H, 7.23. Found: C, 46.25; H, 7.51.

One and nine-hundredths grams of the crude *l*(+)-arginine salt was recrystallized twice from aqueous isopropyl alcohol; $[\alpha]^{25}D$ +57.2° (*c*, 1.747 in water). The purified salt was dissolved in 10 ml. of water and acidified with dilute hydrochloric acid. The crystalline biotin was collected on a filter, washed with water and dried; m. p. 229–231° (micro-block); a mixture of this sample with biotin of natural origin melted without depression; $[\alpha]^{25}D$ +88.8° (*c*, 1.025 in 0.1 *N* sodium hydroxide); yield 0.51 g. (80%). Further purification was accomplished by suspending the biotin in 20 ml. of hot water, adding just enough dilute sodium hydroxide solution to dissolve the solid, then acidifying with hydrochloric acid. The pure crystalline biotin melted at 229–231°; $[\alpha]^{25}D$ +90.4° (*c*, 1.87 in 0.1 *N* sodium hydroxide); yield 0.44 g. (64% overall).

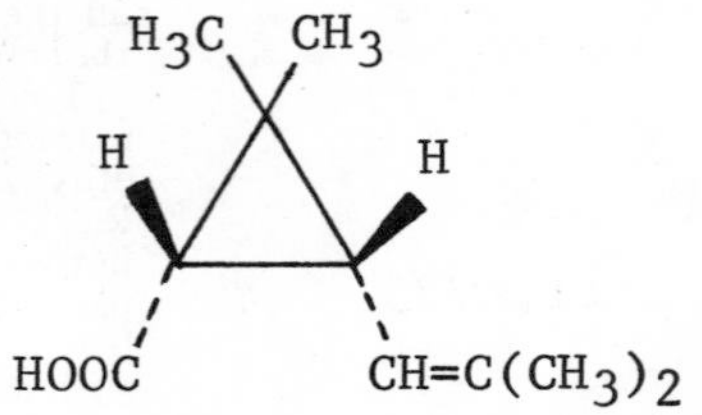

$C_{10}H_{16}O_2$ Ref. 728

Resolution of (±)-cis-*chrysanthemic acid*.—Quinine (32·4 g.) in ethanol (40 ml.) was added to (±)-*cis*-chrysanthemic acid (16·8 g.) in warm ethanol (60 ml.) and the mixture was set aside at room temperature for 3 hours. Filtration of the semi-solid mass of crystals then gave a first fraction of salt (F1, 35·1 g.), m.p. 105–117°, $[\alpha]_D$ − 105·3°. Addition of water (20 ml.) to the filtrate and setting aside overnight gave a second fraction (F2, 6·6 g.), m.p. 112–115°, $[\alpha]_D$ − 108·0°. Evaporation of the filtrate from F2 to 50 ml. gave a third fraction (F3, 4·3 g.), m.p. 109–119°, $[\alpha]_D$ − 104·5°. Further evaporation gave a fourth fraction which was so obviously impure that it was not characterized. By crystallization from 4 : 1-ethanol–water F1 was separated into three fractions : A (24·1 g.), $[\alpha]_D$ − 104·5° ; B (6·2 g.), $[\alpha]_D$ − 110·2° ; and C (4·8 g.), $[\alpha]_D$ − 108·5°. Recrystallization of A from the same solvent gave fractions X (20·4 g.), $[\alpha]_D$ − 103·6°, and Y (2·4 g.), $[\alpha]_D$ − 109·5°. X was pure (+)-cis-*chrysanthemic acid quinine salt*, for on a further crystallization it yielded a first fraction (19·5 g.), $[\alpha]_D$ − 103·5°, and a second fraction (0·8 g.) with the same specific rotation. The pure salt separated from aqueous ethanol in long needles, m.p. 127–135° (decomp.) ; this melting point was not sharpened by recrystallization from ethanol, ethyl acetate, or benzene and was probably due to hydration of the salt (Found : C, 70·7 ; H, 8·35. $C_{10}H_{16}O_2,C_{20}H_{24}O_2N_2,H_2O$ requires C, 70·55 ; H, 8·3%). Decomposition of this salt (19·0 g.) gave pure (+)-cis-*chrysanthemic acid* (6·0 g.), m.p. 40–42°, $[\alpha]_D^{22}$ + 40·8° (*c*, 1·775 in ethanol), $[\alpha]_D^{22}$ + 83·3° (*c*, 1·597 in chloroform). This acid was too soluble in ethyl acetate or in ethanol at − 60° to permit of recrystallization, but a portion was distilled, b.p. 95°/0·1 mm., and had m.p. 41–43°, $[\alpha]_D^{20}$ + 40·7° (*c*, 1·043 in ethanol), $[\alpha]_D^{20}$ + 83·3° (*c*, 1·034 in chloroform) (Found : C, 71·5 ; H, 9·3. $C_{10}H_{16}O_2$ requires C, 71·4 ; H, 9·6%).

Fractions F2, B, and Y were combined and this impure (−)-*cis*- acid quinine salt was crystallized repeatedly from aqueous ethanol, from ethyl acetate, and from benzene. However, the specific rotations of the various fractions remained substantially unaltered between − 108·7° and − 110·0°, so the collected fractions were decomposed to give impure (−)-*cis*-acid (5·1 g.), $[\alpha]_D$ − 25·5° (*c*, 1·525 in ethanol). All the residual fractions, with the exception of F3, were combined and decomposed to give further impure (−)-*cis*-acid (3·6 g.), $[\alpha]_D$ − 23·3°. When this acid (0·5 g.) was dissolved in ethanol (2 ml.) and allowed to evaporate at room temperature, (±)-*cis*-acid, m.p. 114–116°, crystallized and the filtrate after evaporation gave a specimen of acid with $[\alpha]_D$ − 33·6°.

The impure (−)-*cis*-acid (8·2 g.) and (+)-α-phenylethylamine (6·0 g., $[\alpha]_D$ + 34·8° in chloroform) were mixed in ethanol (46 ml.) and set aside for 1 hour. The first fraction of salt (9·1 g.) had m.p. 127–136°, $[\alpha]_D$ + 7·8°, and one crystallization gave pure (−)-cis-*chrysanthemic acid* (+)-*α-phenylethylamine salt* (6·2 g.), m.p. 134–136°, $[\alpha]_D$ + 7·6° (Found : C, 74·8 ; H, 9·4. $C_{10}H_{16}O_2,C_8H_{11}N$ requires C, 74·7 ; H, 9·4%). Decomposition of this salt gave pure (−)-cis-*chrysanthemic acid* (3·2 g.), m.p. 41–43°, $[\alpha]_D^{20}$ − 40·8° (*c*, 1·520 in ethanol), $[\alpha]_D^{19}$ −40·8° (*c*, 1·521 in ethanol), $[\alpha]_D^{19}$ −83·3°(*c*, 1·555 in chloroform) (Found : C, 71·1 ; H, 9·3. $C_{10}H_{16}O_2$ requires C, 71·4 ; H, 9·6%).

$C_{10}H_{16}O_2$ Ref. 728a

cis-Chrysanthemic acid and the derivatives

(+)-*cis-Chrysanthemic acid.* (±)-*cis*-Chrysanthemic acid (250 g) and (+)-1-phenyl-2-*p*-tolylethylamine (314.5 g) ($[\alpha]_D^{20}$ +12.3°, $CHCl_3$) were dissolved in methanol. Evaporation of methanol gave a salt, and this salt was dissolved in hot acetone and allowed to stand overnight at room temperature. Filtration gave a crude salt, and recrystallization from acetone 4 times gave 203 g of pure (+)-*cis*-chrysanthemic acid-(+)-1-phenyl-2-*p*-tolylethylamine salt. Mp 157~158°C, *Anal.* Found: C, 79.21; H, 8.72; N, 3.61. Calcd. for $C_{25}H_{33}O_2N$: C, 79.11; H, 8.76; N, 3.69%.

This salt was decomposed with 5% sodium hydroxide solution and after an optically active amine was extracted with ether, the aqueous solution was acidified with hydrochloric acid. The ether extract was washed with water, dried over magnesium sulfate, and evaporation of ether gave 88 g of pure (+)-*cis*-chrysanthemic acid. Mp 41~42°C. $[\alpha]_D^{19}$ +41.95° (*c*=4.839, EtOH). (lit.[6] $[\alpha]_D^{22}$ +40.8°, EtOH).

(−)-*cis-Chrysanthemic acid.* The filtrate of (+)-*cis*-chrysanthemic acid-(+)-1-phenyl-2-*p*-tolylethylamine salt yielded on decomposition crude (−)-*cis*-chrysanthemic acid. When this crude acid was allowed to stand for a week, oil and crystals were separated, and after crystals were removed, oil was distilled. Bp 110°C (1 mmHg). A small excess of ammonia water (28%) was added to this acid and its ammonium salt was recrystallized 3 times. This salt on decomposition with dilute hydrochloric acid gave pure (−)-*cis*-chrysanthemic acid. Mp 40.5~41.5°C. $[\alpha]_D^{20}$ −41.57° (*c*=4.782, EtOH). (lit.[6] $[\alpha]_D^{20}$ −40.8, EtOH).

6) I.G.M. Campbell and S.H. Harper, J. Sci. Food Agric., 3,189 (1952).

$C_{10}H_{16}O_2$ Ref. 728b

Resolution of (±)-*cis-chrysanthemic acid.* (±)-*cis*-Chrysanthemic acid (16.8 g) and L-lysine (14.6 g) were dissolved in hot methanol (80 ml) and allowed to stand at room temperature for 2 days. Filtration gave crude (+)-*cis*-acid L-lysine salt (12.7 g), mp 204~6°C, $[\alpha]_D^{25}$+18.4° (H_2O). Two recrystallization of the crude salt from 80% (v/v) aqueous methanol gave pure (+)-*cis*-chrysanthemic acid L-lysine salt (7.3 g), mp 205~6°C, $[\alpha]_D^{25}$+20.43° (*c*=3.990, H_2O), *Anal.* Found: C, 61.30; H, 9.45; N, 8.81, Calcd. for $C_{16}H_{30}O_4N_2$: C, 61.12; H, 9.62; N, 8.91%. Yield, 23.2% based on (±)-*cis*-acid. This salt was decomposed with dilute hydrochloric acid to give pure (+)-*cis*-acid (3.4 g), mp 42°C, $[\alpha]_D^{20}$+39.69° (*c*=3.901, EtOH), +82.80° (*c*=4.400, $CHCl_3$), (*lit.* mp 41~3°C, $[\alpha]_D$+40.7° (EtOH), +83.3° ($CHCl_3$)), which was found to be identical with (±)-*cis*-acid on IR spectrum in chloroform. Yield, 20.2% based on (±)-*cis*-acid.

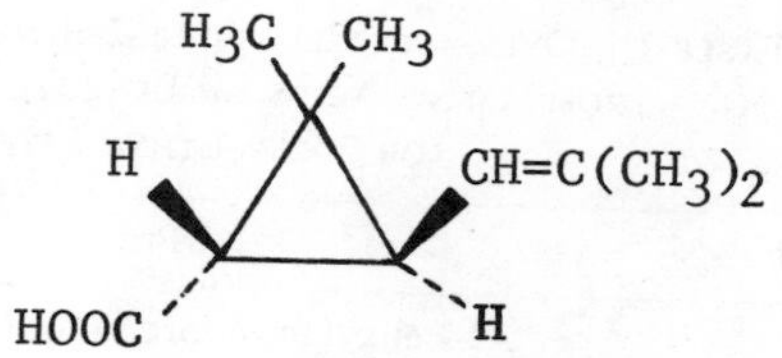

$C_{10}H_{16}O_2$ Ref. 729

Resolution of (±)-trans-*chrysanthemic acid.*—Quinine (32·4 g.) in warm ethanol (30 ml.) was added to a solution of (±)-*trans*-chrysanthemic acid (16·8 g.) in ethanol (25 ml.) and the mixture was filtered and set aside overnight. Filtration then gave a first fraction of salt (F1, 21·8 g.), m.p. 130–160°, $[\alpha]_D$ − 110·1°. Addition of water (25 ml.) to the filtrate from F1 gave, after 24 hours, a second fraction (F2, 19·0 g.), m.p. 80–125° (decomp.), $[\alpha]_D$ − 105·7°. Evaporation of the mother liquor of F2 to 10 ml. gave sticky material which on decomposition yielded *trans*-acid (2·8 g.), $[\alpha]_D$ + 0·8°. Two crystallizations of F1 gave pure (−)-*trans*-chrysanthemic acid quinine salt (12·7 g.), m.p. 159·5–161°, $[\alpha]_D$ − 115·4° (*c*, 1·018 in ethanol). Decomposition of this salt (10·0 g.) by shaking in ether with 2N-hydrochloric acid, washing with water, drying (Na_2SO_4), evaporating, and heating at 80° under reduced pressure gave pure (−)-*trans*-chrysanthemic acid (3·2 g.), whose specific rotations are recorded in Table I. Distillation, b.p. 99–100°/0·2–0·3 mm., gave acid of unchanged specific rotation. Both distilled and undistilled (−)-*trans*-acid crystallized on cooling to − 80° and allowing to regain room temperature. With a mercury-vapour lamp (though not exclusively 5461 Å.) (−)-*trans*-chrysanthemic acid had [α] − 15·3° (*c*, 2·780 in ethanol) and [α] − 27·3° (*c*, 2·930 in chloroform).

Decomposition of F2 gave impure (+)-*trans*-chrysanthemic acid (6·0 g.), $[\alpha]_D$ + 9·9° (*c*, 1·575 in ethanol). On mixing this acid (0·85 g.) with (+)-α-phenylethylamine (0·6 g.) in 50% aqueous ethanol (5 ml.), the salt separated as felted needles (0·5 g.), m.p. 124–126°, $[\alpha]_D$ + 14·4°. One crystallization gave pure (+)-trans-*chrysanthemic acid* (+)-*α-phenylethylamine salt* (0·35 g.), m.p. 126–128°, $[\alpha]_D$ + 14·4°. (−)-trans-*Chrysanthemic acid* (−)-*α-phenylethylamine salt*, prepared from the pure components, had m.p. 126–128°, $[\alpha]_D$ − 14·3°. The above impure (+)-*trans*-acid (5·0 g.) and (−)-α-phenylethylamine (3·6 g., $[\alpha]_D$ − 34·6° in chloroform) were mixed in 50% aqueous ethanol (25 ml.) and set aside overnight. The first fraction (7·0 g.) had $[\alpha]_D$ − 4·9° and after two crystallizations gave pure (+)-trans-*chrysanthemic acid (−)-α-phenylethylamine salt* (3·3 g.), m.p. 135–136°, $[\alpha]_D$ − 3·5° (Found: C, 74·9; H, 9·45. $C_{10}H_{16}O_2,C_8H_{11}N$ requires C, 74·7; H, 9·4%). Decomposition of this salt gave pure (+)-trans-*chrysanthemic acid* (1·8 g.), which crystallized on cooling to − 80° and allowing to regain room temperature, and whose specific rotations are recorded in Table I.

$C_{10}H_{16}O_2$ Ref. 729a

Resolution of (+)-*trans-chrysanthemic acid.* (±)-*trans*-Chrysanthemic acid (17.5 g) and L-lysine (15.6 g) were dissolved in hot methanol (185 ml) and stored in a refrigerator overnight. Filtration gave a first fraction of salt (F₁, 11.5 g), mp 201∼4°C, $[\alpha]_D^{25}$+11.3°(H_2O). The filtrate was evaporated to give crystalline mass (F_2, 20.7 g). Recrystallization of F_2 from 50 ml methanol gave crude salt (F_3, 4.3 g), $[\alpha]_D^{25}$+10.1° (H_2O). The filtrate was evaporated to give crude (−)-*trans*-acid L-lysine salt (15.4 g), which on decomposition yielded crude (−)-*trans*-acid (7.8 g), $[\alpha]_D^{25}$−10.1° (EtOH). Recrystallization of F_3 from 90% (v/v) aqueous methanol gave F_4 (2.6 g), mp 207∼8.5°C, $[\alpha]_D^{22}$+12.0° (H_2O). F_1 and F_4 were combined and recrystallized twice from 90% aqueous methanol to give pure (+)-*trans*-acid L-lysine salt (6.8 g), mp 205∼7°C, $[\alpha]_D^{22}$+12.26° (*c*=4.551, H_2O), *Anal.* Found: C, 60.69; H, 9.64; N, 8.93, Calcd. for $C_{16}H_{30}O_4N_2$: C, 61.12; H, 9.62; N, 8.91%. Yield, 20.8% based on (±)-*trans*-acid. The pure salt was decomposed with dilute hydrochloric acid to give pure (+)-*trans*-acid (3.0 g), mp 18∼9°C, $[\alpha]_D^{22}$+14.52° (*c*=5.008, EtOH), (*lit.*[2] $[\alpha]_D^{23}$+14.4° (EtOH)), which was found to be identical with (±)-*trans*-acid on IR spectrum in chloroform. Yield, 17.1% based on (±)-*trans*-acid.

$C_{10}H_{16}O_2$ Ref. 729b

Resolutions of (±)-*trans-chrysanthemic acid*

Resolution of (±)-*trans-chrysanthemic acid with* L-*2-benzylaminopropanol* (*I*). The amino alcohol (I) (8.3 g) and (±)-*trans*-chrysanthemic acid (8.4 g) were dissolved in 40% aq. EtOH (50 g) and stored in a refrigerator for 3 days. Filtration then gave a crude salt (4.3 g), mp 113∼5°C, which was recrystallized from 30% aq. EtOH (20 g) to yield the salt of (−)-*trans*-acid with the amino alcohol (I) (3.35 g, 40.2%), mp. 115.∼6.5°C, $[\alpha]_D^{24}$ +8.64° (*c*=3.31, EtOH). *Anal.* Found: C, 72.03; H, 9.29; N, 4.28. Calcd. for $C_{20}H_{31}O_3N$: C, 72.03; H, 9.37; H, 4.20%. The pure salt was decomposed with usual manner to give (−)-*trans*-chrysanthemic acid (1.64 g, 39.0%), $[\alpha]_D^{24}$ −14.14° (EtOH), −25.88° ($CHCl_3$).

TABLE I. OVER-ALL YIELDS OF L-2-BENZYLAMINO ALCOHOLS FROM AMINO ACIDS *via* BENZOYL DERIVATIVES OR THEIR ETHYL ESTER

Starting material	Product	Yield %[a]
L-Alanine	L-2-Benzylaminopropanol (I)	43.6
L-Valine	L-2-Benzylamino-3-methylbutanol (II)	66.1
L-Leucine	L-2-Benzylamino-4-methylpentanol (III)	46.7
L-Phenylalanine	L-2-Benzylamino-3-phenylpropanol (IV)	78.2
L-Aspartic acid	L-2-Benzylamino-1,4-butanediol (V)	41.6
L-Glutamic acid	L-2-Benzylamino-1,5-pentanediol (VI)	42.0
L-Lysine	L-2,6-Dibenzylaminohexanol (VII)	52.4

[a] Over-all yield from amino acid.

TABLE II. RESOLUTION OF (±)-*trans*-CHRYSANTHEMIC ACID WITH L-2-BENZYLAMINO ALCOHOLS

Resolving agent L-2-Benzylamino alcohol	Solvent	Product	Yield %	$[\alpha]_D$[a] in CHCL₃
(I)	40% aq. EtOH	(−)-Acid	39.0	−25.9°
(II)	Ether	(+)-Acid	16.0	+24.2°
(III)	isoPr. ether	(+)-Acid	38.6	+22.9°
(IV)	EtOH	(−)-Acid	31.7	−25.9°
(V)		Not capable to resolve		
(VI)		Not capable to resolve		
(VII)		Not capable to resolve		

[a] Lit.[5] $[\alpha]_D$ ($CHCl_3$) of (+)-*trans*-chrysanthemic acid +25.8°

$C_{10}H_{16}O_2$ Ref. 729c

Beispiel 1

16,8 g (±)-trans-Chrysanthemummonocarbonsäure wurden in 120 ml wasserfreiem Äthanol gelöst, und 17,1 g (−)-α-(1-Naphthyl)-äthylamin wurden zur Lösung zugefügt. Die Lösung wurde in einem Kühlschrank stehengelassen. Danach wurden die Kristalle durch Filtrieren gesammelt. Die Kristalle wurden aus 60 ml wasserfreiem Äthanol umkristallisiert, was 10,7 g weiße Kristalle vom F. = 144,5 bis 145,5°C und $[\alpha]_D^{24} = -19{,}90$ (Dimethylsulfoxid) ergab. Die Kristalle wurden in einer 10%igen wäßrigen NaOH-Lösung gelöst. Die ölige Schicht wurde mit Äther extrahiert, und es wurden 5,2 g (−)-α-(1-Naphthyl)-äthylamin gewonnen. Die wäßrige Schicht wurde mit Salzsäure angesäuert und mit Äther extrahiert, was 5,2 g (+)-trans-Chrysanthemummonocarbonsäure ergab. Die Ausbeute betrug 62%. $[\alpha]_D^{25} = +14{,}6$ (Äthanol). Aus der äthanolischen Mutterlauge wurden noch 11,1 g des Amines zurückgewonnen.

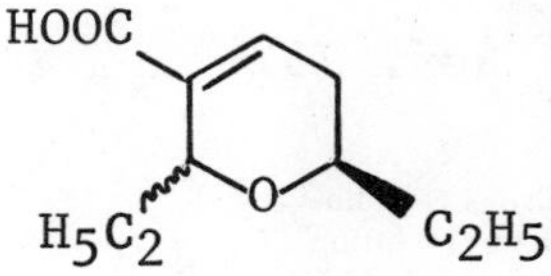

$C_{10}H_{16}O_3$ Ref. 730

Le mode opératoire est le suivant : dans 300^{cm^3} d'acétate de méthyle on dissout 49^g d'acide racémique et 52^g d'éphédrine dextrogyre. Après une nuit la solution se prend en une bouillie cristalline que l'on essore : on obtient 60^g de sel ($[\alpha]_D$ pour l'acide +43°). On redissout ces 60^g de sel dans 300^{cm^3} d'acétate de méthyle et l'on obtient 34^g de sel ($[\alpha]_D$ pour l'acide +155°). Après une nouvelle cristallisation dans 300^{cm^3} d'acétate de méthyle, on recueille 24^g de sel ($[\alpha]_D$ pour l'acide +211°). On fait une nouvelle cristallisation dans 200^{cm^3} d'acétate d'éthyle, ce qui donne 17^g de sel ($[\alpha]_D$ pour l'acide +230°). Enfin, une dernière cristallisation dans 200^{cm^3} d'acétate d'éthyle donne 15^g de sel ($[\alpha]_D$ pour l'acide +234°). Le sel d'éphédrine ainsi obtenu est dissous dans le minimum d'eau, puis additionné d'acide chlorhydrique jusqu'au virage au rouge congo. La solution est extraite à l'éther; la solution éthérée est séchée, puis évaporée à sec; le résidu obtenu est cristallisé dans l'eau. En utilisant l'éphédrine droite, on obtient de la sorte l'acide droit (F. 95°-96° bloc Maquenne), tandis qu'avec l'éphédrine gauche on obtient l'acide gauche.

Analyse de l'acide droit : Trouvé H % 7,60; C % 61,21. Calculé pour $C^8H^{12}O^3$, H % 7,7; C % 61,5.

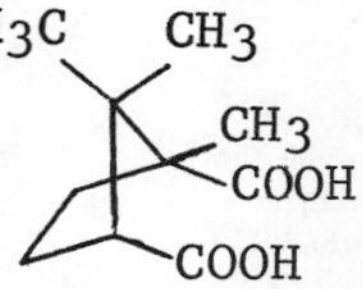

$C_{10}H_{16}O_4$ Ref. 731

Pendant les expériences préliminaires on a constaté que l'acide camphorique racémique forme des sels cristallisant bien avec les bases optiquement actives suivantes: avec la brucine il forme ùn sel acidique (aiguilles f. à 132—135°), au pouvoir rotatoire $[\alpha]_D^{20} = -43{,}75°$ ($c = 0{,}4$; $d = 2$; $\alpha = -0{,}35°$) dans l'éthanol; avec la phényléthylamine lévogyre — un sel acidique (aiguilles longues f. à 178—180°, au pouvoir rotatoire $[\alpha]_D^{20} = -2{,}50°$ ($c = 0{,}4$; $d = 2$; $\alpha = -0{,}02$) dans l'éthanol; avec la quinine — un sel neutre (prismes de l'acetone diluée) et avec la cynchonidine — un sel acidique (aiguilles de l'acétone dilué). Mais ce sont seulement les sels quininique et cynchonidinique qui peuvent être utilisés au dédoublage de l'acide comphorique racémique aux antipodes optiques. La quinine formait avec l'antipode lévogyre un sel moins solluble, qui dédoublait l'acide racémique trés énergiquement. C'est à la troisième cristallisation de l'acétone diluée qu'on pouvait déja obtenir le sel optiquement pur se présentant sous forme de prismes bien distincts, f. à 187—189°, au pouvoir rotatoire $[\alpha]_D^{20} = -150{,}00°$ ($c = 0{,}4$; $d = 2$; $\alpha = -1{,}20°$) dans l'éthanol.

En liant l'acide racémique avec la cynchonidine et faisant la cristallisation fractionnée aussi dans l'acétone dliluée, on obtenu, déja dans la troisième fraction, le sel complètement pur de l'acide dextrogyre, sous forme d'aiguilles bien distinctes f. à 200—202° dont le pouvoir rotatoire était $[\alpha]_D^{20} = -52{,}50°$ ($c = 0{,}4$; $d = 2$; $\alpha = -0{,}42°$) dans l'éthanol.

Les acides camphoriques libres, optiquement actifs, isolés des sels alcaloïdiques, après la cristallisation de l'eau, fondaient à 188,5—189,5° et possédaient l'activité optique $[\alpha]_D^{20} = +48,75°$ ($c = 0,4$, $d = 2$, $\alpha = +0,39°$) et $[\alpha]_D^{20} = -50,00°$ ($c = 0,4$, $d = 2$, $\alpha = -0,40°$) dans l'éthanol.

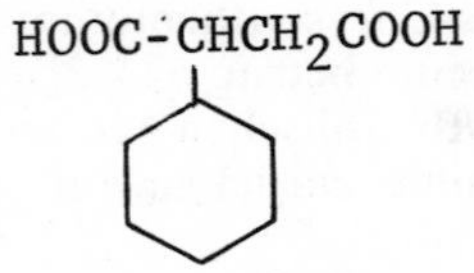

$C_{10}H_{16}O_4$ Ref. 732

Dédoublement optique de l'acide cyclohexylsuccinique

Par hydrogénation de l'acide phénylsuccinique droit, Naps et Johns ont obtenu un acide cyclohexylsuccinique de $[\alpha]_D = +38,3°$ (solution d'acétone), fondant à 95-96° (1). Le dédoublement de l'acide phénylsuccinique a été réalisé par Wren et Williams (2) en utilisant la brucine. Cependant, cette méthode est assez laborieuse quand il s'agit de quantités plus considérables. Nous avons donc essayé de trouver une méthode plus directe pour la préparation de l'acide cyclohexylsuccinique actif. La résolution du racémique est réalisée très facilement à l'aide de la strychnine (sel neutre) et de la quinine (sel acide). Le sel de cinchonidine et le sel neutre de quinine ont donné des résultats moins favorables, et le sel de brucine, cristallisant très bien, a donné un acide complètement inactif. Le pouvoir rotatoire des antipodes obtenus (+ 38.5° et —38.2°) est en accord avec les données de Naps et Johns; l'hydrogénation a donc lieu sans racémisation.

20 g (0,1 mole) de l'acide racémique sont dissous dans un mélange de 150 cm³ d'alcool et 500 cm³ d'eau. La solution est saturée de strychnine à chaud et filtrée. En 24 heures, 45,6 g du sel neutre de strychnine se sont déposés. Le sel est recristallisé plusieurs fois dans de l'alcool dilué. On a décomposé chaque fois un échantillon pesant environ 0,5 g et on a examiné le pouvoir rotatoire de l'acide en solution d'acétone :

Cristallis.	1	2	3	4	5
Eau cm³	500	200	120	100	80
Alcool cm³	150	100	60	50	40
Sel g	45,6	33,4	27,4	23,8	22,4
$[\alpha]_D^{25}$	+16,9°	+33,8°	+37,5°	+38,1°	+38,1°

L'eau-mère de la première cristallisation, saturée de strychnine à chaud, dépose une nouvelle quantité du sel, qui est recristallisée plusieurs fois en utilisant les eaux-mères résultant de la première préparation. On a obtenu 17,5 g du sel; le pouvoir rotatoire de l'acide est $[\alpha]_D^{25} = +38,3°$.

Les deux premières eaux-mères, résultant de la seconde suite de cristallisations du sel de strychnine, ont fourni 8,4 g. d'un acide possédant un pouvoir rotatoire de −36°. On a dissous cet acide (0,042 mole) et 13,6 g de quinine (0,042 mole) dans un mélange de 200 cm³ de méthanol et 300 cm³ d'eau. En 24 heures, 20,4 g. du sel acide de quinine se sont déposés; pour l'acide nous avons trouvé $[\alpha]_D^{25} = -38,0°$. Le pouvoir rotatoire n'est pas augmenté par une recristallisation du sel dans du méthanol dilué.

Acide cyclohexylsuccinique droit

Le sel de strychnine (22 g) est décomposé par du carbonate de sodium et la strychnine est extraite par le chloroforme. L'acide est mis en liberté par l'action de H_2SO_4 et extraction à l'éther. Il est recristallisé dans une très petite quantité d'eau et deux fois dans la ligroïne. Le rendement de l'acide brut est 4,4 g; après les cristallisations on a obtenu 3,0 g. L'acide se présente en petites aiguilles brillantes renfermant de la ligroïne, qui peut être éliminée en chauffant pendant qulques heures à 70-75°. L'acide pur fond à 95,5-96,5°. Il est très soluble dans l'eau et dans la plupart des solvants organiques.

0,1559 g acide : 14,69 cm³ NaOH 0,1060-n
$C_{10}H_{16}O_4$ Poids équiv. calculé 100,1; trouvé 100,1
0,2597 g dissous dans l'acétone à 10,00 cm³ : $\alpha_d^{25} = +2,00°$ (2 dm).
$[\alpha]_D^{25} = +38,5°$; $[M]_D^{25} = +77,1°$.

Le (+)-cyclohexylsuccinate de strychnine se présente en petites plaques renfermant de l'eau de cristallisation.

Acide cyclohexylsuccinique gauche

Le sel acide de quinine (19 g.) est décomposé par H_2SO_4 et l'acide est isolé par extraction à l'éther. Le rendement de l'acide brut est 7,0 g. Il est recristallisé deux fois dans des petits volumes d'eau et deux fois dans la ligroïne. Sauf pour la direction de la rotation il est pratiquement identique à son antipode. Point de fusion 95,5-96,5°.

0,1380 g acide : 12,70 cm³ NaOH 0,1082-n
$C_{10}H_{16}O$ Poids équiv. calculé 100,1; trouvé 100,4.
0,2576 g dissous dans l'acétone à 10,00 cm³ : $\alpha_d^{25} = -1,97°$ (2 dm).
$[\alpha]_D^{25} = -38,2°$; $[M]_D^{25} = -76,5°$.

(1) M. Naps et J. B. Johns : *J. Am, Chem. Soc.* **62**, 2450 (1940).
(2) H. Wren et H. Williams : *J. Chem. Soc.* **109**, 572 (1916).

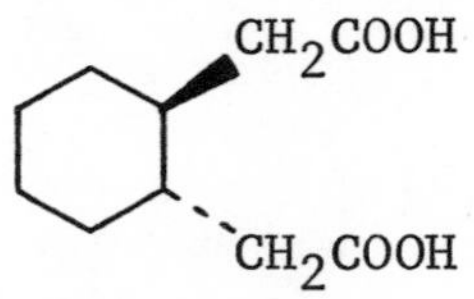

$C_{10}H_{16}O_4$ Ref. 733

Die optischen Antipoden des *trans*-β-Hydrindanons.

25 g d,l-*trans*-Cyclohexan-1,2-diessigsäure werden in 2200 ccm heißem Wasser gelöst und mit 50 g Cinchonidin, das in 600 ccm Äthylalkohol gelöst ist, versetzt und die Lösung filtriert. Beim langsamen Abkühlen krystallisiert das neutrale Cinchonidinsalz der d-Säure in zu großen Büscheln vereinigten, weichen seidigen Nadeln vom Schmelzpunkt 145° aus. Ist die Temperatur auf 25—30° gesunken, saugt man das Salz ab und krystallisiert es aus einem Gemisch von Aceton und Wasser (2:3) 5-mal um. Dann wird durch Ansäuern mit Schwefelsäure die d-Säure in Freiheit gesetzt, im Extraktionsapparat mit Äther extrahiert und aus Wasser umkrystallisiert.

Die Mutterlaugen der ersten Krystallisation werden i. V. eingedampft; das leicht lösliche Cinchonidinsalz der l-Säure krystallisiert langsam. Der großenteils ölig bleibende Rückstand wird mit verdünnter Schwefelsäure angesäuert, die in Freiheit gesetzte unreine l-Säure abfiltriert bzw. extrahiert und von neuem der Spaltung unterworfen. Die bei der zweiten Spaltung aus den öligen Cinchonidinsalzen gewonnene unreine l-Säure wird nun in das Brucinsalz übergeführt; 30 g Säure in 1200 ccm Wasser werden mit 88 g Brucin in 350 ccm Äthylalkohol versetzt. Während mehrerer Wochen krystallisiert das saure Brucinsalz der l-Säure in schönen Krystallen aus, die vollkommen rein sind. Aus ihnen wird durch Ansäuern mit Schwefelsäure, Extraktion mit Äther und Umkrystallisieren aus Wasser die reine l-Säure gewonnen.

Drehung der Säuren: Je 0,5000 g Säure in 10 cm³ absolutem Äthylalkohol, 1 dm-Rohr.

Säure aus dem schwerlöslichen Cinchonidinsalz
$\alpha_D = +2,41°$; $[\alpha]_D^{9,5} = +48,28$,

Säure aus dem Brucinsalz
$\alpha_D = -2,41°$; $[\alpha]_D^{9,5} = -48,28$.

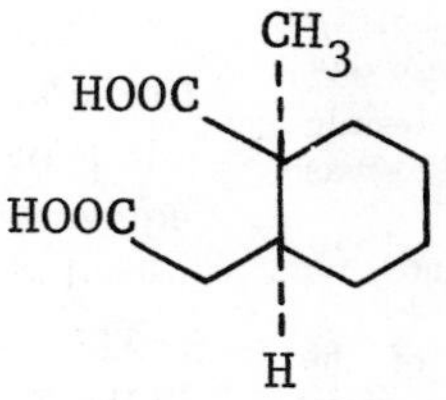

$C_{10}H_{16}O_4$

Ref. 734

(−)-cis-[2-Methyl-2-carboxy-cyclohexyl-(1)]-essigsäure (XVII). 2,151 g racemische cis-[2-Methyl-2-carboxy-cyclohexyl-(1)]-essigsäure[1]) wurden in Aceton mit einer ätherischen Lösung von Diazomethan verestert. Den rohen Diester löste man in 30 cm³ Alkohol, versetzte die Lösung mit 11 cm³ 1-n. Natronlauge und erhitzte 2,5 Std. am Rückfluss. Nach der Aufarbeitung wurden 2,216 g cis-[2-Methyl-2-carbomethoxy-cyclohexyl-(1)]-essigsäure vom Sdp. 187—190°/11 mm (Badtemperatur) erhalten.

Zur Überführung in den (±)-cis-[2-Methyl-2-carbomethoxy-cyclohexyl-(1)]-essigsäure-(−)-menthylester wurden 627 mg des Halbesters mit 2 cm³ Thionylchlorid und einem Tropfen Pyridin versetzt und das Gemisch 10 Min. auf 50—60° erwärmt. Danach dampfte man das Thionylchlorid im Vakuum ab und destillierte das gebildete Säurechlorid im Wasserstrahlvakuum bei 142—155° Badtemperatur.

Das Säurechlorid wurde zusammen mit 445 mg (−)-Menthol in 5 cm³ Pyridin gelöst und nach 24 Std. Stehen bei Zimmertemperatur in üblicher Weise aufgearbeitet, wobei 979 mg ölige neutrale Produkte isoliert wurden.

Zur Trennung in die Diastereomeren wurden 4,535 g dieses Esters in Petroläther-Benzol (10:1) gelöst und an einer Säule aus 270 g Aluminiumoxyd (Aktivität II) chromatographiert: Mit Petroläther-Benzol (10:1, 1000 cm³) wurden 1,393 g Ester eluiert, dessen spezifische Drehung ca. 10° verschieden war von den mit Äther-Methanol (10:1) eluierten letzten Fraktionen. Diese Fraktion wurde mit 3,5 g Kaliumhydroxyd in 12 cm³ Wasser und 12 cm³ Äthanol 10 Std. am Rückfluss verseift und lieferte nach Aufarbeitung und Umkristallisierung aus Aceton-Hexan 410 mg Säure, $[\alpha]_D = -1{,}9°$ (c = 2,46, Aceton). Nach erneuter Umkristallisation, wobei jeweils zuerst das rascher kristallisierende Racemat ausfiel, wurden 95 mg der fast optisch reinen (−)-cis-β-[2-Methyl-2-carboxy-cyclohexyl-(1)]-essigsäure (XVII) gewonnen, $[\alpha]_D = -9°$ (c = 1,86, Aceton). Es gelang ferner, wenn auch erst nach wiederholter chromatographischer Reinigung, aus den Benzolfraktionen nach der Verseifung die (+)-Säure (XVI) rein zu isolieren, $[\alpha]_D = +9°$ (c = 1,0, Aceton), d. h. diejenige Verbindung, die aus (−)-trans-9-Methyl-dekalon-(3) (X) hergestellt worden war.

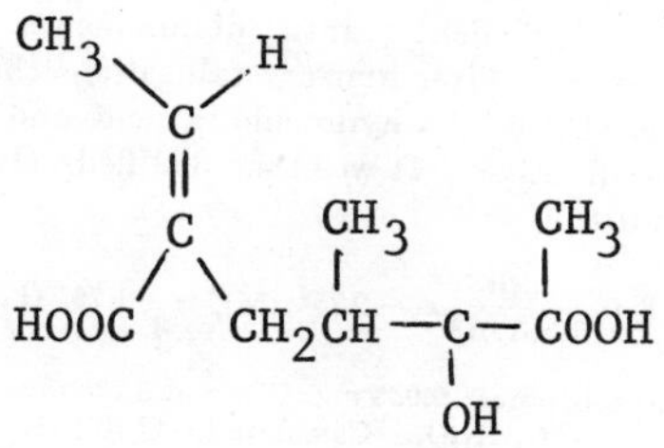

$C_{10}H_{16}O_5$

Ref. 735

Resolution of (±)-Senecic Acid.—A solution of 19.3 mg of *cis*-VA dicarboxylic acid racemate and 52.5 mg of cinchonidine in 8 drops of methanol was heated to boiling and 1 ml of ethyl acetate was added. After standing in the refrigerator for 1 day, this gave 25 mg of crystals: mp 133–137°. This was recrystallized four times to give 9.5 mg of constant mp 136–138° (no depression in melting point when admixed with the cinchonidine salt prepared from senecic acid: mp 136–138°). The infrared (KBr) spectra of the two samples were identical. The synthetic cinchonidine salt was dissolved in dilute hydrochloric acid and extracted with ether. Evaporation of the extract and crystallization from ether–petroleum ether gave crystals, mp 146–148° and no depression resulted when admixed with senecic acid. The infrared (KBr) spectra of the synthetic sample and senecic acid were identical in every respect.

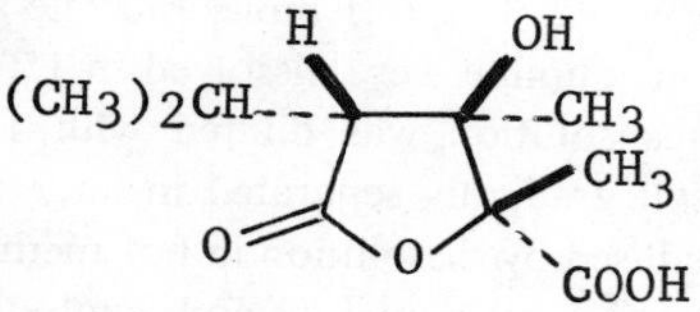

$C_{10}H_{16}O_5$

Ref. 736

Resolution of (±)-Trichodesmic Acid.—A mixture of 21.6 mg of (±)-trichodesmic acid and 29.5 mg of cinchonidine was dissolved in 1.0 ml of methanol by heating. After standing in the refrigerator overnight, the colorless crystals were collected, washed with methanol, and recrystallized from methanol: yield 14 mg, mp 247–248° dec, $[\alpha]_D$ −95.5° (*c* 0.20, ethanol).

Anal. Calcd for $C_{29}H_{38}O_6N_2$: C, 68.21; H, 7.50; N, 5.49. Found: C, 68.36; H, 7.44; N, 5.46.

The filtrate from the above was evaporated under vacuum and the residue was recrystallized from methanol–ethyl acetate to give 16 mg of colorless crystals, mp 226–227° dec, $[\alpha]_D$ −72.5° (*c* 0.40, ethanol).

Anal. Calcd for $C_{29}H_{38}O_6N_2$: C, 68.21; H, 7.50; N, 5.49. Found: C, 68.09; H, 7.61; N, 5.29.

(+)-Trichodesmic Acid.—The cinchonidine salt, mp 247–248° (84 mg), was dissolved in 5 ml of 5% sulfuric acid and extracted five times with ether. The combine extracts were washed with water, dried, and evaporated to dryness under vacuum to give a residue of 32 mg. This was recrystallized from ether–petroleum ethr to give colorless crystals, mp 209–211° dec, $[\alpha]_D$ +2.96° (*c* 1.25, ethanol). A mixture melting point with trichodesmic acid,[14] mp 209–211°, was undepressed. The infrared (KBr) was identical with that published.[4]

Anal. Calcd for $C_{10}H_{16}O_5$: C, 55.54; H, 7.46. Found: C, 55.37; H, 7.42.

(−)-**Trichodesmic Acid.**—Decomposition of 141 mg of the cinchonidine salt, mp 226–227°, as above, gave a residue of 50 mg. Recrystallization from ether–petroleum ether gave 43 mg of colorless crystals, mp 201–203° dec, $[\alpha]_D$ −2.5° (*c* 0.4, ethanol). A mixture melting point with trichodesmic acid (mp 202–204°) was 193–197°. The infrared (KBr) spectrum had the same bands as (+)-trichodesmic acid but of less intensity.

Anal. Calcd for $C_{10}H_{16}O_5$: C, 55.54; H, 7.46. Found: C, 55.78; H, 7.49.

The cinchonidine salt, mp 226–227°, apparently contained a small amount of the diastereomer which would explain the lower melting point and specific rotation of (−)-trichodesmic acid. Because of insufficient material, it was not possible to repeat the resolution.

(4) R. Adams and M. Gianturco, *J. Am. Chem. Soc.*, **78**, 1922 (1956).

(14) This sample was supplied by Professor Yunusov. A sample of trichodesmic acid (insufficient for recrystallization) obtained from Professor Roger Adams in 1959 and stored at room temperature had a melting point of 202–204° which was not depressed on admixture with the synthetic acid. The authors are indebted to Professors Yunusov and Adams for these samples.

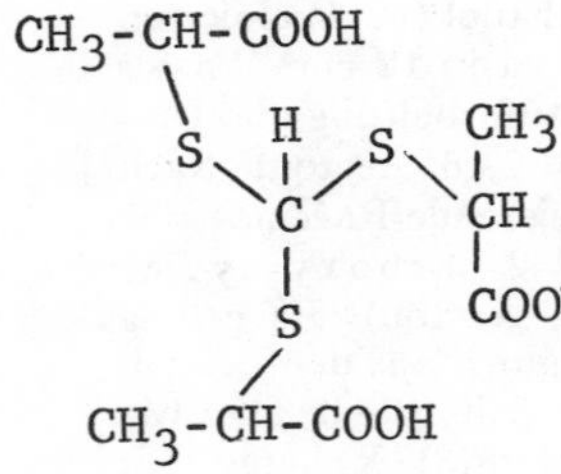

$C_{10}H_{16}O_6S_3$ **Ref. 737**

Laevorotatory acid. 33 g (0.1 mole) racemic acid and 99 g (0.3 mole) anhydrous quinine were dissolved in 150 ml hot methanol and the viscous solution was diluted with 1000 ml hot ethyl acetate. The salt gradually separated in large alum-like crystals. It was recrystallised by dissolution in hot methanol and addition of ethyl acetate (1.5 ml methanol and 8 ml ethyl acetate for 1 g salt). Maximum activity was obtained after two recrystallisations; the yield after two further recrystallisations was 48.1 g. The acid, isolated in the usual way, was recrystallised twice from water and twice from chloroform+ligroin yielding 6.5 g of triangular plates with m.p. 170–171°. [equiv. wt. 109.3; Calc. for $C_{10}H_{16}O_6S_3$ (328.42): equiv. wt. 109.47.] $[\alpha]_D^{25} = -293.7°$; $[M]_D^{25} = -964.5°$ (0.1048 g dissolved in 0.4-N hydrochloric acid to 10.00 ml).

The *tertiary quinine salt* forms large crystals containing ethyl acetate (probably 3 moles), which rapidly evaporates in the air. For analysis the salt was dried to constant weight. (Found: S 7.36; Calc. for $C_{10}H_{16}O_6S_3$, 3 $C_{20}H_{24}N_2O_2$ (1301.64): S 7.39.]

Dextrorotatory acid. The mother liquor from the first crystallisation of the quinine salt yielded 14.8 g acid with $[\alpha]_D^{25} = +197°$. This acid (0.045 mole) and 35 g (0.09 mole) brucine were dissolved in 125 ml water+15 ml acetone. The salt obtained was recrystallised from twice its weight of water to which 15% of acetone had been added. Maximum activity was obtained after one recrystallisation; the yield after a second recrystallisation was 31 g. The acid was liberated and recrystallised as the enantimorph; 30 g salt gave 8.1 g crude product and 5.0 g pure acid, quite similar to the antipode. M.p. 170–171°. Solubility in water at 25°: 28.3 g/l. [Found: equiv. wt. 109.6; Calc. for $C_{10}H_{16}O_6S_3$ (328.42): equiv. wt. 109.47.] $[\alpha]_D^{25} = +293.4°$; $[M]_D^{25} = +963°$ (0.1038 g dissolved in 0.4-N hydrochloric acid to 10.00 ml).

The *secondary brucine salt* formed elongated prisms, weathering in the air. It contains both acetone and water. For analysis it was dried to constant weight over P_2O_5. [Found S 8.61; Calc. for $C_{10}H_{16}O_6S_3$, 2 $C_{23}H_{26}O_4N_2$ (1117.33): S 8.61.]

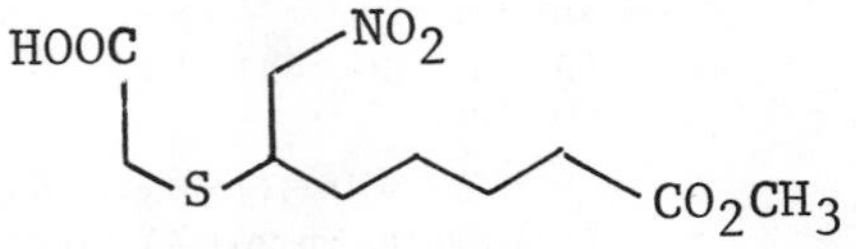

$C_{10}H_{17}NO_6S$ **Ref. 738**

Note:

Treatment of the racemic acid with (+)-α-methylbenzylamine in ethyl acetate followed by dilution with ether gave a salt of the S-enantiomer which upon two recrystallizations from ethyl acetate gave a solid with m.p. 88-89°, $[\alpha]_{365}^{25}$ 155° (c 1.55, $CHCl_3$) with > 97% enantiomeric excess.

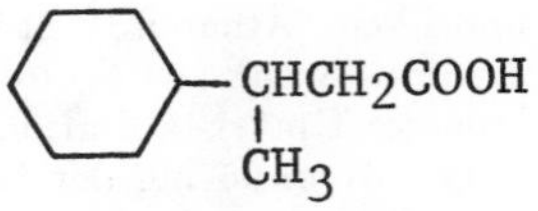

$C_{10}H_{18}O_2$ **Ref. 739**

Resolution of 2-Cyclohexylbutyric Acid-4—The inactive acid was prepared from methylcyclohexylbromomethane and malonic ester.

The acid was dissolved in boiling acetone containing 10 per cent water and an equivalent quantity of quinine was added. This was cooled to −15°. After four crystallizations the quinine salt was decomposed by dilute hydrochloric acid and the organic acid extracted with ether. It was then distilled. B. p. 145° at 4 mm. $D\frac{24}{4} = 1.018$.

$$[\alpha]_D^{24} = \frac{-0.76°}{1 \times 1.018} = -0.75°; [M]_D^{24} = -1.26° \text{ (homogeneous)}$$

3.865 mg. substance: 9.985 mg. CO_2 and 3.736 mg. H_2O

$C_{10}H_{18}O_2$.	Calculated.	C 70.5, H 10.6
170.14	Found.	" 70.5, " 10.8

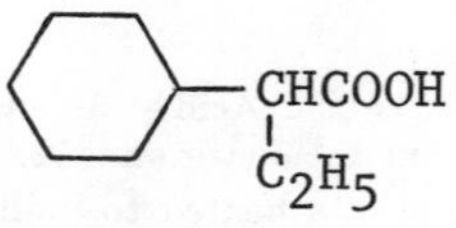

$C_{10}H_{18}O_2$ **Ref. 740**

Resolution of α-Ethylcyclohexaneacetic Acid. To a soln of 578 g (2.0 moles) of dehydroabietylamine dissolved in 4 L of MeOH was added 340 g (2.0 moles) of racemic α-ethylcyclohexaneacetic acid. The stirred soln was slowly dild with 1 l. of H_2O and stored for 16 hr in the refrigerator. The crystals were collected and dried to give 744.7 g of salt, mp 134–141°. Four recrystns from MeOH containing a small quantity of H_2O successively gave the following fractions: 580 g, mp 136–142°; 348 g, mp 138–143°; 209 g, mp 142–145°; and 105 g, mp 143–146°. The final crop (105 g) was added to a mixt of 1 L of satd NaCl and 1 l. of Et_2O. The layers were sepd, and the aqueous layer was washed several times with Et_2O. The aqueous acid was extd with Et_2O. The crude oil resulting from concn of the ext was distd to give 29.4 g of acid: bp 108–110° (0.5 mm); $[M]^{25}_{589}$ -0.085°, $[M]^{25}_{578}$ -0.085°, $[M]^{25}_{546}$ +0.051°, $[M]^{25}_{436}$ +1.5°, and $[M]^{25}_{365}$ +3.08° (*c* 19, heptane). The 29.4-g and 31.4-g fractions were combined.

A small quantity of the other isomer was obtained by concg the filtrate from the first crystn of α-ethylcyclohexaneacetic acid, dehydroabietylamine salt, to dryness giving 141.6 g of solid, mp 128–138°. An attempt to recrystallize the solid from EtOH gave approximately 80 g of EtOH-insol product which was successfully recrystd twice from EtOAc to give the following quantities: 56.6 g, mp 139–144°; 21.6 g, mp 143–146° and mmp with levorotatory salt 134–139°. The 21.6-g fraction was converted to the free acid giving 2 g of distd product: $[M]^{25}_{589}$ -5°, $[M]^{25}_{578}$ -5.3°, $[M]^{25}_{546}$ -6.38°, $[M]^{25}_{436}$ -14.2°, and $[M]^{25}_{365}$ -29.9° (*c* 19, heptane).

layer was acidified with concd HCl and then extd with Et_2O. The combined exts were dried ($MgSO_4$) and concd giving 35.7 g of light yellow oil which was distd to give 31.4 g of oil which solidified: bp 110° (1 mm); and $[M]^{25}_{589}$ -0.27°, $[M]^{25}_{578}$ -0.207°, $[M]^{25}_{546}$ +0.051°, $[M]^{25}_{436}$ +2.48°, and $[M]^{25}_{365}$ 8.9° (*c* 19, heptane).†† The filtrate from the final crop (105 g) of α-ethylcyclohexaneacetic acid, dehydroabietylamine salt, was concd to dryness, and the residue was converted to the free acid by extn with satd Na_2CO_3, followed by washing the aqueous layer with Et_2O. The aqueous layer was acidified with concd HCl, and the α-ethylcyclohexaneacetic

†† Molecular rotations reported[42] for *l*-α-ethylcyclohexaneacetic acid: λ 589.3, *ca.* −1.3°; λ 578.0, −1.267°; λ 546.0, −1.103°; λ 436.0, +1.509°; and λ 365.0 nm, +7.89° (*c* 19, heptane).

(42) A. Rothen and P. A. Levene, *J. Chem. Phys.*, 7, 975 (1939).

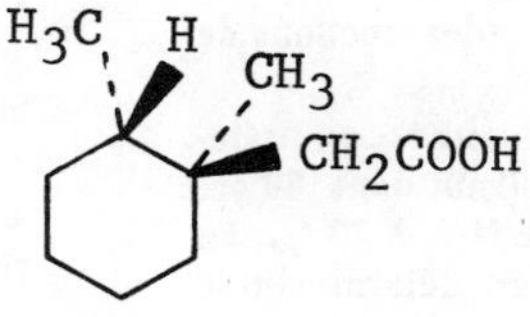

$C_{10}H_{18}O_2$ **Ref. 741**

The acid (0·45 g.) was mixed with 0·894N-sodium hydroxide (1·47 c.c.) and cinchonidine (0·338 g.) with sufficient methyl alcohol to give a clear solution on warming. The *cinchonidine* salt (0·55 g.) crystallised on keeping and was purified by two crystallisations from ethyl acetate, from which it separated in matted needles, m. p. 141—142°, $[\alpha]_{5461}$ −95° in chloroform (*c*, 1·06) (Found: N, 6·0. $C_{10}H_{18}O_2,C_{19}H_{22}ON_2$ requires N, 6·0%). The acid, regenerated from the cinchonidine salt, was converted into the p-*phenylphenacyl* ester, which crystallised from methyl alcohol in leaflets, m.p. 65—67°, $[\alpha]_{5461}$ −6° in ethyl acetate (*c*, 4·0)* (Found: C, 79·4; H, 7·6%). This ester (2 mg.) was mixed in ether with an equal weight of the *d*-ester prepared from hydroxyeremophilone benzoate. After evaporation of the solvent the mixture had m. p. 62—63° both alone and in admixture with the *dl*-ester described above.

The alkaline solution from which the cinchonidine salt had been separated was acidified, and the acid recovered and converted into the p-*phenylphenacyl* ester, which crystallised from methyl alcohol in leaflets, m. p. 62—65°, $[\alpha]_{5461}$ +8° in ethyl acetate (*c*, 3·64)* (Found: C, 79·2; H, 7·6%).

* The rotatory powers of the *p*-phenylphenacyl esters were kindly determined for us by Dr. L. N. Owen. They must be regarded as only very approximate, since they were made in a ½ dm. micro-tube with *ca.* 20 mg. of the ester in 0·5 c.c. of solvent.

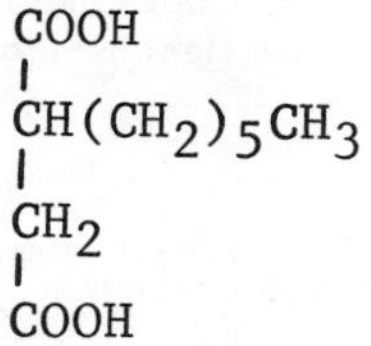

$C_{10}H_{18}O_4$ **Ref. 742**

r-Hexylsuccinic acid (28 grams; prepared according to the directions of Higson and Thorpe, T., 1906, **89**, 1469) and quinine (52·1 grams) were dissolved by gentle warming in ethyl alcohol (560 c.c.). Tufts of small needles slowly separated from the solution at the ordinary temperature, which, when air-dried, weighed 44 grams. The salt was purified by repeated crystallisation from alcohol, 20 c.c. of the solvent being used for each gram of material. The course of the resolution was followed by determining the specific rotation of the acid recovered from the successive filtrates, the observations being made in ethyl-alcoholic solution. In this manner, the following values were found: + 12·97°, + 7·55°, − 2·44°, − 8·7°, − 13·7°, − 17·3°, − 20·2°, − 22·6°, and − 23·6°. The crop which had separated from the final mother liquor weighed 6·8 grams, and, on decomposition with dilute sulphuric acid and extraction with ether, gave 2·4 grams of *l*-hexylsuccinic acid, which melted at 81—83° and had $[\alpha]_D$ − 26·0° in alcohol. It was finally purified by crystallisation from water containing 10 per cent. of alcohol until the specific rotation of successive crops remained constant.

l-*Hexylsuccinic acid* separates from water or very dilute alcohol as a granular powder; it is freely soluble in the hot solvent, sparingly so in the cold, the solutions, however, showing a very pronounced tendency to supersaturation. It melts at 82—83°, whilst the racemic acid melts at 87°:

0·1302 gave 0·2840 CO_2 and 0·1035 H_2O. C=59·5; H=8·8.

$C_{10}H_{18}O_4$ requires C=59·4; H=8·9 per cent.

The specific rotation was determined in the following solvents:

In ethyl-alcoholic solution:

$l=2$, $c=4{\cdot}0025$, $\alpha_D^{15\cdot8}$ − 2·13°, $[\alpha]_D^{15\cdot8}$ − 26·6°.

In acetone solution:

$l=2$, $c=1{\cdot}4875$, α_D − 0·98°, $[\alpha]_D$ − 32·9°.

In benzene solution:

$l=2$, $c=1{\cdot}8775$, α_D + 0·02°, $[\alpha]_D$ + 0·5°.

Attempts to resolve *r*-hexylsuccinic acid by brucine or morphine in aqueous or alcoholic solution, or by strychnine in ethyl-alcoholic solution, were unsuccessful.

$C_{10}H_{18}O_4$ Ref. 742a

Dédoublement de l'acide n hexylsuccinique. — A. Ladenburg(3) a réussi le dédoublement de l'acide méthylsuccinique à l'aide de la strychnine. Nous avons donc essayé de dédoubler l'acide hexylsuccinique par la strychnine, mais l'expérience a fourni un sel qui ne montrait aucun pouvoir rotatoire.

Nous avons alors employé la brucine. Elle a été recristallisée deux fois dans l'alcool ; son pouvoir rotatoire était $\alpha_D = -85°$ en solution dans l'alcool.

Nous avons mis 80,4 gr. d'acide hexylsuccinique racémique dans 600 cc. d'eau distillée et nous avons ajouté peu à peu 156 gr. de brucine, en agitant et en chauffant légèrement jusqu'à ce que toute la brucine soit passée en solution.

En laissant déposer nous avons recueilli différentes fractions de sel de brucine, que nous avons recristallisées.

Nous avons examiné chacune des fractions obtenues, en prélevant 0,7500 gr. de sel de brucine et en le dissolvant dans 30 cc. H_2O à 70° et en ajoutant 2 cc. d'une solution de NH_3 à 20 °/₀, en filtrant la brucine précipitée après une heure et en déterminant le pouvoir rotatoire de la solution.

Voici le résultat obtenu pour les $[\alpha_D]$ des différentes fractions de sels d'ammonium examinées au polarimètre.

$[\alpha_D]$ est calculé suivant la formule $[\alpha_D] = \frac{\alpha . n}{l . p.}$

α = l'angle lu au polarimètre.

l = la longueur du tube, p est le poids de sel d'ammonium correspondant à 0,7500 gr. de sel de brucine et n est le nombre de cc. de solvant employé (H_2O)

Le sel de brucine qui se dépose le premier a donné un sel d'ammonium *droit*.

Première cristallisation :	Fraction	I	II	III		
	$[\alpha_D]$	5°1	3°4	8°1		
Deuxième cristallisation :	Fraction	I	II	III	IV	
	$[\alpha_D]$	27°0	26°0	17°0	— 6°4	
Troisième cristallisation :	Fraction	I	II	III	IV	
	$[\alpha_D]$	27°1	26°5	21°5	— 13°1	
Quatrième cristallisation :	Fraction	I	II	III	IV	V
	$[\alpha_D]$	27°1	27°1	27°0	16°8	3°7

Nous avons réuni toutes les fractions de pouvoir rotatoire constant 27°,1. Nous avons acidifié la solution de sel d'ammonium en ajoutant goutte à goutte une solution de HCl à 10 °/₀ pour ne dépasser la neutralité que d'une goutte.

La solution est alors évaporée jusqu'à petit volume, puis extraite à l'éther ; l'éther est distillé au début, puis quand il ne reste plus qu'un petit volume, on laisse évaporer lentement, de façon que se déposent les cristaux d'acide droit.

Nous avons obtenu 6 gr. d'acide droit pur. P. F. = 83°,2 $[\alpha_D] = + 14°,3$.

Pour retirer l'autre antipode nous avons opéré de la même façon à partir des eaux-mères de cristallisation du sel de brucine.

Nous avons obtenu ainsi 3 gr. d'un acide gauche P.F. = 81°,3 et $[\alpha_D] = - 9°,7$.

Nous avons supposé être en présence de l'eutectique entre l'acide racémique et l'acide gauche. Le diagramme des points de fusion des différents mélanges d'acide droit et gauche a confirmé ce fait car, en corrigeant la courbe de la façon suivante nous avons obtenu une courbe parfaitement symétrique : l'ac. hexylsuccinique gauche obtenu ayant un $[\alpha_D] = - 9°,7$ au lieu de — 14°,3, ne contient que $\frac{100.\ 9°,7}{14°,3} = 68$ °/₀ d'acide hexylsuccinique gauche pur, ce qui correspond à la composition de l'eutectique d'après le diagramme.

Dédoublement de l'acide n propylsuccinique. — Nous avons employé la strychnine par analogie avec le dédoublement de l'acide isopropylsuccinique effectué par J. Braun et W. Reinhard(2) ; 25 gr. d'acide propylsuccinique sont dissous dans 300 cc. H_2O et on ajoute peu à peu en chauffant légèrement 103 gr. de strychnine anhydre, en agitant jusqu'à ce que toute la strychnine soit dissoute.

On recueille le précipité et on examine les différentes fractions de sel de strychnine au polarimètre.

1° cristallisation $[\alpha]_D = + 22°,9$,
2° cristallisation $[\alpha]_D = + 23°,7$,
3° cristallisation $[\alpha]_D = + 24°,6$, qui ne change plus par cristallisation.

Ce sel a été transformé en acide comme précédemment.

Nous avons obtenu 6 gr. d'acide propylsuccinique droit pur P.F. = 93°,9 $\alpha_D = + 9°,6$.

Il a été impossible de retirer l'autre antipode, les eaux-mères étant épaisses et très colorées.

On peut comparer les différents acides alkylsucciniques étudiés dans la littérature, quant aux points de fusion des acides (antipodes et racémiques).

ACIDES	FORMULE	P. F. de l'ac. racémique	P. F. des antipodes	α_D
méthylsucc.	$C_5H_8O_4$	110° (1)	115°	9°88 (2)
éthylsucc.	$C_6H_{10}O_4$	98° (3)	—	—
propylsucc.	$C_7H_{12}O_4$	100°5	93°9	9°6
butylsucc.	$C_8H_{14}O_4$	81°82	—	—
pentylsucc.	$C_9H_{16}O_4$	—	—	—
hexylsucc.	$C_{10}H_{18}O_4$	87°3	83°2	14°3
heptylsucc.	$C_{11}H_{20}O_4$	90° à 96° (4)	—	—

Les acides racémiques forment donc une série inverse avec minimum pour le 4° terme de la série. Il est difficile de tirer une conclusion pour les antipodes, les points de fusion et les α_D de beaucoup de ces acides étant inconnus.

(1) Ch. Hurt et Meinert. Organic Synthesis, vol XII, p. 64 1932.
(2) Ber. D. ch. G. 62, p. 2585. 1928.

(3) Ber. D. chem G 28, p. 1170 (1895).

$HOOC(CH_2)_2$-CH-CH_2COOH
|
$C(CH_3)_3$

$C_{10}H_{18}O_4$ Ref. 743

(3*S*)-3-Tert-butylhexanedioic Acid ((+)-*III*)

Racemic 3-tert-butylhexanedioic acid (35·6 g; 0·176 mol) and brucine (162 g; 0·348 mol) were dissolved in boiling water (5·6 l), filtered while hot and allowed to cool. The salt was filtered and crystallised from water, yield 52·3 g. According to elemental analysis, the salt is a tetrahydrate. For $C_{56}H_{78}N_4O_{16}$ (1063·2) calculated: 63·26% C, 7·39% H, 5·27% N; found: 63·26% C, 7·11% H, 5·45% N. The salt was decomposed with dilute hydrochloric acid and the product extracted continuously with ether. The extract was taken down, the crude residue dissolved in aqueous sodium hydroxide, extracted with ether, and the aqueous layer was acidified and continuously extracted with ether. This procedure afforded 9·9 g of (+)-*III* as a glass which could not be brought to crystallisation, $[\alpha]_D^{20}$ +17·2° (*c* 1, acetone). For $C_{10}H_{18}O_4$ (202·3) calculated: 59·39% C, 8·97% H; found: 59·67% C, 8·95% H.

CH_3
|
$CH_3O_2CCH_2CHCH_2CHCH_2COOH$
|
CH_3

$C_{10}H_{18}O_4$ Ref. 744

Racemic threo *methyl hydrogen 3,5-dimethylpimelate* (S)

Racemic 3,5-dimethylpimelic acid (Q) of m.p. 139.9–140.2° (27.2 g), concentrated hydrochloric acid (5 ml) and methanol (10 ml) were refluxed for 7 hours. The mixture was treated in the manner described in case of the *erythro*-compound. Distillation gave the following fractions:

I	87–115°,	0.6 mm	9.0 g	n_D^{25}	1.4338
II	115°,	0.6	10.0		1.4432
III	115– 82°,	0.6	0.3		1.4440

A residue of solid acid weighing 7.9 g was left in the flask. Fraction I consists of dimethyl ester and fraction II of the mono-methyl ester. The diester and the acid were mixed and refluxed with 5 ml of concentrated hydrochloric acid as described for the *meso*-compound. In this way a total of 17 g of mono-ester was obtained. Redistillation of the ester gave the following fractions:

I	–103°,	0.8 mm	0.24 g	n_D^{25}	1.4370
II	103–112°,	0.8	0.30		1.4420
III	112–122.5°,	0.8	0.36		1.4431
IV	122.5°,	0.8	16.09		1.4431

Fraction IV consists of the desired mono-ester.

(−)-3L, 5D-*Dimethyl-6-methoxycarbonylhexanoic acid* (V)

The racemic mono-ester (S) (16 g) and cinchonidine (23.3 g) were dissolved in warm 45% (v/v) aqueous acetone. The solution was left at room temperature overnight. The crystals were filtered off and the optical activity of the mono-ester, isolated after acidification, determined. The whole procedure was then repeated until the rotation of mono-ester recovered from the mother liquors had the same rotation as mono-ester obtained by decomposing the salt. This was the case after ten crystallizations.

The salt from the tenth crystallization was in the form of rosettes of long needles and had a very beautiful appearance. It was decomposed by means of dilute hydrochloric acid, and the mono-ester was extracted by means of eight portions of ether. The combined ether extracts were washed with water. The wash-water was extracted with ether and, after drying the combined extracts with sodium sulphate, the solvent was distilled off and the residue distilled through a small apparatus of the Claisen type, giving 1.37 g of mono-ester, b.p. 110°, 0.3 mm. (−)-3L,5D-Dimethyl-6-methoxycarbonylhexanoic acid was a slightly viscous, colourless liquid of n_D^{25} 1.4430 and d_4^{25} 1.0414.

Optical rotation

$\alpha_D^{25} - 13.00°$ (undiluted liquid; l, 0.5)
$[\alpha]_D^{25} - 25.0°$ $[M]_D^{25} - 50.5°$

Anal.: Calculated for $C_{10}H_{18}O_4$ (202.2) C, 59.38; H, 8.97 %
Found   C, 59.40;   H, 9.18 %
                                           59.08;   9.05 %

A second resolution starting from 8 g of inactive mono-ester followed an exactly similar course and gave finally 2.6 g salt of the levorotatory enantiomer. The yield was of the order of 20 % of the theoretical.

(−)-3L,5D-Dimethylpimelic acid (Ä)

Levorotatory mono-ester (V, 0.77 g, recovered from the mother liquors of the crystallizations no. 8 and 9 of the resolution just recorded), potassium hydroxide (2 g), water (2 ml) and ethanol (1 ml) were refluxed for 4 hours. After acidification and addition of water, the acid was extracted by means of several portions of ether. The ether extracts were combined, washed with water, and dried by means of sodium sulphate. The solvent was distilled off and the residue crystallized from ethyl acetate. After further crystallization from light petroleum (b.p. 95–100°) the acid melted at 86.0–86.8°. A second crystallization from the same solvent gave (−)-3L,5D-dimethylpimelic acid of m.p. 86.4–87.0° in the form of beautiful, thin needles with a length of several centimetres. On slow cooling of the melt an unstable form melting below 75° was obtained. A sharp melting point could not be obtained because transformation to the higher-melting form occurred as soon as melting began. After melting and solidification had been carried out a number of times, slow crystallization from the melt gave only the high-melting form. The acid may be crystallized from ethyl acetate, high-boiling light petroleum (b.p. about 100°), and cyclohexene, but is too soluble in ether, water, acetonitrile or glacial acetic acid to render these solvents useful for crystallization purposes.

Optical rotation

$\alpha_D^{23} - 2.79°$ (chloroform; l, 1; c, 8.69)
$[\alpha]_D^{23} - 32.1°$ $[M]_D^{23} - 60.4°$

Anal.: Calculated for $C_9H_{16}O_4$ (188.2) C, 57.43; H, 8.57 %
Found   C, 57.74; H, 8.55 %

```
            C2H5
            |
HOOCCH2-CH-CH-CH2COOH
            |
            C2H5
```

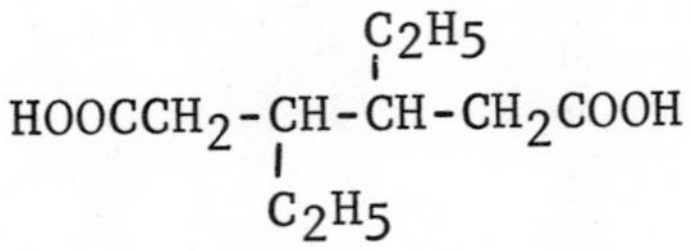

$C_{10}H_{18}O_4$ Ref. 745

(−)-*trans*-3,4-*Diethylcyclopentanone* (21).—Brucine (40 g.) and (±)-3,4-diethyladipic acid [18] (17.3 g.) were dissolved in aqueous ethanol, and the crystals which formed were recrystallised six times from the same solvent. The acid was recovered as usual by treatment of the salt with 2N-hydrochloric acid and ether, and recrystallised from benzene (4.67 g.), m.p. 111.5—112°, $[\alpha]_D^{25} +15° \pm 1.5°$ (c 0.6 in ethanol) {lit.,[18] $[\alpha]_D^{37} +10.4°$ (c 5.54 in ethanol)}.

(18) C.F. Koelsch and C.H. Stratton, J. Am. Chem. Soc., 1944, **66**, 1881.

```
           CH(CH3)2
           |
HOOC-CH-CH-COOH
       |
       CH(CH3)2
```

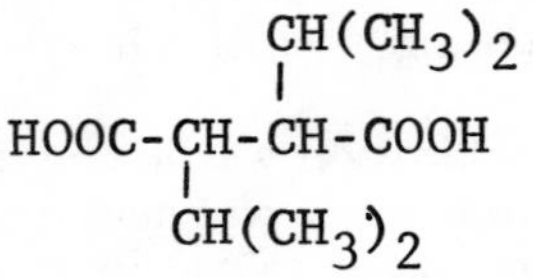

$C_{10}H_{18}O_4$ Ref. 746

Partial resolution of low-melting α,α'-diisopropylsuccinic acid. Low-melting α,α'-diisopropylsuccinic acid, 4.04 g, and brucine, 7.88 g, were dissolved in 250 ml of hot water and the solution filtered warm. After standing for four days at room temperature the crystals formed were filtered off (6.3 g) and recrystallized twice from hot water. The acid was liberated by treatment with dilute hydrochloric acid, dried and recrystallized from ethyl acetate. It showed $[\alpha]_D^{20} = +1.5°$ in absolute ethanolic solution. The rotation was small but significant, as the rotation of the alkaloid is of opposite sign.

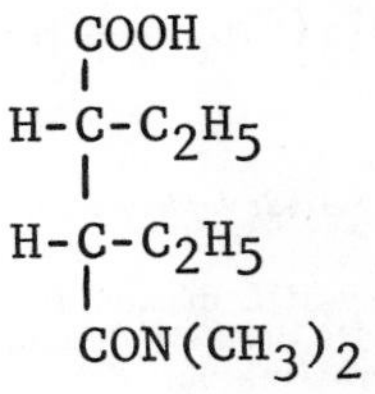

$C_{10}H_{19}NO_3$ Ref. 747

This acid was resolved as follows. The acid (10·0 g.) in ethanol (150 ml.) was titrated to pH 9—10 with quinine methohydroxide.[11] The solvent was removed *in vacuo* (rotatory evaporator) and the residue dried by azeotropic distillation *in vacuo* with absolute ethanol and then ethyl acetate. The resultant product in ethyl acetate (100 ml.), left at 0° overnight, gave white crystals (15·36 g.), m. p. 205—210°. Fractional crystallisation from ethyl acetate–ethanol gave the salt (5·11 g.), m. p. 216—217°, of constant rotation, $[\alpha]_D$ −104° (*c* 1·07 in ethanol). The salt (4·7 g.) in water (150 ml.) was acidified at 0° with aqueous sulphuric acid (3N) to pH 4 in the presence of ether (150 ml.). The ether layer was separated, and the aqueous layer further extracted with ether. The product (1·63 g.) crystallised slowly. Recrystallisation from ether gave (+)-*NN*-dimethyl-*erythro*-diethylsuccinamic acid, m. p. 85—86°, $[\alpha]_D$ +31° (*c* 2·50 in ethanol), $\nu_{max.}$ 1720, 1605 cm.$^{-1}$. The infrared spectrum in chloroform was identical with that of the (±) isomer (see above).

[11] R. T. Major and J. Finkelstein, *J. Amer. Chem. Soc.*, 1941, **63**, 1368.

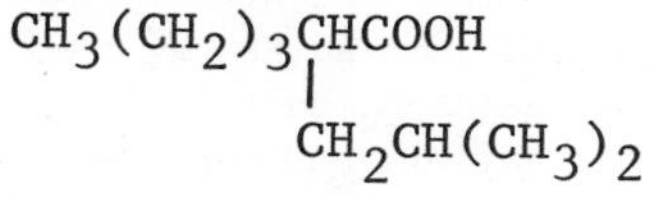

$C_{10}H_{20}O_2$ Ref. 748

d-Butyl-isobutyl-essigsäure.

Zur Darstellung des Brucinsalzes wurden 6 g reine Butyl-iso-butyl-essigsäure mit 14 g trocknem Brucin in 50 ccm heißem Alkohol gelöst und in ein auf 30° erwärmtes Gemisch von 60 ccm Alkohol und 1200 ccm Wasser unter Umschütteln langsam eingegossen. Dabei schied sich ein wenig Salz als Öl ab. Die Mutterlauge wurde abgegossen, der ölige Rückstand in wenig Alkohol gelöst und zur Hauptflüssigkeit wieder zugefügt. Als diese Lösung unter möglichst geringem Druck auf ungefähr ein Liter eingeengt wurde, begann die Krystallisation. Nach längerem Stehen bei 0° betrug die Menge der Krystalle 2.5 g. Die Mutterlauge wurde wieder in derselben Weise bis auf ungefähr 700 ccm eingeengt und ergab dann nach 15-stündigem Stehen im Eisschrank weitere 3 g des Brucinsalzes. Aus der Mutterlauge konnten noch weitere Krystallisationen erhalten werden, die aber nicht mehr so rein waren. Das Salz bildet meist mikroskopisch kleine Prismen; es läßt sich in der vorher angegebenen Weise umkrystallisieren und ist in Alkohol leicht, in Wasser aber schwer löslich. Die zuerst erhaltenen 5.5 g wurden zur Gewinnung der freien Säure mit 25 ccm 2-*n*. Schwefelsäure unter gelindem Erwärmen zersetzt, die Fettsäure ausgeäthert, die ätherische Lösung nochmals mit Schwefelsäure sorgfältig geschüttelt, dann mit Wasser mehrmals gewaschen und schließlich mit Natriumsulfat getrocknet. Nach Verdampfen des Äthers blieb die aktive Säure als farbloses Öl zurück und wurde unter etwas weniger als 1 mm Druck destilliert. Sie ging in der Nähe von 100° über, und die Ausbeute betrug 1.3 g.

Das Präparat drehte bei 21.5° im ½-dcm-Rohr + 2.22° nach rechts und hatte die Dichte 0.8927. Mithin $[\alpha]_D^{21.5} = +4.98°$. Zur weiteren Reinigung der *d*-Säure haben wir sie wieder in das Brucinsalz übergeführt. 1.5 g Säure wurden mit 4 g Brucin in wenig Alkohol gelöst, die Lösung in Wasser gegossen und unter geringem Druck eingeengt. Das auskrystallisierte Salz wurde dann wieder zersetzt und von neuem die Drehung bestimmt. Diese Operation ergab bei Wiederholung langsam steigende Werte. Nach dem 14. Male wurde der Wert der Drehung konstant. Er betrug unter den oben genannten Bedingungen $[\alpha]_D^{21.5} = +5.73°$. Die Analyse dieses Körpers ergab folgende Zahlen:

0.1548 g Sbst.: 0.3946 g CO_2, 0.1646 g H_2O.

$C_{10}H_{20}O_2$ (172.16). Ber. C 69.72, H 11.71.
Gef. » 69.52, » 11.90.

Bei einer zweiten Operation haben wir absichtlich, um die Ausbeute zu erhöhen, größere Mengen des Brucinsalzes auskrystallisieren lassen. Dementsprechend war aber auch zur Erzielung der gleichen Aktivität eine viel öftere Wiederholung der Krystallisation notwendig. Selbstverständlich ist die Ausbeute an dem höchstdrehenden Produkt recht klein. Aus 12 g Racemkörper erhielten wir nach 9-maliger Krystallisation des Brucinsalzes 1.5 g Säure (25 % der Theorie) von $[\alpha]_D^{22} = +5.18°$; dann nach weiteren 3 Krystallisationen 0.5 g Säure von $[\alpha]_D^{22} = +5.73°$.

Ob damit der richtige Wert schon erreicht wurde, ist schwer zu sagen.

Im Folgenden geben wir noch einige Zahlen über das Drehungsvermögen von Lösungen der Säure in Xylol, Alkohol und *n*-Natronlauge.

1. 0.02355 g Sbst., gelöst in Xylol. Gesamtgewicht der Lösung: 0.16615 g. Also Prozentgehalt 14.2. Drehung im ½-dm-Rohr bei 20° = + 0.45° (± 0.03).

2. 0.02240 g Sbst., gelöst in Alkohol. Gesamtgewicht der Lösung: 0.1641 g. Also Prozentgehalt 13.7. Drehung im ½-dm-Rohr bei 20° = + 0.41°.

3. 0.02190 g Sbst., gelöst in *n*-Natronlauge (etwas mehr als 1 Mol.). Gewicht der Lösung: 0.21578 g. Also Prozentgehalt 10.1. Die anfangs etwas trübe Lösung war nach 48 Stunden klar, aber gelb gefärbt. Drehung im ½-dm-Rohr bei 20° = + 0.36° (± 0.03).

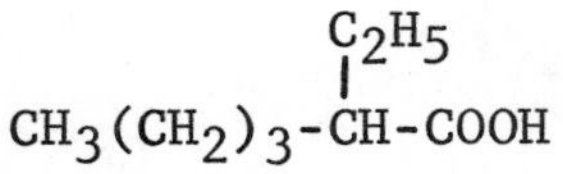

$C_{10}H_{20}O_2$ Ref. 749

(+)-*2-Ethyl-2-methylheptanoic acid* was resolved in 6.1% yield by eight crystallizations of the brucine salt from ethyl acetate: the salt; m.p. 73°; the acid, b.p. 148° (16–17 mm.); α_D^{25} +2.11° (homogeneous, 1 dm.); $[\alpha]_D^{25}$ +2.34°.

The tail fraction gave (−)-*acid*, $[\alpha]_D^{25}$ −1.13°

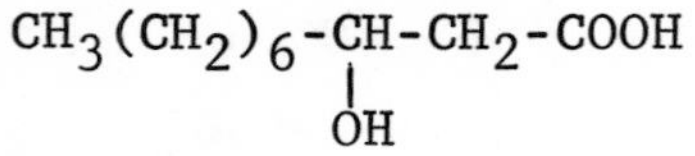

$C_{10}H_{20}O_3$ Ref. 750

Acide hydroxy-3D décanoïque.

L'acide hydroxy-3*DL* décanoïque a été préparé par réduction du céto-3 décanoate de méthyle, $n_D^{19} = 1{,}4418$ (25) par $NaBH_4$; l'hydroxy-acide racémique, F = 54-55,5°, a été dédoublé par recristallisation de son sel de cinchonidine dans l'acétone aqueuse. On obtient l'acide hydroxy-3*D* décanoïque, F=47-48°, $[\alpha]_D = -19{,}0°$ ($CHCl_3$, c=2.00) [litt. (6) : acide *L*, F=48,4°, $[\alpha]_D = +20°$ ($CHCl_3$)].

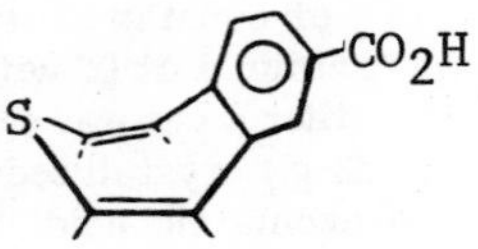

$C_{11}H_8O_2S$ Ref. 751

Zur *Racematspaltung* wurde eine Suspension von 264 mg (0.78 mMol) **3b** in 50 ccm Aceton mit einer Lösung von 316 mg (0.8 mMol) *Brucin* in der gleichen Menge Aceton versetzt. Die beim Erhitzen resultierende Lösung engte man bis auf etwa 20 ccm ein und gab 20 ccm Äthanol hinzu. Beim Abkühlen und Reiben fiel ein kristallines Brucin-Salz aus, welches nochmals aus Aceton/wenig Äthanol umkristallisiert wurde. Ausb. 225 mg vom Schmp. 198—203°. Zur Zerlegung löste man 190 mg dieses Salzes in wenig Aceton, versetzte mit Wasser und 2*n* H_2SO_4, kristallisierte den Niederschlag zweimal aus Benzol um und erhielt 83 mg *linksdrehende Säure* **3b** mit der spezif. Drehung $[\alpha]_{436}^{24}$: −250° (*c* = 0.01 g/ccm in Diäthylenglykoldimethyläther).

$C_{19}H_{12}O_4S$ (336.4) Ber. C 67.84 H 3.60 Gef. C 68.09 H 3.74

Die Mutterlauge des Brucin-Salzes wurde mit verd. Salzsäure angesäuert; der Niederschlag (110 mg) lieferte aus Benzol 30 mg (+)-**3b**; $[\alpha]_{436}^{24}$: +200° (*c* = 0.01 g/ccm in Diäthylenglykoldimethyläther).

In einem zweiten Versuch wurden 956 mg (2.84 mMol) **3b** zusammen mit 1.12 g (2.84 mMol) *Brucin* in 150 ccm heißem Aceton gelöst, anschließend bis auf 30 ccm eingeengt, 30 ccm Äthanol zugegeben und mit dem vorstehend beschriebenen Brucin-Salz angeimpft. Nach der beginnenden Kristallisation fügte man nochmals 40 ccm Aceton zu und beließ über Nacht. Ausb. 744 mg Brucin-Salz vom Schmp. 198—207°. Die analoge Behandlung des Salzes und der Mutterlauge mit verd. Salzsäure führte zu (−)-**3b** und (+)-**3b**, die jedoch noch durch racemische Säure verunreinigt waren. Diese schwerstlösliche Komponente ließ sich durch mehrfaches fraktioniertes Umkristallisieren aus Benzol abtrennen und wurde weiteren Racematspaltungen mit Brucin unterworfen. Auf diesem Wege gelangte man zu 310 mg (−)-**3b**, $[\alpha]_{436}^{20}$: −220°, und 105 mg (+)-**3b**, $[\alpha]_{436}^{20}$: +250° (*c* jeweils 0.01 g/ccm in Diäthylenglykoldimethyläther).

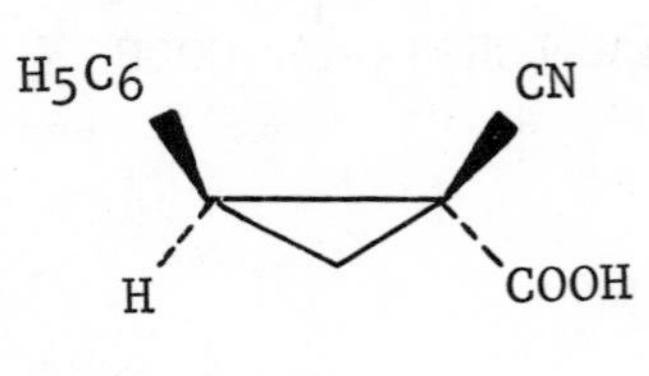

$C_{11}H_9NO_2$ Ref. 752

Resolution of (*E*)-1-Cyano-2-phenylcyclopropanecarboxylic Acid ((*E*)-2). To a dry mixture of 10.0 g (53.5 mmol) of (*E*)-**2** and 21.2 g (53.5 mmol) of brucine (recrystallized from acetone) was added 200 ml of methanol. The mixture was heated until clear, and the solution was stored at 25° and then 0° for several days. The solid that separated was recrystallized 4 times from boiling methanol to give 9.0 g of well-formed hard needles, $[\alpha]^{25}{}_{546}$ −110° (*c* 0.268, $CHCl_3$). The rotation of the salt did not change with another recrystallization to give 7.1 g of needles. The free acid was obtained by shaking 7.1 g of the salt with 50 ml of ether and 50 ml of 1 *M* hydrochloric acid. The aqueous layer was extracted three times more with ether; the combined ether extracts were washed with brine, dried, and evaporated to give 2.2 g of white solid, mp 135–138° (*cf.* 137–138° for racemic acid), $[\alpha]^{25}{}_{546}$ −239° (*c* 0.500, EtOAc). *Anal.* Calcd for $C_{11}H_9NO_2$: C, 70.58; H, 4.85; N, 7.48. Found: C, 70.42; H, 4.89; N, 7.52.

From the brucine salt mother liquors, the positive enantiomer was obtained by evaporating the methanol and recrystallizing the residue four times from hot acetone. From 12 g of residue was obtained 8.4 g of a yellow-white solid, $[\alpha]^{25}{}_{546}$ +74.6° (*c* 0.232, $CHCl_3$). The rotation did not change with another recrystallization to give 6.6 g of yellow-white solid. The 6.6 g of salt was converted to 2.0 g of a white solid, $[\alpha]^{25}{}_{546}$ +239° (*c* 0.575, EtOAc). That the acid obtained was not optically pure was indicated by changes of rotation upon repeated recrystallizations from chloroform (+239°, +241°, +238°, +235°), mp 139–140°. *Anal.* Calcd for $C_{11}H_9NO_2$: C, 70.58; H, 4.85; N, 7.48. Found: C, 70.80; H, 4.94; N, 7.35.

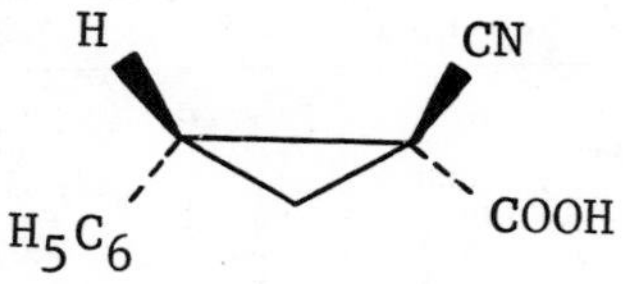

$C_{11}H_9NO_2$ Ref. 753

Resolution of (*Z*)-1-Cyano-2-phenylcyclopropanecarboxylic Acid ((*Z*)-2). To a dry mixture of 1.1 g (5.9 mmol) of (*Z*)-**2** and 1.95 g (5.9 mmol) of quinine was added 50 ml of methanol. The mixture was heated to clarity and stored at 25°. The salt that separated was recrystallized twice to give 1.1 g of soft, long needles, $[\alpha]^{25}_{546}$ $-106°$ (c 0.228, $CHCl_3$). The free acid was obtained in the usual way to give 0.4 g (36%) of an off-white solid, $[\alpha]^{25}_{546}$ $+199°$ (c 0.158, EtOAc).

From the quinine salt mother liquors, the negative enantiomer of (*Z*)-**2** was recovered as 0.7 g (64%) of a yellow solid, $[\alpha]^{25}_{546}$ $-143°$ (c 0.282, EtOAc). These optically impure acids were esterified (see below) without further purification.

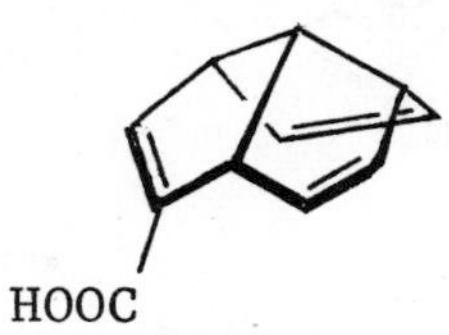

$C_{11}H_{10}O_2$ Ref. 754

Resolution of 12b. A solution of **12b** (1.34 g) in methanol (9.5 ml) and water (2.5 ml) was treated with d-(+)-α-phenethylamine (0.93 g). After several hours at room temperature, the mixture was allowed to stand overnight in a refrigerator. The first crop of crystals was subjected to repeated recrystallization from methanol-water until the optical rotation at 365 nm showed no further change ($[\alpha]_{365}$ $-97.8°$ (c 0.2, ethanol)). Usually three to four recrystallizations sufficed. By this procedure, 347 mg of thick needles was obtained.

Treatment of 235 mg of this (−)-salt with 5% aqueous hydrochloric acid (5 ml) followed by ether extraction (5 × 15 ml), drying, and evaporation provided 130 mg (95%) of optically pure (−)-**12b**; $[\alpha]_D$ $-12.4°$; $[\alpha]_{365}$ $-336°$ (c 0.5, ethanol). After recrystallization from ether-hexane, the white crystals melted at 131-132°; $[\alpha]_D$ $-12.3°$; $[\alpha]_{365}$ $-336°$ (c 0.5, ethanol). The infrared spectrum was identical with that of racemic **12b**.

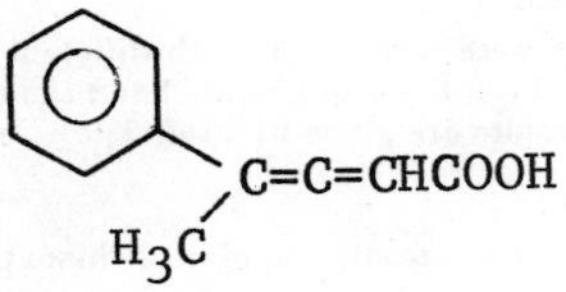

$C_{11}H_{10}O_2$ Ref. 755

Resolution of (±)-4-*Phenylpenta*-2,3-*dienoic Acid*.—(−)-Phenethylamine (1·55 g) in ethyl acetate (5 ml) was added to 4-phenylpenta-2,3-dienoic acid (2·2 g) in warm ethyl acetate (20 ml). After refrigeration (4 days), the crystals were separated and recrystallised to constant rotation from ethyl acetate; m.p. 129·5—130°, $[\alpha]_D^{22}$ $+240°$ (c 1·1 in MeOH) (1·3 g). Decomposition of the salt gave (+)-4-phenylpenta-2,3-dienoic acid (0·6 g, 55%), m.p. 100—101°, $[\alpha]_D^{25}$ $+316°$ (±0·3°) (c 1·1 in EtOH) (Found: C, 75·7; H, 5·8. Calc. for $C_{11}H_{10}O_2$: C, 75·8; H, 5·8%) (lit.,[11] m.p. 101—102°, $[\alpha]_D$ $+318°$), τ −1·2 (1H, s), 2·64 (5H), 4·12 (1H, q, J 3 Hz), and 7·81 (3H, d, J 3 Hz).

Work-up of the mother liquors gave partially resolved (−)-4-phenylpenta-2,3-dienoic acid, m.p. 115°, $[\alpha]_D^{25}$ $-107°$ (±0·3°) (c 1·0 in EtOH) (Found: C, 76·0; H, 5·9%).

[10] A. T. Blomquist, C. H. Liu, and J. C. Bohrer, *J. Amer. Chem. Soc.*, 1952, **74**, 3643.
[11] K. Shingu, S. Hagishita, and M. Nagakawa, *Tetrahedron Letters*, 1967, 4371.
[12] G. Markl, *Chem. Ber.*, 1961, 3005.

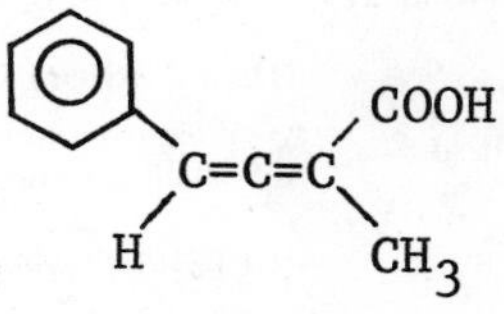

$C_{11}H_{10}O_2$ Ref. 756

*(+)- und (−)-***11**: Nach der allgemeinen Vorschrift wurden aus 4.864 g racem. **11** und 4.130 g Cinchonidin nach vier Umkristallisationen aus Aceton 1.552 g (27.5%) Addukt als farblose Nadeln mit Schmp. 149−150°C (Rotfärbung) erhalten. $[\alpha]^{25}_{589}$ = +60.0°, $[\alpha]^{25}_{589}$ = +535° (c' = 0.004006, C_2H_5OH).

$C_{49}H_{58}N_4O_6$ (799.0) Ber. C 73.66 H 7.32 N 7.01 Gef. C 73.85 H 7.33 N 5.77

Das Filtrat der ersten Kristallisationsfraktion lieferte 2.142 (44%) racem. **11** als farblose Kristalle mit Schmp. 111−113°C (aus Petroläther). — Aus der Mutterlauge dieser Kristallisation wurden 0.123 (2.5%) optisch nicht ganz reines (−)-**11** als farblose Kristalle mit Schmp. 92−95°C (aus Petroläther) gewonnen. $[\alpha]^{25}_{589}$ = −165° (c' = 0.005672, C_2H_5OH). — Aus 1.854 g Komplex wurden 0.397 g (98%) (+)-**11** gewonnen; farblose, grobe Stäbchen mit Schmp. 94−95°C (aus Petroläther). $[\alpha]^{25}_{589}$ = +215° (c' = 0.005740, C_2H_5OH). — CD (C_2H_5OH): λ_{max} ($\Delta\varepsilon$) = 314 (−0.0055), 295 (+0.12), 282 (+0.48), 277 (+0.62), 271 (+0.90), 254 (+4.5), 248 (+5.9), 243 (+7.1), 236 (+7.2), 215 nm (−8.5).

$C_{11}H_{10}O_2$ (174.2) Ber. C 75.84 H 5.79 Gef. C 75.50 H 5.82

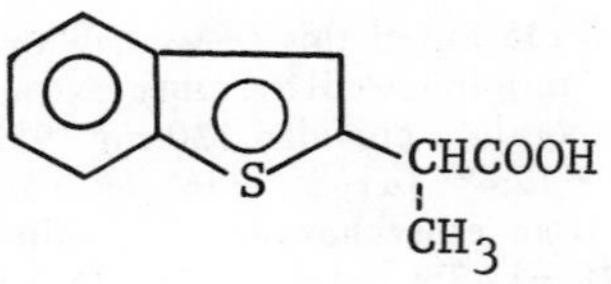

$C_{11}H_{10}O_2S$ Ref. 757

Resolution of the racemic acid

Preliminary experiments were carried out with different bases and solvents. The crystals formed were filtered off, the acid liberated and the optical activity measured in absolute ethanol. The results are given in Table 1.

Table 1. Preliminary tests on the resolution of α-(2-thianaphthenyl)-propionic acid.

Base	$[\alpha]_D$ of acid in abs. ethanol			
	From methanol	From ethanol	From acetone	From ethyl acetate
Cinchonine	oil	oil	oil	+ 2°
Cinchonidine	− 2°	+ 13°	+ 4°	+ 4°
Quinine	± 0	± 0	± 0	± 0
Quinidine	oil	oil	oil	oil
Brucine	− 2°	± 0	− 2°	± 0
Strychnine	oil	oil	oil	oil
Morphine	+ 23°	± 0	+ 20°	+ 26°
Ephedrine	oil	oil	oil	oil
(+)-α-Phenylethylamine	− 2°	− 3°	± 0	± 0
(+)-α-(2-Naphthyl)-ethylamine	− 21°	− 8°	− 5°	− 26°

Table 2. Recrystallization of the morphine salt of α-(2-thianaphthenyl)-propionic acid.

Cryst. no.	Solvent (ml)		Weight of salt (g)	$[\alpha]_D^{25}$ of acid
	Ethanol	Water		
1	600	150	28.1	25°
2	400	100	20.0	30°
3	300	75	16.6	31°
4	250	60	14.5	31°

A mixture of 20.6 g (0.1 mole) of racemic acid and 30.3 (0.1 mole) of morphine was dissolved in a hot mixture of 600 ml of ethanol and 150 ml of water. The salt was filtered off after 24 hours and recrystallized from dilute ethanol. The progress of the resolution was tested on samples from each crystallization, from which the acid was liberated and the optical activity examined in absolute ethanol (Table 2).

The mother liquors from the two first crystallizations were evaporated to dryness at room temperature and 13.5 g acid was isolated with $[\alpha]_D^{25} = -18°$ (absolute ethanol). This acid was added to an ether solution containing about 11.2 g of (+)-α-(2-naphthyl)-ethylamine obtained from 15.6 g of the sulfate (16, 17). The ether was evaporated and the salt recrystallized from dilute ethanol. The progress of the resolution was followed as described above and the results are given in Table 3.

Table 3. Recrystallization of the (+)-α-(2-naphthyl)-ethylamine salt of α-(2-thianaphthenyl)-propionic acid.

Cryst. no.	Solvent (ml)		Weight of salt (g)	$[\alpha]_D^{25}$ of acid
	Ethanol	Water		
1	600	120	20.0	− 27°
2	500	100	16.2	− 31°
3	400	80	14.0	− 31°

(+)-α-(2-Thianaphthenyl)-propionic acid

The acid was liberated from the morphine salt with dilute sulphuric acid and extracted with ether. The ether was evaporated and the crude acid (6.1 g) was recrystallized from a carbon tetrachloride–ligroin mixture. Yield 5.4 g of acid forming small needles with m.p. 134–136°.

Anal. $C_{11}H_{10}O_2S$: Equiv. wt. calc. 206.3; found 207.0.

Optical activity in different solvents

Weighed amounts of the acid were dissolved in different solvents, made up to a volume of 10.00 ml and the rotatory power measured. In the case of water the acid was neutralized with dilute sodium hydroxide to pH 7. The results are given in Table 4.

Table 4. Optical activity of (+)-α-(2-thianaphthenyl)-propionic acid.

Solvent	mg of acid	$2\alpha_D^{25}$	$[\alpha]_D^{25}$	$[M]_D^{25}$
Acetone	100.4	+ 1.04°	+ 51.8°	+ 106.9°
2.2.4-Trimethylpentane	15.5	+ 0.14°	+ 45°	+ 93°
Acetic acid	101.0	+ 0.71°	+ 35.2°	+ 72.6°
Ethanol	105.8	+ 0.66°	+ 31.2°	+ 64.4°
Chloroform	100.8	+ 0.44°	+ 21.8°	+ 45.0°
Benzene	100.8	+ 0.28°	+ 13.9°	+ 28.7°
Water (ion)	100.6	− 0.53°	− 26.3°	− 54.3°

(−)-α-(2-Thianaphthenyl)-propionic acid

The naphthylethylamine salt (13.8 g) was decomposed and the acid isolated in the same way as for the (+)-antipode. 7.5 g of the acid was obtained. It was recrystallized from carbon tetrachloride-ligroin. Yield 7.0 g; m.p. 134–136°.

Anal. $C_{11}H_{10}O_2S$: Equiv. wt. calc. 206.3; found 206.7.

0.1002 g of acid dissolved in absolute ethanol to 10.00 ml: $2\alpha_D^{25} = -0.63°$, $[\alpha]_D^{25} = -31.4°$, $[M]_D^{25} = -64.8°$.

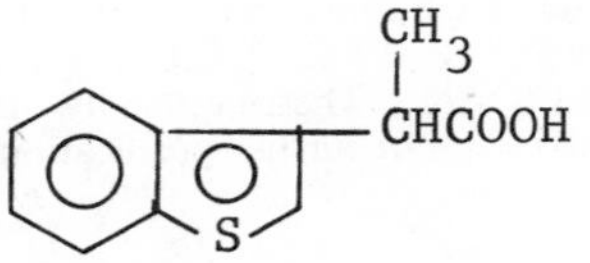

$C_{11}H_{10}O_2S$ Ref. 758

Resolution of the racemic acid

Preliminary experiments were carried out with 0.001 mole of the racemic acid and 0.001 mole of the alkaloids in different solvents. From the crystals formed the acid was liberated and the optical activity measured in absolute ethanol. The results are given in Table 2.

In a hot solution of 160 ml of ethanol and 80 ml of water 8.25 g (0.04 mole) of racemic acid and 11.78 g (0.04 mole) of cinchonidine were dissolved and left to crystallize at room temperature and then in a refrigerator. The salt was filtered off after 24 hours. The acid was isolated from a sample and the optical activity measured in absolute ethanol. The remaining salt was recrystallized from dilute alcohol (Table 3).

The mother liquors from the two first crystallizations were evaporated to dryness at room temperature and 4.55 g acid isolated with $[\alpha]_D^{25} = +52°$ (absolute ethanol). This acid was dissolved with 6.50 cinchonine in 20 ml of boiling ethanol and 5 ml of hot water was added. The salt obtained was recrystallized in the same way as the cinchonide salt (Table 4).

(−)-α-(3-Thianaphthenyl)-propionic acid

The acid was liberated from 6.3 g of the cinchonidine salt with dilute sulphuric acid and extracted with ether. It was recrystallized from ligroin (85–110°). Yield 2.1 g of acid forming small prisms with m.p. 64–66°.

Anal. $C_{11}H_{10}O_2S$: Equiv. wt. calc. 206.3; found 206.8.

Optical activity in different solvents

Weighed amounts of the acid were dissolved in the solvents, made up to a volume of 10.00 ml and the rotatory power determined. The results are given in Table 5.

(+)-α-(3-Thianaphthenyl)-propionic acid

The cinchonine salt was decomposed with dilute sulphuric acid and the organic acid extracted with ether. The yield of the crude acid was 2.0 g and after three recrystallizations from ligroin there remained 1.5 g with m.p. 64–66°.

Table 2. Preliminary tests on the resolution of α-(3-thianaphthenyl)-propionic acid.

Base	$[\alpha]_D^{25}$ of acid in abs. ethanol			
	From methanol	From ethanol	From acetone	From ethyl acetate
Cinchonine	+ 3°	+ 25°	− 14°	—
Cinchonidine	− 43°	− 44°	− 43°	—
Quinine	± 0	− 20°	—	—
Quinidine	± 0	± 0	—	—
Brucine	oil	oil	oil	oil
Strychnine	oil	oil	oil	oil
Morphine	oil	oil	oil	oil
(+)-α-Phenylethylamin	oil	oil	oil	oil
(+)-α-(2-Naphthyl)-ethylamin	+ 10°	+ 11°	—	—

Table 3. Recrystallization of the cinchonidine salt of α-(3-thianaphthenyl)-propionic acid.

Cryst. no.	Solvent (ml)		Weight of salt (g)	$[\alpha]_D^{25}$ of acid
	Ethanol	Water		
1	160	80	10.9	− 54°
2	120	60	8.5	− 77°
3	80	40	7.3	− 80°
4	60	30	6.5	− 80°

Table 4. Recrystallization of the cinchonine salt of α-(3-thianaphthenyl)-propionic acid.

Cryst. no.	Solvent (ml)		Weight of salt (g)	$[\alpha]_D^{25}$ of acid
	Ethanol	Water		
1	20	5	7.3	+ 77°
2	10	3	6.2	+ 78.5°
3	10	3	5.0	+ 78.5°

Table 5. $[\alpha]_D^{25}$ and $[M]_D^{25}$ of (−)-α-(3-thianaphthenyl)-propionic acid.

Solvent	mg of acid	$2\alpha_D^{25}$	$[\alpha]_D^{25}$	$[M]_D^{25}$
Benzene	72.1	− 1.58°	− 109.6°	− 226.1°
2.2.4-Trimethylpentane	17.8	− 0.37°	− 104°	− 215°
Acetic acid	69.4	− 1.39°	− 100.1°	− 206.5°
Acetone	73.2	− 1.35°	− 92.2°	− 190.2°
Chloroform	83.1	− 1.38°	− 83.0°	− 171.2°
Ethanol	75.4	− 1.21°	− 80.2°	− 165.5°
Water (neutr.)	94.1	− 0.02°	− 1°	− 2°

Anal. $C_{11}H_{10}O_2S$: Equiv. wt. calc. 206.3; found 207.1.
0.0756 g acid dissolved in absolute ethanol to 10.00 ml: $2\alpha_D^{25} = +1.21°$, $[\alpha]_D^{25} = +80.0°$, $[M]_D^{25} = +164.8°$.

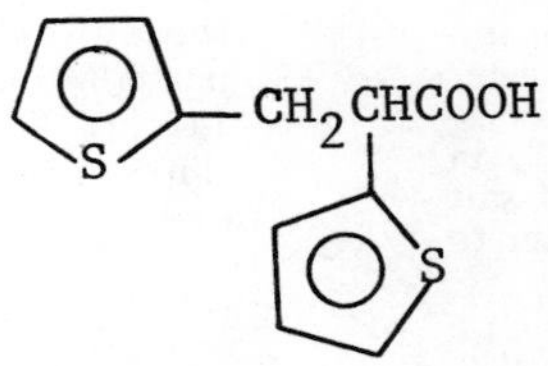

$C_{11}H_{10}O_2S_2$

Ref. 759

Resolution of 2-thenyl-2-thienylacetic acid: 7.5 g (0.031 moles) of optically inactive acid and 5.8 g of *l*-ephedrine dihydrate were dissolved in 200 ml of hot benzene. The solution was allowed to crystallise at room temperature for 12 hours. The precipitate was filtered off, dried and weighed. About 0.5 g was set aside, the acid liberated and the optical activity measured in acetone. The principal part of the salt was recrystallised from benzene and the procedure repeated until the optical activity of the acid remained constant.

Crystallisation No.	1	2	3	4
ml of benzene	200	50	30	30
g of salt obtained	5.1	3.6	3.3	3.1
$[\alpha]_D^{25}$ of acid (acetone)	− 82°	− 88°	− 92°	− 92°

The only alkaloid tried, giving (+)-thenylthienylacetic salt in excess is quinine. This alkaloid loses its resolving power at an optical activity of the acid of about +70°. Thus the specific rotation of the acid was raised from +65° to +71° by five crystallisations from dilute acetone.

Attempts to concentrate the (+)-acid salt in the above-mentioned series by partial evaporation of the solvent were without success.

Attempts to concentrate the (+)-acid by recrystallisation of a moderately dextrorotatory acid from petrol (b.p. 60°–80°) in the usual way and according to a filter-paper method (13) were unsuccessful.

(−)-2-Thenyl-2-thienylacetic acid: 3.1 g of the *l*-ephedrine compound were decomposed with 2-*N* sulphuric acid and the liberated acid was extracted with ether. The ether was removed and the crude acid (1.7 g) was recrystallised from petrol (b.p. 60°–80°). 1.4 g of acid, melting at 79°–81° was obtained as colourless needles or rods.

81.35 mg of acid, 7.64 ml of 0.04475-C NaOH
$C_{11}H_{10}O_2S_2$. Equiv. wt. calc. 238.1, found 237.9.

Weighed amounts of the acid were made up to 10.01 ml with different solvents and the optical activity measured. In the case of water the acid was dissolved in 2.0 ml of ethanol, neutralised with 0.1-C NaOH and finally made up to 10.01 ml with water.

Solvent	mg of acid	$2\alpha_D^{25}$	$[\alpha]_D^{25}$	$[M]_D^{25}$
Acetone	126.9	− 2.434°	− 95.9°	− 228.4°
Ethanol	95.3	− 1.570°	− 82.4°	− 196.2°
Benzene	93.7	− 1.520°	− 81.2°	− 193.2°
Glacial acetic acid	98.1	− 1.688°	− 86.1°	− 204.6°
Water (ion)	89.5	− 1.152°	− 64.4°	− 153.4°

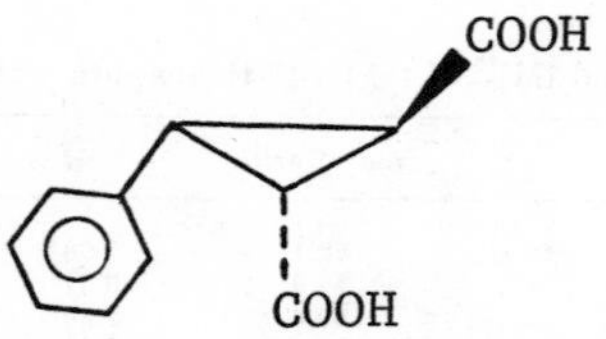

$C_{11}H_{10}O_4$ Ref. 760

Resolution of *trans*-3-Phenylcyclopropane-1,2-dicarboxylic Acid.—To a solution of 0.47 g. of the acid in 3 ml. of acetone was added 1.06 g. of brucine. The solution was warmed until solution was complete. Cooling overnight yielded a negligible amount of solid material. However, it was found that upon the addition of more acetone, a solid did separate. A total of 15 ml. of acetone was added and the mixture allowed to remain in the ice-box for three days. Filtration yielded 0.725 g. of white solid. No attempt was made to purify this further.

Decomposition with dilute acid followed by ether extraction, drying and removal of the ether gave a sample of impure acid which was converted directly to the dimethyl ester with diazomethane. The ester then was purified by recrystallization from hexane, m.p. 78–80°, $[\alpha]^{20}_D$ −20.4°, in methanol. The filtrate from the removal of the solid salt was taken to dryness and worked up in the same manner to give *d*-dimethyl ester, m.p. 76–78°, $[\alpha]^{20}_D$ 42.3°.

Note:

According to author, only partial resolution was achieved.

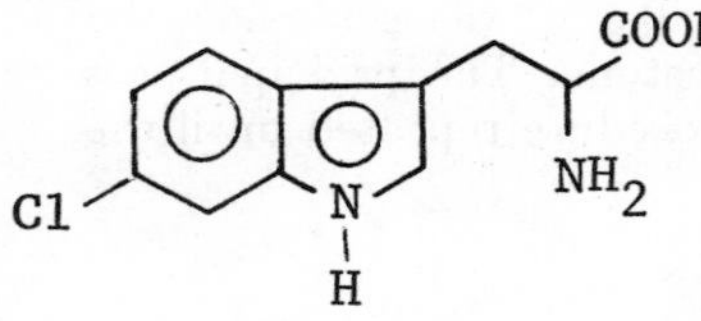

$C_{11}H_{11}ClN_2O_2$ Ref. 761

Resolution of 4:

Racemic-*N*-formyl-6-chlorotryptophan (**4**, 55.8 g, 209 mmol) suspended in ethyl methyl ketone (170 ml) is treated with *S*(−)-α-methyl-*p*-nitrobenzylamine (**1b**; 34.8 g, 209 mmol) and the resulting solution is seeded with the pure salt previously prepared (in lower yield) by a similar process using acetone as the solvent. The mixture is stirred for 4 h at room temperature until the oil that forms initially slowly crystallizes. The product is washed four times with ethyl methyl ketone and dried to give the *S*-α-methyl-*p*-nitrobenzylamine salt of D-(+)-*N*-formyl-6-chlorotryptophan; yield: 48.3 g; $[\alpha]^{23}_D = -17°$ ($c = 1\%$ in methanol). Recrystallization of the salt twice from water (10 ml/g) gives the pure salt; yield: 32.2 g (71%); m.p. 183–184°; $[\alpha]^{23}_D = -28°$ ($c = 1\%$ in methanol). When this salt (32.2 g) is slurried with 1 normal hydrochloric acid (80 ml) for 1 h, D-*N*-formyl-6-chlorotryptophan hydrate is obtained; yield: 20.3 g (96%); m.p. 143–145°; $[\alpha]^{23}_D = -47°$ ($c = 1\%$ in methanol). Hydrolysis of this product with refluxing 2 normal acetic acid (100 ml) for 24 h gives D-6-chlorotryptophan[11]; yield: 12.0 g (71%); m.p. 264° (dec.); $[\alpha]^{23}_D = +28°$ ($c = 1\%$ in methanol), having O.R.D. and C.D. curves similar to those of D-tryptophan.

$C_{11}H_{11}ClN_2O_2$	calc.	C 55.34	H 4.64	N 11.74	Cl 14.85
(238.7)	found	55.39	4.78	11.79	14.74

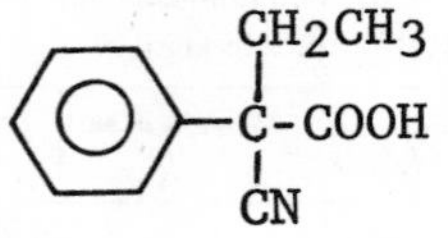

$C_{11}H_{11}NO_2$ Ref. 762

(−)-2-Cyano-2-phenylbutyric Acid.—A modification of the procedure of Cram and Haberfield[13] was used. *dl*-2-Cyano-2-phenylbutyric acid (VII, 100 g) was added to a solution of 100 g of quinine in 200 ml of MeOH. The solution was cooled in an ice bath and stirred vigorously until a precipitate formed. The suspension was then further cooled in a Dry Ice–Me_2CO bath and then filtered. The solid was stirred with *i*-PrOH and filtered, then stirred with Et_2O and filtered again, yielding 87 g of salt, partially melting at 125–135°, with the balance melting at 145–147°, $[\alpha]^{25}_D$ −119° (in MeOH; all rotations are of 1% concentration solution). A 2-g sample was converted to the free acid by dissolving in dilute HCl, extracting the aqueous mixture and Et_2O, and concentrating the Et_2O layer to give 0.8 g of the acid as a colorless oil, n^{26}_D 1.4998, $[\alpha]^{26}_D$ −13° (in $CHCl_3$). The quinine salt was recrystallized three times by rapidly dissolving in $CHCl_3$ and precipitating with *i*-Pr_2O, to give 60 g of salt. Samples from the three recrystallizations gave the following rotations: $[\alpha]^{25}_D$ −120, −121, and −121° (in MeOH). Recrystallization from MeOH did not change the rotation. Hydrolysis gave the acid $[\alpha]^{25}_D$ −19° (in $CHCl_3$), n^{25}_D 1.5139.

(13) D.J. Cram and P. Haberfield, J. Am. Chem. Soc. 83, 2354 (1961)

$C_{11}H_{11}NO_2$ Ref. 762a

Äthyl-phenyl-cyanessigsäure [1A]

(-)Antipode [(-)1A] mit Chinin [32].

100 g razem. [1A] wurden zu einer kalten Lösung von 100 g Chinin in 200 ml Methanol gegeben und bei Raumtemp. möglichst rasch in Lösung gebracht. Nach 48 Std. bei -30° wurden die ausgefallenen Kristalle scharf abgesaugt und zuerst aus 1L und dann aus 0,7L Methanol umkristallisiert. Dabei wurde höchstens auf 30° erwärmt und das Diastereomerensalz zum rascheren Lösen fein gepulvert.

Die nach der letzten Umkristallisation erhaltenen stumpfen Nadeln wurden sofort mit H_2SO_4 zerlegt und lieferten 13,5 g [(-)1A] in farblosen Nadeln vom Schmp. 32°-37°. Nach Lösen in Benzol wurde mehrere Min. im offenen Kölbchen zur Entfernung von Wasser gekocht und nach dem Erkalten mit Petroläther versetzt. Nach 24 Std. bei 4° erhielten wir 12,2 g [(-)1A] in farblosen derben Platten oder Würfeln vom Schmp. 89°-91°, Ausbeute 24,4% d.Th.
(Lit.[32] Schmp.: 41° ; 29% d.Th. ; $[\alpha]^{20}_D$ = - 21°; c = 10%; $CHCl_3$.)

$C_{11}H_{11}NO_2$ (189,2)	Ber.:	C 69,82	H 5,86	N 7,40
	Gef.:	C 69,9	H 5,88	N 7,4

Spezifische Drehung

c = 5,00 , 1 dm Rohr , t = 20° , Chloroform

α_{578} = − 1,170° $[\alpha]^{20}_{578}$ = − 23,4° $[\alpha]^{20}_{D}$ = − 22,1°

α_{546} = − 1,345° $[\alpha]^{20}_{546}$ = − 26,7°

c = 3,00 , 1 dm Rohr , t = 20° , abs. Äthanol

α_{578} = − 0,530° $[\alpha]^{20}_{578}$ = − 17,7° $[\alpha]^{20}_{D}$ = − 17,0°

α_{546} = − 0,595° $[\alpha]^{20}_{546}$ = − 19,8°

α_{436} = − 0,975° $[\alpha]^{20}_{436}$ = − 32,5°

α_{405} = − 1,165° $[\alpha]^{20}_{405}$ = − 38,8°

α_{365} = − 1,505° $[\alpha]^{20}_{365}$ = − 50,2°

Spezifische Drehung

c = 2,77 , 1 dm Rohr , t = 20° , abs. Äthanol

α_{578} = − 1,075° $[\alpha]^{20}_{578}$ = − 38,7° $[\alpha]^{20}_{D}$ = − 37,7°

α_{546} = − 1,230° $[\alpha]^{20}_{546}$ = − 44,3°

α_{436} = − 2,205° $[\alpha]^{20}_{436}$ = − 79,4°

α_{405} = − 2,715° $[\alpha]^{20}_{405}$ = − 97,8°

α_{365} = − 3,730° $[\alpha]^{20}_{365}$ = − 134,4°

(+)Antipode [(+)1A] mit Codein.

52,7 g (+)angereichertes [1A] aus der Mutterlauge des Chinin-ansatzes wurden in 1L Methanol gelöst und zu einer Lösung von 52,7 g Codein in 1L Methanol und 0,85L Wasser gegeben. Nach 48 Std. bei −30° wurden die ausgefallenen farblosen feinen Kristalle scharf abgesaugt und sofort aus 4L 50 proz. Methanol umkristallisiert. Dabei wurde nicht über 30° erwärmt. Die dabei erhaltenen durchscheinenden stumpfen Nadeln wurden sofort mit H_3PO_4 zerlegt und lieferten 15,1 g [(+)1A] vom Schmp. 30°–36°. Die Drehung betrug: $[\alpha]^{20}_{578}$ = + 16,3°

c = 2,00 , 1 dm Rohr , t = 20° , abs. Äthanol

α_{578} = + 0,325°

Wir kristallisierten [(+)1A] aus Benzol/Petroläther um, wie auf S. 74/75 beschrieben und erhielten aus den farblosen leicht zerfließlichen Nadeln fest durchscheinende Platten oder Würfel vom Schmp. 88°–90°.

$C_{11}H_{11}NO_2$ (189,2) Ber.: C 69,82 H 5,86 N 7,4
Gef.: C 69,9 H 5,82 N 7,3

Spezifische Drehung

c = 3,000 , 1 dm Rohr , t = 20° , abs. Äthanol

α_{578} = + 0,515° $[\alpha]^{20}_{578}$ = + 17,2° $[\alpha]^{20}_{D}$ = + 16,5°

α_{546} = + 0,585° $[\alpha]^{20}_{546}$ = + 19,2°

α_{436} = + 0,990° $[\alpha]^{20}_{436}$ = + 33,0°

α_{405} = + 1,170° $[\alpha]^{20}_{405}$ = + 39,0°

α_{365} = + 1,520° $[\alpha]^{20}_{365}$ = + 50,7°

(32) D.J. Cram and P. Haberfield, J. Am. Chem. Soc. 83, 2354 (1961).

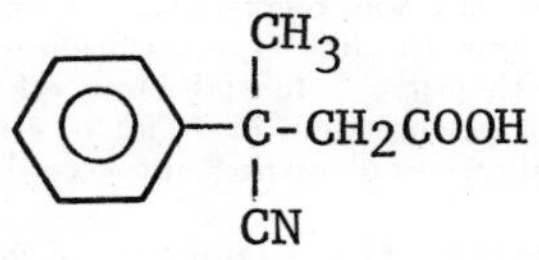

$C_{11}H_{11}NO_2$ Ref. 763

A solution of 250 g. of this crude acid and 250 g. of quinine in 500 ml. of methanol was cooled, and the salt that separated was collected. This material was recrystallized three times from methanol, care being taken never to heat any solutions over 25°. Hydrolysis of samples of salt from these three recrystallizations yielded 2-cyano-2-phenylbutanoic acid of the rotations: $[\alpha]^{25-27}$D −18.85, −21.0 and −21.0° (*c* 10% in $CHCl_3$), respectively. The main batch of salt after the third recrystallization was shaken with a mixture of 200 ml. of dilute sulfuric acid and 400 ml. of ether. The ether solution was washed with water, dried, and evaporated under reduced pressure below 25°. Solvent was removed from the residual oil at 1 mm. and 25°, and the residue crystallized when cooled (white needles), m.p. 41–43° (sealed tube), wt. 37 g., $[\alpha]^{25}$D −21.0° (*c* 10% in $CHCl_3$). After standing at 0° for about one year, this acid lost its hygroscopic character, and its melting point changed to 86–87° (sealed tube). At 0° the compound is optically stable for long periods of time.

Anal. Calcd. for $C_{11}H_{11}NO_2$: C, 69.82; H, 5.85; N, 7.40. Found: C, 69.58; H, 6.03; N, 7.24.

An additional 26 g. of optically pure acid was obtained from recrystallizations of second crops of the quinine salt.

(+)-Ammonium 2-Cyano-2-phenylbutanoate.—Into a solution of 6.5 g. of (−)-2-cyano-2-phenylbutanoic acid in 50 ml. of dry ether was passed ammonia gas. The resulting precipitate of ammonium salt amounted to 6.7 g., m.p. 111° dec., $[\alpha]^{25}$D +29.7° (*c* 10% in CH_3OH).

Anal. Calcd. for $C_{11}H_{14}N_2O_2$: C, 64.06, H, 6.84. Found: C, 63.83; H, 6.83.

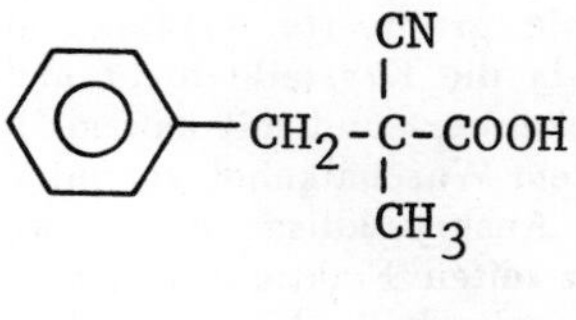

$C_{11}H_{11}NO_2$ Ref. 764

Resolution. Brucine (3·4 g.) was added to a solution of the (±)-acid (1·5 g.) in aqueous acetone (50 c.c. of 75%). The crystals obtained overnight (1·4 g.), m. p. 125—132°, were twice recrystallised from the same solvent, to give needles (0·8 g.), m. p. 135—137°.

The racemic acid (20 g.) and brucine (45 g.) were dissolved in hot aqueous acetone (200 c.c. of 75%), and the cooled solution inoculated with a crystal of the previously prepared salt. After 2 days the crystals were removed by filtration, providing needles, m. p. 135—137° (19 g.). Decomposition gave the (+)-acid (plates; 4·0 g.) which, after recrystallisation from ligroin, had m. p. 87·5—88·5°, $[\alpha]^{19}_{D}$ +25·1° (*l*, 2; *c*, 2·43 in chloroform).

The mother-liquors containing the (−)-acid were concentrated to dryness *in vacuo*. The resulting solid (33 g.) was recrystallised three times from aqueous ethanol (75%), when plate-like crystals (17 g.) of m. p. 127—129° were obtained. Decomposition of the salt with dilute hydrochloric acid gave the (−)-acid (3·0 g.) as platelets, m. p. 88—89° (after recrystallisation from chloroform–ligroin), $[\alpha]^{18.5}_{D}$ −25·7° (*l*, 2; *c*, 2·54 in chloroform).

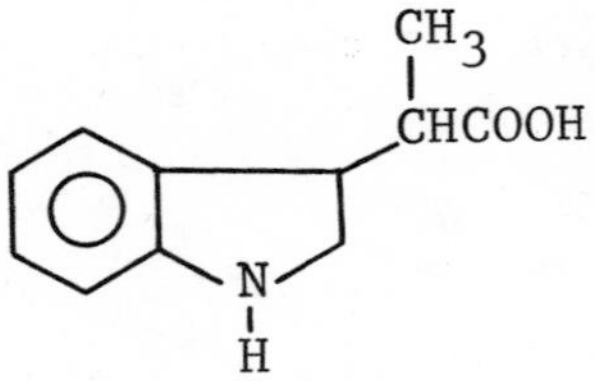

$C_{11}H_{11}NO_2$ Ref. 765

α-(3-Indolyl)-propionic acid (I)

The racemic acid was prepared according to Kögl and Verkaaik (2) in a yield ranging from 30 to 40 %.

Resolution of the racemic acid

Preliminary experiments were carried out with different bases and solvents. From the crystals formed the acid was liberated and the optical activity measured in absolute ethanol (Table 2).

A mixture of 7.6 g (0.04 mole) of racemic acid and 18.7 g (0.04 mole) of brucine was dissolved in 50 ml of hot ethanol and 20 ml of water. The salt was allowed to crystallize in a refrigerator over night and was then filtered off. Some acid was isolated from a sample and its optical activity measured in absolute ethanol. The remaining salt was recrystallized from dilute alcohol (Table 3).

The mother liquor from the first crystallization was evaporated to dryness and 3.5 g acid obtained with $[\alpha]_D^{25} = -46°$ (absolute ethanol). 3.3 g of this acid was dissolved with 5.1 g of cinchonidine in 75 ml of ethyl acetate and 5 ml of methanol. The recrystallizations were carried out according to Table 4.

Table 2. Preliminary tests on the resolution of α-(3-indolyl)-propionic acid.

Base	$[\alpha]_D^{25}$ of acid in abs. ethanol			
	From methanol	From ethanol	From acetone	From ethyl acetate
Cinchonine	oil	− 3°	− 5°	oil
Cinchonidine	− 20°	− 9°	oil	− 26°
Quinine	+ 11°	+ 6°	+ 5°	+ 4°
Quinidine	oil	oil	oil	oil
Brucine	+ 42°	+ 52°	+ 21°	± 0
Strychnine	oil	oil	oil	oil
Morphine	oil	oil	oil	oil
(+)-α-Phenylethylamine	oil	oil	oil	oil
(+)-α-(2-Naphthyl)-ethylamine	+ 8°	+ 10°	+ 3°	+ 5°

Table 3. Recrystallization of the brucine salt of α-(3-indolyl)-propionic acid.

Cryst. no.	Solvent (ml)		Weight of salt (g)	$[\alpha]_D^{25}$ of acid
	Ethanol	Water		
1	50	20	12.3	+ 56°
2	20	5	10.0	+ 69°
3	20	5	9.3	+ 77°
4	20	5	8.7	+ 77°

Table 4. Recrystallization of the cinchonidine salt of α-(3-indolyl)-propionic acid.

Cryst. no.	Solvent (ml)		Weight of salt (g)	$[\alpha]_D^{25}$ of acid
	Ethyl acetate	Methanol		
1	75	5	7.4	− 58°
2	75	5	6.5	− 69°
3	60	4	5.6	− 75°
4	50	3	4.8	− 77°
5	50	3	4.1	− 77°

Table 5. Optical activity of (+)-α-(3-indolyl)-propionic acid.

Solvent	mg of acid	$2\,\alpha_D^{25}$	$[\alpha]_D^{25}$	$[M]_D^{25}$
Benzene	82.9	1.76°	106.2°	200.9°
2,2,4-Trimethylpentane	23.8	0.46°	97°	184°
Acetic acid	83.8	1.61°	96.1°	181.8°
Acetone	81.8	1.51°	92.3°	174.6°
Chloroform	88.7	1.48°	83.4°	157.8°
Ethanol	85.0	1.35°	79.4°	150.2°
Water (ion)	90.7	0.31°	17.1°	32.4°

(+)-α-(3-Indolyl)-propionic acid

The last salt fraction (Table 3) was treated with sulphuric acid and the organic acid extracted with ether. The crude acid (2.4 g) was recrystallized from benzene. Yield 2.2 g with m.p. 138–140°. The equivalent weight could be determined by titration with 0.1-*N* NaOH (phenolphthalein).

Anal. Equiv. wt. calc. 189.2; found 189.5.

Optical activity in different solvents

The acid was dissolved in different solvents to a volume of 10.00 ml and the rotatory power determined. The results are given in Table 5.

(−)-α-(3-Indolyl)-propionic acid

From the last fraction of the cinconidine salt there was obtained 1.4 g acid. It was recrystallized from benzene. Yield 1.2 g with m.p. 138–140°.

Anal. Epuiv. wt. calc. 189.2; found 189.6.

0.0832 g acid dissolved in absolute ethanol to 10.00 ml:
$2\alpha_D^{25} = -1.31°$, $[\alpha]_D^{25} = -78.8°$, $[M]_D^{25} = -149.1°$.

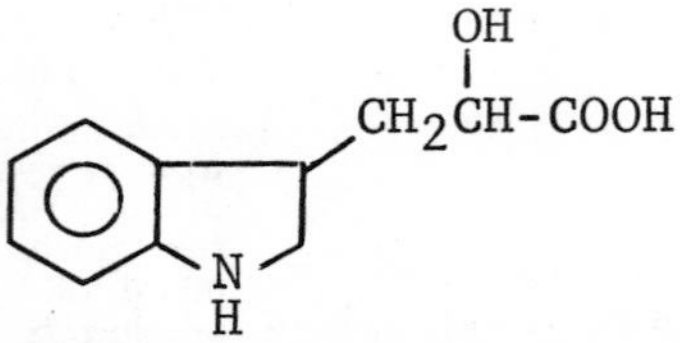

$C_{11}H_{11}NO_3$ Ref. 766

Spaltung der d,l-Indolmilchsäure in ihre Komponenten mittels Chinin. In einem Erlenmeyerschen Kolben von 100 ccm Inhalt wurden 7,0 g d,l-Indolmilchsäure und 10,3 g Chinin in 50 ccm Methylalkohol aufgelöst und in der Kälte freiwillig verdunsten gelassen. Nach 10—15 Tagen schieden sich rosettenförmig gruppierte Krystalle aus, die im Laufe der Zeit langsam wuchsen. Als die Krystalle nicht mehr zunahmen, wurden sie von der Mutterlauge dekantiert und mit kaltem Methanol ausgewaschen. Das Dekantat wurde mit dem Waschalkohol vereinigt, wieder wie oben behandelt und einer zweiten Auskrystallisation unterworfen. Die spezifische Drehung der ersten und zweiten Fraktion betrug $[\alpha]_D = -$ 105,41—104,26.

Da beide Fraktionen, wie angegeben, fast dieselbe Drehung zeigten, wurden sie vereinigt, in verdünnter Salzsäure gelöst und im Extraktionsapparat mit Äther erschöpfend extrahiert. Der Äther wurde bis zu einer kleinen Menge verdunsten gelassen und etwas Petroläther hinzugefügt. Am nächsten Morgen hatten sich derbe, farblose Krystalle nebst feineren, etwas gefärbten ausgeschieden. Die ersteren schmolzen bei 100°, die letzteren bei 97°. Da beide Krystalle aus derselben Substanz zu bestehen schienen, wurden sie zusammen in Wasser gelöst und ihre spezifische Drehung bestimmt.

Substanz 0,2267 g, Lösung 21,1235 g, 1,06 %, Rohr 2 dm.

α + 0,11, $[\alpha]_D$ + 5,18.

Nach der Angabe von Ehrlich und Jacobson[1]) beträgt die spezifische Drehung einer reinen Linksindolmilchsäure $[\alpha]_D = -$ 5,34. Unsere Substanz wurde also als die reine Rechtsindolmilchsäure erkannt.

[1]) Ber. chem. Ges. **44**, 888 (1911).

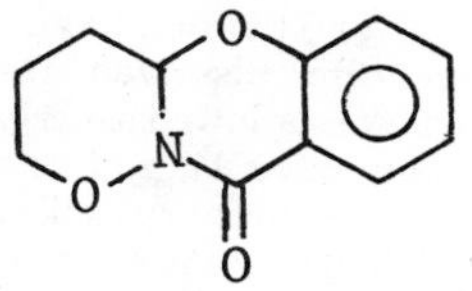

$C_{11}H_{11}NO_3$ Ref. 767

5.4.7. (—)-2,3,4,4a-Tetrahydro-10H-1,2-oxazino[3,2-b][1,3]benzoxazin-10-one (50) and *(+)*-2,3,4,4a-tetrahydro-10H-1,2-oxazino-[3,2-b][1,3]benzoxazin-10-one (51)

10.7 g of 2-(3-chloropropyl)-2,3-dihydro-3-hydroxy-4H-1,3-benzoxazin-4-one (mp. 76—80° C) obtained as described for the preparation of **49** was dissolved in 350 ml of anhydrous Et_2O with warming and 16.2 g of quinine was added and the mixture was stirred for 2 h at room temperature. The precipitate was removed by filtration. The filtrate was set aside and the solid (10.9 g) was slurried with 75 ml of trichloroethylene, filtered and dried. The salt, $[\alpha]^{23}$ —141° (c. 1.0, EtOH), was treated with dilute HCl and extracted with Et_2O. The Et_2O solution was washed successively with dilute HCl and H_2O saturated with NaCl, dried and evaporated to dryness. The solid residue was recrystallized from cyclohexane. The (—)-2-(3-chloropropyl)-2,3-dihydro-3-hydroxy-4H-1,3-benzoxazin-4-one, $[\alpha]^{23}$ —45.6° (c 5.0, EtOH) and mp. 73—75° C, was cyclized with dilute NH_4OH and the crude product was dissolved in hot i-PrOH, and the solution was allowed to cool. Solid was removed and the filtrate was evaporated to dryness. The solid residue was recrystallized from EtOH to give **50**.

The filtrate which was set aside was evaporated to dryness and the residue was treated with 100 ml of Et_2O. The solid was removed by filtration, and the filtrate was washed with dilute HCl and with H_2O saturated with NaCl, dried and evaporated to dryness. The solid residue was recrystallized twice from cyclohexane. The (+)-2-(3-chloropropyl)-2,3-dihydro-3-hydroxy-4H-1,3-benzoxazin-4-one, $[\alpha]^{24}$ +41.4° (c 5.0, EtOH) and mp. 65—69° C, was converted to crude **51** with dilute NH_4OH. For purification, the solid was dissolved in hot i-PrOH and treated as described above for **50**.

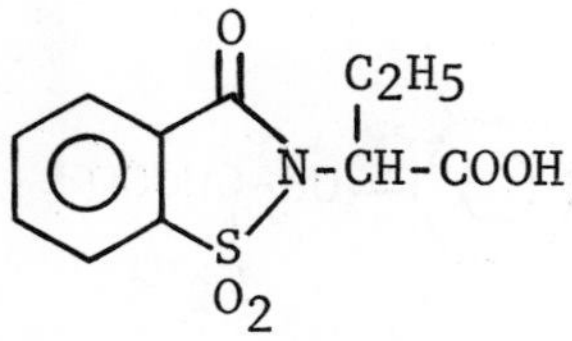

$C_{11}H_{11}NO_5S$ Ref. 768

*(−)-α-Ethyl[3-oxo-1,2-benzisothiazole-2(3*H*)]acetic acid 1,1-dioxide (−2).* Racemic acid (1077 g; 4 mol) and cinchonine (1178 g; 4 mol) were dissolved with heating in a mixture of acetone (16 l) and water (3.2 l). The salt which precipitated on cooling was recrystallized from acetone-water as shown in Table 1. The purification was followed by measurement of the optical rotation of the acid, liberated from 1 g samples of the salt by stirring with 2 N HCl (25 ml), filtering off the solid, washing with much cold water and drying *in vacuo*. The salt was considered optically pure after the sixth crystallization, as three additional crystallizations of a small sample of the salt did not change the optical rotation. The acid was liberated from this preparation and crystallized from methanol. M.p. 139 °C (with a transition point at 134.0−134.5 °C). $[\alpha]_D^{24}$ −61.9±0.1° (*c* 2, acetone).

*(+)-α-Ethyl[3-oxo-1,2-benzisothiazole-2(3*H*)]-acetic acid 1,1-dioxide (+2).* The mother liquor from the first crystallization of the cinchonine salt was evaporated to dryness and the residue was crystallized from 9 l of acetone-water (8:1) to give 409 g salt. The acid liberated from this salt had $[\alpha]_D^{24}$ −6° (*c* 1, acetone). The mother liquor from this second crop of cinchonine salt was evaporated and the acid was liberated from the residue yielding 340 g of product with $[\alpha]_D^{24}$ +29.5° (*c* 1, acetone). This acid (340 g; 1.2 mol) and brucine (500 g; 1.2 mol) were dissolved in hot methanol (95 %; 10 l). On cooling, 465 g of crystalline salt was obtained, which was further recrystallized as shown in Table 2. The purification was followed as described above for the (−)-acid. A small sample of the salt $[\alpha]_D^{24}$ +22.5° was recrystallized four more times from 95 % methanol to a rotation of $[\alpha]_D^{24}$ +25.0°. This rotation remained unchanged after two further crystallizations. The +2 liberated from this final salt had $[\alpha]_D^{24}$ +61.4° (*c* 2, acetone) m.p. 139.5 °C (with a transition point at 133.5−135.0 °C). An attempt to carry out a "triangular" fractional crystallization [6] was unsuccessful, probably because too much decomposition products had accumulated in the mother liquor.

6. Tipson, R. S. In Weissberger, A., Ed., *Technique of Organic Chemistry*, Interscience, New York 1950, Vol. 3, p. 425.

Table 1. **Fractional crystallization of the cinchonine salt of *2* to produce the (−)-form.**

Crystallization No.	1	2	3	4	5	6
Acetone/ml	16000	8500	5000	3000	2000	1010
Water/ml	3200	2200	3000	1600	1500	550
Salt/g	858	502	297	210	106	70
$[\alpha]_D^{24}$ of salt (*c* 1, methanol)/deg.	−108					+105
$[\alpha]_D^{24}$ of acid (*c* 1, acetone)/deg.	−20.7	−40.5	−53.0	−58.0	−59.8	−61.5

Table 2. **Fractional crystallization of the brucine salt of *2* to produce the (+)-form.**

Crystallization No.	1	2	3	4	5	9	10−11
95 % Methanol/ml	10000	5700	3700	2050	1200		
Salt/g	465	300	174	99.5	65.5		
$[\alpha]_D^{24}$ of salt (*c* 2, acetone)/deg.	+13.3	+16.7	+19.5	+21.1	+22.5	+25.5	+25.0
$[\alpha]_D^{24}$ of acid (*c* 1, acetone)/deg.	+37.3	+42.1	+51.2	+54.5	+56.4		+61.4

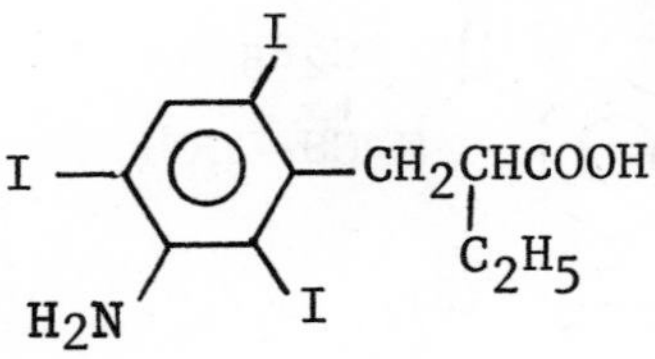

$C_{11}H_{12}I_3NO$ **Ref. 769**

(±)-α-Ethyl-β-(3-amino-2,4,6-triiodophenyl)propionic acid (57.1 g, 0.1 mole) and 12.1 g (0.1 mole) of (−)-α-phenylethylamine were dissolved in 700 ml of boiling 95% EtOH. After decoloration with active charcoal, the solution was kept overnight at room temperature. The (−)-α-phenylethylamine salt was filtered off (29 g), mp 162–163°, $[\alpha]^{20}_D$ −4.0° (*c* 2, EtOH). The mother liquor was used for recovery of the optical isomer. Recrystallization from EtOH gave 10.5 g of crystals, mp 167°, $[\alpha]^{20}_D$ −4.0° (*c* 2, EtOH) (30% over-all).

Anal. Calcd for $C_{19}H_{23}I_3N_2O_2$: C, 32.97; H, 3.25; I, 55.01. Found: C, 33.00; H, 3.29; I, 54.95.

The mother liquor containing the (−)-α-phenethylamine salt was evaporated to dryness. The residue was stirred with 500 ml of 4% NaOH solution and (−)-α-phenethylamine was extracted with four 50-ml portions of ether. After acidification with HCl, 29 g of (+)-α-ethyl-β-(3-amino-2,4,6-triiodophenyl)-propionic acid, mp 75–110°, $[\alpha]^{20}_D$ +3.1 (*c* 2, EtOH), was obtained. This acid (28.6 g, 0.05 mole) and 6.05 g (0.05 mole) of (+)-α-phenethylamine were dissolved in 600 ml of boiling EtOH and the solution was allowed to crystallize at room temperature overnight. This gave 17 g of salt, mp 166–167, $[\alpha]^{20}_D$ +3.6 (*c* 2, EtOH). After further crystallization from 240 ml of EtOH, we obtained 12.2 g of white crystals, mp 167°, $[\alpha]^{20}_D$ +4.1 (*c* 2, EtOH) (35%).

Anal. Calcd for $C_{19}H_{23}I_3N_2O_2$: C, 32.97; H, 3.35; I, 55.01. Found: C, 32.88; H, 3.49; I, 54.75.

(−)-α-Ethyl-β-(3-amino-2,4,6-triiodophenyl)propionic Acid.—The (−)-α-phenethylamine salt (8.2 g) was dissolved in 280 ml of 7% NaOH solution and extracted with three 50-ml portions of ether. The water solution was acidified with HCl, and the acid was filtered, washed (H_2O), and crystallized from 40 ml of AcOH; yield 6 g, mp 162–163°, $[\alpha]^{20}_D$ −5.2 ± 0.1° (*c* 2, EtOH).

Anal. Calcd for $C_{11}H_{12}I_3NO_2$: C, 23.14; I, 66.68. Found: C, 23.15; I, 66.80.

(+)-α-Ethyl-β-(3-amino-2,4,6-triiodophenyl)propionic acid was obtained in the same way as the (−) form, operating on the other enantiomeric salt. From 10 g of salt was obtained 6.9 g of (+) acid, mp 162°, $[\alpha]^{20}_D$ +5.1 ± 0.1° (*c* 2, EtOH).

Anal. Calcd for $C_{11}H_{12}I_3NO_3$: C, 23.14; I, 66.68. Found: C, 23.14; I, 66.70.

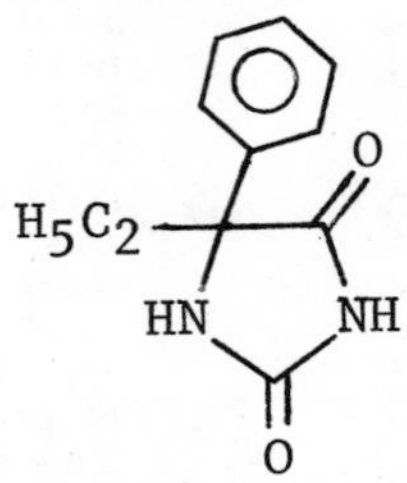

$C_{11}H_{12}N_2O_2$ **Ref. 770**

Resolution of Racemic Phenylethylhydantoin into its Optically Active Components.—Equivalent amounts of 5,5′-phenylethylhydantoin (1 part) m. p. 198°, and of brucine (2.3 parts) are dissolved in absolute alcohol (10–15 parts). On standing, rosets of crystals separate (about 1.6 parts), consisting of the brucine salt of dextro-phenylethylhydantoin. An excess of dilute sulfuric acid is added to a solution of these crystals in absolute alcohol and sufficient water is added to reduce the alcohol content to 10%. The free dextro-phenylethylhydantoin crystallizes in white flaky platelets. They are recrystallized several times from dilute alcohol until free from brucine. The pure substance melts sharply at 237°, and the purest specimens showed a rotation $[\alpha]_D$ +123° in alcoholic solution (0.5514 g. in 25 cc., 2-dm. tube, α_D +5.36°). The specific rotation in alkaline aqueous solution was +169° (0.2521 g. in 25 cc., 2-dm. tube, α_D +3.40°).

Anal. Calcd. for $C_{11}H_{12}O_2N_2$ (204.10): C, 64.67; H, 5.93; N, 13.72. Found: C, 64.09; H, 6.18; N, 13.96.

The mother liquor from the brucine-dextro-hydantoin is carefully acidified until it becomes cloudy. On standing, the levo-phenylethylhydantoin separates in solid crystals. It is purified in the same manner as its enantiomer until its melting point reaches 235–237°; $[\alpha]_D$ −121° (0.2562 g. in 25 cc. alcohol, 2-dm., α_D −2.49°); −167° in aqueous alkali (0.2981 g. in 25 cc., 2-dm., α_D −4.00°).

Anal. Calcd. for $C_{11}H_{12}O_2N_2$ (204.10): C, 64.67; H, 5.93; N, 13.72. Found: C, 64.04; H, 6.11; N, 13.74.

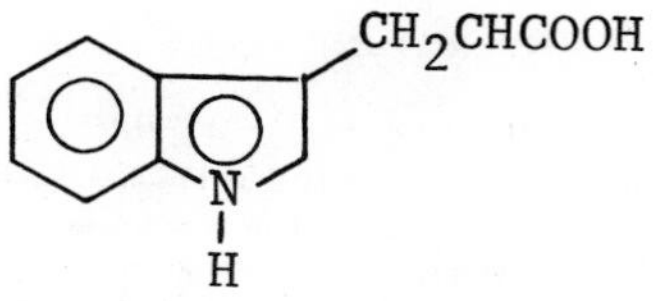

$C_{11}H_{12}N_2O_2$ Ref. 771

Brucine Salt of D- and L-N-Acetyltryptophan.—A mixture of 123 g. of N-acetyl-DL-tryptophan, 208 g. of brucine and 1750 cc. of absolute ethanol was boiled under reflux until solution was effected. After cooling, seeding and storing for twelve hours, the crystalline product was separated and slurried twice with small quantities of ethanol: Weight of dried product, 161 g.; $[\alpha]^{25}D - 16.5 \pm 1°$. Recrystallization from 320 cc. of hot ethanol gave 146 g. of pure brucine salt of N-acetyl-D-tryptophan ($[\alpha]^{25}D - 18.4°$) as was indicated by its constant rotation when subjected to further recrystallization.

The brucine salt of N-acetyl-L-tryptophan was obtained from the resolution mother liquor by concentration of the solution to dryness under reduced pressure, dissolution of the residue in 320 cc. of hot methanol, charcoal treatment of the solution, dilution of the latter with 325 cc. of dry ether, seeding and storage of the mixture for several hours. The crystalline brucine salt was subjected again to the same recrystallization procedure whereby 139 g. of product was obtained; $[\alpha]^{25}D + 1.3 \pm 1°$ (*c*, 1% in water). *Anal.* Calcd. for $C_{36}H_{40}O_7N_4$: C, 67.48; H, 6.29; N, 8.74. Found: C, 67.54; H, 6.13; N, 8.69.

L- and D-Tryptophan.—To a mixture of 49 g. of the recrystallized brucine salt of N-acetyl-L-tryptophan in 170 cc. of water was added 70 cc. of cold 1*N* sodium hydroxide solution. The salt dissolved readily and very soon brucine separated. After cooling in ice for a few hours, the mixture was filtered and the brucine washed with cold water. The combined filtrate and washings were neutralized with hydrochloric acid to *p*H 7.0, concentrated under reduced pressure to 140-cc. volume, treated with charcoal and finally acidified with hydrochloric acid *p*H 3.0. The N-acetyl-L-tryptophan was collected and slurried twice with cold water; weight 16.1 g. (85% yield); $[\alpha]^{25}D + 29°$ (*c*, 1% in H_2O + 1 equivalent NaOH).

A mixture of the product with 160 cc. of 2*N* hydrochloric acid was boiled under reflux for two and one-half hours and the resulting solution was concentrated under reduced pressure to dryness. The residue was dissolved in 40 cc. of hot water and the solution was treated with charcoal. A solution of 7 g. of sodium acetate in 20 cc. of water was added to the product solution and the mixture was stored at 5° for fourteen hours. The product was recrystallized by dissolution in 84 cc. of water containing 2.8 g. of sodium hydroxide, acidifying the warmed solution (at 70°) with 4.5 cc. of acetic acid and storing the mixture at 5° for fourteen hours. The pure L-tryptophan was collected and washed with small amounts of 50% ethanol followed by ethanol and then dry ether; weight, 10.9 g.; 82% yield; $[\alpha]^{25}D -31.90$ (*c*, 1% in water + 1 equivalent of NaOH).

D-Tryptophan was obtained from the brucine salt of its N-acetyl derivative following the same procedure. The over-all yields were slightly better, however.

$C_{11}H_{12}N_2O_2$ Ref. 771a

Preparation of Quinine Diastereoisomers—75.65 gm. of quinine (0.2 mol of hydrated crystals) were added to 49.23 gm. of acetyl-*dl*-tryptophane (0.2 mol) dissolved in 750 cc. of methyl alcohol. The three major fractions of crystals obtained from this solution and from two succeeding concentrations of the mother liquors weighed 33.0 gm., $[\alpha]^{20}_D = -119.5°$; 57 gm., $[\alpha]^{20}_D = -99.7°$; and 12.5 gm., $[\alpha]^{20}_D = -119.4°$, respectively.

Three recrystallizations of the combined first and third fractions from separate 200 cc. portions of methyl alcohol yielded 27 gm. of the acetyl-*d*-tryptophane quinine salt, $[\alpha]^{20}_D = -125.4°$. Further crystallizations produced no change in optical activity. More extensive fractionations of these and other mother liquors yielded an additional 23.8 gm. of the quinine salt of the same optical activity, bringing the total to 89 per cent of the calculated amount. More intensive fractionation, although impractical, would have increased the yield.

After a single recrystallization of the second major fraction from methyl alcohol, a specific rotation of −94° was reached which was not reduced appreciably by three further crystallizations from that solvent. Inasmuch as the acetyl-*l*-tryptophane quinine salt prepared for reference from acetyl-*l*-tryptophane had shown $[\alpha]^{20}_D = -92.2°$, the 30 gm. of salt resulting from the four recrystallizations were dissolved in 75 cc. of ethyl alcohol, the solution was cooled, ether added to incipient opalescence, and the quinine salt allowed to precipitate in the cold. 27.8 gm. of acetyl-*l*-tryptophane quinine salt, $[\alpha]^{20}_D = -91.6°$, were secured. The rotation was not changed by further similar crystallizations. Subsequent fractionation of mother liquors yielded an additional 19.11 gm. of pure acetyl-*l*-tryptophane quinine salt. The combined yield was 82.3 per cent of the theoretical.

Of aid in the fractionation is the difference in appearance of the acetyl-*d*- and the acetyl-*l*-tryptophane quinine salts. The former are very fine crystals which have the appearance of short thick rods; the latter are elongated fibers which mat together and have a tendency to form rosettes.

Isolation of Acetyltryptophane Enantiomorphs—The quinine salts of acetyltryptophane were decomposed as follows: 11.43 gm. (0.2 mol) of the salt were dissolved in 35 cc. of ethyl alcohol, the solution was cooled well, and 1.6 cc. of 16 N sodium hydroxide were added. After thorough mixing, 350 cc. of cold water were added and the flask was cooled in an ice-salt bath to precipitate the bulk of the freed quinine. The precipitate was filtered off. The filtrate was extracted three times with ether to remove the unprecipitated quinine, partially neutralized with 0.9 cc. of 7 N sulfuric acid, concentrated to a small volume *in vacuo*, and acidified. The acetyltryptophane precipitate was recrystallized from water.

From 10.73 gm. of acetyl-*d*-tryptophane quinine salt so treated, 4.5 gm. of acetyl-*d*-tryptophane were obtained, representing 95.2 per cent of the calculated amount. A second run, in which 16.1 gm. of the quinine salt were employed, yielded 6.37 gm. of the acetyl derivative, or 91.7 per cent of the theoretical. The acetyl-*d*-tryptophane melted at 189–190° (corrected) with rapid heating, showed $[\alpha]^{20}_D = -26.2°$, and a nitrogen content of 11.31 per cent (calculated, 11.38).

From 21.46 gm. of the acetyl-*l*-tryptophane quinine salt, 8.7 gm. of acetyl-*l*-tryptophane (93.9 per cent of the theoretical) were obtained. The derivative melted with rapid heating at 189–190° (corrected), gave $[\alpha]^{20}_D = +26.1°$ and nitrogen, 11.32 per cent

[1] Unless otherwise indicated, optical activity measurements were made on 1 per cent solutions in methyl alcohol.

(theoretical, 11.38). A 1 per cent solution in water containing 1 equivalent of sodium hydroxide showed $[\alpha]_D^{20} = +30.3°$.

The active acetyl derivative prepared from *l*-tryptophane by a method essentially the same as that reported by du Vigneaud and Sealock (1932) also melted at 189–190° (corrected) with rapid heating and showed $[\alpha]_D^{20} = +25.1°$ ($[\alpha]_D^{25} = +29.0°$ in 1 per cent solution in water plus 1 equivalent of sodium hydroxide). Both acetyl-*d*-tryptophane and acetyl-*l*-tryptophane crystallize in long, flat rectangles. We have confirmed the observation of du Vigneaud and Sealock that equal-portions of the isomers, mixed and recrystallized from water, form the characteristic glistening rhombic platelets of acetyl-*dl*-tryptophane, m.p. 207–208° (corrected). The same melting point is obtained, however, when equal portions of the dry enantiomorphs are thoroughly mixed and tested without previous recrystallization.

Isolation of d- and l-Tryptophane—5.9 gm. of acetyl-*d*-tryptophane were refluxed with 50 cc. of 2 N sulfuric acid for 2 hours. The liberated tryptophane, isolated through the mercuric salt, weighed 3.2 gm., exhibited $[\alpha]_D^{25} = +32.45°$ (in 0.5 per cent solution in water), melted at 281–282° (corrected) with rapid heating, and showed a nitrogen content of 13.65 per cent (calculated, 13.72).

To avoid the use of the mercuric sulfate reagent, the acetic acid can be distilled off *in vacuo* and the sulfuric acid removed quantitatively by barium hydroxide; 1.46 gm. of acetyl-*d*-tryptophane so treated yielded 0.82 gm. (67.8 per cent of the theoretical) having $[\alpha]_D^{20} = +31.8°$ (in 0.5 per cent solution in water).

By the latter method 2.26 gm. of *l*-tryptophane were obtained from 4.4 gm. of the acetyl-*l*-tryptophane (61.9 per cent). The *l*-tryptophane showed $[\alpha]_D^{20} = -32.1°$ (in 0.5 per cent solution in water), melted at 281–282° (corrected) with rapid heating, and when analyzed showed 13.66 per cent nitrogen (theoretical, 13.72). $[\alpha]_D$ for *l*-tryptophane prepared from protein is usually reported between −28° and −33°.

$C_{11}H_{12}N_2O_2$ Ref. 771b

Resolution of N-Carbobenzoxy-DL-tryptophan. a. Separation of the Diastereoisomeric Salts with Quinine.—The racemic carbobenzoxy derivative (7.6 g.) and 7.6 g. of quinine were heated under reflux for 30 minutes with 100 cc. of acetone, an amount insufficient to dissolve all the salts. After cooling to room temperature the solid was filtered and air-dried; yield 12.2 g., $[\alpha]^{25}D$ −91.02° (*c* 2, methanol). The acetone liquor yielded no more solid. The crystals were twice recrystallized from isopropyl alcohol giving 4.9 g., $[\alpha]^{25}D$ −102.5° (*c* 2, methanol). This was the maximum rotation and represented a recovery of 65% of the less soluble diastereoisomer.

b. N-Carbobenzoxy-D-tryptophan was recovered from the less soluble salt by alkaline decomposition as described above for the salts of carbobenzoxyphenylalanine. The free quinine was removed by filtration and purified by precipitating with alkali from a decolorized solution of the hydrochloride. The aqueous alkaline filtrate from the decomposition was extracted once with chloroform to remove traces of quinine and then acidified with acetic acid. N-Carbobenzoxy-D-tryptophan precipitated as a gelatinous mass. It was filtered by suction and most of the associated water was removed by pressing the filter cake. The dry weight was 3.1 g. (theoretical 3.3 g.), m.p. 136–137°, $[\alpha]^{25}D$ +15.55° (*c* 5, 1 equiv. NaOH). Hanson and Smith[5] reported m.p. 124–126° and $[\alpha]^{29}D$ +15.4°.

c. D-Tryptophan, obtained from the carbobenzoxy derivative by the above-described hydrogenolysis, had $[\alpha]^{25}D$ +31.19° (*c* 1, water). The literature[18] value for L-tryptophan is −32.15°.

d. Impure N-carbobenzoxy-L-tryptophan was obtained from the more soluble quinine salts. It had $[\alpha]^{25}D$ −10.11° (*c* 5, 1 equiv. NaOH). No attempt was made to purify the N-carbobenzoxy-L-tryptophan contained in this material.

***J. Biol. Chem.*,**

(5) H. T. Hanson and E. L. Smith, *ibid.*, **179**, 815 (1949).

(18) M. S. Dunn and L. B. Rockland, *Advances in Protein Chem.*, **3**, 296 (1947).

$C_{11}H_{12}N_2O_2$ Ref. 771c

B. Spaltung des D,L-Methylesters

7,5 g roher D,L-Tryptophanmethylester in 50 cm³ absol. Äthanol werden mit einer auf 50° erwärmten Lösung von 6,4 g Dibenzoyl-D-weinsäure in 150 cm³ abs. Äthanol vereinigt. Nach einigen Stunden beginnen sich Kristallkeime abzuscheiden. Man verdünnt die Lösung mit dem gleichen Volumen (200 cm³) absol. Äther, der in 3–4 Portionen im Verlaufe eines Tages zugesetzt wird. Die Kristallisation des L-Tryptophanmethylester-dibenzoyl-D-hydrogentartrates ist nach insgesamt 24 Stunden beendet.

Ausbeute: 8–9 g; Schmp. 179–181° (korr.); $[\alpha]_D^{20}$: −65,2° (c = 0,52 in Methanol).

$C_{30}H_{28}O_{10}N_2$ (576,4)

Ber.	C 62,50	H 4,86	N 4,86
Gef.	C 62,35	H 5,18	N 4,82

Die Mutterlauge des Spaltansatzes wird im Vakuum auf 20 cm³ eingeengt, unter Kühlung trockener Chlorwasserstoff bis zur deutlich sauren Reaktion eingeleitet und das rohe D-Tryptophanmethylester-hydrochlorid mit viel absol. Äther ausgefällt.

Ausbeute: 4,1 g; Schmp. 211–213° (korr.); $[\alpha]_D^{20}$: −12,2° (c = 4,06 in Methanol).

C. Gewinnung der reinen L- und D-Tryptophanmethylester-hydrochloride

L-Tryptophanmethylester-dibenzoyl-D-hydrogentartrat wird mit der vierfachen Menge absol. Methanol übergossen und Chlorwasserstoff eingeleitet, bis eine klare, stark saure Lösung entstanden ist. Durch Fällung mit viel absol. Äther gewinnt man das rohe L-Esterhydrochlorid.

Ausbeute: 3,6 g; Schmp.: 214–215° (korr.); $[\alpha]_D^{20}$: +14,2° (c = 4,81 in Methanol).

Die so gewonnenen rohen antipodischen Methylesterhydrochloride kristallisiert man aus Methanol-Äther unter Gewinnung der leichter löslichen Anteile um. Man erhält so annähernd 50% der eingesetzten Rohhydrochloride optisch rein.

L-Tryptophanmethylester-hydrochlorid: Schmp.: 216–218° (korr.); $[\alpha]_D^{20}$: +18,2° (c = 3,16 in Methanol).

$C_{12}H_{14}O_2N_2 \cdot HCl$ (254,7)

Ber.	C 56,47	H 5,88	N 10,98
Gef.	C 56,41	H 6,13	N 11,29.

D-Tryptophanmethylester-hydrochlorid: Schmp.: 217–219° (korr.); $[\alpha]_D^{20}$: −17,8° (c = 3,87 in Methanol)

Gef.	C 56,16	H 6,00	N 11,32

Die optischen Reinheitsgrade der Rohesterhydrochloride betrugen somit 77 und 70%.

D. Gewinnung von L- und D-Tryptophan

Zur Aufarbeitung auf L- und D-Tryptophan wurden die Esterhydrochloride, wie bei der D,L-Verbindung beschrieben, mit ammoniakalischem Äther umgesetzt und die freien Ester 15 Minuten mit 0,37 n $Ba(OH)_2$ unter Schütteln bei 20° verseift. Nach Neutralisation mit 0,37 n H_2SO_4 kocht man den Bariumsulfatniederschlag wiederholt mit Wasser aus und saugt unter Aktivkohle-Zugabe ab. Die gesammelten wäßrigen Auszüge werden im Vakuum bis zur Trockne eingedampft und liefern in 70–80proz. Ausbeute bezogen auf das Esterhydrochlorid praktisch reines L- bzw. D-Tryptophan.

L-Tryptophan: Schmp.: 281–283° (korr.); $[\alpha]_D^{21}$: −31,2°[2]) (c = 0,41 in Wasser).

$C_{11}H_{12}O_2N_2$ (204,2)

Ber.	C 64,69	H 5,90	N 13,72
Gef.	C 64,31	H 6,28	N 13,56.

D-Tryptophan: Schmp.: 280–282° (korr.); $[\alpha]_D^{21}$: +30,3° (c = 0,45 in Wasser).

Gef.	C 64,40	H 5,90	N 13,46.

Zur direkten Gewinnung der Antipoden des Tryptophans aus den Rohprodukten des Spaltansatzes wird das L-Estertartrat bzw. D-Esterhydrochlorid wie beschrieben, mit ammoniakalischem Äther zerlegt und der Aminosäureester mit Barytwasser verseift. Die so gewonnenen rohen Aminosäureantipoden besitzen optische Reinheitsgrade von 70–80% und werden aus heißem Wasser umkristallisiert. Hierbei kristallisieren zuerst die racemischen Anteile, nach Einengen und Acetonzugabe die optisch reinen Komponenten.

Ausbeute 50% bezogen auf Rohaminosäure.

L-Tryptophan: Schmp.: 279–280° (korr.); $[\alpha]_D^{20}$: −31,2° (c = 0,39 in Wasser)

Gef. N 13,84

D-Tryptophan: Schmp.: 280–283° (korr.); $[\alpha]_D^{21}$: +31,1° (c = 0,44 in Wasser)

Gef. N 13,52

$C_{11}H_{12}N_2O_2$ Ref. 771d

Resolution of N-acetyl-DL-tryptophan. With 95% L-lysine. N-Acetyl-DL-tryptophan (12.3 g., 0.050 mole) was stirred into a solution of 95% L-lysine (7.3 g., 0.062 mole) in water (18.7 g.) and, when it had dissolved, methanol (50 ml.) was added. The solution was seeded with pure L-lysine N-acetyl-L-tryptophanate (0.5 g., 0.002 mole), then gently stirred for 3.5 hr. at 25°. During this period the solution turned slowly into a thick slurry. This slurry was poured into a 3-cm., coarse-grained, sintered glass funnel and filtered under vacuum. The product, after washing with 75% methanol-water and drying, weighed 4.9 g. (0.013 mole, 27% conversion). It had a specific rotation, $[\alpha]^{22}D$ +17.17 (c, 3.6, water). This compared with $[\alpha]^{22}D$ values of +3.45 (c 2.8, water) for L-lysine N-acetyl-DL-tryptophanate and +20.47 (c 5, water) for L-lysine N-acetyl-L-tryptophanate and indicates that the salt had an optical purity of 90.0%.

With 100% L-lysine. At 25°. Pure L-lysine (11.2 g., 0.095 mole) was dissolved in water (19.8 g.). N-Acetyl-DL-tryptophan (12.3 g., 0.050 mole) was dissolved in the lysine solution, and the L-lysine N-acetyl-DL-tryptophanate solution so formed was diluted with methanol (78 ml.), seeded with L-lysine N-acetyl-L-tryptophanate (*ca.* 1 mg.), then stirred at 25° for 2.25 hr. At the end of this period, the solution was passed through a sintered glass filter, and the solid so obtained was washed with a mixture of methanol (22 ml.) and water (4 ml.). The resolution salt after drying over calcium chloride overnight weighed 6.39 g. (0.0175 mole, 35% conversion). This salt was found to be 90.0% pure L-lysine N-acetyl-L-tryptophanate ($[\alpha]^{22}D$ +16.8, c 4.7, water).

At 45°. To a solution of L-lysine (8.0 g., 0.068 mole) in water (21.8 g.) were added successively N-acetyl-DL-tryptophan (12.3 g., 0.050 mole), methanol (75 ml.), a finely triturated slurry of L-lysine N-acetyl-L-tryptophanate (0.5 g., 0.001 mole) in methanol (10 ml.), and, lastly, an additional amount of methanol (5 ml.). The suspension so obtained was stirred at 45° for 6.67 hr. and then filtered. The filter cake was washed first with 92% methanol–water (25 ml.) and then with 96% methanol–water (25 ml.). The salt obtained in this resolution (6.25 g., 0.0175 mole, 34% conversion) was shown to be 96.0% pure L-lysine N-acetyl-L-tryptophanate ($[\alpha]^{22}D$ +19.27, c 2.9, water).

Isolation of N-acetyl-L-tryptophan from L-lysine N-acetyl-L-tryptophanate. For this separation a sample of L-lysine N-acetyl-L-tryptophanate having a specific rotation, $[\alpha]^{22}D$ +17.2 (c 2.8, water) and an optical purity of 91% was used. A sample (5.8 g., 0.016 mole) of this product was dissolved in 50% methanol–water (100 ml.) and then poured through a bed of washed Dowex 50-8X resin (50 ml., H^+ form, 20–50 mesh). After the L-lysine N-acetyl-L-tryptophanate solution had been run through the bed, the resin was washed with water (50 ml.). After standing overnight, the effluent and the wash solution deposited crystals of N-acetyl-L-tryptophan which were collected by filtration. Concentration of the filtrates by evaporation under vacuum produced an additional quantity of product. The total yield obtained was 2.9 g. (0.012 mole, 79% yield). The material had a specific rotation, $[\alpha]^{29}D$ +21.2 (c 0.85, ethanol) and on optical purity of 91.0%.

An equally effective way of separating the amino acid components of L-lysine N-acetyl-L-tryptophanate was to run a solution of the salt over Dowex 50 (NH_4^+ form) resin. In this case the effluent consisted of a solution of ammonium N-acetyl-L-tryptophanate. N-Acetyl-L-tryptophan was filtered from the solution after acidification.

In both of the above examples, the L-lysine component of the resolution salt was absorbed on the resin and could be recovered by elution by 20% ammonium hydroxide (100 ml. in each case). The ammonium hydroxide eluate was evaporated to remove the ammonia, and the L-lysine reused.

Hydrolysis of N-acetyl-L-tryptophan. N-Acetyl-L-tryptophan (10 g., 0.040 mole) was boiled under reflux with 2 N hydrochloric acid (50 ml.) for 2 hr., then cooled, diluted with an equal volume of methanol, and poured down a 2 × 80 cm. column of Amberlite IR-45 resin (OH^- form). The resin was washed first with methanol (300 ml.) and then with hot water (200 ml.). The effluent and wash solution were mixed and evaporated to dryness at about 50° to give L-tryptophan (7.4 g., 0.036 mole, 89%). The optical purity of this product was 96.0%.

$C_{11}H_{12}N_2O_2$ Ref. 771e

Resolution of N-Acetyl-DL-tryptophan.

N-Acetyl-DL-tryptophan (4.9 g, 0.02 mol) and L-leucinamide (2.9 g, 0.022 mol) were dissolved in 50 ml of ethanol on a water bath at 50°C, and then the solution was let to stand at room temperature for 4 hrs. The crystals formed were filtered, washed with ethanol, and dried in vacuo. Yield, 3.6 g; m.p. 186~187°C. After recrystallization from ethanol 2.3 g of N-acetyl-L-tryptophan salt was yielded; m.p. 187°C; $[\alpha]^{23.5}_D$+20.48° (c, 4: water).

The solution of pure N-acetyl-L-tryptophan salt in 50 ml of water was passed through a column of Diaion SK#1 (H-form). The effluent was concentrated to 100 ml in vacuo, and boiled for 2 hrs with 10 g of barium hydroxide. A calculated amount of sulfuric acid was added to the reaction mixture to remove barium ion. The precipitates were filtered off and washed with hot water. After the combined filtrate was concentrated in vacuo, the residual crops were recrystallized from 60% aqueous ethanol gave 0.8 g of L-tryptophan; $[\alpha]^{20}_D$−32.05° (c, 1: water).

$C_{11}H_{12}N_2O_2$ Ref. 771f

Example 8

24.5 g. of *dl*-N-acetyltryptophan and 13.5 g. of *l*-amphetamine are dissolved with heating in 200 ml. of 95% ethanol, and the solution is allowed to stand for 16 hours at room temperature. The crystals which separate consisting of the *l*-amphetamine salt of *l*-N-acetyltryptophan are collected by filtration and dried; $[\alpha]_D^{26}$=−23.9° (0.5% in ethanol). The crystalline salt is dissolved in water, made alkaline with dilute sodium hydroxide solution, and extracted with small portions of ether. The residual solution is concentrated, acified with dilute hydrochloric acid and allowed to stand overnight with cooling. The resulting crystalline product is isolated by filtration and is dried. The product, *l*-N-acetyltryptophan, has an optical rotation, $[\alpha]_D^{26}$, of −25° (2% in ethanol).

The alcoholic filtrate obtained as described above is concentrated to dryness in vacuo, the residual salt is dissolved in water, and the aqueous solution is subjected to ion exchange and concentration in the same manner as described above. The opposite isomer, *d*-N-acetyltryptophan, is obtained; $[\alpha]_D^{26}$=+25° (2% in ethanol).

$C_{11}H_{12}N_2O_2$ Ref. 771g

Example 2.—Resolution of N-acetyl-DL-tryptophan

10 g. of N-acetyl-DL-tryptophan obtained according to the synthetic process described by Snyder and Smith in "Journal of the American Chemical Society" vol. 66, 1944, page 350, are added to 150 cc. of water while heating at 80° C. 9.5 g. of L(+)-threo-(1-p-nitro phenyl)-2-amino propanediol-(1,3), i. e., an excess of 10%, are added thereto. The warm mixture is cooled, allowed to stand at room temperature for half an hour and filtered off. The resulting crude salt is readily purified by recrystallization from water. The new compound melts on the Maquenne block at 231–232° C.; $[\alpha]_D^{20}$: +26°±2° (c. 1% in water).

For setting free therefrom N-acetyl-L-tryptophan, the crude salt is directly used after filtering off and washing with a small amount of ice water. It is treated with 50 cc. of N/2 hydrochloric acid whereby first solution takes place. Subsequently N-acetyl-L-tryptophan crystallizes from said solution, it is filtered off, washed with ice water, and dried. 4.55 g. of N-acetyl-L-tryptophan, corresponding to a yield of 90%, are obtained. The compound melts at 185–187° C.; $[\alpha]_D^{15}$: +25°±2° (c. 1% in 95% alcohol).

The mother liquors remaining after filtering off the crystallized salt of N-acetyl-L-tryptophan with the resolving agent are acidified with 3 cc. of concentrated hydrochloric acid and the crystalline precipitate is, after crystallisation, filtered off and washed. Thereby 4.5 g. to 4.7 g. of N-acetyl-D-tryptophan are obtained. The yield is about 92%. The compound melts at 185–187° C.; $[\alpha]_D^{15}$: −25°±2° (c. 1% in 95% alcohol).

The resolving agent can be readily recovered from the hydrochloric acid filtrates after removing the precipitated N-acetyl-D- or -L-tryptophans, respectively. Said recovery is effected by rendering said filtrates alkaline and filtering the precipitated resolving agent.

N-acetyl-L-tryptophan is readily saponified to L-tryptophan by proceeding according to the reaction described, for instance, by Shabica and Tishler in "Journal of the American Chemical Society," vol. 71, 1949, page 3251.

N-acetyl-D-tryptophan can be converted into N-acetyl-DL-tryptophan by treating said compound in sodium hydroxide medium with acetic acid anhydride according to the method of Du Vigneaud and Sealock described in "Journal of Biological Chemistry" vol. 96, 1932, page 511.

$C_{11}H_{12}N_2O_2$ Ref. 771h

Example 3.—Resolution of N-formyl-DL-tryptophan

2 g. of N-formyl-DL-tryptophan obtained according to Example 1, are dissolved at 70° C. in 25 cc. of water in the presence of 2 g. of L(+)-threo-(1-p-methyl thiophenyl)-2-amino propanediol-(1,3), (1.84 g. of said base are theoretically required). The salt of N-formyl-L-tryptophan with said base starts to precipitate and crystallize while the solution is hot. The mixture is allowed to cool to 25° C. within one hour, the crystals are filtered off and washed with a small amount of ice water. The salt is recrystallized from water and melts on the Maquenne block at 205° C.±2° C.; $[\alpha]_D^{20}$: +21°±1° (c. 1% in 50% ethanol).

The crude, moist product obtained on filtering off is dissolved in 10 cc. of N hydrochloric acid. Solution takes place quite rapidly whereafter N-formyl-L-tryptophan begins to precipitate and crystallizes. The mixture is allowed to stand for 2 hours, cooled with ice, filtered off, washed with a small amount of ice water and dried. 0.81 g. of N-formyl-L-tryptophan, corresponding to a yield of 81% are obtained. Said compound melts on the Maquenne block at 162° C.; $[\alpha]_D^{20}$: +45°±1° (c. 1% in 95% alcohol). On recrystallization, the product is optically pure: $[\alpha]_D^{20}$=+48°±1 (c.=1%, in 95% ethanol).

The aqueous filtrate obtained after removing by filtering off the salt of N-formyl-L-tryptophan with L(+)-threo-(1-p-methyl thiophenyl)-2-amino propanediol-(1,3), is acidified with 1 cc. of concentrated hydrochloric acid and the acidified solution is allowed to stand overnight. 0.6 g. of optically pure N-formyl-D-tryptophan corresponding to a yield of 60% are obtained. The water-free product has a specific rotatory power of −48°±1° (c. 1% in 95% ethanol).

The various hydrochloric acid mother liquors obtained on decomposing the salts of L- and D-formyl tryptophan, respectively, and filtering off the free L-formyl tryptophan or the free D-formyl tryptophan, respectively, are rendered alkaline by the addition of 5 cc. of commercial ammonia. The alkaline solutions are allowed to stand for 1 hour. 1.8 g. of L(+)-threo-(1-p-methyl thiophenyl)2-amino propanediol-(1,3) precipitate and are collected. The yield on recovery is about 90% of the amount of base employed originally.

$C_{11}H_{12}N_2O_2$ Ref. 771i

Un accès simple au L-tryptophane, à l'aide du L(+) *thréo* 1-*p*-nitrophényl 2-amino propane 1,3-diol, a été décrit dans une récente note (1). Nous venons d'observer qu'un analogue du réactif ci-dessus, le L(+) *thréo* 1-*p*-méthylthiophényl 2-amino propane 1,3-diol est également très favorable à la résolution.

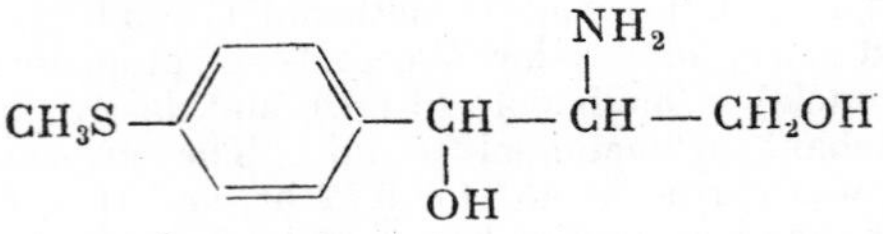

(I)

Le L(+) *thréo* 1-*p*-méthylthiophényl 2-amino propane 1,3-diol, I, a été décrit par Cutler, Stenger et Suter (2). F = 149°, $[\alpha]_D^{20}$ = + 20° ± 1 (*c* = 1 %, alcool à 95°).

On introduit dans 150 cm³ d'eau, à 70°, 9 g de ce réactif puis 10 g de N-acétyl DL-tryptophane (1). On observe une complète dissolution puis une cristallisation. Après refroidissement spontané, essorage et lavage à l'eau, on recristallise dans l'alcool à 50°; F = 224-226°, $[\alpha]_D^{20}$ = + 25° ± 1 (*c* = 1 %, alcool à 50°).

On décompose aussitôt le sel, à température ambiante, par 30 cm³ d'acide chlorhydrique N. La dissolution est suivie d'une cristallisation de N-acétyl L-tryptophane. On refroidit à 0° puis essore, lave à l'eau et sèche. Le rendement est de 4,75 g (95 %) en produit F = 185-187°, $[\alpha]_D^{20}$ = + 25° ± 1 (*c* =1%, alcool à 95°).

Les eaux-mères provenant de la cristallisation précédente sont acidifiées par 3 cm³ d'acide chlorhydrique concentré. On refroidit à 0°, essore, lave à l'eau et sèche. On obtient 4,6 g (92 %) de N-acétyl D-tryptophane, F = 185-187°, $[\alpha]_D^{20}$ = − 25° ± 1 (*c* = 1 %, alcool à 95°).

La suite des opérations est conduite comme il a été précédemment indiqué (1).

Bibliographie.

(1) L. Velluz, G. Amiard et R. Heymès, *Bull. Soc. Chim.*, 1953, 904.

(2) R. A. Cutler, R. J. Stenger et C. M. Suter, *J. Am. Chem. Soc.*, 1952, 74, 5475. Sur l'initiative du Dr. L. Sweet, nous devons à l'obligeance de la Ste Parke-Davis (Detroit) la mise à notre disposition d'un échantillon de L(+) *thréo* 1-*p*-méthylthiophényl 2-amino propane 1,3-diol.

$C_{11}H_{12}N_2O_2$ Ref. 771j

Example 1

A solution of 17 parts d-camphorsulphonic acid in 21 parts water is added to a suspension of 15 parts DL-tryptophan in 40 parts methanol and the mixture is stirred at 20°C. for 6 hours. D-tryptophan d-camphorsulphonate crystals formed during the latter part of this time. Separation of D-tryptophan d-camphorsulphonate is substantially complete after 6 hours and the crystals are then filtered from the solution. The crystals are washed with a little methanol to remove the mother liquor.

Analysis calculated for tryptophan camphorsulphonate:—

C, 57.8; H, 6.4; S, 7.3%

Found:—

C, 58.0; H, 6.2; S, 6.9%.

When the crystals are dissolved in water and analyzed with a polariscope the specific rotation is $[\alpha]_D^{22} = 13.72°$, (C=1.5 in water). Previous calibrations with known isomers have shown that the percent of D-tryptophan d-camphorsulphonate in a mixture with L-tryptophan d-camphorsulphonate is equal to $(\alpha-9.6)/0.044$.

Hence the specific rotation of 13.72° corresponds to 94% D-tryptophan d-camphorsulphonate.

The substantially pure D-tryptophan d-camphorsulphonate is dissolved in aqueous sodium carbonate sufficient to neutralize the camphorsulphonic acid. The solution is partially evaporated and cooled, whereupon D-tryptophan crystallizes out. L-tryptophan is recovered from L-tryptophan d-camphorsulphonate by the same procedure.

$C_{11}H_{12}N_2O_2$ Ref. 771k

IV. Herstellung von D- und L-Tryptophan.

30 g Pankreasenzym[1]) mit einer Aktivität von 13—14 NOVO-Einheiten pro g wurden in 300 cm³ Wasser suspendiert. Man dialysierte die Suspension 15 Stunden gegen fliessendes Wasser und trennte hierauf das Unlösliche durch zentrifugieren ab. Die klare Fermentlösung wurde lyophil getrocknet. Ausbeute 9 g ,,Ferment".

20 g frisch bereiteter, öliger DL-Tryptophanmethylester wurden mit 1 cm³ Wasser und 400 mg ,,Ferment" bei 37° zu einer klaren Lösung verrührt und hierauf bei 37—38° sich selbst überlassen. Das Gemisch begann sich schon nach 1 Stunde zu trüben und war nach 4 Stunden erstarrt. Nach 36-stündiger Reaktion wurde der feste, hellbraune Kuchen mit 50 cm³ Äther verrieben. Man filtrierte und behandelte den ätherunlöslichen Anteil noch 4mal mit je 50 cm³ Äther. Ätherunlösliches 7,9 g. Die Ätherlösungen wurden vereinigt und der Äther vollständig abdestilliert (Badtemperatur unter 40°). Man verrührte das zurückbleibende Öl von neuem mit 5 Gew.-% Wasser und 400 mg ,,Ferment", liess 36 Stunden reagieren und behandelte mit Äther. Ätherunlösliches: 2,1 g. Die Ätherlösungen wurden zweimal mit je 50 cm³ Wasser gewaschen, getrocknet und eingedampft. Es blieben 9,6 g roher D-Ester zurück. Die Ausbeute an rohem L-Tryptophan betrug unter Berücksichtigung des beigemengten Fermenteiweisses 9,2 g.

L-Tryptophan. Das obige ätherunlösliche Reaktionsprodukt wurde in 200 cm³ kochendem Wasser aufgenommen und von ausgeflocktem Fermenteiweiss durch eine dünne Schicht Kohle abfiltriert. Man setzte zur heissen Lösung 12,3 g (1,16 Mol.[2])) Naphtalin-β-sulfosäure und nach erfolgter Auflösung 250 mg Entfärbungskohle zu, filtrierte siedend heiss durch etwas Kohle und wusch mit wenig Wasser nach. Das Nasylat krystallisierte sofort aus. Es wurde nach 1-stündigem Stehen bei 0° abgesaugt und mit wenig Eiswasser gewaschen. Ausbeute 13,5 g. Aus der Mutterlauge wurden nochmals 1,3 g Nasylat von derselben Reinheit gewonnen. Dies gelang allerdings nicht durch blosses Einengen, sondern erst, als die Mutterlauge mit Diäthylamin schwach alkalisch gemacht, im Vakuum zur Trockne verdampft, der Rückstand mit absolutem Alkohol extrahiert und das zurückbleibende Roh-Tryptophan mit Naphtalin-β-sulfosäure behandelt wurde. Man krystallisierte die vereinigten Nasylate nochmals aus Wasser um und erhielt 10,8 g reines L-Salz. Es wurde zur Zerlegung in einer Lösung von 3,2 cm³ Diäthylamin in 155 cm³ absolutem Alkohol suspendiert. Man erwärmte unter gutem Umrühren auf 60° und kühlte dann auf 0° ab. Die Aminosäure wurde abfiltriert, in frischem Alkohol suspendiert, ab-

[1]) Vgl. Fussnote [3]), S. 1911.

[2]) ,,Molgewicht" = 235. Die ursprünglich als Trihydrat vorliegende Säure hatte im Exsikkator durchschnittlich 1,5 Mol Wasser verloren.

[3]) Produkt von *NOVO Terapeutisk Laboratorium A/S*, Kopenhagen; Vertretung für die Schweiz: *Aldepha AG.*, Zürich. Wir möchten der *Aldepha AG.* auch an dieser Stelle für die freundliche Überlassung eines Musters unseren besten Dank aussprechen.

filtriert und mit Alkohol und Äther gewaschen. Ausbeute 4,8 g optisch reines L-Tryptophan. Eine Probe wurde zur Analyse und Drehung aus 66-proz. Alkohol umkrystallisiert und 2 Stunden im Hochvakuum bei 70° getrocknet.

3,991 mg Subst. gaben 9,465 mg CO_2 und 2,200 mg H_2O
2,258 mg Subst. gaben 0,281 cm³ N (27°, 741 mm)

$C_{11}H_{12}O_2N_2$	Ber. C 64,67	H 5,93	N 13,73%
(204,11)	Gef. „ 64,72	„ 6,17	„ 13,81%

$[\alpha]_D^{21} = -32,5^0 \pm 1^0$ (c = 0,5 in Wasser, l = 2 dm).

D-Tryptophan. Die 9,6 g D-Ester wurden zur Verseifung in 0,2-n. methanolischem Bariumhydroxyd (1,5 Mol $Ba(OH)_2$ pro Mol Ester) gelöst und 24 Stunden bei Zimmertemperatur stehen gelassen. Man fügte hierauf die zur Fällung des Bariums berechnete Menge Schwefelsäure hinzu, ersetzte das Methanol nach seiner Entfernung im Vakuum durch ein gleiches Volumen heisses Wasser, kochte auf, liess einige Minuten stehen und filtrierte vom Bariumsulfat ab. Letzteres musste zur Entfernung von adsorbiertem Tryptophan gut mit heissem Wasser ausgewaschen werden. Das Waschwasser wurde mit dem Filtrat vereinigt und im Vakuum zur Trockne verdampft. Ausbeute 8,3 g rohes D-Tryptophan mit der Drehung $[\alpha]_D^{20} = +29,5^0 \pm 2^0$. Die Reinigung wurde wie oben über das Nasylat durchgeführt und gab 5,6 g optisch reines D-Tryptophan. Eine Probe wurde zur Analyse und Drehung aus 66-proz. Alkohol umkrystallisiert und 2 Stunden im Hochvakuum bei 70° getrocknet.

4,085 mg Subst. gaben 9,699 mg CO_2 und 2,139 mg H_2O
2,780 mg Subst. gaben 0,346 cm³ N (26°, 741 mm)

$C_{11}H_{12}O_2N_2$	Ber. C 64,67	H 5,93	N 13,73%
(204,11)	Gef. „ 64,79	„ 5,86	„ 13,85%

$[\alpha]_D^{21} = +32,5^0 \pm 1^0$ (c = 0,5 in Wasser, l = 2 dm).

Die Analysen wurden im Mikroanalytischen Laboratorium der Org.-chem. Anstalt der Universität Basel ausgeführt.

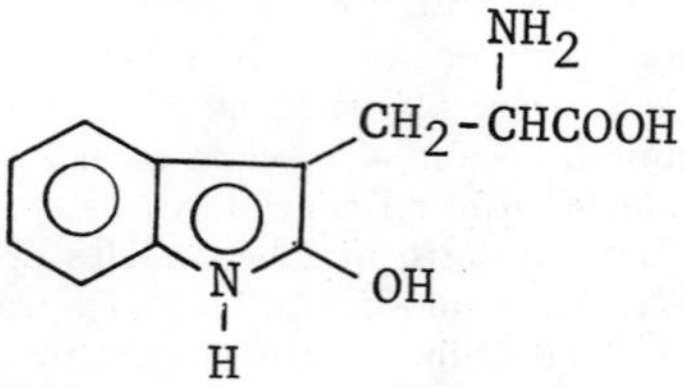

$C_{11}H_{12}N_2O_3$ Ref. 772

N-Acetyl-(+)-α-oxy-tryptophan und *N*-Acetyl-(−)-α-oxy-tryptophan: Man löst 5.24 g *N*-Acetyl-*d,l*-α-oxy-tryptophan und 9.32 g Brucin-tetrahydrat unter Erwärmen in 80 ccm 95-proz. Alkohol und läßt über Nacht bei Raumtemperatur stehen. Die ausgeschiedenen Kristalle werden abfiltriert und getrocknet: 5.86 g, die, dreimal aus 95-proz. Alkohol umkristallisiert, 1.92 g reines Brucin-Salz vom Zersp. 213 bis 213.5° liefern; $[\alpha]_D$: −26.5° (2.0-proz. wäßr. Lösung).

$C_{13}H_{14}O_4N_2 \cdot C_{23}H_{26}O_4N_2$ (656.7) Ber. N 8.54 Gef. N 8.53

1.97 g dieses Brucin-Salzes werden in 150 ccm Wasser gelöst, mit einer konz. wäßr. Lösung von 0.180 g Natriumcarbonat versetzt und das in Freiheit gesetzte Brucin durch viermaliges Ausschütteln mit Chloroform extrahiert. Die wäßr. Lösung wird mit 4 ccm *n*HCl versetzt, zur Trockne verdampft und der feste Rückstand mehrere Male unter Erwärmen mit feuchtem Aceton extrahiert. Der Extrakt wird i. Vak. eingedampft und getrocknet, der Rückstand in 5 ccm Wasser unter Erwärmen gelöst und die Lösung dann i. Vak.-Exsiccator konzentriert. Hierbei scheiden sich die Kristalle des *N*-Acetyl-(−)-α-oxy-tryptophans ab; Ausb. 83.6%. Die Substanz zeigt aus 2 ccm Wasser umkristallisiert den Zersp. 125–125.5°; $[\alpha]_D$: −29.0° (1.91-proz. Lösung in *n*NaOH).

$C_{13}H_{14}O_4N_2 \cdot H_2O$ (280.2) Ber. N 10.0 Gef. N 10.1

Wenn man die bei der Abtrennung des schwer löslichen Brucin-Salzes erhaltene Mutterlauge eindampft und trocknet, erhält man 7.68 g eines pulvrigen Rückstands. Man löst diesen in 150 ccm Wasser, gibt dazu die konz. Lösung von 0.700 g Natriumcarbonat und extrahiert das abgeschiedene Brucin mit Chloroform. Nach Zugabe von 15 ccm *n*HCl wird die Lösung i. Vak. bis auf 20 ccm eingeengt, dann mit einer Spur *N*-Acetyl-*d,l*-Verbindung geimpft und über Nacht stehengelassen. Es scheiden sich 0.56 g Kristalle aus. Wird die Mutterlauge wie im Falle des *N*-Acetyl-(−)-α-oxy-tryptophans behandelt, so steigt die Gesamtausbeute auf 1.09 g kristalliner Substanz. Die Kristalle werden aus 2 ccm Wasser fraktioniert umkristallisiert. Es werden so 0.56 g *N*-Acetyl-(+)-α-oxy-tryptophan vom Zersp. 123–124° erhalten; $[\alpha]_D$: +28.4° (1.99-proz. Lösung in 0.1 *n* NaOH).

$C_{13}H_{14}O_4N_2 \cdot H_2O$ (280.2) Ber. N 10.0 Gef. N 10.1

(+)-α-Oxy-tryptophan und (−)-α-Oxy-tryptophan: Man versetzt 420 mg *N*-Acetyl-(+)-α-oxy-tryptophan mit 2.05 ccm 6*n* H_2SO_4, erwärmt 20 Stdn. auf dem Wasserbad im Kohlendioxyd-Strom, verdünnt dann mit Wasser auf 250 ccm, setzt die ber. Menge von in wenig heißem Wasser gelöstem Bariumhydroxyd zu, filtriert durch

Aktivkohle und konzentriert im Vakuum. Es scheiden sich allmählich nadelförmige Kristalle aus. Beträgt das Gesamtvolumen der Lösung nur noch 5 ccm, so erwärmt man, bis der größte Teil der Kristalle in Lösung gegangen ist, konzentriert weiter im Exsiccator bis auf 2 ccm und filtriert das ausgeschiedene (+)-α-Oxy-tryptophan ab. Farblose Nadeln vom Zersp. 257°; Ausb. 117 mg, $[\alpha]_D$: +40.8° (4.20-proz. Lösung in *n*NaOH).

$C_{11}H_{12}O_3N_2$ (220.2) Ber. C 60.0 H 5.5 N 12.7 Gef. C 60.1 H 5.5 N 12.6

Die Mischprobe mit der aus Phalloidin von H. Wieland und B. Witkop erhaltenen Aminosäure[5]) zeigte keine Schmelzpunkts-Erniedrigung[6]).

(−)-α-Oxy-tryptophan wird aus *N*-Acetyl-(−)-α-oxy-tryptophan nach dem gleichen Verfahren gewonnen. Zersp. 257°; $[\alpha]_D$: −41.3° (1.45-proz. Lösung in *n*NaOH).

$C_{11}H_{12}O_3N_2$ (220.2) Ber. C 60.0 H 5.5 N 12.7 Gef. C 59.8 H 5.4 N 12.5

[6]) Private Mitteilung von Hrn. B. Witkop; den Herren B. Witkop und P. L. Julian möchten wir für die Durchführung der Mischprobe bestens danken.

[5]) A. 548, 171 [1940].

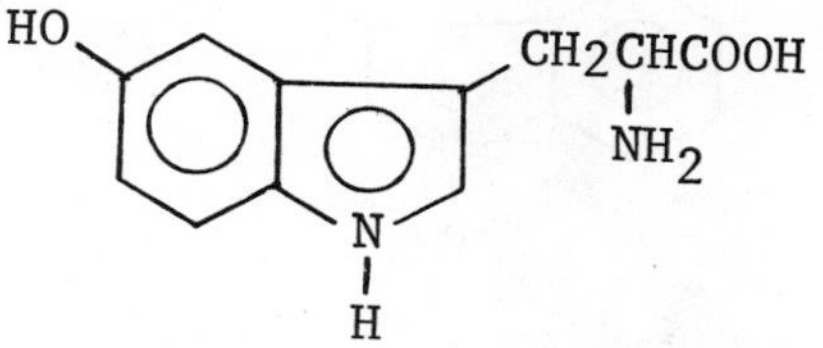

$C_{11}H_{12}N_2O_3$

Ref. 773

Resolution of N-Carbobenzoxy-5-benzyloxy-DL-tryptophan. Carbobenzoxy-5-benzyloxy-DL-tryptophan (47.5 g.) and anhydrous quinine (34.8 g.) were dissolved in 750 ml. of boiling benzene. To promote crystallization, the solution was heated to boiling and cooled three times during a period of 1 hr. The suspension was allowed to stand overnight at room temperature and the crystalline precipitate was collected on a filter, washed with a small volume of benzene and dried *in vacuo;* crop A, 44.4 g., m.p. 136–138°. Crop A was slurried with 500 ml. of benzene, simmered for several minutes and filtered while hot to yield 37.9 g. of white crystalline solid, m.p. 138–140°. The filtrate was cooled to room temperature and yielded an additional 2.0 g., m.p. 137–138°. The two fractions were combined to form crop B, $[\alpha]^{22}_D$ −83.9° (c 1, abs. EtOH). Crop B was simmered with 500 ml. of benzene for 30 min. and filtered to yield crop C; 37.3 g., m.p. 141–142°, $[\alpha]^{22}_D$ −85.8 (c 1, abs. EtOH). Crop C was dissolved in a boiling mixture of 2 l. of benzene and 400 ml. of absolute methanol, filtered, and concentrated *in vacuo* until crystallization commenced. The solution was allowed to stand overnight at room temperature, and was cooled to 10° for 30 min. and filtered to yield 33.8 g. of crop D; $[\alpha]^{22}_D$ −86.9° (c 1, abs. EtOH). Crop D was recrystallized in a similar manner from a boiling mixture of 3 l. of benzene and 60 ml. of abs. EtOH to yield crop E; 30.7 g., m.p. 143–144°, $[\alpha]^{22}_D$ −91.1° (c 1, abs. EtOH). Recrystallization of crop E from a boiling mixture of 4 l. of benzene and 30 ml. of abs. EtOH by the same procedure gave 28.3 g. of crop F; m.p. 168–169° (marked shrinkage at 145°), $[\alpha]^{22}_D$ −92.1 (c 1, abs. EtOH). Recrystallization of crop F in the same manner gave 22.5 g. of crop G, m.p. 145–146°, $[\alpha]^{22}_D$ −94.3° (c 1, abs. EtOH). Further recrystallization did not lead to any further change in the rotation of the salt.

Crop G was suspended in a mixture of 250 ml. of water and 150 ml. of ethyl acetate, and 6*N* HCl was added to the stirred mixture until the *p*H of the aqueous phase was 1.5. The EtOAc was separated and the aqueous phase was extracted two more times with 100 ml. portions of ethyl acetate. The combined ethyl acetate extracts were washed with several small portions of dilute HCl (*p*H 1.5) to remove the last traces of quinine. The extracts were dried over anhydrous Na_2SO_4, treated with charcoal, filtered, and the solvent was removed *in vacuo.* The residue was dissolved in a minimum volume of hot benzene, left at room temperature overnight, and then cooled to 10° for 1 hr. The crystalline product was collected on a filter and dried *in vacuo;* 12.6 g. of N-carbobenzoxy-5-benzyloxy-D-tryptophan (97% yield from salt), m.p. 97–99°, $[\alpha]^{22}_D$ +10.3° (c 1, abs. EtOH). For analysis, a portion was recrystallized three times from hot benzene; m.p. 98–99°.

Anal. Calcd. for $C_{26}H_{24}N_2O_5$: C, 70.25; H, 5.44; N, 6.30. Found: C, 70.28; H, 5.32; N, 6.05.

When the filtrate from crop A was heated to boiling and cooled, further crystallization occurred; 18.8 g. (m.p. 157–158°) of salt was collected. This procedure was repeated on the filtrate and another 11.2 g. (m.p. 156–157°) was obtained. These two fractions were combined as crop H. Crop H was combined with the filtrate from crop B and dissolved in a total volume of 4.75 l. of boiling benzene. The solution was left overnight at room temperature, cooled to 10° for several hours and the crystals which had formed were collected on a filter, washed with cold benzene, and dried *in vacuo;* 7.6 g. of quinine salt was obtained, m.p. 159–160°. The filtrate was concentrated to a volume of 3 l. and a further 20.9 g. of quinine salt, m.p. 157–158°, was thus obtained. These two fractions were combined to form crop I; $[\alpha]^{22}_D$ −35.1° (c 1, abs. EtOH). Crop I was dissolved in 3.6 l. of boiling benzene and crop J was collected after 24 hr. at room temperature; 17.7 g., m.p. 158–159°, $[\alpha]^{22}_D$ −33.3° (c 1, abs. EtOH). Another crop (K) was collected from the same solution after another 24 hr.; 3.0 g., m.p. 157–158°, $[\alpha]^{22}_D$ −32.5° (c 1, abs. EtOH). Crop J was recrystallized from 2.25 l. of boiling benzene to yield 13.7 g. of crop L (m.p. 160–161°), $[\alpha]^{22}_D$ −32.9° (c 1, abs. EtOH), which was combined with crop K and again recrystallized to yield 14.9 g. of crop M; m.p. 160°, $[\alpha]^{22}_D$ −32.4° (c 1, abs. EtOH). Further recrystallization did not lead to a change in the rotation of the salt. Quinine was removed in the manner described for the other isomer; 8.4 g. of *N*-carbobenzoxy-5-benzyloxy-L-tryptophan (98% yield from salt), m.p. 97–99°, $[\alpha]^{22}_D$ −10.1° (c 1, abs. EtOH). For analysis, a small sample was recrystallized from benzene; m.p. 97–99°.

Anal. Calcd. for $C_{26}H_{24}N_2O_5$: C, 70.25; H, 5.44; N, 6.30. Found: C, 70.11; H, 5.33; N, 6.35.

5-Hydroxy-L-tryptophan. Palladium oxide (300 mg.) was added to a solution of 2.0 g. of N-carbobenzoxy-5-benzyloxy-L-tryptophan in 150 ml. of abs. EtOH and 1 ml. of water, and hydrogen was bubbled through the mixture for 3 hr. The precipitate which formed was dissolved by the addition of hot water and the solution was filtered under nitrogen. The filtrate was concentrated to dryness *in vacuo* under nitrogen and washed with ethyl acetate to remove

(20) We are indebted to Drs. J. R. Kimmel and E. L. Smith of this Laboratory for the mercuripapain used in this procedure.

starting material. The remaining solid was dissolved in a minimum volume (about 4 ml.) of hot water under nitrogen, treated with charcoal, filtered under nitrogen, and allowed to crystallize at 5°. 5-Hydroxy-L-tryptophan was recovered as pale pink needles; 0.55 g., m.p. 273° dec., $[\alpha]^{22}_{D}$ −32.5° (c 1, water), $[\alpha]^{22}_{D}$ +16.0° (c 1, 4*N* HCl) (L-tryptophan, $[\alpha]^{23}_{D}$ −31.5° (c 1, water).[21] The filtrate was concentrated to yield an additional 0.11 g. In this manner, a total of 4.7 g. of 5-hydroxy-L-tryptophan (80% theor.) was obtained from 12.0 g. of N-carbobenzoxy-5-benzyloxy-L-tryptophan.

A sample was prepared for analysis by recrystallization from water under nitrogen.

Anal. Calcd. for $C_{11}H_{12}N_2O_3$: C, 59.99; H, 5.49; N, 12.72. Found: C, 59.65; H, 5.44; N, 12.50.

5-Hydroxy-D-tryptophan. N-Carbobenzoxy-5-benzyloxy-D-tryptophan (2.0 g.) was reduced by the procedure described for the L-isomer to yield 0.60 g. (61% yield) of 5-hydroxy-D-tryptophan; m.p. 274° dec., $[\alpha]^{22}_{D}$ +32.2° (c 1, water).

A sample was prepared for analysis by recrystallization from water under nitrogen.

Anal. Calcd. for $C_{11}H_{12}N_2O_3$: C, 59.99; H, 5.49; N, 12.72. Found: C, 59.86; H, 5.62; N, 12.50.

(21) Greenberg, *Chemistry of the Amino Acids and Proteins*, C. C. Thomas, Springfield, Ill., 1945, p. 1177.

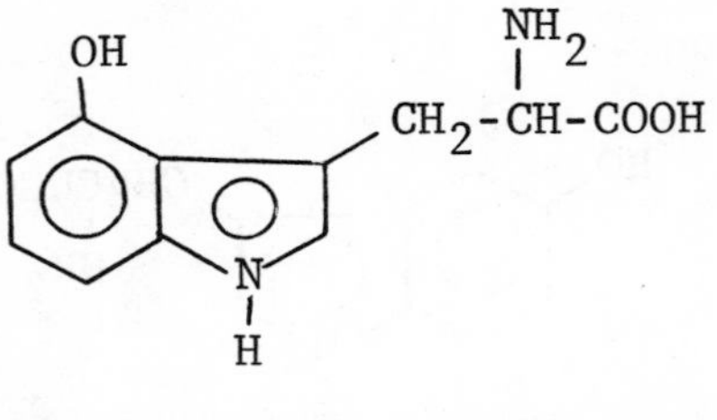

$C_{11}H_{12}N_2O_3$ Ref. 774

Experimentelles. – *Chinidinsalz des N-Benzyloxycarbonyl-5-benzyloxy-L-tryptophans.* Eine Lösung von N-Benzyloxycarbonyl-5-benzyloxy-DL-tryptophan (9,6 g) und 7,2 g Chinidin in 250 ml siedendem Benzol wurde langsam auf Zimmertemperatur abgekühlt und schliesslich 30 Min. in Eiswasser stehengelassen. Der Niederschlag wurde abfiltriert und *in vacuo* bei 45° getrocknet. Ausbeute: 8,5 g. Smp. 135°. Nach zweimaliger Umkristallisation aus 450 ml und 300 ml Benzol wurden 5,5 g (65% d. Th.) Reinprodukt erhalten. Smp. 156°; $[\alpha]^{23}_{D} = +127{,}5°$ ($c = 1$ in Äthanol).

$C_{46}H_{48}O_7N_4$ Ber. N 7,3% Gef. N 7,1%

Chininsalz des N-Benzyloxycarbonyl-5-benzyloxy-D-tryptophans. Die Benzolfiltrate des Chinidinsalzes wurden mit 150 ml Wasser versetzt und unter Umrühren mit 6N HCl auf pH 1,5 eingestellt. Die Benzolphase wurde abgetrennt und die wässerige Schicht mit 100 ml Benzol extrahiert. Die kombinierten Benzolextrakte wurden zweimal mit 75 ml Wasser gewaschen und durch Destillieren auf 250 ml reduziert. Nach Zugabe von 3,4 g Chinin zu der heissen Lösung wurde filtriert, über Nacht bei Zimmertemperatur stehengelassen und der ausgeschiedene Kristallbrei abfiltriert. Der Kristallkuchen, mit etwas kaltem Benzol gewaschen und wie oben getrocknet, lieferte 6 g Produkt. Aus 250 ml Benzol wurden daraus 5 g (60%) reines Chininsalz erhalten. Smp. 144–145°; $[\alpha]^{23}_{D} = -94{,}2°$ ($c = 1$ in Äthanol), (Lit. [2]: $[\alpha]^{22}_{D} = -94{,}3°$).

N-Benzyloxycarbonyl-5-benzyloxy-L-tryptophan. 5 g Chinidinsalz wurden in 50 ml Wasser und 35 ml Äthylacetat aufgeschlämmt und die wässerige Schicht mit 6N HCl auf pH 1,5 eingestellt. Die Äthylacetatschicht wurde abgetrennt und die wässerige Phase noch dreimal mit 25 ml Äthylacetat ausgezogen. Die vereinigten Äthylacetatextrakte wurden mit 20 ml verd. HCl und zweimal mit 20 ml Wasser gewaschen, über Magnesiumsulfat getrocknet, mit etwas Tierkohle behandelt und zur Trockne eingedampft. Der Rückstand wurde in 25 ml heissem Benzol aufgenommen und über Nacht bei Zimmertemperatur stehengelassen. Der kristalline Niederschlag wurde filtriert und im Vakuum über Phosphorpentoxid getrocknet. Ausbeute: 2,8 g (96%). Smp. 99°; $[\alpha]^{21}_{D} = -10{,}3°$ ($c = 1$ in Äthanol), (Lit. [2]: $[\alpha]^{22}_{D} = -10{,}1°$).

N-Benzyloxycarbonyl-5-benzyloxy-D-tryptophan. Die Substanz wurde wie oben angegeben aus ihrem Chininsalz freigesetzt und aus Benzol umkristallisiert. Smp. 99°; $[\alpha]^{25}_{D} = +10{,}15°$ ($c = 1$ in Äthanol), (Lit. [2]: $[\alpha]^{22}_{D} = +10{,}3°$).

5-Hydroxy-L- und -D-tryptophan. 2 g N-Benzyloxycarbonyl-5-benzyloxy-L-tryptophan wurden in 50 ml Äthanol gelöst und die Lösung mit Wasser (ca. 24 ml) versetzt, bis sie sich zu trüben begann. Es wurden nun 0,5 g 10-proz. Palladiumkohle und 0,5 ml konz. HCl hinzugefügt und unter Umschütteln Wasserstoff in die Lösung eingeleitet. Nach etwa 2,5 Std. war die Hydrierung beendet. Das Reaktionsgemisch wurde in einer Stickstoffatmosphäre filtriert, das Filtrat mit 2N NaOH auf pH 5,9 eingestellt und unter vermindertem Druck auf 20 ml eingeengt. Kristallisation wurde durch Kratzen mit einem Glasstab ausgelöst und das Produkt nach zweistündigem Stehen im Kühlschrank in einer Stickstoffatmosphäre filtriert. Die fast farblosen Nädelchen wurden mit etwas Äthanol gewaschen und im Vakuum über KOH und Phosphorpentoxid getrocknet. Ausbeute: 0,7 g (85%) 5-Hydroxy-L-tryptophan. Smp. 273° (Zers.); $[\alpha]^{23}_{D} = -32{,}4°$ ($c = 1$ in Wasser), (Lit. [2]: $[\alpha]^{22}_{D} = -32{,}5°$).

Die D-Form des 5-Hydroxytryptophans wurde auf dieselbe Weise erhalten. Ausbeute: 85%. Smp. 274° (Zers.), $[\alpha]^{24}_{D} = +31{,}5°$ ($c = 1$ in Wasser) (Lit. [2]: $[\alpha]^{22}_{D} = +32{,}2°$).

$C_{11}H_{12}O_3N_2$ Ber. N 12,7% Gef. L-Form N 12,5% D-Form N 12,6%

[1] J. A. Oates & A. Sjoerdsma, Proc. Soc. exp. Biol. Med. *108*, 264 (1961).
[2] A. J. Morris & M. D. Armstrong, J. org. Chemistry *22*, 306 (1957).
[3] E. Fischer & W. A. Jacobs, Ber. deutsch. chem. Ges. *39*, 2942 (1906).

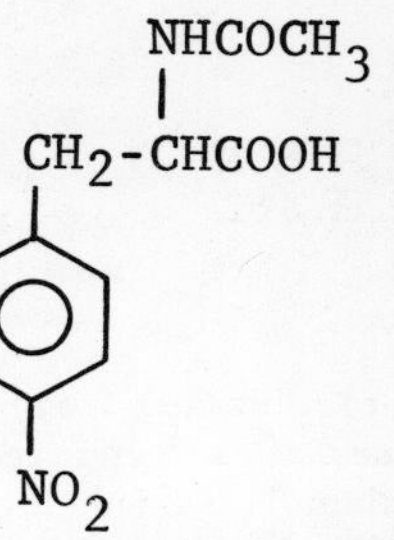

$C_{11}H_{12}N_2O_5$ Ref. 775

To a solution of the DL-acid (23·55 g.) in hot ethanol (200 ml.) was added one of brucine (36·9 g., 1·00 mol.) in hot ethanol (200 ml.), and the mixture left overnight at room temperature. Pale yellow prisms were deposited. The mixture was cooled in ice-water for 1 hr., and the product, m. p. 203—208° (25·35 g., 0·42 mol.) filtered off. Recrystallisation from ethanol gave pale yellow prisms of the *brucine salt*, m. p. 207—209·5°, $[\alpha]_D^{21}$ +19·2° ± 0·5° (*c*, 1·59 in 1 : 1 H_2O–dioxan), of *N*-acetyl-*p*-nitro-L-phenylalanine. Recrystallisation did not significantly affect the rotation or m. p. (Found: C, 61·1; H, 5·85; N, 8·6. $C_{34}H_{38}O_9N_4,H_2O$ requires C, 61·0; H, 6·0; N, 8·4. Found, after drying at 100°/1 mm. for 4 hr.: C, 62·7; H, 6·2; N, 8·6. $C_{34}H_{38}O_9N_4$ requires C, 63·1; H, 5·9; N, 8·7%).

Treatment of an aqueous solution of the salt with ammonia or sodium hydroxide, removal of the brucine by filtration, acidification of the filtrate, and recrystallisation of the product from water yielded colourless prisms of N-*acetyl*-p-*nitro*-L-*phenylalanine*, initial m. p. 170—172°, resolidification, final m. p. 206—209°, $[\alpha]_D^{24}$ +49·7° ± 1° (*c*, 1·55 in EtOH) (Found: C, 52·5; H, 4·8; N, 11·0. $C_{11}H_{12}O_5N_2$ requires C, 52·4; H, 4·8; N, 11·1%).

The ethanolic mother-liquors from the brucine salt were evaporated to dryness under a vacuum, and the residual gum was taken up in hot water. Crystallisation set in on cooling. After an hour at 0°, the product (36·95 g.) was collected and recrystallised from water, yielding yellow prisms of the *brucine salt pentahydrate*, m. p. 98—99°, $[\alpha]_D^{21}$ −36·9° ± 0·5° (*c*, 1·63 in 1 : 1 H_2O–dioxan), of *N*-acetyl-*p*-nitro-D-phenylalanine. Recrystallisation raised the $[\alpha]_D^{21}$ to −37·6° ± 0·5° (*c*, 1·60) (Found, on a sample dried in a vacuum-desiccator over H_2SO_4: C, 55·3; H, 6·6; N, 7·9. $C_{34}H_{38}O_9N_4,5H_2O$ requires C, 55·4; H, 6·6; N, 7·6. Found, on a sample dried to constant wt. at 80° in a high vacuum: C, 62·8; H, 5·65; N, 9·0. $C_{34}H_{38}O_9N_4$ requires C, 63·1; H, 5·9; N, 8·7%). Treatment of an aqueous solution of the salt with aqueous ammonia, removal of the precipitated brucine, acidification with hydrochloric acid, and recrystallisation of the product from water gave colourless prisms of N-*acetyl*-p-*nitro*-D-*phenylalanine*, initial m. p. 170—172°, resolidification, final m. p. 205—206°, $[\alpha]_D^{23}$ −44° ± 0·5° (*c*, 1·45 in EtOH) (Found: C, 52·3; H, 4·8; N, 11·2%).

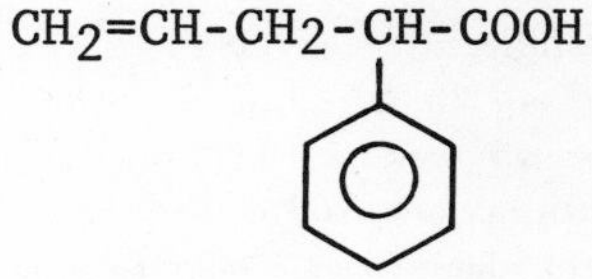

$C_{11}H_{12}O_2$ Ref. 776

The (+)acid was easily obtained according to *Pickard* and *Yates* (loc. cit.[13])) by fractional crystallization of the salts formed by the racemic acid with (—)menthylamine. We did not succeed, however, in isolating from the acetone mother-liquor of the salt (—)bas (+)acid the pure salt with the (—)acid.

Both (+) and (—)acid could be smoothly isolated in a pure form by using (+) and (—)α-phenyl-ethylamine.

To 26.4 g of (*dl*)-allylphenylacetic acid in 200 ml of warm water was added, while stirring intensily, 18.2 g of (+)α-phenyl-ethylamine, whereupon the crystalline salt-mixture separated (42.5 g).

This mixture was first recrystallized from 550 ml of acetone, yielding mainly the salt with the (—)acid (25 g, A).

From the mother liquor (B), consisting for the greater part of the salt with the (+)acid, the acid was recovered after evaporation of the acetone and decomposition of the salt with alkali, the (+)phenyl-ethylamine being extracted with benzene.

After redistillation of the acid, 9.2 g of it were combined with 6.3 g of (—)α-phenyl-ethylamine in 125 ml of boiling ethanol. The salt separating after cooling (13.2 g) was recrystallized twice from absolute ethanol, yielding the salt (—)base (+)acid of constant rotation, $[\alpha]_D^{20}$ = +21.8°, m.p. 162.5—163° C.

10 g of this salt were treated with 2.5 g of NaOH in 40 ml of water. The amine was extracted with benzene, the aqueous layer freed from benzene, traeted with charcoal and acidified. The acid was taken up in benzene, the solution dried over anhydrous Na_2SO_4 and after evaporation of the benzene, the acid was distilled. Yield 5 g, b.p.$_{0.4\ mm}$ 119—120° C; $[\alpha]_D^{20}$ = +104.2°.

Analysis:

Found: C 74.94; H 6.81.
$C_{11}H_{12}O_2$. Calc.: C 75.00; H 6.82.

(—) Allyl-phenylacetic acid.

The 25 g of salt A (see above) obtained from the first recrystallization of the original salt-mixture from acetone, after being recrystallized twice from acetone and subsequently twice from ethanol yielded the pure salt (+)base (—)acid, $[\alpha]_D^{20} = -21.65°$, m.p. 162.5—163° C. The (—)acid was obtained from this salt as described above for the (+)acid. Yield 7.5 g, b.p.$_{9\,mm}$ 153—154° C, $[\alpha]_D^{20} = -104.7°$.

Analysis:

Found: C 74.53; H 6.76.
$C_{11}H_{12}O_2$. Calc.: C 75.00; H 6.82.

$C_{11}H_{12}O_2$ Ref. 776a

29.2 g acid and 20.1 g (−)-α-phenyl-ethylamine were dissolved in 200 ml of hot water. On cooling, 41 g salt were obtained. It was recrystallised from ethanol to constant activity, the rotatory power of the successive crops of the salt being measured in the same solvent. The data are in excellent accordance with those given by VELDSTRA *et al.* (1).

Crystallisation	2	3	4	5	6
ml ethanol	150	75	65	65	40
g salt obtained	22.3	16.3	13.7	10.5	9.0
$[\alpha]_D^{25}$ of the salt	+10.9°	+17.8°	+18.6°	+21.7°	+21.6°

The acid was liberated in the usual way and distilled *in vacuo* passing over at 160° (10 mm Hg).

0.2116 g acid: 11.11 ml 0.1075-*N* NaOH.
Equiv. wt. calc. 176.2, found 177.2.

0.2069 g dissolved in *benzene* to 9.98 ml: $2\alpha_D^{25} = +4.285°$. $[\alpha]_D^{25} = +103.3°$; $[M]_D^{25} = +182.1°$. The same solution was also measured at 20.00°: $2\alpha_D^{20} = +4.395°$. $[\alpha]_D^{20} = +106.0°$; $[M]_D^{20} = +186.8°$. — 0.2126 g dissolved in abs. *ethanol* to 10.00 ml: $2\alpha_D^{25} = +3.84°$. $[\alpha]_D^{25} = +90.3°$; $[M]_D^{25} = +159.1°$. — 0.2114 g dissolved in *acetone* to 10.00 ml: $2\alpha_D^{25} = +4.33°$. $[\alpha]_D^{25} = +102.4°$; $[M]_D^{25} = +180.5°$. — 0.2116 g dissolved in glacial *acetic acid* to 9.98 ml: $2\alpha_D^{25} = +3.88°$. $[\alpha]_D^{25} = +91.5°$; $[M]_D^{25} = +161.2°$.

1. VELDSTRA, H., and VAN DE WESTERINGH, C., Rec. trav. chim. *70*, 1113 (1951).

$C_{11}H_{12}O_2$ Ref. 777

Resolution of VI. Best results were obtained when the reagents were of high quality. The acid was recrystallized to a sharp melting point from hexane (80–81°) and the brucine resolving agent was recrystallized from 1:1 ethanol–water and dried in a vacuum oven at 100° for 24 hr before use.

Powdered, recrystallized, dry brucine (119.7 g, 0.304 mole) was dissolved in 2 l. of reagent acetone in a 4-l. beaker. The acid (53.5 g, 0.304 mole) was dissolved in 250 ml of reagent grade acetone, heated to reflux, and added to the refluxing brucine solution. The beaker was placed on a cork ring and set aside without a cover. After 5 days, small cubic crystals began forming (volume was down to 1 l.). The beaker was covered and the crystals were permitted to grow for another week. The supernatant was transferred to a 1-l. beaker along with a small amount of acetone used to wash the crystals, and this was set aside to crystallize without a cover. The crystals were treated with just enough refluxing acetone to put them into solution in an open 1-l. beaker. Crystal growth began after cooling to ambient and was allowed to continue until half the original volume remained. The supernatant was decanted along with two acetone washes and the crystals were dried to give 70.9 g of brucine acid salt. The supernatant from the first batch of crystals crystallized into a mass of well-defined, cubic crystals after one-third of the solvent evaporated. Recrystallization from acetone with washing and drying gave 73.7 g of brucine acid salt. The supernatant from this recrystallization was combined with the supernatant from the recrystallization of the first batch of crystals and the solvent was evaporated giving what we term "supernatant" brucine acid salt. The three batches of brucine acid salts were each treated separately with 200 ml of 1 *M* HCl solution and 200 ml of hexane and stirred with warming to hasten the neutralization. The two phases were separated and the hexane layer was washed several times with small amounts of water, dried over anhydrous $MgSO_4$, and stripped of solvent to give resolved and "supernatant" acids. Weights, specific rotations, and optical purity are given in Table V.

Table V. Resolution Results

	Weight, g: Brucine salts	Weight, g: Cyclopropane acid	$[\alpha]$,[a] deg: Brucine salts	$[\alpha]$,[a] deg: Cyclopropane acid	Purity,[b] %
(−)-Enantiomer	70.9	21.5	−87.9 ± 0.2	−114.6 ± 0.4	74
(+)-Enantiomer	73.7	20.6	+22.0 ± 0.3	+152.7 ± 0.4	98
Supernatant	40.4	11.3	...	−57.7 ± 0.4	37

[a] Determined as 2% solution in absolute ethanol. [b] These values are based on 100% optically pure acid obtained by a small-scale triangular crystallization of the brucine salt. Constant specific rotations of the recrystallized salt in acetone and finally in absolute ethanol were the criterion of optical purity. The maximum specific rotation of the salt in acetone is $[\alpha]^{25}_D$ − 91.2 ± 0.4°; in absolute ethanol, $[\alpha]^{25}_D$ −91.8 ± 0.4°. Maximum specific rotation of the free acid in absolute ethanol is $[\alpha]^{25}_D$ −155.1 ± 0.4°.

$C_{11}H_{12}O_2$

Ref. 777a

Resolution of (*E*)-2-Methylphenylcyclopropane-2-carboxylic Acids. (*E*)-2-Methylphenylcyclopropane-2-carboxylic acid-*1-d* and its perhydro analog were resolved by the procedure described. Melting points and specific rotations were identical within experimental error for both.

A solution of 30 g (0.18 mol) of ephedrine in 200 ml of hot hexane was added to 29 g (0.16 mol) of the acid also in 200 ml of hot hexane, and the mixture was immediately filtered. The solid obtained was placed in a Soxhlet extractor cup, and the filtrate plus an additional 200 ml of hexane was used to repetitively extract the solid for 3 days. The solid remaining in the cup was dried under vacuum and recrystallized from water. The precipitate which formed on cooling the hexane solution was filtered and recrystallized from anhydrous ether and then water. These two crops of crystals were combined and dried under vacuum to afford 17.7 g (mp 170.5–172°) of ephedrine salt ($[\alpha]$D +71 ± 2° (*c* 2.20, absolute ethanol)). The salt was partially dissolved in 300 ml of hot water and acidified with concentrated HCl. Ether extracts of the cooled solution were dried over $MgSO_4$. Evaporation of the solvent gave 9.3 g (32% of total acid) of (+)-(*E*)-2-methylphenylcyclopropane-2-carboxylic acid, mp 70–73° ($[\alpha]$D +156 ± 2° (*c* 1.53, absolute ethanol). The yield of the (+)-ephedrine salt could be slightly improved, and a sample of pure (−)-ephedrine salt could be obtained from the hexane soluble fraction by tedious triangular recrystallizations from anhydrous ether and water. In this manner, a 37% yield of (+)-ephedrine salt and a 34% yield of (−)-ephedrine salt [mp 119.5–120°; $[\alpha]$D −96 ± 2° (*c* 1.54, absolute ethanol)] were obtained. The literature value for the maximum rotation of this acid is $[\alpha]$D +155.1 ± 0.04° (*c* 2, absolute ethanol).

2,2-Dimethylphenylcyclopropane-*1-d*. The $LiAlH_4$ reduction, tosylation, reduction sequence was carried out as described for the hydrocarbon.[11]

(11) I. Tomoskozi, *Tetrahedron,* **22,** 179 (1966).

(13) C. H. DePuy, F. W. Breitbeil, and K. R. DeBruin, *J. Am. Chem. Soc.,* **88,** 3347 (1966).
We found the resolution with ephedrine to be more convenient than the previously described method using brucine[12] since minor impurities did not cause crystallization difficulties as in the latter case.

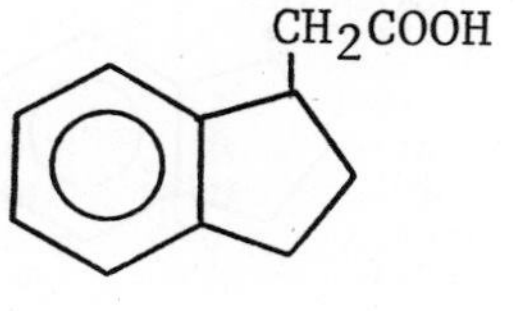

$C_{11}H_{12}O_2$

Ref. 778

Note:

Obtained an acid, m.p. 79.5-80.0°C. $[\alpha]^{25}_D$ = -8.5° (c, 2.58, benzene) using (+)-α-phenylethylamine as resolving agent. Using (-)-α-phenylethylamine as resolving agent, obtained acid, m.p. 78-80° (hexane), $[\alpha]^{25}_D$ = +8.6°C. (c 2.56, benzene).

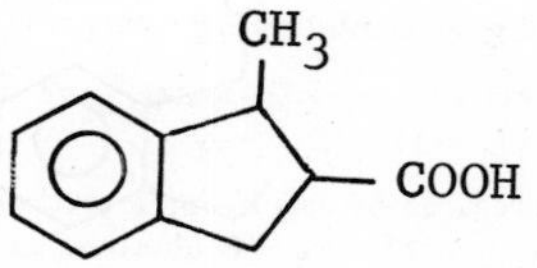

$C_{11}H_{12}O_2$

Ref. 779

l-Menthylamine 2 : 3-*Dihydro-3-methylindene-2-carboxylate,*
$C_{10}H_{11}\cdot CO_2H,C_{10}H_{19}\cdot NH_2$.

Ethereal solutions of the foregoing dihydro-acid and *l*-menthylamine are mixed and the mixture heated for some time on a water-bath. The ether is then distilled off and the viscous mass desiccated, when in the course of a few days it sets to a hard, dark mass. The salt is insoluble in water, but very soluble in all the common organic media except ethyl acetate, from which it can be crystallised. Owing to the dark colour of the substance and the readiness with which resolution takes place when it is crystallised from ethyl acetate, no polarimetric observations of the racemic salt were obtained. After one crystallisation from ethyl acetate, it melted at 152°.

0·2513 gave 9·4 moist nitrogen at 16° and 752 mm. N = 4·32.
$C_{21}H_{33}O_2N$ requires N = 4·22 per cent.

l-Menthylamine d-2 : 3-*Dihydro-3-methylindene-2-carboxylate.*

When the racemic salt described above is crystallised five or six times from ethyl acetate, the *lBdA*-salt is obtained in a pure state, as is shown by consecutive crystallisations giving fractions with a constant rotation. The pure salt crystallises in long, white needles which melt at 170°; it is soluble in most of the ordinary organic media, sparingly so in ethyl acetate and ether, and insoluble in water. The crystals do not lose weight at 100°.

0·2115 gave 0·5931 CO_2 and 0·2500 H_2O. C = 76·47 ; H = 13·13.
0·2083 ,, 7·6 c.c. moist nitrogen at 14° and 758 mm. N = 4·28.
$C_{21}H_{33}O_2N$ requires C = 76·13 ; H = 13·09 ; N = 4·23 per cent.

The following polarimetric observations * on successive fractions were made:

0·3802, made up to 20 c.c. with absolute alcohol, gave α +1·04°, whence $[\alpha]_D$ +27·35° and $[M]_D$ +90·52°.
0·2511, made up to 20 c.c. with absolute alcohol, gave α +0·68°, whence $[\alpha]_D$ +27·08° and $[M]_D$ +89·63°.

d-2 : 3-*Dihydro-3-methylindene-2-carboxylic Acid.*

When the pure *lBdA*-salt is shaken with ether and caustic soda and the alkaline solution separated and acidified, the pure dextrorotatory acid gradually crystallises out in the form of long, flat needles which melt at 86°. The acid is insoluble in water, but very soluble in the ordinary organic media. It does not lose weight when dried at 100°.

0·2968 gave 0·8146 CO_2 and 0·1767 H_2O. C = 74·85 ; H = 6·61.
$C_{11}H_{12}O_2$ requires C = 75·00 ; H = 6·81 per cent.

* All these observations were made in decimetre tubes.

The following determinations of rotatory power were made :

0·3760, made up to 20 c.c. with absolute alcohol, gave α +2·53°, whence $[\alpha]_D$ +67·28° and $[M]_D$ +118·41°.

0·2010, made up to 20 c.c. with absolute alcohol, gave α +1·36°, whence $[\alpha]_D$ +67·66° and $[M]_D$ +119·08°.

0·2472, made up to 20 c.c. with benzene, gave α +1·90°, whence $[\alpha]_D$ +76·86° and $[M]_D$ +135·27°.

0·1959, made up to 20 c.c. with toluene, gave α +1·75°, whence $[\alpha]_D$ +89·33° and $[M]_D$ +157·22°.

Boiled with a large excess of caustic soda or dilute sulphuric acid for four hours, practically no racemisation took place, a small alteration in rotatory power being probably due to some slight decomposition, as the solution deepened considerably in colour.

The *sodium, potassium,* and *barium* salts are soluble in water, the *silver* and *lead* salts insoluble. The *barium* salt crystallises from alcohol in needles and gave the following rotation :

0·4121, made up to 20 c.c. with water, gave α +0·99°, whence $[\alpha]_D$ +24·02° and $[M]_D$ +125·62°.

The *methyl* ester, prepared by saturating a methyl-alcoholic solution of the acid with hydrogen chloride, was obtained as a crystalline solid melting at 68°, which gave the following analytical and polarimetric results.

0·1527 gave 0·4248 CO_2 and 0·1016 H_2O. C = 75·86 ; H = 7·39.
$C_{12}H_{14}O_2$ requires C = 75·78 ; H = 7·36 per cent.

0·2151, made up to 20 c.c. with alcohol, gave α +1·36°, whence $[\alpha]_D$ 63·22° and $[M]_D$ +120·11°.

1·2 : 3-*Dihydro-3-methylindene-2-carboxylic Acid.*

When the mother liquors obtained in the crystallisation of the dextrorotatory acid were worked up, they gave, as the most soluble fraction of the menthylamine salt, a viscid, brown syrup, which, even after prolonged desiccating, would not solidify. The salt was therefore treated with ether and caustic soda solution, and the acid precipitated from the alkaline solution. The acid thus obtained has $[\alpha]_D$ −46·0° in alcoholic solution. On crystallising several times from aqueous alcohol, a small quantity of pure *l*-acid was obtained.

The acid crystallises in long, flat needles and melts at 86°. In general properties, it is exactly similar to the corresponding *d*-acid.

0·1817 gave 0·4990 CO_2 and 0·1120 H_2O. C = 74·89 ; H = 6·84.
$C_{11}H_{12}O_2$ requires C = 75·00 ; H = 6·81 per cent.

The following polarimetric observations were made :

0·2565, made up to 20 c.c. with absolute alcohol, gave α −1·71°, whence $[\alpha]_D$ −66·66° and $[M]_D$ −117·32°.

0·1641, made up to 20 c.c. with benzene, gave α −1·24°, whence $[\alpha]_D$ −75·56° and $[M]_D$ −132·98°.

A mixture made by dissolving equal quantities of the pure *d*- and *l*-acids in alcoholic solution and evaporating the solution to dryness gave, after crystallisation, the racemic acid melting at the same temperature, namely, 82°.

Although indications of the presence of the other theoretically possible isomerides were obtained, sufficient material was not available for their isolation.

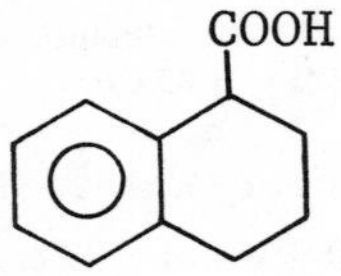

$C_{11}H_{12}O_2$ Ref. 780

Preliminary experiments on resolution. –0.176 g (0.001 moles) of racemic acid and 0.001 moles of the alkaloids were dissolved in dilute ethanol. From the crystals formed the acid was liberated and the optical activity measured in benzene or acetone. The results are given in Table 2.

Table 2.

Base	$[\alpha]_D^{25}$ of acid in benzene	$[\alpha]_D^{25}$ of acid in acetone
Strychnine	oil	—
Cinchonine	+52°	—
Quinine	−51°	—
Brucine	—	+3°
Quinidine	—	±0°
Cinchonidine	+19°	—
(+)-α-Phenylethylamine	oil	—

Optical resolution of 1.2.3.4-tetrahydro-1-naphtoic acid

8.80 g (0.05 moles) of racemic acid and 14.70 g (0.05 moles) of cinchonine were dissolved in 60 ml of hot ethanol (96 %). After adding 40 ml of water the solution was allowed to stand at room temperature for two days. The precipitate was collected (7.15 g) and from a sample (0.25 g) the acid was isolated and the optical activity measured in benzene. The remaining salt was recrystallized from dilute ethanol:

Crystallization	1	2	3	4
ml 96 % ethanol	60	20	10	12
ml water	40	12	10	6
g salt	7.15	5.05	3.85	2.85
$[\alpha]_D^{25}$ of acid	+58.6°	+60.5°	+61.5°	+61.5°

5.25 g (0.03 moles) of 1.2.3.4-tetrahydro-1-naphtoic acid ($[\alpha]_D^{25}$ = −27°, in benzene), isolated from the first mother liquor, and 9.80 g (0.03 moles) of quinine were dissolved in a boiling mixture of 65 ml of ethanol and 50 ml of water. The salt obtained (10.10 g) was recrystallized in the same way as described above:

Crystallization	1	2
ml 96 % ethanol	65	45
ml water	50	30
g salt	10.10	8.70
$[\alpha]_D^{25}$ of acid	−61.5°	−61.5°

(+)-1.2.3.4-Tetrahydro-1-naphtoic acid

2.75 g of the cinchonine salt was decomposed with dilute sulphuric acid and the organic acid extracted with ether. 1.0 g of crude acid was obtained. It was recrystallized twice from petroleum ether (once with charcoal). Yield 0.50 g. The acid forms colourless, glistening plates melting at 55.5–56.5°.

0.0842 g of acid: 9.50 ml of 0.05024-*N* NaOH
Equiv. wt. calc. 176.2, found 176.4
0.1534 g of acid dissolved in benzene to 10.00 ml: $2\alpha_D^{25}$ = +1.95°, $[\alpha]_D^{25}$ = +63.6°; $[M]_D^{25}$ = +112.0°.

(−)-1.2.3.4-Tetrahydro-1-naphtoic acid

The acid was isolated from 8.25 g of the quinine salt in the same way as described for the (+)-acid. The yield of crude acid was 2.80 g. After recrystallization twice from petroleum ether (once with charcoal) 2.10 g of pure acid remained. M.p. 55.5–56.5°. It resembles its optical antipode in all respects.

0.2522 g of acid: 14.21 ml of 0.1006-*N* NaOH
Equiv. wt. calc. 176.2, found 176.4
0.1614 g of acid dissolved in benzene to 10.00 ml: $2\alpha_D^{25}$ = −2.06°, $[\alpha]_D^{25}$ = −63.8°; $[M]_D^{25}$ = −112.4°.

Weighed amounts of acid were made up to 10.00 ml with different solvents and the optical activity measured. The results are given in Table 3.

Table 3.

Solvent	mg of acid	$2\alpha_D^{25}$	$[\alpha]_D^{25}$	$[M]_D^{25}$
Benzene	161.4	−2.06°	−63.8°	−112.4°
2.2.4-Trimethylpentane	155.6	−1.205°	−38.7°	−68.2°
Acetic acid	153.6	−0.715°	−23.3°	−41.0°
Water (neutr.)	154.8	−0.445°	−14.4°	−25.4°
Chloroform	159.5	−0.46°	−14.4°	−25.4°
Ethanol (abs.)	156.5	−0.445°	−14.2°	−25.0°
Acetone	146.6	−0.31°	−10.6°	−18.6°

$C_{11}H_{12}O_2$ Ref. 780a

The racemate was resolved according to *Pickard* and *Yates* (loc.cit. [8])) by means of (—)menthylamine. The amine was obtained by reduction of (—)menthonoxim, and recrystallisation of the hydrochloride from water. (—)Menthylamin-HCl : $[\alpha]_D^{25}$ = —36°.3.

The salts from the racemic acid (22 g in 100 ml of water + 10.5 g of $NaHCO_3$) and menthylamin (23.8 g of the hydrochloride in 75 ml of water) separated as an oil, which soon crystallized. (→ 39 g of the salts). After being recrystallized three times from acetone, the least soluble salt was obtained with a constant rotation, $[\alpha]_D^{20}$ = —30°.7, mp. 105—106° C.

This salt yielded a (—)acid, recrystallized from light petroleum of $[\alpha]_D^{20}$ = = —58°.7 and mp. 48°.5—49°.5 C (a second batch, mp. 51° C).

The data of *Pickard* and *Yates* as to rotation and m.p. differ from ours:

	m.p. salt (—)base (—)acid	$[\alpha]_D$ salt	m.p. (—)acid	$[\alpha]_D$ (—)acid
P. an Y.:	123° C	—29.35°	52.5°	—52.34°
This paper:	105—106° C	—30.7	51°	—58.7°

It may be possible that at that time P. and Y. did not dispose of very pure (—)menthylamine (cf. *Tutin, Kipping* [16])).

Analysis:

Found: C 74.77; H 6.96.
$C_{11}H_{12}O_2$. Calc.: C 75.00; H 6.82.

8) *R. H. Pickard, J. Yates.* J. Chem. Soc. **89**, 1101 (1906).

(+)1,2,3,4-Tetrahydro-α-naphthoic acid.

The crystalline material obtained from the acetone-mother liquor of the salt (—)base (—)acid did not show a sharp m.p., nor did it after recrystallization from acetone. On long standing (2 months) the mother liquor of one of these crystallizations deposited two forms of crystals. The largest and hard crystals could be picked out and washed free from the smaller ones. They proved to be the desired salt (—)base (+)acid. By seeding mother liquor with this material more of it could be obtained: $[\alpha]_D^{20}$ = —12.6°, m.p. 133—134° C.

The (+)acid recovered from this salt, after recrystallization from light petroleum showed a $[\alpha]_D^{20}$ = +59.1°, m.p. 50—51° C.

Also here especially with the m.p. of the salt, our data differ from those given by P. and Y.

	m.p. salt (—)base (+)acid	$[\alpha]_D$ salt	m.p. (+)acid	$[\alpha]_D$ (+)acid
P. an Y.:	78° (not sharp)	12.79°	49—50°	—
This paper:	133—134° C	12.6°	50—51°	+59.1°

Analysis:

Found: C 74.76; H 6.94.
$C_{11}H_{12}O_2$. Calc.: C 75.00; H 6.82.

16) *F. Tutin, F. S. Kipping.* J. Chem. Soc. 85, 69 (1904).

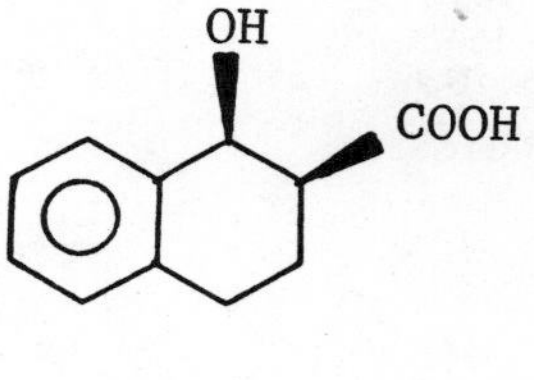

$C_{11}H_{12}O_3$ Ref. 781

Méthyl-2 tétralol-1 *cis* **5** et méthyl-2 tétralone-1 **7**. — L'acide hydroxy-1 tétrahydro-1.2.3.4 naphtoïque-2 *cis* racémique, préparé selon Bernáth ([7]), F 130-130,5°C, est dédoublé par la S-(+)-amphétamine. L'acide optiquement pur, F 105°C, a un pouvoir rotatoire $[\alpha]_D^{22}$ — 119°,9 (c = 2, éthanol).

(7) G. **Bernáth**, P. **Sohár**, K. L. **Láng**, L. **Tornyai** et O. K. **Kovács**, *Acta Chim. Acad. Sc. Hung.*, 64, 1970, p. 81.

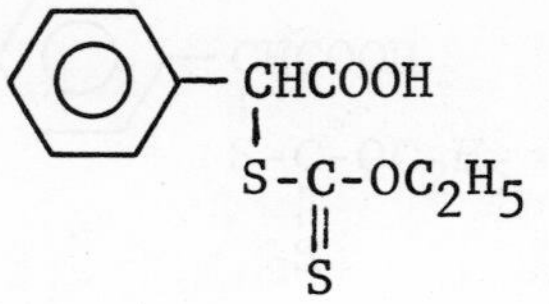

$C_{11}H_{12}O_3S_2$ **Ref. 782**

Resolution of the racemic acid. On preliminary experiments, it was found that both antipodes could be isolated by means of their quinine salts. The brucine salt also crystallised readily, but the acid regenerated from this salt was completely inactive.

The quinine salts were crystallised from a mixture of methanol and water. The salt of the (+)-acid is less soluble and crystallises more readily than that of the antipode. As the separation is partly based upon the different rates of crystallisation, the operations require some care and attention and the results are not strictly reproducible. The following description shows a typical experiment.

25.6 g (0.1 mole) of racemic xantho-phenylacetic acid and 32.4 g (0.1 mole) of anhydrous quinine were dissolved separately in 125 ml methanol at room temperature. The solutions were combined and mixed with 60 ml of water. A few powdered crystals of (+)-salt from an earlier experiment were added and the solution was stirred from time to time with a glass rod. The (+)-salt was gradually deposited as characteristic, shortprismatic brilliant crystals which rapidly sedimented after stirring. After about 25 minutes, the (—)-salt began to separate as loose aggregates of thin needles. The solution was then immediately filtered with suction. The salt fraction obtained (27.2 g) was recrystallised from methanol-water, taking 4 ml of methanol and 1.8 ml of water for each g salt. After each crystallisation, a small sample (about 0.3 g) was set aside, the acid liberated and its rotatory power determined, using ethyl acetate as solvent.

Crystallisation:	1	2	3	4	5
g salt . . .	27.2	22.5	21.1	—	19.6
$[\alpha]_D^{25}$	+178°	+238°	+244°	+242°	+243°

As may be seen from the figures, the acid is optically pure after three crystallisations.

The filtrate from the first crystallisation deposited a voluminous mass of felted crystal needles. The salt (15.2 g) was filtered off after three hours and recrystallised as above, taking for each g salt 4 ml of methanol and 1.5—2 ml of water. The progress of the resolution was followed as described above.

Crystallisation:	1	2	3
g salt . . .	15.2	13.8	12.0
$[\alpha]_D^{25}$	—234°	—243°	—244°

Some further experiments on the isolation of the salt of the (—)-acid may be described.

The combined mother liquors from several expermentsyielded 96 g of quinine salt containing an acid with $[\alpha]_D = -156°$. The salt was recrystallised in the manner described above. No seed-crystals were added and the salt fractions were filtered off after about 24 hours.

Crystallisation:	1	2	3	4	5
g salt . . .	72.0	63.5	54.7	45.7	39.8
$[\alpha]_D^{25}$	—174°	—185°	—227°	—244°	—244°

In another experiment, 51 g of a quinine salt containing an acid with $[\alpha]_D = -150°$ were dissolved in 200 ml of methanol. A suspension of 0.1 g of pure (—)-salt in 65 ml of water was added. The solution was left quite undisturbed for 45 minutes, after which the salt was rapidly filtered of with suction. It was recrystallised as usual.

Crystallisation:	1	2	3
g salt . . .	24.2	18.7	16.5
$[\alpha]_D^{25}$	—202°	—241°	—244°

As seen from these experiments, the resolution is facilitated by seeding the solution.

The quinine salts were decomposed with dilute sulfuric acid, and the active acids were taken up in pure, alcohol-free ether. The ether solutions were evaporated at room temperature. The acids crystallise on seeding; at first they were brought to crystallisation by rubbing a small sample with a drop of water. The recrystallisation met with difficulties as the active acids are very soluble in most organic solvents and have a tendency to separate as oils. The best results are obtained if they are dissolved in so much ligroin that only a very slight turbidity appears on cooling to room temperature. If some seed-crystals are added and the solution is left to stand for some days, finally at —5°, the acids gradually separate as long, lustrous needles.

(—)-*Ethylxanthogene-phenylacetic acid.* 11.8 g of quinine salt gave 5.0 g acid. After one recrystallisation the yield was 3.5 g.

0.2297 g made up to 10.00 ml with ethyl acetate: $2\alpha = -11.25°$.

$[\alpha]_D^{25} = -244.9°$; $[M]_D^{25} = -627.5°$.

The remaining acid was recrystallised once more, yielding 2.15 g with melting point 49.5—51°.

0.2500 g acid made up to 10.00 ml with ethyl acetate: $2\alpha = -12.24°$.

$[\alpha]_D^{25} = -224.8°$; $[M]_D^{25} = -627.5°$.

0.4526 g acid: 11.65 ml 0.1514-N sodium hydroxide.

$C_{11}H_{12}O_3S_2$ (256.8) E. W. calc. 256.8 found 256.6

The quinine salt of the (—)-acid obtained in the course of resolution and described above holds 1 molecule of water.

31.79 mg: 24.67 mg $BaSO_4$. — 0.7922, 1.1151 g: 0.0245, 0.0338 g H_2O.

$C_{11}H_{12}O_3S_2$, $C_{20}H_{24}O_2N_2$, H_2O (598.6) calc. S 10.71 H_2O 3.01

found » 10.66 » 3.09, 3.08.

(+)-*Ethylxanthogene-phenylacetic acid.* The acid was isolated and recrystallised as the antipode.

0.2258 g acid made up to 10.00 ml with ethyl acetate: $2\alpha = +11.08°$.

$[\alpha]_D^{25} = +244.8°$; $[M]_D^{25} = +627.5°$.

After another crystallisation the acid had the melting point 51—52°.

0.2846 g acid made up to 10.00 ml with ethyl acetate: $2\alpha = 11.505°$.

$[\alpha]_D^{25} = +245.2°$; $[M]_D^{25} = +628.5°$.

0.5098 g acid: 13.14 ml 0.1514-N sodium hydroxide.

$C_{11}H_{12}O_3S_2$ (256.8) E. W. calc. 256.8 found 256.0

The quinine salt of the (+)-acid obtained in the course of the resolution holds like that of the antipode 1 molecule of water.

23.52 mg: 18.48 mg $BaSO_4$. — 0.6098, 0.5504 g: 0.0188, 0.0170 g H_2O.

$C_{11}H_{12}O_3S_2$, $C_{20}H_{24}O_2N_2$, H_2O (598.6) calc. S 10.71 H_2O 3.01

found » 10.76 » 3.08, 3.09

The values for the rotatory power of the acid in different solvents are found in table 1. The measurements in ethyl acetate at different concentrations indicate that the influence of the concentration is very insignificant.

Table 1.

$v = 10.00$ ml.

Solvent	g acid	λ = 5893 Å			λ = 5461 Å		
		$2\alpha^{25}$	$[\alpha]^{25}$	$[M]^{25}$	$2\alpha^{25}$	$[\alpha]^{25}$	$[M]^{25}$
Ethyl acetate .	0.1099	+5.395°	+245.4°	+629°	+6.54°	+297.5°	+762.5°
Acetone . . .	0.1058	+5.065°	+239.4°	+613.5°	+6.14°	+290.2°	+744°
Alcohol (abs.) .	0.1115	+5.145°	+230.7°	+591.5°	+6.28°	+281.6°	+722°
Acetic acid . .	0.1114	+5.07°	+227.6°	+583.5°	+6.165°	+276.7°	+709.5°
Chloroform . .	0.1108	+5.81°	+239.6°	+614°	+6.48°	+292.4°	+749.5°
Benzene . . .	0.1050	+4.95°	+235.7°	+604°	+6.035°	+287.4°	+736.5°
Water (neutr.) .	0.1185	+5.14°	+226.4°	+580.5°	+6.27°	+276.2°	+708°

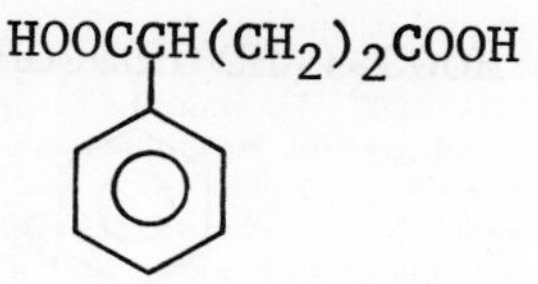

$C_{11}H_{12}O_4$ Ref. 783

Preliminary experiments on resolution.—0.208 g (0.001 moles) of racemic acid and 0.002 moles of the bases were dissolved in ethanol or methanol. The solutions were allowed to stand and in the case of ethanol, water was added gradually until crystals were formed. The crystals were filtered off, the acid liberated and the optical activity measured in absolute ethanol. The results are given in Table 1.

Table 1.

Base	$[\alpha]_D^{25}$ of acid, measured in abs. ethanol	
	From ethanol + water	From methanol
Strychnine	− 3°	—
Cinchonine	+ 20°	+ 27°
Quinine	+ 53°	+ 35°
Brucine	− 5°	− 51°
Quinidine	± 0°	oil
Cinchonidine	+ 20°	− 21°

Optical resolution of α-phenylglutaric acid

33.5 g (0.161 moles) of racemic α-phenylglutaric acid and 104.0 g (0.322 moles) of quinine were dissolved in 800 ml of hot ethanol (96%). The solution was allowed to cool and crystallise for 24 hours in a refrigerator. The precipitate was filtered off, washed with a little cold ethanol, dried and weighed (76.0 g). From a sample (about 0.5 g) the acid was isolated and the optical activity measured in absolute ethanol. The remaining salt was recrystallised from dilute alcohol, and the process repeated until the optical activity of the acid remained constant. The course of the resolution is given in Table 2.

Table 2.

Cryst. No.	Solvent (ml)		Weight of salt (g)	$[\alpha]_D^{25}$ of acid
	Ethanol	Water		
1	800	—	76.0	+ 52°
2	450	—	58.8	+ 72°5
3	450	10	51.5	+ 81°
4	250	15	46.3	+ 83°5
5	250	25	42.0	+ 84°
6	250	25	38.4	+ 86°
7	220	25	35.7	+ 86°
8	200	25	32.8	+ 86°

The mother liquor from the first crystallisation with quinine was evaporated to dryness at room temperature and the acid was liberated from the residue. This acid (13.8 g), which showed an optical activity of about −47° (in abs. ethanol), and 62.0 g of brucine were dissolved in 250 ml hot methanol and the solution was left in a refrigerator for 24 hours. The crystals formed were filtered off and washed with cold methanol. The progress of the resolution was followed as described above and the results are given in Table 3.

Table 3.

Cryst. No.	Methanol (ml)	Weight of salt (g)	$[\alpha]_D^{25}$ of acid
0	—	—	− 47°
1	250	58.6	− 72°
2	150	49.4	− 76°
3	150	47.0	− 77°
4	150	42.6	− 82°
5	150	39.5	− 82°5
6	150	36.7	− 84°
7	150	32.4	− 84°5
8	150	27.2	− 84°5

(+)-α-Phenylglutaric acid

32.8 g of the quinine salt was decomposed with 2 *M* sulphuric acid and the organic acid thoroughly extracted with ether. The ether was removed at room temperature and the crude acid (7.4 g) was recrystallised once from pure formic acid. The acid forms colourless needles or rods with m.p. 129–131°.

0.0498 g of acid: 9.59 ml of 0.0496 *N* NaOH.
Equiv. wt. calc. 104.1, found 104.7.
0.1063 g of acid dissolved in abs. ethanol to 10.00 ml: $2\alpha_D^{25}$ = + 1.83°, $[\alpha]_D^{25}$ = + 86°1; $[M]_D^{25}$ = + 179°0.

Weighed amounts of acid were made up to 10.00 ml with different solvents and the optical activity measured. In the case of water the acid was neutralized with NaOH to pH 7. The results are given in Table 4.

Table 4.

Solvent	mg of acid	$2\alpha_D^{25}$	$[\alpha]_D^{25}$	$[M]_D^{25}$
Acetone	80.5	+ 1°65	+ 102°5	+ 213°5
Ethyl acetate	81.5	+ 1°66	+ 101°8	+ 212°
Acetic acid	81.0	+ 1°51	+ 93°2	+ 194°
Chloroform	84.6	+ 1°47	+ 86°9	+ 181°
Ethanol	106.3	+ 1°83	+ 86°1	+ 179°
Water (ion)	81.6	+ 0°49	+ 30°0	+ 62°5

(−)-α-Phenylglutaric acid

27.2 g of the brucine salt was decomposed in the same way as described for the antipode. 4.5 g of the acid was obtained. It was recrystallised once from formic acid. M.p. 129–131°.

0.1014 g of acid: 9.74 ml of 0.1001 *N* NaOH.
Equiv. wt. calc. 104.1, found 104.0.
0.0889 g of acid dissolved in abs. ethanol to 10.00 ml: $2\alpha_D^{25}$ = − 1°52, $[\alpha]_D^{25}$ = − 85°5; $[M]_D^{25}$ = − 178°0.

$C_{11}H_{12}O_4$ Ref. 783a

Resolution of α-Phenylglutaric Acid (V).[17,28]—It was found that both enantiomers of Va could be obtained by crystallization of the quinine salt; 10.4 g. (0.05 mole) of (±)-V, m.p. 82–84°,[22] and 16 g. (0.05 mole) of quinine were dissolved in a mixture of 100 ml. of ethanol and 100 ml. of water. The solution was left to crystallize, and the salt was purified by repeated crystallization from ethanol. From the last salt fraction, 2.2 g. of the crude (+)-acid, [α]D +77° (*c* 2.20 in abs. ethanol) was obtained as an oil. It was refluxed with 50 ml. of acetyl chloride for half an hour; after removing the excess of acetyl chloride and acetic acid *in vacuo* the residue was crystallized from benzene to give 0.5 g. of (+)-α-phenylglutaric anhydride (Va), m.p. 114.5–116°, [α]D +13° (*c* 3.59 in chloroform). Four grams of the acid, [α]D −66°, obtained from the first filtrate was similarly converted to 0.8 g. of pure (−)-Va, m.p. 114.5–116°, [α]D −12° (*c* 4.20 in chloroform) (lit.[18] m.p. 118–120°, $[\alpha]^{25}_D$ ±41° (ethyl acetate)).[18]

The free enantiomeric acids were obtained by boiling the anhydrides with water and extracting with ether as oils, [α]D +84° (*c* 3.10 in abs. ethanol), −82° (*c* 4.72 in abs. ethanol) (lit.[17] m.p. 129–131°, $[\alpha]^{25}_D$ +86.1°, −85.5° (abs. ethanol)).

(17) L. Westman, *Arkiv Kemi*, **11**, 431 (1957).
(18) L. Westman, *ibid.*, **12**, 161 (1958).

(22) M. F. Ansell and D. H. Hey, *J. Chem. Soc.*, 1683 (1950).

(28) A part of the resolution is by Mr. O. Kimura.

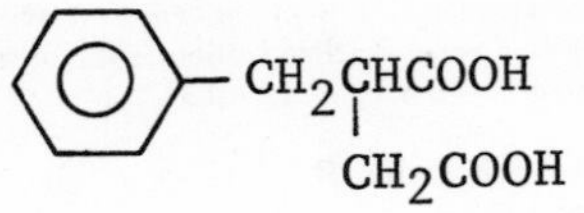

$C_{11}H_{12}O_4$ **Ref. 784**

racem-*Benzylsuccinic* acid was prepared according to FITTIG and RÖDERS (4). After repeated recrystallisations from water it was obtained in the form of glittering rectangular plates with parallel extinction. The melting point is not quite sharp and to a certain extent is dependant on the mode of heating; this is probably due to a slight decomposition (anhydride formation) during the determination. Samples were therefore introduced into the heating bath at different constant temperatures and the melting acid in the capillary was stirred with a thin glass rod. The temperature at which the liquefaction was complete in 30—45 seconds was noted as the melting point. In this way, the m. p. 162.5° was obtained for the racemic acid.

Resolution of the inactive acid. 36 g (0.3 mol) of (+)-phenylethylamine (5) was dissolved in a solution of 31 g (0.15 mol) of inactive acid in 150 ml of hot water. The salt crystallised readily on cooling. After standing for 24 hours, the salt was filtered off and recrystallised several times from hot water. The progress of the resolution was tested on samples (0.25 g) from each crystallisation, from which the acid was isolated and examined in ethyl acetate solution.

Crystallisation	1	2	3	4	5
g water	150	130	130	85	85
g salt obtained .	46.5	31.0	23.1	18.9	14.2
$[\alpha]_D^{25}$ of the acid .	−10.3°	−23.0°	−27.2°	−26.9°	−27.0°

As seen from the figures, the resolution is complete after the third crystallisation. A further crop of salt could be obtained by working up to mother liquors from crystallisations 3—5.

The mother liquor from crystallisation 2 was evaporated to about half its volume and left to crystalline. The salt was filtered off, the mother liquor was combined with the mother liquor from crystallisation 1 and the acid was isolated. In this way, 13.3 g acid with $[\alpha]_D^{25} = +19°$ was obtained. This acid (0.064 mol) was dissolved in a boiling mixture of 100 ml ethanol and 50 ml water, and 35 g (0.105 mol) of strychnine was added. When the strychnine was completely dissolved, the solution was diluted with 100 ml water. On standing, 43.8 g salt was deposited; the acid isolated from a small sample had $[\alpha]_D^{25} = +27.0°$.

The salt was recrystallised from a mixture of 200 ml water and 200 ml ethanol. The yield was 39.2 g and the acid liberated had $[\alpha]_D^{25} = +26.8°$.

(—)-*Benzylsuccinic acid.* The phenylethylamine salt (14.2 g) was decomposed with dilute sulfuric acid, and the benzylsuccinic acid was isolated by extraction with ether. The yield of crude product was 6.6 g and after two recrystallisations from water 5.4 g remained. The acid separates as pointed prisms or needles with parallel extinction. The melting point, determined as described for the inactive compound, was 164.5°.

0.1616 g acid: 14.60 ml 0.1060-N sodium hydroxide.
$C_{11}H_{12}O_4$ (208.2) Equiv. wt. calc. 104.1 found 104.4.

The acid is very soluble in hot water, but sparingly so in cold. The values for the rotatory power in different solvents are given in table 1.

Table 1.

g = weight of acid in g
v = volume of solution in ml
l = length of polarimeter tube in dm.

Solvtnt	g	v	l	α_D^{25}	$[\alpha]_D^{25}$	$[M]_D^{25}$
Ethyl acetate	0.2081	10.00	2	−1.095°	−27.0°	−56.1°
Acetone	0.8205	10.00	2	−1.86°	−29.0°	−60.4°
Alcohol (abs.)	0.8115	10.00	2	−1.80°	−20.9°	−43.5°
Ether	0.2202	10.00	2	−0.805°	−18.3°	−38.1°
Acetic acid	0.2289	10.00	2	−0.875°	−19.5°	−40.6°
Water	0.0536	15.00	3	−0.405°	−38°	−78.5°
Water (neutr.)	0.1535	10.00	2	−1.06°	−34.5°	−71.9°

The (+)-*phenylethylamine salt of the* (—)-*acid* was obtained in the course of the resolution as small prisms or needles. A nitrogen determination on the preparation obtained indicates, that the ratio of amine and acid is not exactly 2:1 as it would be for a secondary salt. On standing over phosphorus pentoxide, no decrease of weight was observed. The salt was not examined more closely.

(+)-*Benzylsuccinic acid.* The strychnine salt (39.2 g) was suspended in 1500 ml of hot water and dissolved by careful addition of hydrochloric acid. The strychnine was precipitated by ammonia in excess, and the remaining solution was evaporated to about 200 ml. After acidification with sulfuric acid, the benzylsuccinic acid was extracted with ether. The yield of crude product was 8.6 g, and after two recrystallisations from water 8.0 g remained. The acid closely resembles its (—)-isomeride in appearance and properties. The melting point is 164.5°.

0.1508 g acid: 13.68 ml 0.1060-N sodium hydroxide.
$C_{11}H_{12}O_4$ (208.2) Equiv. wt. calc. 104.1 found 104.0
0.2814 g made up to 10.00 ml with ethyl acetate: $2\alpha_D^{25} = +1.245°$.

$$[\alpha]_D^{25} = +26.9°; \quad [M]_D^{25} = +56.0°.$$

0.2821 g made up to 10.00 ml with acetone: $2\alpha_D^{25} = +1.84°$.

$$[\alpha]_D^{25} = +28.9°; \quad [M]_D^{25} = +60.1°.$$

A mixed sample containing equal amounts of dextrorotatory and racemic acid showed a distinct depression of the melting point (156°), indicating the existence of an eutectic point.

The *strychnine salt of the* (+)-*acid* was obtained in the course of the resolution in the form of small, rectangular plates with parallel extinction. It is very sparingly soluble. The air-dried salt seems to hold 3 molecules of water.

98.75 mg: 5.124 ml N_2 (19°, 743 mm). — 1.0011 g: 0.0598 g H_2O.
$C_{11}H_{12}O_4$, $2C_{21}H_{22}O_2N_2$, $3H_2O$ (931.1) calc. N 6.02 H_2O 5.80.
found » 5.98 » 5.98.

(4) R. Fittig and P. Röders, Ann. *256*, 95 (1890

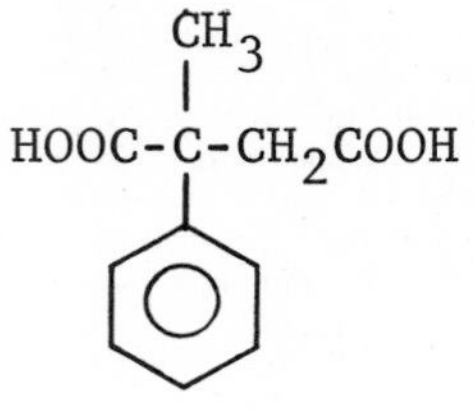

$C_{11}H_{12}O_4$ **Ref. 785**

Acides (+) S et (−) R phenyl-2 methyl-2 succiniques et derives

Dédoublement de l'acide phényl-2 méthyl-2 succinique par la brucine. Mettre en suspension 26 g d'acide phényl-2 méthyl-2 succinique racémique[16] dans 100 ml de $CHCl_3$ et dissoudre 48 g de brucine dans 200 ml de $CHCl_3$. Mélanger les deux solutions. Porter à reflux 20 min. Filtrer, sécher sur Na_2SO_4 et évaporer sous vide. Reprendre la pâte brûnatre résiduelle par 300 ml d'acétonitrile bouillant. Laisser reposer une nuit. On récolte 36 g de cristaux (l'acide résultant présente un pouvoir rotatoire $[\alpha]_D^{25} = +20°$ (c = 4, EtOH)). Ces cristaux, peu solubles dans l'acétonitrile, sont dissous dans 1 l de $CHCl_3$. Concentrer à 250 ml, ajouter 200 ml d'acétonitrile bouillant et abandonner à l'air pendant 8 jours. Filtrer les cristaux qui ont précipité (25 g) et laver à l'éther anhydre.

Acide (+) S phényl-2 méthyl-2 succinique 1. Dissoudre les cristaux précédents dans 250 ml de $CHCl_3$. Ajouter une solution de 4 g de soude dans 20 ml d'eau. Agiter 1/2 h. Décanter et acidifier la couche aqueuse, par HCl concentré. Porter au réfrigérateur. On récolte 8·8 g d'acide (+) S, qui peut être estérifié directement sans autre purification, ou recristallisé dans 40 ml d'eau bouillante. $C_{11}H_{12}O_4$, F = 149°, $[\alpha]_D^{25} = +27°$ (c = 4, EtOH).

Acide (−) R phényl-2 méthyl-2 succinique. La solution-mère dans l'acétonitrile, résulant de la première cristallisation du sel de brucine est évaporée sous vide; le résidu pâteux est repris par 100 ml de $CHCl_3$ et le sel traité comme plus haut. $C_{11}H_{12}O_4$, F = 143°, $[\alpha]_D^{25} = -20°$ (c = 4, EtOH). Ce mélange est utilisé tel quel pour la corrélation chimique.

[16] H. Le Moal, A. Foucaud, R. Carrie, J. Hamelin et C. Sevellec, *Bull. Soc. Chim. Fr.* 579 (1964).

$C_{11}H_{12}O_4$ Ref. 786

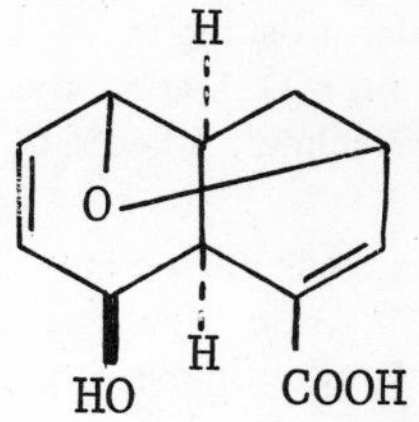

(—)-Säure *IX*

Die heiße Lösung von 6,2 g racemischer Säure *IX* in 50 ml Aceton und 100 ml Methanol wurde mit 14,0 g Tetrahydrat der Brucinbase versetzt. Beim Abkühlen erhielt man 16,0 g rohes Brucinsalz vom Smp. 160°. Zweimaliges Umkristallisieren aus Methanol ergab 5,0 g Diastereomer *A*, Smp. 185°, $[\alpha]_D^{20}$ −17° (1%, Wasser), das in gutentwickelten langen Nadeln kristallisiert.

Für $C_{34}H_{38}N_2O_8$ (602,7) berechnet: 67,76% C, 6,36% H, 4,65% N;
gefunden: 67,76% C, 6,45% H, 4,79% N.

Aus der Mutterlauge wurde beim Einengen und durch wiederholte Kristallisation aus Methanol eine Substanz von unterschiedlicher Kristallform erhalten, die im Gegensatz zum Diastereomer *A* in Wasser ausgezeichnet löslich ist; Smp. 147—150°, $[\alpha]_D^{20}$ −20° (1%, Wasser). Nach den Analysenergebnissen handelt es sich um das Monohydrat des Diastereomers *B*.

Für $C_{34}H_{40}N_2O_9$ (620,7) berechnet: 65,79% C, 6,50% H, 4,51% N;
gefunden: 65,71% C, 6,67% H, 4,59% N.

Die Suspension von 4,5 g Diastereomer *A* des Brucinsalzes in 30 ml Chloroform wurde mit der Lösung von 1,0 g Kaliumhydroxyd in 15 ml Wasser geschüttelt. Die Chloroformlösung des Brucins wurde abgetrennt und die wäßrige Lösung mit verdünnter Salzsäure angesäuert. Es schieden 0,9 g Produkt ab; weitere 0,45 g erhielt man durch Extraktion der Mutterlauge mit Äthylacetat und durch Verdampfen des Extrakts; Smp. 165° (90%iges Methanol), $[\alpha]_D^{20}$ −22 ± 3° (1%, Methanol). Das Produkt ist die linksdrehende Form der Säure *IX*.

Für $C_{11}H_{12}O_4$ (208,2) berechnet: 63,45% C, 5,82% H, 21,62% COOH;
gefunden: 63,21% C, 6,11% H, 21,18% COOH.

$C_{11}H_{12}O_4$ Ref. 786a

Spaltung der (±)-3β,5β-Epoxy-8β-hydroxy-3,4,5,8,9α,10α-hexahydro-1-naphthoesäure (*I*)

Rechtsdrehendes Isomer. Durch Erhitzen eines Gemisches von 20,0 g der (±)-Säure *I*, 16,5 g (—)-Ephedrin, 20 ml Methanol und 300 ml Aceton wurde eine Lösung bereitet, die bereits in der Hitze zu kristallisieren begann. Sie wurde über Nacht bei Raumtemperatur und dann 12 Stunden im Eisschrank bei 0° belassen. Das abgesaugte rohe Salz (18 g) vom Smp. 185° wurde durch Kristallisation aus Methanol–Aceton–Mischung gereinigt. Das erhaltene reine Salz des (—)-Ephedrins mit der (+)-Säure *I* zeigte den Smp. 188° und $[\alpha]_D^{20}$ −40 ± 2° (in Methanol, *c* 1). Für $C_{21}H_{27}NO_5$ (373,4) berechnet: 67,54% C, 7,29% H, 3,75% N; gefunden: 67,79% C, 7,38% H, 3,84% N.

18,0 g dieses Salzes wurden in 35 ml Wasser und 15 ml konz. wäßrigem Ammoniak suspendiert. Das freigesetzte (—)-Ephedrin wurde mit Chloroform extrahiert und durch Abdunsten der Lösung in fast quantitativer Ausbeute zurückerhalten. Die ammoniakalisch–wäßrige Lösung wurde mit Aktivkohle geschüttelt, filtriert und das Filtrat mit Salzsäure angesäuert. Beim Stehen im Eisschrank schieden sich 8,2 g rohe (+)-Säure *I* vom Smp. 149° ab. Durch Extraktion der Mutterlauge mit Äthylacetat und Abdunsten des Extrakts wurden noch weitere 0,3 g des gleichen Produktes erhalten. Das analysenreine Präparat wurde durch Kristallisation aus Wasser gewonnen; Smp. 149—150°, $[\alpha]_D^{20}$ +22 ± 2° (in Methanol, *c* 1). Die Analyse deutet darauf hin, daß die Säure als Monohydrat vorliegt. Für $C_{11}H_{14}O_5$ (226,2) berechnet: 58,40% C, 6,24% H, 19,90% COOH; gefunden: 58,06% C, 6,24% H, 20,15% COOH.

Linksdrehendes Isomer. Das beim Absaugen des (—)-Ephedrinsalzes der (+)-Säure *I* anfallende acetonisch–methanolische Filtrat wurde unter vermindertem Druck zur Trockene abgedunstet und der Rückstand mit 150 ml abs. Äther vermischt; nach Abnutschen wurden 19,0 g rohes Salz vom Smp. 152° erhalten. Durch Kristallisation aus Methanol–Äther-Mischung wurde das analysenreine Salz vom Smp. 155—157° und $[\alpha]_D^{20}$ −16 ± 2° (in Methanol, *c* 1) gewonnen. Für $C_{21}H_{27}NO_5$ (373,4) berechnet: 67,54% C, 7,29% H, 3,75% N; gefunden: 67,28% C, 7,25% H, 3,91% N.

19 g diese Salzes wurden in 35 ml Wasser und 15 ml konz. wäßrigem Ammoniak suspendiert, Ephedrin mit Chloroform extrahiert, die alkalische wäßrige Lösung über Aktivkohle filtriert und das Filtrat mit Salzsäure angesäuert. Beim Stehen in der Kälte schieden sich 6,7 g rohe Säure vom Smp. 149—152° ab. Durch Extraktion mit Äthylacetat und Abdunsten des Extrakts

wurde noch 1,0 g Produkt erhalten, so daß die Gesamtausbeute 7,7 g beträgt. Das analysenreine Produkt wurde durch Umfällen aus der Lösung in verdünntem Ammoniak gewonnen; Smp. 148—149°, $[\alpha]_D^{20}$ —20 ± 2° (in Methanol, *c* 1). Die Analyse deutet auf das Vorliegen des Monohydrats hin. Für $C_{11}H_{14}O_5$ (226,2) berechnet: 58,40% C, 6,24% H, 19,90% COOH; gefunden: 58,26% C, 6,29% H, 20,04% COOH.

In den zitierten Arbeiten[1,2] erhielten wir diese Säure in wasserfreiem Zustand; Smp. 165°, $[\alpha]_D^{20}$ —22 ± 3° (in Methanol, *c* 1). Das nun erhaltene Monohydrat führten wir in das (—)-Lacton vom Smp. 165° über; für dieses Produkt gaben wir früher[1,2] den Smp. 166° an.

1. Novák L., Jílek J. O., Kakáč B., Protiva M.: Tetrahedron Letters *1959*, Nr. 5, 10.
2. Novák L., Jílek J. O., Kakáč B., Ernest I., Protiva M.: diese Zeitschrift *25*, 2196 (1960).

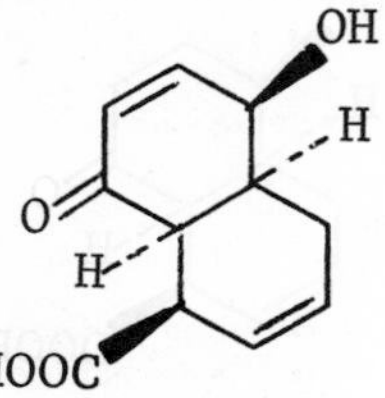

$C_{11}H_{12}O_4$ Ref. 787

EXAMPLE 1

Resolution of dl-1β-carboxy - 5β - hydroxy - 8 - keto-1,4,4aα,5,8,8aα - hexahydronaphthalene by means of quinine and isolation of 1-1β-carboxy-5β-hydroxy-8-keto - 1,4,4aα,5,8,8aα - hexahydronaphthalene (Formula I)

104 g. of dl-1β-carboxy-5β-hydroxy - 8-keto - 1,4,4aα, 5,8,8,aα - hexahydronaphthalene of Formula I, prepared as described by Woodward et al. in "Journ. Am. Chem. Soc.," vol 78, page 2023 (1956), are heated under reflux in 400 cc. of ethanol until complete solution is achieved. 162 g. of quinine free of water of crystallization is added at once to the boiling liquid. The salt of quinine with the levorotatory enantiomorphous compound crystallizes soon thereafter. Boiling without agitation is continued for 15 minutes. The mixture is then placed in ice for two hours and the crystals are filtered off. The filtered crystals are twice made into a paste, each time with 100 cc. of ice-cold ethanol. The pasted crystals are again filtered and the salt of quinine with the levorotatory enantiomorphous compound is dried. The yield amounts to 117 g. The melting point is 225° C.; rotatory power $[\alpha]_D^{20}$=—151°±4°.

The resulting product is suspended in 300 cc. of water and 500 cc. of chloroform. 30 g. of sodium bicarbonate are added thereto and the mixture is stirred for three hours until evolution of carbon dioxide ceases. The resulting emulsion is allowed to settle. The chloroform layer is separated from the aqueous layer and is washed several times with a saturated aqueous solution of sodium bicarbonate. The aqueous layer is extracted three times with chloroform in order to remove therefrom the quinine which may be contained therein. The aqueous layer is combined with the bicarbonate wash waters of the chloroform layer and the combined aqueous solutions are acidified by the addition of 5 N hydrochloric acid to a pH of 2.0. The acidified solution is saturated with sodium chloride and is allowed to stand for one hour. The precipitate is filtered and dried in a vacuum above solid sodium hydroxide at 20° C. 33 g. of the levorotatory compound of Formula I are obtained. The melting point is 210° C.; the rotatory power $[\alpha]_D$=—78°±4°. The crude product contains about 3 g. of sodium chloride. On recrystallization from a mixture of equal parts of alcohol and ether the pure levorotatory enantiomorphous compound is obtained. Its melting point is 210° C.; its optical rotation $[\alpha]_D^{20}$=—85° (concentration: 0.5% in ethanol).

Analysis: $C_{11}H_{12}O_4$=208.21. Calculated: 63.45% C, 5.81% H, 30.74% O. Found: 63.3% C, 5.8% H, 30.5% O.

The aqueous mother liquors remaining after removing the crude product are extracted several times with a mixture of chloroform and ethanol (3:1). The combined extracts are washed with salt-containing water until their pH is 4.0. The washed extracts are then dried over magnesium sulfate and evaporated to dryness in a vacuum. A second batch of the levorotatory enantiomorphous compound is recovered showing a rotatory power

$$[\alpha]_D=-84°$$

after it has been made into a paste by means of a mixture of chloroform and ethanol (3:1). The yield is 3 g.

The mother liquor from the paste of crystals is also evaporated to dryness and the residue is triturated with chloroform. Thereby a third batch of the crystalline enantiomorphous compound is obtained. The total resolution yield, thus, amounts to 80% of the theoretical yield.

EXAMPLE 2

Isolation of d-1β-carboxy-5β-hydroxy-8-keto-1,4,4aα,5,8,8aα-hexahydronaphthalene

The ethanolic mother liquors obtained according to Example 1 after removing the crystals of the quinine salt of the levorotatory isomer of Formula I are evaporated to dryness. The amorphous residue consists of a mixture of the quinine salt of the dextrorotatory enantiomorphous compound and about 10% of the quinine salt of the levorotatory enantiomorphous compound. Said residue is suspended in a mixture of 300 cc. of water and 500 cc. of chloroform with agitation, as described in Example 1. 30 g. of sodium bicarbonate are added thereto. Decomposition of the quinine salt is completed after about 1 hour. The resulting emulsion is allowed to settle. The chloroform layer is removed and washed twice with an aqueous solution of sodium bicarbonate. The aqueous layer is extracted several times with chloroform in order to completely remove the quinine contained therein. The combined aqueous sodium bicarbonate solutions are acidified by the addition of 5 N hydrochloric acid to a pH of 2.0. The acidified solution is allowed to crystallize. The crystals are filtered off after about 1 hour and are dried. 21 g. of a mixture of the racemic compound of Formula I with a small amount of the dextrorotatory enantiomorphous compound are obtained thereby. The optical rotation of said mixture is $[\alpha]_D^{20}$=+22° (concentration: 0.5% in ethanol).

The filtrate remaining after filtering off said mixture of racemate and dextrorotatory compound is saturated with sodium chloride and is extracted several times with a mixture of chloroform and ethanol (3:1). The resulting extract in the organic solvent is washed with salt-containing water until its pH is 4.0, dried over magnesium sulfate, and evaporated to dryness in a vacuum. The residue is made into a paste by means of chloroform and yields a first batch of crystals in the amount of 16 g. This product, which represents the pure dextrorotatory enantiomorphous compound of Formula I has a melting point of 210° C. and an optical rotation

$$[\alpha]_D^{20}=+83°\pm4°$$

The mother liquors yield after evaporation and allowing to stand for several days a second batch of crystals.

The mixture of racemic compound and dextrorotatory enantiomorphous compound in the amount of 21 g. (optical rotation $[\alpha]_D^{20}=+22°$) obtained as described hereinabove is dissolved in as small an amount of warm water as possible, usually about 4 parts by volume. The solution is cooled with ice and the precipitate is filtered off. 11 g. of the racemic compound having a melting point of 202° C. are recovered. They are returned to the resolving process.

The mother liquors obtained thereby are saturated with sodium chloride and yield, by proceeding in the same manner as described hereinabove, a further amount of dextrorotatory enantiomorphous compound.

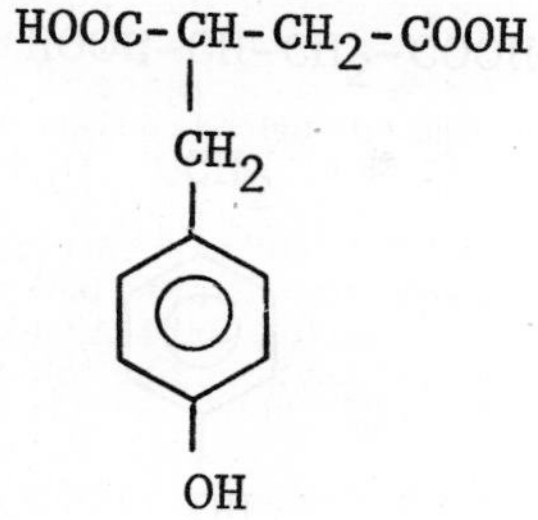

$C_{11}H_{12}O_5$ Ref. 788

2. *Acide p-hydroxybenzyl succinique.*

103 g de *l*-brucine dissous dans 350 cm³ d'éthanol à 50 % sont additionnés à 57 g de racémique. Le mélange est laissé au reflux pendant 2 h. Après refroidissement et repos 24 h, on récupère 110 g de cristaux blancs.

F = 173-174 °C, $[\alpha]_D^{22}$ = — 25,8 °C (C = 0,180 dans alcool à 50 %).

Ces cristaux sont recristallisés dans l'alcool à 50 % (2 150 cm³). Au refroidissement, il se dépose 95 g de cristaux blancs de sel neutre *dAlB*.

Soit $C_{44}H_{52}O_8N_4C_{11}H_{12}O_5$:
F = 175-176 °C, $[\alpha]_D^{20}$ = — 23,4° (C = 0,150 dans l'alcool à 50 %).

On ajoute aux eaux meres une même quantité de brucine, il cristallise alors le sel *lAlB* impur. On recristallise une dizaine de fois ce sel dans le mélange eau-alcool éthylique à 50 % en traitant à chaque fois par du charbon décolorant. On obtient ainsi 1,2 g de diastéréoisomère *lAlB* à 95 % de pureté avec un mauvais rendement.

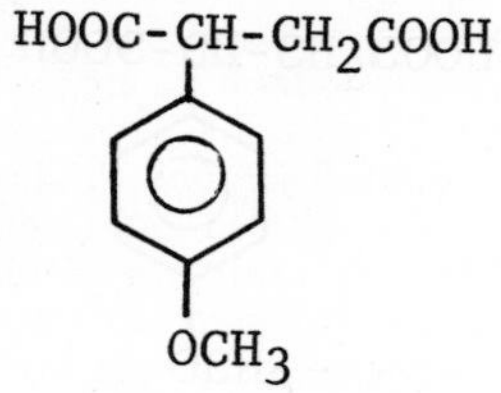

$C_{11}H_{12}O_5$ Ref. 789

***d*-Anisylsuccinic Acid.**—Ten grams of *dl*-anisylsuccinic acid was dissolved in boiling 95% ethyl alcohol and the quantity of brucine (34.6 g.) necessary to form the normal salt was added. The solution was filtered, cooled to 20° and allowed to stand for eighteen hours. On stirring a fine, white crop of crystals melting at 185–190° was obtained. After three recrystallizations from 95% ethanol, the product had a constant melting point of 197–200° and weighed 3.8 g.

Anal. Calcd. for $C_{23}H_{26}O_4N_2 \cdot C_{11}H_{12}O_5$: N, 4.53. Found: N, 4.55.

The dextro acid was regenerated by refluxing the *d*-acid-*l*-brucine salt with 150 cc. of 10% hydrochloric acid solution for two hours. The free acid which crystallized on cooling was filtered off, and the filtrate was then extracted four times with ether. The ether was removed by distillation and the residue was combined with the first crop. The large, glistening needles of *d*-anisylsuccinic acid were recrystallized from hot chlorobenzene containing a small amount of 95% ethanol. The compound melted at 196–197° with sublimation occurring at 191–195°. The sublimate melted sharply at 198.5–199.0°. *Optical rotations.* $[\alpha]^{32}D$ +135.8° (*c* 0.299 in ethanol), $[\alpha]^{32}D$ +161.0° (*c* 0.420 in ethyl acetate), $[\alpha]^{31}D$ +155.4° (*c* 0.869 in acetone), $[\alpha]^{31}_{5461}$ +188.7° (*c* 0.869 in acetone).

***l*-Anisylsuccinic Acid.**—The mother liquor from the *d*-acid-*l*-brucine salt was evaporated to 80 cc. and then diluted with 240 cc. of water, for it was found that the alcoholic solution became sirupy on standing and showed no tendency to crystallize. The aqueous alcoholic solution, which was turbid, was heated until it became clear, then filtered and allowed to cool slowly. A crop of 19 g. of *l*-acid-*l*-brucine salt was filtered and recrystallized from a mixture of water and ethyl alcohol. Thirteen grams of salt melting at 136.5–137.0° was obtained. The analysis indicated that a normal salt containing two molecules of water of crystallization had been formed.

Anal. Calcd. for $2(C_{23}H_{26}O_4N_2) \cdot C_{11}H_{12}O_5 \cdot 2H_2O$: N, 5.35. Found: N, 5.31.

The *l*-anisylsuccinic acid-*l*-brucine salt (13 g.) was hydrolyzed by warming to 58° in 70 cc. of water to which had been added 4 cc. of concentrated hydrochloric acid. Eight-tenths gram of *l*-anisylsuccinic acid crystallized out on cooling: m. p. 196–199°, $[\alpha]^{30}D$ −120.6° (*c* 1.044 in ethanol). After two recrystallizations from hot chlorobenzene containing approximately 2% of ethyl alcohol, a constant specific rotation of $[\alpha]^{29}D$ −122.0° was obtained. Since this value is considerably lower than the activity of the dextro isomeride, ($[\alpha]^{32}D$ (ethanol) + 135.8°), the rotations of the anhydrides prepared from the two isomers were compared: *l*-anisylsuccinic anhydride, $[\alpha]^{29.5}D$ −94.9° (*c* 0.602 in ethanol), *d*-anisylsuccinic anhydride, $[\alpha]^{29.5}D$ +95.2° (*c* 0.598 in ethanol). These data indicate that *l*-anisylsuccinic anhydride can be purified by fractional crystallization more readily than *l*-anisylsuccinic acid. After recrystallization of the two anhydrides, the resolution of *dl*-anisylsuccinic acid was complete.

In contrast to the action of hot water on optically active isomerides of phenylsuccinic acid, it was found the *d*- and *l*-anisylsuccinic acids racemized slowly in boiling water.

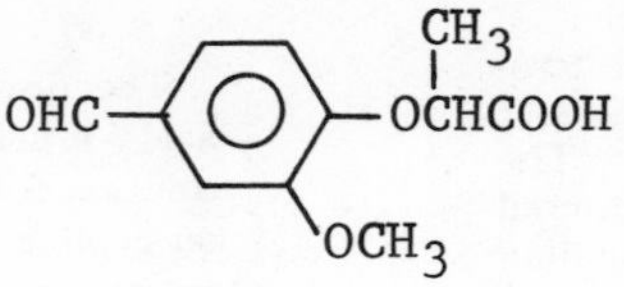

$C_{11}H_{12}O_5$ Ref. 790

Cinchonine salt. Cinchonine (15·16 g.) and the (±)-acid (11·54 g.) were heated under reflux in water (1220 ml.) for 1 hour and filtered. The crystals deposited on cooling were collected and fractionally crystallised from water. The mother-liquor (*M*) was set aside (see below). The final fraction after 5 crystallisations of the *salt* had $[\alpha]_D^{15}$ +77·35° (*c* = 1·049 in alcohol, *l* = 1) (Found: C, 63·5; H, 6·7; N, 5·1. $C_{30}H_{34}O_6N_2,3H_2O$ requires C, 62·9; H, 7·05; N, 4·9%).

(−)-α-(2-*Methoxy-4-formylphenoxy)propionic Acid.*—The final fraction of the cinchonine salt was shaken with excess of sodium carbonate solution (33%). The cinchonine was filtered off, and the aqueous solution extracted 15 times with chloroform to ensure complete removal of the cinchonine. The solution of the sodium salt was acidified with concentrated hydrochloric acid, and the precipitated (−)-*acid* collected, washed, and dried. The aqueous solution, when extracted with ether, gave a further quantity of (−)-acid, which crystallised from benzene–light petroleum in fine colourless needles, m. p. 89—90°; α_D^{15} −0·328° (*l* = 1, *c* = 1·008 in alcohol), $[\alpha]_D^{15}$ −32·5° (Found: C, 58·6; H, 5·4. $C_{11}H_{12}O_5$ requires C, 58·9; H, 5·4%).

(+)-α-(2-*Methoxy-4-formylphenoxy)propionic Acid.*—The mother-liquors (*M*) were concentrated to 100 ml., and the white solid crystallising out removed; when the filtrate was concentrated to 50 ml., no further solid separated. The solution was shaken with excess of sodium carbonate, and the (+)-*acid* recovered as described for the (−)-acid. It crystallised from benzene–light petroleum as fine colourless needles, m. p. 89—90°, α_D^{15} +0·301° (*l* = 1, *c* = 0·912 in alcohol), $[\alpha]_D^{15}$ 33·0° (Found: C, 58·9; H, 5·3%).

A mixture of 0·0715 g. each of the (+)- and the (−)-acid made up to 10 ml. in absolute alcohol was inactive. The (±)-acid recovered from this measurement had m. p. 78—79°, not depressed on admixture with authentic (±)-acid, m. p. 79—80°.

Quinine salt. (*a*) Quinine (12·94 g.) and (±)-acid (8·97 g.) were heated under reflux in water (600 ml.) for 2 hours and filtered. The crystals deposited on cooling were collected and dried. The *salt* was fractionally crystallised from water, the final fraction after 6 crystallisations having $[\alpha]_D^{18}$ −86·8° (*c* = 8·68 in alcohol, *l* = 1) (Found: C, 63·8; H, 6·4. $C_{31}H_{36}O_7N_2,2H_2O$ requires C, 63·7; H, 6·9%). The acid was recovered from the final fraction similarly to the (−)-acid; m. p. 84—85°; α_D^{16} +0·103° (*l* = 1, *c* = 0·76 in alcohol), $[\alpha]_D^{16}$ +13·2° (Found: C, 58·7; H, 5·6%).

(*b*) Quinine (7·088 g.), (±)-acid (11·1984 g.), and sodium carbonate (1·1595 g.) were heated under reflux in water (750 ml.) for 4 hours and concentrated to 500 ml.; the crystals which separated were collected and dried. The salt was fractionally crystallised from water, the final fraction after 6 crystallisations having $[\alpha]_D^{15}$ −80·9° (*c* = 8·18 in absolute alcohol, *l* = 1) (Found: C, 63·9; H, 6·8%). The acid, recovered from the final fraction similarly to the (−)-acid, had m. p. 81—82°; α_D^{15} +0·0825° (*l* = 1, *c* = 1·04 in alcohol), $[\alpha]_D^{15}$ +7·9° (Found: C, 59·3; H, 5·7%).

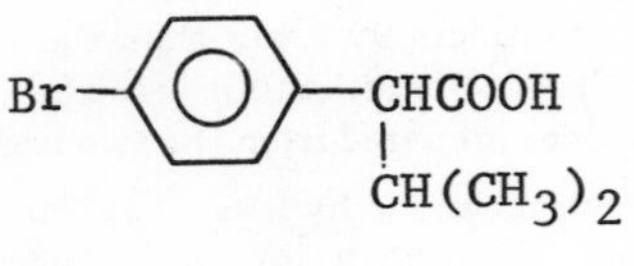

$C_{11}H_{13}BrO_2$ Ref. 791

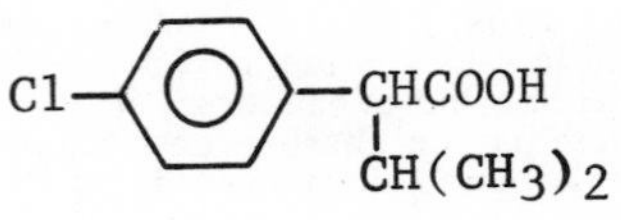

$C_{11}H_{13}ClO_2$ Ref. 792

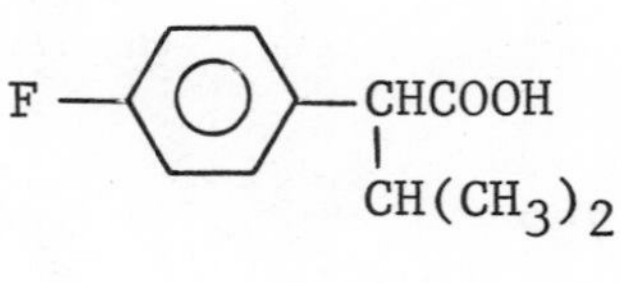

$C_{11}H_{13}FO_2$ Ref. 793

For resolution procedures and physical constants for Refs. 791, 792 and 793 above, see Ref. 905

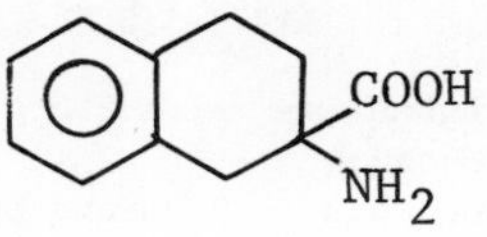

$C_{11}H_{13}NO_2$ Ref. 794

Resolution of Methyl 1,2,3,4-Tetrahydro-2-acetamido-2-naphthoate. To 3.0 g (12.2 mmol) of the methyl ester dissolved in 120 ml of methanol was added 480 ml of water and the pH was adjusted to 7.9 with 0.1 *M* sodium hydroxide. α-Chymotrypsin (400 mg) dissolved in 5 ml of water was added. During 12 hr 60 ml of base was consumed (98% of theoretical) and the reaction stopped. The reaction mixture was extracted with chloroform. The chloroform extracts were washed with water, dried over magnesium sulfate, filtered, and evaporated at reduced pressure to afford 1.5 g (100%) of (−) ester, mp 148–150°. Three recrystallizations from chloroform–hexane gave 1.4 g of (−)-methyl 2-acetamido-1,2,3,4-tetrahydro-2-naphthoate, mp 150.5–151[b], $[\alpha]^{25}D$ −59.6° (*c* 2.75, ethyl acetate).

The aqueous layer was adjusted to pH 2 with concentrated hydrochloric acid and refrigerated overnight. Isolation of the precipitate by filtration gave 900 mg of (+) acid which was recrystallized twice from water. The final yield of (+)-2-acetamido-1,2,3,4-tetrahydro-2-naphthoic acid, mp 259–261°, $[\alpha]^{25}D$ +39.8° (*c* 2, dimethylformamide), was 400 mg.

(−)-2-Acetamido-1,2,3,4-tetrahydro-2-naphthoic acid was obtained by stirring 550 mg of the corresponding methyl ester overnight in 50 ml of 10% sodium hydroxide. The solution was acidified to pH 4 and the crystalline product was removed by filtration and dried at reduced pressure. The yield of (−) acid, mp 263–264°, $[\alpha]^{25}D$ −45.5° (*c* 2, dimethylformamide), was 400 mg.

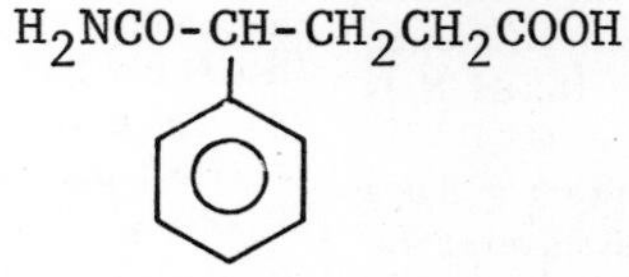

$C_{11}H_{13}NO_3$ Ref. 795

Equimolecular amounts of racemic [III] and D(—)-threo-2-amino-1-(*p*-nitrophenyl)-1,3-propanediol were crystallised six times from *Et*OH until constant rotation was reached. The salt obtained, dissolved in the minimum amount of water and cautiously acidified with conc. HCl, gave crystalline S-[III]: m.p. = 126–8 °C (water); IR (KBr disc), ν_{max}: 3470–3200 (series of bands), 1700, 1640, 1570 cm^{-1}; $[\alpha]_D$ = + 63.5° (*c* 1.02).

Elemental analysis, found % (calcd for $C_{11}H_{13}NO_3$): C 63.41 (63.75), H 6.37 (6.32), N 6.61 (6.76).

A fractional crystallisation of R-[III] recovered from the mother liquor, and showing $[\alpha]_D$ = —40°, allowed the separation of the less soluble racemic compound and of the pure levo isomer having the correct physical constants.

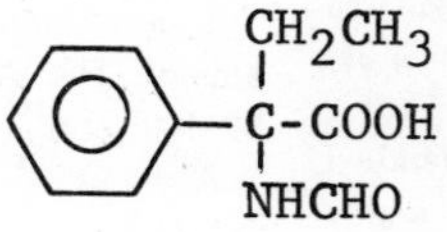

$C_{11}H_{13}NO_3$ Ref. 796

Resolution of N-Formyl-α-ethylphenylglycine(XXX)——The quinine salt of (±)-XXX, prepared by Sobotka's method,[24] was recrystallized from 20% aqueous EtOH five times to give a diastereomeric salt (42.5 g), mp 187—190°, $[\alpha]_D^{25}$ −68.9° (*c*=1.04, EtOH). Recrystallization of the (*S*)(+)-XXX obtained from this salt from aqueous EtOH, gave (*S*)(+)-XXX as colorless prisms (12.96 g), mp 204—205.5° (decomp.), $[\alpha]_D^{26}$ +126° (*c*=1.05, 1N NaOH) (lit.[24] mp 212°, $[\alpha]_D$ +126° (*c*=1.00, aq. alkali)). The (*R*)(−)-XXX recovered from all the mother liquors of recrystallizations of the quinine salt was recrystallized from aqueous EtOH eight times to give (*R*)(−)-XXX as colorless pillars, mp 203—205° (decomp.), $[\alpha]_D^{27}$ −124° (*c*=1.03, 1N NaOH).

24) H. Sobotka, M.F. Holzman and J. Kahm, *J. Am. Chem. Soc.*, **54**, 4697 (1932).

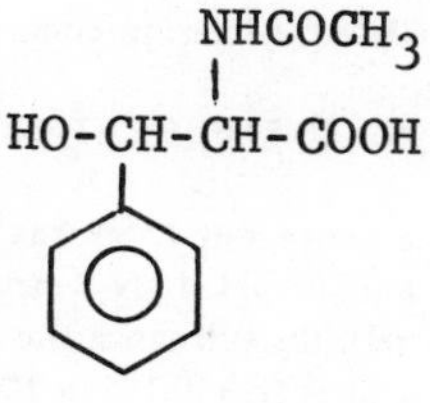

$C_{11}H_{13}NO_4$ Ref. 797

Risoluzione della D,L-treo-N-acetil-β-fenilserina con cinconina. — Per la preparazione della D,L-treo-N-acetil-β-fenilserina oltre all'acetilazione della D,L-treo-β-fenilserina operando secondo Schotten-Baumann con anidride acetica si è impiegato il metodo di O-acetilazione selettiva e successiva migrazione O → N già usato con successo sui fenilaminopropandioli ([14]).

Grammi 1.71 di D,L-treo-β-fenilserina vennero sciolti in 10 cm³ di soluzione al 5,6% di acido cloridrico gassoso in acido acetico glaciale.

Si aggiunsero sotto agitazione 5 cm³ di cloruro di acetile e si lasciò una notte sotto agitazione. Al mattino successivo si filtrò il precipitato formatosi e si lavò con etere etilico. Si pesarono g 2,02 di D,L-treo-O-acetil-β-fenilserina cloridrato a p. f. 156-157° (dec.) che dopo ricristallizzazione da metanolo-etere salì a 160-161° (dec.).

trov.%: HCl 13,92;
per $C_{11}H_{13}O_4N.HCl$ calc. : 14,04.

Grammi 2,59 di D,L-treo-O-acetil-β-fenilserina cloridrato vennero sciolti in 10 cm³ di acqua ed addizionati di 1,68 g di $NaHCO_3$ (2 equivalenti). Il precipitato formatosi dapprima si è lentamente disciolto; vennero allora aggiunti 2 cm³ di acido solforico 2 N ottenendosi g 1,75 di prodotto a p. f. 142-143° (dec.) che per ricristallizzazione dall'acqua salì a 150-151° (dec.). Titolo acidimetrico con NaOH N/10 e fenolftaleina: 99,85%.

Grammi 8,92 di D,L-treo-N-acetil-β-fenilserina si sciolsero a caldo in cm³ 13 di acqua e si addizionarono a caldo ad una soluzione di g 11,76 di cinconina in cm³ 600 di metanolo. Si filtrò da una parte indisciolta e si concentrò a secchezza; il residuo, sciolto in poco cloroformio e fatto evaporare lentamente in capsula, diede un vetro macinabile che fu sciolto in 100 cm³ di acqua calda. Si filtrò da 0,86 g di cinconina separatesi e si fece cristalizzare sotto agitazione; si ottennero g 8,0 a p. f. 89-112° che vennero ricristallizzati da 40 cm³ di acqua filtrando da altra cinconina precipitata. Si ottennero g 5,8 a p. f. 95° (torbido), i quali vennero ricristallizzati al solito modo da 29 cm³ di acqua sempre filtrando via la parte di cinconina separatasi. Vennero pesati g 4,82 a p. f. 96-99° $(\alpha)_D = +0,29°$ (c = 2,0; metanolo) che analizzarono male. Trovato: C=68.03% H=6,93%; per $C_{30}H_{33}O_5N_3$ si calcola: C=69,61; H=6,82.

Grammi 2,0 di questo sale vennero sciolti in 5 cm³ di metanolo ed addizionati di cm³ 5 di NaOH N/1. Si separarono g 1,07 di cinconina e dalle acque madri concentrate ed acidificate al Congo con HCl conc. si ottennero g 0,5 a p. f. 168° (dec.) $(\alpha)_D = -32,2°$ (c = 5,0; NaOH N/1) costituiti da D-treo-N-acetil-β-fenilserina di buona purezza.

Le a.m. della prima cristallizzazione che avevano fornito i g 8,0 a p. f. 89-112° per concentrazione non hanno lasciato cristallizzare alcunchè; diluite con metanolo ed alcalinizzate alla fenolftaleina con NaOH N/1 diedero g 5,7 di cinconina. Le a.m. concentrate ed acidificate al Congo con HCl conc. hanno dato g 2,3 di L-treo-N-acetil-β-fenilserina impura a p. f. 151-158° $(\alpha)_D = +22°$ (c=5,0; NaOH N/1).

(16) C. G. ALBERTI, B. CAMERINO e A. VERCELLONE, questa Gazzetta, 82, 63 (1952).

$C_{11}H_{13}NO_4$ Ref. 797a

Risoluzione della D,L-treo-N-acetil-β-fenilserina con L-treo-1-p.nitro-fenil-2-amino-1,3-propandiolo. — Grammi 89,2 di D,L-treo-N-acetil-β-fenilserina vennero mescolati a g 84,8 di L-treo-1-p.nitrofenil-2-amino-1,3-propandiolo e sciolti a caldo in 200 cm³ di acqua. Dopo trattamento con carboraffina si lasciò cristallizzare per una notte a temperatura ambiente: si ottennero g 77,3 a p. f. 173-177° (dec.) ed $(\alpha)_D = +34°$ (c=2,0; metanolo); le acque madri (a.m. 1) vennero conservate. I 77,3 g furono ricristallizzati nello stesso modo da 3 parti di acqua ottenendosi g 65.88 a p. f. 177-178° (dec.) ed $(\alpha)_D = +39°$ (c=2,0; metanolo). Le acque madri per concentrazione diedero altri 8,35 g a p. f. 170° (dec.) ed $(\alpha)_D = +37,5°$ e delle acque madri (a.m. 2) che vennero lavorate con le a.m. 1 come sarà detto in seguito. Ricristallizzando come al solito i 65,88 g a p. f. 177-178°, si ottennero g 55,0 ancora a p. f. 177-178° ed $(\alpha)_D = +42°$ (c=2,0; metanolo) di L-treo-N-acetil-β-fenilserinato di L-treo-1-p.nitrofenil-2-amino-1,3-propandiolo.

trov.%: C 54,75; H 5,89;
per $C_{20}H_{25}O_8N_3$ calc. : 55,17; 5,79.

Le acque madri unite ai 8,35 g di cui sopra furono concentrate per ottenere altri g 11,0 a p. f. 179° (dec.) ed $(\alpha)_D = +40°$ (c=2,0; metanolo).

Le a.m. 1 + le a.m. 2 per concentrazione non lasciarono cristallizzare alcunchè; si diluì allora ad un rapporto soluto/solvente di 1:5 e si aggiunse soda caustica a 36° Bè fino a pH 11. Cristallizzarono g 43,5 di L-base p. f. 155-160°.

Le a.m. vennero neutralizzate, concentrate ed acidificate al Rosso Congo con HCl conc.: si ebbero g 40,55 di D-treo-N-acetil-β-fenilserina impura a p. f. 156-159° (dec.) ed $(\alpha)_D = -26,8°$ (c=5.0; NaOH N/1) i quali vennero addizionati a g 38,60 di D-treo-1-p.nitrofenil-2-amino-1,3-propandiolo e sciolti con 150 cm³ di acqua a caldo (carboraffina). Dopo una notte si ottennero g 52,5 a p. f. 174-175° (dec.), $(\alpha)_D = -36,5°$ (c=2,0; metanolo) e delle acque madri (a.m. 3) che furono accantonate. I 52,5 g furono ricristallizzati da 3 parti di acqua e diedero g 44,5 di D-treo-N-acetil-β-fenilserinato di D-treo-1-p.nitrofenil-2-amino-1,3-propandiolo ancora a p. f. 174-175° (dec.) $(\alpha)_D = -38°$ (c=2,0; metanolo).

trov.%: C 55,14; H 5.89;
per $C_{20}H_{25}O_8N_3$ calc. : 55,17; 5,79.

Le a.m. vennero concentrate per avere altri g 5,45 a p. f. 176° (dec.) $(\alpha)_D = -36,5°$ (c=2,0; metanolo).

Le acque madri, unite alle a.m. 3, furono portate ad un rapporto soluto: solvente di 1:5 ed alcalinizzate a pH 11 con NaOH di 36 Bè, ottenendosi g 4,7 a p. f. 157-158° di D-base. Le acque madri concentrate ed acidificate al Congo con HCl conc. furono estratte con acetato di etile; per evaporazione dell'estratto si ottennero g 3,7 di D,L-treo-N-acetil-β-fenilserina impura a p. f. 145-146° (dec.) ed $(\alpha)_D = +3,6°$ (c=5,0; NaOH N/1).

Grammi 55,0 di L-treo-N-acetil-β-fenilserinato di L-treo-1-p.nitrofenil-2-amino-1,3-propandiolo furono sciolti in cm³ 275 di acqua e alcalinizzati a pH 11 con NaOH 36° Bè. Si ottennero g 26.46 a p. f. 161-162° di L-base. Le a.m. dopo neutralizzazione furono concentrate ed acidificate al Congo con HCl conc. Si ottennero g 27,4 di L-treo-N-acetil-β-fenilserina a p. f. 166° (dec.) ed $(\alpha)_D = +32,6°$ (c=5,0; NaOH N/1). Dopo ricristallizzazione dall'acqua si ebbero dei prismi aghiformi a p. f. 165-166°, $(\alpha)_D = +34,7° \pm 1°$ (c = 5,0; NaOH N/1) e $+18,0° \pm 3°$ (c = 2,02; metanolo), che diedero un titolo acidimetrico (NaOH N/10 e fenolftaleina) del 100,35%.

In modo identico da g 44.0 di D-treo-N-acetil-β-fenilserinato di D-treo-1-p.nitrofenil-2-amino-1,3-propandiolo si ottennero g 21,0 di D-base a p. f. 159-161° e g 21,9 di D-treo-N-acetil-fenilserina a p. f. 165° (dec.) $(\alpha)_D = -30°$ (c=5,0; NaOH N/1). Dopo ricristallizzazione dall'acqua si ebbero dei prismi aghiformi a p. f. 165-166° (dec.) $(\alpha)_D = -35,7° \pm 1°$ (c = 5,0; NaOH N/1) e $-18,5° \pm 1°$ (c = 2,0; metanolo) il cui titolo acidimetrico è risultato del 99.86%.*

Nel brevetto Svizzero citato (4) viene riportato per la L-treo-N-acetil-β-fenilserina il p. f. di 163-164° (dec.) $(\alpha)_D^{20°} = +16,3°$ (=2,0; metanolo) e per la D-treo-N-acetil-β-fenilserina il p. f. 163-165° (dec.) e $(\alpha)_D^{18°} = -17,1$ (c=2,0; metanolo).

HOOCCH2-CH-COOH
NHCH2C6H5

$C_{11}H_{13}NO_4$ Ref. 798

L-Leucine amide.

(A) From ethyl L-leucinate.

This substance was prepared according to the instructions given for DL-leucine amide[6]. The yield was about 90%; m.p. (after recrystallization from benzene) 99–100°; $[\alpha]_D^{23} = +6°$ ($c = 0.25$ in water).

Anal. Calcd. for $C_6H_{14}N_2O$: C, 55.2; H, 10.8; N, 21.6. Found: C, 55.3; H, 10.6; N, 21.5.

(B) From ethyl L-leucinate hydrochloride.

The ester hydrochloride (25 g) was dissolved in dry methanol (300 ml) and dry gaseous ammonia was passed into the solution until saturation.The reaction mixture was shaken at room temperature for four days. The *p*H was then adjusted to 10 with sodium hydroxide solution and evaporation to dryness *in vacuo* was carried out. The amide was extracted with hot chloroform and after evaporation of the solvent the substance was recrystallized from benzene. Yield 9 g (54%).

Salt of N-benzyl-D-aspartic acid with L-leucine amide.— N-Benzyl-DL-aspartic acid (5.5 g) suspended in ethanol (60 ml) was heated on a water bath with L-leucine amide (3.5 g) until a clear solution resulted. After a short time the salt crystallized in needles united to bundles. After filtration the substance was recrystallized from ethanol, yielding 2.7 g (60%); m.p. 180–182° (sintering commences at about 115°); $[\alpha]_D^{23} = +26°$ ($c = 0.125$ in water).

Anal. Calcd. for $C_{17}H_{27}O_5N_3$: C, 57.6; H, 7.7; N, 11.9. Found: C, 57.2; H, 8.2; N, 11.6.

N-Benzyl-D-aspartic acid. The above salt (2 g) was dissolved in water (2 ml) and acidified with hydrochloric acid to *p*H 4. N-Benzyl-D-aspartic acid precipitated. After addition of ethanol (5 ml), the substance was filtered and dried. Yield 1.24g (98%); m.p. 209°; $[\alpha]_D^{23} = +33.3°$ ($c = 0.125$ in 5% sodium bicarbonate solution).

Salt of N-benzyl-L-aspartic acid and L-leucine amide. After separation of the salt of N-benzyl-D-aspartic acid with L-leucine amide, the second diastereoisomeric salt crystallized. This on recrystallization from ethanol yielded 1.5 g (33.5%); $[\alpha]_D^{25} = -14°$ ($c = 0.125$ in water).

N-Benzyl-L-aspartic acid. Liberation from the salt as for the D-isomer. $[\alpha]_D^{23} = -33.2°$ ($c = 0.125$ in 5% sodium bicarbonate solution); m.p. 209°.

6. YANG, P. S. AND RISING, M. M., 1931, *J. Am. Chem. Soc.*, 53, 3183.

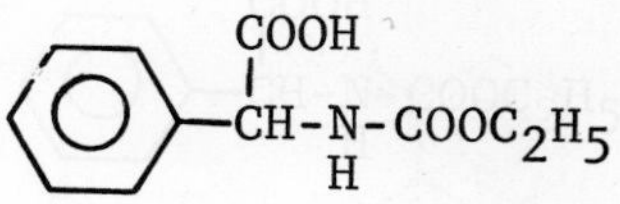

$C_{11}H_{13}NO_4$

Ref. 799

Dédoublement par entraînement de la N-carbéthoxyphénylglycine 5

Dans un quadrucol de 1 l muni d'un agitateur mécanique formé d'un segment de cercle de 10 cm^2 de surface, d'un thermomètre plongeant, d'un réfrigérant, on introduit 178 g de (±) N-carbéthoxyphénylglycine, 16 g d'énantiomère L(+) et 606 g d'acide acétique pur. On chauffe 30 mn dans un bain thermostaté à 70° C puis refroidit en 30 mn à 20° C. La vitesse d'agitation est réglée à 400 t/mn. On amorce la cristallisation par 4 g de cristaux (+) de granulométrie 150 μ et agite 3 h à 20° C. On filtre sur verre fritté (Ø 10 cm, porosité 3) sous légère pression d'azote et récupère le filtrat dans un deuxième quadrucol identique, précédemment taré. On laisse 30 mn sous faible débit d'azote.

Poids de cristaux obtenus (séchés à 40° C sous vide) : 39,6 g, $[\alpha]_D^{20}$ = +145° (c = 1, méthanol), p.o. 91 %.

Poids de solution récupérée : 735,8 g, perte par évaporation : 28,4 g

On lave le quadrucol et le fritté par une quantité d'acide acétique correspondant à la perte. On déduit du pouvoir rotatoire et du poids de cristaux isolés, les quantités de racémique et d'énantiomère (-) ou (+) à ajouter pour se trouver dans les conditions initiales précédentes symétriques. On opère de la même façon en amorçant par l'énantiomère (-) et obtient les résultats du Tableau 3.

Dans toutes les expériences de cristallisation, les compositions des solutions sont déterminées par mesure simultanée de l'indice de réfraction et du pouvoir rotatoire lu qui permettent de déduire la concentration totale en soluté (en g %) et l'excès d'énantiomère (en g %) respectivement, à l'aide de courbes d'étalonnage préalablement établies.

Racémisation de la L(+) N-carbéthoxyphénylglycine 5

On dissout 120 g de soude dans 297 g d'eau et ajoute, à 40°C, 223 g (1 mol) de L(+) N-carbéthoxyphénylglycine. On maintient 5 h à 40°C. Il se forme d'abord deux phases liquides puis le sel de sodium du produit racémique cristallise. On refroidit à 20° C, ajoute 400 g d'eau puis 500 g de toluène et acidifie à pH 0.5. Le produit cristallise après quelques minutes. On refroidit à 5° C, lave par 1500 ml d'eau et sèche à 50° C. On obtient 211 g (Rdt 94,5 %), F = 119° C.

Purification de la D(-) N-carbéthoxyphénylglycine 5

On dissout 113,8 g de D(-) N-carbéthoxyphénylglycine à 89 % de pureté optique dans 228 g d'acétate de n-butyle au reflux, refroidit lentement sous agitation et amorce la cristallisation par 0,1 g de D(-) pur. Après 30 mn à 20° C, on filtre et sèche à 40° C sous vide : 96 g, $[\alpha]_D^{20}$ = -158° (c = 1, méthanol) pureté optique 99,5 %.

D(-) phénylglycine

On dissout 55,7 g (0,25 mol) de D(-) N-carbéthoxyphénylglycine 5 (p.o. 99,5 %) dans 50 g d'acide acétique à 55° C, ajoute 211 g (1,25 mol) d'acide bromhydrique à 48 %, porte à reflux et distille les vapeurs passant entre 34 et 97° C (la température intérieure passant de 114 à 117° C). Après 20 mn environ, le dégagement de CO_2 cesse. On refroidit dans un mélange glace-eau, filtre le précipité formé, lave par 20 ml d'acide acétique froid et sèche à 40° C. On obtient 52,7 g (Rdt 91 %), $[\alpha]_D^{20}$ = -86,5° (c = 1, eau). Ce bromhydrate est repris dans 157 ml d'eau et neutralisé à pH 5,5 par de l'ammoniaque. On filtre les cristaux qu'on lave par 50 ml d'eau glacée. On obtient 31,3 g (Rdt par rapport au bromhydrate 91 %), $[\alpha]_D^{20}$= -157° (c = 1, HCl N), p.o. 99 %.

TABLEAU 3

Dédoublement par entraînement de la N-carbéthoxyphénylglycine 5 effectué sur 800 g de solution sursaturée contenant 178 g de racémique et 16 g d'énantiomère soit une concentration C_o = 24,25 g % et un excès d'énantiomère E_o = 2,0 g %; vitesse d'agitation 400 t/mn, température 20° C, temps de cristallisation 3 h, amorçage avec 4 g de cristaux optiquement purs.

Cycle N°	solution initiale	Cristaux poids (g)	obtenus (*) p.o. (%)	Produits ajoutés à la solution finale (g) (±)	(-)	(+)	AcOH
0	(+)	35,6	(+) 90	35,5	-	0,1	28,5
1	(-)	36,2	(-) 87	35,8	-	0,4	21,4
2	(+)	32,7	(+) 87	29,0	3,7	-	17,9
3	(-)	34,6	(-) 88	32,8	-	1,9	22,6
4	(+)	32,5	(+) 92	30,4	2,0	-	13,9
5	(-)	31,0	(-) 88	-	-	-	-

(*) moins l'amorçage; p.o. = pureté optique.

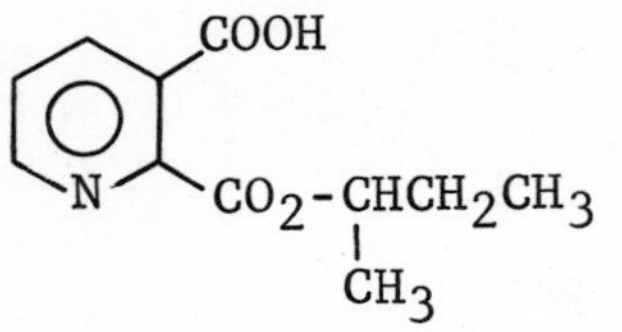

$C_{11}H_{13}NO_4$ Ref. 800

(+)- *and* (−)-sec.-*Butyl 2-Carboxypyridine-3-carboxylate.*—A clear solution of the (±)-3-butyl ester (21 g.) and brucine (37 g.) in warm acetone (100 c.c.) soon deposited crystals of the brucine salt. This (crop A; 46 g.) was removed, washed with acetone, and dissolved in warm chloroform (20 c.c.). Dilution with acetone (40 c.c.) yielded crop B (34 g.). Repetition of the process yielded crop C (28·7 g.) as needles, m. p. 186°. Crop C (25 g.), triturated with acetone (70 c.c.) and 0·5N-sulphuric acid (81 c.c.) and treated as described above, yielded (+)-*sec.*-butyl 2-carboxypyridine-3-carboxylate (6·8 g.) which separated from acetone in needles, m. p. 130—131°, $[\alpha]_D$ +10·9° (*l* 1; *c* 4·60 in $CHCl_3$). The more soluble fraction of the brucine salt contained in the filtrates, which it is essential to concentrate *without heating*, after decomposition as described above, yielded (−)-*sec.*-butyl 2-carboxypyridine-3-carboxylate (7·1 g.) which separated from acetone in needles, m. p. 130—131°, $[\alpha]_D$ −8·6° (*l* 1; *c* 3·86 in $CHCl_3$).

When the filtrates containing the brucine salt were concentrated on the steam-bath before decomposition, the product was the more stable (−)-*sec.*-butyl 3-carboxypyridine-2-carboxylate, m. p. 150—151°, $[\alpha]_D$ −12° in $CHCl_3$.

The brucine salt (1 g.) of *sec.*-butyl 3-carboxypyridine-2-carboxylate was heated at 190° until completely molten, cooled, and decomposed with hydrochloric acid. Crystallisation of the liberated product from cold acetone yielded *sec.*-butyl 2-carboxypyridine-3-carboxylate (0·15 g.), m. p. and mixed m. p. 130—131°.

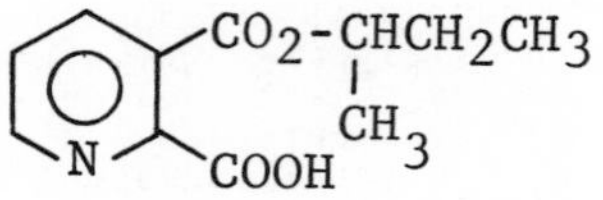

$C_{11}H_{13}NO_4$ Ref. 801

(+)- *and* (−)-sec.-*Butyl 3-Carboxypyridine-2-carboxylate.*—Brucine (49·5 g.) dissolved readily in a warm solution of the (±)-2-butyl ester (28 g.) in acetone (100 c.c.); the solution soon crystallised. Filtration yielded a brucine salt (m. p. 184°; 65 g.; crop A): this was heated for 2 hr. under reflux with acetone (200 c.c.), and the suspension cooled and filtered to give crop B (45·5 g.). This, in turn, was heated under reflux with acetone (100 c.c.) for an hour, cooled, and filtered, to give crop C (36·5 g.), needles, m. p. 186°. Crop C was triturated with acetone (40 c.c.), and the suspension mixed slowly with 0·5N-sulphuric acid (116 c.c.) and shaken. The resulting brucine sulphate was removed and washed with hot acetone (20 c.c.). The aqueous acetone filtrate was slowly evaporated on a current of air, (+)-*sec.*-butyl 3-carboxypyridine-2-carboxylate (10·5 g.) separating as needles, m. p. 150—151°, $[\alpha]_D$ +22·2° (*l* 1; *c* 3·361 in $CHCl_3$).

The combined filtrates containing the more soluble fraction of the brucine salt were concentrated and mixed with 0·5N-sulphuric acid (116 c.c.) and treated as described above. The liberated (−)-*sec.*-butyl 3-carboxypyridine-2-carboxylate (10·2 g.) formed needles, m. p. 150—151°, $[\alpha]_D$ −21·1° (*l* 1; *c* 3·134 in $CHCl_3$).

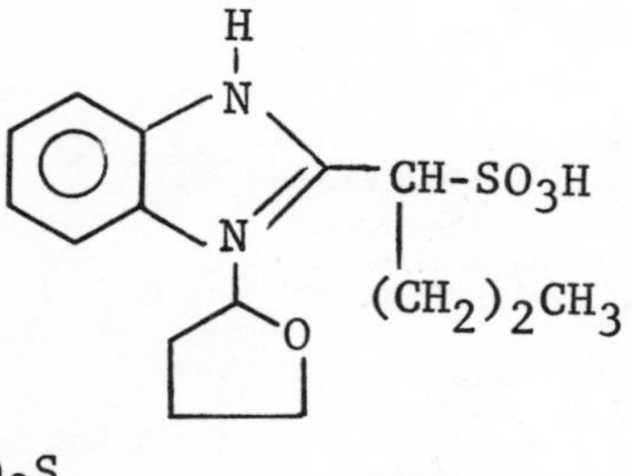

$C_{11}H_{14}N_2O_3S$ Ref. 802

Dedoublement de l'acide d-benzimidazol-2-butyl-x-sulfonique.

L'acide chauffé avec la quantité équimoléculaire de strychnine, en présence de beaucoup d'eau, donne une solution qui, refroidie, dépose le *sel de strychnine* en fines aiguilles.

Substance 0.5611 g ; 0.8988 m. équiv. de baryte.
Trouvé: poids mol. 624.3.
$C_{11}H_{14}O_3N_2S \cdot C_{21}H_{22}O_2N_2 \cdot 2\,H_2O$. Calculé: " " 624.4.

Le sel de baryum, obtenu en décomposant le sel de strychnine par la baryte, est dextrogyre.

Des cristallisations répétées du sel de strychnine dans l'eau bouillante n'ont pas donné le résultat désiré.

On se serait attendu à ce que les eaux-mères des cristallisations successives eussent donné à la longue des sels barytiques dextrogyres, tandis qu'ils sont restés lévogyres.

C'est pour cela que nous avons appliqué la „cristallisation froide" [10]).

On décompose le sel de strychnine par de la baryte et on essore la strychnine, qu'on dissout dans un faible excès d'acide acétique.

Enfin on mélange le benzimidazol-butylsulfonate de baryum avec l'acétate de strychnine en dilution convenable, de sorte qu'une partie du sel de strychnine cristallise lentement.

On répète cette opération plusieurs fois.

A partir de la troisième cristallisation on a examiné le sel de strychnine cristallisé en le transformant en sel barytique. Ainsi on a trouvé les valeurs suivantes de la rotation moléculaire des sels barytiques: 26.6, 34.4, 36.4, 35°.6.

La rotation ne changeant plus, il est probable que le dédoublement est complet.

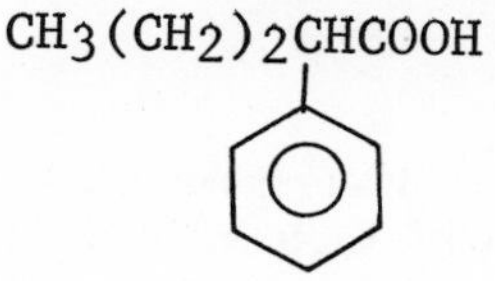

$C_{11}H_{14}O_2$ **Ref. 803**

Preliminary tests on resolution: 0.001 mole of the racemic acids and 0.001 mole of the optically active bases were dissolved in 10 ml of hot ethanol. The solutions were left at room temperature until crystals or oils were formed or until all solvent had disappeared. The crystals were filtered off, and the acid liberated with dilute sulfuric acid and extracted with ether. The ether was removed, the residue dissolved in ethanol and the optical activity measured in a 2 dm tube. The concentration of the acid was determined volumetrically with sodium hydroxide. The results for the three acids are shown in Table 1.

Optical resolution of α-phenylvaleric acid

(+)-*α-Phenylethylamine* (−)-*α-phenylvalerate:* 35.6 g (0.2 moles) of racemic acid and 24.2 g (0.2 moles) of (+)-α-phenylethylamine were dissolved in hot ethanol. After cooling, the crystals were filtered off, dried and weighed. 0.2–0.5 g were set apart, the acid liberated with dilute sulfuric acid, extracted with ether and isolated by evaporation of the solvent. The acid was dissolved in ethanol and the optical activity measured in a 2 dm tube. The concentration of the acid was determined volumetrically with sodium hydroxide. The principal part of the precipitate was recrystallised from ethanol until the liberated acid showed

Table 1.

Base	(I)	(II)	(III)
Strychnine	Oil	Oil	Oil
Brucine	Oil	Oil	Oil
Quinine	±0.0°	Oil	−13°
Quinidine	Oil	Oil	Oil
Cinchonine	±0.0°	−8.4°	Oil
Cinchonidine	±0.0°	Oil	Oil
l-Ephedrine	−13°	Oil	Oil
(+)-α-Phenylethylamine	−33°	−12°	−37°
(+)-Benzedrine			+ 2°

Table 2.

Crystallisation number	(−)-(I)		(+)-(I)	
	g of salt obtained	$[\alpha]_D^{25}$	g of salt obtained	$[\alpha]_D^{25}$
1	26	−35°	18	+38°
2	22	−48°	15	+50°
3	18	−57°	13	+59°
4	14	−60°	12	+62°
5	13	−62°	10	+63°
6	13	−63°	9	+63°
7	12	−63°	8	+63°

constant optical activity. The pure (+)-phenylethylamine (−)-phenylvalerate was obtained as colourless needles. The progress of the resolution is seen in Table 2.

(−)-*α-Phenylethylamine* (+)-*α-phenylvalerate:* The mother liquor from the first crystallisation above was evaporated to dryness and the acid liberated with dilute sulfuric acid and extracted with ether. The ether was removed and the residual acid treated with an equivalent amount of (−)-α-phenylethylamine in hot ethanol. The salt obtained was recrystallised and the optical activity of the liberated acid determined in the same manner as described above for the (−)-acid salt. The results are shown in Table 2.

(+)-*α-Phenylvaleric acid:* (−)-α-Phenylethylamine (+)-α-phenylvalerate was treated with dilute sulfuric acid and the liberated organic acid extracted with ether. The ether was removed and the (+)-phenylvaleric acid was obtained as an almost colourless viscous oil. Attempts to crystallise the acid were without success.

106.4 mg of the acid required 7.70 ml of 0.0776*N* NaOH.

$C_{11}H_{14}O_2$: Equiv. weight calc. 178.2 found 178.0.

90.5 mg of the acid was dissolved in absolute ethanol, made up to 10.00 ml and the optical activity measured.

$2\alpha_D^{25} = +1.147°$ $[\alpha]_D^{25} = +63.4°$ $[M]_D^{25} = +113.0°$

Amide: Colourless crystals. M.p. 106°–108°.

S-Benzylthiuronium salt: Colourless crystals. M.p. 148°–149°.

(−)-*α-Phenylvaleric acid:* (+)-α-Phenylethylamine (−)-α-phenylvalerate was treated with dilute sulfuric acid and the liberated (−)-α-phenylvaleric acid isolated as described for the (+)-acid. Colourless viscous oil.

104.2 mg of the acid required 7.53 ml of 0.0776*N* NaOH.

$C_{11}H_{14}O_2$: Equiv. weight calc. 178.2 found 178.3.

Optical activity in homogenous state: $[\alpha]_D^{20} = -76.2°$ $[M]_D^{20} = -135.5°$.

Optical activity in different solvents: Weighed amounts of the acid were dissolved in the given solvent, made up to a volume of 10.00 ml and the optical activity of the solution was measured in a 2 dm tube. In the case of water the acid was dissolved in 1.0 ml of ethanol, neutralised with 0.0776*N* NaOH and finally made up to 10.00 ml with water. The optical activity was then measured in the usual way. The results are shown in Table 3.

Table 3.

Solvent	mg of acid	$2\alpha_D^{25}$	$[\alpha]_D^{25}$	$[M]_D^{25}$
Acetone	114.1	−2.015°	−88.3°	−157.4°
Benzene	135.5	−2.070°	−76.4°	−136.1°
Acetic acid	126.2	−1.816°	−72.0°	−128.2°
Ether	101.7	−1.342°	−66.0°	−117.6°
Ethanol	153.4	−1.963°	−64.0°	−114.0°
Water (ion)	104.2	−0.06°	− 2.8°	− 5.1°

Amide: Colourless needles. M.p. 106°–108°.

Optical activity in different solvents (*vide ante*) are given in Table 4.

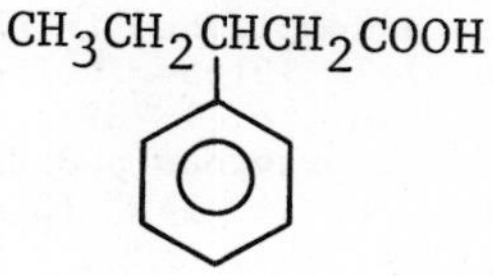

$C_{11}H_{14}O_2$ **Ref. 804**

RESOLUTION OF BETA-PHENYLVALERIC ACID (CVI) WITH QUININE (68):

In 2400 ml. of acetone 213.6 g. (1.20 moles) of Beta-phenylvaleric acid was dissolved. To this was added 453.6 g. (1.20 moles) of quinine, and the resultant mixture was heated to a boil and filtered. After setting in a refrigerator for several days, the white crystalline product was filtered and dried under vacuum. This afforded 415.5 g. of crop I, $[\alpha]_D^{25}$ -82.29° (acetone, 2 dm., c, 2.450).

The acetone was evaporated from the mother liquor on the steam bath and the residue treated with 10% hydrochloric acid. The acidified solution was extracted with benzene and the benzene was distilled at reduced pressure. Distillation afforded 67 g. of acid, m.p. 144-147° (3.7 mm.), $[\alpha]_D^{24}$ +12.76° (benzene, 2 dm., c, 1.865).

Recrystallization of 415 g. of crop I was accomplished from 1350 ml. of hot acetone. After setting in the refrigerator overnight, filtration and vacuum drying, yielded 318 g. of crop II, $[\alpha]_D^{25}$ -85.84° (acetone, 2 dm., c, 1.765). The mother liquor was treated as previously described and afforded 29 g. of acid, b.p. 148-152° (6.3 mm.), $[\alpha]_D^{24}$ +5.24° (benzene, 2 dm., c, 2.119).

Recrystallization of 317.5 g. of crop II from 1100 ml. of acetone was accomplished in the same manner as was done above to yield 278 g. of crop III, $[\alpha]_D^{24}$ -85.54° (acetone, 2 dm., c, 2.210). The mother liquor was treated as previously described and afforded 10.1 g. of acid, b.p. 143-151° (6 mm.), $[\alpha]_D^{23}$ +122° (benzene, 2 dm., c, 2.132).

Recrystallization of 277 g. of crop III from 750 ml. of acetone was accomplished in the same manner as done above to yield 247.5 g. of crop IV, $[\alpha]_D^{24}$ -85.58° (acetone, 2 dm., c, 2.113).

Recrystallization of 247 g. of crop IV from 850

ml. of acetone was accomplished in the same manner as was done above to yield 224 g. of crop V, $[\alpha]_D^{24}$ -88.11° (acetone, 2 dm., c, 1.708). The mother liquor was treated as previously described and afforded 6.4 g. of acid, b.p. 143-148° (5 mm.), $[\alpha]_D^{24}$ -4.01° (benzene 2 dm., c, 1.797).

Recrystallization of 223.5 g. of crop V from 825 ml. of acetone was accomplished in the same manner as was done above to yield 200 g. of crop VI, $[\alpha]_D^{24}$ -87.78° (acetone, 2 dm., c, 1.882). The mother liquor was treated as previously described and afforded 7 g. of acid, b.p. 146-152° (6 mm.), $[\alpha]_D^{23}$ -7.50° (benzene, 2 dm., c, 1.721).

Recrystallization of 199 g. of crop VI from 825 ml. of acetone was accomplished in the same manner as was done above to yield 174.5 g. of crop VII, $[\alpha]_D^{24}$ -87.93° (acetone, 2 dm., c, 1.951). The mother liquor was treated as previously described and afforded 4.8 g. of acid, b.p. 145-147° (3.5 mm.) $[\alpha]_D^{25}$ -11.91° (benzene, 2 dm., c, 2.510).

Recrystallization of 174 g. of crop VII from 824 ml. of acetone was accomplished in the same manner as was done above to yield 151.5 g. of crop VIII $[\alpha]_D^{25}$ -82.82° (acetone, 2 dm., c, 1.787). The mother liquor was treated as previously described and afforded 6.1 g. of acid, b.p. 143-145° (3 mm.), $[\alpha]_D^{25}$ -12.26° (benzene, 2 dm., c, 2.326).

The resolved (l) Beta-phenylvaleric acid was recovered by treatment of 150.5 g. of crop VIII with 10% hydrochloric acid. The resultant solution was extracted with benzene. After the benzene had been distilled at reduced pressure, distillation afforded 49.5 g. of acid (CVI), b.p. 158-158.5° (7.5 mm.), $[\alpha]_D^{24}$ -16.46° (benzene, 2 dm., c, 1.999) [lit. (68) b.p. 140° (4 mm.), $[\alpha]_D^{25}$ -16.5° (benzene)].

(68) P.A. Levene and R.E. Marker, J. Biol. Chem., 93, 763 (1931).

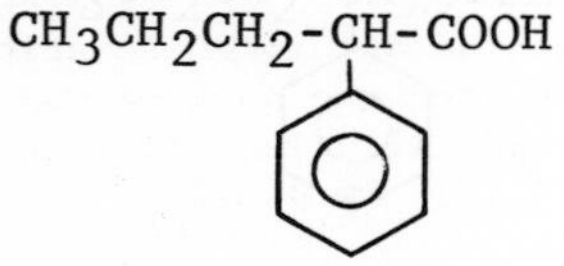

$C_{11}H_{14}O_2$

Ref. 805

Resolution of d,l-Phenylpropylacetic Acid.—54 gm. of the acid were dissolved in a hot mixture of 300 cc. of 95 per cent alcohol and 8600 cc. of water. To this solution were then added 98 gm. of cinchonidine, whereupon the solution was allowed to stand for 36 hours at room temperature. The first fraction of the cinchonidine salt separated in crystalline form and amounted to 57 gm. 2 gm. of the salt were decomposed with concentrated hydrochloric acid. The free acid showed a rotation of

$$[\alpha]_D^{25} = \frac{+2.50° \times 100}{1 \times 16.784} = +14.9°. \quad [M]_D^{25} = +26.5° \text{ (in ether)}.$$

$$[\alpha]_D^{25} = \frac{+2.15° \times 100}{1 \times 16.088} = +13.4°. \quad [M]_D^{24} = +23.8° \text{ (in 75 per cent alcohol)}.$$

From the mother liquor an acid was obtained which showed a rotation of

$$[\alpha]_D^{25} = \frac{-1.27° \times 100}{1 \times 9.000} = -14.1°. \quad [M]_D^{25} = -25.1° \text{ (in ether)}.$$

After two more recrystallizations of the first fraction, an acid was obtained with a rotation of $[\alpha]_D^{25} = +33.1°$ (in ether). Further crystallization did not improve the rotation and the yield on this last fraction was only about 10 per cent.

$C_{11}H_{14}O_2$

Ref. 805a

Resolution of α-*n*-propylphenylacetic acid. The (+)-acid was obtained by crystallization of the (−)-menthylamine salt, according to Pickard and Yates.[11] B.p. 167.5~169° (18 mmHg), $[\alpha]_D$+80.0° (c. 1.70 in benzene) (lit.[11] $[\alpha]_D$+79.05° (c 5.87 in benzene)). The acid (3.2 g), $[\alpha]_D$+28.1° (abs. ethanol), obtained from the mother-liquors of the (−)-menthylamine salt was dissolved with (+)-α-phenylethylamine (2 g) in ethanol (25 ml), the salt precipitated was recrystallized six times from 50% ethanol, and the rotatory power of the acid reached a constant value, $[\alpha]_D$−80.3° (c 1.44 in benzene) (lit.[12] $[\alpha]_D^{25}$−76.4° (c 1.355 in benzene)).

10) M. Delépine and F. Larèze, *Bull. soc. chim. France*, **1955**, 104.
11) R. H. Pickard and J. Yates, *J. Chem. Soc.*, **1909**, 1016.
12) K. Petterson and G. Willdeck, *Arkiv Kemi*, **9**, 333 (1956).

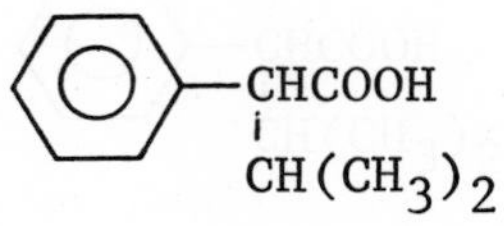

$C_{11}H_{14}O_2$

Ref. 806

Resolution of α-Isopropylphenylacetic Acid.—*dl*-α-Isopropylphenylacetic acid (461 g), (+)-α-phenylethylamine (313 g, α^{27}_D +37.8, neat, *l* = 1), and 63% aqueous ethanol (10 l.) were warmed and allowed to cool slowly. The 353 g of salt which separated was recrystallized four times from 63% aqueous ethanol (6, 5.5, 5, and 4.5 l., respectively) to give 272 g of salt, mp 198–200°, $[\alpha]^{26}_D$ −5.2 (CH_3OH, *c* 2.62). Regeneration with 10% sulfuric acid at 0° gave, after extraction, drying, and distillation, 153.5 g of (−)-α-isopropylphenylacetic acid, bp 132–133° (3.5 mm), which solidified, mp 50.5–51.5°, $[\alpha]^{24}_D$ −62.4 ($CHCl_3$, *c* 4.46). The mother liquors were concentrated, and the partially active (+) acid regenerated in the same way to give 261 g, $[\alpha]^{24}_D$ +36.01 ($CHCl_3$, *c* 3.86). This acid was treated with (−)-α-phenylethylamine (α^{28}_D −36.6, neat, *l* = 1) and the process repeated to obtain the pure acid, $[\alpha]^{25}_D$ +62.5 ($CHCl_3$, *c* 2).

The resolution was also attempted with cinchonidine. α-Isopropylphenylacetic acid (50 g) and cinchonidine (76 g) were heated in 1.5 l. of 60% methanol. The 52 g of salt which was obtained on cooling was recrystallized four times from methanol to give 17 g of salt, mp 136–137.5°, $[\alpha]^{29}_D$ −68.5, which upon regeneration gave an incompletely resolved acid, 5.6 g, $[\alpha]^{27}_D$ +48.7° ($CHCl_3$, *c* 3). Cinchonidine was obviously inferior to α-phenylethylamine for this resolution. Although we were unable to obtain a crystalline cinchonidine salt from the resolved (−)-isopropylphenylacetic acid, the (+) isomer, $[\alpha]^{25}_D$ +62.3 ($CHCl_3$, *c* 3), gave a crystalline cinchonidine salt from which, after several recrystallizations, the original (+) acid was regenerated without significant change in rotation. This gives added evidence that the resolution was complete.

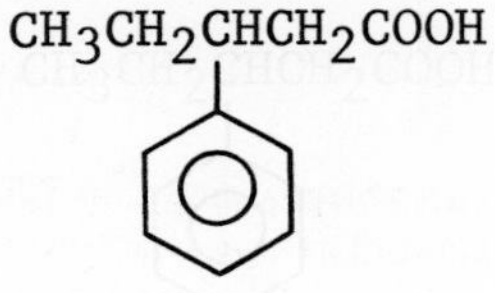

$C_{11}H_{14}O_2$

Ref. 807

(+)-3-phenylpentanoic acid

13.6 g (0.112 moles) of (−)-α-phenylethylamine were added to a hot solution of 20.0 g (0.112 moles) of rac.-3-phenylpentanoic acid in 75 ml of ethanol and 25 ml of water. After standing at room temperature overnight, the salt was collected and dried. It was recrystallized from ethanol and water ten times, and after each recrystallization the acid was liberated from a small amount of the salt and its rotation determined in benzene solution (Table 3).

Table 3. Resolution of 3-phenylpentanoic acid with (−)-α-phenylethylamine.

Crystallization	Ethanol (ml)	Water (ml)	Salt obtained (g)	m.p. of salt (degrees)	$[\alpha]_D^{23}$ of acid (in benzene) (degrees)
1	75	25	21.4	140.0–144.0	+ 5.2
2	30	10	16.8	142.0–146.0	+ 9.8
3	20	7	13.7	142.5–146.5	+12.7
4	21	3	10.9	144.5–149.0	+19.5
5	15	3	8.8	145.0–149.5	+26.7
6	12	3	6.9	145.5–150.0	+34.6
7	7	2	6.0	146.0–150.5	+38.6
8	4	2	5.3	147.0–150.5	+43.1
9	3	2	4.5	147.0–151.0	+44.7
10	3	2	4.1	148.5–151.5	+46.3

(−)-3-phenylpentanoic acid

From the combined mother liquors, (−)-3-phenylpentanoic acid was liberated (14.2 g, 0.080 moles, $[\alpha]_D^{25}$ = −10.9° in benzene.

It was dissolved in 50 ml of ethanol and 20 ml of water, and to the hot solution 9.6 g (0.080 moles) of (+)-α-phenylethylamine were added. The collected salt was recrystallized from ethanol and water (Table 4).

Table 4. **Resolution of 3-phenylpentanoic acid with (+)-α-phenylethylamine.**

Crystallization	Ethanol (ml)	Water (ml)	Salt obtained (g)	m.p. of salt (degrees)	$[\alpha]_D^{23}$ of acid (in benzene) (degrees)
1	50	20	14.8	144.0–146.0	−18.2
2	18	6	12.3	145.5–149.0	−24.1
3	15	5	10.4	146.5–150.5	−30.5
4	13	4	8.6	148.5–151.0	−34.2
5	12	3	7.1	149.0–151.0	−42.3
6	7	2	6.3	149.0–151.5	−45.3

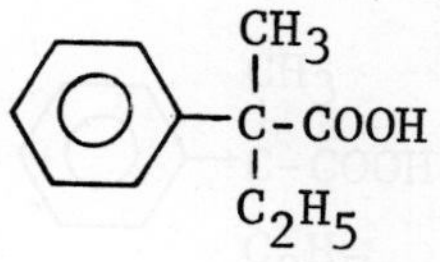

$C_{11}H_{14}O_2$

Ref. 808

racem. 2-Methyl-2-phenyl-buttersäure[22] wurde durch Carboxylierung von *1-Methyl-1-phenyl-propyl-kalium* in 62-proz. Ausb. als farblose Spieße mit Schmp. 56−57° (Lit.[15]: 56.5−57.5°) erhalten. Die Antipodentrennung erfolgte durch fraktionierte Kristallisation des Chininsalzes in verd. Äthanol[15]. Aus 111 g *racem.* Säure erhielt man 54 g Chininsalz mit Schmp. 119−121°, die, in 200 ccm Äther suspendiert, mit 200 ccm 10-proz. Salzsäure kräftig gerührt wurden, bis zwei klare Schichten entstanden. Die wäßr. Phase wurde viermal mit je 100 ccm Äther ausgezogen. Aus den vereinigten Ätherlösungen isolierte man die Säure durch Extraktion mit Natronlauge und Fällen mit konz. Salzsäure. Nach Umlösen aus Petroläther (40−60°) erhielt man 19.8 g farblose glänzende Spieße mit Schmp. 86.5−87° (Lit.[14]: 86−87°) und $[\alpha]_D^{25}$: −28.8° (c = 4.8 in Benzol)*).

In der Literatur finden sich folgende maximalen Drehwerte: $[\alpha]_D^{25}$: −29.5° (c = 4.8 in Benzol)[22] und $[\alpha]_D^{25}$: −28.9° (c = 4.5 in Benzol)[14], so daß unsere Probe 97.8% optisch rein war[22].

$C_{11}H_{14}O_2$ (178.2) Ber. C 74.13 H 7.92 Gef. C 74.45 H 7.68

Die Mutterlaugen der Chininsalz-Kristallisation wurden vereinigt. Wie für die aktive L(−)-Verbindung beschrieben wurde, isolierte man hieraus 67.8 g teilweise aktiver, roher *D(+)-2-Methyl-2-phenyl-buttersäure*, die nach einmaligem Umlösen aus Petroläther (40 bis 60°) 46.6 g Säure mit Schmp. 64−72° und $[\alpha]_D^{25}$: +10.8° (c = 4.8 in Benzol) ergab. Hieraus berechnet sich ein Gehalt von 36.6% D-Form neben Racemat.

[illegible] s. z. B. E. L. ELIEL, Stereochemistry of Carbon Compounds, S. 146ff., McGraw Hill, New York 1962.

14) D. J. CRAM, A. LANGEMAN, J. ALLINGER und K. R. KOPECKY, J. Amer. chem. Soc. **81**,

*) Alle Drehwerte wurden mit einem Zeiss-Winkel-Kreispolarimeter unter Verwendung einer Natriumdampflampe bestimmt. Die angegebenen % optischer Reinheit beziehen sich immer auf den Gehalt an angereichertem Antipoden neben dem Racemat[5].

22) D. J. CRAM und J. D. KNIGHT, J. Amer. chem. Soc. **74**, 5835 [1952].

$C_{11}H_{14}O_2$ Ref. 808a

Resolution of 2-*methyl*-2-*phenylbutyric acid.* To a hot soln of 235 g dehydroabietylamine[10] in 750 ml abs EtOH was added 135 g 2-methyl-2-phenylbutyric acid[9] in 750 ml EtOH. The ppt which formed immediately was kept at 0° for 24 hr before filtering; the crude salt had m.p. 165–175°. Four recrystallizations from 3:1 EtOH–acetone gave 72 g of the salt, m.p. 190–194°. The salt was shaken vigorously with a mixture of dil KOH and ether until all of the solid dissolved, the layers separated, the aqueous layer washed with ether and acidified with conc HCl. Extraction with ether gave 39 g of the (+)-acid, m.p. 78–82°. Recrystallization from pet. ether (b.p. 60–70°) raised the m.p. to 81–82°, $[\alpha]_D^{23}$ +24·5° (*c*, 4.1 in benzene). Recovery of the antipode from the mother liquors of the resolution in the same way gave 85 g of the (−) acid, m.p. 65–68°, $[\alpha]_D$ −9·5° (*c*, 1 in benzene). The optically pure acid is reported[23] to have m.p. 86–87°, $[\alpha]_D^{22}$ 30·2° (*c*, 4·5 in benzene).

[9] D. J. Cram and J. Allinger, *J. Am. Chem. Soc.* **76**, 4516 (1954).
[10] W. G. Gottstein and L. C. Cheney, *J. Org. Chem.* **30**, 2072 (1965).
[23] D. J. Cram and J. D. Knight, *J. Amer. Chem. Soc.*, **74**, 5835 (1952).

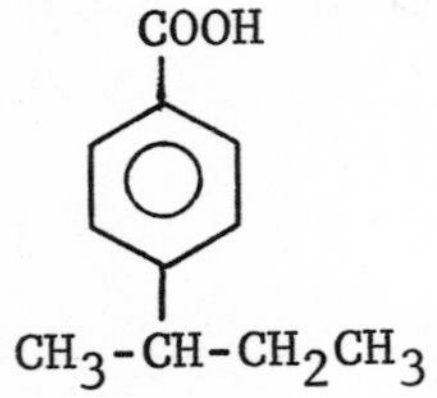

$C_{11}H_{14}O_2$ Ref. 809

Resolution of *p*-*s*-Butylbenzoic Acid by Means of Quinine.—Fifty and seven-tenths grams (0.285 mole) of *p*-*s*-butylbenzoic acid and 92.4 g. (0.285 mole) of anhydrous quinine were placed in a 2-liter Erlenmeyer flask with 1250 cc. of methanol. The material dissolved on heating and the solution was filtered using a hot-water funnel. The filter was washed with 100 cc. of methanol. The solution was placed in the icebox for two days and yielded after this time 83.1 g. of crystals. These were recrystallized from 500 cc. of methanol and yielded 63.4 g. of crystalline salt, $[\alpha]^{24}D$ −129.0° (0.1168 g. in 5 cc. of methanol solution). This salt was recrystallized from 300 cc. of methanol and 40.5 g. of crystalline product was obtained, $[\alpha]^{26}D$ −134.2° (0.1001 g. in 5 cc. methanol solution).

The quinine salt was then decomposed by solution in 100 cc. of glacial acetic acid. The addition of 500 cc. of water precipitated the regenerated *p*-*s*-butylbenzoic acid from the solution. It was removed by filtration and recrystallized from 70% aqueous methanol. The yield was 15 g. (59.2%), m. p. 89–90°. An additional recrystallization gave a product melting at 90.5–91.5°, $[\alpha]^{23}D$ −12.53° (0.0998 g. in 5 cc. methanol solution).

The 15 g. of *p*-*s*-butylbenzoic acid obtained by the above partial resolution was reconverted to the quinine salt (27.0 g. of quinine) and the salt was recrystallized three times from 250 cc. of methanol, $[\alpha]^{26}D$ −138.4° (0.1032 g. in 5 cc. methanol solution). It melted at 184–185°.

Anal. Calcd. for $C_{31}H_{39}N_2O_4$: C, 74.10; H, 7.62. Found: C, 74.14; H, 7.67.

The salt was decomposed in 100 cc. of 20% hydrochloric acid, the mixture was cooled in ice, and the acid separated by filtration. It was then recrystallized from 70% aqueous methanol. The yield was 5.0 g. (19.7%), melting at 88.5–89°, $[\alpha]^{26}D$ −23.5° (0.1031 g. in 5 cc. methanol solution).

Anal. Calcd. for $C_{11}H_{14}O_2$: C, 74.13; H, 7.92. Found: C, 74.02; H, 7.95.

The resolution was repeated several times and in most of these trials the specific rotations of the products approached −23.5°.

By a series of concentrations of the mother liquors from the above resolution and removal of the less-soluble crops of crystals, it was possible to isolate a sample of the quinine salt of $[\alpha]^{27}D$ −121.5° (0.1025 g. in 5 cc. methanol solution). The acid recovered from this had a positive rotation, $[\alpha]^{30}D$ + 18.2° (0.1059 g. in 5 cc. methanol solution).

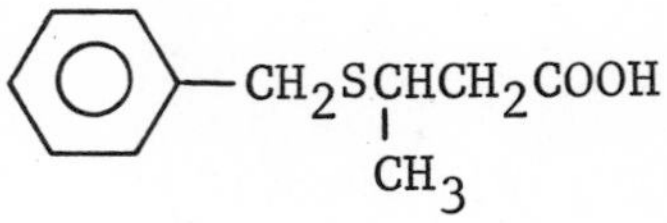

$C_{11}H_{14}O_2S$ Ref. 810

Resolution

(+)*3-Benzylthiobutanoic acid* ((+)*VI*). Racemic VI (21 g, 0.1 mol), dissolved in acetone (50 ml) were mixed with brucine and warmed until all of the base had dissolved. After standing at room temperature overnight, the salt which had precipitated was collected and recrystallized from acetone seven times. The progress of the resolution given in Table IV was followed by liberation of the acid by means of sulfuric acid from a small amount of the salt (0.2 g) after each crystallization. The acid solution was extracted with ether, the etheral solution washed twice with sulfuric acid and five times with water and dried. After evaporation of the solvent, the acid was bulb to bulb distilled under reduced pressure in a microdestillation oven. This could be done without any racemization. After the last crystallization, an acid with $[\alpha]_D^{25}$ = +17.1 ±0.2 (c=1.80, abs. ethanol) was obtained. (All the acid was not liberated but kept as the alkaloid salt.)

(−)*3-Benzylthiobutanoic acid* ((−)*VI*). From the salt obtained after evaporation of the mother liquor from crystallizations 1 and 2 described above, 11 g of acid with $[\alpha]_D^{25}$ = −11 was obtained. To the liquid acid, a crystal of racemic VI was added. After one week the racemic acid which had crystallized was filtered off. The remaining acid (7.6 g) $[\alpha]_D^{25}$ = −14.6 in abs. ethanol, obtained after distillation, was dissolved in ethyl acetate (1 000 ml) and (−)α-(2-naphthyl)ethylamine (6.1 g $[\alpha]_D^{25}$ = −20.1 in ethanol [28]) added. After 24 hours at room temperature, filtration gave 8.4 g of salt with $[\alpha]_D^{25}$ = −15.4 (in abs. ethanol) for the acid liberated.

This salt was recrystallized twice from ethyl acetate (750 ml and 450 ml, respectively). From the last crystallization, 2.5 g salt was obtained with $[\alpha]_D^{25} = -17.0 \pm 0.2$ (c=1.24, abs. ethanol) for the

Table IV

Crystallization no.	Acetone (ml)	Salt obtained (g)	c (g/100 ml)	$[\alpha]_D^{25}$
1	50	48.3	1.45	+ 2.6
2	800	27.5	0.66	+13.0
3	400	21.7	1.03	+14.6
4	400	18.3	0.98	+14.3
5	400	13.9	0.64	+15.9
6	400	10.5	0.68	+17.4
7	500	6.0	1.80	+17.1

acid liberated. Evaporation of the mother liquor gave a further 1.7 g of salt which yielded (−)VI with $[\alpha_D^{25}] = -17.1 \pm 0.2$ (c=1.77, abs. ethanol). A further crystallization of the salt did not increase the optical purity of the acid. The acid was kept as the alkaloid salt.

Anal. Calcd. for $C_{11}H_{14}O_2S$; C, 62.86; H, 6.67; S, 15.24. Found C, 62.49; H, 6.70; S, 15.24.

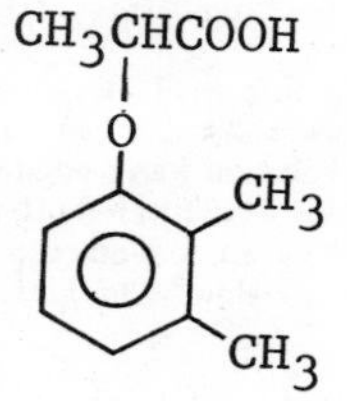

$C_{11}H_{14}O_3$ Ref. 811

Preliminary experiments on resolution were carried out with the common alkaloids in dilute ethanol. The data given below for the rotatory power refer to the free acid in benzene solution.

Cinchonidine gave strongly laevorotatory acid, while the acid from the *quinidine* salt was dextrorotatory. The salts from *strychnine*, *quinine* and *dehydroabietylamine* gave acid of very low activity. The *brucine* and *cinchonine* salts failed to crystallise.

(+)-*2,3-Dimethylphenoxy-propionic acid*. Racemic acid (23.3 g, 0.12 mole) and cinchonidine (35.5 g, 0.12 mole) were dissolved in 280 ml of hot ethanol and 160 ml of hot water were added. After cooling, the solution was seeded with crystals from the preliminary experiments. The salt (13.6 g) was filtered off after 24 hours and washed with 60 ml of 50 % ethanol. On standing in a refrigerator for 24 hours, the filtrate deposited a second crop of salt (6.9 g). The two crops were combined and recrystallised from dilute ethanol; after each crystallisation a small quantity of acid was liberated and its rotatory power determined in benzene, where the (+)-form is *laevorotatory*.

Recrystallisation	1	2	3
ml ethanol	30	40	40
ml water	30	40	40
g salt obtained	18.8	17.8	16.2
$[\alpha]_D^{25}$ of the acid	−66.5°	−66.5°	−66°

The salt from the third recrystallisation gave 6.5 g of crude acid, which was recrystallised from ligroin, dilute formic acid and once more from ligroin. The yield of pure acid was 6.3 g. It was obtained as long, slender needles with m.p. 110.5–111.5°

$C_{11}H_{14}O_3$ (194.22)
Calc. C 68.02 H 7.27 Eq.wt. 194.2
Found C 68.19 H 7.23 Eq.wt. 194.3

0.2607 g acid dissolved in *acetone* to 10.00 ml: $\alpha_D^{25} = +0.27°$ (2 dm). $[\alpha]_D^{25} = +5.2°$; $[M]_D^{25} = +10.1°$. — 0.1899 g acid dissolved in benzene to 10.00 ml: $\alpha_D^{25} = -2.555°$ (2 dm). $[\alpha]_D^{25} = -67.3°$; $[M]_D^{25} = -130.7°$.

(−)-*2,3-Dimethylphenoxy-propionic acid*. The mother liquors from the first two fractions of the cinchonidine salt yielded 14 g acid having $[\alpha]_D^{25} = +24°$ (benzene). This acid (0.072 mole) and 23.3 g (0.072 mole) quinidine were dissolved with heating in 125 ml ethanol + 125 ml water. The salt crystallised readily. It was filtered off after 24 h and recrystallised as described above.

Crystallisation	1	2	3	4
ml ethanol	125	75	75	75
ml water	125	100	100	100
g salt obtained	26.4	24.7	23.4	22.3
$[\alpha]_D^{25}$ of the acid	+65°	+66.5°	+66.5°	—

Table 1.

g = g acid in 10.00 ml solution.

Solvent	g	α_D^{25}(2 dm)	$[\alpha]_D^{25}$	$[M]_D^{25}$
Acetone	0.2030	−0.22°	−5.4°	−10.5°
Ethanol (abs.)	0.1878	−0.06°	−1.6°	− 3.1°
Chloroform	0.2023	−0.07°	−1.7°	− 3.4°
Water (neutr.)	0.2007	+0.545°	+13.6°	+ 26.4°
Benzene	0.1568	+2.11°	+67.3°	+130.7°

The salt from the fourth crystallisation yielded 8.0 g of crude acid. Recrystallisations from ligroin, dilute formic acid and once more from ligroin gave 7.2 g of pure acid, quite similar to the enantiomer. M.p. 110.5–111.5°.

$C_{11}H_{14}O_3$ (194.22)
Calc. C 68.02 H 7.27 Eq.wt. 194.2
Found C 68.20 H 7.25 Eq.wt. 193.9

The rotatory power in various solvents is given in Table 1.

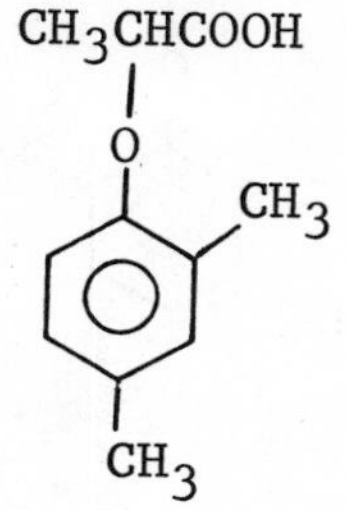

$C_{11}H_{14}O_3$ Ref. 812

Preliminary experiments on resolution were performed with small quantities of acid in dilute ethanol as described above. *Cinchonidine* and *dehydroabietylamine*, which had been found useful for the resolution of 2-chloro-4-methylphenoxy-propionic acid, were first tested. Cinchonidine gave no resolution (even after two recrystallisations of the salt), whereas the dehydroabietylamine salt gave a dextrorotatory acid of high activity. The *quinine* salt gave a practically inactive acid and the salts of *brucine* and *strychnine* could not be brought to crystallisation. The *quinidine* salt did not crystallise readily, but after some experiments with various amounts of ethanol and water, needleshaped crystals were obtained, which could be used for seeding in further experiments. The salt gave a laevorotatory acid of fairly high activity. Dehydroabietylamine and quinidine were thus selected for the final resolution.

(+)-*2,4-Dimethylphenoxy-propionic acid*. Racemic acid (23.3 g, 0.12 mole) was carefully neutralised with ammonia in 270 ml water. Dehydroabietylamine acetate [8] (41.5 g, 0.12 mole) was dissolved in 440 ml hot ethanol and the two solutions were mixed. On standing a crystalline salt separated; it was filtered off after 24 h and washed with dilute ethanol. It was then recrystallised to maximum activity of the acid as described above. The activity was measured in acetone solution.

Crystallisation	1	2	3	4	5
ml ethanol	440	320	300	225	200
ml water	270	330	315	230	210
g salt obtained	21.3	18.3	16.2	12.5	10.3
$[\alpha]_D^{25}$ of the acid	+20°	+27°	+28°	+28.5°	+28°

The salt was decomposed and the acid liberated in the same way as described above for the (+)-2-chloro-4-methylphenoxy-propionic acid. After recrystallisation from dilute formic acid it was obtained as glistening needles with m.p. 83.5–84.5°. The yield of pure acid from 10.3 g salt was 3.9 g (33 % calculated on 23.3 g racemic acid).

$C_{11}H_{14}O_3$ (194.23): calc. equiv.wt. 194.2 C 68.02 H 7.27
found equiv.wt. 193.9 C 68.12, 68.01 H 7.32, 7.34

0.2032 g acid dissolved in *acetone* to 10.00 ml: $\alpha_D^{25} = +1.16°$ (2 dm). $[\alpha]_D^{25} = +28.5°$; $[M]_D^{25} = +55.4°$.

(−)-*2,4-Dimethylphenoxy-propionic acid*. The mother liquor from the first crystallisation of the dehydroabietylamine salt was evaporated to dryness. The salt obtained was decomposed as described above, yielding 14.1 g of a strongly laevorotatory acid. This acid (0.0725 mole) and 23.5 g (0.0725 mole) quinidine were dissolved in a hot mixture of 250 ml ethanol and 100 ml water. After a while, 105 ml water were added. Crystallisation started after about 1 h and after 24 h the salt was filtered off. It was recrystallised to maximum activity of the acid as described above.

Crystallisation	1	2	3	4
ml ethanol	250	190	150	125
ml water	205	160	140	120
g salt obtained	25.6	21.7	19.2	16.4
$[\alpha]_D^{25}$ of the acid	−24°	−28.5°	−28°	−28.5°

The acid was isolated by acidification with dilute sulphuric acid and extraction with ether. The yield after recrystallisation from dilute formic acid was 5.2 g (44.7 % calculated on 23.3 g racemic acid). It was quite similar to the (+)-form. M.p. 83.5–84.5°.

$C_{11}H_{14}O_3$ (194.23): calc. equiv.wt. 194.2 C 68.02 H 7.27
found equiv.wt. 193.4 C 67.97, 68.03 H 7.26, 7.20

Data for the rotatory power in various solvents are given in Table 2.

Table 2

g = g acid dissolved to 10.00 ml.

Solvent	g	α_D^{25} (2 dm)	$[\alpha]_D^{25}$	$[M]_D^{25}$
Acetone	0.2162	−1.24°	−28.7°	−55.7°
Ethanol (abs.)	0.2132	−0.90°	−21.1°	−41.0°
Chloroform	0.2138	−0.69°	−16.1°	−31.3°
Water (neutr.)	0.2153	−0.40°	−9.3°	−18.0°
Benzene	0.1964	+0.96°	+24.4°	+47.4°

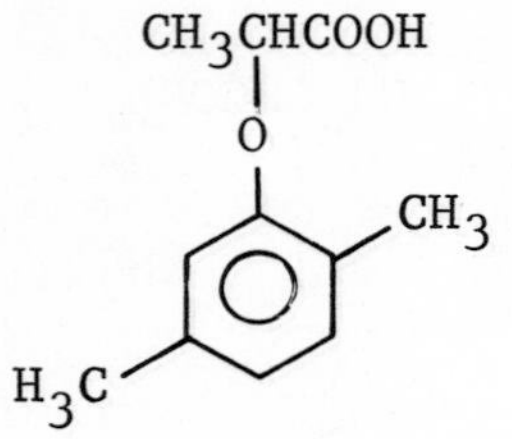

$C_{11}H_{14}O_3$ Ref. 813

Preliminary experiments on resolution were carried out with the common alkaloids and a few synthetic bases. Mixtures of water with methanol, ethanol or acetone were tried as solvents. The *strychnine* salt in 70 % acetone gave an acid with $[\alpha]_D^{25}$ = +22°. Dextrorotatory acid of fairly high activity could also be obtained with *quinine* and (+)-*α-phenyl-ethylamine*, but the results were not reproducible: obviously the resolution was more or less due to different rates of crystallisation, which are influenced by accidental factors. (−)-*β-Phenyl-isopropylamine* in 50 % ethanol gave an acid having $[\alpha]_D^{25}$ = −20°. Laevorotatory acid of lower activity was obtained with *cinchonidine. Cinchonine* and *dehydroabietylamine* gave no resolution and the *brucine* salt failed to crystallise. The final resolution was therefore performed with β-phenyl-isopropylamine and strychnine.

(−)-*2,5-Dimethylphenoxy-propionic acid.* 37.0 g (0.19 mole) of racemic acid and 25.7 g (0.19 mole) of (−)-β-phenyl-isopropylamine were dissolved in 80 ml ethanol + 80 ml water. After 24 h, the salt was filtered off and dried. It was recrystallised three times from dilute ethanol; after each crystallisation, a small sample of the salt was decomposed and the rotatory power of the acid determined.

Crystallisation	1	2	3	4
ml ethanol	80	40	35	35
ml water	80	40	35	35
g salt obtained	16.0	14.2	11.7	11.0
$[\alpha]_D^{25}$ of the acid	−34.5°	−39°	−40°	−40°

The salt (11.0 g) was decomposed with dilute sulphuric acid and the organic acid was extracted with ether. Three recrystallisations from benzene + petrol ether yielded 6.5 g acid as long, slender, glistening needles with m p. 104°.

$C_{11}H_{14}O_3$ (194.22) calc. C 68.02 H 7.27
found C 67.91, 67.91 H 7.24, 7.27

The rotatory power in various solvents is given in Table 1.

Table 1. g = g acid dissolved to 10.00 ml.

Solvent	g	α_D^{25} (2 dm)	$[\alpha]_D^{25}$	$[M]_D^{25}$
Acetone	0.2112	−1.833°	−43.4°	−84.3°
Ethanol (abs.)	0.2017	−1.632°	−40.5°	−78.6°
Chloroform	0.2068	−1.187°	−28.7°	−55.7°
Water (neutr.)	0.2064	−0.783°	−19.0°	−36.8°
Benzene	0.2076	−0.315°	−7.6°	−14.7°

(+)-*2,5-Dimethylphenoxy-propionic acid.* The mother liquor from the first crystallisation of the (−)-β-phenyl-isopropylamine salt was evaporated to dryness. The remaining salt was decomposed, yielding 23.4 g dextrorotatory acid of moderate activity. This acid (0.12 mole) was dissolved with 40 g (0.12 mole) of strychnine in 140 ml acetone + 60 ml water. The salt obtained (38.5 g) was recrystallised 4 times from dilute acetone. The rotatory power of the acid did not change after the second recrystallisation.

The salt obtained after the fourth recrystallisation (19.4 g) was decomposed and the acid liberated as described above. Three recrystallisations from benzene + petrol ether yielded 6.4 g acid as long, slender needles with m.p. 104°.

$C_{11}H_{14}O_3$ (194.22) calc. C 68.02 H 7.27
found C 67.93, 67.98 H 7.28, 7.20

0.2072 g acid dissolved in *acetone* to 10.00 ml: α_D^{25} = +1.801° (2 dm). $[\alpha]_D^{25}$ = +43.5°; $[M]_D^{25}$ = +84.4°.

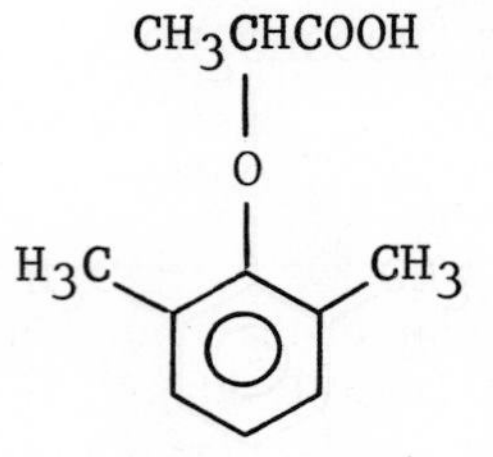

$C_{11}H_{14}O_3$ Ref. 814

Preliminary experiments on resolution were carried out as described above for the 2,3-isomer. *Strychnine* gave a dextrorotatory acid of high activity; it was later found that it was nearly pure (+)-acid. *Cinchonidine* and *dehydroabietylamine* gave laevorotatory acid of low activity and the acid from the *quinine* salt was practically inactive. The *brucine, cinchonine* and *quinidine* salts failed to crystallise. Optically active *β-phenyl-isopropylamine* gave acid of moderate activity, having the same mode of rotation as the base.

(+)-*2,6-Dimethylphenoxy-propionic acid.* Racemic acid (29.1 g, 0.15 mole) and strychnine (50.1 g, 0.15 mole) were dissolved in a hot mixture of 175 ml ethanol and 275 ml water. After cooling, the solution was seeded with crystals from the preliminary experiments. After 24 h, 28.3 g salt were filtered off. Standing for 3 days at room temperature and 6 h in a refrigerator gave a second crop of 9.3 g. The two crops were combined and recrystallised as seen below. The activity of the acid was measured in acetone solution.

Recrystallisation	1	2	3	4	5
ml ethanol	50	40	50	50	40
ml water	100	130	150	150	150
g salt obtained	32.8	30.8	28.7	26.3	25.0
$[\alpha]_D^{25}$ of the acid	+50.5°	+51°	+50.5°	+51°	—

Table 3.

g = g acid in 10.00 ml solution.

Solvent	g	α_D^{25}(2 dm)	$[\alpha]_D^{25}$	$[M]_D^{25}$
Acetone	0.2527	+2.605°	+51.5°	+100.1°
Ethanol (abs.)	0.2080	+1.32°	+31.7°	+61.5°
Chloroform	0.1889	+0.36°	+9.5	+18.5°
Water (neutr.)	0.2124	+0.83°	+19.°	+37.9°
Benzene	0.1878	+0.685°	+18.2	+35.4°

The salt (25.0 g) was decomposed in the ordinary way, yielding 7.8 g crude acid. The salt obtained by evaporation of the mother liquors after recrystallisations 2–5 was recrystallised and yielded a further crop of acid with maximum activity (1.5 g). The total amount of crude acid was recrystallised twice from petrol ether yielding 6.7 g of pure acid. It formed very fine, hairlike needles with m.p. 53–54°.

$C_{11}H_{14}O_3$ (194.22)
Calc. C 68.02 H 7.27 Eq.wt. 194.2
Found C 68.12 H 7.26 Eq.wt. 195.0

The rotatory power in various solvents is given in Table 2.

(−)-*2,6-Dimethylphenoxy-propionic acid.* The mother liquor from the first fraction of the strychnine salt yielded acid with α_D^{25} ~ −40°. Attempts to get maximum activity by recrystallisation of the cinchonidine or dehydroabietylamine salts were not successful. Since the racemic acid is less soluble than the enantiomers in ligroin and petrol ether, isolation of the pure (−)-form from the partially resolved acid was attempted by repeated extractions and recrystallisations from these solvents. The extractions were performed with ice-cold petrol ether. In this way, 7.0 g partially resolved acid yielded 0.54 g pure (−)-acid. It was finally found that resolution via the (−)-β-phenyl-isopropylamine salt gave the best results.

3.2 g (0.0165 mole) acid having $[\alpha]_D^{25}$ = −39° were dissolved with 2.2 g (0.0165 mole) (−)-β-phenyl-isopropylamine in 22 g of hot 96 % ethanol. The salt was recrystallised to constant activity of the acid as seen below.

Crystallisation	1	2	3	4
ml ethanol	22	25	20	15
g salt obtained	4.3	3.5	2.8	2.3
$[\alpha]_D^{25}$ of the acid.	−47.5°	−50°	−51°	−51°

The salt was decomposed in the ordinary way and the acid recrystallised from petrol ether. It was quite similar to the enantiomer. Yield 1.1 g; m.p. 53–54°.

$C_{11}H_{14}O_3$ (194.22)
Calc. C 68.02 H 7.27
Found C 68.14 H 7.33

0.1716 g acid (isolated via the amine salt) was dissolved in *acetone* to 10.00 ml: $\alpha_D^{25} = -1.716°$ (2 dm). $[\alpha]_D^{25} = -51.9°$; $[M]_D^{25} = -101.8°$. — 0.1542 g acid (isolated from partially resolved acid by extraction and recrystallisation) dissolved in *acetone* to 10.00 ml: $\alpha_D^{25} = -1.591°$ (2 dm). $[\alpha]_D^{25} = -51.6°$; $[M]_D^{25} = -101.2°$.

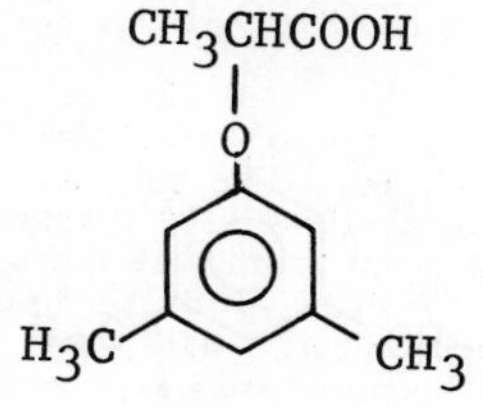

$C_{11}H_{14}O_3$ Ref. 815

Resolution of the acid. Preliminary experiments were performed with the common commercial alkaloids and dehydroabietylamine in dilute ethanol. Some experiments with other solvents were not promising.

Cinchonine gave a salt containing acid with $[\alpha]_D \sim -28°$ (96 % ethanol). The *brucine* salt, which was at first difficult to obtain in a crystalline state, gave an acid with $[\alpha]_D = +27°$. These alkaloids were selected for the final resolution. *Quinine, cinchonidine* and *dehydroabietylamine* gave crystallised salts but no resolution. The salts of *strychnine* and *quinidine* failed to crystallise.

Racemic acid (29.1 g, 0.15 mole) and cinchonine (44.0 g, 0.15 mole) were dissolved in 750 ml ethanol with gentle heating and 750 ml of water were added. The salt crystallised rather slowly and was filtered off after 4 days (fraction 1). After evaporation of about 200 ml solvent, a second crop was obtained (fraction 2). Evaporation of further 200 ml solvent resulted in a third crop (fraction 3). From each fraction, a small sample of the acid was liberated; its melting-point and rotatory power (in 96 % ethanol) were determined.

Fraction	g salt	$[\alpha]_D^{25}$	Melting-point
1	5.7	−32°	118–118.5°
2	12.2	−32°	118–118.5°
3	8.8	−28°	113–116°

The sharp melting-point of the acid from the first two fractions indicated a rather complete resolution. The fractions were recrystallised separately, using 16–17 ml of 50 % ethanol for each gram salt:

Fraction	g salt	$[\alpha]_D^{25}$	Melting-point
1	4.9	−32°	118–119°
2	9.6	−32°	118–119°
3	7.1	−31°	117–118°

The fractions were combined and recrystallised twice from 50 % ethanol as described above, yielding 11.1 g salt. The rotatory power of the acid remained unchanged.

A further crop of pure salt could be obtained from the mother liquors after the recrystallisations.

The mother liquor after the first crystallisation of fraction 3 should contain dextrorotatory acid of fairly high activity. The cinchonine was precipitated with ammonia and filtered off, and the acid was liberated from the filtrate in the usual way. 19.1 g acid having $[\alpha]_D^{25} = +16°$ were obtained. This acid was dissolved with an equivalent amount of brucine in 400 ml of water. The salt separated as an oil, which after 6 h had been transformed into hard, compact crystals. The yield was 33.0 g and the activity of the acid $[\alpha]_D^{25} = +32°$. Recrystallisation from 100 ml of water yielded 30.9 g salt; the activity of the acid remained unchanged.

(−)-3,5-Dimethylphenoxy-propionic acid. The cinchonine salt (11.1 g) was decomposed with dilute sulphuric acid and the organic acid isolated by extraction with ether. The crude acid (4.4 g) was recrystallised twice from 60 ml of ligroin, yielding 4.0 g of long, glistening needles with m.p. 118.5–119.2°.

0.2711 g acid: 13.70 ml 0.1018-*N* NaOH.
$C_{11}H_{14}O_3$ (194.22) Equiv. wt.: calc. 194.2; found 194.5
0.1951 g acid, dissolved in *96 % ethanol* to 10.00 ml: $\alpha_D^{25} = -1.255°$ (2 dm). $[\alpha]_D^{25} = -32.2°$; $[M]_D^{25} = -62.5°$.
0.2427 g acid, dissolved in *acetone* to 10.00 ml: $\alpha_D^{25} = -2.15°$ (2 dm). $[\alpha]_D^{25} = -44.3°$; $[M]_D^{25} = -86.0°$.

(+)-3,5-Dimethylphenoxy-propionic acid. The acid was liberated from the alkaloid salt as described above. 30.9 g brucine salt yielded 8.0 g of crude acid. One recrystallisation from 100 ml ligroin gave 7.6 g of pure acid with m.p. 118.5–119.2°. It formed long, glistening needles, quite similar to the (−)-form.

0.2704 g acid: 13.68 ml 0.1018-*N* NaOH.
$C_{11}H_{14}O_3$ (194.22) Equiv. wt.: calc. 194.2; found 194.2

The rotatory power in various solvents is given in the table below.

Table 1

g = g acid dissolved to 10.00 ml.

Solvent	g	α_D^{25} (2 dm)	$[\alpha]_D^{25}$	$[M]_D^{25}$
Acetone	0.2243	+1.98°	+44.1°	+85.7°
Ethanol (abs.)	0.2297	+1.585°	+34.5°	+67.0°
Ethanol (96 %)	0.1922	+1.24°	+32.3°	+62.6°
Chloroform	0.2120	+0.685°	+16.2°	+31.4°
Water (neutr.)	0.2284	+0.465°	+10.2°	+19.8°
Benzene	0.2136	+0.24°	+5.6°	+10.9°

$C_{11}H_{14}O_3$ Ref. 815a

Preliminary experiments on resolution were carried out on a small scale (0.001 mole acid) in dilute ethanol. The best result was obtained with strychnine, which yielded an acid with $[\alpha]_D^{25} = -42°$. Laevorotatory acid with lower activity (~10°) was obtained from the cinchonine and quinidine salts. Dehydroabietylamine gave an acid with $[\alpha]_D^{25} = +24°$, while cinchonidine gave (+)-acid of lower activity. The acid from the quinine and β-phenyl-isopropylamine salts was racemic and the brucine salt failed to crystallise. Thus, strychnine and dehydroabietylamine were selected for the resolution on a preparative scale.

(−)-*3,4-Dimethylphenoxy-propionic acid.* Racemic acid (23.4 g, 0.12 mole) and strychnine (40.4 g, 0.12 mole) were dissolved in a hot mixture of 100 ml ethanol and 300 ml water. The salt crystallised readily on cooling and was filtered off after standing over-night in a refrigerator. It was recrystallised to maximum activity of the acid; after each crystallisation a small sample of the salt was decomposed and the activity of the acid measured in acetone solution.

The low activity in the first crystallisation is remarkable as compared with the rapid increase in the second and the good results in the preliminary experiments. This may indicate that the resolution is effected by different rates of crystallisation rather than by difference in solubility.

Crystallisation	1	2	3	4	5
ml ethanol	100	70	90	50	50
ml water	300	210	180	100	100
g salt obtained	42.0	31.0	22.4	20.4	18.7
$[\alpha]_D^{25}$ of the acid	−10°	−42°	−46.5°	−47°	−46.5°

The strychnine salt (18.6 g) was decomposed by excess sodium carbonate solution and the alkaloid was taken up in chloroform. The aqueous solution was acidified with sulphuric acid and the organic acid extracted with ether. After drying and evaporation of the ether, the residue was recrystallised three times from ligroin, yielding 5.5 g of pure acid as long, slender, glistening needles with m.p. 75.5–76.5°.

$C_{11}H_{14}O_3$ (194.22)

Calc.	C 68.02	H 7.27
Found	C 68.17, 68.15	H 7.34, 7.21

0.1574 g acid dissolved in *acetone* to 10.00 ml: $\alpha_D^{25} = -0.743°$ (1 dm). $[\alpha]_D^{25} = -47.2°$; $[M]_D^{25} = -91.7°$.

(+)-*3,4-Dimethylphenoxy-propionic acid.* Two experiments were started with moderate quantities (about 6 g) of racemic acid using cinchonidine and dehydroabietylamine, respectively. The latter experiment showed that 96 % ethanol as solvent gives much better resolution than ethanol with 25 or 33 % of water. In the experiment with cinchonidine, the activity of the acid could be raised to about +18° but on further recrystallisation of the salt it remained practically constant. The acid used for the final resolution was a mixture of racemic acid and (+)-acid of moderate activity, regenerated from the experiments described above.

This acid (10.7 g, 0.055 mole), 19.0 g (0.055 mole) dehydroabietylamine acetate and 5.5 g (0.055 mole) potassium bicarbonate were dissolved in 150 ml of 96 % ethanol. After standing in a refrigerator over-night, the salt was filtered off and recrystallised to constant activity of the acid. The course of the resolution was followed as described above.

Crystallisation	1	2	3	4
ml ethanol	150	150	100	100
g salt obtained	19.3	16.5	15.5	13.9
$[\alpha]_D^{25}$ of the acid	+30.5°	+46°	+47°	+47°

The salt was decomposed and the acid liberated as described for the (−)-form. The yield of crude acid was 5.2 g. Three recrystallisations from ligroin gave 3.8 g of pure acid with m.p. 75.5–76.5°. It was quite similar to its enantiomer.

$C_{11}H_{14}O_3$ (194.22)

Calc.	C 68.02	H 7.27
Found	C 68.02, 68.06	H 7.20, 7.22

The values for the rotatory power in various solvents are given in Table 1.

The rotatory power of the pure acids was measured with a Perkin Elmer 141 Polarimeter. For the preliminary experiments and the measurements during the resolutions, a visual instrument of traditional type was used. The melting-point diagrams were recorded by the Kofler method [10] using a hot stage microscope from Leitz.

10. KOFLER, L., and KOFLER, A., Mikromethoden (3 ed.), Innsbruck, 1954.

Table 1.

g = g acid in 10.00 ml solution.

Solvent	g	α_D^{25} (1 dm)	$[\alpha]_D^{25}$	$[M]_D^{25}$
Acetone	0.1480	+0.700°	+47.3°	+91.9°
Ethanol (abs.)	0.1375	+0.481°	+35.0°	+67.9°
Chloroform	0.1530	+0.225°	+14.7°	+28.6°
Water (neutr.)	0.1414	+0.213°	+15.1°	+29.3°
Benzene	0.1375	−0.048°	−3.8°	−7.3°

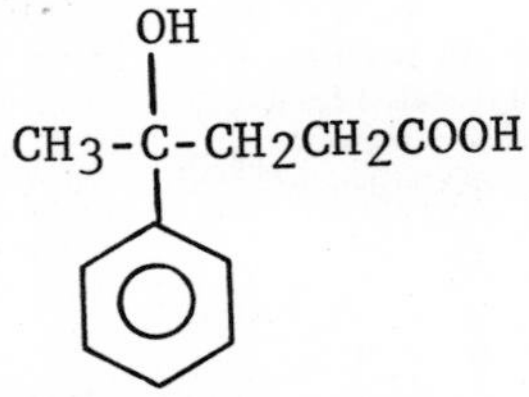

$C_{11}H_{14}O_3$ Ref. 816

Optical Resolution of γ-Hydroxy-γ-phenylvaleric Acid.—The acid (9·7 g., 0·05 g.-mol.) was dissolved in warm water (200 c.c.), and brucine tetrahydrate (23 g., 0·05 g.-mol.) added. The solution was filtered and cooled. Rosettes of needles separated (13·47 g.) which had m. p. 105—110° (decomp.), $[\alpha]_{5780}$ −49·6°, $[\alpha]_{5461}$ −58·2° in chloroform (*c*, 0·998). After five crystallisations from water brucine (+)-γ-hydroxy-γ-phenylvalerate was obtained as sheaves of needles, m. p. 107—108° (decomp.), $[\alpha]_{5780}$ −64·9° ± 0·5°, $[\alpha]_{5461}$ −75·9° ± 0·5° (*c*, 0·995 in chloroform) (Found: C, 64·3; H, 7·3. $C_{34}H_{40}O_7N_2,2\frac{1}{2}H_2O$ requires C, 64·4; H, 7·1%). The salt was treated with excess of warm sodium hydroxide solution, and the brucine removed by three extractions with chloroform. The aqueous layer, freed from chloroform, was acidified with hydrochloric acid at 0° and extracted with ether. The ethereal extract was washed with water and dried (Na_2SO_4). The ether was removed and the residue dissolved in the minimum amount of hot 2·5N-potassium hydroxide. The solution was cooled to 0° and acidified. The γ-hydroxy-γ-phenylvaleric acid which separated had m. p. 122—122·5°, $[\alpha]_{5780}$ +4·8° ± 0·2°, $[\alpha]_{5461}$ +5·7° ± 0·2° (*c*, 2·544 in absolute ethyl alcohol). Dissolution in alkali and reprecipitation by acid did not affect the specific rotation, nor did crystallisation from benzene. The lactone, prepared by repeated distillation with benzene, as described previously, had $[\alpha]_{5780}$ −54·8° ± 0·4°, $[\alpha]_{5461}$ −61·9° ± 0·4° (*c*, 1·214 in absolute ethyl alcohol) (Found: C, 67·9; H, 7·3. Calc. for $C_{11}H_{14}O_3$: C, 68·0; H, 7·3%).

The mother-liquor from which the 13·37 g. of brucine salt had separated was concentrated to 35 c.c., and filtered warm. When the solution was kept, a mixture of needles and plates separated with $[\alpha]_{5461}$ −27·5° (*c*, 1·002 in chloroform). It was freed from brucine and then dissolved afresh in water with the equivalent of brucine. From the solution plates separated, m. p. 95·5—98°, with $[\alpha]_{5461}$ +6·5° ± 0·1° (*c*, 5·062 in chloroform). Recrystallisation from water gave a salt with $[\alpha]_{5461}$ +5·8° ± 0·1° (*c*, 3·441 in chloroform). The brucine was removed and the free acid lactonised. The lactone had $[\alpha]_{5461}$ +61·8° ± 0·5° (*c*, 0·955 in ethyl alcohol) (Found: C, 67·3; H, 7·1. Calc. for $C_{11}H_{14}O_3$: C, 68·0; H, 7·3%).

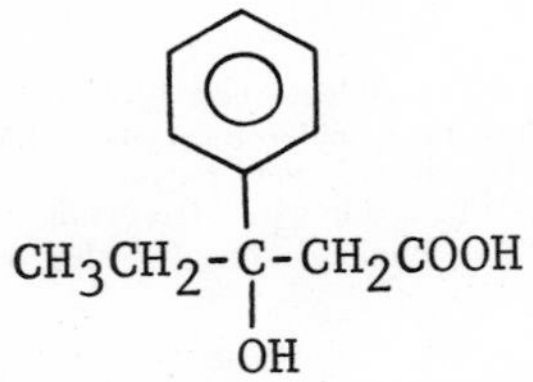

$C_{11}H_{14}O_3$ Ref. 817

Dédoublement de l'acide β-hydroxy β-phénylvalérique par la phényléthylamine

65 g (0·33 mole) d'acide β-hydroxy β-phénylvalérique racémique sont dissous dans 200 cm³ d'éthanol. On y ajoute 40 g (0·33 mole) de (+) phényléthylamine dans 100 cm³ d'éthanol et laisse cristalliser. On obtient 32·75 g de sel. Ce sel recristallisé une fois dans l'éthanol livre 28·7 g qui par traitement à l'acide chlorhydrique fournissent 17·5 g d'acide β-hydroxy β-phénylvalérique $[\alpha]_D^{22}$ = + 27·1° (c = 3, éthanol). Les eaux mères de cristallisation du premier jet de sel (+) sont évaporées à sec et livrent 75 g de sel qui par traitement à l'acide chlorhydrique fournissent 38·7 g d'hydroxyacide partiellement dédoublé $[\alpha]_D^{22}$ = − 13·5° (c = 3, éthanol).

Cet hydroxyacide (0·2 mole) est mis en présence de 24·2 g (0·2 mole) de (−) phényléthylamine. Dans 200 cm³ d'éthanol on obtient 32 g de sel que l'on recristallise dans 130 cm³ d'éthanol. On obtient 28·8 g de sel. Ce sel, traité par l'acide chlorhydrique fournit 17·35 g d'acide β-hydroxy β-phénylvalérique $[\alpha]_D^{22}$ = − 26·2° (c = 3, éthanol). Litt.[30] opt. pur, $[\alpha]_D^{23}$ = 22·0° (c = 3, EtOH), litt.,[33] $[\alpha]_D^{20}$ = 25·9° (c = 3, EtOH).

[30] S. Mitsui et Y. Kudo, *Tetrahedron* **23**, 4271 (1967)

[33] E. B. Dongala, A. Solladié-Cavallo et G. Solladié, *Tetrahedron Letters* 4237 (1972)

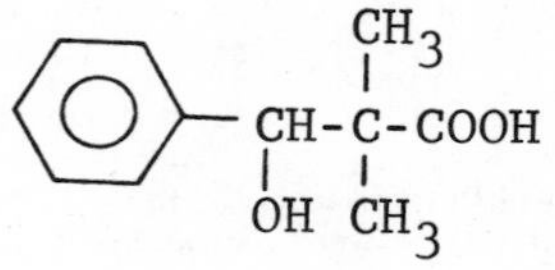

$C_{11}H_{14}O_3$ Ref. 818

Preliminary experiments on resolution

Morphine yielded a crystalline salt, containing excess of (+)-acid. (+)-Phenethylamine yielded a crystalline salt suitable for isolation of the (−)-acid. Quinine, brucine, quinidine and cinchonidine yielded salts containing about equal parts of (+)- and (−)-acid. Strychnine and cinchonine did not give crystalline products.

(+)-β-Oxy-β-phenylpivalic acid

29.1 g inactive acid and 45.5 g morphine were dissolved in 1500 ml hot 25 % ethanol. On cooling the salt crystallized as white prisms. The salt was removed after 4 hours and several recrystallizations were carried out. A sample was taken after each recrystallization, the acid isolated and the melting point and rotatory power in glacial acetic acid were determined. The following series of crystallizations was obtained:

Crystallization	1	2	3	4	5	6
Weight of salt (g)	29.3	23.4	19.9	18.2	16.4	14.2
$[\alpha]_D^{25}$ of the acid	+3.6°	+4.5°	+7.2°	+9.0°	+9.9°	+9.8°
M. p. » »	140°	145°	155°	157°	158°	158°

Analysis 0.3889 g air-dried salt was dried over P_2O_5. Weight reduction: 0.0141 g.

$C_{11}H_{14}O_3$, $C_{17}H_{19}NO_3$, H_2O H_2O calc. 3.62 found 3.63

0.02258 g water-free salt: 0.05759 g CO_2 and 0.01398 g H_2O

$C_{11}H_{14}O_3$, $C_{17}H_{19}NO_3$ calc. C 70.1 H 6.94

found » 69.6 » 6.93

14 g morphine salt yielded 4.1 g acid. The crude product was recrystallized from hot water. 3.6 g acid was collected as white prisms. (M. p. 157—158°)

0.3785 g (+)-acid: 19.20 ml 0.1020-*N* NaOH
E. W. calc 194.2 found 193.3
0.1038 g (+)-acid in glacial *acetic acid*, made up to 10.0 ml:
$3\ \alpha^{25}_{D} = +0.305°$; $[\alpha]^{25}_{D} = +9.8°$; $[M]^{25}_{D} = +19.0°$

(—)-β-Oxy-β-phenylpivalic acid

The acid was liberated from the mother liquor of crystallization 1 above. 15.0 g of this acid, 8.8 g inactive acid and 15.0 g (+)-phenethylamine were dissolved in 1300 ml hot water. After 6 hours the salt was removed. A number of recrystallizations were carried out and from each a sample was taken as above. The following series of crystallizations was obtained:

Crystallization	1	2	3	4	5
Weight of salt (g)	18.0	13.4	10.5	9.0	8.0
$[\alpha]^{25}_{D}$ of the acid	—7.8°	—8.4°	—8.8°	—9.0°	—9.0°
M. p. » » »	155°	156°	157°	158°	158°

Analysis 0.3925 g air-dried salt was dried over P_2O_5. Weight reduction: 0.0210 g.
$C_{11}H_{14}O_3$, $C_8H_{11}N$, H_2O H_2O calc. 5.40 found 5.35
0.02437 g water-free salt: 0.06414 g CO_2 and 0.01751 g H_2O
$C_{11}H_{14}O_3$, $C_8H_{11}N$ calc. C 72.3 H 7.99
found » 71.8 » 8.04

7.7 g salt yielded 4.4 g acid. The crude product was recrystallized from hot water. 4.1 g acid was collected as white prisms closely resembling the other antipode. It melted at 157—158°.

0.3986 g (—)-acid: 20.52 ml 0.1006-*N* NaOH
E. W. calc. 194.2 found 193.1
0.1086 g (—)-acid in glacial *acetic acid*, made up to 10.0 ml:
$3\ \alpha^{25}_{D} = -0.294°$; $[\alpha]^{25}_{D} = -9.0°$; $[M]^{25}_{D} = -17.5°$
0.0763 g (—)-acid in *ethanol*, made up to 10.0 ml:
$3\ \alpha^{25}_{D} = -0.063°$; $[\alpha]^{25}_{D} = -2.8°$; $[M]^{25}_{D} = -5.3°$
0.0983 g (—)-acid in *methanol*, made up to 10.0 ml:
$2\ \alpha^{25}_{D} = -0.103°$; $[\alpha]^{25}_{D} = -5.2°$; $[M]^{25}_{D} = -10.2°$
0.1233 g (—)-acid in *water, neutralized with NaOH*, made up to 10.0 ml:
$2\ \alpha^{25}_{D} = -0.100°$; $[\alpha]^{25}_{D} = -4.1°$; $[M]^{25}_{D} = -7.9°$
0.1021 g (—)-acid in *acetone*, made up to 10.0 ml:
$2\ \alpha^{25}_{D} = -0.080°$; $[\alpha]^{25}_{D} = -3.9°$; $[M]^{25}_{D} = -7.6°$

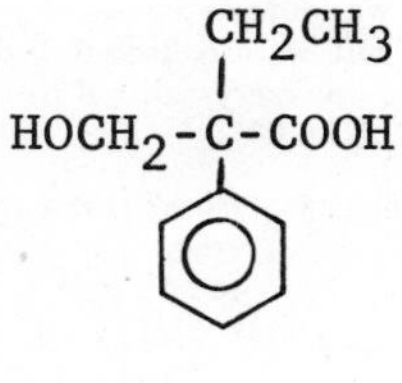

$C_{11}H_{14}O_3$ **Ref. 819**

Acido (+)-α-etil-α-fenil-β-idrossipropionico [(+)-*α-etiltropico*]

Si sciolgono a caldo in cc 290 di etanolo assoluto g 70 di acido DL-α-etil-α-fenil-β-idrossipropionico (XII) e g 116,8 di chinina base. Si aggiungono alla soluzione calda cc 290 di acqua, pure tiepida, e si ottiene soluzione completa. Si lascia raffreddare la miscela a temperatura ambiente per almeno 24 ore permettendo in tal modo la separazione di un abbondante precipitato che viene filtrato su Büchner. Si conservano le acque madri filtrate.

Si hanno così g 79,2 a p.f. 133-134°, $[\alpha]^{20}_{D}$ — 135,6° (c = 2,2 in etanolo). Il precipitato viene ricristallizzato da cc 290 di etanolo e cc 290 di acqua, lasciato 24 ore a temperatura ambiente e poi una notte in ghiacciaia, filtrato, lavato con alcool al 50% e seccato a 80° sotto vuoto. Si ottengono g 68 a p.f. 133-134°, $[\alpha]^{20}_{D}$ — 135,7° (c = 2,1 in etanolo). Il sale di chinina raccolto viene sospeso in cc 300 di acqua e trattato con cc 150 di idrato sodico al 50%. Si estrae la sospensione che ne risulta con cc 1500 di etere e la sospensione acquosa residua dell'estrazione viene acidificata fino al congo con acido cloridrico concentrato ed estratta tre volte con etere (500, 250 e 250 cc). L'estratto etereo viene essiccato su Na_2SO_4. Per evaporazione dell'etere si ottiene un olio, che cristallizza da benzolo-etere di petrolio.

Resa g 19, p.f. 80-81°, $[\alpha]^{20}_{D}$ + 11,50° (c = 2 in etanolo).

Acido (—)-α-etil-α-fenil-β-idrossipropionico

Il filtrato della prima cristallizzazione del sale di chinina dell'acido (+)-α-etiltropico (vedi preparazione precedente) viene evaporato a secco. Si ottiene un olio che non cristallizza. Anche il tentativo di purificazione attraverso il sale di cinconina, non fornisce un prodotto cristallino. L'olio residuo, il sale di chinina dell'acido (—)-α-etiltropico, viene alcalinizzato con idrato sodico al 50% ed estratto con lt 1 di etere. La soluzione alcalina residua viene acidificata al congo con acido cloridrico concentrato ed estratta tre volte con etere solforico (500, 250 e 250 cc). L'estratto etereo viene essiccato su Na_2SO_4 e l'etere evaporato a secco fino ad ottenere un olio che viene sciolto in acqua calda e lasciato una notte in ghiacciaia. Si separa per raffreddamento un olio, dal quale si allontana la soluzione acquosa per decantazione, e che si cristallizza da benzolo-etere di petrolio, ottenendo in tal modo g 5,8 di cristalli a p.f. 80-81°, $[\alpha]^{20}_{D}$ — 11,2° (c = 2 in etanolo). Dalla soluzione decantata si separano per concentrazione sotto vuoto g 23 di acido (—)-α-etiltropico p.f. 78-80°. Anch'essi ricristallizzati da benzolo-etere di petrolio non risultano otticamente unitari bensì frammisti di almeno il 10% di acido (+)-α-etiltropico. Gli spettri I.R. determinati su una sospensione in olio minerale degli antipodi ottici sono identici fra di loro, mentre si differenziano da quello del racemo. L'antipodo (—) ($[\alpha]^{20}_{D}$ — 11,2°) presenta molto leggermente accennate alcune bande presenti solo nello spettro del racemo.

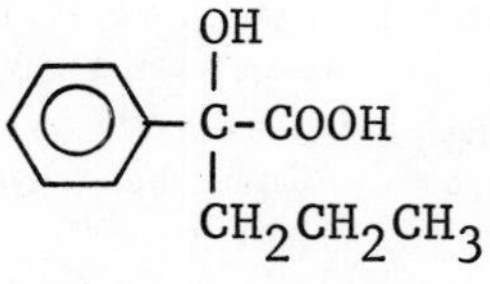

$C_{11}H_{14}O_3$ **Ref. 820**

Note:

a) Amphetamine was used as resolving agent;

b) The following physical constants were reported for the active acid: m.p. 75-81°, $[\alpha]_D$ -21.0° (CH_3OH).

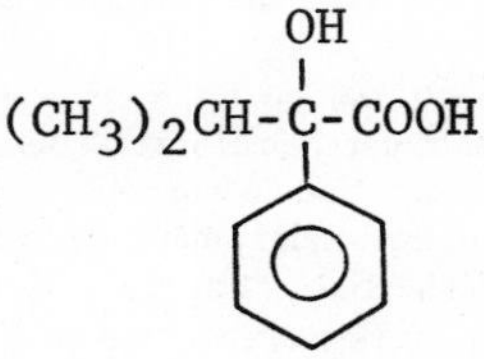

$C_{11}H_{14}O_3$ **Ref. 821**

Das Princip der Methode lässt sich in Kürze, wie folgt, erläutern: Auswahl zweier Alkaloïde A und B derart, dass jedes dieser mit jeder der zu trennenden stereoisomeren Säuren C und C′ Salze von einigermassen verschiedener Löslichkeit liefert. Sodann ist zu beachten, dass die weniger löslichen Salze beider Basen nicht aus einer activen Säure stammen, sondern dass sie Derivate der zwei Isomeren seien. Mit anderen Worten: es soll ein bestimmter Unterschied zwischen der Löslichkeit der Salze AC und AC′, sowie zwischen BC und BC′ bestehen, so dass z. B. die weniger löslichen Salze AC und BC′ sein würden. Aus der Lösung der Salze, die mit A dargestellt sind, wird sich beim Krystallisiren zuerst wesentlich AC abscheiden. Aus dem löslicheren Salze, welches mehr AC′, als AC enthält, wird natürlich mehr von C′, als von C abgeschieden; wird dieses Gemisch mit dem Alkaloïd B behandelt, so wird beim Krystallisiren zuerst mehr BC′, als BC erhalten, woraus ein an C reiches Gemisch der zwei Säuren frei gemacht wird. Durch Wiederholen dieser Operationen gelingt es schliesslich, die beiden activen Säuren rein zu gewinnen.

Die Versuche mit Isopropylphenylglycolsäure sind erfolgreich mittelst Chinin und Cinchonin ausgeführt worden.

Das Chinin giebt mit der genannten Säure ein Salzgemisch, aus welchem durch Umkrystallisiren mit wässerigem Alkohol leicht das Salz der links drehenden Säure (Schmelzp. 204° bis 205°) rein zu gewinnen ist, während das der rechts drehenden Säure nicht ganz frei von ersterem erhalten werden kann. Dagegen lässt sich mittelst Cinchonins das Salz der rechts drehenden Säure (Schmelzp. 201°) rein darstellen. So gelingt es, durch abwechselnde Bereitung von Chinin- und Cinchoninsalzen, Umkrystallisiren derselben, Abscheidung der Säuren etc. die stereoisomeren Säuren vollständig von einander zu trennen.

Diese beiden haben ein viel grösseres Krystallisationsvermögen, als das inactive Gemisch derselben: aus wässrigem Alkohol scheiden sich jene in dicken Tafeln von 153°—154° Schmelzp. aus, während die inactive in Blättchen von 156°—157° Schmelzp. krystallisirt.

Die Löslichkeit in absolutem Alkohol ist folgende:

100 Thle. desselben lösen bei 13°:
47,49 Thle. rechtsdrehende Säure,
47,32 „ linksdrehende Säure,
21,61 „ inactive Säure.

Werden gleiche Theile der beiden activen Säuren in Alkohol gelöst, so scheidet sich die völlig inactive Säure (Schmelzp. 156°—157°) aus. — Das Drehungsvermögen jener in alkoholischer Lösung ist folgendes:

Rechtssäure $[\alpha]_D = 134{,}9°$ (c = 4,0568, l = 2, t = 17°, α = 10,95°.)
Linkssäure $[\alpha]_D = -135°$ (c = 4,0916, l = 2, t = 17°, α = − 11,05°.)

Durch 40stündiges Erhitzen mit Wasser in geschlossenen Röhren auf 180°—200° verlieren die activen Säuren ihr Drehungsvermögen.

Das Chininsalz der linksdrehenden Säure bildet wollige Nadeln von 204°—205° Schmelzp., ziemlich löslich in siedendem, sehr wenig in kaltem Alkohol. 100 Thle. Wasser von 15° lösen 0,09 Thle., 100 Thle. Alkohol 0,54 Thle. des Salzes. Das Drehungsvermögen seiner alkoholischen Lösung ist:

$[\alpha]_D = -118{,}4°$ (c = 0,3800, l = 2, t = 13°, α = − 0,9°).

Das Chininsalz der rechtsdrehenden Säure, aus dieser bereitet, bildet dünne Nädelchen, welche bei 192°—193° schmelzen. 100 Thle. Wasser lösen bei 19° 0,18 Thle., 100 Thle. Alkohol (bei 20°) 1,44 Thle. Sein Drehungsvermögen (in Alkohol) beträgt:

$[\alpha'_D] = -79{,}4°$ (c = 0,9248, l = 2, t = 24°, α = − 1,47°).

Das Cinchoninsalz der rechtsdrehenden Säure krystallisirt in langen, glänzenden Nadeln von 201° Schmelzp., ist in kaltem Wasser wenig, in kochendem merklich löslich. 100 Thle. von 15° lösen 0,17 Thle. Salz, 100 Thl. absoluter Alkohol 3,54 Thle. Das Drehungsvermögen letzterer Lösung ist:

$[\alpha]_D = +136{,}8°$ (c = 2,3014, l = 2, t = 13°, α = + 6,3°).

Das Cinchoninsalz der linksdrehenden Säure, aus dieser bereitet, zeigt geringere Neigung zu krystallisiren, als obiges; es bildet wollige, bei 167° schmelzenden Nadeln.

100 Thle. Wasser lösen bei 19° 0,40 Thle.
100 Thle. absol. Alkohol lösen bei 200° 31,65 Thle.
Rotationsvermögen: $[\alpha]_D = +83{,}4°$ (c = 1,3308, l = 2, t = 24°, α = + 2,22°).

$C_{11}H_{14}O_3$ **Ref. 821a**

Note:
a) Amphetamine was used as resolving agent;
b) The following physical constants were reported for the active acid: m.p. 111-117°, $[\alpha]_D$ -27.0° (CH_3OH).

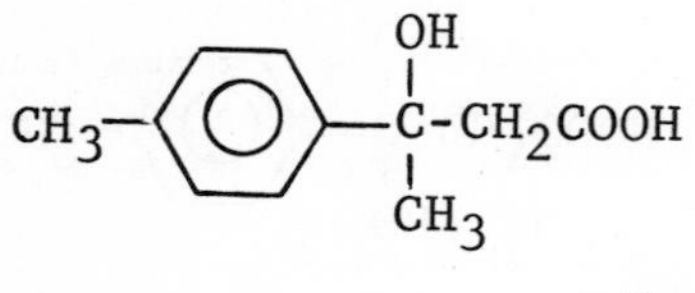

$C_{11}H_{14}O_3$ **Ref. 822**

Resolution of β-(p-Methylphenyl)-β-hydroxybutyric Acid. Brucine (120 g., 0.31 mole) was added to a boiling solution of β-(*p*-methylphenyl)-β-hydroxybutyric acid (65 g., 0.33 mole) in 800 ml. of ethyl acetate. The resulting solution quickly deposited crystals on cooling, m.p. 163–167°. After ten recrystallizations from ethyl acetate the melting point of the brucine salt was found to remain constant at 178–180° dec., $[\alpha]^{23}D$ −17.3° (*c* 0.10, chloroform).

Decomposition of the brucine salt with iced 10% hydrochloric acid, extraction with ether, evaporation of the solvent, and crystallization of the residual oil from benzene–hexane afforded (+)-β-(*p*-methylphenyl)-β-hydroxybutyric acid ((+)-IV) as long needles, m.p. 76–77° (*cf.* racemate, m.p. 72°), $[\alpha]^{23}D$ + 11.9° (*c* 0.10, chloroform).

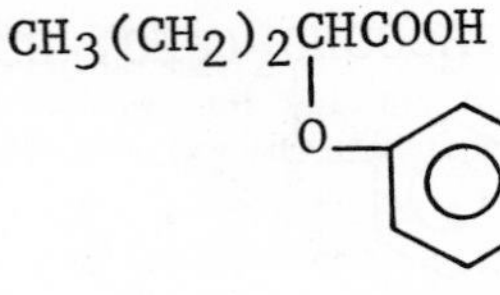

$C_{11}H_{14}O_3$ **Ref. 823**

(+)- *and* (−)-*α-phenoxy-n-valeric acids.*—Most of the optically active bases tried for resolution gave noncrystallizing oils or salts from which only slightly active acids were obtained. Quinine gave some resolution but the best result was obtained with β-phenyl*iso*propylamine, the (−)-form of which gave a salt containing excess of the (+)-acid.

19.4 g (0.10 mole) α-phenoxy-*n*-valeric acid were dissolved in 100 ml 96 % ethanol. After adding 13.5 g (0.10 mole) (−)-β-phenyl*iso*propylamine, water (400 ml) was added and the solution left to crystallize for 6 hours. The salt was collected and recrystallized several times from 20 % ethanol. After each recrystallization $[\alpha]_D$ of the acid was determined, giving the following series: + 8.6°, + 33.2°, + 41.6°, + 44.8°, + 45.8°, + 45.9°, + 45.2°. 6.45 g (20 %) pure (+)-acid-(−)-amine salt were collected. The acid was liberated from the salt and the crude acid (3.9 g) recrystallized from petroleum (b.p. 60–80°). 2.85 g pure (+)-acid were collected as small, colourless needles, m.p. 85.0–86.0°.

80.00 mg (+)-acid: 8.81₅ ml 0.0465-*N* NaOH
$C_{11}H_{14}O_3$ calc. equiv. wt. 194.2
found „ „ 195.2

0.2099 g dissolved in abs. *ethanol* to 20.06 ml: $2\alpha_D^{25}$ = + 0.952°, $[\alpha]_D^{25}$ = + 45.5°; $[M]_D^{25}$ = + 88.4°.

From the first two mother liquors 11.4 g acid with $[\alpha]_D$ = − 23° were obtained. This acid and an equimolecular quantity of (+)-β-phenyl*iso*propylamine (7.9 g) were dissolved in 75 ml of hot 96 % ethanol. Water (300 ml) was added and the solution left to crystallize for 6 hours. The salt was recrystallized several times from 20 % ethanol. The resolution was followed by measuring the rotatory power of the acid of each fraction. The following series was obtained: − 36.2°, − 42.8°, − 44.0°, − 45.6°. 5.55 g (29 %) (−)-acid-(+)-amine salt were collected and from this salt were obtained 3.3 g of the crude (−)-acid. It was recrystallized from petroleum (b.p. 60–80°) yielding 2.4 g pure (−)-acid, m.p. 85.0–86.0°.

78.32 g (−)-acid: 8.62 ml 0.0465-*N* NaOH
$C_{11}H_{14}O_3$ calc. equiv. wt. 194.2
found „ „ 195.3

0.2053 g dissolved in abs. *ethanol* to 20.06 ml: $2\alpha_D^{25}$ = − 0.942°. $[\alpha]_D^{25}$ = − 46.0°; $[M]_D^{25}$ = − 89.4°.
0.2036 g dissolved in *acetone* to 20.06 ml: $2\alpha_D^{25}$ = − 1.161°. $[\alpha]_D^{25}$ = − 57.2°; $[M]_D^{25}$ = − 111.1°.
0.1971 g dissolved in *benzene* to 20.06 ml: $2\alpha_D^{25}$ = − 0.135°. $[\alpha]_D^{25}$ = − 6.9°; $[M]_D^{25}$ = − 13.3°.
0.2049 g dissolved in *chloroform* to 20.06 ml: $2\alpha_D^{25}$ = − 0.665°. $[\alpha]_D^{25}$ = − 32.6°; $[M]_D^{25}$ = − 63.2°.
0.1604 g neutralized with *aqueous NaOH* and made up to 20.06 ml: α_D^{25} = − 0.164°. $[\alpha]_D^{25}$ = − 20.5°; $[M]_D^{25}$ = − 39.8°.

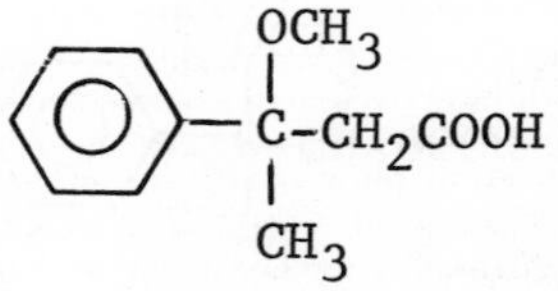

$C_{11}H_{14}O_3$ **Ref. 824**

(+)-3-Methoxy-3-phenylbutanoic Acid. A hot solution of 93.0 g of the acid and 131.0 g of cinchonidine in 3 l. of acetone was filtered and allowed to stand 10 days. The precipitate (95.0 g) was filtered; the filtrate was concentrated and allowed to stand 5 days. The salt which crystallized was added to the previous crop and redissolved (163.0 g) in 3.8 l. of hot acetone. After filtration, concentration of the filtrate, and filtration, 132.0 g of salt was obtained and dissolved in 2.1 l. of acetone. This crop (86.0 g) was filtered after 3 days and then was dissolved in 1.5 l. of acetone. After 1 day, 64 g of salt was recovered and dissolved in 1.3 l. of acetone. After 2 days, 36 g of precipitate was filtered and shaken with 250 ml of ether and 500 ml of 2 *N* hydrochloric acid. After hydrolysis and three recrystallizations from hexane to a constant melting point and rotation, 11.5 g of (+)-X, mp 59.2–60.1°, [α] +35.2° (*c* 0.9, ethanol), was obtained. *Anal.* Calcd for $C_{11}H_{14}O_3$: C, 68.02; H, 7.27. Found: C, 68.13; H, 7.23.

$C_{11}H_{14}O_3$

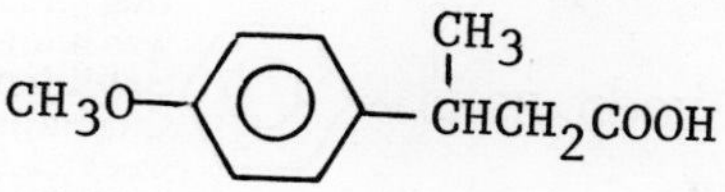

Ref. 825

Resolution of (±)-3-(4-Methoxyphenyl)butyric Acid

The racemic acid[23] was prepared as described by Fort and Leary,[24] and also by hydrogenation (60%) of 3-(4-methoxyphenyl)but-2-enoic acid[24] over Raney nickel as for cinnamic acids;[25] it crystallized from light petroleum (b.p. 60–80°) in prisms, m.p. 63°. Crystallization of the brucine salt from several solvents did not achieve separation of isomeric forms and the (−)-menthyl ester remained as an oil. A solution of quinine (16·2 g, 0·2 mole) and (±)-3-(4-methoxyphenyl)-butyric acid (19·8 g, 0·4 mole) in chloroform (*c.* 40 ml) was diluted with light petroleum (b.p. 60–80°) until the solution became turbid. After 2 days at 0° the supernatant liquid was decanted from a brown oil and allowed to crystallize at 0°. Crystals were collected after 1 week, and after four recrystallizations from chloroform–light petroleum (b.p. 60–80°) the salt formed clusters of needles, m.p. 99–101°, $[\alpha]^{24}_{D}$ −98° (0·5% in EtOH). Concentration of the mother liquors gave material, m.p. 94–95°, $[\alpha]^{24}_{D}$ −110° (0·1% in EtOH), containing an excess of the (−)-acid. The less laevorotatory salt was dissolved in dilute hydrochloric acid and the butyric acid was removed by extraction with ether; crystallization of the acid from light petroleum gave (+)-3-(4-methoxyphenyl)butyric acid (2 g, 10%) in prisms, m.p. 38°, $[\alpha]^{32}_{D}$ +34° (1·4% in EtOH), which gave an *S*-benzylisothiouronium salt, m.p. 161°, $[\alpha]^{30}_{D}$ +33° (0·25% in acetone). The more laevorotatory quinine salt yielded the butyric acid (0·5 g, 2·5%) as a waxy solid, m.p. 32–36°, $[\alpha]^{25}_{D}$ −12° (1·2% in EtOH), containing an excess of the (−)-enantiomorph. The identity of the (+)-acid and its *S*-benzylisothiouronium salt with those obtained by oxidation of the propane was confirmed by mixed m.p. determinations and comparison of infrared spectra.

[24] Fort, A. W., and Leary, R. E., *J. Am. Chem. Soc.*, 1960, **82**, 2494.
[25] Papa, D., Schwenk, E., and Whitman, B., *J. Org. Chem.*, 1942, **7**, 587.

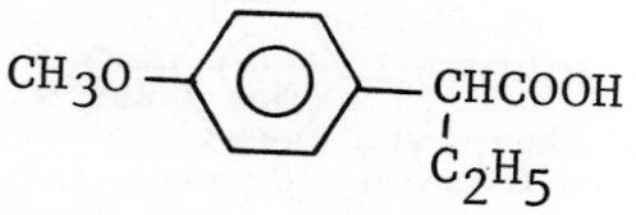

$C_{11}H_{14}O_3$

Ref. 826

Resolution of α-Ethyl-*p*-methoxyphenylacetic Acid.—Fractional crystallization from 80% ethanol of the cinchonidine salt prepared from 605 g of the acid and 940 g of cinchonidine gave a salt, mp 147–148°, $[\alpha]^{24}D$ −62.9 (ethanol, *c* 4), from which the acid was regenerated with 25% sulfuric acid and recrystallized from petroleum ether (bp 55–85°) to give the dextrorotatory acid, mp 85–86°, $[\alpha]^{23}D$ +64.76 (ethanol, *c* 4), 152 g. The product of this acid with (+)-α-phenylethylamine was recrystallized twice from aqueous ethanol to give a salt, $[\alpha]^{24}D$ +14.45° (ethanol, *c* 4), from which the acid was regenerated with essentially unchanged properties, mp 85–86°, $[\alpha]^{23}D$ +64.46 (ethanol, *c* 4).

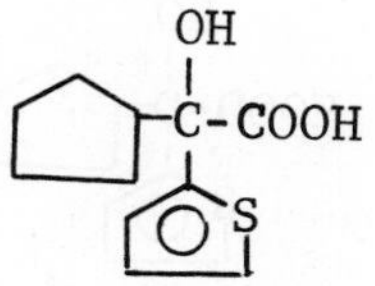

$C_{11}H_{14}O_3S$

Ref. 827

Note:

a) Amphetamine was used as resolving agent;
b) The following physical constants were reported for the active acid: m.p. 95-97°, $[\alpha]_{350}$ -51.0° (CH_3OH), m.p. 94-100°, $[\alpha]_{350}$ +51.3° (CH_3OH).

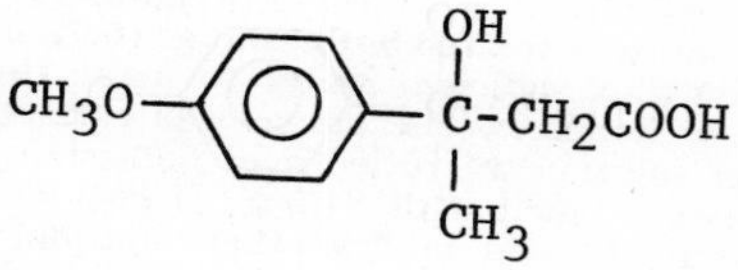

$C_{11}H_{14}O_4$

Ref. 828

Resolution of β-(p-Methoxyphenyl)-β-hydroxybutyric Acid. Brucine (150 g., 0.38 mole) was added to a boiling solution of β-(*p*-methoxyphenyl)-β-hydroxybutyric acid (80 g., 0.38 mole) in 1 l. of ethyl acetate. Upon standing at room temperature crystals were deposited slowly (120 g.), m.p. 128–135°. The melting point of the brucine salt was raised to 133–135° after two recrystallizations from ethyl acetate and remained constant after three further recrystallizations, $[\alpha]^{26}D$ −23.6° (*c* 0.07, chloroform).

The brucine salt was decomposed with *iced* 5% hydrochloric acid in the presence of ether, to avoid any substantial racemization during hydrolysis. Evaporation of the ether extracts afforded (+)-V as a slowly crystallizing oil. By fractional crystallization from benzene–carbon tetrachloride and a few additional recrystallizations to constant rotation, (+)-V was obtained, as rosettes, m.p. 69.5–70.5° (*cf.* racemate, m.p. 80–81°), $[\alpha]^{25}D$ +7.40° (*c* 0.10, chloroform). Hydrolysis of the brucine salt after only two recrystallizations gave acid of relatively high rotation, $[\alpha]^{23}D$ +4.7° (*c* 0.20, chloroform).

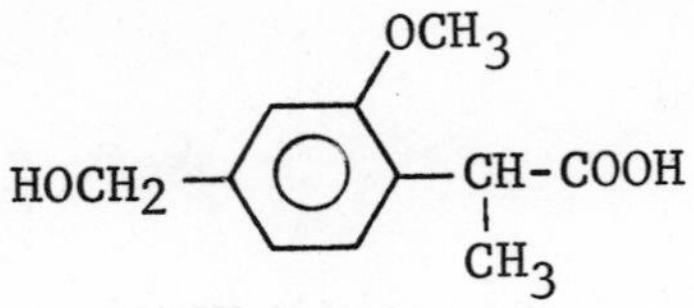

$C_{11}H_{14}O_4$ Ref. 829

Note:

Resolved using quinine in ethyl acetate (salt recrystallized from 2-propanol). There was obtained (+) and (-) acid, m.p. 150°, $[\alpha]^{25}_{578}$ (+) or (-) 41.8° (ethanol, c = 0.9).

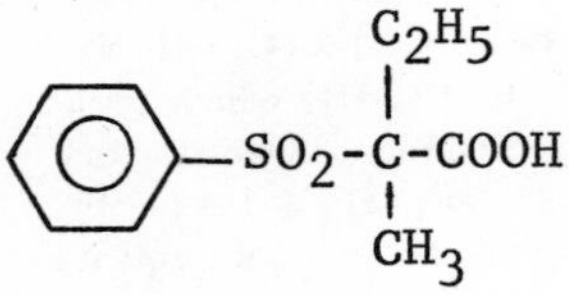

$C_{11}H_{14}O_4S$ Ref. 830

Resolution of *dl*-2-Methyl-2-benzenesulfonylbutyric Acid.—Attempts to resolve the acid using brucine, quinine, strychnine, morphine or the *d*-base from Chloromycetin were unsuccessful. A partial resolution was achieved using *l*-α-phenylethylamine. *l*-α-Phenylethylamine *l*-malate (91.6 g., 0.349 mole) was treated with 28.6 g. (0.72 mole) of sodium hydroxide plus sufficient water to make the mixture convenient to handle. The amine was extracted with ether, the ether solution was dried over potassium hydroxide and then was combined with 84.5 g. of *dl*-2-methyl-2-benzenesulfonylbutyric acid in dry ether solution. Upon mixing the acid and the base, the salt very slowly precipitated. The mixture was allowed to stand overnight to complete the precipitation. An aqueous solution containing 63.9 g. of salt in 28.5 ml. of water was made at room temperature, but, interestingly enough, upon standing at this same temperature this solution slowly precipitated crystals of the salt which had been dissolved in it. The acid recovered from this precipitate showed no optical activity. Precipitation was allowed to continue for several days until an equilibrium condition was achieved. The resulting saturated solution then was placed in a refrigerator and upon standing one week, fairly well formed crystals were observed in low yield. Additional salt was dissolved in the mother liquor at room temperature and the above crystallization procedure repeated. After the total amount of salt had been treated in this manner a second series of crystallizations was made starting with the once-crystallized salt. The entire procedure required several months' time. A yield of 11.4 g. of air-dried salt was obtained.

The salt was treated with 10 ml. of water and 5 ml. of concentrated hydrochloric acid. The mixture was extracted three times with ether and the ether solution was washed and then evaporated to dryness yielding an optically active product. The acid thus obtained was allowed to stand about one day and then was placed in a vacuum desiccator along with a beaker of sulfuric acid. After two days, the mixture crystallized to a soft mass which slowly hardened. The first decarboxylation was made with this acid.

Further purification was achieved by dissolving the above acid in dilute potassium carbonate solution, precipitating with excess sulfuric acid and extracting with benzene using a continuous extractor. The benzene solution, upon standing at room temperature, yielded crystals of acid which showed only slight activity as compared to the fraction in solution. Evaporation of the benzene solution gave a light yellow liquid which very slowly formed a soft solid mass. Treatment of this material with cold benzene resulted in only a partial solution. The insoluble material was set aside since its rotation was $[\alpha]^{22}D$ −0.1° (*c* 9.6 in ethanol). The benzene solution was passed through a column of silicic acid and the benzene evaporated. After a long period of standing the resulting oil crystallized to form a solid with a light yellow color. Treatment of an alcohol solution of this solid with Norit eliminated most of the color; see Table I for the rotations. Upon allowing the alcohol solution to evaporate, long needle-like crystals of the *dl* acid formed first. Evaporation of the residual alcohol under vacuum gave a solid of melting point 86–89°. Several melting points taken during the over-all purification process indicated that the greater the proportion of the *dl*-acid the higher the melting point. The infrared spectrum is essentially indistinguishable from that of the *dl*-acid.

d-2-Methyl-2-benzenesulfonylbutyric acid was prepared from the acid recovered from the mother liquor. An ether solution of this acid was very slowly evaporated resulting in crystallization of the *dl*-acid. The remaining material (1.2 g.), purified by treating both an alcohol solution of the acid and an aqueous solution of the sodium salt with Norit, showed optical activity; $[\alpha]^{27}_{5461}$ +0.81°. No further work was done with the *d*-acid.

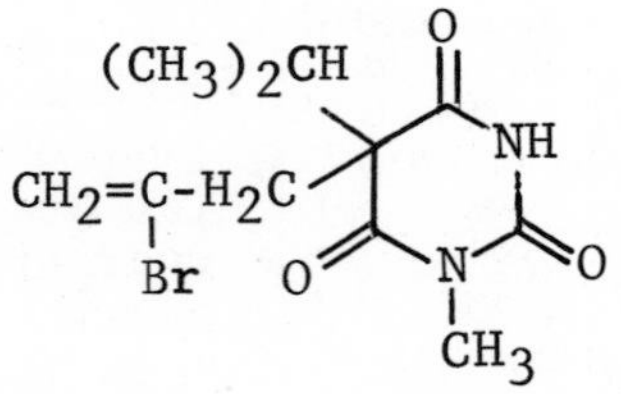

$C_{11}H_{15}BrN_2O_3$ Ref. 831

II. Spaltung von Pronarcon in optische Antipoden

1. Gewinnung der diastereomeren N-Methyl-chininium-pronarcon-salze

12,12 g Pronarcon wurden in 245,0 ml einer titrierten methanol. N-Methyl-chininium-hydroxidlösung gelöst und die Lösung i. Vak. geringfügig eingeengt. Danach schieden sich 10,8 g Kristalle ab, die aus Methanol umkristallisiert wurden. Ausbeute: 8,9 g N-Methyl-chininium-(+)-pronarcon. Nach Trocknen i. Vak. über P_2O_5 Schmp. 238°.

$C_{32}H_{41}BrN_4O_5$ (641,6)	Ber.:	C 59,91	H 6,44
	Gef.:	C 59,86	H 6,53

Aus der Mutterlauge kristallisierten nach weiterem Einengen und Zusatz von abs. Äthanol 6,1 g eines Gemisches. Nach erneutem Einengen und Zusatz von abs. Äthanol und viel Äther fielen 5,0 g feine Kristalle aus, die aus Methanol/Äther umkristallisiert wurden. Ausbeute: 3,9 g N-Methyl-chininium-(—)-pronarcon, Schmp. 220°.

$C_{32}H_{41}BrN_4O_5$ (641,6)	Ber.:	C 59,91	H 6,44
	Gef.:	C 59,70	H 6,38

2. Gewinnung der Pronarcon-Antipoden aus den diastereomeren N-Methyl-chininium-salzen

2.1. (+)-Pronarcon

5,0 g des Salzes vom Schmp. 238° wurden in heißem Methanol gelöst und die Lösung mit 2 n H_2SO_4 und Wasser versetzt. Nach dem Einengen fielen 1,1 g Kristalle an, Schmp. 120°.

$C_{11}H_{15}BrN_2O_3$ (303,2)   Ber.: C 43,58   H 4,99
                                  Gef.: C 43,58   H 4,90

Spezifische Drehung (s. Tab. 1)

a) Messung im Polarimeter mit visueller Ablesung:
   α_D + 0,54°, c = 9,435, 1 dm Rohr, t = 20°, abs. Äthanol.

b) Messung im lichtelektrischen Präzisionspolarimeter:
   c = 4,55, 1 dm Rohr, t = 20°, abs. Äthanol.
   α_{578} + 0,27°, α_{546} + 0,30°, α_{436} + 0,48°, α_{405} + 0,55°, α_{365} + 0,63°.

2.2. (−)-Pronarcon

3,9 g des Salzes vom Schmp. 220° wurden unter Erwärmen in Methanol gelöst und mit 2 n H_2SO_4 und Wasser versetzt. Nadeln, Schmp. 120°. Ausbeute: 1,0 g.

$C_{11}H_{15}BrN_2O_3$ (303,2)   Ber.: C 43,58   H 4,99
                                  Gef.: C 43,86   H 5,16

Spezifische Drehung (s. Tab. 1)

a) Messung im Polarimeter mit visueller Ablesung:
   α_D − 0,50°, c = 9,020, 1 dm Rohr, t = 20°, abs. Äthanol.

b) Messung im lichtelektrischen Präzisionspolarimeter:
   c = 3,88, 1 dm Rohr, t = 20°, abs. Äthanol.
   α_{578} − 0,24°, α_{546} − 0,27°, α_{436} − 0,42°, α_{405} − 0,48°, α_{365} − 0,55°.

$(CH_3)_2CH\text{-}CH\text{-}COOH$
       |
       NHC_6H_5

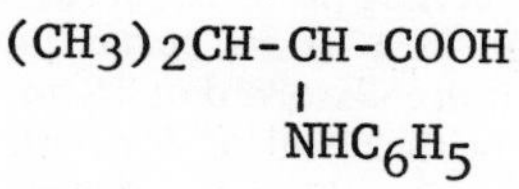

$C_{11}H_{15}NO_2$ Ref. 832

(+)-N-Phenylvaline. A solution of 9.1 g of racemic N-phenylvaline in 35 cc of warm methanol was mixed with the warm solution of 15.3 g of quinine in 5 cc of methanol and 220 cc of acetone. Fine white needles began to form soon after the solution had cooled to room temperature. The crystals were filtered off after about 5 hr and decomposed by shaking with 20 cc of 1 N NaOH solution and 20 cc of chloroform. The aqueous layer was extracted twice with chloroform and then acidified with concentrated hydrochloric acid (pH 4). The precipitate of N-phenylvaline was washed with water and dried; 2.2 g was obtained, $[\alpha]^{25}_D$ +67.0° (3.5% in EtOH). The procedure was repeated by combining the 2.2 g of N-phenylvaline in 7 cc of CH_3OH with the solution of 3.7 g of quinine in 3 cc of CH_3OH and 55 cc of acetone. The salt was filtered after standing overnight at room temperature and decomposed to give 1.32 g of N-phenylvaline, $[\alpha]^{25}_D$ +83.8° (4.3% in EtOH). The highest rotation obtained for this compound by repetitions of this procedure was +86.0°.

Cyclohexenyl-C(CH$_2$CH$_3$)(CN)-COOH

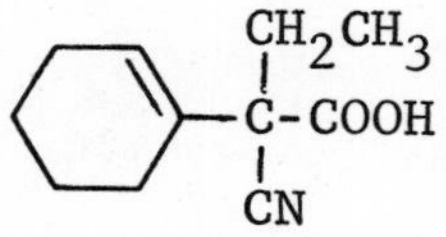

$C_{11}H_{15}NO_2$ Ref. 833

Äthyl-(1'-cyclohexenyl-)-cyanessigsäure [4A]

(+)Antipode [(+)4A] mit Codein

31,7 g (0,1 Mol) Codein wurden unter Erwärmen in 500 ml Methanol und 550 ml Wasser gelöst und zu einer erwärmten Lösung von 19,3 g (0,1 Mol) razem. [4A] in 250 ml Methanol gegeben. Wir ließen den Ansatz bei Raumtemperatur langsam abkühlen. Nach 14 Std. bei 4° erhielten wir farblose Blättchen, die über P_2O_5 getrocknet 21,5 g wogen.

Sie wurden unter Rückfließen in 400 ml Methanol und 180 ml Wasser gelöst und umkristallisiert. Wir erhielten daraus 17 g Diastereomerensalz vom Schmp. 149°-150° (Zers.), welches nach Zerlegen mit H_3PO_4 6 g [(+)4A] in farblosen Kristallen lieferte. Aus Benzol/Petroläther wurden farblose Prismen vom Schmp. 67°-68° erhalten. Ausbeute 62% d.Th.

$C_{11}H_{15}NO_2$ (193,2)   Ber.: C 68,37   H 7,82   N 7,25
                                Gef.: C 68,3   H 7,91   N 7,6

Spezifische Drehung

c = 2,52 , 1 dm Rohr , t = 20° , abs. Äthanol

α_{578} = + 1,350°	$[\alpha]^{20}_{578}$ = + 53,5°	$[\alpha]^{20}_{D}$ = + 50,9°
α_{546} = + 1,560°	$[\alpha]^{20}_{546}$ = + 61,9°	
α_{436} = + 2,885°	$[\alpha]^{20}_{436}$ = + 114,3°	
α_{405} = + 3,600°	$[\alpha]^{20}_{405}$ = + 142,6°	
α_{365} = + 5,060°	$[\alpha]^{20}_{365}$ = + 200,5°	

(-)Antipode [(-)4A] mit Chinin

10 g optisch angereichertes [4A], $[\alpha]^{20}_{D}$ = - 43°, aus der Mutterlauge des Codeinsalzes wurden mit 16 g Chinin in Methanol bei 45° gelöst und die bei - 25° erhaltenen 21 g Diastereomerensalz 2 x aus warmem Methanol umkristallisiert. Dabei wurden 15,5 g farblose stumpfe Nadeln vom Schmp. 135°-137° (Zers.) gewonnen, welche nach Zerlegen mit H_2SO_4 5,2 g [(-)4A] lieferten. Aus Benzol/Petroläther wurden farblose Prismen vom Schmp. 68°-69° erhalten.

$C_{11}H_{15}NO_2$	(193,2)	Ber.: C 68,37	H 7,82	N 7,25
		Gef.: C 68,0	H 7,84	N 7,2

Spezifische Drehung

c = 2,67 , 1 dm Rohr , t = 20° , abs. Äthanol

α_{578} = - 1,440°	$[\alpha]^{20}_{578}$ = - 53,9°	$[\alpha]^{20}_{D}$ = - 51,4°
α_{546} = - 1,655°	$[\alpha]^{20}_{546}$ = - 62,0°	
α_{436} = - 3,050°	$[\alpha]^{20}_{436}$ = -114,2°	
α_{405} = - 3,805°	$[\alpha]^{20}_{405}$ = -142,5°	
α_{365} = - 5,360°	$[\alpha]^{20}_{365}$ = -205,4°	

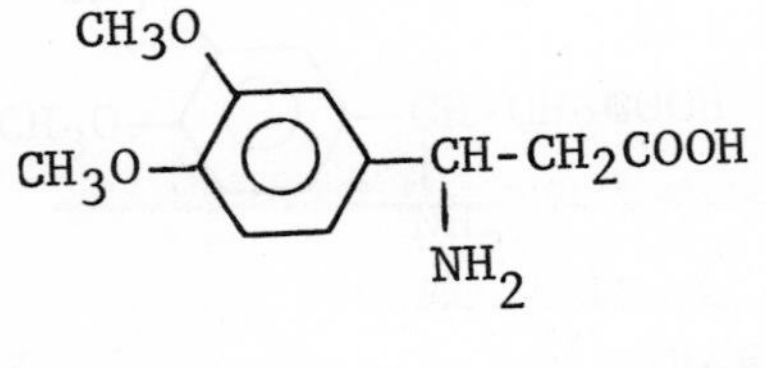

$C_{11}H_{15}NO_4$ Ref. 834

A solution of 1.21 g of (−)-1-phenylethylamine in 2 ml of methanol is added to a solution of 2.67 g of β-acetylamino-β-(3,4-dimethoxy-phenyl)-propionic acid in 80 ml of methanol. On adding ether, crystallisation starts. The crystals which have precipitated are repeatedly recrystallised from ethanol/ether. 1.5 g of the crystals thus obtained are taken up in 8 ml of water and 2.1 ml of 2 N sodium hydroxide solution. The resulting alkaline solution is extracted by shaking with chloroform. Thereafter, it is acidifed with dilute hydrochloric acid and again extracted with chloroform. The chloroform solution is dried and evaporated. The residue thus obtained is washed with water and recrystallised from ethyl acetate. (−)-β-Acetylamino-β-(3,4-dimethoxy-phenyl)-propionic acid is thus obtained.

EXAMPLE 2

27.7 g of (−)-β-acetylamino-β-(3,4-dimethoxy-phenyl)-propionic acid in 280 ml of 1 N hydrochloric acid are heated for 6 hours under reflux. After cooling, the mixture is extracted with methylene chloride. The aqueous acid phase is treated with active charcoal and filtered. The filtrate is evaporated to dryness in vacuo at about 10°C. The residue is recrystallised from 95 percent strength ethanol-ether. The hydrochloride of (+)-β-amino-β-(3,4-dimethoxy-phenyl)-propionic acid of the formula

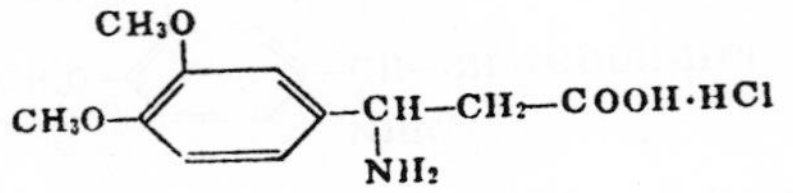

of melting point 204°–206°C; $[\alpha]_D$ = + 6° ± 1°; $[\alpha]_{313m\mu}$ = + 43° ± 1° (c = 1 percent in 0.1 N HCl) is thus obtained. 22.7 g of (+)-β-amino-β-(3,4-dimethoxy-phenyl)-propionic acid hydrochloride are dissolved in 1.2 l of absolute ethanol and treated with 8.7 g of triethylamine. The mixture is stirred for 2 hours at 20°C and the crystals which have precipitated are filtered off. (+)-β-Amino-β-(3,4-dimethoxy-phenyl)-propionic acid of melting point 224°–225°C (decomposition); $[\alpha]_D$ = + 7° ± 1°; $[\alpha]_{313m\mu}$ = + 49° ± 1° (c = 1 percent in 0.1 N HCl) is thus obtained.

The (−)-β-acetylamino-β-(3,4-dimethoxy-phenyl)-propionic acid used as the starting substance can be manufactured as follows:

182 g of (−)-1-phenylethylamine are added to a suspension of 400 g of β-acetylamino-β-(3,4-dimethoxy-phenyl)-propionic acid in 3 l methanol, whereupon a clear solution is produced. 4.5 l of ether are introduced whilst stirring. Thereafter the mixture is allowed to stand at a temperature of 0° to −5°C for about 60 hours. The supernatant solution is then decanted. The crystal mass is stirred with acetone, filtered off and washed with water. The (−)-1-phenylethylamine salt of (−)-β-acetylamino-β-(3,4-dimethoxy-phenyl)-propionic acid, of melting point 165°–167°C, is thus obtained; $[\alpha]_D$ = − 56° ± 1° (c = 1 percent in methanol).

58.4 g of this salt are dissolved in 75 ml of water and rendered alkaline with 28 ml of concentrated sodium hydroxide solution. The mixture is extracted with petroleum ether and the aqueous alkaline phase is rendered acid with 28 ml of concentrated hydrochloric acid. After cooling in an ice bath, the crystals which have precipitated are filtered off and washed with water. (−)-β-Acetylamino-β-(3,4-dimethyoxyphenyl)-propionic acid of melting point 157°–158°C, $[\alpha]_D$ −93° (c = 1 percent in methanol) is obtained. After recrystallisation from water, the melting point rises to 159°–160°C, $[\alpha]_D$ = − 96° ± 1° (c = 1 percent in methanol).

EXAMPLE 3

2.2 g of (+)-β-amino-β-(3,4,-dimethoxy-phenyl)-propionic acid methyl ester hydrochloride ($[\alpha]_D$ = + 14°, c = 1 percent in 0.1 N HCl) in 22 ml of 2 N hydrochloric acid are heated for 2 hours under reflux. Thereafter, the reaction solution is evaporated in vacuo, and the residue is treated with 95 percent strength ethanol and filtered. On adding ether, (+)-β-amino-β-(3,4-dimethoxyphenyl)-propionic acid hydrochloride of the formula

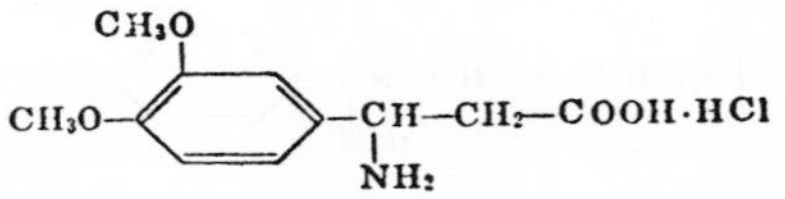

of melting point 202°–203°C (decomposition); $[\alpha]_D$ = + 7° ± 1° (c = 1 percent in 0.1 N HCl), crystallises out.

$C_{11}H_{15}NO_5$

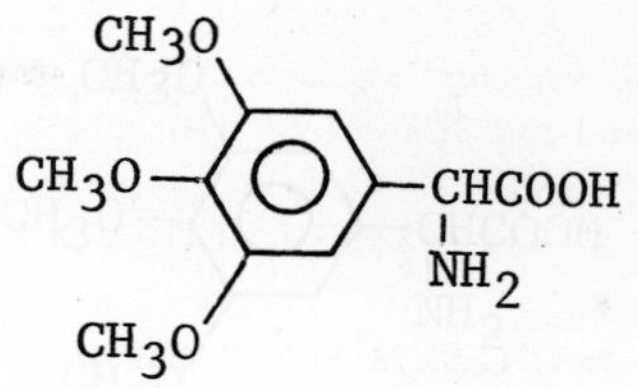

Ref. 835

Dehydroabietylaminsalz der D-*α-Benzyloxycarbonylamino-α-(3,4,5-trimethoxyphenyl)essigsäure* (D-**4**): 20 g (53.4 mmol) DL-α-Benzyloxycarbonylamino-α-(3,4,5-trimethoxyphenyl)-essigsäure (DL-**4**) werden in 200 ml Essigester in der Hitze gelöst und mit 15.2 g (53.4 mmol) Dehydroabietylamin, gelöst in 30 ml Essigester, versetzt. Nach 5 min. Kochen unter Rückfluß läßt man langsam auf Raumtemp. abkühlen und über Nacht bei 5°C stehen. Der weiße Niederschlag wird abgesaugt und mit Essigester gewaschen. Ausbeute 27.6 g (78%); $[\alpha]_{578}^{20} = -2.9°$ ($c = 1$, in Methanol). Nach viermaligem Umkristallisieren aus Methanol unter Zusatz von Wasser erhält man 14.5 g (41.2%) mit $[\alpha]_{578}^{20} = -28.9°$ ($c = 1$, in Methanol).

$C_{39}H_{52}N_2O_7$ (660.8) Ber. C 70.89 H 7.93 N 4.24 Gef. C 71.0 H 8.0 N 4.2

(+)-1-Phenyläthylaminsalz der L-*α-Benzyloxycarbonylamino-α-(3,4,5-trimethoxyphenyl)-essigsäure* (L-**4**): Die Filtrate des Dehydroabietylaminsalzes von D-**4** werden bis zur Trockne eingeengt. Der Rückstand wird in 1 N NaOH aufgenommen, die alkalische Lösung mit Äther ausgeschüttelt, die wäßrige Phase bei 0°C mit konz. Salzsäure bis pH 2 angesäuert und die ausgefallene Säure zweimal mit Essigester extrahiert. Nach Abdestillieren des Lösungsmittels erhält man 10.4 g (27.8 mmol) rohes L-**4**, das in 80 ml Essigester gelöst und mit der Lösung von 3.36 g (27.8 mmol) (+)-1-Phenyläthylamin in 20 ml Essigester versetzt wird. Die Lösung kocht man 5 min unter Rückfluß, gibt 70 ml n-Pentan unter Rühren hinzu und läßt über Nacht auf 0°C abkühlen. Dann wird der Niederschlag abgesaugt, mit n-Pentan gewaschen und viermal aus Essigester umkristallisiert. Ausbeute 7.3 g (53%); $[\alpha]_{578}^{20} = +65.1°$ ($c = 1$, in Methanol).

$C_{27}H_{32}N_2O_7$ (496.6) Ber. C 65.31 H 6.50 N 5.65 Gef. C 65.4 H 6.6 N 5.8

D-*α-Benzyloxycarbonylamino-α-(3,4,5-trimethoxyphenyl)essigsäure* (D-**4**): 14.5 g (22 mmol) des Dehydroabietylaminsalzes von D-**4** werden mit 1 N NaOH in Gegenwart von Äther bis zur vollständigen Auflösung verrührt. Die wäßrig-alkalische Phase wird abgetrennt, mit Essigester überschichtet und unter Umschütteln mit konz. Salzsäure bei Eiskühlung auf pH 2 angesäuert. Die Essigesterphase wird abgetrennt und die wäßrige Schicht nochmals mit Essigester ausgezogen. Die vereinigten Essigesterextrakte werden neutral gewaschen, getrocknet und i. Vak. auf ca. 100 ml eingeengt. Dann wird mit Petroläther versetzt, wobei 6.7 g (81%) D-**4** auskristallisieren; Schmp. 149°C, $[\alpha]_{578}^{20} = -107.7°$ ($c = 1$, in Methanol). — ^{1}H-NMR (CD_3OD): $\delta = 3.71$ (s; 4-CH_3O), 3.77 (s; 3-CH_3O, 5-CH_3O), 5.07 (s; Benzyl-CH_2), 5.22 (s; α-CH), 6.70 (s; 2-H, 6-H), 7.27 ppm (s; 5H, C_6H_5).

$C_{19}H_{21}NO_7$ (375.4) Ber. C 60.79 H 5.64 N 3.74 Gef. C 60.7 H 5.8 N 3.6

·*α-Benzyloxycarbonylamino-α-(3,4,5-trimethoxyphenyl)essigsäure* (L-**4**): 7.5 g (15.1 mmol) des (+)-1-Phenyläthylaminsalzes von L-**4** werden mit einer gesättigten Natriumcarbonat-Lösung in Gegenwart von Äther bis zur vollständigen Auflösung behandelt und die abgetrennte wäßrige Schicht, wie bei der Isolierung von D-**4** beschrieben, aufgearbeitet. Ausbeute 4.5 g (80%); Schmp. 149–150°C, $[\alpha]_{578}^{20} = +108.5°$ ($c = 1$, in Methanol). — ^{1}H-NMR (CD_3OD): $\delta = 3.76$ (s; 4-CH_3O), 3.81 (s; 3-CH_3O, 5-CH_3O), 5.11 (s; Benzyl-CH_2), 5.22 (s; α-CH), 6.74 (s; 2-H, 6-H), 7.34 ppm (s; 5H, C_6H_5).

$C_{19}H_{21}NO_7$ (375.4) Ber. C 60.79 H 5.64 N 3.74 Gef. C 60.8 H 5.8 N 3.2

D-*α-Amino-α-(3,4,5-trimethoxyphenyl)essigsäure* (D-**3**): 5 g (13.3 mmol) D-**4** werden in 30 ml Eisessig suspendiert und mit 30 ml 3.9 N HBr/Eisessig übergossen, die Mischung wird 1 h bei Raumtemp. gerührt und dann das gelartige Gemisch in Äther eingerührt. Das Hydrobromid wird abgesaugt, mit Äther gewaschen und i. Vak. getrocknet, danach in Wasser gelöst, filtriert und das Filtrat bei 0°C mit konz. Ammoniak auf pH 5.1 eingestellt. Nach 2stdg. Stehenlassen bei 0–5°C wird D-**3** abgesaugt, mit eiskaltem Wasser gewaschen, aus Wasser umkristallisiert und bei 50°C i. Vak. getrocknet. Ausbeute 2.4 g (75%); Schmp., 211–212°C, $[\alpha]_{578}^{20} = -147.3°$ ($c = 1$, in Ameisensäure). — ^{1}H-NMR (DCOOD): $\delta = 3.87$–3.89 (2 s; 3,4,5-CH_3O), 5.38 (s; α-CH), 6.9 (s; 2-H, 6-H), 10.62 ppm (s; NH_2).

$C_{11}H_{15}NO_5$ (241.2) Ber. C 54.78 H 6.27 N 5.81 Gef. C 53.8 H 6.6 N 5.6

---

4) *W. Klyne, P. M. Scopes, R. N. Thomas* und *H. Dahn*, Helv. Chim. Acta **54**, 2420 (1971).
5) *M. Legrand* und *R. Viennet*, Bull. Soc. Chim. Fr. **9**, 2798 (1966).
6) *G. Snatzke* und *S. H. Doss*, Tetrahedron **28**, 2539 (1972).

L-α-Amino-α-(3,4,5-trimethoxyphenyl)essigsäure (L-**3**): 4 g (10.7 mmol) L-**4** werden in 30 ml Eisessig suspendiert und mit 30 ml 3.9 N HBr/Eisessig wie voranstehend entacyliert. Ausbeute 2 g (78%); Schmp. 211–212°C, $[\alpha]_{578}^{20} = +146.9°$ (*c* = 1, in Ameisensäure). – ^{1}H-NMR (DCOOD): δ = 3.87–3.88 (2 s; 3,4,5-CH_3O), 5.37 (s; α-CH), 6.89 (s; 2-H, 6-H), 11.1 ppm (s; NH_2).

$C_{11}H_{15}NO_5$ (241.2) Ber. C 54.78 H 6.27 N 5.81 Gef. C 53.9 H 6.6 N 5.9

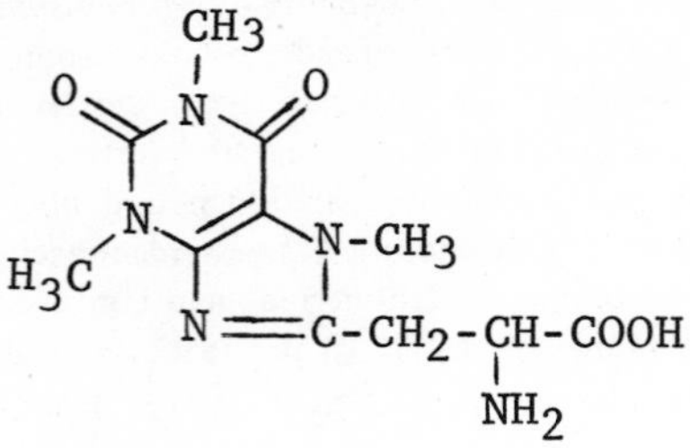

$C_{11}H_{15}N_5O_4$ Ref. 836

Salt of *L*-Threo-1-(p-nitrophenyl)-2-amino-1,3-propanediol and d-N-Carbobenzyloxy-β-(8-caffeinyl)-α-alanine (*L*-III · d-V). A mixture of 6.2 g of racemic N-carbobenzyloxy-β-(8-caffeinyl)-α-alanine and 3.2 g of *L*-threoamine (*L*-III) in 200 ml absolute alcohol was boiled to complete solution and left to stand 4 h at 20°. The crystals which had separated were filtered off and boiled with 60 ml absolute alcohol. 3.7 g (78%) salt (*L*-III · d-V) was obtained with m.p. 190-192°; $[\alpha]_D^{20}$ +26.0° (c 1% in 45% alcohol). The alcoholic filtrate was retained for the isolation of salt (D-III · *l*-V). Found %: C 53.29; H 5.47; N 15.32. $C_{28}H_{33}N_7O_{10}$. Calculated %: C 53.59; H 5.30; N 15.46.

Salt of D-Threo-1-(p-nitrophenyl)-2-amino-1,3-propanediol and *l*-Carbobenzyloxy-β-(8-caffeinyl)-α-alanine (D-III · *l*-V). The filtrate after removal of salt (*L*-III · d-V) was concentrated in vacuum, the residue suspended in ethyl acetate and shaken with 45 ml of 5% HCl. The ethyl acetate solution was washed with water, dried with $MgSO_4$ and concentrated in vacuum. The residue was dissolved by boiling in 80 ml absolute alcohol and run into a hot solution of 1.5 g D-threoamine (D-III) in 60 ml absolute alcohol. After 18 h, crystals of salt (D-III · *l*-V) were filtered off and boiled with 20 ml absolute alcohol. 2.4 g (50%) salt with m.p. 188-190° was obtained; $[\alpha]_D^{20}$ -26° (c 1% in 45% alcohol). Found %: C 53.43; H 5.58; N 15.19. $C_{28}H_{33}N_7O_{10}$. Calculated %: C 53.59; H 5.30; N 15.46.

d-N-Carbobenzyloxy-β-(8-caffeinyl)-α-alanine (d-V). To a suspension of 8.8 g salt (*L*-III · d-V) in 120 ml ethyl acetate was added 95 ml 5% HCl. The mixture was shaken until solution of the residue, the ethyl acetate layer washed with water, shaken with $MgSO_4$ and concentrated at once in vacuum to 10 ml. The precipitate was filtered off and crystallized from absolute alcohol. 3.3 g (57%) d-carbobenzyloxycaffeinylalanine (d-V) with m.p. 188-190° was obtained; $[\alpha]_D^{20}$ +18° (c 1% in 45% alcohol). The compound was dried in vacuum at 100° for analysis. Found %: C 55.10; H 4.83; N 17.03. $C_{19}H_{21}N_5O_6$. Calculated %: C 54.93; H 5.10; N 16.86.

l-N-Carbobenzyloxy-β-(8-caffeinyl)-α-alanine (*l*-V). Analogously to the d-isomer (d-V) 1.5 g (52%) *l*-N-carbobenzyloxy-β-(8-caffeinyl)-α-alanine (*l*-V) was obtained from 4.4 g salt (D-III · *l*-V) and had m.p. 185-186°; $[\alpha]_D^{20}$ -18° (c 1% in 45% alcohol). Found %: C 55.14; H 5.23; N 16.83. $C_{19}H_{21}N_5O_6$. Calculated %: C 54.93; H 5.10; N 16.86.

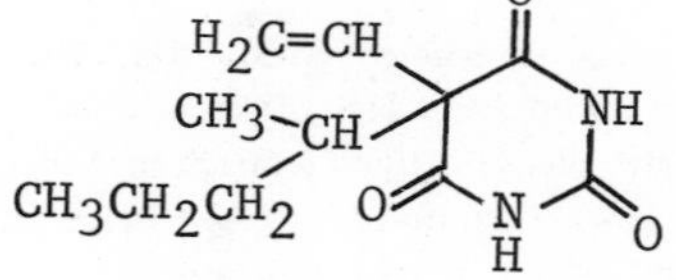

$C_{11}H_{16}N_2O_3$ Ref. 837

IV. Spaltung von Vinylbital in optische Antipoden

1. Gewinnung der diastereomeren N-Methylchininium-vinylbitale

9,7 g Vinylbital wurden in 245,0 ml einer titrierten methanol. N-Methyl-chininium-hydroxidlösung gelöst und das Lösungsmittel bis auf die Hälfte seines Vol. i. Vak. abgezogen. Dann wurde Äther bis zur starken Trübung zugesetzt. Es wurden 7 g eines amorphen Salzes erhalten, das aus Methanol/Äther kristallisierte. Ausbeute: 6,0 g N-Methyl-chininium-(—)-vinylbital. Nach Trocknen i. Vak. über P_2O_5 Schmp. 218°.

$C_{32}H_{42}N_4O_5$ (562,7)	Ber.: C 68,30	H 7,52
	Gef.: C 67,92	H 7,49